S.Chand's

HIGHER ENGINEERING MATHEMATICS

S.Chand's

HIGHER ENGINEERING MATHEMATICS

[For Engineering Students & also useful for Competitive Examinations]

H.K. DASS
M.Sc.
Diploma in Specialist Studies (Mathematics)
University of Hull
England

Er. RAJNISH VERMA
B.E.
Delhi College of Engg.
MBA (FMS) University of Delhi
FIETE, PDBA, NCES
D.G.M., C.M.C. Ltd.

S Chand And Company Limited
(ISO 9001 Certified Company)

S Chand And Company Limited

(ISO 9001 Certified Company)

Head Office: D-92, Sector–2, Noida – 201301, U.P. (India), Ph. 91-120-4682700

Registered Office: A-27, 2nd Floor, Mohan Co-operative Industrial Estate, New Delhi – 110 044, Phone: 011-49731800

www.**schandpublishing.com**; e-mail: **info@schandpublishing.com**

Marketing Offices:

Chennai	:	Ph: 23632120; chennai@schandpublishing.com
Guwahati	:	Ph: 2738811, 2735640; guwahati@schandpublishing.com
Hyderabad	:	Ph: 40186018; hyderabad@schandpublishing.com
Jalandhar	:	Ph: 4645630; jalandhar@schandpublishing.com
Kolkata	:	Ph: 23357458, 23353914; kolkata@schandpublishing.com
Lucknow	:	Ph: 4003633; lucknow@schandpublishing.com
Mumbai	:	Ph: 25000297; mumbai@schandpublishing.com
Patna	:	Ph: 4011400; patna@schandpublishing.com

First Edition 2011
Second Revised Edition 2012
Third Revised Edition 2014
Reprints 2014, 2015, 2018 (Twice), 2019, 2020, 2021, 2023, 2024, 2025 (Twice)

Reprint 2026

ISBN : 978-81-219-3890-7 **Product Code :** H3HEM61MATH10ENAC14O

PRINTED IN INDIA

By Shri Shyamlal Printing Press Private Limited, Plot 20/4, Site-IV, Industrial Area Sahibabad, Ghaziabad – 201 010 and Published by S Chand And Company Limited, A-27, 2nd Floor, Mohan Co-operative Industrial Estate, New Delhi – 110 044.

PREFACE TO THE THIRD REVISED EDITION

The book has received very good response from students and teachers within the country and abroad alike. Its previous edition is exhausted in a very short time.

A chapter on Fuzzy set is included at the end of the book to make it more useful

The misprints which came to my knowledge, have been removed.

We are thankful to the Management Team and the Editorial Department of S. Chand & Company Pvt. Ltd. for all help and support in the publication of this book.

Suggestions for the improvement of the book will be gratefully acknowledged.

D-1/87, Janakpuri
New Delhi-110 058
Tel. 28525078, 32985078, 28521776
Mob. 9350055078
e-mail: hk_dass@yahoo.com

AUTHORS

PREFACE TO THE FIRST EDITION

It gives me great pleasure to present this textbook of **Higher Engineering Mathematics** to the students pursuing various engineering courses.

This book includes the syllabi of various engineering courses and comprises of 65 simple and short chapters to avoid ambiguity. As presently there is not a single textbook which entirely covers so many topics of engineering courses. Therefore an endeavour has been made to cover the syllabus exhaustively and present the subject matter in a systematic and lucid style.

About 2000 solved examples on various topics have been incorporated in the textbook for the better understanding of the students. Care has been taken to systematically grade these examples. While solving the examples, even minor steps have not been missed to take care the difficulty of the students in understanding the subject easily. Most of the examples have been taken from the latest examination papers which should make the students familiar with the standard and trend of questions being set in the examinations.

The author possesses very long and rich experience of more than 50 years in teaching Mathematics to the students preparing examinations of engineering and has first hand experience of the problems and difficulties that students & teachers generally face.

This book should satisfy both average and brilliant students. It would help the students to score good marks in examinations and at the same time would arouse greater intellectual curiosity in them.

We are thankful to the Management Team and the Editorial Department of S. Chand & Company Ltd. for all help and support in the publication of this book.

Suggestions for the improvement of the book will be gratefully acknowledged.

H.K. DASS

D-1/87, Janakpuri
New Delhi-110 058
Tel. 28525078, 32985078, 28521776
Mob. 9350055078
e-mail: hk_dass@yahoo.com

CONTENTS

UNIT – I

UNIT – II

UNIT – IV

UNIT – V

UNIT – VI

UNIT – VIII

UNIT – XII

UNIT – XIII

UNIT – XIV

CHAPTER

1

PARTIAL DIFFERENTIATION

1.1 INTRODUCTION

Area of a rectangle depends upon its length and breadth, hence we can say that area is the function of two variables *i.e.* its length and breadth.

z is called a function of two variables x and y if z has one definite value for every pair of values of x and y. Symbolically, it is written as

$$z = f(x, y)$$

The variable x and y are called independent variables while z is called the dependent variable.

Similarly, we can define z as a function of more than two variables.

Geometrically

Let $$z = f(x, y)$$

where x, y belong to an area A of the xy-plane. For each point (x, y) corresponds a value of z. These values of (x, y, z) form a surface in space.

Hence, the function $z = f(x, y)$ represents a surface.

1.2 LIMIT

The function $f(x, y)$ is said to tend to the limit l as $x \to a$ and $y \to b$ if and only if the limit l is independent of the path followed by the point (x, y) as $x \to a$ and $y \to b$. Then

$$\lim_{\substack{x \to a \\ y \to b}} f(x, y) = l$$

The function $f(x, y)$ in region R is said to tend to the limit l as $x \to a$ and $y \to b$ if and only if corresponding to a positive number ε, there exists another positive number δ such that

$$|f(x, y) - l| < \varepsilon \text{ for } 0 < (x - a)^2 + (y - b)^2 < \delta^2$$

for every point (x, y) in R.

1.3 WORKING RULE TO FIND THE LIMIT

Step 1. Find the value of $f(x, y)$ along $x \to a$ and $y \to b$.

Step 2. Find the value of $f(x, y)$ along $y \to b$ and $x \to a$.

If the values of $f(x, y)$ in step 1 and step 2 remain the same, the limit exists otherwise not.

Step 3. If $a \to 0$ and $b \to 0$, find the limit along $y = mx$ or $y = mx^n$. If the value of the limit does not contain m then limit exists. If it contains m, the limit does not exist.

Note. (*i*) Put $x = 0$ and then $y = 0$ in f. Find its value f_1.

(*ii*) Put $y = 0$ and then $x = 0$ in f. Find the value f_2.

If $f_1 \neq f_2$, limit does not exist.

If $f_1 = f_2$, then

(*iii*) Put $y = mx$ and find the limit f_3.

If $f_1 = f_2 \neq f_3$, then limit does not exist.

If $f_1 = f_2 = f_3$, then

(*iv*) Put $y = mx^n$ and find the limit f_4.
If $f_1 = f_2 = f_3 \neq f_4$, then limit does not exist.
If $f_1 = f_2 = f_3 = f_4$, then limit exists.

Example 1. *If* $f(x, y) = \dfrac{x+y}{2x-y}$, *show that* $\lim_{x\to 0}\left[\lim_{y\to 0} f(x,y)\right] \neq \lim_{y\to 0}\left[\lim_{x\to 0}(x,y)\right]$

Solution. (*i*) $\lim_{x\to 0}\left[\lim_{y\to 0}\dfrac{x+y}{2x-y}\right] = \lim_{x\to 0}\left(\dfrac{x+0}{2x-0}\right) = \dfrac{1}{2}$

(*ii*) $\lim_{y\to 0}\left[\lim_{x\to 0}\dfrac{x+y}{2x-y}\right] = \lim_{y\to 0}\left(\dfrac{0+y}{0-y}\right) = -1$

Hence, $\lim_{x\to 0}\left[\lim_{y\to 0} f(x,y)\right] \neq \lim_{y\to 0}\left[\lim_{x\to 0} f(x,y)\right]$ **Proved.**

Example 2. *Evaluate* $\lim_{\substack{x\to 0\\ y\to 0}} \dfrac{x^2 y}{x^4+y^2}$

Solution. (*i*) $\lim_{\substack{x\to 0\\ y\to 0}} \dfrac{x^2 y}{x^4+y^2} = \lim_{y\to 0}\dfrac{0}{0+y^2} = 0 = f_1$ (say)

(*ii*) $\lim_{\substack{x\to 0\\ y\to 0}} \dfrac{x^2 y}{x^4+y^2} = \lim_{x\to 0}\dfrac{0}{x^4+0} = 0 = f_2$ (say)

Here, $f_1 = f_2$, therefore

(*iii*) Put $y = mx$

$$\lim_{\substack{x\to 0\\ y\to 0}} \frac{x^2 mx}{x^4+m^2x^2} = \lim_{x\to 0}\frac{mx}{x^2+m^2} = 0 = f_3 \quad \text{(say)}$$

Here, $f_1 = f_2 = f_3$, therefore

(*iv*) Put $y = mx^2$

$$\lim_{\substack{x\to 0\\ y\to 0}} \frac{x^2 mx^2}{x^4+m^2x^4} = \frac{m}{1+m^2} = f_4$$

Here, $f_1 = f_2 = f_3 \neq f_4$

Thus, limit does not exist. **Ans.**

Example 3. *Evaluate* $\lim_{\substack{x\to 0\\ y\to 0}} (x^3 + y^3)$.

Solution. (*i*) $\lim_{\substack{x\to 0\\ y\to 0}} (x^3+y^3) = \lim_{y\to 0}(0+y^3) = 0 = f_1$ (say)

(*ii*) $\lim_{\substack{x\to 0\\ y\to 0}} (x^3+y^3) = \lim_{x\to 0}(x^3+0) = 0 = f_2$ (say)

Here, $f_1 = f_2$, therefore

(*iii*) Put $y = mx$

$$\lim_{\substack{x\to 0\\ y\to 0}} (x^3+y^3) = \lim_{x\to 0}\left[\lim_{y\to mx}(x^3+y^3)\right] = \lim_{x\to 0}(x^3+m^3x^3) = 0 = f_3 \quad \text{(say)}$$

Here, $f_1 = f_2 = f_3$, therefore

(*iv*) Put $y = mx^2$

$$\lim_{\substack{x\to 0\\ y\to 0}} (x^3 + y^3) = \lim_{x\to 0}\left[\lim_{y\to mx^2}(x^3+y^3)\right] = \lim_{x\to 0}(x^3 + m^3x^6)$$

$$= \lim_{x\to 0} x^3(1 + m^3x^3) = 0 = f_4 \qquad \text{(say)}$$

Here, $f_1 = f_2 = f_3 = f_4$

Thus, limit exists with value 0. **Ans.**

Example 4. *Evaluate* $\lim_{\substack{x\to 0\\ y\to 0}} \frac{y^2 - x^2}{y^2 + x^2}$, $x \neq 0$, $y \neq 0$

Solution. (*i*) $\lim_{\substack{x\to 0\\ y\to 0}} \left(\frac{y^2 - x^2}{y^2 + x^2}\right) = \lim_{y\to 0}\left(\lim_{x\to 0}\frac{y^2 - x^2}{y^2 + x^2}\right)$

$$= \lim_{y\to 0}\left(\frac{y^2 - 0}{y^2 + 0}\right) = \lim_{y\to 0}\frac{y^2}{y^2} = 1 = f_1 \qquad \text{(say)}$$

(*ii*) $\lim_{\substack{x\to 0\\ y\to 0}} \frac{y^2 - x^2}{y^2 + x^2} = \lim_{x\to 0}\left[\lim_{y\to 0}\frac{y^2 - x^2}{y^2 + x^2}\right]$

$$= \lim_{x\to 0}\left(\frac{0 - x^2}{0 + x^2}\right) = \lim_{x\to 0}\frac{-x^2}{x^2} = -1 = f_2 \qquad \text{(say)}$$

Here $f_1 \neq f_2$

Thus, limit does not exist. **Ans.**

Example 5. *Evaluate* $\lim_{\substack{x\to 0\\ y\to 0}} \frac{x^3 - y^3}{x^2 + y^2}$; $x \neq 0$, $y \neq 0$.

Solution. (*i*) $\lim_{\substack{x\to 0\\ y\to 0}} \frac{x^3 - y^3}{x^2 + y^2} = \lim_{y\to 0}\left[\lim_{x\to 0}\frac{x^3 - y^3}{x^2 + y^2}\right] = \lim_{y\to 0}\frac{0 - y^3}{0 + y^2} = \lim_{y\to 0}(-y) = 0 = f_1$ (say)

(*ii*) $\lim_{\substack{y\to 0\\ x\to 0}} \frac{x^3 - y^3}{x^2 + y^2} = \lim_{x\to 0}\left[\lim_{y\to 0}\frac{x^3 - y^3}{x^2 + y^2}\right] = \lim_{x\to 0}\frac{x^3 - 0}{x^2 + 0} = \lim_{x\to 0}(x) = 0 = f_2$ (say)

Here, $f_1 = f_2$ therefore,

(*iii*) Put $y = mx$

$$\lim_{\substack{x\to 0\\ y\to 0}} \frac{x^3 - y^3}{x^2 + y^2} = \lim_{x\to 0}\left[\lim_{y\to 0}\frac{x^3 - y^3}{x^2 + y^2}\right] = \lim_{x\to 0}\left[\lim_{y\to mx}\frac{x^3 - y^3}{x^2 + y^2}\right]$$

$$= \lim_{x\to 0}\left[\frac{x^3 - m^3x^3}{x^2 + m^2x^2}\right] = \lim_{x\to 0}\frac{x(1 - m^3)}{1 + m^2} = 0 = f_3 \qquad \text{(say)}$$

Here, $f_1 = f_2 = f_3$, therefore

(*iv*) Put $y = mx^2$

$$\lim_{x\to 0}\left[\lim_{y\to 0}\frac{x^3 - y^3}{x^2 + y^2}\right] = \lim_{x\to 0}\left[\lim_{y\to mx^2}\frac{x^3 - y^3}{x^2 + y^2}\right]$$

$$= \lim_{x\to 0}\frac{x^3 - m^3x^6}{x^2 + m^2x^4} = \lim_{x\to 0}\frac{x(1 - m^3x^3)}{1 + m^2x^2} = 0 = f_4 \qquad \text{(say)}$$

Here, $f_1 = f_2 = f_3 = f_4$

Hence, limit exists with value 0. **Ans.**

Example 6. *Evaluate* $\lim\limits_{\substack{x\to 1\\ y\to 2}} \dfrac{x^2+2y}{x+y^2}$. $x \neq 0,\ y \neq 0$

Solution. (*i*) $\lim\limits_{\substack{x\to 1\\ y\to 2}} \dfrac{x^2+2y}{x+y^2} = \lim\limits_{y\to 2}\left[\lim\limits_{x\to 1} \dfrac{x^2+2y}{x+y^2}\right]$

$$= \lim_{y\to 2}\left(\frac{1+2y}{1+y^2}\right) = \frac{1+4}{1+4} = \frac{5}{5} = 1 = f_1 \quad \text{(say)}$$

(*ii*) $\lim\limits_{\substack{y\to 2\\ x\to 1}} \dfrac{x^2+2y}{x+y^2} = \lim\limits_{x\to 1}\left[\lim\limits_{y\to 2} \dfrac{x^2+2y}{x+y^2}\right]$

$$= \lim_{x\to 1} \frac{x^2+4}{x+4} = \frac{1+4}{1+4} = \frac{5}{5} = 1 = f_2 \quad \text{(say)}$$

Here, $f_1 = f_2$

Thus, limit exists. **Ans.**

Example 7. *Evaluate* $\lim\limits_{\substack{x\to\infty\\ y\to 3}} \dfrac{2x-3}{x^3+4y^3}$.

Solution. (*i*) $\lim\limits_{\substack{x\to\infty\\ y\to 3}} \dfrac{2x-3}{x^3+4y^3} = \lim\limits_{y\to 3}\left[\lim\limits_{x\to\infty} \dfrac{2x-3}{x^3+4y^3}\right]$

$$= \lim_{\substack{x\to\infty\\ y\to 3}} \frac{\dfrac{2}{x^2}-\dfrac{3}{x^3}}{1+4\left(\dfrac{y}{x}\right)^3} = \lim_{y\to 3} \frac{0-0}{1+4(0)} = 0 = f_1 \quad \text{(say)}$$

(*ii*) $\lim\limits_{\substack{y\to 3\\ x\to\infty}} \dfrac{2x-3}{x^3+4y^3} = \lim\limits_{x\to\infty}\left[\lim\limits_{y\to 3} \dfrac{2x-3}{x^3+4y^3}\right]$

$$= \lim_{x\to\infty} \frac{2x-3}{x^3+108} = \lim_{x\to\infty} \frac{\dfrac{2}{x^2}-\dfrac{3}{x^3}}{1+\dfrac{108}{x^3}} = \frac{0-0}{1+0} = 0 = f_2 \quad \text{(say)}$$

Here, $f_1 = f_2$.

Hence, the limit exists with value 0. **Ans.**

Example 8. *Evaluate* $\lim\limits_{\substack{x\to\infty\\ y\to 2}} \dfrac{xy+4}{x^2+2y^2}$.

Solution. (*i*) $\lim\limits_{\substack{x\to\infty\\ y\to 2}} \dfrac{xy+4}{x^2+2y^2} = \lim\limits_{x\to\infty}\left[\lim\limits_{y\to 2} \dfrac{xy+4}{x^2+2y^2}\right] = \lim\limits_{x\to\infty} \dfrac{2x+4}{x^2+8}$

$$= \lim_{x\to\infty} \frac{2+\dfrac{4}{x}}{x+\dfrac{8}{x}} = \lim_{x\to\infty} \frac{2}{x} = 0 = f_1 \quad \textbf{Ans.}$$

(*ii*) $\lim\limits_{\substack{x\to\infty\\ y\to 2}} \dfrac{xy+4}{x^2+2y^2} = \lim\limits_{y\to 2}\ \lim\limits_{x\to\infty} \dfrac{\dfrac{y}{x}+\dfrac{4}{x^2}}{1+\dfrac{y^2}{x^2}} = \lim\limits_{y\to 2} \dfrac{0+0}{1+0} = 0 = f_2$

Since $f_1 = f_2$, hence limit exists with value 0. **Ans.**

EXERCISE 1.1

Evaluate the following limits:

1. $\lim\limits_{\substack{x\to 1\\ y\to 2}} \dfrac{2x^2+y^2}{2xy}$ **Ans.** $\dfrac{3}{4}$

2. $\lim\limits_{\substack{x\to 2\\ y\to 3}} \dfrac{x^3+y^2}{x^2-y}$, **Ans.** 17

3. $\lim\limits_{\substack{x\to \infty\\ y\to 3}} \dfrac{2xy-3}{x^3+4y^3}$, **Ans.** 0

4. $\lim\limits_{\substack{x\to 0\\ y\to 0}} \dfrac{xy}{y-x^2}$; $x \neq 0,\ y \neq 0$ **Ans.** Limit does not exist

5. $\lim\limits_{\substack{x\to 0\\ y\to 0}} \dfrac{x-y}{x^2+y^2}$; $x \neq 0,\ y \neq 0$ **Ans.** Limit does not exist

6. $\lim\limits_{\substack{x\to 1\\ y\to 1}} \dfrac{xy-2x}{xy-2y}$ **Ans.** 1

7. $\lim\limits_{\substack{x\to 0\\ y\to 0}} \dfrac{x^3+2y^3}{x^2+4y^2}$ $x \neq 0,\ y \neq 0$ **Ans.** 0

8. $\lim\limits_{\substack{x\to 0\\ y\to 0}} \dfrac{x^2y^3}{x^2+y^2}$ $x \neq 0,\ y \neq 0$ **Ans.** 0

9. $\lim\limits_{\substack{x\to 0\\ y\to 0}} \dfrac{xy+2}{x^2+y^2}$, $x \neq 0,\ y \neq 0$ **Ans.** 0

Choose the correct alternative:

10. The value of $\lim\limits_{(x,\,y)\to(0,\,0)} (x+y)\sin\dfrac{1}{(x+y)}$, $x \neq 0,\ y \neq 0$ is

    (*a*) limit does not exist (*b*) 0
    (*c*) 1 (*d*) –1 **Ans.** (*c*)

11. The value of the $\lim\limits_{(x,\,y)\to(0,\,0)} \dfrac{x+\sqrt{y}}{\sqrt{(x^2+y)}}$, $x \neq 0,\ y \neq 0$ is

    (*a*) limit does not exist (*b*) 0
    (*c*) 1 (*d*) –1 (*AMIETE, June 2010, Dec.* 2007) **Ans.** (*a*)

1.4 CONTINUITY

A function $f(x, y)$ is said to be continuous at a point (a, b) if

$$\lim_{(x,\,y)\to(a,\,b)} f(x, y) = f(a, b)$$

A function is said to be continuous in a domain if it is continuous at every point of the domain.

1.5 WORKING RULE FOR CONTINUITY AT A POINT (*a*, *b*)

Step 1. $f(a, b)$ should be well defined

Step 2. $\lim\limits_{(x,\,y)\to(a,\,b)} f(x, y)$ should exist.

Step 3. $\lim\limits_{(x,\,y)\to(a,\,b)} f(x, y) = f(a, b)$

Example 9. *Test the function* $f(x, y) = \begin{cases} \dfrac{x^3-y^3}{x^2+y^2} & \text{when } x \neq 0,\quad y \neq 0 \\ 0 & \text{when } x = 0,\quad y = 0 \end{cases}$ *for continuity.*

Solution. Step 1. The function is well defined at (0, 0).

Step 2. $$\lim_{(x,\,y)\to(0,\,0)} f(x, y) = \lim_{(x,\,y)\to(0.\,0)} \frac{x^3-y^3}{x^2+y^2} = \lim_{x\to 0}\left[\lim_{y\to mx} \frac{x^3-y^3}{x^2+y^2}\right]$$

$$= \lim_{x\to 0} \frac{x^3-m^3x^3}{x^2+m^2x^2}$$

$$= \lim_{x \to 0} \frac{x(1 - m^3)}{1 + m^2} = 0$$

Thus, limit exists at (0, 0).

Step 3. limit of $f(x)$ at origin = value of the function at origin.

$$\lim_{(x, y) \to (0, 0)} \frac{x^3 - y^3}{x^2 + y^2} = f(0, 0) = 0$$

Hence, the function f is continuous at the origin. **Ans.**

Example 10. *Discuss the continuity of* $f(x, y) = \begin{cases} \dfrac{x}{\sqrt{x^2 + y^2}}, & x \neq 0, \quad y \neq 0 \\ 2, & x = 0, \quad y = 0 \end{cases}$

at the origin.

Solution. Here, we have

$$f(x, y) = \begin{cases} \dfrac{x}{\sqrt{x^2 + y^2}} & x \neq 0, \quad y \neq 0 \\ 2, & x = 0, \quad y = 0 \end{cases}$$

Step 1. The function $f(x, y)$ at (0, 0) is well defined.

Step 2. $$\lim_{(x, y) \to (0, 0)} \frac{x}{\sqrt{x^2 + y^2}} = \lim_{x \to 0} \left[\lim_{y \to mx} \frac{x}{\sqrt{x^2 + y^2}} \right] = \lim_{x \to 0} \frac{x}{\sqrt{x^2 + m^2 x^2}}$$

$$= \lim_{x \to 0} \frac{1}{\sqrt{1 + m^2}}$$

For different values of m the limit f is not unique.

so the $\lim_{(x, y) \to (0, 0)} \dfrac{x}{\sqrt{x^2 + y^2}}$ does not exist.

Hence $f(x, y)$ is not continuous at origin. **Ans.**

1.6 TYPES OF DISCONTINUITY

(*Gujarat Univ. I sem. Jan. 2009*)

1. **First Kind.** $f(x)$ is said to have discontinuity of first kind at the point $x = x_1$ if Right limit $f(x_1 + 0)$ and left limit $f(x_1 - 0)$ exist but are not equal.

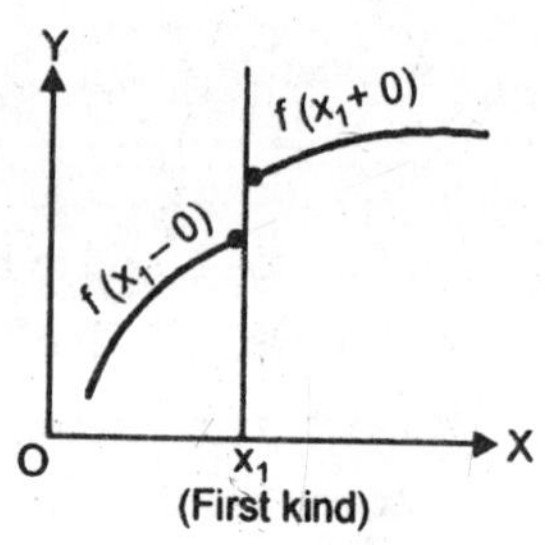

(First kind)

2. **Second Kind.** $f(x)$ is said to have discontinuity of the second kind at $x = x_1$ if neither right limit $f(x_1 + 0)$ exists nor left limit $f(x_1 - 0)$ exists.

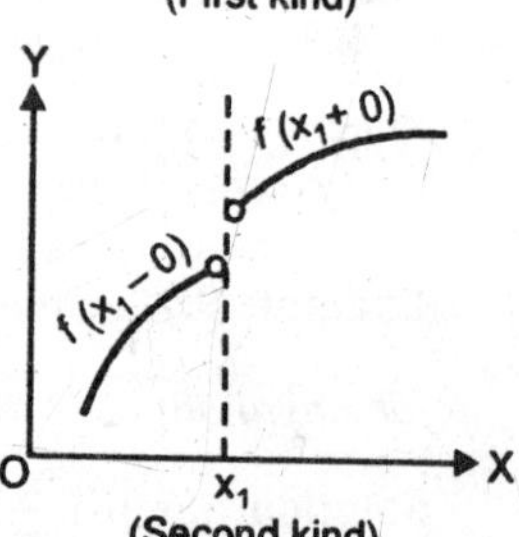

(Second kind)

3. **Third Kind (Mixed discontinuity).** $f(x)$ is said to have mixed discontinuity at the point $x = x_1$ if only one of the two limits right limit $f(x_1 + 0)$ and left limit $f(x_1 - 0)$ exists and not the other.

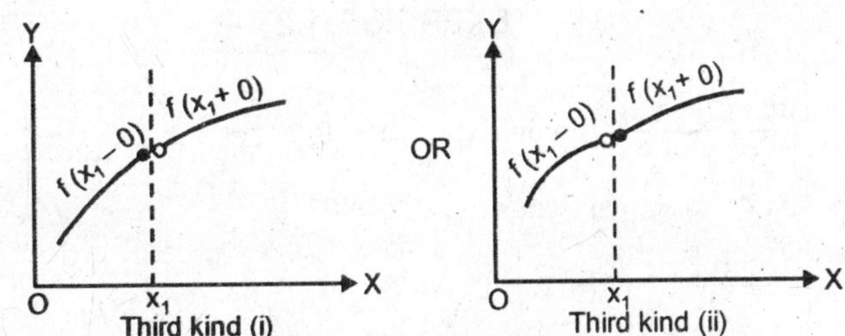

Third kind (i) Third kind (ii)

4. **Fourth Kind (Infinite discontinuity).** $f(x)$ is said to have infinite discontinuinty at the point $x = x_1$ if either one or both limits right limit and left limit $f(x_1 - 0)$ are infinite. If both limits do not exist and if $f(x_1 \pm h)$ oscillates between limits one of which is infinite as $\pm h \to 0$. It is also a point of inifinite discontinuity.

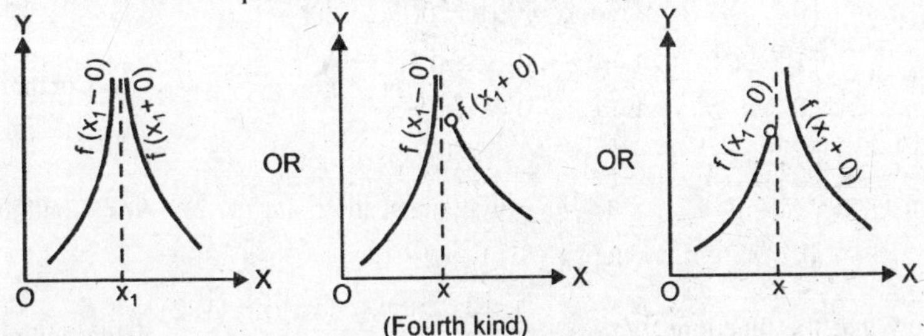

(Fourth kind)

5. **Fifth Kind (Removable discontinuity).** If right limit $f(x_1 + 0)$ is equal to left limit $f(x_1 - 0)$ is not equal to $f(x_1)$, then $f(x)$ is said to have removable discontinuity.

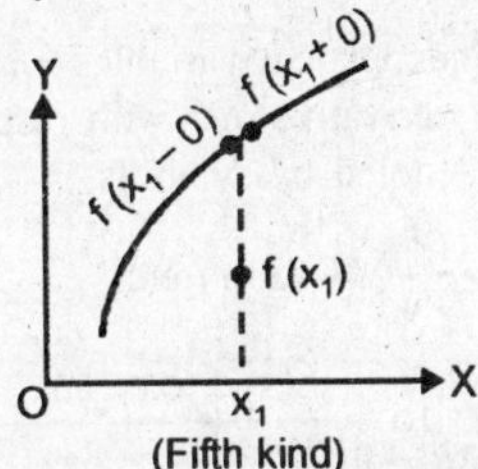

(Fifth kind)

Example 11. *Show that the given function are discontinuous at all the point (2, – 2).*

$$f(x, y) = \begin{cases} \dfrac{x^2 + xy + x + y}{x + y}, & (x, y) \neq (2, -2) \\ 4 & (x, y) = (2, -2) \end{cases}$$

(*A.M.I.E.T.E., June* 2009)

Solution. We have,

$$f(x, y) = \begin{cases} \dfrac{x^2 + xy + x + y}{x + y}, & (x, y) \neq (2, -2) \\ 4 & (x, y) = (2, -2) \end{cases}$$

$$\lim_{(x,y) \to (2,-2)} \frac{x^2 + xy + x + y}{(x + y)} = \lim_{(x,y) \to (2,-2)} \frac{(x + y)(x + 1)}{x + y}$$

$$= \lim_{(x,y) \to (2,-2)} (x + 1) = 2 + 1 = 3$$

Here, $\lim_{(x,y) \to (2,-2)} f(x, y) \neq f(2, -2)$

Since $3 \neq 4$

Hence, $f(x, y)$ is discountinuous at the point (2, –2) **Proved.**

EXERCISE 1.2

Test for continuity:

1. $f(x, y) = \begin{cases} \dfrac{xy(x^2 - y^2)}{x^2 + y^2}, & \text{when } x \neq 0, y \neq 0 \\ 0, & \text{when } x = 0, y = 0 \end{cases}$ at origin. **Ans.** Continuous at origin.

2. $f(x, y) = \begin{cases} \dfrac{x^2 - y^2}{x^2 + y^2}, & \text{when } x \neq 0, y \neq 0 \\ 0; & \text{when } x = 0, y = 0 \end{cases}$ at origin. **Ans.** Not continuous at origin.

3. $f(x, y) = \begin{cases} \dfrac{x^3 y^3}{x^3 + y^3}, & \text{when } x \neq 0, y \neq 0 \\ 0, & \text{when } x = 0, y = 0 \end{cases}$ at origin. **Ans.** Not continuous at origin.

4. $f(x, y) = \begin{cases} x^3 + y^3, & \text{when } x \neq 0, y \neq 0 \\ 0, & \text{when } x = 0, y = 0 \end{cases}$ at origin. **Ans.** Continuous at origin.

5. $f(x, y) = \begin{cases} \dfrac{x^2 + 2y}{x + y^2}, & \\ 1 & \text{when } x = 1, y = 2 \end{cases}$ at the point (1, 2). **Ans.** Continuous at (1, 2).

6. Show that the function $f(x, y) = \begin{cases} 2x^2 + y, & (x, y) \neq (1, 2) \\ 0, & (x, y) = (1, 2) \end{cases}$ is discontinuous at (1, 2)

1.7 PARTIAL DERIVATIVES

Let $z = f(x, y)$ be function of two independent variables x and y. If we keep y constant and x varies then z becomes a function of x only. The derivative of z with respect to x, keeping y as constant is called partial derivative of 'z', w.r.t. 'x' and is denoted by symbols.

$$\frac{\partial z}{\partial x}, \frac{\partial f}{\partial x}, f_x(x, y) \text{ etc.}$$

Then
$$\frac{\partial z}{\partial x} = \lim_{\delta x \to 0} \frac{f(x + \delta x, y) - f(x, y)}{\delta x}$$

The process of finding the partial differential coefficient of z w.r.t. 'x' is that of ordinary differentiation, but with the only difference that we treat y as constant.

Similarly, the partial derivative of 'z' w.r.t. 'y' keeping x as constant is denoted by

$$\frac{\partial z}{\partial y}, \frac{\partial f}{\partial y}, f_y(x, y) \text{ etc.}$$

$$\frac{\partial z}{\partial y} = \lim_{\delta y \to 0} \frac{f(x, y + \delta y) - f(x, y)}{\delta y}$$

Notation.

$$\frac{\partial z}{\partial x} = p, \quad \frac{\partial z}{\partial y} = q, \quad \frac{\partial^2 z}{\partial x^2} = r, \quad \frac{\partial^2 z}{\partial x\, \partial y} = s, \quad \frac{\partial^2 z}{\partial y^2} = t$$

Example 12. *If* $z(x + y) = x^2 + y^2$, *show that*

$$\left(\frac{\partial z}{\partial x} - \frac{\partial z}{\partial y}\right)^2 = 4\left(1 - \frac{\partial z}{\partial x} - \frac{\partial z}{\partial y}\right)$$

Solution.
$$z(x + y) = x^2 + y^2, \quad z = \frac{x^2 + y^2}{x + y}$$

$$\frac{\partial z}{\partial x} = \frac{(x + y)\,2x - (x^2 + y^2).1}{(x + y)^2} = \frac{x^2 + 2xy - y^2}{(x + y)^2}$$

$$\frac{\partial z}{\partial y} = \frac{(x+y)(2y)-(x^2+y^2).1}{(x+y)^2} = \frac{-x^2+2xy+y^2}{(x+y)^2}$$

$$\left(\frac{\partial z}{\partial x}-\frac{\partial z}{\partial y}\right) = \frac{x^2+2xy-y^2}{(x+y)^2} - \frac{-x^2+2xy+y^2}{(x+y)^2}$$

$$= \frac{2x^2-2y^2}{(x+y)^2} = \frac{2(x-y)}{x+y}$$

$$\left(\frac{\partial z}{\partial x}-\frac{\partial z}{\partial y}\right)^2 = \frac{4(x-y)^2}{(x+y)^2} \qquad ...(1)$$

Now,
$$4\left(1-\frac{\partial z}{\partial x}-\frac{\partial z}{\partial y}\right) = 4\left[1-\frac{x^2+2xy-y^2}{(x+y)^2}-\frac{-x^2+2xy+y^2}{(x+y)^2}\right]$$

$$= 4.\frac{x^2+2xy+y^2-x^2-2xy+y^2+x^2-2xy-y^2}{(x+y)^2}$$

$$= 4.\frac{x^2-2xy+y^2}{(x+y)^2} = 4.\frac{(x-y)^2}{(x+y)^2} \qquad ...(2)$$

From (1) and (2), we have

$$\left(\frac{\partial z}{\partial x}-\frac{\partial z}{\partial y}\right)^2 = 4\left(1-\frac{\partial z}{\partial x}-\frac{\partial z}{\partial y}\right)$$ **Proved.**

Example 13. *If* $u = \sin^{-1}\left(\frac{x}{y}\right) + \tan^{-1}\left(\frac{y}{x}\right)$, *then find the value of* $x\frac{\partial u}{\partial x} + y\frac{\partial u}{\partial y}$.

Solution. $u = \sin^{-1}\left(\frac{x}{y}\right) + \tan^{-1}\left(\frac{y}{x}\right)$

$$\frac{\partial u}{\partial x} = \frac{1}{\sqrt{1-\left(\frac{x}{y}\right)^2}}.\frac{1}{y} + \frac{1}{1+\left(\frac{y}{x}\right)^2}.\left(-\frac{y}{x^2}\right) = \frac{1}{\sqrt{y^2-x^2}} - \frac{y}{x^2+y^2}$$

$$x\frac{\partial u}{\partial x} = \frac{x}{\sqrt{y^2-x^2}} - \frac{xy}{x^2+y^2} \qquad ...(1)$$

$$\frac{\partial u}{\partial y} = \frac{1}{\sqrt{1-\left(\frac{x}{y}\right)^2}}\left(-\frac{x}{y^2}\right) + \frac{1}{1+\left(\frac{y}{x}\right)^2}.\frac{1}{x} = -\frac{x}{y\sqrt{y^2-x^2}} + \frac{x}{x^2+y^2}$$

$$y.\frac{\partial u}{\partial y} = -\frac{x}{\sqrt{y^2-x^2}} + \frac{xy}{x^2+y^2} \qquad ...(2)$$

On adding (1) and (2), we have

$$x.\frac{\partial u}{\partial x} + y.\frac{\partial u}{\partial y} = 0$$ **Ans.**

Example 14. *Find* $\frac{\partial u}{\partial r}$ *and* $\frac{\partial u}{\partial \theta}$ *if* $u = e^{r\cos\theta}.\cos(r\sin\theta)$

Solution. $u - e^{r\cos\theta}.\cos(r\sin\theta)$

$$\frac{\partial u}{\partial r} = e^{r\cos\theta}.\ [-\sin(r\sin\theta).\sin\theta] + [\cos\theta.e^{r\cos\theta}].\cos(r\sin\theta)$$

(keeping θ as constant)

$$= e^{r\cos\theta}.[-\sin(r\sin\theta).\sin\theta + \cos(r\sin\theta).\cos\theta]$$

$$= e^{r\cos\theta}.\cos(r\sin\theta + \theta)$$ **Ans.**

$$\frac{\partial u}{\partial\theta} = e^{r\cos\theta}.[-\sin(r\sin\theta).r\cos\theta] + [-r\sin\theta.e^{r\cos\theta}].\cos(r\sin\theta)$$

(keeping r as constant)

$$= -r\,e^{r\cos\theta}.\ [\sin(r\sin\theta).\cos\theta + \sin\theta\cos(r\sin\theta)]$$

$$= -r\,e^{r\cos\theta}.\ \sin(r\sin\theta + \theta)$$ **Ans.**

Example 15. *If* $u = (1 - 2xy + y^2)^{-1/2}$ *prove that,* $x\dfrac{\partial u}{\partial x} - y\dfrac{\partial u}{\partial y} = y^2u^3$.

(A.M.I.E.T.E., June 2010)

Solution. $u = (1 - 2xy + y^2)^{-1/2}$ ...(1)

Differentiating (1) partially w.r.t. 'x', we get

$$\frac{\partial u}{\partial x} = -\frac{1}{2}(1 - 2xy + y^2)^{-3/2}\ (-2y)$$

$$x\frac{\partial u}{\partial x} = xy\ (1 - 2xy + y^2)^{-3/2} \qquad ...(2)$$

Differentiating (1) partially w.r.t. 'y', we get

$$\frac{\partial u}{\partial y} = -\frac{1}{2}(1 - 2xy + y^2)^{-3/2}\ (-2x + 2y)$$

$$y\frac{\partial u}{\partial y} = (xy - y^2)\ (1 - 2xy + y^2)^{-3/2} \qquad ...(3)$$

Subtracting (3) from (2), we get

$$x\frac{\partial u}{\partial x} - y\frac{\partial u}{\partial y} = xy\ (1 - 2xy + y^2)^{-3/2} - (xy - y^2)\ (1 - 2xy + y^2)^{-3/2}$$

$$= y^2\ (1 - 2xy + y^2)^{-3/2} = y^2u^3.$$ **Proved.**

Example 16. *If* $z = e^{ax+by}.f(ax - by)$, *prove that*

$$b\frac{\partial z}{\partial x} + a\frac{\partial z}{\partial y} = 2abz.$$

(A.M.I.E.T.E., Summer 2004)

Solution. $z = e^{ax+by}.f(ax - by)$ ...(1)

Differentiating (1) w.r.t. x, we get

$$\frac{\partial z}{\partial x} = a\,e^{ax+by}.f(ax - by) + e^{ax+by}.a\,f'(ax - by)$$

$$b\frac{\partial z}{\partial x} = a\,b\,e^{ax+by}.f(ax - by) + a\,b\,e^{ax+by}.f'(ax - by) \qquad ...(2)$$

Differentiating (1) w.r.t. 'y', we get

$$\frac{\partial z}{\partial y} = b\,e^{ax+by}.f(ax - by) + e^{ax+by}.(-b)\,f'(ax - by)$$

$$a\frac{\partial z}{\partial y} = a\,b\,e^{ax+by}.f(ax - by) - a\,b\,e^{ax+by}.f'(ax - by) \qquad ...(3)$$

On adding (2) and (3), we get

$$b\frac{\partial z}{\partial x} + a\frac{\partial z}{\partial y} = 2ab\,e^{ax+by}\,f(ax - by) \Rightarrow b\frac{\partial z}{\partial x} + a\frac{\partial z}{\partial y} = 2a\,b\,z$$ **Proved.**

1.8 PARTIAL DERIVATIVES OF HIGHER ORDERS

Let $z = f(x, y)$ then $\frac{\partial z}{\partial x}$ and $\frac{\partial z}{\partial y}$ being the functions of x and y can further be differentiated partially with respect to x and y.

Symbolically $\frac{\partial}{\partial x}\left(\frac{\partial z}{\partial x}\right) = \frac{\partial^2 z}{\partial x^2}$ or $\frac{\partial^2 f}{\partial x^2}$ or f_{xx}

$$\frac{\partial}{\partial y}\left(\frac{\partial z}{\partial x}\right) = \frac{\partial^2 z}{\partial y\,\partial x} \quad \text{or} \quad \frac{\partial^2 f}{\partial y\,\partial x} \quad \text{or} \quad f_{yx}$$

$$\frac{\partial}{\partial x}\left(\frac{\partial z}{\partial y}\right) = \frac{\partial^2 z}{\partial x\,\partial y} \quad \text{or} \quad \frac{\partial^2 f}{\partial x\,\partial y} \quad \text{or} \quad f_{xy}$$

Note. $\frac{\partial^2 z}{\partial y \partial x} = \frac{\partial^2 z}{\partial x \partial y}$

Example 17. *Prove that $y = f(x + at) + g(x - at)$ satisfies*

$$\frac{\partial^2 y}{\partial t^2} = a^2\left(\frac{\partial^2 y}{\partial x^2}\right)$$

where f and g are assumed to be at least twice differentiable and a is any constant.

(*U.P., I Semester, Jan. 2011, A.M.I.E.T.E. June 2009, A.M.I.E., Summer* 2000)

Solution. $y = f(x + at) + g(x - at)$ ...(1)

Differentiating (1) w.r.t. 'x' partially, we get

$$\frac{\partial y}{\partial x} = f'(x + at) + g'(x - at) \Rightarrow \frac{\partial^2 y}{\partial x^2} = f''(x + at) + g''(x - at)$$

Differentiating (1) w.r.t. 't' partially, we get

$$\frac{\partial y}{\partial t} = f'(x + at).a + g'(x - at)(-a) \Rightarrow \frac{\partial^2 y}{\partial t^2} = a^2 f''(x + at) + g''(x - at)\, a^2$$

$$= a^2\,[f''(x + at) + g''(x - at)] = a^2 \frac{\partial^2 y}{\partial x^2}$$

Hence $\frac{\partial^2 y}{\partial t^2} = a^2 \frac{\partial^2 y}{\partial x^2}$ **Proved.**

Example 18. *If $z = \tan(y + ax) + (y - ax)^{3/2}$, then show that $\frac{\partial^2 z}{\partial x^2} = a^2 \frac{\partial^2 z}{\partial y^2}$.* (*M.U.* 2009)

Solution. Here, we have

$$z = \tan(y + ax) + (y - ax)^{3/2} \quad ...(1)$$

Differentiating (1) partially w.r.t. x, we get

$$\frac{\partial z}{\partial x} = a \sec^2(y + ax) - \frac{3}{2} a\,(y - ax)^{1/2}$$

$$\frac{\partial^2 z}{\partial x^2} = 2\,a^2 \sec(y + ax) . \sec(y + ax) \tan(y + ax) + \frac{3}{4} a^2\,(y - ax)^{-\frac{1}{2}}$$

$$\frac{\partial^2 z}{\partial x^2} = 2\,a^2 \sec^2(y + ax) \tan(y + ax) + \frac{3}{4}\, a^2\,(y - ax)^{-1/2}$$

$$= a^2\,[2 \sec^2(y + ax).\tan(y + ax) + \frac{3}{4}\,(y - ax)^{-1/2}] \quad ...(2)$$

Differentiating partially (1) w.r.t. y, we get

$$\frac{\partial z}{\partial y} = \sec^2(y + ax) + \frac{3}{2}(y - ax)^{1/2}$$

$$\frac{\partial^2 z}{\partial y^2} = 2\sec(y + ax).\sec(y + ax)\tan(y + ax) + \frac{3}{4}(y - ax)^{-1/2}$$

$$= 2\sec^2(y + ax)\tan(y + ax) + \frac{3}{4}(y - ax)^{-1/2}$$

$$a^2\frac{\partial^2 z}{\partial y^2} = a^2\left[2\sec^2(y + ax)\tan(y + ax) + \frac{3}{4}(y - ax)^{-1/2}\right]$$

$$= \frac{\partial^2 z}{\partial x^2} \qquad \text{[From (2)] \textbf{Proved.}}$$

Example 19. *If* $u = e^{xyz}$, *find the value of* $\dfrac{\partial^3 u}{\partial x\,\partial y\,\partial z}$. *(A.M.I.E. Winter 2000)*

Solution. $u = e^{xyz}$

$$\frac{\partial u}{\partial z} = e^{xyz}(x\,y) \Rightarrow \frac{\partial^2 u}{\partial y\partial z} = e^{xyz}(x) + e^{xyz}(x\,z)(x\,y) = e^{xyz}(x + x^2 y\,z)$$

$$\frac{\partial^3 u}{\partial x\partial y\partial z} = e^{xyz}(1 + 2x\,y\,z) + e^{xyz}(y\,z).(x + x^2 y\,z)$$

$$= e^{xyz}[1 + 2\,x\,y\,z + x\,y\,z + x^2y^2z^2] = e^{xyz}[1 + 3\,x\,y\,z + x^2y^2z^2] \qquad \textbf{Ans.}$$

Example 20. *If* $z = x\log(x + r) - r$, *where* $r^2 = x^2 + y^2$.

Prove that (i) $\dfrac{\partial^2 z}{\partial x^2} + \dfrac{\partial^2 z}{\partial y^2} = \dfrac{1}{x + r}$ *(ii)* $\dfrac{\partial^3 z}{\partial x^3} = -\left(\dfrac{x}{r^3}\right)$ *(M.U. 2009, 2004, 2002)*

Solution. Here, we have

$$z = x\log(x + r) - r \qquad ...(1)$$

where $r^2 = x^2 + y^2$

$$2r\frac{\partial r}{\partial x} = 2x \Rightarrow \frac{\partial r}{\partial x} = \frac{x}{r}$$

Differentiating (1) partially w.r.t. 'x' we get

$$\frac{\partial z}{\partial x} = \left[\frac{x}{x + r}\left(1 + \frac{\partial r}{\partial x}\right) + \log(x + r).1\right] - \frac{\partial r}{\partial x}$$

$$= \left[\frac{x}{x + r}\left(1 + \frac{x}{r}\right) + \log(x + r)\right] - \frac{x}{r} \qquad \left[\frac{\partial r}{\partial x} = \frac{x}{r}\right]$$

$$= \frac{x}{r} + \log(x + r) - \frac{x}{r} = \log(x + r) \qquad ...(2)$$

Again differentiating (2) partially w.r.t. 'x', we get

$$\frac{\partial^2 z}{\partial x^2} = \frac{1}{x + r}\left(1 + \frac{\partial r}{\partial x}\right) = \frac{1}{x + r}\left(1 + \frac{x}{r}\right) = \frac{1}{r} \qquad ...(3)$$

Differentiating (1) partially w.r.t. 'y', we get

$$\frac{\partial z}{\partial y} = x.\frac{1}{x + r}\left(\frac{\partial r}{\partial y}\right) - \frac{\partial r}{\partial y} = \frac{x}{x + r}.\frac{y}{r} - \frac{y}{r} = \frac{y}{r}\left(\frac{x}{x + r} - 1\right) = -\frac{y}{x + r} \quad \left[\because \frac{\partial r}{\partial y} = \frac{y}{r}\right] \; ...(4)$$

Again differentiating (4) partially w.r.t. 'y', we get

$$\frac{\partial^2 z}{\partial y^2} = -\frac{(x+r)(1) - y\left(\frac{\partial r}{\partial y}\right)}{(x+r)^2} = -\frac{(x+r) - y.\frac{y}{r}}{(x+r)^2}$$

$$= -\frac{r^2 + rx - y^2}{r(x+r)^2} = -\frac{x^2 + y^2 + rx - y^2}{r(x+r)^2}$$

$$= -\frac{rx + x^2}{r(x+r)^2} = -\frac{x(r+x)}{r(x+r)^2} = -\frac{x}{r(x+r)} \quad \text{...(5)}$$

(*i*) Adding (3) and (5), we get

$$\frac{\partial^2 z}{\partial x^2} + \frac{\partial^2 z}{\partial y^2} = \frac{1}{r} - \frac{x}{r(x+r)} = \frac{x + r - x}{r(x+r)} = \frac{1}{x+r} \quad \textbf{Proved.}$$

(*ii*) Differentiating (3) partially w.r.t. '*x*', we get

$$\frac{\partial^3 z}{\partial x^3} = -\frac{1}{r^2}\frac{\partial r}{\partial x} = -\frac{1}{r^2}\left(\frac{x}{r}\right) = -\frac{x}{r^3} \quad \textbf{Proved.}$$

Example 21. *Find the value of n so that the equation* $v = r^n (3\cos^2\theta - 1)$ *satisfies the relation.*

$$\frac{\partial}{\partial r}\left(r^2 \frac{\partial v}{\partial r}\right) + \frac{1}{\sin\theta}.\frac{\partial}{\partial\theta}\left(\sin\theta \frac{\partial v}{\partial\theta}\right) = 0$$ *(Nagpur University, Summer 2005, 2001)*

Solution. $\frac{\partial v}{\partial\theta} = r^n(-6\cos\theta\sin\theta)$

$$\sin\theta \frac{\partial v}{\partial\theta} = -6r^n \cos\theta \sin^2\theta$$

$$\Rightarrow \quad \frac{\partial}{\partial\theta}\left(\sin\theta \frac{\partial v}{\partial\theta}\right) = -6r^n(-\sin^3\theta + 2\sin\theta\cos^2\theta)$$

$$\Rightarrow \quad \frac{1}{\sin\theta}\frac{\partial}{\partial\theta}\left(\sin\theta \frac{\partial v}{\partial\theta}\right) = 6r^n(\sin^2\theta - 2\cos^2\theta) = 6r^n(1 - 3\cos^2\theta) \quad \text{...(1)}$$

Again $\frac{\partial v}{\partial r} = n\,r^{n-1}(3\cos^2\theta - 1)$

$$r^2 \frac{\partial v}{\partial r} = n\,r^{n+1}(3\cos^2\theta - 1)$$

$$\frac{\partial}{\partial r}\left(r^2 \frac{\partial v}{\partial r}\right) = n(n+1)\,r^n(3\cos^2\theta - 1) \quad \text{...(2)}$$

Using the relation $\frac{\partial}{\partial r}\left(r^2 \frac{\partial v}{\partial r}\right) + \frac{1}{\sin\theta}\frac{\partial}{\partial\theta}\left(\sin\theta \frac{\partial v}{\partial\theta}\right) = 0$ ...(3)

Putting the values of (1) and (2) in (3), we get

$$n(n+1)\,r^n(3\cos^2\theta - 1) - 6r^n(3\cos^2\theta - 1) = 0$$

$$r^n(3\cos^2\theta - 1)[n(n+1) - 6] = 0 \quad \Rightarrow \quad n(n+1) - 6 = 0$$

$$\Rightarrow \quad n^2 + n - 6 = 0 \quad \Rightarrow \quad (n+3)(n-2) = 0$$

$$\therefore \quad n = 2, -3 \quad \textbf{Ans.}$$

Example 22. *If* $u = (1 - 2xy + y^2)^{-\frac{1}{2}}$, *prove that* $\dfrac{\partial}{\partial x}\left((1 - x^2)\dfrac{\partial u}{\partial x}\right) + \dfrac{\partial}{\partial y}\left(y^2 \dfrac{\partial u}{\partial y}\right) = 0.$

Solution. We have

$$u = (1 - 2xy + y^2)^{-\frac{1}{2}} \qquad ...(1)$$

$$\frac{\partial u}{\partial x} = -\frac{1}{2}(1 - 2xy + y^2)^{-\frac{3}{2}}(-2y) = \frac{y}{(1 - 2xy + y^2)^{\frac{3}{2}}}$$

$$(1 - x^2)\frac{\partial u}{\partial x} = \frac{(1 - x^2)y}{(1 - 2xy + y^2)^{\frac{3}{2}}}$$

$$\frac{\partial}{\partial x}\left((1 - x^2)\frac{\partial u}{\partial x}\right) = \frac{(1 - 2xy + y^2)^{\frac{3}{2}}(-2xy) - (1 - x^2)\, y\left(\frac{3}{2}(1 - 2xy + y^2)^{\frac{1}{2}}(-2y)\right)}{(1 - 2xy + y^2)^3}$$

Cancelling $(1 - 2xy + y^2)^{\frac{1}{2}}$ from numerator and denominator, we have

$$= \frac{(1 - 2xy + y^2)(-2xy) + 3(1 - x^2)y^2}{(1 - 2xy + y^2)^{\frac{5}{2}}} = \frac{-2xy + 4x^2y^2 - 2xy^3 + 3y^2 - 3x^2y^2}{(1 - 2xy + y^2)^{\frac{5}{2}}}$$

$$= \frac{x^2y^2 - 2xy^3 - 2xy + 3y^2}{(1 - 2xy + y^2)^{\frac{5}{2}}} \qquad ...(2)$$

Differentiating (1) partially w.r.t. y, we get

$$\frac{\partial u}{\partial y} = -\frac{1}{2}(1 - 2xy + y^2)^{-\frac{3}{2}}(-2x + 2y) = \frac{x - y}{(1 - 2xy + y^2)^{\frac{3}{2}}}$$

$$y^2\frac{\partial u}{\partial y} = \frac{xy^2 - y^3}{(1 - 2xy + y^2)^{\frac{3}{2}}}$$

$$\frac{\partial}{\partial y}\left(y^2\frac{\partial u}{\partial y}\right) = \frac{(1 - 2xy + y^2)^{\frac{3}{2}}(2xy - 3y^2) - (xy^2 - y^3)\left(\frac{3}{2}(1 - 2xy + y^2)^{\frac{1}{2}}(-2x + 2y)\right)}{(1 - 2xy + y^2)^3}$$

Dividing numerator and denominator by $(1 - 2xy + y^2)^{\frac{1}{2}}$, we get

$$\frac{\partial}{\partial y}\left(y^2\frac{\partial u}{\partial y}\right) = \frac{(1 - 2xy + y^2)(2xy - 3y^2) + (xy^2 - y^3)\,3\,(x - y)}{(1 - 2xy + y^2)^{\frac{5}{2}}}$$

$$= \frac{(2xy - 4x^2y^2 + 2xy^3 - 3y^2 + 6xy^3 - 3y^4) + 3x^2y^2 - 3xy^3 - 3xy^3 + 3y^4}{(1 - 2xy + y^2)^{\frac{5}{2}}}$$

$$= \frac{-x^2y^2 + 2xy^3 + 2xy - 3y^2}{(1 - 2xy + y^2)^{\frac{5}{2}}} \qquad ...(3)$$

On adding (2) and (3), we get

$$\frac{\partial}{\partial x}(1-x^2)\frac{\partial u}{\partial x}+\frac{\partial}{\partial y}\left(y^2\frac{\partial u}{\partial y}\right) = \frac{x^2y^2-2xy^3-2xy+3y^2}{(1-2xy+y^2)^{\frac{5}{2}}}+\frac{-x^2y^2+2xy^3+2xy-3y^2}{(1-2xy+y^2)^{\frac{5}{2}}}$$

$$= 0$$ **Proved.**

Example 23. *If $x^x\, y^y\, z^z = c$, show that at $x = y = z$.*

$$\frac{\partial^2 z}{\partial x\partial y} = -(x \log ex)^{-1}$$ *(Nagpur University, Summer 2004, 2000)*

Solution. Given $x^x\, y^y\, z^z = c$

Taking log on both sides, we get

$x \log x + y \log y + z \log z = \log c$ ...(1)

Differentiating (1) partially w.r.t. to x, assuming z as a function of x and y, we have

$$\left(x.\frac{1}{x}+\log x\right)+0+\left(z.\frac{1}{z}\frac{\partial z}{\partial x}+\log z\frac{\partial z}{\partial x}\right) = 0$$

$$\Rightarrow \quad (1+\log x)+(1+\log z)\frac{\partial z}{\partial x} = 0 \quad \Rightarrow \quad \frac{\partial z}{\partial x} = -\frac{1+\log x}{1+\log z} \qquad \text{...(2)}$$

Similarly differentiating (1) partially w.r.t. to y, we have

$$\frac{\partial z}{\partial y} = -\frac{1+\log y}{1+\log z} \qquad \text{...(3)}$$

Now differentiating (3) partially w.r.t to x, we get

$$\frac{\partial^2 z}{\partial x\,\partial y} = \frac{\partial}{\partial x}\left(\frac{\partial z}{\partial y}\right) = \frac{\partial}{\partial x}\left[-\frac{1+\log y}{1+\log z}\right] = -(1+\log y).\frac{-1}{(1+\log z)^2}.\frac{1}{z}.\frac{\partial z}{\partial x}$$

Putting the value of $\frac{\partial z}{\partial x}$, we get

$$\frac{\partial^2 z}{\partial x\,\partial y} = \frac{(1+\log y)}{z(1+\log z)^2}\left[-\frac{1+\log x}{1+\log z}\right] = -\frac{(1+\log x)(1+\log y)}{z(1+\log z)^3}$$

$\therefore$ at $x = y = z$, we have,

$$\frac{\partial^2 z}{\partial x\,\partial y} = -\frac{(1+\log x)(1+\log x)}{x(1+\log x)^3} = \frac{-1}{x(1+\log x)}$$

$$= \frac{-1}{x(\log e+\log x)} = \frac{-1}{x\log ex} = -(x \log ex)^{-1}$$ **Proved.**

Example 24. *Prove that if $f(x, y) = \frac{1}{\sqrt{y}}e^{-\frac{(x-a)^2}{4y}}$ then $f_{xy}(x, y) = f_{yx}(x, y)$.*

Solution. $$f(x, y) = \frac{1}{\sqrt{y}}e^{-\frac{(x-a)^2}{4y}} \qquad \text{...(1)}$$

Differentiating $f(x, y)$ partially w.r.t. x, we get

$$f_x(x, y) = \frac{1}{\sqrt{y}}.\frac{[-2(x-a)]}{4y}e^{-\frac{(x-a)^2}{4y}} = \frac{-(x-a)}{2y^{3/2}}e^{-\frac{(x-a)^2}{4y}}$$

Differentiating again partially w.r.t. 'y' by product rule, we have

$$f_{yx}(x, y) = \frac{3(x-a)}{4y^{5/2}}.e^{-\frac{(x-a)^2}{4y}} - \frac{(x-a)^3}{8y^{7/2}}e^{-\frac{(x-a)^2}{4y}}$$

$$= \frac{(x-a)}{8y^{7/2}} . e^{-\frac{(x-a)^2}{4y}} . [6y - (x-a)^2] \quad ...(2)$$

Differentiating (1) partially w.r.t. 'y', we have

$$f_y(x, y) = -\frac{1}{2y^{3/2}} . e^{-\frac{(x-a)^2}{4y}} + \frac{(x-a)^2}{4y^{5/2}} e^{-\frac{(x-a)^2}{4y}}$$

Differentiating again partially w.r.t. 'x', we have

$$f_{xy}(x, y) = \frac{(x-a)}{4y^{5/2}} e^{-\frac{(x-a)^2}{4y}} + \frac{(x-a)}{2y^{5/2}} e^{-\frac{(x-a)^2}{4y}} - \frac{(x-a)^3}{8y^{7/2}} e^{-\frac{(x-a)^2}{4y}}$$

$$= \frac{(x-a)}{8y^{7/2}} e^{-\frac{(x-a)^2}{4y}} [2y + 4y - (x-a)^2]$$

$$= \frac{(x-a)}{8y^{7/2}} e^{-\frac{(x-a)^2}{4y}} [6y - (x-a)^2] \quad ...(3)$$

From (2) and (3), we have

$f_{xy}(x, y) = f_{yx}(x, y)$ **Proved.**

Example 25. *If* $u = x^y$, *show that* $\dfrac{\partial^3 u}{\partial x^2 \partial y} = \dfrac{\partial^3 u}{\partial x \partial y \partial x}$.

Solution. $u = x^y$

$\Rightarrow$ $\log u = \log x^y = y \log x$

Differentiating partially, we have

$$\frac{1}{u} . \frac{\partial u}{\partial x} = \frac{y}{x}, \quad \text{and} \quad \frac{1}{u} . \frac{\partial u}{\partial y} = \log x$$

$$\Rightarrow \quad \frac{\partial u}{\partial x} = u \frac{y}{x}, \quad \text{and} \quad \frac{\partial u}{\partial y} = u \log x$$

$$\frac{\partial^2 u}{\partial y \partial x} = \frac{1}{x} . \left[u + y . \frac{\partial u}{\partial y} \right] = \frac{u}{x} + \frac{y}{x} . \frac{\partial u}{\partial y} = \frac{u}{x} + \frac{uy . \log x}{x} \quad \left(\text{As } \frac{\partial u}{\partial y} = u \log x \right)$$

$$\frac{\partial^3 u}{\partial x \partial y \partial x} = -\frac{u}{x^2} + \frac{1}{x} . \frac{\partial u}{\partial x} + y . \left\{ \frac{x . \left(\frac{\partial u}{\partial x} . \log x + \frac{u}{x} \right) - u \log x}{x^2} \right\}$$

$$= -\frac{u}{x^2} + \frac{1}{x} \frac{\partial u}{\partial x} + \frac{y \log x}{x} \frac{\partial u}{\partial x} + \frac{uy}{x^2} - \frac{uy \log x}{x^2}$$

$$= -\frac{u}{x^2} + \frac{uy}{x^2} + \frac{uy^2 \log x}{x^2} + \frac{uy}{x^2} - \frac{uy \log x}{x^2} \quad \left[\because \frac{\partial u}{\partial x} = \frac{uy}{x} \right]$$

$$= -\frac{u}{x^2} + \frac{2uy}{x^2} + \frac{uy^2 \log x}{x^2} - \frac{uy \log x}{x^2} \quad ...(1)$$

$$\frac{\partial u}{\partial y} = u \log x$$

$$\frac{\partial^2 u}{\partial x \partial y} = \frac{u}{x} + \log x . \frac{\partial u}{\partial x} = \frac{u}{x} + \log x . \frac{uy}{x} \quad \left[\frac{\partial u}{\partial x} = \frac{uy}{x} \right]$$

$$\frac{\partial^3 u}{\partial x^2 \partial y} = -\frac{u}{x^2} + \frac{1}{x} . \frac{\partial u}{\partial x} + y . \frac{x . \left(\frac{u}{x} + \log x . \frac{\partial u}{\partial x} \right) - u \log x \ (1)}{x^2}$$

$$= -\frac{u}{x^2} + \frac{uy}{x^2} + \frac{uy}{x^2} + \frac{y \log x}{x}\frac{\partial u}{\partial x} - \frac{uy \log x}{x^2}$$

$$= -\frac{u}{x^2} + \frac{2uy}{x^2} + \frac{y \log x}{x}\frac{uy}{x} - \frac{uy \log x}{x^2}$$

$$= -\frac{u}{x^2} + \frac{2uy}{x^2} + \frac{uy^2 \log x}{x^2} - \frac{uy \log x}{x^2} \qquad ...(2)$$

From (1) and (2), we get $\dfrac{\partial^3 u}{\partial x^2 \partial y} = \dfrac{\partial^3 u}{\partial x \partial y \partial x}$ **Proved.**

Example 26. *If $u = \log(x^3 + y^3 + z^3 - 3xyz)$, show that*

$$\left(\frac{\partial}{\partial x} + \frac{\partial}{\partial y} + \frac{\partial}{\partial z}\right)^2 u = -\frac{9}{(x+y+z)^2}$$ (*U.P. I Semester, winter 2003*)

Solution. $u = \log(x^3 + y^3 + z^3 - 3xyz)$ ...(1)

Differentiating (1) partially w.r.t. 'x', we get

$$\frac{\partial u}{\partial x} = \frac{3x^2 - 3yz}{x^3 + y^3 + z^3 - 3xyz} \qquad ...(2)$$

Similarly,

$$\frac{\partial u}{\partial y} = \frac{3y^2 - 3zx}{x^3 + y^3 + z^3 - 3xyz} \qquad ...(3)$$

$$\frac{\partial u}{\partial z} = \frac{3z^2 - 3xy}{x^3 + y^3 + z^3 - 3xyz} \qquad ...(4)$$

On adding (2), (3) and (4), we get

$$\frac{\partial u}{\partial x} + \frac{\partial u}{\partial y} + \frac{\partial u}{\partial z} = \frac{3(x^2 + y^2 + z^2 - xy - yz - zx)}{x^3 + y^3 + z^3 - 3xyz}$$

$$= \frac{3(x^2 + y^2 + z^2 - xy - yz - zx)}{(x + y + z)(x^2 + y^2 + z^2 - xy - yz - zx)} = \frac{3}{x + y + z}$$

$$\Rightarrow \left(\frac{\partial}{\partial x} + \frac{\partial}{\partial y} + \frac{\partial}{\partial z}\right) u = \frac{3}{x + y + z}$$

$$\left(\frac{\partial}{\partial x} + \frac{\partial}{\partial y} + \frac{\partial}{\partial z}\right)^2 u = \left(\frac{\partial}{\partial x} + \frac{\partial}{\partial y} + \frac{\partial}{\partial z}\right)\frac{3}{x + y + z}$$

$$= \frac{\partial}{\partial x}\frac{3}{x+y+z} + \frac{\partial}{\partial y}\frac{3}{x+y+z} + \frac{\partial}{\partial z}\frac{3}{x+y+z}$$

$$= -3(x+y+z)^{-2} - 3(x+y+z)^{-2} - 3(x+y+z)^{-2}$$

$$= -\frac{9}{(x+y+z)^2}$$ **Proved.**

Example 27. *If $\dfrac{x^2}{a^2 + u} + \dfrac{y^2}{b^2 + u} + \dfrac{z^2}{c^2 + u} = 1$, show that:*

$$\left(\frac{\partial u}{\partial x}\right)^2 + \left(\frac{\partial u}{\partial y}\right)^2 + \left(\frac{\partial u}{\partial z}\right)^2 = 2\left(x\frac{\partial u}{\partial x} + y\frac{\partial u}{\partial y} + z\frac{\partial u}{\partial z}\right)$$

(*U.P. I Semester, Winter 2002*)

Solution. Given $\dfrac{x^2}{a^2+u} + \dfrac{y^2}{b^2+u} + \dfrac{z^2}{c^2+u} = 1,$ ...(1)

where u is a function of x, y and z.

Differentiating (1) partially with respect to x, we get

$$\frac{2x}{u^2+u} - \left[\frac{x^2}{(a^2+u)^2} + \frac{y^2}{(b^2+u)^2} + \frac{z^2}{(c^2+u)^2}\right]\frac{\partial u}{\partial x} = 0$$

$$\Rightarrow \qquad \frac{\partial u}{\partial x} = \frac{\dfrac{2x}{(a^2+u)}}{\left[\dfrac{x^2}{(a^2+u)^2} + \dfrac{y^2}{(b^2+u)^2} + \dfrac{z^2}{(c^2+u)^2}\right]} \qquad \text{...(2)}$$

Similarly,

$$\frac{\partial u}{\partial y} = \frac{\dfrac{2y}{(b^2+u)}}{\left[\dfrac{x^2}{(a^2+u)^2} + \dfrac{y^2}{(b^2+u)^2} + \dfrac{z^2}{(c^2+u)^2}\right]} \qquad \text{...(3)}$$

$$\frac{\partial u}{\partial z} = \frac{\dfrac{2z}{(c^2+u)}}{\left[\dfrac{x^2}{(a^2+u)^2} + \dfrac{y^2}{(b^2+u)^2} + \dfrac{z^2}{(c^2+u)^2}\right]} \qquad \text{...(4)}$$

Squaring and adding (2), (3) and (4), we get

$$\therefore \qquad \left(\frac{\partial u}{\partial x}\right)^2 + \left(\frac{\partial u}{\partial y}\right)^2 + \left(\frac{\partial u}{\partial z}\right)^2 = \frac{4\left[\dfrac{x^2}{(a^2+u)^2} + \dfrac{y^2}{(b^2+u)^2} + \dfrac{z^2}{(c^2+u)^2}\right]}{\left[\dfrac{x^2}{(a^2+u)^2} + \dfrac{y^2}{(b^2+u)^2} + \dfrac{z^2}{(c^2+u)^2}\right]^2}$$

$$= \frac{4}{\left[\dfrac{x^2}{(a^2+u)^2} + \dfrac{y^2}{(b^2+u)^2} + \dfrac{z^2}{(c^2+u)^2}\right]} \qquad \text{...(5)}$$

Multiplying (1) by x, (2) by y and (3) by z and then on adding, we get

$$\left[x\frac{\partial u}{\partial x} + y\frac{\partial u}{\partial y} + z\frac{\partial u}{\partial z}\right] = \frac{2\left[\dfrac{x^2}{a^2+u} + \dfrac{y^2}{b^2+u} + \dfrac{z^2}{c^2+u}\right]}{\left[\dfrac{x^2}{(a^2+u)^2} + \dfrac{y^2}{(b^2+u)^2} + \dfrac{z^2}{(c^2+u)^2}\right]}$$

$$= \frac{2}{\left[\dfrac{x^2}{(a^2+u)^2} + \dfrac{y^2}{(b^2+u)^2} + \dfrac{z^2}{(c^2+u)^2}\right]} \qquad \text{...(6) [Using (1)]}$$

From (5) and (6), we have

$$\left(\frac{\partial u}{\partial x}\right)^2 + \left(\frac{\partial u}{\partial y}\right)^2 + \left(\frac{\partial u}{\partial z}\right)^2 = 2\left[x\frac{\partial u}{\partial x} + y\frac{\partial u}{\partial y} + z\frac{\partial u}{\partial z}\right]$$ **Proved.**

1.9 WHICH VARIABLE IS TO BE TREATED AS CONSTANT

Let $x = r\cos\theta, \qquad y = r\sin\theta$

To find $\dfrac{\partial r}{\partial x}$, we need a relation between r and x.

$$r = x \sec \theta \qquad ...(1)$$

Differentiating (1) w.r.t. 'x' keeping θ as constant

$$\frac{\partial r}{\partial x} = \sec \theta. \qquad ...(2)$$

Here, we have $$r^2 = x^2 + y^2 \qquad ...(3)$$

Differentiating (3) w.r.t. 'x' keeping y as constant.

$$2r\frac{\partial r}{\partial x} = 2x \quad \text{or} \quad \frac{\partial r}{\partial x} = \frac{x}{r} = \cos \theta \qquad ...(4)$$

From (2), $\frac{\partial r}{\partial x} = \sec \theta$ and from (4), $\frac{\partial r}{\partial x} = \cos \theta$. These two values of $\frac{\partial r}{\partial x}$ make confusion.

To avoid the confusion we use the following notations.

Notation. (*i*) $\left(\frac{\partial r}{\partial x}\right)_\theta$ means the partial derivative of r with respect to x, keeping θ as constant.

From (2), $\left(\frac{\partial r}{\partial x}\right)_\theta = \sec \theta$

(*ii*) $\left(\frac{\partial r}{\partial x}\right)_y$ means the partial derivative of r with respect to x keeping y as constant.

From (4), $\left(\frac{\partial r}{\partial x}\right)_y = \cos \theta$

(*iii*) When no indication is given regarding the variables to be treated as constant

$$\frac{\partial}{\partial x} \text{ means } \left(\frac{\partial}{\partial x}\right)_y, \quad \frac{\partial}{\partial y} \text{ means } \left(\frac{\partial}{\partial y}\right)_x.$$

$$\frac{\partial}{\partial r} \text{ means } \left(\frac{\partial}{\partial r}\right)_\theta, \quad \frac{\partial}{\partial \theta} \text{ means } \left(\frac{\partial}{\partial \theta}\right)_r.$$

Example 28. *If* $x = r \cos \theta$, $y = r \sin \theta$, *find*

(*i*) $\left(\frac{\partial x}{\partial r}\right)_\theta$ (*ii*) $\left(\frac{\partial y}{\partial \theta}\right)_r$ (*iii*) $\left(\frac{\partial r}{\partial x}\right)_y$ (*iv*) $\left(\frac{\partial \theta}{\partial y}\right)_x$

Solution. (*i*) $\left(\frac{\partial x}{\partial r}\right)_\theta$ means the partial derivative of x with respect to r, keeping θ as constant.

$$x = r \cos \theta \Rightarrow \left(\frac{\partial x}{\partial r}\right)_\theta = \cos \theta$$

(*ii*) $\left(\frac{\partial y}{\partial \theta}\right)_r$ means the partial derivative of y with respect to θ, treating r as constant.

$$y = r \sin \theta \Rightarrow \left(\frac{\partial y}{\partial \theta}\right)_r = r \cos \theta$$

(*iii*) $\left(\frac{\partial r}{\partial x}\right)_y$ means the partial derivative of r with respect to x, treating y as constant.

We have to express r as a function of x and y.

$$r = \sqrt{x^2 + y^2} \qquad \text{(From the given equations)}$$

$$\left(\frac{\partial r}{\partial x}\right)_y = \frac{1}{2}\frac{1}{\sqrt{x^2 + y^2}}.2x = \frac{x}{\sqrt{x^2 + y^2}}$$

(*iv*) Before finding $\left(\dfrac{\partial \theta}{\partial y}\right)_x$ we have to express θ in terms of x and y.

$$\theta = \tan^{-1}\frac{y}{x} \qquad \text{(From the given equations)}$$

$$\left(\frac{\partial \theta}{\partial y}\right)_x = \frac{1}{1+\dfrac{y^2}{x^2}}\cdot\frac{1}{x} = \frac{x}{x^2+y^2} \qquad \textbf{Ans.}$$

EXERCISE 1.3

1. If $z^3 - 3yz - 3x = 0$, show that $z\dfrac{\partial z}{\partial x} = \dfrac{\partial z}{\partial y}$ and $z\left[\dfrac{\partial^2 z}{\partial x \partial y} + \left(\dfrac{\partial z}{\partial x}\right)^2\right] = \dfrac{\partial^2 z}{\partial y^2}$.

2. If $z(z^2 + 3x) + 3y = 0$, prove that $\dfrac{\partial^2 z}{\partial x^2} + \dfrac{\partial^2 z}{\partial y^2} = \dfrac{2z(x-1)}{(z^2+x)^3}$.

3. If $z = \log(e^x + e^y)$, show that $rt - s^2 = 0$.

4. If $f(x, y) = x^3y - xy^3$, find $\left[\dfrac{1}{\dfrac{\partial f}{\partial x}} + \dfrac{1}{\dfrac{\partial f}{\partial y}}\right]_{\substack{x=1\\y=2}}$ **Ans.** $-\dfrac{13}{22}$

5. If $\dfrac{1}{u} = \sqrt{x^2+y^2+z^2}$, show that $\dfrac{\partial^2 u}{\partial x^2} + \dfrac{\partial^2 u}{\partial y^2} + \dfrac{\partial^2 u}{\partial z^2} = 0$ az **Ans.** ???

6. Show that the function $u = \text{arc}\tan(y/x)$ satisfies the Laplace equation $\dfrac{\partial^2 u}{\partial x^2} + \dfrac{\partial^2 u}{\partial y^2} = 0$.

7. If $z = y\,f(x^2 - y^2)$ show that $y\dfrac{\partial z}{\partial x} + x\dfrac{\partial z}{\partial y} = \dfrac{xz}{y}$.

8. Show that $\dfrac{\partial^2 z}{\partial x^2} - 2\dfrac{\partial^2 z}{\partial x \partial y} + \dfrac{\partial^2 z}{\partial y^2} = 0$, where $z = x\,.\,f(x+y) + y\,.\,g(x+y)$.

9. If $u(x, y, z) = \log(\tan x + \tan y + \tan z)$, prove that

$\sin 2x\dfrac{\partial u}{\partial x} + \sin 2y\dfrac{\partial u}{\partial y} + \sin 2z\dfrac{\partial u}{\partial z} = 2$ (*U.P. I Semester, Dec. 2006*)

10. If $u = \log(x^2 + y^2) + \tan^{-1}\left(\dfrac{y}{x}\right)$. Show that $\dfrac{\partial^2 u}{\partial x^2} + \dfrac{\partial^2 u}{\partial y^2} = 0$.

11. If $u(x, y, z) = \dfrac{1}{x^2+y^2+z^2}$, find the value of $\dfrac{\partial^2 u}{\partial x^2} + \dfrac{\partial^2 u}{\partial y^2} + \dfrac{\partial^2 u}{\partial z^2}$. **Ans.** $\dfrac{2}{(x^2+y^2+z^2)^2}$

12. If $x = e^{r\cos\theta}\cos(r\sin\theta)$ and $y = e^{r\cos\theta}\sin(r\sin\theta)$

Prove that $\dfrac{\partial x}{\partial r} = \dfrac{1}{r}\dfrac{\partial y}{\partial \theta}$, $\dfrac{\partial y}{\partial r} = -\dfrac{1}{r}\dfrac{\partial x}{\partial \theta}$

Hence deduce that $\dfrac{\partial^2 x}{\partial r^2} + \dfrac{1}{r}\dfrac{\partial x}{\partial r} + \dfrac{1}{r}\dfrac{\partial^2 x}{\partial \theta^2} = 0$

13. If $x = r\cos\theta$, $y = r\sin\theta$, prove that

(*a*) $\dfrac{\partial r}{\partial x} = \dfrac{\partial x}{\partial r}$, $r.\dfrac{\partial \theta}{\partial x} = \dfrac{1}{r}.\dfrac{\partial x}{\partial \theta}$ (*b*) $\dfrac{\partial^2 \theta}{\partial x^2} + \dfrac{\partial^2 \theta}{\partial y^2} = 0$ (*c*) $\dfrac{\partial^2 r}{\partial x^2} + \dfrac{\partial^2 r}{\partial y^2} = \dfrac{1}{r}\left[\left(\dfrac{\partial r}{\partial x}\right)^2 + \left(\dfrac{\partial r}{\partial y}\right)^2\right]$

14. If $z = x^y + y^x$, verify that $\dfrac{\partial^2 z}{\partial x \partial y} = \dfrac{\partial^2 z}{\partial y \partial x}$

15. If $u = f(ax^2 + 2h\,x\,y + by^2)$ and $v = \varphi(ax^2 + 2hx\,y + by^2)$ show that $\dfrac{\partial}{\partial y}\left(u\dfrac{\partial v}{\partial x}\right) = \dfrac{\partial}{\partial x}\left(u\dfrac{\partial v}{\partial y}\right)$.

16. If $u = r^m$, where $r^2 = x^2 + y^2 + z^2$, find the value of $\dfrac{\partial^2 u}{\partial x^2} + \dfrac{\partial^2 u}{\partial y^2} + \dfrac{\partial^2 u}{\partial z^2}$. **Ans.** $m(m+1)r^{m-2}$

17. If $x = \frac{r}{2}(e^{\theta} + e^{-\theta})$, $\quad y = \frac{r}{2}(e^{\theta} - e^{-\theta})$ $\quad$ prove that $\frac{\partial x}{\partial r} = \frac{\partial r}{\partial x}$.

18. If $u(x, t) = a\, e^{-gx} \sin(nt - gx)$, satisfies the equation $\frac{\partial u}{\partial t} = a^2 \frac{\partial^2 u}{\partial x^2}$, show that $ag = \sqrt{\frac{n}{2}}$.

19. If $u = f(x, y)$, $x = r\cos\theta$, $y = r\sin\theta$, then show that

$$\left(\frac{\partial u}{\partial x}\right)^2 + \left(\frac{\partial u}{\partial y}\right)^2 = \left(\frac{\partial u}{\partial r}\right)^2 + \frac{1}{r^2}\left(\frac{\partial u}{\partial \theta}\right)^2$$

20. If $z = (x + y) + (x + y) + \left(\frac{y}{x}\right)$, then prove that $x\left(\frac{\partial^2 z}{\partial x^2} - \frac{\partial^2 z}{\partial y\, \partial x}\right) = y\left[\frac{\partial^2 z}{\partial y^2} - \frac{\partial^2 z}{\partial x\, \partial y}\right]$

(A.M.I.E.T.E., Dec. 2010)

21. Show that the function $f(x, y) = \begin{cases} (x + y)\sin\left(\frac{1}{x + y}\right), & x + y \neq 0 \\ 0, & x + y = 0 \end{cases}$

is continuous at (0, 0) but its partial derivatives of first order do not exist at (0, 0).

(A.M.I.E.T.E., June 2010, Dec. 2007)

Choose the correct objective:

22. If $u = x^2 + y^2$ then the value of $\frac{\partial^2 u}{\partial x \partial y}$ is equal to

(*a*) 0 $\quad$ (*b*) 2 $\quad$ (*c*) $2x + 2y$ $\quad$ (*d*) $y\, x^{y-1}$ *(A.M.I.E.T.E. Dec. 2008)* **Ans.** (*a*)

23. If $u = \log\left(\frac{x^2}{y}\right)$, then the value of $x\frac{\partial u}{\partial x} + y\frac{\partial u}{\partial y}$ *is*

(*a*) $2u$ $\quad$ (*b*) u $\quad$ (*c*) 0 $\quad$ (*d*) 1

(A.M.I.E.T.E. Dec. 2007) **Ans.** (*d*)

24. If $u = x^y$ then the value of $\frac{\partial u}{\partial y}$ is equal to

(*a*) 0 $\quad$ (*b*) $x^y \log(x)$ $\quad$ (*c*) xy^{x-1} $\quad$ (*d*) yx^{y-1}

(A.M.I.E.T.E. Dec. 2007) **Ans.** (*b*)

1.10 HOMOGENEOUS FUNCTION

A function $f(x, y)$ is said to be homogeneous function in which the power of each term is the same.

A function $f(x, y)$ is a homogeneous function of order n, if the degree of each of its terms in x and y is equal to n. Thus

$$a_0\, x^n + a_1 x^{n-1} y + a_2 x^{n-2}\, y^2 + \ldots + a_{n-1}\, xy^{n-1} + a_n\, y^n \qquad \ldots(1)$$

is a homogeneous function of order n.

The polynomial function (1) which can be written as

$$x^n\left[a_0 + a_1\left(\frac{y}{x}\right) + a_2\left(\frac{y}{x}\right)^2 + \ldots + a_{n-1}\left(\frac{y}{x}\right)^{n-1} + a_n\left(\frac{y}{x}\right)^n\right] = x^n\, \varphi\left(\frac{y}{x}\right) \qquad \ldots(2)$$

(*i*) The function $x^3\left[1 + \frac{y}{x} + 3\left(\frac{y}{x}\right)^2 + 5\left(\frac{y}{x}\right)^3\right]$

is a homogeneous function of order 3.

(*ii*) $$\frac{\sqrt{x} + \sqrt{y}}{x^2 + y^2} = \frac{\sqrt{x}\left[1 + \sqrt{\frac{y}{x}}\right]}{x^2\left[1 + \left(\frac{y}{x}\right)^2\right]} = x^{-3/2}.\frac{1 + \sqrt{\frac{y}{x}}}{1 + \left(\frac{y}{x}\right)^2}$$

is a homogeneous function of order $-$ 3/2.

(*iii*) $\sin^{-1}\dfrac{\sqrt{x}+\sqrt{y}}{x^2+y^2}$ is not a homogeneous function as it cannot be written in the form of $x^n f\left(\dfrac{y}{x}\right)$ so that its degree may be pronounced. It is a function of homogeneous expression.

1.11 EULER'S THEOREM ON HOMOGENEOUS FUNCTION

(*A.M.I.E.T.E, June 2009, U.P. I Semester, Dec. 2006*)

Statement. If z is a homogeneous function of x, y of order n, then

$$x.\frac{\partial z}{\partial x}+y.\frac{\partial z}{\partial y} = n\,z$$

Proof. Since z is a homogeneous function of x, y of order n.

$\therefore$ z can be written in the form

$$z = x^n.f\left(\frac{y}{x}\right) \quad ..(1)$$

Differentiating (1) partially w.r.t. 'x', we have

$$\frac{\partial z}{\partial x} = nx^{n-1}.f\left(\frac{y}{x}\right)+x^n.f'\left(\frac{y}{x}\right)\left(-\frac{y}{x^2}\right)$$

$$= nx^{n-1}.f\left(\frac{y}{x}\right)-x^{n-2}y.f'\left(\frac{y}{x}\right)$$

Multiplying both sides by x, we have

$$x\frac{\partial z}{\partial x} = n\,x^n.f\left(\frac{y}{x}\right)-x^{n-1}y.f'\left(\frac{y}{x}\right) \quad ...(2)$$

Differentiating (1) partially w.r.t. 'y', we have

$$\frac{\partial z}{\partial y} = x^n f'\left(\frac{y}{x}\right).\frac{1}{x}$$

Multiplying both sides by y, we get

$$y.\frac{\partial z}{\partial y} = x^{n-1}y.f'\left(\frac{y}{x}\right) \quad ...(3)$$

Adding (2) and (3), we have

$$x.\frac{\partial z}{\partial x}+y.\frac{\partial z}{\partial y} = n.x^n f\left(\frac{y}{x}\right) \Rightarrow x.\frac{\partial z}{\partial x}+y.\frac{\partial z}{\partial y} = n\,z$$

Proved.

Note. If u is a homogeneous function of x, y, z of degree n, then $x\dfrac{\partial u}{\partial x}+y\dfrac{\partial u}{\partial y}+z\dfrac{\partial u}{\partial z} = nu$

Example 29. *Verify Euler's theorem for* $z = \dfrac{x^{1/3}+y^{1/3}}{x^{1/2}+y^{1/2}}$. (U.P. Ist Semester, Dec. 2009)

Solution. Here, we have

$$z = \frac{x^{1/3}+y^{1/3}}{x^{1/2}+y^{1/2}} \quad ..(1)$$

$$= \frac{x^{1/3}\left[1+\left(\frac{y}{x}\right)^{1/3}\right]}{x^{1/2}\left[1+\left(\frac{y}{x}\right)^{1/2}\right]} = x^{-\frac{1}{6}}\phi\left(\frac{y}{x}\right)$$

Thus z is homogeneous function of degree $-\frac{1}{6}$.

By Euler's theorem $x\frac{\partial z}{\partial x}+y\frac{\partial z}{\partial y}=-\frac{1}{6}z$ ...(2)

Differentiating (1) w.r.t. 'x', we get

$$\frac{\partial z}{\partial x}=\frac{\left(x^{\frac{1}{2}}+y^{\frac{1}{2}}\right)\left(\frac{1}{3}x^{-\frac{2}{3}}\right)-\left(x^{\frac{1}{3}}+y^{\frac{1}{3}}\right)\left(\frac{1}{2}x^{-\frac{1}{2}}\right)}{\left(x^{\frac{1}{2}}+y^{\frac{1}{2}}\right)^2}=\frac{\frac{1}{3}x^{\frac{-1}{6}}+\frac{1}{3}x^{-\frac{2}{3}}y^{\frac{1}{2}}-\frac{1}{2}x^{-\frac{1}{6}}-\frac{1}{2}x^{-\frac{1}{2}}y^{\frac{1}{3}}}{\left(x^{\frac{1}{2}}+y^{\frac{1}{2}}\right)^2}$$

$$x\frac{\partial z}{\partial x}=\frac{\frac{1}{3}x^{\frac{5}{6}}+\frac{1}{3}x^{\frac{1}{3}}y^{\frac{1}{2}}-\frac{1}{2}x^{\frac{5}{6}}-\frac{1}{2}x^{\frac{1}{2}}y^{\frac{1}{3}}}{\left(x^{\frac{1}{2}}+y^{\frac{1}{2}}\right)^2} \quad ...(3)$$

$$\frac{\partial z}{\partial y}=\frac{\left(x^{\frac{1}{2}}+y^{\frac{1}{2}}\right)\left(\frac{1}{3}y^{\frac{-2}{3}}\right)-\left(x^{\frac{1}{3}}+y^{\frac{1}{3}}\right)\left(\frac{1}{2}y^{\frac{-1}{2}}\right)}{\left(x^{\frac{1}{2}}+y^{\frac{1}{2}}\right)^2}=\frac{\frac{1}{3}x^{\frac{1}{2}}y^{\frac{-2}{3}}+\frac{1}{3}y^{-\frac{1}{6}}-\frac{1}{2}x^{\frac{1}{3}}y^{-\frac{1}{2}}-\frac{1}{2}y^{\frac{-1}{6}}}{\left(x^{\frac{1}{2}}+y^{\frac{1}{2}}\right)^2}$$

$$y\frac{\partial z}{\partial y}=\frac{\frac{1}{3}x^{\frac{1}{2}}y^{\frac{1}{3}}+\frac{1}{3}y^{\frac{5}{6}}-\frac{1}{2}x^{\frac{1}{3}}y^{\frac{1}{2}}-\frac{1}{2}y^{\frac{5}{6}}}{\left(x^{\frac{1}{2}}+y^{\frac{1}{2}}\right)^2} \quad ...(4)$$

Adding (3) and (4), we get

$$x\frac{\partial z}{\partial x}+y\frac{\partial z}{\partial y}=\frac{\frac{1}{3}x^{\frac{5}{6}}+\frac{1}{3}x^{\frac{1}{3}}y^{\frac{1}{2}}-\frac{1}{2}x^{\frac{5}{6}}-\frac{1}{2}x^{\frac{1}{2}}y^{\frac{1}{3}}+\frac{1}{3}x^{\frac{1}{2}}y^{\frac{1}{3}}+\frac{1}{3}y^{\frac{5}{6}}-\frac{1}{2}x^{\frac{1}{3}}y^{\frac{1}{2}}-\frac{1}{2}y^{\frac{5}{6}}}{\left(x^{\frac{1}{2}}+y^{\frac{1}{2}}\right)^2}$$

$$=\frac{-\frac{1}{6}\left[x^{\frac{5}{6}}+y^{\frac{5}{6}}+x^{\frac{1}{3}}y^{\frac{1}{2}}+x^{\frac{1}{2}}y^{\frac{1}{3}}\right]}{\left(x^{\frac{1}{2}}+y^{\frac{1}{2}}\right)^2}=\frac{-\frac{1}{6}\left[x^{\frac{1}{2}}\left(x^{\frac{1}{3}}+y^{\frac{1}{3}}\right)+y^{\frac{1}{2}}\left(x^{\frac{1}{3}}+y^{\frac{1}{3}}\right)\right]}{\left(x^{\frac{1}{2}}+y^{\frac{1}{2}}\right)^2}$$

$$=\frac{-\frac{1}{6}\left(x^{\frac{1}{2}}+y^{\frac{1}{2}}\right)\left(x^{\frac{1}{3}}+y^{\frac{1}{3}}\right)}{\left(x^{\frac{1}{2}}+y^{\frac{1}{2}}\right)^2}=-\frac{1}{6}\frac{x^{\frac{1}{3}}+y^{\frac{1}{3}}}{x^{\frac{1}{2}}+y^{\frac{1}{2}}}$$

$$x\frac{\partial z}{\partial x}+y\frac{\partial z}{\partial y} = -\frac{1}{6}z \qquad ...(5)$$

From (2) and (5), Euler's theorem is verified. **Verified.**

Example 30. *If* $f(x, y) = \frac{1}{x^2}+\frac{1}{xy}+\frac{\log x - \log y}{x^2+y^2}$, *prove that*

$$x\frac{\partial f}{\partial x}+y\frac{\partial f}{\partial y}+2f = 0.$$

(*A.M.I.E. Summer 2004*)

Solution.

$$f(x, y) = \frac{1}{x^2}+\frac{1}{xy}+\frac{\log x - \log y}{x^2+y^2}$$

$$= \frac{1}{x^2}\left(\frac{y}{x}\right)^0+\frac{1}{x^2}\frac{1}{\left(\frac{y}{x}\right)}-\frac{1}{x^2}\frac{\log\frac{y}{x}}{\left[1+\left(\frac{y}{x}\right)^2\right]}$$

$f(x, y)$ is a homogeneous function of degree – 2.
By Euler's Theorem

$$x\frac{\partial f}{\partial x}+y\frac{\partial f}{\partial y} = -2f \Rightarrow x\frac{\partial f}{\partial x}+y\frac{\partial f}{\partial y}+2f = 0$$

Proved.

Example 31. *Verify Euler's Theorem for* $u = x^2\tan^{-1}\left(\frac{y}{x}\right)-y^2\tan^{-1}\left(\frac{x}{y}\right)$ *and also prove that:*

$$\frac{\partial^2 u}{\partial x\partial y} = \frac{x^2-y^2}{x^2+y^2}.$$

(*M.U. 2008*)

Solution. We have,

$$u = x^2\tan^{-1}\left(\frac{y}{x}\right)-y^2\tan^{-1}\left(\frac{x}{y}\right) \qquad ...(1)$$

Here, u is a homogeneous function of degree two.
By Euler's Theorem, we get

$$x\frac{\partial u}{\partial x}+y\frac{\partial u}{\partial y} = 2u \qquad ...(2)$$

Verification

Differentiating (1) partially w.r.t. x, we get

$$\frac{\partial u}{\partial x} = 2x.\tan^{-1}\left(\frac{y}{x}\right)+x^2.\frac{1}{1+\frac{y^2}{x^2}}\left(-\frac{y}{x^2}\right)\quad y^2\frac{1}{1+\frac{x^2}{y^2}}\left(\frac{1}{y}\right)$$

$$= 2x\tan^{-1}\left(\frac{y}{x}\right)-\frac{x^2y}{x^2+y^2}-\frac{y^3}{(x^2+y^2)}$$

$$= 2x.\tan^{-1}\left(\frac{y}{x}\right)-y\left(\frac{x^2+y^2}{x^2+y^2}\right) = 2x\tan^{-1}\left(\frac{y}{x}\right)-y \qquad ...(3)$$

Differentiating (1) partially w.r.t. 'y', we get

$$\frac{\partial u}{\partial y} = x^2.\frac{1}{1+\frac{y^2}{x^2}}\left(\frac{1}{x}\right)-2y\tan^{-1}\left(\frac{x}{y}\right)-y^2.\frac{1}{1+\frac{x^2}{y^2}}\left(-\frac{x}{y^2}\right)$$

$$= \frac{x^3}{x^2+y^2}-2y\tan^{-1}\left(\frac{x}{y}\right)+\frac{xy^2}{x^2+y^2}$$

$$= -2y \tan^{-1}\left(\frac{x}{y}\right) + x\left(\frac{x^2+y^2}{x^2+y^2}\right) = -2y \tan^{-1}\frac{x}{y} + x \qquad ..(4)$$

Multiplying (3) by x and (4) by y and then adding, we get

$$x\frac{\partial u}{\partial x} + y\frac{\partial u}{\partial y} = 2x^2 \tan^{-1}\left(\frac{y}{x}\right) - xy - 2y^2 \tan^{-1}\left(\frac{x}{y}\right) + xy$$

$$= 2x^2 \tan^{-1}\left(\frac{y}{x}\right) - 2y^2 \tan^{-1}\left(\frac{x}{y}\right) = 2u \qquad ...(5)$$

From (2) and (5), we observe that Euler's theorem is verified.

Now, differentiating (4) again pertially w.r.t. x, we get

$$\frac{\partial^2 u}{\partial x\, \partial y} = -2y \frac{1}{1+\frac{x^2}{y^2}}\left(\frac{1}{y}\right) + 1 = \frac{-2y^2}{x^2+y^2} + 1 = \frac{x^2-y^2}{x^2+y^2} \qquad \textbf{Proved.}$$

Example 32. *Verify Euler's theorem for the function* $u = \sin^{-1}\frac{x}{y} + \tan^{-1}\frac{y}{x}$.

(*A.M.I.E.T.E. Dec., 2010, 2007; U.P. Ist Semester, Dec. 2006*)

Solution.
$$u = x^0 \sin^{-1}\frac{x}{y} + x^0 \tan^{-1}\frac{y}{x} \qquad ...(1)$$

u is a homogeneous function of degree zero so by Euler's Theorem

$$x\frac{\partial u}{\partial x} + y\frac{\partial u}{\partial y} = 0 \qquad ...(2)$$

Verification

Let us find out the value of L.H.S. of (2) by actual differentiation

Differentiating (1) w.r.t. 'x', we get

$$\frac{\partial u}{\partial x} = \frac{1}{\sqrt{1-\frac{x^2}{y^2}}}\left(\frac{1}{y}\right) + \frac{1}{1+\frac{y^2}{x^2}}\left(-\frac{y}{x^2}\right) = \frac{1}{\sqrt{y^2-x^2}} - \frac{y}{x^2+y^2}$$

$$x\frac{\partial u}{\partial x} = \frac{x}{\sqrt{y^2-x^2}} - \frac{xy}{x^2+y^2} \qquad ...(3)$$

Differentiating (1) w.r.t. 'y', we get

$$\frac{\partial u}{\partial y} = \frac{1}{\sqrt{1-\frac{x^2}{y^2}}}\left(-\frac{x}{y^2}\right) + \frac{1}{1+\frac{y^2}{x^2}}\frac{1}{x} = \frac{-x}{y\sqrt{y^2-x^2}} + \frac{x}{x^2+y^2}$$

$$y\frac{\partial u}{\partial y} = \frac{-x}{\sqrt{y^2-x^2}} + \frac{xy}{x^2+y^2} \qquad ...(4)$$

Adding (3) and (4), we get

$$x\frac{\partial u}{\partial x} + y\frac{\partial u}{\partial y} = 0 \qquad ...(5)$$

From (2) and (5) the theorem is verified. **Verified.**

Example 33. *If z be a homogeneous function of degree n, show that*

(*i*) $x.\frac{\partial^2 z}{\partial x^2} + y.\frac{\partial^2 z}{\partial x \partial y} = (n-1)\frac{\partial z}{\partial x}$ (*ii*) $x.\frac{\partial^2 z}{\partial x \partial y} + y.\frac{\partial^2 z}{\partial y^2} = (n-1)\frac{\partial z}{\partial y}$

(*iii*) $x^2.\frac{\partial^2 z}{\partial x^2} + 2xy.\frac{\partial^2 z}{\partial x \partial y} + y^2.\frac{\partial^2 z}{\partial y^2} = n(n-1)z.$

(*A.M.I.E.T.E. June, 2009; Uttarakhand Ist Semester, Dec. 2006*)

Solution. By Euler's Theorem $x.\frac{\partial z}{\partial x} + y\frac{\partial z}{\partial y} = n\,z$ ...(1)

Differentiating (1), partially w.r.t. 'x', we get $\frac{\partial z}{\partial x} + x.\frac{\partial^2 z}{\partial x^2} + y\frac{\partial^2 z}{\partial x \partial y} = n\frac{\partial z}{\partial x}$

$$\Rightarrow \quad x.\frac{\partial^2 z}{\partial x^2} + y\frac{\partial^2 z}{\partial x \partial y} = (n-1)\frac{\partial z}{\partial x}$$ **Proved** (*i*) ...(2)

Differentiating (1), partially w.r.t. 'y', we have $x.\frac{\partial^2 z}{\partial x \partial y} + \frac{\partial z}{\partial y} + y\frac{\partial^2 z}{\partial y^2} = n\frac{\partial z}{\partial y}$

$$\Rightarrow \quad x.\frac{\partial^2 z}{\partial x \partial y} + y.\frac{\partial^2 z}{\partial y^2} = (n-1)\frac{\partial z}{\partial y}$$ **Proved** (*ii*) ...(3)

Multiplying (2) by x, we have $x^2.\frac{\partial^2 z}{\partial x^2} + xy.\frac{\partial^2 z}{\partial x \partial y} = (n-1)\,x\frac{\partial z}{\partial x}$ ...(4)

Multiplying (3) by y, we have $xy.\frac{\partial^2 z}{\partial y \partial x} + y^2.\frac{\partial^2 z}{\partial y^2} = (n-1)\,y\frac{\partial z}{\partial y}$ ...(5)

Adding (4) and (5), we get

$$x^2.\frac{\partial^2 z}{\partial x^2} + 2xy.\frac{\partial^2 z}{\partial x \partial y} + y^2.\frac{\partial^2 z}{\partial y^2} = (n-1)\left(x\frac{\partial z}{\partial x} + y\frac{\partial z}{\partial y}\right)$$

$$= (n-1)\,n\,z$$ [From (1)]

$$= n(n-1)\,z$$ **Proved** (*iii*)

EXERCISE 1.4

1. Verify Euler's theorem in case

(*i*) $f(x, y) = ax^2 + 2hxy + by^2$ (*ii*) $u = (\sqrt{x} + \sqrt{y})\,(x^n + y^n)$,

2. If $v = \frac{x^3 y^3}{x^3 + y^3}$, show that $x.\frac{\partial v}{\partial x} + y.\frac{\partial v}{\partial y} = 3v$.

3. If $z = (x^2 + y^2)/\sqrt{(x+y)}$, prove that $x\frac{\partial z}{\partial x} + y.\frac{\partial z}{\partial x} = \frac{3}{2}z$.

4. If $f(x, y) = x^4 y^2 \sin^{-1}\frac{y}{x}$, then find the value of $x\frac{\partial f}{\partial x} + y\frac{\partial f}{\partial y}$.

(*A.M.I.E.T.E., Winter 2001*) **Ans.** $6\,f(x, y)$

5. State and prove Euler's theorem, and verify for $u = \frac{x}{y} + \frac{y}{z} + \frac{z}{x}$ (*A.M.I.E., Summer 2000*)

6. If $u = \frac{x^2 y^2 z^2}{x^2 + y^2 + z^2} + \cos\frac{xy + yz}{x^2 + y^2 + z^2}$, show that $x\frac{\partial u}{\partial x} + y\frac{\partial u}{\partial y} + z\frac{\partial u}{\partial z} = \frac{4x^2 y^2 z^2}{x^2 + y^2 + z^2}$

7. Verify Euler theorem on homogeneous function when $f(x, y, z) = 3x^2yz + 5xy^2z + 4z^4$

I. Deduction from Euler's theorem

If z is a homogeneous function x, y of degree n and $z = f(u)$, then

$$x\frac{\partial u}{\partial x} + y\frac{\partial u}{\partial y} = n\frac{f(u)}{f'(u)}$$

(*Nagpur University, Winter 2003*)

Proof. Since z is a homogeneous function of x, y of degree n, we have, by Euler's theorem,

$$x\frac{\partial z}{\partial x}+y\frac{\partial z}{\partial y}=nz \qquad \text{... (1)}$$

Now $z=f(u)$, given

$$\therefore \quad \frac{\partial z}{\partial x}=f'(u)\frac{\partial u}{\partial x} \quad \text{and} \quad \frac{\partial z}{\partial y}=f'(u)\frac{\partial u}{\partial y}$$

Substituting in (1), we get

$$x.f'(u)\frac{\partial u}{\partial x}+y\,f'(u)\frac{\partial u}{\partial y}=nf(u) \quad \Rightarrow \quad x\frac{\partial u}{\partial x}+y\frac{\partial u}{\partial y}=n\frac{f(u)}{f'(u)}$$

Note. If $v=f(u)$ where v is a homogeneous function in x, y, z of degree n, then

$$x\frac{\partial u}{\partial x}+y\frac{\partial u}{\partial y}+z\frac{\partial u}{\partial z}=\frac{nf(u)}{f'(u)}$$

Example 34. *If* $u=\log_e\left(\dfrac{x^4+y^4}{x+y}\right)$, *show that* $x\dfrac{\partial u}{\partial x}+y\dfrac{\partial u}{\partial y}=3.$

(*Nagpur University, Summer 2008, Uttarakhand, I Semester 2008*)

Solution. We have, $\quad u=\log_e\left(\dfrac{x^4+y^4}{x+y}\right)$

Here, u is not a homogeneous function but if

$$z=e^u=\frac{x^4+y^4}{x+y}=\frac{x^4\left[1+\left(\dfrac{y}{x}\right)^4\right]}{x\left[1+\left(\dfrac{y}{x}\right)\right]}=x^3\,\varphi\left(\frac{y}{x}\right)$$

Then z is a homogeneous function of degree 3.
By Euler's Deduction formula I

$$x\frac{\partial u}{\partial x}+y\frac{\partial u}{\partial y}=n\frac{f(u)}{f'(u)}=3\frac{e^u}{e^u}=3$$

Proved.

Example 35. *If* $u=\sin^{-1}\left(\dfrac{x^3+y^3+z^3}{ax+by+cz}\right)$, *prove that*

$x\dfrac{\partial u}{\partial x}+y\dfrac{\partial u}{\partial y}+z\dfrac{\partial u}{\partial z}=2\tan u.$ (*U.P., I Semester Winter, 2002, A.M.I.E., Summer 2000*)

Solution. Here, u is not a homogeneous function, we however write

$$v=\sin u=\frac{x^3+y^3+z^3}{ax+by+cz}=\frac{x^3\left[1+\left(\dfrac{y}{x}\right)^3+\left(\dfrac{z}{x}\right)^3\right]}{x\left[a+b\left(\dfrac{y}{x}\right)+c\left(\dfrac{z}{x}\right)\right]}$$

$$=\frac{x^2\left[1+\left(\dfrac{y}{x}\right)^3+\left(\dfrac{z}{x}\right)^3\right]}{a+b\left(\dfrac{y}{x}\right)+c\left(\dfrac{z}{x}\right)}=x^2\,\phi\left(\frac{y}{x},\frac{z}{x}\right)$$

So that v is a homogeneous function of x, y, z of order 2.
By Euler's Theorem,

$$x\frac{\partial u}{\partial x}+y\frac{\partial u}{\partial y}+z\frac{\partial u}{\partial z}=\frac{2f(u)}{f'(u)}=2.\frac{\sin u}{\cos u}$$

$$=2\tan u$$

Proved.

Example 36. *State Euler's Theorem of differential calculus. Hence verify the theorem for the function* $u = \log \dfrac{x^2 + y^2}{xy}$. *(Uttarakhand, I Semester 2008)*

Solution. We have, $\quad u = \log \dfrac{x^2 + y^2}{xy}$

By Euler's Theorem

Here, u is not a homogeneous function but if

$$z = e^u = \frac{x^2 + y^2}{xy} = \frac{x^2\left[1 + \left(\frac{y}{x}\right)^2\right]}{x^2\left(\frac{y}{x}\right)} = \frac{1 + \left(\frac{y}{x}\right)^2}{\left(\frac{y}{x}\right)} = \phi\left(\frac{y}{x}\right)$$

Here z is homogeneous function of degree zero.

By Euler's deduction formula I.

$$x\frac{\partial u}{\partial x} + y\frac{\partial u}{\partial y} = n\frac{f(u)}{f'(u)} = \frac{0\ e^u}{e^u} = 0 \qquad \text{...(1)} \quad \textbf{Ans.}$$

By Direct Differentiation

Now, $\quad u = \log \dfrac{x^2 + y^2}{xy}$

$$\Rightarrow \frac{\partial u}{\partial x} = \frac{1}{\left(\frac{x^2 + y^2}{xy}\right)} \cdot \frac{\partial}{\partial x}\left(\frac{x^2 + y^2}{xy}\right) = \frac{xy}{x^2 + y^2} \frac{\left(xy\frac{\partial}{\partial x}(x^2 + y^2) - (x^2 + y^2)\frac{\partial}{\partial x}(xy)\right)}{(xy)^2}$$

$$= \frac{xy}{x^2 + y^2} \cdot \left[\frac{x\,y\,(2x) - (x^2 + y^2)\,y}{(xy)^2}\right]$$

$$= \frac{1}{xy(x^2 + y^2)}(2x^2y - x^2y - y^3) = \frac{x^2y - y^3}{xy(x^2 + y^2)}$$

$$= \frac{x^2 - y^2}{x(x^2 + y^2)} \quad \Rightarrow \quad x\frac{\partial u}{\partial x} = \frac{x^2 - y^2}{x^2 + y^2} \qquad \text{...(2)}$$

and $$\frac{\partial u}{\partial y} = \frac{1}{\left(\frac{x^2 + y^2}{xy}\right)} \frac{\partial}{\partial y}\left(\frac{x^2 + y^2}{xy}\right) = \frac{xy}{x^2 + y^2} \frac{\left(xy\frac{\partial}{\partial y}(x^2 + y^2) - (x^2 + y^2)\frac{\partial}{\partial y}(xy)\right)}{(xy)^2}$$

$$= \frac{xy}{x^2 + y^2}\left[\frac{x\,y\,(2y) - (x^2 + y^2)\,x}{(xy)^2}\right] = \frac{1}{xy(x^2 + y^2)}(2xy^2 - x^3 - xy^2)$$

$$= \frac{xy^2 - x^3}{xy(x^2 + y^2)} = \frac{y^2 - x^2}{y(x^2 + y^2)} \quad \Rightarrow \quad y\frac{\partial u}{\partial y} = \frac{y^2 - x^2}{x^2 + y^2} \qquad \text{...(3)}$$

Adding (2) and (3), we get

$$x\frac{\partial u}{\partial x} + y\frac{\partial u}{\partial y} = \frac{x^2 - y^2}{x^2 + y^2} + \frac{y^2 - x^2}{x^2 + y^2} = 0 \qquad \text{...(4)}$$

From (1) and (4), we have the same result.

Hence, Euler's Theorem is verified. **Proved.**

II. Deduction: Prove that

$$x^2\frac{\partial^2 u}{\partial x^2}+2xy\frac{\partial^2 u}{\partial x\partial y}+y^2\frac{\partial^2 u}{\partial y^2} = g(u)\,[g'(u)-1]$$ (*Nagpur University, Winter 2003*)

where, $$g(u) = n\frac{f(u)}{f'(u)}$$

Proof. By Euler deduction formula I

$$x\frac{\partial u}{\partial x}+y\frac{\partial u}{\partial y} = n.\frac{f(u)}{f'(u)} \qquad \text{Given } n\frac{f(u)}{f'(u)} = g(u)$$

$$x\frac{\partial u}{\partial x}+y\frac{\partial u}{\partial y} = g(u) \qquad ...(1)$$

Differentiating (1) partially w.r.t. 'x', we have

$$\left(x\frac{\partial^2 u}{\partial x^2}+\frac{\partial u}{\partial x}.1\right)+y\frac{\partial^2 u}{\partial x\partial y} = g'(u)\frac{\partial u}{\partial x}$$

$$\Rightarrow \qquad x\frac{\partial^2 u}{\partial x^2}+y.\frac{\partial^2 u}{\partial x\partial y} = [g'(u)-1]\frac{\partial u}{\partial x} \qquad ...(2)$$

Similarly, on differentiating (1) partially w.r.t. 'y', we have

$$y\frac{\partial^2 u}{\partial y^2}+x.\frac{\partial^2 u}{\partial y\partial x} = [g'(u)-1]\frac{\partial u}{\partial y} \qquad ...(3)$$

Multiplying (2) by x, (3) by y and adding, we get

$$x^2\frac{\partial^2 u}{\partial x^2}+2xy\frac{\partial^2 u}{\partial x\partial y}+y^2\frac{\partial^2 u}{\partial y^2} = [g'(u)-1]\left[x\frac{\partial u}{\partial x}+y\frac{\partial u}{\partial y}\right]$$

$$= [g'(u)-1]g(u) \qquad \text{[From (1)]}$$

$$= g(u)\,[g'(u)-1]$$ **Proved.**

Example 37. *If* $u = \log\left(\frac{x^2+y^2}{\sqrt{x}+\sqrt{y}}\right)$, *find the value of*

(*i*) $x\frac{\partial u}{\partial x}+y\frac{\partial u}{\partial y}$ (*ii*) $x^2\frac{\partial^2 u}{\partial x^2}+2xy\frac{\partial^2 u}{\partial x\partial y}+y^2\frac{\partial^2 u}{\partial y^2}$ (*Nagpur University, Winter 2002*)

Solution. We have, $z = e^u = \frac{x^2+y^2}{\sqrt{x}+\sqrt{y}} = x^{\frac{3}{2}} f\left(\frac{y}{x}\right)$

is homogeneous function of degree $\frac{3}{2}$. $[f(u) = e^u]$

(*i*) By deduction I, we have

$$x\frac{\partial u}{\partial x}+y\frac{\partial u}{\partial y} = n\frac{f(u)}{f'(u)} = \frac{3}{2}\frac{e^u}{e^u} = \frac{3}{2}$$

$$g(u) = \frac{3}{2},\; g'(u) = 0$$

(*ii*) By deduction II, we have

$$x^2\frac{\partial^2 u}{\partial x^2}+2xy\frac{\partial^2 u}{\partial x\partial y}+y^2\frac{\partial^2 u}{\partial y^2} = g(u)\,[g'(u)-1] = \frac{3}{2}[0-1] = -\frac{3}{2}$$ **Ans.**

Example 38. *If* $u = \sin^{-1}\left[\dfrac{x+y}{\sqrt{x}+\sqrt{y}}\right]$

Prove that $x^2\dfrac{\partial^2 u}{\partial x^2} + 2xy\dfrac{\partial^2 u}{\partial x \partial y} + y^2\dfrac{\partial^2 u}{\partial y^2} = \dfrac{-\sin u \cos 2u}{4\cos^3 u}$.

Solution. We have, $u = \sin^{-1}\dfrac{x+y}{\sqrt{x}+\sqrt{y}}$

Let $$z = \sin u = \frac{x+y}{\sqrt{x}+\sqrt{y}} = \frac{x\left[1+\dfrac{y}{x}\right]}{\sqrt{x}\left[1+\sqrt{\dfrac{y}{x}}\right]} = x^{1/2}\,\varphi\left(\frac{y}{x}\right)$$

$$z = f(u) = \sin u$$

z is a homogeneous function of degree $\dfrac{1}{2}$.

By Euler's deduction I

$$x\frac{\partial u}{\partial x} + y\frac{\partial u}{\partial y} = n\frac{f(u)}{f'(u)}$$

$$x\frac{\partial u}{\partial x} + y\frac{\partial u}{\partial y} = \frac{1}{2}\frac{\sin u}{\cos u}$$

$$x\frac{\partial u}{\partial x} + y\frac{\partial u}{\partial y} = \frac{1}{2}\tan u$$

Let $$g(u) = \frac{1}{2}\tan u,\quad g'(u) = \frac{1}{2}\sec^2 u$$

By Euler's deduction II

$$x^2\frac{\partial^2 u}{\partial x^2} + 2xy\frac{\partial^2 u}{\partial x \partial y} + y^2\frac{\partial^2 u}{\partial y^2} = g(u)\,[g'(u)-1] = \frac{1}{2}\tan u\left(\frac{1}{2}\sec^2 u - 1\right)$$

$$= \frac{1}{4}\frac{\sin u}{\cos u}\left(\frac{1}{\cos^2 u} - 2\right) = \frac{1}{4}\frac{\sin u}{\cos^3 u}(1-2\cos^2 u) = \frac{-\sin u\cos 2u}{4\cos^3 u}$$ **Proved.**

Example 39. *If* $u = \tan^{-1}\left(\dfrac{x^3+y^3}{\sqrt{x}+\sqrt{y}}\right)$ *find the value of*

(*i*) $x\dfrac{\partial u}{\partial x} + y\dfrac{\partial u}{\partial y}$ (*ii*) $x^2\dfrac{\partial^2 u}{\partial x^2} + 2xy\dfrac{\partial^2 u}{\partial x \partial y} + y^2\dfrac{\partial^2 u}{\partial y^2}$ (*Nagpur University, Winter 2004*)

Solution. We have, $u = \tan^{-1}\left(\dfrac{x^3+y^3}{\sqrt{x}+\sqrt{y}}\right)$

u is not homogeneous function but we see that

$$z = \tan u = \frac{x^3+y^3}{\sqrt{x}+\sqrt{y}} = \frac{x^3\left(1+\dfrac{y^3}{x^3}\right)}{\sqrt{x}\left(1+\dfrac{\sqrt{y}}{\sqrt{x}}\right)} = x^{\frac{5}{2}}\,\phi\left(\frac{y}{x}\right)$$

$$f(u) = \tan u$$

Now, z is a homogeneous function of x and y of degree $\dfrac{5}{2}$.

(i) By deduction I, we have

$$x\frac{\partial u}{\partial x}+y\frac{\partial u}{\partial y} = n\frac{f(u)}{f'(u)}$$

$$x\frac{\partial u}{\partial x}+y\frac{\partial u}{\partial y} = \frac{5}{2}\frac{\tan u}{\sec^2 u} = \frac{5}{2}\frac{\sin u}{\cos u}\frac{\cos^2 u}{1} = \frac{5}{2}\sin u\cos u = \frac{5}{4}\sin 2u$$

$$g(u) = \frac{5}{4}\sin 2u\,,\quad g'(u) = \frac{5}{2}\cos 2u$$

(ii) By Euler's Deduction II, we have

$$x^2\frac{\partial^2 u}{\partial x^2}+2xy\frac{\partial^2 u}{\partial x\partial y}+y^2\frac{\partial^2 u}{\partial y^2} = g(u)[g'(u)-1]$$

$$= \frac{5}{4}\sin 2u\left[\frac{5}{2}\cos 2u-1\right] = \frac{25}{8}\sin 2u\cos 2u - \frac{5}{4}\sin 2u$$

$$= \frac{25}{16}\sin 4u - \frac{5}{4}\sin 2u$$ **Ans.**

Example 40. *Using Euler's theorem show that if $u = \tan^{-1}(x^2+2y^2)$ then*

(i) $x\frac{\partial u}{\partial x}+y\frac{\partial u}{\partial y} = \sin 2u$ *(ii)* $x^2\frac{\partial^2 u}{\partial x^2}+2xy\frac{\partial^2 u}{\partial x\partial y}+y^2\frac{\partial^2 u}{\partial y^2} = 2\sin u\cos 3u$

Explain whether we can apply Euler's theorem for the function

$$u = f(x, y) = \frac{x^2+y^2+1}{x+1}$$ (*Gujarat, I Semester, Jan. 2009*)

Solution.

(*i*) Here, we have

$$u = \tan^{-1}(x^2+2y^2)$$

u is not homogeneous function.

Let $$z = \tan u = x^2+2y^2$$

z is a homogeneous function of degree 2.

$$x\frac{\partial z}{\partial x}+y\frac{\partial z}{\partial y} = 2z$$

$$x\frac{\partial z}{\partial u}\cdot\frac{\partial u}{\partial x}+y\frac{\partial z}{\partial u}\cdot\frac{\partial u}{\partial y} = 2\tan u$$

$$\Rightarrow\quad x(\sec^2 u)\cdot\frac{\partial u}{\partial x}+y(\sec^2 u)\frac{\partial u}{\partial y} = 2\tan u$$

$$\Rightarrow\quad x\frac{\partial u}{\partial x}+y\frac{\partial u}{\partial y} = \frac{2\tan u}{\sec^2 u}$$

$$\Rightarrow\quad x\frac{\partial u}{\partial x}+y\frac{\partial u}{\partial y} = 2\sin u\cdot\cos u$$

$$\Rightarrow\quad x\frac{\partial u}{\partial x}+y\frac{\partial u}{\partial y} = \sin 2u$$ **Proved.**

(*ii*) By Euler's Theorem of second order

$$x^2\frac{\partial^2 u}{\partial x^2}+2xy\frac{\partial^2 u}{\partial x\partial y}+y^2\frac{\partial^2 u}{\partial y^2} = g(u)[g'(u)-1]$$ $$\left[\begin{array}{l} g(u) = \sin 2u \\ g'(u) = 2\cos 2u \end{array}\right]$$

$$= \sin 2u\,[2\cos 2u - 1] = 2\sin u\cos u\,[2(2\cos^2 u - 1) - 1]$$
$$= 2\sin u\cos u\,[4\cos^2 u - 3]$$
$$= 2\sin u\,[4\cos^3 u - 3\cos u] = 2\sin u\cos 3u \quad \textbf{Ans.}$$

Here $$u = f(x, y) = \frac{x^2 + y^2 + 1}{x + y}$$

But u is not a homogeneous function because 1 in the numerator is of zero degree and other terms of two degree. So, we cannot apply Euler's theorem. For Euler's Theorem the function should be homogeneous.

Example 41. *If* $u = \dfrac{x^3y^3z^3}{x^3 + y^3 + z^3} + \log\left(\dfrac{xy + yz + zx}{x^2 + y^2 + z^2}\right)$ *then prove that:*

$$x\frac{\partial u}{\partial x} + y\frac{\partial u}{\partial y} + z\frac{\partial u}{\partial z} = 6\frac{x^3y^3z^3}{x^3 + y^3 + z^3}$$ (M.U. 2009)

Solution. Here, we have

$$u = \frac{x^3y^3z^3}{x^3 + y^3 + z^3} + \log\left(\frac{xy + yz + zx}{x^2 + y^2 + z^2}\right) = v + w$$

where $$v = \frac{x^3y^3z^3}{x^3 + y^3 + z^3} \quad \text{and} \quad w = \log\frac{xy + yz + zx}{x^2 + y^2 + z^2}$$

v is a homogeneous function of degree 6.

$$x\frac{\partial v}{\partial x} + y\frac{\partial v}{\partial y} + z\frac{\partial v}{\partial z} = 6v \quad \text{...(1)} \quad \text{(By Euler's Theorem)}$$

$$w = \log\frac{xy + yz + zx}{x^2 + y^2 + z^2}$$

Let $$t = e^w = \frac{xy + yz + zx}{x^2 + y^2 + z^2}$$

(t is a homogeneous function of degree zero.)

$$x\frac{\partial w}{\partial x} + y\frac{\partial w}{\partial y} + z\frac{\partial w}{\partial z} = \frac{n\,f(w)}{f'(w)}$$ (By Euler's Theorem)

$$x\frac{\partial w}{\partial x} + y\frac{\partial w}{\partial y} + z\frac{\partial w}{\partial z} = 0 \quad \text{...(2)}$$

Adding (1) and (2), we get

$$x\left(\frac{\partial v}{\partial x} + \frac{\partial w}{\partial x}\right) + y\left(\frac{\partial v}{\partial y} + \frac{\partial w}{\partial y}\right) + z\left(\frac{\partial v}{\partial z} + \frac{\partial w}{\partial z}\right) = 6v + 0$$

$$x\frac{\partial u}{\partial x} + y\frac{\partial u}{\partial y} + z\frac{\partial u}{\partial z} = 6\frac{x^3y^3z^3}{x^3 + y^3 + z^3} \quad \textbf{Proved.}$$

Example 42. *If* $u = x^2\tan^{-1}\left(\dfrac{y}{x}\right) - y^2\tan^{-1}\left(\dfrac{x}{y}\right), x > 0,\ y > 0$ *then evaluate*

$$x^2\frac{\partial^2 u}{\partial x^2} + 2yx\frac{\partial^2 u}{\partial x\partial y} + y^2\frac{\partial^2 u}{\partial y^2}.$$ (A.M.I.E.T.E. Dec 2008)

Solution. Here, we have

$$u = x^2 \tan^{-1}\left(\frac{y}{x}\right) - y^2 \tan^{-1}\left(\frac{x}{y}\right)$$

Let $u = v - w$

where, $v = x^2 \tan^{-1}\left(\frac{y}{x}\right)$ and $w = y^2 \tan^{-1}\left(\frac{x}{y}\right)$

Since, v is a homogeneous function of x, y of degree 2. Therefore

$$x^2 \frac{\partial^2 v}{\partial x^2} + 2yx \frac{\partial^2 v}{\partial x \partial y} + y^2 \frac{\partial^2 v}{\partial y^2} = 2(2-1)v$$

$$= 2v \qquad ...(1)$$

since, w is a homogeneous function of x, y of degree 2. Therefore

$$x^2 \frac{\partial^2 w}{\partial x^2} + 2yx \frac{\partial^2 w}{\partial x \partial y} + y^2 \frac{\partial^2 w}{\partial y^2} = 2(2-1)w = 2w \qquad ...(2)$$

Subtracting (2) from (1), we get

$$x^2\left(\frac{\partial^2 v}{\partial x^2} - \frac{\partial^2 w}{\partial x^2}\right) + 2yx\left(\frac{\partial^2 v}{\partial x \partial y} - \frac{\partial^2 w}{\partial x \partial y}\right) + y^2\left(\frac{\partial^2 v}{\partial y^2} - \frac{\partial^2 w}{\partial y^2}\right) = 2v - 2w = 2(v - w)$$

$$x^2 \frac{\partial^2 u}{\partial x^2} + 2yx \frac{\partial^2 u}{\partial x \partial y} + y^2 \frac{\partial^2 u}{\partial y^2} = 2u \qquad \textbf{Ans.}$$

Example 43. *If* $z = x^n f\left(\frac{y}{x}\right) + y^{-n} \phi\left(\frac{x}{y}\right)$ *then prove that*

$$x^2 \frac{\partial^2 z}{\partial x^2} + 2xy \frac{\partial^2 z}{\partial x \partial y} + y^2 \frac{\partial^2 z}{\partial y^2} + x\frac{\partial z}{\partial x} + y\frac{\partial z}{\partial y} = n^2 z.$$ *(Nagpur University, Summer 2003)*

Solution. Here, we have $\quad z = x^n f\left(\frac{y}{x}\right) + y^{-n} \phi\left(\frac{x}{y}\right)$

Let $\quad z = u + v \qquad ...(1)$

where, $\quad u = x^n f\left(\frac{y}{x}\right)$ and $v = y^{-n} \phi\left(\frac{x}{y}\right)$

Since u is a homogeneous function of x, y of degree n.

$$x\frac{\partial u}{\partial x} + y\frac{\partial u}{\partial y} = nu \qquad ...(2)$$

and $$x^2 \frac{\partial^2 u}{\partial x^2} + 2xy \frac{\partial^2 u}{\partial x \partial y} + y^2 \frac{\partial^2 u}{\partial y^2} = n(n-1)u \qquad ...(3)$$

As v is a homogeneous function of x, y of degree $-n$.

$$x\frac{\partial v}{\partial x} + y\frac{\partial v}{\partial y} = -nv \qquad ...(4)$$

and $$x^2 \frac{\partial^2 v}{\partial x^2} + 2xy \frac{\partial^2 v}{\partial x \partial y} + y^2 \frac{\partial^2 v}{\partial y^2} = -n(-n-1)v = n(n+1)v \qquad ...(5)$$

On adding (2) and (4), we get

$$x\left(\frac{\partial u}{\partial x} + \frac{\partial v}{\partial x}\right) + y\left(\frac{\partial u}{\partial y} + \frac{\partial v}{\partial y}\right) = nu - nv$$

$$x\frac{\partial z}{\partial x} + y\frac{\partial z}{\partial y} = nu - nv \qquad ...(6) \quad \text{[From (1)]}$$

On adding (3) and (5), we get

$$x^2\left(\frac{\partial^2 u}{\partial x^2}+\frac{\partial^2 v}{\partial x^2}\right)+2xy\left(\frac{\partial^2 u}{\partial x\partial y}+\frac{\partial^2 v}{\partial x\partial y}\right)+y^2\left(\frac{\partial^2 u}{\partial y^2}+\frac{\partial^2 v}{\partial y^2}\right)$$

$$= n(n-1)\,u + n\,(n+1)v$$

$$x^2\frac{\partial^2 z}{\partial x^2}+2xy\frac{\partial^2 z}{\partial x\partial y}+y^2\frac{\partial^2 z}{\partial y^2} = n(n-1)u + n\,(n+1)v \qquad ...(7) \quad \text{[From (1)]}$$

$z = u + v$

$\frac{\partial z}{\partial x} = \frac{\partial u}{\partial x}+\frac{\partial v}{\partial x}$

$\frac{\partial^2 z}{\partial x^2} = \frac{\partial^2 u}{\partial x^2}+\frac{\partial^2 v}{\partial x^2}$

On adding (6) and (7), we have

$$x^2\frac{\partial^2 z}{\partial x^2}+2xy\frac{\partial^2 z}{\partial x\partial y}+y^2\frac{\partial^2 z}{\partial y^2}+x\frac{\partial z}{\partial x}+y\frac{\partial z}{\partial y} = n(n-1)u + n\,(n+1)\,v + nu - nv$$

$$= nu\,(n-1+1) + nv\,(n+1-1)$$

$$= n^2u + n^2v = n^2\,(u+v) = n^2z$$

Proved.

Example 44. *If $f(x, y)$ and $\phi(x, y)$ are homogeneous functions of x, y of degree P and q respectively and $u = f(x, y) + \phi(x, y)$, show that*

$$f(x, y) = \frac{1}{P(P-q)}\left[x^2\frac{\partial^2 u}{\partial x^2}+2xy\frac{\partial^2 u}{\partial x\partial y}+y^2\frac{\partial^2 u}{\partial y^2}\right]-\frac{q-1}{P(P-q)}\left[x\frac{\partial u}{\partial x}+y\frac{\partial u}{\partial y}\right]$$

(A.M.I.E.T.E. Winter 2000)

Solution. Since f and ϕ are homogeneous functions of degree P and q respectively, we have

$$x\frac{\partial f}{\partial x}+y\frac{\partial f}{\partial y} = P.f \qquad ...(1)$$

$$x\frac{\partial \phi}{\partial x}+y\frac{\partial \phi}{\partial y} = q.\phi \qquad ...(2)$$

$u = f(x, y) + \phi(x, y)$

$\frac{\partial u}{\partial x} = \frac{\partial f}{\partial x}+\frac{\partial \phi}{\partial x}$

$\frac{\partial u}{\partial y} = \frac{\partial f}{\partial y}+\frac{\partial \phi}{\partial y}$

On adding (1) and (2), we get

$$x\left[\frac{\partial f}{\partial x}+\frac{\partial \phi}{\partial x}\right]+y\left[\frac{\partial f}{\partial y}+\frac{\partial \phi}{\partial y}\right] = P\,f + q\,\phi$$

i.e.,

$$x\frac{\partial u}{\partial x}+y\frac{\partial u}{\partial y} = P\,f + q\,\phi \qquad ...(3)$$

Also $$x^2\frac{\partial^2 f}{\partial x^2}+2xy\frac{\partial^2 f}{\partial x\partial y}+y^2\frac{\partial^2 f}{\partial y^2} = P\,(P-1)\,f \qquad ...(4)$$

And $$x^2\frac{\partial^2 \phi}{\partial x^2}+2xy\frac{\partial^2 \phi}{\partial x\partial y}+y^2\frac{\partial^2 \phi}{\partial y^2} = q\,(q-1)\,\phi \qquad ...(5)$$

On adding (4) and (5), we obtain

$$x^2\left(\frac{\partial^2 f}{\partial x^2}+\frac{\partial^2 \phi}{\partial x^2}\right)+2xy\left(\frac{\partial^2 f}{\partial x\partial y}+\frac{\partial^2 \phi}{\partial x\partial y}\right)+y^2\left(\frac{\partial^2 f}{\partial y^2}+\frac{\partial^2 \phi}{\partial y^2}\right) = P\,(P-1)\,f + q\,(q-1)\,\phi$$

$$x^2\frac{\partial^2 u}{\partial x^2}+2xy\frac{\partial^2 u}{\partial x\partial y}+y^2\frac{\partial^2 u}{\partial y^2} = P\,(P-1)\,f + q\,(q-1)\,\phi$$

$\because\ u = f + \phi$

$\frac{\partial^2 u}{\partial x^2} = \frac{\partial^2 f}{\partial x^2}+\frac{\partial^2 \phi}{\partial x^2}$

$\frac{\partial^2 u}{\partial y^2} = \frac{\partial^2 f}{\partial y^2}+\frac{\partial^2 \phi}{\partial y^2}$

Dividing by $P\,(P-q)$, we get

$$\Rightarrow \frac{1}{P(P-q)}\left[x^2\frac{\partial^2 u}{\partial x^2}+2xy\frac{\partial^2 u}{\partial x\partial y}+y^2\frac{\partial^2 u}{\partial y^2}\right]$$

$$= \frac{1}{P(P-q)}[P(P-1)f + q(q-1)\phi]$$

Subtracting $\frac{q-1}{P(P-q)}\left(x\frac{\partial u}{\partial x}+y\frac{\partial u}{\partial y}\right)$ from both sides, we get

$$\Rightarrow \quad \frac{1}{P(P-q)}\left[x^2\frac{\partial^2 u}{\partial x^2}+2xy\frac{\partial^2 u}{\partial x\partial y}+y^2\frac{\partial^2 u}{\partial y^2}\right]-\frac{(q-1)}{P(P-q)}\left[x\frac{\partial u}{\partial x}+y\frac{\partial u}{\partial y}\right]$$

$$=\frac{1}{P(P-q)}[P(P-1)f+q(q-1)\phi]-\frac{q-1}{P(P-q)}\left[x\frac{\partial u}{\partial x}+y\frac{\partial u}{\partial y}\right]$$

$$=\frac{1}{P(P-q)}[P(P-1)f+q(q-1)\phi-(q-1)[Pf+q\phi]] \qquad \text{[From (3)]}$$

$$=\frac{1}{P(P-q)}[(P^2-P-Pq+P)f+(q^2-q-q^2+q)\phi]$$

$$=\frac{1}{P(P-q)}[(P^2-Pq)f]=\frac{P(P-q)}{P(P-q)}f=f(x,y) \qquad \textbf{Proved.}$$

EXERCISE 1.5

1. If $u=\sin^{-1}\frac{\sqrt{x}-\sqrt{y}}{\sqrt{x}+\sqrt{y}}$, show that $x\frac{\partial u}{\partial x}+y\frac{\partial u}{\partial y}=0$

2. If $u=\cos^{-1}\left(\frac{x+y}{\sqrt{x}+\sqrt{y}}\right)$, show that $x\frac{\partial u}{\partial x}+y\frac{\partial u}{\partial y}+\frac{1}{2}\cot u=0$ *(U.P. I Sem., Dec. 2009)*

3. If $u=\log\frac{x^2+y^2}{x+y}$, prove that $x.\frac{\partial u}{\partial x}+y.\frac{\partial u}{\partial y}=1$. *(U.P. I Semester, Dec. 2008)*

4. If $x=e^u\tan v$, $y=e^u\sec v$, find the value of

$$\left(x.\frac{\partial u}{\partial x}+y.\frac{\partial u}{\partial y}\right).\left(x.\frac{\partial v}{\partial x}+y.\frac{\partial v}{\partial y}\right).$$ *(A.M.I.E., Summer 2001)* **Ans.** 0

[**Hint:** Eliminate u and apply formula I. Again eliminate v and apply the formula]

5. Given $F(u)=V(x, y, z)$ where V is a homogeneous function of x, y, z of degree n, prove that

$$x\frac{\partial u}{\partial x}+y\frac{\partial u}{\partial y}+z\frac{\partial u}{\partial z}=n\frac{F(u)}{F'(u)}$$

6. If $u=\sin^{-1}\left[\frac{x^{1/4}+y^{1/4}}{x^{1/6}+y^{1/6}}\right]$, prove that

$$x^2\frac{\partial^2 u}{\partial x^2}+2xy\frac{\partial^2 u}{\partial x\partial y}+y^2\frac{\partial^2 u}{\partial y^2}=\frac{1}{144}\tan u\,[\tan^2 u-11].$$

7. If $u=\sec^{-1}\left(\frac{x^3-y^3}{x+y}\right)$, show that $x.\frac{\partial u}{\partial x}+y.\frac{\partial u}{\partial y}=2\cot u$, then evaluate

$x^2\frac{\partial^2 u}{\partial x^2}+2xy\frac{\partial^2 u}{\partial x\partial y}+y^2\frac{\partial^2 u}{\partial y^2}$ *(A.M.I.E.T.E., Winter 2001)* **Ans.** $-2\cot u\,(2\,\text{cosec}^2 u+1)$.

8. Find the value of $x^2\frac{\partial^2 u}{\partial x^2}+2xy\frac{\partial^2 u}{\partial x\partial y}+y^2\frac{\partial^2 u}{\partial y^2}$ if $u=\sin^{-1}(x^3+y^3)^{2/5}$

Ans. $\frac{6}{5}\tan u\left[\frac{6}{5}\sec^2 u-1\right]$

9. If $u = \tan^{-1}\left(\dfrac{x^3 + y^3}{x - y}\right)$, prove that

(*i*) $x \dfrac{\partial u}{\partial x} + y \dfrac{\partial u}{\partial y} = \sin 2u$ (*A.M.I.E., Winter 2001*)

(*ii*) $x^2 \dfrac{\partial^2 u}{\partial x^2} + 2xy \dfrac{\partial^2 u}{\partial x\, \partial y} + y^2 \dfrac{\partial^2 u}{\partial y^2} = 2 \cos 3u \sin u$

(*M.U. 2009, Nagpur University, Summer 2002*)

10. If $u = \tan^{-1} \dfrac{\sqrt{x^3 + y^3}}{\sqrt{x} + \sqrt{y}}$, find the value of $x^2 \dfrac{\partial^2 u}{\partial x^2} + 2xy \dfrac{\partial^2 u}{\partial x \partial y} + y^2 \dfrac{\partial^2 u}{\partial y^2}$

Ans. $-2 \sin^3 u \cos u$

11. If $u = f\left(\dfrac{y}{x}\right) + \sqrt{x^2 + y^2}$, find the value of $x^2 \dfrac{\partial^2 u}{\partial x^2} + 2xy \dfrac{\partial^2 u}{\partial x \partial y} + y^2 \dfrac{\partial^2 u}{\partial y^2}$. **Ans.** 0

12. If $z = xy/(x + y)$, find the value of $x^2 \dfrac{\partial^2 z}{\partial x^2} + 2xy \dfrac{\partial^2 z}{\partial x \partial y} + y^2 \dfrac{\partial^2 z}{\partial y^2}$. **Ans.** 0

13. If $u = x\, \phi\left(\dfrac{Y}{X}\right) + \psi\left(\dfrac{Y}{X}\right)$, prove by Euler theorem on homogeneous function that

$$X^2 \frac{\partial^2 u}{\partial x^2} + 2XY \frac{\partial^2 u}{\partial x \partial y} + Y^2 \frac{\partial^2 u}{\partial y^2} = 0.$$

14. If $u = \sin^{-1}\left[\dfrac{x + 2y + 3z}{\sqrt{x^8 + y^8 + z^8}}\right]$, show that

$x \dfrac{\partial u}{\partial x} + y \dfrac{\partial u}{\partial y} + z \dfrac{\partial u}{\partial z} + 3 \tan u = 0.$ (*U.P. I Semester, Winter 2003*)

Tick (✓) the correct answer:

15. If $u = f\left(\dfrac{x}{y}\right)$ then

(*a*) $x \dfrac{\partial u}{\partial x} - y \dfrac{\partial u}{\partial y} = 0$ (*b*) $x \dfrac{\partial u}{\partial x} + y \dfrac{\partial u}{\partial y} = 0$ (*c*) $x \dfrac{\partial u}{\partial x} + y \dfrac{\partial u}{\partial y} = u$ (*d*) $x \dfrac{\partial u}{\partial x} + y \dfrac{\partial u}{\partial y} = 1$

(*U.P., I Semester, Jan. 2011, AMIETE, June 2010*) **Ans.** (*b*)

16. If $z = \sin^{-1} \dfrac{x^2 + y^2}{x + y}$ then $x \dfrac{\partial z}{\partial x} + y \dfrac{\partial z}{\partial y}$ is

(*a*) z (*b*) 2z (*c*) tan z (*d*) sin z

(*AMIETE, June 2010*) **Ans.** (*c*)

CHAPTER

2

TOTAL DIFFERENTIATION

2.1 INTRODUCTION

In partial differentiation of a function of two or more variables, only one variable varies. But in total differentiation, increments are given in all the variables.

2.2 TOTAL DIFFERENTIAL CO-EFFICIENT

Let $z = f(x, y)$ ...(1)

If δx, δy be the increments in x and y respectively, let δz be the corresponding increment in z.

Then $z + \delta z = f(x + \delta x, y + \delta y)$ ...(2)

Subtracting (1) from (2), we have

$$\delta z = f(x + \delta x, y + \delta y) - f(x, y) \qquad ...(3)$$

Adding and subtracting $f(x, y + \delta y)$ on R.H.S. of (3), we have

$$\delta z = f(x + \delta x, y + \delta y) - f(x, y + \delta y) + f(x, y + \delta y) - f(x, y)$$

$$\delta z = \frac{f(x + \delta x, y + \delta y) - f(x, y + \delta y)}{\delta x}\delta x + \frac{f(x, y+\delta y) - f(x, y)}{\delta y}\delta y$$

On taking limit when $\delta x \to 0$ and $\delta y \to 0$

$$dz = \frac{\partial f}{\partial x}dx + \frac{\partial f}{\partial y}dy \qquad ...(4) \quad \textbf{[Remember]}$$

dz is called as the total differential of z.

COROLLARY 1. Differentitiation of composite function

If $z = f(x, y)$

Where $x = \varphi(t)$, $y = \psi(t)$

Here z is composite function of t.

Dividing (4) by dt, we have $\dfrac{dz}{dt} = \dfrac{\partial z}{\partial x}\dfrac{dx}{dt} + \dfrac{\partial z}{\partial y}\dfrac{dy}{dt}$ ...(5) **[Remember]**

Then $\dfrac{dz}{dt}$ is called the total differential co-efficient of z.

COROLLARY 2. Let $z = f(x, y)$

where $x = \varphi(u, v)$, $y = \psi(u, v)$

Then from (5), we obtain

$$\frac{\partial z}{\partial u} = \frac{\partial f}{\partial x}\cdot\frac{\partial x}{\partial u} + \frac{\partial f}{\partial y}\cdot\frac{\partial y}{\partial u} \qquad ...(6)$$

and

$$\frac{\partial z}{\partial v} = \frac{\partial f}{\partial x}\cdot\frac{\partial x}{\partial v} + \frac{\partial f}{\partial y}\cdot\frac{\partial y}{\partial v} \qquad ...(7)$$

Example 1. *If $u = x^3 + y^3$ where, $x = a \cos t$, $y = b \sin t$, find $\frac{du}{dt}$ and verify the result.*

Solution. We have, $u = x^3 + y^3$

$$x = a \cos t$$

$$y = b \sin t$$

$$\frac{du}{dt} = \frac{\partial u}{\partial x}\frac{dx}{dt} + \frac{\partial u}{\partial y}\frac{dy}{dt} = (3x^2)(-a \sin t) + (3y^2)(b \cos t)$$

$$= -3a^3 \cos^2 t \sin t + 3b^3 \sin^2 t \cos t \qquad \text{...(1)}$$

Verification. $u = x^3 + y^3 = a^3 \cos^3 t + b^3 \sin^3 t$

$$\frac{du}{dt} = -3a^3 \cos^2 t \sin t + 3b^3 \sin^2 t \cos t \qquad \text{...(2)}$$

Results (1) and (2) are the same. **Verified.**

Example 2. *Suppose that $u = f(x, y, z)$ and $x = g_1(t)$, $y = g_2(t)$, $z = g_3(t)$. Then write the chain rule for derivative of u w.r.t. to t.*

Express $\partial w/\partial r$ and $\partial w/\partial s$ in terms of r and s if $w = x + 2y + z^2$

$x = r/s, \quad y = r^2 + \log(s), \quad z = 2r.$ *(Gujarat, I Semester, Jan 2009)*

Solution. $u = f(x, y, z)$

$$x = g_1(t) \qquad y = g_2(t) \qquad z = g_3(t)$$

$$\frac{du}{dt} = \frac{\partial u}{\partial x}\frac{\partial g_1}{\partial t} + \frac{\partial u}{\partial y}\frac{\partial g_2}{\partial t} + \frac{\partial u}{\partial z}\frac{\partial g_3}{\partial t}$$

Again $w = x + 2y + z^2$

$$x = \frac{r}{s} \qquad y = r^2 + \log s \qquad z = 2r$$

$$\frac{\partial w}{\partial r} = \frac{\partial w}{\partial x}\frac{\partial x}{\partial r} + \frac{\partial w}{\partial y}\frac{\partial y}{\partial r} + \frac{\partial w}{\partial z}\frac{\partial z}{\partial r} = (1)\left(\frac{1}{s}\right) + (2)(2r) + (2z)(2)$$

$$= \frac{1}{s} + 4r + 4z$$

$$\frac{\partial w}{\partial s} = \frac{\partial w}{\partial x}\frac{\partial x}{\partial s} + \frac{\partial w}{\partial y}\frac{\partial y}{\partial s} + \frac{\partial w}{\partial z}\frac{\partial z}{\partial s}$$

$$= (1)\left(-\frac{r}{s^2}\right) + (2)\left(\frac{1}{s}\right) + (2z)(0) = -\frac{r}{s^2} + \frac{2}{s}$$ **Ans.**

Example 3. *If $z = f(x, y)$, $x = e^u + e^{-v}$, $y = e^{-u} - e^v$. Prove that :*

$$\frac{\partial z}{\partial u} - \frac{\partial z}{\partial v} = x\frac{\partial z}{\partial x} - y\frac{\partial z}{\partial y}$$ *(M. U. 2009)*

Solution. Here we have

$$x = e^u + e^{-v} \qquad \text{...(1)}$$

$$y = e^{-u} - e^v \qquad \text{...(2)}$$

$$\frac{\partial z}{\partial u} = \frac{\partial z}{\partial x}\frac{\partial x}{\partial u} + \frac{\partial z}{\partial y}\frac{\partial y}{\partial u} = \frac{\partial z}{\partial x}(e^u) + \frac{\partial z}{\partial y}(-e^{-u}) \qquad \text{...(3)}$$

$$\frac{\partial z}{\partial v} = \frac{\partial z}{\partial x}\frac{\partial x}{\partial v} + \frac{\partial z}{\partial y}\frac{\partial y}{\partial v} = \frac{\partial z}{\partial x}(-e^{-v}) + \frac{\partial z}{\partial y}(-e^v) \qquad \text{...(4)}$$

Subtracting (4) from (3), we get

$$\frac{\partial z}{\partial u} - \frac{\partial z}{\partial v} = \frac{\partial z}{\partial x}e^u - \frac{\partial z}{\partial y}e^{-u} + \frac{\partial z}{\partial x}e^{-v} + \frac{\partial z}{\partial y}e^v$$

$$= (e^u + e^{-v})\frac{\partial z}{\partial x} - (e^{-u} - e^v)\frac{\partial z}{\partial y}$$

$$= x\frac{\partial z}{\partial x} - y\frac{\partial z}{\partial y}$$ **Proved.**

Example 4. *If $z = f(x, y)$ where $x = e^u \cos v$ and $y = e^u \sin v$, show that*

(*i*) $y\dfrac{\partial z}{\partial u} + x\dfrac{\partial z}{\partial v} = e^{2u}\dfrac{\partial z}{\partial y}$. (*M.U. 2009; Nagpur Univesity 2002*)

(*ii*) $\left(\dfrac{\partial z}{\partial x}\right)^2 + \left(\dfrac{\partial z}{\partial y}\right)^2 = e^{-2u}\left[\left(\dfrac{\partial z}{\partial u}\right)^2 + \left(\dfrac{\partial z}{\partial v}\right)^2\right]$ (*M.U. 2009*)

Solution. (*i*) We have,

$$x = e^u \cos v, \qquad y = e^u \sin v$$

$$\frac{\partial z}{\partial u} = \frac{\partial z}{\partial x}\frac{\partial x}{\partial u} + \frac{\partial z}{\partial y}\frac{\partial y}{\partial u}$$

$$= \frac{\partial z}{\partial x} e^u \cos v + \frac{\partial z}{\partial y} e^u \sin v = x\frac{\partial z}{\partial x} + y\frac{\partial z}{\partial y}$$

$$y\frac{\partial z}{\partial u} = xy\frac{\partial z}{\partial x} + y^2\frac{\partial z}{\partial y} \quad ...(1)$$

And $\dfrac{\partial z}{\partial v} = \dfrac{\partial z}{\partial x}\dfrac{\partial x}{\partial v} + \dfrac{\partial z}{\partial y}\dfrac{\partial y}{\partial v} = \dfrac{\partial z}{\partial x}(-e^u \sin v) + \dfrac{\partial z}{\partial y}(e^u \cos v) = -y\dfrac{\partial z}{\partial x} + x\dfrac{\partial z}{\partial y}$

$$x\frac{\partial z}{\partial v} = -xy\frac{\partial z}{\partial x} + x^2\frac{\partial z}{\partial y} \quad ...(2)$$

On adding (1) and (2), we get

$$y\frac{\partial z}{\partial u} + x\frac{\partial z}{\partial v} = (x^2 + y^2)\frac{\partial z}{\partial y} = (e^{2u}\cos^2 v + e^{2u}\sin^2 v)\frac{\partial z}{\partial y}$$

$$= e^{2u}(\cos^2 v + \sin^2 v)\frac{\partial z}{\partial y} = e^{2u}\frac{\partial z}{\partial y}$$ **Proved.**

(*ii*) $$\frac{\partial z}{\partial u} = \frac{\partial z}{\partial x}\frac{\partial x}{\partial u} + \frac{\partial z}{\partial y}\frac{\partial y}{\partial u}$$

$$= \frac{\partial z}{\partial x}(e^u \cos v) + \frac{\partial z}{\partial y} e^u \sin v$$

$$e^{-u}\frac{\partial z}{\partial u} = \frac{\partial z}{\partial x}\cos v + \frac{\partial z}{\partial y}\sin v$$

$x = e^u \cos v$

$\Rightarrow \dfrac{\partial x}{\partial u} = e^u \cos v$ and $\dfrac{\partial x}{\partial v} = -e^u \sin v$

$y = e^u \sin v$

$\Rightarrow \dfrac{\partial y}{\partial u} = e^u \sin v$ and $\dfrac{\partial y}{\partial v} = e^u \cos v$

On squaring we get

$$e^{-2u}\left(\frac{\partial z}{\partial u}\right)^2 = \left(\frac{\partial z}{\partial x}\right)^2 \cos^2 v + \left(\frac{\partial z}{\partial y}\right)^2 \sin^2 v + 2\left(\frac{\partial z}{\partial x}\right)\left(\frac{\partial z}{\partial y}\right)\sin v \cos v \quad ...(3)$$

Again $$\frac{\partial z}{\partial v} = \frac{\partial z}{\partial x}\frac{\partial x}{\partial v} + \frac{\partial z}{\partial y}\frac{\partial y}{\partial v}$$

$$= \frac{\partial z}{\partial x}(-e^u \sin v) + \frac{\partial z}{\partial y}(e^u \cos v)$$

$$e^{-u}\left(\frac{\partial z}{\partial v}\right) = -\frac{\partial z}{\partial x}\sin v + \frac{\partial z}{\partial y}\cos v$$

On squaring, we get

$$e^{-2u}\left(\frac{\partial z}{\partial v}\right)^2 = \left(\frac{\partial z}{\partial x}\right)^2 \sin^2 v + \left(\frac{\partial z}{\partial y}\right)^2 \cos^2 v - 2\left(\frac{\partial z}{\partial x}\right)\left(\frac{\partial z}{\partial y}\right)\sin v \cos v \quad ...(4)$$

On adding (3) and (4), we get

$$e^{-2u}\left[\left(\frac{\partial z}{\partial u}\right)^2 + \left(\frac{\partial z}{\partial v}\right)^2\right] = \left(\frac{\partial z}{\partial x}\right)^2 (\sin^2 v + \cos^2 v) + \left(\frac{\partial z}{\partial y}\right)^2 (\sin^2 v + \cos^2 v)$$

$$= \left(\frac{\partial z}{\partial x}\right)^2 + \left(\frac{\partial z}{\partial y}\right)^2$$

Proved.

Example 5. *If $w = f(x,y)$, $x = r\cos\theta$, $y = r\sin\theta$, show that*

$$\left(\frac{\partial w}{\partial r}\right)^2 + \frac{1}{r^2}\left(\frac{\partial w}{\partial \theta}\right)^2 = \left(\frac{\partial f}{\partial x}\right)^2 + \left(\frac{\partial f}{\partial y}\right)^2$$

Solution.

$$x = r\cos\theta, \qquad y = r\sin\theta$$

$$\frac{\partial x}{\partial r} = \cos\theta, \qquad \frac{\partial y}{\partial r} = \sin\theta$$

$$\frac{\partial x}{\partial \theta} = -r\sin\theta, \qquad \frac{\partial y}{\partial \theta} = r\cos\theta$$

Now,

$$\frac{\partial w}{\partial r} = \frac{\partial f}{\partial x}\cdot\frac{\partial x}{\partial r} + \frac{\partial f}{\partial y}\cdot\frac{\partial y}{\partial r}$$

$$\frac{\partial w}{\partial r} = \frac{\partial f}{\partial x}\cdot(\cos\theta) + \frac{\partial f}{\partial y}\cdot(\sin\theta) \quad ...(1)$$

$$\frac{\partial w}{\partial \theta} = \frac{\partial f}{\partial x}\cdot\frac{\partial x}{\partial \theta} + \frac{\partial f}{\partial y}\cdot\frac{\partial y}{\partial \theta} = \frac{\partial f}{\partial x}\cdot(-r\sin\theta) + \frac{\partial f}{\partial y}\cdot(r\cos\theta)$$

$$\Rightarrow \quad \frac{1}{r}\frac{\partial w}{\partial \theta} = -\frac{\partial f}{\partial x}\sin\theta + \frac{\partial f}{\partial y}\cos\theta \quad ...(2)$$

Squaring (1) and (2) and adding, we obtain

$$\left(\frac{\partial w}{\partial r}\right)^2 + \frac{1}{r^2}\left(\frac{\partial w}{\partial \theta}\right)^2 = \left(\frac{\partial f}{\partial x}\right)^2 + \left(\frac{\partial f}{\partial y}\right)^2$$

Proved.

Example 6. *If $u = u(y - z,\ z - x,\ x - y)$, prove that*

$$\frac{\partial u}{\partial x} + \frac{\partial u}{\partial y} + \frac{\partial u}{\partial z} = 0$$

(Nagpur University, Winter 2002, U.P., I Sem., Winter 2002, A.M.I.E winter 2001)

Solution. Let $r = y - z, \quad s = z - x, \quad t = x - y$

so that $u = u(r, s, t)$

$$\frac{\partial u}{\partial x} = \frac{\partial u}{\partial r}\frac{\partial r}{\partial x} + \frac{\partial u}{\partial s}\frac{\partial s}{\partial x} + \frac{\partial u}{\partial t}\frac{\partial t}{\partial x}$$

$$= \frac{\partial u}{\partial r}(0) + \frac{\partial u}{\partial s}(-1) + \frac{\partial u}{\partial t}(1) = -\frac{\partial u}{\partial s} + \frac{\partial u}{\partial t} \quad \text{...(1)}$$

$$\frac{\partial u}{\partial y} = \frac{\partial u}{\partial r}\frac{\partial r}{\partial y} + \frac{\partial u}{\partial s}\frac{\partial s}{\partial y} + \frac{\partial u}{\partial t}\frac{\partial t}{\partial y}$$

$$= \frac{\partial u}{\partial r}(1) + \frac{\partial u}{\partial s}(0) + \frac{\partial u}{\partial t}(-1) = \frac{\partial u}{\partial r} - \frac{\partial u}{\partial t} \quad \text{...(2)}$$

$$\frac{\partial u}{\partial z} = \frac{\partial u}{\partial r}\frac{\partial r}{\partial z} + \frac{\partial u}{\partial s}\frac{\partial s}{\partial z} + \frac{\partial u}{\partial t}\frac{\partial t}{\partial z} = \frac{\partial u}{\partial r}(-1) + \frac{\partial u}{\partial s}(1) + \frac{\partial u}{\partial t}(0) = -\frac{\partial u}{\partial r} + \frac{\partial u}{\partial s} \quad \text{...(3)}$$

Adding (1), (2) and (3), we get

$$\frac{\partial u}{\partial x} + \frac{\partial u}{\partial y} + \frac{\partial u}{\partial z} = 0$$ **Proved.**

Example 7. *If* $u = f(e^{y-z}, \ e^{z-x}, \ e^{x-y})$, *then prove that*

$$\frac{\partial u}{\partial x} + \frac{\partial u}{\partial y} + \frac{\partial u}{\partial z} = 0$$ *(M.U. 2008, 2001)*

Solution. We have,

$$u = f(e^{y-z}, \ e^{z-x}, \ e^{x-y})$$

Let $X = e^{y-z}, \quad Y = e^{z-x}, \ Z = e^{x-y}$

$$u = f(X, Y, Z) \quad \text{...(1)}$$

Differentiating partially (1), w.r.t. 'x', we get

$$\Rightarrow \quad \frac{\partial u}{\partial x} = \frac{\partial u}{\partial X}.\frac{\partial X}{\partial x} + \frac{\partial u}{\partial Y}.\frac{\partial Y}{\partial x} + \frac{\partial u}{\partial Z}.\frac{\partial Z}{\partial x} = \frac{\partial u}{\partial X}(0) + \frac{\partial u}{\partial Y}.e^{z-x}(-1) + \frac{\partial u}{\partial Z}.e^{x-y}$$

$$= -\frac{\partial u}{\partial Y}e^{z-x} + \frac{\partial u}{\partial Z}e^{x-y} \quad \text{...(2)}$$

Differentiating (1) partially w.r.t. 'y', we get

$$\frac{\partial u}{\partial y} = \frac{\partial u}{\partial X}.\frac{\partial X}{\partial y} + \frac{\partial u}{\partial Y}.\frac{\partial Y}{\partial y} + \frac{\partial u}{\partial Z}.\frac{\partial Z}{\partial y} = \frac{\partial u}{\partial X}e^{y-z} + \frac{\partial u}{\partial Y}(0) + \frac{\partial u}{\partial Z}e^{x-y}(-1)$$

$$= \frac{\partial u}{\partial X}e^{y-z} - \frac{\partial u}{\partial Z}e^{x-y} \quad \text{...(3)}$$

Differentiating (1) partially w.r.t. 'z', we get

$$\frac{\partial u}{\partial z} = \frac{\partial u}{\partial X}.\frac{\partial X}{\partial z} + \frac{\partial u}{\partial Y}.\frac{\partial Y}{\partial z} + \frac{\partial u}{\partial Z}.\frac{\partial Z}{\partial z} = \frac{\partial u}{\partial X}e^{y-z}(-1) + \frac{\partial u}{\partial Y}e^{z-x}(1) + \frac{\partial u}{\partial Z}(0)$$

$$= -\frac{\partial u}{\partial X}e^{y-z} + \frac{\partial u}{\partial Y}e^{z-x} \quad \text{...(4)}$$

Adding (2), (3) and (4), we get

$$\frac{\partial u}{\partial x} + \frac{\partial u}{\partial y} + \frac{\partial u}{\partial z} = 0$$ **Proved.**

Example 8. *If* $u = u\left(\frac{y-x}{xy}, \frac{z-x}{xz}\right)$, *show that* $x^2\frac{\partial u}{\partial x} + y^2\frac{\partial u}{\partial y} + z^2\frac{\partial u}{\partial z} = 0$

(Nagpur University, Summer 2000, U. P. I. Sem.,

Solution. $u = u\left(\frac{y-x}{xy}, \frac{z-x}{zx}\right) = u(r, s)$

where $r = \frac{y-x}{xy}$, and $s = \frac{z-x}{zx}$

$r = \frac{1}{x} - \frac{1}{y}$ and $s = \frac{1}{x} - \frac{1}{z}$

$\frac{\partial r}{\partial x} = -\frac{1}{x^2}$ and $\frac{\partial s}{\partial x} = -\frac{1}{x^2}$

$\frac{\partial r}{\partial y} = \frac{1}{y^2}$ and $\frac{\partial s}{\partial y} = 0$

$\frac{\partial r}{\partial z} = 0$ and $\frac{\partial s}{\partial z} = \frac{1}{z^2}$

We know that,

$$\frac{\partial u}{\partial x} = \frac{\partial u}{\partial r}\frac{\partial r}{\partial x} + \frac{\partial u}{\partial s}\frac{\partial s}{\partial x} = \frac{\partial u}{\partial r}\left(-\frac{1}{x^2}\right) + \frac{\partial u}{\partial s}\left(-\frac{1}{x^2}\right) = -\frac{1}{x^2}\frac{\partial u}{\partial r} - \frac{1}{x^2}\frac{\partial u}{\partial s}$$

$$\Rightarrow \quad x^2\frac{\partial u}{\partial x} = -\frac{\partial u}{\partial r} - \frac{\partial u}{\partial s} \qquad ...(1)$$

$$\frac{\partial u}{\partial y} = \frac{\partial u}{\partial r}\frac{\partial r}{\partial y} + \frac{\partial u}{\partial s}\frac{\partial s}{\partial y} = \frac{\partial u}{\partial r}\frac{1}{y^2} + \frac{\partial u}{\partial s} \times 0 = \frac{1}{y^2}\frac{\partial u}{\partial r}$$

$$\Rightarrow \quad y^2\frac{\partial u}{\partial y} = \frac{\partial u}{\partial r} \qquad ...(2)$$

$$\frac{\partial u}{\partial z} = \frac{\partial u}{\partial r}\frac{\partial r}{\partial z} + \frac{\partial u}{\partial s}\frac{\partial s}{\partial z} = \frac{\partial u}{\partial r} \times 0 + \frac{\partial u}{\partial s} \times \frac{1}{z^2} = \frac{1}{z^2}\frac{\partial u}{\partial s}$$

$$\Rightarrow \quad z^2\frac{\partial u}{\partial z} = \frac{\partial u}{\partial s} \qquad ...(3)$$

On adding (1), (2) and (3), we get $x^2\frac{\partial u}{\partial x} + y^2\frac{\partial u}{\partial y} + z^2\frac{\partial u}{\partial z} = 0$ **Proved.**

Example 9. *If $u = f(x^2 + 2yz,\ y^2 + 2xz)$ then find the value of*

$$(y^2 - xz)\frac{\partial u}{\partial x} + (x^2 - yz)\frac{\partial u}{\partial y} + (z^2 - xy)\frac{\partial u}{\partial z}$$

(*Nagpur University, Summer 2005, Winter 2001*)

Solution. We have, $u = f(x^2 + 2yz,\ y^2 + 2xz) = f(v, w)$

Where, $v = x^2 + 2yz$, $w = y^2 + 2xz$

$$\frac{\partial u}{\partial x} = \frac{\partial u}{\partial v}\frac{\partial v}{\partial x} + \frac{\partial u}{\partial w}\frac{\partial w}{\partial x} = 2x\frac{\partial u}{\partial v} + 2z\frac{\partial u}{\partial w} \qquad ...(1)$$

$$\frac{\partial u}{\partial y} = \frac{\partial u}{\partial v}\frac{\partial v}{\partial y} + \frac{\partial u}{\partial w}\frac{\partial w}{\partial y} = 2z\frac{\partial u}{\partial v} + 2y\frac{\partial u}{\partial w} \qquad ...(2)$$

$$\frac{\partial u}{\partial z} = \frac{\partial u}{\partial v}\frac{\partial v}{\partial z} + \frac{\partial u}{\partial w}\frac{\partial w}{\partial z} = 2y\frac{\partial u}{\partial v} + 2x\frac{\partial u}{\partial w} \qquad ...(3)$$

$\frac{\partial v}{\partial x} = 2x$, $\frac{\partial v}{\partial y} = 2z$, $\frac{\partial v}{\partial z} = 2y$, $\frac{\partial w}{\partial x} = 2z$, $\frac{\partial w}{\partial y} = 2y$, $\frac{\partial w}{\partial z} = 2x$

Now, $(y^2 - xz)\frac{\partial u}{\partial x} + (x^2 - yz)\frac{\partial u}{\partial y} + (z^2 - xy)\frac{\partial u}{\partial z}$

$$= (y^2 - xz)\,2\left(x\frac{\partial u}{\partial v} + z\frac{\partial u}{\partial w}\right) + (x^2 - yz)\,2\left(z\frac{\partial u}{\partial v} + y\frac{\partial u}{\partial w}\right) + (z^2 - xy)\,2\left(y\frac{\partial u}{\partial v} + x\frac{\partial u}{\partial w}\right)$$

$$= 2(xy^2 - zx^2 + x^2z - yz^2 + yz^2 - xy^2)\frac{\partial u}{\partial v} + 2(y^2z - z^2x + x^2y - y^2z + z^2x - x^2y)\frac{\partial u}{\partial w}$$

$= 0$ **Ans.**

Example 10. *If* $\varphi(cx - az,\ cy - bz) = 0$ *show that* $ap + bq = c$:

$$\text{where } p \equiv \frac{\partial z}{\partial x} \text{ and } q \equiv \frac{\partial z}{\partial y}$$

Solution. We have,

$$\varphi(cx - az,\ cy - bz) = 0$$

$$\varphi(r, s) = 0$$

where $r = cx - az, \qquad s = cy - bz$

$$\frac{\partial r}{\partial x} = c - a\frac{\partial z}{\partial x}, \qquad \frac{\partial s}{\partial x} = -b\frac{\partial z}{\partial x},$$

$$\frac{\partial r}{\partial y} = -a\frac{\partial z}{\partial y}, \qquad \frac{\partial s}{\partial y} = c - b\frac{\partial z}{\partial y}$$

We know that, $$\frac{\partial \varphi}{\partial x} = \frac{\partial \varphi}{\partial r}\frac{\partial r}{\partial x} + \frac{\partial \varphi}{\partial s}\frac{\partial s}{\partial x}$$

$$0 = \frac{\partial \varphi}{\partial r}\left(c - a\frac{\partial z}{\partial x}\right) + \frac{\partial \varphi}{\partial s}\left(-b\frac{\partial z}{\partial x}\right)$$

$$\Rightarrow \quad 0 = c\frac{\partial \varphi}{\partial r} + \frac{\partial z}{\partial x}\left(-a\frac{\partial \varphi}{\partial r} - b\frac{\partial \varphi}{\partial s}\right)$$

$$c\frac{\partial \phi}{\partial r} = \frac{\partial z}{\partial x}\left(a\frac{\partial \phi}{\partial r} + b\frac{\partial \phi}{\partial s}\right)$$

$$\Rightarrow \quad a\frac{\partial z}{\partial x} = \frac{ac\dfrac{\partial \varphi}{\partial r}}{a\dfrac{\partial \varphi}{\partial r} + b\dfrac{\partial \varphi}{\partial s}} \qquad \ldots(1)$$

Again $$\frac{\partial \varphi}{\partial y} = \frac{\partial \varphi}{\partial r}\frac{\partial r}{\partial y} + \frac{\partial \varphi}{\partial s}\frac{\partial s}{\partial y}$$

$$0 = \frac{\partial \varphi}{\partial r}\left(-a\frac{\partial z}{\partial y}\right) + \frac{\partial \varphi}{\partial s}\left(c - b\frac{\partial z}{\partial y}\right)$$

$$0 = c\frac{\partial \varphi}{\partial s} - \frac{\partial z}{\partial y}\left(a\frac{\partial \varphi}{\partial r} + b\frac{\partial \varphi}{\partial s}\right)$$

$$\Rightarrow \quad c\frac{\partial \phi}{\partial s} = \frac{\partial z}{\partial y}\left(a\frac{\partial \phi}{\partial r} + b\frac{\partial \phi}{\partial s}\right)$$

$$\Rightarrow \quad b\frac{\partial z}{\partial y} = \frac{bc\dfrac{\partial \varphi}{\partial s}}{a\dfrac{\partial \varphi}{\partial r} + b\dfrac{\partial \varphi}{\partial s}} \qquad \ldots(2)$$

Adding (1) and (2), we get

$$a\frac{\partial z}{\partial x} + b\frac{\partial z}{\partial y} = \frac{ac\dfrac{\partial \varphi}{\partial r} + bc\dfrac{\partial \phi}{\partial s}}{a\dfrac{\partial \varphi}{\partial r} + b\dfrac{\partial \varphi}{\partial s}} \Rightarrow a\frac{\partial z}{\partial x} + b\frac{\partial z}{\partial y} = c$$ **Proved.**

Example 11. *Transform the equation* $\frac{\partial^2 u}{\partial x^2} + \frac{\partial^2 u}{\partial y^2} = 0$ *into polar co-ordinates.*

Solution. We have,

$$x = r\cos\theta,\ y = r\sin\theta$$

$$r^2 = x^2 + y^2, \qquad \theta = \tan^{-1}\frac{y}{x}$$

$$\frac{\partial u}{\partial x} = \frac{\partial u}{\partial r}\frac{\partial r}{\partial x} + \frac{\partial u}{\partial \theta}\frac{\partial \theta}{\partial x} = \frac{\partial u}{\partial r}\frac{x}{r} + \frac{\partial u}{\partial \theta}\frac{-y}{x^2+y^2} = \frac{\partial u}{\partial r}\cos\theta - \frac{\partial u}{\partial \theta}\frac{\sin\theta}{r}$$

$$\frac{\partial^2 u}{\partial x^2} = \frac{\partial}{\partial x}\left(\frac{\partial u}{\partial x}\right) = \left(\cos\theta\frac{\partial}{\partial r} - \frac{\sin\theta}{r}\frac{\partial}{\partial\theta}\right)\left(\cos\theta\frac{\partial u}{\partial r} - \frac{\sin\theta}{r}\frac{\partial u}{\partial\theta}\right)$$

$$= \cos\theta\frac{\partial}{\partial r}\left(\cos\theta\frac{\partial u}{\partial r} - \frac{\sin\theta}{r}\frac{\partial u}{\partial\theta}\right) - \frac{\sin\theta}{r}\frac{\partial}{\partial\theta}\left(\cos\theta\frac{\partial u}{\partial r} - \frac{\sin\theta}{r}\frac{\partial u}{\partial\theta}\right)$$

$$= \cos\theta\left(\cos\theta\frac{\partial^2 u}{\partial r^2} + \frac{\sin\theta}{r^2}\frac{\partial u}{\partial\theta} - \frac{\sin\theta}{r}\frac{\partial^2 u}{\partial r\,\partial\theta}\right)$$

$$-\frac{\sin\theta}{r}\left(-\sin\theta\frac{\partial u}{\partial r} + \cos\theta\frac{\partial^2 u}{\partial\theta\,\partial r} - \frac{\cos\theta}{r}\frac{\partial u}{\partial\theta} - \frac{\sin\theta}{r}\frac{\partial^2 u}{\partial\theta^2}\right)$$

$$= \cos^2\theta\frac{\partial^2 u}{\partial r^2} + \frac{\sin\theta\cos\theta}{r^2}\frac{\partial u}{\partial\theta} - \frac{\sin\theta\cos\theta}{r}\frac{\partial^2 u}{\partial r\,\partial\theta}$$

$$+\frac{\sin^2\theta}{r}\frac{\partial u}{\partial r} - \frac{\sin\theta\cos\theta}{r}\frac{\partial^2 u}{\partial\theta\,\partial r} + \frac{\sin\theta\cos\theta}{r^2}\frac{\partial u}{\partial\theta} + \frac{\sin^2\theta}{r^2}\frac{\partial^2 u}{\partial\theta^2}$$

$$= \cos^2\theta\frac{\partial^2 u}{\partial r^2} + \frac{2\sin\theta\cos\theta}{r^2}\frac{\partial u}{\partial\theta} - \frac{2\sin\theta\cos\theta}{r}\frac{\partial^2 u}{\partial r\,\partial\theta} + \frac{\sin^2\theta}{r}\frac{\partial u}{\partial r} + \frac{\sin^2\theta}{r^2}\frac{\partial^2 u}{\partial\theta^2} \quad \ldots(1)$$

$$\frac{\partial u}{\partial y} = \frac{\partial u}{\partial r}\frac{\partial r}{\partial y} + \frac{\partial u}{\partial\theta}\frac{\partial\theta}{\partial y} = \frac{\partial u}{\partial r}\frac{y}{r} + \frac{\partial u}{\partial\theta}\frac{x}{x^2+y^2} = \frac{\partial u}{\partial r}\sin\theta + \frac{\partial u}{\partial\theta}\frac{\cos\theta}{r}$$

$$\frac{\partial^2 u}{\partial y^2} = \frac{\partial}{\partial y}\left(\frac{\partial u}{\partial y}\right) = \left(\sin\theta\frac{\partial}{\partial r} + \frac{\cos\theta}{r}\frac{\partial}{\partial\theta}\right)\left(\sin\theta\frac{\partial u}{\partial r} + \frac{\cos\theta}{r}\frac{\partial u}{\partial\theta}\right)$$

$$= \sin\theta\frac{\partial}{\partial r}\left(\sin\theta\frac{\partial u}{\partial r} + \frac{\cos\theta}{r}\frac{\partial u}{\partial\theta}\right) + \frac{\cos\theta}{r}\frac{\partial}{\partial\theta}\left(\sin\theta\frac{\partial u}{\partial r} + \frac{\cos\theta}{r}\frac{\partial u}{\partial\theta}\right)$$

$$= \sin\theta\left[\sin\theta\frac{\partial^2 u}{\partial r^2} - \frac{\cos\theta}{r^2}\frac{\partial u}{\partial\theta} + \frac{\cos\theta}{r}\frac{\partial^2 u}{\partial r\,\partial\theta}\right]$$

$$+\frac{\cos\theta}{r}\left[\cos\theta\frac{\partial u}{\partial r} + \sin\theta\frac{\partial^2 u}{\partial\theta\,\partial r} - \frac{\sin\theta}{r}\frac{\partial u}{\partial\theta} + \frac{\cos\theta}{r}\frac{\partial^2 u}{\partial\theta^2}\right]$$

$$= \sin^2\theta\frac{\partial^2 u}{\partial r^2} - \frac{\sin\theta\cos\theta}{r^2}\frac{\partial u}{r\theta} + \frac{\sin\theta\cos\theta}{r}\frac{\partial^2 u}{\partial r\,\partial\theta}$$

$$+\frac{\cos^2\theta}{r}\frac{\partial u}{\partial r} + \frac{\sin\theta\cos\theta}{r}\frac{\partial^2 u}{\partial\theta\,\partial r} - \frac{\sin\theta\cos\theta}{r^2}\frac{\partial u}{\partial\theta} + \frac{\cos^2\theta}{r^2}\frac{\partial^2 u}{r\theta^2}$$

$$= \sin^2\theta\frac{\partial^2 u}{\partial r^2} - \frac{2\sin\theta\cos\theta}{r^2}\frac{\partial u}{\partial\theta} + \frac{2\sin\theta\cos\theta}{r}\frac{\partial^2 u}{\partial r\,\partial\theta}$$

$$+\frac{\cos^2\theta}{r}\frac{\partial u}{\partial r} + \frac{\cos^2\theta}{r^2}\frac{\partial^2 u}{\partial\theta^2} \quad \ldots(2)$$

Adding (1) and (2), we get

$$\frac{\partial^2 u}{\partial x^2}+\frac{\partial^2 u}{\partial y^2}=(\sin^2\theta+\cos^2\theta)\frac{\partial^2 u}{\partial r^2}+(\sin^2\theta+\cos^2\theta)\frac{1}{r}\frac{\partial u}{\partial r}+(\sin^2\theta+\cos^2\theta)\frac{1}{r^2}\frac{\partial^2 u}{\partial\theta^2}$$

$$=\frac{\partial^2 u}{\partial r^2}+\frac{1}{r}\frac{\partial u}{\partial r}+\frac{1}{r^2}\frac{\partial^2 u}{\partial\theta^2}\qquad \textbf{Ans.}$$

Example 12. *If* $x = r\cos\theta$, $y = r\sin\theta$, *prove that*

$$x\frac{\partial u}{\partial x}+y\frac{\partial u}{\partial y}+y^2\frac{\partial^2 u}{\partial x^2}-2xy\frac{\partial^2 u}{\partial x\,\partial y}+x^2\frac{\partial^2 u}{\partial y^2}=\frac{r\,\partial u}{\partial r}+\frac{\partial^2 u}{\partial\theta^2}$$

(Nagpur University, Summer 2005, Winter 2001)

Solution.

$$\frac{\partial u}{\partial r}=\frac{\partial u}{\partial x}\frac{\partial x}{\partial r}+\frac{\partial u}{\partial y}\frac{\partial y}{\partial r}$$

$$\left[\begin{array}{l|l} x=r\cos\theta & y=r\sin\theta\\ \dfrac{\partial x}{\partial r}=\cos\theta & \dfrac{\partial y}{\partial r}=\sin\theta\\ \dfrac{\partial x}{\partial\theta}=-r\sin\theta & \dfrac{\partial y}{\partial\theta}=r\cos\theta\end{array}\right]$$

$$\Rightarrow\quad \frac{\partial u}{\partial r}=\cos\theta\frac{\partial u}{\partial x}+\sin\theta\frac{\partial u}{\partial y}\qquad\ldots(1)$$

and

$$\frac{\partial u}{\partial\theta}=\frac{\partial u}{\partial x}\frac{\partial x}{\partial\theta}+\frac{\partial u}{\partial y}\frac{\partial y}{\partial\theta}=-r\sin\theta\frac{\partial u}{\partial x}+r\cos\theta\frac{\partial u}{\partial y}$$

$$\Rightarrow\quad \frac{\partial}{\partial\theta}=-r\sin\theta\frac{\partial}{\partial x}+r\cos\theta\frac{\partial}{\partial y}\qquad\therefore\quad \frac{\partial^2 u}{\partial\theta^2}=\frac{\partial}{\partial\theta}\left(\frac{\partial u}{\partial\theta}\right)$$

$$\frac{\partial^2 u}{\partial\theta^2}=\left[-r\sin\theta\frac{\partial}{\partial x}+r\cos\theta\frac{\partial}{\partial y}\right]\left[-r\sin\theta\frac{\partial u}{\partial x}+r\cos\theta\frac{\partial u}{\partial y}\right]$$

$$=-r\sin\theta\frac{\partial}{\partial x}\left[-r\sin\theta\frac{\partial u}{\partial x}+r\cos\theta\frac{\partial u}{\partial y}\right]+r\cos\theta\frac{\partial}{\partial y}\left[-r\sin\theta\frac{\partial u}{\partial x}+r\cos\theta\frac{\partial u}{\partial y}\right]$$

$$=r^2\sin^2\theta\frac{\partial^2 u}{\partial x^2}-r^2\sin\theta\cos\theta\frac{\partial^2 u}{\partial x\,\partial y}-r^2\sin\theta\cos\theta\frac{\partial^2 u}{\partial x\,\partial y}+r^2\cos^2\theta\frac{\partial^2 u}{\partial y^2}$$

$$=r^2\sin^2\theta\frac{\partial^2 u}{\partial x^2}-2r^2\sin\theta\cos\theta\frac{\partial^2 u}{\partial x\,\partial y}+r^2\cos^2\theta\frac{\partial^2 u}{\partial y^2}\qquad\ldots(2)$$

Now, using (1) and (2)

$$r\frac{\partial u}{\partial r}+\frac{\partial^2 u}{\partial\theta^2}=r\left[\cos\theta\frac{\partial u}{\partial x}+\sin\theta\frac{\partial u}{\partial y}\right]+r^2\sin^2\theta\frac{\partial^2 u}{\partial x^2}-2r^2\sin\theta\cos\theta\frac{\partial^2 u}{\partial x\,\partial y}+r^2\cos^2\theta\frac{\partial^2 u}{\partial y^2}$$

$$=r\cos\theta\frac{\partial u}{\partial x}+r\sin\theta\frac{\partial u}{\partial y}+r^2\sin^2\theta\frac{\partial^2 u}{\partial x^2}-2r^2\sin\theta\cos\theta\frac{\partial^2 u}{\partial x\,\partial y}+r^2\cos^2\theta\frac{\partial^2 u}{\partial y^2}$$

$$=x\frac{\partial u}{\partial x}+y\frac{\partial u}{\partial y}+y^2\frac{\partial^2 u}{\partial x^2}-2xy\frac{\partial^2 u}{\partial x\,\partial y}+x^2\frac{\partial^2 u}{\partial y^2}\qquad\textbf{Proved.}$$

Example 13. *If $u = f(x, y)$ where $x = e^r \cos\theta$ and $y = e^r \sin\theta$, then show that*

$$\frac{\partial^2 u}{\partial x^2} + \frac{\partial^2 u}{\partial y^2} = e^{-2r}\left[\frac{\partial^2 u}{\partial r^2} + \frac{\partial^2 u}{\partial \theta^2}\right]$$

(Nagpur University, Summer 2008)

Solution. Here, we have

$$x = e^r \cos\theta, \quad \frac{\partial x}{\partial r} = e^r \cos\theta = x, \quad \frac{\partial x}{\partial \theta} = -e^r \sin\theta = -y$$

$$y = e^r \sin\theta, \quad \frac{\partial y}{\partial r} = e^r \sin\theta = y, \quad \frac{\partial y}{\partial \theta} = e^r \cos\theta = x$$

$$\frac{\partial u}{\partial r} = \frac{\partial u}{\partial x}\frac{\partial x}{\partial r} + \frac{\partial u}{\partial y}\frac{\partial y}{\partial r} = \frac{\partial u}{\partial x}x + \frac{\partial u}{\partial y}y \quad \Rightarrow \quad \frac{\partial}{\partial r} = x\frac{\partial}{\partial x} + y\frac{\partial}{\partial y}$$

$$\frac{\partial^2 u}{\partial r^2} = \frac{\partial}{\partial r}\left(\frac{\partial u}{\partial r}\right) = \left(x\frac{\partial}{\partial x} + y\frac{\partial}{\partial y}\right)\left(x\frac{\partial u}{\partial x} + y\frac{\partial u}{\partial y}\right)$$

$$= x\frac{\partial}{\partial x}\left(x\frac{\partial u}{\partial x}\right) + x\frac{\partial}{\partial x}\left(y\frac{\partial u}{\partial y}\right) + y\frac{\partial}{\partial y}\left(x\frac{\partial u}{\partial x}\right) + y\frac{\partial}{\partial y}\left(y\frac{\partial u}{\partial y}\right)$$

$$= x\left(x\frac{\partial^2 u}{\partial x^2} + \frac{\partial u}{\partial x}\right) + xy\frac{\partial^2 u}{\partial y\,\partial x} + yx\frac{\partial^2 u}{\partial y\,\partial x} + y\left(y\frac{\partial^2 u}{\partial y^2} + \frac{\partial u}{\partial y}\right)$$

$$= x^2\frac{\partial^2 u}{\partial x^2} + x\frac{\partial u}{\partial x} + 2xy\frac{\partial^2 u}{\partial y\,\partial x} + y^2\frac{\partial^2 u}{\partial y^2} + y\frac{\partial u}{\partial y} \qquad \text{...(1)}$$

$$\frac{\partial u}{\partial \theta} = \frac{\partial u}{\partial x}\frac{\partial x}{\partial \theta} + \frac{\partial u}{\partial y}\frac{\partial y}{\partial \theta} = \frac{\partial u}{\partial x}(-y) + \frac{\partial u}{\partial y}x \quad \Rightarrow \quad \frac{\partial}{\partial \theta} = -y\frac{\partial}{\partial x} + x\frac{\partial}{\partial y}$$

$$\frac{\partial^2 u}{\partial \theta^2} = \frac{\partial}{\partial \theta}\left(\frac{\partial y}{\partial \theta}\right) = \left(-y\frac{\partial}{\partial x} + x\frac{\partial}{\partial y}\right)\left(-y\frac{\partial u}{\partial x} + x\frac{\partial u}{\partial y}\right)$$

$$= -y\frac{\partial}{\partial x}\left(-y\frac{\partial u}{\partial x}\right) - y\frac{\partial}{\partial x}\left(x\frac{\partial u}{\partial y}\right) + x\frac{\partial}{\partial y}\left(-y\frac{\partial u}{\partial x}\right) + x\frac{\partial}{\partial y}\left(x\frac{\partial u}{\partial y}\right)$$

$$= y^2\frac{\partial^2 u}{\partial x^2} - y\left(x\frac{\partial^2 u}{\partial x\,\partial y} + \frac{\partial u}{\partial y}\right) + x\left(-y\frac{\partial^2 u}{\partial x\,\partial y} - \frac{\partial u}{\partial x}\right) + x^2\frac{\partial^2 u}{\partial y^2}$$

$$= y^2\frac{\partial^2 u}{\partial x^2} - xy\frac{\partial^2 u}{\partial x\,\partial y} - y\frac{\partial u}{\partial y} - xy\frac{\partial^2 u}{\partial x\,\partial y} - x\frac{\partial u}{dx} + x^2\frac{\partial^2 u}{\partial y^2}$$

$$\frac{\partial^2 u}{\partial \theta^2} = y^2\frac{\partial^2 u}{\partial x^2} - 2xy\frac{\partial^2 u}{\partial x\,\partial y} - y\frac{\partial u}{dy} - x\frac{\partial u}{\partial x} + x^2\frac{\partial^2 u}{\partial y^2} \qquad \text{...(2)}$$

Adding (1) and (2), we get

$$\frac{\partial^2 u}{\partial r^2} + \frac{\partial^2 u}{\partial \theta^2} = x^2\frac{\partial^2 u}{\partial x^2} + x\frac{\partial u}{dx} + 2xy\frac{\partial^2 u}{\partial x\,\partial y} + y^2\frac{\partial^2 u}{\partial y^2} + y\frac{\partial u}{\partial y}$$

$$+ y^2\frac{\partial^2 u}{\partial x^2} - 2xy\frac{\partial^2 u}{\partial x\,\partial y} - y\frac{\partial u}{\partial y} - x\frac{\partial u}{\partial x} + x^2\frac{\partial^2 u}{\partial y^2}$$

$$= (x^2 + y^2)\frac{\partial^2 u}{\partial x^2} + (y^2 + x^2)\frac{\partial^2 u}{\partial y^2} = e^{2r}\frac{\partial^2 u}{\partial x^2} + e^{2r}\frac{\partial^2 u}{\partial y^2}$$

$$\Rightarrow \quad e^{-2r}\left(\frac{\partial^2 u}{\partial r^2} + \frac{\partial^2 u}{\partial \theta^2}\right) = \frac{\partial^2 u}{\partial x^2} + \frac{\partial^2 u}{\partial y^2}$$

Proved.

Example 14. *If $u = f(r)$ and $x = r\cos\theta$, $y = r\sin\theta$, prove that*

$$\frac{\partial^2 u}{\partial x^2}+\frac{\partial^2 u}{\partial y^2}=f''(r)+\frac{1}{r}f'(r).$$

(*Nagpur University, Winter 2004*)

(*A.M.I.E.T.E., Winter 2003, U.P. I Semester, Winter 2005, 2000*)

Solution.

$x = r\cos\theta$
$y = r\sin\theta$

$r^2 = x^2 + y^2$ so that $\dfrac{\partial r}{\partial x}=\dfrac{x}{r}$

$u = f(r)$

$$\frac{\partial u}{\partial x}=\frac{df}{dr}\cdot\frac{\partial r}{\partial x}$$

$$\frac{\partial u}{\partial x}=\frac{df}{dr}\cdot\frac{x}{r}$$

Differentiating again *w.r.t.* x, we get

$$\frac{\partial^2 u}{\partial x^2}=\left(\frac{d^2 f}{dr^2}\cdot\frac{\partial r}{\partial x}\right)\cdot\frac{x}{r}+\frac{df}{dr}\cdot\left[\frac{r.1-x\dfrac{\partial r}{\partial x}}{r^2}\right]=\left(\frac{d^2 f}{dr^2}\frac{x}{r}\right)\cdot\frac{x}{r}+\frac{df}{dr}\cdot\left[\frac{r.1-x\cdot\dfrac{x}{r}}{r^2}\right]$$

$$=\frac{d^2 f}{dr^2}\frac{x^2}{r^2}+\frac{df}{dr}\cdot\frac{r^2-x^2}{r^3}=\frac{d^2 f}{dr^2}\frac{x^2}{r^2}+\frac{df}{dr}\frac{y^2}{r^3} \quad ...(1)$$

Similarly, $$\frac{\partial^2 u}{\partial y^2}=\frac{d^2 f}{dr^2}\frac{y^2}{r^2}+\frac{df}{dr}\frac{x^2}{r^3} \quad ...(2)$$

On adding (1) and (2), we get

$$\frac{\partial^2 u}{\partial x^2}+\frac{\partial^2 u}{\partial y^2}=\frac{d^2 f}{dr^2}\frac{x^2+y^2}{r^2}+\frac{df}{dr}\frac{x^2+y^2}{r^3}=\frac{d^2 f}{dr^2}+\frac{df}{dr}\frac{1}{r}=f''(r)+\frac{1}{r}f'(r)$$

Proved.

Example 15. *A function $f(x, y)$ is rewritten in terms of new variables*

$u = e^x\cos y, \qquad v = e^x\sin y$

Show that (i) $\dfrac{\partial f}{\partial x}=u\dfrac{\partial f}{\partial u}+v\dfrac{\partial f}{\partial v}$ *and (ii)* $\dfrac{\partial f}{\partial y}=-v\dfrac{\partial f}{\partial u}+u\dfrac{\partial f}{\partial v}$

and hence deduce that (iii) $\dfrac{\partial^2 f}{\partial x^2}+\dfrac{\partial^2 f}{\partial y^2}=(u^2+v^2)\left(\dfrac{\partial^2 f}{\partial u^2}+\dfrac{\partial^2 f}{\partial v^2}\right)$

Solution. $u = e^x\cos y,\ \dfrac{\partial u}{\partial x}=e^x\cos y=u,\ \dfrac{\partial u}{\partial y}=-e^x\sin y=-v$

$v = e^x\sin y,\ \dfrac{\partial v}{\partial x}=e^x\sin y=v,\ \dfrac{\partial v}{\partial y}=e^x\cos y=u$

(*i*) We know that $\dfrac{\partial f}{\partial x}=\dfrac{\partial f}{\partial u}\dfrac{\partial u}{\partial x}+\dfrac{\partial f}{\partial v}\dfrac{\partial v}{\partial x}$

$$\frac{\partial f}{\partial x}=\frac{\partial f}{\partial u}u+\frac{\partial f}{\partial v}v \quad ...(1) \textbf{ Proved.}$$

(*ii*) $$\frac{\partial f}{\partial y}=\frac{\partial f}{\partial u}\frac{\partial u}{\partial y}+\frac{\partial f}{\partial v}\frac{\partial v}{\partial y}=\frac{\partial f}{\partial u}.(-v)+\frac{\partial f}{\partial v}u=-v.\frac{\partial f}{\partial u}+u.\frac{\partial f}{\partial v} \quad ...(2) \textbf{ Proved.}$$

(*iii*) $$\frac{\partial^2 f}{\partial x^2}=\frac{\partial}{\partial x}\left(\frac{\partial f}{\partial x}\right)$$

$$=\left(u\cdot\frac{\partial}{\partial u}+v\cdot\frac{\partial}{\partial v}\right)\left(u\cdot\frac{\partial f}{\partial u}+v\cdot\frac{\partial f}{\partial v}\right) \quad \text{[From (1)]}$$

$$= u\frac{\partial}{\partial u}\left(u\frac{\partial f}{\partial u}+v\frac{\partial f}{\partial v}\right)+v\frac{\partial}{\partial v}\left(u\frac{\partial f}{\partial u}+v\frac{\partial f}{\partial v}\right)$$

$$= u\frac{\partial}{\partial u}\left(u\frac{\partial f}{\partial u}\right)+u\frac{\partial}{\partial u}\left(v\frac{\partial f}{\partial v}\right)+v\frac{\partial}{\partial v}\left(u\frac{\partial f}{\partial u}\right)+v\frac{\partial}{\partial v}\left(v\frac{\partial f}{\partial v}\right)$$

$$= u\left(u\frac{\partial^2 f}{\partial u^2}+\frac{\partial f}{\partial u}\right)+u\left(v\frac{\partial^2 f}{\partial u\,\partial v}\right)+v\left(u\frac{\partial^2 f}{\partial v\,\partial u}\right)+v\left(v\frac{\partial^2 f}{\partial v^2}+\frac{\partial f}{\partial v}\right)$$

$$= u^2\frac{\partial^2 f}{\partial u^2}+u\frac{\partial f}{\partial u}+uv\frac{\partial^2 f}{\partial v\,\partial u}+uv\frac{\partial^2 f}{\partial u\,\partial v}+v^2\frac{\partial^2 f}{\partial v^2}+v\frac{\partial f}{\partial v} \qquad ...(3)$$

$$\frac{\partial^2 f}{\partial y^2} = \frac{\partial}{\partial y}\left(\frac{\partial f}{\partial y}\right)=\left(-v\frac{\partial}{\partial u}+u\frac{\partial}{\partial v}\right)\left(-v\frac{\partial f}{\partial u}+u\frac{\partial f}{\partial v}\right) \qquad \text{[From (2)]}$$

$$= -v\frac{\partial}{\partial u}\left(-v\frac{\partial f}{\partial u}+u\frac{\partial f}{\partial v}\right)+u\frac{\partial}{\partial v}\left(-v\frac{\partial f}{\partial u}+u\frac{\partial f}{\partial v}\right)$$

$$= -v\frac{\partial}{\partial u}\left(-v\frac{\partial f}{\partial u}\right)-v\frac{\partial}{\partial u}\left(u\frac{\partial f}{\partial v}\right)+u\frac{\partial}{\partial v}\left(-v\frac{\partial f}{\partial u}\right)+u\frac{\partial}{\partial v}\left(u\frac{\partial f}{\partial v}\right)$$

$$= -v\left(-v\frac{\partial^2 f}{\partial u^2}\right)-v\left(u\frac{\partial^2 f}{\partial u\,\partial v}+\frac{\partial f}{\partial v}\right)+u\left(-v\frac{\partial^2 f}{\partial v\,\partial u}-\frac{\partial f}{\partial u}\right)+u\left(u\frac{\partial^2 f}{\partial v^2}\right)$$

$$= v^2\frac{\partial^2 f}{\partial u^2}-uv\frac{\partial^2 f}{\partial u\,\partial v}-v\frac{\partial f}{\partial u}-uv\frac{\partial^2 f}{\partial u\,\partial v}-u\frac{\partial f}{\partial u}+u^2\frac{\partial^2 f}{\partial v^2} \qquad ...(4)$$

On adding (3) and (4), we obtain

$$\frac{\partial^2 f}{\partial x^2}+\frac{\partial^2 f}{\partial y^2} = u^2\frac{\partial^2 f}{\partial u^2}+v^2\frac{\partial^2 f}{\partial v^2}+v^2\frac{\partial^2 f}{\partial u^2}+u^2\frac{\partial^2 f}{\partial v^2}$$

$$= (u^2+v^2)\frac{\partial^2 f}{\partial u^2}+(u^2+v^2)\frac{\partial^2 f}{\partial v^2} = (u^2+v^2)\left(\frac{\partial^2 f}{\partial u^2}+\frac{\partial^2 f}{\partial v^2}\right) \qquad \textbf{Proved.}$$

Example 16. *If $x + y = 2\,e^{\theta}\cos\phi$ and $x - y = 2\,ie^{\theta}\sin\phi$*

Show that : $\dfrac{\partial^2 V}{\partial \theta^2}+\dfrac{\partial^2 V}{\partial \phi^2} = 4\,x\,y\dfrac{\partial^2 V}{\partial x\,\partial y}$ *(A.M.I.E.T.E., June 2010, Winter 2007)*

(Nagpur University, Summer 2001, Winter 2000, U.P., I Semester, Winter 2001)

Solution. We have, $x + y = 2\,e^{\theta}\cos\phi$ ...(1)

$x - y = 2\,i\,e^{\theta}\sin\phi$ ...(2)

By adding and subtracting equations (1) and (2), we have

$$2\,x = 2e^{\theta}(\cos\phi + i\sin\phi) \Rightarrow x = e^{\theta+i\phi}$$

and $$2\,y = 2e^{\theta}(\cos\phi - i\sin\phi) \Rightarrow y = e^{\theta-i\phi} \qquad ...(3)$$

It is clear that $V = f(x, y)$ and x, y are functions of θ and ϕ. Hence V is a composite function of θ and ϕ.

We want to convert V, θ, ϕ in V, x, y respectively.

From equation (3), we have

$$\frac{\partial x}{\partial \theta} = e^{\theta+i\phi} = x, \quad \frac{\partial x}{\partial \phi} = i\,e^{\theta+i\phi} = i\,x, \quad \frac{\partial y}{\partial \theta} = e^{\theta-i\phi} = y,$$

and $$\frac{\partial y}{\partial \phi} = -ie^{\theta-i\phi} = -iy$$

Now $$\frac{\partial V}{\partial \theta} = \frac{\partial V}{\partial x}\cdot\frac{\partial x}{\partial \theta}+\frac{\partial V}{\partial y}\cdot\frac{\partial y}{\partial \theta} = x\frac{\partial V}{\partial x}+y\frac{\partial V}{\partial y}$$

and
$$\frac{\partial^2 V}{\partial \theta^2} = \frac{\partial}{\partial \theta}\left(\frac{\partial V}{\partial \theta}\right) = \left(x\frac{\partial}{\partial x} + y\frac{\partial}{\partial y}\right)\left(x\frac{\partial V}{\partial x} + y\frac{\partial V}{\partial y}\right)$$
$$= x\frac{\partial}{\partial x}\left[x\frac{\partial V}{\partial x} + y\frac{\partial V}{\partial y}\right] + y\frac{\partial}{\partial y}\left[x\frac{\partial V}{\partial x} + y\frac{\partial V}{\partial y}\right]$$
$$= x\left[x\frac{\partial^2 V}{\partial x^2} + \frac{\partial V}{\partial x} + y\frac{\partial^2 V}{\partial x \partial y}\right] + y\left[x\frac{\partial^2 V}{\partial x \partial y} + y\frac{\partial^2 V}{\partial y^2} + \frac{\partial V}{\partial y}\right]$$
$$= x^2\frac{\partial^2 V}{\partial x^2} + y^2\frac{\partial^2 V}{\partial y^2} + 2xy\frac{\partial^2 V}{\partial x \partial y} + \left(x\frac{\partial V}{\partial x} + y\frac{\partial V}{\partial y}\right) \quad ...(4)$$

Again
$$\frac{\partial V}{\partial \phi} = \frac{\partial V}{\partial x}\frac{\partial x}{\partial \phi} + \frac{\partial V}{\partial y}\frac{\partial y}{\partial \phi} = i\left[x\frac{\partial V}{\partial x} - y\frac{\partial V}{\partial y}\right]$$
$$\frac{\partial}{\partial \phi} = i\left[x\frac{\partial}{\partial x} - y\frac{\partial}{\partial y}\right]$$
$$\frac{\partial^2 V}{\partial \phi^2} = \frac{\partial}{\partial \phi}\left(\frac{\partial V}{\partial \phi}\right) = i\left(x\frac{\partial}{\partial x} - y\frac{\partial}{\partial y}\right) i\left(x\frac{\partial V}{\partial x} - y\frac{\partial V}{\partial y}\right)$$
$$= ix\frac{\partial}{\partial x}.i\left[x\frac{\partial V}{\partial x} - y\frac{\partial V}{\partial y}\right] - iy\frac{\partial}{\partial y}.i\left[x\frac{\partial V}{\partial x} - y\frac{\partial V}{\partial y}\right]$$
$$= -x\frac{\partial}{\partial x}\left[x\frac{\partial V}{\partial x} - y\frac{\partial V}{\partial y}\right] + y\frac{\partial}{\partial y}\left[x\frac{\partial V}{\partial x} - y\frac{\partial V}{\partial y}\right]$$
$$= -x\left[\frac{\partial V}{\partial x} + x\frac{\partial^2 V}{\partial x^2} - y\frac{\partial^2 V}{\partial x \partial y}\right] + y\left[x\frac{\partial^2 V}{\partial x \partial y} - y\frac{\partial^2 V}{\partial y^2} - \frac{\partial V}{\partial y}\right]$$
$$= -\left[x\frac{\partial V}{\partial x} + y\frac{\partial V}{\partial y}\right] + 2xy\frac{\partial^2 V}{\partial x \partial y} - \left[x^2\frac{\partial^2 V}{\partial x^2} + y^2\frac{\partial^2 V}{\partial y^2}\right] \quad ...(5)$$

Adding (4) and (5), we get
$$\frac{\partial^2 V}{\partial \theta^2} + \frac{\partial^2 V}{\partial \phi^2} = 4xy\frac{\partial^2 V}{\partial x \partial y}$$
Proved.

EXERCISE 2.1

1. If $z = u^2 + v^2$ and $u = at^2$, $v = 2\,at$, find $\frac{dz}{dt}$. **Ans.** $4\,a^2 t\,(t^2 + 2)$

2. If $z = \sin^{-1}(x - y)$, $x = 3\,t$, $y = 4\,t^3$; show that $\frac{dz}{dt} = \frac{3}{\sqrt{1-t^2}}$.

3. If $w = f(u, v)$, where $u = x + y$ and $v = x - y$, show that $\frac{dw}{dx} + \frac{dw}{dy} = 2\frac{dw}{du}$

4. If $u = xe^y z$, where $y = \sqrt{a^2 - x^2}$, $z = \sin^3 x$. Find $\frac{du}{dx}$ **Ans.** $e^y z\left(1 - \frac{x^2}{y} + 3x\cot x\right)$

5. If $u = x^2 + y^2 + z^2 - 2\,xyz = 1$, show that $\frac{dx}{\sqrt{1-x^2}} + \frac{dy}{\sqrt{1-y^2}} + \frac{dz}{\sqrt{1-z^2}} = 0$

[**Hint.** $du = \frac{\partial u}{\partial x}dx + \frac{\partial u}{\partial y}dy + \frac{\partial u}{\partial z}dz = 0$

$= 2\,(x - yz)\,dx + 2\,(y - zx)\,dy + 2\,(z - xy)\,dz = 0$

But $x^2 + y^2 + z^2 - 2\,xyz = 1$, $\Rightarrow y^2 - 2xyz = 1 - x^2 - z^2$

$y^2 - 2xyz + x^2z^2 = 1 + x^2z^2 - x^2 - z^2$ $\Rightarrow (y - xz)^2 = (1 - x^2)(1 - z^2)$]

6. If $z = z(u, v)$, $u = x^2 - 2xy - y^2$ and $v = y$. Show that

$$(x+y)\frac{\partial z}{\partial x} + (x-y)\frac{\partial z}{\partial y} = 0 \text{ is equivalent to } \frac{\partial z}{\partial v} = 0$$

7. If $V = f(2x - 3y,\ 3y - 4z,\ 4z - 2x)$, compute the value of $6V_x + 4V_y + 3V_z$.
(*U.P., I Semester., Dec. 2008*) **Ans.** = 0

8. If $z = \sqrt{x^2 + y^2}$ and $x^3 + y^3 + 3axy = 5a^2$, find the value of $\frac{dz}{dx}$, when $x = y = a$. **Ans.** 0

9. By changing the independent variables x and t to u and v by means of the relationships
$u = x - at$, $v = x + at$

Show that $a^2 \frac{\partial^2 y}{\partial x^2} - \frac{\partial^2 y}{\partial t^2} = 4a^2 \frac{\partial^2 y}{\partial u\, \partial v}$

10. If $x^2 = au + bv$, $y = au - bv$, prove that $\left(\frac{du}{dx}\right)_y \cdot \left(\frac{dx}{du}\right)_v = \frac{1}{2} = \left(\frac{\partial v}{\partial y}\right)_x \cdot \left(\frac{\partial y}{\partial v}\right)_u$

11. If $z = f(x, y)$ where $x = uv$, $y = \frac{u+v}{u-v}$, show that $2x\frac{\partial z}{dx} = u\frac{\partial z}{du} + v\frac{\partial z}{dv}$.

12. If $u = x\cos\frac{y}{z}$, $x = 3r^2 + 2s$, $y = 4r - 2s^3$, $z = 2r^2 - 3s^2$ find $\frac{\partial u}{\partial r}, \frac{\partial u}{\partial s}$.

Ans. $\frac{\partial u}{\partial r} = 6r\cos\frac{y}{z} - \frac{4x}{z}\sin\frac{y}{z} + \frac{4xyr}{z^2}\sin\frac{y}{z}$, $\frac{\partial u}{\partial s} = 2\cos\frac{y}{z} + \frac{6xs^2}{z}\sin\frac{y}{z} - \frac{6xys}{z^2}\sin\frac{y}{z}$

13. Find $\frac{\partial w}{\partial \theta}$ and $\frac{\partial w}{\partial \phi}$ given that $w(x, y, z) = f(x^2 + y^2 + z^2)$ where $x = r\cos\theta\,.\cos\phi$,

$y = r\cos\theta\,.\sin\phi$, $z = r\sin\theta$. **Ans.** $\frac{\partial w}{\partial \phi} = 0$, $\frac{\partial w}{\partial \phi} = 0$.

14. If $x = \frac{\cos\theta}{u}$, $y = \frac{\sin\theta}{u}$ and $z = f(x, y)$, then show that

$$\frac{\partial^2 z}{\partial x^2} + \frac{\partial^2 z}{\partial y^2} = u^4\frac{\partial^2 z}{\partial u^2} + u^3\frac{\partial z}{\partial u} + u^4\frac{\partial^2 z}{\partial \theta^2}$$

15. If $x = z\sin^{-1}\frac{y}{x}$ where $x = 3r^2 + 2s$, $y = 4r - 2s^3$, $z = 2r^2 - 3s^2$, find $\frac{\partial u}{\partial r}, \frac{\partial u}{\partial s}$

Ans. $\frac{\partial u}{\partial r} = -\frac{6ryz}{x\sqrt{x^2 - y^2}} + \frac{4z}{\sqrt{x^2 - y^2}} + 4r\sin\frac{y}{x}$, $\frac{\partial u}{\partial s} = \frac{2yz}{x\sqrt{x^2 - y^2}} - \frac{6s^2 z}{\sqrt{x^2 - y^2}} - 6s\sin\frac{y}{x}$

16. If $z = \frac{\sin u}{\cos v}$, where $u = \frac{\cos y}{\sin x}$, $v = \frac{\cos x}{\sin y}$, find $\frac{\partial z}{\partial x}$.

Ans. $\frac{(u\cot x\cos u\sin y + z\sin v\sin x)}{(\cos v\sin y)}$

17. If v is a function of u, v, where $u = x - y$ and $v = xy$ prove that

$$x\frac{\partial^2 v}{\partial x^2} + y\frac{\partial^2 v}{\partial y^2} = (x+y)\left(\frac{\partial^2 v}{\partial u^2} + xy\frac{\partial^2 v}{\partial v^2}\right)$$

18. If $z = f(x, y)$, where $x = r\cos\theta$, $y = r\sin\theta$, prove that

$$\left(\frac{\partial z}{\partial x}\right)^2 + \left(\frac{\partial z}{\partial y}\right)^2 = \left(\frac{\partial z}{\partial r}\right)^2 + \frac{1}{r^2}\left(\frac{\partial z}{\partial \theta}\right)^2$$

and $$\frac{\partial^2 z}{\partial x^2} + \frac{\partial^2 z}{\partial y^2} = \frac{\partial^2 z}{\partial r^2} + \frac{1}{r^2}\left(\frac{\partial^2 z}{\partial \theta^2}\right) + \frac{1}{r}\left(\frac{\partial z}{\partial r}\right)$$

19. If $x = r \sin\theta \cos\phi$, $y = r \sin\theta \sin\phi$, $z = r\cos\theta$, $v = v(x, y, z)$, prove that

$$\left(\frac{\partial v}{\partial x}\right)^2 + \left(\frac{\partial v}{\partial y}\right)^2 + \left(\frac{\partial v}{\partial z}\right)^2 = \left(\frac{\partial v}{\partial r}\right)^2 + \left(\frac{1}{r}\frac{\partial v}{\partial \theta}\right)^2 + \left(\frac{1}{r\sin\theta}\frac{\partial v}{\partial \varphi}\right)^2$$

20. If v be a potential function such that $v = v(r)$ and $r^2 = x^2 + y^2 + z^2$, show that

$$\frac{\partial^2 v}{\partial x^2} + \frac{\partial^2 v}{\partial y^2} + \frac{\partial^2 v}{\partial z^2} = \frac{d^2 v}{dr^2} + \frac{2}{r}\frac{dv}{dr}$$

21. If $x = u + v + w$, $y = vw + wu + uv$, $z = uvw$ and F is a function of x, y, z, then show that

$$u\frac{\partial F}{\partial u} + v\frac{\partial F}{\partial v} + w\frac{\partial F}{\partial w} = x\frac{\partial F}{\partial x} + 2y\frac{\partial F}{\partial y} + 3z\frac{\partial F}{\partial z}$$

22. If $u = x + ay$ and $v = x + by$, transform the equation

$2\frac{\partial^2 z}{\partial x^2} - 5\frac{\partial^2 z}{\partial x\,\partial y} + 3\frac{\partial^2 z}{\partial y^2} = 0$ into the equation $\frac{\partial^2 z}{\partial u\,\partial v} = 0$, find the values of a and b.

(A.M.I.E.T.E., Summer 2000) **Ans.** $\left(a = 1, b = \frac{2}{3}\right), \left(a = \frac{2}{3}, b = 1\right)$

2.3 IMPORTANT DEDUCTIONS

Let $\qquad z = f(x, y)$, then

$$dz = \frac{\partial f}{\partial x}dx + \frac{\partial f}{\partial y}dy$$

If $\qquad z = 0,\ dz = 0$

$$0 = \frac{\partial f}{\partial x}dx + \frac{\partial f}{\partial y}dy \Rightarrow \frac{\partial f}{\partial y}\cdot\frac{dy}{dx} = -\frac{\partial f}{\partial x}$$

$$\Rightarrow \boxed{\frac{dy}{dx} = -\frac{\frac{\partial f}{\partial x}}{\frac{\partial f}{\partial y}}}$$ **[Remember]** ...(1)

We can find $\frac{d^2 y}{dx^2}$ by differentiating (1).

Let $\qquad \frac{\partial f}{\partial x} = p,\quad \frac{\partial f}{\partial y} = q,\quad \frac{\partial^2 f}{\partial x^2} = r,\quad \frac{\partial^2 f}{\partial x\,\partial y} = s,\quad \frac{\partial^2 f}{\partial y^2} = t$

From (1), $\frac{dy}{dx} = -\frac{p}{q}$.

On differentiating again, we obtain $\frac{d^2 y}{dx^2} = -\frac{q\frac{dp}{dx} - p\frac{dq}{dx}}{q^2}$ ...(2)

But $$\frac{dp}{dx} = \frac{\partial p}{\partial x} + \frac{\partial p}{\partial y}\frac{dy}{dx}$$

$$\Rightarrow \frac{dp}{dx} = \frac{\partial}{\partial x}\left(\frac{\partial f}{\partial x}\right) + \frac{\partial}{\partial y}\left(\frac{\partial f}{\partial x}\right)\cdot\frac{dy}{dx}$$

$$\frac{dp}{dx} = \frac{\partial^2 f}{\partial x^2} + \frac{\partial^2 f}{\partial y\,\partial x}\frac{dy}{dx} = \frac{\partial^2 f}{\partial x^2} + \frac{\partial^2 f}{\partial y\,\partial x}\left(\frac{-\frac{\partial f}{\partial x}}{\frac{\partial f}{\partial y}}\right) = r - s\frac{p}{q} = \frac{qr - ps}{q}$$

$$\frac{dq}{dx} = \frac{\partial q}{\partial x} + \frac{\partial q}{\partial y}\frac{dy}{dx} = \frac{\partial}{\partial x}\left(\frac{\partial f}{\partial y}\right) + \frac{\partial}{\partial y}\left(\frac{\partial f}{\partial y}\right)\left(\frac{-p}{q}\right)$$

$$= \frac{\partial^2 f}{\partial x \partial y} - \frac{\partial^2 f}{\partial y^2}\frac{p}{q} = s - \frac{tp}{q} = \frac{qs - tp}{q}$$

$\frac{\partial z}{\partial x} = p$, $\frac{\partial z}{\partial y} = q$, $\frac{\partial^2 z}{\partial x^2} = r$, $\frac{\partial^2 z}{\partial x \partial y} = s$

Making substitutions in (2), we obtain

$$\frac{d^2 y}{dx^2} = -\frac{q\,\frac{qr - ps}{q} - p\,\frac{qs - tp}{q}}{q^2} = -\frac{q^2 r - pqs - pqs + p^2 t}{q^3}$$

$$\Rightarrow \boxed{\frac{d^2 y}{dx^2} = -\frac{q^2 r - 2pqs + p^2 t}{q^3}}$$

Example 17. *If $x^3 + 3x^2y + 6xy^2 + y^3 = 1$, find $\frac{dy}{dx}$.*

Solution. Let $f(x, y) = x^3 + 3x^2y + 6xy^2 + y^3 - 1 = 0$

$$\frac{\partial f}{\partial x} = 3x^2 + 6xy + 6y^2 \quad \Rightarrow \quad \frac{\partial f}{\partial y} = 3x^2 + 12xy + 3y^2$$

$$\frac{dy}{dx} = -\frac{\frac{\partial f}{\partial x}}{\frac{\partial f}{\partial y}} = -\frac{3x^2 + 6xy + 6y^2}{3x^2 + 12xy + 3y^2} = -\frac{x^2 + 2xy + 2y^2}{x^2 + 4xy + y^2}$$ **Ans.**

Second method.

$$x^3 + 3x^2y + 6xy^2 + y^3 - 1 = 0.$$

On differentiation, we have

$$3x^2 + \left(3x^2\frac{dy}{dx} + 6xy\right) + \left(6x \,.\, 2y\frac{dy}{dx} + 6y^2\right) + 3y^2\frac{dy}{dx} = 0$$

$$\left(3x^2 + 12xy + 3y^2\right)\frac{dy}{dx} = -3x^2 - 6xy - 6y^2$$

$$\frac{dy}{dx} = \frac{-3x^2 - 6xy - 6y^2}{3x^2 + 12xy + 3y^2} = -\frac{x^2 + 2xy + 2y^2}{x^2 + 4xy + y^2}$$ **Ans.**

Example 18. *If $y^3 - 3ax^2 + x^3 = 0$, then prove that*

$$\frac{d^2 y}{dx^2} = -\frac{2a^2x^2}{y^5}$$

Solution. Let $f(x, y) = y^3 - 3ax^2 + x^3$ ...(1)

$$p = \frac{\partial f}{\partial x} = -6ax + 3x^2, \quad q = \frac{\partial f}{\partial y} = 3y^2$$

$$r = \frac{\partial^2 f}{\partial x^2} = -6a + 6x, \quad s = \frac{\partial^2 f}{\partial x \partial y} = 0, \quad t = \frac{\partial^2 f}{\partial y^2} = 6y$$

$$\frac{d^2 y}{dx^2} = -\frac{q^2 r - 2pqs + p^2 t}{q^3}$$ ...(2) [Art 2.3]

Putting the values of p, q, r, s and t in (2), we get

$$\frac{d^2 y}{dx^2} = -\frac{(3y^2)^2(-6a + 6x) - 2(-6ax + 3x^2)(3y^2)(0) + (-6ax + 3x^2)^2(6y)}{(3y^2)^3}$$

$$= -\frac{54y^4(-a + x) + 54(-2ax + x^2)^2 y}{27y^6}$$

$$\frac{d^2 y}{d x^2} = -\frac{2 y^3(-a+x) + 2(4a^2 x^2 + x^4 - 4 a x^3)}{y^5}$$

Putting the value of $y^3 = 3ax^2 - x^3$ from (1) in (3), we get

$$\frac{d^2 y}{d x^2} = -\frac{2(3 a x^2 - x^3)(-a+x) + 2(4a^2 x^2 + x^4 - 4 a x^3)}{y^5}$$

$$= -\frac{-6a^2x^2 + 6ax^3 + 2ax^3 - 2x^4 + 8a^2x^2 + 2x^4 - 8ax^3}{y^5}$$

$$\frac{d^2 y}{d x^2} = -\frac{2a^2 x^2}{y^5}$$ **Proved.**

Example 19. *If* $x^3 + y^3 - 3axy = 0$, *find* $\frac{dy}{dx}$ *and* $\frac{d^2y}{dx^2}$.

Solution. Let $f(x, y) = x^3 + y^3 - 3axy = 0$

$$p = \frac{\partial f}{\partial x} = 3x^2 - 3ay, \quad q = \frac{\partial f}{\partial y} = 3y^2 - 3ax$$

$$r = \frac{\partial^2 f}{\partial x^2} = 6x, \quad s = \frac{\partial^2 f}{\partial x \partial y} = -3a, \quad t = \frac{\partial^2 f}{\partial y^2} = 6y$$

$$\frac{dy}{dx} = -\frac{\frac{\partial f}{\partial x}}{\frac{\partial f}{\partial y}} = -\frac{3x^2 - 3ay}{3y^2 - 3ax} = \frac{ay - x^2}{y^2 - ax}$$

$$\frac{d^2 y}{d x^2} = -\frac{q^2 r - 2pqs + p^2 t}{q^3} \quad \text{...(1) [Art. 2.3]}$$

Putting the values of p, q, r, s and t in (1), we get

$$\frac{d^2 y}{d x^2} = -\frac{(3y^2 - 3ax)^2 6x - 2(3x^2 - 3ay)(3y^2 - 3ax)(-3a) + (3x^2 - 3ay)^2(6y)}{(3y^2 - 3ax)^3}$$

$$= -\frac{2x(y^2 - ax)^2 + 2a(x^2 - ay)(y^2 - ax) + 2y(x^2 - ay)^2}{(y^2 - ax)^3}$$

$$\frac{d^2 y}{d x^2} = \frac{2a^3 xy}{(ax - y^2)^3}$$ **Ans.**

Example 20. *Find* $\frac{dy}{dx}$ *when* $(\cos x)^y = (\sin y)^x$

Solution. Given equation can be written as :

$(\cos x)^y - (\sin y)^x = 0$

Here $f(x, y) = (\cos x)^y - (\sin y)^x = 0$

$$\frac{\partial f}{\partial x} = y(\cos x)^{y-1}(-\sin x) - (\sin y)^x \log \sin y$$

$$= -[y \sin x (\cos x)^{y-1} + (\sin y)^x \log \sin y]$$

$$\frac{\partial f}{\partial y} = (\cos x)^y \log \cos x - x(\sin y)^{x-1} \cos y$$

$$\frac{dy}{dx} = -\frac{\frac{\partial f}{\partial x}}{\frac{\partial f}{\partial y}} \quad \text{[Art. 2.3]}$$

$$\frac{dy}{dx} = \frac{y \sin x (\cos x)^{y-1} + (\sin y)^x \log \sin y}{(\cos x)^y \log \cos x - x(\sin y)^{x-1} \cos y} \qquad ...(1)$$

In (1), put $(\cos x)^y$ for $(\sin y)^x$

$$\frac{dy}{dx} = \frac{y \sin x (\cos x)^{y-1} + (\cos x)^y \log \sin y}{(\cos x)^y \log \cos x - \frac{x (\cos x)^y}{\sin y} . \cos y}$$

$$= \frac{(\cos x)^y \ [y \tan x + \log \sin y]}{(\cos x)^y \ [\log \cos x - x \cot y]} = \frac{y \tan x + \log \sin y}{\log \cos x - x \cot y} \qquad \textbf{Ans.}$$

Example 21. *If $f(x, y) = 0$ and $\phi (y, z) = 0$, show that*

$$\frac{\partial f}{\partial y} . \frac{\partial \varphi}{\partial z} . \frac{dz}{dx} = \frac{\partial f}{\partial x} . \frac{\partial \varphi}{\partial y} .$$

Solution. $f(x, y) = 0$ ...(1)

$\varphi (y, z) = 0$ ...(2)

Differentiating (1) w.r.t. x, we get

$$0 = \frac{\partial f}{\partial x} + \frac{\partial f}{\partial y} . \frac{dy}{dx} \Rightarrow \frac{dy}{dx} = -\frac{\frac{\partial f}{\partial x}}{\frac{\partial f}{\partial y}} \qquad ...(3)$$

Differentiating (2) w.r.t. 'y', we get

$$0 = \frac{\partial \phi}{\partial y} + \frac{\partial \phi}{\partial z} . \frac{dz}{dy} \Rightarrow \frac{dz}{dy} = -\frac{\frac{\partial \varphi}{\partial y}}{\frac{\partial \varphi}{\partial z}} \qquad ...(4)$$

Multiplying (3) and (4), we get

$$\frac{dy}{dx} \times \frac{dz}{dy} = \left(-\frac{\frac{\partial f}{\partial x}}{\frac{\partial f}{\partial y}} \right) \left(-\frac{\frac{\partial \varphi}{\partial y}}{\frac{\partial \varphi}{\partial z}} \right) \Rightarrow \frac{dz}{dx} = \frac{\frac{\partial f}{\partial x} \times \frac{\partial \varphi}{\partial y}}{\frac{\partial f}{\partial y} \times \frac{\partial \varphi}{\partial z}} \Rightarrow \frac{\partial f}{\partial y} . \frac{\partial \varphi}{\partial z} . \frac{dz}{dx} = \frac{\partial f}{\partial x} \frac{\partial \varphi}{\partial y}$$

Proved.

EXERCISE 2.2

Find $\frac{dy}{dx}$ in the following cases :

1. $x \sin (x - y) - (x + y) = 0$ **Ans.** $[y + x^2 \cos (x - y) / [x + x^2 \cos (x - y)]$
2. $x^y = y^x$ **Ans.** $y (y - x \log y) / x (x - y \log x)$
3. Find $\frac{d^2 y}{dx^2}$ If $ax^2 + 2 hxy + by^2 = 1$, **Ans.** $\frac{h^2 - ab}{(hx + by)^3}$
4. If $u = x^2 y + y^2 z + z^2 x$ and if z is defined implicitly as a function of x and y by the equation $x^2 + yz + z^3 = 0$

   find $\frac{\partial u}{\partial x}$, where u is considered as a function of x and y alone.

   Ans. $\frac{\partial u}{\partial x} = 2xy + z^2 - (y^2 + 2zx) \left(\frac{2x}{y + 3z^2} \right)$
5. Find $\frac{dy}{dx}$, if $\tan^{-1} \frac{x}{y} + y^3 + 1 = 0$ **Ans.** $\frac{y}{x - 3x^2 y^2 - 3y^4}$
6. If $f(x, y, z) = 0$, prove that

   $$\left(\frac{dx}{dy} \right)_z \cdot \left(\frac{dy}{dz} \right)_x \cdot \left(\frac{dz}{dx} \right)_y = -1$$

   $\left[\textbf{Hint}. \left(\frac{dx}{dy} \right)_z = -\frac{\partial f}{\partial y} \div \frac{\partial f}{\partial x} \right]$

2.4 TYPICAL CASES

Example 22. *If* $x = f(u, v)$, $y = \phi(u, v)$, *find* $\frac{\partial u}{\partial x}, \frac{\partial u}{\partial y}, \frac{\partial v}{\partial x}, \frac{\partial v}{\partial y}$.

Solution.

$$x = f(u, v) \quad \ldots(1)$$

$$y = \phi(u, v) \quad \ldots(2)$$

Differentiating (1), (2) w.r.t. x (treating y as constant), we obtain

$$1 = \frac{\partial f}{\partial u}\cdot\frac{\partial u}{\partial x} + \frac{\partial f}{\partial v}\cdot\frac{\partial v}{\partial x} \quad \ldots(3) \qquad \Rightarrow \qquad 0 = \frac{\partial \varphi}{\partial u}\cdot\frac{\partial u}{\partial x} + \frac{\partial \varphi}{\partial v}\cdot\frac{\partial v}{\partial x} \quad \ldots(4)$$

Solving the equations (3) and (4) for $\frac{\partial u}{\partial x}$ and $\frac{\partial v}{\partial x}$, we obtain

$$\frac{\partial u}{\partial x} = \frac{\frac{\partial \varphi}{\partial v}}{\frac{\partial f}{\partial u}\cdot\frac{\partial \varphi}{\partial v} - \frac{\partial f}{\partial v}\cdot\frac{\partial \varphi}{\partial u}} \quad \textbf{Ans.} \qquad \Rightarrow \qquad \frac{\partial v}{\partial x} = \frac{-\frac{\partial \varphi}{\partial u}}{\frac{\partial f}{\partial u}\cdot\frac{\partial \varphi}{\partial v} - \frac{\partial f}{\partial v}\cdot\frac{\partial \varphi}{\partial u}} \quad \textbf{Ans.}$$

Similarly, differentiating (1) and (2) w.r.t. y, we get

$$0 = \frac{\partial f}{\partial u}\cdot\frac{\partial u}{\partial y} + \frac{\partial f}{\partial v}\cdot\frac{\partial v}{\partial y} \quad \ldots(5) \qquad \Rightarrow \qquad 1 = \frac{\partial \varphi}{\partial u}\cdot\frac{\partial u}{\partial y} + \frac{\partial \varphi}{\partial v}\cdot\frac{\partial v}{\partial y} \quad \ldots(6)$$

Solving the equations (5) and (6) for $\frac{\partial u}{\partial y}$ and $\frac{\partial v}{\partial y}$, we obtain

$$\frac{\partial u}{\partial y} = \frac{-\frac{\partial f}{\partial v}}{\frac{\partial f}{\partial u}\cdot\frac{\partial \varphi}{\partial v} - \frac{\partial f}{\partial v}\cdot\frac{\partial \varphi}{\partial u}} \quad \textbf{Ans.} \qquad \Rightarrow \qquad \frac{\partial v}{\partial y} = \frac{\frac{\partial f}{\partial u}}{\frac{\partial f}{\partial u}\cdot\frac{\partial \varphi}{\partial v} - \frac{\partial f}{\partial v}\cdot\frac{\partial \varphi}{\partial u}} \quad \textbf{Ans.}$$

Example 23. *If* $x = u^2 - v^2$ *and* $y = uv$, *find*

$\frac{\partial u}{\partial x}, \quad \frac{\partial u}{\partial y}, \quad \frac{\partial v}{\partial x}$ *and* $\frac{\partial v}{\partial y}$ *(Nagpur University, Winter 2003)*

Solution. Here, we have

$$x = u^2 - v^2 \quad \ldots(1)$$

$$y = uv \quad \ldots(2)$$

Differentiating (1) w.r.t. x, we get

$$\Rightarrow \qquad 1 = 2u\frac{\partial u}{\partial x} - 2v\frac{\partial v}{\partial x} \quad \ldots(3)$$

Similarly differentiating (2) w.r.t. x, we get

$$0 = v\frac{\partial u}{\partial x} + u\frac{\partial v}{\partial x} \quad \ldots(4)$$

By solving (3) and (4), we get

$$\frac{\partial u}{\partial x} = \frac{u}{2u^2 + 2v^2}, \qquad \frac{\partial v}{\partial x} = -\frac{v}{2u^2 + 2v^2} \quad \textbf{Ans.}$$

On differentiating (1) and (2) w.r.t. y, we get

$$0 = 2u\frac{\partial u}{\partial y} - 2v\frac{\partial v}{\partial y} \quad \ldots(5)$$

$$1 = v\frac{\partial u}{\partial y} + u\frac{\partial v}{\partial y} \quad \ldots(6)$$

On solving (5) and (6), we get

$$\frac{\partial v}{\partial y} = \frac{u}{u^2 + v^2}, \quad \text{and} \quad \frac{\partial u}{\partial y} = \frac{v}{u^2 + v^2} \quad \textbf{Ans.}$$

Example 24. *If* $x^2 - y^2 + u^2 + 2v^2 = 1,\ x^2 + y^2 - u^2 - v^2 = 2,$

$$\text{find } \frac{\partial u}{\partial x} \text{ and } \frac{\partial v}{\partial x}.$$

Solution.
$$x^2 - y^2 + u^2 + 2v^2 = 1 \quad \text{...(1)}$$
$$x^2 + y^2 - u^2 - v^2 = 2 \quad \text{...(2)}$$

Differentiating (1) partially w.r.t. 'x', we get

$$2x + 2u.\frac{\partial u}{\partial x} + 4v.\frac{\partial v}{\partial x} = 0$$

$$\Rightarrow \qquad u\frac{\partial u}{\partial x} + 2v.\frac{\partial v}{\partial x} = -x \quad \text{...(3)}$$

Differentiating (2) partially w.r.t. 'x', we get

$$2x - 2u\frac{\partial u}{\partial x} - 2v.\frac{\partial v}{\partial x} = 0 \ \Rightarrow \ u\frac{\partial u}{\partial x} + v.\frac{\partial v}{\partial x} = x \quad \text{...(4)}$$

On subtracting (4) from (3), we get

$$v\frac{\partial v}{\partial x} = -2x \ \Rightarrow \ \frac{\partial v}{\partial x} = -\frac{2x}{v}$$

Similarly, we get
$$\frac{\partial u}{\partial x} = \frac{3x}{u}$$
Ans.

2.5 GEOMETRICAL INTERPRETATION OF $\frac{\partial z}{\partial x}$ AND $\frac{\partial z}{\partial y}$ *(Gujarat, I Semester, Jan. 2009)*

Let $z = f(x, y)$ be a surface S.

Let $y = k$ be a plane parallel to XZ – plane, passing through $P\ (x, k, z)$ cutting the surface $z = f(x, y)$ along the curve APB.

This section APB is a plane curve whose equations are

$$z = f(x, y)$$
$$y = k$$

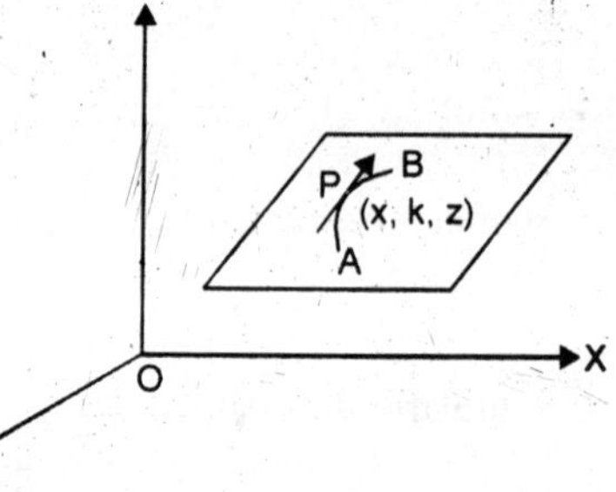

The slope of the tangent to this curve is given by $\frac{\partial z}{\partial x}$.

Similarly, $\frac{\partial z}{\partial y}$ is the slope of the tangent to the curve of intersection of the surface $z = f(x, y)$ with a plane parallel to YZ-plane.

2.6 TANGENT PLANE TO A SURFACE

Let $f(x, y, z) = 0$ be the equation of a surface S. Now we wish to find out the equation of a tangent plane to S at the point $P\ (x_1, y_1, z_1)$.

Let $Q\ (x_1 + \delta x_1,\ y_1 + \delta y_1,\ z_1 + \delta z_1)$ be a neighbouring point to P. Let the arc PQ be δs and the chord PQ be δc.

The direction cosines of PQ are

$$\frac{\delta x}{\delta c}, \quad \frac{\delta y}{\delta c}, \quad \frac{\delta z}{\delta c}$$

$$\Rightarrow \qquad \frac{\delta x}{\delta s}.\frac{\delta s}{\delta c}, \quad \frac{\delta y}{\delta s}.\frac{\delta s}{\delta c}, \quad \frac{\delta z}{\delta s}.\frac{\delta s}{\delta c}$$

The $\delta s \to 0$, $Q \to P$ and PQ tends to a tangent line PT. The direction cosines of PT are

$$\frac{dx}{ds}, \quad \frac{dy}{ds}, \quad \frac{dz}{ds} \qquad ...(1)$$

Differentiating F (x, y, z) = 0 w.r.t. 's', we get

$$\frac{\partial F}{\partial x}\frac{dx}{ds}+\frac{\partial F}{\partial y}\frac{dy}{ds}+\frac{\partial F}{\partial z}\frac{dz}{ds} = 0 \qquad ...(2)$$

From (1) and (2) it is clear that the tangent whose direction cosines are $\frac{dx}{ds}, \frac{dy}{ds}, \frac{dz}{ds}$ is perpendicular to a line having direction ratios

$$\frac{\partial F}{\partial x}, \quad \frac{\partial F}{\partial y}, \quad \frac{\partial F}{\partial z} \qquad ...(3)$$

There are a number of tangent lines at P to the curves joining P and Q. All these tangents will be perpendicular to the line having direction ratios as given by (3).
Hence all these tangent lines will lie in a plane known as tangent plane.

Equation of tangent plane

$$(x-x_1)\frac{\partial F}{\partial x}+(y-y_1)\frac{\partial F}{\partial y}+(z-z_1)\frac{\partial F}{\partial z} = 0$$

Equation of the normal to the plane.

$$\frac{x-x_1}{\frac{\partial F}{\partial x}}=\frac{y-y_1}{\frac{\partial F}{\partial y}}=\frac{z-z_1}{\frac{\partial F}{\partial z}}$$

Example 25. *Find the equation of the tangent plane and normal line to the surface* $x^2 + 2y^2 + 3z^2 = 12$ *at (1, 2, – 1).* *(AMIETE, Dec. 2010)*

Solution. $F(x, y, z) = x^2 + 2y^2 + 3z^2 - 12$

$$\frac{\partial F}{\partial x} = 2x, \quad \frac{\partial F}{\partial y} = 4y, \quad \frac{\partial F}{\partial z} = 6z$$

At the point (1, 2, – 1) $\quad \frac{\partial F}{\partial x} = 2, \quad \frac{\partial F}{\partial y} = 8, \quad \frac{\partial F}{\partial z} = -6$

Hence the equation of the tangent plane at (1, 2, – 1) is

$$2(x-1) + 8(y-2) - 6(z+1) = 0$$

$$\Rightarrow \quad 2x + 8y - 6z = 24 \quad \Rightarrow \quad x + 4y - 3z = 12$$

Equation of normal is $\frac{x-1}{2}=\frac{y-2}{8}=\frac{z+1}{-6}$

$$\Rightarrow \quad \frac{x-1}{1}=\frac{y-2}{4}=\frac{z+1}{-3}$$ **Ans.**

Example 26. *Find the equations of tangent plane and the normal line to the surface* $2x^2 + y^2 + 2z = 3$ *at the point (2, 1, – 3).*

Solution. $F(x, y, z) = 2x^2 + y^2 + 2z - 3 = 0$

$$\Rightarrow \quad \frac{\partial F}{\partial x} = 4x, \quad \frac{\partial F}{\partial y} = 2y, \quad \frac{\partial F}{\partial z} = 2$$

At the point (2, 1, – 3)

$$\frac{\partial F}{\partial x} = 8, \quad \frac{\partial F}{\partial y} = 2, \quad \frac{\partial F}{\partial z} = 2$$

Hence, the equation of the tangent plane at (2, 1, – 3) is

$$8(x-2)+2(y-1)+2(z+3)=0$$

$$\Rightarrow \quad 8x+2y+2z-12=0$$

$$\Rightarrow \quad 4x+y+z-6=0$$ **Ans.**

Equation of normal is

$$\frac{x-2}{8}=\frac{y-1}{2}=\frac{z+3}{2}$$

$$\Rightarrow \quad \frac{x-2}{4}=\frac{y-1}{1}=\frac{z+3}{1}$$ **Ans.**

Example 27. *Show that the surface $x^2-2yz+y^3=4$ is perpendicular to any number of the family of surfaces $x^2+1=(2-4a)y^2+az^2$ at the point of intersection* (1, – 1, 2).

Solution.

$$f(x,y,z)=x^2-2yz+y^3-4=0 \quad ...(1)$$

$$F(x,y,z)=x^2+1-(2-4a)y^2-az^2=0 \quad ...(2)$$

$$\frac{\partial f}{\partial x}=2x, \quad \frac{\partial f}{\partial y}=-2z+3y^2, \quad \frac{\partial f}{\partial z}=-2y$$

Direction ratios to the normal of the tangent plane to (1) are

$$2x, -2z+3y^2, -2y$$

DRs at the point (1, – 1, 2) are 2, – 1, 2.

Now differentiating (2), we get

$$\frac{\partial F}{\partial x}=2x, \quad \frac{\partial F}{\partial y}=-2(2-4a)y, \quad \frac{\partial F}{\partial z}=-2az.$$

Direction ratios to the normal of the tangent plane to (2) are

$$2x, (-4+8a)y, -2az.$$

DRs at the point (1, – 1, 2) are 2, 4 – 8*a*, – 4*a*

Now

$$l_1 l_2+m_1 m_2+n_1 n_2 = (2)(2)+(-1)(4-8a)+2(-4a)$$

$$= 4-4+8a-8a=0.$$

Hence the given surfaces are perpendicular at (1, – 1, 2). **Ans.**

EXERCISE 2.3

1. Find the equation of tangent plane and the normal line to the surface

   $xyz=6$ at (1, 2, 3). **Ans.** $6x+3y+2z=18, \ \frac{x-1}{6}=\frac{y-2}{3}=\frac{z-3}{2}$

2. Find the equations of the tangent plane and the normal to the surface $z^2=4(1+x^2+y^2)$ at (2, 2, 6). **Ans.** $4x+4y-3z+2=0, \ \frac{x-2}{4}=\frac{y-2}{4}=\frac{z-6}{-3}$

3. Find the equations of the tangent plane and the normal to the surface

   $\frac{x^2}{2}-\frac{y^2}{3}=z$ at (2, 3, – 1) **Ans.** $2x-2y-z+1=0, \ \frac{x-2}{2}=\frac{y-3}{-2}=\frac{z+6}{-1}$

4. Show that the plane $3x+12y-6z-17=0$, touch the conicoid $3x^2-6y^2+9z^2+17=0$. Find also the point of contact. **Ans.** $\left(1, 2, \frac{2}{3}\right)$

5. Show that the plane $ax+by+cz+d=0$ touches the surface $px^2+qy^2+2z=0$, if $\frac{a^2}{p}+\frac{b^2}{q}+2cd=0$.

CHAPTER 3

MAXIMA AND MINIMA OF FUNCTIONS (TWO VARIABLES)

3.1 MAXIMUM VALUE

A function $f(x, y)$ is said to have a maximum value at $x = a$, $y = b$, if there exists a small neighbourhood of (a, b) such that,

$$f(a, b) > f(a + h, b + k)$$

Minimum Value. A function $f(x, y)$ is said to have a minimum value for $x = a$, $y = b$, if there exists a small neighbourhood of (a, b) such that.

$$f(a, b) < f(a + h, b + k)$$

The maximum and minimum values of a function are also called extreme or extremum values of the function.

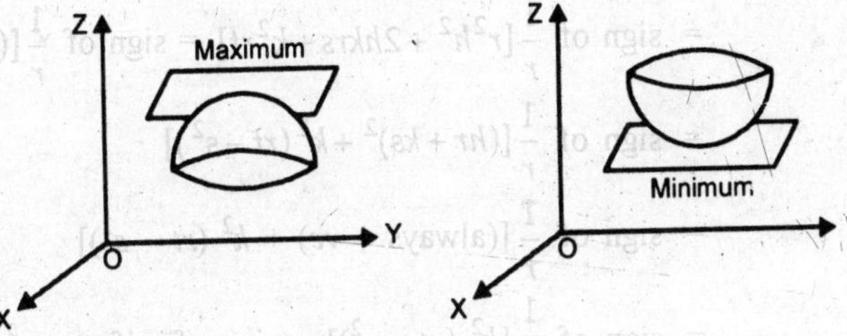

Saddle Point or Minimax

It is a point where function is neither maximum nor minimum.

Geometrically such a surface (looks like the leather seat on the back of a horse) forms a ridge rising in one direction and falling in another direction.

3.2 CONDITIONS FOR EXTREMUM VALUES

If $f(a + h, b + k) - f(a, b)$ remains of the same sign for all values (positive or negative) of h, k then $f(a, b)$ is said to be extremum value of $f(x, y)$ at (a, b)

(*i*) If $f(a + h, b + k) - f(a, b) < 0$, then $f(a, b)$ is maximum.

(*ii*) If $f(a + h, b + k) - f(a, b) > o$, then $f(a, b)$ is minimum.

By Taylor's Theorem

$$f(a+h, b+k) = f(a, b) + \left(h\frac{\partial f}{\partial x} + k\frac{\partial f}{\partial y}\right)_{(a,b)} + \frac{1}{2!}\left[h^2\frac{\partial^2 f}{\partial x^2} + 2hk\frac{\partial^2 f}{\partial x \partial y} + k^2\frac{\partial^2 f}{\partial y^2}\right] + \ldots$$

$$\Rightarrow \quad f(a + h, b + k) - f(a, b) = \left(h\frac{\partial f}{\partial x} + k\frac{\partial f}{\partial y}\right)_{(a,b)}$$

$$+ \frac{1}{2!}\left[h^2\frac{\partial^2 f}{\partial x^2} + 2hk\frac{\partial^2 f}{\partial x \partial y} + k^2\frac{\partial^2 f}{\partial y^2}\right] + \ldots \quad \ldots(1)$$

$$\Rightarrow \quad f(a + h, b + k) - f(a, b) = \left(h\frac{\partial f}{\partial x} + k\frac{\partial f}{\partial y}\right)_{(a,b)} \quad \ldots(2)$$

For small values of h, k, the second and higher order terms are still smaller and hence may be neglected.

The sign of L.H.S. of (2) is governed by $h\frac{\partial f}{\partial x}+k\frac{\partial f}{\partial y}$ which may be positive or negative depending on h, k.

Hence, the necessary condition for (a, b) to be a maximum or minimum is that

$$\left(h\frac{\partial f}{\partial x}+k\frac{\partial f}{\partial y}\right) = 0 \Rightarrow \frac{\partial f}{\partial x}=0, \quad \frac{\partial f}{\partial y}=0$$

By solving the equations, we get, point $x = a$, $y = b$ which may be maximum or minimum value.

Then from (1)

$$f(a+h, b+k) - f(a, b) = \frac{1}{2!}\left[h^2\frac{\partial^2 f}{\partial x^2}+2hk\frac{\partial^2 f}{\partial x\partial y}+k^2\frac{\partial^2 f}{\partial y^2}\right]$$

$$= \frac{1}{2!}[h^2r+2hks+k^2t] \qquad \text{...(3)}$$

where $r = \frac{\partial^2 f}{\partial x^2}, s = \frac{\partial^2 f}{\partial x\partial y}, t = \frac{\partial^2 f}{\partial y^2}$ at (a, b)

Now the sign of L.H.S. of (3) is sign of $[rh^2 + 2hks + k^2 t]$

$= \text{sign of } \frac{1}{r}[r^2h^2+2hkrs+k^2rt] = \text{sign of } \frac{1}{r}[(r^2h^2+2hkrs+k^2s^2)+(-k^2s^2+k^2rt)]$

$= \text{sign of } \frac{1}{r}[(hr+ks)^2+k^2(rt-s^2)]$

$= \text{sign of } \frac{1}{r}[(\text{always} + ve) + k^2(rt-s^2)]$ $\qquad [(hr+ks)^2 = +ve]$

$= \text{sign of } \frac{1}{r}[k^2(rt-s^2)] = \text{sign of } r \text{ if } rt - s^2 > 0$

Hence, if $rt - s^2 > 0$, then $f(x, y)$ has a maximum or minimum at (a, b) according as $r < 0$ or $r > 0$.

Note: (*i*) If $rt - s^2 < 0$, then L.H.S. will change with h and k hence there is no maximum or minimum at (a, b), *i.e.*, it is a saddle point.

(*ii*) If $rt - s^2 = 0$, then $rh^2 + 2shk + tk^2 = \frac{1}{r}[(rh+sk)^2 + k^2(rt-s^2)]$

$= \frac{1}{r}(rh+sk)^2$ which is zero for values of h, k, such that

$\Rightarrow \quad \frac{h}{k} = -\frac{s}{r}$

This is, therefore, a doubtful case, further investigation is required.

3.3 WORKING RULE TO FIND EXTREMUM VALUES

(*i*) Differentiate $f(x, y)$ and find out

$$\frac{\partial f}{\partial x}, \frac{\partial f}{\partial y}, \frac{\partial^2 f}{\partial x^2}, \frac{\partial^2 f}{\partial x\partial y}, \frac{\partial^2 f}{\partial y^2}$$

(*ii*) Put $\frac{\partial f}{\partial x} = 0$ and $\frac{\partial f}{\partial y} = 0$ and solve these equations for x and y. Let (a, b) be the values of (x, y).

(*iii*) Evaluate $r = \frac{\partial^2 f}{\partial x^2}, s = \frac{\partial^2 f}{\partial x\partial y}, t = \frac{\partial^2 f}{\partial y^2}$ for these values (a, b).

(iv) If $rt - s^2 > 0$ and

(a) $r < 0$, then $f(x, y)$ has a maximum value.

(b) $r > 0$, then $f(x, y)$ has a minimum value.

(v) If $rt - s^2 < 0$, then $f(x, y)$ has no extremum value at the point (a, b).

(iv) If $rt - s^2 = 0$, then the case is doubtful and needs further investigation.

Note: The point (a, b), which are the roots of $\frac{\partial f}{\partial x} = 0$ and $\frac{\partial f}{\partial y} = 0$ are called stationary points.

Example 1. *Find the absolute maximum and minimum values of*

$$f(x, y) = 2 + 2x + 2y - x^2 - y^2$$

on triangular plane in the first quadrant, bounded by the lines $x = 0$, $y = 0$ and $y = 9 - x$. *(Gujarat, I Semester, Jan. 2009)*

Solution. We have, $f(x, y) = 2 + 2x + 2y - x^2 - y^2$

$$\frac{\partial f}{\partial x} = 2 - 2x, \qquad \frac{\partial f}{\partial y} = 2 - 2y$$

$$\frac{\partial^2 f}{\partial x^2} = -2, \qquad \frac{\partial^2 f}{\partial x \partial y} = 0, \qquad \frac{\partial^2 f}{\partial x^2} = -2$$

For maxima and minima,

$$\frac{\partial f}{\partial x} = 0 \quad \Rightarrow \quad 2 - 2x = 0 \quad \Rightarrow \quad x = 1$$

and $$\frac{\partial f}{\partial y} = 0 \quad \Rightarrow \quad 2 - 2y = 0 \quad \Rightarrow \quad y = 1$$

At (1, 1) $\qquad rt - s^2 = (-2)(-2) - 0 = +4.$

Here $\qquad r = \frac{\partial^2 f}{\partial x^2} = -2 = -\text{ve}$

Hence $f(x, y)$ is maximum at (1, 1).

Maximum value of $f(x, y) = 2 + 2 + 2 - 1 - 1 = 4.$ **Ans.**

Example 2. *Examine the function $f(x, y) = y^2 + 4xy + 3x^2 + x^3$ for extreme values.* *(M.U. 2008)*

Solution. We have, $f(x, y) = y^2 + 4xy + 3x^2 + x^3$

$$p = \frac{\partial f}{\partial x} = 4y + 6x + 3x^2 \qquad q = \frac{\partial f}{\partial y} = 2y + 4x$$

$$r = \frac{\partial^2 f}{\partial x^2} = 6 + 6x \qquad s = \frac{\partial^2 f}{\partial x \partial y} = 4$$

$$t = \frac{\partial^2 f}{\partial y^2} = 2$$

For maxima or minima

$$\frac{\partial f}{\partial x} = 0 \quad \text{and} \quad \frac{\partial f}{\partial y} = 0$$

$\Rightarrow \quad 4y + 6x + 3x^2 = 0 \quad$ ...(1) $\qquad 2y + 4x = 0$

$\Rightarrow \quad y = -2x \quad$ ...(2)

Putting the value of y from (2) in (1), we get

$$4(-2x) + 6x + 3x^2 = 0$$

$\Rightarrow \quad 3x^2 + 6x - 8x = 0$

$\Rightarrow \quad 3x^2 - 2x = 0 \quad \Rightarrow \quad x(3x - 2) = 0$

$$\Rightarrow \quad x = 0 \text{ or } x = \frac{2}{3}$$

when $x = 0$ then $y = 0$

when $x = \frac{2}{3}$ then $y = -2\left(\frac{2}{3}\right) = -\frac{4}{3}$

Thus, the stationary points are (0, 0) and $\left(\frac{2}{3}, \frac{-4}{3}\right)$,

	(0, 0)	$\left(\frac{2}{3}, \frac{-4}{3}\right)$
$r = 6 + 6x$	6	10
$s = 4$	4	4
$t = 2$	2	2
$rt - s^2$	-4	$+4$

At (0, 0) there is no extremum value, since $rt - s^2 < 0$;

At $\left(\frac{2}{3}, \frac{-4}{3}\right)$, $rt - s^2 > 0$, $r > 0$.

Therefore $\left(\frac{2}{3}, \frac{-4}{3}\right)$ is a point of minimum value.

The minimum value of $f\left(\frac{2}{3}, -\frac{4}{3}\right) = \left(\frac{-4}{3}\right)^2 + 4\left(\frac{2}{3}\right)\left(-\frac{4}{3}\right) + 3\left(\frac{2}{3}\right)^2 + \left(\frac{2}{3}\right)^3$

$$\Rightarrow \quad f\left(\frac{2}{3}, \frac{-4}{3}\right) = \frac{16}{9} - \frac{32}{9} + \frac{12}{9} + \frac{8}{27} = \frac{8}{27} - \frac{4}{9} = -\frac{4}{27}$$ **Ans.**

Example 3. *Show that the minimum value of* $u = xy + a^3\left(\frac{1}{x} + \frac{1}{y}\right)$ *is* $3a^2$.

(*M. U.* 2002)

Solution. We have,

$$f(x, y) = xy + a^3\left(\frac{1}{x} + \frac{1}{y}\right)$$

$$p = \frac{\partial f}{\partial x} = y - \frac{a^3}{x^2} \qquad q = \frac{\partial f}{\partial y} = x - \frac{a^3}{y^2}$$

$$r = \frac{\partial^2 f}{\partial x^2} = \frac{2a^3}{x^3} \qquad s = \frac{\partial^2 f}{\partial x \partial y} = 1$$

$$t = \frac{\partial^2 f}{\partial y^2} = \frac{2a^3}{y^3}$$

For maxima and minima

$$\frac{\partial f}{\partial x} = 0 \quad \text{and} \quad \frac{\partial f}{\partial y} = 0$$

$$\Rightarrow \quad y - \frac{a^3}{x^2} = 0 \qquad \Rightarrow \quad x - \frac{a^3}{y^2} = 0$$

$$\Rightarrow \quad x^2 y = a^3 \quad \ldots(1) \qquad \Rightarrow \quad x = \frac{a^3}{y^2} \quad \ldots(2)$$

Putting the value of x from (2) in (1), we get

$$\left(\frac{a^3}{y^2}\right)^2 y = a^3 \Rightarrow \frac{a^6}{y^3} = a^3 \Rightarrow y = a \quad \ldots(3)$$

On putting the value of y from (3) in (1), we get

$$x^2\, a = a^3 \Rightarrow x^2 = a^2 \Rightarrow x = \pm a$$

Thus, the stationary pairs are (a, a) and $(-a, a)$

	(a, a)	$(-a, a)$
$r = \dfrac{2a^3}{x^3}$	2	-2
$s = 1$	1	1
$t = \dfrac{2a^3}{y^3}$	2	2
$rt - s^2$	$+3$	-5

At $(-a, a)$, $rt - s^2 = -ve$

Hence, we reject this pair.

At (a, a), $r = +ve$, $rt - s^2 = +ve$

Hence, $f(x, y)$ is minimum at (a, a).

$$\text{Minimum value} = a^2 + a^3\left(\frac{1}{a}+\frac{1}{a}\right) = 3a^2$$ **Ans.**

Example 4. *Examine $f(x, y) = x^3 + y^3 - 3axy$ for maximum and minimum values.* *(GBTU, Dec. 2012. M.U. 2004, 2003; U.P. I sem. Dec. 2004)*

Solution. We have, $f(x, y) = x^3 + y^3 - 3axy$

$$p = \frac{\partial f}{\partial x} = 3x^2 - 3ay, \quad q = \frac{\partial f}{\partial y} = 3y^2 - 3ax$$

$$r = \frac{\partial^2 f}{\partial x^2} = 6x, \quad s = \frac{\partial^2 f}{\partial x \partial y} = -3a, \quad t = \frac{\partial^2 f}{\partial y^2} = 6y$$

For maxima and minima

$$\frac{\partial f}{\partial x} = 0 \text{ and } \frac{\partial f}{\partial y} = 0.$$

$$3x^2 - 3ay = 0 \qquad\qquad 3y^2 - 3ax = 0$$

$$\Rightarrow x^2 = ay \Rightarrow y = \frac{x^2}{a} \quad ...(1) \qquad \Rightarrow y^2 = ax \quad ...(2)$$

Putting the value of y from (1) in (2), we get

$$x^4 = a^3 x \Rightarrow x(x^3 - a^3) = 0$$

$$\Rightarrow x(x - a)(x^2 + ax + a^2) = 0$$

$$x = 0, a$$

Putting $x = 0$ in (1), we get $y = 0$

Putting $x = a$ in (1), we get $y = a$,

	$(0, 0)$	(a, a)
r	0	$6a$
s	$-3a$	$-3a$
t	0	$6a$
$rt - s^2$	$-9a^2 < 0$	$27a^2 > 0$

At (0, 0) there is no extremum value, since $rt - s^2 < 0$.

At (a, a), $rt - s^2 > 0$, $r > 0$

Therefore (a, a) is a point of minimum value.

The minimum value of $f(a, a) = a^3 + a^3 - 3a^3 = -a^3$ **Ans.**

Example 5. *Show that the function*

$$f(x, y) = x^3 + y^3 - 63(x + y) + 12xy$$

is maximum at (– 7, – 7) *and minimum at* (3, 3.)

Solution. We have, $f(x, y) = x^3 + y^3 - 63(x + y) + 12xy$ ...(1)

$$\frac{\partial f}{\partial x} = 3x^2 - 63 + 12y, \qquad \frac{\partial f}{\partial y} = 3y^2 - 63 + 12x$$

$$\frac{\partial^2 f}{\partial x^2} = 6x, \quad \frac{\partial^2 f}{\partial x \partial y} = 12, \qquad \frac{\partial^2 f}{\partial y^2} = 6y$$

For extremum, we have

$$p = \frac{\partial f}{\partial x} = 3x^2 - 63 + 12y = 0 \quad \Rightarrow x^2 + 4y - 21 = 0 \quad ...(2)$$

$$q = \frac{\partial f}{\partial x} = 3y^2 - 63 + 12x = 0 \quad \Rightarrow y^2 + 4x - 21 = 0 \quad ...(3)$$

$$r = \frac{\partial^2 f}{\partial x^2} = 6x \qquad s = \frac{\partial^2 f}{\partial x \partial y} = 12 \qquad t = \frac{\partial^2 f}{\partial y^2} = 6y$$

We have to solve (2) and(3) for x, y.

On subtracting (3) from (2), we have

$$x^2 - y^2 - 4(x - y) = 0 \quad \Rightarrow \quad (x - y)(x + y - 4) = 0$$

$$x = y \text{ and } x + y = 4 \quad ...(4)$$

If $x = y$ then (2) becomes, $x^2 + 4x - 21 = 0$, $\quad (x + 7)(x - 3) = 0$

$$x = -7, \text{ and } x = 3$$

$$y = -7, \text{ and } y = 3$$

Two stationary points are (–7, –7) and (3, 3)

On solving (2) and (4), we get

$$x^2 + 4(4 - x) - 21 = 0 \Rightarrow x^2 - 4x - 5 = 0$$

$$\Rightarrow \quad (x - 5)(x + 1) = 0$$

$$x = -1, x = 5$$

$$y = 5, y = -1$$

Two more stationary points are (–1, 5) and (5, –1)

Hence four possible extremum points of $f(x, y)$ are (– 7, – 7), (3, 3), (– 1, 5) and (5, – 1) may be.

	(– 7, – 7)	(3, 3)	(– 1, 5)	(5, – 1)
$r = 6x$	– 42	+ 18	– 6	30
$s = 12$	12	12	12	12
$t = 6y$	– 42	18	30	– 6
$rt - s^2$	+ 1620	+ 180	– 324	– 324

At (– 7, – 7)

$$r = -\text{ve}, \text{ and } rt - s^2 = +\text{ve}$$

Hence, $f(x, y)$ is maximum at (– 7, – 7).

At (3, 3)

$$r = +\text{ve}, \text{ and } rt - s^2 = +\text{ve}$$

Hence $f(x, y)$ is minimum at (3, 3). **Proved.**

Example 6. *Find the points on the surface $z^2 = xy + 1$ nearest to the origin.*

(*M.U.* 2002, 2001; *Nagpur University, Summer* 2005, *Winter* 2004, 2002)

Solution. Let the point on the surface $z^2 = xy + 1$ be (x, y, z). ...(1)

$\therefore$ Its distance from origin is $r = \sqrt{x^2+y^2+z^2}$.

Let $u = r^2 = x^2 + y^2 + z^2$ ...(2)

We have to find the values of x, y, z for which 'u' is minimum.

Put the value of z^2 from (1) in (2) to get the equation in two variables only.

$$u = x^2 + y^2 + xy + 1$$

For maximum or minimum, we have

$$\frac{\partial u}{\partial x} = 2x + y = 0 \quad ...(3) \qquad \text{and} \qquad \frac{\partial u}{\partial y} = 2y + x = 0 \quad ...(4)$$

Solving equations (3) and (4), we get $x = 0$, $y = 0$

On putting $x = 0$, $y = 0$ in $z^2 = xy + 1$, we get

$$z^2 = 0 + 1 = 1 \qquad \therefore z = \pm 1$$

Now, $\frac{\partial^2 u}{\partial x^2} = 2, \quad \frac{\partial^2 u}{\partial y^2} = 2, \quad \frac{\partial^2 u}{\partial x\, \partial y} = 1$

$$r = 2, \qquad t = 2, \qquad s = 1$$

$$rt - s^2 = 2 \times 2 - (1)^2 = +3$$

But $r = +2$, so u in minimum for $(0, 0, \pm 1)$ **Ans.**

Example 7. *Divide* 120 *into three parts so that the sum of their products taken two at a time shall be maximum.*

Solution. Let x, y, z be the number whose sum is 120.

i.e., $x + y + z = 120 \Rightarrow z = 120 - x - y$ ...(1)

Let $f = xy + yz + zx$

$\Rightarrow$ $f = xy + y(120 - x - y) + x(120 - x - y)$ [Using (1)]

$\Rightarrow$ $f = xy + 120y - xy - y^2 + 120x - x^2 - xy$

$\Rightarrow$ $f = 120x + 120y - xy - x^2 - y^2$

$$p = \frac{\partial f}{\partial x} = 120 - y - 2x \qquad q = \frac{\partial f}{\partial y} = 120 - x - 2y$$

$$r = \frac{\partial^2 f}{\partial x^2} = -2 \qquad s = \frac{\partial^2 f}{\partial x\, \partial y} = -1$$

$$t = \frac{\partial^2 f}{\partial y^2} = -2$$

For maxima and minima

$$\frac{\partial f}{\partial x} = 0 \qquad \text{and} \qquad \frac{\partial f}{\partial y} = 0$$

$\Rightarrow$ $120 - y - 2x = 0$ | $\Rightarrow$ $120 - x - 2y = 0$ ...(3)

$\Rightarrow$ $y = 120 - 2x$ ...(2)

Putting the value of y from (2) in (3), we get

$120 - x - 2(120 - 2x) = 0$

$\Rightarrow$ $120 - x - 240 + 4x = 0 \Rightarrow 3x = 120 \Rightarrow x = 40$

Putting the value of x in (1), we get

$$y = 120 - 2(40) = 120 - 80 = 40$$

Thus, the stationary pair is (40, 40).

	(40, 40)
$r = -2$	-2
$s = -1$	-1
$t = -2$	-2
$rt - s^2$	$+3$

At (40, 40), $r = -$ ve and $rt - s^2 = +$ ve
Hence, f is maximum at (40, 40).
Putting $x = 40$, $y = 40$ in (1), we get

$$40 + 40 + z = 120 \Rightarrow z = 40$$

Hence, f is maximum at $x = 40$, $y = 40$ and $z = 40$. **Ans.**

Example 8. *Divide 24 into three parts such that continued product of first, square of second and cube of third is a maximum.* (*Nagpur University, Summer 2005, Winter 2001*)

Solution. Let, 24 be divided into x, y, z then $x + y + z = 24 \Rightarrow z = 24 - x - y$
and $f(x, y, z) = x^3 y^2 z = x^3 y^2 (24 - x - y)$

$$\frac{\partial f}{\partial x} = 72 x^2 y^2 - 4 x^3 y^2 - 3 x^2 y^3$$

$$\frac{\partial f}{\partial y} = 48 x^3 y - 2 x^4 y - 3 x^3 y^2$$

For a maximum value of f, $\frac{\partial f}{\partial x} = 0$ and $\frac{\partial f}{\partial y} = 0$.

Hence, we have $x^2y^2 (72 - 4x - 3y) = 0$

$$\Rightarrow \quad 72 - 4x - 3y = 0 \quad ...(1)$$

and $x^3 y (48 - 2x - 3y) = 0$

$$\Rightarrow \quad 48 - 2x - 3y = 0 \quad ...(2)$$

By (2) × 2 – (1), we get $24 - 3y = 0 \Rightarrow y = 8$

Hence, $x = \frac{48 - 24}{2} = 12$

	$x = 12, y = 8$
$r = \frac{\partial^2 f}{\partial x^2} = 144 xy^2 - 12x^2 y^2 - 6xy^3$	$144.(12).(8)^2 - 12 (12)^2 8^2 - 6 (12) 8^3 = 12(64)(-48)$ $= -36864$
$t = \frac{\partial^2 f}{\partial y^2} = 48 x^3 - 2x^4 - 6x^3 y$	$48.(12)^3 - 2 (12)^4 - 6(12)^3 8 = (12)^3 (-24)$ $= -41472$
$s = \frac{\partial^2 f}{\partial x \partial y} = 144x^2 y - 8x^3 y - 9x^2 y^2$	$144.(12)^2 (8) - 8(12)^3 8 - 9(12)^2 (8)^2 = 144(8)(-24)$ $= -27648$
$rt - s^2$	$(-36864)(-41472) - (-27648)^2 = 764411904 > 0$

Since $rt - s^2 > 0$ and $r < 0$, so f is maximum at (12, 8)
Putting $x = 12$, $y = 8$ in $x + y + z = 24$, we get $z = 4$.
Hence, the division is 12, 8 and 4. **Ans.**

Example 9. *A rectangular box, open at the top, is to have a volume of 32 c.c. Find the dimensions of the box requiring least material for its construction.*
(*A.M.I.E.T.E., June 2009, M.U. 2009; U.P. I semester Dec. 2005, A.M.I.E Summer 2001*)

Solution. Let l, b and h be the length, breadth, and height of the box respectively and S its surface area and V the volume.

$$V = 32 \text{ c.c.}$$

$$\Rightarrow \quad l\,b\,h = 32 \text{ or } b = \frac{32}{lh}$$

$$S = 2(l + b)h + l\,b \quad ...(1)$$

Putting the value of b in (1), we get

$$S = 2\left(l + \frac{32}{lh}\right)h + l\left(\frac{32}{lh}\right)$$

$$S = 2\,l\,h + \frac{64}{l} + \frac{32}{h} \qquad \ldots(2)$$

Differentiating (2) partially w.r.t. l, we get

$$\frac{\partial S}{\partial l} = 2\,h - \frac{64}{l^2} \qquad \ldots(3)$$

Differentiating (2) partially w.r.t. h, we get

$$\frac{\partial S}{\partial h} = 2\,l - \frac{32}{h^2} \qquad \ldots(4)$$

For maximum and minimum S, we get

$$\frac{\partial S}{\partial l} = 0 \Rightarrow 2\,h - \frac{64}{l^2} = 0 \Rightarrow h = \frac{32}{l^2} \qquad \ldots(5)$$

$$\frac{\partial S}{\partial h} = 0 \Rightarrow 2\,l - \frac{32}{h^2} = 0 \Rightarrow l = \frac{16}{h^2} \qquad \ldots(6)$$

From (5) and (6), $l = 4$, $h = 2$ and $b = 4$

$$\frac{\partial^2 S}{\partial l^2} = \frac{128}{l^3} = \frac{128}{64} = 2$$

$$\frac{\partial^2 S}{\partial l\,\partial h} = 2$$

$$\frac{\partial^2 S}{\partial h^2} = \frac{64}{h^3} = \frac{64}{8} = 8$$

$$\frac{\partial^2 S}{\partial l^2}\cdot\frac{\partial^2 S}{\partial h^2} - \left(\frac{\partial^2 S}{\partial l\,\partial h}\right)^2 = (2)\,(8) - (2)^2 = +\,12$$

$\dfrac{\partial^2 S}{\partial l^2} = +\,2$, so S is minimum for $l = 4$ cm, $b = 4$ cm, $h = 2$ cm **Ans.**

Example 10. *In a plane triangle ABC, find the maximum value of cos A cos B cos C.* (*Nagpur University Summer* 2000)

Solution.

$$\cos A \cos B \cos C = \cos A \cos B \cos [\pi - (A + B)]$$
$$= -\cos A \cos B \cos (A + B)$$

Let

$$f(A, B) = -\cos A \cos B \cos (A + B)$$

$$\frac{\partial f}{\partial A} = -\cos B\,[-\sin A \cos (A + B) - \cos A \sin (A + B)]$$
$$= \cos B \sin [A + (A + B)]$$
$$= \cos B \sin (2A + B)$$

$$\frac{\partial f}{\partial B} = -\cos A\,[-\sin B \cos (A + B) - \cos B \sin (A + B)]$$
$$= \cos A \sin [B + (A + B)] = \cos A \sin (A + 2B)$$

$$r = \frac{\partial^2 f}{\partial A^2} = 2 \cos B \cos (2A + B)$$

$$s = \frac{\partial^2 f}{\partial A \partial B} = -\sin B \sin (2A + B) + \cos B \cos (2A + B)$$
$$= \cos [B + (2A + B)] = \cos (2A + 2B)$$

$$t = \frac{\partial^2 f}{\partial B^2} = 2\cos A\cos(A+2B)$$

For maxima and minima

$$\frac{\partial f}{\partial A} = 0 \text{ and } \frac{\partial f}{\partial B} = 0$$

$$\cos B\sin(2A+B) = 0 \qquad ...(1)$$

$$\cos A\sin(A+2B) = 0 \qquad ...(2)$$

From (1) if $\cos B = 0$, then $B = \pi/2$

$$\cos A\sin(A+2B) = 0$$

From (2), $\cos A\sin(A+\pi) = 0 \Rightarrow \cos A(-\sin A) = 0$

$\Rightarrow$ either $\cos A = 0$, *i.e.* $A = \pi/2$ which is not possible.

$\sin A = 0$ *i.e.* $A = 0$ or π, which is not possible.

$$\left[\begin{array}{l} \because\ A+B+C=\pi \\ \frac{\pi}{2}+\frac{\pi}{2}+C=\pi \Rightarrow C=0 \end{array}\right]$$

From (1) if $\cos B \neq 0$, similarly $\cos A \neq 0$.

From (1), $\sin(2A+B) = 0 \Rightarrow 2A+B = \pi$ ...(3)

From (2), $\sin(A+2B) = 0 \Rightarrow A+2B = \pi$ ...(4)

Solving (3) and (4), we get $A = B = \pi/3$

$$r = 2\cos\frac{\pi}{3}\cos\pi = 2\times\frac{1}{2}(-1) = -1$$

$$s = \cos\frac{4\pi}{3} = -\frac{1}{2}$$

$$t = 2\cos\frac{\pi}{3}\cos\pi = 2\left(\frac{1}{2}\right)(-1) = -1$$

$$rt - s^2 = 1 - \frac{1}{4} = \frac{3}{4} = +\text{ ve}$$

Also, $r = -1 = -$ ve

$\Rightarrow$ $f(A, B)$ is maximum at $A = B = \frac{\pi}{3}$

$$\text{Maximum value} = f(\pi/3, \pi/3) = \cos\frac{\pi}{3}\cos\frac{\pi}{3}\cos\frac{\pi}{3}$$

$$= \left(\frac{1}{2}\right)\left(\frac{1}{2}\right)\left(\frac{1}{2}\right) = \frac{1}{8}.$$

Ans.

Example 11. *Prove that if the perimeter of a triangle is constant, its area is maximum when the triangle is equilateral.*

Solution. Let a, b, c, be the sides of a triangle whose perimeter $2s$ is constant.

Then $2s = a+b+c \Rightarrow c = 2s-a-b$ ...(1)

$$\Delta = \sqrt{s(s-a)(s-b)(s-c)} = \sqrt{s(s-a)(s-b)(a+b-s)} \qquad \text{[Using (1)]}$$

Let $z = \Delta^2 = s(s-a)(s-b)(a+b-s) = f(a, b)$

$$\frac{\partial f}{\partial a} = s(s-b)\frac{\partial}{\partial a}[(s-a)(a+b-s)]$$

$$= s(s-b)[-(a+b-s)+(s-a)] = s(s-b)(2s-2a-b)$$

$$\frac{\partial f}{\partial b} = s(s-a)\frac{\partial}{\partial b}[(s-b)(a+b-s)]$$

$$= s(s-a)[-(a+b-s)+(s-b)] = s(s-a)(2s-a-2b)$$

Now, $\dfrac{\partial f}{\partial a} = 0$ and $\dfrac{\partial f}{\partial b} = 0$

$\Rightarrow\ s(s-b)(2s-2a-b) = 0$ and $s(s-a)(2s-a-2b) = 0$

$\Rightarrow\ (s-b)(2s-2a-b) = 0$ ...(2) $\Rightarrow$ $(s-a)(2s-a-2b) = 0$ ...(3) $[\because s \neq 0]$

From (2), $s = b$ or $2s = 2a + b$

When $s = b$, from (3), $(b-a)(-a) = 0 \Rightarrow b = a$ $[\because a \neq 0]$

When $2s = 2a + b$, from (3), $\dfrac{b}{2}(a-b) = 0$ or $a = b$ $[\because b \neq 0]$

If we express z as a function of b and c, we similarly get $b = c$

$$\therefore \quad a = b = c = \frac{2s}{3}$$

$$\frac{\partial^2 f}{\partial a^2} = -2s(s-b) \quad \Rightarrow \quad \frac{\partial^2 f}{\partial b^2} = -2s(s-a)$$

$$\frac{\partial^2 f}{\partial a\,\partial b} = s[-(2s-a-2b)-(s-a)] = s(2a+2b-3s)$$

$$\frac{\partial^2 f}{\partial a^2} = -2s\left(\frac{s}{3}\right) = -\frac{2s^2}{3} < 0$$

$$\frac{\partial^2 f}{\partial a\,\partial b} = s\left(\frac{4s}{3}+\frac{4s}{3}-3s\right) = s\left(-\frac{s}{3}\right) = -\frac{s^2}{3}$$

$$\frac{\partial^2 f}{\partial b^2} = -2s\left(\frac{s}{3}\right) = -\frac{2s^2}{3}$$

$$\frac{\partial^2 f}{\partial a^2}\frac{\partial^2 f}{\partial b^2} - \left(\frac{\partial^2 f}{\partial a \partial b}\right)^2 = \frac{4s^4}{9} - \frac{s^4}{9} = \frac{s^4}{3} > 0. \text{ Also } \frac{\partial^2 f}{\partial a^2} < 0$$

Hence, Δ is maximum when $a = b = c = \dfrac{2s}{3}$ *i.e.* when the triangle is equilateral. **Proved.**

Example 12. *Show that the diameter of the right circular cylinder of greatest curved surface which can be inscribed in a given cone is equal to the radius of the cone.*

(R.G.P.V. Bhopal, April 2010)

Solution. Let R be the radius of the cone ABC and H be the height of the cone.

A cylinder PQRS of radius r and height h is inscribed in the cone.

In Δ APT and Δ PBQ

$$\frac{AT}{PQ} = \frac{PT}{BQ} \Rightarrow \frac{H-h}{h} = \frac{r}{R-r} \Rightarrow \left(\frac{H}{h}-1\right) = \frac{r}{R-r} \Rightarrow \frac{H}{h} = \frac{r}{R-r}+1$$

$$\frac{H}{h} = \frac{r+R-r}{R-r} \Rightarrow \frac{H}{h} = \frac{R}{R-r} \Rightarrow h = \frac{R-r}{R}H \quad ...(1)$$

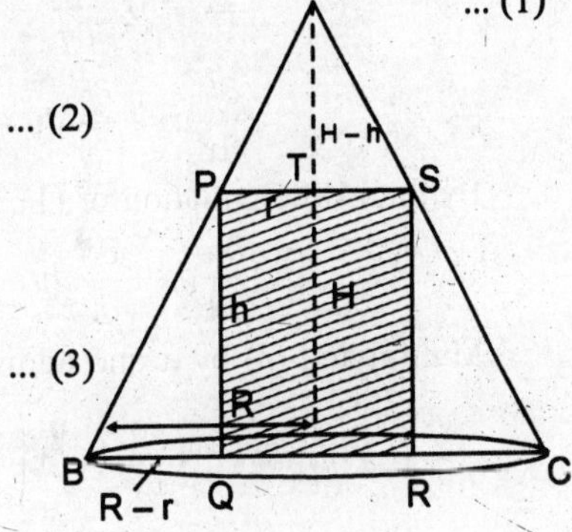

Let the curved surface of the cylinder be S.

$$S = 2\pi r h \quad ...(2)$$

Putting the value of h from (1) in (2), we get

$$S = 2\pi r \frac{R-r}{R} H$$

$$= \frac{2\pi H}{R}(rR - r^2) \quad ...(3)$$

Differentiating (3) w.r.t. 'r', we get

$$\frac{\partial S}{\partial r} = \frac{2\pi H}{R}(R-2r)$$

$$\frac{\partial^2 S}{\partial r^2} = \frac{2\pi H}{R}(-2) = -\frac{4\pi H}{R} = -Ve$$

For maximum curved surface area $\frac{\partial S}{\partial r} = 0$

$$\Rightarrow \frac{2\pi H}{R}(R - 2r) = 0 \Rightarrow R - 2r = 0 \Rightarrow R = 2r \text{ and } \frac{\partial^2 S}{\partial r^2} = -Ve$$

Hence, for maximum 'S', Diameter of the cylinder = Radius of the cone. **Proved.**

EXERCISE 3.1

Find the stationary points of the following functions

1. $f(x, y) = x^3 y^2 (1 - x - y)$ [*A.M.I.E., Summer 2004*] **Ans.** $\left(\frac{1}{2}, \frac{1}{3}\right)$, Maximum
2. $f(x, y) = x^3 + 3xy^2 - 15x^2 - 15y^3 + 72x.$ (*M.U. 2007, 2005, 2004*) **Ans.** (6, 0), (4 0)
3. $f(x, y) = x^2 + 2xy + 2y^2 + 2x + 3y$ such that $x^2 - y = 1.$ **Ans.** $\left(-\frac{3}{4}, -\frac{7}{16}\right), -\frac{155}{128}.$
4. $f(x, y) = xy\, e^{-(2x + 3y)}$ (*A.M.I.E., Winter 2000*) **Ans.** (3, 2)
5. Find the extreme value of the function $f(x, y) = x^2 + y^2 + xy + x - 4y + 5.$ State whether this value is a relative maximum or a relative minimum.
   Ans. Minimum value of $f(x, y)$ at $(-2, 3) = -2.$
6. Find the values of x and y for which $x^2 + y^2 + 6x + 12$ has a minimum value and find this minimum value. **Ans.** $(-3, 0)$, 3.
7. In a plane triangle ABC, find the maximum value of $\cos A \cos B \cos C$.
8. Find a point within a triangle such that the sum of the square of its distances from the three angular points is a minimum.
9. Find the maximum and minimum values of $x^3 + 3xy^2 - 3x^2 - 3y^2 + 4.$ (*M.U. 2006*)
   Ans. Maximum at (0, 0) and minimum (2,0)
10. Test the function $f(x, y) = (x^2 + y^2)\, e^{-(x^2 + y^2)}$ for maxima and minima for limits not on the circle $x^2 + y^2 = 1.$ (*U.P. Q. Bank 2001*)

Tick (✓) the correct answer in the following :

11. One of the stationary values of the function $f(x, y) = x^4 + y^4 - 2x^2 + 4xy - 2y^2$ is
    (*a*) $(\sqrt{2}, -\sqrt{2})$ (*b*) $(2, -2)$ (*c*) $(\sqrt{2}, \sqrt{2})$ (*d*) $(-2, 2)$
    (*AMIETE, June 2009*)

3.4 LAGRANGE METHOD OF UNDETERMINED MULTIPLIERS

Let $f(x, y, z)$ be a function of three variables x, y, z and the variables be connected by the relation.

$$\phi(x, y, z) = 0 \quad \text{...(1)}$$

$\Rightarrow f(x, y, z)$ to have stationary values,

$$\frac{\partial f}{\partial x} = 0, \quad \frac{\partial f}{\partial y} = 0, \quad \frac{\partial f}{\partial z} = 0$$

$$\Rightarrow \quad \frac{\partial f}{\partial x}dx + \frac{\partial f}{\partial y}dy + \frac{\partial f}{\partial z}dz = 0 \quad \text{...(2)}$$

By total differentiation of (1), we get

$$\frac{\partial \phi}{\partial x}dx + \frac{\partial \phi}{\partial y}dy + \frac{\partial \phi}{\partial z}dz = 0 \quad \text{...(3)}$$

Multiplying (3) by λ and adding to (2), we get

$$\left(\frac{\partial f}{\partial x}dx + \lambda\frac{\partial \phi}{\partial x}dx\right) + \left(\frac{\partial f}{\partial y}dy + \lambda\frac{\partial \phi}{\partial y}dy\right) + \left(\frac{\partial f}{\partial z}dz + \lambda\frac{\partial \phi}{\partial z}dz\right) = 0$$

$$\left(\frac{\partial f}{\partial x}+\lambda\frac{\partial \phi}{\partial x}\right)dx+\left(\frac{\partial f}{\partial y}+\lambda\frac{\partial \phi}{\partial y}\right)dy+\left(\frac{\partial f}{\partial z}+\lambda\frac{\partial \phi}{\partial z}\right)dz = 0$$

This equation will hold good if

$$\frac{\partial f}{\partial x}+\lambda\frac{\partial \phi}{\partial x} = 0 \qquad ...(4)$$

$$\frac{\partial f}{\partial y}+\lambda\frac{\partial \phi}{\partial y} = 0 \qquad ...(5)$$

$$\frac{\partial f}{\partial z}+\lambda\frac{\partial \phi}{\partial z} = 0 \qquad ...(6)$$

On solving (1), (4), (5), (6), we can find the values of x, y, z and λ for which $f(x, y, z)$ has stationary value.

Draw Back in Lagrange method is that the nature of stationary point cannot be determined.

Example 13. *The shape of a hole pored by a drill is a cone surmounted by cylinder. If the cylinder be of height h and radius r and the semi-vertical angle of the cone be* α, *where* $\tan\alpha = \frac{h}{r}$ *show that for a total height H of the hole, the volume removed is maximum if* $h = H(\sqrt{7}+1)/6$. *(R.G.P.V., Bhopal I sem. 2003)*

Solution. Let $ABCD$ be the given cylinder of height 'h' and radius 'r' and DPC be the cone of course, of radius r.

Now, since α is the semi-vertical angle of the cone.

$$\therefore \qquad \tan\alpha = \frac{PC}{OP} = \frac{r}{OP} \qquad ...(1)$$

but, given that $\tan\alpha = \frac{h}{r}$ ...(2)

From (1) and (2), we have $\frac{h}{r} = \frac{r}{OP}$

$$\Rightarrow \qquad OP = \frac{r^2}{h} \qquad ...(3)$$

Total height of the hole = H

$$\Rightarrow \qquad H = h + OP \Rightarrow OP = H - h \qquad ...(4)$$

From (3) and (4)

$$\frac{r^2}{h} = H - h \qquad ...(5)$$

Again, let $\phi = \text{H} - h - \frac{r^2}{h}$ ...(6)

In drilling a hole, the volume of the removed portion

$$\frac{\partial\phi}{\partial r} = -\frac{2r}{h}, \qquad \frac{\partial\phi}{\partial h} = -1 + \frac{r^2}{h^2}$$

V = Volume of the cylinder + Volume of the cone.

$$= \pi r^2 h + \frac{1}{3}\pi r^2 (OP) = \pi r^2 h + \frac{1}{3}\pi r^2 . \frac{r^2}{h} \qquad \text{[From (3)]}$$

$$V = \pi r^2 h + \frac{\pi r^4}{3h},$$

$$\frac{\partial V}{\partial r} = 2\pi rh + \frac{4\pi r^3}{3h} \qquad ...(7)$$

By Legrange Method

$$\frac{\partial V}{\partial r} + \lambda \frac{\partial \phi}{\partial r} = 0 \quad \Rightarrow \quad 2\pi r h + \frac{4\pi r^3}{3h} + \lambda\left(\frac{-2r}{h}\right) = 0 \qquad ...(8)$$

$$\frac{\partial V}{\partial h} + \lambda \frac{\partial \phi}{\partial h} = 0 \quad \Rightarrow \quad \pi r^2 - \frac{\pi r^4}{3h^2} + \lambda\left(-1 + \frac{r^2}{h^2}\right) = 0 \qquad ...(9)$$

Multiplying (8) by r and (9) by $2h$, we get

$$2\pi r^2 h + \frac{4\pi r^4}{3h} + \lambda\left(\frac{-2r^2}{h}\right) = 0 \quad \Rightarrow \quad 2\pi r^2 h - \frac{2\pi r^4}{3h} + 2\lambda\left(-h + \frac{r^2}{h}\right) = 0$$

On subtracting, we get

$$\frac{6\pi r^4}{3h} + \lambda\left(\frac{-2r^2}{h} + 2h - \frac{2r^2}{h}\right) = 0 \Rightarrow \frac{6\pi r^4}{3h} + \lambda\left(\frac{-4r^2}{h} + 2h\right) = 0$$

$$2\pi r^4 + \lambda(-4r^2 + 2h^2) = 0 \Rightarrow \pi r^4 + \lambda(-2r^2 + h^2) = 0$$

$$\Rightarrow \qquad \lambda = \frac{\pi r^4}{-h^2 + 2r^2}$$

Putting the value of λ in (8), we get

$$2\pi r h + \frac{4\pi r^3}{3h} + \left(\frac{\pi r^4}{-h^2 + 2r^2}\right)\left(-\frac{2r}{h}\right) = 0$$

$$\Rightarrow \qquad h + \frac{2r^2}{3h} + \frac{r^4}{h(h^2 - 2r^2)} = 0 \qquad \left[\frac{r^2}{h} = H - h\right]$$

$$h + \frac{2}{3}(H - h) + \frac{h^2(H-h)^2}{h[h^2 - 2h(H-h)]} = 0$$

$$h + \frac{2}{3}(H - h) + \frac{(H^2 + h^2 - 2hH)}{h - 2H + 2h} = 0$$

$$h + \frac{2}{3}(H - h) + \frac{H^2 + h^2 - 2hH}{3h - 2H} = 0$$

$$3h^2 - 2Hh + \frac{2}{3}(H - h)(3h - 2H) + H^2 + h^2 - 2hH = 0$$

$$9h^2 - 6Hh + 6Hh - 4H^2 - 6h^2 + 4Hh + 3H^2 + 3h^2 - 6hH = 0$$

$$6h^2 - 2hH - H^2 = 0$$

$$h = \frac{2H \pm \sqrt{4H^2 + 24H^2}}{12}$$

$$h = \frac{H \pm H\sqrt{7}}{6} = H\,\frac{[\sqrt{7} + 1]}{6} \text{ (–ve is not possible)}$$

Proved.

Example 14. *Find the minimum value of $x^2 + y^2 + z^2$ subject to the condition $xyz = a^3$.*

(A.M.I.T.E., June 2009)

Solution. Let

$$f(x, y) = x^2 + y^2 + z^2 \text{ and} \qquad ...(1)$$

$$\phi(x, y) = xyz - a^3 \qquad ...(2)$$

By Lagrange's method

$$\frac{\partial f}{\partial x} + \lambda \frac{\partial \phi}{\partial x} = 0 \quad \Rightarrow \quad 2x + \lambda(yz) = 0 \qquad ...(3)$$

$$\frac{\partial f}{\partial y}+\lambda\frac{\partial \phi}{\partial y}=0 \quad \Rightarrow \quad 2y+\lambda\,(xz)=0 \qquad ...(4)$$

$$\frac{\partial f}{\partial z}+\lambda\frac{\partial \phi}{\partial z}=0 \quad \Rightarrow \quad 2z+\lambda\,(xy)=0 \qquad ...(5)$$

On multiplying (3) by x, (4) by y and (5) by z, we get

$$2x^2+\lambda\,(xyz)=0 \qquad ...(6)$$
$$2y^2+\lambda\,(xyz)=0 \qquad ...(7)$$
$$2z^2+\lambda\,(xyz)=0 \qquad ...(8)$$

On subtracting (7) from (6), we get

$$2x^2-2y^2=0 \quad \Rightarrow \quad x=y$$

On subtracting (8) from (7), we get

$$2y^2-2z^2=0 \quad \Rightarrow \quad y=z$$

so $x=y=z$

Now, putting the value of y and z in term of x in (2), we get

$$(x)\,(x)\,(x)=a^3 \quad \Rightarrow \quad x^3=a^3 \quad \Rightarrow \quad x=a$$

$$\Rightarrow \qquad y=a,\ z=a$$

Putting the values of x, y, z in (1), we get

$$f(x,y)=a^2+a^2+a^2=3a^2$$

Hence, the minimum value of $f(x, y)=3a^2$ **Ans.**

Example 15. *Find the point upon the plane $ax+by+cz=p$ at which the function*

$$f=x^2+y^2+z^2$$

has a minimum value and find this minimum f. (*Nagpur University, Winter* 2000)

Solution. We have, $\quad f=x^2+y^2+z^2 \qquad ...(1)$

$$ax+by+cz=p \quad \Rightarrow \quad \phi=ax+by+cz-p \qquad ...(2)$$

$$\frac{\partial f}{\partial x}+\lambda\frac{\partial \varphi}{\partial x}=0 \quad \Rightarrow \quad 2x+\lambda\,a=0 \quad \Rightarrow \quad x=\frac{-\lambda\,a}{2}$$

$$\frac{\partial f}{\partial y}+\lambda\frac{\partial \varphi}{\partial y}=0 \quad \Rightarrow \quad 2y+\lambda\,b=0 \quad \Rightarrow \quad y=\frac{-\lambda\,b}{2}$$

$$\frac{\partial f}{\partial z}+\lambda\frac{\partial \varphi}{\partial z}=0 \quad \Rightarrow \quad 2\,z+\lambda\,c=0 \quad \Rightarrow \quad z=\frac{-\lambda\,c}{2}$$

Substituting the values of x, y, z in (2), we get

$$a\left(\frac{-\lambda\,a}{2}\right)+b\left(\frac{-\lambda\,b}{2}\right)+c\left(\frac{-\lambda\,c}{2}\right)=p$$

$$\lambda\,(a^2+b^2+c^2)=-2p \quad \Rightarrow \quad \lambda=\frac{-2p}{a^2+b^2+c^2}$$

$$\therefore \qquad x=\frac{a\,p}{a^2+b^2+c^2},\quad y=\frac{b\,p}{a^2+b^2+c^2},\quad z=\frac{c\,p}{a^2+b^2+c^2}$$

The minimum value of

$$f=\frac{a^2p^2}{(a^2+b^2+c^2)^2}+\frac{b^2p^2}{(a^2+b^2+c^2)^2}+\frac{c^2p^2}{(a^2+b^2+c^2)^2}$$

$$=\frac{p^2(a^2+b^2+c^2)}{(a^2+b^2+c^2)^2}=\frac{p^2}{a^2+b^2+c^2}$$ **Ans.**

Example 16. *Find the maximum value of $u=x^p\,y^q\,z^r$ when the variables x, y, z are subject to the condition $ax+by+cz=p+q+r$.*

Solution. Here, we have $\quad u=x^p\,y^q\,z^r \qquad ...(1)$

If $\quad \log u=p\log x+q\log y+r\log z \qquad ...(2)$

$$\frac{1}{u}\frac{\partial u}{\partial x} = \frac{p}{x} \quad \Rightarrow \quad \frac{\partial u}{\partial x} = \frac{pu}{x}$$

$$\frac{1}{u}\frac{\partial u}{\partial y} = \frac{q}{y} \quad \Rightarrow \quad \frac{\partial u}{\partial y} = \frac{qu}{y}$$

$$\frac{1}{u}\frac{\partial u}{\partial z} = \frac{r}{z} \quad \Rightarrow \quad \frac{\partial u}{\partial z} = \frac{ru}{z}$$

$$ax + by + cz = p + q + r$$

$$\phi(x, y, z) = ax + by + cz - p - q - r$$

$$\frac{\partial \phi}{\partial x} = a, \qquad \frac{\partial \phi}{\partial y} = b, \qquad \frac{\partial \phi}{\partial z} = c$$

Lagranges equations are

$$\frac{\partial u}{\partial x} + \lambda\frac{\partial \phi}{\partial x} = 0 \Rightarrow \frac{pu}{x} + \lambda a = 0 \quad \Rightarrow \quad x = -\frac{pu}{\lambda a}$$

$$\frac{\partial u}{\partial y} + \lambda\frac{\partial \phi}{\partial y} = 0 \Rightarrow \frac{qu}{y} + \lambda b = 0 \quad \Rightarrow \quad y = -\frac{qu}{\lambda b}$$

$$\frac{\partial u}{\partial z} + \lambda\frac{\partial \phi}{\partial z} = 0 \Rightarrow \frac{ru}{z} + \lambda c = 0 \quad \Rightarrow \quad z = -\frac{ru}{\lambda c}$$

Putting in $ax + by + cz = p + q + r$ we have

$$-\frac{pu}{\lambda} - \frac{qu}{\lambda} - \frac{ru}{\lambda} = p + q + r$$

$$-\frac{u}{\lambda}(p+q+r) = p + q + r, \quad \Rightarrow \quad -\frac{u}{\lambda} = 1, \quad \Rightarrow \quad \lambda = -u$$

$$x = -\frac{pu}{\lambda a} = \frac{-pu}{-ua} = \frac{p}{a} \qquad y = -\frac{qu}{\lambda b} = \frac{-qu}{-ub} = \frac{q}{b}$$

$$z = -\frac{ru}{\lambda c} = \frac{-ru}{-uc} = \frac{r}{c}$$

Putting in (1), we have

$$\text{Maximum value of} \quad u = \left(\frac{p}{a}\right)^p \left(\frac{q}{b}\right)^q \left(\frac{r}{c}\right)^r$$ **Ans.**

Example 17. *If $xyz = 8$, find the value of x, y, z for which $u = \dfrac{5xyz}{x + 2y + 4z}$ is maximum.*

(K. University Dec. 2008)

Solution. Here, we have

$$u = \frac{5xyz}{x + 2y + 4z} \qquad \text{... (1)}$$

$$xyz = 8 \quad \Rightarrow \quad xyz - 8 = 0 \quad \text{Let } \phi = xyz - 8 = 0 \qquad \text{... (2)}$$

$$\frac{\partial u}{\partial x} = \frac{(x+2y+4z)\,5yz - 5xyz\,(1)}{(x+2y+4z)^2} = \frac{5yz\,(x+2y+4z-x)}{(x+2y+4z)^2}$$

$$= \frac{10y^2z + 20yz^2}{(x+2y+4z)^2}$$

Similarly $$\frac{\partial u}{\partial y} = \frac{5x^2z + 20xz^2}{(x+2y+4z)^2}, \quad \text{and} \quad \frac{\partial u}{\partial z} = \frac{5x^2y + 10xy^2}{(x+2y+4z)^2}$$

Lagranges equations are

$$\frac{\partial u}{\partial x}+\lambda\frac{\partial \phi}{\partial x}=0 \quad \Rightarrow \quad \frac{10y^2z+20yz^2}{(x+2y+4z)^2}+\lambda yz=0 \qquad \ldots(3)$$

$$\frac{\partial u}{\partial y}+\lambda\frac{\partial \phi}{\partial y}=0 \quad \Rightarrow \quad \frac{5x^2z+20xz^2}{(x+2y+4z)^2}+\lambda xz=0 \qquad \ldots(4)$$

$$\frac{\partial u}{\partial z}+\lambda\frac{\partial \phi}{\partial z}=0 \quad \Rightarrow \quad \frac{5x^2y+10xy^2}{(x+2y+4z)^2}+\lambda xy=0 \qquad \ldots(5)$$

On multiplying (3) by x and (4) by y and then subtracting, we get

$$\frac{10xy^2z+20xyz^2}{(x+2y+4z)^2}-\frac{5x^2yz+20xyz^2}{(x+2y+4z)^2}=0 \Rightarrow 10xy^2z+20xyz^2-5x^2yz-20xyz^2=0$$

$$\Rightarrow \quad 10xy^2z-5x^2yz=0, \quad \Rightarrow \quad 2y=x \qquad \ldots(6)$$

On multiplying (4) by y and (5) by z and then subtracting, we get

$$\frac{5x^2yz+20xyz^2}{(x+2y+4z)^2}-\frac{5x^2yz+10xy^2z}{(x+2y+4z)^2}=0 \Rightarrow 5x^2yz+20xyz^2-5x^2yz-10xy^2z=0$$

$$\Rightarrow \quad 20xyz^2-10xy^2z=0 \quad \Rightarrow \quad 2z=y$$

But $x=2y=2(2z)=4z$ $\Rightarrow$ $x=4z$

On putting $x=4z$, $y=2z$ in (2), we get

$$(4z)(2z)z=8 \quad \Rightarrow \quad 8z^3=8 \quad \Rightarrow \quad z^3=1 \quad \Rightarrow \quad z=1$$

$$x=4, \quad y=2, \quad z=1$$ **Ans.**

Example 18. *Show that the stationary value of* $u=x^m y^n z^p$

where $x+y+z=a$ *is* $\dfrac{m^m n^n p^p \, a^{m+n+p}}{(m+n+p)^{m+n+p}}$ *(Nagpur Universtiy, Summer 2003)*

Solution. Here, we have $u=x^m y^n z^p$ *and* ..(1)

$$\phi=x+y+z-a \qquad \ldots(2)$$

$$\frac{\partial u}{\partial x}=mx^{m-1}y^n z^p, \qquad \frac{\partial \phi}{\partial x}=1$$

$$\frac{\partial u}{\partial y}=n\,x^m y^{n-1}z^p, \qquad \frac{\partial \phi}{\partial y}=1$$

$$\frac{\partial u}{\partial z}=p\,x^m y^n z^{p-1}, \qquad \frac{\partial \phi}{\partial z}=1$$

By Lagrange's Method

$$\frac{\partial u}{\partial x}+\lambda\frac{\partial \phi}{\partial x}=0 \quad \Rightarrow \quad mx^{m-1}y^n z^p+\lambda=0 \qquad \ldots(3)$$

$$\frac{\partial u}{\partial y}+\lambda\frac{\partial \phi}{\partial y}=0 \quad \Rightarrow \quad n\,x^m y^{n-1}z^p+\lambda=0 \qquad \ldots(4)$$

$$\frac{\partial u}{\partial z}+\lambda\frac{\partial \phi}{\partial z}=0 \quad \Rightarrow \quad p\,x^m y^n z^{p-1}+\lambda=0 \qquad \ldots(5)$$

Multiplying (3), (4), (5) by x, y, z respectively and adding, we get

$$x^m y^n z^p(m+n+p)+\lambda(x+y+z)=0$$

$$\Rightarrow \quad x^m y^n z^p(m+n+p)+\lambda a=0 \qquad \text{[From (2)]}$$

$$\Rightarrow \quad \lambda=-\frac{x^m y^n z^p(m+n+p)}{a}$$

Putting the value of λ in (3), we get

$$m\,x^{m-1}\,y^n\,z^p - \frac{x^m y^n z^p (m+n+p)}{a} = 0 \Rightarrow m - x\,\frac{(m+n+p)}{a} = 0$$

$$\Rightarrow \qquad \frac{m}{x} = \frac{m+n+p}{a} \qquad \Rightarrow \quad x = \frac{am}{m+n+p}$$

Similarly, $\quad y = \dfrac{an}{m+n+p} \quad$ and $\quad z = \dfrac{ap}{m+n+p}$

Putting the values of x, y and z in (1), we get

$$u = \left(\frac{am}{m+n+p}\right)^m \cdot \left(\frac{an}{m+n+p}\right)^n \cdot \left(\frac{ap}{m+n+p}\right)^p$$

$$u = \frac{m^m n^n p^p a^{m+n+p}}{(m+n+p)^{m+n+p}}$$ **Proved.**

Example 19. *Show that the rectangular solid of maximum volume that can be inscribed in a sphere is a cube.*

Solution. Let $2x$, $2y$, $2z$ be the length, breadth and height of the rectangular solid.
Let R be the radius of the sphere.

Volume of solid $V = 8\,x\,.\,y\,.\,z$ ...(1)

$$x^2 + y^2 + z^2 = R^2 \qquad ....(2)$$

$$\Rightarrow \qquad \phi\,(x, y, z) = x^2 + y^2 + z^2 - R^2 = 0$$

$$\frac{\partial V}{\partial x} + \lambda \frac{\partial \phi}{\partial x} = 0 \Rightarrow 8yz + \lambda\,(2x) = 0 \qquad ...(3)$$

$$\frac{\partial V}{\partial y} + \lambda \frac{\partial \phi}{\partial y} = 0 \Rightarrow 8x\,z + \lambda\,(2\,y) = 0 \qquad ...(4)$$

$$\frac{\partial V}{\partial z} + \lambda \frac{\partial \phi}{\partial z} = 0 \Rightarrow 8xy + \lambda\,(2\,z) = 0 \qquad ...(5)$$

From (3) $\quad 2\,\lambda\,x = -\,8y\,z \quad \Rightarrow \quad 2\,\lambda\,x^2 = -\,8x\,y\,z$

From (4) $\quad 2\,\lambda\,y = -\,8x\,z \quad \Rightarrow \quad 2\,\lambda\,y^2 = -\,8x\,y\,z$

From (5) $\quad 2\,\lambda\,z = -\,8x\,y \quad \Rightarrow \quad 2\,\lambda\,z^2 = -\,8\,x\,y\,z$

$$2\,\lambda\,x^2 = 2\,\lambda\,y^2 = 2\,\lambda\,z^2$$

$$\Rightarrow \qquad x^2 = y^2 = z^2 \qquad \Rightarrow \quad x = y = z$$

Hence rectangular solid is a cube. **Proved.**

Example 20. *If x, y, z are the length of the perpendiculars dropped any point P to the three sides of a triangle of constant area A, find the minimum value of $x^2 + y^2 + z^2$.*

(*Nagpur University, Summer 2004*)

Solution. Let x, y, z are the length of the perpendiculars dropped from point P to the three sides a, b and c of a triangle.

Area (Δ *ABC*) = Area (Δ *PBC*) + Area (Δ *PAC*) + Area (Δ *PAB*)

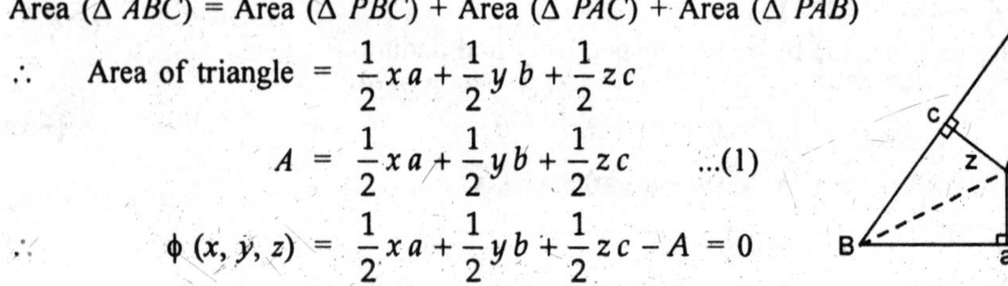

$$\therefore \quad \text{Area of triangle} = \frac{1}{2}x\,a + \frac{1}{2}y\,b + \frac{1}{2}z\,c$$

$$A = \frac{1}{2}x\,a + \frac{1}{2}y\,b + \frac{1}{2}z\,c \qquad ...(1)$$

$$\therefore \quad \phi\,(x, y, z) = \frac{1}{2}x\,a + \frac{1}{2}y\,b + \frac{1}{2}z\,c - A = 0$$

$$F(x, y, z) = x^2 + y^2 + z^2$$

$\therefore$ Lagranges multiplier method

$$\frac{\partial F}{\partial x}+\lambda\frac{\partial \phi}{\partial x} = 0 \quad \Rightarrow \quad 2x+\lambda\frac{1}{2}a = 0 \qquad ...(2)$$

$$\frac{\partial F}{\partial y}+\lambda\frac{\partial \phi}{\partial y} = 0 \quad \Rightarrow \quad 2y+\lambda\frac{1}{2}b = 0 \qquad ...(3)$$

$$\frac{\partial F}{\partial z}+\lambda\frac{\partial \phi}{\partial z} = 0 \quad \Rightarrow \quad 2z+\lambda\frac{1}{2}c = 0 \qquad ...(4)$$

Multiplying (2) by x, (3) by y and (4) by z and adding, we get

$$2(x^2+y^2+z^2)+\lambda\left[\frac{1}{2}xa+\frac{1}{2}yb+\frac{1}{2}zc\right] = 0$$

$$\therefore \quad 2F + \lambda A = 0 \quad \Rightarrow \quad \lambda = \frac{-2F}{A}$$

From (2), $2x-\frac{2F}{A}.\frac{1}{2}a = 0 \quad \Rightarrow \quad x = \frac{aF}{2A}$

From (3), $2y-\frac{2F}{A}.\frac{1}{2}b = 0 \quad \Rightarrow \quad y = \frac{bF}{2A}$

From (4), $2z-\frac{2F}{A}.\frac{1}{2}c = 0 \quad \Rightarrow \quad z = \frac{cF}{2A}$

Putting the values of x, y, z in (1), we get

$$A = \frac{1}{2}\frac{aF}{2A}.a + \frac{1}{2}\frac{bF}{2A}.b + \frac{1}{2}\frac{cF}{2A}.c = \frac{F}{4A}[a^2+b^2+c^2]$$

$$\Rightarrow \quad 4A^2 = F(a^2+b^2+c^2) \quad \Rightarrow \quad F = \frac{4A^2}{a^2+b^2+c^2}$$

Hence, $x^2 + y^2 + z^2 = \dfrac{4A^2}{a^2+b^2+c^2}$ **Ans.**

Example 21. *Find the dimension of rectangular box of maximum capacity whose surface area is given when (a) box is open at the top (b) box is closed.* *(U.P., I Semester, Dec 2008)*

Solution. Let x, y, z be the length, breadth and height of rectangular box respectively. Then

volume, $V = xyz \Rightarrow \frac{\partial V}{\partial x} = yz, \frac{\partial V}{\partial y} = xz, \frac{\partial V}{\partial z} = xy$

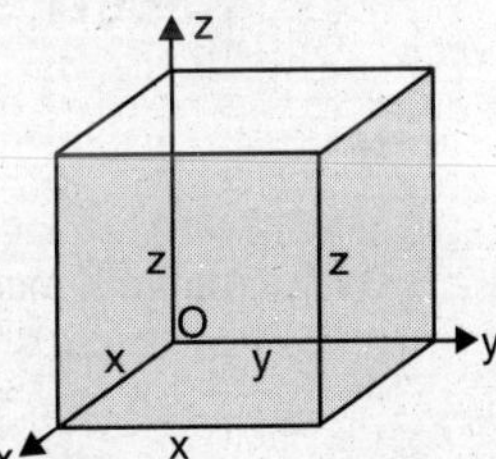

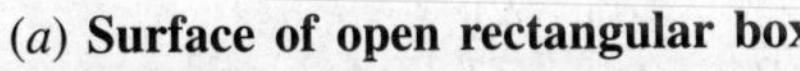

(*a*) **Surface of open rectangular box**

Surface area, $S_1 = xy + 2(yz + zx)$

$$\Rightarrow \quad \frac{\partial S_1}{\partial x} = y+2z, \frac{\partial S_1}{\partial y} = x+2z, \frac{\partial S_1}{\partial z} = 2(x+y)$$

By Lagrange's method

$$\frac{\partial V}{\partial x}+\lambda\frac{\partial S_1}{\partial x} = 0 \Rightarrow yz + [y + 2z]\lambda = 0 \qquad ...(1)$$

$$\frac{\partial V}{\partial y}+\lambda\frac{\partial S_1}{\partial y} = 0 \Rightarrow xz + [x + 2z]\lambda = 0 \qquad ...(2)$$

$$\frac{\partial V}{\partial z}+\lambda\frac{\partial S_1}{\partial z} = 0 \Rightarrow xy + [2y + 2x]\lambda = 0 \qquad ...(3)$$

Multiplying (1) by x, (2) by y and subtracting, we get

$$2z(x-y)\lambda = 0 \Rightarrow x = y \qquad ....(4)$$

Multiplying (2) by y and (3) by z and then subtracting, we get

$$x\,(y - 2z)\,\lambda = 0 \Rightarrow y = 2z \qquad \text{....(5)}$$

From (4) and (5), we get

$$x = y = 2z$$

$\Rightarrow$ Length = Breadth = 2 × Height.

(*b*) **Surface of closed rectangular box:**

Surface area $S_2 = 2\,(xy + yz + zx) \Rightarrow \dfrac{\partial S_2}{\partial x} = 2\,(y + z), \dfrac{\partial S_2}{\partial y} = 2\,(x + z), \dfrac{\partial S_2}{\partial z} = 2\,(x + y)$

By Lagrange's method, we have

$$\frac{\partial V}{\partial x} + \lambda\frac{\partial S_2}{\partial x} = 0 \Rightarrow yz + 2\,(y + z)\,\lambda = 0 \qquad \text{...(6)}$$

$$\frac{\partial V}{\partial y} + \lambda\frac{\partial S_2}{\partial y} = 0 \Rightarrow xz + 2\,(x + z)\,\lambda = 0 \qquad \text{...(7)}$$

$$\frac{\partial V}{\partial z} + \lambda\frac{\partial S_2}{\partial z} = 0 \Rightarrow xy + 2\,(y + x)\,\lambda = 0 \qquad \text{...(8)}$$

Multiplying (6) by x and (7) by y and then subtracting, we get

$$2z\,(x - y)\,\lambda = 0 \Rightarrow x = y \qquad \text{...(9)}$$

Multiplying (7) by y and (8) by z and then subtracting, we get

$$2x\,(y - z)\,\lambda = 0 \Rightarrow y = z \qquad \text{...(10)}$$

From (9) and (10), we get

$$x = y = z$$

Thus, Length = Breadth = Height. **Ans.**

Example 22. *Find the area of a greatest rectangle that can be inscribed in an ellipse*

$$\frac{x^2}{a^2} + \frac{y^2}{b^2} = 1$$

(*Nagpur University, Winter 2001*)

Solution. Let, $ABCD$ be the rectangle.

Let the co-ordinates of point A be (x, y).

$$AB = 2x,\; BC = 2y$$

$$\text{Area} = A = (2x)\,(2y) = 4\,xy, \quad \frac{\partial A}{\partial x} = 4y, \quad \frac{\partial A}{\partial y} = 4x \qquad \text{...(1)}$$

$$\phi = \frac{x^2}{a^2} + \frac{y^2}{b^2} - 1 = 0, \quad \frac{\partial \phi}{\partial x} = \frac{2x}{a^2}, \quad \frac{\partial \phi}{\partial y} = \frac{2y}{b^2}$$

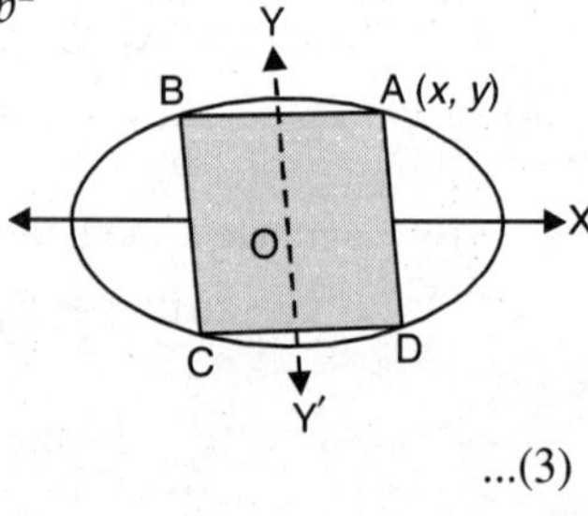

By Lagrange's method

$$\frac{\partial A}{\partial x} + \lambda\frac{\partial \phi}{\partial x} = 0$$

$$\Rightarrow 4y + \lambda\frac{2x}{a^2} = 0 \Rightarrow \lambda = -\frac{2ya^2}{x} \qquad \text{...(2)}$$

$$\frac{\partial A}{\partial y} + \lambda\frac{\partial \phi}{\partial y} = 0 \Rightarrow 4x + \lambda\frac{2y}{b^2} = 0 \qquad \text{...(3)}$$

Putting the value of λ from (2) in (3), we get

$$4x - \frac{2ya^2}{x}\left(\frac{2y}{b^2}\right) = 0 \qquad \text{[From (2)]}$$

$$\Rightarrow \qquad x^2 - \frac{y^2a^2}{b^2} = 0 \qquad \Rightarrow \qquad \frac{x^2}{a^2} = \frac{y^2}{b^2} \qquad ...(4)$$

Putting the value of $\frac{y^2}{b^2}$ from (4) in $\frac{x^2}{a^2}+\frac{y^2}{b^2}=1$, we get

$$\frac{x^2}{a^2}+\frac{x^2}{a^2} = 1 \qquad \Rightarrow \frac{2x^2}{a^2}=1 \qquad \Rightarrow \qquad x = \pm\frac{a}{\sqrt{2}}$$

Similarly, $\qquad y = \pm\frac{b}{\sqrt{2}}$

$$\text{Area} = 4xy = 4\left(\frac{a}{\sqrt{2}}\right)\left(\frac{b}{\sqrt{2}}\right) = 2ab$$

Hence, area of greatest rectangle inscribed in the ellipse = $2ab$ **Ans.**

Example 23. *Use the method of the Lagrange's multipliers to find the volume of the largest rectangular parallelopiped that can be inscribed in the ellipsoid* $\frac{x^2}{a^2}+\frac{y^2}{b^2}+\frac{z^2}{c^2}=1.$

(*Nagpur Univesity, Summer 2008, Winter 2003*)
(*A.M.I.E.T.E., Summer 2004, U.P., I Semester, Winter 2002, 2000*)

Solution. Here, we have $\qquad \frac{x^2}{a^2}+\frac{y^2}{b^2}+\frac{z^2}{c^2} = 1$

$$\Rightarrow \qquad \phi(x, y, z) = \frac{x^2}{a^2}+\frac{y^2}{b^2}+\frac{z^2}{c^2}-1=0 \qquad ...(1)$$

Let $2x$, $2y$, $2z$ be the length, breadth and height of the rectangular parallelopiped inscribed in the ellipsoid.

$$V = (2x)(2y)(2z) = 8\,xyz$$

$$\frac{\partial V}{\partial x} = 8\,yz; \qquad \frac{\partial V}{\partial y}=8xz, \qquad \frac{\partial V}{\partial z}=8xy$$

$$\frac{\partial \phi}{\partial x} = \frac{2x}{a^2}, \qquad \frac{\partial \phi}{\partial y}=\frac{2y}{b^2}, \qquad \frac{\partial \phi}{\partial z}=\frac{2z}{c^2}$$

Lagrange's equations are

$$\frac{\partial V}{\partial x}+\lambda\frac{\partial \phi}{\partial x} = 0 \qquad \Rightarrow \qquad 8\,yz+\lambda\,\frac{2x}{a^2}=0 \qquad ...(1)$$

$$\frac{\partial V}{\partial y}+\lambda\frac{\partial \phi}{\partial y} = 0 \qquad \Rightarrow \qquad 8\,xz+\lambda\,\frac{2y}{b^2}=0 \qquad ...(2)$$

$$\frac{\partial V}{\partial z}+\lambda\frac{\partial \phi}{\partial z} = 0 \qquad \Rightarrow \qquad 8\,xy+\lambda\,\frac{2z}{c^2}=0 \qquad ...(3)$$

Multiplying (1), (2) and (3) by x, y, z respectively and adding, we get

$$24xyz+2\lambda\left[\frac{x^2}{a^2}+\frac{y^2}{b^2}+\frac{z^2}{c^2}\right] = 0 \qquad \left[\frac{x^2}{a^2}+\frac{y^2}{b^2}+\frac{z^2}{c^2}=1\right]$$

$$\Rightarrow \qquad 24\,xyz + 2\,\lambda\,(1) = 0 \qquad \Rightarrow \qquad \lambda = -12\,xyz$$

Putting the value of λ in (1), we get

$$8\,yz + (-12\,xyz)\,\frac{2x}{a^2} = 0 \qquad \Rightarrow \qquad 1-\frac{3x^2}{a^2} = 0$$

$$\Rightarrow \qquad x = \frac{a}{\sqrt{3}}$$

Similarly from (2) and (3), we have

$$y = \frac{b}{\sqrt{3}}, \quad z = \frac{c}{\sqrt{3}}$$

Volume of the largest rectangular parallelopiped = 8 xyz

$$= 8\left(\frac{a}{\sqrt{3}}\right)\left(\frac{b}{\sqrt{3}}\right)\left(\frac{c}{\sqrt{3}}\right) = \frac{8\,abc}{3\sqrt{3}}$$ **Ans.**

Example 24. *The pressure P at any point (x, y, z) in space is $P = 400\,xyz^2$. Find the highest pressure at the surface of a unit sphere $x^2 + y^2 + z^2 = 1$.* (*Gujarat, I Semester, Jan. 2009*)

Solution. We have, $P = 400\,x\,y\,z^2$

$$x^2 + y^2 + z^2 = 1, \qquad \phi\,(x, y, z) = x^2 + y^2 + z^2 - 1$$

$$\frac{\partial P}{\partial x} + \lambda\frac{\partial \phi}{\partial x} = 0 \quad \Rightarrow \quad 400\,y\,z^2 + \lambda\,(2\,x) = 0 \qquad ...(1)$$

$$\frac{\partial P}{\partial y} + \lambda\frac{\partial \phi}{\partial y} = 0 \quad \Rightarrow \quad 400\,x\,z^2 + \lambda\,(2\,y) = 0 \qquad ...(2)$$

$$\frac{\partial P}{\partial z} + \lambda\frac{\partial \phi}{\partial z} = 0 \quad \Rightarrow \quad 800\,x\,y\,z + \lambda\,(2\,z) = 0 \qquad ...(3)$$

Multiplying (1) by x, (2) by y and (3) by z and adding, we get

$$1600\,x\,y\,z^2 + 2\lambda\,(x^2 + y^2 + z^2) = 0$$

$$1600\,x\,y\,z^2 + 2\lambda\,(1) = 0 \qquad (x^2 + y^2 + z^2 = 1)$$

$$\Rightarrow \qquad \lambda = -\,800\,x\,y\,z^2$$

Putting the value of λ in (l), we get

$$400\,y\,z^2 + 2\,x\,(-\,800\,x\,y\,z^2) = 0 \Rightarrow 1 - 4\,x^2 = 0 \Rightarrow x = \pm\frac{1}{2}$$

Similarly, $$y = \pm\frac{1}{2}$$

On putting the value of λ in (3), we get

$$800\,x\,y\,z - 1600\,x\,y\,z^3 = 0$$

$$1 - 2\,z^2 = 0 \quad \Rightarrow \quad z = \pm\frac{1}{\sqrt{2}}$$

On putting the values of x, y, z in P, we get

$$P = 400 \times \frac{1}{2}\times\frac{1}{2}\times\frac{1}{2} = 50$$ **Ans.**

Example 25. *A scope probe in the shape of ellipsoid $4x^2 + y^2 + 4z^2 = 16$ enters the earth atmosphere and its surface begins to heat. After one hour the temperature at the point (x, y, z) on the surface is $T\,(x, y, z) = 8x^2 + 4yz - 16z + 600$. Find the hottest point on the probe surface.* (*Nagpur University, Summer 2001*)

Solution. Given temperature on the surface (x, y, z) is

$$T\,(x, y, z) = 8x^2 + 4yz - 16z + 600 \qquad ...(1)$$

and ellipsoid is $$4x^2 + y^2 + 4z^2 = 16$$

Let $$\phi = 4x^2 + y^2 + 4z^2 - 16 = 0 \qquad ...(2)$$

$$\frac{\partial T}{\partial x} = 16x, \quad \text{and} \quad \frac{\partial \phi}{\partial x} = 8x$$

$$\frac{\partial T}{\partial y} = 4z, \quad \text{and} \quad \frac{\partial \phi}{\partial y} = 2y$$

$$\frac{\partial T}{\partial z} = 4y - 16, \quad \text{and} \quad \frac{\partial \phi}{\partial z} = 8z$$

By Lagrange's method

$$\frac{\partial T}{\partial x} + \lambda \frac{\partial \phi}{\partial x} = 0 \quad \Rightarrow \quad 16x + 8\lambda x = 0 \quad \text{...(3)}$$

$$\frac{\partial T}{\partial y} + \lambda \frac{\partial \phi}{\partial y} = 0 \quad \Rightarrow \quad 4z + 2\lambda y = 0 \quad \text{...(4)}$$

$$\frac{\partial T}{\partial z} + \lambda \frac{\partial \phi}{\partial z} = 0 \quad \Rightarrow \quad 4y - 16 + 8\lambda z = 0 \quad \text{...(5)}$$

From (3), $\lambda = -2$

Putting the value of λ in (4) and (5), we get

$$4z + 2(-2)y = 0 \quad \Rightarrow \quad z - y = 0 \quad \text{...(6)}$$

and $$4y - 16 + 8(-2)z = 0 \quad \Rightarrow \quad y - 4z = 4 \quad \text{...(7)}$$

Adding (6) and (7), we get

$$-3z = 4 \text{ and } z = -\frac{4}{3}$$

From (6), $y = z = -\frac{4}{3}$

On putting the values of y and z in (2), we get

$$4x^2 + \left(-\frac{4}{3}\right)^2 + 4\left(-\frac{4}{3}\right)^2 - 16 = 0 \quad \Rightarrow \quad x^2 = \frac{16}{9} \Rightarrow x = \pm\frac{4}{3}$$

Hence, the hottest points on the probe surface is $\left(\pm\frac{4}{3}, \frac{-4}{3}, \frac{-4}{3}\right)$ **Ans.**

Example 26. *A tent of a given volume has a square base of side 2a, has its four-side vertical of length b and is surmounted by a regular pyramid of height h. Find the values of a and b in terms of h such that the canvas required for its construction is minimum.*

Solution. Let V be the volume and S be the surface of the tent.

$$V = 4a^2b + \frac{1}{3}(4a^2)h \qquad \left[\text{Volume of pyramid} = \frac{1}{3}\text{ Area of the base} \times \text{height}\right]$$

$$S = 8ab + 4a\sqrt{a^2 + h^2} \qquad \left[\text{Surface Area of pyramid} = \frac{1}{2}\text{ perimeter} \times \text{slant height}\right]$$

$$\frac{\partial S}{\partial a} + \lambda \frac{\partial V}{\partial a} = 0$$

$$\Rightarrow \quad 8b + 4\sqrt{a^2 + h^2} + \frac{4a^2}{\sqrt{a^2 + h^2}} + \lambda\left[8ab + \frac{8ah}{3}\right] = 0 \quad \text{...(1)}$$

$$\frac{\partial S}{\partial b} + \lambda \frac{\partial V}{\partial b} = 0 \quad \Rightarrow \quad 8a + 4\lambda a^2 = 0 \quad \text{...(2)}$$

$$\frac{\partial S}{\partial h} + \lambda \frac{\partial V}{\partial h} = 0 \quad \Rightarrow \quad \frac{4ah}{\sqrt{a^2 + h^2}} + \frac{4}{3}\lambda a^2 = 0 \quad \text{...(3)}$$

From (2) $$\lambda a + 2 = 0 \quad \Rightarrow \quad \lambda a = -2 \quad \text{...(4)}$$

From (3) $$\underline{12ah + 4\lambda a^2\sqrt{a^2 + h^2}} = 0$$

$$\Rightarrow \quad 3h + \lambda a\sqrt{a^2 + h^2} = 0 \quad \text{...(5)}$$

Substituting the value of λ a from (4) in (5), we get

$$3h - 2\sqrt{a^2 + h^2} = 0 \Rightarrow 9h^2 = 4a^2 + 4h^2 \Rightarrow 4a^2 = 5h^2$$

$$a = \frac{\sqrt{5}}{2}h$$

Substituting $\lambda\, a = -2$ and $a = \frac{\sqrt{5}}{2}h$ in (1) and simplifying, we get

$$8b + 4\sqrt{\frac{5h^2}{4} + h^2} + \frac{5h^2}{\sqrt{\frac{5h^2}{4} + h^2}} - 2\left[8b + \frac{8h}{3}\right] = 0$$

$$\Rightarrow \quad 8b + 6h + \frac{10h}{3} - 16b - \frac{16h}{3} = 0$$

$$\Rightarrow \quad -8b + 4h = 0 \Rightarrow b = \frac{h}{2}.$$

Thus, when $a = \frac{\sqrt{5}}{2}h$ and $b = \frac{h}{2}$ we get the stationary value of S. **Ans.**

Example 27. *Find the maximum and minimum distances of the point* (3, 4, 12) *from the sphere* $x^2 + y^2 + z^2 = 1$. *(AMIETE, June 2010)*

Solution. Let the co-ordinates of the given point be (x, y, z), then its distance (D) from (3, 4, 12).

$$D = \sqrt{(x-3)^2 + (y-4)^2 + (z-12)^2}$$

$$\Rightarrow \quad F(x, y, z) = (x-3)^2 + (y-4)^2 + (z-12)^2$$

$$x^2 + y^2 + z^2 = 1$$

$$\phi(x, y, z) = x^2 + y^2 + z^2 - 1$$

$$\frac{\partial F}{\partial x} + \lambda \frac{\partial \phi}{\partial x} = 2(x-3) + 2\lambda x = 0 \quad \text{...(1)}$$

$$\frac{\partial F}{\partial y} + \lambda \frac{\partial \phi}{\partial y} = 2(y-4) + 2\lambda y = 0 \quad \text{...(2)}$$

$$\frac{\partial F}{\partial z} + \lambda \frac{\partial \phi}{\partial z} = 2(z-12) + 2\lambda z = 0 \quad \text{...(3)}$$

Multiplying (1) by x, (2) by y and (3) by z and adding, we get

$$(x^2 + y^2 + z^2) - 3x - 4y - 12z + \lambda(x^2 + y^2 + z^2) = 0$$

$$1 - 3x - 4y - 12z + \lambda = 0 \quad \text{...(4)}$$

From (1) $$x = \frac{3}{1+\lambda} \quad \text{...(5)}$$

From (2) $$y = \frac{4}{1+\lambda} \quad \text{...(6)}$$

From (3) $$z = \frac{12}{1+\lambda} \quad \text{...(7)}$$

Putting these values of x, y, z in (4),we have

$$1+\lambda - \frac{9}{1+\lambda} - \frac{16}{1+\lambda} - \frac{144}{1+\lambda} = 0 \quad \Rightarrow \quad (1+\lambda)^2 = 169$$

$$\Rightarrow \quad 1 + \lambda = \pm 13$$

Putting the value of $1 + \lambda$ in (5), (6) and (7) we have the points

$$\left(\frac{3}{13}, \frac{4}{13}, \frac{12}{13}\right) \text{ and } \left(\frac{-3}{13}, \frac{-4}{13}, \frac{-12}{13}\right).$$

The minimum distance $= \sqrt{\left(3-\frac{3}{13}\right)^2+\left(4-\frac{4}{13}\right)^2+\left(12-\frac{12}{13}\right)^2} = 12$

The maximum distance $= \sqrt{\left(3+\frac{3}{13}\right)^2+\left(4+\frac{4}{13}\right)^2+\left(12+\frac{12}{13}\right)^2} = 14$ **Ans.**

Example 28. *Use the method of Lagrange's multipliers to find the extreme values of $f(x, y, z) = 2x + 3y + z$ subject to $x^2 + y^2 = 5$ and $x + z = 1$.*

(*A.M.I.E.T.E, June 2010, Dec. 2007, Uttarakhand, I Semester, Dec. 2006*)

Solution. Let
$$f(x, y, z) = 2x + 3y + z \qquad ...(1)$$
$$\phi(x, y) = x^2 + y^2 - 5 \qquad ...(2)$$
$$\psi(x, z) = x + z - 1 \qquad ...(3)$$

Lagranges Multipliers Equations are

$$\frac{\partial f}{\partial x}+\lambda\frac{\partial \phi}{\partial x}+\mu\frac{\partial \psi}{\partial x} = 0 \Rightarrow 2 + \lambda(2x) + \mu(1) = 0 \qquad ...(4)$$

$$\frac{\partial f}{\partial y}+\lambda\frac{\partial \phi}{\partial y}+\mu\frac{\partial \psi}{\partial y} = 0 \Rightarrow 3 + \lambda(2y) + \mu(0) = 0 \qquad ...(5)$$

$$\frac{\partial f}{\partial z}+\lambda\frac{\partial \phi}{\partial z}+\mu\frac{\partial \psi}{\partial z} = 0 \Rightarrow 1 + \lambda(0) + \mu(1) = 0 \Rightarrow \mu = -1 \qquad ...(6)$$

Putting the value of μ in (4) and (5), we get

$$2 + 2\lambda x - 1 = 0 \Rightarrow 2\lambda x = -1, \Rightarrow x = -\frac{1}{2\lambda}$$

$$3 + 2\lambda y = 0 \Rightarrow 2\lambda y = -3, \Rightarrow y = -\frac{3}{2\lambda}$$

Putting the values of x, y in $x^2 + y^2 = 5$, we get

$$\frac{1}{4\lambda^2}+\frac{9}{4\lambda^2}=5 \Rightarrow \frac{10}{4\lambda^2} = 5 \Rightarrow 2\lambda^2 = 1$$

$$\Rightarrow \lambda^2 = \frac{1}{2} \Rightarrow \lambda = \pm\frac{1}{\sqrt{2}}$$

We know that
$$x = -\frac{1}{2\lambda} = \pm\frac{\sqrt{2}}{2} = \pm\frac{1}{\sqrt{2}}$$

$$y = -\frac{3}{2\lambda} = \pm\frac{3\sqrt{2}}{2} = \pm\frac{3}{\sqrt{2}}$$

From (3), $x + z = 1 \Rightarrow z = 1 - x = 1 \mp \frac{1}{\sqrt{2}}$

Putting $x = \frac{1}{\sqrt{2}}, y = \frac{3}{\sqrt{2}}$ and $z = 1 - \frac{1}{\sqrt{2}}$ in (1), we get

$$f = \frac{2}{\sqrt{2}} + \frac{9}{\sqrt{2}} + 1 - \frac{1}{\sqrt{2}} = \frac{10}{\sqrt{2}} + 1 = 5\sqrt{2} + 1$$

Putting $x = -\frac{1}{\sqrt{2}}, y = -\frac{3}{\sqrt{2}}$ and $z = 1 + \frac{1}{\sqrt{2}}$ in (1), we get

$$f = 2\left(-\frac{1}{\sqrt{2}}\right)+3\left(-\frac{3}{\sqrt{2}}\right)+\left(1+\frac{1}{\sqrt{2}}\right) = -\frac{2}{\sqrt{2}}-\frac{9}{\sqrt{2}}+1+\frac{1}{\sqrt{2}}$$
$$= 1-5\sqrt{2}$$ **Ans.**

Example 29. *A torpedo has the shape of a cylinder with conical ends. For given surface area, show that the dimensions which give maximum volume are,* $l = h = \dfrac{2}{\sqrt{5}}r$, *where l is the length of the cylinder, r its radius and h the altitude of the cone.*

(A.M.I.E.T.E. Summer 2000)

Solution. Let V be the volume enclosed by the torpedo and S its surface. Then

$$V = \text{Volume of the cylinder} + \text{Volume of two cones}$$

$$= \pi r^2 l + 2.\frac{1}{3}\pi r^2 h$$

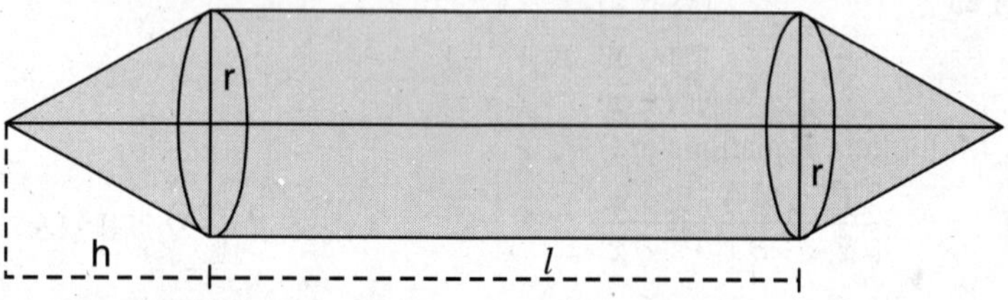

r r h l

S = Surface of the cylinder + 2 (surface of the cone)

= (Circumference of the base × height) + 2 ($\frac{1}{2}$ × perimeter of the base × slant height)

$$\phi = S = 2\pi r l + 2\pi r\sqrt{(r^2+h^2)} \qquad \text{...(1)}$$

Let $$F(l, h, r) = V = \pi r^2 l + \frac{2}{3}\pi r^2 h \qquad \text{...(2)}$$

By Lagrange's method

$$\frac{\partial F}{\partial l} + \lambda\frac{\partial \phi}{\partial l} = 0 \quad\Rightarrow\quad \pi r^2 + 2\pi r\lambda = 0 \quad\Rightarrow\quad \lambda = -\frac{r}{2} \qquad \text{...(3)}$$

$$\frac{\partial F}{\partial h} + \lambda\frac{\partial \phi}{\partial h} = 0 \quad\Rightarrow\quad \frac{2}{3}\pi r^2 + \frac{2\pi r h\lambda}{\sqrt{(r^2+h^2)}} = 0 \qquad \text{...(4)}$$

and $$\frac{\partial F}{\partial r} + \lambda\frac{\partial \phi}{\partial r} = 0 \Rightarrow 2\pi r l + \frac{4}{3}\pi r h + \lambda\left[2\pi l + 2\pi\sqrt{(r^2+h^2)} + \frac{2\pi r^2}{\sqrt{(r^2+h^2)}}\right] = 0 \qquad \text{...(5)}$$

Putting the value of λ in (4), we get

$$\frac{2}{3}\pi r^2 + \frac{2\pi r h}{\sqrt{r^2+h^2}}\left(-\frac{r}{2}\right) = 0 \;\Rightarrow\; \frac{2}{3} - \frac{h}{\sqrt{r^2+h^2}} = 0$$

$$\Rightarrow \quad \frac{2}{3} = \frac{h}{\sqrt{r^2+h^2}} \;\Rightarrow\; \frac{4}{9} = \frac{h^2}{r^2+h^2} \;\Rightarrow 9h^2 = 4r^2 + 4h^2 \;\Rightarrow\; 5h^2 = 4r^2$$

$$h = \frac{2}{\sqrt{5}}r$$

Putting $\lambda = -\dfrac{r}{2}$ and $h = \dfrac{2r}{\sqrt{5}}$ in (5), we get

$$2rl + \frac{4}{3}r\left(\frac{2}{\sqrt{5}}r\right) - \frac{r}{2}\left(2l + 2\sqrt{r^2 + \frac{4}{5}r^2} + \frac{2r^2}{\sqrt{r^2 + \frac{4r^2}{5}}}\right) = 0$$

$$2l + \frac{8r}{3\sqrt{5}} - l - \sqrt{\frac{9r^2}{5}} - r^2\sqrt{\frac{5}{9r^2}} = 0 \quad\Rightarrow\quad l = \frac{-8r}{3\sqrt{5}} + \frac{3r}{\sqrt{5}} + \frac{\sqrt{5}}{3}r$$

$$l = \left(\frac{-8}{3\sqrt{5}} + \frac{3}{\sqrt{5}} + \frac{\sqrt{5}}{3}\right) r \qquad \Rightarrow \quad l = \frac{2r}{\sqrt{5}}$$

Hence $\quad l = h = \dfrac{2r}{\sqrt{5}}$. **Proved.**

Example 30. *If $u = ax^2 + by^2 + cz^2$ where $x^2 + y^2 + z^2 = 1$ and $lx + my + nz = 0$ prove that stationary values of 'u' satisfy the equation*

$$\frac{l^2}{a-u} + \frac{m^2}{b-u} + \frac{n^2}{c-u} = 0$$

Solution. We have, $\quad u = ax^2 + by^2 + cz^2 \qquad$ ...(1)

Let $\quad \phi = x^2 + y^2 + z^2 - 1 \qquad$ ...(2)

$\psi = lx + my + nz \qquad$ ...(3)

$$\frac{\partial u}{\partial x} = 2\,a\,x, \quad \frac{\partial u}{\partial y} = 2\,b\,y, \quad \frac{\partial u}{\partial z} = 2\,c\,z$$

$$\frac{\partial \phi}{\partial x} = 2\,x, \quad \frac{\partial \phi}{\partial y} = 2\,y, \quad \frac{\partial \phi}{\partial z} = 2\,z$$

$$\frac{\partial \psi}{\partial x} = l, \quad \frac{\partial \psi}{\partial y} = m, \quad \frac{\partial \psi}{\partial z} = n$$

By Lagrange's method

$$\frac{\partial u}{\partial x} + \lambda_1 \frac{\partial \phi}{\partial x} + \lambda_2 \frac{\partial \psi}{\partial x} = 0, \qquad 2\,a\,x + 2\,x\,\lambda_1 + \lambda_2\,l = 0 \qquad ...(4)$$

$$\frac{\partial u}{\partial y} + \lambda_1 \frac{\partial \phi}{\partial y} + \lambda_2 \frac{\partial \psi}{\partial y} = 0, \qquad 2\,b\,y + 2\,y\,\lambda_1 + \lambda_2\,m = 0 \qquad ...(5)$$

$$\frac{\partial u}{\partial z} + \lambda_1 \frac{\partial \phi}{\partial z} + \lambda_2 \frac{\partial \psi}{\partial z} = 0, \qquad 2\,c\,z + 2\,z\,\lambda_1 + \lambda_2\,n = 0 \qquad ...(6)$$

Multiplying (4), (5) and (6) by x, y and z respectively and adding, we get

$$(2\,ax^2 + 2\,by^2 + 2\,cz^2) + (2\,x^2 + 2\,y^2 + 2\,z^2)\,\lambda_1 + (lx + my + nz)\,\lambda_2 = 0$$

$$2u + 2\lambda_1 = 0, \qquad \lambda_1 = -u$$

Putting the value of λ_1 in (4), (5) and (6), we get

$$2\,a\,x - 2\,x\,u + \lambda_2\,l = 0, \qquad x = \frac{-\lambda_2 l}{2(a-u)}$$

$$2\,b\,y - 2\,y\,u + \lambda_2\,m = 0, \qquad y = \frac{-\lambda_2\,m}{2(b-u)}$$

$$2\,c\,z - 2\,z\,u + \lambda_2\,n = 0, \qquad z = \frac{-\lambda_2\,n}{2(c-u)}$$

Putting the values of x, y, z in (3), we get

$$\frac{-\lambda_2\,l^2}{2(a-u)} + \frac{-\lambda_2\,m^2}{2(b-u)} + \frac{-\lambda_2 n^2}{2(c-u)} = 0$$

$$\frac{l^2}{a-u} + \frac{m^2}{b-u} + \frac{n^2}{c-u} = 0 \qquad \textbf{Proved.}$$

EXERCISE 3.2

1. Show that the greatest value of $x^m y^n$ where x and y are positive and $x + y = a$ is $\dfrac{m^m . n^n . a^{m+n}}{(m+n)^{m+n}}$, where a is constant.
2. Find the absolute maximum and minimum values of the function $f(x, y) = 3x^2 + y^2 - x$ over the region $2x^2 + y^2 \le 1$. *(A.M.I.E.T.E., Dec. 2008)*
3. Using Lagrange's method (of multipliers), find the critical (stationary values) of the function $f(x, y, z) = x^2 + y^2 + z^2$, given that $z^2 = xy + 1$. **Ans.** (0, 0, – 1), (0, 0, 1).
4. The sum, of three numbers is constant. Prove that their product is a maximum when they are equal.
5. Using the method of Lagrange's multipliers, find the largest product of the numbers x, y and z when $x^2 + y^2 + z^2 = 9$. **Ans.** $3\sqrt{3}$
6. Find a point in the plane $x + 2y + 3z = 13$ nearest to the point (1, 1, 1) using the method of Lagrange's multipliers. **Ans.** $\left(\dfrac{3}{2}, 2, \dfrac{5}{2}\right)$
7. Using the Lagrange's method (of multipliers), find the shortest distance from the point (1, 2, 2) to the sphere $x^2 + y^2 = 36$. **Ans.** 3
8. Find the shortest and the longest distances from the point (1, 2, – 1) to the $x^2 + y^2 + z^2 = 24$. *(U.P. I Semester, Dec 2009)* **Ans.** $\sqrt{6}, 3\sqrt{6}$
9. The sum of the surfaces of a sphere and a cube is given. Show that when the sum of the volumes is least, the diameter of the sphere is equal to the edge of the cube.
10. If $u = \dfrac{a^3}{x^2} + \dfrac{b^3}{y^2} + \dfrac{c^3}{z^2}$, where $x + y + z = 1$, prove that the stationary value of u is given by
$$x = \frac{a}{a+b+c},\quad y = \frac{b}{a+b+c},\quad z = \frac{c}{a+b+c}$$
11. If $u = a^3x^2 + b^3y^2 + c^2z^2$ where $x^{-1} + y^{-1} + z^{-1} = 1$, show that the stationary value of u is given by
$$x = \frac{\Sigma a}{a},\quad y = \frac{\Sigma b}{b},\quad z = \frac{\Sigma c}{c}$$ *(AMIETE, Dec. 2009)*
12. Find maximum value of the expression $\sum_{i=1}^{n} a_i x_i$ with $\sum_{i=1}^{n} x_i^2 = 1$, where $a_1, a_2, a_3 \ldots a_n$ are positive constants. **Ans.** $(a_1^2 + a_2^2 + \ldots\ldots + a_n^2)^{\frac{1}{2}}$
13. If r is the distance of a point on conic $ax^2 + by^2 + cz^2 = 1$, $lx + my + nz = 0$ from origin, then the stationary values of r are given by the equation.
$$\frac{l^2}{1-ar^2} + \frac{m^2}{1-br^2} + \frac{n^2}{1-cr^2} = 0$$ *(A.M.I.E.T.E., Winter 2002)*
14. If x and y satisfy the relation $ax^2 + by^2 = ab$, prove that the extreme values of function $u = x^2 + xy + y^2$ are given by the roots of the equation $4(u - a)(u - b) = ab$ *(A.M.I.E.T.E., Winter 2000)*
15. Use the Lagranges method of undetermined multipliers to find the minimum value of $x^2 + y^2 + z^2$ subject to the conditions $x + y + z = 1$, $xyz + 1 = 0$.
16. The temperature 'T' at any point (xyz) in space is $T(xyz) = kxyz^2$ where k is constant. Find the highest temperature on the surface of the sphere $x^2 + y^2 + z^2 = a^2$. **Ans.** $\dfrac{ka^4}{8}$
17. Prove that the stationary values of $u = \dfrac{x^2}{a^4} + \dfrac{y^2}{b^4} + \dfrac{z^2}{c^4}$, where $lx + my + nz = 0$ and $\dfrac{x^2}{a^2} + \dfrac{y^2}{b^2} + \dfrac{z^2}{c^2} = 1$ are the roots of the equation : $\dfrac{l^2 a^4}{1-a^2u} + \dfrac{m^2 b^4}{1-b^2u} + \dfrac{n^2 c^4}{1-c^2u} = 0$

CHAPTER

4

Errors

4.1 ERROR DETERMINATION

We know that $\lim_{\partial x \to 0} \frac{\delta y}{\delta x} = \frac{dy}{dx}$

$$\frac{\partial y}{\partial x} = \frac{dy}{dx} \text{ approximately} \quad \Rightarrow \quad \delta y = \left(\frac{dy}{dx}\right).\delta x \text{ approximately}$$

Definition.

(*i*) *δx is known as absolute error in x.*

(*ii*) $\frac{\delta x}{x}$ *is known as relative error in x.*

(*iii*) $\left(\frac{\delta x}{x}\right) \times 100$ is known as *percentage error* in *x*.

Example 1. *Find the percentage error in the area of a rectangle when an error of +1 percent is made in measuring its length and breadth* (*R.G.P.V., Bhopal, June 2008, 2004*)

Solution. Let *A* be the area and *l*, *b*, be the length and breadth respectively.

Area of the rectangle, $A = lb$

Taking log of both the sides, we have

$$\log A = \log l + \log b \qquad \text{... (1)}$$

Differentiating equation (1), we get

$$\frac{\delta A}{A} = \frac{\delta l}{l} + \frac{\delta b}{b}$$

$$\Rightarrow \quad 100\frac{\delta A}{A} = 100\frac{\delta l}{l} + 100\frac{\delta b}{b} = +1 + 1 = +2$$

Hence, percentage error in area = + 2 **Ans.**

Example 2. *The power dissipated in a resistor is given by* $P = \frac{E^2}{R}$. *Find by using calculus the approximate percentage change in P when E is increased by 3% and R is decreased by 2%.* (*A.M.I.E. Summer 2001*)

Solution. Here, we have $\frac{100\,\delta E}{E} = 3\%$ and $\frac{100\,\delta R}{R} = -2\%$

$$P = \frac{E^2}{R} \quad \Rightarrow \quad \log P = 2 \log E - \log R$$

On differentiating, we get

$$\frac{\delta P}{P} = \frac{2}{E}\delta E - \frac{\delta R}{R} \quad \Rightarrow \quad 100\,\frac{\delta P}{P} = 2 \times \frac{100\,\delta E}{E} - \frac{100\,\delta R}{R}$$

$$100\frac{\delta P}{P} = 2\,(3) - (-\,2) = 8$$

Percentage change in P = 8% **Ans.**

Example 3. *The focal length of a mirror is found from the formula* $\frac{2}{f} = \frac{1}{v} - \frac{1}{u}$. *Find the percentage error in f if u and v are both in error by 2% each.*

Solution. *Here, we have*

$$\frac{2}{f} = \frac{1}{v} - \frac{1}{u} \qquad ...(1)$$

$$\frac{100\,\delta v}{v} = 2\%, \qquad \frac{100\,\delta u}{u} = 2\%$$

Taking the differential of (1), we get $-\frac{2}{f^2}\delta f = -\frac{1}{v^2}\delta v + \frac{1}{u^2}\delta u$

$$-\frac{2}{f}\left(\frac{100}{f}\right)\delta f = -\frac{1}{v}\left(\frac{100}{v}\right)\delta v + \frac{1}{u}\left(\frac{100}{u}\right)\delta u = -\frac{1}{v}.2 + \frac{1}{u}.2 = -2\left(\frac{l}{v} - \frac{1}{u}\right)$$

$$= -\frac{4}{f} \qquad \left[\because \frac{1}{v} - \frac{1}{u} = \frac{2}{f}\right]$$

$$\Rightarrow \qquad \frac{100}{f}.\delta f = 2$$

Hence, the percentage error in f = 2% **Ans.**

Example 4. *In determining the specific gravity by the formula* $S = \frac{A}{A - w}$ *where A is the weight in air and w is the weight in water; A can be read within 0.01 gm and w within 0.02 gm. Find approximately the maximum error in S if the readings are A = 1.1gm, w = 0.6gm. Find also the maximum relative error.* *(Q. Bank U.P.T.U.2001)*

Solution. We have, $S = \frac{A}{A - w}$

Differentiating, we get

$$\delta S = \frac{(A - w)\,\delta A - A\,(\delta A - \delta w)}{(A - w)^2} \qquad ...(1)$$

Maximum error in S can be obtained if we take $\delta A = -\,0.01$ and $\delta w = 0.02$

$\therefore$ From (1), $$\delta S = \frac{(1.1 - 0.6)\,(-0.01) - (1.1)\,(-0.01 - 0.02)}{(1.1 - 0.6)^2} = 0.112$$

Maximum relative error in

$$S = \frac{(\delta S)_{max}}{S} = \frac{0.112}{\left(\frac{1.1}{1.1 - 0.6}\right)} = 0.05091. \qquad \textbf{Ans.}$$

Example 5. *What error in the common logarithm of a number will be produced by an error of 1% in the number.* *(U.P., Q. Bank, I Semester, 2001)*

Solution. Let x be the number.

and $$y = \log_{10} x \qquad ...(1)$$

On differentiating both sides of (1), we get

$$\delta y = \frac{\delta x}{x}\log_e 10$$

$$y = (\log_{10} x \log_e 10)\log_{10} e = \log_e x \log_{10} e$$

On differentiating (1) w.r.t. 'x', we get

$$\frac{\delta y}{y} = \frac{\delta x}{x}\log_{10} e$$

$$\frac{\delta y}{y} = \frac{1}{100}\left(100\frac{\delta x}{x}\right)\log_{10} e \qquad \left[\text{Given } 100\frac{\delta x}{x} = 1\%\right]$$

$$= \frac{1}{100}(1)\log_{10} e = \frac{1}{100}(0.4343) \qquad [\log_{10} e = 0.4343]$$

$$= 0.004343\%$$

Ans.

Example 6. *The power 'P' required to propel a steamer of length 'l' at a speed 'u' is given by $P = \lambda u^3 l^3$ where λ is constant. If u is increased by 3% and l is decreased by 1%, find the corresponding increase in 'P'.* (*U.P., I Semester, Dec. 2009*)

Solution. We have,

$$P = \lambda u^3 l^3$$

On taking log on both sides, we get

$$\log P = \log \lambda + 3 \log u + 3 \log l$$

On differentiating, we get

$$\frac{\delta P}{P} = 0 + 3\frac{\delta u}{u} + 3\frac{\delta l}{l}$$

$$\Rightarrow \quad \frac{100\,\delta P}{P} = 3\times\frac{100\,\delta u}{u} + 3\times 100\frac{\delta l}{l} \qquad \left[\text{Given } \frac{100\,\delta u}{u} = 3,\ \frac{100\,\delta l}{l} = -1\right]$$

$$= 3\,(3) + 3\,(-1) = 9 - 3 = 6\%$$

Ans.

Example 7. *Find the percentage error in the area of an ellipse if one percent error is made in measuring the major and minor axis.* (*K. University Dec. 2008*)

Solution. Here we have the equation of an ellipse is

$$\frac{x^2}{a^2} + \frac{y^2}{b^2} = 1 \qquad \dots (1)$$

Differentiating equation (1), we get

$$\frac{2x\,\delta x}{a^2} + \frac{2y\,\delta y}{b^2} = 0$$

Area of ellipse (A) = $\pi\, a\, b$.

$\log A = \log \pi + \log a + \log b$

Differentiating, we get

$$\frac{\delta A}{A} = 0 + \frac{2\delta a}{2a} + \frac{2\delta b}{2b} \qquad \left[\begin{array}{l}\text{Major axis} = 2a\\ \text{Minor axis} = 2b\end{array}\right.$$

$$100\frac{\delta A}{A} = 100\frac{(2\delta a)}{2a} + \frac{100\,(2\delta b)}{2b} = 1 + 1$$

$$= 2\%$$

Ans.

Example 8. *The range R of a projectile which starts with a velocity v at an elevation α is given by $R = \dfrac{v^2 \sin 2\alpha}{g}$. Find the percentage error in R due to an error of 1% in v and an error of 0.5% in α.* (*K. University Dec. 2009*)

Solution. Here, we have

$$R = \frac{v^2 \sin 2\alpha}{g}. \qquad \dots (1)$$

Taking log of both sides of (1), we get

$$\log R = 2 \log v + \log \sin 2\alpha - \log g \qquad \dots (2)$$

Differentiating (2), we get

$$\frac{\delta R}{R} = \frac{2\delta v}{v} + \frac{2 \cos 2\alpha . \delta\alpha}{\sin 2\alpha} - \frac{\delta g}{g}$$

$$100 \frac{\delta R}{R} = 100 . \frac{2\delta v}{v} + 100 \frac{2 \alpha \cos 2\alpha . \delta\alpha}{\sin 2\alpha} - 100 \frac{\delta g}{g}$$

$$= 2(1) + 2\alpha \cot 2\alpha (0.5)$$

$$= 2 + \alpha \cot 2\alpha$$ **Ans.**

Example 9. *The period T of a simple pendulum is*

$$T = 2\pi \sqrt{\frac{l}{g}}$$

Find the maximum error in T due to possible errors upto 1% in l and 2% in g.

(AMITE, Dec. 2010, Uttarakhand, I Semester, 2009, Dec. 2006)

Solution. Here, we have $T = 2\pi \sqrt{\frac{l}{g}}$ ... (1)

$$100 \times \frac{\delta l}{l} = 1, \qquad 100 \times \frac{\delta g}{g} = 2$$

Taking log of both sides of (1), we get

$$\Rightarrow \qquad \log T = \log 2\pi + \frac{1}{2} \log l - \frac{1}{2} \log g$$

Differentiating, we get

$$\frac{\delta T}{T} = 0 + \frac{1}{2} \frac{\delta l}{l} - \frac{1}{2} \frac{\delta g}{g}$$

$$100 \times \left(\frac{\delta T}{T}\right) = \frac{1}{2} \times \left[100 \times \left(\frac{\delta l}{l}\right) - 100 \times \left(\frac{\delta g}{g}\right)\right]$$

$$100 \times \left(\frac{\delta T}{T}\right) = \frac{1}{2}[1 \pm 2] = \frac{3}{2}$$

Maximum error in $T = 1.5\%$ **Ans.**

Example 10. *The time T of a pendulum of length l under certain conditions is given by* $T = 2\pi\sqrt{\frac{l}{g'}}$ *where* $g' = g\left(\frac{r}{r+h}\right)^2$. *Find the error in T due to error p% and q% in h and l respectively. (g and r are constants)*

Solution. We have,

$$\frac{100\, \delta h}{h} = \text{P and } \frac{100\, \delta l}{l} = q$$

$$T = 2\pi\sqrt{\frac{l}{g'}} \qquad \dots(1) \qquad \left[g' = g\left(\frac{r}{r+h}\right)^2\right]$$

On putting the value of g' in (1), we get

$$T = 2\pi \sqrt{\frac{l}{g\left(\frac{r}{r+h}\right)^2}} = 2\pi \sqrt{\left(\frac{l}{g}\right) . \left(\frac{r+h}{r}\right)}$$

Taking log of both the sides, we get

$$\log T = \log 2\pi + \frac{1}{2}\log l - \frac{1}{2}\log g + \log (r+h) - \log r$$

$$\therefore \quad \frac{1}{T}\delta T = 0 + \frac{1}{2}\frac{\delta l}{l} - 0 + \frac{1}{r+h}\delta h - 0$$

$$\Rightarrow \quad \frac{100\,\delta T}{T} = \frac{1}{2}\left(100\,\frac{\delta l}{l}\right) + \frac{1}{r+h}h\left(100\,\frac{\delta h}{h}\right)$$

$$\Rightarrow \quad \frac{100\,\delta T}{T} = \frac{1}{2}.q + \frac{h}{r+h}.p$$

Hence percentage error in T is $\frac{1}{2}q + \frac{hp}{r+h}\%$ **Ans.**

Example 11. *A balloon is in the form of right circular cylinder of radius 1.5 m and length 4 m and is surmounted by hemispherical ends. If the radius is increased by 0.01 m and the length by 0.05 m, find the percentage change in the volume of the balloon.*

(*U.P. I Semester Dec., 2005, Comp. 2002*)

Solution. Radius of the cylinder (r) = 1.5 m

Length of the cylinder (h) = 4 m

Volume of balloon = Volume of cylinder + Volume of two hemispherical ends

1.5 m

4 m

$$\text{Volume } (V) = \pi r^2 h + \frac{2}{3}\pi r^3 + \frac{2}{3}\pi r^3 = \pi r^2 h + \frac{4}{3}\pi r^3$$

Differentiating, we get

$$\delta V = \pi\, 2r\,\delta r\,.h + \pi r^2.\delta h + \frac{4}{3}\pi\, 3r^2\,.\delta r$$

$$\frac{\delta V}{V} = \frac{\pi r[2\,\delta r\,.h + r.\delta h + 4r\,\delta r]}{\pi r^2 h + \frac{4}{3}\pi r^3} = \frac{2.\delta r.h + r.\delta h + 4r.\delta r}{rh + \frac{4}{3}r^2}$$

$$= \frac{2\times 0.01\times 4 + 1.5\times 0.05 + 4\times 1.5\times 0.01}{1.5\times 4 + \frac{4}{3}(1.5)^2} = \frac{0.08 + 0.075 + 0.06}{6+3} = \frac{0.215}{9}$$

$$100\frac{\delta V}{V} = \frac{100\times 0.215}{9} = \frac{21.5}{9} = 2.389\ \%$$ **Ans.**

Example 12. *In estimating the number of bricks in a pile which is measured to be (5m × 10m × 5m), count of bricks in taken as 100 bricks per m^3. Find the error in the cost when the tape is stretched 2% beyond its standard length. The cost of bricks is Rs. 2,000 per thousand bricks.*

Solution. Volume $V = x\,y\,z$

$$\log V = \log x + \log y + \log z$$

Differentiating, we get

$$\frac{\delta V}{V} = \frac{\delta x}{x} + \frac{\delta y}{y} + \frac{\delta z}{z}$$

$$100\,\frac{\delta V}{V} = \frac{100\,\delta x}{x} + \frac{100\,\delta y}{y} + \frac{100\,\delta z}{z} = 2 + 2 + 2 = 6$$

$$\delta V = \frac{6V}{100} = \frac{6(5\times 10\times 5)}{100} = 15 \text{ cubic metre.}$$

Number of bricks in $\delta V = 15 \times 100 = 1500$

Error in cost $= \dfrac{1500 \times 2000}{1000} = 3000$

This error in cost, a loss to the seller of brick = Rs. 3000 **Ans.**

Example 13. *With the usual meaning for a, b, c and s, if Δ be the area of a triangle, prove that the error in Δ resulting from a small error in the measurement of c is given by*

$$\delta\,\Delta = \frac{\Delta}{4}\left[\frac{1}{s}+\frac{1}{s-a}+\frac{1}{s-b}-\frac{1}{s-c}\right]\delta\,c.$$

Solution. We know that

$$\Delta = \sqrt{s(s-a)(s-b)(s-c)}$$

$$\log\Delta = \frac{1}{2}\left[\log s+\log(s-a)+\log(s-b)+\log(s-c)\right] \quad \left[\begin{array}{l}\because \Delta = \frac{1}{2}bc\sin A\\ b = a\cos C + c\cos A\end{array}\right]$$

$$\frac{1}{\Delta}\frac{\delta\Delta}{\delta c} = \frac{1}{2}\left[\frac{1}{s}\frac{\delta s}{\delta c}+\frac{1}{s-a}\frac{\delta(s-a)}{\delta c}+\frac{1}{s-b}\frac{\delta(s-b)}{\delta c}+\frac{1}{s-c}\frac{\delta(s-c)}{\delta c}\right]$$

$$\left[s = \frac{1}{2}(a+b+c),\ \frac{\delta s}{\delta c}=\frac{1}{2} \quad \text{and } s-c = \frac{1}{2}(a+b-c),\ \frac{d}{dc}(s-c) = \frac{ds}{dc}-1 = \frac{1}{2}-1 = -\frac{1}{2}\right]$$

Putting the values of $\dfrac{\delta s}{\delta c}, \dfrac{\delta(s-a)}{\delta c}, \dfrac{\delta(s-b)}{\delta c}$ and $\dfrac{\delta(s-c)}{\delta c}$, we get

$$\Rightarrow \quad \frac{1}{\Delta}\frac{\delta\Delta}{\delta c} = \frac{1}{2}\left[\frac{1}{2}\frac{1}{s}+\frac{1}{2}\frac{1}{s-a}+\frac{1}{2}\frac{1}{s-b}-\frac{1}{2}\frac{1}{s-c}\right]$$

$$\Rightarrow \quad \delta\,\Delta = \frac{\Delta}{4}\left[\frac{1}{s}+\frac{1}{s-a}+\frac{1}{s-b}-\frac{1}{s-c}\right]\delta\,c$$ **Proved.**

Example 14. *The angles of a triangle are calculated from the sides a, b, c. If small changes δa, δb, δc are made in the sides, show that approximately*

$$\delta A = \frac{a}{2\Delta}\left[\delta a - \delta b.\cos C - \delta c.\cos B\right],$$

where Δ is the area of the triangle and A, B, C are the angles opposite to a, b, c respectively. Verify that $\delta A + \delta B + \delta C = 0$

(A.M.I.E.T.E., June 2010, Dec. 2007, Summer 2001
U.P.,I Sem,. Winter 2001)

Solution. We know that

$$\cos A = \frac{b^2+c^2-a^2}{2bc} \Rightarrow a^2 = b^2+c^2-2\,b\,c\cos A \quad ...(1)$$

Differentiating both sides of (1), we get

$$2a\,\delta a = 2\,b\,\delta\,b + 2\,c\,\delta\,c - 2\,b\,\delta\,c\cos A - 2\,\delta\,b\,c\cos A + 2\,b\,c\sin A\,\delta\,A$$ (approx.)

$$a\,\delta\,a = b\,\delta b + c\,\delta c - b\,\delta c\cos A - \delta b\,c\cos A + b\,c\sin A\,\delta A$$

$$\Rightarrow bc\sin A\,\delta\,A = a\,\delta a - (b - c\cos A)\,\delta b - (c - b\cos A)\,\delta c$$

$$\Rightarrow \quad 2\,\Delta\,\delta\,A = a\,\delta a - (a\cos C + c\cos A - c\cos A)\,\delta b - (a\cos B + b\cos A - b\cos A)\,\delta c$$

$$2\,\Delta\,\delta\,A = a\,\delta\,a - a\,\delta\,b\cos C - a\,\delta\,c\cos B$$

$$= a\,(\delta\,a - \delta\,b\cos C - \delta\,c\cos B)$$

$$\Rightarrow \qquad \delta A = \frac{a}{2\Delta}[\delta a - \delta b \cdot \cos C - \delta c \cdot \cos B] \qquad \text{...(2) Proved.}$$

Similarly,

$$\delta B = \frac{b}{2\Delta}[\delta b - \delta c \cdot \cos A - \delta a \cdot \cos C] \qquad \text{...(3)}$$

$$\delta C = \frac{c}{2\Delta}[\delta c - \delta a \cdot \cos B - \delta b \cdot \cos A] \qquad \text{...(4)}$$

On adding (2), (3) and (4), we get

$$[\delta A + \delta B + \delta C] = \frac{1}{2\Delta}\ [(a - b\cos C - c\cos B)\ \delta a + (b - a\cos C - c\cos A)\ \delta b + (c - a\cos B - b\cos A)\ \delta c]$$

$$= \frac{1}{2\Delta}\left[\{a - (b\cos C + c\cos B)\}\ \delta a + \{b - (a\cos C + c\cos A)\}\ \delta b + \{c - (a\cos B + b\cos A)\}\ \delta c\right]$$

$$= \frac{1}{2\Delta}\left[(a-a)\,\delta a + (b-b)\,\delta b + (c-c)\,\delta c\right] = 0 \qquad [\because b\cos C + c\cos B = a] \quad \textbf{Verified.}$$

Example 15. *Find the possible percentage error in computing the parallel resistance r of three resistances r_1, r_2, r_3 from the formula $\frac{1}{r} = \frac{1}{r_1} + \frac{1}{r_2} + \frac{1}{r_3}$, if r_1, r_2 and r_3 are each in error by plus 1.2%.*

Solution. Here,

$$\frac{1}{r} = \frac{1}{r_1} + \frac{1}{r_2} + \frac{1}{r_3}, \qquad \text{...(1)}$$

Differentiating, we get

$$-\frac{1}{r^2}dr = -\frac{1}{r_1^2}dr_1 - \frac{1}{r_2^2}dr_2 - \frac{1}{r_3^2}dr_3$$

$$\Rightarrow \quad \frac{1}{r}\left(\frac{100\,dr}{r}\right) = \frac{1}{r_1}\left(\frac{100\,dr_1}{r_1}\right) + \frac{1}{r_2}\left(\frac{100\,dr_2}{r_2}\right) + \frac{1}{r_3}\left(\frac{100\,dr_3}{r_3}\right)$$

$$= \frac{1}{r_1}(1.2) + \frac{1}{r_2}(1.2) + \frac{1}{r_3}(1.2) = (1.2)\left[\frac{1}{r_1} + \frac{1}{r_2} + \frac{1}{r_3}\right]$$

$$= 1.2\left(\frac{1}{r}\right) \qquad \text{[From (1)]}$$

$$\Rightarrow \qquad \frac{100\,dr}{r} = 1.2\ \% \qquad \textbf{Ans.}$$

Example 16. *The height h and semi-vertical angle α of a cone are measured, and from there A, the total area of the cone, including the base, is calculated. If h and α are in error by small quantities δh and δα respectively, find the corresponding error in the area. Show further that, if $\alpha = \frac{\pi}{6}$, an error of + 1 percent in h will be approximately compensated by an error of – 19.8' in α.* (*A.M.I.E.T.E. Summer 2003*)

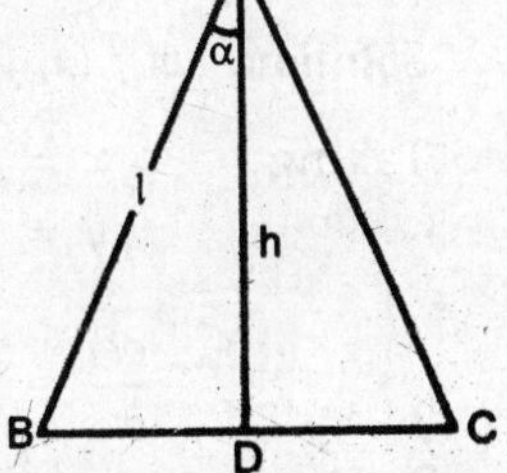

Solution. Let l be the slant height of the cone and r its radius.

$$l = h\sec\alpha$$

$$r = h\tan\alpha$$

$$A = \pi r^2 + \pi r l$$

$$= \pi h^2\tan^2\alpha + \pi(h\tan\alpha)(h\sec\alpha)$$

$$= \pi h^2[\tan^2\alpha + \tan\alpha\sec\alpha]$$

$$\delta A = 2\pi h\,\delta h\,[\tan^2\alpha + \tan\alpha\sec\alpha]$$
$$+ \pi h^2 [2\tan\alpha\sec^2\alpha\,\delta\alpha + \sec^2\alpha\,.\,\delta\alpha\,.\,\sec\alpha + \tan\alpha\sec\alpha\tan\alpha\,\delta\alpha]$$
$$\delta A = 2\pi h[\tan\alpha + \sec\alpha]\tan\alpha\,.\,\delta h + \pi.h^2\,[2\tan\alpha\sec\alpha + \sec^2\alpha + \tan^2\alpha]\,.\,\sec\alpha\,.\,\delta\alpha$$
$$\delta A = 2\pi h\,[\tan\alpha + \sec\alpha]\tan\alpha\,.\,\delta h + \pi h^2\,[\tan\alpha + \sec\alpha]^2\sec\alpha.\,\delta\alpha$$
$$\delta A = \pi h^2\,[\tan\alpha + \sec\alpha]\;[2\tan\alpha\frac{\delta h}{h} + (\tan\alpha + \sec\alpha)\sec\alpha\,.\,\delta\alpha]$$

On putting $\delta A = 0,\quad \alpha = \frac{\pi}{6},\quad \frac{\delta h}{h}\times 100 = 1,$ we get

$$0 = \pi h^2\left[\tan\frac{\pi}{6} + \sec\frac{\pi}{6}\right]\left[\left(2\tan\frac{\pi}{6}\right)\frac{1}{100} + \left(\tan\frac{\pi}{6} + \sec\frac{\pi}{6}\right)\sec\frac{\pi}{6}\,\delta\alpha\right]$$
$$0 = \left(2\tan\frac{\pi}{6}\right)\frac{1}{100} + \left(\tan\frac{\pi}{6} + \sec\frac{\pi}{6}\right)\sec\frac{\pi}{6}\,\delta\alpha$$
$$0 = \left[\frac{2}{\sqrt{3}}\frac{1}{100} + \left(\frac{1}{\sqrt{3}} + \frac{2}{\sqrt{3}}\right)\frac{2}{\sqrt{3}}\,\delta\alpha\right]$$
$$\frac{2}{\sqrt{3}}\frac{1}{100} = -\left(\frac{1}{\sqrt{3}} + \frac{2}{\sqrt{3}}\right)\frac{2}{\sqrt{3}}\,\delta\alpha$$
$$\frac{1}{100} = -\left(\frac{1}{\sqrt{3}} + \frac{2}{\sqrt{3}}\right)\delta\alpha \Rightarrow \frac{1}{100} = -\frac{3}{\sqrt{3}}\delta\alpha \Rightarrow \delta\alpha = -\frac{1}{100\sqrt{3}}$$
$$\delta\alpha = -\frac{1}{100\sqrt{3}}\frac{180}{\pi}\text{degree} = \left(-\frac{9}{5\sqrt{3}\pi}\right)60\text{ minutes} = -19.8\text{ minutes}\qquad \textbf{Ans.}$$

Example 17. *Compute an approximate value of* $(1.04)^{3.01}$.

Solution. Let $f(x, y) = x^y$

We have $\qquad \frac{\partial f}{\partial x} = y\,x^{y-1},\qquad \frac{\partial f}{\partial y} = x^y\log x \qquad ...(1)$

Here $\quad$ Let $\quad x = 1,\quad \delta x = 0.04$; $\quad y = 3,\quad \delta y = 0.01 \qquad ...(2)$

Now $\qquad df = \frac{\partial f}{\partial x}dx + \frac{\partial f}{\partial y}dy$

$\qquad = y\,x^{y-1}\,dx + x^y\log x\,dy \qquad ...(3)$ [Using (1)]

Substituting the values from (2) in (3), we get

$$df = (3)\,(1)\,(0.04) + (1)^3\log(1)\,(0.01) = 0.12$$
$$(1.04)^{3.01} = f(1, 3) + df = 1 + 0.12 = 1.12\qquad \textbf{Ans.}$$

Example 18. *Find* $[(3.82)^2 + 2(2.1)^3]^{\frac{1}{5}}$

Solution. Let $f(x, y) = \left(x^2 + 2y^3\right)^{\frac{1}{5}}$

Taking $\quad x = 4,\quad \delta x = 3.82 - 4 = -0.18$

$\qquad\quad y = 2,\quad \delta y = 2.1 - 2 = 0.1$

$$\frac{\partial f}{\partial x} = \frac{1}{5}\left[x^2 + 2y^3\right]^{-\frac{4}{5}}(2x) = \frac{2\,(4)}{5}\left[16 + 2(2)^3\right]^{-\frac{4}{5}} = \frac{8}{5}\left(\frac{1}{16}\right) = \frac{1}{10}$$

$$\frac{\partial f}{\partial y} = \frac{1}{5}\left[x^2+2y^3\right]^{-\frac{4}{5}}\left(6y^2\right)=\frac{6}{5}(2)^2\left[16+2(2)^3\right]^{-\frac{4}{5}}=\frac{24}{5}\times\left(\frac{1}{16}\right)=\frac{3}{10}$$

By total differentiation,

$$df=\frac{\partial f}{\partial x}dx+\frac{\partial f}{\partial y}dy = \frac{1}{10}(-0.18)+\frac{3}{10}(0.1)=-0.018+0.03=0.012$$

$$\left[(3.82)^2+2(2.1)^3\right]^{\frac{1}{5}} = f(4, 2) + df$$

$$= [(4)^2+2(2)^3]^{\frac{1}{5}}+0.012=2+0.012=2.012 \quad \textbf{Ans.}$$

Example 19. *Find approximate value of* $[(0.98)^2 + (2.01)^2 + (1.94)^2]^{1/2}$.

Solution. Let $f(x, y, z) = (x^2 + y^2 + z^2)^{1/2}$

Taking $x = 1, y = 2$ and $z = 2$ so that

$dx = -0.02, dy = 0.01\ dz = -0.06$

From (1), $\frac{\partial f}{\partial x} = x(x^2+y^2+z^2)^{-1/2}, \frac{\partial f}{\partial y}=y(x^2+y^2+z^2)^{-1/2},$

$$\frac{\partial f}{\partial z} = z(x^2+y^2+z^2)^{-1/2}$$

Now, $df=\frac{\partial f}{\partial x}dx+\frac{\partial f}{\partial y}dy+\frac{\partial f}{\partial z}dz$ [By total differentiation]

$$= \frac{\partial f}{\partial x}(dx+dy+dz)$$

$$= (x^2+y^2+z^2)^{-1/2}(x\,dx+y\,dy+z\,dz)$$

$$= (1+4+4)^{-\frac{1}{2}}[1(-0.02)+2(0.01)+2(-0.06)]$$

$$= \frac{1}{3}(-0.02+0.02-0.12)=-0.04$$

$\therefore \quad [(0.98)^2 + (2.01)^2 + (1.94)^2]^{1/2} = f(1, 2, 2) + df = 3 + (-0.04) = 2.96.$ **Ans.**

EXERCISE 4.1

1. If the density ρ of a body be inferred from its weights W, ω in air and water respectively, show that the relative error in ρ due to errors δ W, δ ω in W, ω respectively is $\frac{\delta\rho}{\rho}=\frac{-\omega}{W-\omega}\cdot\frac{\delta W}{W}+\frac{\delta\omega}{W-\omega}$

2. The time T of a complete oscillation of a simple pendulum of length 'L' is governed by the equation $T = 2\pi\sqrt{\frac{L}{g}}$, g is constant, find the approximate error in the calculated value of T corresponding to an error of 2% in the value of L. (*U.P., I semester, Dec. 2008*) **Ans.** 1%

3. The period of oscillation of a pendulum is computed by the formula

$$T = 2\pi\sqrt{\frac{l}{g}}$$

Show that the percentage error in $T=\frac{1}{2}$ [% error in l – % error in g]

If l = 6 cm and relative error in g is equal to $\frac{1}{160}$, find the error in the determination of T. (Given g = 981 cm/sec^2) **Ans.** – 00153

4. The indicated horse power I of an engine is calculated from the formula.

$$I = PLAN/33000$$

where $A = \frac{\pi}{4} d^2$. Assuming that errors of r per cent may have been made in measuring P, L, N and d. Find the greatest possible error in I. **Ans.** $5\ r\ \%$

5. The dimensions of a cone are radius 4 cm, height 6 cm. What is the error in its volume if the scale used in taking the measurement is short by 0.01 cm per cm. ...(2)
Ans. $0.96\ \pi\ \text{cm}^3$

6. The work that must be done to propel a ship of displacement D for a distance s in time t is proportional to $s^2 D^{2/3} t^2$.
Find approximately the percentage increase of work necessary when the displacement is increased by 1% the time is diminished by 1% and the distance is increased by 3%. **Ans.** $\frac{14}{3}\%$

7. The power P required to propel a ship of length l moving with a velocity V is given by $P = kV^3 l^2$.
Find the percentage increase in power if increase in velocity is 3% and increase in length is 4%.
Ans. 17%

8. In estimating the cost of a pile of bricks measured as 2 m × 15 m × 1.2 m, the tape is stretched 1% beyond the standard length if the count is 450 bricks to 1 m^3 and bricks cost Rs. 1300 per 1000, find the approximate error in the cost. **Ans.** Rs. 631.80

9. In estimating the cost of a pile of bricks measured as 6' × 50' × 4', the tape is stretched 1% beyond the standard length. If the count is 12 bricks to ft^3, and bricks cost Rs. 100 per 1000, find the approximate error in the cost. **Ans.** 720 bricks, Rs. 25.20 (*U.P. I Sem. Dec. 2004*)

10. The sides of a triangle are measured as 12 cm and 15 cm and the angle included between them as 60°. If the lengths can be measured within 1% accuracy while the angle can be measured within 2% accuracy. Find the percentage error in determining (i) area of the triangle (ii) length of opposite side of the triangle. (*A.M.I.E.T.E., Winter 2002*) **Ans.** ???

11. In determining the specific gravity by the formula $S = \frac{A}{A - w}$, where A is the weight in air and w is the weight in water. A can be read within 0.01 gm and w within 0.02 gm. Find approximately the maximum error in S if the reading are A = 1.1 gm and w = 0.6 gm. **Ans.** – 0.112

12. The voltage V across a resistor is measured with error h, and the resistance R is measured with an error k. Show that the error in calculating the power $W(V, R) = \frac{V^2}{R}$ generated in the resistor is $\frac{V}{R^2}(2Rh - Vk.)$. If V can be measured to an accuracy of 0.5 *p.c.* and R to an accuracy of 1 *p.c.*, what is the approximate possible percentage error in W? **Ans.** Zero per cent

13. Find the percentage error in calculating the area of ellipse $\frac{x^2}{a^2} + \frac{y^2}{b^2} = 1$, when error of + 1% is made in measuring the major and minor axes. (*A.M.I.E.T.E., June 2009*) **Ans.** 2%

14. If $f = x^2\, y^2\, z^{\frac{1}{10}}$, find the approximate value of f, when x = 1.99, y = 3.01 and z = 0.98.
Ans. 107.784

15. Find approximate value of $\sqrt{(0.98)^2 + (2.01)^2 + (1.94)^2}$ **Ans.** 2.96

CHAPTER

5

JACOBIANS

5.1 JACOBIANS

If u and v are functions of the two independent variables x and y, then the determinant

$$\begin{vmatrix} \frac{\partial u}{\partial x} & \frac{\partial u}{\partial y} \\ \frac{\partial v}{\partial x} & \frac{\partial v}{\partial y} \end{vmatrix}$$

is called the Jacobian of u, v with respect to x, y and is written as

$$\frac{\partial(u, v)}{\partial(x, y)} \text{ or } J\left(\frac{u, v}{x, y}\right)$$

Similarly the Jacobian of u, v, w with respect to x, y, z is

$$\frac{\partial(u, v, w)}{\partial(x, y, z)} = \begin{vmatrix} \frac{\partial u}{\partial x} & \frac{\partial u}{\partial y} & \frac{\partial u}{\partial z} \\ \frac{\partial v}{\partial x} & \frac{\partial v}{\partial y} & \frac{\partial v}{\partial z} \\ \frac{\partial w}{\partial x} & \frac{\partial w}{\partial y} & \frac{\partial w}{\partial z} \end{vmatrix}$$

Example 1. *If* $x = r\cos\theta$, $y = r\sin\theta$; *evaluate* $\frac{\partial(x, y)}{\partial(r, \theta)}$ *and* $\frac{\partial(r, \theta)}{\partial(x, y)}$

Solution. $x = r\cos\theta, \qquad y = r\sin\theta$

$$\frac{\partial x}{\partial r} = \cos\theta \qquad \frac{\partial y}{\partial r} = \sin\theta$$

$$\frac{\partial x}{\partial \theta} = -r\sin\theta, \qquad \frac{\partial y}{\partial \theta} = r\cos\theta$$

$$\frac{\partial(x, y)}{\partial(r, \theta)} = \begin{vmatrix} \frac{\partial x}{\partial r} & \frac{\partial x}{\partial \theta} \\ \frac{\partial y}{\partial r} & \frac{\partial y}{\partial \theta} \end{vmatrix} = \begin{vmatrix} \cos\theta & -r\sin\theta \\ \sin\theta & r\cos\theta \end{vmatrix}$$

$$= r\cos^2\theta + r\sin^2\theta = r(\cos^2\theta + \sin^2\theta) = r$$

Now $\quad r^2 = x^2 + y^2, \qquad \theta = \tan^{-1}\frac{y}{x}$

$$\frac{\partial r}{\partial x} = \frac{x}{r}, \qquad \frac{\partial \theta}{\partial x} = \frac{-y}{x^2 + y^2} = -\frac{y}{r^2}$$

$$\frac{\partial r}{\partial y}=\frac{y}{r} \qquad \frac{\partial\theta}{\partial y}=\frac{x}{x^2+y^2}=\frac{x}{r^2}$$

$$\frac{\partial(r,\theta)}{\partial(x,y)}=\begin{vmatrix}\frac{\partial r}{\partial x} & \frac{\partial r}{\partial y}\\ \frac{\partial\theta}{\partial x} & \frac{\partial\theta}{\partial y}\end{vmatrix}=\begin{vmatrix}\frac{x}{r} & \frac{y}{r}\\ -\frac{y}{r^2} & \frac{x}{r^2}\end{vmatrix}=\frac{x^2}{r^3}+\frac{y^2}{r^3}=\frac{x^2+y^2}{r^3}=\frac{r^2}{r^3}=\frac{1}{r}$$ **Ans.**

Note : $\frac{\partial(x,y)}{\partial(r,\theta)}\times\frac{\partial(r,\theta)}{\partial(x,y)}=r\cdot\frac{1}{r}=1.$

Example 2. *Find the Jacobian* $J\left(\frac{u,v}{x,y}\right)$ *for* $u=e^x\sin y$ *and*

$v=x\log\sin y$ *(U.P.I semester, Dec 2009)*

Solution. Here, we have

$$u=e^x\sin y \qquad v=x\log\sin y$$

$$\frac{\partial u}{\partial x}=e^x\sin y \qquad \frac{\partial v}{\partial x}=\log\sin y$$

$$\frac{\partial u}{\partial y}=e^x\cos y \qquad \frac{\partial v}{\partial y}=\frac{x\cos y}{\sin y}$$

$$J\left(\frac{u,v}{x,y}\right)=\begin{vmatrix}\frac{\partial u}{\partial x} & \frac{\partial u}{\partial y}\\ \frac{\partial v}{\partial x} & \frac{\partial v}{\partial y}\end{vmatrix}=\begin{vmatrix}e^x\sin y & e^x\cos y\\ \log\sin y & \frac{x\cos y}{\sin y}\end{vmatrix}$$

$$=e^x x\cos y-e^x\cos y\log\sin y=e^x\cos y\,[x-\log\sin y]$$ **Ans.**

Example 3. *If* $x=a\cosh\xi\cos\eta,\ y=a\sinh\xi\sin\eta$, *show that*

$$\frac{\partial(x,y)}{\partial(\xi,\eta)}=\frac{a^2}{2}(\cosh 2\xi-\cos 2\eta)$$

Solution. $x=a\cosh\xi\cos\eta$

$y=a\sinh\xi\sin\eta$

$$\frac{\partial(x,y)}{\partial(\xi,\eta)}=\begin{vmatrix}\frac{\partial x}{\partial\xi} & \frac{\partial x}{\partial\eta}\\ \frac{\partial y}{\partial\xi} & \frac{\partial y}{\partial\eta}\end{vmatrix}=\begin{vmatrix}a\sinh\xi\cos\eta & -a\cosh\xi\sin\eta\\ a\cosh\xi\sin\eta & a\sinh\xi\cos\eta\end{vmatrix}$$

$$=a^2\begin{vmatrix}\sinh\xi\cos\eta & -\cosh\xi\sin\eta\\ \cosh\xi\sin\eta & \sinh\xi\cos\eta\end{vmatrix}$$

$$=a^2\,[\sinh^2\xi\cos^2\eta+\cosh^2\xi\sin^2\eta]$$

$$=a^2\,[\sinh^2\xi(1-\sin^2\eta)+(1+\sinh^2\xi)\sin^2\eta]$$

$$=a^2\,[\sinh^2\xi-\sinh^2\xi\sin^2\eta+\sin^2\eta+\sinh^2\xi\sin^2\eta]$$

$$=a^2\,[\sinh^2\xi+\sin^2\eta]=\frac{a^2}{2}\,[\cosh 2\xi-1+1-\cos 2\eta]$$

$$=\frac{a^2}{2}\,[\cosh 2\xi-\cos 2\eta]$$ **Proved.**

Example 4. *If* $y_1 = \frac{x_2 x_3}{x_1},\ y_2 = \frac{x_3 x_1}{x_2},\ y_3 = \frac{x_1 x_2}{x_3}$.

Show that the Jacobian of y_1, y_2, y_3 with respect to x_1, x_2, x_3 is 4.

(UP. I Sem. 2004, Comp.2002,A.M.I.E.,Winter 2001, Summer 2002, 2000)

Solution. We have, $y_1 = \frac{x_2 x_3}{x_1},\ y_2 = \frac{x_3 x_1}{x_2},\ y_3 = \frac{x_1 x_2}{x_3}$.

$$\frac{\partial(y_1, y_2, y_3)}{\partial(x_1, x_2, x_3)} = \begin{vmatrix} \frac{\partial y_1}{\partial x_1} & \frac{\partial y_1}{\partial x_2} & \frac{\partial y_1}{\partial x_3} \\ \frac{\partial y_2}{\partial x_1} & \frac{\partial y_2}{\partial x_2} & \frac{\partial y_2}{\partial x_3} \\ \frac{\partial y_3}{\partial x_1} & \frac{\partial y_3}{\partial x_2} & \frac{\partial y_3}{\partial x_3} \end{vmatrix} = \begin{vmatrix} -\frac{x_2 x_3}{x_1^2} & \frac{x_3}{x_1} & \frac{x_2}{x_1} \\ \frac{x_3}{x_2} & -\frac{x_3 x_1}{x_2^2} & \frac{x_1}{x_2} \\ \frac{x_2}{x_3} & \frac{x_1}{x_3} & -\frac{x_1 x_2}{x_3^2} \end{vmatrix}$$

$$= \frac{1}{x_1^2\, x_2^2\, x_3^2} \begin{vmatrix} -x_2 x_3 & x_3 x_1 & x_1 x_2 \\ x_2 x_3 & -x_3 x_1 & x_1 x_2 \\ x_2 x_3 & x_3 x_1 & -x_1 x_2 \end{vmatrix} = \frac{x_1^2\, x_2^2\, x_3^2}{x_1^2\, x_2^2\, x_3^2} \begin{vmatrix} -1 & 1 & 1 \\ 1 & -1 & 1 \\ 1 & 1 & -1 \end{vmatrix}$$

$$= -1(1-1) - 1(-1-1) + 1(1+1)$$

$$= 0 + 2 + 2 = 4$$ **Proved.**

Example 5. *If* $x = r \sin\theta \cos\phi$
$y = r \sin\theta \sin\phi$
$z = r \cos\theta,$

show that $\frac{\partial(x, y, z)}{\partial(r, \theta\ \phi)} = r^2 \sin\theta$ *(U.P, I, Sem. Winter2002)*

Solution.

$x = r \sin\theta\ \cos\phi,\qquad y = r \sin\theta\ \sin\phi,\qquad z = r\cos\theta$

$\frac{\partial x}{\partial r} = \sin\theta \cos\phi,\qquad \frac{\partial y}{\partial r} = \sin\theta \sin\phi,\qquad \frac{\partial z}{\partial r} = \cos\theta$

$\frac{\partial x}{\partial \theta} = r\cos\theta \cos\phi,\qquad \frac{\partial y}{\partial \theta} = r\cos\theta \sin\phi,\qquad \frac{\partial z}{\partial \theta} = -r\sin\theta$

$\frac{\partial x}{\partial \phi} = -r \sin\theta \sin\phi,\qquad \frac{\partial y}{\partial \phi} = r\sin\theta \cos\phi,\qquad \frac{\partial z}{\partial \phi} = 0$

$$\frac{\partial(x, y, z)}{\partial(r, \theta\ \varphi)} = \begin{vmatrix} \frac{\partial x}{\partial r} & \frac{\partial x}{\partial \theta} & \frac{\partial x}{\partial \varphi} \\ \frac{\partial y}{\partial r} & \frac{\partial y}{\partial \theta} & \frac{\partial y}{\partial \varphi} \\ \frac{\partial z}{\partial r} & \frac{\partial z}{\partial \theta} & \frac{\partial z}{\partial \varphi} \end{vmatrix} = \begin{vmatrix} \sin\theta\cos\varphi & r\cos\theta\cos\varphi & -r\sin\theta\sin\varphi \\ \sin\theta\sin\varphi & r\cos\theta\sin\varphi & r\sin\theta\cos\varphi \\ \cos\theta & -r\sin\theta & 0 \end{vmatrix}$$

$$= r^2 \sin\theta \begin{vmatrix} \sin\theta\cos\varphi & \cos\theta\cos\varphi & -\sin\varphi \\ \sin\theta\sin\varphi & \cos\theta\sin\varphi & \cos\varphi \\ \cos\theta & -\sin\theta & 0 \end{vmatrix}$$

$= r^2 \sin\theta\ [\sin\theta \cos\phi\ (0 + \sin\theta \cos\phi) - \cos\theta \cos\phi\ (0 - \cos\theta \cos\phi) - \sin\phi\ (-\sin^2\theta \sin\phi - \cos^2\theta \sin\phi)]$

$= r^2 \sin\theta\ [\sin^2\theta \cos^2\phi + \cos^2\theta \cos^2\phi + \sin^2\theta \sin^2\phi + \cos^2\theta \sin^2\phi]$

$= r^2 \sin\theta\ [(\sin^2\theta + \cos^2\theta)\cos^2\phi + (\sin^2\theta + \cos^2\theta) \sin^2\phi]$

$= r^2 \sin\theta\ [\cos^2\phi + \sin^2\phi] = r^2 \sin\theta$ **Proved.**

Theorem. *If the relations connecting u_i's and x_i's are of the form.*

$$u_1 = f_1(x_1)$$
$$u_2 = f_2(x_1, x_2)$$
$$\ldots\ldots\ldots\ldots$$
$$\ldots\ldots\ldots\ldots$$
$$u_n = f_n(x_1, x_2, \ldots\ldots x_n)$$

then
$$\frac{\partial(u_1, u_2, \ldots\ldots, u_n)}{\partial(x_1, x_2, \ldots\ldots, x_n)} = \frac{\partial u_1}{\partial x_1} \cdot \frac{\partial u_2}{\partial x_2} \cdot \frac{\partial u_3}{\partial x_3} \ldots\ldots \frac{\partial u_n}{\partial x_n}$$

Proof. We know that,

$$\frac{\partial(u_1, u_2, \ldots\ldots, u_n)}{\partial(x_1, x_2, \ldots\ldots, x_n)} = \begin{vmatrix} \frac{\partial u_1}{\partial x_1} & \frac{\partial u_1}{\partial x_2} & \frac{\partial u_1}{\partial x_3} & \ldots & \frac{\partial u_1}{\partial x_n} \\ \frac{\partial u_2}{\partial x_1} & \frac{\partial u_2}{\partial x_2} & \frac{\partial u_2}{\partial x_3} & \ldots & \frac{\partial u_2}{\partial x_n} \\ \ldots & \ldots & \ldots & \ldots & \ldots \\ \ldots & \ldots & \ldots & \ldots & \ldots \\ \frac{\partial u_n}{\partial x_1} & \frac{\partial u_n}{\partial x_2} & \frac{\partial u_n}{\partial x_3} & \ldots & \frac{\partial u_n}{\partial x_n} \end{vmatrix}$$

u_1 is not function of $x_2, x_3, \ldots\ldots x_n$. Therefore,

$$\frac{\partial u_1}{\partial x_2} = 0, \frac{\partial u_1}{\partial x_3} = 0, \ldots\ldots \frac{\partial u_1}{\partial x_n} = 0$$

u_2 is not function of $x_3, x_4, x_5 \ldots\ldots x_n$. Therefore,

$$\frac{\partial u_2}{\partial x_3} = 0, \frac{\partial u_2}{\partial x_4} = 0, \ldots\ldots \frac{\partial u_2}{\partial x_n} = 0$$

u_3 is not function of $x_4, x_5, x_6 \ldots\ldots x_n$. Therefore,

$$\frac{\partial u_3}{\partial x_4} = 0, \frac{\partial u_3}{\partial x_5} = 0, \ldots\ldots \frac{\partial u_3}{\partial x_n} = 0$$

u_n is function of $x_1, x_2, \ldots\ldots x_n$ therefore all $\frac{\partial u_n}{\partial x_1}, \frac{\partial u_n}{\partial x_2}, \ldots\ldots \frac{\partial u_n}{\partial x_n}$ will exist.

Putting these values in (1), we get

$$\frac{\partial(u_1, u_2, u_3 \ldots\ldots, u_n)}{\partial(x_1, x_2, x_3 \ldots\ldots, x_n)} = \begin{vmatrix} \frac{\partial u_1}{\partial x_1} & 0 & 0 & \ldots & 0 \\ \frac{\partial u_2}{\partial x_1} & \frac{\partial u_2}{\partial x_2} & 0 & \ldots & 0 \\ \frac{\partial u_3}{\partial x_1} & \frac{\partial u_3}{\partial x_2} & \frac{\partial u_3}{\partial x_3} & \ldots & 0 \\ \ldots & \ldots & \ldots & \ldots & \ldots \\ \ldots & \ldots & \ldots & \ldots & \ldots \\ \frac{\partial u_n}{\partial x_1} & \frac{\partial u_n}{\partial x_2} & \frac{\partial u_n}{\partial x_3} & \ldots & \frac{\partial u_n}{\partial x_n} \end{vmatrix}$$

Expanding the determinant in terms of first row, we get

$$\frac{\partial(u_1, u_2, \ldots\ldots\ldots\ldots, u_n)}{\partial(x_1, x_2, \ldots\ldots\ldots\ldots, x_n)} = \frac{\partial u_1}{\partial x_1} \cdot \frac{\partial u_2}{\partial x_2} \cdot \frac{\partial u_3}{\partial x_3} \cdots \frac{\partial u_n}{\partial x_n}.$$ **Proved.**

Example 6. *If* $y_1 = \cos x_1$, $y_2 = \sin x_1 \cos x_2$ and $y_3 = \sin x_1 \sin x_2 \cos\cos x_3$, *then show that*

$$\frac{\partial(y_1, y_2, y_3)}{\partial(x_1, x_2, x_3)} = -\sin^3 x_1 \sin^2 x_2 \sin x_3.$$

Solution.
$$\frac{\partial(y_1, y_2, y_3)}{\partial(x_1, x_2, x_3)} = \frac{\partial y_1}{\partial x_1} \cdot \frac{\partial y_2}{\partial x_2} \cdot \frac{\partial y_3}{\partial x_3}$$
$$= (-\sin x_1)(-\sin x_1 \sin x_2)\ (-\sin x_1 \sin x_2 \sin x_3)$$
$$= -\sin^3 x_1 \sin^2 x_2 \sin x_3.$$ **Ans.**

Example 7. *If* $y_1 = 1 - x_1$, $y_2 = x_1(1 - x_2)$, $y_3 = x_1 x_2 (1 - x_3)$................

$\ldots\ldots\ldots y_n = x_1 x_2 x_3 \ldots\ldots\ldots x_{n-1}(1 - x_n)$

then show that

$$\frac{\partial(y_1, y_2, \ldots\ldots y_n)}{\partial(x_1, x_2, \ldots\ldots\ldots x_n)} = (-1)^n x_1^{n-1} x_2^{n-2} \ldots\ldots\ldots\ldots\ldots x_{n-1}.$$

Solution. we know that,

$$\frac{\partial(y_1, y_2, \ldots\ldots y_n)}{\partial(x_1, x_2, \ldots\ldots\ldots x_n)} = \frac{\partial y_1}{\partial x_1} \cdot \frac{\partial y_2}{\partial x_2} \ldots\ldots \frac{\partial y_n}{\partial x_n}.$$
$$= (-1)(-x_1)(-x_1 x_2)\ldots.(-x_1 x_2 \ldots\ldots\ldots\ldots x_{n-1})$$
$$= (-1)^n x_1^{n-1} x_2^{n-2} \ldots\ldots\ldots\ldots x_{n-1}$$ **Proved.**

Example 8. *If* $u_1 = \frac{x_1}{x_n}$, $u_2 = \frac{x_2}{x_n}$, $u_3 = \frac{x_3}{x_n}, \ldots,$ $u_{n-1} = \frac{x_{n-1}}{x_n}$ *and*
$x_1^2 + x_2^2 + x_3^2 + \ldots + x_{n-1}^2 + x_n^2 = 1$, *find the value of the Jacobian* $\frac{\partial(u_1, u_2, u_3, \ldots, u_{n-1})}{\partial(x_1, x_2, x_3, \ldots, x_{n-1})}$

Solution. Here, we have $u_1 = \frac{x_1}{x_n} \Rightarrow \frac{\partial u_1}{\partial x_1} = \frac{x_n - x_1 . \frac{\partial x_n}{\partial x_1}}{x_n^2}$ ...(1)

Differentiating $x_1^2 + x_2^2 + x_3^2 + \ldots + x_n^2 = 1$ partially w.r.t x_1, we get

$$2x_1 + 2x_n \frac{\partial x_n}{\partial x_1} = 0 \Rightarrow \frac{\partial x_n}{\partial x_1} = -\frac{x_1}{x_n} \quad \ldots(2)$$

Putting the value of $\frac{\partial x_n}{\partial x_1}$ from (2) in (1), we get

$$\frac{\partial u_1}{\partial x_1} = \frac{x_n - x_1\left(\frac{-x_1}{x_n}\right)}{x_n^2} = \frac{x_n^2 + x_1^2}{x_n^3}$$

Similarly,
$$\frac{\partial u_1}{\partial x_2} = -\frac{x_1}{x_n^2} \cdot \frac{\partial x_n}{\partial x_2} = -\frac{x_1}{x_n^2} \cdot \left(\frac{-x_2}{x_n}\right) = \frac{x_1 x_2}{x_n^3}$$

$$\frac{\partial(u_1, u_2, u_3, \ldots, u_{n-1})}{\partial(x_1, x_2, x_3, \ldots, x_{n-1})} = \begin{vmatrix} \frac{x_n^2+x_1^2}{x_n^3} & \frac{x_1x_2}{x_n^3} & \frac{x_1x_3}{x_n^3} & \ldots\ldots & \frac{x_1x_{n-1}}{x_n^3} \\ \frac{x_2x_1}{x_n^3} & \frac{x_n^2+x_2^2}{x_n^3} & \frac{x_2x_3}{x_n^3} & \ldots\ldots & \frac{x_2x_{n-1}}{x_n^3} \\ \frac{x_3x_1}{x_n^3} & \frac{x_3x_2}{x_n^3} & \frac{x_n^2+x_3^2}{x_n^3} & \ldots\ldots & \frac{x_3x_{n-1}}{x_n^3} \\ \ldots\ldots & \ldots\ldots & \ldots\ldots & \ldots\ldots & \ldots\ldots \\ \ldots\ldots & \ldots\ldots & \ldots\ldots & \ldots\ldots & \ldots\ldots \\ \frac{x_{n-1}x_1}{x_n^3} & \frac{x_{n-1}x_2}{x_n^3} & \frac{x_{n-1}x_3}{x_n^3} & \ldots\ldots & \frac{x_n^2+x_{n-1}^2}{x_n^3} \end{vmatrix}$$

Taking $\frac{1}{x_n^3}$ common from each of $(n-1)$ columns and we get

$$= \left(\frac{1}{x_n^3}\right)^{n-1} \begin{vmatrix} x_n^2+x_1^2 & x_1x_2 & x_1x_3 & \ldots & x_1x_{n-1} \\ x_2x_1 & x_n^2+x_2^2 & x_2x_3 & \ldots & x_2x_{n-1} \\ x_3x_1 & x_3x_2 & x_n^2+x_3^2 & \ldots & x_3x_{n-1} \\ \ldots & \ldots & \ldots & \ldots & \ldots \\ \ldots & \ldots & \ldots & \ldots & \ldots \\ x_{n-1}x_1 & x_{n-1}x_2 & x_{n-1}x_3 & \ldots & x_n^2+x_{n-1}^2 \end{vmatrix}$$

Multiplying I column by x_1 and dividing the determinant by x_1, we get

$$= \frac{1}{x_1 x_n^{3n-3}} \begin{vmatrix} x_1(x_n^2+x_1^2) & x_1x_2 & x_1x_3 & \ldots & x_1x_{n-1} \\ x_2x_1^2 & x_n^2+x_2^2 & x_1x_3 & \ldots & x_2x_{n-1} \\ x_3x_1^2 & x_3x_2 & x_n^2+x_3^2 & \ldots & x_3x_{n-1} \\ \ldots & \ldots & \ldots & \ldots & \ldots \\ x_{n-1}x_1^2 & x_{n-1}x_2 & x_{n-1}x_3 & \ldots & x_n^2+x_{n-1}^2 \end{vmatrix}$$

Multiplying c_2 by x_2, c_3 by x_3 and so on and then adding to 1 column, we get

$$= \frac{1}{x_1 x_n^{3n-3}} \begin{vmatrix} x_1(x_n^2+x_1^2)+x_1x_2^2+x_1x_3^2+\ldots+x_1x_{n-1}^2 & x_1x_2 & x_1x_3 & \ldots & x_1x_{n-1} \\ x_2x_1^2+x_2(x_n^2+x_2^2)+x_2x_3^2+\ldots+x_2x_{n-1}^2 & x_n^2+x_2^2 & x_2x_3 & \ldots & x_2x_{n-1} \\ x_3x_1^2+x_2^2x_3+x_3(x_n^2+x_3^2)+\ldots+x_3x_{n-1}^2 & x_3x_2 & x_n^2+x_3^2 & \ldots & x_3x_{n-1} \\ \ldots & \ldots & \ldots & \ldots & \ldots \\ \ldots & \ldots & \ldots & \ldots & \ldots \\ x_{n-1}x_1^2+x_2^2x_{n-1}+\ldots+x_{n-1}x_{n-1}^2 & x_{n-1}x_2 & x_{n-1}x_3 & \ldots & x_n^2+x_{n-1}^2 \end{vmatrix}$$

$$= \frac{1}{x_1 x_n^{3n-3}} \begin{vmatrix} x_1(x_n^2+x_1^2+x_2^2+x_3^2\ldots+x_{n-1}^2) & x_1x_2 & x_1x_3 & \ldots & x_1x_{n-1} \\ x_2(x_1^2+x_n^2+x_2^2+x_3^2+\ldots x_{n-1}^2) & x_n^2+x_2^2 & x_2x_3 & \ldots & x_2x_{n-1} \\ x_3(x_1^2+x_2^2+x_n^2+x_3^2+\ldots x_{n-1}^2) & x_3x_2 & x_n^2+x_3^2 & \ldots & x_3x_{n-1} \\ x_{n-1}(x_1^2+x_2^2+x_3^2+\ldots+x_n^2+x_{n-1}^2) & x_{n-1}x_2 & x_{n-1}x_3 & \ldots & x_n^2+x_{n-1}^2 \end{vmatrix}$$

Apply $= x_1^2 + x_2^2 + x_3^2 + ... + x_{n-1}^2 + x_n^2 = 1$

$$J = \frac{1}{x_1 x_n^{3n-3}} \begin{vmatrix} x_1 & x_1x_2 & x_1x_3 & \cdots & x_1x_{n-1} \\ x_2 & x_n^2 + x_2^2 & x_2x_3 & \cdots & x_2x_{n-1} \\ x_3 & x_3x_2 & x_n^2 + x_3^2 & \cdots & x_3x_{n-1} \\ \cdots & \cdots & \cdots & \cdots & \cdots \\ \cdots & \cdots & \cdots & \cdots & \cdots \\ x_{n-1} & x_{n-1}x_2 & x_{n-1}x_3 & \cdots & x_n^2 + x_{n-1}^2 \end{vmatrix}$$

Taking x_1 common from R_1 and then applying $R_2 \to R_2 - x_2R_1$, $R_3 \to R_3 - x_3R_1$, ..., $R_{n-1} \to R_{n-1} - x_{n-1} R_1$, we get

$$J = \frac{1}{x_n^{3n-3}} \begin{vmatrix} 1 & x_2 & x_3 & \cdots & x_{n-1} \\ 0 & x_n^2 & 0 & \cdots & 0 \\ 0 & 0 & x_n^2 & \cdots & 0 \\ \cdots & \cdots & \cdots & \cdots & \cdots \\ 0 & 0 & 0 & \cdots & x_n^2 \end{vmatrix}$$

$$= \frac{1}{x_n^{3n-3}} (x_n^2)^{n-2} = \frac{1}{(x_n)^{n+1}}.$$ **Ans.**

EXERCISE 5.1

1. If $u = x^2$, $v = y^2$, find $\dfrac{\partial(u,v)}{\partial(x,y)}$ **Ans.** $4xy$
2. If $u = \dfrac{y-x}{1+xy}$ and $v = \tan^{-1}y - \tan^{-1}x$, find $\dfrac{\partial(u,v)}{\partial(x,y)}$ **Ans.** 0
3. If $u = xyz$, $v = xy + yz + zx$, $w = x + y + z$, compute $\dfrac{\partial(u,v,w)}{\partial(x, y\ z)}$. **Ans.** $(x-y)(y-z)(z-x)$
4. If $x = r\cos\theta$, $y = r\sin\theta$, $z = z$ find $\dfrac{\partial(x,y,z)}{\partial(r,\theta\ z)}$.
5. If $u_1 = \dfrac{x_1}{x_n}$, $u_2 = \dfrac{x_2}{x_n}$ ....$u_{n-1} = \dfrac{x_{n-1}}{x_n}$ and $x_1^2 + x_2^2 + x_3^2 + ... + x_n^2 = 1$ *find* $\dfrac{\partial(u_1, u_2, ...u_{n-1})}{\partial(x_1, x_2, ...x_{n-1})}$ **Ans.** $\dfrac{1}{x_n^{n-1}}$
6. Find the value of $\dfrac{\partial(y_1, y_2, y_3)}{\partial(x_1, x_2, x_3)}$, $y_1 = (1-x_1)$, $y_2 = x_1(1-x_2)$, $y_3 = x_1x_2(1-x_3)$ **Ans.** $-x_1^2x_2$
7. If $u = \dfrac{x}{y-z}$, $v = \dfrac{y}{z-x}$, $w = \dfrac{z}{x-y}$ then show that $\dfrac{\partial(u,v,w)}{\partial(x, y, z)} = 0$.
8. Fill in the blanks :

   (*i*) If $x = r\cos\theta$, $y = r\sin\theta$, then the value of Jacobians $\dfrac{\partial(x,y)}{\partial(r,\theta)}$ is...

   (*ii*) If $u = x(1-y)$ $v = xy$, then the value of the Jacobian $\dfrac{\partial(u,v)}{\partial(x,y)} = ...$ **Ans.** (*i*) r, (*ii*) x

5.2 PROPERTIES OF JACOBIANS

(1) First Property

If u and v are the functions of x and y, then

$$\frac{\partial(u,v)}{\partial(x,y)} \times \frac{\partial(x,y)}{\partial(u,v)} = 1$$

(A.M.I.E.T.E., Summer 2009, 2005)

Proof. *Let* $\quad u = f(x, y) \qquad \ldots(1)$

$\qquad v = \phi(x, y) \qquad \ldots(2)$

$$\frac{\partial(u,v)}{\partial(x,y)} \times \frac{\partial(x,y)}{\partial(u,v)} = \begin{vmatrix} \frac{\partial u}{\partial x} & \frac{\partial u}{\partial y} \\ \frac{\partial v}{\partial x} & \frac{\partial v}{\partial y} \end{vmatrix} \times \begin{vmatrix} \frac{\partial x}{\partial u} & \frac{\partial x}{\partial v} \\ \frac{\partial y}{\partial u} & \frac{\partial y}{\partial v} \end{vmatrix}$$

On interchanging the rows and columns of second determinant

$$= \begin{vmatrix} \frac{\partial u}{\partial x} & \frac{\partial u}{\partial y} \\ \frac{\partial v}{\partial x} & \frac{\partial v}{\partial y} \end{vmatrix} \times \begin{vmatrix} \frac{\partial x}{\partial u} & \frac{\partial y}{\partial u} \\ \frac{\partial x}{\partial v} & \frac{\partial y}{\partial v} \end{vmatrix} = \begin{vmatrix} \frac{\partial u}{\partial x}\cdot\frac{\partial x}{\partial u} + \frac{\partial u}{\partial y}\cdot\frac{\partial y}{\partial u} & \frac{\partial u}{\partial x}\cdot\frac{\partial x}{\partial v} + \frac{\partial u}{\partial y}\cdot\frac{\partial y}{\partial v} \\ \frac{\partial v}{\partial x}\cdot\frac{\partial x}{\partial u} + \frac{\partial v}{\partial y}\cdot\frac{\partial y}{\partial u} & \frac{\partial v}{\partial x}\cdot\frac{\partial x}{\partial v} + \frac{\partial v}{\partial y}\cdot\frac{\partial y}{\partial v} \end{vmatrix} \qquad \ldots(3)$$

On differentiating (1) and (2) w.r.t. u and v, we get

$$\left.\begin{aligned} \frac{\partial u}{\partial u} &= 1 = \frac{\partial u}{\partial x}\cdot\frac{\partial x}{\partial u} + \frac{\partial u}{\partial y}\cdot\frac{\partial y}{\partial u} \\ \frac{\partial u}{\partial v} &= 0 = \frac{\partial u}{\partial x}\cdot\frac{\partial x}{\partial v} + \frac{\partial u}{\partial y}\cdot\frac{\partial y}{\partial v} \\ \frac{\partial v}{\partial v} &= 1 = \frac{\partial v}{\partial x}\cdot\frac{\partial x}{\partial v} + \frac{\partial v}{\partial y}\cdot\frac{\partial y}{\partial v} \\ \frac{\partial v}{\partial u} &= 0 = \frac{\partial v}{\partial x}\cdot\frac{\partial x}{\partial u} + \frac{\partial v}{\partial y}\cdot\frac{\partial y}{\partial u} \end{aligned}\right\} \qquad \ldots(4)$$

On making substitutions from (4) in (3), we get

$$\frac{\partial(u,v)}{\partial(x,y)} \times \frac{\partial(x,y)}{\partial(u,v)} = \begin{vmatrix} 1 & 0 \\ 0 & 1 \end{vmatrix} = 1$$

Proved.

Example 9. *If* $x = uv, \quad y = \frac{u+v}{u-v}$, *find* $\frac{\partial(u,v)}{\partial(x,y)}$.

Solution. Here it is easy to find $\frac{\partial x}{\partial u}, \frac{\partial x}{\partial v}, \frac{\partial y}{\partial u}, \frac{\partial y}{\partial v}$. But to find $\frac{\partial u}{\partial x}, \frac{\partial u}{\partial y}, \frac{\partial v}{\partial x}, \frac{\partial v}{\partial y}$ is comparatively difficult. So we first find $\frac{\partial(x,y)}{\partial(u,v)}$.

$$\frac{\partial(x,y)}{\partial(u,v)} = \begin{vmatrix} \frac{\partial x}{\partial u} & \frac{\partial x}{\partial v} \\ \frac{\partial y}{\partial u} & \frac{\partial y}{\partial v} \end{vmatrix} = \begin{vmatrix} v & u \\ \frac{-2v}{(u-v)^2} & \frac{2u}{(u-v)^2} \end{vmatrix}$$

$$= \frac{uv}{(u-v)^2}\begin{vmatrix} 1 & 1 \\ -2 & 2 \end{vmatrix} = \frac{uv}{(u-v)^2}(2+2) = \frac{4uv}{(u-v)^2}$$

But $$\frac{\partial(u,v)}{\partial(x,y)}\times\frac{\partial(x,y)}{\partial(u,v)}=1$$

$$\frac{\partial(u,v)}{\partial(x,y)}\times\frac{4uv}{(u-v)^2}=1\Rightarrow\frac{\partial(u,v)}{\partial(x,y)}=\frac{(u-v)^2}{4uv}$$ **Ans.**

Example 10. *If* $u = xyz$, $v = x^2 + y^2 + z^2$, $w = x + y + z$, *find* $J=\frac{\partial(x,y,z)}{\partial(u,v,w)}$

Solution. Since u, v, w are explicitly given, so first we evaluate

$$J'=\frac{\partial(u,v,w)}{\partial(x,y,z)}$$

$$J'=\begin{vmatrix}\frac{\partial u}{\partial x} & \frac{\partial u}{\partial y} & \frac{\partial u}{\partial z}\\ \frac{\partial v}{\partial x} & \frac{\partial v}{\partial y} & \frac{\partial v}{\partial z}\\ \frac{\partial w}{\partial x} & \frac{\partial w}{\partial y} & \frac{\partial w}{\partial z}\end{vmatrix}=\begin{vmatrix}yz & zx & xy\\ 2x & 2y & 2z\\ 1 & 1 & 1\end{vmatrix}$$

$= yz\,(2y - 2z) - zx\,(2x - 2z) + xy\,(2x - 2y) = 2\,[yz\,(y - z) - zx\,(x - z) + xy\,(x - y)]$

$= 2\,[x^2y - x^2z - xy^2 + xz^2 + y^2z - yz^2] = 2\,[x^2\,(y - z) - x\,(y^2 - z^2) + yz\,(y - z)]$

$= 2\,(y - z)\,[x^2 - x\,(y + z) + yz] = 2\,(y - z)\,[y\,(z - x) - x\,(z - x)]$

$= 2\,(y - z)\,(z - x)\,(y - x) = -2\,(x - y)\,(y - z)(z - x)$

Hence, by $JJ' = 1$, we have

$$J=\frac{\partial(x,y,z)}{\partial(u,v,w)}=\frac{-1}{2(x-y)(y-z)(z-x)}$$ **Ans.**

EXERCISE 5.2

1. Given $u = x^2 - y^2$, $v = 2xy$, calculate $\frac{\partial(x,y)}{\partial(u,v)}$. **Ans.** $\frac{1}{4(x^2+y^2)}$

2. If $x = uv$, $y=\frac{u+v}{u-v}$, find $\frac{\partial(u,v)}{\partial(x,y)}$. **Ans.** $\frac{(u-v)^2}{4uv}$

3. If $x = r\sin\theta\cos\phi$, $y = r\sin\theta\sin\phi$, $z = r\cos\theta$, find $\frac{\partial(r,\theta\,\varphi)}{\partial(x,y,z)}$. **Ans.** $\frac{1}{r^2\sin\theta}$

4. If $x = u\,(1 - v)$, $y = uv$, verify that $\frac{\partial(x,y)}{\partial(u,v)}\times\frac{\partial(u,v)}{\partial(x,y)}=1$ *(A.M.I.E. Summer 2001)*

5. For the transformation $x = e^u\cos v$, $y = e^u\sin v$, prove that $\frac{\partial(x,y)}{\partial(u,v)}\times\frac{\partial(u,v)}{\partial(x,y)}=1$

6. Verify $JJ'=1$, if $x = uv$, $y=\frac{u}{v}$. **7.** Verify $JJ' = 1$, if $x = e^v\sec u$, $y = e^v\tan u$.

8. Verify $JJ' = 1$, if $x = \sin\theta\cos\phi$, $y = \sin\theta\sin\phi$

(2) Second Property

If u, v are the functions of r, s where r, s are functions of x, y, then

$$\frac{\partial(u,v)}{\partial(x,y)}=\frac{\partial(u,v)}{\partial(r,s)}\times\frac{\partial(r,s)}{\partial(x,y)}$$

Proof. $\dfrac{\partial(u, v)}{\partial(r, s)} \times \dfrac{\partial(r, s)}{\partial(x, y)} = \begin{vmatrix} \dfrac{\partial u}{\partial r} & \dfrac{\partial u}{\partial s} \\ \dfrac{\partial v}{\partial r} & \dfrac{\partial v}{\partial s} \end{vmatrix} \times \begin{vmatrix} \dfrac{\partial r}{\partial x} & \dfrac{\partial r}{\partial y} \\ \dfrac{\partial s}{\partial x} & \dfrac{\partial s}{\partial y} \end{vmatrix}$

On interchanging the columns and rows in second determinant

$$= \begin{vmatrix} \dfrac{\partial u}{\partial r} & \dfrac{\partial u}{\partial s} \\ \dfrac{\partial v}{\partial r} & \dfrac{\partial v}{\partial s} \end{vmatrix} \times \begin{vmatrix} \dfrac{\partial r}{\partial x} & \dfrac{\partial s}{\partial x} \\ \dfrac{\partial r}{\partial y} & \dfrac{\partial s}{\partial y} \end{vmatrix} = \begin{vmatrix} \dfrac{\partial u}{\partial r}\cdot\dfrac{\partial r}{\partial x} + \dfrac{\partial u}{\partial s}\cdot\dfrac{\partial s}{\partial x} & \dfrac{\partial u}{\partial r}\cdot\dfrac{\partial r}{\partial y} + \dfrac{\partial u}{\partial s}\cdot\dfrac{\partial s}{\partial y} \\ \dfrac{\partial v}{\partial r}\cdot\dfrac{\partial r}{\partial x} + \dfrac{\partial v}{\partial s}\cdot\dfrac{\partial s}{\partial y} & \dfrac{\partial v}{\partial r}\cdot\dfrac{\partial r}{\partial y} + \dfrac{\partial v}{\partial s}\cdot\dfrac{\partial s}{\partial y} \end{vmatrix}$$

$$= \begin{vmatrix} \dfrac{\partial u}{\partial x} & \dfrac{\partial u}{\partial y} \\ \dfrac{\partial v}{\partial x} & \dfrac{\partial v}{\partial y} \end{vmatrix} = \frac{\partial(u, v)}{\partial(x, y)}$$

Ans.

Similarly, $\dfrac{\partial(u, v, w)}{\partial(x, y, z)} = \dfrac{\partial(u, v, w)}{\partial(r, s, t)} \times \dfrac{\partial(r, s, t)}{\partial(x, y, z)}$

Example 11. *Find the value of the Jacobian* $\dfrac{\partial(u, v)}{\partial(r, \theta)}$, *where* $u = x^2 - y^2$, $v = 2xy$ *and*

$x = r\cos\theta$, $y = r\sin\theta$.

Solution. $u = x^2 - y^2$

$v = 2xy$

$$\frac{\partial(u, v)}{\partial(x, y)} = \begin{vmatrix} \dfrac{\partial u}{\partial x} & \dfrac{\partial u}{\partial y} \\ \dfrac{\partial v}{\partial x} & \dfrac{\partial v}{\partial y} \end{vmatrix} = \begin{vmatrix} 2x & -2y \\ 2y & 2x \end{vmatrix} = 4\left(x^2 + y^2\right) = 4r^2$$

$$\frac{\partial(x, y)}{\partial(r, \theta)} = \begin{vmatrix} \dfrac{\partial x}{\partial r} & \dfrac{\partial x}{\partial \theta} \\ \dfrac{\partial y}{\partial r} & \dfrac{\partial y}{\partial \theta} \end{vmatrix} = \begin{vmatrix} \cos\theta & -r\sin\theta \\ \sin\theta & r\cos\theta \end{vmatrix} = r\cos^2\theta + r\sin^2\theta = r$$

$$\frac{\partial(u, v)}{\partial(r, \theta)} = \frac{\partial(u, v)}{\partial(x, y)} \times \frac{\partial(x, y)}{\partial(r, \theta)} = 4r^2 \cdot r = 4r^3$$

Ans.

Example 12. *If* $x = \sqrt{vw}$, $y = \sqrt{wu}$, $z = \sqrt{uv}$ and

$u = r\sin\theta\cos\phi$, $v = r\sin\theta\sin\phi$, $w = r\cos\theta$. Calculate $\dfrac{\partial(x, y, z)}{\partial(r, \theta, \phi)}$

(*Gujarat, I Semester, Jan 2009*)

Solution. We have,

$x = \sqrt{vw}$, $\quad y = \sqrt{wu}$, $\quad z = \sqrt{uv}$

$u = r\sin\theta\cos\phi$, $\quad v = r\sin\theta\cos\phi$, $\quad w = r\cos\theta$

$$\frac{\partial(x, y, z)}{\partial(r, \theta, \phi)} = \frac{\partial(x, y, z)}{\partial(u, v, w)} \frac{\partial(u, v, w)}{\partial(r, \theta, \phi)}$$

$$\frac{\partial(x,y,z)}{\partial(u,v,w)} = \begin{vmatrix} \frac{\partial x}{\partial u} & \frac{\partial x}{\partial v} & \frac{\partial x}{\partial w} \\ \frac{\partial y}{\partial u} & \frac{\partial y}{\partial v} & \frac{\partial y}{\partial w} \\ \frac{\partial z}{\partial u} & \frac{\partial z}{\partial v} & \frac{\partial z}{\partial w} \end{vmatrix} = \begin{vmatrix} 0 & \frac{\sqrt{w}}{2\sqrt{v}} & \frac{\sqrt{v}}{2\sqrt{w}} \\ \frac{\sqrt{w}}{2\sqrt{u}} & 0 & \frac{\sqrt{u}}{2\sqrt{w}} \\ \frac{\sqrt{v}}{2\sqrt{u}} & \frac{\sqrt{u}}{2\sqrt{v}} & 0 \end{vmatrix} = \frac{1}{8}\begin{vmatrix} 0 & \sqrt{\frac{w}{v}} & \sqrt{\frac{v}{w}} \\ \sqrt{\frac{w}{v}} & 0 & \sqrt{\frac{u}{w}} \\ \sqrt{\frac{v}{u}} & \sqrt{\frac{u}{v}} & 0 \end{vmatrix}$$

$$= \frac{1}{8}\left[-\frac{\sqrt{w}}{v}\left(0\frac{\sqrt{uv}}{uw}\right) + \sqrt{\frac{u}{w}}\left(\frac{\sqrt{wu}}{uv}\right)\right] = \frac{1}{8}[1+1]$$

$$\frac{\partial(u,v,w)}{\partial(r,\theta,\phi)} = \begin{vmatrix} \frac{\partial u}{\partial r} & \frac{\partial x}{\partial \theta} & \frac{\partial u}{\partial \phi} \\ \frac{\partial v}{\partial r} & \frac{\partial y}{\partial \theta} & \frac{\partial v}{\partial \phi} \\ \frac{\partial w}{\partial r} & \frac{\partial w}{\partial \theta} & \frac{\partial w}{\partial \phi} \end{vmatrix} = \begin{vmatrix} \sin\theta\cos\phi & r\cos\theta\cos\phi & -\sin\phi \\ \sin\theta\sin\phi & \cos\theta\sin\phi & \cos\phi \\ \cos\phi & -\sin\theta & 0 \end{vmatrix}$$

$$= r^2\sin\theta[\sin\theta\cos\phi\ (0+\sin\theta\cos\phi) - \cos\theta\cos\phi\ (0-\cos\phi\cos\theta)$$
$$-\sin\phi(-\sin^2\theta\sin\phi - \cos^2\theta\ \sin\phi)]$$

$$= r^2\sin\theta[\sin\theta\cos^2\phi + \cos^2\theta\cos^2\phi + \sin^2\theta - \sin^2\phi + \cos^2\theta\sin^2\phi]$$

$$= r^2\sin\theta\ [\sin^2\theta + \cos^2\theta)\cos^2\phi + (\sin^2\phi + \cos^2 q)\sin^2\theta$$

$$= r^2\sin\theta[\cos^2\phi + \sin^2\phi] = r^2\sin\theta$$

Putting the values of $\frac{\partial(x,y,z)}{\partial(u,v,w)}$ and $\frac{\partial(u,v,w)}{\partial(r,\theta,\phi)}$ from (2) and (3) in (1), we get

$$\frac{\partial(x,y,z)}{\partial(r,\theta,\phi)} = \frac{1}{4}\cdot r^2\sin\theta$$

Example 13. *Verify the chain rule for Jacobians if $x = u$, $y = u \tan v$, $z = w$.*

(*U.P;I Semester, Dec 2008*)

Solution. Here, we have

$$x = u \quad \Rightarrow \quad u = x$$

$$y = u\tan v \quad \Rightarrow \quad v = \tan^{-1}\frac{y}{u} \quad \Rightarrow \quad \frac{\partial v}{\partial y} = \frac{1}{1+\frac{y^2}{u^2}}\cdot\frac{1}{u}$$

$$z = w \quad \Rightarrow \quad w = z$$

Let
$$f_1 = x - u$$
$$f_2 = y - u\tan v$$
$$f_3 = z - w$$

We have to prove chain rule

$$\frac{\partial(f_1,f_2,f_3)}{\partial(x,y,z)} = \frac{\partial(f_1,f_2,f_3)}{\partial(u,v,w)}\cdot\frac{\partial(u,v,w)}{\partial(x,y,z)} \quad ...(1)$$

Now, $$\frac{\partial(f_1,f_2,f_3)}{\partial(x,y,z)}=\begin{vmatrix}\frac{\partial f_1}{\partial x} & \frac{\partial f_1}{\partial y} & \frac{\partial f_1}{\partial z}\\ \frac{\partial f_2}{\partial x} & \frac{\partial f_2}{\partial y} & \frac{\partial f_2}{\partial z}\\ \frac{\partial f_3}{\partial x} & \frac{\partial f_3}{\partial y} & \frac{\partial f_3}{\partial z}\end{vmatrix}=\begin{vmatrix}1 & 0 & 0\\ 0 & 1 & 0\\ 0 & 0 & 1\end{vmatrix}=1 \qquad ...(2)$$

$$\frac{\partial(f_1,f_2,f_3)}{\partial(u,v,w)}=\begin{vmatrix}\frac{\partial f_1}{\partial u} & \frac{\partial f_1}{\partial v} & \frac{\partial f_1}{\partial w}\\ \frac{\partial f_2}{\partial u} & \frac{\partial f_2}{\partial v} & \frac{\partial f_2}{\partial w}\\ \frac{\partial f_3}{\partial u} & \frac{\partial f_3}{\partial v} & \frac{\partial f_3}{\partial w}\end{vmatrix}=\begin{vmatrix}-1 & 0 & 0\\ -\tan v & -u\sec^2 v & 0\\ 0 & 0 & 1\end{vmatrix}=u\sec^2 v$$

$$\frac{\partial(u,v,w)}{\partial(x,y,z)}=\begin{vmatrix}\frac{\partial u}{\partial x} & \frac{\partial u}{\partial y} & \frac{\partial u}{\partial z}\\ \frac{\partial v}{\partial x} & \frac{\partial v}{\partial y} & \frac{\partial v}{\partial z}\\ \frac{\partial w}{\partial x} & \frac{\partial w}{\partial y} & \frac{\partial w}{\partial z}\end{vmatrix}=\begin{vmatrix}1 & 0 & 0\\ 0 & \frac{u}{u^2+y^2} & 0\\ 0 & 0 & 1\end{vmatrix}=\frac{u}{u^2+y^2}$$

$$\frac{\partial(f_1,f_2,f_3)}{\partial(u,v,w)}\times\frac{\partial(u,v,w)}{\partial(x,y,z)}=(u\sec^2 v)\left(\frac{u}{u^2+y^2}\right)=\frac{u^2\sec^2 v}{u^2+u^2\tan^2 v}$$

$$=\frac{u^2(1+\tan^2 v)}{u^2(1+\tan^2 v)}=1 \qquad ...(3)$$

From (1), (2) and (3), we have

$$\frac{\partial(f_1,f_2,f_3)}{\partial(x,y,z)}=\frac{\partial(f_1,f_2,f_3)}{\partial(u,v,w)}\cdot\frac{\partial(u,v,w)}{\partial(x,y,z)}$$

Hence, chain rule is verified. **Proved.**

EXERCISE 5.3

1. If $u = e^x \cos y$, $v = e^x \sin y$, where $x = lr + sm$, $y = mr - sl$, verify

$$\frac{\partial(u,\ v)}{\partial(x,\ y)}\times\frac{\partial(x,\ y)}{\partial(r,\ s)}=\frac{\partial(u,\ v)}{\partial(r,\ s)}$$

2. If $u = x\left(1-r^2\right)^{-\frac{1}{2}}$, $v = y\left(1-r^2\right)^{-\frac{1}{2}}$

$w = z\left(1-r^2\right)^{-\frac{1}{2}}$ where $r^2 = x^2 + y^2 + z^2$

Show that $\dfrac{\partial(u,\ v,\ w)}{\partial(x,\ y,\ z)}=\left(1-r^2\right)^{-\frac{5}{2}}$

3. If $u = x+y+z$, $u^2 v=y+z$, $u^3w=z$, show that $\dfrac{\partial(u,\ v,\ w)}{\partial(x,\ y,\ z)}=u^{-5}$

[**Hint.** Put $r= x+y+z$, $s= y+z$, $t =z$

$u=r$ $u^2v = s$ $u^3w = t$

$$\left.\frac{\partial(u, v, w)}{\partial(x, y, z)}=\frac{\partial(u, v, w)}{\partial(r, s, t)}\times\frac{\partial(r, s, t)}{\partial(x, y, z)}=\frac{1}{\left(\frac{\partial(r, s, t)}{\partial(u, v, w)}\right)}\times\frac{\partial(r, s, t)}{\partial(x, y, z)}\right]$$

4. If $u = x + y + z$, $uv = y + z$, $uvw = z$. Evaluate $\frac{\partial(x, y, z)}{\partial(u, v, w)}$. *(K.U. Dec. 2009)* **Ans.** $u^2 v$

5. If $u^3+v^3 = x+y$, $u^2+v^2=x^3+y^3$, show that $\frac{\partial(u, v)}{\partial(x,y)}=\frac{1}{2}\frac{(y^2-x^2)}{uv(u-v)}$.

(3) Third Property

If functions u, v, w of three independent variables x, y, z are not independent, then $\frac{\partial(u, v, w)}{\partial(x, y, z)} = 0$.

Proof. As u, v, w are not independent, then $f(u, v, w) = 0$ ...(1)

Differentiating (1) w.r.t. x, y, z, we get

$$\frac{\partial f}{\partial u}\cdot\frac{\partial u}{\partial x}+\frac{\partial f}{\partial v}\cdot\frac{\partial v}{\partial x}+\frac{\partial f}{\partial w}\cdot\frac{\partial w}{\partial x}=0 \qquad ...(2)$$

$$\frac{\partial f}{\partial u}\cdot\frac{\partial u}{\partial y}+\frac{\partial f}{\partial v}\cdot\frac{\partial v}{\partial y}+\frac{\partial f}{\partial w}\cdot\frac{\partial w}{\partial y}=0 \qquad ...(3)$$

$$\frac{\partial f}{\partial u}\cdot\frac{\partial u}{\partial z}+\frac{\partial f}{\partial v}\cdot\frac{\partial v}{\partial z}+\frac{\partial f}{\partial w}\cdot\frac{\partial w}{\partial z}=0 \qquad ...(4)$$

Eliminating $\frac{\partial f}{\partial u}, \frac{\partial f}{\partial v}, \frac{\partial f}{\partial w}$ from (2), (3) and (4), we have

$$\begin{vmatrix} \frac{\partial u}{\partial x} & \frac{\partial v}{\partial x} & \frac{\partial w}{\partial x} \\ \frac{\partial u}{\partial y} & \frac{\partial v}{\partial y} & \frac{\partial w}{\partial y} \\ \frac{\partial u}{\partial z} & \frac{\partial v}{\partial z} & \frac{\partial w}{\partial z} \end{vmatrix}=0$$

On interchanging rows and columns, we get

$$\Rightarrow \begin{vmatrix} \frac{\partial u}{\partial x} & \frac{\partial u}{\partial y} & \frac{\partial u}{\partial z} \\ \frac{\partial v}{\partial x} & \frac{\partial v}{\partial y} & \frac{\partial v}{\partial z} \\ \frac{\partial w}{\partial x} & \frac{\partial w}{\partial y} & \frac{\partial w}{\partial z} \end{vmatrix}=0 \Rightarrow \frac{\partial(u, v, w)}{\partial(x, y, z)}=0$$

Proved.

Example 14. *If* $u=\sin^{-1}x+\sin^{-1}y$, $v= x\sqrt{1-y^2}+y\sqrt{1-x^2}$ *find* $\frac{\partial(u, v)}{\partial(x, y)}$. *Is* u, v *functionally related ? If so, find the relationship.* *(Q. bank U. P .T. U. 2001)*

Solution. Here, we have

$$u = \sin^{-1}x+\sin^{-1}y \qquad ....(1) \text{ (Given)}$$

$$\frac{\partial u}{\partial x}=\frac{1}{\sqrt{1-x^2}} \text{ and } \frac{\partial u}{\partial y}=\frac{1}{\sqrt{1-y^2}}$$

Also

$$v=x\sqrt{1-y^2}+y\sqrt{1-x^2} \qquad ... (2) \text{ (Given)}$$

$$\frac{\partial v}{\partial x} = \sqrt{1-y^2} - \frac{xy}{\sqrt{1-x^2}}, \qquad \frac{\partial v}{\partial y} = \frac{-xy}{\sqrt{1-y^2}} + \sqrt{1-x^2}$$

Now,
$$\frac{\partial (u, v)}{\partial (x, y)} = \begin{vmatrix} \frac{\partial u}{\partial x} & \frac{\partial u}{\partial y} \\ \frac{\partial v}{\partial x} & \frac{\partial v}{\partial y} \end{vmatrix} = \begin{vmatrix} \frac{1}{\sqrt{1-x^2}} & \frac{1}{\sqrt{1-y^2}} \\ \sqrt{1-y^2} - \frac{xy}{\sqrt{1-x^2}} & \frac{-xy}{\sqrt{1-y^2}} + \sqrt{1-x^2} \end{vmatrix}$$

$$= \frac{-xy}{\sqrt{(1-x^2)(1-y^2)}} + 1 - 1 + \frac{xy}{\sqrt{(1-x^2)(1-y^2)}} = 0$$

By the third property of Jacobian u and v are functionally related.

Let $\sin^{-1} x = \alpha$, $\sin \alpha = x$ and $\cos \alpha = \sqrt{1 - \sin^2 \alpha} = \sqrt{1-x^2}$

$\sin^{-1} y = \beta$, $\sin \beta = y$ and $\cos \beta = \sqrt{1-\sin^2 \beta} = \sqrt{1-y^2}$

$\sin(\alpha + \beta) = \sin \alpha \cos \beta + \cos \alpha \sin \beta$

$$= x\sqrt{1-y^2} + y\sqrt{1-x^2}$$

$$\alpha + \beta = \sin^{-1}\left[x\sqrt{1-y^2} + y\sqrt{1-x^2}\right]$$

$$\sin^{-1} x + \sin^{-1} y = \sin^{-1}\left[x\sqrt{1-y^2} + y\sqrt{1-x^2}\right]$$

We know that

$$u = \sin^{-1} x + \sin^{-1} y$$

$$= \sin^{-1}\left[x\sqrt{1-y^2} + y\sqrt{1-x^2}\right] = \sin^{-1} v$$

$$\Rightarrow \qquad \sin u = v$$

This is the required relation between u and v. **Ans.**

Example 15. *If $u = xy + yz + zx$, $v = x^2 + y^2 + z^2$ and $w = x + y + z$, determine whether there is a functional relationship between u, v, w and if so, find it.* (*Uttarakhand, Ist Sem. Dec. 2009*)

Solution. $u = xy + yz + zx, \quad v = x^2 + y^2 + z^2, \quad w = x + y + z$

$$\frac{\partial(u, v, w)}{\partial(x, y, z)} = \begin{vmatrix} \frac{\partial u}{\partial x} & \frac{\partial u}{\partial y} & \frac{\partial u}{\partial z} \\ \frac{\partial v}{\partial x} & \frac{\partial v}{\partial y} & \frac{\partial v}{\partial z} \\ \frac{\partial w}{\partial x} & \frac{\partial w}{\partial y} & \frac{\partial w}{\partial z} \end{vmatrix} = \begin{vmatrix} y+z & z+x & x+y \\ 2x & 2y & 2z \\ 1 & 1 & 1 \end{vmatrix}$$

$$= 2\begin{vmatrix} y+z & z+x & x+y \\ x & y & z \\ 1 & 1 & 1 \end{vmatrix} = 2\begin{vmatrix} x+y+z & x+y+z & x+y+z \\ x & y & z \\ 1 & 1 & 1 \end{vmatrix} \quad R_1 + R_2$$

$$= 2(x+y+z)\begin{vmatrix} 1 & 1 & 1 \\ x & y & z \\ 1 & 1 & 1 \end{vmatrix} = 0 \qquad (R_1 = R_3)$$

Hence the functional relationship exists between, u, v and w.

Now $w^2 = (x + y + z)^2 = x^2 + y^2 + z^2 + 2(xy + yz + zx)$

$$w^2 = v + 2u$$

$w^2 - v - 2u = 0$ which is the required relationship. **Ans.**

Example 16. *If* $u = x+2y+z,\ \ v = x-2y+3z,\ \ w = 2xy - xz + 4yz - 2z^2$
Show that they are functionally related, and find the relation between them.

(Nagpur University, Winter 2000)

Solution. $u = x+2y+z \qquad \therefore \quad \frac{\partial u}{\partial x} = 1, \quad \frac{\partial u}{\partial y} = 2, \quad \frac{\partial u}{\partial z} = 1$

$$v = x-2y+3z \qquad \therefore \quad \frac{\partial v}{\partial x} = 1, \quad \frac{\partial v}{\partial y} = -2, \quad \frac{\partial v}{\partial z} = 3$$

$$w = 2xy - xz + 4yz - 2z^2 \ \therefore \ \frac{\partial w}{\partial x} = 2y - z, \frac{\partial w}{\partial y} = 2x + 4z, \frac{\partial w}{\partial z} = -x + 4y - 4z$$

We know that $\frac{\partial(u,v,w)}{\partial(x,y,z)} = \begin{vmatrix} \frac{\partial u}{\partial x} & \frac{\partial u}{\partial y} & \frac{\partial u}{\partial z} \\ \frac{\partial v}{\partial x} & \frac{\partial v}{\partial y} & \frac{\partial v}{\partial z} \\ \frac{\partial w}{\partial x} & \frac{\partial w}{\partial y} & \frac{\partial w}{\partial z} \end{vmatrix} = \begin{vmatrix} 1 & 2 & 1 \\ 1 & -2 & 3 \\ 2y-z & 2x+4z & -x+4y-4z \end{vmatrix}$

$$= 1\{-2(-x+4y-4z) - 3(2x+4z)\} - 2\{(-x+4y-4z) - 3(2y-z) + 1(2x+4z + 2(2y-z)\}$$
$$= 2x - 8y + 8z - 6x - 12z + 2x - 8y + 8z + 12y - 6z + 2x + 4z + 4y - 2z$$
$$= 6x - 6x - 8y + 8y + 12z - 12z = 0.$$

$\because \quad \frac{\partial(u,v,w)}{\partial(x,y,z)} = 0$, the functions u, v, w are functionally related.

Now we have to find the relation between u, v and w clearly

$$u + v = 2x + 4z = 2(x + 2z)$$
$$u - v = 4y - 2z = 2(2y - z)$$
$$(u+v)\,(u-v) = 4(x+2z)\,(2y-z)$$
$$= 4(2xy - xz + 4yz - 2z^2)$$

$\therefore \qquad u^2 - v^2 = 4\,w$ is the required relationship. **Ans.**

Example 17. *If* $u = 3x+2y-z,\ \ v = x-2y+z.$

$w = x(x+2y-z)$, *show that they are functionally related and find relation.*

(Nagpur University, Summer 2008)

Solution. Here, we have

$$u = 3x + 2y - z, \qquad \frac{\partial u}{\partial x} = 3, \qquad \frac{\partial u}{\partial y} = 2, \qquad \frac{\partial u}{\partial z} = -1$$

$$v = x - 2y + z, \qquad \frac{\partial v}{\partial x} = 1, \qquad \frac{\partial v}{\partial y} = -2, \qquad \frac{\partial v}{\partial z} = 1$$

$$w = x^2 + 2xy - xz, \qquad \frac{\partial w}{\partial x} = 2x + 2y - z, \qquad \frac{\partial w}{\partial y} = 2x, \qquad \frac{\partial w}{\partial z} = -x$$

$$\frac{\partial(u,v,w)}{\partial(x,y,z)} = \begin{vmatrix} \frac{\partial u}{\partial x} & \frac{\partial u}{\partial y} & \frac{\partial u}{\partial z} \\ \frac{\partial v}{\partial x} & \frac{\partial v}{\partial y} & \frac{\partial v}{\partial z} \\ \frac{\partial w}{\partial x} & \frac{\partial w}{\partial y} & \frac{\partial w}{\partial z} \end{vmatrix} = \begin{vmatrix} 3 & 2 & -1 \\ 1 & -2 & 1 \\ 2x+2y-z & 2x & -x \end{vmatrix}$$

$$= 3(2x-2x) - 2(-x-2x-2y+z) - 1\,(2x+4x+4y-2z)$$

$$= 3(0) - 2(-3x-2y+z) - 6x - 4y + 2z = 0 + 6x + 4y - 2z - 6x - 4y + 2z$$

$$= 0$$

Hence the functional relationship exists between u, v and w.
Now, we have to find the relation between u, v and w.

$$u + v = (3x+2y-z) + (x-2y+z) = 4x$$

$$u - v = (3x+2y-z) - (x-2y+z) = 2x+4y-2z$$

$$\Rightarrow \quad (u+v)\,(u-v) = 4x(2x+4y-2z)$$

$$\Rightarrow \quad u^2 - v^2 = 8x(x+2y-z)$$

$$\Rightarrow \quad u^2 - v^2 = 8w$$

Ans.

Example 18. *Show that the functions :*

$u = x + y + z,$

$v = x^2 + y^2 + z^2 - 2xy - 2xz - 2yz$ *and*

$w = x^3 + y^3 + z^3 - 3xyz$ *are functionally related.*

Find the relation between them. *(Nagpur University, Winter 2001)*

Solution.
$$\frac{\partial(u,v,w)}{\partial(x,y,z)} = \begin{vmatrix} \frac{\partial u}{\partial x} & \frac{\partial u}{\partial y} & \frac{\partial u}{\partial z} \\ \frac{\partial v}{\partial x} & \frac{\partial v}{\partial y} & \frac{\partial v}{\partial z} \\ \frac{\partial w}{\partial x} & \frac{\partial w}{\partial y} & \frac{\partial w}{\partial z} \end{vmatrix}$$

$$= \begin{vmatrix} 1 & 1 & 1 \\ 2x-2y-2z & 2y-2x-2z & 2z-2y-2x \\ 3x^2-3yz & 3y^2-3xz & 3z^2-3xy \end{vmatrix}$$

$$= \begin{vmatrix} 1 & 0 & 0 \\ 2(x-y-z) & 4(y-x) & 4(z-x) \\ 3(x^2-yz) & 3[y^2-x^2+z(y-x)] & 3[z^2-x^2+y(z-x)] \end{vmatrix} \quad \begin{matrix} C_2 \to C_2 - C_1 \\ C_3 \to C_3 - C_1 \end{matrix}$$

$$= 4(y-x)3[z^2-x^2+y(z-x)] - 4(z-x)\,3[(y^2-x^2)+z(y-x)]$$

$$= 12(y-x)\,(z-x)\,(x+y+z) - 12(z-x)\,(y-x)\,(x+y+z) = 0$$

Hence, u, v & w are functionally related.

$$w = x^3 + y^3 + z^3 - 3xyz$$

$$= (x+y+z)\,(x^2+y^2+z^2-xy-yz-zx)$$

$$= (x+y+z)\,[x^2+y^2+z^2-2(xy+yz+zx)+xy+yz+zx]$$

$$= u\left\{v + \frac{1}{4}[(x+y+z)^2 - (x^2+y^2+z^2-2xy-2yz-2zx)]\right\}$$

$$= u\left\{v + \frac{1}{4}(u^2 - v)\right\} = \frac{u^3}{4} + \frac{3}{4}uv$$

$\Rightarrow$ $4w = u^3 + 3uv$ is the required relationship. **Ans.**

Example 19. *Verify whether the following functions are functionally dependent, and if so, find the relation between them.*

$$u = \frac{x+y}{1-xy}, \qquad v = \tan^{-1}x + \tan^{-1}y.$$

Solution. $$\frac{\partial(u, v)}{\partial(x, y)} = \begin{vmatrix} \frac{\partial u}{\partial x} & \frac{\partial u}{\partial y} \\ \frac{\partial v}{\partial x} & \frac{\partial v}{\partial y} \end{vmatrix} = \begin{vmatrix} \frac{1+y^2}{(1-xy)^2} & \frac{1+x^2}{(1-xy)^2} \\ \frac{1}{1+x^2} & \frac{1}{1+y^2} \end{vmatrix} = \frac{1}{(1-xy)^2} - \frac{1}{(1-xy)^2} = 0$$

Hence, u, v are functionally related.

$$\tan^{-1}x + \tan^{-1}y = \tan^{-1}\frac{x+y}{1-xy}$$

$$v = \tan^{-1}u$$

$\Rightarrow$ $u = \tan v$. **Ans.**

Example 20. *If* $u = \dfrac{x+y}{x-y}$ *and* $v = \dfrac{xy}{(x-y)^2}$, *find* $\dfrac{\partial(u,v)}{\partial(x,y)}$.

Are u and v functionally related? If so, find the relationship.

(*Nagpur University, Summer 2005, 2004, Winter 2004*)

Solution. We have, $u = \dfrac{x+y}{x-y}$

$$\therefore \quad \frac{\partial u}{\partial x} = \frac{(x-y)-(x+y)}{(x-y)^2} = \frac{-2y}{(x-y)^2}, \text{ and } \frac{\partial u}{\partial y} = \frac{2x}{(x-y)^2}$$

Also $$v = \frac{xy}{(x-y)^2}$$

Hence, $$\frac{\partial v}{\partial x} = \frac{(x-y)^2 \cdot y - xy(2)(x-y)}{(x-y)^4} = -\frac{y(x+y)}{(x-y)^3}$$

$$\frac{\partial v}{\partial y} = \frac{(x-y)^2 \cdot x - xy\,(2)\,(x-y)\,(-1)}{(x-y)^4} = \frac{(x^3 - 2x^2y + xy^2) + 2x^2y - 2xy^2}{(x-y)^4} = \frac{x(x+y)}{(x-y)^3}$$

Now, $$\frac{\partial(u,v)}{\partial(x,y)} = \begin{vmatrix} \frac{\partial u}{\partial x} & \frac{\partial u}{\partial y} \\ \frac{\partial v}{\partial x} & \frac{\partial v}{\partial y} \end{vmatrix} \begin{vmatrix} \frac{-2y}{(x-y)^2} & \frac{2x}{(x-y)^2} \\ \frac{-y(x+y)}{(x-y)^3} & \frac{x(x+y)}{(x-y)^3} \end{vmatrix} = \frac{-2xy(x+y)}{(x-y)^5} + \frac{2xy(x+y)}{(x-y)^5} = 0$$

Since Jacobian of u, v with respect to x, y is zero therefore u and v are functionally related.

Now, $$u^2 = \frac{(x+y)^2}{(x-y)^2} \text{ and } 4v = \frac{4xy}{(x-y)^2}$$

$$\therefore \quad u^2 - 4v = \frac{(x+y)^2 - 4xy}{(x-y)^2} = \frac{x^2 + 2xy + y^2 - 4xy}{(x-y)^2}$$

$$\Rightarrow \quad u^2 - 4v = \frac{(x-y)^2}{(x-y)^2} = 1 \quad \Rightarrow \quad u^2 = 1 + 4v$$

This is the required relationship between u and v. **Ans.**

Example 21. *Show that $ax^2 + 2h\,xy + by^2$ and $Ax^2 + 2Hxy + By^2$ are independent unless :*

$$\frac{a}{A} = \frac{b}{B} = \frac{h}{H}$$

Solution. Let $u = ax^2 + 2hxy + by^2$, $\frac{\partial u}{\partial x} = 2ax + 2hy$, $\frac{\partial u}{\partial y} = 2hx + 2by$

$$v = Ax^2 + 2Hxy + By^2 \quad \frac{\partial v}{\partial x} = 2Ax + 2Hy, \quad \frac{\partial v}{\partial y} = 2Hx + 2By$$

u and v will not be independent if there exists a relationship between them and in that case $\frac{\partial(u,v)}{\partial(x,y)}$ should vanish identically.

i.e.
$$\frac{\partial(u,v)}{\partial(x,y)} = \begin{vmatrix} 2(ax+hy) & 2(hx+by) \\ 2(Ax+Hy) & 2(Hx+By) \end{vmatrix} = 0$$

$$\Rightarrow \quad (ax + hy)\,(Hx + By) - (hx + by)\,(Ax + Hy) = 0$$

$$\Rightarrow \quad (aH - Ah)x^2 + (aB - Ab)\,xy + (Bh - bH)\,y^2 = 0$$

Now, the variables x and y are independent and as such the coefficients of x^2 and y^2 should vanish separately.

$$\therefore \quad aH - Ah = 0 \quad \Rightarrow \quad \frac{a}{A} = \frac{h}{H}$$

and
$$Bh - bH = 0 \quad \Rightarrow \quad \frac{h}{H} = \frac{b}{B}$$

Hence, $\frac{a}{A} = \frac{h}{H} = \frac{b}{B}$ and these conditions also make the coefficient of xy zero.

Hence, $\frac{a}{A} = \frac{b}{B} = \frac{h}{H}$ are the required conditions. **Proved.**

EXERCISE 5.4

1. Verify whether $u = \frac{x-y}{x+y}$, $v = \frac{x+y}{x}$ are functionally dependent, and if so, find the relation between them. **Ans.** $u = \frac{2-v}{v}$

2. Determine functional dependence and find relation between
$$u = \frac{x-y}{x+y}, \quad v = \frac{xy}{(x+y)^2}$$
Ans. $4v = 1 - u^2$

3. Are $x + y - z$, $x - y + z$, $x^2 + y^2 + z^2 - 2yz$ functionally dependent ? If so, find a relation between them. **Ans.** $u^2 + v^2 = 2w$

4. If $u = x + y + z$, $v = x^2 + y^2 + z^2$, $w = x^3 + y^3 + z^3 - 3xyz$, prove that u, v, w are not independent and find the relation between them. **Ans.** $2w = u\,(3v - u^2)$

5. Are the following two functions of x, y, z functionally dependent ? If so find the relation between them.

$$u = \frac{x-y}{x+z}, \quad v = \frac{x+z}{y+z}$$ **Ans.** $v = \frac{1}{1-u}$

6. Determine whether $u = \sin^{-1} x + \sin^{-1} y$ and $v = x\sqrt{1-y^2} + y\sqrt{1-x^2}$ are functionally dependent.
(*A.M.I.E., Winter 2001*) **Ans.** $v = \sin u$

7. If $u = \frac{x+v}{z}$, $v = \frac{y+z}{x}$, $w = \frac{y(x+y+z)}{xz}$, show that u, v, w are not indpendent and find the relation between them. **Ans.** $u^2 - 4v = 1$

5.3 JACOBIAN OF IMPLICIT FUNCTIONS

The variables x, y, u, v are connected by implicit functions

$$f_1(x, y, u, v) = 0 \quad \text{... (1)}$$

$$f_2(x, y, u, v) = 0 \quad \text{... (2)}$$

where u, v are implicit functions of x, y.

Differentiating (1) and (2) w.r.t. x, y, we get

$$\frac{\partial f_1}{\partial x} + \frac{\partial f_1}{\partial u}\frac{\partial u}{\partial x} + \frac{\partial f_1}{\partial v}\frac{\partial v}{\partial x} = 0 \quad \text{... (3)}$$

$$\frac{\partial f_1}{\partial y} + \frac{\partial f_1}{\partial u}\frac{\partial u}{\partial y} + \frac{\partial f_1}{\partial v}\frac{\partial v}{\partial y} = 0 \quad \text{... (4)}$$

$$\frac{\partial f_2}{\partial x} + \frac{\partial f_2}{\partial u}\frac{\partial u}{\partial x} + \frac{\partial f_2}{\partial v}\frac{\partial v}{\partial x} = 0 \quad \text{...(5)}$$

$$\frac{\partial f_2}{\partial y} + \frac{\partial f_2}{\partial u}\frac{\partial u}{\partial y} + \frac{\partial f_2}{\partial v}\frac{\partial v}{\partial y} = 0 \quad \text{...(6)}$$

Now we have

$$\frac{\partial(f_1, f_2)}{\partial(u,v)} \times \frac{\partial(u,v)}{\partial(x,y)} = \begin{vmatrix} \frac{\partial f_1}{\partial u} & \frac{\partial f_1}{\partial v} \\ \frac{\partial f_2}{\partial u} & \frac{\partial f_2}{\partial v} \end{vmatrix} \times \begin{vmatrix} \frac{\partial u}{\partial x} & \frac{\partial u}{\partial y} \\ \frac{\partial v}{\partial x} & \frac{\partial v}{\partial y} \end{vmatrix} = \begin{vmatrix} \frac{\partial f_1}{\partial u}\frac{\partial u}{\partial x} + \frac{\partial f_1}{\partial v}\frac{\partial v}{\partial x} & \frac{\partial f_1}{\partial u}\frac{\partial u}{\partial y} + \frac{\partial f_1}{\partial v}\frac{\partial v}{\partial y} \\ \frac{\partial f_2}{\partial u}\frac{\partial u}{\partial x} + \frac{\partial f_2}{\partial v}\frac{\partial v}{\partial x} & \frac{\partial f_2}{\partial u}\frac{\partial u}{\partial y} + \frac{\partial f_2}{\partial v}\frac{\partial v}{\partial y} \end{vmatrix}$$

$$= \begin{vmatrix} -\frac{\partial f_1}{\partial x} & -\frac{\partial f_1}{\partial y} \\ -\frac{\partial f_2}{\partial x} & -\frac{\partial f_2}{\partial y} \end{vmatrix} \quad \text{[From (3), (4), (5) and (6)]}$$

$$= (-1)^2 \frac{\partial(f_1, f_2)}{\partial(x,y)}$$

$$\frac{\partial(u,v)}{\partial(x,y)} = (-1)^2 \frac{\partial(f_1, f_2)/\partial(x,y)}{\partial(f_1, f_2)/\partial(u,v)}$$

In general, the variables $x_1, x_2, \ldots x_n$ are connected with $u_1, u_2, \ldots u_n$ implicitly as $f_1(x_1, x_2, \ldots x_n, u_1, u_2, \ldots u_n) = 0$, $f_2(x_1, x_2, \ldots x_n, u_1, u_2, \ldots u_n) = 0$ ... ... $f_n(x_1, x_2, \ldots x_n, u_1, u_2, \ldots u_n) = 0$

Then we have

$$\frac{\partial(u_1, u_2, \ldots u_n)}{\partial(x_1, x_2, \ldots x_n)} = (-1)^n \frac{\partial(f_1, f_2 \ldots f_n)/\partial(x_1, x_2, \ldots x_n)}{\partial(f_1, f_2 \ldots f_n)/\partial(u_1, u_2, \ldots u_n)}$$

Example 22. *If $x^2 + y^2 + u^2 - v^2 = 0$ and $uv + xy = 0$, prove that* $\dfrac{\partial(u,v)}{\partial(x,y)} = \dfrac{x^2-y^2}{u^2+v^2}$

Solution. Let $\quad f_1 = x^2 + y^2 + u^2 - v^2 \quad , f_2 = uv + xy$

$$\frac{\partial(f_1,f_2)}{\partial(x,y)} = \begin{vmatrix} \frac{\partial f_1}{\partial x} & \frac{\partial f_1}{\partial y} \\ \frac{\partial f_2}{\partial x} & \frac{\partial f_2}{\partial y} \end{vmatrix} = \begin{vmatrix} 2x & 2y \\ y & x \end{vmatrix} = 2(x^2 - y^2)$$

$$\frac{\partial(f_1,f_2)}{\partial(u,v)} = \begin{vmatrix} \frac{\partial f_1}{\partial u} & \frac{\partial f_1}{\partial v} \\ \frac{\partial f_2}{\partial u} & \frac{\partial f_2}{\partial v} \end{vmatrix} = \begin{vmatrix} 2u & -2v \\ v & u \end{vmatrix} = 2(u^2 + v^2)$$

But $$\frac{\partial(u,v)}{\partial(x,y)} = (-1)^2 \frac{\frac{\partial(f_1,f_2)}{\partial(x,y)}}{\frac{\partial(f_1,f_2)}{\partial(u,v)}} = \frac{2(x^2-y^2)}{2(u^2+v^2)} = \frac{x^2-y^2}{u^2+v^2}$$ **Proved.**

Example 23. *If $u^3 + v + w = x + y^2 + z^2$, $u + v^3 + w = x^2 + y + z^2$, $u + v + w^3 = x^2 + y^2 + z$,*

Prove that $$\frac{\partial(u,v,w)}{\partial(x,y,z)} = \frac{1-4(xy+yz+zx)+16xyz}{2-3(u^2+v^2+w^2)+27u^2v^2w^2}$$

Solution. Let

$$f_1 = u^3 + v + w - x - y^2 - z^2$$
$$f_2 = u + v^3 + w - x^2 - y - z^2$$
$$f_3 = u + v + w^3 - x^2 - y^2 - z$$

Now, $$\frac{\partial(f_1,f_2,f_3)}{\partial(x,y,z)} = \begin{vmatrix} -1 & -2y & -2z \\ -2x & -1 & -2z \\ -2x & -2y & -1 \end{vmatrix} = -1+4\,(yz + zx + xy) - 16xyz$$

and $$\frac{\partial(f_1,f_2,f_3)}{\partial(u,v,w)} = \begin{vmatrix} 3u^2 & 1 & 1 \\ 1 & 3v^2 & 1 \\ 1 & 1 & 3w^2 \end{vmatrix} = 2 - 3(u^2 + v^2 + w^2) + 27u^2v^2w^2$$

$$\frac{\partial(u,v,w)}{\partial(x,y,z)} = (-1)^3 \frac{\partial(f_1,f_2,f_3)/\partial(x,y,z)}{\partial(f_1,f_2,f_3)/\partial(u,v,w)} = \frac{1-4(yz+zx+xy)+16xyz}{2-3(u^2+v^2+w^2)+27u^2v^2w^2}$$ **Proved.**

Example 24. *If $x + y + z = u$, $y + z = uv$, $z = uvw$, show that* $\dfrac{\partial(x,y,z)}{\partial(u,v,z)} = u^2 v$

Solution. Let

$$f_1 = x + y + z - u$$
$$f_2 = y + z - uv$$
$$f_3 = z - uvw$$

$$\frac{\partial(f_1,f_2,f_3)}{\partial(x,y,z)} = \begin{vmatrix} \frac{\partial f_1}{\partial x} & \frac{\partial f_1}{\partial y} & \frac{\partial f_1}{\partial z} \\ \frac{\partial f_2}{\partial x} & \frac{\partial f_2}{\partial y} & \frac{\partial f_2}{\partial z} \\ \frac{\partial f_3}{\partial x} & \frac{\partial f_3}{\partial y} & \frac{\partial f_3}{\partial z} \end{vmatrix} = \begin{vmatrix} 1 & 1 & 1 \\ 0 & 1 & 1 \\ 0 & 0 & 1 \end{vmatrix} = 1$$

$$\frac{\partial(f_1,f_2,f_3)}{\partial(u,v,w)}=\begin{vmatrix}\frac{\partial f_1}{\partial u} & \frac{\partial f_1}{\partial v} & \frac{\partial f_1}{\partial w}\\ \frac{\partial f_2}{\partial u} & \frac{\partial f_2}{\partial v} & \frac{\partial f_2}{\partial w}\\ \frac{\partial f_3}{\partial u} & \frac{\partial f_3}{\partial v} & \frac{\partial f_3}{\partial w}\end{vmatrix}=\begin{vmatrix}-1 & 0 & 0\\ -v & -u & 0\\ -vw & -uv & -uv\end{vmatrix}=-u^2v$$

But $$\frac{\partial(x,y,z)}{\partial(u,v,w)}=(-1)^3\frac{\frac{\partial(f_1,f_2,f_3)}{\partial(u,v,w)}}{\frac{\partial(f_1,f_2,f_3)}{\partial(x,y,z)}}=-\frac{-u^2v}{1}=u^2v$$ **Proved.**

Exalmple 25. *If u,v,w are the roots of the equation* $(\lambda-x)^3+(\lambda-y)^3+(\lambda-z)^3=0$ *in* λ, *find* $\frac{\partial(u,v,w)}{\partial(x,y,z)}$. *(U.P. I Semester, Winter 2001)*

Solution. We have $(\lambda-x)^3+(\lambda-y)^3+(\lambda-z)^3=0$

$\Rightarrow \quad 3\lambda^3-3(x+y+z)\lambda^2+3(x^2+y^2+z^2)\lambda-(x^3+y^3+z^3)=0$

Sum of the roots $=u+v+w=x+y+z$ ...(1)

Product of the roots $=uv+vw+wu=x^2+y^2+z^2$ ...(2)

$$u\,v\,w=\frac{1}{3}(x^3+y^3+z^3) \quad ...(3)$$

Equations (1),(2) and (3) can be rewritten as

$$f_1=u+v+w-x-y-z$$
$$f_2=uv+vw+wu-x^2-y^2-z^2$$
$$f_3=uvw-\frac{1}{3}(x^3+y^3+z^3)$$

$$\frac{\partial(f_1,f_2,f_3)}{\partial(x,y,z)}=\begin{vmatrix}\frac{\partial f_1}{\partial x} & \frac{\partial f_1}{\partial y} & \frac{\partial f_1}{\partial z}\\ \frac{\partial f_2}{\partial x} & \frac{\partial f_2}{\partial y} & \frac{\partial f_2}{\partial z}\\ \frac{\partial f_3}{\partial x} & \frac{\partial f_3}{\partial y} & \frac{\partial f_3}{\partial z}\end{vmatrix}=\begin{vmatrix}-1 & -1 & -1\\ -2x & -2y & -2z\\ -x^2 & -y^2 & -z^2\end{vmatrix}$$

$$=(-1)(-2)(-1)\begin{vmatrix}1 & 1 & 1\\ x & y & z\\ x^2 & y^2 & z^2\end{vmatrix}=-2\begin{vmatrix}0 & 0 & 1\\ x-y & y-z & z\\ x^2-y^2 & y^2-z^2 & z^2\end{vmatrix}\begin{matrix}C_1\to C_1-C_2\\ C_2\to C_2-C_3\end{matrix}$$

$$=-2(x-y)(y-z)\begin{vmatrix}0 & 0 & 1\\ 1 & 1 & z\\ x+y & y+z & z^2\end{vmatrix}=-2(x-y)(y-z)(y+z-x-y)=-2(x-y)(y-z)(z-x)$$

$$\frac{\partial(f_1,f_2,f_3)}{\partial(u,v,w)}=\begin{vmatrix}\frac{\partial f_1}{\partial u} & \frac{\partial f_1}{\partial v} & \frac{\partial f_1}{\partial w}\\ \frac{\partial f_2}{\partial u} & \frac{\partial f_2}{\partial v} & \frac{\partial f_2}{\partial w}\\ \frac{\partial f_3}{\partial u} & \frac{\partial f_3}{\partial v} & \frac{\partial f_3}{\partial w}\end{vmatrix}=\begin{vmatrix}1 & 1 & 1\\ v+w & u+w & u+v\\ vw & wu & uv\end{vmatrix}=\begin{vmatrix}0 & 0 & 1\\ v-u & w-v & u+v\\ w(v-u) & u(w-v) & uv\end{vmatrix}$$

$$= (v-u)(w-v)\begin{vmatrix} 0 & 0 & 1 \\ 1 & 1 & u+v \\ w & u & uv \end{vmatrix} \begin{aligned} &= (v-u)(w-v)(u-w) \\ &= -(u-v)(v-w)(w-u) \end{aligned}$$

$$\frac{\partial(u,v,w)}{\partial(x,y,z)} = (-1)^3 \frac{\dfrac{\partial(f_1,f_2,f_3)}{\partial(x,y,z)}}{\dfrac{\partial(f_1,f_2,f_3)}{\partial(u,v,w)}} = -\frac{-2(x-y)(y-z)(z-x)}{-(u-v)(v-w)(w-u)} = \frac{-2(x-y)(y-z)(z-x)}{(u-v)(v-w)(w-u)} \quad \textbf{Ans}$$

EXERCISE 5.5

1. If $u^3+v^3 = x+y = u^2+v^2 = x^3+y^3$, then prove that $\dfrac{\partial(u,v)}{\partial(x,y)} = \dfrac{1}{2}\dfrac{y^2-x^2}{2uv(u-v)}$

2. If $u^3+v^3+w^3 = x+y+z$,

$u^2+v^2+w^2 = x^3+y^3+z^3$,

$u+v+w = x^2+y^2+z^2$,

show that $\dfrac{\partial(u,v,w)}{\partial(x,y,z)} = \dfrac{(x-y)(y-z)(z-x)}{(u-v)(v-w)(w-u)}$

3. If $u = \dfrac{x}{\sqrt{1-r^2}}, v = \dfrac{y}{\sqrt{1-r^2}}, w = \dfrac{z}{\sqrt{1-r^2}}$, show that

$\dfrac{\partial(u,v,w)}{\partial(x,y,z)} = \dfrac{1}{(1-r^2)^{5/2}}$ where $r^2 = x^2+y^2+z^2$

4. If $u_1 = x_1+x_2+x_3+x_4,\ u_1u_2 = x_2+x_3+x_4,\ u_1u_2u_3 = x_3+x_4,\ u_1u_2u_3u_4 = x_4$

show that $\dfrac{\partial(x_1,x_2,x_3,x_4)}{\partial(u_1,u_2,u_3,u_4)} = u_1^3.u_2^2.u_3$

5. If u, v, w are the roots of the equation in λ and $\dfrac{x}{a+\lambda}+\dfrac{y}{b+\lambda}+\dfrac{z}{c+\lambda} = 1$, then find $\dfrac{\partial(x,y,z)}{\partial(u,v,w)}$.

Ans. $\dfrac{(u-v)(v-w)(w-u)}{(a-b)(b-c)(c-a)}$

5.4 PARTIAL DERIVATIVES OF IMPLICIT FUNCTIONS BY JACOBIAN

Given $f_1(x,y,u,v) = 0, f_2(x,y,u,v) = 0$

$$\frac{\partial f_1}{\partial u}\frac{\partial u}{\partial x} + \frac{\partial f_1}{\partial v}\frac{\partial v}{\partial x} + \frac{\partial f_1}{\partial x}.1 = 0 \quad ...(1)$$

$$\frac{\partial f_2}{\partial u}\frac{\partial u}{\partial x} + \frac{\partial f_2}{\partial v}\frac{\partial v}{\partial x} + \frac{\partial f_2}{\partial x}.1 = 0 \quad ...(2)$$

Solving (1) and (2), we get

$$\frac{\dfrac{\partial u}{\partial x}}{\dfrac{\partial f_1}{\partial v}.\dfrac{\partial f_2}{\partial x} - \dfrac{\partial f_1}{\partial x}.\dfrac{\partial f_2}{\partial v}} = \frac{\dfrac{\partial v}{\partial x}}{\dfrac{\partial f_1}{\partial x}.\dfrac{\partial f_2}{\partial u} - \dfrac{\partial f_1}{\partial u}.\dfrac{\partial f_2}{\partial x}} = \frac{1}{\dfrac{\partial f_1}{\partial u}.\dfrac{\partial f_2}{\partial v} - \dfrac{\partial f_1}{\partial v}.\dfrac{\partial f_2}{\partial u}}$$

$$\frac{\partial u}{\partial x} = \frac{\dfrac{\partial f_1}{\partial v}.\dfrac{\partial f_2}{\partial x} - \dfrac{\partial f_1}{\partial x}.\dfrac{\partial f_2}{\partial v}}{\dfrac{\partial f_1}{\partial u}.\dfrac{\partial f_2}{\partial v} - \dfrac{\partial f_1}{\partial v}.\dfrac{\partial f_2}{\partial u}} = -\frac{\dfrac{\partial(f_1,f_2)}{\partial(x,v)}}{\dfrac{\partial(f_1,f_2)}{\partial(u,v)}}$$

$$\frac{\partial v}{\partial x}=\frac{\dfrac{\partial f_1}{\partial x}\cdot\dfrac{\partial f_2}{\partial u}-\dfrac{\partial f_1}{\partial u}\cdot\dfrac{\partial f_2}{\partial x}}{\dfrac{\partial f_1}{\partial u}\cdot\dfrac{\partial f_2}{\partial v}-\dfrac{\partial f_1}{\partial v}\cdot\dfrac{\partial f_2}{\partial u}}=-\frac{\dfrac{\partial(f_1,f_1)}{\partial(x,u)}}{\dfrac{\partial(f_1,f_2)}{\partial(v,u)}}$$

and if, $f_1(x, y, z, u, v, w) = 0, f_2(x, y, z, u, v, w) = 0,$
$f_3(x, y, z, u, v, w) = 0$

$$\frac{\partial x}{\partial u}=-\frac{\partial(f_1,f_2,f_3)/\partial(u,y,z)}{\partial(f_1,f_2,f_3)/\partial(x,y,z)}$$

and so on

Note. First we write the Jacobian in the denominator and then we write the Jacobian in the numerator by replacing x by u.

Example 26. *Use Jacobians to find* $\left(\dfrac{\partial u}{\partial x}\right)_v$ *if:*

$$u^2+xv^2-xy=0 \textit{ and } u^2+xyv+v^2=0$$

Solution. Let $f_1=u^2+xv^2-xy,\ f_2=u^2+xyv+v^2$

$$\frac{\partial(f_1,f_2)}{\partial(u,v)}=\begin{vmatrix}\dfrac{\partial f_1}{\partial u} & \dfrac{\partial f_1}{\partial v}\\ \dfrac{\partial f_2}{\partial u} & \dfrac{\partial f_2}{\partial v}\end{vmatrix}=\begin{vmatrix}2u & 2xv\\ 2u & xy+2v\end{vmatrix}=2uxy+4uv-4uxv$$

$$\frac{\partial(f_1,f_2)}{\partial(x,v)}=\begin{vmatrix}\dfrac{\partial f_1}{\partial x} & \dfrac{\partial f_1}{\partial v}\\ \dfrac{\partial f_2}{\partial x} & \dfrac{\partial f_2}{\partial v}\end{vmatrix}=\begin{vmatrix}v^2-y & 2xv\\ yv & xy+2v\end{vmatrix}=xyv^2+2v^3-xy^2-2yv-2xyv^2$$

$$=-xyv^2+2v^3-xy^2-2yv$$

$$\frac{\partial u}{\partial x}=-\frac{\partial(f_1,f_2)/\partial(x,v)}{\partial(f_1,f_2)/\partial(u,v)}$$

$$=\frac{xyv^2-2v^3+xy^2+2yv}{2uxy+4uv-4xuv}$$ **Ans.**

Example 27. *If* $u=x+y^2,\ v=y+z^2,\ w=z+x^2$*, prove that*

(*i*) $\dfrac{\partial x}{\partial u}=-\dfrac{1}{1+8xyz}$

(*ii*) *Also find* $\dfrac{\partial^2 x}{\partial u^2}$.

Solution. (*i*) Here, $f_1=u-x-y^2,\ f_2=v-y-z^2,\ f_3=w-z-x^2.$

Now $$\frac{\partial(f_1,f_2,f_3)}{\partial(u,y,z)}=\begin{vmatrix}1 & -2y & 0\\ 0 & -1 & -2z\\ 0 & 0 & -1\end{vmatrix}=1;$$

$$\frac{\partial(f_1, f_2, f_3)}{\partial(x, y, z)} = \begin{vmatrix} -1 & -2y & 0 \\ 0 & -1 & -2z \\ -2x & 0 & -1 \end{vmatrix} = -1(1+0) + 2y\,(0 - 4zx) = -1 - 8xyz$$

$$\frac{\partial x}{\partial u} = -\frac{\partial(f_1, f_2, f_3)/\partial(u, y, z)}{\partial(f_1, f_2, f_3)/\partial(x, y, z)}$$

$$\frac{\partial x}{\partial u} = -\left(\frac{1}{-1 - 8xyz}\right) = \frac{1}{1 + 8xyz}$$ **Ans.**

(*ii*)
$$\frac{\partial^2 x}{\partial u^2} = \frac{\partial}{\partial u}\left(\frac{\partial x}{\partial u}\right) = \frac{\partial}{\partial u}\left(\frac{1}{1 + 8xyz}\right) = -\frac{1}{(1 + 8xyz)^2}\frac{\partial}{\partial u}(1 + 8xyz)$$

$$= \frac{-1}{(1 + 8xyz)^2}\left[0 + 8\left(\frac{\partial x}{\partial u}yz + \frac{\partial y}{\partial u}zx + \frac{\partial z}{\partial u}xy\right)\right]$$

$$= \frac{-8}{(1 + 8xyz)^2}\left[\frac{\partial x}{\partial u}yz + \frac{\partial y}{\partial u}zx + \frac{\partial z}{\partial u}xy\right] \qquad ...(1)$$

We have $\dfrac{\partial(f_1, f_2, f_3)}{\partial(x, u, z)} = \begin{vmatrix} -1 & 1 & 0 \\ 0 & 0 & -2z \\ -2x & 0 & -1 \end{vmatrix} = 4zx\,;$

$$\frac{\partial(f_1, f_2, f_3)}{\partial(x, y, u)} = \begin{vmatrix} -1 & -2y & 1 \\ 0 & -1 & 0 \\ -2x & 0 & 0 \end{vmatrix} = -2x.$$

Now, $\dfrac{\partial y}{\partial u} = -\dfrac{\partial(f_1, f_2, f_3)/\partial(x, u, z)}{\partial(f_1, f_2, f_3)/\partial(x, y, z)}$

$\therefore$ $\dfrac{\partial y}{\partial u} = -\dfrac{4zx}{-1 - 8xyz} = \dfrac{4zx}{1 + 8xyz}$

$$\frac{\partial z}{\partial u} = -\frac{\partial(f_1, f_2, f_3)/\partial(x, y, u)}{\partial(f_1, f_2, f_3)/\partial(x, y, u)}$$

$\therefore$ $\dfrac{\partial z}{\partial u} = -\dfrac{-2x}{-1 - 8xyz} = \dfrac{-2x}{1 + 8xyz}$

Substituting in (1), we have

$$\frac{\partial^2 x}{\partial u^2} = \frac{-8}{(1 + 8xyz)^2}\left[\frac{yz}{1 + 8xyz} + \frac{4z^2x^2}{1 + 8xyz} + \frac{-2x^2y}{1 + 8xyz}\right]$$

$$= -\frac{-8(yz + 4z^2x^2 - 2x^2y)}{(1 + 8xyz)^2}$$ **Ans.**

Example 28. *Given,* $x = u + v + w,\ y = u^2 + v^2 + w^2,\ z = u^3 + v^3 + w^3$

show that $\dfrac{\partial u}{\partial x} = \dfrac{vw}{(u - v)(u - w)}$

Solution. Let
$$f_1 = u + v + w - x = 0$$
$$f_2 = u^2 + v^2 + w^2 - y = 0$$
$$f_3 = u^3 + v^3 + w^3 - z = 0$$

$$\frac{\partial u}{\partial x} = -\frac{\partial(f_1 f_2 f_3)/\partial(x,v,w)}{\partial(f_1 f_2 f_3)/\partial(u,v,w)} \quad ...(1)$$

$$\frac{\partial(f_1, f_2, f_3)}{\partial(x,v,w)} = \begin{vmatrix} -1 & 1 & 1 \\ 0 & 2v & 2w \\ 0 & 3v^2 & 3w^2 \end{vmatrix} = -6vw\begin{vmatrix} 1 & 1 \\ v & w \end{vmatrix}$$

$$= 6vw(v-w) \quad ...(2)$$

$$\frac{\partial(f_1, f_2, f_3)}{\partial(u,v,w)} = \begin{vmatrix} 1 & 1 & 1 \\ 2u & 2v & 2w \\ 3u^2 & 3v^2 & 3w^2 \end{vmatrix} = 6\,(v-u)\,(w-u)\,(w-v) \quad ...(3)$$

Thus from (1),(2) and (3), we get

$$\frac{\partial u}{\partial x} = -\frac{6vw(v-w)}{6(v-u)(w-u)(w-v)} = \frac{vw}{(u-v)(u-w)}$$ **Proved.**

EXERCISE 5.6

1. If $u^2 + xv^2 - uxy = 0, v^2 - xy^2 + 2uv + u^2 = 0$, find $\frac{\partial u}{\partial x}$. **Ans.** $-\frac{(v^2-uy)(u+v)+xvy^2}{(u+v)(2u-xy-2xv)}$

2. If $x = u + e^{-v}\sin u, y = v + e^{-v}\cos u$, find $\frac{\partial u}{\partial y}, \frac{\partial v}{\partial x}$. **Ans.** $\frac{\partial u}{\partial y} = \frac{\partial v}{\partial x} = \frac{e^{-v}\sin u}{1-e^{-3v}}$

3. If $x = u^2 - v^2, y = 2uv$, find $\frac{\partial u}{\partial x}; \frac{\partial u}{\partial y}; \frac{\partial v}{\partial x}; \frac{\partial v}{\partial y}$ and $\frac{\partial(u,v)}{\partial(x,y)}$.

   Ans. $\frac{u}{2(u^2+v^3)}; \frac{v}{2(u^2+v^3)}; \frac{-v}{2(u^2+v^2)}; \frac{u}{2(u^2+v^2)}; \frac{1}{4(u^2+v^2)}$

4. If $u^3 + xv^2 - uy = 0, u^2 + xyv + v^2 = 0$, find $\frac{\partial u}{\partial x}$ **Ans.** $\frac{xyv^2 - 2v^3}{3xyu^2 - xy^2 + 6u^2v - 2vy - 4xuv}$

5. If $u^2 + xv^2 = x + y, v^2 + yu^2 = x - y$, find $\frac{\partial u}{\partial x}, \frac{\partial v}{\partial y}$ **Ans.** $\frac{1-x-v^2}{2u(1-xy)}, \frac{1+y+u^2}{-2v(1-xy)}$

6. If $u = xyz, v = x^2 + y^2 + z^2, w = x + y + z$ find $\frac{\partial x}{\partial u}$. **Ans.** $\frac{1}{(x-y)(x-z)}$

7. If $u = x^2 + y^2 + z^2, v = xyz$, find $\frac{\partial x}{\partial u}$. **Ans.** $\frac{x}{2(2x^2-y^2)}$

Choose the Correct answer:

1. The Jacobian $\frac{\partial(u,v)}{\partial(x,y)}$ for the function $u = e^x \sin y, \; v = (x + \log\sin y)$ is

   (*a*) 1 (*b*) $\sin x \sin y - xy\cos x\cos y$

   (*c*) 0 (*c*) $\frac{e^x}{x}$ (*U.P.I. Semester, Dec. 2008*) **Ans.** (*c*)

CHAPTER 6

TAYLOR'S SERIES FOR FUNCTIONS OF TWO VARIABLES

6.1 TAYLOR'S SERIES OF TWO VARIABLES

If $f(x, y)$ and all its partial derivatives upto the nth order are finite and continuous for all points (x, y), where

$$a \le x \le a+h,\ b \le y \le b+k$$

Then $f(a+h, b+k) = f(a,b)+\left(h\frac{\partial}{\partial x}+k\frac{\partial}{\partial y}\right)f+\frac{1}{2!}\left(h\frac{\partial}{\partial x}+k\frac{\partial}{\partial y}\right)^2 f+\frac{1}{3!}\left(h\frac{\partial}{\partial x}+k\frac{\partial}{\partial y}\right)^3 f+\ldots$

Proof. Suppose that $f(x+h, y+k)$ is a function of one variable only, say x where y is assumed as constant. Expanding by Taylor's Theorem for one variable, we have

$$f(x+\delta x, y+\delta y) = f(x, y+\delta y)+\delta x\frac{\partial}{\partial x}f(x, y+\delta y)+\frac{(\delta x)^2}{2!}\frac{\partial^2}{\partial x^2}f(x, y+\delta y)+\ldots\ldots$$

Now expanding for y, we get

$$= \left[f(x,y)+\delta y\frac{\partial}{\partial y}f(x,y)+\frac{(\delta y)^2}{2!}\frac{\partial^2}{\partial y^2}f(x,y)+\ldots\right]+\delta x\cdot\frac{\partial}{\partial x}\left[f(x,y)+\delta y\frac{\partial}{\partial y}f(x,y)+\ldots\ldots\right]$$

$$+\frac{(\delta x)^2}{2!}\frac{\partial^2}{\partial x^2}\left[f(x,y)+\delta y\frac{\partial}{\partial y}f(x,y)+\ldots\right]+\ldots$$

$$= \left[f(x,y)+\delta y\frac{\partial}{\partial y}f(x,y)+\frac{(\delta y)^2}{2!}\cdot\frac{\partial^2}{\partial y^2}f(x,y)+\ldots\right]+$$

$$+\delta x\left[\frac{\partial f(x,y)}{\partial x}+\delta y\cdot\frac{\partial^2}{\partial x\,\partial y}f(x,y)\right]+\frac{(\delta x)^2}{2!}\left[\frac{\partial^2}{\partial x^2}f(x,y)+\ldots\right]+\ldots$$

$$= f(x,y)+\left[\delta x\frac{\partial f(x,y)}{\partial x}+\delta y\cdot\frac{\partial f(x,y)}{\partial y}\right]+\frac{1}{2!}\left[(\delta x)^2\frac{\partial^2 f(x,y)}{\partial x^2}+2\,\delta x\cdot\delta y\cdot\frac{\partial^2 f(x,y)}{\partial x\,\partial y}\right.$$

$$\left.+(\delta y)^2\cdot\frac{\partial^2 f(x,y)}{\partial y^2}\right]+\ldots$$

$$\Rightarrow f(a+h, b+k) = f(a,b)+\left[h\frac{\partial f}{\partial x}+k\frac{\partial f}{\partial y}\right]+\frac{1}{2!}\left[h^2\frac{\partial^2 f}{\partial x^2}+2hk\frac{\partial^2 f}{\partial x\,\partial y}+k^2\frac{\partial^2 f}{\partial y^2}\right]+\ldots$$

$$\Rightarrow \quad f(a+h, b+k) = f(a,b) + \left[h\frac{\partial}{\partial x} + k\frac{\partial}{\partial y}\right] f + \frac{1}{2!}\left[h\frac{\partial}{\partial x} + k\frac{\partial}{\partial y}\right]^2 f + \ldots\ldots$$

On putting $a = 0, b = 0, h = x, k = y$, we get

$$f(x,y) = f(0,0) + \left(x\frac{\partial f}{\partial x} + y\frac{\partial f}{\partial y}\right) + \frac{1}{2!}\left(x^2\frac{\partial^2 f}{\partial x^2} + 2xy\frac{\partial^2 f}{\partial x\,\partial y} + y^2\frac{\partial^2 f}{\partial y^2}\right) + \ldots\ldots$$

Example 1. *Expand e^x. $\sin y$ in powers of x and y, $x = 0$, $y = 0$ as far as terms of third degree.*

Solution.

		$x = 0, y = 0$
$f(x, y)$	$e^x \sin y,$	0
$f_x(x, y)$	$e^x \sin y,$	0
$f_y(x, y)$	$e^x \cos y,$	1
$f_{xx}(x, y)$	$e^x \sin y,$	0
$f_{xy}(x, y)$	$e^x \cos y,$	1
$f_{yy}(x, y)$	$-e^x \sin y,$	0
$f_{xxx}(x, y)$	$e^x \sin y,$	0
$f_{xxy}(x, y)$	$e^x \cos y,$	1
$f_{xyy}(x, y)$	$-e^x \sin y,$	0
$f_{yyy}(x, y)$	$-e^x \cos y,$	-1

By Taylor's theorem

$$f(x,y) = f(0,0) + \left(x\frac{\partial}{\partial x} + y\frac{\partial}{\partial y}\right) f(0,0) + \frac{1}{2!}\left(x\frac{\partial}{\partial x} + y\frac{\partial}{\partial y}\right)^2 f(0,0)$$

$$+ \frac{1}{3!}\left(x\frac{\partial}{\partial x} + y\frac{\partial}{\partial y}\right)^3 f(0,0) + \ldots$$

$$= f(0,0) + x f_x(0,0) + y f_y(0,0) + \frac{x^2}{2!} f_{xx}(0,0) + \frac{2xy}{2!} f_{xy}(0,0) + \frac{y^2}{2!} f_{yy}(0,0)$$

$$+ \frac{1}{3!}x^3 f_{xxx}(0,0) + \frac{3x^2 y}{3!} f_{xxy}(0,0) + \frac{3}{3!} x y^2 f_{xyy}(0,0) + \frac{1}{3!} y^3 f_{yyy}(0,0) + \ldots$$

$$e^x \sin y = 0 + x(0) + y(1) + \frac{x^2}{2}(0) + xy(1) + \frac{y^2}{2}(0) + \frac{x^3}{6}(0) + \frac{3x^2 y}{6}(1)$$

$$+ \frac{3xy^2}{6}(0) + \frac{y^3}{6}(-1) + \ldots\ldots$$

$$= y + xy + \frac{x^2 y}{2} - \frac{y^3}{6} + \ldots\ldots$$ **Ans.**

Example 2. *Find the expansion for cos x cos y in powers of x, y upto fourth order terms.*

Solution.

		$x = 0, y = 0$
$f(x, y)$	$\cos x \cos y,$	1
f_x	$-\sin x \cos y,$	0
f_y	$-\cos x \sin y,$	0
f_{xx}	$-\cos x \cos y,$	-1
f_{xy}	$\sin x \sin y,$	0
f_{yy}	$-\cos x \cos y,$	-1
f_{xxx}	$\sin x \cos y,$	0
f_{xxy}	$\cos x \sin y,$	0
f_{xyy}	$\sin x \cos y,$	0
f_{yyy}	$\cos x \sin y,$	0
f_{xxxx}	$\cos x \cos y,$	1
f_{xxxy}	$-\sin x \sin y,$	0
f_{xxyy}	$\cos x \cos y,$	1
f_{xyyy}	$-\sin x \sin y,$	0
f_{yyyy}	$\cos x \cos y,$	1

By Taylor's Series

$$f(x, y) = f(0, 0) + x f_x(0, 0) + y f_y(0, 0) + \frac{1}{2!}\Big[x^2 f_x^2(0, 0)$$

$$+ 2xy f_{xy}(0, 0) + y^2 f_{yy}(0, 0)\Big] + \frac{1}{3!}\Big[x^3 f_x^3(0, 0) + 3x^2 y f_{xy}^{2}(0, 0)$$

$$+ 3xy^2 f_{xy}^{2}(0, 0) + y^3 f_y^3(0, 0)\Big] + \frac{1}{4!}\Big[x^4 f_x^4(0, 0) + 4x^3 y f_{xy}^{3}(0, 0)$$

$$+ 6x^2y^2 f_{xy}^{22}(0, 0) + 4xy^3 f_{xy}^{3}(0, 0) + y^4 f_y^4(0, 0)\Big] + \ldots$$

$$\cos x \cos y = 1 + 0 + 0 + \frac{1}{2}(-x^2 + 0 - y^2) + \frac{1}{6}(0 + 0 + 0 + 0)$$

$$+ \frac{1}{24}(x^4 + 0 + 6x^2y^2 + 0 + y^4)$$

$$= 1 - \frac{x^2}{2} - \frac{y^2}{2} + \frac{x^4}{24} + \frac{x^2y^2}{4} + \frac{y^2}{24} + \ldots$$ **Ans.**

Example 3. *Find the first six terms of the expansion of the function e^x log (1 + y) in a Taylor's series in the neighbourhood of the point* (0, 0).

Solution.

		$x=0, y=0$
$f(x, y)$	$e^x \log(1+y)$	0
$\dfrac{\partial f}{\partial x}$	$e^x \log(1+y)$	0
$\dfrac{\partial f}{\partial y}$	$\dfrac{e^x}{1+y}$	1
$\dfrac{\partial^2 f}{\partial x^2}$	$e^x \log(1+y)$	0
$\dfrac{\partial^2 f}{\partial y^2}$	$-\dfrac{e^x}{(1+y)^2}$	-1
$\dfrac{\partial^2 f}{\partial x \partial y}$	$\dfrac{e^x}{(1+y)}$	1

Taylor's series is

$$f(x, y) = f(0,0) + \left(x\frac{\partial f}{\partial x} + y\frac{\partial f}{\partial y}\right) + \frac{1}{2!}\left(x^2\frac{\partial^2 f}{\partial x^2} + 2xy\frac{\partial^2 f}{\partial x \partial y} + y^2\frac{\partial^2 f}{\partial y^2}\right) + \ldots$$

$$\Rightarrow \quad e^x \log(1+y) = 0 + (x \times 0 + y \times 1) + \frac{1}{2!}[x^2 \times (0) + 2xy \times 1 + y^2 \times (-1)] + \ldots$$

$$\Rightarrow \quad e^x \log(1+y) = y + xy - \frac{y^2}{2}$$ **Ans.**

EXERCISE 6.1

1. Expand $e^x \cos y$ at (0, 0) upto three terms. **Ans.** $1 + x + \frac{1}{2}(x^2 - y^2) + \ldots\ldots\ldots$
2. Expand $z = e^{2x} \cos 3y$ in power series of x and y upto quadratic terms. *(AMIE Summer 2004)*
   Ans. $1 + 2x + 2x^2 - \frac{9}{2}y^2 + \ldots$
3. Show that $e^y \log(1+x) = x + xy - \dfrac{x^2}{2}$ approximately.
4. Verify $\sin(x+y) = x + y - \dfrac{(x+y)^3}{3} + \ldots\ldots$

Example 4. *Expand* $\sin(xy)$ *in powers of* $(x-1)$ *and* $\left(y - \frac{\pi}{2}\right)$ *as far as the terms of second degree.* *(Nagpur University, Summer 2003)*

Solution. We have, $f(x, y) = \sin(xy)$

Here
$$\left[\begin{array}{l} a + h = x \text{ and } h = x - 1 \\ \Rightarrow a + (x-1) = x \Rightarrow a = 1 \end{array}\right.$$
$$\left[\begin{array}{l} b + k = y \text{ and } k = y - \dfrac{\pi}{2} \\ \Rightarrow b + y - \dfrac{\pi}{2} = y \Rightarrow b = \dfrac{\pi}{2} \end{array}\right.$$

		$x = 1, y = \frac{\pi}{2}$
$f(x, y)$	$\sin(xy)$	1
$f_x(x, y)$	$y\cos(xy)$,	0
$f_y(x, y)$	$x\cos(xy)$,	0
$f_{xx}(x, y)$	$-y^2\sin(xy)$,	$-\frac{\pi^2}{4}$
$f_{xy}(x, y)$	$\cos(xy) - xy\sin(xy)$,	$-\frac{\pi}{2}$
$f_{yy}(x, y)$	$-x^2\sin(xy)$,	-1

By Taylor's theorem for a function of two variables, we have

$$f(a+h, b+k) = f(a, b) + hf_x(a, b) + kf_y(a, b) + \frac{1}{2!}\left\{h^2 f_{xx}(a, b) + 2hkf_{xy}(a, b) + k^2 f_{yy}(a, b)\right\}$$

$$\Rightarrow f(x, y) = f\left(1, \frac{\pi}{2}\right) + (x-1) f_x\left(1, \frac{\pi}{2}\right) + \left(y - \frac{\pi}{2}\right) f_y\left(1, \frac{\pi}{2}\right)$$

$$+ \frac{1}{2!}\left\{(x-1)^2 f_{xx}\left(1, \frac{\pi}{2}\right) + 2(x-1)\left(y - \frac{\pi}{2}\right) f_{xy}\left(1, \frac{\pi}{2}\right) + \left(y - \frac{\pi}{2}\right)^2 f_{yy}\left(1, \frac{\pi}{2}\right)\right\}$$

$$\Rightarrow \sin(xy) = 1 + (x-1).0 + \left(y - \frac{\pi}{2}\right).0 +$$

$$\frac{1}{2!}\left\{(x-1)^2\left(-\frac{\pi^2}{4}\right) + 2(x-1)\left(y - \frac{\pi}{2}\right)\left(-\frac{\pi}{2}\right) + \left(y - \frac{\pi}{2}\right)^2(-1)\right\} + \ldots$$

$$\Rightarrow \sin(xy) = 1 - \frac{\pi^2}{8}(x-1)^2 - \frac{\pi}{2}(x-1)\left(y - \frac{\pi}{2}\right) - \frac{1}{2}\left(y - \frac{\pi}{2}\right)^2 + \ldots$$ **Ans.**

Example 5. *Expand $e^x . \cos y$ in powers of x and $\left[y - \frac{\pi}{2}\right]$ upto terms of degree 3.*

(*Nagpur University, Summer 2002*)

Solution. Here, we have

$$\left[\begin{array}{l} a + h = x \text{ and } h = x \\ \Rightarrow a + x = x \Rightarrow a = 0 \end{array}\right. \Rightarrow \left[\begin{array}{l} b + k = y \text{ and } k = y - \frac{\pi}{2} \\ \Rightarrow b + y - \frac{\pi}{2} = y \Rightarrow b = \frac{\pi}{2} \end{array}\right]$$

$$f(x+h, y+k) = f(x, y) + \left(h\frac{\partial}{\partial x} + k\frac{\partial}{\partial y}\right) f + \frac{1}{2!}\left(h\frac{\partial}{\partial x} + k\frac{\partial}{\partial y}\right)^2 f$$

$$+ \frac{1}{3!}\left(h\frac{\partial}{\partial x} + k\frac{\partial}{\partial y}\right)^3 f + \ldots \quad \ldots(1)$$

		$x = 0, y = \frac{\pi}{2}$
$f(x, y)$	$e^x \cos y$	0
$\frac{\partial f}{\partial x}$	$e^x \cos y$	0
$\frac{\partial f}{\partial y}$	$-e^x \sin y$	-1
$\frac{\partial^2 f}{\partial x^2}$	$e^x \cos y$	0
$\frac{\partial^2 f}{\partial x \partial y}$	$-e^x \sin y$	-1
$\frac{\partial^2 f}{\partial y^2}$	$-e^x \cos y$	0
$\frac{\partial^3 f}{\partial x^3}$	$e^x \cos y$	0
$\frac{\partial^3 f}{\partial x^2 \partial y}$	$-e^x \sin y$	-1
$\frac{\partial^3 f}{\partial y^2 \partial x}$	$-e^x \cos y$	0
$\frac{\partial^3 f}{\partial y^3}$	$e^x \sin y$	1

Putting these values in (1), we get

$$e^x \cos y = 0 + x\,.\,0 + \left(y - \frac{\pi}{2}\right)(-1) +$$

$$\frac{1}{2!}\left[x^2\,.\,0 + 2x\left(y - \frac{\pi}{2}\right)(-1) + \left(y - \frac{\pi}{2}\right)^2 .\,0\right] +$$

$$\frac{1}{3!}\left[x^3\,.\,0 + 3x^2\left(y - \frac{\pi}{2}\right)(-1) + 3x\left(y - \frac{\pi}{2}\right)^2 .\,0 + \left(y - \frac{\pi}{2}\right)^3 .\,1\right] + \ldots$$

$$= -\left(y - \frac{\pi}{2}\right) - x\left(y - \frac{\pi}{2}\right) - \frac{1}{2}x^2\left(y - \frac{\pi}{2}\right) + \frac{1}{6}\left(y - \frac{\pi}{2}\right)^3 + \ldots$$

Ans.

Example 6. *Expand e^x cos y near the point $\left(1, \frac{\pi}{4}\right)$ by Taylor's Theorem.*

(U.P., I Semester Dec. 2007)

Solution.
$$f(x + h, y + k) = f(x, y) + \left[h\frac{\partial}{\partial x} + k\frac{\partial}{\partial y}\right] f + \frac{1}{2!}\left(h\frac{\partial}{\partial x} + k\frac{\partial}{\partial y}\right)^2 f + \frac{1}{3!}\left(h\frac{\partial}{\partial x} + k\frac{\partial}{\partial y}\right)^3 f + \ldots$$

$$e^x \cos y = f(x, y) = f\left[1+(x-1), \frac{\pi}{4}+\left(y-\frac{\pi}{4}\right)\right]$$

where $\quad h = x-1,\ k = y - \dfrac{\pi}{4}$

$$= f\left(1+h, \frac{\pi}{4}+k\right)$$

$\Rightarrow \quad x = 1,\ y = \dfrac{\pi}{4}$

$f(x, y)$	$e^x \cos y$	$\dfrac{e}{\sqrt{2}}$
$\dfrac{\partial f}{\partial x}$	$e^x \cos y,$	$\dfrac{e}{\sqrt{2}}$
$\dfrac{\partial f}{\partial y}$	$-e^x \sin y,$	$\dfrac{-e}{\sqrt{2}}$
$\dfrac{\partial^2 f}{\partial x^2}$	$e^x \cos y,$	$\dfrac{e}{\sqrt{2}}$
$\dfrac{\partial^2 f}{\partial y^2}$	$-e^x \cos y,$	$\dfrac{-e}{\sqrt{2}}$
$\dfrac{\partial^2 f}{\partial x\, \partial y}$	$-1\, e^x \sin y,$	$\dfrac{-e}{\sqrt{2}}$

Putting these values in Taylor's Theorem, we get

$$e^x \cos y = \frac{e}{\sqrt{2}} + \left[(x-1)\frac{e}{\sqrt{2}} + \left(y-\frac{\pi}{4}\right)\left(\frac{-e}{\sqrt{2}}\right)\right]$$

$$+\frac{1}{2!}\left[(x-1)^2 \frac{e}{\sqrt{2}} + 2(x-1)\left(y-\frac{\pi}{4}\right)\left(\frac{-e}{\sqrt{2}}\right) + \left(y-\frac{\pi}{4}\right)^2\left(\frac{-e}{\sqrt{2}}\right)\right] + \ldots$$

$$= \frac{e}{\sqrt{2}}\left[1+(x-1)-\left(y-\frac{\pi}{4}\right)+\frac{(x-1)^2}{2}-(x-1)\left(y-\frac{\pi}{4}\right)-\left(y-\frac{\pi}{4}\right)^2+\ldots\right]$$ **Ans.**

Example 7. *If $f(x, y) = \tan^{-1}(x y)$, compute an approximate value of $f(0.9, -1.2)$.*

Solution. We have,

$$f(x, y) = \tan^{-1}(x y)$$

Let us expand $f(x, y)$ near the point $(1, -1)$

$$f(0.9, -1.2) = f(1-0.1, -1-0.2)$$

$$= f(1,-1) + \left[(-0.1)\frac{\partial f}{\partial x} + (-0.2)\frac{\partial f}{\partial y}\right] + \frac{1}{2!}\left[(-0.1)^2 \frac{\partial^2 f}{\partial x^2}\right.$$

$$\left. + 2(-0.1)(-0.2)\frac{\partial^2 f}{\partial x\, \partial y} + (-0.2)^2 \frac{\partial^2 f}{\partial y^2}\right] + \ldots \qquad \ldots(1)$$

Substituting the values of $\dfrac{\partial f}{\partial x}, \dfrac{\partial f}{\partial y}$ etc. in (1), we get

$$f(0.9, -1.2) = -\frac{\pi}{4} + (-0.1)\left(-\frac{1}{2}\right) + (-0.2)\left(\frac{1}{2}\right) + \frac{1}{2}\left[(-0.1)^2\left(\frac{1}{2}\right)\right.$$

$$\left. + 2(-0.1)(-0.2)\, 0 + (-0.2)^2\left(\frac{1}{2}\right)\right] + \ldots$$

$$= -\frac{22}{28} + 0.05 - 0.1 + \frac{1}{2}(0.005 + 0.02)$$

$$= -0.786 + 0.05 - 0.1 + 0.0125 = -0.8235$$ **Ans.**

		$x = 1, y = -1$
$f(x, y)$	$\tan^{-1}(xy)$	$-\dfrac{\pi}{4}$
$\dfrac{\partial f}{\partial x}$	$\dfrac{y}{1+x^2y^2}$,	$-\dfrac{1}{2}$
$\dfrac{\partial f}{\partial y}$	$\dfrac{x}{1+x^2y^2}$,	$\dfrac{1}{2}$
$\dfrac{\partial^2 f}{\partial x^2}$	$-\dfrac{(2x)y}{(1+x^2y^2)^2}$,	$\dfrac{1}{2}$
$\dfrac{\partial^2 f}{\partial y\,\partial x}$	$\dfrac{1+x^2y^2-x(2x\,y^2)}{(1+x^2y^2)^2}=\dfrac{1-x^2y^2}{(1+x^2y^2)^2}$	0
$\dfrac{\partial^2 f}{\partial y^2}$	$\dfrac{-x\,(2x^2y)}{(1+x^2y^2)^2}$,	$\dfrac{1}{2}$

Example 8. *Obtain Taylor's expansion of* $\tan^{-1}\dfrac{y}{x}$ *about (1, 1) upto and including the second degree terms. Hence compute f (1. 1, 0.9).* (*U.P., I Sem. Winter 2005, 2002*)

Solution.

		$x = 1, y = 1$
$f(x, y)$	$\tan^{-1}\dfrac{y}{x}$	$\dfrac{\pi}{4}$
$\dfrac{\partial f}{\partial x}$	$\dfrac{1}{1+\dfrac{y^2}{x^2}}\left(-\dfrac{y}{x^2}\right)=\dfrac{-y}{x^2+y^2}$,	$-\dfrac{1}{2}$
$\dfrac{\partial f}{\partial y}$	$\dfrac{1}{1+\dfrac{y^2}{x^2}}\left(\dfrac{1}{x}\right)=\dfrac{x}{x^2+y^2}$,	$\dfrac{1}{2}$
$\dfrac{\partial^2 f}{\partial x^2}$	$\dfrac{y\,(2x)}{(x^2+y^2)^2}=\dfrac{2xy}{(x^2+y^2)^2}$,	$\dfrac{1}{2}$
$\dfrac{\partial^2 f}{\partial y^2}$	$\dfrac{-x\,(2y)}{(x^2+y^2)^2}=\dfrac{-2xy}{(x^2+y^2)^2}$,	$-\dfrac{1}{2}$
$\dfrac{\partial^2 f}{\partial y\,\partial x}$	$\dfrac{(x^2+y^2)-(x)(2x)}{(x^2+y^2)^2}=\dfrac{y^2-x^2}{(x^2+y^2)^2}$,	0

By Taylor's Theorem

$$f(x, y) = f(a, b)+\left[(x-a)\frac{\partial f}{\partial x}+(y-b)\frac{\partial f}{\partial y}\right]+\frac{1}{2!}\left[(x-a)^2\frac{\partial^2 f}{\partial x^2}\right.$$
$$\left.+2\,(x-a)\,(y-b)\frac{\partial^2 f}{\partial x\partial y}+(y-b)^2\frac{\partial^2 f}{\partial y^2}\right]+\ldots$$

Here, $a = 1, b = 1$

$$\tan^{-1}\frac{y}{x} = \frac{\pi}{4}+(x-1)\left(-\frac{1}{2}\right)+(y-1)\frac{1}{2}+\frac{1}{2!}\left[(x-1)^2\left(\frac{1}{2}\right)\right.$$
$$\left.+2\,(x-1)\,(y-1)\,(0)+(y-1)^2\left(-\frac{1}{2}\right)\right]+\ldots$$

$$\tan^{-1}\frac{y}{x} = \frac{\pi}{4} - \frac{1}{2}(x-1) + \frac{1}{2}(y-1) + \frac{1}{4}(x-1)^2 - \frac{1}{4}(y-1)^2 + .. \qquad ...(1)$$

Putting $(x-1) = 1.1 - 1 = 0.1$, $(y-1) = 0.9 - 1 = -0.1$ in (1), we get

$$f(1.1, 0.9) = \frac{\pi}{4} - \frac{1}{2}(0.1) - \frac{1}{2}(-0.1) + \frac{1}{4}(0.1)^2 - \frac{1}{4}(-0.1)^2$$

$$= 0.786 - 0.05 + 0.05 + 0.0025 - 0.0025 = 0.786 \qquad \textbf{Ans.}$$

Example 9. *Expand* $\dfrac{(x+h)(y+k)}{x+h+y+k}$ *in powers of h, k upto and inclusive of the second degree terms.* *(A.M.I.E.T.E., Summer 2001)*

Solution. $f(x+h, y+k) = \dfrac{(x+h)(y+k)}{x+h+y+k}$

$$f(x, y) = \frac{xy}{x+y}$$

$$\frac{\partial f}{\partial x} = \frac{(x+y)\,y - xy}{(x+y)^2} = \frac{y^2}{(x+y)^2}$$

$$\frac{\partial f}{\partial y} = \frac{(x+y)\,x - xy}{(x+y)^2} = \frac{x^2}{(x+y)^2} \quad \Rightarrow \quad \frac{\partial^2 f}{\partial x^2} = \frac{-2y^2}{(x+y)^3}$$

$$\frac{\partial^2 f}{\partial y \partial x} = \frac{(x+y)^2\, 2x - 2(x+y)\,x^2}{(x+y)^4} = \frac{(x+y)\,2x - 2x^2}{(x+y)^3} = \frac{2xy}{(x+y)^3}$$

$$\frac{\partial^2 f}{\partial y^2} = \frac{-2x^2}{(x+y)^3}$$

$$f(x+h, y+k) = f(x, y) + \left(h\frac{\partial}{\partial x} + k\frac{\partial}{\partial y}\right) f(x, y) + \frac{1}{2!}\left(h\frac{\partial}{\partial x} + k\frac{\partial}{\partial y}\right)^2 f(x, y) + ...$$

$$\frac{(x+h)(y+k)}{x+h+y+k} = \frac{xy}{x+y} + h\frac{y^2}{(x+y)^2} + k\frac{x^2}{(x+y)^2}$$

$$+ \frac{h^2}{2!}\frac{(-2y^2)}{(x+y)^3} + \frac{1}{2!}2hk\frac{2xy}{(x+y)^3} + \frac{1}{2!}k^2\frac{(-2x^2)}{(x+y)^3} + ...$$

$$= \frac{xy}{x+y} + \frac{hy^2}{(x+y)^2} + \frac{kx^2}{(x+y)^2} - \frac{h^2y^2}{(x+y)^3} + \frac{2hkxy}{(x+y)^3} - \frac{k^2x^2}{(x+y)^3} + ... \qquad \textbf{Ans.}$$

Example 10. *Expand* $x^2y + 3y - 2$ *in powers of* $x - 1$ *and* $y + 2$ *using Taylor's Theorem.* *(A.M.I.E.T.E., Winter 2003, A.M.I.E., Summer 2004, 2003)*

Solution. $f(x, y) = x^2y + 3y - 2$

Here $a + h = x$ and $h = x - 1$, so $a = 1$

$b + k = y$ and $k = y + 2$ so $b = -2$

Now Taylor's Theorem is

$$f(a+h, b+k) = f(a, b) + \left(h\frac{\partial f}{\partial x} + k\frac{\partial f}{\partial y}\right)_{(a,b)} + \frac{1}{2!}\left[h^2\frac{\partial^2 f}{\partial x^2} + 2hk\frac{\partial^2 f}{\partial x \partial y} + k^2\frac{\partial^2 f}{\partial y^2}\right]_{(a,b)}$$

$$+ \frac{1}{3!}\left(h^3\frac{\partial^3 f}{\partial x^3} + 3h^2k\frac{\partial^3 f}{\partial x^2 \partial y} + 3hk^2\frac{\partial^2 f}{\partial x \partial y^2} + k^3\frac{\partial^3 f}{\partial y^3}\right)_{(a,b)} + ...$$

		$x = 1, y = -2$
$f(x, y)$	$x^2y + 3y - 2,$	-10
$f_x(x, y)$	$2xy,$	-4
$f_y(x, y)$	$x^2 + 3,$	4
$f_{xx}(x, y)$	$2y,$	-4
$f_{xy}(x, y)$	$2x,$	2
$f_{yy}(x, y)$	$0,$	0
$f_{xxx}(x, y)$	$0,$	0
$f_{xxy}(x, y)$	$2,$	2
$f_{xyy}(x, y)$	$0,$	0
$f_{yyy}(x, y)$	$0,$	0

Putting the values of $f(a, b)$ etc. in Taylor's Theorem, we get

$$x^2y + 3y - 2 = -10 + [(x-1)(-4) + (y+2)(4)]$$
$$+\frac{1}{2!}[(x-1)^2(-4) + 2(x-1)(y+2)(2) + (y+2)^2(0)]$$
$$+\frac{1}{3!}[(x-1)^3(0) + 3(x-1)^2(y+2)(2) + 3(x-1)(y+2)^2(0) + (y+2)^3(0)]$$

$$x^2y + 3y - 2 = -10 - 4(x-1) + 4(y+2) - 2(x-1)^2 + 2(x-1)(y+2) + (x-1)^2(y+2)$$ **Ans.**

Example 11. *Expand x^y in powers of $(x - 1)$ and $(y - 1)$ upto the third degree terms.*

(U.P., I Sem. Winter 2003)

Solution. $f(x, y) = x^y$

Here $\quad a + h = x$, and $h = x - 1 \Rightarrow a = 1$

$\quad b + k = y$, and $k = y - 1 \Rightarrow b = 1$

Now Taylor's series is

$$f(a+h, b+k) = f(a, b) + \left(h\frac{\partial f}{\partial x} + k\frac{\partial f}{\partial y}\right) + \frac{1}{2!}\left(h^2\frac{\partial^2 f}{\partial x^2} + 2hk\frac{\partial^2 f}{\partial x\partial y} + k^2\frac{\partial^2 f}{\partial y^2}\right)$$
$$+\frac{1}{3!}\left[h^3\frac{\partial^3 f}{\partial x^3} + 3h^2k\frac{\partial^3 f}{\partial x\partial x\partial y} + 3hk^2\frac{\partial^3 f}{\partial x\,\partial y\partial y} + k^3\frac{\partial^2 f}{\partial y^3}\right]$$

$$x^y = 1 + (x-1) + 0 + \frac{1}{2!}[0 + 2(x-1)(y-1) + 0] + \frac{1}{3!}[0 + 3(x-1)^2(y-1) + 0 + 0]$$
$$= 1 + (x-1) + (x-1)(y-1) + \frac{1}{2}(x-1)^2(y-1)$$ **Ans.**

		$x = 1, y = 1$
$f(x, y)$	x^y	1
$f_x(x, y)$	$y\,x^{y-1}$	1
$f_y(x, y)$	$x^y \log x$	0
$f_{xx}(x, y)$	$y\,(y-1)\,x^{y-2}$	0
$f_{yx}(x, y)$	$x^y \cdot \frac{1}{x} + y\,x^{y-1} \log x = x^{y-1} + y\,x^{y-1} \log x$	1
$f_{yy}(x, y)$	$x^y (\log x)^2$	0
$f_{xxx}(x, y)$	$y\,(y-1)\,(y-2)\,x^{y-3}$	0
$f_{xxy}(x, y)$	$(y-1)\,x^{y-2} + y\,(y-1)\,x^{y-2} \log x + y\,x^{y-1} \cdot \frac{1}{x}$ $= (y-1)\,x^{y-2} + y\,(y-1)\,x^{y-2} \log x + y\,x^{y-2}$	1
$f_{xyy}(x, y)$	$yx^{y-1} (\log x)^2 + x^y \frac{2 \log x}{x}$ $= y\,x^{y-1} (\log x)^2 + 2x^{y-1} \log x$	0
$f_{yyy}(x, y)$	$x^y (\log x)^3$	0

EXERCISE 6.2

1. Expand e^{xy} at (1, 1) upto three terms.

Ans. $e\,[1 + (x-1) + (y-1) + \frac{1}{2!}[(x-1)^2 + 4\,(x-1)\,(y-1) + (y-1)^2]$

2. Expand y^x at (1, 1) upto second term **Ans.** $1 + (y-1) + (x-1)\,(y-1) + \ldots\ldots.$
3. Expand $e^{ax} \sin by$ in powers of x and y as far as the terms of third degree.

Ans. $by + abxy + \frac{1}{3!}\left(3a^2\, bx^2 y - b^3 y^3\right) + \ldots.$

4. Expand $(x^2 y + \sin y + e^x)$ in powers of $(x-1)$ and $(x-\pi)$.

Ans. $\pi + e + (x-1)\,(2\pi + e) + \frac{1}{2}\,(x-1)^2\,(2\pi + e) + 2\,(x-1)\,(y-\pi).$

5. Expand $(1 + x + y^2)^{1/2}$ at (1, 0). **Ans.** $\sqrt{2}\left[1 + \frac{x-1}{4} - \frac{(x-1)^2}{32} + \frac{y^2}{4} + \ldots\right]$

6. Obtain the linearised form $T(x, y)$ of the function $f(x, y) = x^2 - xy + \frac{1}{2}\,y^2 + 3$ at the point (3, 2), using the Taylor's series expansion. Find the maximum error in magnitude in the approximation $f(x, y) \approx T(x, y)$ over the rectangle R: $|x - 3| < 0.1$, $|y - 2| < 0.1$.

Ans. $8 + 4\,(x-3) - (y-2)$., Error 0.04.

7. Expand $\sin\,(x+h)\,(y+k)$ by Taylor's Theorem.

Ans. $\sin xy + h\,(x+y) \cos xy + hk \cos xy - \frac{1}{2} h^2\,(x+y)^2 \sin xy + \ldots$

Fill in the blank:

8. If $f(x) = f(0) + kf_1(0) + \frac{k^2}{2!} f_2(\theta k),\ 0 < \theta < 1$ then the value of θ when $k = 1$ and $f(x) = (1-x)^{3/2}$ is given as .... *(U.P. Ist Semester, Dec 2008)*

CHAPTER 7

DOUBLE INTEGRALS

7.1 DOUBLE INTEGRATION

We know that

$$\int_a^b f(x)\,dx = \lim_{\substack{n\to\infty \\ \delta x\to 0}} [f(x_1)\,\delta x_1 + f(x_2)\,\delta x_2 + f(x_3)\,\delta x_3 + \ldots + f(x_n)\,\delta x_n]$$

Let us consider a function $f(x, y)$ of two variable x and y defined in the finite region A of xy-plane. Divide the region A into elementary areas.

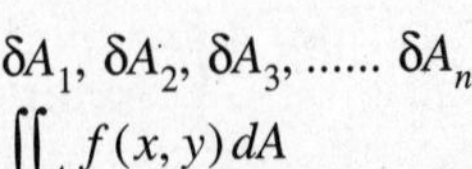

$$\delta A_1,\ \delta A_2,\ \delta A_3,\ \ldots\ldots\ \delta A_n$$

Then $$\iint_A f(x, y)\,dA$$

$$= \lim_{\substack{n\to\infty \\ \delta A\to 0}} [f(x_1, x_1)\,\delta A_1 + f(x_2, y_2)\,\delta A_2 + \ldots.. + f(x_n, y_n)\,\delta A_n]$$

7.2 EVALUATION OF DOUBLE INTEGRAL

Double integral over region A may be evaluated by two successive integrations.

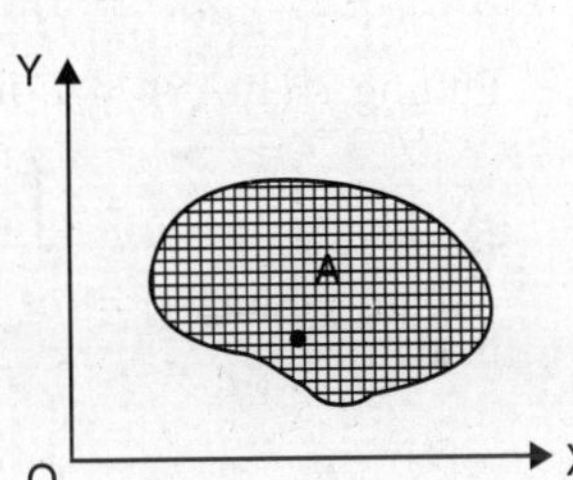
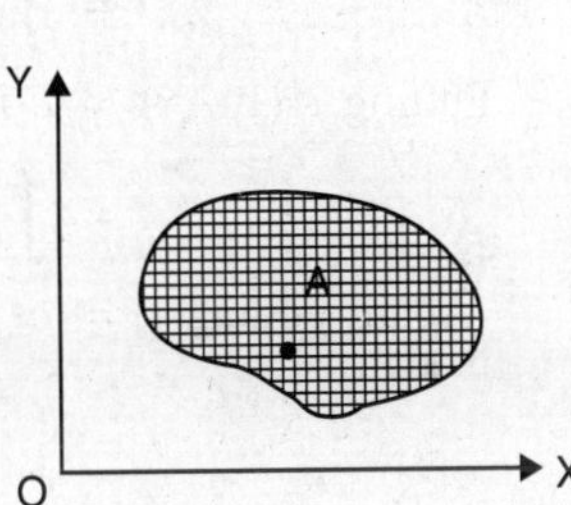

If A is described as $f_1(x) \le y \le f_2(x)$ $[y_1 \le y \le y_2]$

and $$a \le x \le b,$$

Then $$\iint_A f(x, y)\,dA = \int_a^b \int_{y_1}^{y_2} f(x, y)\,dx\,dy$$

(1) First Method

$$\iint_A f(x, y)\,dA = \int_a^b \boxed{\int_{y_1}^{y_2} f(x, y)\,dy}\,dx$$

$f(x, y)$ is first integratred with respect to y treating x as constant between the limits a and b.

In the region we take an elementary area $\delta x \delta y$. Then integration w.r.t y (x keeping constant). converts small rectangle $\delta x \delta y$ into a strip PQ ($y\,\delta x$). While the integration of the result w.r.t. x corresponding to the sliding to the strip PQ, from AD to BC covering the while region $ABCD$.

Second method

$$\iint_A f(x, y)\,dxdy = \int_c^d \boxed{\int_{x_1}^{x_2} f(x, y)\,dx}\,dy$$

Here $f(x,y)$ is first integrated w.r.t x keeping y constant between the limits x_1 and x_2 and then the resulting expression is integrated with respect to y between the limits c and d

Take a small area $\delta x \delta y$. The integration w.r.t. x between the limits x_1, x_2 keeping y fixed indicates that integration is done, along PQ. Then the integration of result w.r.t y corresponds to sliding the strips PQ from BC to AD covering the whole region $ABCD$.

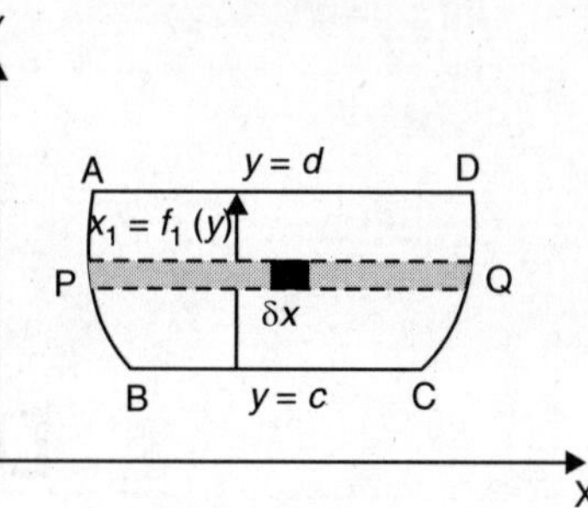

Note. For constant limits, it does not matter whether we first integrate w.r.t x and then w.r.t y or vice versa.

Example 1. *Evaluate* $\int_0^1 \int_0^x (x^2+y^2)\, dA$, *where dA indicates small area in xy-plane.*

(*Gujarat, I Semester, Jan. 2009*)

Solution. Let $I = \int_0^1 \int_0^x (x^2+y^2)\, dy\, dx = \int_0^1 \left[x^2 y + \frac{y^3}{3}\right]_0^x dx$

$$= \int_0^1 \left[x^2 (x-0) + \frac{1}{3}(x^3-0)\right] dx = \int_0^1 \left[x^3 + \frac{x^3}{3}\right] dx$$

$$= \int_0^1 \frac{4}{3} x^3\, dx = \frac{4}{3}\left[\frac{x^4}{4}\right]_0^1 = \frac{1}{3}[1-0] = \frac{1}{3} \text{ sq. units.} \qquad \textbf{Ans.}$$

Example 2. *Evaluate* $\int_{-1}^{1} \int_0^{1-x} x^{1/3}\, y^{-1/2}\, (1-x-y)^{1/2}\, dy\, dx$. (*M.U., II Semester 2002*)

Solution. Here, we have

$$I = \int_{-1}^{1} \int_0^{1-x} x^{1/3}\, y^{-1/2}\, (1-x-y)^{1/2}\, dy\, dx \qquad ...(1)$$

Putting $(1-x) = c$ in (1), we get

$$I = \int_{-1}^{1} x^{1/3}\, dx \int_0^c y^{-1/2}\, (c-y)^{1/2}\, dy \qquad ...(2)$$

Again putting $y = ct \Rightarrow dy = c\, dt$ in (2), we get

$$I = \int_{-1}^{1} x^{\frac{1}{3}}\, dx \int_0^1 c^{-\frac{1}{2}}\, t^{-\frac{1}{2}}\, (c - ct)^{\frac{1}{2}}\, c\, dt$$

$$= \int_{-1}^{1} x^{1/3}\, dx \int_0^1 c^{-1/2}\, t^{-1/2}\, c^{1/2}\, (1-t)^{1/2}\, c\, dt$$

$$= \int_{-1}^{1} c\, x^{1/3}\, dx \int_0^c t^{-1/2} (1-t)^{1/2}\, dt = \int_{-1}^{1} c\, x^{1/3}\, dx \int_0^1 t^{1/2-1}\, (1-t)^{3/2-1}\, dt$$

$$= \int_{-1}^{1} c\, x^{\frac{1}{3}}\, dx\, \beta\left(\frac{1}{2}, \frac{3}{2}\right) \qquad \left[\int_0^1 x^{l-1} (1-x)^{m-1}\, dx = \beta(l, m)\right]$$

$$= \int_{-1}^{1} c\, x^{1/3}\, dx \frac{\left\lceil\frac{1}{2}\right.\left\lceil\frac{3}{2}\right.}{\left\lceil\frac{1}{2}+\frac{3}{2}\right.} = \int_{-1}^{1} c\, x^{1/3}\, dx \frac{\left\lceil\frac{1}{2}\right. \cdot \frac{1}{2}\left\lceil\frac{1}{2}\right.}{\lceil 2} = \int_{-1}^{1} c\, x^{1/3}\, dx \frac{\sqrt{\pi}\,\frac{1}{2}\sqrt{\pi}}{1}$$

$$= \int_{-1}^{1} c\, x^{1/3} \frac{\pi}{2}\, dx = \frac{\pi}{2} \int_{-1}^{1} x^{1/3} . c\, dx$$

Putting the value of c, we get

$$I = \frac{\pi}{2}\int_{-1}^{1} x^{1/3}(1-x)\,dx = \frac{\pi}{2}\int_{-1}^{1}(x^{1/3}-x^{4/3})\,dx = \frac{\pi}{2}\left[\frac{x^{4/3}}{\frac{4}{3}} - \frac{x^{7/3}}{\frac{7}{3}}\right]_{-1}^{1}$$

$$= \frac{\pi}{2}\left[\frac{3}{4}(1) - \frac{3}{7}(1) - \frac{3}{4}(-1) + \frac{3}{7}(-1)\right] = \frac{\pi}{2}\left[\frac{9}{14}\right] = \frac{9\pi}{28} \qquad \textbf{Ans.}$$

Example 3. *Evaluate* $\iint_R (x+y)\,dy\,dx$, *R is the region bounded by* $x = 0$, $x = 2$, $y = x$, $y = x + 2$. *(Gujarat, I Semester, Jan. 2009)*

Solution. Let $I = \iint_R (x+y)\,dy\,dx$

The limits are $x = 0$, $x = 2$, $y = x$ and $y = x + 2$

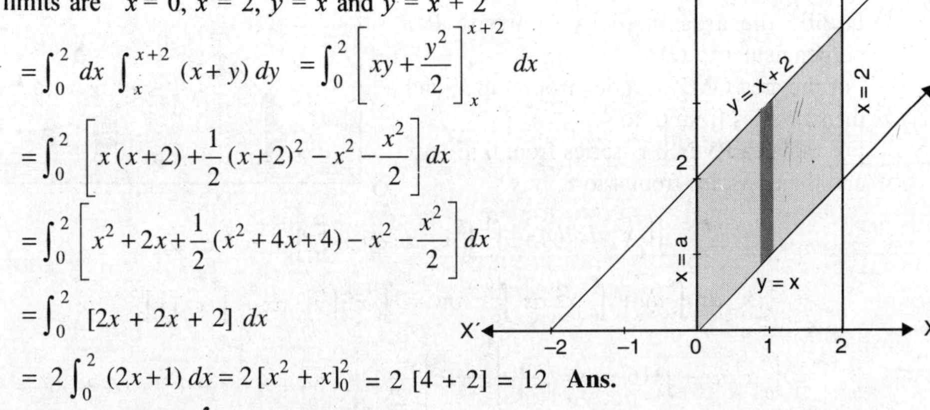

$$I = \int_0^2 dx \int_x^{x+2}(x+y)\,dy = \int_0^2\left[xy + \frac{y^2}{2}\right]_x^{x+2} dx$$

$$= \int_0^2\left[x(x+2) + \frac{1}{2}(x+2)^2 - x^2 - \frac{x^2}{2}\right]dx$$

$$= \int_0^2\left[x^2 + 2x + \frac{1}{2}(x^2+4x+4) - x^2 - \frac{x^2}{2}\right]dx$$

$$= \int_0^2 [2x + 2x + 2]\,dx$$

$$= 2\int_0^2 (2x+1)\,dx = 2[x^2 + x]_0^2 = 2\,[4 + 2] = 12 \quad \textbf{Ans.}$$

Example 4. *Evaluate* $\iint_R xy\,dx\,dy$

where R is the quadrant of the circle $x^2 + y^2 = a^2$ *where* $x \geq 0$ *and* $y \geq 0$.

(A.M.I.E.T.E, Summer 2004, 1999)

Solution. Let the region of integration be the first quadrant of the circle OAB.

$$\iint_R xy\,dx\,dy \;(x^2+y^2=a^2,\; y=\sqrt{a^2-x^2})$$

First we integrate w.r.t. y and then w.r.t. x.

The limits for y are 0 and $\sqrt{a^2-x^2}$ and for x, 0 to a.

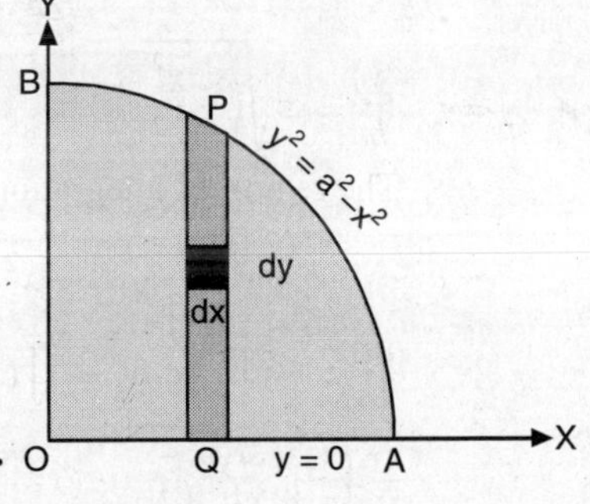

$$= \int_0^a x\,dx\int_0^{\sqrt{a^2-x^2}} y\,dy = \int_0^a x\,dx\left[\frac{y^2}{2}\right]_0^{\sqrt{a^2-x^2}}$$

$$= \frac{1}{2}\int_0^a x(a^2-x^2)\,dx = \frac{1}{2}\left[\frac{a^2x^2}{2} - \frac{x^4}{4}\right]_0^a = \frac{a^4}{8} \quad \textbf{Ans.}$$

Example 5. *Evaluate* $\iint_s \sqrt{xy - y^2}\,dy\,dx$,

where S is a triangle with vertices (0, 0), (10, 1) and (1, 1).

Solution. Let the verties of a triangle OBA be (0, 0) (10, 1) and (1, 1).

Equation of OA is $x = y$.

Equation of OB is $x = 10\,y$.

The region of ΔOBA, given by the limits

$y \leq x \leq 10\,y$ and $0 \leq y \leq 1$.

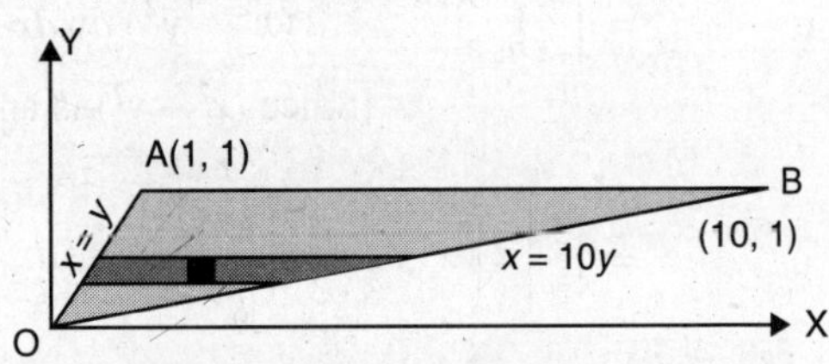

$$\iint_s \sqrt{xy-y^2}\,dy\,dx = \int_0^1 dy\int_y^{10y}(xy-y^2)^{1/2}\,dx$$

$$= \int_0^1 dy \left[\frac{2}{3}\frac{1}{y}(xy - y^2)^{3/2}\right]_y^{10y} = \int_0^1 \frac{2}{3}\frac{1}{y}(9y^2)^{3/2}\, dy = 18\int_0^1 y^2\, dy$$

$$= 18\left[\frac{y^3}{3}\right]_0^1 = \frac{18}{3} = 6 \qquad \textbf{Ans.}$$

Example 6. *Evaluate* $\iint_A x^2\, dx\, dy$, *where A is the region in the first quadrant bounded by the hyperbola xy = 16 and the lines y = x, y = 0 and x = 8.* (*A.M.I.E., Summer 2001*)

Solution. The line *OP*, $y = x$ and the curve *PS*, $xy = 16$ intersect at (4, 4).
The line *SN*, $x = 8$ intersects the hyperbola at *S* (8, 2). $y = 0$ is *x*-axis.
The area *A* is shown shaded.

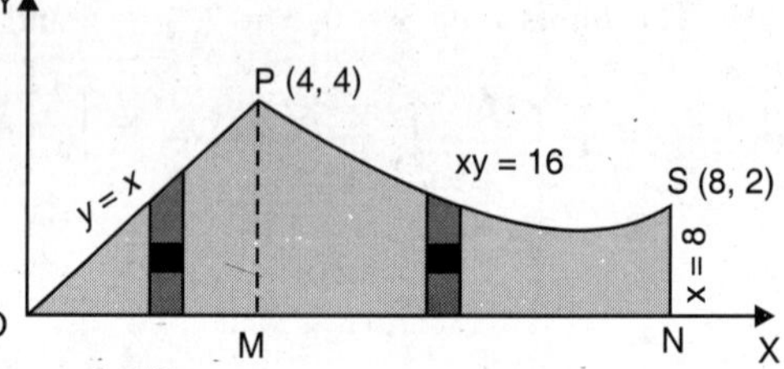

Divide the area in to two part by *PM* perpendicular to *OX*.
For the area OMP, *y* varies from 0 to *x*, and then *x* varies from 0 to 4.
For the area PMNS, *y* -series from 0 to 16/*x* and then *x* varies from 4 to 8.

$$\therefore \quad \iint_A x^2\, dx\, dy = \int_0^4\int_0^x x^2\, dx\, dy + \int_4^8\int_0^{16/x} x^2\, dx\, dy$$

$$= \int_0^4 x^2\, dx \int_0^x dy + \int_4^8 x^2\, dx \int_0^{16/x} dy = \int_0^4 x^2\,[y]_0^x\, dx + \int_4^8 x^2\,[y]_0^{16/x}\, dx$$

$$= \int_0^4 x^3\, dx + \int_4^8 16\, x\, dx = \left[\frac{x^4}{4}\right]_0^4 + 16\left[\frac{x^2}{2}\right]_4^8 = 64 + 8\,(8^2 - 4^2) = 64 + 384 = 448. \ \textbf{Ans.}$$

Example 7. *Evaluate* $\iint (x + y)^2\, dx\, dy$ *over the area bounded by the ellipse* $\frac{x^2}{a^2} + \frac{y^2}{b^2} = 1$

(*U.P. Ist Semester Compartment 2004*)

Solution. For the ellipse $\frac{x^2}{a^2} + \frac{y^2}{b^2} = 1$

$$\Rightarrow \quad \frac{y}{b} = \pm\sqrt{1 - \frac{x^2}{a^2}} \quad \Rightarrow \quad y = \pm\frac{b}{a}\sqrt{a^2 - x^2}$$

$\therefore$ The region of integration can be expressed as

$$-a \le x \le a \text{ and } -\frac{b}{a}\sqrt{a^2 - x^2} \le y \le \frac{b}{a}\sqrt{a^2 - x^2}$$

$$\therefore \quad \iint (x + y)^2\, dx\, dy = \iint (x^2 + y^2 + 2xy)\, dx\, dy$$

$$= \int_{-a}^{a}\int_{(-b/a)\sqrt{a^2 - x^2}}^{b/a\sqrt{a^2 - x^2}} (x^2 + y^2 + 2xy)\, dx\, dy$$

$$= \int_{-a}^{a}\int_{(-b/a)\sqrt{a^2 - x^2}}^{b/a\sqrt{a^2 - x^2}} (x^2 + y^2)\, dx\, dy + \int_{-a}^{a}\int_{(-b/a)\sqrt{a^2 - x^2}}^{b/a\sqrt{a^2 - x^2}} 2xy\, dy\, dx$$

$$= \int_{-a}^{a}\int_0^{b/a\sqrt{a^2 - x^2}} 2\,(x^2 + y^2)\, dy\, dx + 0$$

[Since $(x^2 + y^2)$ is an even function of *y* and $2xy$ is an odd function of *y*]

$$= \int_{-a}^{a}\left[2\left(x^2 y + \frac{y^3}{3}\right)\right]_0^{\left(\frac{b}{a}\right)\sqrt{a^2 - x^2}} dx$$

$$= 2\int_{-a}^{a}\left[x^2 \times \frac{b}{a}\sqrt{a^2-x^2} + \frac{1}{3}\frac{b^3}{a^3}(a^2-x^2)^{3/2}\right]dx$$

$$= 4\int_0^a\left[\frac{b}{a}x^2\sqrt{a^2-x^2} + \frac{b^3}{3a^3}(a^2-x^2)^{3/2}\right]dx$$

[On putting $x = a \sin\theta$ and $dx = a\cos\theta\, d\theta$]

$$= 4\int_0^{\frac{\pi}{2}}\left(\frac{b}{a}.a^2\sin^2\theta.a\cos\theta + \frac{b^3}{3a^3}a^3\cos^3\theta\right)\times a\cos\theta\, d\theta$$

$$= 4\int_0^{\frac{\pi}{2}}\left(a^3 b\sin^2\theta\cos^2\theta + \frac{ab^3}{3}\cos^4\theta\right)d\theta = 4\left[a^3 b.\frac{1}{4}.\frac{1}{2}.\frac{\pi}{2} + \frac{ab^3}{3}.\frac{3}{4}.\frac{1}{2}.\frac{\pi}{2}\right]$$

$$= \frac{\pi}{4}(a^3 b + ab^3) = \frac{\pi}{4}ab(a^2+b^2)$$ **Ans.**

Example 8. *Evaluate* $\iint (x^2+y^2)\,dx\,dy$ *throughout the area enclosed by the curves* $y = 4x$, $x + y = 3$, $y = 0$ *and* $y = 2$.

Solution. Let OC represent $y = 4x$; BD, $x + y = 3$; OB, $y = 0$, and CD, $y = 2$. The given integral is to be evaluated over the area A of the trapezium $OCDB$.

Area $OCDB$ consists of area OCE, area $ECDF$ and area FDB.

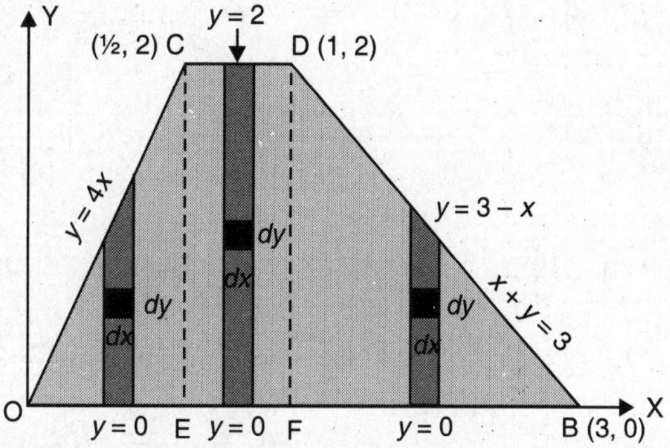

The co-ordinates of C, D and B are $\left(\frac{1}{2}, 2\right)$ (1, 2) and (3, 0) respectively.

$$\therefore \quad \iint_A (x^2+y^2)\,dy\,dx$$

$$= \iint_{OCE}(x^2+y^2)\,dy\,dx + \iint_{ECDE}(x^2+y^2)\,dy\,dx + \iint_{FDB}(x^2+y^2)\,dy\,dx$$

$$= \underbrace{\int_0^{1/2}dx\int_0^{4x}(x^2+y^2)\,dy}_{I_1} + \underbrace{\int_{1/2}^{1}dx\int_0^{2}(x^2+y^2)\,dy}_{I_2} + \underbrace{\int_1^3 dx\int_0^{3-x}(x^2+y^2)\,dy}_{I_3}$$

Now, $$I_1 = \int_0^{1/2}dx\int_0^{4x}(x^2+y^2)\,dy = \int_0^{1/2}\left[x^2y + \frac{y^3}{3}\right]_0^{4x}dx = \int_0^{1/2}\frac{76}{3}x^3\,dx$$

$$= \frac{76}{3}\int_0^{1/2}x^3\,dx = \frac{76}{3}\left[\frac{x^4}{4}\right]_0^{1/2} = \frac{76}{3}\left[\frac{1}{4}.\frac{1}{16}\right] = \frac{19}{48}$$

$$I_2 = \int_{1/2}^{1}dx\int_{1/2}^{1}(x^2+y^2)\,dy = \int_{1/2}^{1}\left[x^2y + \frac{y^3}{3}\right]_0^2 dx = \int_{1/2}^{1}\left(2x^2 + \frac{8}{3}\right)dx$$

$$= \left[\frac{2x^3}{3} + \frac{8}{3}x\right]_{1/2}^{1} = \left[\left(\frac{2}{3}+\frac{8}{3}\right) - \left(\frac{2}{3}.\frac{1}{8} + \frac{8}{3}.\frac{1}{2}\right)\right] = \frac{23}{12}$$

$$I_3 = \int_1^3 dx\int_0^{3-x}(x^2+y^2)\,dy = \int_1^3\left[x^2y + \frac{y^3}{3}\right]_0^{3-x}dx = \int_0^3\left[x^2(3-x) + \frac{(3-x)^3}{3}\right]dx$$

$$= \int_1^3\left[3x^2 - x^3 + \frac{(3-x)^3}{3}\right]dx = \left[x^3 - \frac{x^4}{4} - \frac{(3-x)^4}{3}\right]_1^3$$

$$= \left[27 - \frac{81}{4} - 0 - 1 + \frac{1}{4} + \frac{16}{12}\right] = \frac{22}{3}$$

$$\therefore \quad \int_A \int (x^2 + y^2)\, dy\, dx = I_1 + I_2 + I_3 = \frac{19}{48} + \frac{23}{12} + \frac{22}{3} = \frac{463}{48} = 9\frac{31}{48}. \qquad \textbf{Ans.}$$

EXERCISE 7.1

Evaluate

1. $\int_0^2 \int_0^{x^2} e^{\frac{y}{x}}\, dy\, dx$ **Ans.** $e^2 - 1$

2. $\int_0^a \int_0^{\sqrt{ay}} xy\, dx\, dy$ **Ans.** $\frac{a^4}{6}$

3. $\int_0^a \int_0^{\sqrt{a^2 - y^2}} dx\, dy$ **Ans.** $\frac{\pi a^2}{4}$

4. $\int_0^1 \int_{y^2}^{y} (1 + xy^2)\, dx\, dy$ **Ans.** $\frac{41}{210}$

5. $\int_0^{2a} \int_0^{\sqrt{2ax - x}} xy\, dy\, dx$ **Ans.** $\frac{2a^4}{3}$

6. $\int_0^{2a} \int_0^{\sqrt{2ax - x^2}} x^2\, dy\, dx$ **Ans.** $\frac{5\pi a^4}{8}$

7. $\int_0^{a} \int_0^{\sqrt{a^2 - x^2}} \sqrt{a^2 - x^2 - y^2}\, dy\, dx$ **Ans.** $\frac{\pi a^3}{4}$

8. $\int_0^{1} \int_0^{\sqrt{\frac{1}{2}(1 - y^2)}} \frac{dx\, dy}{\sqrt{1 - x^2 - y^2}}$ **Ans.** $\frac{\pi}{4}$

9. $\int_0^{a} \int_0^{\sqrt{a^2 - x^2}} \frac{dx\, dy}{(1 + e^y)\sqrt{a^2 - x^2 - y^2}}$ **Ans.** $\frac{\pi}{2} \log \frac{2e^a}{1 + e^a}$

10. $\int_0^{a} \int_y^{a} \frac{x\, dx\, dy}{\sqrt{x^2 + y^2}}$ **Ans.** $\frac{a^2}{2} \log(\sqrt{2} + 1)$

11 $\int_{x=0}^{1} \int_{y=0}^{2} (x^2 + 3xy^2)\, dx\, dy$ *(A.M.I.E.T.E., June 2009)* **Ans.** $\frac{14}{3}$

12. $\iint_A (5 - 2x - y)\, dx\, dy$, where A is given by $y = 0$, $x + 2y = 3$, $x = y^2$. **Ans.** $\frac{217}{60}$

13. $\iint_A xy\, dx\, dy$, where A is given by $x^2 + y^2 - 2x = 0$, $y^2 = 2x$, $y = x$. **Ans.** $\frac{7}{12}$

14. $\iint_A \sqrt{4x^2 - y^2}\, dx\, dy$, where A is the triangle given by $y = 0$, $y = x$ and $x = 1$. **Ans.** $\frac{1}{3}\left(\frac{\pi}{3} + \frac{\sqrt{3}}{2}\right)$

15. $\iint_R x^2\, dx\, dy$, where R is the two-dimensional region bounded by the curves $y = x$ and $y = x^2$. **Ans.** $\frac{1}{20}$

16. $\iint_A \sqrt{xy(1 + x - y)}\, dx\, dy$ where A is the area bounded by $x = 0$, $y = 0$ and $x + y = 1$. **Ans.** $\frac{2\pi}{105}$

7.3 EVALUATION OF DOUBLE INTEGRALS IN POLAR CO-ORDINATES

We have to evalaute $\int_{\theta_1}^{\theta_2} \int_{r_1(\theta)}^{r_2(\theta)} f(r, \theta)\, dr\, d\theta$ over the region bounded by the staight lines $\theta = \theta_1$ and $\theta = \theta_2$ and the curves $r = r_1$ (q) and $r = r_2$ (θ). We first integrate with respect to r between the limits $r = r_1(\theta)$ and $r = r_2(\theta)$ and taking θ as constant. Then the resulting expression is integrated with respect to θ between the limits $\theta = \theta_1$ and $\theta = \theta_2$.

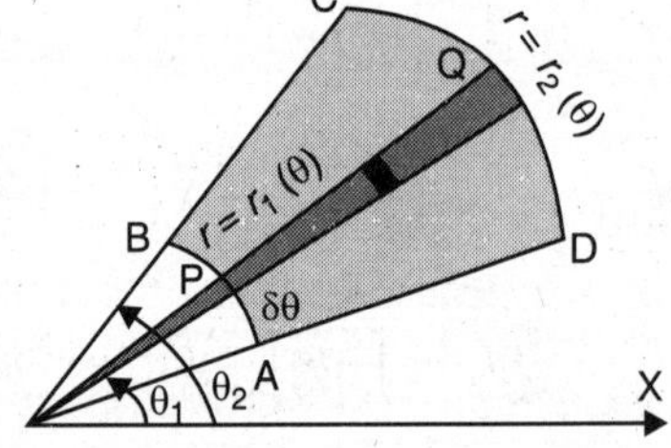

The area of integration is $ABCD$. On integrating first with respect to r, the strip extends from P to Q and the integration with respect to θ means the rotation ot this strip PQ from AD to BC.

Example 9. *Evaluate* $\int_0^{\pi/2}\left[\int_0^{a\cos\theta} r\sqrt{a^2-r^2}\,dr\right]d\theta.$

Solution. $I = \int_0^{\pi/2}\left[\int_0^{a\cos\theta} r\sqrt{a^2-r^2}\,dr\right]d\theta. = \int_0^{\frac{\pi}{2}}\left[\int_0^{a\cos\theta} -\frac{1}{2}(a^2-r^2)^{\frac{1}{2}}(-2r)\,dr\right]d\theta.$

$$= \int_0^{\pi/2}\left[-\frac{1}{2}\cdot\frac{(a^2-r^2)^{3/2}}{3/2}\right]_0^{a\cos\theta} d\theta. = -\frac{1}{3}\int_0^{\frac{\pi}{2}}\left[(a^2-a^2\cos^2\theta)^{\frac{3}{2}}-(a^2)^{\frac{3}{2}}\right]d\theta.$$

$$= -\frac{1}{3}\int_0^{\frac{\pi}{2}}\left[a^2(1-\cos^2\theta)^{\frac{3}{2}}-a^3\right]d\theta.$$

$$= -\frac{1}{3}\int_0^{\pi/2}(a^3\sin^3\theta-a^3)\,d\theta = -\frac{a^3}{3}\left[\frac{2}{3}-\frac{\pi}{2}\right] = \frac{a^3}{18}(3\pi-4).$$ **Ans.**

Example 10. *Evaluate* $\int_0^{\frac{\pi}{2}}\int_{a(1-\cos\theta)}^{a} r^2\,dr\,d\theta$

θ = π/2, r = a, θ = 0, O, r = a(1 − cos θ)

Solution. $\int_0^{\frac{\pi}{2}} d\theta\int_{a(1-\cos\theta)}^{a} r^2\,dr$

$$= \int_0^{\frac{\pi}{2}} d\theta\left[\frac{r^3}{3}\right]_{a(1-\cos\theta)}^{a} = \int_0^{\frac{\pi}{2}} d\theta\left(\frac{a^3}{3}-\frac{a^3(1-\cos\theta)^3}{3}\right)$$

$$= \frac{a^3}{3}\int_0^{\frac{\pi}{2}}[1-(1-\cos\theta)^3]\,d\theta = \frac{a^3}{3}\int_0^{\frac{\pi}{2}}[1-(1-3\cos\theta+3\cos^2\theta-\cos^3\theta]\,d\theta$$

$$= \frac{a^3}{3}\int_0^{\frac{\pi}{2}}(3\cos\theta-3\cos^2\theta+\cos^3\theta)\,d\theta$$

$$= \frac{a^3}{3}\left[[3\sin\theta]_0^{\frac{\pi}{2}}-3\frac{1}{2}\frac{\pi}{2}+\frac{2}{3.1}\right] = \frac{a^3}{3}\left[3-\frac{3\pi}{4}+\frac{2}{3}\right] = \frac{a^3}{36}[44-9\pi]$$ **Ans.**

Example 11. *Evaluate* $\iint r^3\,dr\,d\theta$, *over the area bounded between the circles* $r = 2\cos\theta$ *and* $r = 4\cos\theta$. *(K. University, Dec. 2008)*

Solution. Here, we have

$r = 2\cos\theta$ circle ... (1)

$r = 4\cos\theta$ circle ... (2)

Let $I = \iint r^3\,dr\,d\theta$ ... (3)

Y, r = 4 cos θ, r = 2 cos θ, O, θ, X, Y′

$$= \int_{-\frac{\pi}{2}}^{\frac{\pi}{2}} d\theta\int_{2\cos\theta}^{4\cos\theta} r^3\,dr$$

$$= \int_{-\frac{\pi}{2}}^{\frac{\pi}{2}} d\theta\left[\frac{r^4}{4}\right]_{2\cos\theta}^{4\cos\theta}$$

$$= \frac{1}{4}\int_{-\frac{\pi}{2}}^{\frac{\pi}{2}} d\theta\,(256\cos^4\theta-16\cos^4\theta) = \frac{240}{4}\int_{-\frac{\pi}{2}}^{\frac{\pi}{2}}\cos^4\theta\,d\theta$$

$$= 120\int_0^{\frac{\pi}{2}}\cos^4\theta\,d\theta$$ [$\cos^4\theta$ is an even function]

$$= 120 \frac{\Gamma\left(\frac{4+1}{2}\right)\Gamma\left(\frac{0+1}{2}\right)}{2\,\Gamma\left(\frac{4+1+0+1}{2}\right)} = 60\,\frac{\Gamma\left(\frac{5}{2}\right)\Gamma\left(\frac{1}{2}\right)}{\Gamma 3}$$

$$= 60\left[\frac{3}{2}\left(\frac{1}{2}\right)\Gamma\left(\frac{1}{2}\right)\right]\Gamma\left(\frac{1}{2}\right) = \frac{45}{2}(\sqrt{\pi})(\sqrt{\pi}) = \frac{45\pi}{2}$$ **Ans.**

Example 12. *Transform the integral to cartesian form and hence evaluate* $\int_0^{\pi}\int_0^{a} r^3 \sin\theta\cos\theta\, dr\, d\theta.$ *(M.U., II Semester 2000)*

Solution. Here, we have

$$\int_0^{\pi}\int_0^{a} r^3 \sin\theta\cos\theta\, dr\, d\theta \quad \ldots(1)$$

Here the region *i.e.*, semicircle *ABC* of integration is bounded by $r = 0$, *i.e.*, x-axis.

$r = a$ *i.e.*, circle, $\theta = 0$ and $\theta = \pi$ *i.e.*, x-axis in the second quadrant.

$$\int\int (r\sin\theta)(r\cos\theta)(r\, d\theta\, dr)$$

Putting $x = r\cos\theta$, $y = r\sin\theta$, $dx\, dy = r\, d\theta\, dr$ in (1), we get

$$\int_{-a}^{a}\int_0^{\sqrt{a^2-x^2}} xy\, dy\, dx = \int_{-a}^{a} x\, dx \int_0^{\sqrt{a^2-x^2}} y\, dy$$

$$= \int_{-a}^{a} x\, dx \left[\frac{y^2}{2}\right]_0^{\sqrt{a^2-x^2}} = \int_{-a}^{a} x\, dx\, \frac{(a^2-x^2)}{2}$$

$$= \frac{1}{2}\int_{-a}^{a} (a^2 x - x^3)\, dx = 0$$ **Ans.** $\left[\text{Since } f(x) \text{ is odd function } \int_{-a}^{a} f(x)\, dx = 0\right]$

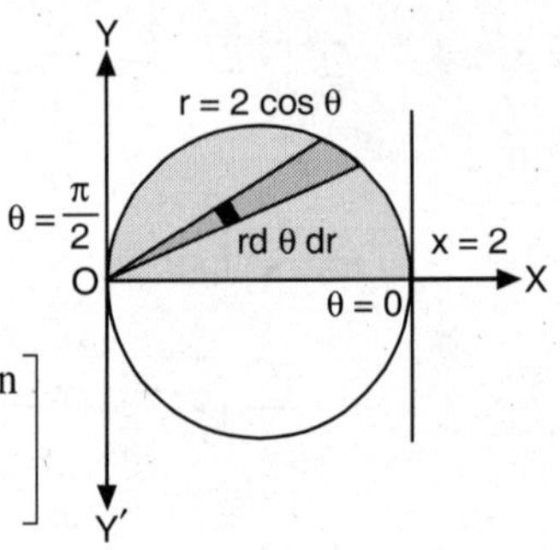

Example 13. *Evaluate* $\int_0^2\int_0^{\sqrt{2x-x^2}} (x^2+y^2)\, dy\, dx$

Solution. $\int_0^2\int_0^{\sqrt{2x-x^2}} (x^2+y^2)\, dy\, dx$

Limits of $\quad y = \sqrt{2x-x^2} \Rightarrow y^2 = 2x - x^2 \Rightarrow x^2 + y^2 - 2x = 0 \quad \ldots(1)$

(1) represents a circle whose centre is (1, 0) and radius = 1.

Lower limit of y is 0 *i.e.*, x-axis.

Region of integration is upper half circle.

Let us convert (1) into polar co-ordinate by putting

$$x = r\cos\theta,\ y = r\sin\theta$$
$$r^2 - 2r\cos\theta = 0 \Rightarrow r = 2\cos\theta$$

Limits of r are 0 to $2\cos\theta$

Limits of θ are 0 to $\frac{\pi}{2}$

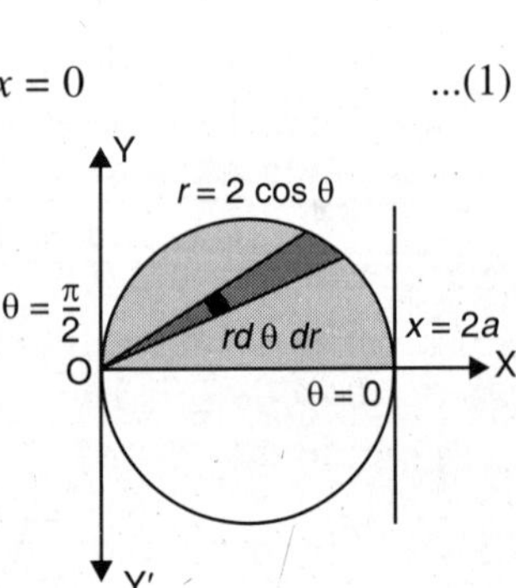

$$\int_0^2\int_0^{\sqrt{2x-x^2}} (x^2+y^2)\, dy\, dx = \int_0^{\frac{\pi}{2}}\int_0^{2\cos\theta} r^2 (r\, d\theta\, dr) = \int_0^{\frac{\pi}{2}} d\theta \int_0^{2\cos\theta} r^3\, dr = \int_0^{\frac{\pi}{2}} d\theta \left[\frac{r^4}{4}\right]_0^{2\cos\theta}$$

$$= 4\int_0^{\frac{\pi}{2}} \cos^4\theta\, d\theta = 4\times\frac{3\times 1\times\pi}{4\times 2\times 2} = \frac{3\pi}{4}$$ **Ans.**

Example 14. *Evaluate* $\int_0^2 \int_0^{\sqrt{2x-x^2}} \frac{x\,dy\,dx}{\sqrt{x^2+y^2}}$ *by changing to polar coordinates.*

Solution. In the given integral, y varies from 0 to $\sqrt{2x-x^2}$ and x varies from 0 to 2.

$$y = \sqrt{2x - x^2}$$

$$\Rightarrow \quad y^2 = 2x - x^2$$

$$\Rightarrow \quad x^2 + y^2 = 2x$$

In polar co-ordinates, we have $r^2 = 2r\cos\theta \quad \Rightarrow \quad \mathbf{r = 2\cos\theta}$.

$\therefore$ For the region of integarion, r varies from 0 to $2\cos\theta$ and θ varies from 0 to $\frac{\pi}{2}$.

In the given integral, replacing x by $r\cos\theta$, y by $r\sin\theta$, $dy\,dx$ by $r\,dr\,d\theta$, we have

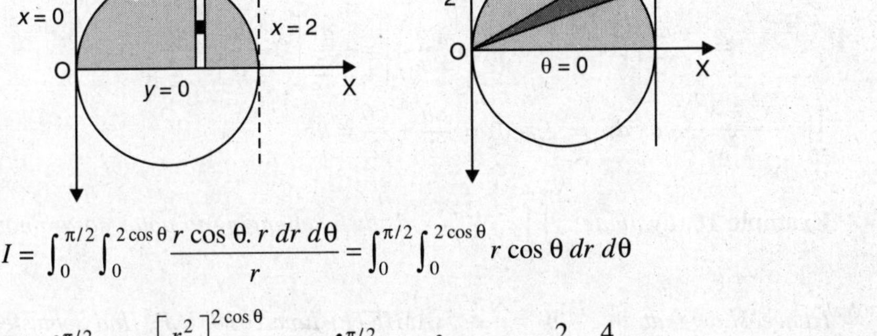

$$I = \int_0^{\pi/2} \int_0^{2\cos\theta} \frac{r\cos\theta . r\,dr\,d\theta}{r} = \int_0^{\pi/2} \int_0^{2\cos\theta} r\cos\theta\,dr\,d\theta$$

$$= \int_0^{\pi/2} \cos\theta \left[\frac{r^2}{2}\right]_0^{2\cos\theta} d\theta = \int_0^{\pi/2} 2\cos^3\theta\,d\theta = 2.\frac{2}{3} = \frac{4}{3}. \qquad \textbf{Ans.}$$

Example 15. *Evaluate* $\iint \frac{(x^2+y^2)^2}{x^2y^2}\,dx\,dy$ *over the area common to* $x^2 + y^2 = ax$ *and* $x^2 + y^2 = by$, *a, b > 0.* *(M.U. II Semester 2008, 2003, 2002)*

Solution. The boundary of area of integration are

$$x^2 + y^2 = ax \quad \Rightarrow \quad r^2 = ar\cos\theta \quad \Rightarrow \quad r = a\cos\theta$$

and $$x^2 + y^2 = by \quad \Rightarrow \quad r^2 = br\sin\theta \quad \Rightarrow \quad r = b\sin\theta$$

The region of integration is bounded by $r = a\cos\theta$ and $r = b\sin\theta$.

Point of intersection is given by

$$ra\cos\theta = rb\sin\theta$$

$$\tan\theta = \frac{a}{b} \quad \Rightarrow \quad \theta = \tan^{-1}\frac{a}{b}$$

And $$\frac{(x^2+y^2)^2}{x^2y^2} = \frac{r^4}{r^4\sin^2\theta\cos^2\theta} = \frac{1}{\sin^2\theta\cos^2\theta}$$

$$dx\,dy = r\,dr\,d\theta$$

$$\iint \frac{(x^2+y^2)^2}{x^2y^2}\,dx\,dy = \int_0^{\tan^{-1}\frac{a}{b}} \int_0^{b\sin\theta} \frac{1}{\sin^2\theta\cos^2\theta}\,r\,dr\,d\theta + \int_{\tan^{-1}\frac{a}{b}}^{\frac{\pi}{2}} \int_0^{a\cos\theta} \frac{1}{\sin^2\theta\cos^2\theta}\,r\,dr\,d\theta$$

$$= I_1 + I_2 \text{ (say)}$$

$$I_1 = \int_0^{\tan^{-1}\frac{a}{b}} \int_0^{b\sin\theta} \frac{1}{\sin^2\theta\cos^2\theta}\, r\,dr\,d\theta = \int_0^{\tan^{-1}\frac{a}{b}} \frac{1}{\sin^2\theta\cos^2\theta}\,d\theta \left[\frac{r^2}{2}\right]_0^{b\sin\theta}$$

$$= \int_0^{\tan^{-1}\frac{a}{b}} \frac{1}{\sin^2\theta\cos^2\theta}\left(\frac{b^2\sin^2\theta}{2}\right)d\theta = \frac{1}{2}b^2\int_0^{\tan^{-1}\frac{a}{b}}\sec^2\theta\,d\theta = \frac{b^2}{2}[\tan\theta]_0^{\tan^{-1}\frac{a}{b}}$$

$$= \frac{1}{2}b^2\tan\left(\tan^{-1}\frac{a}{b}\right) = \frac{1}{2}b^2\frac{a}{b} = \frac{ab}{2}.$$

$$I_2 = \int_{\tan^{-1}\frac{a}{b}}^{\frac{\pi}{2}} \int_0^{r=a\cos\theta} \frac{d\theta}{\sin^2\theta\cos^2\theta}\, r\,dr$$

$$= \int_{\tan^{-1}\frac{a}{b}}^{\frac{\pi}{2}} \frac{d\theta}{\sin^2\theta\cos^2\theta}\left[\frac{r^2}{2}\right]_0^{a\cos\theta} = \int_{\tan^{-1}\frac{a}{b}}^{\frac{\pi}{2}} \frac{d\theta}{\sin^2\theta\cos^2\theta}\left(\frac{a^2\cos^2\theta}{2}\right)$$

$$= \frac{a^2}{2}\int_{\tan^{-1}\frac{a}{b}}^{\frac{\pi}{2}} \frac{d\theta}{\sin^2\theta} = \frac{a^2}{2}\int_{\tan^{-1}\frac{a}{b}}^{\frac{\pi}{2}} \text{cosec}^2\theta\,d\theta = \frac{a^2}{2}[-\cot\theta]_{\tan^{-1}\frac{a}{b}}^{\frac{\pi}{2}}$$

$$= -\frac{a^2}{2}\left[\cot\frac{\pi}{2} - \cot\left(\tan^{-1}\frac{a}{b}\right)\right] = -\frac{a^2}{2}\left[0 - \frac{b}{a}\right] = \frac{ab}{2}$$

$$\iint \frac{(x^2+y^2)^2}{x^2y^2}\,dx\,dy = I_1 + I_2 = \frac{ab}{2} + \frac{ab}{2} = ab$$ **Ans.**

Example 16. *Evaluate :* $\int_0^\infty\int_0^\infty e^{-(x^2+y^2)}\,dx\,dy$ *by changing to polar co-ordinates.*

Hence, show that $\int_0^\infty e^{-x^2}\,dx = \frac{\sqrt{\pi}}{2}$ *(AMIETE, June 2010, U.P., IInd Semester, Summer 2002)*

Solution. From the limit of integration, we find that the first integration is along a vertical strip extending from $y = 0$ to $y = \infty$. The strip slides from $x = 0$ and goes to $x = \infty$. Thus the region of integration is the whole of first quadrant.

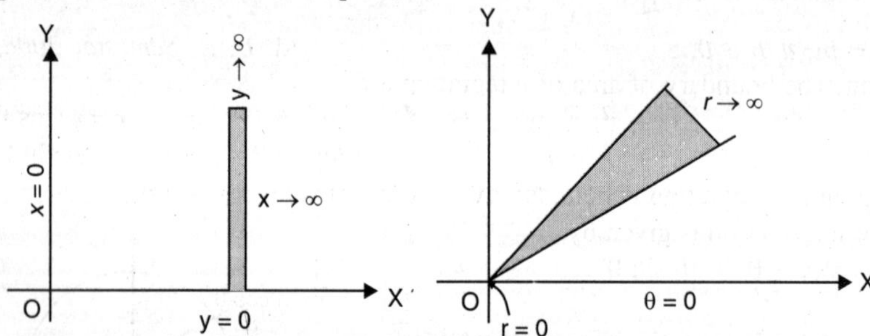

This region can be covered by radial strips extending from $r = 0$ to $r = \infty$. The strip starts from $\theta = 0$ and goes upto $\theta = \pi/2$.

Hence, $$\int_0^\infty\int_0^\infty e^{-(x^2+y^2)}\,dx\,dy = \int_0^{\pi/2}\int_0^\infty e^{-r^2}\,r\,dr\,d\theta$$

$$= -\frac{1}{2}\int_0^{\pi/2}\int_0^\infty(-2r)\,e^{-r^2}\,dr\,d\theta = -\frac{1}{2}\int_0^{\pi/2}[e^{-r^2}]_0^\infty\,d\theta$$

$$= -\frac{1}{2}\int_0^{\pi/2}(0-1)\,d\theta = \frac{1}{2}\int_0^{\pi/2}1\,d\theta = \frac{\pi}{4}$$ **Ans.**

Let $$I = \int_0^\infty e^{-x^2}\,dx \qquad ...(1)$$

Also, $$I = \int_0^{\infty} e^{-y^2}\, dy$$ ...(2) [Property of definite integrals]

Multiplying (1) and (2), we get

$$I^2 = \int_0^{\infty}\int_0^{\infty} e^{-(x^2+y^2)}\, dx\, dy = \frac{\pi}{4}$$ [As obtained above]

$$\Rightarrow \quad I = \sqrt{\frac{\pi}{4}} = \frac{\sqrt{\pi}}{2}$$ **Proved.**

EXERCISE 7.2

Evaluate the folloing:

1. $\int_0^{\pi}\int_0^{a(1-\cos\theta)} 2\pi r^2 \sin\theta\, d\theta\, dr$ **Ans.** $\frac{8}{3}\pi a^3$

2. $\int_0^{\pi}\int_0^{a(1+\cos\theta)} r^2 \cos\theta\, dr\, d\theta$ **Ans.** $\frac{5}{8}\pi a^3$

3. $\iint_A \frac{r\, dr\, d\theta}{\sqrt{r^2+a^2}}$ where A is a loop of $r^2 = a^2\cos 2\theta$ **Ans.** $2a - \frac{\pi a}{2}$

4. $\iint_A r^2 \sin\theta\, d\theta\, dr$ where A is $r = 2a\cos\theta$ above initial line. (*A.M.I.E. Winter 2001*) **Ans.** $\frac{2a^3}{3}$

5. Calculate the integral $\iint \frac{(x-y)^2}{x^2+y^2}\, dx\, dy$ over the circle $x^2+y^2 \le 1$. **Ans.** $\pi - 2$

6. $\iint (x^2+y^2)\, x\, dx\, dy$ over the positive quadrant of the circle $x^2+y^2 = a^2$ by changing to polar coordinates. **Ans.** $\frac{a^2}{5}$

7. $\iint_R \sqrt{x^2+y^2}\, dx\, dy$ by changing to polar coordinates, R is the region in the *xy*-plane bounded by the circles $x^2+y^2 = 4$ (*AMIETE, Dec. 2009*) **Ans.** $\frac{38\pi}{3}$

8. Convert into polar coordinates

$$\int_0^{2a}\int_0^{2ax-x^2} dx\, dy$$ **Ans.** $\int_0^{\pi/2}\int_0^{2a\cos\theta} r\, d\theta\, dr$

9. $\iint r^3\, dr\, d\theta$, over the area bounded between the circles $r = 2b\cos\theta$ and $r = 2b\cos\theta$. **Ans.** $\frac{3\pi}{2}(a^4 - b^4)$

10. $\iint r\sin\theta\, dr\, d\theta$ over the area of the cardiod $r = a(1+\cos\theta)$ above the initial line. **Ans.** $\frac{5}{8}\pi a^3$

11. $\int\int_A x^2\, dr\, d\theta$, where A is the area between the circles $r = a\cos\theta$ and $r = 2a\cos\theta$. **Ans.** $\frac{28a^3}{9}$

12. Transform the integral $\int_0^1\int_0^x f(x,y)\, dy\, dx$ to the integral in polar co-ordinates.

Ans. $\int_0^{\pi/4}\int_0^{\sec\theta} f(r,\theta)\, r\, d\theta\, dr$

7.4 CHANGE OF ORDER OF INTEGRATION

On changing the order of integration, the limits of integration change. To find the new limits, we draw the rough sketch of the region of integration.

Some of the problems connected with double integrals, which seem to be complicated, can be made easy to handle by a change in the order of integration.

Example 17. *Evaluate* $\int_0^a\int_y^a \frac{x}{x^2+y^2}\, dx\, dy$ *by changing the order of integration.*

(*AMIETE, June 2010, Nagpur University, Summer 2008*)

Solution. Here we have

$$I = \int_0^a \int_y^a \frac{x}{x^2 + y^2}\, dx\, dy$$

Here $x = a, x = y, y = 0$ and $y = a$

The area of integration is OAB.

On changing the order of integration Lower limit of $y = 0$ and upper limit is $y = x$.

Lower limit of $x = 0$ and upper limit is $x = a$.

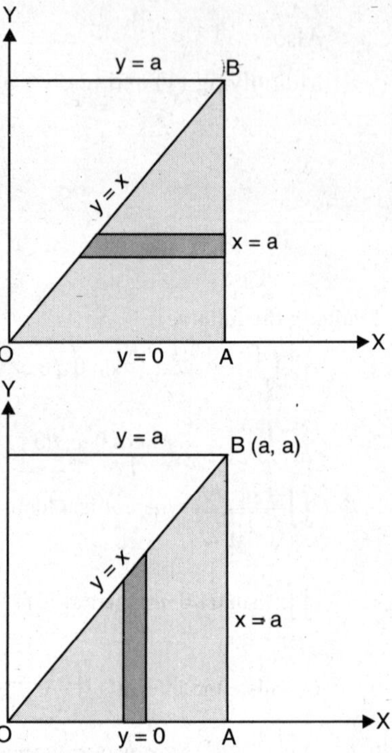

$$I = \int_0^a x dx \int_0^{y=x} \frac{1}{x^2 + y^2} dy$$

$$= \int_0^a x dx \left[\frac{1}{x} \tan^{-1} \frac{y}{x}\right]_0^{y=x}$$

$$= \int_0^a \frac{x}{x} dx \left(\tan^{-1} \frac{x}{x} - \tan^{-1} 0\right)$$

$$= \int_0^a dx \left(\frac{\pi}{4}\right) \quad = \frac{\pi}{4}[x]_0^a = \frac{a\pi}{4} \quad \textbf{Ans.}$$

Example 18. *Change the order of integration in*

$I = \int_0^1 \int_{x^2}^{2-x} xy\, dx\, dy$ *and hence evaluate the same.*

(A.M.I.E.T.E., June 2010, 2009, U.P. I Sem., Dec., 2004)

Solution. $I = \int_0^1 \int_{x^2}^{2-x} xy\, dx\, dy$

The region of integration is shown by shaded portion in the figure bounded by parabola $y = x^2$ and the line $y = 2 - x$.

The point of intersection of the parabola $y = x^2$ and the line $y = 2 - x$ is B (1, 1).

In the figure below (left) we have taken a strip parallel to y-axis and the order of integration is

$$\int_0^1 x\, dx \int_{x^2}^{2-x} y\, dy$$

In the second figure above we have taken a strip parallel to x-axis in the area OBC and second strip in the area ABC. The limits of x in the area OBC are 0 and $\sqrt{y}$ and the limits of x in the area ABC are 0 and $2 - y$.

$$= \int_0^1 y\, dy \int_0^{\sqrt{y}} x\, dx + \int_1^2 y\, dx \int_0^{2-y} x\, dx = \int_0^1 y\, dy \left[\frac{x^2}{2}\right]_0^{\sqrt{y}} + \int_0^{\sqrt{y}} y\, dy \left[\frac{x^2}{2}\right]_0^{2-y}$$

$$= \frac{1}{2}\int_0^1 y^2\,dy + \frac{1}{2}\int_1^2 y(2-y)^2\,dy = \frac{1}{2}\left[\frac{y^3}{3}\right]_0^1 + \frac{1}{2}\int_1^2 (4y - 4y^2 + y^3)\,dy$$

$$= \frac{1}{6} + \frac{1}{2}\left[2y^2 - \frac{4}{3}y^3 + \frac{y^4}{4}\right]_1^2 = \frac{1}{6} + \frac{1}{2}\left[8 - \frac{32}{3} + 4 - 2 + \frac{4}{3} - \frac{1}{4}\right]$$

$$= \frac{1}{6} + \frac{1}{2}\left[\frac{96 - 128 + 48 - 24 + 16 - 3}{12}\right] = \frac{1}{6} + \frac{5}{24} = \frac{9}{24} = \frac{3}{8}$$ **Ans.**

Example 19. *Evaluate the integral* $\int_0^\infty \int_0^x x \exp\left(-\frac{x^2}{y}\right) dx\,dy$ *by changing the order of integration* (*U.P. I Semester Dec., 2005*)

Solution. Limits are given

$y = 0$ and $y = x$

$x = 0$ and $x = \infty$

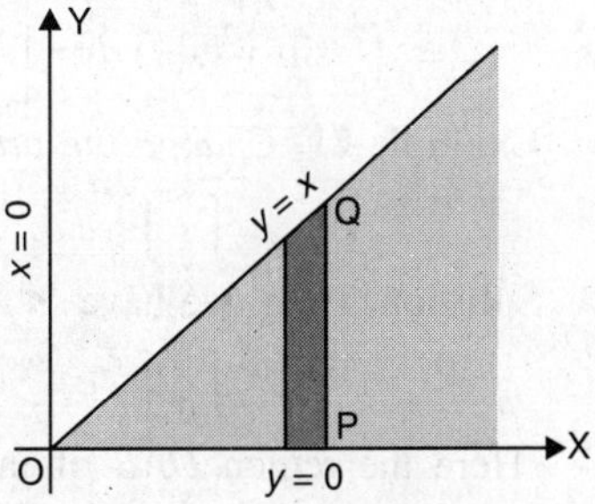

Here, the elementary strip PQ extends from $y = 0$ to $y = x$ and this vertical strip slides from $x = 0$ to $x = \infty$.

The region of integration is shown by shaded portion in the figure bounded by $y = 0$, $y = x$, $x = 0$ and $x = \infty$.

On changing the order of integration, we first integrate with respect to x along a horizontal strip RS which extends from $x = y$ to $x = \infty$ and this horizontal strip slides from $y = 0$ to $y = \infty$ to cover the given region of integration.

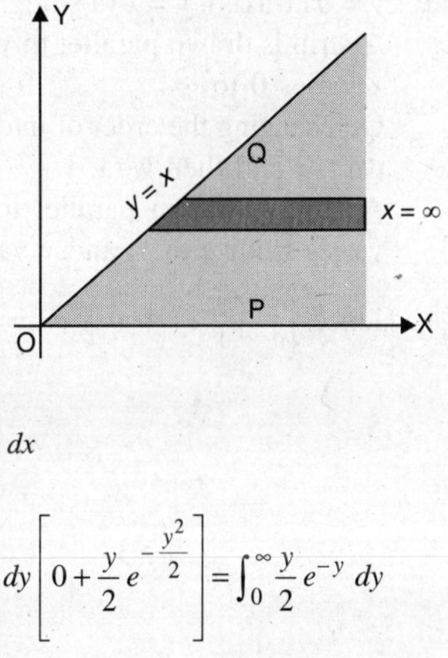

New limits :

$x = y$ and $x = \infty$

$y = 0$ and $y = \infty$

We first integrate with respect to x.

Thus,

$$\int_0^\infty dy \int_y^\infty x e^{-\frac{x^2}{y}}\,dx = \int_0^\infty dy \int_y^\infty -\frac{y}{2}\left(-\frac{2x}{y}e^{-\frac{x^2}{y}}\right) dx$$

$$= \int_0^\infty dy\left[-\frac{y}{2}e^{-\frac{x^2}{y}}\right]_y^\infty = \int_0^\infty dy\left[0 + \frac{y}{2}e^{-\frac{y^2}{2}}\right] = \int_0^\infty \frac{y}{2}e^{-y}\,dy$$

$$= \left[\frac{y}{2}(-e^{-y}) - \left(\frac{1}{2}\right)(e^{-y})\right]_0^\infty$$ (Integrating by parts)

$$= \left[(0-0) - \left(0 - \frac{1}{2}\right)\right] = \frac{1}{2}$$ **Ans.**

Example 20. *Evaluate the double integral.*

$\int_0^1 \int_x^1 \sin(y^2)\,dy\,dx$ (*B.P.U.T.; I Semester 2008*)

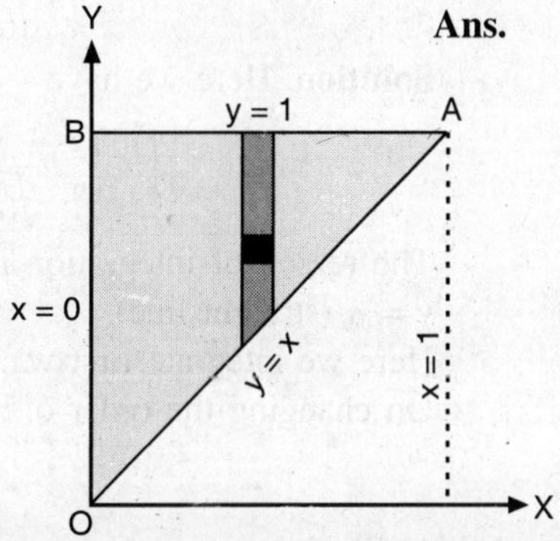

Solution. Here, we have $\int_0^1 \int_x^1 \sin(y^2)\,dy\,dx$

The region OAB integration is bounded by the straight lines $y = x$, $x = 0$ and $y = 1$. A strip is drawn parallel to y axis. y varies from x to 1 and x varies from 0 to 1.

In the given problem the integration is first w.r.t. y and then w.r.t. x. But by this way the evaluation of integral is difficult.

So we change the order of integration. Now we will integrate first w.r.t. x and then w.r.t. y. Here we draw a strip parallel to x axis. On this strip x varies from 0 to y and y varies from 0 to 1.

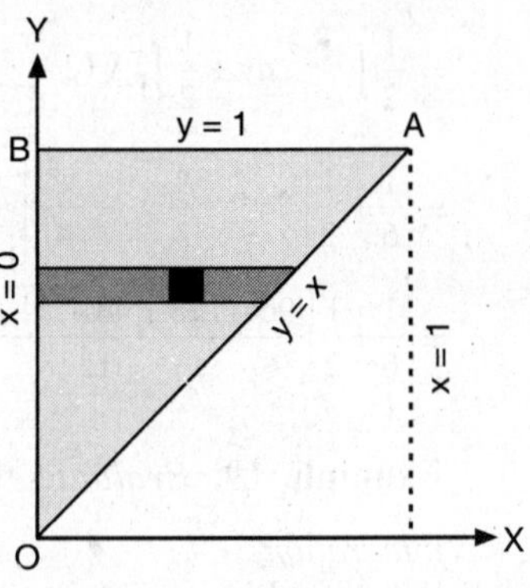

Hence, $\int_0^1 dx \int_x^1 \sin(y^2)\, dy$

$$= \int_0^1 \sin(y^2)\, dy \int_0^y dx = \int_0^1 \sin(y^2)\, dy\, [x]_0^y$$

$$= \int_0^1 \sin y^2 . (y)\, dy = \left[\frac{\cos y^2}{2}\right]_0^1 = \frac{\cos 1}{2} - \frac{1}{2} \quad \textbf{Ans.}$$

Example 21. *Change the order of the integration*

$$\int_0^\infty \int_0^x e^{-xy}\, y\, dy\, dx$$

(*B.P.U.T.; I Semester 2008*)

Solution. Here, we have

$$\int_0^\infty \int_0^x e^{-xy}\, y\, dy\, dx$$

Here the region OAB of integration is bounded by $y = 0$ (x-axis), $y = x$ (a straight line), $x = 0$, *i.e.*, y axis.

A strip is drawn parallel to y-axis, y varies 0 to x and x varies 0 to ∞.

On changing the order of integration, first we integrate w.r.t. x and then w.r.t. y.

A strip is drawn parallel to x-axis. On this strip x varies from y to ∞ and y varies from 0 to ∞.

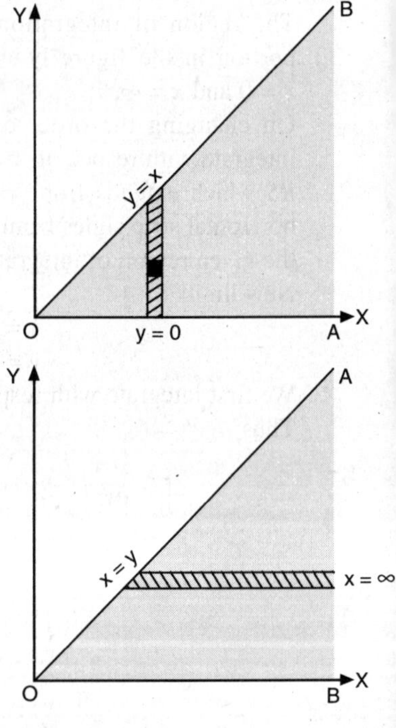

Hence $\int_0^\infty \int_0^x e^{-xy}\, y\, dy\, dx = \int_0^\infty y\, dy \int_y^\infty e^{-xy} dx$

$$= \int_0^\infty y\, dy \left(\frac{e^{-xy}}{-y}\right)_y^\infty$$

$$= \int_0^\infty \frac{y\, dy}{-y} [0 - e^{y^2}]$$

$$= \int_0^\infty e^{-y^2}\, dy = \frac{1}{2}\sqrt{\pi} \quad \textbf{Ans.}$$

Example 22. *Change the order of integration and evaluate:*

$$\int_0^a \int_0^y \frac{dx\, dy}{\sqrt{(a^2 + x^2)(a - y)(y - x)}}$$

(*M.U., II Semester 2004, 2003*)

Solution. Here we have

$$\int_0^a \int_0^y \frac{dx\, dy}{\sqrt{(a^2 + x^2)(a - y)(y - x)}}$$

The region of integration is bounded by $x = 0$ (y axis) $x = y$ (straight line), $y = 0$ (x-axis), $y = a$ (straight line).

Here we integrate first w.r.t. x and then y.

On changing the order of integration we have to integrate first w.r.t. y and then x.

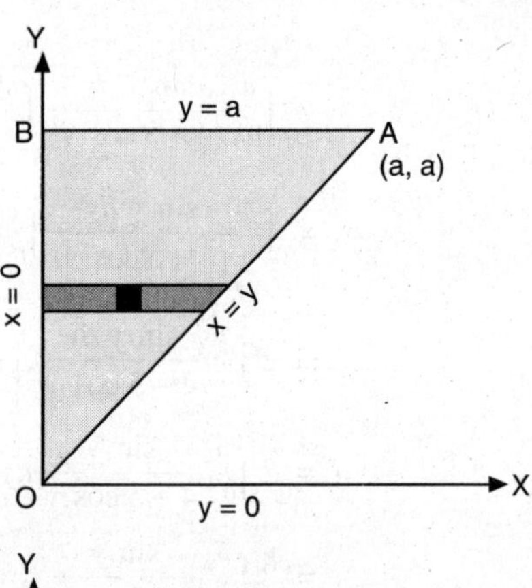

$$\text{Given integral} = \int_0^a \int_0^y \frac{dx\,dy}{\sqrt{(a^2+x^2)(a-y)(y-x)}}$$

$$= \int_0^a \int_x^a \frac{dy\,dx}{\sqrt{(a^2+x^2)(a-y)(y-x)}}$$

Put $y - x = t^2$, $dy = 2t\,dt$

When $y = x$, $t = 0$, when $y = a$, $t = \sqrt{a-x}$

$$\therefore \qquad I = \int_0^a \int_0^{\sqrt{a-x}} \frac{dx}{\sqrt{a^2+x^2}} \cdot \frac{2t\,dt}{\sqrt{(a-x)-t^2}.t}$$

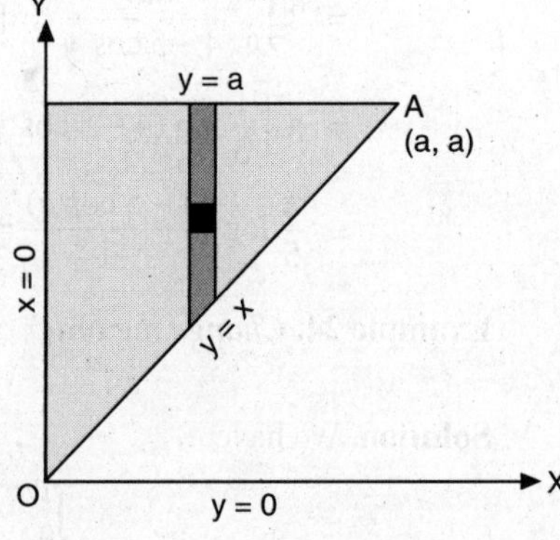

$$= 2\int_0^a \frac{dx}{\sqrt{a^2+x^2}} \left[\sin^{-1}\frac{t}{\sqrt{a-x}}\right]_0^{\sqrt{a-x}}$$

$$= 2\int_0^a \frac{1}{\sqrt{a^2+x^2}} [\sin^{-1} 1 - \sin^{-1} 0]\,dx$$

$$= 2\int_0^a \frac{1}{\sqrt{a^2+x^2}} \cdot \frac{\pi}{2}\,dx$$

$$= 2\left(\frac{\pi}{2}\right)\int_0^a \frac{1}{\sqrt{a^2+x^2}}\,dx$$

$$= \pi\left[\log\left(x+\sqrt{a^2+x^2}\right)\right]_0^a$$

$$= \pi\left[\log\left(a+\sqrt{a^2+a^2}\right) - \log\left(0+\sqrt{a^2+0}\right)\right] = \pi\,[\log\,(a+\sqrt{2}a) - \log a]$$

$$= \pi\,[\log a\,(1+\sqrt{2}) - \log a] = \pi \log\,(1+\sqrt{2})$$ **Ans.**

Example 23. *Change the order of integration and evaluate*

$$\int_0^a \int_0^x \frac{\sin y\,dy\,dx}{\sqrt{[(a-x)(x-y)](4-5\cos y)^2}}$$ *(M.U., II Semester 2002)*

Solution. The limits of y are 0 and x, that of x are 0 and a. The area OAB of integration is bounded by $y = 0$, $y = x$ and $x = 0$ and $x = a$. The given function is integrated first w.r.t. y and then x.

The strip is drawn parallel to y-axis and varies $y = 0$ and $y = x$ and x varies from $x = 0$ and $x = a$.

On changing the order of integration we integrate first w.r.t. x and then y.

On the strip parallel to x-axis, x varies from $x = y$ to $x = a$ and y varies from $y = 0$ to $y = a$.

On changing the order of integration.

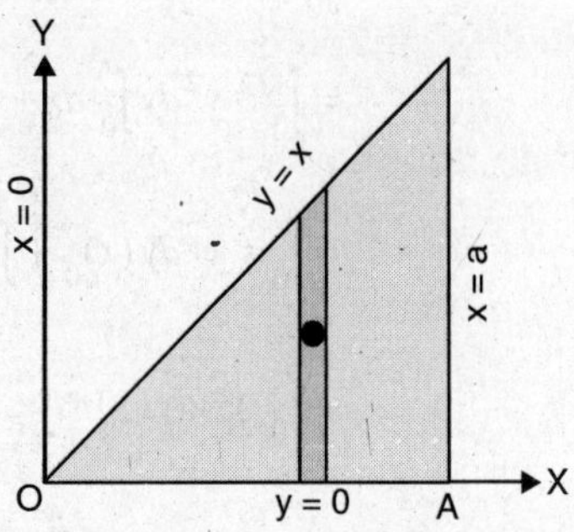

The given integral

$$= \int_{y=0}^{y=a} \int_{x=y}^{x=a} \frac{\sin y}{(4-5\cos y)} \cdot \frac{dx}{\sqrt{[(a-x)(x-y)]}}\,dy$$

On putting $x - y = t^2$, $dx = 2\,t\,dt$ in (1), we get

$$= \int_0^a \int_0^{\sqrt{a-y}} \frac{\sin y}{(4-5\cos y)} \cdot \frac{2t\,dt}{\sqrt{[(a-y-t^2)]\,t}}\,dy$$

$$= \int_0^a \frac{\sin y\, dy}{(4-5\cos y)} \int_0^{\sqrt{a-y}} \frac{2t\, dt}{\sqrt{[(a-y)-t^2]}\,t}$$

$$\left[\begin{array}{l}\text{Put } x-y=t^2 \ \therefore\ dx = 2t\, dt \\ \text{When } x = y, \text{ then } t = 0; \\ \text{when } x = a, \text{ then } t = \sqrt{a-y}\end{array}\right]$$

$$= \int_0^a \frac{\sin y\, dy}{(4-5\cos y)} \int_0^{\sqrt{a-y}} \frac{2dt}{\sqrt{(a-y)-t^2}}$$

$$= 2\int_0^a \frac{\sin y\, dy}{4-5\cos y} \cdot \left[\sin^{-1}\left(\frac{t}{\sqrt{a-y}}\right)\right]_0^{\sqrt{a-y}} dy$$

$$= 2\int_0^a \frac{\sin y}{4-5\cos y}\, dy \left(\sin^{-1}\frac{\sqrt{a-y}}{\sqrt{a-y}} - \sin^{-1} 0\right)$$

$$= 2\int_0^a \frac{\sin y}{4-5\cos y}\, dy\, [\sin^{-1}(1)] = 2\,.\frac{\pi}{2}\int_0^a \frac{\sin y\, dy}{4-5\cos y}$$

$$= \pi\left[\frac{1}{5}\log(4-5\cos y)\right]_0^a = \frac{\pi}{5}[\log(4-5\cos a) - \log(-1)]$$

$$= \frac{\pi}{5}\log\frac{(4-5\cos a)}{-1} = \frac{\pi}{5}\log(5\cos a - 4)$$ **Ans.**

Example 24. *Change the order of integration and evaluate* $\int_0^{a/\sqrt{2}} \int_x^{\sqrt{a^2-x^2}} y^2\, d A$

(*Gujarat, I Semester, Jan,. 2009*)

Solution. We have,

$$\int_0^{a/\sqrt{2}} \int_x^{\sqrt{a^2-x^2}} y^2 d\, A$$

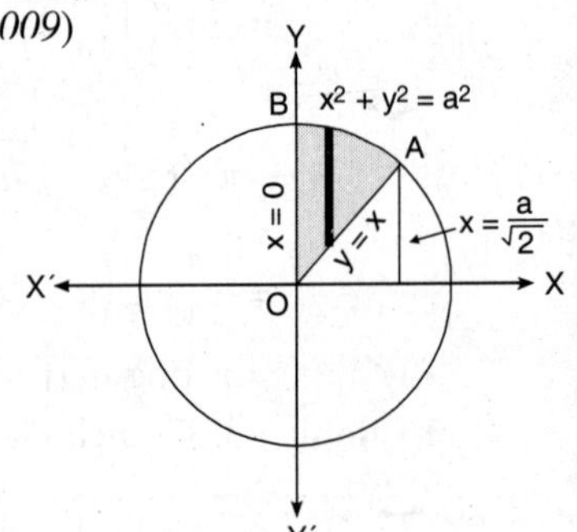

Here the limits are

$$x = 0$$
$$x = \frac{a}{\sqrt{2}}$$
$$y = x$$
$$y = \sqrt{a^2 - x^2} \Rightarrow x^2 + y^2 = a^2$$

Area of integration is shaded area OAB in the figure.

On changing the order of integration we integrate first w.r.t. 'x' and then w.r.t. 'y'

In this way we have to divided the area of integration in two parts OAC and ABC.

$$\int_0^{\frac{a}{\sqrt{2}}} d\,x \int_x^{\sqrt{a^2-x^2}} y^2 dy$$

$$= \int_0^{\frac{a}{\sqrt{2}}} y^2 dy \int_0^y dx + \int_{\frac{a}{\sqrt{2}}}^a y^2 dy \int_0^{\sqrt{a^2-y^2}} dx$$

$$= \int_0^{\frac{a}{\sqrt{2}}} y^2 dy\, (x)_0^y + \int_{\frac{a}{\sqrt{2}}}^a y^2 dy (x)_0^{\sqrt{a^2-y^2}}$$

$$= \int_0^{\frac{a}{2}} y^2\, dy\, (y) + \int_{\frac{a}{\sqrt{2}}}^a y^2\, dy \sqrt{a^2-y^2}$$ (Put $y = a\sin\theta$)

$$= \int_0^{\frac{a}{\sqrt{2}}} y^3\, dy + \int_{\frac{\pi}{4}}^{\frac{\pi}{2}} a^2 \sin^2\theta\, (a\cos\theta)\, a\cos\theta\, d\,\theta = \left[\frac{y^4}{4}\right]_0^{\frac{a}{\sqrt{2}}} + a^4 \int_{\frac{\pi}{4}}^{\frac{\pi}{2}} \sin^2\theta \cos^2\theta\, d\,\theta$$

$$= \frac{\frac{a^4}{4}}{4} + a^4 \int_{\frac{\pi}{4}}^{\frac{\pi}{2}} \sin^2\theta(1-\sin^2\theta)\, d\,\theta = \frac{a^4}{16} + a^4\left(\frac{\pi}{32}\right) = a^4\left(\frac{\pi+2}{32}\right)$$ **Ans.**

Example 25. *Change the order of integration*

$$\int_0^1 \int_{\sqrt{2x-x^2}}^{1+\sqrt{1-x^2}} f(x, y)\,dy\,dx \qquad \textit{(M.U. II Semester 2009)}$$

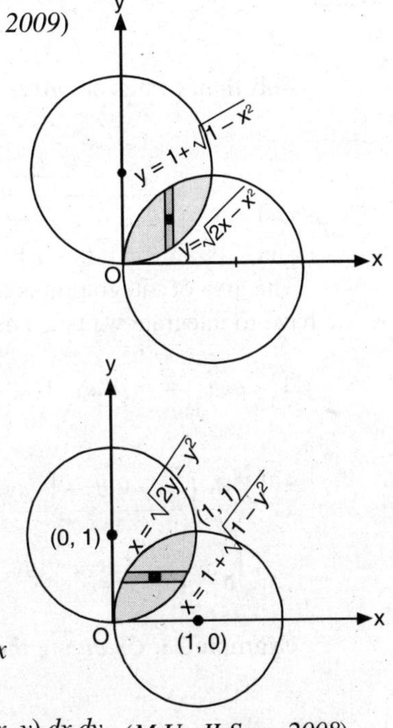

Solution. Here, we have

$$\int_0^1 \int_{\sqrt{2x-x^2}}^{1+\sqrt{1-x^2}} f(x, y)\,dy\,dx$$

In the given integral problem it is integrated first w.r.t. y and then x.

The limits of x are 0 and 1.

The limits of y are $\sqrt{2x - x^2}$ (circle) and $1 + \sqrt{1 - x^2}$ (circle).

The shaded region of integration is bounded by two circles.

On changing the order of integration we integrate first w.r.t. 'x' and then 'y'.

The limits of y are 0 and 1.

The limits of x are $1 + \sqrt{1 - y^2}$ and $\sqrt{2y - y^2}$.

Thus $\displaystyle\int_0^1 dx \int_{\sqrt{2x-x^2}}^{1+\sqrt{1-x^2}} f(x, y)\,dy = \int_0^1 dy \int_{\sqrt{2y-y^2}}^{1+\sqrt{1-y^2}} f(x, y)\,dx$

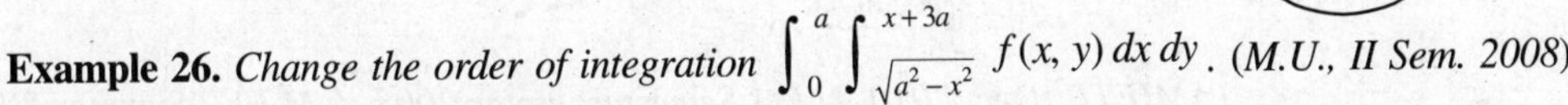

Example 26. *Change the order of integration* $\displaystyle\int_0^a \int_{\sqrt{a^2-x^2}}^{x+3a} f(x, y)\,dx\,dy$. *(M.U., II Sem. 2008)*

Solution. Here we have $\displaystyle\int_0^a \int_{\sqrt{a^2-x^2}}^{x+3a} f(x, y)\,dx\,dy$ ...(1)

Here we have integrated (1) first w.r.t. 'y' and then x.

The limits of y are $\sqrt{a^2 - x^2}$ (circle) and $x + 3a$ and the limits of x are 0 and a. The shaded portion *ABCDA* of the region of the integration is bounded by $y = \sqrt{a^2 - x^2}$ (circle), $y = x + 3a$ (straight line) $x = 0$ (y-axis) and $x = a$ (a straight line).

On changing the order of integration we have to integrate (1) w.r.t. to x first and then y.

For this way we have to divide the region *ABCDA* of integration into three parts *AFD*, *DFGC* and *BCG*.

One part is *AFDA* in which the limits of x are $\sqrt{a^2 - y^2}$ and $x = a$ and the limits of y are 0 and a.

Second part is *FGCD* in which the limits of $x = 0$ and $x = a$ and the limits of y are a and $3a$.

In third part *BCGB* the limits of x are $x = y - 3a$ and $x = a$ and the limits of y are $3a$ and $4a$.

Hence, $\displaystyle\int_0^a \int_{\sqrt{a^2-x^2}}^{x+3a} f(x, y)\,dy\,dx$

$$= \int_0^a \int_{\sqrt{a^2-y^2}}^{a} f(x, y)\,dx\,dy + \int_a^{3a} \int_0^{a} f(x, y)\,dx\,dy + \int_{3a}^{4a} \int_{y-3a}^{a} f(x, y)\,dx\,dy \qquad \textbf{Ans.}$$

Example 27. *Change the order of integration in the double integral*

$$\int_0^{2a}\int_{\sqrt{2ax-x^2}}^{\sqrt{2ax}} V\,dx\,dy$$

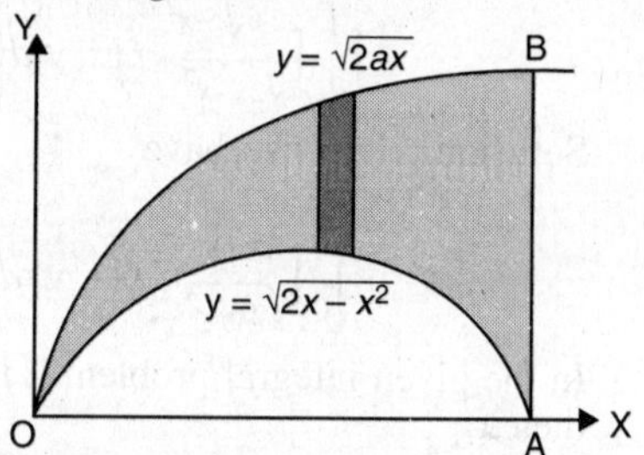

Solution. Limits are given as

$$x = 0,\ x = 2a$$

$$y = \sqrt{2ax}$$

and $y = \sqrt{2ax - x^2} \Rightarrow y^2 = 2ax$

and $(x-a)^2 + y^2 = a^2$

The area of integration is the shaded portion *OAB*. On changing the order of integration first we have to integrate w.r.t. *x*, The area of integration has three portions *BCE*, *ODE* and *ACD*.

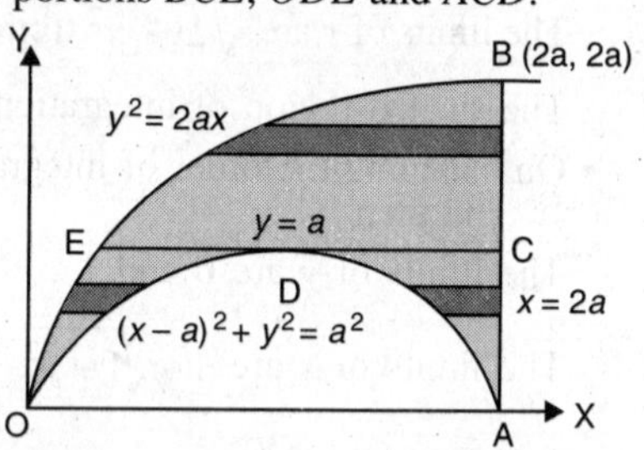

$$\int_0^{2a} dx\int_{\sqrt{2ax-x^2}}^{\sqrt{2ax}} V\,dy$$

$$= \int_0^{2a} dy\int_{y^2/2a}^{2a} V\,dx + \int_0^{a} dy\int_{y^2/2a}^{a+\sqrt{a^2+y^2}} V\,dx$$

$$+ \int_0^{a} dy\int_{a+\sqrt{a^2-y^2}}^{2a} V\,dx$$ **Ans.**

Example 28. *Changing the order of integration of* $\int_0^\infty\int_0^\infty e^{-xy}\sin nx\,dx\,dy$ *show that*

$$\int_0^\infty \frac{\sin nx}{x}dx = \frac{\pi}{2}.$$

(AMIETE, Dec. 2010, U.P. I Semester winter 2003, A.M.I.E., Summer 2000)

Solution. The region of integration is bounded by $x = 0, x = \infty, y = 0, y = \infty$, *i.e.*, first quadrant.

$$\int_0^\infty\int_0^\infty e^{-xy}\sin nx\,dx\,dy = \int_0^\infty dy\int_0^\infty e^{-xy}\sin nx\,dx$$

$$= \int_0^\infty dy\left[\frac{e^{-xy}}{n^2+y^2}\{-y\sin x - n\cos nx\}\right]_0^\infty$$

$$= \int_0^\infty dy\left[0+\frac{n}{n^2+y^2}\right] = \int_0^\infty \frac{n}{n^2+y^2}dy = \left[\tan^{-1} y\right]_0^\infty = \frac{\pi}{2} \quad ...(1)$$

On changing the order of integration

$$\int_0^\infty\int_0^\infty e^{-xy}\sin nx\,dx\,dy = \int_0^\infty \sin nx\,dx\int_0^\infty e^{-xy}\,dy$$

$$= \int_0^\infty \sin nx\,dx\left[\frac{e^{-xy}}{-x}\right]_0^\infty = \int_0^\infty \frac{\sin nx\,dx}{x}\left[-\frac{1}{e^{xy}}\right]_0^\infty$$

$$= \int_0^\infty \frac{\sin nx}{x}dx[-0+1] = \int_0^\infty \frac{\sin nx}{x}dx \quad ...(2)$$

From (1) and (2), $\int_0^\infty \frac{\sin nx}{x}dx = \frac{\pi}{2}$ **Proved.**

Example 29. *Change order of integration and hence evaluate:*

$$\int_0^a\int_{\sqrt{ax}}^a \frac{y^2\,dx\,dy}{\sqrt{y^4-a^2x^2}}$$

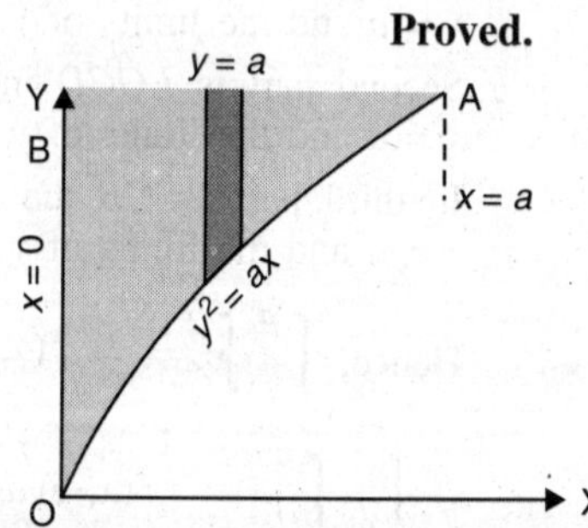

Solution. The given limits show that the area of integration lies between $y^2 = ax$, $y = a$, $x = 0$ and $x = a$. We can consider it as lying between $y = 0$, $y = a$, $x = 0$ and $x = y^2/a$ by changing the order of integration. Hence, the given integral.

$$\int_{x=0}^{a}\int_{y=\sqrt{ax}}^{a}\frac{y^2\,dx\,dy}{\sqrt{y^4-a^2x^2}}=\int_{y=0}^{a}\int_{x=0}^{\frac{y^2}{a}}\frac{y^2\,dy\,dx}{\sqrt{y^4-a^2x^2}}$$

$$=\frac{1}{a}\int_0^a\int_0^{\frac{y^2}{a}}\frac{y^2\,dy\,dx}{\sqrt{\left(\frac{y^2}{a}\right)^2-x^2}}\qquad=\frac{1}{a}\int_a^a y^2\left[\sin^{-1}\left(\frac{ax}{y^2}\right)\right]_0^{y^2/a}dy$$

$$=\frac{1}{a}\int_0^a y^2\left[\sin^{-1}(1)-\sin^{-1}(0)\right]dy=\frac{\pi}{2a}\int_0^a y^2\,dy=\frac{\pi}{2a}\left(\frac{y^3}{3}\right)_0^a=\frac{\pi}{6a}(a^3)=\frac{\pi a^2}{6}.\qquad\textbf{Ans.}$$

Example 30. *Evaluate* $\int_0^a\int_0^x\frac{f'(y)\,dy\,dx}{[(a-x)(x-y)]^{1/2}}$

Solution. Let $I=\int_0^a\int_0^x\frac{f'(y)\,dy\,dx}{[(a-x)(x-y)]^{1/2}}$

Here the limits are $x=0$, $x=a$ and $y=0$, $y=x$. Evidently the region of integration is *OABO*.

By changing the order of integration, we have

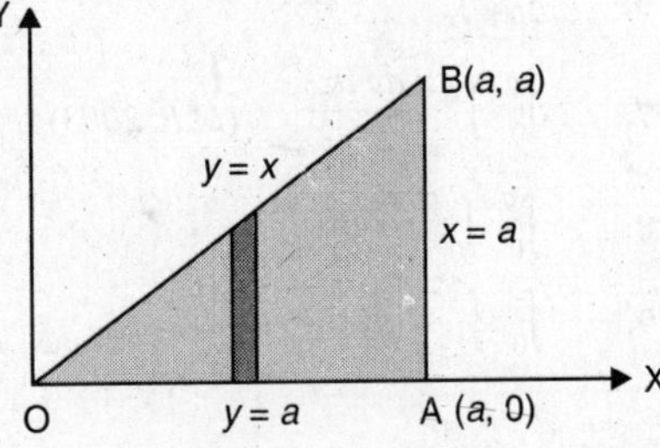

$$I=\int_0^a\int_y^a\frac{f'(y)\,dy\,dx}{[(a-x)(x-y)]^{1/2}}$$

$$=\int_0^a f'(y)\,dy\int_y^a\frac{dx}{\sqrt{(a-x)(x-y)}}\quad\ldots(1)$$

Let us find the values of $(a-x)$ and $(x-y)$ for (1)

Putting $x=a\cos^2\theta+y\sin^2\theta$

We have $\quad a-x=a-a\cos^2\theta-y\sin^2\theta$

$\qquad=a(1-\cos^2\theta)-y\sin^2\theta$

$\qquad a-x=a\sin^2\theta-y\sin^2\theta=(a-y)\sin^2\theta \qquad\ldots(2)$

$\Rightarrow\quad -dx=2(a-y)\sin\theta\cos\theta\,d\theta$, keeping y constant.

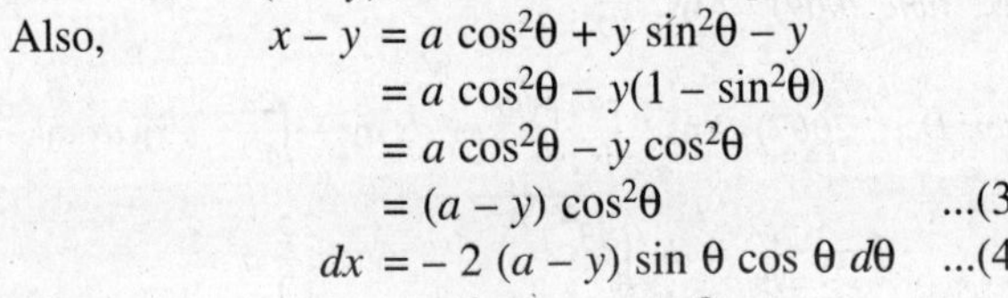

Also, $\quad x-y=a\cos^2\theta+y\sin^2\theta-y$

$\qquad=a\cos^2\theta-y(1-\sin^2\theta)$

$\qquad=a\cos^2\theta-y\cos^2\theta$

$\qquad=(a-y)\cos^2\theta \qquad\ldots(3)$

$\qquad dx=-2(a-y)\sin\theta\cos\theta\,d\theta \qquad\ldots(4)$

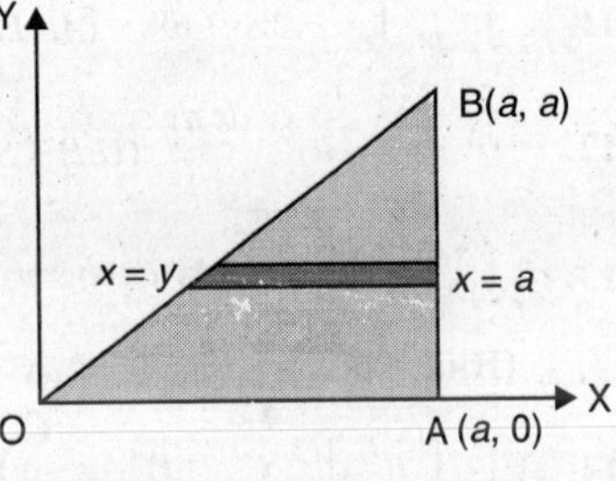

when $\quad x=y$ then $x-y=0$

Upper limt $\quad x=a$

$\qquad x-y=(a-y)\cos^2\theta$

$\qquad a-y=(a-y)\cos^2\theta$

$\Rightarrow\quad \cos^2\theta=1\Rightarrow\theta=0$

lower limit $\quad x=y$

$\qquad x-y=(a-y)\cos^2\theta$

$\qquad x-x=(a-y)\cos^2\theta\ \Rightarrow\ 0=\cos^2\theta\ \Rightarrow\ \theta=\frac{\pi}{2}$

Putting the values of $a-x$, $x-y$ and dx from (2), (3), and (4) respectively in (1), we get

$$I=\int_0^a f'(y)\,dy\int_{\frac{\pi}{2}}^{0}\frac{-2(a-y)\sin\theta\cos\theta}{\sqrt{(a-y)\sin^2\theta.(a-y)\cos^2\theta}}\,d\theta$$

$$=\int_0^a f'(y)\,dy\int_{\frac{\pi}{2}}^{0}\frac{-2(a-y)\sin\theta\cos\theta}{(a-y)\sin\theta\cos\theta}\,d\theta$$

$$=-2\int_0^a f'(y)\,dy\int_{\frac{\pi}{2}}^{0}d\theta=2\int_0^a f'(y)\,[\theta]_0^{\pi/2}=2\int_0^a f'(y)\,dy.\frac{\pi}{2}=2[f(y)]_0^a.\frac{\pi}{2}=[f(a)-f(0)]\,\pi$$

EXERCISE 7.3

Change the order of integration and hence evaluate the following:

1. $\int_0^a \int_0^x \frac{\cos y\, dy}{\sqrt{(a-x)(a-y)}}\, dx$ **Ans.** (a) $\int_0^a dy \int_y^a \frac{\cos y\, dx}{\sqrt{(a-x)(a-y)}}$ (b) $2 \sin a$.

2. $\int_0^{2a} \int_{\frac{x^2}{4a}}^{3a-x} (x^2+y^2)\, dy\, dx$ **Ans.** (a) $\int_0^a dy \int_0^{2\sqrt{ay}} (x^2+y^2)\, dx + \int_a^{3a} dy \int_0^{3a-y} (x^2+y^2)\, dx$ (b) $\frac{314\, a^4}{35}$.

3. $\int_0^1 \int_{x^2}^{x} (x^2+y^2)^{-1/2}\, dy\, dx$ **Ans.** $\int_0^1 dy \int_y^{\sqrt{y}} (x^2+y^2)^{-1/2}\, dx$.

4. $\int_0^a \int_{\sqrt{a^2-y^2}}^{y+a} f(x, y)\, dx\, dy$ (*A.M.I.E.T.E., Summer 2000*)

Ans. $\int_0^a dx \int_{\sqrt{a^2-x^2}}^{a} f(x, y)\, dy + \int_a^{2a} dx \int_{x-a}^{a} f(x, y)\, dy$

5. $\int_{-a}^{a} \int_0^{\sqrt{a^2-y^2}} f(x, y)\, dx\, dy$ **Ans.** $\int_0^a dx \int_{-\sqrt{a^2-x^2}}^{\sqrt{a^2-x^2}} f(x, y)\, dy$

6. $\int_0^1 \int_x^{2-x} \frac{x}{y}\, dy\, dx$ **Ans.** $\int_0^a \frac{dy}{y} \int_0^y x dx + \int_1^2 \frac{dy}{y} \int_0^{2-y} x dx$; $\log \frac{4}{e}$

7. $\int_0^b \int_y^a \frac{x\, dy\, dx}{x^2+y^2}$ (*M.P. 2003*) **Ans.** $\frac{\pi ab}{4}$

8. $\int_0^a \int_0^{bx/a} x\, dy\, dx$ **Ans.** (a) $\int_0^b dy \int_{ay/b}^{a} x\, dx$ (b) $\frac{1}{3} a^2 b$

9. $\int_0^5 \int_{2-x}^{2+x} f(x, y)\, dx\, dy$ (*A.M.I.E.T.E. Winter 1999*)

Ans. $\int_0^2 dy \int_{2-y}^{5} f(x, y)\, dx + \int_2^7 dy \int_{y-2}^{5} f(x, y)\, dx$

10. $\int_0^\infty \int_{-y}^{y} (y^2-x^2)\, e^{-y}\, dx\, dy$ **Ans.** $\int_{-\infty}^{\infty} dx \int_{-x}^{x} (y^2-x^2)\, e^{-y}\, dy$, 8 (*AMIE., Summer 2000*)

11. $\int_{y=0}^{1} \int_{x=\sqrt{y}}^{2-y} yx\, dx\, dy$ (*A.M.I.E.T.E., June 2009*) **Ans.** $\frac{7}{12}$

12. $\int_0^a \int_{\frac{x^2}{a}}^{2a-x} xy\, dx\, dy$ (*U.P. I Semester, Dec., 2007*) **Ans.** $\int_0^a \int_0^{\sqrt{ay}} xy\, dx\, dy + \int_0^{2a-y} xy\, dx\, dy$, $\frac{3a^2}{8}$

13. $\int_0^a \int_{a-\sqrt{a^2-y^2}}^{a+\sqrt{a^2-y^2}} xy\, dx\, dy$ **Ans.** $\int_0^{2a} x\, dx \int_0^{\sqrt{a^2-(x-a)^2}} y\, dy$, $\frac{2}{3} a^4$

[**Hint:** Put $x = a \sin^2 \theta \Rightarrow dx = 2\, a \sin \theta \cos \theta\, d\,\theta$]

14. $\int_0^1 \int_{-1}^{1-y} x^{1/3}\, y^{-1/2}\, (1-x-y)^{1/2}\, dx\, dy$ **Ans.** $\int_{-1}^{1} x^{\frac{1}{3}}\, dx \int_0^{1-x} y^{-\frac{1}{2}} (1-x-y)^{\frac{1}{2}}\, dy$, $-\frac{3\pi}{7}$

15. $\int_0^{2a} dx \int_0^{\frac{x^2}{4a}} (x+y)^3\, dy$ **Ans.** $\int_0^a dy \int_{\sqrt{4a\,y}}^{2a} (x+y)^3\, dx$

16. $\int_0^1 \int_0^y (x^2+y^2)\, dx\, dy + \int_1^2 \int_0^{2-y} (x^2+y^2)\, dx\, dy$ (*A.M.I.E. Winter 2000*)

Ans. $\int_0^1 dx \int_x^{2-x} (x^2+y^2)\, dy$, $\frac{5}{3}$

17. $\int_0^a \int_0^{\sqrt{a^2-x^2}} y^2 \sqrt{x^2+y^2}\, dx\, dy$ by changing into polar coordinates. **Ans.** $\frac{\pi\, a^5}{20}$

(*U.P., I Semester, Dec. 2007, A.M.I.E., Summer 2001*)

18. $\int_0^1 \int_1^2 \frac{1}{x^2+y^2}\, dx\, dy + \int_0^2 \int_y^2 \frac{1}{x^2+y^2}\, dx\, dy = \int_R \int \frac{1}{x^2+y^2}\, dy\, dx$

Recognise the region R of integration on the R.H.S. and then evaluate the integral on the right in the order indicated. (*AMIETE, Dec. 2004*)

Ans. Region R is $x = 0$, $x = y$, $y = 1$ and $y = 2$, $\frac{\pi}{4} \log 2$.

19. Express as single integral and evaluate :

$$\int_0^{\frac{a}{\sqrt{2}}} \int_0^x x\,dx\,dy + \int_{\frac{a}{\sqrt{2}}}^a \int_0^{\sqrt{a^2-x^2}} x\,dx\,dy$$

Ans. $\int_0^{\frac{a}{\sqrt{2}}} dy \int_y^{\sqrt{a^2-y^2}} x\,dx, \frac{5a^3}{6\sqrt{2}}$

20. Express as single integral and evaluate :

$$\int_0^1 \int_0^y (x^2+y^2)\,dx\,dy + \int_1^2 \int_0^{2-y} (x^2+y^2)\,dx\,dy$$

Ans. $\int_0^1 dx \int_x^{2-x} (x^2+y^2)\,dy, \frac{5}{3}$

21. If $f(x, y)\,dx\,dy$, where R is the circle $x^2 + y^2 = a^2$, is R equivalent to the repeated integral.

(AMIE winter 2001) [**Ans.** $\int_0^{2\pi}\int_0^1 (r, \theta)\, r\,dr\,d\theta.$]

7.5 CHANGE OF VARIABLES

Sometimes the problems of double integration can be solved easily by change of independent variables. Let the double integral as be $\iint_R f(x, y)\,dx\,dy$. It is to be changed by the new variables u, v.

The relation cf x, y with u, v are given as $x = f(u, v), y = \Psi(u, v)$. Then tne double integratio 's converted into.

$\iint_{R'} f\{\phi(u, v), \Psi(u, v)\}\,|J|\,du\,dv$, where

$$dx\,dy = |J|\,du\,dv = \frac{\partial(x, y)}{\partial(u, v)}\,du\,dv = \begin{vmatrix} \frac{\partial x}{\partial u} & \frac{\partial x}{\partial v} \\ \frac{\partial y}{\partial u} & \frac{\partial y}{\partial v} \end{vmatrix} du\,dv$$

Example 31. *Evaluate* $\iint_R (x+y)^2\,dx\,dy$, *where R is the parallelogram in the xy-plane with vertices (1, 0), (3, 1), (2, 2), (0, 1), using the transformation u = x + y and v = x – 2y.*

(U.P., I Semester, 2003)

Solution. The region of integration is a parallelogram $ABCD$, where A (1, 0), B (3, 1), C (2, 2) and D (0, 1) in xy-plane.

The new region of integration is a rectangle $A'B'C'D'$ in uv-plane

xy-plane	$A \equiv (x, y)$ $A \equiv (1, 0)$	$B \equiv (x, y)$ $B \equiv (3, 1)$	$C \equiv (x, y)$ $C \equiv (2, 2)$	$D \equiv (x, y)$ $D \equiv (0, 1)$
uv-plane	$A' \equiv (u, v)$ $A' \equiv (x+y, x-2y)$ $A' \equiv (1+0, 1-2\times 0)$ $A' \equiv (1, 1)$	$B' \equiv (u, v)$ $B' \equiv (x+y, x-2y)$ $B' \equiv (3+1, 3-2\times 1)$ $B' \equiv (4, 1)$	$C' \equiv (u, v)$ $C' \equiv (u, v)$ $C' \equiv (2+2, 2-2\times 2)$ $C' \equiv (4, -2)$	$D' \equiv (u, v)$ $D' \equiv (0+1, 0-2\times 1)$ $D' \equiv (1, -2)$

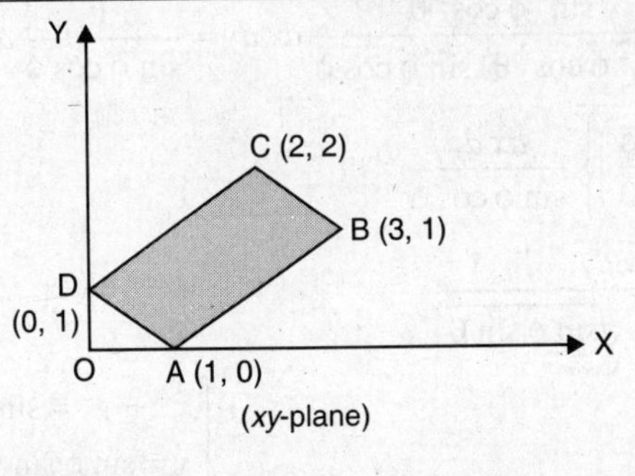

(*xy*-plane)

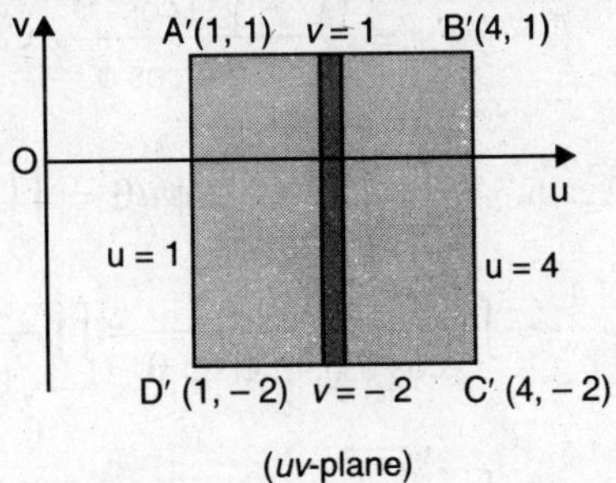

(*uv*-plane)

and $\left\{\begin{matrix} u = x + y \\ v = x - 2y \end{matrix}\right\} \Rightarrow x = \frac{1}{3}(2u + v)$, and $y = \frac{1}{3}(u - v)$

$$J = \frac{\partial(x, y)}{\partial(u, v)} = \begin{vmatrix} \frac{\partial x}{\partial u} & \frac{\partial x}{\partial v} \\ \frac{\partial y}{\partial u} & \frac{\partial y}{\partial v} \end{vmatrix} = \begin{vmatrix} \frac{2}{3} & \frac{1}{3} \\ \frac{1}{3} & -\frac{1}{3} \end{vmatrix} = -\frac{1}{3}$$

$$dx\, dy = |J|\, du\, dv = \frac{1}{3}\, du\, dv$$

$$\iint_R (x+y)^2\, dx\, dy = \int_{-2}^{1}\int_{1}^{4} u^2 . \frac{1}{3}\, du\, dv = \int_{-2}^{1} \frac{1}{3}\left[\frac{u^3}{3}\right]_1^4 dv = \int_{-2}^{1} 7\, dv = 7\,[v]_{-2}^{1} = 7 \times 3 = 21 \textbf{ Ans.}$$

Example 32. *Transform* $\int_0^{\pi/2}\int_0^{\pi/2} \sqrt{\frac{\sin\phi}{\sin\theta}}\, d\phi\, d\theta$ *by the substitution* $x = \sin\phi\cos\theta$, $y = \sin\phi \sin\theta$ *and show that its value is* π.

Solution. $x = \sin\phi\cos\theta$, $y = \sin\phi\sin\theta$.

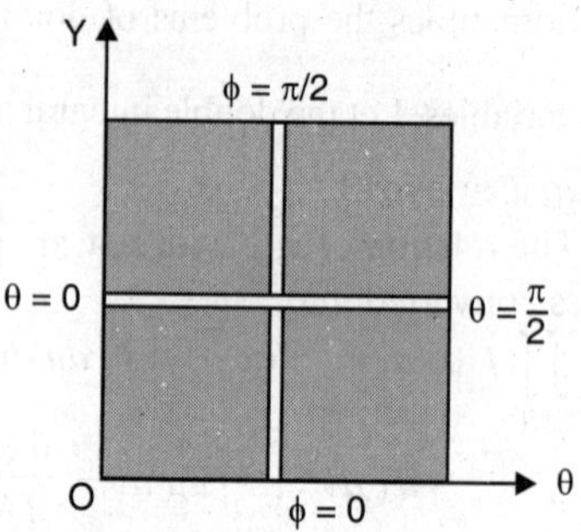

$\Rightarrow \quad x^2 + y^2 = \sin^2\phi$ and $\frac{y}{x} = \tan\theta$

limit of θ are 0 are $\frac{\pi}{2}$.

Also, the limits of ϕ are 0 and $\frac{\pi}{2}$.

$$x^2 + y^2 = \sin^2\phi = \sin^2\frac{\pi}{2} = 1$$

limits of x are 0 to $\sqrt{1-y^2}$

limits of y are 0 to 1.

$$\text{Now,}\quad d\phi\, d\theta = \frac{\partial(\phi, \theta)}{\partial(x, y)}\, dx\, dy = \begin{vmatrix} \frac{\partial\phi}{\partial x} & \frac{\partial\theta}{\partial x} \\ \frac{\partial\phi}{\partial y} & \frac{\partial\theta}{\partial y} \end{vmatrix} dx\, dy$$

$$= \begin{vmatrix} \frac{x}{\sin\phi\cos\phi} & \frac{-y\cos^2\theta}{x^2} \\ \frac{y}{\sin\phi\cos\phi} & \frac{\cos^2\theta}{x} \end{vmatrix} dx\, dy = \left[\frac{\cos^2\theta}{\sin\phi\cos\phi} + \frac{y^2\cos^2\theta}{x^2\sin\phi\cos\phi}\right] dx\, dy$$

$$= \frac{(x^2+y^2)\cos^2\theta}{x^2\sin\phi\cos\phi}\, dx\, dy = \frac{\sin^2\phi\cos^2\theta}{(\sin^2\phi\cos^2\theta)\sin\phi\cos\phi}\, dx\, dy = \frac{1}{\sin\phi\cos\phi}\, dx\, dy$$

$$\text{Again,}\quad \iint \sqrt{\left(\frac{\sin\phi}{\sin\theta}\right)}\, d\phi\, d\theta = \iint \sqrt{\left\{\left(\frac{\sin\phi}{\sin\theta}\right)\right\}}\, \frac{dx\, dy}{\sin\phi\cos\phi}$$

$$= \iint \frac{dx\, dy}{\cos\phi\sqrt{\sin\phi\sin\theta}} = \iint \frac{dx\, dy}{\sqrt{1-\sin^2\phi}\,\sqrt{\sin\phi\sin\theta}}$$

$$= \int_0^1\int_0^{\sqrt{1-y^2}} \frac{1}{\sqrt{1-(x^2+y^2)}\,\sqrt{y}}\, dx\, dy \qquad \left[\begin{matrix} x^2+y^2 = \sin^2\phi \\ y = \sin\phi\sin\theta \end{matrix}\right]$$

$$I = \int_0^1\int_0^{\sqrt{1-y^2}} \frac{dx\, dy}{\sqrt{(y)}\,\sqrt{\{(1-y^2)-x^2}} = \int_0^1 \frac{dy}{\sqrt{y}} \int_0^{\sqrt{1-y^2}} \frac{dx}{\sqrt{\{(1-y^2)-x^2\}}},$$

$$= \int_0^1 \frac{1}{\sqrt{(y)}} \left\{ \sin^{-1} \frac{x}{\sqrt{(1-y^2)}} \right\}_0^{\sqrt{(1-y^2)}} dy \;=\; \int_0^1 \frac{1}{\sqrt{(y)}} \{\sin^{-1} 1 - \sin^{-1} 0\}\, dy$$

$$= \frac{\pi}{2} \{2\sqrt{(y)}\}_0^1 = \pi$$ **Proved.**

Example 33. *Using the transformation $x + y = u$, $y = uv$ show that*

$$\int_0^1 \int_0^{1-x} e^{y/(x+y)}\, dy\, dx = \frac{1}{2}(e-1)$$

Solution. Since, $x = u(1 - v)$, $y = uv$

$$\therefore \quad J = \frac{\partial(x, y)}{\partial(u, v)} = \begin{vmatrix} \frac{\partial x}{\partial u} & \frac{\partial x}{\partial v} \\ \frac{\partial y}{\partial u} & \frac{\partial y}{\partial v} \end{vmatrix} = \begin{vmatrix} 1-v & -u \\ v & u \end{vmatrix}$$

$$= u - uv + uv = u$$

$$\therefore \quad dx\, dy = |J|\, du\, dv = u\, du\, dv$$

Also
$$x = 0 \Rightarrow u(1 - v) = 0$$
$$\Rightarrow u = 0,\ v = 1$$
$$y = 0 \Rightarrow uv = 0$$
$$\Rightarrow u = 0,\ v = 0$$
$$x + y = \Rightarrow u = 1$$

Hence, the limits of u are 0 to 1 and the limits of v are 0 to 1. The area of integration is $O'PQR$ in uv-plance.

$$\therefore \quad \int_0^1 \int_0^{1-x} e^{y(x+y)}\, dy\, dx = \int_0^1 \int_0^1 e^{uv/u} |J|\, du\, dv$$

$$= \int_0^1 \int_0^1 u e^v\, du\, dv = \left(\frac{u^2}{2}\right)_0^1 (e^v)_0^1 = \frac{1}{2}(e-1).$$ **Proved.**

Example 34. *Using the transformation $x + y = u$, $y = uv$, show that*

$\iint [xy(1 - x - y)]^{1/2}\, dx\, dy = \frac{2\pi}{105}$, *integration being taken over the area of the tringle bounded by the lines $x = 0$, $y = 0$, $x + y = 1$.*

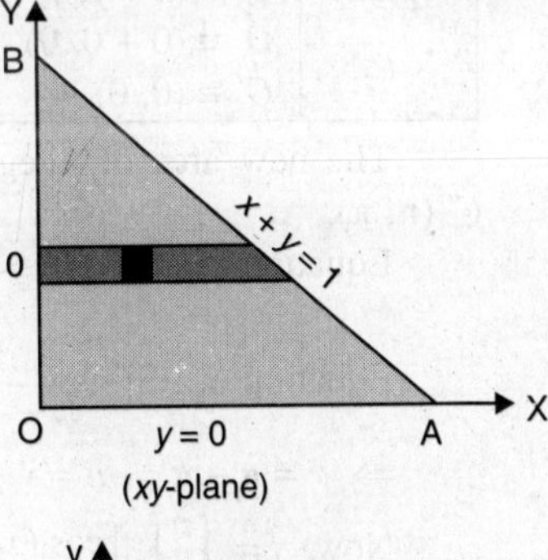

(*xy*-plane)

Solution. $\iint [xy(1 - x - y)]^{1/2}\, dx\, dy$

$$x + y = u \text{ or } x = u - y = u - uv,$$

$$dx\, dy = \frac{\partial(x, y)}{\partial(u, v)}\, du\, dv = \begin{vmatrix} \frac{\partial x}{\partial u} & \frac{\partial x}{\partial v} \\ \frac{\partial y}{\partial u} & \frac{\partial y}{\partial v} \end{vmatrix} du\, dv$$

$$dx\, dy = \begin{vmatrix} 1-v & -u \\ v & u \end{vmatrix} du\, dv = u\, du\, dv.$$

$$x = 0 \quad \Rightarrow \quad u(1 - v) = 0$$
$$\Rightarrow \quad u = 0,\ v = 1$$
$$y = 0 \quad \Rightarrow \quad uv = 0$$
$$\Rightarrow \quad u = 0,\ v = 0$$
$$x + y = 1 \quad \Rightarrow \quad u = 1$$

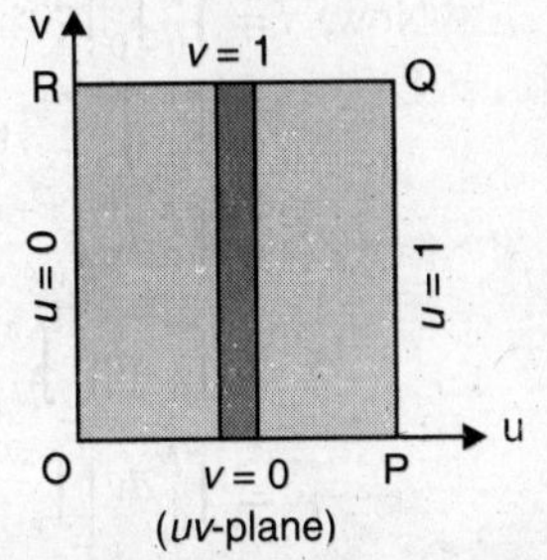

(*uv*-plane)

Hence, the limits of u are from 0 to 1 and the limits of v are from 0 to 1.
The area of integration is a square $OPQR$ in uv-plane.
On putting $x = u - uv$, $y = uv$, $dx\, dy = u\, du\, dv$ in (1), we get

$$\iint (u - uv)^{1/2}\, (uv)^{1/2}\, (1 - v)^{1/2}\, u\, du\, dv$$

$$= \int_0^1 u^2\, (1-u)^{1/2}\, du \int_0^1 v^{1/2}\, (1-v)^{1/2}\, dv = \frac{\Gamma 3}{\Gamma \frac{9}{2}} \times \frac{\Gamma\frac{3}{2}\, \Gamma\frac{3}{2}\, \Gamma\frac{3}{2}}{\Gamma 5}$$

$$= \frac{2.\Gamma\frac{3}{2}}{\frac{7}{2}\cdot\frac{5}{2}\cdot\frac{3}{2}\cdot\Gamma\frac{3}{2}} \times \frac{\frac{1}{2}\Gamma\frac{1}{2}\cdot\frac{1}{2}\Gamma\frac{1}{2}}{2} = \frac{1}{\frac{7}{2}\cdot\frac{5}{2}\cdot\frac{3}{2}} \times \frac{\frac{1}{2}\sqrt{\pi}.\frac{1}{2}\sqrt{\pi}}{1} = \frac{2\pi}{105}$$ **Ans.**

Example 35. *Using the transformation* $x + y = u$, $y = v$, *evaluate* $\int_0^{\pi}\int_0^{\pi} |\cos(x+y)|\, dx\, dy$.

Solution. Let $$I = \int_0^{\pi}\int_0^{\pi} |\cos(x+y)|\, dx\, dy$$

$$\left.\begin{matrix} x + y = u \\ y = v \end{matrix}\right\} \Rightarrow \begin{matrix} x = u - v \\ y = v \end{matrix}$$

$$J = \frac{\partial(x, y)}{\partial(u, v)} = \begin{vmatrix} \frac{\partial x}{\partial u} & \frac{\partial x}{\partial v} \\ \frac{\partial y}{\partial u} & \frac{\partial y}{\partial v} \end{vmatrix} = \begin{vmatrix} 1 & -1 \\ 0 & 1 \end{vmatrix} = 1$$

$$dx\, dy = |J|\, du\, dv = du\, dv$$

The area of integration is a square OABC in xy-plane, where O is (0, 0), $A \equiv (\pi, 0)$, $B \equiv (\pi, \pi)$, $C \equiv (0, \pi)$)

xy-plane –	$O \equiv (x, y)$ $O \equiv (1, 0)$ $O' \equiv (u, v)$	$A \equiv (x, y)$ $A \equiv (\pi, 0)$ $A' \equiv (u, v)$	$B \equiv (x, y)$ $B \equiv (2, 2)$ $B' \equiv (u, v)$	$C \equiv (x, y)$ $C \equiv (0, 1)$ $C' \equiv (u, v)$
uv-plane	$O' \equiv (x + y, y)$ $O' \equiv (0 + 0, 0)$ $O' \equiv (0, 0)$	$A' \equiv (x + y, y)$ $A' \equiv (\pi + 0, 0)$ $A' \equiv (\pi, 0)$	$B' \equiv (x + y, y)$ $B' \equiv (\pi + \pi, \pi)$ $B' \equiv (2\pi, \pi)$	$C' \equiv (x + y, y)$ $C' \equiv (0 + \pi, \pi)$ $C' \equiv (\pi, \pi)$

The new area of integration is $O'A'B'C'$ in uv-plane where $O'(0, 0)$, $A'(\pi, 0)$, $B'(2\pi, \pi)$, $C'(\pi, \pi)$.

Equation of $O'C'$ is $v = u$

Equation of $A'B'$ is $v - 0 = \dfrac{0 - \pi}{\pi - 2\pi}(u - \pi)$

$\Rightarrow v = u - \pi \Rightarrow u = v + \pi$

Now, $I = \int_0^{\pi}\int_0^{\pi} |\cos(x+y)|\, dx\, dy$

$$= \int_0^{\pi}\int_v^{\pi+v} |\cos u|\, du\, dv$$

$$= \int_0^{\pi} dv \int_v^{\pi+v} |\cos u|\, du$$

$$= \int_0^{\pi} dv \left[\int_v^{\pi/2} |\cos u|\, du + \int_{\pi/2}^{\pi} |\cos u|\, du + \int_{\pi}^{\pi+v} |\cos u|\, du\right]$$

$$= \int_0^{\pi} dv \left[\int_v^{\pi/2} \cos u\, du + \int_{\pi/2}^{\pi} -\cos u\, du + \int_{\pi}^{\pi+v} -\cos u\, du\right]$$

$$= \int_0^\pi dv\left[(\sin u)_v^{\pi/2} - (\sin u)_{\pi/2}^{\pi} - (\sin u)_{\pi}^{\pi+v}\right]$$

$$= \int_0^\pi dv\,[1 - \sin v) - (0-1) - |\sin(\pi+v) - 0|]$$

$$= \int_0^\pi dv\,[(1-\sin v) + 1 - \{-\sin v - 0|]$$

$$= \int_0^\pi dv\,[1 - \sin v + 1 + \sin v]$$

$$= \int_0^\pi dv\,(2) = 2\int_0^\pi dv = 2(v)_0^\pi = 2(\pi - 0) = 2\pi$$

Interval	cos *u*	| cos *u*|
$\left(v, \frac{\pi}{2}\right)$	+ ve	cos *u*
$\left(\frac{\pi}{2}, \pi\right)$	– ve	– cos *u*
$(\pi, \pi+v)$	– ve	– cos *u*

Example 36. *Evaluate* $\iint \sqrt{\dfrac{a^2b^2 - b^2x^2 - a^2y^2}{a^2b^2 + b^2x^2 + a^2y^2}}\,dx\,dy$ *where R is the region bounded by the ellipse* $\dfrac{x^2}{a^2} + \dfrac{y^2}{b^2} = 1$*, using polar coordinates.* (M.U. II Semester, 2009)

Solution. The region of integration is bounded by the ellipse $\dfrac{x^2}{a^2} + \dfrac{y^2}{b^2} = 1$.

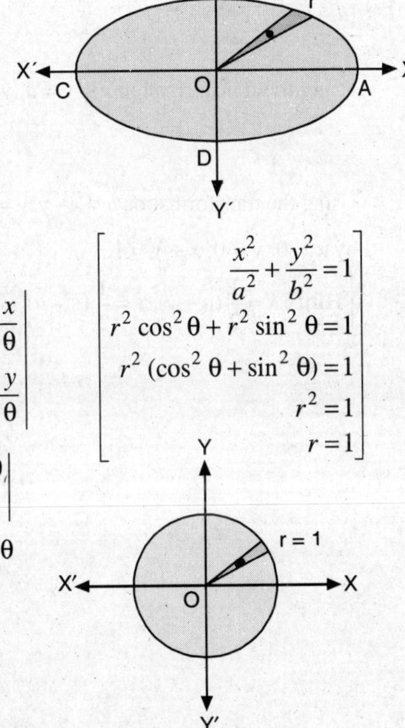

Let us substitute $\quad \dfrac{x}{a} = r\cos\theta$

$\Rightarrow \quad x = a\,r\cos\theta$

$\dfrac{y}{b} = r\sin\theta$

$\Rightarrow \quad y = b\,r\sin\theta$

$$\left[\begin{aligned} \frac{x^2}{a^2} + \frac{y^2}{b^2} &= 1 \\ r^2\cos^2\theta + r^2\sin^2\theta &= 1 \\ r^2(\cos^2\theta + \sin^2\theta) &= 1 \\ r^2 &= 1 \\ r &= 1 \end{aligned}\right]$$

$$J = \frac{\partial(x, y)}{\partial(r, \theta)} = \begin{vmatrix} \dfrac{\partial x}{\partial r} & \dfrac{\partial x}{\partial \theta} \\ \dfrac{\partial y}{\partial r} & \dfrac{\partial y}{\partial \theta} \end{vmatrix}$$

$$= \begin{vmatrix} a\cos\theta & -ar\sin\theta \\ b\sin\theta & br\cos\theta \end{vmatrix}$$

$$= abr\cos^2\theta + abr\sin^2\theta$$

$$= abr(\cos^2\theta + \sin^2\theta)$$

$$= abr$$

$$dx\,dy = J\,dr\,d\theta$$

$$= abr\,dr\,d\theta$$

$$\iint \sqrt{\left[\frac{a^2b^2 - b^2x^2 - a^2y^2}{a^2b^2 + b^2x^2 + a^2y^2}\right]}\,dx\,dy = 4\int_0^{\frac{\pi}{2}}\int_0^1 \sqrt{\left[\frac{a^2b^2 - b^2a^2r^2\cos^2\theta - a^2b^2r^2\sin^2\theta}{a^2b^2 + b^2a^2r^2\cos^2\theta + a^2b^2r^2\sin^2\theta}\right]}\,abr\,dr\,d\theta$$

$$= 4\int_0^{\frac{\pi}{2}}\int_0^1 \sqrt{\frac{a^2b^2 - a^2b^2r^2(\cos^2\theta + \sin^2\theta)}{a^2b^2 + a^2b^2r^2(\cos^2\theta + \sin^2\theta)}}\,abr\,dr\,d\theta$$

$$= 4\int_0^{\frac{\pi}{2}}\int_0^1 \sqrt{\frac{a^2b^2 - a^2b^2r^2}{a^2b^2 + a^2b^2r^2}}\,abr\,dr\,d\theta$$

$$= 4\int_0^{\frac{\pi}{2}}\int_0^1 \sqrt{\frac{a^2b^2(1-r^2)}{a^2b^2(1+r^2)}}\, abr\, dr\, d\theta = 4\int_0^{\frac{\pi}{2}}\int_0^1 \sqrt{\frac{1-r^2}{1+r^2}}\, abr\, dr\, d\theta$$

$$= 4\int_0^{\frac{\pi}{2}}\int_0^1 \sqrt{\frac{(1-r^2)\,(1-r^2)}{(1+r^2)\,(1-r^2)}}\, abr\, dr\, d\theta = 4ab\int_0^{\frac{\pi}{2}}\int_0^1 \frac{1-r^2}{\sqrt{1-r^4}}\, r\, dr\, d\theta \;\; [\text{Put } r^2 = \sin t]$$

$$= 4ab\int_0^{\frac{\pi}{2}}\int_0^{\frac{\pi}{2}} \frac{1-\sin t}{\cos t}\cdot\frac{1}{2}\cos t\, dt\, d\theta = 2ab\int_0^{\frac{\pi}{2}}\int_0^{\frac{\pi}{2}} (1-\sin t)\, dt\, d\theta$$

$$= 2ab\int_0^{\frac{\pi}{2}} [t+\cos t]_0^{\pi/2}\, d\theta = 2ab\int_0^{\frac{\pi}{2}} \left[\frac{\pi}{2}-1\right] d\theta = 2ab\left(\frac{\pi}{2}-1\right)\int_0^{\frac{\pi}{2}} d\theta$$

$$= 2ab\left(\frac{\pi}{2}-1\right)[\theta]_0^{\pi/2} = \pi ab\left(\frac{\pi}{2}-1\right).$$ **Ans.**

EXERCISE 7.4

1. Evaluate $\int_0^{\infty}\int_0^{\infty} e^{-(x+y)} \sin\left(\frac{\pi y}{x+y}\right) dx\, dy$ by means of the transformation $u = x + y$, $v = y$ from (x, y) to (u, v) **Ans.** $\frac{1}{\pi}$

2. Using the transformation $x + y = u$, $y = uv$, show that $\int_0^1\int_0^{1-x} \frac{y}{e^{x+y}}\, dy\, dx = \frac{1}{2}(e-1)$

(A.M.I.E. Winter 2001)

3. Using the transformation $u = x - y$, $v = x + y$, prove that $\iint_R \cos\frac{x-y}{x+y}\, dx\, dy = \frac{1}{2}\sin 1$ where R is bounded by $x = 0$, $y = 0$, $x + y = 1$

$\left[\textbf{Hint}: x = \frac{1}{2}(u+v),\ y = \frac{1}{2}(v-u) \text{ so that } |J| = \frac{1}{2}\right]$

APPLICATION OF THE DOUBLE INTEGRALS

(AREA, CENTRE OF GRAVITY, MASS, VOLUME)

8.1 INTRODUCTION

In this chapter, we will study how to find out area, centre of gravity, mass of lamina and the volume by revolving the area.

8.2 AREA IN CARTESIAN CO-ORDINATES

Let the curves AB and CD be $y_1 = f_1\,(x)$ and $y_2 = f_2\,(x)$.

Let the ordinates AD and BC be $x = a$ and $x = b$.

So the area enclosed by the two curves $y_1 = f_1\,(x)$ and $y_2 = f_2(x)$ and $x = a$ and $x = b$ is $ABCD$.

Let $P(x, y)$ and $Q(x + \delta x, y + \delta y)$ be two neighbouring points, thent the area of the small rectangle $PQ = \delta x.\ \delta y$.

Area of the vertical strip $= \lim\limits_{\partial y \to 0} \sum\limits_{y_1}^{y_2} \delta x\, \delta y = \delta x \int_{y_1}^{y_2} dy\, \delta x$ the width of the strip is constant throughout.

If we add all the strips from $x = a$ to $x = b$, we get

The area $ABCD = \lim\limits_{\partial x \to 0} \sum\limits_{a}^{b} \delta x \int_{y_1}^{y_2} dy = \int_a^b dx \int_{y_1}^{y_2} dy$

$$\boxed{\textbf{Area} = \int_a^b \int_{y_1}^{y_2} dx\, dy}$$

Example 1. *Find the area bounded by the parabola $y^2 = 4ax$ and its latus rectum.*

Solution. Required area = 2 (area (ASL)

$$= 2\int_0^a \int_0^{2\sqrt{ax}} dy\, dx$$

$$= 2\int_0^a 2\sqrt{ax}\, dx$$

$$= 4\sqrt{a}\left(\frac{x^{3/2}}{3/2}\right)_0^a = \frac{8a^2}{3}$$

Example 2. *Find the area between the parabolas $y^2 = 4\,ax$ and $x^2 = 4\,ay$.*

Solution.

$$y^2 = 4ax \qquad ...(1)$$

$$x^2 = 4ay \qquad ...(2)$$

On solving the equations (1) and (2) we get the point of intersection $(4a, 4a)$.

Divide the area into horizontal strips of width δy, x varies from $P, \frac{y^2}{4a}$ to $Q, \sqrt{4ay}$ and then y varies from $O(y = 0)$ to A $(y = 4a)$.

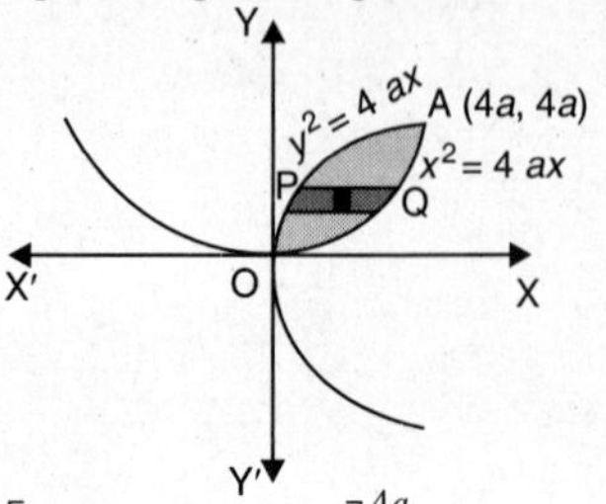

$\therefore$ The required area $= \int_0^{4a} dy \int_{y^2/4a}^{\sqrt{4ay}} dx$

$$= \int_0^{4a} dy\,[x]_{y^2/4a}^{\sqrt{4ay}} = \int_0^{4a} dy \left[\sqrt{4ay} - \frac{y^2}{4a}\right] = \left[\sqrt{4a}\,\frac{y^{3/2}}{\frac{3}{2}} - \frac{y^3}{12\,a}\right]_0^{4a}$$

$$= \left[\frac{4\sqrt{a}}{3}(4a)^{3/2} - \frac{(4a)^3}{12a}\right] = \left[\frac{32}{3}a^2 - \frac{16}{3}a^2\right] = \frac{16}{3}a^2 \qquad \textbf{Ans.}$$

Example 3. *Find by double integration the area enclosed by the pair of curves*

$$y = 2 - x \text{ and } y^2 = 2(2 - x)$$

Solution.

$$y = 2 - x$$
$$y^2 = 2\,(2 - x)$$

On solving the equations (1) and (2), we get the points of intersection (2, 0) and (0, 2).

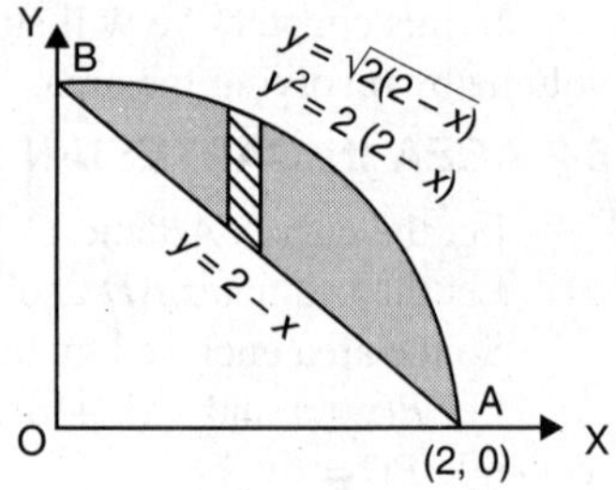

$$A = \int\int dx\,dy$$

$$\text{The required area} = \int_0^2 dx \int_{2-x}^{\sqrt{2(2-x)}} dy$$

$$= \int_0^2 dx\,[y]_{2-x}^{\sqrt{2(2-x)}} = \int_0^2 dx\,[\sqrt{4-2x} - 2 + x] = \left[\frac{2}{3\times -2}(4-2x)^{3/2} - 2x + \frac{x^2}{2}\right]_0^2$$

$$= \left[-\frac{1}{3}(4-2x)^{3/2} - 2x + \frac{x^2}{2}\right]_0^2 = \left(-4 + \frac{4}{2}\right) + \frac{8}{3} = \frac{2}{3} \qquad \textbf{Ans.}$$

Example 4. *By double integration, find the whole area of the curve $a^2x^2 = y^3(2a - y)$.*

(U.P., II Semester, Summer 2001)

Solution. $\qquad a^2x^2 = y^3(2a - y) \qquad ...(1)$

The area enclosed by the curve (1) is the shaded portion of the figure

Take a horizontal strip extending from $x = 0$ to $x = \frac{y^{3/2}\sqrt{(2a-y)}}{a}$.

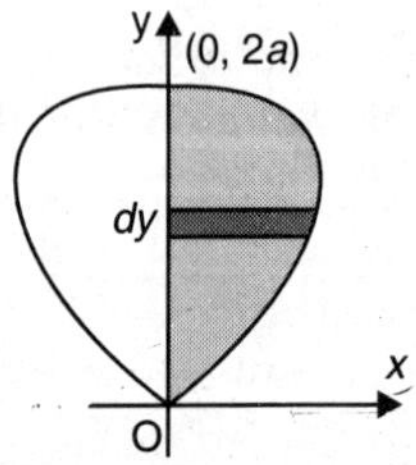

The required area

$$= \int\int dx\,dy = 2\int_0 dy \int_0^{\frac{y^{3/2}\sqrt{(2a-y)}}{a}} dx = 2\int_0^{2a} dy\,[x]_0^{\frac{y^{3/2}\sqrt{(2a-y)}}{a}}$$

$$= 2\int_0^{2a} dy\,\frac{y^{3/2}\sqrt{(2a-y)}}{a} = \frac{2}{a}\int_0^{2a} y^{3/2}\sqrt{2a-y}\,dy,$$

Put $y = 2a\sin^2\theta$ so that $dy = 4a\sin\theta\cos\theta\,d\theta$

$$\text{Area} = \frac{2}{a}\int_0^{2a} y^{3/2}\sqrt{2a-y}\,dy = \frac{2}{a}\int_0^{\frac{\pi}{2}}(2a\sin^2\theta)^{\frac{3}{2}}\sqrt{2a - 2a\sin^2\theta}\,[4a\sin\theta\cos\theta\,d\theta]$$

$$= \frac{2}{a}(2a)^{3/2}\sqrt{2a}.(4a)\int_0^{\frac{\pi}{2}}\sin^3\theta\cos\theta.\sin\theta\cos\theta\,d\theta$$

$$= 32a^2 \int_0^{\frac{\pi}{2}} \sin^4\theta \cos^2\theta \, d\theta = 32a^2 \frac{\Gamma\left(\frac{5}{2}\right)\Gamma\left(\frac{3}{2}\right)}{2\Gamma(4)}$$

$$= 32a^2 \frac{\frac{3}{2}\times\frac{1}{2}\times\Gamma\left(\frac{1}{2}\right)\times\frac{1}{2}\Gamma\left(\frac{1}{2}\right)}{2\times 3\times 2\times 1} = 32a^2 \frac{\frac{3}{2}\times\frac{1}{2}\sqrt{\pi}\times\frac{1}{2}\sqrt{\pi}}{12} = \pi a^2 \quad \textbf{Ans.}$$

Example 5. *Find the mass of the area bounded by the curves* $y^2 = x$ *and* $y = x^3$, *if* $\rho = \mu(x^2 + y^2)$.

(*Nagpur University, Summer 2008*)

Solution. Here we have

$y^2 = x$ and $y = x^3$

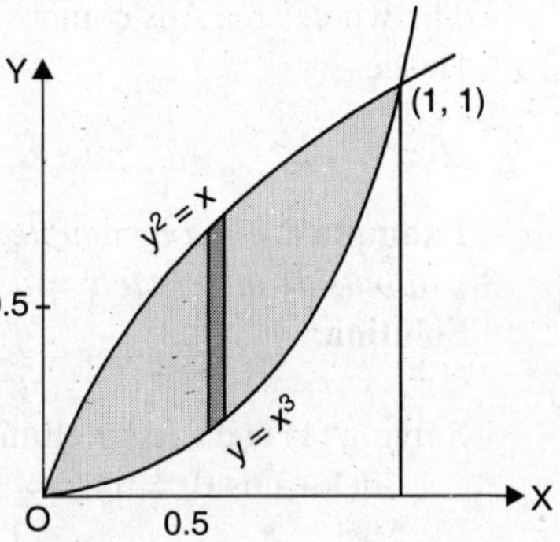

Area bounded by the curves $= \iint dx\,dy$

Mass of the area

$$= \iint \mu(x^2 + y^2)\,dx\,dy = \mu \int_0^1 \int_{x^3}^{\sqrt{x}} (x^2 + y^2)\,dx\,dy$$

$$= \mu \int_0^1 \int_{x^3}^{\sqrt{x}} (x^2 + y^2)\,dy = \mu \int_0^1 \left[x^2 y + \frac{y^3}{3}\right]_{x^3}^{\sqrt{x}}$$

$$= \mu \int_0^1 \left[\left(x^{\frac{5}{2}} + \frac{x^{\frac{3}{2}}}{3}\right) - \left(x^5 + \frac{x^9}{3}\right)\right] = \mu \left[\frac{2}{7}x^{\frac{7}{2}} + \frac{2}{15}x^{\frac{5}{2}} - \frac{x^6}{6} - \frac{x^{10}}{30}\right]_0^1 = \mu\left[\frac{2}{7} + \frac{2}{15} - \frac{1}{6} - \frac{1}{30}\right]$$

$$= \frac{\mu}{210}[60 + 28 - 35 - 7] = \frac{46}{210}\mu = \frac{23}{105}\mu \qquad \textbf{Ans}$$

EXERCISE 8.1

Use double integration in the following questions:

1. Find the area bounded by $y = x - 2$ and $y^2 = 2x + 4$. **Ans.** 18.
2. Find the area between the circle $x^2 + y^2 = a^2$ and the line $x + y = a$ in the first quadrant. **Ans.** $(\pi - 2)a^2/4$
3. Find the area of a plate in the form of quadrant of the ellipse $\frac{x^2}{a^2} + \frac{y^2}{b^2} = 1$. **Ans.** $\frac{\pi ab}{4}$
4. Find the area included between the curves $y^2 = 4a(x + a)$ and $y^2 = 4b(b - x)$. **Ans.** $\frac{8\sqrt{ab}}{3}$ (*A.M.I.E.T.E., Summer 2001*)
5. Find the area bounded by (*a*) $y^2 = 4 - x$ and $y^2 = x$. **Ans.** $\frac{16\sqrt{2}}{3}$

   (*b*) $x - 2y + 4 = 0$, $x + y - 5 = 0$, $y = 0$ (*A.M.I.E., Winter 2001*) **Ans.** $\frac{27}{2}$
6. Find the area enclosed by the leminscate $r^2 = a^2 \cos 2\theta$. **Ans.** a^2
7. Find the area common to the circles $x^2 + y^2 = a^2$ and $x^2 + y^2 = 2ax$. **Ans.** $\left[\frac{\pi}{3} - \frac{\sqrt{3}}{4}\right]a^2$
8. Find the area included between the curves $y = x^2 - 6x + 3$ and $y = 2x + 9$.

   (*A.M.I.E., Summer 2001*) **Ans.** $\frac{88\sqrt{22}}{3}$
9. Determine the area of region bounded by the curves $xy = 2$, $4y = x^2$, $y = 4$. **Ans.** $\frac{28}{3} - 4\log 2$

   (*U.P. I Semester 2003*)

8.3 AREA IN POLAR CO-ORDINATES

$$\text{Area} = \iint r\, d\theta\, dr$$

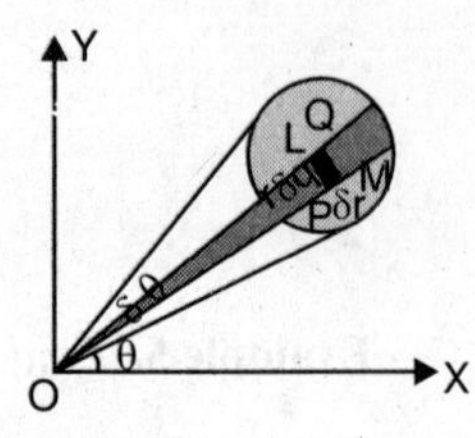

Let us consider the area enclosed by the curve $r = f(\theta)$.
Let $P(r, \theta)$, $Q(r + \delta r, \theta + \delta\theta)$ be two neighbouring points.
Draw ares PL and QM, radii r and $r + \delta r$.

$$PL = r\delta\theta,\ PM = \delta r$$

Area of rectangle $PLQM = PL \times PM$

$$= (r\delta\theta)\,(d\delta r) = r\,\delta\theta\,\delta r.$$

The whole area A is composed of such small rectangles.
Hence,

$$A = \lim_{\substack{\partial r \to 0 \\ \delta\theta \to 0}} \sum \sum r\,\delta\theta.\,\delta r = \iint r\, d\theta\, dr$$

Example 6. *Find by double integration, the area lying inside the cardioid $r = a\,(1 + \cos\theta)$ and outside the circle $r = a$.* *(Nagpur University, Winter 2000)*

Solution.
$$r = a\,(1 + \cos\theta) \qquad \ldots(1)$$
$$r = a \qquad \ldots(2)$$

Solving (1) and (2), by eliminating r, we get

$$a(1 + \cos\theta) = a \quad \Rightarrow \quad 1 + \cos\theta = 1$$

$$\cos\theta = 0 \quad \Rightarrow \quad \theta = -\frac{\pi}{2} \text{ or } \frac{\pi}{2}$$

limits of r are a and $a(1 + \cos\theta)$

limits of θ are $-\dfrac{\pi}{2}$ to $\dfrac{\pi}{2}$

Required area = Area ABCDA

$$= \int_{-\pi/2}^{\pi/2} \int_{r \text{ for circle}}^{\text{for cardioid}} r\, d\theta\, dr$$

$$= \int_{-\pi/2}^{\pi/2} \int_{a}^{a(1+\cos\theta)} r\, d\theta\, dr \qquad = \int_{-\pi/2}^{\pi/2} \left(\frac{r^2}{2}\right)_a^{a(1+\cos\theta)} d\theta$$

$$= \frac{a^2}{2} \int_{-\pi/2}^{\pi/2} [(1 + \cos\theta)^2 - 1]\, d\theta \qquad = \frac{a^2}{2} \int_{-\pi/2}^{\pi/2} (\cos^2\theta + 2\cos\theta)\, d\theta$$

$$= a^2 \int_0^{\pi/2} (\cos^2\theta + 2\cos\theta)\, d\theta \qquad = a^2 \left[\int_0^{\pi/2} \cos^2\theta\, d\theta + 2\int_0^{\pi/2} \cos\theta\, d\theta\right]$$

$$= a^2 \left[\frac{\pi}{4} + 2\,(\sin\theta)_0^{\pi/2}\right] = a^2 \left[\frac{\pi}{4} + 2\right] \qquad = \frac{a^2}{4}(\pi + 8) \qquad \textbf{Ans.}$$

Example 7. *Find by double integration, the area lying inside the circle $r = a\sin\theta$ and outside the cardioid $r = a\,(1 - \cos\theta)$.*

Solution. We have,

$$r = a\sin\theta \qquad \ldots(1)$$
$$r = a\,(1 - \cos\theta) \qquad \ldots(2)$$

Solving (1) and (2) by eliminating r, we have

$$\sin\theta = 1 - \cos\theta \Rightarrow \sin\theta + \cos\theta = 1$$

Squaring above, we get

$$\sin^2\theta + \cos^2\theta + 2\sin\theta\cos\theta = 1$$

$$\Rightarrow \quad 1 + \sin 2\theta = 1 \Rightarrow \sin 2\theta = 0 \Rightarrow 2\theta = 0 \text{ or } \pi = \theta = 0 \text{ or } \frac{\pi}{2}$$

The required area is shaded portion in the fig.

Limits of r are $a(1 - \cos\theta)$ and $a \sin\theta$, limits of θ are 0 and $\frac{\pi}{2}$

$$\text{Required area} = \int_0^{\frac{\pi}{2}} \int_{a(1-\cos\theta)}^{a\sin\theta} r\,dr\,d\theta$$

$$= \int_0^{\frac{\pi}{2}} \left[\frac{r^2}{2}\right]_{a(1-\cos\theta)}^{a\sin\theta} d\theta = \frac{1}{2}\int_0^{\pi/2} a^2\,[\sin^2\theta - (1-\cos\theta)^2]\,d\theta$$

$$= \frac{a^2}{2}\int_0^{\pi/2} (\sin^2\theta - 1 - \cos^2\theta + 2\cos\theta]\,d\theta$$

$$= \frac{a^2}{2}\int_0^{\pi/2} (-2\cos^2\theta + 2\cos\theta)\,d\theta$$

$$= \frac{a^2}{2}\left[\int_0^{\pi/2} -2\cos^2\theta\,d\theta + \int_0^{\pi/2} 2\cos\theta\,d\theta\right]$$

$$= \frac{a^2}{2}\left[\left(-2.\frac{\pi}{4}\right) + 2\,(\sin\theta)_0^{\pi/2}\right]$$

$$= \frac{a^2}{2}\left[-\frac{\pi}{2} + 2\left(\sin\frac{\pi}{2} - \sin 0\right)\right] = \frac{a^2}{2}\left[-\frac{\pi}{2} + 2\right] = a^2\left(1 - \frac{\pi}{4}\right) \quad \textbf{Ans.}$$

Example 8. *Find by double integration, the area lying inside a cardioid $r = 1 + \cos\theta$ and outside the parabola $r\,(1 + \cos\theta) = 1$.*

Solutio. We have,

$$r = 1 + \cos\theta \quad ...(1)$$

$$r\,(1 + \cos\theta) = 1 \quad ...(2)$$

Solving (1) and (2), we get

$$(1 + \cos\theta)\,(1 + \cos\theta) = 1$$

$$(1 + \cos\theta)^2 = 1$$

$$1 + \cos\theta = 1$$

$$\cos\theta = 0 \Rightarrow \theta = \pm\frac{\pi}{2}$$

limits of r are $1 + \cos\theta$ and $\frac{1}{1+\cos\theta}$

limits of θ are $-\frac{\pi}{2}$ to $\frac{\pi}{2}$.

Required area = Area *ADCBA* (Shaded portion)

$$= \int_{-\pi/2}^{\pi/2} \int_{\frac{1}{1+\cos\theta}}^{1+\cos\theta} r\,d\theta\,dr = \int_{\frac{-\pi}{2}}^{\frac{\pi}{2}} \left(\frac{r^2}{2}\right)_{\frac{1}{1+\cos\theta}}^{1+\cos\theta} d\theta = \frac{1}{2}\int_{-\pi/2}^{\pi/2} \left[(1+\cos\theta)^2 - \frac{1}{(1+\cos\theta)^2}\right] d\theta$$

$$= \frac{1}{2}\int_{-\pi/2}^{\pi/2} \left[(1+\cos^2\theta + 2\cos\theta) - \frac{1}{\left(2\cos^2\frac{\theta}{2}\right)^2}\right] d\theta$$

$$= 2 \times \frac{1}{2}\int_0^{\pi/2} \left[(1+\cos^2\theta + 2\cos\theta) - \frac{1}{4}\sec^4\frac{\pi}{2}\right] d\theta$$

$$= \int_0^{\pi/2}\left[(1+\cos^2\theta+2\cos\theta)-\frac{1}{4}\left(1+\tan^2\frac{\theta}{2}\right)\sec^2\frac{\theta}{2}\right]d\theta$$

$$= \int_0^{\pi/2}\left[\left(1+\frac{1+\cos 2\theta}{2}+2\cos\theta\right)-\frac{1}{4}\left(1+\tan^2\frac{\pi}{2}\right)\sec^2\frac{\pi}{2}\right]d\theta$$

$$= \int_0^{\pi/2}\left[1+\frac{1}{2}+\frac{\cos 2\theta}{2}+2\cos\theta-\frac{1}{4}\left(\sec^2\frac{\theta}{2}+\tan^2\frac{\theta}{2}\times\sec^2\frac{\theta}{2}\right)\right]d\theta$$

$$= \left[\theta+\frac{\theta}{2}+\frac{\sin 2\theta}{4}+2\sin\theta-\frac{1}{4}\left(2\tan\frac{\theta}{2}+\frac{2}{3}\tan^3\frac{\theta}{2}\right)\right]_0^{\frac{\pi}{2}}$$

$$= \left[\frac{\pi}{2}+\frac{\pi}{4}+0+2\sin\frac{\pi}{2}-\frac{1}{2}\tan\frac{\pi}{4}-\frac{1}{6}\tan^3\frac{\pi}{4}\right] = \left[\frac{3\pi}{4}+2-\frac{1}{2}-\frac{1}{6}\right] = \left[\frac{3\pi}{4}+\frac{4}{3}\right] \qquad \textbf{Ans.}$$

Example 9. *Find by double integration the area included between the curves $r = a\,(\sec\theta + \cos\theta)$ and its asymptotes.*

Solution. We have,

Equation of curve is, $r = a\,(\sec\theta + \cos\theta)$ ...(1)

Equation of asymptotes CD is, $r = a\sec\theta$ ...(2)

limits of r are $a\,(\sec\theta + \cos\theta)$ and $a\sec\theta$

limits of θ are $-\dfrac{\pi}{2}$ and $\dfrac{\pi}{2}$

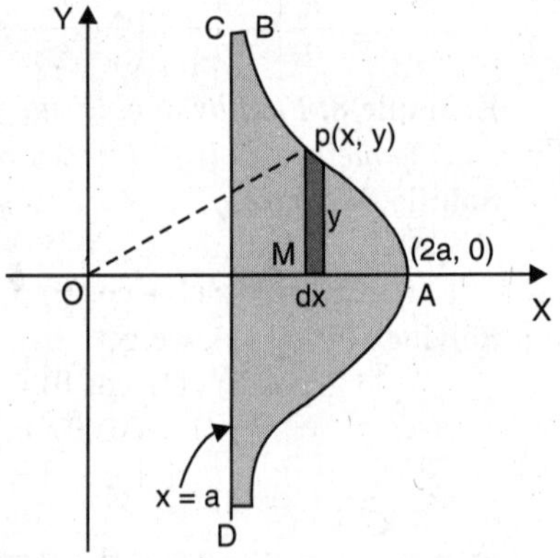

Required area = Shaded portion of the figure

$$= \int_{-\pi/2}^{\pi/2}\int_{a\sec\theta}^{a(\sec\theta+\cos\theta)} r\,dr\,d\theta$$

$$= \int_{-\pi/2}^{\pi/2}\left(\frac{r^2}{2}\right)_{a\sec\theta}^{a(\sec\theta+\cos\theta)} d\theta$$

$$= \frac{a^2}{2}\int_{-\pi/2}^{\pi/2}[(\sec\theta+\cos\theta)^2-\sec^2\theta]\,d\theta$$

$$= \frac{a^2}{2}\int_{-\pi/2}^{\pi/2}(\cos^2\theta+2)\,d\theta = \frac{2a^2}{2}\int_0^{\pi/2}(\cos^2\theta+2)\,d\theta$$

$$= a^2\left(\frac{1}{2}.\frac{\pi}{2}+2.\frac{\pi}{2}\right) = a^2\left(\frac{\pi}{4}+\pi\right) = \frac{5\pi}{4}a^2. \qquad \textbf{Ans.}$$

Example 10. *Find the area included between the curve $x = a(\theta - \sin\theta)$, $y = a\,(1 - \cos\theta)$ and its base.*

Solution. When $\theta = 0$

$$x = a\,(0 - \sin 0) = a \times 0 = 0$$
$$y = a\,(1 - \cos 0) = a\,(1 - 1) = a \times 0 = 0$$

When $\theta = 2\pi$

$$x = a\,(2\pi - \sin 2\pi) = a\,(2\pi - 0) = 2a\pi$$
$$y = a\,(1 - \cos 2\pi) = a(1 - 1) = a \times 0 = 0$$

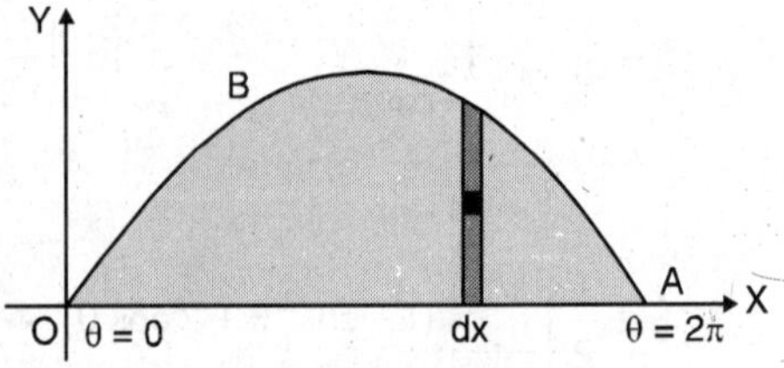

Therefore, the limits for θ are from 0 to 2π.

Required area = area *OBAO*

$$= \int_0^{2\pi} y\frac{dx}{d\theta}\,d\theta = \int_0^{2\pi} a(1-\cos\theta)\,a\,(1-\cos\theta)\,d\theta$$

$$= a^2\int_0^{2\pi}(1-\cos\theta)^2\,d\theta = 4a^2\int_0^{2\pi}\sin^4\frac{\theta}{2}\,d\theta$$

$$= 8a^2 \int_0^{\pi} \sin^4 \phi \, d\phi \qquad \left[\begin{array}{l} \text{Put } \dfrac{\theta}{2} = \phi \\ \Rightarrow d\theta = 2d\phi \end{array}\right]$$

$$= 8a^2 . 2\int_0^{\pi/2} \sin^4 \phi \, d\phi = 16a^2 \cdot \frac{3}{4} \cdot \frac{1}{2} \cdot \frac{\pi}{2} = 3a^2\pi \qquad \textbf{Ans.}$$

EXERCISE 8.2

1. Find the area of cardioid $r = a(1 + \cos\theta)$. **Ans.** $\dfrac{3\pi a^2}{2}$
2. Find the area of the curve $r^2 = a^2 \cos 2\theta$. **Ans.** a^2
3. Find the area enclosed by the curve $r = 2\,a \cos\theta$ **Ans.** πa^2
4. Find the area enclosed by the curve $r = 3 + 2 \cos\theta$. **Ans.** $11\,\pi$
5. Find the area enclosed by the curve

   $r^3 = a^2 \cos^2\theta + b^2 \sin^2\theta$. **Ans.** $\dfrac{\pi}{2}(a^2 + b^2)$
6. Show that the area of the region included between the cardioides $r = a(1 + \cos\theta)$ and $r = a\,(1 - \cos\theta)$ is $\dfrac{a^2}{2}(3\pi - 8)$.
7. Find the area outside the circle $r = 2$ and inside the cardioid $r = 2(1 + \cos\theta)$. **Ans.** $(\pi + 8)$
8. Find the area inside the circle $r = 2a\cos\theta$ and outside the circle $r = a$. **Ans.** $2a^2\left(\dfrac{\pi}{3} + \dfrac{\sqrt{3}}{4}\right)$
9. Find the area inside the circle $r = 4 \sin\theta$ and outside the lemniscate $r^2 = 8 \cos 2\theta$.

   Ans. $\left(\dfrac{8}{3}\pi + 4\sqrt{3} - 4\right)$

8.4 VOLUME OF SOLID BY ROTATION OF AN AREA (DOUBLE INTEGRAL)

When the area enclosed by a curve $y = f(x)$ is revolved about an axis, a solid is generated, we have to find out the volume of solid generated.

Volume of the solid generated about x-axis

$$= \int_a^b \int_{y_1(x)}^{y_2(x)} 2\pi\, PQ\, dx\, dy$$

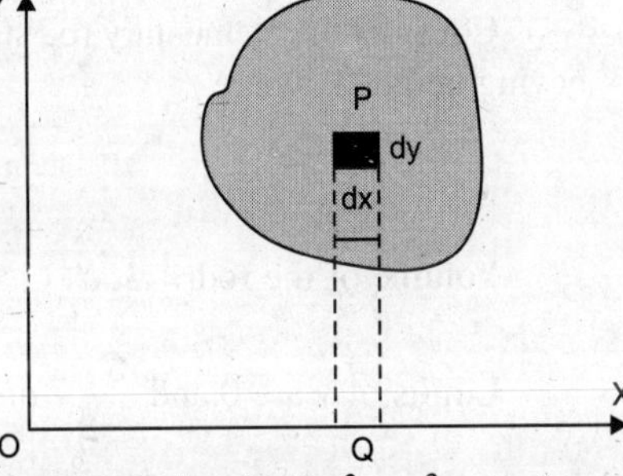

Example 11. *Find the volume of the torus generated by revolving the circle $x^2 + y^2 = 4$ about the line $x = 3$.*

Solution. $x^2 + y^2 = 4$

$$V = \int\int (2\pi\, PQ)\, dx\, dy = 2\pi \int\int (3 - x)\, dx\, dy$$

$$= 2\pi \int_{-2}^{+2} dx \int_{-\sqrt{4-x^2}}^{+\sqrt{4-x^2}} (3 - x)\, dy$$

$$= 2\pi \int_{-2}^{+2} dx\, (3y - xy)_{-\sqrt{4-x^2}}^{+\sqrt{4-x^2}}$$

$$= 2\pi \int_{-2}^{+2} dx\, [3\sqrt{4-x^2} - x\sqrt{4-x^2} + 3\sqrt{4-x^2} - x\sqrt{4-x^2}]$$

$$= 4\pi [3\sqrt{4-x^2} - x\sqrt{4-x^2}]\, dx = 4\pi \left[3\frac{x}{2}\sqrt{4-x^2} + 3 \times \frac{4}{2}\sin^{-1}\frac{x}{2} + \frac{1}{3}(4-x^2)^{3/2}\right]_{-2}^{2}$$

$$= 4\pi\left[6\times\frac{\pi}{2}+6\times\frac{\pi}{2}\right] = 24\,\pi^2$$ **Ans.**

Example 12. *Calculate by double integration the volume generated by the revolution of the cardioid $r = a\,(1 - \cos\theta)$ about its axis.* *(AMIETE, June 2010)*

Solution. $r = a\,(1 - \cos\theta)$

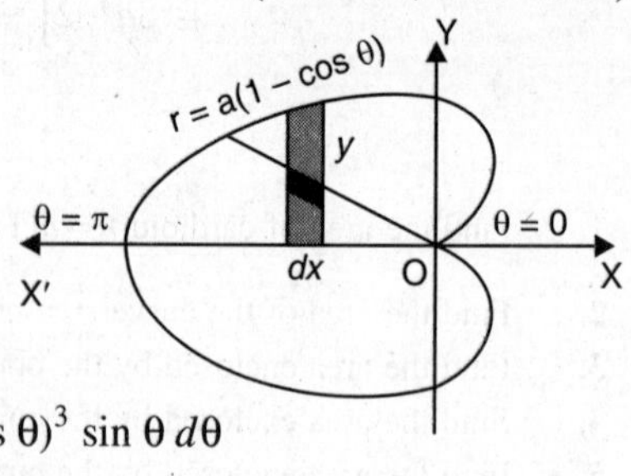

$$V = 2\pi\int\int y\,dx\,dy \Rightarrow V = 2\pi\int\int (r\,d\theta\,dr)\,y$$

$$= 2\pi\int d\theta\int r\,dr\,(r\sin\theta)$$

$$= 2\pi\int_0^{\pi}\sin\theta\,d\theta\int_0^{a(1-\cos\theta)} r^2\,dr$$

$$= 2\pi\int_0^{\pi}\sin\theta\,d\theta\left[\frac{r^3}{3}\right]_0^{a(1-\cos\theta)} = \frac{2\pi}{3}\int_0^{\pi} a^3\,(1-\cos\theta)^3\sin\theta\,d\theta$$

$$= \frac{2\pi a^3}{3}\left[\frac{(1-\cos\theta)^4}{4}\right]_0^{\pi} = \frac{2\pi a^3}{12}[16] = \frac{8}{3}\pi a^3$$ **Ans.**

Example 13. *A pyramid is bounded by the three co-ordinate planes and the plane $x + 2y + 3z = 6$. Compute this volume by double integration.*

Solution. $x + 2y + 3z = 6$ ...(1)

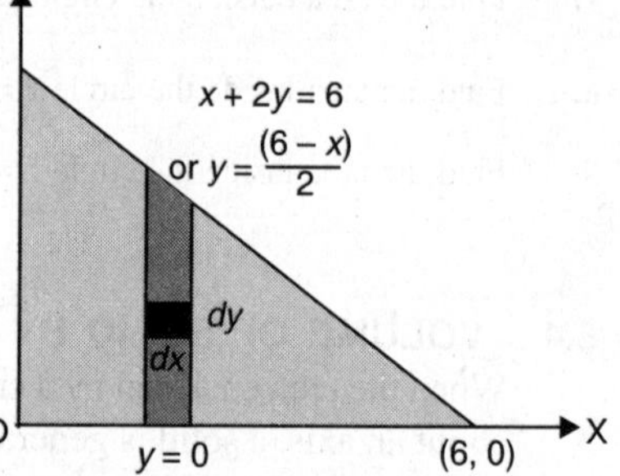

$x = 0,\ y = 0,\ z = 0$ are co-ordinate planes.

The line of intersection of plane (1) and xy plane ($z = 0$) is

$$x + 2y = 6 \qquad ...(2)$$

The base of the pyramid may be taken to be the triangle bounded by x-axis, y-axis and the line (2).

An elementary area on the base is $dx\,dy$.

Consider the elementary rod standing on this area and having height z, where

$$3z = 6 - x - 2y \text{ or } z = \frac{6-x-2y}{3}$$

Volume of the rod = $dx\,dy,\ z$, Limits for z are 0 and $\frac{6-x-2y}{3}$.

Limits of y are 0 and $\frac{6-x}{2}$ and limits of x are 0 and 6.

$$\text{Required volume} = \int_0^6\int_0^{\frac{6-x}{2}} z\,dx\,dy = \int_0^6 dx\int_0^{\frac{6-x}{2}}\frac{6-x-2y}{3}\,dy$$

$$= \frac{1}{3}\int_0^6 dx\,\left(6x - xy - y^2\right)_0^{\frac{6-x}{2}} = \frac{1}{3}\int_0^6\left(\frac{6(6-x)}{2} - \frac{x(6-x)}{2} - \left(\frac{6-x}{2}\right)^2\right)dx$$

$$= \frac{1}{3}\int_0^6\left(\frac{36-6x}{2} - \frac{6x-x^2}{2} - \frac{36+x^2-12x}{4}\right)dx$$

$$= \frac{1}{12}\int_0^6 (72 - 12x - 12x + 2x^2 - 36 - x^2 + 12x)\,dx$$

$$= \frac{1}{12}\int_0^6 (x^2 - 12x + 36)\,dx = \frac{1}{12}\left[\frac{x^3}{3} - \frac{12x^2}{2} + 36x\right]_0^6$$

$$= \frac{1}{12}[72 - 216 + 216] = 6$$ **Ans.**

EXERCISE 8.3

1. Find the volume of the sphere $x^2 + y^2 + z^2 = a^2$ by revolving area of the circle $x^2 + y^2 = a^2$. **Ans.** $\frac{4}{3}\pi a^3$

8.5 CENTRE OF GRAVITY

$$\bar{x} = \frac{\int\int \rho\, x\, dx\, dy}{\int\int \rho\, dx\, dy}, \quad \bar{y} = \frac{\int\int \rho\, y\, dx\, dy}{\int\int \rho\, dx\, dy}$$

Example 14. *Find the position of the C.G. of a semi-circular lamina of radius a if its density varies as the square of the distance from the diameter.* (*AMIETE, Dec. 2010*)

Solution. Let the bounding diameter be as the x-axis and a line perpendicular to the diameter and passing through the centre is y-axis. Equation of the circle is $x^2 + y^2 = a^2$. By symmetry $\bar{x} = 0$.

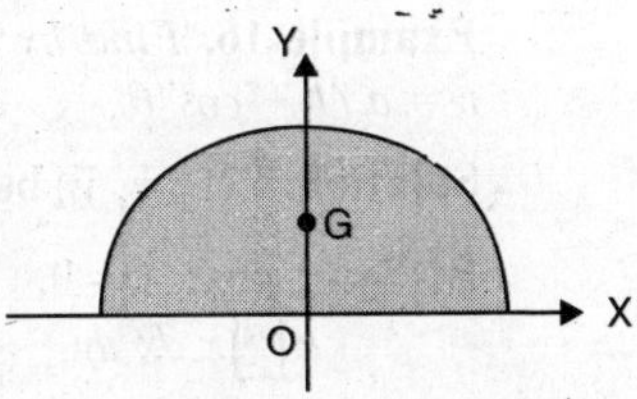

$$\bar{y} = \frac{\int\int y\, \rho\, dx\, dy}{\int\int \rho\, dx\, dy} = \frac{\int\int (\lambda\, y^2)\, y\, dx\, dy}{\int\int (\lambda\, y^2)\, dx\, dy} = \frac{\int_{-a}^{a} dx \int_0^{\sqrt{a^2 - x^2}} y^3\, dy}{\int_{-a}^{a} dx \int_0^{\sqrt{a^2 - x^2}} y^2\, dy}$$

Y, G, O, X

Put $x = a \sin\theta$

$$= \frac{\int_{-a}^{a} dx \left[\frac{y^4}{4}\right]_0^{\sqrt{a^2 - x^2}}}{\int_{-a}^{a} dx \left(\frac{y^3}{3}\right)_0^{\sqrt{a^2 - x^2}}} = \frac{3\int_{-a}^{a} (a^2 - x^2)^2\, dx}{4\int_{-a}^{a} (a^2 - x^2)^{3/2}\, dx}$$

$$= \frac{3\int_{-\frac{\pi}{2}}^{\frac{\pi}{2}} (a^2 - a^2 \sin^2\theta)^2\, a\cos\theta\, d\theta}{4\int_{-\frac{\pi}{2}}^{\frac{\pi}{2}} (a^2 - a^2\sin^2\theta)^{3/2}\, a\cos\theta\, d\theta} = \frac{3\int_{-\frac{\pi}{2}}^{\frac{\pi}{2}} a^5 \cos^5\theta\, d\theta}{4\int_{-\frac{\pi}{2}}^{\frac{\pi}{4}} a^4 \cos^4\theta\, d\theta}$$

$$= \frac{3a}{4}\, \frac{\frac{4\times 2}{5\times 3}}{\frac{3\times 1}{4\times 2}\frac{\pi}{2}} = \left(\frac{3a}{4}\right)\left(\frac{8}{15}\right)\left(\frac{16}{3\pi}\right) = \frac{32a}{15\pi}$$

Hence C.G. is $\left(0, \frac{32a}{15\pi}\right)$ **Ans.**

Example 15. *Find C.G. of the area in the positive quadrant of the curve* $x^{2/3} + y^{2/3} = a^{2/3}$.

Solution. For C.G. of area; $\bar{x} = \frac{\int\int x\, dx\, dy}{\int\int dx\, dy}, \quad \bar{y} = \frac{\int\int y\, dx\, dy}{\int\int dx\, dy}$

$$\bar{x} = \frac{\int_0^a x\, dx \int_0^{(a^{2/3} - x^{2/3})^{3/2}} dy}{\int_0^a dx \int_0^{(a^{2/3} - x^{2/3})^{3/2}} dy} = \frac{\int_0^a x\, dx\, [y]_0^{(a^{2/3} - x^{2/3})^{3/2}}}{\int_0^a dx\, [y]_0^{(a^{23} - x^{2/3})^{3/2}}}$$ [Put $x = a\cos^3\theta$]

$$= \frac{\int_0^a x\, dx\, (a^{2/3} - x^{2/3})^{3/2}}{\int_0^a dx\, (a^{2/3} - x^{2/3})^{3/2}} = \frac{\int_{\frac{\pi}{2}}^{0} a\cos^3\theta\, (a^{2/3} - a^{2/3}\cos^2\theta)^{3/2}\, (-3a\cos^2\theta\sin\theta\, d\theta)}{\int_{\frac{\pi}{2}}^{0} (a^{2/3} - a^{2/3}\cos^2\theta)^{3/2}\, (-3a\cos^2\theta\sin\theta\, d\theta)}$$

$$= \frac{\int_0^{\frac{\pi}{2}} 3a^3 \cos^3\theta \sin^3\theta \cos^2\theta \sin\theta \, d\theta}{\int_0^{\frac{\pi}{2}} 3a^2 \sin^3\theta \cos^2\theta \sin\theta \, d\theta} = \frac{a\int_0^{\frac{\pi}{2}} \sin^4\theta \cos^5\theta \, d\theta}{\int_0^{\frac{\pi}{2}} \sin^4\theta \cos^2\theta \, d\theta} = \frac{\dfrac{\Gamma\left(\frac{5}{2}\right)\Gamma\left(\frac{6}{2}\right)}{2\,\Gamma\left(\frac{4}{2}\right)}a}{\dfrac{\Gamma\left(\frac{5}{2}\right)\Gamma\left(\frac{3}{2}\right)}{2\,\Gamma(4)}}$$

$$= \frac{\Gamma(3)\,\Gamma(4)\, a}{\Gamma\left(\frac{3}{2}\right)\Gamma\left(\frac{11}{2}\right)} = \frac{(2)\,(6)\, a}{\frac{1}{2}\cdot\frac{9}{2}\cdot\frac{7}{2}\cdot\frac{5}{2}\cdot\frac{3}{2}\cdot\frac{1}{2}\,\pi} = \frac{256\, a}{315\,\pi}, \quad \text{Similarly,} \quad \bar{y} = \frac{256\, a}{315\,\pi}$$

Hence, C.G. of the area is $\left(\dfrac{256\,a}{315\,\pi}, \dfrac{256\,a}{315\,\pi}\right)$.

Example 16. *Find by double integration, the centre of gravity of the area of the cardioid $r = a\,(1 + \cos\theta)$.*

Solution. Let $(\bar{x}, \bar{y})$ be the C.G. the cardioid

By Symmetry, $\bar{y} = 0$.

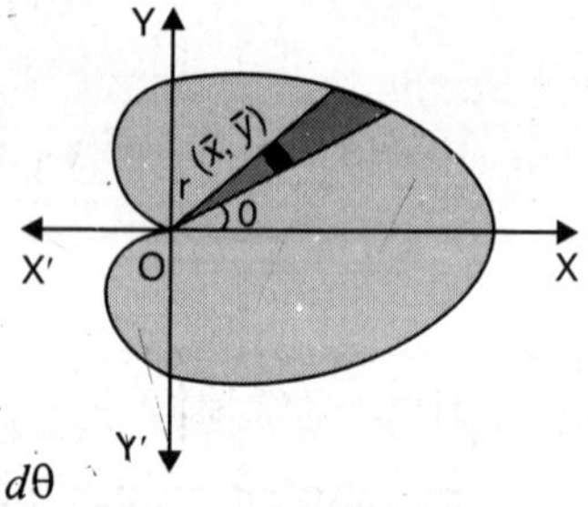

$$\bar{x} = \frac{\iint_A x \, dx \, dy}{\iint_A dx \, dy} = \frac{\iint_A x \, dx \, dy}{\iint_A dx \, dy}$$

$$= \frac{\int_{-\pi}^{\pi}\int_0^{a(1+\cos\theta)} (r\cos\theta)\,(r\,d\theta\,dr)}{\int_{-\pi}^{\pi}\int_0^{a(1+\cos\theta)} r\,d\theta\,dr} = \frac{\int_{-\pi}^{\pi}\cos\theta\,d\theta\int_0^{a(1+\cos\theta)} r^2\,dr}{\int_{-\pi}^{\pi}d\theta\int_0^{a(1+\cos\theta)} r\,dr}$$

$$= \frac{\int_{-\pi}^{\pi}\cos\theta\,d\theta\left[\frac{r^3}{3}\right]_0^{a\,(a+\cos\theta)}}{\int_{-\pi}^{\pi}d\theta\left(\frac{r^2}{2}\right)_0^{a\,(1+\cos\theta)}} = \frac{\int_{-\pi}^{\pi}\cos\theta\,d\theta.\,\frac{a^3}{3}(1+\cos\theta)^3}{\int_{-\pi}^{\pi}d\theta\,\frac{a^2}{2}(1+\cos\theta)^2}$$

$$= \frac{\frac{a^3}{3}\int_{-\pi}^{\pi}\left(2\cos^2\frac{\theta}{2}-1\right)\left(1+2\cos^2\frac{\theta}{2}-1\right)^3 d\theta}{\frac{a^2}{2}\int_{-\pi}^{\pi}\left(1+2\cos^2\frac{\theta}{2}-1\right)d\theta}$$

$$= \frac{a^3}{3}\int_{-\pi}^{\pi}\left(2\cos^2\frac{\theta}{2}-1\right)\left(8\cos^6\frac{\theta}{2}\right)d\theta \div \frac{a^2}{2}\int_{-\pi}^{\pi}4\cos^4\frac{\theta}{2}\,d\theta$$

$$= \frac{8a^3}{3}\int_{-\pi}^{\pi}\left(2\cos^8\frac{\theta}{2}-\cos^6\frac{\theta}{2}\right)d\theta \div 2a^2\int_{-\pi}^{\pi}\cos^4\frac{\theta}{2}\,d\theta$$

$$= \frac{2\times 8a^3}{3}\int_0^{\pi}\left(2\cos^8\frac{\theta}{2}-\cos^6\frac{\theta}{2}\right)d\theta \div 4a^2\int_0^{\pi}\cos^4\frac{\theta}{2}\,d\theta$$

$$= \frac{16a^3}{3}\int_0^{\pi/2}(2\cos^8 t - \cos^6 t)\,(2\,dt) \div 4a^2\int_0^{\pi/2}\cos^4 t\,(2\,dt)$$

$$= \frac{32\,a^3}{3}\left[\frac{2\times 7\times 5\times 3\times 1}{8\times 6\times 4\times 2}\,\frac{\pi}{2} - \frac{5\times 3\times 1}{6\times 4\times 2}\,\frac{\pi}{2}\right] \div 8a^2\left(\frac{3\times 1}{4\times 2}\,\frac{\pi}{2}\right)$$

$$= \frac{32a^3}{3}\left(\frac{35\pi}{128}-\frac{5\pi}{32}\right) \div 8a^2\left(\frac{3\pi}{16}\right) = \frac{8a^3}{3}\times\frac{15\pi}{128}\times\frac{16}{8a^2\times 3\pi} = \frac{5a}{24}$$

Ans.

Example 17. *OA is the diameter of semiciruclar disc, the density at any point varies its distance from O. Find the position of entire of gravity given given that OA = a.*

Solution. $\rho \propto r \Rightarrow \rho = k\,r$, where k is constant.

where κ is constant.

r varies from $r = 0$ to $r = a\cos\theta$ and θ varies from $\theta = 0$ to $\theta = \dfrac{\pi}{2}$.

If $(\bar{x}, \bar{y})$ are the coordinates of centre of gravity, then

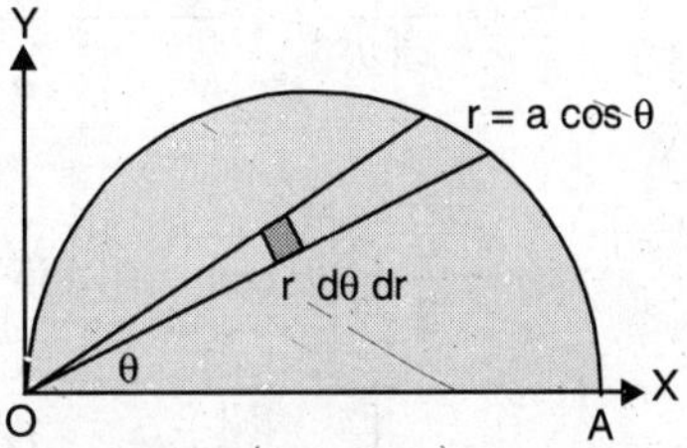

Y, r = a cos θ, r dθ dr, θ, O, A, X

$$\bar{x} = \frac{\iint x\rho\, dA}{\iint \rho\, dA} = \frac{\int_0^{\frac{\pi}{2}}\int_0^{a\cos\theta}(r\cos\theta)(k\,r)(r\,d\theta\,dr)}{\int_0^{\frac{\pi}{2}}\int_0^{a\cos\theta}(k\,r)(r\,d\theta\,dr)}$$

$$= \frac{\int_0^{\frac{\pi}{2}}\cos\theta\,d\theta\int_0^{a\cos\theta} r^3\,dr}{\int_0^{a\cos\theta} r^2\,dr} = \frac{\int_0^{\frac{\pi}{2}}\cos\theta\,d\theta\left[\frac{r^4}{4}\right]_0^{a\cos\theta}}{\int_0^{\frac{\pi}{2}}\left[\frac{r^3}{3}\right]_0^{a\cos\theta}} = \frac{\frac{1}{4}\int_0^{\frac{\pi}{2}}\cos\theta\,a\,\theta(a^4\cos^4\theta)}{\frac{1}{3}\int_0^{\frac{\pi}{2}} d\theta(a^3\cos^3\theta)}$$

$$= 3\frac{a^4}{4a^3}\frac{\int_0^{\frac{\pi}{2}}\cos^5\theta\,d\theta}{\int_0^{\frac{\pi}{2}}\cos^3\theta\,d\theta} = \frac{3a}{4}\left[\frac{\frac{4\times 2}{5\times 3}}{\frac{2}{3}}\right] = \frac{3a}{4}\left(\frac{4}{5}\right) = \frac{3a}{4}$$

$$\bar{y} = \frac{\iint y\rho\,dA}{\iint \rho\,dA} = \frac{\int_0^{\frac{\pi}{2}}\int_0^{a\cos\theta}(r\sin\theta)k\,r(r\,d\theta\,dr)}{\int_0^{\frac{\pi}{2}}\int_0^{a\cos\theta}(kr)\,r\,d\theta\,dr} \qquad \rho = k\,r$$

$$= \frac{k\int_0^{\frac{\pi}{2}}\sin\theta\,d\theta\int_0^{a\cos\theta} r^3\,dr}{k\int_0^{\frac{\pi}{2}}d\theta\int_0^{a\cos\theta} r^2\,dr} = \frac{\int_0^{\frac{\pi}{2}}\sin\theta\,d\theta\left[\frac{r^4}{4}\right]_0^{a\cos\theta}}{\int_0^{\frac{\pi}{2}}d\theta\left[\frac{r^3}{3}\right]_0^{a\cos\theta}}$$

$$= \frac{\frac{1}{4}\int_0^{\frac{\pi}{2}}\sin\theta\,d\theta(a^4\cos^4\theta)}{\frac{1}{3}\int_0^{\frac{\pi}{2}}d\theta(a^3\cos^3\theta)} = \frac{3a^4\int_0^{\frac{\pi}{2}}\cos^4\theta(\sin\theta\,d\theta)}{4a^3\int_0^{\frac{\pi}{2}}\cos^3\theta\,d\theta}$$

$$= \frac{3a}{4}\frac{\left[-\frac{\cos^5\theta}{5}\right]_0^{\frac{\pi}{2}}}{\frac{2}{3}} = \frac{3}{4}\times\frac{3}{2}\,a\left[-0+\frac{1}{5}\right] = \frac{9a}{40}$$ Hence C.G is at $\left(\frac{3a}{5}, \frac{9a}{40}\right)$. **Ans.**

.6 CENTRE OF GRAVITY OF AN ARC

Example 18. *Find the C.G. of the arc of the curve*

$x = a(\theta + \sin\theta)$, $y = a(1 - \cos\theta)$ in the positive quadrant.

Solution. We know that, $\bar{x} = \dfrac{\int x\,ds}{\int ds}$, $\bar{y} = \dfrac{\int y\,ds}{\int ds}$

Now, $ds = \sqrt{\left(\frac{dx}{d\theta}\right)^2 + \left(\frac{dy}{d\theta}\right)^2}\, d\theta$

$$= \sqrt{\{a^2 (1+\cos\theta)^2 + a^2 \sin^2\theta\}}\, d\theta = a\sqrt{1+2\cos\theta+\cos^2\theta+\sin^2\theta}\, d\theta$$

$$= a\sqrt{1+2\cos\theta+1}\, d\theta = a\sqrt{2(1+\cos\theta)}\, d\theta = a\sqrt{4\cos^2\frac{\theta}{2}}\, d\theta = 2a\cos\frac{\theta}{2}\, d\theta$$

$$\bar{x} = \frac{\int x\, dx}{\int ds} = \frac{\int_0^\pi a(\theta+\sin\theta)\, 2a\cos\frac{\theta}{2}\, d\theta}{\int_0^\pi 2a\cos\frac{\theta}{2}\, d\theta} = \frac{a\int_0^\pi \left(\theta + 2\sin\frac{\theta}{2}\cos\frac{\theta}{2}\, d\theta\right)}{\left[2\sin\frac{\pi}{2}\right]_0^\pi}$$

$$= \frac{a}{2}\int_0^\pi \left[\theta\cos\frac{\theta}{2} + 2\sin\frac{\theta}{2}\cos^2\frac{\theta}{2}\right] d\theta = \frac{a}{2}\int_0^{\frac{\pi}{2}} (2t\cos t + 2\sin t\cos^2 t)\, 2\, dt$$

$$= 2a\left[t\sin t + \cos t - \frac{\cos^3 t}{3}\right]_0^{\frac{\pi}{2}} = 2a\left[\frac{\pi}{2} - 1 + \frac{1}{3}\right] = a\left[\pi - \frac{4}{3}\right]$$

$$\bar{y} = \frac{\int y\, ds}{\int ds} = \frac{\int_0^\pi a(1-\cos\theta)\, 2a\cos\frac{\theta}{2}\, d\theta}{\int_0^\pi 2a\cos\frac{\theta}{2}\, d\theta} = \frac{a\int_0^\pi 2\sin^2\frac{\theta}{2}\cos\frac{\theta}{2}\, d\theta}{\int_0^\pi \cos\frac{\theta}{2}\, d\theta}$$

$$= \frac{a\, r\left[\sin^3\frac{\theta}{2}\right]_0^\pi}{3\left[2\sin\frac{\theta}{2}\right]_0^\pi} = \frac{4a}{3\times 2} = \frac{2a}{3}$$ Hence, C.G. of the arc is $\left[a\left(\pi - \frac{4}{3}\right), \frac{2a}{3}\right]$ **Ans.**

EXERCISE 8.4

1. Find the centre of gravity of the area bounded by the parabola $y^2 = x$ and the line $x + y = 2$. **Ans.** $\left(\frac{8}{5}, -\frac{1}{2}\right)$

2. Find the centroid of the tetrahedron bounded by the coordinate planes and the plane $x + y + z = 1$, the density at any point varying as its distance from the face $z = 0$. **Ans.** $\left(\frac{1}{5}, \frac{1}{5}, \frac{2}{5}\right)$

3. Find the centroid of the area enclosed by the parabola $y^2 = 4\, ax$, the axis of x and latus rectum. **Ans.** $\left(\frac{3a}{20}, \frac{3a}{16}\right)$

4. Find the centroid of the loop of curve $r^2 = a^2 \cos 2\theta$. **Ans.** $\left(\frac{\pi a\sqrt{2}}{8}, 0\right)$

5. Find the centroid of solid formed by revolving about the x-axis that part of the area of the ellipse $\frac{x^2}{a^2} + \frac{y^2}{b^2} = 1$ which lies in the first quadrant. **Ans.** $\left(\frac{3a}{8}, 0\right)$

6. Find the average density of the sphere of radius a whose density at a distance r from the centre of the sphere is $\rho = \rho_0\left[1 + k\frac{r^3}{a^3}\right]$. $\rho_0\left(1 + \frac{k}{2}\right)$

7. The density at a point on a circular lamina varies as the distance from a point O on the circumference. Show that the C.G. divides the diameter through O in the ratio 3 : 2.

TRIPLE INTEGRATION

9.1 INTRODUCTION

In this chapter we will learn triple integration. Before that we will discuss the following coordinate systems.

1. Cartesian coordinates
2. Spherical coordinates
3. Cylindrical coordinates

1. Cartesian Coordinates

Take a point O in the space draw three mutually perpendicular lines through O. O is known as origin and these three lines are known as x-axis, y-axis, z-axis. There are three coordinate planes.

1. *xy-plane* : The plane passing through x-axis and y-axis is known as xy-plane.

2. *yz-plane* : The plane passing through y-axis and z-axis is known as yz-plane.

3. *zx-plane* : The plane passing through z-axis and x-axis is known as zx-plane.

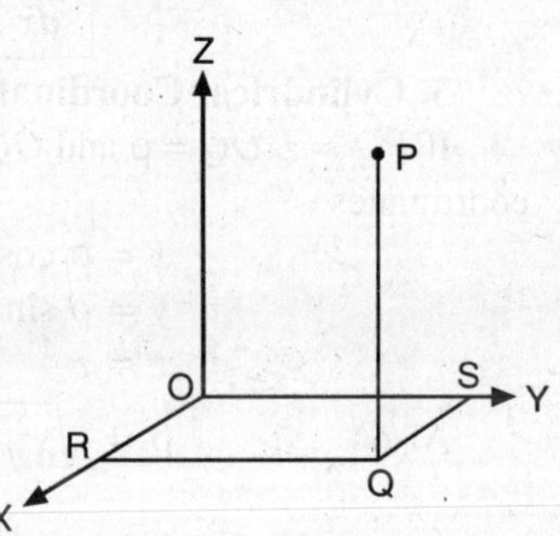

Consider a point P in the space draw perpendicular PQ to $x\,y$-plane. PQ is known as z-coordinate; from Q draw perpendicular lines QR and QS to x-axis and y-axis respectively. QS is x-coordinate, QR is y-coordinate and small element of the solid in cartesian coordinates is $dx\ dy\ dz$.

2. Spherical Coordinates

In the adjoining figure join OP, OP is denoted by r, θ is called the angle between OP *i.e.* r and z-axis.

ϕ is the angle between x-axis and OQ.

Then the coordinates (r, θ, ϕ) of a point P are known as spherical coordinates.

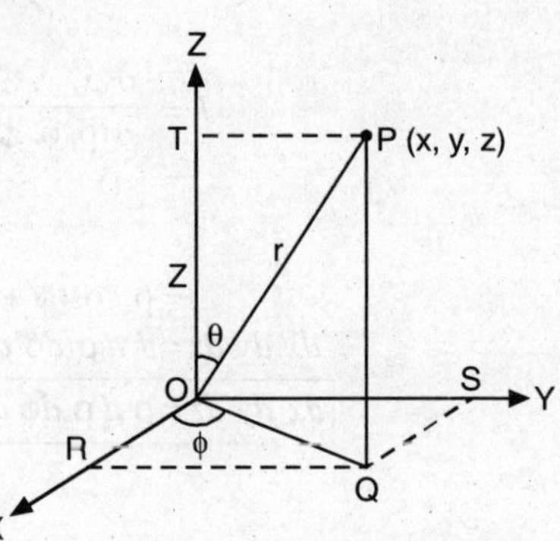

$$Z = PQ = OT = OP\cos\theta$$
$$= r\cos\theta$$
$$PT = OP\sin\theta$$
$$= r\sin\theta$$
$$\Rightarrow \qquad OQ = r\sin\theta$$

In right angled triangle QRO, $\angle R$ is right angle.

$$x = OR = OQ \cos\phi = r \sin\theta \cos\phi$$

Again in right angled triangle QSO

$$y = OQ \sin\phi = r \sin\theta \sin\phi$$

$$\boxed{\begin{aligned} \mathbf{x} &= \mathbf{r \sin\theta \cos\Phi} \\ \mathbf{y} &= \mathbf{r \sin\theta \sin\Phi} \\ \mathbf{z} &= \mathbf{r \cos\theta} \end{aligned}}$$

Also, $$r^2 = x^2 + y^2 + z^2$$

$$\Phi = \tan^{-1}\frac{y}{x}$$

$$\theta = \tan^{-1}\left(\frac{\sqrt{x^2+y^2}}{z}\right)$$

If $$J = \frac{\partial(x, y, z)}{\partial(r, \theta, \phi)}$$

$x = r \sin\theta \cos\phi$
$y = r \sin\theta \sin\phi$
$z = r \cos\theta$

$$= \begin{vmatrix} \frac{\partial x}{\partial r} & \frac{\partial x}{\partial \theta} & \frac{\partial x}{\partial \phi} \\ \frac{\partial y}{\partial r} & \frac{\partial y}{\partial \theta} & \frac{\partial y}{\partial \phi} \\ \frac{\partial z}{\partial r} & \frac{\partial z}{\partial \theta} & \frac{\partial z}{\partial \phi} \end{vmatrix} = r^2 \sin\theta$$

$$dx\,dy\,dz = J\,dr\,d\theta\,d\phi$$

$$\boxed{\mathbf{dx\,dy\,dz = r^2 \sin\theta\,dr\,d\theta\,d\phi}}$$

3. Cylindrical Coordinate System

If $PQ = z$, $OQ = \rho$ and OQ makes angle Φ with the x-axis then (z, ρ, Φ) are called cylindrical coordinates.

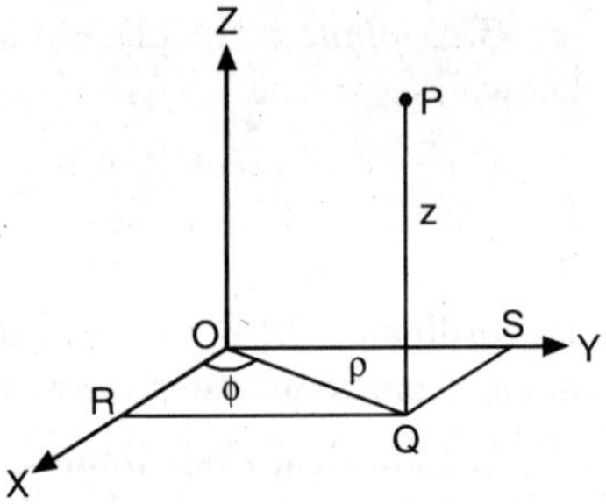

$$x = \rho \cos\Phi$$
$$y = \rho \sin\Phi$$
$$z = z$$

Also $$\rho = \sqrt{x^2 + y^2}$$

$$\Phi = \tan^{-1}\left(\frac{y}{x}\right)$$

$$z = z$$

$$J = \frac{\partial(x, y, z)}{\partial(\rho, \phi, z)} = \begin{vmatrix} \frac{\partial x}{\partial \rho} & \frac{\partial x}{\partial \phi} & \frac{\partial x}{\partial z} \\ \frac{\partial y}{\partial \rho} & \frac{\partial y}{\partial \phi} & \frac{\partial y}{\partial z} \\ \frac{\partial z}{\partial \rho} & \frac{\partial z}{\partial \phi} & \frac{\partial z}{\partial z} \end{vmatrix} = \begin{vmatrix} \cos\phi & -\rho\sin\phi & 0 \\ \sin\phi & \rho\cos\phi & 0 \\ 0 & 0 & 1 \end{vmatrix}$$

$$= \rho\cos^2\phi + \rho\sin^2\phi = \rho(\cos^2\phi + \sin^2\phi) = \rho$$

$$dx\,dy\,dz = J\,d\rho\,d\phi\,dz$$

$$\boxed{\mathbf{dx\,dy\,dz = \rho\,d\rho\,d\phi\,dz}}$$

9.2 TRIPLE INTEGRATION

Let a function $f(x, y, z)$ be a continuous at every point of a finite region S of three dimensional space. Consider n sub-spaces $\delta S_1, \delta S_2, \delta S_3, \ldots \delta S_n$ of the space S.

If (x_r, y_r, z_r) be a point in the rth subspace.

The limit of the sum $\sum_{r=1}^{n} f(x_r, y_r, z_r)\, \delta S_r$, as $n \to \infty$, $\delta S_r \to 0$ is known as the triple integral of $f(x, y, z)$ over the space S.

Symbolically, it is denoted by

$$\iiint_S f(x, y, z)\, dS$$

It can be calculated as $\int_{x_1}^{x_2}\int_{y_1}^{y_2}\int_{z_1}^{z_2} f(x, y, z)\; dx\, dy\, dz$. First we integrate with respect to z treating x, y as constant between the limits z_1 and z_2. The resulting expression (function of x, y) is integrated with respect to y keeping x as constant between the limits y_1 and y_2. At the end we integrate the resulting expression (function of x only) within the limits x_1 and x_2.

$$\boxed{\int_{x_1=a}^{x_2=b} \Psi(x)\, dx \;\boxed{\int_{y_1=\phi_1(x)}^{y_2=\phi_2(x)} \phi(x, y)\, dy \;\boxed{\int_{z_1=f_1(x, y)}^{z_2=f_2(x, y)} f(x, y, z)\, dz}}}$$

First we integrate from inner most integral w.r.t. z, then we integrate with respect to y and finally the outer most with respect to x.

But the above order of integration is immaterial provided the limits change accordingly.

Example 1. *Evaluate* $\iiint_R (x + y + z)\; dx\, dy\, dz$, *where* $R : 0 \le x \le 1,\ 1 \le y \le 2,\ 2 \le z \le 3$.

Solution. $\int_0^1 dx \int_1^2 dy \int_2^3 (x + y + z)\, dz = \int_0^1 dx \int_1^2 dy \left[\frac{(x + y + z)^2}{2}\right]_2^3$

$$= \frac{1}{2}\int_0^1 dx \int_1^2 dy\, [(x + y + 3)^2 - (x + y + 2)^2] = \frac{1}{2}\int_0^1 dx \int_1^2 (2x + 2y + 5)\,.1.dy$$

$$= \frac{1}{2}\int_0^1 dx \left[\frac{(2x + 2y + 5)^2}{4}\right]_1^2 = \frac{1}{8}\int_0^1 dx\, [(2x + 4 + 5)^2 - (2x + 2 + 5)^2]$$

$$= \frac{1}{8}\int_0^1 (4x + 16)\,.\,2\, dx = \int_0^1 (x + 4)\, dx = \left[\frac{x^2}{2} + 4x\right]_0^1 = \frac{1}{2} + 4 = \frac{9}{2} \qquad \textbf{Ans.}$$

Example 2. *Evaluate the integral :* $\int_0^{\log 2}\int_0^x\int_0^{x + \log y} e^{x + y + z}\, dz\, dy\, dx$.

Solution. $\int_0^{\log 2}\int_0^x\int_0^{x + \log y} e^{x + y + z}\, dz\, dy\, dx$.

$$= \int_0^{\log 2} e^x dx \int_0^x e^y dy \int_0^{x + \log y} e^z dz = \int_0^{\log 2} e^x\, dx \int_0^x e^y dy (e^z)_0^{x + \log y}$$

$$= \int_0^{\log 2} e^x dx \int_0^x e^y dy\, (e^{x + \log y} - 1) = \int_0^{\log 2} e^x\, dx \int_0^x e^y dy\, (e^{\log y}\,.\,e^x - 1)$$

$$= \int_0^{\log 2} e^x dx \int_0^x e^y\, (y e^x - 1)\, dy = \int_0^{\log 2} e^x dx \left[(ye^x - 1)\, e^y - \int e^x\,.\,e^y\, dy\right]_0^x$$

$$= \int_0^{\log 2} e^x dx \left[(ye^x - 1)\, e^y - e^{x + y}\right]_0^x = \int_0^{\log 2} e^x dx\, [(xe^x - 1)\, e^x - e^{2x} + 1 + e^x]$$

$$= \int_0^{\log 2} e^x dx\, [xe^{2x} - e^x - e^{2x} + 1 + e^x] = \int_0^{\log 2} (xe^{3x} - e^{3x} + e^x)\, dx$$

$$= \left[x\frac{e^{3x}}{3} - \int 1.\frac{e^{3x}}{3}\,dx - \frac{e^{3x}}{3} + e^x\right]_0^{\log 2} = \left[\frac{x}{3}e^{3x} - \frac{e^{3x}}{9} - \frac{e^{3x}}{3} + e^x\right]_0^{\log 2}$$

$$= \frac{\log 2}{3}e^{3\log 2} - \frac{e^{3\log 2}}{9} - \frac{e^{3\log 2}}{3} + e^{\log 2} + \frac{1}{9} + \frac{1}{3} - 1$$

$$= \frac{\log 2}{3}e^{\log 2^3} - \frac{e^{\log 2^3}}{9} - \frac{e^{\log 2^3}}{3} + e^{\log 2} + \frac{1}{9} + \frac{1}{3} - 1$$

$$= \frac{8}{3}\log 2 - \frac{8}{9} - \frac{8}{3} + 2 + \frac{1}{9} + \frac{1}{3} - 1 = \frac{8}{3}\log 2 - \frac{19}{9}$$ **Ans.**

Example 3. *Evaluate* $\int_0^{\log 2}\int_0^x\int_0^{x+y} e^{x+y+z}\,dx\,dy\,dz.$

(M.U. II Semester, 2005, 2003, 2002)

Solution. $I = \int_0^{\log 2}\int_0^x e^{x+y}\left[e^z\right]_0^{x+y} dx\,dy$

$$= \int_0^{\log 2}\int_0^x e^{x+y}(e^{x+y} - 1)\,dx\,dy = \int_0^{\log 2}\int_0^x \left[e^{2(x+y)} - e^{(x+y)}\right] dx\,dy$$

$$= \int_0^{\log 2}\left[e^{2x}.\frac{e^{2y}}{2} - e^x.e^y\right]_0^x dx = \int_0^{\log 2}\left(\frac{e^{4x}}{2} - e^{2x} - \frac{e^{2x}}{2} + e^x\right)dx$$

$$= \left[\frac{e^{4x}}{8} - \frac{e^{2x}}{2} - \frac{e^{2x}}{4} + e^x\right]_0^{\log 2} = \left[\frac{e^{4\log 2}}{8} - \frac{e^{2\log 2}}{2} - \frac{e^{2\log 2}}{4} + e^{\log 2}\right] - \left(\frac{1}{8} - \frac{1}{2} - \frac{1}{4} + 1\right)$$

$$= \left(\frac{e^{\log 16}}{8} - \frac{e^{\log 4}}{2} - \frac{e^{\log 4}}{4} + e^{\log 2}\right) - \left(\frac{1}{8} - \frac{1}{2} - \frac{1}{4} + 1\right)$$

$$= \left(\frac{16}{8} - \frac{4}{2} - \frac{4}{4} + 2\right) - \left(\frac{1}{8} - \frac{1}{2} - \frac{1}{4} + 1\right) = \frac{5}{8}$$ **Ans.**

Example 4. *Evaluate* $\iiint_R (x^2 + y^2 + z^2)\,dx\,dy\,dz$

where R denotes the region bounded by x = 0, y = 0, z = 0 and x + y + z = a, (a > 0)

Solution. $\iiint_R (x^2 + y^2 + z^2)\,dx\,dy\,dz$

$x + y + z = a$ or $z = a - x - y$

Upper limit of $z = a - x - y$

On x-y plane, $x + y + z = a$ becomes $x + y = a$ as shown in the figure.

Upper limit of $y = a - x$

Upper limit of $x = a$

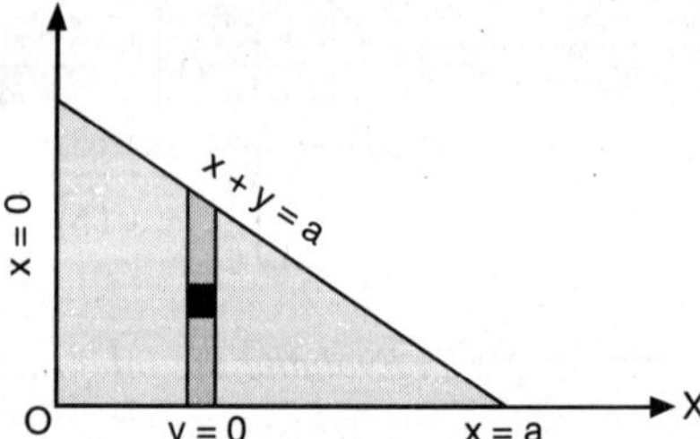

$$= \int_{x=0}^{a} dx \int_{y=0}^{a-x} dy \int_{z=0}^{a-x-y} (x^2 + y^2 + z^2)\,dz = \int_0^a dx \int_0^{a-x} dy \left(x^2 z + y^2 z + \frac{z^3}{3}\right)_0^{a-x-y}$$

$$= \int_0^a dx \int_0^{a-x} dy \left[x^2(a - x - y) + y^2(a - x - y) + \frac{(a - x - y)^3}{3}\right]$$

$$= \int_0^a dx \int_0^{a-x} \left[x^2(a - x) - x^2 y + (a - x)\,y^2 - y^3 + \frac{(a - x - y)^3}{3}\right] dy$$

$$= \int_0^a dx \left[x^2(a - x)\,y - \frac{x^2 y^2}{2} + (a - x)\frac{y^3}{3} - \frac{y^4}{4} - \frac{(a - x - y)^4}{12}\right]_0^{a-x}$$

$$= \int_0^a dx\left[x^2(a-x)^2 - \frac{x^2}{2}(a-x)^2 + (a-x)\frac{(a-x)^3}{3} - \frac{(a-x)^4}{4} + \frac{(a-x)^4}{12}\right]$$

$$= \int_0^a \left[\frac{x^2}{2}(a-x)^2 + \frac{(a-x)^4}{6}\right]dx = \int_0^a \left[\frac{1}{2}(a^2x^2 - 2ax^3 + x^4) + \frac{(a-x)^4}{6}\right]dx$$

$$= \left[\frac{1}{2}a^2\frac{x^3}{3} - \frac{ax^4}{4} + \frac{x^5}{10} - \frac{(a-x)^5}{30}\right]_0^a = \frac{a^5}{6} - \frac{a^5}{4} + \frac{a^5}{10} + \frac{a^5}{30} = \frac{a^5}{20} \qquad \textbf{Ans.}$$

Example 5. *Compute* $\iiint \frac{dx\,dy\,dz}{(x+y+z+1)^3}$ *if the region of integration is bounded by the coordinate planes and the plane* $x + y + z = 1$. (*M.U., II Semester 2007, 2006*)

Solution. Let the given region be R, then R is expressed as

$$0 \le z \le 1 - x - y, \quad 0 \le y \le 1 - x, \quad 0 \le x \le 1.$$

$$\iiint_R \frac{dx\,dy\,dz}{(x+y+z+1)^3} = \int_0^1 dx \int_0^{1-x} dy \int_0^{1-x-y} \frac{dz}{(x+y+z+1)^3}$$

$$= \int_0^1 dx \int_0^{1-x} dy \left[\frac{1}{-2(x+y+z+1)^2}\right]_0^{1-x-y}$$

$$= -\frac{1}{2}\int_0^1 dx \int_0^{1-x} dy \left[\frac{1}{(x+y+1-x-y+1)^2} - \frac{1}{(x+y+1)^2}\right]$$

$$= -\frac{1}{2}\int_0^1 dx \int_0^{1-x} \left[\frac{1}{4} - \frac{1}{(x+y+1)^2}\right]dy = -\frac{1}{2}\int_0^1 dx \left[\frac{y}{4} + \frac{1}{x+y+1}\right]_0^{1-x}$$

$$= -\frac{1}{2}\int_0^1 dx \left[\frac{1-x}{4} + \frac{1}{x+1+1-x} - \frac{1}{x+1}\right] = -\frac{1}{2}\int_0^1 \left[\frac{1-x}{4} + \frac{1}{2} - \frac{1}{x+1}\right]dx$$

$$= -\frac{1}{2}\left[-\frac{(1-x)^2}{8} + \frac{x}{2} - \log(x+1)\right]_0^1 = -\frac{1}{2}\left[\frac{1}{2} - \log 2 + \frac{1}{8}\right] = -\frac{1}{2}\left[\frac{5}{8} - \log 2\right]$$

$$= \frac{1}{2}\log 2 - \frac{5}{16} \qquad \textbf{Ans.}$$

Example 6. *Compute* $\iiint_V x^2\,dx\,dy\,dz$ *over volume of tetrahedron bounded by*

$x = 0,\ \ y = 0,\ \ z = 0$ *and* $\frac{x}{a} + \frac{y}{b} + \frac{z}{c} = 1.$ (*M.U. II Semester, 2008*)

Solution. Here, we have

$$I = \iiint_V x^2\,dx\,dy\,dz \qquad ...(1)$$

Where V is bounded by $x = 0$, $y = 0$, $z = 0$ and $\frac{x}{a} + \frac{y}{b} + \frac{z}{c} = 1$

Putting $\frac{x}{a} = u$, $\frac{y}{b} = v$, $\frac{z}{c} = w$ so that $dx = a\,du$, $dy = b\,dv$, $dz = c\,dw$ in (1), we get

$$I = \int_0^1 \int_0^{1-u} \int_0^{1-u-v} a^2u^2\ (a\,du)\ (b\,dv)\ (c\,dw)$$

$$= a^3bc \int_0^1 u^2du \int_0^{1-u} dv \int_0^{1-u-v} dw = a^3bc \int_0^1 u^2\,du \int_0^{1-u} dv\,[w]_0^{1-u-v}$$

$$= a^3bc \int_0^1 u^2du \int_0^{1-u} (1-u-v)\,dv = a^3bc \int_0^1 u^2du \left[v - uv - \frac{v^2}{2}\right]_0^{1-u}$$

$$= a^3bc \int_0^1 u^2\left[1 - u - u(1-u) - \frac{(1-u)^2}{2}\right]du$$

$$= a^3bc\int_0^1 u^2\left[1-u-u+u^2-\frac{1}{2}-\frac{u^2}{2}+u\right]du$$

$$= a^3bc\int_0^1 u^2\left(\frac{1}{2}-u+\frac{1}{2}u^2\right)du = a^3bc\int_0^1\left(\frac{u^2}{2}-u^3+\frac{u^4}{2}\right)du$$

$$= a^3bc\left[\frac{1}{2}.\frac{u^3}{3}-\frac{u^4}{4}+\frac{1}{2}.\frac{u^5}{5}\right]_0^1 = a^3bc\left[\frac{1}{6}-\frac{1}{4}+\frac{1}{10}\right] = a^3bc\left(\frac{1}{60}\right)=\frac{a^3bc}{60}$$ **Ans.**

Example 7. *Evaluate* $\iiint x^2yz\ dx\ dy\ dz$ *throughout the volume bounded by the planes* $x = 0$, $y = 0$, $z = 0$, $\frac{x}{a}+\frac{y}{b}+\frac{z}{c} = 1.$ (*M.U. II Semester* 2003, 2002, 2001)

Solution. Here, we have

$$I = \iiint x^2yz\ dx\ dy\ dz \quad ...(1)$$

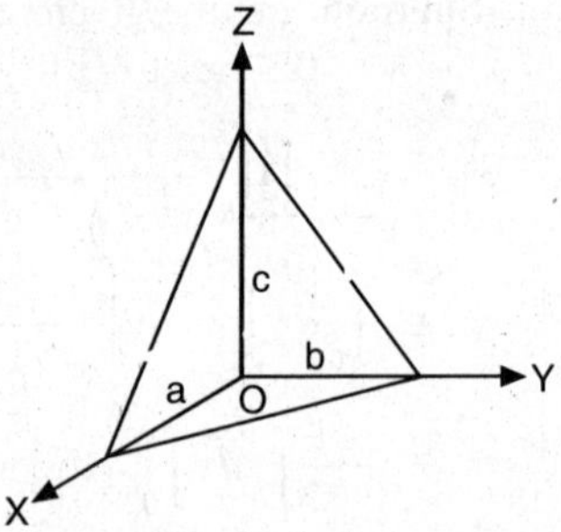

Putting $x = au$, $y = bv$, $z = cw$

$dx = a\ du$, $dy = b\ dv$, $dz = c\ dw$ in (1), we get

$$I = \iiint a^2bc\,u^2vw\,a\,bc\ du\ dv\ dw$$

Limits are for $u = 0, 1$ for $v = 0, 1 - u$ and for $w = 0, 1 - u - v$

$$u + v + w = 1$$

$$I = \int_{u=o}^1\int_{v=0}^{1-u}\int_{w=0}^{1-u-v} a^3b^2c^2u^2vw\ du\ dv\ dw = \int_0^1\int_0^{1-u} a^3b^2c^2u^2v\left[\frac{w^2}{2}\right]_0^{1-u-v} du\ dv$$

$$= \frac{a^3b^2c^2}{2}\int_0^1\int_0^{1-u} u^2v(1-u-v)^2\ du\ dv$$

$$= \frac{a^3b^2c^2}{2}\int_0^1\int_0^{1-u} u^2v\left[(1-u)^2-2(1-u)v+v^2\right]du\ dv$$

$$= \frac{a^3b^2c^2}{2}\int_0^1\int_0^{1-u} u^2[(1-u)^2v-2(1-u)v^2+v^3]\,du\ dv$$

$$= \frac{a^3b^2c^2}{2}\int_0^1 u^2\left[(1-u)^2\frac{v^2}{2}-2(1-u)\frac{v^3}{3}+\frac{v^4}{4}\right]_0^{1-u} du$$

$$= \frac{a^3b^2c^2}{2}\int_0^1 u^2\left[\frac{(1-u)^4}{2}-\frac{2(1-u)^4}{3}+\frac{(1-u)^4}{4}\right]du$$

$$= \frac{a^3b^2c^2}{2}\int_0^1 \frac{u^2(1-u)^4}{12}du = \frac{a^3b^2c^2}{24}\int_0^1 u^{3-1}(1-u)^{5-1}du$$

$$= \frac{a^3b^2c^2}{24}\beta(3,5) = \frac{a^3b^2c^2}{24}.\frac{\lceil 3\ \lceil 5}{\lceil 8} = \frac{a^3b^2c^2}{24}.\left(\frac{2!\,4!}{7!}\right) = \frac{a^3b^2c^2}{2520}.$$ **Ans.**

9.3 INTEGRATION BY CHANGE OF CARTESIAN COORDINATES INTO SPHERICAL COORDINATES

Sometime it becomes easy to integrate by changing the cartesian coordinates into spherical coordinates.

The relations between the cartesian and spherical polar co-ordinates of a point are given by the relations

$$x = r\sin\theta\cos\phi$$
$$y = r\sin\theta\sin\phi$$
$$z = r\cos\theta$$

$$dx\,dy\,dz = |J|\,dr\,d\theta\,d\phi$$
$$= r^2 \sin\theta\,dr\,d\theta\,d\phi$$

Note. 1. Spherical coordinates are very useful if the expression $x^2 + y^2 + z^2$ is involved in the problem.

2. In a sphere $x^2 + y^2 + z^2 = a^2$ the limits of r are 0 and a and limits of θ are 0, π and that of ϕ are 0 and 2π.

Example 8. *Evaluate the integral* $\iiint (x^2 + y^2 + z^2)\,dx\,dy\,dz$ *taken over the volume enclosed by the sphere* $x^2 + y^2 + z^2 = 1$.

Solution. Let us convert the given integral into spherical polar co-ordinates. By putting

$$x = r\sin\theta\cos\phi\ ;\quad y = r\sin\theta\sin\phi\ ;\quad z = r\cos\theta$$

$$\iiint (x^2 + y^2 + z^2)\,dx\,dy\,dz = \int_0^{2\pi}\int_0^{\pi}\int_0^{1} r^2(r^2 \sin\theta\,d\theta\,d\,\phi\,dr)$$

$$= \int_0^{2\pi} d\phi \int_0^{\pi} \sin\theta\,d\theta \int_0^1 r^4 dr = \int_0^{2\pi} d\phi \int_0^{\pi} \sin\theta\,d\theta \left(\frac{r^5}{5}\right)_0^1 = \frac{1}{5}\int_0^{2\pi} d\phi\,[-\cos\theta]_0^{\pi} = \frac{2}{5}\int_0^{2\pi} d\phi$$

$$= \frac{2}{5}(\phi)_0^{2\pi} = \frac{4\pi}{5}$$ **Ans.**

Example 9. *Evaluate* $\iiint (x^2 + y^2 + z^2)\,dx\,dy\,dz$ *over the first octant of the sphere* $x^2 + y^2 + z^2 = a^2$. *(M.U. II Semester 2007)*

Solution. Here, we have

$$I = \iiint (x^2 + y^2 + z^2)\,dx\,dy\,dz \qquad ...(1)$$

Putting $x = r\sin\theta\cos\phi$, $y = r\sin\theta\sin\phi$, $z = r\cos\theta$ and $dx\,dy\,dz = r^2\sin\theta\,dr\,d\theta\,d\phi$ in (1), we get

Limits of r are 0, a for θ are 0, $\frac{\pi}{2}$ for ϕ are 0, $\frac{\pi}{2}$.

$$I = \int_0^{\frac{\pi}{2}}\int_0^{\frac{\pi}{2}}\int_0^{a} r^2 . r^2 \sin\theta\,dr\,d\theta\,d\phi = \int_0^{\frac{\pi}{2}} d\phi \int_0^{\frac{\pi}{2}} \sin\theta\,d\theta \int_0^a r^4\,dr$$

$$\left(\begin{aligned} x^2 + y^2 + z^2 &= r^2\sin^2\theta\cos^2\phi + r^2\sin^2\theta\sin^2\phi + r^2\cos^2\theta \\ &= r^2\sin^2\theta + r^2\cos^2\theta = r^2 \end{aligned}\right)$$

$$= [\phi]_0^{\pi/2}\,[-\cos\theta]_0^{\pi/2}\left[\frac{r^5}{5}\right]_0^a = \frac{\pi}{2}.(1).\frac{a^5}{5} = \pi.\frac{a^5}{10}.$$ **Ans.**

Example 10. *Evaluate* $\iiint \frac{dx\,dy\,dz}{x^2 + y^2 + z^2}$ *throughout the volume of the sphere* $x^2 + y^2 + z^2 = a^2$.

(M.U. II Semester 2002, 2001)

Solution. Here, we have

$$I = \iiint \frac{dx\,dy\,dz}{x^2 + y^2 + z^2} \qquad ...(1)$$

Putting $x = r\sin\theta\cos\phi$, $y = r\sin\theta\sin\phi$, $z = r\cos\theta$ and $dx\,dy\,dz = r^2\sin\theta\,dr\,d\theta\,d\phi$ in (1), we get

The limits of r are 0 and a, for θ are 0 and $\frac{\pi}{2}$ for ϕ are 0 and $\frac{\pi}{2}$ in first octant.

$$I = 8\int_0^{\frac{\pi}{2}}\int_0^{\frac{\pi}{2}}\int_0^{a} \frac{r^2\sin\theta\,dr\,d\theta\,d\phi}{r^2} \qquad [\text{Sphere } x^2 + y^2 + z^2 \text{ lies in 8 quadrants}]$$

$$I = 8\int_0^{\frac{\pi}{2}} d\phi \int_0^{\frac{\pi}{2}} \sin\theta\, d\theta \int_0^a dr = 8[\phi]_0^{\pi/2}[-\cos\theta]_0^{\pi/2}[r]_0^a = 8\left(\frac{\pi}{2}-0\right)(0+1)(a+0)$$

$$= 8\frac{\pi}{2}.1.a = 4\pi a$$ **Ans.**

Example 11. *Evaluate* $\iiint \frac{z^2 dx\, dy\, dz}{x^2+y^2+z^2}$ *over the volume of the sphere* $x^2 + y^2 + z^2 = 2.$

(M.U. II Semester 2005, 2004)

Solution. Here, we have

$$I = \iiint \frac{z^2 dx\, dy\, dz}{x^2+y^2+z^2} \qquad ...(1)$$

Putting $x = r\sin\theta\cos\phi$, $y = r\sin\theta\sin\phi$, $z = r\cos\theta$, $dx\,dy\,dz = r^2\sin\theta\, dr\, d\theta\, d\phi$ in (1), we get

[The limits r, θ and ϕ over the first octant of $x^2 + y^2 + z^2 = r^2$ are $0, \sqrt{2}$; $0, \frac{\pi}{2}$ and $0, \frac{\pi}{2}$].

$$I = 8\int_0^{\frac{\pi}{2}}\int_0^{\frac{\pi}{2}}\int_0^{\sqrt{2}} \frac{r^4\cos^2\theta\sin\theta}{r^2} dr\, d\theta\, d\phi = 8\int_0^{\frac{\pi}{2}} d\phi \int_0^{\frac{\pi}{2}} \cos^2\theta\sin\theta\, d\theta . \int_0^{\sqrt{2}} r^2 dr$$

$$= 8[\phi]_0^{\pi/2}\left[-\frac{\cos^3\theta}{3}\right]_0^{\frac{\pi}{2}}\left[\frac{r^3}{3}\right]_0^{\sqrt{2}} = 8\frac{\pi}{2}.\frac{1}{3}.\frac{2\sqrt{2}}{3} = \frac{8\pi\sqrt{2}}{9}.$$ **Ans.**

Example 12. *Evaluate* $\iiint xyz\ dx\, dy\, dz$ *over the positive octant of the sphere* $x^2 + y^2 + z^2 = a^2$.

(M.U. II Semester, 2002)

Solution. Here, we have

$$I = \iiint xyz\, dx\, dy\, dz \text{ over the first quadrant} \qquad ...(1)$$

Putting $x = r\sin\theta\cos\phi$, $y = r\sin\theta\sin\phi$, $z = r\cos\theta$, $dx\,dy\,dz = r^2\sin\theta\, dr\, d\theta\, d\phi$ in (1), we get

$$I = \int_0^{\frac{\pi}{2}}\int_0^{\frac{\pi}{2}}\int_0^a (r\sin\theta\cos\phi)(r\sin\theta\sin\phi)(r\cos\theta)(r^2\sin\theta\, d\theta\, dr\, d\phi)$$

$$I = \int_0^{\frac{\pi}{2}}\int_0^{\frac{\pi}{2}}\int_0^a r^5\sin^3\theta\cos\theta\sin\phi\cos\phi\, dr\, d\theta\, d\phi$$

$$= \int_0^{\frac{\pi}{2}} \sin\phi\cos\phi\, d\phi \int_0^{\frac{\pi}{2}} \sin^3\theta\cos\theta\, d\theta \int_0^a r^5 dr$$

$$= \left[\frac{\sin^2\phi}{2}\right]_0^{\frac{\pi}{2}}\left[\frac{\sin^4\theta}{4}\right]_0^{\frac{\pi}{2}}\left[\frac{r^6}{6}\right]_0^a = \frac{1}{2}.\frac{1}{4}.\frac{a^6}{6} = \frac{a^6}{48}.$$ **Ans.**

In first octant of $x^2 + y^2 + z^2 = a^2$
Limits of $r = 0$ and a
for $\theta = 0$ and $\frac{\pi}{2}$
for $\phi = 0$ and $\frac{\pi}{2}$.

Example 13. *Evaluate* $\iiint \sqrt{x^2+y^2}\ dx\, dy\, dz$ *over the volume bounded by the right circular cone* $x^2 + y^2 = z^2$, $z > 0$ *and the planes* $z = 0$ *and* $z = 1$. *(M.U. II Semester, 2004, 2002)*

Solution. Here, we have

$$I = \iiint \sqrt{x^2+y^2}\ dx\, dy\, dz \qquad ...(1)$$

Putting $x = r\cos\theta$, $y = r\sin\theta$, $z = z$ (cylindrical coordinates),
$dx\,dy\,dz = r.dr\, d\theta\, dz$ in (1), we get
[Now limits for r are 0 to 1 for θ are 0 to 2π for z are r to 1.]

$$I = \int_0^1 r^2 dr \int_0^{2\pi} d\theta \int_r^1 dz$$

$$= \int_0^1 r^2 dr\, [\theta]_0^{2\pi}\, [z]_r^1$$

$$= \int_0^1 r^2 . 2\pi . (1-r)\, dr$$

$$= 2\pi \left[\frac{r^3}{3} - \frac{r^4}{4}\right]_0^1 = \frac{2\pi}{12} = \frac{\pi}{6} \qquad \textbf{Ans.}$$

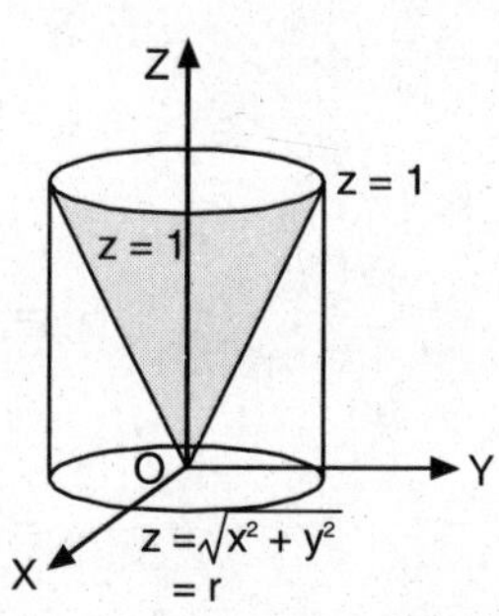

Example 14. *Evaluate* $\iiint xyz\,(x^2 + y^2 + z^2)\; dx\, dy\, dz$ *over the first octant of the sphere* $x^2 + y^2 + z^2 = a^2$. *(M.U. II Semester 2009, 2005, 2004, 2002)*

Solution. Here, we have

$$I = \iiint xyz\,(x^2 + y^2 + z^2)\; dx\, dy\, dz \qquad ...(1)$$

Putting $x = r \sin\theta \cos\phi$, $y = r \sin\theta \sin\phi$, $z = r\cos\theta$, $dx\, dy\, dz = r^2 \sin\theta\, dr\, d\theta\, d\phi$ in (1), we get

$$\left(\begin{aligned} x^2 + y^2 + z^2 &= r^2 \sin^2\theta \cos^2\phi + r^2 \sin^2\theta \sin^2\phi + r^2\cos^2\theta \\ &= r^2 \sin^2\theta + r^2 \cos^2\theta = r^2 \end{aligned}\right)$$

$$\left[\text{Limits of } r, \theta \text{ and } \phi \text{ in the first octant are } 0, a\ ;\ 0, \frac{\pi}{2} \text{ and } 0, \frac{\pi}{2}\right]$$

$$I = \int_0^{\frac{\pi}{2}} \int_0^{\frac{\pi}{2}} \int_0^a r^3 \sin^2\theta \cos\theta \sin\phi \cos\phi\, r^2 . r^2 \sin\theta\, dr\, d\theta\, d\phi$$

$$= \int_0^{\frac{\pi}{2}} \sin\phi \cos\phi\, d\phi \int_0^{\frac{\pi}{2}} \sin^3\theta \cos\theta\, d\theta \int_0^a r^7 dr$$

$$= \left[\frac{\sin^2\phi}{2}\right]_0^{\frac{\pi}{2}} \left[\frac{\sin^4\theta}{4}\right]_0^{\frac{\pi}{2}} \left[\frac{r^8}{8}\right]_0^a = \frac{1}{2}.\frac{1}{4}.\frac{a^8}{8} = \frac{a^8}{64} \qquad \textbf{Ans.}$$

Example 15. *Evaluate* $\iiint (x^2y^2 + y^2z^2 + z^2x^2)\; dx\, dy\, dz$ *over the volume of the sphere* $x^2 + y^2 + z^2 = a^2$. *(M.U. II Semester 2003, 2002)*

Solution. Here, we have

$$I = \iiint (x^2y^2 + y^2z^2 + z^2x^2)\; dx\, dy\, dz \qquad ...(1)$$

Putting $x = r \sin\theta \cos\phi$, $y = r \sin\theta \sin\phi$, $z = r\cos\theta$, $dx\, dy\, dz = r^2 \sin\theta\, dr\, d\theta\, d\phi$, in (1), we get

$$\left[\text{In first octant the limits of } r \text{ are } 0, a \text{ for } \theta \text{ are } 0, \frac{\pi}{2} \text{ for } \phi \text{ are } 0 \text{ and } \frac{\pi}{2}\right]$$

$$I = 8 \int_0^{\frac{\pi}{2}} \int_0^{\frac{\pi}{2}} \int_0^a (r^4 \sin^4\theta \sin^2\phi \cos^2\phi + r^4 \sin^2\theta \cos^2\theta \sin^2\phi + r^4 \sin^2\theta \cos^2\theta \cos^2\phi) . r^2 \sin\theta\, dr\, d\theta\, d\phi$$

$$= 8 \int_0^{\frac{\pi}{2}} \int_0^{\frac{\pi}{2}} \int_0^a r^6 (\sin^4\theta \sin^2\phi \cos^2\phi + \sin^2\theta \cos^2\theta) \sin\theta\, dr\, d\theta\, d\phi$$

$$= 8 \int_0^{\frac{\pi}{2}} \int_0^{\frac{\pi}{2}} (\sin^4\theta \sin^2\phi \cos^2\phi + \sin^2\theta \cos^2\theta) . \sin\theta\, d\theta\, d\phi \int_0^a r^6 dr$$

$$= \frac{8a^7}{7}\left[\int_0^{\frac{\pi}{2}}\int_0^{\frac{\pi}{2}} \sin^5\theta \sin^2\phi \cos^2\phi \, d\theta \, d\phi + \int_0^{\frac{\pi}{2}}\int_0^{\frac{\pi}{2}} \sin^3\theta \cos^2\theta \, d\theta \, d\phi\right]$$

$$= \frac{8a^7}{7}\left[\int_0^{\frac{\pi}{2}} \sin^5\theta \, d\theta \int_0^{\frac{\pi}{2}} \sin^2\phi \cos^2\phi \, d\phi + \int_0^{\frac{\pi}{2}} \sin^3\theta \cos^2\theta \, d\theta \int_0^{\frac{\pi}{2}} d\phi\right]$$

$$= \frac{8a^7}{7}\left[\frac{1}{2}\cdot\frac{2!\left|\frac{1}{2}\right.}{\frac{5}{2}\cdot\frac{3}{2}\cdot\frac{1}{2}\left|\frac{1}{2}\right.}\cdot\frac{1}{2}\cdot\frac{\frac{1}{2}\left|\frac{1}{2}\right.\cdot\frac{1}{2}\left|\frac{1}{2}\right.}{2!} + \frac{1}{2}\cdot\frac{1!.\frac{1}{2}\left|\frac{1}{2}\right.}{\frac{5}{2}\cdot\frac{3}{2}\cdot\frac{1}{2}\left|\frac{1}{2}\right.}\cdot\frac{\pi}{2}\right]$$

$$= \frac{8a^7}{7}\left[\frac{1}{2}\cdot\frac{2!}{\frac{5}{2}\cdot\frac{3}{2}\cdot\frac{1}{2}}\cdot\frac{1}{2}\cdot\frac{\frac{1}{2}\cdot\sqrt{\pi}\cdot\frac{1}{2}\cdot\sqrt{\pi}}{2!} + \frac{1}{2}\cdot\frac{1!.\frac{1}{2}}{\frac{5}{2}\cdot\frac{3}{2}\cdot\frac{1}{2}}\cdot\frac{\pi}{2}\right]$$

$$= \frac{8a^7}{7}\left[\frac{8}{15}\cdot\frac{1}{16}\cdot\pi + \frac{1}{15}\cdot\pi\right] = \frac{8a^7}{7}\left[\frac{1}{30}+\frac{1}{15}\right]\pi = \frac{8a^7}{7}\cdot\frac{1}{10}\pi = \frac{4a^7\pi}{35} \quad \textbf{Ans.}$$

EXERCISE 9.1

Evaluate the following :

1. $\int_{-1}^{1}\int_{-2}^{2}\int_{-3}^{3} dx\,dy\,dz$ *(M.U., II Semester 2002)* **Ans.** 48

2. $\int_0^4\int_0^x\int_0^{x+y} z\,dz\,dy\,dx$ *(R.G.P.V. Bhopal I Sem. 2003)* **Ans.** 70

3. $\int_1^2\int_0^1\int_{-1}^1 (x^2+y^2+z^2)\,dx\,dy\,dz$ **Ans.** 6

4. $\int_0^1\int_0^1\int_0^1 (x^2+y^2+z^2)\,dz\,dy\,dx$ *(AMIETE, June 2006)* **Ans.** 1

5. $\int_{-1}^1\int_0^z\int_{x-z}^{x+z} (x-y+z)\,dx\,dy\,dz$ *(AMIETE, Summer 2004)* **Ans.** 0

6. $\iiint_R (x-y-z)\;dx\,dy\,dz,$ where $R: 1 \le x \le 2;\; 2 \le y \le 3;\; 1 \le z \le 3$ **Ans.** 2

$\int_0^2\int_1^3\int_1^2 xy^2z\,dx\,dy\,dz$ *(AMIETE, Dec. 2007)* **Ans.** 26 **8.** $\int_0^1 dx\int_0^2 dy\int_1^2 x^2yz\,dz$ **Ans.** 1

9. $\iiint x^2yz\;dx\,dy\,dz$ throughout the volume bounded by $x = 0$, $y = 0$, $z = 0$, $x + y + z = 1$.

(M.U. II Semester, 2003) **Ans.** $\frac{1}{2520}$

10. $\int_0^1\int_0^{1-x}\int_0^{1-x^2-y^2} dz\,dy\,dx$ **Ans.** $\frac{1}{3}$

11. $\int_1^e\int_1^{\log y}\int_1^{e^x} \log z\,dz\,dx\,dy$ **Ans.** $\frac{1}{2}(e^2-8e+13)$

12. $\iiint_T y\,dx\,dy\,dz,$ where T is the region bounded by the surfaces $x = y^2$, $x = y + 2$, $4z = x^2 + y^2$ and $z = y + 3$. *(AMIETE Dec. 2008)* **Ans.** $\frac{92}{15}$

13. $\int_0^2\int_0^x\int_0^{2x+2y} e^{x+y+z}\,dz\,dy\,dx$ *(M.U. II Semester, 2003)*

Ans. $\frac{1}{3}\left[\frac{e^{12}}{6}-\frac{e^6}{3}-\frac{1}{6}+\frac{1}{3}\right]-\frac{1}{2}[e^4-1]+[e^2-1]$

14. $\iiint (x+y+z)\ dx\ dy\ dz$ over the tetrahedron bounded by the planes $x = 0$, $y = 0$, $z = 0$ and $x + y + z = 1$. **Ans.** $\frac{1}{8}$

15. $\int_0^a \int_0^{a-x} \int_0^{a-x-y} x^2\ dx\ dy\ dz$ **Ans.** $\frac{a^5}{60}$

16. $\int_{-2}^{2} \int_{-\sqrt{(4-x^2)/2}}^{\sqrt{(4-x^2)/2}} \int_{x^2+3y^2}^{8-x^2-y^2} dz\ dy\ dx$ **Ans.** $8\sqrt{2}\pi$

17. $\int_{-1}^{1} \int_0^{z} \int_{x-z}^{x+z} (x+y+z)\ dz\ dx\ dy$ *(M.U. II Semester, 2000, 02)* **Ans.** 0

18. $\int_0^2 \int_0^y \int_{x-y}^{x+y} (x+y+z)\ dx\ dy\ dz$ *(M.U. II Semester 2004)* **Ans.** 16

19. $\iiint \sqrt{1-\frac{x^2}{a^2}-\frac{y^2}{b^2}-\frac{z^2}{c^2}}\ dx\ dy\ dz$ throughout the volume of the ellipsoid $\frac{x^2}{a^2}+\frac{y^2}{b^2}+\frac{z^2}{c^2} = 1$.

Ans. $\frac{\pi^2}{4} abc$

20. $\iiint \sqrt{\frac{x^2}{a^2}+\frac{y^2}{b^2}+\frac{z^2}{c^2}}\ dx\ dy\ dz$ over the volume of the ellipsoid $\frac{x^2}{a^2}+\frac{y^2}{b^2}+\frac{z^2}{c^2} = 1$. **Ans.** $\frac{4\pi}{3} abc$

21. $\iiint x^{l-1} y^{m-1} z^{n-1}\ dx\ dy\ dz$
throughout the volume of the tetrahedron $x \geq 0$, $y \geq 0$, $z \geq 0$, $x + y + z \leq 1$.

Ans. $\frac{1}{(l+m+n)} \cdot \frac{\overline{|l}\ \overline{|m}\ \overline{|n}}{\overline{|l+m+n}}$

22. $\iiint \frac{dx\ dy\ dz}{\sqrt{1-x^2-y^2-z^2}}$ taken throughout the volume of the sphere $x^2 + y^2 + z^2 = 1$, lying in the first octant. **Ans.** $\frac{\pi^2}{8}$

23. $\int_0^{\pi} 2d\theta \int_0^{a(1+\cos\theta)} r\ dr \int_0^h \left[1-\frac{r}{a(1+\cos\theta)}\right] dz$ **Ans.** $\frac{\pi a^2}{2} h$

24. $\int_0^{\pi/2} \int_0^{a \sin\theta} \int_0^{(a^2-r^2)/a} r\ d\theta\ dr\ dz$ **Ans.** $\frac{5a^3}{64}$

25. $\iiint z^2\ dx\ dy\ dz$ over the volume common to the sphere $x^2 + y^2 + z^2 = a^2$ and the cylinder $x^2 + y^2 + z^2 = ax$. **Ans.** $\frac{2a^5\pi}{15}$

26. $\iiint_V \frac{dx\ dy\ dz}{(1+x^2+y^2+z^2)^2}$ where V is the volume in the first octant. **Ans.** $\frac{\pi^2}{8}$

27. $\iiint \frac{dx\ dy\ dz}{(x^2+y^2+z^2)^{3/2}}$ over the volume bounded by the spheres $x^2 + y^2 + z^2 = 16$ and $x^2 + y^2 + z^2 = 25$. *(M.U. II Semester, 2001, 03)* **Ans.** $4\pi \log (5/4)$

28. $\iiint_T z^2\ dx\ dy\ dz$ over the volume bounded by the cylinder $x^2 + y^2 = a^2$ and the paraboloid $x^2 + y^2 = z$ and the plane $z = 0$. **Ans.** $\frac{\pi a^8}{12}$

29. $\iiint_T z\ dx\ dy\ dz$, where T is region bounded by the cone $x^2 \tan^2 \alpha + y^2 \tan^2 \beta = z^2$ and the plane $z = 0$ to $z = h$ in the first octant. *(AMIETE, Dec. 2009*

CHAPTER 10

APPLICATION OF TRIPLE INTEGRATION

10.1 INTRODUCTION

In this chapter we will discuss how to find out volume, surface area, mass, C.G., moment of Inertia of solids and centre of pressure of fluids.

10.2 VOLUME = $\iiint dx\, dy\, dz$.

The elementary volume δv is δx . δy . δz and therefore the volume of the whole solid is obtained by evaluating the triple integral.

$$\delta V = \delta x\ \delta y\ \delta z$$

$$\boxed{V = \iiint dx\, dy\, dz.}$$

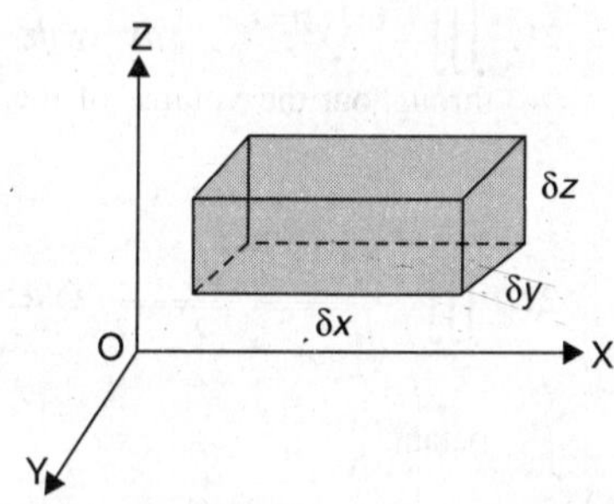

Note : (*i*) Mass = volume × density

$$= \iiint \rho\, dx\, dy\, dz \text{ if } \rho \text{ is the density.}$$

(*ii*) In cylindrical co-ordinaties, we have

$$\boxed{V = \iiint_V r\, dr\, d\varphi\, dz}$$

(*iii*) In spherical polar co-ordinates, we have

$$\boxed{V = \iiint_V r^2 \sin\theta dr\, d\theta\, d\varphi}$$

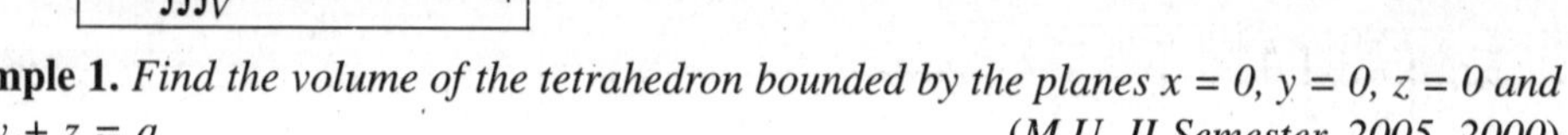

Example 1. *Find the volume of the tetrahedron bounded by the planes x = 0, y = 0, z = 0 and x + y + z = a.* (*M.U. II Semester, 2005, 2000*)

Solution. Here, we have a solid which is bounded by $x = 0$, $y = 0$, $z = 0$ and $x + y + z = a$ planes.

The limits of z are 0 and $a - x - y$, the limits of y are 0 and $1 - x$, the limits of x are 0 and a.

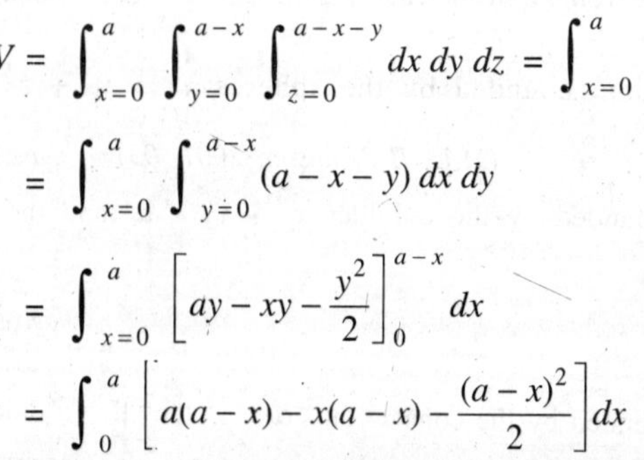

$$V = \int_{x=0}^{a} \int_{y=0}^{a-x} \int_{z=0}^{a-x-y} dx\, dy\, dz = \int_{x=0}^{a} \int_{y=0}^{a-x} [z]_0^{a-x-y}\, dx\, dy$$

$$= \int_{x=0}^{a} \int_{y=0}^{a-x} (a - x - y)\, dx\, dy$$

$$= \int_{x=0}^{a} \left[ay - xy - \frac{y^2}{2}\right]_0^{a-x} dx$$

$$= \int_0^a \left[a(a-x) - x(a-x) - \frac{(a-x)^2}{2}\right] dx$$

$$= \int_0^a \left[a^2 - ax - ax + x^2 - \frac{a^2}{2} + ax - \frac{x^2}{2} \right] dx$$

$$= \int_0^a \left(\frac{a^2}{2} - ax + \frac{x^2}{2} \right) dx$$

$$= \left[\frac{a^2}{2} . x - \frac{ax^2}{2} + \frac{x^3}{6} \right]_0^a = a^3 \left(\frac{1}{2} - \frac{1}{2} + \frac{1}{6} \right) = \frac{a^3}{6}. \quad \textbf{Ans.}$$

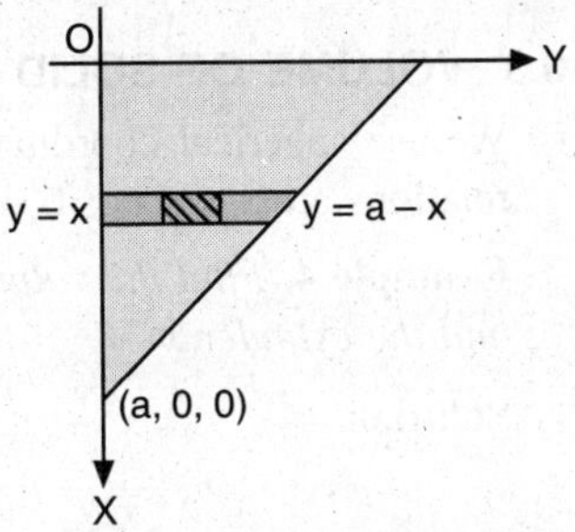

O
Y
y = x
y = a − x
(a, 0, 0)
X

Example 2. *Find the volume of the cylindrical column standing on the area common to the parabolas $y^2 = x$, $x^2 = y$ and cut off by the surface $z = 12 + y - x^2$. (U.P., II Sem., Summer 2001)*

Solution. We have,

$y^2 = x$
$x^2 = y$
$z = 12 + y - x^2$

$$V = \int_0^1 dx \int_{x^2}^{\sqrt{x}} dy \int_0^{12 + y - x^2} dz = \int_0^1 dx \int_{x^2}^{\sqrt{x}} (12 + y - x^2)\, dy$$

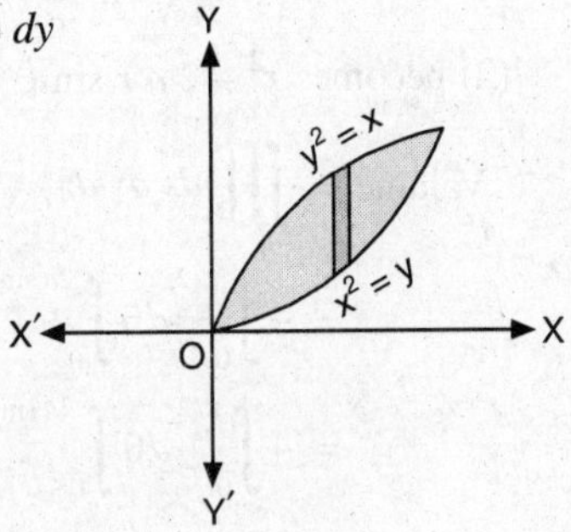

$$= \int_0^1 dx \left(12y + \frac{y^2}{2} - x^2 y \right)_{x^2}^{\sqrt{x}}$$

$$= \int_0^1 \left(12\sqrt{x} + \frac{x}{2} - x^{5/2} - 12x^2 - \frac{x^4}{2} + x^4 \right) dx$$

$$= \left[\frac{2}{3} \times 12 x^{3/2} + \frac{x^2}{4} - \frac{2}{7} x^{7/2} - 4x^3 - \frac{x^5}{10} + \frac{x^5}{5} \right]_0^1$$

$$= 8 + \frac{1}{4} - \frac{2}{7} - 4 - \frac{1}{10} + \frac{1}{5} = 4 + \frac{1}{4} - \frac{2}{7} - \frac{1}{10} + \frac{1}{5} = \frac{560 + 35 - 40 - 14 + 28}{140} = \frac{569}{140} \quad \textbf{Ans.}$$

Y
$y^2 = x$
$x^2 = y$
X′
O
X
Y′

Example 3. *A triangular prism is formed by planes whose equations are $ay = bx$, $y = 0$ and $x = a$. Find the volume of the prism between the planes $z = 0$ and surface $z = c + xy$.*

(M.U. II Semester 2000; U.P., Ist Semester, 2009 (C.O) 2003)

Solution. Required volume $= \displaystyle\int_0^a \int_0^{\frac{bx}{a}} \int_0^{c + xy} dz\, dy\, dx$

$$= \int_0^a \int_0^{\frac{bx}{a}} (c + xy)\, dy\, dx$$

$$= \int_0^a \left(cy + \frac{xy^2}{2} \right)_0^{\frac{bx}{a}} dx$$

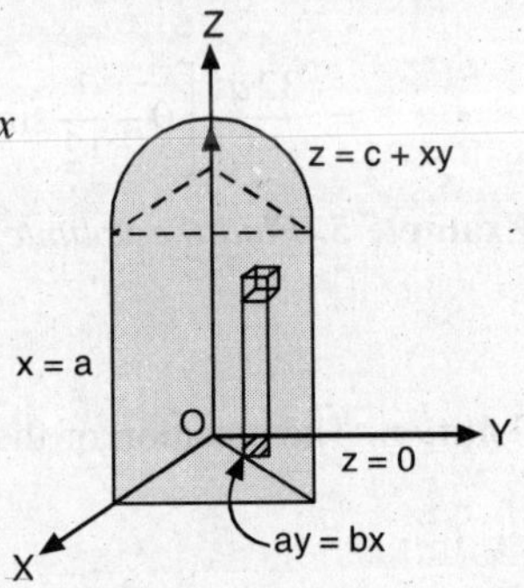

$$= \int_0^a \left(\frac{cbx}{a} + \frac{b^2}{2a^2} x^3 \right) dx = \frac{bc}{a} \left(\frac{x^2}{2} \right)_0^a + \frac{b^2}{2a^2} \left(\frac{x^4}{4} \right)_0^a$$

$$= \frac{abc}{2} + \frac{b^2 a^2}{8} = \frac{ab}{8}(4c + ab) \quad \textbf{Ans.}$$

Z
z = c + xy
x = a
O
Y
z = 0
ay = bx
X

10.3 VOLUME OF SOLID BOUNDED BY SPHERE OR BY CYLINDER

We use spherical coordinates (r, θ, ϕ) and the cylindrical coordinates are (ρ, ϕ, z) and the relations are $x = \rho \cos \phi$, $y = \rho \sin \phi$.

Example 4. *Find the volume of a solid bounded by the spherical surface $x^2 + y^2 + z^2 = 4a^2$ and the cylinder $x^2 + y^2 - 2ay = 0$.*

Solution.
$$x^2 + y^2 + z^2 = 4a^2 \qquad ...(1)$$
$$x^2 + y^2 - 2ay = 0 \qquad ...(2)$$

Considering the section in the positive quadrant of the xy-plane and taking z to be positive (that is volume above the xy-plane) and changing to polar co-ordinates, (1) becomes

$$r^2 + z^2 = 4a^2 \quad \Rightarrow \quad z^2 = 4a^2 - r^2$$

$$\therefore \qquad z = \sqrt{4a^2 - r^2}$$

(2) becomes $r^2 - 2ar \sin\theta = 0 \Rightarrow r = 2a \sin\theta$

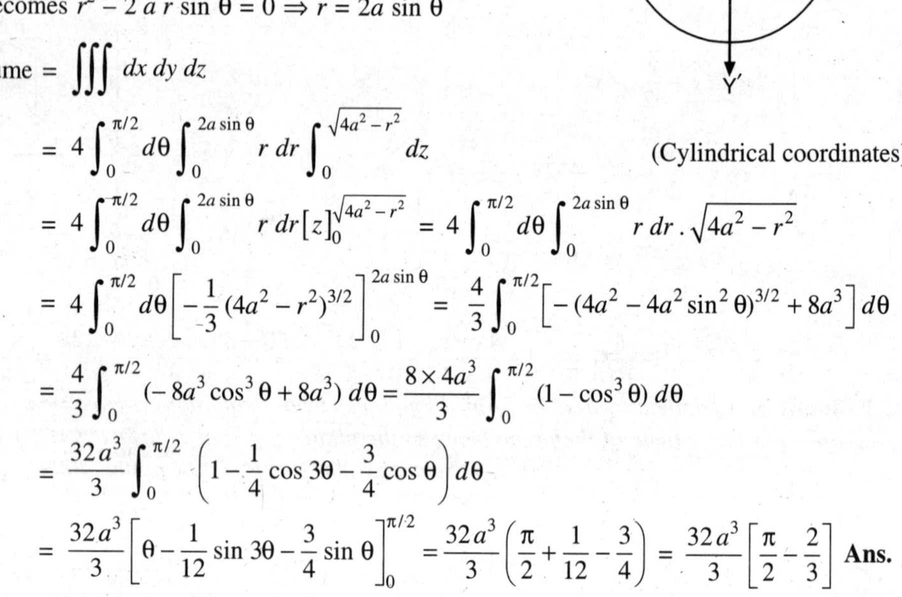

$$\text{Volume} = \iiint dx\, dy\, dz$$

$$= 4\int_0^{\pi/2} d\theta \int_0^{2a\sin\theta} r\, dr \int_0^{\sqrt{4a^2-r^2}} dz \qquad \text{(Cylindrical coordinates)}$$

$$= 4\int_0^{\pi/2} d\theta \int_0^{2a\sin\theta} r\, dr [z]_0^{\sqrt{4a^2-r^2}} = 4\int_0^{\pi/2} d\theta \int_0^{2a\sin\theta} r\, dr \,.\, \sqrt{4a^2 - r^2}$$

$$= 4\int_0^{\pi/2} d\theta \left[-\frac{1}{3}(4a^2 - r^2)^{3/2}\right]_0^{2a\sin\theta} = \frac{4}{3}\int_0^{\pi/2} \left[-(4a^2 - 4a^2\sin^2\theta)^{3/2} + 8a^3\right] d\theta$$

$$= \frac{4}{3}\int_0^{\pi/2} (-8a^3\cos^3\theta + 8a^3)\, d\theta = \frac{8 \times 4a^3}{3}\int_0^{\pi/2} (1 - \cos^3\theta)\, d\theta$$

$$= \frac{32a^3}{3}\int_0^{\pi/2} \left(1 - \frac{1}{4}\cos 3\theta - \frac{3}{4}\cos\theta\right) d\theta$$

$$= \frac{32a^3}{3}\left[\theta - \frac{1}{12}\sin 3\theta - \frac{3}{4}\sin\theta\right]_0^{\pi/2} = \frac{32a^3}{3}\left(\frac{\pi}{2} + \frac{1}{12} - \frac{3}{4}\right) = \frac{32a^3}{3}\left[\frac{\pi}{2} - \frac{2}{3}\right] \quad \textbf{Ans.}$$

Example 5. *Find the volume enclosed by the solid*

$$\left(\frac{x}{a}\right)^{2/3} + \left(\frac{y}{b}\right)^{2/3} + \left(\frac{z}{c}\right)^{2/3} = 1$$

Solution. The equation of the solid is

$$\left(\frac{x}{a}\right)^{2/3} + \left(\frac{y}{b}\right)^{2/3} + \left(\frac{z}{c}\right)^{2/3} = 1$$

Putting
$$\left(\frac{x}{a}\right)^{1/3} = u \quad \Rightarrow \quad x = au^3 \quad \Rightarrow \quad dx = 3au^2\, du$$
$$\left(\frac{y}{b}\right)^{1/3} = v \quad \Rightarrow \quad y = bv^3 \quad \Rightarrow \quad dy = 3bv^2\, dv$$
$$\left(\frac{z}{c}\right)^{1/3} = w \quad \Rightarrow \quad z = cw^3 \quad \Rightarrow \quad dz = 3cw^2\, dw$$

The equation of the solid becomes

$$u^2 + v^2 + w^2 = 1 \quad ...(1)$$

$$V = \iiint dx\, dy\, dz \quad ...(2)$$

On putting the values of dx, dy and dz in (2), we get

$$V = \iiint 27abc\, u^2v^2w^2 du\, dv\, dw \quad ..(3)$$

(1) represents a sphere.
Let us use spherical coordinates.

$$u = r \sin\theta \cos\phi, \qquad v = r\sin\theta\sin\phi,$$

$$w = r\cos\theta, \qquad du\, dv\, dw = r^2 \sin\theta\, dr\, d\theta\, d\phi$$

On substituting spherical coordinates in (3), we have

$$V = 27abc \, .\, 8 \int_{r=0}^{1} \int_{\phi=0}^{\pi/2} \int_{\theta=0}^{\pi/2} r^2 \sin^2\theta \cos^2\phi \, .\, r^2 \sin^2\theta \sin^2\phi \, .\, r^2 \cos^2\theta \, .\, r^2 \sin\theta\, dr\, d\theta\, d\phi$$

$$= 216\, abc \int_{r=0}^{1} r^8\, dr \int_{\phi=0}^{\pi/2} \sin^2\phi \cos^2\phi\, d\phi \int_{\theta=0}^{\pi/2} \sin^5\theta \cos^2\theta\, d\theta$$

$$= 216abc \left[\frac{r^9}{9}\right]_0^1 . \left(\frac{\sqrt{\frac{3}{2}}\sqrt{\frac{3}{2}}}{2\sqrt{3}}\right)\left(\frac{\sqrt{3}\sqrt{\frac{3}{2}}}{2\sqrt{\frac{9}{2}}}\right) = 24\, abc . \frac{1}{2} . \frac{\sqrt{\frac{3}{2}}\sqrt{\frac{3}{2}}}{\sqrt{3}} . \frac{1}{2} . \frac{\sqrt{3}\sqrt{\frac{3}{2}}}{\sqrt{\frac{9}{2}}}$$

$$= 6abc . \frac{\left[\left(\frac{1}{2}\right)\sqrt{\frac{1}{2}}\right]^2}{2!} . \frac{2!\sqrt{\frac{3}{2}}}{\left(\frac{7}{2}\right)\left(\frac{5}{2}\right)\frac{3}{2}\sqrt{\frac{3}{2}}} = 6abc . \frac{1}{4} . \pi \frac{1}{\left(\frac{7}{2}\right)\left(\frac{5}{2}\right)\left(\frac{3}{2}\right)} = \frac{4}{35} abc\, \pi$$

Ans.

Example 6. *Find the volume bounded above by the sphere $x^2 + y^2 + z^2 = a^2$ and below by the cone $x^2 + y^2 = z^2$.* (*U.P. II Semester 2002*)

Solution. The equation of the sphere is $x^2 + y^2 + z^2 = a^2$ ...(1)

and that of the cone is $x^2 + y^2 = z^2$ ...(2)

In polar coordinates $x = r \sin\theta \cos\phi$, $y = r\sin\theta\sin\phi$, $z = r\cos\theta$

The equation (1) in polar co-ordinates is

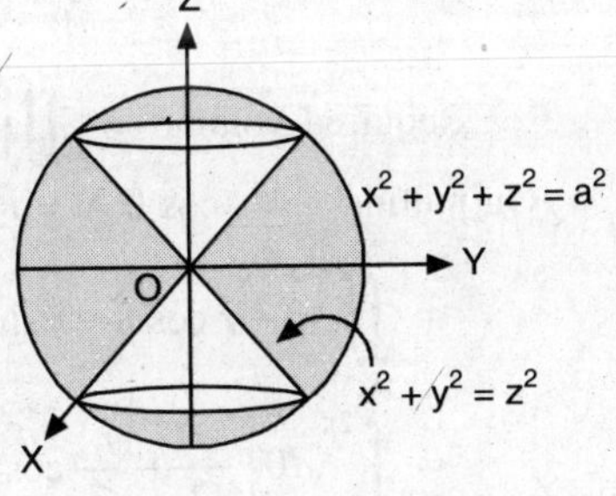

$$(r\sin\theta\cos\phi)^2 + (r\sin\theta\sin\phi)^2 + (r\cos\theta)^2 = a^2$$

$$\Rightarrow \quad r^2\sin^2\theta\cos^2\phi + r^2\sin^2\theta\sin^2\phi + r^2\cos^2\theta = a^2$$

$$\Rightarrow \quad r^2\sin^2\theta(\cos^2\phi + \sin^2\phi) + r^2\cos^2\theta = a^2$$

$$\Rightarrow \quad r^2\sin^2\theta + r^2\cos^2\theta = a^2$$

$$\Rightarrow \quad r^2(\sin^2\theta + \cos^2\theta) = a^2$$

$$\Rightarrow \quad r^2 = a^2 \Rightarrow r = a$$

The equation (2) in polar co-ordinates is

$$(r\sin\theta\cos\phi)^2 + (r\sin\theta\sin\phi)^2 = (r\cos\theta)^2$$

$$\Rightarrow \quad r^2\sin^2\theta(\cos^2\phi + \sin^2\phi) = r^2\cos^2\theta \Rightarrow r^2\sin^2\theta = r^2\cos^2\theta$$

$$\Rightarrow \quad \tan^2\theta = 1 \Rightarrow \tan\theta = 1 \Rightarrow \quad \theta = \pm\frac{\pi}{4}$$

Thus equations (1) and (2) in polar coordinates are respectively,

$$r = a \quad \text{and} \quad \theta = \pm\frac{\pi}{4}$$

The volume in the first octant is one fourth only.

Limits in the first octant : r varies 0 to a, θ from 0 to $\frac{\pi}{4}$ and ϕ from 0 to $\frac{\pi}{2}$.

The required volume lies between $x^2 + y^2 + z^2 = a^2$ and $x^2 + y^2 = z^2$.

$$V = 4\int_0^{\frac{\pi}{2}}\int_0^{\frac{\pi}{4}}\int_0^a r^2 \sin\theta \, dr\, d\theta\, d\phi = 4\int_0^{\frac{\pi}{2}} d\phi \int_0^{\frac{\pi}{4}} \sin\theta \, d\theta \left[\frac{r^3}{3}\right]_0^a$$

$$= 4\int_0^{\frac{\pi}{2}} d\phi \int_0^{\frac{\pi}{4}} \sin\theta\, d\theta . \frac{a^3}{3} = \frac{4a^3}{3}\int_0^{\frac{\pi}{2}} d\phi \,[-\cos\theta]_0^{\frac{\pi}{4}} = \frac{4a^3}{3}(\phi)_0^{\frac{\pi}{2}}\left[-\frac{1}{\sqrt{2}}+1\right]$$

$$= \frac{2}{3}\pi a^3\left(1-\frac{1}{\sqrt{2}}\right)$$ **Ans.**

10.4 VOLUME OF SOLID BOUNDED BY CYLINDER OR CONE

We use cylindrical coordinates (r, θ, z).

Example 7. *Find the volume of the solid bounded by the parabolic $y^2 + z^2 = 4x$ and the plane $x = 5$.*

Solution. $y^2 + z^2 = 4x$, $x = 5$

$$V = \int_0^5 dx \int_{-2\sqrt{x}}^{2\sqrt{x}} dy \int_{-\sqrt{4x-y^2}}^{\sqrt{4x-y^2}} dz = 4\int_0^5 dx \int_0^{2\sqrt{x}} dy \int_0^{\sqrt{4x-y^2}} dz$$

$$= 4\int_0^5 dx \int_0^{2\sqrt{x}} dy\,[z]_0^{\sqrt{4x-y^2}} = 4\int_0^5 dx \int_0^{2\sqrt{x}} dy\,\sqrt{4x-y^2}$$

$$= 4\int_0^5 dx\left[\frac{y}{2}\sqrt{4x-y^2}+\frac{4x}{2}\sin^{-1}\frac{y}{2\sqrt{x}}\right]_0^{2\sqrt{x}} = 4\int_0^5\left[0+2x\left(\frac{\pi}{2}\right)\right]dx = 4\pi\int_0^5 x\,dx$$

$$= 4\pi\left[\frac{x^2}{2}\right]_0^5 = 50\pi$$ **Ans.**

Example 8. *Calculate the volume of the solid bounded by the following surfaces :*

$$z = 0, \quad x^2 + y^2 = 1, \quad x + y + z = 3$$

Solution.

$$x^2 + y^2 = 1 \quad ...(1)$$
$$x + y + z = 3 \quad ...(2)$$
$$z = 0 \quad ...(3)$$

$$\text{Required Volume} = \iiint dx\,dy\,dz = \iint dx\,dy\,[z]_0^{3-x-y} = \iint (3-x-y)\,dx\,dy$$

On putting $x = r\cos\theta$, $y = r\sin\theta$, $dx\,dy = r\,d\theta\,dr$, we get

$$= \iint (3 - r\cos\theta - r\sin\theta)\, r\,d\theta\,dr = \int_0^{2\pi} d\theta \int_0^1 (3r - r^2\cos\theta - r^2\sin\theta)\,dr$$

$$= \int_0^{2\pi} d\theta\left(\frac{3r^2}{2}-\frac{r^3}{3}\cos\theta-\frac{r^3}{3}\sin\theta\right)_0^1 = \int_0^{2\pi}\left(\frac{3}{2}-\frac{1}{3}\cos\theta-\frac{1}{3}\sin\theta\right)d\theta$$

$$= \left[\frac{3}{2}\theta-\frac{1}{3}\sin\theta+\frac{1}{3}\cos\theta\right]_0^{2\pi} = 3\pi-\frac{1}{3}\sin 2\pi+\frac{1}{3}\cos 2\pi-\frac{1}{3} = 3\pi$$ **Ans.**

Example 9. *Find the volume bounded by the cylinder $x^2 + y^2 = 4$ and the planes $y + z = 4$ and $z = 0$.*

Solution. $x^2 + y^2 = 4 \Rightarrow y = \pm\sqrt{4-x^2}$

$$y + z = 4 \quad \Rightarrow \quad z = 4 - y \text{ and } z = 0$$

x varies from -2 to $+2$.

$$V = \iiint dx\, dy\, dz$$

$$= \int_{-2}^{2} dx \int_{-\sqrt{4-x^2}}^{\sqrt{4-x^2}} dy \int_{0}^{4-y} dz$$

$$= \int_{-2}^{2} dx \int_{-\sqrt{4-x^2}}^{\sqrt{4-x^2}} dy\, [z]_0^{4-y}$$

$$= \int_{-2}^{2} dx \int_{-\sqrt{4-x^2}}^{\sqrt{4-x^2}} dy\, (4-y) = \int_{-2}^{2} dx \left[4y - \frac{y^2}{2}\right]_{-\sqrt{4-x^2}}^{\sqrt{4-x^2}}$$

$$= \int_{-2}^{2} dx \left[4\sqrt{4-x^2} - \frac{1}{2}(4-x^2) + 4\sqrt{4-x^2} + \frac{1}{2}(4-x^2)\right]$$

$$= 8\int_{-2}^{2} \sqrt{4-x^2}\, dx = 8\left[\frac{x}{2}\sqrt{4-x^2} + \frac{4}{2}\sin^{-1}\frac{x}{2}\right]_{-2}^{2} = 16\pi$$ **Ans.**

Example 10. *Find the volume in the first octant bounded by the cylinder $x^2 + y^2 = 2$ and the planes $z = x + y$, $y = x$, $z = 0$ and $x = 0$.* *(M.U. II Semester 2005)*

Solution. Here, we have the solid bounded by

$$x^2 + y^2 = 2 \text{ (cylinder)}$$

$$(\text{or} \quad r^2 = 2)$$

$$z = x + y \quad \Rightarrow \quad z = r(\cos\theta + \sin\theta) \qquad \text{(plane)}$$

$$y = x \quad \Rightarrow \quad r\sin\theta = r\cos\theta \qquad \text{(plane)}$$

$$\Rightarrow \quad \tan\theta = 1 \quad \Rightarrow \quad \theta = \frac{\pi}{4}$$

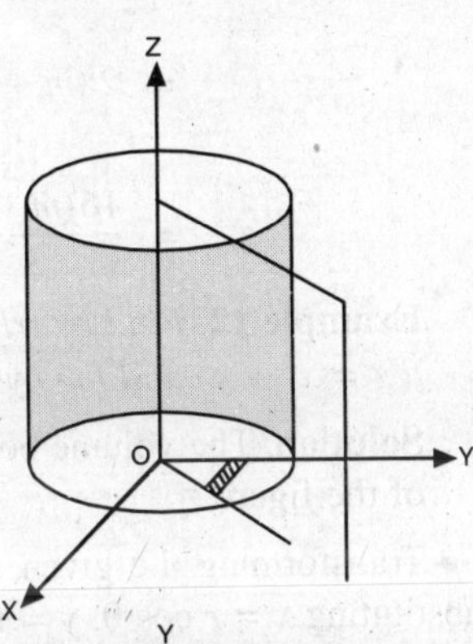

$$x = 0 \Rightarrow r\cos\theta = 0 \Rightarrow \cos\theta = 0 \Rightarrow \theta = \frac{\pi}{2}$$

z varies from 0 to $r(\cos\theta + \sin\theta)$

r varies from 0 to $\sqrt{2}$

θ varies from $\frac{\pi}{4}$ to $\frac{\pi}{2}$

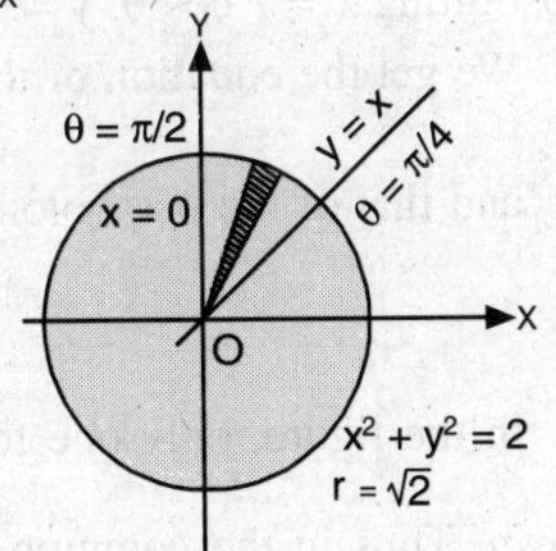

$$\therefore \quad V = \int_{\theta=\pi/4}^{\pi/2} \int_{r=0}^{\sqrt{2}} \int_{z=0}^{r(\cos\theta+\sin\theta)} r\, dr\, d\theta\, dz$$

$$= \int_{\theta=\pi/4}^{\pi/2} \int_{r=0}^{\sqrt{2}} r[z]_0^{r(\cos\theta+\sin\theta)}\, dr\, d\theta$$

$$= \int_{\theta=\pi/4}^{\pi/2} \int_{r=0}^{\sqrt{2}} r^2(\cos\theta+\sin\theta)\, dr\, d\theta$$

$$= \int_{\theta=\pi/4}^{\pi/2} (\cos\theta+\sin\theta)\left[\frac{r^3}{3}\right]_0^{\sqrt{2}} d\theta = \frac{2\sqrt{2}}{3}\int_{\theta=\pi/4}^{\pi/2} (\cos\theta+\sin\theta)\, d\theta$$

$$= \frac{2\sqrt{2}}{3}[\sin\theta - \cos\theta]_{\pi/4}^{\pi/2} = \frac{2\sqrt{2}}{3}\left[(1-0) - \left(\frac{1}{\sqrt{2}} - \frac{1}{\sqrt{2}}\right)\right] = \frac{2\sqrt{2}}{3}$$ **Ans.**

Example 11. *Show that the volume of the wedge intercepted between the cylinder $x^2 + y^2 = 2ax$ and planes $z = mx$, $z = nx$ is $\pi(m - n)\, a^3$.* (*M.U. II Semester, 2000*)

Solution. The equation of the cylinder is $x^2 + y^2 = 2\,a\,x$

we convert the cartesian coordinates into cylindrical coordinates.

$$x = r\cos\theta$$
$$y = r\sin\theta$$
$$x^2 + y^2 = 2ax \Rightarrow r^2 = 2ar\cos\theta$$
$$\Rightarrow \qquad r = 2a\cos\theta$$

r varies from 0 to $2a\cos\theta$

θ varies from $-\dfrac{\pi}{2}$ to $\dfrac{\pi}{2}$

and z varies from $z = nx$ $(z = nr\cos\theta)$ to $z = m\,x$ $(z = m\,r\cos\theta)$

$$V = 2\int_{\theta=0}^{\pi/2}\int_{r=0}^{2a\cos\theta}\int_{z=nr\cos\theta}^{mr\cos\theta} r\,dr\,d\theta\,dz$$

$$= 2\int_{\theta=0}^{\pi/2}\int_{r=0}^{2a\cos\theta} r\left[z\right]_{nr\cos\theta}^{mr\cos\theta} dr\,d\theta$$

$$= 2\int_{\theta=0}^{\pi/2}\int_{r=0}^{2a\cos\theta} r\,.(m-n)\,r\cos\theta\,dr\,d\theta$$

$$= 2\,(m-n)\int_{\theta=0}^{\pi/2}\int_{r=0}^{2a\cos\theta} r^2\cos\theta\,dr\,d\theta$$

$$= 2\,(m-n)\int_{\theta=0}^{\pi/2}\left[\frac{r^3}{3}\right]_0^{2a\cos\theta}\cos\theta\,d\theta = 2\,(m-n)\int_{\theta=0}^{\pi/2}\frac{8a^3}{3}\cos^3\theta\cos\theta\,d\theta$$

$$= \frac{16(m-n)}{3}a^3\int_{\theta=0}^{\pi/2}\cos^4\theta\,d\theta = \frac{16(m-n)}{3}.a^3.\frac{3}{4}.\frac{1}{2}.\frac{\pi}{2} = (m-n)\pi a^3 \quad \textbf{Ans.}$$

Example 12. *Find by triple integration, the volume of the region bounded by the paraboloid $a\,z = x^2 + y^2$ and the cylinder $x^2 + y^2 = R^2$.* (*U.P. I Semester Dec. 2008*)

Solution. The volume bounded by the paraboloid and the cylinder is shown as shaded portion of the figure.

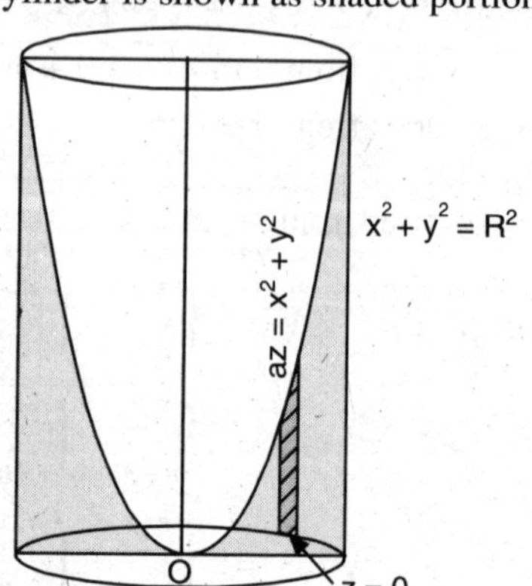

Transforming the given equations to the polar form, by substituting $x = r\cos\theta$, $y = r\sin\theta$

We get the equation of the cylinder

$$x^2 + y^2 = R^2 \text{ as } r = R$$

and that of the paraboloid as

$$a\,z = r^2 \Rightarrow z = \frac{r^2}{a}$$

In the figure, only one fourth of the common volume is shown. Thus in the common region, z varies from 0 to $\dfrac{r^2}{a}$ and r and ϕ vary on the circle $r = R$ (In the xy or r, θ plane).

The variation of r is from 0 to R and that of ϕ is 0 to $\dfrac{\pi}{2}$.

$$\text{Required volume } V = 4\int_0^{\pi/2}\int_0^R\int_0^{\frac{r^2}{a}} r\,dz\,dr\,d\theta \qquad \text{(cylindrical coordinates)}$$

$$= 4\int_0^{\pi/2}\int_0^R r\left[z\right]_0^{r^2/a} dr\, d\theta = 4\int_0^{\pi/2}\int_0^R r\frac{r^2}{a}\, dr\, d\theta = 4\int_0^{\pi/2}\left[\frac{r^4}{4a}\right]_0^R d\theta$$

$$= \frac{1}{a}\int_0^{\pi/2} R^4\, d\theta = \frac{R^4}{a}\cdot\frac{\pi}{2} = \frac{\pi R^4}{2a}$$ **Ans.**

Example 13. *A cylindrical hole of radius b is bored through a sphere of radius a. Find the volume of the remaining solid.* *(M.U. II Semester 2004)*

Solution. Let the equation of the sphere be

$$x^2 + y^2 + z^2 = a^2$$

Now, we will solve this problem using cylindrical coordinates

$$x = r\cos\theta$$
$$y = r\sin\theta$$
$$z = z$$

Limits of z are 0 and $\sqrt{a^2-(x^2+y^2)}$ *i.e.*, $\sqrt{a^2-r^2}$

Limits of r are a and b.

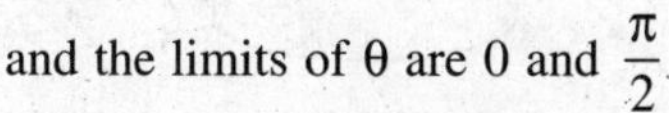

and the limits of θ are 0 and $\frac{\pi}{2}$

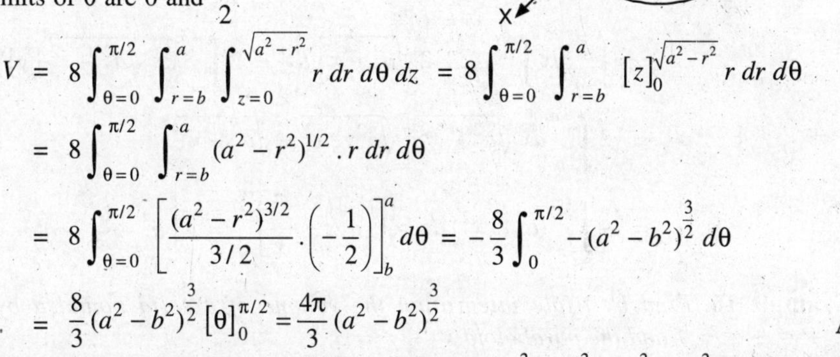

$$V = 8\int_{\theta=0}^{\pi/2}\int_{r=b}^{a}\int_{z=0}^{\sqrt{a^2-r^2}} r\, dr\, d\theta\, dz = 8\int_{\theta=0}^{\pi/2}\int_{r=b}^{a}\left[z\right]_0^{\sqrt{a^2-r^2}} r\, dr\, d\theta$$

$$= 8\int_{\theta=0}^{\pi/2}\int_{r=b}^{a}(a^2-r^2)^{1/2}\cdot r\, dr\, d\theta$$

$$= 8\int_{\theta=0}^{\pi/2}\left[\frac{(a^2-r^2)^{3/2}}{3/2}\cdot\left(-\frac{1}{2}\right)\right]_b^a d\theta = -\frac{8}{3}\int_0^{\pi/2} -(a^2-b^2)^{\frac{3}{2}}\, d\theta$$

$$= \frac{8}{3}(a^2-b^2)^{\frac{3}{2}}\left[\theta\right]_0^{\pi/2} = \frac{4\pi}{3}(a^2-b^2)^{\frac{3}{2}}$$ **Ans.**

Example 14. *Find the volume cut off from the sphere $x^2+y^2+z^2=a^2$ by the cylinder $x^2+y^2=ax$.*

Solution. Cylindrical co-ordinates

$$x = r\cos\theta\ ;\ y = r\sin\theta$$

$$x^2+y^2+z^2 = a^2 \Rightarrow r^2+z^2 = a^2$$

$$x^2+y^2 = ax \Rightarrow r^2 = ar\cos\theta \Rightarrow r = a\cos\theta$$

$$\text{Volume} = \iiint dx\, dy\, dz = \iiint (r\, d\theta\, dr)\, dz$$

$$= \int_{-\pi/2}^{\pi/2} d\theta\int_0^{a\cos\theta} r\, dr\int_{-\sqrt{a^2-r^2}}^{\sqrt{a^2-r^2}} dz$$

$$= 4\int_0^{\pi/2} d\theta\int_0^{a\cos\theta} r\, dr\int_0^{\sqrt{a^2-r^2}} dz$$

$$= 4\int_0^{\pi/2} d\theta\int_0^{a\cos\theta} r\, dr\left[z\right]_0^{\sqrt{a^2-r^2}} = 4\int_0^{\pi/2} d\theta\int_0^{a\cos\theta} r\, dr\sqrt{a^2-r^2}$$

$$= 4\int_0^{\pi/2} d\theta\left[\left(-\frac{1}{2}\right)\frac{2}{3}\left(a^2-r^2\right)^{\frac{3}{2}}\right]_0^{a\cos\theta}$$

$$= -\frac{4}{3}\int_0^{\frac{\pi}{2}} d\theta\left[\left(a^2 - a^2\cos^2\theta\right)^{\frac{3}{2}} - a^3\right] = -\frac{4}{3}\int_0^{\pi/2} d\theta\left[a^3\left(1-\cos^2\theta\right)^{3/2} - a^3\right]$$

$$= \frac{4}{3}\int_0^{\pi/2}(a^3 - a^3\sin^3\theta)\,d\theta = \frac{4a^3}{3}\int_0^{\pi/2}(1-\sin^3\theta)\,d\theta = \frac{4a^3}{3}\left[\frac{\pi}{2}-\frac{2}{3}\right]$$ **Ans.**

Example 15. *Find the volume common to the cylinders*
$x^2 + y^2 = a^2$ *and* $x^2 + z^2 = a^2$ (*M.U. II Semester 2004*)

Solution. $x^2 + y^2 = a^2$; $x^2 + z^2 = a^2$

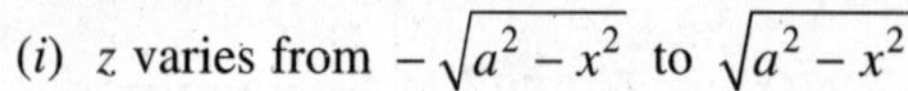

(*i*) z varies from $-\sqrt{a^2 - x^2}$ to $\sqrt{a^2 - x^2}$

(*ii*) y varies from $-\sqrt{a^2 - x^2}$ to $\sqrt{a^2 - x^2}$

(*ii*) x varies from $-a$ to a.

$$\text{Required volume} = \int_{-a}^{+a} dx \int_{-\sqrt{a^2-x^2}}^{\sqrt{a^2-x^2}} dy \int_{-\sqrt{a^2-x^2}}^{\sqrt{a^2-x^2}} dz$$

$$= \int_{-a}^{+a} dx \int_{-\sqrt{a^2-x^2}}^{\sqrt{a^2-x^2}} dy\,[z]_{-\sqrt{a^2-x^2}}^{\sqrt{a^2-x^2}}$$

$$= \int_{-a}^{+a} dx \int_{-\sqrt{a^2-x^2}}^{\sqrt{a^2-x^2}} 2\sqrt{a^2-x^2}\,dy = 2\int_{-a}^{+a} dx\,\sqrt{a^2-x^2}\,[y]_{-\sqrt{a^2-x^2}}^{\sqrt{a^2-x^2}}$$

$$= 2\int_{-a}^{+a} dx\,\sqrt{a^2-x^2}\left[\sqrt{a^2-x^2}+\sqrt{a^2-x^2}\right]$$

$$= 4\int_{-a}^{+a}(a^2-x^2)\,dx = 4\left[a^2x - \frac{x^3}{3}\right]_{-a}^{+a} = 4\left[a^3 - \frac{a^3}{3} + a^3 - \frac{a^3}{3}\right] = \frac{16a^3}{3}$$ **Ans.**

Example 16. *Find by triple integration the volume of a solid bounded by the sphere* $x^2 + y^2 + z^2 = 4$ *and the paraboloid* $x^2 + y^2 = 3z$.

Solution. $x^2 + y^2 + z^2 = 4 \Rightarrow z = \sqrt{4 - x^2 - y^2}$ ...(1)

and $x^2 + y^2 = 3z \quad \Rightarrow \quad z = \dfrac{x^2+y^2}{3}$ ...(2)

Limits of z are $\sqrt{4 - x^2 - y^2}$ and $\dfrac{x^2+y^2}{3}$

Area of cross-section at the intersection of (1) and (2) is given by

$$4 - x^2 - y^2 = \left(\frac{x^2+y^2}{3}\right)^2$$

$\Rightarrow \quad (x^2+y^2)^2 + 9(x^2+y^2) - 36 = 0$

$\Rightarrow \quad (x^2+y^2-3)(x^2+y^2+12) = 0$

$\Rightarrow \quad x^2 + y^2 = 3$, a circle in the area of cross-section

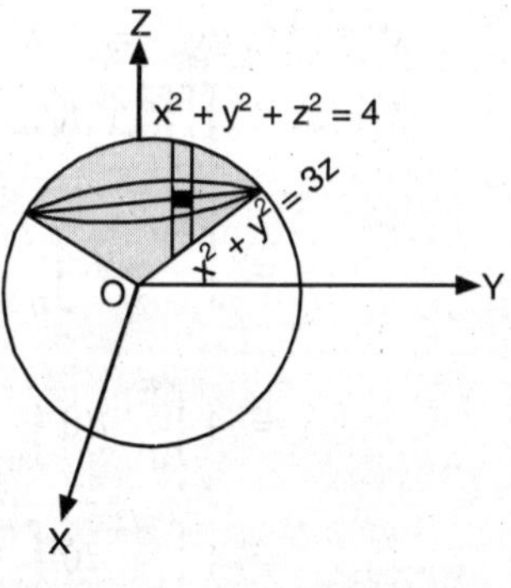

$$V = \iiint dx\,dy\,dz = \iint dx\,dy\int_{\frac{x^2+y^2}{3}}^{\sqrt{4-x^2-y^2}} dz$$

$$= \iint dx\,dy\,[z]_{\frac{x^2+y^2}{3}}^{\sqrt{4-x^2-y^2}} = \iint dx\,dy\left[\sqrt{4-x^2-y^2} - \frac{x^2+y^2}{3}\right]$$

$(x^2 + y^2 = r^2,\ dx\,dy = r\,d\theta\,dr)$

Limits of r are 0 to $\sqrt{3}$ and that of θ are from 0 to 2π.

$$V = \int_0^{2\pi}\int_0^{\sqrt{3}}\left(\sqrt{4-r^2}-\frac{r^2}{3}\right) r\,d\theta\,dr = \int_0^{2\pi} d\theta \int_0^{\sqrt{3}}\left[r\left(4-r^2\right)^{\frac{1}{2}}-\frac{r^3}{3}\right]dr$$

$$= (\theta)_0^{2\pi}\left[-\frac{1}{2}\times\frac{2}{3}\left(4-r^2\right)^{\frac{3}{2}}-\frac{r^4}{12}\right]_0^{\sqrt{3}} = 2\pi\left[-\frac{1}{3}-\frac{3}{4}+\frac{8}{3}\right]=\frac{19\pi}{6} \qquad \textbf{Ans.}$$

10.5 VOLUME BOUNDED BY A PARABOLOID

When a volume is bounded by a paraboloid it is convenient to use cartesian coordinates again.

Example 17. *Find the volume cut off from the paraboloid*

$$x^2+\frac{y^2}{4}+z = 1 \text{ by the plane } z = 0. \qquad \textit{(M.U. II Semester 2005)}$$

Solution. We have

$$x^2+\frac{y^2}{4}+z = 1 \quad \text{(Paraboloid)} \qquad ...(1)$$

$$z = 0 \quad (x\text{-}y \text{ plane}) \qquad ...(2)$$

z varies from 0 to $1-x^2-\dfrac{y^2}{4}$

y varies from $-2\sqrt{1-x^2}$ to $2\sqrt{1-x^2}$

x varies from -1 to 1.

$$V = \iiint dx\,dy\,dz$$

$$= \int_{-1}^{1} dx\int_{-2\sqrt{1-x^2}}^{2\sqrt{1-x^2}} dy\int_0^{\left(1-x^2-\frac{y^2}{4}\right)} dz$$

$$= \int_{-1}^{1}\int_{-2\sqrt{1-x^2}}^{2\sqrt{1-x^2}}\left(1-x^2-\frac{y^2}{4}\right)dx\,dy$$

$$= 4\int_0^1\int_0^{2\sqrt{1-x^2}}\left(1-x^2-\frac{y^2}{4}\right)dx\,dy$$

$$= 4\int_0^1\left[(1-x^2)\,y-\frac{y^3}{12}\right]_0^{2\sqrt{1-x^2}} .\,dx$$

$$= 4\int_0^1\left[(1-x^2).\,2\sqrt{1-x^2}-\frac{8}{12}(1-x^2)^{3/2}\right]d\,x$$

$$= 4\int_0^1\left[2(1-x^2)^{3/2}-\frac{2}{3}(1-x^2)^{3/2}\right]d\,x$$

On putting $x = \sin\theta$, we get

$$V = 4\int_0^1\frac{4}{3}(1-x^2)^{3/2}\,dx = \frac{16}{3}\int_0^{\pi/2}(-\sin^2\theta)^{3/2}\cos\theta\,d\theta$$

$$= \frac{16}{3}\int_0^{\pi/2}\cos^4\theta\,d\theta = \frac{16}{3}.\frac{3}{4}.\frac{1}{2}.\frac{\pi}{2} = \pi \qquad \textbf{Ans.}$$

Example 18. *Find the volume bounded by the paraboloid $x^2+y^2 = az$ and the cylinder $x^2+y^2 = a^2$.* *(M.U. II Semester 2007)*

Solution. The required solid is bounded by a cylinder

$$x^2 + y^2 = a^2 \quad \Rightarrow \quad r^2 = a^2$$

and the paraboloid $x^2 + y^2 = a\,z \Rightarrow r^2 = az \Rightarrow z = \dfrac{r^2}{a}$

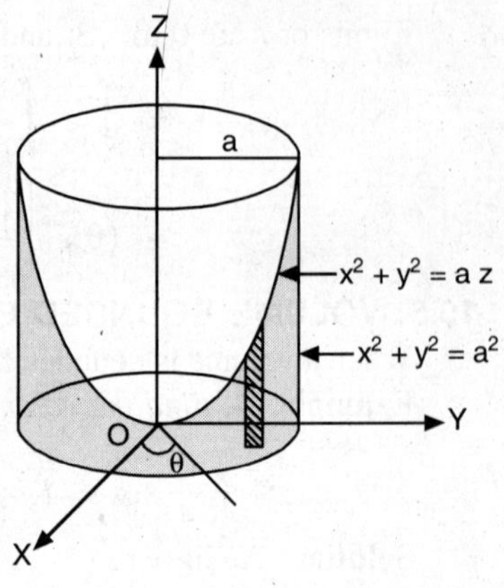

z varies from 0 to $\dfrac{r^2}{a}$

r varies from 0 to a

and θ varies from 0 to $\dfrac{\pi}{2}$

and the solid lies in four octants.

$$V = 4\int_{\theta=0}^{\pi/2}\int_{r=0}^{a}\int_{z=0}^{r^2/a} r\,dr\,d\theta\,dz$$

$$= 4\int_{\theta=0}^{\pi/2}\int_{r=0}^{a} r\left[z\right]_{z=0}^{r^2/a} dr\,d\theta = 4\int_{\theta=0}^{\pi/2}\int_{r=0}^{a}\frac{r^3}{a}\,dr\,d\theta = \frac{4}{a}\int_{\theta=0}^{\pi/2}\left[\frac{r^4}{4}\right]_0^a d\theta$$

$$= \frac{4}{a}\int_{\theta=0}^{\pi/2}\left[\frac{a^4}{4}\right]d\theta = a^3\int_0^{\pi/2} d\theta = a^3\left[\theta\right]_0^{\pi/2} = \frac{\pi}{2}a^3 \qquad \textbf{Ans.}$$

Example 19. *Find the volume bounded by the surfaces*

$$z = 4 - x^2 - \frac{1}{4}y^2 \quad \text{and} \quad z = 3x^2 + \frac{y^2}{4} \qquad \textit{(M.U. II Semester, 2003)}$$

Solution. Here, we have

$$z + x^2 + \frac{y^2}{4} = 4 \quad \text{(Paraboloid)} \quad \ldots(1)$$

$$z = 3x^2 + \frac{y^2}{4} \quad \text{(Paraboloid)} \quad \ldots(2)$$

Let us find out the equation of the section intersected by (1) and (2).

Solving them, we get

$$4 - x^2 - \frac{1}{4}y^2 = 3x^2 + \frac{y^2}{4}$$

$$\Rightarrow \quad 4x^2 + \frac{y^2}{2} = 4 \quad i.e. \quad x^2 + \frac{y^2}{8} = 1$$

The cross-section is an ellipse in four octants

$$V = 4\int_{x=0}^{1}\int_{y=0}^{\sqrt{8(1-x^2)}}\int_{3x^2+(y^2/4)}^{4-x^2-(y^2/4)} dz\,dy\,dx$$

$$= 4\int_{x=0}^{1}\int_{y=0}^{\sqrt{8(1-x^2)}}\left(4 - x^2 - \frac{y^2}{4} - 3x^2 - \frac{y^2}{4}\right)dy\,dx$$

$$= 4\int_{x=0}^{1}\int_{y=0}^{\sqrt{8(1-x^2)}}\left(4 - 4x^2 - \frac{y^2}{2}\right)dy\,dx$$

$$= 4\int_{x=0}^{1}\left[4y - 4x^2y - \frac{y^3}{6}\right]_0^{\sqrt{8(1-x^2)}} dx = 4\int_0^1\left[4(1-x^2)\,y - \frac{y^3}{6}\right]_0^{\sqrt{8(1-x^2)}} dx$$

$$= 4\int_0^1 \left[4(1-x^2).\sqrt{8}.\sqrt{1-x^2} - \frac{1}{6}8\sqrt{8}\,(1-x^2)^{3/2}\right]dx = 4\int_0^1 \left(4\sqrt{8} - \frac{4\sqrt{8}}{3}\right)(1-x^2)^{3/2}dx$$

$$= \frac{64\sqrt{2}}{3}\int_0^1 (1-x^2)^{3/2}\,dx \qquad [\text{Put } x = \sin\theta \;\Rightarrow\; dx = \cos\theta\, d\theta]$$

$$= \frac{64\sqrt{2}}{3}\int_0^{\pi/2} \cos^4\theta\, d\theta = \frac{64\sqrt{2}}{3}.\frac{3}{4}.\frac{1}{2}.\frac{\pi}{2} = 4\sqrt{2}\,.\pi$$ **Ans.**

Example 20. *Find the volume enclosed between the cylinders $x^2 + y^2 = a\,x$, and $z^2 = a\,x$.*

Solution. Here, we have $x^2 + y^2 = ax$ ...(1)

$z^2 = ax$ ...(2)

$$V = \iiint dx\, dy\, dz$$

$$= \int_0^a dx \int_{-\sqrt{ax-x^2}}^{\sqrt{ax-x^2}} dy \int_{-\sqrt{ax}}^{\sqrt{ax}} dz = 2\int_0^a dx \int_{-\sqrt{a^2-x^2}}^{\sqrt{ax-x^2}} dy \int_0^{\sqrt{ax}} dz$$

$$= 2\int_0^a dx \int_{-\sqrt{ax-x^2}}^{\sqrt{ax-x^2}} dy\,(z)_0^{\sqrt{ax}} = 2\int_0^a dx \int_{-\sqrt{ax-x^2}}^{\sqrt{ax-x^2}} dy\,\sqrt{ax} = 2\int_0^a \sqrt{ax}\,dx\,[y]_{-\sqrt{ax-x^2}}^{\sqrt{ax-x^2}}$$

$$= 2\int_0^a \sqrt{ax}\,dx\left(2\sqrt{ax-x^2}\right) = 4\sqrt{a}\int_0^a x\sqrt{a-x}\,dx$$

Putting $x = a\sin^2\theta$ so that $dx = 2a\sin\theta\cos\theta\, d\theta$, we get

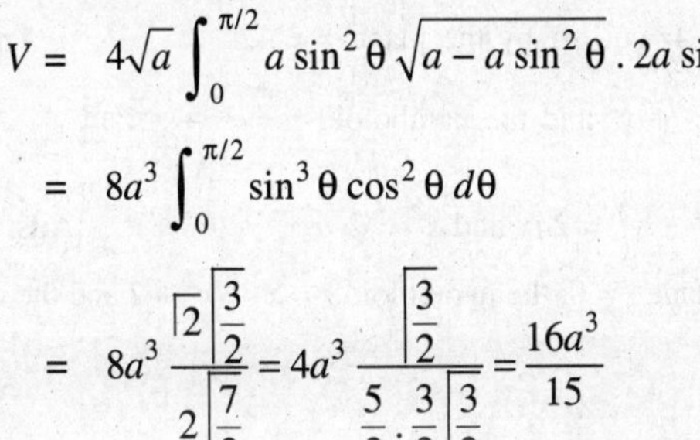

$$V = 4\sqrt{a}\int_0^{\pi/2} a\sin^2\theta\sqrt{a - a\sin^2\theta}\,.\,2a\sin\theta\cos\theta\, d\theta$$

$$= 8a^3\int_0^{\pi/2} \sin^3\theta\cos^2\theta\, d\theta$$

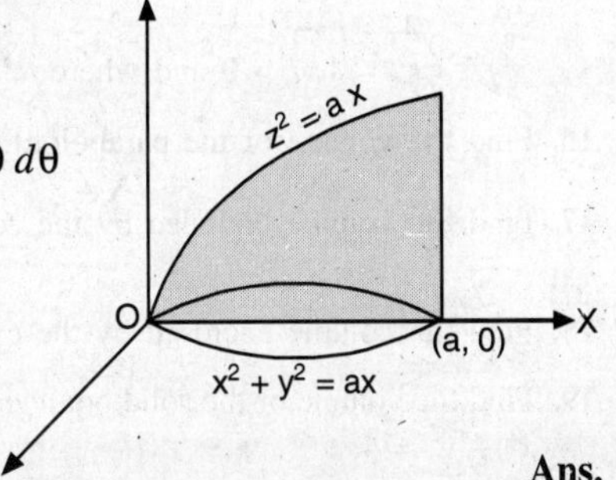

$$= 8a^3\frac{\boxed{2}\,\boxed{\tfrac{3}{2}}}{2\,\boxed{\tfrac{7}{2}}} = 4a^3\frac{\boxed{\tfrac{3}{2}}}{\tfrac{5}{2}.\tfrac{3}{2}\boxed{\tfrac{3}{2}}} = \frac{16a^3}{15}$$ **Ans.**

EXERCISE 10.1

1. Find the volume bounded by the coordinate planes and the plane.

$\frac{x}{a} + \frac{y}{b} + \frac{z}{c} = 1$ *(MTU, 2012)* **Ans.** $\frac{abc}{6}$

2. Find the volume bounded by the cylinders $y^2 = x$ and $x^2 = y$ between the planes $z = 0$ and $x + y + z = 2$. **Ans.** $\frac{11}{30}$

3. Find the volume bounded by the co-ordinate planes and the plane.

$l\,x + m\,y + n\,z = 1$ *(A.M.I.E.T.E. Winter 2001)* **Ans.** $\frac{1}{6\,l\,m\,n}$

4. Find the volume of the sphere $x^2 + y^2 + z^2 = a^2$ by triple integration. *(AMIETE, June 2009)* **Ans.** $\frac{4}{3}\pi a^3$

5. Find the volume of the ellipsoid $\frac{x^2}{a^2} + \frac{y^2}{b^2} + \frac{z^2}{c^2} = 1$ **Ans.** $\frac{4\pi\, a\, b\, c}{3}$

6. Find the volume bounded by the cylinder $x^2 + y^2 = a^2$ and the planes $y + z = 2a$ and $z = 0$. *(M.U. II Semester 2000, 02, 06)* **Ans.** $2\pi a^3$

7. Find the volume bounded by the cylinder $x^2 + y^2 = a^2$ and the planes $z = 0$ and $y + z = b$.
Ans. $\pi a^2 b$

8. Find the volume of the region bounded by $z = x^2 + y^2$, $z = 0$, $x = -a$, $x = a$ and $y = -a$, $y = a$.
Ans. $\frac{8}{3}a^4$

9. Find the volume enclosed by the cylinder $x^2 + y^2 = 9$ and the planes $x + z = 5$ and $z = 0$.
Ans. $45\pi - 36$

10. Compute the volume of the solid bounded by $x^2 + y^2 = z$, $z = 2x$.(*A.M.I.E., Summer 2000*)
Ans. 2π

11. Find the volume cut from the paraboloid $4z = x^2 + y^2$ by plane $z = 4$.
(*U.P. I Semester, Dec. 2005*) **Ans.** 32π

12. By using triple integration find the volume cut off from the sphere $x^2 + y^2 + z^2 = 16$ by the plane $z = 0$ and the cylinder $x^2 + y^2 = 4x$.
Ans. $\frac{64}{9}(3\pi - 4)$

13. The sphere $x^2 + y^2 + z^2 = a^2$ is pierced by the cylinder $x^2 + y^2 = a^2(x^2 - y^2)$.

Prove that the volume of the sphere that lies inside the cylinder is $\frac{8}{3}\left[\frac{\pi}{4} + \frac{5}{3} - \frac{4\sqrt{2}}{3}\right]a^3$.

14. Find the volume of the solid bounded by the surfaces $z = 0$, $3z = x^2 + y^2$ and $x^2 + y^2 = 9$.
(*A.M.I.E.T.E., Summer 2005*) **Ans.** $\frac{27\pi}{2}$

15. Obtain the volume bounded by the surface $z = c\left(1 - \frac{x}{a}\right)\left(1 - \frac{y}{b}\right)$ and a quadrant of the elliptic cylinder $\frac{x^2}{a^2} + \frac{y^2}{b^2} = 1$, $z > 0$ and where $a, b > 0$.
A.M.I.E.T.E., Dec. 2005) **Ans.** $\pi\, abc$

16. Find the volume of the paraboloid $x^2 + y^2 = 4z$ cut off by the plane $z = 4$.
Ans. 32π

17. Find the volume bounded by the cone $z^2 = x^2 + y^2$ and the paraboloid $z = x^2 + y^2$.
Ans. $\frac{\pi}{6}$

18. Find the volume enclosed by the cylinders $x^2 + y^2 = 2ax$ and $z^2 = 2ax$.
Ans. $\frac{128a^3}{15}$

19. Find the volume of the solid bounded by the plane $z = 0$, the paraboloid $z = x^2 + y^2 + 2$ and the cylinder $x^2 + y^2 = 4$.
Ans. 16π

20. The triple integral $\iiint dx\,dy\,dz$ gives

(*a*) Volume of region (*b*) Surface area of region T
(*c*) Area of region *T* (*d*) Density of region *T*. (*A.M.I.E.T.E., 2002*) **Ans.** (*a*)

10.6 SURFACE AREA

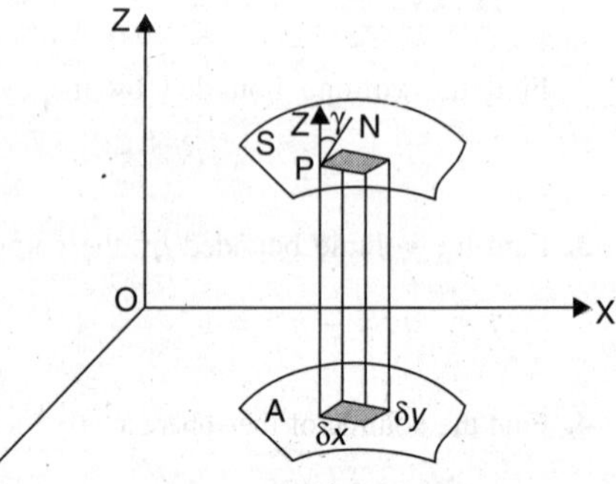

Let $z = f(x,y)$ be the surface S. Let its projection on the x-y plane be the region A. Consider an element $\delta x.\ \delta y$ in the region A. Erect a cylinder on the element $\delta x.\ \delta y$ having its generator parallel to OZ and meeting the surface S in an element of area δs.

$\therefore \qquad \delta x\ \delta y = \delta s \cos\gamma,$

Where γ is the angle between the xy-plane and the tangent plane to S at P, *i.e.*, it is the angle between the Z-axis and the normal to S at P.

The direction cosines of the normal to the surface $F(x, y, z) = 0$ are proportional to

$$\frac{\partial F}{\partial x}, \frac{\partial F}{\partial y}, \frac{\partial F}{\partial z}$$

$\therefore$ The direction of the normal to S $[F = f(x, y) - z]$ are proportional to $-\frac{\partial z}{\partial x}, -\frac{\partial z}{\partial y}, 1$ and those of the Z-axis are 0, 0 , 1.

$$\text{Direction cosines} = \frac{-\frac{\partial z}{\partial x}}{\sqrt{\left(\frac{\partial z}{\partial x}\right)^2+\left(\frac{\partial z}{\partial y}\right)^2+1}}, \frac{-\frac{\partial z}{\partial y}}{\sqrt{\left(\frac{\partial z}{\partial x}\right)^2+\left(\frac{\partial z}{\partial y}\right)^2+1}}, \frac{1}{\sqrt{\left(\frac{\partial z}{\partial x}\right)^2+\left(\frac{\partial z}{\partial y}\right)^2+1}},$$

Hence
$$\cos\gamma = \frac{1}{\sqrt{\left[\left(\frac{\partial z}{\partial x}\right)^2+\left(\frac{\partial z}{\partial y}\right)^2+1\right]}} \quad (\cos\theta = l_1 l_2 + m_1 m_2 + n_1 n_2)$$

$$\delta S = \frac{\delta x\, \delta y}{\cos\gamma} = \sqrt{\left[\left(\frac{\partial z}{\partial x}\right)^2+\left(\frac{\partial z}{\partial y}\right)^2+1\right]}\,\delta x\,\delta y\,; \quad S = \iint_A \sqrt{\left[\left(\frac{\partial z}{\partial x}\right)^2+\left(\frac{\partial z}{\partial y}\right)^2+1\right]}\,dx\,dy$$

Example 21. *Evolute* $\iint_s 6xy\,ds$, *where S is the portion of the plane* $x + y + z = 1$ *that lies infront of the yz-plane.* *(Gujarat, I Semester, Jan. 2009)*

Solution. Here, we nave

$$I = \iint_s 6xy\,ds$$

$$x + y + z = 1,\ z = 1 - x - y \qquad \ldots (1)$$

$$\Rightarrow \quad \frac{\partial z}{\partial x} = -1 \quad \Rightarrow \quad \frac{\partial z}{\partial y} = -1$$

The direction ratio of the normal to the plane (1) are $-\frac{\partial z}{\partial x}, = -1$

Direction cosines of the normal to the surface

$$\frac{-\frac{\partial z}{\partial x}}{\sqrt{\left(\frac{\partial z}{\partial x}\right)^2+\left(\frac{\partial z}{\partial y}\right)^2+1}}, \frac{-\frac{\partial z}{\partial y}}{\sqrt{\left(\frac{\partial z}{\partial x}\right)^2+\left(\frac{\partial z}{\partial y}\right)^2+1}}, \frac{1}{\sqrt{\left(\frac{\partial z}{\partial x}\right)^2+\left(\frac{\partial z}{\partial y}\right)^2+1}}$$

Direction cosines of the normal (x-axis) to the plane 1, 0, 0

$$\cos\alpha = \frac{-\frac{\partial z}{\partial x}}{\sqrt{\left(\frac{\partial z}{\partial x}\right)^2+\left(\frac{\partial z}{\partial y}\right)^2+1}} \Rightarrow \delta s = \frac{\delta x\,\delta y}{\cos\alpha} = \frac{\sqrt{\left(\frac{\partial z}{\partial x}\right)^2+\left(\frac{\partial z}{\partial y}\right)^2+1}}{-\frac{\partial z}{\partial x}}\,\delta x\,\delta y$$

$$I = \iint 6xy\,ds = 6\int x dx \int y \frac{\sqrt{\left(\frac{\partial z}{\partial x}\right)^2+\left(\frac{\partial z}{\partial s}\right)^2+1}}{-\frac{\partial z}{\partial x}}\,dy$$

$$= 6\int_0^1 x dx \int_0^{1-x} \frac{\sqrt{1+1+1}}{-1}\,dy = 6\int_0^1 dx \left[\frac{y^2}{2}\right]_0^{1-x} (-\sqrt{3})$$

$$= -3\sqrt{3}\int_0^{1-x} x dx\,(1-x)^2 = -3\sqrt{3}\int_0^1 (x^3 - 2x^2 + x)\,dx$$

$$= -3\sqrt{3}\left(\frac{x^4}{4}-\frac{2x^3}{3}+\frac{x^2}{2}\right)_0^1 = -3\sqrt{3}\left[\frac{1}{4}-\frac{2}{3}+\frac{1}{2}\right] = -\frac{3\sqrt{3}}{12} = -\frac{\sqrt{3}}{4}$$

$$\text{Surface Area} = \frac{\sqrt{3}}{4}$$ **Ans.**

Example 22. *Find the surface area of the cylinder $x^2 + z^2 = 4$ inside the cylinder $x^2 + y^2 = 4$.*

Solution. $x^2 + y^2 = 4$

$$2x + 2z\frac{\partial z}{\partial x} = 0 \text{ or } \frac{\partial z}{\partial x} = -\frac{x}{z},\ \frac{\partial z}{\partial y} = 0$$

$$\left(\frac{\partial z}{\partial x}\right)^2 + \left(\frac{\partial z}{\partial y}\right)^2 + 1 = \frac{x^2}{z^2} + 1 = \frac{x^2+z^2}{z^2} = \frac{4}{4-x^2}$$

Hence, the required surface area

$$= 8\int_0^2\int_0^{\sqrt{4-x^2}} \sqrt{\left[\left(\frac{\partial z}{\partial x}\right)^2 + \left(\frac{\partial z}{\partial y}\right)^2 + 1\right]}\, dx\, dy$$

$$= 8\int_0^2\int_0^{\sqrt{4-x^2}} \frac{2}{\sqrt{4-x^2}}\, dx\, dy = 16\int_0^2 \frac{1}{\sqrt{4-x^2}}[y]_0^{\sqrt{4-x^2}}\, dx = 16\int_0^2 \frac{1}{\sqrt{4-x^2}}[\sqrt{4-x^2}]\, dx$$

$$= 16\int_0^2 dx = 16\,(x)_0^2 = 32$$ **Ans.**

Example 23. *Find the surface area of the sphere $x^2 + y^2 + z^2 = 9$ lying inside the cylinder $x^2 + y^2 = 3y$.*

Solution. $x^2 + y^2 + z^2 = 9$

$$2x + 2z\frac{\partial z}{\partial x} = 0, \quad \frac{\partial z}{\partial x} = -\frac{x}{z}$$

$$2x + 2z\frac{\partial z}{\partial y} = 0, \quad \frac{\partial z}{\partial y} = -\frac{y}{z}$$

$$\left[\left(\frac{\partial z}{\partial x}\right)^2 + \left(\frac{\partial z}{\partial y}\right)^2 + 1\right] = \frac{x^2}{z^2} + \frac{y^2}{z^2} + 1 = \frac{x^2+y^2+z^2}{z^2} = \frac{9}{9-x^2-y^2} = \frac{9}{9-r^2} \quad \begin{bmatrix} x = r\cos\theta \\ y = r\sin\theta \end{bmatrix}$$

$$x^2 + y^2 = 3y \text{ or } r^2 = 3r\sin\theta \text{ or } r = 3\sin\theta.$$

Hence, the required surface area

$$= \iint \sqrt{\left[\left(\frac{\partial z}{\partial x}\right)^2 + \left(\frac{\partial z}{\partial y}\right)^2 + 1\right]}\, dx\, dy = 4\int_0^{\pi/2}\int_0^{3\sin\theta} \frac{3}{\sqrt{9-r^2}}\, r\, d\theta\, dr = 12\int_0^{\pi/2} d\theta \int_0^{3\sin\theta} \frac{r\, dr}{\sqrt{9-r^2}}$$

$$= 12\int_0^{\pi/2} d\theta\,[-\sqrt{9-r^2}\,]_0^{3\sin\theta} = 12\int_0^{\pi/2} [-\sqrt{9-9\sin^2\theta} + 3]\, d\theta$$

$$= 36\int_0^{\pi/2} (-\cos\theta + 1)\, d\theta = 36\,(-\sin\theta + \theta)_0^{\pi/2} = 36\left(-1 + \frac{\pi}{2}\right) = 18\,(\pi - 2)$$ **Ans.**

Example 24. *Find the surface area of the section of the cylinder $x^2 + y^2 = a^2$ made by the plane $x + y + z = a$.*

Solution.

$$x^2 + y^2 = a^2 \quad \text{... (1)}$$

$$x + y + z = a \quad \text{... (2)}$$

The projection of the surface area on xy-plane is a circle

$$x^2 + y^2 = a^2$$

$$1 + \frac{\partial z}{\partial x} = 0 \quad \text{or} \quad \frac{\partial z}{\partial x} = -1$$

$$1 + \frac{\partial z}{\partial y} = 0 \quad \text{or} \quad \frac{\partial z}{\partial y} = -1$$

$$\sqrt{\left(\frac{\partial z}{\partial x}\right)^2+\left(\frac{\partial z}{\partial y}\right)^2+1} = \sqrt{(-1)^2+(-1)^2+1} = \sqrt{3}$$

Hence the required surface area

$$= 4\int_0^a\int_0^{\sqrt{a^2-x^2}}\sqrt{\left(\frac{\partial z}{\partial x}\right)^2+\left(\frac{\partial z}{\partial y}\right)^2+1}\;dx\,dy = 4\int_0^a\int_0^{\sqrt{a^2-x^2}}\sqrt{3}\,dx\cdot dy$$

$$= 4\sqrt{3}\int_0^a [y]_0^{\sqrt{a^2-x^2}}\,dx = 4\sqrt{3}\int_0^a \sqrt{a^2-x^2}\,dx$$

$$= 4\sqrt{3}\left[\frac{x}{2}\sqrt{a^2-x^2}+\frac{a^2}{2}\sin^{-1}\frac{x}{a}\right]_0^a = 4\sqrt{3}\left[0+\frac{a^2}{2}\frac{\pi}{2}\right] = 4\sqrt{3}\left(\frac{a^2\pi}{4}\right) = \sqrt{3}\,\pi\,a^2 \quad \textbf{Ans.}$$

Example 25. *Find the area of that part of the surface of the paraboloid of the paraboloid $y^2 + z^2 = 2\,ax$, which lies between the cylinder, $y^2 = ax$ and the plane $x = a$.*

Solution.
$$y^2 + z^2 = 2\,ax \quad \text{... (1)}$$
$$y^2 = ax \quad \text{... (2)}$$
$$x = a \quad \text{... (3)}$$

Differentiating (1), we get

$$2z\frac{\partial z}{\partial x} = 2\,a, \quad \frac{\partial z}{\partial x} = \frac{a}{z}$$

$$2y + 2\,z\frac{\partial z}{\partial y} = 0, \quad \frac{\partial z}{\partial y} = -\frac{y}{z}$$

$$\left(\frac{\partial z}{\partial x}\right)^2+\left(\frac{\partial z}{\partial y}\right)^2+1 = \frac{a^2}{z^2}+\frac{y^2}{z^2}+1 = \frac{a^2+y^2}{z^2}+1 \qquad \left[\begin{array}{l} y^2+z^2 = 2\,ax \\ z^2 = 2\,ax-y^2 \end{array}\right]$$

$$= \frac{a^2+y^2}{2\,a\,x-y^2}+1 = \frac{a^2+y^2+2\,a\,x-y^2}{2\,a\,x-y^2} = \frac{a^2+2\,a\,x}{2\,a\,x-y^2}$$

$$S = \int_0^a\int_{-\sqrt{ax}}^{\sqrt{ax}}\sqrt{\left(\frac{\partial z}{\partial x}\right)^2+\left(\frac{\partial z}{\partial y}\right)^2+1}\;dx\,dy = \int_0^a\int_{-\sqrt{ax}}^{\sqrt{ax}}\sqrt{\frac{a^2+2\,ax}{2\,ax-y^2}}\;dx\,dy \qquad \left[\begin{array}{l} y^2 = ax \\ y = \pm\sqrt{ax} \end{array}\right]$$

$$= \sqrt{a}\int_0^a\int_{-\sqrt{ax}}^{\sqrt{ax}}\sqrt{\frac{a+2\,x}{2\,ax-y^2}}\;dx\,dy = \sqrt{a}\int_0^a\sqrt{a+2\,x}\,dx\int_{-\sqrt{ax}}^{\sqrt{ax}}\frac{1}{\sqrt{2\,ax+y^2}}\,dy$$

$$= \sqrt{a}\int_0^a\sqrt{a+2\,x}\,dx\left[\sin^{-1}\frac{y}{\sqrt{2\,ax}}\right]_{-\sqrt{ax}}^{\sqrt{ax}}$$

$$= \sqrt{a}\int_0^a\sqrt{a+2\,x}\,dx\left[\sin^{-1}\frac{1}{\sqrt{2}}-\sin^{-1}\left(-\frac{1}{\sqrt{2}}\right)\right] = \sqrt{a}\int_0^a\sqrt{a+2\,x}\,dx\left[\frac{\pi}{4}-\left(\frac{\pi}{4}\right)\right]$$

$$= \sqrt{a}\,\frac{\pi}{2}\int_0^a\sqrt{a+2\,x}\,dx = \frac{\pi}{2}\cdot\frac{\sqrt{a}}{2}\cdot\frac{2}{3}[(a+2x)^{3/2}]_0^a$$

$$= \frac{\pi\sqrt{a}}{6}[(3a)^{3/2}-a^{3/2}] = \frac{\pi\,a^2}{6}[3\sqrt{3}-1] \qquad \textbf{Ans.}$$

EXERCISE 10.2

1. Find the surface area of sphere $x^2 + y^2 + z^2 = 16$. **Ans.** 64π
2. Find the surface area of the portion of the cylinder $x^2 + y^2 = 4y$ lying inside the sphere $x^2 + y^2 + z^2 = 16$. **Ans.** 64.
3. Show that the area of surfaces $cz = xy$ intercepted by the cylinder $x^2 + y^2 = b^2$ is $\iint_A \frac{\sqrt{c^2 + x^2 + y^2}}{c}\, dx\, dy$, where A is the area of the circle $x^2 + y^2 = b^2$, $z = 0$

   Ans. $\frac{2}{3}\frac{\pi}{c}\left[(c^2 + b^2)^{\frac{1}{2}} - c^2\right]$
4. Find the area of the portion of the sphere $x^2 + y^2 + z^2 = a^2$ lying inside the cylinder $x^2 + y^2 = ax$. **Ans.** $2(\pi - 2)a^2$
5. Find the area of the surface of the cone $z^2 = 3(x^2 + y^2)$ cut out by the paraboloid $z = x^2 + y^2$ using surface integral. **Ans.** 6π

10.7 CALCULATION OF MASS

We have,

Volume $= \iint_V dx\, dy\, dz$ Density $= \rho = f(x, y, z)$

[Density = Mass per unit volume] Mass = Volume × Density

Mass $= \iint_V dx\, dy\, dz$ $\boxed{\text{Mass} = \iint_V f(x, y, z)\, dx\, dy\, dz}$

Example 26. Find the mass of a plate which is formed by the co-ordinate planes and the plane $\frac{x}{a} + \frac{y}{b} + \frac{z}{c} = 1$, the density is given by $\rho = k\,x\,y\,z$. *(U.P., I Semester, Dec., 2003)*

Solution. The plate is bounded by the planes $x = 0$, $y = 0$, $z = 0$ and $\frac{x}{a} + \frac{y}{b} + \frac{z}{c} = 1$.

$$\text{Mass} = \iiint dx\, dy\, dz\, \rho = \int_0^c \int_0^{b\left(1-\frac{z}{c}\right)} \int_0^{a\left(1-\frac{y}{b}-\frac{z}{c}\right)} dx\, dy\, dz\, (k\, xyz)$$

$$= k\int_0^c z\, dz \int_0^{b\left(1-\frac{z}{c}\right)} y\, dy \int_0^{a\left(1-\frac{y}{b}-\frac{z}{c}\right)} x\, dx = k\int_0^c z\, dz \int_0^{b\left(1-\frac{z}{c}\right)} y\, dy \left(\frac{x^2}{2}\right)_0^{a\left(1-\frac{y}{b}-\frac{z}{c}\right)}$$

$$= k\int_0^c z\, dz \int_0^{b\left(1-\frac{z}{c}\right)} y\, dy\, \frac{a^2}{2}\left(1-\frac{y}{b}-\frac{z}{c}\right)^2 = \frac{k\,a^2}{2}\int_0^c z\, dz \int_0^{b\left(1-\frac{z}{c}\right)} y\left[\left(1-\frac{z}{c}\right)-\frac{y}{b}\right]^2 dy$$

$$= \frac{k\,a^2}{2}\int_0^c z\, dz \int_0^{b\left(1-\frac{z}{c}\right)} \left[y\left(1-\frac{z}{c}\right)^2 + \frac{y^3}{b^2} - \frac{2y^2}{b}\left(1-\frac{z}{c}\right)\right] dy$$

$$= \frac{k\,a^2}{2}\int_0^c z\, dz \left[\frac{y^2}{2}\left(1-\frac{z}{c}\right)^2 + \frac{y^4}{4b^2} - \frac{2y^3}{3b}\left(1-\frac{z}{c}\right)\right]_0^{b\left(1-\frac{z}{c}\right)}$$

$$= \frac{k\,a^2}{2}\int_0^c z\, dz \left[\frac{b^2}{2}\left(1-\frac{z}{c}\right)^4 + \frac{b^4}{4b^2}\left(1-\frac{z}{c}\right)^4 - \frac{2}{3}\cdot\frac{b^3}{b}\left(1-\frac{z}{c}\right)^4\right]$$

$$= \frac{k\,a^2}{2}\int_0^c z\left[\frac{b^2}{2} + \frac{b^2}{4} - \frac{2b^2}{3}\right]\left(1-\frac{z}{c}\right)^4 dz = \frac{k\,a^2\,b^2}{2\ \ 12}\int_0^c \left(1-\frac{z}{c}\right)^4 dz \qquad [\text{Put } z = c\sin^2\theta]$$

$$= \frac{k\,a^2\,b^2\,c^2}{12}\int_0^{\frac{\pi}{2}} c\sin^2\theta\,(1-\sin^2\theta)^4\,(2c\sin\theta\cos\theta\, d\theta)$$

$$= \frac{k^2 a^2 b^2 c^2}{12} \int_0^{\pi/2} \sin^2 \theta (\cos^8 \theta) \sin\theta \cos\theta \, d\theta = \frac{k^2 a^2 b^2 c^2}{12} \int_0^{\pi/2} \sin^3\theta \cos^9\theta \, d\theta$$

$$= \frac{k^2 a^2 b^2 c^2}{12} \frac{\left\lceil\frac{3+1}{2}\right. \left\lceil\frac{9+1}{2}\right.}{2\left\lceil\frac{3+9+2}{2}\right.} = \frac{k\, a^2 b^2 c^2}{12} \cdot \frac{\lceil 2 \lceil 5}{2\lceil 7} = \frac{k\, a^2 b^2 c^2}{12} \frac{(1)(\lceil 5)}{2\times 6\times 5 \lceil 5} = \frac{k\, a^2 b^2 c^2}{720} \quad \textbf{Ans.}$$

Example 27. *Find the mass of a plate in the shape of the curve* $\left(\frac{x}{a}\right)^{2/3} + \left(\frac{y}{b}\right)^{2/3} = 1$, *the density being given by* $\rho = \mu\, xy$. *(Nagpur University, Summer 2003)*

Solution. Let the required mass be M which is four times the mass in the first quadrant.

From the equation of the curve $\left(\frac{y}{b}\right)^{2/3} = 1 - \left(\frac{x}{a}\right)^{2/3} \Rightarrow y = b\left[1 - \left(\frac{x}{a}\right)^{2/3}\right]^{3/2} = y_1$

For the region OAB, x varies from 0 to a and y varies from 0 to y_1.

$$\therefore M = 4\int_0^a \int_0^{y_1} \rho \, dy\, dx = 4\int_0^a \int_0^{y_1} \mu\, xy\, dy\, dx$$

$$= 4\int_0^a \mu x \cdot \left[\frac{y^2}{2}\right]_0^{y_1} dx = 2\mu \int_0^a x y_1^2 \, dx$$

$$= 2\mu \int_0^a x b^2 \left[1 - \left(\frac{x}{a}\right)^{2/3}\right]^3 dx$$

Y
B (0, b)
y = y₁
y = 0
A′
(–a, 0)
A (a, 0)
B′ (0, b)

Put $x = a \sin 3\,\theta$, then $dx = 3a \sin^2\theta \cos\theta \, d\theta$

When $x = 0$, $\theta = 0$; when $x = a$, $\theta = \frac{\pi}{2}$

$$\therefore M = 2\mu b^2 \int_0^{\pi/2} a \sin^3\theta (1 - \sin^2\theta)^3 \,.\, 3a \sin^2\theta \cos\theta \, d\theta$$

$$= 6\mu a^2 b^2 \int_0^{\pi/2} \sin^5\theta \cos^7\theta \, d\theta = 6\mu a^2 b^2 \cdot \frac{4\cdot 2\cdot 6\cdot 4\cdot 2}{12\cdot 10\cdot 8\cdot 6\cdot 4\cdot 2} = \frac{\mu a^2 b^2}{20} \quad \textbf{Ans.}$$

Example 28. *Find the mass of a lamina in the form of the cardioid r = a (1 + cos θ) whose density at any point varies as the square of its distance from the initial line.*

(Nagpur University, Winter 2005)

Solution. Let the required mass be M which is twice the mass above the initial line.

Since the distance of any point (r, θ) from the initial line is $r \sin\theta$, the density at (r, θ) is given by $r = m\,(r \sin\theta)^2 = \mu\, r^2 \sin^2\theta$.

For the region above the initial line, θ varies from 0 to π and r varies from 0 to $a(1 + \cos\theta)$.

r = a (1 + cos θ)
r
θ = π
θ
θ = 0
O
A
X

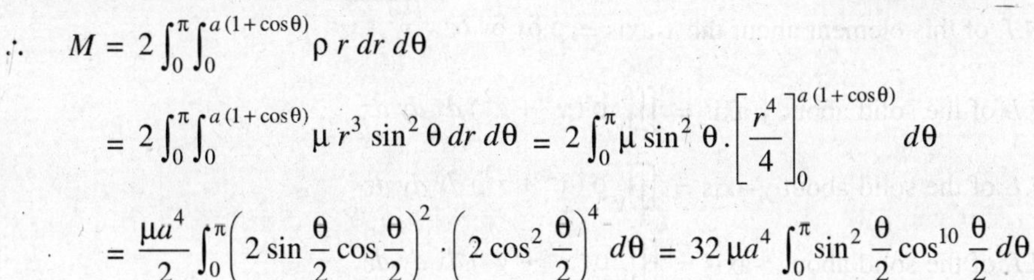

$$\therefore \quad M = 2\int_0^{\pi} \int_0^{a(1+\cos\theta)} \rho\, r\, dr\, d\theta$$

$$= 2\int_0^{\pi} \int_0^{a(1+\cos\theta)} \mu\, r^3 \sin^2\theta \, dr\, d\theta = 2\int_0^{\pi} \mu \sin^2\theta \,.\left[\frac{r^4}{4}\right]_0^{a(1+\cos\theta)} d\theta$$

$$= \frac{\mu a^4}{2} \int_0^{\pi} \left(2\sin\frac{\theta}{2}\cos\frac{\theta}{2}\right)^2 \cdot \left(2\cos^2\frac{\theta}{2}\right)^4 d\theta = 32\,\mu a^4 \int_0^{\pi} \sin^2\frac{\theta}{2} \cos^{10}\frac{\theta}{2}\, d\theta$$

Put $\frac{\theta}{2} = t$, then $d\theta = 2dt$; when $\theta = 0$, $t = 0$; when $\theta = \pi$, $t = \frac{\pi}{2}$

$$\therefore \quad M = 32\,\mu a^4 \int_0^{\pi} 2\sin^2 t \cos^{10} t\, dt = 64\,\mu\, a^4 \cdot \frac{1.9.7.5\,3.1}{12.10.8.6.4.2} \cdot \frac{\pi}{2} = \frac{21}{32}\mu\pi a^4 \qquad \textbf{Ans.}$$

10.8 CENTRE OF GRAVITY

$$\bar{x} = \frac{\iiint x\,\rho\, dx\, dy\, dz}{\iiint \rho\, dx\, dy\, dz},\ \bar{y} = \frac{\iiint y\,\rho\, dx\, dy\, dz}{\iiint \rho\, dx\, dy\, dz},\ \bar{z} = \frac{\iiint z\,\rho\, dx\, dy\, dz}{\iiint \rho\, dx\, dy\, dz}$$

Example 29. *Find the co-ordinates of the centre of gravity of the positive octant of the sphere $x^2 + y^2 + z^2 = a^2$, density being given = k xyz.*

Solution. $$\bar{x} = \frac{\iiint_V x\,\rho\, dx\, dy\, dz}{\iiint_V \rho\, dx\, dy\, dz} = \frac{\iiint z\,\rho\, dx\, dy\, dz}{\iiint \rho\, dx\, dy\, dz} = \frac{\iiint_V x^2 yz\, dx\, dy\, dz}{\iiint_V xyz\, dx\, dy\, dz}$$

Converting into polar co-ordinates, $x = r\sin\theta\cos\phi$, $y = r\sin\theta\sin\phi$, $z = r\cos\theta$,
$dx\, dy\, dz = r^2 \sin\theta\, dr\, d\theta\, d\phi$

$$\bar{x} = \frac{\int_0^{\pi/2}\int_0^{\pi/2}\int_0^a (r\sin\theta\cos\phi)^2 (r\sin\theta\sin\phi)(r\cos\theta)(r^2\sin\theta\, dr\, d\theta\, d\phi)}{\int_0^{\pi/2}\int_0^{\pi/2}\int_0^a (r\sin\theta\cos\phi)(r\sin\theta\sin\phi)(r\cos\theta)(r^2\sin\theta\, dr\, d\theta\, d\phi)}$$

$$= \frac{\int_0^{\pi/2}\int_0^{\pi/2}\int_0^a r^6 \sin^4\theta\cos\theta\sin\phi\cos^2\phi\, dr\, d\theta\, d\phi}{\int_0^{\pi/2}\int_0^{\pi/2}\int_0^a r^5 \sin^3\theta\cos\theta\sin\phi\cos\phi\, dr\, d\theta\, d\phi}$$

$$= \frac{\int_0^{\pi/2} \sin\phi\cos^2\phi\, d\phi \int_0^{\pi/2}\sin^4\theta\cos\theta\, d\theta \int_0^a r^6\, dr}{\int_0^{\pi/2} \sin\phi\cos\phi\, d\phi \int_0^{\pi/2}\sin^3\theta\cos\theta\, d\theta \int_0^a r^5\, dr}$$

$$= \frac{\left[-\frac{\cos^3\phi}{3}\right]_0^{\pi/2}\left[\frac{\sin^5\theta}{5}\right]_0^{\pi/2}\left[\frac{r^7}{7}\right]_0^a}{-\left[\frac{\cos^2\phi}{2}\right]_0^{\pi/2}\left[\frac{\sin^4\theta}{4}\right]_0^{\pi/2}\left[\frac{r^6}{6}\right]_0^a} = \frac{\left(\frac{1}{3}\right)\left(\frac{1}{5}\right)\left(\frac{a^7}{7}\right)}{\left(\frac{1}{2}\right)\left(\frac{1}{4}\right)\left(\frac{a^6}{6}\right)} = \frac{16\,a}{35}$$

Similarly, $\bar{y} = \bar{z} = \frac{16\,a}{35}$

Hence, C.G. is $\left(\frac{16\,a}{35}, \frac{16\,a}{35}, \frac{16\,a}{35}\right)$ **Ans.**

10.9 MOMENT OF INERTIA OF A SOLID

Let the mass of an element of a solid of volume V be $\rho\ \delta x\, \delta y\, \delta z$.

Perpendicular distance of this element from the x-axis $= \sqrt{y^2 + z^2}$

M.I. of this element about the x-axis $= \rho\ \delta x\, \delta y\, \delta z\, \sqrt{y^2 + z^2}$

M.I. of the solid about x-axis $= \iiint_V \rho\,(y^2 + z^2)\, dx\, dy\, dz$

M.I. of the solid about y-axis $= \iiint_V \rho\,(x^2 + z^2)\, dx\, dy\, dz$

M.I. of the solid about z-axis $= \iiint_V \rho\,(x^2 + y^2)\, dx\, dy\, dz$

The Perpendicular Axes Theorem

If I_{ox} and I_{oy} be the moments of inertia of a lamina about x-axis and y-axis respectively and I_{oz} be the moment of inertia of the lamina about an axis perpendicular to the lamina and passing through the point of intersection of the axes OX and OY.

$$I_{OZ} = I_{OX} + I_{OY}$$

The Parallel Axes Theorem

M.I. of a lamina about an axis in the plane of the lamina equals the sum of the moment of inertia about a parallel centroidal axis in the plane of lamina together with the product of the mass of the lamina and square of the distance between the two axes.

$$I_{AB} = I_{XX} + M\bar{y}^2$$

Example 30. *Find M.I. of a sphere about diameter.*

Solution. Let a circular disc of δx thickness be perpendicular to the given diameter XX' at a distance x from it.

$$\text{The radius of the disc} = \sqrt{a^2 - x^2}$$

$$\text{Mass of the disc} = \rho\,\pi\,(a^2 - x^2)$$

Moment of inertia of the disc about a diameter perpendicular on it

$$= \frac{1}{2} MR^2 = \frac{1}{2}[\rho\,\pi\,(a^2 - x^2)]\,(a^2 - x^2) = \frac{1}{2}\rho\,\pi\,(a^2 - x^2)^2$$

$$\text{M.I. of the sphere} = \int_{-a}^{a} \frac{1}{2}\rho\,\pi\,(a^2 - x^2)^2\,dx = 2\left(\frac{1}{2}\rho\,\pi\right)\int_0^a [a^4 - 2a^2x^2 + x^4]\,dx$$

$$= \rho\,\pi\left[a^4 x - \frac{2a^2x^3}{3} + \frac{x^5}{5}\right]_0^a = \rho\,\pi\left[a^5 - \frac{2a^5}{3} + \frac{a^5}{5}\right]$$

$$= \frac{8}{15}\pi\,\rho\,a^5 = \frac{2}{5}\left(\frac{4\pi}{3}a^3\rho\right)a^2 = \frac{2}{5}M\,a^2 \qquad \textbf{Ans.}$$

Example 31. *The mass of a solid right circular cylinder of radius a and height h is M. Find the moment of inertia of the cylinder about (i) its axis (ii) a line through its centre of gravity perpendicular to its axis (iii) any diameter through its base.*

Solution. *To find M.I. about OX. Consider a disc at a distance x from O at the base.*

M.I. *of the about OX,*

$$= \frac{(\pi a^2 \rho\,dx)\,a^2}{2} = \frac{\pi\rho a^4\,dx}{2}$$

(*i*) M.I. of the cylinder about OX

$$\int_0^h \frac{\pi\rho a^4 dx}{2} = \frac{\pi\rho a^4}{2}(x)_0^h = \frac{\pi\rho a^4 h}{2} = (\pi a^2 h)\,\rho\cdot\frac{a^2}{2} = \frac{M a^2}{2}$$

(*ii*) *M.I.* of the disc about a line through *C.G.* and perpendicular to *OX*.

$$I_{OX} + I_{OY} = I_{OZ}$$

$$I_{OX} + I_{OX} = I_{OZ}$$

$$I_{OX} = \frac{1}{2} I_{OZ}$$

M.I. of the disc about a line through

$$C.G. = \frac{1}{2}\left(\frac{M a^2}{2}\right) = \frac{M a^2}{4}$$

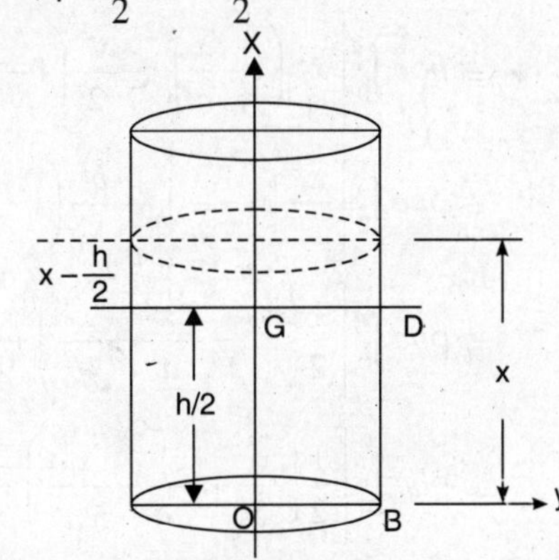

M.I. of the disc about the diameter $= \left(\dfrac{\pi a^2 \rho\, dx}{4}\right) a^2$

M.I. of the disc about line $GD = \dfrac{\pi a^2 \rho\, dx}{4} + (\pi a^2 \rho\, dx)\left(x - \dfrac{h}{2}\right)^2$

Hence, *M.I.* of cylinder about $GD = \int_0^h \dfrac{\pi a^2 \rho}{4} dx + \int_0^h (\pi a^2 \rho\, dx)\left(x - \dfrac{h}{2}\right)^2$

$$= \frac{\pi a^2 \rho}{4} (x)_0^h + \left[\frac{\pi a^2 \rho}{4}\left(x - \frac{h}{2}\right)^3\right]_0^h = \frac{\pi a^2 \rho\, h}{4} + \left[\frac{\pi a^2 \rho}{3}\left(\frac{h}{2}\right)^3 + \frac{\pi a^2 \rho}{3}\left(\frac{h}{2}\right)^3\right]$$

$$= \frac{\pi a^2 \rho\, h}{4} + \frac{\pi a^2 \rho\, h^3}{12} = \frac{M a^2}{4} + \frac{M h^2}{12}$$

(*iii*) *M.I.* of cylinder about line *OB* (through) base

$$I_{OB} = I_G + M\left(\frac{h}{2}\right)^2 = \frac{M a^2}{4} + \frac{M h^2}{12} + \frac{M h^2}{4} = \frac{M a^2}{4} + \frac{M h^2}{3}$$ **Ans.**

Example 32. *Find the moment of inertia and radius of gyration about z-axis of the region in the first octant bounded by* $\dfrac{x}{a} + \dfrac{y}{b} + \dfrac{z}{c} = 1$.

Solution. Let *r* be the density.

M.I. of tetrahedron about *z*-axis

$$= \iiint (\rho\, dx\, dy\, dz)\,(x^2 + y^2)$$

$$= \rho \int_0^a dx \int_0^{b\left(1-\frac{x}{a}\right)} (x^2 + y^2)\, dy \int_0^{c\left(1-\frac{x}{a}-\frac{y}{b}\right)} dz = \rho \int_0^a dx \int_0^{b\left(1-\frac{x}{a}\right)} (x^2 + y^2)\, dy\, (z)_0^{c\left(1-\frac{x}{a}-\frac{y}{b}\right)}$$

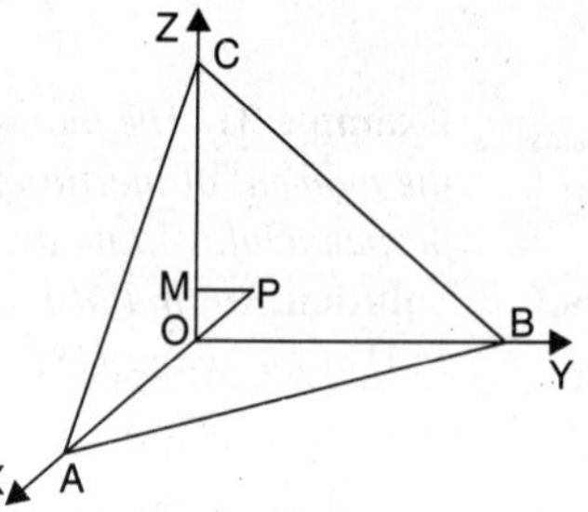

$$= \rho \int_0^a dx \int_0^{b\left(1-\frac{x}{a}\right)} (x^2 + y^2)\, dy\; c\, c\left(1 - \frac{x}{a} - \frac{y}{b}\right)$$

$$= \rho c \int_0^a dx \int_0^{b\left(1-\frac{x}{a}\right)} \left[x^2\left(1 - \frac{x}{a}\right) - \frac{x^2 y}{b} + y^2\left(1 - \frac{x}{a}\right) - \frac{y^3}{b}\right] dy$$

$$= \rho c \int_0^a dx \left[x^2\left(1 - \frac{x}{a}\right) y - \frac{x^2 y^2}{2b} + \frac{y^3}{3}\left(1 - \frac{x}{a}\right) - \frac{y^4}{4b}\right]_0^{b\left(1-\frac{x}{a}\right)}$$

$$= \rho c \int_0^a dx \left[x^2\left(1 - \frac{x}{a}\right) b\left(1 - \frac{x}{a}\right) - \frac{x^2}{2b} b^2\left(1 - \frac{x}{a}\right)^2 + \frac{b^3}{3}\left(1 - \frac{x}{a}\right)^3\left(1 - \frac{x}{a}\right) - \frac{b^4}{4b}\left(1 - \frac{x}{a}\right)^4\right]$$

$$= b\rho c \int_0^a \left[x^2\left(1 - \frac{x}{a}\right)^2 - \frac{x^2}{2}\left(1 - \frac{x}{a}\right)^2 - \frac{b^2}{3}\left(1 - \frac{x}{a}\right)^4 - \frac{b^2}{4}\left(1 - \frac{x}{a}\right)^4\right] dx$$

$$= \rho bc \int_0^a \left[\frac{x^2}{2}\left(1 - \frac{x}{a}\right)^2 + \frac{b^2}{12}\left(1 - \frac{x}{a}\right)^4\right] dx$$

$$= \rho bc \int_0^a \left[\frac{1}{2}\left(x^2 - \frac{2x^3}{a} + \frac{x^4}{a^2}\right) + \frac{b^2}{12}\left(1 - \frac{4x}{a} + \frac{6x^2}{a^2} - \frac{4x^3}{a^3} + \frac{x^4}{a^4}\right)\right] dx$$

$$= \rho bc \int_0^a \left[\frac{1}{2}\left(\frac{x^3}{3} - \frac{x^4}{2a} + \frac{x^5}{5a^2}\right) + \frac{b^2}{12}\left(x - \frac{2x^2}{a} + \frac{6x^2}{a^2} - \frac{4x^3}{a^3} + \frac{x^4}{a^4}\right)\right]_0^a dx$$

$$= \rho bc \int_0^a \left[\frac{1}{2}\left(\frac{a^3}{3} - \frac{a^3}{2} + \frac{a^3}{5}\right) + \frac{b^2}{12}\left(a - 2a + 2a - a + \frac{a}{5}\right)\right]$$

$$= \rho bc \left[\frac{a^3}{60} + \frac{ab^2}{60}\right] = \rho \frac{abc}{60}(a^2 + b^2)$$

$$\text{Radius of gyration} = \sqrt{\frac{M.I.}{Mass}} = \frac{\sqrt{\frac{\rho abc}{60}(a^2 + b^2)}}{\frac{\rho abc}{6}} = \sqrt{\frac{1}{10}(a^2 + b^2)}$$ **Ans.**

Example 33. *A solid body of density ρ is in the shape of the solid formed by revolution of the cardioid r = a (1 + cos θ) about the initial line. Show that its moment of inertia about a straight line through the pole perpendicular to the initial line is* $\left(\frac{352}{105}\right)\pi \rho a^5$.

(U.P., II Semester Summer 2001)

Solution. $r = a(1 + \cos\theta)$

Consider an elementary area $r\, d\theta\, dr$. This area when revolved about OX generates a circular ring of radius $r \sin\theta$.

Its mass $= (2\pi r \sin\theta)(r\, d\theta\, dr\, \rho)$.

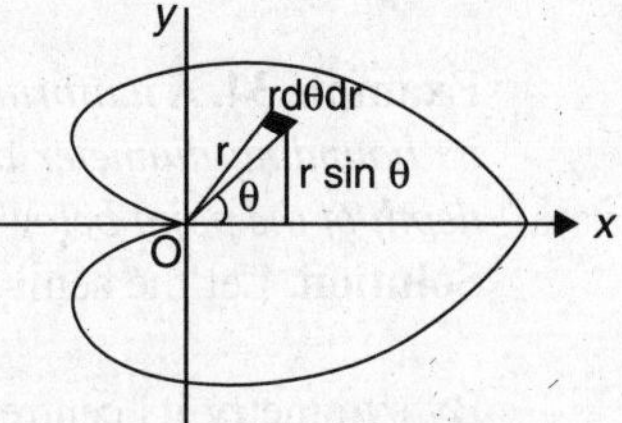

($M.I.$ of a ring about a diameter $= \frac{Ma^2}{2}$)

M.I. of the ring about a diameter parallel to $OY = (2\pi r \sin\theta)(r\, d\theta\, dr) P \left(\frac{r^2 \sin^2\theta}{2}\right)$

M.I. of the ring about OY = M.I. of the ring about a diameter parallel to OY + mass of the ring $(r\cos\theta)^2$

$$= (2\pi r \sin\theta\, r\, d\theta\, dr\, \rho)\left(\frac{r^2 \sin^2\theta}{2} + r^2\cos^2\theta\right)$$

So M.I. of the solid generated by revolution about OY

$$= 2\pi\rho \int_0^\pi \int_a^{a(1+\cos\theta)} r^4 \sin\theta \left(\frac{\sin^2\theta}{2} + \cos^2\theta\right) d\theta\, dr$$

$$= \pi\rho \int_0^\pi \sin\theta\,(1 + \cos^2\theta)\, d\theta \int_a^{a(1+\cos\theta)} r^4\, dr = \pi\rho \int_0^\pi \sin\theta\,(1+\cos^2\theta)\, d\theta \left(\frac{r^5}{5}\right)_0^{a(1+\cos\theta)}$$

$$= \pi\rho \int_0^\pi \sin\theta\,(1+\cos^2\theta)\, d\theta \frac{a^5(1+\cos\theta)^5}{5} = \pi\rho \frac{a^5}{5}\int_0^\pi (1+\cos^2\theta)(1+\cos\theta)^5 \sin\theta\, d\theta$$

$$= \frac{\pi\rho a^5}{5}\int_0^\pi \left[1 + \left(2\cos^2\frac{\theta}{2} - 1\right)^2\right]\left(2\cos^2\frac{\theta}{5}\right)^5 2\sin\frac{\theta}{2}\cos\frac{\theta}{2}\, d\theta$$

Putting $\frac{\theta}{2} = t,\ d\theta = 2\,dt = \frac{265\pi\rho a^5}{5}\int_0^{\pi/2} [\cos^{11} t \sin t + 2\cos^{15} t \sin t - 2\cos^{13} t \sin t]\, dt$

$$= \frac{265\pi\rho a^5}{5}\left[-\frac{\cos^{12} t}{12} - 2\frac{\cos^{16} t}{16} + 2\frac{\cos^{14} t}{14}\right]_0^{\pi/2} = \frac{352\pi\rho a^5}{105}$$ **Proved.**

10.10 CENTRE OF PRESSURE

The centre of pressure of a plane area immersed in a fluid is the point at which the resultant force acts on the area.

Consider a plane area A immersed vertically in a homogeneous liquid. Let x-axis be the line of intersection of the plane with the free surface. Any line in this plane and perpendicular to x-axis is the y-axis.

Let P be the pressure at the point (x, y). Then the pressure on elementary area $\delta x\, \delta y$ is $P\, \delta x\, \delta y$.

Let $(\bar{x}, \bar{y})$ be the centre of pressure. Taking moment about y-axis.

$$\bar{x} \cdot \iint_A P\, dx\, dy = \iint_A Px\, dx\, dy$$

$$\bar{x} = \frac{\iint_A Px\, dx\, dy}{\iint_A P\, dx\, dy}$$

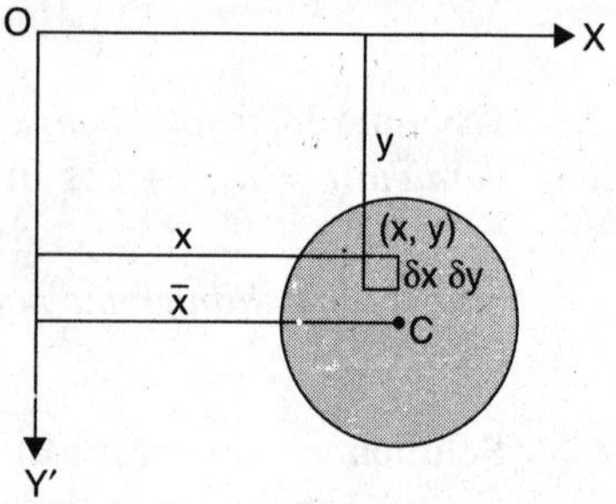

Similarly,
$$\bar{y} = \frac{\iint_A Py\, dx\, dy}{\iint_A P\, dx\, dy}$$

Example 34. *A uniform semi-circular lamina is immersed in a fluid with its plane vertical and its bounding diameter on the free surface. If the density at any point of the fluid varies as the depth of the point below the free surface, find the position of the centre of pressure of the lamina.*

Solution. Let the semi-circular lamina be

$$x^2 + y^2 = a^2$$

By symmetry its centre of pressure lies on OY.

Let ky be the density of the fluid.

$$\bar{y} = \frac{\iint_A Py\, dx\, dy}{\iint_A P\, dx\, dy} = \frac{\iint_A (\rho y)\, y\, dx\, dy}{\iint_A (\rho y)\, dx\, dy} \qquad (\because \rho = ky)$$

$$= \frac{\iint_A (ky \cdot y)\, y\, dx\, dy}{\iint_A (ky \cdot y)\, dx\, dy} = \frac{\iint_A y^3\, dx\, dy}{\iint_A y^2\, dx\, dy} = \frac{\int_{-a}^{a} dx \int_0^{\sqrt{a^2 - x^2}} y^3\, dy}{\int_{-a}^{a} dx \int_0^{\sqrt{a^2 - x^2}} y^2\, dy}$$

$$= \frac{\int_{-a}^{a} dx \left[\frac{y^4}{4}\right]_0^{\sqrt{a^2 - x^2}}}{\int_{-a}^{a} dx \left[\frac{y^3}{3}\right]_0^{\sqrt{a^2 - x^2}}} = \frac{3}{4} \frac{\int_{-a}^{a} dx\, (a^2 - x^2)^2}{\int_{-a}^{a} dx\, (a^2 - x^2)^{3/2}}$$

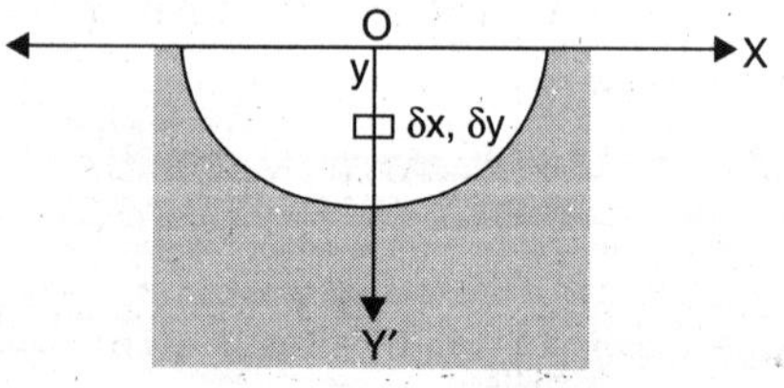

$$= \frac{3}{4} \frac{\int_{-\pi/2}^{\pi/2} (a \cos\theta\, d\theta)\, (a^2 - a^2 \sin^2\theta)^2}{\int_{-\pi/2}^{\pi/2} (a \cos\theta\, d\theta)\, (a^2 - a^2 \sin^2\theta)^{3/2}} \qquad (\text{Put } x = a \sin\theta)$$

$$= \frac{3a}{4} \frac{\int_{-\pi/2}^{\pi/2} \cos^5\theta\, d\theta}{\int_{-\pi/2}^{\pi/2} \cos^4\theta\, d\theta}$$

$$= \frac{3a}{4} \frac{2\int_0^{\pi/2} \cos^5\theta\, d\theta}{2\int_0^{\pi/2} \cos^4\theta\, d\theta} = \frac{3a}{4} \frac{\frac{4 \times 2}{5 \times 3}}{\frac{3 \times 1}{4 \times 2} \frac{\pi}{2}} = \frac{32\, a}{15\, \pi}$$

Ans.

EXERCISE 10.3

1. Find the mass of the solid bounded by the ellipsoid $\frac{x^2}{a^2}+\frac{y^2}{b^2}+\frac{z^2}{c^2}=1$ and the co-ordinate planes, where the density at any point $P(x, y, z)$ is $k\,xyz$. **Ans.** P
2. If the density at a point varies as the square of the distance of the point from XOY plane, find the mass of the volume common to the sphere $x^2 + y^2 + z^2 = a^2$ and cylinder $x^2 + y^2 = ax$. **Ans.** $\frac{4k}{15}a^5\left(\frac{\pi}{2}-\frac{8}{15}\right)$
3. Find the mass of the plate in the form of one loop of leminscate $r^2 = a^2 \sin 2\theta$, where $\rho = k r^2$. **Ans.** $\frac{k\pi a^4}{16}$
4. Find the mass of the plate which is inside the circle $r = 2a\cos\theta$ and outside the circle $r = a$, if the density varies as the distance from the pole.
5. Find the mass of a lamina in the form of the cardioid $r = a(1+\cos\theta)$ whose density at any point varies as the square of its distance from the initial line. **Ans.** $\frac{21\pi k a^4}{32}$
6. Find the centroid of the region in the first octant bounded by $\frac{x}{a}+\frac{y}{b}+\frac{z}{c}=1$. **Ans.** $\left(\frac{a}{4}, \frac{b}{4}, \frac{c}{4}\right)$
7. Find the centroid of the region bounded by $z = 4 - x^2 - y^2$ and xy-plane. **Ans.** $\left(0, 0, \frac{4}{3}\right)$
8. Find the position of $C.G.$ of the volume intercepted between the parallelepiped $x^2+y^2 = a(a-z)$ and the plane $z = 0$. **Ans.** $\left(0, 0, \frac{a}{3}\right)$
9. A solid is cut off the cylinder $x^2 + y^2 = a^2$ by the plane $z = 0$ and that part of the olane $z = mx$ for which z is positive. The density of the solid cut off at any point varies as the height of the point above plane $z = 0$. Find $C.G$ of the solid. **Ans.** $\bar{z} = \frac{64\,ma}{45\,\pi}$
10. If an area is bounded by two concentric semi-circles with their common bounding diameter in a free surface, prove that the depth of the centre of pressure is $\frac{3\pi}{16}\frac{(a+b)(a^2+b^2)}{a^2+ab+b^2}$
11. An ellipse $\frac{x^2}{a^2}+\frac{y^2}{b^2}=1$ is immersed vertically in a fluid with its major axis horizontal. If its centre be at depth h, find the depth of its centre of pressure. **Ans.** $h+\frac{b^2}{4h}$
12. A horizontal boiler has a flat bottom and its ends are plane and semi-circular. If it is just full of water, show that the depth of centre of pressure of either end is 0.7 × total depth approximately.
13. A quadrant of a circle of radius a is just immersed vertically in a homogeneous liquid with one edge in the surface. Determine the co-ordinates of the centre of pressure. **Ans.** $\left(\frac{3a}{8}, \frac{3\pi a}{16}\right)$
14. Find the product of inertia of an equilateral triangle about two perpendicular axes in its plane at a vertex, one of the axes being along a side.
15. Find the $M.I.$ of a right circular cylinder of radius a and height h about axis if density varies as distance from the axis. **Ans.** $\frac{2}{5}k\pi a^5 h$

16. Compute the moment of inertia of a right circular cone whose altitude is h and base radius r, about (*i*) the axis of symmetry (*ii*) the diameter of the base. **Ans.** (*i*) $\dfrac{\pi h r^4}{10}$ (*ii*) $\dfrac{\pi h r^2}{60}(2h^2+3r^2)$

17. Find the moment of inertia for the area of the cardioid $r = a(1 - \cos\theta)$ relative to the pole.

Ans. $\dfrac{35\pi a^4}{16}$

18. Find the M.I. about the line $\theta = \dfrac{\pi}{2}$ of the area enclosed by $r = a(1 + \cos\theta)$.

19. Find the moment of inertia of the uniform solid in the form of octant of the ellipsoid $\dfrac{x^2}{a^2}+\dfrac{y^2}{b^2}+\dfrac{z^2}{c^2}=1$ about OX **Ans.** $\dfrac{M}{5}(b^2+c^2)$

20. Prove that the moment of inertia of the area included between the curves $y^2 = 4ax$ and $x^2 = 4ay$ about the x-axis is $\dfrac{144}{35} M a^2$, where M is the mass of area included between the curves.

21. A solid body of density p is the shape of solid formed by revolution of the cardioid $r = a(1 + \cos\theta)$ about the initial line. Show that its moment of inertia about a straight line through the pole perpendicular to the initial line is $\left(\dfrac{352}{105}\right)\pi l a^5$. (*U. P. II Semester, Summer 2001*)

22. Find the product of inertia of a disc in the form of a quadrant of a circle of radius '*a*' about bounding radii. (*U. P. II Semester, Summer 2002*) **Ans.** $\rho\dfrac{a^4}{4}$

23. Show that the principal axes at the origin of the triangle enclosed by $x = 0, y = 0, \dfrac{x}{a}+\dfrac{y}{b}=1$ are inclined at angles α and $\alpha+\dfrac{\pi}{2}$ to the x-axis, where $a = \dfrac{1}{2}\tan^{-1}\left(\dfrac{ab}{a^2-b^2}\right)$

(*U.P. II Semester Summer* 2001)

Choose the correct answer:

24. The triple integral $\iiint_T dx\,dy\,dz$ gives

(*i*) Volume of region T (*ii*) Surface area of region T
(*ii*) Area of region T (*iv*) Density of region T. (*A.M.I.E.T.E. 2002*)

Ans. (*i*)

25. The volume of the solid under the surface $az = x^2 + y^2$ and whose base R is the circle $x^2 + y^2 = a^2$ is given as

(*i*) $\dfrac{\pi}{2a}$ (*ii*) $\dfrac{\pi a^3}{2}$ **Ans.** (*ii*)

(*iii*) $\dfrac{4}{3}\pi a^3$ (*iv*) None of the above. [*U.P., I. Sem. Dec. 2008*]

DIFFERENTIAL EQUATIONS OF FIRST ORDER

11.1 DEFINITION

An equation which involves differential co-efficient is called a differential equation.

For example,

1. $\frac{dy}{dx} = \frac{1+x^2}{1-y^2}$ 2. $\frac{d^2y}{dx^2} = 2\frac{dy}{dx} - 8y = 0$ 3. $\left[1+\left(\frac{dy}{dx}\right)^2\right]^{3/2} = k\frac{d^2y}{dx^2}$

4. $x\frac{\partial u}{\partial x} + y\frac{\partial u}{\partial y} = nu,$ 5. $\frac{\partial^2 z}{\partial x \partial y} = \frac{\partial z}{\partial y}$

There are two types of differential equations :

(1) Ordinary Differential Equation

A differential equation involving derivatives with respect to a single independent variable is called an ordinary differential equation.

(2) Partial Differential Equation

A differential equation involving partial derivatives with respect to more than one independent variable is called a partial differential equation.

11.2 ORDER AND DEGREE OF A DIFFERENTIAL EQUATION

The *order* of a differential equation is the order of the highest differential co-efficient present in the equation. Consider

1. $L\frac{d^2q}{dt^2} + R\frac{dq}{dt} + \frac{q}{c} = E\sin wt.$ 2. $\cos x\frac{d^2y}{dx^2} + \sin x\left(\frac{dy}{dx}\right)^2 + 8y = \tan x$

3. $\left[1+\left(\frac{dy}{dx}\right)^2\right]^3 = \left(\frac{d^2y}{dx^2}\right)^2$

The order of the above equations is 2.

The degree of a differential equation is the degree of the highest derivative after removing the radical sign and fraction.

The *degree* of the equation (1) and (2) is 1. The degree of the equation (3) is 2.

11.3 FORMATION OF DIFFERENTIAL EQUATIONS

The differential equations can be formed by differentiating the ordinary equation and eliminating the arbitrary constants.

Example 1. *Form the differential equation by eliminating arbitrary constants, in the following cases and also write down the order of the differential equations obtained.*

(*a*) $y = A x + A^2$ (*b*) $y = A \cos x + B \sin x$ (*c*) $y^2 = Ax^2 + Bx + C$.

(*R.G.P.V. Bhopal, June 2008*)

Solution. (*a*) $y = Ax + A^2$ ... (1)

On differentiation $\frac{dy}{dx} = A$

Putting the value of A in (1), we get $y = x\frac{dy}{dx} + \left(\frac{dy}{dx}\right)^2$ **Ans.**

On eliminating one constant A we get the differential equation of order 1.

(*b*) $y = A \cos x + B \sin x$

On differentiation $\frac{dy}{dx} = -A\sin x + B\cos x$

Again differentiating

$$\frac{d^2y}{dx^2} = -A\cos x - B\sin x \Rightarrow \frac{d^2y}{dx^2} = -(A\cos x + B\sin x)$$

$$\Rightarrow \quad \frac{d^2y}{dx^2} = -y \Rightarrow \frac{d^2y}{dx^2} + y = 0 \qquad \textbf{Ans.}$$

This is differential equation of order 2 obtained by eliminating two constants A and B.

(*c*) $y^2 = Ax^2 + Bx + C$

On differentiation $2y\frac{dy}{dx} = 2Ax + B$

Again differentiating $2y\frac{d^2y}{dx^2} + 2\left(\frac{dy}{dx}\right)^2 = 2A$

On differentiating again $y\frac{d^3y}{dx^3} + \frac{dy}{dx}\frac{d^2y}{dx^2} + 2\frac{dy}{dx}\frac{d^2y}{dx^2} = 0 \Rightarrow y\frac{d^3y}{dx^3} + 3\frac{dy}{dx}\frac{d^2y}{dx^2} = 0$ **Ans.**

This is the differential equation of order 3, obtained by eliminating three constants A, B, C.

Example 2. *Determine the differential equation whose set of independent solution is* $\{e^x, xe^x, x^2 e^x\}$ (*U.P., II Semester, Summer 2002*)

Solution. Here, we have

$$y = c_1 e^x + c_2 x e^x + c_3 x^2 e^x \qquad \text{... (1)}$$

Differentiating both sides, we get

$$y' = c_1 e^x + c_2 x e^x + c_2 e^x + c_3 x^2 e^x + c_3 2x e^x$$

$$= (c_1 e^x + c_2 x e^x + c_3 x^2 e^x) + c_2 e^x + c_3 2x e^x$$

$$= y + c_2 e^x + c_3 2x e^x \qquad \text{[Using (1)]}$$

$$\Rightarrow \quad y' - y = c_2 e^x + c_3 2x e^x \qquad \text{... (2)}$$

Again, differentiating both sides, we get

$$\Rightarrow \quad y'' - y' = c_2 e^x + 2c_3 e^x + 2x c_3 e^x$$

$$\Rightarrow \quad y'' - y' = (c_2 e^x + 2x c_3 e^x) + 2c_3 e^x$$

$$\Rightarrow \quad y'' - y' = y' - y + 2c_3 e^x \qquad \text{[Using (2)]}$$

$$\Rightarrow \quad y'' - 2y' + y = 2c_3 e^x \qquad \text{... (3)}$$

Finally, on differentiating both sides, we get

$$\Rightarrow \quad y''' - 2y'' + y' = 2c_3 e^x$$

$\Rightarrow \quad y''' - 2y'' + y' = y'' - 2y' + y$ [Using (3)]

$\Rightarrow \quad y''' - 2y'' - y'' + y' + 2y' - y = 0$

$\Rightarrow \quad y''' - 3y'' + 3y' - y = 0$

$\Rightarrow \quad (D-1)^3 y = 0$ **Ans.**

Example 3. *By the elimination of the constants A and B obtain the differential equation of which $xy = Ae^x + Be^{-x} + x^2$ is the solution.* *(U.P.B. Pharma (C.O.) 2005)*

Solution. We have $xy = Ae^x + Be^{-x} + x^2$ .... (1)

On differentiating (1), we get $x\frac{dy}{dx} + y = Ae^x - Be^{-x} + 2x$ ... (2)

Again differentiating (2), we get $x\frac{d^2y}{dx^2} + \frac{dy}{dx} + \frac{dy}{dx} = Ae^x + Be^{-x} + 2$ ... (3)

From (1), we have $Ae^x + Be^{-x} = xy - x^2$ ... (4)

Putting the value of $Ae^x + Be^{-x}$ From (4) in (3), we have

$$x\frac{d^2y}{dx^2} + 2\frac{dy}{dx} = xy - x^2 + 2, \quad x\frac{d^2y}{dx^2} + 2\frac{dy}{dx} - xy = 2 - x^2$$ **Ans.**

EXERCISE 11.1

1. Write the order and the degree of the following differential equations.

(*i*) $\frac{d^2y}{dx^2} + a^2x = 0;$ (*ii*) $\left[1 + \left(\frac{dy}{dx}\right)^2\right]^{\frac{3}{2}} = \frac{d^2y}{dx^2};$ (*iii*) $x^2\left(\frac{d^2y}{dx^2}\right)^3 + y\left(\frac{dy}{dx}\right)^4 + y^4 = 0.$

Ans. (*i*) 2,1 (*ii*) 2,2 (*iii*) 2,3

2. Give an example of each of the following type of differential equations.

(*i*) A linear-differential equation of second order and first degree **Ans.** *Q*, 1, (*i*)

(*ii*) A non-linear differential equation of second order and second degree **Ans.** *Q*, 1 (*ii*)

(*iii*) Second order and third degree. **Ans.** *Q* 1 (*iii*)

3. Obtain the differential equation of which $y^2 = 4a(x + a)$ is a solution.

Ans. $y^2\left(\frac{dy}{dx}\right)^2 + 2xy\frac{dy}{dx} - y^2 = 0$

4. Obtain the differential equation associated with the primitive $Ax^2 + By^2 = 1$.

Ans. $xy\frac{d^2y}{dx^2} + x\left(\frac{dy}{dx}\right)^2 - y\frac{dy}{dx} = 0$

5. Find the differential equation corresponding to

$y = a\,e^{3x} + be^x$. **Ans.** $\frac{d^2y}{dx^2} - 4\frac{dy}{dx} + 3y = 0$

6. By the elimination of constants *A* and *B*, find the differential equation of which

$y = e^x(A\cos x + B\sin x)$ is a solution. **Ans.** $\frac{d^2y}{dx^2} - 2\frac{dy}{dx} + 2y = 0$

7. Find the differential equation whose solution is $y = a\cos(x = 3)$. *(A.M.I.E., Summer 2000)*

Ans. $\frac{dy}{dx} = -\tan(x + 3)$

8. Show that set of function $\left\{x, \frac{1}{x}\right\}$ forms a basis of the differential equation $x^2y'' + xy' - y = 0$.

Obtain a particular solution when $y(1) = 1, y(1) = 2$. **Ans.** $y = \frac{3x}{2} - \frac{1}{2x}$

11.4 SOLUTION OF A DIFFERENTIAL EQUATION

In the example 1(*b*), $y = A \cos x + B \sin x$, on eliminating A and B we get the differential equation

$$\frac{d^2y}{dx^2} + y = 0$$

$y = A \cos x + B \sin x$ is called the solution of the differential equation $\frac{d^2y}{dx^2} + y = 0$.

The order of the differential equation $\frac{d^2y}{dx^2} + y = 0$ is two and the solution $y = A \cos x + B \sin x$ contains two arbitrary constants. The number of arbitrary constants in the solution is equal to the order of the differential equation.

An equation containing dependent variable (y) and independent variable (x) and free from derivative, which satisfies the differential equation, is called the solution (primative) of the differential equation.

11.5 GEOMETRICAL MEANING OF THE DIFFERENTIAL EQUATION OF FIRST ORDER AND FIRST DEGREE

Let $$f\left(x,\ y,\ \frac{dy}{dx}\right) = 0 \qquad \ldots (1)$$

be a differential equation of the first order and first degree.

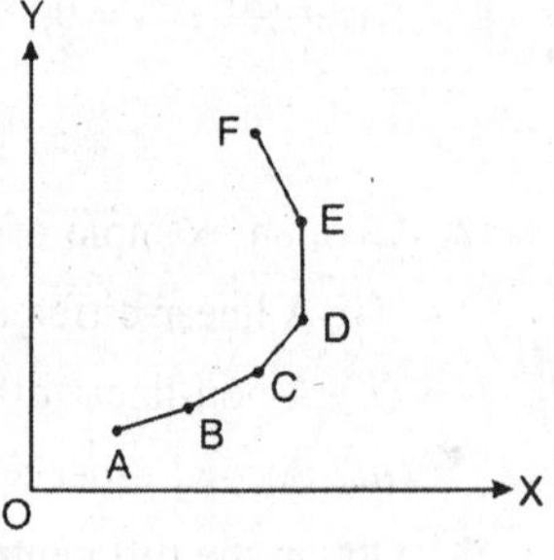

It is known that a direction of a curve at a particular point is given by the tangent line at that point and the slope of the tangent is $\frac{dy}{dx}$ at that point. Let $A\ (x_0\ ,\ y_0)$ be any initial point. From (1), we can find $\frac{dy}{dx}$ at $A\ (x_0\ ,\ y_0)$.

With the help of $\frac{dy}{dx}$ at $A\ (x_0, y_0)$ draw the tangent at the point A. On the tangent line take a neighbouring point $B\ (x_1, y_1)$.

Find $\frac{dy}{dx}$ at the point $B\ (x_1, y_1)$ from equation (1) and draw the tangent at B with the help of $\frac{dy}{dx}$ at (x_1, y_1).

Take a neighbouring point $C\ (x_2, y_2)$ on this tangent and in this way draw another tangent at the point C. Similarly draw, some more tangents by taking the neighbouring points on them. They form a smooth curve *i.e.* $y = f_1\ (x)$ which is the solution (1).

Again we take another starting point $A'\ (x_0', y_0')$. We can draw another curve starting from A'. In this way we can draw a number of curves.

The given differential equation represents a family of curves.

11.6 DIFFERENTIAL EQUATIONS OF THE FIRST ORDER AND FIRST DEGREE

We will discuss the standard methods of solving the differential equations of the following types:

(*i*) Equations solvable by separation of the variables. (*ii*) Homogeneous equations.
(*iii*) Linear equations of the first order. (*iv*) Exact differential equations.

11.7 VARIABLES SEPARABLE

If a differential equation can be written in the form

$$f(y)\,dy = \phi(x)\,dx$$

We say that variables are separable, y on left hand side and x on right hand side.

We get the solution by integrating both sides.

Working Rule:

Step 1. Separate the variables as $f(y)\,dy = \phi(x)\,dx$

Step 2. Integrate both sides as $\int f(y)\,dy = \int \phi(x)\,dx$

Step 3. Add an arbitrary constant C on R.H.S.

Example 4. *Solve* : $\dfrac{dy}{dx} = \dfrac{x(2\log x + 1)}{\sin y + y\cos y}$ *(U.P., II, 2008, U.P.B Pharm (C.O.) 2005)*

Solution. We have, $\dfrac{dy}{dx} = \dfrac{x(2\log x + 1)}{\sin y + y\cos y}$

Separating the variables, we get

$$(\sin y + y\cos y)\,dy = \{x\,(2\log x + 1)\}\,dx$$

Integrating both the sides, we get $\int (\sin y + y\cos y)dy = \int \{x(2\log x + 1)\}\,dx + C$

$$-\cos y + y\sin y - \int (1)\cdot \sin y\,dy = 2\int \log x \cdot x\,dx + \int x\,dx + C$$

$$\Rightarrow \quad -\cos y + y\sin y + \cos y = 2\left[\log x \cdot \frac{x^2}{2} - \int \frac{1}{x}\cdot\frac{x^2}{2}dx\right] + \frac{x^2}{2} + C$$

$$\Rightarrow \quad y\sin y = 2\log x \cdot \frac{x^2}{2} - \int x\,dx + \frac{x^2}{2} + C$$

$$\Rightarrow \quad y\sin y = 2\log x \cdot \frac{x^2}{2} - \frac{x^2}{2} + \frac{x^2}{2} + C$$

$$\Rightarrow \quad y\sin y = x^2\log x + C$$ **Ans.**

Example 5. *Solve the differential equation.*

$$x^4\frac{dy}{dx} + x^3y = -\sec(xy).$$ *(A.M.I.E.T.E., Winter 2003)*

Solution. $x^4\dfrac{dy}{dx} + x^3y = -\sec(xy) \quad \Rightarrow \quad x^3\left(x\dfrac{dy}{dx} + y\right) = -\sec xy$

Put $v = xy, \quad \dfrac{dv}{dx} = x\dfrac{dy}{dx} + y \quad \Rightarrow \quad x^3\dfrac{dv}{dx} = -\sec v$

$\Rightarrow \quad \dfrac{dv}{\sec v} = -\dfrac{dx}{x^3} \quad \Rightarrow \quad \int \cos v\,dv = -\int \dfrac{dx}{x^3} + c$

$\Rightarrow \quad \sin v = \dfrac{1}{2x^2} + c \quad \Rightarrow \quad \sin xy = \dfrac{1}{2x^2} + c$ **Ans.**

Example 6. *Solve* : $\cos(x + y)dy = dx$

Solution. $\cos(x + y)\,dy = dx \quad \Rightarrow \quad \dfrac{dy}{dx} = \sec(x + y)$

On putting $x + y = z$

So that $$1+\frac{dy}{dx}=\frac{dz}{dx} \Rightarrow \frac{dy}{dx}=\frac{dz}{dx}-1$$

$$\frac{dz}{dx}-1=\sec z \Rightarrow \frac{dz}{dx}=1+\sec z$$

Separating the variables, we get

$$\frac{dz}{1+\sec z}=dx$$

On integrating,

$$\int\frac{\cos z}{\cos z+1}dz=\int dx \Rightarrow \int\left[1-\frac{1}{\cos z+1}\right]dz=x+C$$

$$\int\left(1-\frac{1}{2\cos^2\frac{z}{2}-1+1}\right)dz=x+C$$

$$\int\left(1-\frac{1}{2}\sec^2\frac{z}{2}\right)dz=x+C \Rightarrow z-\tan\frac{z}{2}=x+C$$

$$x+y-\tan\frac{x+y}{2}=x+C$$

$$y-\tan\frac{x+y}{2}=C$$ **Ans.**

Example 7. *Solve the equation.*

$$(2x^2+3y^2-7)\,x\,dx-(3x^2+2y^2-8)\,y\,dy=0$$ *(U.P. II Semester, Summer 2005)*

Solution. We have

$$(2x^2+3y^2-7)\,x\,dx-(3x^2+2y^2-8)\,y\,dy=0$$

Re-arranging (1), we get $\dfrac{x\,dx}{y\,dy}=\dfrac{3x^2+2y^2-8}{2x^2+3y^2-7}$

Applying componendo and dividendo rule, we get

$$\frac{x\,dx+y\,dy}{x\,dx-y\,dy}=\frac{5x^2+5y^2-15}{x^2-y^2-1} \Rightarrow \frac{x\,dx+y\,dy}{x^2+y^2-3}=5\left(\frac{x\,dx-y\,dy}{x^2-y^2-1}\right)$$

Multiplying by 2 both the sides, we get

$$\Rightarrow \left(\frac{2x\,dx+2y\,dy}{x^2+y^2-3}\right)=5\left(\frac{2x\,dx-2y\,dy}{x^2-y^2-1}\right)$$

Integrating both sides, we get

$$\log(x^2+y^2-3)=5\log(x^2-y^2-1)+\log C$$

$$\Rightarrow x^2+y^2-3=C(x^2-y^2-1)^5$$ **Ans.**

where C is arbitrary constant of integration.

EXERCISE 11.2

Solve the following differential equations :

1. $\dfrac{dx}{x}=\tan y\cdot dy$ **Ans.** $x\cos y=C$

2. $\dfrac{dy}{dx}=\dfrac{\sqrt{1-y^2}}{\sqrt{1-x^2}}$ **Ans.** $\sin^{-1}y=\sin^{-1}x+C$

3. $y(1+x^2)^{1/2}dy+x\sqrt{1+y^2}\,dx=0$ **Ans.** $\sqrt{1+y^2}+\sqrt{1+x^2}=C$

4. $\sec^2 x\tan y\,dx+\sec^2 y\tan x\,dy=0$ **Ans.** $\tan x\tan y=C$

5. $(1 + x^2)\, dy - x\, y\, dx = 0$ **Ans.** $y^2 = C\,(1 + x^2)$

6. $(e^y + 1) \cos x\, dx + e^y \sin x\, dy = 0$ **Ans.** $(e^y + 1) \sin x = C$

7. $3\, e^x \tan y\, dx + (1 - e^x)\, sec^2\, y\, dy = 0$ **Ans.** $(1 - e^x)^3 = C \tan y$

8. $(e^y + 2) \sin x\, dx - e^y \cos x\, dy = 0$ **Ans.** $(e^y + 2) \cos x = C$

9. $\dfrac{dy}{dx} = e^{x-y} + x^2 e^{-y}$ **Ans.** $e^y = e^x + \dfrac{x^3}{3} + C$

10. $\dfrac{dy}{dx} = 1 + \tan(y - x)$ [Put $y - x = z$] **Ans.** $\sin(y - x) = e^{x+c}$

11. $(4x + y)^2 \dfrac{dx}{dy} = 1$ **Ans.** $\tan^{-1} \dfrac{4x + y}{2} = 2x + C$

12. $\dfrac{dy}{dx} = (4x + y + 1)^2$ [**Hint.** Put $4x + y + 1 = z$] **Ans.** $\tan^{-1} \dfrac{4x + y + 1}{2} = 2x + C$

11.8 HOMOGENEOUS DIFFERENTIAL EQUATIONS

A differential equation of the form $\dfrac{dy}{dx} = \dfrac{f(x, y)}{\phi(x, y)}$

is called a homogeneous equation if each term of $f(x, y)$ and $\phi(x, y)$ is of the same degree *i.e.*,

$$\frac{dy}{dx} = \frac{3xy + y^2}{3x^2 + xy}$$

In such case we put $\boxed{y = vx. \text{ and } \dfrac{dy}{dx} = v + x\dfrac{dv}{dx}}$

The reduced equation involves v and x only. This new differential equation can be solved by *variables separable* method.

Working Rule

Step 1. Put $y = vx$ so that $\dfrac{dy}{dx} = v + x\dfrac{dv}{dx}$ **Step 2.** Separate the variables.

Step 3. Integrate both the sides. **Step 4.** Put $v = \dfrac{y}{x}$ and simplify.

Example 8. *Solve the following differential equation*

$$(2xy + x^2)\, y = 3y^2 + 2xy \qquad \text{(A.M.I.E.T.E. Dec. 2006)}$$

Solution. We have, $(2xy + x^2)\dfrac{dy}{dx} = 3y^2 + 2xy \Rightarrow \dfrac{dy}{dx} = \dfrac{3y^2 + 2xy}{2xy + x^2}$

Put $y = vx$ so that $\dfrac{dy}{dx} = v + x\dfrac{dv}{dx}$

On substituting, the given equation becomes $v + x\dfrac{dv}{dx} = \dfrac{3v^2x^2 + 2vx^2}{2vx^2 + x^2} = \dfrac{3v^2 + 2v}{2v + 1}$

$$\Rightarrow \quad x\frac{dv}{dx} = \frac{3v^2 + 2v - 2v^2 - v}{2v + 1} \qquad \Rightarrow \quad x\frac{dv}{dx} = \frac{v^2 + v}{2v + 1} \quad \Rightarrow \quad \left(\frac{2v + 1}{v^2 + v}\right) dv = \frac{dx}{x}$$

$$\Rightarrow \quad \int \left(\frac{2v + 1}{v^2 + v}\right) dv = \int \frac{dx}{x} \qquad \Rightarrow \quad \log(v^2 + v) \log x + \log c$$

$$\Rightarrow \quad v^2 + v = cx \qquad \Rightarrow \quad \frac{y^2}{x^2} + \frac{y}{x} = cx$$

$$\Rightarrow \quad y^2 + xy = cx^3$$

Example 9. *Solve the equation* :

$$\frac{dy}{dx}=\frac{y}{x}+x\sin\frac{y}{x}$$

Solution. $$\frac{dy}{dx}=\frac{y}{x}+x\sin\frac{y}{x} \qquad \ldots (1)$$

Put $y = vx$ in (1) so that $\frac{dy}{dx}=v+x\frac{dv}{dx}$

$$v+x\frac{dv}{dx}=v+x\sin v$$

$$\Rightarrow \quad x\frac{dv}{dx}=x\sin v \quad \Rightarrow \quad \frac{dv}{dx}=\sin v$$

Separating the variable, we get

$$\Rightarrow \quad \frac{dv}{\sin v}=dx \quad \Rightarrow \quad \int \operatorname{cosec} v\, dv=\int dx+C$$

$$\log\tan\frac{v}{2}=x+C \quad \Rightarrow \quad \log\tan\frac{y}{2x}=x+C$$ **Ans.**

EXERCISE 11.3

Solve the following differential equations:

1. $(y^2 - xy)\, dx + x^2 dy = 0$ **Ans.** $\frac{x}{y}=\log x+C$
2. $(x^2 - y^2)\, dx+2xy\, dy = 0$ *(AMIETE, June 2009)* **Ans.** $x^2 + y^2 = ax$
3. $x(y-x)\frac{dy}{dx}=y(y+x).$ **Ans.** $\frac{y}{x}-\log xy=a$
4. $x(x - y)\, dy + y^2 dx = 0$ *(U.P. B. Pharm (C.O.) 2005)* **Ans.** $y = x \log Cy$
5. $\frac{dy}{dx}+\frac{x-2y}{2x-y}=0$ **Ans.** $y - x = C(x + y)^3$
6. $\frac{dy}{dx}=\tan\frac{y}{x}+\frac{y}{x}$ **Ans.** $\sin\frac{y}{x}=Cx$
7. $\frac{dy}{dx}=\frac{3xy+y^2}{3x^2}$ **Ans.** $3x + y\log x + Cy = 0$
8. $\frac{dy}{dx}=\frac{x^2-2y^2}{2xy}$ **Ans.** $4y^2 - x^2 = \frac{C}{x^2}$
9. $(x^2 + y^2)\, dy = xy\, dx$ **Ans.** $-\frac{x^2}{2y^2}+\log y=C$
10. $x^2y\, dx - (x^3 + y^3)\, dy = 0$ **Ans.** $\frac{-x^3}{3y^3}+\log y=C$
11. $(y^2 + 2xy)\, dx + (2x^2 + 3xy)\, dy = 0$ *(AMIETE, Summer 2004)* **Ans.** $xy^2(x + y) = C$
12. $(2xy^2 - x^3)\, dy + (y^3 - 2yx^2)\, dx = 0$ **Ans.** $y^2(y^2 - x^2) = Cx^{-2}$
13. $(x^3 - 3xy^2)\, dx + (y^3 - 3x^2y)\, dy = 0,\ y(0) = 1$ **Ans.** $x^4 - 6x^2y^2 + y^4 = 1$
14. $2xy^2\, dy - (x^3 + 2y^3)\, dx = 0$ **Ans.** $2y^3 = 3x^3\log x + 3x^3 + C$
15. $x\sin\frac{y}{x}dy=\left(y\sin\frac{y}{x}-x\right)dx$ **Ans.** $\cos\frac{y}{x}=\log x+C$
16. $\left\{x\cos\frac{y}{x}+y\sin\frac{y}{x}\right\}y-\left\{y\sin\frac{y}{x}-x\cos\frac{y}{x}\right\}x\frac{dy}{dx}=0$ **Ans.** $xy\cos\frac{y}{x}=a$
17. $\frac{dy}{dx}=\frac{y+\sqrt{x^2+y^2}}{x}$ **Ans.** $y+\sqrt{x^2+y^2}=Cx^2$

18. $x\dfrac{dy}{dx} = y(\log y - \log x + 1)$ *(AMIETE, Summer 2004)* **Ans.** $\log\dfrac{y}{x} = Cx$

19. $xy\log\dfrac{x}{y}dx + \left(y^2 - x^2\log\dfrac{x}{y}\right)dy = 0$ given that $y(1) = 0$

Ans. $\dfrac{x^2}{2y^2}\log\dfrac{x}{y} - \dfrac{x^2}{4y^2} + \log y = 1 - \dfrac{3}{4e^2}$

20. $(1 + e^{\frac{x}{y}})\,dx + e^{\frac{x}{y}}\left(1 - \dfrac{x}{y}\right)dy = 0$ *(AMIETE, June 2009)* **Ans.** $e^{\frac{x}{y}} + \dfrac{x}{y} = e^{-y} + C$

11.9 EQUATIONS REDUCIBLE TO HOMOGENEOUS FORM

Case I. $\dfrac{a}{A} + \dfrac{b}{B}$

The equations of the form

$$\boxed{\frac{dy}{dx} = \frac{ax + by + c}{Ax + By + C}}$$

can be reduced to the homogeneous form by the substitution if $\dfrac{a}{A} + \dfrac{b}{B}$

$x = X + h,\ y = Y + k$ (h,k being constants)

$$\therefore \quad \frac{dy}{dx} = \frac{dY}{dX}$$

The given differential equation reduces to

$$\frac{dY}{dX} = \frac{a(X+h) + b(Y+k) + c}{A(X+h) + B(Y+k) + C} = \frac{aX + bY + ah + bk + c}{AX + BY + Ah + Bk + C}$$

Choose h, k so that $\quad ah + bk + c = 0$

$\quad Ah + Kk + C = 0$

Then the given equation becomes homogeneous $\dfrac{dY}{dX} = \dfrac{aX + bY}{AX + BY}$

Case II. If $\dfrac{a}{A} = \dfrac{b}{B}$ then the value of h, k will not be finite.

$$\frac{a}{A} = \frac{b}{B} = \frac{1}{m} \quad \text{(say)}$$

$$A = am,\ B = bm$$

The given equation becomes $\dfrac{dy}{dx} = \dfrac{ax + by + c}{m(ax + by) + c}$

Now put $ax + by = z$ and apply the method of variables separable.

Example 10. *Solve* : $\dfrac{dy}{dx} = \dfrac{x + 2y - 3}{2x + y - 3}$

Solution. Put $x = X + h, \quad y = Y + k.$

The given equation reduces to

$$\therefore \quad \frac{dY}{dX} = \frac{(X+h) + 2(Y+k) - 3}{2(X+h) + (Y+k) - 3} \qquad \left(\frac{1}{2} \neq \frac{2}{1}\right)$$

$$= \frac{X + 2Y + (h + 2k - 3)}{2X + Y + (2h + k - 3)} \qquad \text{.... (1)}$$

Now choose h and k so that $h + 2k - 3 = 0$, $2h + k - 3 = 0$
Solving these equations we get $h = k = 1$

$$\therefore \qquad \frac{dY}{dX} = \frac{X + 2Y}{2X + Y} \qquad \dots (2)$$

Put $Y = vX$, so that $\dfrac{dY}{dX} = v + X\dfrac{dv}{dX}$

The equation (2) is transformed as

$$v + X\frac{dv}{dX} = \frac{X + 2vX}{2X + vX} = \frac{1 + 2v}{2 + v}$$

$$X\frac{dv}{dX} = \frac{1 + 2v}{2 + v} - v = \frac{1 - v^2}{2 + v} \qquad \Rightarrow \qquad \left(\frac{2 + v}{1 - v^2}\right) dv = \frac{dX}{X}$$

$$\Rightarrow \qquad \frac{1}{2}\frac{1}{(1 + v)} dv + \frac{3}{2}\frac{1}{1 - v} dv = \frac{dX}{X} \qquad \text{(Partial fractions)}$$

On integrating, we have

$$\frac{1}{2}\log(1 + v) - \frac{3}{2}\log(1 - v) = \log X + \log C$$

$$\Rightarrow \qquad \log\frac{1 + v}{(1 - v)^3} = \log C^2 X^2 \qquad \Rightarrow \qquad \frac{1 + v}{(1 - v)^3} = C^2 X^2$$

$$\frac{1 + \dfrac{Y}{X}}{\left(1 - \dfrac{Y}{X}\right)^3} = C^2 X^2 \qquad \Rightarrow \qquad \frac{X + Y}{(X - Y)^3} = C^2 \quad \text{or} \quad X + Y = C^2 (X - Y)^3$$

Put $\quad X = x - 1$ and $Y = y - 1 \qquad \Rightarrow \qquad x + y - 2 = a\,(x - y)^3$ **Ans.**

Example 11. *Solve* : $(x + 2y)\,(dx - dy) = dx + dy$

Solution. $(x + 2y)\,(dx - dy) = dx + dy \quad \Rightarrow \quad (x + 2y - 1)\,dx - (x + 2y + 1)\,dy = 0$

$$\Rightarrow \qquad \frac{dy}{dx} = \frac{x + 2y - 1}{x + 2y + 1} \qquad \dots (1)$$

Hence $\qquad \dfrac{a}{A} = \dfrac{b}{B} \quad i.e., \left(\dfrac{1}{1} = \dfrac{2}{2}\right)$ (Case of failure)

Now put $x + 2y = z$ so that $1 + 2\dfrac{dy}{dx} = \dfrac{dz}{dx}$

Equation (1) becomes

$$\frac{1}{2}\frac{dz}{dx} - \frac{1}{2} = \frac{z - 1}{z + 1} \qquad \Rightarrow \qquad \frac{dz}{dx} = 2\frac{(z - 1)}{z + 1} + 1 = \frac{3z - 1}{z + 1}$$

$$\Rightarrow \qquad \frac{z + 1}{3z - 1} dz = dx \qquad \Rightarrow \qquad \left(\frac{1}{3} + \frac{4}{3}\frac{1}{3z - 1}\right) dz = dx$$

On integrating, $\qquad \dfrac{z}{3} + \dfrac{4}{9}\log(3z - 1) = x + C$

$$3z + 4\log(3z - 1) = 9x + 9C$$

$$\Rightarrow \qquad 3(x + 2y) + 4\log(3x + 6y - 1) = 9x + 9C$$

$$3x - 3y + a = 2\log(3x + 6y - 1) \qquad \textbf{Ans.}$$

EXERCISE 11.4

Solve the following differential equations :

1. $\frac{dy}{dx}=\frac{2x+9y-20}{6x+2y-10}$ **Ans.** $(2x-y)^2 = C(x+2y-5)$

2. $\frac{dy}{dx}=\frac{y-x+1}{y+x+5}$ **Ans.** $\log[(y+3)^2+(x+2)^2]+2\tan^{-1}\frac{y+3}{x+2}=a$

3. $\frac{dy}{dx}=\frac{x-y-2}{x+y+6}$ **Ans.** $(y+4)^2+2(x+2)(y+4)-(x+2)^2=a^2$

4. $\frac{dy}{dx}=\frac{y+x-2}{y-x-4}$ *(AMIETE, Dec. 2009)* **Ans.** $-(y-3)^2+2(x+1)(y-3)+(x+1)^2=a$

5. $\frac{dy}{dx}=\frac{2x-5y+3}{2x+4y-6}$ **Ans.** $(x-4y+3)(2x+y-3)=a$

6. $(2x+y+1)\,dx+(4x+2y-1)\,dy=0$ **Ans.** $2(2x+y)+\log(2x+y-1)=3x+C$

7. $(x-y-2)\,dx-(2x-2y-3)\,dy=0$ **Ans.** $\log(x-y-1)=x-2y+C$ *(U.P. B. Pharm (C.O.) 2005)*

8. $(6x-4y+1)\,dy-(3x-2y+1)\,dx=0$ *(A.M.I.E.T. E., Dec. 2006)*
Ans. $4x-8y-\log(12x-xy+1)=c$

9. $\frac{dy}{dx}=-\frac{3y-2x+7}{7y-3x+3}$ *(A.M.I.E.T.E., Summer 2004)* **Ans.** $(x+y-1)^5(x-y-1)^2=1$

10. $\frac{dy}{dx}=\frac{2y-x-4}{y-3x+3}$ *(AMIETE, Dec. 2010)*
ns. $X^2-5XY+Y^2=c\left[\frac{2Y+(-5+\sqrt{21})X}{2Y-(5+\sqrt{21})X}\right]^{\frac{1}{\sqrt{21}}}$, $X=x-2$, $Y=y-3$

11.10 LINEAR DIFFERENTIAL EQUATIONS

A differential equation of the form

$$\boxed{\frac{dy}{dx}+Py=Q} \qquad \dots(1)$$

is called a linear differential equation, where P and Q, are functions of x (but not of y) or constants.

In such case, multiply both sides of (1) by $e^{\int P dx}$

$$e^{\int Pdx}\left(\frac{dy}{dx}+Py\right)=Q e^{\int Pdx} \qquad \dots(2)$$

The left hand side of (2) is

$$\frac{d}{dx}\left[y.e^{\int Pdx}\right]$$

(2) becomes

$$\frac{d}{dx}\left[y.e^{\int P dx}\right]=Q.e^{\int P dx}$$

Integrating both sides, we get

$$y.e^{\int P dx}=\int Q.e^{\int P dx}dx+C$$

This is the required solution.

Note. $e^{\int P dx}$ is called the integrating factor.

Solution is

$$\boxed{y\times[I.F.]=\int Q\,[I.F.]\,dx+C}$$

Working Rule

Step 1. Convert the given equation to the standard form of linear differential equation

i.e. $$\frac{dy}{dx} + Py = Q$$

Step 2. Find the integrating factor i.e. I.F. = $e^{\int P dx}$

Step 3. Then the solution is $y(I.F.) = \int Q\,(I.F.)dx + C$

Example 12. *Solve*: $$(x+1)\frac{dy}{dx} - y = e^x (x+1)^2$$ (*A.M.I.E.T.E., Summer 2002*)

Solution. $$\frac{dy}{dx} - \frac{y}{x+1} = e^x (x+1)$$

$$\text{Integrating factor } = e^{-\int \frac{dx}{x+1}} = e^{-\log(x+1)} = e^{\log(x+1)^{-1}} = \frac{1}{x+1}$$

The solution is $$y.\frac{1}{x+1} = \int e^x.(x+1).\frac{1}{x+1}dx = \int e^x dx$$

$$\frac{y}{x+1} = e^x + C$$ **Ans.**

Example 13. *Solve a differential equation*

$$(x^3 - x)\frac{dy}{dx} - (3x^2 - 1)y = x^5 - 2x^3 + x.$$ (*Nagpur University, Summer 2008*)

Solution. We have $(x^3 - x)\frac{dy}{dx} - (3x^2 - 1)y = x^5 - 2x^3 + x$

$$\Rightarrow \quad \frac{dy}{dx} - \frac{3x^2 - 1}{x^3 - x}y = \frac{x^5 - 2x^3 + x}{x^3 - x} \quad \Rightarrow \quad \frac{dy}{dx} - \frac{3x^2 - 1}{x^3 - x}y = x^2 - 1$$

$$\text{I.F.} = e^{\int -\frac{3x^2-1}{x^3-x}dx} = e^{-\log(x^3 - x)} = e^{\log(x^3-x)^{-1}} = \frac{1}{x^3 - x}$$

Its solution is

$$y(\text{I.F.}) = \int Q(I.F.)\,dx + C \quad \Rightarrow \quad y\left(\frac{1}{x^3 - x}\right) = \int \frac{x^2 - 1}{x^3 - x}dx + C$$

$$\Rightarrow \quad \frac{y}{x^3 - x} = \int \frac{x^2 - 1}{x(x^2 - 1)}dx + C \quad \Rightarrow \quad \frac{y}{x^3 - x} = \int \frac{1}{x}dx + C$$

$$\Rightarrow \quad \frac{y}{x^3 - x} = \log x + C \quad \Rightarrow \quad y = (x^3 - x)\log x + (x^3 - x)\,C$$ **Ans.**

Example 14. *Solve* $\sin x\frac{dy}{dx} + 2y = \tan^3\left(\frac{x}{2}\right)$ (*Nagpur University, Summer 2004*)

Solution. Given equation : $\sin x\frac{dy}{dx} + 2y = \tan^3\frac{x}{2} \Rightarrow \frac{dy}{dx} + \frac{2}{\sin x}y = \frac{\tan^3\frac{x}{2}}{\sin x}$

This is linear form of $\frac{dy}{dx} + Py = Q$

$$\therefore \quad P = \frac{2}{\sin x} \text{ and } Q = \frac{\tan^3\frac{x}{2}}{\sin x}$$

$$\therefore \quad \text{I.F.} = e^{\int P dx} = e^{\int \frac{2}{\sin x}dx} = e^{2\int \operatorname{cosec} x\,dx} = e^{2\log\tan\frac{x}{2}} = \tan^2\frac{x}{2}$$

$\therefore$ Solution is $\quad y.(\text{I.F.}) = \int \text{I.F.}(Q\,dx) + C$

$$y\tan^2\frac{x}{2} = \int \tan^2\frac{x}{2}\cdot\frac{\tan^3\frac{x}{2}}{2\sin\frac{x}{2}\cdot\cos\frac{x}{2}} + C = \frac{1}{2}\int\frac{\tan^4\frac{x}{2}}{\cos^2\frac{x}{2}}dx + C$$

$$= \frac{1}{2}\int \tan^4\frac{x}{2}.\sec^2\frac{x}{2}\,dx + C \qquad \text{... (1)}$$

Putting $\tan\frac{x}{2} = t$ so that $\frac{1}{2}\sec^2\frac{x}{2}dx = dt$ on R.H.S. (1), we get

$$y.\tan^2\frac{x}{2} = \frac{1}{2}\int t^4(2dt) + C \quad\Rightarrow\quad y\tan^2\frac{x}{2} = \frac{t^5}{5} + C$$

$$y\tan^2\frac{x}{2} = \frac{\tan^5\frac{x}{2}}{5} + C \qquad \textbf{Ans.}$$

EXERCISE 11.5

Solve the following differential equations:

1. $\frac{dy}{dx} + \frac{1}{x}y = x^3 - 3$ **Ans.** $xy = \frac{x^5}{5} - \frac{3x^2}{2} + C$
2. $(2y - 3x)\,dx + x\,dy = 0$ **Ans.** $y\,x^2 = x^3 + C$
3. $\frac{dy}{dx} + y\cot x = \cos x$ **Ans.** $y\sin x = \frac{\sin^2 x}{2} + C$
4. $\frac{dy}{dx} + y\sec x = \tan x$ **Ans.** $y = \frac{C - x}{\sec x + \tan x} + 1$
5. $\cos^2 x\frac{dy}{dx} + y = \tan x$ **Ans.** $y = \tan x - 1 + Ce^{-\tan x}$
6. $(x + a)\frac{dy}{dx} - 3y = (x + a)^5$ **Ans.** $2y = (x + a)^5 + 2C\,(x + a)^3$
7. $x\cos x\frac{dy}{dx} + y(x\sin x + \cos x) = 1$ **Ans.** $x\,y = \sin x + C\cos x$
8. $x\log x\frac{dy}{dx} + y = 2\log x$ **Ans.** $y\log x = (\log x)^2 + C$
9. $x\frac{dy}{dx} + 2y = x^2\log x$ **Ans.** $yx^2 = \frac{x^4}{4}\log x - \frac{x^4}{16} + C$
10. $dr + (2r\cot\theta + \sin 2\theta)\,d\theta = 0$ **Ans.** $r\sin^2\theta = \frac{-\sin^4\theta}{2} + C$
11. $\frac{dy}{dx} + y\cos x = \frac{1}{2}\sin 2x$ **Ans.** $y = \sin x - 1 + Ce^{-\sin x}$
12. $(1 - x^2)\frac{dy}{dx} + 2xy = x(1 - x^2)^{1/2}$ **Ans.** $y = \sqrt{1 - x^2} + C(1 - x^2)$
13. $\sec x\frac{dy}{dx} = y + \sin x$ *(A.M.I.E.T.E., Dec 2005)* **Ans.** $y = -\sin x - 1 + ce^{\sin x}$
14. $y' + y\tan x = \cos x, y(0) = 0$ *(A.M.I.E.T.E., June 2006)* **Ans.** $y = x\cos x$
15. Solve $(1 + y^2)\,dx = (\tan^{-1} y - x)\,dy$ *(AMIETE, Dec. 2009)* **Ans.** $x = -\tan^{-1} y - 1 + ce^{\tan^{-1} y}$
16. Find the value of α so that e^2 is an integrating factor of differential equation $x(1 - y)\,dx - dy = 0$. *(A.M.I.E.T.E., Summer 2005)* **Ans.** $\alpha = \frac{1}{2}$

17. Slove the differential equation $\cot 3x \dfrac{dy}{dx} - 3y = \cos 3x + \sin 3x,\ 0 < x < \dfrac{\pi}{2}$.

(*AMIETE, Dec. 2009*) **Ans.** $y \cos 3x = \dfrac{1}{12}[6x - \sin 6x - \cos 6x]$

18. The value of α so that $e^{\alpha y^2}$ is an integrating factor of the differential equation

$(e^{\frac{-y^2}{2}} - xy)\, dy - dx = 0$ is (*A.M.I.E.T.E. Dec., 2005*)

(*a*) -1 (*b*) 1 (*c*) $\dfrac{1}{2}$ (*d*) $-\dfrac{1}{2}$ **Ans.** (c)

19. The solution of the differential equation $(y + x)^2 \dfrac{dy}{dx} = a^2$ is given by

(*a*) $y + x = a \tan\left(\dfrac{y-c}{a}\right)$ (*b*) $y - x = \tan\left(\dfrac{y-c}{a}\right)$

(*c*) $y - x = a \tan(y - c)$ (*d*) $a(y - x) = \tan\left(y - \dfrac{c}{a}\right)$ **Ans.** (*a*)

(*AMIETE, June 2010*)

11.11 EQUATIONS REDUCIBLE TO THE LINEAR FORM (BERNOULLI EQUATION)

The equation of the form

$$\frac{dy}{dx} + Py = Qy^n \quad \text{...(1)}$$

where **P** and **Q** are constants or functions of x can be reduced to the linear form on dividing by y^n and substituting $\dfrac{1}{y^{n-1}} = z$

On dividing bothsides of (1) by y^n, we get

$$\frac{1}{y^n}\frac{dy}{dx} + \frac{1}{y^{n-1}} P = Q \quad \text{...(2)}$$

Put $\dfrac{1}{y^{n-1}} = z$, so that $\dfrac{(1-n)}{y^n}\dfrac{dy}{dx} = \dfrac{dz}{dx} \quad \Rightarrow \quad \dfrac{1}{y^n}\dfrac{dy}{dx} = \dfrac{dz}{1-n}$

$\therefore$ (2) becomes $\dfrac{1}{1-n}\dfrac{dz}{dx} + Pz = Q$ or $\dfrac{dz}{dx} + P(1-n)z = Q(1-n)$

which is a linear equation and can be solved easily by the previous method discussed in article 11.10 on page 217.

Example 15. *Solve* $x^2 dy + y(x + y)\, dx = 0$ (*U.P. II Semester Summer 2006*)

Solution. We have, $x^2\, dy + y\,(x + y)\, dx = 0$

$$\Rightarrow \quad \frac{dy}{dx} + \frac{y}{x} = -\frac{y^2}{x^2} \quad \Rightarrow \quad \frac{1}{y^2}\frac{dy}{dx} + \frac{1}{xy} = -\frac{1}{x^2}$$

Put $-\dfrac{1}{y} = z$ so that $\dfrac{1}{y^2}\dfrac{dy}{dx} = \dfrac{dz}{dx}$

The given equation reduces to a linear differential equation in z.

$$\frac{dz}{dx} - \frac{z}{x} = -\frac{1}{x^2}$$

$$\textbf{I.F.} = e^{-\int \frac{1}{x} dx} = e^{-\log x} = e^{\log 1/x} = \frac{1}{x}.$$

Hence the solution is

$$z.\frac{1}{x} = \int -\frac{1}{x^2}.\frac{1}{x}\,dx + C \quad \Rightarrow \quad \frac{z}{x} = \int -x^{-3}\,dx + C$$

$$\Rightarrow \quad -\frac{1}{xy} = -\frac{x^{-2}}{-2} + C \quad \Rightarrow \quad \frac{1}{xy} = -\frac{1}{2x^2} - C \qquad \textbf{Ans.}$$

Example 16. *Solve:* $x\frac{dy}{dx} + y\log y = xy\ e^x$ *(A.M.I.E., Summer 2000)*

Solution. $x\frac{dy}{dx} + y\log y = xy\,e^x$

Dividing by xy, we get

$$\frac{1}{y}\frac{dy}{dx} + \frac{1}{x}\log y = e^x \qquad ...(1)$$

Put $\log y = z$, so that $\frac{1}{y}\frac{dy}{dx} = \frac{dz}{dx}$

Equation (1) becomes, $\frac{dz}{dx} + \frac{z}{x} = e^x$

$$\text{I.F.} = e^{\int \frac{1}{x}dx} = e^{\log x} = x$$

Solution is $$z\,x = \int x\,e^x\,d\,x + C$$

$$z\,x = x\,e^x - e^x + C$$

$$\Rightarrow \quad x\log y = x\,e^x - e^x + C \qquad \textbf{Ans.}$$

Example 17. *Solve:* $\frac{dy}{dx} - \frac{\tan y}{1+x} = (1+x)e^x\sec y.$ *(Nagpur University, Summer 2000)*

Solution. $\frac{dy}{dx} - \frac{\tan y}{1+x} = (1+x)e^x\sec y$

$$\Rightarrow \quad \cos y\frac{dy}{dx} - \frac{\sin y}{1+x} = (1+x)\,e^x \qquad ...(1)$$

Put $\sin y = z$, so that $\cos y\frac{dy}{dx} = \frac{dz}{dx}$

(1) becomes $\frac{dz}{dx} - \frac{z}{1+x} = (1+x)e^x$

$$I.F. = e^{-\int \frac{1}{1+x}dx} = e^{-\log(1+x)} = e^{\log 1/(1+x)} = \frac{1}{1+x}$$

Solution is $$z.\frac{1}{1+x} = \int (1+x)e^x.\frac{1}{1+x}\,dx + C = \int e^x\,dx + C$$

$$\frac{\sin y}{1+x} = e^x + C \qquad \textbf{Ans}$$

Example 18. *Solve:* $\tan y\frac{dy}{dx} + \tan x = \cos y\cos^2 x$ *(Nagpur University, Summer 2000)*

Solution. $\tan y\frac{dy}{dx} + \tan x = \cos y\cos^2 x$

$$\sec y\tan y\frac{dy}{dx} + \sec y\,\tan x = \cos^2 x$$

Writing $\qquad z = \sec y$, so that $\dfrac{dz}{dx} = \sec y \tan y \dfrac{dy}{dx}$

The equation becomes $\dfrac{dz}{dx} + z \tan x = \cos^2 x$

$$\text{I.F.} = e^{\int \tan x\,dx} = e^{\log \sec x} = \sec x$$

$\therefore$ The solution of the equation is

$$z \sec x = \int \cos^2 x \sec x \, dx + C$$

$$\sec y \sec x = \int \cos x \, dx + C = \sin x + C$$

$$\sec y = (\sin x + C) \cos x$$ **Ans.**

Example 19. *Solve differential equation*

$$\frac{dy}{dx} + \frac{1}{x} \tan y = \frac{1}{x^2} \tan y \sin y$$ (*Nagpur University, Summer 2000*)

Solution. We have, $\dfrac{dy}{dx} + \dfrac{1}{x} \tan y = \dfrac{1}{x^2} \tan y \sin y \quad \Rightarrow \quad \dfrac{1}{\tan y \sin y} \dfrac{dy}{dx} + \dfrac{1}{x} \dfrac{1}{\sin y} = \dfrac{1}{x^2}$

$$\Rightarrow \quad \cot y \operatorname{cosec} y \frac{dy}{dx} + \frac{1}{x} \operatorname{cosec} y = \frac{1}{x^2} \qquad \text{... (1)}$$

Putting $\operatorname{cosec} y = z$, and $-\operatorname{cosec} y \cot y \dfrac{dy}{dx} = \dfrac{dz}{dx}$ in (1), we get

$$-\frac{dz}{dx} + \frac{1}{x} z = \frac{1}{x^2}$$

$$\frac{dz}{dx} - \frac{1}{x} z = -\frac{1}{x^2}$$

$$\text{I.F.} = e^{-\int \frac{1}{x} dx} = e^{-\log x} = e^{\log x^{-1}} = x^{-1} = \frac{1}{x}$$

Its solution is $\qquad z\,(\text{I.F.}) = \int Q (\text{I.F.})\, dx + C$

$$z \cdot \frac{1}{x} = \int \left(\frac{-1}{x^2} \right) \frac{1}{x} dx + C = -\int \frac{1}{x^3} dx + C = -\frac{x^{-2}}{-2} + C$$

$$z = \frac{1}{2x^2} + C x \quad \Rightarrow \quad \operatorname{cosec} y = \frac{1}{2x^2} + C x$$ **Ans.**

Example 20. $\quad x \left[\dfrac{dx}{dy} + y \right] = 1 - y$ (*Nagpur University, Summer 2004*)

Solution. $\quad x \left(\dfrac{dy}{dx} + y \right) = (1 - y)$

$$\Rightarrow \quad \frac{dy}{dx} + y = \frac{1}{x} - \frac{y}{x} \quad \Rightarrow \quad \frac{dy}{dx} + \left(1 + \frac{1}{x} \right) y = \frac{1}{x}$$

which is in linear form of $\dfrac{dy}{dx} + Py = Q$.

$$\therefore \quad P = \left(1 + \frac{1}{x} \right), \qquad Q = \frac{1}{x}$$

$$\text{I.F.} = e^{\int P dx} = e^{\int \left(1 + \frac{1}{x}\right) dx} = e^{x + \log x} = e^x . e^{\log x} = e^x . x = x e^x$$

Its solution is

$$y(\text{I.F.}) = \int \text{I.F.}(Q\,dx) + C$$

$$y(x.e^x) = \int (x.e^x) \times \frac{1}{x}\,dx + C \Rightarrow y(x.e^x) = \int e^x\,dx + C$$

$$y\,(x\,.\,e^x) = e^x + C$$

$\therefore$ $$y = \frac{1}{x} + \frac{C}{x}e^{-x}$$ **Ans.**

Example 21. *Solve the differential equation.*

$$y \log y\,dx + (x - \log y)\,dy = 0$$ *(Uttarakhand II Semester, June 2007)*

Solution. We have,

$$y \log y\,dx + (x - \log y)\,dy = 0$$

$$\Rightarrow \quad \frac{dx}{dy} = \frac{-x + \log y}{y \log y} \quad \Rightarrow \quad \frac{dx}{dy} = \frac{-x}{y \log y} + \frac{\log y}{y \log y}$$

$$\Rightarrow \quad \frac{dx}{dy} + \frac{x}{y \log y} = \frac{1}{y}$$

$$\text{I.F.} = e^{\int \frac{1}{y \log y} dy} = e^{\log(\log y)} = \log y$$

Its solution is $$x.\log y = \int \frac{1}{y}(\log y)\,dy$$

$$x.\log y = \frac{(\log y)^2}{2} + C$$ **Ans.**

Example 22. $$y e^y dx = (y^3 + 2x e^y) dy$$ *(Nagpur University, Winter 2003)*

Solution. $$y e^y dx = (y^3 + 2x e^y) dy \quad \Rightarrow \quad \frac{dx}{dy} - \frac{2x}{y} = \frac{y^2}{e^y}$$

which is linear in x

$\therefore$ $$P = \frac{-2}{y} \quad \text{and} \quad Q = \frac{y^2}{e^y}$$

$$\text{I.F.} = e^{\int P\,dy} = e^{\int \frac{-2}{y} dy} = e^{-2 \log y} = e^{\log y^{-2}} = y^{-2} = \frac{1}{y^2}$$

Its solution is $$x(\text{I.F.}) = \int Q.(\text{I.F.}\,dy) + C \quad \Rightarrow \quad x.\frac{1}{y^2} = \int \frac{y^2}{e^y} \times \frac{1}{y^2}\,dy + C$$

$$\frac{x}{y^2} = \int e^{-y}.dy + C \quad \Rightarrow \quad \frac{x}{y^2} = -e^{-y} + C$$

$\therefore$ $$\frac{x}{y^2} + e^{-y} = C$$ **Ans.**

Example 23. *Solve* $\dfrac{dy}{dx} = \dfrac{y}{2y \log y + y - x}$ *(Nagpur University, Summer 2003)*

Solution. $$\frac{dx}{dy} = \frac{2y \log y + y - x}{y}$$

$$\Rightarrow \quad \frac{dx}{dy} + \frac{1}{y}x = 1 + 2 \log y$$

Which is of the form $\frac{dx}{dy} + Px = Q$

Here $P = \frac{1}{y}$ and $Q = 1 + 2\log y$

$$\text{I.F.} = e^{\int P\,dy} = e^{\int \frac{1}{y} dy} = e^{\log y} = y$$

Its solution is

$$x.(I.F.) = \int Q(I.F.)\,dy + C \quad \Rightarrow \quad xy = \int (1 + 2\log y)\,y\,dy + C$$

$$\Rightarrow \quad xy = \int (y + 2y\log y)\,dy + C = \frac{y^2}{2} + 2\left[\log y.\frac{y^2}{2} - \int \frac{1}{y}.\frac{y^2}{2}\,dy\right] + C$$

$$= \frac{y^2}{2} + 2\left[\frac{y^2}{2}\log y - \frac{1}{4}y^2\right] + C = y^2 \log y + C$$

$$\Rightarrow \quad x = y\log y + \frac{C}{y}$$ **Ans.**

Example 24. *Solve* : $\frac{dy}{dx} = \frac{y+1}{(y+2)\,e^y - x}$ *(Nagpur University, Winter 2004)*

Solution. $\frac{dx}{dy} = \frac{(y+2)\,e^y}{y+1} - \frac{x}{y+1}$

$$\Rightarrow \quad \frac{dx}{dy} + \frac{x}{y+1} = \frac{(y+2)\,e^y}{y+1}$$

which is linear in x

Here $P = \frac{1}{y+1}$ and $Q = \frac{y+2}{y+1}e^y$

$$\text{I.F.} = e^{\int P\,dy} = e^{\int \frac{1}{y+1} dy} = e^{\log(y+1)} = y+1$$

Its solution is

$$x(\text{I.F.}) = \int Q(\text{I.F.}dy) + C$$

$$\Rightarrow \quad x(y+1) = \int \frac{y+2}{y+1}e^y(y+1)\,dy + C$$

$$\Rightarrow \quad x(y+1) = \int (y+2)e^y\,dy + C = (y+2)e^y - \int \frac{d}{dy}(y+2).e^y\,dy + C$$

$$\Rightarrow \quad x(y+1) = (y+2)e^y - \int e^y\,dy + C$$

$$\Rightarrow \quad x(y+1) = (y+2)\,e^y - e^y + C$$

$$\Rightarrow \quad x(y+1) = (y+1)\,e^y + C \quad \Rightarrow \quad x = e^y + \frac{C}{y+1}$$ **Ans.**

Example 25. *Solve*: $(1 + y^2)\,dx = (\tan^{-1} y - x)\,dy$.

(AMIETE, June 2010, 2004, R.G.P.V., Bhopal, April 2010, June 2008, U.P. (B. Pharm) 2005)

Solution. $(1 + y^2)\,dx = (\tan^{-1} y - x)\,dy$

$$\frac{dx}{dy} = \frac{\tan^{-1} y - x}{1+y^2} \quad \Rightarrow \quad \frac{dx}{dy} + \frac{x}{1+y^2} = \frac{\tan^{-1} y}{1+y^2}$$

This is a linear differential equation.

$$\text{I.F.} = e^{\int \frac{1}{1+y^2} dy} = e^{\tan^{-1} y}$$

Its solution is $\quad x.e^{\tan^{-1} y} = \int e^{\tan^{-1} y} \frac{\tan^{-1} y}{1+y^2} dy + C$

Put $\quad \tan^{-1} y = t$ on R.H.S., so that $\frac{1}{1+y^2} dy = dt$

$$x.e^{\tan^{-1} y} = \int e^t .t\, dt + C = t.e^t - e^t + C = e^{\tan^{-1} y}\left(\tan^{-1} y - 1\right) + C$$

$$x = \left(\tan^{-1} y - 1\right) + Ce^{-\tan^{-1} y} \qquad \textbf{Ans.}$$

Example 26. *Solve* : $r \sin\theta - \frac{dr}{d\theta}\cos\theta = r^2$ *(Nagpur University, Summer 2005)*

Solution. The given equation can be written as $-\frac{dr}{d\theta}\cos\theta + r\sin\theta = r^2$ ... (1)

Dividing (1) by $r^2 \cos\theta$, we get $-r^{-2}\frac{dr}{d\theta} + r^{-1}\tan\theta = \sec\theta$ ... (2)

Putting $\quad r^{-1} = v$ so that $-r^{-2}\frac{dr}{d\theta} = \frac{dv}{d\theta}$ in (2), we get

$$\frac{dv}{d\theta} + v\tan\theta = \sec\theta$$

$$\text{I.F.} = e^{\int \tan\theta\, d\theta} = e^{\log \sec\theta} = \sec\theta$$

Solution is $\quad v\sec\theta = \int \sec\theta,\ \sec\theta + C \quad \Rightarrow \quad v\sec\theta = \int \sec^2\theta\, d\theta + C$

$$\frac{\sec\theta}{r} = \tan\theta + C \quad \Rightarrow \quad r^{-1} = (\sin\theta + C\cos\theta,$$

$$\therefore \quad r = \frac{1}{\sin\theta + C\cos\theta} \qquad \textbf{Ans.}$$

EXERCISE 11.6

Solve the following differential equations:

1. $\frac{1}{y^2}\frac{dy}{dx} - \frac{1}{y} = 2xe^{-x}$ **Ans.** $e^x + x^2y + Cy = 0$
2. $3\frac{dy}{dx} + 3\frac{y}{x} = 2x^4y^4$ **Ans.** $\frac{1}{y^3} = x^5 + Cx^3$
3. $\frac{dy}{dx} = y\tan x - y^2 \sec x$ **Ans.** $\sec x = (\tan x + C)\, y$
4. $\frac{dy}{dx} = 2y\tan x + y^2\tan^2 x$, if $y = 1$ at $x = 0$ **Ans.** $\frac{1}{y}\sec^2 x = -\frac{\tan^3 x}{3} + 1$
5. $\frac{dy}{dx} + \tan x \tan y = \cos x \sec y$ **Ans.** $\sin y \sec x = x + C$
6. $dy + y\tan x \,.\, dx = y^2 \sec x \,.\, dx$ **Ans.** $y\,(x + C) + \cos x = 0$
7. $(x^2y^2 + xy)\, y\, dx + (x^2y^2 - 1)\, x\, dy = 0$ **Ans.** $x\,y = \log C\,y$
8. $(x^2 + y^2 + x)\, dx + xy\, dy = 0$ **Ans.** $x^2y^2 = -\frac{x^4}{2} - \frac{2x^3}{3} + C$

9. $\dfrac{dy}{dx}+y=3e^x y^3$ **Ans.** $\dfrac{1}{y^2}=6e^x+Ce^{2x}$

10. $(x-y^2)\,dx+2\,x\,y\,dy=0$ **Ans.** $\dfrac{y^2}{x}+\log x=C$

11. $e^y\left(\dfrac{dy}{dx}+1\right)=e^x$ **Ans.** $e^{x+y}=\dfrac{e^{2x}}{2}+C$

12. $x^2y-x^3\dfrac{dy}{dx}=y^4\cos x$ **Ans.** $x^3=y^3(3\sin x-C)$

13. $3\dfrac{dy}{dx}+\dfrac{2}{x+1}.y=\dfrac{x^2}{y^2}$ **Ans.** $y^3(x+1)^2=\dfrac{x^5}{5}+\dfrac{x^4}{2}+\dfrac{x^3}{3}+C$

14. $\cos x\dfrac{dy}{dx}+4y\sin x=4\sqrt{y}\sec x$ **Ans.** $\sqrt{y}\sec^2 x=2\left[\tan x+\dfrac{\tan^3 x}{3}\right]+C$

15. $\dfrac{dy}{dx}+x\sin 2y=x^3\cos^2 y$ **Ans.** $\tan y=\dfrac{1}{2}(x^2-1)+Ce^{-x^2}$

16. $\dfrac{1}{1+y^2}\dfrac{dy}{dx}+2x\tan^{-1}y=x^3$ **Ans.** $\tan^{-1}y=\dfrac{1}{2}(x^2-1)+Ce^{-x^2}$

17. $e^{-y}\sec^2 y\,dy=dx+x\,dy$ **Ans.** $x\,e^y=\tan y+C$

18. $(x+y+1)\dfrac{dy}{dx}=1$ **Ans.** $x+y+2=C\,e^y$

19. $\dfrac{dy}{dx}=\dfrac{y^3}{e^{2x}+y^2}$ **Ans.** $e^{-2x}y^2+2\log y+C=0$

20. $dx-xy(1+xy^2)\,dy=0$ **Ans.** $-\dfrac{1}{x}=y^2-2+Ce^{-y^2/2}$

21. $\dfrac{dy}{dx}+\dfrac{y}{x}\log y=\dfrac{y}{x^2}(\log y)^2$ *(A.M.I.E.T.E., Summer 2004, 2003, Winter 2003, 2001)*

Ans. $\dfrac{1}{x\log y}=\dfrac{1}{2x^2}+C$

22. $3\dfrac{dy}{dx}+xy=xy^{-2}$ *(A.M.I.E.T.E., June 2009)* **Ans.** $y^3=1+Ce^{-x^2/2}$

27. $x\dfrac{dy}{dx}+y=x^3y^6$ *(AMIETE, June 2010)* **Ans.** $\dfrac{1}{y^5x^5}=\dfrac{5}{2x^2}+C$

23. General solution of linear differential equation of first order $\dfrac{dx}{dy}+Px=Q$ (where P and Q are constants or functions of y) is

(*a*) $ye^{\int P.dx}=\int Qe^{\int P.dx}\,dx+c$ (*b*) $xe^{\int P.dy}=\int Qe^{\int P.dy}\,dy+c$

(*c*) $y=\int Qe^{\int P.dx}\,dx+c$ (*d*) $x=\int Qe^{\int P.dy}\,dy+c$ *(AMIETE, June, 2010)* **Ans.** (*b*)

11.12 EXACT DIFFERENTIAL EQUATION

An exact differential equation is formed by directly differentiating its primitive (solution) without any other process

$$Mdx+Ndy=0$$

is said to be an exact differential equation if it satisfies the following condition

$$\boxed{\frac{\partial M}{\partial y}=\frac{\partial N}{\partial x}}$$

where $\frac{\partial M}{\partial y}$ denotes the differential co-efficient of M with respect to y keeping x constant and $\frac{\partial N}{\partial x}$, the differential co-efficient of N with respect to x, keeping y constant.

Method for Solving Exact Differential Equations

Step I. Integrate M w.r.t. x keeping y constant

Step II. Integrate w.r.t. y, only those terms of N which do not contain x.

Step III. Result of I + Result of II = Constant.

Example 27. *Solve* :

$$(5x^4 + 3x^2y^2 - 2xy^3)\,dx + (2x^3y - 3x^2y^2 - 5y^4)\,dy = 0$$

Solution. Here, $M = 5x^4 + 3x^2y^2 - 2xy^3$, $\quad N = 2x^3y - 3x^2y^2 - 5y^4$

$$\frac{\partial M}{\partial y} = 6x^2y - 6xy^2, \quad \frac{\partial N}{\partial x} = 6x^2y - 6xy^2$$

Since, $\frac{\partial M}{\partial y} = \frac{\partial N}{\partial x}$, the given equation is exact.

Now $\int M\,dx + \int(\text{terms of } N \text{ is not containing } x)\,dy = C$ (y constant)

$$\int\left(5x^4 + 3x^2y^2 - 2xy^3\right)dx + \int -5y^4\,dy = C$$

$$\Rightarrow \quad x^5 + x^3y^2 - x^2y^3 - y^5 = C \qquad \textbf{Ans.}$$

Example 28. *Solve*: $\left\{2xy\cos x^2 - 2xy + 1\right\}dx + \left\{\sin x^2 - x^2 + 3\right\}dy = 0$

(Nagpur University, Summer 2000)

Solution. Here we have

$$\left\{2xy\cos x^2 - 2xy + 1\right\}dx + \left\{\sin x^2 - x^2 + 3\right\}dy = 0 \qquad \dots(1)$$

$$M\,dx + N\,dy = 0 \qquad \dots(2)$$

Comparing (1) and (2), we get

$$M = 2xy\cos x^2 - 2xy + 1 \quad \Rightarrow \quad \frac{\partial M}{\partial y} = 2x\cos x^2 - 2x$$

$$N = \sin x^2 - x^2 + 3 \quad \Rightarrow \quad \frac{\partial N}{\partial x} = 2x\cos x^2 - 2x$$

Here, $\therefore \quad \frac{\partial M}{\partial y} = \frac{\partial N}{\partial x}$

So the given differential equation is exact differential equation.

Hence solution is $\int\limits_{y \text{ as const}} M\,dx + \int(\text{terms of } N \text{ not containing } x)\,dy = C$

$$\int(2xy\cos x^2 - 2xy + 1)\,dx + \int 3\,dy = C$$

$$\Rightarrow \quad \int[y(2x\cos x^2) - y(2x) + 1]\,dx + 3\int dy = C$$

$$\Rightarrow \quad y\int 2x\cos x^2\,dx - y\int 2x\,dx + \int 1\,dx + 3\int y\,dy = C$$

Put $x^2 = t$ so that $2x\,dx = dt$

$$y\int\cos t\,dt - 2y\frac{x^2}{2} + x + 3y = C$$

$$\Rightarrow \quad y\sin t - x^2y + x + 3y = C$$

$$y\sin x^2 - yx^2 + x + 3y = C \qquad \textbf{Ans.}$$

Example 29. *Solve* :

$$(1+e^{x/y})+e^{x/y}\left(1-\frac{x}{y}\right)\frac{dy}{dx}=0$$

(*Nagpur University, Summer 2008, A.M.I.E.T.E. June, 2009*)

Solution. We have,

$$\left(1+e^{\frac{x}{y}}\right)+e^{\frac{x}{y}}\left(1-\frac{x}{y}\right)\frac{dy}{dx}=0 \quad \Rightarrow \quad \left(1+e^{\frac{x}{y}}\right)dx+\left(e^{\frac{x}{y}}-e^{\frac{x}{y}}\frac{x}{y}\right)dy=0$$

$$M=1+e^{\frac{x}{y}} \quad \Rightarrow \quad \frac{\partial M}{\partial y}=-\frac{x}{y^2}e^{\frac{x}{y}}$$

$$N=e^{\frac{x}{y}}-e^{\frac{x}{y}}\frac{x}{y} \quad \Rightarrow \quad \frac{\partial N}{\partial x}=\frac{1}{y}e^{\frac{x}{y}}-\frac{1}{y}e^{\frac{x}{y}}-\frac{x}{y^2}e^{\frac{x}{y}}=-\frac{x}{y^2}e^{\frac{x}{y}}$$

$$\Rightarrow \quad \frac{\partial M}{\partial y}=\frac{\partial N}{\partial x}$$

$\therefore$ Given equation is exact.

Its solution is $\int\left(1+e^{\frac{x}{y}}\right)dx+\int(\text{terms of } N \text{ not containing } x)\,dy=C$

$$\Rightarrow \quad \int\left(1+e^{\frac{x}{y}}\right)dx+\int 0\,dy=C \quad \Rightarrow \quad x+ye^{\frac{x}{y}}=C$$ **Ans.**

Example 30. *Solve* : $x\,dx+y\,dy=\dfrac{a^2(x\,dy-y\,dx)}{x^2+y^2}$ (*U.P. Second Semester Summer 2005*)

Solution. We have, $x\,dx+y\,dy=\dfrac{a^2(x\,dy-y\,dx)}{x^2+y^2}$

$$\Rightarrow \quad \left(x+\frac{a^2y}{x^2+y^2}\right)dx+\left(y-\frac{(a^2x)}{x^2+y^2}\right)dy=0$$

Here, $$M=x+\frac{a^2y}{x^2+y^2}, \qquad N=y-\frac{a^2x}{x^2+y^2}$$

Now, $$\frac{\partial M}{\partial y}=\frac{a^2\left(x^2-y^2\right)}{\left(x^2+y^2\right)^2}, \quad \frac{\partial N}{\partial x}=\frac{a^2\left(x^2-y^2\right)}{\left(x^2+y^2\right)^2}$$

Since, $$\frac{\partial M}{\partial y}=\frac{\partial N}{\partial x}$$

Therefore equation is exact. Hence,

$$\int\left(x+\frac{a^2y}{x^2+y^2}\right)dx+\int y\,dy=C$$

$$\Rightarrow \quad \frac{x^2}{2}+a^2y.\frac{1}{y}\tan^{-1}\left(\frac{x}{y}\right)+\frac{y^2}{2}=C$$

$$\Rightarrow \quad \left(\frac{x^2+y^2}{2}\right)+a^2\tan^{-1}\left(\frac{x}{y}\right)=C$$ **Ans.**

Example 31. *Solve*: $[1+\log(x\,y)]\,dx+\left[1+\dfrac{x}{y}\right]dy=0$ (*Nagpur University, Winter 2003*)

Solution. $[1+\log x\,y]\,dx+\left[1+\dfrac{x}{y}\right]dy=0$

$$\therefore \qquad [1+\log x+\log y]\,dx+\left[1+\frac{x}{y}\right]dy=0$$

which is in the form

$$M\,dx+N\,dy=0$$

$$M=[1+\log x+\log y] \quad \text{and} \quad N=1+\frac{x}{y}$$

$$\Rightarrow \quad \frac{\partial M}{\partial y}=\frac{1}{y} \quad \text{and} \quad \Rightarrow \quad \frac{\partial N}{\partial x}=\frac{1}{y}$$

$$\Rightarrow \quad \frac{\partial M}{\partial y}=\frac{\partial N}{\partial x}$$

Hence the given differential equation is exact.

$\therefore$ Solution is $\displaystyle\int_{y\text{ constant}} M\,dx+\int N\,(\text{terms not containing } x)\,dy=C$

y constant

$$\therefore \qquad \int(1+\log x+\log y)\,dx+\int dy=C$$

$$\Rightarrow \qquad x+\int\log x\,dx+\int\log y\,dx+y=C \qquad \ldots(1)$$

Now, $\displaystyle\int\log x\,dx=\int\log x.(1)\,dx=(\log x)x-\int\left[\frac{d}{dx}(\log x)x\right]dx=x\log x-\int\frac{1}{x}.x\,dx$

$$=x\log x-\int dx=x\log x-x=x[\log x-1]$$

$\therefore$ Equation (1) becomes $\Rightarrow$ $x+x\log x-x+x\log y+y=C$

$x\,[\log x+\log y]+y=C \Rightarrow x\log xy+y=C$ **Ans.**

Example 32. *Find the value of* λ, *for the differential equation*

$(xy^2+\lambda x^2y)dx+(x+y)x^2dy=0$ *is exact*

Solve the equation for this value of λ.

(*Uttarakhad, II Summer 2010, Nagpur University, Summer 2002*)

Solution. Here $(xy^2+\lambda x^2 y)\,dx+(x+y)x^2dy=0$ ... (1)

which is of the form $M\,dx+N\,dx=0$

Where $M=xy^2+\lambda\, x^2y$ and $N=(x+y)\,x^2=x^3+x^2y$

$$\frac{\partial M}{\partial y}=2xy+\lambda x^2, \qquad \frac{\partial N}{\partial x}=3x^2+2xy$$

Condition of to be exact is

$$\frac{\partial M}{\partial y}=\frac{\partial N}{\partial x}$$

$$2xy+\lambda x^2=3x^2+2xy$$

$$\Rightarrow \qquad \lambda x^2 = 3x^2 \qquad \Rightarrow \qquad \lambda = 3$$

If $\lambda = 3$ then (1) becomes an excat differential equation.

$$(xy^2 + 3x^2y)\, dx + (x + y)\, x^2\, dy = 0$$

Its solution is given by

$$\underset{(y\,=\,\text{constant})}{\int M\, dx} + \int (\text{terms of } N \text{ not containing } x)\, dy = \text{constant}$$

$$\Rightarrow \qquad \int (xy^2 + 3x^2 y)\, dx + \int 0\, dy = C$$

$$\Rightarrow \qquad \frac{x^2y^2}{2} + \frac{3x^3y}{3} = C \qquad \Rightarrow \qquad \frac{x^2y^2}{2} + x^3y = C$$

$$x^2y^2 + 2x^3y = C_1$$ **Ans.**

EXERCISE 11.7

Solve the following differential equation (1 – 12).

1. $(x + y - 10)\, dx + (x - y - 2)\, dy = 0$ **Ans.** $\frac{x^2}{2} + xy - 10x - \frac{y^2}{2} - 2y = C$
2. $(y^2 - x^2)\, dx + 2xy\, dy = 0$ **Ans.** $\frac{x^3}{3} = xy^2 + C$
3. $\left(1 + 3e^{x/y}\right) dx + 3e^{x/y}\left(1 - \frac{x}{y}\right) dy = 0$ *(R.G.P.V. Bhopal, Winter 2010)* **Ans.** $x + 3y\, e^{x/y} = C$
4. $(2x - y)\, dx = (x - y)\, dy$ **Ans.** $xy = x^2 + \frac{y^2}{2} + C$
5. $(y \sec^2 x + \sec x \tan x)\, dx + (\tan x + 2y)\, dy = 0$ **Ans.** $y \tan x + \sec x + y^2 = C$
6. $(ax + hy + g)\, dx + (hx + by + f)\, dy = 0$ **Ans.** $ax^2 + 2h\, xy + by^2 + 2gx + 2fy + C = 0$
7. $(x^4 - 2xy^2 + y^4)\, dx - (2x^2y - 4xy^3 + \sin y)\, dy = 0$ **Ans.** $\frac{x^5}{5} - x^2y^2 + xy^4 + \cos y = C$
8. $(2xy + e^y)\, dx + (x^2 + xe^y)\, dy = 0$ **Ans.** $x^2y + xe^y = C$
9. $(x^2 + 2ye^{2x})\, dy + (2xy + 2y^2e^{2x})\, dx = 0$ **Ans.** $x^2y + y^2e^{2x} = C$
10. $\left[y\left(1 + \frac{1}{x}\right) + \cos y\right] dx + (x + \log x - x \sin y)\, dy = 0$ *(M.D.U., 2010)*
    Ans. $y\,(x + \log x) + x \cos y = C$
11. $(x^3 - 3xy^2)\, dx + (y^3 - 3x^2y)\, dy = 0,\ y(0) = 1$ **Ans.** $x^4 - 6x^2y^2 + y^4 = 1$
12. The differential equation $M(x, y)\, dx + N(x, y)\, dy = 0$ is an exact differential equation if
    (*a*) $\frac{\partial M}{\partial y} + \frac{\partial N}{\partial x} = 0$ (*b*) $\frac{\partial M}{\partial y} - \frac{\partial N}{\partial x} = 0$ (*c*) $\frac{\partial M}{\partial y} \times \frac{\partial N}{\partial x} = 1$ (*d*) None of the above
    (A.M.I.E.T.E. Dec. 2010, Dec 2006) **Ans.** (*b*)

11.13 EQUATIONS REDUCIBLE TO THE EXACT EQUATIONS

Sometimes a differential equation which is not exact may become so, on multiplication by a suitable function known as the integrating factor.

Rule 1. If $\dfrac{\frac{\partial M}{\partial y} - \frac{\partial N}{\partial x}}{N}$ is a function of x alone, say $f(x)$, then $\text{I.F.} = e^{\int f(x)\, dx}$

Example 33. *Solve* $(2x \log x - xy)\, dy + 2y\, dx = 0$ ... (1)

Solution. $M = 2y, \qquad N = 2x \log x - xy$

$$\frac{\partial M}{\partial y} = 2, \qquad \frac{\partial N}{\partial x} = 2(1+\log x) - y$$

Here, $$\frac{\frac{\partial M}{\partial y} - \frac{\partial N}{\partial x}}{N} = \frac{2-2-2\log x + y}{2x\log x - xy} = \frac{-(2\log x - y)}{x(2\log x - y)} = -\frac{1}{x} = f(x)$$

$$\text{I.F.} = e^{\int f(x)dx} = e^{\int -\frac{1}{x}dx} = e^{-\log x} = e^{\log x^{-1}} = x^{-1} = \frac{1}{x}$$

On multiplying the given differential equation (1) by $\frac{1}{x}$, we get

$$\frac{2y}{x}dx + (2\log x - y)dy = 0 \qquad \Rightarrow \qquad \int \frac{2y}{x}dx + \int -y\, dy = c$$

$$\Rightarrow \qquad 2y\log x - \frac{1}{2}y^2 = c$$ **Ans.**

EXERCISE 11.8

Solve the following differential equations:

1. $(y \log y)\, dx + (x - \log y)\, dy = 0$ **Ans.** $2x \log y = c + (\log y)^2$
2. $\left(y + \frac{1}{3}y^3 + \frac{1}{2}x^2\right)dx + \frac{1}{4}(1+y^2)\,x\, dy = 0$ **Ans.** $\frac{yx^4}{4} + \frac{y^3x^4}{12} + \frac{x^6}{12} = c$
3. $(y - 2x^3)\, dx - x(1 - xy)\, dy = 0$ **Ans.** $-\frac{y}{x} - x^2 + \frac{y^2}{2} = c$
4. $(x \sec^2 y - x^2 \cos y)\, dy = (\tan y - 3x^4)\, dx$ **Ans.** $-\frac{1}{x}\tan y - x^3 + \sin y = c$
5. $(x - y^2)\, dx + 2xy\, dy = 0$ **Ans.** $y^2 = cx - x \log x$

Rule II. If $\dfrac{\frac{\partial N}{\partial x} - \frac{\partial M}{\partial y}}{M}$ is a function of y alone, say $f(y)$, then

$$\text{I.F.} = e^{\int f(y)dy}$$

Example 34. *Solve* $(y^4 + 2y)\, dx + (xy^3 + 2y^4 - 4x)\, dy = 0$

Solution. Here $M = y^4 + 2y; \qquad N = xy^3 + 2y^4 - 4x$ ...(1)

$$\therefore \qquad \frac{\partial M}{\partial y} = 4y^3 + 2; \qquad \frac{\partial N}{\partial x} = y^3 - 4$$

$$\therefore \qquad \frac{\frac{\partial N}{\partial x} - \frac{\partial M}{\partial y}}{M} = \frac{(y^3-4)-(4y^3+2)}{y^4+2y} = \frac{-3(y^3+2)}{y(y^3+2)} = -\frac{3}{y} = f(y)$$

$$\text{I.F.} = e^{\int f(y)dy} = e^{\int -\frac{3}{y}dy} = e^{-3\log y} = e^{\log y^{-3}} = y^{-3} = \frac{1}{y^3}$$

On multiplying the given equation (1) by $\frac{1}{y^3}$ we get the exact differential equation.

$$\left(y+\frac{2}{y^2}\right)dx+\left(x+2y-\frac{4x}{y^3}\right)dy=0$$

$$\int\left(y+\frac{2}{y^2}\right)dx+\int 2y\,dy=c \quad\Rightarrow\quad x\left(y+\frac{2}{y^2}\right)+y^2=c \qquad \textbf{Ans.}$$

EXERCISE 11.9

Solve the following differential equations:

1. $(3x^2y^4+2xy)\,dx+(2x^3y^3-x^2)\,dy=0$ **Ans.** $x^3y^2+\dfrac{x^2}{y}=c$

2. $(xy^3+y)\,dx+2(x^2y^2+x+y^4)\,dy=0$ **Ans.** $\dfrac{x^2y^4}{2}+xy^2+\dfrac{y^6}{3}=c$

3. $y(x^2y+e^x)dx-e^x dy=0$ **Ans.** $\dfrac{x^3}{3}+\dfrac{e^x}{y}=c$

4. $(2x^4y^4e^y+2xy^3+y)\,dx+(x^2y^4e^y-x^2y^2-3x)\,dy=0$ **Ans.** $x^2e^y+\dfrac{x^2}{y}+\dfrac{x}{y^3}=c$

Rule III. If M is of the form $M=y\,f_1(xy)$ and N is of the form $N=x\,f_2(xy)$

Then $\qquad \text{I.F.}=\dfrac{1}{M.x-N.y}$

Example 35. *Solve* $y\,(xy+2x^2y^2)\,dx+x\,(xy-x^2y^2)\,dy=0$

Solution. $\qquad y\,(xy+2x^2y^2)\,dx+x\,(xy-x^2y^2)\,dy=0 \qquad \ldots(1)$

Dividing (1) by xy, we get

$$y\,(1+2xy)\,dx+x\,(1-xy)\,dy=0 \qquad \ldots(2)$$

$$M=y\,f_1(xy),\quad N=x\,f_2(xy)$$

$$\text{I.F.}=\frac{1}{Mx-Ny}=\frac{1}{xy(1+2xy)-xy(1-xy)}=\frac{1}{3x^2y^2}$$

On multiplying (2) by $\dfrac{1}{3x^2y^2}$, we have an exact differential equation

$$\left(\frac{1}{3x^2y}+\frac{2}{3x}\right)dx+\left(\frac{1}{3xy^2}-\frac{1}{3y}\right)dy=0 \quad\Rightarrow\quad \int\left(\frac{1}{3x^2y}+\frac{2}{3x}\right)dx+\int-\frac{1}{3y}\,dy=c$$

$$\Rightarrow\quad -\frac{1}{3xy}+\frac{2}{3}\log x-\frac{1}{3}\log y=c \quad\Rightarrow\quad -\frac{1}{xy}+2\log x-\log y=b \qquad \textbf{Ans.}$$

EXERCISE 11.10

Solve the following differential equations

1. $(y-xy^2)\,dx-(x+x^2y)\,dy=0$ **Ans.** $\log\left(\dfrac{x}{y}\right)-xy=A$

2. $y\,(1+xy)\,dx+x(1-xy)\,dy=0$ **Ans.** $xy\log\left(\dfrac{y}{x}\right)=cxy-1$

3. $y\,(1+xy)\,dx+x\,(1+xy+x^2y^2)\,dy=0$ **Ans.** $\dfrac{1}{2x^2y^2}+\dfrac{1}{xy}-\log y=c$

4. $(xy \sin xy + \cos xy)\, y\, dx + (xy \sin xy - \cos xy)\, x\, dy = 0$ **Ans.** $y \cos xy = cx$

Rule IV. For of this type of $x^m y^n (ay\,dx + bx\,dy) + x^{m'} y^{n'} (a'\,y\,dx + b'\,x\,dy) = 0,$ the integrating factor is $x^h y^k$.

where $$\frac{m+h+1}{a} = \frac{n+k+1}{b}, \quad \text{and} \quad \frac{m'+h+1}{a'} = \frac{n'+k+1}{b'}$$

Example 36. *Solve* $(y^3 - 2x^2y)\, dx + (2xy^2 - x^3)\, dy = 0$

Solution. $$(y^3 - 2x^2y)\, dx + (2xy^2 - x^3)\, dy = 0$$

$$y^2\,(ydx + 2xdy) + x^2\,(-2ydx - xdy) = 0$$

Here $m = 0,\ h = 2,\ a = 1,\ b = 2, \quad m' = 2,\ n' = 0,\ a' = -2,\ b' = -1$

$$\frac{0+h+1}{1} = \frac{2+k+1}{2} \text{ and } \frac{2+h+1}{-2} = \frac{0+k+1}{-1}$$

$$\Rightarrow \quad 2h + 2 = 2 + k + 1 \text{ and } h + 3 = 2k + 2$$

$$\Rightarrow \quad 2h - k = 1 \text{ and } h - 2k = -1$$

On solving $h = k = 1$. Integrating Factor $= xy$

Multiplying the given equation by xy, we get

$$(xy^4 - 2x^3y^2)\, dx + (2x^2y^3 - x^4y)\, dy = 0$$

which is an exact differential equation.

$$\int (xy^4 - 2x^3y^2)dx = C \quad \Rightarrow \quad \frac{x^2y^4}{2} - \frac{2x^4y^2}{4} = C$$

$$\Rightarrow \quad x^2y^4 - x^4y^2 = C' \quad \Rightarrow \quad x^2y^2(y^2 - x^2) = C'$$ **Ans.**

Example 37. Solve $(3y - 2xy^3)\, dx + (4x - 3x^2y^2)\, dy = 0.$ *(U.P., II Semester, June 2007)*

Solution. $$(3y - 2xy^3)\, dx + (4x - 3x^2y^2)\, dy = 0$$

$$\Rightarrow \quad (3y\, dx + 4x\, dy) + xy^2(-2y\, dx - 3x\, dy) = 0 \quad ...(1)$$

Comparing the coefficients of (1) with

$x^m y^n (a\,y\,dx + b\,x\,dy) + x^{m'} y^{n'} (a'\,y\,dx + b'\,x\,dy) = 0,$ we get

$m = 0,\ n = 0,\ a = 3,\ b = 4$

$m' = 1,\ n' = 2,\ a' = -2,\ b' = -3$

To find the integrating factor $x^h y^k$

$$\frac{m+h+1}{a} = \frac{n+k+1}{b} \text{ and } \frac{m'+h+1}{a'} = \frac{n'+k+1}{b'}$$

$$\frac{0+h+1}{3} = \frac{0+k+1}{4} \text{ and } \frac{1+h+1}{-2} = \frac{2+k+1}{-3}$$

$$\Rightarrow \quad \frac{h+1}{3} = \frac{k+1}{4} \text{ and } \frac{h+2}{2} = \frac{k+3}{3} \Rightarrow 4h - 3k + 1 = 0 \quad ...(2)$$

and $$3h - 2k = 0 \quad \Rightarrow \quad h = \frac{2k}{3} \quad ...(3)$$

Putting the value of h from (3) in (2), we get

$$\frac{8k}{3} - 3k + 1 = 0 \quad \Rightarrow \quad -\frac{k}{3} + 1 = 0 \quad \Rightarrow \quad k = 3$$

Putting $k = 3$ in (2), we get $h = \frac{2k}{3} = \frac{2\times 3}{3} = 2$

$$\text{I.F.} = x^h y^k = x^2 y^3$$

On multiplying the given differential equation by x^2y^3, we get

$$x^2y^3\,(3y - 2xy^3)dx + x^2y^3(4x - 3x^2y^2)\,dy = 0$$

$$(3x^2y^4 - 2x^3y^6)\,dx + (4x^3y^3 - 3x^4y^5)\,dy = 0$$

This is the exact differential equation.

Its solution is $\int (3x^2y^4 - 2x^3y^6)dx = 0 \quad \Rightarrow \quad x^3y^4 - \frac{x^4}{2}y^6 = C$ **Ans.**

EXERCISE 11.11

Solve the following differential equations.

1. $(2y\,dx + 3x\,dy) + 2xy\,(3y\,dx + 4x\,dy) = 0$ **Ans.** $x^2y^3\,(1 + 2xy) = c$

2. $(y^2 + 2yx^2)\,dx + (2x^3 - xy)\,dy = 0$ **Ans.** $4(xy)^{1/2} - \frac{2}{3}\left(\frac{y}{x}\right)^{3/2} = c$

3. $(3x + 2y^2)y\,dx + 2x\,(2x + 3y^2)\,dy = 0$ **Ans.** $x^2y^4\,(x + y^2) = c$

4. $(2x^2y^2 + y)\,dx - (x^3y - 3x)\,dy = 0$ **Ans.** $\frac{7}{5}x^{10/7}y^{-5/7} - \frac{7}{4}x^{-4/7}y^{-12/7} = c$

5. $x\,(3y\,dx + 2x\,dy) + 8y^4\,(y\,dx + 3x\,dy) = 0$ **Ans.** $x^3y^2 + 4x^2y^6 = c$

Rule V.

If the given equation $M\,dx + N\,dy = 0$ is homogeneous equation and $Mx + Ny \neq 0$, then $\frac{1}{Mx + Ny}$ is an integrating factor.

Example 38. *Solve* $\frac{dy}{dx} = \frac{x^3 + y^3}{xy^2}$

Solution. $(x^3 + y^3)\,dx - (xy^2)\,dy = 0$ ... (1)

Here $M = x^3 + y^3, \qquad N = -xy^2$

$$\text{I.F.} = \frac{1}{Mx + Ny} = \frac{1}{x(x^3 + y^3) - xy^2(y)} = \frac{1}{x^4}$$

Multiplying (1) by $\frac{1}{x^4}$ we get $\frac{1}{x^4}(x^3 + y^3)dx + \frac{1}{x^4}(-xy^2)dy = 0$

$\Rightarrow \left(\frac{1}{x} + \frac{y^3}{x^4}\right)dx - \frac{y^2}{x^3}dy = 0$, which is an exact differential equation.

$$\int\left(\frac{1}{x} + \frac{y^3}{x^4}\right)dx = c \quad \Rightarrow \quad \log x - \frac{y^3}{3x^3} = c$$ **Ans.**

EXERCISE 11.12

Solve the following differential equations:

1. $x^2y\,dx - (x^3 + y^3)\,dy = 0$ **Ans.** $-\frac{x^3}{3y^3} + \log y = c$

2. $(y^3 - 3xy^2)\,dx + (2x^2y - xy^2)\,dy = 0$ **Ans.** $\frac{y}{x} + 3\log x - 2\log y = c$

3. $(x^2y - 2xy^2)\,dx - (x^3 - 3x^2y)\,dy = 0$ **Ans.** $\frac{x}{y} - 2\log x + 3\log y = c$

4. $(y^3 - 2yx^2)\,dx + (2xy^2 - x^3)\,dy = 0$ **Ans.** $x^2y^4 - x^4y^2 = c$

11.14 DIFFERENTIAL EQUATIONS REDUCIBLE TO EXACT FORM (BY INSPECTION)

The following differentials, which commonly occur, help in selecting the suitable integrating factor.

(*i*) $y\,dx + x\,dy = d\,[xy]$

(*ii*) $\frac{x\,dy - y\,dx}{x^2} = d\left[\frac{y}{x}\right]$

(*iii*) $\frac{y\,dx - x\,dy}{y^2} = d\left[\frac{x}{y}\right]$

(*iv*) $\frac{x\,dy - y\,dx}{x^2 + y^2} = d\left[\tan^{-1}\left(\frac{y}{x}\right)\right]$

(*v*) $\frac{x\,dy - y\,dx}{xy} = d\left[\log\left(\frac{y}{x}\right)\right]$

(*vi*) $\frac{x\,dx - y\,dy}{x^2 + y^2} = d\left[\frac{1}{2}\log\left(x^2 + y^2\right)\right]$

(*vii*) $\frac{x\,dy - y\,dx}{x^2 - y^2} = d\left[\frac{1}{2}\log\frac{x+y}{x-y}\right]$

(*viii*) $\frac{x\,dy + y\,dx}{x^2y^2} = d\left[-\frac{1}{xy}\right]$

11.15 EQUATIONS OF FIRST ORDER AND HIGHER DEGREE

The differential equations will involve $\frac{dy}{dx}$ in higher degree and $\frac{dy}{dx}$ will be denoted by p. The differential equation will be of the form $f(x, y, p) = 0$.

Case 1. Equations solvable for p.

Example 39. *Solve* : $x^2 = 1 + p^2$

Solution. $x^2 = 1 + p^2 \Rightarrow p^2 = x^2 - 1$

$$\Rightarrow \quad p = \pm\sqrt{x^2 - 1} \quad \Rightarrow \quad \frac{dy}{dx} = \pm\sqrt{x^2 - 1} \quad \Rightarrow \quad dy = \pm\sqrt{x^2 - 1}\,dx$$

which gives on integration $y = \pm\frac{x}{2}\sqrt{x^2 - 1} \mp \frac{1}{2}\log\left(x + \sqrt{x^2 - 1}\right) + c$ **Ans.**

Case II. Equations solvable for y.

(*i*) Differentiate the given equation w.r.t. "x".

(*ii*) Eliminate p from the given equation and the equation obtained as above.

(*iii*) The eliminant is the required solution.

Example 40. *Solve*: $y = (x - a)\,p - p^2$.

Solution. $y = (x - a)\,p - p^2$ ... (1)

Differentiating (1) w.r.t. "x" we obtain

$$\frac{dy}{dx} = p + (x - a)\frac{dp}{dx} - 2p\frac{dp}{dx}$$

$$p = p + (x - a)\frac{dp}{dx} - 2p\frac{dp}{dx}$$

$$\Rightarrow \quad 0 = (x - a)\frac{dp}{dx} - 2p\frac{dp}{dx}$$

$$\Rightarrow \quad 0 = \frac{dp}{dx}[x - a - 2p] \quad \Rightarrow \quad \frac{dp}{dx} = 0$$

On integration, we get $p = c$.

Putting the value of p in (1), we get

$$y = (x - a)\, c - c^2$$ **Ans.**

Case III. Equations solvable for x

(*i*) Differentiate the given equation w.r.t. "y".
(*ii*) Solve the equation obtained as in (1) for p.
(*iii*) Eliminate p, by putting the value of p in the given equation.
(*iv*) The eliminant is the required solution.

Example 41. *Solve:* $y = 2px + yp^2$

Solution. $$y = 2px + yp^2 \quad \ldots (1)$$

$$\Rightarrow \quad 2px = y - yp^2 \quad \Rightarrow \quad 2x = \frac{y}{p} - yp \quad \ldots (2)$$

Differentiating (2) w.r.t. "y" we get

$$2\frac{dx}{dy} = \frac{1}{p} - \frac{y}{p^2}\frac{dp}{dy} - p - y\frac{dp}{dy}$$

$$\Rightarrow \quad \frac{2}{p} = \frac{1}{p} - \frac{y}{p^2}\frac{dp}{dy} - p - y\frac{dp}{dy} \quad \Rightarrow \quad \frac{1}{p} + p = -\frac{y}{p^2}\frac{dp}{dy} - y\frac{dp}{dy}$$

$$\Rightarrow \quad \frac{1}{p} + p = -y\left(\frac{1}{p^2} + 1\right)\frac{dp}{dy} \quad \Rightarrow \quad \frac{1+p^2}{p} = -y\frac{1+p^2}{p^2}\frac{dp}{dy}$$

$$\Rightarrow \quad 1 = -\frac{y}{p}\frac{dp}{dy} \Rightarrow -\frac{dy}{y} = \frac{dp}{p} \quad \Rightarrow \quad -\log y = \log p + \log c'$$

$$\Rightarrow \quad \log p\,y = \log c \Rightarrow \quad p\,y = c \quad \Rightarrow \quad p = \frac{c}{y}$$

Putting the value of p in (1), we get

$$y = 2\left(\frac{c}{y}\right)x + y\left(\frac{c^2}{y^2}\right) \quad \Rightarrow \quad y^2 = 2\,cx + c^2$$

$$\Rightarrow \quad y^2 = c(2x + c)$$ **Ans.**

Class IV. Clairaut's Equation.

The equation $y = px + f(p)$ is known as Clairaut's equation. ... (1)

Differentiating (1) w.r.t. "x", we get

$$\frac{dy}{dx} = p + x\frac{dp}{dx} + f'(p)\frac{dp}{dx}$$

$$\Rightarrow \quad p = p + x\frac{dp}{dx} + f'(p)\frac{dp}{dx} \quad \Rightarrow \quad 0 = x\frac{dp}{dx} + f'(p)\frac{dp}{dx}$$

$$\Rightarrow \quad [x + f'(p)]\frac{dp}{dx} = 0 \quad \Rightarrow \quad \frac{dp}{dx} = 0 \quad \Rightarrow \quad p = a \quad \text{(constant)}$$

Putting the value of p in (1), we have

$$y = ax + f(a)$$

which is the required solution.

Method. In the Clairaut's equation, on replacing p by a (constant), we get the solution of the equation.

Example 42. *Solve* : $p = \log(p\,x - y)$

Solution. $p = \log(p\,x - y)$ or $e^p = p\,x - y$ or $y = p\,x - e^p$

Which is Clairaut's equation.

Hence its solution is $y = a\,x - e^a$ **Ans.**

EXERCISE 11.13

Solve the following differential equations.

1. $xp^2 + x = 2yp$ **Ans.** $2cy = c^2x^2 + 1$
2. $x(1 + p^2) = 1$ **Ans.** $y - c = \sqrt{(x - x^2)} - \tan^{-1}\sqrt{\frac{1-x}{x}}$
3. $x^2p^2 + xyp - 6y^2 = 0$ **Ans.** $y = \frac{c}{x^3}, y = c_1x^2$
4. $\frac{dy}{dx} - \frac{dx}{dy} = \frac{x}{y} - \frac{y}{x}$ **Ans.** $xy = c,\ x^2 - y^2 = c$
5. $y = px + p^3$ **Ans.** $y = ax + a^3$
6. $x^2(y - px) = yp^2$ **Ans.** $y^2 = cx^2 + c^2$

11.16 ORTHOGONAL TRAJECTORIES

Two families of curves are such that every curve of either family cuts each curve of the other family at right angles. They are called orthogonal trajectories of each other.

Orthogonal trajectories are very useful in engineering problems.

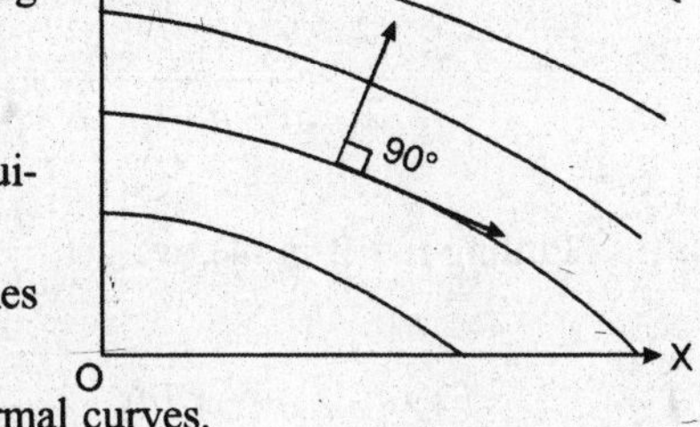

For example:

(*i*) The path of an electric field is perpendicular to equipotential curves.

(*ii*) In fluid flow, the stream lines and equipotential lines are orthogonal trajectories.

(*iii*) The lines of heat flow is perpendicular to isothermal curves.

Working rule to find orthogonal trajectories of curves

Step 1. By differenciating the equaton of curves find the differential equations in the form $f\left(x, y, \frac{dy}{dx}\right) = 0$

Step 2. Replace $\frac{dy}{dx}$ by $-\frac{dx}{dy}$ $(M_1 . M_2 = -1)$

Step 3. Solve the differential equation of the orthogonal trajectories *i.e.*, $f\left(x, y - \frac{dx}{dy}\right) = 0$

Self-orthogonal. A given family of curves is said to be 'self-orthogonal' if the family of orthogonal trajectory is the same as the given family of curves.

Example 43. Find the orthogonal trajectories of the family of curves $xy = c$.

Solution. Here, we have

$$xy = c \qquad \text{... (1)}$$

Differentiating (1), w.r.t., "x", we get

$$y + x\frac{dy}{dx} = 0 \quad \Rightarrow \quad \frac{dy}{dx} = -\frac{y}{x}$$

On replacing $\frac{dy}{dx}$ by $-\frac{dx}{dy}$, we get

$$\Rightarrow \quad -\frac{dx}{dy} = -\frac{y}{x} \quad \Rightarrow \quad \frac{dy}{dx} = \frac{x}{y}$$

$$y\,dy = x\,dx \qquad \text{... (2)}$$

Integrating (2), we get $\frac{y^2}{2} = \frac{x^2}{2} + c$

$\Rightarrow \quad y^2 - x^2 = 2c$ **Ans.**

Example 44. Find the orthogonal trajectories of $x^p + cy^p = 1$, p = constant.

Solution. Here we have $x^p + cy^p = 1$ ... (1)

Differentiating (1), we get $px^{p-1} + pcy^{p-1}\frac{dy}{dx} = 0 \Rightarrow \frac{dy}{dx} = -\frac{x^{p-1}}{cy^{p-1}}$ ... (2)

Putting the value of $\frac{1}{c}$ in (2), we get $\frac{dy}{dx} = -\frac{x^{p-1}}{y^{p-1}}\frac{y^p}{1-x^p} \Rightarrow \frac{dy}{dx} = -\frac{x^{p-1}y}{1-x^p}$ ... (3)

Replacing $\frac{dy}{dx}$ by $-\frac{dx}{dy}$ for orthogonal trajectory in (3), we get

$$-\frac{dx}{dy} = -\frac{x^{p-1}y}{1-x^p} \Rightarrow \frac{dy}{dx} = \frac{1-x^p}{x^{p-1}y} \quad \text{... (4)}$$

$$\Rightarrow \quad y\,dy = \frac{1-x^p}{x^{p-1}}dx \quad \Rightarrow \quad \int y\,dy = \int x^{1-p}dx - \int x\,dx$$

$$\frac{y^2}{2} = \frac{x^{1-p+1}}{1-p+1} - \frac{x^2}{2} + c \Rightarrow y^2 = \frac{2x^{2-p}}{2-p} - x^2 + 2c$$ **Ans.**

Putting $p = 2$ in (4), we get $\frac{dy}{dx} = \frac{1-x^2}{xy} \Rightarrow y\,dy = \frac{1-x^2}{x}dx$

$$y\,dy = \left(\frac{1}{x} - x\right)dx \quad \Rightarrow \quad \frac{y^2}{2} = \log x - \frac{x^2}{2} + \log c$$

$$\log x + \log c = \frac{x^2+y^2}{2} \quad \Rightarrow \quad 2\log x + 2\log c = x^2 + y^2$$

$$x^2c^2 = e^{x^2+y^2}$$

$$c_1x^2 = e^{x^2+y^2} \quad [c_1 = c^2]$$ **Ans.**

Example 45. *Show that the family of porabolas $y^2 = 2cx + c^2$ is "self-orthogonal."*

Solution. Here we have

$$y^2 = 2cx + c^2 \quad \text{... (1)}$$

Differentiating (1), we get $2y\frac{dy}{dx} = 2c \quad \Rightarrow \quad c = y\frac{dy}{dx}$

Putting the value of c in (1), we have $y^2 = 2\left(y\frac{dy}{dx}\right)x + \left(y\frac{dy}{dx}\right)^2$ ... (2)

Putting $\frac{dy}{dx} = p$ in (2), we get

$$y^2 = 2ypx + y^2p^2 \quad \text{... (3)}$$

This is differential equation of give n family of parabolas.

For orthogonal trajectories we put $-\frac{1}{p}$ for p in (3)

$$y^2 = 2y\left(-\frac{1}{p}\right)x + y^2\left(-\frac{1}{p}\right)^2 \quad \Rightarrow \quad y^2 = -\frac{2yx}{p} + \frac{y^2}{p^2}$$

$$\Rightarrow \qquad y^2p^2 = -2pyx + y^2$$

Rewriting, we get

$$\Rightarrow \qquad y^2 = 2ypx + y^2p^2$$

Which is same as equation (3). Thus (2) is D.E. for the given family and its orthogonal trajectories.

Hence, the given family is self-orthogonal. **Proved.**

Example 46. *Show that the system of confocal conics*

$$\frac{x^2}{a^2+\lambda}+\frac{y^2}{b^2+\lambda}=1$$

Where λ is a parameter, is self orthogonal.

Solution. Here we have $\dfrac{x^2}{a^2+\lambda}+\dfrac{y^2}{b^2+\lambda}=1$ ... (1)

Differentiating (1), we get

$$\frac{2x}{a^2+\lambda}+\frac{2y}{b^2+\lambda}\frac{dy}{dx}=0 \qquad \left(\text{Put } \frac{dy}{dx}=p\right)$$

$$\Rightarrow \qquad \frac{x}{a^2+\lambda}+\frac{yp}{b^2+\lambda}=0$$

$$\Rightarrow \qquad x(b^2+\lambda)+py(a^2+\lambda)=0 \quad \Rightarrow \quad \lambda(x+py) = -b^2x-a^2yp$$

$$\Rightarrow \qquad \lambda=\frac{-(b^2x+a^2yp)}{x+py}$$

Now $\quad a^2+\lambda = a^2-\dfrac{b^2x+a^2yp}{x+py}=\dfrac{a^2x+a^2py-b^2x-a^2yp}{x+py}=\dfrac{(a^2-b^2)x}{x+py}$

Again $\quad b^2+\lambda = b^2-\dfrac{b^2x+a^2yp}{x+py}=\dfrac{b^2x+b^2py-b^2x-a^2yp}{x+py}=\dfrac{-(a^2-b^2)yp}{x+py}$

Eliminating λ by putting the value of $a^2+\lambda$ and $b^2+\lambda$ in (1), we get

$$\frac{x^2(x+py)}{(a^2-b^2)x}+\frac{y^2(x+py)}{-(a^2-b^2)yp}=1 \quad \Rightarrow \quad \frac{x(x+py)}{(a^2-b^2)}-\frac{y(x+py)}{(a^2-b^2)p}=1$$

$$\frac{x+py}{a^2-b^2}\left[x-\frac{y}{p}\right]=1 \Rightarrow \frac{(x+py)\left(x-\dfrac{y}{p}\right)}{a^2-b^2}=1 \Rightarrow (x+py)\left(x-\frac{y}{p}\right)=a^2-b^2 \qquad \text{... (2)}$$

Equation (2) is the differential equation of (1),
To get the differential equation of orthogonal trajectory

Replace p by $-\dfrac{1}{p}$ in (2) $\left(x-\dfrac{1}{p}y\right)(x+py)=a^2-b^2$ ...(3)

Equation (3) is the same as eq. (2).

Thus the differential equation of the family of the orthogonal trajectory is the same as the differential equation of the family of the given curves.

Hence it is a self orthogonal family of curves. **Ans.**

EXERCISE 11.14

Find the orthogonal trajectories of the following family of curves:

1. $y^2 = cx^3$ **Ans.** $(x+1)^2 + y^2 = a^2$ 2. $x^2 - y^2 = cx$ **Ans.** $y(y^2 + 3x^2) = c$
3. $x^2 - y^2 = c$ **Ans.** $xy = c$
4. $(a + x)y^2 = x^2(3a - x)$ **Ans.** $(x^2 + y^2)^5 = cy^3(5x^2 + y^2)$
5. $y = ce^{-2x} + 3x$, passing through the point (0, 3)

   Ans. $9x - 3y + 5 = -4e^{6(3-y)}$
6. $16x^2 + y^2 = c$ **Ans.** $y^{16} = kx$
7. $y = \tan x + c$ **Ans.** $2x + 4y + \sin 2x = k$
8. $y = ax^2$ **Ans.** $x^2 + 2y^2 = c$
9. $x^2 + (y - c)^2 = c^2$ **Ans.** $x^2 + y^2 = cx$
10. $x^2 + y^2 + 2gx + 2fy + c = 0$ **Ans.** $x^2 + y^2 + 2fy - c = 0$
11. Family of parabolas through origin and focii on y-axis.

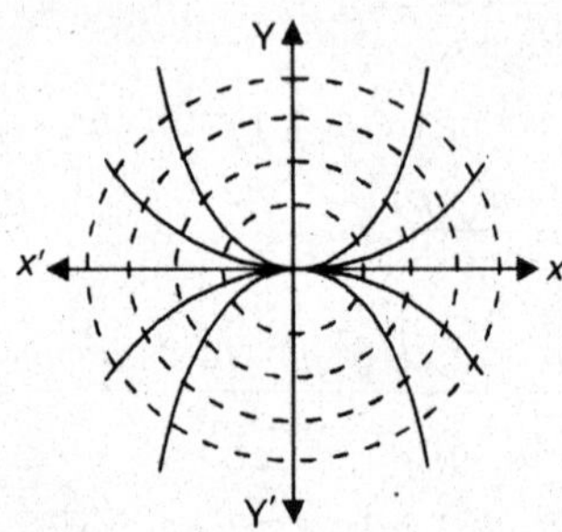

Ans. Ellipses with centre at the origin and focii on x-axis.

12. Show that the system of rectangular hyperbola $x^2 - y^2 = c^2$ and $xy = c^2$ are mutually orthogonal trajectories.
13. Show that the family of curves $y^2 = 4c(c + x)$ is self orthogonal.

11.17 POLAR EQUATION OF THE FAMILY OF CURVES

Let the polar equation of the family of curves be $f(r, \theta, c) = 0$ ... (1)

Working Rule

Step 1. On differentiating and eliminating the arbitrary constant c between (1) and $f'(r, \theta, c) = 0$ we get the differential equation of (1) *i.e.*,

$$F\left(r, \theta, \frac{dr}{d\theta}\right) = 0 \qquad \text{... (2)}$$

Step 2. Replace $\frac{dr}{d\theta}$ by $-r^2 \frac{d\theta}{dr}$ in (2). Here we will get the differential equation of orthogonal trajectory *i.e.*,

$$F\left(r, \theta - r^2 \frac{d\theta}{dr}\right) = 0 \qquad \text{... (3)}$$

Step 3. Integrating (3) to get the equation of the orthogonal trajectory.

Example 47. *Find the orthogonal trajectory of the cardioids $r = a(1 - \cos\theta)$.*

Solution. We have, $r = a(1 - \cos\theta)$ ... (1)

Differentiating (1) w.r.t. θ, we get $\frac{dr}{d\theta} = a\sin\theta$ ... (2)

Dividing (2) by (1) to eliminate a, we get

$$\frac{1}{r}\frac{dr}{d\theta} = \frac{\sin\theta}{1-\cos\theta} = \frac{2\sin\frac{\theta}{2}\cos\frac{\theta}{2}}{1-1+2\sin^2\frac{\theta}{2}} = \cot\frac{\theta}{2} \quad ... (3)$$

which is the differential equation of (1).

Replacing $\frac{dr}{d\theta}$ by $-r^2\frac{d\theta}{dr}$ in (3), we get $\frac{1}{r}\left(-r^2\frac{d\theta}{dr}\right) = \cot\frac{\theta}{2}$

$$r\frac{d\theta}{dr} = -\cot\frac{\theta}{2}$$

Separating the variables we get $\frac{dr}{r} = -\tan\frac{\theta}{2}\,d\theta$ ... (4)

Integrating (4), we get $\log r = 2\log\cos\frac{\theta}{2} + \log c \ = \log c\cos^2\frac{\theta}{2}$

$\Rightarrow \qquad r = c\cos^2\frac{\theta}{2} \qquad \Rightarrow \qquad r = \frac{c}{2}(1+\cos\theta)$

Which is the required trajectory. **Ans.**

Example 48. *Find the orthogonal trajectory the family of curves*

$$r^2 = c\sin 2\theta$$

Solution. We have

$$r^2 = c\sin 2\theta \quad ... (1)$$

Differentiating (1), we get $2r\frac{dr}{d\theta} = 2c\cos 2\theta$ ... (2)

Dividing (2) by (1), to eliminate 'c' we get $\frac{2}{r}\frac{dr}{d\theta} = 2\cot 2\theta$ ... (3)

Replacing $\frac{dr}{d\theta}$ by $-r^2\frac{d\theta}{dr}$ in (3), we have $\frac{2}{r}\left(-r^2\frac{d\theta}{dr}\right) = 2\cot 2\theta$

$$-2r\frac{d\theta}{dr} = 2\cot 2\theta \quad ... (4)$$

Separating the variables of (4), we obtain $\frac{dr}{r} = -\tan 2\theta\, d\theta$ ... (5)

Integrating (5), we get $\log r = \frac{1}{2}\log\cos 2\theta + \log c$

$$2\log r = \log c\cos\theta$$

$$r^2 = c\cos 2\theta$$

which is the required trajectory **Ans.**

Example 49. *Find the orthogonal trajectory of the family of curves*

$$r = c(\sec\theta + \tan\theta)$$

Solution. We have $r = c(\sec\theta + \tan\theta)$ ... (1)

Differentiating (1) w.r.t. 'θ' we get $\frac{dr}{d\theta} = c(\sec\theta\tan\theta + \sec^2\theta)$ ...(2)

$$\frac{dr}{d\theta} = c\sec\theta\,(\tan\theta + \sec\theta)$$

Dividing (2) by (1), we get $\frac{1}{r}\frac{dr}{d\theta} = \sec\theta$ ... (3)

Separating the variables of (3), we have $\frac{1}{r}\frac{dr}{d\theta} \equiv \sec\theta$ ... (4)

Replacing $\frac{dr}{d\theta}$ by $-r^2\frac{d\theta}{dr}$, we obtain

$$\frac{1}{r}\left(-r^2\frac{d\theta}{dr}\right) = \sec\theta \quad\Rightarrow\quad -r\frac{d\theta}{dr} = \sec\theta \qquad \text{... (5)}$$

Separating the variables of (5), we obtain $\frac{dr}{r} = -\cos\theta\, d\theta$ ... (6)

Integrating (6), we get $\log r = -\sin\theta + c \quad\Rightarrow\quad r = c'e^{-\sin\theta}$

which is the required orthogonal trajectory. **Ans.**

EXERCISE 11.15

Find the orthogonal trajectory of the following families of the curves:

1. $r = ce^{\theta}$ **Ans.** $r = ke^{-\theta}$
2. $r = c\theta^2$ **Ans.** $r = ke^{-\frac{\theta^2}{4}}$
3. $r = a(1+\cos\theta)$ **Ans.** $r = c(1-\cos\theta)$
4. $r^n \sin n\theta = a^n$ **Ans.** $r^n \cos n\theta = c^n$
5. $r = a\cos^2\theta$ **Ans.** $r^2 = c\sin\theta$
6. $r = 2a(\sin\theta + \cos\theta)$ **Ans.** $r = 2c(\sin\theta - \cos\theta)$
7. $r = c(1+\sin^2\theta)$ **Ans.** $r^2 = k\cos\theta.\cot\theta$
8. $r = \frac{a}{1+2\cos\theta}$ **Ans.** $r^2\sin^3\theta = (1+\cos\theta)$

LINEAR DIFFERENTIAL EQUATIONS OF SECOND ORDER

12.1 LINEAR DIFFERENTIAL EQUATIONS

If the degree of the dependent variable and all derivatives is one, such differential equations are called *linear differential equations e.g.*

(1) $\frac{d^2y}{dx^2}+5\frac{dy}{dx}+6y=x^2+x+1$ (2) $2\frac{d^2x}{dt^2}-\frac{dx}{dt}-3x=f(t)$

12.2 NON LINEAR DIFFERENTIAL EQUATIONS

If the degree of the dependent variable and / or its derivatives are of greater than 1 such differential equations are called one-linear differential equations.

(1) $\frac{d^2y}{dx^2}+\frac{dy}{dx}+y^2=\sin x$ (2) $\frac{d^2y}{dx^2}+2\left(\frac{dy}{dx}\right)^2+y^2=e^x$ (3) $\left(\frac{d^2x}{dt^2}\right)^2+\frac{dx}{dt}+x=f(t)$

The order of a differential equation is the highest order of the derivative involved. All the above differential equations are of second order.

Fourier and Laplace transforms are mathematical tools to solve the differential equations.

12.3 LINEAR DIFFERENTIAL EQUATIONS OF SECOND ORDER WITH CONSTANT COEFFICIENTS

The general form of the linear differential equation of second order is

$$\frac{d^2y}{dx^2}+P\frac{dy}{dx}+Qy=R$$

where P and Q are constants and R is a function of x or constant.

Differential operator. Symbol D stands for the operation of differential *i.e.*,

$$Dy=\frac{dy}{dx},\quad D^2y=\frac{d^2y}{dx^2}$$

$\frac{1}{D}$ stands for the operation of integration.

$\frac{1}{D^2}$ stands for the operation of integration twice.

$\frac{d^2y}{dx^2}+P\frac{dy}{dx}+Qy=R$ can be written in the operator form.

$$D^2y+P\,Dy+Qy=R \quad\Rightarrow\quad (D^2+PD+Q)\,y=R$$

12.4 COMPLETE SOLUTION = COMPLEMENTARY FUNCTION + PARTICULAR INTEGRAL

Let us consider a linear differential equation of the first order

$$\frac{dy}{dx}+Py=Q \qquad ...(1)$$

Its solution is $\quad ye^{\int Pdx}=\int(Q\,e^{\int Pdx})\,dx+C$

$$\Rightarrow \qquad y=Ce^{-\int Pdx}+e^{-\int Pdx}\int(Qe^{\int Pdx})\,dx$$

$$\Rightarrow \qquad y=cu+v\text{(say)} \qquad ...(2)$$

where $u=e^{-\int Pdx}$ and $v=e^{-\int Pdx}\int Q\,e^{\int Pdx}\,dx$

(*i*) Now differentiating $u=e^{-\int Pdx}$ w.r.t. x, we get $\frac{du}{dx}=-Pe^{-\int Pdx}=-Pu$

$$\Rightarrow \qquad \frac{du}{dx}+Pu=0 \qquad \Rightarrow \qquad \frac{d(cu)}{dx}+P(cu)=0$$

which shows that $y = c.u$ is the solution of $\frac{dy}{dx}+Py=0$

(*ii*) Differentiating $v=e^{-\int Pdx}\int(Qe^{\int Pdx}\,dx$ with respect to x, we get

$$\frac{dv}{dx}=-Pe^{\int Pdx}\int(Qe^{\int Pdx})\,dx+e^{-\int Pdx}Qe^{\int Pdx} \qquad \Rightarrow \qquad \frac{dv}{dx}=-Pv+Q$$

$\Rightarrow \qquad \frac{dv}{dx}+Pv=Q$ which shows that $y = v$ is the solution of $\boxed{\frac{dy}{dx}+Py=Q}$

Solution of the differential equation (1) is (2) consisting of two parts *i.e. cu* and *v*. *cu* is the solution of the differential equation whose R.H.S. is zero. *cu* is known as *complementary function.* Second part of (2) is *v* free from any arbitrary constant and is known as *particular integral.*

Complete Solution = Complementary Function + Particular Integral.

$$\Rightarrow \qquad \boxed{y=C.F.+P.I.}$$

12.5 METHOD FOR FINDING THE COMPLEMENTARY FUNCTION

(1) In finding the complementary function, R.H.S. of the given equation is replaced by zero.

(2) Let $y = C_1\, e^{mx}$ be the C.F. of

$$\frac{d^2y}{dx^2}+P\frac{dy}{dx}+Qy=0 \qquad ...(1)$$

Putting the values of y, $\frac{dy}{dx}$ and $\frac{d^2y}{dx^2}$ in (1) then $C_1e^{mx}\,(m^2 + Pm + Q) = 0$

$\Rightarrow \qquad m^2 + Pm + Q = 0$. It is called **Auxiliary equation.**

(3) Solve the auxiliary equation :

Case I : Roots, Real and Different. If m_1 and m_2 are the roots, then the C.F. is

$$y=C_1e^{m_1x}+C_2e^{m_2x}$$

Case II : Roots, Real and Equal. If both the roots are m_1, m_1 then the C.F. is

$$y=(C_1+C_2x)\,e^{m_1x}$$

Proof. Equation (1) can be written as

$$(D - m_1)(D - m_1)y = 0 \qquad \text{... (2)}$$

Replacing $(D - m_1)y = v$ in (2), we get

$$(D - m_1)v = 0 \qquad \text{... (3)}$$

$$\frac{dv}{dx} - m_1 v = 0 \quad \Rightarrow \quad \frac{dv}{v} = m_1 dx \quad \Rightarrow \quad \log v = m_1 x + \log c_2 \quad \Rightarrow \quad v = c_2 e^{m_1 x}$$

$$v = c_2 e^{m_1 x}$$

From (3) $\quad (D-1)y = c_2 e^{m_1 x}$

This is the linear differential equation.

$$\text{I.F.} = e^{-m_1 \int dx} = e^{-m_1 x}$$

Solution is

$$y.e^{-m_1 x} = \int (c_2 e^{m_1 x})(e^{-m_1 x})\, dx + c_1 = \int c_2\, dx + c_1 = c_2 x + c_1$$

$$y = (c_2 x + c_1) e^{m_1 x}$$

$$\text{C.F.} = (c_1 + c_2 x)\, e^{m_1 x}$$

Example 1. *Solve:* $\dfrac{d^2 y}{dx^2} - 8\dfrac{dy}{dx} + 15y = 0.$

Solution. Given equation can be written as

$$(D^2 - 8D + 15)\, y = 0$$

Here auxiliary equation is $m^2 - 8m + 15 = 0$

$\Rightarrow \quad (m - 3)(m - 5) = 0 \qquad \therefore\ m = 3, 5$

Hence, the required solution is

$$y = C_1\, e^{3x} + C_2 e^{5x}$$ **Ans.**

Example 2. *Solve:* $\dfrac{d^2 y}{dx^2} - 6\dfrac{dy}{dx} + 9y = 0$

Solution. Given equation can be written as

$$(D^2 - 6D + 9)\, y = 0$$

A.E. is $m^2 - 6m + 9 = 0 \quad \Rightarrow \quad (m - 3)^2 = 0 \quad \Rightarrow \quad m = 3, 3$

Hence, the required solution is

$$y = (C_1 + C_2 x)\, e^{3x}$$ **Ans.**

Example 3. *Solve:* $\dfrac{d^2 y}{dx^2} + 4\dfrac{dy}{dx} + 5y = 0,$

$$y = 2 \text{ and } \frac{dy}{dx} = \frac{d^2 y}{dx^2} \text{ when } x = 0.$$

Solution. Here the auxiliary equation is

$$m^2 + 4m + 5 = 0$$

Its root are $-2 \pm i$

The complementary function is

$$y = e^{-2x}\,(A \cos x + B \sin x) \qquad \text{...(1)}$$

On putting $y = 2$ and $x = 0$ in (1), we get

$$2 = A$$

On putting $A = 2$ in (1), we have

$$y = e^{-2x}\,[2\cos x + B\sin x] \qquad ...(2)$$

On differentiating (2), we get

$$\frac{dy}{dx} = e^{-2x}[-2\sin x + B\cos x] - 2e^{-2x}[2\cos x + B\sin x]$$

$$= e^{-2x}\,[(-2B-2)\sin x + (B-4)\cos x]$$

$$\frac{d^2y}{dx^2} = e^{-2x}[(-2B-2)\cos x - (B-4)\sin x]$$

$$-2e^{-2x}\,[(-2B-2)\sin x + (B-4)\cos x]$$

$$= e^{-2x}\,[(-4B+6)\cos x + (3B+8)\sin x]$$

But $\qquad \dfrac{dy}{dx} = \dfrac{d^2y}{dx^2}$

$$e^{-2x}\,[(-2B-2)\sin x + (B-4)\cos x] = e^{-2x}\,[(-4B+6)\cos x + (3B+8)\sin x]$$

On putting $x = 0$, we get

$$B - 4 = -4B + 6 \qquad \Rightarrow \qquad B = 2$$

(2) becomes, $\qquad y = e^{-2x}\,[2\cos x + 2\sin x]$

$$y = 2e^{-2x}\,[\sin x + \cos x] \qquad \textbf{Ans.}$$

Example 4. *The general solution of the diffcrential equation*

$$\frac{d^5y}{dx^5} - \frac{d^3y}{dx^3} = 0 \text{ is given by} \qquad (U.P.\ II\ Semester,\ 2009)$$

Solution. Here, we have

$$\frac{d^5y}{dx^5} - \frac{d^3y}{dx^3} = 0$$

or $\qquad D^5y - D^3y = 0 \quad \Rightarrow \quad (D^5 - D^3)y = 0 \quad \Rightarrow \quad D^3(D^2 - 1)y = 0$

A.E. is $m^3(m^2 - 1) = 0 \quad \Rightarrow \quad m = 0, 0, 0, 1, -1$

Here the solution is

$$y = (C_1 + C_2x + C_3x^2) + C_3e^x + C_5e^{-x} \qquad \textbf{Ans.}$$

Case III: Roots Imaginary. If the roots are $\alpha \pm i\beta$, then the solution will be

$$y = C_1e^{(\alpha+i\beta)x} + C_2e^{(\alpha-i\beta)x} = e^{\alpha x}.[C_1e^{i\beta x} + C_2e^{-i\beta x}]$$

$$= e^{\alpha x}[C_1(\cos\beta x + i\sin\beta x) + C_2(\cos\beta x - i\sin\beta x)]$$

$$= e^{\alpha x}[(C_1 + C_2)\cos\beta x + i(C_1 - C_2)\sin\beta x]$$

$$= e^{\alpha x}[A\cos\beta x + B\sin\beta x]$$

EXERCISE 12.1

Solve the following equations :

1. $\dfrac{d^2y}{dx^2} - 3\dfrac{dy}{dx} + 2y = 0$ **Ans.** $y = C_1e^x + C_2e^{2x}$ 2. $\dfrac{d^2y}{dx^2} + \dfrac{dy}{dx} - 30y = 0$ **Ans.** $y = C_1e^{5x} + C_2e^{-6x}$

3. $\dfrac{d^2y}{dx^2} - 8\dfrac{dy}{dx} + 16y = 0$ **Ans.** $y = (C_1 + C_2x)\,e^{4x}$

4. $\frac{d^2y}{dx^2}+\mu^2 y=0$ **Ans.** $y=C_1\cos\mu x+C_2\sin\mu x$

5. $(D^2+2D+2)y=0,\ y(0)=0,\ y'(0)=1$ (*A.M.I.E.T.E., June 2006*) **Ans.** $y=e^{-x}\sin x$

6. $\frac{d^3y}{dx^3}-2\frac{d^2y}{dx^2}+4\frac{dy}{dx}-8y=0$ **Ans.** $y=C_1e^{2x}+C_2\cos 2x+C_3\sin 2x$

7. $\frac{d^4y}{dx^4}-32\frac{d^2y}{dx^2}+256=0$ (*A.M.I.E.T.E., Dec. 2004*) **Ans.** $y=(C_1+x)\cos 4x+(C^3+C^4x)\sin 4x$

8. $\frac{d^4y}{dx^4}-4\frac{d^3y}{dx^3}+8\frac{d^2y}{dx^2}-8\frac{dy}{dx}+4y=0$ **Ans.** $y=e^x[(C_1+C_2x)\cos x+(C_3+C_4x)\sin x]$

9. $\frac{d^4y}{dx^4}+\frac{d^2y}{dx^2}=0,\ y(0)=y'(0)=y''(0)=0,\ y'''(0)=1$ **Ans.** $y=x-\sin x$

10. The equation for the bending of a strut is EI $\frac{d^2y}{dx^2}+Py=0$

If $y=0$ when $x=0$, and $y=a$ when $x=\frac{1}{2}$, find y. **Ans.** $y=\dfrac{a\sin\sqrt{\frac{P}{EI}}\,x}{\sin\sqrt{\frac{P}{EI}}\,\frac{1}{2}}$

11. $\frac{d^3y}{dx^3}+6\frac{d^2y}{dx^2}+12\frac{dy}{dx}+8y=0,\ y(0)=0,$ and $y'(0)=0$ and $y''(0)=2$

(*A.M.I.E.T.E. Dec. 2008*) **Ans.** $y=x^2e^{-2x}$

12. $\frac{d^3y}{dx^3}+\frac{d^2y}{dx^2}+\frac{4dy}{dx}+4y=0,\ y(0)=0,\ y'(0)=0,\ y''(0)=-5,$ **Ans.** $y=-e^x+\cos 2x-\frac{1}{2}\sin 2x$

13. $(D^8+6D^6-32D^2)y=0$ (*A.M.I.E.T.E., Summer 2005*)

Ans. $y=C_1+C_2x+C_3e^{\sqrt{2}x}+C_4e^{-\sqrt{2}x}+C_5\cos 2x+C_6\sin 2x$

14. Show that non-trivial solutions of the boundary value problem $y^{(iv)}-w^4y=0,\ y(0)=0=y''(0),$ $y(L)=0,\ y''(L)=0$ are $y(x)=\sum_{n=1}^{\infty}D_n\sin\left(\frac{n\pi x}{L}\right)$ where D_n are constants.

(*A.M.I.E.T.E. Dec. 2005*)

15. Solve the initial value problem $y'''+6y''+11y'+6y=0,\ y(0)=0,\ y'(0)=1,\ y''(0)=-1.$

(*A.M.I.E.T.E., Dec. 2006*) **Ans.** $y=2e^{-x}-3e^{-2x}+e^{-3x}$.

16. Let y_1, y_2 be two linearly independent solutions of the differential equation $yy''-(y')^2=0$. Then, $c_1y_1+c_2y_2$, where c_1, c_2 are constants is a solution of this differential equation for
(*a*) $c_1=c_2=0$ only. (*b*) $c_1=0$ or $c_2=0$ (*c*) no value of c_1, c_2. (*d*) all real c_1, c_2

(*A.M.I.E.T.E., Dec. 2004*)

17. The solution of the differential equation $\frac{d^2y}{dx^2}+y=0$ satisfying the initial conditions $y(0)=1$, $y\left(\frac{\pi}{2}\right)=2$ is

(*a*) $y=2\cos(x)+\sin(x)$ (*b*) $y=\cos(x)+2\sin(x)$
(*c*) $y=\cos(x)+\sin(x)$ (*d*) $y=2\cos(x)+2\sin(x)$ (*AMIETE, Dec. 2009*) **Ans.** (*b*)

18. Find the complementary function of $(D-2)^2 = 8\,(e^{2x} + \sin 2x - x^2)$

(a) $(C_1 + C_2 e^{2x})\,x$ (b) $(C_1 + C_2 x)\,e^{2x}$

(c) $(C_1 x + C_2 x^2)\,e^{2x}$ (d) $(C_1 x + C_2 ex^2)\,.\,2^{2x}$ (*AMIETE, Dec. 2010*) **Ans.** (b)

19. Solution of $\dfrac{d^3y}{dx^3} + 2\dfrac{d^2y}{dx^2} + \dfrac{dy}{dx} = 0$ is

(a) $y = C_1 e^x + C_2 e^{2x}$ (b) $y = C_1 + (C_2 + C_3 x)\,e^{-x}$

(c) $y = (C_1 + C_2 x + C_3 x^2)\,e^{-x}$ (d) $y = C_1 + C_2 e^{-x}$ (*AMIETE, June 2009*) **Ans.** (b)

12.6 RULES TO FIND PARTICULAR INTEGRAL

(i) $\dfrac{1}{f(D)}e^{ax} = \dfrac{1}{f(a)}e^{ax}$

If $f(a) = 0$ then $\dfrac{1}{f(D)}\cdot e^{ax} = x\cdot\dfrac{1}{f'(a)}\cdot e^{ax}$

If $f'(a) = 0$ then $\dfrac{1}{f(D)}\cdot e^{ax} = x^2\dfrac{1}{f''(a)}\cdot e^{ax}$

(ii) $\dfrac{1}{f(D)}x^n = [f(D)]^{-1}x^n$ Expand $[f(D)]^{-1}$ and then operate.

(iii) $\dfrac{1}{f(D^2)}\sin ax = \dfrac{1}{f(-a^2)}\sin ax$ and $\dfrac{1}{f(D^2)}\cos ax = \dfrac{1}{f(-a^2)}\cos ax$

If $f(-a^2) = 0$ then $\dfrac{1}{f(D^2)}\sin ax = x\cdot\dfrac{1}{f'(-a^2)}\cdot\sin ax$

(iv) $\dfrac{1}{f(D)}e^{ax}\cdot\phi(x) = e^{ax}\cdot\dfrac{1}{f(D+a)}\phi(x)$

(v) $\dfrac{1}{D+a}\phi(x) = e^{-ax}\int e^{ax}\cdot\phi(x)\,dx$

12.7 $\boxed{\dfrac{1}{f(D)}e^{ax} = \dfrac{1}{f(a)}e^{ax}}$

We know that, $D.e^{ax} = a.e^{ax}$, $D^2e^{ax} = a^2.e^{ax}$,..............., $D^n e^{ax} = a^n e^{ax}$

Let $f(D)\,e^{ax} = (D^n + K_1 D^{n-1} + \ldots + K_n)\,e^{ax} = (a^n + K_1 a^{n-1} + \ldots + K_n)e^{ax} = f(a)\,e^{ax}$.

Operating both sides by $\dfrac{1}{f(D)}$

$$\frac{1}{f(D)}\cdot f(D)e^{ax} = \frac{1}{f(D)}\cdot f(a)e^{ax}$$

$$\Rightarrow \qquad e^{ax} = f(a)\frac{1}{f(D)}\cdot e^{ax} \qquad \Rightarrow \frac{1}{f(D)}e^{ax} = \frac{1}{f(a)}e^{ax}$$

If $f(a) = 0$, then the above rule fails.

Then $\dfrac{1}{f(D)}e^{ax} = x\cdot\dfrac{1}{f'(D)}e^{ax} = x\dfrac{1}{f'(a)}e^{ax} \;\Rightarrow\; \boxed{\dfrac{1}{f(D)}e^{ax} = x\,.\,\dfrac{1}{f'(a)}e^{ax}}$

If $f'(a) = 0$ then $\boxed{\dfrac{1}{f(D)}e^{ax} = x^2\dfrac{1}{f''(a)}e^{ax}}$

Example 5. *Solve the differential equation*

$$\frac{d^2x}{dt^2}+\frac{g}{t}x=\frac{g}{l}L$$

where g, l, L are constants subject to the conditions,

$$x = a,\ \frac{dx}{dt}=0 \ \text{at } t = 0.$$

Solution. We have, $\frac{d^2x}{dt^2}+\frac{g}{t}x=\frac{g}{l}L \quad \Rightarrow \quad \left(D^2+\frac{g}{l}\right)x=\frac{g}{l}L$

A.E. is $m^2+\frac{g}{l}=0 \quad \Rightarrow \quad m=\pm\, i\sqrt{\frac{g}{l}}$

$$\text{C.F.} = C_1\cos\sqrt{\frac{g}{l}}\,t+C_2\sin\sqrt{\frac{g}{l}}\,t$$

$$\text{P.I.} = \frac{1}{D^2+\frac{g}{l}}\cdot\frac{g}{l}L=\frac{g}{l}L\frac{1}{D^2+\frac{g}{l}}e^{0t}=\frac{g}{l}L\frac{1}{0+\frac{g}{l}}=L \qquad [D=0]$$

$\therefore$ General solution is = C.F. + P.I.

$$x=C_1\cos\left(\sqrt{\frac{g}{l}}\right)t+C_2\sin\left(\sqrt{\frac{g}{l}}\right)t+L \qquad \text{...(1)}$$

$$\frac{dx}{dt}=-C_1\sqrt{\frac{g}{l}}\sin\left(\sqrt{\frac{g}{l}}\right)t+C_2\sqrt{\frac{g}{l}}\cos\left(\sqrt{\frac{g}{l}}\right)t$$

Put $t = 0$ and $\frac{dx}{dt}=0$

$$0=C_2\sqrt{\frac{g}{l}} \qquad \therefore \quad C_2=0$$

(1) becomes $x=C_1\cos\sqrt{\frac{g}{l}}\,t+L$ ...(2)

Put $x = a$ and $t = 0$ in (2), we get

$$a=C_1+L \quad \text{or} \quad C_1=a-L$$

On putting the value of C_1 in (2), we get $x=(a-L)\cos\left(\sqrt{\frac{g}{l}}\right)t+L$ **Ans.**

Example 6. *Solve* : $\frac{d^2y}{dx^2}+6\frac{dy}{dx}+9y=5e^{3x}$

Solution. $(D^2 + 6D + 9)y = 5e^{3x}$

Auxiliary equation is $m^2 + 6m + 9 = 0 \quad \Rightarrow \quad (m + 3)^2 = 0 \quad \Rightarrow \quad m = -3, -3,$

$$\text{C.F.} = (C_1 + C_2x)\, e^{-3x}$$

$$\text{P.I.} = \frac{1}{D^2+6D+9}.5.e^{3x}=5\frac{e^{3x}}{(3)^2+6(3)+9}=\frac{5e^{3x}}{36}$$

The complete solution is $y=(C_1+C_2x)e^{-3x}+\frac{5e^{3x}}{36}$ **Ans.**

Example 7. *Solve :* $\frac{d^2y}{dx^2}-6\frac{dy}{dx}+9y=6e^{3x}+7e^{-2x}-\log 2$

Solution. $(D^2-6D+9)y=6e^{3x}+7e^{-2x}-\log 2$

A.E. is $(m^2-6m+9)=0 \Rightarrow (m-3)^2=0, \Rightarrow m=3,3$

$$\text{C.F.}=(C_1+C_2x)e^{3x}$$

$$\text{P.I.}=\frac{1}{D^2-6D+9}6e^{3x}+\frac{1}{D^2-6D+9}7e^{-2x}+\frac{1}{D^2-6D+9}(-\log 2)$$

$$=x\frac{1}{2D-6}6e^{3x}+\frac{1}{4+12+9}7e^{-2x}-\log 2\frac{1}{D^2-6D+9}e^{0x}$$

$$=x^2\frac{1}{2}\cdot 6\cdot e^{3x}+\frac{7}{25}e^{-2x}-\log 2\left(\frac{1}{9}\right)=3x^2e^{3x}+\frac{7}{25}e^{-2x}-\frac{1}{9}\log 2$$

Complete solution is $y=(C_1+C_2\,x)e^{3x}+3x^2e^{3x}+\frac{7}{25}e^{-2x}-\frac{1}{9}\log 2$ **Ans.**

EXERCISE 12.2

Solve the following differential equations:

1. $[D^2+5D+6]\,[y]=e^x$ **Ans.** $C_2\,e^{-2x}+C_2e^{-3x}+\frac{e^x}{12}$

2. $\frac{d^2y}{dx^2}-3\frac{dy}{dx}+2y=e^{3x}$ **Ans.** $C_1e^x+C_2e^{2x}+\frac{e^{3x}}{2}$ *(A.M.I.E.T.E. June 2010, 2007)*

3. $(D^3+2D^2-D-2)\,y=e^x$ **Ans.** $C_1e^x+C_2e^{-x}+C_3e^{-2x}+\frac{x}{6}e^x$

4. $\frac{d^2y}{dx^2}+2\frac{dy}{dx}+2y=\sinh x$ **Ans.** $e^{-x}[C_1\cos x+C_2\sin x]+\frac{e^x}{10}-\frac{e^{-x}}{2}$

5. $\frac{d^2y}{dx^2}+4\frac{dy}{dx}+5y=-2\cosh x$ **Ans.** $e^{-2x}(C_1\cos x+C_2\sin x)-\frac{1}{10}e^x-\frac{e^{-x}}{2}$

6. $(D^3-2D^2-5D+6)\,y=e^{3x}$ **Ans.** $C_1e^x+C_2e^{-2x}+C_3e^{3x}+\frac{x.e^{3x}}{10}$

7. $\frac{d^3y}{dx^3}-\frac{d^2y}{dx^2}+4\frac{dy}{dx}-4y=e^x$ **Ans.** $C_1e^x+C_2\cos 2x+C_3\sin 2x+\frac{x\,e^x}{5}$

8. $\frac{d^2y}{dx^2}-6\frac{dy}{dx}+9y=e^{3x}$ **Ans.** $(C_1+C_2x)e^{3x}+\frac{x^2}{2}e^{3x}$

9. $\frac{d^3y}{dx^3}+3\frac{d^2y}{dx^2}+3\frac{dy}{dx}+y=e^{-x}$ **Ans.** $(C_1+C_2x+C_3x^2)e^{-x}+\frac{x^3}{6}e^{-x}$

10. $\frac{d^2y}{dx^2}-\frac{dy}{dx}-6y=e^x\cosh 2x$ **Ans.** $C_1e^{3x}+C_2e^{-2x}+\frac{1}{10}xe^{3x}-\frac{1}{8}e^{-x}$

11. $(D-2)\,(D+1)^2\,y=e^{2x}+e^x$ **Ans.** $C_1e^{2x}+(C_2+C_3x)e^{-x}+\frac{x}{9}e^{2x}-\frac{e^x}{4}$

12. $(D-1)^3\,y=16\,e^{3x}$ **Ans.** $(C_1+C_2x+C_3\,x^2)\,e^x+2e^{3x}$

13. The particular integral (PI) of differential equation $[D^2+5D+6]\,y=e^x$ is

(*a*) $\frac{e^x+x}{12}$ (*b*) $\frac{e^x-x}{12}$ (*c*) $\frac{e^{-x}}{12}$ (*d*) $\frac{e^x}{12}$ *(AMIETE, June 2010)* **Ans.** (*d*)

12.8 $\boxed{\dfrac{1}{f(D)}x^n = [f(D)]^{-1}x^n.}$

Expand $[f(D)]^{-1}$ by the Binomial theorem in ascending powers of D as far as the result of operation on x^n is zero.

Example 8. *Solve the differential equation* $\dfrac{d^2y}{dx^2}+a^2y=\dfrac{a^2R}{p}(l-x)$

where a, R, p and l are constants subject to the conditions $y = 0$, $\dfrac{dy}{dx}=0$ *at* $x = 0$.

Solution. $\dfrac{d^2y}{dx^2}+a^2y=\dfrac{a^2}{p}R(l-x) \quad \Rightarrow \quad (D^2+a^2)y=\dfrac{a^2}{p}R(l-x)$

A.E. is $m^2 + a^2 = 0 \Rightarrow m=\pm ia$

C.F. $= C_1 \cos ax + C_2 \sin ax$

$$\text{P.I.} = \frac{1}{D^2+a^2}\frac{a^2}{p}R(l-x) = \frac{a^2R}{p}\frac{1}{a^2}\left[\frac{1}{1+\frac{D^2}{a^2}}\right](l-x)=\frac{R}{p}\left[1+\frac{D^2}{a^2}\right]^{-1}(l-x)$$

$$= \frac{R}{p}\left[1-\frac{D^2}{a^2}\right](l-x)=\frac{R}{p}(l-x)$$

$$y = C_1\cos ax+C_2\sin ax+\frac{R}{p}(l-x) \qquad \text{...(1)}$$

On putting $y = 0$, and $x = 0$ in (1), we get $0 = C_1+\dfrac{R}{p}l \Rightarrow \quad C_1=-\dfrac{Rl}{p}$

On differentiating (1), we get $\dfrac{dy}{dx}=-a\,C_1\sin ax+a\,C_2\cos ax-\dfrac{R}{p}$ ...(2)

On putting $\dfrac{dy}{dx}=0$ and $x = 0$ in (2), we have

$$0 = a\,C_2-\frac{R}{p} \quad \Rightarrow \quad C_2=\frac{R}{a.p}$$

On putting the values of C_1 and C_2 in (1), we get

$$y=-\frac{R}{p}l\cos a.x+\frac{R}{a.p}\sin a.x+\frac{R}{p}(l-x) \Rightarrow y=\frac{R}{p}\left[\frac{\sin ax}{a}-l\cos a.x+l-x\right] \qquad \textbf{Ans.}$$

EXERCISE 12.3

Solve the following equations :

1. $(D^2 + 5D + 4)\,y = 3 - 2x$ **Ans.** $C_1e^{-x}+C_2e^{-4x}+\frac{1}{8}(11-4x)$
2. $\dfrac{d^2y}{dx^2}+2\dfrac{dy}{dx}+y=x$ **Ans.** $(C_1 + C_2 x)\,e^{-x} + x - 2$
3. $(2D^2 + 3D + 4)\,y = x^2 - 2x$ **Ans.** $e^{-\frac{3}{4}x}[A\cos\frac{\sqrt{23}}{4}x+B\sin\frac{\sqrt{23}}{4}x]+\frac{1}{32}[8x^2-28x+13]$
4. $(D^2 - 4D + 3)\,y = x^3$ **Ans.** $C_1e^x+C_2e^{3x}+\frac{1}{27}(9x^3+36x^2+78x+80)$.

5. $5\frac{d^3y}{dx^3}-\frac{d^2y}{dx^2}-6\frac{dy}{dx}=1+x^2.$ **Ans.** $A+Be^{-2x}+Ce^{3x}-\frac{1}{36}\left(2x^3-x^2+\frac{25}{3}x\right)$

6. $\frac{d^4y}{dx^4}+4y=x^4$ **Ans.** $e^x(C_1\cos x+C_2\sin x)+e^{-x}(C_3\cos x+C_4\sin x)+\frac{1}{4}(x^4-6)$

7. $\frac{d^2y}{dx^2}+2p\frac{dy}{dx}+(p^2+q^2)y=e^{cx}+p.q\,x^2$

Ans. $e^{-px}[C_1\cos qx+C_2\sin qx]+\frac{e^{Cx}}{(p+C)^2+q^2}+\frac{pq}{p^2+q^2}\left[x^2-\frac{4px}{p^2+q^2}+\frac{6p^2-2q^2}{(p^2+q^2)^2}\right]$

8. $D^2(D^2+4)y=96x^2$ **Ans.** $C_1+C_2x+C_3\cos 2x+C_4\sin 2x+2x^2(x^2-3)$

12.9 $\boxed{\frac{1}{f(D^2)}\sin ax=\frac{\sin ax}{f(-a^2)}}$ $\boxed{\frac{1}{f(D^2)}\cdot\cos ax=\frac{\cos ax}{f(-a^2)}}$

$D(\sin ax)=a.\cos ax$, $D^2(\sin ax)=D(a\cos ax)=-a^2.\sin ax$

$D^4(\sin ax)=D^2.D^2(\sin ax)=D^2(-a^2\sin ax)=(-a^2)^2\sin ax$

$(D^2)^n\sin ax=(-a^2)^n\sin ax$

Hence, $f(D^2)\sin ax=f(-a^2)\sin ax$

$$\frac{1}{f(D^2)}\cdot f(D^2)\sin ax=\frac{1}{f(D^2)}.f(-a^2).\sin ax$$

$$\sin ax=f(-a^2)\frac{1}{f(D^2)}\sin ax \Rightarrow \frac{1}{f(D^2)}\cdot\sin ax=\frac{\sin ax}{f(-a^2)}$$

Similarly, $$\frac{1}{f(D^2)}\cos ax=\frac{\cos ax}{f(-a^2)}$$

If $f(-a^2)=0$ then above rule fails.

$$\frac{1}{f(D^2)}\sin ax=x\frac{\sin ax}{f'(-a^2)}$$

If $f'(-a^2)=0$ then, $\frac{1}{f(D^2)}\sin ax=x^2\frac{\sin ax}{f''(-a^2)}$

Example 9. *Solve* : $(D^2+4)y=\cos 2x$

(R.G.P.V., Bhopal June, 2008, A.M.I.E.T.E. Dec 2008)

Solution. $(D^2+4)y=\cos 2x$

Auxiliary equation is $m^2+4=0$

$m=\pm 2i$, C.F. $=A\cos 2x+B\sin 2x$

$$\text{P.I.}=\frac{1}{D^2+4}\cos 2x=x.\frac{1}{2D}\cos 2x=\frac{x}{2}\left(\frac{1}{2}\sin 2x\right)=\frac{x}{4}\sin 2x$$

Complete solution is $y=A\cos 2x+B\sin 2x+\frac{x}{4}\sin 2x$ **Ans.**

Example 10. *Solve* : $\frac{d^3y}{dx^3}-3\frac{d^2y}{dx^3}+4\frac{dy}{dx}-2y=e^x+\cos x$ *(U.P., II Semester, Summer* 2006, 2001)

Solution. Given $(D^3-3D^2+4D-2)y=e^x+\cos x$

A.E. is $m^3 - 3m^2 + 4m - 2 = 0$

$\Rightarrow$ $(m-1)(m^2 - 2m + 2) = 0$, *i.e.*, $m = 1,\ 1 \pm i$

$\therefore$ $$\text{C.F.} = C_1 e^x + e^x (C_2 \cos x + C_3 \sin x)$$

$$\text{P.I.} = \frac{1}{(D-1)(D^2-2D+2)} e^x + \frac{1}{D^3 - 3D^2 + 4D - 2} \cos x$$

$$= \frac{1}{(D-1)(1-2+2)} e^x + \frac{1}{(-1)D - 3(-1) + 4D - 2} \cos x$$

$$= \frac{1}{(D-1)} e^x + \frac{1}{3D+1} \cos x = x \frac{1}{1} e^x + \frac{3D-1}{9D^2-1} \cos x$$

$$= e^x . x + \frac{(-3\sin x - \cos x)}{-9-1} = e^x . x + \frac{1}{10}(3\sin x + \cos x)$$

Hence, complete solution is

$$y = C_1 e^x + e^x (C_2 \cos x + C_3 \sin x) + x\, e^x + \frac{1}{10}(3 \sin x + \cos x) \qquad \textbf{Ans.}$$

Example 11. *Solve* : $(D^4 - 3D^2 - 4)y = 5 \sin 2x - e^{-2x}$ (*Nagpur University, Summer* 2001)

Solution. Auxiliary equation is

$m^4 - 3m^2 - 4 = 0$

$(m^2 + 1)(m^2 - 4) = 0$

$\Rightarrow$ $m^2 + 1 = 0 \Rightarrow m^2 - 4 = 0 \Rightarrow m = \pm i \Rightarrow m = \pm 2$

$\therefore$ $$\text{C.F.} = C_1 e^{2x} + C_2 e^{-2x} + C_3 \cos x + C_4 \sin x$$

$$\text{P.I.} = \frac{1}{D^4 - 3D^2 - 4}(5 \sin 2x - e^{-2x}) = 5.\frac{1}{D^4 - 3D^2 - 4} \sin 2x - \frac{1}{D^4 - 3D^2 - 4} e^{-2x}$$

$$= 5 \frac{1}{(-2^2)^2 - 3(-2^2) - 4} \sin 2x - \frac{1}{16 - 12 - 4} e^{-2x} \qquad \text{(Rule fails)}$$

$$= \frac{5}{24} \sin 2x - x \frac{1}{4D^3 - 6D} e^{-2x}$$

$$= \frac{5}{24} \sin 2x - x \frac{1}{4(-2)^3 - 6(-2)} e^{-2x} = \frac{5}{24} \sin 2x + \frac{x\, e^{-2x}}{20} \qquad \textbf{Ans.}$$

Example 12. *Solve* : $(D^3 + 1)y = \cos^2\left(\frac{x}{2}\right) + e^{-x}$ (*Nagpur University, Summer* 2004)

Solution. $$(D^3 + 1)y = \cos^2\left(\frac{x}{2}\right) + e^{-x}$$

A.E. is $m^3 + 1 = 0$

$$(m + 1)(m^2 - m + 1) = 0 \qquad \Rightarrow \qquad m = -1$$

or $$m = \frac{-(-1) \pm \sqrt{1-4}}{2} = \frac{1 \pm i\sqrt{3}}{2} \qquad \Rightarrow \qquad m = \frac{1}{2} \pm i \frac{\sqrt{3}}{2}$$

$\therefore$ $$\text{C.F.} = C_1 e^{-x} + e^{\frac{x}{2}} \left[C_2 \cos \frac{\sqrt{3}}{2} x + C_3 \sin \frac{\sqrt{3}}{2} x \right]$$

$$\text{P.I.} = \frac{1}{D^3+1}\left[\cos^2\left(\frac{x}{2}\right) + e^{-x}\right] = \frac{1}{D^3+1} \cos^2\left(\frac{x}{2}\right) + \frac{1}{D^3+1} e^{-x} \qquad [\text{Put } D = -1]$$

$$= \frac{1}{D^3+1}\left(\frac{1+\cos x}{2}\right)+\frac{1}{3D^2+1}e^{-x}$$

$$= \frac{1}{2}\frac{1}{D^3+1}e^{0x}+\frac{1}{2}\frac{1}{D^3+1}\cos x+\frac{1}{3(-1)^2+1}e^{-x} = \frac{1}{2}+\frac{1}{2}\frac{1}{-D+1}\cos x+\frac{1}{4}e^{-x}$$

$$= \frac{1}{2}-\frac{1}{2}\frac{(D+1)\cos x}{(D-1)(D+1)}+\frac{1}{4}e^{-x} = \frac{1}{2}-\frac{1}{2}\frac{(-\sin x+\cos x)}{(D^2-1)}+\frac{1}{4}e^{-x}$$

$$= \frac{1}{2}+\frac{1}{2}\frac{\sin x}{(D^2-1)}-\frac{1}{2}\frac{1}{(D^2-1)}\cos x+\frac{1}{4}e^{-x}$$

Put $D^2 = -1$

$$= \frac{1}{2}+\frac{1}{2}\frac{\sin x}{(-1-1)}-\frac{1}{2}\frac{1}{(-1-1)}\cos x+\frac{1}{4}e^{-x} = \frac{1}{2}-\frac{\sin x}{4}+\frac{\cos x}{4}+\frac{1}{4}e^{-x}$$

$$\text{P.I.} = \frac{1}{2}+\frac{1}{4}(\cos x-\sin x+e^{-x})$$

Hence, the complete solution is

$$y = C_1 e^{-x}+e^{\frac{x}{2}}\left[C_2\cos\frac{\sqrt{3}}{2}x+C_3\sin\frac{\sqrt{3}}{2}x\right]+\frac{1}{2}+\frac{1}{4}(\cos x-\sin x+e^{-x})$$ **Ans.**

Example 13. *Solve the differential equation*

$$\frac{d^2y}{dx^2}-2\frac{dy}{dx}+2y = \sinh x+\sin\sqrt{2}\,x.$$ (*Nagpur University, Winter* 2001)

Solution. A.E. is $m^2 - 2m + 2 = 0$

$$\therefore \quad m = \frac{2\pm\sqrt{4-8}}{2} = \frac{2\pm 2i}{2} = 1\pm i$$

$$\text{C.F.} = e^x(C_1\cos x + C_2\sin x) \qquad \left(\sinh x = \frac{e^x-e^{-x}}{2}\right)$$

$$\text{P.I.} = \frac{1}{D^2-2D+2}\sinh x+\frac{1}{D^2-2D-2}\sin\sqrt{2}x$$

$$\text{P.I.} = \frac{1}{D^2-2D+2}\left(\frac{e^x-e^{-x}}{2}\right)+\frac{1}{D^2-2D+2}\sin\sqrt{2}x$$

$$= \frac{1}{2}\left[\frac{1}{1-2+2}e^x-\frac{1}{1+2+2}e^{-x}\right]+\frac{1}{-2-2D+2}\sin\sqrt{2}x$$

$$= \frac{1}{2}e^x-\frac{1}{10}e^{-x}+\frac{1}{2\sqrt{2}}\cos\sqrt{2}x \qquad \left(\frac{1}{D}\sin\sqrt{2}x = \int\sin\sqrt{2}\,x\,dx\right)$$

Hence the solution is,

$$y = e^x(C_1\cos x+C_2\sin x)+\frac{1}{2}e^x-\frac{1}{10}e^{-x}+\frac{1}{2\sqrt{2}}\cos\sqrt{2}x.$$ **Ans.**

EXERCISE 12.4

Solve the following differential equations :

1. $\frac{d^2y}{dx^2}+6y = \sin 4x$ **Ans.** $C_1\cos\sqrt{6}x+C_2\sin\sqrt{6}x-\frac{1}{10}\sin 4x$

2. $\frac{d^2x}{dt^2}+2\frac{dx}{dt}+3x=\sin t$ **Ans.** $e^{-t}[A\cos\sqrt{2}t+B\sin\sqrt{2}t]-\frac{1}{4}(\cos t-\sin t)$

3. $\frac{d^2x}{dt^2}+2\frac{dx}{dt}+5x=\sin 2t$, given that when $t = 0$, $x = 3$ and $\frac{dx}{dt}=0$

Ans. $e^{-t}\left[\frac{55}{17}\cos 2t+\frac{53}{34}\sin 2t\right]-\frac{1}{17}(4\cos 2t-\sin 2t)$

4. $\frac{d^2y}{dx^2}-7\frac{dy}{dx}+6y=2\sin 3x$, given that $y = 1$, $\frac{dy}{dx}=0$ when $x = 0$.

Ans. $-\frac{13}{75}e^{6x}+\frac{27}{25}e^{x}+\frac{1}{75}(7\cos 3x-\sin 3x)$

5. $(D^3 + 1)\,y = 2\cos^2 x$

Ans. $C_1e^{-x}+e^{\frac{1}{2}x}\left(C_2\cos\frac{\sqrt{3}}{2}x+C_3\sin\frac{\sqrt{3}}{2}x\right)+1+\frac{1}{65}(-8\sin 2x+\cos 2x)$

6. $(D^2 + a^2)\,y = \sin ax$ *(A.M.I.E.T.E., June 2009)* (**Ans.** $C_1\cos ax+C_2\sin ax-\frac{x}{2a}\cos ax$

7. $(D^4 + 2a^2D^2 + a^4)\,y = 8\cos ax$ **Ans.** $(C_1+C_2x+C_3\cos ax+C_4\sin ax)-\frac{x^2}{a^2}\cos ax$

8. $\frac{d^2y}{dx^2}+3\frac{dy}{dx}+2y=\sin 2x$ *(A.M.I.E.T.E., Summer* 2002)

Ans. $C_1e^{-x}+C_2e^{-2x}-\frac{1}{20}(3\cos 2x+\sin 2x)$

9. $\frac{d^2y}{dx^2}+y=\sin 3x\cos 2x$ **Ans.** $C_1\cos x+C_2\sin x+\frac{1}{48}[-\sin 5x-12x\cos x]$

10. $\frac{d^2y}{dx^2}-2\frac{dy}{dx}-3y=2e^{2x}+10\sin 3x$ given that $y(0) = 2$ and $y'(0) = 4$

Ans. $\frac{29}{12}e^{3x}-\frac{1}{12}e^{-x}-\frac{2}{3}e^{2x}+\frac{1}{3}[\cos 3x-2\sin 3x]$

11. $\frac{d^2y}{dx^2}+3\frac{dy}{dx}+2y=4\cos^2 x$ *(R.G.P.V., Bhopal, I Semester, June 2007)*

Ans. $C_1e^{-x}+C_2e^{-2x}-e^{2x}+\frac{1}{10}(3\sin 2x-\cos 2x)+1$

12. $\frac{d^2y}{dx^2}-2\frac{dy}{dx}+3y=\cos x+x^2$

Ans. $e^{x}[C_1\cos\sqrt{2}\,x+C_2\sin\sqrt{2}\,x]+\frac{1}{4}(\cos x-\sin x)+\frac{1}{3}(x^2+\frac{4}{3}x+\frac{2}{9})$

13. $(D^3-3D^2+4D-2)y=e^x+\cos x$

Ans. $(C_1+C_2\cos x+C_3\sin x)e^x+\frac{1}{10}(3\sin x+\cos x)$

14. $(D^3 - 4D^2 + 13D)\,y = 1 + \cos 2x$

Ans. $C_1+e^{2x}(C_2\cos 3x+C_3\sin 3x)+\frac{1}{290}(9\sin 2x+8\cos 2x)+\frac{x}{13}$

15. $(D^2 - 4D + 4)\,y = e^{2x} + x^3 + \cos 2x$

Ans. $(C_1+C_2x)e^{2x}+\frac{1}{2}x^2e^{2x}+\frac{1}{8}(2x^3+6x^2+9x+6)-\frac{1}{8}\sin 2x$

16. $\frac{d^2y}{dx^2}+n^2y=h\sin px$ $(P\neq n)$

where h, p and n are constants satisfying the conditions

$y = a$, $\frac{dy}{dx}=b$ for $x = 0$ **Ans.** $a\cos nx+\left(\frac{b}{n}-\frac{ph}{n(n^2-p^2)}\right)\sin nx+\frac{h\sin px}{(n^2-p^2)}$

17. $y''+y'-2y=-6\sin 2x-18\cos 2x$, $y(0) = 2$, $y'(0) = 2$ **Ans.** $-e^{-2x}+3\cos 2x$

12.10 $\boxed{\dfrac{1}{f(D)}.e^{ax}\cdot\phi(x)=e^{ax}.\dfrac{1}{f(D+a)}.\phi(x)}$

$$D[e^{ax}\phi(x)]=e^{ax}D\phi(x)+ae^{ax}\phi(x)=e^{ax}(D+a)\phi(x)$$

$$D^2[e^{ax}\phi(x)]=D[e^{ax}(D+a)\phi(x)]=e^{ax}(D^2+aD)\phi(x)+ae^{ax}(D+a)\phi(x)$$

$$=e^{ax}(D^2+2a\,D+a^2)\phi(x)=e^{ax}(D+a)^2\,\phi(x)$$

Similarly, $\quad D^n[e^{ax}\phi(x)]=e^{ax}(D+a)^n\phi(x)$

$$f(D)[e^{ax}\,\phi(x)]=e^{ax}f(D+a)\phi(x)$$

$$e^{ax}\,\phi(x)=\frac{1}{f(D)}.[e^{ax}f(D+a)\phi(x)] \qquad ...(1)$$

Put $f(D+a)\phi(x)=X$, so that $\phi(x)=\dfrac{1}{f(D+a)}.X$

Substituting these values in (1), we get

$$e^{ax}\frac{1}{f(D+a)}X=\frac{1}{f(D)}[e^{ax}.X] \;\Rightarrow\; \frac{1}{f(D)}[e^{ax}.\phi(x)]=e^{ax}\frac{1}{f(D+a)}\phi(x)$$

Example 14. *Obtain the general solution of the differential equation*

$y''-2y'+2y=x+e^x\cos x.$ (*U.P. II Semester Summer*, 2002)

Solution. $\quad y''-2y'+2y=x+e^x\cos x$

A.E. is $\quad m^2-2m+2=0 \quad\Rightarrow m=1\pm i$

C.F. $=e^x\,(A\cos x+B\sin x)$

$$\text{P.I.}=\frac{1}{D^2-2D+2}x+\frac{1}{D^2-2D+2}e^x\cos x$$

where $\quad I_1=\dfrac{1}{D^2-2D+2}x=\dfrac{1}{2\left[1-D+\dfrac{D^2}{2}\right]}x=\dfrac{1}{2\left[1-\left(D-\dfrac{D^2}{2}\right)\right]}x$

$$=\frac{1}{2}\left[1-\left(D-\frac{D^2}{2}\right)\right]^{-1}x=\frac{1}{2}\left[1+\left(D-\frac{D^2}{2}\right)+\ldots\right]x=\frac{1}{2}\left[x+Dx-\frac{D^2}{2}x+\ldots\right]=\frac{1}{2}[x+1]$$

and, $\quad I_2=\dfrac{1}{D^2-2D+2}e^x\cos x=e^x\dfrac{1}{(D+1)^2-2(D+1)+2}\cos x=e^x\dfrac{1}{D^2+1}\cos x=e^x.x\dfrac{1}{2D}\cos x$

$$\left[\text{If } f(-a^2)=0,\text{ then } \frac{1}{f(D^2)}\cos ax=x\frac{1}{f'(-a^2)}\cos ax\right]$$

$$=\frac{1}{2}xe^x\sin x$$

y = C.F. + P.I.

$$=e^x(A\cos x+B\sin x)+\frac{1}{2}(x+1)+\frac{1}{2}xe^x\sin x. \qquad \textbf{Ans.}$$

Example 15. *Solve* : $(D^2 - 4D + 4)\, y = x^3 e^{2x}$

Solution. $(D^2 - 4D + 4)\, y = x^3 e^{2x}$

A.E. is $m^2 - 4m + 4 = 0 \quad \Rightarrow \quad (m-2)^2 = 0 \quad \Rightarrow \quad m = 2, 2$

C.F. $= (C_1 + C_2 x)\, e^{2x}$

$$\text{P.I.} = \frac{1}{D^2 - 4D + 4} x^3 \cdot e^{2x} = e^{2x} \frac{1}{(D+2)^2 - 4(D+2) + 4} x^3$$

$$= e^{2x} \frac{1}{D^2} x^3 = e^{2x} . \frac{1}{D}\left(\frac{x^4}{4}\right) = e^{2x} \cdot \frac{x^5}{20}$$

The complete solution is $y = (C_1 + C_2 x) e^{2x} + e^{2x} . \dfrac{x^5}{20}$ **Ans.**

Example 16. *Solve* :

$$\frac{d^4 y}{dx^4} - y = \cos x . \cosh x$$ (*Nagpur University, Summer* 2003)

Solution. We have, $(D^4 - 1)\, y = \cos x \cosh x$

A.E. is $m^4 - 1 = 0 \Rightarrow (m^2 - 1)(m^2 + 1) = 0 \quad \Rightarrow \quad m = \pm 1, \pm i$

C.F. $= C_1 e^x + C_2 e^{-x} + C_3 \cos x + C_4 \sin x$

$$\text{P.I.} = \frac{1}{D^4 - 1} \cos x \cosh x = \frac{1}{D^4 - 1} \cos x \left(\frac{e^x + e^{-x}}{2}\right)$$

$$= \frac{1}{2}\left[\frac{1}{D^4 - 1} e^x \cos x + \frac{1}{D^4 - 1} e^{-x} \cos x\right] = \frac{1}{2}\left[e^x \frac{1}{(D+1)^4 - 1} \cos x + e^{-x} \frac{1}{(D-1)^4 - 1} \cos x\right]$$

$$= \frac{1}{2}\left[e^x \frac{1}{D^4 + 4D^3 + 6D^2 + 4D} \cos x + e^{-x} \frac{1}{D^4 - 4D^3 + 6D^2 - 4D} \cos x\right]$$

$$= \frac{1}{2}\left[e^x \frac{1}{(-1)^2 + 4D(-1) + 6(-1) + 4D} \cos x + e^{-x} \frac{1}{(-1)^2 - 4D(-1) + 6(-1) - 4D} \cos x\right]$$

$$= \frac{1}{2}\left[e^x \frac{1}{-5} \cos x + e^{-x} \frac{1}{-5} \cos x\right] = -\frac{1}{5}\left(\frac{e^x + e^{-x}}{2}\right) \cos x = -\frac{1}{5} \cosh x \cos x$$

Hence, the complete solution is

$$y = C_1 e^x + C_2 e^{-x} + C_3 \cos x + C_4 \sin x - \frac{1}{5} \cos x \cosh x$$ **Ans.**

Example 17. *Solve the differential equation* :

$$\frac{d^3 y}{dx^3} - 7\frac{d^2 y}{dx^2} + 10\frac{dy}{dx} = e^{2x} \sin x$$ (*AMIETE, June 2010, Nagpur University, Summer* 2005)

Solution. $$\frac{d^3 y}{dx^3} - 7\frac{d^2 y}{dx^2} + 10\frac{dy}{dx} = e^{2x} \sin x$$

$\Rightarrow \quad D^3 y - 7D^2 y + 10\, Dy = e^{2x} \sin x$

A.E. is

$m^3 - 7m^2 + 10\, m = 0 \quad \Rightarrow \quad (m-2)(m^2 - 5m) = 0$

$\Rightarrow \quad m\,(m-2)(m-5) = 0 \quad \Rightarrow \quad m = 0, 2, 5$

C.F $= C_1 e^{0x} + C_2 e^{2x} + C_3 e^{5x}$

$$\text{P.I.} = \frac{1}{D^3 - 7D^2 + 10D} e^{2x} \sin x = e^{2x} \frac{1}{(D+2)^3 - 7(D+2)^2 + 10(D+2)} . \sin x$$

$$= e^{2x} \frac{1}{D^3+6D^2+12D+8-7D^2-28D-28+10D+20} . \sin x$$

$$= e^{2x} \frac{1}{D^3-D^2-6D} \sin x \quad = e^{2x} \frac{1}{(-1^2)\,D-(-1^2)-6D} \sin x$$

$$= e^{2x} \frac{1}{-D+1-6D} \sin x \ = e^{2x} \frac{1}{1-7D} \sin x \ = e^{2x} \frac{1+7D}{1-49D^2} \sin x \ = e^{2x} \frac{1+7D}{1-49(-1^2)} \sin x$$

$$= e^{2x} \frac{1+7D}{50} \sin x \ = \frac{e^{2x}}{50} (\sin x + 7\cos x)$$

Complete solution is

$$y = \text{C.F.} + \text{P.I.}$$

$$\Rightarrow \qquad y = C_1 + C_2\, e^{2x} + C_3\, e^{5x} + \frac{e^{2x}}{50} (\sin x + 7\cos x)$$ **Ans.**

Example 18. *A body executes damped forced vibrations given by the equation*

$$\frac{d^2x}{dt^2} + 2k\frac{dx}{dt} + b^2 x = e^{-kt} \sin \omega t.$$

Solve the differential equation for both the cases when $\omega^2 \neq b^2 - k^2$ *and when* $\omega^2 = b^2 - k^2$.

(*U.P., II Semester, Summer* 2002)

Solution. The given equation is $(D^2 + 2kD + b^2)\, x = e^{-kt} \sin \omega\, t,$ ...(1)

which is a linear differential equation with constant coefficients.

A.E. is $m^2 + 2km + b^2 = 0$ or $m = \dfrac{-2k \pm \sqrt{(4k^2 - 4b^2)}}{2} = -k \pm \sqrt{(k^2 - b^2)}$

As the given problem is on vibration, we must have $k^2 < b^2$

$$m = -k \pm \sqrt{-(b^2-k^2)} = -k \pm i\sqrt{(b^2-k^2)}$$

$$\text{C.F.} = e^{-kt}\left\{C_1 \cos\sqrt{(b^2-k^2)}\,t + C_2 \sin\sqrt{(b^2-k^2)}\,t\right\}$$

$$\text{P.I.} = \frac{1}{D^2+2kD+b^2} e^{-kt} \sin \omega t = e^{-kt} \frac{1}{(D-k)^2 + 2k(D-k) + b^2} \sin \omega t$$

$$= e^{-kt} \frac{1}{D^2 + (b^2-k^2)} \sin \omega t = e^{-kt} \frac{1}{-\omega^2 + (b^2-k^2)} \sin \omega t, \text{ if } \omega^2 \neq b^2 - k^2 \quad \text{... (2)}$$

If $\boldsymbol{\omega^2 = b^2 - k^2}$**, then** $\text{P.I.} = e^{-kt} . t \dfrac{1}{2D} \sin \omega t = e^{-kt}\left(-\dfrac{t}{2\omega} \cos \omega t\right),$ ...(3)

Case. I. If $\omega^2 \neq b^2 - k^2$, the complete solution of (1) is

$$x = e^{-kt}\left\{C_1 \cos\sqrt{(b^2-k^2)}\,t + C_2 \sin\sqrt{(b^2-k^2)}\,t\right\} + \frac{e^{-kt}}{(b^2-k^2)-\omega^2} \sin \omega t \quad \text{[From (2)]}$$

Case II. If $\omega^2 = b^2 - k^2$, the complete solution of (1) is

$$x = e^{-kt}\{C_1 \cos \omega t + C_2 \sin \omega t\} - \frac{e^{-kt}\, t \cos \omega t}{2\omega}$$ [From (3)] **Ans.**

Example 19. *Solve* $(D^2 + 6D + 9)y = \dfrac{e^{-3x}}{x^3}$.

(*Nagpur University, Summer 2002, A.M.I.E.T.E., June 2009*)

Solution A.E. is $m^2 + 6m + 9 = 0$

$(m + 3)^2 = 0 \qquad \therefore \qquad m = -3, -3$

C.F. = $(C_1 + C_2x)\, e^{-3x}$

$$\text{P.I.} = \frac{1}{D^2+6D+9}\frac{e^{-3x}}{x^3} = e^{-3x}\frac{1}{(D-3)^2+6(D-3)+9}\frac{1}{x^3}$$

$$= e^{-3x}\frac{1}{D^2-6D+9+6D-18+9}\frac{1}{x^3} \qquad = e^{-3x}\frac{1}{D^2}(x^{-3})$$

$$= e^{-3x}\frac{1}{D}\left(\frac{x^{-2}}{-2}\right) = e^{-3x}\frac{x^{-1}}{(-2)(-1)} = \frac{e^{-3x}x^{-1}}{2} = \frac{e^{-3x}}{2x}$$

Hence, the solution is $y = (C_1 + C_2x)e^{-3x} + \dfrac{e^{-3x}}{2x}$ **Ans.**

Example 20. *Solve* $(D^2 - 4D + 3)\, y = 2x\, e^{2x} + 3e^x \cos 2x$

Solution. The auxiliary equation is

$m^2 - 4\, m + 3 = 0$ which gives $m = 1, 3$

C.F. = $C_1e^x + C_2e^{3x}$

$$\text{P.I.} = \frac{1}{D^2-4D+3}2xe^{3x} + \frac{1}{D^2-4D+3}3e^x\cos 2x$$

$$= 2e^{3x}.\frac{1}{(D+3)^2-4(D+3)+3}x + 3e^x.\frac{1}{(D+1)^2-4(D+1)+3}\cos 2x$$

$$= 2e^{3x}.\frac{1}{D^2+2D}x + 3e^x.\frac{1}{D^2-2D}\cos 2x = 2e^{3x}.\frac{1}{2D(1+D/2)}x + 3e^x.\frac{1}{-4-2D}\cos 2x$$

$$= e^{3x}.\frac{1}{D}\left(1+\frac{D}{2}\right)^{-1}x - \frac{3e^x}{2}.\frac{1}{2+D}\cos 2x = e^{3x}.\frac{1}{D}\left(1-\frac{D}{2}+\frac{D^2}{4}\ldots\right)x - \frac{3e^x}{2}.\frac{2-D}{4-D^2}\cos 2x$$

$$= e^{3x}.\left(\frac{1}{D}-\frac{1}{2}+\frac{D}{4}\ldots\right)x - \frac{3e^x}{2}.\frac{2-D}{4+4}\cos 2x = e^{3x}.\left(\frac{x^2}{2}-\frac{x}{2}+\frac{1}{4}\right) - \frac{3e^x}{16}(2\cos 2x + 2\sin 2x)$$

The complete solution is

$$y = C_1e^x + C_2e^{3x} + e^{3x}\left(\frac{x^2}{2}-\frac{x}{2}\right) - \frac{3e^x}{8}(\cos 2x + \sin 2x)$$

The term $\dfrac{e^{3x}}{4}$ has been omitted from the P.I., since C_2e^{3x} is present in the C.F. **Ans.**

Example 21. *Find the complete solution of* $\dfrac{d^2y}{dx^2} - 3\dfrac{dy}{dx} + 2y = xe^{3x} + \sin 2x$

(*U.P. II Semester* 2003)

Solution. The auxiliary equation is

$$m^2 - 3m + 2 = 0 \quad \Rightarrow \quad m^2 - 2m - m + 2 = 0$$

$$\Rightarrow \quad (m-2)(m-1) = 0 \quad \Rightarrow \quad m = 1, 2$$

C.F. = $C_1e^x + C_2e^{2x}$

$$\text{P.I.} = \frac{1}{D^2-3D+2}(xe^{3x} + \sin 2x) = \frac{1}{D^2-3D+2}xe^{3x} + \frac{1}{D^2-3D+2}\sin 2x$$

$$= e^{3x}\frac{1}{(D+3)^2-3(D+3)+2}x + \frac{1}{-4-3D+2}\sin 2x$$

$$= e^{3x}\frac{1}{D^2+6D+9-3D-9+2}x + \frac{1}{-3D-2}\sin 2x = e^{3x}\frac{1}{D^2+3D+2}x - \frac{1}{3D+2}\sin 2x$$

$$= \frac{e^{3x}}{2}\left[1+\left(\frac{3D+D^2}{2}\right)\right]^{-1} x - \frac{(3D-2)}{9D^2-4}\sin 2x = \frac{e^{3x}}{2}\left[1-\left(\frac{3D+D^2}{2}\right)+\ldots\right]x - \frac{(3D-2)}{9(-4)-4}\sin 2x$$

$$= \frac{e^{3x}}{2}\left[x-\left(\frac{3D+D^2}{2}\right)x+\ldots\right] - \frac{(3D-2)}{-36-4}\sin 2x = \frac{e^{3x}}{2}\left[x-\frac{3}{2}\right]+\frac{3D-2}{40}\sin 2x$$

$$\Rightarrow \quad \text{P.I.} = \frac{e^{3x}}{4}(2x-3)+\frac{1}{40}(6\cos 2x - 2\sin 2x) = \frac{e^{3x}}{4}(2x-3)+\frac{3}{20}\cos 2x - \frac{1}{20}\sin 2x$$

The complete solution is

$$y = \text{C.F.} + \text{P.I.}$$

$$\Rightarrow \quad y = c_1 e^x + c_2 e^{2x} + \frac{e^{3x}}{4}(2x-3)+\frac{3}{20}\cos 2x - \frac{1}{20}\sin 2x \qquad \textbf{Ans.}$$

Example 22. *Solve* : $\frac{d^2y}{dx^2} - 2\frac{dy}{dx} + y = xe^x \sin x$

(*R.G.P.V., Bhopal, 2001, Nagpur University, Winter* 2002)

Solution. The given equation is $(D^2 - 2D + 1)\, y = xe^x \sin x$

A.E. is $m^2 - 2m + 1 = 0 \qquad \therefore\ m = 1, 1$

C.F. $= (C_1 + C_2 x)\, e^x$

$$\text{P.I.} = \frac{1}{(D-1)^2}e^x . x\sin x = e^x \frac{1}{(D+1-1)^2} x\sin x = e^x \frac{1}{D^2} x\sin x = e^x . \frac{1}{D}\int x\sin x\, dx$$

Integrating by parts

$$= e^x \frac{1}{D}[x(-\cos x) - \int(-\cos x)\, dx] = e^x . \frac{1}{D}(-x\cos x + \sin x)$$

$$= e^x \int(-x\cos x + \sin x)\, dx = e^x\left\{-x\sin x + \int 1.\sin x\, dx - \cos x\right\}$$

$$= e^x\,[-x\sin x - \cos x - \cos x] = -e^x(x\sin x + 2\cos x)$$

Hence, the complete solution is

$$y = (C_1 + C_2 x)\, e^x - e^x (x \sin x + 2\cos x). \qquad \textbf{Ans.}$$

Example 23. *Solve* $(D^2 + 5D + 6)\, y = e^{-2x}\sec^2 x\,(1 + 2\tan x)$ (*A.M.I.E.T.E., Summer 2003*)

Solution. $(D^2 + 5D + 6)\, y = e^{-2x}\sec^2 x\,(1 + 2\tan x)$

Auxiliary Equation is $m^2 + 5m + 6 = 0$

$$\Rightarrow (m+2)(m+3) = 0 \Rightarrow m = -2, \text{ and } m = -3$$

Hence, complementry function (C.F.) $= C_1 e^{-2x} + C_2 e^{-3x}$

$$\text{P.I.} = \frac{1}{D^2+5D+6} e - 2x\sec 2x(1+2\tan x) = e^{-2x}\frac{1}{(D-2)^2+5(D-2)+6}\sec^2 x(1+2\tan x)$$

$$= e^{-2x}\frac{1}{D^2-4D+4+5D-10+6}\sec^2 x(1+2\tan x)$$

$$= e^{-2x}\frac{1}{D^2+D}\sec^2 x(1+2\tan x) = e^{-2x}\left[\frac{\sec^2 x}{D^2+D}+\frac{2\tan x\sec^2 x}{D^2+D}\right]$$

$$= e^{-2x}\frac{1}{D(D+1)}\sec^2 + \frac{1}{D(D+1)}2\tan x\sec^2 x$$

$$= e^{-2x}\left[\left(\frac{1}{D}-\frac{1}{D+1}\right)\sec^2 x + \right]\left(\frac{1}{D}-\frac{1}{D+1}\right)2\tan x\sec^2 x$$

$$= e^{-2x}\left[\frac{1}{D}\sec^2 x - \frac{1}{D+1}\sec^2 x + \frac{1}{D}2\tan x\sec^2 x - \frac{1}{D+1}2\tan x\sec^2 x\right]$$

$$= e^{-2x}\left[\tan x - e^{-x}\int ex.\sec 2x\,dx + \tan^2 x - e^{-x}\int 2e^x\tan x\sec 2x\,dx\right]$$

Now, $= e^{-2x}\int e^x\sec^2 x\,dx = e^x\sec^2 x - \int e^x.2\sec x\sec x\tan x.dx$

$$= e^x\sec^2 x - 2\int ex\sec^2 x.\tan x\,dx$$

$$\therefore \text{ P.I.} = e^{-2x}\left[\tan x^{-e} - x.e^x\sec^2 x + 2e^{-x}\int e^x\sec x\tan x\,dx + \tan^2 x - 2e^{-x}\int e^x\sec^2 x\tan x\,dx\right]$$

$$= e^{-2x}[\tan x - \sec^2 x + \tan^2 x] = e^{-2x}[\tan x - (\sec^2 x - \tan 2x] = e^{-2x}(\tan x - 1)$$

$\therefore$ Complete solution is

$$\Rightarrow \ y = C.F. + P.I. = C_1e^{-2x} + C_2e^{-3x} + e^{-2x}(\tan x - 1)$$

Example 24. *Solve the differential equation* $(D^2 - 4D + 4)\,y = 8x^2\,e^{2x}\sin 2x$

(U.P. II Semester, Summer 2008, Uttrakhand 2007, 2005, 2004; Nagpur University June 2008)

Solution. $(D^2 - 4D + 4)\,y = 8x^2\,e^{2x}\sin 2x$

A.E. is $(m^2 - 4m + 4) = 0 \quad \Rightarrow \quad (m-2)^2 = 0 \quad \Rightarrow \quad m = 2, 2$

C.F. $= (C_1 + C_2x)\,e^{2x}$

$$\text{P.I.} = \frac{1}{D^2 - 4D + 4}8x^2e^{2x}\sin 2x = 8\frac{1}{(D-2)^2}x^2e^{2x}\sin 2x$$

$$= 8e^{2x}\frac{1}{(D-2+2)^2}x^2\sin 2x = 8e^{2x}\frac{1}{D^2}x^2\sin 2x$$

$$= 8e^{2x}\frac{1}{D}\left[x^2\frac{(-\cos 2x)}{2} - 2x\left(-\frac{\sin 2x}{4}\right) + 2\frac{\cos 2x}{8}\right] = 8e^{2x}\frac{1}{D}\left[-\frac{x^2}{2}\cos 2x + \frac{x\sin 2x}{2} + \frac{\cos 2x}{4}\right]$$

$$= 8e^{2x}\left[-\frac{x^2}{2}\left(\frac{\sin 2x}{2}\right) - \left(\frac{-2x}{2}\right)\left(-\frac{\cos 2x}{4}\right) + (-1)\left(-\frac{\sin 2x}{8}\right) + \frac{x}{2}\left(-\frac{\cos 2x}{2}\right) - \left(\frac{1}{2}\right)\left(-\frac{\sin 2x}{4}\right) + \frac{\sin 2x}{8}\right]$$

$$= e^{2x}[-2x^2\sin 2x - 2x\cos 2x + \sin 2x - 2x\cos 2x + \sin 2x + \sin 2x]$$

$$= e^{2x}[-2x^2\sin 2x - 4x\cos 2x + 3\sin 2x] = -e^{2x}[4x\cos 2x + (2x^2 - 3)\sin 2x]$$

Complete solution is, y = C.F. + P.I.

$$y = (C_1 + C_2x)\,e^{2x} - e^{2x}[4x\cos 2x + (2x^2 - 3)\sin 2x]$$ **Ans.**

EXERCISE 12.5

Solve the following equations :

1. $(D^2 - 5D + 6)\,y = e^x\sin x$ **Ans.** $y = C_1e^{2x} + C_2e^{3x} + \dfrac{e^x}{10}(3\cos x + \sin x)$

2. $\dfrac{d^2y}{dx^2} - 7\dfrac{dy}{dx} + 10y = e^{2x}\sin x$ **Ans.** $y = C_1e^{2x} + C_2e^{5x} + \dfrac{e^{2x}}{10}(3\cos x - \sin x)$

3. $\dfrac{d^3y}{dx^3} - 2\dfrac{dy}{dx} + 4y = e^x\cos x$ **Ans.** $y = C_1e^{-2x} + e^x(C_2\cos x + C_3\sin x) + \dfrac{xe^x}{20}(3\sin x - \cos x)$

4. $(D^2 - 4D + 3)\,y = 2xe^{3x} + 3e^{3x}\cos 2x$

   Ans. $y = C_1e^x + C_2e^{3x} + \dfrac{1}{2}e^{3x}(x^2 - x) + \dfrac{3}{8}e^{3x}(\sin 2x - \cos 2x)$

5. $\dfrac{d^2y}{dx^2}+2\dfrac{dy}{dx}+y=\dfrac{e^{-x}}{x^2}$ **Ans.** $y=(C_1+C_2x)\,e^{-x}-e^{-x}\log x$

6. $(D^2-4)\,y=x^2\,e^{3x}$ **Ans.** $y=C_1e^{2x}+C_2e^{-2x}+\dfrac{e^{3x}}{5}\left[x^2-\dfrac{12x}{5}+\dfrac{62}{25}\right]$

7. $(D^2-3D+2)\,y=2x^2\,e^{4x}+5e^{3x}$ **Ans.** $y=C_1e^{x}+C_2e^{2x}+\dfrac{e^{4x}}{54}[18x^2-30x+19]+\dfrac{5}{2}e^{3x}$

8. $\dfrac{d^2y}{dx^2}-4y=x\sinh x$ **Ans.** $y=C_1e^{2x}+C_2e^{-2x}-\dfrac{x}{3}\sinh x-\dfrac{2}{9}\cosh x$

9. $\dfrac{d^2y}{dt^2}+2h\dfrac{dy}{dt}+(h^2+p^2)y=ke^{-ht}\cos pt$ **Ans.** $y=e^{-ht}[A\cos pt+B\sin pt]+\dfrac{k}{2p}te^{-ht}\sin pt$

12.11 TO FIND THE VALUE OF $\dfrac{1}{f(D)}x^n\sin ax$.

Now $\dfrac{1}{f(D)}x^n(\cos ax+i\sin ax)=\dfrac{1}{f(D)}x^n e^{iax}=e^{iax}\dfrac{1}{f(D+ia)}x^n$

$$\boxed{\frac{1}{f(D)}\cdot x^n\sin ax=\text{Imaginary part of } e^{iax}\frac{1}{f(D+ia)}\cdot x^n}$$

$$\boxed{\frac{1}{f(D)}\cdot x^n\cos ax=\text{Real part of } e^{iax}\frac{1}{f(D+ia)}\cdot x^n}$$

Example 25. *Solve* $\dfrac{d^2y}{dx^2}-2\dfrac{dy}{dx}+y=x\sin x$

Solution. Auxiliary equation is $m^2-2m+1=0$ or $m=1,1$

C.F. $=(C_1+C_2x)\,e^x$

P.I. $=\dfrac{1}{D^2-2D+1}x\cdot\sin x\quad(e^{ix}=\cos x+i\sin x)$

$=$ Imaginary part of $\dfrac{1}{D^2-2D+1}x(\cos x+i\sin x)$ $=$ Imaginary part of $\dfrac{1}{D^2-2D+1}x\cdot e^{ix}$

$=$ Imaginary part of $e^{ix}\dfrac{1}{(D+i)^2-2(D+i)+1}\cdot x$

$=$ Imaginary part of $e^{ix}\dfrac{1}{D^2-2(1-i)D-2i}\cdot x$

$=$ Imaginary part of $e^{ix}\dfrac{1}{-2i}\left[1-(1+i)D-\dfrac{1}{2i}D^2\right]^{-1}\cdot x$

$=$ Imaginary part of $(\cos x+i\sin x)\left(\dfrac{i}{2}\right)[1+(1+i)D]x$

$=$ Imaginary part of $\dfrac{1}{2}(i\cos x-\sin x)\,[x+1+i]$

P.I. $=\dfrac{1}{2}x\cos x+\dfrac{1}{2}\cos x-\dfrac{1}{2}\sin x$

Complete solution is $y=(C_1+C_2x)e^x+\dfrac{1}{2}(x\cos x+\cos x-\sin x)$ **Ans.**

EXERCISE 12.6

Solve the following differential equations :

1. $(D^2 + 4)y = 3x \sin x$ **Ans.** $C_1 \cos 2x + C_2 \sin 2x + x \sin x - \frac{2}{3} \cos x$

2. $\frac{d^2y}{dx^2} - y = x \sin 3x + \cos x$ **Ans.** $C_1e^x + C_2e^{-x} - \frac{1}{10}\left[\frac{3}{5}\cos 3x + x \sin 3x + 5\cos x\right]$

3. $\frac{d^2y}{dx^2} - y = x \sin x + e^x + x^2e^x$ **Ans.** $C_1e^x + C_2e^{-x} - \frac{1}{2}[x\sin x + \cos x] + \frac{x}{12}e^x(2x^2 - 3x + 9)$

4. $(D^4 + 2D^2 + 1)\, y = x^2 \cos x$

Ans. $(C_1 + C_2x)\cos x + (C_3 + C_4x)\sin x + \frac{1}{12}x^3 \sin x - \frac{1}{48}(x^4 - 9x^2)\cos x$

12.12 GENERAL METHOD OF FINDING THE PARTICULAR INTEGRAL OF ANY FUNCTION ϕ (*x*)

$$\text{P.I.} = \frac{1}{D-a}\phi(x) = y \qquad ...(1)$$

or $$(D-a)\frac{1}{D-a}\cdot\phi(x) = (D-a)\cdot y$$

$$\phi(x) = (D-a)y \qquad \text{or} \qquad \phi(x) = Dy - ay$$

$\frac{dy}{dx} - ay = \phi(x)$ which is the linear differential equation.

Its solution is $ye^{-\int a\,dx} = \int e^{-\int a\,dx}\cdot\phi(x)\,dx$ or $ye^{-ax} = \int e^{-ax}\cdot\phi(x)\,dx$

$$y = e^{ax}\int e^{-ax}\cdot\phi(x)\,dx$$

$$\boxed{\frac{1}{D-a}\cdot\phi(x) = e^{ax}\int e^{-ax}\cdot\phi(x)\,dx}$$

Example 26. *Solve* $\frac{d^2y}{dx^2} + 9y = \sec 3x.$

Solution. Auxiliary equation is $m^2 + 9 = 0$ or $m = \pm 3i$,

$$\text{C.F.} = C_1 \cos 3x + C_2 \sin 3x$$

$$\text{P.I.} = \frac{1}{D^2+9}\cdot\sec 3x = \frac{1}{(D+3i)\,(D-3i)}\cdot\sec 3x = \frac{1}{6i}\left[\frac{1}{D-3i} - \frac{1}{D+3i}\right]\cdot\sec 3x$$

$$= \frac{1}{6i}\cdot\frac{1}{D-3i}\cdot\sec 3x - \frac{1}{6i}\frac{1}{D+3i}\cdot\sec 3x \qquad ...(1)$$

Now, $\frac{1}{D-3i}\sec 3x = e^{3ix}\int e^{-3ix}\sec 3x\,dx$ $\left[\frac{1}{D-a}\phi(x) = e^{ax}\int e^{-ax}\phi(x)\,dx\right]$

$$= e^{3ix}\int\frac{\cos 3x - i\sin 3x}{\cos 3x}dx = e^{3ix}\int(1 - i\tan 3x)\,dx = e^{3ix}(x + \frac{i}{3}\log\cos 3x)$$

Changing i to $-i$, we have $\frac{1}{D+3i}\sec 3x = e^{-3ix}(x - \frac{i}{3}\log\cos 3x)$

Putting these values in (1), we get

$$\text{P.I.} = \frac{1}{6i}\left[e^{3ix}\left(x + \frac{i}{3}\log\cos 3x\right) - e^{-3ix}\left(x - \frac{i}{3}\log\cos 3x\right)\right]$$

$$= \frac{x}{6i}e^{3ix} + \frac{e^{3ix}\log\cos 3x}{18} - \frac{xe^{-3ix}}{6i} + \frac{e^{-3ix}}{18}\log\cos 3x$$

$$= \frac{x}{3}\frac{e^{3ix} - e^{-3ix}}{2i} + \frac{1}{9}\cdot\frac{e^{3ix} + e^{-3ix}}{2}\log\cos 3x = \frac{x}{3}\sin 3x + \frac{1}{9}\cos 3x \cdot \log\cos 3x$$

Hence, complete solution is $y = C_1 \cos 3x + C_2 \sin 3x + \frac{x}{3}\sin 3x + \frac{1}{9}\cos 3x \cdot \log\cos 3x$ **Ans.**

Example 27. *Solve* $\frac{d^2y}{dx^2} + 2\frac{dy}{dx} + 2y = e^{-x}\sec^3 x.$ *(Nagpur, Winter 2000)*

Solution. Here we have

$$\frac{d^2y}{dx^2} + 2\frac{dy}{dx} + 2y = e^{-x}\sec^3 x$$

$$(D^2 + 2D + 2)y = e^{-x}\sec^3 x$$

A.E. is $m^2 + 2m + 2 = 0 \qquad \Rightarrow m = \frac{-2 \pm \sqrt{4-8}}{2} = -1 + i$

$C.F. = e^{-x}(C_1 \cos x + C_2 \sin x)$

$$\text{P.I.} = \frac{1}{D^2 + 2D + 2}e^{-x}\sec^3 x = e^{-x}\frac{1}{(D-1)^2 + 2(D-1) + 2}\sec^{3x}$$

$$= e^{-x}\frac{1}{D^2 + 1}\sec^{3x} = \frac{1}{(D+i)(D-i)}\sec^3 x = e^{-x}\frac{1}{2i}\left[\frac{1}{D-i} - \frac{1}{D+i}\right]\sec^3 x \qquad ...(1)$$

Now $\frac{1}{D-i}\sec^3 x = e^{ix}\int e^{-ix}\sec^3 x\, dx \qquad \left[\frac{1}{D-a}\phi(x) = e^{ax}\int e^{-ax}\phi(x)\, dx\right]$

$$= e^{ix}\int(\cos x - i\sin x)\sec^3 x\, dx = e^{ix}\int[\sec^2 x - i\tan x\sec^2 x]dx = e^{ix}\left[\tan x - \frac{i\tan^2 x}{2}\right] \quad ...(2)$$

Similarly $\frac{1}{D+i}\sec^3 x = e^{-ix}\left[\tan x + i\frac{\tan^2 x}{2}\right]$ ...(3) [changing i to $-i$]

Putting the values from (2) and (3) in (1), we get

$$P.I. = \frac{e^{-x}}{2i}\left[e^{ix}\left(\tan x - \frac{i\tan^2 x}{2}\right) - e^{-ix}\left(\tan x + i\frac{\tan^2 x}{2}\right)\right]$$

$$= e^{-x}\left[\tan x\frac{e^{ix} - e^{-ix}}{2i} - i\frac{\tan^2 x}{2}\left(\frac{e^{ix} + e^{-ix}}{2i}\right)\right] = e^{-x}\left[\tan x\sin x - \frac{\tan^2 x}{2}\cos x\right]$$

$$= e^{-x}\left[\tan x . \sin x - \frac{\tan x}{2}\frac{\sin x}{\cos x}\cos x\right] = e^{-x}\left[\tan x\sin x - \frac{\tan x}{2}\sin x\right] = e^{-x}\left(\frac{1}{2}\tan x\sin x\right)$$

Complete solution = C.F. + P.I.

$$= e^{-x}(c_1\cos x + c_2\sin x) + \frac{e^{-x}}{2}\tan x\sin x = e^{-x}\left[C_1\cos x + C_2\sin x + \frac{\sin x\tan x}{2}\right] \textbf{ Ans.}$$

Example 28. *Solve* $\frac{d^2y}{dx^2} - 2\frac{dy}{dx} + 2y = e^x\tan x.$

Solution. Here we have $\frac{d^2y}{dx^2} - 2\frac{dy}{dx} + 2y = e^x\tan x$

$$(D^2 - 2D + 2)y = e^x\tan x$$

A.E. is $m^2 - 2m + 2 = 0 \qquad \Rightarrow m = \frac{2 \pm \sqrt{4-8}}{2} = 1 \pm i$

$C.F. = e^x(C_1\cos x + C_2\sin x)$

$$P.I. = \frac{1}{D^2 - 2D + 2} e^{-x} \tan x = e^x \frac{1}{(D+1)^2 - 2(D+1) + 2} \tan x$$

$$= e^x \frac{1}{D^2 + 1} \tan x = e^x \frac{1}{(D+i)(D-i)} \tan x = \frac{e^x}{2i}\left[\frac{1}{D-i} - \frac{1}{D+i}\right] \tan x \qquad ...(1)$$

Now $\frac{1}{D-i} \tan x = e^{ix} \int e^{-ix} \tan x \, dx$

$$= e^{ix} \int (\cos x - i \sin x) \tan x \, dx = e^{ix} \int \left(\sin x - i \frac{\sin^2 x}{\cos x} \right) dx$$

$$= e^{ix} \int \left[\sin x - \frac{i(1 - \cos^2 x)}{\cos x} \right] dx = e^{ix} \int (\sin x - i \sec x + i \cos x] dx$$

$$= e^{ix}[-\cos x + i \sin x - i \log(\sec x + \tan x)] \qquad ...(2)$$

Similarly $\quad \frac{1}{D+i} \tan x = e^{-ix}[-\cos x - i \sin x + i \log(\sec x + \tan x)] \qquad ...(3)$

On putting the values from (2) and (3) in (1), we get

$$P.I. = \frac{e^x}{2i}[e^{ix}(-\cos x + i \sin x - i \log(\sec x + \tan x) - e^{-ix}(-\cos x - i \sin x + i \log(\sec x + \tan x)]$$

$$= e^x \left[-\cos x \frac{e^{ix} - e^{-ix}}{2i} + \sin x \frac{e^{ix} + e^{-ix}}{2} - \log(\sec x + \tan x) \frac{e^{ix} + e^{-ix}}{2} \right]$$

$$= e^x[-\cos x \sin x + \sin x \cos x - \cos x \log(\sec x + \tan x)]$$

$$= - e^x \cos x \log(\sec x + \tan x)$$

Complete solution = C.F. + P.I.

$$= e^x (C_1 \cos x + C_2 \sin x) - e^x \cos x \log(\sec x + \tan x) \qquad \textbf{Ans.}$$

EXERCISE 12.7

Solve the following differential equations :

1. $\frac{d^2 y}{dx^2} + a^2 y = \sec ax$ *(R.G.P.V., Bhopal April, 2010)*

Ans. $C_1 \cos ax + C_2 \sin ax + \frac{x}{a} \sin ax + \frac{1}{a^2} \cos ax \cdot \log \cos ax$

2. $\frac{d^2 y}{dx^2} + y = \text{cosec}\, x$ **Ans.** $C_1 \cos x + C_2 \sin x - x \cos x + \sin x \log \sin x$

3. $(D^2 + 4) y = \tan 2x$ **Ans.** $C_1 \cos 2x + C_2 \sin 2x - \frac{1}{4} \cos 2x \log(\sec 2x + \tan 2x)$

4. $\frac{d^2 y}{dx^2} + y = (x - \cot x)$ *(A.M.I.E. Winter 2002)*

Ans. $C_1 \cos x + C_2 \sin x - x \cos^2 x - \sin x \log(\text{cosec}\, x - \cot x)$

CAUCHY – EULER EQUATIONS, METHOD OF VARIATION OF PARAMETERS

13.1 CAUCHY EULER HOMOGENEOUS LINEAR EQUATIONS

$$a_n x_n \frac{d^n y}{dx^n} + a_{n-1} x^{n-1} \frac{d^{n-1} y}{d x^{n-1}} + \dots + a_0 y = \phi(x) \qquad \dots (1)$$

where $a_0, a_1, a_2, \dots$ are constants, is called a homogeneous equation.

Put $\quad x = e^z, \quad z = \log_e x, \quad \frac{d}{dz} \equiv D$

$$\frac{dy}{dx} = \frac{dy}{dz} \cdot \frac{dz}{dx} = \frac{1}{x}\frac{dy}{dz} \Rightarrow x\frac{dy}{dx} = \frac{dy}{dz} \Rightarrow x\frac{dy}{dx} = Dy$$

Again,
$$\frac{d^2y}{dx^2} = \frac{d}{dx}\left(\frac{dy}{dx}\right) = \frac{d}{dx}\left(\frac{1}{x}\frac{dy}{dz}\right) = -\frac{1}{x^2}\frac{dy}{dz} + \frac{1}{x}\frac{d^2y}{dz^2}\frac{dz}{dx}$$

$$= -\frac{1}{x^2}\frac{dy}{dz} + \frac{1}{x}\frac{d^2y}{dz^2}\frac{1}{x} = \frac{1}{x^2}\left(\frac{d^2y}{dz^2} - \frac{dy}{dz}\right) = \frac{1}{x^2}(D^2 - D)\, y$$

$$x^2 \frac{d^2y}{dx^2} = (D^2 - D)\, y$$

or
$$\boxed{\mathbf{x^2 \frac{d^2y}{dx^2} = D(D-1)\, y}}$$

Similarly.
$$\boxed{\mathbf{x^3 \frac{d^3y}{dx^3} = D(D-1)(D-2)\, y}}$$

The substitution of these values in (1) reduces the given homogeneous equation to a differential equation with constant coefficients.

Example 1. *Solve:* $\quad x^2 \frac{d^2y}{dx^2} - 2x\frac{dy}{dx} - 4y = x^4 \qquad$ (*A.M.I.E. Summer* 2000)

Solution. We have, $\quad x^2 \frac{d^2y}{dx^2} - 2x\frac{dy}{dx} - 4y = x^4 \qquad \dots (1)$

Putting $x = e^z$, $\quad D \equiv \frac{d}{dz}$, $x\frac{dy}{dx} = Dy$, $x^2\frac{d^2y}{dx^2} = D(D-1)\, y$ in (1), we get

$$D(D-1)\, y - 2\,Dy - 4y = e^{4z} \quad \text{or} \quad (D^2 - 3D - 4)\, y = e^{4z}$$

A.E. is $\quad m^2 - 3m - 4 = 0 \quad \Rightarrow \quad (m-4)(m+1) = 0 \quad \Rightarrow \quad m = -1, 4$

$$\text{C.F.} = C_1 e^{-z} + C_2 e^{4z}$$

$$\text{P.I.} = \frac{1}{D^2 - 3D - 4} e^{4z} \qquad \text{[Rule Fails]}$$

$$= z \frac{1}{2D-3} e^{4z} = z \frac{1}{2(4)-3} e^{4z} = \frac{z e^{4z}}{5}$$

Thus, the complete solution is given by

$$y = C_1 e^{-z} + C_2 e^{4z} + \frac{z e^{4z}}{5} \quad \Rightarrow \quad y = \frac{C_1}{x} + C_2 x^4 + \frac{1}{5} x^4 \log x \quad \textbf{Ans.}$$

Example 2. *Solve:* $\quad x^2 \dfrac{d^2y}{dx^2} + 4x \dfrac{dy}{dx} + 2y = e^x$

(U.P., II Semester 2005; Nagpur University, Summer 2001)

Solution. Given equation is $x^2 \dfrac{d^2y}{dx^2} + 4x \dfrac{dy}{dx} + 2y = e^x$ ...(1)

To solve (1), we put $x = e^z$ or $z = \log x$ and $D \equiv \dfrac{d}{dz}$

$$x \frac{dy}{dx} = Dy \quad \text{and} \quad x^2 \frac{d^2y}{dx^2} = D(D-1)\, y$$

Substituting these values in (1), it reduces

$$[\,D(D-1) + 4D + 2\,]\, y = e^{e^z} \quad \Rightarrow \quad (D^2 + 3D + 2)\, y = e^{e^z}$$

$\therefore$ It's *A.E.* is $m^2 + 3m + 2 = 0$

$\therefore$ $(m+1)(m+2) = 0 \quad \therefore \quad m = -1, -2$

$\therefore$ $\text{C.F.} = C_1 e^{-z} + C_2 e^{-2z} = \dfrac{C_1}{x} + \dfrac{C_2}{x^2}$

$$\text{P.I.} = \frac{1}{D^2 + 3D + 2} e^{e^z} = \frac{1}{(D+1)(D+2)} e^{e^z}$$

$$= \left(\frac{1}{D+1} - \frac{1}{D+2}\right) e^{e^z} = \frac{1}{D+1} e^{e^z} - \frac{1}{D+2} e^{e^z}$$

$$= e^{-z} \int e^{e^z} . e^z\, dz - e^{-2z} \int e^{e^z} . e^{2z}\, dz \qquad \left[\text{Since } \frac{1}{D-a} X = e^{ax} \int X e^{-ax}\, dx\right]$$

Put $\quad e^z = t$ so that $e^z dz = dt$

$$\text{P.I.} = e^{-z} \int e^t\, dt - e^{-2z} \int e^t . t\, dt = e^{-z} e^t - e^{-2z} (te^t - e^t)$$

$$= e^{-z} e^{e^z} - e^{-2z} (e^z e^{e^z} - e^{e^z})$$

$$= e^{-z} e^{e^z} - e^{-z} e^{e^z} + e^{-2z} e^{e^z} = e^{-2z} e^{e^z} = x^{-2} e^x = \frac{e^x}{x^2} \qquad (\because x = e^z)$$

Hence, the C.S. of (1) is $\; y = \text{C.F.} + \text{P.I.}$

$$y = \frac{C_1}{x} + \frac{C_2}{x^2} + \frac{e^x}{x^2} \qquad \textbf{Ans.}$$

Example 3. *Solve* $(x^3 D^3 + x^2 D^2 - 2)\, y = x - \dfrac{1}{x^3}$ *(Nagpur University, Summer* 2000)

Solution. Put $x = e^z$, $z = \log x$

Let $\quad D_1 \equiv \dfrac{d}{dz}$

Then $\quad x^2 \dfrac{d^2y}{dx^2} = D_1 (D_1 - 1)\, y \quad \Rightarrow \quad x^3 \dfrac{d^3y}{dx^3} = D_1 (D_1 - 1)(D_1 - 2)\, y$

Substituting in the given equation, we get

$$[\, D_1 (D_1 - 1)(D_1 - 2) + D_1 (D_1 - 1) - 2\,]\, y = e^z + e^{-3z}$$

$$\Rightarrow \quad [\, D_1^3 - 3D_1^2 + 2D_1 + D_1^2 - D_1 - 2\,]\, y = e^z + e^{-3z}$$

$$\Rightarrow \quad (\, D_1^3 - 2D_1^2 + D_1 - 2\,)\, y = e^z + e^{-3z}$$

A.E. is $m^3 - 2m^2 + m - 2 = 0$

i.e. $(m - 2)(m^2 + 1) = 0 \qquad$ *i.e.* $m = 2, \quad \pm\, i.$

$$\text{C.F.} = C_1 e^{2z} + C_2 \cos z + C_3 \sin z$$

$$\text{P.I.} = \frac{1}{D_1^3 - 2D_1^2 + D_1 - 2} e^z + \frac{1}{D_1^3 - 2D_1^2 + D_1 - 2} e^{-3z}$$

$$= \frac{1}{1 - 2 + 1 - 2} e^z + \frac{1}{-27 - 18 - 3 - 2} e^{-3z} \qquad \begin{bmatrix} D_1 = 1 \\ D_1 = -3 \end{bmatrix}$$

$$= -\frac{e^z}{2} - \frac{1}{50} e^{-3z}$$

$$\therefore \quad y = C_1 e^{2z} + C_2 \cos z + C_3 \sin z - \frac{1}{2} e^z - \frac{1}{50} e^{-3z}$$

$$= C_1 x^2 + C_2 \cos (\log x) + C_3 \sin (\log x) - \frac{1}{2} x - \frac{1}{50} \cdot \frac{1}{x^3}$$ **Ans.**

Example 4. *Solve* $x^2 \dfrac{d^2y}{dx^2} + x \dfrac{dy}{dx} + y = \sin (\log x^2)$ *(Nagpur University, Summer* 2005)

Solution. We have, $\quad x^2 \dfrac{d^2y}{dx^2} + x \dfrac{dy}{dx} + y = \sin (\log x^2) \qquad \ldots (1)$

Let $x = e^z$, $\quad$ so that $z = \log x$, $\quad D \equiv \dfrac{d}{dz}$

(1) becomes

$$D(D - 1)\, y + Dy + y = \sin (2z) \quad \Rightarrow \quad (D^2 + 1)\, y = \sin 2z$$

A.E. is $m^2 + 1 = 0 \qquad$ or $\qquad m = \pm\, i$

$$C.F. = C_1 \cos z + C_2 \sin z$$

$$\text{P.I} = \frac{1}{D^2 + 1} \sin 2z = \frac{1}{-4 + 1} \sin 2z = -\frac{1}{3} \sin 2z$$

$$y = \text{C.F.} + \text{P.I.} = C_1 \cos z + C_2 \sin z - \frac{1}{3} \sin 2z$$

$$= C_1 \cos (\log x) + C_2 \sin (\log x) - \frac{1}{3} \sin (\log x^2)$$ **Ans.**

Example 5. *Solve* $x^3 \dfrac{d^3y}{dx^3} + 3x^2 . \dfrac{d^2y}{dx^2} + x\dfrac{dy}{dx} + y = x + \log x$

(*AMIETE, June 2010, U.P., II Semester, Summer,* 2001)

Solution. Putting $x = e^z$ or $z = \log x$ and denoting $\dfrac{d}{dz}$ by D the equation becomes

$$[D(D-1)(D-2) + 3\,D(D-1) + D + 1\,]\,y = e^z + z$$

$$\Rightarrow \qquad [\,D^3 + 1\,]\,y = e^z + z$$

$$\therefore \qquad \text{A.E. is} \qquad m^3 + 1 = 0$$

$$\Rightarrow \qquad (m+1)(m^2 - m + 1) = 0 \quad \Rightarrow \quad m = -1, \frac{1 \pm i\sqrt{3}}{2}$$

$$\text{C.F.} = C_1\, e^{-z} + e^{\frac{1}{2}z}\left\{C_2 \cos\frac{\sqrt{3}}{2}z + C_3 \sin\frac{\sqrt{3}}{2}z\right\}$$

$$\text{P.I.} = \frac{1}{D^3+1}\{e^z + z\}$$

$$= \frac{1}{D^3+1}e^z + \frac{1}{D^3+1}z = \frac{e^z}{1+1} + (1+D^3)^{-1}z$$

$$= \frac{1}{2}e^z + (1 - D^3 + \ldots)\,z = \frac{1}{2}e^z + z.$$

$\therefore$ Complete solution is

$$y = C_1\, e^{-z} + e^{z/2}\left\{C_2 \cos\frac{\sqrt{3}}{2}z + C_3 \sin\frac{\sqrt{3}}{2}z\right\} + \frac{1}{2}e^z + z$$

$$\Rightarrow \qquad y = C_1\, x^{-1} + \sqrt{x}\left\{C_2 \cos\left(\frac{\sqrt{3}}{2}\log x\right) + C_3 \sin\left(\frac{\sqrt{3}}{2}\log x\right)\right\} + \frac{1}{2}x + \log x \qquad \textbf{Ans.}$$

Example 6. *Solve:* $x^2 \dfrac{d^3y}{dx^3} + 3x\dfrac{d^2y}{dx^2} + \dfrac{dy}{dx} = x^2 \log x$ (*Nagpur University, Summer* 2003)

Solution. We have, $x^3 \dfrac{d^3y}{dx^3} + 3x^2 \dfrac{d^2y}{dx^2} + x\dfrac{dy}{dx} = x^3 \log x$

Let $x = e^z$ so that $z = \log x$, $D \equiv \dfrac{d}{dz}$

The equation becomes after substitution

$$[D\,(D-1)\,(D-2) + 3D\,(D-1) + D\,]\,y = z\,e^{3z} \qquad \Rightarrow \qquad D^3y = ze^{3z}$$

Auxiliary equation is $m^3 = 0 \Rightarrow \qquad m = 0, 0, 0.$

C.F. $= C_1 + C_2\, z + C_3\, z^2 = C_1 + C_2 \log x + C_3 (\log x)^2$

$$\text{P.I.} = \frac{1}{D^3}\,.\,z\,e^{3z} = e^{3z}\,.\,\frac{1}{(D+3)^3}.z$$

$$= e^{3z}\,\frac{1}{27}\left(1 + \frac{D}{3}\right)^{-3} z = \frac{e^{3z}}{27}(1-D)\,z = \frac{e^{3z}}{27}(z-1) = \frac{x^3}{27}(\log x - 1)$$

Complete solution is $\qquad y = C_1 + C_2 \log x + C_3 (\log x)^2 + \dfrac{x^3}{27}(\log x - 1) \qquad$ **Ans.**

Example 7. *Solve the differential equation*

$$x^2\frac{d^2y}{dx^2}-3x\frac{dy}{dx}+5y=x\log x.$$ (*Nagpur University, Winter* 2002, *Summer* 2000)

Solution. Putting $x=e^z$ or $z=\log x$ and $D\equiv\dfrac{d}{dz}$

On putting $x\dfrac{dy}{dx}=Dy$ and $x^2\dfrac{d^2y}{dx^2}=D(D-1)\,y$ in the given equation, we get

$$[\,D(D-1)-3D+5\,]\,y=ze^z$$

i.e. $(D^2-4D+5)\,y=ze^z$

It's A.E. is $m^2-4m+5=0$

$$\therefore\quad m=\frac{4\pm\sqrt{16-20}}{2}=\frac{4\pm 2i}{2}=2\pm i$$

$$\text{C.F.}=e^{2z}\,(C_1\cos z+C_2\sin z)$$

$$\text{P.I.}=\frac{1}{D^2-4D+5}\,ze^z$$

$$=e^z\cdot\frac{1}{(D+1)^2-4(D+1)+5}\,z \qquad \text{(by replacing } D \text{ by } D+1)$$

$$=e^z\cdot\frac{1}{D^2-2D+2}\,z=\frac{e^z}{2}\cdot\frac{1}{1+\left(\dfrac{D^2-2D}{2}\right)}\,z$$

$$=\frac{e^z}{2}\left[1+\left(\frac{D^2-2D}{2}\right)\right]^{-1}z=\frac{e^z}{2}\left[1-\left(\frac{D^2-2D}{2}\right)+\ldots\right]z$$

$$=\frac{e^z}{2}[\,z+Dz\,]=\frac{e^z}{2}(\,z+1).$$

Hence, the solution is

$$y=e^{2z}[C_1\cos z+C_2\sin z]+\frac{e^z}{2}(z+1)$$

$$y=x^2\,[\,C_1\cos(\log x)+C_2\sin(\log x)\,]+\frac{x}{2}(1+\log x)$$ **Ans.**

Example 8. *Solve the homogeneous linear differential equation.*

$$x^2\frac{d^2y}{dx^2}+x\frac{dy}{dx}+y=(\log x)\sin(\log x)$$

(*Nagpur University, Winter* 2002, *U.P. II Semester, Summer* 2002)

Solution. Since given equation is homogeneous,

Put $x=e^z \Rightarrow \log x=z$

Also, $x\dfrac{dy}{dx}=Dy,\quad x^2\dfrac{d^2y}{dx^2}=D(D-1)\,y,\quad D\equiv\dfrac{d}{dz}$

The transformed equation is

$$D(D-1)\,y+Dy+y=z\sin z$$

$$(D^2-D+D+1)\,y=z\sin z$$

$$(D^2+1)\,y=z\sin z$$

A.E. is $\quad m^2 + 1 = 0 \Rightarrow \quad m = \pm i$

$$\text{C.F.} = C_1 \cos z + C_2 \sin z$$

$$P.I. = \frac{1}{D^2+1} z \sin z$$

$$= \text{Imaginary part of } \frac{1}{D^2+1} z(\cos z + i \sin z)$$

$$= \text{Imaginary part of } \frac{1}{D^2+1} ze^{iz} \qquad = \text{Imaginary part of } e^{iz} \frac{1}{(D+i)^2+1} z$$

$$= \text{Imaginary part of } e^{iz} \frac{1}{D^2+2iD-1+1} z \quad = \text{Imaginary part of } e^{iz} \frac{1}{D^2+2iD} z$$

$$= \text{Imaginary part of } e^{iz} \frac{1}{2iD} \frac{1}{\left(1+\frac{D}{2i}\right)} z \qquad = \text{Imaginary part of } e^{iz} \frac{1}{2iD}\left(1-\frac{D}{2i}\right) z$$

$$= \text{Imaginary part of } e^{iz} \frac{1}{2iD}\left(z-\frac{1}{2i}\right) \qquad = \text{Imaginary part of } e^{iz} \frac{1}{2i}\left(\frac{z^2}{2}-\frac{z}{2i}\right)$$

$$= \text{Imaginary part of } \frac{1}{2i} (\cos z + i \sin z)\left(\frac{z^2}{2}-\frac{z}{2i}\right)$$

$$= \text{Imaginary part of } (\cos z + i \sin z)\left(\frac{z^2}{4i}+\frac{z}{4}\right)$$

$$= \text{Imaginary part of } (\cos z + i \sin z)\left(-i\frac{z^2}{4}+\frac{z}{4}\right) \quad = -\frac{z^2}{4}\cos z + \frac{z}{4}\sin z$$

Complete solution is, y = C.F. + P.I.

$$y = C_1 \cos z + C_2 \sin z - \frac{z^2}{4}\cos z + \frac{z}{4}\sin z$$

$$y = C_1 \cos(\log x) + C_2 \sin(\log x) - \frac{1}{4}(\log x)^2 \cos(\log x) + \frac{1}{4}(\log x)\sin(\log x) \quad \textbf{Ans.}$$

Example 9. *The radial displacement in a rotating disc at a distance r from the axis is given by* $r^2\frac{d^2u}{dr^2} + r\frac{du}{dr} - u + kr^3 = 0,$ *where k is a constant. Solve the equation under the conditions u = 0 when r = 0, u = 0 when r = a.* (*Nagpur University, Summer* 2008)

Solution. Here, we have

$$r^2\frac{d^2u}{dr^2} + r\frac{du}{dr} - u + kr^3 = 0 \qquad \text{... (1)}$$

On putting $r = e^z$, $\quad r\frac{du}{dr} = Dz, \quad r^2\frac{d^2u}{dr^2} = D(D-1)z \quad$ in (1), we get

$$D(D-1)u + Du - u = -ke^{3z} \qquad \left[D \equiv \frac{d}{dz}\right]$$

$$\Rightarrow \qquad (D^2 - D + D - 1)\, u = -ke^{3z} \quad \Rightarrow \quad (D^2 - 1)u = -k\, e^{3z}$$

A.E. is $\quad m^2 - 1 = 0 \quad \Rightarrow \quad m = \pm 1$

$$\text{C.F.} = C_1 e^z + C_2 e^{-z} = C_1\, r + \frac{C_2}{r}$$

$$\text{P.I.} = \frac{1}{D^2 - 1}(-ke^{3z}) = -k\, \frac{1}{(3)^2 - 1}\, e^{3z} = -\frac{k}{8}\, e^{3z} = \frac{-k}{8}\, r$$

C.S. = C.F. + P.I.

$$u = C_1\, r + \frac{C_2}{r} - \frac{k}{8}\, r^3 \qquad \text{... (2)}$$

Putting $\quad u = 0, \quad r = 0 \quad$ in (2), we get

$$0 = C_2$$

Putting $C_2 = 0$ in (2), we get

$$u = C_1\, r - \frac{k}{8}\, r^3 \qquad \text{... (3)}$$

Putting $u = 0$, $r = a$ in (3), we get

$$0 = C_1\, a - \frac{k}{8}\, a^3 \quad \Rightarrow \quad C_1 = \frac{k}{8}\, a^2$$

Putting $\quad C_1 = \frac{k}{8}\, a^2$ in (3), we get

$$u = \frac{k}{8}\, a^2 r - \frac{k}{8}\, r^3$$

$$u = \frac{kr}{8}\,(a^2 - r^2)$$

Ans.

13.2 LEGENDRE'S HOMOGENEOUS DIFFERENTIAL EQUATIONS

A linear differential equation of the form

$$(a + bx)^n\, \frac{d^n y}{dx^n} + a_1\, (a + bx)^{n-1}\, \frac{d^{n-1} y}{dx^{n-1}} + \ldots + a_n y = X \qquad \text{... (1)}$$

where a, b, a_1, a_2, ... a_n are constants and X is a function of x, is called Legendre's linear equation.

Equation (1) can be reduced to linear differential equation with constant coefficients by the substitution

$$a + bx = e^z \qquad \Rightarrow \qquad z = \log\,(a + bx)$$

so that $\qquad \dfrac{dy}{dx} = \dfrac{dy}{dz} \cdot \dfrac{dz}{dx} = \dfrac{b}{a + bx} \cdot \dfrac{dy}{dz}$

$\Rightarrow \qquad (a + bx)\dfrac{dy}{dx} = b\,\dfrac{dy}{dz} = b\, Dy, \quad D \equiv \dfrac{d}{dz} \quad \Rightarrow \quad (a + bx)\, \dfrac{dy}{dx} = b\, Dy$

where

Again $\qquad \dfrac{d^2 y}{dx^2} = \dfrac{d}{dx}\left(\dfrac{dy}{dx}\right) = \dfrac{d}{dx}\left(\dfrac{b}{a + bx} \cdot \dfrac{dy}{dz}\right)$

$$= -\frac{b^2}{(a + bx)^2}\, \frac{dy}{dz} + \frac{b}{(a + bx)} \cdot \frac{d^2 y}{dz^2} \cdot \frac{dz}{dx}$$

$$= -\frac{b^2}{(a+bx)^2}\frac{dy}{dz} + \frac{b}{(a+bx)}\cdot\frac{d^2y}{dz^2}\cdot\frac{b}{(a+bx)}$$

$$\Rightarrow \quad (a+bx)^2\frac{d^2y}{dx^2} = -b^2\frac{dy}{dz} + b^2\frac{d^2y}{dz^2}$$

$$= b^2\left(\frac{d^2y}{dz^2} - \frac{dy}{dz}\right) = b^2(D^2y - Dy) = b^2 D(D-1)y$$

$$\Rightarrow \quad (a+bx)^2\frac{d^2y}{dx^2} = b^2 D(D-1)$$

Similarly, $(a+bx)^3\dfrac{d^3y}{dx^3} = b^3 D(D-1)(D-2)y$

....

$$(a+bx)^n\frac{d^ny}{dx^n} = b^n D(D-1)(D-2)\ldots\ldots(D-n+1)y$$

Substituting these values in equation (1), we get a linear differential equation with constant coefficients, which can be solved by the method given in the previous section.

Example 10. *Solve* $(1+x)^2\dfrac{d^2y}{dx^2} + (1+x)\dfrac{dy}{dx} + y = \sin 2\{\log(1+x)\}$

Solution. We have, $(1+x)^2\dfrac{d^2y}{dx^2} + (1+x)\dfrac{dy}{dx} + y = \sin 2\{\log(1+x)\}$

Put $1+x = e^z$ or $\log(1+x) = z$

$(1+x)\dfrac{dy}{dx} = Dy$ and $(1+x)^2\dfrac{d^2y}{dx^2} = D(D-1)y$, where $D \equiv \dfrac{d}{dz}$

Putting these values in the given differential equation, we get

$D(D-1)y + Dy + y = \sin 2z$ or $(D^2 - D + D + 1)y = \sin 2z$

$(D^2+1)y = \sin 2z$

A.E. is $m^2 + 1 = 0 \Rightarrow m = \pm i$

C.F. $= A\cos z + B\sin z$

$$\text{P.I.} = \frac{1}{D^2+1}\sin 2z = \frac{1}{-4+1}\sin 2z = -\frac{1}{3}\sin 2z$$

Now, complete solution is $y =$ C.F. + P.I.

$$\Rightarrow \quad y = A\cos z + B\sin z - \frac{1}{3}\sin 2z$$

$$\Rightarrow \quad y = A\cos\{\log(1+x)\} + B\sin\{\log(1+x)\} - \frac{1}{3}\sin 2\{\log(1+x)\}$$ **Ans.**

Example 11. *Solve:* $(3x+2)^2\dfrac{d^2y}{dx^2} + 3(3x+2)\dfrac{dy}{dx} - 36y = 3x^2 + 4x + 1.$

(Uttarakhand II, Semester, June 2007)

Solution. Let $3x + 2 = e^z \quad \Rightarrow z = \log(3x+2)$ $\quad \left[x = \dfrac{e^z - 2}{3}\right]$

So that $(3x+2)\dfrac{dy}{dx} = 3Dy$ and $(3x+2)^2\dfrac{d^2y}{dx^2} = 9D(D-1)y$ where $D \equiv \dfrac{d}{dz}$

Putting these values in the given differential equation, we get

$$9D(D-1)\,y + 9Dy - 36y = 3\left(\frac{e^z-2}{3}\right)^2 + 4\left(\frac{e^z-2}{3}\right) + 1$$

$$\therefore \quad (9D^2-36)\,y = \frac{1}{3}(e^{2z} - 4e^z + 4) + \frac{4}{3}e^z - \frac{8}{3} + 1 = \frac{e^{2z}}{3} - \frac{1}{3}$$

A.E. is $\quad 9m^2 - 36 = 0$

$$m^2 - 4 = 0 \qquad \therefore \quad m = \pm 2$$

$$\text{C.F.} = C_1 e^{2z} + C_2 e^{-2z}$$

$$\text{P.I.} = \frac{1}{9D^2-36}\left[\frac{e^{2z}}{3} - \frac{1}{3}\right] = \frac{1}{27}\frac{1}{D^2-4}e^{2z} - \frac{1}{3}\frac{1}{9D^2-36}e^{0z}$$

$$= \frac{1}{27}z\frac{1}{2D}e^{2z} - \frac{1}{3}\frac{1}{0-36} = \frac{1}{27}z\left(\frac{e^{2z}}{4}\right) + \frac{1}{108} = \frac{1}{108}[ze^{2z}+1]$$

Complete Solution is

$$y = C.F. + P.I. = C_1 e^{2z} + C_2 e^{-2z} + \frac{1}{108}(ze^{2z}+1)$$

$$y = C_1(3x+2)^2 + \frac{C_2}{(3x+2)^2} + \frac{1}{108}[(3x+2)^2 \log(3x+2) + 1] \qquad \textbf{Ans.}$$

EXERCISE 13.1

Solve the following differential equations:

1. $x^2\dfrac{d^2y}{dx^2} - 4x\dfrac{dy}{dx} + 6y = \dfrac{42}{x^4}$ **Ans.** $C_1 x^2 + C_2 x^3 + \dfrac{1}{x^4}$

2. $(x^2D^2 - 3xD + 4)\,y = 2x^2$ **Ans.** $(C_1 + C_2 \log x)\,x^2 + x^2(\log x)^2$

3. $x^2\dfrac{d^2y}{dx^2} - x\dfrac{dy}{dx} + y = \log x$ *(AMIETE, June 2010)* **Ans.** $(C_1 + C_2\log x)\,x + \log x + 2$

4. $\dfrac{d^2y}{dx^2} + \dfrac{1}{x}\dfrac{dy}{dx} = \dfrac{12\log x}{x^2}$ **Ans.** $C_1 + C_2\log x + 2(\log x)^3$

5. $(x^2D^2 - xD - 3)\,y = x^2\log x$ (*A.M.I.E. Winter* 2001, *Summer* 2001)

   Ans. $\dfrac{C_1}{x} + C_2x^3 - \dfrac{x^2}{3}\left(\log x + \dfrac{2}{3}\right)$

6. $x^2\dfrac{d^2y}{dx^2} - 2x\dfrac{dy}{dx} + 2y = x^2 + \sin(5\log x)$

   Ans. $c_1x + c_2x^2 + x^2\log x + \dfrac{1}{754}[15\cos(5\log x) - 23\sin(5\log x)]$

7. $x^2\dfrac{d^2y}{dx^2} - 3x\dfrac{dy}{dx} + y = \log x\dfrac{\sin(\log x)+1}{x}$ (*AMIETE, Dec. 2009*)

   Ans. $y + C_1x^{2+\sqrt{3}} + C_2x^{2-\sqrt{3}} + \dfrac{1}{x}\left[\dfrac{382}{61}\cos\log x + \dfrac{54}{61}\sin(\log x) + 6\log x\cos(\log x) + \right.$

   $\left. 5\log x\sin(\log x)\right] + \dfrac{1}{6x}$

8. $(1+x)^2 \frac{d^2y}{dx^2} + (1+x)\frac{dy}{dx} + y = 2 \sin \log (1+x)$

Ans. $y = C_1 \cos \log (1+x) + C_2 \sin \log (1+x) - \log (1+x) \cos \log (1+x)$

9. Which of the basis of solutions are for the differential equation $x^2 \frac{d^2y}{dx^2} - x\frac{dy}{dx} + y = 0$

(*a*) $x, x\, I_n\, x,$ (*b*) $I_n\, x, e^x$ (*c*) $\frac{1}{x}, \frac{1}{x^2},$ (*d*) $\frac{1}{x^2} e^x, x\, I_n\, x$

(*A.M.I.E., Winter* 2001) **Ans.** (*a*)

10. The general solution of $x^2 \frac{d^2y}{dx^2} - 5x\frac{dy}{dx} + 9y = 0$ is

(*a*) $(C_1 + C_2 x)\, e^{3x}$ (*b*) $(C_1 + C_{2n}\, x)\, x^3$ (*c*) $(C_1 + C_2 x)\, x^3$ (*d*) $(C_1 + C_2 l_n\, x)\, e^{x^3}$

(*AMIETE, Dec. 2009*) **Ans.** (*b*)

11. To transform $x\frac{d^2y}{dx^2} + \frac{dy}{dx} = \frac{1}{x}$ into a linear differential equation with constant coefficients, the required substitution is

(*a*) $x = \sin t$ (b) $x = t^2 + 1$ (*c*) $x = \log t$ (*d*) $x = e^t$

(*AMIETE, June 2010*) **Ans.** (*d*)

13.3 METHOD OF VARIATION OF PARAMETERS

To find particular integral of

$$\frac{d^2y}{dx^2} + b\frac{dy}{dx} + c\,y = X \qquad \text{... (1)}$$

Let complementary function $= Ay_1 + By_2$, so that y_1 and y_2 satisfy

$$\frac{d^2y}{dx^2} + b\frac{dy}{dx} + c\,y = 0 \qquad \text{... (2)}$$

Let us assume particular integral $y = uy_1 + vy_2$, ... (3)

where u and v are unknown functions of x.

Differentiation (3) *w.r.t.* x, we have $y' = uy_1' + vy_2' + u'y_1 + v'y_2$ assuming that u, v satisfy the equation

$$u'y_1 + v'y_2 = 0 \qquad \text{... (4)}$$

then $$y' = uy_1' + vy_2' \qquad \text{... (5)}$$

Differentiating (5) *w.r.t.x*, we have $y'' = uy_1'' + u'y_1' + vy_2'' + v'y_2'$

Substituting the values of y, y' and y'' in (1), we get

$$(uy_1'' + u'y_1' + vy_2'' + v'\,y_2') + b\,(uy_1' + vy_2') + c\,(uy_1 + vy_2) = X$$

$$\Rightarrow \quad u(y_1'' + by_1' + cy_1) + v(y_2'' + by_2' + cy_2) + (u'\,y_1' + v'\,y_2') = X \qquad \text{... (6)}$$

y_1 and y_2 will satisfy equation (1)

$$\therefore \quad y_1'' + by_1' + cy_1 = 0 \qquad \text{... (7)}$$

and $$y_2'' + by_2' + cy_2 = 0 \qquad \text{... (8)}$$

Putting the values of expressions from (7) and (8) in (6), we get

$$\Rightarrow \quad u'\,y_1' + v'y_2' = X \qquad \text{... (9)}$$

Solving (4) and (9), we get

$$u' = \begin{vmatrix} 0 & y_2 \\ X & y_2' \end{vmatrix} \div \begin{vmatrix} y_1 & y_2 \\ y_1' & y_2' \end{vmatrix} = \frac{-y_2 X}{y_1 y_2' - y_1' y_2}$$

$$v' = \begin{vmatrix} y_1 & 0 \\ y_1' & X \end{vmatrix} \div \begin{vmatrix} y_1 & y_2 \\ y_1' & y_2' \end{vmatrix} = \frac{y_1 X}{y_1 y_2' - y_1' y_2}$$

$$\boxed{u = \int \frac{-y_2 X}{y_1 y_2' - y_1' y_2} dx}$$

$$\boxed{v = \int \frac{y_1 X}{y_1 y_2' - y_1' y_2} dx}$$

General solution = complementary function + particular integral.

Working Rule

Step 1. Find out the *C.F. i.e.*, $A y_1 + B y_2$

Step 2. Particular integral = $u y_1 + v y_2$

Step 3. Find u and v by the formulae

$$u = \int \frac{-y_2 X}{y_1 y_2' - y_1' y_2} dx, \quad v = \int \frac{y_1 X}{y_1 y_2' - y_1' y_2} dx$$

Example 12. *Solve* $\frac{d^2 y}{dx^2} + y = \operatorname{cosec} x.$ (*Nagpur University, Summer* 2005)

Solution. $(D^2 + 1) y = \operatorname{cosec} x$

A.E. is $m^2 + 1 = 0 \Rightarrow m = \pm i$

$$\text{C.F.} = A \cos x + B \sin x$$

Here $y_1 = \cos x, \quad y_2 = \sin x$

$$\text{P.I.} = y_1 u + y_2 v$$

where $u = \int \frac{-y_2 . \operatorname{cosec} x \, dx}{y_1 . y'_2 - y'_1 . y_2} = \int \frac{-\sin x . \operatorname{cosec} x \, dx}{\cos x (\cos x) - (-\sin x)(\sin x)}$

$$= \int \frac{-\sin x . \frac{1}{\sin x} dx}{\cos^2 x + \sin^2 x} = -\int dx = -x$$

$$v = \int \frac{y_1 . X \, dx}{y_1 . y'_2 - y'_1 y_2} = \int \frac{\cos x . \operatorname{cosec} x \, dx}{\cos x (\cos x) - (-\sin x)(\sin x)}$$

$$= \int \frac{\cos x . \frac{1}{\sin x}}{\cos^2 x + \sin^2 x} dx = \int \frac{\cot x \, dx}{1} = \log \sin x$$

$$P.I. = uy_1 + vy_2 = -x \cos x + \sin x (\log \sin x)$$

General solution = C.F. + P.I.

$$y = A \cos x + B \sin x - x \cos x + \sin x .(\log \sin x)$$ **Ans.**

Example 13. *Apply the mehtod of variation of parameters to solve*

$$\frac{d^2y}{dx^2} + y = \tan x \qquad \text{(A. M. I. E. T. E., Dec. 2010, Winter 2001, Summer 2000)}$$

Solution. We have, $\frac{d^2y}{dx^2} + y = \tan x$

$$(D^2 + 1)y = \tan x$$

A.E. is $\quad m^2 = -1 \quad$ or $\quad m = \pm i$

C. F. $\quad y = A\cos x + B\sin x$

Here, $\quad y_1 = \cos x, \qquad y_2 = \sin x$

$$y_1 \cdot y'_2 - y'_1 \cdot y_2 = \cos x(\cos x) - (-\sin x)\sin x = \cos^2 x + \sin^2 x = 1$$

P. I. $\quad = u.\, y_1 + v.\, y_2$ where

$$u = \int \frac{-y_2 \tan x}{y_1 \cdot y'_2 - y'_1 \cdot y_2}\, dx = -\int \frac{\sin x \tan x}{1}\, dx$$

$$= -\int \frac{\sin^2 x}{\cos x}\, dx = -\int \frac{1-\cos^2 x}{\cos x}\, dx$$

$$= \int (\cos x - \sec x)\, dx = \sin x - \log(\sec x + \tan x)$$

$$v = \int \frac{y_1 \tan x}{y_1 \cdot y'_2 - y'_1 \cdot y_2}\, dx = \int \frac{\cos x \cdot \tan x}{1}\, dx = \int \sin x\, dx = -\cos x$$

P. I. $= u.\, y_1 + v.\, y_2$

$$= [\sin x - \log(\sec x + \tan x)]\cos x - \cos x \sin x$$

$$= -\cos x \log(\sec x + \tan x)$$

Complete solution is

$$y = A\cos x + B\sin x - \cos x\log(\sec x + \tan x)$$ **Ans.**

Example 14. *Use variation of parameters method to solve* $y'' + y = \sec x$

(*Nagpur University, Winter,* 2002, 2001, *U. P. Second Semester* 2002, *AMIETE, June 2010,* 2004)

Solution. We have, $\frac{d^2y}{dx^2} + y = \sec x$ ... (1)

A. E. is $\quad (m^2 + 1) = 0 \quad \Rightarrow \quad m^2 = -1 \Rightarrow \quad m = \pm i$

Complementary Function of (1) is

$$\text{C. F.} = A\cos x + B\sin x$$

Here $\quad y_1 = \cos x, \qquad y_2 = \sin x$

P. I. $= uy_1 + vy_2$ ... (2)

where $\quad u = \int \frac{-y_2 \sec x}{y_1 \cdot y'_2 - y'_1 \cdot y_2}\, dx \qquad \left[\begin{aligned} y_1 y'_2 - y_1' y_2 &= \cos x \cos x - (-\sin x)\sin x \\ &= \cos^2 x + \sin^2 x = 1 \end{aligned}\right]$

On putting the values of y_2 and $y_1 y_2' - y_1' y_2$, we get

$$u = \int \frac{-\sin x \sec x}{1}\, dx = -\int \tan x\, dx = \log \cos x$$

$$v = \int \frac{y_1 \sec x}{y_1 y_2' - y_1' y_2} dx$$

On putting the values of y_1 and $y_1 y_2' - y_1' y_2$, we get

$$v = \int \frac{\cos x \cdot \sec x}{1} dx = \int dx = x$$

Putting the values of u and v in (2), we get

$$P.\ I. = \cos x \cdot \log \cos x + x \sin x$$

Complete solution is $y = C.\ F. + P.\ I.$

$\Rightarrow \quad y = A \cos x + B \sin x + \cos x \log \cos x + x \sin x$ **Ans.**

Example 15. *Obtain general solution of the differential equation* $x^2 y'' + xy' - y = x^3 e^x$.

(*Nagpur University, Summer* 2004, *U. P. II Semester, Summer* 2002)

Solution. The given differential equation is $x^2 y'' + xy' - y = x^3 e^x$

$$\Rightarrow \quad x^2 \frac{d^2 y}{dx^2} + x \frac{dy}{dx} - y = x^3 e^x \quad \ldots (1)$$

Putting $x = e^z \quad \Rightarrow \quad D = \frac{d}{dz},\ x \frac{dy}{dx} = Dy, \quad x^2 \frac{d^2 y}{dx^2} = D(D-1)\, y,$ in (1), we get

$$D(D-1)\, y + D\, y - y = e^{3z} e^{e^z} \quad \Rightarrow \quad (D^2 - 1) y = e^{3z} e^{e^z}$$

A. E. is $m^2 - 1 = 0 \quad \Rightarrow \quad m = \pm 1$

$\therefore \quad$ C. F. $= c_1 e^z + c_2 e^{-z}$

$$= uy_1 + vy_2, \text{ where } y_1 = e^{-z},\ y = \left[y_1 = x_1\ y_2 = \frac{1}{x} \right]$$

$$\text{P.I.} = uy_1 + vy_2$$

Also $u = -\int \frac{y_2 z}{y_1 y_2' - y_1' y_2} dz = -\int \frac{e^{-z} . e^{3z} . e^{e^z}}{e^z(-e^{-z}) - e^z(e^{-z})} dz = -\int \frac{e^{2z} e^{e^z}}{-1-1} dz$

$$\left[x = e^z,\ dx = e^z dz \quad dz = \frac{dx}{e^z} = \frac{dx}{x} \right.$$

$$= \frac{1}{2} \int e^{2z} e^{e^z} dz = \int x^2 e^x \frac{dx}{x} = \frac{1}{2} \int x e^x dx$$

$$= \frac{1}{2} \left[xe^x - (1)\, e^x \right] = \frac{1}{2} (xe^x - e^x)$$

and $v = \int \frac{y_1 z}{y_1 y_2' - y_1' y_2} dz = \int \frac{e^z . e^{3z} . e^{e^z}}{e^z(-e^{-z}) - e^z(e^{-z})} dz = \int \frac{e^{4z} e^{e^z}}{-1-1} dz = \int \frac{x^4 e^x}{-2} \frac{dx}{x} = -\frac{1}{2} \int x^3 e^x dx$

$$= -\frac{1}{2} [x^3 e^x - 3x^2 e^x + 6x e^x - 6e^x]$$

P.I. $= uy_1 + vy_2 = \frac{1}{2} (xe^x - e^x)\, x - \frac{1}{2} (x^3 e^x - 3x^2 e^x + 6xe^x - 6e^x) \frac{1}{x}$

$$= \frac{1}{2} \left[x^2 - x - x^2 + 3x - 6 + \frac{6}{x} \right] e^x = \frac{1}{2} \left(2x - 6 + \frac{6}{x} \right) e^x = \left(x - 3 + \frac{3}{x} \right) e^x$$

Complete solution = C.F. + P.I.

$$y = (c_1 e^z + c_2 e^{-z}) + \left(x - 3 + \frac{3}{x} \right) e^x$$

$$= c_1 x + \frac{c_2}{x} + \left(x - 3 + \frac{3}{x} \right) e^x$$ **Ans.**

$$= \frac{e^x}{2}\left[x^4 - 3x^3 + 6x^2 - 6x - x^4 + 5x^3 - 20x^2 + 60x - 120 + \frac{120}{x}\right]$$

$$= \frac{e^x}{2}\left[2x^3 - 14x^2 + 54x - 120 + \frac{120}{x}\right]$$

$y = C.F. + P.I.$

$$= C_1 x + \frac{C_2}{x} + \left(x^3 - 7x^2 + 27x - 60 + \frac{60}{x}\right) e^x$$ **Ans.**

Example 16. *Solve by method of variation of parameters:*

$$\frac{d^2y}{dx^2} - y = \frac{2}{1+e^x}$$ (*Uttarakhand, II Semester, June* 2007, *A.M.I.E.T.E., Summer* 2001)

(*Nagpur University, Summer* 2001)

Solution. $$\frac{d^2y}{dx^2} - y = \frac{2}{1+e^x}$$

A. E. is $(m^2 - 1) = 0$

$m^2 = 1, \quad m = \pm 1$

$C.F. = C_1 e^x + C_2 e^{-x}$

$\therefore \quad P.I. = uy_1 + vy_2$

Here, $y_1 = e^x, \quad y_2 = e^{-x}$

and $y_1 \cdot y'_2 - y'_1 \cdot y_2 = -e^x \cdot e^{-x} - e^x . e^{-x} = -2$

$$u = \int \frac{-y_2 X}{y_1 . y'_2 - y'_1 \cdot y_2}\, dx = -\int \frac{e^{-x}}{-2} \times \frac{2}{1+e^x}\, dx$$

$$= \int \frac{e^{-x}}{1+e^x}\, dx = \int \frac{dx}{e^x(1+e^x)} = \int \left(\frac{1}{e^x} - \frac{1}{1+e^x}\right) dx$$

$$= \int e^{-x}\, dx - \int \frac{e^{-x}}{e^{-x}+1}\, dx = -e^{-x} + \log(e^{-x}+1)$$

$$v = \int \frac{y_1 X}{y_1 . y'_2 - y'_1 . y_2}\, dx = \int \frac{e^x}{-2} \frac{2}{1+e^x}\, dx = -\int \frac{e^x}{1+e^x}\, dx = -\log(1+e^x)$$

$$\text{P.I.} = u.\, y_1 + v.\, y_2 = [-e^{-x} + \log(e^{-x}+1)]\, e^x - e^{-x} \log(1+e^x)$$

$$= -1 + e^x \log(e^{-x}+1) - e^{-x} \log(e^x+1)$$

Complete solution $= y = C_1 e^x + C_2 e^{-x} - 1 + e^x \log(e^{-x}+1) - e^{-x} \log(e^x+1)$ **Ans.**

Example 17. *Apply the method of variation of parameters to solve*

$$\frac{d^2y}{dx^2} - 3\frac{dy}{dx} + 2y = \frac{e^x}{1+e^x}$$ (*U.P. II Semester Summer* 2005)

Solution. Given equation is $\dfrac{d^2y}{dx^2} - 3\dfrac{dy}{dx} + 2y = \dfrac{e^x}{1+e^x}$

Auxiliary equation is $m^2 - 3m + 2 = 0 \Rightarrow (m-1)(m-2) = 0 \Rightarrow m = 1, 2$

C. F. $= C_1 e^x + C_2 e^{2x}$

$= C_1 y_1 + C_2 y_2$, Here $y_1 = e^x$, $y_2 = e^{2x}$

P. I. $= uy_1 + vy_2$,

$$u = \int \frac{-y_2 X\, dx}{y_1 y_2' - y_2 y_1'} = \int \frac{-e^{2x} \cdot \frac{e^x}{1+e^x}}{e^x (2e^{2x}) - e^{2x}(e^x)} dx = \int \frac{\frac{-e^{3x}}{1+e^x}}{2e^{3x} - e^{3x}} dx$$

$$= \int \frac{-e^{3x}}{e^{3x}(1+e^x)} dx = -\int \frac{1}{1+e^x} dx = -\int \frac{e^{-x}}{e^{-x}+1} \cdot dx \cdot \qquad \text{[Dividing by } e^x\text{]}$$

$$= \log (e^{-x} + 1)$$

Now, $$v = \int \frac{y_1 X\, dx}{y_1 y_2' - y_2 y_1'} = \int \frac{e^x \left(\frac{e^x}{1+e^x}\right) dx}{e^x (2e^{2x}) - e^{2x}(e^x)} = \int \frac{\frac{e^{2x}}{1+e^x}}{2e^{3x} - e^{3x}} dx = \int \frac{e^{2x}}{e^{3x}(1+e^x)} dx$$

$$= \int \frac{1}{e^x (1+e^x)} dx = \int \left(\frac{1}{e^x} - \frac{1}{1+e^x}\right) dx \qquad \text{[By Partial fraction]}$$

$$= \int \left(e^{-x} - \frac{e^{-x}}{e^{-x}+1}\right) dx = -e^{-x} + \log (e^{-x} + 1)$$

P.I. $= uy_1 + vy_2$

P.I. $= e^x \log (e^{-x} + 1) + e^{2x} \{-e^{-x} + \log (e^{-x} + 1)\}$

P.I. $= e^x \log (e^{-x} + 1) - e^x + e^{2x} \log (e^{-x} + 1)$

Complete solution is

$y =$ C. F. + P. I.

$= C_1 e^x + C_2 e^{2x} + e^x \log (e^{-x} + 1) - e^x + e^{2x} \log (e^{-x} + 1)$ **Ans.**

Example 18. *Solve by method of variation of parameters.*

$$\frac{d^2 y}{dx^2} - y = \left(1 + \frac{1}{e^x}\right)^{-2}$$

(*Nagpur University, Summer* 2000)

Solution. $$\frac{d^2 y}{dx^2} - y = \left(\frac{e^x + 1}{e^x}\right)^{-2} = \frac{e^{2x}}{(e^x + 1)^2}$$

A. E. is $m^2 - 1 = 0$ $\therefore$ $m = \pm 1$

C. F. $= C_1 e^{-x} + C_2 e^x$

Let P. I. $= uy_1 + vy_2$, where $y_1 = e^{-x}$, $y_2 = e^x$

$y_1 y_2' - y_1' y_2 = e^{-x} \cdot e^x + e^{-x} \cdot e^x = 2$

$$u = \int \frac{-y_2 X}{y_1 y_2' - y_1' y_2} dx = \frac{1}{2} \int -e^x \cdot \frac{e^{2x}}{(1+e^x)^2} dx = -\frac{1}{2} \int \frac{e^{2x}}{(1+e^x)^2} e^x dx$$

Putting $t = 1 + e^x$, $dt = e^x\, dx$, we get

$$u = -\frac{1}{2} \int (t-1)^2 \frac{dt}{t^2}$$

$$u=-\frac{1}{2}\int\left(1-\frac{2}{t}+t^{-2}\right)dt=-\frac{1}{2}(t-2\log t-t^{-1})$$

$$u=-\frac{1}{2}(1+e^x)+\log(1+e^x)+\frac{1}{2}(1+e^x)^{-1}$$

Now, $$v=\int\frac{y_1X}{y_1y_2'-y_1'y_2}\,dx=\frac{1}{2}\int\frac{e^{-x}.e^{2x}}{(1+e^x)^2}\,dx=\frac{1}{2}\int\frac{e^x\,dx}{(1+e^x)^2}$$

$$=\frac{1}{2}\int\frac{1}{t^2}\,dt=-\frac{1}{2}\frac{1}{t}=-\frac{1}{2}(1+e^x)^{-1}$$ where $t=1+e^x$

P.I. $=uy_1+vy_2$

$$\text{P.I.}=e^{-x}\left[-\frac{1}{2}(1+e^x)+\log(1+e^x)+\frac{1}{2}(1+e^x)^{-1}\right]-\frac{1}{2}e^x(1+e^x)^{-1}$$

$$=-\frac{1}{2}(1+e^x)^{-1}\{e^x-e^{-x}\}+e^{-x}\left(-\frac{1}{2}\right)[(1+e^x)+e^{-x}\log(1+e^x)]$$

$$=-(1+e^x)^{-1}\sinh x-\frac{1}{2}e^{-x}(1+e^x)+e^{-x}\log(1+e^x)$$

Hence the solution is

$$y=C_1e^{-x}+C_2e^x-(1+e^x)^{-1}\sinh x-\frac{1}{2}e^{-x}(1+e^x)+e^{-x}\log(1+e^x)$$ **Ans.**

Example 19. *Solve by method of variation of parameters*

$$\frac{d^2y}{dx^2}-2\frac{dy}{dx}=e^x\sin x$$ (*U.P. II Semester,* 2003)

Solution. $$\frac{d^2y}{dx^2}-2\frac{dy}{dx}=e^x\sin x$$

$\Rightarrow$ $$(D^2-2D)y=e^x\sin x$$

A. E. is $m^2-2m=0 \Rightarrow m(m-2)=0 \Rightarrow m=0,2$

C. F. $=C_1+C_2e^{2x}$

P. I. $=uy_1+vy_2$ where, $y_1=1$, $y_2=e^{2x}$

$\therefore$ $$u=\int\frac{-y_2X\,dx}{y_1y_2'-y_1'y_2}=\int\frac{-e^{2x}.e^x\sin x}{1(2e^{2x})-0(e^{2x})}\,dx$$

$$=-\frac{1}{2}\int e^x\sin x\,dx \qquad \left[\int e^{ax}\sin bx\,dx=\frac{e^{ax}}{a^2+b^2}(a\sin bx-b\cos bx)\right]$$

$$=-\frac{1}{2}\frac{e^x}{1+1}\{\sin x-\cos x\}=-\frac{e^x}{4}(\sin x-\cos x)$$

$$v=\int\frac{y_1X\,dx}{y_1y_2'-y_1'y_2}=\int\frac{1.e^x\sin x\,dx}{1(2e^{2x})-0(e^{2x})}=\frac{1}{2}\int e^{-x}\sin x\,dx$$

$$=\frac{1}{2}\frac{e^{-x}}{(-1)^2+(1)^2}[-\sin x-\cos x]$$

$$=-\frac{1}{2}\frac{e^{-x}}{2}(\sin x+\cos x)=-\frac{e^{-x}}{4}(\sin x+\cos x)$$

$$\text{P. I.} = uy_1 + vy_2 = -\frac{e^x}{4}(\sin x - \cos x)\,1 - \frac{e^{-x}}{4}(\sin x + \cos x)\,e^{2x}$$

$$= -\frac{e^x}{4}[\sin x - \cos x + \sin x + \cos x] = -\frac{e^x}{2}\sin x$$

The complete solution is

$$y = C.\,F. + P.\,I.$$

$$\Rightarrow \quad y = C_1 + C_2 e^{2x} - \frac{e^x}{2}\sin x$$ **Ans.**

Example 20. *Solve :* $\dfrac{d^2y}{dx^2} + 3\dfrac{dy}{dx} + 2y = \sin(e^x)$ (*Nagpur University, Winter* 2003)

Solution. The given equation can be written as

$$(D^2 + 3D + 2)y = \sin e^x$$

A. E. is $(m^2 + 3m + 2) = 0 \quad \Rightarrow \quad (m + 1)(m + 2) = 0$

$m = -1, -2,$

$\therefore \quad \text{C. F.} = C_1 e^{-x} + C_2 e^{-2x}$

$\text{P. I.} = uy_1 + vy_2$

where $y_1 = e^{-x}, \quad y_2 = e^{-2x}$

$y_1' = -e^{-x}, \quad y_2' = -2e^{-2x}$

$$y_1y_2' - y_2y_1' = e^{-x}(-2e^{-2x}) - e^{-2x}(-e^{-x}) = -2e^{-3x} + e^{-3x} = -e^{-3x}$$

$$u = \int \frac{-y_2 X}{y_1y'_2 - y'_1 y_2}\,dx = \int \frac{-e^{-2x}}{-e^{-3x}}\sin(e^x)\,dx$$

$$= \int e^x \sin(e^x)\,dx = -\cos(e^x)$$

$$= \int \frac{y_1 X}{y_1y'_2 - y'_1 y_2}\,dx = \int \frac{e^{-x}\sin(e^x)\,dx}{-e^{-3x}}$$

$$= -\int e^{2x}\sin(e^x)\,dx = -\int t\sin t\,dt \qquad [t = e^x \text{ so that } dt = e^x\,dx]$$

$$= t\cos t - \sin t = e^x\cos(e^x) - \sin(e^x)$$

Putting the values of u, v, y_1 and y_2 in (1), we get

$$\text{P.I.} = e^{-x}[-\cos(e^x)] + e^{-2x}[e^x\cos(e^x) - \sin(e^x)] = -e^{-2x}\sin(e^x)$$

The solution is $y = C_1e^{-x} + C_2e^{-2x} - e^{-2x}\sin(e^x).$ **Ans.**

Example 21. *Solve by method of variation of parameters:*

$$\frac{d^2y}{dx^2} + y = (x - \cot x)$$ (*Nagpur University, Winter* 2001)

Solution. Here A. E. is $m^2 + 1 = 0 \qquad \therefore \qquad m = \pm i$

$$C.\,F. = C_1\cos x + C_2\sin x = C_1y_1 + C_2y_2$$

where $y_1 = \cos x, \qquad y_2 = \sin x$

$y'_1 = -\sin x, \qquad y'_2 = \cos x$

$$y_1y'_2 - y_2y'_1 = \cos x\,.\cos x + \sin x.\,\sin x = 1$$

Let $\quad$ P. I. $= uy_1 + vy_2$ $\quad$ ... (1)

Where
$$u = \int \frac{-y_2 X}{y_1y_2' - y_2y'_1}\,dx = \int \frac{-\sin x(x - \cot x)}{1}\,dx$$
$$= \int \cos x\,dx - \int x \sin x\,dx = \sin x - \{x(-\cos x) + \sin x\} = x\cos x$$
$$v = \int \frac{y_1 X}{y_1y'_2 - y_2y'_1}\,dx = \int \cos x\,(x - \cot x)\,dx$$
$$= \int x\cos x\,dx - \int \frac{\cos^2 x}{\sin x}\,dx = \int x\cos x\,dx - \int \frac{1 - \sin^2 x}{\sin x}\,dx$$
$$= \int x\cos x\,dx - \int \operatorname{cosec} x\,dx + \int \sin x\,dx$$
$$= x\sin x + \cos x - \log(\operatorname{cosec} x - \cot x) - \cos x$$
$$= x\sin x - \log(\operatorname{cosec} x - \cot x)$$

Putting the values of u, v, y_1, y_2 in (1), we get
$$\text{P. I.} = \cos x.\,x\cos x + \sin x\,\{x\sin x - \log(\operatorname{cosec} x - \cot x)\}$$
$$= x\cos^2 x + x\sin^2 x - \sin x\log(\operatorname{cosec} x - \cot x)$$
$$= x - \sin x\log(\operatorname{cosec} x - \cot x)$$

Hence, complete solution is
$$y = C_1\cos x + C_2\sin x + x - \sin x\log(\operatorname{cosec} x - \cot x) \qquad \textbf{Ans.}$$

Example 22. *Solve* $\dfrac{d^2y}{dx^2} + 2\dfrac{dy}{dx} + 2y = e^{-x}\sec^3 x.$ $\quad$ (*Nagpur University, Winter* 2000)

Solution. A. E. is $m^2 + 2m + 2 = 0$
$$m = \frac{-2 + \sqrt{4 - 8}}{2} = \frac{-2 \pm 2i}{2} = -1 + i$$

C. F. $= e^{-x}(C_1\cos x + C_2\sin x) = C_1y_1 + C_2y_2$

where $y_1 = e^{-x}\cos x \quad \Rightarrow \quad y_1' = -e^{-x}\sin x - e^{-x}\cos x$

$y_2 = e^{-x}\sin x \quad \Rightarrow \quad y'_2 = -e^{-x}\sin x + e^{-x}\cos x$
$$y_1y'_2 - y'_1y_2 = e^{-x}\cos x(-e^{-x}\sin x + e^{-x}\cos x) - (-e^{-x}\cos x - e^{-x}\sin x)\,e^{-x}\sin x$$
$$= e^{-2x}(\sin^2 x + \cos^2 x - \sin x\cos x + \sin x\cos x)$$
$$= e^{-2x}(\sin^2 x + \cos^2 x) = e^{-2x}$$

Let P.I. $= uy_1 + vy_2$

where
$$u = -\int \frac{y_2 X}{y_1y'_2 - y'_1y_2}\,dx = -\int \frac{e^{-x}\sin x.\,e^{-x}\sec^3 x}{e^{-2x}}\,dx$$
$$= -\int \sin x.\sec^3 x\,dx = -\int \tan x\cdot\sec^2 x\,dx = -\frac{\tan^2 x}{2}$$
$$v = \int \frac{y_1 X}{y_1y'_2 - y'_1y_2}\,dx = \int \frac{e^{-x}\cos x.\,e^{-x}\sec^3 x}{e^{-2x}}\,dx = \int \cos x.\sec^3 x\,dx = \int \sec^2 x\,dx = \tan x$$

P. I. $= uy_1 + vy_2$
$$= \frac{-\tan^2 x}{2}e^{-x}\cos x + \tan x.\,e^{-x}\sin x = e^{-x}\left[\frac{-\tan^2 x}{2}\cos x + \tan x.\sin x\right]$$

$$= e^{-x}\left[\frac{-\sin x\,.\,\tan x}{2} + \tan x\,.\,\sin x\right] = \frac{1}{2}e^{-x}\sin x\tan x$$

Complete solution is

$$y = C.\,F. + P.\,I.$$

$$= e^{-x}(C_1\cos x + C_2\sin x) + \frac{1}{2}e^{-x}\sin x\tan x$$ **Ans.**

Example 23. *Apply the method of variation of parameters to solve*

$$\frac{d^2y}{dx^2} - y = e^{-x}\sin(e^{-x}) + \cos(e^{-x})$$ (*Nagpur University, Summer* 2002)

Solution. The auxiliary equation is $m^2 - 1 = 0$

$$\therefore \quad m = \pm 1$$

$$\therefore \quad C.\,F. = C_1 e^x + C_2 e^{-x} = C_1 y_1 + C_2 y_2$$

where $y_1 = e^x$ and $y_2 = e^{-x}$

Let P. I. $= uy_1 + vy_2$

where $$u = \int \frac{-y_2 X}{y_1 y'_2 - y'_1 y_2}\,dx = \int -\frac{e^{-x}\{e^{-x}\sin(e^{-x}) + \cos(e^{-x})\}}{e^x(-e^{-x}) - e^x e^{-x}}\,dx$$

i.e. $$u = \frac{1}{2}\int [e^{-2x}\sin(e^{-x}) + e^{-x}\cos(e^{-x})]\,dx \quad (\text{Put } e^{-x} = t,\ -e^{-x}dx = dt)$$

$$= -\frac{1}{2}\int (t\sin t + \cos t)\,dt = -\frac{1}{2}\{-t\cos t + \sin t + \sin t\}$$

$$= \frac{1}{2}[t\cos t - 2\sin t] = \frac{1}{2}[e^{-x}\cos(e^{-x}) - 2\sin(e^{-x})]$$

$$v = \int \frac{y_1 X}{y_1 y'_2 - y'_1 y_2}\,dx = \int \frac{e^x\{e^{-x}\sin(e^{-x}) + \cos(e^{-x})\}}{e^x(-e^{-x}) - e^x e^{-x}}\,dx$$

$$= -\frac{1}{2}\int \{\sin(e^{-x}) + e^x\cos(e^{-x})\}\,dx$$

$$\left[e^{-x} = t \text{ and } -e^{-x}dx = dt \text{ i.e. } dx = -\frac{dt}{t}\right]$$

$$= \frac{1}{2}\int\left(\frac{\sin t}{t} + \frac{\cos t}{t^2}\right)dt = \frac{1}{2}\left\{\frac{-\cos t}{t} - \int\frac{\cos t}{t^2}\,dt + \int\frac{\cos t}{t}\,dt\right\}$$

$$= -\frac{1}{2}\frac{\cos t}{t} = -\frac{1}{2}e^x\cos(e^{-x})$$

$$P.\,I. = e^x\left\{\frac{1}{2}e^{-x}\cos(e^{-x}) - \sin(e^{-x})\right\} - \frac{1}{2}e^{-x}e^x\cos(e^{-x})$$

$$= -e^x\sin(e^{-x})$$

$\therefore$ Required solution is y = C. F. + P. I.

i.e. $$y = C_1e^x + C_2e^{-x} - e^x\sin(e^{-x})$$ **Ans.**

Example 24. *Solve* $(D^2 + 2D + 1)\,y = 4e^{-x}\log x$

by method of variation of parameters. (*Nagpur University, Winter* 2004)

Solution. $(D^2 + 2D + 1)\,y = 4e^{-x}\log x$

A. E. is $m^2 + 2m + 1 = 0$

$$(m+1)^2 = 0$$

$$m = -1, -1$$

$$\therefore \quad C.F. = (C_1 + C_2 x)\, e^{-x} \quad \Rightarrow C.F. = C_1 e^{-x} + C_2 x e^{-x} = C_1 y_1 + C_2 y_2$$

$$y_1 = e^{-x} \quad \Rightarrow \quad y'_1 = -e^{-x}$$

$$y_2 = xe^{-x} \quad \Rightarrow \quad y'_2 = -xe^{-x} + e^{-x}$$

$$y_1 y'_2 - y'_1 y_2 = e^{-x}(-xe^{-x} + e^{-x}) + e^{-x}(x e^{-x})$$

$$= -xe^{-2x} + e^{-2x} + xe^{-2x} = e^{-2x}$$

Let P. I. $= uy_1 + vy_2$ ... (1)

where $u = -\int \dfrac{y_2 X}{y_1 y'_2 - y'_1 y_2}\, dx$

$$u = -\int \frac{xe^{-x}\, 4e^{-x} \log x}{e^{-2x}}\, dx = -4\int x \log x\, dx = -4\left[\log x \frac{x^2}{2} - \int \frac{1}{x}\frac{x^2}{2}\, dx\right]$$

$$= -4\left[\frac{x^2}{2}\log x - \frac{1}{2}\frac{x^2}{2}\right] = -2x^2 \log x + x^2$$

$$u = x^2(1 - 2\log x)$$

and $v = \int \dfrac{y_1 X}{y_1 y'_2 - y'_1 y_2}\, dx$

$$v = \int \frac{e^{-x}\,.\,4e^{-x} \log x}{e^{-2x}}\, dx = 4\int \log x\, dx = 4\int 1\cdot \log x\, dx = 4\left[(\log x)\, x - \int \frac{1}{x}\,.\,x\, dx\right]$$

$$= 4[x \log x - x]$$

$$v = 4x(\log x - 1)$$

Putting the values of u, v, y_1 and y_2 in (1), we get

$$\text{P. I.} = x^2(1 - 2\log x)\, e^{-x} + 4x(\log x - 1)x\, e^{-x} = x^2 e^{-x}[1 - 2\log x + 4\log x - 4]$$

$$\text{P.I.} = x^2 e^{-x}[2\log x - 3]$$

C. S. = C. F. + P. I.

$$y = (C_1 + C_2 x)\, e^{-x} + x^2 e^{-x}[2\log x - 3]$$ **Ans.**

EXERCISE 13.2

Solve the following equations by variation of parameters method.

1. $\dfrac{d^2y}{dx^2} - 4y = e^{2x}$ **Ans.** $y = C_1 e^{2x} + C_2 e^{-2x} + \dfrac{x}{4}e^{2x} - \dfrac{e^{2x}}{16}$

2. $\dfrac{d^2y}{dx^2} + y = \sin x$ **Ans.** $y = C_1 \cos x + C_2 \sin x - \dfrac{x}{2}\cos x + \dfrac{1}{4}\sin x$

3. $\dfrac{d^2y}{dx^2} - 3\dfrac{dy}{dx} + 2y = \sin x$ **Ans.** $y = C_1 e^{x} + C_2 e^{2x} + \dfrac{1}{10}(3\cos x + \sin x)$

4. $\dfrac{d^2y}{dx^2} + y = \sec x \tan x$ **Ans.** $y = C_1 \cos x + C_2 \sin x + x\cos x + \sin x \log \sec x - \sin x$

5. $y'' - 6y' + 9y = \dfrac{e^{3x}}{x^2}$ *(AMIETE, June 2010, 2009)* **Ans.** $y = (C_1 + x\,C_2)\, e^{3x} - e^{3x}\log x$

Simultaneous Linear Differential Equations

14.1 SIMULTANEOUS DIFFERENTIAL EQUATIONS

If two or more dependent variables are functions of a single independent variable, the equations involving their deriavatives are called simultaneous equations, *e.g.*

$$\frac{dx}{dt} + 4y = t$$

$$\frac{dy}{dt} + 2x = e^t$$

The method of solving these equations is based on the process of elimination, as we solve algebraic simultaneous equations.

Example 1. *The equations of motions of a particle are given by*

$$\frac{dx}{dt} + \omega y = 0$$

$$\frac{dy}{dt} - \omega x = 0$$

Find the path of the particle and show that it is a circle.

(*R.G.P.V. Bhopal, Feb. 2006, U.P. II Semester summer* 2009)

Solution. On putting $\frac{d}{dt} \equiv D$ in the equations, we have

$$Dx + \omega y = 0 \quad \text{...(1)}$$

$$-\omega x + Dy = 0 \quad \text{...(2)}$$

On multiplying (1) by w and (2) by D, we get

$$\omega Dx + \omega^2 y = 0 \quad \text{...(3)}$$

$$-\omega Dx + D^2 y = 0 \quad \text{...(4)}$$

On adding (3) and (4), we obtain

$$\omega^2 y + D^2 y = 0 \quad \Rightarrow \quad (D^2 + \omega^2)\, y = 0 \quad \text{...(5)}$$

Now, we have to solve (5) to get the value of y.

A.E. is $m^2 + \omega^2 = 0 \quad \Rightarrow \quad m^2 = -\omega^2 \quad \Rightarrow \quad m = \pm i\omega$

$$\therefore \quad y = A \cos \omega t + B \sin \omega t \quad \text{...(6)}$$

$\Rightarrow Dy = -A\omega \sin \omega t + B\omega \cos wt$

On putting the value of Dy in (2), we get

$-\omega x - A\omega \sin \omega t + B\omega \cos \omega t = 0$

$$\Rightarrow \quad \omega x = -A\omega \sin \omega t + B\omega \cos \omega t$$

$$\Rightarrow \qquad x = -A \sin \omega t + B \cos \omega t \qquad ...(7)$$

On squaring (6) and (7) and adding , we get

$$x^2 + y^2 = A^2(\cos^2\omega t + \sin^2\omega t) + B^2 (\cos^2\omega t + \sin^2\omega t)$$

$$\Rightarrow \qquad x^2 + y^2 = A^2 + B^2$$

This is the equation of circle. **Proved.**

Example 2. *Solve the following differential equation*

$$\frac{dx}{dt} = y + 1, \quad \frac{dy}{dt} = x + 1$$

(U.P. II Semester, 2009)

Solution. Here, we have

$$Dx - y = 1 \qquad ...(1)$$

$$-x + Dy = 1 \qquad ...(2)$$

Multiplying (1) by D, we get

$$D^2x - Dy = D.1 \qquad ...(3)$$

Adding (2) and (3), we get

$$(D^2 - 1)\, x = 1 + D.1$$

$$\Rightarrow \qquad (D^2 - 1)\, x = 1 \text{ or } (D^2 - 1)x = e^0 \qquad [D.\,(1) = 0]$$

A.E. is $\quad m^2 - 1 = 0 \;\Rightarrow\; m = \pm 1$

$$\therefore \qquad \text{C.F.} = c_1 e^t + c_2 e^{-t}$$

$$\text{P.I.} = \frac{1}{D^2 - 1} . e^0 = \frac{1}{0 - 1} e^0 = -1$$

$$\therefore \qquad x = \text{C.F.} + \text{P.I.} = c_1 e^t + c_2 e^{-t} - 1$$

From (1), $y = \dfrac{dx}{dt} - 1$

$$\Rightarrow \qquad y = \frac{d}{dt}(c_1 e^t + c_2 e^{-t} - 1) - 1$$

$$\Rightarrow \qquad \left.\begin{aligned} y &= c_1 e^t - c_2 e^{-t} - 1 \\ \text{and} \quad x &= c_1 e^t + c_2 e^{-t} - 1 \end{aligned}\right\} \qquad \textbf{Ans.}$$

Example 3. *Solve*:

$$\frac{dx}{dt} + y = \sin t$$

$$\frac{dy}{dt} + x = \cos t \qquad \text{where } y(0) = 0, \quad x(0) = 2$$

(R.G.P.V., Bhopal, I Semester, April, 2010 June 2007)

Solution. We have,

$$\frac{dx}{dt} + y = \sin t \quad \Rightarrow \quad Dx + y = \sin t \qquad ... (1)$$

$$\frac{dy}{dt} + x = \cos t \quad \Rightarrow \quad Dy + x = \cos t \qquad ... (2)$$

Multiplying (2) by D, we get

$$D^2 y + Dx = D \cos t$$

$$D^2 y + Dx = -\sin t \qquad ... (3)$$

Subtracting (1) from (3), we have

$$D^2 y - y = -2 \sin t$$

$$\Rightarrow \qquad (D^2 - 1)\, y = -2 \sin t$$

A.E. is $m^2 - 1 = 0$ $\Rightarrow$ $m^2 = 1$ $\Rightarrow$ $m = \pm 1$

C.F. $= C_1 e^t + C_2 e^{-t}$

P.I. $= \dfrac{1}{D^2 - 1}(-2 \sin t)$

$\Rightarrow$ P.I. $= \dfrac{1}{-1-1}(-2 \sin t) = \sin t$

Complete solution = C.F. + P.I.

$$y = C_1 e^t + C_2 e^{-t} + \sin t \qquad \dots (4)$$

Putting $y = 0$ and $t = 0$ in (4), we get

$$0 = C_1 + C_2 \qquad \text{or} \qquad C_2 = -C_1$$

On putting $C_2 = -C_1$ in (4), we get

$$y = C_1 e^t - C_1 e^{-t} + \sin t$$

On putting the value of y in (2), we get

$$D(C_1 e^t - C_1 e^{-t} + \sin t) + x = \cos t$$

$$C_1 e^t + C_1 e^{-t} + \cos t + x = \cos t$$

$$x = -C_1 e^t - C_1 e^{-t} \qquad \dots(5)$$

On putting $x = 2$, $t = 0$ in (5), we get

$$2 = -C_1 - C_1 \qquad \Rightarrow \qquad C_1 = -1$$

Putting the value of C_1 in (5) and (4), we have

$$x = e^t + e^{-t}$$

$$y = -e^t + e^{-t} + \sin t$$

Which is the required solution. **Ans.**

Example 4. *Solve:* $\dfrac{dx}{dt} + 4x + 3y = t$

$$\frac{dy}{dt} + 2x + 5y = e^t$$

[*U.P. II Semester, 2006*]

Solution. Here, we have

$$(D+4)x + 3y = t \qquad \dots(1)$$

$$2x + (D+5)y = e^t \qquad \dots(2) \quad \left(D \equiv \frac{d}{dt}\right)$$

To eliminate y, operating (1) by $(D + 5)$ and multiplying (2) by 3 then subtracting, we get

$$(D+5)(D+4)x + 3(D+5)y - 3(2x) - 3(D+5)y = (D+5)t - 3e^t$$

$$[(D+4)(D+5) - 6]x = (D+5)t - 3e^t$$

$$(D^2 + 9D + 14)x = 1 + 5t - 3e^t$$

Auxiliary equation is

$$m^2 + 9m + 14 = 0 \Rightarrow m = -2, -7$$

$\therefore$ C.F. $= c_1 e^{-2t} + c_2 e^{-7t}$

$$\text{P.I.} = \frac{1}{D^2 + 9D + 14}(1 + 5t - 3e^t)$$

$$= \frac{1}{D^2+9D+14} e^{0t} + 5 \frac{1}{D^2+9D+14} t - 3 \frac{1}{D^2+9D+14} e^t$$

$$= \frac{1}{0^2+9(0)+14} e^{0t} + 5 . \frac{1}{14\left(1+\frac{9D}{14}+\frac{D^2}{14}\right)} t - 3 \frac{1}{1^2+9(1)+14} e^t \quad ... (5)$$

$$= \frac{1}{14} + \frac{5}{14}\left[1+\left(\frac{9D}{14}+\frac{D^2}{14}\right)\right]^{-1} t - \frac{1}{8} e^t = \frac{1}{14} + \frac{5}{14}\left[1-\left(\frac{9D}{14}+\frac{D}{14}\right)+\cdots\right] t - \frac{1}{8} e^t$$

$$= \frac{1}{14} + \frac{5}{14}\left(t - \frac{9}{14}\right) - \frac{1}{8} e^t = \frac{5}{14} t - \frac{31}{196} - \frac{1}{8} e^t$$

$$x = c_1 e^{-2t} + c_2 e^{-7t} + \frac{5}{14} t - \frac{31}{196} - \frac{1}{8} e^t$$

$$3y = t - \frac{dx}{dt} - 4x \quad \text{[From (1)]}$$

$$= t - \frac{d}{dt}\left[c_1 e^{-2t} + c_2 e^{-7t} + \frac{5}{14} t - \frac{31}{196} - \frac{1}{8} e^t\right] - 4\left[c_1 e^{-2t} + c_2 e^{-7t} + \frac{5}{14} t - \frac{31}{196} - \frac{e^t}{8}\right]$$

$$3y = t + 2c_1 e^{-2t} + 7c_2 e^{-7t} - \frac{5}{14} + \frac{1}{8} e^t - 4c_1 e^{-2t} - 4c_2 e^{-7t} - \frac{10}{7} t + \frac{31}{49} + \frac{1}{2} e^t$$

$$\therefore y = \frac{1}{3}\left[-2c_1 e^{-2t} + 3c_2 e^{-7t} - \frac{3}{7} t + \frac{27}{98} + \frac{5}{8} e^t\right]$$

Hence,
$$x = c_1 e^{-2t} + c_2 e^{-7t} + \frac{5}{14} t - \frac{31}{196} - \frac{1}{8} e^t$$

$$y = -\frac{2}{3} c_1 e^{-2t} + c_2 e^{-7t} - \frac{1}{7} t + \frac{9}{98} + \frac{5}{24} e^t$$ **Ans.**

Example 5. *Solve:* $\frac{dx}{dt} + 5x + y = e^t, \quad \frac{dy}{dt} + x + 5y = e^{5t}.$ *(R.G.P.V. Bhopal, 2003)*

Solution. Here, we have

$$(D+5)\,x + y = e^t \quad ...(1)$$

$$x + (D+5)y = e^{5t} \quad ...(2) \left(D \equiv \frac{d}{dt}\right)$$

Multiplying (1) by $(D + 5)$, we get

$$(D+5)^2 x + (D+5)y = (D+5)\, e^t \quad ...(3)$$

Subtracting (3) form (2), we get

$$\left\{1-(D+5)^2\right\} x = e^{5t} - (D+5)e^t$$

$$\Rightarrow \quad [\,1 - D^2 - 10D - 25]x = e^{5t} - e^t - 5e^t$$

$$\Rightarrow \quad (D^2 + 10D + 24)x = 6e^t - e^{5t}$$

Auxiliary equation is

$$m^2 + 10m + 24 = 0 \quad \Rightarrow \quad (m+4)(m+6) = 0$$

$$\Rightarrow \quad m = -4, -6$$

$$\therefore \quad \text{C.F.} = c_1 e^{-4t} + c_2 e^{-6t}$$

$$\text{P.I.} = \frac{1}{D^2 + 10D + 24}(6e^t - e^{5t})$$

$$= \frac{6}{1^2 + 10(1) + 24} e^t - \frac{1}{(5)^2 + 10(5) + 24} e^{5t} = \frac{6e^t}{35} - \frac{e^{5t}}{99}$$

Thus $\quad x = \text{C.F.} + \text{P.I.} = c_1 e^{-4t} + c_2 e^{-6t} + \dfrac{6e^t}{35} - \dfrac{e^{5t}}{99}$

From (1),

$$y = e^t - (D+5)x = e^t - \frac{dx}{dt} - 5x$$

$$= e^t - \frac{d}{dt}\left(c_1 e^{-4t} + c_2 e^{-6t} + \frac{6e^t}{35} - \frac{e^{5t}}{99}\right) - 5\left(c_1 e^{-4t} + c_2 e^{-6t} + \frac{6e^t}{35} - \frac{e^{5t}}{99}\right)$$

$$= e^t + 4c_1 e^{-4t} + 6c_2 e^{-6t} - \frac{6e^t}{35} + \frac{5e^{5t}}{99} - 5c_1 e^{-4t} - 5c_2 e^{-6t} - \frac{30e^t}{35} + \frac{5e^{5t}}{99}$$

$$y = -\frac{1}{35} e^t - c_1 e^{-4t} + \frac{10}{99} e^{5t} + c_2 e^{-6t}$$

$$x = c_1 e^{-6t} + c_2 e^{-4t} + \frac{6e^t}{35} - \frac{e^{5t}}{99}$$ **Ans.**

Example 6. *Solve the following system of differential equations*

$$Dx + Dy + 3x = \sin t \text{ and}$$

$$Dx + y - x = \cos t$$ *(U.P. II Semester, Summer* 2003)

Solution. We have, $(D+3)\,x + Dy = \sin t$ ...(1)

$$(D-1)\,x + y = \cos t \quad \text{...(2)}$$

Operating (2) by D, we get

$$D\,(D-1)\,x + Dy = -\sin t \quad \text{...(3)}$$

Subtracting (1) from (3), we get

$$\{D\,(D-1) - (D+3)\}\,x = -2\sin t$$

$$\Rightarrow \quad \{D^2 - D - D - 3\}\,x = -2\sin t$$

$$\Rightarrow \quad (D^2 - 2D - 3)\,x = -2\sin t$$

A.E. is $\quad m^2 - 2m - 3 = 0 \quad \Rightarrow (m+1)(m-3) = 0$

$$\Rightarrow \quad (D-3)(D+1) = 0 \quad \Rightarrow m = 3, -1$$

$$\therefore \quad C.F. = c_1\, e^{3t} + c_2\, e^{-t}$$

$$P.I. = \frac{1}{D^2 - 2D - 3}(-2\sin t) = -2\,\frac{1}{(-1) - 2D - 3}\sin t$$

$$= 2 \cdot \frac{1}{2\,(D+2)}\sin t = \frac{(D-2)}{D^2 - 4}\sin t$$

$$= \frac{D-2}{-1-4}\sin t = \frac{\cos t - 2\sin t}{-5} = \frac{1}{5}(2\sin t - \cos t)$$

$$x = \text{C.F.} + \text{P.I.} = c_1 e^{3t} + c_2 e^{-t} + \frac{1}{5}(2\sin t - \cos t) \qquad ...(4)$$

From (2), we get

$$(D-1)\,x + y = \cos t$$

$$\Rightarrow (D-1)\left\{c_1 e^{3t} + c_2 e^{-t} + \frac{1}{5}(2\sin t - \cos t)\right\} + y = \cos \text{t}$$

$$\Rightarrow y = \cos t - D\left\{c_1 e^{3t} + c_2 e^{-t} + \frac{1}{5}(2\sin t - \cos t)\right\} + \left\{c_1 e^{3t} + c_2 e^{-t} + \frac{1}{5}(2\sin t - \cos t)\right\}$$

$$= \cos t - 3c_1 e^{3t} + c_2 e^{-t} - \frac{1}{5}(2\cos t + \sin t) + c_1 e^{3t} + c_2 e^{-t} + \frac{1}{5}(2\sin t - \cos t)$$

$$= \cos t - 2c_1 e^{3t} + 2c_2 e^{-t} - \frac{1}{5}[3\cos t - \sin t]$$

$$= \frac{2}{5}\cos t + \frac{1}{5}\sin t + 2c_2 e^{-t} - 2c_1 e^{3t}$$

Here, $y = \frac{1}{5}(2\cos t + \sin t) - 2c_1 e^{3t} + 2c_2 e^{-t}$ ...(5)

(4) and (5) are the required solutions **Ans.**

Example 7. *Solve the simultaneous equations*

$$\frac{dx}{dt} + \frac{dy}{dt} - 2y = 2\cos t - 7\sin t \qquad ...(1)$$

$$\frac{dx}{dt} - \frac{dy}{dt} + 2x = 4\cos t - 3\sin t \qquad ...(2)$$

(*U.P., B. Pharm* 2005, *II Semester, Summer* 2001)

Solution. $Dx + (D-2)\,y = 2\cos t - 7\sin t$ ...(3)

$(D+2)\,x - Dy = 4\cos t - 3\sin t$ ...(4)

Operating (3) by D (4) by $(D-2)$, we get

$$\Rightarrow \begin{array}{l} D^2x + D(D-2)\,y = -2\sin t - 7\cos t \qquad ...(5) \\ (D^2-4)\,x - D(D-2)\,y = (D-2)\,4\cos t - (D-2)\,3\sin t \qquad ...(6) \end{array}$$

On adding (5) and (6), we get

$$(D^2 + D^2 - 4)\,x = -2\sin t - 7\cos t - 4\sin t - 8\cos t - 3\cos t + 6\sin t$$

$$\Rightarrow (2D^2 - 4)\,x = -18\cos t$$

$$\Rightarrow (D^2 - 2)\,x = -9\cos t$$

A.E. is $m^2 - 2 = 0 \Rightarrow m = \pm\sqrt{2}$, C.F. $= c_1 e^{\sqrt{2}t} + c_2 e^{-\sqrt{2}t}$

$$P.I. = \frac{1}{D^2-2}(-9\cos t) = \frac{-9}{-1-2}\cos t = 3\cos \text{t}$$

$$x = c_1 e^{\sqrt{2}t} + c_2 e^{-\sqrt{2}t} + 3\cos t \qquad ...(7)$$

Putting the value of x in (2), we get

$$\frac{dy}{dt} = \sqrt{2}\,c_1 e^{\sqrt{2}t} - \sqrt{2}\,c_2 e^{-\sqrt{2}t} - 3\sin t + 2c_1 e^{\sqrt{2}t} + 2c_2 e^{-\sqrt{2}t} + 6\cos t - 4\cos t + 3\sin t$$

$$\frac{dy}{dt} = (2+\sqrt{2})\,c_1 e^{\sqrt{2}t} + (2-\sqrt{2})\,c_2 e^{-\sqrt{2}t} + 2\cos t$$

On integrating, we get

$$\Rightarrow y = (\sqrt{2}+1)\, c_1 e^{\sqrt{2}t} - (\sqrt{2}-1)\, c_2 e^{-\sqrt{2}t} + 2\sin t + c_3 \qquad ...(8)$$

Relations (7) and (8) are the required solutions **Ans.**

Example 8. *Solve* $\frac{dx}{dt} = 2y,\ \frac{dy}{dt} = 2z,\ \frac{dz}{dt} = 2x$ (*Uttarakhand, II Semester, June* 2007)

Solution. Here, we have

$$\frac{dx}{dt} = 2y \quad \Rightarrow \quad Dx = 2y \qquad ...(1)$$

$$\frac{dy}{dt} = 2z \quad \Rightarrow \quad Dy = 2z \qquad ...(2)$$

$$\frac{dz}{dt} = 2x \quad \Rightarrow \quad Dz = 2x \qquad ...(3)$$

From (1), we have $\frac{dx}{dt} = 2y$

$$\Rightarrow \quad \frac{d^2x}{dt^2} = \frac{2\,dy}{dt} = 2\,(2z) = 4z \qquad \left[\text{Using (2), } \frac{dy}{dt} = 2z\right]$$

$$\frac{d^3x}{dt^3} = 4\frac{dz}{dt} = 4\,(2x) = 8x \qquad \left[\text{Using (3), } \frac{dz}{dt} = 2x\right]$$

$$\Rightarrow \quad \frac{d^3x}{dt^3} - 8x = 0 \quad \Rightarrow \quad (D^3 - 8)\,x = 0$$

A.E. is $m^3 - 8 = 0 \Rightarrow (m-2)(m^2+2m+4) = 0$

$\Rightarrow \quad m - 2 = 0 \Rightarrow m = 2$

or $m^2 + 2m + 4 = 0 \Rightarrow m = \frac{-2 \pm \sqrt{4-16}}{2} = \frac{-2 \pm i\sqrt{12}}{2} = -1 \pm i\sqrt{3}$

So the C.F. of x is

$$x = C_1 e^{2t} + e^{-t}\,(A\cos\sqrt{3}t + B\sin\sqrt{3}t) \qquad ...(4)$$

$$[A = C_2\cos\alpha,\ B = C_2\sin\alpha] \qquad \left[\begin{array}{l}\tan\alpha = \frac{B}{A} \\ \alpha = \tan^{-1}\left(\frac{B}{A}\right)\end{array}\right]$$

$$x = C_1 e^{2t} + e^{-t}\,[C_2\cos\alpha\cos\sqrt{3}t + C_2\sin\alpha\sin\sqrt{3}t]$$

$$x = C_1 e^{2t} + e^{-t} C_2\cos(\sqrt{3}t - \alpha) = C_1 e^{2t} + C_2 e^{-t}\cos(\sqrt{3}t - \alpha)$$

From (3), we have $\frac{dz}{dt} = 2x$

$$\Rightarrow \frac{dz}{dt} = 2C_1 e^{2t} + 2C_2 e^{-t}\cos(\sqrt{3}t - \alpha) \qquad [\text{On putting the value of } x]$$

$$z = C_1 e^{2t} + 2C_2 \frac{e^{-t}}{\sqrt{1+3}}\cos(\sqrt{3}t - \alpha - \beta) \qquad \left[\int e^{ax}\cos bx\,dx = \frac{e^{ax}}{\sqrt{a^2+b^2}}\cos(bx - \beta)\right]$$

$$\Rightarrow z = C_1 e^{2t} + 2C_2 \frac{e^{-t}}{\sqrt{1+3}}\cos\left[\sqrt{3}t - \alpha - \frac{2\pi}{3}\right] \qquad \left[\beta = \tan^{-1}\frac{\sqrt{3}}{-1} = \frac{2\pi}{3}\right]$$

$$\Rightarrow \quad z = C_1 e^{2t} + C_2 e^{-t} \cos\left(\sqrt{3}\,t - \alpha + \frac{4\pi}{3}\right) \qquad ...(5) \qquad \left[-\frac{2\pi}{3} = \frac{4\pi}{3}\right]$$

From (2), we have $\frac{dy}{dx} = 2z$

$$\Rightarrow \quad \frac{dy}{dt} = 2C_1 e^{2t} + 2C_2 e^{-t} \cos\left(\sqrt{3}\,t - \alpha + \frac{4\pi}{3}\right) \qquad \text{[On putting the value of } z\text{]}$$

$$\Rightarrow \quad y = \int 2C_1 e^{2t}\, dt + 2C_2 \int e^{-t} \cos\left(\sqrt{3}\,t - \alpha + \frac{4\pi}{3}\right) dt \qquad \left(\gamma = \tan^{-1}\frac{\sqrt{3}}{-1} = \frac{2\pi}{3}\right)$$

$$y = C_1 e^{2t} + 2C_2 \frac{e^{-x}}{\sqrt{1+3}} \cos\left(\sqrt{3}\,t - \alpha + \frac{4\pi}{3} - \gamma\right)$$

$$\Rightarrow \quad y = C_1 e^{2t} + 2C_2 \frac{e^{-t}}{\sqrt{1+3}} \cos\left(\sqrt{3}\,t - \alpha + \frac{4\pi}{3} - \frac{2\pi}{3}\right)$$

$$y = C_1 e^{2t} + C_2 e^{-t} \cos\left(\sqrt{3}\,t - \alpha + \frac{2\pi}{3}\right) \qquad ...(6)$$

Relations (4), (5) and (6) are the required solutions. **Ans.**

Example 9. *Solve the following simultaneous equations :*

$$\frac{d^2x}{dt^2} - 3x - 4y = 0, \quad \frac{d^2y}{dt^2} + x + y = 0 \qquad (\textit{U.P. II Semester, Summer } 2005)$$

Solution. We have, $\frac{d^2x}{dt^2} - 3x - 4y = 0$

$$\frac{d^2y}{dt^2} + x + y = 0$$

$$(D^2 - 3)x - 4y = 0 \qquad ...(1)$$

$$x + (D^2 + 1)\, y = 0 \qquad ...(2)$$

Operating equation (2) by $(D^2 - 3)$, we get

$$(D^2 - 3)x + (D^2 - 3)(D^2 + 1)\, y = 0 \qquad ...(3)$$

Subtracting (3) from (1), we get

$$-4y - (D^2 - 3)(D^2 + 1)\, y = 0 = \Rightarrow -4y - (D^4 - 2D^2 - 3)\, y = 0$$

$$\Rightarrow \quad (D^4 - 2D^2 - 3 + 4)\, y = 0 \quad \Rightarrow (D^4 - 2D^2 + 1)\, y = 0$$

$$\Rightarrow \quad (D^2 - 1)^2\, y = 0$$

A.E. is $\quad (m^2 - 1)^2 = 0 \quad \Rightarrow \quad (m^2 - 1) = 0 \Rightarrow \quad m = \pm 1$

$$y = (c_1 + c_2 t)e^t + (c_3 + c_4 t)e^{-t} \qquad ...(4)$$

From (2), we have

$$x = -(D^2 + 1)y$$

$$= -D^2 y - y$$

$$= -D^2 [(c_1 + c_2 t)\, e^t + (c_3 + c_4 t)e^{-t}] - [(c_1 + c_2 t)e^t + (c_3 + c_4 t)e^{-t}]$$

$$= -D\, [\{(c_1 + c_2 t)e^t + c_2 e^t\} + \{(c_3 + c_4 t)(-e^{-t}) + c_4 e^{-t}\}] - [(c_1 + c_2 t)e^t + (c_3 + c_4 t)e^{-t}]$$

$$= -[(c_1 + c_2 t)\, e^t + c_2 e^t + c_2 e^t + (c_3 + c_4 t)(e^{-t}) - c_4 e^{-t} - c_4 e^{-t}] - [(c_1 + c_2 t)\, e^t + (c_3 + c_4 t)e^{-t}]$$

$$= -[(c_1 + c_2 t + 2c_2 + c_1 + c_2 t)\, e^t + (c_3 + c_4 t - 2c_4 + c_3 + c_4\, t)\, e^{-t}]$$

$$= -[(2c_1 + 2c_2 + 2c_2 t)e^t + (2c_3 - 2c_4 + 2c_4 t)e^{-t}] \qquad ...(5)$$

Relations (4) and (5) are the required solutions. **Ans.**

Example 10. *A mechanical system with two degrees of freedom satisfies the equations:*

$$2\frac{d^2x}{dt^2} + 3\frac{dy}{dt} = 4$$

$$2\frac{d^2y}{dt^2} - 3\frac{dx}{dt} = 0$$

Obtain expressions for x and y in terms of t, given x, y $\frac{dx}{dt}, \frac{dy}{dt}$ *all vanish at* $t = 0$

(R.G.P.V., Bhopal, I Sem. 2003)

Solution. $2D^2x + 3Dy = 4$ ... (1)

$$-3Dx + 2D^2y = 0 \quad \text{... (2)}$$

Multiplying (1) by 3 and (2) by $2D$ We get

$$6D^2x + 9Dy = 12 \quad \text{...(3)}$$

$$-6D^2x + 4D^3y = 0 \quad \text{...(4)}$$

Adding (3) and (4), we have

$$4D^3y + 9Dy = 12$$

$$\Rightarrow (4D^3 + 9D)\,y = 12$$

A.E. $\Rightarrow 4m^3 + 9m = 0$

$$\Rightarrow m\,(4m^2 + 9) = 0$$

$$m = 0,\ m = \frac{3}{2}i\,;\ -\frac{3}{2}i$$

$$\therefore\ y = c_1 + \left(c_2 \cos\frac{3}{2}t + c_3 \sin\frac{3}{2}t\right)$$

$$\text{P.I.} = \frac{12}{4D^3 + 9D}e^0 = t\cdot\frac{12}{12\,D^2 + 9}e^0 \qquad \Rightarrow \text{P.I.} = t\frac{12}{9} = \frac{12t}{9} = \frac{4t}{3}$$

$\therefore$ G.S = C.F. + P.I.

$$\Rightarrow y = c_1 + c_2\cos\frac{3}{2}t + c_3\sin\frac{3}{2}t + \frac{4t}{3}$$

$$y = 0,\ t = 0 \Rightarrow c_1 + c_2 = 0 \Rightarrow c_2 = -c_1$$

$$\therefore\quad y = c_1 - c_1\cos\frac{3}{2}t + c_3\sin\frac{3}{2}t + \frac{4}{3}t$$

$$\frac{dy}{dt} = \frac{3}{2}c_1\sin\frac{3}{2}t + \frac{3}{2}c_3\cos\frac{3}{2}t + \frac{4}{3}$$

$$t = 0,\ \frac{dy}{dt} = 0 \Rightarrow 0 = \frac{3}{2}c_3 + \frac{4}{3} \Rightarrow c_3 = -\frac{8}{9}$$

$$\text{Now } y = c_1 - c_1\cos\frac{3}{2}t - \frac{8}{9}\sin\frac{3}{2}t + \frac{4}{3}t$$

Putting the value of y in (1), we get

$$\frac{2d^2x}{dt^2} + \frac{3\,d}{dt}\left(c_1 - c_1\cos\frac{3}{2}t - \frac{8}{9}\sin\frac{3}{2}t + \frac{4t}{3}\right) = 4$$

$$\Rightarrow 2\frac{d^2x}{dt^2} + 3\left[0 + c_1\frac{3}{2}\sin\frac{3}{2}t - \frac{3}{2}\times\frac{8}{9}\cos\frac{3}{2}t + \frac{4}{3}\right] = 4$$

$$\Rightarrow 2\frac{d^2x}{dt^2} + \frac{9}{2}c_1\sin\frac{3}{2}t - 4\cos\frac{3}{2}t + 4 = 4$$

$$\Rightarrow 2\frac{d^2x}{dt^2} = -\frac{9}{2}c_1\sin\frac{3}{2}t + 4\cos\frac{3}{2}t$$

Integrating, we get

$$\therefore \quad 2\frac{dx}{dt} = \frac{9}{2}c_1 \times \frac{2}{3}\cos\frac{3}{2}t + 4 \times \frac{2}{3}\sin\frac{3}{2}t + c_4$$

$$\therefore \ t = 0, \frac{dx}{dt} = 0 \Rightarrow 0 = 3c_1 + c_4$$

$$\Rightarrow c_4 = -3c_1$$

Now, $2\frac{dx}{dt} = 3c_1 \cos\frac{3}{2}t + \frac{8}{3}\sin\frac{3}{2}t - 3c_1$

Again integrating we have

$$\therefore \quad 2x = 3c_1 \times \frac{2}{3}\sin\frac{3}{2}t - \frac{8}{3} \times \frac{2}{3}\cos\frac{3}{2}t - 3c_1 t + c_5$$

$$t = 0,\ x = 0 \Rightarrow 0 = -\frac{16}{9} + c_5 \Rightarrow c_5 = \frac{16}{9}$$

$$\therefore \quad 2x = 2c_1 \sin\frac{3}{2}t - \frac{16}{9}\cos\frac{3t}{2} - 3c_1 t + \frac{16}{9}$$

$$\Rightarrow x = c_1 \sin\frac{3}{2}t - \frac{8}{9}\cos\frac{3}{2}t - \frac{3}{2}c_1 t + \frac{8}{9}$$

Hence, $x = c_1 \sin\frac{3}{2}t - \frac{8}{9}\cos\frac{3}{2}t - \frac{3}{2}c_1 t + \frac{8}{9}$

$$y = c_1 - c_1 \cos\frac{3}{2}t - \frac{8}{9}\sin\frac{3}{2}t + \frac{4}{3}t \qquad \text{Ans}$$

Example 11. *Solve* : $\frac{d^2x}{dt^2} + \frac{dy}{dt} + 3x = e^{-t},$

$$\frac{d^2y}{dt^2} - 4\frac{dx}{dt} + 3y = \sin 2t$$ (*U.P., II Semester, June* 2007)

Solution. We have, $D^2x + Dy + 3x = e^{-t} \quad \Rightarrow \quad (D^2 + 3)x + Dy = e^{-t}$ ...(1)

$D^2y - 4Dx + 3y = \sin 2t \quad \Rightarrow \quad -4Dx + (D^2 + 3)y = \sin 2t$ ...(2)

To eliminate y operating (1) by $(D^2 + 3)$ and (2) by D, we get

$(D^2 + 3)^2 x + D(D^2 + 3)y = (D^2 + 3)e^{-t}$

$-4D^2 x + D(D^2 + 3)y = D \sin 2t$

$(D^4 + 6D^2 + 9)x + D(D^2 + 3)y = e^{-t} + 3e^{-t}$ ...(3)

$-4D^2x + D(D^2 + 3)y = 2\cos 2t$ ...(4)

Subtracting (4) from (3), we get

$(D^4 + 10D^2 + 9)x = 4e^{-t} - 2\cos 2t$

A.E. is $m^4 + 10m^2 + 9 = 0 \quad \Rightarrow (m^2 + 1)(m^2 + 9) = 0 \quad \Rightarrow m = \pm i, m = \pm 3$

C.F. $= C_1 \cos t + C_2 \sin t + C_3 \cos 3t + C_4 \sin 3t$

$$\text{P.I.} = \frac{1}{D^4 + 10D^2 + 9}4e^{-t} - \frac{1}{D^4 + 10D^2 + 9}(2\cos 2t)$$

$$= \frac{4}{1+10+9}e^{-t} - \frac{1}{(-4)^2+10(-4)+9}(2\cos 2t) = \frac{e^{-t}}{5} + \frac{2}{15}\cos 2t$$

$$x = C_1 \cos t + C_2 \sin t + C_3 \cos 3t + C_4 \sin 3t + \frac{e^{-t}}{5} + \frac{2}{15}\cos 2t$$

Putting the value of x in (2), we get

$$-4D\left[C_1 \cos t + C_2 \sin t + C_3 \cos 3t + C_4 \sin 3t + \frac{e^{-t}}{5} + \frac{2}{15}\cos 2t\right] + (D^2+3)\,y = \sin 2t$$

$$\Rightarrow \quad -4\left(-C_1 \sin t + C_2 \cos t - 3C_3 \sin 3t + 3C_4 \cos 3t - \frac{e^{-t}}{5} - \frac{4}{15}\sin 2t\right) + (D^2+3)\,y = \sin 2t$$

$$\Rightarrow \quad (D^2+3)\,y = \sin 2t - 4C_1 \sin t + 4C_2 \cos t - 12\,C_3 \sin 3t + 12C_4 \cos 3t - \frac{4}{5}e^{-t} - \frac{16}{15}\sin 2t$$

A.E. is $m^2 + 3 = 0 \Rightarrow m = \pm i\sqrt{3}$; C.F. $= C_1 \cos\sqrt{3}\,t + C_2 \sin\sqrt{3}\,t$

$$\text{P.I.} = \frac{1}{D^2+3}\left[\sin 2t - 4C_1 \sin t + 4C_2 \cos t - 12C_3 \sin 3t + 12C_4 \cos 3t - \frac{4}{5}e^{-t} - \frac{16}{15}\sin 2t\right]$$

$$= \frac{1}{D^2+3}\left(-\frac{1}{15}\sin 2t\right) + \frac{1}{D^2+3}(-4C_1 \sin t) + \frac{1}{D^2+3}4\,C_2 \cos t$$
$$+ \frac{1}{D^2+3}(-12C_3 \sin 3t) + \frac{1}{D^2+3}(12C_4 \cos 3t) + \frac{1}{D^2+3}\left(\frac{-4}{5}e^{-t}\right)$$

$$= \frac{1}{-4+3}\left(-\frac{1}{15}\sin 2t\right) + \frac{1}{-1+3}(-4C_1 \sin t) + \frac{1}{-1+3}4C_2 \cos t + \frac{1}{-9+3}(-12C_3 \sin 3t)$$
$$+ \frac{1}{-9+3}(12\,C_4 \cos 3t) + \frac{1}{1+3}\left(\frac{-4}{5}e^{-t}\right)$$

$$= \frac{1}{15}\sin 2t - 2C_1 \sin t + 2C_2 \cos t + 2C_3 \sin 3t - 2C_4 \cos 3t - \frac{1}{5}e^{-t}$$

$y =$ C.F. + P.I.

$$y = C_1 \cos\sqrt{3}\,t + C_2 \sin\sqrt{3}t + \frac{1}{15}\sin 2t - 2\,C_1 \sin t + 2C_2 \cos t + 2C_3 \sin 3t$$
$$- 2C_4 \cos 3t - \frac{1}{5}e^{-t}$$

$$x = C_1 \cos t + C_2 \sin t + C_3 \cos 3t + C_4 \sin 3t + \frac{e^{-t}}{5} + \frac{2}{15}\cos 2t$$ **Ans.**

Example 12. *Solve the simultaneous differential equations*

$$\frac{d^2x}{dt^2} - 4\frac{dx}{dt} + 4x = y,$$

$$\frac{d^2y}{dt^2} + 4\frac{dy}{dt} + 4y = 25x + 16e^t$$ *(U.P. II Semester, 2001)*

Solution. Here, we have

$$(D^2 - 4D + 4)x - y = 0 \qquad ...(1)$$

$$-25x + (D^2 + 4D + 4)y = 16e^t \qquad ...(2) \quad \left(D \equiv \frac{d}{dt}\right)$$

Operating (1) by $D^2 + 4D + 4$ and adding to (2), we get

$$\Rightarrow \quad (D^2+4D+4)(D^2-4D+4)\,x - (D^2+4D+4)\,y - 25x + (D^2+4D+4)\,y = 16e^t$$

$$(D^2-4D+4)(D^2+4D+4)\,x - 25x = 16e^t \quad \Rightarrow \quad (D^4 - 8D^2 - 9)x = 16e^t$$

Auxiliary equation is

$$m^4 - 8m^2 - 9 = 0 \quad \Rightarrow (m^2 - 9)(m^2 + 1) = 0 \Rightarrow m = \pm i, \pm 3$$

$$\text{C.F.} = c_1 e^{3t} + c_2 e^{-3t} + c_3 \cos t + c_4 \sin t, \quad \text{P.I.} = \frac{1}{D^4 - 8D^2 - 9}(16e^t) = -e^t$$

$$x = \text{C.F.} + \text{P.I.}$$

$$x = c_1 e^{3t} + c_2 e^{-3t} + c_3 \cos t + c_4 \sin t - e^t$$

From (1), $y = \dfrac{d^2x}{dt^2} - 4\dfrac{dx}{dt} + 4x$

$$= \frac{d^2}{dt^2}\left(c_1 e^{3t} + c_2 e^{-3t} + c_3 \cos t + c_4 \sin t - e^t\right) - 4\frac{d}{dt}\left(c_1 e^{3t} + c_2 e^{-3t} + c_3 \cos t + c_4 \sin t - e^t\right)$$
$$+ 4\left(c_1 e^{3t} + c_2 e^{-3t} + c_3 \cos t + c_4 \sin t - e^t\right)$$

$$= \frac{d}{dt}\left[3c_1 e^{3t} - 3c_2 e^{-3t} + c_3(-\sin t) + c_4 \cos t - e^t\right]$$
$$- 4\left[3c_1 e^{3t} - 3c_2 e^{-3t} + c_3(-\sin t) + c_4 \cos t - e^t\right]$$
$$+ 4\left[c_1 e^{3t} + c_2 e^{-3t} + c_3 \cos t + c_4 \sin t - e^t\right]$$

$$= 9c_1 e^{3t} + 9c_2 e^{-3t} - c_3 \cos t - c_4 \sin t - e^t +$$
$$\left[-12c_1 e^{3t} + 12c_2 e^{-3t} + 4c_3 \sin t - 4c_4 \cos t + 4e^t\right]$$
$$+ \left[4c_1 e^{3t} + 4c_2 e^{-3t} + 4c_3 \cos t + 4c_4 \sin t - 4e^t\right]$$

$$y = c_1 e^{3t} + 25c_2 e^{-3t} + (3c_3 - 4c_4)\cos t + (4c_3 + 3c_4)\sin t - e^t$$

$$x = c_1 e^{3t} + c_2 e^{-3t} + c_3 \cos t + c_4 \sin t - e^t \quad \text{[From (3)]}$$

Ans.

EXERCISE 14.1

Solve the following simultaneous equations:

1. $\dfrac{dx}{dt} + 2x - 3y = 0$, $\dfrac{dy}{dt} - 3x + 2y = 0$ **Ans.** $x = c_1 e^t - c_2 e^{-5t}$, $y = c_1 e^t + c_2 e^{-5t}$

2. $\dfrac{d^2y}{dt^2} = x$, $\dfrac{d^2x}{dt^2} = y$ **Ans.** $x = c_1 e^t + c_2 e^{-t} + (c_3 \cos t + c_4 \sin t)$

$y = c_1 e^t + c_2 e^{-t} - (c_3 \cos t + c_4 \sin t)$

3. $\dfrac{dx}{dt} + 5x - 2y = t$, $\dfrac{dy}{dt} + 2x + y = 0$ **Ans.** $x = -\dfrac{1}{27}(1 + 6t)e^{-3t} + \dfrac{1}{27}(1 + 3t)$

So that $x = y = 0$ when $t = 0$ (*AMIETE, June 2009, U.P., II Semester, June* 2008)

Ans. $y = -\dfrac{2}{27}(2 + 3t)e^{-3t} + \dfrac{2}{27}(2 - 3t)$

4. $\dfrac{dx}{dt} - y = t$, $\dfrac{dy}{dt} = t^2 - x$ **Ans.** $x = c_1 \cos t + c_2 \sin t + t^2 - 1$; $y = -c_1 \sin t + c_2 \cos t + t$

5. $\dfrac{dx}{dt} + 2y + \sin t = 0$

$\dfrac{dy}{dt} - 2x - \cos t = 0$ **Ans.** $x = c_1 \cos 2t + c_2 \sin 2t - \cos t$; $y = c_1 \sin 2t - c_2 \cos 2t - \sin$

6. $4\frac{dx}{dt} - \frac{dy}{dt} + 3x = \sin t$

$\frac{dx}{dt} + y = \cos t$ **Ans.** $x = c_1 e^{-t} + c_2 e^{-3t}, y = c_1 e^{-t} + 3 c_2 e^{-3t} + \cos t$

7. $\frac{dy}{dx} = x$ and $\frac{dx}{dt} = y + e^{2t}$ **Ans.** $x = C_1 e^t + C_2 e^{-t} + \frac{2}{3} e^{2t}, y = C_1 e^t - C_2 e^{-t} + \frac{1}{3} e^{2t}$

8. $\frac{dx}{dt} = y + t, \quad \frac{dy}{dx} = -2x + 3y + 1$ **Ans.** $x = c_1 e^t + \frac{1}{2} c_2 e^{2t} - \frac{3}{2} t - \frac{5}{4}, y = c_1 e^t + c_2 e^{2t} - t - \frac{3}{2}$

9. $t\frac{dx}{dt} + y = 0, \quad t\frac{dy}{dx} + x = 0$ **Ans.** $x = c_1 t + c_2 t^{-1}, y = c_2 t^{-1} - c_1 t$

given $x(1) = 1$ and $y(-1) = 0$

10. $\frac{dx}{dt} + y = \sin t, \quad \frac{dy}{dt} + x = \cos t,$ given that $x = 2, y = 0$ when $t = 0$ *(U.P., II Semester, 2004)*

Ans. $x = e^t + e^{-t}, y = \sin t - e^t + e^{-t}$

11. $(D - 1)\, x + Dy = 2t + 1$

$(2D + 1)\, x + 2\, Dy = t$ **Ans.** $x = -t - \frac{2}{3}, y = \frac{t^2}{2} + \frac{4}{3} t + C$

12. $\frac{dx}{dt} + \frac{2}{t}(x - y) = 1,$ *(U.P., II Semester, Summer (C.O.) 2005)*

$\frac{dy}{dt} + \frac{1}{t}(x + 5y) = t$ **Ans.** $x = At^{-4} + Bt^{-3} + \frac{t^2}{15} + \frac{3y}{10}, \quad y = -At^{-4} - \frac{1}{2} Bt^{-3} + \frac{2t^2}{15} - \frac{t}{20}$

13. $(D^2 - 1)x + 8Dy = 16e^t$ and $Dx + 3(D^2 + 1)y = 0$ *(Q. Bank U.P.T.U. 2001)*

Ans. $y = c_1 \cos \frac{t}{\sqrt{3}} + c_2 \sin \frac{t}{\sqrt{3}} + c_3 \cosh \sqrt{3}\, t + c_4 \sin h \sqrt{3}\, t + 2e^t$

$x = \sqrt{3}\, c_1 \sin \frac{t}{\sqrt{3}} - \sqrt{3}\, c_2 \cos \frac{t}{\sqrt{3}} - 3\sqrt{3}\, c_3 \sinh \sqrt{3}\, t - 3\sqrt{3}\, c_4 \cosh \sqrt{3}\, t - 6e^t - 3t.$

14. $\frac{dx}{dt} + \frac{2}{t}(x - y) = 1,$ *(U.P. II Semester, 2005)*

$\frac{dy}{dt} + \frac{1}{t}(x + 5y) = t.$ **Ans.** $x = At^{-4} + Bt^{-3} + \frac{t^2}{15} + \frac{3t}{10}, \quad y = -At^{-4} - \frac{1}{2} Bt^{-3} + \frac{2t^2}{15} - \frac{t}{20}$

DIFFERENTIAL EQUATION OF OTHER TYPES

15.1 INTRODUCTION

In this chapter we have to solve the following eight different types of equations:

1. $\dfrac{d^n y}{dx^n} = f(x)$.   2. $\dfrac{d^n y}{dx^n} = f(y)$.
3. Equation which do not contain y directly.
4. Equations that do not contain x directly.
5. (*i*) Equations whose one part (u) of C.F. is given
   (*ii*) Not a single part u or v of C.F of the equation is given
6. Normal form (removal of first derivative).
7. By changing the independent variable of the different equation.
8. By variation of parameters.

15.2 EQUATION OF THE TYPE $\dfrac{d^n y}{dx^n} = f(x)$

This type of exact differential equations are solved by successive integration.

Example 1. *Solve* $\dfrac{d^2 y}{dx^2} = x^2 \sin x$.

Solution. We have $\dfrac{d^2 y}{dx^2} = x^2 \sin x$ ...(1)

Integrating the differential equation (1), we get

$$\frac{dy}{dx} = x^2(-\cos x) - (2x)(-\sin x) + (2)(\cos x) + c_1$$

$$\frac{dy}{dx} = x^2 \cos x + 2x \sin x + 2 \cos x + c_1$$

Integrating again, we have $y = [(-x^2)(\sin x) - (-2x)(-\cos x) + (-2)(-\sin x)]$

$$+ [(2x)(-\cos x) - 2(-\sin x)] + 2 \sin x + c_1 x + c_2$$

$$= -x^2 \sin x - 4x \cos x + 6 \sin x + c_1 x + c_2$$ **Ans.**

Example 2. *Solve* $\dfrac{d^3 y}{dx^3} = x + \log x$.

Solution. We have, $\dfrac{d^3 y}{dx^3} = x + \log x$ ...(1)

integrating the differential equation (1), we get $\dfrac{d^2 y}{dx^2} = \dfrac{x^2}{2} + (\log x)(x) - \int \dfrac{1}{x} \cdot x\, dx + c_1$

$$\frac{d^2y}{dx^2} = \frac{x^2}{2} + x\log x - x + c_1 \qquad ...(2)$$

Again integrating (2), we have,

$$\frac{dy}{dx} = \frac{x^3}{6} + (\log x)\left(\frac{x^2}{2}\right) - \int \frac{1}{x}\cdot\frac{x^2}{2}\,dx - \frac{x^2}{2} + c_1 x + c_2$$

$$\frac{dy}{dx} = \frac{x^3}{6} + \frac{x^2}{2}\log x - \frac{x^2}{4} - \frac{x^2}{2} + c_1 x + c_2 \qquad ...(3)$$

Again integrating (3), we obtain

$$y = \frac{x^4}{24} + (\log x)\frac{x^3}{6} - \int \frac{1}{x}\frac{x^3}{6}\,dx - \frac{x^3}{12} - \frac{x^3}{6} + c_1\frac{x^2}{2} + c_2 x + c_3$$

$$\Rightarrow \quad y = \frac{x^4}{24} + \frac{x^3}{6}\log x - \frac{x^3}{18} - \frac{x^3}{12} - \frac{x^3}{6} + \frac{c_1 x^2}{2} + c_2 x + c_3$$

$$\Rightarrow \quad y = \frac{x^4}{24} + \frac{x^3}{6}\log x - \frac{11x^3}{36} + c_1\frac{x^2}{2} + c_2 x + c_3 \qquad \textbf{Ans.}$$

EXERCISE 15.1

Solve the following differential equations:

1. $\dfrac{d^5y}{dx^5} = x$ **Ans.** $y = \dfrac{x^6}{720} + \dfrac{c_1 x^4}{24} + \dfrac{c_2 x^3}{6} + \dfrac{c_3 x^2}{2} + c_4 x + c_5$
2. $\dfrac{d^2y}{dx^2} = x e^x$ **Ans.** $y = (x-2)e^x + c_1 x + c_2$
3. $\dfrac{d^4y}{dx^4} = x + e^{-x} - \cos x$ **Ans.** $y = \dfrac{x^5}{120} + e^{-x} - \cos x + c_1\dfrac{x^3}{6}$
4. $x^2\dfrac{d^2y}{dx^2} = \log x$ **Ans.** $y = -\dfrac{1}{2}(\log x)^2 + \log x - c_1 x + c_2$
5. $\dfrac{d^3y}{dx^3} = \log x$ **Ans.** $y = \dfrac{1}{36}[6x^3\log x - 11x^3 + c_1 x^2 + c_2 x + c_3]$
6. $\dfrac{d^3y}{dx^3} = \sin^2 x$ **Ans.** $y = \dfrac{x^3}{12} + \dfrac{\sin 2x}{16} + \dfrac{c_1 x^2}{2} + c_2 x + c_3$

15.3 EQUATION OF THE TYPE $\boxed{\dfrac{d^n y}{dx^n} = f(y)}$

Multiplying by $2\dfrac{dy}{dx}$, we get $2\dfrac{dy}{dx}\dfrac{d^2y}{dx^2} = 2f(y)\dfrac{dy}{dx}$ ...(1)

Integrating (1), we have $\left(\dfrac{dy}{dx}\right)^2 = 2\int f(y)\,dy + c = \phi(y)$ (say)

$$\frac{dy}{dx} = \sqrt{\phi(y)} \Rightarrow \frac{dy}{\sqrt{\phi(y)}} = dx \quad \Rightarrow \quad \int \frac{dy}{\sqrt{\phi(y)}} = x + c$$

Example 3. *Solve* $\dfrac{d^2y}{dx^2} = \sqrt{y}$, *under the condition* $y = 1.\ \dfrac{dy}{dx} = \dfrac{2}{\sqrt{3}}$ *at* $x = 0$

Solution. We have $\dfrac{d^2y}{dx^2} = \sqrt{y}$ ...(1)

Multiplying (1) by $2\frac{dy}{dx}$, we get $2\frac{dy}{dx}\frac{d^2y}{dx^2} = 2\sqrt{y}\frac{dy}{dx}$ ...(2)

Integrating (2), we get $\left(\frac{dy}{dx}\right)^2 = \frac{4}{3}y^{3/2} + c_1$ ...(3)

On putting $y = 1$ and $\frac{dy}{dx} = \frac{2}{\sqrt{3}}$, we have $c_1 = 0$

Equation (3) becomes $\left(\frac{dy}{dx}\right)^2 = \frac{4}{3}y^{3/2}$ or $\frac{dy}{dx} = \frac{2}{\sqrt{3}}y^{3/4}$ or $y^{-3/4}\,dy = \frac{2}{\sqrt{3}}dx$

Again integrating $\frac{y^{1/4}}{\frac{1}{4}} = \frac{2}{\sqrt{3}}x + c_2 \quad \Rightarrow 4y^{\frac{1}{4}} = \frac{2}{\sqrt{3}}x + c_2$ ...(4)

On putting $x = 0$, $y = 1$, we get $c_2 = 4$

(4) becomes $$4\,y^{1/4} = \frac{2}{\sqrt{3}}x + 4$$ **Ans.**

Example 4. *Solve* $\frac{d^2y}{dx^2} = \sec^2 y \tan y$ *under the condition* $y = 0$ *and* $\frac{dy}{dx} = 1$ *when* $x = 0$.

Solution. $$\frac{d^2y}{dx^2} = \sec^2 y \tan y \quad \Rightarrow \quad 2\frac{dy}{dx}\frac{d^2y}{dx^2} = 2\sec^2 y \tan y\frac{dy}{dx}$$

$$\int 2\frac{dy}{dx}\frac{d^2y}{dx^2} = \int 2\sec^2 y \tan y\frac{dy}{dx}$$

$$\left(\frac{dy}{dx}\right)^2 = \tan^2 y + c_1 \quad \text{or} \quad \frac{dy}{dx} = \sqrt{\tan^2 y + c_1}$$

On putting $y = 0$, and $\frac{dy}{dx} = 1$, we get $c_1 = 1$

Now, $$\frac{dy}{dx} = \sqrt{\tan^2 y + 1} = \sec y$$

$\Rightarrow$ $$\cos y\,dy = dx$$

On integrating we get $\sin y = x + c$

On putting $y = 0$, $x = 0$, we have $c = 0$

$$\sin y = x \Rightarrow y = \sin^{-1} x$$ **Ans.**

Example 5. *Solve* $\frac{d^2y}{dx^2} = 2(y^3 + y)$, *under the condition* $y = 0$, $\frac{dy}{dx} = 1$ *when* $x = 0$.

(*U.P., II Semester, Summer 2003*)

Solution. $$\frac{d^2y}{dx^2} = 2(y^3 + y) \text{ or } 2\frac{dy}{dx}\frac{d^2y}{dx^2} = 4(y^3 + y)\frac{dy}{dx}$$

Integrating, we get

$$\left(\frac{dy}{dx}\right)^2 = 4\left(\frac{y^4}{4} + \frac{y^2}{2}\right) + c_1 = y^4 + 2y^2 + c_1 \quad ...(1)$$

On putting $y = 0$ and $\frac{dy}{dx} = 1$ in (1), we get $1 = c_1$

Equation (1) becomes $\left(\frac{dy}{dx}\right)^2 = y^4 + 2y^2 + 1 = (y^2 + 1)^2$

$$\frac{dy}{dx} = y^2 + 1 \text{ or } \frac{dy}{1+y^2} = dx$$

Again integraiting, we get $\tan^{-1} y = x + c_2$ ...(2)

On putting $y = 0$ and $x = 0$ in (2), we have $0 = c_2$

Equation (2) is reduced to $\tan^{-1} y = x \Rightarrow y = \tan x$ **Ans.**

Example 6. *A motion is governed by* $\frac{d^2x}{dt^2} = 36x^{-2}$, *given that at* $t = 0$, $x = 8$ *and* $\frac{dx}{dt} = 0$, *find the displacement at any time t.*

Solution. We have $\frac{d^2x}{dt^2} = 36x^{-2} \Rightarrow 2\frac{d^2x}{dt^2}\frac{dx}{dt} = 2 \times 36x^{-2}\frac{dx}{dt}$ ...(1)

Integrating (1), we hae $\left(\frac{dx}{dt}\right)^2 = -72x^{-1} + c_1$ ...(2)

Putting $x = 8$ and $\frac{dx}{dt} = 0$ in (2), we get $0 = -\frac{72}{8} + c_1$ or $c_1 = 9$

(2) becomes $\left(\frac{dx}{dt}\right)^2 = -\frac{72}{x} + 9$ or $\left(\frac{dx}{dt}\right)^2 = \frac{-72+9x}{x} \Rightarrow \frac{dx}{dt} = 3\sqrt{\frac{(x-8)}{x}}$

$$\Rightarrow \int \frac{\sqrt{x}\,dx}{\sqrt{x-8}} = 3\int dt + c_2 \Rightarrow \int \frac{x\,dx}{\sqrt{x^2-8x}} = 3t + c_2$$

$$\frac{1}{2}\int \frac{2x-8+8}{\sqrt{x^2-8x}}\,dx = 3t + c_2$$

$$\frac{1}{2}\int \frac{2x-8}{\sqrt{x^2-8x}}\,dx + 4\int \frac{1}{\sqrt{(x-4)^2-(4)^2}}\,dx = 3t + c_2$$

$$\sqrt{x^2-8x} + 4\cosh^{-1}\frac{x-4}{4} = 3t + c_2 \quad ...(3)$$

On putting $x = 8$ and $t = 0$ in (3), we get $c_2 = 0$

(3) becomes $\sqrt{x^2-8x} + 4\cosh^{-1}\frac{x-4}{4} = 3t$ **Ans.**

EXERCISE 15.2

1. $y^3\frac{d^2y}{dx^2} = a$ **Ans.** $c_1 y^2 = (c_1 x + c_2)^2$

2. $e^{2y}\frac{d^2y}{dx^2} = 1$ **Ans.** $c_1 e^y = \cosh(c_1 x + c_2)$

3. $\sin^3 y\frac{d^2y}{dx^2} = \cos y$ **Ans.** $\sin[(x+c_2)\sqrt{(1+c_1)}] + \sqrt{\left(\frac{1+c_1}{c_1}\right)}\cos y = 0$

4. A particle is acted upon by a force $\mu\left(x+\frac{a^4}{x^3}\right)$ per unit mass towards the origin where x is the distance from the origin at time t. If it starts that it will arrive at the origin in time $\frac{\pi}{4\sqrt{\mu}}$.

5. In the case of a stretched elastic string which has one end fixed and a particle of mass m attached to the other end, the equation of motion is

$$\frac{d^2s}{dt^2} = -\frac{mg}{e}(\bar{s} - l)$$

where l is the natural length of the string and e its elongwith due to a weight mg. Find s and v determining the constants, so that $s = s_0$ at the time $t = 0$ and $v = 0$ when $t = 0$.

Ans. $v = -\sqrt{\left(\frac{g}{e}\right)}\,[(s_0 - l)^2 - (s - l)^2]^{1/2},\ s - l = (s_0 - l)\cos\left[\sqrt{\left(\frac{g}{e}\right)}.t\right]$

15.4 EQUATIONS WHICH DO NOT CONTAIN 'y' DIRECTLY

The equation which do not contain y directly, can be written

$$f\left(\frac{d^n y}{dx^n}, \frac{d^{n-1} y}{dx^{n-1}}, \ldots\ldots, \frac{dy}{dx}, x\right) = 0 \qquad \ldots(1)$$

On substituting $\frac{dy}{dx} = P$ *i.e.*, $\frac{d^2 y}{dx^2} = \frac{dP}{dx}, \frac{d^3 y}{dx^3} = \frac{d^2 P}{dx^2}$ etc. in (1), we get $f\left(\frac{d^{n-1} P}{dx^{n-1}}, \ldots\ldots., P, x\right) = 0$

Example 7. *Solve* $\frac{d^2 y}{dx^2} + \frac{dy}{dx} + \left(\frac{dy}{dx}\right)^3 = 0$

Solution. On putting $\frac{dy}{dx} = P$ and $\frac{d^2 y}{dx^2} = \frac{dP}{dx}$, equation (1) becomes

$$\frac{dP}{dx} + P + P^3 = 0 \text{ or } \frac{dP}{dx} + P(1 + P^2) = 0$$

$$\frac{dP}{dx} = -P(1 + P^2) \text{ or } \frac{dP}{P(1 + P^2)} = -dx \Rightarrow \left(\frac{1}{P} - \frac{P}{1 + P^2}\right) dP = -dx$$

On integrating, we have

$$\log P - \frac{1}{2}\log(1 + P^2) = -x + c_1 \text{ or } \log\frac{P}{\sqrt{1 + P^2}} = -x + c_1$$

$$\frac{P}{\sqrt{1 + P^2}} = e^{-x + c_1} \text{ or } \frac{P^2}{1 + P^2} = a^2 e^{-2x} \Rightarrow P^2 = (1 + P^2)\, a^2 e^{-2x}$$

$$\Rightarrow P^2(1 - a^2 e^{-2x}) = a^2 e^{-2x} \Rightarrow P = \frac{a e^{-x}}{\sqrt{1 - a^2 e^{-2x}}} \Rightarrow \frac{dy}{dx} = \frac{a e^{-x}}{\sqrt{1 - a^2 e^{-2x}}}$$

$$dy = \frac{a e^{-x}}{\sqrt{1 - a^2 e^{-2x}}}\, dx$$

On integration, we get $\quad y = -\sin^{-1}(a e^{-x}) + b$ **Ans.**

Example 8. *Solve* $\frac{d^2 y}{dx^2} = \left[1 - \left(\frac{dy}{dx}\right)^2\right]^{1/2}$ (*U.P. Second Sem., 2002*)

Solution. We have, $\frac{d^2 y}{dx^2} = \left[1 - \left(\frac{dy}{dx}\right)^2\right]^{1/2}$ ... (1)

Putting $P = \frac{dy}{dx} \Rightarrow \frac{dP}{dx} = \frac{d^2 y}{dx^2}$ in (1), we get $\frac{dP}{dx} = \sqrt{1 - P^2} \Rightarrow \frac{dP}{\sqrt{1 - P^2}} = dx$

On integrating, we have

$$\sin^{-1} P = x + c \quad \Rightarrow \quad P = \sin(x + c)$$

$$\frac{dy}{dx} = \sin(x + c)$$

On integrating, we have $\quad y = -\cos(x + c) + c_1$ **Ans.**

Example 9. *Solve* $x\dfrac{d^2y}{dx^2}+x\left(\dfrac{dy}{dx}\right)^2-\dfrac{dy}{dx}=0$ *(U.P. II Semester, 2010)*

Solution. On putting $\dfrac{dy}{dx}=P$ and $\dfrac{d^2y}{dx^2}=\dfrac{dP}{dx}$ in the given equation, we get

$$x\frac{dP}{dx}+xP^2-P=0\Rightarrow\frac{1}{P^2}\frac{dP}{dx}-\frac{1}{P}\frac{1}{x}=-1 \qquad ...(1)$$

Again putting $\dfrac{1}{P}=z$ so that $-\dfrac{1}{P^2}\dfrac{dP}{dx}=\dfrac{dz}{dx}$

Equation (1) becomes $\dfrac{-dz}{dx}-\dfrac{z}{x}=-1\Rightarrow\dfrac{dz}{dx}+\dfrac{z}{x}=1$

$$\text{I.F.}=e^{\int\frac{1}{x}dx}=e^{\log x}=x$$

Hence, solution is $\quad zx=\int x\,dx+C_1$ or $zx=\dfrac{x^2}{2}+C_1$ or $\dfrac{1}{P}x=\dfrac{x^2}{2}+C$

$$\Rightarrow \quad \frac{x}{P}=\frac{x^2+2C}{2}\Rightarrow P=\frac{2x}{x^2+2c_1}\Rightarrow\frac{dy}{dx}=\frac{2x}{x^2+2c_1}\Rightarrow dy=\frac{2x}{x^2+2c_1}dx$$

On integrating, we have $\quad y=\log(x^2+2c_1)+c_2$ **Ans.**

EXERCISE 15.3

Solving the following differential equations:

1. $(1+x^2)\dfrac{d^2y}{dx^2}+x\dfrac{dy}{dx}+ax=0$ **Ans** $y=c_2-ax+c_1\log[x+\sqrt{(1+x^2)}]$

2. $(1+x^2)\dfrac{d^2y}{dx^2}+1+\left(\dfrac{dy}{dx}\right)^2=0$ **Ans.** $y=-\dfrac{x}{k}+\dfrac{1+k^2}{k^2}\log(1+kx)+a$

3. $\dfrac{d^4y}{dx^4}-\cot x\dfrac{d^3y}{dx^3}=0$ **Ans.** $y=c_1\cos x+c_2x^2+c_3x+c_4$

4. $2x\dfrac{d^3y}{dx^3}\cdot\dfrac{d^2y}{dx^2}=\left[\dfrac{d^2y}{dx^2}\right]^2-a^2$ **Ans.** $15c_1^2y=4(c_1x+a^2)^{5/2}+c_2x+c_3$

5. $e^{x^2/2}\left[x\dfrac{d^2y}{dx^2}-\dfrac{dy}{dx}\right]=x^3$ **Ans.** $y=e^{-x^2/2}+c_1\dfrac{x^2}{2}+c_2$

15.5 EQUATIONS THAT DO NOT CONTAIN 'x' DIRECTLY

The equations that do not contain x directly are of the form

$$f\left(\frac{d^ny}{dx^n},\frac{d^{n-1}y}{dx^{n-1}},......\frac{dy}{dx},y\right)=0 \qquad ...(1)$$

On substituting $\dfrac{dy}{dx}=P,\dfrac{d^2y}{dx^2}=\dfrac{dP}{dx}=\dfrac{dP}{dx}\cdot\dfrac{dy}{dx}=\dfrac{dP}{dy}P$ in the equation (1), we get

$$\left[\frac{dP^{n-1}}{dy^{n-1}},.....P,y\right]=0 \qquad ...(2)$$

Equation (2) is solved for P. Let

$$P=f_1(y)\Rightarrow\frac{dy}{dx}=f_1(y)\text{ or }\frac{dy}{f_1(y)}=dx$$

$$\Rightarrow \qquad \int\frac{dy}{f_1(y)}=x+c$$

Example 10. *Solve* $y\frac{d^2y}{dx^2}+\left(\frac{dy}{dx}\right)^2=\frac{dy}{dx}$ ...(1)

Solution. Put $\frac{dy}{dx}=P, \frac{d^2y}{dx^2}=\frac{dP}{dx}=\frac{dP}{dy}\cdot\frac{dy}{dx}=P\frac{dP}{dy}$ in equation (1)

$$yP\frac{dP}{dy}+P^2=P \Rightarrow y\frac{dP}{dy}=1-P$$

$$\Rightarrow \quad \frac{dp}{1-P}=\frac{dy}{y} \Rightarrow -\log(1-P)=\log y+\log c_1$$

$$\Rightarrow \quad \frac{1}{1-P}=c_1 y \Rightarrow P=1-\frac{1}{c_1 y} \text{ or } \frac{dy}{dx}=\frac{c_1 y-1}{c_1 y}$$

$$\Rightarrow \quad \frac{c_1 y}{c_1 y-1}dy=dx \Rightarrow \left(1+\frac{1}{c_1 y-1}\right)dy=dx$$

$$y+\frac{1}{c_1}\log(c_1 y-1)=x+c_1$$ **Ans.**

Example 11. *Solve* $y\frac{d^2y}{dx^2}+\left(\frac{dy}{dx}\right)^2=y^2$ ...(1)

Solution. Put $\frac{dy}{dx}=P, \frac{d^2y}{dx^2}=\frac{dP}{dx}=\frac{dP}{dy}\frac{dy}{dx}=P\frac{dP}{dy}$ in (1)

$$yP\frac{dP}{dy}+P^2=y^2 \text{ or } P\frac{dP}{dy}+\frac{P^2}{y}=y \quad ...(2)$$

Put $P^2=z$ or $2P\frac{dP}{dy}=\frac{dz}{dy}$ in (2), $\frac{1}{2}\frac{dz}{dy}+\frac{z}{y}=y$ or $\frac{dz}{dy}+\frac{2z}{y}=2y$

$$\text{I.F.}=e^{\int\frac{2}{y}dy}=e^{2\log y}=e^{\log y^2}=y^2$$

Hence, the solution is $\quad zy^2=\int 2y.(y^2)\,dy+c$

$$\Rightarrow \quad P^2y^2=\frac{y^4}{2}+c$$

$$\Rightarrow \quad 2P^2y^2=y^4+k \text{ or } \sqrt{2}\,yP=\sqrt{y^4+k} \qquad [\text{Put } 2c=k]$$

$$\Rightarrow \quad \sqrt{2}\,y\frac{dy}{dx}=\sqrt{y^4+k} \text{ or } \sqrt{2}\frac{y\,dy}{\sqrt{y^4+k}}=dx$$

$$\Rightarrow \quad \frac{1}{\sqrt{2}}\frac{dt}{\sqrt{t^2+k}}=dx \quad [\text{Put } y^2=t,\ 2y\,dy=dt] \Rightarrow \frac{1}{\sqrt{2}}\sin h^{-1}\frac{t}{\sqrt{k}}=x+c$$

$$\Rightarrow \quad \sin h^{-1}\frac{y^2}{\sqrt{k}}=\sqrt{2}\,x+c \text{ or } y^2=\sqrt{k}\sin h(\sqrt{2}\,x+c)$$ **Ans.**

EXERCISE 15.4

Solve the following differential equations:

1. $\frac{d^2y}{dx^2}+\left(\frac{dy}{dx}\right)^2=1$ **Ans.** $y^2=x^2+ax+b$
2. $y\frac{d^2y}{dx^2}-\left(\frac{dy}{dx}\right)^2+2\frac{dy}{dx}=0$ **Ans.** $cy+2=d\,e^{cx^a}$
3. $2y\frac{d^2y}{dx^2}-3\left(\frac{dy}{dx}\right)^2-4y^2=0$ **Ans.** $y=a\sec^2(x+b)$
4. $y\frac{d^2y}{dx^2}+\frac{dy}{dx}+\left(\frac{dy}{dx}\right)^3=0$ **Ans.** $y=a-\sin^{-1}(b\,e^{-x})$

5. $y\dfrac{d^2y}{dx^2}=\left\{\dfrac{dy}{dx}\right\}^2\left[1-\dfrac{dy}{dx}\cos y+y\dfrac{dy}{dx}\sin y\right]$ **Ans.** $x=c_1+c_2\log y+\sin y$

6. $y\dfrac{d^2y}{dx^2}-\left(\dfrac{dy}{dx}\right)^2=y^2\log y$ **Ans.** $\log y=b.\,e^x+a\,e^{-x}$

15.6 EQUATION WHOSE ONE SOLUTION IS KNOWN

If $y = u$ is given solution belonging to the complementary function of the differential equation. Let the other solution be $y = v$. Then $y = u.\,v$ is complete solution of the differential equation.

Let $\dfrac{d^2y}{dx^2}+P\dfrac{dy}{dx}+Qy=R$ ....(1), be the differential equation and u is the solution included in the complementary function of (1)

$$\therefore \quad \frac{d^2u}{dx^2}+P\frac{du}{dx}+Qu=0 \quad ...(2)$$

$$y=u.\,v$$

$$\frac{dy}{dx}=v\frac{du}{dx}+u\frac{dv}{dx}$$

$$\frac{d^2y}{dx^2}=v\frac{d^2u}{dx^2}+2\frac{dv}{dx}\frac{du}{dx}+u\frac{d^2v}{dx^2}$$

Substituting the values of $y,\dfrac{dy}{dx},\dfrac{d^2y}{dx^2}$ in (1), we get

$$v\frac{d^2u}{dx^2}+2\frac{dv}{dx}\frac{du}{dx}+u\frac{d^2v}{dx^2}+P\left(v\frac{du}{dx}+u\frac{dv}{dx}\right)+Qu.v=R$$

On arranging

$$\Rightarrow v\left[\frac{d^2u}{dx^2}+P\frac{du}{dx}+Qu\right]+u\left[\frac{d^2v}{dx^2}+P\frac{dv}{dx}\right]+2\frac{du}{dx}\cdot\frac{dv}{dx}=R$$

The first bracket is zero by virtue of relation (2), and the remaing is divided by u.

$$\frac{d^2v}{dx^2}+P\frac{dv}{dx}+\frac{2}{u}\frac{du}{dx}\frac{dv}{dx}=\frac{R}{u}$$

$$\Rightarrow \quad \frac{d^2v}{dx^2}+\left[P+\frac{2}{u}\frac{du}{dx}\right]\frac{dv}{dx}=\frac{R}{u} \quad ...(3)$$

Let $\dfrac{dv}{dx}=z$, so that $\dfrac{d^2v}{dx^2}=\dfrac{dz}{dx}$

Equation (3) becomes

$$\boxed{\frac{dz}{dx}+\left[P+\frac{2}{u}\frac{du}{dx}\right]z=\frac{R}{u}}$$

This is the linear differential equation of first order and can be solved (z can be found), which will contain one constant.

On integration $z=\dfrac{dv}{dx}$, we can get v.

Having found v, the solution is $y = uv$.

Note: Rule to find out the integral belonging to the complementary function

Rule	Condition	u
1	$1 + P + Q \;= 0$	e^x
2	$1 - P + Q \;= 0$	e^{-x}
3	$1 + \dfrac{P}{a} + \dfrac{Q}{a^2} \;= 0$	e^{ax}
4	$P + Qx \;= 0$	x
5	$2 + 2Px + Qx^2 \;= 0$	x^2
6	$n(n-1) + Pnx + Qx^2 \;= 0$	x^n

Example 12. *Solve $y'' - 4xy' + (4x^2 - 2)y = 0$ given that $y = e^{x^2}$ is an integral included in the complementary function.* (*U.P., II Semester, 2004*)

Solution. $y'' - 4xy' + (4x^2 - 2)y = 0$ ...(1)

On putting $y = v.e^{x^2}$ in (1), the reduced equation as in the article 15.6.

$$\frac{d^2v}{dx^2} + \left[P + \frac{2}{u}\frac{du}{dx}\right]\frac{dv}{dx} = 0 \qquad [P = -4x,\ Q = 4x^2 - 2,\ R = 0]$$

$$\Rightarrow \quad \frac{d^2v}{dx^2} + \left[-4x + \frac{2}{e^{x^2}}(2x\,e^{x^2})\right]\frac{dv}{dx} = 0$$

$$\Rightarrow \quad \frac{d^2v}{dx^2} + [-4x + 4x]\frac{dv}{dx} = 0 \quad \Rightarrow \quad \frac{d^2v}{dx^2} = 0 \Rightarrow \frac{dv}{dx} = c, \Rightarrow v = c_1x + c_2$$

$$\therefore \quad y = uv \qquad [u = e^{x^2}]$$

$$y = e^{x^2}(c_1x + c_2)$$ **Ans.**

Example 13. *Solve* $x\dfrac{d^2y}{dx^2} - (2x-1)\dfrac{dy}{dx} + (x-1)y = 0$

given that $y = e^x$ is an integral included in the complementary function.

Solution. $x\dfrac{d^2y}{dx^2} - (2x-1)\dfrac{dy}{dx} + (x-1)\,y = 0$

$$\Rightarrow \quad \frac{d^2y}{dx^2} - \frac{2x-1}{x}\frac{dy}{dx} + \frac{x-1}{x}y = 0 \qquad [1 + P + Q = 0] \qquad ...(1)$$

By putting $y = ve^x$ in (1),we get the reduced equation as in the article 15.6.

$$\frac{d^2v}{dx^2} + \left[P + \frac{2}{u}\frac{du}{dx}\right]\frac{dv}{dx} = 0 \qquad ...(2)$$

Putting $u = e^x$ and $\dfrac{dv}{dx} = z$ in (2), we get $\dfrac{dz}{dx} + \left[-\dfrac{2x-1}{x} + \dfrac{2}{e^x}e^x\right]z = 0$

$$\Rightarrow \quad \frac{dz}{dx} + \frac{-2x+1+2x}{x}z = 0 \quad \Rightarrow \quad \frac{dz}{dx} + \frac{z}{x} = 0$$

$$\Rightarrow \quad \frac{dz}{z} = -\frac{dx}{x} \Rightarrow \log z = -\log x + \log c_1$$

$$\Rightarrow \quad z = \frac{c_1}{x} \text{ or } \frac{dv}{dx} = \frac{c_1}{x} \text{ or } dv = c_1\frac{dx}{x} \Rightarrow v = c_1\log x + c_2$$

$$y = u.\,v = e^x(c_1\log x + c_2)$$ **Ans.**

Example 14. *Solve* $x^2 \dfrac{d^2y}{dx^2} - 2x[1+x]\dfrac{dy}{dx} + 2(1+x)y = x^3$

Solution. $x^2 \dfrac{d^2y}{dx^2} - 2x(1+x)\dfrac{dy}{dx} + 2(1+x)y = x^3$

$$\Rightarrow \quad \frac{d^2y}{dx^2} - \frac{2x(1+x)}{x^2}\frac{dy}{dx} + \frac{2(1+x)y}{x^2} = x \qquad \ldots(1)$$

Here $\qquad P + Qx = -\dfrac{2x(1+x)}{x^2} + \dfrac{2(1+x)}{x^2}x = 0$

Hence $y = x$ is a solution of the C.F. and the other solution is v.

Putting $y = vx$ in (1), we get the reduced equation as in article 15.6

$$\frac{d^2v}{dx^2} + \left\{P + \frac{2}{u}\frac{du}{dx}\right\}\frac{du}{dx} = \frac{x}{u}$$

$$\frac{d^2v}{dx^2} + \left[\frac{-2x(1+x)}{x^2} + \frac{2}{x}(1)\right]\frac{dv}{dx} = \frac{x}{x}$$

$$\Rightarrow \quad \frac{d^2v}{dx^2} - 2\frac{dv}{dx} = 1 \Rightarrow \frac{dz}{dx} - 2z = 1 \qquad \left[\frac{dv}{dx} = z\right]$$

which is a linear differential equation of first order and $I.F. = e^{\int -2\,dx} = e^{-2x}$

Its solution is $z\,e^{-2x} = \int e^{-2x}\,dx + c_1$

$$z\,e^{-2x} = \frac{e^{-2x}}{-2} + c_1 \quad \text{or} \quad z = \frac{-1}{2} + c_1 e^{2x}$$

$$\Rightarrow \quad \frac{dv}{dx} = -\frac{1}{2} + c_1 e^{2x} \quad \text{or} \quad dv = \left(-\frac{1}{2} + c_1 e^{2x}\right)dx \quad \Rightarrow \quad v = \frac{-x}{2} + \frac{c_1}{2}e^{2x} + c_2$$

$$y = uv = x\left(\frac{-x}{2} + \frac{c_1}{2}e^{2x} + c_2\right) \qquad \textbf{Ans.}$$

Example 15. *Verify that* $y = e^2x$ *is a solution of* $(2x+1)y'' - 4(x+1)y' + 4y = 0$. *Hence find the general solution.*

Solution. We have

$$(2x+1)\frac{d^2y}{dx^2} - 4(x+1)\frac{dy}{dx} + 4y = 0 \qquad \ldots(1)$$

$$y = e^{2x}, \quad y' = 2e^{2x}, \quad y'' = 4e^{2x}$$

Substiting the values of y, y' and y'' in (1), we get

$$(2x+1)\,4e^{2x} - 4(x+1)\,2e^{2x} + 4e^{2x} = 0$$

or $\qquad [8x + 4 - 8x - 8 + 4]\,e^{2x} = 0 \qquad \Rightarrow \qquad 0 = 0$

Thus $\qquad y_1 = e^{2x}$ is a solution

Equation (1) in the standard form is

$$y'' - \frac{4(x+1)}{(2x+1)}y' + \frac{4}{(2x+1)}y = 0$$

So $\qquad P(x) = -\dfrac{4(x+1)}{(2x+1)}$

Then $\qquad \omega(x) = \dfrac{1}{y_1^2}e^{-\int P\,dx}$

Now $$\int -P\,dx = -\int -\frac{4(x+1)}{(2x+1)}\,dx = \int\left(\frac{4x+2}{2x+1}+\frac{2}{2x+1}\right)dx$$
$$= 2x + \ln(2x+1)$$

Then $$\omega = \frac{1}{(e^{2x})^2}\,e^{2x+\ln(2x+1)} = \frac{e^{2x}}{(e^{2x})^2}.(2x+1)$$

$$\omega(x) = \frac{2x+1}{e^{2x}}$$

Now $v(x) = \int \omega(x)\,dx = \int \frac{2x+1}{e^{2x}}\,dx$

Integrating by parts $v(x) = (2x+1)\dfrac{e^{-2x}}{-2} - 2.\dfrac{e^{-2x}}{4}$

The required second solution

$$y_2(x) = y_1(x)\,v(x) = e^{2x}\left[-\frac{2x+1}{2}.\frac{1}{e^{2x}} - \frac{1}{2}\frac{1}{e^{2x}}\right] = -x-1 = -(x+1)$$

Then the general solution is

$y(x) = c_1 y_1(x) + c_2 y_2(x) = c_1 e^{2x} - c_2(x+1)$ **Ans.**

Example 16. *Solve $x^2 y'' - (x^2+2x)y' + (x+2)y = x^3 e^x$ given that $y = x$ is a solution.*

Solution. $x^2 y'' - (x^2+2x)y' + (x+2)y = x^3 e^x$

$$\Rightarrow \quad y'' - \frac{x^2+2x}{x^2}y' + \frac{x+2}{x^2}y = xe^x \qquad ...(1)$$

On putting $y = vx$ in (1), the reduced equation as in the article 20.6.

$$\frac{d^2v}{dx^2} + \left\{P + \frac{2}{u}\frac{du}{dx}\right\}\frac{dv}{dx} = \frac{R}{u} \Rightarrow \frac{d^2v}{dx^2} + \left[-\frac{x^2+2x}{x^2} + \frac{2}{x}(1)\right]\frac{dv}{dx} = \frac{xe^x}{x}$$

$$\Rightarrow \quad \frac{d^2v}{dx^2} - \frac{dv}{dx} = e^x \Rightarrow \frac{dz}{dx} - z = e^x \qquad \left(z = \frac{dv}{dx}\right)$$

which is a linear differential equation

$$I.F. = e^{-\int dx} = e^{-x} \quad \Rightarrow \quad z\,e^{-x} = \int e^x.e^{-x}\,dx + c$$

$$z\,e^{-x} = x + c \text{ or } z = e^x.x + c\,e^x \Rightarrow \frac{dv}{dx} = e^x.x + c\,e^x$$

$$v = x.e^x - e^x + c\,e^x + c_1 \Rightarrow v = (x - 1 + c)\,e^x + c_1$$

$y = vx = (x^2 - x + cx)\,e^x + c_1 x$ **Ans.**

Example 17. *Solve* $(x+2)\dfrac{d^2y}{dx^2} - (2x+5)\dfrac{dy}{dx} + 2y = (x+1)e^x$

Solution. $$\frac{d^2y}{dx^2} - \frac{2x+5}{x+2}\frac{dy}{dx} + \frac{2y}{x+2} = \frac{(x+1)\,e^x}{x+2} \qquad ...(1)$$

Here $P = \dfrac{2x+5}{x+2},\ Q = \dfrac{2}{x+2},\ R = \dfrac{(x+1)e^x}{x+2}$

$$1 + \frac{P}{2} + \frac{Q}{4} = 0 \quad \Rightarrow \quad 1 - \frac{2x+5}{2x+4} + \frac{2}{4x+8} = 0 \qquad \left(1 + \frac{P}{a} + \frac{Q}{a^2} = 0,\ \text{choosing } a = 2\right)$$

Hence $y = e^{2x}$ is a part of C.F.

Putting $y = e^{2x}v$ in (1), the reduced equation as in the article 15.6.

$$\frac{d^2v}{dx^2}+\left[P+\frac{2}{u}\frac{du}{dx}\right]\frac{dv}{dx} = \frac{(x+1)\,e^x}{u\,(x+2)}$$

On putting the values of P, Q and R, we get

$$\Rightarrow \quad \frac{d^2v}{dx^2}+\left[-\frac{2x+5}{x+2}+\frac{2}{e^{2x}}2e^{2x}\right]\frac{dv}{dx} = \frac{(x+1)\,e^x}{e^{2x}(x+2)}$$

$$\Rightarrow \quad \frac{d^2v}{dx^2}+\left[-\frac{2x+5}{x+2}+4\right]\frac{dv}{dx} = \frac{x+1}{x+2}e^{-x}$$

$$\Rightarrow \quad \frac{dz}{dx}+\frac{2x+3}{x+2}z = \frac{x+1}{x+2}e^{-x} \qquad \left(\frac{dv}{dx}=z\right)$$

which is a linear differential equation,

$$I.F. = e^{\int\frac{2x+3}{x+2}dx} = e^{\int\left(2-\frac{1}{x+2}\right)dx} = e^{2x-\log(x+2)} = \frac{e^{2x}}{x+2}$$

Its solution is $\quad z.\dfrac{e^{2x}}{x+2} = \displaystyle\int\frac{e^{2x}}{x+2}\,\frac{x+1}{x+2}e^{-x}\,dx+c$

$$= \int\frac{e^x\,(x+1)}{(x+2)^2}dx+c = \int e^x\left[\frac{1}{x+2}-\frac{1}{(x+2)^2}\right]dx+c = \int\frac{e^x dx}{x+2}-\int\frac{e^x dx}{(x+2)^2}+c$$

$$= \frac{e^x}{x+2}+\int\frac{e^x dx}{(x+2)^2}-\int\frac{e^x\,dx}{(x+2)^2}+c = \frac{e^x}{x+2}+c$$

$$\Rightarrow \quad z = e^{-x}+c\,(x+2)e^{-2x} \Rightarrow \frac{dv}{dx} = e^{-x}+c(x+2)e^{-2x}$$

$$v = \int e^{-x}\,dx + c\int(x+2)e^{-2x}dx + c_1 = -e^{-x}+c\left[\frac{(x+2)\,e^{-2x}}{-2}-\frac{e^{-2x}}{4}\right]+c_1$$

$$= -e^{-x}-\frac{ce^{-2x}}{4}[2x+5]+c_1$$

$y = u.v$

$$y = e^{2x}\left[-e^{-x}-\frac{ce^{-2x}}{4}(2x+5)+c_1\right] \Rightarrow y = -e^x+\frac{c}{4}(2x+5)+c_1\,e^{2x}$$ **Ans.**

EXERCISE 15.5

Solve the following differential equations:

1. $(3-x)\dfrac{d^2y}{dx^2}-(9-4x)\dfrac{dy}{dx}+(6-3x)\,y=0$, given $y = e^x$ is a solution.

Ans. $y = \dfrac{c_1}{8}e^{3x}\,(4x^3-42x^2+150x-183)+c_2e^x$

2. $x\dfrac{d^2y}{dx^2}-\dfrac{dy}{dx}+(1-x)y=x^2\,e^{-x}$ given $y = e^x$ is an integral included in C.F.

Ans. $y = c_2\,e^x+c_1\,(2x+1)\,e^{-x}-\dfrac{1}{4}(2x^2+2x+1)e^{-x}$

3. $(1-x^2)\dfrac{d^2y}{dx^2}+x\dfrac{dy}{dx}-y=x(1-x^2)^{3/2}$, given $y = x$ is part of C.F.

Ans. $y = -\dfrac{x}{9}(1-x^2)^{3/2}-c_1\,[\sqrt{(1-x^2)}+x\sin^{-1}x]+c_2x.$

4. $\sin^2 x\dfrac{d^2y}{dx^2}=2y$, given that $y = \cot x$ is a solution **Ans.** $cy = 1 + (c_1 - x)\cot x$

5. $\dfrac{d^2y}{dx^2}-x^2\dfrac{dy}{dx}+xy=x$, given $y = x$ is a part of C.F. **Ans.** $y = 1+c_1x\displaystyle\int\frac{1}{x^2}e^{\frac{x^3}{3}}dx+c_2x$

6. $(x \sin x + \cos x)\dfrac{d^2y}{dx^2} - x\cos x\dfrac{dy}{dx} + y\cos x = 0$ given $y = x$ is solution.

Ans. $y = c_2 x - c_1 \cos x$

7. $x^2\dfrac{d^2y}{dx^2} + x\dfrac{dy}{dx} - y = 0$, given that $y = x + \dfrac{1}{x}$ is one integral. **Ans.** $y = c_2\left(x + \dfrac{1}{x}\right) + \dfrac{c_1}{x}$

8. $x^2\dfrac{d^2y}{dx^2} + 2x\dfrac{dy}{dx} - 12y = x^3 \log x$ (*U.P., II Semester 2004*)

[**Hint.** $(n(n-1) + pnx + Qx^2 = 0)$, n = 3, satisfies this equation. Put $y = vx^3, \dfrac{dv}{dx} = z$]

Ans. $y = \left(c_1 x^3 + \dfrac{c_2}{x^4}\right) + \dfrac{x^3}{98}\log x\,(7\log x - 2)$

15.7 NORMAL FORM (REMOVAL OF FIRST DERIVATIVE)

Consider the differential equation $\dfrac{d^2y}{dx^2} + P\dfrac{dy}{dx} + Qy = R$ ...(1)

Put $y = uv$ where v is not an integral solution of C.F.

$$\frac{dy}{dx} = v\frac{du}{dx} + u\frac{du}{dx}$$

$$\frac{d^2y}{dx^2} = u\frac{d^2v}{dx^2} + 2\frac{du}{dx}\frac{dv}{dx} + v\frac{d^2u}{dx^2}$$

On putting the values of $y, \dfrac{dy}{dx}, \dfrac{d^2y}{dx^2}$ in (1) we get

$$\left(u\frac{d^2v}{dx^2} + 2\frac{dv}{dx}\frac{du}{dx} + v\frac{d^2u}{dx^2}\right) + P\left(u\frac{du}{dx} + v\frac{du}{dx}\right) + Q.uv = R$$

$$\Rightarrow \quad v\frac{d^2u}{dx^2} + \frac{du}{dx}\left(Pv + 2\frac{dv}{dx}\right) + u\left(\frac{d^2v}{dx^2} + P\frac{dv}{dx} + Q.v\right) = R$$

$$\Rightarrow \quad \frac{d^2u}{dx^2} + \frac{du}{dx}\left(P + \frac{2}{u}\frac{dv}{dx}\right) + \frac{u}{v}\left(\frac{d^2v}{dx^2} + P\frac{dv}{dx} + Q.v\right) = \frac{R}{v} \quad ...(2)$$

Here in the last bracket on L.H.S. is not zero $y =\ v$ is not a part of C.F.

Here we shall remove the first derivative.

$$P + \frac{2}{v}\frac{dv}{dx} = 0 \text{ or } \frac{dv}{v} = -\frac{1}{2}P\,dx \text{ or } \log v = \frac{-1}{2}\int P\,dx$$

$$v = e^{-\frac{1}{2}\int P\,dx}$$

In (2) we have to find out the value of the last bracket *i.e.*, $\dfrac{d^2v}{dx^2} + P\dfrac{dv}{dx} + Qv$

$$\frac{dv}{dx} = -\frac{P}{2}e^{-\frac{1}{2}\int P\,dx} = -\frac{1}{2}Pv \qquad \left[\therefore v = e^{-1/2\int p\,dx}\right]$$

$$\frac{d^2v}{dx^2} = -\frac{1}{2}\frac{dP}{dx}v - \frac{P}{2}\frac{dv}{dx} = -\frac{1}{2}\frac{dP}{dx}v - \frac{P}{2}\left(-\frac{1}{2}Pv\right) = -\frac{1}{2}\frac{dP}{dx}v + \frac{1}{4}P^2v$$

$$\therefore \frac{d^2v}{dx^2} + P\frac{dv}{dx} + Qv = -\frac{1}{2}\frac{dP}{dx}v + \frac{1}{4}P^2v + P\left(-\frac{1}{2}Pv\right) + Qv = v\left[Q - \frac{1}{2}\frac{dP}{dx} - \frac{1}{4}P^2\right]$$

Equation (1) is transformed as

$$\frac{d^2u}{dx^2} + \frac{u}{v}v\left\{Q - \frac{1}{2}\frac{dP}{dx} - \frac{P^2}{4}\right\} = \frac{R}{v}$$

$$\Rightarrow \quad \frac{d^2u}{dx^2} + u\left\{Q - \frac{1}{2}\frac{dP}{dx} - \frac{P^2}{4}\right\} = R e^{\frac{1}{2}\int P\,dx}$$

$$\frac{d^2u}{dx^2} + Q_1 u = R_1 \qquad \text{where } Q_1 = \left[Q - \frac{1}{2}\frac{dP}{dx} - \frac{P^2}{4}\right]$$

$$R_1 = R e^{\frac{1}{2}\int P\,dx} \text{ or } \frac{R}{v}$$

$$y = uv \quad \text{and} \quad v = e^{-\frac{1}{2}\int P\,dx}$$ **Ans.**

Example 18. *Solve* $\frac{d}{dx}\left[\cos^2 x \frac{dy}{dx}\right] + \cos^2 x.\, y = 0$

Solution. We have, $\frac{d}{dx}\left(\cos^2 x \frac{dy}{dx}\right) + \cos^2 x.\, y = 0$

$$\Rightarrow \frac{d^2y}{dx^2}\cos^2 x - 2\cos x \sin x \frac{dy}{dx} + (\cos^2 x)y = 0 \Rightarrow \frac{d^2y}{dx^2} - 2\tan x.\frac{dy}{dx} + y = 0$$

Here, $P = -2\tan x,\ Q = 1,\ R = 0$

$$Q_1 = Q - \frac{1}{2}\frac{dP}{dx} - \frac{P^2}{4} = 1 - \frac{1}{2}(-2\sec^2 x) - \frac{4\tan^2 x}{4}$$

$$= 1 + \sec^2 x - \tan^2 x = 1 + 1 = 2$$

$$R_1 = R e^{\frac{1}{2}\int P\,dx} = 0$$

$$v = e^{-\frac{1}{2}\int P\,dx} = e^{-\frac{1}{2}\int(-2\tan x)dx} = e^{\int \tan x\,dx} = e^{\log \sec x} = \sec x$$

Normal equation is

$$\frac{d^2u}{dx^2} + Q_1 u = R_1$$

$$\frac{d^2u}{dx^2} + 2u = 0 \quad \text{or} \quad (D_2 + 2)\,u = 0 \quad \Rightarrow \quad D = \pm i\sqrt{2}$$

$$u = c_1 \cos\sqrt{2}\,x + c_2 \sin\sqrt{2}\,x$$

$$y = u.v$$

$$= [c_1\cos\sqrt{2}\,x + c_2\sin\sqrt{2}x]\sec x$$ **Ans.**

Example 19. *Solve* $x^2\frac{d^2y}{dx^2} - 2(x^2 + x)\frac{dy}{dx} + (x^2 + 2x + 2)y = 0$

Solution. We have, $\frac{d^2y}{dx^2} - \frac{2(x^2+x)}{x^2}\frac{dy}{dx} + \left(\frac{x^2+2x+2}{x^2}\right)y = 0$ ...(1)

Here $p = -2\left(1 + \frac{1}{x}\right),\ Q = \frac{x^2+2x+2}{x^2},\ R = 0$

In order to remove the first derivative, we put $y = u.v$ in (1) to get the normal equation

$$\frac{d^2v}{dx^2} + Q_1 v = R_1$$ ...(2)

where $v = e^{-\frac{1}{2}\int p\,dx} = e^{-\frac{1}{2}\int -2\left(1+\frac{1}{x}\right)dx} = e^{\int\left(1+\frac{1}{x}\right)dx} = e^x . e^{\log x} = x\,e^x$

$$Q_1 = Q - \frac{1}{2}\frac{dp}{dx} - \frac{p^2}{4} = \frac{x^2+2x+2}{x^2} - \frac{1}{2}\left(\frac{2}{x^2}\right) - \frac{4}{4}\left(1+\frac{1}{x}\right)^2$$

$$= 1 + \frac{2}{x} + \frac{2}{x^2} - \frac{1}{x^2} - 1 - \frac{1}{x^2} - \frac{2}{x}$$

$$R_1 = R e^{\frac{1}{2}\int p\,dx} = 0$$

On putting the values of Q_1 and R_1 in (2), we get

$$\frac{d^2u}{dx^2} + 0(u) = 0 \Rightarrow \frac{d^2u}{dx^2} = 0$$

$$\frac{du}{dx} = c_1 \Rightarrow u = c_1x + c_2$$

$\therefore$ $\quad y = u.v = (c_1x + c_2)\,x\,e^x$ **Ans.**

Example 20. *Solve* $\frac{d^2y}{dx^2} - 4x\frac{dy}{dx} + (4x^2-1)y = -3e^{x^2}\sin 2x$ *(U.P. II Semester, (C.O.) 2004)*

Solution. We have, $\frac{d^2y}{dx^2} = 4x\frac{dy}{dx} + (4x^2-1)\,y = -3e^{x^2}\sin 2x$ ...(1)

Here $p = -4x$, $Q = 4x^2 - 1$, $R = -3e^{x^2}\sin 2x$

In order to remove the first derivative, $v = e^{-\frac{1}{2}\int p\,dx} = e^{-\frac{1}{2}\int -4x\,dx} = e^{2\int x\,dx} = e^{x^2}$

On putting $y = uv$, the normal equation is $\frac{d^2u}{dx^2} + Q_1 u = R_1$ ...(2)

where $Q_1 = Q - \frac{1}{2}\frac{dp}{dx} - \frac{p^2}{4} = (4x^2-1) - \frac{1}{2}(-4) - \frac{16x^2}{4} = 4x^2 - 1 + 2 - 4x^2 = 1$

$$R_1 = \frac{R}{v} = \frac{-3e^{x^2}\sin 2x}{e^{x^2}} = -3\sin 2x$$

Equation (2) becomes $\frac{d^2u}{dx^2} + u = -3\sin 2x$

$(D^2 + 1)u = -3\sin 2x$

A.E. is $m^2 + 1 = 0 \Rightarrow m = \pm i \Rightarrow C.F. = c_1\cos x + c_2\sin x$

$$P.I. = \frac{1}{D^2+1}(-3\sin 2x) = \frac{-3\sin 2x}{-4+1} = \sin 2.$$

$$u = c_1\cos x + c_2\sin x + \sin 2x$$

$$y = u.\,v = (c_1\cos x + c_2\sin x + \sin 2x)e^{x^2}$$ **Ans.**

EXERCISE 15.6

Solve the following differential equations:

1. $\frac{d^2y}{dx^2} - 2\tan x.\,y - 5y = 0$ **Ans.** $y = (a\,e^{2x} + e^{-3x})\sec x$

2. $\frac{d^2y}{dx^2} - 4x\frac{dy}{dx} + (4x^2-3)\,y = e^{x^2}$ **Ans.** $y = (c_1e^x + c_2e^{-x} - 1)$

3. $\frac{d^2y}{dx^2} - 2x\frac{dy}{dx} + (x^2+2)y = e^{\frac{1}{2}(x^2+2x)}$ **Ans.** $y = (c_1\cos\sqrt{3}x + c_2\sin\sqrt{3}\,x)\,e^{\frac{x^2}{2}} + \frac{1}{4}e^x.e^{\frac{x^2}{2}}$

4. $\frac{d^2y}{dx^2} - 2x\frac{dy}{dx} + \left(n^2 + \frac{2}{x^2}\right)y = 0$ **Ans.** $y = (c_1\cos nx + c_2\sin nx)x$

5. $\frac{d^2y}{dx^2} + \frac{2}{x}\frac{dy}{dx} - n^2y = 0$ **Ans.** $y = (c_1e^{nx} + c_2 + e^{-nx})\frac{1}{x}$

6. $\dfrac{d^2y}{dx^2}+\dfrac{1}{x^{\frac{1}{3}}}\dfrac{dy}{dx}+\left(\dfrac{1}{4x^{2/3}}-\dfrac{1}{x^{4/3}}-\dfrac{6}{x^2}\right)y=0$ **Ans.** $y=(c_1x^3+c_2x^{-2})e^{-\frac{3}{4}x^{\frac{2}{3}}}$

7. $\dfrac{d^2y}{dx^2}-\dfrac{1}{\sqrt{x}}\dfrac{dy}{dx}+\dfrac{y}{4x^2}(-8+\sqrt{x}+x)=0$ **Ans.** $y=(c_1x^2+c_2x^{-1})\,e^{\sqrt{x}}$

15.8 METHOD OF SOLVING LINEAR DIFFERENTIAL EQUATIONS BY CHANGING THE INDEPENDENT VARIABLE

Consider, $$\frac{d^2y}{dx^2}+P\frac{dy}{dx}+Qy=R \qquad ...(1)$$

Let us change the independent variable x to z and $z=f(x)$

$$\frac{dy}{dx}=\frac{dy}{dz}\frac{dz}{dx} \Rightarrow \frac{d^2y}{dx^2}=\frac{d^2y}{dz^2}\left(\frac{dz}{dx}\right)^2+\frac{dy}{dz}\frac{d^2z}{dx^2}$$

Putting the values of $\frac{dy}{dx}$ and $\frac{d^2y}{dx^2}$ in (1), we get

$$\left(\frac{d^2y}{dz^2}\left(\frac{dz}{dx}\right)^2+\frac{dy}{dz}\frac{d^2z}{dx^2}\right)+P\left(\frac{dy}{dz}\frac{d^2z}{dx^2}\right)+Qy=R$$

$$\Rightarrow \frac{d^2y}{dz^2}\left(\frac{dz}{dx}\right)^2+\left(P\frac{dz}{dx}+\frac{d^2z}{dx^2}\right)\frac{dy}{dz}+Qy=R$$

$$\Rightarrow \frac{d^2y}{dz^2}+\frac{P\left(\frac{dz}{dx}+\frac{d^2z}{dx^2}\right)}{\left(\frac{dz}{dx}\right)^2}\frac{dy}{dz}+\frac{Qy}{\left(\frac{dz}{dx}\right)^2}=\frac{R}{\left(\frac{dz}{dx}\right)^2} \Rightarrow \frac{d^2y}{dz^2}+P_1\frac{dy}{dz}+Q_1y=R \qquad ...(2)$$

where $$P_1=\frac{P\left(\frac{dz}{dx}+\frac{d^2z}{dx^2}\right)}{\left(\frac{dz}{dx}\right)^2},\quad Q_1=\frac{Q}{\left(\frac{dz}{dx}\right)^2},\quad R_1=\frac{R}{\left(\frac{dz}{dx}\right)^2}$$

Equation (2) is solved either by taking $P_1=0$ or $Q_1=a$ constant.

Equation (2) can be solved by by two methods, by taking

First Method, $P_1=0$

Second Method, $Q=$ constant

Working Rule

Step 1. Coefficient of $\frac{d^2y}{dx^2}$ should be made as 1 if it is not so.

Step 2. To get P, Q and R, compare the given differential equation with the standard form $y''+Py'+Qy=R$.

Step 3. Find P_1, Q_1 and R_1 by the following formulae.

$$P_1=\frac{\frac{d^2y}{dx^2}+p\frac{dz}{dx}}{\left(\frac{dz}{dx}\right)^2},\quad R_1=\frac{R}{\left(\frac{dz}{dx}\right)^2}$$

Step 4. Find out the value of z by taking

First Method, $P_1=0$ **Second Method.** $Q_1=$ constant

Step 5. We get a reduced equation $\frac{d^2y}{dz^2} + P_1 \frac{dy}{dz} + Q_1 y = R_1$

On solving this equation we can find out the value of y in terms of z.
Then write down the solution in terms of x by replacing the value of z.

Example 21. *Solve* $\frac{d^2y}{dx^2} + \cot x \frac{dy}{dx} + 4\, y \operatorname{cosec}^2 x = 0$

Solution. We have, $\frac{d^2y}{dx^2} + \cot x \frac{dy}{dx} + 4\, y \operatorname{cosec}^2 x = 0$ ...(1)

Here, $P = \cot x$, $Q = 4 \operatorname{cosec}^2 x$ and $R = 0$

Changing the independent variable from x to z, the equation becomes

$$\frac{d^2y}{dx^2} + P_1 \frac{dy}{dz} + Q_1 y = 0 \quad ...(2)$$

where
$$P_1 = \frac{P\frac{dz}{dx} + \frac{d^2z}{dx^2}}{\left(\frac{dz}{dx}\right)^2}, \quad Q_1 = \frac{Q}{\left(\frac{dz}{dx}\right)^2}$$

Case I. Let us take $P_1 = 0$

$$\frac{p\frac{dz}{dx} + \frac{d^2z}{dx^2}}{\left(\frac{dz}{dx}\right)^2} = 0 \text{ or } P\frac{dz}{dx} + \frac{d^2z}{dx^2} = 0 \Rightarrow \frac{d^2z}{dx^2} + \cot x \frac{dz}{dx} = 0 \quad ...(3)$$

Put
$$\frac{dz}{dx} = v, \quad \frac{d^2z}{dx^2} = \frac{dv}{dx}$$

(3) becomes $\frac{dv}{dx} + (\cot x)\, v = 0 \Rightarrow \frac{dv}{v} = -\cot x.\, dx$

$\Rightarrow$
$$\log v = -\log \sin x + \log c = \log c \log c \operatorname{cosec} x \Rightarrow v = c \operatorname{cosec} x$$

$$\frac{dz}{dx} = c \operatorname{cosec} x \text{ or } dz = (c \operatorname{cosec} x)\, dx \Rightarrow z = c \log \tan \frac{x}{2}$$

Case II.

Now let us take $Q_1 =$ Constant.

$$Q_1 = \frac{Q}{\left(\frac{dz}{dx}\right)^2} = \frac{4 \operatorname{cosec}^2 x}{c^2 \operatorname{cosec}^2 x} = \frac{4}{c^2} \text{ which is constant}$$

Hence the equation (2) reduces to

$$\frac{d^2y}{dz^2} + 0\frac{dy}{dz} + \frac{4}{c^2} y = 0 \text{ or } \frac{d^2y}{dz^2} + \frac{4}{c^2} y = 0 \quad \left[\because P_1 = 0, Q_1 = \frac{4}{c^2}\right]$$

$\Rightarrow$ $\left(D^2 + \frac{4}{c^2}\right) y = 0$, A.E. is $m^2 + \frac{4}{c^2} = 0 \Rightarrow m = \pm i \frac{2}{c}$

$$\text{C.F.} = c_1 \cos \frac{2z}{c} + c_2 \sin \frac{2z}{c} \quad \left(z = c \log \tan \frac{x}{2}\right)$$

$\Rightarrow$
$$y = c_1 \cos\left(2 \log \tan \frac{x}{2}\right) + c_2 \sin\left(2 \log \tan \frac{x}{2}\right)$$ **Ans.**

Example 22. *Solve* $x^6 \frac{d^2y}{dx^2} + 3x^5 \frac{dy}{dx} + a^2 y = \frac{1}{x^2}$

Solution. We have, $\frac{d^2y}{dx^2} + \frac{3}{x}\frac{dy}{dx} + a^2 \frac{y}{x^6} = \frac{1}{x^8}$ ...(1)

Here $P = \frac{3}{x}$ and $Q = \frac{a^2}{x^6}$

On changing the independent variable x to z, the equation (1) is reduced to

$$\frac{d^2y}{dz^2} + P_1 \frac{dy}{dz} + Q_1 y = R_1 \quad ...(2)$$

Using Second Method

Let $Q_1 = a_2$ (constant) $\quad Q_1 = \frac{Q}{\left(\frac{dz}{dx}\right)^2} = \frac{a^2}{x^6\left(\frac{dz}{dx}\right)^2}$ = constant = a_2 (say)

$$\therefore \quad x^6\left(\frac{dz}{dx}\right)^2 = 1 \Rightarrow x^3\left(\frac{dz}{dx}\right) = 1 \Rightarrow \frac{dz}{dx} = \frac{1}{x^3} \Rightarrow dz = \frac{dx}{x^3} \Rightarrow z = \frac{x^{-2}}{-2} + c$$

On differentiating twice, we have $\frac{d^2z}{dx^2} = \frac{-3}{x^4}$

$$P_1 = \frac{P\frac{dz}{dx} + \frac{d^2z}{dx^2}}{\left(\frac{dz}{dx}\right)^2} = \frac{\frac{3}{x}\cdot\frac{1}{x^3} + \left(\frac{-3}{x^4}\right)}{\left(\frac{1}{x^3}\right)^2} = 0 \Rightarrow R_1 = \frac{R}{\left(\frac{dz}{dx}\right)^2} = \frac{\frac{1}{x^8}}{\frac{1}{x^6}} = \frac{1}{x^2} = -2z$$

On putting the values of P_1, Q_1 and R_1 in (2), we get

$$\frac{d^2y}{dz^2} + a^2 y = -2z \quad \Rightarrow \quad (D^2 + a^2)\, y = -2z$$

A.E. is $m^2 + a^2 = 0$, $\quad m = \pm\, i\, a$, $\quad \Rightarrow \quad$ C.F. $= c_1 \cos az + c_2 \sin az$

$$\text{P.I.} = \frac{1}{D^2 + a^2}(-2z) = \frac{1}{a^2}\left[1 + \frac{D^2}{a^2}\right]^{-1}(-2z) = \frac{1}{a^2}\left[1 - \frac{D^2}{a^2}\right](-2z) = \frac{-2z}{a^2} = \frac{1}{a^2x^2}$$

$$y = \text{C.F.} + \text{P.I.}$$

$$y = c_1 \cos\frac{a}{2x^2} - c_2 \sin\frac{a}{2x^2} + \frac{1}{a^2x^2}$$ **Ans.**

Example 23. *Solve* $\frac{d^2y}{dx^2} - \frac{1}{x}\frac{dy}{dx} + 4x^2 y = x^4$ *(U.P., I Semester Summer 2003, 2002)*

Solution. We have, $\frac{d^2y}{dx^2} - \frac{1}{x}\frac{dy}{dx} + 4x^2 y = x^4$ ...(1)

Hence $P = -\frac{1}{x}$, $Q = 4x^2$, $R = x^4$

On changing the independent variable x to z, the equation (1) is transformed as

$$\frac{d^2y}{dz^2} + P_1 \frac{dy}{dz} + Q_1 y = R_1 \quad ...(2)$$

Using Second Method

Let $\quad Q_1 = 1$ (constant)

but $$Q_1 = \frac{Q}{\left(\frac{dz}{dx}\right)^2} = \frac{4x^2}{\left(\frac{dz}{dx}\right)^2} = \text{constant} = 1 \quad \text{(say)}$$

$$\left(\frac{dz}{dx}\right)^2 = 4x^2 \Rightarrow \frac{dz}{dx} = 2x$$

$$\Rightarrow \quad dz = 2x\,dx \Rightarrow z = x^2 + c \Rightarrow x^2 = z - c \qquad ...(3)$$

$$P_1 = \frac{P\left(\frac{dz}{dx}\right) + \frac{d^2z}{dx^2}}{\left(\frac{dz}{dx}\right)^2} = \frac{-\frac{1}{x}(2x) + 2}{(2x)^2} = 0$$

$$R_1 = \frac{R}{\left(\frac{dz}{dx}\right)^2} = \frac{x^4}{4x^2} = \frac{x^2}{4} = \frac{z-c}{4} \qquad \text{[Using (3)]}$$

On putting the value of P_1, Q_1 and R_1 in (2), we get

$$\frac{d^2y}{dz^2} + (0)\frac{dy}{dz} + (1)\,y = \frac{z-c}{4} \Rightarrow \frac{d^2y}{dz^2} + y = \frac{z-c}{4} \Rightarrow (D^2+1)\,y = \frac{z-c}{4}$$

A.E. is $m^2 + 1 = 0$, $\Rightarrow m = \pm i$

$$\text{C.F.} = c_1 \cos z + c_2 \sin z$$

$$\text{P.I.} = \frac{1}{D^2+1}\left(\frac{z-c}{4}\right) = (1+D^2)^{-1}\frac{z-c}{4} = (1-D^2)\frac{z-c}{4} = \frac{z-c}{4}$$

Now complete solution = C.F. + P.I.

$$\Rightarrow \quad y = c_1 \cos z + c_2 \sin z + \frac{z-c}{4} \Rightarrow y = c_1 \cos x^2 + c_2 \sin x^2 + \frac{x^2}{4} \qquad \textbf{Ans.}$$

Example 24. *Solve the following differential equation by changing the independent variable*

$$x\frac{d^2y}{dx^2} + (4x^2-1)\frac{dy}{dx} + 4x^3y = 2x^3 \qquad \textit{(U.P., II Semester, Summer 2006)}$$

Solution. We have

$$x\frac{d^2y}{dx^2} + (4x^2-1)\frac{dy}{dx} + 4x^3y = 2x^3$$

$$\Rightarrow \quad \frac{d^2y}{dx^2} + \left(\frac{4x^2-1}{x}\right)\frac{dy}{dx} + 4x^2y = 2x^2 \qquad ...(1)$$

Here, $P = \dfrac{4x^2-1}{x}$, $Q = 4x^2$ and $R = 2x^2$

Changing the independent variable from x to z, the equation becomes

$$\frac{d^2y}{dz^2} + P_1\frac{dy}{dz} + Q_1y = R_1 \qquad ...(2)$$

where $$P_1 = \frac{P\frac{dz}{dx} + \frac{d^2z}{dx^2}}{\left(\frac{dz}{dx}\right)^2}, \quad Q_1 = \frac{Q}{\left(\frac{dz}{dx}\right)^2}, \quad R_1 = \frac{R}{\left(\frac{dz}{dx}\right)^2}$$

Using Second Method

Choose $Q_1 = 1$ (constant) therefore $1 = \dfrac{Q}{\left(\dfrac{dz}{dx}\right)^2}$

$$\Rightarrow \quad \left(\frac{dz}{dx}\right)^2 = 4x^2 \qquad [\because Q = 4x^2]$$

$$\Rightarrow \quad \frac{dz}{dx} = 2x \qquad \Rightarrow dz = 2x\,dx \quad \Rightarrow \quad z = x^2$$

$$\frac{d^2z}{dx^2} = 2$$

and $P_1 = \dfrac{P\left(\dfrac{dz}{dx}\right) + \dfrac{d^2z}{dx^2}}{\left(\dfrac{dz}{dx}\right)^2} = \dfrac{\left(\dfrac{4x^2-1}{x}\right)(2x)+2}{(2x)^2} \quad \Rightarrow P_1 = \dfrac{8x^2-2+2}{4x^2} = 2$

$$R_1 = \frac{R}{\left(\dfrac{dz}{dx}\right)^2} = \frac{2x^2}{4x^2} = \frac{1}{2}$$

The equation (2) is transformed to

$$\frac{d^2y}{dz^2} + 2\frac{dy}{dz} + y = \frac{1}{2}$$

A.E. is $\quad m^2 + 2m + 1 = 0$

$\Rightarrow \quad (m+1)^2 = 0 \quad \Rightarrow \quad m = -1, -1$

$$\text{C.F.} = (C_1 + C_2 z)\, e^{-z}$$

$$\text{P.I.} = \frac{1}{D^2+2D+1}\left(\frac{1}{2}\right) = \frac{1}{0+0+1}\left(\frac{1}{2}\right) = \frac{1}{2}$$

Complete solution is

$$y = C.F. + P.I. = (c_1 + c_1 z)e^{-z} + \frac{1}{2}$$

$$\Rightarrow \quad y = (c_1 + c_2x^2)\, e^{-x^2} + \frac{1}{2} \qquad \textbf{Ans.}$$

EXERCISE 15.7

Solve the following differential equations:

1. $x^4\dfrac{d^2y}{dx^2} + 2x^3\dfrac{dy}{dx} + a^2y = 0$ **Ans.** $y = c_1\cos\dfrac{a}{x} + c_2\sin\dfrac{a}{x}$

2. $\cos x\dfrac{d^2y}{dx^2} + \dfrac{dy}{dx}\sin x - 2y\cos^3 x = 2\cos^5 x$ **Ans.** $y = c_1e^{\sqrt{2}\sin x} + c_2e^{-\sqrt{2}\sin x} + \sin^2 x$

3. $\dfrac{d^2y}{dx^2} + \tan x\dfrac{dy}{dx} + y\cos^2 x = 0$ **Ans.** $y = c_1\cos(\sin x) + c_2\sin(\sin x)$

4. $x\dfrac{d^2y}{dx^2} - \dfrac{dy}{dx} - 4x^3y = 8x^2\sin x^2$ **Ans.** $y = c_1e^{x^2} + c_2e^{\cos x} + \dfrac{1}{6}e^{-\cos x}$

5. $\dfrac{d^2y}{dx^2} + (3\sin x - \cot x)\dfrac{dy}{dx} + 2y\sin^2 x = e^{-\cos x}\sin^2 x$ **Ans.** $y = c_1e^{2\cos x} + c_2e^{\cos x} + \dfrac{1}{6}e^{-\cos x}$

6. $\dfrac{d^2y}{dx^2} + (\tan x - 1)^2\dfrac{dy}{dx} - n(n-1)y\sec^4 x = 0$ **Ans.** $y = C_1e^{-n\tan x} + C_2e^{(n-1)\tan x}$

7. $\dfrac{d^2y}{dx^2} - \cot x \dfrac{dy}{dx} - y \sin^2 x = \cos x - \cos^3 x$ **Ans.** $y = C_1 e^{-\cos x} + C_2 e^{\cos x} - \cos x$

8. $\dfrac{d^2y}{dx^2} + (3 \sin x - \cot x) \dfrac{dy}{dx} + 2y \sin^2 x = e^{-\cos x} \sin^2 x$ **Ans.** $y = C_1 e^{\cos x} + C_2 e^{2\cos x} + \dfrac{1}{6} e^{-\cos x}$

15.9 METHOD OF VARIATION OF PARAMETERS

Here, the method to find out C.F. is different from the methods discussed earlier. Now, we will find C.F. new methods and then will apply the method of variation of parameters as discussed in Article 13.3 on page 275.

Example 25. *By the method of variation of parameters, solve the differential equation*

$$\frac{d^2y}{dx^2} + (1 - \cot x) \frac{dy}{dx} - y \cot x = \sin^2 x. \qquad \textit{(U.P., II Semester, Summer 2002)}$$

Solution. First we shall find the C.F. of the given equation *i.e.,* the solution of the equation.

$$\frac{d^2y}{dx^2} + (1 - \cot x) \frac{dy}{dx} - y \cot x = 0 \qquad \left[\frac{d^2y}{dx^2} + P\frac{dy}{dx} + Q\,y = 0\right] \qquad ...(1)$$

Here $P = 1 - \cot x$, $Q = -\cot x$

$\therefore \quad 1 - P + Q = 1 - (1 - \cot x) - \cot x = 0$, $\therefore$ $y = e^{-x}$ is apart of the C.F.

Putting $y = ve^{-x}$, $\qquad \dfrac{dy}{dx} = -ve^{-x} \dfrac{dy}{dx}$

$$\frac{d^2y}{dx^2} = e^{-x} \frac{d^2v}{dx^2} - 2e^{-x}\frac{dv}{dx} + ve^{-x} \text{ the equation (1) reduces to}$$

$$\frac{d^2v}{dx^2} - (1 + \cot x)\frac{dv}{dx} = 0 \Rightarrow \frac{dp}{dx} - (1 + \cot x)p = 0. \quad \text{where } p = \frac{dv}{dx}$$

$$\Rightarrow \qquad \frac{dp}{p} = (1 + \cot x)\, dx$$

Integrating, we get $\qquad \log p = x + \log \sin x + \log c_1 \Rightarrow \log\left[\dfrac{p}{c_1 \sin x}\right] = x$

$$\Rightarrow \qquad \frac{P}{c_1 \sin x} = e^x \Rightarrow P = c_1\, e^x \sin x$$

$$\therefore \qquad p = \frac{dv}{dx} = c_1\, e^x \sin x$$

$$\Rightarrow \qquad v = c_1 \int e^x \sin x\, dx = c_1 . \frac{1}{2} e^x (\sin x - \cos x) + c_2$$

$\therefore$ The solution of (1) i.e., C.F. of the given equation is

$$y = ve^{-x} = c_1 . \frac{1}{2} (\sin x - \cos x) + c_2 e^{-x}$$

Let $y = A(\sin x - \cos x) + Be^{-x}$ be the complete solution of the given equation where A and B are functions of x, so chosen that the given equation will be satisfied ...(2)

$$\therefore \quad \frac{dy}{dx} = A(\cos x + \sin x) - Be^{-x} + \frac{dA}{dx} (\sin x - \cos x) + \frac{dB}{dx} e^{-x}.$$

Let us choose A and B such that

$$\frac{dA}{dx} (\sin x - \cos x) + \frac{dB}{dx} e^{-x} = 0 \qquad ...(3)$$

$$\therefore \quad \frac{dy}{dx} = A(\cos x + \sin x) - Be^{-x}$$

and $\dfrac{d^2y}{dx^2} = \dfrac{dA}{dx}(\cos x + \sin x) - \dfrac{dB}{dx}e^{-x} + A(-\sin x + \cos x) + Be^{-x}$

Putting these values of $\dfrac{d^2y}{dx^2}, \dfrac{dy}{dx}$ and y in the given equation, we get

$$\frac{dA}{dx}(\cos x + \sin x) - \frac{dB}{dx}e^{-x} = \sin^2 x. \qquad ...(4)$$

Solving (3) and (4), we get $A = -\dfrac{1}{2}\cos x + c_1$

and $B = \dfrac{1}{4}\int e^x (\sin 2x - 1 + \cos 2x)\,dx + c_2 = \dfrac{1}{4}\int e^x \sin 2x\,dx - \dfrac{1}{4}\int e^x dx + \dfrac{1}{4}\int e^x \cos 2x\,dx$

$$= \frac{1}{4}.\frac{e^x}{5}(\sin 2x - 2\cos 2x) - \frac{e^x}{4} + \frac{1}{4}.\frac{e^x}{5}(\cos 2x + 2\sin 2x) + c_2$$

$$= \frac{e^x}{20}(3\sin 2x - \cos 2x) - \frac{e^x}{4} + c_2.$$

Putting the values of A and B in (2), the general solution of the given equation is

$$y = \left(-\frac{1}{2}\cos x + c_1\right)(\sin x - \cos x) + \left\{\frac{e^x}{20}(3\sin 2x - \cos 2x) - \frac{e^x}{4} + c_2\right\}e^{-x}$$

$$= -\frac{1}{2}\cos x \sin x + \frac{1}{2}\cos^2 x + c_1 \sin x - c_1 \cos x + \frac{3}{20}\sin 2x - \frac{1}{20}\cos 2x - \frac{1}{4} + c_2 e^{-x}$$

$$= -\frac{1}{4}(\sin 2x) + \frac{1}{4}(\cos 2x + 1) + c_1(\sin x - \cos x) + \frac{3}{20}\sin 2x - \frac{1}{20}\cos 2x - \frac{1}{4} + c_2 e^{-x}$$

$$y = c_1(\sin x - \cos x) + c_2 e^{-x} - \frac{1}{10}(\sin 2x - 2\cos 2x)$$ **Ans.**

15.10 METHOD OF UNDETERMINED COEFFICIENTS

Let $y = A(\sin x - \cos x) + Be^{-x}$ be the complete solution of the given equation where A and B are function of x, so chosen that the given equation will be satisfied.

$$\therefore \quad \frac{dy}{dx} = A(\cos x + \sin x) - Be^{-x} + \frac{dA}{dx}(\sin x - \cos x) + \frac{dB}{dx}e^{-x}. \qquad ...(2)$$

Let us choose A and B such that

$$\frac{dA}{dx}(\sin x - \cos x) + \frac{dB}{dx}e^{-x} = 0. \qquad ...(3)$$

Now (2) becomes

$$\therefore \quad \frac{dy}{dx} = A(\cos x + \sin x) - Be^{-x}$$

and $\dfrac{d^2y}{dx^2} = \dfrac{dA}{dx}(\cos x + \sin x) - \dfrac{dB}{dx}e^{-x} + A(-\sin x + \cos x) + Be^{-x}$

Putting these values of $\dfrac{d^2y}{dx^2}, \dfrac{dy}{dx}$ and y in the given equation, we get

$$\frac{dA}{dx}(\cos x + \sin x) - \frac{dB}{dx}e^{-x} + A(-\sin x + \cos x) + Be^{-x} + (1 - \cot x)$$
$$[A(\cos x + \sin x) - B e^{-x}] - [A(\sin x - \cos x) + Be^{-x}]\cot x = \sin^2 x$$

$$\Rightarrow \quad \frac{dA}{dx}(\cos x + \sin x) - \frac{dB}{dx}e^{-x} = \sin^2 x$$

Solving (3) and (4), we get

$$\frac{dA}{dx}=\frac{1}{2}\sin x \quad \text{and} \quad \frac{dB}{dx} = \frac{1}{2}e^x(\sin x\cos x-\sin^2 x) = \frac{e^x}{4}(\sin 2x+\cos 2x-1)$$

Integrating these, we get

$$A = -\frac{1}{2}\cos x+C_1$$

and

$$B = \frac{1}{4}\int e^x(\sin 2x-1+\cos 2x)\,dx = \frac{1}{4}\int e^x\sin 2x\,dx-\frac{1}{4}\int e^x\,dx+\frac{1}{4}\int e^x\cos 2x\,dx$$

$$= \frac{1}{4}.\frac{e^x}{5}(\sin 2x-2\cos 2x)-\frac{e^x}{4}+\frac{1}{4}.\frac{e^x}{5}(\cos 2x+2\sin 2x)+C_3$$

$$= \frac{e^x}{20}(3\sin 2x-\cos 2x)-\frac{e^x}{4}+C_2$$

Putting the values of A and B in (2), the general solution of the given equation is

$$y = \left(-\frac{1}{2}\cos x+C_1\right)(\sin x-\cos x)+\left\{\frac{e^x}{20}(3\sin 2x-\cos 2x)-\frac{e^x}{4}+C_2\right\}e^{-x}$$

$$= -\frac{1}{2}\cos x\sin x+\frac{1}{2}\cos^2 x+C_1\sin x-C_1\cos x+\frac{3}{20}\sin 2x-\frac{1}{20}\cos x-\frac{1}{4}+C_2e^{-x}$$

$$= -\frac{1}{4}(\sin 2x)+\frac{1}{4}(\cos 2x+1)+C_1(\sin x-\cos x)+\frac{3}{20}\sin 2x-\frac{1}{20}\cos 2x-\frac{1}{4}+C_2e^{-x}$$

$$y = C_1(\sin x-\cos x)+C_2e^{-x}-\frac{1}{10}(\sin 2x-2\cos 2x)$$ **Ans.**

Example 26. *Solve by the method of variation of parameters:*

$$x^2\frac{d^2y}{dx^2}-2x(1+x)\frac{dy}{dx}+2(1+x)\,y=x^3.$$

Solution. Dividing the given equation by x^2, we get

$$\frac{d^2y}{dx^2}-\frac{2(1+x)}{x}\frac{dy}{dx}+\frac{2(1+x)}{x^2}y=x. \qquad ...(1)$$

Firstly, we shall find the C.F. of (1) *i.e.*, the complete solution of the following differential equation

$$\frac{d^2y}{dx^2}-\frac{2(1+x)}{x}\frac{dy}{dx}+\frac{2(1+x)}{x^2}y=0 \qquad ...(2)$$

Here $\qquad P=-\frac{2(1+x)}{x},\ Q=\frac{2(1+x)}{x^2},\ R=0$

Clearly, $\qquad P+Qx=-\frac{2(1+x)}{x}+\frac{2(1+x)}{x}=0$

$\therefore \qquad y=x$ is part of C.F

Hence, $\qquad y=x$ here

Let $v=vx$ be complete primitive of.(2). So putting $y=vx$ in (2), the reduced equation is

$$\frac{d^2v}{dx^2}+\left(\frac{2}{u}\frac{du}{dx}+P\right)\frac{dv}{dx} = \frac{R}{u}$$

i.e., $$\frac{d^2y}{dx^2}+\left\{\frac{2}{x}.1-\frac{2(1+x)}{x}\right\}\frac{dv}{dx} = 0$$

$$\Rightarrow \qquad \frac{d^2v}{dx^2}-\frac{2dv}{dx} = 0 \Rightarrow (D^2-2D)v=0 \qquad [D\equiv d/dx]$$

$$\Rightarrow \qquad D(D-2)\,v = 0$$

$$v = C_1 + C_2 e^{2x}$$

Thus from $y = vx$, the complete solution of (2) *i.e.*, the C.F. of equation (1) is given by

$$y = C_1 x + C_2 x e^{2x} \Rightarrow C.F. = C_1 y_1 + C_2 y_2$$

Method of Undetermined Coefficients

Now suppose that the complete solution of (1) is given by

$$y = Ax + B\,x\,e^{2x} \qquad ...(4)$$

where A and B are functions of x.

$$\therefore \quad \frac{dy}{dx} = A + Be^{2x} + 2Bxe^{2x} + \frac{dA}{dx}.x + \frac{dB}{dx}xe^{2x} = A + Be^{2x} + 2Bxe^{2x} \qquad ...(5)$$

$$\text{where} \qquad \frac{dA}{dx} + \frac{dB}{dx}.e^{2x} = 0 \qquad ...(6)$$

Differentiating (5), we have

$$\frac{d^2y}{dx^2} = \frac{dA}{dx} + \left(\frac{dB}{dx}e^{2x} + 2Be^{2x}\right) + \left(2\frac{dB}{dx}xe^{2x} + 2Be^{2x} + 4Be^{2x}\right)$$

$$= \left(\frac{dA}{dx} + e^{2x}\frac{dB}{dx}\right) + \left(4Be^{2x} + 4Bxe^{2x} + 2xe^{2x}\frac{dB}{dx}\right)$$

$$= 0 + 4Be^{2x} + 4Bxe^{2x} + 2xe^{2x}\frac{dB}{dx} \qquad ...(7) \text{ [Using (6)]}$$

Subtituting the values of $\frac{d^2y}{dx^2}, \frac{dy}{dx}$ and y from (7), (5) and (4) respectively in (1), we get

$$\frac{dA}{dx} + e^{2x}(2x+1)\frac{dB}{dx} = x \qquad ...(8)$$

Solving (6) and (8), we get

$$\frac{dA}{dx} = -\frac{1}{2}, \qquad \frac{dB}{dx} = \frac{1}{2}e^{-2x}$$

$$\text{Integrating} \qquad A = -\frac{1}{2}x + C_1, \qquad B = -\frac{1}{4}e^{-2x} + C_2$$

Substituting values of A and B in (4), the complete solution of (1) is given by

$$y = \left(-\frac{1}{2}x + c_1\right)x\left(-\frac{1}{4}e^{-2x} + c_2\right)xe^{2x} = C_1x + C_2xe^{2x} - \frac{1}{2}x^2 - \frac{1}{4}x.$$

EXERCISE 15.8

Solve the following differential equations by the method of variation of parameters:

1. $(1-x)\frac{d^2y}{dx^2} + x\frac{dy}{dx} - y = (1-x)^2.$ **Ans.** $y = C_1e^x + C_2x + x^2 + x + 1$

2. $(1-x^2)\frac{d^2y}{dx} - 4x\frac{dy}{dx} - (1+x^2)y = x$ **Ans.** $y = \frac{1}{1-x^2}(C_1\cos x + C_2\sin x + x)$

3. $(1-x)\frac{d^2y}{dx^2} + x\frac{dy}{dx} - y = x(1-x^2)^{\frac{3}{2}}$ **Ans.** $y = C_1(\sqrt{1-x^2} + x\sin^{-1}x) + C_2x - \frac{x}{9}(1-x^2)^{3/2}$

4. $x\frac{d^2y}{dx^2} - (2x+1)\frac{dy}{dx} + (x+1)y = (x^2+x-1)e^{2x}$ **Ans.** $y = C_1x + C_2x^{-1}x + x\int e^x x^{-3}(x^2 - 2x + 2)\,dx$

5. $x\frac{d^2y}{dx^2} - 2(x+1)\frac{dy}{dx} + (x+2)y = (x-2)e^x$ **Ans.** $y = \frac{C_1}{3}x^3e^x + C_2e^x + \left(x - \frac{x^2}{2}\right)e^x.$

6. $x^2\frac{d^2y}{dx^2} + \frac{dy}{dx} - y = x^2e^x$ **Ans.** $y = C_1x + \frac{C_2}{x} + e^x, (x > 0)$

APPLICATIONS TO DIFFERENTIAL EQUATIONS

16.1 INTRODUCTION

In this chapter, we shall study the application of differential equations to various physical problems.

16.2 ELECTRICAL CIRCUIT

We will consider circuits made up of

(*i*) Voltage source which may be a battery or a generator.

(*ii*) Resistance, inductance and capacitance.

Table of Elements, Symbols and Units

	Element	*Symbol*	*Unit*
1.	Charge	q	coulomb
2.	Current	i	ampere
3.	Resistance,	+ R − (→ i) R	ohm
4.	Inductance,	+ L − (→ i) L	henry
5.	Capacitance,	+ C − (→ i) C	farad
6.	Electromotive force or voltage (constant)	+ E − (→ i) constant V	volt
7.	Variable voltage	+ E − (→ i) variable V	volt

The formation of differential equation for an electric circuit depends upon the following laws.

(*i*) $i = \dfrac{dq}{dt}$, (*ii*) Voltage drop across resistance $R = Ri$

(*iii*) Voltage drop across inductance $L = L.\dfrac{di}{dt}$

(*iv*) Voltage drop across capacitance $C = \dfrac{q}{C}$

Kirchhoff's laws

I. Voltage law. The algebraic sum of the voltage drop around any closed circuit is equal to the resultant electromotive force in the circuit.

II. Current law. At a junction or node, current coming is equal to current going.

(*i*) **L - R series circuit.** Let i be the current flowing in the circuit containing resistance R and inductance L in series, with voltage source E, at any time t.

By voltage law $\quad Ri + L\dfrac{di}{dt} = \text{E} \quad \Rightarrow \quad \dfrac{di}{dt} + \dfrac{R}{L}i = \dfrac{E}{L}$ ...(1) (*M.U. II Semester, 2009*)

This is the linear differential equation

$$\text{I.F.} = e^{\int \frac{R}{L} dt} = e^{\frac{R}{L}t}$$

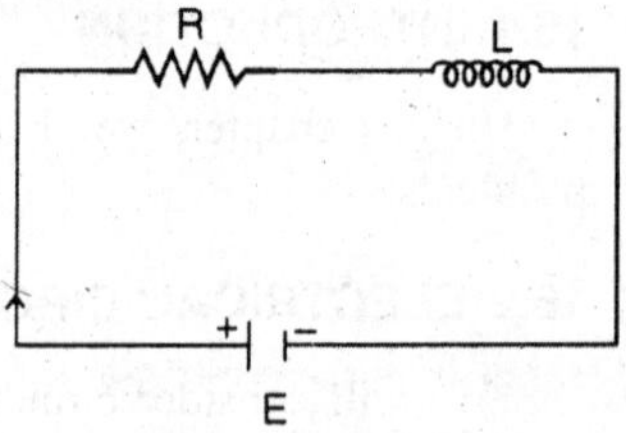

Its solution is $\quad i.e^{\frac{R}{L}t} = \int \dfrac{E}{L} e^{\frac{R}{L}t} dt + A$

$\Rightarrow \quad i.e^{\frac{R}{L}t} = \dfrac{E}{L} \times \dfrac{L}{R} e^{\frac{R}{L}t} + A$

$\Rightarrow \quad i = \dfrac{E}{R} + Ae^{-\frac{Rt}{L}}$ ...(2)

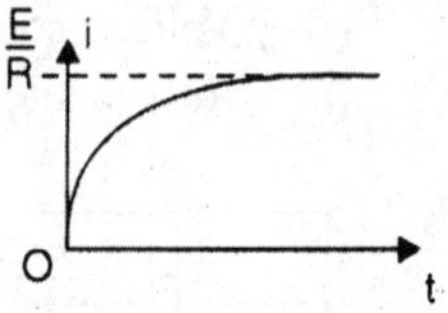

At $t = 0$, $\quad i = 0 \Rightarrow A = -\dfrac{E}{R}$

Thus, (2) becomes $i = \dfrac{E}{R}\left[1 - e^{-\frac{Rt}{L}}\right]$

R L i E $\frac{E}{R}$ i O t

(*ii*) **C-R series circuit.** Let i be current in the circuit containing resistance R, L, and capacitance C in series with voltage source E, at any time t.

By voltage law

$$Ri + \frac{q}{C} = E \qquad \left[i = \frac{dq}{dt}\right]$$

$$\Rightarrow \quad R\frac{dq}{dt} + \frac{q}{C} = E$$

Example 1. *An inductance of 2 henries and a resistance of 20 ohms are connected in series with an e.m.f. E volts. If the current is zero when t = 0, find the current at the end of 0.01 sec if E = 100 Volts.* (*U.P., II Semester, June 2008*)

Solution. Differential equation of the above circuit is as

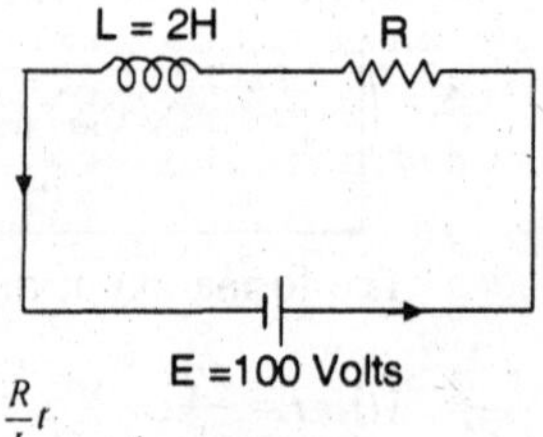

L = 2H R E =100 Volts

1st case: $\quad L\dfrac{di}{dt} + Ri = E \Rightarrow \dfrac{di}{dt} + \dfrac{R}{L}i = \dfrac{E}{L}$

$$\text{I.F.} = e^{\int \frac{R}{L} dt} = e^{\frac{Rt}{L}}$$

Its solution is $\quad ie^{\frac{Rt}{L}} = \dfrac{E}{L}\int e^{\frac{Rt}{L}} dt \quad \Rightarrow \quad ie^{\frac{Rt}{L}} = \dfrac{E}{L}\dfrac{L}{R}e^{\frac{R}{L}t} + A$

$\Rightarrow \quad ie^{\frac{Rt}{L}} = \dfrac{E}{R}e^{\frac{R}{L}t} + A$ ...(1)

Putting $t = 0$, $i = 0$; in (1), we get $0 = \dfrac{E}{R} + A \Rightarrow A = -\dfrac{E}{R}$

Putting the value of A in (1), we get

$$i e^{\frac{Rt}{L}} = \frac{E}{R} e^{\frac{Rt}{L}} - \frac{E}{R} \quad \Rightarrow \quad i = \frac{E}{R} - \frac{E}{R} e^{-\frac{R}{L}t}$$

$$i = \frac{E}{R}\left[1 - e^{-\frac{R}{L}t}\right] \qquad \text{...(2)}$$

On putting the values of E, R and L in (2), we get

$$i = \frac{100}{20}\left[1 - e^{-\frac{20}{2}t}\right] = 5\,[1 - e^{-10t}]$$

$$= 5\,[1 - e^{-10 \times 0.01}] = 5\,[1 - e^{-0.1}] = 5\left[1 - \frac{1}{e^{0.1}}\right] \text{ at } [t = 0.01 \text{ sec}]$$

$$= 0.475 \text{ Approx.}$$ **Ans.**

Example 2. *Solve the equation* $L\dfrac{di}{dt} + Ri = E_0 \sin wt$

where L, R and E_0 *are constants and discuss the case when t increases indefinitely.*

Solution. $L\dfrac{di}{dt} + Ri = E_0 \sin wt$

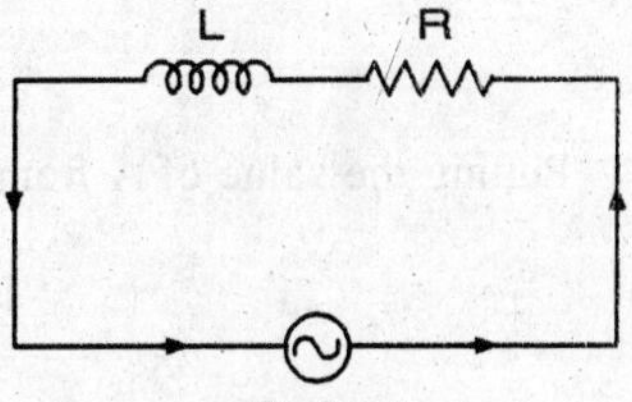

L R

E_0 sin wt

$$\Rightarrow \quad \frac{di}{dt} + \frac{R}{L} i = \frac{E_0}{L} \sin wt$$

$$\text{I.F.} = e^{\int \frac{R}{L} dt} = e^{\frac{R}{L}t}$$

Solution is
$$i.e^{\frac{R}{L}t} = \frac{E_0}{L} \int e^{\frac{R}{L}t} \sin wt\, dt + A$$

$$\left[\int e^{ax} \sin bx\, dx = \frac{e^{ax}}{\sqrt{a^2 + b^2}} \sin\left(bx - \tan^{-1}\frac{b}{a}\right)\right]$$

$$\Rightarrow \quad i.e^{\frac{R}{L}t} = \frac{E_0}{L} \frac{e^{\frac{R}{L}t}}{\sqrt{\frac{R^2}{L^2} + w^2}} \sin\left(wt - \tan^{-1}\frac{Lw}{R}\right) + A$$

$$i = \frac{E_0}{\sqrt{R^2 + L^2 w^2}} \sin\left(wt - \tan^{-1}\frac{Lw}{R}\right) + Ae^{-\frac{R}{L}t}$$

As t increases indefinitely, then $Ae^{-\frac{Rt}{L}}$ tends to zero.

so
$$i = \frac{E_0}{\sqrt{R^2 + L^2 w^2}} \sin\left(wt - \tan^{-1}\frac{Lw}{R}\right)$$ **Ans.**

Example 3. *A condenser of capacity C farads with* V_0 *is discharged through a resistance R ohms. Show that if q coulomb is the charge on the condenser, i ampere the current and V the voltage at time t.*

$$q = CV,\ V = Ri \text{ and } i = \frac{dq}{dt}, \text{ hence show that } V = V_0 e^{-\frac{t}{RC}}$$

Solution.

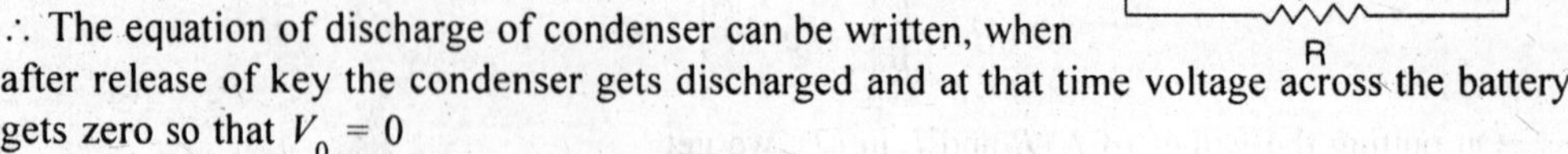

Voltage across $R = Ri$

Voltage drop across capacitance $= \dfrac{q}{C}$

$\therefore$ The equation of discharge of condenser can be written, when after release of key the condenser gets discharged and at that time voltage across the battery gets zero so that $V_0 = 0$

The differential equation of the above circuit is

$$Ri + \frac{q}{C} = 0 \quad \Rightarrow \quad R\frac{dq}{dt} + \frac{q}{C} = 0 \qquad \left(\text{as } i = \frac{dq}{dt}\right)$$

$$\Rightarrow \quad \frac{dq}{dt} + \frac{q}{RC} = 0 \quad \Rightarrow \quad \frac{dq}{dt} = -\frac{q}{RC} \Rightarrow \frac{dq}{q} = -\frac{1}{RC}dt$$

Integrating both sides, we get

$$\int \frac{dq}{q} = -\frac{1}{RC}\int dt \quad \Rightarrow \quad \log q = -\frac{1}{RC}t + A \qquad ...(1)$$

But at $t = 0$, the charge at the condenser is q_0 such that

$$\log q_0 = -\frac{1}{RC}(0) + A \quad \Rightarrow \quad A = \log q_0 \qquad ...(2)$$

Putting the value of A from (2) in (1), we have

$$\log q = -\frac{1}{RC}t + \log q_0 \quad \Rightarrow \quad \log q - \log q_0 = -\frac{1}{RC}t$$

$$\Rightarrow \quad \log\frac{q}{q_0} = -\frac{1}{RC}t \quad \Rightarrow \quad \frac{q}{q_0} = e^{-\frac{t}{RC}}$$

$$\Rightarrow \quad q = q_0\, e^{-\frac{t}{RC}} \qquad ...(3)$$

Dividing both side of (3) by C, we get

$$\frac{q}{C} = \frac{q_0}{C}e^{-\frac{t}{RC}} \quad \Rightarrow \quad V = V_0 e^{-\frac{t}{RC}} \qquad \left[\text{as } \frac{q}{C} = V\right] \quad \textbf{Proved.}$$

Example 4. *The equations of electromotive force in terms of current i for an electrical circuit having resistance R and a condenser of capacity C, in series, is* $E = Ri + \int \frac{i}{C}dt$. *Find the current i at any time t, when* $E = E_0 \sin wt$. *(U.P. II Semester, Summer 2006)*

Solution. We have,

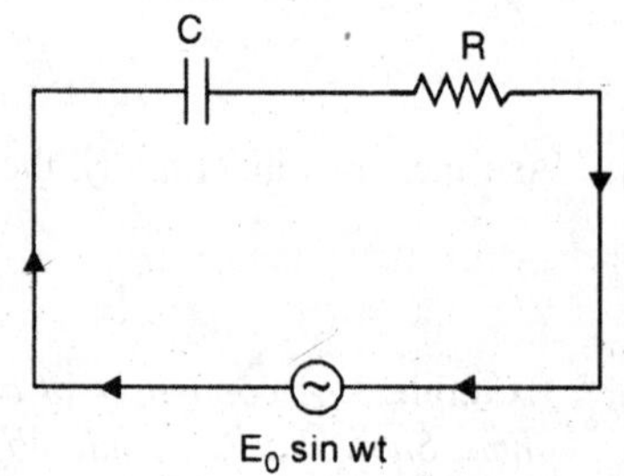

$$Ri + \int \frac{i}{C}\,dt = E_0 \sin wt$$

Differentiating both the sides, we get

$$\frac{R\,di}{dt} + \frac{i}{C} = E_0\, w \cos wt$$

$$\Rightarrow \quad \frac{di}{dt} + \frac{i}{RC} = \frac{E_0}{R} w \cos wt$$

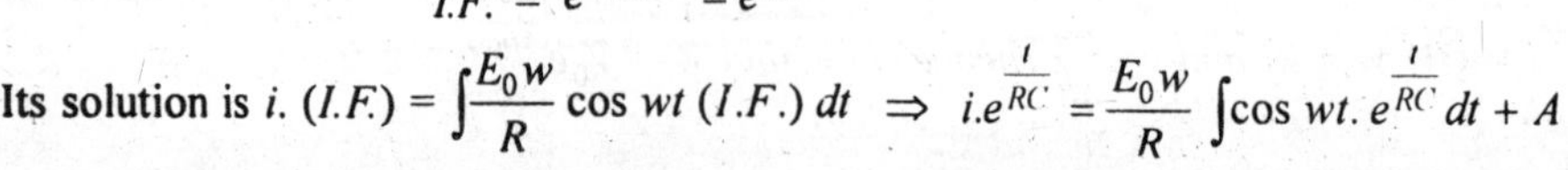

$$I.F. = e^{\int \frac{1}{RC}dt} = e^{\frac{t}{RC}}$$

Its solution is $i.\,(I.F.) = \int \frac{E_0 w}{R} \cos wt\,(I.F.)\,dt \Rightarrow i.e^{\frac{t}{RC}} = \frac{E_0 w}{R}\int \cos wt.\, e^{\frac{t}{RC}}\,dt + A$

$$\left[\int e^{ax}\cos bx\,dx = \frac{e^{ax}}{a^2+b^2}[a\cos bx + b\sin bx]\right]$$

$$\Rightarrow \quad i.e^{\frac{t}{RC}} = \frac{E_0 w}{R}\frac{e^{\frac{t}{RC}}}{\frac{1}{R^2C^2}+w^2}\left[\frac{1}{RC}\cos wt + w\sin wt\right] + A$$

$$= \frac{E_0 w}{R}\frac{R^2C^2\, e^{\frac{t}{RC}}}{1+w^2R^2C^2}\left[\frac{1}{RC}\cos wt + w\sin wt\right] + A$$

$$i = E_0 w.\frac{RC^2}{1+w^2R^2C^2}\left[\frac{1}{RC}\cos wt + w\sin wt\right] + Ae^{-\frac{t}{RC}}$$

$$i = E_0 w.\frac{C}{1+w^2R^2C^2}[\cos wt + w\,RC\sin wt] + Ae^{-\frac{t}{RC}} \quad \textbf{Ans.}$$

EXERCISE 16.1

1. A coil having a resistance of 15 ohms and an inductance of 10 henries is connected to 90 volts supply. Determine the value of current after 2 seconds. ($e^{-3} = 0.05$) **Ans.** 5.985 amp.
2. A resistance of 70 ohms, an inductance of 0.80 henry are connected in series with a battery of 10 volts. Determine the expression for current as a function of time after $t = 0$.

   Ans. $i = \frac{1}{7}\left(1 - e^{-\frac{175}{2}t}\right)$
3. A circuit consists of resistance R ohms and a condenser of C farads connected to a constant e.m.f. E; if $\frac{q}{C}$ is the voltage of the condenser at time t after closing the circuit Show that $\frac{q}{C} = E - Ri$ and hence show that the voltage at time t is $E\left(1 - e^{-\frac{t}{CR}}\right)$.
4. Show that the current $i = \frac{q}{CR}e^{-\frac{t}{R.C.}}$ during the discharge of a condenser of charge Q coulomb through a resistance R ohms.
5. A condenser of capacity C farads with voltage v_0 is discharged through a resistance R ohms. Show that if q coulomb is the charge on the condenser, i ampere the current and v the voltage at time t.

   $q = Cv$, $v = Ri$ and $i = -\frac{dq}{dt}$, hence show that $v = v_0\, e^{-\frac{1}{Rc}}$.
6. Solve $L\frac{di}{dt} + Ri = E\cos wt$ **Ans.** $i = \frac{E}{L^2w^2 + R^2}\left(R\cos wt + Lw\sin wt - \text{Re}^{-\frac{Rt}{L}}\right)$
7. A circuit consists of a resistance R ohms and an inductance of L henry connected to a generator of $E\cos(wt + \alpha)$ voltage. Find the current in the circuit. ($i = 0$, when $t = 0$).

   Ans. $i = \frac{E}{\sqrt{R^2+L^2w^2}}\cos\left[wt + \alpha - \tan^{-1}\frac{Lw}{R}\right] - \frac{E}{\sqrt{R^2+L^2w^2}}.e^{-\frac{R}{L}t}\cos\left[\alpha - \tan^{-1}\frac{Lw}{R}\right]$

16.3 SECOND ORDER DIFFERENTIAL EQUATION

We have already discussed $R - L$ and $R - L - C$ electric circuits. Here we want to do circuit problems involving second order differential equations.

Example 5. *The damped LCR circuit is governed by the equation*

$$L\frac{d^2q}{dt^2}+R\frac{dq}{dt}+\left(\frac{1}{C}\right)q=0$$

where L, R, C are positive constants. Find the conditions under which the circuit is overdamped, underdamped and critically damped. Find also the critical resistance.

(*U.P. II Semester, Summer 2005*)

Solution. Given equation is

$$L\frac{d^2q}{dt^2}+R\frac{dq}{dt}+\left(\frac{1}{C}\right)q=0 \Rightarrow \frac{d^2q}{dt^2}+\frac{R}{L}\frac{dq}{dt}+\left(\frac{1}{LC}\right)q=0 \qquad ...(1)$$

Let $\frac{R}{L}=2p$ and $\frac{1}{LC}=w^2$

Thus equation (1) becomes

$$\frac{d^2q}{dt^2}+2p\frac{dq}{dt}+w^2q=0$$

Its auxiliary equation is

$$m^2+2pm+w^2=0$$

$$\Rightarrow \qquad m=-p\pm\sqrt{p^2-w^2}$$

Case 1. When $p > w$, roots are real and distinct solution of equation (1) is

$$q=Ae^{(-p+\sqrt{p^2-w^2})t}+Be^{(-p-\sqrt{p^2-w^2})t}$$

In this case q is always positive, this is a condition of over damping.

Thus if $p > w$

$$\frac{R}{2L}>\frac{1}{\sqrt{LC}}$$

$$R>2\sqrt{\frac{L}{C}}$$

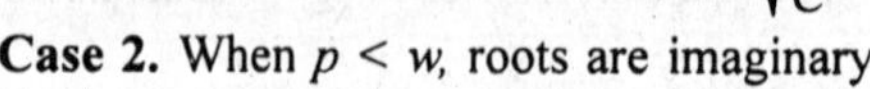

Case 2. When $p < w$, roots are imaginary

$$q=e^{-pt}\left(A\cos\sqrt{w^2-p^2}+B\sin\sqrt{w^2-p^2}\right)$$

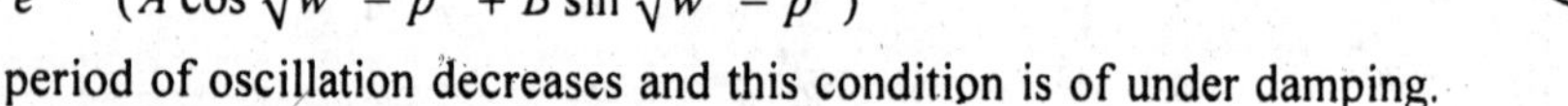

period of oscillation decreases and this condition is of under damping.

Case 3. When $p = w$, roots are equal $q=(A+Bt)e^{-pt}$,

This is a condition of critically damped.

Critical resistance is given by

$$p=w$$

$$\Rightarrow \qquad \frac{R}{2L}=\frac{1}{\sqrt{LC}}$$

$$R=2\sqrt{\frac{L}{C}} \quad \textbf{Ans.}$$

Example 6. *A circuit consists of resistance of 5ohms, inductance of 0.05 Henrys and capacitance of 4×10^{-4} farads. If $q(0)=0$, $i(0)=0$ find $q(t)$ and $i(t)$, when an emf of 110 volts is applied.*

(*M.D.U., 2010*)

Solution.

$$L\frac{d^2Q}{dt^2}+R\frac{dQ}{dt}+\left(\frac{1}{C}\right)Q=110 \quad \text{... (1)}$$

$$\Rightarrow \quad \frac{d^2Q}{dt^2}+\frac{R}{L}\frac{dQ}{dt}+\frac{1}{LC}Q=\frac{110}{L}$$

L C R

E = 110 volt

Let $\frac{R}{L}=2p$ and $\frac{1}{LC}=w^2$

Thus equation is

$$\frac{d^2Q}{dt^2}+2p\frac{dQ}{dt}+w^2Q=\frac{110}{0.05} \qquad [L = 0.05]$$

$$\Rightarrow \quad \frac{d^2Q}{dt^2}+2p\frac{dQ}{dt}+w^2Q=2200$$

Its auxiliary equation is

$m^2 + 2\,p\,m + w^2 = 0$

$$m=-p\pm\sqrt{p^2-w^2} \quad \text{... (2)}$$

Here, we have

$R = 5$ ohms, $L = 0.05$ Henrys, $C = 4 \times 10^{-4}$ farads

$$\therefore \quad 2p=\frac{R}{L}=\frac{5}{0.05}=100 \quad \Rightarrow \quad p=50$$

$$w^2=\frac{1}{LC}=\frac{1}{0.05\times 4\times 10^{-4}}=50000$$

Putting the values of p and w in (2), we get

$$m=-50\pm\sqrt{(50)^2-50000}=-50\pm\sqrt{2500-50000}$$

$$\Rightarrow \quad m=-50\pm\sqrt{-47500}=-50\pm 50\sqrt{-19}=-50\pm 50\sqrt{19}\,i$$

$$\text{C.F.}=e^{-50t}\left(A\cos 50\sqrt{19}\,t+B\sin 50\sqrt{19}\,t\right) \quad \text{... (3)}$$

$$\text{P.I.}=\frac{1}{D^2+2PD+w^2}2200$$

$$=\frac{1}{D^2+100\,D+50000}2200 \qquad [D=0]$$

$$=\frac{2200}{50000}=\frac{22}{500}=\frac{11}{250}$$

Complete solution = C.F. + P.I.

$$Q=e^{-50t}\left[A\cos 50\sqrt{19}\,t+B\sin 50\sqrt{19}\,t\right]+\frac{11}{250} \quad \text{... (4)}$$

On putting $Q = 0,\quad t = 0$ in (4), we get

$$Q = A + \frac{11}{250} \qquad \Rightarrow \qquad A = -\frac{11}{250}$$

On differentiating (4), we get

$$i = \frac{dQ}{dt} = -50\, e^{-50t}\ [A \cos 50\sqrt{19}\, t + B \sin 50\sqrt{19}\, t] +$$

$$e^{-50t}\ [-50\sqrt{19}\ A \sin 50\sqrt{19}\, t + 50\sqrt{19}\ B \cos 50\sqrt{19}\, t] \qquad \ldots (5)$$

On putting $i = 0,\quad t = 0$ in (5), we get

$$0 = -50\, A + 50\sqrt{19}\ B \qquad \ldots (6)$$

On putting $A = -\frac{11}{250}$ in (6), we get

$$0 = -50\left(-\frac{11}{250}\right) + 50\sqrt{19}\ B \qquad \Rightarrow \qquad B = -\frac{11}{5 \times 50\sqrt{19}}$$

$$B = -\frac{11}{250\sqrt{19}}$$

On putting the values of A and B in (4), we get

$$Q = e^{-50t}\left[-\frac{11}{250}\cos 50\sqrt{19}\, t - \frac{11}{250\sqrt{19}} \sin 50\sqrt{19}\, t\right] + \frac{11}{250}$$

On putting the values of A and B in (5), we get

$$i = -50\, e^{-50t}\left[\left(-\frac{11}{250}\right)\cos 50\sqrt{19}\, t - \frac{11}{250\sqrt{19}} \sin 50\sqrt{19}\, t\right] +$$

$$e^{-50t}\left[-50\sqrt{19}\,.\left(\frac{-11}{250}\right)\sin 50\sqrt{19}\, t + 50\sqrt{19}\left(\frac{-11}{250\sqrt{19}}\right)\cos 50\sqrt{19}\, t\right]$$

$$= e^{-50t}\left[\left(\frac{11}{5} - \frac{11}{5}\right)\cos 50\sqrt{19}\, t + \left(\frac{11}{5\sqrt{19}} + \frac{11\sqrt{19}}{5}\right)\sin 50\sqrt{19}\, t\right]$$

$$\Rightarrow \qquad i = e^{-50t}\,\frac{11 + 11 \times 19}{5 \times 19}\sin 50\sqrt{19}\, t$$

$$\Rightarrow \qquad i = e^{-50t}\,\frac{44}{\sqrt{19}}\sin 50\sqrt{19}\, t = \frac{44}{\sqrt{19}}\, e^{-50t} \sin 50\sqrt{19}\, t$$ **Ans.**

Example 7. *An electric circuit consists of an inductance 0.1 henry, a resistance of 20 ohms and a condenser of capacitance 25 microfarads. Find the charge q and the current i at time t, given the initial conditions q = 0.05 coulombs, i = 0 when t = 0*

Solution. The differential equation of the above given circuit can be written as

$$L\frac{di}{dt} + Ri + \frac{q}{C} = 0 \quad \Rightarrow \quad L\frac{d^2q}{dt^2} + R\frac{dq}{dt} + \frac{q}{C} = 0 \qquad \left[i = \frac{dq}{dt}\right]$$

$$\Rightarrow \qquad \frac{d^2q}{dt^2} + \frac{R}{L}\frac{dq}{dt} + \frac{q}{LC} = 0$$

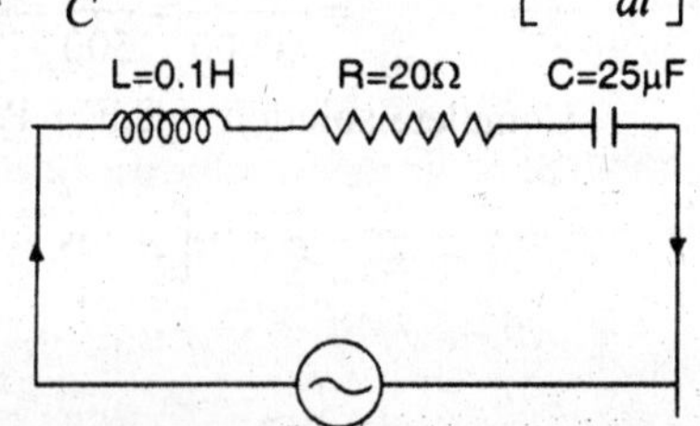

First, we will solve the equation and then put the values of R, L and C. For convenience we put

$$\frac{R}{L} = 2b, b = \frac{R}{2L} = \frac{20}{2 \times 0.1} = 100$$

Let $\frac{1}{LC} = k^2$

$\Rightarrow$ $$k = \sqrt{\frac{1}{LC}} = \sqrt{\frac{1}{0.1 \times 25 \times 10^{-6}}} = \sqrt{\frac{10^7}{25}} = 632.5 \geq 100$$

Our equation reduces to

$$\frac{d^2q}{dt^2} + 2b\frac{dq}{dt} + k^2 q = 0$$

A.E. is $m^2 + 2b\,m + k^2 = 0$

So that $$m = \frac{-2b \pm \sqrt{4b^2 - 4k^2}}{2} = -b \pm \sqrt{b^2 - k^2} = -b \pm j\sqrt{k^2 - b^2}$$

C.F. is $q = e^{-bt}[A \cos\sqrt{k^2 - b^2}\, t + B \sin\sqrt{k^2 - b^2} t]$ ...(1)

On putting $q = 0.05$ and $t = 0$ in (1), we get $0.05 = A$

On differentiating (1), we get

$$\frac{dq}{dt} = -be^{-bt}[A \cos\sqrt{k^2 - b^2}\, t + B \sin\sqrt{k^2 - b^2} t]$$

$$+ e^{-bt}[-A\sqrt{k^2 - b^2} \sin\sqrt{k^2 - b^2} t + B\sqrt{k^2 - b^2} \cos\sqrt{k^2 - b^2} t] \quad ...(2)$$

On putting $\frac{dq}{dt} = 0$ and $t = 0$ in (2), we get

$$0 = -bA + B\sqrt{k^2 - b^2} \Rightarrow B = \frac{bA}{\sqrt{k^2 - b^2}} = \frac{0.05\, b}{\sqrt{k^2 - b^2}}$$

Substituting the values of A and B in (1), we have

$$q = e^{-bt}[0.05 \cos\sqrt{k^2 - b^2} t + \frac{0.05\, b}{\sqrt{k^2 - b^2}} \sin\sqrt{k^2 - b^2}\, t] \quad ...(3)$$

Now, $$\sqrt{k^2 - b^2} = \sqrt{\frac{10^7}{25} - (100)^2} = \sqrt{400000 - 10000} = \sqrt{390000} = 624.5$$

On putting these values in (3), we have

$$q = e^{-100t}[0.05 \cos 624.5t + \frac{0.05 \times 100}{624.5} \sin 624.5t]$$

$\Rightarrow$ $$q = e^{-100\,t}[0.05 \cos 624.5t + 0.008 \sin 624.5t] \quad ...(4)$$ **Ans.**

On differentiating (4), we have

$$\frac{dq}{dt} = -100\, e^{-100\,t}[0.05 \cos 624.5t + 0.008 \sin 624.5t]$$

$$+ e^{-100\,t}[-0.05 \times 624.5 \sin 624.5t + 0.008 \times 624.5 \cos 624.5t]$$

$\Rightarrow$ $$i = e^{-100\,t}[(-5 + 4.996) \cos 624.5t - (0.8 + 31.225) \sin 624.5t]$$

$$= e^{-100\,t}[-0.004 \cos 624.5t - 32.025 \sin 624.5t]$$

$$= -32\, e^{-100\,t} \sin 624.5t. \text{ approximately}$$ **Ans.**

Example 8. *An alternating e.m.f. E sin wt is applied to an inductance L and capacitance C in series. Show that the current in the circuit is* $\frac{Ew}{(n^2 - w^2)L}$ *(cos wt – cos nt), where* $n^2 = \frac{1}{LC}$.

(U.P. II Semester, June 2010, 2009)

Solution. The differential equation for the above circuit is

$$L\frac{d^2q}{dt^2} + \frac{q}{C} = E \sin wt$$

$$\Rightarrow \quad \frac{d^2q}{dt^2} + \frac{q}{LC} = \frac{E}{L}\sin wt$$

$$\Rightarrow \quad \left(D^2 + \frac{1}{LC}\right)q = \frac{E}{L}\sin wt$$

A.E. is $m^2 + \frac{1}{LC} = 0 \Rightarrow m^2 + n^2 = 0 \Rightarrow m = \pm i\, n$ $\qquad \left(\because \frac{1}{LC} = n^2\right)$

$$\text{C.F.} = A \cos nt + B \sin nt$$

$$\text{P.I.} = \frac{1}{D^2 + n^2}\frac{E}{L}\sin wt$$

$$\Rightarrow \quad \text{P.I.} = \frac{1}{-w^2 + n^2}\frac{E}{L}\sin wt$$

Complete solution is $\quad q = A \cos n\,t + B \sin n\,t + \frac{E}{(n^2 - w^2)\,L}\sin wt \qquad ...(1)$

On putting $q = 0$, $t = 0$ in (1), we get

$$0 = A$$

On putting the value of A in (1), we get

$$q = B \sin nt + \frac{E}{(n^2 - w^2)L}\sin wt \qquad ...(2)$$

On differentiating (2) w.r.t., 't', we get

$$\frac{dq}{dt} = B\,n \cos nt + \frac{Ew}{(n^2 - w^2)\,L}\cos wt$$

$$\Rightarrow \quad i = B\,n \cos nt + \frac{Ew}{(n^2 - w^2)\,L}\cos wt \qquad ...(3)$$

On putting $i = 0$, $t = 0$ in (3), we get

$$0 = Bn + \frac{Ew}{(n^2 - w^2)L} \quad \Rightarrow \quad B = -\frac{Ew}{n(n^2 - w^2)L}$$

Putting the value of B in (3), we get

$$i = -\frac{Ewn}{n(n^2 - w^2)L}\cos nt + \frac{Ew}{(n^2 - w^2)L}\cos wt$$

$$\Rightarrow \quad i = \frac{Ew}{(n^2 - w^2)L}(\cos wt - \cos nt)$$ **Proved.**

Example 9. *For an electric circuit with circuit constants, L, R, C the charge q on a plate condenser is given by*

$$L\frac{d^2q}{dt^2} + R\frac{dq}{dt} + \frac{q}{C} = E \text{ and the current by } i = \frac{dq}{dt}$$

Let L = 1 henry, $C = 10^{-4}$ *farad, R = 100 ohms, E = 100 volts,*

Suppose that no charge present and no current is flowing at time t = 0, when the e.m.f. is applied. Determine q and i at any time t.

Solution. The differential equation is

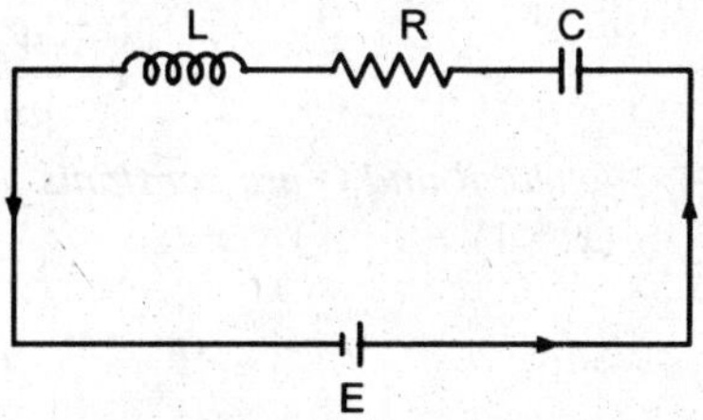

$$L\frac{d^2q}{dt^2}+R\frac{dq}{dt}+\frac{q}{C}=E$$

$$\Rightarrow \quad \frac{d^2q}{dt^2}+\frac{R}{L}\frac{dq}{dt}+\frac{q}{LC}=\frac{E}{L}$$

L R C E

Putting $\frac{R}{L}=2b$ and $\frac{1}{LC}=k^2$ in (1), we have

$$\frac{d^2q}{dt^2}+2b\frac{dq}{dt}+k^2q=\frac{E}{L} \qquad ...(1)$$

This equation is exactly identical, we have

$$\Rightarrow \quad q=\frac{E}{k^2L}+e^{-bt}[A\cos\sqrt{k^2-b^2}\,t+B\sin\sqrt{k^2-b^2}\,t] \qquad ...(2)$$

On putting $q=0$ and $t=0$ in (1), we get

$$0=\frac{E}{k^2L}+A \quad\Rightarrow\quad A=-\frac{E}{k^2L}$$

Differentiating (2), we have

$$\frac{dq}{dt}=-be^{-bt}[A\cos\sqrt{k^2-b^2}\,t+B\sin\sqrt{k^2-b^2}\,t]$$

$$+e^{-bt}[-A\sqrt{k^2-b^2}\sin\sqrt{k^2-b^2}\,t+B\sqrt{k^2-b^2}\cos\sqrt{k^2-b^2}\,t] \qquad ...(3)$$

On putting $\frac{dq}{dt}=0$ and $t=0$ in (3), we have

$$0=-bA+B\sqrt{k^2-b^2} \quad\Rightarrow\quad B=\frac{bA}{\sqrt{k^2-b^2}}=\frac{-\frac{bE}{k^2L}}{\sqrt{k^2-b^2}}=-\frac{bE}{k^2L\sqrt{k^2-b^2}}$$

Substituting the values of A and B in (2), we have

$$q=\frac{E}{k^2L}+e^{-bt}\left[-\frac{E}{k^2L}\cos\sqrt{k^2-b^2}\,t-\frac{bE}{k^2L\sqrt{k^2-b^2}}\sin\sqrt{k^2-b^2}\,t\right]$$

$$=\frac{E}{k^2L}\left[1-e^{-bt}\left(\cos\sqrt{k^2-b^2}\,t+\frac{b}{\sqrt{k^2-b^2}}\sin\sqrt{k^2-b^2}\,t\right)\right] \qquad ...(4)$$

Now,

$$\frac{E}{k^2L}=\frac{E}{\frac{1}{LC}.L}=EC=100\times10^{-4}=\frac{1}{100}$$

$$b=\frac{R}{2L}=\frac{100}{2\times1}=50$$

$$\sqrt{k^2-b^2}=\sqrt{\frac{1}{LC}-(50)^2}=\sqrt{\frac{1}{10^{-4}}-(50)^2}=\sqrt{10000-2500}=\sqrt{7500}=50\sqrt{3}$$

On putting these values in (4), we get

$$q=\frac{1}{100}\left[1-e^{-50t}(\cos 50\sqrt{3}\,t+\frac{1}{\sqrt{3}}\sin 50\sqrt{3}\,t)\right] \qquad \textbf{Ans.}$$

Example 10. *The voltage V and the current i at a distance x from the sending end of the transmission line satisfy the equations.*

$$-\frac{dV}{dx} = Ri, \quad -\frac{di}{dx} = GV$$

where R and G are constants. If $V = V_0$ at the sending end ($x = 0$) and $V = 0$ at receiving end ($x = 1$).

Show that $$V = V_0 \left\{\frac{\sinh n(l - x)}{\sinh nl}\right\}, \text{ when } n^2 = RG$$

Solution. We have, $$-\frac{dV}{dx} = Ri \qquad \ldots(1)$$

$$-\frac{di}{dx} = GV \qquad \ldots(2)$$

When $x = 0$, $V = V_0$; When $x = l$, $V = 0$

Putting the value of i from (1) in (2), we get

$$-\frac{d}{dx}\left(-\frac{dV}{dx}\frac{1}{R}\right) = GV \Rightarrow \frac{d^2V}{dx^2} = RGV$$

$$\Rightarrow \quad \frac{d^2V}{dx^2} - (RG)V = 0 \Rightarrow (D^2 - RG)V = 0 \qquad (RG = n^2)$$

A.E. is $m^2 - n^2 = 0$, $m = \pm n$

$$\therefore \quad V = A\,e^{nx} + B\,e^{-nx} \qquad \ldots(3)$$

Now, we have to find out the values of A and B with the help of given conditions.

On putting $x = 0$ and $V = V_0$ in (3), we get

$$V_0 = A + B \qquad \ldots(4)$$

On putting $x = 1$ and $V = 0$ in (3), we get

$$0 = A\,e^{nl} + B\,e^{-nl} \qquad \ldots(5)$$

On solving (4) and (5), we have

$$A = \frac{V_0}{1 - e^{2nl}}, \quad B = \frac{-V_0\,e^{2nl}}{1 - e^{2nl}}$$

Substituting the values of A and B in (3), we have

$$V = \frac{V_0\,e^{nx}}{1 - e^{2nl}} - \frac{V_0\,e^{2nl}\,e^{-nx}}{1 - e^{2nl}} = \frac{V_0\,[e^{nx} - e^{2nl - nx}]}{1 - e^{2nl}}$$

$$= \frac{V_0\,[e^{(nl - nx)} - e^{-(nl - nx)}]}{e^{nl} - e^{-nl}} = V_0 \left\{\frac{\sinh n(l - x)}{\sinh nl}\right\} \qquad \textbf{Proved.}$$

EXERCISE 16.2

1. For an electric circuit with circuit constants L, R, C the charge q on the plate of the condensor is given by :

$$L\frac{d^2q}{dt^2} + R\frac{dq}{dt} + \frac{q}{C} = 0$$

Find q at any time t.

Discuss the case when R is negligible and show that q is oscillatory. Calculate its period and frequency.

$$\textbf{Ans. } q = e^{-\frac{R}{2L}t}\left[A\cos\frac{\sqrt{4CL - R^2C^2}}{2LC}t + B\sin\frac{\sqrt{4LC - R^2C^2}}{2LC}t\right]$$

$$q = A\cos\frac{1}{\sqrt{LC}}t + B\sin\frac{1}{\sqrt{LC}}t,\ \text{Period} = 2\pi\sqrt{LC},\ \text{frequency} = \frac{1}{2\pi\sqrt{LC}}$$

2. A condenser of capacity C is discharged through an iductance L and a resistance R in series and the charge q at any time t is given by

$$L\frac{d^2q}{dt^2} + R\frac{dq}{dt} + \frac{q}{C} = 0$$

If $L = 10$ milli henry, $R = 200$ ohms, $C = 0.1\ \mu\ F$ and also when $t = 0$, charge $q = 0.01$ coulomb and current $\frac{dq}{dt} = 0$. Find the value of q at any time t and find the frequency of the circuit if the discharge is oscillatory. **Ans.** $q = e^{-10000\,t}\,[\,0.01\cos 3\times 10^4\,t + 0.33\times 10^{-2}\sin 3\times 10^4\,t]$

$$\text{frequency} = \frac{3\times 10^4}{2\pi}$$

3. A condenser of capacity C is discharged through L and a resistance R in series and the charge q at any time t is given by the equation

$$L\frac{d^2q}{dt^2} + R\frac{dq}{dt} + \frac{q}{C} = 0$$

If $L = 0.5$ henry, $R = 300$ ohms, $C = 2\times 10^{-6}$ farad and also when $t = 0$, charge $q = 0.01$ and current $\frac{dq}{dt} = 0$, find the value of q in terms of t. **Ans.** $q = e^{-200\,t}\,[0.01\cos 100\sqrt{91}\,t + 0.0031\sin 100\sqrt{91}\,t]$

4. A 10^{-3} farad capacitor is connected in series with 0.05 henry inductor and 10 ohms resistor. Initially, the current in the circuit is zero and the charge on the capacitor is also zero. If the e.m.f. is 50 sin 200 t. Find the charge t seconds after the circuit is closed.

Ans. $q = 0.125\,e^{-100\,t}\,[\cos 100\,t + \sin 100\,t] - 25\cos 200\,t + 0.125\sin 200\,t$

5. For an electric circuit with circuit constants L, R, C, the charge q on a plate on the condenser is given by

$$L\frac{d^2q}{dt^2} + R\frac{dq}{dt} + \frac{q}{C} = E\sin wt$$

and the current $i = \frac{dq}{dt}$. The circuit is tuned to resonance so that $w^2 = \frac{1}{LC}$

If $R^2 = \frac{4L}{C}$ and $q = i = 0$ at $t = 0$, show that

$$q = \frac{E}{Rw}\left[-\cos wt + e^{-\frac{Rt}{2L}}\left(\cos pt + \frac{R}{2LP}\sin pt\right)\right] \qquad \text{where } p^2 = \frac{1}{LC} - \frac{R^2}{4L^2}$$

6. If the charge on one of the coatings of a layden jar be q when a force $E\cos pt$ acts in the circuit connecting the coatings and the circuit contains inductance L. Resistance R and capacitance C satisfying the differential equation:

$$\left(LD^2 + RD + \frac{1}{C}\right)q = E\cos pt$$

$D = \frac{d}{dt}$, find an expression for the charge given

$q = \frac{dq}{dt} = 0$ when $t = 0$.

7. An e.m.f. $E\sin pt$ is applied at $t = 0$ to a circuit containing a condenser C and inductance L in series. The current x satisfies the equation

$$L\frac{dx}{dt} + \frac{1}{C}\int x\,dt = E \sin pt$$

If $p^2 = \frac{1}{LC}$, and initially the current x and the charge q are zero, show that the current in the circuit at time t is given by

$$x = \frac{E}{2L} t \sin pt, \text{ where } x = -\frac{dq}{dt}.$$

8. An L-C-R circuit has $R = 180$ ohm, $C = \frac{1}{280}$ farad, $L = 20$ henries and an applied voltage $E(t) = 10 \sin t$. Assuming that no charge is present but an initial current of i ampere is flowing at $t = 0$ when the voltage is first applied, find q and $i = \frac{dq}{dt}$ at any time t. q is given by the differential equation

$$L\frac{d^2q}{dt^2} + R\frac{dq}{dt} + \frac{q}{C} = E(t)$$

Ans. $q = \frac{11}{50}e^{-2t} - \frac{101}{500}e^{-7t} + \frac{1}{1000}[26 \sin t - 18 \cos t]$

16.4 MECHANICAL ENGINEERING PROBLEMS

Rectilinear Motion

When a body moves in a straight line the motion is called rectilinear motion. If x be the distance of the body at any time t from starting point then we have its velocity v given by

$$\text{velocity} = v = \frac{dx}{dt} \qquad \boxed{v = \frac{dx}{dt}}$$

If the acceleration of the body be 'a' then

$$\text{Acceleration} = a = \frac{dv}{dt} = \frac{d^2x}{dt^2}$$

Since $\qquad a = \frac{dv}{dt} = \frac{dv}{dx}\cdot\frac{dx}{dt} = v\frac{dv}{dx}$

$$\boxed{a = \frac{d^2x}{dt^2}}$$

$$\boxed{a = v\frac{dv}{dx}}$$

If mass of a body is m and the body is moving with acceleration a by a force F acting on it, then

$$\boxed{F = ma}$$

$$\boxed{F = m\frac{d^2x}{dt^2}}$$

$$\boxed{F = mv\frac{dv}{dx}}$$

Example 11. *A moving body is opposed by a force per unit mass of value cx and resistance per unit mass of value bv^2 where x and v are the displacement and velocity of the particle at that instant. Find the velocity of the particle in terms of x, if it starts from rest.*

Solution. By Newton's second law of motion, the equation of motion of the body is

$$v\frac{dv}{dx} = -cx - bv^2 \Rightarrow v\frac{dv}{dx} + bv^2 = -cx \qquad ...(1)$$

Putting $v^2 = z,\ 2v\dfrac{dv}{dx} = \dfrac{dz}{dx}$, in (1), we get

$$\frac{1}{2}\frac{dz}{dx} + bz = -cx \Rightarrow \frac{dz}{dx} + 2bz = -2cx$$

$$\text{I.F.} = e^{\int 2bd} = e^{2bx}$$

Its solution is $z.\, e^{2bx} = \int -2cx\, e^{2bx}\, dx + A$

$$= -2c\left[\frac{x\, e^{2bx}}{2b} - \int \frac{e^{2bx}}{2b}\, dx\right] + A'$$

$$= -\frac{c}{b}\, x e^{2bx} + \frac{c}{b}\frac{e^{2bx}}{2b} + A'$$

$\Rightarrow$ $$z = -\frac{c}{b}x + \frac{c}{2b^2} + A'\, e^{-2bx}$$

$$v^2 = -\frac{cx}{b} + \frac{c}{2b^2} + A'\, e^{-2bx} \qquad ...(2)$$

Putting $v = 0$ and $x = 0$ in (2), we have

$\therefore$ $$0 = \frac{c}{2b^2} + A' \qquad \Rightarrow \qquad A' = -\frac{c}{2b^2}$$

$\therefore$ Equation (2) becomes $v^2 = -\dfrac{cx}{b} + \dfrac{c}{2b^2} - \dfrac{c}{2b^2}\, e^{-2bx}$ **Ans.**

Example 12. *A particle of mass m moves in a straight line under the action of force mn^2x which is always directed towards a fixed point O on the line. Determine the displacement x(t) if the resistance to the motion is 2 λmnv given that initially $x = 0$, $\dfrac{dx}{dt} = 0$, $(0 < \lambda < 1)$.*

(U.P. II Sem 2010)

Solution. Equation of motion is

$$mv\frac{dv}{dx} = -mn^2x - 2\lambda mnv \qquad ...(1)$$

$$\frac{dv}{dx} = -\frac{(n^2x + 2\lambda nv)}{v}$$

Let $v = Zx$

$$\frac{dv}{dx} = Z + x\frac{dZ}{dx}$$

Putting the value of v, $\dfrac{dv}{dx}$ in (1), we get

$$Z + x\frac{dZ}{dx} = -\left(\frac{n^2x + 2\lambda nZx}{Zx}\right) = -\left(\frac{n^2 + 2\lambda nZ}{Z}\right)$$

$$x\frac{dZ}{dx} = \frac{-n^2 - 2\lambda nZ - Z^2}{Z}$$

$$\int \frac{Z\, dZ}{n^2 + 2\lambda nZ + Z^2} = -\int \frac{dx}{x}$$

$$\Rightarrow \quad \int \frac{Z\,dZ}{z^2 + 2\lambda n Z + n^2} = -\int \frac{dx}{x}$$

$$\Rightarrow \quad \frac{1}{2}\int \frac{2Z + 2n\lambda - 2n\lambda}{z^2 + 2\lambda nZ + n^2}\,dZ = -\int \frac{dx}{x}$$

$$\Rightarrow \quad \frac{1}{2}\int \frac{2Z + 2\lambda n}{Z^2 + 2\lambda nZ + n^2}\,dz - \int \frac{\lambda n}{Z^2 + 2\lambda nZ + n^2}\,dZ = -\int \frac{dx}{x}$$

$$\Rightarrow \quad \frac{1}{2}\log(Z^2 + 2\lambda nZ + n^2) - \int \frac{\lambda n}{Z^2 + 2\lambda nZ + \lambda^2 n^2 + n^2 - \lambda^2 n^2}\,dZ = -\log x + C$$

$$\Rightarrow \quad \frac{1}{2}\log(Z^2 + 2\lambda nZ + n^2) - \lambda n\int \frac{1}{(Z + \lambda n)^2 + n^2(1 - \lambda^2)}\,dZ = -\log x + C$$

$$\Rightarrow \quad \frac{1}{2}\log(Z^2 + 2\lambda nZ + n^2) - \lambda n\int \frac{1}{(Z + \lambda n)^2 + \left[n\sqrt{1 - \lambda^2}\right]^2}\,dZ = -\log x + C$$

$$\Rightarrow \quad \frac{1}{2}\log(Z^2 + 2\lambda nZ + n^2) + \log x - \frac{\lambda n}{n\sqrt{1 - \lambda^2}}\tan^{-1}\left(\frac{Z + \lambda n}{n\sqrt{1 - \lambda^2}}\right) = C$$

$$\Rightarrow \quad \log x\sqrt{Z^2 + 2\lambda nZ + n^2} - \frac{\lambda}{\sqrt{1 - \lambda^2}}\tan^{-1}\left(\frac{\frac{v}{x} + \lambda n}{n\sqrt{1 - \lambda^2}}\right) = C$$

$$\Rightarrow \quad \log x\sqrt{\frac{v^2}{x^2} + 2\lambda n\left(\frac{v}{x}\right) + n^2} - \frac{\lambda}{\sqrt{1 - \lambda^2}}\tan^{-1}\left(\frac{\frac{v}{x} + \lambda n}{n\sqrt{1 - \lambda^2}}\right) = C$$

$$\Rightarrow \quad \log\sqrt{v^2 + 2\lambda nvx + n^2x^2} - \frac{\lambda}{\sqrt{1 - \lambda^2}}\tan^{-1}\left(\frac{v + \lambda nx}{nx\sqrt{1 - \lambda^2}}\right) = C \qquad \text{....(2)}$$

On putting $v = 0$ and $x = 0$ in (2) we get

$$\log 0 - \frac{\lambda}{\sqrt{1 - \lambda^2}}\tan^{-1} 0 = C \quad \Rightarrow \quad C = 0$$

Putting $C = 0$ in (2), we get

$$\log\sqrt{v^2 + 2\lambda nvx + n^2x^2} - \frac{\lambda}{\sqrt{1 - \lambda^2}}\tan^{-1}\left(\frac{v + \lambda nx}{nx\sqrt{1 - \lambda^2}}\right) = 0$$

Ans.

16.5 VERTICAL MOTION

Example 13. *A body falling vertically under gravity encounters resistance of the atmosphere. If the resistance varies as the velocity, show that the equation of motion is given by*

$$\frac{du}{dt} = g - ku$$

where u is the velocity, k is a constant and g is the acceleration due to gravity. Show that as t increases, u approaches the value g/k. Also, if $u = \frac{dx}{dt}$ *where x is the distance fallen by the body from rest in time t, show that*

$$x = \frac{gt}{k} - \frac{g}{k^2}(1 - e^{-kt})$$

Solution. Let the mass of the falling body be unity.

$$\text{Acceleration} = \frac{du}{dt}$$

$$\text{Force acting downward} = 1.\frac{du}{dt} = \frac{du}{dt}$$

$$\text{Force of resistance} = ku$$

$$\text{Net force acting downward} = g - ku \quad \Rightarrow \quad \frac{du}{dt} = g - ku \qquad \text{...(1)}$$ **Proved.**

$$\Rightarrow \qquad \frac{du}{g - ku} = dt$$

Integrating, we get $\displaystyle\int \frac{du}{g - ku} = \int dt$

$$\Rightarrow \qquad t = -\frac{1}{k}\log(g - ku) + \log A = \log(g - ku)^{-1/k} A$$

$$A(g - ku)^{-1/k} = e^t \quad \Rightarrow \quad (g - ku) = A^k e^{-kt}$$

$$\Rightarrow \qquad u = \frac{g}{k} - \frac{A^k}{k} e^{-kt}$$

If t increases very large then $\dfrac{A^k}{k} e^{-kt} = 0$

$$\Rightarrow \qquad u = \frac{g}{k} \qquad \text{when } t \to \infty$$ **Proved.**

Given $\qquad u = \dfrac{dx}{dt} \quad \Rightarrow \quad \dfrac{du}{dt} = \dfrac{d^2x}{dt^2}$

Putting the values of $\dfrac{du}{dt}$ and u in (1), we get

$$\frac{d^2x}{dt^2} + k\frac{dx}{dt} = g \quad \Rightarrow \quad (D^2 + kD)\,x = g$$

$$\text{A.E. is} \quad m(m + k) = 0 \quad \Rightarrow \quad m = 0,\ m = -k$$

$$\text{C.F.} = A_1 + A_2 e^{-kt}$$

$$\text{P.I.} = \frac{1}{D^2 + kD} g = t\frac{1}{2D + k} g$$

$$= \frac{t}{k}\frac{1}{\left(1 + \frac{2D}{k}\right)} g = \frac{t}{k}\left(1 + \frac{2D}{k}\right)^{-1} g = \frac{t}{k}\left(1 - \frac{2D}{k}\right) g = \frac{t\,g}{k}$$

$$x = A_1 + A_2 e^{-kt} + \frac{gt}{k} \qquad \text{...(2)}$$

Putting the values of $t = 0$ and $x = 0$ in (2), we get

$$0 = A_1 + A_2 \quad \Rightarrow \quad A_2 = -A_1$$

(2) becomes $\qquad x = A_1 - A_1 e^{-kt} + \dfrac{gt}{k} \qquad \text{...(3)}$

On differentiating (3), we get $\dfrac{dx}{dt} = A_1 k e^{-kt} + \dfrac{g}{k} \qquad \text{...(4)}$

On putting $\frac{dx}{dt} = 0$, when $t = 0$ in (4), we get $0 = A_1 k + \frac{g}{k} \Rightarrow A_1 = -\frac{g}{k^2}$

Putting the value of A_1 in (3), we get

$$x = -\frac{g}{k^2} + \frac{g}{k^2} e^{-kt} + \frac{gt}{k} \Rightarrow x = \frac{gt}{k} - \frac{g}{k^2}(1 - e^{-kt}) \qquad \textbf{Proved.}$$

Example 14. *A particle falls under gravity in a resisting medium whose resistance varies with velocity. Find the relation between distance and velocity if initially the particle starts from rest.* (*U.P., II Semester, Summer 2003*)

Solution. By Newton's second law of motion, the equation of motion of the body is

$$m \frac{V\,dV}{dx} = m\,g - m\,k\,V$$

$$\Rightarrow \qquad \frac{V\,dV}{dx} = g - k\,V$$

$$\Rightarrow \qquad \frac{V\,dV}{g - k\,V} = dx \Rightarrow \frac{-dV}{k} + \frac{g}{k}\frac{dV}{g - k\,V} = d\,x$$

Integrating, we get

$$-\frac{V}{k} + \frac{g}{k}\left(-\frac{1}{k}\right) \log\,(g + k\,V) = x + A$$

$$-\frac{V}{k} - \frac{g}{k^2} \log\,(g - kV) = x + A \qquad \text{...(1)}$$

Initially, $x = 0,\ V = 0$

$$-\frac{g}{k^2} \log g = A$$

(1) becomes

$$-\frac{V}{k} - \frac{g}{k^2} \log\,(g - k\,V) = x - \frac{g}{k^2} \log g$$

$$\Rightarrow \qquad -\frac{V}{k} - \frac{g}{k^2} \log \frac{g - k\,V}{g} = x \qquad \textbf{Ans.}$$

Example 15. *The acceleration and velocity of a body falling in the air approximately satisfy the equation :*

Acceleration $= g - kv^2$, *where v is the velocity of the body at any time t, and g, k are constants. Find the distance traversed as a function of the time, if the body falls from rest.*

Show that value of v will never exceed $\sqrt{\frac{g}{k}}$.

Solution Acceleration $= g - k\,v^2 \Rightarrow \frac{dv}{dt} = g - k\,v^2 \Rightarrow \frac{dv}{g - k\,v^2} = dt.$

$$\Rightarrow \qquad \frac{1}{2\sqrt{g}}\left[\frac{1}{\sqrt{g} + \sqrt{k}.v} + \frac{1}{\sqrt{g} - \sqrt{k}.v}\right] dv = dt$$

On integrating, we get

$$\frac{1}{2\sqrt{g}}\frac{1}{\sqrt{k}} \log\,(\sqrt{g} + \sqrt{k}.v) - \frac{1}{2\sqrt{gk}} \log\,(\sqrt{g} - \sqrt{k}.v) = t + A$$

$$\Rightarrow \qquad \frac{1}{2\sqrt{gk}} \log \frac{\sqrt{g} + \sqrt{k}.v}{\sqrt{g} - \sqrt{k}.v} = t + A \qquad \text{...(1)}$$

On putting $t = 0$, $v = 0$ in (1), we get $\frac{1}{2\sqrt{gk}} \log 1 = 0 + A \Rightarrow A = 0$

Equation (1) becomes $\frac{1}{2\sqrt{gk}} \log \frac{\sqrt{g}+\sqrt{k}.v}{\sqrt{g}-\sqrt{k}.v} = t \quad \Rightarrow \quad \log \frac{\sqrt{g}+\sqrt{k}.v}{\sqrt{g}-\sqrt{k}.v} = 2\sqrt{gk}\, t$

$$\Rightarrow \qquad \frac{\sqrt{g}+\sqrt{k}.v}{\sqrt{g}-\sqrt{k}.v} = e^{2\sqrt{gk}\,t}$$

By componendo and dividendo, we have

$$\frac{\sqrt{k}.v}{\sqrt{g}} = \frac{e^{2\sqrt{gk}t}-1}{e^{2\sqrt{gk}t}+1} = \frac{e^{\sqrt{gk}\,t}-e^{-\sqrt{gk}\,t}}{e^{\sqrt{gk}\,t}+e^{-\sqrt{gk}\,t}} = \tan h\sqrt{gk}\, t$$

$$\Rightarrow \qquad v = \sqrt{\frac{g}{k}} \tan h\sqrt{gk}\, t$$

Whatever the value of t may be $\tanh\sqrt{gk}\, t \le 1$.

Hence the value of v will never exceed $\sqrt{\frac{g}{k}}$. **Proved.**

$$\frac{dx}{dt} = \sqrt{\frac{g}{k}} \tanh\sqrt{gk}\, t$$

Integrating again, we get $\quad x = \sqrt{\frac{g}{k}} \int \tanh\sqrt{gk}\, t\, dt$

$$= \frac{1}{k} \log\cosh\sqrt{gk}\, t + B$$

when $t = 0$, $x = 0$ then $B = 0$

$$\therefore \qquad x = \frac{1}{k} \log\cosh\sqrt{gk}\, t$$ **Ans.**

EXERCISE 16.3

1. A moving body is opposed by a force proportional to the displacement and by a resistance proportional to the square of velocity. Prove that the velocity is given by

$V^2 = ae - \frac{cx}{b} + \frac{c}{ab^2}$ **Hint.** Equation of motion is $m\frac{VdV}{dx} = -K_1x - K_2V^2$

2. A particle of mass m is projected vertically upward with an initial velocity v_0. The resisting force at any time is K times the velocity. Formulate the differential equation of motion and show that the distance s covered by the particle at any time t is given by

$$s = \left(\frac{g}{K^2} + \frac{v_0}{K}\right)(1 - e^{-Kt}) - \frac{g}{K}t$$

3. A particle falls in a vertical line under gravity (supposed constant) and the force of air resistance to its motion is proportional to its velocity. Show that its velocity cannot exceed a particular limit.

Ans. $V = \frac{g}{K}$

4. A body falling from rest is subjected to a force of gravity and an air resistance of $\frac{n^2}{g}$ times the square of velocity. Show that the distance travelled by the body in t seconds in $\frac{g}{n^2} \log\cosh nt$.

5. A body of mass m, falling from rest is subject to the force of gravity and an air resistance proportional to the square of the velocity Kv^2. If it falls through a distance x and possesses a velocity v, at the instant, prove that

$$\frac{2kx}{m} = \log\left(\frac{a^2}{a^2 - v^2}\right) \quad \text{where} \quad \frac{mg}{k} = a^2$$ *(A.M.I.E.T.E., June 2009)*

16.6 VERTICAL ELASTIC STRING

Let an elastic string OA of length l be attached to a fixed point O and a particle of mass m be attached at the other end A. When a mass m hangs freely then it pulls the string OA and it stays at B, in the position of equilibrium. The tension T in the string balances the mass mg hanging at A.

a
A
mg B
x
P
C

By Hooke's Law

$$\frac{\text{Stress } (T)}{\text{Strain}} = \text{Constant} = \text{Modulus of Elasticity } (E)$$

and
$$\text{Strain} = \frac{\text{Extension in length}}{\text{Original length}} = \frac{a}{l}$$

$$T = \frac{Ea}{l} \quad \text{or} \quad mg = \frac{Ea}{l} \qquad \ldots(1)$$

The string is further pulled down to a point C and then released. Then the particle at the lower end C will make motion up and down between B and C. Let the particle be at P at any time t, where $BP = x$.

The down ward force $= mg$

Upward force $=$ Tension T

$$T = E\left(\frac{a+x}{l}\right) \qquad \ldots(2) \quad (\text{Extension} = a + x)$$

$$\text{Resultant down force} = mg - T = mg - E\left(\frac{a+x}{l}\right)$$

$$= mg - \frac{Ea}{l} - \frac{Ex}{l} = mg - mg - \frac{Ex}{l} = \frac{-Ex}{l}$$

$$\text{Downward Acceleration} = -\frac{Ex}{ml}$$

Equation of motion is $\dfrac{d^2x}{dt^2} = -\dfrac{Ex}{ml}$, which is a S.H.M.

Its time period $T = \dfrac{2\pi}{\sqrt{\dfrac{E}{ml}}} = 2\pi\sqrt{\dfrac{ml}{E}}$

16.7 HORIZONTAL ELASTIC STRING

If one end of the elastic string be fixed at O on a table. The other end A of the elastic string of length l is attached to a particle of mass m.

The string is stretched to a point B and then released. The particle comes into motion. Let the particle be at a distance x from A at any time t. The weight mg of the particle is acting downward and is balanced by the reaction R of the table.

The only force acting upon the particle is the tension of the string.

←l→← x →

B′ A′ O A P B →X

← a →

By Hook's Law $\frac{\text{stress}}{\text{strain}}$ = constant of elasticity

$$T = E\frac{x}{l}$$ (x = Extension of the length of the string)

(E = Modulus of elasticity)

Equation of motion is $m\frac{d^2x}{dt^2} = -\frac{Ex}{l}$

$$\Rightarrow \qquad \frac{d^2x}{dt^2} = -\left(\frac{E}{ml}\right)x \qquad ...(1)$$

The motion of the equation is S.H.M.

On multiplying (1) by $\frac{2\,dx}{dt}$, we get

$$\frac{2\,d^2x}{dt^2}\frac{dx}{dt} = -\frac{2\,Ex}{ml}\frac{dx}{dt}$$

On integrating, we get $\left(\frac{dx}{dt}\right)^2 = \frac{-Ex^2}{ml} + A \Rightarrow v^2 = \frac{-Ex^2}{ml} + A$ ....(2)

If the velocity of the particle is zero at S, amplitude $AB = a$. The particle moves from A to B and back B to A will be a S.H.M. The particle moves towards O. Then it moves with uniform velocity upto A'.

On putting $v = 0$, $x = a$, we get $0 = \frac{-Ea^2}{ml} + A \Rightarrow A = \frac{Ea^2}{ml}$

$$v^2 = \frac{E}{ml}(a^2 - x^2)$$

At A, $x = 0$, $v = \sqrt{\left(\frac{E}{\lambda_1}\right)}\,a$. This is the maximum velocity. The particle moves from A to A' with this velocity. After that the string again stretches and motion becomes S.H.M.

Periodic Time of S.H.M. (from A to B, B to A, A' to B', B' to A') + time taken by the particle from A to A' and A' to A with constant velocity $\sqrt{\left(\frac{E}{l}\right)}\,a$.

$$= \frac{2\pi}{\sqrt{\left(\frac{E}{lm}\right)}} + \frac{4l}{\sqrt{\left(\frac{E}{lm}\right)}\,a} = \sqrt{\left(\frac{2lm}{E}\right)}\left(\pi + \frac{2l}{a}\right)$$

Example 16. *A light elastic string of original length l is hung by one end to the other end are tied successively particles of masses m, m'. If t_1 and t_2 be the periods of small oscillations corresponding to these weights are c_1, c_2 the statical extensions, prove that*

$$g(t_1^2 - t_2^2) = 4\,\pi^2\,(c_1 - c_2)$$

Solution. $$mg = T_1 \qquad ...(1)$$

$$mg = E\frac{c_1}{l} \qquad ...(2)$$

$$m'g = E\frac{c_2}{l} \qquad ...(3)$$

Equation of motion of first particle, $m\dfrac{d^2x}{dt^2} = mg - E\dfrac{(x+c_1)}{l}$

$$\Rightarrow \qquad m\frac{d^2x}{dt^2} = mg - E\frac{c_1}{l} - \frac{Ex}{l} = -E\frac{x}{l} \qquad \left[\text{From (2)}, mg = E\frac{c_1}{l}\right]$$

Motion of S.H.M. with $t_1 = \dfrac{2\pi}{\sqrt{\dfrac{E}{lm}}}$ similarly, $t_2 = \dfrac{2\pi}{\sqrt{\dfrac{E}{lm'}}}$

$$\therefore \qquad t_1^2 - t_2^2 = 4\pi^2\frac{l}{E}(m - m') = 4\pi^2\left(\frac{c_1}{g} - \frac{c_2}{g}\right) \qquad \text{[Using (2) and (3)]}$$

$$\Rightarrow \qquad g(t_1^2 - t_2^2) = 4\pi^2(c_1 - c_2)$$ **Ans.**

16.8 SIMPLE HARMONIC MOTION

If a particle moves in a straight line such that its acceleration is always directed towards a fixed point in the line and is proportional to the distance of the particle from the fixed point.

Let O be the fixed point in the line $A'A$. Let the position of the particle be P at time t. Since the acceleration is always directed towards O, *i.e.*, the acceleration is in the direction opposite to that in which x increases, then

$$\frac{d^2x}{dt^2} = -\mu^2 x \qquad ...(1) \quad (\mu \text{ is constant}, OP = x)$$

$$D^2x = -\mu^2 x \Rightarrow (D^2 + \mu^2)x = 0$$

A.E. is $\qquad m^2 + \mu^2 = 0, \quad m = \pm i\mu$

$$x = A\cos\mu t + B\sin\mu t \qquad ...(2)$$

The solution of (1) is

$$\text{Velocity of particle at } P = \frac{dx}{dt} = -A\mu\sin\mu t + B\mu\cos\mu t \qquad ...(3)$$

At A, velocity $= \dfrac{dx}{dt} = 0, x = a, t = 0$

Putting $x = a, t = 0$ in (2), we get $a = A$

Putting $\dfrac{dx}{dt} = 0, t = 0$ in (3), we get $0 = 0 + B\mu \Rightarrow B = 0$

Equation (2) becomes $x = a\cos\mu t$

Equation (3) becomes $\qquad \dfrac{dx}{dt} = -a\mu\sin\mu t$

$$\text{velocity} = -a\mu\sqrt{1-\cos^2\mu t} = -a\mu\sqrt{1-\left(\frac{x}{a}\right)^2} \qquad (x = a\cos\mu t)$$

$$= -\mu\sqrt{a^2 - x^2} \qquad ...(4)$$

At A, $x = a$ and $v = 0$

At O, $x = 0$, and acceleration $= 0$, velocity is maximum.

O is called the centre of motion or the mean position.

Amplitude. In S.H.M. the distance from the centre to the position of maximum displacement is called the amplitude of the motion. OA is the maximum distance and is called the amplitude.

From (4), $$-\frac{dx}{\sqrt{a^2 - x^2}} = \mu\, dt$$

Integrating, we get $$\cos^{-1}\frac{x}{a} = \mu\, t + A \qquad \text{...(5)}$$

Putting $t = 0$, $x = a$ in (5), we get

$$0 = 0 + A \quad \Rightarrow \quad A = 0$$

On putting the value of A, (5) becomes, $\cos^{-1}\frac{x}{a} = \mu\, t \quad \Rightarrow \quad x = a \cos \mu\, t$

Particle will reach O in time t_1,

$$0 = a \cos \mu\, t_1 \quad \Rightarrow \quad 0 = \cos \mu\, t_1$$

$$\Rightarrow \quad \cos\frac{\pi}{2} = \cos \mu\, t_1, \quad \Rightarrow \quad \frac{\pi}{2} = \mu\, t_1 \quad \Rightarrow \quad t_1 = \frac{\pi}{2\mu}$$

Time period: In a S.H.M., the time taken to make a complete oscillation is called time period.

Frequency: The number of complete oscillations per second is called the frequency of motion. If η is the frequency and T is the time period.

$$\text{Time period} = T = 4\left(\frac{\pi}{2\mu}\right) = \frac{2\pi}{\mu}$$

$$\text{Frequency} = n = \frac{1}{T} = \frac{\mu}{2\pi}$$

(*i*) The equation of S.H.M. is $\dfrac{d^2x}{dt^2} = -\,\omega^2 x$

(*ii*) The velocity v at a distance x from the centre at time t is

$$v^2 = \omega^2\,(a^2 - x^2)$$

$x = a \cos \omega t$, where a is the amplitude and ω is the angular velocity.

(*iii*) Maximum acceleration $= \omega^2 a$ (At the extreme point)

(*iv*) Maximum velocity $= \omega\, a$ (at the centre)

(*v*) Time period $= T = \dfrac{2\pi}{\omega}$

Example 17. *A particle moves with S.H.M. of period 12 secs, travels 8 cm from the position of rest in 2 secs. Find the amplitude, the maximum velocity and the velocity at the end of 2 secs.*

Solution. $$T = \frac{2\pi}{\sqrt{\mu}} = 12 \Rightarrow \sqrt{\mu} = \frac{\pi}{6}$$

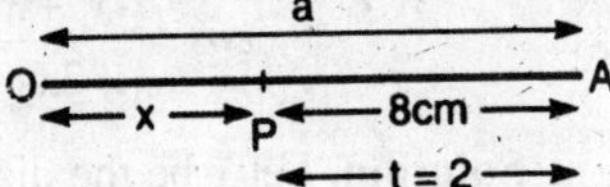

Let a be the amplitude OA

$$AP = 8 \text{ cm}, \; OP = x = a - 8, \; t = 2 \text{ secs.}$$

We know that $$x = a \cos \sqrt{\mu}\, t$$

$$a - 8 = a \cos 2\sqrt{\mu} \qquad \left(\sqrt{\mu} = \frac{\pi}{6}\right)$$

$$= a\cos\frac{\pi}{3} = \frac{a}{2} \quad \Rightarrow \quad a = 16$$

Maximum velocity $= \sqrt{\mu}\, a = \frac{\pi}{6} \times 16 = \frac{8\pi}{3} = 4.619$ cm/sec.

Velocity v at the end of two seconds $= \sqrt{\mu}\sqrt{a^2 - x^2} = \frac{\pi}{6}\sqrt{256 - 64}$

$$= \frac{\pi}{6}\sqrt{192} = \frac{4\pi\sqrt{3}}{3} \text{ cm/ sec.}$$ **Ans.**

Example 18. *A particle is performing a simple harmonic motion of period T about a centre O and it passes through a point P where OP = b with velocity v in the direction OP. Prove that the time which elapses before it returns to P is*

$$\frac{T}{\pi}\tan^{-1}\frac{vT}{2\pi b}.$$

A' — O — P — A

Solution. We have to find time taken from P to A and then A to P

$t = 2$ (time from A to P)

$$= 2\int_0^t dt = 2\int_a^p \frac{dx}{\sqrt{\mu}\sqrt{a^2 - x^2}} \text{ (Ignoring } - ve \text{ sign)} \left(\frac{dx}{dt} = -\sqrt{\mu}\sqrt{a^2 - x^2}\right)$$

$$= \frac{2}{\sqrt{\mu}}\left[\cos^{-1}\frac{x}{a}\right]_a^b = \frac{2}{\sqrt{\mu}}\left[\cos^{-1}\frac{b}{a} - \cos^{-1}\frac{a}{b}\right] = \frac{2}{\sqrt{\mu}}\cos^{-1}\frac{b}{a}$$

$$\Rightarrow \quad t = \frac{2}{\sqrt{\mu}}\tan^{-1}\left(\frac{\sqrt{a^2 - b^2}}{b}\right)$$

(Right triangle: hypotenuse a, base b, perpendicular $\sqrt{a^2 - b^2}$, angle θ)

$$\left[\begin{aligned} v^2 &= \mu(a^2 - b^2) \\ \Rightarrow \quad v &= \sqrt{\mu}\sqrt{(a^2 - b^2)} \\ \Rightarrow \quad \frac{v}{\sqrt{\mu}} &= \sqrt{a^2 - b^2} \end{aligned}\right]$$

$$= \frac{2}{\sqrt{\mu}}\tan^{-1}\left(\frac{v}{b\sqrt{\mu}}\right)$$

$$= \frac{2}{\frac{2\pi}{T}}\tan^{-1}\left[\frac{v}{b\left(\frac{2\pi}{T}\right)}\right] \quad \left[T = \frac{2\pi}{\sqrt{\mu}} \Rightarrow \sqrt{\mu} = \frac{2\pi}{T}\right] = \frac{T}{\pi}\tan^{-1}\left[\frac{vT}{2\pi b}\right]$$ **Proved.**

Example 19. *At the end of three successive seconds, the distances of a point moving with S.H.M. from its mean position are x_1, x_2, x_3. Show that the time of complete oscillation is*

$$\frac{2\pi}{\cos^{-1}\left(\frac{x_1 + x_3}{2x_2}\right)}$$

Solution. Let x be the distance from the centre at any time t,

$$x = a\sin\sqrt{\mu}\, t$$

$$x_1 = a\sin\sqrt{\mu}\, t,\ x_2 = a\sin\sqrt{\mu}\,(t+1),\ x_3 = a\sin\sqrt{\mu}\,(t+2)$$

$$\Rightarrow \quad x_1 + x_3 = a[\sin\sqrt{\mu}\, t + \sin\sqrt{\mu}\,(t+2)]$$

$$= 2a \sin \sqrt{\mu}\,(t+1) \cos \sqrt{\mu} = 2x_2 . \cos \sqrt{\mu}$$

$$\Rightarrow \quad \frac{x_1 + x_3}{2x_2} = \cos \sqrt{\mu} \quad \Rightarrow \quad \sqrt{\mu} = \cos^{-1} . \frac{x_1 + x_3}{2x_2}$$

$$\text{Time period} = \frac{2\pi}{\sqrt{\mu}} = \frac{2\pi}{\cos^{-1} \dfrac{x_1 + x_3}{2x_2}}$$ **Proved.**

Example 20. *A point moves in a straight line towards a centre of force* $\mu/(\text{distance})^3$, *starting from rest at a distance 'a' from the centre of force; show that the time of reaching a point distant b from the centre of force is* $(a/\sqrt{\mu})\sqrt{(a^2 - b^2)}$, *and that its velocity is* $\dfrac{\sqrt{\mu}}{ab}\sqrt{(a^2 - b^2)}$.

(*U.P., II Semester, Summer 2001*)

Solution. Let a point move from A towards the centre of force O.

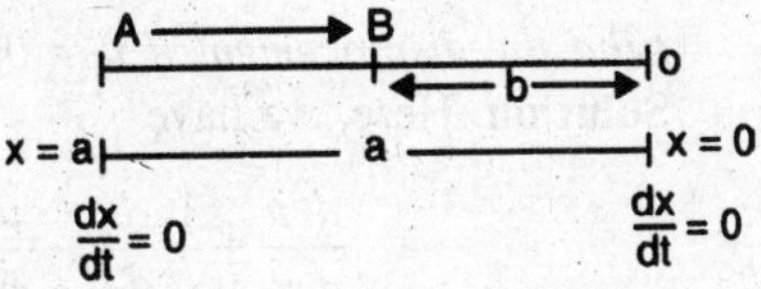

$$\therefore \quad \frac{m\mu}{x^3} = -m\frac{d^2x}{dt^2} \Rightarrow \frac{d^2x}{dt^2} = -\frac{\mu}{x^3}$$

A → B, b, O; $x = a$, a, $x = 0$; $\frac{dx}{dt} = 0$, $\frac{dx}{dt} = 0$

$$\Rightarrow \quad 2\frac{dx}{dt} . \frac{d^2x}{dt^2} = -2\frac{\mu}{x^3} . \frac{dx}{dt}$$

Integrating, we get

$$\left(\frac{dx}{dt}\right)^2 = -2 . \mu \frac{x^{-2}}{-2} + C$$

$$\Rightarrow \quad V^2 = \frac{\mu}{x^2} + C \qquad \text{...(1)}$$

At A, $V = 0$ and $x = a$, $\quad \therefore \ 0 = \dfrac{\mu}{a^2} + C \quad \Rightarrow C = -\dfrac{\mu}{a^2}$

On putting the value of C, (1) becomes

$$V^2 = \frac{\mu}{x^2} - \frac{\mu}{a^2} = \mu\left(\frac{a^2 - x^2}{x^2 a^2}\right) \qquad \text{...(2)}$$

Therefore velocity when $x = b$ is given by

$$V^2 = \mu\left(\frac{a^2 - b^2}{a^2 b^2}\right) \quad \Rightarrow \quad V = \pm\sqrt{\mu}\,\frac{\sqrt{a^2 - b^2}}{ab}$$

$$\Rightarrow \quad V = -\sqrt{\mu}\,\frac{\sqrt{a^2 - b^2}}{ab}$$

From (2), $$\left(\frac{dx}{dt}\right)^2 = \mu\frac{(a^2 - x^2)}{x^2 a^2} \quad \Rightarrow \quad \frac{dx}{dt} = -\sqrt{\mu}\,\frac{\sqrt{a^2 - x^2}}{xa}$$

$$dt = -\frac{1}{\sqrt{\mu}}\frac{xa}{\sqrt{a^2 - x^2}}\,dx$$

Integrating, we get

$$t = -\frac{1}{\sqrt{\mu}}\int\frac{xa\,dx}{\sqrt{a^2 - x^2}}$$

Let $\quad a^2 - x^2 = z^2 \Rightarrow -2x\,dx = 2z\,dz$

$$t = \frac{a}{\sqrt{\mu}} \int \frac{z\,dz}{z} = \frac{a}{\sqrt{\mu}} \int dz = \frac{a}{\sqrt{\mu}} z + C_1 = \frac{a}{\sqrt{\mu}} \sqrt{a^2 - x^2} + C_1 \quad ...(3)$$

At A, $t = 0$, $x = a$,

On putting $t = 0$, $x = a$ in (3), we get $0 = 0 + C_1 \Rightarrow C_1 = 0$

On putting the value of C_1 in (3), we have

$$t = \frac{a}{\sqrt{\mu}} \sqrt{a^2 - x^2}$$

At B, $x = b$, $t = \frac{a}{\sqrt{\mu}} \sqrt{a^2 - b^2}$ **Proved.**

Example 21. *The radial displacement 'u' in a rotating disc at a distance r from the axis is given by*

$$\frac{d^2u}{dr^2} + \frac{1}{r}\frac{du}{dr} - \frac{u}{r^2} + kr = 0$$

Find the displacement if $u = 0$ at $r = 0$ and at $r = a$. (*M.U. II Semester, 2008*)

Solution. Here, we have

$$\frac{d^2u}{dr^2} + \frac{1}{r}\frac{du}{dr} - \frac{u}{r^2} + kr = 0$$

$$\Rightarrow \quad r^2 \frac{d^2u}{dr^2} + r\frac{du}{dr} - u = -k r^3 \quad ...(1)$$

This is the homogeneous equation.

On putting $z = \log r \Rightarrow r = e^z$ and $r^2 \frac{d^2u}{dr^2} = D(D-1)u$, $r\frac{du}{dr} = Du$ in (1), we get

$$D(D-1)u + Du - u = -k e^{3z}$$

$$\Rightarrow \quad (D^2 - D + D - 1)u = -k e^{3z}$$

$$\Rightarrow \quad (D^2 - 1)u = -k e^{3z}$$

A.E. is $m^2 - 1 = 0 \Rightarrow m = 1, m = -1$

$$\text{C.F.} = A e^z + B e^{-z}$$

$$\text{P.I.} = \frac{1}{D^2 - 1}(-k e^{3z}) = -k \frac{1}{(3)^2 - 1} e^{3z} = -\frac{k}{8} e^{3z}$$

Complete solution is u = C.F. + P.I.

$$\Rightarrow \quad u = A e^z + B e^{-z} - \frac{k}{8} e^{3z}$$

$$\Rightarrow \quad u = A r + \frac{B}{r} - \frac{k}{8} r^3$$ **Ans.**

EXERCISE 16.4

1. At what distance from the centre the velocity in a S.H.M. will be one fourth of the maximum?

Ans. $x = \pm \frac{\sqrt{15}\, a}{4}$

2. A particle moving in a straight line with S.H.M. has velocities 3m/sec. and 2m/sec. respectively when it is at distances 1 metre and 1.3 metre from the centre of its path. Find its period and acceleration at the greatest distance from the centre of motion.

3. A particle moves with S.H.M. If, when at a distance of 3 and 4 cm from the centre of the path, its velocities are 8 and 6 cm per sec. respectively. Find its period, maximum velocity and acceleration when at its greatest distance from the centre. **Ans.** π secs, 10 cm/sec, 20 cm/sec^2.

4. A point executes S.H.M. such that in two of its positions, the velocities are u, v and the corresponding accelerations α, β. Show that the distance between the positions is $\dfrac{v^2 - u^2}{\alpha + \beta}$, and find the amplitude of the motion.

5. A particle of mass m is oscillating in a straight line about a centre of force, O towards which when at a distance r the force is $m\,n^2\,r$ and a is the amplitude of oscillation, when at a distance $\dfrac{a\sqrt{3}}{2}$ from O, the particle receives a blow in the direction of motion which generates a velocity na. If this velocity be away from O, show that the new amplitude is $a\sqrt{3}$.

6. The speed v of the point P which moves in a line is given by the relation $v^2 = a + 2bx - cx^2$ where x is the distance of the point P from a fixed point on the path, and a, b, c are constants the motion is simple harmonic if c is positive; determine the period and the amplitude of the motion.

Ans. $T = \dfrac{2\pi}{\sqrt{c}}$, Amplitude $= \dfrac{\sqrt{b^2 + ab}}{c}$

7. In the case of a stretched elastic string which has one end fixed and a particle of mass m attached to the other end, the equation of motion is

$$\frac{d^2 s}{dt^2} = -\frac{mg}{e}(s - l)$$

where l is the natural length of the string and e its elongation due to a weight mg. Find s and v determining the constants, so that $s = s_0$ at the time $t = 0$ and $v = 0$ when $t = 0$.

Ans. $v = -\sqrt{\left(\dfrac{g}{e}\right)}\,[(s_0 - l)^2 - (s - l)^2]^{1/2}$, $s - l = (s_0 - l)\cos\left[\sqrt{\left(\dfrac{g}{e}\right)}\,.\,t\right]$

16.9 THE SIMPLE PENDULUM

A particle of mass m suspended vertically by a light inextensible string oscillating under gravity constitutes a simple pendulum.

Let l be the length of the string, O be the fixed point. A be the initial position of the bob. In the displaced position P at any time t, then forces acting on the bob are

(*i*) weight mg acting vertically downward

(*ii*) the tension T in the string.

$$mg\cos\theta = T$$

$$\text{Restoring force} = m\frac{d^2x}{dt^2} = -mg\sin\theta \quad (x = AP)$$

$$\Rightarrow \quad \frac{d^2x}{dt^2} = -g\sin\theta$$

$$= -g\theta \quad (\sin\theta = \theta \text{ if } \theta \text{ is small})$$

$$= -g\frac{x}{l}$$

$$\Rightarrow \quad D^2x + \frac{g}{l}x = 0 \;\Rightarrow\; \left(D^2 + \frac{g}{l}\right)x = 0$$

A.E. is $m^2 + \dfrac{g}{l} = 0 \;\Rightarrow\; m^2 = \dfrac{-g}{l}$, $m = \pm i\sqrt{\dfrac{g}{l}}$

$\therefore \qquad x = c_1 \cos \sqrt{\frac{g}{l}}\, t + c_2 \sin \sqrt{\frac{g}{l}}\, t$

Thus, the motion of the bob is simple harmonic and

$$\text{Period of Oscillation} = \frac{2\pi}{\sqrt{\frac{g}{l}}} = 2\pi\sqrt{\frac{l}{g}}$$

Note. The motion of the bob from one end to the other end is half oscillation and is called a **beat** or **swing.**

In a second pendulum, the time of one beat is one second. The number of beats in a day

$$= 24 \times 60 \times 60 = 86400.$$

16.10 OSCILLATIONS OF A SPRING

(*a*) **Free oscillations.** Let a spring be fixed at O and a mass m is suspended from the lower end A.

Let S (= AB) be the elongation produced by the mass m hanging. B is called the position of static equilibrium and S is called the static extension.

We choose the downward direction as the positive direction and regard forces which act down as positive and upward forces as negative.

If k be the restoring force per unit stretch of the spring due to elasticity.

Equilibrium at $B \quad mg = ks$

Let the mass m be displaced through a further distance x (= BP) from the equilibrium position B.

Weight mg is acting downward

Restoring force $k(s + x)$ is acting upward.

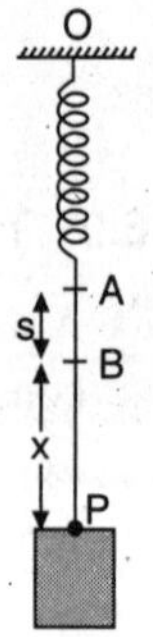

O A s B x P

Equation of motion is $\quad m\frac{d^2x}{dt^2} = mg - K(s + x)$

$$= mg - ks - kx \qquad (mg = ks)$$

$$= mg - mg - kx = -kx$$

$$\Rightarrow \qquad \frac{d^2x}{dt^2} = -\frac{k}{m}x \quad \Rightarrow \quad \frac{d^2x}{dt^2} + \frac{k}{m}x = 0$$

$$\Rightarrow \qquad \frac{d^2x}{dt^2} + \omega^2 x = 0 \qquad \left(\omega^2 = \frac{k}{m}\right)$$

Example 22. *A spring of negligible weight hangs vertically. A mass m is attached to the other end. If the mass is moving with velocity u when the spring is unstretched, find the velocity v as a function of the stretch x.*

Solution. If x is the increase in length of the spring when velocity of the mass m is v, then the equation of motion is

$$m.v\frac{dv}{dx} = mg - T$$

where $T = kx$, by Hooke's law, k being Young's modulus

$$\therefore \qquad mv\frac{dv}{dx} = mg - kx \quad \Rightarrow \quad mv\,dv = (mg - kx)\,dx$$

Integrating, we get $$m\frac{v^2}{2} = mgx - k\frac{x^2}{2} + c$$

Now, $v = u$ when $x = 0 \Rightarrow c = m\dfrac{u^2}{2}$

$$\therefore \quad m\frac{v^2}{2} = mgx - k\frac{x^2}{2} + m\frac{u^2}{2} \Rightarrow mv^2 = 2\,mgx - kx^2 + mu^2$$

$$\Rightarrow \quad v^2 = 2gx + u^2 - \frac{k}{m}x^2$$ **Ans.**

Example 23. *A spring for which stiffness k = 700 Newtons/m hangs in a vertical position with its upper end fixed. A mass of 7 kg is attached to the lower end. After coming to rest, the mass is pulled down 0.05 m and released. Discuss the resulting motion of the mass, neglecting air resistance.*

Solution. The equation of motion is $m\dfrac{d^2x}{dt^2} = -kx$

$$\Rightarrow \quad 7\frac{d^2x}{dt^2} = -700\,x \quad \Rightarrow \quad \frac{d^2x}{dt^2} + 100x = 0 \quad \Rightarrow \quad D^2x + 100\,x = 0$$

A.E. is $m^2 + 100 = 0 \Rightarrow m = \pm i\,10$

Its solution is $$x = C_1 \cos 10\,t + C_2 \sin 10t \qquad \text{...(1)}$$

Putting $t = 0$, $x = 0.05$ in (1), we get $0.05 = C_1$

Differentiating (1), we have $$\frac{dx}{dt} = -10\,C_1 \sin 10\,t + 10\,C_2 \cos 10\,t \qquad \text{...(2)}$$

On putting $v = \dfrac{dx}{dt} = 0$ and $t = 0$ in (2), $0 = 10\,C_2 \Rightarrow C_2 = 0$

On substituting the values of C_1 and C_2 in (1), we obtain
$$x = 0.05 \cos 10\,t$$

This is S.H.M. $$\text{Period} = \frac{2\pi}{\omega} = \frac{2\pi}{10} = 0.628 \text{ secs.}$$

$$\text{Frequency} = \frac{10}{2\pi} = 1.59 \text{ cycle/sec}$$

$$\text{Amplitude} = 0.50\,m$$ **Ans.**

Example 24. *A mass M suspended from the end of a helical spring is subjected to a periodic force f = F sin ω t in the direction of its length. The force f is measured positive vertically downwards and at zero time M is at rest. If the spring stiffness is S, prove that the displacement of M at time t from the commencement of motion is given by*

$$x = \frac{F}{M\,(P^2 - \omega^2)}\left[\sin \omega t - \frac{\omega}{P}\sin pt\right] \textit{ where } P^2 = \frac{S}{M}$$

and damping effects are neglected.

Solution. Let x be the displacement from the equilibrium position, the equation of motion is

$$M\frac{d^2x}{dt^2} = -S\,x + F \sin \omega\,t$$

$$\frac{d^2x}{dt^2} + \frac{S}{M}x = \frac{F}{M}\sin \omega\,t \quad \Rightarrow \quad \frac{d^2x}{dt^2} + P^2x = \frac{F}{M}\sin \omega\,t \qquad \left(\because \frac{S}{M} = P^2\right)$$

$$\Rightarrow \qquad (D^2 + P^2) = \frac{F}{M} \sin wt$$

A.E. is $m^2 + P^2 = 0$, $m = \pm iP$

$$\therefore \qquad \text{C.F.} = (c_1 \cos pt + c_2 \sin pt)$$

$$\text{P.I.} = \frac{1}{D^2 + P^2} \frac{F}{M} \sin \omega t = \frac{F}{M} \frac{1}{-\omega^2 + P^2} \sin \omega t$$

$$\therefore \qquad x = c_1 \cos pt + c_2 \sin pt + \frac{F}{M} \frac{1}{P^2 - \omega^2} \sin \omega t \qquad \text{...(1)}$$

Putting $t = 0$ and $x = 0$ in (1), we get, $0 = c_1$

Equation (1) becomes $\quad x = c_2 \sin pt + \dfrac{F}{M} \dfrac{1}{P^2 - \omega^2} \sin \omega t \qquad$ ...(2)

Differentiating (2), we obtain $\dfrac{dx}{dt} = c_2\, p \cos pt + \dfrac{F}{M} \dfrac{\omega}{P^2 - \omega^2} \cos \omega t \qquad$ ...(3)

Putting $\dfrac{dx}{dt} = 0$ and $t = 0$ in (3), we have

$$0 = c_2 P + \frac{F}{M} \frac{\omega}{P^2 - \omega^2} \Rightarrow c_2 = -\frac{F\omega}{PM\,(P^2 - \omega^2)}$$

On substituting the value of c_2 in (2), we get

$$x = -\frac{\omega}{P} \frac{F}{M\,(P^2 - \omega^2)} \sin pt + \frac{F}{M} \frac{1}{P^2 - \omega^2} \sin \omega t$$

$$= \frac{F}{M(P^2 - \omega^2)} \left[\sin \omega t - \frac{\omega}{P} \sin pt \right] \qquad \textbf{Proved.}$$

(*b*) Damped Free Oscillation. If the motion of the mass m be opposed by some resistance, proportional to the velocity $\left(= K_1 \dfrac{dx}{dt} \right)$, the oscillations are said to be damped.

The equation of motion is $m \dfrac{d^2x}{dt^2} = mg - k(x + s) - k_1 \dfrac{dx}{dt} \qquad [mg = ks]$

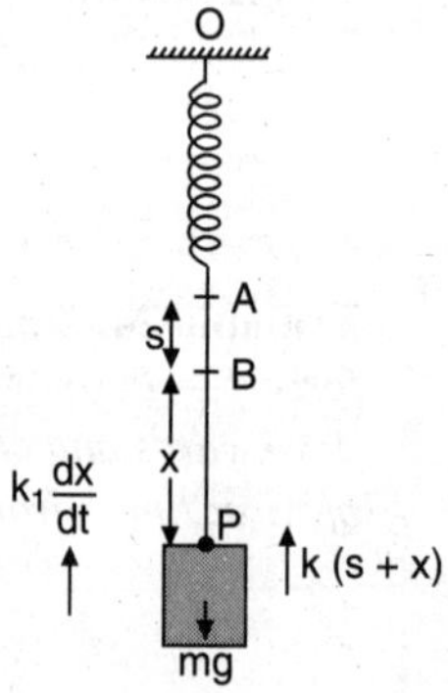

$$\Rightarrow m \frac{d^2x}{dt^2} = mg - mg - kx - k_1 \frac{dx}{dt} \Rightarrow m \frac{d^2x}{dt^2} = -kx - k_1 \frac{dx}{dt}$$

$$\Rightarrow \qquad \frac{d^2x}{dt^2} + \frac{k_1}{m} \frac{dx}{dt} + \frac{k}{m} x = 0$$

On putting $\dfrac{k_1}{m} = 2\lambda$, $\dfrac{k}{m} = \mu^2$, we get

$$\Rightarrow \qquad \frac{d^2x}{dt^2} + 2\lambda \frac{dx}{dt} + \mu^2 x = 0 \quad \Rightarrow \quad (D^2 + 2\lambda D + \mu^2)x = 0$$

A.E. is $\qquad m^2 + 2\lambda m + \mu^2 = 0 \qquad$ ...(1)

$$\Rightarrow \qquad m = \frac{-2\lambda \pm \sqrt{4\lambda^2 - 4\mu^2}}{2} \quad \Rightarrow \quad m = -\lambda \pm \sqrt{\lambda^2 - \mu^2} \qquad \text{...(2)}$$

Case I. If $\lambda > \mu$

Then m is real and distinct.

Hence, the solution of (1) is $x = c_1 e^{[-\lambda+\sqrt{\lambda^2-\mu^2}]t} + c_2 e^{[-\lambda-\sqrt{\lambda^2-\mu^2}]t}$ ...(3)

From (3), it is obvious that x is positive and decreases to zero as $t \Rightarrow \infty$.

The restoring force, in this case, so great that the motion is non-oscillatory and is known as **over damped** or **dead beat motion.**

Case II. If $\lambda = \mu$

From (2), $m = -\lambda \pm \sqrt{\mu^2 - \mu^2} \Rightarrow m = -\lambda$

The roots of (1) are real and equal.

$\therefore \quad x = (c_1 + c_2 t)e^{-\lambda t}$

Here also x is non-oscillatory and $x \Rightarrow 0$ as $t \Rightarrow \infty$. This is the critical damping. There is no oscillatory term in the solution hence motion becomes a periodic or non-oscillatory.

Case III. If $\lambda < \mu$

From (2), $m = -\lambda \pm \sqrt{\lambda^2 - \mu^2}$

$m = -\lambda \pm i\sqrt{\mu^2 - \lambda^2}$

$\therefore \quad x = e^{-\lambda t}[c_1 \cos\sqrt{\mu^2 - \lambda^2}\, t + c_2 \sin\sqrt{\mu^2 - \lambda^2}\, t]$

$x = e^{-\lambda t} r \sin(\sqrt{\mu^2 - \lambda^2}\, t + \alpha) \quad [c_1 = r \sin\alpha,\ c_2 = r\cos\alpha]$

Hence, the motion is oscillatory.

The periodic time of the oscillation $= T = \dfrac{2\pi}{\sqrt{\mu^2 - \lambda^2}}$ which is greater than the periodic time in case of free oscillations which is $\dfrac{2\pi}{\mu}$. Thus, the effect of damping is to increase the periodic time of oscillation.

Example 25. *A spring fixed at the upper end supports a weight of 980 gm at its lower end. The spring stretches $\frac{1}{2}$ cm under a load of 10 gm and the resistance (in gm ωt) to the motion of the weight is numerically equal to $\frac{1}{10}$ of the speed of the weight in cm/sec. The weight is pulled down $\frac{1}{4}$ cm below its equilibrium position and then released. Find the expression for the distance of weight from its equilibrium position at time t during its first upward motion. Also find the time t it takes the damping factor to drop to $\frac{1}{10}$ of its initial value.*

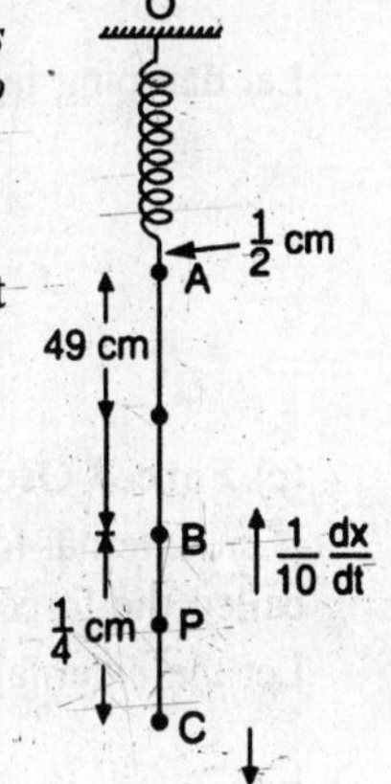

Solution. Let OA be a spring fixed at O and a load of 10 gm is attached at A. The spring is stretched by $\frac{1}{2}$ cm.

$$mg = T_0 \Rightarrow 10 = T_0 = k.\frac{1}{2} \Rightarrow k = 20 \text{ gm/cm}$$

Let B be the equilibrium after attaching a weight 980 gm at A.

$$mg = kx \text{ where } x = AB$$

$$980 = 20x \Rightarrow x = AB = \frac{980}{20} = 49 \text{ cm}$$

After static equilibrium the weight is pulled down to C and released. $\left(BC = \frac{1}{4} \text{ cm}\right)$. After release, the weight be at P after time t.

$$BP = x$$

$$T = K \,.\, AP = 20\,(49 + x) = 980 + 20x$$

Equation of motion is $m\dfrac{d^2x}{dt^2} = \omega - T - \dfrac{1}{10}\dfrac{dx}{dt}$ $\quad\left(\text{Resistance} = \dfrac{1}{10}\dfrac{dx}{dt}\right)$

$$\frac{980}{g}\frac{d^2x}{dt^2} = 980 - (980 + 20x) - \frac{1}{10}\frac{dx}{dt} \qquad [g = 980 \text{ cm/sec}^2]$$

$$\Rightarrow \quad 10\frac{d^2x}{dt^2} + \frac{dx}{dt} + 200x = 0 \Rightarrow 10\,D^2x + Dx + 200\,x = 0$$

A.E. is $10\,m^2 + m + 200 = 0$

$$\Rightarrow \quad m = \frac{-1 \pm \sqrt{1 - 8000}}{20} = \frac{-1 \pm i\,89.4}{20}$$

$$\Rightarrow \quad m = -\,0.05 \pm i\,4.5$$

$\therefore$ C.F. $\quad x = e^{-0.05\,t}\,[c_1 \cos 4.5\,t + c_2 \sin 4.5\,t]$ ...(1)

On putting $t = 0$ and $x = \frac{1}{4}$ in (1), we get $\frac{1}{4} = c_1$

On differentiating (1), we get

$$\frac{dx}{dt} = -\,0.05\,e^{-0.05\,t}[c_1 \cos 4.5\,t + c_2 \sin 4.5\,t] + e^{-\,0.05t}[-\,4.5\,c_1 \sin 4.5t + 4.5\,c_2 \cos 4.5\,t] \quad ..(2)$$

On putting $\frac{dx}{dt} = 0$ and $t = 0$ in (2), we have

$$0 = -\,0.05\,c_1 + 4.5\,c_2 \Rightarrow c_2 = \frac{0.05}{4.5}\,c_1 = \frac{0.05}{4.5}\left(\frac{1}{4}\right) = 0.0028$$

On substituting the values of c_1 and c_2 in (1), we obtain

$$x = e^{-0.05\,t}[0.25 \cos 4.5t + 0.0028 \sin 4.5\,t]$$

Damping factor $= b\,e^{-0.05t}$ (b = constant of proportionality)

Initial value of damping factor $= b\,e^0 = b$

Let damping factor after time t be $\frac{b}{10}$.

$$\frac{b}{10} = b\,e^{-\,0.05t} \Rightarrow e^{\frac{t}{20}} = 10 \Rightarrow \frac{t}{20} = \log_e 10$$

$$\Rightarrow \quad t = 20 \log_e 10 \Rightarrow t = 20\frac{\log_{10} 10}{\log_{10} e} = \frac{20}{\log_{10} e} = 20 \times 2.3 = 46 \text{ secs.}$$ **Ans.**

(c) Forced Oscillations (without damping)

If an external force is applied on the point of support of the spring, it oscillates. The motion is called the forced oscillatory motion.

Let the external force be $q \cos nt$.

Equation of motion is $m\dfrac{d^2x}{dt^2} = mg - ks - kx + q \cos nt$ $\quad (mg = ks)$

$$\Rightarrow \quad m\frac{d^2x}{dt^2} = -kx + q\cos nt \Rightarrow \frac{d^2x}{dt^2} = -\frac{k}{m}x + \frac{q}{m}\cos nt \qquad \text{...(1)}$$

Let $\frac{k}{m} = \mu^2$ and $\frac{q}{m} = e$, then (1) becomes

$$\frac{d^2x}{dt^2} = -\mu^2 x + e\cos nt$$

$$\Rightarrow \quad \frac{d^2x}{dt^2} + \mu^2 x = e\cos nt$$

$$\Rightarrow \quad (D^2 + \mu^2)x = e\cos nt \qquad \text{...(2)}$$

A.E. is $m^2 + \mu^2 = 0 \Rightarrow m = \pm i\mu$

$$\text{C.F.} = c_1\cos\mu t + c_2\sin\mu t$$

$$\text{P.I.} = \frac{1}{D^2+\mu^2}e\cos nt$$

Case (*a*). If $\mu \neq n$, $\quad \text{P.I.} = e\frac{1}{-n^2+\mu^2}\cos nt$

Complete solution of (1) is $\quad x = c_1\cos\mu t + c_2\sin\mu t + \frac{e}{\mu^2-n^2}\cos nt \qquad \begin{bmatrix} c_1 = A\cos\alpha \\ c_2 = A\sin\alpha \end{bmatrix}$

$$x = A\cos(\mu t + \alpha) + \frac{e}{\mu^2-n^2}\cos nt \qquad \text{...(3)}$$

Equation (2) shows that the motion is the resultant of two oscillatory motions, *i.e.*, the first due to $A\cos(\mu t+\alpha)$ gives free oscillation of period $\frac{2\pi}{\mu}$ and the second due to $\frac{e}{\mu^2-n^2}\cos nt$ gives forced oscillations of period $\frac{2\pi}{n}$. If μ is large, then the frequency of free oscillations is very high, then the amplitude $\frac{e}{\mu^2-n^2}$ of forced oscillations is small.

Case (*b*). If $\mu = n$ $\quad \text{P.I.} = \frac{1}{D^2+\mu^2}e\cos nt = e.t\frac{1}{2D}\cos nt = \frac{et}{2}\int\cos nt\,dt$

$$\text{P.I.} = \frac{et}{2}\left(\frac{\sin nt}{n}\right)$$

$$x = c_1\cos\mu t + c_2\sin\mu t + \frac{et\sin nt}{2n}$$

$$= c_1\cos\mu t + c_2\sin\mu t + e\frac{t}{2\mu}\sin\mu t \qquad (n=\mu)$$

$$= c_1\cos\mu t + \left(c_2 + \frac{et}{2\mu}\right)\sin\mu t$$

Let $c_1 = r\sin\phi$ and $\left(c_2 + \frac{et}{2\mu}\right) = r\cos\phi$ $\qquad \left[\begin{array}{l} r = \sqrt{c_1^2 + \left(c_2+\frac{et}{2\mu}\right)^2} \\ \phi = \tan^{-1}\left(\frac{c_1}{c_2+\frac{et}{2\mu}}\right) \end{array} \right]$

$$\Rightarrow \quad x = r\sin\phi\cos\mu t + r\cos\phi\sin\mu t$$

$$= r\sin(\mu t + \phi)$$

The period of oscillation $= \frac{2\pi}{\mu}$

Amplitude $= \sqrt{c_1^2 + \left(c_2 + \frac{et}{2\mu}\right)}$ and it increases as t increases.

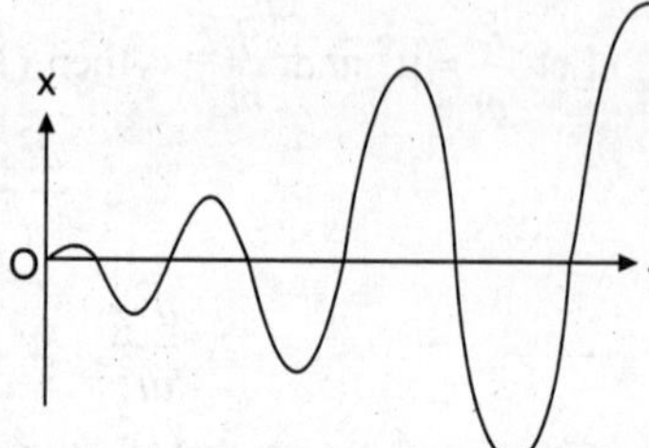

After long time, the amplitude of the oscillation may become abnormally large causing over strain and consequently break down the system. But it does not happen as there is always some resistance in the system.

Resonance. If the frequency due to external periodic force becomes equal to the natural frequency of the system, the phenomenon is known as *resonance.*

In designing a machine or structure, occurrence of the resonance is always to be avoided so that the system may not break down. While marching over a bridge, the soldiers avoid that their steps may not be in rythm with the natural frequency of the bridge. Resonance may cause the collapse the bridge.

(*d*) **Forced Oscillations (with damping)**

If there is an additional damping force, proportional to velocity, then equation of motion is

$$m\frac{d^2x}{dt^2} = mg - ks - kx - k_1\frac{dx}{dt} + q\cos nt \qquad (mg = ks)$$

$$\Rightarrow \quad m\frac{d^2x}{dt^2} = -kx - k_1\frac{dx}{dt} + q\cos nt$$

$$\Rightarrow \quad \frac{d^2x}{dt^2} = -\frac{k}{m}x - \frac{k_1}{m}\frac{dx}{dt} + \frac{q}{m}\cos nt \qquad \ldots(1)$$

On putting $\frac{k}{m} = \mu^2, \frac{k_1}{m} = 2\lambda, \frac{q}{m} = e$ in equation (1), we get

$$\Rightarrow \quad \frac{d^2x}{dt^2} = -\mu^2 x - 2\lambda\frac{dx}{dt} + e\cos nt$$

$$\Rightarrow \quad \frac{d^2x}{dt^2} + 2\lambda\frac{dx}{dt} + \mu^2 x = e\cos nt \qquad \ldots(2)$$

$$\Rightarrow \quad (D^2 + 2\lambda D + \mu^2)x = e\cos nt$$

A.E. is $\quad m^2 + 2\lambda m + \mu^2 = 0 \Rightarrow m = \frac{-2\lambda \pm \sqrt{4\lambda^2 - 4\mu^2}}{2}$

$$m = -\lambda \pm \sqrt{\lambda^2 - \mu^2}$$

$$\text{C.F.} = c_1 e^{[-\lambda + \sqrt{\lambda^2-\mu^2}]t} + c_2 e^{[-\lambda - \sqrt{\lambda^2-\mu^2}]t} = e^{-\lambda t}\left[c_1 e^{\sqrt{\lambda^2-\mu^2}t} + c_2 e^{-\sqrt{\lambda^2-\mu^2}t}\right]$$

$$\text{P.I.} = \frac{1}{D^2 + 2\lambda D + \mu^2} e\cos nt = e\frac{1}{-n^2 + 2\lambda D + \mu^2}\cos nt$$

$$= e\frac{1}{(\mu^2 - n^2) + 2\lambda D}\cos nt = e\frac{(\mu^2 - n^2) - 2\lambda D}{(\mu^2 - n^2)^2 - 4\lambda^2 D^2}\cos nt$$

$$= e\frac{(\mu^2 - n^2) - 2\lambda D}{(\mu^2 - n^2)^2 - 4\lambda^2(-n^2)}\cos nt = e\frac{(\mu^2 - n^2)\cos nt + 2n\lambda\sin nt}{(\mu^2 - n^2)^2 + 4\lambda^2 n^2}$$

Put $\mu^2 - n^2 = r\cos\phi$ and $2n\lambda = r\sin\phi$

$$\left[\begin{array}{l} r = \sqrt{(\mu^2 - n^2)^2 + 4n^2\lambda^2} \\ \phi = \tan^{-1}\dfrac{2n\lambda}{\mu^2 - n^2} \end{array}\right]$$

$$\text{P.I.} = e\frac{r\cos\phi\cos nt + r\sin\phi\sin nt}{(\mu^2 - n^2)^2 + 4\lambda^2 n^2}$$

$$= e\frac{r\cos(nt - \phi)}{r^2} = \frac{e}{r}\cos(nt - \phi)$$

Complete solution is x = C.F. + P.I.

$$x = e^{-\lambda t}[c_1 e^{\sqrt{\lambda^2 - \mu^2}\,t} + c_2 e^{-\sqrt{\lambda^2 - \mu^2}\,t}] + \frac{e}{r}\cos(nt - \phi) \qquad ...(3)$$

C.F. in (3) represents free oscillations of the system, which die out as $t \to \infty$, due to $e^{-\lambda t}$.

$\frac{e}{r}\cos(nt - \phi)$ represents the forced oscillation.

Its constant amplitude $= \frac{e}{r} = \frac{e}{\sqrt{(\mu^2 - n^2)^2 + 4n^2\lambda^2}}$

Its period $= \frac{2\pi}{n}$ which is the same as that of impressed force. As t increases, the free oscillations (given in the C.F.) die out while the forced oscillations persist giving the steady state of motion.

Example 26. *A body executes damped forced vibrations given e the equation*

$\frac{d^2x}{dt^2} + 2k\frac{dx}{dt} + b^2x = e^{-kt}\sin\omega t.$ *Solve the equation for both the cases, when* $w^2 \neq b^2 - k^2$ *and when* $w^2 = b^2 - k^2$. (*Uttarakhand, II Semester, June 2007*)

Solution. We have,

$$\frac{d^2x}{dt^2} + 2k\frac{dx}{dt} + b^2x = e^{-kt}\sin\omega t$$

$$\Rightarrow \quad (D^2 + 2kD + b^2)x = e^{-kt}\sin\omega t \qquad ...(1)$$

Which is a linear differential equation with constant coefficients.

A.E. is $m^2 + 2km + b^2 = 0 \quad \Rightarrow \quad m = \frac{-2k \pm \sqrt{4k^2 - 4b^2}}{2} = -k \pm \sqrt{k^2 - b^2}$

As the given problem is on vibrations, we must have $k^2 < b^2$

$$\therefore \quad m = -k \pm \sqrt{\{-(b^2 - k^2)\}} = -k \pm i\sqrt{(b^2 - k^2)}$$

$$\therefore \Rightarrow \quad \text{C.F.} = e^{-kt}[C_1\cos\sqrt{b^2 - k^2}\,t + C_2\sin\sqrt{b^2 - k^2}\,t]$$

$$= e^{-kt}A\cos\{\sqrt{b^2 - k^2}\,t + \beta\}$$

where A and B are arbitrary constants.

$$\text{P.I.} = \frac{1}{D^2 + 2kD + b^2}e^{-kt}\sin\omega t = e^{-kt}\frac{1}{(D-k)^2 + 2k(D-k) + b^2}\sin\omega t$$

$$= e^{-kt}\frac{1}{D^2 + (b^2 - k^2)}\sin\omega t = e^{-kt}\frac{1}{-\omega^2 + (b^2 - k^2)}\sin\omega t \quad \text{if } \omega^2 \neq b^2 - k^2$$

When $\omega^2 - (b^2 - k^2) = 0$

$$\text{P.I.} = e^{-kt}\,t\frac{1}{2D}\sin\omega t = e^{-kt}\left(-\frac{t}{2\omega}\cos\omega t\right),$$

Case I. If $w^2 \neq b^2 - k^2$, the complete solution of (1) is

$$x = Ae^{-kt} \cos \{\sqrt{(b^2 - k^2)}\, t + \beta\} + \frac{e^{-kt}}{(b^2 - k^2) - \omega^2} \sin \omega t$$

Case II. If $\omega^2 = b^2 - k^2$, the complete solution of (1) is

$$x = Ae^{-kt} \cos (\sqrt{b^2 - k^2}\, t + \beta) + \frac{-e^{-kt}\, t \cos wt}{2\omega} \qquad (b^2 - k^2 = \omega^2)$$

$$x = Ae^{-kt} \cos (\omega t + \beta) - \frac{e^{-kt}\, t \cos \omega t}{2\omega}$$

Ans.

Example 27. *A spring of negligible weight which stretches 1 inch under tension of 2 lb is fixed at one end and is attached to a weight of w lb at the other. It is found that resonance occurs when an axial periodic force 2 cos 2t lb acts on the weight. Show that when the free vibrations have died out, the forced vibrations are given by x = ct sin 2t and find values of w and c.* *(Uttarakhand, II Semester, June 2007)*

Solution. When a weight of 2 lb is attached to A, spring stretches by $\frac{1}{12}$ *ft.*

Stress = k strain

$$\therefore \quad 2 = k.\frac{1}{12} \quad \Rightarrow \quad k = 24 \text{ lb/ft.}$$

Let B be the position of the weight ω attached to A then,
Stress = k strain

$$\omega = k \times AB \;\Rightarrow\; \omega = 24\, AB \;\Rightarrow\; AB = \frac{\omega}{24} \text{ ft.}$$

At any time t, let the weight T_p be at P where $BP = x$.

A
B
X
T
P
2 cos 2t

Tension at P,
$$T_P = k \times AP = 24\left(\frac{\omega}{24} + x\right) \qquad [AP = AB + BP]$$

$$T_P = \omega + 24x$$

Its equation of motion is

$$\frac{\omega}{g} \cdot \frac{d^2x}{dt^2} = -T + \omega + 2 \cos 2t \qquad \text{[Mass} \times \text{acceleration = Force]}$$

$$= -\omega - 24x + \omega + 2 \cos 2t \qquad [T = \omega + 24x]$$

$$\Rightarrow \quad \omega \frac{d^2x}{dt^2} + 24gx = 2g \cos 2t \qquad \ldots(1)$$

The phenomenon of resonance occurs when the period of free oscillations is equal to the period of forced oscillations,

From (1),
$$\frac{d^2x}{dt^2} + \mu^2 x = \frac{2g}{\omega} \cos 2t \qquad \ldots(2) \left[\text{where } \mu^2 = \frac{24g}{\omega}\right]$$

The period of free oscillations is found as $\frac{2\pi}{\mu}$

and the period of the force $\left(\frac{2g}{\omega}\right) \cos 2t$ is π.

$$\therefore \quad \frac{2\pi}{\mu} = \pi \quad \Rightarrow \quad \mu = 2$$

On putting the value of μ in $\mu^2 = \frac{24g}{\omega}$, we get $4 = \frac{24g}{\omega} \Rightarrow \omega = 6g$

On putting the values of μ and w in (2), we get

$$\frac{d^2x}{dt^2}+4x=\frac{1}{3}\cos 2t \qquad ...(3)\ [\because\ \omega = 6g \text{ and } \mu = 2]$$

We know that the free oscillations are given by the C.F. and the forced oscillations are given by P.I. Thus, when the free oscillations have died out, the forced oscillations are given by the P.I. of (3).

$$\text{P.I.} = \frac{1}{3}\left(\frac{1}{D^2+4}\cos 2t\right)=\frac{1}{3}t.\frac{1}{2D}\cos 2t=\frac{t}{12}\sin 2t$$

Hence, $$c=\frac{1}{12}.$$ **Ans.**

EXERCISE 16.5

1. A mass of 30 kg is attached to a spring for which k = 750 Newton/m and brought to rest. Find the position of the mass at time t if a force equal to 20 sin 2 t is applied to it.

Ans. $x = -\ 0.013 \sin 5t + 0.032 \sin 2t$

2. A body weighing 4.9 kg is hung from a spring. A pull of 10 kg will stretch the spring to 5 cm. The body is pulled down 6 cm below the static equilibrium position and then released. Find the displacement of the body from its equilibrium position at time t seconds, the maximum velocity and the period of oscillation. **Ans.** 0.06 cos 20 t m; 1.2 m/sec; 0.314 sec.

3. A spring is such that it would stretch 72 mm by a mass of 15 kg. A mass of 30 kg is attached and brought to rest. The resistance of the medium is numerically equal to $20\frac{dx}{dt}$ Newtons. Find the equation of the motion of the weight if it is pulled down 140 mm and given an upward velocity 3m/sec.

Ans. $x = e^{-2t}\,(0.14 \cos 8t - 0.34 \sin 8t)$

4. A body weighing 10 kg is hung from a spring. A pull of 20 kg wt. will stretch the spring to 10 cm. The body is pulled down to 20 cm below the static equilibrium position and then released. Find the displacement of the body from its equilibrium position at time t sec, the maximum velocity and the period of oscillation. **Ans.** x = 0.2 cos 14 t, Max. Vel = 2.8 m/sec, period of oscillation = 0.45 sec

5. A particle of mass m is attached to one end of a light spring of modulus λ, the other end being fixed and the spring vertical. Prove that the velocity of the particle when it has traversed a distance 'a' is

$$\sqrt{2ag-\frac{\lambda}{m}a^2}$$

16.11 BEAM

A bar whose length is much greater than its cross-section and its thickness is called a *beam*.

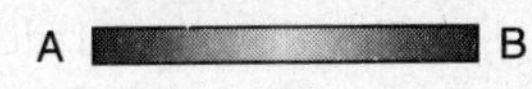

Supported beam. If a beam may just rest on a support like a knife edge is called a *supported beam.*

Fixed beam. If one or both ends of a beam are firmly fixed then it is called *fixed beam.*

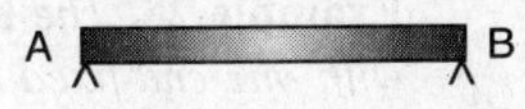

Cantilever. If one end of a beam is fixed and the other end is loaded, it is called a *cantilever.*

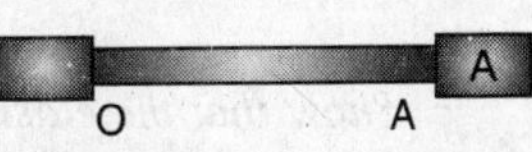

Bending of Beam. Let a beam be fixed at one end and the other end is loaded. Then the upper surface is elongated and therefore under tension and the lower surface is shortened so under compression.

Neutral Surface. In between the lower and upper surface there is a surface which is neither stretched nor compressed. It is known as a *neutral surface.*

Bending Moment. Whenever a beam is loaded it deflects from its original position. If M is the bending moment of the forces acting on it, then

$$M=\frac{EI}{R} \qquad \text{...(1)}$$

where E = Modulus of elasticity of the beam,

I = Moment of inertia of the cross-section of beam about neutral axis.

R = Radius of curvature of the curved beam

$$R=\frac{\left[1+\left(\frac{dy}{dx}\right)^2\right]^{3/2}}{\frac{d^2y}{dx^2}}=\frac{1}{\frac{d^2y}{dx^2}} \qquad \left[\text{neglecting } \frac{dy}{dx}\right]$$

Thus equation (1) becomes $M = EI\,\dfrac{d^2y}{dx^2}$

Boundary Conditions

(*i*) *End freely supported.* At the freely supported end there will be no deflection and no bending moment.

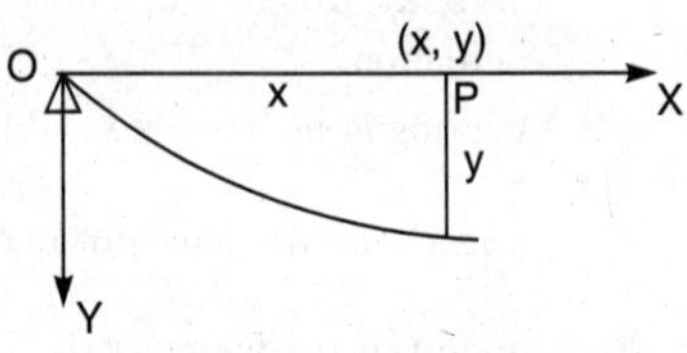

$$y = 0,\quad \frac{d^2y}{dx^2}=0$$

(*ii*) *Fixed end horizontally.* Deflection and slope of the beam are zero.

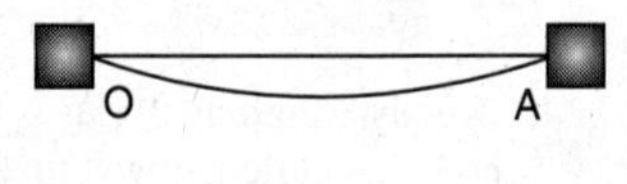

$$y = 0,\quad \frac{dy}{dx}=0$$

(*iii*) *Perfectly free end.* At the free end there is no bending moment or shear force.

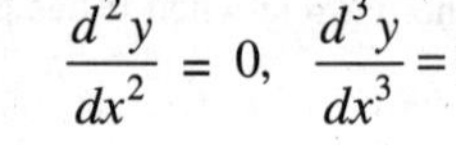

$$\frac{d^2y}{dx^2}=0,\quad \frac{d^3y}{dx^3}=0$$

Convention of signs

The sign of the moment about NN' on the left NN' is positive if anticlockwise and negative if clockwise.

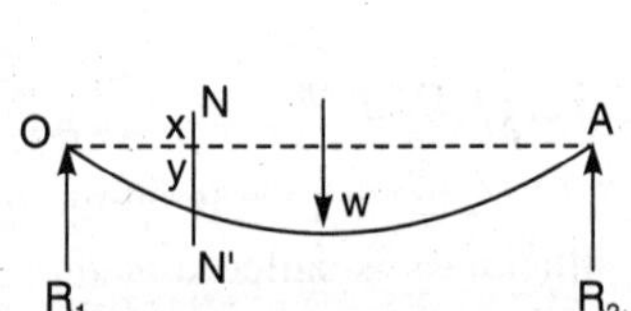

The downward deflection is positive and length x on right-side is also positive. Slope $\dfrac{dy}{dx}$ is positive if downward.

Example 28. *The differential equation satisfied by a beam uniformly loaded (W kg/metre), with one end fixed and the second end subjected to tensile force P, is given by*

$$E.I.\frac{d^2y}{dx^2}=Py-\frac{1}{2}Wx^2$$

Show that the elastic curve for the beam with conditions

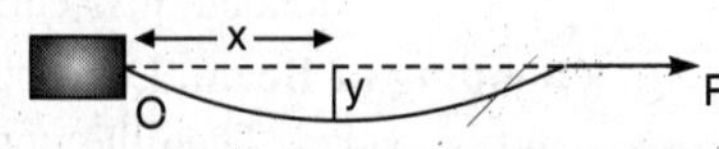

$$y = 0=\frac{dy}{dx} \text{ at } x=0. \textit{ is given by}$$

$$y=\frac{W}{Pn^2}\;- \cosh nx)+\frac{Wx^2}{2P} \quad \textit{where } n^2=\frac{P}{EI}$$

Solution. We have, $E.I.\dfrac{d^2y}{dx^2} = Py - \dfrac{1}{2}W.x^2$ ...(1)

$$\Rightarrow \quad \frac{d^2y}{dx^2} - \frac{P}{E.I.}y = -\frac{W}{2\,E.I.}x^2 \quad \Rightarrow \quad \left(D^2 - \frac{P}{E.I.}\right)y = -\frac{W}{2\,E.I.}x^2$$

A.E. is $m^2 - \dfrac{P}{E.I.} = 0 \Rightarrow m^2 = \dfrac{P}{E.I.} = n^2 \Rightarrow m = \pm n$ $\quad \left(n^2 = \dfrac{P}{EI}\right)$

C.F. $= c_1\,e^{nx} + c_2\,e^{-nx}$

$$\text{P.I.} = \frac{1}{D^2 - \frac{P}{E.I.}}\left(-\frac{W}{2\,E.I.}\right)x^2 = -\frac{W}{2\,E.I.}\frac{1}{D^2 - n^2}.x^2$$

$$= \frac{W}{2n^2.\,E.I.}\left(1 - \frac{D^2}{n^2}\right)^{-1}.x^2 = \frac{W}{2n^2\,E.I.}\left(1 + \frac{D^2}{n^2}\right).x^2 = \frac{W}{2n^2\,E.I.}\left(x^2 + \frac{2}{n^2}\right)$$

$$\therefore \quad y = c_1\,e^{nx} + c_2 e^{-nx} + \frac{W}{2n^2\,E.I.}\left(x^2 + \frac{2}{n^2}\right) \qquad ...(2)$$

Differentiating (2) w.r.t. x, we get

$$\frac{dy}{dx} = n c_1\,e^{nx} - n c_2\,e^{-nx} + \frac{W}{2n^2\,E.I.}(2x) \qquad ...(3)$$

Putting $x = 0$, $\dfrac{dy}{dx} = 0$ in (3), we get

$$0 = n\,c_1 - n\,c_2 \quad \Rightarrow \quad c_1 = c_2$$

Putting $x = 0$, $y = 0$ in (2), we get

$$0 = c_1 + c_2 + \frac{W}{2n^2\,E.I.}\,\frac{2}{n^2} \quad \Rightarrow \quad 0 = c_1 + c_2 + \frac{W}{n^4\,E.I.} \qquad ...(4)$$

Putting $c_1 = c_2$ in (4), we get $0 = 2c_1 + \dfrac{W}{n^4\,E.I.} \Rightarrow c_1 = -\dfrac{W}{2n^4\,E.I.}$,

Now, $\quad n^2 = \dfrac{P}{E.I.} \Rightarrow n^2\,E.I. = P$

$$\Rightarrow \quad c_1 = c_2 = -\frac{W}{2n^2 P}$$

Putting the values of c_1 and c_2 in (2), we get

$$y = \frac{-W}{2n^2 P}(e^{nx} + e^{-nx}) + \frac{W}{2P}\left(x^2 + \frac{2}{n^2}\right)$$

$$y = \frac{-W}{n^2 P}\cosh nx + \frac{W}{2P}x^2 + \frac{W}{Pn^2}$$

$$y = \frac{W}{P.n^2}(1 - \cosh nx) + \frac{W\,x^2}{2P}$$ **Ans.**

EXERCISE 16.6

1. A beam of length l and of uniform cross-section has the differential equation of its elastic curve as

$$E.I.\frac{d^2y}{dx^2}=\frac{w}{2}\left(\frac{l^2}{4}-x^2\right)$$

where E is the modulus of elasticity, I is the moment of inertia of the cross-section, w is weight per unit length and x is measured from the centre of span.

If at $x=0, \frac{dy}{dx}=0$. Prove that the equation of the elastic curve is

$$y=\frac{1}{2}\cdot\frac{2}{E.I.}\left(\frac{l^3.x^2}{8}-\frac{x^4}{12}\right)+\frac{5w.l^4}{384\,E.I.}$$

2. A horizontal tie rod of length l is freely pinned at each end. It carries a uniform load w kg per unit length and has a horizontal pull P. Find the central deflection and the maximum bending moment, taking the origin at one of its ends. **Ans.** $\frac{w}{a}\left(\sec h\frac{al}{2}-1\right)$ where $a^2=\frac{P}{EI}$

3. A light horizontal strut AB is freely pinned at A and B. It is under the action of equal and opposite compressive forces P at its ends and it carries a load W at its centre. Then for $0<x<\frac{l}{2}$,

$$EI\frac{d^2y}{dx^2}+Py+\frac{1}{2}Wx=0$$

Also $y=0$ at $x=0$ and $\frac{dy}{dx}=0$ at $x=\frac{l}{2}$. Prove that $y=\frac{W}{2P}\left(\frac{\sin ax}{a\cos\frac{al}{2}}-x\right)$, where $a^2=\frac{P}{EI}$

4. A horizontal tie-rod of length $2l$ with concentrated load W at its centre and ends freely hinged satisfies the differential equation $EI\frac{d^2y}{dx^2}=Py-\frac{W}{2}x$. With conditions $x=0$, $y=0$ and $x=l, \frac{dy}{dx}=0$. Prove that the deflection and bending moment M at the centre ($x=l$) are given by $\delta=\frac{W}{2Pn}(nl-\tan nl)$ and $M=-\frac{W}{2n}\tan h\,nl$, where $n^2\,EI=P$.

16.12 PROJECTILE

Example 29. *A particle is projected with velocity u making an angle α with the horizontal. Neglecting air resistance, show that the equation of its path is the parabola.*

$$y=x\tan\alpha-\frac{gx^2}{2u^2\cos^2\alpha}.$$

Find the time of flight, the greatest height attained and the range on the horizontal plane.

Solution. If a particle of mass m is projected from 0 with velocity u in a direction making an angle α with the horizontal.

Let horizontal line and vertical line be taken as x-axis and y-axis. Let $P(x, y)$ be the position of the particle at time t.

Horizontal component of $u = u\cos\alpha$

Vertical component of $u = u\sin\alpha$

The force acting on the particle = weight = mg

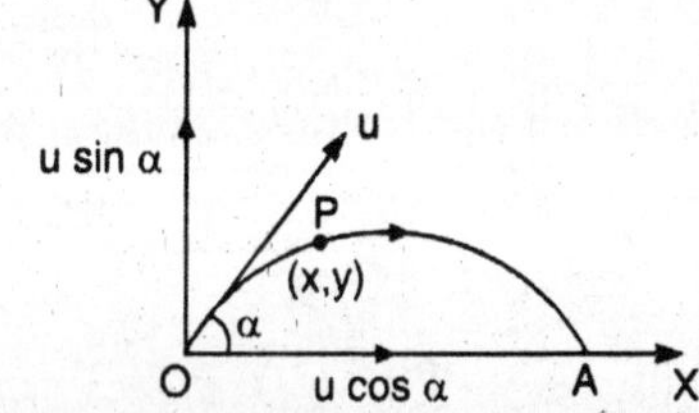

The equations of motion are

Horizontal components: $m\dfrac{d^2x}{dt^2} = 0 \quad \Rightarrow \quad \dfrac{d^2x}{dt^2} = 0$ ...(1)

Vertical components: $m\dfrac{d^2y}{dx^2} = -mg \quad \Rightarrow \quad \dfrac{d^2y}{dx^2} = -g$ ...(2)

Integrating (1), we have $\dfrac{dx}{dt} = c_1$ ...(3)

Initially $\dfrac{dx}{dt} = u\cos\alpha$, time $t = 0$, putting in (3), we get

$$u\cos\alpha = c_1$$

On putting the value of c_1 in (3), we get

$$\frac{dx}{dt} = u\cos\alpha \quad ...(4)$$

Integrating (4), we have $x = (u\cos\alpha)\,t + c_2$ ...(5)

Putting (initial condition) $t = 0$, $\alpha = 0$, we get $c_2 = 0$

(5) becomes, $x = (u\cos\alpha)\,t$ ...(6)

Integrating (2), we have $\dfrac{dy}{dt} = -gt + c_3$ ...(7)

Putting (initial condition) $t = 0$, $\dfrac{dy}{dt} = u\sin\alpha$ in (7), we get

$$u\sin\alpha = c_3$$

(7) becomes $\dfrac{dy}{dt} = -gt + u\sin\alpha$ ...(8)

Integrating (8), we get $y = -g\dfrac{t^2}{2} + (u\sin\alpha)\,t + c_4$ ...(9)

Putting (initial condition), $t = 0$, $y = 0$, we obtain $0 = c_4$

(9) becomes, $y = -\dfrac{gt^2}{2} + (u\sin\alpha)\,t$ ...(10)

Equation (6) and (10) give the position of the particle at any time t. The equation of the path described by the particle is obtained by eliminating t, between the equations (6) and (10),

From (6), $t = \dfrac{x}{u\cos\alpha}$

Substituting this value of t in (10), we get

$$y = (u\sin\alpha)\frac{x}{u\cos\alpha} - \frac{g}{2}\left(\frac{x^2}{u^2\cos^2\alpha}\right)$$

$$y = x\tan\alpha - \frac{gx^2}{2u^2\cos^2\alpha}$$

which is the equation of the projectile, the path is parabola.

Time of flight. At the point A, $y = 0$

On putting $y = 0$ in (10), we get

$$0 = -\frac{gt^2}{2} + (u \sin \alpha)t \quad \Rightarrow \quad t = \frac{2u \sin \alpha}{g}$$

Greatest Height

At the highest point, the vertical component becomes zero, *i.e.*, $\frac{dy}{dt} = 0$

Putting $\frac{dy}{dt} = 0$ in (8), we get $0 = -gt + u \sin \alpha \quad \Rightarrow \quad t = \frac{u \sin \alpha}{g}$

For the highest point, we substitute the value of t in (10), we get

$$y = -\frac{g}{2}\left(\frac{u \sin \alpha}{g}\right)^2 + u \sin \alpha\left(\frac{u \sin \alpha}{g}\right) = -\frac{u^2 \sin^2 \alpha}{2g} + \frac{u^2 \sin^2 \alpha}{g} = \frac{u^2 \sin^2 \alpha}{2g}$$

Range, *i.e.*, OA is the horizontal distance covered during flight.

Putting the value of $t = \frac{2u \sin \alpha}{g}$ in (6), we get

$$\text{Range} = (u \cos \alpha)\left(\frac{2u \sin \alpha}{g}\right) = \frac{u^2 \sin 2\alpha}{g}$$

$$\text{Maximum range} = \frac{u^2}{g} \text{ if } \sin 2\alpha = 1 \quad \Rightarrow \quad \alpha = \frac{\pi}{4}$$ **Ans.**

Example 30. *The equation of motion under certain conditions are*

$$m\frac{d^2x}{dt^2} + eh\frac{dy}{dt} = eE \qquad \text{...(1)}$$

$$m\frac{d^2y}{dt^2} - eh\frac{dx}{dt} = 0 \qquad \text{..(2)}$$

With condition $x = \frac{dx}{dt} = y = \frac{dy}{dt} = 0$ *when* $t = 0$, *find the path of electron.*

Solution. Multiplying (2) by k and adding to (1), we get

$$m\frac{d^2x}{dt^2} + mk\frac{d^2y}{dt^2} + eh\frac{dy}{dt} - ehk\frac{dx}{dt} = eE$$

$$\Rightarrow \quad m\frac{d^2}{dt^2}(x + ky) - ehk\frac{d}{dt}\left(-\frac{y}{k} + x\right) = e.\ E \qquad \text{...(3)}$$

Let us choose k such that $x + ky = x - \frac{y}{k}$

$$\Rightarrow \quad k = -\frac{1}{k} \quad \Rightarrow \quad k^2 = -1 \Rightarrow k = \pm i$$

Putting $x + ky = u$ in (3), we have

$$m\frac{d^2u}{dt^2} - ehk\frac{du}{dt} = eE \quad \Rightarrow \quad \frac{d^2u}{dt^2} - wk\frac{du}{dt} = \frac{eE}{m} \qquad \left(w = \frac{eh}{m}\right)$$

$$\Rightarrow \quad D^2 u - wk\, Du = \frac{eE}{m}$$

A.E. is $m^2 - wkm = 0 \quad \Rightarrow \quad m(m - wk) = 0 \quad \Rightarrow \quad m = 0$ or $m = wk$

$$\text{C.F.} = c_1 + c_2\, e^{wkt}$$

$$\text{P.I.} = \frac{1}{D^2 - wkD}\frac{eE}{m} = \frac{eE}{m}\frac{1}{D^2 - wkD}e^{0t}$$

$$= \frac{eEt}{m}\frac{1}{2D - wk}e^{0t} = \frac{wEt}{h}\frac{1}{0 - wk} = -\frac{Et}{hk} \qquad \left(w = \frac{eh}{m}\right)$$

The complete solution is

$$u = c_1 + c_2\, e^{wkt} - \frac{Et}{h.k} \text{ or } x + ky = c_1 + c_2\, e^{wkt} - \frac{Et}{hk} \qquad ...(4)$$

Putting the value of k, *i.e.*, $i, -i$ in (4), we get

$$x + iy = c_1 + c_2\, e^{iwt} - \frac{Et}{ih} \qquad ...(5)$$

$$x - iy = c_3 + c_4\, e^{-iwt} + \frac{Et}{ih} \qquad ...(6)$$

Differentiating (5) and (6), we get

$$\frac{dx}{dt} + i\frac{dy}{dt} = c_2\, iwe^{iwt} + \frac{iE}{h} \qquad ...(7)$$

$$\frac{dx}{dt} - i\frac{dy}{dt} = -iwc_4\, e^{-iwt} - \frac{iE}{h} \qquad ...(8)$$

Initial conditions $\qquad x = y = \dfrac{dx}{dt} = \dfrac{dy}{dt} = 0$ when $t = 0$

Putting these values in (5), (6), (7) and (8), we get

$$0 = c_1 + c_2 \Rightarrow c_2 = -c_1$$

$$0 = c_3 + c_4 \Rightarrow c_4 = -c_3$$

$$0 = iwc_2 + \frac{iE}{h} \Rightarrow c_2 = -\frac{E}{wh}$$

$$0 = -iwc_4 - \frac{iE}{h} \Rightarrow c_4 = -\frac{E}{wh}$$

On substituting the values of c_1, c_2, c_3 and c_4 in (5) and (6), we get

$$x + iy = \frac{E}{wh} - \frac{E}{hw}e^{iwt} + i\frac{Et}{h} \qquad ...(9)$$

$$x - iy = \frac{E}{wh} - \frac{E}{wh}e^{-iwt} - i\frac{Et}{h} \qquad ...(10)$$

On adding (9) and (10), we get

$$2x = \frac{2E}{hw} - \frac{E}{hw}(e^{iwt} + e^{-iwt})$$

$$x = \frac{E}{hw} - \frac{E}{hw}\cos wt$$

$$x = \frac{E}{hw}(1 - \cos wt)$$

Subtracting (10) from (9), we obtain

$$2iy = -\frac{E}{hw}(e^{iwt} - e^{-iwt}) + \frac{2iEt}{h}$$

$$\Rightarrow \qquad y = -\frac{E}{wh}\left(\frac{e^{iwt} - e^{-iwt}}{2i}\right) + \frac{Et}{h} = -\frac{E}{wh}\sin wt + \frac{Et}{h} \Rightarrow y = \frac{E}{hw}(wt - \sin wt) \textbf{ Ans.}$$

Example 31. *Assuming that a spherical rain drop evaporates at rate proportional to its surface area and if its radius originally is 3 mm and one hour later has been reduced to 2 mm, find an expression for the radius of the rain drop at any time.*

Solution. Evaporation $\propto$ surface area

$$\frac{dV}{dt} \propto S \quad \Rightarrow \quad \frac{dV}{dt} = kS \qquad \ldots(1)$$

$$V = \frac{4}{3}\pi r^3 \quad \Rightarrow \quad \frac{dV}{dt} = 4\pi r^2 \frac{dr}{dt} \qquad \ldots(2)$$

Putting the value of $\frac{dV}{dt}$ from (1) in (2), we have

$$4\pi r^2 \frac{dr}{dt} = kS \Rightarrow S\frac{dr}{dt} = kS \quad \Rightarrow \quad \frac{dr}{dt} = k$$

$$r = kt + c \qquad \ldots(3) \; [S = 4\pi r^2]$$

Putting $t = 0$, $r = 3$ in (3), we get $3 = c$

(3) becomes $\quad r = kt + 3 \qquad \ldots(4)$

Putting $t = 1$ and $r = 2$ in (4), we get $2 = k + 3 \;\Rightarrow\; k = -1$

(4) becomes $\quad r = -t + 3$ **Ans.**

EXERCISE 16.7

1. The current i_1 and i_2 in mesh are given by the differential equations

$$\frac{di_1}{dt} - wi_2 = a \cos pt, \frac{di_2}{dt} + wi_1 = a \sin pt$$

Find the currents i_1 and i_2 if $i_1 = i_2 = 0$ at $t = 0$. **Ans.** $i_1 = \frac{a}{P+w} \sin pt, i_2 = \frac{a}{P+w} \cos pt$

2. A particle moving in a plane is subjected to a force directed towards a fixed point O and proportional to the distance of the particle from O. Show that the differential equations of motion are of the form $\frac{d^2x}{dt^2} = -k^2 x, \frac{d^2y}{dt^2} = -k^2 y.$ Find the cartesian equation of the path of the particle if $x = 1$, $y = 0$, $\frac{dx}{dt} = 0$ and $\frac{dy}{dt} = 2$ when $t = 0$. **Ans.** $4x^2 + k^2 y^2 = 4$

3. A projectile of mass m is fired into the air with initial velocity v_0 at an angle θ with the ground. Neglecting all forces except gravity and the resistance of air, the assumed proportional to velocity. Find the position of the projectile at time t.

$$x = \frac{1}{k}(v_0 \cos \theta)(1 - e^{kt}), \quad y = \frac{1}{k}\left\{\frac{g}{k} + v_0 \sin \theta\,(1 - e^{-kt}) - gt\right\}$$

4. A particle of unit mass is projected with velocity u at an inclination α above the horizon in a medium whose resistance is k times the velocity. Show that its direction will again make an angle α with the horizon after a time $\frac{1}{k}\log\left\{1 + \frac{2ku}{g}\sin \alpha\right\}$.

5. An inclined plane makes angle α with the horizontal. A projectile is launched from the bottom of the inclined plane with speed v in a direction making an angle β with the horizontal. Set up the differential equations and find

(*i*) the range on the incline (*ii*) the maximum range up the incline.

Ans. (*i*) $\frac{2u^2 \sin(\alpha - \beta)\cos \alpha}{g \cos^2 \beta}$ (*ii*) $u^2 g(1 + \sin \beta)$

DETERMINANTS

17.1 INTRODUCTION

With the help of determinants, we can solve a system of simultaneous equations by Cramer's Rule. Determinants are also used in calculating inverse of a square matrix.

17.2 DETERMINANT AS ELIMINANT

Consider the following three equations having three unknowns, x, y and z.

$$a_1 x + b_1 y + c_1 z = 0 \quad \text{...(1)}$$
$$a_2 x + b_2 y + c_2 z = 0 \quad \text{...(2)}$$
$$a_3 x + b_3 y + c_3 z = 0 \quad \text{...(3)}$$

From (2) and (3) by cross-multiplication; we get

$$\frac{x}{b_2 c_3 - b_3 c_2} = \frac{y}{a_3 c_2 - a_2 c_3} = \frac{z}{a_2 b_3 - a_3 b_2} = k \text{ (say)}$$

$$x = (b_2 c_3 - b_3 c_2)\, k$$
$$y = (a_3 c_2 - a_2 c_3)\, k$$

and $\quad z = (a_2 b_3 - a_3 b_2)\, k$

Substituting the values of x, y and z in (1), we get the eliminant

$$a_1 (b_2c_3 - b_3c_2)\, k + b_1 (a_3c_2 - a_2c_3)\, k + c_1 (a_2b_3 - a_3b_2)\, k = 0$$

or $\quad a_1 (b_2c_3 - b_3c_2) - b_1 (a_2c_3 - a_3c_2) + c_1 (a_2b_3 - a_3b_2) = 0 \quad$ ...(4)

From (1), (2) and (3) by suppressing x, y, z the remaining can be written in the determinant as

$$\begin{vmatrix} a_1 & b_1 & c_1 \\ a_2 & b_2 & c_2 \\ a_3 & b_3 & c_3 \end{vmatrix} = 0 \quad \text{...(5)}$$

This is the determinant of third order.

As (4) and (5) both are the eliminant of the same equations.

$$\therefore \quad \begin{vmatrix} a_1 & b_1 & c_1 \\ a_2 & b_2 & c_2 \\ a_3 & b_3 & c_3 \end{vmatrix} = a_1 (b_2c_3 - b_3c_2) - b_1 (a_2c_3 - a_3c_2) + c_1 (a_2b_3 - a_3b_2) = 0$$

or $$\begin{vmatrix} a_1 & b_1 & c_1 \\ a_2 & b_2 & c_2 \\ a_3 & b_3 & c_3 \end{vmatrix} = a_1 \begin{vmatrix} b_2 & c_2 \\ b_3 & c_3 \end{vmatrix} - b_1 \begin{vmatrix} a_2 & c_2 \\ a_3 & c_3 \end{vmatrix} + c_1 \begin{vmatrix} a_2 & b_2 \\ a_3 & b_3 \end{vmatrix}$$

17.3. MINOR

The minor of an element is defined as a determinant obtained by deleting the row and column containing the element.

Thus the minors of a_1, b_1 and c_1 are respectively.

$$\begin{vmatrix} b_2 & c_2 \\ b_3 & c_3 \end{vmatrix}, \begin{vmatrix} a_2 & c_2 \\ a_3 & c_3 \end{vmatrix} \text{ and } \begin{vmatrix} a_2 & b_2 \\ a_3 & b_3 \end{vmatrix}$$

Thus $\begin{vmatrix} a_1 & b_1 & c_1 \\ a_2 & b_2 & c_2 \\ a_3 & b_3 & c_3 \end{vmatrix} = a_1$ (minor of a_1) $- b_1$ (minor of b_1) $+ c_1$ (minor of c_1).

17.4. COFACTOR

Cofactor = $(-1)^{r+c}$ Minor

where r is the number of rows of the element and c is the number of columns of the element.

The cofactor of any element of ith row and jth column is

$(-1)^{i+j}$ minor

Thus the cofactor of $a_1 = (-1)^{1+1}(b_2c_3 - b_3c_2) = +(b_2c_3 - b_3c_2)$

The cofactor of $b_1 = (-1)^{1+2}(a_2c_3 - a_3c_2) = -(a_2c_3 - a_3c_2)$

The cofactor of $c_1 = (-1)^{1+3}(a_2b_3 - a_3b_2) = +(a_2b_3 - a_3b_2)$

The determinant $= a_1$ (cofactor of a_1) $+ a_2$ (cofactor of a_2) $+ a_3$ (cofactor of a_3).

Example 1. *Write down the minors and co-factors of each element and also evaluate the determinant:*

$$\begin{vmatrix} 1 & 3 & -2 \\ 4 & -5 & 6 \\ 3 & 5 & 2 \end{vmatrix}$$

Solution. M_{11} = Minor of element (1) $= \begin{vmatrix} 1 & 3 & -2 \\ 4 & -5 & 6 \\ 3 & 5 & 2 \end{vmatrix}$

By eleminating the row and column of (1), the remaining is minor of (1)

$$= \begin{vmatrix} -5 & 6 \\ 5 & 2 \end{vmatrix} = (-5) \times 2 - (6 \times 5) = -10 - 30 = -40$$

Cofactor of element (1) $= A_{11} = (-1)^{1+1} M_{11} = (-1)^2(-40) = -40$

M_{12} = Minor of element (3)

By eleminating the row and column of (3), we get

$$= \begin{vmatrix} 1 & 3 & -2 \\ 4 & -5 & 6 \\ 3 & 5 & 2 \end{vmatrix} = \begin{vmatrix} 4 & 6 \\ 3 & 2 \end{vmatrix} = (4 \times 2) - (3 \times 6) = 8 - 18 = -10$$

$\Rightarrow$ Cofactor of element (–2) $= A_{12} = (-1)^{1+2}(-10) = 10$

M_{13} = Minor of element (– 2)

$$= \begin{vmatrix} 1 & 3 & -2 \\ 4 & -5 & 6 \\ 3 & 5 & 2 \end{vmatrix} = \begin{vmatrix} 4 & -5 \\ 3 & 5 \end{vmatrix} = (4 \times 5) - (-5) \times 3 = 20 + 15 = 35$$

$\Rightarrow$ Cofactor of element (– 2) = $A_{13} = (-1)^{1+3} M_{13} = (-1)^4\, 35 = 35$

M_{21} = Minor of element (4)

$$= \begin{vmatrix} 1 & 3 & -2 \\ 4 & -5 & 6 \\ 3 & 5 & 2 \end{vmatrix} = \begin{vmatrix} 3 & -2 \\ 5 & 2 \end{vmatrix} = (3\times 2)-(-2)\times 5 = 6+10 = 16$$

$\Rightarrow$ Cofactor of element (4) = $A_{21} = (-1)^{2+1} M_{21} = (-1)^{2+1}(16) = -16$

M_{22} = Minor of element (– 5)

$$= \begin{vmatrix} 1 & 3 & -2 \\ 4 & -5 & 6 \\ 3 & 5 & 2 \end{vmatrix} = \begin{vmatrix} 1 & -2 \\ 3 & 2 \end{vmatrix} = (1\times 2)-(-2)\times 3 = 2+6 = 8$$

$\Rightarrow$ Cofactor of element (– 5) = $A_{22} = (-1)^{2+2} M_{22} = (-1)^{2+2}(8) = 8$

M_{23} = Minor of element (6)

$$= \begin{vmatrix} 1 & 3 & -2 \\ 4 & -5 & 6 \\ 3 & 5 & 2 \end{vmatrix} = \begin{vmatrix} 1 & 3 \\ 3 & 5 \end{vmatrix} = (1\times 5)-3\times 3 = 5-9 = -4$$

$\Rightarrow$ Cofactor of element (6) = $A_{23} = (-1)^{2+3} M_{23} = (-1)^{2+3}(-4) = 4$

M_{31} = Minor of element (3)

$$= \begin{vmatrix} 1 & 3 & -2 \\ 4 & -5 & 6 \\ 3 & 5 & 2 \end{vmatrix} = \begin{vmatrix} 3 & -2 \\ -5 & 6 \end{vmatrix} = (3\times 6)-(-2)\times(-5) = 18-10 = 8$$

$\Rightarrow$ Cofactor of element (3) = $A_{31} = (-1)^{3+1} M_{31} = (-1)^{3+1}\, 8 = 8$

M_{32} = Minor of element (5)

$$= \begin{vmatrix} 1 & 3 & -2 \\ 4 & -5 & 6 \\ 3 & 5 & 2 \end{vmatrix} = \begin{vmatrix} 1 & -2 \\ 4 & 6 \end{vmatrix} = (1\times 6)-(-2)\times 4 = 6+8 = 14$$

$\Rightarrow$ Cofactor of element (5) = $A_{32} = (-1)^{3+2} M_{32} = (-1)^{3+2}\, 14 = -14$

M_{33} = Minor of element (2)

$$= \begin{vmatrix} 1 & 3 & -2 \\ 4 & -5 & 6 \\ 3 & 5 & 2 \end{vmatrix} = \begin{vmatrix} 1 & 3 \\ 4 & -5 \end{vmatrix} = 1\times(-5)-(4\times 3) = -5-12 = -17$$

Cofactor of element (2) = $A_{33} = (-1)^{3+3} M_{33} = (-1)^{3+3}(-17) = -17.$

$$\begin{vmatrix} 1 & 3 & -2 \\ 4 & -5 & 6 \\ 3 & 5 & 2 \end{vmatrix} = 1 \times (\text{cofactor of } 1) + 3 \times (\text{cofactor of } 3) + (-2) \times [\text{cofactor of } (-2)].$$

$= 1 \times (-40) + 3 \times (10) + (-2) \times (35) = -40 + 30 - 70 = -80$ **Ans.**

Example 2. *Evaluate the determinants* :

$$\begin{vmatrix} 3 & -1 & -2 \\ 0 & 0 & -1 \\ 3 & -5 & 0 \end{vmatrix}$$

Solution. We have, two zero entries in the second row. So, expanding along 2nd row:

$$\begin{vmatrix} 3 & -1 & -2 \\ 0 & 0 & -1 \\ 3 & -5 & 0 \end{vmatrix} = -0\begin{vmatrix} -1 & -2 \\ -5 & 0 \end{vmatrix} + 0\begin{vmatrix} 3 & -2 \\ 3 & 0 \end{vmatrix} - (-1)\begin{vmatrix} 3 & -1 \\ 3 & -5 \end{vmatrix}$$

$= -0 + 0 + 1(-15 + 3) = -12$ **Ans.**

Example 3. *Prove that the determinant* $\begin{vmatrix} x & \sin\theta & \cos\theta \\ -\sin\theta & -x & 1 \\ \cos\theta & 1 & x \end{vmatrix}$ *is independent of* θ.

Solution. We have,

$$\begin{vmatrix} x & \sin\theta & \cos\theta \\ -\sin\theta & -x & 1 \\ \cos\theta & 1 & x \end{vmatrix} = x\begin{vmatrix} -x & 1 \\ 1 & x \end{vmatrix} - \sin\theta\begin{vmatrix} -\sin\theta & 1 \\ \cos\theta & x \end{vmatrix} + \cos\theta\begin{vmatrix} -\sin\theta & -x \\ \cos\theta & 1 \end{vmatrix}$$

$$= x(-x^2-1) - \sin\theta\,(-x\sin\theta - \cos\theta) \; + \cos\theta\,(-\sin\theta + x\cos\theta)$$

$$= -x^3 - x + x\sin^2\theta + \sin\theta\cos\theta - \sin\theta\cos\theta + x\cos^2\theta$$

$$= -x^3 - x + x\,(\sin^2\theta + \cos^2\theta) \; = -x^3 - x + x$$

Thus, the determinant is independent of θ. **Proved.**

Example 4. *Evaluate the determinant* $\begin{vmatrix} 1 & 0 & 4 \\ 3 & 5 & -1 \\ 0 & 1 & 2 \end{vmatrix}$.

(*i*) *With the help of second row,* (*ii*) *with the help of third column.*

Solution.

(*i*) $\begin{vmatrix} 1 & 0 & 4 \\ 3 & 5 & -1 \\ 0 & 1 & 2 \end{vmatrix}$ = 3 × (cofactor of 3) + 5 × (cofactor of 5) + (– 1) (cofactor of – 1).

$$= 3 \times (-1)^{2+1}\begin{vmatrix} 0 & 4 \\ 1 & 2 \end{vmatrix} + 5 \times (-1)^{2+2}\begin{vmatrix} 1 & 4 \\ 0 & 2 \end{vmatrix} + (-1) \times (-1)^{2+3}\begin{vmatrix} 1 & 0 \\ 0 & 1 \end{vmatrix}$$

$$= -3 \times (0-4) + 5\,(2-0) + (1-0) = 12 + 10 + 1 = 23$$ **Ans.**

(*ii*) $\begin{vmatrix} 1 & 0 & 4 \\ 3 & 5 & -1 \\ 0 & 1 & 2 \end{vmatrix}$ = 4 × (cofactor of 4) + (– 1) (cofactor of (– 1)) + 2 × (cofactor of 2)

$$= 4 \times (-1)^{1+3}\begin{vmatrix} 3 & 5 \\ 0 & 1 \end{vmatrix} + (-1)(-1)^{2+3}\begin{vmatrix} 1 & 0 \\ 0 & 1 \end{vmatrix} + 2 \times (-1)^{3+3}\begin{vmatrix} 1 & 0 \\ 3 & 5 \end{vmatrix}$$

$$= 4 \times (3-0) + (1-0) + 2\,(5-0) = 12 + 1 + 10 = 23$$ **Ans.**

EXERCISE 17.1

Write the minors and co-factors of each element of the following determinants and also evaluate the determinant in each case :

1. $\begin{vmatrix} 42 & 1 & 6 \\ 28 & 7 & 4 \\ 14 & 3 & 2 \end{vmatrix}$

Ans. $M_{11} = 2$, $M_{12} = 0$, $M_{13} = -14$, $M_{21} = -16$, $M_{22} = 0$
$M_{23} = 112$, $M_{31} = -38$, $M_{32} = 0$, $M_{33} = 266$
$A_{11} = 2$, $A_{12} = 0$, $A_{13} = -14$, $A_{21} = 16$, $A_{22} = 0$
$A_{23} = -112$, $A_{31} = -38$, $A_{32} = 0$, $A_{33} = 266$ $|A| = 0$

2. $\begin{vmatrix} 1 & a & bc \\ 1 & b & ca \\ 1 & c & ab \end{vmatrix}$

Ans. $M_{11} = (ab^2 - ac^2)$, $M_{12} = (ab - ac)$, $M_{13} = (c - b)$, $M_{21} = a^2b - bc^2$

$M_{22}=(ab-bc)$, $M_{23}=(c-a)$, $M_{31}=(ca^2-cb^2)$, $M_{32}=ca-bc$, $M_{33}=(b-a)$,
$A_{11}=(ab^2-ac^2)$, $A_{12}=(ac-ab)$, $A_{13}=(c-b)$, $A_{21}=bc^2-a^2b$
$A_{22}=(ab-bc)$, $A_{23}=(a-c)$, $A_{31}=(ca^2-cb^2)$, $A_{32}=(bc-ca)$, $A_{33}=(b-a)$
$|A|=(a-b)(b-c)(c-a)$.

Expand the following determinants :

3. $\begin{vmatrix} 2 & -3 & 4 \\ 5 & 1 & -6 \\ -7 & 8 & -9 \end{vmatrix}$ **Ans.** $|A|=5$

4. $\begin{vmatrix} 5 & 0 & 7 \\ 8 & -6 & -4 \\ 2 & 3 & 9 \end{vmatrix}$ **Ans.** $|A|=42$

5. $\begin{vmatrix} a & h & g \\ h & b & f \\ g & f & c \end{vmatrix}$ **Ans.** $|A|=abc+2fgh-af^2-bg^2-ch^2$

Expand the following determinants by two methods :

(*i*) along the-third row. (*ii*) along the-third column.

6. $\begin{vmatrix} 1 & -3 & 2 \\ 4 & -1 & 2 \\ 3 & 5 & 2 \end{vmatrix}$ **Ans.** $|A|=40$

7. $\begin{vmatrix} 3 & -2 & 4 \\ 1 & 2 & 1 \\ 0 & 1 & -1 \end{vmatrix}$ **Ans.** $|A|=-7$

8. $\begin{vmatrix} 2 & 3 & -2 \\ 1 & 2 & 3 \\ -2 & 1 & -3 \end{vmatrix}$ **Ans.** $|A|=-37$

17.5. PROPERTIES OF DETERMINANTS

Property (*i*). *The value of a determinant remains unaltered; if the rows are interchanged into columns (or the columns into rows).*

Verification. Let $\Delta=\begin{vmatrix} a_1 & a_2 & a_3 \\ b_1 & b_2 & b_3 \\ c_1 & c_2 & c_3 \end{vmatrix}$

Expanding along first row, we get

$$\Delta=a_1\begin{vmatrix} b_2 & b_3 \\ c_2 & c_3 \end{vmatrix}-a_2\begin{vmatrix} b_1 & b_3 \\ c_1 & c_3 \end{vmatrix}+a_3\begin{vmatrix} b_1 & b_2 \\ c_1 & c_2 \end{vmatrix}$$

$$=a_1(b_2c_3-b_3c_2)-a_2(b_1c_3-b_3c_1)+a_3(b_1c_2-b_2c_1) \quad \text{...(1)}$$

By interchanging the rows and columns of Δ, we get the determinant

$$\Delta_1=\begin{vmatrix} a_1 & b_1 & c_1 \\ a_2 & b_2 & c_2 \\ a_3 & b_3 & c_3 \end{vmatrix}$$

$$=a_1(b_2c_3-b_3c_2)-b_1(a_2c_3-a_3c_2)+c_1(a_2b_3-a_3b_2)$$

$$=a_1b_2c_3-a_1b_3c_2-a_2b_1c_3+a_3b_1c_2+a_2b_3c_1-a_3b_2c_1$$

$$=(a_1b_2c_3-a_1b_3c_2)-(a_2b_1c_3-a_2b_3c_1)+(a_3b_1c_2-a_3b_2c_1)$$

$$=a_1(b_2c_3-b_3c_2)-a_2(b_1c_3-b_3c_1)+a_3(b_1c_2-b_2c_1) \quad \text{...(2)}$$

From (1) and (2), we have

$$\Delta=\Delta_1.$$

It follows: The value of determinant remains unaltered, if the rows are interchanged into columns (or the columns into rows). **Proved.**

Property (*ii*). *If two rows (or two columns) of a determinant are interchanged, the sign of the value of the determinant changes.*

Verification. Let $\Delta = \begin{vmatrix} a_1 & a_2 & a_3 \\ b_1 & b_2 & b_3 \\ c_1 & c_2 & c_3 \end{vmatrix}$

$$= a_1 (b_2 c_3 - b_3 c_2) - a_2 (b_1 c_3 - b_3 c_1) + a_3 (b_1 c_2 - b_2 c_1) \quad \text{... (1)}$$

Interchanging first and third rows, the new determinant obtained is given by

$$\Delta_1 = \begin{vmatrix} c_1 & c_2 & c_3 \\ b_1 & b_2 & b_3 \\ a_1 & a_2 & a_3 \end{vmatrix}$$

Expanding along third row, we get

$$\Delta_1 = a_1 (c_2 b_3 - b_2 c_3) - a_2 (c_1 b_3 - c_3 b_1) + a_3 (b_2 c_1 - b_1 c_2)$$

$$= -[a_1 (b_2 c_3 - b_3 c_2) - a_2 (b_1 c_3 - b_3 c_1) + a_3 (b_1 c_2 - b_2 c_1)] \quad \text{... (2)}$$

From (1) and (2), we have

$$\Delta_1 = -\Delta$$

Hence, property (ii) is verified. **Proved.**

Example 5. *Verify property (ii) for* $\begin{vmatrix} 1 & 2 & 3 \\ 4 & 0 & 6 \\ 7 & 8 & 9 \end{vmatrix}$

Solution. Let $\Delta = \begin{vmatrix} 1 & 2 & 3 \\ 4 & 0 & 6 \\ 7 & 8 & 9 \end{vmatrix} = 1(0-48) - 2(36-42) + 3(32-0)$

$$= -48 + 12 + 96 = 60$$

Interchanging second the third rows, we have

$$\Delta_1 = \begin{vmatrix} 1 & 2 & 3 \\ 7 & 8 & 9 \\ 4 & 0 & 6 \end{vmatrix} = 1(48-0) - 2(42-36) + 3(0-32)$$

$$= 48 - 12 - 96 = -60$$

Thus, $\Delta_1 = \Delta$

Hence property (*ii*) is verified. **Verified.**

Property (*iii*). *If two rows (or columns) of a determinant are identical, the value of the determinant is zero.*

Verification. Let $\Delta = \begin{vmatrix} a_1 & b_1 & c_1 \\ a_1 & b_1 & c_1 \\ a_3 & b_3 & c_3 \end{vmatrix}$, so that the first two rows are identical.

By interchanging the first two rows, we get the same determinant D.

By property (*ii*), on interchanging the rows, the sign of the determinant changes.

or $\Delta = -\Delta$ or $2\Delta = 0$ or $\Delta = 0$ **Proved.**

Example 6. *Evaluate* : $\Delta = \begin{vmatrix} 2 & 3 & 4 \\ 2 & 3 & 4 \\ 1 & 5 & 6 \end{vmatrix}$

Solution. Expanding along first row, we have

$$\Delta = 2(18-20) - 3(12-4) + 4(10-3)$$

$$= 2 \times (-2) - 3(8) + 4(7) = -4 - 24 + 28 = 0$$

Here, R_1 and R_2 are identical. **Verified.**

Property (*iv*). *If the elements of any row (or column) of a determinant be each multiplied by the same number, the determinant is multiplied by that number.*

Verification. Let $\Delta = \begin{vmatrix} a_1 & b_1 & c_1 \\ a_2 & b_2 & c_2 \\ a_3 & b_3 & c_3 \end{vmatrix}$

and Δ_1 be the determinant obtained by multiplying the elements of the first row by k. Then

$$\Delta_1 = \begin{vmatrix} ka_1 & kb_1 & kc_1 \\ a_2 & b_2 & c_2 \\ a_3 & b_3 & c_3 \end{vmatrix}$$

$$= ka_1 (b_2c_3 - b_3c_2) - kb_1 (a_2c_3 - a_3c_2) + kc_1 (a_2b_3 - a_3b_2)$$
$$= k [a_1 (b_2c_3 - b_3c_2) - b_1 (a_2c_3 - a_3c_2) + c_1 (a_2b_3 - a_3b_2)]$$
$$= k \begin{vmatrix} a_1 & b_1 & c_1 \\ a_2 & b_2 & c_2 \\ a_3 & b_3 & c_3 \end{vmatrix} = k\,\Delta.$$

Hence, $\begin{vmatrix} ka_1 & kb_1 & kc_1 \\ a_2 & b_2 & c_2 \\ a_3 & b_3 & c_3 \end{vmatrix} = k \begin{vmatrix} a_1 & b_1 & c_1 \\ a_2 & b_2 & c_2 \\ a_3 & b_3 & c_3 \end{vmatrix}$ **Proved.**

Example 7. *Verify the property (iv) by*

$$\Delta = \begin{vmatrix} 2 & 5 & 8 \\ 3 & 7 & 1 \\ 2 & 0 & 2 \end{vmatrix}$$

Solution. $\Delta = \begin{vmatrix} 2 & 5 & 8 \\ 3 & 7 & 1 \\ 2 & 0 & 2 \end{vmatrix} = 2\,(14-0) - 5\,(6-2) + 8\,(0-14) = 28 - 20 - 112 = -104$

Multiplying the first column by 5, we get

$$\Delta_1 = \begin{vmatrix} 10 & 5 & 8 \\ 15 & 7 & 1 \\ 10 & 0 & 2 \end{vmatrix} = 10\,(14-0) - 5\,(30-10) + 8\,(0-70)$$

$$= 140 - 100 - 560 = -520 = 5\,(-104)$$

$$\Delta_1 = 5\,\Delta$$

Property (*iv*) is verified. **Verified.**

Property (*v*). *The value of the determinant remains unaltered if to the elements of one row (or column) be added any constant multiple of the corresponding elements of any other row (or column) respectively.*

Verification. Let $\Delta = \begin{vmatrix} a_1 & b_1 & c_1 \\ a_2 & b_2 & c_2 \\ a_3 & b_3 & c_3 \end{vmatrix}$

On multiplying the second column by l and the third column by m and adding to the first column, we get

$$\Delta' = \begin{vmatrix} a_1 + lb_1 + mc_1 & b_1 & c_1 \\ a_2 + lb_2 + mc_2 & b_2 & c_2 \\ a_3 + lb_3 + mc_3 & b_3 & c_3 \end{vmatrix} = \begin{vmatrix} a_1 & b_1 & c_1 \\ a_2 & b_2 & c_2 \\ a_3 & b_3 & c_3 \end{vmatrix} + l \begin{vmatrix} b_1 & b_1 & c_1 \\ b_2 & b_2 & c_2 \\ b_3 & b_3 & c_3 \end{vmatrix} + m \begin{vmatrix} c_1 & b_1 & c_1 \\ c_2 & b_2 & c_2 \\ c_3 & b_3 & c_3 \end{vmatrix}$$

$$= \Delta + 0 + 0 \qquad \text{(Since columns are identical)}$$

$$= \Delta$$

Proved.

Example 8. *Verify the property (v) by*

$$\Delta = \begin{vmatrix} 1 & 2 & 4 \\ 3 & 2 & 5 \\ 0 & 4 & 6 \end{vmatrix}$$

Solution. $\Delta = \begin{vmatrix} 1 & 2 & 4 \\ 3 & 2 & 5 \\ 0 & 4 & 6 \end{vmatrix} = 1\,(12-20) - 2\,(18-0) + 4\,(12-0) = -8 - 36 + 48 = 4$

On multiplying the second column by 5 and third column by 6 and adding to the first column, we get

$$\Delta_1 = \begin{vmatrix} 1+10+24 & 2 & 4 \\ 3+10+30 & 2 & 5 \\ 0+20+36 & 4 & 6 \end{vmatrix} = \begin{vmatrix} 35 & 2 & 4 \\ 43 & 2 & 5 \\ 56 & 4 & 6 \end{vmatrix} = 35\,(12-20) - 2\,(258-280) + 4\,(172-112)$$

$$= 35\,(-8) - 2\,(-22) + 4\,(60) = -280 + 44 + 240 = 284 - 280 = 4$$

$$\Delta_1 = \Delta$$

Verified.

Example 9. *Show that*

$$\Delta = \begin{vmatrix} b-c & c-a & a-b \\ c-a & a-b & b-c \\ a-b & b-c & c-a \end{vmatrix} = 0$$

Solution. Let $\quad \Delta = \begin{vmatrix} b-c & c-a & a-b \\ c-a & a-b & b-c \\ a-b & b-c & c-a \end{vmatrix}$

Applying $C_1 \to C_1 + C_2 + C_3$, we get

$$\Delta = \begin{vmatrix} 0 & c-a & a-b \\ 0 & a-b & b-c \\ 0 & b-c & c-a \end{vmatrix} = 0 \qquad [\because\ C_1 \text{ consists of all zeroes.}]$$

Proved.

Example 10. *Without expanding, evaluate the determinant*

$$\begin{vmatrix} \sin\alpha & \cos\alpha & \sin(\alpha+\delta) \\ \sin\beta & \cos\beta & \sin(\beta+\delta) \\ \sin\gamma & \cos\gamma & \sin(\gamma+\delta) \end{vmatrix}$$

Solution. Let $\quad \Delta = \begin{vmatrix} \sin\alpha & \cos\alpha & \sin(\alpha+\delta) \\ \sin\beta & \cos\beta & \sin(\beta+\delta) \\ \sin\gamma & \cos\gamma & \sin(\gamma+\delta) \end{vmatrix}$

$$\Rightarrow \quad \Delta = \begin{vmatrix} \sin\alpha & \cos\alpha & \sin\alpha\cos\delta + \cos\alpha\sin\delta \\ \sin\beta & \cos\beta & \sin\beta\cos\delta + \cos\beta\sin\delta \\ \sin\gamma & \cos\gamma & \sin\gamma\cos\delta + \cos\gamma\sin\delta \end{vmatrix}$$

$$[\because \sin(A+B) = \sin A \cos B + \cos A \sin B]$$

$$\Rightarrow \quad \Delta = \begin{vmatrix} \sin\alpha & \cos\alpha & 0 \\ \sin\beta & \cos\beta & 0 \\ \sin\gamma & \cos\gamma & 0 \end{vmatrix} \quad [\text{Applying } C_3 \to C_3 - \cos\delta.C_1 - \sin\delta.C_2]$$

$$\Rightarrow \quad \Delta = 0 \quad [\because C_3 \text{ consists of all zeroes.}]$$ **Ans.**

Example 11. *By using property of determinants prove that :*

$$\begin{vmatrix} 1 & x & x^2 \\ x^2 & 1 & x \\ x & x^2 & 1 \end{vmatrix} = (1-x^3)^2$$

Solution. L.H.S. $= \begin{vmatrix} 1 & x & x^2 \\ x^2 & 1 & x \\ x & x^2 & 1 \end{vmatrix} = \begin{vmatrix} 1+x+x^2 & x & x^2 \\ 1+x+x^2 & 1 & x \\ 1+x+x^2 & x^2 & 1 \end{vmatrix}$ [Applying $C_1 \to C_1 + C_2 + C_3$]

$$= (1+x+x^2)\begin{vmatrix} 1 & x & x^2 \\ 1 & 1 & x \\ 1 & x^2 & 1 \end{vmatrix}$$

$$= (1+x+x^2)\begin{vmatrix} 1 & x & x^2 \\ 0 & 1-x & x-x^2 \\ 0 & x^2-x & 1-x^2 \end{vmatrix} \quad [\text{Applying } R_2 \to R_2 - R_1 \text{ and } R_3 \to R_3 - R_1]$$

$$= (1+x+x^2)(1)\{(1-x)(1-x^2) - (x^2-x)(x-x^2)\}$$

$$= (1+x+x^2)(1-x)^2\{1+x+x^2\} = \{(1-x)(1+x+x^2)\}^2 = (1-x^3)^2 = \text{R.H.S.}$$

Proved.

Example 12. *Using properties of determinants, prove that*

$$\begin{vmatrix} x & y & z \\ x^2 & y^2 & z^2 \\ x^3 & y^3 & z^3 \end{vmatrix} = xyz(x-y)(y-z)(z-x).$$

Solution. Let $\Delta = \begin{vmatrix} x & y & z \\ x^2 & y^2 & z^2 \\ x^3 & y^3 & z^3 \end{vmatrix} = xyz\begin{vmatrix} 1 & 1 & 1 \\ x & y & z \\ x^2 & y^2 & z^2 \end{vmatrix}$

Operate : $C_1 \to C_1 - C_2$; $C_2 \to C_2 - C_3$.

$$\Delta = xyz\begin{vmatrix} 0 & 0 & 1 \\ x-y & y-z & z \\ x^2-y^2 & y^2-z^2 & z^2 \end{vmatrix} = xyz\begin{vmatrix} x-y & y-z \\ x^2-y^2 & y^2-z^2 \end{vmatrix} \quad (\text{On expanding by } R_1)$$

$$= xyz(x-y)(y-z)\begin{vmatrix} 1 & 1 \\ x+y & y+z \end{vmatrix} = xyz(x-y)(y-z)(z-x).$$ **Proved.**

Example 13. *Using the properties of determinants, show that*

$$\begin{vmatrix} a+x & y & z \\ x & a+y & z \\ x & y & a+z \end{vmatrix} = a^2(a+x+y+z).$$

Solution. Let $\Delta = \begin{vmatrix} a+x & y & z \\ x & a+y & z \\ x & y & a+z \end{vmatrix}$

Operate : $C_1 \to C_1 + C_2 + C_3$

$$\Delta = \begin{vmatrix} a+x+y+z & y & z \\ a+x+y+z & a+y & z \\ a+x+y+z & y & a+z \end{vmatrix}$$

Taking $(a + x + y + z)$ common from Ist column, we get

$$= (a+x+y+z)\begin{vmatrix} 1 & y & z \\ 1 & a+y & z \\ 1 & y & a+z \end{vmatrix}$$

Operate : $R_2 \to R_2 - R_1$, $R_3 \to R_3 - R_1$

$$= (a+x+y+z)\begin{vmatrix} 1 & y & z \\ 0 & a & 0 \\ 0 & 0 & a \end{vmatrix} = (a+x+y+z)\times 1\begin{vmatrix} a & 0 \\ 0 & a \end{vmatrix} \qquad \text{[Expanding along } C_1\text{]}$$

$$= (a + x + y + z)\, a^2 = a^2(a + x + y + z)$$ **Proved.**

Example 14. *If w is the one of the imaginary cube roots of unity, find the value of the determinant:*

$$\begin{vmatrix} 1 & \omega & \omega^2 \\ \omega & \omega^2 & 1 \\ \omega^2 & 1 & \omega \end{vmatrix}$$

Solution. The given determinant $= \begin{vmatrix} 1 & \omega & \omega^2 \\ \omega & \omega^2 & 1 \\ \omega^2 & 1 & \omega \end{vmatrix}$

By $R_1 \to R_1 + R_2 + R_3$, we get

$$= \begin{vmatrix} 1+\omega+\omega^2 & 1+\omega+\omega^2 & 1+\omega+\omega^2 \\ \omega & \omega^2 & 1 \\ \omega^2 & 1 & \omega \end{vmatrix} = \begin{vmatrix} 0 & 0 & 0 \\ \omega & \omega^2 & 1 \\ \omega^2 & 1 & \omega \end{vmatrix} \qquad [\because\ 1 + \omega + \omega^2 = 0]$$

$= 0$ (Since each entry in R_1 is zero.) **Ans.**

Example 15. *Without expanding the determinant, show that $(a + b + c)$ is a factor of the determinant* $\begin{vmatrix} a & b & c \\ b & c & a \\ c & a & b \end{vmatrix}$.

Solution. Let $\Delta = \begin{vmatrix} a & b & c \\ b & c & a \\ c & a & b \end{vmatrix}$ Operate : $C_1 \to C_1 + C_2 + C_3$

$$\Rightarrow \qquad \Delta = \begin{vmatrix} a+b+c & b & c \\ a+b+c & c & a \\ a+b+c & a & b \end{vmatrix} = (a+b+c)\begin{vmatrix} 1 & b & c \\ 1 & c & a \\ 1 & a & b \end{vmatrix}$$

$\Rightarrow$ $(a + b + c)$ is a factor of Δ. **Proved.**

Example 16. *Without expanding the determinant, prove that* $\begin{vmatrix} 1 & a & b+c \\ 1 & b & c+a \\ 1 & c & a+b \end{vmatrix} = 0.$

Solution. Let
$$\Delta = \begin{vmatrix} 1 & a & b+c \\ 1 & b & c+a \\ 1 & c & a+b \end{vmatrix}$$

Operate : $C_3 \to C_3 + C_2$.

$$\therefore \qquad \Delta = \begin{vmatrix} 1 & a & a+b+c \\ 1 & b & a+b+c \\ 1 & c & a+b+c \end{vmatrix} = (a+b+c)\begin{vmatrix} 1 & a & 1 \\ 1 & b & 1 \\ 1 & c & 1 \end{vmatrix}$$

$= 0$ ($\because$ C_1 and C_3 are identical). **Proved.**

Example 17. *Without expanding the determinant, prove that*

$$\begin{vmatrix} \frac{1}{a} & a^2 & bc \\ \frac{1}{b} & b^2 & ca \\ \frac{1}{c} & c^2 & ab \end{vmatrix} = 0$$

Solution. Let
$$\Delta = \begin{vmatrix} \frac{1}{a} & a^2 & bc \\ \frac{1}{b} & b^2 & ca \\ \frac{1}{c} & c^2 & ab \end{vmatrix}$$

Multiply R_1 by a, R_2 by b and R_3 by c.

$$\Delta = \frac{1}{abc}\begin{vmatrix} 1 & a^3 & abc \\ 1 & b^3 & abc \\ 1 & c^3 & abc \end{vmatrix} = \frac{1}{abc}.abc\begin{vmatrix} 1 & a^3 & 1 \\ 1 & b^3 & 1 \\ 1 & c^3 & 1 \end{vmatrix} = 1 \times 0 = 0.$$

(Since C_1 and C_3 are identical) **Proved.**

Example 18. *Using properties of determinants, prove that :*

$$\begin{vmatrix} 1 & a & a^3 \\ 1 & b & b^3 \\ 1 & c & c^3 \end{vmatrix} = (a-b)(b-c)(c-a)(a+b+c)$$

Solution. Let
$$\Delta = \begin{vmatrix} 1 & a & a^3 \\ 1 & b & b^3 \\ 1 & c & c^3 \end{vmatrix}$$

Operate : $R_1 \to R_1 - R_2$; $R_2 \to R_2 - R_3$

$$\Delta = \begin{vmatrix} 0 & a-b & a^3-b^3 \\ 0 & b-c & b^3-c^3 \\ 1 & c & c^3 \end{vmatrix} = 1.\begin{vmatrix} a-b & a^3-b^3 \\ b-c & b^3-c^3 \end{vmatrix} \qquad \text{(Expand along } C_1\text{)}$$

$$= (a-b)(b-c)\begin{vmatrix} 1 & a^2+ab+b^2 \\ 1 & b^2+bc+c^2 \end{vmatrix}$$

Operate : $R_1 \to R_1 - R_2$

$$\Delta = (a-b)(b-c)\begin{vmatrix} 0 & (a^2-c^2)+(ab-bc) \\ 1 & b^2+bc+c^2 \end{vmatrix}$$

$$= (a-b).(b-c).(-1)[(a^2-c^2)+b(a-c)]$$

$$= (a-b).(b-c)(c-a)(a+b+c).$$

Proved.

[**Note :** It can also be proved by factor Theorem easily]

Example 19. *Evaluate*

$$\begin{vmatrix} a-b-c & 2a & 2a \\ 2b & b-c-a & 2b \\ 2c & 2c & c-a-b \end{vmatrix}$$

Solution. By $R_1 \to R_1 + R_2 + R_3$, we get

$$\begin{vmatrix} a+b+c & a+b+c & a+b+c \\ 2b & b-c-a & 2b \\ 2c & 2c & c-a-b \end{vmatrix} = (a+b+c)\begin{vmatrix} 1 & 1 & 1 \\ 2b & b-c-a & 2b \\ 2c & 2c & c-a-b \end{vmatrix}$$

$$= (a+b+c)\begin{vmatrix} 1 & 0 & 0 \\ 2b & -(a+b+c) & 0 \\ 2c & 0 & -(a+b+c) \end{vmatrix} \begin{matrix} \\ C_2 - C_1 \\ C_3 - C_1 \end{matrix}$$

On expanding by first row $= (a+b+c)(a+b+c)^2 = (a+b+c)^3$. **Ans.**

Example 20. *By using properties of determinants prove that:*

$$\begin{vmatrix} 1+a^2-b^2 & 2ab & -2b \\ 2ab & 1-a^2+b^2 & 2a \\ 2b & -2a & 1-a^2-b^2 \end{vmatrix} = (1+a^2+b^2)^3$$

Solution. Let $\Delta = \begin{vmatrix} 1+a^2-b^2 & 2ab & -2b \\ 2ab & 1-a^2+b^2 & 2a \\ 2b & -2a & 1-a^2-b^2 \end{vmatrix}$

Applying $C_1 \to C_1 - bC_3$, $\quad C_2 \to C_2 + a\,C_3$, we get

$$\Delta = \begin{vmatrix} 1+a^2+b^2 & 0 & -2b \\ 0 & 1+a^2+b^2 & 2a \\ b(1+a^2+b^2) & -a(1+a^2+b^2) & 1-a^2-b^2 \end{vmatrix}$$

Taking $(1 + a^2 + b^2)$ common from C_1 and C_2, we get

$$\Delta = (1+a^2+b^2)^2 \begin{vmatrix} 1 & 0 & -2b \\ 0 & 1 & 2a \\ b & -a & 1-a^2-b^2 \end{vmatrix}$$

Applying $C_3 \to C_3 + 2b\, C_1$, we get

$$\Delta = (1+a^2+b^2)^2 \begin{vmatrix} 1 & 0 & 0 \\ 0 & 1 & 2a \\ b & -a & 1-a^2+b^2 \end{vmatrix}$$

Expanding along R_1, we get

$$\Delta = (1+a^2+b^2)^2 \begin{vmatrix} 1 & 2a \\ -a & 1-a^2+b^2 \end{vmatrix} = (1 + a^2 + b^2)^2 (1 - a^2 + b^2 + 2a^2)$$

$$= (1 + a^2 + b^2)^3$$ **Proved.**

Example 21. *Prove that*

$$\begin{vmatrix} \alpha & \beta & \gamma \\ \alpha^2 & \beta^2 & \gamma^2 \\ \beta+\gamma & \gamma+\alpha & \alpha+\beta \end{vmatrix} = (\alpha - \beta)(\beta - \gamma)(\gamma - \alpha)(\alpha + \beta + \gamma)$$

Solution. Let $\Delta = \begin{vmatrix} \alpha & \beta & \gamma \\ \alpha^2 & \beta^2 & \gamma^2 \\ \beta+\gamma & \gamma+\alpha & \alpha+\beta \end{vmatrix}$.

Applying $R_3 \to R_1 + R_3$, we get

$$\Delta = \begin{vmatrix} \alpha & \beta & \gamma \\ \alpha^2 & \beta^2 & \gamma^2 \\ \alpha+\beta+\gamma & \alpha+\beta+\gamma & \alpha+\beta+\gamma \end{vmatrix}$$

$$= (\alpha + \beta + \gamma) \begin{vmatrix} \alpha & \beta & \gamma \\ \alpha^2 & \beta^2 & \gamma^2 \\ 1 & 1 & 1 \end{vmatrix}$$ [Taking out $(\alpha + \beta + \gamma)$ common from R_3]

$$= (\alpha + \beta + \gamma) \begin{vmatrix} \alpha & \beta-\alpha & \gamma-\alpha \\ \alpha^2 & \beta^2-\alpha^2 & \gamma^2-\alpha^2 \\ 1 & 0 & 0 \end{vmatrix}$$ Applying $C_2 \to C_2 - C_1$, $C_3 \to C_3 - C_1$

$$= (\alpha + \beta + \gamma)(\beta - \alpha)(\gamma - \alpha) \begin{vmatrix} \alpha & 1 & 1 \\ \alpha^2 & \beta+\alpha & \gamma+\alpha \\ 1 & 0 & 0 \end{vmatrix}$$

$$= (\alpha + \beta + \gamma)(\beta - \alpha)(\gamma - \alpha).1 \begin{vmatrix} 1 & 1 \\ \beta+\alpha & \gamma+\alpha \end{vmatrix}$$ [Expanding along R_3]

$$= (\alpha + \beta + \gamma)(\beta - \alpha)(\gamma - \alpha)(\gamma + \alpha - \beta - \alpha)$$

$$= (\alpha + \beta + \gamma)(\beta - \gamma)(\gamma - \alpha)(\alpha - \beta)$$ **Proved.**

Example 22. *Prove that*

$$\begin{vmatrix} -a^2 & ab & ac \\ ba & -b^2 & bc \\ ac & bc & -c^2 \end{vmatrix} = 4\,a^2b^2c^2$$

Solution. Let $\Delta = \begin{vmatrix} -a^2 & ab & ac \\ ba & -b^2 & bc \\ ac & bc & -c^2 \end{vmatrix}$

Taking a, b, c common from C_1, C_2 and C_3 respectively, we get

$$\Delta = abc \begin{vmatrix} -a & a & a \\ b & -b & b \\ c & c & -c \end{vmatrix}$$

$$= a^2b^2c^2 \begin{vmatrix} -1 & 1 & 1 \\ 1 & -1 & 1 \\ 1 & 1 & -1 \end{vmatrix} \quad \text{[Taking } a, b, c \text{ common from } R_1, R_2 \text{ and } R_3 \text{ respectively]}$$

$$= a^2b^2c^2 \begin{vmatrix} -1 & 0 & 0 \\ 1 & 0 & 2 \\ 1 & 2 & 0 \end{vmatrix} \quad [\text{Applying } C_2 \to C_2 + C_1,\ C_3 \to C_3 + C_1]$$

$$= a^2b^2c^2\,(-1) \begin{vmatrix} 0 & 2 \\ 2 & 0 \end{vmatrix}$$

$$= a^2b^2c^2\,(-1)\,(0-4) = 4a^2\,b^2\,c^2 \quad [\text{Expanding along } R_1]$$ **Proved.**

Example 23. *Using properties of determinants, prove the following:*

$$\begin{vmatrix} a & b & c \\ a-b & b-c & c-a \\ b+c & c+a & a+b \end{vmatrix} = a^3 + b^3 + c^3 - 3abc.$$

Solution. We have,

$$\begin{vmatrix} a & b & c \\ a-b & b-c & c-a \\ b+c & c+a & a+b \end{vmatrix} = \begin{vmatrix} a+b+c & b & c \\ a-b+b-c+c-a & b-c & c-a \\ b+c+c+a+a+b & c+a & a+b \end{vmatrix}$$

By applying $C_1 \to C_1 + C_2 + C_3$

$$= \begin{vmatrix} a+b+c & b & c \\ 0 & b-c & c-a \\ 2(a+b+c) & c+a & a+b \end{vmatrix} = \begin{vmatrix} a+b+c & b & c \\ 0 & b-c & c-a \\ 0 & c+a-2b & a+b-2c \end{vmatrix} \quad \text{By } R_3 \to R_3 - 2R_1$$

Expanding along C_1, we get

$= (a + b + c)\ \{(b - c)\,(a + b - 2c) - (c - a)\,(c + a - 2b)\}$

$= (a + b + c)\ \{(ab + b^2 - 2bc - ac - bc + 2c^2) - (c^2 + ac - 2bc - ac - a^2 + 2ab)\}$

$= (a + b + c)\ \{ab + b^2 - 2bc - ac - bc + 2c^2 - c^2 - ac + 2bc + ac + a^2 - 2ab\}$

$= (a + b + c)\,(a^2 + b^2 + c^2 - ab - bc - ca) = a^3 + b^3 + c^3 - 3abc$ **Proved.**

Example 24. *If* $\begin{vmatrix} a & a^2 & a^3 - 1 \\ b & b^2 & b^3 - 1 \\ c & c^2 & c^3 - 1 \end{vmatrix} = 0$, *prove that* $abc = 1$.

Solution.

$$\begin{vmatrix} a & a^2 & a^3-1 \\ b & b^2 & b^3-1 \\ c & c^2 & c^3-1 \end{vmatrix} = 0 \Rightarrow \begin{vmatrix} a & a^2 & a^3 \\ b & b^2 & b^3 \\ c & c^2 & c^3 \end{vmatrix} + \begin{vmatrix} a & a^2 & -1 \\ b & b^2 & -1 \\ c & c^2 & -1 \end{vmatrix} = 0$$

$$\Rightarrow \quad abc \begin{vmatrix} 1 & a & a^2 \\ 1 & b & b^2 \\ 1 & c & c^2 \end{vmatrix} - \begin{vmatrix} a & a^2 & 1 \\ b & b^2 & 1 \\ c & c^2 & 1 \end{vmatrix} = 0$$

(Taking out common a, b, c from R_1, R_2 and R_3 from 1st determinant)

$$\Rightarrow \quad abc \begin{vmatrix} 1 & a & a^2 \\ 1 & b & b^2 \\ 1 & c & c^2 \end{vmatrix} + \begin{vmatrix} a & 1 & a^2 \\ b & 1 & b^2 \\ c & 1 & c^2 \end{vmatrix} = 0 \quad \text{(Interchanging } C_2 \text{ and } C_3\text{)}$$

$$\Rightarrow \quad abc \begin{vmatrix} 1 & a & a^2 \\ 1 & b & b^2 \\ 1 & c & c^2 \end{vmatrix} - \begin{vmatrix} 1 & a & a^2 \\ 1 & b & b^2 \\ 1 & c & c^2 \end{vmatrix} = 0$$

(Interchanging C_1 and C_2 of the second determinant)

$$\Rightarrow \quad (abc-1) \begin{vmatrix} 1 & a & a^2 \\ 1 & b & b^2 \\ 1 & c & c^2 \end{vmatrix} = 0 \quad \Rightarrow \quad abc - 1 = 0 \quad \Rightarrow \quad abc = 1$$

Proved.

Example 25. *Show that*

$$\begin{vmatrix} b+c & c+a & a+b \\ q+r & r+p & p+q \\ y+z & z+x & x+y \end{vmatrix} = 2 \begin{vmatrix} a & b & c \\ p & q & r \\ x & y & z \end{vmatrix}$$

Solution. The above determinant can be expressed as the sum of 8 determinants as given below:

$$\begin{vmatrix} b+c & c+a & a+b \\ q+r & r+p & p+q \\ y+z & z+x & x+y \end{vmatrix} = \begin{vmatrix} b & c & a \\ q & r & p \\ y & z & x \end{vmatrix} + \begin{vmatrix} b & a & a \\ q & p & p \\ y & x & x \end{vmatrix} + \begin{vmatrix} b & c & b \\ q & r & q \\ y & z & y \end{vmatrix} + \begin{vmatrix} b & a & b \\ q & p & q \\ y & x & y \end{vmatrix}$$

$$+ \begin{vmatrix} c & c & a \\ r & r & p \\ z & z & x \end{vmatrix} + \begin{vmatrix} c & a & a \\ r & p & p \\ z & x & x \end{vmatrix} + \begin{vmatrix} c & c & b \\ r & r & q \\ z & z & y \end{vmatrix} + \begin{vmatrix} c & a & b \\ r & p & q \\ z & x & y \end{vmatrix}$$

$$= \begin{vmatrix} b & c & a \\ q & r & p \\ y & z & x \end{vmatrix} + 0 + 0 + 0 + 0 + 0 + 0 + \begin{vmatrix} c & a & b \\ r & p & q \\ z & x & y \end{vmatrix}$$

$$= (-1)^2 \begin{vmatrix} a & b & c \\ p & q & r \\ x & y & z \end{vmatrix} + (-1)^2 \begin{vmatrix} a & b & c \\ p & q & r \\ x & y & z \end{vmatrix} = 2 \begin{vmatrix} a & b & c \\ p & q & r \\ x & y & z \end{vmatrix}$$

Proved.

Example 26. *If a, b, c are in A.P; then find the determinant:*

$$\begin{vmatrix} x+2 & x+3 & x+2a \\ x+3 & x+4 & x+2b \\ x+4 & x+5 & x+2c \end{vmatrix}$$

Solution. Applying $R_1 \to R_1 + R_3 - 2R_2$ to the given determinant, we have

$$\begin{vmatrix} (x+2)+(x+4)-2(x+3) & (x+3)+(x+5)-2(x+4) & (x+2a)+(x+2c)-2(x+2b) \\ x+3 & x+4 & x+2b \\ x+4 & x+5 & x+2c \end{vmatrix}$$

$$= \begin{vmatrix} 0 & 0 & 2a+2c-4b \\ x+3 & x+4 & x+2b \\ x+4 & x+5 & x+2c \end{vmatrix}$$

$$= \begin{vmatrix} 0 & 0 & 0 \\ x+3 & x+4 & x+2b \\ x+4 & x+5 & x+2c \end{vmatrix} = 0 \qquad \begin{bmatrix} \because & a, b, c \text{ are in A.P.} \\ \Rightarrow & 2b = a+c \\ \Rightarrow & 2a+2c = 4b \end{bmatrix}$$

[$\because$ R_1 consists of all zeroes.] **Ans.**

Example 27. *Prove that*

$$\begin{vmatrix} 2\alpha & \alpha+\beta & \alpha+\gamma \\ \beta+\alpha & 2\beta & \beta+\gamma \\ \gamma+\alpha & \gamma+\beta & 2\gamma \end{vmatrix} = 0$$

Solution. Given determinant $= \begin{vmatrix} \alpha+\alpha & \alpha+\beta & \alpha+\gamma \\ \beta+\alpha & \beta+\beta & \beta+\gamma \\ \gamma+\alpha & \gamma+\beta & \gamma+\gamma \end{vmatrix}$

The above determinant can be expressed as the sum of 8 determinants.
Each of the 8 determinants has either two identical columns or identical rows.
$\therefore$ Each of the resulting determinant is zero. Hence the result. **Proved.**

Example 28. *Using properties of determinants, prove that:*

$$\begin{vmatrix} 3a & -a+b & -a+c \\ -b+a & 3b & -b+c \\ -c+a & -c+b & 3c \end{vmatrix} = 3(a+b+c)(ab+bc+ca)$$

Solution.

$$\begin{vmatrix} 3a & -a+b & -a+c \\ -b+a & 3b & -b+c \\ -c+a & -c+b & 3c \end{vmatrix} = \begin{vmatrix} a+b+c & -a+b & -a+c \\ a+b+c & 3b & -b+c \\ a+b+c & -c+b & 3c \end{vmatrix} \quad (\text{Applying } C_1 \to C_1 + C_2 + C_3)$$

$$= (a+b+c)\begin{vmatrix} 1 & -a+b & -a+c \\ 1 & 3b & -b+c \\ 1 & -c+b & 3c \end{vmatrix} = (a+b+c)\begin{vmatrix} 1 & -a+b & -a+c \\ 0 & a+2b & a-b \\ 0 & a-c & a+2c \end{vmatrix} \begin{matrix} \\ R_2 \to R_2 - R_1 \\ R_3 \to R_3 - R_1 \end{matrix}$$

$= (a + b + c) .1.\{(a + 2b)(a + 2c) - (a - c)(a - b)\}$ [Expanding along C_1]
$= (a + b + c) \{(a^2 + 2ac + 2ab + 4bc) - (a^2 - ab - ac + bc)\}$
$= (a + b + c)(3ab + 3bc + 3ca) = 3(a + b + c)(ab + bc + ca)$ **Proved.**

Example 29. *Show that $x = -(a + b + c)$ is one root of the equation:*

$$\begin{vmatrix} x+a & b & c \\ b & x+c & a \\ c & a & x+b \end{vmatrix} = 0$$

and solve the equation completely.

Solution. By $C_1 \to C_1 + C_2 + C_3$, we get $\begin{vmatrix} x+a+b+c & b & c \\ x+a+b+c & x+c & a \\ x+a+b+c & a & x+b \end{vmatrix} = 0$

$$\Rightarrow \quad (x+a+b+c)\begin{vmatrix} 1 & b & c \\ 1 & x+c & a \\ 1 & a & x+b \end{vmatrix} = 0$$

$$\Rightarrow \quad (x+a+b+c)\begin{vmatrix} 1 & b & c \\ 0 & x-b+c & a-c \\ 0 & a-b & x+b-c \end{vmatrix} = 0, \; R_2 \to R_2 - R_1;\; R_3 \to R_3 - R_1$$

On expanding by first column, we get

$$(x+a+b+c)\,[(x-b+c)(x+b-c) - (a-b)(a-c)] = 0$$

$$\Rightarrow \quad (x+a+b+c)\,[x^2 - (b-c)^2 - (a^2 - ac - ab + bc)] = 0$$

$$\Rightarrow \quad (x+a+b+c)\,(x^2 - b^2 - c^2 + 2bc - a^2 + ac + ab - bc] = 0$$

$$\Rightarrow \quad (x+a+b+c)\,(x^2 - a^2 - b^2 - c^2 + ab + bc + ca) = 0$$

Either $x + a + b + c = 0 \;\Rightarrow\; x = -(a+b+c)$

or $\quad x^2 - a^2 - b^2 - c^2 + ab + bc + ca = 0$

$$\Rightarrow \quad x = \pm\sqrt{a^2 + b^2 + c^2 - ab - bc - ca}$$

Hence, $x = -(a + b + c)$ is one root of the given equation. **Proved.**

Example 30. *Find the value of*

$$\begin{vmatrix} (b+c)^2 & a^2 & a^2 \\ b^2 & (c+a)^2 & b^2 \\ c^2 & c^2 & (a+b)^2 \end{vmatrix}$$

Solution. By $C_1 \to C_1 - C_3, C_2 \to C_2 - C_3$, we get

$$\begin{vmatrix} (b+c)^2 - a^2 & a^2 - a^2 & a^2 \\ b^2 - b^2 & (c+a)^2 - b^2 & b^2 \\ c^2 - (a+b)^2 & c^2 - (a+b)^2 & (a+b)^2 \end{vmatrix}$$

$$= \begin{vmatrix} (a+b+c)(b+c-a) & 0 & a^2 \\ 0 & (a+b+c)(c+a-b) & b^2 \\ (a+b+c)(c-a-b) & (a+b+c)(c-a-b) & (a+b)^2 \end{vmatrix}$$

On taking out $(a + b + c)$ as common from 1st and 2nd columns, we get

$$= (a+b+c)^2\begin{vmatrix} b+c-a & 0 & a^2 \\ 0 & c+a-b & b^2 \\ c-a-b & c-a-b & (a+b)^2 \end{vmatrix}$$

$$= (a+b+c)^2 \begin{vmatrix} -a+b+c & 0 & a^2 \\ 0 & a-b+c & b^2 \\ -2b & -2a & 2ab \end{vmatrix} \quad R_3 \to R_3 - (R_1 + R_2)$$

$$= -2(a+b+c)^2 \begin{vmatrix} -a+b+c & 0 & a^2 \\ 0 & a-b+c & b^2 \\ b & a & -ab \end{vmatrix}$$

On expanding by first row, we get

$$= -2\,(a+b+c)^2\,[(-a+b+c)\,\{-ab\,(a-b+c) - ab^2\} + a^2\,\{0 - b\,(a-b+c)\}]$$
$$= -2\,(a+b+c)^2\,[(-a+b+c)\,(-a^2b - abc) - a^2b\,(a-b+c)]$$
$$= -2ab\,(a+b+c)^2\,[(-a+b+c)\,(-a-c) - a\,(a-b+c)]$$
$$= -2ab\,(a+b+c)^2\,[a^2 + ac - ab - bc - ac - c^2 - a^2 + ab - ac]$$
$$= -2ab\,(a+b+c)^2\,(-bc - ac - c^2) = 2abc\,(a+b+c)^2\,(b+a+c)$$
$$= \;\; 2abc\,(a+b+c)^3.$$

Ans.

Example 31. *Using properties of determinants, solve for x :*

$$\begin{vmatrix} a+x & a-x & a-x \\ a-x & a+x & a-x \\ a-x & a-x & a+x \end{vmatrix} = 0$$

Solution. Given that,

$$\begin{vmatrix} a+x & a-x & a-x \\ a-x & a+x & a-x \\ a-x & a-x & a+x \end{vmatrix} = 0$$

Applying $C_1 \to C_1 + C_2 + C_3$, we get

$$\begin{vmatrix} 3a-x & a-x & a-x \\ 3a-x & a+x & a-x \\ 3a-x & a-x & a+x \end{vmatrix} = 0$$

$$\Rightarrow \qquad (3a-x) \begin{vmatrix} 1 & a-x & a-x \\ 1 & a+x & a-x \\ 1 & a-x & a+x \end{vmatrix} = 0$$

Now applying $R_2 \to R_2 - R_1$ and $R_3 \to R_3 - R_1$, we get

$$\Rightarrow \qquad (3a-x) \begin{vmatrix} 1 & a-x & a-x \\ 0 & 2x & 0 \\ 0 & 0 & 2x \end{vmatrix} = 0$$

Expanding along C_1, we get

$\Rightarrow$ $(3a - x)\,(4x^2 - 0) = 0$

$\Rightarrow$ $4x^2\,(3a - x) = 0$ $\Rightarrow$ If $4x^2 = 0$, then $x = 0$

$\Rightarrow$ If $3a - x = 0$, then $x = 3a$ Hence, $x = 0$ or $3a$ **Ans.**

Example 32. *Using properties of determinants, prove the following*

$$\begin{vmatrix} 1+a & 1 & 1 \\ 1 & 1+b & 1 \\ 1 & 1 & 1+c \end{vmatrix} = abc\left(\frac{1}{a} + \frac{1}{b} + \frac{1}{c} + 1\right)$$

Solution. Let $$\Delta = \begin{vmatrix} 1+a & 1 & 1 \\ 1 & 1+b & 1 \\ 1 & 1 & 1+c \end{vmatrix}$$

Taking a, b and c common from R_1, R_2 and R_3 rows respectively.

$$\Delta = abc \begin{vmatrix} \frac{1+a}{a} & \frac{1}{a} & \frac{1}{a} \\ \frac{1}{b} & \frac{1+b}{b} & \frac{1}{b} \\ \frac{1}{c} & \frac{1}{c} & \frac{1+c}{c} \end{vmatrix} = abc \begin{vmatrix} 1+\frac{1}{a} & \frac{1}{a} & \frac{1}{a} \\ \frac{1}{b} & 1+\frac{1}{b} & \frac{1}{b} \\ \frac{1}{c} & \frac{1}{c} & 1+\frac{1}{c} \end{vmatrix}$$

$$= abc \begin{bmatrix} 1+\frac{1}{a}+\frac{1}{b}+\frac{1}{c} & 1+\frac{1}{a}+\frac{1}{b}+\frac{1}{c} & 1+\frac{1}{a}+\frac{1}{b}+\frac{1}{c} \\ \frac{1}{b} & 1+\frac{1}{b} & \frac{1}{b} \\ \frac{1}{c} & \frac{1}{c} & 1+\frac{1}{c} \end{bmatrix} \quad R_1 \to R_1 + R_2 + R_3$$

Taking $\left(1+\frac{1}{a}+\frac{1}{b}+\frac{1}{c}\right)$ common from R_1, we get

$$\Delta = abc\left(1+\frac{1}{a}+\frac{1}{b}+\frac{1}{c}\right) \begin{vmatrix} 1 & 1 & 1 \\ \frac{1}{b} & 1+\frac{1}{b} & \frac{1}{b} \\ \frac{1}{c} & \frac{1}{c} & 1+\frac{1}{c} \end{vmatrix}$$

Operate : $C_2 \to C_2 - C_1$; $C_3 \to C_3 - C_1$

$$\Delta = abc\left(\frac{1}{a}+\frac{1}{b}+\frac{1}{c}+1\right) \begin{vmatrix} 1 & 0 & 0 \\ \frac{1}{b} & 1 & 0 \\ \frac{1}{c} & 0 & 1 \end{vmatrix} = abc\left(\frac{1}{a}+\frac{1}{b}+\frac{1}{c}+1\right)$$

(On-expanding by R_1). **Proved.**

Example 33. *Prove that :*

$$\begin{vmatrix} a & a^2 & bc \\ b & b^2 & ac \\ c & c^2 & ab \end{vmatrix} = (a-b)(b-c)(c-a)(ab+bc+ac).$$

Solution. Let $$\Delta = \begin{vmatrix} a & a^2 & bc \\ b & b^2 & ac \\ c & c^2 & ab \end{vmatrix}$$

$$= \frac{1}{abc} \begin{vmatrix} a^2 & a^3 & abc \\ b^2 & b^3 & abc \\ c^2 & c^3 & abc \end{vmatrix} = \frac{1}{abc}.abc \begin{vmatrix} a^2 & a^3 & 1 \\ b^2 & b^3 & 1 \\ c^2 & c^3 & 1 \end{vmatrix}$$

$$\therefore \quad \Delta = \begin{vmatrix} a^2 - b^2 & a^3 - b^3 & 0 \\ b^2 - c^2 & b^3 - c^3 & 0 \\ c^2 & c^3 & 1 \end{vmatrix} \begin{matrix} R_1 \to R_1 - R_2 \\ R_2 \to R_2 - R_3 \\ \end{matrix}$$

$$= (a-b)(b-c) \begin{vmatrix} a+b & a^2 + ab + b^2 & 0 \\ b+c & b^2 + bc + c^2 & 0 \\ c^2 & c^3 & 1 \end{vmatrix}$$

Expand by C_3

$$= (a-b)(b-c).1 \begin{vmatrix} a+b & a^2 + ab + b^2 \\ b+c & b^2 + bc + c^2 \end{vmatrix}$$

$$= (a-b)(b-c) \begin{vmatrix} a+b & a^2 + ab + b^2 \\ c-a & bc + c^2 - a^2 - ab \end{vmatrix} \quad R_2 \to R_2 - R_1$$

$$= (a-b)(b-c) \begin{vmatrix} a+b & a^2 + ab + b^2 \\ c-a & b(c-a) + (c^2 - a^2) \end{vmatrix}$$

$$= (a-b)(b-c)(c-a) \begin{vmatrix} a+b & a^2 + ab + b^2 \\ 1 & b+c+a \end{vmatrix}$$

$$= (a-b)(b-c)(c-a)\,[(a+b)(a+b+c) - 1 \,.\, (a^2 + ab + b^2)]$$

$$= (a-b)(b-c)(c-a)(ab + bc + ac).$$ **Proved.**

EXERCISE 17.2

Expand the following determinants, using properties of the determinants :

1. $\begin{vmatrix} 1 & 3 & 7 \\ 4 & 9 & 1 \\ 2 & 7 & 6 \end{vmatrix}$ **Ans.** 51.

2. $\begin{vmatrix} x & a & a \\ a & x & a \\ a & a & x \end{vmatrix}$ **Ans.** $(x + 2a)(x - a)^2$

3. Show that $\begin{vmatrix} 0 & x-a & x-b \\ x-a & 0 & x-c \\ x-b & x-c & 0 \end{vmatrix} = 2(x-a)(x-b)(x-c).$

4. $\begin{vmatrix} \frac{1}{a} & a & bc \\ \frac{1}{b} & b & ca \\ \frac{1}{c} & c & ab \end{vmatrix} = 0$

5. $\begin{vmatrix} x+y & y+z & z+x \\ z & x & y \\ 1 & 1 & 1 \end{vmatrix} = 0.$

6. $\begin{vmatrix} x+4 & 2x & 2x \\ 2x & x+4 & 2x \\ 2x & 2x & x+4 \end{vmatrix} = (5x+4)(4-x)^2$

7. $\begin{vmatrix} a+b+2c & a & b \\ c & b+c+2a & b \\ c & a & c+a+2b \end{vmatrix} = 2(a+b+c)^3$

8. $\begin{vmatrix} 1 & x+y & x^2 + y^2 \\ 1 & y+z & y^2 + z^2 \\ 1 & z+x & z^2 + x^2 \end{vmatrix} = (x-y)(y-z)(z-x).$

9. $\begin{vmatrix} b^2c^2 & bc & b+c \\ c^2a^2 & ca & c+a \\ a^2b^2 & ab & a+b \end{vmatrix} = 0$

10. $\begin{vmatrix} 1 & a & a^2 - bc \\ 1 & b & b^2 - ca \\ 1 & c & c^2 - ab \end{vmatrix} = 0.$

11. $\begin{vmatrix} a & a+b & a+b+c \\ 2a & 3a+2b & 4a+3b+2c \\ 3a & 6a+3b & 10a+6b+3c \end{vmatrix} = a^3.$

12. $\begin{vmatrix} 1 & a & bc \\ 1 & b & ca \\ 1 & c & ab \end{vmatrix} = (a-b)(b-c)(c-a).$

13. $\begin{vmatrix} 1 & a & a^2 \\ 1 & b & b^2 \\ 1 & c & c^2 \end{vmatrix} = \begin{vmatrix} 1 & a & bc \\ 1 & b & ca \\ 1 & c & ab \end{vmatrix}$

14. $\begin{vmatrix} a+b+c & -c & -b \\ -c & a+b+c & -a \\ -b & -a & a+b+c \end{vmatrix} = 2(a+b)(b+c)(c+a).$

15. $\begin{vmatrix} a^2 & bc & ac+c^2 \\ a^2+ab & b^2 & ac \\ ab & b^2+bc & c^2 \end{vmatrix} = 4a^2b^2c^2$

16. $\begin{vmatrix} 3a & -a+b & -a+c \\ -b+a & 3b & -b+c \\ -c+a & -c+b & 3c \end{vmatrix} = 3\,(a+b+c)\,(ab+bc+ca)$

17.6 FACTOR THEOREM

If the elements of a determinant are polynomials in a variable x and if the substitution $x = a$ makes two rows (or columns) identical then $(x - a)$ is a factor of the determinant.

When two rows are identical, the value of the determinant is zero. The expansion of a determinant being polynomial in x vanishes on putting $x = a$, then $x - a$ is its factor by the Remainder theorem.

Example 34. *Show that*

$$\begin{vmatrix} 1 & 1 & 1 \\ x & y & z \\ x^2 & y^2 & z^2 \end{vmatrix} = (x-y)(y-z)(z-x)$$

Solution. If we put $x = y$, $y = z$ *and* $z = x$ then in each case two columns become identical and the determinant vanishes.

$\therefore$ $(x - y)$, $(y - z)$, and $(z - x)$ are the factors.

Since the determinant is of third degree, the other factor can be numerical only k (say).

$$\begin{vmatrix} 1 & 1 & 1 \\ x & y & z \\ x^2 & y^2 & z^2 \end{vmatrix} = k(x-y)(y-z)(z-x) \qquad \text{... (1)}$$

This leading term (product of the elements of the diagonal elements) L.H.S. of (1) is yz^2 and in the expansion of R.H.S. *i.e.* $k\,(x - y)\,(y - z)\,(z - x)$ we get kyz^2

Equating the coefficient of yz^2 on both sides of (1), we have

$$k = 1$$

Hence, the expansion $= (x - y)\,(y - z)\,(z - x)$. **Proved.**

Example 35. *Using properties of determinants, prove that*

$$\begin{vmatrix} x & x^2 & 1+px^3 \\ y & y^2 & 1+py^3 \\ z & z^2 & 1+pz^3 \end{vmatrix} = (1+pxyz)(x-y)(y-z)(z-x), \text{ where } p \text{ is any scalar.}$$

Solution. We have,

$$\begin{vmatrix} x & x^2 & 1+px^3 \\ y & y^2 & 1+py^3 \\ z & z^2 & 1+pz^3 \end{vmatrix} = \begin{vmatrix} x & x^2 & 1 \\ y & y^2 & 1 \\ z & z^2 & 1 \end{vmatrix} + \begin{vmatrix} x & x^2 & px^3 \\ y & y^2 & py^3 \\ z & z^2 & pz^3 \end{vmatrix} = \begin{vmatrix} x & x^2 & 1 \\ y & y^2 & 1 \\ z & z^2 & 1 \end{vmatrix} + pxyz \begin{vmatrix} 1 & x & x^2 \\ 1 & y & y^2 \\ 1 & z & z^2 \end{vmatrix}$$

$$= \begin{vmatrix} 1 & x & x^2 \\ 1 & y & y^2 \\ 1 & z & z^2 \end{vmatrix} + pxyz \begin{vmatrix} 1 & x & x^2 \\ 1 & y & y^2 \\ 1 & z & z^2 \end{vmatrix} = (1+pxyz) \begin{vmatrix} 1 & x & x^2 \\ 1 & y & y^2 \\ 1 & z & z^2 \end{vmatrix} \quad \text{... (1)}$$

If we put $x = y$ then two rows become identical and the determinant vanishes.

$\Rightarrow$ $(x - y)$ is a factor.

If we put $y = z$ then two rows become identical and the determinant vanishes.

$\Rightarrow$ $(y - z)$ is a factor.

If we put $z = x$, then two rows become identical and the determinant vanishes.

$\Rightarrow$ $(z - x)$ is also a factor.

Since the determinant is of the third degree, the other factor can be numerical k,

$$\begin{vmatrix} 1 & x & x^2 \\ 1 & y & y^2 \\ 1 & z & z^2 \end{vmatrix} = k(x-y)(y-z)(z-x) \quad \text{...(2)}$$

This leading term (product of the diagonal elements) L.H.S of (2) is yz^2 and in the expansion of R.H.S. i.e., $k(x - y)(y - z)(z - x)$ we get kyz^2.

Equating the coefficients of yz^2, we have $k = 1$

$$\text{Hence, } \begin{vmatrix} 1 & x & x^2 \\ 1 & y & y^2 \\ 1 & z & z^2 \end{vmatrix} = (x-y)(y-z)(z-x) \quad \text{... (3)}$$

From (1) and (3), we have

$$\text{The given determinant} = \begin{vmatrix} x & x^2 & 1+px^3 \\ y & y^2 & 1+py^3 \\ z & z^2 & 1+pz^3 \end{vmatrix} = (1+pxyz)(x-y)(y-z)(z-x)$$ **Proved.**

Example 36. *Factorize*

$$\Delta = \begin{vmatrix} 1 & 1 & 1 \\ a^2 & b^2 & c^2 \\ a^3 & b^3 & c^3 \end{vmatrix}$$

Solution. Putting $a = b$, $C_1 = C_2$ and hence $\Delta = 0$.

$\therefore$ $a - b$ is a factor of Δ.

Similarly, $b - c$, $c - a$ are also factors of Δ.

$\therefore$ $(a - b)(b - c)(c - a)$ is a third degree factor of Δ which itself is of the fifth degree as is judged from the leading term b^2c^3.

$\therefore$ The remaining factor must be of the second degree. As Δ is symmetrical in a, b, c the remaining factor must, therefore, be of the form

$$k(a^2 + b^2 + c^2) + l(ab + bc + ca)$$

$\therefore \quad \Delta = (a - b)(b - c)(c - a)\{k(a^2 + b^2 + c^2) + l(ab + bc + ca)\}$

If $k \neq 0$, we shall get terms like a^4b, b^4c etc. which do not occur in Δ. Hence, k must be zero.

$\therefore \quad \Delta = (a - b)(b - c)(c - a)\{0 + l(ab + bc + ca)\}$

or $\quad \Delta = l(a - b)(b - c)(c - a)(ab + bc + ca)$

The leading term in $\quad \Delta = b^2c^3$

The corresponding term on R.H.S. $= l\, b^2c^3$

$\therefore \quad l = 1$

Hence, $\Delta = (a - b)(b - c)(c - a)(ab + bc + ca)$. **Ans.**

EXERCISE 17.3

1. Evaluate, without expanding

$$\begin{vmatrix} a & a^2 & 1 + a^3 \\ b & b^2 & 1 + b^3 \\ c & c^2 & 1 + c^3 \end{vmatrix}$$

Ans. $(a - b)(b - c)(c - a)(1 + abc)$

2. Solve the equation $\begin{vmatrix} x^3 - a^3 & x^2 & x \\ b^3 - a^3 & b^2 & b \\ c^3 - a^3 & c^2 & c \end{vmatrix} = 0, b \neq c, c \neq 0, b \neq 0.$ **Ans.** $x = \dfrac{a^3}{bc}$, $x = b$, $x = c$

3. Without expanding, show that

$$\Delta = \begin{vmatrix} (a - x)^2 & (a - y)^2 & (a - z)^2 \\ (b - x)^2 & (b - y)^2 & (b - z)^2 \\ (c - x)^2 & (c - y)^2 & (c - z)^2 \end{vmatrix} = 2(a - b)(b - c)(c - a)(x - y)(y - z)(z - x).$$

4. Show (without expanding) that

$$\begin{vmatrix} bc & a^2 & a^2 \\ b^2 & ca & b^2 \\ c^2 & c^2 & ab \end{vmatrix} = \begin{vmatrix} bc & ab & ca \\ ab & ca & bc \\ ca & bc & ab \end{vmatrix}$$

$$= -\frac{1}{2}(ab + bc + ca)[(ab - bc)^2 + (bc - ca)^2 + (ca - ab)^2]$$

17.7 SPECIAL TYPES OF DETERMINANTS

(i) Ortho-symmetric Determinant. If every element of the leading diagonal is the same and the equidistant elements from the diagonal are equal, then the determinant is said to be ortho-symmetric determinant.

$$\begin{vmatrix} a & h & g \\ h & a & f \\ g & f & a \end{vmatrix}$$

(ii) Skew-Symmetric Determinant. If the elements of the leading diagonal are all zero and every other element is equal to its conjugate with sign changed, the determinant is said to be Skew-symmetric.

$$\begin{vmatrix} 0 & -a & -b \\ a & 0 & -c \\ b & c & 0 \end{vmatrix}$$

Property 1. *A Skew-symmetric determinant of odd order vanishes.*

Example 37. *Prove that*

$$\begin{vmatrix} 0 & -a & -b \\ a & 0 & -c \\ b & c & 0 \end{vmatrix} = 0$$

Solution. Taking out (– 1) common from each of the three columns

$$\Delta = (-1)^3 \begin{vmatrix} 0 & a & b \\ -a & 0 & c \\ -b & -c & 0 \end{vmatrix}$$

Changing rows into columns

$$\Delta = (-1)^3 \begin{vmatrix} 0 & -a & -b \\ a & 0 & -c \\ b & c & 0 \end{vmatrix} = (-1)^3\, \Delta = -\Delta \quad \text{or} \quad 2\Delta = 0 \quad \text{or} \quad \Delta = 0$$

17.8 APPLICATION OF DETERMINANTS

Area of triangle. We know that the area of a triangle, whose vertices are (x_1, y_1), (x_2, y_2) and (x_3, y_3) is given by

$$\Delta = \frac{1}{2}\left[x_1 (y_2 - y_3) - x_2 (y_1 - y_3) + x_3 (y_1 - y_2)\right]$$

$$= \frac{1}{2}\left[x_1 \begin{vmatrix} y_2 & 1 \\ y_3 & 1 \end{vmatrix} - x_2 \begin{vmatrix} y_1 & 1 \\ y_3 & 1 \end{vmatrix} + x_3 \begin{vmatrix} y_1 & 1 \\ y_2 & 1 \end{vmatrix}\right] = \frac{1}{2}\begin{vmatrix} x_1 & y_1 & 1 \\ x_2 & y_2 & 1 \\ x_3 & y_3 & 1 \end{vmatrix}$$

Note. Since area is always a positive quantity, therefore we always take the absolute value of the determinant for the area.

Condition of collinearity of three points. Let $A\,(x_1, y_1)$, $B\,(x_2, y_2)$ and $C\,(x_3, y_3)$ be three points. Then, A, B, C are collinear

⇔ area of triangle $ABC = 0$

$$\Leftrightarrow \frac{1}{2}\begin{vmatrix} x_1 & y_1 & 1 \\ x_2 & y_2 & 1 \\ x_3 & y_3 & 1 \end{vmatrix} = 0 \Leftrightarrow \begin{vmatrix} x_1 & y_1 & 1 \\ x_2 & y_2 & 1 \\ x_3 & y_3 & 1 \end{vmatrix} = 0$$

Proved.

Example 38. *Using determinants, find the area of the triangle with vertices (– 2, –3), (3, 2) and (–1, –8).*

Solution. The area of the given triangle $= \frac{1}{2}\begin{vmatrix} -2 & -3 & 1 \\ 3 & 2 & 1 \\ -1 & -8 & 1 \end{vmatrix}$

$$= \frac{1}{2}\begin{vmatrix} -2 & -3 & 1 \\ 5 & 5 & 0 \\ 1 & -5 & 0 \end{vmatrix} \qquad \begin{matrix} R_2 \to R_2 - R_1 \\ R_3 \to R_3 - R_1 \end{matrix}$$

Expand by C_3 we get

$$= \frac{1}{2}.1.\begin{vmatrix} 5 & 5 \\ 1 & -5 \end{vmatrix} = \frac{1}{2}(-25 - 5) = \frac{|-30|}{2} = 15 \text{ sq. units}$$

Ans.

Example 39. *If area of triangle is 35 sq. units with verties (2, –6), (5, 4) and (k, 4). Then find k.*

Solution. Let the vertices of triangle be A (2, –6), B (5, 4) and C (k, 4). Since the area of the triangle ABC is 35 sq. units, we have

$$\frac{1}{2}\begin{vmatrix} 2 & -6 & 1 \\ 5 & 4 & 1 \\ k & 4 & 1 \end{vmatrix} = \pm 35$$

$$\Rightarrow \quad \frac{1}{2}\begin{vmatrix} 2 & -6 & 1 \\ 3 & 10 & 0 \\ k-2 & 10 & 0 \end{vmatrix} = \pm 35 \qquad [\text{Applying } R_2 \to R_2 - R_1 \text{ and } R_3 \to R_3 - R_1]$$

$$\Rightarrow \quad \frac{1}{2}\begin{vmatrix} 3 & 10 \\ k-2 & 10 \end{vmatrix} = \pm 35 \Rightarrow \quad \frac{1}{2}\{30 - 10(k-2)\} = \pm 35 \qquad [\text{Exanding along } C_3]$$

$$\Rightarrow \quad 30 - 10k + 20 = \pm 70$$

$$\Rightarrow \quad 10k = 50 \mp 70 \quad \Rightarrow \quad k = 12 \quad \text{or} \quad k = -2$$

Ans.

Example 40. *Show that points $A(a, b+c), B(b, c+a), C(c, a+b)$ are collinear.*

Solution. The area of the triangle formed by the given points:

$$= \frac{1}{2}\begin{vmatrix} a & b+c & 1 \\ b & c+a & 1 \\ c & a+b & 1 \end{vmatrix}$$

Operate : $R_2 \to R_2 - R_1$, $R_3 \to R_3 - R_1$

$$= \frac{1}{2}\begin{vmatrix} a & b+c & 1 \\ b-a & a-b & 0 \\ c-a & a-c & 0 \end{vmatrix} = \frac{1}{2}(1)\{(b-a)(a-c) - (a-b)(c-a)\} \quad [\text{Expanding along } C_3]$$

$$= \frac{1}{2}[ab - bc - a^2 + ac - ac + a^2 + bc - ab] = \frac{1}{2}[0] = 0$$

Hence, the given points are collinear. **Proved.**

Example 41. *Using determinants, show that the points (11, 7), (5, 5) and (– 1, 3) are collinear.*

Solution. The area of the triangle formed by the given points

$$= \frac{1}{2}\begin{vmatrix} 11 & 7 & 1 \\ 5 & 5 & 1 \\ -1 & 3 & 1 \end{vmatrix}$$

Operate : $R_1 \to R_1 - R_2$; $R_2 \to R_2 - R_3$

$$= \frac{1}{2}\begin{vmatrix} 6 & 2 & 0 \\ 6 & 2 & 0 \\ -1 & 3 & 1 \end{vmatrix} = \frac{1}{2}.0 = 0. \qquad (\because \ R_1 \text{ and } R_2 \text{ are identical})$$

Hence, the given points are collinear. **Proved.**

Example 42. *Using determinants, find the area of the triangle whose vertices are (1, 4) (2, 3) and (– 5, – 3). Are the given points collinear?*

Solution. Area of the required triangle

$$= \frac{1}{2}\begin{vmatrix} 1 & 4 & 1 \\ 2 & 3 & 1 \\ -5 & -3 & 1 \end{vmatrix} = \frac{1}{2}[1(3+3) - 4(2+5) + 1(-6+15)] = \frac{1}{2}[6 - 28 + 9] = \frac{13}{2} \neq 0$$

Hence, the given points are not collinear. **Ans.**

Example 43. *Find the equation of line joining A (1, 2) and B (3, 6) using determinants.*

Solution. Let $P(x, y)$ be any point on AB. Then, area of triangle ABP is zero. So,

$$\frac{1}{2}\begin{vmatrix} 1 & 2 & 1 \\ 3 & 6 & 1 \\ x & y & 1 \end{vmatrix} = 0$$

$$\Rightarrow \frac{1}{2}\begin{vmatrix} 1 & 2 & 1 \\ 2 & 4 & 0 \\ x-1 & y-2 & 0 \end{vmatrix} = 0 \quad [\text{Applying } R_2 \to R_2 - R_1 \text{ and } R_3 \to R_3 - R_1]$$

$$\Rightarrow \frac{1}{2}(1)\{2(y-2) - 4(x-1)\} = 0 \qquad [\text{Expanding along } C_3]$$

$$\Rightarrow y - 2 - 2x + 2 = 0 \Rightarrow y = 2x$$ **Ans.**

EXERCISE 17.4

1. Using determinants, find the area of the triangle with vertices (2,– 7), (1, 3), (10, 8). **Ans.** $A = \frac{95}{2}$
2. Using determinants, show that the points (3, 8), (– 4, 2) and (10, 14) are collinear.
3. Using determinants, find the area of the triangle whose vertices are (– 2, 4), (2, – 6) and (5, 4). Are the given points collinear ? **Ans.** Area = 35, not collinear
4. Using determinants, find the area of the triangle whose vertices are (– 1, – 3), (2, 4) and (3, – 1). Are the given points collinear ? **Ans.** Area = 11, not collinear
5. Using determinants, find the area of the triangle whose vertices are (1, – 1), (2, 4) and (– 3, 5). Are the given points collinear ? **Ans.** Area = 13, not collinear
6. Find the value of α, so that the points (1, – 5), (– 4, 5) and (α, 7) are collinear. **Ans.** α = – 5
7. Find the value of x, if the area of triangle is 35 square cms with vertices (x, 4), (2, – 6), (5, 4). **Ans.** $x = -2, 12$
8. Using determinants find the value of k, so that the points (k, 2 – 2k), (– k + 1, 2k) and (– 4 – k, 6 – 2 k) may be collinear. **Ans.** $k = -1, \frac{1}{2}$
9. If the points (x, – 2), (5, 2) and (8, 8) are collinear, find x using determinants. **Ans.** $x = 3$
10. If the points (3, – 2), (x, 2) and (8, 8) are collinear, find x using determinants. **Ans.** $x = 5$

17.9 SOLUTION OF SIMULTANEOUS LINEAR EQUATIONS BY DETERMINANTS (CRAMER'S RULE)

Let us solve the following equations.

$$a_1 x + b_1 y + c_1 z = d_1$$
$$a_2 x + b_2 y + c_2 z = d_2$$
$$a_3 x + b_3 y + c_3 z = d_3$$

Let
$$D = \begin{vmatrix} a_1 & b_1 & c_1 \\ a_2 & b_2 & c_2 \\ a_3 & b_3 & c_3 \end{vmatrix} \text{ or } x\,D = \begin{vmatrix} a_1x & b_1 & c_1 \\ a_2x & b_2 & c_2 \\ a_3x & b_3 & c_3 \end{vmatrix}$$

Multiplying the 2nd column by y and 3rd column by z and adding to the 1st column, we get

$$x\,D_1 = \begin{vmatrix} a_1x + b_1y + c_1z & b_1 & c_1 \\ a_2x + b_2y + c_2z & b_2 & c_2 \\ a_3x + b_3y + c_3z & b_3 & c_3 \end{vmatrix}, \quad x\,D_1 = \begin{vmatrix} d_1 & b_1 & c_1 \\ d_2 & b_2 & c_2 \\ d_3 & b_3 & c_3 \end{vmatrix} \Rightarrow x = \frac{\begin{vmatrix} d_1 & b_1 & c_1 \\ d_2 & b_2 & c_2 \\ d_3 & b_1 & c_3 \end{vmatrix}}{\begin{vmatrix} a_1 & b_1 & c_1 \\ a_2 & b_2 & c_2 \\ a_3 & b_3 & c_3 \end{vmatrix}} = \frac{D_1}{D}$$

Similarly, $y = \frac{D_2}{D} = \frac{\begin{vmatrix} a_1 & d_1 & c_1 \\ a_2 & d_2 & c_2 \\ a_3 & d_3 & c_3 \end{vmatrix}}{\begin{vmatrix} a_1 & b_1 & c_1 \\ a_2 & b_2 & c_2 \\ a_3 & b_3 & c_3 \end{vmatrix}} \Rightarrow z = \frac{D_3}{D} = \frac{\begin{vmatrix} a_1 & b_1 & d_1 \\ a_2 & b_2 & d_2 \\ a_3 & b_3 & d_3 \end{vmatrix}}{\begin{vmatrix} a_1 & b_1 & c_1 \\ a_2 & b_2 & c_2 \\ a_3 & b_3 & c_3 \end{vmatrix}}$

Thus, $x = \frac{D_1}{D},\ y = \frac{D_2}{D},\ z = \frac{D_3}{D}$ **Ans.**

Example 44. *Solve the following system of equations using Cramer's rule :*

$$5x - 7y + z = 11$$
$$6x - 8y - z = 15$$
$$3x + 2y - 6z = 7$$

Solution. The given equations are

$$5x - 7y + z = 11$$
$$6x - 8y - z = 15$$
$$3x + 2y - 6z = 7$$

Here, $D = \begin{vmatrix} 5 & -7 & 1 \\ 6 & -8 & -1 \\ 3 & 2 & -6 \end{vmatrix} = 5\,(48 + 2) + 7\,(-36 + 3) + 1\,(12 + 24) = 55$

$$D_1 = \begin{vmatrix} 11 & -7 & 1 \\ 15 & -8 & -1 \\ 7 & 2 & -6 \end{vmatrix} = 11\,(48 + 2) + 7\,(-90 + 7) + 1\,(30 + 56) = 55$$

$$D_2 = \begin{vmatrix} 5 & 11 & 1 \\ 6 & 15 & -1 \\ 3 & 7 & -6 \end{vmatrix} = 5\,(-90 + 7) - 11\,(-36 + 3) + 1\,(42 - 45) = -55$$

$$D_3 = \begin{vmatrix} 5 & -7 & 11 \\ 6 & -8 & 15 \\ 3 & 2 & 7 \end{vmatrix} = 5\,(-56 - 30) + 7\,(42 - 45) + 11\,(12 + 24) = -55$$ **Proved.**

By Cramer's Rule

$$x = \frac{D_1}{D} = \frac{55}{55} = 1, \quad y = \frac{D_2}{D} = \frac{-55}{55} = -1, \quad z = \frac{D_3}{D} = \frac{-55}{55} = -1$$

Hence, $x = 1, y = -1, z = -1$ **Ans.**

Example 45. *Solve, by determinants, the following set of simultaneous equations :*

$$5x - 6y + 4z = 15$$
$$7x + 4y - 3z = 19$$
$$2x + y + 6z = 46$$

Solution. $D = \begin{vmatrix} 5 & -6 & 4 \\ 7 & 4 & -3 \\ 2 & 1 & 6 \end{vmatrix} = 419 \qquad D_1 = \begin{vmatrix} 15 & -6 & 4 \\ 19 & 4 & -3 \\ 46 & 1 & 6 \end{vmatrix} = 1257$

$$D_2 = \begin{vmatrix} 5 & 15 & 4 \\ 7 & 19 & -3 \\ 2 & 46 & 6 \end{vmatrix} = 1676 \qquad D_3 = \begin{vmatrix} 5 & -6 & 15 \\ 7 & 4 & 19 \\ 2 & 1 & 46 \end{vmatrix} = 2514$$

By Cramer's Rule

$$x = \frac{D_1}{D} = \frac{1257}{419} = 3 \qquad y = \frac{D_2}{D} = \frac{1676}{419} = 4. \qquad z = \frac{D_3}{D} = \frac{2514}{419} = 6.$$

Hence $x = 3, y = 4, z = 6$ **Ans.**

Example 46. *Solve, using Cramer's rule*

$$3x - 2y + 4z = 5$$
$$x + y + 3z = 2$$
$$-x + 2y - z = 1$$

Solution.
$$D = \begin{vmatrix} 3 & -2 & 4 \\ 1 & 1 & 3 \\ -1 & 2 & -1 \end{vmatrix} = -5 \qquad D_1 = \begin{vmatrix} 5 & -2 & 4 \\ 2 & 1 & 3 \\ 1 & 2 & -1 \end{vmatrix} = -33$$

$$D_2 = \begin{vmatrix} 3 & 5 & 4 \\ 1 & 2 & 3 \\ -1 & 1 & -1 \end{vmatrix} = -13 \qquad D_3 = \begin{vmatrix} 3 & -2 & 5 \\ 1 & 1 & 2 \\ -1 & 2 & 1 \end{vmatrix} = 12$$

By Cramer's Rule

$$x = \frac{D_1}{D} = \frac{-33}{-5} = \frac{33}{5} \qquad y = \frac{D_2}{D} = \frac{-13}{-5} = \frac{13}{5} \qquad z = \frac{D_3}{D} = \frac{12}{-5} = \frac{-12}{5}$$

Hence, $x = \frac{33}{5},\ y = \frac{13}{5},\ z = \frac{-12}{5}$ **Ans.**

Example 47. *Solve the following system of equations by using determinants :*

$$x + y + z = 1$$
$$ax + by + cz = k$$
$$a^2x + b^2y + c^2z = k^2$$

Solution. We have,

$$D = \begin{vmatrix} 1 & 1 & 1 \\ a & b & c \\ a^2 & b^2 & c^2 \end{vmatrix} = \begin{vmatrix} 1 & 0 & 0 \\ a & b-a & c-a \\ a^2 & b^2-a^2 & c^2-a^2 \end{vmatrix} \quad \text{[Applying } C_2 \to C_2 - C_1 \text{ and } C_3 \to C_3 - C_1\text{]}$$

$$= (b-a)(c-a) \begin{vmatrix} 1 & 0 & 0 \\ a & 1 & 1 \\ a^2 & b+a & c+a \end{vmatrix}$$

$$= (b-a)(c-a).1. \begin{vmatrix} 1 & 1 \\ b+a & c+a \end{vmatrix} \quad \text{[Expanding along } R_1\text{]}$$

$$= (b-a)(c-a)(c+a-b-a) = (b-c)(c-a)(a-b) \quad ...(1)$$

$$D_1 = \begin{vmatrix} 1 & 1 & 1 \\ k & b & c \\ k^2 & b^2 & c^2 \end{vmatrix} = (b-c)(c-k)(k-b) \quad \text{[Replacing } a \text{ by } k \text{ in (1)]}$$

$$D_2 = \begin{vmatrix} 1 & 1 & 1 \\ a & k & c \\ a^2 & k^2 & c^2 \end{vmatrix} = (k-c)(c-a)(a-k) \qquad \text{[Replacing } b \text{ by } k \text{ in (1)]}$$

and $$D_3 = \begin{vmatrix} 1 & 1 & 1 \\ a & b & k \\ a^2 & b^2 & k^2 \end{vmatrix} = (a-b)(b-k)(k-a) \qquad \text{[Replacing } c \text{ by } k \text{ in (1)]}$$

$$\therefore \quad x = \frac{D_1}{D} = \frac{(b-c)(c-k)(k-b)}{(b-c)(c-a)(a-b)} = \frac{(c-k)(k-b)}{(c-a)(a-b)},$$

$$y = \frac{D_2}{D} = \frac{(k-c)(c-a)(a-k)}{(b-c)(c-a)(a-b)} = \frac{(k-c)(a-k)}{(b-c)(a-b)},$$

and $$z = \frac{D_3}{D} = \frac{(a-b)(b-k)(k-a)}{(a-b)(b-c)(c-a)} = \frac{(b-k)(k-a)}{(b-c)(c-a)}$$

Hence, $x = \dfrac{(c-k)(k-b)}{(c-a)(a-b)}$, $y = \dfrac{(k-c)(a-k)}{(b-c)(a-b)}$ and $z = \dfrac{(b-k)(k-a)}{(b-c)(c-a)}$ **Ans.**

EXERCISE 17.6

Using Cramer's Rule, solve the following system of equations :

1. $2x - 3y = 7$, $7x - 3y = 10$ **Ans.** $x = \frac{3}{5}, y = -\frac{29}{15}$

2. $2x + y = 1$, $x - 2y = 8$ **Ans.** $x = 2, y = -3$

3. $2x + 3y = 10$, $x + 6y = 4.$ **Ans.** $x = \frac{16}{3}, y = -\frac{2}{9}$

4. $5x + 2y = 3$, $3x + 2y = 5.$ **Ans.** $x = -1, y = 4$

5. $7x - 2y = -7$, $2x - y = 1.$ **Ans.** $x = -3, y = -7$

6. $x - 2y = 4$, $-3x + 5y = -7$ **Ans.** $x = -6, y = -5$

7. $x - 4y - z = 11$
$2x - 5y + 2z = 39$
$-3x + 2y + z = 1.$ **Ans.** $x = -1, y = -5, z = 8$

8. $x + 3y - 2z = 5$
$2x + y + 4z = 8$
$6x + y - 3z = 5$ **Ans.** $x = 1, y = 2, z = 1$

9. $x + 2y + 5z = 23$
$3x + y + 4z = 26.$
$6x + y + 7z = 47$ **Ans.** $x = 4, y = 2, z = 3$

10. $x + y + z = 1$
$3x + 5y + 6z = 4$
$9x + 2y - 36x = 17$ **Ans.** $x = \frac{1}{3}, y = 1, z = -\frac{1}{3}$

11. $2y - z = 0$
$x + 3y = -4$
$3x + 4y = 3$ **Ans.** $x = 5, y = -3, z = -6$

12. $x + y + z = -1$
$x + 2y + 3z = -4$
$x + 3y + 4z = -6$ **Ans.** $x = 1, y = -1, z = -1$

13. $x + y + z = 1$ **Ans.** $x = \dfrac{(2-k)(3-k)}{2}$, $y = \dfrac{(1-k)(3-k)}{-1}$, $z = \dfrac{(1-k)(2-k)}{2}$

$x + 2y + 3z = k$
$1^2x + 2^2y + 3^2z = k^2$

14. Show that there are three real values of λ for which the equations:

$$(a-\lambda)x + by + cz = 0$$
$$bx + (c-\lambda)y + az = 0$$
$$cx + ay + (b-\lambda)z = 0$$

are simultaneously true, and that the product of these values of λ is $\begin{vmatrix} a & b & c \\ b & c & a \\ c & a & b \end{vmatrix}$

17.10 RULE FOR MULTIPLICATION OF TWO DETERMINANTS

Multiply the elements of the first row of Δ_1 with the corresponding elements of the first, the second and the third row of Δ_2 respectively.

Their respective sums form the elements of the first row of $\Delta_1\Delta_2$. Similarly multiply the elements of the second row of Δ_1 with the corresponding elements of first, second and third row of the Δ_2 to from the second row of $\Delta_1\Delta_2$ and so on.

Example 48. *Find the product*

$$\begin{vmatrix} a_1 & b_1 & c_1 \\ a_2 & b_2 & c_2 \\ a_3 & b_3 & c_3 \end{vmatrix} \times \begin{vmatrix} \alpha_1 & \beta_1 & \gamma_1 \\ \alpha_2 & \beta_2 & \gamma_2 \\ \alpha_3 & \beta_3 & \gamma_3 \end{vmatrix}$$

Solution. Produce of the given determinants

$$= \begin{vmatrix} a_1\alpha_1 + b_1\beta_1 + c_1\gamma_1 & a_1\alpha_2 + b_1\beta_2 + c_1\gamma_2 & a_1\alpha_3 + b_1\beta_3 + c_1\gamma_3 \\ a_2\alpha_1 + b_2\beta_1 + c_2\gamma_1 & a_2\alpha_2 + b_2\beta_2 + c_2\gamma_2 & a_2\alpha_3 + b_2\beta_3 + c_2\gamma_3 \\ a_3\alpha_1 + b_3\beta_1 + c_3\gamma_1 & a_3\alpha_2 + b_3\beta_2 + c_3\gamma_2 & a_3\alpha_2 + b_3\beta_3 + c_3\gamma_3 \end{vmatrix}$$ **Ans.**

Example 49. *Find*

$$\begin{vmatrix} a & b & c \\ b & c & a \\ c & a & b \end{vmatrix} \times \begin{vmatrix} -a & c & b \\ -b & a & c \\ -c & b & a \end{vmatrix}$$ *and hence show that*

$$= \begin{vmatrix} 2bc - a^2 & c^2 & b^2 \\ c^2 & 2ca - b^2 & a^2 \\ b^2 & a^2 & 2ab -- c^2 \end{vmatrix} = (a^3 + b^3 + c^3 - 3abc)^2$$

Solution. Product of the given determinants

$$= \begin{vmatrix} -a^2 + bc + bc & -ab + ab + c^2 & -ac + b^2 + ac \\ -ab + c^2 + ab & -b^2 + ac + ac & -bc + bc + a^2 \\ -ca + ca + b^2 & -bc + a^2 + bc & -c^2 + ab + ab \end{vmatrix} = \begin{vmatrix} 2bc - a^2 & c^2 & b^2 \\ c^2 & 2ca - b^2 & a^2 \\ b^2 & a^2 & 2ab - c^2 \end{vmatrix}$$

Now $\begin{vmatrix} -a & c & b \\ -b & a & c \\ -c & b & a \end{vmatrix} = (-1)^2 \begin{vmatrix} a & b & c \\ b & c & a \\ c & a & b \end{vmatrix} = a(bc - a^2) - b(b^2 - ac) + c(ab - c^2)$

$$= -(a^3 + b^3 + c^3 - 3abc)$$

Product $= (a^3 + b^3 + c^3 - 3abc)^2$ **Proved**

Example 50. *Prove that the determinant*

$$\begin{vmatrix} 2b_1+c_1 & c_1+3a_1 & 2a_1+3b_1 \\ 2b_2+c_2 & c_2+3a_2 & 2a_2+3b_2 \\ 2b_3+c_3 & c_3+3a_3 & 2a_3+3b_3 \end{vmatrix}$$

is a multiple of the determinant $\begin{vmatrix} a_1 & b_1 & c_1 \\ a_2 & b_2 & c_2 \\ a_3 & b_3 & c_3 \end{vmatrix}$ *and find the other factor.*

Solution. $\begin{vmatrix} 2b_1+c_1 & c_1+3a_1 & 2a_1+3b_1 \\ 2b_2+c_2 & c_2+3a_2 & 2a_2+3b_2 \\ 2b_3+c_3 & c_3+3a_3 & 2a_3+3b_3 \end{vmatrix} = \begin{vmatrix} a_1 & b_1 & c_1 \\ a_2 & b_2 & c_2 \\ a_3 & b_3 & c_3 \end{vmatrix} \times \begin{vmatrix} 0 & 2 & 1 \\ 3 & 0 & 1 \\ 2 & 3 & 0 \end{vmatrix}$ **Ans.**

Example 51. *Prove that* $\begin{vmatrix} 1 & \cos(\beta-\alpha) & \cos(\gamma-\alpha) \\ \cos(\alpha-\beta) & 1 & \cos(\gamma-\beta) \\ \cos(\alpha-\gamma) & \cos(\beta-\gamma) & 1 \end{vmatrix} = 0$

Solution. $\begin{vmatrix} \cos\alpha & \sin\alpha & 0 \\ \cos\beta & \sin\beta & 0 \\ \cos\gamma & \sin\gamma & 0 \end{vmatrix} \times \begin{vmatrix} \cos\alpha & \sin\alpha & 0 \\ \cos\beta & \sin\beta & 0 \\ \cos\gamma & \sin\gamma & 0 \end{vmatrix} = 0$

$$\Rightarrow \begin{vmatrix} \cos^2\alpha+\sin^2\alpha & \cos\alpha\cos\beta+\sin\alpha\sin\beta & \cos\alpha\cos\gamma+\sin\alpha\sin\gamma \\ \cos\beta\cos\alpha+\sin\beta\sin\alpha & \cos^2\beta+\sin^2\beta & \cos\beta\cos\gamma+\sin\beta\sin\gamma \\ \cos\gamma\cos\alpha+\sin\gamma\sin\alpha & \cos\gamma\cos\beta+\sin\gamma\sin\beta & \cos^2\gamma+\sin^2\gamma \end{vmatrix} = 0$$

The above determinant can be split into eight determinants and each determinants having identical column is zero. **Proved.**

ALGEBRA OF MATRICES

18.1 DEFINITION

Let us consider a set of simultaneous equations,

$$x + 2y + 3z + 5t = 0$$
$$4x + 2y + 5z + 7t = 0$$
$$3x + 4y + 2z + 6t = 0.$$

Now we write down the coefficients of x, y, z, t of the above equations and enclose them within brackets and then we get

$$A = \begin{bmatrix} 1 & 2 & 3 & 5 \\ 4 & 2 & 5 & 7 \\ 3 & 4 & 2 & 6 \end{bmatrix}$$

The above system of numbers, arranged in a rectangular array in rows and columns and bounded by the brackets, is called a matrix.

It has got 3 rows and 4 columns and in all $3 \times 4 = 12$ elements. It is termed as 3×4 matrix, to be read as [3 by 4 matrix]. In the double subscripts of an element, the first subscript determines the row and the second subscript determines the column in which the element lies, a_{ij} lies in the ith row and jth column.

18.2 VARIOUS TYPES OF MATRICES

(*i*) **Row Matrix.** If a matrix has only one row and any number of columns, it is called a *Row matrix*, *e.g.*,

$$[2\ 7\ 3\ 9]$$

(*b*) **Column Matrix.** A matrix, having one column and any number of rows, is called a *Column matrix*, *e.g.*, $\begin{bmatrix} 1 \\ 2 \\ 3 \end{bmatrix}$

(*c*) **Null Matrix or Zero Matrix.** Any matrix, in which all the elements are zeros, is called a *Zero matrix* or *Null matrix e.g.*,

$$\begin{bmatrix} 0 & 0 & 0 & 0 \\ 0 & 0 & 0 & 0 \end{bmatrix}$$

(*d*) **Square Matrix.** A matrix, in which the number of rows is equal to the number of columns, is called a square matrix *e.g.*,

$$\begin{bmatrix} 2 & 5 \\ 1 & 4 \end{bmatrix}$$

(*e*) **Diagonal Matrix.** A square matrix is called a diagonal matrix, if all its non-diagonal elements are zero *e.g.*,

$$\begin{bmatrix} 1 & 0 & 0 \\ 0 & 3 & 0 \\ 0 & 0 & 4 \end{bmatrix}$$

(*f*) **Scalar matrix.** A diagonal matrix in which all the diagonal elements are equal to a scalar, say (k) is called a scalar matrix.

For example;

$$\begin{bmatrix} 2 & 0 & 0 \\ 0 & 2 & 0 \\ 0 & 0 & 2 \end{bmatrix}, \begin{bmatrix} -6 & 0 & 0 & 0 \\ 0 & -6 & 0 & 0 \\ 0 & 0 & -6 & 0 \\ 0 & 0 & 0 & -6 \end{bmatrix}$$

i.e., $A = [a_{ij}]_{n \times n}$ is a scalar matrix if $a_{ij} = \begin{cases} 0, & \text{when } i \neq j \\ k, & \text{when } i = j \end{cases}$

(*g*) **Unit or Identity Matrix.** A square matrix is called a unit matrix if all the diagonal elements are unity and non-diagonal elements are zero *e.g.*,

$$\begin{bmatrix} 1 & 0 & 0 \\ 0 & 1 & 0 \\ 0 & 0 & 1 \end{bmatrix}, \begin{bmatrix} 1 & 0 \\ 0 & 1 \end{bmatrix}$$

(*h*) **Symmetric Matrix.** A square matrix will be called symmetric, if for all values of i and j, $a_{ij} = a_{ji}$ *i.e.*, $A' = A$

$$e.g., \begin{bmatrix} a & h & g \\ h & b & f \\ g & f & c \end{bmatrix}$$

(*i*) **Skew Symmetric Matrix.** A square matrix is called skew symmetric matrix, if

(1) $a_{ij} = -a_{ji}$ for all values of i and j, or $A' = -A$

(2) All diagonal elements are zero, *e.g.*,

$$\begin{bmatrix} 0 & -h & -g \\ h & 0 & -f \\ g & f & 0 \end{bmatrix}$$

(*j*) **Triangular Matrix.** (Echelon form) A square matrix, all of whose elements below the leading diagonal are zero, is called an *upper triangular matrix.* A square matrix, all of whose elements above the leading diagonal are zero, is called *a lower triangular matrix* *e.g.*,

$$\begin{bmatrix} 1 & 3 & 2 \\ 0 & 4 & 1 \\ 0 & 0 & 6 \end{bmatrix} \qquad \begin{bmatrix} 2 & 0 & 0 \\ 4 & 1 & 0 \\ 5 & 6 & 7 \end{bmatrix}$$

Upper triangular matrix | Lower triangular matrix

(*k*) **Transpose of a Matrix.** If in a given matrix A, we interchange the rows and the corresponding columns, the new matrix obtained is called the transpose of the matrix A and is denoted by A' or A^T *e.g.*,

$$A = \begin{bmatrix} 2 & 3 & 4 \\ 1 & 0 & 5 \\ 6 & 7 & 8 \end{bmatrix}, \ A' = \begin{bmatrix} 2 & 1 & 6 \\ 3 & 0 & 7 \\ 4 & 5 & 8 \end{bmatrix}$$

(*l*) **Orthogonal Matrix.** A square matrix A is called an orthogonal matrix if the product of the matrix A and the transpose matrix A' is an identity matrix *e.g.*,

$$A.\, A' = I$$

if $|A| = 1$, matrix A is proper.

(*m*) **Conjugate of a Matrix**

Let $$A = \begin{bmatrix} 1+i & 2-3i & 4 \\ 7+2i & -i & 3-2i \end{bmatrix}$$

Conjugate of matrix A is $\overline{A}$

$$\overline{A} = \begin{bmatrix} 1-i & 2+3i & 4 \\ 7-2i & i & 3+2i \end{bmatrix}$$

(*n*) **Matrix A^{θ}.** Transpose of the conjugate of a matrix A is denoted by A^{θ}.

Let $$A = \begin{bmatrix} 1+i & 2-3i & 4 \\ 7+2i & -i & 3-2i \end{bmatrix}$$

$$\overline{A} = \begin{bmatrix} 1-i & 2+3i & 4 \\ 7-2i & +i & 3+2i \end{bmatrix}$$

$$(\overline{A})' = \begin{bmatrix} 1-i & 7-2i \\ 2+3i & i \\ 4 & 3+2i \end{bmatrix}$$

$$A^{\theta} = \begin{bmatrix} 1-i & 7-2i \\ 2+3i & i \\ 4 & 3+2i \end{bmatrix}$$

(*o*) **Unitary Matrix.** A square matrix A is said to be unitary if

$$A^{\theta} A = I$$

e.g. $$A = \begin{bmatrix} \frac{1+i}{2} & \frac{-1+i}{2} \\ \frac{1+i}{2} & \frac{1-i}{2} \end{bmatrix}, \quad A^{\theta} = \begin{bmatrix} \frac{1-i}{2} & \frac{1-i}{2} \\ \frac{-1-i}{2} & \frac{1+i}{2} \end{bmatrix}, \quad A \cdot A^{\theta} = I$$

(*p*) **Hermitian Matrix.** A square matrix $A = (a_{ij})$ is called Hermitian matrix, if every *i-jth* element of A is equal to conjugate complex *j-i*th element of A.

In other words, $$a_{ij} = \overline{a}_{ji}$$

e.g. $$\begin{bmatrix} 1 & 2+3i & 3+i \\ 2-3i & 2 & 1-2i \\ 3-i & 1+2i & 5 \end{bmatrix}$$

Necessary and sufficient condition for a matrix A to be Hermitian is that $A = A^{\theta}$ *i.e.* conjugate transpose of A

$$\Rightarrow \qquad A = (\overline{A})'.$$

(*q*) **Skew Hermitian Matrix.** A square matrix $A = (a_{ij})$ will be called a Skew Hermitian matrix if every *i-j*th element of A is equal to negative conjugate complex of *j-i*th element of A.

In other words, $$a_{ij} = -\overline{a}_{ji}$$

All the elements in the principal diagonal will be of the form

$$a_{ii} = -\bar{a}_{ii} \qquad \text{or} \qquad a_{ii} + \bar{a}_{ii} = 0$$

If $\qquad a_{ii} = a + ib \qquad$ then $\quad \bar{a}_{ii} = a - ib$

$$(a + ib) + (a - ib) = 0 \qquad \Rightarrow \qquad 2a = 0 \Rightarrow a = 0$$

So, a_{ii} is pure imaginary $\Rightarrow a_{ii} = 0$.

Hence, all the diagonal elements of a Skew Hermitian Matrix are either zeros or pure imaginary.

$$e.g. \qquad \begin{bmatrix} i & 2-3i & 4+5i \\ -(2+3i) & 0 & 2i \\ -(4-5i) & 2i & -3i \end{bmatrix}$$

The necessary and sufficient condition for a matrix A to be Skew Hermitian is that

$$A^{\theta} = -A$$

$$(\bar{A})' = -A$$

(*r*) **Idempotent Matrix.** A matrix, such that $A^2 = A$ is called Idempotent Matrix.

$$e.g.\ A = \begin{bmatrix} 2 & -2 & -4 \\ -1 & 3 & 4 \\ 1 & -2 & -3 \end{bmatrix}, A^2 = \begin{bmatrix} 2 & -2 & -4 \\ -1 & 3 & 4 \\ 1 & -2 & -3 \end{bmatrix}\begin{bmatrix} 2 & -2 & -4 \\ -1 & 3 & 4 \\ 1 & -2 & -3 \end{bmatrix} = \begin{bmatrix} 2 & -2 & -4 \\ -1 & 3 & 4 \\ 1 & -2 & -3 \end{bmatrix} = A$$

(*s*) **Periodic Matrix.** A matrix A will be called a Periodic Matrix, if

$$A^{k+1} = A$$

where k is a +ve integer. If k is the least + ve integer, for which $A^{k+1} = A$, then k is said to be the period of A. If we choose $k = 1$, we get $A^2 = A$ and we call it to be idempotent matrix.

(*t*) **Nilpotent Matrix.** A matrix will be called a Nilpotent matrix, if $A^k = 0$ (null matrix) where k is a +ve integer ; if however k is the least +ve integer for which $A^k = 0$, then k is the *index* of the nilpotent matrix.

$$e.g.,\ A = \begin{bmatrix} ab & b^2 \\ -a^2 & -ab \end{bmatrix}, A^2 = \begin{bmatrix} ab & b^2 \\ -a^2 & -ab \end{bmatrix}\begin{bmatrix} ab & b^2 \\ -a^2 & -ab \end{bmatrix} = \begin{bmatrix} 0 & 0 \\ 0 & 0 \end{bmatrix} = 0$$

A is nilpotent matrix whose index is 2.

(*u*) **Involuntary Matrix.** A matrix A will be called an Involuntary matrix, if $A^2 = I$ (unit matrix). Since $I^2 = I$ always $\quad \therefore$ Unit matrix is involuntary.

(*v*) **Equal Matrices.** Two matrices are said to be equal if

(*i*) They are of the same order.

(*ii*) The elements in the corresponding positions are equal.

Thus if $\qquad A = \begin{bmatrix} 2 & 3 \\ 1 & 4 \end{bmatrix}, B = \begin{bmatrix} 2 & 3 \\ 1 & 4 \end{bmatrix}$

Here $\qquad A = B$

(*w*) **Singular Matrix.** If the determinant of the matrix is zero, then the matrix is known as singular matrix e.g. $A = \begin{bmatrix} 1 & 2 \\ 3 & 6 \end{bmatrix}$ is singular matrix, because $|A| = 6 - 6 = 0$.

Example 1. *Find the values of x, y, z and 'a' which satisfy the matrix equation.*

$$\begin{bmatrix} x+3 & 2y+x \\ z-1 & 4a-6 \end{bmatrix} = \begin{bmatrix} 0 & -7 \\ 3 & 2a \end{bmatrix}$$

Solution. As the given matrices are equal, so their corresponding elements are equal.

$$x + 3 = 0 \quad \Rightarrow \quad x = -3 \qquad ...(1)$$
$$2y + x = -7 \qquad ...(2)$$
$$z - 1 = 3 \quad \Rightarrow \quad z = 4 \qquad ...(3)$$
$$4a - 6 = 2a \quad \Rightarrow \quad a = 3 \qquad ...(4)$$

Putting the value of $x = -3$ from (1) into (2), we have

$$2y - 3 = -7 \quad \Rightarrow \quad y = -2$$

Hence, $\quad x = -3, \quad y = -2, \quad z = 4, \quad a = 3$ **Ans.**

18.3 ADDITION OF MATRICES

If A and B be two matrices of the same order, then their sum, $A + B$ is defined as the matrix, each element of which is the sum of the corresponding elements of A and B.

Thus if $\quad A = \begin{bmatrix} 4 & 2 & 5 \\ 1 & 3 & -6 \end{bmatrix}, \quad B = \begin{bmatrix} 1 & 0 & 2 \\ 3 & 1 & 4 \end{bmatrix}$

then $\quad A + B = \begin{bmatrix} 4+1 & 2+0 & 5+2 \\ 1+3 & 3+1 & -6+4 \end{bmatrix} = \begin{bmatrix} 5 & 2 & 7 \\ 4 & 4 & -2 \end{bmatrix}$

If $\quad A = [a_{ij}], \; B = [b_{ij}] \;$ then $\; A + B = [a_{ij} + b_{ij}]$

Example 2. *Show that any square matrix can be expressed as the sum of two matrices, one symmetric and the other anti-symmetric.*

Solution. Let A be a given square matrix.

Then $\quad A = \frac{1}{2}(A + A') + \frac{1}{2}(A - A')$

Now, $\quad (A + A')' = A' + A = A + A'$.

$\therefore \quad A + A'$ is a symmetric matrix.

Also, $\quad (A - A')' = A' - A = -(A - A')$

$\therefore \quad A - A'$ or $\frac{1}{2}$ (A – A′) is an anti-symmetric matrix.

$$\boxed{\mathbf{A = \frac{1}{2}(A + A') + \frac{1}{2}(A - A')}}$$

Square matrix = Symmetric matrix + Anti-symmetric matrix **Proved.**

Example 3. *Write matrix A given below as the sum of a symmetric and a skew symmetric matrix.*

$$A = \begin{pmatrix} 1 & 2 & 4 \\ -2 & 5 & 3 \\ -1 & 6 & 3 \end{pmatrix}$$

Solution. $A = \begin{bmatrix} 1 & 2 & 4 \\ -2 & 5 & 3 \\ -1 & 6 & 3 \end{bmatrix}$ On transposing, we get $A' = \begin{bmatrix} 1 & -2 & -1 \\ 2 & 5 & 6 \\ 4 & 3 & 3 \end{bmatrix}$

On adding A and A′, we have

$$A + A' = \begin{bmatrix} 1 & 2 & 4 \\ -2 & 5 & 3 \\ -1 & 6 & 3 \end{bmatrix} + \begin{bmatrix} 1 & -2 & -1 \\ 2 & 5 & 6 \\ 4 & 3 & 3 \end{bmatrix} = \begin{bmatrix} 2 & 0 & 3 \\ 0 & 10 & 9 \\ 3 & 9 & 6 \end{bmatrix} \qquad ...(1)$$

On subtracting A' from A, we get

$$A - A' = \begin{bmatrix} 1 & 2 & 4 \\ -2 & 5 & 3 \\ -1 & 6 & 3 \end{bmatrix} - \begin{bmatrix} 1 & -2 & -1 \\ 2 & 5 & 6 \\ 4 & 3 & 3 \end{bmatrix} = \begin{bmatrix} 0 & 4 & 5 \\ -4 & 0 & -3 \\ -5 & 3 & 0 \end{bmatrix} \qquad ...(2)$$

On adding (1) and (2), we have

$$2A = \begin{bmatrix} 2 & 0 & 3 \\ 0 & 10 & 9 \\ 3 & 9 & 6 \end{bmatrix} + \begin{bmatrix} 0 & 4 & 5 \\ -4 & 0 & -3 \\ -5 & 3 & 0 \end{bmatrix}$$

$$A = \begin{bmatrix} 1 & 0 & \frac{3}{2} \\ 0 & 5 & \frac{9}{2} \\ \frac{3}{2} & \frac{9}{2} & 3 \end{bmatrix} + \begin{bmatrix} 0 & 2 & \frac{5}{2} \\ -2 & 0 & -\frac{3}{2} \\ \frac{-5}{2} & \frac{3}{2} & 0 \end{bmatrix}$$

A = [Symmetric matrix] + [Skew symmetric matrix.] **Ans.**

Example 4. *Express* $A = \begin{bmatrix} 1 & -2 & -3 \\ 3 & 0 & 5 \\ 5 & 6 & 1 \end{bmatrix}$ *as the sum of a lower triangular matrix and upper triangular matrix.*

Solution. Let $A = L + U$

$$\begin{bmatrix} 1 & -2 & -3 \\ 3 & 0 & 5 \\ 5 & 6 & 1 \end{bmatrix} = \begin{bmatrix} a & 0 & 0 \\ b & c & 0 \\ d & e & f \end{bmatrix} + \begin{bmatrix} 1 & p & q \\ 0 & 1 & r \\ 0 & 0 & 1 \end{bmatrix}$$

$$\begin{bmatrix} 1 & -2 & -3 \\ 3 & 0 & 5 \\ 5 & 6 & 1 \end{bmatrix} = \begin{bmatrix} a+1 & 0+p & 0+q \\ b+0 & c+1 & 0+r \\ d+0 & e+0 & f+1 \end{bmatrix}$$

Equating the corresponding elements on both the sides, we get

$$a + 1 = 1 \qquad p = -2 \qquad q = -3$$
$$b = 3 \qquad c + 1 = 0 \qquad r = 5$$
$$d = 5 \qquad e = 6 \qquad f + 1 = 1$$

On solving these equations, we get

$$a = 0 \qquad p = -2 \qquad q = -3$$
$$b = 3 \qquad c = -1 \qquad r = 5$$
$$d = 5 \qquad e = 6 \qquad f = 0$$

Hence, $L = \begin{bmatrix} 0 & 0 & 0 \\ 3 & -1 & 0 \\ 5 & 6 & 0 \end{bmatrix}$ and $U = \begin{bmatrix} 1 & -2 & -3 \\ 0 & 1 & 5 \\ 0 & 0 & 1 \end{bmatrix}$ **Ans.**

18.4 PROPERTIES OF MATRIX ADDITION

Only matrices of the same order can be added or subtracted.

(*i*) **Commutative Law.** $A + B = B + A$.

(*ii*) **Associative law.** $A + (B + C) = (A + B) + C$.

18.5 SUBTRACTION OF MATRICES

The difference of two matrices is a matrix, each element of which is obtained by subtracting the elements of the second matrix from the corresponding element of the first.

$$A - B = [a_{ij} - b_{ij}]$$

Thus
$$\begin{bmatrix} 8 & 6 & 4 \\ 1 & 2 & 0 \end{bmatrix} - \begin{bmatrix} 3 & 5 & 1 \\ 7 & 6 & 2 \end{bmatrix}$$

$$= \begin{bmatrix} 8-3 & 6-5 & 4-1 \\ 1-7 & 2-6 & 0-2 \end{bmatrix} = \begin{bmatrix} 5 & 1 & 3 \\ -6 & -4 & -2 \end{bmatrix}$$ **Ans.**

18.6 SCALAR MULTIPLE OF A MATRIX

If a matrix is multiplied by a scalar quantity k, then each element is multiplied by k, *i.e.*

$$A = \begin{bmatrix} 2 & 3 & 4 \\ 4 & 5 & 6 \\ 6 & 7 & 9 \end{bmatrix}$$

$$3A = 3\begin{bmatrix} 2 & 3 & 4 \\ 4 & 5 & 6 \\ 6 & 7 & 9 \end{bmatrix} = \begin{bmatrix} 3\times 2 & 3\times 3 & 3\times 4 \\ 3\times 4 & 3\times 5 & 3\times 6 \\ 3\times 6 & 3\times 7 & 3\times 9 \end{bmatrix} = \begin{bmatrix} 6 & 9 & 12 \\ 12 & 15 & 18 \\ 18 & 21 & 27 \end{bmatrix}$$

EXERCISE 18.1

1. (*i*) If $A = \begin{bmatrix} -1 & 7 & 1 \\ 2 & 3 & 4 \\ 5 & 0 & 5 \end{bmatrix}$, represent it as A = B + C where B is a symmetric and C is a skew-symmetric matrix.

   (*b*) Express $\begin{bmatrix} 1 & 2 & 0 \\ 3 & 7 & 1 \\ 5 & 9 & 3 \end{bmatrix}$ as a sum of symmetric and skew-symmetric matrix.

   Ans. (*i*) $A = \begin{bmatrix} -1 & \frac{9}{2} & 3 \\ \frac{9}{2} & 3 & 2 \\ 3 & 2 & 5 \end{bmatrix} + \begin{bmatrix} 0 & \frac{5}{2} & -2 \\ \frac{-5}{2} & 0 & 2 \\ 2 & -2 & 0 \end{bmatrix}$ (*b*) $A = \begin{bmatrix} 1 & \frac{5}{2} & \frac{5}{2} \\ \frac{5}{2} & 7 & 5 \\ \frac{5}{2} & 5 & 3 \end{bmatrix} + \begin{bmatrix} 0 & -\frac{1}{2} & -\frac{5}{2} \\ \frac{1}{2} & 0 & -4 \\ \frac{5}{2} & 4 & 0 \end{bmatrix}$

2. Matrices A and B are such that

   $$3A - 2B = \begin{bmatrix} 2 & 1 \\ -2 & -1 \end{bmatrix} \text{ and } -4A + B = \begin{bmatrix} -1 & 2 \\ -4 & 3 \end{bmatrix}$$

   Find A and B. **Ans.** $A = \begin{bmatrix} 0 & -1 \\ 2 & -1 \end{bmatrix}$, $B = \begin{bmatrix} -1 & -2 \\ 4 & -1 \end{bmatrix}$

3. Given $3\begin{bmatrix} x & y \\ z & w \end{bmatrix} = \begin{bmatrix} x & 6 \\ -1 & 2w \end{bmatrix} + \begin{bmatrix} 4 & x+y \\ z+w & 3 \end{bmatrix}$

   Find x, y, z and w. **Ans.** $x = 2,\ y = 4,\ z = 1,\ w = 3$

4. If $A = \begin{bmatrix} 0 & 2 & 0 \\ 1 & 0 & 3 \\ 1 & 1 & 2 \end{bmatrix}$, $B = \begin{bmatrix} 1 & 2 & 1 \\ 2 & 1 & 0 \\ 0 & 0 & 3 \end{bmatrix}$ **Ans.** (*i*) $= \begin{bmatrix} 3 & 10 & 3 \\ 8 & 3 & 6 \\ 2 & 2 & 13 \end{bmatrix}$, (*ii*) $= \begin{bmatrix} -4 & -2 & -4 \\ -5 & -4 & 9 \\ 3 & 3 & -6 \end{bmatrix}$

   Find (*i*) $2A + 3B$ (*ii*) $3A - 4B$.

18.7 MULTIPLICATION

The product of two matrices A and B is only possible if the number of columns in A is equal to the number of rows in B.

Let $A = [a_{ij}]$ be an $m \times n$ matrix and $B = [b_{ij}]$ be an $n \times p$ matrix. Then the product AB of these matrices is an $m \times p$ matrix $C = [c_{ij}]$ where

$$c_{ij} = a_{i1}\, b_{1j} + a_{i2}\, b_{2j} + a_{i3}\, b_{3j} + \ldots + a_{in}\, b_{nj}$$

18.8 $(AB)' = B'A'$

If A and B are two matrices conformal for product AB, then show that $(AB)' = B'A'$, where dash represents transpose of a matrix.

Solution. Let $A = (a_{ij})$ be an $m \times n$ matrix and $B = (b_{ij})$ be $n \times p$ matrix.

Since AB is $m \times p$ matrix, $(AB)'$ is a $p \times m$ matrix.

Further B' *is* $p \times n$ matrix and A' an $n \times m$ matrix and therefore $B'\,A'$ is a $p \times m$ matrix.

Then $(AB)'$ and $B'\,A'$ are matrices of the same order.

Now the (j, i)th element of $(AB)' = (i, j)$th element of $(AB) = \sum_{k=1}^{n} a_{ik}b_{kj}$ ...(1)

Also the jth row of B' is $b_{1j}, b_{2j} \ldots b_{nj}$, and ith column of A' is $a_{i1}, a_{i2}, a_{i3} \ldots a_{in}$.

$\therefore$ (j, i)th element of $B'A' = \sum_{k=1}^{n} b_{kj}a_{ik}$ ...(2)

From (1) and (2), we have (j, i)th element of $(AB)' = (j, i)$ th element of $B'A'$.

As the matrices $(AB)'$ and $B'A'$ are of the same order and their corresponding elements are equal, we have $(AB)' = B'A'$. **Proved.**

18.9 PROPERTIES OF MATRIX MULTIPLICATION

1. Multiplication of matrices is not commutative.

$$AB \neq BA$$

2. Matrix multiplication is associative, if conformability is assured.

$$A\,(BC) = (AB)\,C$$

3. Matrix multiplication is distributive with respect to addition.

$$A\,(B + C) = AB + AC$$

4. Multiplication of matrix A by unit matrix.

$$AI = IA = A$$

5. Multiplicative inverse of a matrix exists if $|A| \neq 0$.

$$A \,.\, A^{-1} = A^{-1} \,.\, A = I$$

6. If A is a square then $A \times A = A^2$, $A \times A \times A = A^3$.

7. $A^0 = I$

8. $I^n = I$, where n is positive integer.

Example 5. *If* $A = \begin{bmatrix} 0 & 1 & 2 \\ 1 & 2 & 3 \\ 2 & 3 & 4 \end{bmatrix}$ *and* $B = \begin{bmatrix} 1 & -2 \\ -1 & 0 \\ 2 & -1 \end{bmatrix}$

obtain the product AB and explain why BA is not defined.

Solution. The number of columns in A is 3 and the number of rows in B is also 3, therefore the product AB is defined.

$$AB = \begin{bmatrix} 0 & 1 & 2 \\ 1 & 2 & 3 \\ 2 & 3 & 4 \end{bmatrix} \begin{matrix} R_1 \\ R_2 \\ R_3 \end{matrix} \times \overset{\begin{matrix} C_1 & C_2 \end{matrix}}{\begin{bmatrix} 1 & -2 \\ -1 & 0 \\ 2 & -1 \end{bmatrix}} = \begin{bmatrix} R_1 C_1 & R_1 C_2 \\ R_2 C_1 & R_2 C_2 \\ R_3 C_1 & R_3 C_2 \end{bmatrix}$$

R_1, R_2, R_3 are rows of A and C_1, C_2 are columns of B.

$$= \begin{bmatrix} \boxed{0\ 1\ 2}\boxed{\begin{matrix} 1 \\ -1 \\ 2 \end{matrix}} & \boxed{0\ 1\ 2}\boxed{\begin{matrix} -2 \\ 0 \\ -1 \end{matrix}} \\ \boxed{1\ 2\ 3}\boxed{\begin{matrix} 1 \\ -1 \\ 2 \end{matrix}} & \boxed{1\ 2\ 3}\boxed{\begin{matrix} -2 \\ 0 \\ -1 \end{matrix}} \\ \boxed{2\ 3\ 4}\boxed{\begin{matrix} 1 \\ -1 \\ 2 \end{matrix}} & \boxed{2\ 3\ 4}\boxed{\begin{matrix} -2 \\ 0 \\ -1 \end{matrix}} \end{bmatrix}$$

For convenience of multiplication, we write the columns in horizontal rectangles.

$$= \begin{bmatrix} \boxed{0 \quad 1 \quad 2} & \boxed{0 \quad 1 \quad 2} \\ \boxed{1 \quad -1 \quad 2} & \boxed{-2 \quad 0 \quad -1} \\ \boxed{1 \quad 1 \quad 3} & \boxed{1 \quad 2 \quad 3} \\ \boxed{1 \quad -1 \quad 2} & \boxed{-2 \quad 0 \quad -1} \\ \boxed{2 \quad 3 \quad 4} & \boxed{2 \quad 3 \quad 4} \\ \boxed{1 \quad -1 \quad 2} & \boxed{-2 \quad 0 \quad -1} \end{bmatrix}$$

$$= \begin{bmatrix} 0 \times 1 + 1 \times (-1) + 2 \times 2 & 0 \times (-2) + 1 \times 0 + 2 \times (-1) \\ 1 \times 1 + 2 \times (-1) + 3 \times 2 & 1 \times (-2) + 2 \times 0 + 3 \times (-1) \\ 2 \times 1 + 3 \times (-1) + 4 \times 2 & 2 \times (-2) + 3 \times 0 + 4 \times (-1) \end{bmatrix}$$

$$= \begin{bmatrix} 0 - 1 + 4 & 0 + 0 - 2 \\ 1 - 2 + 6 & -2 + 0 - 3 \\ 2 - 3 + 8 & -4 + 0 - 4 \end{bmatrix} = \begin{bmatrix} 3 & -2 \\ 5 & -5 \\ 7 & -8 \end{bmatrix}$$ **Ans.**

Since, the number of columns of B is (2) $\neq$ the number of rows of A is 3, BA is not defined.

Example 6. *If* $A = \begin{bmatrix} 1 & -2 & 3 \\ 2 & 3 & -1 \\ -3 & 1 & 2 \end{bmatrix}$ *and* $B = \begin{bmatrix} 1 & 0 & 2 \\ 0 & 1 & 2 \\ 1 & 2 & 0 \end{bmatrix}$

from the products AB and BA, and show that $AB \neq BA$.

Solution. Here, $$AB = \begin{bmatrix} 1 & -2 & 3 \\ 2 & 3 & -1 \\ -3 & 1 & 2 \end{bmatrix} \begin{bmatrix} 1 & 0 & 2 \\ 0 & 1 & 2 \\ 1 & 2 & 0 \end{bmatrix}$$

$$= \begin{bmatrix} 1 - 0 + 3 & 0 - 2 + 6 & 2 - 4 + 0 \\ 2 + 0 - 1 & 0 + 3 - 2 & 4 + 6 - 0 \\ -3 + 0 + 2 & 0 + 1 + 4 & -6 + 2 + 0 \end{bmatrix} = \begin{bmatrix} 4 & 4 & -2 \\ 1 & 1 & 10 \\ -1 & 5 & -4 \end{bmatrix}$$

$$BA = \begin{bmatrix}1 & 0 & 2\\0 & 1 & 2\\1 & 2 & 0\end{bmatrix}\begin{bmatrix}1 & -2 & 3\\2 & 3 & -1\\-3 & 1 & 2\end{bmatrix} = \begin{bmatrix}1+0-6 & -2+0+2 & 3-0+4\\0+2-6 & 0+3+2 & 0-1+4\\1+4+0 & -2+6+0 & 3-2+0\end{bmatrix} = \begin{bmatrix}-5 & 0 & 7\\-4 & 5 & 3\\5 & 4 & 1\end{bmatrix}$$

$AB \neq BA$ **Proved.**

Example 7. *If* $A = \begin{bmatrix}1 & 2\\-2 & 3\end{bmatrix}$, $B = \begin{bmatrix}2 & 1\\2 & 3\end{bmatrix}$ *and* $C = \begin{bmatrix}-3 & 1\\2 & 0\end{bmatrix}$

Verify that $(AB)\,C = A\,(BC)$ *and* $A\,(B + C) = AB + AC$.

Solution. We have, $AB = \begin{bmatrix}1 & 2\\-2 & 3\end{bmatrix} \times \begin{bmatrix}2 & 1\\2 & 3\end{bmatrix}$

$$= \begin{bmatrix}(1)(2)+(2)(2) & (1)(1)+(2)(3)\\(-2)(2)+(3)(2) & (-2)(1)+(3)(3)\end{bmatrix} = \begin{bmatrix}6 & 7\\2 & 7\end{bmatrix}$$

$$BC = \begin{bmatrix}2 & 1\\2 & 3\end{bmatrix} \times \begin{bmatrix}-3 & 1\\2 & 0\end{bmatrix} = \begin{bmatrix}-6+2 & 2+0\\-6+6 & 2+0\end{bmatrix} = \begin{bmatrix}-4 & 2\\0 & 2\end{bmatrix}$$

$$AC = \begin{bmatrix}1 & 2\\-2 & 3\end{bmatrix} \times \begin{bmatrix}-3 & 1\\2 & 0\end{bmatrix} = \begin{bmatrix}-3+4 & 1+0\\6+6 & -2+0\end{bmatrix} = \begin{bmatrix}1 & 1\\12 & -2\end{bmatrix}$$

$$B + C = \begin{bmatrix}2+(-3) & 1+1\\2+2 & 3+0\end{bmatrix} = \begin{bmatrix}-1 & 2\\4 & 3\end{bmatrix}$$

(*i*) $$(AB)\,C = \begin{bmatrix}6 & 7\\2 & 7\end{bmatrix} \times \begin{bmatrix}-3 & 1\\2 & 0\end{bmatrix} = \begin{bmatrix}-18+14 & 6+0\\-6+14 & 2+0\end{bmatrix} = \begin{bmatrix}-4 & 6\\8 & 2\end{bmatrix} \quad ...(1)$$

and $$A\,(BC) = \begin{bmatrix}1 & 2\\-2 & 3\end{bmatrix}\begin{bmatrix}-4 & 2\\0 & 2\end{bmatrix} = \begin{bmatrix}-4+0 & 2+4\\8+0 & -4+6\end{bmatrix} = \begin{bmatrix}-4 & 6\\8 & 2\end{bmatrix} \quad ...(2)$$

Thus from (1) and (2), we get

$$(AB)\,C = A\,(BC)$$

(*ii*) $$A\,(B + C) = \begin{bmatrix}1 & 2\\-2 & 3\end{bmatrix}\begin{bmatrix}-1 & 2\\4 & 3\end{bmatrix}$$

$$= \begin{bmatrix}-1+8 & 2+6\\2+12 & -4+9\end{bmatrix} = \begin{bmatrix}7 & 8\\14 & 5\end{bmatrix} \quad ...(3)$$

$$AB + AC = \begin{bmatrix}6+1 & 7+1\\2+12 & 7-2\end{bmatrix} = \begin{bmatrix}7 & 8\\14 & 5\end{bmatrix} \quad ...(4)$$

Thus from (3) and (4), we get

$A\,(B + C) = AB + AC$ **Verified.**

Example 8. *If* $A = \begin{pmatrix}1 & 2 & 2\\2 & 1 & 2\\2 & 2 & 1\end{pmatrix}$ *show that* $A^2 - 4A - 5I = 0$ *where* I, 0 *are the unit matrix and the null matrix of order 3 respectively. Use this result to find* A^{-1}. (*A.M.I.E., Summer 2004*)

Solution. Here, we have $A = \begin{bmatrix}1 & 2 & 2\\2 & 1 & 2\\2 & 2 & 1\end{bmatrix}$

$$A^2 = \begin{bmatrix} 1 & 2 & 2 \\ 2 & 1 & 2 \\ 2 & 2 & 1 \end{bmatrix}\begin{bmatrix} 1 & 2 & 2 \\ 2 & 1 & 2 \\ 2 & 2 & 1 \end{bmatrix} = \begin{bmatrix} 9 & 8 & 8 \\ 8 & 9 & 8 \\ 8 & 8 & 9 \end{bmatrix}$$

$$A^2 - 4A - 5I = \begin{bmatrix} 9 & 8 & 8 \\ 8 & 9 & 8 \\ 8 & 8 & 9 \end{bmatrix} - 4\begin{bmatrix} 1 & 2 & 2 \\ 2 & 1 & 2 \\ 2 & 2 & 1 \end{bmatrix} - 5\begin{bmatrix} 1 & 0 & 0 \\ 0 & 1 & 0 \\ 0 & 0 & 1 \end{bmatrix}$$

$$A^2 - 4A - 5I = \begin{bmatrix} 9-4-5 & 8-8-0 & 8-8-0 \\ 8-8-0 & 9-4-5 & 8-8-0 \\ 8-8-0 & 8-8-0 & 9-4-5 \end{bmatrix} = \begin{bmatrix} 0 & 0 & 0 \\ 0 & 0 & 0 \\ 0 & 0 & 0 \end{bmatrix}$$

$$A^2 - 4A - 5I = 0 \quad \Rightarrow \quad 5I = A^2 - 4A$$

Multiplying by A^{-1}, we get

$$5A^{-1} = A - 4I$$

$$= \begin{bmatrix} 1 & 2 & 2 \\ 2 & 1 & 2 \\ 2 & 2 & 1 \end{bmatrix} - 4\begin{bmatrix} 1 & 0 & 0 \\ 0 & 1 & 0 \\ 0 & 0 & 1 \end{bmatrix} = \begin{bmatrix} -3 & 2 & 2 \\ 2 & -3 & 2 \\ 2 & 2 & -3 \end{bmatrix}$$

$$A^{-1} = \frac{1}{5}\begin{bmatrix} -3 & 2 & 2 \\ 2 & -3 & 2 \\ 2 & 2 & -3 \end{bmatrix}$$ **Ans.**

Example 9. *Show by means of an example that in matrices $AB = 0$ does not necessarily mean that either $A = 0$ or $B = 0$, where 0 stands for the null matrix.*

Solution. Let $$A = \begin{bmatrix} 1 & -1 & 1 \\ -3 & 2 & -1 \\ -2 & 1 & 0 \end{bmatrix}, \quad B = \begin{bmatrix} 1 & 2 & 3 \\ 2 & 4 & 6 \\ 1 & 2 & 3 \end{bmatrix}$$

$$AB = \begin{bmatrix} 1-2+1 & 2-4+2 & 3-6+3 \\ -3+4-1 & -6+8-2 & -9+12-3 \\ -2+2+0 & -4+4+0 & -6+6+0 \end{bmatrix} = \begin{bmatrix} 0 & 0 & 0 \\ 0 & 0 & 0 \\ 0 & 0 & 0 \end{bmatrix}$$

$$AB = 0.$$

But here neither $A = 0$ nor $B = 0$. **Proved.**

Example 10. *If $AB = AC$, it is not necessarily true that $B = C$ i.e. like ordinary algebra, the equal matrices in the identity cannot be cancelled.*

Solution. Let $$AB = \begin{bmatrix} 1 & -3 & 2 \\ 2 & 1 & -3 \\ 4 & -3 & -1 \end{bmatrix}\begin{bmatrix} 1 & 4 & 1 & 0 \\ 2 & 1 & 1 & 1 \\ 1 & -2 & 1 & 2 \end{bmatrix} = \begin{bmatrix} -3 & -3 & 0 & 1 \\ 1 & 15 & 0 & -5 \\ -3 & 15 & 0 & -5 \end{bmatrix}$$

$$AC = \begin{bmatrix} 1 & -3 & 2 \\ 2 & 1 & -3 \\ 4 & -3 & -1 \end{bmatrix}\begin{bmatrix} 2 & 1 & -1 & -2 \\ 3 & -2 & -1 & -1 \\ 2 & -5 & -1 & 0 \end{bmatrix} = \begin{bmatrix} -3 & -3 & 0 & 1 \\ 1 & 15 & 0 & -5 \\ -3 & 15 & 0 & -5 \end{bmatrix}$$ **Proved.**

Here, $AB = AC$. But $B \neq C$.

Example 11. *Represent each of the transformations*

$$x_1 = 3y_1 + 2y_2, \quad y_1 = z_1 + 2z_2 \quad \text{and} \quad x_2 = -y_1 + 4y_2, \; y_2 = 3z_1$$

by the use of matrices and find the composite transformation which expresses x_1, x_2 in terms of z_1, z_2.

Solution. The equations in the matrix form are

$$\begin{bmatrix} x_1 \\ x_2 \end{bmatrix} = \begin{bmatrix} 3 & 2 \\ -1 & 4 \end{bmatrix} \begin{bmatrix} y_1 \\ y_2 \end{bmatrix} \qquad ...(1)$$

$$\begin{bmatrix} y_1 \\ y_2 \end{bmatrix} = \begin{bmatrix} 1 & 2 \\ 3 & 0 \end{bmatrix} \begin{bmatrix} z_1 \\ z_2 \end{bmatrix} \qquad ...(2)$$

Substituting the values of y_1, y_2 in (1), we get

$$\begin{bmatrix} x_1 \\ x_2 \end{bmatrix} = \begin{bmatrix} 3 & 2 \\ -1 & 4 \end{bmatrix} \begin{bmatrix} 1 & 2 \\ 3 & 0 \end{bmatrix} \begin{bmatrix} z_1 \\ z_2 \end{bmatrix} = \begin{bmatrix} 9 & 6 \\ 11 & -2 \end{bmatrix} \begin{bmatrix} z_1 \\ z_2 \end{bmatrix} = \begin{bmatrix} 9z_1 + 6z_2 \\ 11z_1 - 2z_2 \end{bmatrix}$$

$$x_1 = 9z_1 + 6z_2, \quad x_2 = 11z_1 - 2z_2$$ **Ans.**

Example 12. *Prove that the product of two matrices*

$$\begin{bmatrix} \cos^2 \theta & \cos \theta \sin \theta \\ \cos \theta \sin \theta & \sin^2 \theta \end{bmatrix} \text{ and } \begin{bmatrix} \cos^2 \phi & \cos \phi \sin \phi \\ \cos \phi \sin \phi & \sin^2 \phi \end{bmatrix}$$

is zero when θ and ϕ differ by an odd multiple of $\frac{\pi}{2}$.

Solution. $= \begin{bmatrix} \cos^2 \theta & \cos \theta \sin \theta \\ \cos \theta \sin \theta & \sin^2 \theta \end{bmatrix} \times \begin{bmatrix} \cos^2 \phi & \cos \phi \sin \phi \\ \cos \phi \sin \phi & \sin^2 \phi \end{bmatrix}$

$$= \begin{bmatrix} \cos^2 \theta \cos^2 \phi + \cos \theta \sin \theta \cos \phi \sin \phi & \cos^2 \theta \cos \phi \sin \phi + \cos \theta \sin \theta \sin^2 \phi \\ \cos \theta \sin \theta \cos^2 \phi + \sin^2 \theta \cos \phi \sin \phi & \cos \theta \sin \theta \cos \phi \sin \phi + \sin^2 \theta \sin^2 \phi \end{bmatrix}$$

$$= \begin{bmatrix} \cos \theta \cos \phi (\cos \theta \cos \phi + \sin \theta \sin \phi) & \cos \theta \sin \phi (\cos \theta \cos \phi + \sin \theta \sin \phi) \\ \sin \theta \cos \phi (\cos \theta \cos \phi + \sin \theta \sin \phi) & \sin \theta \sin \phi (\cos \theta \cos \phi + \sin \theta \sin \phi) \end{bmatrix}$$

$$= \begin{bmatrix} \cos \theta \cos \phi \cos (\theta - \phi) & \cos \theta \sin \phi \cos (\theta - \phi) \\ \sin \theta \cos \phi \cos (\theta - \phi) & \sin \theta \sin \phi \cos (\theta - \phi) \end{bmatrix}$$

Given $\qquad \theta - \phi = (2n + 1) \frac{\pi}{2}$

$$\cos (\theta - \phi) = \cos (2n + 1) \frac{\pi}{2} = 0$$

$\therefore$ The product $= \begin{bmatrix} 0 & 0 \\ 0 & 0 \end{bmatrix} = 0.$ **Proved.**

Example 13. *Verify that*

$$A = \frac{1}{3} \begin{bmatrix} 1 & 2 & 2 \\ 2 & 1 & -2 \\ -2 & 2 & -1 \end{bmatrix} \text{ is orthogonal.}$$

Solution. $A = \frac{1}{3} \begin{bmatrix} 1 & 2 & 2 \\ 2 & 1 & -2 \\ -2 & 2 & -1 \end{bmatrix} \quad \therefore \quad A' = \frac{1}{3} \begin{bmatrix} 1 & 2 & -2 \\ 2 & 1 & 2 \\ 2 & -2 & -1 \end{bmatrix}$

$$AA' = \frac{1}{9} \begin{bmatrix} 1 & 2 & 2 \\ 2 & 1 & -2 \\ -2 & 2 & -1 \end{bmatrix} \begin{bmatrix} 1 & 2 & -2 \\ 2 & 1 & 2 \\ 2 & -2 & -1 \end{bmatrix} = \frac{1}{9} \begin{bmatrix} 9 & 0 & 0 \\ 0 & 9 & 0 \\ 0 & 0 & 9 \end{bmatrix} \begin{bmatrix} 1 & 0 & 0 \\ 0 & 1 & 0 \\ 0 & 0 & 1 \end{bmatrix} = I$$

Hence, A is an orthogonal matrix. **Verified.**

Example 14. *Determine the values of* α, β, γ *when*

$$\begin{bmatrix} 0 & 2\beta & \gamma \\ \alpha & \beta & -\gamma \\ \alpha & -\beta & \gamma \end{bmatrix} \text{ is orthogonal.}$$

Solution. Let $A = \begin{bmatrix} 0 & 2\beta & \gamma \\ \alpha & \beta & -\gamma \\ \alpha & -\beta & \gamma \end{bmatrix}$

On transposing A, we have

$$A' = \begin{bmatrix} 0 & \alpha & \alpha \\ 2\beta & \beta & -\beta \\ \gamma & -\gamma & \gamma \end{bmatrix}$$

If A is orthogonal, then $AA' = I$

$$\begin{bmatrix} 0 & 2\beta & \gamma \\ \alpha & \beta & -\gamma \\ \alpha & -\beta & \gamma \end{bmatrix} \begin{bmatrix} 0 & \alpha & \alpha \\ 2\beta & \beta & -\beta \\ \gamma & -\gamma & \gamma \end{bmatrix} = \begin{bmatrix} 1 & 0 & 0 \\ 0 & 1 & 0 \\ 0 & 0 & 1 \end{bmatrix}$$

$$\Rightarrow \quad \begin{bmatrix} 4\beta^2+\gamma^2 & 2\beta^2-\gamma^2 & -2\beta^2+\gamma^2 \\ 2\beta^2-\gamma^2 & \alpha^2+\beta^2+\gamma^2 & \alpha^2-\beta^2-\gamma^2 \\ -2\beta^2+\gamma^2 & \alpha^2-\beta^2-\gamma^2 & \alpha^2+\beta^2+\gamma^2 \end{bmatrix} = \begin{bmatrix} 1 & 0 & 0 \\ 0 & 1 & 0 \\ 0 & 0 & 1 \end{bmatrix}$$

Equating the corresponding elements, we have

$$\left.\begin{matrix} 4\beta^2+\gamma^2 = 1 \\ 2\beta^2-\gamma^2 = 0 \end{matrix}\right\} \Rightarrow \beta = \pm\frac{1}{\sqrt{6}}, \gamma = \pm\frac{1}{\sqrt{3}}$$

But $\alpha^2 + \beta^2 + \gamma^2 = 1$ as $\beta = \pm\frac{1}{\sqrt{6}}, \gamma = \pm\frac{1}{\sqrt{3}}, \alpha = \pm\frac{1}{\sqrt{2}}$ **Ans.**

Example 15. *Prove that*

$$(AB)^n = A^n \cdot B^n, \text{ if } A \cdot B = B \cdot A$$

Solution.

$$(AB)^1 = AB = (A) \cdot (B)$$

$$(AB)^2 = (AB) \cdot (AB) = (ABA) \cdot B = \{ A\,(AB) \} \cdot B$$
$$= (A^2B) \cdot B = A^2\,(B \cdot B) = A^2 \cdot B^2$$

Suppose that

$$(AB)^n = A^n \cdot B^n$$

$$(AB)^{n+1} = (AB)^n \cdot (AB) = (A^n \cdot B^n) \cdot (AB) = A^n \cdot (B^nA) \cdot B$$
$$= A^n \cdot (B^{n-1} \cdot BA) \cdot B = A^n \cdot (B^{n-1} \cdot AB) \cdot B$$
$$= A^n \cdot (B^{n-2} \cdot B \cdot AB) \cdot B = A^n \cdot (B^{n-2} \cdot AB \cdot B) \cdot B$$
$$= A^n \cdot (B^{n-2} \cdot AB^2) \cdot B, \text{ continuing the process } n \text{ times.}$$
$$= A^n \cdot (A \cdot B^n) \cdot B = A^n \cdot (A \cdot B^{n+1}) = A^{n+1} \cdot B^{n+1}$$

Hence, taking the above to be true for $n = n$, we have shown that it is true for $n = n + 1$ and also it was true for $n = 1, 2, ...$ so it is universally true. **Proved.**

EXERCISE 18.2

1. Compute AB, if

$A = \begin{bmatrix} 1 & 2 & 3 \\ 4 & 5 & 6 \end{bmatrix}$ and $B = \begin{bmatrix} 2 & 5 & 3 \\ 3 & 6 & 4 \\ 4 & 7 & 5 \end{bmatrix}$ **Ans.** $\begin{bmatrix} 20 & 38 & 26 \\ 47 & 92 & 62 \end{bmatrix}$

2. If $A = \begin{bmatrix} 1 & 3 & 0 \\ -1 & 2 & 1 \\ 0 & 0 & 2 \end{bmatrix}$, $B = \begin{bmatrix} 2 & 3 & 4 \\ 1 & 2 & 3 \\ -1 & 1 & 2 \end{bmatrix}$. From the product AB and BA. Show that $AB \neq BA$.

3. If $A = \begin{bmatrix} 0 & 1 & 0 \\ 0 & 0 & 1 \\ 0 & 0 & 0 \end{bmatrix}$, $B = \begin{bmatrix} 0 & 0 & 0 \\ 1 & 0 & 0 \\ 0 & 1 & 0 \end{bmatrix}$

(*i*) Calculate AB and BA. Hence evaluate $A^2 B + B^2 A$

(*ii*) Show that for any number k, $(A + kB^2)^3 = KI$, where I is the unit matrix.

4. If $A = \begin{bmatrix} 0 & 1 \\ -1 & 0 \end{bmatrix}$ choose α and β so that $(\alpha I + \beta A)^2 = A$ **Ans.** $\alpha = \beta = \pm \frac{1}{\sqrt{2}}$

5. Write the following transformation in matrix form :

$$x_1 = \frac{\sqrt{3}}{2} y_1 + \frac{1}{2} y_2 \; ; \; x_2 = -\frac{1}{2} y_1 + \frac{\sqrt{3}}{2} y_2$$

Hence, find the transformation in matrix form which expresses y_1, y_2 in terms of x_1, x_2.

Ans. $y_1 = \frac{\sqrt{3}}{2} x_1 - \frac{1}{2} x_2$, $y_2 = \frac{1}{2} x_1 + \frac{\sqrt{3}}{2} x_2$

6. If $A = \begin{bmatrix} 0 & -\tan\frac{\alpha}{2} \\ \tan\frac{\alpha}{2} & 0 \end{bmatrix}$ and I is a unit matrix, show that $I + A = (I - A)\begin{bmatrix} \cos\alpha & -\sin\alpha \\ \sin\alpha & \cos\alpha \end{bmatrix}$

7. If $f(x) = x^3 - 20x + 8$, find $f(A)$ where $A = \begin{bmatrix} 1 & 1 & 3 \\ 1 & 3 & -3 \\ -2 & -4 & -4 \end{bmatrix}$ **Ans.** 0

8. Show that $\begin{bmatrix} \cos\theta & -\sin\theta \\ \sin\theta & \cos\theta \end{bmatrix} = \begin{bmatrix} 1 & -\tan\frac{\theta}{2} \\ \tan\frac{\theta}{2} & 1 \end{bmatrix} \begin{bmatrix} 1 & \tan\frac{\theta}{2} \\ -\tan\frac{\theta}{2} & 1 \end{bmatrix}^{-1}$

9. If $A = \begin{bmatrix} 3 & -3 & 4 \\ 2 & -3 & 4 \\ 0 & -1 & 1 \end{bmatrix}$ then show that $A^3 = A^{-1}$.

10. Verify whether the matrix $A = \frac{1}{3}\begin{bmatrix} 2 & 2 & 1 \\ 2 & -1 & 2 \\ -1 & 2 & 2 \end{bmatrix}$ is orthogonal.

11. Show that $\begin{bmatrix} \cos\phi & 0 & \sin\phi \\ \sin\theta\sin\phi & \cos\theta & -\sin\theta\cos\phi \\ -\cos\theta\sin\phi & \sin\theta & \cos\theta\cos\phi \end{bmatrix}$ is an orthogonal matrix.

12. Show that $A = \begin{bmatrix} \cos\theta & 0 & \sin\theta \\ 0 & 1 & 0 \\ -\sin\theta & 0 & \cos\theta \end{bmatrix}$ is an orthogonal matrix. *(A.M.I.E., Summer 2004)*

13. If A and B are square matrices of the same order, explain in general

(*i*) $(A + B)^2 \neq A^2 + 2AB + B^2$ (*ii*) $(A - B)^2 \neq A^2 - 2AB + B^2$ (*iii*) $(A + B)(A - B) \neq A^2 - B^2$

18.10 ADJOINT OF A SQUARE MATRIX

Let the determinant of the square matrix A be $|A|$.

If $A = \begin{bmatrix} a_1 & a_2 & a_3 \\ b_1 & b_2 & b_3 \\ c_1 & c_2 & c_3 \end{bmatrix}$, Than $|A| = \begin{vmatrix} a_1 & a_2 & a_3 \\ b_1 & b_2 & b_3 \\ c_1 & c_2 & c_3 \end{vmatrix}$

The matrix formed by the co-factors of the elements in

$$|A| \text{ is } \begin{bmatrix} A_1 & A_2 & A_3 \\ B_1 & B_2 & B_3 \\ C_1 & C_2 & C_3 \end{bmatrix}.$$

where $A_1 = \begin{vmatrix} b_2 & b_3 \\ c_2 & c_3 \end{vmatrix} = b_2c_3 - b_3c_2,$ $\quad A_2 = -\begin{vmatrix} b_1 & b_3 \\ c_1 & c_3 \end{vmatrix} = -b_1c_3 + b_3c_1$

$$A_3 = \begin{vmatrix} b_1 & b_2 \\ c_1 & c_2 \end{vmatrix} = b_1c_2 - b_2c_1, \qquad B_1 = -\begin{vmatrix} a_2 & a_3 \\ c_2 & c_3 \end{vmatrix} = -a_2c_3 + a_3c_2$$

$$B_2 = \begin{vmatrix} a_1 & a_3 \\ c_1 & c_3 \end{vmatrix} = a_1c_3 - a_3c_1, \qquad B_3 = -\begin{vmatrix} a_1 & a_2 \\ c_1 & c_2 \end{vmatrix} = -a_1c_2 + a_2c_1$$

$$C_1 = \begin{vmatrix} a_2 & a_3 \\ b_2 & b_3 \end{vmatrix} = a_2b_3 - a_3b_2, \qquad C_2 = -\begin{vmatrix} a_1 & a_3 \\ b_1 & b_3 \end{vmatrix} = -a_1b_3 + a_3b_1$$

$$C_3 = \begin{vmatrix} a_1 & a_2 \\ b_1 & b_2 \end{vmatrix} = a_1b_2 - a_2b_1$$

Then the transpose of the matrix of co-factors

$$\begin{bmatrix} A_1 & B_1 & C_1 \\ A_2 & B_2 & C_2 \\ A_3 & B_3 & C_3 \end{bmatrix}$$

is called the adjoint of the matrix A and is written as adj A.

18.11 MATHEMATICAL INDUCTION

By mathematical induction we can prove results for all positive integers. If the result to be proved for the positive integer n then we apply the following method.

Working Rule:

Step 1. Verify the result for $n = 1$

Step 2. Assume the result to be true for $n = k$ and then prove that it is true for $n = k + 1$.

Explanation. By step 1, the result is true for $n = k = 1$

By step 2, the result is true for $n = k + 1 = 1 + 1 = 2$ $\quad (k = 1)$

Again, the result is also true for $n = k + 1 = 2 + 1 = 3$ $\quad (k = 2)$

Similarly, the result is also true for $n = k + 1 = 3 + 1 = 4$ $\quad (k = 3)$

Hence, in this way the result is true for all positive integer n.

Example 16. *By mathematical induction,*

$$\textit{if } A = \begin{bmatrix} \cos\alpha & \sin\alpha \\ -\sin\alpha & \cos\alpha \end{bmatrix}, \textit{ show that } A^n = \begin{bmatrix} \cos n\alpha & \sin n\alpha \\ -\sin n\alpha & \cos n\alpha \end{bmatrix}$$

Where n is a positive integer.

Solution. We prove the result by mathematical induction :

$$A^n = \begin{bmatrix} \cos n\alpha & \sin n\alpha \\ -\sin n\alpha & \cos n\alpha \end{bmatrix}$$

Let us verify the result for $n = 1$.

$$A^1 = \begin{bmatrix} \cos 1\alpha & \sin 1\alpha \\ -\sin 1\alpha & \cos 1\alpha \end{bmatrix} = \begin{bmatrix} \cos\alpha & \sin\alpha \\ -\sin\alpha & \cos\alpha \end{bmatrix} = A \qquad \text{[Given]}$$

The result is true when $n = 1$.

Let us assume that the result is true for any positive integer k.

$$A^k = \begin{bmatrix} \cos k\alpha & \sin k\alpha \\ -\sin k\alpha & \cos k\alpha \end{bmatrix}$$

Now, $$A^{k+1} = A^k.A = \begin{bmatrix} \cos k\alpha & \sin k\alpha \\ -\sin k\alpha & \cos k\alpha \end{bmatrix} \begin{bmatrix} \cos\alpha & \sin\alpha \\ -\sin\alpha & \cos\alpha \end{bmatrix}$$

$$= \begin{bmatrix} \cos k\alpha\cos\alpha - \sin k\alpha\sin\alpha & \cos k\alpha\sin\alpha + \sin k\alpha\cos\alpha \\ -\sin k\alpha\cos\alpha - \cos k\alpha\sin\alpha & -\sin k\alpha\sin\alpha + \cos k\alpha\cos\alpha \end{bmatrix}$$

$$= \begin{bmatrix} \cos(k\alpha+\alpha) & \sin(k\alpha+\alpha) \\ -\sin(k\alpha+\alpha) & \cos(k\alpha+\alpha) \end{bmatrix} = \begin{bmatrix} \cos(k+1)\alpha & \sin(k+1)\alpha \\ -\sin(k+1)\alpha & \cos(k+1)\alpha \end{bmatrix}$$

The result is true for $n = k + 1$.

Hence, by mathematical induction the result is true for all positive integer n. **Proved.**

Example 17. *Factorise the matrix* $A = \begin{bmatrix} 5 & -2 & 1 \\ 7 & 1 & -5 \\ 3 & 7 & 4 \end{bmatrix}$ *into the form LU, where L is lower triangular and U is upper triangular matrix.*

Solution. Let $A = LU$

$$\Rightarrow \begin{bmatrix} 5 & -2 & 1 \\ 7 & 1 & -5 \\ 3 & 7 & 4 \end{bmatrix} = \begin{bmatrix} l_{11} & 0 & 0 \\ l_{21} & l_{22} & 0 \\ l_{31} & l_{32} & l_{33} \end{bmatrix} \begin{bmatrix} 1 & u_{12} & u_{13} \\ 0 & 1 & u_{23} \\ 0 & 0 & 1 \end{bmatrix} \qquad \ldots (1)$$

$$\Rightarrow \begin{bmatrix} 5 & -2 & 1 \\ 7 & 1 & -5 \\ 3 & 7 & 4 \end{bmatrix} = \begin{bmatrix} l_{11} & l_{11}u_{12} & l_{11}u_{13} \\ l_{21} & l_{21}u_{12} + l_{22} & l_{21}u_{13} + l_{22}u_{23} \\ l_{31} & l_{31}u_{12} + l_{32} & l_{31}u_{13} + l_{32}u_{23} + l_{33} \end{bmatrix}$$

Equating the corresponding elements of equal matrices, we get

$\Rightarrow \quad l_{11} = 5 \qquad l_{11}u_{12} = -2 \qquad l_{11}u_{13} = 1$

$l_{21} = 7 \qquad l_{21}u_{12} + l_{22} = 1 \qquad l_{21}u_{13} + l_{22}u_{23} = -5$

$l_{31} = 3 \qquad l_{31}u_{12} + l_{32} = 7 \qquad l_{31}u_{13} + l_{32}u_{23} + l_{33} = 4$

Let us solve the above equations along first column.

$l_{11} = 5$

$l_{21} = 7$

$l_{31} = 3$

Let us solve along first row.

$l_{11}u_{12} = -2 \qquad \Rightarrow 5\,u_{12} = -2 \qquad \Rightarrow u_{12} = -\frac{2}{5}$

$l_{11}u_{13} = 1 \qquad \Rightarrow 5\,u_{13} = 1 \qquad \Rightarrow u_{13} = \frac{1}{5}$

Let us solve along second column.

$$l_{21}\, u_{12} + l_{22} = 1 \quad \Rightarrow 7\left(-\frac{2}{5}\right) + l_{22} = 1 \quad \Rightarrow l_{22} = 1 + \frac{14}{5} = \frac{19}{5}$$

$$l_{31}\, u_{12} + l_{32} = 7 \quad \Rightarrow 3\left(-\frac{2}{5}\right) + l_{32} = 7 \quad \Rightarrow l_{32} = 7 + \frac{6}{5} = \frac{41}{5}$$

Let us solve along second row,

$$l_{21}u_{13} + l_{22}u_{23} = -5 \quad \Rightarrow 7\left(\frac{1}{5}\right) + \frac{19}{5}u_{23} = -5 \quad \Rightarrow u_{23} = \left(-5 - \frac{7}{5}\right)\frac{5}{19} = -\frac{32}{19}$$

Let us solve along third column,

$$l_{31}u_{13} + l_{32}u_{23} + l_{33} = 4 \Rightarrow 3\left(\frac{1}{5}\right) + \left(\frac{41}{5}\right)\left(-\frac{32}{19}\right) + l_{33} = 4 \Rightarrow l_{33} = 4 - \frac{3}{5} + \frac{1312}{95} = \frac{327}{19}$$

Putting the values of $l_{11}, l_{21}, l_{22}, l_{31}, l_{32}, l_{33}, u_{12}, u_{13}, u_{23}$ in (1), we get

$$\begin{bmatrix} 5 & -2 & 1 \\ 7 & 1 & -5 \\ 3 & 7 & 4 \end{bmatrix} = \begin{bmatrix} 5 & 0 & 0 \\ 7 & \frac{19}{5} & 0 \\ 3 & \frac{41}{5} & \frac{327}{19} \end{bmatrix} \begin{bmatrix} 1 & -\frac{2}{5} & \frac{1}{5} \\ 0 & 1 & -\frac{32}{19} \\ 0 & 0 & 1 \end{bmatrix}$$ **Ans.**

18.12 PROPERTY OF ADJOINT MATRIX

The product of a matrix A and its adjoint is equal to unit matrix multiplied by the determinant A.

Proof. If A be a square matrix, then (Adjoint A) $\cdot A = A \cdot$ (Adjoint A) $= |A| \cdot I$

Let
$$A = \begin{bmatrix} a_1 & a_2 & a_3 \\ b_1 & b_2 & b_3 \\ c_1 & c_2 & c_3 \end{bmatrix} \text{ and adj} . A = \begin{bmatrix} A_1 & B_1 & C_1 \\ A_2 & B_2 & C_2 \\ A_3 & B_3 & C_3 \end{bmatrix}$$

$$A . (\text{adj}. A) = \begin{bmatrix} a_1 & a_2 & a_3 \\ b_1 & b_2 & b_3 \\ c_1 & c_2 & c_3 \end{bmatrix} \times \begin{bmatrix} A_1 & B_1 & C_1 \\ A_2 & B_2 & C_2 \\ A_3 & B_3 & C_3 \end{bmatrix}$$

$$= \begin{bmatrix} a_1 A_1 + a_2 A_2 + a_3 A_3 & a_1 B_1 + a_2 B_2 + a_3 B_3 & a_1 C_1 + a_2 C_2 + a_3 C_3 \\ b_1 A_1 + b_2 A_2 + b_3 A_3 & b_1 B_1 + b_2 B_2 + b_3 B_3 & b_1 C_1 + b_2 C_2 + b_3 C_3 \\ c_1 A_1 + c_2 A_2 + c_3 A_3 & c_1 B_1 + c_2 B_2 + c_3 B_3 & c_1 C_1 + c_2 C_2 + c_3 C_3 \end{bmatrix}$$

$$= \begin{bmatrix} |A| & 0 & 0 \\ 0 & |A| & 0 \\ 0 & 0 & |A| \end{bmatrix} = |A| \begin{bmatrix} 1 & 0 & 0 \\ 0 & 1 & 0 \\ 0 & 0 & 1 \end{bmatrix} = |A|\, I$$ *(A.M.I.E., Summer 2004)*

18.13 INVERSE OF A MATRIX

If A and B are two square matrices of the same order, such that

$$AB = BA = I \qquad (I = \text{unit matrix})$$

then B is called the inverse of A *i.e.* $B = A^{-1}$ and A is the inverse of B.

Condition for a square matrix A to possess an inverse is that matrix A is non-singular, i.e., $|A| \neq 0$

If A is a square matrix and B be its inverse, then $AB = I$

Taking determinant of both sides, we get

$$|AB| = |I| \text{ or } |A||B| = I$$

From this relation it is clear that $|A| \neq 0$

i.e. the matrix A is non-singular.

To find the inverse matrix with the help of adjoint matrix

We know that $\quad A \cdot (Adj.\ A) = |A|\, I$

$$\Rightarrow \quad A \cdot \frac{1}{|A|}(A\,dj.\ A) = I \qquad \text{[Provided } |A| \neq 0] \qquad ...(1)$$

and $$A \cdot A^{-1} = I \qquad ...(2)$$

From (1) and (2), we have

$$\therefore \quad \boxed{A^{-1} = \frac{1}{|A|}(Adj.\ A)}$$

Example 18. *If* $A = \begin{bmatrix} 3 & -3 & 4 \\ 2 & -3 & 4 \\ 0 & -1 & 1 \end{bmatrix}$, *find* A^{-1}. *(A.M.I.E. Summer 2004)*

Solution. $A = \begin{bmatrix} 3 & -3 & 4 \\ 2 & -3 & 4 \\ 0 & -1 & 1 \end{bmatrix}$

$$|A| = 3(-3+4) + 3(2-0) + 4(-2-0) = 3 + 6 - 8 = 1$$

The co-factors of elements of various rows of $|A|$ are

$$\begin{bmatrix} (-3+4) & (-2-0) & (-2-0) \\ (3-4) & (3-0) & (3-0) \\ (-12+12) & (-12+8) & (-9+6) \end{bmatrix}$$

Therefore, the matrix formed by the co-factors of $|A|$ is

$$\begin{bmatrix} 1 & -2 & -2 \\ -1 & 3 & 3 \\ 0 & -4 & -3 \end{bmatrix}, \quad Adj.\ A = \begin{bmatrix} 1 & -1 & 0 \\ -2 & 3 & -4 \\ -2 & 3 & -3 \end{bmatrix}$$

$$\therefore \quad A^{-1} = \frac{1}{|A|}Adj.\ A = \frac{1}{1}\begin{bmatrix} 1 & -1 & 0 \\ -2 & 3 & -4 \\ -2 & 3 & -3 \end{bmatrix} = \begin{bmatrix} 1 & -1 & 0 \\ -2 & 3 & -4 \\ -2 & 3 & -3 \end{bmatrix} \qquad \textbf{Ans.}$$

Example 19. *If* $A = \frac{1}{9}\begin{bmatrix} -8 & 1 & 4 \\ 4 & 4 & 7 \\ 1 & -8 & 4 \end{bmatrix}$, *prove that* $A^{-1} = A'$, A' *being the transpose of* A.

(A.M.I.E., Winter 2000)

Solution. We have, $\quad A = \frac{1}{9}\begin{bmatrix} -8 & 1 & 4 \\ 4 & 4 & 7 \\ 1 & -8 & 4 \end{bmatrix}, \quad A' = \frac{1}{9}\begin{bmatrix} -8 & 4 & 1 \\ 1 & 4 & -8 \\ 4 & 7 & 4 \end{bmatrix}$

$$AA' = \frac{1}{9}\begin{bmatrix} -8 & 1 & 4 \\ 4 & 4 & 7 \\ 1 & -8 & 4 \end{bmatrix} \cdot \frac{1}{9}\begin{bmatrix} -8 & 4 & 1 \\ 1 & 4 & -8 \\ 4 & 7 & 4 \end{bmatrix}$$

$$= \frac{1}{81}\begin{bmatrix} 64+1+16 & -32+4+28 & -8-8+16 \\ -32+4+28 & 16+16+49 & 4-32+28 \\ -8-8+16 & 4-32+28 & 1+64+16 \end{bmatrix}$$

$$= \frac{1}{81}\begin{bmatrix} 81 & 0 & 0 \\ 0 & 81 & 0 \\ 0 & 0 & 81 \end{bmatrix} = \begin{bmatrix} 1 & 0 & 0 \\ 0 & 1 & 0 \\ 0 & 0 & 1 \end{bmatrix} \text{ or } AA' = I$$

$$A' = A^{-1}$$ **Proved.**

Example 20. *If A and B are non-singular matrices of the same order then,*

$$(AB)^{-1} = B^{-1} \cdot A^{-1}$$

Hence prove that $(A^{-1})^m = (A^m)^{-1}$ *for any positive integer m.*

Solution. We know that,

$$(AB) \cdot (B^{-1} A^{-1}) = [(AB) B^{-1}] \cdot A^{-1} = [A (BB^{-1}] \cdot A^{-1}$$

$$= [AI] A^{-1} = A \cdot A^{-1} = I$$

Also, $$B^{-1} A^{-1} \cdot (AB) = B^{-1}[A^{-1} \cdot (AB)] = B^{-1} [(A^{-1} A) \cdot B]$$

$$= B^{-1} [I . B] = B^{-1} \cdot B = I$$

By definition of the inverse of a matrix, $B^{-1} A^{-1}$ is inverse of AB.

$\Rightarrow$ $$B^{-1} A^{-1} = (AB)^{-1}$$ **Proved.**

$$(A^m)^{-1} = [A \cdot A^{m-1}]^{-1} = (A^{m-1})^{-1} A^{-1}$$

$$= (A \cdot A^{m-2})^{-1} \cdot A^{-1} = [(A^{m-2})^{-1} \cdot A^{-1}] \cdot A^{-1} = (A^{m-2})^{-1} (A^{-1})^2$$

$$= (A \cdot A^{m-3})^{-1} \cdot (A^{-1})^2 = [(A^{m-3})^{-1} \cdot A^{-1}] (A^{-1})^2 = (A^{m-3})^{-1} (A^{-1})^3$$

$$= A^{-1} (A^{-1})^{m-1} = (A^{-1})^m$$ **Proved.**

Example 21. *Find A satisfying the Matrix equation.*

$$\begin{bmatrix} 2 & 1 \\ 3 & 2 \end{bmatrix} A \begin{bmatrix} -3 & 2 \\ 5 & -3 \end{bmatrix} = \begin{bmatrix} -2 & 4 \\ 3 & -1 \end{bmatrix}$$

Solution. $$\begin{bmatrix} 2 & 1 \\ 3 & 2 \end{bmatrix} A \begin{bmatrix} -3 & 2 \\ 5 & -3 \end{bmatrix} = \begin{bmatrix} -2 & 4 \\ 3 & -1 \end{bmatrix}$$

Both sides of the equation are pre-multiplied by the inverse of $\begin{bmatrix} 2 & 1 \\ 3 & 2 \end{bmatrix}$ *i.e.,* $\begin{bmatrix} 2 & -1 \\ -3 & 2 \end{bmatrix}$

$$\begin{bmatrix} 2 & -1 \\ -3 & 2 \end{bmatrix}\begin{bmatrix} 2 & 1 \\ 3 & 2 \end{bmatrix} A \begin{bmatrix} -3 & 2 \\ 5 & -3 \end{bmatrix} = \begin{bmatrix} 2 & -1 \\ -3 & 2 \end{bmatrix}\begin{bmatrix} -2 & 4 \\ 3 & -1 \end{bmatrix}$$

$$\begin{bmatrix} 1 & 0 \\ 0 & 1 \end{bmatrix} A \begin{bmatrix} -3 & 2 \\ 5 & -3 \end{bmatrix} = \begin{bmatrix} -7 & 9 \\ 12 & -14 \end{bmatrix}$$

$$A \begin{bmatrix} -3 & 2 \\ 5 & -3 \end{bmatrix} = \begin{bmatrix} -7 & 9 \\ 12 & -14 \end{bmatrix}$$

Again both sides are post-multiplied by the inverse of $\begin{bmatrix} -3 & 2 \\ 5 & -3 \end{bmatrix}$ *i.e.* $\begin{bmatrix} 3 & 2 \\ 5 & 3 \end{bmatrix}$

$$A \begin{bmatrix} -3 & 2 \\ 5 & -3 \end{bmatrix}\begin{bmatrix} 3 & 2 \\ 5 & 3 \end{bmatrix} = \begin{bmatrix} -7 & 9 \\ 12 & -14 \end{bmatrix}\begin{bmatrix} 3 & 2 \\ 5 & 3 \end{bmatrix}$$

$$A \begin{bmatrix} 1 & 0 \\ 0 & 1 \end{bmatrix} = \begin{bmatrix} 24 & 13 \\ -34 & -18 \end{bmatrix} \Rightarrow A = \begin{bmatrix} 24 & 13 \\ -34 & -18 \end{bmatrix}$$ **Ans.**

EXERCISE 18.3

Find the adjoint and inverse of the following matrices: (1 - 3)

1. $\begin{bmatrix} 2 & 5 & 3 \\ 3 & 1 & 2 \\ 1 & 2 & 1 \end{bmatrix}$ **Ans.** $\frac{1}{4}\begin{bmatrix} -3 & 1 & 7 \\ -1 & -1 & 5 \\ 5 & 1 & -13 \end{bmatrix}$ **2.** $\begin{bmatrix} 1 & 1 & 2 \\ 1 & 9 & 3 \\ 1 & 4 & 2 \end{bmatrix}$ **Ans.** $-\frac{1}{3}\begin{bmatrix} 6 & 6 & -15 \\ 1 & 0 & -1 \\ -5 & -3 & 8 \end{bmatrix}$

3. $\begin{bmatrix} 1 & 0 & -1 \\ 3 & 4 & 5 \\ 0 & -6 & -7 \end{bmatrix}$ **Ans.** $\frac{1}{20}\begin{bmatrix} 2 & 6 & 4 \\ 21 & -7 & -8 \\ -18 & 6 & 4 \end{bmatrix}$

4. If $A = \begin{bmatrix} 3 & -4 \\ 1 & -1 \end{bmatrix}$, then show that $A^n = \begin{bmatrix} 1+2n & -4n \\ n & 1-2n \end{bmatrix}$

5. If $A = \begin{bmatrix} 1 & 1 & -2 \\ -1 & 2 & 1 \\ 0 & 1 & -1 \end{bmatrix}, P = \begin{bmatrix} 1 & 1 & 3 \\ 0 & 3 & 2 \\ 1 & 1 & 1 \end{bmatrix}$, show that $P^{-1} AP = \begin{bmatrix} -1 & 0 & 0 \\ 0 & 2 & 0 \\ 0 & 0 & 1 \end{bmatrix}$

6. If $A = \begin{bmatrix} 1 & 1 & 1 \\ 1 & 2 & 3 \\ 1 & 4 & 9 \end{bmatrix}, B = \begin{bmatrix} 2 & 5 & 3 \\ 3 & 1 & 2 \\ 1 & 2 & 1 \end{bmatrix}$, show that $(AB)^{-1} = B^{-1} A^{-1}$.

7. Given the matrix $A = \begin{bmatrix} 3 & 2 & 2 \\ 1 & 3 & 1 \\ 5 & 3 & 4 \end{bmatrix}$ compute det (A), A^{-1} and the matrix B such that $AB = \begin{bmatrix} 3 & 4 & 2 \\ 1 & 6 & 1 \\ 5 & 6 & 4 \end{bmatrix}$

Also compute BA. Is $AB = BA$?

Ans. $5, \frac{1}{5}\begin{bmatrix} 9 & -2 & -4 \\ 1 & 2 & -1 \\ -12 & 1 & 7 \end{bmatrix}. B = \begin{bmatrix} 1 & 0 & 0 \\ 0 & 2 & 0 \\ 0 & 0 & 1 \end{bmatrix}, AB \neq BA$

8. Find the condition of k such that the matrix

$A = \begin{bmatrix} 1 & 3 & 4 \\ 3 & k & 6 \\ -1 & 5 & 1 \end{bmatrix}$ has an inverse. Obtain A^{-1} for $k = 1$. **Ans.** $k \neq -\frac{3}{5}, A^{-1} = \frac{1}{8}\begin{bmatrix} -29 & 17 & 14 \\ -9 & 5 & 6 \\ 16 & -8 & -8 \end{bmatrix}$

9. Prove that $(A^{-1})^T = (A^T)^{-1}$.

10. If $A\begin{bmatrix} 0 & 1 \\ 2 & -1 \end{bmatrix} = \begin{bmatrix} 2 & 1 \\ -1 & 0 \end{bmatrix}$ where $A = \begin{bmatrix} a & b \\ c & d \end{bmatrix}$, then A is

(*a*) $\begin{bmatrix} 2 & 1 \\ 0 & 0 \end{bmatrix}$ (*b*) $\begin{bmatrix} 0 & 1 \\ 2 & -1 \end{bmatrix}$ (*c*) $\begin{bmatrix} 2 & 1 \\ -1 & 0 \end{bmatrix}$ (*d*) $\begin{bmatrix} 2 & 1 \\ -\frac{1}{2} & -\frac{1}{2} \end{bmatrix}$ (*AMIETE, June 2010*) **Ans.** (*d*)

18.14 ELEMENTARY TRANSFORMATIONS

Any one of the following operations on a matrix is called an elementary transformation.

1. Interchanging any two rows (or columns). This transformation is indicated by R_{ij}, if the ith and jth rows are interchanged.
2. Multiplication of the elements of any row R_i (or column) by a non-zero scalar quantity k is denoted by $(k.R_i)$.
3. Addition of constant multiplication of the elements of any row R_j to the corresponding elements of any other row R_j is denoted by $(R_i + kR_j)$.

If a matrix B is obtained from a matrix A by one or more E-operations, then B is said to be equivalent to A. The symbol ~ is used for equivalence.

$$i.e., A \sim B.$$

Example 22. *Reduce the following matrix to upper triangular form (Echelon form) :*

$$\begin{bmatrix} 1 & 2 & 3 \\ 2 & 5 & 7 \\ 3 & 1 & 2 \end{bmatrix}$$

Solution. *Upper triangular matrix.* If in a square matrix, all the elements below the principal diagonal are zero, the matrix is called an upper triangular matrix.

$$\begin{bmatrix} 1 & 2 & 3 \\ 2 & 5 & 7 \\ 3 & 1 & 2 \end{bmatrix} \sim \begin{bmatrix} 1 & 2 & 3 \\ 0 & 1 & 1 \\ 0 & -5 & -7 \end{bmatrix} \begin{matrix} R_2 \to R_2 - 2R_1 \\ R_3 \to R_3 - 3R_1 \end{matrix} \sim \begin{bmatrix} 1 & 2 & 3 \\ 0 & 1 & 1 \\ 0 & 0 & -2 \end{bmatrix} R_3 \to R_3 + 5R_2$$ **Ans.**

Example 23. *Transform* $\begin{bmatrix} 1 & 3 & 3 \\ 2 & 4 & 10 \\ 3 & 8 & 4 \end{bmatrix}$ *into a unit matrix.* *(Q. Bank U.P., 2001)*

Solution. $$\begin{bmatrix} 1 & 3 & 3 \\ 2 & 4 & 10 \\ 3 & 8 & 4 \end{bmatrix} \sim \begin{bmatrix} 1 & 3 & 3 \\ 0 & -2 & 4 \\ 0 & -1 & -5 \end{bmatrix} \begin{matrix} R_2 \to R_2 - 2R_1 \\ R_3 \to R_3 - 3R_1 \end{matrix}$$

$$\sim \begin{bmatrix} 1 & 3 & 3 \\ 0 & 1 & -2 \\ 0 & -1 & -5 \end{bmatrix} R_2 \to -\frac{1}{2}R_2 \sim \begin{bmatrix} 1 & 0 & 9 \\ 0 & 1 & -2 \\ 0 & 0 & -7 \end{bmatrix} \begin{matrix} R_1 \to R_1 - 3R_2 \\ \\ R_3 \to R_3 + R_2 \end{matrix}$$

$$\sim \begin{bmatrix} 1 & 0 & 9 \\ 0 & 1 & -2 \\ 0 & 0 & 1 \end{bmatrix} R_3 \to -\frac{1}{7}R_3 \sim \begin{bmatrix} 1 & 0 & 0 \\ 0 & 1 & 0 \\ 0 & 0 & 1 \end{bmatrix} \begin{matrix} R_1 \to R_1 - 9R_3 \\ R_2 \to R_2 + 2R_3 \end{matrix}$$

18.15 ELEMENTARY MATRICES

A matrix obtained from a unit matrix by a single elementary transformation is called elementary matrix.

$$I = \begin{bmatrix} 1 & 0 & 0 \\ 0 & 1 & 0 \\ 0 & 0 & 1 \end{bmatrix}$$

Consider the matrix obtained by $R_2 + 3R_1$

$$\begin{bmatrix} 1 & 0 & 0 \\ 3 & 1 & 0 \\ 0 & 0 & 1 \end{bmatrix}$$ is called the elementary matrix.

18.16 THEOREM

Every elementary row transformation of a matrix can be affected by pre-multiplication with the corresponding elementary matrix.

Consider the matrix $$A = \begin{bmatrix} 2 & 3 & 4 \\ 5 & 6 & 7 \\ 3 & 5 & 9 \end{bmatrix}$$

Let us apply row transformation $R_3 + 4\,R_1$ and we get a matrix B.

$$B = \begin{bmatrix} 2 & 3 & 4 \\ 5 & 6 & 7 \\ 11 & 17 & 25 \end{bmatrix}$$

Now we shall show that pre-multiplication of A by corresponding elementary matrix $R_3 + 4\,R_1$ will give us B.

Now, if $I = \begin{bmatrix} 1 & 0 & 0 \\ 0 & 1 & 0 \\ 0 & 0 & 1 \end{bmatrix}$ then, Elementary matrix $= \begin{bmatrix} 1 & 0 & 0 \\ 0 & 1 & 0 \\ 4 & 0 & 1 \end{bmatrix}_{(R_3 + 4R_1)}$

$$\therefore \quad \text{Elementary matrix} \times A = \begin{bmatrix} 1 & 0 & 0 \\ 0 & 1 & 0 \\ 4 & 0 & 1 \end{bmatrix} \times \begin{bmatrix} 2 & 3 & 4 \\ 5 & 6 & 7 \\ 3 & 5 & 9 \end{bmatrix} = \begin{bmatrix} 2 & 3 & 4 \\ 5 & 6 & 7 \\ 11 & 17 & 25 \end{bmatrix} = B$$

Similarly, we can show that every elementary column transformation of a matrix can be affecte by post-multiplication with the corresponding elementary matrix.

18.17 TO COMPUTE THE INVERSE OF A MATRIX FROM ELEMENTARY MATRICES (Gauss-jordan Method)

If A is reduced to I by elementary transformation then

$$PA = I \qquad \text{where} \qquad P = P_n P_{n-1} \ldots P_2 P_1$$

$$\therefore \qquad P = A^{-1} \qquad = \text{Elementary matrix.}$$

Working rule. Write $A = IA$. Perform elementary row transformation on A of the left side and on I of the right hand side so that A is reduced to I and I of right hand side is reduced to P getting $I = PA$.

Then P is the inverse of A.

18.18 THE INVERSE OF A SYMMETRIC MATRIX

The elementary transformations are to be transformed so that the property of being symmetric is preserved. This requires that the transformations occur in pairs, a row transformation must be followed immediately by the same column transformation.

Example 24. *Find the inverse of the following matrix employing elementary transformations:*

$$\begin{bmatrix} 3 & -3 & 4 \\ 2 & -3 & 4 \\ 0 & -1 & 1 \end{bmatrix}$$

(U.P., I Semester, Compartment 2002)

Solution. The given matrix is $A = \begin{bmatrix} 3 & -3 & 4 \\ 2 & -3 & 4 \\ 0 & -1 & 1 \end{bmatrix}$

$$\begin{bmatrix} 3 & -3 & 4 \\ 2 & -3 & 4 \\ 0 & -1 & 1 \end{bmatrix} = \begin{bmatrix} 1 & 0 & 0 \\ 0 & 1 & 0 \\ 0 & 0 & 1 \end{bmatrix} A \quad \Rightarrow \quad \begin{bmatrix} 1 & -1 & \frac{4}{3} \\ 2 & -3 & 4 \\ 0 & -1 & 1 \end{bmatrix} = \begin{bmatrix} \frac{1}{3} & 0 & 0 \\ 0 & 1 & 0 \\ 0 & 0 & 1 \end{bmatrix} A \quad R_1 \to \frac{R_1}{3}$$

$$\Rightarrow \begin{bmatrix} 1 & -1 & \frac{4}{3} \\ 0 & -1 & \frac{4}{3} \\ 0 & -1 & 1 \end{bmatrix} = \begin{bmatrix} \frac{1}{3} & 0 & 0 \\ -\frac{2}{3} & 1 & 0 \\ 0 & 0 & 1 \end{bmatrix} A \quad R_2 \to R_2 - 2R_1 \qquad \Rightarrow \begin{bmatrix} 1 & -1 & \frac{4}{3} \\ 0 & 1 & -\frac{4}{3} \\ 0 & -1 & 1 \end{bmatrix} = \begin{bmatrix} \frac{1}{3} & 0 & 0 \\ \frac{2}{3} & -1 & 0 \\ 0 & 0 & 1 \end{bmatrix} A \quad R_2 \to -R_2$$

$$\Rightarrow \begin{bmatrix} 1 & -1 & \frac{4}{3} \\ 0 & 1 & -\frac{4}{3} \\ 0 & 0 & -\frac{1}{3} \end{bmatrix} = \begin{bmatrix} \frac{1}{3} & 0 & 0 \\ \frac{2}{3} & -1 & 0 \\ \frac{2}{3} & -1 & 1 \end{bmatrix} A \quad R_3 \to R_3 + R_2 \qquad \Rightarrow \begin{bmatrix} 1 & -1 & \frac{4}{3} \\ 0 & 1 & -\frac{4}{3} \\ 0 & 0 & 1 \end{bmatrix} = \begin{bmatrix} \frac{1}{3} & 0 & 0 \\ \frac{2}{3} & -1 & 0 \\ -2 & 3 & -3 \end{bmatrix} A \quad R_3 \to -3R_3$$

$$\Rightarrow \begin{bmatrix} 1 & -1 & 0 \\ 0 & 1 & 0 \\ 0 & 0 & 1 \end{bmatrix} = \begin{bmatrix} 3 & -4 & 4 \\ -2 & 3 & -4 \\ -2 & 3 & -3 \end{bmatrix} A \quad \begin{matrix} R_1 \to R_1 - \frac{4}{3}R_3 \\ R_2 \to R_2 + \frac{4}{3}R_3 \end{matrix}$$

$$\Rightarrow \begin{bmatrix} 1 & 0 & 0 \\ 0 & 1 & 0 \\ 0 & 0 & 1 \end{bmatrix} = \begin{bmatrix} 1 & -1 & 0 \\ -2 & 3 & -4 \\ -2 & 3 & -3 \end{bmatrix} A \quad R_1 \to R_1 + R_2$$

Hence, $$A^{-1} = \begin{bmatrix} 1 & -1 & 0 \\ -2 & 3 & -4 \\ -2 & 3 & -3 \end{bmatrix}$$ **Ans.**

Example 25. *Find by elementary row transformation the inverse of the matrix.*

$$\begin{bmatrix} 0 & 1 & 2 \\ 1 & 2 & 3 \\ 3 & 1 & 1 \end{bmatrix}$$

(U.P., I Semester, Winter 2003, 2000)

Solution. Let $A = \begin{bmatrix} 0 & 1 & 2 \\ 1 & 2 & 3 \\ 3 & 1 & 1 \end{bmatrix}$

Elementary row transformation, which will reduce $A = IA$ to $I = PA$, then matrix P will be the inverse of matrix A.

$$\begin{bmatrix} 0 & 1 & 2 \\ 1 & 2 & 3 \\ 3 & 1 & 1 \end{bmatrix} = \begin{bmatrix} 1 & 0 & 0 \\ 0 & 1 & 0 \\ 0 & 0 & 1 \end{bmatrix} A \qquad \Rightarrow \begin{bmatrix} 1 & 2 & 3 \\ 0 & 1 & 2 \\ 3 & 1 & 1 \end{bmatrix} = \begin{bmatrix} 0 & 1 & 0 \\ 1 & 0 & 0 \\ 0 & 0 & 1 \end{bmatrix} A \quad R_1 \leftrightarrow R_2$$

$$\Rightarrow \begin{bmatrix} 1 & 2 & 3 \\ 0 & 1 & 2 \\ 0 & -5 & -8 \end{bmatrix} = \begin{bmatrix} 0 & 1 & 0 \\ 1 & 0 & 0 \\ 0 & -3 & 1 \end{bmatrix} A \quad R_3 \to R_3 - 3R_1 \qquad \Rightarrow \begin{bmatrix} 1 & 2 & 3 \\ 0 & 1 & 2 \\ 0 & 0 & 2 \end{bmatrix} = \begin{bmatrix} 0 & 1 & 0 \\ 1 & 0 & 0 \\ 5 & -3 & 1 \end{bmatrix} A \quad R_3 \to R_3 + 5R_2$$

$$\begin{bmatrix} 1 & 2 & 3 \\ 0 & 1 & 2 \\ 0 & 0 & 1 \end{bmatrix} = \begin{bmatrix} 0 & 1 & 0 \\ 1 & 0 & 0 \\ \frac{5}{2} & -\frac{3}{2} & \frac{1}{2} \end{bmatrix} A \quad R_3 \to \frac{1}{2} R_3 \qquad \Rightarrow \begin{bmatrix} 1 & 2 & 0 \\ 0 & 1 & 0 \\ 0 & 0 & 1 \end{bmatrix} = \begin{bmatrix} -\frac{15}{2} & \frac{11}{2} & -\frac{3}{2} \\ -4 & 3 & -1 \\ \frac{5}{2} & -\frac{3}{2} & \frac{1}{2} \end{bmatrix} A \quad \begin{matrix} R_1 \to R_1 - 3R_3 \\ R_2 \to R_2 - 2R_3, \end{matrix}$$

$$\Rightarrow \begin{bmatrix} 1 & 0 & 0 \\ 0 & 1 & 0 \\ 0 & 0 & 1 \end{bmatrix} = \begin{bmatrix} \frac{1}{2} & -\frac{1}{2} & \frac{1}{2} \\ -4 & 3 & -1 \\ \frac{5}{2} & -\frac{3}{2} & \frac{1}{2} \end{bmatrix} A \quad R_1 \to R_1 - 2R_2$$

$$I = PA \Rightarrow P = A^{-1}$$

$$A^{-1} = \begin{bmatrix} \frac{1}{2} & -\frac{1}{2} & \frac{1}{2} \\ -4 & 3 & -1 \\ \frac{5}{2} & \frac{-3}{2} & \frac{1}{2} \end{bmatrix}$$

Ans.

Example 26. *Find the inverse of the matrix M by applying elementary transformations*

$$\begin{bmatrix} 0 & 2 & 1 & 3 \\ 1 & 1 & -1 & -2 \\ 1 & 2 & 0 & 1 \\ -1 & 1 & 2 & 6 \end{bmatrix}$$

[U.P.T.U.(C.O.) 2003]

Solution. Here, we have $A = \begin{bmatrix} 0 & 2 & 1 & 3 \\ 1 & 1 & -1 & -2 \\ 1 & 2 & 0 & 1 \\ -1 & 1 & 2 & 6 \end{bmatrix}$

Let

$$\begin{bmatrix} 0 & 2 & 1 & 3 \\ 1 & 1 & -1 & -2 \\ 1 & 2 & 0 & 1 \\ -1 & 1 & 2 & 6 \end{bmatrix} \sim \begin{bmatrix} 1 & 0 & 0 & 0 \\ 0 & 1 & 0 & 0 \\ 0 & 0 & 1 & 0 \\ 0 & 0 & 0 & 1 \end{bmatrix} A$$

$$\begin{bmatrix} 1 & 1 & -1 & -2 \\ 0 & 2 & 1 & 3 \\ 1 & 2 & 0 & 1 \\ -1 & 1 & 2 & 6 \end{bmatrix} \sim \begin{bmatrix} 0 & 1 & 0 & 0 \\ 1 & 0 & 0 & 0 \\ 0 & 0 & 1 & 0 \\ 0 & 0 & 0 & 1 \end{bmatrix} A \quad R_1 \leftrightarrow R_2$$

$$\begin{bmatrix} 1 & 1 & -1 & -2 \\ 0 & 2 & 1 & 3 \\ 0 & 1 & 1 & 3 \\ 0 & 2 & 1 & 4 \end{bmatrix} \sim \begin{bmatrix} 0 & 1 & 0 & 0 \\ 1 & 0 & 0 & 0 \\ 0 & -1 & 1 & 0 \\ 0 & 1 & 0 & 1 \end{bmatrix} A \quad \begin{matrix} R_3 \to R_3 - R_1 \\ R_4 \to R_4 + R_1 \end{matrix}$$

$$\begin{bmatrix} 1 & 1 & -1 & -2 \\ 0 & 1 & 1 & 3 \\ 0 & 2 & 1 & 3 \\ 0 & 2 & 1 & 4 \end{bmatrix} = \begin{bmatrix} 0 & 1 & 0 & 0 \\ 0 & -1 & 1 & 0 \\ 1 & 0 & 0 & 0 \\ 0 & 1 & 0 & 1 \end{bmatrix} A \quad R_3 \leftrightarrow R_2$$

$$\begin{bmatrix} 1 & 1 & -1 & -2 \\ 0 & 1 & 1 & 3 \\ 0 & 0 & -1 & -3 \\ 0 & 0 & -1 & -2 \end{bmatrix} = \begin{bmatrix} 0 & 1 & 0 & 0 \\ 0 & -1 & 1 & 0 \\ 1 & 2 & -2 & 0 \\ 0 & 3 & -2 & 1 \end{bmatrix} A \quad \begin{matrix} R_3 \to R_3 - 2R_2 \\ R_4 \to R_4 - 2R_2 \end{matrix}$$

$$\begin{bmatrix} 1 & 1 & -1 & -2 \\ 0 & 1 & 1 & 3 \\ 0 & 0 & -1 & -3 \\ 0 & 0 & 0 & 1 \end{bmatrix} = \begin{bmatrix} 0 & 1 & 0 & 0 \\ 0 & -1 & 1 & 0 \\ 1 & 2 & -2 & 0 \\ -1 & 1 & 0 & 1 \end{bmatrix} A \quad R_4 \to R_4 - R_3$$

$$\begin{bmatrix} 1 & 1 & -1 & -2 \\ 0 & 1 & 1 & 3 \\ 0 & 0 & 1 & 3 \\ 0 & 0 & 0 & 1 \end{bmatrix} = \begin{bmatrix} 0 & 1 & 0 & 0 \\ 0 & -1 & 1 & 0 \\ -1 & -2 & 2 & 0 \\ -1 & 1 & 0 & 1 \end{bmatrix} A \quad R_3 \to - R_3$$

$$\begin{bmatrix} 1 & 1 & -1 & 0 \\ 0 & 1 & 1 & 0 \\ 0 & 0 & 1 & 0 \\ 0 & 0 & 0 & 1 \end{bmatrix} = \begin{bmatrix} -2 & 3 & 0 & 2 \\ 3 & -4 & 1 & -3 \\ 2 & -5 & 2 & -3 \\ -1 & 1 & 0 & 1 \end{bmatrix} A \quad \begin{matrix} R_1 \to R_1 + 2R_4 \\ R_2 \to R_2 - 3R_4 \\ R_3 \to R_3 - 3R_4 \end{matrix}$$

$$\begin{bmatrix} 1 & 1 & 0 & 0 \\ 0 & 1 & 0 & 0 \\ 0 & 0 & 1 & 0 \\ 0 & 0 & 0 & 1 \end{bmatrix} = \begin{bmatrix} 0 & -2 & 2 & -1 \\ 1 & 1 & -1 & 0 \\ 2 & -5 & 2 & -3 \\ -1 & 1 & 0 & 1 \end{bmatrix} A \quad \begin{matrix} R_1 \to R_1 + R_3 \\ R_2 \to R_2 - R_3 \end{matrix}$$

$$\begin{bmatrix} 1 & 0 & 0 & 0 \\ 0 & 1 & 0 & 0 \\ 0 & 0 & 1 & 0 \\ 0 & 0 & 0 & 1 \end{bmatrix} = \begin{bmatrix} -1 & -3 & 3 & -1 \\ 1 & 1 & -1 & 0 \\ 2 & -5 & 2 & -3 \\ -1 & 1 & 0 & 1 \end{bmatrix} A \quad R_1 \to R_1 - R_2$$

$$I = A^{-1} A$$

Hence,
$$A^{-1} = \begin{bmatrix} -1 & -3 & 3 & -1 \\ 1 & 1 & -1 & 0 \\ 2 & -5 & 2 & -3 \\ -1 & 1 & 0 & 1 \end{bmatrix}$$
Ans.

EXERCISE 18.4

Reduce the matrices to triangular form:

$A = \begin{bmatrix} 1 & 2 & 3 \\ 2 & 5 & 7 \\ 3 & 1 & 2 \end{bmatrix}$ **Ans.** $\begin{bmatrix} 1 & 2 & 3 \\ 0 & 1 & 1 \\ 0 & 0 & -2 \end{bmatrix}$

2. $\begin{bmatrix} 3 & 1 & 4 \\ 1 & 2 & -5 \\ 0 & 1 & 5 \end{bmatrix}$ **Ans.** $\begin{bmatrix} 0 & 1 & 4 \\ 0 & 5 & -19 \\ 0 & 0 & 22 \end{bmatrix}$

Find the inverse of the following matrices:

3. $\begin{bmatrix} 1 & 3 & 3 \\ 1 & 4 & 3 \\ 1 & 3 & 4 \end{bmatrix}$ **Ans.** $\begin{bmatrix} 7 & -3 & -3 \\ -1 & 1 & 0 \\ -1 & 0 & 1 \end{bmatrix}$

4. $\begin{bmatrix} 1 & -1 & 1 \\ 4 & 1 & 0 \\ 8 & 1 & 1 \end{bmatrix}$ **Ans.** $\begin{bmatrix} 1 & 2 & -1 \\ -4 & -7 & 4 \\ -4 & -9 & 5 \end{bmatrix}$

5. Use elementary row operations to find inverse of $A = \begin{bmatrix} 1 & 1 & 3 \\ 1 & 3 & -3 \\ -2 & -4 & -4 \end{bmatrix}$ **Ans.** $\frac{1}{4}\begin{bmatrix} 12 & 4 & 6 \\ -5 & -1 & -3 \\ -1 & -1 & -1 \end{bmatrix}$

(AMIETE, June 2010)

6. $\begin{bmatrix} 2 & 1 & -1 & 2 \\ 1 & 3 & 2 & -3 \\ -1 & 2 & 1 & -1 \\ 2 & -3 & -1 & 4 \end{bmatrix}$ **Ans.** $\frac{1}{18}\begin{bmatrix} 2 & 5 & -7 & 1 \\ 5 & -1 & 5 & -2 \\ -7 & 5 & 11 & 10 \\ 1 & -2 & 10 & 5 \end{bmatrix}$

7. $\begin{bmatrix} 1 & 2 & 3 & 1 \\ 1 & 3 & 3 & 2 \\ 2 & 4 & 3 & 3 \\ 1 & 1 & 1 & 1 \end{bmatrix}$ *(Q. Bank U.P. II Semester 2001)* **Ans.** $\begin{bmatrix} 1 & -2 & 1 & 0 \\ 1 & -2 & 2 & -3 \\ 0 & 1 & -1 & 1 \\ -2 & 3 & -2 & 3 \end{bmatrix}$

8. $\begin{bmatrix} 2 & 1 & -1 & 2 \\ 1 & 3 & 2 & -3 \\ -1 & 2 & 1 & -1 \\ 2 & -3 & -1 & 4 \end{bmatrix}$ **Ans.** $\frac{1}{18}\begin{bmatrix} 2 & 5 & -7 & 1 \\ 5 & -1 & 5 & -2 \\ -7 & 5 & 11 & 10 \\ 1 & -2 & 10 & 5 \end{bmatrix}$

9. $\begin{bmatrix} 2 & -6 & -2 & -3 \\ 5 & -13 & -4 & -7 \\ -1 & 4 & 1 & 2 \\ 0 & 1 & 0 & 1 \end{bmatrix}$ **Ans.** $\begin{bmatrix} -2 & 1 & 0 & 1 \\ 1 & 0 & 2 & -1 \\ -4 & 1 & -3 & 1 \\ -1 & 0 & -2 & 2 \end{bmatrix}$

10. $\begin{bmatrix} 1 & 3 & 3 & 2 & 1 \\ 1 & 4 & 3 & 3 & -1 \\ 1 & 3 & 4 & 1 & 1 \\ 1 & 1 & 1 & 1 & -1 \\ 1 & -2 & -1 & 2 & 2 \end{bmatrix}$ **Ans.** $\frac{1}{15}\begin{bmatrix} 30 & -20 & -15 & 25 & -5 \\ 30 & -11 & -18 & 7 & -8 \\ -30 & 12 & 21 & -9 & 6 \\ -15 & 12 & 6 & -9 & 6 \\ 15 & -7 & -6 & -1 & -1 \end{bmatrix}$

11. If X, Y are non-singular matrices and $B = \begin{bmatrix} X & O \\ O & Y \end{bmatrix}$, show that $B^{-1} = \begin{bmatrix} X^{-1} & O \\ O & Y^{-1} \end{bmatrix}$ where O is a null matrix.

CHAPTER 19

RANK OF MATRIX

19.1 RANK OF A MATRIX

The rank of a matrix is said to be r if

(*a*) It has at least one non-zero minor of order r.

(*b*) Every minor of A of order higher than r is zero.

Note: (*i*) Non-zero row is that row in which all the elements are not zero.

(*ii*) The rank of the product matrix AB of two matrices A and B is less than the rank of either of the matrices A and B.

(*iii*) Corresponding to every matrix A of rank r, there exist non-singular matrices P and Q such that $\quad PAQ = \begin{bmatrix} I_r & 0 \\ 0 & 0 \end{bmatrix}$

19.2 NORMAL FORM (CANONICAL FORM)

By performing elementary transformation, any non-zero matrix A can be reduced to one of the following four forms, called the Normal form of A :

(*i*) I_r $\qquad$ (*ii*) $[I_r \;\; 0]$ $\qquad$ (*iii*) $\begin{bmatrix} I_r \\ 0 \end{bmatrix}$ $\qquad$ (*iv*) $\begin{bmatrix} I_r & 0 \\ 0 & 0 \end{bmatrix}$

The number r so obtained is called the rank of A and we write $\rho(A) = r$. The form $\begin{bmatrix} I_r & 0 \\ 0 & 0 \end{bmatrix}$ is called first canonical form of A. Since both row and column transformations may be used here, the element 1 of the first row obtained can be moved in the first column. Then both the first row and first column can be cleared of other non-zero elements. Similarly, the element 1 of the second row can be brought into the second column, and so on.

Example 1. *Find the rank of the following matrix by reducing it to normal form*

$$A = \begin{bmatrix} 1 & 2 & -1 & 3 \\ 4 & 1 & 2 & 1 \\ 3 & -1 & 1 & 2 \\ 1 & 2 & 0 & 1 \end{bmatrix} \qquad \text{(U.P. I Sem., Com. 2002, Winter 2001)}$$

Solution.

$$\begin{bmatrix} 1 & 2 & -1 & 3 \\ 4 & 1 & 2 & 1 \\ 3 & -1 & 1 & 2 \\ 1 & 2 & 0 & 1 \end{bmatrix} \sim \begin{bmatrix} 1 & 2 & -1 & 3 \\ 0 & -7 & 6 & -11 \\ 0 & -7 & 4 & -7 \\ 0 & 0 & 1 & -2 \end{bmatrix} \begin{matrix} \\ R_2 \to R_2 - 4R_1 \\ R_3 \to R_3 - 3R_1 \\ R_4 \to R_4 - R_1 \end{matrix}$$

$$\sim \begin{bmatrix} 1 & 0 & 0 & 0 \\ 0 & -7 & 6 & -11 \\ 0 & -7 & 4 & -7 \\ 0 & 0 & 1 & -2 \end{bmatrix} \sim \begin{bmatrix} 1 & 0 & 0 & 0 \\ 0 & -7 & 6 & -11 \\ 0 & 0 & -2 & 4 \\ 0 & 0 & 1 & -2 \end{bmatrix} R_3 \to R_3 - R_2$$

$$C_2 \to C_2 - 2C_1, C_3 \to C_3 + C_1, C_4 \to C_4 - 3C_1$$

$$\sim \begin{bmatrix} 1 & 0 & 0 & 0 \\ 0 & -7 & 0 & 0 \\ 0 & 0 & -2 & 4 \\ 0 & 0 & 1 & -2 \end{bmatrix} \sim \begin{bmatrix} 1 & 0 & 0 & 0 \\ 0 & -7 & 0 & 0 \\ 0 & 0 & -2 & 4 \\ 0 & 0 & 0 & 0 \end{bmatrix} R_4 \to R_4 + \frac{1}{2} R_3$$

$$C_3 \to C_3 + \frac{6}{7} C_2, C_4 \to C_4 - \frac{11}{7} C_2,$$

$$C_4 \to C_4 + 2C_3 \sim \begin{bmatrix} 1 & 0 & 0 & 0 \\ 0 & -7 & 0 & 0 \\ 0 & 0 & -2 & 0 \\ 0 & 0 & 0 & 0 \end{bmatrix} \sim \begin{bmatrix} 1 & 0 & 0 & 0 \\ 0 & 1 & 0 & 0 \\ 0 & 0 & 1 & 0 \\ 0 & 0 & 0 & 0 \end{bmatrix} \begin{matrix} \\ R_2 \to -1/7\, R_2 \\ R_3 \to -1/2\, R_3 \\ \\ \end{matrix}$$

Rank of $A = 3$ **Ans.**

Example 2. *For which value of 'b' the rank of the matrix*

$$A = \begin{bmatrix} 1 & 5 & 4 \\ 0 & 3 & 2 \\ b & 13 & 10 \end{bmatrix} \text{ is 2,}$$

(U.P., I Semester, 2008)

Solution. Here, we have

$$A = \begin{bmatrix} 1 & 5 & 4 \\ 0 & 3 & 2 \\ b & 13 & 10 \end{bmatrix} \sim \begin{bmatrix} 1 & 5 & 4 \\ 0 & 3 & 2 \\ 0 & 13-5b & 10-4b \end{bmatrix} R_3 \to R_3 - bR_1$$

$$\sim \begin{bmatrix} 1 & 0 & 0 \\ 0 & 3 & 2 \\ 0 & 13-5b & 10-4b \end{bmatrix} \begin{matrix} C_2 \to C_2 - 5C_1 \\ C_3 \to C_3 - 4C_1 \end{matrix} \sim \begin{bmatrix} 1 & 0 & 0 \\ 0 & 3 & 2 \\ 0 & 0 & \frac{2(2-b)}{3} \end{bmatrix} R_3 \to R_3 - \frac{13-5b}{3} R_2$$

If rank of A is 2, than $\frac{2(2-b)}{3}$ must be zero.

i.e; $\frac{2(2-b)}{3} = 0 \qquad \Rightarrow 2 - b = 0 \qquad \Rightarrow b = 2$ **Ans.**

Example 3. *Reduce the matrix to normal form and find its rank.*

$$\begin{bmatrix} 2 & 3 & 4 & 5 \\ 3 & 4 & 5 & 6 \\ 4 & 5 & 6 & 7 \\ 9 & 10 & 11 & 12 \end{bmatrix}$$

(R.G.P.V., Bhopal, I Sem. April 2009, 2003)

Solution. $$\begin{bmatrix} 2 & 3 & 4 & 5 \\ 3 & 4 & 5 & 6 \\ 4 & 5 & 6 & 7 \\ 9 & 10 & 11 & 12 \end{bmatrix} \sim \begin{bmatrix} 2 & 3 & 4 & 5 \\ 0 & -\frac{1}{2} & -1 & \frac{-3}{2} \\ 0 & -1 & -2 & -3 \\ 0 & \frac{-7}{2} & -7 & \frac{-21}{2} \end{bmatrix} \begin{matrix} R_2 \to R_2 - \frac{3}{2} R_1 \\ R_3 \to R_3 - 2R_1 \\ R_4 \to R_4 - \frac{9}{2} R_1 \end{matrix}$$

$$\sim\begin{bmatrix} 2 & 0 & 0 & 0 \\ 0 & -\frac{1}{2} & -1 & \frac{-3}{2} \\ 0 & -1 & -2 & -3 \\ 0 & \frac{-7}{2} & -7 & \frac{-21}{2} \end{bmatrix} \begin{matrix} C_2 \to C_2 - \frac{3}{2}C_1 \\ C_3 \to C_3 - 2C_1 \\ C_4 \to C_4 - \frac{5}{2}C_1 \end{matrix} \sim \begin{bmatrix} 1 & 0 & 0 & 0 \\ 0 & 1 & 2 & 3 \\ 0 & 1 & 2 & 3 \\ 0 & 1 & 2 & 3 \end{bmatrix} \begin{matrix} R_1 \to \frac{1}{2}R_1 \\ R_2 \to -2R_2 \\ R_3 \to -R_3 \\ R_4 \to \frac{-2}{7}R_4 \end{matrix}$$

$$\sim\begin{bmatrix} 1 & 0 & 0 & 0 \\ 0 & 1 & 2 & 3 \\ 0 & 0 & 0 & 0 \\ 0 & 0 & 0 & 0 \end{bmatrix} \begin{matrix} \\ \\ R_3 \to R_3 - R_2 \\ R_4 \to R_4 - R_2 \end{matrix} \sim \begin{bmatrix} 1 & 0 & 0 & 0 \\ 0 & 1 & 0 & 0 \\ 0 & 0 & 0 & 0 \\ 0 & 0 & 0 & 0 \end{bmatrix} \begin{matrix} C_3 \to C_3 - 2C_2 \\ C_4 \to C_4 - 3C_2 \end{matrix} = \begin{bmatrix} I_2 & 0 & 0 \\ 0 & 0 & 0 \\ 0 & 0 & 0 \end{bmatrix}$$

Hence its rank = 2 **Ans.**

Example 4. *Find the rank of the matrix.*

$$A = \begin{bmatrix} 1 & 3 & 4 & 2 \\ 2 & -1 & 3 & 2 \\ 3 & -5 & 2 & 2 \\ 6 & -3 & 8 & 6 \end{bmatrix},$$ *by reducing it to normal form.*

(Uttarakhand, I semester, Dec. 2006)

Solution. We have,
$$A = \begin{bmatrix} \textcircled{1} & 3 & 4 & 2 \\ 2 & -1 & 3 & 2 \\ 3 & -5 & 2 & 2 \\ 6 & -3 & 8 & 6 \end{bmatrix} = \begin{bmatrix} 1 & 3 & 4 & 2 \\ 0 & -7 & -5 & -2 \\ 0 & -14 & -10 & -4 \\ 0 & -21 & -16 & -6 \end{bmatrix} \begin{matrix} \\ R_2 \to R_2 - 2R_1 \\ R_3 \to R_3 - 3R_1 \\ R_4 \to R_4 - 6R_1 \end{matrix}$$

$$= \begin{bmatrix} 1 & 0 & 0 & 0 \\ 0 & \textcircled{-7} & -5 & -2 \\ 0 & -14 & -10 & -4 \\ 0 & -21 & -16 & -6 \end{bmatrix} \begin{matrix} C_2 \to C_2 - 3C_1 \\ C_3 \to C_3 - 4C_1 \\ C_4 \to C_4 - 2C_1 \end{matrix}$$

$$= \begin{bmatrix} 1 & 0 & 0 & 0 \\ 0 & -7 & -5 & -2 \\ 0 & 0 & 0 & 0 \\ 0 & 0 & -1 & 0 \end{bmatrix} \begin{matrix} R_3 \to R_3 - 2R_2 \\ R_4 \to R_4 - 3R_2 \end{matrix} = \begin{bmatrix} 1 & 0 & 0 & 0 \\ 0 & -7 & -5 & -2 \\ 0 & 0 & -1 & 0 \\ 0 & 0 & 0 & 0 \end{bmatrix} R_3 \leftrightarrow R_4$$

$$= \begin{bmatrix} 1 & 0 & 0 & 0 \\ 0 & -7 & 0 & 0 \\ 0 & 0 & -1 & 0 \\ 0 & 0 & 0 & 0 \end{bmatrix} \begin{matrix} C_3 \to C_3 - \frac{5}{7}C_2 \\ C_4 \to C_4 - \frac{2}{7}C_2 \end{matrix} = \begin{bmatrix} 1 & 0 & 0 & 0 \\ 0 & 1 & 0 & 0 \\ 0 & 0 & 1 & 0 \\ 0 & 0 & 0 & 0 \end{bmatrix} \begin{matrix} R_2 \to -\frac{1}{7}R_2 \\ R_3 \to -R_3 \end{matrix}$$

$$= \begin{bmatrix} I_3 & 0 \\ 0 & 0 \end{bmatrix}$$ which is normal form.

Hence, Rank (A) = 3. **Ans.**

Example 5. *Reduce the matrix A to its normal form, when*

$$A = \begin{bmatrix} 1 & 2 & -1 & 4 \\ 2 & 4 & 3 & 4 \\ 1 & 2 & 3 & 4 \\ -1 & -2 & 6 & -7 \end{bmatrix}$$

Hence, find the rank of A. *(U.P., I Semester, Dec. 2004, Winter 2001)*

Solution. The given matrix is $A = \begin{bmatrix} 1 & 2 & -1 & 4 \\ 2 & 4 & 3 & 4 \\ 1 & 2 & 3 & 4 \\ -1 & -2 & 6 & -7 \end{bmatrix}$

$$\sim \begin{bmatrix} 1 & 2 & -1 & 4 \\ 0 & 0 & 5 & -4 \\ 0 & 0 & 4 & 0 \\ 0 & 0 & 5 & -3 \end{bmatrix} \begin{matrix} \\ R_2 \to R_2 - 2R_1 \\ R_3 \to R_3 - R_1 \\ R_4 \to R_4 + R_1 \end{matrix} \sim \begin{bmatrix} 1 & 0 & 0 & 0 \\ 0 & 0 & 5 & -4 \\ 0 & 0 & 4 & 0 \\ 0 & 0 & 5 & -3 \end{bmatrix} \begin{matrix} C_2 \to C_2 - 2C_1 \\ C_3 \to C_3 + C_1 \\ C_4 \to C_4 - 4C_1 \end{matrix}$$

$$\sim \begin{bmatrix} 1 & 0 & 0 & 0 \\ 0 & 5 & 0 & -4 \\ 0 & 4 & 0 & 0 \\ 0 & 5 & 0 & -3 \end{bmatrix} C_3 \leftrightarrow C_2 \sim \begin{bmatrix} 1 & 0 & 0 & 0 \\ 0 & 5 & 0 & -4 \\ 0 & 0 & 0 & \frac{16}{5} \\ 0 & 0 & 0 & 1 \end{bmatrix} \begin{matrix} \\ \\ R_3 \to R_3 - \frac{4}{5}R_2 \\ R_4 \to R_4 - R_2 \end{matrix}$$

$$\sim \begin{bmatrix} 1 & 0 & 0 & 0 \\ 0 & 5 & -4 & 0 \\ 0 & 0 & \frac{16}{5} & 0 \\ 0 & 0 & 1 & 0 \end{bmatrix} C_4 \leftrightarrow C_3 \sim \begin{bmatrix} 1 & 0 & 0 & 0 \\ 0 & 5 & 0 & 0 \\ 0 & 0 & \frac{16}{5} & 0 \\ 0 & 0 & 0 & 0 \end{bmatrix} \begin{matrix} \\ R_2 \to R_2 + \frac{5}{4}R_3 \\ \\ R_4 \to R_4 - \frac{5}{16}R_3 \end{matrix}$$

$$\sim \begin{bmatrix} 1 & 0 & 0 & 0 \\ 0 & 1 & 0 & 0 \\ 0 & 0 & 1 & 0 \\ 0 & 0 & 0 & 0 \end{bmatrix} \begin{matrix} R_2 \to 1/5\, R_2 \\ R_3 \to 5/16\, R_3 \end{matrix} \sim \begin{bmatrix} I_3 & 0 \\ 0 & 0 \end{bmatrix}$$

Which is the required normal form.

And since, the non-zero rows are 3 hence, the rank of the given matrix is 3. **Ans.**

Example 6. *Find non-singular matrices P, Q so that PAQ is a normal form where*

$$A = \begin{bmatrix} 2 & 1 & -3 & -6 \\ 3 & -3 & 1 & 2 \\ 1 & 1 & 1 & 2 \end{bmatrix}$$ *(R.G.P.V., Bhopal, April, 2010, U.P., I Sem. Winter 2002)*

and hence find its rank.

Solution. Order of A is 3×4

Total number of rows in $A = 3$; $\therefore$ Consider unit matrix I_3.

Total number of columns in $A = 4$

Hence, consider unit matrix I_4,

$$\therefore \quad A_{3 \times 4} = I_3\, A\, I_4$$

$$\begin{bmatrix} 2 & 1 & -3 & -6 \\ 3 & -3 & 1 & 2 \\ 1 & 1 & 1 & 2 \end{bmatrix} = \begin{bmatrix} 1 & 0 & 0 \\ 0 & 1 & 0 \\ 0 & 0 & 1 \end{bmatrix} A \begin{bmatrix} 1 & 0 & 0 & 0 \\ 0 & 1 & 0 & 0 \\ 0 & 0 & 1 & 0 \\ 0 & 0 & 0 & 1 \end{bmatrix}$$

$$\begin{bmatrix} 1 & 1 & 1 & 2 \\ 3 & -3 & 1 & 2 \\ 2 & 1 & -3 & -6 \end{bmatrix} = \begin{bmatrix} 0 & 0 & 1 \\ 0 & 1 & 0 \\ 1 & 0 & 0 \end{bmatrix} A \begin{bmatrix} 1 & 0 & 0 & 0 \\ 0 & 1 & 0 & 0 \\ 0 & 0 & 1 & 0 \\ 0 & 0 & 0 & 1 \end{bmatrix} R_1 \leftrightarrow R_3$$

$$\begin{bmatrix} 1 & 1 & 1 & 2 \\ 0 & -6 & -2 & -4 \\ 0 & -1 & -5 & -10 \end{bmatrix} = \begin{bmatrix} 0 & 0 & 1 \\ 0 & 1 & -3 \\ 1 & 0 & -2 \end{bmatrix} A \begin{bmatrix} 1 & 0 & 0 & 0 \\ 0 & 1 & 0 & 0 \\ 0 & 0 & 1 & 0 \\ 0 & 0 & 0 & 1 \end{bmatrix} \begin{matrix} R_2 \to R_2 - 3R_1 \\ R_3 \to R_3 - 2R_1 \end{matrix}$$

$$C_2 \to C_2 - C_1,\ C_3 \to C_3 - C_1,\ C_4 \to C_4 - 2C_1$$

$$\begin{bmatrix} 1 & 0 & 0 & 0 \\ 0 & -6 & -2 & -4 \\ 0 & -1 & -5 & -10 \end{bmatrix} = \begin{bmatrix} 0 & 0 & 1 \\ 0 & 1 & -3 \\ 1 & 0 & -2 \end{bmatrix} A \begin{bmatrix} 1 & -1 & -1 & -2 \\ 0 & 1 & 0 & 0 \\ 0 & 0 & 1 & 0 \\ 0 & 0 & 0 & 1 \end{bmatrix}$$

$$\begin{bmatrix} 1 & 0 & 0 & 0 \\ 0 & 6 & 2 & 4 \\ 0 & 1 & 5 & 10 \end{bmatrix} = \begin{bmatrix} 0 & 0 & 1 \\ 0 & -1 & 3 \\ -1 & 0 & 2 \end{bmatrix} A \begin{bmatrix} 1 & -1 & -1 & -2 \\ 0 & 1 & 0 & 0 \\ 0 & 0 & 1 & 0 \\ 0 & 0 & 0 & 1 \end{bmatrix} \begin{matrix} R_2 \to (-1)R_2 \\ R_3 \to (-1)R_3 \end{matrix}$$

$$\begin{bmatrix} 1 & 0 & 0 & 0 \\ 0 & 1 & 5 & 10 \\ 0 & 6 & 2 & 4 \end{bmatrix} = \begin{bmatrix} 0 & 0 & 1 \\ -1 & 0 & 2 \\ 0 & -1 & 3 \end{bmatrix} A \begin{bmatrix} 1 & -1 & -1 & -2 \\ 0 & 1 & 0 & 0 \\ 0 & 0 & 1 & 0 \\ 0 & 0 & 0 & 1 \end{bmatrix} R_2 \leftrightarrow R_3$$

$$\begin{bmatrix} 1 & 0 & 0 & 0 \\ 0 & 1 & 5 & 10 \\ 0 & 0 & -28 & -56 \end{bmatrix} = \begin{bmatrix} 0 & 0 & 1 \\ -1 & 0 & 2 \\ 6 & -1 & -9 \end{bmatrix} A \begin{bmatrix} 1 & -1 & -1 & -2 \\ 0 & 1 & 0 & 0 \\ 0 & 0 & 1 & 0 \\ 0 & 0 & 0 & 1 \end{bmatrix} R_3 \to R_3 - 6R_2$$

$$C_3 \to C_3 - 5C_2, \qquad C_4 \to C_4 - 10C_2$$

$$\begin{bmatrix} 1 & 0 & 0 & 0 \\ 0 & 1 & 0 & 0 \\ 0 & 0 & -28 & -56 \end{bmatrix} = \begin{bmatrix} 0 & 0 & 1 \\ -1 & 0 & 2 \\ 6 & -1 & -9 \end{bmatrix} A \begin{bmatrix} 1 & -1 & 4 & 8 \\ 0 & 1 & -5 & -10 \\ 0 & 0 & 1 & 0 \\ 0 & 0 & 0 & 1 \end{bmatrix}$$

$$\begin{bmatrix} 1 & 0 & 0 & 0 \\ 0 & 1 & 0 & 0 \\ 0 & 0 & 1 & 2 \end{bmatrix} = \begin{bmatrix} 0 & 0 & 1 \\ -1 & 0 & 2 \\ -\frac{6}{28} & \frac{1}{28} & \frac{9}{28} \end{bmatrix} A \begin{bmatrix} 1 & -1 & 4 & 8 \\ 0 & 1 & -5 & -10 \\ 0 & 0 & 1 & 0 \\ 0 & 0 & 0 & 1 \end{bmatrix} R_3 \to -\frac{1}{28} R_3$$

$$\begin{bmatrix} 1 & 0 & 0 & 0 \\ 0 & 1 & 0 & 0 \\ 0 & 0 & 1 & 0 \end{bmatrix} = \begin{bmatrix} 0 & 0 & 1 \\ -1 & 0 & 2 \\ -\frac{6}{28} & \frac{1}{28} & \frac{9}{28} \end{bmatrix} A \begin{bmatrix} 1 & -1 & 4 & 0 \\ 0 & 1 & -5 & 0 \\ 0 & 0 & 1 & -2 \\ 0 & 0 & 0 & 1 \end{bmatrix} C_4 \to C_4 - 2C_3$$

$$N = PAQ$$

$$P = \begin{bmatrix} 0 & 0 & 1 \\ -1 & 0 & 2 \\ -\frac{3}{14} & \frac{1}{28} & \frac{9}{28} \end{bmatrix}, Q = \begin{bmatrix} 1 & -1 & 4 & 0 \\ 0 & 1 & -5 & 0 \\ 0 & 0 & 1 & -2 \\ 0 & 0 & 0 & 1 \end{bmatrix}$$ **Ans.**

Note. P and Q are not unique.

Normal form of the given matrix is $\begin{bmatrix} 1 & 0 & 0 & 0 \\ 0 & 1 & 0 & 0 \\ 0 & 0 & 1 & 0 \end{bmatrix}$

The number of non zero rows in the normal matrix = 3
Hence Rank = 3 **Ans.**

Example 7. *If* $A = \begin{bmatrix} 3 & -3 & 4 \\ 2 & -3 & 4 \\ 0 & -1 & 1 \end{bmatrix}$, *Find two non singular matrices P and Q such that* $PAQ = I$. *Hence find* A^{-1}.

Solution. $A_{3\times 3} = I_3 A I_3$

$$\begin{bmatrix} 3 & -3 & 4 \\ 2 & -3 & 4 \\ 0 & -1 & 1 \end{bmatrix} = \begin{bmatrix} 1 & 0 & 0 \\ 0 & 1 & 0 \\ 0 & 0 & 1 \end{bmatrix} A \begin{bmatrix} 1 & 0 & 0 \\ 0 & 1 & 0 \\ 0 & 0 & 1 \end{bmatrix}$$

$$\begin{bmatrix} 1 & 0 & 0 \\ 2 & -3 & 4 \\ 0 & -1 & 1 \end{bmatrix} = \begin{bmatrix} 1 & -1 & 0 \\ 0 & 1 & 0 \\ 0 & 0 & 1 \end{bmatrix} A \begin{bmatrix} 1 & 0 & 0 \\ 0 & 1 & 0 \\ 0 & 0 & 1 \end{bmatrix} R_1 \to R_1 - R_2$$

$$\begin{bmatrix} 1 & 0 & 0 \\ 0 & -3 & 4 \\ 0 & -1 & 1 \end{bmatrix} = \begin{bmatrix} 1 & -1 & 0 \\ -2 & 3 & 0 \\ 0 & 0 & 1 \end{bmatrix} A \begin{bmatrix} 1 & 0 & 0 \\ 0 & 1 & 0 \\ 0 & 0 & 1 \end{bmatrix} R_3 \to R_2 - 2R_1$$

$$\begin{bmatrix} 1 & 0 & 0 \\ 0 & 3 & 4 \\ 0 & 1 & 1 \end{bmatrix} = \begin{bmatrix} 1 & -1 & 0 \\ -2 & 3 & 0 \\ 0 & 0 & 1 \end{bmatrix} A \begin{bmatrix} 1 & 0 & 0 \\ 0 & -1 & 0 \\ 0 & 0 & 1 \end{bmatrix} C_2 \to -C_2$$

$$\begin{bmatrix} & & 0 \\ 0 & 1 & 1 \\ 0 & 3 & 4 \end{bmatrix} = \begin{bmatrix} 1 & -1 & 0 \\ 0 & 0 & 1 \\ -2 & 3 & 0 \end{bmatrix} A \begin{bmatrix} 1 & 0 & 0 \\ 0 & -1 & 0 \\ 0 & 0 & 1 \end{bmatrix} R_2 \leftrightarrow R_3$$

$$\begin{bmatrix} 1 & 0 & 0 \\ 0 & 1 & 1 \\ 0 & 0 & 1 \end{bmatrix} = \begin{bmatrix} 1 & -1 & 0 \\ 0 & 0 & 1 \\ -2 & 3 & -3 \end{bmatrix} A \begin{bmatrix} 1 & 0 & 0 \\ 0 & -1 & 0 \\ 0 & 0 & 1 \end{bmatrix} R_3 \to R_3 - 3R_2$$

$$\begin{bmatrix} 1 & 0 & 0 \\ 0 & 1 & 0 \\ 0 & 0 & 1 \end{bmatrix} = \begin{bmatrix} 1 & -1 & 0 \\ 0 & 0 & 1 \\ -2 & 3 & -3 \end{bmatrix} A \begin{bmatrix} 1 & 0 & 0 \\ 0 & -1 & 1 \\ 0 & 0 & 1 \end{bmatrix} C_3 \to C_3 - C_2$$

$$I_3 = PAQ$$

$$A^{-1} = QP, \qquad A^{-1} = \begin{bmatrix} 1 & 0 & 0 \\ 0 & -1 & 1 \\ 0 & 0 & 1 \end{bmatrix} \begin{bmatrix} 1 & -1 & 0 \\ 0 & 0 & 1 \\ -2 & 3 & -3 \end{bmatrix} \qquad \begin{bmatrix} I = PAQ \\ P^{-1} = AQ \\ P^{-1}Q^{-1} = A \\ (P^{-1}Q^{-1})^{-1} = A^{-1} \\ QP = A^{-1} \end{bmatrix}$$

$$\Rightarrow \qquad A^{-1} = \begin{bmatrix} 1 & -1 & 0 \\ -2 & 3 & -4 \\ -2 & 3 & -3 \end{bmatrix}$$ **Ans.**

19.3 RANK OF MATRIX BY TRIANGULAR FORM

Rank = Number of non-zero row in upper triangular matrix.

Note. Non-zero row is that row which does not contain all the elements as zero.

Example 8. *Find the rank of the matrix*

$$\begin{bmatrix} 1 & 2 & 3 & 2 \\ 2 & 3 & 5 & 1 \\ 1 & 3 & 4 & 5 \end{bmatrix}$$ *(U.P., I Semester, Winter 2003, 2000)*

Solution.
$$\begin{bmatrix} 1 & 2 & 3 & 2 \\ 2 & 3 & 5 & 1 \\ 1 & 3 & 4 & 5 \end{bmatrix} \sim \begin{bmatrix} 1 & 2 & 3 & 2 \\ 0 & -1 & -1 & -3 \\ 0 & 1 & 1 & 3 \end{bmatrix} \begin{matrix} \\ R_2 \to R_2 - 2R_1 \\ R_3 \to R_3 - R_1 \end{matrix}$$

$$\sim \begin{bmatrix} 1 & 2 & 3 & 2 \\ 0 & -1 & -1 & -3 \\ 0 & 0 & 0 & 0 \end{bmatrix} \begin{matrix} \\ \\ R_3 \to R_3 + R_2 \end{matrix}$$

Rank = Number of non zero rows = 2. **Ans.**

Example 9. *Find the rank of the matrix*

$$\begin{bmatrix} -1 & 2 & 3 & -2 \\ 2 & -5 & 1 & 2 \\ 3 & -8 & 5 & 2 \\ 5 & -12 & -1 & 6 \end{bmatrix}$$

Solution.
$$\begin{bmatrix} -1 & 2 & 3 & -2 \\ 2 & -5 & 1 & 2 \\ 3 & -8 & 5 & 2 \\ 5 & -12 & -1 & 6 \end{bmatrix} \sim \begin{bmatrix} -1 & 2 & 3 & -2 \\ 0 & -1 & 7 & -2 \\ 0 & -2 & 14 & -4 \\ 0 & -2 & 14 & -4 \end{bmatrix} \begin{matrix} \\ R_2 \to R_2 + 2R_1 \\ R_3 \to R_3 + 3R_1 \\ R_4 \to R_4 + 5R_1 \end{matrix}$$

$$\sim \begin{bmatrix} -1 & 2 & 3 & -2 \\ 0 & -1 & 7 & -2 \\ 0 & 0 & 0 & 0 \\ 0 & 0 & 0 & 0 \end{bmatrix} \begin{matrix} \\ R_3 \to R_3 - 2R_2 \\ R_4 \to R_4 - 2R_2 \\ \\ \end{matrix}$$

Here the 4th order and 3rd order minors are zero. But a minor of second order

$$\begin{vmatrix} 3 & -2 \\ 7 & -2 \end{vmatrix} = -6 + 14 = 8 \neq 0$$

Rank = Number of non-zero rows = 2. **Ans.**

Example 10. *Find the rank of matrix*

$$\begin{bmatrix} 2 & 3 & -2 & 4 \\ 3 & -2 & 1 & 2 \\ 3 & 2 & 3 & 4 \\ -2 & 4 & 0 & 5 \end{bmatrix}$$

(*U.P., I Semester, Dec., 2006*)

Solution. Multiplying R_1 by $\frac{1}{2}$, we get 1 as pivotal element

$$\sim \begin{bmatrix} \textcircled{1} & \frac{3}{2} & -1 & 2 \\ 3 & -2 & 1 & 2 \\ 3 & 2 & 3 & 4 \\ -2 & 4 & 0 & 5 \end{bmatrix}$$

$$\sim \begin{bmatrix} \textcircled{1} & \frac{3}{2} & -1 & 2 \\ 0 & -\frac{13}{2} & 4 & -4 \\ 0 & -\frac{5}{2} & 6 & -2 \\ 0 & 7 & -2 & 9 \end{bmatrix} \begin{matrix} \\ R_2 \to R_2 - 3R_1 \\ R_3 \to R_3 - 3R_1 \\ R_4 \to R_4 + 2R_1 \end{matrix} \sim \begin{bmatrix} 1 & 0 & 0 & 0 \\ 0 & \textcircled{1} & -\frac{8}{13} & \frac{8}{13} \\ 0 & -\frac{5}{2} & 6 & -2 \\ 0 & 7 & -2 & 9 \end{bmatrix} \begin{matrix} R_2 \to -\frac{2}{13} R_2 \\ C_2 \to C_2 - \frac{3}{2} C_1 \\ C_3 \to C_3 + C_1 \\ C_4 \to C_4 - 2C_1 \end{matrix}$$

$$\sim \begin{bmatrix} 1 & 0 & 0 & 0 \\ 0 & \textcircled{1} & -\frac{8}{13} & \frac{8}{13} \\ 0 & 0 & \frac{58}{13} & -\frac{6}{13} \\ 0 & 0 & \frac{30}{13} & \frac{61}{13} \end{bmatrix} \begin{matrix} \\ \\ R_3 \to R_3 + \frac{5}{2} R_2 \\ R_4 \to R_4 - 7\,R_2 \end{matrix} \sim \begin{bmatrix} 1 & 0 & 0 & 0 \\ 0 & 1 & 0 & 0 \\ 0 & 0 & \frac{58}{13} & -\frac{6}{13} \\ 0 & 0 & \frac{30}{13} & \frac{61}{13} \end{bmatrix} \begin{matrix} \\ \\ C_3 \to C_3 + \frac{8}{13} C_2 \\ C_4 \to C_4 - \frac{8}{13} C_2 \end{matrix}$$

$$\sim \begin{bmatrix} 1 & 0 & 0 & 0 \\ 0 & 1 & 0 & 0 \\ 0 & 0 & \textcircled{1} & -\frac{3}{29} \\ 0 & 0 & \frac{30}{13} & \frac{61}{13} \end{bmatrix} R_3 \to \frac{13}{58} R_3 \sim \begin{bmatrix} 1 & 0 & 0 & 0 \\ 0 & 1 & 0 & 0 \\ 0 & 0 & \textcircled{1} & -\frac{3}{29} \\ 0 & 0 & 0 & \frac{143}{29} \end{bmatrix} R_4 \to R_4 - \frac{30}{13} R_3$$

$$\sim \begin{bmatrix} 1 & 0 & 0 & 0 \\ 0 & 1 & 0 & 0 \\ 0 & 0 & 1 & 0 \\ 0 & 0 & 0 & \frac{143}{29} \end{bmatrix} C_4 \to C_4 + \frac{3}{29} C_3 \sim \begin{bmatrix} 1 & 0 & 0 & 0 \\ 0 & 1 & 0 & 0 \\ 0 & 0 & 1 & 0 \\ 0 & 0 & 0 & 1 \end{bmatrix} R_4 \to \frac{29}{143} R_4$$

$$= I_4$$

Hence, the rank of the given matrix = 4 **Ans.**

Example 11. *Use elementary transformation to reduce the following matrix A to triangular from and hence find the rank of A.*

$$A = \begin{bmatrix} 2 & 3 & -1 & -1 \\ 1 & -1 & -2 & -4 \\ 3 & 1 & 3 & -2 \\ 6 & 3 & 0 & -7 \end{bmatrix}$$

(R.G.P.V., Bhopal, June 2007, Winter 2003, U.P., I Semester, Dec. 2005)

Solution. We have,

$$A = \begin{bmatrix} 2 & 3 & -1 & -1 \\ 1 & -1 & -2 & -4 \\ 3 & 1 & 3 & -2 \\ 6 & 3 & 0 & -7 \end{bmatrix} \approx \begin{bmatrix} 1 & -1 & -2 & -4 \\ 2 & 3 & -1 & -1 \\ 3 & 1 & 3 & -2 \\ 6 & 3 & 0 & -7 \end{bmatrix} R_1 \longleftrightarrow R_2$$

$$\approx \begin{bmatrix} 1 & -1 & -2 & -4 \\ 0 & 5 & 3 & 7 \\ 0 & 4 & 9 & 10 \\ 0 & 9 & 12 & 17 \end{bmatrix} \begin{matrix} \\ R_2 \to R_2 - 2R_1 \\ R_3 \to R_3 - 3R_1 \\ R_4 \to R_4 - 6R_1 \end{matrix} \approx \begin{bmatrix} 1 & -1 & -2 & -4 \\ 0 & 5 & 3 & 7 \\ 0 & 0 & 33/5 & 22/5 \\ 0 & 0 & 33/5 & 22/5 \end{bmatrix} \begin{matrix} \\ \\ R_3 \to R_3 - 4/5\, R_2 \\ R_4 \to R_4 - 9/5\, R_2 \end{matrix}$$

$$\approx \begin{bmatrix} 1 & -1 & -2 & -4 \\ 0 & 5 & 3 & 7 \\ 0 & 0 & 33/5 & 22/5 \\ 0 & 0 & 0 & 0 \end{bmatrix} R_4 \to R_4 - R_3$$

$R(A)$ = Number of non-zero rows.

$\Rightarrow \quad R(A) = 3$ **Ans.**

Example 12. *Prove that the points (x_1, y_1), (x_2, y_2) and (x_3, y_3) are collinear if and only if the rank of the matrix* $\begin{bmatrix} x_1 & y_1 & 1 \\ x_2 & y_2 & 1 \\ x_3 & y_3 & 1 \end{bmatrix}$ *is less than three.*

Solution. Necessary condition.

Since the points (x_1, y_1), (x_2, y_2) and (x_3, y_3) are collinear. Therefore, the area of the triangle formed by these points is zero.

$$\therefore \quad \frac{1}{2}\begin{bmatrix} x_1 & y_1 & 1 \\ x_2 & y_2 & 1 \\ x_3 & y_3 & 1 \end{bmatrix} = 0 \quad \Rightarrow \quad \begin{bmatrix} x_1 & y_1 & 1 \\ x_2 & y_2 & 1 \\ x_3 & y_3 & 1 \end{bmatrix} = 0$$

$\Rightarrow$ The rank of matrix $\begin{bmatrix} x_1 & y_1 & 1 \\ x_2 & y_2 & 1 \\ x_3 & y_3 & 1 \end{bmatrix}$ is less than 3.

Given, the three points are collinear and we have proved that the rank of matrix is less than 3.

Hence, the condition is necessary.

Sufficient condition.

Given : The rank of the following matrix is less than 3.

$$\text{Rank of } \begin{bmatrix} x_1 & y_1 & 1 \\ x_2 & y_2 & 1 \\ x_3 & y_3 & 1 \end{bmatrix} \leq 3 \quad \Rightarrow \quad \begin{vmatrix} x_1 & y_1 & 1 \\ x_2 & y_2 & 1 \\ x_3 & y_3 & 1 \end{vmatrix} = 0 \quad \Rightarrow \quad \frac{1}{2}\begin{vmatrix} x_1 & y_1 & 1 \\ x_2 & y_2 & 1 \\ x_3 & y_3 & 1 \end{vmatrix} = 0$$

Thus, the three points (x_1, y_1), (x_2, y_2) and (x_3, y_3) are collinear. Given, the rank of matrix is less than 3 and we have proved that the points are collinear.

Hence, the condition is sufficient. **Proved.**

Theorem

The rank of the product matrix AB of two matrices A and B is less than the rank of the either of the matrices A and B.

Proof. Let r_1 and r_2 be the ranks of the matrices A and B respectively.

Since r_1 is the rank of the matrix A, therefore

$$A \sim \begin{bmatrix} Ir_1 & 0 \\ 0 & 0 \end{bmatrix} \qquad \text{... (1)}$$

Where Ir_1 is the unit matrix of order r_1 and contains r_1 rows.

Post multiplying (1) by B, we get

$$AB \sim \begin{bmatrix} Ir_1 & 0 \\ 0 & 0 \end{bmatrix} B$$

But $\begin{bmatrix} Ir_1 & 0 \\ 0 & 0 \end{bmatrix}$ B can have r_1 non-zero rows at the most.

$$\text{Rank of } AB = \text{Rank of } \begin{bmatrix} Ir_1 & 0 \\ 0 & 0 \end{bmatrix} B$$

$$\text{Rank of } AB = \text{Rank } \begin{bmatrix} Ir_1 & 0 \\ 0 & 0 \end{bmatrix} B \leq r_1$$

Rank of $AB \leq$ Rank of A

Similarly we can prove that,

Rank of $AB \leq$ Rank of B. **Proved.**

EXERCISE 19.1

Find the rank of the following matrices:

1. $\begin{bmatrix} 1 & 2 & 3 \\ 2 & 4 & 7 \\ 3 & 6 & 10 \end{bmatrix}$ **Ans.** 2

2. $\begin{bmatrix} 1 & 2 & 1 \\ -1 & 0 & 2 \\ 2 & 1 & -3 \end{bmatrix}$ **Ans.** 3

3. $\begin{bmatrix} 0 & 1 & 2 & -2 \\ 4 & 0 & 2 & 6 \\ 2 & 1 & 3 & 1 \end{bmatrix}$ **Ans.** 2

4. $\begin{bmatrix} 2 & 4 & 3 & -2 \\ -3 & -2 & -1 & 4 \\ 6 & -1 & 7 & 2 \end{bmatrix}$ **Ans.** 3

5. $\begin{bmatrix} 3 & 4 & 1 & 1 \\ 2 & 4 & 3 & 6 \\ -1 & -2 & 6 & 4 \\ 1 & -1 & 2 & -3 \end{bmatrix}$ **Ans.** 4

6. $\begin{bmatrix} 1 & 4 & 3 & -2 & 1 \\ -2 & -3 & -1 & 4 & 3 \\ -1 & 6 & 7 & 2 & 9 \\ -3 & 3 & 6 & 6 & 12 \end{bmatrix}$ **Ans.** 2

Reduce the following matrices to Echelon form and find out the rank:

7. $\begin{bmatrix} 1 & 1 & 2 \\ 1 & 2 & 2 \\ 2 & 2 & 3 \end{bmatrix}$ **Ans.** $\begin{bmatrix} 1 & 0 & 0 \\ 0 & 1 & 0 \\ 0 & 0 & 1 \end{bmatrix}$, Rank = 3 **8.** $\begin{bmatrix} 1 & 2 & 3 & 0 \\ 2 & 4 & 3 & 2 \\ 3 & 2 & 1 & 3 \\ 6 & 8 & 7 & 5 \end{bmatrix}$ **Ans.** $\begin{bmatrix} I_3 & 0 \\ 0 & 0 \end{bmatrix}$, Rank = 3

9. $\begin{bmatrix} 3 & 2 & 5 & 7 & 12 \\ 1 & 1 & 2 & 3 & 5 \\ 3 & 3 & 6 & 9 & 15 \end{bmatrix}$ **Ans.** $\begin{bmatrix} I_2 & 0 \\ 0 & 0 \end{bmatrix}$, Rank = 2 **10.** $\begin{bmatrix} 2 & -4 & 3 & 1 & 0 \\ 1 & -2 & 1 & -4 & 2 \\ 0 & 1 & -1 & 3 & 1 \\ 4 & -7 & 4 & -4 & 5 \end{bmatrix}$ **Ans.** $\begin{bmatrix} I_3 & 0 \\ 0 & 0 \end{bmatrix}$, Rank = 3

Using elementary transformations, reduce the following matrices to the canonical form (or row-reduced Echelon form):

11. $A = \begin{bmatrix} 0 & 0 & 0 & 0 & 0 \\ 0 & 1 & 2 & 3 & 4 \\ 0 & 2 & 3 & 4 & 1 \\ 0 & 3 & 4 & 1 & 2 \end{bmatrix}$ **Ans.** $\begin{bmatrix} 1 & 0 & 0 & 0 \\ 0 & 1 & 0 & 0 \\ 0 & 0 & 1 & 0 \end{bmatrix}$ **12.** $A = \begin{bmatrix} 0 & 4 & -12 & 8 & 9 \\ 0 & 2 & -6 & 2 & 5 \\ 0 & 1 & -3 & 6 & 4 \\ 0 & -8 & 24 & 3 & 1 \end{bmatrix}$ **Ans.** $\begin{bmatrix} 1 & 0 & 0 & 0 \\ 0 & 1 & 0 & 0 \\ 0 & 0 & 1 & 0 \end{bmatrix}$

Using elementary transformations, reduce the following matrices to the normal form:

13. $A = \begin{bmatrix} 1 & 2 & 0 & -1 \\ 3 & 4 & 1 & 2 \\ -2 & 3 & 2 & 5 \end{bmatrix}$ **Ans.** $\begin{bmatrix} 1 & 0 & 0 & 0 \\ 0 & 1 & 0 & 0 \\ 0 & 0 & 1 & 0 \end{bmatrix}$ **14.** $A = \begin{bmatrix} 1 & 2 & 3 & 4 \\ 3 & 4 & 1 & 2 \\ 4 & 3 & 1 & 2 \end{bmatrix}$ **Ans.** $\begin{bmatrix} 1 & 0 & 0 & 0 \\ 0 & 1 & 0 & 0 \\ 0 & 0 & 1 & 0 \end{bmatrix}$

Obtain a matrix N in the normal form equivalent to

15. $A = \begin{bmatrix} 0 & 0 & 0 & 0 & 0 \\ 0 & 4 & 5 & 0 & 0 \\ 0 & 9 & 1 & -1 & 2 \\ 0 & 10 & 0 & 1 & 11 \end{bmatrix}$

Hence find non-singular matrices P and Q such that $PAQ = N$.

16. $\begin{bmatrix} 1 & -3 & 1 & 2 \\ 0 & 1 & 2 & 3 \\ 3 & 4 & 1 & -2 \end{bmatrix}$

Find the rank of the following matrix by reducing it into normal form:

17. $A = \begin{bmatrix} 1 & 3 & 2 & 5 & 1 \\ 2 & 2 & -1 & 6 & 3 \\ 1 & 1 & 2 & 3 & -1 \\ 0 & 2 & 5 & 2 & -3 \end{bmatrix}$ **Ans.** 4 **18.** $A = \begin{bmatrix} 1 & 2 & 3 & 1 \\ 1 & 3 & 3 & 2 \\ 2 & 4 & 3 & 3 \\ 1 & 1 & 1 & 1 \end{bmatrix}$ **Ans.** 3

Choose the correct answer:

19. Rank of matrix $A = \begin{bmatrix} 1 & 2 & 3 \\ 1 & 4 & 2 \\ 2 & 6 & 5 \end{bmatrix}$ is

(*a*) 0 (*b*) 1 (*c*) 3 (*d*) 2 (*AMIETE, June 2009*) **Ans.** (*d*)

20. For which value of 'b' the rank of the matrix $A = \begin{bmatrix} 1 & 5 & 4 \\ 0 & 3 & 2 \\ b & 13 & 10 \end{bmatrix}$ is

(*a*) 1 (*b*) 2 (*c*) 3 (*d*) 0 (*AMIETE, Dec. 2009*) **Ans.** (*b*)

CONSISTENCY OF LINEAR SYSTEM OF EQUATIONS AND THEIR SOLUTION (LINEAR DEPENDENCE)

20.1 SOLUTION OF SIMULTANEOUS EQUATIONS

The matrix of the coefficients of x, y, z is reduced into Echelon form by elementary row transformations. At the end of the row transformation the value of z is calculated from the last equation and value of y and the value of x are calculated by the backward substitution.

Example 1. *Solve the following equations*

$$x - y + 2z = 3, \quad x + 2y + 3z = 5, \quad 3x - 4y - 5z = -13$$

Solution. In the matrix form, the equations are written in the following form.

$$\begin{bmatrix} 1 & -1 & 2 \\ 1 & 2 & 3 \\ 3 & -4 & -5 \end{bmatrix}\begin{bmatrix} x \\ y \\ z \end{bmatrix} = \begin{bmatrix} 3 \\ 5 \\ -13 \end{bmatrix} \text{ or } \begin{bmatrix} 1 & -1 & 2 \\ 0 & 3 & 1 \\ 0 & -1 & -11 \end{bmatrix}\begin{bmatrix} x \\ y \\ z \end{bmatrix} = \begin{bmatrix} 3 \\ 2 \\ -22 \end{bmatrix} \begin{matrix} R_2 \to R_2 - R_1 \\ R_3 \to R_3 - 3R_1 \end{matrix}$$

$$\begin{bmatrix} 1 & -1 & 2 \\ 0 & 3 & 1 \\ 0 & 0 & -\frac{32}{3} \end{bmatrix}\begin{bmatrix} x \\ y \\ z \end{bmatrix} = \begin{bmatrix} 3 \\ 2 \\ -\frac{64}{3} \end{bmatrix} R_3 \to R_3 + \frac{1}{3}R_2$$

$$x - y + 2z = 3 \qquad \ldots(1)$$

$$3y + z = 2 \qquad \ldots(2)$$

$$\frac{-32}{3}z = \frac{-64}{3} \Rightarrow z = 2$$

Putting the value of z in (2), we get

$$3y + 2 = 2 \Rightarrow y = 0$$

Putting the value of y, z in (1), we get

$$x - 0 + 4 = 3 \Rightarrow x = -1$$

$$x = -1, y = 0, z = 2 \qquad \textbf{Ans.}$$

Example 2. *Find all the solutions of the system of equations*

$$x_1 + 2x_2 - x_3 = 1, \; 3x_1 - 2x_2 + 2x_3 = 2, \; 7x_1 - 2x_2 + 3x_3 = 5$$

Solution. $$\begin{bmatrix} 1 & 2 & -1 \\ 3 & -2 & 2 \\ 7 & -2 & 3 \end{bmatrix}\begin{bmatrix} x_1 \\ x_2 \\ x_3 \end{bmatrix} = \begin{bmatrix} 1 \\ 2 \\ 5 \end{bmatrix}$$

$$R_2 \to R_2 - 3R_1,\ R_3 \to R_3 - 7R_1$$

$$\begin{bmatrix} 1 & 2 & -1 \\ 0 & -8 & 5 \\ 0 & -16 & 10 \end{bmatrix}\begin{bmatrix} x_1 \\ x_2 \\ x_3 \end{bmatrix} = \begin{bmatrix} 1 \\ -1 \\ -2 \end{bmatrix},\ \begin{bmatrix} 1 & 2 & -1 \\ 0 & -8 & 5 \\ 0 & 0 & 0 \end{bmatrix}\begin{bmatrix} x_1 \\ x_2 \\ x_3 \end{bmatrix} = \begin{bmatrix} 1 \\ -1 \\ 0 \end{bmatrix} R_3 \to R_3 - 2R_2$$

$$x_1 + 2x_2 - x_3 = 1 \qquad \text{...(1)}$$

$$-8x_2 + 5x_3 = -1 \qquad \text{...(2)}$$

Let $x_3 = k$

Putting $x_3 = k$ in (2), we get

$$-8x_2 + 5k = -1 \Rightarrow x_2 = \frac{1}{8}(5k+1)$$

Substituting the values of x_3, x_1 in (1), we get

$$x_1 + \frac{1}{4}(5k+1) - k = 1$$

$$\therefore \qquad x_1 = 1 + k - \frac{5k}{4} - \frac{1}{4} = -\frac{k}{4} + \frac{3}{4}$$

$$\therefore \qquad x_1 = -\frac{k}{4} + \frac{3}{4},\ x_2 = \frac{5k}{8} + \frac{1}{8},\ x_3 = k$$

The equations have infinite solution. **Ans.**

Example 3. *Express the following system of equations in matrix form and solve them by the elimination method (Gauss Jordan Method)*

$$2x_1 + x_2 + 2x_3 + x_4 = 6$$
$$6x_1 - 6x_2 + 6x_3 + 12x_4 = 36$$
$$4x_1 + 3x_2 + 3x_3 - 3x_4 = -1$$
$$2x_1 + 2x_2 - x_3 + x_4 = 10$$

Solution. The equations are expressed in matrix form as

$$\begin{bmatrix} 2 & 1 & 2 & 1 \\ 6 & -6 & 6 & 12 \\ 4 & 3 & 3 & -3 \\ 2 & 2 & -1 & 1 \end{bmatrix}\begin{bmatrix} x_1 \\ x_2 \\ x_3 \\ x_4 \end{bmatrix} = \begin{bmatrix} 6 \\ 36 \\ -1 \\ 10 \end{bmatrix}$$

$$\Rightarrow \begin{bmatrix} 2 & 1 & 2 & 1 \\ 0 & -9 & 0 & 9 \\ 0 & 1 & -1 & -5 \\ 0 & 1 & -3 & 0 \end{bmatrix}\begin{bmatrix} x_1 \\ x_2 \\ x_3 \\ x_4 \end{bmatrix} = \begin{bmatrix} 6 \\ 18 \\ -13 \\ 4 \end{bmatrix} \begin{matrix} R_2 \to R_2 - 3R_1 \\ R_3 \to R_3 - 2R_1 \\ R_4 \to R_4 - R_1 \end{matrix}$$

$$\begin{bmatrix} 2 & 1 & 2 & 1 \\ 0 & 1 & 0 & -1 \\ 0 & 1 & -1 & -5 \\ 0 & 1 & -3 & 0 \end{bmatrix}\begin{bmatrix} x_1 \\ x_2 \\ x_3 \\ x_4 \end{bmatrix} = \begin{bmatrix} 6 \\ -2 \\ -13 \\ 4 \end{bmatrix} R_2 \to \frac{R_2}{-9}$$

$$\begin{bmatrix} 2 & 1 & 2 & 1 \\ 0 & 1 & 0 & -1 \\ 0 & 0 & -1 & -4 \\ 0 & 0 & -3 & 1 \end{bmatrix}\begin{bmatrix} x_1 \\ x_2 \\ x_3 \\ x_4 \end{bmatrix} = \begin{bmatrix} 6 \\ -2 \\ -11 \\ 6 \end{bmatrix} \begin{matrix} R_3 \to R_3 - R_2 \\ R_4 \to R_4 - R_2 \end{matrix}$$

$$\begin{bmatrix} 2 & 1 & 2 & 1 \\ 0 & 1 & 0 & -1 \\ 0 & 0 & -1 & -4 \\ 0 & 0 & 0 & 13 \end{bmatrix} \begin{bmatrix} x_1 \\ x_2 \\ x_3 \\ x_4 \end{bmatrix} = \begin{bmatrix} 6 \\ -2 \\ -11 \\ 39 \end{bmatrix} \quad R_4 \to R_4 - 3R_3$$

$2x_1 + x_2 + 2x_3 + x_4 = 6$ ...(1)

$x_2 - x_4 = -2$ ...(2)

$-x_3 - 4x_4 = -11$ ...(3)

$13x_4 = 39 \Rightarrow x_4 = 3$

Putting the value of x_4 in (3), we get

$$-x_3 - 12 = -11 \Rightarrow x_3 = -1$$

Putting the value of x_4 in (2), we get

$$x_2 - 3 = -2 \Rightarrow x_2 = 1$$

Substituting the values of x_4, x_3 and x_2 in (1), we get

$$2x_1 + 1 - 2 + 3 = 6 \text{ or } 2x_1 = 4 \Rightarrow x_1 = 2$$

$\therefore$ $x_1 = 2, x_2 = 1, x_3 = -1, x_4 = 3$ **Ans.**

Example 4. *Find the general solution of the system of equations:*

$$3x_1 + 2x_3 + 2x_4 = 0$$
$$-x_1 + 7x_2 + 4x_3 + 9x_4 = 0$$
$$7x_1 - 7x_2 - 5x_4 = 0$$

Solution. The system of equations in the matrix form is expressed as

$$\begin{bmatrix} 3 & 0 & 2 & 2 \\ -1 & 7 & 4 & 9 \\ 7 & -7 & 0 & -5 \end{bmatrix} \begin{bmatrix} x_1 \\ x_2 \\ x_3 \\ x_4 \end{bmatrix} = \begin{bmatrix} 0 \\ 0 \\ 0 \end{bmatrix}$$

$$\begin{bmatrix} -1 & 7 & 4 & 9 \\ 3 & 0 & 2 & 2 \\ 7 & -7 & 0 & -5 \end{bmatrix} \begin{bmatrix} x_1 \\ x_2 \\ x_3 \\ x_4 \end{bmatrix} = \begin{bmatrix} 0 \\ 0 \\ 0 \end{bmatrix} \quad R_1 \leftrightarrow R_2$$

$$\begin{bmatrix} -1 & 7 & 4 & 9 \\ 0 & 21 & 14 & 29 \\ 0 & 42 & 28 & 58 \end{bmatrix} \begin{bmatrix} x_1 \\ x_2 \\ x_3 \\ x_4 \end{bmatrix} = \begin{bmatrix} 0 \\ 0 \\ 0 \end{bmatrix} \quad \begin{matrix} R_2 \to R_2 + 3R_1 \\ R_3 \to R_3 + 7R_1 \end{matrix}$$

$$\begin{bmatrix} -1 & 7 & 4 & 9 \\ 0 & 21 & 14 & 29 \\ 0 & 0 & 0 & 0 \end{bmatrix} \begin{bmatrix} x_1 \\ x_2 \\ x_3 \\ x_4 \end{bmatrix} = \begin{bmatrix} 0 \\ 0 \\ 0 \end{bmatrix} \quad R_3 \to R_3 - 2R_2$$

$-x_1 + 7x_2 + 4x_3 + 9x_4 = 0$ ...(1)

$21x_2 + 14x_3 + 29x_4 = 0$ ...(2)

Let $x_4 = a, x_3 = b$

From (2), $21x_2 + 14b + 29a = 0$ or $x_2 = -\dfrac{2b}{3} - \dfrac{29a}{21}$

From (1), $-x_1 + 7\left(-\dfrac{2b}{3} - \dfrac{29a}{21}\right) + 4b + 9a = 0$

$$x_1 = -\frac{2a}{3} - \frac{2b}{3}$$

$$x_1 = -\frac{2}{3}(a+b),\ x_2 = -\frac{1}{21}(29a+14b)$$

$$x_3 = b,\ x_4 = a$$ **Ans.**

20.2 TYPES OF LINEAR EQUATIONS

(1) Consistent. A system of equations is said to be *consistent,* if they have one or more solution *i.e.*

$x + 2y = 4$	$x + 2y = 4$
$3x + 2y = 2$	$3x + 6y = 12$
Unique solution	Infinite solution

(2) Inconsistent. If a system of equation has no solution, it is said to be inconsistent *i.e.*

$$x + 2y = 4$$
$$3x + 6y = 5$$

20.3 CONSISTENCY OF A SYSTEM OF LINEAR EQUATIONS

$$a_{11}x_1 + a_{12}x_2 + \ldots a_{1n}x_n = b_1$$
$$a_{21}x_1 + a_{22}x_2 + \ldots a_{2n}x_n = b_2$$
$$\ldots\ldots\ldots\ldots\ldots\ldots\ldots\ldots\ldots\ldots$$
$$a_{m1}x_1 + a_{m2}x_2 + \ldots a_{mn}x_n = b_m$$

$$\Rightarrow \begin{bmatrix} a_{11} & a_{12} \ldots\ldots & a_{1n} \\ a_{21} & a_{22} \ldots\ldots & a_{2n} \\ \ldots\ldots & \ldots\ldots & \ldots\ldots \\ a_{m1} & a_{m2} \ldots\ldots & a_{mn} \end{bmatrix} \begin{bmatrix} x_1 \\ x_2 \\ \ldots \\ x_m \end{bmatrix} = \begin{bmatrix} b_1 \\ b_2 \\ \ldots \\ b_m \end{bmatrix}$$

$$\Rightarrow \quad AX = B$$

and $C = [A, B] = \begin{bmatrix} a_{11} & a_{12} \ldots\ldots & a_{1n} & b_1 \\ a_{21} & a_{22} \ldots\ldots & a_{2n} & b_2 \\ \ldots\ldots & \ldots\ldots & \ldots\ldots & \ldots \\ a_{m1} & a_{m2} \ldots\ldots & a_{mn} & b_m \end{bmatrix}$

is called the **augmented** matrix.

$$\boxed{[A:B] = C}$$

(a) Consistent equations. If Rank A = Rank C

(*i*) *Unique solution:* Rank A = Rank $C = n$ where n = number of unknown.

(*ii*) *Infinite solution:* Rank A = Rank $C = r,\ r < n$

(b) Inconsistent equations. If Rank $A \neq$ Rank C.

In Brief :

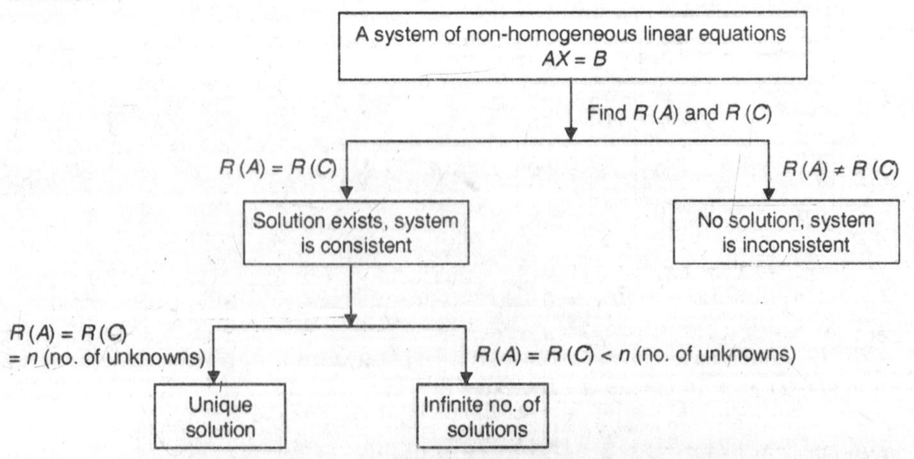

Example 5. *Show that the equations*

$$2x + 6y = -11,\ 6x + 20y - 6z = -3,\ 6y - 18z = -1$$

are not consistent.

Solution. Augmented matrix $C = [A, B]$

$$= \begin{bmatrix} 2 & 6 & 0 & : & -11 \\ 6 & 20 & -6 & : & -3 \\ 0 & 6 & -18 & : & -1 \end{bmatrix} \sim \begin{bmatrix} 2 & 6 & 0 & : & -11 \\ 0 & 2 & -6 & : & 30 \\ 0 & 6 & -18 & : & -1 \end{bmatrix} R_2 \to R_2 - 3R_1$$

$$\sim \begin{bmatrix} 2 & 6 & 0 & : & -11 \\ 0 & 2 & -6 & : & 30 \\ 0 & 0 & 0 & : & -91 \end{bmatrix} R_3 \to R_3 - 3\,R_2$$

The rank of C is 3 and the rank of A is 2.

Rank of $A \neq$ Rank of C. The equations are not consistent. **Ans.**

Example 6. *Test the consistency and hence solve the following set of equation.*

$$x_1 + 2x_2 + x_3 = 2$$
$$3x_1 + x_2 - 2x_3 = 1$$
$$4x_1 - 3x_2 - x_3 = 3$$
$$2x_1 + 4x_2 + 2x_3 = 4$$

(U.P., I Semester, Compartment 2002)

Solution. The given set of equations is written in the matrix form:

$$\begin{bmatrix} 1 & 2 & 1 \\ 3 & 1 & -2 \\ 4 & -3 & -1 \\ 2 & 4 & 2 \end{bmatrix} \begin{bmatrix} x_1 \\ x_2 \\ x_3 \end{bmatrix} = \begin{bmatrix} 2 \\ 1 \\ 3 \\ 4 \end{bmatrix}$$

$$AX = B$$

Here, we have augmented matrix $C = [A : B] \sim \begin{bmatrix} 1 & 2 & 1 & : & 2 \\ 3 & 1 & -2 & : & 1 \\ 4 & -3 & -1 & : & 3 \\ 2 & 4 & 2 & : & 4 \end{bmatrix}$

$$\sim \begin{bmatrix} 1 & 2 & 1 & : & 2 \\ 0 & -5 & -5 & : & -5 \\ 0 & -11 & -5 & : & -5 \\ 0 & 0 & 0 & : & 0 \end{bmatrix} \begin{matrix} R_2 \to R_2 - 3\,R_1 \\ R_3 \to R_3 - 4\,R_1 \\ R_4 \to R_4 - 2\,R_1 \end{matrix} \sim \begin{bmatrix} 1 & 2 & 1 & : & 2 \\ 0 & 1 & 1 & : & 1 \\ 0 & -11 & -5 & : & -5 \\ 0 & 0 & 0 & : & 0 \end{bmatrix} R_2 \to -\frac{1}{5} R_2$$

$$\sim \begin{bmatrix} 1 & 2 & 1 & : & 2 \\ 0 & 1 & 1 & : & 1 \\ 0 & 0 & 6 & : & 6 \\ 0 & 0 & 0 & : & 0 \end{bmatrix} R_3 \to R_3 + 11\,R_2 \sim \begin{bmatrix} 1 & 2 & 1 & : & 2 \\ 0 & 1 & 1 & : & 1 \\ 0 & 0 & 1 & : & 1 \\ 0 & 0 & 0 & : & 0 \end{bmatrix} R_3 \to \frac{1}{6} R_3$$

Number of non-zero rows = Rank of matrix.

$$\Rightarrow \qquad R(C) = R(A) = 3$$

Hence, the given system is consistent and possesses a unique solution. In matrix form the system reduces to

$$\begin{bmatrix} 1 & 2 & 1 \\ 0 & 1 & 1 \\ 0 & 0 & 1 \\ 0 & 0 & 0 \end{bmatrix} \begin{bmatrix} x_1 \\ x_2 \\ x_3 \end{bmatrix} = \begin{bmatrix} 2 \\ 1 \\ 1 \\ 0 \end{bmatrix}$$

$$x_1 + 2x_2 + x_3 = 2 \qquad ...(1)$$
$$x_2 + x_3 = 1 \qquad ...(2)$$
$$x_3 = 1$$

From (2), $x_2 + 1 = 1 \Rightarrow x_2 = 0$

From (1), $x_1 + 0 + 1 = 2 \Rightarrow x_1 = 1$

Hence, $x_1 = 1,\ x_2 = 0$ and $x_3 = 1$ **Ans.**

Example 7. *Test for consistency and solve :*

$$5x + 3y + 7z = 4,\ 3x + 26y + 2z = 9,\ 7x + 2y + 10z = 5$$

Solution. The augmented matrix $C = [A, B]$ (*R.G. P.V. Bhopal I. Sem. April 2009-08-03*)

$$\begin{bmatrix} 5 & 3 & 7 & : & 4 \\ 3 & 26 & 2 & : & 9 \\ 7 & 2 & 10 & : & 5 \end{bmatrix} \sim \begin{bmatrix} 1 & \frac{3}{5} & \frac{7}{5} & : & \frac{4}{5} \\ 3 & 26 & 2 & : & 9 \\ 7 & 2 & 10 & : & 5 \end{bmatrix} R_1 \to \frac{1}{5} R_1$$

$$\sim \begin{bmatrix} 1 & \frac{3}{5} & \frac{7}{5} & : & \frac{4}{5} \\ 0 & \frac{121}{5} & -\frac{11}{5} & : & \frac{33}{5} \\ 0 & -\frac{11}{5} & \frac{1}{5} & : & -\frac{3}{5} \end{bmatrix} \begin{matrix} \\ R_2 \to R_2 - 3R_1 \\ R_3 \to R_3 - 7R_1 \end{matrix} \sim \begin{bmatrix} 1 & \frac{3}{5} & \frac{7}{5} & : & \frac{4}{5} \\ 0 & \frac{121}{5} & -\frac{11}{5} & : & \frac{33}{5} \\ 0 & 0 & 0 & : & 0 \end{bmatrix} R_3 \to R_3 + \frac{1}{11} R_2$$

Rank of $A = 2 =$ Rank of C

Hence, the equations are consistent. But the rank is less than 3 *i.e.* number of unknows. So its solutions are infinite.

$$\begin{bmatrix} 1 & \frac{3}{5} & \frac{7}{5} \\ 0 & \frac{121}{5} & -\frac{11}{5} \\ 0 & 0 & 0 \end{bmatrix} \begin{bmatrix} x \\ y \\ z \end{bmatrix} = \begin{bmatrix} \frac{4}{5} \\ \frac{33}{5} \\ 0 \end{bmatrix}$$

$$x + \frac{3}{5} y + \frac{7}{5} z = \frac{4}{5}$$

$$\frac{121}{5} y - \frac{11z}{5} = \frac{33}{5} \quad \text{or } 11y - z = 3$$

Let $z = k$ then $11y - k = 3$ or $y = \frac{3}{11} + \frac{k}{11}$

$$x + \frac{3}{5}\left[\frac{3}{11} + \frac{k}{11}\right] + \frac{7}{5} k = \frac{4}{5} \quad \text{or } x = -\frac{16}{11} k + \frac{7}{11}$$ **Ans.**

Example 8. *Test the consistency of following system of linear equations and hence find the solution.*

$$4x_1 - x_2 = 12$$
$$-x_1 + 5x_2 - 2x_3 = 0$$
$$-2x_2 + 4x_3 = -8$$

(*U.P., I semester Dec. 2005*)

Solution. The given equation in the matrix form is

$$\begin{bmatrix} 4 & -1 & 0 \\ -1 & 5 & -2 \\ 0 & -2 & 4 \end{bmatrix} \begin{bmatrix} x_1 \\ x_2 \\ x_3 \end{bmatrix} = \begin{bmatrix} 12 \\ 0 \\ -8 \end{bmatrix}$$

$$AX = B$$

where, $A = \begin{bmatrix} 4 & -1 & 0 \\ -1 & 5 & -2 \\ 0 & -2 & 4 \end{bmatrix}, X = \begin{bmatrix} x_1 \\ x_2 \\ x_3 \end{bmatrix}, B = \begin{bmatrix} 12 \\ 0 \\ -8 \end{bmatrix}$

$C = [A, B]$

$$C = \begin{bmatrix} 4 & -1 & 0 & : & 12 \\ -1 & 5 & -2 & : & 0 \\ 0 & -2 & 4 & : & -8 \end{bmatrix} \sim \begin{bmatrix} -1 & 5 & -2 & : & 0 \\ 4 & -1 & 0 & : & 12 \\ 0 & -2 & 4 & : & -8 \end{bmatrix} R_1 \leftrightarrow R_2$$

$$\sim \begin{bmatrix} -1 & 5 & -2 & : & 0 \\ 0 & 19 & -8 & : & 12 \\ 0 & -2 & 4 & : & -8 \end{bmatrix} R_2 \to R_2 + 4R_1$$

$$\sim \begin{bmatrix} -1 & 5 & -2 & : & 0 \\ 0 & 19 & -8 & : & 12 \\ 0 & 0 & \frac{60}{19} & : & \frac{-128}{19} \end{bmatrix} R_3 \to R_3 + \frac{2}{19} R_2$$

Here, rank of A is 3 and Rank of C is also 3.

$$R(A) = R(C) = 3$$

Hence, the equations are consistent with unique solution.

$$\begin{bmatrix} -1 & 5 & -2 \\ 0 & 19 & -8 \\ 0 & 0 & \frac{60}{19} \end{bmatrix} \begin{bmatrix} x_1 \\ x_2 \\ x_3 \end{bmatrix} = \begin{bmatrix} 0 \\ 12 \\ \frac{-128}{19} \end{bmatrix}$$

$$-x_1 + 5x_2 - 2x_3 = 0 \qquad \text{...(1)}$$

$$19x_2 - 8x_3 = 12 \qquad \text{...(2)}$$

$$\frac{60}{19} x_3 = \frac{-128}{19} \quad \Rightarrow \quad x_3 = -\frac{128}{19} \times \frac{19}{60} \quad \Rightarrow \quad x_3 = \frac{-32}{15}$$

On putting the value of x_3 in (2), we get

$$19x_2 - 8\left(\frac{-32}{15}\right) = 12 \quad \Rightarrow \quad 19x_2 = 12 - \frac{256}{15} = \frac{-76}{15}$$

$$\Rightarrow \quad x_2 = \frac{-76}{15 \times 19} = -\frac{4}{15}$$

On putting the values of x_2 and x_3 in (1), we get

$$-x_1 + 5\left(-\frac{4}{15}\right) - 2\left(\frac{-32}{15}\right) = 0$$

$$\Rightarrow \quad -x_1 = \frac{20}{15} - \frac{64}{15} = \frac{-44}{15} \Rightarrow x_1 = \frac{44}{15}$$

Hence, $x_1 = \frac{44}{15}, x_2 = \frac{-4}{15}$ and $x_3 = \frac{-32}{15}$ **Ans.**

Example 9. *Test for consistency the following system of equations and, if consistent, solve them.*

$$x_1 + 2x_2 - x_3 = 3$$
$$3x_1 - x_2 + 2x_3 = 1$$
$$2x_1 - 2x_2 + 3x_3 = 2$$
$$x_1 - x_2 + x_3 = -1$$

(*U.P. I Semester, Winter 2002*)

Solution. The augmented matrix $C = [A, B]$

$$\begin{bmatrix} 1 & 2 & -1 & : & 3 \\ 3 & -i & 2 & : & 1 \\ 2 & -2 & 3 & : & 2 \\ 1 & -1 & 1 & : & -1 \end{bmatrix} \sim \begin{bmatrix} 1 & 2 & -1 & : & 3 \\ 0 & -7 & 5 & : & -8 \\ 0 & -6 & 5 & : & -4 \\ 0 & -3 & 2 & : & -4 \end{bmatrix} \begin{matrix} \\ R_2 \to R_2 - 3R_1 \\ R_3 \to R_3 - 2R_1 \\ R_4 \to R_4 - R_1 \end{matrix}$$

$$\sim \begin{bmatrix} 1 & 2 & -1 & : & 3 \\ 0 & -7 & 5 & : & -8 \\ 0 & 0 & \frac{5}{7} & : & \frac{20}{7} \\ 0 & 0 & \frac{-1}{7} & : & \frac{-4}{7} \end{bmatrix} \begin{matrix} \\ \\ R_3 \to R_3 - \frac{6}{7}R_2 \\ R_4 \to R_4 - \frac{3}{7}R_2 \end{matrix} \sim \begin{bmatrix} 1 & 2 & -1 & : & 3 \\ 0 & -7 & 5 & : & -8 \\ 0 & 0 & \frac{5}{7} & : & \frac{20}{7} \\ 0 & 0 & 0 & : & 0 \end{bmatrix} R_4 \to R_4 + \frac{1}{5}R_3$$

Rank of $C = 3 =$ Rank of A

Hence, the system of equations is consistent with unique solution.

Now, $$\begin{bmatrix} 1 & 2 & -1 \\ 0 & -7 & 5 \\ 0 & 0 & \frac{5}{7} \end{bmatrix} \begin{bmatrix} x_1 \\ x_2 \\ x_3 \end{bmatrix} = \begin{bmatrix} 3 \\ -8 \\ \frac{20}{7} \end{bmatrix}$$

$$x_1 + 2x_2 - x_3 = 3 \qquad ...(1)$$
$$-7x_2 + 5x_3 = -8 \qquad ...(2)$$
$$\frac{5}{7}x_3 = \frac{20}{7} \Rightarrow x_3 = 4$$

Form (2), $-7x_2 + 5 \times 4 = -8 \Rightarrow -7x_2 = -28 \quad \Rightarrow x_2 = 4$

Form (1), $x_1 + 2 \times 4 - 4 = 3 \quad \Rightarrow x_1 = 3 - 8 + 4 = -1$

Hence, $x_1 = -1, \; x_2 = 4, \; x_3 = 4$ **Ans.**

Example 10. *Discuss the consistency of the following system of equations*

$2x + 3y + 4z = 11, \; x + 5y + 7z = 15, \; 3x + 11y + 13z = 25.$

If found consistent, solve it. (*A.M.I.E.T.E., Winter 2001*)

Solution. The augmented matrix $C = [A, B]$

$$\begin{bmatrix} 2 & 3 & 4 & 11 \\ 1 & 5 & 7 & 15 \\ 3 & 11 & 13 & 25 \end{bmatrix} \sim \begin{bmatrix} 1 & 5 & 7 & 15 \\ 2 & 3 & 4 & 11 \\ 3 & 11 & 13 & 25 \end{bmatrix} R_1 \leftrightarrow R_2$$

$$R_2 \to R_2 - 2R_1, \; R_3 \to R_3 - 3R_1, \quad R_2 \to -\frac{1}{7}R_2, \; R_3 \to -\frac{1}{4}R_3, \quad R_3 \to R_3 - R_2$$

$$\begin{bmatrix} 1 & 5 & 7 & 15 \\ 0 & -7 & -10 & -19 \\ 0 & -4 & -8 & -20 \end{bmatrix} \sim \begin{bmatrix} 1 & 5 & 7 & 15 \\ 0 & 1 & \frac{10}{7} & \frac{19}{7} \\ 0 & 1 & 2 & 5 \end{bmatrix} \sim \begin{bmatrix} 1 & 5 & 7 & 15 \\ 0 & 1 & \frac{10}{7} & \frac{19}{7} \\ 0 & 0 & \frac{4}{7} & \frac{16}{7} \end{bmatrix}$$

Rank of C = 3 = Rank of A

Hence, the system of equations is consistent with unique solution.

Now,
$$\begin{bmatrix} 1 & 5 & 7 \\ 0 & 1 & \frac{10}{7} \\ 0 & 0 & \frac{4}{7} \end{bmatrix} \begin{bmatrix} x \\ y \\ z \end{bmatrix} = \begin{bmatrix} 15 \\ \frac{19}{7} \\ \frac{16}{7} \end{bmatrix}$$

$$\Rightarrow \quad x + 5y + 7z = 15 \qquad \ldots(1)$$

$$y + \frac{10z}{7} = \frac{19}{7} \qquad \ldots(2)$$

$$\frac{4z}{7} = \frac{16}{7} \Rightarrow z = 4$$

From (2), $\quad y + \frac{10}{7} \times 4 = \frac{19}{7} \Rightarrow y = -3$

From (1), $\quad x + 5(-3) + 7(4) = 15 \Rightarrow x = 2$

$$x = 2,\ y = -3,\ z = 4$$ **Ans.**

Example 11. Test for the consistency of the following system of equations :

$$x_1 + 2x_2 + 3x_3 + 4x_4 = 5$$
$$6x_1 + 7x_2 + 8x_3 + 9x_4 = 10$$
$$11x_1 + 12x_2 + 13x_3 + 14x_4 = 15$$
$$16x_1 + 17x_2 + 18x_3 + 19x_4 = 20$$
$$21x_1 + 22x_2 + 23x_3 + 24x_4 = 25$$

Solution. The given equations are written in the matrix form.

$$\begin{bmatrix} 1 & 2 & 3 & 4 \\ 6 & 7 & 8 & 9 \\ 11 & 12 & 13 & 14 \\ 16 & 17 & 18 & 19 \\ 21 & 22 & 23 & 24 \end{bmatrix} \begin{bmatrix} x_1 \\ x_2 \\ x_3 \\ x_4 \end{bmatrix} = \begin{bmatrix} 5 \\ 10 \\ 15 \\ 20 \\ 25 \end{bmatrix}$$

$AX = B$

$$C = [A : B] = \begin{bmatrix} 1 & 2 & 3 & 4 & : & 5 \\ 6 & 7 & 8 & 9 & : & 10 \\ 11 & 12 & 13 & 14 & : & 15 \\ 16 & 17 & 18 & 19 & : & 20 \\ 21 & 22 & 23 & 24 & : & 25 \end{bmatrix} = \begin{bmatrix} 1 & 2 & 3 & 4 & 5 \\ 0 & 5 & 10 & 15 & 20 \\ 0 & 10 & 20 & 30 & 40 \\ 0 & 15 & 30 & 45 & 60 \\ 0 & 20 & 40 & 60 & 80 \end{bmatrix} \begin{matrix} \\ R_2 \to R_2 - 6R_1 \\ R_3 \to R_3 - 11R_1 \\ R_4 \to R_4 - 16R_1 \\ R_5 \to R_5 - 21R_1 \end{matrix}$$

$$= \begin{bmatrix} 1 & 2 & 3 & 4 & 5 \\ 0 & 5 & 10 & 15 & 20 \\ 0 & 0 & 0 & 0 & 0 \\ 0 & 0 & 0 & 0 & 0 \\ 0 & 0 & 0 & 0 & 0 \end{bmatrix} \begin{matrix} \\ \\ R_3 \to R_3 - 2R_2 \\ R_4 \to R_4 - 3R_2 \\ R_5 \to R_5 - 4R_2 \end{matrix}$$

Number of non zero rows is only 2.

So Rank (A) = Rank (C) = 2

Since Rank (A) = Rank (C) < Number of unknows.

The given system of equations is consistent and has infinite number of solutions. **Ans.**

Example 12. *For what values of k, the equations $x + y + z = 1$, $2x + y + 4z = k$, $4x + y + 10z = k^2$ has a solution?* *(Q. Bank U.P. T.U. 2001)*

Solution. Here, we have

$$x + y + z = 1$$
$$2x + y + 4z = k$$
$$4x + y + 10z = k^2$$

$$\Rightarrow \quad \begin{bmatrix} 1 & 1 & 1 \\ 2 & 1 & 4 \\ 4 & 1 & 10 \end{bmatrix}\begin{bmatrix} x \\ y \\ z \end{bmatrix} = \begin{bmatrix} 1 \\ k \\ k^2 \end{bmatrix}$$

$$AX = B$$

$$C = [A : B] = \begin{bmatrix} 1 & 1 & 1 & : & 1 \\ 2 & 1 & 4 & : & k \\ 4 & 1 & 10 & : & k^2 \end{bmatrix} \sim \begin{bmatrix} 1 & 1 & 1 & : & 1 \\ 0 & -1 & 2 & : & k-2 \\ 0 & -3 & 6 & : & k^2-4 \end{bmatrix} \begin{matrix} \\ R_2 \to R_2 - 2R_1 \\ R_3 \to R_3 - 4R_1 \end{matrix}$$

$$\sim \begin{bmatrix} 1 & 1 & 1 & : & 1 \\ 0 & -1 & 2 & : & k-2 \\ 0 & 0 & 0 & : & k^2-3k+2 \end{bmatrix} R_3 \to R_3 - 3R_2 \quad \Rightarrow \begin{bmatrix} 1 & 1 & 1 \\ 0 & -1 & 2 \\ 0 & 0 & 0 \end{bmatrix}\begin{bmatrix} x \\ y \\ z \end{bmatrix} = \begin{bmatrix} 1 \\ k-2 \\ k^2-3k+2 \end{bmatrix}$$

If the given system has solutions, then $R(A) = R(C)$ But $R(A) = 2$

$R(C) = 2$ if $k^2 - 3k + 2 = 0 \Rightarrow (k - 1)(k - 2) = 0 \Rightarrow k = 1, k = 2$

Case I. When $k = 1$, we have

$$x + y + z = 1 \qquad ...(1)$$
$$-y + 2z = 1 - 2 = -1 \qquad ...(2)$$

Let $z = \lambda$

Putting the value of $z = \lambda$ in (2), we have

$$-y + 2\lambda = -1 \Rightarrow y = 2\lambda + 1$$

Putting the values of y and z in (1), we have

$$x + (2\lambda + 1) + \lambda = 1 \quad \Rightarrow \quad x = -3\lambda$$

Hence solution is

$$x = -3\lambda$$
$$y = 2\lambda + 1$$
$$z = \lambda$$

(λ is an arbitray constant)

Case II. When $k = 2$, we have

$$x + y + z = 1 \qquad ...(3)$$
$$-y + 2z = 4 - 6 + 2 \Rightarrow -y + 2z = 0 \qquad ...(4)$$

Let $z = c$

Putting the value of $z = c$ in (4), we have

$$-y + 2c = 0 \Rightarrow y = 2c$$

Putting the values of y and z in (1), we have

$$x + 2c + c = 1 \Rightarrow x = -3c + 1$$

Hence the solution is

$x = 1 - 3c$, $y = 2c$, $z = c$, where c is an arbitrary constant. **Ans.**

Example 13. *Investigate the values of λ and μ so that the equations:*

$$2x + 3y + 5z = 9$$
$$7x + 3y - 2z = 8$$
$$2x + 3y + \lambda z = \mu$$

have (i) no solution (ii) a unique solution

(iii) an infinite number of solutions. *(R.G.P.V. Bhopal, I Semester, June 2007)*

Solution. Here, we have,

$$2x + 3y + 5z = 9$$
$$7x + 3y - 2z = 8$$
$$2x + 3y + \lambda z = \mu$$

The above equations are written in the matrix form

$$\begin{bmatrix} 2 & 3 & 5 \\ 7 & 3 & -2 \\ 2 & 3 & \lambda \end{bmatrix}\begin{bmatrix} x \\ y \\ z \end{bmatrix} = \begin{bmatrix} 9 \\ 8 \\ \mu \end{bmatrix}$$
$$AX = B$$

$$C = [A : B] = \begin{bmatrix} 2 & 3 & 5 & : & 9 \\ 7 & 3 & -2 & : & 8 \\ 2 & 3 & \lambda & : & \mu \end{bmatrix} = \begin{bmatrix} 2 & 3 & 5 & : & 9 \\ 0 & -\frac{15}{2} & -\frac{39}{2} & : & -\frac{47}{2} \\ 0 & 0 & \lambda - 5 & : & \mu - 9 \end{bmatrix} \begin{matrix} \\ R_2 \to R_2 - \frac{7}{2} R_1 \\ R_3 \to R_3 - R_1 \end{matrix}$$

(*i*) **No solution.** Rank $(A) \neq$ Rank (C)

$\lambda - 5 = 0$ or $\lambda = 5$ and $\mu - 9 \neq 0$ $\Rightarrow$ $\mu \neq 9$

(*ii*) **A unique solution.** Rank $(A) = R(C) =$ Number of unknowns

$\lambda - 5 \neq 0$ $\Rightarrow$ $\lambda \neq 5$ and $\mu \neq 9$

(*iii*) **An infinite number of solutions.** Rank $(A) =$ Rank $(C) = 2$

$\lambda - 5 = 0$ and $\mu - 9 = 0$

$\lambda = 5$ and $\mu = 9$ **Ans.**

Example 14. *Determine for what values of* λ *and* μ *the following equations have* (*i*) *no solution,* (*ii*) *a unique solution;* (*iii*) *infinite number of solutions.*

$x + y + z = 6,\ x + 2y + 3z = 10,\ x + 2y + \lambda z = \mu$ (*U.P., I Sem. Winter 2001*)

Solution.
$$\begin{bmatrix} 1 & 1 & 1 \\ 1 & 2 & 3 \\ 1 & 2 & \lambda \end{bmatrix}\begin{bmatrix} x \\ y \\ z \end{bmatrix} = \begin{bmatrix} 6 \\ 10 \\ \mu \end{bmatrix}$$
$$AX = B$$

$$C = (A, B) = \begin{bmatrix} 1 & 1 & 1 & \cdot & 6 \\ 1 & 2 & 3 & \cdot & 10 \\ 1 & 2 & \lambda & \cdot & \mu \end{bmatrix} \sim \begin{bmatrix} 1 & 1 & 1 & \cdot & 6 \\ 0 & 1 & 2 & \cdot & 4 \\ 0 & 1 & \lambda - 1 & \cdot & \mu - 6 \end{bmatrix} \begin{matrix} \\ R_2 \to R_2 - R_1 \\ R_3 \to R_3 - R_1 \end{matrix}$$

$$\sim \begin{bmatrix} 1 & 1 & i & \cdot & 6 \\ 0 & 1 & 2 & \cdot & 4 \\ 0 & 0 & \lambda - 3 & \cdot & \mu - 10 \end{bmatrix} R_3 \to R_3 - R_2$$

(*i*) There is no solution if $R(A) \neq R(C)$

i.e. $\lambda - 3 = 0$ or $\lambda = 3$ and $\mu - 10 \neq 0$ or $\mu \neq 10$

(*ii*) There is a unique solution if $R(A) = R(C) = 3$

i.e. $\lambda - 3 \neq 0$ or $\lambda \neq 3$, μ may have any value.

(*iii*) There are infinite solutions if $R(A) = R(C) = 2$

$\lambda - 3 = 0$ or $\lambda = 3$ and $\mu - 10 = 0$ or $\mu = 10$ **Ans.**

Example 15. *Find for what values of* λ *and* μ *the system of linear equations:*

$$x + y + z = 6$$
$$x + 2y + 5z = 10$$
$$2x + 3y + \lambda z = \mu$$

has (*i*) *a unique solution* (*ii*) *no solution*

(*iii*) *infinite solutions. Also find the solution for* $\lambda = 2$ *and* $\mu = 8$.

(*Uttarakhand, 1st semester, Dec. 2006*)

Solution. $\begin{bmatrix} 1 & 1 & 1 \\ 1 & 2 & 5 \\ 2 & 3 & \lambda \end{bmatrix}\begin{bmatrix} x \\ y \\ z \end{bmatrix} = \begin{bmatrix} 6 \\ 10 \\ \mu \end{bmatrix}$

$$AX = B$$

$$C = (A, B) = \begin{bmatrix} 1 & 1 & 1 & : & 6 \\ 1 & 2 & 5 & : & 10 \\ 2 & 3 & \lambda & : & \mu \end{bmatrix} \sim \begin{bmatrix} 1 & 1 & 1 & : & 6 \\ 0 & 1 & 4 & : & 4 \\ 0 & 1 & \lambda - 2 & : & \mu - 12 \end{bmatrix} \begin{matrix} \\ R_2 \to R_2 - R_1 \\ R_3 \to R_3 - 2R_1 \end{matrix}$$

$$\sim \begin{bmatrix} 1 & 1 & 1 & : & 6 \\ 0 & 1 & 4 & : & 4 \\ 0 & 0 & \lambda - 6 & : & \mu - 16 \end{bmatrix} R_3 \to R_3 - R_2 \qquad \text{...(1)}$$

(*i*) **A unique solution**

If $R(A) = R(C) = 3$

then $\lambda - 6 \neq 0 \Rightarrow \lambda \neq 6$ and $\mu - 16 \neq 0 \Rightarrow \mu \neq 16$

(*ii*) **No solutions**

If $R(A) \neq R(C)$, then $R(A) = 2$ and $R(C) = 3$

$\lambda - 6 = 0 \Rightarrow \lambda = 6$ and $\mu - 16 \neq 0 \Rightarrow \mu \neq 16$

(*iii*) **Infinite solutions**

If $R(A) = R(C) = 2$

then $\lambda - 6 = 0$ and $\mu - 16 = 0$

$\Rightarrow \quad \lambda = 6$ and $\mu = 16$

(*iv*) Putting $\lambda = 2$ and $\mu = 8$ in (1), we get

$$\begin{bmatrix} 1 & 1 & 1 & : & 6 \\ 0 & 1 & 4 & : & 4 \\ 0 & 0 & -4 & : & -8 \end{bmatrix} \Rightarrow \begin{bmatrix} 1 & 1 & 1 \\ 0 & 1 & 4 \\ 0 & 0 & -4 \end{bmatrix}\begin{bmatrix} x \\ y \\ z \end{bmatrix} = \begin{bmatrix} 6 \\ 4 \\ -8 \end{bmatrix}$$

$$x + y + z = 6$$
$$y + 4z = 4$$
$$-4z = -8 \qquad \Rightarrow \qquad z = 2$$

Putting $z = 2$ in (3), we get

$$y + 8 = 4 \qquad \Rightarrow \qquad y = -4$$

Putting $y = -4$, $z = 2$ in (2), we get

$$x - 4 + 2 = 6 \qquad \Rightarrow \qquad x = 8$$

Hence, $\quad x = 8, \; y = -4, \; z = 2$ **Ans.**

Example 16. *Show that the equations*

$$-2x + y + z = a$$
$$x - 2y + z = b$$
$$x + y - 2z = c$$

have no solution unless $a + b + c = 0$. In which case they have infinitely many solutions? Find these solutions when $a = 1$, $b = 1$ and $c = -2$.

Solution. Augmented matrix,

$$C = [A : B] = \begin{bmatrix} -2 & 1 & 1 & : & a \\ 1 & -2 & 1 & : & b \\ 1 & 1 & -2 & : & c \end{bmatrix} \qquad [\text{Rank } (A) = 2]$$

$$\sim \begin{bmatrix} ① & 1 & -2 & : & c \\ 1 & -2 & 1 & : & b \\ -2 & 1 & 1 & : & a \end{bmatrix} R_1 \leftrightarrow R_3 \qquad \sim \begin{bmatrix} 1 & ① & -2 & : & c \\ 0 & -3 & 3 & : & b-c \\ 0 & 3 & -3 & : & a+2c \end{bmatrix} \begin{matrix} \\ R_2 \to R_2 - R_1 \\ R_3 \to R_2 + 2R_1 \end{matrix}$$

$$\sim \begin{bmatrix} 1 & 1 & -2 & : & c \\ 0 & -3 & 3 & : & b-c \\ 0 & 0 & 0 & : & a+b+c \end{bmatrix} R_3 \to R_3 + R_2 \qquad \text{... (1)}$$

Case I. If $a + b + c \neq 0$

Rank of $C = 3$.

But Rank of $A = 2$

$\Rightarrow$ $R(C) \neq R(A)$ where A is the coefficient matrix.

Hence, the system being inconsistent, have no solution.

Case II. If $a + b + c = 0$

Rank of $C = 2$ and $R(A) = 2$

$\Rightarrow$ $R(C) = R(A)$

Hence, the system has infinite number of solutions.

Case III. On putting $a = 1$, $b = 1$ and $c = -2$ in (1), we get

$$\begin{bmatrix} 1 & 1 & -2 & : & -2 \\ 0 & -3 & 3 & : & 3 \\ 0 & 0 & 0 & : & 0 \end{bmatrix}$$

$$x + y - 2z = -2 \qquad \text{...(2)}$$

$$-3y + 3z = 3 \qquad \text{...(3)}$$

Let $z = k$, k being an arbitrary constant.

From (3) $-3y + 3k = 3 \quad \Rightarrow \quad y = k - 1$

Putting $y = k - 1$ and $z = k$ in (2), we get

$$x + (k-1) - 2k = -2 \quad \Rightarrow \quad x = k - 1$$

Hence, the solutions are $x = k - 1, \ y = k - 1, \ z = k$ **Ans.**

Example 17. *Find for what values of k the set of equations*

$$2x - 3y + 6z - 5t = 3, \ y - 4z + t = 1, \ 4x - 5y + 8z - 9t = k$$

has (i) no solution (ii) infinite number of solutions. (*A.M.I.E.T.E., Summer 2004*)

Solution. The augmented matrix $C = [A, B]$

$$\begin{bmatrix} 2 & -3 & 6 & -5 & \cdot & 3 \\ 0 & 1 & -4 & 1 & \cdot & 1 \\ 4 & -5 & 8 & -9 & \cdot & k \end{bmatrix} \sim \overset{R_3 \to R_3 - 2R_1}{\begin{bmatrix} 2 & -3 & 6 & -5 & \cdot & 3 \\ 0 & 1 & -4 & 1 & \cdot & 1 \\ 0 & 1 & -4 & 1 & \cdot & k-6 \end{bmatrix}}$$

$$\sim \begin{bmatrix} 2 & -3 & 6 & -5 & \cdot & 3 \\ 0 & 1 & -4 & 1 & \cdot & 1 \\ 0 & 0 & 0 & 0 & \cdot & k-7 \end{bmatrix} R_3 \to R_3 - R_2$$

(*i*) There is no solution if $R(A) \neq R(C)$

$k - 7 \neq 0$ or $k \neq 7$, $R(A) = 2$ and $R(C) = 3$.

(*ii*) There are infinite solutions if $R(A) = R(C) = 2$

$k - 7 = 0 \Rightarrow k = 7$ **Ans.**

$$\begin{bmatrix} 2 & -3 & 6 & -5 \\ 0 & 1 & -4 & 1 \end{bmatrix} \begin{bmatrix} x \\ y \\ z \\ t \end{bmatrix} = \begin{bmatrix} 3 \\ 1 \end{bmatrix}$$

$$2x - 3y + 6z - 5t = 3 \qquad ...(1)$$
$$y - 4z + t = 1 \qquad ...(2)$$

Let $t = k_1$ and $z = k_2$.

From (2), $y - 4k_2 + k_1 = 1$ or $y = 1 + 4k_2 - k_1$

From (l), $2x - 3 - 12k_2 + 3k_1 + 6k_2 - 5k_1 = 3$

$\Rightarrow$ $$2x = 6 + 6k_2 + 2k_1 \Rightarrow x = 3 + 3k_2 + k_1$$
$$y = 1 + 4k_2 - k_1 \Rightarrow z = k_2,\ t = k_1$$ **Ans.**

20.4. HOMOGENEOUS EQUATIONS

For a system of homogeneous linear equations $AX = O$

(*i*) $X = O$ is always a solution. This solution in which each unknown has the value zero is called the **Null Solution** or the **Trivial solution.** Thus a homogeneous system is always consistent.

A system of homogeneous linear equations has either the trivial solution or an infinite number of solutions.

(*ii*) If $R(A)$ = number of unknowns, the system has only the trivial solution.

(*iii*) If $R(A)$ < number of unknowns, the system has an infinite number of non-trivial solutions.

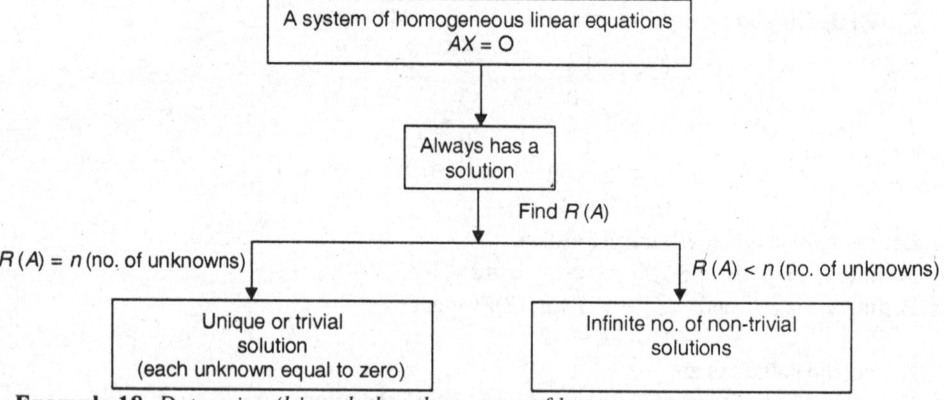

Example 18. *Determine 'b' such that the system of homogeneous equations*

$$2x + y + 2z = 0;$$
$$x + y + 3z = 0;$$
$$4x + 3y + bz = 0$$

has (i) Trivial solution

(ii) Non-Trivial solution . Find the Non-Trivial solution using matrix method.

(*U.P., I Sem Dec 2008*)

Solution. Here, we have

$$2x + y + 2z = 0$$
$$x + y + 3z = 0$$
$$4x + 3y + bz = 0$$

(*i*) **For trivial solution:** We know that $x = 0$, $y = 0$ and $z = 0$. So, b can have any value.

(*ii*) **For non-trivial solution:** The given equations are written in the matrix form as :

$$\begin{bmatrix} 2 & 1 & 2 \\ 1 & 1 & 3 \\ 4 & 3 & b \end{bmatrix} \begin{bmatrix} x \\ y \\ z \end{bmatrix} = \begin{bmatrix} 0 \\ 0 \\ 0 \end{bmatrix}$$

$$AX = B$$

$$R_1 \leftrightarrow R_2, \qquad R_2 \to R_2 - 2R_1,\ R_3 \to R_3 - 4R_1, \qquad R_3 \to R_3 - R_2$$

$$C = \begin{bmatrix} 2 & 1 & 2 & : & 0 \\ 1 & 1 & 3 & : & 0 \\ 4 & 3 & b & : & 0 \end{bmatrix} \sim \begin{bmatrix} 1 & 1 & 3 & : & 0 \\ 2 & 1 & 2 & : & 0 \\ 4 & 3 & b & : & 0 \end{bmatrix} \sim \begin{bmatrix} 1 & 1 & 3 & : & 0 \\ 0 & -1 & -4 & : & 0 \\ 0 & -1 & b-12 & : & 0 \end{bmatrix} \sim \begin{bmatrix} 1 & 1 & 3 & : & 0 \\ 0 & -1 & -4 & : & 0 \\ 0 & 0 & b-8 & : & 0 \end{bmatrix}$$

For non trivial solution or infinite solutions $R(C) = R(A) = 2 <$ Number of unknowns

$$b - 8 = 0, \quad b = 8$$ **Ans.**

Example 19. *Find the values of k such that the system of equations*

$x + ky + 3z = 0, \quad 4x + 3y + kz = 0, \quad 2x + y + 2z = 0$

has non-trivial solution.

Solution. The set of equations is written in the form of matrices

$$\begin{bmatrix} 1 & k & 3 \\ 4 & 3 & k \\ 2 & 1 & 2 \end{bmatrix} \begin{bmatrix} x \\ y \\ z \end{bmatrix} = \begin{bmatrix} 0 \\ 0 \\ 0 \end{bmatrix}, \quad AX = B, \; C = [A : B] = \begin{bmatrix} 1 & k & 3 & : & 0 \\ 4 & 3 & k & : & 0 \\ 2 & 1 & 2 & : & 0 \end{bmatrix}$$

On interchanging first and third rows, we have

$$\begin{bmatrix} 2 & 1 & 2 & : & 0 \\ 4 & 3 & k & : & 0 \\ 1 & k & 3 & : & 0 \end{bmatrix}$$

$$R_2 \to R_2 - 2R_1, \quad R_3 \to R_3 - \frac{1}{2}R_1 \qquad R_3 \to R_3 - \left(k - \frac{1}{2}\right) R_2$$

$$\begin{bmatrix} 2 & 1 & 2 & : & 0 \\ 0 & 1 & k-4 & : & 0 \\ 0 & k-\frac{1}{2} & 2 & : & 0 \end{bmatrix} \sim \begin{bmatrix} 2 & 1 & 2 & : & 0 \\ 0 & 1 & k-4 & : & 0 \\ 0 & 0 & 2-\left(k-\frac{1}{2}\right)(k-4) & : & 0 \end{bmatrix}$$

For a non-trivial solution or for infinite solution, $R(A) = R(C) = 2$

so $$2 - \left(k - \frac{1}{2}\right)(k-4) = 0 \Rightarrow 2 - k^2 + 4k + \frac{k}{2} - 2 = 0$$

$$\Rightarrow \quad -k^2 + \frac{9}{2}k = 0 \Rightarrow k\left(-k + \frac{9}{2}\right) = 0 \Rightarrow k = \frac{9}{2}, k = 0$$ **Ans.**

20.5 CRAMER'S RULE

$$a_1x + b_1y + c_1z = d_1$$
$$a_2x + b_2y + c_2z = d_2$$
$$a_3x + b_3y + c_3z = d_3$$

then $$D = \begin{vmatrix} a_1 & b_1 & c_1 \\ a_2 & b_2 & c_2 \\ a_3 & b_3 & c_3 \end{vmatrix}, \qquad D_1 = \begin{vmatrix} d_1 & b_1 & c_1 \\ d_2 & b_2 & c_2 \\ d_3 & b_3 & c_3 \end{vmatrix}$$

$$D_2 = \begin{vmatrix} a_1 & d_1 & c_1 \\ a_2 & d_2 & c_2 \\ a_3 & d_3 & c_3 \end{vmatrix}, \qquad D_3 = \begin{vmatrix} a_1 & b_1 & d_1 \\ a_2 & b_2 & d_2 \\ a_3 & b_3 & d_3 \end{vmatrix}$$

$$x = \frac{D_1}{D}, \quad y = \frac{D_2}{D}, \quad z = \frac{D_3}{D}$$

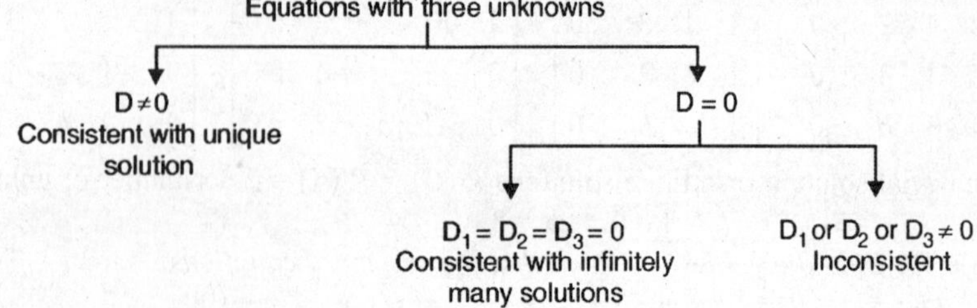

Example 20. *Show that the homogeneous system of equations*

$$x + y\cos\gamma + z\cos\beta = 0;\quad x\cos\gamma + y + z\cos\alpha = 0;\quad x\cos\beta + y\cos\alpha + z = 0$$

has non-trivial solution if $\alpha + \beta + \gamma = 0$. *(Q. Bank U.P.T.U. 2001)*

Solution. If the system has only non-trivial solutions , then

$$\begin{vmatrix} 1 & \cos\gamma & \cos\beta \\ \cos\gamma & 1 & \cos\alpha \\ \cos\beta & \cos\alpha & 1 \end{vmatrix} = 0$$

$$\Rightarrow \quad 1 - \cos^2\alpha + \cos\gamma(\cos\alpha\cos\beta - \cos\gamma) + \cos\beta(\cos\gamma\cos\alpha - \cos\beta) = 0$$

$$\Rightarrow \quad \sin^2\alpha - \cos^2\beta - \cos^2\gamma + 2\cos\alpha\cos\beta\cos\gamma = 0$$

$$\Rightarrow \quad -(\cos^2\beta - \sin^2\alpha) - \cos^2\gamma + 2\cos\alpha\cos\beta\cos\gamma = 0$$

$$\Rightarrow \quad -\cos(\alpha + \beta)\cos(\beta - \alpha) - \cos^2\gamma + 2\cos\alpha\cos\beta\cos\gamma = 0$$

[if $\alpha + \beta + \gamma = 0$]

$$\Rightarrow \quad -\cos(-\gamma)\cos(\beta - \alpha) - \cos^2\gamma + 2\cos\alpha\cos\beta\cos\gamma = 0$$

$$\Rightarrow \quad -\cos\gamma[\cos(\beta - \alpha) + \cos(\beta + \alpha)] + 2\cos\alpha\cos\beta\cos\gamma = 0$$

$$\Rightarrow \quad -2\cos\beta\cos\alpha\cos\gamma + 2\cos\alpha\cos\beta\cos\gamma = 0$$

which is true.

Hence, the given homogeneous system of equations has non-trivial solution if $\alpha + \beta + \gamma = 0$.

Proved.

Example 21. *Find values of λ for which the following system of equations is consistent and has non-trivial solutions. Solve equations for all such values of λ.*

$$(\lambda - 1)x + (3\lambda + 1)y + 2\lambda z = 0$$

$$(\lambda - 1)x + (4\lambda - 2)y + (\lambda + 3)z = 0$$

$$2x + (3\lambda + 1)y + 3(\lambda - 1)z = 0$$

(A.M.I.E.T.E., Summer 2010, 2001)

Solution.

$$\begin{bmatrix} (\lambda-1) & (3\lambda+1) & 2\lambda \\ (\lambda-1) & (4\lambda-2) & (\lambda+3) \\ 2 & (3\lambda+1) & (3\lambda-3) \end{bmatrix} \begin{bmatrix} x \\ y \\ z \end{bmatrix} = \begin{bmatrix} 0 \\ 0 \\ 0 \end{bmatrix} \quad \ldots(1)$$

$$AX = 0$$

For infinite solutions, $|A| = 0$

$$\begin{vmatrix} \lambda-1 & 3\lambda+1 & 2\lambda \\ \lambda-1 & 4\lambda-2 & \lambda+3 \\ 2 & 3\lambda+1 & 3\lambda-3 \end{vmatrix} = 0, \quad \begin{vmatrix} 0 & -\lambda+3 & \lambda-3 \\ \lambda-1 & 4\lambda-2 & \lambda+3 \\ 2 & 3\lambda+1 & 3\lambda-3 \end{vmatrix} \begin{matrix} R_1 \to R_1 - R_2 \\ = 0, \\ \end{matrix}$$

$$\begin{vmatrix} 0 & 0 & \lambda-3 \\ \lambda-1 & 5\lambda+1 & \lambda+3 \\ 2 & 6\lambda-2 & 3\lambda-3 \end{vmatrix} = 0, \quad C_2 \to C_2 + C_3$$

$$(\lambda - 3)[(\lambda - 1)(6\lambda - 2) - 2(5\lambda + 1)] = 0$$

$$[6\lambda^2 - 8\lambda + 2 - 10\lambda - 2] = 0 \text{ or } 6\lambda^2 - 18\lambda = 0 \text{ or } 6\lambda(\lambda - 3) = 0, \lambda = 3$$

On putting $\lambda = 3$ in (1), we get

$$\begin{bmatrix} 2 & 10 & 6 \\ 2 & 10 & 6 \\ 2 & 10 & 6 \end{bmatrix}\begin{bmatrix} x \\ y \\ z \end{bmatrix} = \begin{bmatrix} 0 \\ 0 \\ 0 \end{bmatrix} \Rightarrow \begin{bmatrix} 2 & 10 & 6 \\ 0 & 0 & 0 \\ 0 & 0 & 0 \end{bmatrix}\begin{bmatrix} x \\ y \\ z \end{bmatrix} = \begin{bmatrix} 0 \\ 0 \\ 0 \end{bmatrix}$$

$$2x + 10y + 6z = 0 \Rightarrow x + 5y + 3z = 0$$

Let $x = k_1, y = k_2, 3z = -k_1 - 5k_2 \Rightarrow z = \dfrac{-k_1}{3} - \dfrac{5k_2}{3}$ **Ans.**

EXERCISE 20.1

Test the consistency of the following equations and solve them if possible.

1. $3x + 3y + 2z = 1,\ \ x + 2y = 4,\ \ 10y + 3z = -2,\ \ 2x - 3y - z = 5$

   Ans. Consistent, $x = 2,\ y = 1,\ z = -4$ *(R.G.P.V. Bhopal 1st Sem 2001)*

2. $x_1 - x_2 + x_3 - x_4 + x_5 = 1,\qquad 2x_1 - x_2 + 3x_3 + 4x_5 = 2,$

   $3x_1 - 2x_2 + 2x_3 + x_4 + x_5 = 1,\qquad x_1 + x_3 + 2x_4 + x_5 = 0$ *(A.M.I.E.T.E., Winter 2003)*

   Ans. $x_1 = -3k_1 + k_2 - 1, x_2 = -3k_1 - 1, x_3 = k_1 - 2k_2 + 1, x_4 = k_1, x_4 = k_1, x_5 = k_2 1$

3. Find the value of k for which the following system of equations is consistent.

   $3x_1 - 2x_2 + 2x_3 = 3,\ x_1 + kx_2 - 3x_3 = 0,\ 4x_1 + x_2 + 2x_3 = 7$ **Ans.** $k = \dfrac{1}{4}$

4. Find the value of λ for which the system of equations

   $x + y + 4z = 1, x + 2y - 2z = 1, \lambda x + y + z = 1$

   will have a unique solution. *(A.M.I.E., Winter 2000)* **Ans.** $\lambda \neq \dfrac{7}{10}$

5. Determine the values of a and b for which the system $\begin{bmatrix} 3 & -2 & 1 \\ 5 & -8 & 9 \\ 2 & 1 & a \end{bmatrix}\begin{bmatrix} x \\ y \\ z \end{bmatrix} = \begin{bmatrix} b \\ 3 \\ -1 \end{bmatrix}$

   (*i*) has a unique solution, (*ii*) has no solution and, (*iii*) has infinitely many solutions.

   Ans. (*i*) $a \neq -3$, (*ii*) $a = -3, b \neq \dfrac{1}{3}$, (*iii*) $a = -3, b = \dfrac{1}{3}$

6. Choose λ that makes the following system of linear equations consistent and find the general solution of the system for that λ.

   $x + y - z + t = 2, 2y + 4z + 2t = 3, x + 2y + z + 2t = \lambda$

   Ans. $\lambda = \dfrac{7}{2}, x = \dfrac{1}{2} + 3k_2, y = \dfrac{3}{2} - 2k_2 - k_1, z = k_2, t = k_1$

7. Show that the equations

   $3x + 4y + 5z = a, 4x + 5y + 6z = b, 5x + 6y + 7z = c$

   don't have a solution unless $a + c = 2b$.

   Solve the equations when $a = b = c = -1$ (*MTU, Dec. 2012*) **Ans.** $x = k + 1, y = -2k - 1, z = k$

8. Find the values of k, such that the system of equations

   $4x_1 + 9x_2 + x_3 = 0,\ kx_1 + 3x_2 + kx_3 = 0, x_1 + 4x_2 + 2x_3 = 0$

   has non-trivial solution. Hence, find the solution of the system.

   Ans. $k = 1, x_1 = 2\lambda, x_2 = -\lambda, x_3 = \lambda$

9. Find values of λ for which the following system of equations has a non-trivial solution.

   $3x_1 + x_2 - \lambda x_3 = 0, 2x_1 + 4x_2 + \lambda x_3 = 0, 8x_1 - 4x_2 - 6x_3 = 0$ **Ans.** $\lambda = 1$

10. Find value of λ so that the following system of homogeneous equations have exactly two linearly independent solutions

    $\lambda x_1 - x_2 - x_3 = 0,\ -x_1 + \lambda x_2 - x_3 = 0,\ -x_1 - x_2 + \lambda x_3 = 0,$ **Ans.** $\lambda = -1$

11. Find the values of k for which the following system of equations has a non-trivial solution.

    $(3k - 8)x + 3y + 3z = 0, 3x + (3k - 8)y + 3z = 0, 3x + 3y + (3k - 8)z = 0$ *(AMIETE, June 2010)*

    Ans. $k = \dfrac{2}{3}, \dfrac{11}{3}$

12. Solve the homogeneous system of equations :
$4x + 3y - z = 0,\ 3x + 4y + z = 0,\ x - y - 2z = 0,\ 5x + y - 4z = 0$ **Ans.** $x = k, y = -k, z = k$

13. If $A = \begin{bmatrix} -1 & 2 & 1 \\ 3 & -1 & 2 \\ 0 & 1 & \lambda \end{bmatrix}$ **Ans.** (*i*) $\lambda \neq 1$, (*ii*) $\lambda = 1$

find the values of λ for which equation $AX = 0$ has (*i*) a unique solution, (*ii*) more than one solution.

14. Show that the following system of equations:

$$x + 2y - 2u = 0,\quad 2x - y - u = 0,\quad x + 2z - u = 0,\quad 4x - y + 3z - u = 0$$

do not have a non-trivial solution.

15. Determine the values of λ and μ such that the following system has (*i*) no solution (*ii*) a unique solution (*iii*) infinite number of solutions:

$$2x - 5y + 2z = 8,\quad 2x + 4y + 6z = 5,\quad x + 2y + \lambda z = \mu$$

Ans. (*i*) $\lambda = 3,\ \mu \neq \frac{5}{2}$ (*ii*) $\lambda \neq 3$, (*iii*) $\lambda = 3,\ \mu = \frac{5}{2}$

16. Test the following system of equations for consistency. If possible, solve for non-trivial solutions.

$$3x + 4y - z - 6t = 0,\ 2x + 3y + 2z - 3t = 0,\ 2x + y - 14z - 9t = 0,\ x + 3y + 13z + 3t = 0$$

(*A.M.I.E.T.E., Winter 2000*) **Ans.** $x = 11k_1 + 6k_2,\ y = -8k_1 - 3k_2,\ z = k_1\ t = k_2$

17. Given the following system of equations

$$2x - 2y + 5z + 3z = 0,\ 4x - y + z + w = 0,\ 3x - 2y + 3z + 4w = 0,\ x - 3y + 7z + 6w = 0$$

Reduce the coefficient matrix A into Echelon form and find the rank utilising the property of rank, test the given system of equation for consistency and if possible find the solution of the given system.

(*A.M.I.E.T.E., Summer 2001*) **Ans.** $x = 5k,\ y = 36k,\ z = 7k,\ w = 9k$

18. Find the values of λ for which the equations

$$(2 - \lambda)x + 2y + 3 = 0,\quad 2x + (4 - \lambda)y + 7 = 0,\quad 2x + 5y + (6 - \lambda) = 0$$

are consistent and find the values of x and y corresponding to each of these values of λ.

(*R.G.P.V, Bhopal I sem. 2003, 2001*) **Ans.** $\lambda = 1, -1, 12$.

20.6 VECTORS

A n-tuple is a set of n similar things. If the place of every members of a set is fixed then it is called an *ordered* set. Any ordered n-tuple of numbers is called a *n-vector*. Thus the coordinates of a point in space is called 3-vector (x, y, z). The members of a set are called the components of a vector so x, y, z in a 3-vector are called components.

$x_1, x_2, x_3 .... x_n$ are the components of a n-vector $X = (x_1, x_2, x_3 .... x_n)$.

Each row of a matrix is a vector and each column of the matrix is also a vector.

20.7 LINEAR DEPENDENCE AND INDEPENDENCE OF VECTORS

Vectors (matrices) $X_1, X_2, ... X_n$ are said to be dependent if

(1) all the vectors (row or column matrices) are of the same order.

(2) n scalars $\lambda_1, \lambda_2, ... \lambda_n$ (not all zero) exist such that

$$\lambda_1 X_1 + \lambda_2 X_2 + \lambda_3 X_3 + ... + \lambda_n X_n = 0$$

Otherwise they are linearly independent.

Remember: If in a set of vectors, any vector of the set is the combination of the remaining vectors, then the vectors are called dependent vectors.

Example 22. *Examine the following vectors for linear dependence and find the relation if it exists.*

$X_1 = (1, 2, 4),\ X_2 = (2, -1, 3),\ X_3 = (0, 1, 2),\ X_4 = (-3, 7, 2)$ (*U.P., I Sem. Winter* 2002)

Solution. Consider the matrix equation

$$\lambda_1 X_1 + \lambda_2 X_2 + \lambda_3 X_3 + \lambda_4 X_4 = 0$$

$$\Rightarrow \lambda_1 (1, 2, 4) + \lambda_2 (2, -1, 3) + \lambda_3 (0, 1, 2) + \lambda_4 (-3, 7, 2) = 0$$

$$\lambda_1 + 2\lambda_2 + 0\lambda_3 - 3\lambda_4 = 0$$
$$2\lambda_1 - \lambda_2 + \lambda_3 + 7\lambda_4 = 0$$
$$4\lambda_1 + 3\lambda_2 + 2\lambda_3 + 2\lambda_4 = 0$$

This is the homogeneous system

$$\begin{bmatrix} 1 & 2 & 0 & -3 \\ 2 & -1 & 1 & 7 \\ 4 & 3 & 2 & 2 \end{bmatrix} \begin{bmatrix} \lambda_1 \\ \lambda_2 \\ \lambda_3 \\ \lambda_4 \end{bmatrix} = \begin{bmatrix} 0 \\ 0 \\ 0 \end{bmatrix} \quad \text{or } A\,\lambda = 0$$

$$\begin{bmatrix} 1 & 2 & 0 & -3 \\ 0 & -5 & 1 & 13 \\ 0 & -5 & 2 & 14 \end{bmatrix} \begin{bmatrix} \lambda_1 \\ \lambda_2 \\ \lambda_3 \\ \lambda_4 \end{bmatrix} = \begin{bmatrix} 0 \\ 0 \\ 0 \end{bmatrix} \qquad \begin{matrix} R_2 \to R_2 - 2R_1 \\ R_3 \to R_3 - 4R_1 \end{matrix}$$

$$\begin{bmatrix} 1 & 2 & 0 & -3 \\ 0 & -5 & 1 & 13 \\ 0 & 0 & 1 & 1 \end{bmatrix} \begin{bmatrix} \lambda_1 \\ \lambda_2 \\ \lambda_3 \\ \lambda_4 \end{bmatrix} = \begin{bmatrix} 0 \\ 0 \\ 0 \end{bmatrix} \qquad R_3 \to R_3 - R_2$$

$$\lambda_1 + 2\,\lambda_2 - 3\,\lambda_4 = 0$$
$$-5\,\lambda_2 + \lambda_3 + 13\,\lambda_4 = 0$$
$$\lambda_3 + \lambda_4 = 0$$

Let $\quad \lambda_4 = t,\ \lambda_3 + t = 0,\ \lambda_3 = -t$

$$-5\lambda_2 - t + 13\,t = 0,\ \lambda_2 = \frac{12\,t}{5}$$

$$\lambda_1 + \frac{24\,t}{5} - 3t = 0 \text{ or } \lambda_1 = \frac{-9\,t}{5}$$

Hence, the given vectors are linearly dependent.

Substituting the values of λ in (1), we get

$$-\frac{9\,t\,X_1}{5} + \frac{12\,t}{5} X_2 - t\,X_3 + t\,X_4 = 0 \Rightarrow -\frac{9\,X_1}{5} + \frac{12\,X_2}{5} - X_3 + X_4 = 0$$

$\Rightarrow \quad 9\,X_1 - 12\,X_2 + 5\,X_3 - 5\,X_4 = 0$ **Ans.**

Example 23. *Define linear dependence and independence of vectors.*

Examine for linear dependence [1, 0, 2, 1], [3, 1, 2, 1], [4, 6, 2, –4], [–6, 0, –3, –4] and find the relation between them, if possible.

Solution. Consider the matrix equation

$$\lambda_1 X_1 + \lambda_2 X_2 + \lambda_3 X_3 + \lambda_4 X_4 = 0 \qquad \text{...(1)}$$

$$\lambda_1\,(1, 0, 2, 1) + \lambda_2\,(3, 1, 2, 1) + \lambda_3\,(4, 6, 2, -4) + \lambda_4\,(-6, 0, -3, -4) = 0$$

$$\lambda_1 + 3\,\lambda_2 + 4\,\lambda_3 - 6\,\lambda_4 = 0$$
$$0\,\lambda_1 + \lambda_2 + 6\,\lambda_3 + 0\,\lambda_4 = 0$$
$$2\,\lambda_1 + 2\,\lambda_2 + 2\,\lambda_3 - 3\,\lambda_4 = 0$$
$$\lambda_1 + \lambda_2 - 4\,\lambda_3 - 4\,\lambda_4 = 0$$

$$\begin{bmatrix} 1 & 3 & 4 & -6 \\ 0 & 1 & 6 & 0 \\ 2 & 2 & 2 & -3 \\ 1 & 1 & -4 & -4 \end{bmatrix} \begin{bmatrix} \lambda_1 \\ \lambda_2 \\ \lambda_3 \\ \lambda_4 \end{bmatrix} = \begin{bmatrix} 0 \\ 0 \\ 0 \\ 0 \end{bmatrix}$$

$$\begin{bmatrix} 1 & 3 & 4 & -6 \\ 0 & 1 & 6 & 0 \\ 0 & -4 & -6 & 9 \\ 0 & -2 & -8 & 2 \end{bmatrix} \begin{bmatrix} \lambda_1 \\ \lambda_2 \\ \lambda_3 \\ \lambda_4 \end{bmatrix} = \begin{bmatrix} 0 \\ 0 \\ 0 \\ 0 \end{bmatrix} \quad \begin{matrix} R_3 \to R_3 - 2R_1 \\ R_4 \to R_4 - R_1 \end{matrix}$$

$$\begin{bmatrix} 1 & 3 & 4 & -6 \\ 0 & 1 & 6 & 0 \\ 0 & 0 & 18 & 9 \\ 0 & 0 & 4 & 2 \end{bmatrix} \begin{bmatrix} \lambda_1 \\ \lambda_2 \\ \lambda_3 \\ \lambda_4 \end{bmatrix} = \begin{bmatrix} 0 \\ 0 \\ 0 \\ 0 \end{bmatrix} \quad \begin{matrix} R_3 \to R_3 + 4R_2 \\ R_4 \to R_4 + 2R_2 \end{matrix}$$

$$\begin{bmatrix} 1 & 3 & 4 & -6 \\ 0 & 1 & 6 & 0 \\ 0 & 0 & 18 & 9 \\ 0 & 0 & 0 & 0 \end{bmatrix} \begin{bmatrix} \lambda_1 \\ \lambda_2 \\ \lambda_3 \\ \lambda_4 \end{bmatrix} = \begin{bmatrix} 0 \\ 0 \\ 0 \\ 0 \end{bmatrix} \quad R_4 \to R_4 - \frac{2}{9}R_3$$

$$\lambda_1 + 3\lambda_2 + 4\lambda_3 - 6\lambda_4 = 0$$
$$\lambda_2 + 6\lambda_3 = 0$$
$$18\lambda_3 + 9\lambda_4 = 0$$

Let $\lambda_4 = t$, $\quad 18\lambda_3 + 9t = 0$ or $\lambda_3 = \dfrac{-t}{2}$

$$\lambda_2 - 3t = 0 \text{ or } \lambda_2 = 3t$$
$$\lambda_1 + 9t - 2t - 6t = 0$$
$$\lambda_1 = -t$$

Substituting the values of λ_1, λ_2, λ_3 and λ_4 in (1), we get

$$tX_1 + 3tX_2 - \frac{t}{2}X_3 + tX_4 = 0 \quad \text{or} \quad 2X_1 - 6X_2 + X_3 - 2X_4 = 0$$ **Ans.**

Example 24. *Show that row vectors of the matrix*

$$\begin{bmatrix} 1 & 2 & -2 \\ -1 & 3 & 0 \\ 0 & -2 & 1 \end{bmatrix}$$ *are linearly independent.* (*U.P., I Sem, Dec 2009*)

Solution. Here, we have three vectors

$$X_1 = (1, 2, -2)'$$
$$X_2 = (-1, 3, 0)'$$
$$X_3 = (0, -2, 1)'$$

$$X_1 = \begin{bmatrix} 1 \\ 2 \\ -2 \end{bmatrix}, X_2 = \begin{bmatrix} -1 \\ 3 \\ 0 \end{bmatrix}, X_3 = \begin{bmatrix} 0 \\ -2 \\ 1 \end{bmatrix}$$

Consider the equation

$$\lambda_1 X_1 + \lambda_2 X_2 + \lambda_3 X_3 = 0 \qquad \ldots(1)$$

$$\lambda_1 \begin{bmatrix} 1 \\ 2 \\ -2 \end{bmatrix} + \lambda_2 \begin{bmatrix} -1 \\ 3 \\ 0 \end{bmatrix} + \lambda_3 \begin{bmatrix} 0 \\ -2 \\ 1 \end{bmatrix} = \begin{bmatrix} 0 \\ 0 \\ 0 \end{bmatrix}$$

$$\lambda_1 - \lambda_2 + 0\lambda_3 = 0$$
$$2\lambda_1 + 3\lambda_2 - 2\lambda_3 = 0$$

$$-2\lambda_1 + 0\lambda_2 + \lambda_3 = 0$$

which is the system of homogeneous equations

$$\begin{bmatrix} 1 & -1 & 0 \\ 2 & 3 & -2 \\ -2 & 0 & 1 \end{bmatrix}\begin{bmatrix} \lambda_1 \\ \lambda_2 \\ \lambda_3 \end{bmatrix} = \begin{bmatrix} 0 \\ 0 \\ 0 \end{bmatrix} \Rightarrow \begin{bmatrix} 1 & -1 & 0 \\ 0 & 5 & -2 \\ 0 & -2 & 1 \end{bmatrix}\begin{bmatrix} \lambda_1 \\ \lambda_2 \\ \lambda_3 \end{bmatrix} = \begin{bmatrix} 0 \\ 0 \\ 0 \end{bmatrix} \begin{matrix} \\ R_2 \to R_2 - 2R_1 \\ R_3 \to R_3 + 2R_1 \end{matrix}$$

$$\begin{bmatrix} 1 & -1 & 0 \\ 0 & 5 & -2 \\ 0 & 0 & \frac{1}{5} \end{bmatrix}\begin{bmatrix} \lambda_1 \\ \lambda_2 \\ \lambda_3 \end{bmatrix} = \begin{bmatrix} 0 \\ 0 \\ 0 \end{bmatrix} R_3 \to R_3 + \frac{2}{5}R_2$$

$$\lambda_1 - \lambda_2 = 0 \qquad \text{...(2)}$$

$$5\lambda_2 - 2\lambda_3 = 0 \qquad \text{...(3)}$$

$$\frac{1}{5}\lambda_3 = 0 \Rightarrow \lambda_3 \qquad \text{....(4)}$$

Putting the value of λ_3 in (3), we get

$$5\lambda_2 - 2(0) = 0 \Rightarrow \lambda_2 = 0$$

Putting the value of λ_2 in (2), we get

$$\lambda_1 - 0 = 0 \Rightarrow \lambda_1 = 0$$

Thus non zero values of $\lambda_1, \lambda_2, \lambda_3$ do not exist which can satisfy (1). Hence by definition the given system of vectors is linearly independent. **Proved.**

20.8 LINEARLY DEPENDENCE AND INDEPENDENCE OF VECTORS BY RANK METHOD

1. If the rank of the matrix of the given vectors is equal to number of vectors, then the vectors are linearly independent.
2. If the rank of the matrix of the given vectors is less than the number of vectors, then the vectors are linearly dependent.

Example 25. *Show using a matrix that the set of vectors*

X = [1, 2,– 3, 4], Y = [3, – 1, 2, 1] , Z = [1,–5 ,8 –7] is linearly dependent.

Solution. Here, we have

$$X = [1, 2, -3, 4], \quad Y = [3, -1, 2, 1], \quad Z = [1, -5, 8, -7]$$

Let us form a matrix of the above vectors

$$\begin{bmatrix} 1 & 2 & -3 & 4 \\ 3 & -1 & 2 & 1 \\ 1 & -5 & 8 & -7 \end{bmatrix} \sim \begin{bmatrix} 1 & 2 & -3 & 4 \\ 0 & -7 & 11 & -11 \\ 0 & -7 & 11 & -11 \end{bmatrix} \begin{matrix} \\ R_2 \to R_2 - 3R_1 \\ R_3 \to R_3 - R_1 \end{matrix} \sim \begin{bmatrix} 1 & 2 & -3 & 4 \\ 0 & -7 & 11 & -11 \\ 0 & 0 & 0 & 0 \end{bmatrix} R_3 \to R_3 - R_2$$

Here the rank of the matrix = 2 < Number of vectors

Hence, vectors are linearly dependent. **Proved.**

Example 26. *Show using a matrix that the set of vectors : [2, 5, 2, –3], [3, 6, 5, 2], [4, 5, 14, 14], [5, 10, 8, 4] is linearly independent.*

Solution. Here, the given vectors are

[2, 5, 2, –3], [3, 6, 5, 2], [4, 5, 14, 14], [5, 10, 8, 4]

Let us form a matrix of the above vectors :

$$\begin{bmatrix} 2 & 5 & 2 & -3 \\ 3 & 6 & 5 & 2 \\ 4 & 5 & 14 & 14 \\ 5 & 10 & 8 & 4 \end{bmatrix} \sim \begin{bmatrix} 2 & 5 & 2 & -3 \\ 1 & 1 & 3 & 5 \\ 1 & -1 & 9 & 12 \\ 1 & 5 & -6 & -10 \end{bmatrix} \begin{matrix} \\ R_2 \to R_2 - R_1 \\ R_3 \to R_3 - R_2 \\ R_4 \to R_4 - R_3 \end{matrix}$$

$$\sim\begin{bmatrix}1 & 1 & 3 & 5\\ 2 & 5 & 2 & -3\\ 1 & -1 & 9 & 12\\ 1 & 5 & -6 & -10\end{bmatrix} \begin{matrix}R_1 \leftrightarrow R_2\\ R_2 \leftrightarrow R_1\\ \\ \end{matrix} \sim\begin{bmatrix}1 & 1 & 3 & 5\\ 0 & 3 & -4 & -13\\ 0 & -2 & 6 & 7\\ 0 & 4 & -9 & -15\end{bmatrix} \begin{matrix}\\ R_2 \to R_2 - 2R_1\\ R_3 \to R_3 - R_1\\ R_4 \to R_4 - R_1\end{matrix}$$

$$\sim\begin{bmatrix}1 & 1 & 3 & 5\\ 0 & 3 & -4 & -13\\ 0 & 0 & \frac{10}{3} & \frac{-5}{3}\\ 0 & 0 & \frac{-11}{3} & \frac{7}{3}\end{bmatrix} \begin{matrix}\\ \\ R_3 \to R_3 + \frac{2}{3}R_2\\ R_4 \to R_4 - \frac{4}{3}R_2\end{matrix} \sim\begin{bmatrix}1 & 1 & 3 & 5\\ 0 & 3 & -4 & -13\\ 0 & 0 & \frac{10}{3} & \frac{-5}{3}\\ 0 & 0 & 0 & \frac{1}{2}\end{bmatrix} R_4 \to R_4 + \frac{11}{10}R_3$$

Here, the rank of the matrix = 4 = Number of vectors

Hence, the vectors are linearly independent. **Proved.**

EXERCISE 20.2

Examine the following system of vectors for linear dependence. If dependent, find the relation between them.

1. $X_1 = (1, -1, 1), X_2 = (2, 1, 1), X_3 = (3, 0, 2)$. **Ans.** Dependent, $X_1 + X_2 - X_3 = 0$
2. $X_1 = (1, 2, 3), X_2 = (2, -2, 6)$. **Ans.** Independent
3. $X_1 = (3, 1, -4), X_2 = (2, 2, -3), X_3 = (0, -4, 1)$. **Ans.** Dependent, $2X_1 - 3X_2 - X_3 = 0$
4. $X_1 = (1, 1, 1, 3), X_2 = (1, 2, 3, 4), X_3 = (2, 3, 4, 7)$. **Ans.** Dependent, $X_1 + X_2 - X_3 = 0$
5. $X_1 = (1, 1, -1, 1), X_2 = (1, -1, 2, -1), X_3 = (3, 1, 0, 1)$. **Ans.** Dependent, $2X_1 + X_2 - X_3 = 0$
6. $X_1 = (1, -1, 2, 0), X_2 = (2, 1, 1, 1), X_3 = (3, -1, 2, -1), X_4 = (3, 0, 3, 1)$.
   Ans. Dependent, $X_1 + X_2 - X_4 = 0$
7. Show that the column vectors of following matrix A are linearly independent:
   $$A = \begin{bmatrix}1 & 0 & 0\\ 6 & 2 & 1\\ 4 & 3 & 2\end{bmatrix}$$
8. Show that the vectors $x_1 = (2, 3, 1, -1), x_2 = (2, 3, 1, -2), x_3 = (4, 6, 2, 1)$ are linearly dependent. Express one of the vectors as linear combination of the others.
9. Find whether or not the following set of vectors are linearly dependent or independent:
   (*i*) (1, –2), (2, 1), (3, 2) (*ii*) (1, 1, 1, 1), (0, 1, 1, 1), (0, 0, 1, 1), (0, 0, 0, 1).
   Ans. (*i*) Dependent (*ii*) Independent
10. Show that the vectors $x_1 = (a_1, b_1), x_2 = (a_2, b_2)$ are linearly dependent if $a_1 b_2 - a_2 b_1 = 0$.

20.9 ANOTHER METHOD (ADJOINT METHOD) TO SOLVE LINEAR EQUATIONS

Let the equations be

$$a_1 x + a_2 y + a_3 z = d_1$$
$$b_1 x + b_2 y + b_3 z = d_2$$
$$c_1 x + c_2 y + c_3 z = d_3$$

We write the above equations in the matrix form

$$\begin{bmatrix}a_1 x + a_2 y + a_3 z\\ b_1 x + b_2 y + b_3 z\\ c_1 x + c_2 y + c_3 z\end{bmatrix} = \begin{bmatrix}d_1\\ d_2\\ d_3\end{bmatrix} \quad \text{or} \quad \begin{bmatrix}a_1 & a_2 & a_3\\ b_1 & b_2 & b_3\\ c_1 & c_2 & c_3\end{bmatrix}\begin{bmatrix}x\\ y\\ z\end{bmatrix} = \begin{bmatrix}d_1\\ d_2\\ d_3\end{bmatrix}$$

$$AX = B \qquad ...(1)$$

where $A = \begin{bmatrix}a_1 & a_2 & a_3\\ b_1 & b_2 & b_3\\ c_1 & c_2 & c_3\end{bmatrix}, X = \begin{bmatrix}x\\ y\\ z\end{bmatrix}$ and $B = \begin{bmatrix}d_1\\ d_2\\ d_3\end{bmatrix}$

Multiplying (1) by A^{-1}.

$$A^{-1}AX = A^{-1}B \text{ or } IX = A^{-1}B \text{ or } X = A^{-1}B.$$

Example 27. *Solve, with the help of matrices, the simultaneous equations*

$x + y + z = 3,\ x + 2y + 3z = 4,\ x + 4y + 9z = 6$ (*A.M.I.E., Summer* 2004, 2003)

Solution. The given equations in the matrix form are written as below:

$$\begin{bmatrix} 1 & 1 & 1 \\ 1 & 2 & 3 \\ 1 & 4 & 9 \end{bmatrix}\begin{bmatrix} x \\ y \\ z \end{bmatrix} = \begin{bmatrix} 3 \\ 4 \\ 6 \end{bmatrix}$$

$$AX = B$$

where $A = \begin{bmatrix} 1 & 1 & 1 \\ 1 & 2 & 3 \\ 1 & 4 & 9 \end{bmatrix}, X = \begin{bmatrix} x \\ y \\ z \end{bmatrix}, B = \begin{bmatrix} 3 \\ 4 \\ 6 \end{bmatrix}$

Now we have to find out the A^{-1}.

$$|A| = 1 \times 6 + 1 \times (-6) + 1 \times 2 = 6 - 6 + 2 = 2$$

$$\text{Matrix of co-factors} = \begin{bmatrix} 6 & -6 & 2 \\ -5 & 8 & -3 \\ 1 & -2 & 1 \end{bmatrix}, \textit{Adjoint } A = \begin{bmatrix} 6 & -5 & 1 \\ -6 & 8 & -2 \\ 2 & -3 & 1 \end{bmatrix}$$

$$A^{-1} = \frac{1}{|A|} \textit{Adjoint } A = \frac{1}{2}\begin{bmatrix} 6 & -5 & 1 \\ -6 & 8 & -2 \\ 2 & -3 & 1 \end{bmatrix}$$

$$X = A^{-1}B = \frac{1}{2}\begin{bmatrix} 6 & -5 & 1 \\ -6 & 8 & -2 \\ 2 & -3 & 1 \end{bmatrix}\begin{bmatrix} 3 \\ 4 \\ 6 \end{bmatrix}$$

$$\Rightarrow \quad \begin{bmatrix} x \\ y \\ z \end{bmatrix} = \frac{1}{2}\begin{bmatrix} 18-20+6 \\ -18+32-12 \\ 6-12+6 \end{bmatrix} = \frac{1}{2}\begin{bmatrix} 4 \\ 2 \\ 0 \end{bmatrix} = \begin{bmatrix} 2 \\ 1 \\ 0 \end{bmatrix}$$

$x = 2, y = 1, z = 0$ **Ans.**

Example 28. *Given the matrices*

$$A \equiv \begin{bmatrix} 1 & 2 & 3 \\ 3 & -1 & 1 \\ 4 & 2 & 1 \end{bmatrix}, X \equiv \begin{bmatrix} x \\ y \\ z \end{bmatrix} \textit{ and } C \equiv \begin{bmatrix} 1 \\ 2 \\ 3 \end{bmatrix}$$

Write down the linear equations given by AX = C and solve for x, y, z by the matrix method.

Solution. $AX = C$

$$\begin{bmatrix} 1 & 2 & 3 \\ 3 & -1 & 1 \\ 4 & 2 & 1 \end{bmatrix}\begin{bmatrix} x \\ y \\ z \end{bmatrix} = \begin{bmatrix} 1 \\ 2 \\ 3 \end{bmatrix}$$

$$X = A^{-1}.C$$

$$\begin{bmatrix} x \\ y \\ z \end{bmatrix} = \begin{bmatrix} 1 & 2 & 3 \\ 3 & -1 & 1 \\ 4 & 2 & 1 \end{bmatrix}^{-1}\begin{bmatrix} 1 \\ 2 \\ 3 \end{bmatrix}$$

$$\text{Matrix of co-factors of } A = \begin{bmatrix} -3 & 1 & 10 \\ 4 & -11 & 6 \\ 5 & 8 & -7 \end{bmatrix}$$

$$|A| = 1(-3) + 2(1) + 3(10) = -3 + 2 + 30 = 29$$

$$\text{Adj. } A = \begin{bmatrix} -3 & 4 & 5 \\ 1 & -11 & 8 \\ 10 & 6 & -7 \end{bmatrix}$$

$$A^{-1} = \frac{1}{|A|} Adj.\, A = \frac{1}{29}\begin{bmatrix} -3 & 4 & 5 \\ 1 & -11 & 8 \\ 10 & 6 & -7 \end{bmatrix}$$

$$X = A^{-1} C$$

$$\begin{bmatrix} x \\ y \\ z \end{bmatrix} = \frac{1}{29}\begin{bmatrix} -3 & 4 & 5 \\ 1 & -11 & 8 \\ 10 & 6 & -7 \end{bmatrix}\begin{bmatrix} 1 \\ 2 \\ 3 \end{bmatrix}$$

$$\begin{bmatrix} x \\ y \\ z \end{bmatrix} = \frac{1}{29}\begin{bmatrix} -3 & +8 & +15 \\ 1 & -22 & +24 \\ 10 & +12 & -21 \end{bmatrix} = \frac{1}{29}\begin{bmatrix} 20 \\ 3 \\ 1 \end{bmatrix} = \begin{bmatrix} \frac{20}{29} \\ \frac{3}{29} \\ \frac{1}{29} \end{bmatrix}$$

Hence, $x = \frac{20}{29}, y = \frac{3}{29}, z = \frac{1}{29}$ **Ans.**

Example 29. *By the method of matrix, inversion, solve the system.*

$$\begin{bmatrix} 1 & 1 & 1 \\ 2 & 5 & 7 \\ 2 & 1 & -1 \end{bmatrix}\begin{bmatrix} x & u \\ y & v \\ z & w \end{bmatrix} = \begin{bmatrix} 9 & 2 \\ 52 & 15 \\ 0 & -1 \end{bmatrix}$$

Solution. $\begin{bmatrix} 1 & 1 & 1 \\ 2 & 5 & 7 \\ 2 & 1 & -1 \end{bmatrix}\begin{bmatrix} x & u \\ y & v \\ z & w \end{bmatrix} = \begin{bmatrix} 9 & 2 \\ 52 & 15 \\ 0 & -1 \end{bmatrix}$

$$\begin{bmatrix} x & u \\ y & v \\ z & w \end{bmatrix} = \begin{bmatrix} 1 & 1 & 1 \\ 2 & 5 & 7 \\ 2 & 1 & -1 \end{bmatrix}^{-1}\begin{bmatrix} 9 & 2 \\ 52 & 15 \\ 0 & -1 \end{bmatrix}$$

$$= \frac{-1}{4}\begin{bmatrix} -12 & 2 & 2 \\ 16 & -3 & -5 \\ -8 & 1 & 3 \end{bmatrix}\begin{bmatrix} 9 & 2 \\ 52 & 15 \\ 0 & -1 \end{bmatrix} = \frac{-1}{4}\begin{bmatrix} -4 & 4 \\ -12 & -8 \\ -20 & -4 \end{bmatrix} = \begin{bmatrix} 1 & -1 \\ 3 & 2 \\ 5 & 1 \end{bmatrix}$$

$x = 1,\quad u = -1$
$y = 3,\quad v = 2$
$z = 5,\quad w = 1$ **Ans.**

EXERCISE 20.3

Solve the following equations

1. $3x + y + 2z = 3,\ 2x - 3y - z = -3,\ x + 2y + z = 4$ (*A.M.I.E. Winter* 2001)
   Ans. $x = 1,\ y = 2, z = -1$
2. $x + 2y + 3z = 1,\ 2x + 3y + 8z = 2,\ x + y + z = 3$ **Ans.** $x = \frac{9}{2}, y = -1, z = -\frac{1}{2}$
3. $4x + 2y - z = 9,\ x - y + 3z = -4,\ 2x + z = 1$ **Ans.** $x = 1, y = 2, z = -1$
4. $5x + 3y + 3z = 48,\ 2x + 6y - 3z = 18,\ 8x - 3y + 2z = 21$ **Ans.** $x = 3, y = 5, z = 6$
5. $x + y + z = 6,\ x - y + 2z = 5,\ 3x + y + z = 8$ **Ans.** $x = 1, y = 2, z = 3$

6. $x + 2y + 3z = 1,\ 3x - 2y + z = 2,\ 4x + 2y + z = 3$ **Ans.** $x = \frac{7}{10}, y = \frac{3}{40}, z = \frac{1}{20}$

7. $9x + 4y + 3z = -1,\ 5x + y + 2z = 1,\ 7x + 3y + 4z = 1$ **Ans.** $x = 0, y = -1, z = 1$

8. $x + y + z = 8,\ x - y + 2z = 6,\ 9x + 5y - 7z = 14$ **Ans.** $x = 5, y = \frac{5}{3}, z = \frac{4}{3}$

9. $3x + 2y + 4z = 7,\ 2x + y + z = 4,\ x + 3y + 5z = 2$ **Ans.** $x = \frac{9}{4}, y = -\frac{9}{8}, z = \frac{5}{8}$

10. Represent each of the transformations
$x_1 = 3y_1 + 2y_2,\ x_2 = -y_1 + 4y_2$ and $y_1 = z_1 + 2z_2,\ y_2 = -3z_1$
by the use of matrices, find the composite transformation which expresses x_1, x_2 in terms of z_1, z_2. **Ans.** $x_1 = -3z_1 + 6z_2,\ x_2 = -13z_1 - 2z_2$

20.10 PARTITIONING OF MATRICES

Sub matrix. A matrix obtained by deleting some of the rows and columns of a matrix A is said to be sub matrix.

For example, $A = \begin{bmatrix} 4 & 1 & 0 \\ 5 & 2 & 1 \\ 6 & 3 & 4 \end{bmatrix}$, then $\begin{bmatrix} 4 & 1 \\ 5 & 2 \end{bmatrix}, \begin{bmatrix} 5 & 2 \\ 6 & 3 \end{bmatrix}, \begin{bmatrix} 1 & 0 \\ 2 & 1 \end{bmatrix}$ are the sub matrices.

Partitioning: A matrix may be subdivided into sub matrices by drawing lines parallel to its rows and columns. These sub matrices may be considered as the elements of the original matrix.

For example,
$$A = \begin{bmatrix} 2 & 1 & \vdots & 0 & 4 & 1 \\ 1 & 0 & \vdots & 2 & 3 & 4 \\ \cdots & \cdots & \vdots & \cdots & \cdots & \cdots \\ 4 & 5 & \vdots & 1 & 6 & 5 \end{bmatrix}$$

$$A_{11} = \begin{bmatrix} 2 & 1 \\ 1 & 0 \end{bmatrix}, \quad A_{12} = \begin{bmatrix} 0 & 4 & 1 \\ 2 & 3 & 4 \end{bmatrix}$$
$$A_{21} = [4\ \ 5], \quad A_{22} = [1\ \ 6\ \ 5]$$

Then we may write
$$A = \begin{bmatrix} A_{11} & A_{12} \\ A_{21} & A_{22} \end{bmatrix}$$

So, the matrix is partitioned. The dotted lines divide the matrix into sub-matrices. $A_{11}, A_{12}, A_{21}, A_{22}$ are the sub-matrices but behave like elements of the original matrix A. The matrix A can be partitioned in several ways.

Addition by submatrices: Let A and B be two matrices of the same order and are partitioned identically.

For example;
$$A = \begin{bmatrix} 2 & 3 & 4 & \vdots & 5 \\ 0 & 1 & 2 & \vdots & 3 \\ \cdots & \cdots & \cdots & \vdots & \cdots \\ 3 & 4 & 5 & \vdots & 6 \\ \cdots & \cdots & \cdots & \vdots & \cdots \\ 4 & 5 & 0 & \vdots & 1 \end{bmatrix} \qquad B = \begin{bmatrix} 3 & 1 & 4 & \vdots & 6 \\ 2 & 1 & 0 & \vdots & 4 \\ \cdots & \cdots & \cdots & \vdots & \cdots \\ 4 & 5 & 1 & \vdots & 2 \\ \cdots & \cdots & \cdots & \vdots & \cdots \\ 1 & 3 & 4 & \vdots & 5 \end{bmatrix}$$

$$A = \begin{bmatrix} A_{11} & A_{12} \\ A_{21} & A_{22} \\ A_{31} & A_{32} \end{bmatrix}, \quad B = \begin{bmatrix} B_{11} & B_{12} \\ B_{21} & B_{22} \\ B_{31} & B_{32} \end{bmatrix}$$

$$A + B = \begin{bmatrix} A_{11} + B_{11} & A_{12} + B_{12} \\ A_{21} + B_{21} & A_{22} + B_{22} \\ A_{31} + B_{31} & A_{32} + B_{32} \end{bmatrix}$$

20.11 MULTIPLICATION BY SUB-MATRICES

Two matrices A and B, which are conformable to the product AB are partitioned in such a way that the columns of A partitioned in the same way as the rows of B are partitioned. But the rows of A and columns of B can be partitioned in any way.

For example, Here A is a 3×4 matrix and B is 4×3 matrix.

$$A = \begin{bmatrix} 1 & 2 & 3 & \vdots & 4 \\ 0 & 1 & 2 & \vdots & 3 \\ 1 & 4 & 1 & \vdots & 2 \end{bmatrix} \text{ and } B = \begin{bmatrix} 4 & 5 & 6 \\ 3 & 2 & 1 \\ 1 & 0 & 4 \\ \cdots & \cdots & \cdots \\ 2 & 5 & 3 \end{bmatrix}$$

The partitioning of the columns of A is the same as the partitioning of the rows of B. Here, A is partitioned after third column, B has been partitioned after third row.

Example 30. *If C and D are two non-singular matrices, show that if*

$$A = \begin{bmatrix} C & 0 \\ 0 & D \end{bmatrix}, \quad then \; A^{-1} = \begin{bmatrix} C^{-1} & 0 \\ 0 & D^{-1} \end{bmatrix}$$

Solution. Let $$A^{-1} = \begin{bmatrix} E & F \\ G & H \end{bmatrix} \qquad ...(1)$$

Then $$AA^{-1} = \begin{bmatrix} C & 0 \\ 0 & D \end{bmatrix} \begin{bmatrix} E & F \\ G & H \end{bmatrix} = \begin{bmatrix} CE + 0G & CF + 0H \\ 0E + DG & 0F + DH \end{bmatrix}$$

So that $$\begin{bmatrix} CE + 0G & CF + 0H \\ 0E + DG & 0F + DH \end{bmatrix} = \begin{bmatrix} I & 0 \\ 0 & I \end{bmatrix}$$

$$CE + 0G = I \Rightarrow CE = I$$
$$CF + 0H = 0 \Rightarrow CF = 0$$
$$0E + DG = 0 \Rightarrow DG = 0$$
$$0F + DH = I \Rightarrow DH = I$$

Since, C is non singular and $CF = 0$, $\quad \therefore F = 0$

$$CE = I \Rightarrow E = C^{-1}$$

Similarly, D is non singular and $DG = 0 \Rightarrow G = 0$ and $DH = I \Rightarrow H = D^{-1}$

Putting these values in (1), we get

$$A^{-1} = \begin{bmatrix} C^{-1} & 0 \\ 0 & D^{-1} \end{bmatrix}$$ **Proved.**

Inverse By Partitioning: Let the matrix B be the inverse of the matrix A. Matrices A and B are partitioned as

$$A = \begin{bmatrix} A_{11} & A_{12} \\ A_{21} & A_{22} \end{bmatrix}, \quad B = \begin{bmatrix} B_{11} & B_{12} \\ B_{21} & B_{22} \end{bmatrix}$$

Since, $$AB = BA = I$$

$$\begin{bmatrix} A_{11} & A_{12} \\ A_{21} & A_{22} \end{bmatrix}\begin{bmatrix} B_{11} & B_{12} \\ B_{21} & B_{22} \end{bmatrix} = \begin{bmatrix} B_{11} & B_{12} \\ B_{21} & B_{22} \end{bmatrix}\begin{bmatrix} A_{11} & A_{12} \\ A_{21} & A_{22} \end{bmatrix} = \begin{bmatrix} I & 0 \\ 0 & I \end{bmatrix}$$

$$\begin{bmatrix} A_{11} B_{11} + A_{12} B_{21} & A_{11} B_{12} + A_{12} B_{22} \\ A_{21} B_{11} + A_{22} B_{21} & A_{21} B_{12} + A_{22} B_{22} \end{bmatrix} = \begin{bmatrix} B_{11} A_{11} + B_{12} A_{21} & B_{11} A_{12} + B_{12} A_{22} \\ B_{21} A_{11} + B_{22} A_{21} & B_{21} A_{12} + B_{22} A_{22} \end{bmatrix} = \begin{bmatrix} I & 0 \\ 0 & I \end{bmatrix}$$

Let us solve the equations for B_{11}, B_{12}, B_{21} and B_{22}.

Let, $\quad B_{22} = M^{-1}$

From (2), $\quad B_{12} = -A_{11}^{-1}(A_{22} B_{22}) = -(A_{11}^{-1} A_{22}) M^{-1}$

From (3), $\quad B_{21} = -(B_{22} A_{21}) A_{11}^{-1} = -M^{-1}(A_{21} A_{11}^{-1})$

From (1), $\quad B_{11} = A_{11}^{-1} - A_{11}^{-1}(A_{12} B_{21}) = A_{11}^{-1} - (A_{11}^{-1} A_{12}) B_{21}$

$$= A_{11}^{-1} + (A_{11}^{-1} A_{12}) M^{-1} (A_{21} A_{11}^{-1})$$

Here $\quad M = A_{22} - A_{21}(A_{11}^{-1} A_{22})$

Note: A is usually taken of order $n - 1$.

Example 31. *Find the inverse of the following matrix by partitioning*

$$\begin{bmatrix} 1 & 3 & 3 \\ 1 & 4 & 3 \\ 1 & 3 & 4 \end{bmatrix}$$

Solution. Let the matrix be partitioned into four submatrices as follows:

Let $\quad A = \begin{bmatrix} 1 & 3 & 3 \\ 1 & 4 & 3 \\ 1 & 3 & 4 \end{bmatrix}$

$$A_{11} = \begin{bmatrix} 1 & 3 \\ 1 & 4 \end{bmatrix}; \; A_{12} = \begin{bmatrix} 3 \\ 3 \end{bmatrix}$$

$$A_{21} = [1 \quad 3]; \; A_{22} = [4]$$

We have to find $A^{-1} = \begin{bmatrix} B_{11} & B_{12} \\ B_{21} & B_{22} \end{bmatrix}$ where

$$B_{11} = A_{11}^{-1} + (A_{11}^{-1} A_{12})(M^{-1})(A_{21} A_{11}^{-1})$$

$$B_{21} = -M^{-1}(A_{21} A_{11}^{-1})$$

$$B_{12} = -A_{11}^{-1} A_{12} M^{-1}; \; B_{22} = M^{-1}$$

and $\quad M = A_{22} - A_{21}(A_{11}^{-1} A_{12})$

Now

$$A_{11}^{-1} = \begin{bmatrix} 4 & -3 \\ -1 & 1 \end{bmatrix}; \; A_{11}^{-1} A_{12} = \begin{bmatrix} 4 & -3 \\ -1 & 1 \end{bmatrix}\begin{bmatrix} 3 \\ 3 \end{bmatrix} = \begin{bmatrix} 3 \\ 0 \end{bmatrix}$$

$$A_{21} A_{11}^{-1} = [1 \quad 3]\begin{bmatrix} 4 & -3 \\ -1 & 1 \end{bmatrix} = [1 \quad 0]$$

$$M = [4] - [1 - 3]\begin{bmatrix} 3 \\ 0 \end{bmatrix} = [4] - [3] = [1]$$

$$M^{-1} = [1]$$

$$B_{11} = \begin{bmatrix} 4 & -3 \\ -1 & 1 \end{bmatrix} + \begin{bmatrix} 3 \\ 0 \end{bmatrix}[1 \quad 0] = \begin{bmatrix} 4 & -3 \\ -1 & 1 \end{bmatrix} + \begin{bmatrix} 3 & 0 \\ 0 & 0 \end{bmatrix} \Rightarrow B_{11} = \begin{bmatrix} 7 & -3 \\ -1 & 1 \end{bmatrix}$$

$$B_{21} = -[1][1 \quad 0] = -[1 \quad 0]$$

$$B_{12} = -\begin{bmatrix} 3 \\ 0 \end{bmatrix}$$

$$B_{22} = [1]$$

$$A^{-1} = \begin{bmatrix} B_{11} & B_{12} \\ B_{21} & B_{22} \end{bmatrix} = \begin{bmatrix} 7 & -3 & -3 \\ -1 & 1 & 0 \\ -1 & 0 & 1 \end{bmatrix}$$ **Ans.**

Example 32. *Find the inverse of* $A = \begin{bmatrix} 1 & 2 & 3 & 1 \\ 1 & 3 & 3 & 2 \\ 2 & 4 & 3 & 3 \\ 1 & 1 & 1 & 1 \end{bmatrix}$ *by partitioning.*

Solution. (*a*) Take $G_3 = \begin{bmatrix} 1 & 2 & 3 \\ 1 & 3 & 3 \\ 2 & 4 & 3 \end{bmatrix}$ and partition so that

$$A_{11} = \begin{bmatrix} 1 & 2 \\ 1 & 3 \end{bmatrix}, A_{12} = \begin{bmatrix} 3 \\ 3 \end{bmatrix}, A_{21} = \begin{bmatrix} 2 & 4 \end{bmatrix}, \text{ and } A_{22} = [3]$$

Now,
$$A_{11}^{-1} = \begin{bmatrix} 3 & -2 \\ -1 & 1 \end{bmatrix}, \; A_{11}^{-1} A_{12} = \begin{bmatrix} 3 & -2 \\ -1 & 1 \end{bmatrix}\begin{bmatrix} 3 \\ 3 \end{bmatrix} = \begin{bmatrix} 3 \\ 0 \end{bmatrix},$$

$$A_{21} A_{11}^{-1} = [2 \quad 4]\begin{bmatrix} 3 & -2 \\ -1 & 1 \end{bmatrix} = [2 \quad 0]$$

$$M = A_{22} - A_{21}(A_{11}^{-1} A_{12}) = [3] - [2 \quad 4]\begin{bmatrix} 3 \\ 0 \end{bmatrix} = [-3], \text{ And } M^{-1} = [-1/3]$$

Then

$$B_{11} = A_{11}^{-1} + (A_{11}^{-1} A_{12}) M^{-1}(A_{21} A_{11}^{-1}) = \begin{bmatrix} 3 & -2 \\ -1 & 1 \end{bmatrix} + \begin{bmatrix} 3 \\ 0 \end{bmatrix}\begin{bmatrix} -\frac{1}{3} \end{bmatrix}[2 \quad 0] = \begin{bmatrix} 3 & -2 \\ -1 & 1 \end{bmatrix} - \begin{bmatrix} 2 & 0 \\ 0 & 0 \end{bmatrix}$$

$$= \frac{1}{3}\begin{bmatrix} 3 & -6 \\ -3 & 3 \end{bmatrix}$$

$$B_{12} = -(A_{11}^{-1} A_{12}) M^{-1} = \frac{1}{3}\begin{bmatrix} 3 \\ 0 \end{bmatrix}$$

$$B_{21} = -M^{-1}(A_{21} A_{11}^{-1}) = \frac{1}{3}[2 \quad 0]$$

$$B_{22} = M^{-1} = \begin{bmatrix} -\frac{1}{3} \end{bmatrix}$$

and
$$G_3^{-1} = \begin{bmatrix} B_{11} & B_{12} \\ B_{21} & B_{22} \end{bmatrix} = \frac{1}{3}\begin{bmatrix} 3 & -6 & 3 \\ -3 & 3 & 0 \\ 2 & 0 & -1 \end{bmatrix}$$

(*b*) Partition A so that $A_{11} = \begin{bmatrix} 1 & 2 & 3 \\ 1 & 3 & 3 \\ 2 & 4 & 3 \end{bmatrix}$, $A_{12} = \begin{bmatrix} 1 \\ 2 \\ 3 \end{bmatrix}$, $A_{21} = \begin{bmatrix} 1 & 1 & 1 \end{bmatrix}$, and $A_{22} = [1]$.

Now, $A_{11}^{-1} = \frac{1}{3}\begin{bmatrix} 3 & -6 & 3 \\ -3 & 3 & 0 \\ 2 & 0 & -1 \end{bmatrix}$, $A_{11}^{-1} A_{12} = \frac{1}{3}\begin{bmatrix} 0 \\ 3 \\ -1 \end{bmatrix}$, $A_{21} A_{11}^{-1} = \frac{1}{3}[2 \quad -3 \quad 2]$

$$M = [1] - [1 \quad 1 \quad 1]\left(\frac{1}{3}\right)\begin{bmatrix} 0 \\ 3 \\ -1 \end{bmatrix} = \left[\frac{1}{3}\right] \cdot \text{ and } M^{-1} = [3]$$

Then
$$B_{11} = \frac{1}{3}\begin{bmatrix} 3 & -6 & 3 \\ -3 & 3 & 0 \\ 2 & 0 & -1 \end{bmatrix} + \frac{1}{3}\begin{bmatrix} 0 \\ 3 \\ -1 \end{bmatrix}[3]\frac{1}{3}[2 \; -3 \; 2]$$

$$= \frac{1}{3}\begin{bmatrix} 3 & -6 & 3 \\ -3 & 3 & 0 \\ 2 & 0 & -1 \end{bmatrix} + \frac{1}{3}\begin{bmatrix} 0 & 0 & 0 \\ 6 & -9 & 6 \\ -2 & 3 & -2 \end{bmatrix} = \begin{bmatrix} 1 & -2 & 1 \\ 1 & -2 & 2 \\ 0 & 1 & -1 \end{bmatrix}$$

$$B_{12} = \begin{bmatrix} 0 \\ -3 \\ 1 \end{bmatrix}, \; B_{21} = [-2 \quad 3 \quad -2], \; B_{22} = [3]; \; A^{-1} = \begin{bmatrix} B_{11} & B_{12} \\ B_{21} & B_{22} \end{bmatrix} = \begin{bmatrix} 1 & -2 & 1 & 0 \\ 1 & -2 & 2 & -3 \\ 0 & 1 & -1 & 1 \\ -2 & 3 & -2 & 3 \end{bmatrix}$$ **Ans.**

EXERCISE 20.4

1. Compute A + B using partitioning

$A = \begin{bmatrix} 4 & 1 & 0 & 5 \\ 6 & 7 & 8 & 1 \\ 0 & 2 & 1 & 1 \\ 1 & 2 & 0 & 1 \end{bmatrix}$, $B = \begin{bmatrix} 3 & 2 & 1 & 1 \\ 1 & 0 & 1 & 1 \\ 2 & 1 & 2 & 1 \\ 0 & 1 & 2 & 3 \end{bmatrix}$ **Ans.** $\begin{bmatrix} 7 & 3 & 1 & 6 \\ 7 & 7 & 9 & 2 \\ 2 & 3 & 3 & 2 \\ 1 & 3 & 2 & 4 \end{bmatrix}$

2. Compute AB using partitioning

$A = \begin{bmatrix} 1 & 2 & 0 & 1 \\ 4 & 1 & 3 & 2 \\ 2 & 1 & 3 & 0 \end{bmatrix}$, $B = \begin{bmatrix} 2 & 3 & 1 \\ 0 & 1 & 4 \\ 4 & 1 & 2 \\ 2 & 1 & 2 \end{bmatrix}$ **Ans.** $\begin{bmatrix} 4 & 6 & 11 \\ 24 & 18 & 18 \\ 16 & 10 & 12 \end{bmatrix}$

3. Find the inverse of $\begin{bmatrix} A & B \\ C & 0 \end{bmatrix}$ where B, C are non-singular. **Ans.** $\begin{bmatrix} 0 & C^{-1} \\ B^{-1} & -B^{-1}AC^{-1} \end{bmatrix}$

Find the inverse of the following metrices by partitioning:

4. $\begin{bmatrix} 2 & 1 & -1 \\ 1 & 3 & 2 \\ -1 & 2 & 1 \end{bmatrix}$ **Ans.** $\frac{1}{10}\begin{bmatrix} 1 & 3 & -5 \\ 3 & -1 & 5 \\ -5 & 5 & -5 \end{bmatrix}$ **5.** $\begin{bmatrix} 1 & 2 & -1 \\ -1 & 1 & 2 \\ 2 & -1 & 1 \end{bmatrix}$ **Ans.** $\frac{1}{14}\begin{bmatrix} 3 & -1 & 5 \\ 5 & 3 & -1 \\ -1 & 5 & 3 \end{bmatrix}$

6. $\begin{bmatrix} 2 & 3 & 4 \\ 4 & 3 & 1 \\ 1 & 2 & 4 \end{bmatrix}$ **Ans.** $\frac{1}{5}\begin{bmatrix} -10 & 4 & 9 \\ 15 & -4 & -14 \\ -5 & 1 & 6 \end{bmatrix}$ **7.** $\begin{bmatrix} 1 & 2 & 3 \\ 2 & 4 & 5 \\ 3 & 5 & 6 \end{bmatrix}$ **Ans.** $\begin{bmatrix} 1 & -3 & 2 \\ -3 & 3 & -1 \\ 2 & -1 & 0 \end{bmatrix}$

8. $\begin{bmatrix} 3 & 4 & 2 & 7 \\ 2 & 3 & 3 & 2 \\ 52 & 7 & 3 & 9 \\ 2 & 3 & 2 & 3 \end{bmatrix}$ **Ans.** $\frac{1}{2}\begin{bmatrix} -1 & 11 & 7 & -26 \\ -1 & -7 & -3 & 16 \\ 1 & 1 & -1 & 0 \\ 1 & -1 & -1 & 2 \end{bmatrix}$

Choose the correct answer:

9. If $3x + 2y + z - 0$, $x + 4y + z = 0$, $2x + y + 4z = 0$, be a system of equations then

(*i*) System is inconsistent (*ii*) it has only trivial solution

(*iii*) it can be reduced to a single equation thus solution does not exist

(*iv*) Determinant of the coefficient matrix is zero. (*AMIETE, June 2010*) **Ans.** (*ii*)

Eigen Values, Eigen Vector, Cayley Hamilton Theorem, Diagonalisation

(Complex and Unitary Matrices, Application)

21.1 INTRODUCTION

Eigen values and eigen vectors are used in the study of ordinary differential equations, analysing population growth and finding powers of matrices.

21.2 EIGEN VALUES

Let $$\begin{bmatrix} a_{11} & a_{12} & a_{13} & \cdots & a_{1n} \\ a_{21} & a_{22} & a_{23} & \cdots & a_{2n} \\ a_{31} & a_{32} & a_{33} & \cdots & a_{3n} \\ \cdots & \cdots & \cdots & \cdots & \cdots \\ a_{n1} & a_{n2} & a_{n3} & \cdots & a_{nn} \end{bmatrix} \begin{bmatrix} x_1 \\ x_2 \\ x_3 \\ \vdots \\ x_n \end{bmatrix} = \begin{bmatrix} y_1 \\ y_2 \\ y_3 \\ \vdots \\ y_n \end{bmatrix}$$

$$AX = Y \qquad \text{...(1)}$$

Where A is the matrix , X is the column vector and Y is also column vector.

Here column vector X is transformed into the column vector Y by means of the square matrix A.

Let X be a such vector which transforms into λX by means of the transformation (1). Suppose the linear transformation $Y = AX$ transforms X into a scalar multiple of itself i.e. λX.

$$AX = Y = \lambda X$$

$$AX - \lambda IX = 0$$

$$(A - \lambda I)X = 0 \qquad \text{...(2)}$$

Thus the unknown scalar λ is known as an eigen value of the matrix A and the corresponding non zero vector X as **eigen vector.**

The eigen values are also called characteristic values or proper values or latent values.

Let $$A = \begin{bmatrix} 2 & 2 & 1 \\ 1 & 3 & 1 \\ 1 & 2 & 2 \end{bmatrix}$$

$$A - \lambda I = \begin{bmatrix} 2 & 2 & 1 \\ 1 & 3 & 1 \\ 1 & 2 & 2 \end{bmatrix} - \lambda \begin{bmatrix} 1 & 0 & 0 \\ 0 & 1 & 0 \\ 0 & 0 & 1 \end{bmatrix} = \begin{bmatrix} 2-\lambda & 2 & 1 \\ 1 & 3-\lambda & 1 \\ 1 & 2 & 2-\lambda \end{bmatrix} \quad \text{characteristic matrix}$$

(*b*) **Characteristic Polynomial:** The determinant $| A - \lambda I |$ when expanded will give a polynomial, which we call as characteristic polynomial of matrix A.

For example; $\begin{vmatrix} 2-\lambda & 2 & 1 \\ 1 & 3-\lambda & 1 \\ 1 & 2 & 2-\lambda \end{vmatrix}$

$$= (2-\lambda)(6-5\lambda+\lambda^2-2) - 2(2-\lambda-1) + 1(2-3+\lambda)$$
$$= -\lambda^3 + 7\lambda^2 - 11\lambda + 5$$

(*c*) **Characteristic Equation:** The equation $|A - \lambda I| = 0$ is called the characteristic equation of the matrix A *e.g.*

$$\lambda^3 - 7\lambda^2 + 11\lambda - 5 = 0$$

(*d*) **Characteristic Roots or Eigen Values:** The roots of characteristic equation $|A - \lambda I| = 0$ are called characteristic roots of matrix A. *e.g.*

$$\lambda^3 - 7\lambda^2 + 11\lambda - 5 = 0$$

$\Rightarrow$ $(\lambda-1)(\lambda-1)(\lambda-5) = 0$ $\therefore \lambda = 1, 1, 5$

Characteristic roots are 1, 1, 5.

Some Important Properties of Eigen Values

(*AMIETE, Dec. 2009*)

(1) Any square matrix A and its transpose A' have the same eigen values.

Note. The sum of the elements on the principal diagonal of a matrix is called the **trace** of the matrix.

(2) The sum of the eigen values of a matrix is equal to the **trace** of the matrix.

(3) The product of the eigen values of a matrix A is equal to the **determinant** of A.

(4) If $\lambda_1, \lambda_2, \ldots \lambda_n$ are the eigen values of A, then the eigen values of

(*i*) kA are $k\lambda_1, k\lambda_2, \ldots\ldots, k\lambda_n$ (*ii*) A^m are $\lambda_1^m, \lambda_2^m, \ldots\ldots\ldots, \lambda_n^m$

(*iii*) A^{-1} are $\dfrac{1}{\lambda_1}, \dfrac{1}{\lambda_2}, \ldots\ldots, \dfrac{1}{\lambda_n}$.

Example 1. *Find the characteristic roots of the matrix* $\begin{bmatrix} 6 & -2 & 2 \\ -2 & 3 & -1 \\ 2 & -1 & 3 \end{bmatrix}$

Solution. The characteristic equation of the given matrix is

$$\begin{vmatrix} 6-\lambda & -2 & 2 \\ -2 & 3-\lambda & -1 \\ 2 & -1 & 3-\lambda \end{vmatrix} = 0$$

$\Rightarrow$ $(6-\lambda)(9-6\lambda+\lambda^2-1) + 2(-6+2\lambda+2) + 2(2-6+2\lambda) = 0$

$\Rightarrow$ $-\lambda^3 + 12\lambda^2 - 36\lambda + 32 = 0$

By trial, $\lambda = 2$ is a root of this equation.

$\Rightarrow$ $(\lambda-2)(\lambda^2-10\lambda+16) = 0 \Rightarrow (\lambda-2)(\lambda-2)(\lambda-8) = 0$

$\Rightarrow$ $\lambda = 2, 2, 8$ are the characteristic roots or Eigen values. **Ans.**

Example 2. *Find the eigen values of the matrix :*

$$\begin{bmatrix} 2 & 1 & 1 \\ 1 & 2 & 1 \\ 0 & 0 & 1 \end{bmatrix}$$

(*R.G.P.V. Bhopal, I Semester, June 2007*)

Solution. Let $A = \begin{bmatrix} 2 & 1 & 1 \\ 1 & 2 & 1 \\ 0 & 0 & 1 \end{bmatrix}$

The characteristic equation of A is

$$|A-\lambda I|=0$$

$$\begin{vmatrix} 2-\lambda & 1 & 1 \\ 1 & 2-\lambda & 1 \\ 0 & 0 & 1-\lambda \end{vmatrix}=0$$

Expanding the determinant with the help of third row, we have

$\Rightarrow \quad (1-\lambda)\left[(2-\lambda)^2-1\right]=0 \quad \Rightarrow \quad (1-\lambda)\left(\lambda^2-4\lambda+4-1\right)=0$

$\Rightarrow \quad (1-\lambda)\left(\lambda^2-4\lambda+3\right)=0 \quad \Rightarrow \quad (1-\lambda)(\lambda-3)(\lambda-1)=0$

$\Rightarrow \quad \lambda=1,\ 1,\ 3$

The eigen values of the given matrix are 1, 1 and 3. **Ans.**

Example 3. *The matrix A is defined as* $A=\begin{bmatrix} 1 & 2 & -3 \\ 0 & 3 & 2 \\ 0 & 0 & -2 \end{bmatrix}$

Find the eigen values of $3A^3+5A^2-6A+2I$.

Solution. $|A-\lambda I|=0$

$$\begin{vmatrix} 1-\lambda & 2 & -3 \\ 0 & 3-\lambda & 2 \\ 0 & 0 & -2-\lambda \end{vmatrix}=0$$

$\Rightarrow \quad (1-\lambda)(3-\lambda)(-2-\lambda)=0$ or $\lambda=1,\ 3,\ -2$

Eigen values of $A^3 = 1, 27, -8$; Eigen values of $A^2 = 1, 9, 4$

Eigen values of $A = 1, 3, -2$; Eigen values of $I = 1, 1, 1$

$\therefore$ Eigen values of $3A^3+5A^2-6A+2I$

First eigen value $= 3(1)^3+5(1)^2-6(1)+2(1) = 4$

Second eigen value $= 3(27)+5(9)-6(3)+2(1) = 110$

Third eigen value $= 3(-8)+5(4)-6(-2)+2(1) = 10$

Required eigen values are 4, 110, 10 **Ans.**

Example 4. *Show that for any square matrix A, the product of all the eigen values of A is equal to det (A), and the sum of all the eigen values of A is equal to the sum of the diagonal elements.* (*U.P., I Semester, Winter* 2003)

Solution. Let $A=\begin{bmatrix} a_{11} & a_{12} & a_{13} \\ a_{21} & a_{22} & a_{23} \\ a_{31} & a_{32} & a_{33} \end{bmatrix}$, $|A-\lambda I| = \begin{vmatrix} a_{11}-\lambda & a_{12} & a_{13} \\ a_{21} & a_{22}-\lambda & a_{23} \\ a_{31} & a_{32} & a_{33}-\lambda \end{vmatrix}$

$|A-\lambda I|$

$= (a_{11}-\lambda)\left[(a_{22}-\lambda)(a_{33}-\lambda)-a_{32}a_{23}\right]-a_{12}\left[a_{21}(a_{33}-\lambda)-a_{31}a_{23}\right]+a_{13}\left[a_{21}a_{32}-a_{31}(a_{22}-\lambda)\right]$

$= (a_{11}-\lambda)\left[a_{22}a_{33}-(a_{22}+a_{33})\lambda+\lambda^2-a_{32}a_{23}\right]-a_{12}\left[a_{21}a_{33}-a_{21}\lambda-a_{31}a_{23}\right]+a_{13}(a_{21}a_{32}-a_{31}a_{22}+a_{31}\lambda)$

$= a_{11}a_{22}a_{33}+(-a_{11}a_{22}-a_{11}a_{33})\lambda+a_{11}\lambda^2-a_{11}a_{32}a_{23}+(-a_{22}a_{33}+a_{32}a_{23})\lambda+(a_{22}+a_{33})\lambda^2-\lambda^3-a_{12}a_{21}a_{33}+a_{12}a_{31}a_{23}+a_{12}a_{21}\lambda+a_{13}a_{21}a_{32}-a_{13}a_{31}a_{22}+a_{13}a_{31}\lambda$

$= -\lambda^3+\lambda^2(a_{11}+a_{22}+a_{33})+\lambda(-a_{11}a_{22}-a_{11}a_{33}+a_{12}a_{21}-a_{22}a_{33}+a_{23}a_{32}+a_{13}a_{31})$
$\quad -\left[a_{11}(a_{22}a_{33}-a_{23}a_{32})-a_{12}(a_{21}a_{33}-a_{23}a_{31})+a_{13}(a_{21}a_{32}-a_{31}a_{22})\right] \quad \ldots(1)$

If $\lambda_1, \lambda_2, \lambda_3$ be the roots of the equation (1) then
Sum of the roots $= \lambda_1 + \lambda_2 + \lambda_3 = a_{11} + a_{22} + a_{33}$ = Sum of the diagonal elements.
Product of the roots

$$= \lambda_1\lambda_2\lambda_3 = [a_{11}(a_{22}a_{33} - a_{23}a_{32}) - a_{12}(a_{21}a_{33} - a_{23}a_{31}) + a_{13}(a_{21}a_{32} - a_{31}a_{22})]$$

= Determinant A. **Proved.**

Example 5. *Let λ be an eigen value of a matrix A. Then prove that*

(*i*) $\lambda + k$ is an eigen value of $A + kI$

(*ii*) $k\lambda$ is an eigen value of kA. (*Gujarat, II Semester, June 2009*)

Solution. Here, A has eigen value λ. $\Rightarrow \quad |A - \lambda I| = 0$ ... (1)

(*ii*) Adding and subtracting kI from (1) we get

$$|A + kI - \lambda I - kI| = 0$$

$\Rightarrow \quad |(A + kI) - (\lambda + k) I| = 0 \Rightarrow \quad A + kI$ has $\lambda + k$ eigen value.

(*i*) Multiplying (1), by k, we get

$$k|A - \lambda I| = 0 \qquad \Rightarrow |kA - k\lambda I| = 0$$

$\Rightarrow \quad kA$ has eigen value $k\lambda$. **Proved.**

Example 6. *If $\lambda_1, \lambda_2, \ldots \lambda_n$ are the eigen values of A, find the eigen values of the martrix $(A - \lambda I)^2$.*

Solution. $(A - \lambda I)^2 = A^2 - 2\lambda AI + \lambda^2 I^2 = A^2 - 2\lambda A + \lambda^2 I$.

Eigen values of A^2 are $\lambda_1^2, \lambda_2^2, \lambda_3^3 \ldots \lambda_n^2$
Eigen values of $2\lambda A$ are $2\lambda\lambda_1, \quad 2\lambda\lambda_2, \quad 2\lambda\lambda_3 \ldots 2\lambda\lambda_n$.
Eigen values of $\lambda^2 I$ are λ^2.
$\therefore$ Eigen values of $A^2 - 2\lambda A + \lambda^2 I$

$$\lambda_1^2 - 2\lambda\lambda_1 + \lambda^2, \quad \lambda_2^2 - 2\lambda\lambda_2 + \lambda^2, \quad \lambda_3^2 - 2\lambda\lambda_3 + \lambda^2 \ldots\ldots$$

$\Rightarrow \quad (\lambda_1 - \lambda)^2, \quad (\lambda_2 - \lambda)^2, \quad (\lambda_3 - \lambda)^2, \ldots (\lambda_n - \lambda)^2$ **Ans.**

Example 7. *Prove that a matrix A and its transpose A′ have the same characteristic roots.*

Solution. Characteristic equation of matrix A is

$$|A - \lambda I| = 0 \qquad \ldots (1)$$

Characteristic equation of matrix A' is

$$|A' - \lambda I| = 0 \qquad \ldots (2)$$

Clearly both (1) and (2) are same, as we know that

$$|A| = |A'|$$

i.e., a determinant remains unchanged when rows be changed into columns and columns into rows. **Proved.**

Example 8. *If A and P be square matrices of the same type and if P be invertible, show that the matrices A and $P^{-1}AP$ have the same characteristic roots.*

Solution. Let us put $B = P^{-1}AP$ and we will show that characteristic equations for both A and B are the same and hence they have the same characteristic roots.

$$B - \lambda I = P^{-1}AP - \lambda I = P^{-1}AP - P^{-1}\lambda IP = P^{-1}(A - \lambda I)P$$

$$\therefore \quad |B - \lambda I| = |P^{-1}(A - \lambda I)P| = |P^{-1}||A - \lambda I||P|$$

$$= |A - \lambda I||P^{-1}||P| = |A - \lambda I||P^{-1}P|$$

$$= |A - \lambda I||I| = |A - \lambda I| \text{ as } |I| = 1$$

Thus the matrices A and B have the same characteristic equations and hence the same characteristic roots. **Proved.**

Example 9. *If A and B be two square invertible matrices, then prove that AB and BA have the same characteristic roots.*

Solution. Now $AB = IAB = B^{-1} B (AB) = B^{-1} (BA) B$ ...(1)

But by Ex. 8, matrices BA and $B^{-1} (BA) B$ have same characteristic roots or matrices BA and AB by (1) have same characteristic roots. **Proved.**

Example 10. *If A and B be n rowed square matrices and if A be invertible, show that the matrices $A^{-1} B$ and BA^{-1} have the same characteristics roots.*

Solution. $A^{-1} B = A^{-1} BI = A^{-1} B (A^{-1}A) = A^{-1} (BA^{-1}) A.$ ...(1)

But by Ex. 8, matrices BA^{-1} and $A^{-1} (BA^{-1})A$ have same characteristic roots or matrices BA^{-1} and $A^{-1} B$ by (1) have same characteristic roots. **Proved.**

Example 11. *Show that 0 is a characteristic root of a matrix, if and only if, the matrix is singular.*

Solution. Characteristic equation of matrix A is given by

$$|A - \lambda I| = 0$$

If $\lambda = 0$, then from above it follows that $|A| = 0$ *i.e.* Matrix A is singular.

Again if Matrix A is singular *i.e.*, $|A| = 0$ then

$$|A - \lambda I| = 0 \Rightarrow |A| - \lambda |I| = 0, 0 - \lambda \cdot 1 = 0 \Rightarrow \lambda = 0.$$ **Proved.**

Example 12. *Show that characteristic roots of a triangular matrix are just the diagonal elements of the matrix.*

Solution. Let us consider the triangular matrix.

$$A = \begin{bmatrix} a_{11} & 0 & 0 & 0 \\ a_{21} & a_{22} & 0 & 0 \\ a_{31} & a_{32} & a_{33} & 0 \\ a_{41} & a_{42} & a_{43} & a_{44} \end{bmatrix}$$

Characteristic equation is $|A - \lambda I| = 0$

or
$$\begin{vmatrix} a_{11}-\lambda & 0 & 0 & 0 \\ a_{21} & a_{22}-\lambda & 0 & 0 \\ a_{31} & a_{32} & a_{33}-\lambda & 0 \\ a_{41} & a_{42} & a_{43} & a_{44}-\lambda \end{vmatrix} = 0$$

On expansion it gives

$$(a_{11} - \lambda)(a_{22} - \lambda)(a_{33} - \lambda)(a_{44} - \lambda) = 0$$

$$\therefore \quad \lambda = a_{11}, \quad a_{22}, \quad a_{33}, \quad a_{44}$$

which are diagonal elements of matrix A. **Proved.**

Example 13. *If λ is an eigen value of an orthogonal matrix, then $\frac{1}{\lambda}$ is also eigen value.*

[**Hint:** $AA' = I$ if λ is the eigen value of A, then $\lambda^2 = 1, \quad \lambda = \frac{1}{\lambda}$]

Example 14. *Find the eigen values of the orthogonal matrix.*

$$B = \frac{1}{3}\begin{bmatrix} 1 & 2 & 2 \\ 2 & 1 & -2 \\ 2 & -2 & 1 \end{bmatrix}$$

Solution. The characteristic equation of

$$A = \begin{bmatrix} 1 & 2 & 2 \\ 2 & 1 & -2 \\ 2 & -2 & 1 \end{bmatrix} \quad \text{is} \quad \begin{vmatrix} 1-\lambda & 2 & 2 \\ 2 & 1-\lambda & -2 \\ 2 & -2 & 1-\lambda \end{vmatrix} = 0$$

$\Rightarrow \quad (1-\lambda)[(1-\lambda)(1-\lambda)-4]-2[2(1-\lambda)+4]+2[-4-2(1-\lambda)]=0$

$\Rightarrow \quad (1-\lambda)(1-2\lambda+\lambda^2-4)-2(2-2\lambda+4)+2(-4-2+2\lambda)=0$

$\Rightarrow \quad \lambda^3-3\lambda^2-9\lambda+27=0$

$\Rightarrow \quad (\lambda-3)^2(\lambda+3)=0$

The eigen values of A are 3, 3, –3, so the eigen values of $B=\frac{1}{3}A$ are 1, 1, –1.

Note. If $\lambda = 1$ is an eigen value of B then its reciprocal $\frac{1}{\lambda}=\frac{1}{1}=1$ is also an eigen value of B. **Ans.**

EXERCISE 21.1

Show that, for any square matrix A.

1. If λ be an eigen value of a non singular matrix A, show that $\frac{|A|}{\lambda}$ is an eigen value of the matrix adj A.

2. There are infinitely many eigen vectors corresponding to a single eigen value.

3. Find the product of the eigen values of the matrix $\begin{bmatrix} 3 & -3 & 3 \\ 2 & 1 & 1 \\ 1 & 5 & 6 \end{bmatrix}$ **Ans.** 18

4. Find the sum of the eigen values of the matrix $\begin{bmatrix} 3 & 2 & 1 \\ 1 & 3 & 2 \\ 4 & 1 & 5 \end{bmatrix}$ **Ans.** 11

5. Find the eigen value of the inverse of the matrix $\begin{bmatrix} 4 & 6 & 6 \\ 1 & 3 & 2 \\ -1 & -4 & -3 \end{bmatrix}$ **Ans.** $-1, 1, \frac{1}{4}$

6. Find the eigen values of the square of the matrix $\begin{bmatrix} 1 & 0 & -1 \\ 1 & 2 & 1 \\ 2 & 2 & 3 \end{bmatrix}$ **Ans.** 1, 4, 9

7. Find the eigen values of the matrix $\begin{bmatrix} 3 & 1 & 4 \\ 0 & 2 & 6 \\ 0 & 0 & 5 \end{bmatrix}^3$ **Ans.** 8, 27, 125

8. The sum and product of the eigen values of the matrix $A=\begin{bmatrix} 2 & 2 & 1 \\ 1 & 3 & 1 \\ 1 & 2 & 2 \end{bmatrix}$ are respectively

(*a*) 7 and 7 (*b*) 7 and 5 (*c*) 7 and 6 (*d*) 7 and 8 (*AMIETE, June 2010*) **Ans.** (*b*)

21.3 CAYLEY-HAMILTON THEOREM

Satement. Every square matrix satisfies its own characteristic equation.

If $|A-\lambda I|=(-1)^n(\lambda^n+a_1\lambda^{n-1}+a_2\lambda^{n-2}+\cdots+a_n)$ be the characteristic polynomial of $n \times n$ matrix $A=(a_{ij})$, then the matrix equation

$$X^n+a_1X^{n-1}+a_2X^{n-2}+\ldots+a_nI=0 \text{ is satisfied by } X=A \text{ i.e.,}$$

$$A^n+a_1A^{n-1}+a_2A^{n-2}+\ldots+a_nI=0$$

Proof. Since the elements of the matrix $A-\lambda I$ are at most of the first degree in λ, the elements of adj. $(A-\lambda I)$ are at most degree $(n-1)$ in λ. Thus, adj. $(A-\lambda I)$ may be written as a matrix polynomial in λ, given by

$$Adj(A-\lambda I)=B_0\lambda^{n-1}+B_1\lambda^{n-2}+\ldots+B_{n-1}$$

where $B_0, B_1, \ldots, B_{n-1}$ are $n \times n$ matrices, their elements being polynomial in λ.

We know that

$$(A-\lambda I)\,Adj(A-\lambda I) = |A-\lambda I|\,I$$

$$(A-\lambda I)\left(B_0\lambda^{n-1} + B_1\lambda^{n-2} + \ldots + B_{n-1}\right) = (-1)^n\left(\lambda^n + a_1\lambda^{n-1} + \ldots + a_n\right)I$$

Equating coefficient of like power of λ on both sides, we get

$$-IB_0 = (-1)^n I$$

$$AB_0 - IB_1 = (-1)^n a_1 I$$

$$AB_1 - IB_2 = (-1)^n a_2 I$$

....

$$AB_{n-1} = (-1)^n a_n I$$

On multiplying the equation by $A^n, A^{n-1}, \ldots, I$ respectively and adding, we obtain

$$0 = (-1)^n\left[A^n + a_1 A^{n-1} + \ldots + a_n I\right]$$

Thus $\qquad A^n + a_1 A^{n-1} + \ldots + a_n I = 0$

for example, Let A be square matrix and if

$$\lambda^3 - 2\lambda^2 + 3\lambda - 4 = 0 \qquad \ldots(1)$$

be its characteristic equation, then according to Cayley Hamilton Theorem (1) is satisfied by A.

$$A^3 - 2A^2 + 3A - 4I = 0 \qquad \ldots(2)$$

We can find out A^{-1} from (2). On premultiplying (2) by A^{-1}, we get

$$A^2 - 2A + 3I - 4A^{-1} = 0$$

$$A^{-1} = \frac{1}{4}\left[A^2 - 2A + 3I\right]$$

Example 15. *Verify Cayley-Hamilton theorem for the matrix*

$$A = \begin{pmatrix} 1 & 2 \\ 2 & -1 \end{pmatrix}$$ *and hence find A^{-1}.* (U.P.,I Sem., Dec 2008)

Solution. The characteristic equation of the matrix is $|A - \lambda I| = 0$

$$\begin{vmatrix} 1-\lambda & 2 \\ 2 & -1-\lambda \end{vmatrix} = 0$$

$$(1-\lambda)(-1-\lambda) - 4 = 0 \Rightarrow -1 + \lambda^2 - 4 = 0 \Rightarrow \lambda^2 - 5 = 0$$

By Cayley-Hamilton Theorem,

$$A^2 - 5I = 0 \qquad \ldots(1)$$

Now, $\quad A^2 = A.A = \begin{bmatrix} 1 & 2 \\ 2 & -1 \end{bmatrix}\begin{bmatrix} 1 & 2 \\ 2 & -1 \end{bmatrix} = \begin{bmatrix} 5 & 0 \\ 0 & 5 \end{bmatrix}$

$$A^2 - 5I = \begin{bmatrix} 5 & 0 \\ 0 & 5 \end{bmatrix} - 5\begin{bmatrix} 1 & 0 \\ 0 & 1 \end{bmatrix} = \begin{bmatrix} 5 & 0 \\ 0 & 5 \end{bmatrix} + \begin{bmatrix} -5 & 0 \\ 0 & -5 \end{bmatrix} = \begin{bmatrix} 0 & 0 \\ 0 & 0 \end{bmatrix} = 0 \qquad \ldots(2)$$

From (1) and (2), Cayley-Hamilton theorem is verified.

Again from (1), we have

$$A^2 - 5I = 0$$

Multiplying by A^{-1}, we get

$$A - 5A^{-1} = 0 \Rightarrow A^{-1} = \frac{1}{5}A \Rightarrow A^{-1} = \frac{1}{5}\begin{bmatrix} 1 & 2 \\ 2 & -1 \end{bmatrix} = \begin{bmatrix} \frac{1}{5} & \frac{2}{5} \\ \frac{2}{5} & -\frac{1}{5} \end{bmatrix}$$ **Ans**

Example 16. *Find the characteristic equation of the matrix A.*

$$A=\begin{bmatrix}4 & 3 & 1\\ 2 & 1 & -2\\ 1 & 2 & 1\end{bmatrix}$$

Hence find A^{-1}. *(R.G.P.V., Bhopal, Feb. 2006)*

Solution Characteristic equation is

$$\begin{vmatrix}4-\lambda & 3 & 1\\ 2 & 1-\lambda & -2\\ 1 & 2 & 1-\lambda\end{vmatrix}=0$$

$$(4-\lambda)\left[1+\lambda^2-2\lambda+4\right]-3(2-2\lambda+2)+1\cdot(4-1+\lambda)=0$$

$$\Rightarrow \quad (4-\lambda)(\lambda^2-2\lambda+5)-3(-2\lambda+4)+(3+\lambda)=0$$

$$\Rightarrow \quad 4\lambda^2-8\lambda+20-\lambda^3+2\lambda^2-5\lambda+6\lambda-12+3+\lambda=0$$

$$\Rightarrow \quad -\lambda^3+6\lambda^2-6\lambda+11=0 \quad \text{or} \quad \lambda^3-6\lambda^2+6\lambda-11=0$$

By Cayley-Hamilton Theorem

$$A^3-6A^2+6A-11I=0 \qquad ...(1)$$

Multiplying (1) by A^{-1}, we get

$$A^2-6A+6I-11A^{-1}=0 \quad \text{or} \quad 11A^{-1}=A^2-6A+6I$$

$$11A^{-1}=\begin{bmatrix}4 & 3 & 1\\ 2 & 1 & -2\\ 1 & 2 & 1\end{bmatrix}\begin{bmatrix}4 & 3 & 1\\ 2 & 1 & -2\\ 1 & 2 & 1\end{bmatrix}-6\begin{bmatrix}4 & 3 & 1\\ 2 & 1 & -2\\ 1 & 2 & 1\end{bmatrix}+6\begin{bmatrix}1 & 0 & 0\\ 0 & 1 & 0\\ 0 & 0 & 1\end{bmatrix}$$

$$=\begin{bmatrix}23 & 17 & -1\\ 8 & 3 & -2\\ 9 & 7 & -2\end{bmatrix}+\begin{bmatrix}-24 & -18 & -6\\ -12 & -6 & 12\\ -6 & -12 & -6\end{bmatrix}+\begin{bmatrix}6 & 0 & 0\\ 0 & 6 & 0\\ 0 & 0 & 6\end{bmatrix}=\begin{bmatrix}5 & -1 & -7\\ -4 & 3 & 10\\ 3 & -5 & -2\end{bmatrix}$$

$$A^{-1}=\frac{1}{11}\begin{bmatrix}5 & -1 & -7\\ -4 & 3 & 10\\ 3 & -5 & -2\end{bmatrix}$$ **Ans.**

Example 17. *Find the characteristic equation of the matrix* $A=\begin{bmatrix}2 & 1 & 1\\ 0 & 1 & 0\\ 1 & 1 & 2\end{bmatrix}$

Verify Cayley Hamilton Theorem and hence prove that :

$$A^8-5A^7+7A^6-3A^5+A^4-5A^3+8A^2-2A+I=\begin{bmatrix}8 & 5 & 5\\ 0 & 3 & 0\\ 5 & 5 & 8\end{bmatrix}$$

(Gujarat, II Semester, June 2009)

Solution. Characteristic equation of the matrix A is

$$\begin{vmatrix}2-\lambda & 1 & 1\\ 0 & 1-\lambda & 0\\ 1 & 1 & 2-\lambda\end{vmatrix}=0$$

$$\Rightarrow \quad (2-\lambda)[(1-\lambda)(2-\lambda)]-1(0)+1(0-1+\lambda)=0$$

$$\Rightarrow \qquad \lambda^3 - 5\lambda^2 + 7\lambda - 3 = 0$$

According to Cayley-Hamilton Theorem

$$A^3 - 5A^2 + 7A - 3I = 0 \qquad ...(1)$$

We have to verify the equation (1).

$$A^2 = \begin{bmatrix} 2 & 1 & 1 \\ 0 & 1 & 0 \\ 1 & 1 & 2 \end{bmatrix}\begin{bmatrix} 2 & 1 & 1 \\ 0 & 1 & 0 \\ 1 & 1 & 2 \end{bmatrix} = \begin{bmatrix} 5 & 4 & 4 \\ 0 & 1 & 0 \\ 4 & 4 & 5 \end{bmatrix}$$

$$A^3 = A^2.A = \begin{bmatrix} 2 & 1 & 1 \\ 0 & 1 & 0 \\ 1 & 1 & 2 \end{bmatrix}\begin{bmatrix} 5 & 4 & 4 \\ 0 & 1 & 0 \\ 4 & 4 & 5 \end{bmatrix} = \begin{bmatrix} 14 & 13 & 13 \\ 0 & 1 & 0 \\ 13 & 13 & 14 \end{bmatrix}$$

$$A^3 - 5A^2 + 7A - 3I = \begin{bmatrix} 14 & 13 & 13 \\ 0 & 1 & 0 \\ 13 & 13 & 14 \end{bmatrix} - 5\begin{bmatrix} 5 & 4 & 4 \\ 0 & 1 & 0 \\ 4 & 4 & 5 \end{bmatrix} + 7\begin{bmatrix} 2 & 1 & 1 \\ 0 & 1 & 0 \\ 1 & 1 & 2 \end{bmatrix} - 3\begin{bmatrix} 1 & 0 & 0 \\ 0 & 1 & 0 \\ 0 & 0 & 1 \end{bmatrix}$$

$$= \begin{bmatrix} 14-25+14-3 & 13-20+7+0 & 13-20+7+0 \\ 0+0+0+0 & 1-5+7-3 & 0-0+0-0 \\ 13-20+7+0 & 13-20+7-0 & 14-25+14-3 \end{bmatrix} = \begin{bmatrix} 0 & 0 & 0 \\ 0 & 0 & 0 \\ 0 & 0 & 0 \end{bmatrix} = 0$$

Hence Cayley Hamilton Theorem is verified.

Now, $A^8 - 5A^7 + 7A^6 - 3A^5 + A^4 - 5A^3 + 8A^2 - 2A + I$

$$= A^5\left(A^3 - 5A^2 + 7A - 3I\right) + A\left(A^3 - 5A^2 + 7A - 3I\right) + A^2 + A + I$$

$$= A^5 \times O + A \times O + A^2 + A + I \;\; = A^2 + A + I$$

$$= \begin{bmatrix} 5 & 4 & 4 \\ 0 & 1 & 0 \\ 4 & 4 & 5 \end{bmatrix} + \begin{bmatrix} 2 & 1 & 1 \\ 0 & 1 & 0 \\ 1 & 1 & 2 \end{bmatrix} + \begin{bmatrix} 1 & 0 & 0 \\ 0 & 1 & 0 \\ 0 & 0 & 1 \end{bmatrix}$$

$$= \begin{bmatrix} 5+2+1 & 4+1+0 & 4+1+0 \\ 0+0+0 & 1+1+1 & 0+0+0 \\ 4+1+0 & 4+1+0 & 5+2+1 \end{bmatrix} = \begin{bmatrix} 8 & 5 & 5 \\ 0 & 3 & 0 \\ 5 & 5 & 8 \end{bmatrix}$$

Proved.

21.4 POWER OF MATRIX (by Cayley Hamilton Theorem)

Any positive integral power A^m of matrix A is linearly expressible in terms of those of lower degree, where m is a positive integer and n is the degree of characteristic equation such that $m > n$.

Example 18. *Find A^4 with the help of Cayley Hamilton Theorem, if*

$$A = \begin{bmatrix} 1 & 0 & -1 \\ 1 & 2 & 1 \\ 2 & 2 & 3 \end{bmatrix}$$

Solution. Here, we have

$$A = \begin{bmatrix} 1 & 0 & -1 \\ 1 & 2 & 1 \\ 2 & 2 & 3 \end{bmatrix}$$

Characteristic equation of the matrix A is

$$\begin{vmatrix} 1-\lambda & 0 & -1 \\ 1 & 2-\lambda & 1 \\ 2 & 2 & 3-\lambda \end{vmatrix} = 0 \quad \Rightarrow \quad \lambda^3 - 6\lambda^2 - 11\lambda - 6 = 0 \quad \Rightarrow \quad (\lambda-1)(\lambda-2)(\lambda-3) = 0$$

Eigen values of A are 1, 2, 3.

Let $\lambda^4 = \left(\lambda^3 - 6\lambda^2 - 11\lambda - 6\right)Q(\lambda) + \left(a\lambda^2 + b\lambda + c\right) = 0$ ...(1)

(where $Q(\lambda)$ is quotient)

Put $\lambda = 1$ in (1), $(1)^4 = a + b + c \quad \Rightarrow \quad a + b + c = 1$ ...(2)

Put $\lambda = 2$ in (1), $(2)^4 = 4a + 2b + c \quad \Rightarrow \quad 4a + 2b + c = 16$ ... (3)

Put $\lambda = 3$ in (1), $(3)^4 = 9a + 3b + c \quad \Rightarrow \quad 9a + 3b + c = 81$ ... (4)

Solving (2), (3) and (4), we get

$a = 25, \quad b = -60, \quad c = 36$

Replacing λ by matrix A in (1), we get

$$A^4 = \left(A^3 - 6A^2 + 11A - 6\right)Q(A) + \left(aA^2 + bA + c\right)$$

$$= O + aA^2 + bA + cI = 25A^2 - 60A + 36I$$

$$= 25\begin{bmatrix} 1 & 0 & -1 \\ 1 & 2 & 1 \\ 2 & 2 & 3 \end{bmatrix}\begin{bmatrix} 1 & 0 & -1 \\ 1 & 2 & 1 \\ 2 & 2 & 3 \end{bmatrix} + (-60)\begin{bmatrix} 1 & 0 & -1 \\ 1 & 2 & 1 \\ 2 & 2 & 3 \end{bmatrix} + 36\begin{bmatrix} 1 & 0 & 0 \\ 0 & 1 & 0 \\ 0 & 0 & 1 \end{bmatrix}$$

$$= \begin{bmatrix} -25 & -50 & -100 \\ 125 & 150 & 100 \\ 250 & 250 & 225 \end{bmatrix} + \begin{bmatrix} -60 & 0 & 60 \\ -60 & -120 & -60 \\ -120 & -120 & -180 \end{bmatrix} + \begin{bmatrix} 36 & 0 & 0 \\ 0 & 36 & 0 \\ 0 & 0 & 36 \end{bmatrix}$$

$$= \begin{bmatrix} -25-60+36 & -50+0+0 & -100+60+0 \\ 125-60+0 & 150-120+36 & 100-60+0 \\ 250-120+0 & 250-120+0 & 225-180+36 \end{bmatrix} = \begin{bmatrix} -49 & -50 & -40 \\ 65 & 66 & 40 \\ 130 & 130 & 81 \end{bmatrix}$$

(It is also solved by diagonalization method on page 496 Example 38.)

EXERCISE 21.2

1. Find the characteristic polynomial of the matrix

$$A = \begin{bmatrix} 3 & 1 & 1 \\ -1 & 5 & -1 \\ 1 & -1 & 3 \end{bmatrix}$$

Verify Cayley-Hamilton Theorem for this matrix. Hence find A^{-1}. **Ans.** $A^{-1} = \frac{1}{20}\begin{bmatrix} 7 & -2 & -3 \\ 1 & 4 & 1 \\ -2 & 2 & 8 \end{bmatrix}$

2. Use Cayley-Hamilton Theorem to find the inverse of the matrix

$$\begin{bmatrix} \cos\theta & \sin\theta \\ -\sin\theta & \cos\theta \end{bmatrix}$$

Ans. $\begin{bmatrix} \cos\theta & -\sin\theta \\ \sin\theta & \cos\theta \end{bmatrix}$

3. Using Cayley-Hamilton Theorem, find A^{-1}, given that

$$A = \begin{bmatrix} 2 & -1 & 3 \\ 1 & 0 & 2 \\ 4 & -2 & 1 \end{bmatrix}$$

Ans. $-\frac{1}{5}\begin{bmatrix} 4 & -5 & -2 \\ 7 & -10 & -1 \\ -2 & 0 & 1 \end{bmatrix}$

4. Find the characteristic equation of the matrix

$$A = \begin{bmatrix} 1 & 3 & 7 \\ 4 & 2 & 3 \\ 1 & 2 & 1 \end{bmatrix}$$

(R.G.P.V., Bhopal, Summer 2004)

and show that the equation is also satisfied by A. **Ans.** $\lambda^3 - 4\lambda^2 - 20\lambda - 35 = 0$

5. Using, Cayley-Hamilton Theorem obtain the inverse of the matrix

$$\begin{bmatrix} 1 & 1 & 3 \\ 1 & 3 & -3 \\ -2 & -4 & -4 \end{bmatrix}$$ (*R.G.P.V. Bhopal, I Sem., 2003*) **Ans.** $\frac{1}{8}\begin{bmatrix} 24 & 8 & 12 \\ -10 & -2 & -6 \\ -2 & -2 & -2 \end{bmatrix}$

Show that the matrix $A = \begin{bmatrix} 1 & -2 & 2 \\ 1 & 2 & 3 \\ 0 & -1 & 2 \end{bmatrix}$

satisfies its characteristic equation. Hence find A^{-1}. **Ans.** $\frac{1}{9}\begin{bmatrix} 7 & 2 & -10 \\ -2 & 2 & -1 \\ -1 & 1 & 4 \end{bmatrix}$

7. Verify Cayley-Hamilton Theorem for the matrix

$$A = \begin{bmatrix} 1 & 1 & 2 \\ 3 & 1 & 1 \\ 2 & 3 & 1 \end{bmatrix}$$

Hence evaluate A^{-1}. **Ans.** $\frac{1}{11}\begin{bmatrix} -2 & 5 & -1 \\ -1 & -3 & 5 \\ 7 & -1 & -2 \end{bmatrix}$

8. Verify Cayley-Hamilton theorem for the matrix

$$A = \begin{bmatrix} 1 & -2 & 3 \\ 2 & 4 & -2 \\ -1 & 1 & 2 \end{bmatrix}$$

9. Find adj. A by using Cayley-Hamilton thmeorem where A is given by

$$A = \begin{bmatrix} 1 & 2 & 1 \\ 0 & 1 & -1 \\ 3 & -1 & 1 \end{bmatrix}$$ (*R.G.P.V., Bhopal, April 2010*) **Ans.** $\begin{bmatrix} 0 & -3 & -3 \\ -3 & -2 & 1 \\ -3 & 7 & 1 \end{bmatrix}$

10. If a matrix $A = \begin{bmatrix} 1 & 0 & 0 \\ 0 & -1 & 0 \\ 1 & 0 & 1 \end{bmatrix}$, find the matrix A^{32}, using Cayley Hamilton Theorem. **Ans.** $\begin{vmatrix} 1 & 0 & 0 \\ 0 & 1 & 0 \\ 32 & 0 & 1 \end{vmatrix}$

21.5 CHARACTERISTIC VECTORS OR EIGEN VECTORS

As we have discussed in Art 21.2,

A column vector X is transformed into column vector Y by means of a square matrix A. Now we want to multiply the column vector X by a scalar quantity λ so that we can find the same transformed column vector Y.

i.e., $AX = \lambda X$

X is known as eigen vector.

Example 19. *Show that the vector (1, 1, 2) is an eigen vector of the matrix*

$$A = \begin{bmatrix} 3 & 1 & -1 \\ 2 & 2 & -1 \\ 2 & 2 & 0 \end{bmatrix}$$ *corresponding to the eigen value 2.*

Solution. Let $X = (1, 1, 2)$.

Now, $$AX = \begin{bmatrix} 3 & 1 & -1 \\ 2 & 2 & -1 \\ 2 & 2 & 0 \end{bmatrix}\begin{bmatrix} 1 \\ 1 \\ 2 \end{bmatrix} = \begin{bmatrix} 3+1-2 \\ 2+2-2 \\ 2+2+0 \end{bmatrix} = \begin{bmatrix} 2 \\ 2 \\ 4 \end{bmatrix} = 2\begin{bmatrix} 1 \\ 1 \\ 2 \end{bmatrix} = 2X$$

Corresponding to each characteristic root λ, we have a corresponding non-zero vector X which satisfies the equation $[A - \lambda I]X = 0$. The non-zero vector X is called characteristic vector or Eigen vector.

21.6 PROPERTIES OF EIGEN VECTORS

1. The eigen vector X of a matrix A is not unique.
2. If $\lambda_1, \lambda_2, \ldots, \lambda_n$ be distinct eigen values of an $n \times n$ matrix then corresponding eigen vectors $X_1, X_2, \ldots, X_n$ form a linearly independent set.
3. If two or more eigen values are equal it may or may not be possible to get linearly independent eigen vectors corresponding to the equal roots.
4. Two eigen vectors X_1 and X_2 are called orthogonal vectors if $X_1' X_2 = 0$.
5. Eigen vectors of a symmetric matrix corresponding to different eigen values are orthogonal.

Normalised form of vectors. To find normalised form of $\begin{bmatrix} a \\ b \\ c \end{bmatrix}$, we divide each element by $\sqrt{a^2+b^2+c^2}$.

For example, normalised form of $\begin{bmatrix} 1 \\ 2 \\ 2 \end{bmatrix}$ is $\begin{bmatrix} 1/3 \\ 2/3 \\ 2/3 \end{bmatrix}$ $\qquad \left[\sqrt{1^2+2^2+2^2} = 3\right]$

21.7 ORTHOGONAL VECTORS

Two vectors X and Y are said to be orthogonal if $X_1^T X_2 = X_2^T X_1 = 0$.

Example 20. *Determine whether the eigen vectors of the matrix*

$$A = \begin{bmatrix} 1 & 0 & -1 \\ 1 & 2 & 1 \\ 2 & 2 & 3 \end{bmatrix}$$

are orthogonal.

Solution. Characteristic equation is

$$\begin{vmatrix} 1-\lambda & 0 & -1 \\ 1 & 2-\lambda & 1 \\ 2 & 2 & 3-\lambda \end{vmatrix} = 0$$

$$\Rightarrow \quad (1-\lambda)[(2-\lambda)(3-\lambda)-2]-0-1[2-2(2-\lambda)]=0$$

$$\Rightarrow \quad (1-\lambda)(6-5\lambda+\lambda^2-2)-(2-4+2\lambda)=0 \Rightarrow (\lambda-1)(\lambda^2-5\lambda+4)+2(\lambda-1)=0$$

$$\Rightarrow \quad (1-\lambda)(\lambda^2-5\lambda+4)-2(\lambda-1)=0 \qquad \Rightarrow (\lambda-1)[\lambda^2-5\lambda+4+2]=0$$

$$\Rightarrow (\lambda-1)(\lambda^2-5\lambda+6)=0 \qquad \Rightarrow (\lambda-1)(\lambda-2)(\lambda-3)=0$$

So, $\lambda = 1, 2, 3$ are three distinct eigen values of A.

For $\lambda = 1$

$$\begin{bmatrix} 1-\lambda & 0 & -1 \\ 1 & 2-\lambda & 1 \\ 2 & 2 & 3-\lambda \end{bmatrix}\begin{bmatrix} x_1 \\ x_2 \\ x_3 \end{bmatrix} = \begin{bmatrix} 0 \\ 0 \\ 0 \end{bmatrix}$$

$$\Rightarrow \begin{bmatrix} 1-1 & 0 & -1 \\ 1 & 2-1 & 1 \\ 2 & 2 & 3-1 \end{bmatrix}\begin{bmatrix} x_1 \\ x_2 \\ x_3 \end{bmatrix} = \begin{bmatrix} 0 \\ 0 \\ 0 \end{bmatrix} \Rightarrow \begin{bmatrix} 0 & 0 & -1 \\ 1 & 1 & 1 \\ 2 & 2 & 2 \end{bmatrix}\begin{bmatrix} x_1 \\ x_2 \\ x_3 \end{bmatrix} = \begin{bmatrix} 0 \\ 0 \\ 0 \end{bmatrix}$$

$$\Rightarrow \qquad x_3 = 0$$

$$x_1 + x_2 + x_3 = 0 \Rightarrow x_2 = -x_3 - x_1$$

Let $x_1 = k$ then $x_2 = 0 - k = -k$

$$X_1 = \begin{bmatrix} k \\ -k \\ 0 \end{bmatrix} \Rightarrow X_1 = k\begin{bmatrix} 1 \\ -1 \\ 0 \end{bmatrix}$$

For λ = 2

$$\begin{bmatrix} 1-2 & 0 & -1 \\ 1 & 2-2 & 1 \\ 2 & 2 & 3-2 \end{bmatrix}\begin{bmatrix} x_1 \\ x_2 \\ x_3 \end{bmatrix} = \begin{bmatrix} 0 \\ 0 \\ 0 \end{bmatrix} \Rightarrow \begin{bmatrix} -1 & 0 & -1 \\ 1 & 0 & 1 \\ 2 & 2 & 1 \end{bmatrix}\begin{bmatrix} x_1 \\ x_2 \\ x_3 \end{bmatrix} = \begin{bmatrix} 0 \\ 0 \\ 0 \end{bmatrix}$$

$$\Rightarrow \quad \begin{aligned} x_1 + 0x_2 + x_3 &= 0 \\ 2x_1 + 2x_2 + x_3 &= 0 \end{aligned}$$

$$\frac{x_1}{0-2} = \frac{x_2}{2-1} = \frac{x_3}{2-0} = k$$

$$\Rightarrow \quad x_1 = 2k, \quad x_2 = -k, \quad x_3 = -2k$$

$$X_2 = k\begin{bmatrix} 2 \\ -1 \\ -2 \end{bmatrix}$$

For λ = 3

$$\begin{bmatrix} 1-3 & 0 & -1 \\ 1 & 2-3 & 1 \\ 2 & 2 & 3-3 \end{bmatrix}\begin{bmatrix} x_1 \\ x_2 \\ x_3 \end{bmatrix} = \begin{bmatrix} 0 \\ 0 \\ 0 \end{bmatrix} \Rightarrow \begin{bmatrix} -2 & 0 & -1 \\ 1 & -1 & 1 \\ 2 & 2 & 0 \end{bmatrix}\begin{bmatrix} x_1 \\ x_2 \\ x_3 \end{bmatrix} = \begin{bmatrix} 0 \\ 0 \\ 0 \end{bmatrix}$$

$$\left.\begin{aligned} -2x_1 + 0x_2 - x_3 &= 0 \\ x_1 - x_2 + x_3 &= 0 \end{aligned}\right\} \Rightarrow \frac{x_1}{0-1} = \frac{x_2}{-1+2} = \frac{x_3}{2-0} = k$$

$$\Rightarrow \quad x_1 = k, \quad x_2 = -k, \quad x_3 = -2k$$

$$X_3 = \begin{bmatrix} x_1 \\ x_2 \\ x_3 \end{bmatrix} = \begin{bmatrix} k \\ -k \\ -2k \end{bmatrix} = k\begin{bmatrix} 1 \\ -1 \\ -2 \end{bmatrix}$$

$$X'_1X_2 = [1, -1, 0]\begin{bmatrix} 2 \\ -1 \\ -2 \end{bmatrix} = 3, \quad X'_2X_3 = [2, -1, -2]\begin{bmatrix} 1 \\ -1 \\ -2 \end{bmatrix} = 7, \quad X'_3X_1 = [1, -1, -2]\begin{bmatrix} 1 \\ -1 \\ 0 \end{bmatrix} = 2$$

Since $X_1^T X_2 = 3 \neq 0, \quad X_2^T X_3 = 7 \neq 0, \quad X_3^T X_1 = 2 \neq 0$

Thus, there are three distinct eigen vectors. So X_1, X_2, X_3 are not orthogonal eigen vectors.

21.8 NON-SYMMETRIC MATRICES WITH NON-REPEATED EIGEN VALUES

Example 21. *Show that if $\lambda_1, \lambda_2, \ldots\ldots, \lambda_n$ be the eigen values of the matrix A, then A^n has the eigen values $\lambda^n{}_1, \lambda^n{}_2, \ldots\ldots\lambda^n{}_n$.*

Solution. Let λ be an eigen value of the matrix A.

Therefore, $$AX = \lambda X \qquad \ldots(1)$$

By premultiplying both sides of (1) by A^{n-1}, we get

$$A^{n-1}(AX) = A^{n-1}(\lambda X) \quad \Rightarrow \quad A^n X = \lambda(A^{n-1}X) \qquad \ldots(2)$$

But $\quad A^2X = A(AX) = A(\lambda X)$

$$= \lambda(AX) = \lambda(\lambda X) = \lambda^2 X \qquad \text{(From (1) } AX = \lambda X\text{)}$$

$$A^3X = A(A^2X) = \lambda(\lambda^2 X) = \lambda^3 X$$

Similarly, $A^4 X = \lambda^4 X$

....

....

$$A^n X = \lambda^n X$$

$\Rightarrow$ λ^n is an eigen value of A^n.

Hence, if $\lambda_1, \lambda_2, ...\lambda_n$ be the eigen values of A, then $\lambda_1^n, \lambda_2^n, \lambda_3^n, ...\lambda_n^n$ be the eigen values of A^n.

Example 22. *If λ be an eigen value of matrix A (non-zero matrix), show that λ^{-1} is an eigen value of A^{-1}.*

Solution. We have, λ is an eigen value of matrix A.

$$AX = \lambda X \qquad ... (1)$$

where X is eigen vector

Premultiplying both sides of (1) by A^{-1}, we get

$$A^{-1}(AX) = A^{-1}(\lambda X) \quad \Rightarrow \quad (A^{-1}A)X = \lambda(A^{-1}X)$$

$$\Rightarrow \quad IX = \lambda(A^{-1}X) \quad \Rightarrow \quad X = \lambda(A^{-1}X)$$

$$\Rightarrow \quad \frac{1}{\lambda}X = A^{-1}X \quad \Rightarrow \quad A^{-1}X = \lambda^{-1}X$$

Hence, λ^{-1} is an eigen value of A^{-1}. **Proved.**

Example 23. *Find the eigen value and corresponding eigen vectors of the matrix*

$$A = \begin{pmatrix} -5 & 2 \\ 2 & -2 \end{pmatrix}$$

(U.P.I Sem., Dec 2008)

Solution. $|A - \lambda I| = 0$

$$\Rightarrow \begin{vmatrix} -5-\lambda & 2 \\ 2 & -2-\lambda \end{vmatrix} = 0 \Rightarrow (-5-\lambda)(-2-\lambda) - 4 = 0$$

$$\Rightarrow \lambda^2 + 7\lambda + 10 - 4 = 0 \Rightarrow \lambda^2 + 7\lambda + 6 = 0$$

$$(\lambda + 1)(\lambda + 6) = 0 \Rightarrow \lambda = -1, -6$$

The eigen values of the given matrix are – 1 and – 6.

(*i*) When $\lambda = -1$, the corresponding eigen vectors are given by

$$\begin{bmatrix} -5+1 & 2 \\ 2 & -2+1 \end{bmatrix}\begin{bmatrix} x_1 \\ x_2 \end{bmatrix} = \begin{bmatrix} 0 \\ 0 \end{bmatrix} \Rightarrow \begin{bmatrix} -4 & 2 \\ 2 & -1 \end{bmatrix}\begin{bmatrix} x_1 \\ x_2 \end{bmatrix} = \begin{bmatrix} 0 \\ 0 \end{bmatrix}$$

$$\Rightarrow 2x_1 - x_2 = 0 \Rightarrow x_1 = \frac{1}{2}x_2$$

Let $x_1 = k$, then $x_2 = 2k$, Hence, eigen vector $X_1 = \begin{bmatrix} k \\ 2k \end{bmatrix}$

(*ii*) When $\lambda = -6$, the corresponding eigen vectors are given by

$$\begin{bmatrix} -5+6 & 2 \\ 2 & -2+6 \end{bmatrix}\begin{bmatrix} x_1 \\ x_2 \end{bmatrix} = \begin{bmatrix} 0 \\ 0 \end{bmatrix} \Rightarrow \begin{bmatrix} 1 & 2 \\ 2 & 4 \end{bmatrix}\begin{bmatrix} x_1 \\ x_2 \end{bmatrix} = \begin{bmatrix} 0 \\ 0 \end{bmatrix}$$

$$\Rightarrow x_1 + 2x_2 = 0 \qquad \Rightarrow x_1 = -2x_2$$

Let $x_1 = k_1$, then $x_2 = -\frac{1}{2}k_1$

Hence eigen vector $X_2 = \begin{bmatrix} k_1 \\ -\frac{k_1}{2} \end{bmatrix}$ or $\begin{bmatrix} 2k_1 \\ -k_1 \end{bmatrix}$

Hence eigen vectors are $\begin{bmatrix} k \\ 2k \end{bmatrix}$ and $\begin{bmatrix} 2k_1 \\ -k_1 \end{bmatrix}$ **Ans.**

Example 24. *Find the eigen values and eigen vectors of matrix* $A = \begin{bmatrix} 3 & 1 & 4 \\ 0 & 2 & 6 \\ 0 & 0 & 5 \end{bmatrix}$

(AMIETE, June 2010, 2009)

Solution. $|A - \lambda I| = \begin{vmatrix} 3-\lambda & 1 & 4 \\ 0 & 2-\lambda & 6 \\ 0 & 0 & 5-\lambda \end{vmatrix} = (3-\lambda)(2-\lambda)(5-\lambda)$

Hence the characteristic equation of matrix A is given by

$$|A - \lambda I| = 0 \quad \Rightarrow \quad (3-\lambda)(2-\lambda)(5-\lambda) = 0$$

$$\therefore \quad \lambda = 2, 3, 5.$$

Thus the eigen values of matrix A are 2, 3, 5.

The eigen vectors of the matrix A corresponding to the eigen value λ is given by the non-zero solution of the equation $(A - \lambda I)X = 0$

or $$\begin{bmatrix} 3-\lambda & 1 & 4 \\ 0 & 2-\lambda & 6 \\ 0 & 0 & 5-\lambda \end{bmatrix} \begin{bmatrix} x_1 \\ x_2 \\ x_3 \end{bmatrix} = \begin{bmatrix} 0 \\ 0 \\ 0 \end{bmatrix} \quad \ldots (1)$$

When $\lambda = 2$, the corresponding eigen vector is given by

$$\begin{bmatrix} 3-2 & 1 & 4 \\ 0 & 2-2 & 6 \\ 0 & 0 & 5-2 \end{bmatrix} \begin{bmatrix} x_1 \\ x_2 \\ x_3 \end{bmatrix} = \begin{bmatrix} 0 \\ 0 \\ 0 \end{bmatrix}$$

$$\Rightarrow \quad \begin{bmatrix} 1 & 1 & 4 \\ 0 & 0 & 6 \\ 0 & 0 & 3 \end{bmatrix} \begin{bmatrix} x_1 \\ x_2 \\ x_3 \end{bmatrix} = \begin{bmatrix} 0 \\ 0 \\ 0 \end{bmatrix}$$

$$\Rightarrow \quad \begin{aligned} x_1 + x_2 + 4x_3 &= 0 \\ 0x_1 + 0x_2 + 6x_3 &= 0 \end{aligned}$$

$$\frac{x_1}{6-0} = \frac{x_2}{0-6} = \frac{x_3}{0-0} = k \quad \Rightarrow \quad \frac{x_1}{1} = \frac{x_2}{-1} = \frac{x_3}{0} = k \quad \Rightarrow x_1 = k,\ x_2 = -k,\ x_3 = 0$$

Hence $X_1 = \begin{bmatrix} k \\ -k \\ 0 \end{bmatrix} = k \begin{bmatrix} 1 \\ -1 \\ 0 \end{bmatrix}$ can be taken as an eigen vector of A corresponding to the eigen value $\lambda = 2$

When $\lambda = 3$, substituting in (1), the corresponding eigen vector is given by

$$\begin{bmatrix} 3-3 & 1 & 4 \\ 0 & 2-3 & 6 \\ 0 & 0 & 5-3 \end{bmatrix} \begin{bmatrix} x_1 \\ x_2 \\ x_3 \end{bmatrix} = \begin{bmatrix} 0 \\ 0 \\ 0 \end{bmatrix} \quad \Rightarrow \quad \begin{bmatrix} 0 & 1 & 4 \\ 0 & -1 & 6 \\ 0 & 0 & 2 \end{bmatrix} \begin{bmatrix} x_1 \\ x_2 \\ x_3 \end{bmatrix} = \begin{bmatrix} 0 \\ 0 \\ 0 \end{bmatrix}$$

$$\begin{aligned} 0x_1 + x_2 + 4x_3 &= 0 \\ 0x_1 - x_2 + 6x_3 &= 0 \end{aligned}$$

$$\frac{x_1}{6+4} = \frac{x_2}{0-0} = \frac{x_3}{0-0} \quad \Rightarrow \quad \frac{x_1}{10} = \frac{x_2}{0} = \frac{x_3}{0} = \frac{k}{10}$$

$$x_1 = k,\ x_2 = 0,\ x_3 = 0$$

Hence, $X_2 = \begin{bmatrix} k \\ 0 \\ 0 \end{bmatrix} = k\begin{bmatrix} 1 \\ 0 \\ 0 \end{bmatrix}$ can be taken as an eigen vector of A corresponding to the eigen value $\lambda = 3$.

When $\lambda = 5$.

Again, when $\lambda = 5$, substituting in (1), the corresponding eigen vector is given by

$$\begin{bmatrix} 3-5 & 1 & 4 \\ 0 & 2-5 & 6 \\ 0 & 0 & 5-5 \end{bmatrix}\begin{bmatrix} x_1 \\ x_2 \\ x_3 \end{bmatrix} = \begin{bmatrix} 0 \\ 0 \\ 0 \end{bmatrix} \Rightarrow \begin{bmatrix} -2 & 1 & 4 \\ 0 & -3 & 6 \\ 0 & 0 & 0 \end{bmatrix}\begin{bmatrix} x_1 \\ x_2 \\ x_3 \end{bmatrix} = \begin{bmatrix} 0 \\ 0 \\ 0 \end{bmatrix}$$

$$-2x_1 + x_2 + 4x_3 = 0$$
$$-3x_2 + 6x_3 = 0$$

By cross-multiplication method, we have

$$\frac{x_1}{6+12} = \frac{x_2}{0+12} = \frac{x_3}{6-0} \Rightarrow \frac{x_1}{18} = \frac{x_2}{12} = \frac{x_3}{6} \Rightarrow \frac{x_1}{3} = \frac{x_2}{2} = \frac{x_3}{1} = k$$

$x_1 = 3k, \quad x_2 = 2k, \quad x_3 = k$

Hence, $X_3 = \begin{bmatrix} 3k \\ 2k \\ k \end{bmatrix} = k\begin{bmatrix} 3 \\ 2 \\ 1 \end{bmatrix}$ can be taken as an eigen vector of A corresponding to the eigen value $\lambda = 5$. **Ans.**

EXERCISE 21.3

Non-symmetric matrix with different eigen values:

Find the eigen values and the corresponding eigen vectors for the following matrices:

1. $\begin{bmatrix} 4 & 2 & -2 \\ -5 & 3 & 2 \\ -2 & 4 & 1 \end{bmatrix}$ **Ans.** $1, 2, 5; \begin{bmatrix} 2 \\ 1 \\ 4 \end{bmatrix}, \begin{bmatrix} 1 \\ 1 \\ 2 \end{bmatrix}, \begin{bmatrix} 0 \\ 1 \\ 0 \end{bmatrix}$

2. $\begin{bmatrix} 2 & -2 & 3 \\ 1 & 1 & 1 \\ 1 & 3 & -1 \end{bmatrix}$ **Ans.** $-2, 1, 3; \begin{bmatrix} 11 \\ 1 \\ 14 \end{bmatrix}, \begin{bmatrix} -1 \\ 1 \\ 1 \end{bmatrix}, \begin{bmatrix} 1 \\ 1 \\ 1 \end{bmatrix}$

3. $\begin{bmatrix} -9 & 2 & 6 \\ 5 & 0 & -3 \\ -16 & 4 & 11 \end{bmatrix}$ **Ans.** $-1, 1, 2; \begin{bmatrix} 2 \\ -1 \\ 3 \end{bmatrix}, \begin{bmatrix} 1 \\ -1 \\ 2 \end{bmatrix}, \begin{bmatrix} 2 \\ -1 \\ 4 \end{bmatrix}$

4. $\begin{bmatrix} 4 & 6 & 6 \\ 1 & 3 & 2 \\ -1 & -4 & -3 \end{bmatrix}$ **Ans.** $-1, 1, 4; \begin{bmatrix} -6 \\ -2 \\ 7 \end{bmatrix}, \begin{bmatrix} 0 \\ 1 \\ -1 \end{bmatrix}, \begin{bmatrix} 3 \\ 1 \\ -1 \end{bmatrix}$

21.9 NON-SYMMETRIC MATRIX WITH REPEATED EIGEN VALUES

Example 25. *Find the eigen values and eigen vectors of the matrix:*

$$\begin{bmatrix} 2 & 1 & 1 \\ 1 & 2 & 1 \\ 0 & 0 & 1 \end{bmatrix}$$

(*R.G.P.V. Bhopal, June 2004*)

Solution. We have, $A = \begin{bmatrix} 2 & 1 & 1 \\ 1 & 2 & 1 \\ 0 & 0 & 1 \end{bmatrix}$.

Characteristic equation of A is $|A-\lambda I|=0$

$$\Rightarrow \begin{vmatrix} 2-\lambda & 1 & 1 \\ 1 & 2-\lambda & 1 \\ 0 & 0 & 1-\lambda \end{vmatrix}=0$$

On expanding the determinant by the third row, we get

$$\Rightarrow (1-\lambda)\{(2-\lambda)(2-\lambda)-1\}=0 \quad \Rightarrow \quad (1-\lambda)\{(2-\lambda)^2-1\}=0$$

$$\Rightarrow (1-\lambda)(2-\lambda+1)(2-\lambda-1)=0 \quad \Rightarrow \quad (1-\lambda)(3-\lambda)(1-\lambda)=0$$

$$\Rightarrow \lambda=1,1,3$$

when $\lambda = 1$

$$\begin{bmatrix} 2-1 & 1 & 1 \\ 1 & 2-1 & 1 \\ 0 & 0 & 1-1 \end{bmatrix}\begin{bmatrix} x \\ y \\ z \end{bmatrix}=\begin{bmatrix} 0 \\ 0 \\ 0 \end{bmatrix} \quad \Rightarrow \quad \begin{bmatrix} 1 & 1 & 1 \\ 1 & 1 & 1 \\ 0 & 0 & 0 \end{bmatrix}\begin{bmatrix} x \\ y \\ z \end{bmatrix}=\begin{bmatrix} 0 \\ 0 \\ 0 \end{bmatrix}.$$

$$\begin{bmatrix} 1 & 1 & 1 \\ 0 & 0 & 0 \\ 0 & 0 & 0 \end{bmatrix}\begin{bmatrix} x \\ y \\ z \end{bmatrix}=\begin{bmatrix} 0 \\ 0 \\ 0 \end{bmatrix} R_2 \to R_2-R_1 \quad \Rightarrow \quad x+y+z=0$$

Let $x = k_1$ and $y = k_2$

$$k_1 + k_2 + z = 0 \quad \Rightarrow \quad z = -(k_1 + k_2)$$

$$X_1 = \begin{bmatrix} k_1 \\ k_2 \\ -(k_1+k_2) \end{bmatrix} = k\begin{bmatrix} 1 \\ 1 \\ -2 \end{bmatrix} \qquad [\text{If } k_1 = k_2 = k]$$

$$\text{Again } \lambda=1, \quad X_2 = \begin{bmatrix} 1 \\ 0 \\ -1 \end{bmatrix} \qquad [\text{Again if } k_1 = 1,\ k_2 = 0,\ -(k_1 + k_2) = -1]$$

when $\lambda = 3$

$$\begin{bmatrix} 2-3 & 1 & 1 \\ 1 & 2-3 & 1 \\ 0 & 0 & 1-3 \end{bmatrix}\begin{bmatrix} x \\ y \\ z \end{bmatrix}=\begin{bmatrix} 0 \\ 0 \\ 0 \end{bmatrix} \quad \Rightarrow \quad \begin{bmatrix} -1 & 1 & 1 \\ 1 & -1 & 1 \\ 0 & 0 & -2 \end{bmatrix}\begin{bmatrix} x \\ y \\ z \end{bmatrix}=\begin{bmatrix} 0 \\ 0 \\ 0 \end{bmatrix}$$

$$\begin{bmatrix} -1 & 1 & 1 \\ 0 & 0 & 2 \\ 0 & 0 & -2 \end{bmatrix}\begin{bmatrix} x \\ y \\ z \end{bmatrix}=\begin{bmatrix} 0 \\ 0 \\ 0 \end{bmatrix} \quad R_2 \to R_2 + R_1$$

$$-x+y+z=0$$

$$2z = 0 \Rightarrow z = 0$$

$$-x+y+0=0 \Rightarrow x=y=k \text{ (say)}$$

$$X_3 = \begin{bmatrix} k \\ k \\ 0 \end{bmatrix} = k\begin{bmatrix} 1 \\ 1 \\ 0 \end{bmatrix}$$

Ans.

Example 26. *Find all the Eigen values and Eigen vectors of the matrix*

$$A = \begin{bmatrix} -2 & 2 & -3 \\ 2 & 1 & -6 \\ -1 & -2 & 0 \end{bmatrix}$$

(AMIETE, Dec. 2009)

Solution. Characteristic equation of A is

$$\begin{vmatrix} -2-\lambda & 2 & -3 \\ 2 & 1-\lambda & -6 \\ -1 & -2 & 0-\lambda \end{vmatrix} = 0$$

$\Rightarrow \quad (-2-\lambda)[-\lambda+\lambda^2-12]-2(-2\lambda-6)-3(-4+1-\lambda)=0$

$\lambda^3+\lambda^2-21\lambda-45=0$ .... (1)

By trial: If $\lambda=-3$, then $-27+9+63-45=0$, so $(\lambda+3)$ is one factor of (1).

The remaining factors are obtained on dividing (1) by $\lambda+3$.

−3	1	1	−21	−45
		−3	6	45
	1	−2	−15	0

$\lambda^2-2\lambda-15=0 \quad \Rightarrow (\lambda-5)(\lambda+3)=0$

$\Rightarrow (\lambda+3)(\lambda+3)(\lambda-5)=0 \Rightarrow \lambda=5,-3,-3$

To find the eigen vectors for corresponding eigen values, we will consider the matrix equation

$$(A-\lambda I)X=0 \quad i.e., \quad \begin{bmatrix} -2-\lambda & 2 & -3 \\ 2 & 1-\lambda & -6 \\ -1 & -2 & 0-\lambda \end{bmatrix}\begin{bmatrix} x \\ y \\ z \end{bmatrix} = \begin{bmatrix} 0 \\ 0 \\ 0 \end{bmatrix} \quad \text{... (2)}$$

On putting $\lambda=5$ in eq. (2), it becomes $\begin{bmatrix} -7 & 2 & -3 \\ 2 & -4 & -6 \\ -1 & -2 & -5 \end{bmatrix}\begin{bmatrix} x \\ y \\ z \end{bmatrix} = \begin{bmatrix} 0 \\ 0 \\ 0 \end{bmatrix}$

We have $\quad -7x+2y-3z=0,$
$\quad 2x-4y-6z=0$

$$\frac{x}{-12-12}=\frac{y}{-6-42}=\frac{z}{28-4} \quad \text{or} \quad \frac{x}{-24}=\frac{y}{-48}=\frac{z}{24} \quad \text{or} \quad \frac{x}{1}=\frac{y}{2}=\frac{z}{-1}=k$$

$x=k, \quad y=2k, \quad z=-k$

Hence, the eigen vector $X_1=\begin{bmatrix} k \\ 2k \\ -k \end{bmatrix}=k\begin{bmatrix} 1 \\ 2 \\ -1 \end{bmatrix}$

Put $\lambda=-3$ in eq. (2), it becomes $\begin{bmatrix} 1 & 2 & -3 \\ 2 & 4 & -6 \\ -1 & -2 & 3 \end{bmatrix}\begin{bmatrix} x \\ y \\ z \end{bmatrix} = \begin{bmatrix} 0 \\ 0 \\ 0 \end{bmatrix}$

We have $\quad x+2y-3z=0,$
$\quad 2x+4y-6z=0,$
$\quad -x-2y+3z=0$

Here first, second and third equations are the same.

Let $x=k_1, y=k_2$ then $z=\frac{1}{3}(k_1+2k_2)$

Hence, the eigen vector is $\begin{bmatrix} k_1 \\ k_2 \\ \frac{1}{3}(k_1+2k_2) \end{bmatrix}$

Let $k_1=0, k_2=3$, Hence $X_2=\begin{bmatrix} 0 \\ 3 \\ 2 \end{bmatrix}$

Since the matrix is non-symmetric, the corresponding eigen vectors X_2 and X_3 must be linearly independent. This can be done by choosing

$$k_1 = 3,\ k_2 = 0, \text{ and Hence } X_3 = \begin{bmatrix} 3 \\ 0 \\ 1 \end{bmatrix}$$

$$\text{Hence, } X_1 = \begin{bmatrix} 1 \\ 2 \\ -1 \end{bmatrix}, \quad X_2 = \begin{bmatrix} 0 \\ 3 \\ 2 \end{bmatrix}, \quad X_3 = \begin{bmatrix} 3 \\ 0 \\ 1 \end{bmatrix}.$$

Ans.

EXERCISE 21.4

Non-symmetric matrices with repeated eigen values
Find the eigen values and eigen vectors of the following matrices:

1. $\begin{bmatrix} 2 & -2 & 2 \\ 1 & 1 & 1 \\ 1 & 3 & -1 \end{bmatrix}$ **Ans.** $-2, 2, 2;\ \begin{bmatrix} -4 \\ -1 \\ 7 \end{bmatrix}, \begin{bmatrix} 0 \\ 1 \\ 1 \end{bmatrix}$

2. $\begin{bmatrix} 2 & 2 & 1 \\ 1 & 3 & 1 \\ 1 & 2 & 2 \end{bmatrix}$ **Ans.** $1, 1, 5;\ \begin{bmatrix} 1 \\ 2 \\ -5 \end{bmatrix}, \begin{bmatrix} 1 \\ 1 \\ 1 \end{bmatrix}$

3. $\begin{bmatrix} 2 & 1 & 1 \\ 2 & 3 & 2 \\ 3 & 3 & 4 \end{bmatrix}$ **Ans.** $1, 1, 7;\ \begin{bmatrix} 0 \\ 1 \\ -1 \end{bmatrix}, \begin{bmatrix} 1 \\ 0 \\ -1 \end{bmatrix}, \begin{bmatrix} 1 \\ 2 \\ 3 \end{bmatrix}$

4. $\begin{bmatrix} -9 & 4 & 4 \\ -8 & 3 & 4 \\ -16 & 8 & 7 \end{bmatrix}$ **Ans.** $-1, -1, 3;\ \begin{bmatrix} 0 \\ 1 \\ -1 \end{bmatrix}, \begin{bmatrix} 1 \\ 1 \\ 1 \end{bmatrix}, \begin{bmatrix} 1 \\ 1 \\ 2 \end{bmatrix}$

5. $\begin{bmatrix} 1 & 1 & 0 \\ 0 & 1 & 0 \\ 0 & 0 & 1 \end{bmatrix}$ (*AMIETE, Dec. 2010*) **Ans.** $1, 1, 1,\ \begin{bmatrix} 1 \\ 0 \\ 1 \end{bmatrix}$

21.10 SYMMETRIC MATRICES WITH NON REPEATED EIGEN VALUES

Example 27. *Find the eigen values and the corresponding eigen vectors of the matrix*

$$\begin{bmatrix} -2 & 5 & 4 \\ 5 & 7 & 5 \\ 4 & 5 & -2 \end{bmatrix}$$

Solution. $|A - \lambda I| = 0$

$$\begin{vmatrix} -2-\lambda & 5 & 4 \\ 5 & 7-\lambda & 5 \\ 4 & 5 & -2-\lambda \end{vmatrix} = 0 \quad \Rightarrow \quad \lambda^3 - 3\lambda^2 - 90\lambda - 216 = 0$$

By trial: Take $\lambda = -3$, then $-27 - 27 + 270 - 216 = 0$

By synthetic division

$$\begin{array}{r|rrrr} -3 & 1 & -3 & -90 & -216 \\ & & -3 & 18 & 216 \\ \hline & 1 & -6 & -72 & 0 \end{array}$$

$$\lambda^2 - 6\lambda - 72 = 0 \Rightarrow (\lambda - 12)(\lambda + 6) = 0 \Rightarrow \lambda = -3, -6, 12$$

Matrix equation for eigen vectors $[A - \lambda I]\, X = 0$

$$\begin{bmatrix} -2-\lambda & 5 & 4 \\ 5 & 7-\lambda & 5 \\ 4 & 5 & -2-\lambda \end{bmatrix} \begin{bmatrix} x \\ y \\ z \end{bmatrix} = \begin{bmatrix} 0 \\ 0 \\ 0 \end{bmatrix} \qquad \ldots(1)$$

Eigen Vector

On putting $\lambda = -3$ in (1), it will become

$$\begin{bmatrix} 1 & 5 & 4 \\ 5 & 10 & 5 \\ 4 & 5 & 1 \end{bmatrix} \begin{bmatrix} x \\ y \\ z \end{bmatrix} = \begin{bmatrix} 0 \\ 0 \\ 0 \end{bmatrix} \quad \Rightarrow \quad \begin{cases} x + 5y + 4z = 0 \\ 5x + 10y + 5z = 0 \end{cases}$$

$$\frac{x}{25-40} = \frac{y}{20-5} = \frac{z}{10-25} \qquad \text{or} \qquad \frac{x}{1} = \frac{y}{-1} = \frac{z}{1}$$

Eigen vector $X_1 = \begin{bmatrix} 1 \\ -1 \\ 1 \end{bmatrix}$.

Eigen vector corresponding to eigen value $\lambda = -6$.

Equation (1) becomes

$$\begin{bmatrix} 4 & 5 & 4 \\ 5 & 13 & 5 \\ 4 & 5 & 4 \end{bmatrix} \begin{bmatrix} x \\ y \\ z \end{bmatrix} = \begin{bmatrix} 0 \\ 0 \\ 0 \end{bmatrix} \quad \text{or} \quad \begin{cases} 4x + 5y + 4z = 0 \\ 5x + 13y + 5z = 0 \end{cases}$$

$$\frac{x}{25-52} = \frac{y}{20-20} = \frac{z}{52-25} \qquad \text{or} \qquad \frac{x}{-1} = \frac{y}{0} = \frac{z}{1}$$

eigen vector $X_2 = \begin{bmatrix} -1 \\ 0 \\ 1 \end{bmatrix}$

Eigen vector corresponding to eigen value $\lambda = 12$.

Equation (1) becomes

$$\begin{bmatrix} -14 & 5 & 4 \\ 5 & -5 & 5 \\ 4 & 5 & -14 \end{bmatrix} \begin{bmatrix} x \\ y \\ z \end{bmatrix} = \begin{bmatrix} 0 \\ 0 \\ 0 \end{bmatrix} \quad \text{or} \quad \begin{cases} -14x + 5y + 4z = 0 \\ 5x - 5y + 5z = 0 \end{cases}$$

$$\frac{x}{25+20} = \frac{y}{20+70} = \frac{z}{70-25} \qquad \text{or} \qquad \frac{x}{1} = \frac{y}{2} = \frac{z}{1}$$

Eigen vector $X_3 = \begin{bmatrix} 1 \\ 2 \\ 1 \end{bmatrix}$ **Ans.**

Example 28. *Find the eigen values, eigen vectors the modal matrix given below.*

$$\begin{bmatrix} 1 & 0 & 0 \\ 0 & 3 & -1 \\ 0 & -1 & 3 \end{bmatrix}$$

(R.G.P.V. Bhopal, I Sem., 2003)

Solution. The characteristic equation of the given matrix is

$$\begin{vmatrix} 1-\lambda & 0 & 0 \\ 0 & 3-\lambda & -1 \\ 0 & -1 & 3-\lambda \end{vmatrix} = 0$$

$$\Rightarrow \quad (1-\lambda)\{(3-\lambda)^2 - 1\} = 0 \quad \Rightarrow \quad (1-\lambda)(3-\lambda+1)(3-\lambda-1) = 0$$

$$\Rightarrow \quad (1-\lambda)(4-\lambda)(2-\lambda) = 0 \quad \Rightarrow \quad \lambda = 1, 2, 4$$

Eigen vectors

When λ = 1,

$$\begin{bmatrix} 0 & 0 & 0 \\ 0 & 2 & -1 \\ 0 & -1 & 2 \end{bmatrix}\begin{bmatrix} x_1 \\ x_2 \\ x_3 \end{bmatrix} = \begin{bmatrix} 0 \\ 0 \\ 0 \end{bmatrix} \sim \begin{bmatrix} 0 & 0 & 0 \\ 0 & 2 & -1 \\ 0 & 0 & \frac{3}{2} \end{bmatrix}\begin{bmatrix} x_1 \\ x_2 \\ x_3 \end{bmatrix} = \begin{bmatrix} 0 \\ 0 \\ 0 \end{bmatrix} \quad R_3 \to R_3 + \frac{1}{2}R_2$$

$\Rightarrow \quad 2x_2 - x_3 = 0$ ... (1)

$\frac{3}{2}x_3 = 0 \Rightarrow x_3 = 0$ ... (2)

Putting $x_3 = 0$ from (2) in (1), we get $2x_2 - 0 = 0 \Rightarrow x_2 = 0$

$$\text{Eigen Vector} = \begin{bmatrix} 1 \\ 0 \\ 0 \end{bmatrix}$$

When λ = 2,

$$\begin{bmatrix} -1 & 0 & 0 \\ 0 & 1 & -1 \\ 0 & -1 & 1 \end{bmatrix}\begin{bmatrix} x_1 \\ x_2 \\ x_3 \end{bmatrix} = \begin{bmatrix} 0 \\ 0 \\ 0 \end{bmatrix}$$

$$\Rightarrow \begin{bmatrix} 1 & 0 & 0 \\ 0 & 1 & -1 \\ 0 & 0 & 0 \end{bmatrix}\begin{bmatrix} x_1 \\ x_2 \\ x_3 \end{bmatrix} = \begin{bmatrix} 0 \\ 0 \\ 0 \end{bmatrix} \quad \begin{matrix} R_1 \to -R_1 \\ R_3 \to R_3 + R_2 \end{matrix}$$

$x_1 = 0$

$x_2 - x_3 = 0 \Rightarrow x_2 = x_3$

$$\text{Eigen vector} = \begin{bmatrix} 0 \\ 1 \\ 1 \end{bmatrix}$$

When λ = 4,

$$\begin{bmatrix} -3 & 0 & 0 \\ 0 & -1 & -1 \\ 0 & -1 & -1 \end{bmatrix}\begin{bmatrix} x_1 \\ x_2 \\ x_3 \end{bmatrix} = \begin{bmatrix} 0 \\ 0 \\ 0 \end{bmatrix}$$

$$-3x_1 = 0$$

$$-x_2 - x_3 = 0$$

$$x_2 = -x_3$$

$$\text{Eigen Vector} = \begin{bmatrix} 0 \\ 1 \\ -1 \end{bmatrix}$$

$$\text{Modal matrix} = \begin{bmatrix} 1 & 0 & 0 \\ 0 & 1 & 1 \\ 0 & 1 & -1 \end{bmatrix}$$ **Ans.**

EXERCISE 21.5

Symmetric matrices with non-repeated eigen values

Find the eigen values and eigen vectors of the following matrices:

1. $\begin{bmatrix} 5 & 0 & 1 \\ 0 & -2 & 0 \\ 1 & 0 & 5 \end{bmatrix}$ **Ans.** $-2, 4, 6;$ $\begin{bmatrix} 0 \\ 1 \\ 0 \end{bmatrix}, \begin{bmatrix} 1 \\ 0 \\ -1 \end{bmatrix}, \begin{bmatrix} 1 \\ 0 \\ 1 \end{bmatrix}$ **2.** $\begin{bmatrix} 3 & -1 & 1 \\ -1 & 5 & -1 \\ 1 & -1 & 3 \end{bmatrix}$ **Ans.** $2, 3, 6;$ $\begin{bmatrix} -1 \\ 0 \\ 1 \end{bmatrix}, \begin{bmatrix} 1 \\ 1 \\ 1 \end{bmatrix}, \begin{bmatrix} 1 \\ -2 \\ 1 \end{bmatrix}$

3. $\begin{bmatrix} 8 & -6 & 2 \\ -6 & 7 & -4 \\ 2 & -4 & 3 \end{bmatrix}$ (*U.P., I Semester, Jan 2011*) **Ans.** $0, 3, 15;$ $\begin{bmatrix} 1 \\ 2 \\ 2 \end{bmatrix}, \begin{bmatrix} 2 \\ 1 \\ -2 \end{bmatrix}, \begin{bmatrix} 2 \\ -2 \\ 1 \end{bmatrix}$

4. $\begin{bmatrix} 2 & 4 & -6 \\ 4 & 2 & -6 \\ -6 & -6 & -15 \end{bmatrix}$ **Ans.** $-2, 9, -18;$ $\begin{bmatrix} 1 \\ -1 \\ 0 \end{bmatrix}, \begin{bmatrix} 2 \\ 2 \\ -1 \end{bmatrix}, \begin{bmatrix} 1 \\ 1 \\ 4 \end{bmatrix}$ **5.** $\begin{bmatrix} 1 & 1 & 3 \\ 1 & 5 & 1 \\ 3 & 1 & 1 \end{bmatrix}$ **Ans.** $-2, 3, 6;$ $\begin{bmatrix} -1 \\ 0 \\ 1 \end{bmatrix}, \begin{bmatrix} 1 \\ -1 \\ 1 \end{bmatrix}, \begin{bmatrix} 1 \\ 2 \\ 1 \end{bmatrix}$

(*AMIETE, June 2009*)

21.11 SYMMETRIC MATRICES WITH REPEATED EIGEN VALUES

Example 29. *Find all the eigen values and eigen vectors of the matrix*

$$\begin{bmatrix} 2 & -1 & 1 \\ -1 & 2 & -1 \\ 1 & -1 & 2 \end{bmatrix}$$

Solution. The characteristic equation is $\begin{vmatrix} 2-\lambda & -1 & 1 \\ -1 & 2-\lambda & -1 \\ 1 & -1 & 2-\lambda \end{vmatrix} = 0$

$$\Rightarrow \quad (2-\lambda)[(2-\lambda)^2 - 1] + 1[-2+\lambda+1] + 1[1-2+\lambda] = 0$$

$$\Rightarrow \quad (2-\lambda)(4-4\lambda+\lambda^2-1) + (\lambda-1) + \lambda - 1 = 0$$

$$\Rightarrow \quad 8 - 8\lambda + 2\lambda^2 - 2 - 4\lambda + 4\lambda^2 - \lambda^3 + \lambda + 2\lambda - 2 = 0$$

$$\Rightarrow \quad -\lambda^3 + 6\lambda^2 - 9\lambda + 4 = 0$$

$$\Rightarrow \quad \lambda^3 - 6\lambda^2 + 9\lambda - 4 = 0 \qquad \dots (1)$$

On putting $\lambda = 1$ in (1), the equation (1) is satisfied. So $\lambda - 1$ is one factor of the equation (1). The other factor $(\lambda^2 - 5\lambda + 4)$ is got on dividing (1) by $\lambda - 1$.

$$\Rightarrow \quad (\lambda-1)(\lambda^2 - 5\lambda + 4) = 0 \text{ or } (\lambda-1)(\lambda-1)(\lambda-4) = 0 \quad \Rightarrow \lambda = 1, 1, 4$$

The eigen values are 1, 1, 4.

When $\lambda = 4$ $\begin{pmatrix} 2-4 & -1 & 1 \\ -1 & 2-4 & -1 \\ 1 & -1 & 2-4 \end{pmatrix} \begin{pmatrix} x_1 \\ x_2 \\ x_3 \end{pmatrix} = \begin{bmatrix} 0 \\ 0 \\ 0 \end{bmatrix} \Rightarrow \begin{pmatrix} -2 & -1 & 1 \\ -1 & -2 & -1 \\ 1 & -1 & -2 \end{pmatrix} \begin{pmatrix} x_1 \\ x_2 \\ x_3 \end{pmatrix} = \begin{bmatrix} 0 \\ 0 \\ 0 \end{bmatrix}$

$$-2x_1 - x_2 + x_3 = 0$$

$$x_1 - x_2 - 2x_3 = 0$$

$$\Rightarrow \quad \frac{x_1}{2+1} = \frac{x_2}{1-4} = \frac{x_3}{2+1} \quad \Rightarrow \quad \frac{x_1}{1} = \frac{x_2}{-1} = \frac{x_3}{1} = k$$

$$x_1 = k, \quad x_2 = -k, \quad x_3 = k$$

$$X_1 = \begin{bmatrix} k \\ -k \\ k \end{bmatrix} = k \begin{bmatrix} 1 \\ -1 \\ 1 \end{bmatrix} \quad \text{or} \quad X_1 = \begin{bmatrix} 1 \\ -1 \\ 1 \end{bmatrix}$$

When $\lambda = 1$ $\quad \begin{pmatrix} 2-1 & -1 & 1 \\ -1 & 2-1 & -1 \\ 1 & -1 & 2-1 \end{pmatrix}\begin{pmatrix} x_1 \\ x_2 \\ x_3 \end{pmatrix} = 0$

$$\Rightarrow \begin{pmatrix} 1 & -1 & 1 \\ -1 & 1 & -1 \\ 1 & -1 & 1 \end{pmatrix}\begin{pmatrix} x_1 \\ x_2 \\ x_3 \end{pmatrix} = 0 \Rightarrow \begin{pmatrix} 1 & -1 & 1 \\ 0 & 0 & 0 \\ 0 & 0 & 0 \end{pmatrix}\begin{pmatrix} x_1 \\ x_2 \\ x_3 \end{pmatrix} = 0, \; \begin{matrix} R_2 \to R_2 + R_1 \\ R_3 \to R_3 - R_1 \end{matrix}$$

$$x_1 - x_2 + x_3 = 0$$

Let $x_1 = k_1$ and $x_2 = k_2$

$$k_1 - k_2 + x_3 = 0 \quad \text{or} \quad x_3 = k_2 - k_1$$

$$X_2 = \begin{bmatrix} k_1 \\ k_2 \\ k_2 - k_1 \end{bmatrix} \Rightarrow X_2 = \begin{bmatrix} 1 \\ 1 \\ 0 \end{bmatrix} \qquad \begin{bmatrix} k_1 = 1 \\ k_2 = 1 \end{bmatrix}$$

Let $X_3 = \begin{bmatrix} l \\ m \\ n \end{bmatrix}$

As X_3 is orthogonal to X_1 since the given matrix is symmetric

$$[1, -1, 1]\begin{bmatrix} l \\ m \\ n \end{bmatrix} = 0 \quad \text{or} \quad l - m + n = 0 \qquad \ldots (2)$$

As X_3 is orthogonal to X_2 since the given matrix is symmetric

$$[1, 1, 0]\begin{bmatrix} l \\ m \\ n \end{bmatrix} = 0 \quad \text{or} \quad l + m + 0 = 0 \qquad \ldots (3)$$

Solving (2) and (3), we get $\dfrac{l}{0-1} = \dfrac{m}{1-0} = \dfrac{n}{1+1} \Rightarrow \dfrac{l}{-1} = \dfrac{m}{1} = \dfrac{n}{2}$

$$X_3 = \begin{bmatrix} -1 \\ 1 \\ 2 \end{bmatrix}$$

Ans.

EXERCISE 21.6

Symmetric matrices with repeated eigen values

Find the eigen values and the corresponding eigen vectors of the following matrices:

1. $\begin{bmatrix} 1 & 2 & 3 \\ 2 & 4 & 6 \\ 3 & 6 & 9 \end{bmatrix}$ **Ans.** 0, 0, 14; $\begin{bmatrix} -2 \\ 1 \\ 0 \end{bmatrix}, \begin{bmatrix} 3 \\ 6 \\ -5 \end{bmatrix}, \begin{bmatrix} 1 \\ 2 \\ 3 \end{bmatrix}$

2. $\begin{bmatrix} 2 & 0 & 1 \\ 0 & 3 & 0 \\ 1 & 0 & 2 \end{bmatrix}$ **Ans.** 1, 3, 3; $\begin{bmatrix} 1 \\ 0 \\ -1 \end{bmatrix}, \begin{bmatrix} 1 \\ 1 \\ 1 \end{bmatrix}, \begin{bmatrix} 1 \\ -2 \\ 1 \end{bmatrix}$

3. $\begin{bmatrix} 6 & -2 & 2 \\ -2 & 3 & -1 \\ 2 & -1 & 3 \end{bmatrix}$ **Ans.** 8, 2, 2; $\begin{bmatrix} 2 \\ -1 \\ 1 \end{bmatrix}, \begin{bmatrix} 1 \\ 0 \\ -2 \end{bmatrix}, \begin{bmatrix} 1 \\ 2 \\ 0 \end{bmatrix}$

4. $\begin{bmatrix} 6 & -3 & 3 \\ -3 & 6 & -3 \\ 3 & -3 & 6 \end{bmatrix}$ **Ans.** 3, 3, 12

21.12 MATRIX HAVING ONLY ONE LINEARLY INDEPENDENT EIGEN VECTOR

Example 30. *Find the eigen values and eigen vectors of*

$$A=\begin{bmatrix}-3 & -7 & -5\\ 2 & 4 & 3\\ 1 & 2 & 2\end{bmatrix}$$

has less than three linearly independent eigen vectors. It is possible to obtain a similarity transformation that will diagonalise this matrix.

Solution. The characteristic equation of the given matrix is

$$|A-\lambda I|=0$$

$$\Rightarrow \quad \begin{vmatrix}-3-\lambda & -7 & -5\\ 2 & 4-\lambda & 3\\ 1 & 2 & 2-\lambda\end{vmatrix}=0$$

$$\Rightarrow \quad (-3-\lambda)[(4-\lambda)(2-\lambda)-6]+7[2(2-\lambda)-3]-5[4-(4-\lambda)]=0$$

$$\Rightarrow \quad \lambda^3-3\lambda^2+3\lambda-1=0 \Rightarrow (\lambda-1)^3=1 \quad \Rightarrow \quad \lambda=1,\ 1,\ 1$$

Eigen values of the given matrix A are 1, 1, 1. Eigen vector when $\lambda=1$

$$\begin{bmatrix}-3-1 & -7 & -5\\ 2 & 4-1 & 3\\ 1 & 2 & 2-1\end{bmatrix}\begin{bmatrix}x_1\\ x_2\\ x_3\end{bmatrix}=\begin{bmatrix}0\\ 0\\ 0\end{bmatrix} \Rightarrow \begin{bmatrix}-4 & -7 & -5\\ 2 & 3 & 3\\ 1 & 2 & 1\end{bmatrix}\begin{bmatrix}x_1\\ x_2\\ x_3\end{bmatrix}=\begin{bmatrix}0\\ 0\\ 0\end{bmatrix}$$

$$\Rightarrow \quad -4x_1-7x_2-5x_3=0 \qquad \text{... (1)}$$

$$2x_1+3x_2+3x_3=0 \qquad \text{... (2)}$$

$$\frac{x_1}{-21+15}=\frac{x_2}{-10+12}=\frac{x_3}{-12+14}$$

$$\Rightarrow \quad \frac{x_1}{-6}=\frac{x_2}{2}=\frac{x_3}{2}=k \quad \text{(say)}$$

Thus, $x_1=-6k$, $x_2=2k$ and $x_3=2k$

$$X=\begin{bmatrix}x_1\\ x_2\\ x_3\end{bmatrix}=\begin{bmatrix}-6k\\ 2k\\ 2k\end{bmatrix}=2k\begin{bmatrix}-3\\ 1\\ 1\end{bmatrix}$$

All the eigen vectors are same and hence linearly independent. **Ans.**

21.13 MATRIX HAVING ONLY TWO EIGEN VECTORS

Example 31. *Find the eigen values and eigen vectors of* $A=\begin{bmatrix}3 & 10 & 5\\ -2 & -3 & -4\\ 3 & 5 & 7\end{bmatrix}$

has less than three linearly independent eigen vectors. Is it possible to obtain a similarity transformation that will diagonalise this matrix?

Solution. The characteristic equation of the given matrix A is

$$|A-\lambda I|=0$$

$$\Rightarrow \quad \begin{vmatrix}3-\lambda & 10 & 5\\ -2 & -3-\lambda & -4\\ 3 & 5 & 7-\lambda\end{vmatrix}=0$$

$$\Rightarrow \quad (3-\lambda)[(-3-\lambda)(7-\lambda)+20]-10[-2(7-\lambda)+12]+5[-10-3(-3-\lambda)]=0$$

$$\Rightarrow \quad (3-\lambda)[-21+3\lambda-7\lambda+\lambda^2+20]-10[-14+2\lambda+12]+5[-10+9+3\lambda]=0$$

$$\Rightarrow \quad (3-\lambda)(\lambda^2-4\lambda-1)-10(2\lambda-2)+5(3\lambda-1)=0$$

$$\Rightarrow \quad \lambda^3-7\lambda^2+16\lambda-12=0 \quad \Rightarrow \quad (\lambda-3)(\lambda-2)(\lambda-2)=0 \quad \Rightarrow \quad \lambda=3,2,2$$

Eigen values of the given matrix A are 3, 2, 2.

Eigen vector, when $\lambda = 3$

$$\begin{bmatrix} 3-3 & 10 & 5 \\ -2 & -3-3 & -4 \\ 3 & 5 & 7-3 \end{bmatrix}\begin{bmatrix} x_1 \\ x_2 \\ x_3 \end{bmatrix}=\begin{bmatrix} 0 \\ 0 \\ 0 \end{bmatrix} \Rightarrow \begin{bmatrix} 0 & 10 & 5 \\ -2 & -6 & -4 \\ 3 & 5 & 4 \end{bmatrix}\begin{bmatrix} x_1 \\ x_2 \\ x_3 \end{bmatrix}=\begin{bmatrix} 0 \\ 0 \\ 0 \end{bmatrix}$$

$$-2x_1-6x_2-4x_3=0 \qquad \text{... (1)}$$

$$3x_1+5x_2+4x_3=0 \qquad \text{... (2)}$$

Solving (1) and (2) by cross multiplication method, we have

$$\frac{x_1}{-24+20}=\frac{x_2}{-12+8}=\frac{x_3}{-10+18}$$

$$\Rightarrow \quad \frac{x_1}{-4}=\frac{x_2}{-4}=\frac{x_3}{8}=k \text{ (say)}$$

Thus, $x_1 = -4k$, $x_2 = -4k$ and $x_3 = 8k$

$$X=\begin{bmatrix} x_1 \\ x_2 \\ x_3 \end{bmatrix}=\begin{bmatrix} -4k \\ -4k \\ 8k \end{bmatrix}=4k\begin{bmatrix} -1 \\ -1 \\ 2 \end{bmatrix}$$

Eigen vector when $\lambda = 2$

$$\begin{bmatrix} 3-2 & 10 & 5 \\ -2 & -3-2 & -4 \\ 3 & 5 & 7-2 \end{bmatrix}\begin{bmatrix} x_1 \\ x_2 \\ x_3 \end{bmatrix}=\begin{bmatrix} 0 \\ 0 \\ 0 \end{bmatrix} \Rightarrow \begin{bmatrix} 1 & 10 & 5 \\ -2 & -5 & -4 \\ 3 & 5 & 5 \end{bmatrix}\begin{bmatrix} x_1 \\ x_2 \\ x_3 \end{bmatrix}=\begin{bmatrix} 0 \\ 0 \\ 0 \end{bmatrix}$$

$$\Rightarrow \quad x_1+10x_2+5x_3=0 \qquad \text{... (3)}$$

$$-2x_1-5x_2-4x_3=0 \qquad \text{... (4)}$$

Solving (3) and (4) by cross multiplication method, we have

$$\frac{x_1}{-40+25}=\frac{x_2}{-10+4}=\frac{x_3}{-5+20} \quad \Rightarrow \quad \frac{x_1}{-15}=\frac{x_2}{-6}=\frac{x_3}{15}=k \text{ (say)}$$

$$\Rightarrow \quad x_1=-15k \qquad x_2=-6k, \; x_3=15k$$

$$X=\begin{bmatrix} x_1 \\ x_2 \\ x_3 \end{bmatrix}=\begin{bmatrix} -15k \\ -6k \\ 15k \end{bmatrix}=3k\begin{bmatrix} -5 \\ -2 \\ 5 \end{bmatrix}$$

We get one eigen vector corresponding to repeated root $\lambda_2 = 2 = \lambda_3$.

Eigen vectors corresponding to $\lambda_2 = 2 = \lambda_3$ are not linearly independent. Similarity transformation is not possible. **Ans.**

21.14 COMPLEX EIGEN VALUES

Example 32. *Show that if* $0 < \theta < \pi$, *then* $A=\begin{bmatrix} \cos\theta & -\sin\theta \\ \sin\theta & \cos\theta \end{bmatrix}$ *has no real eigen values and consequently no eigen vector.* *(Gujarat, II Semester, June 2009)*

Solution. The characteristic equation of A is $\begin{vmatrix} \cos\theta-\lambda & -\sin\theta \\ \sin\theta & \cos\theta-\lambda \end{vmatrix}=0$

$\Rightarrow \quad (\cos\theta - \lambda)^2 + \sin^2\theta = 0$

$\Rightarrow \quad \cos^2\theta - 2\lambda\cos\theta + \lambda^2 + \sin^2\theta = 0$

$\Rightarrow \quad \lambda^2 - 2\lambda\cos\theta + 1 = 0$

$$\Rightarrow \quad \lambda = \frac{2\cos\theta \pm \sqrt{4\cos^2\theta - 4}}{2} = \frac{2\cos\theta \pm 2i\sqrt{1-\cos^2\theta}}{2} = \cos\theta \pm i\sin\theta$$

Hence, the given matrix A has no real eigen values and consequently no eigen vector. **Proved.**

Example 33. *If a matrix A is non-singular. Then $\lambda = 0$ is not its eigen value.*

Solution. Since matrix A is non-singular then $|A| \neq 0$

$$\Rightarrow \quad |A - 0I| \neq 0$$

Hence $\lambda = 0$ is not its eigen value. **Proved.**

21.15 ALGEBRAIC MULTIPLICITY

Algebraic multiplicity of an eigen value is the number of times of repetition of an eigen value.

It is denoted by $\text{mult}_a(\lambda)$.

For example, the eigen values of a matrix $\begin{bmatrix} -2 & 2 & -3 \\ 2 & 1 & -6 \\ -1 & -2 & 0 \end{bmatrix}$ are $-3, -3, 5$.

The $\text{mult}_a(-3) = 2$ and $\text{mult}_a(5) = 1$

21.16 GEOMETRIC MULTIPLICITY

Geometric multiplicity of an eigen value is the number of linearly independent eigen vectors corresponding to λ.

It is denoted by $\text{Mult}_g(\lambda)$

In example 30, two linearly independent eigen vectors corresponding to

$$\lambda = -3 \text{ are } \begin{bmatrix} 0 \\ 3 \\ 2 \end{bmatrix} \text{ and } \begin{bmatrix} 3 \\ 0 \\ 1 \end{bmatrix}$$

so the $\text{mult}_g(-3) = 2$

And the eigen vector corresponding to $\lambda = 5$ is $\begin{bmatrix} 1 \\ 2 \\ -1 \end{bmatrix}$ so the $\text{mult}_g(5) = 1$.

21.17 REGULAR EIGEN VALUE

If the algebraic multiplicity and geometric multiplicity of an eigen value are equal, then the eigen value is called *regular*.

Example 34. *Find the algebraic multiplicity and geometric multiplicity of an eigen value of the matrix $A = \begin{bmatrix} 3 & 10 & 5 \\ -2 & -3 & -4 \\ 3 & 5 & 7 \end{bmatrix}$ and show geometric multiplicity cannot be greater than algebraic multiplicity.*

Solution. The characteristic equation of the given matrix is

$$\begin{vmatrix} 3-\lambda & 10 & 5 \\ -2 & -3-\lambda & -4 \\ 3 & 5 & 7-\lambda \end{vmatrix} = 0$$

$$\Rightarrow \qquad \lambda^3 - 7\lambda^2 + 16\lambda - 12 = 0$$

$$\Rightarrow \qquad (\lambda - 2)(\lambda - 2)(\lambda - 3) = 0$$

$$\Rightarrow \qquad \lambda = 2, 2, 3$$

Therefore 2 is a multiple eigen value repeating 2 times. So Algebraic Multiplicity of 2 is 2.

$$\text{Mult}_a(2) = 2. \qquad \text{...(A)}$$

We shall find the eigen vector corresponding to the eigen value 2.

$$X = \begin{bmatrix} 3-2 & 10 & 5 \\ -2 & -3-2 & -4 \\ 3 & 5 & 7-2 \end{bmatrix} \begin{bmatrix} x_1 \\ x_2 \\ x_3 \end{bmatrix} = \begin{bmatrix} 0 \\ 0 \\ 0 \end{bmatrix}$$

$$\Rightarrow \qquad \begin{bmatrix} 1 & 10 & 5 \\ -2 & -5 & -4 \\ 3 & 5 & 5 \end{bmatrix} \begin{bmatrix} x_1 \\ x_2 \\ x_3 \end{bmatrix} = \begin{bmatrix} 0 \\ 0 \\ 0 \end{bmatrix}$$

$$x_1 + 10x_2 + 5x_3 = 0 \qquad \text{...(1)}$$

$$-2x_1 - 5x_2 - 4x_3 = 0 \qquad \text{...(2)}$$

Solving (1) and (2) by cross multiplication method, we have

$$\frac{x_1}{-40+25} = \frac{x_2}{-10+4} = \frac{x_3}{-5+20}$$

$$\Rightarrow \qquad \frac{x_1}{-15} = \frac{x_2}{-6} = \frac{x_3}{15} = k \text{ (say)}$$

Thus $x_1 = -15\,k, \quad x_2 = -6\,k, \quad x_3 = -15\,k.$

$$X = \begin{bmatrix} x_1 \\ x_2 \\ x_3 \end{bmatrix} = \begin{bmatrix} -15k \\ -6k \\ 15k \end{bmatrix} = 3k \begin{bmatrix} -5 \\ -2 \\ 5 \end{bmatrix}$$

Here the linearly independent eigen vector is 1.

So the, geometric multiplicity of eigen value 2 is 1

$$\text{Mult}_g(2) = 1 \qquad \text{...(B)}$$

Hence from (*A*) and (*B*) **Ans.**

Geometric multiplicity < Algebraic multiplicity

Notes: (1) If the values of x_1, x_2, x_3 are in terms of k (one independent value), then there is only one linearly independent eigen vector. So the geometric multiplicity is 1.

(2) If the values of x_1, x_2, x_3 are in terms of k_1, k_2 (two independent values, then there are two linearly independent eigen vectors. So the geometric multiplicity is 2.

EXERCISE 21.7

From the following matrices; find eigen value, Algebraic multiplicity, Geometric multiplicity.

1. $\begin{bmatrix} -2 & -1 \\ 5 & 4 \end{bmatrix}$ **Ans.** $\lambda = -1$, $\text{Mult}_a(-1) = 1$, $\text{Mult}_g(-1) = 1$
$\lambda = 3$, $\text{Mult}_a(3) = 1$, $\text{Mult}_g(3) = 1$

2. $\begin{bmatrix} 1 & 1 & 1 \\ 0 & 1 & 1 \\ 0 & 0 & 1 \end{bmatrix}$ **Ans.** $\lambda = 1$, $\text{Mult}_a(1) = 3$, $\text{Mult}_g(1) = 1$

3. $\begin{bmatrix} 0 & 0 & 1 \\ 1 & 0 & -3 \\ 0 & 1 & 3 \end{bmatrix}$ **Ans.** $\lambda = 1$, $\text{Mult}_a(1) = 3$, $\text{Mult}_g(1) = 1$

4. $\begin{bmatrix} 1 & 2 & 3 \\ 0 & 2 & 3 \\ 0 & 0 & 2 \end{bmatrix}$ **Ans.** $\lambda = 2$, $Mult_a(2) = 2$, $Mult_g(2) = 1$

5. $\begin{bmatrix} 2 & 2 & 1 \\ 1 & 3 & 1 \\ 1 & 2 & 2 \end{bmatrix}$ **Ans.** $\lambda = 5$, $Mult_a(5) = 1$, $Mult_g(5) = 1$

6. $\begin{bmatrix} 1 & 2 & 2 \\ 0 & 2 & 1 \\ -1 & 2 & 2 \end{bmatrix}$ Ans. $\lambda = 1$, $Mult_a(1) = 1$, $Mult_g(1) = 1$
$\lambda = 2$, $Mult_a(2) = 2$, $Mult_g(2) = 1$

7. $\begin{bmatrix} 5 & 4 & -4 \\ 4 & 5 & -4 \\ -1 & -1 & 2 \end{bmatrix}$ **Ans.** $\lambda = 1$, $Mult_a(1) = 2$, $Mult_g(1) = 2$
$\lambda = 10$, $Mult_a(10) = 1$, $Mult_g(10) = 1$

8. $\begin{bmatrix} 1 & 0 & 0 & 0 \\ 0 & 1 & 0 & 0 \\ 0 & 0 & 1 & 1 \\ 0 & 0 & 0 & 1 \end{bmatrix}$ **Ans.** $\lambda = 1$, $Mult_a(1) = 4$, $Mult_g(1) = 3$

21.18 SIMILARITY TRANSFORMATION

Let A and B be two square matrices of order n. Then B is said to be similar to A if there exists a non-singular matrix P such that

$$B = P^{-1} AP \quad ...(1)$$

Equation (1) is called a similar transformation.

21.19 DIAGONALISATION OF A MATRIX

Diagonalisation of a matrix A is the process of reduction of A to a diagonal form 'D'. If A is related to D by a similarity transformation such that $D = P^{-1} AP$ then A is reduced to the diagonal matrix D through *modal matrix P*. D is also called *spectral matrix* of A.

21.20 ORTHOGONAL TRANSFORMATION OF A SYMMETRIC MATRIX TO DIAGONAL FORM

Let A be a symmetric matrix, then

$$A \,.\, A' = I \quad ...(1)$$

and

$$A \,.\, A^{-1} = I \quad ...(2)$$

From (1) and (2), we have $A^{-1} = A'$

We know that, diagonalisation transformation of a symmetric matrix is

$$P^{-1}AP = D$$

If we normalize each eigen vector and use them to form the normalized modal matirx N then N is an orthogonal matrix.

Then, $N' AN = D$

Transforming A into D by means of the transformation $N' AN = D$ is called as orthogonal transformation.

Note. To normalize eigen vector divide each element of the vector by the square root of the sum of the squares of all the elements of the vector.

21.21 THEOREM ON DIAGONALIZATION OF A MATRIX

Theorem. *If a square matrix A of order n has n linearly independent eigen vectors, then a matrix P can be found such that P^{-1} AP is a diagonal matrix.*

Proof. We shall prove the theorem for a matrix of order 3. The proof can be easily extended to matrices of higher order.

Let
$$A = \begin{bmatrix} a_1 & b_1 & c_1 \\ a_2 & b_2 & c_2 \\ a_3 & b_3 & c_3 \end{bmatrix}$$

and let $\lambda_1, \lambda_2, \lambda_3$ be its eigen values and X_1, X_2, X_3 the corresponding eigen vectors, where

$$X_1 = \begin{bmatrix} x_1 \\ y_1 \\ z_1 \end{bmatrix}, \qquad X_2 = \begin{bmatrix} x_2 \\ y_2 \\ z_2 \end{bmatrix}, \qquad X_3 = \begin{bmatrix} x_3 \\ y_3 \\ z_3 \end{bmatrix}$$

For the eigen value λ_1, the eigen vector is given by

$$\left.\begin{aligned} (a_1 - \lambda_1)x_1 + b_1 y_1 + c_1 z_1 &= 0 \\ a_2 x_1 + (b_2 - \lambda_1) y_1 + c_2 z_1 &= 0 \\ a_3 x_1 + b_3 y_1 + (c_3 - \lambda_1) z_1 &= 0 \end{aligned}\right\} \quad \text{...(1)}$$

$\therefore$ We have

$$\left.\begin{aligned} a_1 x_1 + b_1 y_1 + c_1 z_1 &= \lambda_1 x_1 \\ a_2 x_1 + b_2 y_1 + c_2 z_1 &= \lambda_1 y_1 \\ a_3 x_1 + b_3 y_1 + c_3 z_1 &= \lambda_1 z_1 \end{aligned}\right\} \quad \text{...(2)}$$

Similarly, for λ_2 and λ_3, we have

$$\left.\begin{aligned} a_1 x_2 + b_1 y_2 + c_1 z_2 &= \lambda_2 x_2 \\ a_2 x_2 + b_2 y_2 + c_2 z_2 &= \lambda_2 y_2 \\ a_3 x_2 + b_3 y_2 + c_3 z_2 &= \lambda_2 z_2 \end{aligned}\right\} \quad \text{...(3)}$$

and
$$\left.\begin{aligned} a_1 x_3 + b_1 y_3 + c_1 z_3 &= \lambda_3 x_3 \\ a_2 x_3 + b_2 y_3 + c_2 z_3 &= \lambda_3 y_3 \\ a_3 x_3 + b_3 y_3 + c_3 z_3 &= \lambda_3 z_3 \end{aligned}\right\} \quad \text{...(4)}$$

We consider the matrix
$$P = \begin{bmatrix} x_1 & x_2 & x_3 \\ y_1 & y_2 & y_3 \\ z_1 & z_2 & z_3 \end{bmatrix}$$

whose columns are the eigen vectors of A.

Then
$$AP = \begin{bmatrix} a_1 & b_1 & c_1 \\ a_2 & b_2 & c_2 \\ a_3 & b_3 & c_3 \end{bmatrix} \begin{bmatrix} x_1 & x_2 & x_3 \\ y_1 & y_2 & y_3 \\ z_1 & z_2 & z_3 \end{bmatrix}$$

$$= \begin{pmatrix} a_1 x_1 + b_1 y_1 + c_1 z_1 & a_1 x_2 + b_1 y_2 + c_1 z_2 & a_1 x_3 + b_1 y_3 + c_1 z_3 \\ a_2 x_1 + b_2 y_1 + c_2 z_1 & a_2 x_2 + b_2 y_2 + c_2 z_2 & a_2 x_3 + b_2 y_3 + c_2 z_3 \\ a_3 x_1 + b_3 y_1 + c_3 z_1 & a_3 x_2 + b_3 y_2 + c_3 z_2 & a_3 x_3 + b_3 y_3 + c_3 z_3 \end{pmatrix}$$

$$= \begin{pmatrix} \lambda_1 x_1 & \lambda_2 x_2 & \lambda_3 x_3 \\ \lambda_1 y_1 & \lambda_2 y_2 & \lambda_3 y_3 \\ \lambda_1 z_1 & \lambda_2 z_2 & \lambda_3 z_3 \end{pmatrix} \qquad \text{[Using results (2), (3) and (4)]}$$

$$= \begin{pmatrix} x_1 & x_2 & x_3 \\ y_1 & y_2 & y_3 \\ z_1 & z_2 & z_3 \end{pmatrix} \begin{pmatrix} \lambda_1 & 0 & 0 \\ 0 & \lambda_2 & 0 \\ 0 & 0 & \lambda_3 \end{pmatrix} = PD$$

where D is the Diagonal matrix $\begin{pmatrix} \lambda_1 & 0 & 0 \\ 0 & \lambda_2 & 0 \\ 0 & 0 & \lambda_3 \end{pmatrix}$.

$$\therefore \quad AP = PD$$

$$\Rightarrow \quad P^{-1} AP = P^{-1} PD = D$$

Notes 1. The square matrix P, which diagonalises A, is found by grouping the eigen vectors of A into square-matrix and the resulting diagonal matrix has the eigen values of A as its diagonal elements.

2. The transformation of a matrix A to $P^{-1} AP$ is known as a *similarity transformation.*

3. The reduction of A to a diagonal matrix is, obviously, a particular case of similarity transformation.

4. The matrix P which diagonalises A is called the *modal matrix* of A and the resulting diagonal matrix D is known as the *spectra matrix* of A.

Example 35. *Find a matrix P which diagonalizes the matrix*

$A = \begin{bmatrix} 4 & 1 \\ 2 & 3 \end{bmatrix}$, *verify $P^{-1} AP = D$ where D is the diagonal matrix.* (*U.P., I Semester, Dec. 2008*)

Solution. The characteristic equation of matrix A is

$$\begin{vmatrix} 4-\lambda & 1 \\ 2 & 3-\lambda \end{vmatrix} = 0 \quad \Rightarrow \quad (4-\lambda)(3-\lambda) - 2 = 0$$

$$\Rightarrow \lambda^2 - 7\lambda + 12 - 2 = 0 \quad \Rightarrow \quad \lambda^2 - 7\lambda + 10 = 0$$

$$\Rightarrow (\lambda - 2)(\lambda - 5) = 0 \quad \Rightarrow \quad \lambda = 2, \lambda = 5$$

Eigen values are 2 and 5.

(*i*) When $\lambda = 2$, eigen vectors are given by the matrix equation

$$\begin{bmatrix} 4-2 & 1 \\ 2 & 3-2 \end{bmatrix}\begin{bmatrix} x_1 \\ x_2 \end{bmatrix} = \begin{bmatrix} 0 \\ 0 \end{bmatrix} \Rightarrow \begin{bmatrix} 2 & 1 \\ 2 & 1 \end{bmatrix}\begin{bmatrix} x_1 \\ x_2 \end{bmatrix} = \begin{bmatrix} 0 \\ 0 \end{bmatrix}$$

$$\Rightarrow \quad 2x_1 + x_2 = 0 \Rightarrow x_2 = -2x_1$$

Let $\quad x_1 = k, x_2 = -2k$

Hence, the eigen vector $\quad X_1 = \begin{bmatrix} k \\ -2k \end{bmatrix}$ or $\begin{bmatrix} 1 \\ -2 \end{bmatrix}$

(ii) When $\lambda = 5$, eigen vectors are given by the matrix equation

$$\begin{bmatrix} 4-5 & 1 \\ 2 & 3-5 \end{bmatrix}\begin{bmatrix} x_1 \\ x_2 \end{bmatrix} = \begin{bmatrix} 0 \\ 0 \end{bmatrix} \Rightarrow \begin{bmatrix} -1 & 1 \\ 2 & -2 \end{bmatrix}\begin{bmatrix} x_1 \\ x_2 \end{bmatrix} = \begin{bmatrix} 0 \\ 0 \end{bmatrix}$$

$$\Rightarrow \quad -x_1 + x_2 = 0 \Rightarrow x_1 = x_2$$

Let $x_1 = k$, then $x_2 = k$

Hence, the eigen vector $\quad X_2 = \begin{bmatrix} k \\ k \end{bmatrix}$ or $\begin{bmatrix} 1 \\ 1 \end{bmatrix}$

Modal matrix $\quad P = \begin{bmatrix} 1 & 1 \\ -2 & 1 \end{bmatrix} \quad \Rightarrow P^{-1} = \frac{1}{3}\begin{bmatrix} 1 & -1 \\ 2 & 1 \end{bmatrix}$

For diagonalization

$$D = P^{-1} AP = \frac{1}{3}\begin{bmatrix} 1 & -1 \\ 2 & 1 \end{bmatrix}\begin{bmatrix} 4 & 1 \\ 2 & 3 \end{bmatrix}\begin{bmatrix} 1 & 1 \\ -2 & 1 \end{bmatrix}$$

$$= \frac{1}{3}\begin{bmatrix} 1 & -1 \\ 2 & 1 \end{bmatrix}\begin{bmatrix} 2 & 5 \\ -4 & 5 \end{bmatrix} = \frac{1}{3}\begin{bmatrix} 6 & 0 \\ 0 & 15 \end{bmatrix} = \begin{bmatrix} 2 & 0 \\ 0 & 5 \end{bmatrix}$$ **Verified.**

Example 36. *Let* $A = \begin{bmatrix} 6 & -2 & 2 \\ -2 & 3 & -1 \\ 2 & -1 & 3 \end{bmatrix}$ *Find matrix P such that $P^{-1}AP$ is diagonal matrix.*

Solution. The characteristic equation of the matrix A is

$$\begin{vmatrix} 6-\lambda & -2 & 2 \\ -2 & 3-\lambda & -1 \\ 2 & -1 & 3-\lambda \end{vmatrix} = 0$$

$$\Rightarrow \quad (6-\lambda)[9+\lambda^2-6\lambda-1]+2[-6+2\lambda+2]+2[2-6+2\lambda]=0$$

$$\Rightarrow \quad (6-\lambda)(\lambda^2-6\lambda+8)-8+4\lambda-8+4\lambda=0$$

$$\Rightarrow \quad 6\lambda^2-36\lambda+48-\lambda^3+6\lambda^2-8\lambda-16+8\lambda=0$$

$$\Rightarrow -\lambda^3+12\lambda^2-36\lambda+32=0 \quad \Rightarrow \quad \lambda^3-12\lambda^2+36\lambda-32=0$$

$$\Rightarrow \quad (\lambda-2)^2(\lambda-8)=0 \quad \Rightarrow \quad \lambda = 2, 2, 8$$

Eigen vector for $\lambda = 2$

$$\begin{bmatrix} 4 & -2 & 2 \\ -2 & 1 & -1 \\ 2 & -1 & 1 \end{bmatrix}\begin{bmatrix} x_1 \\ x_2 \\ x_3 \end{bmatrix} = \begin{bmatrix} 0 \\ 0 \\ 0 \end{bmatrix} \text{ or } \begin{bmatrix} 2 & -1 & 1 \\ -2 & 1 & -1 \\ 2 & -1 & 1 \end{bmatrix}\begin{bmatrix} x_1 \\ x_2 \\ x_3 \end{bmatrix} = \begin{bmatrix} 0 \\ 0 \\ 0 \end{bmatrix} \quad \begin{matrix} R_2 \to R_1 + R_2 \\ R_3 \to R_2 + R_3 \end{matrix}$$

$$\begin{bmatrix} 2 & -1 & 1 \\ 0 & 0 & 0 \\ 0 & 0 & 0 \end{bmatrix}\begin{bmatrix} x_1 \\ x_2 \\ x_3 \end{bmatrix} = \begin{bmatrix} 0 \\ 0 \\ 0 \end{bmatrix} \text{ or } 2x_1 - x_2 + x_3 = 0$$

This equation is satisfied by $x_1 = 0, x_2 = 1, x_3 = 1$

$$X_1 = \begin{bmatrix} 0 \\ 1 \\ 1 \end{bmatrix}$$

and again $\quad x_1 = 1, x_2 = 3, x_3 = 1.$

$$X_2 = \begin{bmatrix} 1 \\ 3 \\ 1 \end{bmatrix}$$

Eigen vector for $\lambda = 8$

$$\begin{bmatrix} -2 & -2 & 2 \\ -2 & -5 & -1 \\ 2 & -1 & -5 \end{bmatrix}\begin{bmatrix} x_1 \\ x_2 \\ x_3 \end{bmatrix} = \begin{bmatrix} 0 \\ 0 \\ 0 \end{bmatrix}$$

$$-2x_1 - 2x_2 + 2x_3 = 0$$

$$-2x_1 - 5x_2 - x_3 = 0$$

$$\frac{x_1}{2+10} = \frac{x_2}{-4-2} = \frac{x_3}{10-4} \Rightarrow \frac{x_1}{12} = \frac{x_2}{-6} = \frac{x_3}{6} \Rightarrow \frac{x_1}{2} = \frac{x_2}{-1} = \frac{x_3}{1}$$

$$X_3 = \begin{bmatrix} 2 \\ -1 \\ 1 \end{bmatrix}$$

$$\therefore \qquad P = \begin{bmatrix} 0 & 1 & 2 \\ 1 & 3 & -1 \\ 1 & 1 & 1 \end{bmatrix}, \quad P^{-1} = -\frac{1}{6}\begin{bmatrix} 4 & 1 & -7 \\ -2 & -2 & 2 \\ -2 & 1 & -1 \end{bmatrix}$$

Now $P^{-1}AP = -\frac{1}{6}\begin{bmatrix} 4 & 1 & -7 \\ -2 & -2 & 2 \\ -2 & 1 & -1 \end{bmatrix}\begin{bmatrix} 6 & -2 & 2 \\ -2 & 3 & -1 \\ 2 & -1 & 3 \end{bmatrix}\begin{bmatrix} 0 & 1 & 2 \\ 1 & 3 & -1 \\ 1 & 1 & 1 \end{bmatrix} = \begin{bmatrix} 2 & 0 & 0 \\ 0 & 2 & 0 \\ 0 & 0 & 8 \end{bmatrix}$ **Ans.**

Example 37. *The matrix* $A = \begin{bmatrix} a & h \\ h & b \end{bmatrix}$ *is transformed to the diagonal form* $D = T^{-1}AT$, *where* $T = \begin{bmatrix} \cos\theta & \sin\theta \\ -\sin\theta & \cos\theta \end{bmatrix}$. *Find the value of* θ *which gives this diagonal transformation.*

Solution. $T = \begin{bmatrix} \cos\theta & \sin\theta \\ -\sin\theta & \cos\theta \end{bmatrix}$ $\therefore$ $T^{-1} = \begin{bmatrix} \cos\theta & -\sin\theta \\ \sin\theta & \cos\theta \end{bmatrix}$

Now $T^{-1}AT = \begin{bmatrix} \cos\theta & -\sin\theta \\ \sin\theta & \cos\theta \end{bmatrix}\begin{bmatrix} a & h \\ h & b \end{bmatrix}\begin{bmatrix} \cos\theta & \sin\theta \\ -\sin\theta & \cos\theta \end{bmatrix}$

$$= \begin{bmatrix} a\cos\theta - h\sin\theta & h\cos\theta - b\sin\theta \\ a\sin\theta + h\cos\theta & h\sin\theta + b\cos\theta \end{bmatrix}\begin{bmatrix} \cos\theta & \sin\theta \\ -\sin\theta & \cos\theta \end{bmatrix}$$

$$= \begin{bmatrix} a\cos^2\theta - 2h\sin\theta\cos\theta + b\sin^2\theta & (a-b)\sin\theta\cos\theta - h\sin^2\theta + h\cos^2\theta \\ (a-b)\sin\theta\cos\theta + h\cos^2\theta - h\sin^2\theta & a\sin^2\theta + 2h\sin\theta\cos\theta + b\cos^2\theta \end{bmatrix}$$

$$= \begin{bmatrix} a\cos^2\theta - h\sin 2\theta + b\sin^2\theta & (a-b)\sin\theta\cos\theta + h\cos 2\theta \\ (a-b)\sin\theta\cos\theta + h\cos 2\theta & a\sin^2\theta + h\sin 2\theta + b\cos^2\theta \end{bmatrix}$$

$= \begin{bmatrix} d_1 & 0 \\ 0 & d_2 \end{bmatrix}$ being diagonal matrix .

$$\therefore \qquad (a-b)\sin\theta\cos\theta + h\cos 2\theta = 0$$

$$\Rightarrow \qquad \frac{a-b}{2}\sin 2\theta + h\cos 2\theta = 0 \qquad \Rightarrow \frac{a-b}{2}\sin 2\theta = -h\cos 2\theta$$

$$\Rightarrow \qquad \tan 2\theta = \frac{2h}{b-a} \qquad \Rightarrow \qquad \theta = \frac{1}{2}\tan^{-1}\frac{2h}{b-a}$$ **Ans.**

EXERCISE 21.8

1. Find the matrix B which transforms the matrix

$A = \begin{bmatrix} 8 & -8 & -2 \\ 4 & -3 & -2 \\ 3 & -4 & 1 \end{bmatrix}$ to a diagonal matrix. **Ans.** $B = \begin{bmatrix} 4 & 3 & 2 \\ 3 & 2 & 1 \\ 2 & 1 & 1 \end{bmatrix}$

2. For the matrix $A = \begin{bmatrix} 4 & 1 & 0 \\ 1 & 4 & 1 \\ 0 & 1 & 4 \end{bmatrix}$, determine a matrix P such that $P^{-1}AP$ is diagonal matrix.

Ans. $P = \begin{bmatrix} -1 & 1 & 1 \\ 0 & -\sqrt{2} & \sqrt{2} \\ 1 & 1 & 1 \end{bmatrix}$

3. Determine the eigen values and the corresponding eigen vectors of the matrix $A = \begin{bmatrix} 5 & 7 & -5 \\ 0 & 4 & -1 \\ 2 & 8 & -3 \end{bmatrix}$

Hence find the matrix P such that $P^{-1}AP$ is diagonal matrix. **Ans.** $P = \begin{bmatrix} 2 & 1 & -1 \\ 1 & 1 & 1 \\ 3 & 2 & 1 \end{bmatrix}$

4. Reduce the following matrix A into a diagonal matrix

$$A = \begin{bmatrix} 8 & -6 & 2 \\ -6 & 7 & -4 \\ 2 & -4 & 3 \end{bmatrix}$$

Ans. $\begin{bmatrix} 0 & 0 & 0 \\ 0 & 3 & 0 \\ 0 & 0 & 15 \end{bmatrix}$

21.22 POWERS OF A MATRIX (By diagonalisation)

We can obtain powers of a matrix by using diagonalisation.

We know that $D = P^{-1}AP$

Where A is the square matrix and P is a non-singular matrix.

$$D^2 = (P^{-1}AP)(P^{-1}AP) = P^{-1}A(PP^{-1})AP = P^{-1}A^2P$$

Similarly $D^3 = P^{-1}A^3P$

In general $D^n = P^{-1}A^nP$ ...(1)

Pre-multiply (1) by P and post-multiply by P^{-1}

$$PD^nP^{-1} = P(P^{-1}A^nP)P^{-1}$$
$$= (PP^{-1})A^n(PP^{-1})$$
$$= A^n$$

Procedure: (1) Find eigen values for a square matrix A.

(2) Find eigen vectors to get the modal matrix P.

(3) Find the diagonal matrix D, by the formula $D = P^{-1}AP$

(4) Obtain A^n by the formula $A^n = PD^nP^{-1}$.

Example 38. *Find a matrix P which transform the matrix* $A = \begin{bmatrix} 1 & 0 & -1 \\ 1 & 2 & 1 \\ 2 & 2 & 3 \end{bmatrix}$ *to diagonal form. Hence* A^4.

Solution. Characteristic equation of the matrix A is

$$\begin{vmatrix} 1-\lambda & 0 & -1 \\ 1 & 2-\lambda & 1 \\ 2 & 2 & 3-\lambda \end{vmatrix} = 0$$

or $\lambda^3 - 6\lambda^2 + 11\lambda - 6 = 0$

or $(\lambda-1)(\lambda-2)(\lambda-3) = 0$

$\Rightarrow \lambda = 1, 2, 3$

For $\lambda = 1$, eigen vector is given by

$$\begin{bmatrix} 1-1 & 0 & -1 \\ 1 & 2-1 & 1 \\ 2 & 2 & 3-1 \end{bmatrix}\begin{bmatrix} x_1 \\ x_2 \\ x_3 \end{bmatrix} = \begin{bmatrix} 0 \\ 0 \\ 0 \end{bmatrix} \Rightarrow \begin{bmatrix} 0 & 0 & -1 \\ 1 & 1 & 1 \\ 2 & 2 & 2 \end{bmatrix}\begin{bmatrix} x_1 \\ x_2 \\ x_3 \end{bmatrix} = \begin{bmatrix} 0 \\ 0 \\ 0 \end{bmatrix}$$

$$\left.\begin{array}{r} 0x_1 + 0x_2 - x_3 = 0 \\ x_1 + x_2 + x_3 = 0 \end{array}\right] \Rightarrow \frac{x_1}{0+1} = \frac{x_2}{-1+0} = \frac{x_3}{0} \text{ or } x_1 = 1, x_2 = -1, x_3 = 0$$

Eigen vector is [1, –1, 0].

For $\lambda = 2$, eigen vector is given by

$$\begin{bmatrix} 1-2 & 0 & -1 \\ 1 & 2-2 & 1 \\ 2 & 2 & 3-2 \end{bmatrix}\begin{bmatrix} x_1 \\ x_2 \\ x_3 \end{bmatrix} = \begin{bmatrix} 0 \\ 0 \\ 0 \end{bmatrix} \Rightarrow \begin{bmatrix} -1 & 0 & -1 \\ 1 & 0 & 1 \\ 2 & 2 & 1 \end{bmatrix}\begin{bmatrix} x_1 \\ x_2 \\ x_3 \end{bmatrix} = \begin{bmatrix} 0 \\ 0 \\ 0 \end{bmatrix}$$

$$\Rightarrow \quad \begin{bmatrix} 0 & 0 & 0 \\ 1 & 0 & 1 \\ 2 & 2 & 1 \end{bmatrix}\begin{bmatrix} x_1 \\ x_2 \\ x_3 \end{bmatrix} = \begin{bmatrix} 0 \\ 0 \\ 0 \end{bmatrix} \quad R_1 \to R_1 + R_2$$

$$\left.\begin{aligned} x_1 + 0x_2 + x_3 &= 0 \\ 2x_1 + 2x_2 + x_3 &= 0 \end{aligned}\right]$$

$$\Rightarrow \quad \frac{x_1}{0-2} = \frac{x_2}{2-1} = \frac{x_3}{2-0} \quad \Rightarrow \quad x_1 = -2, \quad x_2 = 1, \quad x_3 = 2$$

Eigen vector is [–2, 1, 2].

For $\lambda = 3$, eigen vector is given by

$$\begin{bmatrix} 1-3 & 0 & -1 \\ 1 & 2-3 & 1 \\ 2 & 2 & 3-3 \end{bmatrix}\begin{bmatrix} x_1 \\ x_2 \\ x_3 \end{bmatrix} = \begin{bmatrix} 0 \\ 0 \\ 0 \end{bmatrix} \Rightarrow \begin{bmatrix} -2 & 0 & -1 \\ 1 & -1 & 1 \\ 2 & 2 & 0 \end{bmatrix}\begin{bmatrix} x_1 \\ x_2 \\ x_3 \end{bmatrix} = \begin{bmatrix} 0 \\ 0 \\ 0 \end{bmatrix}$$

$$\left.\begin{aligned} -2x_1 + 0x_2 - x_3 &= 0 \\ x_1 - x_2 + x_3 &= 0 \end{aligned}\right]$$

$$\Rightarrow \quad \frac{x_1}{0-1} = \frac{x_2}{-1+2} = \frac{x_3}{2-0} \quad \Rightarrow \quad x_1 = -1, \quad x_2 = 1, \quad x_3 = 2$$

Eigen vector is [–1, 1, 2].

Modal matrix $P = \begin{bmatrix} 1 & -2 & -1 \\ -1 & 1 & 1 \\ 0 & 2 & 2 \end{bmatrix}$ and $P^{-1} = -\frac{1}{2}\begin{bmatrix} 0 & 2 & -1 \\ 2 & 2 & 0 \\ -2 & -2 & -1 \end{bmatrix}$

Now $P^{-1}AP = \begin{bmatrix} 0 & -1 & \frac{1}{2} \\ -1 & -1 & 0 \\ 1 & 1 & \frac{1}{2} \end{bmatrix}\begin{bmatrix} 1 & 0 & -1 \\ 1 & 2 & 1 \\ 2 & 2 & 3 \end{bmatrix}\begin{bmatrix} 1 & -2 & -1 \\ -1 & 1 & 1 \\ 0 & 2 & 2 \end{bmatrix} = \begin{bmatrix} 1 & 0 & 0 \\ 0 & 2 & 0 \\ 0 & 0 & 3 \end{bmatrix} = D$

$$A^4 = PD^4P^{-1} = \begin{bmatrix} 1 & -2 & -1 \\ -1 & 1 & 1 \\ 0 & 2 & 2 \end{bmatrix}\begin{bmatrix} 1 & 0 & 0 \\ 0 & 16 & 0 \\ 0 & 0 & 81 \end{bmatrix}\begin{bmatrix} 0 & -1 & \frac{1}{2} \\ -1 & -1 & 0 \\ 1 & 1 & \frac{1}{2} \end{bmatrix} = \begin{bmatrix} -49 & -50 & -40 \\ 65 & 66 & 40 \\ 130 & 130 & 81 \end{bmatrix} \quad \textbf{Ans.}$$

EXERCISE 21.9

Find a matrix P which transforms the following matrices to diagonal form. Hence calculate the power matrix.

1. If $A = \begin{bmatrix} 1 & 1 & 3 \\ 1 & 5 & 1 \\ 3 & 1 & 1 \end{bmatrix}$, calculate A^4. **Ans.** $\begin{bmatrix} 251 & 405 & 235 \\ 405 & 891 & 405 \\ 235 & 405 & 251 \end{bmatrix}$

2. If $A = \begin{bmatrix} 3 & -1 & 1 \\ -1 & 5 & -1 \\ 1 & -1 & 3 \end{bmatrix}$, calculate A^4. **Ans.** $\begin{bmatrix} 251 & -405 & 235 \\ -405 & 891 & -405 \\ 235 & -405 & 251 \end{bmatrix}$

3. If $A=\begin{bmatrix}2 & -1 & 1\\ -1 & 2 & -1\\ 1 & -1 & 2\end{bmatrix}$, calculate A^6. **Ans.** $\begin{bmatrix}1366 & -1365 & 1365\\ -1365 & 1366 & -1365\\ 1365 & -1365 & 1366\end{bmatrix}$

4. If $A=\begin{bmatrix}1 & 1 & 1\\ 0 & 2 & 1\\ -4 & 4 & 3\end{bmatrix}$, calculate A^8. **Ans.** $\begin{bmatrix}-12099 & 12355 & 6305\\ -12100 & 12356 & 6305\\ -13120 & 13120 & 6561\end{bmatrix}$

5. Show that the matrix A is diagonalisable $A=\begin{bmatrix}3 & 1 & -1\\ -2 & 1 & 2\\ 0 & 1 & 2\end{bmatrix}$. If so obtain the matrix P such that P^{-1} AP is a diagonal matrix. *(AMIETE, June 2010)*

21.23 SYLVESTER THEOREM

Let $P(A) = C_0 A^n + C_1 A^{n-1} + C_2 A^{n-2} + \dots + C_{n-1} A + C_n I$

and $|\lambda I - A| = f(\lambda)$ and Adjoint matrix of $[\lambda I - A] = [f(\lambda)]$

$$z(\lambda) = \frac{[f(\lambda)]}{f'(\lambda)} = \frac{\text{Adjoint matrix of } [\lambda I - A]}{f'(\lambda)}$$

Then according to Sylvester's theorem

$$P(A) = P(\lambda_1).Z(\lambda_1) + P(\lambda_2).Z(\lambda_2) + P(\lambda_3).Z(\lambda_3) + \dots$$

$$= \sum_{r=1}^{n} P(\lambda_r).Z(\lambda_r)$$

Example 39. *If* $A=\begin{bmatrix}2 & 0\\ 0 & 1\end{bmatrix}$, *find* A^{100}.

Solution. $f(\lambda) = |\lambda I - A| = \begin{vmatrix}\lambda & 0\\ 0 & \lambda\end{vmatrix} - \begin{vmatrix}2 & 0\\ 0 & 1\end{vmatrix} = \begin{vmatrix}\lambda-2 & 0\\ 0 & \lambda-1\end{vmatrix} = 0$

$\Rightarrow$ $f(\lambda) = (\lambda-2)(\lambda-1) = 0$ or $\lambda_1 = 1,\ \lambda_2 = 2$

$f(\lambda) = \lambda^2 - 3\lambda + 2,\ f'(\lambda) = 2\lambda - 3$

$f'(2) = 4 - 3 = 1, f'(1) = 2 - 3 = -1$

$$[f(\lambda)] = \text{Adjoint matrix of the matrix } [\lambda I - A] = \begin{bmatrix}\lambda-1 & 0\\ 0 & \lambda-2\end{bmatrix}$$

$$Z(\lambda_1) = Z(1) = \frac{[f(1)]}{f'(1)} = \frac{1}{-1}\begin{bmatrix}0 & 0\\ 0 & -1\end{bmatrix} = \begin{bmatrix}0 & 0\\ 0 & 1\end{bmatrix}$$

$$Z(\lambda_2) = Z(2) = \frac{[f(2)]}{f'(2)} = \frac{1}{1}\begin{bmatrix}1 & 0\\ 0 & 0\end{bmatrix}$$

By Sylvester theorem $P(A) = P(\lambda_1).Z(\lambda_1) + P(\lambda_2).Z(\lambda_2)$

$$A^{100} = P(\lambda_1)Z(\lambda_1) + P(\lambda_2)Z(\lambda_2)$$

$$= \lambda_1^{100}\begin{bmatrix}0 & 0\\ 0 & 1\end{bmatrix} + \lambda_2^{100}\begin{bmatrix}1 & 0\\ 0 & 0\end{bmatrix} = 1^{100}\begin{bmatrix}0 & 0\\ 0 & 1\end{bmatrix} + 2^{100}\begin{bmatrix}1 & 0\\ 0 & 0\end{bmatrix}$$

$$= \begin{bmatrix}0 & 0\\ 0 & 1\end{bmatrix} + \begin{bmatrix}2^{100} & 0\\ 0 & 0\end{bmatrix} = \begin{bmatrix}2^{100} & 0\\ 0 & 1\end{bmatrix}$$ **Ans.**

Example 40. *If* $A=\begin{bmatrix}1 & 0\\ 0 & 3\end{bmatrix}$, *find* A^{50}.

Solution. $f(\lambda)=|\lambda I-A|=\begin{vmatrix}\lambda & 0\\ 0 & \lambda\end{vmatrix}-\begin{vmatrix}1 & 0\\ 0 & 3\end{vmatrix}=\begin{vmatrix}\lambda-1 & 0\\ 0 & \lambda-3\end{vmatrix}=0$

$\Rightarrow \quad f(\lambda)=(\lambda-1)(\lambda-3)=0$ or $\lambda_1=1,\ \lambda_2=3$

$f(\lambda)=\lambda^2-4\lambda+3,\ f'(\lambda)=2\lambda-4$

$f'(1)=2-4=-2,\ f'(3)=6-4=2$

$[f(\lambda)]$ = Adjoint matrix of the matrix $[\lambda I-A]=\begin{bmatrix}\lambda-3 & 0\\ 0 & \lambda-1\end{bmatrix}$

$$Z(\lambda_1)=Z(1)=\frac{[f(1)]}{f'(1)}=\frac{1}{-2}\begin{bmatrix}-2 & 0\\ 0 & 0\end{bmatrix}=\begin{bmatrix}1 & 0\\ 0 & 0\end{bmatrix}$$

$$Z(\lambda_2)=Z(3)=\frac{[f(3)]}{f'(3)}=\frac{1}{2}\begin{bmatrix}0 & 0\\ 0 & 2\end{bmatrix}=\begin{bmatrix}0 & 0\\ 0 & 1\end{bmatrix}$$

By Sylvester theorem $P(A)=P_1(\lambda_1).Z(\lambda_1)+P(\lambda_2).Z(\lambda_2)$

$$A^{50}=P(\lambda_1)Z(\lambda_1)+P(\lambda_2)Z(\lambda_2)=\lambda_1^{50}\begin{bmatrix}1 & 0\\ 0 & 0\end{bmatrix}+\lambda_2^{50}\begin{bmatrix}0 & 0\\ 0 & 1\end{bmatrix}$$

$$=1^{50}\begin{bmatrix}1 & 0\\ 0 & 0\end{bmatrix}+3^{50}\begin{bmatrix}0 & 0\\ 0 & 1\end{bmatrix}=\begin{bmatrix}1 & 0\\ 0 & 0\end{bmatrix}+\begin{bmatrix}0 & 0\\ 0 & 3^{50}\end{bmatrix}=\begin{bmatrix}1 & 0\\ 0 & 3^{50}\end{bmatrix}$$ **Ans.**

EXERCISE 21.10

1. Verify Sylvesters theorem for A^3, where $A=\begin{bmatrix}1 & 4\\ 3 & 2\end{bmatrix}$

Use sylvesters theorem in solving the following:

2. Given $A=\begin{bmatrix}1 & 0\\ 0 & 3\end{bmatrix}$, find A^{256}. **Ans.** $\begin{bmatrix}1 & 0\\ 0 & 3^{256}\end{bmatrix}$

3. Given $A=\begin{bmatrix}\lambda_1 & 0\\ 0 & \lambda_2\end{bmatrix}$, show that $e^A=\begin{bmatrix}e^{\lambda_1} & 0\\ 0 & e^{\lambda_2}\end{bmatrix}$.

4. Given $A=\begin{bmatrix}-1 & 3\\ 1 & 1\end{bmatrix}$, show that $2\sin A=|\sin 2|\,A$.

5. Prove that $3\tan A=A\tan(3)$ where $A=\begin{bmatrix}-1 & 4\\ 2 & 1\end{bmatrix}$

6. Prove that $\sin^2 A+\cos^2 A=1$, where $A=\begin{bmatrix}1 & 2\\ -1 & 4\end{bmatrix}$

7. Given $A=\begin{bmatrix}1 & -2 & -3\\ 0 & 2 & 0\\ 0 & 0 & 3\end{bmatrix}$, find A^{-1}. **Ans.** $\begin{bmatrix}1 & -1 & 1\\ 0 & -\frac{1}{2} & 0\\ 0 & 0 & \frac{1}{3}\end{bmatrix}$

8. Given $A=\begin{bmatrix}1 & 20 & 0\\ -1 & 7 & 1\\ 3 & 0 & -2\end{bmatrix}$, find $\tan A$.

Ans. $\dfrac{\tan 1}{2}\begin{bmatrix}-18 & 60 & 20\\ 0 & 0 & 0\\ -18 & 60 & 20\end{bmatrix}+\dfrac{\tan 2}{-1}\begin{bmatrix}-20 & 80 & 20\\ -1 & 4 & 1\\ -15 & 60 & 15\end{bmatrix}+\dfrac{\tan 3}{2}\begin{bmatrix}-20 & 100 & 20\\ -2 & 10 & 2\\ -12 & 60 & 12\end{bmatrix}$

21.24 COMPLEX MATRICES

Conjugate of a Complex Number

$z = x + iy$ is called a complex number where $\sqrt{-1} = i$, x, y are real numbers. $\bar{z} = x - iy$ is called the conjugate of the complex number z, e.g.,

Complex number	Conjugate number
$2 + 3i$	$2 - 3i$
$-4 - 5i$	$-4 + 5i$
$6i$	$-6i$
2	2

Conjugate of a matrix. The matrix formed by replacing the elements of a matrix by their respective conjugate numbers is called the conjugate of A and is denoted by $\bar{A}$.

$$A = (a_{ij})_{m\times n}, \text{ then } \bar{A} = (\bar{a}_{ij})_{m\times n}$$

Example

$$\text{If } A = \begin{bmatrix} 3+4i & 2-i & 4 \\ i & 2 & -3i \end{bmatrix} \text{ then } \bar{A} = \begin{bmatrix} 3-4i & 2+i & 4 \\ -i & 2 & 3i \end{bmatrix}$$

21.25 THEOREM

If A and B be two matrices and their conjugate matrices are $\bar{A}$ and $\bar{B}$ respectively, then

(*i*) $(\bar{\bar{A}}) = A$ (*ii*) $(\overline{A+B}) = \bar{A} + \bar{B}$ (*iii*) $(\overline{kA}) = \bar{k}\bar{A}$ (*iv*) $(\overline{AB}) = \bar{A}\bar{B}$

Proof. Let $A = [a_{ij}]_{m\times n}$, then

$\bar{A} = [\bar{a}_{ij}]_{m\times n}$ where $\bar{a}_{ij}$ is the conjugate complex of a_{ij}.

The (i, j) th element of $(\bar{\bar{A}})$ = the conjugate complex of the (i, j)th element of $\bar{A}$

= the conjugate complex of $\bar{a}_{ij}$

= a_{ij} = the (i, j)th element of A.

Hence $(\bar{\bar{A}}) = A$. **Proved.**

(*ii*) Let $A = [a_{ij}]_{m\times n}$ and $B = [b_{ij}]_{m\times n}$

$\bar{A} = [\bar{a}_{ij}]_{m\times n}$ and $\bar{B} = [\bar{b}_{ij}]_{m\times n}$

(i, j) th element of $(\overline{A+B})$ = conjugate complex of (i, j) th element of $(A + B)$

= conjugate complex of $(a_{ij} + b_{ij})$

$= \overline{(a_{ij} + b_{ij})} = \bar{a}_{ij} + \bar{b}_{ij}$

= (i, j)th element of $\bar{A}$ + (i, j)th element of $\bar{B}$

= (i, j)th element of $(\bar{A} + \bar{B})$

Hence, $(\overline{A+B}) = \bar{A} + \bar{B}$ **Proved.**

(*iii*) Let $A = [a_{ij}]_{m\times n}$, let k be any complex number.

The (i, j)th element of $(\overline{kA})$ = conjugate complex of the (i, j)th element of kA

= conjugate complex of ka_{ij}

$= \overline{ka_{ij}} = \bar{k}\cdot\bar{a}_{ij}$

$= \bar{k}\cdot(i, j)$th element of $\bar{A} = (i, j)$th element of $\bar{k}.\ \bar{A}$

Hence, $\overline{kA} = \bar{k}\cdot\bar{A}$ **Proved.**

(*iv*) Let $\quad A = [a_{ij}]_{m \times n},\ B = [b_{ij}]_{n \times p}$

Then $\quad \overline{A} = [\overline{a_{ij}}]_{m\times n},\ \overline{B} = [\overline{b_{ij}}]_{n\times p}$

The (i, j)th element of $(\overline{AB})$ = conjugate complex of (i, j)th element of AB

$$= \text{conjugate complex of } \sum_{j=1}^{n} a_{ij}\, b_{jk} = \left(\sum_{j=1}^{n} \overline{a_{ij} b_{jk}}\right) = \sum_{j=1}^{n} \overline{a_{ij}} \cdot \overline{b_{jk}}$$

$= (i, j)$th element of $\overline{A} \cdot \overline{B}$

Hence, $\quad (\overline{AB}) = \overline{A} \cdot \overline{B}$ **Proved.**

21.26 TRANSPOSE OF CONJUGATE OF A MATRIX

The transpose of a conjugate of a matrix A is denoted by A^θ or A^*.

$$(\overline{A})' = A^\theta$$

The (i, j)th element of A^θ = (j, i)th element of $\overline{A}$
= conjugate complex of (j, i)th element of A.

Example 41. *If* $A = \begin{bmatrix} 2+3i & 1-2i & 2+4i \\ 3-4i & 4+3i & 2-6i \\ 5 & 5+6i & 3 \end{bmatrix}$, *find* A^θ

Solution. We have, $A = \begin{bmatrix} 2+3i & 1-2i & 2+4i \\ 3-4i & 4+3i & 2-6i \\ 5 & 5+6i & 3 \end{bmatrix} \Rightarrow \overline{A} = \begin{bmatrix} 2-3i & 1+2i & 2-4i \\ 3+4i & 4-3i & 2+6i \\ 5 & 5-6i & 3 \end{bmatrix}$

$$A^\theta = (\overline{A})' = \begin{bmatrix} 2-3i & 3+4i & 5 \\ 1+2i & 4-3i & 5-6i \\ 2-4i & 2+6i & 3 \end{bmatrix}$$ **Ans.**

EXERCISE 21.11

1. If the matrix $A = \begin{bmatrix} 1+i & 3-5i \\ 2i & 5 \end{bmatrix}$, find (*i*) $\overline{A}$ (*ii*) $(\overline{A})'$ (*iii*) A^θ (*iv*) $(A^\theta)^\theta$

Ans. (*i*) $\overline{A} = \begin{bmatrix} 1-i & 3+5i \\ -2i & 5 \end{bmatrix}$ (*ii*) $(\overline{A})' = \begin{bmatrix} 1-i & -2i \\ 3+5i & 5 \end{bmatrix}$

(*iii*) $A^\theta = \begin{bmatrix} 1-i & -2i \\ 3+5i & 5 \end{bmatrix}$ (*iv*) $(A^\theta)^\theta = \begin{bmatrix} 1+i & 3-5i \\ 2i & 5 \end{bmatrix}$

21.27 HERMITIAN MATRIX

Definition. A square matrix $A = [a_{ij}]$ is said to be Hermitian if the (i, j)th element of A, *i.e.*,

$$a_{ij} = \overline{a}_{ji} \text{ for all } i \text{ and } j.$$

For example, $\begin{bmatrix} 2 & 3+4i \\ 3-4i & 1 \end{bmatrix}, \begin{bmatrix} a & b-id \\ b+id & c \end{bmatrix}$

Hence all the elements of the principal diagonal are real.

A necessary and sufficient condition for a matrix A to be Hermitian is that $A = A^\theta$.

Example 42. *Prove that the following*

(*i*) $(A^\theta)^\theta = A$ (*ii*) $(A+B)^\theta = A^\theta + B^\theta$ (*iii*) $(kA)^\theta = \overline{k}\, A^\theta$ (*iv*) $(AB)^\theta = B^\theta \cdot A^\theta$

where A^θ *and* B^θ *be the transposed conjugates of A and B respectively, A and B being conformable to multiplication.*

Solution.

(*i*) $(A^\theta)^\theta = [\{\overline{(\overline{A})'}\}]' = [\overline{\overline{A}}] = A$ as $\{(A)'\}' = \overline{A}$

(*ii*) $(A+B)^\theta = (\overline{A+B})' = (\overline{A}+\overline{B})'$

$= (\overline{A})' + (\overline{B})' = A^\theta + B^\theta$

(*iii*) $(kA)^\theta = (\overline{kA})' = (\overline{k}\,\overline{A})' = \overline{k}\,(\overline{A})' = \overline{k}\,A^\theta$

(*iv*) $(AB)^\theta = (\overline{AB})' = (\overline{A}\cdot\overline{B})' = (\overline{B})'\cdot(\overline{A})' = B^\theta \cdot A^\theta$ **Proved.**

Example 43. *Prove that matrix* $A = \begin{bmatrix} 1 & 1-i & 2 \\ 1+i & 3 & i \\ 2 & -i & 0 \end{bmatrix}$ *is Hermitian.*

Solution. $\overline{A} = \begin{bmatrix} 1 & 1+i & 2 \\ 1-i & 3 & -i \\ 2 & i & 0 \end{bmatrix} \Rightarrow (\overline{A})' = \begin{bmatrix} 1 & 1-i & 2 \\ 1+i & 3 & i \\ 2 & -i & 0 \end{bmatrix}$

$\Rightarrow \quad A^\theta = A \quad \Rightarrow A$ is Hermitian matrix. **Proved.**

Example 44. *Show that* $A = \begin{bmatrix} -i & 3+2i & -2-i \\ -3+2i & 0 & 3-4i \\ 2-i & -3-4i & -2i \end{bmatrix}$ *is Skew-Hermitian matrix.*

Solution. $\overline{A} = \begin{bmatrix} i & 3-2i & -2+i \\ -3-2i & 0 & 3+4i \\ 2+i & -3+4i & 2i \end{bmatrix}$

$(\overline{A})' = \begin{bmatrix} i & -3-2i & 2+i \\ 3-2i & 0 & -3+4i \\ -2+i & 3+4i & 2i \end{bmatrix}$

$\Rightarrow \quad A^\theta = \begin{bmatrix} i & -3-2i & 2+i \\ 3-2i & 0 & -3+4i \\ -2+i & 3+4i & 2i \end{bmatrix}$ $[\because A^\theta = (\overline{A})']$

$= -\begin{bmatrix} -i & 3+2i & -2-i \\ -3+2i & 0 & 3-4i \\ 2-i & -3-4i & -2i \end{bmatrix} = -A$

$A^\theta = -A \Rightarrow A$ is Skew-Hermitian matrix. **Proved.**

Example 45. *Show that the matrix $B^\theta AB$ is Hermitian or Skew-Hermitian according as A is Hermitian or Skew-Hermitian.*

Solution. (*i*) Let A be Hermitian $\Rightarrow A^\theta = A$

Now $(B^\theta A B)^\theta = (AB)^\theta\,(B^\theta)^\theta$

$= B^\theta \cdot A^\theta \cdot B$

$= B^\theta \cdot A \cdot B \quad (A^\theta = A)$

Hence, $A^\theta AB$ is Hermitian.

(*ii*) Let A be Skew-Hermitian $\Rightarrow A^\theta = -A$

Now, $$(B^{\theta}AB)^{\theta} = (AB)^{\theta}\cdot(B^{\theta})^{\theta}$$
$$= B^{\theta}\cdot A^{\theta}\cdot B$$
$$= -B^{\theta}A\cdot B \qquad (A^{\theta} = -A)$$

Hence, $B^{\theta}\ AB$ is Skew-Hermitian. **Proved.**

21.28 THE CHARACTERISTIC ROOTS OF A HERMITIAN MATRIX ARE ALL REAL

Solution.

We know that matrix A is Hermitian if

$$A^{\theta} = A \text{ i.e., where } A^{\theta} = \left(\overline{A'}\right) \text{ or } \left(\overline{A}\right)'$$

Also $(\lambda A)^{\theta} = \overline{\lambda}A^{\theta}$ and $(AB)^{\theta} = B^{\theta}A^{\theta}$.

If λ is a characteristic root of matrix A then $AX = \lambda X$. ...(1)

$\therefore$ $(AX)^{\theta} = (\lambda X)^{\theta}$ or $X^{\theta}A^{\theta} = \overline{\lambda}X^{\theta}$.

But A is Hermitian. $\therefore A^{\theta} = A$.

$\therefore$ $X^{\theta}A = \overline{\lambda}X^{\theta}$ $\quad\therefore\quad X^{\theta}AX = \overline{\lambda}X^{\theta}X$ ...(2)

Again from (1) $X^{\theta}AX = X^{\theta}\lambda X = \lambda X^{\theta}X$ ...(3)

Hence from (2) and (3) we conclude that $\overline{\lambda} = \lambda$ showing that λ is real.

Deduction 1. *From above we conclude that characteristic roots of real symmetric matrix are all real, as in this case, real symmetric matrix will be Hermitian.*

For symmetric, we know that $A' = A$. $\left(\overline{A'}\right) = \overline{A}$

or $A^{\theta} = A$ $\quad\therefore\ \overline{A} = A$ as A is real. Rest as above.

21.29 SKEW-HERMITIAN MATRIX

Definition. A square matrix $A = (a_{ij})$ is said to be Skew-Hermitian matrix if the (i, j)th element of A is equal to the negative of the conjugate complex of the (j, i)th element of A, *i.e.*, $a_{ij} = -\overline{a}_{ji}$ for all i and j.

If A is a Skew-Hermitian matrix, then

$$a_{ii} = -\overline{a}_{ii}$$
$$a_{ii} + \overline{a}_{ii} = 0$$

Obviously, a_{ii} is either a pure imaginary number or must be zero.

For example, $\begin{bmatrix} 0 & -3+4i \\ 3+4i & 0 \end{bmatrix}$ and $\begin{bmatrix} 0 & a-ib \\ -a-ib & 0 \end{bmatrix}$ are Skew-Hermitian matrixes.

A necessary and sufficient condition for a matrix A to be Skew-Hermitian is that $A^{\theta} = -A$.

Deduction 2. *Characteristic roots of a skew Hermitian matrix is either zero or a pure imaginary numbers.*

If A is skew Hermitian, then iA is Hermitian.

Also λ be a characteristic root of A then $AX = \lambda X$.

$\therefore$ $(i\,.A)\,X = (i\lambda)\,X.$

Above shows that $i\lambda$ is characteristic root of matrix iA, which is Hermitian and hence $i\lambda$ should be real, which will be possible if λ is either pure imaginary or zero.

Example 46. *Show that every square matrix can be expressed as $R + iS$ uniquely where R and S are Hermitian matrices.*

Solution. Let A be any square matrix. It can be rewritten as

$$A = \left\{\frac{1}{2}(A + A^{\theta})\right\} + i\left\{\frac{1}{2i}(A - A^{\theta})\right\} = R + iS$$

where $R = \frac{1}{2}(A + A^\theta),\ S = \frac{1}{2i}(A - A^\theta)$

Now we have to show that R and S are Hermitian matrices.

$$R^\theta = \frac{1}{2}(A + A^\theta)^\theta = \frac{1}{2}[A^\theta + (A^\theta)^\theta] = \frac{1}{2}(A^\theta + A) = \frac{1}{2}(A + A^\theta) = R$$

Thus R is Hermitian matrix.

Now, $$S^\theta = \left[\frac{1}{2i}(A - A^\theta)\right]^\theta = -\frac{1}{2i}(A - A^\theta)^\theta$$

$$= -\frac{1}{2i}[A^\theta - (A^\theta)^\theta] = \frac{-1}{2i}(A^\theta - A) = \frac{1}{2i}(A - A^\theta) = S$$

Thus S is a Hermitian matrix.

Hence $A = R + iS$, where R and S are Hermitian matrices.

Now, we have to show its **uniqueness**.

Let $A = P + iQ$ be another expression, where P and Q are Hermitian matrices, *i.e.*,

$$P^\theta = P,\ Q^\theta = Q$$

Then $$A^\theta = (P + iQ)^\theta = P^\theta + (iQ)^\theta = P^\theta - iQ^\theta = P - iQ$$

$$A = P + iQ \text{ and } A^\theta = P - iQ$$

$\Rightarrow$ $$P = \frac{1}{2}(A + A^\theta) = R \text{ and } Q = \frac{1}{2i}(A - A^\theta) = S$$

Hence $A = R + iS$ is the unique expression, where R and S are Hermitian matrices. **Proved.**

Example 47. *Express the matrix* $A = \begin{bmatrix} i & 2-3i & 4+5i \\ 6+i & 0 & 4-5i \\ -i & 2-i & 2+i \end{bmatrix}$ *as a sum of Hermitian and Skew Hermitian matrix.* *(U.P.I Sem Dec, 2009)*

Solution. Here, we have

$$A = \begin{bmatrix} i & 2-3i & 4+5i \\ 6+i & 0 & 4-5i \\ -i & 2-i & 2+i \end{bmatrix} \quad \ldots (1)$$

$$\bar{A} = \begin{bmatrix} -i & 2+3i & 4-5i \\ 6-i & 0 & 4+5i \\ i & 2+i & 2-i \end{bmatrix}$$

$$(\bar{A})' = \begin{bmatrix} -i & 6-i & i \\ 2+3i & 0 & 2+i \\ 4-5i & 4+5i & 2-i \end{bmatrix}$$

$$A^\theta = \begin{bmatrix} -i & 6-i & i \\ 2+3i & 0 & 2+i \\ 4-5i & 4+5i & 2-i \end{bmatrix} \quad \ldots (2)$$

On adding (1) & (2), we get

$$A + A^\theta = \begin{bmatrix} 0 & 8-4i & 4+6i \\ 8+4i & 0 & 6-4i \\ 4-6i & 6+4i & 4 \end{bmatrix}$$

Let
$$R = \frac{1}{2}\left[A+A^{\theta}\right] = \begin{bmatrix} 0 & 4-2i & 2+3i \\ 4+2i & 0 & 3-2i \\ 2-3i & 3+2i & 2 \end{bmatrix} \quad \text{... (3)}$$

On subtracting (2) from (1), we get

$$A - A^{\theta} = \begin{bmatrix} 2i & -4-2i & 4+4i \\ 4-2i & 0 & 2-6i \\ -4+4i & -2-6i & 2i \end{bmatrix}$$

$$\frac{1}{2}\left(A - A^{\theta}\right) = \begin{bmatrix} i & -2-i & 2+2i \\ 2-i & 0 & 1-3i \\ -2+2i & -1-3i & i \end{bmatrix} \quad \text{... (4)}$$

From (3) and (4), we have

$$A = \underbrace{\begin{bmatrix} 0 & 4-2i & 2+3i \\ 4+2i & 0 & 3-2i \\ 2-3i & 3+2i & 2 \end{bmatrix}}_{\text{Hermitian matrix}} + \underbrace{\begin{bmatrix} i & -2-i & 2+2i \\ 2-i & 0 & 1-3i \\ -2+2i & -1-3i & i \end{bmatrix}}_{\text{Skew-Hermitian matrix}}$$

Example 48. *Express the matrix* $A = \begin{bmatrix} 1+i & 2 & 5-5i \\ 2i & 2+i & 4+2i \\ -1+i & -4 & 7 \end{bmatrix}$ *as the sum of Hermitian matrix and Skew-Hermitian matrix.*

Solution. $$A = \begin{bmatrix} 1+i & 2 & 5-5i \\ 2i & 2+i & 4+2i \\ -1+i & -4 & 7 \end{bmatrix} \Rightarrow \overline{A} = \begin{bmatrix} 1-i & 2 & 5+5i \\ -2i & 2-i & 4-2i \\ -1-i & -4 & 7 \end{bmatrix} \quad \text{...(1)}$$

$$(\overline{A})' = \begin{bmatrix} 1-i & -2i & -1-i \\ 2 & 2-i & -4 \\ 5+5i & 4-2i & 7 \end{bmatrix} \Rightarrow A^{\theta} = \begin{bmatrix} 1-i & -2i & -1-i \\ 2 & 2-i & -4 \\ 5+5i & 4-2i & 7 \end{bmatrix} \quad \text{...(2)}$$

On adding (1) and (2), we get

$$A + A^{\theta} = \begin{bmatrix} 2 & 2-2i & 4-6i \\ 2+2i & 4 & 2i \\ 4+6i & -2i & 14 \end{bmatrix}$$

Let
$$R = \frac{1}{2}(A + A^{\theta}) = \begin{bmatrix} 1 & 1-i & 2-3i \\ 1+i & 2 & i \\ 2+3i & -i & 7 \end{bmatrix} \quad \text{...(3)}$$

On subtracting (2) from (1), we get

$$A - A^{\theta} = \begin{bmatrix} 2i & 2+2i & 6-4i \\ -2+2i & 2i & 8+2i \\ -6-4i & -8+2i & 0 \end{bmatrix}$$

Let
$$S = \frac{1}{2}(A - A^{\theta}) = \begin{bmatrix} i & 1+i & 3-2i \\ -1+i & i & 4+i \\ -3-2i & -4+i & 0 \end{bmatrix} \quad \text{...(4)}$$

From (3) and (4), we have

$$A = \begin{bmatrix} 1 & 1-i & 2-3i \\ 1+i & 2 & i \\ 2+3i & -i & 7 \end{bmatrix} + \begin{bmatrix} i & 1+i & 3-2i \\ -1+i & i & 4+i \\ -3-2i & -4+i & 0 \end{bmatrix}$$ **Ans.**

Hermitian matrix Skew-Hermitian matrix

Example 49. *For any square matrix, if* $AA^\theta = I$ *show that* $A^\theta A = I$.

Solution. $AA^\theta = I$ (given)

So A is invertible.

Let B be another matrix such that

$$AB = BA = I \quad ...(1)$$

Now $B = BI = B(AA^\theta)$ $(AA^\theta = I)$

$= (BA)\,A^\theta$

$= IA^\theta = A^\theta$ [Using (1)]

We know that $BA = I$ [From (1)]

Putting the value of B from (2) in (1), we get

$\Rightarrow$ $A^\theta A = I$ **Proved.**

21.30 CHARACTERISTIC ROOTS OF A SKEW-HERMITIAN MATRIX IS EITHER ZERO OR PURELY AN IMAGINARY NUMBER

[*U.P. (C.O.) 2003*]

Since A is a skew-Hermitian matrix:

$\therefore$ $i\,A$ is Hermitian matrix.

Let λ be a characteristic root of A.

Then, $AX = \lambda X \Rightarrow (iA)\,X = (i\lambda)\,X$

$\Rightarrow i\lambda$ is a characteristic root of matrix iA.

But $i\lambda$ is a characteristic root of Hermitian matrix.

Therefore, $i\lambda$ should be real.

Hence, λ is either zero or purely imaginary. **Proved.**

21.31 PERIODIC MATRIX

A square matrix is said to be periodic, if $A^{k+1} = A$, where k is a positive integer. If k is the least positive integer for which $A^{k+1} = A$, then A is said to be of period k.

21.32 IDEMPOTENT MATRIX

A square matrix is said to be idempotent provided $A^2 = A$.

21.33 PROVE THAT THE EIGEN VALUES OF AN IDEMPOTENT MATRIX ARE EITHER ZERO OR UNITY

(*R.G.P.V., Bhopal, I Semester, June* 2007)

Solution. Let A be an idempotent matrix.

$\therefore$ $A^2 = A$

Let λ be a characteristic root of A and the corresponding vector be X. Hence $X \neq 0$ and

$$AX = \lambda X \quad ...(1)$$

$\Rightarrow$ $A(AX) = A(\lambda X) = \lambda(AX)$

$\Rightarrow$ $(AA)X = \lambda(\lambda X)$ [$\because$ From (1), $AX = \lambda X$]

$\Rightarrow$ $A^2X = \lambda^2 X$

$\Rightarrow$ $AX = \lambda^2 X$ [$\because A^2 = A$]

$\Rightarrow$ $\lambda X = \lambda^2 X$ [From (1), $AX = \lambda X$]

$\Rightarrow \quad (\lambda^2 - \lambda)X = 0 \qquad \Rightarrow \qquad \lambda^2 - \lambda = 0$

$\Rightarrow \quad \lambda(\lambda - 1) = 0 \qquad \Rightarrow \qquad \lambda = 0, 1 \qquad [\because X \neq 0]$

Hence, the eigen values of an idempotent matrix are either zero or unity. **Proved.**

Example 50. *Determine all the idempotent diagonal matrices of order n.*

Solution. Let A = diag. $[d_1, d_2, d_3, \ldots d_n]$ be an idempotent matrix of order n.

Here, for the matrix 'A' to be idempotent $A^2 = A$

$$\begin{bmatrix} d_1 & 0 & 0........0 \\ 0 & d_2 & 0........0 \\ 0 & 0 & d_3.......0 \\ 0 & 0 & 0........d_n \end{bmatrix}\begin{bmatrix} d_1 & 0 & 0........0 \\ 0 & d_2 & 0........0 \\ 0 & 0 & d_3.......0 \\ 0 & 0 & 0........d_n \end{bmatrix} = \begin{bmatrix} d_1 & 0 & 0........0 \\ 0 & d_2 & 0........0 \\ 0 & 0 & d_3.......0 \\ 0 & 0 & 0........d_n \end{bmatrix}$$

$$\Rightarrow \quad \begin{bmatrix} d_1^2 & 0 & 0........0 \\ 0 & d_2^2 & 0........0 \\ 0 & 0 & d_3^2.......0 \\ 0 & 0 & 0........d_n^2 \end{bmatrix} = \begin{bmatrix} d_1 & 0 & 0........0 \\ 0 & d_2 & 0........0 \\ 0 & 0 & d_3.......0 \\ 0 & 0 & 0........d_n \end{bmatrix}$$

$\therefore \qquad d_1^2 = d_1; \ d_2^2 = d_2 .... d_n^2 = d_n$

i.e., $\qquad d_1 = 0, 1; \ d_2 = 0, 1; \ d_3 = 0, 1 \ .... \ d_n = 0, 1.$

Hence diag. $[d_1, d_2, d_3 \ldots d_n]$, is the required idempotent matrix where

$d_1 = d_2 = d_3 = \ldots d_n = 0$ or 1. **Ans.**

EXERCISE 21.12

1. Which of the following matrices are Hermitian:

$(a) \begin{bmatrix} 1 & 2+i & 3-i \\ 2+i & 2 & 4-i \\ 3+i & 4+i & 3 \end{bmatrix} \qquad (b) \begin{bmatrix} 2i & 3 & 1 \\ 4 & -1 & 6 \\ 3 & 7 & 2i \end{bmatrix}$

$(c) \begin{bmatrix} 4 & 2-i & 5+2i \\ 2+i & 1 & 2-5i \\ 5-2i & 2+5i & 2 \end{bmatrix} \qquad (d) \begin{bmatrix} 0 & i & 3 \\ -7 & 0 & 5i \\ 3i & 1 & 0 \end{bmatrix}$ **Ans.** (*c*)

2. Which of the following matrices are Skew-Hermitian:

$(a) \begin{bmatrix} 2i & -3 & 4 \\ 3 & 3i & -5 \\ -4 & 5 & 4i \end{bmatrix} \qquad (b) \begin{bmatrix} 3i & -1 & 2 \\ 1 & 2i & -6 \\ 4 & 6 & -3i \end{bmatrix}$

$(c) \begin{bmatrix} 0 & 1-i & 2+3i \\ -1-i & 0 & 6i \\ -2+3i & 6i & 4i \end{bmatrix} \qquad (d) \begin{bmatrix} 1 & 3 & 7+i \\ 3i & -i & 6 \\ 7-i & 8 & 0 \end{bmatrix}$ **Ans.** (*a*), (*c*)

3. Give an example of a matrix which is Skew-symmetric but not Skew-Hermitian.

Ans. $\begin{bmatrix} 0 & 2+3i \\ -2-3i & 0 \end{bmatrix}$

4. If A be a Hermitian matrix, show that iA is Skew-Hermitian. Also show that if B be a Skew-Hermitian matrix, then iB must be Hermitian.

5. If A and B are Hermitian matrices, then show that $AB + BA$ is Hermitian and $AB - BA$ is Skew-Hermitian.

6. If A is any square matrix, show that $A + A^{\theta}$ is Hermitian.

7. If $H = \begin{bmatrix} 3 & 5+2i & -3 \\ 5-2i & 7 & 4i \\ -3 & -4i & 5 \end{bmatrix}$, show that H is a Hermitian matrix.

Verify that iH is a Skew-Hermitian matrix.

8. Show that for any complex square matrix A,

(*i*) $(A + A^*)$ is a Hermitian matrix, where $A^* = \overline{A}^T$

(*ii*) $(A - A^*)$ is Skew-Hermitian matrix.

(*iii*) AA^* and A^*A are Hermitian matrices.

9. Show that any complex square matrix can be uniquely expressed as the sum of a Hermitian matrix and a Skew-Hermitian matrix.

10. Express $A = \begin{bmatrix} i & 2-3i & 4+5i \\ 6+i & 0 & 4-5i \\ -i & 2-i & 2+i \end{bmatrix}$ as the sum of Hermitian and Skew-Hermitian matrices.

11. Prove that the latent roots of a Hermitian matrix are all real.

12. If A = $\begin{bmatrix} 2+i & 3 & -1+3i \\ -5 & i & 4-2i \end{bmatrix}$ show that AA^* is a Hermitian matrix; where A^* is the conjugate transpose of A. (*AMIETE, June 2010*)

21.34 UNITARY MATRIX

A square matrix A is said to be unitary matrix if

$$A \cdot A^{\theta} = A^{\theta} A = I$$

Example 51. *If A is a unitary matrix, show that A^T is also unitary.*

Solution. $A \cdot A^{\theta} = A^{\theta} A = I$, since A is a unitary matrix.

$$(AA^{\theta})^{\theta} = (A^{\theta} A)^{\theta} = I^{\theta} \qquad (I^{\theta} = I)$$

$$(AA^{\theta})^{\theta} = (A^{\theta} A)^{\theta} = I$$

$$(A^{\theta})^{\theta} A^{\theta} = A^{\theta} (A^{\theta})^{\theta} = I$$

$$AA^{\theta} = A^{\theta} A = I \qquad [\text{since } (A^{\theta})^{\theta} = A]$$

$$(AA^{\theta})^T = (A^{\theta} A)^T = (I)^T$$

$$(A^{\theta})^T A^T = A^T (A^{\theta})^T = I$$

$$(A^T)^{\theta} \cdot A^T = A^T (A^T)^{\theta} = I$$

Hence, A^T is a unitary matrix. **Proved.**

Example 52. *If A is a unitary matrix, show that A^{-1} is also unitary.*

Solution. $AA^{\theta} = A^{\theta} A = I$, since A is a unitary matrix.

$$(AA^{\theta})^{-1} = (A^{\theta} \cdot A)^{-1} = (I)^{-1} \quad \text{taking inverse}$$

$$(A^{\theta})^{-1} \cdot A^{-1} = A^{-1} (A^{\theta})^{-1} = I$$

$$(A^{-1})^{\theta} \cdot A^{-1} = A^{-1} (A^{-1})^{\theta} = I$$

Hence, A^{-1} is a unitary matrix. **Proved.**

Example 53. *If A and B are two unitary matrices, show that AB is a unitary matrix.*

Solution. $A \cdot A^{\theta} = A^{\theta} A = I$ since A is a unitary matrix. ...(1)

Similarly, $$B \cdot B^\theta = B^\theta B = I \qquad ...(2)$$

Now,
$$\begin{aligned}(AB)(AB)^\theta &= (AB)(B^\theta \cdot A^\theta)\\ &= A(BB^\theta)\cdot A^\theta\\ &= A\,I\,A^\theta && \text{[From (2)]}\\ &= AA^\theta = I && \text{[From (1)]}\end{aligned}$$

Again,
$$\begin{aligned}(AB)^\theta \cdot (AB) &= (B^\theta \cdot A^\theta)\,(AB)\\ &= B^\theta (A^\theta A)\, B && \text{[From (1)]}\\ &= B^\theta\, I\, B\\ &= B^\theta B\\ &= I && \text{[From (2)]}\end{aligned}$$

Hence, AB is a unitary matrix. **Proved.**

Example 54. *Prove that the matrix* $\frac{1}{\sqrt{3}}\begin{bmatrix} 1 & 1+i \\ 1-i & -1 \end{bmatrix}$ *is unitary.*

Solution. Let $A = \frac{1}{\sqrt{3}}\begin{bmatrix} 1 & 1+i \\ 1-i & -1 \end{bmatrix}$

$$A^\theta = \frac{1}{\sqrt{3}}\begin{bmatrix} 1 & 1+i \\ 1-i & -1 \end{bmatrix}$$

$$A^\theta \cdot A = \frac{1}{\sqrt{3}}\begin{bmatrix} 1 & 1+i \\ 1-i & -1 \end{bmatrix} \times \frac{1}{\sqrt{3}}\begin{bmatrix} 1 & 1+i \\ 1-i & -1 \end{bmatrix}$$

$$= \frac{1}{3}\begin{bmatrix} 1+(1+1) & (1+i)-(1+i) \\ (1-i)-1(1-i) & (1+1)+1 \end{bmatrix} = \frac{1}{3}\begin{bmatrix} 3 & 0 \\ 0 & 3 \end{bmatrix} = \begin{bmatrix} 1 & 0 \\ 0 & 1 \end{bmatrix} = I$$

Hence, A is a unitary matrix. **Proved.**

Example 55. *Show that the matrix* $A = \begin{bmatrix} \alpha + i\gamma & -\beta + i\delta \\ \beta + i\delta & \alpha - i\gamma \end{bmatrix}$ *is a unitary matrix if*

$$\alpha^2 + \beta^2 + \gamma^2 + \delta^2 = 1$$ (*U.P., I Semester, Dec.* 2005)

Solution. We have,
$$A = \begin{bmatrix} \alpha + i\gamma & -\beta + i\delta \\ \beta + i\delta & \alpha - i\gamma \end{bmatrix}$$
$$A^\theta = \begin{bmatrix} \alpha - i\gamma & \beta - i\delta \\ -\beta - i\delta & \alpha + i\gamma \end{bmatrix}$$

We know that, a square matrix A is said to be unitary if $A\,A^\theta = I$

$$\begin{bmatrix} \alpha + i\gamma & -\beta + i\delta \\ \beta + i\delta & \alpha - i\gamma \end{bmatrix}\begin{bmatrix} \alpha - i\gamma & \beta - i\delta \\ -\beta - i\delta & \alpha + i\gamma \end{bmatrix} = \begin{bmatrix} 1 & 0 \\ 0 & 1 \end{bmatrix}$$

$$\begin{bmatrix} \alpha^2+\gamma^2+\beta^2+\delta^2 & \alpha\beta - i\alpha\delta + i\beta\gamma + \gamma\delta - \alpha\beta - i\beta\gamma + i\alpha\delta - \delta\gamma \\ \alpha\beta - i\beta\gamma + i\alpha\delta + \gamma\delta - \alpha\beta - i\alpha\delta + i\beta\gamma - \delta\gamma & \beta^2+\delta^2+\alpha^2+\gamma^2 \end{bmatrix} = \begin{bmatrix} 1 & 0 \\ 0 & 1 \end{bmatrix}$$

$$\Rightarrow \quad \begin{bmatrix} \alpha^2+\beta^2+\gamma^2+\delta^2 & 0 \\ 0 & \alpha^2+\beta^2+\gamma^2+\delta^2 \end{bmatrix} = \begin{bmatrix} 1 & 0 \\ 0 & 1 \end{bmatrix}$$

$$\Rightarrow \quad \alpha^2+\beta^2+\gamma^2+\delta^2 = 1$$ **Proved.**

Example 56. *Define a unitary matrix. If* $N = \begin{bmatrix} 0 & 1+2i \\ -1+2i & 0 \end{bmatrix}$ *is a matrix, then show that* $(I - N)(I + N)^{-1}$ *is a unitary matrix, where I is an identity matrix.*

(*U.P., I Semester, Winter* 2000)

Solution. Unitary matrix: A square matrix 'A' is said to be unitary if $A^\theta A = I$, where $A^\theta = (\overline{A})^T$ and I is an identity matrix.

we have $$N = \begin{bmatrix} 0 & 1+2i \\ -1+2i & 0 \end{bmatrix}$$

$$I - N = \begin{bmatrix} 1 & 0 \\ 0 & 1 \end{bmatrix} - \begin{bmatrix} 0 & 1+2i \\ -1+2i & 0 \end{bmatrix} = \begin{bmatrix} 1 & -1+2i \\ 1-2i & 1 \end{bmatrix} \quad ...(1)$$

Now we have to find $(I + N)^{-1}$

$$I + N = \begin{bmatrix} 1 & 0 \\ 0 & 1 \end{bmatrix} + \begin{bmatrix} 0 & 1+2i \\ -1+2i & 0 \end{bmatrix} = \begin{bmatrix} 1 & 1+2i \\ -1+2i & 1 \end{bmatrix}$$

$$|I + N| = 1 - (-1 - 4) = 6$$

$$\text{Adj. } (I + N) = \begin{bmatrix} 1 & -1-2i \\ 1-2i & 1 \end{bmatrix}$$

$$(I + N)^{-1} = \frac{Adj(I+N)}{|I+N|} = \frac{1}{6}\begin{bmatrix} 1 & -1-2i \\ 1-2i & 1 \end{bmatrix} \quad ...(2)$$

For unitary matrix, $A^\theta A = I$

From (1) and (2), we get

$$\therefore \ (I - N)(I + N)^{-1} = \frac{1}{6}\begin{bmatrix} 1 & -1-2i \\ 1-2i & 1 \end{bmatrix}\begin{bmatrix} 1 & -1-2i \\ 1-2i & 1 \end{bmatrix} = \frac{1}{6}\begin{bmatrix} -4 & -2-4i \\ 2-4i & -4 \end{bmatrix} = B \text{ (say)}$$

Now $$(\overline{B})^T = \frac{1}{6}\begin{bmatrix} -4 & 2+4i \\ -2+4i & -4 \end{bmatrix}$$

$$\therefore \quad (\overline{B})^T B = \frac{1}{36}\begin{bmatrix} -4 & 2+4i \\ -2+4i & -4 \end{bmatrix}\begin{bmatrix} -4 & -2-4i \\ 2-4i & -4 \end{bmatrix} = \frac{1}{36}\begin{bmatrix} 36 & 0 \\ 0 & 36 \end{bmatrix} = I.$$

Hence the result. **Proved.**

21.35. THE MODULUS OF EACH CHARACTERISTIC ROOT OF A UNITARY MATRIX IS UNITY.

(*U.P., I Semester, Compartment* 2002)

Solution. Suppose A is a unitary matrix. Then

$$A^\theta A = I.$$

Let λ be a characteristic root of A. Then

$$AX = \lambda X \quad ...(1)$$

Taking conjugate transpose of both sides of (1), we get

$$(AX)^\theta = \overline{\lambda} X^\theta \quad ...(2)$$

$$\Rightarrow \quad X^\theta A^\theta = \overline{\lambda} X^\theta$$

From (1) and (2), we have

$$(X^\theta A^\theta)(AX) = \overline{\lambda}\lambda X^\theta X$$

$$\Rightarrow \quad X^\theta (A^\theta A) X = \overline{\lambda}\lambda X^\theta X$$

$$\Rightarrow \quad X^{\theta} IX = \bar{\lambda}\lambda X^{\theta} X \qquad (\because A^{\theta}.A = I)$$

$$\Rightarrow \quad X^{\theta} X = \bar{\lambda}\lambda X^{\theta} X$$

$$\Rightarrow \quad X^{\theta} X(\bar{\lambda}\lambda - 1) = 0 \qquad ...(3)$$

Since, $X^{\theta}\ X \neq 0$ therefore (3) gives

$$\lambda\bar{\lambda} - 1 = 0. \text{ or } \lambda\bar{\lambda} = 1 \quad \text{or} \quad |\lambda|^2 = 1 \quad \Rightarrow \quad |\lambda| = 1$$ **Proved.**

EXERCISE 21.13

1. Show that the matrix $A = \frac{1}{\sqrt{2}}\begin{bmatrix} 1 & i \\ -i & -1 \end{bmatrix}$ is unitary.

2. Prove that a real matrix is unitary if it is orthogonal.
3. Prove that the following matrix is unitary:

$$\begin{bmatrix} \frac{1}{2}(1+i) & \frac{1}{2}(-1+i) \\ \frac{1}{2}(1+i) & \frac{1}{2}(1-i) \end{bmatrix}$$

4. Show that $U = \frac{1}{\sqrt{3}}\begin{bmatrix} 1 & 1 & 1 \\ 1 & \omega & \omega^2 \\ 1 & \omega^2 & \omega \end{bmatrix}$ is a unitary matrix, where ω is the complex cube root of unity.

5. Prove that the latent roots of a unitary matrix have unit modulus.
6. Verify that the matrix

$$A = \frac{1}{2}\begin{bmatrix} 1+i & 1-i \\ 1-i & 1+i \end{bmatrix}$$

has eigen values with unit modulus.

Tick (✓) the correct answer:

7. If λ is an eigen value of the matrix 'M' then for the matrix $(M - \lambda I)$, which of the following statement (*s*) is/are coorrect ?
   (*i*) Skew symmetric (*ii*) Non singular (*iii*) Singular (*iv*) None of these **Ans.** (*iii*)
   (*U.P., I Sem. Dec. 2009*)

8. A square matrix A is idempotent if :
   (*i*) $A' = A$ (*ii*) $A' = -A$ (*iii*) $A^2 = A$ (*iv*) $A^2 = I$ **Ans.** (*iii*)
   (*R.G.P.V. Bhopal, I Semester June,* 2007)

9. If a square matrix U such that $\overline{U}' = U^{-1}$ then U is
   (*i*) Orthogonal (*ii*) Unitary (*iii*) Symmetric (*iv*) Hermitian **Ans.** (*ii*)
   (*R.G.P.V. Bhopal, I Semester June,* 2007)

10. If λ is an eigen value of a non-singular matrix A then the eigen value of A^{-1} is
    (*i*) $1/\lambda$ (*ii*) λ (*iii*) $-\lambda$ (*iv*) $-1/\lambda$
    (*AMIETE, June 2010*) **Ans.** (*i*)

CHAPTER

22

REVIEW OF VECTOR ALGEBRA

22.1 VECTORS

A vector is a quantity having both magnitude and direction such as force, velocity acceleration, displacement etc.

22.2 ADDITION OF VECTORS

Let $\vec{a}$ and $\vec{b}$ be two given vectors

$\vec{OA} = \vec{a}$ and $\vec{AB} = \vec{b}$ then vector $\vec{OB}$ is called the sum of $\vec{a}$ and $\vec{b}$.

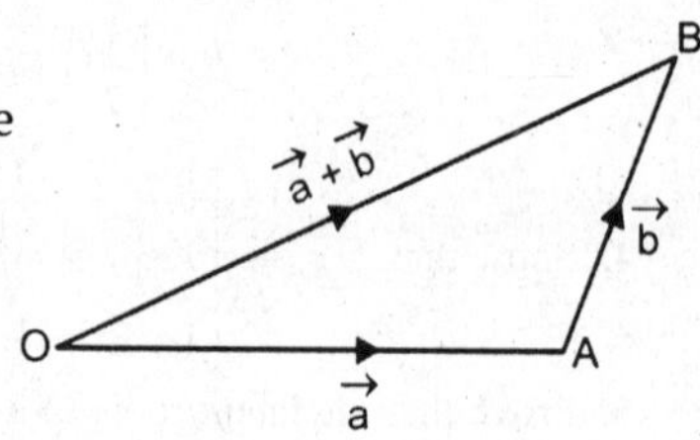

Symbolically

$$\overrightarrow{OA} + \overrightarrow{AB} = \overrightarrow{OB}$$

$$\vec{a} + \vec{b} = \overrightarrow{OB}$$

22.3 RECTANGULAR RESOLUTION OF A VECTOR

Let OX, OY, OZ be the three rectangular axes. Let $\hat{i}, \hat{j}, \hat{k}$ be three unit vectors and parallel to three axes.

If $\vec{OP} = \hat{n}$ and the co-ordinates of P be (x, y, z)

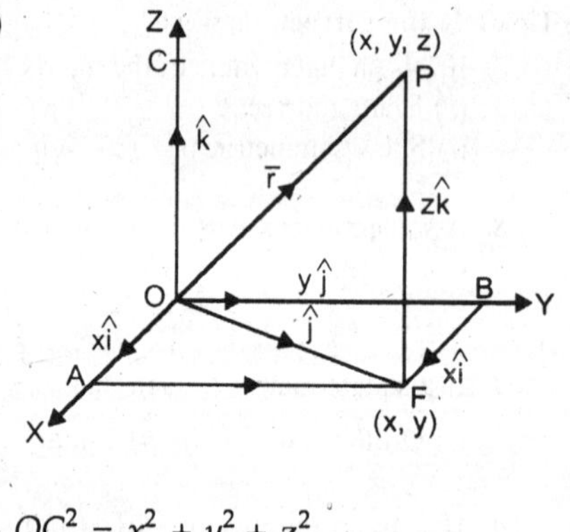

$$\vec{OA} = x\hat{i}, \quad \vec{OB} = y\hat{j} \quad \text{and} \quad \vec{OC} = z\hat{k}$$

$$\vec{OP} = \vec{OF} + \vec{FP}$$

$$\Rightarrow \quad \vec{OP} = (\vec{OA} + \vec{AF}) + \vec{FP}$$

$$\Rightarrow \quad \vec{OP} = \vec{OA} + \vec{OB} + \vec{OC}$$

$$\vec{r} = x\hat{i} + y\hat{j} + z\hat{k}$$

$$\Rightarrow \quad OP^2 = OF^2 + FP^2$$

$$= (OA^2 + AF^2) + FP^2 = OA^2 + OB^2 + OC^2 = x^2 + y^2 + z^2$$

$$OP = \sqrt{x^2 + y^2 + z^2}$$

$$|\vec{r}| = \sqrt{x^2 + y^2 + z^2}$$

22.4 UNIT VECTOR

Let a vector be $x\hat{i} + y\hat{j} + z\hat{k}$.

$$\text{Unit vector} = \frac{x\hat{i} + y\hat{j} + z\hat{k}}{\sqrt{x^2 + y^2 + z^2}}$$

Example 1. *If $\vec{a}$ and $\vec{b}$ be two unit vectors and α be the angle between them, then find the value of α such that $\vec{a} + \vec{b}$ is a unit vector.* (Nagpur, University, Winter 2001)

Solution. Let $\vec{OA} = \vec{a}$ be a unit vector and $\vec{AB} = \vec{b}$ is another unit vector and α be the angle between $\vec{a}$ and $\vec{b}$.

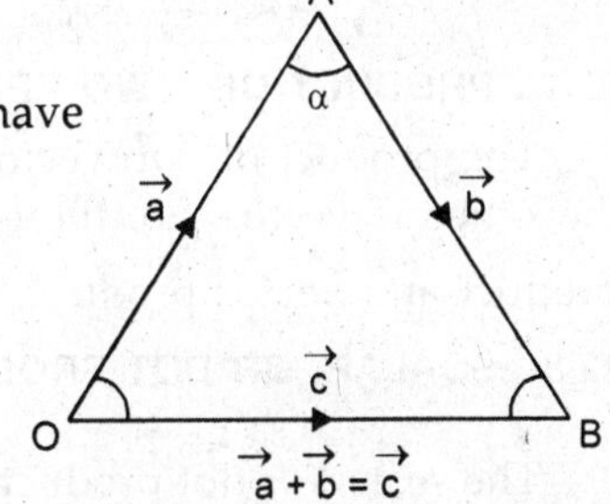

If $\vec{OB} = \vec{c} = \vec{a} + \vec{b}$ is also a unit vector then, we have

$$|\vec{OA}| = 1$$

$$|\vec{OB}| = 1$$

$$|\vec{OB}| = 1$$

OAB is an equilateral triangle.

Hence each angle of $\Delta\, OAB$ is $\dfrac{\pi}{3}$ **Ans.**

22.5 POSITION VECTOR OF A POINT

The position vector of a point A with respect to origin O is the vector $\vec{OA}$ which is used to specify the position of A w.r.t. O.

To find $\overrightarrow{AB}$ if the position vectors of the point A and point B are given.

If the position vectors of A and B are $\vec{a}$ and $\vec{b}$. Let the origin be O.

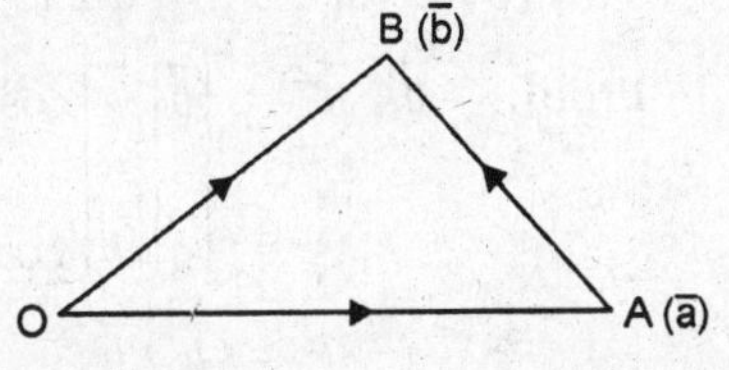

Then $\quad \vec{OA} = \vec{a}, \quad \vec{OB} = \vec{b}$

$$\vec{OA} + \vec{AB} = \vec{OB}$$

$$\vec{AB} = \vec{OB} - \vec{OA}$$

$$\Rightarrow \quad \vec{AB} = \vec{b} - \vec{a}$$

$\vec{AB}$ = **Position vector of B – Position vector of A**

Example 2. *If A and B are (3, 4, 5) and (6, 8, 9), find $\vec{AB}$.*

Solution. $\vec{AB}$ = Position vector of B – Position vector of A

$$= (6\hat{i} + 8\hat{j} + 9\hat{k}) - (3\hat{i} + 4\hat{j} + 5\hat{k})$$

$$= 3\hat{i} + 4\hat{j} + 4\hat{k}$$ **Ans.**

22.6 RATIO FORMULA

To find the position vector of the point which divides the line joining two given points.

Let A and B be two points and a point C divides AB in the ratio of $m : n$.

Let O be the origin, then

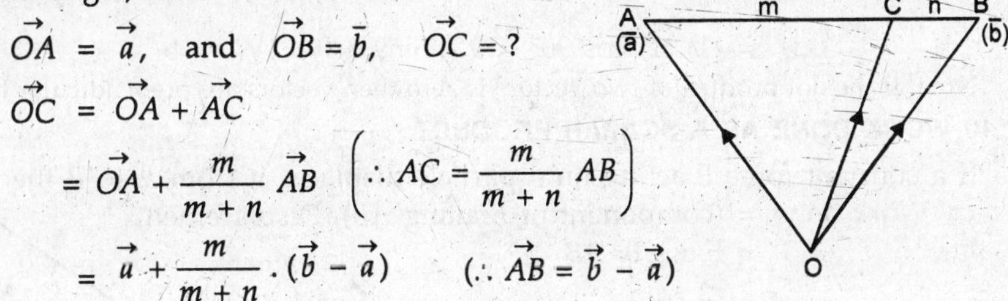

$$\vec{OA} = \vec{a}, \quad \text{and} \quad \vec{OB} = \bar{b}, \quad \vec{OC} = ?$$

$$\vec{OC} = \vec{OA} + \vec{AC}$$

$$= \vec{OA} + \frac{m}{m+n}\vec{AB} \qquad \left(\therefore AC = \frac{m}{m+n} AB\right)$$

$$= \vec{a} + \frac{m}{m+n}\cdot(\vec{b} - \vec{a}) \qquad (\therefore \vec{AB} = \vec{b} - \vec{a})$$

$$\vec{OC} = \frac{m\vec{b} + n\vec{a}}{m + n}$$

Cor. If $m = n = 1$, then C will be the mid-point, and

$$\vec{OC} = \frac{\vec{a} + \vec{b}}{2}$$

22.7 PRODUCT OF TWO VECTORS

The product of two vectors results in two different ways, the one is a number and the other is vector. So, there are two types of product of two vectors, namely scalar product and vector product. They are written as $\vec{a} \cdot \vec{b}$ and $\vec{a} \times \vec{b}$.

22.8 SCALAR, OR DOT PRODUCT

The scalar, or dot product of two vectors $\vec{a}$ and $\vec{b}$ is defined to be $|\vec{a}|\,|\vec{b}|\cos\theta$ *i.e.*, scalar where θ is the angle between $\vec{a}$ and $\vec{b}$.

Symbolically, $\vec{\mathbf{a}} \cdot \vec{\mathbf{b}} = |\vec{\mathbf{a}}|\,|\vec{\mathbf{b}}| \cos\theta$

Due to a dot between $\vec{a}$ and $\vec{b}$ this product is also called dot product.

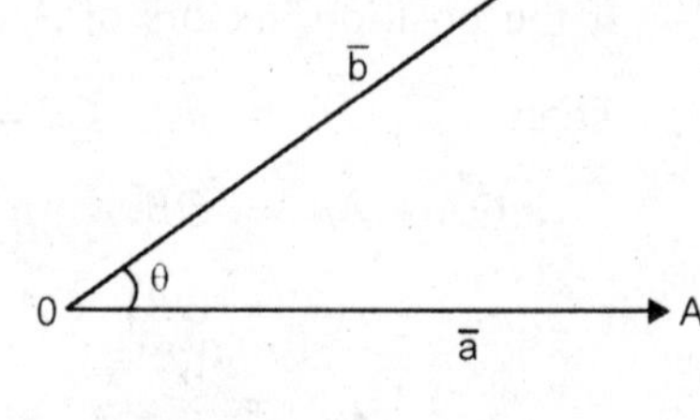

The scalar product is commutative

To Prove. $\vec{a} \cdot \vec{b} = \vec{b} \cdot \vec{a}$

Proof. $\vec{b} \cdot \vec{a} = |\vec{b}|\,|\vec{a}| \cos(-\theta)$

$= |\vec{a}|\,|\vec{b}| \cos\theta$

$= \vec{a} \cdot \vec{b}$ **Proved.**

Geometrical interpretation. The scalar product of two vectors is the product of one vector and the length of the projection of the other in the direction of the first.

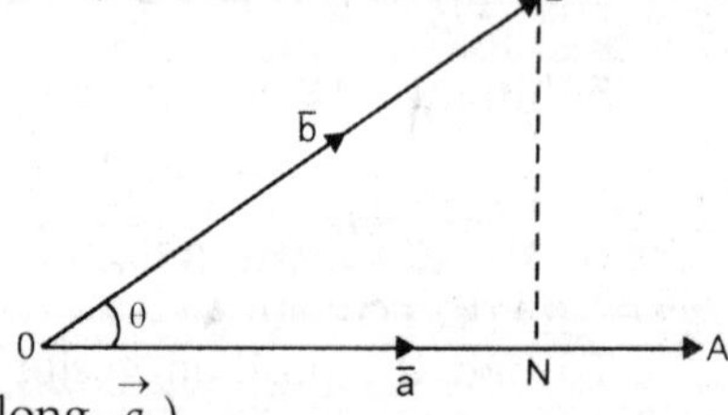

Let $\vec{OA} = \vec{a}$ and $\vec{OB} = \vec{b}$

then $\vec{a} \cdot \vec{b} = (OA) \cdot (OB) \cos\theta$

$= OA \cdot OB \cdot \dfrac{ON}{OB}$

$= OA \cdot ON$

$=$ (Length of $\vec{a}$) (projection of $\vec{b}$ along $\vec{a}$)

22.9 USEFUL RESULTS

$\hat{i} \cdot \hat{i} = (1)(1)\cos 0° = 1$ Similarly, $\hat{j} \cdot \hat{j} = 1$, $\hat{k} \cdot \hat{k} = 1$

$\hat{i} \cdot \hat{j} = (1)(1)\cos 90° = 0$ Similarly, $\hat{j} \cdot \hat{k} = 0$, $\hat{k} \cdot \hat{i} = 0$

Note. If the dot product of two vectors is zero then vectors are prependicular to each other.

22.10 WORK DONE AS A SCALAR PRODUCT

If a constant force F acting on a particle displaces it from A to B then,

Work done $=$ (component of F along AB). Displacement

$= F\cos\theta.\ AB$

$= \vec{F} \cdot \vec{AB}$

Work done = Force . Displacement

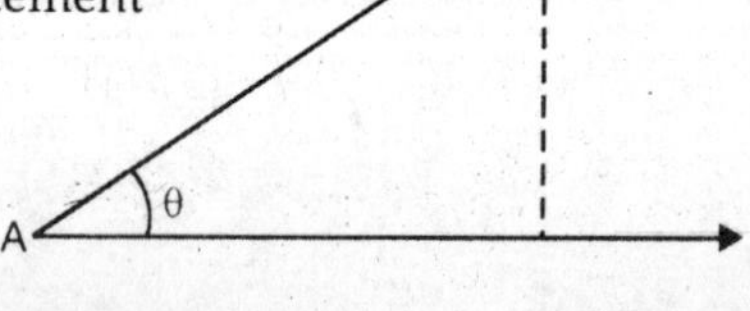

22.11 VECTOR PRODUCT OR CROSS PRODUCT

1. The vector, or cross product of two vectors $\vec{a}$ and $\vec{b}$ is defined to be a vector such that

(*i*) Its magnitude is $|\vec{a}||\vec{b}| \sin \theta$, where θ is the angle between $\vec{a}$ and $\vec{b}$.

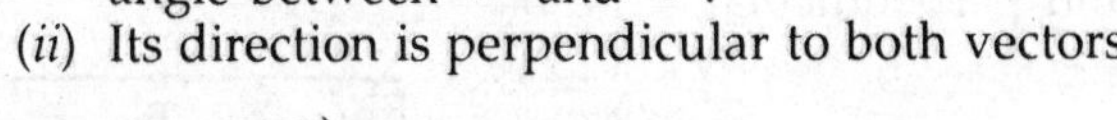

(*ii*) Its direction is perpendicular to both vectors $\vec{a}$ and $\vec{b}$.

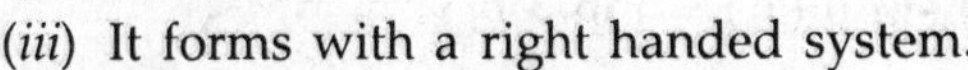

(*iii*) It forms with a right handed system.

Let $\hat{\eta}$ be a unit vector perpendicular to both the vectors $\vec{a}$ and $\vec{b}$

$$\vec{a} \times \vec{b} = |\vec{a}||\vec{b}| \sin \theta . \hat{\eta}$$

2. Useful results

Since $\hat{i}, \hat{j}, \hat{k}$ are three mutually perpendicular unit vectors, then

$$\hat{i} \times \hat{i} = \hat{j} \times \hat{j} = \hat{k} \times \hat{k} = 0$$

$$\hat{i} \times \hat{j} = -\hat{j} \times \hat{i} = \hat{k} \qquad \hat{j} \times \hat{i} = -\hat{i} \times \hat{j}$$

$$\hat{j} \times \hat{k} = -\hat{k} \times \hat{j} = \hat{i} \qquad \text{and} \qquad \hat{k} \times \hat{j} = -\hat{j} \times \hat{k}$$

$$\hat{k} \times \hat{i} = -\hat{i} \times \hat{k} = \hat{j} \qquad \hat{i} \times \hat{k} = -\hat{k} \times \hat{i}$$

22.12 VECTOR PRODUCT EXPRESSED AS A DETERMINANT

If
$$\vec{a} = a_1 \hat{i} + a_2 \hat{j} + a_3 \hat{k}$$
$$\vec{b} = b_1 \hat{i} + b_2 \hat{j} + b_3 \hat{k}$$

$$\vec{a} \times \vec{b} = (a_1 \hat{i} + a_2 \hat{j} + a_3 \hat{k}) \times (b_1 \hat{i} + b_2 \hat{j} + b_3 \hat{k})$$

$$= a_1 b_1 (\hat{i} \times \hat{i}) + a_1 b_2 (\hat{i} \times \hat{j}) + a_1 b_3 (\hat{i} \times \hat{k}) + a_2 b_1 (\hat{j} \times \hat{i}) + a_2 b_2 (\hat{j} \times \hat{j})$$
$$+ a_2 b_3 (\hat{j} \times \hat{k}) + a_3 b_1 (\hat{k} \times \hat{i}) + a_3 b_2 (\hat{k} \times \hat{j}) + a_3 b_3 (\hat{k} \times \hat{k})$$

$$= a_1 b_2 \hat{k} - a_1 b_3 \hat{j} - a_2 b_1 \hat{k} + a_2 b_3 \hat{i} + a_3 b_1 \hat{j} - a_3 b_2 \hat{i}$$

$$= (a_2 b_3 - a_3 b_2) \hat{i} - (a_1 b_3 - a_3 b_1) \hat{j} + (a_1 b_2 - a_2 b_1) \hat{k}$$

$$= \begin{vmatrix} \hat{i} & \hat{j} & \hat{k} \\ a_1 & a_2 & a_3 \\ b_1 & b_2 & b_3 \end{vmatrix}$$

22.13 AREA OF PARALLELOGRAM

Example 3. *Find the area of a parallelogram whose adjacent sides are i – 2j + 3 k and 2i + j – 4k.*

Solution. Vector area of || gm $= \begin{vmatrix} \hat{i} & \hat{j} & \hat{k} \\ 1 & -2 & 3 \\ 2 & 1 & -4 \end{vmatrix}$

$$= (8-3)\hat{i} - (-4-6)\hat{j} + (1+4)\hat{k} = 5\hat{i} + 10\hat{j} + 5\hat{k}$$

Area of parallelogram $= \sqrt{(5)^2 + (10)^2 + (5)^2} = 5\sqrt{6}$ **Ans.**

22.14 MOMENT OF A FORCE

Let a force F $(\overrightarrow{PQ})$ act at a point P.

Moment of $\vec{F}$ about O

= Product of force F and perpendicular distance $(ON.\hat{\eta})$

$$= (PQ)\,(ON)(\hat{\eta}) = (PQ)\,(OP)\sin\theta\,(\hat{\eta}) = \overrightarrow{OP} \times \overrightarrow{PQ}$$

$$\Rightarrow \quad \vec{M} = \vec{r} \times \vec{F}$$

22.15 ANGULAR VELOCITY

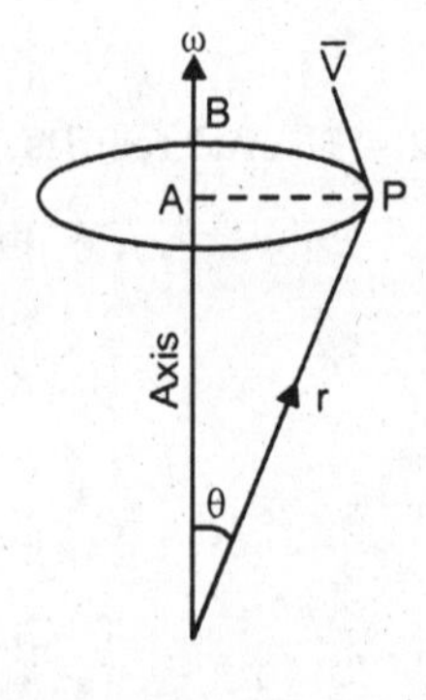

Let a rigid body be rotating about the axis OA with the angular velocity ω which is a vector and its magnitude is ω radians per second and its direction is parallel to the axis of rotation OA.

Let P be any point on the body such that $\overline{OP} = \vec{r}$ and $\angle AOP = \theta$ and $AP \perp OA$. Let the velocity of P be V.

Let $\hat{\eta}$ be a unit vector perpendicular to $\vec{\omega}$ and $\vec{r}$.

$$\vec{\omega} \times \vec{r} = (\omega\, r \sin\theta)\;\hat{\eta} = (\omega\, AP)\;\hat{\eta} = (\text{Speed of } P)\;\hat{\eta}$$

$$= \text{Velocity of } P \perp \text{ to } \vec{\omega} \text{ and } r$$

Hence $\boxed{\vec{V} = \vec{\omega} \times \vec{r}}$

22.16 SCALAR TRIPLE PRODUCT

Let $\vec{a}, \vec{b}, \vec{c}$ be three vectors then their dot product is written as $\vec{a}\,.\,(\vec{b} \times \vec{c})$ or $[\vec{a}\;\vec{b}\;\vec{c}]$.

If $\vec{a} = a_1\hat{i} + a_2\hat{j} + a_3\hat{k},\; \vec{b} = b_1\hat{i} + b_2\hat{j} + b_3\hat{k}$, and $\vec{c} = c_1\hat{i} + c_2\hat{j} + c_3\hat{k}$

$$\vec{a}\,.(\vec{b} \times \vec{c}) = (a_1\hat{i} + a_2\hat{j} + a_3\hat{k}).[(b_1\hat{i} + b_2\hat{j} + b_3\hat{k}) \times (c_1\hat{i} + c_2\hat{j} + c_3\hat{k})]$$

$$= (a_1\hat{i} + a_2\hat{j} + a_3\hat{k}).[(b_2c_3 - b_3c_2)\hat{i} + (b_3c_1 - b_1c_3)\hat{j} + (b_1c_2 - b_2c_1)\hat{k}]$$

$$= a_1(b_2c_3 - b_3c_2) + a_2(b_3c_1 - b_1c_3) + a_3(b_1c_2 - b_2c_1)$$

$$= \begin{vmatrix} a_1 & a_2 & a_3 \\ b_1 & b_2 & b_3 \\ c_1 & c_2 & c_3 \end{vmatrix}$$

Similarly, $\vec{b}\,.\,(\vec{c} \times \vec{a})$ and $\vec{c}\,.\,(\vec{a} \times \vec{b})$ have the same value.

$$\therefore \quad \vec{a}\,.\,(\vec{b} \times \vec{c}) = \vec{b}\,.\,(\vec{c} \times \vec{a}) = \vec{c}\,.\,(\vec{a} \times \vec{b})$$

The value of the product depends upon the cyclic order of the vector, but is independent of the position of the dot and cross. These may be interchanged.

The value of the product changes if the order is non-cyclic.

Note. $\vec{a} \times (\vec{b}\,.\,\vec{c})$ **and** $(\vec{a}\,.\,\vec{b}) \times \vec{c}$ **are meaningless.**

22.17 GEOMETRICAL INTERPRETATION

The scalar triple product $\vec{a}\cdot(\vec{b}\times\vec{c})$ represents the volume of the parallelopiped having $\vec{a}, \vec{b}, \vec{c}$ as its co-terminous edges.

$\vec{a}\cdot(\vec{b}\times\vec{c}) = \vec{a}\,.$ Area of || gm $OBDC\ \hat{\eta}$

= Area of || gm $OBDC$ × perpendicular distance between the parallel faces $OBDC$ and $AEFG$.

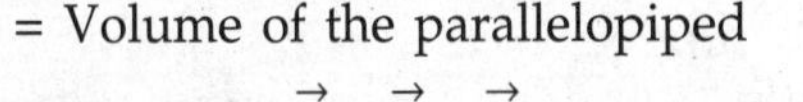

= Volume of the parallelopiped

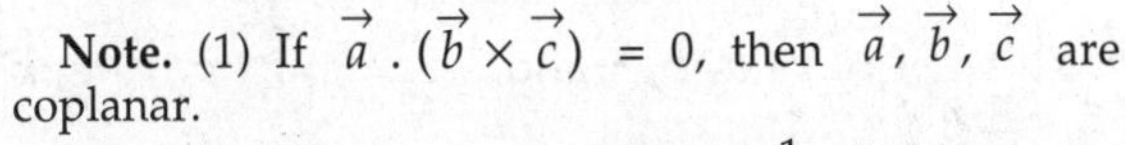

Note. (1) If $\vec{a}\cdot(\vec{b}\times\vec{c}) = 0$, then $\vec{a}, \vec{b}, \vec{c}$ are coplanar.

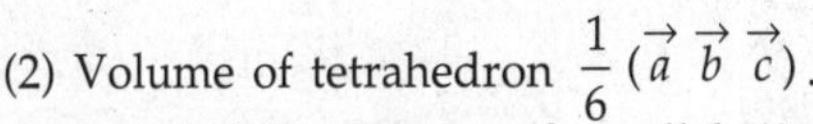

(2) Volume of tetrahedron $\frac{1}{6}(\vec{a}\ \vec{b}\ \vec{c})$.

Example 4. *Find the volume of parallelopiped if*

$$\vec{a} = -3\hat{i} + 7\hat{j} + 5\hat{k},\quad \vec{b} = -3\hat{i} + 7\hat{j} - 3\hat{k}, \text{ and } \vec{c} = 7\hat{i} - 5\hat{j} - 3\hat{k}$$

are the three co-terminous edges of the parallelopiped.

Solution.

$$\text{Volume} = \vec{a}\cdot(\vec{b}\times\vec{c})$$

$$= \begin{vmatrix} -3 & 7 & 5 \\ -3 & 7 & -3 \\ 7 & -5 & -3 \end{vmatrix} = -3(-21-15) - 7(9+21) + 5(15-49)$$

$$= 108 - 210 - 170 = -272$$

Volume = 272 cube units. **Ans.**

Example 5. *Show that the volume of the tetrahedron having* $\vec{A}+\vec{B}, \vec{B}+\vec{C}, \vec{C}+\vec{A}$ *as concurrent edges is twice the volume of the tetrahendron having* $\vec{A}, \vec{B}, \vec{C}$ *as concurrent edges.*

Solution. Volume of tetrahendron $= \frac{1}{6}(\vec{A}+\vec{B})\cdot[(\vec{B}+\vec{C})\times(\vec{C}+\vec{A})]$

$$= \frac{1}{6}(\vec{A}+\vec{B})\cdot[\vec{B}\times\vec{C} + \vec{B}\times\vec{A} + \vec{C}\times\vec{C} + \vec{C}\times\vec{A}] \qquad [\vec{C}\times\vec{C} = 0]$$

$$= \frac{1}{6}(\vec{A}+\vec{B})\cdot(\vec{B}\times\vec{C} + \vec{B}\times\vec{A} + \vec{C}\times\vec{A})$$

$$= \frac{1}{6}[\vec{A}\cdot(\vec{B}\times\vec{C}) + \vec{A}\cdot(\vec{B}\times\vec{A}) + \vec{A}\cdot(\vec{C}\times\vec{A}) + \vec{B}\cdot(\vec{B}\times\vec{C}) + \vec{B}\cdot(\vec{B}\times\vec{A}) + \vec{B}\cdot(\vec{C}\times\vec{A})]$$

$$= \frac{1}{6}[\vec{A}\cdot(\vec{B}\times\vec{C}) + \vec{B}\cdot(\vec{C}\times\vec{A})] = \frac{1}{3}\vec{A}\cdot(\vec{B}\times\vec{C})$$

$$= 2\times\frac{1}{6}[\vec{A}\ \vec{B}\ \vec{C}]$$

= 2 Volume of tetrahedron having $\vec{A}, \vec{B}, \vec{C}$, as concurrent edges. **Proved.**

Example 6. *Using vectors, establish Cramer's rule for solving these following linear equations*

$$a_1 x + b_1 y + c_1 z = d_1,\ a_2 x + b_2 y + c_2 z = d_2,\ a_3 x + b_3 y + c_3 z = d_3$$

Solution. Here, we have, $a_1 x + b_1 y + c_1 z = d_1$ ...(1)

$a_2 x + b_2 y + c_2 z = d_2$ ...(2)

$a_3 x + b_3 y + c_3 z = d_3$ ...(3)

Multiplying equations (1), (2) and (3) by $\hat{i}, \hat{j}$ and $\hat{k}$ respectively, we get

$$(a_1 x + b_1 y + c_1 z)\ \hat{i} = d_1 \hat{i} \qquad ...(4)$$
$$(a_2 x + b_2 y + c_2 z)\ \hat{j} = d_2 \hat{j} \qquad ...(5)$$
$$(a_3 x + b_3 y + c_3 z)\ \hat{k} = d_3 \hat{k} \qquad ...(6)$$

Adding (4), (5) and (6), we have

$$(a_1 \hat{i} + a_2 \hat{j} + a_3 \hat{k})\, x + (b_1 \hat{i} + b_2 \hat{j} + b_3 \hat{k})\, y + (c_1 \hat{i} + c_2 \hat{j} + c_3 \hat{k})\, z = d_1 \hat{i} + d_2 \hat{j} + d_3 \hat{k}$$

$$(\vec{a}\, x + \vec{b}\, y + \vec{c}\, z) = \bar{d} \qquad ...(7)$$

$$[\vec{a} = a_1 \hat{i} + a_2 \hat{j} + a_3 \hat{k},\ \vec{b} = b_1 \hat{i} + b_2 \hat{j} + b_3 \hat{k}, \text{ and } \vec{c} = c_1 \hat{i} + c_2 \hat{j} + c_3 \hat{k}]$$

Take scalar product of (7) with $(\vec{b} \times \vec{c})$

$$\vec{a} \cdot (\vec{b} \times \vec{c})\, x + \vec{b} \cdot (\vec{b} \times \vec{c})\, y + \vec{c} \cdot (\vec{b} \times \vec{c})\, z = \bar{d} \cdot (\vec{b} \times \vec{c})$$

$$\Rightarrow \qquad (\vec{a}\, \vec{b}\, \vec{c})\, x = (\vec{d}\, \vec{b}\, \vec{c}) \qquad \Rightarrow x = \frac{(\vec{d}\, \vec{b}\, \vec{c})}{(\vec{a}\, \vec{b}\, \vec{c})}$$

Similarly, $(\vec{b}\, \vec{c}\, \vec{a})\, y = (\vec{d}\, \vec{c}\, \vec{a}) \qquad \Rightarrow y = \dfrac{(\vec{d}\, \vec{c}\, \vec{a})}{(\vec{b}\, \vec{c}\, \vec{a})}$

$$(\vec{c}\, \vec{a}\, \vec{b})\, z = (\vec{d}\, \vec{a}\, \vec{b}) \qquad \Rightarrow z = \frac{(\vec{d}\, \vec{a}\, \vec{b})}{(\vec{c}\, \vec{a}\, \vec{b})}$$

Thus $\quad x = \dfrac{(\vec{d}\, \vec{b}\, \vec{c})}{(\vec{a}\, \vec{b}\, \vec{c})}, \qquad y = \dfrac{(\vec{d}\, \vec{c}\, \vec{a})}{(\vec{b}\, \vec{c}\, \vec{a})}, \qquad z = \dfrac{(\vec{d}\, \vec{a}\, \vec{b})}{(\vec{c}\, \vec{a}\, \vec{b})}$ **Proved.**

EXERCISE 22.1

1. Find the volume of the parallelopiped with adjacent sides.

$$\overline{OA} = 3\hat{i} - \hat{j}, \quad \overline{OB} = \hat{j} + 2\hat{k}, \quad \text{and} \quad \overline{OC} = \hat{i} + 5\hat{j} + 4\hat{k}$$

extending from the origin of co-ordinates O. **Ans.** 20

2. Find the volume of the tetrahedron whose vertices are the points A (2, –1, –3), B (4, 1, 3) C (3, 2, –1) and D (1, 4, 2). **Ans.** $7\dfrac{1}{3}$

3. Choose y in order that the vectors $\vec{a} = 7\hat{i} + y\hat{j} + \hat{k}$, $\vec{b} = 3\hat{i} + 2\hat{j} + \hat{k}$, $\vec{c} = 5\hat{i} + 3\hat{j} + \hat{k}$ are linearly dependent. **Ans.** $y = 4$

4. Prove that

$$[\vec{a} + \vec{b},\ \vec{b} + \vec{c},\ \vec{c} + \vec{a}] = 2\, [\vec{a}\, \vec{b}\, \vec{c}]$$

22.18 COPLANARITY QUESTIONS

Example 7. *Find the volume of tetrahedron having vertices*

$$(-\hat{j} - \hat{k}), \quad (4\hat{i} + 5\hat{j} + q\hat{k}), \quad (3\hat{i} + 9\hat{j} + 4\hat{k}) \text{ and } 4(-\hat{i} + \hat{j} + \hat{k}).$$

Also find the value of q for which these four points are coplanar.

(Nagpur University, Summer 2004, 2003, 2002)

Solution. Let $\vec{A} = -\hat{j} - \hat{k}$, $\bar{B} = 4\hat{i} + 5\hat{j} + q\hat{k}$, $\bar{C} = 3\hat{i} + 9\hat{j} + 4\hat{k}$, $\bar{D} = 4(-\hat{i} + \hat{j} + \hat{k})$

$$\overline{AB} = \vec{B} - \vec{A} = 4\hat{i} + 5\hat{j} + q\hat{k} - (-\hat{j} - \hat{k}) = 4\hat{i} + 6\hat{j} + (q+1)\hat{k}$$

$$\overline{AC} = \vec{C} - \vec{A} = (3\hat{i} + 9\hat{j} + 4\hat{k}) - (-\hat{j} - \hat{k}) = 3\hat{i} + 10\hat{j} + 5\hat{k}$$

$$\overline{AD} = \vec{D} - \vec{A} = 4(-\hat{i} + \hat{j} + \hat{k}) - (-\hat{j} - \hat{k}) = -4\hat{i} + 5\hat{j} + 5\hat{k}$$

$$\text{Volume of the tetrahedron} = \frac{1}{6}[\overline{AB}\ \overline{AC}\ \overline{AD}]$$

$$= \frac{1}{6}\begin{vmatrix} 4 & 6 & q+1 \\ 3 & 10 & 5 \\ -4 & 5 & 5 \end{vmatrix} = \frac{1}{6}\{4(50-25) - 6(15+20) + (q+1)(15+40)\}$$

$$= \frac{1}{6}\{100 - 210 + 55(q+1)\} = \frac{1}{6}(-110 + 55 + 55q)$$

$$= \frac{1}{6}(-55 + 55q) = \frac{55}{6}(q-1)$$

If four points A, B, C and D are coplanar, then $(\overline{AB}\ \overline{AC}\ \overline{AD}) = 0$
i.e., Volume of the tetrahedron = 0

$$\Rightarrow \quad \frac{55}{6}(q-1) = 0 \quad \Rightarrow \quad q = 1$$ **Ans.**

Example 8. *Find m so that the vectors*
$2\hat{i} - 4\hat{j} + 5\hat{k};\ \hat{i} - m\hat{j} + \hat{k}$, and $3\hat{i} + 2\hat{j} - 5\hat{k}$ *are coplanar.*
(*Nagpur University, Winter 2003*)

Solution. Let
$$\vec{a} = 2\hat{i} - 4\hat{j} + 5\hat{k}$$
$$\vec{b} = \hat{i} - m\hat{j} + \hat{k}$$
$$\vec{c} = 3\hat{i} + 2\hat{j} - 5\hat{k}$$

$\vec{a}$, $\vec{b}$ and $\vec{c}$ are coplanar if $\vec{a}\cdot(\vec{b}\times\vec{c}) = 0$

$$\vec{a}\cdot(\vec{b}\times\vec{c}) = \begin{vmatrix} 2 & -4 & 5 \\ 1 & -m & 1 \\ 3 & 2 & -5 \end{vmatrix} = 0$$

$$0 = 2(5m - 2) + 4(-5 - 3) + 5(2 + 3m)$$
$$0 = 10m - 4 - 32 + 10 + 15m$$
$$0 = 25m - 26$$
$$25m = 26$$

$$\therefore \quad m = \frac{26}{25}.$$ **Ans.**

Example 9. *Show that the vectors*

$$(5\vec{a} + 6\vec{b} + 7\vec{c}),\ (7\vec{a} - 8\vec{b} + 9\vec{c}) \text{ and } (3\vec{a} + 20\vec{b} + 5\vec{c})$$

are coplanar, $\vec{a}, \vec{b}, \vec{c}$ *being three non-collinear vectors.*
(*Nagpur University, Summer 2003*)

Solution. Let
$$\vec{\alpha} = 5\bar{a} + 6\bar{b} + 7\bar{c}$$
$$\vec{\beta} = 7\bar{a} - 8\bar{b} + 9\bar{c}$$
$$\vec{\gamma} = 3\bar{a} + 20\bar{b} + 5\bar{c}$$

$$\vec{\alpha}\cdot(\vec{\beta}\times\vec{\gamma}) = \begin{vmatrix} 5 & 6 & 7 \\ 7 & -8 & 9 \\ 3 & 20 & 5 \end{vmatrix}$$
$$= 5(-40 - 180) - 6(35 - 27) + 7(140 + 24)$$

$$= -5 \times 220 - 6 \times 8 + 7 \times 164$$
$$= -1100 - 48 + 1148 = 0$$

Hence $\vec{\alpha}, \vec{\beta}$ and $\vec{\gamma}$ are coplanar. **Proved.**

Example 10. *If four points whose position vectors are $\bar{a}, \bar{b}, \bar{c}, \bar{d}$ are coplanar, show that*

$$[\vec{a}\ \vec{b}\ \vec{c}] = [\vec{a}\ \vec{d}\ \vec{b}] + [\vec{a}\ \vec{d}\ \vec{c}] + [\vec{d}\ \vec{b}\ \vec{c}]$$ *(Nagpur University, Summer 2005)*

Solution. Let A, B, C, D be four points whose position vectors are $\vec{a}, \vec{b}, \vec{c}, \vec{d}$.

$$\vec{AD} = \vec{d} - \vec{a}, \quad \vec{BD} = \vec{d} - \vec{b} \quad \text{and} \quad \vec{CD} = \vec{d} - \vec{c}$$

If $\vec{AD}, \vec{BD}, \vec{CD}$ are coplanar, then

$$\vec{AD} \cdot (\vec{BD} \times \vec{CD}) = 0$$

$$\Rightarrow \quad (\vec{d} - \vec{a}) \cdot [(\vec{d} - \vec{b}) \times (\vec{d} - \vec{c})] = 0$$

$$\Rightarrow (\vec{d} - \vec{a}) \cdot [\vec{d} \times \vec{d} - \vec{d} \times \vec{c} - \vec{b} \times \vec{d} + \vec{b} \times \vec{c}] = 0$$

$$\Rightarrow \quad (\vec{d} - \vec{a}) \cdot [-\vec{d} \times \vec{c} - \vec{b} \times \vec{d} + \vec{b} \times \vec{c}] = 0$$

$$\Rightarrow \ -\vec{d} \cdot (\vec{d} \times \vec{c}) - \vec{d} \cdot (\vec{b} \times \vec{d}) + \vec{d} \cdot (\vec{b} \times \vec{c}) + \vec{a} \cdot (\vec{d} \times \vec{c}) + \vec{a} \cdot (\vec{b} \times \vec{a}) - \vec{a} \cdot (\vec{b} \times \vec{c}) = 0$$

$$\Rightarrow -0 + 0 + [\vec{d}\ \vec{b}\ \vec{c}] + [\vec{a}\ \vec{d}\ \vec{c}] + [\vec{a}\ \vec{b}\ \vec{d}] - [\vec{a}\ \vec{b}\ \vec{c}] = 0$$

$$\Rightarrow \quad [\vec{a}\ \vec{b}\ \vec{c}] = [\vec{a}\ \vec{b}\ \vec{d}] + [\vec{a}\ \vec{d}\ \vec{c}] + [\vec{d}\ \vec{b}\ \vec{c}]$$ **Proved.**

Example 11. *Assuming that $\vec{a} \cdot (\vec{b} \times \vec{c}) \neq 0$ and $\vec{d} = x\vec{a} + y\vec{b} + z\vec{c}$. Find the values of x, y and z.* *(Nagpur University, Winter 2002)*

Solution. We have, $\vec{d} = x\vec{a} + y\vec{b} + z\vec{c}$ ...(1)

Taking dot product of (1) with $\vec{b} \times \vec{c}$, we get

$$[\vec{d}\ \vec{b}\ \vec{c}] = \vec{d} \cdot (\vec{b} \times \vec{c}) = (x\vec{a} + y\vec{b} + z\vec{c}) \cdot (\vec{b} \times \vec{c})$$

$$= x\{\vec{a} \cdot (\vec{b} \times \vec{c})\} + y\{\vec{b} \cdot (\vec{b} \times \vec{x})\} + z\{\vec{c} \cdot (\vec{b} \times \vec{c})\} = x[\vec{a}\ \vec{b}\ \vec{c}] + 0 + 0$$

$$\therefore \quad x = \frac{[\vec{d}\ \vec{b}\ \vec{c}]}{[\vec{a}\ \vec{b}\ \vec{c}]}$$

Similarly taking dot product of (1) with $(\vec{c} \times \vec{a})$ and $(\vec{a} \times \vec{b})$ Separately, we get

$$y = \frac{[\vec{d}\ \vec{c}\ \vec{a}]}{[\vec{a}\ \vec{b}\ \vec{c}]} \text{ and } z = \frac{[\vec{d}\ \vec{a}\ \vec{b}]}{[\vec{a}\ \vec{b}\ \vec{c}]}$$ **Ans.**

EXERCISE 22.2

1. Determine λ such that

   $\bar{a} = \hat{i} + \hat{j} + \hat{k}, \bar{b} = 2\hat{i} - 4\hat{k}$, and $\bar{c} = \hat{i} + \lambda\hat{j} + 3\hat{k}$ are coplanar. **Ans.** $\lambda = 5/3$

2. Show that the four points

   $-6\hat{i} + 3\hat{j} + 2\hat{k}, 3\hat{i} - 2\hat{j} + 4\hat{k}, 5\hat{i} + 7\hat{j} + 3\hat{k}$ and $-13\hat{i} + 17\hat{j} - \hat{k}$ are coplanar.

3. Find the constant a such that the vectors

   $2\hat{i} - \hat{j} + \hat{k}, \hat{i} + 2\hat{j} - 3\hat{k}$, and $3\hat{i} + a\hat{j} + 5\hat{k}$ are coplanar. **Ans.** -4

4. Prove that four points

   $4\hat{i} + 5\hat{j} + \hat{k}, -(\hat{j} + \hat{k}), 3\hat{i} + 9\hat{j} + 4\hat{k}, 4(-\hat{i} + \hat{j} + \hat{k})$ are coplanar.

5. If the vectors $\vec{a}, \vec{b}$ and $\vec{c}$ are coplanar, show that

$$\begin{vmatrix} \vec{a} & \vec{b} & \vec{c} \\ \vec{a}\cdot\vec{a} & \vec{a}\cdot\vec{b} & \vec{a}\cdot\vec{c} \\ \vec{b}\cdot\vec{a} & \vec{b}\cdot\vec{b} & \vec{b}\cdot\vec{c} \end{vmatrix} = 0$$

22.19 VECTOR PRODUCT OF THREE VECTORS

(A.M.I.E.T.E., Summer, 2004, 2000)

Let $\vec{a}$, $\vec{b}$ and $\vec{c}$ be three vectors then their vector product is written as $\vec{a} \times (\vec{b} \times \vec{c})$.

Let $\vec{a} = a_1\hat{i} + a_2\hat{j} + a_3\hat{k}$,

$\vec{b} = b_1\hat{i} + b_2\hat{j} + b_3\hat{k}$,

$\vec{c} = c_1\hat{i} + c_2\hat{j} + c_3\hat{k}$

$$\vec{a} \times (\vec{b} \times \vec{c}) = (a_1\hat{i} + a_2\hat{j} + a_3\hat{k}) \times (b_1\hat{i} + b_2\hat{j} + b_3\hat{k}) \times (c_1\hat{i} + c_2\hat{j} + c_3\hat{k})$$

$$= (a_1\hat{i} + a_2\hat{j} + a_3\hat{k}) \times [(b_2c_3 - b_3c_2)\hat{i} + (b_3c_1 - b_1c_3)\hat{j} + (b_1c_2 - b_2c_1)\hat{k}]$$

$$= [a_2(b_1c_2 - b_2c_1) - a_3(b_3c_1 - b_1c_3)]\hat{i} + [a_3(b_2c_3 - b_3c_2) - a_1(b_1c_2 - b_2c_1)]\hat{j} + [a_1(b_3c_1 - b_1c_3) - a_2(b_2c_3 - b_3c_2)\hat{k}]$$

$$= (a_1c_1 + a_2c_2 + a_3c_3)(b_1\hat{i} + b_2\hat{j} + b_3\hat{k}) - (a_1b_1 + a_2b_2 + a_3b_3)(c_1\hat{i} + c_2\hat{j} + c_3\hat{k})$$

$$= (\vec{a}\cdot\vec{c})\vec{b} - (\vec{a}\cdot\vec{b})\vec{c}.$$ **Ans.**

Example 12. *Let* $\vec{a} = \hat{i} + \hat{j} - \hat{k}$, $\vec{b} = \hat{i} - \hat{j} + \hat{k}$, $\vec{c} = \hat{i} - \hat{j} - \hat{k}$.

Find the vector $\vec{a} \times (\vec{b} \times \vec{c})$.

Solution. $\vec{a} \times (\vec{b} \times \vec{c}) = (\vec{a}\cdot\vec{c})\vec{b} - (\vec{a}\cdot\vec{b})\vec{c}$

$$= [(\hat{i} + \hat{j} - \hat{k})\cdot(\hat{i} - \hat{j} - \hat{k})](\hat{i} - \hat{j} + \hat{k}) - [(\hat{i} + \hat{j} - \hat{k})\cdot(\hat{i} - \hat{j} + \hat{k})](\hat{i} - \hat{j} - \hat{k})$$

$$= (1 - 1 + 1)(\hat{i} - \hat{j} + \hat{k}) - [(1 - 1 - 1)(\hat{i} - \hat{j} - \hat{k})]$$

$$= (\hat{i} - \hat{j} + \hat{k}) + (\hat{i} - \hat{j} - \hat{k}) = 2\hat{i} - 2\hat{j}$$ **Ans**

Example 13. *Prove that :*

$\vec{a} \times (\vec{b} \times \vec{c}) + \vec{b} \times (\vec{c} \times \vec{a}) + \vec{c} \times (\vec{a} \times \vec{b}) = 0$ *(Nagpur University, Winter 2008)*

Solution. Here, we have

$$\vec{a} \times (\vec{b} \times \vec{c}) + \vec{b} \times (\vec{c} \times \vec{a}) + \vec{c} \times (\vec{a} \times \vec{b})$$

$$= [(\vec{a}\cdot\vec{c})\vec{b} - (\vec{a}\cdot\vec{b})\vec{c}] + [(\vec{b}\cdot\vec{a})\vec{c} - (\vec{b}\cdot\vec{c})\vec{a}] + [(\vec{c}\cdot\vec{b})\vec{b} - (\vec{c}\cdot\vec{a})\vec{b}]$$

$$= [(\vec{b}\cdot\vec{a})\vec{c} - (\vec{a}\cdot\vec{b})\vec{c}] + [(\vec{c}\cdot\vec{b})\vec{a} - (\vec{b}\cdot\vec{c})\vec{a}] + [(\vec{a}\cdot\vec{c})\vec{b} - (\vec{c}\cdot\vec{a})\vec{b}]$$

$$= [(\vec{a}\cdot\vec{b})\vec{c} - (\vec{a}\cdot\vec{b})\vec{c}] + [(\vec{b}\cdot\vec{c})\vec{a} - (\vec{b}\cdot\vec{c})\vec{a}] + [(\vec{c}\cdot\vec{a})\vec{b} - (\vec{c}\cdot\vec{a})\vec{b}]$$

$$= 0 + 0 + 0 = 0$$ **Proved.**

Example 14. *Prove that :*

$\hat{i} \times (\vec{a} \times \hat{i}) + \hat{j} \times (\vec{a} \times \hat{j}) + \hat{k} \times (\vec{a} \times \hat{k}) = 2\vec{a}$ *(Nagpur University, Winter 2003)*

Solution. Let $\vec{a} = a_1\hat{i} + a_2\hat{j} + a_3\hat{k}$

Now, L.H.S. $= \hat{i} \times (\vec{a} \times \hat{i}) + \hat{j} \times (\vec{a} \times \hat{j}) + \hat{k} \times (\vec{a} \times \hat{k})$

$$= \hat{i} \times \left[(a_1 \hat{i} + a_2 \hat{j} + a_3 \hat{k}) \times \hat{i}\right] + \hat{j} \times \left[(a_1 \hat{i} + a_2 \hat{j} + a_3 \hat{k}) \times \hat{j}\right] + \hat{k} \times \left[(a_1 \hat{i} + a_2 \hat{j} + a_3 \hat{k}) \times \hat{k}\right]$$

$$= \hat{i} \times \left[a_1(\hat{i} \times \hat{i}) + a_2(\hat{j} \times \hat{i}) + a_3(\hat{k} \times \hat{i})\right] + \hat{j} \times \left[a_1(\hat{i} \times \hat{j}) + a_2(\hat{j} \times \hat{j}) + a_3(\hat{k} \times \hat{j})\right] + \hat{k} \times \left[a_1(\hat{i} \times \hat{k}) + a_2(\hat{j} \times \hat{k}) + a_3(\hat{k} \times \hat{k})\right]$$

$$= \hat{i} \times \left[0 - a_2 \hat{k} + a_3 \hat{j}\right] + \hat{j} \times \left[a_1 \hat{k} + 0 - a_3 \hat{i}\right] + \hat{k} \times \left[-a_1 \hat{j} + a_2 \hat{i} + 0\right]$$

$$= -a_2(\hat{i} \times \hat{k}) + a_3(\hat{i} \times \hat{j}) + a_1(\hat{j} \times \hat{k}) - a_3(\hat{j} \times \hat{i}) - a_1(\hat{k} \times \hat{j}) + a_2(\hat{k} \times \hat{i})$$

$$= a_2 \hat{j} + a_3 \hat{k} + a_1 \hat{i} + a_3 \hat{k} + a_1 \hat{i} + a_2 \hat{j} = 2a_1 \hat{i} + 2a_2 \hat{j} + 2a_3 \hat{k}$$

$$= 2(a_1 \hat{i} + a_2 \hat{j} + a_3 \hat{k}) = 2\vec{a}$$

Proved.

Example 15. *Show that for any scalar* λ, *the vectors* $\vec{x}, \vec{y}$ *given by*

$$\vec{x} = \lambda \vec{a} + \frac{q(\vec{a} \times \vec{b})}{a^2}, \quad \vec{y} = \frac{(1 - p\lambda)}{q}\vec{a} - \frac{p(\vec{a} \times \vec{b})}{a^2}$$ *satisfy the equations*

$$p\vec{x} + q\vec{y} = \vec{a} \text{ and } \vec{x} \times \vec{y} = \vec{b}.$$ *(Nagpur University, Winter 2004)*

Solution. The given equations are

$$p\vec{x} + q\vec{y} = \vec{a} \qquad ...(1)$$

$$\vec{x} \times \vec{y} = \vec{b} \qquad ...(2)$$

Multiplying equation (1) vectorially by $\vec{x}$, we get

$$\vec{x} \times (p\vec{x} + q\vec{y}) = \vec{x} \times \vec{a}$$

$$p(\vec{x} \times \vec{x}) + q(\vec{x} \times \vec{y}) = \vec{x} \times \vec{a}$$

$$q \times (\vec{x} \times \vec{y}) = \vec{x} \times \vec{a}, \qquad \text{as } \vec{x} \times \vec{x} = 0$$

$$\vec{x} \times \vec{a} = q\vec{b}, \qquad [\text{From (2) } \vec{x} \times \vec{y} = \vec{b}] \quad ...(3)$$

Multiplying (3) vectorially by $\vec{a}$, we have

$$\vec{a} \times (\vec{x} \times \vec{a}) = \vec{a} \times q\vec{b}$$

$$(\vec{a} \cdot \vec{a})\vec{x} - (\vec{a} \cdot \vec{x})\vec{a} = q(\vec{a} \times \vec{b})$$

$$a^2 \vec{x} - (\vec{a} \cdot \vec{x})\vec{a} = q(\vec{a} \times \vec{b}) \quad \Rightarrow \quad a^2 \vec{x} = (\vec{a} \cdot \vec{x})\vec{a} + q(\vec{a} \times \vec{b})$$

$$\vec{x} = \frac{(\vec{a} \cdot \vec{x})\vec{a}}{a^2} + \frac{q(\vec{a} \times \vec{b})}{a^2}$$

$$\vec{x} = \lambda a + \frac{q(\vec{a} \times \vec{b})}{a^2} \qquad \text{where } \lambda = \frac{\vec{a} \cdot \vec{x}}{a^2}$$

Substituting the value of $\vec{x}$ in (1), we get $p\left\{\lambda \vec{a} + \dfrac{q(\vec{a} \times \vec{b})}{a^2}\right\} + q\vec{y} = \vec{a}$

$$q\vec{y} = \vec{a} - p\left\{\lambda \vec{a} + \frac{q(\vec{a} \times \vec{b})}{a^2}\right\}$$

$$\vec{y} = \frac{(1 - p\lambda)\,\vec{a}}{q} - \frac{p\,(\vec{a} \times \vec{b})}{a^2}$$ **Ans.**

EXERCISE 22.3

1. Show that $\vec{a} \times (\vec{b} \times \vec{a}) = (\vec{a} \times \vec{b}) \times \vec{a}$
2. Write the correct answer

(*a*) $(\vec{A} \times \vec{B}) \times \vec{C}$ lies in the plane of

(*i*) $\vec{A}$ and $\vec{B}$ (*ii*) $\vec{B}$ and $\vec{C}$ (*iii*) $\vec{C}$ and $\vec{A}$ **Ans.** (*ii*)

(*b*) The value of $\vec{a} \,.\, (\vec{b} + \vec{c}) \times (\vec{a} + \vec{b} + \vec{c})$ is

(*i*) Zero (*ii*) $[\vec{a}\,,\, \vec{b}\,,\, \vec{c}] + [\vec{b}\,,\, \vec{c}\,,\, \vec{a}]$ (*iii*) $[\vec{a}\,,\, \vec{b}\,,\, \vec{c}]$ (*iv*) None of these **Ans.** (*ii*)

22.20 SCALAR PRODUCT OF FOUR VECTORS

Prove the identity

$$(\vec{a} \times \vec{b}) \,.\, (\vec{c} \times \vec{d}) = (\vec{a} \,.\, \vec{c})\,(\vec{b} \,.\, \vec{d}) - (\vec{a} \,.\, \vec{d})\,(\vec{b} \,.\, \vec{c})$$

Proof. $(\vec{a} \times \vec{b}) \,.\, (\vec{c} \times \vec{d}) = (\vec{a} \times \vec{b}) \,.\, \vec{r}$

$= \vec{a} \,.\, (\vec{b} \times \vec{r})$ dot and cross can be interchanged. Put $\vec{c} \times \vec{d} = \vec{r}$

$= \vec{a} \,.\, [\vec{b} \times (\vec{c} \times \vec{d})] = \vec{a} \,.\, [(\vec{b} \,.\, \vec{d})\, \vec{c} - (\vec{b} \,.\, \vec{c})\, \vec{d}]$

$= (\vec{a} \,.\, \vec{c})\,(\vec{b} \,.\, \vec{d}) - (\vec{a} \,.\, \vec{d})\,(\vec{b} \,.\, \vec{c})$

$$= \begin{vmatrix} \vec{a} \,.\, \vec{c} & \vec{a} \,.\, \vec{d} \\ \vec{b} \,.\, \vec{c} & \vec{b} \,.\, \vec{d} \end{vmatrix}$$ **Proved.**

EXERCISE 22.4

1. If $\vec{a} = 2i + 3j - k$, $\vec{b} = -i + 2j - 4k$, $\vec{c} = i + j + k$, find $(\vec{a} \times \vec{b}) \,.\, (\vec{a} \times \vec{c})$. **Ans.** –74
2. Prove that $(\vec{a} \times \vec{b}) \,.\, (\vec{a} \times \vec{c}) = a^2 (\vec{b} \,.\, \vec{c}) - (\vec{a} \,.\, \vec{b})\,(\vec{a} \,.\, \vec{c})$.

22.21 VECTOR PRODUCT OF FOUR VECTORS

Let $\vec{a}, \vec{b}, \vec{c}$ and $\vec{d}$ be four vectors then their vector product is written as

$$(\vec{a} \times \vec{b}) \times (\vec{c} \times \vec{d})$$

Now, $(\vec{a} \times \vec{b}) \times (\vec{c} \times \vec{d}) = \vec{r} \times (\vec{c} \times \vec{d})$ [Put $\vec{a} \times \vec{b} = \vec{r}$]

$= (\vec{r} \,.\, \vec{d})\, \vec{c} - (\vec{r} \,.\, \vec{c})\, \vec{d}$

$= [(\vec{a} \times \vec{b}) \,.\, \vec{d}]\, \vec{c} - [(\vec{a} \times \vec{b}) \,.\, \vec{c}]\, \vec{d}$

$= [\vec{a}\ \vec{b}\ \vec{d}]\, \vec{c} - [\vec{a}\ \vec{b}\ \vec{c}]\, \vec{d}$

$\therefore$ $(\vec{a} \times \vec{b}) \times (\vec{c} \times \vec{d})$ lies in the plane of $\vec{c}$ and $\vec{d}$. ...(1)

Again, $(\vec{a} \times \vec{b}) \times (\vec{c} \times \vec{d}) = (\vec{a} \times \vec{b}) \times \vec{s}$ [Put = $\vec{c} \times \vec{d} = \vec{s}$]

$= - \vec{s} \times (\vec{a} \times \vec{b}) = -(\vec{s} \,.\, \vec{b})\, \vec{a} + (\vec{s} \,.\, \vec{a})\, \vec{b}$

$= -[(\vec{c} \times \vec{d}) \,.\, \vec{b}]\, \vec{a} + [(\vec{c} \times \vec{d}) \,.\, \vec{a}]\, \vec{b} = -[(\vec{b}\ \vec{c}\ \vec{d}]\, \vec{a} + [\vec{a}\ \vec{c}\ \vec{d}]\, \vec{b}$

$\therefore \quad (\vec{a} \times \vec{b}) \times (\vec{c} \times \vec{d})$ lies in the plane of $\vec{a}$ and $\vec{b}$. ...(2)

Geometrical interpretation : From (1) and (2) we conclude that $(\vec{a} \times \vec{b}) \times (\vec{c} \times \vec{d})$ is a vector parallel to the line of intersection of the plane containing $\vec{a}$, $\vec{b}$ and plane containing $\vec{c}$, $\vec{d}$.

Example 16. *Show that*

$$(\vec{B} \times \vec{C}) \times (\vec{A} \times \vec{D}) + (\vec{C} \times \vec{A}) \times (\vec{B} \times \vec{D}) + (\vec{A} \times \vec{B}) \times (\vec{C} \times \vec{D}) = -2(\vec{A}\ \vec{B}\ \vec{C})\vec{D}$$

Solution. L.H.S. $= (\vec{B} \times \vec{C}) \times (\vec{A} \times \vec{D}) + (\vec{C} \times \vec{A}) \times (\vec{B} \times \vec{D}) + (\vec{A} \times \vec{B}) \times (\vec{C} \times \vec{D})$

$= [(\vec{B}\ \vec{C}\ \vec{D})\vec{A} - (\vec{B}\ \vec{C}\ \vec{A})\vec{D}] + [(\vec{C}\ \vec{A}\ \vec{D})\vec{B} - (\vec{C}\ \vec{A}\ \vec{B})\vec{D}] + [(-\vec{B}\ \vec{C}\ \vec{D})\vec{A} + (\vec{A}\ \vec{C}\ \vec{D})\vec{B}]$

$= (\vec{B}\ \vec{C}\ \vec{D})\vec{A} - (\vec{B}\ \vec{C}\ \vec{D})\vec{A} + (\vec{C}\ \vec{A}\ \vec{D})\vec{B} + (\vec{A}\ \vec{C}\ \vec{D})\vec{B} - (\vec{B}\ \vec{C}\ \vec{A})\vec{D} - (\vec{C}\ \vec{A}\ \vec{B})\vec{D}$

$= -(\vec{A}\ \vec{C}\ \vec{D})\vec{B} + (\vec{A}\ \vec{C}\ \vec{D})\vec{B} - (\vec{A}\ \vec{B}\ \vec{C})\vec{D} - (\vec{A}\ \vec{B}\ \vec{C})\vec{D}$

$= -2(\vec{A}\ \vec{B}\ \vec{C})\vec{D}$ = R.H.S. **Proved.**

Example 17. *Prove that*

$$(\vec{a} \times \vec{b}) \cdot \left\{(\vec{b} \times \vec{c}) \times (\vec{c} \times \vec{a})\right\} = [\vec{a} \cdot (\vec{b} \times \vec{c})]^2$$ *(Nagpur University, Summer 2003)*

Solution. L.H.S. $= (\vec{a} \times \vec{b}) \cdot \left\{(\vec{b} \times \vec{c}) \times (\vec{c} \times \vec{a})\right\}$ ...(1)

By applying the formula of vector triple product on $(\vec{b} \times \vec{c}) \times (\vec{c} \times \vec{a})$ in (1), we get

$$\text{L.H.S.} = (\vec{a} \times \vec{b}) \cdot \left[\left\{(\vec{b}\ \vec{c}\ \vec{a})\vec{c} - (\vec{b}\ \vec{c}\ \vec{c})\vec{a}\right\}\right]$$

$$= (\vec{a} \times \vec{b}) \cdot [(\vec{a}\ \vec{b}\ \vec{c})\vec{c} - 0(\vec{a})] = (\vec{a}\ \vec{b}\ \vec{c})(\vec{a} \times \vec{b}) \cdot \vec{c}$$

$$= [\vec{a}\,(\vec{b} \times \vec{c})][\vec{a} \cdot (\vec{b} \times \vec{c})] = [\vec{a} \cdot (\vec{b} \times \vec{c})]^2 = \text{R.H.S.}$$ **Proved.**

EXERCISE 22.5

Show that:

1. $(\vec{b} \times \vec{c}) \times (\vec{c} \times \vec{a}) = \vec{c}\,(\vec{a}\ \vec{b}\ \vec{c})$ when $(\vec{a}\ \vec{b}\ \vec{c})$ stands for scalar triple product.
2. $[\vec{b} \times \vec{c},\quad \vec{c} \times \vec{a},\quad \vec{a} \times \vec{b})] = [\vec{a}\ \vec{b}\ \vec{c}]$
3. $\vec{d}\,[\vec{a} \times \{\vec{b} \times (\vec{c} \times \vec{d})\}] = [(\vec{b} \cdot \vec{d})[\vec{a} \cdot (\vec{c} \times \vec{d})]$
4. $\vec{a}\,[\vec{a} \times [\vec{a} \times (\vec{a} \times \vec{b})] = a^2\,(\vec{b} \times \vec{a})$
5. $[(\vec{a} \times \vec{b}) \times (\vec{a} \times \vec{c})] \cdot \vec{d} = (\vec{a} \cdot \vec{d})[\vec{a}\ \vec{b}\ \vec{c}]$
6. $2a^2 = \left|\vec{a} \times \hat{i}\right|^2 + \left|\vec{a} \times \hat{j}\right|^2 + \left|\vec{a} \times \hat{k}\right|^2$
7. $\vec{a} \times \vec{b} = [(\hat{i} \times \vec{a}) \cdot \vec{b}]\,\hat{i} + [(\hat{j} \times \vec{a}) \cdot \vec{b}]\,\hat{j} + [(\hat{k} \times \vec{a}) \cdot \vec{b}]\,\hat{k}$
8. $\vec{p} \times [(\vec{a} \times \vec{q}) \times (\vec{b} \times \vec{r})] + \vec{q} \times [(\vec{a} \times \vec{r}) \times (\vec{b} \times \vec{p})] + \vec{r} \times [(\vec{a} \times \vec{p}) \times (\vec{b} \times \vec{q})] = 0$

DIFFERENTIATION OF VECTORS

(POINT FUNCTION, GRADIENT, DIVERGENCE AND CURL OF A VECTOR AND THEIR PHYSICAL INTERPRETATIONS)

23.1 VECTOR FUNCTION

If vector r is a function of a scalar variable t, then we write

$$\vec{r} = \vec{r}(t)$$

If a particle is moving along a curved path then the position vector $\vec{r}$ of the particle is a function of t. If the component of $f(t)$ along x-axis, y-axis, z-axis are $f_1(t), f_2(t), f_3(t)$ respectively. Then,

$$\overrightarrow{f(t)} = f_1(t)\hat{i} + f_2(t)\hat{j} + f_3(t)\hat{k}$$

23.2 DIFFERENTIATION OF VECTORS

Let O be the origin and P be the position of a moving particle at time t.

Let $\overrightarrow{OP} = \vec{r}$

Let Q be the position of the particle at the time $t + \delta t$ and the position vector of Q is $\overrightarrow{OQ} = \vec{r} + \delta\vec{r}$

$$\overrightarrow{PQ} = \overrightarrow{OQ} - \overrightarrow{OP}$$

$$= (\vec{r} + \delta\vec{r}) - \vec{r} = \delta\vec{r}$$

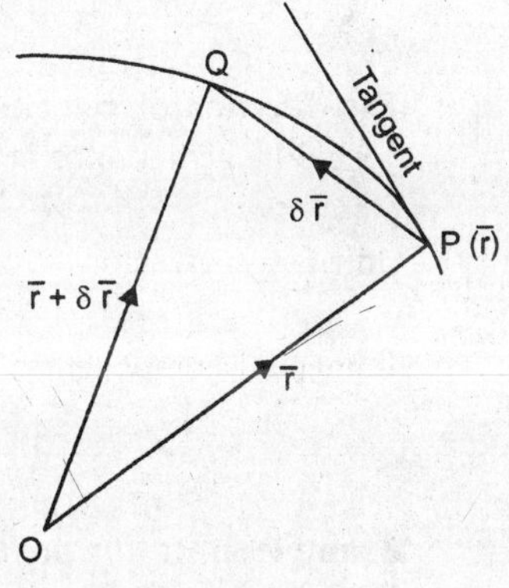

$\dfrac{\delta\vec{r}}{\delta t}$ is a vector. As $\delta t \to 0$, Q tends to P and the chord becomes the tangent at P.

We define $\dfrac{d\vec{r}}{dt} = \lim\limits_{\delta t \to 0} \dfrac{\delta\vec{r}}{\delta t}$, then

$\dfrac{d\vec{r}}{dt}$ is a vector in the direction of the *tangent* at P.

$\dfrac{d\vec{r}}{dt}$ is also called the differential coefficient of $\vec{r}$ with respect to 't'.

Similarly, $\dfrac{d^2\vec{r}}{dt^2}$ is the second order derivative of $\vec{r}$.

$\dfrac{d\vec{r}}{dt}$ gives the velocity of the particle at P, which is along the tangent to its path. Also $\dfrac{d^2\vec{r}}{dt^2}$ gives the *acceleration* of the particle at P.

23.3 FORMULAE OF DIFFERENTIATION

(i) $\frac{d}{dt}(\vec{F}+\vec{G})=\frac{d\vec{F}}{dt}+\frac{d\vec{G}}{dt}$ (ii) $\frac{d}{dt}(\vec{F}\phi)=\frac{d\vec{F}}{dt}\phi+\vec{F}\frac{d\phi}{dt}$ *(U.P. I semester, Dec. 2005)*

(iii) $\frac{d}{dt}(\vec{F}.\vec{G})=\vec{F}.\frac{d\vec{G}}{dt}+\frac{d\vec{F}}{dt}.\vec{G}$ (iv) $\frac{d}{dt}(\vec{F}\times\vec{G})=\vec{F}\times\frac{d\vec{G}}{dt}+\frac{d\vec{F}}{dt}\times\vec{G}$

(v) $\frac{d}{dt}[\vec{a}\,\vec{b}\,\vec{c}]=\left[\frac{d\vec{a}}{dt}\vec{b}\,\vec{c}\right]+\left[\vec{a}\frac{d\vec{b}}{dt}\vec{c}\right]+\left[\vec{a}\,\vec{b}\frac{d\vec{c}}{dt}\right]$

(vi) $\frac{d}{dt}[\vec{a}\times(\vec{b}\times\vec{c})]=\frac{d\vec{a}}{dt}\times(\vec{b}\times\vec{c})+\vec{a}\times\left(\frac{d\vec{b}}{dt}\times\vec{c}\right)+\vec{a}\times\left(\vec{b}\times\frac{d\vec{c}}{dt}\right)$

The order of the functions $\vec{F}, \vec{G}$ is not to be changed.

Example 1. *A particle moves along the curve $\bar{r}=(t^3-4t)\hat{i}+(t^2+4t)\hat{j}+(8t^2-3t^3)\hat{k}$, where t is the time. Find the magnitude of the tangential components of its acceleration at t = 2.*

(Nagpur University, Summer 2005)

Solution. We have, $\vec{r} = (t^3-4t)\hat{i}+(t^2+4t)\hat{j}+(8t^2-3t^3)\hat{k}$

$$\text{Velocity} = \frac{d\vec{r}}{dt}=(3t^2-4)\hat{i}+(2t+4)\hat{j}+(16t-9t^2)\hat{k}$$

At $t = 2$, Velocity $= 8\hat{i}+8\hat{j}-4\hat{k}$

$$\text{Acceleration} = \vec{a} = \frac{d^2\vec{r}}{dt^2}=6t\hat{i}+2\hat{j}+(16-18t)\hat{k}$$

At $t = 2$ $\vec{a}=12\hat{i}+2\hat{j}-20\hat{k}$

The direction of velocity is along tangent.
So the tangent vector is velocity.

$$\text{Unit tangent vector,} \quad \hat{T}=\frac{\vec{v}}{|v|}=\frac{8\hat{i}+8\hat{j}-4\hat{k}}{\sqrt{64+64+16}}=\frac{8\hat{i}+8\hat{j}-4\hat{k}}{12}=\frac{2\hat{i}+2\hat{j}-\hat{k}}{3}$$

Tangential component of acceleration, $a_t = \bar{a}.\hat{T}$

$$= (12\hat{i}+2\hat{j}-20\hat{k}).\frac{2\hat{i}+2\hat{j}-\hat{k}}{3} = \frac{24+4+20}{3}=\frac{48}{3}=16 \textbf{ Ans.}$$

Example 2. *At any point of the curve x = 3 cos t, y = 3 sin t, z = 4 t, find*
(i) Tangent vector (ii) Unit tangent vector *(R.G.P.V., Bhopal, II Semester June 2007)*

Solution. We have, $x = 3\cos t$, $y = 3\sin t$, $z = 4t$

$$\vec{r} = x\hat{i}+y\hat{j}+z\hat{k} \Rightarrow \vec{r} = (3\cos t)\hat{i}+(3\sin t)\hat{j}+(4t)\hat{k}$$

(i) $$\frac{d\vec{r}}{dt} = (-3\sin t)\hat{i}+(3\cos t)\hat{j}+4\hat{k}$$

which is the required tangent vector.

Magnitude of tangent vector $= \sqrt{(-3\sin t)^2+(3\cos t)^2+(4)^2}=5.$

(ii) Unit tangent vector $= \frac{1}{5}(-3\sin t\,\hat{i}+3\cos t\,\hat{j}+4\hat{k})$ **Ans.**

Example 3. *Show that* $\frac{d}{dt}\left[\vec{u}\,\frac{d\vec{u}}{dt}\,\frac{d^2\vec{u}}{dt^2}\right] = \left[\bar{u}\,\frac{d\vec{u}}{dt}\,\frac{d^3\vec{u}}{dt^3}\right]$

Solution. We know that $\frac{d}{dt}(x\,y\,z) = \frac{dx}{dt}yz + x\frac{dy}{dt}z + xy\frac{dz}{dt}$

$$\frac{d}{dt}\left[\vec{u}\,\frac{d\vec{u}}{dt}\,\frac{d^2\vec{u}}{dt^2}\right] = \left[\frac{d\vec{u}}{dt}\,\frac{d\vec{u}}{dt}\,\frac{d^2\vec{u}}{dt}\right] + \left[\bar{u}\,\frac{d^2\vec{u}}{dt^2}\,\frac{d^2\vec{u}}{dt^2}\right] + \left[u\,\frac{d\vec{u}}{dt}\,\frac{d^3\vec{u}}{dt^3}\right]$$

$$= 0 + 0 + \left[\vec{u}\,\frac{d\vec{u}}{dt}\,\frac{d^3\vec{u}}{dt^3}\right] = \left[\vec{u}\,\frac{d\vec{u}}{dt}\,\frac{d^3\vec{u}}{dt^3}\right]$$ **Proved.**

Example 4. *If* $\frac{d\vec{a}}{dt} = \vec{u}\times\vec{a}$ *and* $\frac{d\vec{b}}{dt} = \vec{u}\times\vec{b}$ *then prove that* $\frac{d}{dt}[\vec{a}\times\vec{b}] = \vec{u}\times(\vec{a}\times\vec{b})$

(M.U. 2009)

Solution. We have,

$$\frac{d}{dt}[\vec{a}\times\vec{b}] = \vec{a}\times\frac{d\vec{b}}{dt} + \frac{d\vec{a}}{dt}\times\vec{b} = \vec{a}\times(\vec{u}\times\vec{b}) + (\vec{u}\times\vec{a})\times\vec{b}$$

$$= \vec{a}\times(\vec{u}\times\vec{b}) - \vec{b}\times(\vec{u}\times\vec{a})$$

$$= (\vec{a}.\vec{b})\,\vec{u} - (\vec{a}.\vec{u})\,\vec{b} - [(\vec{b}.\vec{a})\,\vec{u} - (\vec{b}.\vec{u})\,\vec{a}]$$

(Vector triple product)

$$= (\vec{a}.\vec{b})\,\vec{u} - (\vec{u}.\vec{a})\,\vec{b} - (\vec{a}.\vec{b})\,\vec{u} + (\vec{u}.\vec{b})\,\vec{a}$$

$$= (\vec{u}.\vec{b})\,\vec{a} - (\vec{u}.\vec{a})\,\vec{b}$$

$$= \vec{u}\times(\vec{a}\times\vec{b})$$ **Proved.**

Example 5. *Find the angle between the surface* $x^2 + y^2 + z^2 = 9$ and $z = x^2 + y^2 - 3$ at (2, –1, 2). *(M.D.U. Dec. 2009)*

Solution. Here, we have

$$x^2 + y^2 + z^2 = 9 \quad ...(1)$$

$$z = x^2 + y^2 - 3 \quad ...(2)$$

Normal to (1) $\eta_1 = \nabla(x^2 + y^2 + z^2 - 9)$

$$= \left(\hat{i}\frac{\partial}{\partial x} + \hat{j}\frac{\partial}{\partial y} + \hat{k}\frac{\partial}{\partial z}\right)(x^2 + y^2 + z^2 - 9) = 2\,x\hat{i} + 2\,y\hat{j} + 2\,z\hat{k}$$

Normal to (1) at (2, – 1, 2), $\eta_1 = 4\hat{i} - 2\hat{j} + 4\hat{k}$ ...(3)

Normal to (2), $\eta_2 = \nabla(z - x^2 - y^2 + 3)$

$$= \left(\hat{i}\frac{\partial}{\partial x} + \hat{j}\frac{\partial}{\partial y} + \hat{k}\frac{\partial}{\partial z}\right)(z - x^2 - y^2 + 3) = -2\,x\hat{i} - 2\,y\hat{j} + \hat{k}$$

Normal to (2) at (2, – 1, 2), $\eta_2 = -4\hat{i} + 2\hat{j} + \hat{k}$ ...(4)

$$\eta_1.\eta_2 = |\eta_1||\eta_2|\cos\theta$$

$$\cos\theta = \frac{\eta_1.\eta_2}{|\eta_1||\eta_2|} = \frac{(4\hat{i} - 2\hat{j} + 4\hat{k}).(-4\hat{i} + 2\hat{j} + \hat{k})}{|4\hat{i} - 2\hat{j} + 4\hat{k}||-4\hat{i} + 2\hat{j} + \hat{k}|} = \frac{-16 - 4 + 4}{\sqrt{16+4+16}\,\sqrt{16+4+1}}$$

$$= \frac{-16}{6\sqrt{21}} = \frac{-8}{3\sqrt{21}}$$

$$\theta = \cos^{-1}\left(\frac{-8}{3\sqrt{21}}\right)$$

Hence the angle between (1) and (2) $\cos^{-1}\left(\frac{-8}{3\sqrt{21}}\right)$ **Ans**

EXERCISE 23.1

1. The coordinates of a moving particle are given by $x = 4t - \frac{t^2}{2}$ and $y = 3 + 6t - \frac{t^3}{6}$. Find the velocity and acceleration of the particle when $t = 2$ secs. **Ans.** 4.47, 2.2
2. A particle moves along the curve
$$x = 2t^2, \quad y = t^2 - 4t \quad \text{and} \quad z = 3t - 5$$
where t is the time. Find the components of its velocity and acceleration at time $t = 1$, in the direction $\hat{i} - 3\hat{j} + 2\hat{k}$. *(Nagpur, Summer 2001)* **Ans.** $\frac{8\sqrt{14}}{7}, -\frac{\sqrt{14}}{7}$
3. Find the unit tangent and unit normal vector at $t = 2$ on the curve $x = t^2 - 1$, $y = 4t - 3$, $z = 2t^2 - 6t$ where t is any variable. **Ans.** $\frac{1}{3}(2\hat{i} + 2\hat{j} + \hat{k})$, $\frac{1}{3\sqrt{5}}(2\hat{i} + 2\hat{k})$
4. Prove that $\frac{d}{dt}(\vec{F} \times \vec{G}) = \vec{F} \times \frac{d\vec{G}}{dt} + \frac{d\vec{F}}{dt} \times \vec{G}$
5. Find the angle between the tangents to the curve $\bar{r} = t^2\hat{i} + 2t\hat{j} - t^3\hat{k}$, at the points $t = \pm 1$. **Ans.** $\cos^{-1}\left(\frac{9}{17}\right)$
6. If the surface $5x^2 - 2byz = 9x$ be orthogonal to the surface $4x^2y + z^3 = 4$ at the point $(1, -1, 2)$ then b is equal to
(*a*) 0 (*b*) 1 (*c*) 2 (*d*) 3 *(AMIETE, Dec. 2009)* **Ans.** (*b*)

23.4 SCALAR AND VECTOR POINT FUNCTIONS

Point function. A variable quantity whose value at any point in a region of space depends upon the position of the point, is called a *point function*. There are two types of point functions.

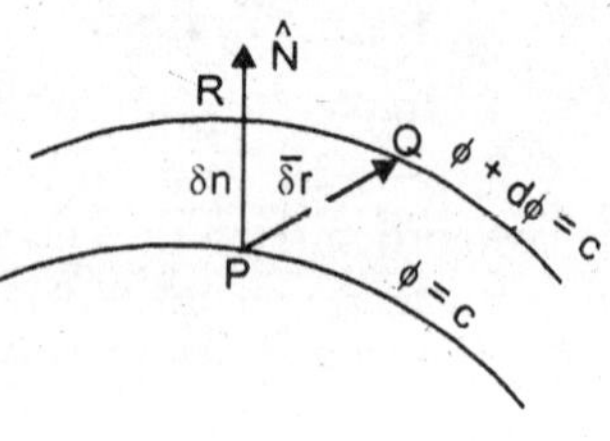

(*i*) **Scalar point function.** If to each point P (x, y, z) of a region R in space there corresponds a unique scalar $f(P)$, then f is called a scalar point function. *For example*, the temperature distribution in a heated body, density of a body and potential due to gravity are the examples of a scalar point function.

(*ii*) **Vector point function.** If to each point P (x, y, z) of a region R in space there corresponds a unique vector $f(P)$, then f is called a *vector point function*. The velocity of a moving fluid, gravitational force are the examples of vector point function.

(U.P., I Semester, Winter 2000)

Vector Differential Operator Del *i.e.* ∇

The vector differential operator Del is denoted by ∇. It is defined as

$$\nabla = \hat{i}\frac{\partial}{\partial x} + \hat{j}\frac{\partial}{\partial y} + \hat{k}\frac{\partial}{\partial z}$$

23.5 GRADIENT OF A SCALAR FUNCTION

If ϕ (x, y, z) be a scalar function then $\hat{i}\frac{\partial \phi}{\partial x} + \hat{j}\frac{\partial \phi}{\partial y} + \hat{k}\frac{\partial \phi}{\partial z}$ is called the gradient of the scalar function ϕ.

And is denoted by grad ϕ.

Thus, $$\text{grad } \phi = \hat{i}\frac{\partial \phi}{\partial x} + \hat{j}\frac{\partial \phi}{\partial y} + \hat{k}\frac{\partial \phi}{\partial z}$$

$$\text{gard } \phi = \left(\hat{i}\frac{\partial}{\partial x} + \hat{j}\frac{\partial}{\partial y} + \hat{k}\frac{\partial}{\partial z}\right)\phi(x, y, z)$$

$$\text{gard } \phi = \nabla \phi \qquad (\nabla \text{ is read del or nebla})$$

23.6 GEOMETRICAL MEANING OF GRADIENT, NORMAL

(*U.P. Ist Semester, Dec 2006*)

If a surface $\phi(x, y, z) = c$ passes through a point P. The value of the function at each point on the surface is the same as at P. Then such a surface is called a *level surface* through P. *For example*, If $\phi(x, y, z)$ represents potential at the point P, then *equipotential surface* $\phi(x, y, z) = c$ is a *level surface*.

Two level surfaces can not intersect.

Let the level surface pass through the point P at which the value of the function is ϕ. Consider another level surface passing through Q, where the value of the function is $\phi + d\phi$.

Let $\bar{r}$ and $\bar{r} + \delta\bar{r}$ be the position vector of P and Q then $\overrightarrow{PQ} = \delta\bar{r}$

$$\nabla\phi . d\bar{r} = \left(\hat{i}\frac{\partial\phi}{\partial x} + \hat{j}\frac{\partial\phi}{\partial y} + \hat{k}\frac{\partial\phi}{\partial z}\right).(\hat{i}\,dx + \hat{j}\,dy + \hat{k}\,dz)$$

$$= \frac{\partial\phi}{\partial x}dx + \frac{\partial\phi}{\partial y}dy + \frac{\partial\phi}{\partial z}dz = d\phi \qquad ...(1)$$

If Q lies on the level surface of P, then $d\phi = 0$

Equation (1) becomes $\nabla\phi . dr = 0$. Then $\overline{\nabla}\phi$ is $\perp$ to $d\bar{r}$ (tangent).

Hence, $\nabla\phi$ is **normal** to the surface $\phi(x, y, z) = c$

Let $\nabla\phi = |\nabla\phi|\,\hat{N}$, where $\hat{N}$ is a unit normal vector. Let δn be the perpendicular distance between two level surfaces through P and R. Then the rate of change of ϕ in the direction of the normal to the surface through P is $\frac{\partial\phi}{\partial n}$.

$$\frac{d\phi}{dn} = \lim_{\delta n\to 0}\frac{\delta\phi}{\delta n} = \lim_{\delta n\to 0}\frac{\nabla\phi . d\vec{r}}{\delta n}$$

$$= \lim_{\delta n\to 0}\frac{|\nabla\phi|\,\hat{N}.d\vec{r}}{\delta n} \qquad \left\{\begin{aligned}\hat{N}.\vec{\delta r} &= |\vec{N}|\,|\vec{\delta r}|\cos\theta \\ &= |\vec{\delta r}|\cos\theta = \delta n\end{aligned}\right\}$$

$$= \lim_{\delta n\to 0}\frac{|\nabla\phi|\,\delta n}{\delta n} = |\nabla\phi|$$

$$\therefore \qquad |\nabla\phi| = \frac{\partial\phi}{\partial n}$$

Hence, gradient ϕ is a vector normal to the surface $\phi = c$ and has a magnitude equal to the rate of change of ϕ along this normal.

23.7 NORMAL AND DIRECTIONAL DERIVATIVE

(*i*) **Normal.** If $\phi(x, y, z) = c$ represents a family of surfaces for different values of the constant c. On differentiating ϕ, we get $d\phi = 0$

But $\qquad d\phi = \nabla\phi . d\vec{r}$ so $\nabla\phi . dr = 0$

The scalar product of two vectors $\nabla\phi$ and $d\vec{r}$ being zero, $\nabla\phi$ and $d\vec{r}$ are perpendicular to each other. $d\vec{r}$ is in the direction of tangent to the given surface.

Thus $\nabla\phi$ is a vector *normal* to the surface $\phi(x, y, z) = c$.

(*ii*) **Directional derivative.** The component of $\nabla\phi$ in the direction of a vector $\vec{d}$ is equal to $\nabla\phi.\hat{d}$ and is called the directional derivative of ϕ in the direction of $\vec{d}$.

$$\frac{\partial\phi}{\partial r} = \lim_{\delta r \to 0} \frac{\delta\phi}{\delta r} \qquad \text{where, } \delta r = PQ$$

$\frac{\partial\phi}{\partial r}$ is called the *directional derivative* of ϕ at P in the direction of PQ.

Let a unit vector along PQ be $\hat{N}'$

$$\frac{\delta n}{\delta r} = \cos\theta \Rightarrow \delta r = \frac{\delta n}{\cos\theta} = \frac{\delta n}{\hat{N}\cdot\hat{N}'} \qquad \ldots(1)$$

Now $$\frac{\partial\phi}{\partial r} = \lim_{\delta r \to 0}\left[\frac{\delta\phi}{\frac{\delta n}{\hat{N}.\hat{N}'}}\right] = \hat{N}.\hat{N}'\frac{\partial\phi}{\partial n} \qquad \left[\text{From (1), } \delta r = \frac{\delta n}{\hat{N}.\hat{N}'}\right]$$

$$= \hat{N}'.\hat{N}\,|\nabla\phi| \quad = \hat{N}'.\nabla\phi \qquad (\because \hat{N}\,|\nabla\phi| = \nabla\phi)$$

Hence, $\frac{\partial\phi}{\partial r}$, directional derivative is the component of $\nabla\phi$ in the direction $\hat{N}'$

$$\frac{\partial\phi}{\partial r} = \hat{N}'.\nabla\phi = |\nabla\phi|\cos\theta \le |\nabla\phi|$$

Hence, $\nabla\phi$ is the maximum rate of change of ϕ.

Example 6. *For the vector field* (*i*) $\vec{A} = m\hat{i}$ *and* (*ii*) $\vec{A} = m\vec{r}$. *Find* $\nabla.\vec{A}$ *and* $\nabla\times\vec{A}$. *Draw the sketch in each case.* (*Gujarat, I Semester, Jan. 2009*)

Solution. (*i*) Vector $\vec{A} = m\hat{i}$ is represented in the figure.

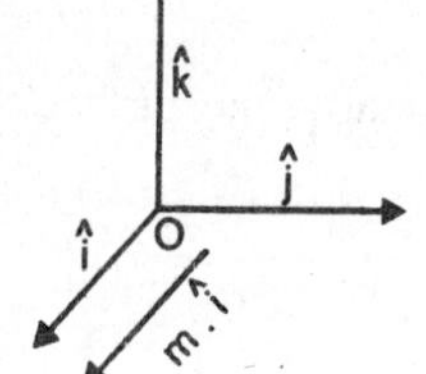

(*ii*) $\vec{A} = m\vec{r}$ is represented in the figure.

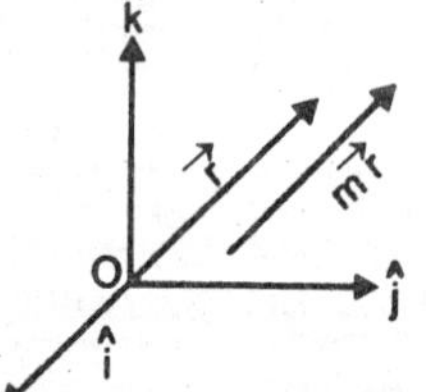

(*iii*) $$\nabla.\vec{A} = \left(\hat{i}\frac{\partial}{\partial x} + \hat{j}\frac{\partial}{\partial y} + \hat{k}\frac{\partial}{\partial z}\right).(x\hat{i} + y\hat{j} + z\hat{k}) = 1 + 1 + 1 = 3$$

$\nabla.\vec{A} = 3$ is represented on the number line at 3.

(*iv*) $$\nabla\times\vec{A} = \left(\hat{i}\frac{\partial}{\partial x} + \hat{j}\frac{\partial}{\partial y} + \hat{k}\frac{\partial}{\partial z}\right)\times(x\hat{i} + y\hat{j} + z\hat{k})$$

$$= \begin{vmatrix} \hat{i} & \hat{j} & \hat{k} \\ \frac{\partial}{\partial x} & \frac{\partial}{\partial y} & \frac{\partial}{\partial z} \\ x & y & z \end{vmatrix} = 0$$

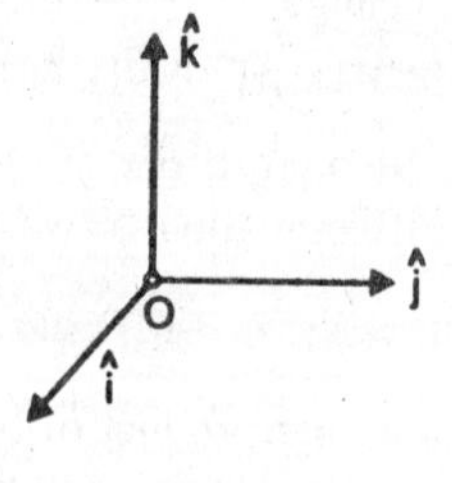

are represented in the adjoining figure.

Example 7. *If $\phi = 3x^2y - y^3z^2$; find grad ϕ at the point (1, –2, –1).*

(AMIETE, June 2009, U.P., I Semester, Dec. 2006)

Solution.

$$\text{grad } \phi = \nabla\phi$$

$$= \left(\hat{i}\frac{\partial}{\partial x} + \hat{j}\frac{\partial}{\partial y} + \hat{k}\frac{\partial}{\partial z}\right)(3x^2y - y^3z^2)$$

$$= \hat{i}\frac{\partial}{\partial x}(3x^2y - y^3z^2) + \hat{j}\frac{\partial}{\partial y}(3x^2y - y^3z^2) + \hat{k}\frac{\partial}{\partial z}(3x^2y - y^3z^2)$$

$$= \hat{i}\,(6xy) + \hat{j}\,(3x^2 - 3y^2z^2) + \hat{k}\,(-2y^3z)$$

$$\text{grad } \phi \text{ at } (1, -2, -1) = \hat{i}\,(6)(1)(-2) + \hat{j}\,[(3)(1) - 3(4)(1)] + \hat{k}\,(-2)(-8)(-1)$$

$$= -12\hat{i} - 9\hat{j} - 16\hat{k}$$

Ans.

Example 8. *If $u = x + y + z$, $v = x^2 + y^2 + z^2$, $w = yz + zx + xy$ prove that grad u, grad v and grad w are coplanar vectors.* *[U.P., I Semester, 2001]*

Solution. We have,

$$\text{grad } u = \left(\hat{i}\frac{\partial}{\partial x} + \hat{j}\frac{\partial}{\partial y} + \hat{k}\frac{\partial}{\partial z}\right)(x + y + z) = \hat{i} + \hat{j} + \hat{k}$$

$$\text{grad } v = \left(\hat{i}\frac{\partial}{\partial x} + \hat{j}\frac{\partial}{\partial y} + \hat{k}\frac{\partial}{\partial z}\right)(x^2 + y^2 + z^2) = 2x\,\hat{i} + 2y\,\hat{j} + 2z\,\hat{k}$$

$$\text{grad } w = \left(\hat{i}\frac{\partial}{\partial x} + \hat{j}\frac{\partial}{\partial y} + \hat{k}\frac{\partial}{\partial z}\right)(yz + zx + xy) = \hat{i}(z + y) + \hat{j}(z + x) + \hat{k}(y + x)$$

[For vectors to be coplanar, their scalar triple product is 0]

$$\text{Now, grad } u.(\text{grad } v \times \text{grad } w) = \begin{vmatrix} 1 & 1 & 1 \\ 2x & 2y & 2z \\ z + y & z + x & y + x \end{vmatrix} = 2\begin{vmatrix} 1 & 1 & 1 \\ x & y & z \\ z + y & z + x & y + x \end{vmatrix}$$

$$= 2\begin{vmatrix} 1 & 1 & 1 \\ x + y + z & x + y + z & x + y + z \\ z + y & z + x & y + x \end{vmatrix} \quad [\text{Applying } R_2 \to R_2 + R_3]$$

$$= 2(x + y + z)\begin{vmatrix} 1 & 1 & 1 \\ 1 & 1 & 1 \\ y + z & z + x & x + y \end{vmatrix} = 0$$

Since the scalar product of grad u, grad v and grad w are zero, hence these vectors are coplanar vectors. **Proved.**

Example 9. *Find the directional derivative of $x^2y^2z^2$ at the point (1, 1, –1) in the direction of the tangent to the curve $x = e^t$, $y = \sin 2t + 1$, $z = 1 - \cos t$ at $t = 0$.*

(Nagpur University, Summer 2005)

Solution. Let $\phi = x^2\,y^2\,z^2$

Directional Derivative of ϕ

$$= \nabla\phi = \left(\hat{i}\frac{\partial}{\partial x} + \hat{j}\frac{\partial}{\partial y} + \hat{k}\frac{\partial}{\partial z}\right)(x^2y^2z^2)$$

$$\nabla\phi = 2xy^2z^2\,\hat{i} + 2yx^2z^2\,\hat{j} + 2zx^2y^2\,\hat{k}$$

Directional Derivative of ϕ at (1, 1, –1)

$$= 2(1)(1)^2(-1)^2\,\hat{i} + 2(1)(1)^2\,(-1)^2\,\hat{j} + 2(-1)(1)^2(1)^2\,\hat{k}$$

$$= 2\hat{i} + 2\hat{j} - 2\hat{k} \qquad ...(1)$$

$$\vec{r} = x\hat{i} + y\hat{j} + z\hat{k} = e^t\hat{i} + (\sin 2t + 1)\hat{j} + (1 - \cos t)\hat{k}$$

Tangent vector, $\vec{T} = \dfrac{d\vec{r}}{dt} = e^t\hat{i} + 2\cos 2t\,\hat{j} + \sin t\,\hat{k}$

$$\text{Tangent(at } t = 0) = e^0\hat{i} + 2(\cos 0)\hat{j} + (\sin 0)\hat{k} = \hat{i} + 2\hat{j} \qquad ...(2)$$

Required directional derivative along tangent $= (2\hat{i} + 2\hat{j} - 2\hat{k})\dfrac{(\hat{i} + 2\hat{j})}{\sqrt{1+4}}$

[From (1), (2)]

$$= \frac{2+4+0}{\sqrt{5}} = \frac{6}{\sqrt{5}}$$ **Ans.**

Example 10. *Find the unit normal to the surface $xy^3z^2 = 4$ at $(-1, -1, 2)$.* (*M.U. 2008*)

Solution. Let $\phi(x, y, z) = xy^3z^2 - 4$

We know that $\nabla\phi$ is the vector normal to the surface $\phi(x, y, z) = c$.

$$\text{Normal vector} = \nabla\phi = \hat{i}\frac{\partial\phi}{\partial x} + \hat{j}\frac{\partial\phi}{\partial y} + \hat{k}\frac{\partial\phi}{\partial z}$$

Now $= \hat{i}\dfrac{\partial}{\partial x}(xy^3z^2) + \hat{j}\dfrac{\partial}{\partial y}(xy^3z^2) + \hat{k}\dfrac{\partial}{\partial z}(xy^3z^2)$

$\Rightarrow$ Normal vector $= y^3z^2\hat{i} + 3xy^2z^2\hat{j} + 2xy^3z\hat{k}$

Normal vector at $(-1, -1, 2) = -4\hat{i} - 12\hat{j} + 4\hat{k}$

Unit vector normal to the surface at $(-1, -1, 2)$.

$$= \frac{\nabla\phi}{|\nabla\phi|} = \frac{-4\hat{i} - 12\hat{j} + 4\hat{k}}{\sqrt{16+144+16}} = -\frac{1}{\sqrt{11}}(\hat{i} + 3\hat{j} - \hat{k})$$ **Ans.**

Example 11. *Find the rate of change of $\phi = xyz$ in the direction normal to the surface $x^2y + y^2x + yz^2 = 3$ at the point $(1, 1, 1)$.* (*Nagpur University, Summer 2001*)

Solution. Rate of change of $\phi = \Delta\,\phi$

$$= \left(\hat{i}\frac{\partial}{\partial x} + \hat{j}\frac{\partial}{\partial y} + \hat{k}\frac{\partial}{\partial z}\right)(x\,y\,z) = \hat{i}\,yz + \hat{j}\,xz + \hat{k}\,xy$$

Rate of change of ϕ at $(1, 1, 1) = (\hat{i} + \hat{j} + \hat{k})$

Normal to the surface $\Psi = x^2y + y^2x + yz^2 - 3$ is given as -

$$\nabla\psi = \left(\hat{i}\frac{\partial}{\partial x} + \hat{j}\frac{\partial}{\partial y} + \hat{k}\frac{\partial}{\partial z}\right)(x^2y + y^2x + yz^2 - 3)$$

$$= \hat{i}(2xy + y^2) + \hat{j}(x^2 + 2xy + z^2) + \hat{k}\,2yz$$

$$(\nabla\psi)_{(1,1,1)} = 3\hat{i} + 4\hat{j} + 2\hat{k}$$

$$\text{Unit normal} = \frac{3\hat{i} + 4\hat{j} + 2\hat{k}}{\sqrt{9+16+4}}$$

Required rate of change of $\phi = (\hat{i} + \hat{j} + \hat{k}).\dfrac{(3\hat{i} + 4\hat{j} + 2\hat{k})}{\sqrt{9+16+4}} = \dfrac{3+4+2}{\sqrt{29}} = \dfrac{9}{\sqrt{29}}$ **Ans.**

Example 12. *Find the constants m and n such that the surface m x^2 – 2nyz = (m + 4)x will be orthogonal to the surface $4x^2y + z^3 = 4$ at the point (1, –1, 2).*

(*M.D.U. Dec. 2009, Nagpur University, Summer 2002*)

Solution. The point P (1, –1, 2) lies on both surfaces. As this point lies in

$$mx^2 - 2nyz = (m + 4)x, \text{ so we have}$$

$$m - 2n(-2) = (m + 4)$$

$$\Rightarrow \quad m + 4n = m + 4 \quad \Rightarrow \quad n = 1$$

$$\therefore \quad \text{Let } \phi_1 = mx^2 - 2yz - (m + 4)x \text{ and } \phi_2 = 4x^2y + z^3 - 4$$

$$\text{Normal to } \phi_1 = \nabla\phi_1$$

$$= \left(\hat{i}\frac{\partial}{\partial x} + \hat{j}\frac{\partial}{\partial y} + \hat{k}\frac{\partial}{\partial z}\right)[mx^2 - 2yz - (m+4)x]$$

$$= \hat{i}(2mx - m - 4) - 2z\,\hat{j} - 2y\,\hat{k}$$

$$\text{Normal to } \phi_1 \text{ at } (1, -1, 2) = \hat{i}(2m - m - 4) - 4\,\hat{j} + 2\,\hat{k} = (m - 4)\,\hat{i} - 4\,\hat{j} + 2\,\hat{k}$$

$$\text{Normal to } \phi_2 = \nabla\phi_2$$

$$= \left(\hat{i}\frac{\partial}{\partial x} + \hat{j}\frac{\partial}{\partial y} + \hat{k}\frac{\partial}{\partial z}\right)(4x^2y + z^3 - 4) = \hat{i}\,8xy + 4x^2\,\hat{j} + 3z^2\,\hat{k}$$

$$\text{Normal to } \phi_2 \text{ at } (1, -1, 2) = -8\,\hat{i} + 4\,\hat{j} + 12\,\hat{k}$$

Sinec ϕ_1 and ϕ_2 are orthogonal, then normals are perpendicular to each other.

$$\nabla\phi_1 \cdot \nabla\phi_2 = 0$$

$$\Rightarrow \quad [(m - 4)\,\hat{i} - 4\,\hat{j} + 2\,\hat{k}].[-8\,\hat{i} + 4\,\hat{j} + 12\,\hat{k}] = 0$$

$$\Rightarrow \quad -8\,(m - 4) - 16 + 24 = 0$$

$$\Rightarrow \quad m - 4 = -2 + 3 \quad \Rightarrow \quad m = 5$$

Hence $m = 5$, $n = 1$ **Ans.**

Example 13. *Find the values of constants λ and μ so that the surfaces $\lambda x^2 - \mu yz = (\lambda + 2)\,x$, $4x^2y + z^3 = 4$ intersect orthogonally at the point (1, – 1, 2).*

(*AMIETE, II Sem., Dec. 2010, June 2009*)

Solution. Here, we have

$$\lambda x^2 - \mu yz = (\lambda + 2)\,x \qquad \text{...(1)}$$

$$4x^2\,y + z^3 = 4 \qquad \text{...(2)}$$

$$\text{Normal to the surface (1), } = \nabla\left[\lambda x^2 - \mu yz - (\lambda + 2)x\right]$$

$$= \left[\hat{i}\frac{\partial}{\partial x} + \hat{j}\frac{\partial}{\partial y} + \hat{k}\frac{\partial}{\partial z}\right]\left[\lambda x^2 - \mu yz - (\lambda + 2)x\right]$$

$$= \hat{i}\,(2\lambda x - \lambda - 2) + \hat{j}\,(-\mu z) + \hat{k}\,(-\mu y)$$

$$\textbf{Normal at (1, –1, 2)} = \hat{i}\;(2\lambda - \lambda - 2) - \hat{j}\;(-2\mu) + \hat{k}\,\mu \qquad \text{...(3)}$$

$$= \hat{i}\;(\lambda - 2) + \hat{i}\;(2\mu) + \hat{k}\,\mu$$

Normal at the surface (2)

$$= \left(\hat{i}\frac{\partial}{\partial x} + \hat{j}\frac{\partial}{\partial y} + \hat{k}\frac{\partial}{\partial z}\right)(4x^2y + z^3 - 4)$$

$$= \hat{i}\;(8xy) + \hat{j}\;(4x^2) + \hat{k}\;(3z^2)$$

$$\textbf{Normal at the point (1, –1, 2)} = -8\,\hat{i} + 4\,\hat{j} + 12\,\hat{k} \qquad \text{...(4)}$$

Since (3) and (4) are orthogonal so

$$\left[\hat{i}\,(\lambda-2)+\hat{j}\,(2\mu)+\hat{k}\,\mu\right].\left[-8\hat{i}+4\hat{j}+12\hat{k}\right]=0$$

$$-8\,(\lambda-2)+4(2\mu)+12\mu=0 \quad\Rightarrow\quad -8\lambda+16+8\mu+12\mu=0$$

$$-8\lambda+20\mu+16=0 \quad\Rightarrow\quad 4(-2\lambda+5\mu+4)=0$$

$$-2\lambda+5\mu+4=0 \quad\Rightarrow\quad 2\lambda-5\mu=4 \qquad ...(5)$$

Point (1, – 1, 2) will satisfy (1)

$$\therefore \qquad \lambda(1)^2-\mu(-1)(2)=(\lambda+2)\,(1)\Rightarrow\ \lambda+2\mu=\lambda+2\ \Rightarrow\ \mu=1$$

Putting $\mu = 1$ in (5), we get

$$2\lambda-5=4 \quad\Rightarrow\quad \lambda=\frac{9}{2}$$

Hence $\lambda=\frac{9}{2}$ and $\mu=1$ **Ans.**

Example 14 *Find the angle between the surfaces $x^2+y^2+z^2=9$ and $z=x^2+y^2-3$ at the point (2, –1, 2).* *(Nagpur University, Summer 2002)*

Solution. Normal on the surface $(x^2+y^2+z^2-9=0)$

$$\nabla\phi = \left(\hat{i}\frac{\partial}{\partial x}+\hat{j}\frac{\partial}{\partial y}+\hat{k}\frac{\partial}{\partial z}\right)(x^2+y^2+z^2-9)=(2x\,\hat{i}+2y\,\hat{j}+2z\,\hat{k})$$

Normal at the point (2, –1, 2) = $4\hat{i}-2\hat{j}+4\hat{k}$ ...(1)

Normal on the surface $(z=x^2+y^2-3) = \left(\hat{i}\frac{\partial}{\partial x}+\hat{j}\frac{\partial}{\partial y}+\hat{k}\frac{\partial}{\partial z}\right)(x^2+y^2-z-3)$

$$= 2x\,\hat{i}+2y\,\hat{j}-\hat{k}$$

Normal at the point (2, –1, 2) = $4\hat{i}-2\hat{j}-\hat{k}$ ...(2)

Let θ be the angle between normals (1) and (2).

$$(4\hat{i}-2\hat{j}+4\hat{k}).(4\hat{i}-2\hat{j}-\hat{k}) = \sqrt{16+4+16}\,\sqrt{16+4+1}\,\cos\theta$$

$$16+4-4 = 6\sqrt{21}\cos\theta \quad\Rightarrow\quad 16 = 6\sqrt{21}\cos\theta$$

$$\Rightarrow \qquad \cos\theta = \frac{8}{3\sqrt{21}} \quad\Rightarrow\quad \theta=\cos^{-1}\frac{8}{3\sqrt{21}}$$ **Ans.**

Example 15. *Find the directional derivative of $\frac{1}{r}$ in the direction $\bar{r}$ where $\bar{r}=x\,\hat{i}+y\,\hat{j}+z\,\hat{k}$.*

(Nagpur University, Summer 2004, U.P., I Semester, Winter 2005, 2002)

Solution. Here, $\phi\,(x, y, z) = \frac{1}{r}=\frac{1}{\sqrt{x^2+y^2+z^2}}=(x^2+y^2+z^2)^{-\frac{1}{2}}$

$$\text{Now } \nabla\left(\frac{1}{r}\right) = \left(\hat{i}\frac{\partial}{\partial x}+\hat{j}\frac{\partial}{\partial y}+\hat{k}\frac{\partial}{\partial z}\right)(x^2+y^2+z^2)^{-\frac{1}{2}}$$

$$= \frac{\partial}{\partial x}(x^2+y^2+z^2)^{-\frac{1}{2}}\,\hat{i}+\frac{\partial}{\partial y}(x^2+y^2+z^2)^{-\frac{1}{2}}\,\hat{j}+\frac{\partial}{\partial z}(x^2+y^2+z^2)^{-\frac{1}{2}}\,\hat{k}$$

$$= \left\{-\frac{1}{2}(x^2+y^2+z^2)^{-\frac{3}{2}}\,2x\right\}\hat{i}+\left(-\frac{1}{2}(x^2+y^2+z^2)^{-\frac{3}{2}}\,2y\right)\hat{j}+\left\{-\frac{1}{2}(x^2+y^2+z^2)^{-\frac{3}{2}}\,2z\right\}\hat{k}$$

$$= \frac{-(x\hat{i} + y\hat{j} + z\hat{k})}{(x^2 + y^2 + z^2)^{3/2}} \quad ...(1)$$

and $\hat{r}$ = unit vector in the direction of $x\hat{i} + y\hat{j} + z\hat{k}$

$$= \frac{x\hat{i} + y\hat{j} + z\hat{k}}{\sqrt{x^2 + y^2 + z^2}} \quad ...(2)$$

So, the required directional derivative

$$= \nabla\phi.\hat{r} = -\frac{x\hat{i} + y\hat{j} + z\hat{k}}{(x^2 + y^2 + z^2)^{3/2}} \cdot \frac{x\hat{i} + y\hat{j} + z\hat{k}}{(x^2 + y^2 + z^2)^{1/2}} = \frac{x^2 + y^2 + z^2}{(x^2 + y^2 + z^2)^2} \quad \text{[From (1), (2)]}$$

$$= \frac{1}{x^2 + y^2 + z^2} = \frac{1}{r^2}$$ **Ans.**

Example 16. *Find the direction in which the directional derivative of* $\phi(x, y) = \dfrac{x^2 + y^2}{xy}$ *at (1, 1) is zero and hence find out component of velocity of the vector* $\bar{r} = (t^3 + 1)\hat{i} + t^2\hat{j}$ *in the same direction at t = 1.* (*Nagpur University, Winter 2000*)

Solution. Directional derivative $= \nabla\phi = \left(\hat{i}\dfrac{\partial}{\partial x} + \hat{j}\dfrac{\partial}{\partial y} + \hat{k}\dfrac{\partial}{\partial z}\right)\left(\dfrac{x^2 + y^2}{xy}\right)$

$$= \hat{i}\left[\frac{x\,y.2x - (x^2 + y^2)\,y}{x^2y^2}\right] + \hat{j}\left[\frac{xy.2y - x(y^2 + x^2)}{x^2y^2}\right]$$

$$= \hat{i}\left[\frac{x^2y - y^3}{x^2y^2}\right] + \hat{j}\left[\frac{xy^2 - x^3}{x^2y^2}\right]$$

Directional Derivative at (1, 1) $= \hat{i}\,0 + \hat{j}\,0 = 0$

Since $(\nabla\phi)_{(1,\,1)} = 0$, the directional derivative of ϕ at (1, 1) is zero in any direction.

Again $\bar{r} = (t^3 + 1)\hat{i} + t^2\hat{j}$

Velocity, $\bar{v} = \dfrac{d\bar{r}}{dt} = 3t^2\,\hat{i} + 2t\,\hat{j}$

Velocity at $t = 1$ is $= 3\hat{i} + 2\hat{j}$

The component of velocity in the same direction of velocity

$$= (3\hat{i} + 2\hat{j}).\left(\frac{3\hat{i} + 2\hat{j}}{\sqrt{9 + 4}}\right) = \frac{9 + 4}{\sqrt{13}} = \sqrt{13}$$ **Ans.**

Example 17. *Find the directional derivative of* $\phi(x, y, z) = x^2yz + 4xz^2$ *at (1, –2, 1) ir the direction of* $2\hat{i} - \hat{j} - 2\hat{k}$. *Find the greatest rate of increase of* ϕ.

(*Uttarakhand, I Semester, Dec. 2006*)

Solution. Here, $\phi(x, y, z) = x^2yz + 4xz^2$

Now, $$\nabla\phi = \left(\hat{i}\frac{\partial}{\partial x} + \hat{j}\frac{\partial}{\partial y} + \hat{k}\frac{\partial}{\partial z}\right)(x^2yz + 4xz^2)$$

$$= (2xyz + 4z^2)\hat{i} + (x^2z)\hat{j} + (x^2y + 8xz)\hat{k}$$

$$\nabla\phi \text{ at } (1, -2, 1) = \{2(1)(-2)(1) + 4(1)^2\}\hat{i} + (1\times 1)\hat{j} + \{1(-2) + 8(1)(1)\}\hat{k}$$

$$= (-4+4)\hat{i} + \hat{j} + (-2+8)\hat{k} = \hat{j} + 6\hat{k}$$

Let $\hat{a}$ = unit vector $= \dfrac{2\hat{i} - \hat{j} - 2\hat{k}}{\sqrt{4+1+4}} = \dfrac{1}{3}(2\hat{i} - \hat{j} - 2\hat{k})$

So, the required directional derivative at (1, –2, 1)

$$= \nabla\phi.\hat{a} = (\hat{j} + 6\hat{k}).\frac{1}{3}(2\hat{i} - \hat{j} - 2\hat{k}) = \frac{1}{3}(-1-12) = \frac{-13}{3}$$

Greatest rate of increase of $\phi = \left|\hat{j} + 6\hat{k}\right| = \sqrt{1+36}$

$$= \sqrt{37}$$ **Ans.**

Example 18. *Find the directional derivative of the function $\phi = x^2 - y^2 + 2z^2$ at the point P (1, 2, 3) in the direction of the line PQ where Q is the point (5, 0, 4).*

(AMIETE, Dec. 2010, Nagpur University, Summer 2008, U.P., I Sem., Winter 2000)

Solution. Directional derivative $= \bar{\nabla}\phi$

$$= \left(\hat{i}\frac{\partial}{\partial x} + \hat{j}\frac{\partial}{\partial y} + \hat{k}\frac{\partial}{\partial z}\right)(x^2 - y^2 + 2z^2) = 2x\hat{i} - 2y\hat{j} + 4z\hat{k}$$

Directional Derivative at the point P (1, 2, 3) $= 2\hat{i} - 4\hat{j} + 12\hat{k}$ ...(1)

$$\overline{PQ} = \overline{Q} - \overline{P} = (5, 0, 4) - (1, 2, 3) = (4, -2, 1) \quad ...(2)$$

Directional Derivative along $PQ = (2\hat{i} - 4\hat{j} + 12\hat{k}).\dfrac{(4\hat{i} - 2\hat{j} + \hat{k})}{\sqrt{16+4+1}}$ [From (1) and (2)]

$$= \frac{8+8+12}{\sqrt{21}} = \frac{28}{\sqrt{21}}$$ **Ans.**

Example 19. *Find the directional derivative of $\phi = 4\,e^{2x-y+z}$ at the point (1, 1, –1) in the directional towards the point (–3, 5, 6).* *(Nagpur University, Winter 2003, Summer 2000)*

Solution. Directional derivative $= \nabla\phi$

$$= \left(\hat{i}\frac{\partial}{\partial x} + \hat{j}\frac{\partial}{\partial y} + \hat{k}\frac{\partial}{\partial z}\right)4\,e^{2x-y+z}$$

$$= 4[\hat{i}\,2e^{2x-y+z} - \hat{j}\,e^{2x-y+z} + \hat{k}\,e^{2x-y+z}] = 4[2\hat{i} - \hat{j} + \hat{k}]\,e^{2x-y+z}$$

Directional Derivative at (1, 1, –1)

$$= 4[2\hat{i} - \hat{j} + \hat{k}]\,e^{2-1-1} = 4[2\hat{i} - \hat{j} + \hat{k}] \quad ...(1)$$

Direction of Directional Derivative

$$= (-3\hat{i} + 5\hat{j} + 6\hat{k}) - (\hat{i} + \hat{j} - \hat{k}) = -4\hat{i} + 4\hat{j} + 7\hat{k} \quad ...(2)$$

Directional Derivative in the direction of $(-4\hat{i} + 4\hat{j} + 7\hat{k})$

$$= \left|(8\hat{i} - 4\hat{j} + 4\hat{k}).\frac{(-4\hat{i} + 4\hat{j} + 7\hat{k})}{\sqrt{16+16+49}}\right|$$ [From (1) and (2)]

$$= \left|\frac{1}{9}[-32 - 16 + 28]\right| = \left|-\frac{20}{9}\right| = \frac{20}{9}$$ **Ans.**

Example 20. *For the function* $\phi(x, y) = \frac{x}{x^2+y^2}$, *find the magnitude of the directional derivative along a line making an angle 30° with the positive x-axis at (0, 2).*

(*A.M.I.E.T.E., Winter 2002*)

Solution. Directional derivative $= \vec{\nabla}\phi$

$$= \left(\hat{i}\frac{\partial}{\partial x}+\hat{j}\frac{\partial}{\partial y}+\hat{k}\frac{\partial}{\partial z}\right)\frac{x}{x^2+y^2} = \hat{i}\left(\frac{1}{x^2+y^2}-\frac{x(2x)}{(x^2+y^2)^2}\right)-\hat{j}\frac{x(2y)}{(x^2+y^2)^2}$$

$$= \hat{i}\frac{y^2-x^2}{(x^2+y^2)^2}-\hat{j}\frac{2xy}{(x^2+y^2)^2}$$

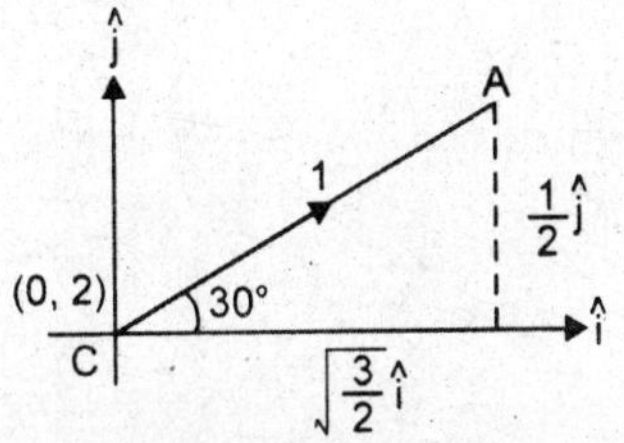

Directional derivative at the point (0, 2)

$$= \hat{i}\frac{4-0}{(0+4)^2}-\hat{j}\frac{2(0)(2)}{(0+4)^2}=\frac{\hat{i}}{4}$$

A
1
$\frac{1}{2}\hat{j}$
(0, 2)
30°
C
$\frac{\sqrt{3}}{2}\hat{i}$

Directional derivative at the point (0, 2) in the direction $\vec{CA}$ *i.e.* $\left(\frac{\sqrt{3}}{2}\hat{i}+\frac{1}{2}\hat{j}\right)$

$$= \frac{\hat{i}}{4}\cdot\left(\frac{\sqrt{3}}{2}\hat{i}+\frac{1}{2}\hat{j}\right)$$

$$\left\{\vec{CA}=\vec{OB}+\vec{BA}=\hat{i}\cos 30°+\hat{j}\sin 30° = \left(\frac{\sqrt{3}}{2}\hat{i}+\frac{1}{2}\hat{j}\right)\right\}$$

$$= \frac{\sqrt{3}}{8}$$ **Ans.**

Example 21. *Find the directional derivative of* $\vec{V}^2$, *where* $\vec{V} = xy^2\hat{i}+zy^2\hat{j}+xz^2\hat{k}$, *at the point (2, 0, 3) in the direction of the outward normal to the sphere* $x^2+y^2+z^2=14$ *at the point (3, 2, 1).*

(*A.M.I.E.T.E., Dec. 2007*)

Solution. $V^2 = \vec{V}.\vec{V}$

$$= (xy^2\hat{i}+zy^2\hat{j}+xz^2\hat{k}).(xy^2\hat{i}+zy^2\hat{j}+xz^2\hat{k}) = x^2y^4+z^2y^4+x^2z^4$$

Directional derivative $= \vec{\nabla}V^2$

$$= \left(\hat{i}\frac{\partial}{\partial x}+\hat{j}\frac{\partial}{\partial y}+\hat{k}\frac{\partial}{\partial z}\right)(x^2y^4+z^2y^4+x^2z^4)$$

$$= (2xy^4+2xz^4)\hat{i}+(4x^2y^3+4y^3z^2)\hat{j}+(2y^4z+4x^2z^3)\hat{k}$$

Directional derivative at (2, 0, 3) $= (0+2\times2\times81)\hat{i}+(0+0)\hat{j}+(0+4\times4\times27)\hat{k}$

$$= 324\hat{i}+432\hat{k} = 108(3\hat{i}+4\hat{k}) \quad ...(1)$$

Normal to $x^2+y^2+z^2-14 = \nabla\phi$

$$= \left(\hat{i}\frac{\partial}{\partial x}+\hat{j}\frac{\partial}{\partial y}+\hat{k}\frac{\partial}{\partial z}\right)(x^2+y^2+z^2-14)$$

$$= (2x\hat{i}+2y\hat{j}+2z\hat{k})$$

Normal vector at (3, 2, 1) $= 6\hat{i}+4\hat{j}+2\hat{k}$ ...(2)

Unit normal vector $= \frac{6\hat{i}+4\hat{j}+2\hat{k}}{\sqrt{36+16+4}} = \frac{2(3\hat{i}+2\hat{j}+\hat{k})}{2\sqrt{14}} = \frac{3\hat{i}+2\hat{j}+\hat{k}}{\sqrt{14}}$ [From (1), (2)]

Directional derivative along the normal $= 108(3\hat{i} + 4\hat{k}).\dfrac{3\hat{i}+2\hat{j}+\hat{k}}{\sqrt{14}}$.

$$= \frac{108 \times (9+4)}{\sqrt{14}} = \frac{1404}{\sqrt{14}} \qquad \textbf{Ans.}$$

Example 22. *Find the directional derivative of $\nabla(\nabla f)$ at the point $(1, -2, 1)$ in the direction of the normal to the surface $xy^2z = 3x + z^2$, where $f = 2x^3y^2z^4$.* (U.P., I Semester, Dec 2008)

Solution. Here, we have

$$f = 2x^3 y^2 z^4$$

$$\nabla f = \left(\hat{i}\frac{\partial}{\partial x} + \hat{j}\frac{\partial}{\partial y} + \hat{k}\frac{\partial}{\partial z}\right)(2x^3y^2z^4) = 6x^2y^2z^4\hat{i} + 4x^3yz^4\hat{j} + 8x^3y^2z^3\hat{k}$$

$$\nabla(\nabla f) = \left(\hat{i}\frac{\partial}{\partial x} + \hat{j}\frac{\partial}{\partial y} + \hat{k}\frac{\partial}{\partial z}\right)(6x^2y^2z^4\hat{i} + 4x^3yz^4\hat{j} + 8x^3y^2z^3\hat{k})$$

$$= 12xy^2z^4 + 4x^3z^4 + 24x^3y^2z^2$$

Directional derivative of $\nabla(\nabla f)$

$$= \left(\hat{i}\frac{\partial}{\partial x} + \hat{j}\frac{\partial}{\partial y} + \hat{k}\frac{\partial}{\partial z}\right)(12xy^2z^4 + 4x^3z^4 + 24x^3y^2z^2)$$

$$= (12y^2z^4 + 12x^2z^4 + 72x^2y^2z^2)\hat{i} + (24xyz^4 + 48x^3yz^2)\hat{j} + (48xy^2z^3 + 16x^3z^3 + 48x^3y^2z)\hat{k}$$

Directional derivative at $(1, -2, 1) = (48 + 12 + 288)\hat{i} + (-48 - 96)\hat{j} + (192 + 16 + 192)\hat{k}$

$$= 348\hat{i} - 144\hat{j} + 400\hat{k}$$

Normal to$(xy^2z - 3x - z^2) = \nabla(xy^2z - 3x - z^2)$

$$= \left(\hat{i}\frac{\partial}{\partial x} + \hat{j}\frac{\partial}{\partial y} + \hat{k}\frac{\partial}{\partial z}\right)(xy^2z - 3x - z^2)$$

$$= (y^2z - 3)\hat{i} + (2xyz)\hat{j} + (xy^2 - 2z)\hat{k}$$

Normal at$(1, -2, 1) = \hat{i} - 4\hat{j} + 2\hat{k}$

Unit Normal Vector $= \dfrac{\hat{i} - 4\hat{j} + 2\hat{k}}{\sqrt{1+16+4}} = \dfrac{1}{\sqrt{21}}(\hat{i} - 4\hat{j} + 2\hat{k})$

Directional derivative in the direction of normal

$$= (348\hat{i} - 144\hat{j} + 400\hat{k})\frac{1}{\sqrt{21}}(\hat{i} - 4\hat{j} + 2\hat{k})$$

$$= \frac{1}{\sqrt{21}}(348 + 576 + 800) = \frac{1724}{\sqrt{21}} \qquad \textbf{Ans.}$$

Example 23. *If the directional derivative of $\phi = a\,x^2\,y + b\,y^2\,z + c\,z^2\,x$ at the point $(1, 1, 1)$ has maximum magnitude 15 in the direction parallel to the line $\dfrac{x-1}{2} = \dfrac{y-3}{-2} = \dfrac{z}{1}$, find the values of a, b and c.* (U.P. I semester, Winter 2001)

Solution. Given $\qquad \phi = a\,x^2\,y + b\,y^2\,z + c\,z^2\,x$

$$\therefore \qquad \bar{\nabla}\phi = \left(\hat{i}\frac{\partial}{\partial x} + \hat{j}\frac{\partial}{\partial y} + \hat{k}\frac{\partial}{\partial z}\right)(a\,x^2\,y + b\,y^2\,z + c\,z^2\,x)$$

$$= \hat{i}(2a\,x\,y + c\,z^2) + \hat{j}(a\,x^2 + 2b\,y\,z) + \hat{k}(b\,y^2 + 2c\,z\,x)$$

$$\bar{\nabla}\phi \text{ at the point } (1, 1, 1) = \hat{i}(2a+c)+\hat{j}(a+2b)+\hat{k}(b+2c) \quad ...(1)$$

We know that the maximum value of the directional derivative is in the direction of $\bar{\nabla}\phi$.

$$i.e.\ |\nabla\phi| = 15 \Rightarrow (2a+c)^2 + (2b+a)^2 + (2c+b)^2 = (15)^2$$

But, the directional derivative is given to be maximum parallel to the line

$$\frac{x-1}{2} = \frac{y-3}{-2} = \frac{z}{1} \text{ i.e., parallel to the vector } 2\hat{i}-2\hat{j}+\hat{k}. \quad ...(2)$$

On comparing the coefficients of (1) and (2)

$$\Rightarrow \quad \frac{2a+c}{2} = \frac{2b+a}{-2} = \frac{2c+b}{1}$$

$$\Rightarrow \quad 2a+c = -2b-a \Rightarrow 3a+2b+c=0 \quad ...(3)$$

and $\quad 2b+a = -2(2c+b)$

$$\Rightarrow \quad 2b+a = -4c-2b \Rightarrow a+4b+4c=0 \quad ...(4)$$

Rewriting (3) and (4), we have

$$\left.\begin{aligned} 3a+2b+c=0 \\ a+4b+4c=0 \end{aligned}\right\} \Rightarrow \frac{a}{4} = \frac{b}{-11} = \frac{c}{10} = k \text{ (say)}$$

$$\Rightarrow \quad a = 4k, \quad b=-11k \text{ and } c = 10k.$$

Now, we have

$$(2a+c)^2 + (2b+a)^2 + (2c+b)^2 = (15)^2$$

$$\Rightarrow \quad (8k+10k)^2 + (-22k+4k)^2 + (20k-11k)^2 = (15)^2$$

$$\Rightarrow \quad k = \pm\frac{5}{9}$$

$$\Rightarrow \quad a = \pm\frac{20}{9}, \quad b = \pm\frac{55}{9} \text{ and } c = \pm\frac{50}{9}$$ **Ans.**

Example 24. *If $\bar{r} = x\hat{i}+y\hat{j}+z\hat{k}$, show that :*

(i) grad $r = \frac{\vec{r}}{r}$ *(ii) grad* $\left(\frac{1}{r}\right) = -\frac{\vec{r}}{r^3}$. *(Nagpur University, Summer 2002)*

Solution. (*i*) $\bar{r} = x\hat{i}+y\hat{j}+z\hat{k} \Rightarrow r = \sqrt{x^2+y^2+z^2} \Rightarrow r^2 = x^2+y^2+z^2$

$$\therefore \quad 2r\frac{\partial r}{\partial x} = 2x \Rightarrow \frac{\partial r}{\partial x} = \frac{x}{r}$$

Similarly, $\frac{\partial r}{\partial y} = \frac{y}{r}$ and $\frac{\partial r}{\partial z} = \frac{z}{r}$

$$\text{grad } r = \nabla r = \left(\hat{i}\frac{\partial}{\partial x}+\hat{j}\frac{\partial}{\partial y}+\hat{k}\frac{\partial}{\partial z}\right)r = \hat{i}\frac{\partial r}{\partial x}+\hat{j}\frac{\partial r}{\partial y}+\hat{k}\frac{\partial r}{\partial z}$$

$$= \hat{i}\frac{x}{r}+\hat{j}\frac{y}{r}+\hat{k}\frac{z}{r} = \frac{x\hat{i}+y\hat{j}+z\hat{k}}{r} = \frac{\bar{r}}{r}$$ **Proved.**

$$(ii) \text{ grad}\left(\frac{1}{r}\right) = \nabla\left(\frac{1}{r}\right) = \left(\hat{i}\frac{\partial}{\partial x}+\hat{j}\frac{\partial}{\partial y}+\hat{k}\frac{\partial}{\partial z}\right)\left(\frac{1}{r}\right) = \hat{i}\frac{\partial}{\partial x}\left(\frac{1}{r}\right)+\hat{j}\frac{\partial}{\partial y}\left(\frac{1}{r}\right)+\hat{k}\frac{\partial}{\partial z}\left(\frac{1}{r}\right)$$

$$= \hat{i}\left(-\frac{1}{r^2}\frac{\partial r}{\partial x}\right)+\hat{j}\left(\frac{-1}{r^2}\frac{\partial r}{\partial y}\right)+\hat{k}\left(-\frac{1}{r^2}\frac{\partial r}{\partial z}\right)$$

$$= \hat{i}\left(-\frac{1}{r^2}\frac{x}{r}\right)+\hat{j}\left(-\frac{1}{r^2}\frac{y}{r}\right)+\hat{k}\left(-\frac{1}{r^2}\frac{z}{r}\right) = -\frac{x\hat{i}+y\hat{j}+z\hat{k}}{r^3} = -\frac{\bar{r}}{r^3}$$ **Proved.**

Example 25. *Prove that* $\nabla^2 f(r) = f''(r) + \frac{2}{r} f'(r)$. *(K. University, Dec. 2008)*

Solution.

$$\nabla f(r) = \left(i\frac{\partial}{\partial x} + j\frac{\partial}{\partial y} + k\frac{\partial}{\partial z}\right) f(r)$$

$$\left[r^2 = x^2 + y^2 + z^2 \Rightarrow 2r\frac{\partial r}{\partial x} = 2x \Rightarrow \frac{\partial r}{\partial x} = \frac{x}{r}, \quad \frac{\partial r}{\partial y} = \frac{y}{r} \text{ and } \frac{\partial r}{\partial z} = \frac{z}{r}\right]$$

$$= i f'(r)\frac{\partial r}{\partial x} + j f'(r)\frac{\partial r}{\partial y} + k f'(r)\frac{\partial r}{\partial z} = f'(r)\left[i\frac{x}{r} + j\frac{y}{r} + k\frac{z}{r}\right]$$

$$= f'(r)\frac{xi + yj + zk}{r}$$

$$\nabla^2 f(r) = \nabla[\nabla f(r)] = \left(i\frac{\partial}{\partial x} + j\frac{\partial}{\partial y} + k\frac{\partial}{\partial z}\right)\left[f'(r)\frac{xi + yj + zk}{r}\right]$$

$$= \frac{\partial}{\partial x}\left[f'(r)\frac{x}{r}\right] + \frac{\partial}{\partial y}\left[f'(r)\frac{y}{r}\right] + \frac{\partial}{\partial z}\left[f'(r)\frac{z}{r}\right]$$

$$= \left(f''(r)\frac{\partial r}{\partial x}\right)\left(\frac{x}{r}\right) + f'(r)\frac{r1 - x\frac{\partial r}{\partial x}}{r^2} + \left(f''(r)\frac{\partial r}{\partial y}\right)\left(\frac{y}{r}\right) + f'(r)\frac{r.1 - y\frac{\partial r}{\partial y}}{r^2} +$$

$$\left(f''(r)\frac{\partial r}{\partial z}\right)\left(\frac{z}{r}\right) + f'(r)\frac{r.1 - z\frac{\partial z}{\partial r}}{r^2}$$

$$= \left(f''(r)\frac{x}{r}\right)\left(\frac{x}{r}\right) + f'(r)\frac{r - \frac{x^2}{r}}{r^2} + \left(f''(r)\frac{y}{r}\right)\left(\frac{y}{r}\right) + f'(r)\frac{r - \frac{y^2}{r}}{r^2} + \left(f''(r)\frac{z}{r}\right)\left(\frac{z}{r}\right) + f'(r)\frac{r - \frac{z^2}{r}}{r^2}$$

$$= \left(f''(r)\frac{x}{r}\right)\left(\frac{x}{r}\right) + f'(r)\frac{r^2 - x^2}{r^3} + \left(f''(r)\frac{y}{r}\right) + f'(r)\frac{r^2 - y^2}{r^3} + \left(f''(r)\frac{z}{r}\right)\left(\frac{z}{r}\right) + f'(r)\frac{r^2 - z^2}{r^3}$$

$$= f''(r)\frac{x^2}{r^2} + f'(r)\frac{y^2 + z^2}{r^3} + f''(r)\frac{y^2}{r^2} + f'(r)\frac{x^2 + z^2}{r^3} + f''(r)\frac{z^2}{r^2} + f'(r)\frac{x^2 + y^2}{r^3}$$

$$= f''(r)\left[\frac{x^2}{r^2} + \frac{y^2}{r^2} + \frac{z^2}{r^2}\right] + f'(r)\left[\frac{y^2 + z^2}{r^3} + \frac{z^2 + x^2}{r^3} + \frac{x^2 + y^2}{r^3}\right]$$

$$= f''(r)\frac{x^2 + y^2 + x^2}{r^2} + f'(r)\frac{2(x^2 + y^2 + z^2)}{r^3} = f''(r)\frac{r^2}{r^2} + f'(r)\frac{2r^2}{r^3}$$

$$= f''(r) + f'(r)\frac{2}{r}$$

Ans.

EXERCISE 23.1

1. Evaluate grad ϕ if $\phi = \log(x^2 + y^2 + z^2)$ **Ans.** $\frac{2(x\hat{i} + y\hat{j} + z\hat{k})}{x^2 + y^2 + z^2}$

2. Find a unit normal vector to the surface $x^2 + y^2 + z^2 = 5$ at the point (0, 1, 2). **Ans.** $\frac{1}{\sqrt{5}}(\hat{j} + 2\hat{k})$ *(AMIETE, June 2010)*

3. Calculate the directional derivative of the function $\phi(x, y, z) = xy^2 + yz^3$ at the point (1, –1, 1) in the direction of (3, 1, –1) *(A.M.I.E.T.E. Winter 2009, 2000)* **Ans.** $\frac{5}{\sqrt{11}}$

4. Find the direction in which the directional derivative of $f(x, y) = (x^2 - y^2)/xy$ at (1, 1) is zero.

(*Nagpur Winter 2000*) **Ans.** $\frac{\hat{i}+\hat{j}}{\sqrt{2}}$

5. Find the directional derivative of the scalar function of $(x, y, z) = xyz$ in the direction of the outer normal to the surface $z = xy$ at the point (3, 1, 3). **Ans.** $\frac{27}{\sqrt{11}}$

6. The temperature of the points in space is given by $T(x, y, z) = x^2 + y^2 - z$. A mosquito located at (1, 1, 2) desires to fly in such a direction that it will get warm as soon as possible. In what direction should it move? **Ans.** $\frac{1}{3}(2\hat{i}+2\hat{j}-\hat{k})$

7. If $\phi(x, y, z) = 3xz^2y - y^3z^2$, find grad ϕ at the point (1, –2, –1) **Ans.** $-(16\hat{i}+9\hat{j}+4\hat{k})$

8. Find a unit vector normal to the surface $x^2y + 2xz = 4$ at the point (2, –2, 3).

Ans. $\frac{1}{3}(-\hat{i}+2\hat{j}+2\hat{k})$

9. What is the greatest rate of increase of the function $u = xyz^2$ at the point (1, 0, 3)? **Ans.** 9

10. If θ is the acute angle between the surfaces $xyz^2 = 3x + z^2$ and $3x^2 - y^2 + 2z = 1$ at the point (1, –2, 1) show that $\cos\theta = 3/7\sqrt{6}$.

11. Find the values of constants a, b, c so that the maximum value of the directional directive of $\phi = axy^2 + byz + cz^2x^3$ at (1, 2, –1) has a maximum magnitude 64 in the direction parallel to the axis of z. **Ans.** $a = b$, $b = 24$, $c = -8$

12. The position vector of a particle at time t is $R = \cos(t-1)\, i + \sinh(t-1)\, j + at^2k$. If at $t = 1$, the acceleration of the particle be perpendicular to its position vector, then a is equal to

(*a*) 0 (*b*) 1 (*c*) $\frac{1}{2}$ (*d*) $\frac{1}{\sqrt{2}}$ (*AMIETE, Dec. 2009*) **Ans.** (*d*)

23.8 DIVERGENCE OF A VECTOR FUNCTION

The divergence of a vector point function $\vec{F}$ is denoted by *div F* and is defined as below.

Let $\vec{F} = F_1\hat{i} + F_2\hat{j} + F_3\hat{k}$

$$\text{div}\,\vec{F} = \vec{\nabla}.\vec{F} = \left(\hat{i}\frac{\partial}{\partial x} + \hat{j}\frac{\partial}{\partial y} + \hat{k}\frac{\partial}{\partial z}\right)(\hat{i}F_1 + \hat{j}F_2 + \hat{k}F_3) = \frac{\partial F_1}{\partial x} + \frac{\partial F_2}{\partial y} + \frac{\partial F_3}{\partial z}$$

It is evident that div F is scalar function.

23.9 PHYSICAL INTERPRETATION OF DIVERGENCE

Let us consider the case of a fluid flow. Consider a small rectangular parallelopiped of dimensions dx, dy, dz parallel to x, y and z axes respectively.

Let $\vec{V} = V_x\hat{i} + V_y\hat{j} + V_z\hat{k}$ be the velocity of the fluid at $P(x, y, z)$.

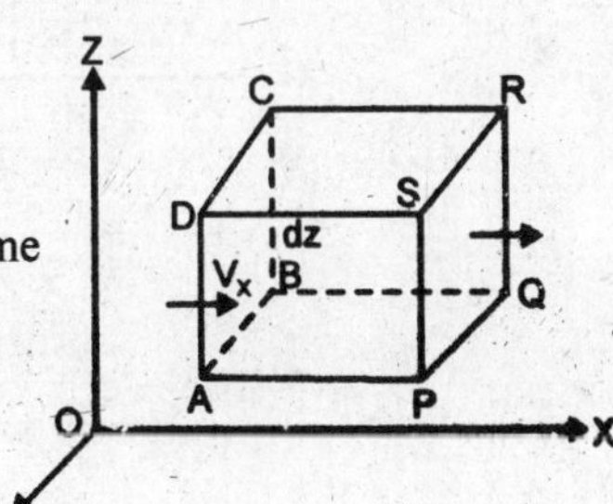

$\therefore$ Mass of fluid flowing in through the face $ABCD$ in unit time

$= \text{Velocity} \times \text{Area of the face} = V_x\,(dy\,dz)$

Mass of fluid flowing out across the face $PQRS$ per unit time

$$= V_x(x + dx)\,(dy\,dz)$$

$$= \left(V_x + \frac{\partial V_x}{\partial x}dx\right)(dy\,dz)$$

Net decrease in mass of fluid in the parallelopiped

corresponding to the flow along x-axis per unit time

$$= V_x\, dy\, dz - \left(V_x + \frac{\partial V_x}{\partial x} dx\right) dy\, dz$$

$$= -\frac{\partial V_x}{\partial x} dx\, dy\, dz \qquad \text{(Minus sign shows decrease)}$$

Similarly, the decrease in mass of fluid to the flow along y-axis $= \dfrac{\partial V_y}{\partial y} dx\, dy\, dz$

and the decrease in mass of fluid to the flow along z-axis $= \dfrac{\partial V_z}{\partial z} dx\, dy\, dz$

Total decrease of the amount of fluid per unit time $= \left(\dfrac{\partial V_x}{\partial x} + \dfrac{\partial V_y}{\partial y} + \dfrac{\partial V_z}{\partial z}\right) dx\, dy\, dz$

Thus the rate of loss of fluid per unit volume $= \dfrac{\partial V_x}{\partial x} + \dfrac{\partial V_y}{\partial y} + \dfrac{\partial V_z}{\partial z}$

$$= \left(\hat{i}\frac{\partial}{\partial x} + \hat{j}\frac{\partial}{\partial y} + \hat{k}\frac{\partial}{\partial z}\right).(\hat{i} V_x + \hat{j} V_y + \hat{k} V_z) = \overline{\nabla}.\overline{V} = \text{div}\, \overline{V}$$

If the fluid is compressible, there can be no gain or loss in the volume element. Hence

$$\text{div}\, \vec{V} = 0 \qquad ...(1)$$

and V is called a *Solenoidal* vector function.

Equation (1) is also called the *equation of continuity or conservation of mass.*

Example 26. *If* $\vec{v} = \dfrac{x\hat{i} + y\hat{j} + z\hat{k}}{\sqrt{x^2 + y^2 + z^2}}$, *find the value of div* $\overline{v}$.

(*U.P., I Semester, Winter 2000*)

Solution. We have, $\vec{v} = \dfrac{x\hat{i} + y\hat{j} + z\hat{k}}{\sqrt{x^2 + y^2 + z^2}}$

$$\text{div}\, \vec{v} = \vec{\nabla}.\vec{v} = \left(\hat{i}\frac{\partial}{\partial x} + \hat{j}\frac{\partial}{\partial y} + \hat{k}\frac{\partial}{\partial z}\right).\left(\frac{x\hat{i} + y\hat{j} + z\hat{k}}{(x^2 + y^2 + z^2)^{1/2}}\right)$$

$$= \frac{\partial}{\partial x}\frac{x}{(x^2 + y^2 + z^2)^{1/2}} + \frac{\partial}{\partial y}\frac{y}{(x^2 + y^2 + z^2)^{1/2}} + \frac{\partial}{\partial z}\frac{z}{(x^2 + y^2 + z^2)^{1/2}}$$

$$= \frac{\left[(x^2 + y^2 + z^2)^{1/2} - x.\frac{1}{2}(x^2 + y^2 + z^2)^{-\frac{1}{2}}.2x\right]}{(x^2 + y^2 + z^2)}$$

$$+ \frac{\left[(x^2 + y^2 + z^2)^{\frac{1}{2}} - y.\frac{1}{2}(x^2 + y^2 + z^2)^{-\frac{1}{2}} \times 2y\right]}{(x^2 + y^2 + z^2)} + \frac{\left[(x^2 + y^2 + z^2)^{1/2} - z.\frac{1}{2}(x^2 + y^2 + z^2)^{-1/2}.2z\right]}{(x^2 + y^2 + z^2)}$$

$$= \frac{(x^2 + y^2 + z^2) - x^2}{(x^2 + y^2 + z^2)^{3/2}} + \frac{(x^2 + y^2 + z^2) - y^2}{(x^2 + y^2 + z^2)^{3/2}} + \frac{(x^2 + y^2 + z^2) - z^2}{(x^2 + y^2 + z^2)^{3/2}}$$

$$= \frac{y^2 + z^2 + x^2 + z^2 + x^2 + y^2}{(x^2 + y^2 + z^2)^{3/2}} = \frac{2(x^2 + y^2 + z^2)}{(x^2 + y^2 + z^2)^{3/2}} = \frac{2}{\sqrt{(x^2 + y^2 + z^2)}} \qquad \textbf{Ans.}$$

Example 27. *If $u = x^2 + y^2 + z^2$, and $\overline{r} = x\hat{i} + y\hat{j} + z\hat{k}$, then find div $(u\overline{r})$ in terms of u.*

(A.M.I.E.T.E., Summer 2004)

Solution. $$\text{div}\,(u\vec{r}) = \left(\hat{i}\frac{\partial}{\partial x} + \hat{j}\frac{\partial}{\partial y} + \hat{k}\frac{\partial}{\partial z}\right).[(x^2+y^2+z^2)(x\hat{i}+y\hat{j}+z\hat{k})]$$

$$= \left(\hat{i}\frac{\partial}{\partial x} + \hat{j}\frac{\partial}{\partial y} + \hat{k}\frac{\partial}{\partial z}\right).[(x^2+y^2+z^2)x\hat{i}+(x^2+y^2+z^2)y\hat{j}+(x^2+y^2+z^2)z\hat{k}]$$

$$= \frac{\partial}{\partial x}(x^3+xy^2+xz^2) + \frac{\partial}{\partial y}(x^2y+y^3+yz^2) + \frac{\partial}{\partial z}(x^2z+y^2z+z^3)$$

$$= (3x^2+y^2+z^2) + (x^2+3y^2+z^2) + (x^2+y^2+3z^2) = 5(x^2+y^2+z^2) = 5u$$ **Ans.**

Example 28. *Find the value of n for which the vector $r^n\vec{r}$ is solenoidal, where $\overline{r} = x\hat{i} + y\hat{j} + z\hat{k}$.*

Solution. Divergence $$\vec{F} = \vec{\nabla}.\vec{F} = \vec{\nabla}.r^n\vec{r} = \nabla.(x^2+y^2+z^2)^{n/2}(x\hat{i}+y\hat{j}+z\hat{k})$$

$$= \left[\hat{i}\frac{\partial}{\partial x} + \hat{j}\frac{\partial}{\partial y} + \hat{k}\frac{\partial}{\partial z}\right].[(x^2+y^2+z^2)^{n/2}x\hat{i}+(x^2+y^2+z^2)^{n/2}y\hat{j}+(x^2+y^2+z^2)^{n/2}z\hat{k}]$$

$$= \frac{n}{2}(x^2+y^2+z^2)^{n/2-1}(2x^2) + (x^2+y^2+z^2)^{n/2} + \frac{n}{2}(x^2+y^2+z^2)^{n/2-1}(2y^2)$$
$$+ (x^2+y^2+z^2)^{n/2} + \frac{n}{2}(x^2+y^2+z^2)^{n/2-1}(2z^2) + (x^2+y^2+z^2)^{n/2}$$

$$= n(x^2+y^2+z^2)^{n/2-1}(x^2+y^2+z^2) + 3(x^2+y^2+z^2)^{n/2}$$
$$= n(x^2+y^2+z^2)^{n/2} + 3(x^2+y^2+z^2)^{n/2} = (n+3)(x^2+y^2+z^2)^{n/2}$$

If $r^n\vec{r}$ is solenoidal, then $(n+3)(x^2+y^2+z^2)^{n/2} = 0$ or $n+3=0$ or $n=-3$. **Ans.**

Example 29. *Show that* $\nabla\left[\frac{(\vec{a}.\vec{r})}{r^n}\right] = \frac{\vec{a}}{r^n} - \frac{n(\vec{a}.\vec{r})\vec{r}}{r^{n+2}}$. *(M.U. 2005)*

Solution. We have, $$\frac{\vec{a}.\vec{r}}{r^n} = \frac{(a_1\hat{i}+a_2\hat{j}+a_3\hat{k}).(x\hat{i}+y\hat{j}+z\hat{k})}{r^n} = \frac{a_1x+a_2y+a_3z}{r^n}$$

Let $$\phi = \frac{\vec{a}.\vec{r}}{r^n} = \frac{a_1x+a_2y+a_3z}{r^n}$$

$$\therefore \quad \frac{\partial\phi}{\partial x} = \frac{r^n.a_1 - (a_1x+a_2y+a_3z)\,nr^{n-1}(\partial r/\partial x)}{r^{2n}}$$

But $r^2 = x^2+y^2+z^2 \Rightarrow 2r\frac{\partial r}{\partial x} = 2x \Rightarrow \frac{\partial r}{\partial x} = \frac{x}{r}$

$$\therefore \quad \frac{\partial\phi}{\partial x} = \frac{a_1r^n - (a_1x+a_2y+a_3z).n\,r^{n-2}.x}{r^{2n}} = \frac{a_1}{r^n} - \frac{n(a_1x+a_2y+a_3z)x}{r^{n+2}}$$

$$\therefore \quad \nabla\phi = \frac{\partial\phi}{\partial x}\hat{i} + \frac{\partial\phi}{\partial y}\hat{j} + \frac{\partial\phi}{\partial z}\hat{k}$$

$$= \frac{1}{r^n}(a_1\hat{i}+a_2\hat{j}+a_3\hat{k}) - \frac{n}{r^{n+2}}[(a_1x+a_2y+a_3z)(x\hat{i}+y\hat{j}+z\hat{k})]$$

$$= \frac{\overline{a}}{r^n} - \frac{n}{r^{n+2}}(\overline{a}.\overline{r})\overline{r}$$

Example 30. *Let $\vec{r} = x\hat{i} + y\hat{j} + z\hat{k}, r = |\vec{r}|$ and $\vec{a}$ is a constant vector. Find the value of*

$$div\left(\frac{\vec{a}\times\vec{r}}{r^n}\right)$$

Solution. Let

$$\vec{a} = a_1\hat{i} + a_2\hat{j} + a_3\hat{k}$$

$$\vec{a}\times\vec{r} = (a_1\hat{i} + a_2\hat{j} + a_3\hat{k})\times(x\hat{i} + y\hat{j} + z\hat{k})$$

$$= \begin{vmatrix} \hat{i} & \hat{j} & \hat{k} \\ a_1 & a_2 & a_3 \\ x & y & z \end{vmatrix} = (a_2 z - a_3 y)\hat{i} - (a_1 z - a_3 x)\hat{j} + (a_1 y - a_2 x)\hat{k}$$

$$\frac{\vec{a}\times\vec{r}}{|\vec{r}|^n} = \frac{(a_2 z - a_3 y)\hat{i} - (a_1 z - a_3 x)\hat{j} + (a_1 y - a_2 x)\hat{k}}{(x^2+y^2+z^2)^{n/2}}$$

$$\text{div}\left(\frac{\vec{a}\times\vec{r}}{|\vec{r}|^n}\right) = \vec{\nabla}.\frac{\vec{a}\times\vec{r}}{|\vec{r}|^n}$$

$$= \left(\hat{i}\frac{\partial}{\partial x} + \hat{j}\frac{\partial}{\partial y} + \hat{k}\frac{\partial}{\partial z}\right).\frac{(a_2 z - a_3 y)\hat{i} - (a_1 z - a_3 x)\hat{j} + (a_1 y - a_2 x)\hat{k}}{(x^2+y^2+z^2)^{n/2}}$$

$$= \frac{\partial}{\partial x}\frac{a_2 z - a_3 y}{(x^2+y^2+z^2)^{n/2}} - \frac{\partial}{\partial y}\frac{a_1 z - a_3 x}{(x^2+y^2+z^2)^{n/2}} + \frac{\partial}{\partial z}\frac{(a_1 y - a_2 x)}{(x^2+y^2+z^2)^{n/2}}$$

$$-\frac{n}{2}\frac{(a_2 z - a_3 y)\,2x}{(x^2+y^2+z^2)^{\frac{n+2}{2}}} + \frac{n}{2}\frac{(a_1 z - a_3 x)\,2y}{(x^2+y^2+z^2)^{\frac{n+2}{2}}} - \frac{n}{2}\frac{(a_1 y - a_2 x)\,2z}{(x^2+y^2+z^2)^{\frac{n+2}{2}}}$$

$$= -\frac{n}{(x^2+y^2+z^2)^{\frac{n+2}{2}}}[(a_2 z - a_3 y)x - (a_1 z - a_3 x)y + (a_1 y - a_2 x)z]$$

$$= -\frac{n}{(x^2+y^2+z^2)^{\frac{n+2}{2}}}[a_2 zx - a_3 xy - a_1 yz + a_3 xy + a_1 yz - a_2 zx] = 0$$ **Ans.**

Example 31. *Find the directional derivative of div $(\vec{u})$ at the point (1, 2, 2) in the direction of the outer normal of the sphere $x^2 + y^2 + z^2 = 9$ for $\vec{u} = x^4\hat{i} + y^4\hat{j} + z^4\hat{k}$.*

Solution. div $(\vec{u}) = \nabla.\vec{u}$

$$= \left(\hat{i}\frac{\partial}{\partial x} + \hat{j}\frac{\partial}{\partial y} + \hat{k}\frac{\partial}{\partial z}\right).(x^4\hat{i} + y^4\hat{j} + z^4\hat{k}) = 4x^3 + 4y^3 + 4z^3$$

Outer normal of the sphere $= \nabla(x^2 + y^2 + z^2 - 9)$

$$= \left(\hat{i}\frac{\partial}{\partial x} + \hat{j}\frac{\partial}{\partial y} + \hat{k}\frac{\partial}{\partial z}\right)(x^2 + y^2 + z^2 - 9) = 2x\hat{i} + 2y\hat{j} + 2z\hat{k}$$

Outer normal of the sphere at (1, 2, 2) $= 2\hat{i} + 4\hat{j} + 4\hat{k}$ ...(1)

Directional derivative $= \vec{\nabla}(4x^3 + 4y^3 + 4z^3)$

$$= \left(\hat{i}\frac{\partial}{\partial x} + \hat{j}\frac{\partial}{\partial y} + \hat{k}\frac{\partial}{\partial z}\right)(4x^3 + 4y^3 + 4z^3) = 12x^2\hat{i} + 12y^2\hat{j} + 12z^2\hat{k}$$

Directional derivative at (1, 2, 2) $= 12\hat{i} + 48\hat{j} + 48\hat{k}$ ...(2)

Directional derivative along the outer normal $= (12\hat{i} + 48\hat{j} + 48\hat{k}).\dfrac{2\hat{i}+4\hat{j}+4\hat{k}}{\sqrt{4+16+16}}$ [From (1), (2)]

$$= \frac{24+192+192}{6} = 68$$ **Ans.**

Example 32. *Show that div (grad r^n) = $n\,(n+1)r^{n-2}$, where*

$$r = \sqrt{x^2+y^2+z^2}$$

Hence, show that $\nabla^2\left(\dfrac{1}{r}\right) = 0.$ *(AMIETE, 2010, U.P. I Semester, Dec. 2004*

Solution. $\text{grad }(r^n) = \hat{i}\dfrac{\partial}{\partial x}r^n + \hat{j}\dfrac{\partial}{\partial y}r^n + \hat{k}\dfrac{\partial}{\partial z}r^n$ by definition

$$= \hat{i}\,n\,r^{n-1}\frac{\partial r}{\partial x} + \hat{j}\,n\,r^{n-1}\frac{\partial r}{\partial y} + \hat{k}.n\,r^{n-1}\frac{\partial r}{\partial z} = n\,r^{n-1}\left[\hat{i}\frac{\partial r}{\partial x} + \hat{j}\frac{\partial r}{\partial y} + \hat{k}\frac{\partial r}{\partial z}\right]$$

$$= n\,r^{n-1}\left[\hat{i}\left(\frac{x}{r}\right) + \hat{j}\left(\frac{y}{r}\right) + \hat{k}\left(\frac{z}{r}\right)\right] = nr^{n-2}(x\hat{i}+y\hat{j}+z\hat{k}) = nr^{n-2}\bar{r}.$$

$$\left[\because r^2 = x^2+y^2+z^2 \Rightarrow 2r\frac{\partial r}{\partial x} = 2x \Rightarrow \frac{\partial r}{\partial x} = \frac{x}{r} \text{ etc.}\right]$$

Thus, $\text{grad }(r^n) = n\,r^{n-2}\,x\hat{i} + n\,r^{n-2}\,y\hat{j} + n\,r^{n-2}\,z\hat{k}$ ...(1)

$\therefore$ $\text{div grad } r^n = \text{div }[n\,r^{n-2}\,x\hat{i} + n\,r^{n-2}\,y\hat{j} + n\,r^{n-2}\,z\hat{k}]$

$$= \left(\hat{i}\frac{\partial}{\partial x} + \hat{j}\frac{\partial}{\partial y} + \hat{k}\frac{\partial}{\partial z}\right).(nr^{n-2}x\hat{i} + nr^{n-2}y\hat{j} + nr^{n-2}z\hat{k})$$ [From (1)]

$$= \frac{\partial}{\partial x}(n\,r^{n-2}x) + \frac{\partial}{\partial y}(n\,r^{n-2}y) + \frac{\partial}{\partial z}(n\,r^{n-2}z)$$ (By definition)

$$= \left(n\,r^{n-2} + nx\,(n-2)\,r^{n-3}\frac{\partial r}{\partial x}\right) + \left(n\,r^{n-2} + ny\,(n-2)\,r^{n-3}\frac{\partial r}{\partial y}\right) + \left(n\,r^{n-2} + nz\,(n-2)\,r^{n-3}\frac{\partial r}{\partial z}\right)$$

$$= 3n\,r^{n-2} + n(n-2)r^{n-3}\left[x\frac{\partial r}{\partial x} + y\frac{\partial r}{\partial y} + z\frac{\partial r}{\partial z}\right]$$

$$= 3n\,r^{n-2} + n(n-2)r^{n-3}\left[x\left(\frac{x}{r}\right) + y\left(\frac{y}{r}\right) + z\left(\frac{z}{r}\right)\right]$$

$$\left[\because r^2 = x^2+y^2+z^2 \Rightarrow 2r\frac{\partial r}{\partial x} = 2x \Rightarrow \frac{\partial r}{\partial x} = \frac{x}{r} \text{ etc.}\right]$$

$$= 3nr^{n-2} + n\,(n-2)r^{n-4}\,[x^2+y^2+z^2]$$

$$= 3nr^{n-2} + n\,(n-2)\,r^{n-4}.r^2 \qquad (\because r^2 = x^2+y^2+z^2)$$

$$= r^{n-2}\,[3n + n^2 - 2n] = r^{n-2}\,(n^2+n) = n(n+1)\,r^{n-2}$$

If we put $n = -1$

$\text{div grad }(r^{-1}) = -1\,(-1+1)\,r^{-1-2}$

$\Rightarrow \quad \nabla^2\left(\dfrac{1}{r}\right) = 0$

Ques. If $\vec{r} = x\hat{i} + y\hat{j} + z\hat{k}$, and $r = |r|$ find div $\left(\dfrac{\boldsymbol{r}}{r^2}\right)$. *(U.P. I Sem., Dec. 2006)* **Ans.** $\dfrac{1}{r^2}$

EXERCISE 23.3

1. If $r = x\hat{i} + y\hat{j} + z\hat{k}$ and $r = |\vec{r}|$, show that (*i*) div $\left(\dfrac{\vec{r}}{|\bar{r}|^3}\right) = 0,$

   (*ii*) div $(r\,\phi) = 3\phi + r$ grad ϕ.

2. Show that the vector $V = (x+3y)\hat{i} + (y-3z)\hat{j} + (x-2z)\hat{k}$ is solenoidal.

   (*R.G.P.V., Bhopal, Dec. 2003*)

3. Show that $\nabla.(\phi\, A) = \nabla\phi.A + \phi(\nabla.A)$

4. If ρ, ϕ, z are cylindrical coordinates, show that grad (log ρ) and grad ϕ are solenoidal vectors.

5. Obtain the expression for $\nabla^2 f$ in spherical coordinates from their corresponding expression in orthogonal curvilinear coordinates.

Prove the following:

6. (*a*) $\nabla.(\nabla\phi) = \nabla^2\phi$.

7. $\vec{\nabla} \times \dfrac{(\vec{A} \times \vec{R})}{r^n} = \dfrac{(2-n)\vec{A}}{r^n} + \dfrac{n(\vec{A}.\vec{R})\vec{R}}{r^{n+2}}, r = |\vec{R}|$

8. div $(f\,\nabla\, g)$ – div $(g\,\nabla\, f) = f\,\nabla^2 g - g\,\nabla^2 f$

23.10 CURL

(*U.P., I semester, Dec. 2006*)

The curl of a vector point function F is defined as below

$$\text{curl } \vec{F} = \vec{\nabla} \times \vec{F} \qquad (\vec{F} = F_1\hat{i} + F_2\hat{j} + F_3\hat{k})$$

$$= \left(\hat{i}\frac{\partial}{\partial x} + \hat{j}\frac{\partial}{\partial y} + \hat{k}\frac{\partial}{\partial z}\right) \times (F_1\hat{i} + F_2\hat{j} + F_3\hat{k})$$

$$= \begin{vmatrix} \hat{i} & \hat{j} & \hat{k} \\ \dfrac{\partial}{\partial x} & \dfrac{\partial}{\partial y} & \dfrac{\partial}{\partial z} \\ F_1 & F_2 & F_3 \end{vmatrix} = \hat{i}\left(\frac{\partial F_3}{\partial y} - \frac{\partial F_2}{\partial z}\right) - \hat{j}\left(\frac{\partial F_3}{\partial x} - \frac{\partial F_1}{\partial z}\right) + \hat{k}\left(\frac{\partial F_2}{\partial x} - \frac{\partial F_1}{\partial y}\right)$$

Curl $\vec{F}$ is a vector quantity.

23.11 PHYSICAL MEANING OF CURL

(*M.D.U., Dec. 2009, U.P. I Semester, Winter 2009, 2000*)

We know that $\vec{V} = \vec{\omega} \times \vec{r}$, where ω is the angular velocity, $\vec{V}$ *is the linear velocity and* $\vec{r}$ *is the position vector of a point on the rotating body.*

$$\text{Curl } \vec{V} = \vec{\nabla} \times \vec{V} \qquad \begin{bmatrix} \vec{\omega} = \omega_1\hat{i} + \omega_2\hat{j} + \omega_3\hat{k} \\ \vec{r} = x\hat{i} + y\hat{j} + z\hat{k} \end{bmatrix}$$

$$= \vec{\nabla} \times (\vec{\omega} \times \vec{r}) = \vec{\nabla} \times [(\omega_1\hat{i} + \omega_2\hat{j} + \omega_3\hat{k}) \times (x\hat{i} + y\hat{j} + z\hat{k})]$$

$$= \vec{\nabla} \times \begin{vmatrix} \hat{i} & \hat{j} & \hat{k} \\ \omega_1 & \omega_2 & \omega_3 \\ x & y & z \end{vmatrix} = \vec{\nabla} \times [(\omega_2 z - \omega_3 y)\hat{i} - (\omega_1 z - \omega_3 x)\hat{j} + (\omega_1 y - \omega_2 x)\hat{k}]$$

$$= \left(\hat{i}\frac{\partial}{\partial x} + \hat{j}\frac{\partial}{\partial y} + \hat{k}\frac{\partial}{\partial z}\right) \times [(\omega_2 z - \omega_3 y)\hat{i} - (\omega_1 z - \omega_3 x)\hat{j} + (\omega_1 y - \omega_2 x)\hat{k}]$$

$$= \begin{vmatrix} \hat{i} & \hat{j} & \hat{k} \\ \frac{\partial}{\partial x} & \frac{\partial}{\partial y} & \frac{\partial}{\partial z} \\ \omega_2 z - \omega_3 y & \omega_3 x - \omega_1 z & \omega_1 y - \omega_2 x \end{vmatrix}$$

$$= (\omega_1 + \omega_1)\hat{i} - (-\omega_2 - \omega_2)\hat{j} + (\omega_3 + \omega_3)\hat{k} = 2(\omega_1 \hat{i} + \omega_2 \hat{j} + \omega_3 \hat{k}) = 2\omega$$

Curl $\vec{V} = 2\omega$ which shows that curl of a vector field is connected with rotational properties of the vector field and justifies the name *rotation* used for curl.

If Curl $\bar{F} = 0$, the field F is termed as *irrotational.*

Example 33. *Find the divergence and curl of* $\vec{v} = (xyz)\hat{i} + (3x^2y)\hat{j} + (xz^2 - y^2z)\hat{k}$ *at (2, –1, 1)* *(Nagpur University, Summer 2003)*

Solution. Here, we have

$$\vec{v} = (x\,y\,z)\hat{i} + (3x^2y)\hat{j} + (xz^2 - y^2z)\hat{k}$$

$$\text{Div. } \vec{v} = \nabla . \vec{v}$$

$$\text{Div } \vec{v} = \frac{\partial}{\partial x}(x\,y\,z) + \frac{\partial}{\partial y}(3x^2y) + \frac{\partial}{\partial z}(xz^2 - y^2z)$$

$$= yz + 3x^2 + 2x\,z - y^2 \qquad = -1 + 12 + 4 - 1 = 14 \text{ at } (2, -1, 1)$$

$$\text{Curl } \bar{v} = \begin{vmatrix} \hat{i} & \hat{j} & \hat{k} \\ \frac{\partial}{\partial x} & \frac{\partial}{\partial y} & \frac{\partial}{\partial z} \\ xyz & 3x^2y & xz^2 - y^2z \end{vmatrix} = -2yz\,\hat{i} - (z^2 - xy)\hat{j} + (6xy - xz)\hat{k}$$

$$= -2yz\,\hat{i} + (xy - z^2)\hat{j} + (6xy - xz)\hat{k}$$

Curl at (2, –1, 1)

$$= -2(-1)(1)\,\hat{i} + \{(2)(-1) - 1\}\hat{j} + \{6(2)(-1) - 2(1)\}\hat{k}$$

$$= 2\hat{i} - 3\hat{j} - 14\hat{k}$$ **Ans.**

Example 34. *If* $\vec{V} = \dfrac{x\hat{i} + y\hat{j} + z\hat{k}}{\sqrt{x^2 + y^2 + z^2}}$, *find the value of curl* $\vec{V}$. *(U.P., I Semester, Winter 2000)*

Solution. $\text{Curl } \vec{V} = \vec{\nabla} \times \vec{V}$

$$= \left(\hat{i}\frac{\partial}{\partial x} + \hat{j}\frac{\partial}{\partial y} + \hat{k}\frac{\partial}{\partial z}\right) \times \left(\frac{x\hat{i} + y\hat{j} + z\hat{k}}{(x^2 + y^2 + z^2)^{1/2}}\right)$$

$$= \begin{vmatrix} \hat{i} & \hat{j} & \hat{k} \\ \frac{\partial}{\partial x} & \frac{\partial}{\partial y} & \frac{\partial}{\partial z} \\ \frac{x}{(x^2 + y^2 + z^2)^{1/2}} & \frac{y}{(x^2 + y^2 + z^2)^{1/2}} & \frac{z}{(x^2 + y^2 + z^2)^{1/2}} \end{vmatrix}$$

$$= \hat{i}\left[\frac{\partial}{\partial y}\left(\frac{z}{(x^2+y^2+z^2)^{1/2}}\right)-\frac{\partial}{\partial z}\left(\frac{y}{(x^2+y^2+z^2)^{1/2}}\right)\right]-\hat{j}\left[\frac{\partial}{\partial x}\left(\frac{z}{(x^2+y^2+z^2)^{1/2}}\right)\right.$$

$$\left.-\frac{\partial}{\partial z}\left(\frac{x}{(x^2+y^2+z^2)^{1/2}}\right)\right]+\hat{k}\left[\frac{\partial}{\partial x}\left(\frac{y}{(x^2+y^2+z^2)^{1/2}}\right)-\frac{\partial}{\partial y}\left(\frac{x}{(x^2+y^2+z^2)^{1/2}}\right)\right]$$

$$= \hat{i}\left[\frac{-yz}{(x^2+y^2+z^2)^{3/2}}+\frac{y.z}{(x^2+y^2+z^2)^{3/2}}\right]-\hat{j}\left[\frac{-zx}{(x^2+y^2+z^2)^{3/2}}+\frac{zx}{(x^2+y^2+z^2)^{3/2}}\right]$$

$$+\hat{k}\left[\frac{-xy}{(x^2+y^2+z^2)^{3/2}}+\frac{xy}{(x^2+y^2+z^2)^{3/2}}\right]=0 \qquad \textbf{Ans.}$$

Example 35. *Prove that* $(y^2-z^2+3yz-2x)\hat{i}+(3xz+2xy)\hat{j}+(3xy-2xz+2z)\hat{k}$ *is both solenoidal and irrotational.* *(U.P., I Sem, Dec. 2008)*

Solution. Let $\vec{F}=(y^2-z^2+3yz-2x)\hat{i}+(3xz+2xy)\hat{j}+(3xy-2xz+2z)\hat{k}$

For solenoidal, we have to prove $\vec{\nabla}.\vec{F}=0$.

Now, $\vec{\nabla}.\vec{F}=\left[\hat{i}\frac{\partial}{\partial x}+\hat{j}\frac{\partial}{\partial y}+\hat{k}\frac{\partial}{\partial z}\right]\cdot\left[(y^2-z^2+3yz-2x)\hat{i}+(3xz+2xy)\hat{j}+(3xy-2xz+2z)\hat{k}\right]$

$$= -2+2x-2x+2=0$$

Thus, $\vec{F}$ is solenoidal. For irrotational, we have to prove Curl $\overline{F}=0$.

Now, $\quad$ Curl $\vec{F}=\begin{vmatrix}\hat{i} & \hat{j} & \hat{k}\\ \frac{\partial}{\partial x} & \frac{\partial}{\partial y} & \frac{\partial}{\partial z}\\ y^2-z^2+3yz-2x & 3xz+2xy & 3xy-2xz+2z\end{vmatrix}$

$$= (3x-3x)\hat{i}-(-2z+3y-3y+2z)\hat{j}+(3z+2y-2y-3z)\hat{k}$$

$$= 0\hat{i}+0\hat{j}+0\hat{k}=0$$

Thus, $\vec{F}$ is irrotational.

Hence, $\vec{F}$ is both solenoidal and irrotational. **Proved.**

Example 36. *Determine the constants a and b such that the curl of vector*

$\overline{A}=(2xy+3yz)\hat{i}+(x^2+axz-4z^2)\hat{j}-(3xy+byz)\hat{k}$ *is zero.*

(U.P. I Semester, Dec 2008)

Solution. $\quad$ Curl $A=\left(\hat{i}\frac{\partial}{\partial x}+\hat{j}\frac{\partial}{\partial y}+\hat{k}\frac{\partial}{\partial z}\right)\times[(2xy+3yz)\hat{i}+(x^2+axz-4z^2)\hat{j}-(3xy+byz)\hat{k}]$

$$=\begin{vmatrix}\hat{i} & \hat{j} & \hat{k}\\ \frac{\partial}{\partial x} & \frac{\partial}{\partial y} & \frac{\partial}{\partial z}\\ 2xy+3yz & x^2+axz-4z^2 & -3xy-byz\end{vmatrix}$$

$$= [-3x - bz - ax + 8z]\hat{i} - [-3y - 3y]\hat{j} + [2x + az - 2x - 3z]\hat{k}$$

$$= [-x(3+a) + z(8-b)]\hat{i} + 6y\hat{j} + z(-3+a)\hat{k}$$

$$= 0 \qquad \text{(given)}$$

i.e., $3 + a = 0$ and $8 - b = 0$, $\quad -3 + a = 0 \Rightarrow a = 3$

$a = -3, 3 \qquad b = 8$ **Ans.**

Example 37. *If a vector field is given by*

$\vec{F} = (x^2 - y^2 + x)\hat{i} - (2xy + y)\hat{j}$. *Is this field irrotational ? If so, find its scalar potential.*

(*U.P. I Semester, Dec 2009*)

Solution. Here, we have

$$\vec{F} = (x^2 - y^2 + x)\hat{i} - (2xy + y)\hat{j}$$

$$\text{Curl } F = \nabla \times \vec{F}$$

$$= \left(\hat{i}\frac{\partial}{\partial x} + \hat{j}\frac{\partial}{\partial y} + \hat{k}\frac{\partial}{\partial z}\right) \times (x^2 - y^2 + x)\hat{i} - (2xy + y)\hat{j}$$

$$= \begin{vmatrix} \hat{i} & \hat{j} & \hat{k} \\ \dfrac{\partial}{\partial x} & \dfrac{\partial}{\partial y} & \dfrac{\partial}{\partial z} \\ x^2 - y^2 + x & -2xy - y & 0 \end{vmatrix} = \hat{i}(0-0) - \hat{j}(0-0) + \hat{k}(-2y + 2y) = 0$$

Hence, vector field $\vec{F}$ is irrotational.

To find the scalar potential function ϕ

$$\vec{F} = \nabla \phi$$

$$d\phi = \frac{\partial \phi}{\partial x}dx + \frac{\partial \phi}{\partial y}dy + \frac{\partial \phi}{\partial z}dz = \left|\hat{i}\frac{\partial \phi}{\partial x} + \hat{j}\frac{\partial \phi}{\partial y} + \hat{k}\frac{\partial \phi}{\partial z}\right| \cdot (\hat{i}\,dx + \hat{j}\,dy + \hat{k}\,dz)$$

$$= \left(\hat{i}\frac{\partial}{\partial x} + \hat{j}\frac{\partial}{\partial y} + \hat{k}\frac{\partial}{\partial z}\right)\phi \cdot (\vec{d}\,\vec{r}) = \nabla\phi \cdot \vec{d}\,\vec{r} = \vec{F} \cdot \vec{d}\,\vec{r}$$

$$= [(x^2 - y^2 + x)\hat{i} - (2xy + y)\hat{j}] \cdot (\hat{i}\,dx + \hat{j}\,dy + \hat{k}\,dz)$$

$$= (x^2 - y^2 + x)dx - (2xy + y)dy.$$

$$\phi = \int [(x^2 - y^2 + x)dx - (2xy + y)dy] + c$$

$$= \int \left[x^2\,dx + x\,dx - y\,dy - y^2\,dx - 2\,xy\,dy\right] + c = \frac{x^3}{3} + \frac{x^2}{2} - \frac{y^2}{2} - xy^2 +$$

Hence, the scalar potential is $\dfrac{x^3}{3} + \dfrac{x^2}{2} - \dfrac{y^2}{2} - xy^2 + c$ **Ans.**

Example 38. *Find the scalar potential function f for* $\vec{A} = y^2\hat{i} + 2xy\hat{j} - z^2\hat{k}$.

(*Gujarat, I Semester, Jan. 2009*)

Solution. We have,

$$\vec{A} = y^2\hat{i} + 2xy\hat{j} - z^2\hat{k}$$

$$\text{Curl } \vec{A} = \nabla \times \vec{A} = \left(\hat{i}\frac{\partial}{\partial x} + \hat{j}\frac{\partial}{\partial y} + \hat{k}\frac{\partial}{\partial z}\right) \times (y^2\hat{i} + 2xy\hat{j} - z^2\hat{k})$$

$$= \begin{vmatrix} \hat{i} & \hat{j} & \hat{k} \\ \frac{\partial}{\partial x} & \frac{\partial}{\partial y} & \frac{\partial}{\partial z} \\ y^2 & 2xy & -z^2 \end{vmatrix} = \hat{i}(0) - \hat{j}(0) + \hat{k}(2y - 2y) = 0$$

Hence, $\vec{A}$ is irrotational. To find the scalar potential function f.

$$\vec{A} = \nabla f$$

$$df = \frac{\partial f}{\partial x} dx + \frac{\partial f}{\partial y} dy + \frac{\partial f}{\partial z} dz = \left(\hat{i}\frac{\partial f}{\partial x} + \hat{j}\frac{\partial f}{\partial y} + \hat{k}\frac{\partial f}{\partial z}\right).(\hat{i}\, dx + \hat{j}\, dy + \hat{k}\, dz)$$

$$= \left(\hat{i}\frac{\partial}{\partial x} + \hat{j}\frac{\partial}{\partial y} + \hat{k}\frac{\partial}{\partial z}\right) f.dr = \nabla f.d\vec{r}$$

$$= \vec{A}.dr \qquad (A = \nabla f)$$

$$= (y^2\hat{i} + 2xy\hat{j} - z^2\hat{k}).(\hat{i}\, dx + \hat{j}\, dy + \hat{k}\, dz)$$

$$= y^2\, dx + 2xy\, dy - z^2\, dz = d\,(xy^2) - z^2\, dz$$

$$f = \int d\,(xy^2) - \int z^2\, dz = xy^2 - \frac{z^3}{3} + C$$ **Ans.**

Example 39. *A vector field is given by* $\vec{A} = (x^2 + xy^2)\hat{i} + (y^2 + x^2y)\hat{j}$. *Show that the field is irrotational and find the scalar potential.*(*Nagpur Univeristy, Summer 2003, Winter 2002*)

Solution. $\vec{A}$ is irrotational if curl $\vec{A} = 0$

$$\text{Curl } \vec{A} = \nabla \times \vec{A} = \begin{vmatrix} \hat{i} & \hat{j} & \hat{k} \\ \frac{\partial}{\partial x} & \frac{\partial}{\partial y} & \frac{\partial}{\partial z} \\ x^2 + xy^2 & y^2 + x^2y & 0 \end{vmatrix} = \hat{i}(0-0) - \hat{j}(0-0) + \hat{k}(2xy - 2xy) = 0$$

Hence, $\vec{A}$ is irrotational. If ϕ is the scalar potential, then

$$\vec{A} = \text{grad } \phi$$

$$d\phi = \frac{\partial \phi}{\partial x} dx + \frac{\partial \phi}{\partial y} dy + \frac{\partial \phi}{\partial z} dz \qquad \text{[Total differential coefficient]}$$

$$= \left(\hat{i}\frac{\partial \phi}{\partial x} + \hat{j}\frac{\partial \phi}{\partial y} + \hat{k}\frac{\partial \phi}{\partial z}\right).(\hat{i}\, dx + \hat{j}\, dy + \hat{k}\, dz) = \text{grad } \phi\,.\, dr$$

$$= \vec{A}.dr = [(x^2 + xy^2)\hat{i} + (y^2 + x^2y)\hat{j}].(\hat{i}\, dx + \hat{j}\, dy + \hat{k}\, dz)$$

$$= (x^2 + xy^2)\, dx + (y^2 + x^2y)\, dy = x^2\, dx + y^2\, dy + (x\, dx)y^2 + (x^2)\,(y\, dy)$$

$$\phi = \int x^2 dx + \int y^2 dy + \int [(x\, dx)\, y^2 + (x^2)\,(y\, dy)] = \frac{x^3}{3} + \frac{y^3}{3} + \frac{x^2y^2}{2} + c$$ **Ans.**

Example 40. *Show that* $\vec{V}(x, y, z) = 2xyz\hat{i} + (x^2z + 2y)\hat{j} + x^2y\hat{k}$ *is irrotational and find a scalar function u(x, y, z) such that* $\vec{V}$ = *grad (u).*

Solution. $\vec{V}(x, y, z) = 2xyz\hat{i} + (x^2z + 2y)\hat{j} + x^2y\hat{k}$

$$\text{Curl } \vec{V} = \left(\hat{i}\frac{\partial}{\partial x}+\hat{j}\frac{\partial}{\partial y}+\hat{k}\frac{\partial}{\partial z}\right)\times[2xyz\,\hat{i}+(x^2z+2y)\,\hat{j}+x^2y\,\hat{k}]$$

$$= \begin{vmatrix} \hat{i} & \hat{j} & \hat{k} \\ \frac{\partial}{\partial x} & \frac{\partial}{\partial y} & \frac{\partial}{\partial z} \\ 2xyz & x^2z+2y & x^2y \end{vmatrix}$$

$$= (x^2-x^2)\hat{i}-(2xy-2xy)\hat{j}+(2xz-2xz)\hat{k}=0$$

Hence, $\vec{V}(x, y, z)$ is irrotational.

To find corresponding scalar function u, consider the following relations given

$$\vec{V} = \text{grad }(u)$$

or $$\vec{V} = \vec{\nabla}(u) \quad ...(1)$$

$$du = \frac{\partial u}{\partial x}dx+\frac{\partial u}{\partial y}dy+\frac{\partial u}{\partial z}dz \quad \text{(Total differential coefficient)}$$

$$= \left(\hat{i}\frac{\partial u}{\partial x}+\hat{j}\frac{\partial u}{\partial y}+\hat{k}\frac{\partial u}{\partial z}\right).(\hat{i}\,dx+\hat{j}\,dy+\hat{k}\,dz)$$

$$= \vec{\nabla}u.d\vec{r}=\vec{V}.d\vec{r} \quad \text{[From (1)]}$$

$$= [2xyz\,\hat{i}+(x^2z+2y)\,\hat{j}+x^2y\,\hat{k}].(\hat{i}\,dx+\hat{j}\,dy+\hat{k}\,dz)$$

$$= 2xyz\,dx+(x^2z+2y)\,dy+x^2y\,dz$$

$$= y(2xz\,dx+x^2\,dz)+(x^2z)\,dy+2y\,dy$$

$$= [yd(x^2z)+(x^2z)\,dy]+2y\,dy = d(x^2yz)+2y\,dy$$

Integrating, we get $u = x^2yz+y^2$ **Ans.**

Example 41. *A fluid motion is given by* $\vec{v}=(y+z)\hat{i}+(z+x)\hat{j}+(x+y)\hat{k}$. *Show that the motion is irrotational and hence find the velocity potential.*

(*AMIETE, Dec. 2007, Uttarakhand, I Semester 2006; U.P., I Semester, Winter 2003*)

Solution. $\text{Curl } \vec{v} = \nabla\times\vec{v}$

$$= \left(\hat{i}\frac{\partial}{\partial x}+\hat{j}\frac{\partial}{\partial y}+\hat{k}\frac{\partial}{\partial z}\right)\times[(y+z)\hat{i}+(z+x)\hat{j}+(x+y)\hat{k}]$$

$$= \begin{vmatrix} \hat{i} & \hat{j} & \hat{k} \\ \frac{\partial}{\partial x} & \frac{\partial}{\partial y} & \frac{\partial}{\partial z} \\ y+z & z+x & x+y \end{vmatrix} = (1-1)\hat{i}-(1-1)\hat{j}+(1-1)\hat{k}=0$$

Hence, $\vec{v}$ is irrotational.

To find the corresponding velocity potential ϕ, consider the following relation.

$$\bar{v} = \nabla\phi$$

$$d\phi = \frac{\partial\phi}{\partial x}dx+\frac{\partial\phi}{\partial y}dy+\frac{\partial\phi}{\partial z}dz \quad \text{[Total Differential coefficient]}$$

$$= \left(\hat{i}\frac{\partial\phi}{\partial x}+\hat{j}\frac{\partial\phi}{\partial y}+\hat{k}\frac{\partial\phi}{\partial z}\right).(\hat{i}\,dx+\hat{j}\,dy+\hat{k}\,dz) = \left(\hat{i}\frac{\partial}{\partial x}+\hat{j}\frac{\partial}{\partial y}+\hat{k}\frac{\partial}{\partial z}\right)\phi.d\vec{r} = \nabla\phi.d\vec{r} = \vec{v}.d\vec{r}$$

$$= [(y+z)\hat{i}+(z+x)\hat{j}+(x+y)\hat{k}].(\hat{i}\,dx+\hat{j}\,dy+\hat{k}\,dz)$$
$$= (y+z)\,dx+(z+x)\,dy+(x+y)\,dz$$
$$= y\,dx+z\,dx+z\,dy+x\,dy+x\,dz+y\,dz$$
$$\phi = \int(y\,dx+x\,dy)+\int(z\,dy+y\,dz)+\int(z\,dx+x\,dz)$$
$$\phi = xy+yz+zx+c$$

Velocity potential $= xy+yz+zx+c$ **Ans.**

Example 42. *A fluid motion is given by* $\vec{v} = (y\sin z-\sin x)\hat{i}+(x\sin z+2yz)\hat{j}+(xy\cos z+y^2)\hat{k}$ *is the motion irrotational? If so, find the velocity potential.*

Solution. Curl $\vec{v} = \vec{\nabla}\times\vec{v}$

$$= \left(\hat{i}\frac{\partial}{\partial x}+\hat{j}\frac{\partial}{\partial y}+\hat{k}\frac{\partial}{\partial z}\right)\times(y\sin z-\sin x)\hat{i}+(x\sin z+2yz)\hat{j}+(xy\cos z+y^2)\hat{k}$$

$$= \begin{vmatrix}\hat{i} & \hat{j} & \hat{k}\\ \frac{\partial}{\partial x} & \frac{\partial}{\partial y} & \frac{\partial}{\partial z}\\ y\sin z-\sin x & x\sin z+2yz & xy\cos z+y^2\end{vmatrix}$$

$$= (x\cos z+2y-x\cos z-2y)\hat{i}-[y\cos z-y\cos z]\hat{j}+(\sin z-\sin z)\hat{k} = 0$$

Hence, the motion is irrotational.

So, $\bar{v} = \bar{\nabla}\phi$ where ϕ is called velocity potential.

$$d\phi = \frac{\partial\phi}{\partial x}dx+\frac{\partial\phi}{\partial y}dy+\frac{\partial\phi}{\partial z}dz \qquad \text{[Total differential coefficient]}$$

$$= \left(\hat{i}\frac{\partial\phi}{\partial x}+\hat{j}\frac{\partial\phi}{\partial y}+\hat{k}\frac{\partial\phi}{\partial z}\right).(\hat{i}\,dx+\hat{j}\,dy+\hat{k}\,dz) = \vec{\nabla}\phi.d\vec{r} = \vec{v}.d\vec{r}$$

$$= [(y\sin z-\sin x)\hat{i}+(x\sin z+2yz)\hat{j}+(xy\cos z+y^2)\hat{k}].[\hat{i}\,dx+\hat{j}\,dy+\hat{k}\,dz]$$
$$= (y\sin z-\sin x)\,dx+(x\sin z+2\,y\,z)\,dy+(x\,y\cos z+y^2)\,dz$$
$$= (y\sin z\,dx+x\,dy\sin z+x\,y\cos z\,dz)-\sin x\,dx+(2\,y\,z\,dy+y^2\,dz)$$
$$= d\,(x\,y\sin z)+d\,(\cos x)+d\,(y^2\,z)$$
$$\phi = \int d\,(xy\sin z)+\int d\,(\cos x)+\int d(y^2z)$$
$$\phi = xy\sin z+\cos x+y^2z+c$$

Hence, Velocity potential $= xy\sin z+\cos x+y^2z+c.$ **Ans.**

Example 43. *Prove that* $\vec{F} = r^2\vec{r}$ *is conservative and find the scalar potential* ϕ *such that*

$$\vec{F} = \nabla\phi.$$ (*Nagpur University, Summer 2004*)

Solution. Given $\vec{F} = r^2\vec{r} = r^2(x\hat{i}+y\hat{j}+z\hat{k}) = r^2\,x\hat{i}+r^2\,y\hat{j}+r^2\,z\hat{k}$

Consider $\vec{\nabla}\times\vec{F} = \begin{vmatrix}\hat{i} & \hat{j} & \hat{k}\\ \frac{\partial}{\partial x} & \frac{\partial}{\partial y} & \frac{\partial}{\partial z}\\ r^2x & r^2y & r^2z\end{vmatrix}$

$$= \hat{i}\left[\frac{\partial}{\partial y}r^2 z - \frac{\partial}{\partial z}r^2 y\right] - \hat{j}\left[\frac{\partial}{\partial x}r^2 z - \frac{\partial}{\partial z}r^2 x\right] + \hat{k}\left[\frac{\partial}{\partial x}r^2 y - \frac{\partial}{\partial y}r^2 x\right]$$

$$= \hat{i}\left[2rz\frac{\partial r}{\partial y} - 2ry\frac{\partial r}{\partial z}\right] - \hat{j}\left[2rz\frac{\partial r}{\partial x} - 2rx\frac{\partial r}{\partial z}\right] + \hat{k}\left[2ry\frac{\partial r}{\partial x} - 2rx\frac{\partial r}{\partial y}\right]$$

$$\left[\text{But } r^2 = x^2 + y^2 + z^2, \frac{\partial r}{\partial x} = \frac{x}{r}, \ \frac{\partial r}{\partial y} = \frac{y}{r}, \ \frac{\partial r}{\partial z} = \frac{z}{r}\right]$$

$$= \hat{i}\left[2rz\frac{y}{r} - 2ry\frac{z}{r}\right] - \hat{j}\left[2rz\frac{x}{r} - 2rx\frac{z}{r}\right] + \hat{k}\left[2ry\frac{x}{r} - 2rx\frac{y}{r}\right]$$

$$= \hat{i}(2yz - 2yz) - \hat{j}(2zx - 2zx) + \hat{k}(2xy - 2xy) = 0\hat{i} - 0\hat{j} + 0\hat{k} = 0$$

$$\therefore \quad \nabla \times \vec{F} = 0$$

$\therefore$ $\vec{F}$ is irrotational $\quad\therefore\quad$ F is conservative.

Consider scalar potential ϕ such that $\vec{F} = \nabla\phi$.

$$d\phi = \frac{\partial \phi}{\partial x}dx + \frac{\partial \phi}{\partial y}dy + \frac{\partial \phi}{\partial z}dz \qquad \text{[Total differential coefficient]}$$

$$= \left(\hat{i}\frac{\partial \phi}{\partial x} + \hat{j}\frac{\partial \phi}{\partial y} + \hat{k}\frac{\partial \phi}{\partial z}\right).(\hat{i}\,dx + \hat{j}\,dy + \hat{k}\,dz)$$

$$= \left(\hat{i}\frac{\partial}{\partial x} + \hat{j}\frac{\partial}{\partial y} + \hat{k}\frac{\partial}{\partial z}\right)\phi.(\hat{i}\,dx + \hat{j}\,dy + \hat{k}\,dz) = \nabla\phi.(\hat{i}\,dx + \hat{j}\,dy + \hat{k}\,dz)$$

$$= \vec{F}.(\hat{i}\,dx + \hat{j}\,dy + \hat{k}\,dz) \qquad = r^2\vec{r}.(\hat{i}\,dx + \hat{j}\,dy + \hat{k}\,dz) \qquad (\nabla\phi = \vec{F})$$

$$= (x^2 + y^2 + z^2)(\hat{i}x + \hat{j}y + \hat{k}z).(\hat{i}\,dx + \hat{j}\,dy + \hat{k}\,dz)$$

$$= (x^2 + y^2 + z^2)(x\,dx + y\,dy + z\,dz)$$

$$= x^3\,dx + y^3\,dy + z^3\,dz + (x\,dx)\,y^2 + (x^2)(y\,dy) + (x\,dx)z^2 + z^2(y\,dy) + x^2(z\,dz) + y^2(z\,dz)$$

$$\phi = \int x^3\,dx + \int y^3\,dy + \int z^3\,dz + \int[(x\,dx)y^2 + (y\,dy)x^2] + \int[(x\,dx)z^2 + (z\,dz)x^2] + \int[(y\,dy)z^2 + (z\,dz)y^2]$$

$$= \frac{x^4}{4} + \frac{y^4}{4} + \frac{z^4}{4} + \frac{1}{2}x^2y^2 + \frac{1}{2}x^2z^2 + \frac{1}{2}y^2z^2 + c$$

$$= \frac{1}{4}(x^4 + y^4 + z^4 + 2x^2y^2 + 2x^2z^2 + 2y^2z^2) + c$$ **Ans.**

Example 44. *Show that the vector field* $\vec{F} = \dfrac{\vec{r}}{|\vec{r}|^3}$ *is irrotational as well as solenoidal. Find the scalar potential.*

(*Nagpur University, Summer 2008, 2001, U.P. I Semester Dec. 2005, 2001*)

Solution. $$F = \frac{\vec{r}}{|\vec{r}|^3} = \frac{x\hat{i} + y\hat{j} + z\hat{k}}{(x^2 + y^2 + z^2)^{3/2}}$$

$$\text{Curl } \vec{F} = \vec{\nabla} \times \vec{F} = \left(\hat{i}\frac{\partial}{\partial x} + \hat{j}\frac{\partial}{\partial y} + \hat{k}\frac{\partial}{\partial z}\right) \times \left(\frac{x\hat{i} + y\hat{j} + z\hat{k}}{(x^2 + y^2 + z^2)^{3/2}}\right)$$

$$= \begin{vmatrix} \hat{i} & \hat{j} & \hat{k} \\ \dfrac{\partial}{\partial x} & \dfrac{\partial}{\partial y} & \dfrac{\partial}{\partial z} \\ \dfrac{x}{(x^2+y^2+z^2)^{3/2}} & \dfrac{y}{(x^2+y^2+z^2)^{3/2}} & \dfrac{z}{(x^2+y^2+z^2)^{3/2}} \end{vmatrix}$$

$$= \hat{i}\left[\frac{-3}{2}\frac{2yz}{(x^2+y^2+z^2)^{5/2}} + \frac{3}{2}\frac{2yz}{(x^2+y^2+z^2)^{5/2}}\right]$$

$$-\hat{j}\left[\frac{-3}{2}\frac{2xz}{(x^2+y^2+z^2)^{5/2}} - \left(-\frac{3}{2}\right)\frac{2xz}{(x^2+y^2+z^2)^{5/2}}\right]$$

$$+\hat{k}\left[-\frac{3}{2}\frac{2xy}{(x^2+y^2+z^2)^{5/2}} - \left(-\frac{3}{2}\right)\frac{2xy}{(x^2+y^2+z^2)^{5/2}}\right]$$

$$= 0$$

Hence, $\vec{F}$ is irrotational.

$\Rightarrow \quad \vec{F} = \vec{\nabla}\phi$, where ϕ is called scalar potential

$$d\phi = \frac{\partial\phi}{\partial x}dx + \frac{\partial\phi}{\partial y}dy + \frac{\partial\phi}{\partial z}dz \qquad \text{[Total differential coefficient]}$$

$$= \left(\hat{i}\frac{\partial\phi}{\partial x} + \hat{j}\frac{\partial\phi}{\partial y} + \hat{k}\frac{\partial\phi}{\partial z}\right).(\hat{i}\,dx + \hat{j}\,dy + \hat{k}\,dz) = \vec{\nabla}\phi.\vec{dr} = \vec{F}.\vec{dr}$$

$$= \frac{x\hat{i} + y\hat{j} + z\hat{k}}{(x^2+y^2+z^2)^{3/2}}.(\hat{i}\,dx + \hat{j}\,dy + \hat{k}\,dz) = \frac{x\,dx + y\,dy + z\,dz}{(x^2+y^2+z^2)^{3/2}}$$

$$\phi = \frac{1}{2}\int\frac{2x\,dx + 2y\,dy + 2z\,dz}{(x^2+y^2+z^2)^{3/2}}$$

$$= \frac{1}{2}\left(-\frac{2}{1}\right)(x^2+y^2+z^2)^{-\frac{1}{2}} = -\frac{1}{(x^2+y^2+z^2)^{\frac{1}{2}}} = -\frac{1}{|\vec{r}|} \qquad \textbf{Ans.}$$

Now, $\quad \text{Div}\,\vec{F} = \vec{\nabla}.\vec{F}$

$$= \left(\hat{i}\frac{\partial}{\partial x} + \hat{j}\frac{\partial}{\partial y} + \hat{k}\frac{\partial}{\partial z}\right).\frac{x\hat{i} + y\hat{j} + z\hat{k}}{(x^2+y^2+z^2)^{3/2}}$$

$$= \frac{\partial}{\partial x}\frac{x}{(x^2+y^2+z^2)^{3/2}} + \frac{\partial}{\partial y}\frac{y}{(x^2+y^2+z^2)^{3/2}} + \frac{\partial}{\partial z}\frac{z}{(x^2+y^2+z^2)^{3/2}}$$

$$= \frac{(x^2+y^2+z^2)^{3/2}\,(1) - x\left(\frac{3}{2}\right)(x^2+y^2+z^2)^{1/2}\,(2x)}{(x^2+y^2+z^2)^3}$$

$$+ \frac{(x^2+y^2+z^2)^{3/2}\,(1) - y\left(\frac{3}{2}\right)(x^2+y^2+z^2)^{1/2}\,(2y)}{(x^2+y^2+z^2)^3}$$

$$+ \frac{(x^2+y^2+z^2)^{3/2}\,(1) - z\left(\frac{3}{2}\right)(x^2+y^2+z^2)^{1/2}\,(2z)}{(x^2+y^2+z^2)^3}$$

$$= \frac{(x^2+y^2+z^2)^{1/2}}{(x^2+y^2+z^2)^3}\,[x^2+y^2+z^2-3x^2+x^2+y^2+z^2-3y^2+x^2+y^2+z^2-3z^2]$$

$$= 0$$

Hence, $\vec{F}$ is solenoidal. **Proved.**

Example 45. *Given the vector field* $\vec{V} = (x^2 - y^2 + 2xz)\hat{i} + (xz - xy + yz)\hat{j} + (z^2 + x^2)\hat{k}$ *find curl V. Show that the vectors given by curl V at P_0 (1, 2, –3) and P_1 (2, 3, 12) are orthogonal.*

Solution. $\overline{\text{Curl}\,\vec{V}} = \nabla \times \vec{V}$

$$= \left(\hat{i}\frac{\partial}{\partial x} + \hat{j}\frac{\partial}{\partial y} + \hat{k}\frac{\partial}{\partial z}\right) \times [(x^2 - y^2 + 2xz)\hat{i} + (xz - xy + yz)\hat{j} + (z^2 + x^2)\hat{k}]$$

$$\text{curl}\,\vec{V} = \begin{vmatrix} \hat{i} & \hat{j} & \hat{k} \\ \frac{\partial}{\partial x} & \frac{\partial}{\partial y} & \frac{\partial}{\partial z} \\ x^2 - y^2 + 2xz & xz - xy + yz & z^2 + x^2 \end{vmatrix}$$

$$= -(x+y)\hat{i} - (2x - 2x)\hat{j} + (z - y + 2y)\hat{k} = -(x+y)\hat{i} + (y+z)\hat{k}$$

$$\text{curl}\,\vec{V} \text{ at } P_0\,(1, 2, -3) = -(1+2)\hat{i} + (2-3)\hat{k} = -3\hat{i} - \hat{k}$$

$$\text{curl}\,\vec{V} \text{ at } P_1\,(2, 3, 12) = -(2+3)\hat{i} + (3+12)\hat{k} = -5\hat{i} + 15\hat{k}$$

The curl $\vec{V}$ at (1, 2, –3) and (2, 3, 12) are perpendicular since

$$(-3\hat{i} - \hat{k}).(-5\hat{i} + 15\hat{k}) = 15 - 15 = 0\backslash$$ **Proved.**

Example 46. *Find the constants a, b, c, so that*

$$\vec{F} = (x + 2y + az)\hat{i} + (bx - 3y - z)\hat{j} + (4x + cy + 2z)\hat{k} \qquad ...(1)$$

is irrotational and hence find function ϕ such that $\vec{F} = \nabla\phi$.

(Nagpur University, Summer 2005, Winter 2000; R.G.P.V., Bhopal 2009)

Solution. We have,

$$\therefore \qquad \nabla \times \vec{F} = \begin{vmatrix} \hat{i} & \hat{j} & \hat{k} \\ \frac{\partial}{\partial x} & \frac{\partial}{\partial y} & \frac{\partial}{\partial z} \\ (x + 2y + az) & (bx - 3y - z) & (4x + cy + 2z) \end{vmatrix}$$

$$= (c+1)\hat{i} - (4-a)\hat{j} + (b-2)\hat{k}$$

As $\vec{F}$ is irrotational, $\nabla \times \vec{F} = \vec{0}$

i.e., $(c+1)\hat{i} - (4-a)\hat{j} + (b-2)\hat{k} = 0\hat{i} + 0\hat{j} + 0\hat{k}$

$\therefore$ $c + 1 = 0, \quad 4 - a = 0 \quad$ and $\quad b - 2 = 0$

i.e., $a = 4, \quad b = 2, \quad c = -1$

Putting the values of *a*, *b*, *c* in (1), we get

$$\vec{F} = (x + 2y + 4z)\hat{i} + (2x - 3y - z)\hat{j} + (4x - y + 2z)\hat{k}$$

Now we have to find ϕ such that $\vec{F} = \nabla\phi$

We know that

$$d\phi = \frac{\partial \phi}{\partial x}dx + \frac{\partial \phi}{\partial y}dy + \frac{\partial \phi}{\partial z}dz \qquad \text{[Total differential coefficient]}$$

$$= \left(\hat{i}\frac{\partial \phi}{\partial x} + \hat{j}\frac{\partial \phi}{\partial y} + \hat{k}\frac{\partial \phi}{\partial z}\right).(\hat{i}\,dx + \hat{j}\,dy + \hat{k}\,dz)$$

$$= \left(\hat{i}\frac{\partial}{\partial x} + \hat{j}\frac{\partial}{\partial y} + \hat{k}\frac{\partial}{\partial z}\right)\phi.(\hat{i}\,dx + \hat{j}\,dy + \hat{k}\,dz) = \nabla\phi.(\hat{i}\,dx + \hat{j}\,dy + \hat{k}\,dz)$$

$$= \bar{F}.(\hat{i}\,dx + \hat{j}\,dy + \hat{k}\,dz)$$

$$= [(x+2y+4z)\hat{i} + (2x-3y-z)\hat{j} + (4x-y+2z)\hat{k}].(\hat{i}\,dx + \hat{j}\,dy + \hat{k}\,dz)$$

$$= (x+2y+4z)\,dx + (2x-3y-z)\,dy + (4x-y+2z)\,dz$$

$$= x\,dx - 3y\,dy + 2z\,dz + (2y\,dx + 2x\,dy) + (4z\,dx + 4x\,dz) + (-z\,dy - y\,dz)$$

$$\phi = \int x\,dx - 3\int y\,dy + 2\int z\,dz + \int(2y\,dx + 2x\,dy) + \int(4z\,dx + 4x\,dz) - \int(z\,dy + y\,dz)$$

$$= \frac{x^2}{2} - \frac{3y^2}{2} + z^2 + 2xy + 4zx - yz + c \qquad \textbf{Ans.}$$

Example 47. *Let* $\vec{V}\,(x, y, z)$ *be a differentiable vector function and* $\phi(x, y, z)$ *be a scalar function. Derive an expression for div* $(\phi\vec{V})$ *in terms of* $\phi\,.\,\vec{V}$, *div* $\vec{V}$ *and* $\nabla\phi$.

(*U.P. I Semester, Winter 2003*)

Solution. Let $\vec{V} = V_1\hat{i} + V_2\hat{j} + V_3\hat{k}$

$$\text{div}\,(\phi\vec{V}) = \vec{\nabla}.(\phi\vec{F})$$

$$= \left(\hat{i}\frac{\partial}{\partial x} + \hat{j}\frac{\partial}{\partial y} + \hat{k}\frac{\partial}{\partial z}\right).[\phi V_1\hat{i} + \phi V_2\hat{j} + \phi V_3\hat{k}] = \frac{\partial}{\partial x}(\phi V_1) + \frac{\partial}{\partial y}(\phi V_2) + \frac{\partial}{\partial z}(\phi V_3)$$

$$= \left(\phi\frac{\partial V_1}{\partial x} + \frac{\partial \phi}{\partial x}V_1\right) + \left(\phi\frac{\partial V_2}{\partial y} + \frac{\partial \phi}{\partial y}V_2\right) + \left(\phi\frac{\partial V_3}{\partial z} + \frac{\partial \phi}{\partial z}V_3\right)$$

$$= \phi\left(\frac{\partial V_1}{\partial x} + \frac{\partial V_2}{\partial y} + \frac{\partial V_3}{\partial z}\right) + \left(\frac{\partial \phi}{\partial x}V_1 + \frac{\partial \phi}{\partial y}V_2 + \frac{\partial \phi}{\partial z}V_3\right)$$

$$= \phi\left(\hat{i}\frac{\partial}{\partial x} + \hat{j}\frac{\partial}{\partial y} + \hat{k}\frac{\partial}{\partial z}\right).(V_1\hat{i} + V_2\hat{j} + V_3\hat{k}) + \left(\hat{i}\frac{\partial \phi}{\partial x} + \hat{j}\frac{\partial \phi}{\partial y} + \hat{k}\frac{\partial \phi}{\partial z}\right).(V_1\hat{i} + V_2\hat{j} + V_3\hat{k})$$

$$= \phi(\nabla.\vec{V}) + (\vec{\nabla}\phi).\vec{V} = \phi\,(\text{div}\,\vec{V}) + (\text{grad}\,\phi).\,\vec{V} \qquad \textbf{Ans.}$$

Example 48. *If* $\vec{A}$ *is a constant vector and* $\vec{R} = x\hat{i} + y\hat{j} + z\hat{k}$, *then prove that*

$$\text{Curl}\left[\left(\vec{A}\,.\,\vec{R}\right)\vec{A}\right] = \vec{A}\times\vec{R}$$

(*K. University, Dec. 2009*)

Solution. Let $\vec{A} = A_1\hat{i} + A_2\hat{j} + A_3\hat{k}$, $\qquad \vec{R} = x\hat{i} + y\hat{j} + z\hat{k}$

$$\vec{A}\,.\,\vec{R} = (A_1\hat{i} + A_2\hat{j} + A_3\hat{k})\,.\,(x\hat{i} + y\hat{j} + z\hat{k}) = A_1x + A_2y + A_3z$$

$$[\vec{A}\,.\,\vec{R}]\,\vec{R} = (A_1x + A_2y + A_3z)\,(x\hat{i} + y\hat{j} + z\hat{k})$$

$$= (A_1x^2 + A_2xy + A_3zx)\,\hat{i} + (A_1xy + A_2y^2 + A_3yz)\,\hat{j} + (A_1xz + A_2yz + A_3z^2)\,\hat{k}$$

$$\text{Curl}\left[(\vec{A}\,.\,\vec{R})\,\vec{R}\right] = \begin{vmatrix} \hat{i} & \hat{j} & \hat{k} \\ \frac{\partial}{\partial x} & \frac{\partial}{\partial y} & \frac{\partial}{\partial z} \\ A_1x^2 + A_2xy + A_3zx & A_2xy + A_2y^2 + A_3yz & A_1xz + A_2yz + A_3z^2 \end{vmatrix}$$

$$= (A_2 z - A_3 y)\ \hat{i} - (A_1 z - A_3 x)\ \hat{j} + [A_1 y - A_2 x]\ \hat{k} \quad \ldots (1)$$

$$\text{L.H.S.} = \vec{A} \times \vec{R}$$

$$= (A_1 \hat{i} + A_2 \hat{j} + A_3 k) \times (x\hat{i} + y\hat{j} + z\hat{k})$$

$$= \begin{vmatrix} \hat{i} & \hat{j} & k \\ A_1 & A_2 & A_3 \\ x & y & z \end{vmatrix}$$

$$= (A_2 z - A_3 y)\ \hat{i} - (A_1 z - A_3 x)\ \hat{j} + (A_1 y - A_2 x)\ \hat{k}$$

$$= \text{R.H.S.} \qquad \text{[From (1)]}$$

Example 49. *Suppose that $\vec{U}, \vec{V}$ and f are continuously differentiable fields then Prove that, div $(\vec{U} \times \vec{V}) = \vec{V}.\text{curl}\, \vec{U} - \vec{U}.\text{curl}\, \vec{V}$.* *(GBTU, 2012, M.U. 2003, 2005)*

Solution. Let

$$\vec{U} = u_1 \hat{i} + u_2 \hat{j} + u_3 \hat{k},\ \vec{V} = v_1 \hat{i} + v_2 \hat{j} + v_3 \hat{k}$$

$$\vec{U} \times \vec{V} = \begin{vmatrix} \hat{i} & \hat{j} & \hat{k} \\ u_1 & u_2 & u_3 \\ v_1 & v_2 & v_3 \end{vmatrix}$$

$$= (u_2 v_3 - u_3 v_2)\hat{i} - (u_1 v_3 - u_3 v_1)\hat{j} + (u_1 v_2 - u_2 v_1)\hat{k}$$

$$\text{div } (\vec{U} \times \vec{V}) = \left(\hat{i}\frac{\partial}{\partial x} + \hat{j}\frac{\partial}{\partial y} + \hat{k}\frac{\partial}{\partial z}\right).[(u_2 v_3 - u_3 v_2)\hat{i} - (u_1 v_3 - u_3 v_1)\hat{j} + (u_1 v_2 - u_2 v_1)\hat{k}]$$

$$= \frac{\partial}{\partial x}(u_2 v_3 - u_3 v_2) + \frac{\partial}{\partial y}(-u_1 v_3 + u_3 v_1) + \frac{\partial}{\partial z}(u_1 v_2 - u_2 v_1)$$

$$= \left[u_2 \frac{\partial v_3}{\partial x} + v_3 \frac{\partial u_2}{\partial x} - u_3 \frac{\partial v_2}{\partial x} - v_2 \frac{\partial u_3}{\partial x}\right] + \left[-u_1 \frac{\partial v_3}{\partial y} - v_3 \frac{\partial u_1}{\partial y} + u_3 \frac{\partial v_1}{\partial y} + v_1 \frac{\partial u_3}{\partial y}\right]$$
$$+ \left[u_1 \frac{\partial v_2}{\partial z} + v_2 \frac{\partial u_1}{\partial z} - u_2 \frac{\partial v_1}{\partial z} - v_1 \frac{\partial u_2}{\partial z}\right]$$

$$= v_1\left(\frac{\partial u_3}{\partial y} - \frac{\partial u_2}{\partial z}\right) + v_2\left(-\frac{\partial u_3}{\partial x} + \frac{\partial u_1}{\partial z}\right) + v_3\left(\frac{\partial u_2}{\partial x} - \frac{\partial u_1}{\partial y}\right)$$
$$+ u_1\left(-\frac{\partial v_3}{\partial y} + \frac{\partial v_2}{\partial z}\right) + u_2\left(\frac{\partial v_3}{\partial x} - \frac{\partial v_1}{\partial z}\right) + u_3\left(\frac{\partial v_1}{\partial y} - \frac{\partial v_2}{\partial x}\right)$$

$$= (v_1 \hat{i} + v_2 \hat{j} + v_3 \hat{k}).\left[\hat{i}\left(\frac{\partial u_3}{\partial y} - \frac{\partial u_2}{\partial z}\right) + \hat{j}\left(\frac{\partial u_1}{\partial z} - \frac{\partial u_3}{\partial x}\right) + \hat{k}\left(\frac{\partial u_2}{\partial x} - \frac{\partial u_1}{\partial y}\right)\right]$$
$$-(u_1 \hat{i} + u_2 \hat{j} + u_3 \hat{k}).\left[\hat{i}\left(-\frac{\partial v_2}{\partial z} + \frac{\partial v_3}{\partial y}\right) + \hat{j}\left(-\frac{\partial v_3}{\partial x} + \frac{\partial v_1}{\partial z}\right) + \hat{k}\left(-\frac{\partial v_1}{\partial y} + \frac{\partial v_2}{\partial x}\right)\right]$$

$$= V.(\vec{\nabla} \times \vec{U}) - \vec{U}.(\vec{\nabla} \times V) = \vec{V}.\text{curl}\, \vec{U} - \vec{U}.\text{curl}\, \vec{V}$$ **Proved.**

Example 50. *Prove that*

$$\vec{\nabla} \times (\vec{F} \times \vec{G}) = \vec{F}(\vec{\nabla}.\vec{G}) - \vec{G}(\vec{\nabla}.\vec{F}) + (\vec{G}.\vec{\nabla})\vec{F} - (\vec{F}.\vec{\nabla})\vec{G}$$ *(M.U. 2004, 2005)*

Solution.

$$\vec{\nabla} \times (\vec{F} \times \vec{G}) = \Sigma \hat{i} \times \frac{\partial}{\partial x}(\vec{F} \times \vec{G})$$

$$= \Sigma \hat{i} \times \left(\frac{\partial F}{\partial x} \times \vec{G} + \vec{F} \times \frac{\partial G}{\partial x}\right) = \Sigma \hat{i} \times \left(\frac{\partial F}{\partial x} \times \vec{G}\right) + \Sigma \hat{i} \times \left(\vec{F} \times \frac{\partial G}{\partial x}\right)$$

$$= \Sigma\left[(\hat{i}.\vec{G})\frac{\partial F}{\partial x} - \left(\hat{i}\frac{\partial F}{\partial x}\right)\vec{G}\right] + \Sigma\left[\left(\hat{i}\frac{\partial G}{\partial x}\right)\vec{F} - (\hat{i}.\vec{F})\frac{\partial G}{\partial x}\right]$$

$$= \Sigma(\vec{G}.\hat{i})\frac{\partial F}{\partial x} - \vec{G}\Sigma\left(\hat{i}.\frac{\partial F}{\partial x}\right) + \vec{F}\Sigma\left(\hat{i}.\frac{\partial G}{\partial x}\right) - \Sigma(\vec{F}.\hat{i})\frac{\partial G}{\partial x}$$

$$= \vec{F}\left(\Sigma\hat{i}\frac{\partial G}{\partial x}\right) - \vec{G}\Sigma\left(\hat{i}.\frac{\partial F}{\partial x}\right) + \Sigma(\vec{G}.\hat{i})\frac{\partial F}{\partial x} - \Sigma(\vec{F}.\hat{i})\frac{\partial G}{\partial x}$$

$$= \vec{F}(\vec{\nabla}\vec{G}) - \vec{G}(\vec{\nabla}.\vec{F}) + (\vec{G}.\vec{\nabla})\vec{F} - (\vec{F}.\vec{\nabla})\vec{G}$$

Proved.

Example 51. *Prove that, for every field $\vec{V}$; div curl $\vec{V} = 0$.*

(*Nagpur University, Summer 2004; AMIETE, Sem II, June 2010*)

Solution. Let $V = V_1\hat{i} + V_2\hat{j} + V_3\hat{k}$

$$\text{div}\ (\overline{\text{curl}\ \vec{V}}) = \vec{\nabla}.(\vec{\nabla}\times\vec{V})$$

$$= \vec{\nabla}.\begin{vmatrix} \hat{i} & \hat{j} & \hat{k} \\ \frac{\partial}{\partial x} & \frac{\partial}{\partial y} & \frac{\partial}{\partial z} \\ V_1 & V_2 & V_3 \end{vmatrix}$$

$$= \left(\hat{i}\frac{\partial}{\partial x} + \hat{j}\frac{\partial}{\partial y} + \hat{k}\frac{\partial}{\partial z}\right).\left[\hat{i}\left(\frac{\partial V_3}{\partial y} - \frac{\partial V_2}{\partial z}\right) - \hat{j}\left(\frac{\partial V_3}{\partial x} - \frac{\partial V_1}{\partial z}\right) + \hat{k}\left(\frac{\partial V_2}{\partial x} - \frac{\partial V_1}{\partial y}\right)\right]$$

$$= \frac{\partial}{\partial x}\left(\frac{\partial V_3}{\partial y} - \frac{\partial V_2}{\partial z}\right) - \frac{\partial}{\partial y}\left(\frac{\partial V_3}{\partial x} - \frac{\partial V_1}{\partial z}\right) + \frac{\partial}{\partial z}\left(\frac{\partial V_2}{\partial x} - \frac{\partial V_1}{\partial y}\right)$$

$$= \frac{\partial^2 V_3}{\partial x\,\partial y} - \frac{\partial^2 V_2}{\partial x\,\partial z} - \frac{\partial^2 V_3}{\partial y\,\partial x} + \frac{\partial^2 V_1}{\partial y\,\partial z} + \frac{\partial^2 V_2}{\partial z\,\partial x} - \frac{\partial^2 V_1}{\partial z\,\partial y}$$

$$= \left(\frac{\partial^2 V_1}{\partial y\,\partial z} - \frac{\partial^2 V_1}{\partial z\,\partial y}\right) + \left(\frac{\partial^2 V_2}{\partial z\,\partial x} - \frac{\partial^2 V_2}{\partial x\,\partial z}\right) + \left(\frac{\partial^2 V_3}{\partial x\,\partial y} - \frac{\partial^2 V_3}{\partial y\,\partial x}\right)$$

$$= 0$$

Ans.

Example 52. *If $\vec{a}$ is a constant vector, show that*

$$\vec{a}\times(\vec{\nabla}\times\vec{r}) = \vec{\nabla}(\vec{a}.\vec{r}) - (\vec{a}.\vec{\nabla})\vec{r}.$$

(*U.P., Ist Semester, Dec. 2007*)

Solution. $\vec{a} = a_1\hat{i} + a_2\hat{j} + a_3\hat{k},\quad \vec{r} = r_1\hat{i} + r_2\hat{j} + r_3\hat{k}$

$$\vec{\nabla}\times\vec{r} = \begin{vmatrix} \hat{i} & \hat{j} & \hat{k} \\ \frac{\partial}{\partial x} & \frac{\partial}{\partial y} & \frac{\partial}{\partial z} \\ r_1 & r_2 & r_3 \end{vmatrix} = \left(\frac{\partial r_3}{\partial y} - \frac{\partial r_2}{\partial z}\right)\hat{i} - \left(\frac{\partial r_3}{\partial x} - \frac{\partial r_1}{\partial z}\right)\hat{j} + \left(\frac{\partial r_2}{\partial x} - \frac{\partial r_1}{\partial y}\right)\hat{k}$$

$$\vec{a}\times(\vec{\nabla}\times\vec{r}) = \begin{vmatrix} \hat{i} & \hat{j} & \hat{k} \\ a_1 & a_2 & a_3 \\ \frac{\partial r_3}{\partial y} - \frac{\partial r_2}{\partial z} & -\frac{\partial r_3}{\partial x} + \frac{\partial r_1}{\partial z} & \frac{\partial r_2}{\partial x} - \frac{\partial r_1}{\partial y} \end{vmatrix}$$

$$= \left[\left(a_2\frac{\partial r_2}{\partial x} - a_2\frac{\partial r_1}{\partial y}\right) - \left(-a_3\frac{\partial r_3}{\partial x} + a_3\frac{\partial r_1}{\partial z}\right)\right]\hat{i} - \left[a_1\frac{\partial r_2}{\partial x} - a_1\frac{\partial r_1}{\partial y} - a_3\frac{\partial r_3}{\partial y} + a_3\frac{\partial r_2}{\partial z}\right]\hat{j}$$

$$+\left[-a_1\frac{\partial r_3}{\partial x} + a_1\frac{\partial r_1}{\partial z} - a_2\frac{\partial r_3}{\partial y} + a_2\frac{\partial r_2}{\partial z}\right]\hat{k}$$

$$= \left[\left(a_1\hat{i}\frac{\partial r_1}{\partial x} + a_2\hat{i}\frac{\partial r_2}{\partial x} + a_3\hat{i}\frac{\partial r_3}{\partial x}\right) + \left(a_1\hat{j}\frac{\partial r_1}{\partial y} + a_2\hat{j}\frac{\partial r_2}{\partial y} + a_3\hat{j}\frac{\partial r_3}{\partial y}\right)\right.$$

$$\left.+\left(a_1\hat{k}\frac{\partial r_1}{\partial z} + a_2\hat{k}\frac{\partial r_2}{\partial z} + a_3\hat{k}\frac{\partial r_3}{\partial z}\right)\right] - \left[\left(a_1\hat{i}\frac{\partial r_1}{\partial x} + a_1\hat{j}\frac{\partial r_2}{\partial x} + a_1\hat{k}\frac{\partial r_3}{\partial x}\right)\right.$$

$$\left.+\left(a_2\hat{i}\frac{\partial r_1}{\partial y} + a_2\hat{j}\frac{\partial r_2}{\partial y} + a_2\hat{k}\frac{\partial r_3}{\partial y}\right) + \left(a_3\hat{i}\frac{\partial r_1}{\partial z} + a_3\hat{j}\frac{\partial r_2}{\partial z} + a_3\hat{k}\frac{\partial r_3}{\partial z}\right)\right]$$

$$= \left(\hat{i}\frac{\partial}{\partial x} + \hat{j}\frac{\partial}{\partial y} + \hat{k}\frac{\partial}{\partial z}\right)(a_1r_1 + a_2r_2 + a_3r_3) - \left[a_1\frac{\partial}{\partial x} + a_2\frac{\partial}{\partial y} + a_3\frac{\partial}{\partial z}\right](r_1\hat{i} + r_2\hat{j} + r_3\hat{k})$$

$$= \nabla(\vec{a}.\vec{r}) - (\vec{a}.\vec{\nabla})\vec{r}$$

Proved.

Example 53. *If r is the distance of a point (x, y, z) from the origin, prove that* $Curl\left(k \times grad\frac{1}{r}\right) + grad\left(k.grad\frac{1}{r}\right) = 0$, *where k is the unit vector in the direction OZ.*

(U.P., I Semester, Winter 2000)

Solution. $r^2 = (x-0)^2 + (y-0)^2 + (z-0)^2 = x^2 + y^2 + z^2$

$$\Rightarrow \quad \frac{1}{r} = (x^2 + y^2 + z^2)^{-1/2}$$

$$\text{grad}\ \frac{1}{r} = \vec{\nabla}\frac{1}{r} = \left(\hat{i}\frac{\partial}{\partial x} + \hat{j}\frac{\partial}{\partial y} + \hat{k}\frac{\partial}{\partial z}\right)(x^2 + y^2 + z^2)^{-1/2}$$

$$= -\frac{1}{2}(x^2 + y^2 + z^2)^{-3/2}\,(2x\hat{i} + 2y\hat{j} + 2z\hat{k})$$

$$= -(x^2 + y^2 + z^2)^{-3/2}\,(x\hat{i} + y\hat{j} + z\hat{k})$$

$$k \times \text{grad}\ \frac{1}{r} = k \times [-(x^2 + y^2 + z^2)^{-3/2}\,(x\hat{i} + y\hat{j} + z\hat{k})]$$

$$= -(x^2 + y^2 + z^2)^{-3/2}\,(x\hat{j} - y\hat{i})$$

$$\text{curl}\left(k \times \text{grad}\frac{1}{r}\right) = \vec{\nabla} \times \left(k \times \text{grad}\frac{1}{r}\right)$$

$$= \left(\hat{i}\frac{\partial}{\partial x} + \hat{j}\frac{\partial}{\partial y} + \hat{k}\frac{\partial}{\partial z}\right) \times [-(x^2 + y^2 + z^2)^{-3/2}\,(x\hat{j} - y\hat{i})]$$

$$= \begin{vmatrix} \hat{i} & \hat{j} & \hat{k} \\ \frac{\partial}{\partial x} & \frac{\partial}{\partial y} & \frac{\partial}{\partial z} \\ \frac{y}{(x^2+y^2+z^2)^{3/2}} & \frac{-x}{(x^2+y^2+z^2)^{3/2}} & 0 \end{vmatrix}$$

$$= -\left(-\frac{3}{2}\right)\frac{(-x)\,(2z)}{(x^2+y^2+z^2)^{5/2}}\hat{i} + -\frac{3}{2}\frac{y\,(2z)}{(x^2+y^2+z^2)^{5/2}}\hat{j} + \left[-\frac{3}{2}\frac{(-x)(2x)}{(x^2+y^2+z^2)^{5/2}}\right.$$

$$\left.- \frac{1}{(x^2+y^2+z^2)^{3/2}} - \frac{(-3/2)\,(y)\,(2y)}{(x^2+y^2+z^2)^{5/2}} - \frac{1}{(x^2+y^2+z^2)^{3/2}}\right]\hat{k}$$

$$= \frac{-3xz}{(x^2+y^2+z^2)^{5/2}}\hat{i} - \frac{3yz}{(x^2+y^2+z^2)^{5/2}}\hat{j} + \frac{(3x^2-x^2-y^2-z^2+3y^2-x^2-y^2-z^2)}{(x^2+y^2+z^2)^{5/2}}\hat{k}$$

$$= \frac{-3xz\,\hat{i} - 3yz\,\hat{j} + (x^2+y^2-2z^2)\hat{k}}{(x^2+y^2+z^2)^{5/2}} \quad ...(1)$$

$$k \,.\, \text{grad}\,\frac{1}{r} = k.[-(x^2+y^2+z^2)^{-3/2}\,(x\hat{i} + y\hat{j} + z\hat{k})] = \frac{-z}{(x^2+y^2+z^2)^{3/2}}$$

$$\text{grad}\left(k.\text{grad}\,\frac{1}{r}\right) = \left(\hat{i}\frac{\partial}{\partial x} + \hat{j}\frac{\partial}{\partial y} + \hat{k}\frac{\partial}{\partial z}\right)\frac{-z}{(x^2+y^2+z^2)^{3/2}}$$

$$= -\frac{3}{2}\frac{\hat{i}(-z)(2x)}{(x^2+y^2+z^2)^{5/2}} + -\frac{3}{2}\frac{\hat{j}(-z)(2y)}{(x^2+y^2+z^2)^{5/2}}$$

$$+\left[-\frac{3}{2}\frac{(-z)(2z)}{(x^2+y^2+z^2)^{5/2}} - \frac{1}{(x^2+y^2+z^2)^{3/2}}\right]\hat{k}$$

$$= \frac{3xz\,\hat{i} + 3yz\,\hat{j} + (3z^2-x^2-y^2-z^2)\hat{k}}{(x^2+y^2+z^2)^{5/2}} = \frac{3xz\,\hat{i} + 3yz\,\hat{j} - (x^2+y^2-2z^2)\hat{k}}{(x^2+y^2+z^2)^{5/2}} \quad ...(2)$$

Adding (1) and (2), we get

$$\text{Curl}\left(k \times \text{grad}\,\frac{1}{r}\right) + \text{grad}\left(k.\text{grad}\,\frac{1}{r}\right) = 0$$ **Proved.**

Example 54. *Prove that* $\nabla \times \left(\dfrac{\vec{a} \times \vec{r}}{r^n}\right) = \dfrac{(2-n)\,\vec{a}}{r^n} + \dfrac{n(\vec{a}\,.\,\vec{r})\,\vec{r}}{r^{n+2}}$.

(*M.U. 2009, 2005, 2003, 2002; AMIETE, II Sem. June 2010*)

Solution. We have,

$$\frac{\vec{a} \times \vec{r}}{r^n} = \frac{1}{r^n}\begin{vmatrix} \hat{i} & \hat{j} & \hat{k} \\ a_1 & a_2 & a_3 \\ x & y & z \end{vmatrix}$$

$$= \frac{1}{r^n}(a_2 z - a_3 y)\,\hat{i} + \frac{1}{r^n}(a_3 x - a_1 z)\,\hat{j} + \frac{1}{r^n}(a_1 y - a_2 x)\,\hat{k}$$

$$\nabla \times \frac{(\vec{a} \times \vec{r})}{r^n} = \begin{vmatrix} \hat{i} & \hat{j} & \hat{k} \\ \dfrac{\partial}{\partial x} & \dfrac{\partial}{\partial y} & \dfrac{\partial}{\partial z} \\ \dfrac{a_2 z - a_3 y}{r^n} & \dfrac{a_3 x - a_1 z}{r^n} & \dfrac{a_1 y - a_2 x}{r^n} \end{vmatrix}$$

$$= \hat{i}\left[\frac{\partial}{\partial y}\left(\frac{a_1 y - a_2 x}{r^n}\right) - \frac{\partial}{\partial z}\left(\frac{a_3 x - a_1 z}{r^n}\right)\right] - \hat{j}\left[\frac{\partial}{\partial x}\left(\frac{a_1 y - a_2 x}{r^n}\right) - \frac{\partial}{\partial z}\left(\frac{a_2 z - a_3 y}{r^n}\right)\right]$$

$$+ \hat{k}\left[\frac{\partial}{\partial x}\left(\frac{a_3 x - a_1 z}{r^n}\right) - \frac{\partial}{\partial y}\left(\frac{a_2 z - a_3 y}{r^n}\right)\right]$$

Now, $\quad r^2 = x^2 + y^2 + z^2 \Rightarrow 2r\dfrac{\partial r}{\partial x} = 2x \quad \Rightarrow \quad \dfrac{\partial r}{\partial x} = \dfrac{x}{r}$

Similarly, $\quad \dfrac{\partial r}{\partial y} = \dfrac{y}{r}, \qquad \dfrac{\partial r}{\partial z} = \dfrac{z}{r}$

$$\therefore\ \nabla\times\left(\frac{\vec{a}\times\vec{r}}{r^n}\right) = \hat{i}\left[\left\{-nr^{-n-1}.\left(\frac{y}{r}\right)(a_1y-a_2x)+\frac{1}{r^n}a_1\right\}\right.$$

$$\left.-\left\{-nr^{-n-1}\left(\frac{z}{r}\right)(a_3x-a_1z)+\frac{1}{r^n}(-a_1)\right\}\right] + \text{two similar terms}$$

$$= \hat{i}\left[-\frac{n}{r^{n+2}}(a_1y^2-a_2xy)+\frac{a_1}{r^n}+\frac{n}{r^{n+2}}(a_3xz-a_1z^2)+\frac{a_1}{r^n}\right] + \text{two similar terms}$$

$$= \hat{i}\left[\frac{2a_1}{r^n}-\frac{n}{r^{n+2}}a_1(y^2+z^2)+\frac{n}{r^{n+2}}(a_2xy+a_3xz)\right] + \text{two similar terms}$$

Adding and subtracting $\frac{n}{r^{n+2}}a_1x^2$ to third and from second term, we get

$$\vec{\nabla}\times\left(\frac{\vec{a}\times\vec{r}}{r^n}\right) = \hat{i}\left[\frac{2a_1}{r^n}-\frac{na_1}{r^{n+2}}(x^2+y^2+z^2)+\frac{n}{r^{n+2}}(a_1x^2+a_2xy+a_3xz)\right] + \text{two similar terms}$$

$$= \hat{i}\left[\frac{2a_1}{r^n}-\frac{na_1}{r^{n+2}}r^2+\frac{n}{r^{n+2}}x(a_1x+a_2y+a_3z)\right] + \text{two similar terms}$$

$$= \hat{i}\left[\frac{2a_1}{r^n}-\frac{na_1}{r^n}+\frac{n}{r^{n+2}}x(a_1x+a_2y+a_3z)\right] + \hat{j}\left[\frac{2a_2}{r^n}-\frac{na_2}{r^n}+\frac{n}{r^{n+2}}y(a_2y+a_3z+a_1x)\right]$$

$$+\hat{k}\left[\frac{2a_3}{r^n}-\frac{na_3}{r^n}+\frac{n}{r^{n+2}}z(a_3z+a_1x+a_2y)\right]$$

$$= \frac{2}{r^n}(a_1\hat{i}+a_2\hat{j}+a_3\hat{k})-\frac{n}{r^n}(a_1\hat{i}+a_2\hat{j}+a_3\hat{k}) + \frac{n}{r^{n+2}}(a_1x+a_2y+a_3z)(x\hat{i}+y\hat{j}+z\hat{k})$$

$$= \frac{2-n}{r^n}(a_1\hat{i}+a_2\hat{j}+a_3\hat{k})+\frac{n}{r^{n+2}}(a_1x+a_2y+a_3z)(x\hat{i}+y\hat{j}+z\hat{k})$$

$$= \frac{2-n}{r^n}\vec{a}+\frac{n}{r^{n+2}}(\vec{a}.\vec{r})\vec{r}$$

Proved.

Example 55. *If f and g are two scalar point functions, prove that*

div $(f\nabla g) = f\nabla^2 g + \nabla f\,\nabla g.$ *(U.P., I Semester, compartment, Winter 2001)*

Solution. We have, $\nabla g = \frac{\partial g}{\partial x}\hat{i}+\frac{\partial g}{\partial y}\hat{j}+\frac{\partial g}{\partial z}\hat{k}$

$$\Rightarrow\quad f\nabla g = f\frac{\partial g}{\partial x}\hat{i}+f\frac{\partial g}{\partial y}\hat{j}+f\frac{\partial g}{\partial z}\hat{k}$$

$$\Rightarrow\quad \text{div}\,(f\nabla g) = \frac{\partial}{\partial x}\left(f\frac{\partial g}{\partial x}\right)+\frac{\partial}{\partial y}\left(f\frac{\partial g}{\partial y}\right)+\frac{\partial}{\partial z}\left(f\frac{\partial g}{\partial z}\right)$$

$$= f\left(\frac{\partial^2 g}{\partial x^2}+\frac{\partial^2 g}{\partial y^2}+\frac{\partial^2 g}{\partial z^2}\right)+\left(\frac{\partial f}{\partial x}\frac{\partial g}{\partial x}+\frac{\partial f}{\partial y}\frac{\partial g}{\partial y}+\frac{\partial f}{\partial z}\frac{\partial g}{\partial z}\right)$$

$$= f\left(\frac{\partial^2}{\partial x^2}+\frac{\partial^2}{\partial y^2}+\frac{\partial^2}{\partial z^2}\right)g+\left(\frac{\partial f}{\partial x}\hat{i}+\frac{\partial f}{\partial y}\hat{j}+\frac{\partial f}{\partial z}\hat{k}\right).\left(\frac{\partial g}{\partial x}\hat{i}+\frac{\partial g}{\partial y}\hat{j}+\frac{\partial g}{\partial z}\hat{k}\right)$$

$$= f\nabla^2 g + \nabla f.\nabla g$$

Proved.

Example 56. *Establish the reation curl curl* $\vec{f} = \vec{\nabla}\,div\,\vec{f} - \nabla^2\vec{f}$.

(U.P., I Semester, Compartment 2002)

Solution. Let $\vec{f} = f_1\hat{i} + f_2\hat{j} + f_3\hat{k}$, then by definition,

$$\text{Curl } \vec{f} = \begin{vmatrix} \hat{i} & \hat{j} & \hat{k} \\ \frac{\partial}{\partial x} & \frac{\partial}{\partial y} & \frac{\partial}{\partial z} \\ f_1 & f_2 & f_3 \end{vmatrix} = \left(\frac{\partial f_3}{\partial y} - \frac{\partial f_2}{\partial z}\right)\hat{i} + \left(\frac{\partial f_1}{\partial z} - \frac{\partial f_3}{\partial x}\right)\hat{j} + \left(\frac{\partial f_2}{\partial x} - \frac{\partial f_1}{\partial y}\right)\hat{k}$$

$$\therefore \text{ Curl curl } \vec{f} = \begin{vmatrix} \hat{i} & \hat{j} & \hat{k} \\ \frac{\partial}{\partial x} & \frac{\partial}{\partial y} & \frac{\partial}{\partial z} \\ \frac{\partial f_3}{\partial y} - \frac{\partial f_2}{\partial z} & \frac{\partial f_1}{\partial z} - \frac{\partial f_3}{\partial x} & \frac{\partial f_2}{\partial x} - \frac{\partial f_1}{\partial y} \end{vmatrix} = \left[\frac{\partial}{\partial y}\left(\frac{\partial f_2}{\partial x} - \frac{\partial f_1}{\partial y}\right) - \frac{\partial}{\partial z}\left(\frac{\partial f_1}{\partial z} - \frac{\partial f_3}{\partial x}\right)\right]\hat{i}$$

$$-\left[\frac{\partial}{\partial x}\left(\frac{\partial f_2}{\partial x} - \frac{\partial f_1}{\partial y}\right) - \frac{\partial}{\partial z}\left(\frac{\partial f_3}{\partial y} - \frac{\partial f_2}{\partial z}\right)\right]\hat{j} + \left[\frac{\partial}{\partial x}\left(\frac{\partial f_1}{\partial z} - \frac{\partial f_3}{\partial x}\right) - \frac{\partial}{\partial y}\left(\frac{\partial f_3}{\partial y} - \frac{\partial f_2}{\partial z}\right)\right]\hat{k}$$

$$= \left[\frac{\partial^2 f_2}{\partial y \partial x} - \frac{\partial^2 f_1}{\partial y^2} - \frac{\partial^2 f_1}{\partial z^2} + \frac{\partial^2 f_3}{\partial z \partial x}\right]\hat{i}$$

$$-\left[\frac{\partial^2 f_2}{\partial x^2} - \frac{\partial^2 f_1}{\partial x \partial y} - \frac{\partial^2 f_3}{\partial z \partial y} + \frac{\partial^2 f_2}{\partial z^2}\right]\hat{j} + \left[\frac{\partial^2 f_1}{\partial x \partial z} - \frac{\partial^2 f_3}{\partial x^2} - \frac{\partial^2 f_3}{\partial y^2} + \frac{\partial^2 f_2}{\partial y \partial z}\right]\hat{k}$$

$$= \left[\frac{\partial^2 f_1}{\partial x^2} + \frac{\partial^2 f_2}{\partial y \partial x} + \frac{\partial^2 f_3}{\partial z \partial x} - \left(\frac{\partial^2 f_1}{\partial x^2} + \frac{\partial^2 f_1}{\partial y^2} + \frac{\partial^2 f_1}{\partial z^2}\right)\right]\hat{i}$$

$$-\left[\frac{\partial^2 f_1}{\partial x \partial z} + \frac{\partial^2 f_2}{\partial y^2} + \frac{\partial^2 f_3}{\partial x \partial y} - \left(\frac{\partial^2 f_2}{\partial x^2} + \frac{\partial^2 f_2}{\partial y^2} + \frac{\partial^2 f_2}{\partial z^2}\right)\right]\hat{j}$$

$$+\left[\frac{\partial^2 f_1}{\partial x \partial z} - \frac{\partial^2 f_1}{\partial y^2} + \frac{\partial^2 f_2}{\partial y \partial z} - \left(\frac{\partial^2 f_3}{\partial x^2} + \frac{\partial^2 f_3}{\partial y^2} + \frac{\partial^2 f_3}{\partial z^2}\right)\right]\hat{k}$$

$$= \left[\frac{\partial}{\partial x}\left(\frac{\partial f_1}{\partial x} + \frac{\partial f_2}{\partial y} + \frac{\partial f_3}{\partial z}\right) - \left(\frac{\partial^2}{\partial x^2} + \frac{\partial^2}{\partial y^2} + \frac{\partial^2}{\partial z^2}\right)f_1\right]\hat{i}$$

$$+\left[\frac{\partial}{\partial y}\left(\frac{\partial f_2}{\partial x} + \frac{\partial f_2}{\partial y} + \frac{\partial f_2}{\partial z}\right) - \left(\frac{\partial^2}{\partial x^2} + \frac{\partial^2}{\partial y^2} + \frac{\partial^2}{\partial z^2}\right)f_2\right]\hat{j}$$

$$+\left[\frac{\partial}{\partial z}\left(\frac{\partial f_3}{\partial x} + \frac{\partial f_3}{\partial y} + \frac{\partial f_3}{\partial z}\right) - \left(\frac{\partial^2}{\partial x^2} + \frac{\partial^2}{\partial y^2} + \frac{\partial^2}{\partial z^2}\right)f_3\right]\hat{k}$$

$$= \left(\hat{i}\frac{\partial}{\partial x} + \hat{j}\frac{\partial}{\partial y} + \hat{k}\frac{\partial}{\partial z}\right)(\nabla f_1 + \nabla f_2 + \nabla f_3)$$

$$= \frac{\partial}{\partial x}(\text{div } \vec{f} - \nabla^2 f_1)\hat{i} + \frac{\partial}{\partial y}(\text{div } \vec{f} - \nabla^2 f_2)\hat{j} + \frac{\partial}{\partial z}(\text{div } \vec{f} - \nabla^2 f_3)\hat{k}$$

$$= \left(\hat{i}\frac{\partial}{\partial x} + \hat{j}\frac{\partial}{\partial y} + \hat{k}\frac{\partial}{\partial z}\right)\text{div } \vec{f} - \nabla^2[f_1\hat{i} + f_2\hat{j} + f_3\hat{k}] = \nabla(\nabla.\vec{f}) - \nabla^2\vec{f}$$

$$= \text{grad div } \vec{f} - \nabla^2\vec{f}$$

Proved.

Example 57. *For a solenoidol vector $\vec{F}$, show that curl curl curl curl $\vec{F} = \nabla^4 \vec{F}$.*

(M.D.U., Dec. 2009)

Solution. Since vector $\vec{F}$ is solenoidal, so div $\vec{F} = 0$ ... (1)

We know that curl curl $\vec{F}$ = grad div $(\vec{F} - \nabla^2 \vec{F})$ ... (2)

Using (1) in (2), grad div $\vec{F}$ = grad (0) = 0 ... (3)

On putting the value of grad div F in (2), we get

curl curl $\vec{F} = -\nabla^2 \vec{F}$ ... (4)

Now, curl curl curl curl $\vec{F}$ = curl curl $(-\nabla^2 \vec{F})$ [Using (4)]

$= -$ curl curl $(\nabla^2 \vec{F}) = -$ [grad div $(\nabla^2 \vec{F}) - \nabla^2 (\nabla^2 \vec{F})$] [Using (2)]

$= -$ grad $(\nabla \cdot \nabla^2 \vec{F}) + \nabla^2 (\nabla^2 \vec{F}) = -$ grad $(\nabla^2 \nabla \cdot \vec{F}) + \nabla^4 \vec{F}$ $[\nabla \cdot \vec{F} = 0]$

$= 0 + \nabla^4 \vec{F} = \nabla^4 \vec{F}$ [Using (1)] **Proved.**

EXERCISE 23.4

1. Find the divergence and curl of the vector field $V = (x^2 - y^2)\hat{i} + 2xy\hat{j} + (y^2 - xy)\hat{k}$.

Ans. Divergence = $4x$, Curl = $(2y - x)\hat{i} + y\hat{j} + 4y\hat{k}$

2. If a is constant vector and r is the radius vector, prove that

(*i*) $\nabla(\vec{a} \cdot \vec{r}) = \vec{a}$ (*ii*) div $(\vec{r} \times \vec{a}) = 0$ (*iii*) curl $(\vec{r} \times \vec{a}) = -2\vec{a}$

where $\vec{r} = x\hat{i} + y\hat{j} + z\hat{k}$ and $\vec{a} = a_1\hat{i} + a_2\hat{j} + a_3\hat{k}$.

3. Prove that:

$\nabla(A.B) = (A.\nabla)B + (B.\nabla)A + A \times (\nabla \times B) + B \times (\nabla \times A)$ *(R.G.P.V. Bhopal, June 2004)*

4. If $F = (x + y + 1)\hat{i} + \hat{j} - (x + y)\hat{k}$, show that F.curl $F = 0$.

(R.G.P.V. Bhopal, Feb. 2006, June 2004)

Prove that

5. $\vec{\nabla} \times (\phi \vec{F}) = (\vec{\nabla}\phi) \times \vec{F} + \phi(\vec{\nabla} \times \vec{F})$

6. $\nabla.(\vec{F} \times \vec{G}) = \vec{G}.(\nabla \times \vec{F}) - \vec{F}.(\nabla \times \vec{G})$

7. Prove that curl $(\vec{a} \times \vec{r}) = 2\vec{a}$

8. Evaluate curl grad, rm, where $\vec{r} = |\vec{r}| = |x\hat{i} + y\hat{j} + z\hat{k})$ **Ans.** 0

9. Find div $\vec{F}$ and curl F where F = grad $(x^3 + y^3 + z^3 - 3xyz)$. *(R.G.P.V. Bhopal Dec. 2003)*

Ans. div $\vec{F} = 6(x + y + z)$, curl $\vec{F} = 0$

10. Find out values of a, b, c for which $\vec{v} = (x + y + az)\hat{i} + (bx + 3y - z)\hat{j} + (3x + cy + z)\hat{k}$ is irrotational.

Ans. $a = 3$, $b = 1$, $c = -1$

11. Determine the constants a, b, c, so that $\vec{F} = (x + 2y + az)\hat{i} + (bx - 3y - z)\hat{j} + (4x + cy + 2z)\hat{k}$ is irrotational. Hence find the scalar potential ϕ such that $\vec{F}$ = grad ϕ.

(R.G.P.V. Bhopal, Feb. 2005) **Ans.** $a = 4$, $b = 2$, $c = 1$

Potential $\phi = \left(\dfrac{x^2}{2} - \dfrac{3y^2}{2} + z^2 + 2xy - yz + 4zx\right)$

Choose the correct alternative:

12. The magnitude of the vector drawn in a direction perpendicular to the surface $x^2 + 2y^2 + z^2 = 7$ at the point (1, –1, 2) is

(*i*) $\frac{2}{3}$ (*ii*) $\frac{3}{2}$ (*iii*) 3 (*iv*) 6 (*A.M.I.E.T.E., Summer 2000*) **Ans.** (*iv*)

13. If $u = x^2 - y^2 + z^2$ and $\bar{V} = x\hat{i} + y\hat{j} + z\hat{k}$ then $\nabla \cdot (u\bar{V})$ is equal to

(*i*) $5u$ (*ii*) $5|\vec{V}|$ (*iii*) $5(u - |\vec{V}|)$ (*iv*) $5(u - |\vec{V}|)$ (*A.M.I.E.T.E., June 2007*) **Ans.** (*i*)

14. A unit normal to $x^2 + y^2 + z^2 = 5$ at (0, 1, 2) is equal to

(*i*) $\frac{1}{\sqrt{5}}(\hat{i} + \hat{j} + \hat{k})$ (*ii*) $\frac{1}{\sqrt{5}}(\hat{i} + \hat{j} - \hat{k})$ (*iii*) $\frac{1}{\sqrt{5}}(\hat{j} + 2\hat{k})$ (*iv*) $\frac{1}{\sqrt{5}}(\hat{i} - \hat{j} + \hat{k})$

(*A.M.I.E.T.E., Dec. 2008*) **Ans.** (*iii*)

15. The directional derivative of $\phi = x\,y\,z$ at the point (1, 1, 1) in the direction $\hat{i}$ is:

(*i*) –1 (*ii*) $-\frac{1}{3}$ (*iii*) 1 (*iv*) $\frac{1}{3}$ **Ans.** (*iii*)

(*R.G.P.V. Bhopal, II Sem., June 2007*)

16. If $\vec{r} = x\hat{i} + y\hat{j} + z\hat{k}$ and $r = |\vec{r}|$ then $\nabla\phi(r)$ is:

(*i*) $\phi'(r)\,\vec{r}$ (*ii*) $\frac{\phi(r)\,\vec{r}}{r}$ (*iii*) $\frac{\phi'(r)\,\vec{r}}{r}$ (*iv*) None of these **Ans.** (*iii*)

(*R.G.P.V. Bhopal, II Semester, Feb. 2006*)

17. If $\vec{r} = x\hat{i} + y\hat{j} + z\hat{k}$ is position vector, then value of $\nabla(\log r)$ is (*U.P., I Sem, Dec 2008*)

(*i*) $\frac{\vec{r}}{r}$ (*ii*) $\frac{\vec{r}}{r^2}$ (*iii*) $-\frac{\vec{r}}{r^3}$ (*iv*) none of the above. **Ans.** (*ii*)

18. If $\vec{r} = x\hat{i} + y\hat{j} + z\hat{k}$ and $|\vec{r}| = r$, then div $\vec{r}$ is:

(*i*) 2 (*ii*) 3 (*iii*) –3 (*iv*) –2 **Ans.** (*ii*)

(*R.G.P.V. Bhopal, II Semester, Feb. 2006*)

19. If $\vec{V} = xy^2\hat{i} + 2yx^2z\hat{j} - 3yz^2\hat{k}$ then curl $\vec{V}$ at point (1, –1, 1) is

(*i*) $-(\hat{j} + 2\hat{k})$ (*ii*) $(\hat{i} + 3\hat{k})$ (*iii*) $-(\hat{i} + 2\hat{k})$ (*iv*) $(\hat{i} + 2\hat{j} + \hat{k})$

(*R.G.P.V. Bhopal, II Semester, Feb 2006*) **Ans.** (*iii*)

20. If $\vec{A}$ is such that $\nabla \times \vec{A} = 0$ then $\vec{A}$ is called

(*i*) Irrotational (*ii*) Solenoidal (*iii*) Rotational (*iv*) None of these

(*A.M.I.E.T.E., Dec. 2008*) **Ans.** (*i*)

21. If $\vec{F}$ is a conservative force field, then the value of curl $\vec{F}$ is

(*i*) 0 (*ii*) 1 (*iii*) $\overline{\nabla F}$ (*iv*) –1 (*A.M.I.E.T.E., June 2007*) **Ans.** (*i*)

22. If $\nabla^2\,[(1 - x)(1 - 2x)]$ is equal to

(*i*) 2 (*ii*) 3 (*iii*) 4 (*iv*) 6 (*A.M.I.E.T.E., Dec. 2009*) **Ans.** (*iii*)

23. If $\vec{R} = xi + yj + zk$ and $\vec{A}$ is a constant vector, curl $(\vec{A} \times \vec{R})$ is equal to

(*i*) $\vec{R}$ (*ii*) $2\vec{R}$ (*iii*) $\vec{A}$ (*iv*) $2\vec{A}$ (*A.M.I.E.T.E., Dec. 2009*) **Ans.** (*iv*)

24. If r is the distance of a point (x, y, z) from the origin, the value of the expression $\hat{j} \times \text{grad}\,\frac{1}{2}$ equals

(*i*) $(x^2 + y^2 + z^2)^{-\frac{3}{2}}(\hat{j}z - \hat{k}x)$ (*ii*) $(x^2 + y^2 + z^2)^{-\frac{3}{2}}(\hat{j}z - \hat{i}z)$

(*iii*) zero (*iv*) $(x^2 + y^2 + z^2)^{-\frac{3}{2}}(\hat{j}y - \hat{k}x)$

(*AMIETE, Dec. 2010*) **Ans.** (*ii*)

CHAPTER 24

VECTOR INTEGRATION

24.1 LINE INTEGRAL

Let $\vec{F}(x, y, z)$ be a vector function and a curve AB.

Line integral of a vector function $\vec{F}$ along the curve AB is defined as integral of the component of $\vec{F}$ along the tangent to the curve AB.

Component of $\vec{F}$ along a tangent PT at P

$$= \text{Dot product of } \vec{F} \text{ and unit vector along PT}$$

$$= \vec{F} \cdot \frac{\vec{dr}}{ds} \left(\frac{\vec{dr}}{ds} \text{ is a unit vector along tangent PT} \right)$$

Line integral $= \sum \vec{F} \cdot \dfrac{\vec{dr}}{ds}$ from A to B along the curve

$$\therefore \text{ Line integral} = \int_c \left(\vec{F} \cdot \frac{\vec{dr}}{ds} \right) ds = \int_c \vec{F} \cdot \vec{dr}$$

Note (1) Work. If $\overline{F}$ represents the variable force acting on a particle along arc AB, then the total work done $= \int_A^B \vec{F} \cdot \vec{dr}$

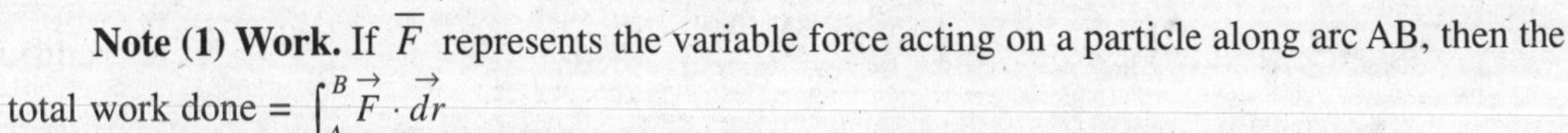

(2) Circulation. If $\vec{v}$ represents the velocity of a liquid then $\oint_c \vec{V} \cdot \vec{dr}$ is called the circulation of V round the closed curve c.

If the circulation of V round every closed curve is zero then V is said to be irrotational there.

(3) When the path of integration is a closed curve then notation of integration is $\oint$ in place of $\int$.

Example 1. *If a force $\vec{F} = 2x^2 y\hat{i} + 3xy\hat{j}$ displaces a particle in the xy-plane from (0, 0) to (1, 4) along a curve $y = 4x^2$. Find the work done.*

Solution. Work done $= \int_c \vec{F} . \vec{dr}$ $\qquad \left[\begin{array}{l} \vec{r} = x\hat{i} + y\hat{j} \\ \vec{dr} = dx\hat{i} + dy\,\hat{j} \end{array} \right]$

$$= \int_c (2x^2 y\hat{i} + 3xy\,\hat{j}) . (dx\hat{i} + dy\,\hat{j})$$

$$= \int_c (2x^2 y\, dx + 3xy\, dy)$$

Putting the values of y and dy, we get $\left(\begin{matrix} y = 4x^2 \\ dy = 8x\,dx \end{matrix}\right)$

$$= \int_0^1 [2x^2 (4x^2)\,dx + 3x(4x^2)\,8x\,dx]$$

$$= 104 \int_0^1 x^4\,dx = 104\left(\frac{x^5}{5}\right)_0^1 = \frac{104}{5} \qquad \textbf{Ans.}$$

Example 2. *Evaluate* $\int_C \vec{F}\cdot\vec{dr}$ *where* $\vec{F} = x^2\hat{i} + xy\hat{j}$ *and C is the boundary of the square in the plane z = 0 and bounded by the lines x = 0, y = 0, x = a and y = a.*

(*Nagpur University, Summer 2001*)

Solution. $\int_C \vec{F}\cdot\vec{d}\,r = \int_{OA} \vec{F}\cdot\vec{dr} + \int_{AB} \vec{F}\cdot\vec{dr} + \int_{BC} \vec{F}\cdot\vec{dr} + \int_{CO} \vec{F}\cdot\vec{dr}$

Here $\vec{r} = x\hat{i} + y\hat{j}, \quad \vec{d}\,r = dx\hat{i} + dy\hat{j}, \quad \vec{F} = x^2\hat{i} + xy\,\hat{j}$

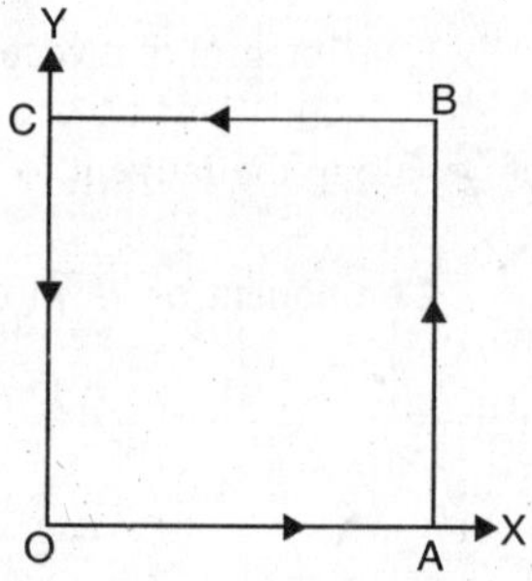

$$\vec{F}\cdot\vec{dr} = x^2dx + xydy \qquad ...(1)$$

On OA, $y = 0$ $\qquad \therefore \vec{F}\cdot\vec{dr} = x^2dx$

$$\int_{OA} \vec{F}\cdot\vec{dr} = \int_0^a x^2dx = \left[\frac{x^3}{3}\right]_0^a = \frac{a^3}{3} \qquad ...(2)$$

On AB, $x = a$ $\qquad \therefore dx = 0$
(1) becomes

$$\therefore \vec{F}\cdot\vec{dr} = aydy$$

$$\int_{Ab} \vec{F}\cdot\vec{dr} = \int_0^a aydy = a\left[\frac{y^2}{2}\right]_0^a = \frac{a^3}{2} \qquad ...(3)$$

On BC, $y = a$ $\qquad \therefore dy = 0$

$\Rightarrow$ (1) becomes $\qquad \vec{F}\cdot\vec{dr} = x^2dx$

$$\int_{BC} \vec{F}\cdot\vec{dr} = \int_a^0 x^2dx = \left[\frac{x^3}{3}\right]_a^0 = \frac{-a^3}{3} \qquad ...(4)$$

On CO, $x = 0$, $\qquad \therefore \vec{F}\cdot\vec{dr} = 0$
(1) becomes

$$\int_{CO} \vec{F}\cdot\vec{dr} = 0 \qquad ...(5)$$

On adding (2), (3), (4) and (5), we get $\int_C \vec{F}\cdot\vec{dr} = \frac{a^3}{3} + \frac{a^3}{2} - \frac{a^3}{3} + 0 = \frac{a^3}{2}$ **Ans.**

Example 3. *A vector field is given by*

$\vec{F} = (2y+3)\,\hat{i} + xz\hat{j} + (yz - x)\,\hat{k}$. *Evaluate* $\int_C \vec{F}\cdot\vec{dr}$ *along the path c is x = 2t, y = t, z = t*^{3} *from t = 0 to t = 1.*

(*Nagpur University, Winter 2003*)

Solution. $\int_C \vec{F}\cdot\vec{dr} = \int_C (2y+3)\,dx + (xz)\,dy + (yz - x)\,dz$

$$\left[\begin{matrix} \text{Since } x = 2t & y = t & z = t^3 \\ \therefore \frac{dx}{dt} = 2 & \frac{dy}{dt} = 1 & \frac{dz}{dt} = 3t^2 \end{matrix}\right]$$

$$= \int_0^1 (2t+3)(2\,dt) + (2t)(t^3)\,dt + (t^4 - 2t)(3t^2\,dt) = \int_0^1 (4t + 6 + 2t^4 + 3t^6 - 6t^3)\,dt$$

$$= \left[4\frac{t^2}{2} + 6t + \frac{2}{5}t^5 + \frac{3}{7}t^7 - \frac{6}{4}t^4 \right]_0^1 = \left[2t^2 + 6t + \frac{2}{5}t^5 + \frac{3}{7}t^7 - \frac{3}{2}t^4 \right]_0^1$$

$$= 2 + 6 + \frac{2}{5} + \frac{3}{7} - \frac{3}{2} = 7.32857.$$ **Ans.**

Example 4. *If* $\bar{F} = 2y\hat{i} - z\hat{j} + x\hat{k}$, *evaluate* $\int_C \vec{F} \times \vec{dr}$ *along the curve*

$x = \cos t,\ y = \sin t,\ z = 2\cos t$ *from* $t = 0$ *to* $t = \frac{\pi}{2}$. *(Nagpur University, winter 2002)*

Solution. We have, $\vec{r} = x\hat{i} + y\hat{j} + z\hat{k}$

$$\vec{dr} = dx\,\hat{i} + dy\,\hat{j} + dz\,\hat{k}$$

$$\vec{F} \times \vec{dr} = \begin{vmatrix} \hat{i} & \hat{j} & \hat{k} \\ 2y & -z & x \\ dx & dy & dz \end{vmatrix}$$

$$= (-zdz - xdy)\,\hat{i} - (2y\,dz - xdx)\,\hat{j} + (2ydy + zdx)\,\hat{k}$$

$$= [-2\cos t\,(-2\sin t)\,dt - \cos t\,(\cos t)\,dt]\,\hat{i}$$

$$-[\,2\sin t\,(-2\sin t)\,dt - \cos t\,(-\sin t)\,dt\,]\,\hat{j}$$

$$+\ [\,2\sin t\,(\cos t)\,dt + 2\cos t\,(-\sin t)\,dt\,]\,\hat{k}$$

$$= [\,(4\cos t\sin t - \cos^2 t)\,\hat{i} + (4\sin^2 t - \cos t\sin t)\,\hat{j}\,]\,dt$$

$$\therefore \int_C \vec{F} \times \vec{dr} = \int_0^{\frac{\pi}{2}} [\,(4\cos t\sin t - \cos^2 t)\,\hat{i} + (4\sin^2 t - \cos t\sin t)\,\hat{j}\,]\,dt$$

$$= \int_0^{\frac{\pi}{2}} \left[\left\{ 2\sin 2t - \frac{\cos 2t + 1}{2} \right\} \hat{i} \right] dt + \int_0^{\frac{\pi}{2}} \left[\left\{ 2(1 - \cos 2t) - \frac{1}{2}\sin 2t \right\} \hat{j} \right] dt$$

$$= \left[-\cos 2t - \frac{1}{4}\sin 2t - \frac{1}{2}t \right]_0^{\frac{\pi}{2}} \hat{i} + \left[2t - \sin 2t + \frac{1}{4}\cos 2t \right]_0^{\frac{\pi}{2}} \hat{j}$$

$$= \left[-\cos\pi - \frac{1}{4}\sin\pi - \frac{1}{2}\left(\frac{\pi}{2}\right) + \cos 0 + \frac{1}{4}\sin 0 + \frac{1}{2}(0) \right] \hat{i} +$$

$$\left[\pi - \sin\pi + \frac{1}{4}\cos\pi - 0 + \sin 0 - \frac{1}{4}\cos 0 \right] \hat{j}$$

$$= \left[1 - 0 - \frac{\pi}{4} + 1 + 0 \right] \hat{i} + \left[\pi - 0 - \frac{1}{4} + 0 - \frac{1}{4} \right] \hat{j} = \left(2 - \frac{\pi}{4} \right) \hat{i} + \left(\pi - \frac{1}{2} \right) \hat{j}$$ **Ans.**

Example 5. *The acceleration of a particle at time t is given by*

$$\vec{a} = 18\cos 3t\,\hat{i} - 8\sin 2t\,\hat{j} + 6t\,\hat{k}.$$

If the velocity $\vec{v}$ *and displacement* $\vec{r}$ *be zero at* $t = 0$, *find* $\vec{v}$ *and* $\vec{r}$ *at any point t.*

Solution. Here, $\vec{a} = \dfrac{d^2\vec{r}}{dt^2} = 18\cos 3t\,\hat{i} - 8\sin 2t\,\hat{j} + 6t\,\hat{k}.$

On integrating, we have

$$\vec{v} = \frac{d\vec{r}}{dt} = \hat{i}\int 18\cos 3t\, dt + \hat{j}\int -8\sin 2t\, dt + \hat{k}\int 6t\, dt$$

$$\Rightarrow \quad \vec{v} = 6\sin 3t\,\hat{i} + 4\cos 2t\,\hat{j} + 3t^2\,\hat{k} + \vec{c} \qquad ...(1)$$

At $t = 0,\ \vec{v} = \vec{0}$

Putting $t = 0$ and $\vec{v} = 0$ in (1), we get

$$\vec{0} = 4\hat{j} + \vec{c} \ \Rightarrow\ \vec{c} = -4\hat{j}$$

$$\therefore \quad \vec{v} = \frac{d\vec{r}}{dt} = 6\sin 3t\,\hat{i} + 4(\cos 2t - 1)\,\hat{j} + 3t^2\hat{k}$$

Again integrating, we have

$$\vec{r} = \hat{i}\int 6\sin 3t\, dt + \hat{j}\int 4(\cos 2t - 1)\, dt + \hat{k}\int 3t^2\, dt$$

$$\Rightarrow \quad \vec{r} = -2\cos 3t\,\hat{i} + (2\sin 2t - 4t)\hat{j} + t^3\,\hat{k} + \vec{c_1} \qquad ...(2)$$

At, $t = 0,\ \vec{r} = 0$

Putting $t = 0$ and $\vec{r} = 0$ in (2), we get

$$\therefore \quad \vec{0} = -2\hat{i} + \vec{C_1} \ \Rightarrow\ \vec{C_1} = 2\hat{i}$$

Hence, $\vec{r} = 2(1 - \cos 3t)\,\hat{i} + 2(\sin 2t - 2t)\hat{j} + t^3\hat{k}$ **Ans.**

Example 6. *If* $\vec{A} = (3x^2 + 6y)\,\hat{i} - 14yz\hat{j} + 20xz^2\,\hat{k}$, *evaluate the line integral* $\oint \vec{A}.d\vec{r}$ *from* (0, 0, 0) *to* (1, 1, 1) *along the curve C.*

$x = t,\ y = t^2,\ z = t^3.$ *(Uttarakhand, I Semester, Dec. 2006)*

Solution. We have,

$$\int_C \vec{A}.d\vec{r} = \int_C [(3x^2 + 6y)\,\hat{i} - 14yz\hat{j} + 20xz^2\hat{k}].[\hat{i}\,dx + \hat{j}\,dy + \hat{k}\,dz]$$

$$= \int_C [(3x^2 + 6y)\,dx - 14yz\,dy + 20xz^2 dz]$$

If $x = t,\ y = t^2,\ z = t^3$, then points (0, 0, 0) and (1, 1, 1) correspond to $t = 0$ and $t = 1$ respectively.

$$\text{Now, } \int_C \vec{A}.d\vec{r} = \int_{t=0}^{t=1} [(3t^2 + 6t^2)\,d(t) - 14t^2\,t^3\,d(t^2) + 20t\,(t^3)^2\,d(t^3)]$$

$$= \int_{t=0}^{t=1} [9t^2\,dt - 14t^5 . 2t\,dt + 20\,t^7 . 3\,t^2\,dt] = \int_0^1 (9t^2 - 28t^6 + 60t^9)\,dt$$

$$= \left[9\left(\frac{t^3}{3}\right) - 28\left(\frac{t^7}{7}\right) + 60\left(\frac{t^{10}}{10}\right)\right]_0^1 = 3 - 4 + 6 = 5$$ **Ans.**

Example 7. *Compute* $\int_c \vec{F}.\vec{dr}$, *where* $\vec{F} = \dfrac{\hat{i}y - \hat{j}x}{x^2 + y^2}$ *and c is the circle* $x^2 + y^2 = 1$ *traversed counter clockwise.*

Solution. $\vec{r} = \hat{i}\,x + \hat{j}\,y + \hat{k}\,z,\ d\bar{r} = \hat{i}\,dx + \hat{j}\,dy + \hat{k}\,dz$

$$\int_c \vec{F}.\vec{dr} = \int_c \frac{\hat{i}y - \hat{j}x}{x^2 + y^2} \cdot (\hat{i}dx + \hat{j}dy + \hat{k}dz)$$

$$= \int_c \frac{ydx - xdy}{x^2 + y^2} = \int_c (ydx - xdy) \qquad ...(1)\ [\because x^2 + y^2 = 1]$$

Parametric equation of the circle are $x = \cos\theta$, $y = \sin\theta$.
Putting $x = \cos\theta$, $y = \sin\theta$, $dx = -\sin\theta\, d\theta$, $dy = \cos\theta\, d\theta$ in (1), we get

$$\int_C \vec{F}\, d\vec{r} = \int_0^{2\pi} \sin\theta(-\sin\theta\, d\theta) - \cos\theta(\cos\theta\, d\theta)$$

$$= -\int_0^{2\pi}(\sin^2\theta + \cos^2\theta)\, d\theta = -\int_0^{2\pi} d\theta = -(\theta)_0^{2\pi} = -2\pi \qquad \textbf{Ans.}$$

Example 8. *Show that the vector field* $\vec{F} = 2x(y^2+z^3)\hat{i} + 2x^2y\hat{j} + 3x^2z^2\hat{k}$ *is conservative. Find its scalar potential and the work done in moving a particle from* (–1, 2, 1) to (2, 3, 4).
(A.M.I.E.T.E. June 2010, 2009)

Solution. Here, we have

$$\vec{F} = 2x(y^2+z^3)\,\hat{i} + 2x^2y\,\hat{j} + 3x^2z^2\hat{k}$$

$$\text{Curl } \vec{F} = \nabla\times\vec{F}$$

$$= \begin{vmatrix} \hat{i} & \hat{j} & \hat{k} \\ \dfrac{\partial}{\partial x} & \dfrac{\partial}{\partial y} & \dfrac{\partial}{\partial z} \\ 2x(y^2+z^3) & 2x^2y & 3x^2z^2 \end{vmatrix} = (0-0)i - (6xz^2 - 6xz^2)\hat{j} + (4xy - 4xy)\hat{k} = 0$$

Hence, vector field $\vec{F}$ is irrotational.
To find the scalar potential function ϕ

$$\vec{F} = \vec{\nabla}\phi$$

$$d\phi = \frac{\partial\phi}{\partial x}dx + \frac{\partial\phi}{\partial y}dy + \frac{\partial\phi}{\partial z}dz = \left(\hat{i}\frac{\partial\phi}{\partial x} + \hat{j}\frac{\partial\phi}{\partial y} + \hat{k}\frac{\partial\phi}{\partial z}\right).\left(\hat{i}dx + \hat{j}dy + \hat{k}dz\right)$$

$$= \left(\hat{i}\frac{\partial}{\partial x} + \hat{j}\frac{\partial}{\partial y} + \hat{k}\frac{\partial}{\partial z}\right)\phi.\left(d\vec{r}\right) = \nabla\phi.d\vec{r} = \vec{F}.d\vec{r}$$

$$= \left[2x(y^2+z^3)\hat{i} + 2x^2y\hat{j} + 3x^2z^2\hat{k}\right](\hat{i}dx + \hat{j}dy + \hat{k}dz)$$

$$= 2x(y^2+z^3)\,dx + 2x^2y\,dy + 3x^2z^2dz$$

$$\phi = \int\left[2x(y^2+z^3)dx + 2x^2ydy + 3x^2z^2dz\right] + C$$

$$\int(2xy^2dx + 2x^2ydy) + (2xz^3dx + 3x^2z^2dz) + C = x^2y^2 + x^2z^3 + C$$

Hence, the scalar potential is $x^2y^2 + x^2z^3 + C$
Now, for conservative field

$$\text{Work done} = \int_{(-1,2,1)}^{(2,3,4)} \vec{F}.d\vec{r} = \int_{(-1,2,1)}^{(2,3,4)} d\phi = [\phi]_{(-1,2,1)}^{(2,3,4)} = \left[x^2y^2 + x^2z^3 + c\right]_{(-1,2,1)}^{(2,3,4)}$$

$$= (36 + 256) - (2 - 1) = 291 \qquad \textbf{Ans.}$$

Example 9. *A vector field is given by* $\vec{F} = (\sin y)\,\hat{i} + x\,(1+\cos y)\,\hat{j}$. *Evaluate the line integral over a circular path* $x^2 + y^2 = a^2$, $z = 0$. *(Nagpur University, Winter 2001)*
Solution. We have,

$$\text{Work done} = \int_C \vec{F}.d\vec{r}$$

$$= \int_C [(\sin y)\hat{i} + x(1+\cos y)\hat{j}].[dx\hat{i} + dy\hat{j}] \quad (\because\ z = 0 \text{ hence } dz = 0)$$

$$\Rightarrow \quad \int_C \vec{F}.\vec{dr} = \int_C \sin y\, dx + x(1+\cos y)\, dy = \int_C (\sin y\, dx + x \cos y\, dy + x\, dy)$$

$$= \int_C d(x \sin y) + \int_C x\, dy$$

(where d is differential operator).

The parametric equations of given path

$x^2 + y^2 = a^2$ are $x = a\cos\theta$, $y = a\sin\theta$,

Where θ varies form 0 to 2π

$$\therefore \quad \int_C \vec{F}.\vec{dr} = \int_0^{2\pi} d[a\cos\theta \sin(a\sin\theta)] + \int_0^{2\pi} a\cos\theta \,.\, a\cos\theta\, d\theta$$

$$= \int_0^{2\pi} d[a\cos\theta\sin(a\sin\theta)] + \int_0^{2\pi} a^2\cos^2\theta\,.\,d\theta$$

$$= [a\cos\theta\sin(a\sin\theta)]_0^{2\pi} + \int_0^{2\pi} a^2\cos^2\theta\, d\theta$$

$$= 0 + a^2\int_0^{2\pi}\left(\frac{1+\cos 2\theta}{2}\right)d\theta = \frac{a^2}{2}\left[\theta + \frac{\sin 2\theta}{2}\right]_0^{2\pi}$$

$$= \frac{a^2}{2}.2\pi = \pi a^2$$

Ans.

Example 10. *Determine whether the line integral* $\int (2xyz^2)\,dx + (x^2z^2 + z\cos yz)\,dy + (2x^2yz + y\cos yz)\,dz$ *is independent of the path of integration ? If so, then evaluate it from (1, 0, 1) to* $\left(0, \frac{\pi}{2}, 1\right)$.

Solution. $\int_c (2xyz^2)\,dx + (x^2z^2 + z\cos yz)\,dy + (2x^2yz + y\cos yz)\,dz$

$$= \int_c [(2xyz^2\hat{i}) + (x^2z^2 + z\cos yz)\hat{j} + (2x^2yz + y\cos yz)\hat{k}].(\hat{i}dx + \hat{j}dy + \hat{k}dz)$$

$$= \int_c \vec{F}\cdot\vec{dr}$$

This integral is independent of path of integration if

$$\vec{F} = \nabla\phi \Rightarrow \nabla\times\vec{F} = 0$$

$$\nabla\times\overline{F} = \begin{vmatrix} \hat{i} & \hat{j} & \hat{k} \\ \frac{\partial}{\partial x} & \frac{\partial}{\partial y} & \frac{\partial}{\partial z} \\ 2xyz^2 & x^2z^2 + z\cos yz & 2x^2yz + y\cos yz \end{vmatrix}$$

$= (2x^2z + \cos yz - yz\sin yz - 2x^2z - \cos yz + yz\sin yz)\,\hat{i} - (4xyz - 4xyz)\hat{j} + (2xz^2 - 2xz^2)\,\hat{k}$
$= 0$

Hence, the line integral is independent of path.

$$d\phi = \frac{\partial\phi}{\partial x}dx + \frac{\partial\phi}{\partial y}dy + \frac{\partial\phi}{\partial z}dz \qquad \text{(Total differentiation)}$$

$$= \left(\hat{i}\frac{\partial\phi}{\partial x} + \hat{j}\frac{\partial\phi}{\partial y} + \hat{k}\frac{\partial\phi}{\partial z}\right)\cdot(\hat{i}dx + \hat{j}dy + \hat{k}dz) = \nabla\phi\cdot dr = \vec{F}\cdot\vec{d}r$$

$= [(2xyz^2)\,\hat{i} + (x^2z^2 + z\cos yz)\,\hat{j} + (2x^2yz + y\cos yz)\,\hat{k}].(\hat{i}dx + \hat{j}dy + \hat{k}dz)$
$= 2xyz^2\,dx + (x^2z^2 + z\cos yz)\,dy + (2x^2yz + y\cos yz)\,dz$
$= [(2x\,dx)\,yz^2 + x^2\,(dy)\,z^2 + x^2y\,(2z\,dz)] + [(\cos yz\,dy)\,z + (\cos yz\,dz)\,y]$

$$= d(x^2yz^2) + d(\sin yz)$$

$$\phi = \int d(x^2yz^2) + \int d(\sin yz) = x^2yz^2 + \sin yz$$

$$[\phi]_A^B = \phi(B) - \phi(A)$$

$$= [x^2yz^2 + \sin yz]_{(0, \frac{\pi}{2}, 1)} - [x^2yz^2 + \sin yz]_{(1, 0, 1)} = \left[0 + \sin\left(\frac{\pi}{2} \times 1\right)\right] - [0 + 0]$$

$$= 1$$ **Ans.**

EXERCISE 24.1

1. Find the work done by a force $y\hat{i} + x\hat{j}$ which displaces a particle from origin to a point $(\hat{i} + \hat{j})$. **Ans.** 1
2. Find the work done when a force $\overline{F} = (x^2 - y^2 + x)\,\hat{i} - (2xy + y)\,\hat{j}$ moves a particle from origin to (1, 1) along a parabola $y^2 = x$. **Ans.** $\frac{2}{3}$
3. Show that $\vec{V} = (2xy + z^3)\,\hat{i} + x^2\,\hat{j} + 3xz^2\,\hat{k}$ is a conservative field. Find its scalar potential ϕ such that $\vec{V}$ = grad ϕ. Find the work done by the force $\vec{V}$ in moving a particle from (1, – 2, 1) to (3, 1, 4). **Ans.** $x^2y + xz^3$, 202
4. Show that the line integral $\int_c (2xy + 3)\,dx + (x^2 - 4z)\,dy - 4y\,dz$ where c is any path joining (0, 0, 0) to (1, – 1, 3) does not depend on the path c and evaluate the line integral. **Ans.** 14
5. Find the work done in moving a particle once round the ellipse $\frac{x^2}{25} + \frac{y^2}{16} = 1$, $z = 0$, under the field of force given by $F = (2x - y + z)\,\hat{i} + (x + y - z^2)\,\hat{j} + (3x - 2y + 4z)\,\hat{k}$. Is the field of force conservative? (*A.M.I.E.T.E., Winter 2000*) **Ans.** 40π
6. If $\vec{\nabla}\phi = (y^2 - 2xyz^3)\,\hat{i} + (3 + 2xy - x^2z^3)\,\hat{j} + (z^3 - 3x^2yz^2)\,\hat{k}$, find ϕ. **Ans.** $3y + \frac{z^4}{4} + xy^2 - x^2yz^3$
7. $\int_C \vec{R}\,.\,d\vec{R}$ is independent of the path joining any two point if it is. (*A.M.I.E.T.E., June 2010*)
(*i*) irrotational field (*ii*) solenoidal field (*iii*) rotational field (*iv*) vector field. **Ans.** (*i*)

24.2 SURFACE INTEGRAL

A surface $r = f(u, v)$ is called smooth if $f\,(u, v)$ posses continous first order partial derivative.

Let $\vec{F}$ be a vector function and S be the given surface.

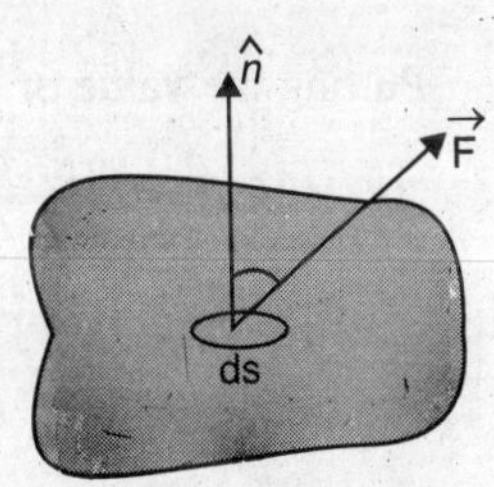

Surface integral of a vector function $\vec{F}$ over the surface S is defined as the integral of the components of $\vec{F}$ along the normal to the surface.

Component of $\vec{F}$ along the normal

$= \vec{F}\,.\,\hat{n}$, where n is the unit normal vector to an element ds and

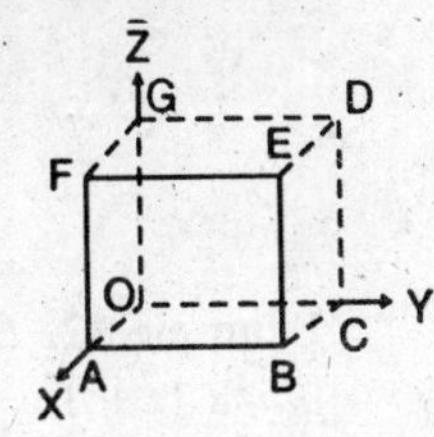

$$\hat{n} = \frac{\text{grad } f}{|\text{grad } f|} \qquad ds = \frac{dx\,dy}{(\hat{n}\cdot\hat{k})}$$

Surface integral of F over S

$$= \Sigma\,\vec{F}\cdot\hat{n} \qquad = \iint_S (\vec{F}\cdot\hat{n})\,ds$$

Note. (1) Flux = $\iint_S (\vec{F}\cdot\hat{n})\,ds$ where, $\overline{F}$ represents the velocity of a liquid.

f $\iint_S (\vec{F}\cdot\hat{n})\,ds = 0$, then $\vec{F}$ is said to be a *solenoidal* vector point function.

Example 11. *Evaluate* $\iint_S \vec{A}.\hat{n}\, ds$ *where* $\vec{A}=(x+y^2)\,\hat{i}-2x\hat{j}+2yz\hat{k}$ *and S is the surface of the plane* $2x+y+2z=6$ *in the first octant.* (*Nagpur University, Summer 2000*)

Solution. A vector normal to the surface "S" is given by

$$\nabla(2x+y+2z)=\left(\hat{i}\frac{\partial}{\partial x}+\hat{j}\frac{\partial}{\partial y}+\hat{k}\frac{\partial}{\partial z}\right)(2x+y+2z)=2\hat{i}+\hat{j}+2\hat{k}$$

And $\hat{n}$ = a unit vector normal to surface S

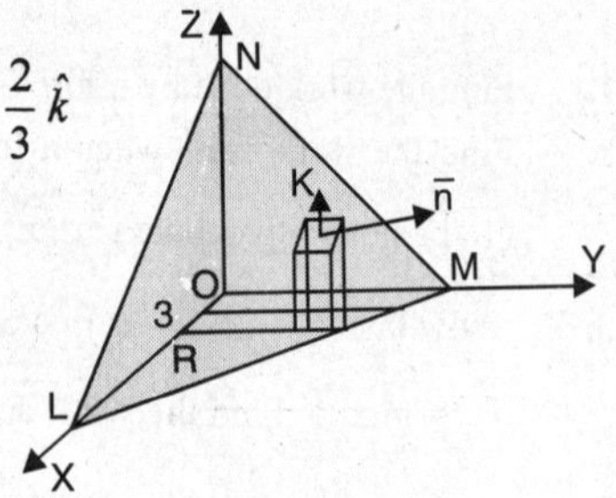

$$=\frac{2\hat{i}+\hat{j}+2\hat{k}}{\sqrt{4+1+4}}=\frac{2}{3}\hat{i}+\frac{1}{3}\hat{j}+\frac{2}{3}\hat{k}$$

$$\hat{k}.\hat{n}=\hat{k}.\left(\frac{2}{3}\hat{i}+\frac{1}{3}\hat{j}+\frac{2}{3}\hat{k}\right)=\frac{2}{3}$$

$$\therefore \qquad \iint_S \bar{A}.\hat{n}\,ds=\iint_R \bar{A}.\hat{n}\frac{dx\,dy}{\hat{k}.\bar{n}}$$

Where R is the projection of S.

Now, $\vec{A}.\hat{n}=[(x+y^2)\,\hat{i}-2x\hat{j}+2yz\hat{k}].\left(\frac{2}{3}\hat{i}+\frac{1}{3}\hat{j}+\frac{2}{3}\hat{k}\right)$

$$=\frac{2}{3}(x+y^2)-\frac{2}{3}x+\frac{4}{3}yz=\frac{2}{3}y^2+\frac{4}{3}yz \qquad ...(1)$$

Putting the value of z in (1), we get

$$\vec{A}.\hat{n}=\frac{2}{3}y^2+\frac{4}{3}y\left(\frac{6-2x-y}{2}\right) \qquad \left(\because \text{ on the plane } 2x+y+2z=6,\ z=\frac{(6-2x-y)}{2}\right)$$

$$\vec{A}.\hat{n}=\frac{2}{3}y(y+6-2x-y)=\frac{4}{3}y(3-x) \qquad ...(2)$$

Hence, $$\iint_S \vec{A}.\hat{n}\,ds=\iint_R \bar{A}.\bar{n}\frac{dx\,dy}{|\hat{k}.\bar{n}|} \qquad ...(3)$$

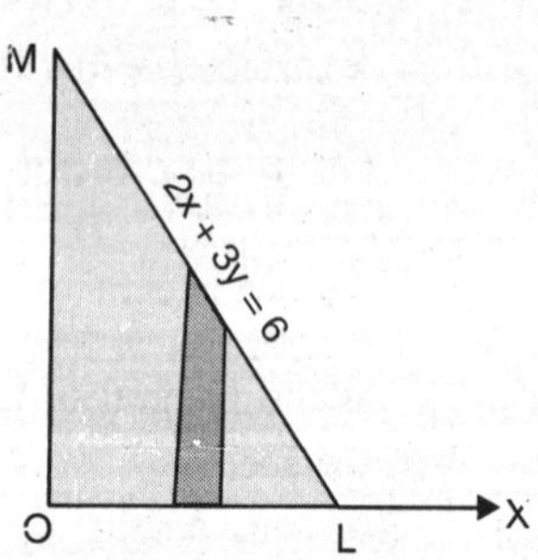

Putting the value of $\vec{A}.\hat{n}$ from (2) in (3), we get

$$\iint_S \vec{A}.\hat{n}\,ds=\iint_R \frac{4}{3}y(3-x).\frac{3}{2}\,dx\,dy=\int_0^3\int_0^{6-2x}2y(3-x)\,dy\,dx$$

$$=\int_0^3 2(3-x)\left[\frac{y^2}{2}\right]_0^{6-2x}dx$$

$$=\int_0^3(3-x)(6-2x)^2\,dx=4\int_0^3(3-x)^3\,dx$$

$$=4.\left[\frac{(3-x)^4}{4(-1)}\right]_0^3=-(0-81)=81 \qquad \textbf{Ans.}$$

Example 12. *Evaluate* $\iint_S \vec{A}.\hat{n}\,dS$, *where* $\vec{A}=18z\hat{i}-12\hat{j}+3y\hat{k}$ *and S is the part of the plane* $2x+3y+6z=12$ *included in the first octant.* (*Uttarakhand, I semester, Dec. 2006*)

Solution. Here, $\vec{A}=18z\hat{i}-12\hat{j}+3y\hat{k}$

Given surface $f(x,y,z)=2x+3y+6z-12$

$$\text{Normal vector}=\nabla f=\left(\hat{i}\frac{\partial}{\partial x}+\hat{j}\frac{\partial}{\partial y}+\hat{k}\frac{\partial}{\partial z}\right)(2x+3y+6z-12)=2\hat{i}+3\hat{j}+6\hat{k}$$

$\hat{n}$ = unit normal vector at any point (x, y, z) of $2x + 3y + 6z = 12$

$$= \frac{2\hat{i} + 3\hat{j} + 6\hat{k}}{\sqrt{4+9+36}} = \frac{1}{7}(2\hat{i} + 3\hat{j} + 6\hat{k})$$

$$dS = \frac{dx\,dy}{\hat{n}.\hat{k}} = \frac{dx\,dy}{\frac{1}{7}(2\hat{i} + 3\hat{j} + 6\hat{k}).\hat{k}} = \frac{dx dy}{\frac{6}{7}} = \frac{7}{6}\,dx\,dy$$

Now, $$\iint \vec{A}.\hat{n}\,dS = \iint (18z\hat{i} - 12\hat{j} + 3y\hat{k}).\frac{1}{7}(2\hat{i} + 3\hat{j} + 6\hat{k})\frac{7}{6}\,dx\,dy$$

$$= \iint (36z - 36 + 18y)\frac{dx\,dy}{6} = \iint (6z - 6 + 3y)\,dx\,dy$$

Putting the value of $6z = 12 - 2x - 3y$, we get

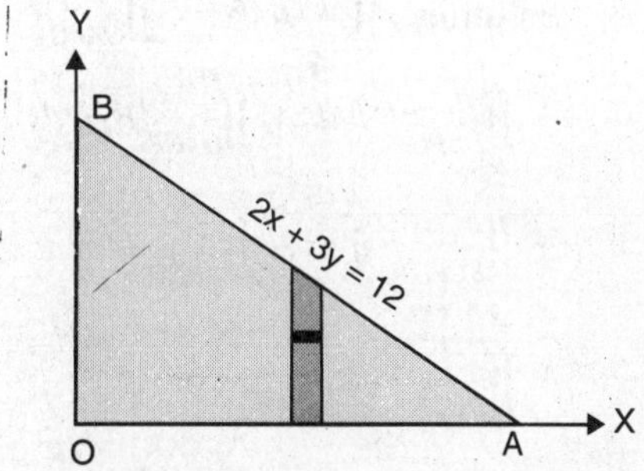

$$= \int_0^6 \int_0^{\frac{1}{3}(12-2x)} (12 - 2x - 3y - 6 + 3y)\,dx\,dy$$

$$= \int_0^6 \int_0^{\frac{1}{3}(12-2x)} (6 - 2x)\,dx\,dy$$

$$= \int_0^6 (6 - 2x)\,dx \int_0^{\frac{1}{3}(12-2x)} dy$$

$$= \int_0^6 (6 - 2x)\,dx\,(y)_0^{\frac{1}{3}(12-2x)}$$

$$= \int_0^6 (6 - 2x)\frac{1}{3}(12 - 2x)\,dx = \frac{1}{3}\int_0^6 (4x^2 - 36x + 72)\,dx$$

$$= \frac{1}{3}\left[\frac{4x^3}{3} - 18x^2 + 72x\right]_0^6 = \frac{1}{3}[4\times 36\times 2 - 18\times 36 + 72\times 6] = \frac{72}{3}[4 - 9 + 6] = 24 \textbf{ Ans.}$$

Example 13. *Evaluate* $\iint_S (yz\hat{i} + zx\hat{j} + xy\hat{k})\cdot\vec{ds}$ *where S is the surface of the sphere* $x^2 + y^2 + z^2 = a^2$ *in the first octant.* (U.P., I Semester, Dec. 2004)

Solution. Here, $\phi = x^2 + y^2 + z^2 - a^2$

Vector normal to the surface $= \nabla\phi = \hat{i}\frac{\partial\phi}{\partial x} + \hat{j}\frac{\partial\phi}{\partial y} + \hat{k}\frac{\partial\phi}{\partial z}$

$$= \left(\hat{i}\frac{\partial}{\partial x} + \hat{j}\frac{\partial}{\partial y} + \hat{k}\frac{\partial}{\partial z}\right)(x^2 + y^2 + z^2 - a^2) = 2x\hat{i} + 2y\hat{j} + 2z\hat{k}$$

$$\hat{n} = \frac{\nabla\phi}{|\nabla\phi|} = \frac{2x\hat{i} + 2y\hat{j} + 2z\hat{k}}{\sqrt{4x^2 + 4y^2 + 4z^2}} = \frac{x\hat{i} + y\hat{j} + z\hat{k}}{\sqrt{x^2 + y^2 + z^2}}$$

$$= \frac{x\hat{i} + y\hat{j} + z\hat{k}}{a} \qquad [\because x^2 + y^2 + z^2 = a^2]$$

Here. $$\vec{F} = yz\hat{i} + zx\hat{j} + xy\hat{k}$$

$$\vec{F}\cdot\hat{n} = (yz\hat{i} + zx\hat{j} + xy\hat{k})\cdot\left(\frac{x\hat{i} + y\hat{j} + z\hat{k}}{a}\right) = \frac{3xyz}{a}$$

Now, $$\iint_S F\cdot\hat{n}\,ds = \iint_S (\vec{F}\cdot\hat{n})\frac{dx\,dy}{|\hat{k}.\hat{n}|} = \int_0^a \int_0^{\sqrt{a^2-x^2}} \frac{3xyz\,dx\,dy}{a\left(\frac{z}{a}\right)}$$

$$= 3\int_0^a \int_0^{\sqrt{a^2-x^2}} xy\,dy\,dx = 3\int_0^a x\left(\frac{y^2}{2}\right)_0^{\sqrt{a^2-x^2}} dx$$

$$= \frac{3}{2}\int_0^a x\,(a^2-x^2)\,dx = \frac{3}{2}\left(\frac{a^2x^2}{2}-\frac{x^4}{4}\right)_0^a = \frac{3}{2}\left(\frac{a^4}{2}-\frac{a^4}{4}\right) = \frac{3a^4}{8}.$$ **Ans.**

Example 14. *Show that* $\iint_S \vec{F}\cdot\hat{n}\,ds = \frac{3}{2}$, *where* $\vec{F} = 4\,xz\hat{i} - y^2\hat{j} + yz\,\hat{k}$ *and S is the surface of the cube bounded by the planes,*
$x=0,\ x=1,\ y=0,\ y=1,\ z=0,\ z=1.$

S.No.	Surface	Outward normal	ds	
1	OABC	$-k$	$dx\,dy$	$z=0$
2	DEFG	k	$dx\,dy$	$z=1$
3	OAGF	$-j$	$dx\,dz$	$y=0$
4	BCED	j	$dx\,dz$	$y=1$
5	ABDG	i	$dy\,dz$	$x=1$
6	OCEF	$-i$	$dy\,dz$	$x=0$

Solution. $\iint_S \vec{F}\cdot\hat{n}\,ds = \iint_{OABC}\vec{F}\cdot\hat{n}\,ds$

$+ \iint_{DEFG}\vec{F}\cdot\hat{n}\,ds + \iint_{OAGF}\vec{F}\cdot\hat{n}\,ds$

$+ \iint_{BCED}\vec{F}\cdot\hat{n}\,ds + \iint_{ABDG}\vec{F}\cdot\hat{n}\,ds$

$+ \iint_{OCEF}\vec{F}\cdot\hat{n}\,ds$ ...(1)

Now, $\iint_{OABC}\vec{F}\cdot n\,ds = \iint_{OABC}(4\,xz\hat{i} - y^2\,\hat{j} + yz\,\hat{k})\,(-k)\,dx\,dy = \int_0^1\int_0^1 -yz\,dx\,dy = 0$ (as $z=0$)

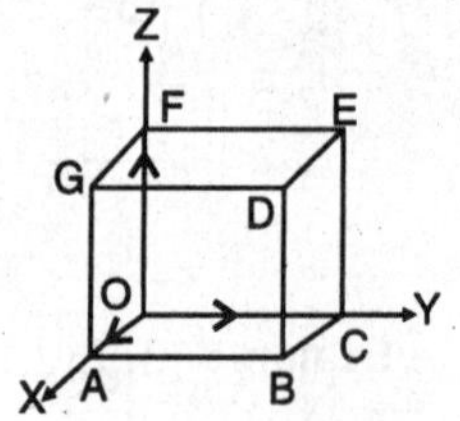

$\iint_{DEFG}(4\,xz\hat{i} - y^2\,\hat{j} + yz\,\hat{k})\cdot\hat{k}\,dx\,dy$

$$= \iint_{DEFG} yz\,dx\,dy = \int_0^1\int_0^1 y\,(1)\,dx\,dy \quad (\text{as } z=1)$$

$$= \int_0^1 dx\left[\frac{y^2}{2}\right]_0^1 = [x]_0^1\,\frac{1}{2} = \frac{1}{2}$$

Z, F, E, G, D, O, Y, C, X, A, B

$\iint_{OAGF}(4\,xz\,\hat{i} - y^2\hat{j} + yz\,\hat{k})\cdot(-j)\,dx\,dz = \iint_{OAGF} y^2\,dx\,dz = 0$ (as $y=0$)

$\iint_{BCED}(4\,xz\,\hat{i} - y^2\hat{j} + yz\,\hat{k})\cdot\hat{j}\,dx\,dz = \iint_{BCED}(-y^2)\,dx\,dz$

$$= -\int_0^1 dx\int_0^1 dz = -(x)_0^1\,(z)_0^1 = -1 \quad (\text{as } y=1)$$

$\iint_{ABDG}(4\,xz\hat{i} - y^2\hat{j} + yz\hat{k})\cdot\hat{i}\,dy\,dz = \iint 4\,xz\,dy\,dz = \int_0^1\int_0^1 4\,(1)\,z\,dy\,dz$ (as $x=1$)

$$= 4\,(y)_0^1\left(\frac{z^2}{2}\right)_0^1 = 4\,(1)\left(\frac{1}{2}\right) = 2$$

$\iint_{OCEF}(4\,xz\,\hat{i} - y^2\hat{j} + yz\,\hat{k})\,(-\hat{i})\,dy\,dz = \int_0^1\int_0^1 -4\,xz\,dy\,dz = 0$ (as $x=0$)

On putting these values in (1), we get

$\iint_S \overline{F}\cdot\hat{n}\,ds = 0 + \frac{1}{2} + 0 - 1 + 2 + 0 = \frac{3}{2}$ **Proved.**

EXERCISE 24.2

1. Evaluate $\iint_S \vec{A}.\hat{n}\, ds$, where $\vec{A} = z\hat{i} + x\hat{j} - 3y^2 z\hat{k}$ and S is the surface of the cylinder $x^2 + y^2 = 16$ included in the first octant between $z = 0$ and $z = 5$. **Ans.** 90

2. If $\vec{r} = t\hat{i} - t^2\hat{j} + (t-1)\hat{k}$ and $\vec{S} = 2t^2\hat{i} + 6t\hat{k}$, evaluate $\int_0^2 \vec{r} \cdot \vec{S}\, dt$. **Ans.** 12

3. Evaluate $\iint_S \vec{F} \cdot \hat{n}\, ds$, where, $F = 2yx\hat{i} - yz\hat{j} + x^2\hat{k}$ over the surface S of the cube bounded by the coordinate planes and planes $x = a$, $y = a$ and $z = a$. **Ans.** $\frac{1}{2}a^4$

4. If $\vec{F} = 2y\hat{i} - 3\hat{j} + x^2\hat{k}$ and S is the surface of the parabolic cylinder $y^2 = 8x$ in the first octant bounded by the planes $y = 4$, and $z = 6$, then evaluate $\iint_S \vec{F} \cdot \hat{n}\, dS$. **Ans.** 132

24.3 VOLUME INTEGRAL

Let $\vec{F}$ be a vector point function and volume V enclosed by a closed surface.

The volume integral $= \iiint_V \vec{F}\, dv$

Example 15. *If $\vec{F} = 2z\hat{i} - x\hat{j} + y\hat{k}$, evaluate $\iiint_V \vec{F}\, dv$ where, v is the region bounded by the surfaces*

$$x = 0,\ y = 0,\ x = 2,\ y = 4,\ z = x^2,\ z = 2.$$

Solution.
$$\iiint_V \vec{F}\, dv = \iiint (2z\hat{i} - x\hat{j} + y\hat{k})\, dx\, dy\, dz$$

$$= \int_0^2 dx \int_0^4 dy \int_{x^2}^2 (2z\hat{i} - x\hat{j} + y\hat{k})\, dz = \int_0^2 dx \int_0^4 dy\, [z^2\hat{i} - xz\hat{j} + yz\hat{k}]_{x^2}^2$$

$$= \int_0^2 dx \int_0^4 dy\, [4\hat{i} - 2x\hat{j} + 2y\hat{k} - x^4\hat{i} + x^3\hat{j} - x^2 y\hat{k}]$$

$$= \int_0^2 dx \left[4y\hat{i} - 2xy\hat{j} + y^2\hat{k} - x^4 y\hat{i} + x^3 y\hat{j} - \frac{x^2 y^2}{2}\hat{k}\right]_0^4$$

$$= \int_0^2 (16\hat{i} - 8x\hat{j} + 16\hat{k} - 4x^4\hat{i} + 4x^3\hat{j} - 8x^2\hat{k})\, dx$$

$$= \left[16x\hat{i} - 4x^2\hat{j} + 16x\hat{k} - \frac{4x^5}{5}\hat{i} + x^4\hat{j} - \frac{8x^3}{3}\hat{k}\right]_0^2$$

$$= 32\hat{i} - 16\hat{j} + 32\hat{k} - \frac{128}{5}\hat{i} + 16\hat{j} - \frac{64}{3}\hat{k} = \frac{32\hat{i}}{5} + \frac{32\hat{k}}{3} = \frac{32}{15}(3\hat{i} + 5\hat{k})$$ **Ans.**

EXERCISE 24.3

1. If $\vec{F} = (2x^2 - 3z)\hat{i} - 2xy\hat{j} - 4x\hat{k}$, then evaluate $\iiint_V \nabla \vec{F}\, dV$, where V is bounded by the plane $x = 0$, $y = 0$, $z = 0$ and $2x + 2y + z = 4$. **Ans.** $\frac{8}{3}$

2. Evaluate $\iiint_V \phi\, dV$, where $\phi = 45\, x^2 y$ and V is the closed region bounded by the planes $4x + 2y + z = 8$, $x = 0$, $y = 0$, $z = 0$ **Ans.** 128

3. If $\vec{F} = (2x^2 - 3z)\,\hat{i} - 2xy\,\hat{j} - 4x\hat{k}$, then evaluate $\iiint_V \nabla \times \vec{F}\, dV$, where V is the closed region bounded by the planes $x = 0, y = 0, z = 0$ and $2x + 2y + z = 4$. **Ans.** $\frac{8}{3}(\hat{j} - \hat{k})$

4. Evaluate $\iiint_V (2x + y)\, dV$, where V is closed region bounded by the cylinder $z = 4 - x^2$ and the planes $x = 0, y = 0, y = 2$ and $z = 0$. **Ans.** $\frac{80}{3}$

5. If $\vec{F} = 2xz\hat{i} - x\hat{j} + y^2\hat{k}$, evaluate $\iiint \vec{F}\, dV$ over the region bounded by the surfaces $x = 0$, $y = 0$, $y = 6$ and $z = x^2$, $z = 4$. **Ans.** $(16\hat{i} - 3\hat{j} + 48\hat{k})$

24.4 GREEN'S THEOREM (For a plane)

Statement. *If $\phi(x, y), \psi(x, y), \frac{\partial \phi}{\partial y}$ and $\frac{\partial \psi}{\partial x}$ be continuous functions over a region R bounded by simple closed curve C in x – y plane, then* (*MTU, Dec. 2012*)

$$\oint_C (\phi\, dx + \psi\, dy) = \iint_R \left(\frac{\partial \psi}{\partial x} - \frac{\partial \phi}{\partial y}\right) dx\, dy$$ (*AMIETE, June 2010, U.P., I Semester, Dec. 2007*)

Proof. Let the curve C be divided into two curves C_1 (ABC) and C_1 (CDA).
Let the equation of the curve C_1 (ABC) be $y = y_1(x)$ and equation of the curve C_2 (CDA) be $y = y_2(x)$.
Let us see the value of

$$\iint_R \frac{\partial \phi}{\partial y} dx\, dy = \int_{x=a}^{x=c}\left[\int_{y=y_1(x)}^{y=y_2(x)} \frac{\partial \phi}{\partial y} dy\right] dx = \int_a^c [\phi(x, y)]_{y=y_1(x)}^{y=y_2(x)} dx$$

$$= \int_a^c [\phi(x, y_2) - \phi(x, y_1)]\, dx = -\int_c^a \phi(x, y_2)\, dx - \int_a^c \phi(x, y_1)\, dx$$

$$= -\left[\int_c^a \phi(x, y_2)\, dx + \int_a^c \phi(x, y_1)\, dx\right]$$

$$= -\left[\int_{c_2} \phi(x, y)\, dx + \int_{c_1} \phi(x, y)\, dx\right] = -\oint_c \phi(x, y)\, dx$$

Thus, $$\oint_c \phi\, dx = -\iint_R \frac{\partial \phi}{\partial y} dx\, dy \qquad ...(1)$$

Similarly, it can be shown that

$$\oint_c \psi\, dy = \iint \frac{\partial \psi}{\partial x} dx\, dy \qquad ...(2)$$

On adding (1) and (2), we get

$$\oint (\phi\, dx + \psi\, dy) = \iint_R \left(\frac{\partial \psi}{\partial x} - \frac{\partial \phi}{\partial y}\right) dx\, dy$$ **Proved.**

Note. Green's Theorem in vector form

$$\int_c \vec{F} \cdot d\vec{r} = \iint_R (\nabla \times \vec{F}) \cdot \hat{k}\, dR$$

where, $\vec{F} = \phi\,\hat{i} + \psi\,\hat{j}$, $\bar{r} = x\hat{i} + y\hat{j}$, $\hat{k}$ is a unit vector along z-axis and $dR = dx\, dy$.

Example 16. *A vector field $\vec{F}$ is given by $\vec{F} = \sin y\hat{i} + x(1 + \cos y)\,\hat{j}$.*

Evaluate the line integral $\int_C \vec{F} \cdot d\vec{r}$ where C is the circular path given by $x^2 + y^2 = a^2$.

Solution. $\vec{F} = \sin y\hat{i} + x(1 + \cos y)\,\hat{j}$

$$\int_C \vec{F}\cdot \vec{dr} = \int_C [\sin y\hat{i} + x(1+\cos y)\,\hat{j}]\cdot(\hat{i}dx + \hat{j}dy) = \int_C \sin y\, dx + x(1+\cos y)\, dy$$

On applying Green's Theorem, we have

$$\oint_c (\phi\, dx + \psi\, dy) = \iint_s \left(\frac{\partial \psi}{\partial x} - \frac{\partial \phi}{\partial y}\right) dx\, dy$$

$$= \iint_s [(1+\cos y) - \cos y]\, dx\, dy$$

where s is the circular plane surface of radius a.

$$= \iint_s dx\, dy = \text{Area of circle} = \pi a^2. \quad \textbf{Ans.}$$

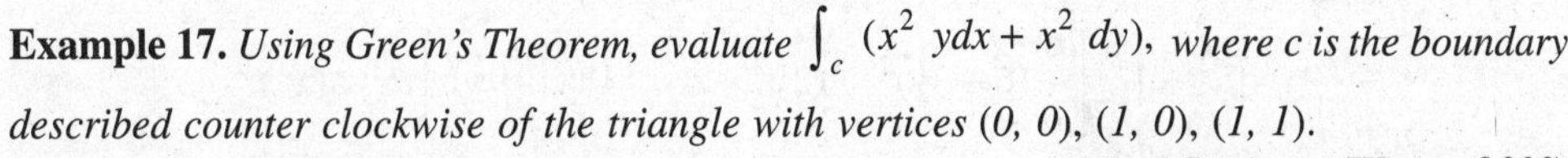

Example 17. *Using Green's Theorem, evaluate* $\int_c (x^2\, y dx + x^2\, dy)$, *where c is the boundary described counter clockwise of the triangle with vertices (0, 0), (1, 0), (1, 1).*

(*U.P., I Semester, Winter 2003*)

Solution. By Green's Theorem, we have

$$\int_c (\phi\, dx + \psi\, dy) = \iint_R \left(\frac{\partial \psi}{\partial x} - \frac{\partial \phi}{\partial y}\right) dx\, dy$$

$$\int_c (x^2 y\, dx + x^2\, dy) = \iint_R (2x - x^2)\, dx\, dy$$

$$= \int_0^1 (2x - x^2)\, dx \int_0^x dy = \int_0^1 (2x - x^2)\, dx\, [y]_0^x$$

$$= \int_0^1 (2x - x^2)(x)\, dx = \int_0^1 (2x^2 - x^3)\, dx = \left(\frac{2x^3}{3} - \frac{x^4}{4}\right)_0^1$$

$$= \left(\frac{2}{3} - \frac{1}{4}\right) = \frac{5}{12} \quad \textbf{Ans.}$$

Example 18. *State and verify Green's Theorem in the plane for* $\oint (3x^2 - 8y^2)\, dx + (4y - 6xy)\, dy$ *where C is the boundary of the region bounded by* $x \geq 0$, $y \leq 0$ and $2x - 3y = 6$.

(*Uttarakhand, I Semester, Dec. 2006*)

Solution. Statement: See Article 24.4 on page 576.

Here the closed curve C consists of straight lines OB, BA and AO, where coordinates of A and B are (3, 0) and (0, – 2) respectively. Let R be the region bounded by C.

Then by Green's Theorem in plane, we have

$$\oint [(3x^2 - 8y^2)\, dx + (4y - 6xy)\, dy]$$

$$= \iint_R \left[\frac{\partial}{\partial x}(4y - 6xy) - \frac{\partial}{\partial y}(3x^2 - 8y^2)\right] dx\, dy \quad ...(1)$$

$$= \iint_R (-6y + 16y)\, dx\, dy = \iint_R 10y\, dx\, dy$$

$$= 10\int_0^3 dx \int_{\frac{1}{3}(2x-6)}^{0} y\, dy = 10\int_0^3 dx \left[\frac{y^2}{2}\right]_{\frac{1}{3}(2x-6)}^{0} = -\frac{5}{9}\int_0^3 dx\, (2x-6)^2$$

$$= -\frac{5}{9}\left[\frac{(2x-6)^3}{3\times 2}\right]_0^3 = -\frac{5}{54}(0+6)^3 = -\frac{5}{54}(216) = -20 \quad ...(2)$$

Now we evaluate L.H.S. of (1) along OB, BA and AO.

Along OB, $x = 0$, $dx = 0$ and y varies form 0 to –2.

Along BA, $x = \frac{1}{2}(6+3y)$, $dx = \frac{3}{2}\, dy$ and y varies from –2 to 0.

and along AO, $y = 0$, $dy = 0$ and x varies from 3 to 0.

L.H.S. of (1) $= \oint [(3x^2 - 8y^2)\, dx + (4y - 6xy)\, dy]$

$$= \int_{OB} [(3x^2 - 8y^2)\, dx + (4y - 6xy)\, dy] + \int_{BA} [(3x^2 - 8y^2)\, dx + (4x - 6xy)\, dy]$$

$$+ \int_{AO} [(3x^2 - 8y^2)\, dx + (4y - 6xy)\, dy]$$

$$= \int_0^{-2} 4y\, dy + \int_{-2}^{0} \left[\frac{3}{4}(6+3y)^2 - 8y^2\right]\left(\frac{3}{2}dy\right) + [4y - 3(6+3y)\, y]\, dy + \int_3^0 3x^2\, dx$$

$$= [2y^2]_0^{-2} + \int_{-2}^{0} \left[\frac{9}{8}(6+3y)^2 - 12y^2 + 4y - 18y - 9y^2\right] dy + (x^3)_3^0$$

$$= 2[4] + \int_{-2}^{0} \left[\frac{9}{8}(6+3y)^2 - 21y^2 - 14y\right] dy + (0 - 27)$$

$$= 8 + \left[\frac{9}{8}\frac{(6+3y)^3}{3\times 3} - 7y^3 - 7y^2\right]_{-2}^{0} - 27 = -19 + \left[\frac{216}{8} + 7(-2)^3 + 7(-2)^2\right]$$

$$= -19 + 27 - 56 + 28 = -20 \qquad \text{...(3)}$$

With the help of (2) and (3), we find that (1) is true and so Green's Theorem is verified.

Example 19. *Verify Green's Theorem in the plane for*

$$\oint_C (3x^2 - 8y^2)\, dx + (4y - 6xy)\, dy$$

Where C is the boundary of the region defined by $y = \sqrt{x}$, and $y = x^2$ *(K.University, Dec. 2008)*

Solution. Here we have,

$$\oint_C (3x^2 - 8y^2)\, dx + (4y - 6xy)\, dy$$

By Green's Theorem, we have

$$\int_C (\phi\, dx + \psi\, dy) = \iint_S \left(\frac{\partial \psi}{\partial x} - \frac{\partial \phi}{\partial y}\right) dx\, dy$$

$$= \int_0^1 \int_{x^2}^{\sqrt{x}} \left[\frac{\partial}{\partial x}(4y - 6xy) - \frac{\partial}{\partial y}(3x^2 - 8y^2)\right] dx\, dy$$

$$= \int_0^1 \int_{x^2}^{\sqrt{x}} (-6y + 16y)\, dx\, dy = 10 \int_0^1 \int_{x^2}^{\sqrt{x}} y\, dx\, dy = 10 \int_0^1 dx \left(\frac{y^2}{2}\right)_{x^2}^{\sqrt{x}} = \frac{10}{2}\int_0^1 dx\left(x - x^4\right)$$

$$= 5\left(\frac{x^2}{2} - \frac{x^5}{5}\right)_0^1 = 5\left(\frac{1}{2} - \frac{1}{5}\right) = 5\left(\frac{3}{10}\right) = \frac{3}{2}$$

Ans.

Example 20. *Apply Green's Theorem to evaluate* $\int_C [(2x^2 - y^2)\, dx + (x^2 + y^2)\, dy]$, *where C is the boundary of the area enclosed by the x-axis and the upper half of circle* $x^2 + y^2 = a^2$.

(M.D.U. Dec. 2009, U.P., I Sem., Dec. 2004)

Solution. $\int_C [(2x^2 - y^2)\, dx + (x^2 + y^2)\, dy]$

By Green's Theorem, we've $\int_C (\phi\, dx + \psi\, dy) = \iint_S \left(\frac{\partial \psi}{\partial x} - \frac{\partial \phi}{\partial y}\right) dx\, dy$

$$= \int_{-a}^{a}\int_0^{\sqrt{a^2-x^2}} \left[\frac{\partial}{\partial x}(x^2+y^2) - \frac{\partial}{\partial y}(2x^2-y^2)\right] dx\, dy$$

$$= \int_{-a}^{a}\int_0^{\sqrt{a^2-x^2}} (2x+2y)\, dx\, dy = 2\int_{-a}^{a} dx \int_0^{\sqrt{a^2-x^2}} (x+y)\, dy$$

$$= 2\int_{-a}^{a} dx \left(xy + \frac{y^2}{2}\right)_0^{\sqrt{a^2-x^2}} = 2\int_{-a}^{a}\left(x\sqrt{a^2-x^2} + \frac{a^2-x^2}{2}\right) dx$$

Y, $y = \sqrt{a^2 - x^2}$, X′, O, X

$$= 2\int_{-a}^{a} x\sqrt{a^2-x^2}\, dx + \int_{-a}^{a} (a^2-x^2)\, dx \qquad \left[\begin{aligned}\int_{-a}^{a} f(x)\,dx &= 2\int_0^a f(x)\,dx, && f \text{ is even}\\ &= 0, && f \text{ is odd}\end{aligned}\right]$$

$$= 0 + 2\int_0^a (a^2-x^2)\, dx = 2\left(a^2x - \frac{x^3}{3}\right)_0^a = 2\left(a^3 - \frac{a^3}{3}\right) = \frac{4a^3}{3} \qquad \textbf{Ans.}$$

Example 21. *Evaluate* $\oint_C -\frac{y}{x^2+y^2}\, dx + \frac{x}{x^2+y^2}\, dy$, *where* $C = C_1 \cup C_2$ *with* $C_1 : x^2 + y^2 = 1$ *and* $C_2 : x = \pm 2,\ y = \pm 2.$ *(Gujarat, I Semester, Jan 2009)*

Solution. $\oint_C -\frac{y}{x^2+y^2}\, dx + \frac{x}{x^2+y^2}\, dy$

$$= \iint \left(\frac{\partial}{\partial x}\frac{x}{x^2+y^2} + \frac{\partial}{\partial y}\frac{y}{x^2+y^2}\right) dx\, dy$$

$$= \iint \left[\frac{(x^2+y^2)1 - 2x(x)}{(x^2+y^2)^2} + \frac{(x^2+y^2)1 - 2y(y)}{(x^2+y^2)^2}\right] dx\, dy$$

$$= \iint \left[\frac{x^2+y^2-2x^2}{(x^2+y^2)^2} + \frac{x^2+y^2-2y^2}{(x^2+y^2)^2}\right] dx\, dy$$

Y, y = 2, $x^2+y^2=1$, X′, O, X, x = −2, x = 2, y = −2, Y′

$$= \iint \left[\frac{y^2-x^2}{(x^2+y^2)^2} + \frac{x^2-y^2}{(x^2+y^2)^2}\right] dx\, dy = \iint \frac{0}{(x^2+y^2)^2}\, dx\, dy = 0 \qquad \textbf{Ans.}$$

24.5 AREA OF THE PLANE REGION BY GREEN'S THEOREM

Proof. We know that

$$\int_C M dx + N dy = \iint_A \left(\frac{\partial N}{\partial x} - \frac{\partial M}{\partial y}\right) dx\, dy \qquad \text{...(1)}$$

On putting $N = x \left(\frac{\partial N}{\partial x} = 1\right)$ and $M = -y \left(\frac{\partial M}{\partial y} = 1\right)$ in (1), we get

$$\int_C -y\, dx + x\, dy = \iint_A [1-(-1)]\, dx\, dy = 2\iint dx\, dy = 2A$$

$$\text{Area} = \frac{1}{2}\int_C (x\, dy - y\, dx)$$

Example 22. *Using Green's theorem, find the area of the region in the first quadrant bounded by the curves*

$$y = x,\ y = \frac{1}{x},\ y = \frac{x}{4}$$ *(U.P. I, Semester, Dec. 2008)*

Solution. By Green's Theorem Area A of the region bounded by a closed curve C is given by

$$A = \frac{1}{2}\oint_C (xdy - ydx)$$

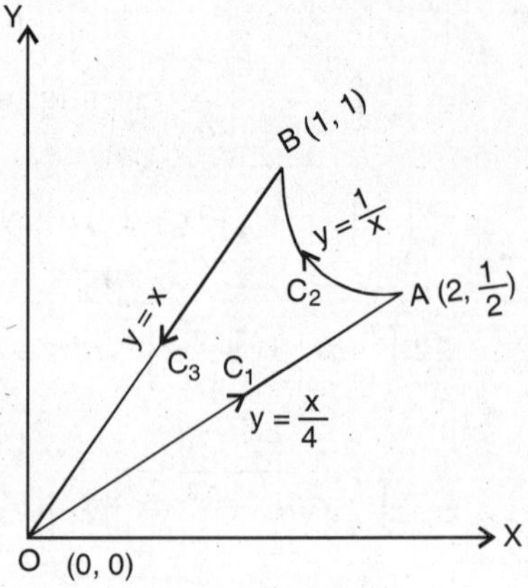

Here, C consists of the curves $C_1 : y = \frac{x}{4}$, $C_2 : y = \frac{1}{x}$

and $C_3 : y = x$ So

$$\left[A = \frac{1}{2}\oint_C = \frac{1}{2}\left[\int_{C_1} + \int_{C_2} + \int_{C_3}\right] = \frac{1}{2}(I_1 + I_2 + I_3)\right]$$

Along $\quad C_1 : y = \frac{x}{4}, dy = \frac{1}{4}dx, x : 0 \text{ to } 2$

$$I_1 = \int_{C_1}(xdy - ydx) = \int_{C_1}\left(x\frac{1}{4}dx - \frac{x}{4}dx\right) = 0$$

Along $\quad C_2 : y = \frac{1}{x}, dy = -\frac{1}{x^2}dx, x : 2 \text{ to } 1$

$$I_2 = \int_{C_2}(xdy - ydx) = \int_2^1\left[x\left(-\frac{1}{x^2}\right)dx - \frac{1}{2}dx\right] = [-2\log x]_2^1 = 2\log 2$$

Along $\quad C_3 : y = x, dy = dx \,;\, x : 1 \text{ to } 0 \,;$

$$I_3 = \int_{C_3}(xdy - ydx) = \int(xdx - xdx) = 0$$

$$A = \frac{1}{2}(I_1 + I_2 + I_3) = \frac{1}{2}(0 + 2\log 2 + 0) = \log 2$$ **Ans.**

EXERCISE 24.4

1. Evaluate $\int_c [(3x^2 - 6yz)\,dx + (2y + 3xz)\,dy + (1 - 4xyz^2)\,dz]$ from (0, 0, 0) to (1, 1, 1) along the path c given by the straight line from (0, 0, 0) to (0, 0, 1) then to (0, 1, 1) and then to (1, 1, 1).

2. Verify Green's Theorem in plane for $\int_C (x^2 + 2xy)\,dx + (y^2 + x^3 y)\,dy$, where c is a square with the vertices P (0, 0), Q (1, 0), R (1, 1) and S (0, 1). **Ans.** $-\frac{1}{2}$

3. Verify Green's Theorem for $\int_c (x^2 - 2xy)\,dx + (x^2 y + 3)\,dy$ around the boundary c of the region $y^2 = 8x$ and $x = 2$.

4. Use Green's Theorem in a plane to evaluate the integral $\int_c [(2x^2 - y^2)\,dx + (x^2 + y^2)\,dy]$, where c is the boundary in the xy-plane of the area enclosed by the x-axis and the semi-circle $x^2 + y^2 = 1$ in the upper half xy-plane. **Ans.** $\frac{4}{3}$

5. Apply Green's Theoem to evaluate $\int_c [(y - \sin x)\,dy + \cos x\,dy]$, where c is the plane triangle enclosed by the lines $y = 0$, $x = \frac{\pi}{2}$ and $y = \frac{2x}{\pi}$. **Ans.** $-\frac{\pi^2 + 8}{4\pi}$

6. Either directly or by Green's Therorem, evaluate the line integral $\int_c e^{-x}(\cos y\,dx - \sin y\,dy)$, where c is the rectangle with vertices (0, 0), (π, 0,), $\left(\pi, \frac{\pi}{2}\right)$ and $\left(0, \frac{\pi}{2}\right)$. **Ans.** $2\,(1 - e^{-\pi})$

(*AMIETE II Sem June 2010*)

7. Verify the Green's Theorem to evaluate the line integral $\int_c (2y^2\,dx + 3x\,dy)$, where c is the boundary of the closed region bounded by $y = x$ and $y = x^2$.

(*U.P., I Semester, Dec. 20005, AMIETE Summer 2004, Winter 2001*) **Ans.** $\frac{27}{4}$

8. Evaluate $\iint_S \bar{F}\,.\,\hat{n}.ds$, where $\vec{F} = xy\,\hat{i} - x^2\hat{j} + (x+z)\,\hat{k}$ and s is the region of the plane $2x + 2y + z = 6$ in the first octant. (*A.M.I.E.T.E., Summer 2004, Winter 2001*) **Ans.** $\frac{27}{4}$

9. Verify Green's Theorem for $\int_C \left[(xy+y^2)\,dx + x^2 dy\right]$ where C is the boundary by $y = x$ and $y = x^2$. (*AMIETE, June 2010*) **Ans.** $-\frac{1}{20}$

24.6 STOKE'S THEOREM (Relation between Line Integral and Surface Integral)

(*Uttarakhand, I Sem. 2008, U.P., Ist Semester, Dec. 2006*)

Statement. Surface integral of the component of curl $\overline{F}$ along the normal to the surface S, taken over the surface S bounded by curve C is equal to the line integral of the vector point function $\vec{F}$ taken along the closed curve C.

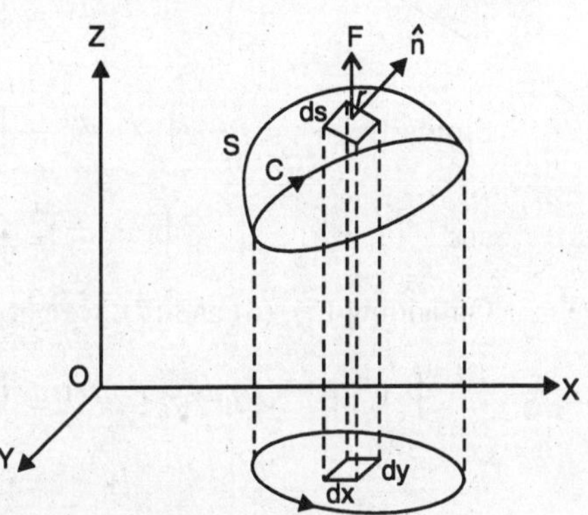

Mathematically

$$\oint \vec{F}\,.\,d\,\vec{r} = \iint_S \text{curl}\,\vec{F}\cdot\hat{n}\,ds$$

where $\hat{n} = \cos\alpha\,\hat{i} + \cos\beta\,\hat{j} + \cos\gamma\,\hat{k}$ is a unit external normal to any surface ds,

Proof. Let

$$\vec{r} = x\hat{i} + y\hat{j} + z\hat{k}$$

$$d\vec{r} = \hat{i}\,dx + \hat{j}\,dy + \hat{k}\,dz$$

$$F = F_1\,\hat{i} + F_2\,\hat{j} + F_3\,\hat{k}$$

On putting the values of $\vec{F}, d\,\vec{r}$ in the statement of the theorem

$$\oint_c (F_1\,\hat{i} + F_2\,\hat{j} + F_3\,\hat{k})\cdot(\hat{i}\,dx + \hat{j}\,dy + \hat{k}\,dz)$$

$$= \iint_S \left(i\frac{\partial}{\partial x} + j\frac{\partial}{\partial y} + k\frac{\partial}{\partial z}\right) \times (F_1\,\hat{i} + F_2\,\hat{j} + F3\,\hat{k}).\,(\cos\alpha\,\hat{i} + \cos\beta\,\hat{j} + \cos\gamma\,\hat{k})\,ds$$

$$\oint (F_1\,dx + F_2\,dy + F_3\,dz) = \iint_S \left[\left(\frac{\partial F_3}{\partial y} - \frac{\partial F_2}{\partial z}\right)\hat{i} + \left(\frac{\partial F_1}{\partial z} - \frac{\partial F_3}{\partial x}\right)\hat{j} + \left(\frac{\partial F_2}{\partial x} - \frac{\partial F_1}{\partial y}\right)\hat{k}\right].$$

$$(\hat{i}\cos\alpha + \hat{j}\cos\beta + \hat{k}\cos\gamma)\,ds$$

$$= \iint_S \left[\left(\frac{\partial F_3}{\partial y} - \frac{\partial F_2}{\partial z}\right)\cos\alpha + \left(\frac{\partial F_1}{\partial z} - \frac{\partial F_3}{\partial x}\right)\cos\beta + \left(\frac{\partial F_2}{\partial x} - \frac{\partial F_1}{\partial y}\right)\cos\gamma\right] ds \quad ...(1)$$

Let us first prove

$$\oint_c F_1\,dx = \iint_S \left[\left(\frac{\partial F_1}{\partial z}\cos\beta - \frac{\partial F_1}{\partial y}\cos\gamma\right)\right] ds \quad ...(2)$$

Let the equation of the surface S be $z = g\,(x, y)$. The projection of the surface on $x - y$ plane is region R.

$$\oint_c F_1\,(x, y, z)\,dx = \oint_c F_1\,[x, y, g(x, y)]\,dx$$

$$= -\iint_R \frac{\partial}{\partial y} F_1\,(x, y, g)\,dx\,dy \quad \text{[By Green's Theorem]}$$

$$= -\iint_R \left(\frac{\partial F_1}{\partial y} + \frac{\partial F_1}{\partial z}\frac{\partial g}{\partial y}\right) dx\,dy \quad ...(3)$$

The direction consines of the normal to the surface $z = g(x, y)$ are given by

$$\frac{\cos\alpha}{\frac{-\partial g}{\partial x}} = \frac{\cos\beta}{\frac{-\partial g}{\partial y}} = \frac{\cos\gamma}{1}$$

And $dx\,dy$ = projection of ds on the xy-plane = $ds \cos \gamma$

Putting the values of ds in R.H.S. of (2)

$$\iint_S \left(\frac{\partial F_1}{\partial z} \cos \beta - \frac{\partial F_1}{\partial y} \cos \gamma \right) ds = \iint_R \left(\frac{\partial F_1}{\partial z} \cos \beta - \frac{\partial F_1}{\partial y} \cos \gamma \right) \frac{dx\,dy}{\cos \gamma}$$

$$= \iint_R \left(\frac{\partial F_1}{\partial z} \frac{\cos \beta}{\cos \gamma} - \frac{\partial F_1}{\partial y} \right) dx\,dy = \iint_R \left(\frac{\partial F_1}{\partial z} \left(-\frac{\partial g}{\partial y} \right) - \frac{\partial F_1}{\partial y} \right) dx\,dy$$

$$= -\iint_R \left(\frac{\partial F_1}{\partial y} + \frac{\partial F_1}{\partial z} \frac{\partial g}{\partial y} \right) dx\,dy \qquad ...(4)$$

From (3) and (4), we get

$$\oint_c F_1\,dx = \iint_S \left(\frac{\partial F_1}{\partial z} \cos \beta - \frac{\partial F_1}{\partial y} \cos \gamma \right) ds \qquad ...(5)$$

Similarly,
$$\oint_c F_2\,dy = \iint_S \left(\frac{\partial F_2}{\partial x} \cos \gamma - \frac{\partial F_2}{\partial z} \cos \alpha \right) ds \qquad ...(6)$$

and
$$\oint_c F_3\,dz = \iint_S \left(\frac{\partial F_3}{\partial y} \cos \alpha - \frac{\partial F_3}{\partial x} \cos \beta \right) ds \qquad ...(7)$$

On adding (5), (6) and (7), we get

$$\oint_c (F_1\,dx + F_2\,dy + F_3\,dz) = \iint_S \left(\frac{\partial F_1}{\partial z} \cos \beta - \frac{\partial F_1}{\partial y} \cos \gamma + \frac{\partial F_2}{\partial x} \cos \gamma - \frac{\partial F_2}{\partial z} \cos \alpha \right.$$
$$\left. + \frac{\partial F_3}{\partial y} \cos \alpha - \frac{\partial F_3}{\partial x} \cos \beta \right) ds$$ **Proved.**

24.7 ANOTHER METHOD OF PROVING STOKE'S THEOREM

The circulation of vector F around a closed curve C is equal to the flux of the curve of the vector through the surface S bounded by the curve C.

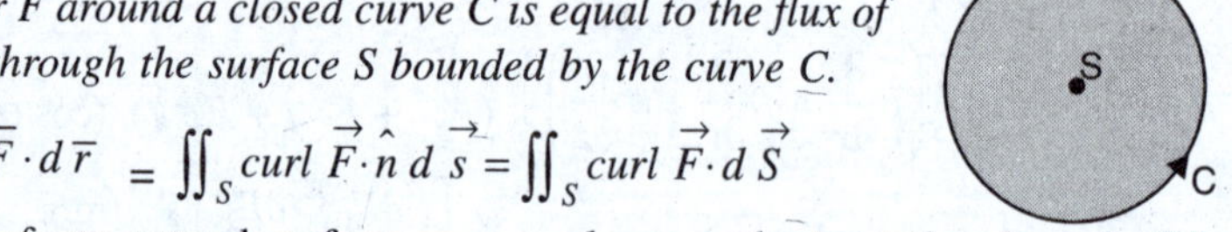

$$\oint_c \overline{F} \cdot d\,\overline{r} = \iint_S curl\ \vec{F} \cdot \hat{n}\,d\,\vec{s} = \iint_S curl\ \vec{F} \cdot d\,\vec{S}$$

Proof : The projection of any curved surface over xy-plane can be treated as kernal of the surface integral over actual surface

Now, $$\iint_S (\nabla \times \vec{F}) \cdot \hat{k}\,d\,\vec{S} = \iint_S (\nabla \times \vec{F}) \cdot (\vec{i} \times \vec{j})\,dx\,dy \qquad [\hat{k} = \hat{i} \times \hat{j}]$$

$$= \iint_S [(\nabla \cdot \hat{i})\,(\vec{F} \cdot \hat{j}) - (\nabla \cdot \hat{j})\,(\vec{F} \cdot \hat{i})]\,dx\,dy = \iint_S \left[\frac{\partial}{\partial x}(F_y) - \frac{\partial}{\partial y}(F_x) \right] dx\,dy$$

$$= \iint_S [F_x\,dx + F_y\,dy] \text{ [By Green's theorem]}$$

$$= \iint_S [\hat{i}\,F_x + \hat{j}\,F_y] \cdot (\hat{i}\,dx + \hat{j}\,dy) = \oint_c \vec{F} \,.\, \vec{dr}$$

$$\iint_S curl\ \vec{F} \cdot \hat{n}\,dS = \oint_c \vec{F} \,.\, \vec{d}\,r.$$

where, $\vec{F} = F_x\,\hat{i} + F_y\,\hat{j} + F_z\,\hat{k}$ and $\vec{dr} = dx\,\hat{i} + dy\,\hat{j} + dz\,\hat{k}$

Example 23. *Evaluate by Strokes theorem* $\oint_C (yz\,dx + zx\,dy + xy\,dz)$ *where C is the curve* $x^2 + y^2 = 1,\ z = y^2$. *(M.D.U., Dec 2009)*

Solution. Here we have $\oint yz\,dx + zx\,dy + xy\,dz$

$$= \int (yz\hat{i} + zx\hat{j} + xy\hat{k}).\,(\hat{i}dx + \hat{j}dy + kdz)$$

$$= \oint F.dx \qquad \text{Curl } \overline{F} = \begin{vmatrix} \hat{i} & \hat{j} & \hat{k} \\ \frac{\partial}{\partial x} & \frac{\partial}{\partial y} & \frac{\partial}{\partial z} \\ yz & zx & xy \end{vmatrix}$$

$$= \int curl\ F.\eta\ ds \qquad = (x - x)\ \hat{i} + (y - y)\ \hat{j} + (z - z)\ \hat{k}$$

$$= 0 \quad = 0$$ **Ans.**

Example 24. *Using Stoke's theorem or otherwise, evaluate*

$$\int_c [(2x - y)\, dx - yz^2 dy - y^2 z\, dz]$$

where c is the circle $x^2 + y^2 = 1$, corresponding to the surface of sphere of unit radius.

(U.P., I Semester, Winter 2001)

Solution. $\int_c [(2x - y)\, dx - yz^2 dy - y^2 z\, dz]$

$$= \int_c [(2x - y)\,\hat{i} - yz^2\,\hat{j} - y^2 z\,\hat{k}] \cdot (\hat{i}\, dx + \hat{j}\, dy + \hat{k}\, dz)$$

By Stoke's theorem $\oint \vec{F} \cdot d\vec{r} = \iint_S \text{Curl } \vec{F} \cdot \hat{n}\, ds$ ...(1)

$$\text{Curl } \vec{F} = \nabla \times \vec{F} = \begin{vmatrix} \hat{i} & \hat{j} & \hat{k} \\ \frac{\partial}{\partial x} & \frac{\partial}{\partial y} & \frac{\partial}{\partial z} \\ 2x - y & -yz^2 & -y^2 z \end{vmatrix}$$

$$= (-2yz + 2yz)\,\hat{i} - (0 - 0)\,\hat{j} + (0 + 1)\,\hat{k} = \hat{k}$$

Putting the value of curl $\vec{F}$ in (1), we get

$$\oint \vec{f} \cdot d\vec{r} = \iint \hat{k} \cdot \hat{n}\, ds = \iint \hat{k} \cdot \hat{n} \frac{dx\, dy}{\hat{n} \cdot \hat{k}} = \iint dx\, dy = \text{Area of the circle} = \pi \quad \left[\because ds = \frac{dx\, dy}{(\hat{n} \cdot \hat{k})}\right]$$

Example 25. *Evaluate* $\int_C \vec{F} \cdot d\vec{r}$, *where* $F(x, y, z) = -y^2\hat{i} + x\hat{j} + z^2\hat{k}$ *and C is the curve of intersection of the plane y + z = 2 and the cylinder $x^2 + y^2 = 1$. (Gujarat, I sem. Jan. 2009)*

Solution. $\oint_C \vec{F} \cdot d\vec{r} = \iint_S \text{curl } \vec{F} \cdot \hat{n}\, ds = \iint_S \text{curl } (-y^2\,\hat{i} + x\,\hat{j} + z^2\,\hat{k})\,\hat{n}\, ds$ ...(1)

$$F(x, y, z) = -y^2\,\hat{i} + x\,\hat{j} + z^2\,\hat{k}$$ *(By Stoke's Theorem)*

$$\text{Curl } \vec{F} = \begin{vmatrix} \hat{i} & \hat{j} & \hat{k} \\ \frac{\partial}{\partial x} & \frac{\partial}{\partial y} & \frac{\partial}{\partial z} \\ -y^2 & x & z^2 \end{vmatrix}$$

$$= \hat{i}\,(0 - 0) - \hat{j}\,(0 - 0) + \hat{k}\,(1 + 2y) = (1 + 2y)\,\hat{k}$$

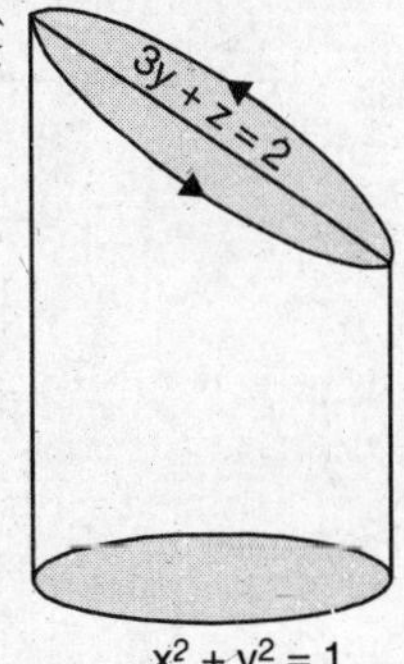

Normal vector $= \nabla \vec{F}$

$$= \left(\hat{i}\frac{\partial}{\partial x} + \hat{j}\frac{\partial}{\partial y} + \hat{k}\frac{\partial}{\partial z}\right)(y + z - 2) = \hat{j} + \hat{k}$$

Unit normal vector $\hat{n} = \dfrac{\hat{j} + \hat{k}}{\sqrt{2}}$

$$ds = \frac{dx\, dy}{\hat{\eta} \cdot \hat{k}}$$

On putting the values of curl $\vec{F}, \hat{n}$ and ds in (1), we get

$$\int_C \vec{F}\cdot d\vec{r} = \iint_S (1+2y)\,\hat{k}\cdot\frac{\hat{j}+\hat{k}}{\sqrt{2}}\,\frac{dx\,dy}{\left(\dfrac{\hat{j}+\hat{k}}{\sqrt{2}}\right)\cdot\hat{k}}$$

$$= \iint \frac{1+2y}{\sqrt{2}}\,\frac{dx\,dy}{\dfrac{1}{\sqrt{2}}} = \iint (1+2y)\,dx\,dy = \int_0^{2\pi}\int_0^1 (1+2r\sin\theta)\,r\,d\theta\,dr$$

$$= \int_0^{2\pi}\int_0^1 (r+2r^2\sin\theta)\,d\theta\,dr$$

$$= \int_0^{2\pi} d\theta\left[\frac{r^2}{2}+\frac{2r^3}{3}\sin\theta\right]_0^1 = \int_0^{2\pi}\left[\frac{1}{2}+\frac{2}{3}\sin\theta\right]d\theta$$

$$= \left[\frac{\theta}{2}-\frac{2}{3}\cos\theta\right]_0^{2\pi} = \left(\pi-\frac{2}{3}-0+\frac{2}{3}\right) = \pi \qquad \textbf{Ans.}$$

Example 26. *Apply Stoke's Theorem to find the value of*

$$\int_c (y\,dx + z\,dy + x\,dz)$$

where c is the curve of intersection of $x^2+y^2+z^2=a^2$ and $x+z=a$. (*Nagpur, Summer 2001*)

Solution. $\int_c (y\,dx + z\,dy + x\,dz)$

$$= \int_c (y\hat{i}+z\hat{j}+x\hat{k})\cdot(\hat{i}\,dx+\hat{j}\,dy+\hat{k}\,dz) = \oint_C (y\hat{i}+z\hat{j}+x\hat{k})\cdot d\vec{r}$$

$$= \iint_S \text{curl}\,(y\hat{i}+z\hat{j}+x\hat{k})\cdot\hat{n}\,ds \qquad \text{(By Stoke's Theorem)}$$

$$= \iint_S \left(\hat{i}\frac{\partial}{\partial x}+\hat{j}\frac{\partial}{\partial y}+\hat{k}\frac{\partial}{\partial z}\right)\times(y\hat{i}+z\hat{j}+x\hat{k})\cdot\hat{n}\,ds = \iint_S -(\hat{i}+\hat{j}+\hat{k})\cdot\hat{n}\,ds \quad ...(1)$$

where S is the circle formed by the intersection of $x^2+y^2+z^2=a^2$ and $x+z=a$.

$$\hat{n} = \frac{\nabla\phi}{|\nabla\phi|} = \frac{\left(\hat{i}\dfrac{\partial}{\partial x}+\hat{j}\dfrac{\partial}{\partial y}+\hat{k}\dfrac{\partial}{\partial z}\right)(x+z-a)}{|\nabla\phi|} = \frac{\hat{i}+\hat{k}}{\sqrt{1+1}}$$

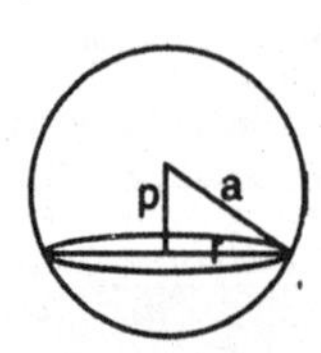

$$\therefore \qquad \hat{n} = \frac{\hat{i}}{\sqrt{2}}+\frac{\hat{k}}{\sqrt{2}}$$

Putting the value of $\hat{n}$ in (1), we have

$$= \iint_S -(\hat{i}+\hat{j}+\hat{k})\cdot\left(\frac{\hat{i}}{\sqrt{2}}+\frac{\hat{k}}{\sqrt{2}}\right)ds$$

$$= \iint_S -\left(\frac{1}{\sqrt{2}}+\frac{1}{\sqrt{2}}\right)ds \qquad \left[\text{Use } r^2 = R^2 - p^2 = a^2 - \frac{a^2}{2} = \frac{a^2}{2}\right]$$

$$= \frac{-2}{\sqrt{2}}\iint_S ds = \frac{-2}{\sqrt{2}}\pi\left(\frac{a}{\sqrt{2}}\right)^2 = -\frac{\pi a^2}{\sqrt{2}} \qquad \textbf{Ans.}$$

Example 27. *Use Stoke's Theorem to evaluate* $\int_c \vec{v} \cdot \vec{dr}$, *where* $\vec{v} = y^2\hat{i} + xy\hat{j} + xz\hat{k}$, *and c is the bounding curve of the hemisphere* $x^2 + y^2 + z^2 = 9$, $z > 0$, *oriented in the positive direction.*

Solution. By Stoke's theorem

$$\int_c \vec{v} \cdot \vec{dr} = \iint_S (\text{curl } \vec{v}) \cdot \hat{n}\, ds = \iint_S (\nabla \times \vec{v}) \cdot \hat{n}\, ds$$

$$\nabla \times \vec{v} = \begin{vmatrix} \hat{i} & \hat{j} & \hat{k} \\ \frac{\partial}{\partial x} & \frac{\partial}{\partial y} & \frac{\partial}{\partial z} \\ y^2 & xy & xz \end{vmatrix} = (0-0)\,\hat{i} - (z-0)\,\hat{j} + (y-2y)\,\hat{k} = -z\hat{j} - y\hat{k}$$

$$\hat{n} = \frac{\nabla \phi}{|\nabla \phi|} = \frac{\left(i\frac{\partial}{\partial x} + j\frac{\partial}{\partial y} + k\frac{\partial}{\partial z}\right)(x^2+y^2+z^2-9)}{|\nabla \phi|}$$

$$= \frac{2x\hat{i} + 2y\hat{j} + 2z\hat{k}}{\sqrt{4x^2+4y^2+4z^2}} = \frac{x\hat{i}+y\hat{j}+z\hat{k}}{\sqrt{x^2+y^2+z^2}} = \frac{x\hat{i}+y\hat{j}+z\hat{k}}{3}$$

$$(\nabla \times \vec{v}) \cdot \hat{n} = (-z\hat{j} - y\hat{k}) \cdot \frac{x\hat{i}+y\hat{j}+z\hat{k}}{3} = \frac{-yz-yz}{3} = \frac{-2yz}{3}$$

$$\hat{n} \cdot \hat{k}\, ds = dx\, dy \Rightarrow \frac{x\hat{i}+y\hat{j}+z\hat{k}}{3} . \hat{k}\ dx = dx\, dy \Rightarrow \frac{z}{3} ds = dx\, dy$$

$$\therefore \qquad ds = \frac{3}{z} dx\, dy$$

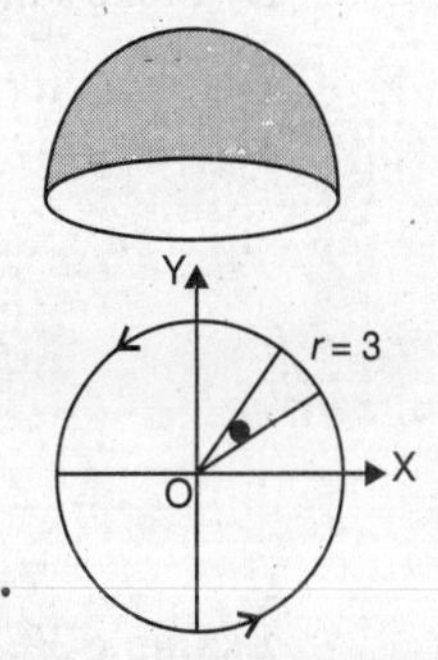

$$\iint_S (\nabla \times \vec{v}) \cdot \hat{n}\, ds = \iint \left(\frac{-2yz}{3}\right)\left(\frac{3}{z} dx\, dy\right) = -\iint 2y\, dx\, dy$$

$$= -\iint 2r \sin\theta\, r\, d\theta\, dr = -2 \int_0^{2\pi} \sin\theta\, d\theta \int_0^3 r^2\, dr$$

$$= -2\,(-\cos\theta)_0^{2\pi} \cdot \left[\frac{r^3}{3}\right]_0^3 = -2\,(-1+1)\,9 = 0 \quad \textbf{Ans.}$$

Example 28. *Evaluate the surface integral* $\iint_S \text{curl } \vec{F} . \hat{n}\, dS$ *by transforming it into a line integral, S being that part of the surface of the paraboloid* $z = 1 - x^2 - y^2$ *for which* $z \geq 0$ and $\vec{F} = y\,\hat{i} + z\,\hat{j} + x\,\hat{k}$.

(K. University, Dec. 2008)

Solution. $\qquad \vec{\nabla} \times \vec{F} = \begin{vmatrix} \hat{i} & \hat{j} & \hat{k} \\ \frac{\partial}{\partial x} & \frac{\partial}{\partial y} & \frac{\partial}{\partial z} \\ y & z & x \end{vmatrix} = -\hat{i} - \hat{j} - \hat{k}$

Obviously $\qquad \hat{n} = \hat{k}.$

Therefore $\qquad (\nabla \times \vec{F}) \cdot \hat{n} = (-\hat{i} - \hat{j} - \hat{k}) . \hat{k} = -1$

Hence $\iint_S (\nabla \times \vec{F}) \cdot \hat{n}\, ds = \iint_S (-1)\, dx\, dy = -\iint_S dx\, dy$

$= -\pi\,(1)^2 = -\pi.$ (Area of circle $= \pi r^2$) **Ans.**

Example 29. *Evaluate* $\oint_C \vec{F} \cdot \vec{dr}$ *by Stoke's Theorem, where* $\vec{F} = y^2\,\hat{i} + x^2\,\hat{j} - (x+z)\,\hat{k}$ *and C is the boundary of triangle with vertices at (0, 0, 0), (1, 0, 0) and (1, 1, 0).*

(*U.P., I Semester, Winter 2000*)

Solution. We have, curl $\vec{F} = \nabla \times \vec{F}$

$$= \begin{vmatrix} \hat{i} & \hat{j} & \hat{k} \\ \dfrac{\partial}{\partial x} & \dfrac{\partial}{\partial y} & \dfrac{\partial}{\partial z} \\ y^2 & x^2 & -(x+z) \end{vmatrix} = 0.\,\hat{i} + \hat{j}\,2\,(x-y)\,\hat{k}.$$

We observe that z co-ordinate of each vertex of the triangle is zero.

Therefore, the triangle lies in the xy-plane.

$\therefore \quad \hat{n} = \hat{k}$

$\therefore \quad$ curl $\vec{F} \cdot \hat{n} = [\hat{j} + 2\,(x-y)\,\hat{k}]\,.\,\hat{k} = 2\,(x-y).$

In the figure, only xy-plane is considered.

The equation of the line OB is $y = x$

By Stoke's theorem, we have

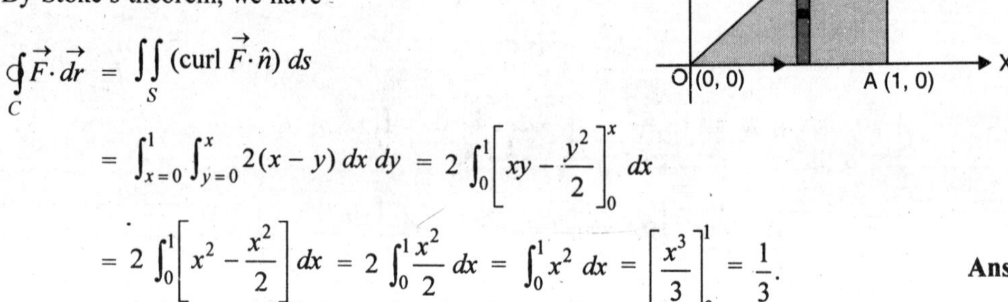

$$\oint_C \vec{F} \cdot \vec{dr} = \iint_S (\text{curl } \vec{F} \cdot \hat{n})\, ds$$

$$= \int_{x=0}^{1} \int_{y=0}^{x} 2(x-y)\, dx\, dy = 2 \int_0^1 \left[xy - \frac{y^2}{2} \right]_0^x dx$$

$$= 2 \int_0^1 \left[x^2 - \frac{x^2}{2} \right] dx = 2 \int_0^1 \frac{x^2}{2}\, dx = \int_0^1 x^2\, dx = \left[\frac{x^3}{3} \right]_0^1 = \frac{1}{3}. \qquad \textbf{Ans.}$$

Example 30. *Use the Stoke's Theorem to evaluate*

$$\int_C [(x+2y)\, dx + (x-z)\, dy + (y-z)\, dz]$$

where c is the boundary of the triangle with vertices (2, 0, 0), (0, 3, 0) and (0, 0, 6) oriented in the anti-clockwise direction.

Solution.

$$\int_C \vec{F} \cdot \vec{dr} = \int_C [(x+2y)\, dx + (x-z)\, dy + (y-z)\, dz]$$

$$= \int_C [(x+2y)\,\hat{i} + (x-z)\,\hat{j} + (y-z)\,\hat{k}]\,.\,[\hat{i}\, dx + \hat{j} dy + \hat{k} dz]$$

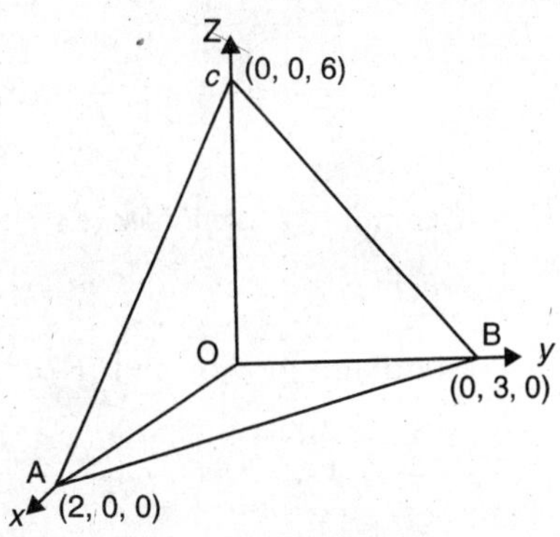

$$\therefore \quad \vec{F} = (x+2y)\,\hat{i} + (x-z)\,\hat{j} + (y-z)\,\hat{k}$$

$$\nabla \times \vec{F} = \begin{vmatrix} \hat{i} & \hat{j} & \hat{k} \\ \dfrac{\partial}{\partial x} & \dfrac{\partial}{\partial y} & \dfrac{\partial}{\partial z} \\ x+2y & x-z & y-z \end{vmatrix}$$

$$= (1+1)\,\hat{i} - (0-0)\,\hat{j} + (1-2)\,\hat{k} = 2\hat{i} - \hat{k}$$

S is the surface of the plane $\dfrac{x}{2} + \dfrac{y}{3} + \dfrac{z}{6} = 1$,

$\hat{n}$ is the normal to the plane ABC.

$$\text{Normal Vector} = \nabla\phi = \left[\hat{i}\frac{\partial}{\partial x} + \hat{j}\frac{\partial}{\partial y} + \hat{k}\frac{\partial}{\partial z}\right]\left[\frac{x}{2} + \frac{y}{3} + \frac{z}{6} - 1\right]$$

$$= \frac{\hat{i}}{2} + \frac{\hat{j}}{3} + \frac{\hat{k}}{6} = \frac{1}{6}(3\hat{i} + 2\hat{j} + \hat{k})$$

$$\hat{n} = \frac{\frac{1}{6}(3\hat{i} + 2\hat{j} + \hat{k})}{\frac{1}{6}\sqrt{9+4+1}} = \frac{1}{\sqrt{14}}(3\hat{i} + 2\hat{j} + \hat{k})$$

$$(\nabla \times \vec{F})\cdot\hat{n} = (2\hat{i} - \hat{k}).\frac{1}{\sqrt{14}}(3\hat{i} + 2\hat{j} + \hat{k}) = \frac{1}{\sqrt{14}}(6-1) = \frac{5}{\sqrt{14}}$$

$$\int_C \vec{F}\cdot d\vec{r} = \iint_S (\text{curl } \vec{F})\cdot\hat{n}\, ds$$

$$= \iint_s \frac{5}{\sqrt{14}}\, ds = \frac{5}{\sqrt{14}}\iint_R \frac{dx\, dy}{\hat{k}.\frac{1}{\sqrt{14}}(3\hat{i} + 2\hat{j} + \hat{k})} = 5\iint_R dx\, dy$$

where R is the projection of S on the $x\,y$-plane *i.e.* triangle OAB.

$$= 5.\text{ Area of triangle } OAB = \frac{5}{2}(2\times 3) = 15$$ **Ans.**

Example 31. *Evaluate* $\oint_C \vec{F}\cdot d\vec{r}$ *by Stoke's Theorem, where* $\vec{F} = (x^2 + y^2)\,\hat{i} - 2\,xy\,\hat{j}$ *and C is the boundary of the rectangle* $x = \pm\, a$, $y = 0$ *and* $y = b$. (*U.P., I Semester, Winter 2002*)

Solution. Since the z co-ordinate of each vertex of the given rectangle is zero, hence the given rectangle must lie in the xy-plane.

Here, the co-ordinates of A, B, C and D are $(a, 0)$, (a, b), $(-a, b)$ and $(-a, 0)$ respectively.

$$\therefore \quad \text{Curl } \overline{F} = \begin{vmatrix} \hat{i} & \hat{j} & \hat{k} \\ \dfrac{\partial}{\partial x} & \dfrac{\partial}{\partial y} & \dfrac{\partial}{\partial z} \\ x^2 + y^2 & -2xy & 0 \end{vmatrix} = -4\,y\,k$$

Here, $\hat{n} = \hat{k}$, so by Stoke's theorem, we've

$$\oint_C \vec{F}\cdot d\vec{r} = \iint_S \text{curl } \overline{F}\cdot\hat{n}\, d\, s$$

$$= \iint_S (-4\,y\hat{k})\cdot(\hat{k})\, d\,x\, d\,y = -4\int_{x=-a}^{a}\int_{y=0}^{b} y\, dx\, dy$$

$$= -4\int_{-a}^{a}\left[\frac{y^2}{2}\right]_0^b dx = -2b^2\int_{-a}^{a} dx = -4ab^2$$ **Ans.**

Example 32. *Apply Stoke's Theorem to calculate* $\int_c 4\,y\,dx + 2\,z\,dy + 6\,y\,dz$ *where c is the curve of intersection of* $x^2 + y^2 + z^2 = 6\,z$ *and* $z = x + 3$.

Solution.
$$\int_c \vec{F}\cdot\vec{dr} = \int_c 4\,y\,dx + 2\,z\,dy + 6\,y\,dz$$
$$= \int_c (4\,y\hat{i} + 2\,z\hat{j} + 6\,y\hat{k})\cdot(\hat{i}dx + \hat{j}dy + \hat{k}dz)$$
$$\vec{F} = 4\,y\hat{i} + 2\,z\hat{j} + 6\,y\hat{k}$$
$$\nabla\times\vec{F} = \begin{vmatrix}\hat{i} & \hat{j} & \hat{k}\\ \frac{\partial}{\partial x} & \frac{\partial}{\partial y} & \frac{\partial}{\partial z}\\ 4y & 2z & 6y\end{vmatrix} = (6-2)\,\hat{i} - (0-0)\,\hat{j} + (0-4)\,\hat{k} = 4\hat{i} - 4\,\hat{k}$$

S is the surface of the circle $x^2 + y^2 + z^2 = 6z$, $z = x + 3$, $\hat{n}$ is normal to the plane $x - z + 3 = 0$

$$\hat{n} = \frac{\nabla\phi}{|\nabla\phi|} = \frac{\left(\hat{i}\frac{\partial}{\partial x} + \hat{j}\frac{\partial}{\partial y} + \hat{k}\frac{\partial}{\partial z}\right)(x - z + 3)}{|\nabla\phi|} = \frac{\hat{i} - \hat{k}}{\sqrt{1+1}} = \frac{\hat{i} - \hat{k}}{\sqrt{2}}$$

$$(\nabla\times F)\cdot\hat{n} = (4\,\hat{i} - 4\,\hat{k})\cdot\frac{\hat{i} - \hat{k}}{\sqrt{2}} = \frac{4+4}{\sqrt{2}} = 4\sqrt{2}$$

$$\int_c \vec{F}\cdot\vec{dr} = \iint_S (\text{curl } F)\cdot\hat{n}\,ds = \iint_S 4\sqrt{2}\,(dx\,dz) = 4\sqrt{2}\ (\text{area of circle})$$

Centre of the sphere $x^2 + y^2 + (z - 3)^2 = 9$, (0, 0, 3) lies on the plane $z = x + 3$. It means that the given circle is a great circle of sphere, where radius of the circle is equal to the radius of the sphere.

Radius of circle = 3, Area = $\pi\,(3)^2 = 9\,\pi$

$$\iint_S (\nabla\times F)\cdot\hat{n}\,ds = 4\sqrt{2}(9\pi) = 36\sqrt{2}\,\pi$$ **Ans.**

Example 33 *Verify Stoke's Theorem for the function* $\overline{F} = z\hat{i} + x\,\hat{j} + y\,\hat{k}$, *where C is the unit circle in xy-plane bounding the hemisphere* $z = \sqrt{(1 - x^2 - y^2)}$. *(U.P., I Semester Comp. 2002)*

Solution. Here
$$\overline{F} = z\hat{i} + x\,\hat{j} + y\,\hat{k}. \quad ...(1)$$

Also,
$$\overline{r} = x\hat{i} + y\,\hat{j} + z\,\hat{k} \;\Rightarrow\; \overline{dr} = d\,x\hat{i} + dy\,\hat{j} + dz\,\hat{k}.$$

$\therefore$
$$\overline{F}\cdot\overline{dr} = z\,dx + x\,dy + y\,dz.$$

$\therefore$
$$\oint_C \overline{F}\cdot\overline{dr} = \oint_C (z\,dx + x\,dy + y\,dz). \quad ...(2)$$

On the circle C, $x^2 + y^2 = 1$, $z = 0$ on the xy-plane. Hence on C, we have $z = 0$ so that $dz = 0$. Hence (2) reduces to

$$\oint_C \overline{F}\cdot\overline{dr} = \oint_C x\,dy. \quad ...(3)$$

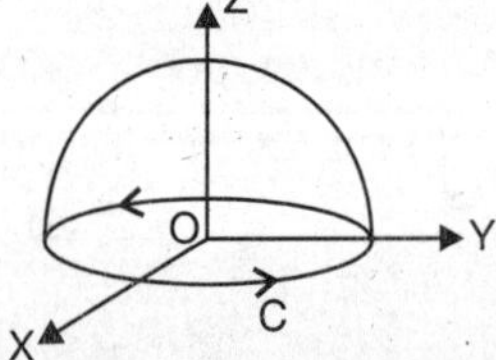

Now the parametric equations of C, *i.e.*, $x^2 + y^2 = 1$ are
$$x = \cos\phi,\ y = \sin\phi. \quad ...(4)$$

Using (4), (3) reduces to $\oint_C \vec{F}\cdot d\vec{r} = \int_{\phi=0}^{2\pi} \cos\phi \cos\phi \, d\phi = \int_0^{2\pi} \frac{1+\cos 2\phi}{2} d\phi$

$$= \frac{1}{2}\left[\phi + \frac{\sin 2\phi}{2}\right]_0^{2\pi} = \pi \quad ...(5)$$

Let $P(x, y, z)$ be any point on the surface of the hemisphere $x^2 + y^2 + z^2 = 1$, O origin is the centre of the sphere.

Radius $= \text{OP} = x\hat{i} + y\hat{j} + z\hat{k}$

Normal $= x\hat{i} + y\hat{j} + z\hat{k}$

$$\hat{n} = \frac{x\hat{i} + y\hat{j} + z\hat{k}}{\sqrt{x^2 + y^2 + z^2}} = x\hat{i} + y\hat{j} + z\hat{k}$$

(Radius is $\perp$ to tangent *i.e.* Radius is normal)

$x = \sin\theta\cos\phi,\; y = \sin\theta\sin\phi,\; z = \cos\theta$ ...(6)

$\hat{n} = \sin\theta\cos\phi\,\hat{i} + \sin\theta\sin\phi\,\hat{j} + \cos\theta\,\hat{k}$

Also, $$\text{Curl } \vec{F} = \begin{vmatrix} \hat{i} & \hat{j} & \hat{k} \\ \partial/\partial x & \partial/\partial y & \partial/\partial z \\ z & x & y \end{vmatrix} = \hat{i} + \hat{j} + \hat{k} \quad ...(7)$$

$$\text{Curl } \vec{F}\cdot\hat{n} = (\hat{i}+\hat{j}+\hat{k}).(\sin\theta\cos\phi\,\hat{i} + \sin\theta\sin\phi\,\hat{j} + \sin\theta\,\hat{k})$$
$$= \sin\theta\cos\phi + \sin\theta\sin\phi + \cos\theta$$

$$\therefore \quad \iint_s \text{Curl } \vec{F}\cdot\hat{n}\, dS = \int_{\theta=0}^{\pi/2}\int_{\phi=0}^{2\pi} (\hat{i}+\hat{j}+\hat{k})$$

$$.(\sin\theta\cos\phi\,\hat{i} + \sin\theta\sin\phi\,\hat{j} + \cos\theta\,\hat{k})\sin\theta\, d\theta\, d\phi$$

$$= \int_{\theta=0}^{\pi/2} \sin\theta\, d\theta \int_{\phi=0}^{2\pi} (\sin\theta\cos\phi + \sin\theta\sin\phi + \cos\theta)\, d\phi$$

[$\because$ dS = Elementary area on hemisphere = $\sin\theta\, d\theta\, d\phi$]

$$= \int_0^{\pi/2} \sin\theta\, d\theta\, [\sin\theta\sin\phi + \sin\theta(-\cos\phi) + \phi\cos\theta]_0^{2\pi} = \int_0^{\pi/2} \sin\theta\, d\theta$$

$$= \int_0^{\pi/2} (0 + 0 + 2\pi\sin\theta\cos\theta)\, d\theta = \pi\int_0^{\pi/2} \sin 2\theta\, d\theta = \pi\left[-\frac{\cos 2\theta}{2}\right]_0^{\pi/2}$$

$= -(\pi/2)\,[-1-1] = \pi.$

From (5) and (8), $\oint_C \vec{F}\cdot d\vec{r} = \iint_S \text{curl } \overline{F}\cdot\hat{n}\, dS$, which verifies Stokes's theorem.

Example 34. *Verify Stoke's theorem for the vector field* $\overline{F} = (2x - y)\hat{i} - yz^2\hat{j} - y^2 z\hat{k}$ *over the upper half of the surface* $x^2 + y^2 + z^2 = 1$ *bounded by its projection on xy- plane.*

(*Nagpur University, Summer 2001*)

Solution. Let S be the upper half surface of the sphere $x^2 + y^2 + z^2 = 1$. The boundary C or S is a circle in the xy plane of radius unity and centre O. The equation of C are $x^2 + y^2 = 1$, $z = 0$ whose parametric form is

$x = \cos t,\; y = \sin t,\; z = 0,\; 0 < t < 2\pi$

$$\int_C \vec{F}\cdot d\vec{r} = \int_C [(2x-y)\hat{i} - yz^2\hat{j} - y^2 z\hat{k}].[\hat{i}\,dx + \hat{j}\,dy + \hat{k}\,dz]$$

$$= \int_C [(2x-y)\,dx - yz^2\,dy - y^2 z\,dz]$$

$$= \int_C (2x-y)\,dx, \text{ since on } C,\ z = 0 \text{ and } 2z = 0$$

$$= \int_0^{2\pi} (2\cos t - \sin t)\frac{dx}{dt}\,dt = \int_0^{2\pi} (2\cos t - \sin t)(-\sin t)\,dt$$

$$= \int_0^{2\pi} (-\sin 2t + \sin^2 t)\, dt = \int_0^{2\pi} \left(-\sin 2t + \frac{1-\cos 2t}{2}\right) dt$$

$$= \left[\frac{\cos 2t}{2} + \frac{t}{2} - \frac{\sin 2t}{4}\right]_0^{2\pi} = \frac{1}{2} + \pi - \frac{1}{2} = \pi \qquad ...(1)$$

$$\text{Curl } \vec{F} = \begin{vmatrix} \hat{i} & \hat{j} & \hat{k} \\ \frac{\partial}{\partial x} & \frac{\partial}{\partial y} & \frac{\partial}{\partial z} \\ 2x - y & -yz^2 & -y^2 z \end{vmatrix} = (-2yz + 2yz)\,\hat{i} + (0-0)\,\hat{j} + (0+1)\,\hat{k} = \hat{k}$$

$$\text{Curl } \vec{F}.\hat{n} = \hat{k}.\hat{n} = \hat{n}.\hat{k}$$

$$\iint_S Curl\ \vec{F}.\hat{n}\, ds = \iint_S \hat{n}.\hat{k}\, ds = \iint_R \hat{n}.\hat{k}.\frac{dx}{\hat{n}}.\frac{dy}{\hat{k}}$$

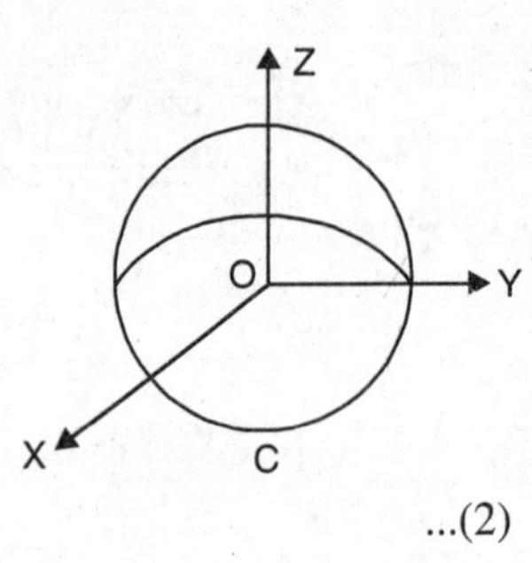

Where R is the projection of S on xy-plane.

$$= \int_{-1}^{1} \int_{-\sqrt{1-x^2}}^{\sqrt{1-x^2}} dx\, dy$$

$$= \int_{-1}^{1} 2\sqrt{1-x^2}\, dx = 4\int_0^1 \sqrt{1-x^2}\, dx$$

$$= 4\left[\frac{x}{2}\sqrt{1-x^2} + \frac{1}{2}\sin^{-1} x\right]_0^1 = 4\left[\frac{1}{2}.\frac{\pi}{2}\right] = \pi \qquad ...(2)$$

From (1) and (2), we have

$\therefore \quad \int_C \vec{F}.\vec{dr} = \iint \text{Curl } \overline{F}.\hat{n}\, ds$ which is the Stoke's theorem. **Ans.**

Example 35 *Verify Stoke's Theorem for*

$$\vec{F} = (x^2 + y - 4)\,\hat{i} + 3\,xy\hat{j} + (2\,xz + z^2)\,\hat{k}$$

over the surface of hemisphere $x^2 + y^2 + z^2 = 16$ *above the xy-plane.*

Solution. $\int_c \vec{F}\cdot\vec{dr}$, where c is the boundary of the circle $x^2 + y^2 + z^2 = 16$ (bounding the hemispherical surface)

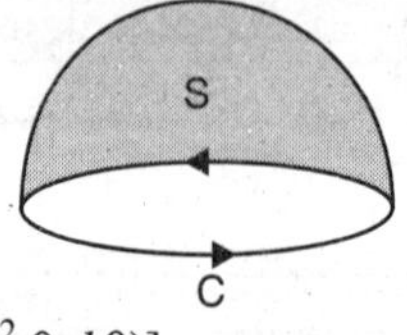

$$= \int_c [(x^2 + y - 4)\hat{i} + 3xy\hat{j} + (2xz + z^2)\hat{k}]\cdot(\hat{i}dx + \hat{j}dy)$$

$$= \int_c [(x^2 + y - 4)\, dx + 3\,xy\, dy)]$$

Putting $x = 4\cos\theta,\ y = 4\sin\theta,\ dx = -4\sin\theta\, d\theta,\ dy = 4\cos\theta\, d\theta$

$$= \int_0^{2\pi} [(16\cos^2\theta + 4\sin\theta - 4)(-4\sin\theta\, d\theta) + (192\sin\theta\cos^2\theta\, d\theta)]$$

$$= 16\int_0^{2\pi} [-4\cos^2\theta\sin\theta - \sin^2\theta + \sin\theta + 12\sin\theta\cos^2\theta]\, d\theta$$

$$= 16\int_0^{2\pi} (8\sin\theta\cos^2\theta - \sin^2\theta + \sin\theta)\, d\theta$$

$$= -16\int_0^{2\pi} \sin^2\theta\, d\theta \qquad \left\{\begin{aligned} &\int_0^{2\pi} \sin^n\theta\cos\theta\, d\theta = 0 \\ &\int_0^{2\pi} \cos^n\theta\sin\theta\, d\theta = 0 \end{aligned}\right\}$$

$$= -16\times 4\int_0^{\frac{\pi}{2}} \sin^2\theta\, d\theta = -64\left(\frac{1}{2}\frac{\pi}{2}\right) = -16\,\pi.$$

To evaluate surface integral $\nabla\times\vec{F} = \begin{vmatrix} \hat{i} & \hat{j} & \hat{k} \\ \frac{\partial}{\partial x} & \frac{\partial}{\partial y} & \frac{\partial}{\partial z} \\ x^2 + y - 4 & 3xy & 2xz + z^2 \end{vmatrix}$

$$= (0-0)\,\hat{i} - (2z-0)\,\hat{j} + (3y-1)\,\hat{k} = -2z\hat{j} + (3y-1)\,\hat{k}$$

$$\hat{n} = \frac{\nabla \phi}{|\nabla \phi|} = \frac{\left(\hat{i}\frac{\partial}{\partial x} + \hat{j}\frac{\partial}{\partial y} + \hat{k}\frac{\partial}{\partial z}\right)(x^2+y^2+z^2-16)}{|\nabla \phi|}$$

$$= \frac{2x\hat{i} + 2y\hat{j} + 2z\hat{k}}{\sqrt{4x^2+4y^2+4z^2}} = \frac{x\hat{i} + y\hat{j} + z\hat{k}}{\sqrt{x^2+y^2+z^2}} = \frac{x\hat{i} + y\hat{j} + z\hat{k}}{4}$$

$$(\nabla \times \vec{F})\cdot\hat{n} = [-2z\hat{j} + (3y-1)\,\hat{k}]\cdot\frac{x\hat{i} + y\hat{j} + z\hat{k}}{4} = \frac{-2yz + (3y-1)z}{4}$$

$$\hat{k}\cdot\hat{n}\cdot ds = dx\,dy \Rightarrow \frac{x\hat{i} + y\hat{j} + z\hat{k}}{4}\,.\,k\,ds = dx\,dy \Rightarrow \frac{z}{4}\,ds = dx\,dy$$

$$\therefore \quad ds = \frac{4}{z}\,dx\,dy$$

$$\iint(\nabla\times F)\cdot\hat{n}\,ds = \iint\frac{-2yz + (3y-1)z}{4}\left(\frac{4}{z}\,dx\,dy\right)$$

$$= \iint[-2y + (3y-1)]\,dx\,dy = \iint(y-1)\,dx\,dy$$

On putting $x = r\cos\theta$, $y = r\sin\theta$, $dx\,dy = r\,d\theta\,dr$, we get

$$= \iint(r\sin\theta - 1)\,r\,d\theta\,dr = \int d\theta\int(r^2\sin\theta - r)\,dr$$

$$= \int_0^{2\pi} d\theta\left(\frac{r^3}{3}\sin\theta - \frac{r^2}{2}\right)_0^4 = \int_0^{2\pi} d\theta\left(\frac{64}{3}\sin\theta - 8\right)$$

$$= \left(-\frac{64}{3}\cos\theta - 8\theta\right)_0^{2\pi} = \frac{-64}{3} - 16\pi + \frac{64}{3} = -16\pi$$

The line integral is equal to the surface integral, hence Stoke's Theorem is verified. **Proved.**

Example 36. *Verify Stoke's theorem for a vector field defined by* $\vec{F} = (x^2 - y^2)\,\hat{i} + 2xy\,\hat{j}$ *in the rectangular in xy-plane bounded by lines* $x = 0$, $x = a$, $y = 0$, $y = b$.

(Nagpur University, Summer 2000)

Solution. Here we have to verify Stoke's theorem $\int_C \vec{F}\,.\,\vec{dr} = \iint_S (\nabla\times\vec{F})\,.\,\hat{n}\,ds$

Where 'C' be the boundary of rectangle (*ABCD*) and *S* be the surface enclosed by curve *C*.

$$\vec{F} = (x^2 - y^2)\,\hat{i} + (2xy)\,\hat{j}$$

$$\vec{F}\,.\,\vec{dr} = [(x^2-y^2)\,\hat{i} + 2xy\,\hat{j}]\,.\,[\hat{i}\,dx + \hat{j}\,dy]$$

$$\Rightarrow \quad \vec{F}\,.\,\vec{dr} = (x^2 + y^2)\,dx + 2xy\,dy \qquad \ldots(1)$$

Now,

$$\int_C \vec{F}\,.\,\vec{dr} = \int_{OA} F\,.\,\vec{dr} + \int_{AB}\vec{F}\,.\,\vec{dr} + \int_{BC}\vec{F}\,.\,\vec{dr} + \int_{CO}\vec{F}\,.\,\vec{dr} \qquad \ldots(2)$$

Along *OA*, put $y = 0$ so that $k\,dy = 0$ in (1) and $\vec{F}\,.\,\vec{dr} = x^2\,dx$,
Where *x* is from 0 to *a*.

$$\therefore \quad \int_{OA}\vec{F}\,.\,\vec{dr} = \int_0^a x^2 dx = \left[\frac{x^3}{3}\right]_0^a = \frac{a^3}{3} \qquad \ldots(3)$$

Along *AB*, put $x = a$ so that $dx = 0$ in (1), we get $\vec{F}\,.\,\vec{dr} = 2ay\,dy$
Where *y* is from 0 to *b*.

$$\therefore \quad \int_{AB}\vec{F}\,.\,\vec{dr} = \int_0^b 2ay\,dy = [ay^2]_0^b = ab^2 \qquad \ldots(4)$$

Along BC, put $y = b$ and $dy = 0$ in (1) we get $\bar{F}\,.\,\bar{dr} = (x^2 - b^2)\,dx$, where x is from a to 0.

$$\therefore \qquad \int_{BC} \vec{F}\,.\,\vec{dr} = \int_a^0 (x^2 - b^2)\,dx = \left[\frac{x^3}{3} - b^2 x\right]_a^0 = \frac{-a^3}{3} + b^2 a \qquad ...(5)$$

Along CO, put $x = 0$ and $dx = 0$ in (1), we get $\vec{F}\,.\,\vec{dr} = 0$

$$\therefore \qquad \int_{CO} \vec{F}\,.\,\vec{dr} = 0 \qquad ...(6)$$

Putting the values of integrals (3), (4), (5) and (6) in (2), we get

$$\int_C \vec{F}\,.\,\vec{dr} = \frac{a^3}{3} + ab^2 - \frac{a^3}{3} + ab^2 + 0 = 2ab^2 \qquad ...(7)$$

Now we have to evaluate R.H.S. of Stoke's Theorem *i.e.* $\iint_S (\nabla \times \vec{F})\,.\,\hat{n}\,ds$

We have,

$$\nabla \times \vec{F} = \begin{vmatrix} \hat{i} & \hat{j} & \hat{k} \\ \dfrac{\partial}{\partial x} & \dfrac{\partial}{\partial y} & \dfrac{\partial}{\partial z} \\ x^2 - y^2 & 2xy & 0 \end{vmatrix} = (2y + 2y)\,\hat{k} = 4y\,\hat{k}$$

Also the unit vector normal to the surface S in outward direction is $\hat{n} = k$

($\because$ z-axis is normal to surface S)

Also in xy-plane $ds = dx\,dy$

$$\therefore \qquad \iint_S (\nabla \times \vec{F})\,.\,\hat{n}.\,ds = \iint_R 4y\,\hat{k}\,.\,\hat{k}\,dx\,dy = \iint_R 4y\,dx\,dy.$$

Where R be the region of the surface S.

Consider a strip parallel to y-axis. This strip starts on line y = 0 (*i.e.* x-axis) and end on the line $y = b$, We move this strip from $x = 0$ (y-axis) to $x = a$ to cover complete region R.

$$\therefore \qquad \iint_S (\nabla \times \vec{F})\,.\,\hat{n}.\,ds = \int_0^a \left[\int_0^b 4y\,dy\right] dx = \int_0^a [2y^2]_0^b\,dx$$

$$= \int_0^a 2b^2\,dx = 2b^2\,[x]_0^a = 2ab^2 \qquad ...(8)$$

$\therefore$ From (7) and (8), we get

$\int_C \vec{F}\,.\,\vec{dr} = \iint_S (\nabla \times \bar{F})\,.\,\hat{n}\,ds$ and hence the Stoke's theorem is verified.

Example 37. *Verify Stoke's Theorem for the function*

$$\vec{F} = x^2\hat{i} - xy\hat{j}$$

integrated round the square in the plane $z = 0$ and bounded by the lines

$$x = 0,\ y = 0,\ x = a,\ y = a.$$

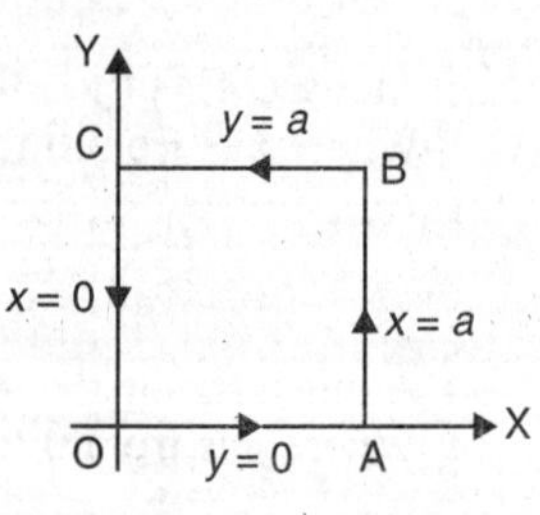

Solution. We have, $\vec{F} = x^2\hat{i} - xy\hat{j}$

$$\nabla \times \vec{F} = \begin{vmatrix} \hat{i} & \hat{j} & \hat{k} \\ \dfrac{\partial}{\partial x} & \dfrac{\partial}{\partial y} & \dfrac{\partial}{\partial z} \\ x^2 & -xy & 0 \end{vmatrix}$$

$$= (0 - 0)\,\hat{i} - (0 - 0)\,\hat{j} + (-y - 0)\,\hat{k} = -y\hat{k} \qquad (\hat{n} \perp \text{ to } xy \text{ plane } i.e.\ \hat{k})$$

$$\iint_S (\nabla \times \vec{F}) \cdot \hat{n}\, ds = \iint_S (-yk) \cdot k\, dx\, dy$$

$$= \int_0^a dx \int_0^a -y dy = \int_0^a dx \left[-\frac{y^2}{2}\right]_0^a = -\frac{a^2}{2}(x)_0^a = -\frac{a^3}{2} \qquad ...(1)$$

To obtain line integral

$$\int_C \vec{F} \cdot \vec{dr} = \int_C (x^2 \hat{i} - xy\hat{j}) \cdot (\hat{i}\, dx + \hat{j}\, dy) = \int_C (x^2\, dx - xy\, dy)$$

where c is the path $OABCO$ as shown in the figure.

Also, $$\int_C \vec{F} \cdot \vec{dr} = \int_{OABCO} \vec{F} \cdot dr = \int_{OA} \vec{F} \cdot dr + \int_{AB} \vec{F} \cdot dr + \int_{BC} \vec{F} \cdot dr + \int_{CO} \vec{F} \cdot dr \qquad ...(2)$$

Along OA, $y = 0$, $dy = 0$

line	Eq. of line		Lower limit	Upper limit
OA	$y = 0$	$dy = 0$	$x = 0$	$x = a$
AB	$x = a$	$dx = 0$	$y = 0$	$y = a$
BC	$y = a$	$dy = 0$	$x = a$	$x = 0$
CO	$x = 0$	$dx = 0$	$y = a$	$y = 0$

$$\int_{OA} \vec{F} \cdot \vec{dr} = \int_{OA} (x^2 dx - xydy)$$

$$= \int_0^a x^2\, dx = \left[\frac{x^3}{3}\right]_0^a = \frac{a^3}{3}$$

Along AB, $x = a$, $dx = 0$

$$\int_{AB} \vec{F} \cdot \vec{dr} = \int_{AB} (x^2\, dx - x\, y\, d\, y)$$

$$= \int_0^a -a\, y\, d\, y = -a\left[\frac{y^2}{2}\right]_0^a = -\frac{a^3}{2}$$

Along BC, $y = a$, $dy = 0$

$$\int_{BC} \vec{F} \cdot \vec{dr} = \int_{BC} (x^2 dx - xy\, dy) = \int_a^0 x^2 dx = \left[\frac{x^3}{3}\right]_a^0 = -\frac{a^3}{3}$$

Along CO, $x = 0$, $dx = 0$

$$\int_{CO} \vec{F} \cdot \vec{dr} = \int_{CO} (x^2 dx - xy\, dy) = 0$$

Putting the values of these integrals in (2), we have

$$\int_C \vec{F} \cdot \vec{dr} = \frac{a^3}{3} - \frac{a^3}{2} - \frac{a^3}{3} + 0 = -\frac{a^3}{2} \qquad ...(3)$$

From (1) and (3), $\iint_S (\vec{\nabla} \times \vec{F}) \cdot \hat{n}\, ds = \int_C \vec{F} \cdot \vec{dr}$

Hence, Stoke's Theorem is verified. **Ans.**

Example 38. *Verify Stoke's Theorem for* $\vec{F} = (x + y)\,\hat{i} + (2x - z)\,\hat{j} + (y + z)\,\hat{k}$ *for the surface of a triangular lamina with vertices* (2, 0, 0), (0, 3, 0) *and* (0, 0, 6).

(*Nagpur University 2004, K. U. Dec. 2009, 2008, A.M.I.E.T.E., Summer 2000*)

Solution. Here the path of integration c consists of the straight lines AB, BC, CA where the co-ordinates of A, B, C and (2, 0, 0), (0, 3, 0) and (0, 0, 6) respectively. Let S be the plane surface of triangle ABC bounded by C. Let $\hat{n}$ be unit normal vector to surface S. Then by Stoke's Theorem, we must have

$$\oint_c \vec{F} \cdot \vec{dr} = \iint_s \text{curl}\, \vec{F} \cdot \hat{n}\, ds \qquad ...(1)$$

L.H.S. of (1)= $\int_{ABC}^{c} \vec{F}\cdot\vec{dr} = \int_{AB} \vec{F}\cdot\vec{dr} + \int_{BC} \vec{F}\cdot\vec{dr} + \int_{CA} \vec{F}\cdot\vec{dr}$

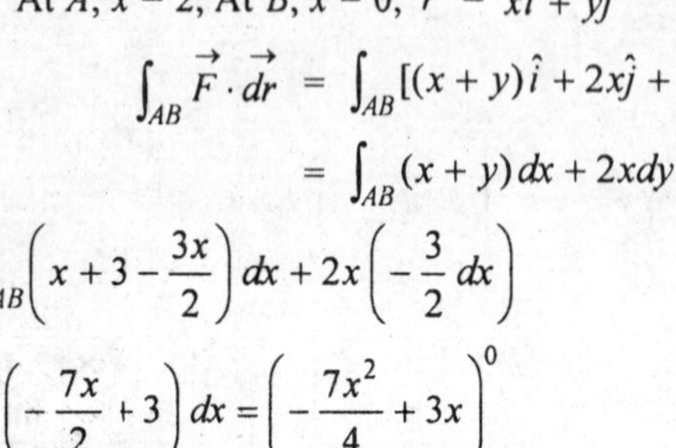

Along line AB, $z = 0$, equation of AB is $\frac{x}{2} + \frac{y}{3} = 1$

$\Rightarrow \quad y = \frac{3}{2}(2 - x),\ dy = -\frac{3}{2}\,dx$

At A, $x = 2$, At B, $x = 0$, $\vec{r} = x\hat{i} + y\hat{j}$

$$\int_{AB} \vec{F}\cdot\vec{dr} = \int_{AB} [(x + y)\hat{i} + 2x\hat{j} + y\hat{k}]\cdot(\hat{i}dx + \hat{j}dy)$$

$$= \int_{AB} (x + y)\,dx + 2x\,dy$$

$$= \int_{AB} \left(x + 3 - \frac{3x}{2}\right) dx + 2x\left(-\frac{3}{2}\,dx\right)$$

$$= \int_{2}^{0} \left(-\frac{7x}{2} + 3\right) dx = \left(-\frac{7x^2}{4} + 3x\right)_{2}^{0}$$

$= (7 - 6) = +1$

line	Eq. of line		Lower limit	Upper limit
AB	$\frac{x}{2} + \frac{y}{3} = 1$ $z = 0$	$dy = -\frac{3}{2}\,dx$	At A $x = 2$	At B $x = 0$
BC	$\frac{y}{3} + \frac{z}{6} = 1$ $x = 0$	$dz = -2dy$	At B $y = 3$	At C $y = 0$
CA	$\frac{x}{2} + \frac{z}{6} = 1$ $y = 0$	$dz = -3dx$	At C $x = 0$	At A $x = 2$

Along line BC, $x = 0$, Equation of BC is $\frac{y}{3} + \frac{z}{6} = 1$ or $z = 6 - 2y$, $dz = -2dy$

At B, $y = 3$, At C, $y = 0$, $\vec{r} = y\hat{j} + z\hat{k}$

$$\int_{BC} \vec{F}\cdot\vec{dr} = \int_{BC} [yi + zj + (y + z)k]\cdot(jdy + kdz) = \int_{BC} -z\,dy + (y + z)\,dz$$

$$= \int_{3}^{0} (-6 + 2y)\,dy + (y + 6 - 2y)(-2dy)$$

$$= \int_{3}^{0} (4y - 18)\,dy = (2y^2 - 18y)_{3}^{0} = 36$$

Along line CA, $y = 0$, Eq. of CA, $\frac{x}{2} + \frac{z}{6} = 1$ or $z = 6 - 3x$, $dz = -3dx$

At C, $x = 0$, at A, $x = 2$, $\vec{r} = x\hat{i} + z\hat{k}$

$$\int_{CA} \vec{F}\cdot\vec{dr} = \int_{CA} [x\hat{i} + (2x - z)\,\hat{j} + z\hat{k}]\cdot[dx\hat{i} + dz\hat{k}] = \int_{CA} (x\,dx + z\,dz)$$

$$= \int_{0}^{2} x\,dx + (6 - 3x)(-3dx) = \int_{0}^{2} (10x - 18)\,dx = [5x^2 - 18x]_{0}^{2} = -16$$

$$\text{L.H.S. of (1)} = \int_{ABC} \vec{F}\cdot\vec{dr} = \int_{AB} \vec{F}\cdot\vec{dr} + \int_{BC} \vec{F}\cdot\vec{dr} + \int_{CA} \vec{F}\cdot\vec{dr} = 1 + 36 - 16 = 21 \quad \text{...(2)}$$

$$\text{Curl } \overline{F} = \nabla\times\overline{F} = \left(\hat{i}\frac{\partial}{\partial x} + \hat{j}\frac{\partial}{\partial y} + \hat{k}\frac{\partial}{\partial z}\right)\times[(x + y)\,\hat{i} + (2x - z)\,\hat{j} + (y + z)\,k]$$

$$= \begin{vmatrix} \hat{i} & \hat{j} & \hat{k} \\ \frac{\partial}{\partial x} & \frac{\partial}{\partial y} & \frac{\partial}{\partial z} \\ x + y & 2x - z & y + z \end{vmatrix} = (1 + 1)\,\hat{i} - (0 - 0)\,\hat{j} + (2 - 1)\,\hat{k} = 2\hat{i} + \hat{k}$$

Equation of the plane of ABC is $\frac{x}{2} + \frac{y}{3} + \frac{z}{6} = 1$

Normal to the plane ABC is

$$\nabla\phi = \left(\hat{i}\frac{\partial}{\partial x}+\hat{j}\frac{\partial}{\partial y}+\hat{k}\frac{\partial}{\partial z}\right)\left(\frac{x}{2}+\frac{y}{3}+\frac{z}{6}-1\right) = \frac{\hat{i}}{2}+\frac{\hat{j}}{3}+\frac{\hat{k}}{6}$$

$$\text{Unit Normal Vector} = \frac{\frac{\hat{i}}{2}+\frac{\hat{j}}{3}+\frac{\hat{k}}{6}}{\sqrt{\frac{1}{4}+\frac{1}{9}+\frac{1}{36}}}$$

$$\hat{n} = \frac{1}{\sqrt{14}}(3\hat{i}+2\hat{j}+\hat{k})$$

$$\text{R.H.S. of (1)} = \iint_s \text{curl } \vec{F}\cdot\hat{n}\,ds = \iint_s (2\hat{i}+\hat{k})\cdot\frac{1}{\sqrt{4}}(3\hat{i}+2\hat{j}+\hat{k})\frac{dx\,dy}{\frac{1}{\sqrt{14}}(3\hat{i}+2\hat{j}+\hat{k}).\hat{k}}$$

$$= \iint_s \frac{(6+1)}{\sqrt{14}}\frac{dx\,dy}{\frac{1}{\sqrt{14}}} = 7\iint dx\,dy = 7 \text{ Area of } \Delta \text{ OAB}$$

$$= 7\left(\frac{1}{2}\times 2\times 3\right) = 21 \qquad ...(3)$$

with the help of (2) and (3) we find (1) is true and so Stoke's *Theorem* is verified.

Example 39. *Verify Stoke's Theorem for*

$$\vec{F} = (y-z+2)\,\hat{i} + (yz+4)\,\hat{j} - (xz)\,\hat{k}$$

over the surface of a cube x = 0, y = 0, z = 0, x = 2, y = 2, z = 2 above the XOY plane (open the bottom).

Solution. Consider the surface of the cube as shown in the figure. Bounding path is *OABCO* shown by arrows.

$$\int_c \vec{F}\cdot\vec{d}r = \int[(y-z+2)\,\hat{i}+(yz+4)\,\hat{j}-(xz)\,\hat{k}]\cdot(\hat{i}dx+\hat{j}dy+\hat{k}dz)$$

$$= \int_c (y-z+2)\,dx + (yz+4)\,dy - xz\,dz$$

$$\int_c \vec{F}\cdot\vec{dr} = \int_{OA}\vec{F}\cdot\vec{dr} + \int_{AB}\vec{F}\cdot\vec{dr} + \int_{BC}\vec{F}\cdot\vec{dr} + \int_{CO}\vec{F}\cdot\vec{dr} \qquad ...(1)$$

(1) Along *OA*, $y = 0$, $dy = 0$, $z = 0$, $dz = 0$

$$\int_{OA}\vec{F}\cdot\vec{dr} = \int_0^2 2\,dx = [2x]_0^2 = 4$$

(2) Along *AB*, $x = 2$, $dx = 0$, $z = 0$, $dz = 0$

$$\int_{AB}\vec{F}\cdot\vec{dr} = \int_0^2 4\,dy = 4\,(y)_0^2 = 8$$

(3) Along *BC*, $y = 2$, $dy = 0$, $z = 0$, $dz = 0$

$$\int_{BC}\vec{F}\cdot\vec{dr} = \int_0^2 (2-0+2)\,dx = (4x)_2^0 = -8$$

(4) Along *CO*, $x = 0$, $dx = 0$, $z = 0$, $dz = 0$

$$\int_{CO}\vec{F}\cdot\vec{dr} = \int(y-0+2)\times 0+(0+4)\,dy - 0$$

	Line	Equ. of line		Lower limit	Upper limit	$\bar{F}.\bar{dr}$
1	*OA*	$y=0$, $z=0$	$dy=0$, $dz=0$	$x=0$	$x=2$	$2\,dx$
2	*AB*	$x=2$, $z=0$	$dx=0$, $dz=0$	$y=0$	$y=2$	$4\,dy$
3	*BC*	$y=2$, $z=0$	$dy=0$, $dz=0$	$x=2$	$x=0$	$4\,dx$
4	*CO*	$x=0$, $z=0$	$dx=0$, $dz=0$	$y=2$	$y=0$	$4\,dy$

$$= 4 \int dy = 4\,(y)_2^0 = -8$$

On putting the values of these integrals in (1), we get

$$\int_c \vec{F}\cdot\vec{dr} = 4 + 8 - 8 - 8 = -4$$

To obtain surface integral

$$\nabla \times \vec{F} = \begin{vmatrix} \hat{i} & \hat{j} & \hat{k} \\ \dfrac{\partial}{\partial x} & \dfrac{\partial}{\partial y} & \dfrac{\partial}{\partial z} \\ y - z + 2 & yz + 4 & -xz \end{vmatrix}$$

$$= (0 - y)\,\hat{i} - (-z + 1)\,\hat{j} + (0 - 1)\,\hat{k} = -y\hat{i} + (z - 1)\,\hat{j} - \hat{k}$$

Here we have to integrate over the five surfaces, *ABDE, OCGF, BCGD, OAEF, DEFG.*
Over the surface *ABDE* ($x = 2$), $\hat{n} = i$, $ds = dy\,dz$

$$\iint (\nabla \times \vec{F})\cdot\hat{n}\,ds = \iint [-yi + (z-1)\,j - k]\cdot i\,dx\,dz = \iint -y\,dy\,dz$$

$$= \iint_R \left[F_3\,(x, y, z)\right]_{z = f_1(x,y)}^{z = f_2(x,y)} dx\,dy$$

$$= -\int_0^2 y\,dy \int_0^2 dz = -\left[\frac{y^2}{2}\right]_0^2 [z]_0^2 = -4$$

	Surface	Outward normal	ds	
1	*ABDE*	i	$dy\,dz$	$x = 2$
2	*OCGF*	$-i$	$dy\,dz$	$x = 0$
3	*BCGD*	j	$dx\,dz$	$y = 2$
4	*OAEF*	$-j$	$dx\,dz$	$y = 0$
5	*DEFG*	k	$dx\,dy$	$z = 2$

Over the surface OCGF ($x = 0$), $\hat{n} = -i$, $ds = dy\,dz$

$$\iint (\nabla \times \vec{F})\cdot\hat{n}\,ds = \iint [-y\hat{i} + (z-1)\,\hat{j} - \hat{k}]\cdot(-\hat{i})\,dy\,dz$$

$$= \iint y\,dy\,dz = \int_0^2 y\,dy \int_0^2 dz = 2\left[\frac{y^2}{2}\right]_0^2 = 4$$

(3) Over the surface *BCGD*, ($y = 2$), $\hat{n} = j$, $ds = dx\,dz$

$$\iint (\nabla \times \vec{F})\cdot\hat{n}\,ds = \iint [-y\hat{i} + (z-1)\,\hat{j} - \hat{k}]\cdot\hat{j}\,dx\,dz$$

$$= -\iint (z-1)\,dx\,dz = \int_0^2 dx \int_0^2 (z-1)\,dz = -(x)_0^2\left(\frac{z^2}{2} - z\right)_0^2 = 0$$

(4) Over the surface *OAEF*, ($y = 0$), $\hat{n} = -\hat{j}$, $ds = dx\,dz$

$$\iint (\nabla \times \vec{F})\cdot\hat{n}\,ds = \iint [-y\hat{i} + (z-1)\hat{j} - \hat{k}]\cdot(-\hat{j})\,dx\,dz$$

$$= -\iint (z-1)\,dx\,dz = -\int_0^2 dx \int_0^2 (z-1)\,dz = -(x)_0^2\left(\frac{z^2}{2} - z\right)_0^2 = 0$$

(5) Over the surface *DEFG*, ($z = 2$), $\hat{n} = k$, $ds = dx\,dy$

$$\iint (\nabla \times \vec{F})\cdot\hat{n}\,ds = \iint [-y\hat{i} + (z-1)\hat{j} - \hat{k}]\cdot\hat{k}\,dx\,dy = -\iint dx\,dy$$

$$= -\int_0^2 dx \int_0^2 dy = -[x]_0^2\,[y]_0^2 = -4$$

Total surface integral $= -4 + 4 + 0 + 0 - 4 = -4$

Thus $\iint_S \text{curl}\,\vec{F}\cdot\hat{n}\,ds = \int_c \vec{F}\cdot\vec{dr} = -4$

which verifies Stoke's Theorem. **Ans.**

EXERCISE 24.5

1. Use the Stoke's Theorem to evaluate $\int_C y^2 dx + xy\, dy + xz\, dz$,

where C is the bounding curve of the hemisphere $x^2 + y^2 + z^2 = 1$, $z \geq 0$, oriented in the positive direction. **Ans.** 0

2. Evaluate the integral for $\int_C y^2 dx + z^2 dy + x^2 dz$, where C is the triangular closed path joining the points (0, 0, 0), (0, a, 0) and (0, 0, a) by transforming the integral to surface integral using Stoke's Theorem.

Ans. $\frac{a^3}{3}$.

3. Verify Stoke's Theorem for $\vec{A} = 3y\hat{i} - xz\hat{j} + yz^2\hat{k}$, where S is the surface of the paraboloid $2z = x^2 + y^2$ bounded by $z = 2$ and c is its boundary traversed in the clockwise direction. **Ans.** -20π

4. Evaluate $\int_C \vec{F} \cdot d\vec{R}$ where $\vec{F} = y\hat{i} + xz^3\hat{j} - zy^3\hat{k}$, C is the circl $x^2 + y^2 = 4$, $z = 1.5$ **Ans.** $\frac{19}{2}\pi$

5. Verify Stoke's Theorem for the vector field

$$\vec{F} = (2y + z)\,\hat{i} + (x - z)\,\hat{j} + (y - x)\,\hat{k}$$

over the portion of the plane $x + y + z = 1$ cut off by the co-ordinate planes.

6. Evaluate $\int_C \vec{F} \cdot dr$ by Stoke's Theorem for $\vec{F} = yz\,\hat{i} + zx\,\hat{j} + xy\,k$ and C is the curve of intersection of $x^2 + y^2 = 1$ and $y = z^2$. **Ans.** 0

7. If $\vec{F} = (x - z)\,\hat{i} + (x^3 + yz)\,\hat{j} + 3xy^2\,\hat{k}$ and S is the surface of the cone $z = a - \sqrt{(x^2 + y^2)}$ above the xy-plane, show that $\iint_S \text{curl}\,\vec{F} \cdot dS = 3\pi a^4 / 4$.

8. If $\vec{F} = 3y\hat{i} - xy\hat{j} + yz2\hat{k}$ and S is the surface of the paraboloid $2z = x^2 + y^2$ bounded by $z = 2$, show by using Stoke's Theorem that $\iint_S (\nabla \times \vec{F}) \cdot dS = 20\pi$.

0. If $\vec{F} = (y^2 + z^2 - x^2)\,\hat{i} + (z^2 + x^2 - y^2)\,\hat{j} + (x^2 + y^2 - z^2)\,\hat{k}$, evaluate $\int \text{curl}\,\vec{F} \cdot \hat{n}\, ds$ integrated over the portion of the surface $x^2 + y^2 - 2ax + az = 0$ above the plane $z = 0$ and verify Stoke's Theorem; where $\hat{n}$ is unit vector normal to the surface. *(A.M.I.E.T.E., Winter 20002)* **Ans.** $2\pi a^3$

10. Evaluate by using Stoke's Theorem $\int_C [\sin z\, dx - \cos x\, dy + \sin y\, dz]$ where C is the boundary of rectangle $0 \leq x \leq \pi$, $0 \leq y \leq 1$, $z = 3$. *(AMIETE, June 2010)* **Ans.** 2

24.8 GAUSS'S THEOREM OF DIVERGENCE

(Relation between surface integral and volume integral)

(GBTU, Dec. 2012, U.P., Ist Semester, Jan., 2011, Dec, 2006)

Statement. The surface integral of the normal component of a vector function F taken around a closed surface S is equal to the integral of the divergence of F taken over the volume V enclosed by the surface S.

Mathematically

$$\iint_S \vec{F} . \hat{n}\, ds = \iiint_V div\, \vec{F} dw$$

Proof. Let $\vec{F} = F_1\hat{i} + F_2\hat{j} + F_3\hat{k}$.

Putting the values of $\vec{F}, \hat{n}$ in the statement of the divergence theorem, we have

$$\iint_S F_1\,\hat{i} + F_2\,\hat{j} + F_3\,\hat{k}\cdot\hat{n}\,ds = \iiint_V \left(\hat{i}\frac{\partial}{\partial x} + \hat{j}\frac{\partial}{\partial y} + \hat{k}\frac{\partial}{\partial z}\right)(F_1\,\hat{i} + F_2\,\hat{j} + F_3\,\hat{k})\,dx\,dy\,dz.$$

$$= \iiint_V \left(\frac{\partial F_1}{\partial x} + \frac{\partial F_2}{\partial y} + \frac{\partial F_3}{\partial z}\right) dx\,dy\,dz \qquad ...(1)$$

We require to prove (1).

Let us first evaluate $\iiint_V \frac{\partial F_3}{\partial z}\,dx\,dy\,dz$.

$$\iiint_V \frac{\partial F_3}{\partial z}\,dx\,dy\,dz = \iint_R \left[\int_{z=f_1(x,y)}^{z=f_2(x,y)} \frac{\partial F_3}{\partial z}\,dz\right] dx\,dy$$

$$= \iint_R [F_3(x, y, f_2) - F_3(x, y, f_1)]\,dx\,dy \qquad ...(2)$$

For the upper part of the surface *i.e.* S_2, we have

$$dx\,dy = ds_2 \cos r_2 = \hat{n}_2 . \hat{k}\;ds_2$$

Again for the lower part of the surface *i.e.* S_1, we have,

$$dx\,dy = -\cos r_1,\; ds_1 = \hat{n}_1 .\;\hat{k}\,ds_1$$

$$\iint_R F_3(x, y, f_2)\,dx\,dy = \iint_{S_2} F_3\,\hat{n}_2\cdot\hat{k}\,ds_2$$

and $\qquad \iint_R F_3(x, y, f_1)\,dx\,dy = -\iint_{S_1} F_3\,\hat{n}_1\cdot\hat{k}\,ds_1$

Putting these values in (2), we have

$$\iiint_V \frac{\partial F_3}{\partial z}\,dv = \iint_{S_2} F_3\,\hat{n}_2\cdot\hat{k}\,ds_2 + \iint_{S_1} F_3\,\hat{n}_1\cdot\hat{k}\,ds_1 = \iint_S F_3\,\hat{n}\cdot\hat{k}\,ds \qquad ...(3)$$

Similarly, it can be shown that

$$\iiint_V \frac{\partial F_2}{\partial y}\,dv = \iint_S F_2\,\hat{n}\cdot\hat{j}\,ds \qquad ...(4)$$

$$\iiint_V \frac{\partial F_1}{\partial x}\,dv = \iint_S F_1\,\hat{n}\cdot\hat{i}\,ds \qquad ...(5)$$

Adding (3), (4) & (5), we have

$$\iiint \left(\frac{\partial F_1}{\partial x} + \frac{\partial F_2}{\partial y} + \frac{\partial F_3}{\partial z}\right) dv$$

$$= \iint_S (F_1\,\hat{i} + F_2\,\hat{j} + F_3\,\hat{k})\cdot\hat{n}\cdot ds$$

$$\Rightarrow \iiint_V (\nabla\cdot\overline{F})\,dv = \iint_S \overline{F}\cdot\hat{n}\cdot ds$$ **Proved.**

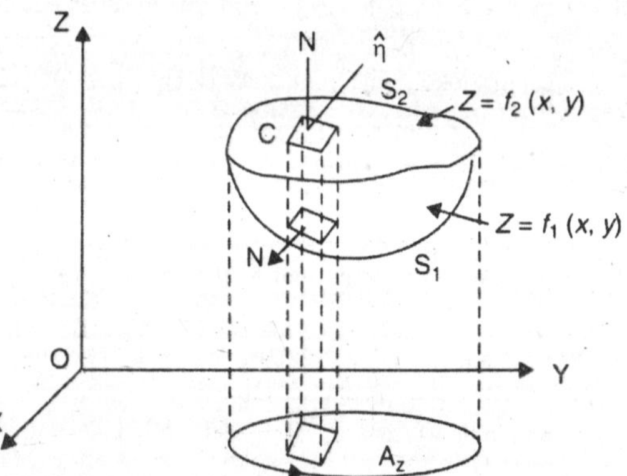

Example 40. *State Gauss's Divergence theorem* $\iint_S \vec{F}.\hat{n}\,ds = \iiint Div\,\vec{F}\,dv$ *where S is the surface of the sphere* $x^2 + y^2 + z^2 = 16$ *and* $\vec{F} = 3x\,\hat{i} + 4y\,\hat{j} + 5z\,\hat{k}$.

(*Nagpur University, Winter 2004*)

Solution. Statement of Gauss's Divergence theorem is given in Art 24.8 on page 597.
Thus by Gauss's divergence theorem,

$$\iint_S \vec{F}.\hat{n}\,ds = \iint_v \int \nabla.\vec{F}\,dv \quad \text{Here } \vec{F} = 3x\,\hat{i} + 4y\,\hat{j} + 5z\,\hat{k}$$

$$\nabla . \vec{F} = \left(\hat{i}\frac{\partial}{\partial x} + \hat{j}\frac{\partial}{\partial y} + \hat{k}\frac{\partial}{\partial z}\right).(3x\hat{i} + 4y\hat{j} + 5z\hat{k})$$

$$\nabla . \vec{F} = 3 + 4 + 5 = 14$$

Putting the value of ∇ . F, we get

$$\iint_S \vec{F} . \hat{n}\, ds = \iint_v \int 14 . dv \qquad \text{where } v \text{ is volume of a sphere}$$

$$= 14\, v$$

$$= 14 \frac{4}{3}\pi (4)^3 = \frac{3584\pi}{3} \qquad \textbf{Ans.}$$

Example 41. *Evaluate* $\iint_S \vec{F} . \hat{n}\, ds$ *where* $\vec{F} = 4xz\,\hat{i} - y^2\,\hat{j} + yz\,\hat{k}$ *and S is the surface of the cube bounded by* $x = 0,\ x = 1,\ y = 0,\ y = 1,\ z = 0,\ z = 1.$

(U.P., Ist Semester, 2009, Nagpur University, Winter 2003)

Solution. By Divergence theorem,

$$\iint_S \vec{F} . \hat{n}\, ds = \iint_v \int (\nabla . \vec{F})\, dv$$

$$= \iiint_v \left(\hat{i}\frac{\partial}{\partial x} + \hat{j}\frac{\partial}{\partial y} + \hat{k}\frac{\partial}{\partial z}\right).(4xz\,\hat{i} - y^2\,\hat{j} + yz\,\hat{k})\, dv$$

$$= \iint_v \int \left[\frac{\partial}{\partial x}(4xz) + \frac{\partial}{\partial y}(-y^2) + \frac{\partial}{\partial z}(yz)\right] dx\, dy\, dz$$

$$= \iint_v \int (4z - 2y + y)\, dx\, dy\, dz$$

$$= \iint_v \int (4z - y)\, dx\, dy\, dz = \int_0^1 \int_0^1 \left(\frac{4z^2}{2} - yz\right)_0^1 dx\, dy$$

$$= \int_0^1 \int_0^1 (2z^2 - yz)_0^1\, dx\, dy = \int_0^1 \int_0^1 (2 - y)\, dx\, dy$$

$$= \int_0^1 \left(2y - \frac{y^2}{2}\right)_0^1 dx = \frac{3}{2}\int_0^1 dx = \frac{3}{2}[x]_0^1 = \frac{3}{2}(1) = \frac{3}{2} \qquad \textbf{Ans.}$$

Note: This question is directly solved as on example 14 on Page 574.

Example 42. *Find* $\iint \vec{F}\cdot\hat{n}\cdot ds,$ *where* $\vec{F} = (2x + 3z)\,\hat{i} - (xz + y)\,\hat{j} + (y^2 + 2z)\,\hat{k}$ *and S is the surface of the sphere having centre (3, – 1, 2) and radius 3.*

(AMIETE, Dec. 2010, U.P., I Semester, Winter 2005, 2000)

Solution. Let *V* be the volume enclosed by the surface *S*.
By Divergence theorem, we've

$$\iint_S \vec{F}\cdot\hat{n}\cdot ds = \iiint_V div\, \vec{F}\, dv.$$

Now, $$div\, \vec{F} = \frac{\partial}{\partial x}(2x + 3z) + \frac{\partial}{\partial y}[-(xz + y)] + \frac{\partial}{\partial z}(y^2 + 2z) = 2 - 1 + 2 = 3$$

$$\therefore \quad \iint_S \vec{F}\cdot\hat{n}\cdot ds = \iiint_V 3\, dv = 3\iiint_V dv = 3V.$$

Again V is the volume of a sphere of radius 3. Therefore

$$V = \frac{4}{3}\pi r^3 = \frac{4}{3}\pi (3)^3 = 36\,\pi.$$

$$\therefore \quad \iint_S \vec{F}\cdot\hat{n}\cdot ds = 3V = 3 \times 36\,\pi = 108\,\pi \qquad \textbf{Ans.}$$

Example 43. *Use Divergence Theorem to evaluate* $\iint_S \vec{A}\cdot\vec{ds}$,

where $\vec{A} = x^3\hat{i} + y^3\hat{j} + z^3\hat{k}$ *and S is the surface of the sphere* $x^2 + y^2 + z^2 = a^2$.

(*AMIETE, Dec. 2009*)

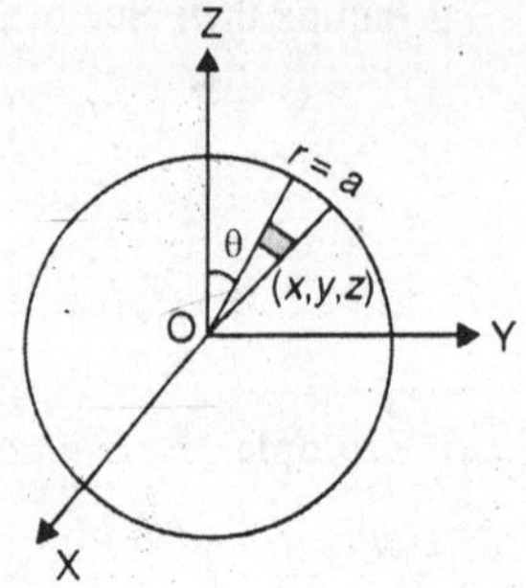

Solution. $\iint_S \vec{A}\cdot\vec{ds} = \iiint_v \text{div}\,\vec{A}\,dV$

$$= \iiint_v \left(\hat{i}\frac{\partial}{\partial x} + \hat{j}\frac{\partial}{\partial y} + \hat{k}\frac{\partial}{\partial z}\right)\cdot(x^3\hat{i} + y^3\hat{j} + z^3\hat{k})\,dV$$

$$= \iiint_v (3x^2 + 3y^2 + 3z^2)\,dV = 3\iiint_v (x^2 + y^2 + z^2)\,dV$$

On putting $x = r\sin\theta\cos\phi$, $y = r\sin\theta\sin\phi$, $z = r\cos\theta$, we get

$$= 3\iiint_V r^2\,(r^2\sin\theta\,dr\,d\theta\,d\phi) = 3\times 8\int_0^{\frac{\pi}{2}} d\phi \int_0^{\frac{\pi}{2}} \sin\theta\,d\theta \int_0^a r^4\,dr$$

$$= 24\,(\phi)_0^{\frac{\pi}{2}}\,(-\cos\theta)_0^{\frac{\pi}{2}}\left(\frac{r^5}{5}\right)_0^a = 24\left(\frac{\pi}{2}\right)(-0+1)\left(\frac{a^5}{5}\right) = \frac{12\pi a^5}{5}$$ **Ans.**

Example 44. *Use divergence Theorem to show that*

$$\iint_S \nabla(x^2 + y^2 + z^2)\,d\vec{s} = 6\,V$$

where S is any closed surface enclosing volume V. (*U.P., I Semester, Winter 2002*)

Solution. Here $\nabla\,(x^2 + y^2 + z^2) = \left(\hat{i}\frac{\partial}{\partial x} + \hat{j}\frac{\partial}{\partial y} + \hat{k}\frac{\partial}{\partial z}\right)\cdot(x^2 + y^2 + z^2)$

$$= 2x\,\hat{i} + 2y\,\hat{j} + 2z\,\hat{k} = 2\,(x\,\hat{i} + y\,\hat{j} + z\,\hat{k})$$

$$\therefore \quad \iint_S \nabla(x^2 + y^2 + z^2)\cdot ds = \iint_S \nabla(x^2 + y^2 + z^2)\cdot\hat{n}\,ds$$

$\hat{n}$ being outward drawn unit normal vector to S

$$= \iint_S 2\,(x\hat{i} + y\,\hat{j} + z\,\hat{k})\cdot\hat{n}\,ds$$

$$= 2\iiint_V div\,(x\hat{i} + y\,\hat{j} + z\,\hat{k})\,dv \qquad \text{...(1)}$$

(By Divergence Theorem)
(V being volume enclosed by *S*)

Now, $\quad$ div. $(x\,\hat{i} + y\,\hat{j} + z\hat{k}) = \left(\hat{i}\frac{\partial}{\partial x} + \hat{j}\frac{\partial}{\partial y} + \hat{k}\frac{\partial}{\partial z}\right)\cdot(x\hat{i} + y\,\hat{j} + z\,\hat{k})$

$$= \frac{\partial x}{\partial x} + \frac{\partial y}{\partial y} + \frac{\partial z}{\partial z} = 3 \qquad \text{...(2)}$$

From (1) & (2), we have

$$\iint \nabla(x^2 + y^2 + z^2)\cdot dS = 2\iiint_V 3\,dv = 6\iiint_V dv = 6\,V$$ **Proved.**

Example 45. *Evaluate* $\iint_S (y^2z^2\hat{i} + z^2x^2\hat{j} + z^2y^2\hat{k})\,\hat{n}\,dS$, *where S is the part of the sphere* $x^2 + y^2 + z^2 = 1$ *above the xy-plane and bounded by this plane.*

Solution. Let V be the volume enclosed by the surface S. Then by divergence Theorem, we have

$$\iint_S (y^2z^2\hat{i} + z^2x^2\hat{j} + z^2y^2\hat{k})\,\hat{n}\,dS = \iiint_V div\,(y^2z^2\hat{i} + z^2x^2\hat{j} + z^2y^2\hat{k})\,dV$$

$$= \iiint_V \left[\frac{\partial}{\partial x}(y^2z^2) + \frac{\partial}{\partial y}(z^2x^2) + \frac{\partial}{\partial z}(z^2y^2)\right] dV = \iint_V 2z\,y^2\,dV = 2\iint_V zy^2\,dV$$

Changing to spherical polar coordinates by putting
$x = r \sin\theta \cos\phi, \quad y = r \sin\theta \sin\phi, \quad z = r\cos\theta, \; dV = r^2 \sin\theta \, dr\, d\theta\, d\phi$

To cover V, the limits of r will be 0 to 1, those of θ will be 0 to $\frac{\pi}{2}$ and those of ϕ will be 0 to 2π.

$$\therefore \quad 2\iiint_V zy^2\, dV = 2\int_0^{2\pi}\int_0^{\pi/2}\int_0^1 (r\cos\theta)(r^2\sin^2\theta\sin^2\phi)\, r^2\sin\theta\cdot dr\, d\theta\, d\phi$$

$$= 2\int_0^{2\pi}\int_0^{\pi/2}\int_0^1 r^5\sin^3\theta\cos\theta\sin^2\phi\, dr\, d\theta\, d\phi$$

$$= 2\int_0^{2\pi}\int_0^{\frac{\pi}{2}} \sin^3\theta\cos\theta\sin^2\phi\left[\frac{r^6}{6}\right]_0^1 d\theta\, d\phi$$

$$= \frac{2}{6}\int_0^{2\pi}\sin^2\phi\cdot\frac{2}{4.2}\,d\phi = \frac{1}{12}\int_0^{2\pi}\sin^2\phi\, d\phi = \frac{\pi}{12} \qquad \textbf{Ans.}$$

Example 46. *Use Divergence Theorem to evaluate* $\iint_S \vec{F}\cdot d\vec{S}$ *where* $\vec{F} = 4x\hat{i} - 2y^2\hat{j} + z^2\hat{k}$ *and S is the surface bounding the region* $x^2 + y^2 = 4$, $z = 0$ *and* $z = 3$.

(A.M.I.E.T.E., Summer 2003, 2001)

Solution. By Divergence Theorem,

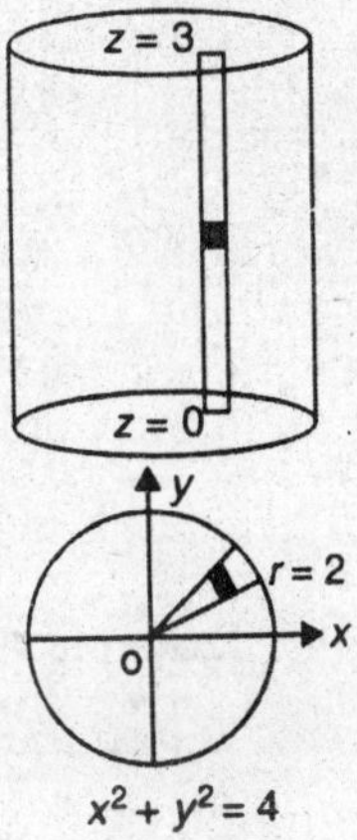

$$\iint_S \vec{F}\cdot dS = \iiint_V div\,\vec{F}\,dV$$

$$= \iiint_V\left(\hat{i}\frac{\partial}{\partial x} + \hat{j}\frac{\partial}{\partial y} + \hat{k}\frac{\partial}{\partial z}\right)\cdot(4x\hat{i} - 2y^2\hat{j} + z^2\hat{k})\,dV$$

$$= \iiint_V (4 - 4y + 2z)\,dx\,dy\,dz$$

$$= \iint dx\,dy\int_0^3 (4 - 4y + 2z)dz = \iint dx\,dy\,[4z - 4yz + z^2]_0^3$$

$$= \iint (12 - 12y + 9)\,dx\,dy = \iint (21 - 12y)\,dx\,dy$$

Let us put $x = r\cos\theta$, $y = r\sin\theta$

$$= \iint (21 - 12r\sin\theta)\,r\,d\theta\,dr = \int_0^{2\pi} d\theta\int_0^2 (21r - 12r^2\sin\theta)\,dr$$

$$= \int_0^{2\pi} d\theta\left[\frac{21r^2}{2} - 4r^3\sin\theta\right]_0^2 = \int_0^{2\pi} d\theta\,(42 - 32\sin\theta) = (42\theta + 32\cos\theta)_0^{2\pi}$$

$$= 84\pi + 32 - 32 = 84\pi \qquad \textbf{Ans.}$$

Example 47. *Apply the Divergence Theorem to compute* $\iint \vec{u}\cdot\hat{n}\,ds$, *where s is the surface of the cylinder* $x^2 + y^2 = a^2$ *bounded by the planes* $z = 0$, $z = b$ *and where* $u = \hat{i}x - \hat{j}y + \hat{k}z$.

Solution. By Gauss's Divergence Theorem

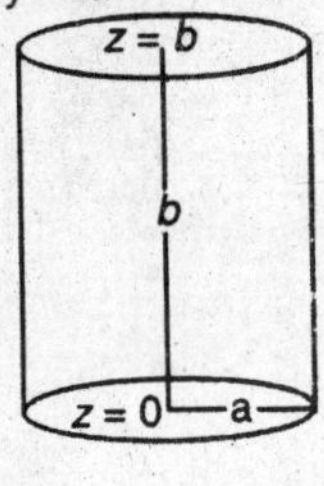

$$\iint \vec{u}\cdot\hat{n}ds = \iiint_V (\nabla\cdot\vec{u})\,dv$$

$$= \iiint_V\left(\hat{i}\frac{\partial}{\partial x} + \hat{j}\frac{\partial}{\partial y} + \hat{k}\frac{\partial}{\partial z}\right)\cdot(\hat{i}x - \hat{j}y + \hat{k}z)dv$$

$$= \iiint_V\left(\frac{\partial x}{\partial x} - \frac{\partial y}{\partial y} + \frac{\partial z}{\partial z}\right)dv = \iiint_V (1 - 1 + 1)\,dv$$

$$= \iiint_V dv = \iiint_V dx\,dy\,dz = \text{Volume of the cylinder} = \pi a^2 b \qquad \textbf{Ans.}$$

Example 48. *Apply Divergence Theorem to evaluate* $\iiint_V \vec{F}\,.\,\hat{n}\,ds$, *where* $\vec{F} = 4x^3\hat{i} - x^2y\,\hat{j} + x^2z\hat{k}$ *and S is the surface of the cylinder* $x^2 + y^2 = a^2$ *bounded by the planes* $z = 0$ *and* $z = b$. *(U.P. Ist Semester, Dec. 2006)*

Solution. We have,

$$\vec{F} = 4x^3\hat{i} - x^2y\,\hat{j} + x^2z\hat{k}$$

$$\therefore \quad div\,\overline{F} = \left(\hat{i}\frac{\partial}{\partial x} + \hat{j}\frac{\partial}{\partial y} + \hat{k}\frac{\partial}{\partial z}\right)\cdot(4x^3\hat{i} - x^2y\hat{j} + x^2z\hat{k})$$

$$= \frac{\partial}{\partial x}(4x^3) + \frac{\partial}{\partial y}(-x^2y) + \frac{\partial}{\partial z}(x^2z) = 12x^2 - x^2 + x^2 = 12\,x^2$$

$$\text{Now,}\quad \iiint_V div\,\vec{F}\,dV = 12\int_{x=-a}^{a}\int_{y=-\sqrt{a^2-x^2}}^{\sqrt{a^2-x^2}}\int_{z=0}^{b} x^2\,dz\,dy\,dx$$

$$= 12\int_{x=-a}^{a}\int_{y=-\sqrt{a^2-x^2}}^{\sqrt{a^2-x^2}} x^2\,(z)_0^b\,dy\,dx = 12\,b\int_{-a}^{a} x^2\,(y)_{-\sqrt{a^2-x^2}}^{\sqrt{a^2-x^2}}\,dx$$

$$= 12\,b\int_{-a}^{a} x^2.2\sqrt{a^2-x^2}\,dx = 24\,b\int_{-a}^{a} x^2\sqrt{a^2-x^2}\,dx$$

$$= 48\,b\int_0^a x^2\sqrt{a^2-x^2}\,dx \qquad [\text{Put } x = a\sin\theta,\ dx = a\cos\theta\,d\theta]$$

$$= 48\,b\int_0^{\pi/2} a^2\sin^2\theta\; a\cos\theta\; a\cos\theta\,d\theta$$

$$= 48\,ba^4\int_0^{\pi/2}\sin^2\theta\cdot\cos^2\theta\,d\theta = 48\,ba^4\,\frac{\Gamma\frac{3}{2}\,\Gamma\frac{3}{2}}{2\,\Gamma 3}$$

$$= 48\,ba^4\,\frac{\frac{1}{2}\sqrt{\pi}\cdot\frac{1}{2}\sqrt{\pi}}{2\cdot 2} = 3\,b\,a^4\pi \qquad \textbf{Ans.}$$

Example 49. *Evaluate surface integral* $\iint \vec{F}\cdot\hat{n}\,ds$, *where* $\vec{F} = (x^2 + y^2 + z^2)\,(\hat{i} + \hat{j} + \hat{k})$, *S is the surface of the tetrahedron* $x = 0$, $y = 0$, $z = 0$, $x + y + z = 2$ *and n is the unit normal in the outward direction to the closed surface S.*

Solution. By Divergence theorem

$$\iint_S \vec{F}\cdot\hat{n}\,ds = \iiint_V \text{div}\,\vec{F}\cdot dv$$

where S is the surface of tetrahedron $x = 0$, $y = 0$, $z = 0$, $x + y + z = 2$

$$= \iiint_V\left(\hat{i}\frac{\partial}{\partial x} + \hat{j}\frac{\partial}{\partial y} + \hat{k}\frac{\partial}{\partial z}\right)\cdot(x^2 + y^2 + z^2)\,(\hat{i} + \hat{j} + \hat{k})\,dv$$

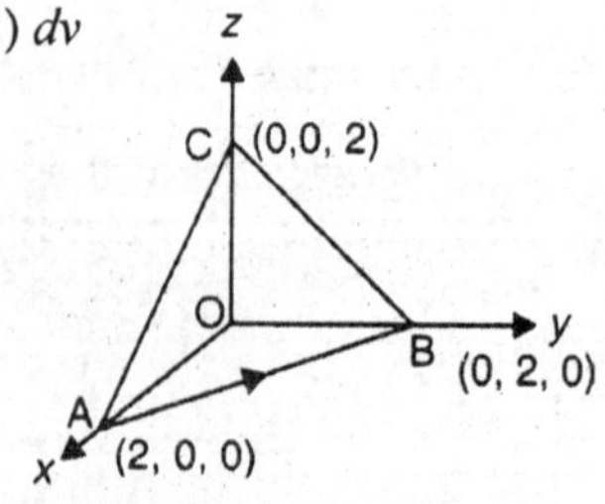

$$= \iiint_V (2x + 2y + 2z)\,dv$$

$$= 2\iiint_V (x + y + z)\,dx\,dy\,dz$$

$$= 2\int_0^2 dx\int_0^{2-x} dy\int_0^{2-x-y}(x + y + z)\,dz$$

$$= 2\int_0^2 dx\int_0^{2-x} dy\left(xz + yz + \frac{z^2}{2}\right)_0^{2-x-y}$$

$$= 2\int_0^2 dx \int_0^{2-x} dy \left(2x - x^2 - xy + 2y - xy - y^2 + \frac{(2-x-y)^2}{2} \right]$$

$$= 2\int_0^2 dx \left[2xy - x^2 y - x y^2 + y^2 - \frac{y^3}{3} - \frac{(2-x-y)^3}{6} \right]_0^{2-x}$$

$$= 2\int_0^2 dx \left[2x(2-x) - x^2(2-x) - x(2-x)^2 + (2-x)^2 - \frac{(2-x)^3}{3} + \frac{(2-x)^3}{6} \right]$$

$$= 2\int_0^2 \left(4x - 2x^2 - 2x^2 + x^3 - 4x + 4x^2 - x^3 + (2-x)^2 - \frac{(2-x)^3}{3} + \frac{(2-x)^3}{6} \right]$$

$$= 2\left[2x^2 - \frac{4x^3}{3} + \frac{x^4}{4} - 2x^2 + \frac{4x^3}{3} - \frac{x^4}{4} - \frac{(2-x)^3}{3} + \frac{(2-x)^4}{12} - \frac{(2-x)^4}{24} \right]_0^2$$

$$= 2\left[-\frac{(2-x)^3}{3} + \frac{(2-x)^4}{12} - \frac{(2-x)^4}{24} \right]_0^2 = 2\left[\frac{8}{3} - \frac{16}{12} + \frac{16}{24} \right] = 4$$ **Ans.**

Example 50. *Use the Divergence Theorem to evaluate*

$$\iint_S (x\, dy\, dz + y\, dz\, dx + z\, dx\, dy)$$

where S is the portion of the plane $x + 2y + 3z = 6$ *which lies in the first Octant.* (*U.P., I Semester, Winter 2003*)

Solution. $\iint_S (f_1\, dy\, dz + f_2\, dx\, dz + f_3\, dx\, dy)$

$$= \iiint_V \left(\frac{\partial f_1}{\partial x} + \frac{\partial f_2}{\partial y} + \frac{\partial f_3}{\partial z} \right) dx\, dy\, dz$$

where S is a closed surface bounding a volume V.

$$\therefore \quad \iint_S (x\, dy\, dz + y\, dz\, dx + z\, dx\, dy)$$

$$= \iiint_V \left[\frac{\partial x}{\partial x} + \frac{\partial y}{\partial y} + \frac{\partial z}{\partial z} \right] dx\, dy\, dz$$

$$= \iiint_V (1+1+1)\, dx\, dy\, dz = 3 \iiint_V dx\, dy\, dz$$

= 3 (Volume of tetrahedron $OABC$)

$$= 3\,[(\tfrac{1}{3} \text{ Area of the base } \Delta\, OAB) \times \text{height } OC]$$

$$= 3\left[\frac{1}{3}\left(\frac{1}{2} \times 6 \times 3 \right) \times 2 \right] = 18$$ **Ans.**

Example 51. *Use Divergence Theorem to evaluate :* $\iint (x\, dy\, dz + y\, dz\, dx + z\, dx\, dy)$ *over the surface of a sphere radius a.* (*K. University, Dec. 2009*)

Solution. Here, we have

$$\iint_S [x\, dy\, dz + y\, dx\, dz + z\, dx\, dy]$$

$$= \iiint_V \left(\frac{\partial f_1}{\partial x} + \frac{\partial f_2}{\partial y} + \frac{\partial f_3}{\partial z} \right) dx\, dy\, dz = \iiint_V \left(\frac{\partial x}{\partial x} + \frac{\partial y}{\partial y} + \frac{\partial z}{\partial z} \right) dx\, dy\, dz$$

$$= \iiint_V (1 + 1 + 1)\, dx\, dy\, dz = 3 \text{ (volume of the sphere)}$$

$$= 3\left(\frac{4}{3} \pi a^3 \right) = 4\pi a^3$$ **Ans.**

Example 52. *Using the divergence theorem, evaluate the surface integral* $\iint_S (yz\, dy\, dz + zx\, dz\, dx + xy\, dy\, dx)$ *where* $S : x^2 + y^2 + z^2 = 4.$

(*AMIETE, Dec. 2010, UP, I Sem., Dec 2008*)

Solution. $\iint_S (f_1\, dy\, dz + f_2\, dx\, dz + f_3\, dx\, dy)$

$$= \iiint_V \left(\frac{\partial f_1}{\partial x} + \frac{\partial f_2}{\partial y} + \frac{\partial f_3}{\partial z} \right) dx\, dy\, dz$$

where S is closed surface bounding a volume V.

$\therefore \quad \iint_S (yz\, dy\, dz + zx\, dx\, dz + xy\, dx\, dy)$

$$= \iiint_V \left(\frac{\partial (yz)}{\partial x} + \frac{\partial (zx)}{\partial y} + \frac{\partial (xy)}{\partial z} \right) dx\, dy\, dz = \iiint_V (0 + 0 + 0)\, dx\, dy\, dz$$

$= 0$ **Ans.**

Example 53. *Evaluate* $\iint_S xz^2\, dy\, dz + (x^2 y - z^3)\, dz\, dx + (2xy + y^2 z)\, dx\, dy$ *where S is the surface of hemispherical region bounded by*

$$z = \sqrt{a^2 - x^2 - y^2} \text{ and } z = 0.$$

Solution. $\iint_S (f_1\, dy\, dz + f_2\, dz\, dx + f_3\, dx\, dy) = \iiint_V \left(\frac{\partial f_1}{\partial x} + \frac{\partial f_2}{\partial y} + \frac{\partial f_3}{\partial z} \right) dx\, dy\, dz$

where S is a closed surface bounding a volume V.

$\therefore \quad \iint_S xz^2\, dy\, dz + (x^2 y - z^3)\, dz\, dx + (2xy + y^2 z)\, dx\, dy$

$$= \iiint_V \left[\frac{\partial}{\partial x}(xz^2) + \frac{\partial}{\partial y}(x^2 y - z^3) + \frac{\partial}{\partial z}(2xy + y^2 z) \right] dx\, dy\, dz$$

(Here V is the volume of hemisphere)

$$= \iiint_V (z^2 + x^2 + y^2)\, dx\, dy\, dz$$

Let $x = r \sin\theta \cos\phi,\ y = r \sin\theta \sin\phi,\ z = r\cos\theta$

$$= \iiint r^2 (r^2 \sin\theta\, dr\, d\theta\, d\phi) = \int_0^{2\pi} d\phi \int_0^{\frac{\pi}{2}} \sin\theta\, d\theta \int_0^a r^4\, dr$$

$$= (\phi)_0^{2\pi} (-\cos\theta)_0^{\pi/2} \left(\frac{r^5}{5} \right)_0^a = 2\pi(-0 + 1)\frac{a^5}{5} = \frac{2\pi a^5}{5}$$ **Ans.**

Example 54. *Evaluate* $\iint_S \vec{F} \cdot \hat{n}\, ds$ *over the entire surface of the region above the xy-plane bounded by the cone* $z^2 = x^2 + y^2$ *and the plane* $z = 4$, *if* $F = 4xz\hat{i} + xyz^2\hat{j} + 3z\hat{k}$.

Solution. If V is the volume enclosed by S, then V is bounded by the surfaces $z = 0$, $z = 4$, $z^2 = x^2 + y^2$.

By divergence theorem, we have

$$\iint_S \vec{F} \cdot \hat{n}\, ds = \iiint_V \operatorname{div} \vec{F}\, dx\, dy\, dz$$

$$= \iiint_V \left[\frac{\partial}{\partial x}(4xz) + \frac{\partial}{\partial y}(xyz^2) + \frac{\partial}{\partial z}(3z) \right] dx\, dy\, dz$$

$$= \iiint_V (4z + xz^2 + 3)\, dx\, dy\, dz$$

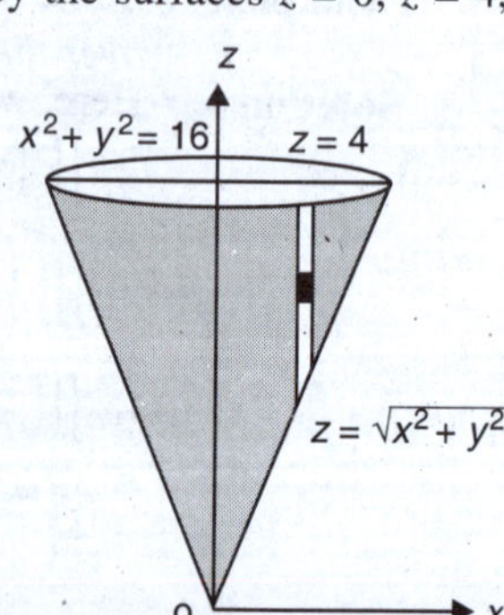

Limits of z are $\sqrt{x^2 + y^2}$ and 4.

$$\iiint_{\sqrt{x^2+y^2}}^{4}(4z+xz^2+3)\,dz\,dy\,dx = \iint\left[2z^2+\frac{xz^3}{3}+3z\right]_{\sqrt{x^2+y^2}}^{4} dy\,dx$$

$$= \iint\left[\left(32+\frac{64x}{3}+12\right)-\{2(x^2+y^2)+x(x^2+y^2)^{3/2}+3\sqrt{x^2+y^2}\}\right]dy\,dx$$

$$= \iint\left(44+\frac{64x}{3}-2(x^2+y^2)-x(x^2+y^2)^{3/2}-3\sqrt{x^2+y^2}\right)dy\,dx$$

Putting $x = r\cos\theta$ and $y = r\sin\theta$, we have

$$= \iint\left(44+\frac{64r\cos\theta}{3}-2r^2-r\cos\theta\, r^3-3r\right)r\,d\theta\,dr$$

Limits of r are 0 to 4.
and limits of θ are 0 to 2π.

$$= \int_0^{2\pi}\int_0^4\left(44r+\frac{64r^2\cos\theta}{3}-2r^3-r^5\cos\theta-3r^2\right)d\theta\,dr$$

$$= \int_0^{2\pi}\left[22r^2+\frac{64\times r^3\cos\theta}{9}-\frac{r^4}{2}-\frac{r^6}{6}\cos\theta-r^3\right]_0^4 d\theta$$

$$= \int_0^{2\pi}\left[22(4)^2+\frac{64\times(4)^3\cos\theta}{9}-\frac{(4)^4}{2}-\frac{(4)^6}{6}\cos\theta-(4)^3\right]d\theta$$

$$= \int_0^{2\pi}\left[352+\frac{64\times 64}{9}\cos\theta-128-\frac{(4)^6}{6}\cos\theta-64\right]d\theta$$

$$= \int_0^{2\pi}\left[160+\left(\frac{64\times 64}{9}-\frac{(4)^6}{6}\right)\cos\theta\right]d\theta$$

$$= \left[160\,\theta+\left(\frac{64\times 64}{9}-\frac{(4)^6}{6}\right)\sin\theta\right]_0^{2\pi} = 160\,(2\pi)+\left(\frac{64\times 64}{9}-\frac{(4)^6}{6}\right)\sin 2\pi$$

$= 320\,\pi$ **Ans.**

Example 55. *The vector field $\vec{F} = x^2\hat{i}+z\hat{j}+yz\hat{k}$ is defined over the volume of the cuboid given by $0 \le x \le a$, $0 \le y \le b$, $0 \le z \le c$, enclosing the surface S. Evaluate the surface integral*

$$\iint_S \vec{F}\cdot\vec{ds}$$

(*U.P., I Semester, Winter 2001*)

Solution. By Divergence Theorem, we have

$$\iint_S (x^2\hat{i}+z\hat{j}+yz\hat{k})\cdot ds = \iiint_V div\,(x^2\hat{i}+z\hat{j}+yz\hat{k})\,dv,$$

where V is the volume of the cuboid enclosing the surface S.

$$= \iiint_V\left(\hat{i}\frac{\partial}{\partial x}+\hat{j}\frac{\partial}{\partial y}+\hat{k}\frac{\partial}{\partial z}\right)\cdot(x^2\hat{i}+z\hat{j}+yz\hat{k})\,dv$$

$$= \iiint_V\left\{\frac{\partial}{\partial x}(x^2)+\frac{\partial}{\partial y}(z)+\frac{\partial}{\partial z}(yz)\right\}dx\,dy\,dz$$

$$= \int_{x=0}^{a}\int_{y=0}^{b}\int_{z=0}^{c}(2x+y)\,dx\,dy\,dz = \int_0^a dx\int_0^b dy\int_0^c (2x+y)\,dz$$

$$= \int_0^a dx\int_0^b [2xz+yz]_0^c\,dy = \int_0^a dx\int_0^b (2xc+yc)\,dy$$

$$= c\int_0^a dx \int_0^b (2x+y)\,dy = c\int_0^a \left[2xy + \frac{y^2}{2}\right]_0^b dx = c\int_0^a \left(2bx + \frac{b^2}{2}\right) dx$$

$$= c\left[\frac{2bx^2}{2} + \frac{b^2 x}{2}\right]_0^a = c\left[a^2 b + \frac{ab^2}{2}\right] = abc\left(a + \frac{b}{2}\right) \qquad \textbf{Ans.}$$

Example 56. *Verify the divergence Theorem for the function* $\overline{F} = 2x^2yi - y^2j + 4xz^2k$ *taken over the region in the first octant bounded by* $y^2 + z^2 = 9$ *and* $x = 2$.

Solution. $\iiint_V \nabla\cdot\vec{F}\,d$ $\qquad \iiint \left(\hat{i}\frac{\partial}{\partial x} + \hat{j}\frac{\partial}{\partial y} + \hat{k}\frac{\partial}{\partial z}\right)\cdot(2x^2y\hat{i} - y^2\hat{j} + 4xz^2k)\,dV$.

$$= \iiint (4xy - 2y + 8xz)\,dx\,dy\,dz = \int_0^2 dx \int_0^3 dy \int_0^{\sqrt{9-y^2}} (4xy - 2y + 8xz)\,dz$$

$$= \int_0^2 dx \int_0^3 dy\,(4xyz - 2yz + 4xz^2)_0^{\sqrt{9-y^2}}$$

$$= \int_0^2 dx \int_0^3 [4xy\sqrt{9-y^2} - 2y\sqrt{9-y^2} + 4x(9-y^2)]\,dy$$

$$= \int_0^2 dx \left[-\frac{4x}{2}\frac{2}{3}(9-y^2)^{3/2} + \frac{2}{3}(9-y^2)^{3/2} + 36xy - \frac{4xy^3}{3}\right]_0^3$$

$$= \int_0^2 (0 + 0 + 108x - 36x + 36x - 18)\,dx = \int_0^2 (108x - 18)\,dx = \left[108\frac{x^2}{2} - 18x\right]_0^2$$

$$= 216 - 36 = 180 \qquad \ldots(1)$$

Here $\iint_S \vec{F}\cdot\hat{n}\,ds = \iint_{OABC} \vec{F}\cdot\hat{n}\,ds + \iint_{OCE} \vec{F}\cdot\hat{n}\,ds + \iint_{OADE} \vec{F}\cdot\hat{n}\,ds + \iint_{ABD} \vec{F}\cdot\hat{n}\,ds + \iint_{BDEC} \vec{F}\cdot\hat{n}\,ds$

$$\iint_{BDEC} \overline{F}\cdot\hat{n}\,ds = \iint_{BDEC} (2x^2y\hat{i} - y^2\hat{j} + 4xz^2\hat{k})\,.\,\hat{n}\,ds$$

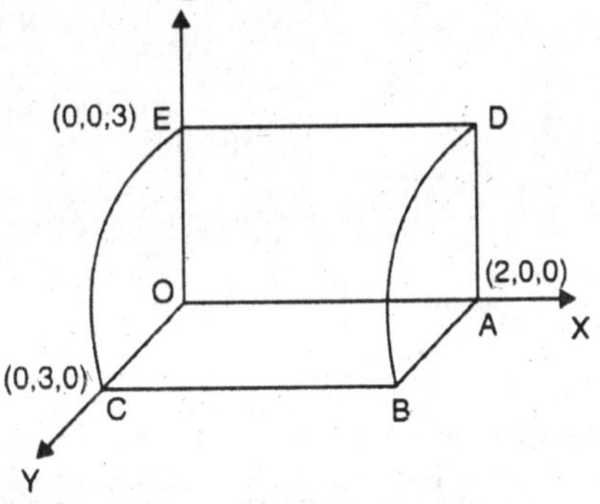

Normal vector

$$= \nabla\phi = \left(\hat{i}\frac{\partial}{\partial x} + \hat{j}\frac{\partial}{\partial y} + \hat{k}\frac{\partial}{\partial z}\right)(y^2 + z^2 - 9)$$

$$= 2y\hat{j} + 2z\hat{k}$$

Unit normal vector $= \hat{n} = \dfrac{2y\hat{j} + 2z\hat{k}}{\sqrt{4y^2 + 4z^2}} = \dfrac{y\hat{j} + z\hat{k}}{\sqrt{y^2 + z^2}}$

$$= \frac{y\hat{j} + z\hat{k}}{\sqrt{9}} = \frac{y\hat{j} + z\hat{k}}{3}$$

Z, (0,0,3) E, D, O, (2,0,0), A, X, (0,3,0), C, B, Y

$$\iint_{BDEC} (2x^2y\hat{i} - y^2\hat{j} + 4xz^2k)\cdot\frac{y\hat{j} + zk}{3}\,ds = \frac{1}{3}\iint_{BDEC} (-y^3 + 4xz^3)\,ds$$

$$\left[dx\,dy = ds\,(\hat{n}\cdot k) = ds\left(\frac{y\hat{j} + z\hat{k}}{3}\cdot\hat{k}\right) = ds\,\frac{z}{3} \text{ or } ds = \frac{dx\,dy}{\frac{z}{3}}\right]$$

$$= \frac{1}{3}\iint_{BDEC} (-y^3 + 4xz^3)\frac{dxdy}{\frac{z}{3}} = \int_0^2 dx \int_0^3 \left(-\frac{y^3}{z} + 4xz^2\right) dy \qquad \begin{pmatrix} y = 3\sin\theta, \\ z = 3\cos\theta \end{pmatrix}$$

$$= \int_0^2 dx \int_0^{\frac{\pi}{2}} \left[\frac{-27\sin^3\theta}{3\cos\theta} + 4x(9\cos^2\theta)\right]$$

$$= \int_0^2 dx\left(-27\times\frac{2}{3}+108\,x\times\frac{2}{3}\right) = \int_0^2 (-18+72\,x)\,dx$$

$$= \left[-18x+36x^2\right]_0^2 = 108 \qquad ...(2)$$

$$\iint_{OABC} \vec{F}\cdot\hat{n}\,ds = \iint_{OABC} (2x^2y\hat{i}-y^2\hat{j}+4xz^2\hat{k})\cdot(-\hat{k})\,ds$$

$$= \iint_{OABC} 4xz^2\,ds = 0 \qquad ...(3) \text{ because in } OABC \text{ } xy\text{-plane, } z=0$$

$$\iint_{OADE} \vec{F}\cdot\hat{n}\,ds = \iint_{OADE} (2x^2y\hat{i}-y^2\hat{j}+4xz^2\hat{k})\cdot(-\hat{j})\,ds = \iint_{OADE} y^2\,ds = 0 \qquad ...(4)$$

because in $OADE$ xz-plane, $y = 0$

$$\iint_{OCE} \vec{F}\cdot\hat{n}\,ds = \iint_{OCE} (2x^2y\hat{i}-y^2\hat{j}+4xz^2\hat{k})\cdot(-\hat{i})\,ds = \iint_{OCE} -2x^2y\,ds = 0 \qquad ...(5)$$

because in OCE yz-plane, $x = 0$

$$\iint_{ABD} \vec{F}\cdot\hat{n}\,ds = \iint_{ABD} (2x^2y\hat{i}-y^2\hat{j}+4xz^2\hat{k})\cdot(\hat{i})\,ds = \iint_{ABD} 2x^2y\,ds$$

$$= \iint 2x^2y\,dy\,dz = \int_0^3 dz\int_0^{\sqrt{9-z^2}} 2\,(2)^2\,y\,dy \qquad \text{because in } ABD \text{ plane, } x=2$$

$$= 8\int_0^3 dz\left[\frac{y^2}{2}\right]_0^{\sqrt{9-z^2}} = 4\int_0^3 dz\,(9-z^2) = 4\left[9z-\frac{z^3}{3}\right]_0^3 = 4\,[27-9] = 72 \qquad ...(6)$$

On adding (2), (3), (4), (5) and (6), we get

$$\iint_S \vec{F}\cdot\hat{n}\,ds = 108+0+0+0+72 = 180 \qquad ...(7)$$

From (1) and (7), we have $\iiint_V \nabla\cdot\vec{F}\,dV = \iint_S \vec{F}\cdot\hat{n}\,ds$

Hence the theorem is verified.

Example 57. *Verify the Gauss divergence Theorem for*

$\vec{F} = (x^2-yz)\,\hat{i} + (y^2-zx)\,\hat{j} + (z^2-xy)\,\hat{k}$ *taken over the rectangular parallelopiped* $0\le x\le a,\ 0\le y\le b,\ 0\le z\le c.$ *(U.P., I Semester, Compartment 2002)*

Solution. We have

$$\text{div }\vec{F} = \nabla\cdot\vec{F} = \left(\hat{i}\frac{\partial}{\partial x}+\hat{j}\frac{\partial}{\partial y}+\hat{k}\frac{\partial}{\partial z}\right)\cdot[(x^2-yz)\,\hat{i}+(y^2-zx)\,\hat{j}+(z^2-xy)\,\hat{k}]$$

$$= \frac{\partial}{\partial x}(x^2-yz)+\frac{\partial}{\partial y}(y^2-zx)+\frac{\partial}{\partial z}(z^2-xy) = 2x+2y+2z$$

$$\therefore \text{ Volume integral} = \iiint_V \nabla\cdot F\,dV = \iiint_V 2\,(x+y+z)\,dV$$

$$= 2\int_{x=0}^a\int_{y=0}^b\int_{z=0}^c (x+y+z)\,dx\,dy\,dz = 2\int_0^a dx\int_0^b dy\int_0^c (x+y+z)\,dz$$

$$= 2\int_0^a dx\int_0^b dy\left(xz+yz+\frac{z^2}{2}\right)_0^c = 2\int_0^a dx\int_0^b dy\left(cx+cy+\frac{c^2}{2}\right)$$

$$= 2\int_0^a dx\left(cxy+c\frac{y^2}{2}+\frac{c^2y}{2}\right)_0^b = 2\int_0^a dx\left(bcx+\frac{b^2c}{2}+\frac{bc^2}{2}\right)$$

$$= 2\left[\frac{bc\,x^2}{2} + \frac{b^2\,cx}{2} + \frac{bc^2 x}{2}\right]_0^a = [a^2bc + ab^2c + abc^2]$$

$$= abc\,(a + b + c) \qquad \ldots(A)$$

To evaluate $\iint_S \vec{F}\cdot\hat{n}\,ds$, where S consists of six plane surfaces.

$$\iint_S \vec{F}\cdot\hat{n}\,ds = \iint_{OABC} \vec{F}\cdot\hat{n}\,ds + \iint_{DEFG} \vec{F}\cdot\hat{n}\,ds + \iint_{OAFG} \vec{F}\cdot\hat{n}\,as$$

$$+ \iint_{BCDE} \vec{F}\cdot\hat{n}\,ds + \iint_{ABEF} \vec{F}\cdot\hat{n}\,ds + \iint_{OCDG} \vec{F}\cdot\hat{n}\,ds$$

$$\iint_{OABC} \vec{F}\cdot\hat{n}\,ds = \iint_{OABC} \{(x^2 - yz)\hat{i} + (y^2 - xz)\hat{j} + (z^2 - xy)\hat{k}\}\,(-\hat{k})\,dx\,dy$$

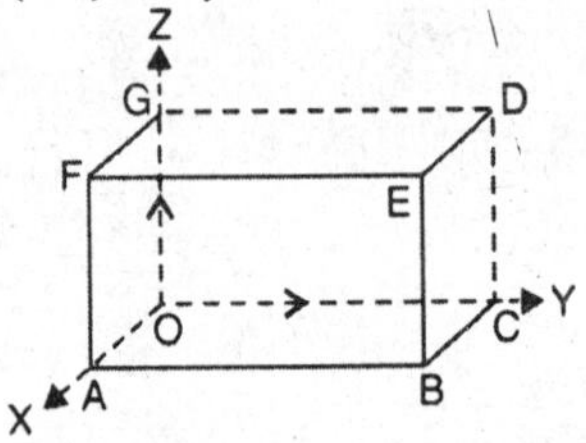

$$= -\int_0^a\int_0^b (z^2 - xy)\,dx\,dy$$

$$= -\int_0^a\int_0^b (0 - xy)\,dx\,dy = \frac{a^2\,b^2}{4} \qquad \ldots(1)$$

$$\iint_{DEFG} \vec{F}\cdot\hat{n}\,ds = \iint_{DEFG} \{(x^2 - yz)\hat{i} + (y^2 - xz)\hat{j} + (z^2 - xy)\hat{k}\}\,(\hat{k})\,dx\,dy$$

$$= \int_0^a\int_0^b (z^2 - xy)\,dx\,dy = \int_0^a\int_0^b (c^2 - xy)\,dx\,dy$$

$$= \int_0^a \left[c^2 y - \frac{xy^2}{2}\right]_0^b dx = \int_0^a \left(c^2 b - \frac{x\,b^2}{2}\right) dx$$

$$= \left[c^2 bx - \frac{x^2\,b^2}{4}\right]_0^a = abc^2 - \frac{a^2\,b^2}{4} \qquad \ldots(2)$$

S.No.	Surface	Outward normal	ds	
1	OABC	$-k$	$dx\,dy$	$z = 0$
2	DEFG	k	$dx\,dy$	$z = c$
3	OAFG	$-j$	$dx\,dz$	$y = 0$
4	BCDE	j	$dx\,dz$	$y = b$
5	ABEF	i	$dy\,dz$	$x = a$
6	OCDG	$-i$	$dy\,dz$	$x = 0$

$$\iint_{OAFG} \vec{F}\cdot\hat{n}\,ds = \iint_{OAFG} \{(x^2 - yz)\hat{i} + (y^2 - zx)\hat{j} + (z^2 - xy)\hat{k}\}\,(-\hat{j})\,dx\,dz$$

$$= -\iint_{OAFG} (y^2 - zx)\,dx\,dz$$

$$= -\int_0^a dx \int_0^c (0 - zx)\,dz = \int_0^a dx\left(\frac{x\,z^2}{2}\right)_0^c = \int_0^a \frac{x\,c^2}{2}\,dx = \left[\frac{x^2\,c^2}{4}\right]_0^a = \frac{a^2\,c^2}{4} \qquad \ldots(3)$$

$$\iint_{BCDE} \vec{F}\cdot\hat{n}\,ds = \iint \{(x^2 - y\,z)\hat{i} + (y^2 - zx)\,\hat{j} + (z^2 - xy)\hat{k}\}\cdot\hat{j}\,dx\,dz = \iint_{BCDE} (y^2 - xz)\,dx\,dz$$

$$= -\int_0^a dx \int_0^c (b^2 - x\,z)\,dz = \int_0^a \left(b^2\,z - \frac{x\,z^2}{2}\right)_0^c dx = \int_0^a \left(b^2\,c - \frac{x\,c^2}{2}\right) dx$$

$$= \left[b^2\,c\,x - \frac{x^2\,c^2}{4}\right]_0^a = a\,b^2\,c - \frac{a^2 c^2}{4} \qquad \ldots(4)$$

$$\iint_{ABEF} \vec{F}\cdot\hat{n}\,ds = \iint_{ABEF} \{(x^2 - y\,z)\hat{i} + (y^2 - xz)\,\hat{j} + (z^2 - x\,y)\,\hat{k}\}\cdot\hat{i}\,dy\,dz$$

$$= \iint_{ABEF} (x^2 - y\,z)\,dy\,dz = \int_0^b dy \int_0^c (a^2 - y\,z)\,d\,z = \int_0^b dy\left(a^2 z - \frac{y\,z^2}{2}\right)_0^c$$

$$= \int_0^b \left(a^2 c - \frac{y c^2}{2} \right) dy = \left[a^2 c y - \frac{y^2 c^2}{4} \right]_0^b = a^2 b c - \frac{b^2 c^2}{4} \qquad \text{...(5)}$$

$$\iint_{OCDG} \vec{F} \cdot \hat{n}\, ds = \iint_{OCDG} \{(x^2 - yz)\hat{i} + (y^2 - zx)\hat{j} + (z^2 - xy)\hat{k}\} \cdot (-\hat{i})\, dy\, dz$$

$$= \int_0^b \int_0^c (x^2 - yz)\, dy\, dz = -\int_0^b dy \int_0^c (-yz)\, dz = -\int_0^b dy \left[\frac{-y z^2}{2} \right]_0^c$$

$$= \int_0^b \frac{y c^2}{2} dy = \left[\frac{y^2 c^2}{4} \right]_0^b = \frac{b^2 c^2}{4} \qquad \text{...(6)}$$

Adding (1), (2), (3), (4), (5) and (6), we get

$$\iint \vec{F} \cdot \hat{n}\, ds = \left(\frac{a^2 b^2}{4} \right) + \left(abc^2 - \frac{a^2 b^2}{4} \right) + \left(\frac{a^2 c^2}{4} \right) + \left(a b^2 c - \frac{a^2 c^2}{4} \right) + \left(\frac{b^2 c^2}{4} \right) + \left(a^2 b c - \frac{b^2 c^2}{4} \right)$$

$$= abc^2 + ab^2c + a^2bc$$

$$= abc\,(a + b + c) \qquad \text{...(B)}$$

From (A) and (B), Gauss divergence Theorem is verified. **Verified.**

Example 58. *Verify Divergence Theorem, given that $\vec{F} = 4xz\hat{i} - y^2\hat{j} + yz\,\hat{k}$ and S is the surface of the cube bounded by the planes x = 0, x = 1, y = 0, y = 1, z = 0, z = 1.*

Solution. $\nabla \cdot \vec{F} = \left(\hat{i} \frac{\partial}{\partial x} + \hat{j} \frac{\partial}{\partial y} + \hat{k} \frac{\partial}{\partial z} \right) \cdot (4\,zx\hat{i} - y^2\hat{j} + yz\hat{k})$

$$= 4z - 2y + y$$

$$= 4z - y$$

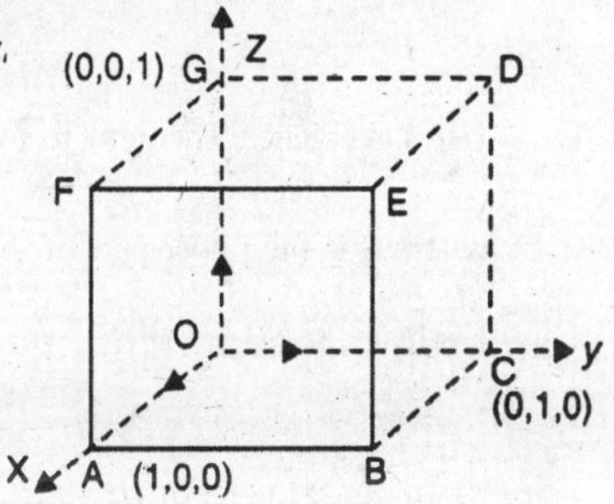

$$\text{Volume Integral} = \iiint \nabla \cdot \vec{F}\, dv$$

$$= \iiint (4z - y)\, dx\, dy\, dz$$

$$= \int_0^1 dx \int_0^1 dy \int_0^1 (4z - y)\, dz$$

$$= \int_0^1 dx \int_0^1 dy\, (2z^2 - yz)_0^1 = \int_0^1 dx \int_0^1 dy\, (2 - y)$$

$$= \int_0^1 dx \left(2y - \frac{y^2}{2} \right)_0^1 = \int_0^1 dx \left(2 - \frac{1}{2} \right) = \frac{3}{2} \int_0^1 dx = \frac{3}{2} (x)_0^1 = \frac{3}{2} \qquad \text{...(1)}$$

To evaluate $\iint_S \vec{F} \cdot \hat{n}\, ds$, where S consists of six plane surfaces.

Over the face $OABC$, $z = 0$, $dz = 0$, $\hat{n} = -\hat{k}$, $ds = dx\, dy$

$$\iint \vec{F} \cdot \hat{n}\, ds = \int_0^1 \int_0^1 (-y^2\hat{j}) \cdot (-\hat{k})\, dx\, dy = 0$$

Over the face $BCDE$, $y = 1$, $dy = 0$

$$\iint \vec{F} \cdot \hat{n}\, ds = \int_0^1 \int_0^1 (4\,xz\hat{i} - \hat{j} + z\hat{k}) \cdot (\hat{j})\, dx\, dz$$

$$\hat{n} = \hat{j},\ ds = dx\, dz \quad = \int_0^1 \int_0^1 -\, dx dz$$

S.No.				$\hat{n}$	ds
1	OABC	$z = 0$	$dz = 0$	$-k$	$dx\,dy$
2	BCDE	$y = 1$	$dy = 0$	j	$dx\,dz$
3	DEFG	$z = 1$	$dz = 0$	k	$dx\,dy$
4	OCDG	$x = 0$	$dx = 0$	$-i$	$dy\,dz$
5	AOGF	$x = 0$	$dy = 0$	$-j$	$dx\,dz$
6	ABEF	$x = 1$	$dx = 0$	i	$dy\,dz$

$$= -\int_0^1 dx \int_0^1 dz = -(x)_0^1 (z)_0^1 = -(1)(1) = -1$$

Over the face *DEFG*, $z = 1$, $dz = 0$, $\hat{n} = \hat{k}$, $ds = dx\, dy$

$$\iint \vec{F}\cdot\hat{n}\, ds = \int_0^1\int_0^1 [4x(1) - y^2\hat{j} + y(1)\hat{k}]\cdot(\hat{k})\, dx\, dy$$

$$= \int_0^1\int_0^1 y\, dx\, dy = \int_0^1 dx \int_0^1 y\, dy = (x)_0^1 \left(\frac{y^2}{2}\right)_0^1 = \frac{1}{2}$$

Over the face *OCDG*, $x = 0$, $dx = 0$, $\hat{n} = -\hat{i}$, $ds = dy\, dz$

$$\iint \vec{F}\cdot\hat{n}\, ds = \int_0^1\int_0^1 (0\hat{i} - y^2\hat{j} + yz\hat{k})\cdot(-\hat{i})\, dy\, dz = 0$$

Over the face *AOGF*, $y = 0$, $dy = 0$, $\hat{n} = -\hat{j}$, $ds = dx\, dz$

$$\iint \vec{F}\cdot\hat{n}\, ds = \int_0^1\int_0^1 (4xz\hat{i})\cdot(-\hat{j})\, dx\, dz = 0$$

Over the face *ABEF*, $x = 1$, $dx = 0$, $\hat{n} = \hat{i}$, $ds = dy\, dz$

$$\iint \vec{F}\cdot\hat{n}\, ds = \int_0^1\int_0^1 [(4z\hat{i} - y^2\hat{j} + yz\hat{k})\cdot(\hat{i})]\, dy\, dz = \int_0^1\int_0^1 4z\, dy\, dz$$

$$= \int_0^1 dy \int_0^1 4z\, dz = \int_0^1 dy\, (2z^2)_0^1 = 2\int_0^1 dy = 2(y)_0^1 = 2$$

On adding we see that over the whole surface $\iint \vec{F}\cdot\hat{n}\, ds = \left(0 - 1 + \frac{1}{2} + 0 + 0 + 2\right) = \frac{3}{2}$ ...(2)

From (1) and (2), we have $\iiint_V \nabla\cdot\vec{F}\, dv = \iint_S \vec{F}\cdot\hat{n}\, ds$ **Verified.**

EXERCISE 24.6

1. Use Divergence Theorem to evaluate $\iint_s (y^2z^2\hat{i} + z^2x^2\hat{j} + x^2y^2\hat{k}).\overline{ds}$,

where S is the upper part of the sphere $x^2 + y^2 + z^2 = 9$ above xy- plane. **Ans.** $\frac{243\pi}{8}$

2. Evaluate $\iint_S (\nabla \times \vec{F}).\, ds$, where S is the surface of the paraboloid $x^2 + y^2 + z = 4$ above the xy-plane and $\vec{F} = (x^2 + y - 4)\,\hat{i} + 3\,xy\hat{j} + (2\,xz + z^2)\,\hat{k}$. **Ans.** -4π

3. Evaluate $\iint_s [xz^2\, dy\, dz + (x^2y - z^3)\, dz\, dx + (2xy + y^2z)\, dx\, dy]$, where S is the surface enclosing a region bounded by hemisphere $x^2 + y^2 + z^2 = 4$ above XY-plane. **Ans.** $\frac{64\pi}{5}$

4. Verify Divergence Theorem for $\vec{F} = x^2\hat{i} + z\hat{j} + yz\hat{k}$, taken over the cube bounded by $x = 0, x = 1, y = 0, y = 1, z = 0, z = 1$. **Ans.** $\frac{3}{2}$

5. Evaluate $\iint_S (2xy\,\hat{i} + yz^2\hat{j} + xz\,\hat{k})\cdot\vec{ds}$ over the surface of the region bounded by

$x = 0, y = 0, y = 3, z = 0$ and $x + 2z = 6$ **Ans.** $\frac{351}{2}$

6. Verify Divergence Theorem for $\vec{F} = (x + y^2)\,\hat{i} - 2\,x\hat{j} + 2\,yz\hat{k}$ and the volume of a tetrahedron bounded by co-ordinate planes and the plane $2x + y + 2z = 6$.

(*Nagpur, Winter 2000, A.M.I.E.T.E.. Winter 2000*)

7. Verify Divergence Theorem for the function $\vec{F} = y\hat{i} + x\hat{j} + z^2\,\hat{k}$ over the region bounded by $x^2 + y^2 = 9$, $z = 0$ and $z = 2$.

8. Use the Divergence Theorem to evaluate $\iint_s x^3\, dy\, dz + x^2y\, dz\, dx + x^2z\, dx\, dy$,

where S is the surface of the region bounded by the closed cylinder

$x^2 + y^2 = a^2$, $(0 \le z \le b)$ and $z = 0$, $z = b$. **Ans.** $\frac{5\pi a^4 b}{4}$

9. Evaluate the integral $\iint_S (z^2 - x)\, dy\, dz - xy\, dx\, dz + 3z\, dx\, dy$, where S is the surface of closed region bounded by $z = 4 - y^2$ and planes $x = 0, x = 3, z = 0$ by transforming it with the help of Divergence Theorem to a triple integral. **Ans.** 16

10. Evaluate $\iint_S \frac{ds}{\sqrt{a^2x^2 + b^2y^2 + c^2z^2}}$ over the closed surface of the ellipsoid $ax^2 + by^2 + cz^2 = 1$ by applying Divergence Theorem. *(MTU, Dec. 2012)* **Ans.** $\frac{4\pi}{\sqrt{(abc)}}$

11. Apply Divergence Theorem to evaluate $\iint (l x^2 + m y^2 + n z^2)\, ds$ taken over the sphere $(x - a)^2 + (y - b)^2 + (z - c)^2 = r^2$, l, m, n being the direction cosines of the external normal to the sphere. *(AMIETE June 2010, 2009)* **Ans.** $\frac{8\pi}{3}(a + b + c)\, r^3$

12. Show that $\iiint_V (u \nabla \cdot \vec{V} + \nabla u \cdot \vec{V})\, dv = \iint_S u \vec{V} \cdot ds.$

13. If $E = \text{grad}\, \phi$ and $\nabla^2 \phi = 4\pi\rho$, prove that $\iint_S \vec{E} \cdot \vec{n}\, ds = -4\pi \iiint_V \rho\, dv$ where $\vec{n}$ is the outward unit normal vector, while dS and dV are respectively surface and volume elements.

Pick up the correct option from the following:

14. If $\vec{F}$ is the velocity of a fluid particle then $\int_C \vec{F} . \vec{dr}$ represents.

(*a*) Work done (*b*) Circulation (*c*) Flux (*d*) Conservative field.

(U.P. Ist Semester, Dec 2009) **Ans.** (*b*)

15. If $\vec{f} = ax\,\vec{i} + by\,\vec{j} + cz\,\vec{k}$, a, b, c, constants, then $\iint f . dS$ where S is the surface of a unit sphere is

(*a*) $\frac{\pi}{3}(a + b + c)$ (*b*) $\frac{4}{3}\pi(a + b + c)$ (*c*) $2\pi(a + b + c)$ (*d*) $\pi(a + b + c)$

(U.P., Ist Semester, 2009) **Ans.** (*b*)

16. A force field $\vec{F}$ is said to be conservative if

(*a*) Curl $\vec{F} = 0$ (*b*) grad $\bar{F} = 0$ (*c*) Div $\bar{F} = 0$ (*d*) Curl (grad $\bar{F}$) = 0

(AMIETE, Dec. 2006) **Ans.** (*a*)

17. The line integral $\int_c x^2\, dx + y^2\, dy$, where C is the boundary of the region $x^2 + y^2 < a^2$ equals

(*a*) 0, (*b*) a (*c*) πa^2 (*d*) $\frac{1}{2}\pi a^2$

(AMIETE, Dec. 2006) **Ans.** (*b*)

Complex Numbers

25.1 INTRODUCTION

We have learnt the complex numbers in the previous class. Here we will review the complex number. In this chapter we will learn how to add, subtract, multiply and divide complex numbers.

25.2 COMPLEX NUMBERS

A number of the form $a + i\,b$ is called a complex number when a and b are real numbers and $i = \sqrt{-1}$. We call 'a' the real part and 'b' the imaginary part of the complex number $a + ib$. If $a = 0$ the number $i\,b$ is said to be purely imaginary, if $b = 0$ the number a is real.

A complex number $x + iy$ is denoted by z.

25.3 GEOMETRICAL REPRESENTATION OF IMAGINARY NUMBERS

Let OA be positive numbers which is represented by x and OA' by $-x$.

And $\qquad -x = (i)^2\, x = i\,(ix)$ is on OX'.

It means that the multiplication of the real number x by i twice amounts to the rotation of OA through two right angles to reach OA'.

Thus, it means that multiplication of x by i is equivalent to the rotation of x through one right angle to reach OA''.

Hence, y-axis is known as imaginary axis.

Multiplication by i rotates its direction through a right angle.

25.4 ARGAND DIAGRAM

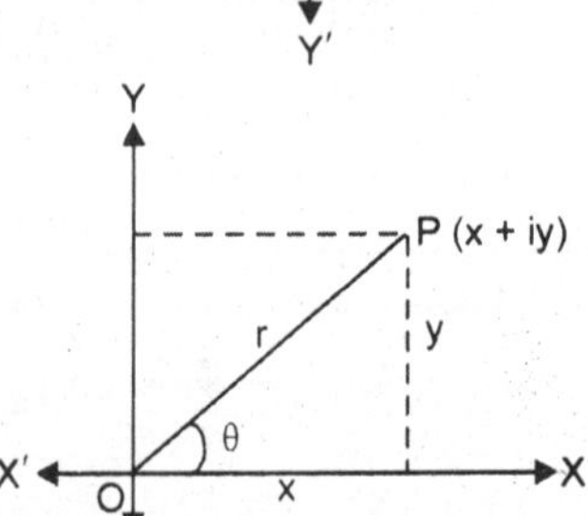

Mathematician Argand represented a complex number in a diagram known as Argand diagram. A complex number $x + iy$ can be represented by a point P whose co-ordinate are (x, y). The axis of x is called the real axis and the axis of y the imaginary axis. The distance OP is the **modulus** and the angle, OP makes with the x-axis, is the **argument** of $x + iy$.

25.5 EQUAL COMPLEX NUMBERS

If two complex numbers $a + ib$ and $c + id$ are equal, prove that

$$a = c \quad \text{and} \quad b = d$$

Solution. We have, $\qquad a + ib = c + id \;\Rightarrow\; a - c = i(d - b)$

$$(a - c)^2 = -(d - b)^2 \;\Rightarrow\; (a - c)^2 + (d - b)^2 = 0$$

Here sum of two positive numbers is zero. This is only possible if each number is zero.

i.e., $(a - c)^2 = 0 \;\Rightarrow\; a = c$ and $(d - b)^2 = 0 \;\Rightarrow\; b = d$ **Ans.**

25.6 ADDITION OF COMPLEX NUMBERS

Let $a + ib$ and $c + id$ be two complex numbers, then

$$(a + ib) + (c + id) = (a + c) + i\,(b + d)$$

Procedure. In addition of complex numbers we add real parts with real parts and imaginary parts with imaginary parts.

Example 1. *Add the following complex numbers:*

$$z_1 = 2 + \frac{3}{2}i, \quad z_2 = -5 + \frac{7}{4}i, \quad z_3 = \frac{5}{4} - \frac{8}{3}i, \quad z_4 = -\frac{11}{2} - i$$

Solution.
$$z_1 + z_2 + z_3 + z_4 = \left(2 + \frac{3}{2}i\right) + \left(-5 + \frac{7}{4}i\right) + \left(\frac{5}{4} - \frac{8}{3}i\right) + \left(-\frac{11}{2} - i\right)$$

$$= \left(2 - 5 + \frac{5}{4} - \frac{11}{2}\right) + \left(\frac{3}{2} + \frac{7}{4} - \frac{8}{3} - 1\right)i$$

$$= \left(\frac{8 - 20 + 5 - 22}{4}\right) + \left(\frac{18 + 21 - 32 - 12}{12}\right)i$$

$$= -\frac{29}{4} - \frac{5}{12}i \qquad \textbf{Ans.}$$

25.7 ADDITION OF COMPLEX NUMBERS BY GEOMETRY

Let $z_1 = x_1 + i\,y_1$ and $z_2 = x_2 + i\,y_2$ be two complex numbers represented by the points P and Q on the Argand diagram.

Complete the parallelogram $OPRQ$.

Draw PK, RM, QL, perpendiculars on OX.

Also draw $PN \perp$ to RM.

$$OM = OK + KM = OK + OL = x_1 + x_2$$

and
$$RM = MN + NR = KP + LQ = y_1 + y_2$$

$\therefore$ The co-ordinates of R are $(x_1 + x_2, y_1 + y_2)$ and it represents the complex number.

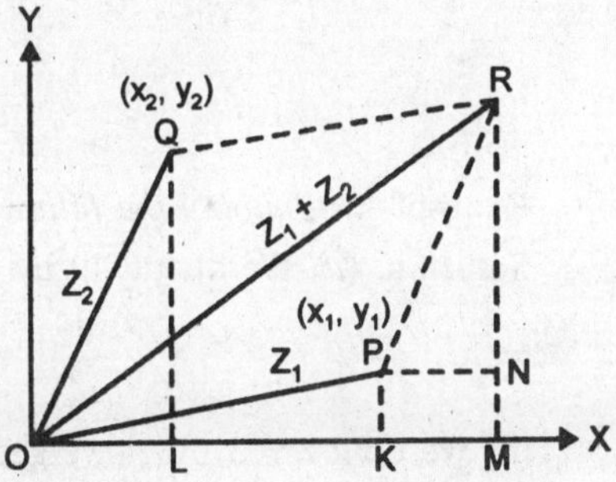

$$(x_1 + x_2) + i\,(y_1 + y_2) = (x_1 + iy_1) + (x_2 + iy_2)$$

Thus the sum of two complex numbers is represented by the extremity of the diagonal of the parallelogram formed by $OP\ (z_1)$ and $OQ\ (z_2)$ as adjacent sides.

$$|z_1 + z_2| = OR \quad \text{and} \quad \text{amp}\,(z_1 + z_2) = \angle ROM.$$

25.8 SUBTRACTION

$$(a + ib) - (c + id) = (a - c) + i(b - d)$$

Procedure. In subtraction of complex numbers we subtract real parts from real parts and imaginary parts from imaginary parts.

Example 2. *Subtract* $z_1 = \frac{3}{4} - \frac{7}{3}i$ *from* $z_2 = \frac{-5}{3} + \frac{11}{5}i$.

Solution.
$$z_2 - z_1 = \left(\frac{-5}{3} + \frac{11}{5}i\right) - \left(\frac{3}{4} - \frac{7}{3}i\right) = \left(\frac{-5}{3} - \frac{3}{4}\right) + \left(\frac{11}{5} + \frac{7}{3}\right)i$$

$$= \left(\frac{-20 - 9}{12}\right) + \left(\frac{33 + 35}{15}\right)i = \frac{-29}{12} + \frac{68}{15}i \qquad \textbf{Ans.}$$

SUBTRACTION OF COMPLEX NUMBERS BY GEOMETRY.

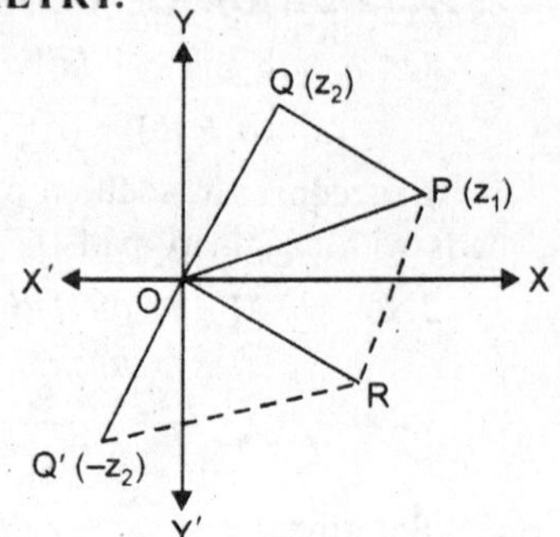

Let P and Q represent two complex numbers

$$z_1 = x_1 + iy_1 \text{ and } z_2 = x_2 + iy_2.$$

Then $$z_1 - z_2 = z_1 + (-z_2)$$

$z_1 - z_2$ means the addition of z_1 and $-z_2$.

$-z_2$ is represented by OQ' formed by producing OQ to OQ' such that $OQ = OQ'$.

Complete the parallelogram $OPRQ'$, then the sum of z_1 and $-z_2$ represented by OR.

25.9 POWERS OF *i*

Some time we need various powers of i.

We know that $i = \sqrt{-1}$.

On squaring both sides, we get

$$i^2 = -1$$

Multiplying by i both sides, we get

$$i^3 = -i$$

Again, $$i^4 = (i^3)(i) = (-i)(i) = -(i^2) = -(-1) = 1$$

$$i^5 = (i^4)(i) = (1)(i) = i$$

$$i^6 = (i^4)(i^2) = (1)(-1) = -1$$

$$i^7 = (i^4)(i^3) = 1(-i) = -i$$

$$i^8 = (i^4)(i^4) = (1)(1) = 1.$$

Example 3. *Simplify the following:* (*a*) i^{49}, (*b*) i^{103}.

Solution. (*a*) We divide 49 by 4 and we get

$$49 = 4 \times 12 + 1$$

$$i^{49} = i^{4 \times 12 + 1} = (i^4)^{12}(i^1) = (1)^{12}(i) = i$$

(*b*) we divide 103 by 4, we get

$$103 = 4 \times 25 + 3$$

$$i^{103} = i^{4 \times 25 + 3} = (i^4)^{25}(i^3) = (1)^{25}(-i) = -i$$ **Ans.**

25.10 MULTIPLICATION

$$(a + ib) \times (c + id) = ac - bd + i(ad + bc)$$

Proof. $$(a + ib) \times (c + id) = ac + iad + ibc + i^2bd$$

$$= ac + i(ad + bc) + (-1)bd \qquad [\because i^2 = -1]$$

$$= (ac - bd) + (ad + bc)i$$

Example 4. *Multiply 3 + 4i by 7 – 3i.*

Solution. Let $z_1 = 3 + 4i$ and $z_2 = 7 - 3i$

$$z_1.z_2 = (3 + 4i).(7 - 3i)$$

$$= 21 - 9i + 28i - 12i^2$$

$$= 21 - 9i + 28i - 12(-1) \qquad [\because i^2 = -1]$$

$$= 21 - 9i + 28i + 12$$

$$= 33 + 19i$$ **Ans.**

Multiplication of complex numbers (Polar form) :

Let $z_1 = r_1(\cos \theta_1 + i \sin \theta_1)$ and $z_2 = r_2(\cos \theta_2 + i \sin \theta_.)$

$$x_1 = r_1 \cos \theta_1, \quad y_1 = r_1 \sin \theta_1$$

$$x_2 = r_2 \cos \theta_2, \quad y_2 = r_2 \sin \theta_2$$

$$z_1 = x_1 + iy_1 = r_1(\cos \theta_1 + i \sin \theta_1) \qquad |z_1| = r_1$$

$$z_2 = x_2 + iy_2 = r_2(\cos \theta_2 + i \sin \theta_2) \qquad |z_2| = r_2$$

$$z_1 . z_2 = r_1 r_2(\cos \theta_1 + i \sin \theta_1)(\cos \theta_2 + i \sin \theta_2)$$

$$= r_1 r_2 [\cos \theta_1 \cos \theta_2 - \sin \theta_1 \sin \theta_2 + i(\sin \theta_1 \cos \theta_2 + \cos \theta_1 \sin \theta_2)]$$

$$= r_1 r_1[\cos (\theta_1 + \theta_2) + i \sin (\theta_1 + \theta_2)], \qquad |z_1 z_2| = r_1 r_2$$

The modulus of the product of two complex numbers is the product of their moduli and the argument of the product is the sum of their arguments.

Graphical method

Let P, Q represent the complex numbers.

$$z_1 = x_1 + iy_1$$

$$= r_1(\cos \theta_1 + i \sin \theta_1)$$

$$z_2 = x_2 + iy_2$$

$$= r_2(\cos \theta_2 + i \sin \theta_2)$$

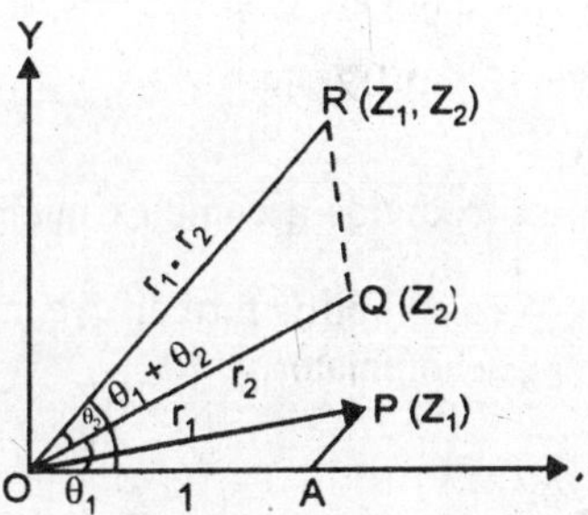

Cut off $OA = 1$ along x-axis. Construct $\Delta\ ORQ$ on OQ similar to $\Delta\ OAP$.

So that $\dfrac{OR}{OP} = \dfrac{OQ}{OA} \Rightarrow \dfrac{OR}{OP} = \dfrac{OQ}{1} \Rightarrow OR = OP . OQ = r_1 r_2$

$$\angle XOR = \angle AOQ + \angle QOR = \theta_2 + \theta_1$$

Hence the product of two complex numbers z_1, z_2 is represented by the point R, such that

(*i*) $|z_1 . z_2| = |z_1|.|z_2|$ (*ii*) $\text{Arg } (z_1 . z_2) = \text{Arg } (z_1) + \text{Arg } (z_2)$

25.11 *i* (IOTA) AS AN OPERATOR

Multipliction of a complex number by i.

Let $$z = x + iy = r(\cos \theta + i \sin \theta)$$

$$i = 0 + i . 1 = \left[\cos \frac{\pi}{2} + i \sin \frac{\pi}{2}\right]$$

$$i . z = r(\cos \theta + i \sin \theta) . \left[\cos \frac{\pi}{2} + i \sin \frac{\pi}{2}\right]$$

$$= r\left[\cos\left(\theta + \frac{\pi}{2}\right) + i \sin\left(\theta + \frac{\pi}{2}\right)\right]$$

Hence a complex number multiplied by i results :

The rotation of the complex number by $\dfrac{\pi}{2}$ in anticlockwise direction without change in magnitude.

25.12 CONJUGATE OF A COMPLEX NUMBER

Two complex numbers which differ only in the sign of imaginary parts are called conjugate of each other.

A pair of complex number $a + ib$ and $a - ib$ are said to be conjugate of each other.

Theorem. Show that the sum and product of a complex number and its conjugate complex are both real.

Proof. Let $x + iy$ be a complex number and $x - iy$ its conjugate complex.

$$\text{Sum} = (x + iy) + (x - iy) = 2x \qquad \text{(Real)}$$

$$\text{Product} = (x + iy).(x - iy) = x^2 + y^2. \qquad \text{(Real)}$$

Proved.

Note. Let a complex number be z. Then the conjugate complex number is denoted by $\bar{z}$.

Example 5. *Find out the conjugate of a complex number $7 + 6i$.*

Solution. Let $z = 7 + 6i$

To find conjugate complex number of $7 + 6i$ we change the sign of imaginary number.

Conjugate of $z = \bar{z} = 7 - 6i$ **Ans.**

25.13 DIVISION

To divide a complex number $a + ib$ by $c + id$, we write it as $\dfrac{a + ib}{c + id}$.

To simplify further, we multiply the numerator and denominator by the conjugate of the denominator.

$$\frac{a + ib}{c + id} = \frac{(a + ib)}{(c + id)} \times \frac{(c - id)}{(c - id)} = \frac{ac - iad + ibc - i^2 bd}{(c)^2 - (id)^2}$$

$$= \frac{ac - i(ad - bc) + bd}{c^2 - d^2 i^2} \qquad [\because i^2 = -1]$$

$$= \frac{ac + bd}{c^2 + d^2} + \frac{bc - ad}{c^2 + d^2} i$$

Example 6. *Divide $1 + i$ by $3 + 4i$.*

Solution.

$$\frac{1 + i}{3 + 4i} = \frac{1 + i}{3 + 4i} \times \frac{3 - 4i}{3 - 4i}$$

$$= \frac{3 - 4i + 3i - 4i^2}{9 - 16i^2}$$

$$= \frac{3 - i + 4}{9 + 16} = \frac{7}{25} - \frac{1}{25} i$$

Ans.

DIVISION (By Algebra)

Let $z_1 = r_1(\cos \theta_1 + i \sin \theta_1)$ and $z_2 = r_2(\cos \theta_2 + i \sin \theta_2)$

$$\frac{z_1}{z_2} = \frac{r_1(\cos \theta_1 + i \sin \theta_1)}{r_2(\cos \theta_2 + i \sin \theta_2)} = \frac{r_1(\cos \theta_1 + i \sin \theta_1)(\cos \theta_2 - i \sin \theta_2)}{r_2(\cos \theta_2 + i \sin \theta_2)(\cos \theta_2 - i \sin \theta_2)}$$

$$= \frac{r_1[(\cos \theta_1 \cos \theta_2 + \sin \theta_1 \sin \theta_2) + i(\sin \theta_1 \cos \theta_2 - \sin \theta_2 \cos \theta_1)]}{r_2(\cos^2 \theta_2 + \sin^2 \theta_2)}$$

$$= \frac{r_1}{r_2}[\cos (\theta_1 - \theta_2) + i \sin (\theta_1 - \theta_2)]$$

The modulus of the quotient of two complex numbers is the quotient of their moduli, and the argument of the quotient is the difference of their arguments.

25.14 DIVISION OF COMPLEX NUMBERS BY GEOMETRY

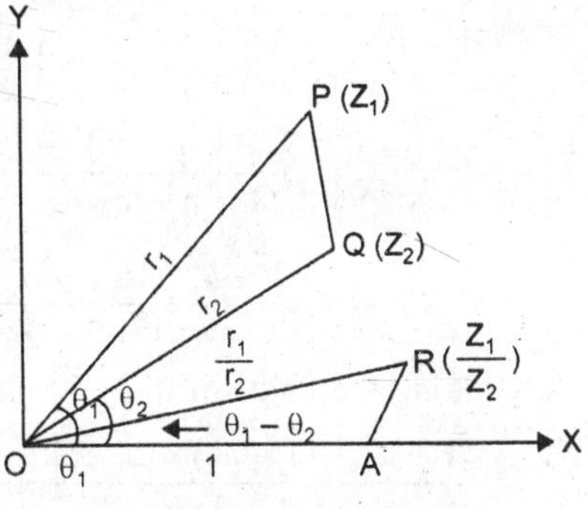

Let P and Q represent the complex numbers.

$$z_1 = x_1 + i\,y_1 = r_1(\cos\theta_1 + i\sin\theta_1)$$
$$z_2 = x_2 + i\,y_2 = r_2(\cos\theta_2 + i\sin\theta_2)$$

Cut off $OA = 1$, construct $\Delta\, OAR$ on OA similar to $\Delta\, OQP$

So that $\dfrac{OR}{OA} = \dfrac{OP}{OQ} \Rightarrow \dfrac{OR}{1} = \dfrac{OP}{OQ}$

$$OR = \frac{OP}{OQ} = \frac{r_1}{r_2}$$

$$\angle AOR = \angle QOP = \angle AOP - \angle AOQ = \theta_1 - \theta_2$$

$\therefore$ R represents the number $\dfrac{r_1}{r_2}[\cos(\theta_1 - \theta_2) + i\sin(\theta_1 - \theta_2)]$

Hence the complex number $\dfrac{z_1}{z_2}$ is represented by the point R.

(*i*) $\left|\dfrac{z_1}{z_2}\right| = \dfrac{|z_1|}{|z_2|}$ (*ii*) Arg. $\left(\dfrac{z_1}{z_2}\right)$ = Arg. (z_1) – Arg. (z_2).

Example 7. *Express* $\dfrac{(6+i).(2-i)}{(4+3i).(1-2i)}$ *in the form of a + ib.*

Solution. $\dfrac{(6+i).(2-i)}{(4+3i).(1-2i)} = \dfrac{12+1+i(2-6)}{4+6+i(3-8)} = \dfrac{13-4i}{10-5i}$

$$= \frac{(13-4i)(10+5i)}{(10-5i)(10+5i)} = \frac{150+25i}{100+25} = \frac{6+i}{5} = \frac{6}{5} + \frac{1}{5}i. \quad \textbf{Ans.}$$

Example 8. *If* $a = \cos\theta + i\sin\theta$, *prove that* $1 + a + a^2 = (1 + 2\cos\theta)(\cos\theta + i\sin\theta)$.

Solution. Here we have $a = \cos\theta + i\sin\theta$

$$\begin{aligned}1 + a + a^2 &= 1 + (\cos\theta + i\sin\theta) + (\cos\theta + i\sin\theta)^2\\ &= 1 + \cos\theta + i\sin\theta + \cos^2\theta + 2i\sin\theta\cos\theta - \sin^2\theta\\ &= (\cos\theta + i\sin\theta) + (1 - \sin^2\theta) + \cos^2\theta + 2i\sin\theta\cos\theta\\ &= (\cos\theta + i\sin\theta) + \cos^2\theta + \cos^2\theta + 2i\sin\theta\cos\theta\\ &= (\cos\theta + i\sin\theta) + 2\cos^2\theta + 2i\sin\theta\cos\theta\\ &= (\cos\theta + i\sin\theta) + 2\cos\theta(\cos\theta + i\sin\theta)\\ &= (\cos\theta + i\sin\theta)(1 + 2\cos\theta)\end{aligned}$$

Proved.

Example 9. *Solve for* θ *such that the expression* $\dfrac{3 + 2i\sin\theta}{1 - 2i\sin\theta}$ *is imaginary.*

Solution. $\dfrac{3 + 2i\sin\theta}{1 - 2i\sin\theta} = \dfrac{(3 + 2i\sin\theta)(1 + 2i\sin\theta)}{(1 - 2i\sin\theta)(1 + 2i\sin\theta)} = \dfrac{3 - 4\sin^2\theta + 8i\sin\theta}{1 + 4\sin^2\theta}$

If $3 - 4\sin^2\theta = 0$ then $\dfrac{3 - 4\sin^2\theta + 8i\sin\theta}{1 + 4\sin^2\theta}$ = purely imaginary.

$$\sin^2\theta = \frac{3}{4} \quad \text{or} \quad \sin\theta = \frac{\sqrt{3}}{2} \quad \text{or} \quad \theta = \frac{\pi}{3} \qquad \textbf{Ans.}$$

Example 10. *If* $a^2 + b^2 + c^2 = 1$ *and* $b + ic = (1 + a)z$, *prove that* $\dfrac{a + ib}{1 + c} = \dfrac{1 + iz}{1 - iz}$.

Solution. Here, we have $b + ic = (1 + a)z \Rightarrow z = \dfrac{b + ic}{1 + a}$

$$\frac{1+iz}{1-iz} = \frac{1+i\dfrac{b+ic}{1+a}}{1-i\dfrac{b+ic}{1+a}} = \frac{1+a+ib-c}{1+a-ib+c}$$

$$= \frac{[(1+a+ib)-c]}{(1+a+c-ib)} \times \frac{(1+a+ib+c)}{(1+a+c+ib)} = \frac{(1+a+ib)^2-c^2}{(1+a+c)^2+b^2}$$

$$= \frac{1+a^2-b^2+2a+2ib+2iab-c^2}{1+a^2+c^2+2a+2c+2ac+b^2} = \frac{1+a^2-b^2-c^2+2a+2ib+2iab}{1+(a^2+b^2+c^2)+2a+2c+2ac}$$

Putting the value of $a^2+b^2+c^2=1$ in the above, we get

$$= \frac{1+a^2-(1-a^2)+2a+2ib+2iab}{1+1+2a+2c+2ac} = \frac{2(a^2+a+ib+iab)}{2(1+a+c+ac)} = \frac{2(1+a)(a+ib)}{2(1+a)(1+c)} = \frac{a+ib}{1+c}$$

Proved.

Example 11. *If $z = \cos\theta + i\sin\theta$, prove that*

(*a*) $\dfrac{2}{1+z} = 1 - i\tan\dfrac{\theta}{2}$ (*b*) $\dfrac{1+z}{1-z} = i\cot\dfrac{\theta}{2}$

Solution. Here, we have $z = \cos\theta + i\sin\theta$

(*a*) $$\frac{2}{1+z} = \frac{2}{1+(\cos\theta+i\sin\theta)} = \frac{2}{(1+\cos\theta)+i\sin\theta} \times \frac{(1+\cos\theta)-i\sin\theta}{(1+\cos\theta)-i\sin\theta}$$

$$= \frac{2[(1+\cos\theta)-i\sin\theta]}{(1+\cos\theta)^2+\sin^2\theta}$$

$$= \frac{2[(1+\cos\theta)-i\sin\theta]}{2(1+\cos\theta)} = 1 - \frac{i\sin\theta}{1+\cos\theta}$$

$$= 1 - i\frac{2\sin\left(\frac{\theta}{2}\right)\cos\left(\frac{\theta}{2}\right)}{2\cos^2\left(\frac{\theta}{2}\right)} = 1 - i\tan\left(\frac{\theta}{2}\right)$$

$$(1+\cos\theta)^2+\sin^2\theta$$
$$= 1+\cos^2\theta+2\cos\theta+\sin^2\theta$$
$$= 1+(\sin^2\theta+\cos^2\theta)+2\cos\theta$$
$$= 1+1+2\cos\theta$$
$$= 2+2\cos\theta$$
$$= 2(1+\cos\theta)$$

Proved.

(*b*) $$\frac{1+z}{1-z} = \frac{(1+\cos\theta)+i\sin\theta}{(1-\cos\theta)-i\sin\theta} = \frac{2\cos^2\frac{\theta}{2}+2i\sin\frac{\theta}{2}\cos\frac{\theta}{2}}{2\sin^2\frac{\theta}{2}-2i\sin\frac{\theta}{2}\cos\frac{\theta}{2}}$$

$$= \frac{\cos\frac{\theta}{2}}{\sin\frac{\theta}{2}}\cdot\left(\frac{\cos\frac{\theta}{2}+i\sin\frac{\theta}{2}}{\sin\frac{\theta}{2}-i\cos\frac{\theta}{2}}\right) = \cot\frac{\theta}{2}\left(\frac{\cos\frac{\theta}{2}+i\sin\frac{\theta}{2}}{\sin\frac{\theta}{2}-i\cos\frac{\theta}{2}}\right)$$

Now, $$\frac{\cos\frac{\theta}{2}+i\sin\frac{\theta}{2}}{\sin\frac{\theta}{2}-i\cos\frac{\theta}{2}} = \left(\frac{\cos\frac{\theta}{2}+i\sin\frac{\theta}{2}}{\sin\frac{\theta}{2}-i\cos\frac{\theta}{2}}\right)\left(\frac{\sin\frac{\theta}{2}+i\cos\frac{\theta}{2}}{\sin\frac{\theta}{2}+i\cos\frac{\theta}{2}}\right)$$

$$= \frac{\cos\frac{\theta}{2}\sin\frac{\theta}{2}+i\cos^2\frac{\theta}{2}+i\sin^2\frac{\theta}{2}-\sin\frac{\theta}{2}\cos\frac{\theta}{2}}{\sin^2\frac{\theta}{2}+\cos^2\frac{\theta}{2}} = \frac{i\left(\cos^2\frac{\theta}{2}+\sin^2\frac{\theta}{2}\right)}{\left(\cos^2\frac{\theta}{2}+\sin^2\frac{\theta}{2}\right)} = i$$

Thus, $$\frac{1+z}{1-z} = i\cot\frac{\theta}{2}$$

Proved.

Example 12. *If $x = \cos\theta + i\sin\theta$, $y = \cos\phi + i\sin\phi$, prove that*

$$\frac{x-y}{x+y} = i\tan\left(\frac{\theta-\phi}{2}\right)$$ (*M.U. 2008*)

Solution. We have,

$$\frac{x-y}{x+y} = \frac{(\cos\theta + i\sin\theta) - (\cos\phi + i\sin\phi)}{(\cos\theta + i\sin\theta) + (\cos\phi + i\sin\phi)}$$

$$= \frac{(\cos\theta - \cos\phi) + i(\sin\theta - \sin\phi)}{(\cos\theta + \cos\phi) + i(\sin\theta + \sin\phi)}$$

$$= \frac{\left[-2\sin\left(\frac{\theta+\phi}{2}\right)\sin\left(\frac{\theta-\phi}{2}\right) + 2i\cos\left(\frac{\theta+\phi}{2}\right)\sin\left(\frac{\theta-\phi}{2}\right)\right]}{\left[2\cos\left(\frac{\theta+\phi}{2}\right)\cos\left(\frac{\theta-\phi}{2}\right) + 2i\sin\left(\frac{\theta+\phi}{2}\right)\cos\left(\frac{\theta-\phi}{2}\right)\right]}$$

$$= \frac{2i\sin\left(\frac{\theta-\phi}{2}\right)\left[\cos\left(\frac{\theta+\phi}{2}\right) + i\sin\left(\frac{\theta+\phi}{2}\right)\right]}{2\cos\left(\frac{\theta-\phi}{2}\right)\left[\cos\left(\frac{\theta+\phi}{2}\right) + i\sin\left(\frac{\theta+\phi}{2}\right)\right]} = i\tan\left(\frac{\theta-\phi}{2}\right)$$ **Proved.**

EXERCISE 25.1

1. If $z = 1 + i$, find (*i*) z^2 (*ii*) $\frac{1}{z}$ and plot them on the Argand diagram. **Ans.** (*i*) $2i$, (*ii*) $\frac{1}{2} - \frac{i}{2}$

Express the following in the form $a + ib$, where a and b are real (2 – 4):

2. $\dfrac{2-3i}{4-i}$ **Ans.** $\frac{11}{17} - \frac{10}{17}i$ **3.** $\dfrac{(3+4i)(2+i)}{1+i}$ **Ans.** $\frac{13}{2} + \frac{9}{2}i$

4. $\dfrac{(1+2i)^3}{(1+i)(2-i)}$ **Ans.** $-\frac{7}{2} + \frac{1}{2}i$

5. The points A, B, C represent the complex numbers z_1, z_2, z_3 respectively, and G is the centroid of the triangle ABC, if $4z_1 + z_2 + z_3 = 0$, show that the origin is the mid-point of AG.

6. $ABCD$ is a parallelogram on the Argand plane. The affixes of A, B, C are $8 + 5i$, $-7 - 5i$, $-5 + 5i$, respectively. Find the affix of D. **Ans.** $10 + 15i$

7. If z_1, z_2, z_3 are three complex numbers and

$$a_1 = z_1 + z_2 + z_3$$
$$b_1 = z_1 + \omega z_2 + \omega^2 z_3$$
$$c_1 = z_1 + \omega^2 z_2 + \omega z_3$$

show that $|a_1|^2 + |b_1|^2 + |c_1|^2 = 3\{|z_1|^2 + |z_2|^2 + |z_3|^2\}$

where ω, ω^2 are cube roots of unity.

8. Find the complex conjugate of $\dfrac{2+3i}{1-i}$. **Ans.** $-\frac{1}{2} - \frac{5}{2}i$

9. If $x + iy = \dfrac{1}{a+ib}$, prove that $(x^2 + y^2)(a^2 + b^2) = 1$

10. Find the value of $x^2 - 6x + 13$, when $x = 3 + 2i$. **Ans.** 0

11. If $\alpha - i\beta = \dfrac{1}{a-ib}$, prove that $(\alpha^2 + \beta^2)(a^2 + b^2) = 1$. (*M.U. 2008*)

12. If $\dfrac{1}{\alpha + i\beta} + \dfrac{1}{a + ib} = 1$, where α, β, a, b are real, express b in terms of α, β.

Ans. $\dfrac{-\beta}{\alpha^2 + \beta^2 - 2\alpha + 1}$

13. If $(x + iy)^{1/3} = a + ib$, then show that $4(a^2 - b^2) = \dfrac{x}{a} + \dfrac{y}{b}$.

14. If $(x + iy)^3 = u + iv$, then show that $\dfrac{u}{x} + \dfrac{v}{y} = 4(x^2 - y^2)$.

15. Find the values of x and y, if $\dfrac{(1+i)x - 2i}{3+i} + \dfrac{(2-3i)y + i}{3-i} = i$. **Ans.** $x = 3$ and $y = -1$

16. If $a + ib = \dfrac{(x+i)^2}{2x^2+1}$, prove that $a^2 + b^2 = \dfrac{(x^2+1)^2}{(2x^2+1)^2}$.

25.15 MODULUS AND ARGUMENT

Let $x + iy$ be a complex number.

Putting $x = r\cos\theta$ and $y = r\sin\theta$ so that $r = \sqrt{x^2 + y^2}$

$$\cos\theta = \frac{x}{\sqrt{x^2+y^2}} \quad \text{and} \quad \sin\theta = \frac{y}{\sqrt{x^2+y^2}}$$

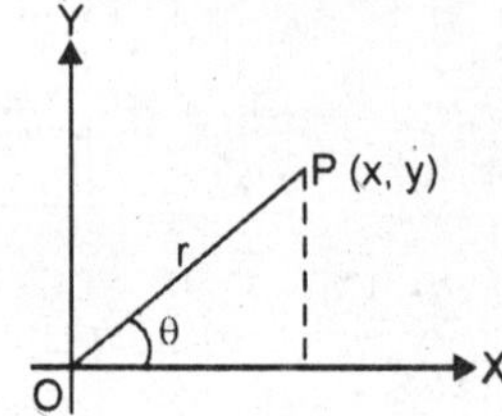

the positive value of the root being taken.

Then r called the *modulus* or absolute value of the complex number $x + iy$ and is denoted by $|x + iy|$.

The angle θ is called the *argument* or *amplitude* of the complex number $x + iy$ and is denoted by arg. $(x + iy)$.

It is clear that θ will have infinite number of values differing by multiples of 2π. The values of θ lying in the range $-\pi < \theta \le \pi$ $[(0 < \theta < \pi)$ or $(-\pi < \theta < 0)]$ is called the *principal value* of the argument.

The principal value of θ is written either between 0 and π or between 0 and $-\pi$.

A complex number $x + iy$ is denoted by a single letter z. The number $x - iy$ (conjugate) is denoted by $\bar{z}$. The complex number in polar form is $r(\cos\theta + i\sin\theta)$.

Modulus of z is denoted by $|z|$ and $|z|^2 = x^2 + y^2$.

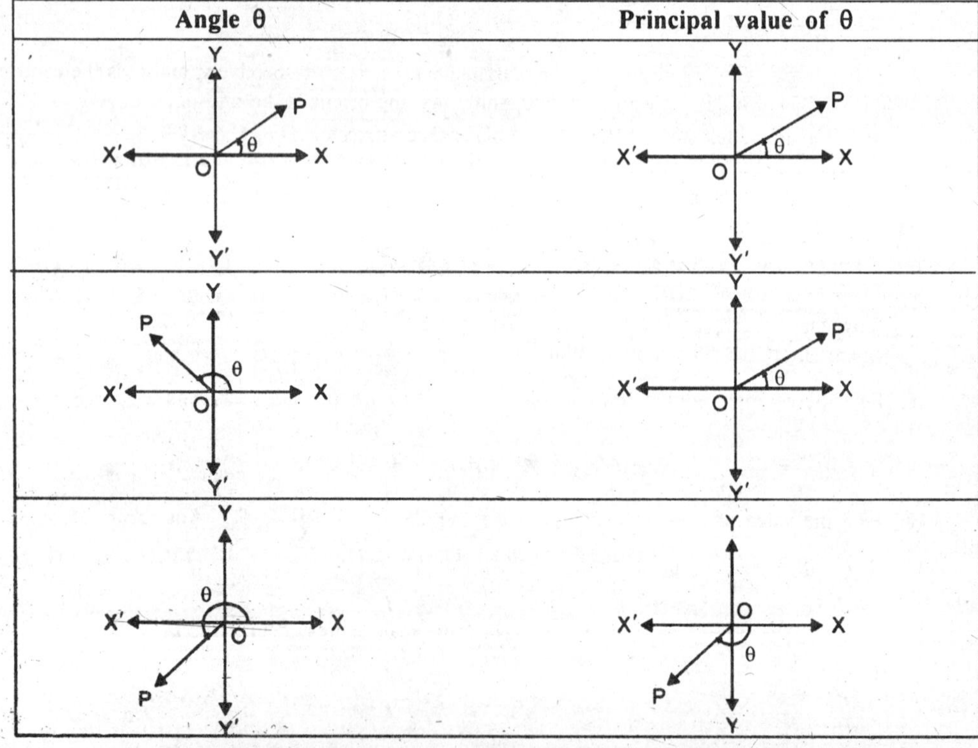

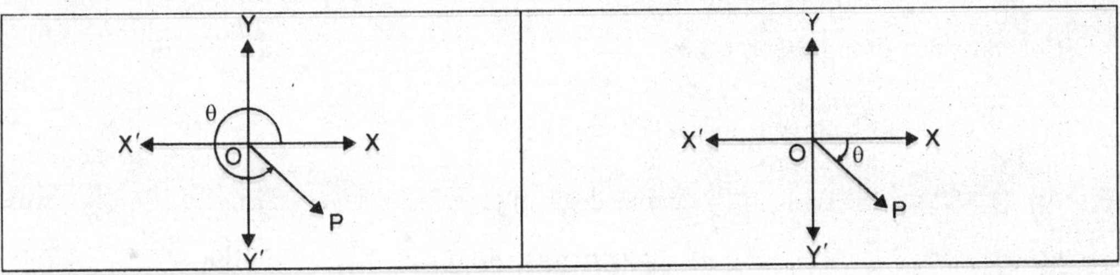

For example (*i*) the principal value of 240° is –120°.

(*ii*) the principal value of 330° is –30°.

Example 13. *Find the modulus and principal argument of the complex number*

$$\frac{1+2i}{1-(1-i)^2}.$$

Solution. $$\frac{1+2i}{1-(1-i)^2} = \frac{1+2i}{1-(1-1-2i)} = \frac{1+2i}{1+2i} = 1 = 1+0i$$

$$\therefore \quad \left|\frac{1+2i}{1-(1-i)^2}\right| = |1+0i| = \sqrt{1^2} = 1$$ **Ans.**

$$\text{Principal argument of } \frac{1+2i}{1-(1-i)^2} = \text{Principal argument of } 1+0i$$

$$= \tan^{-1}\frac{0}{1} = \tan^{-1} 0 = 0°.$$

Hence modulus = 1 and principal argument = 0°. **Ans.**

Example 14. *Find the modulus and principal argument of the complex number :*

$$1 + \cos\alpha + i\sin\alpha. \qquad \left(0 < \alpha < \frac{\pi}{2}\right)$$

Solution. Let $(1+\cos\alpha) + i\sin\alpha = r(\cos\theta + i\sin\theta)$

Equating real and imaginary parts, we get

$$1 + \cos\alpha = r\cos\theta \qquad \text{...(1)}$$

And $$\sin\alpha = r\sin\theta \qquad \text{...(2)}$$

Squaring and adding (1) and (2), we get

$$r^2(\cos^2\theta + \sin^2\theta) = (1+\cos\alpha)^2 + (\sin\alpha)^2$$

$$\Rightarrow \quad r^2 = 1 + \cos^2\alpha + 2\cos\alpha + \sin^2\alpha = 1 + 2\cos\alpha + 1$$

$$= 2(1+\cos\alpha) = 2\left(1 + 2\cos^2\frac{\alpha}{2} - 1\right) = 4\cos^2\frac{\alpha}{2}$$

$$\Rightarrow \quad r = 2\cos\frac{\alpha}{2}$$

From (1), we have, $$\cos\theta = \frac{1+\cos\alpha}{r} = \frac{1 + 2\cos^2\frac{\alpha}{2} - 1}{2\cos\frac{\alpha}{2}} = \cos\frac{\alpha}{2} \qquad \text{...(3)}$$

From (2), we have, $$\sin\theta = \frac{\sin\alpha}{r} = \frac{\sin\alpha}{2\cos\frac{\alpha}{2}} = \frac{2\sin\frac{\alpha}{2}\cos\frac{\alpha}{2}}{2\cos\frac{\alpha}{2}} = \sin\frac{\alpha}{2} \qquad \text{...(4)}$$

$$\text{Argument} = \tan^{-1}\frac{\sin\alpha}{1+\cos\alpha} = \tan^{-1}\frac{2\sin\frac{\alpha}{2}\cos\frac{\alpha}{2}}{1 + 2\cos^2\frac{\alpha}{2} - 1} = \tan^{-1}\tan\frac{\alpha}{2} = \frac{\alpha}{2}$$

General value of argument $= 2\pi k + \frac{\alpha}{2}$

$\theta = \frac{\alpha}{2}$ satisfied both equations, (1) and (2),

Arg $(1 + \cos\alpha + i\sin\alpha) = \frac{\alpha}{2}$ and modulus of $(1 + \cos\alpha + i\sin\alpha) = r = 2\cos\frac{\alpha}{2}$ **Ans.**

EXERCISE 25.2

Find the modulus and principal argument of the following complex numbers:

1. $-\sqrt{3} - i$ **Ans.** $2, -\frac{5\pi}{6}$

2. $\frac{(1+i)^2}{1-i}$ **Ans.** $\sqrt{2}, \frac{3\pi}{4}$

3. $\sqrt{\left(\frac{1+i}{1-i}\right)}$ **Ans.** $1, \frac{\pi}{4}$

4. $\tan\alpha - i$ **Ans.** $\sec\alpha, -\left(\frac{\pi}{2} - \alpha\right)$

5. $1 - \cos\alpha + i\sin\alpha$ **Ans.** $2\sin\frac{\alpha}{2}, \frac{\pi - \alpha}{2}$

6. $(4 + 2i)(-3 + \sqrt{2}i)$ **Ans.** $2\sqrt{55}, \tan^{-1}\left(\frac{3 - 2\sqrt{2}}{6 + \sqrt{2}}\right)$

Find the modulus of the following complex numbers :

7. $\overline{(7 + i^2)} + (6 - i) - (4 - 3i^3)$ **Ans.** $4\sqrt{5}$

8. $\overline{(5 - 6i)} - (5 + 6i) + (8 - i)$ **Ans.** $\sqrt{185}$

9. $(8 - i^3) - (7i^2 + 5) + \overline{(9 - i)}$ **Ans.** $\sqrt{365}$

10. $(5 + 6i^{11}) + (8i^3 + i^5) + (i^2 - i^4)$ **Ans.** $\sqrt{178}$

11. If arg. $(z + 2i) = \frac{\pi}{4}$ and arg. $(z - 2i) = \frac{3\pi}{4}$, find z. **Ans.** $z = 2$

Example 15. *If* $z_1 = \cos\alpha + i\sin\alpha$, $z_2 = \cos\beta + i\sin\beta$ *show that*

$$\frac{1}{2i}\left(\frac{z_1}{z_2} - \frac{z_2}{z_1}\right) = \sin(\alpha - \beta)$$

Solution. We have

$$\frac{z_1}{z_2} = \frac{\cos\alpha + i\sin\alpha}{\cos\beta + i\sin\beta} = \frac{\cos\alpha + i\sin\alpha}{\cos\beta + i\sin\beta} \times \frac{\cos\beta - i\sin\beta}{\cos\beta - i\sin\beta}$$

$$= \frac{(\cos\alpha\cos\beta + \sin\alpha\sin\beta) + i(\sin\alpha\cos\beta - \cos\alpha\sin\beta)}{\cos^2\beta + \sin^2\beta}$$

$$= \cos(\alpha - \beta) + i\sin(\alpha - \beta) \quad \ldots(1)$$

$$\frac{z_2}{z_1} = \frac{1}{\cos(\alpha - \beta) + i\sin(\alpha - \beta)} \times \frac{\cos(\alpha - \beta) - i\sin(\alpha - \beta)}{\cos(\alpha - \beta) - i\sin(\alpha - \beta)}$$

$$= \frac{\cos(\alpha - \beta) - i\sin(\alpha - \beta)}{\cos^2(\alpha - \beta) + \sin^2(\alpha - \beta)} = \cos(\alpha - \beta) - i\sin(\alpha - \beta) \quad \ldots(2)$$

Subtracting (2) from (1), we get

$$\frac{z_1}{z_2} - \frac{z_2}{z_1} = 2i\sin(\alpha - \beta)$$ **Proved.**

Example 16. *If* z_1 *and* z_2 *are any two complex numbers, prove that*

$$|z_1 + z_2|^2 + |z_1 - z_2|^2 = 2[|z_1|^2 + |z_2|^2]$$

Solution. Let

$$z_1 = x_1 + iy_1$$

$$z_2 = x_2 + iy_2$$

$$|z_1 + z_2|^2 = |(x_1 + iy_1) + (x_2 + iy_2)|^2$$

$$= |(x_1 + x_2) + i(y_1 + y_2)|^2$$
$$= (x_1 + x_2)^2 + (y_1 + y_2)^2 \qquad ...(1)$$

Similarly $\quad |z_1 - z_2|^2 = (x_1 - x_2)^2 + (y_1 - y_2)^2 \qquad ...(2)$

and $\quad |z_1|^2 = x_1^2 + y_1^2 \qquad ...(3)$

$$|z_2|^2 = x_2^2 + y_2^2 \qquad ...(4)$$

$$\text{L.H.S.} = |z_1 + z_2|^2 + |z_1 - z_2|^2 = (x_1 + x_2)^2 + (y_1 + y_2)^2 + (x_1 - x_2)^2 + (y_1 - y_2)^2$$

[Using (1) and (2)]

$$= x_1^2 + 2x_1x_2 + x_2^2 + y_1^2 + 2y_1y_2 + y_2^2$$
$$+ x_1^2 - 2x_1x_2 + x_2^2 + y_1^2 - 2y_1y_2 + y_2^2$$
$$= 2[x_1^2 + x_2^2 + y_1^2 + y_2^2] = 2[(x_1^2 + y_1^2) + (x_2^2 + y_2^2)] \qquad ...(5)$$
$$= 2[|z_1|^2 + |z_2|^2] = \text{R.H.S.}$$

Proved.

Example 17. *If z_1 and z_2 are two complex numbers such that $|z_1 + z_2| = |z_1 - z_2|$, prove that*

$$arg.\ z_1 - arg.\ z_2 = \frac{\pi}{2}$$

(M.U. 2002, 2007)

Solution. Let
$$z_1 = x_1 + iy_1$$
$$z_2 = x_2 + iy_2$$

Given that $\quad |z_1 + z_2| = |z_1 - z_2|$

$$\Rightarrow \quad |(x_1 + iy_1) + (x_2 + iy_2)| = |(x_1 + iy_1) - (x_2 + iy_2)|$$
$$\Rightarrow \quad |(x_1 + x_2) + i(y_1 + y_2)| = |(x_1 - x_2) + (y_1 - y_2)i|$$
$$\Rightarrow \quad (x_1 + x_2)^2 + (y_1 + y_2)^2 = (x_1 - x_2)^2 + (y_1 - y_2)^2$$
$$\Rightarrow x_1^2 + x_2^2 + 2x_1x_2 + y_1^2 + y_2^2 + 2y_1y_2 = x_1^2 + x_2^2 - 2x_1x_2 + y_1^2 + y_2^2 - 2y_1y_2$$
$$\Rightarrow \quad 2x_1x_2 + 2y_1y_2 = -2x_1x_2 - 2y_1y_2$$
$$\Rightarrow \quad 4x_1x_2 + 4y_1y_2 = 0$$
$$\Rightarrow \quad x_1x_2 + y_1y_2 = 0 \qquad ...(1)$$

Now, $\quad \text{arg. } z_1 - \text{arg. } z_2 = \tan^{-1}\left(\frac{y_1}{x_1}\right) - \tan^{-1}\left(\frac{y_2}{x_2}\right)$

$$= \tan^{-1}\left[\frac{\frac{y_1}{x_1} - \frac{y_2}{x_2}}{1 + \left(\frac{y_1}{x_1}\right)\left(\frac{y_2}{x_2}\right)}\right] = \tan^{-1}\left(\frac{x_2y_1 - x_1y_2}{x_1x_2 + y_1y_2}\right)$$

$$= \tan^{-1}\left(\frac{x_2y_1 - x_1y_2}{0}\right) = \tan^{-1}\infty = \frac{\pi}{2} \qquad \text{[Using (1)]}$$

$$\text{arg. } z_1 - \text{arg. } z_2 = \frac{\pi}{2}$$

Proved.

Example 18. *Find the complex number z if arg $(z + 1) = \frac{\pi}{6}$ and arg $(z - 1) = \frac{2\pi}{3}$.*

(M.U. 2009, 2000, 01, 02, 03)

Solution. Let $\quad z = x + iy \qquad ...(1)$

$\therefore \quad z + 1 = (x + 1) + iy$

We also given that

$$\text{Arg } (z + 1) = \tan^{-1}\left(\frac{y}{x + 1}\right) = \frac{\pi}{6}$$

$\therefore \qquad \dfrac{y}{x+1} = \tan 30° = \dfrac{1}{\sqrt{3}}$

$\therefore \qquad \sqrt{3}y = x + 1 \qquad$ ...(2)

Now $\qquad z - 1 = (x - 1) + iy \qquad$ [From (1)]

and $\qquad \tan^{-1}\left(\dfrac{y}{x-1}\right) = \dfrac{2\pi}{3} \Rightarrow \dfrac{y}{x-1} = \tan 120°$

$\Rightarrow \qquad \dfrac{y}{x-1} = -\cot 30° = -\sqrt{3}$

$\therefore \qquad -y = \sqrt{3}x - \sqrt{3}$

$\Rightarrow \qquad -\sqrt{3}y = 3x - 3 \qquad$ ...(3)

Adding (2) and (3), we get

$$0 = 4x - 2 \quad \Rightarrow \quad 4x = 2 \Rightarrow x = \frac{1}{2}$$

Putting $x = \dfrac{1}{2}$ in (2), we get

$$\sqrt{3}y = \frac{1}{2} + 1 \Rightarrow \sqrt{3}y = \frac{3}{2} \Rightarrow y = \frac{\sqrt{3}}{2}$$

Putting the values of x and y in (1), we get

$$z = \frac{1}{2} + \frac{\sqrt{3}}{2}i$$ **Ans.**

Example 19. *Prove that*

(*i*) $|z_1 + z_2| \le |z_1| + |z_2|$ $\qquad$ (*ii*) $|z_1 - z_2| \ge |z_1| - |z_2|$

Solution. (*a*) (By Geometry) Let $z_1 = x_1 + iy_1$ and $z_2 = x_2 + iy_2$ be the two complex numbers shown in the figure

$$|z_1| = OP, \quad |z_2| = OQ$$

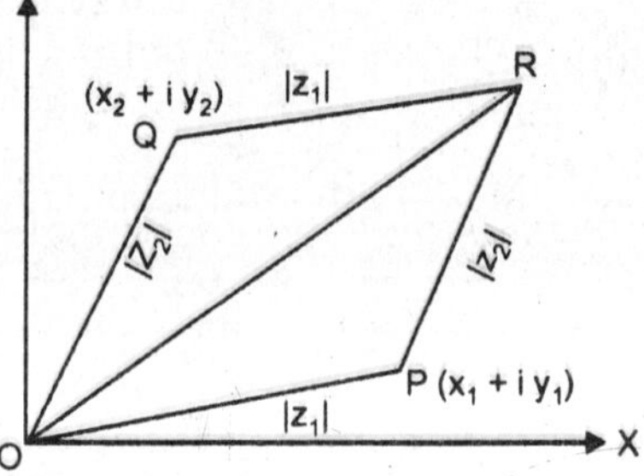

(*i*) Since in a traingle any side is less than the sum of the other two.

In Δ OPR, $\quad OR < OP + PR$, $OR < OP + OQ$

$\Rightarrow \qquad |z_1 + z_2| < |z_1| + |z_2|$

$OR = OP + PR$ if O, P, R are collinear.

or $\qquad |z_1 + z_2| = |z_1| + |z_2|$

(*ii*) Again, any side of a triangle is greater than the difference between the other two, we have

In Δ OPR

$OR > OP - PR, \quad \Rightarrow \quad OR > OP - OQ$

$$|z_1 - z_2| > |z_1| - |z_2|$$ **Proved.**

(*b*) By Algebra. $\qquad z_1 + z_2 = x_1 + iy_1 + x_2 + iy_2 = (x_1 + x_2) + i(y_1 + y_2)$

(*i*) $\qquad |z_1 + z_2|^2 = (x_1 + x_2)^2 + (y_1 + y_2)^2$

$= x_1^2 + x_2^2 + 2x_1x_2 + y_1^2 + y_2^2 + 2y_1y_2 = (x_1^2 + y_1^2) + (x_2^2 + y_2^2) + 2(x_1x_2 + y_1y_2)$

$= (x_1^2 + y_1^2) + (x_2^2 + y_2^2) + 2\sqrt{(x_1x_2 + y_1y_2)^2}$

$= |z_1|^2 + |z_2|^2 + 2\sqrt{x_1^2x_2^2 + y_1^2y_2^2 + 2x_1x_2y_1y_2}$

$[\because (x_1y_2 - x_2y_1)^2 \ge 0$ or $x_1^2\,y_2^2 + x_2^2\,y_1^2 \ge 2x_1x_2y_1y_2]$

$$|z_1 + z_2|^2 \le |z_1|^2 + |z_2|^2 + 2\sqrt{x_1^2x_2^2 + y_1^2y_2^2 + x_1^2y_2^2 + x_2^2y_1^2}$$

$$\le |z_1|^2 + |z_2|^2 + 2\sqrt{(x_1^2 + y_1^2)(x_2^2 + y_2^2)}$$

$$\le |z_1|^2 + |z_2|^2 + 2|z_1||z_2|$$

$$\le \{|z_1| + |z_2|\}^2$$

$$|z_1 + z_2| \le |z_1| + |z_2|$$

(*ii*)

$$|z_1| = |(z_1 - z_2) + z_2| \le |z_1 - z_2| + |z_2|$$

$$|z_1| - |z_2| \le |z_1 - z_2|$$

$$|z_1 - z_2| \ge |z_1| - |z_2|$$

Proved.

EXERCISE 25.3

1. If $z = x + iy$, prove that $\left(\frac{z}{\bar{z}} + \frac{\bar{z}}{z}\right) = 2\left(\frac{x^2 - y^2}{x^2 + y^2}\right)$.

2. If $z = a\cos\theta + ia\sin\theta$, prove that $\left(\frac{z}{\bar{z}} + \frac{\bar{z}}{z}\right) = 2\cos 2\theta$.

3. Prove that $\left|\frac{z - 1}{\bar{z} - 1}\right| = 1$.

4. Let $z_1 = 2 - i$, $z_2 = -2 + i$, find

   (*i*) $\text{Re}\left[\frac{z_1 z_2}{\bar{z}_1}\right]$ (*ii*) $\text{Im}\left[\frac{1}{z_1 \bar{z}_2}\right]$ **Ans.** (*i*) $-\frac{2}{5}$, (*ii*) 0

5. If $|z| = 1$, prove that $\frac{z-1}{z+1}$ $(z \ne 1)$ is a pure imaginary number, what will you conclude, if $z = 1$? **Ans.** If $z = 1$, $\frac{z-1}{z+1} = 0$, which is purely real.

25.16 POLAR FORM

Polar form of a complex number as we have discussed above

$$x = r\cos\theta \quad \text{and} \quad y = r\sin\theta$$

$$\Rightarrow \quad x + iy = r(\cos\theta + i\sin\theta)$$

$$= r\,e^{i\theta} \quad \text{(Exponential form)} \qquad (e^{i\theta} = \cos\theta + i\sin\theta)$$

Procedure. To convert $x + iy$ into polar.

We write

$$x = r\cos\theta$$

$$y = r\sin\theta$$

On solving these equations, we get the value of θ which satisfy both the equations and $r = \sqrt{x^2 + y^2}$.

25.17 TYPES OF COMPLEX NUMBERS

1. Cartesian form : $x + iy$
2. Polar form : $r(\cos\theta + i\sin\theta)$
3. Exponential form : $re^{i\theta}$

Example 20. *Express in polar form* : $1 - \sqrt{2} + i$

Solution. Let $(1 - \sqrt{2}) + i = r(\cos\theta + i\sin\theta)$

$$\therefore \quad 1 - \sqrt{2} = r\cos\theta \qquad ...(1)$$

$$1 = r\sin\theta \qquad ...(2)$$

Squaring and adding (1) and (2), we get

$$r^2(\cos^2\theta + \sin^2\theta) = (1 - \sqrt{2})^2 + 1^2$$

$$\Rightarrow \qquad r^2 = 1 - 2\sqrt{2} + 2 + 1$$

$$\Rightarrow \qquad r = \sqrt{4 - 2\sqrt{2}}$$

Putting the value of r in (1) and (2), we get

$$\cos\theta = \frac{1-\sqrt{2}}{\sqrt{4-2\sqrt{2}}} \quad \text{and} \quad \sin\theta = \frac{1}{\sqrt{4-2\sqrt{2}}}$$

Hence, the polar form is $\sqrt{4-2\sqrt{2}}\left\{\frac{1-\sqrt{2}}{\sqrt{4-2\sqrt{2}}} + i\frac{1}{\sqrt{4-2\sqrt{2}}}\right\}$ **Ans.**

Example 21. *Find the smallest positive integer n for which*

$$\left(\frac{1+i}{1-i}\right)^n = 1.$$ *(Nagpur University, Winter 2004)*

Solution. $$\left[\frac{1+i}{1-i}\right]^n = 1$$

$$\left[\frac{1+i}{1-i} \times \frac{1+i}{1+i}\right]^n = 1 \Rightarrow \left(\frac{1-1+2i}{1+1}\right)^n = 1$$

$$(i)^n = 1 = (i)^4 \quad \Rightarrow \quad n = 4$$ **Ans.**

Example 22. *If* $i^{\alpha+i\beta} = \alpha + i\beta$, *prove that* $\alpha^2 + \beta^2 = e^{-(4n+1)\pi\beta}$

(Nagpur University, Summer 2003)

Solution. We have, $\alpha + i\beta = i^{(\alpha+i\beta)} = e^{\log i^{\alpha+i\beta}}$

$$\alpha + i\beta = e^{(\alpha+i\beta)\log i} = e^{(\alpha+i\beta)(\log i + 2n\pi i)}$$

$$= e^{(\alpha+i\beta)[\log(\cos\pi/2 + i\sin\pi/2) + 2n\pi i]}$$

$$\Rightarrow \qquad \alpha + i\beta = e^{(\alpha+i\beta)[\log e^{i\pi/2} + 2n\pi i]} = e^{(\alpha+i\beta)[i\pi/2 + 2n\pi i]}$$

$$= e^{i\alpha(\pi/2+2n\pi) - \beta(\pi/2+2n\pi)} = e^{-\beta\pi(2n+1/2)} \times e^{\pi\alpha(2n+1/2)i}$$

$$\Rightarrow \qquad \alpha + i\beta = e^{-\pi\beta(4n+1)/2}\left[\cos\left[\pi\alpha\frac{(4n+1)}{2}\right] + i\sin\left[\pi\alpha\frac{(4n+1)}{2}\right]\right]$$

Equating real and imaginary parts, we get

$$\alpha = e^{-\pi\beta(4n+1)/2}.\cos\left\{\frac{1}{2}\pi\alpha\,(4n+1)\right\} \qquad ...(1)$$

and

$$\beta = e^{-\pi\beta(4n+1)/2}.\sin\left\{\frac{1}{2}\pi\alpha\,(4n+1)\right\} \qquad ...(2)$$

Squaring and adding (1) and (2), we get

$$\alpha^2 + \beta^2 = e^{-\pi\beta(4n+1)}.\left[\cos^2\left\{\frac{1}{2}\pi\,\alpha\,(4n+1)\right\} + \sin^2\left\{\frac{1}{2}\pi\,\alpha\,(4n+1)\right\}\right]$$

$$\therefore \qquad \alpha^2 + \beta^2 = e^{-\pi\beta(4n+1)}$$

Hence the result. **Proved.**

EXERCISE 25.4

Express the following complex numbers into polar form :

1. $\dfrac{1+i}{1-i}$ **Ans.** $\cos\dfrac{\pi}{2} + i\sin\dfrac{\pi}{2}$
2. $\dfrac{-35+5i}{4\sqrt{2}+3\sqrt{2}i}$ **Ans.** $5\left(\cos\dfrac{3\pi}{4} + i\sin\dfrac{3\pi}{4}\right)$
3. $\dfrac{3(-4-\sqrt{3}+4\sqrt{3}i-i)}{8+2i}$ **Ans.** $3\left(\cos\dfrac{2\pi}{3} + i\sin\dfrac{2\pi}{3}\right)$
4. $\dfrac{2+6\sqrt{3}i}{5+\sqrt{3}i}$ **Ans.** $2\left(\cos\dfrac{\pi}{3} + i\sin\dfrac{\pi}{3}\right)$

5. $\dfrac{2+3i}{3-7i}$ **Ans.** $r = \sqrt{754}$, $\theta = \tan^{-1}\left(-\dfrac{23}{15}\right)$ **6.** $\left(\dfrac{4-5i}{2+3i}\right)\cdot\left(\dfrac{3+2i}{7+i}\right)$ **Ans.** 0.905, $\theta = \tan^{-1}(-7.2)$

7. $\dfrac{(2+5i)(-3+i)}{(1-2i)^2}$ **Ans.** $\dfrac{\sqrt{290}}{5}$, $\tan^{-1}\left(-\dfrac{1}{17}\right)$ **8.** $\dfrac{1+7i}{(2-i)^2}$ **Ans.** $\sqrt{2}\left(\cos\dfrac{3\pi}{4} + i\sin\dfrac{3\pi}{4}\right)$

9. $\dfrac{1+3i}{1-2i}$ **Ans.** $\sqrt{2}\left(\cos\dfrac{3\pi}{4} + i\sin\dfrac{3\pi}{4}\right)$ **10.** $\dfrac{i-1}{\cos\dfrac{\pi}{3} + i\sin\dfrac{\pi}{3}}$ **Ans.** $\sqrt{2}\left(\cos\dfrac{5\pi}{12} + i\sin\dfrac{5\pi}{12}\right)$

25.18 SQUARE ROOT OF A COMPLEX NUMBER

Let $a + ib$ be a complex number and its square root is $x + iy$.

i.e., $\sqrt{a+ib} = x + iy$ ...(1)

where x and $y \in R$.

Squaring both sides of (1), we get

$$a + ib = (x+iy)^2$$

$$\Rightarrow \quad a + ib = x^2 + i^2y^2 + i\,2xy$$

$$\Rightarrow \quad a + ib = (x^2 - y^2) + i\,2xy \qquad [\because i^2 = -1] \qquad ...(2)$$

Equating real and imaginary parts of (2), we get

$$x^2 - y^2 = a \qquad ...(3)$$

and $$2xy = b \qquad ...(4)$$

Also, we know that

$$(x^2+y^2)^2 = (x^2-y^2)^2 + 4x^2y^2$$

$$\Rightarrow \quad (x^2+y^2)^2 = a^2 + b^2 \qquad \text{[Using (3) and (4)]}$$

$$\Rightarrow \quad x^2 + y^2 = \sqrt{a^2+b^2} \qquad ...(5)$$

Adding (3) and (5), we get

$$2x^2 = a + \sqrt{a^2+b^2} \quad \Rightarrow \quad x = \sqrt{\frac{a+\sqrt{a^2+b^2}}{2}}$$

Example 23. *Find the square root of the complex number $5 + 12i$.*

Solution. Let $\sqrt{5+12i} = x + iy$ ...(1)

Squaring both sides of (1), we get $5 + 12i = (x+iy)^2 = (x^2 - y^2) + i\,2xy$ ...(2)

Equating real and imaginary parts of (2), we get

$$x^2 - y^2 = 5 \qquad ...(3)$$

and $$2xy = 12 \qquad ...(4)$$

Now, $$x^2 + y^2 = \sqrt{(x^2-y^2)^2 + 4x^2y^2} = \sqrt{(5)^2 + (12)^2}$$

$$= \sqrt{25+144} = \sqrt{169} = 13$$

$$\Rightarrow \quad x^2 + y^2 = 13 \qquad ...(5)$$

Adding (3) and (5), we get $2x^2 = 5 + 13 = 18 \Rightarrow x = \sqrt{\dfrac{18}{2}} = \sqrt{9} = \pm 3$

Subtracting (3) from (5), we get $2y^2 = 13 - 5 = 8 \Rightarrow y = \sqrt{\dfrac{8}{2}} = \sqrt{4} = \pm 2$

Since, xy is positive, so x and y are of same sign. Hence, $x = \pm 3$, $y = \pm 2$

$\therefore \quad \sqrt{5+12i} = \pm 3 \pm 2i$ *i.e.* $(3+2i)$ or $-(3+2i)$ **Ans.**

Example 24. *Find the square root of $-4 - 3i$.*

Solution. Let $\sqrt{-4-3i} = x + iy$ ...(1)

Squaring both sides of (1), we get

$$-4 - 3i = (x+iy)^2 = (x^2 - y^2) + i\,2xy \qquad ...(2)$$

Equating real and imaginary parts of (2), we get

$$x^2 - y^2 = -4 \quad ...(3)$$

And $$2xy = -3 \quad ...(4)$$

Now, $$x^2 + y^2 = \sqrt{(x^2 - y^2)^2 + 4x^2y^2} = \sqrt{16 + 9} = \sqrt{25} = \pm 5$$

$$\Rightarrow \quad x^2 + y^2 = 5 \quad ...(5) \quad (\because x^2 + y^2 \geq 0)$$

Adding (3) and (5), we get

$$2x^2 = 5 - 4 = 1 \Rightarrow x = \sqrt{\frac{1}{2}} = \pm\frac{1}{\sqrt{2}}$$

Subtracting (3) from (5), we get

$$2y^2 = 5 + 4 = 9 \Rightarrow y = \sqrt{\frac{9}{2}} = \pm\frac{3}{\sqrt{2}}$$

Since, xy is negative, so x and y will be of different signs. Hence, $x = \pm\frac{1}{\sqrt{2}}$, $y = \mp\frac{3}{\sqrt{2}}$

$$\therefore \quad \sqrt{-4 - 3i} = \pm\left(\frac{1}{\sqrt{2}} - \frac{3}{\sqrt{2}}i\right)$$ **Ans.**

Example 25. *Prove that if the sum and product of two complex numbers are real then the two numbers must be either real or conjugate.* *(M.U. 2008)*

Solution. Let z_1 and z_2 be the two complex numbers.

We are given that $z_1 + z_2 = a$ (real)

and $z_1 . z_2 = b$ (real)

If sum and product of the roots of a quadratic equation are given. Then the equation becomes

$$x^2 - (\text{sum of the roots})\, x + \text{product of the roots} = 0$$

$$x^2 - ax + b = 0$$

$$\text{Root} = x = \frac{a \pm \sqrt{a^2 - 4b}}{2}$$

Case I. If $a^2 > 4b$ Then both the roots are real

Case II. If $a^2 < 4b$

Then $$\text{one root} = \frac{a}{2} + i\frac{\sqrt{4b - a^2}}{2}$$

$$\text{Second root} = \frac{a}{2} - i\frac{\sqrt{4b - a^2}}{2}$$

These roots are conjugate to each other. **Proved.**

EXERCISE 25.5

Find the square root of the following :

1. $1 + i$ **Ans.** $\left\{\pm\sqrt{\frac{\sqrt{2}+1}{2}} \pm \sqrt{\frac{\sqrt{2}-1}{2}}i\right\}$

2. $1 - i$ **Ans.** $\left\{\pm\sqrt{\frac{\sqrt{2}+1}{2}} \mp \sqrt{\frac{\sqrt{2}-1}{2}}i\right\}$

3. i **Ans.** $\left\{\pm\frac{1}{\sqrt{2}} \pm \frac{1}{\sqrt{2}}i\right\}$

4. $15 - 8i$ **Ans.** $1 - 4i, -1 + 4i$

5. $-2 + 2\sqrt{3}i$ **Ans.** $\pm(1 + \sqrt{3}i)$

6. $3 + 4\sqrt{7}i$ **Ans.** $\pm(\sqrt{7} + 2i)$

7. $\frac{2 + 3i}{5 - 4i} + \frac{2 - 3i}{5 + 4i}$ **Ans.** $\pm\frac{2}{\sqrt{41}}i$

8. $x^2 - 1 + i\,2x$ **Ans.** $\pm(x + i)$

9. $3 - 4i$ **Ans.** $\pm(2 - i)$

25.19 EXPONENTIAL AND CIRCULAR FUNCTIONS OF COMPLEX VARIABLES

Proof. $\boxed{\cos\theta + i\sin\theta = e^{i\theta}}$

$$e^z = 1 + z + \frac{z^2}{2!} + \frac{z^3}{3!} + \frac{z^4}{4!} + \ldots \quad \ldots(1)$$

$$\sin z = z - \frac{z^3}{3!} + \frac{z^5}{5!} - \frac{z^7}{7!} + \ldots \quad \ldots(2)$$

$$\cos z = 1 - \frac{z^2}{2!} + \frac{z^4}{4!} - \frac{z^6}{6!} + \ldots \quad \ldots(3)$$

From (2) and (3), we have

$$\cos z + i\sin z = \left(1 - \frac{z^2}{2!} + \frac{z^4}{4!} - \frac{z^6}{6!} + \ldots\right) + i\left(z - \frac{z^3}{3!} + \frac{z^5}{5!} - \ldots\right)$$

$$= 1 + \frac{(iz)^1}{1!} + \frac{(iz)^2}{2!} + \frac{(iz)^3}{3!} + \ldots = e^{iz}$$

Therefore, $\cos z + i\sin z = e^{iz}$ ...(4)

Similarly, $\cos z - i\sin z = e^{-iz}$ ...(5)

From (4) and (5), we have

$$\cos z = \frac{e^{iz} + e^{-iz}}{2} \quad \ldots(6)$$

$$\sin z = \frac{e^{iz} - e^{-iz}}{2i} \quad \ldots(7)$$

25.20 DE MOIVRE'S THEOREM (By Exponential Function)

$$(\cos\theta + i\sin\theta)^n = \cos n\theta + i\sin n\theta$$

Proof. We know that

$$e^{i\theta} = \cos\theta + i\sin\theta$$

$$(e^{i\theta})^n = (\cos\theta + i\sin\theta)^n$$

$$e^{in\theta} = (\cos\theta + i\sin\theta)^n$$

$$(\cos n\theta + i\sin n\theta) = (\cos\theta + i\sin\theta)^n$$ **Proved.**

If n is a fraction, then $\cos n\theta + i\sin n\theta$ is one of the values of $(\cos\theta + i\sin\theta)^n$

25.21 DE MOIVRE'S THEOREM (BY INDUCTION)

Statement: For any rational number n the value or one of the values of

$$(\cos\theta + i\sin\theta)^n = \cos n\theta + i\sin n\theta$$

Proof. Case I. Let n be a non-negative integer. By actual multiplication,

$$(\cos\theta_1 + i\sin\theta_1)(\cos\theta_2 + i\sin\theta_2) = (\cos\theta_1\cos\theta_2 - \sin\theta_1\sin\theta_2) + i(\cos\theta_1\sin\theta_2 + \sin\theta_1\cos\theta_2)$$

$$= \cos(\theta_1 + \theta_2) + i\sin(\theta_1 + \theta_2) \quad \ldots(1)$$

Similarly we can prove that

$$(\cos\theta_1 + i\sin\theta_1)(\cos\theta_2 + i\sin\theta_2)(\cos\theta_3 + i\sin\theta_3)$$
$$= \cos(\theta_1 + \theta_2 + \theta_3) + i\sin(\theta_1 + \theta_2 + \theta_3)$$

Continuing in this way, we can prove that

$$(\cos\theta_1 + i\sin\theta_1)(\cos\theta_2 + i\sin\theta_2)\ldots(\cos\theta_n + i\sin\theta_n)$$
$$= \cos(\theta_1 + \theta_2 \ldots + \theta_n) + i\sin(\theta_1 + \theta_2 + \ldots + \theta_n)$$

Putting $\theta_1 = \theta_2 = \theta_3 = \ldots \theta_n = \theta$, we get

$$(\cos\theta + i\sin\theta)^n = (\cos n\theta + i\sin n\theta)$$

Case II. Let n be a negative integer, say $n = -m$ where m is a positive integer. Then,

$$(\cos\theta + i\sin\theta)^n = (\cos\theta + i\sin\theta)^{-m}$$

$$= \frac{1}{(\cos\theta + i\sin\theta)^m} = \frac{1}{(\cos m\theta + i\sin m\theta)} \quad \text{[By case I]}$$

$$= \frac{1}{(\cos m\theta + i \sin m\theta)} \cdot \frac{(\cos m\theta - i \sin m\theta)}{(\cos m\theta - i \sin m\theta)} = \frac{\cos m\theta - i \sin m\theta}{\cos^2 m\theta + \sin^2 m\theta}$$

$= \cos m\,\theta - i \sin m\,\theta$ $\quad [\because \cos^2 m\,\theta + \sin^2 m\,\theta = 1]$

$= \cos(-m\,\theta) + i\sin(-m\,\theta) = \cos n\,\theta + i \sin n\,\theta$

Hence, the theorem is true for negative integers also.

Case III. Let n be a proper fraction $\frac{p}{q}$ where p and q are integers. Without loss of generality we can select q to be positive integer, p may be a positive or negative integer.

Since q is a positive integer

Now, $\left(\cos\frac{\theta}{q} + i\sin\frac{\theta}{q}\right)^q = \cos q \cdot \frac{\theta}{q} + i \sin q \cdot \frac{\theta}{q}$ [By case I]

$= \cos\theta + i\sin\theta$

Taking the q th root of both sides, we get

$$(\cos\theta + i\sin\theta)^{\frac{1}{q}} = \cos\frac{\theta}{q} + i\sin\frac{\theta}{q}$$

Raising both sides to the power p,

$$(\cos\theta + i\sin\theta)^{\frac{p}{q}} = \left(\cos\frac{\theta}{q} + i\sin\frac{\theta}{q}\right)^p = \cos p.\frac{\theta}{q} + i\sin p.\frac{\theta}{q} \quad \text{[By case I and II]}$$

Hence, one of the values of $(\cos\theta + i\sin\theta)^n$ is $\cos n\,\theta + i\sin n\,\theta$ when n is a proper fraction.

Thus, the theorem is true for all rational values of n.

Example 26. *Express* $\frac{(\cos\theta + i\sin\theta)^8}{(\sin\theta + i\cos\theta)^4}$ *in the form* $(x + iy)$.

Solution.
$$\frac{(\cos\theta + i\sin\theta)^8}{(\sin\theta + i\cos\theta)^4} = \frac{(\cos\theta + i\sin\theta)^8}{(i)^4\left(\cos\theta + \frac{1}{i}\sin\theta\right)^4}$$

$$= \frac{(\cos\theta + i\sin\theta)^8}{(\cos\theta - i\sin\theta)^4} = \frac{(\cos\theta + i\sin\theta)^8}{[\cos(-\theta) + i\sin(-\theta)]^4}$$

$$= \frac{(\cos\theta + i\sin\theta)^8}{[(\cos\theta + i\sin\theta)^{-1}]^4} = \frac{(\cos\theta + i\sin\theta)^8}{(\cos\theta + i\sin\theta)^{-4}} = (\cos\theta + i\sin\theta)^{12}$$

$= \cos 12\,\theta + i\sin 12\,\theta$ **Ans.**

Example 27. *Prove that* $(1 + \cos\theta + i\sin\theta)^n + (1 + \cos\theta - i\sin\theta)^n = 2^{n+1}\cos^n\frac{\theta}{2}\cos\frac{n\,\theta}{2}$ *where n is an integer.*

Solution. L.H.S. $= (1 + \cos\theta + i\sin\theta)^n + (1 + \cos\theta - i\sin\theta)^n$

$$= \left[1 + 2\cos^2\frac{\theta}{2} - 1 + 2\,i\sin\frac{\theta}{2}\cos\frac{\theta}{2}\right]^n + \left[1 + 2\cos^2\frac{\theta}{2} - 1 - i\,2\sin\frac{\theta}{2}\cos\frac{\theta}{2}\right]^n$$

$$= \left[2\cos^2\frac{\theta}{2} + 2\,i\sin\frac{\theta}{2}\cos\frac{\theta}{2}\right]^n + \left[2\cos^2\frac{\theta}{2} - 2\,i\sin\frac{\theta}{2}\cos\frac{\theta}{2}\right]^n$$

$$= \left(2\cos\frac{\theta}{2}\right)^n\left[\cos\frac{\theta}{2} + i\sin\frac{\theta}{2}\right]^n + \left(2\cos\frac{\theta}{2}\right)^n\left[\cos\frac{\theta}{2} - i\sin\frac{\theta}{2}\right]^n$$

$$= 2^n\cos^n\frac{\theta}{2}\left[\cos\frac{n\,\theta}{2} + i\sin\frac{n\,\theta}{2}\right] + 2^n\cos^n\frac{\theta}{2}\left[\cos\frac{n\,\theta}{2} - i\sin\frac{n\,\theta}{2}\right]$$

$$= 2^n \cos^n \frac{\theta}{2}\left[\cos\frac{n\theta}{2} + i\sin\frac{n\theta}{2} + \cos\frac{n\theta}{2} - i\sin\frac{n\theta}{2}\right]$$

$$= 2^n \cos^n \frac{\theta}{2}\left(2\cos\frac{n\theta}{2}\right) = 2^{n+1}\cos^n\frac{\theta}{2}\cos\frac{n\theta}{2} = \text{R.H.S.}$$ **Proved.**

Example 28. *Evaluate* $\left(\dfrac{1+\sin\alpha + i\cos\alpha}{1+\sin\alpha - i\cos\alpha}\right)^n$ *(M.U. 2001, 2004, 2005)*

Solution. We know that,

$$1 = \sin^2\alpha + \cos^2\alpha$$

$$\Rightarrow \quad 1 = \sin^2\alpha - i^2\cos^2\alpha$$

$$\Rightarrow \quad 1 = (\sin\alpha + i\cos\alpha)(\sin\alpha - i\cos\alpha) \qquad ...(1)$$

Adding $\sin\alpha + i\cos\alpha$ both sides of (1), we get

$$1 + \sin\alpha + i\cos\alpha = (\sin\alpha + i\cos\alpha)(\sin\alpha - i\cos\alpha) + (\sin\alpha + i\cos\alpha)$$

$$= (\sin\alpha + i\cos\alpha)(\sin\alpha - i\cos\alpha + 1)$$

$$\therefore \quad \frac{1+\sin\alpha + i\cos\alpha}{1+\sin\alpha - i\cos\alpha} = \sin\alpha + i\cos\alpha$$

$$= \cos\left(\frac{\pi}{2}-\alpha\right) + i\sin\left(\frac{\pi}{2}-\alpha\right) \qquad ...(2)$$

$$\Rightarrow \quad \left(\frac{1+\sin\alpha + i\cos\alpha}{1+\sin\alpha - i\cos\alpha}\right)^n = \left\{\cos\left(\frac{\pi}{2}-\alpha\right) + i\sin\left(\frac{\pi}{2}-\alpha\right)\right\}^n$$

$$= \cos n\left(\frac{\pi}{2}-\alpha\right) + i\sin n\left(\frac{\pi}{2}-\alpha\right)$$ **Ans.**

Example 29. *If* $2\cos\theta = x + \dfrac{1}{x}$ *and* $2\cos\phi = y + \dfrac{1}{y}$, *then prove that*

$x^p . y^q + \dfrac{1}{x^p . y^q} = 2\cos(p\theta + q\phi).$ *(Nagpur University, Summer 2000)*

Solution. We have,

$$x + \frac{1}{x} = 2\cos\theta \quad \Rightarrow \quad x^2 - 2x\cos\theta + 1 = 0$$

$$\Rightarrow \quad x = \frac{2\cos\theta \pm \sqrt{4\cos^2\theta - 4}}{2} = \cos\theta \pm \sqrt{-\sin^2\theta}$$

Putting i^2 for -1 and considering the positive sign, we get

$$x = \cos\theta + i\sin\theta \text{ and similarly, } y = \cos\phi + i\sin\phi$$

Now, $$x^p = (\cos\theta + i\sin\theta)^p = \cos p\theta + i\sin p\theta$$

and $$y^q = (\cos\phi + i\sin\phi)^q = \cos q\phi + i\sin q\phi$$

(by De-Moivre's theorem)

$$\therefore \quad x^p . y^q = (\cos p\theta + i\sin p\theta)(\cos q\phi + i\sin q\phi)$$

$$= \cos(p\theta + q\phi) + i\sin(p\theta + q\phi) \qquad ...(1)$$

Also $$\frac{1}{x^p . y^q} = [\cos(p\theta + q\phi) + i\sin(p\theta + q\phi)]^{-1}$$

$$= \cos(p\theta + q\phi) - i\sin(p\theta + q\phi) \qquad ...(2)$$

Adding (1) and (2), we get

$$\therefore \quad x^p . y^q + \frac{1}{x^p . y^q} = 2\cos(p\theta + q\phi)$$ **Proved.**

Example 30. *Prove that the general value of* θ *which satisfies the equation*

$(\cos\theta + i\sin\theta).(\cos 2\theta + i\sin 2\theta) \ldots (\cos n\theta + i\sin n\theta) = 1$ *is* $\dfrac{4m\pi}{n(n+1)}$, *where m is any integer.*

Solution. $(\cos\theta + i\sin\theta)(\cos 2\theta + i\sin 2\theta)\ldots(\cos n\theta + i\sin n\theta) = 1$

$$(\cos\theta + i\sin\theta)(\cos\theta + i\sin\theta)^2\ldots(\cos\theta + i\sin\theta)^n = 1$$

$$(\cos\theta + i\sin\theta)^{1+2\ldots+n} = 1$$

$$(\cos\theta + i\sin\theta)^{\frac{n(n+1)}{2}} = (\cos 2m\pi + i\sin 2m\pi)$$

$$\cos\frac{n(n+1)}{2}\theta + i\sin\frac{n(n+1)}{2}\theta = \cos 2m\pi + i\sin 2m\pi$$

$$\frac{n(n+1)}{2}\theta = 2m\pi \Rightarrow \theta = \frac{4m\pi}{n(n+1)}$$ **Proved.**

Example 31. *If* $(a_1 + ib_1)\,.\,(a_2 + ib_2)\ldots(a_n + ib_n) = A + iB$, *then prove that*

(i) $\tan^{-1}\frac{b_1}{a_1} + \tan^{-1}\frac{b_2}{a_2} + \ldots + \tan^{-1}\frac{b_n}{a_n} = \tan^{-1}\frac{B}{A}$

(ii) $(a_1^2 + b_1^2)(a_2^2 + b_2^2)\ \ldots\ (a_n^2 + b_n^2) = A^2 + B^2$

Solution. Let $a_1 = r_1\cos\alpha_1,\quad b_1 = r_1\sin\alpha_1$

$a_2 = r_2\cos\alpha_2,\quad b_2 = r_2\sin\alpha_2$

....

$a_n = r_n\cos\alpha_n,\quad b_n = r_n\sin\alpha_n$

$A = R\cos\theta,\quad B = R\sin\theta,$

$$(a_1 + ib_1).(a_2 + ib_2)\ldots(a_n + ib_n) = A + iB \qquad \text{(Given)}$$

$$r_1(\cos\alpha_1 + i\sin\alpha_1)\,r_2(\cos\alpha_2 + i\sin\alpha_2)\ldots r_n(\cos\alpha_n + i\sin\alpha_n) = R(\cos\theta + i\sin\theta)$$

$$r_1 r_2\ldots r_n[\cos(\alpha_1 + \alpha_2 + \ldots\alpha_n) + i\sin(\alpha_1 + \alpha_2 + \ldots\alpha_n)] = R(\cos\theta + i\sin\theta)$$

$\therefore\quad r_1 r_2 \ldots\ldots\ldots\ldots r_n = R$

$\Rightarrow\quad (a_1^2 + b_1^2)(a_2^2 + b_2^2)\ldots(a_n^2 + b_n^2) = A^2 + B^2$

And $\quad \alpha_1 + \alpha_2 + \ldots + \alpha_n = \theta$

$$\Rightarrow \tan^{-1}\frac{b_1}{a_1} + \tan^{-1}\frac{b_2}{a_2} + \ldots. + \tan^{-1}\frac{b_n}{a_n} = \tan^{-1}\frac{B}{A}$$ **Proved.**

Example 32. *If* $\cos\alpha + \cos\beta + \cos\gamma = 0 = \sin\alpha + \sin\beta + \sin\gamma$, *then prove that*

(i) $\sin^2\alpha + \sin^2\beta + \sin^2\gamma = \cos^2\alpha + \cos^2\beta + \cos^2\gamma = \frac{3}{2}$

(ii) $\cos 2\alpha + \cos 2\beta + \cos 2\gamma = 0$

(iii) $\cos(\alpha + \beta) + \cos(\beta + \gamma) + \cos(\gamma + \alpha) = 0$

(iv) $\sin(\alpha + \beta) + \sin(\beta + \gamma) + \sin(\gamma + \alpha) = 0$ *(M.U. 2009)*

Solution. Here, we have

$$(\cos\alpha + \cos\beta + \cos\gamma) + i(\sin\alpha + \sin\beta + \sin\gamma) = 0$$

$$(\cos\alpha + i\sin\alpha) + (\cos\beta + i\sin\beta) + (\cos\gamma + i\sin\gamma) = 0$$

$$\therefore\quad a + b + c = 0 \text{ say} \qquad \ldots(1)$$

where, $a = \cos\alpha + i\sin\alpha$, $b = \cos\beta + i\sin\beta$ and $c = \cos\gamma + i\sin\gamma$

Also we can write

$$(\cos\alpha - i\sin\alpha) + (\cos\beta - i\sin\beta) + (\cos\gamma - i\sin\gamma) = 0$$

$$\Rightarrow\quad (\cos\alpha + i\sin\alpha)^{-1} + (\cos\beta + i\sin\beta)^{-1} + (\cos\gamma + i\sin\gamma)^{-1} = 0$$

$$\Rightarrow\quad \frac{1}{a} + \frac{1}{b} + \frac{1}{c} = 0$$

$$\Rightarrow \quad \frac{bc + ca + ab}{abc} = 0 \Rightarrow ab + bc + ca = 0 \quad \text{...(2)}$$

But $\quad (a + b + c)^2 = (a^2 + b^2 + c^2) + 2(ab + bc + ca)$

$$0 = (a^2 + b^2 + c^2) + 0 \quad \text{[From (1) and (2)]}$$

$$\Rightarrow \quad a^2 + b^2 + c^2 = 0$$

$$\Rightarrow \quad (\cos\alpha + i\sin\alpha)^2 + (\cos\beta + i\sin\beta)^2 + (\cos\gamma + i\sin\gamma)^2 = 0$$

$$\Rightarrow \quad (\cos 2\alpha + i\sin 2\alpha) + (\cos 2\beta + i\sin 2\beta) + (\cos 2\gamma + i\sin 2\gamma) = 0$$

$$\Rightarrow \quad (\cos 2\alpha + \cos 2\beta + \cos 2\gamma) + i(\sin 2\alpha + \sin 2\beta + \sin 2\gamma) = 0$$

$$\Rightarrow \quad \cos 2\alpha + \cos 2\beta + \cos 2\gamma = 0. \quad \text{...(3)}$$

$$\Rightarrow \quad 2\cos^2\alpha - 1 + 2\cos^2\beta - 1 + 2\cos^2\gamma - 1 = 0$$

$$\Rightarrow \quad \cos^2\alpha + \cos^2\beta + \cos^2\gamma = \frac{3}{2} \quad \text{...(4)}$$

Further $\quad 1 - \sin^2\alpha + 1 - \sin^2\beta + 1 - \sin^2\gamma = \frac{3}{2}$

$$\Rightarrow \quad \sin^2\alpha + \sin^2\beta + \sin^2\gamma = \frac{3}{2} \quad \text{..(5)}$$

Again consider $\quad ab + bc + ca = 0 \quad$ [From (2)]

$$\Rightarrow \quad (\cos\alpha + i\sin\alpha)(\cos\beta + i\sin\beta) + (\cos\beta + i\sin\beta)(\cos\gamma + i\sin\gamma) + (\cos\gamma + i\sin\gamma)(\cos\alpha + i\sin\alpha) = 0$$

$$\Rightarrow \quad [\cos(\alpha + \beta) + i\sin(\alpha + \beta)] + [\cos(\beta + \gamma) + i\sin(\beta + \gamma)] + [\cos(\gamma + \alpha) + i\sin(\gamma + \alpha)] = 0$$

Equating real and imaginary parts, we get

$$\cos(\alpha + \beta) + \cos(\beta + \gamma) + \cos(\gamma + \alpha) = 0$$

$$\sin(\alpha + \beta) + \sin(\beta + \gamma) + \sin(\gamma + \alpha) = 0$$

Proved.

EXERCISE 25.6

1. If n is a positive integer show that $(a + ib)^n + (a - ib)^n = 2r^n \cos n\theta$ where $r^2 = a^2 + b^2$ and $\theta = \tan^{-1}\left(\frac{b}{a}\right)$. Hence deduce that $\left(1 + i\sqrt{3}\right)^8 + \left(1 - i\sqrt{3}\right)^8 = -2^8$.

2. If n be a positive integer, prove that $(1 + i)^n + (1 - i)^n = 2^{\frac{n+2}{2}} \cos\frac{n\pi}{4}$

3. Show that $(a + ib)^{m/n} + (a - ib)^{m/n} = 2(a^2 + b^2)^{\frac{m}{2n}} \cos\left(\frac{m}{n}\tan^{-1}\frac{b}{a}\right)$.

4. If $P = \cos\theta + i\sin\theta$, $q = \cos\phi + i\sin\phi$, show that

(*i*) $\frac{P - q}{P + q} = i\tan\frac{\theta - \phi}{2}$ $\quad$ (*ii*) $\frac{(P + q)(Pq - 1)}{(P - q)(Pq + 1)} = \frac{\sin\theta + \sin\phi}{\sin\theta - \sin\phi}$

5. If $x = \cos\theta + i\sin\theta$, show that (*i*) $x^m + \frac{1}{x^m} = 2\cos m\theta$ $\quad$ (*ii*) $x^m - \frac{1}{x^m} = 2i\sin m\theta$.

6. Prove that $\tanh(\log\sqrt{3}) = \frac{1}{2}$

7. Prove that $[\sin(\alpha + \theta) - e^{i\alpha}\sin\theta]^n = \sin^n\alpha\, e^{-in\theta}$

8. If $x + \frac{1}{x} = 2\cos\theta$, $y + \frac{1}{y} = 2\cos\phi$, $z + \frac{1}{z} = 2\cos\psi$, show that

$$xyz + \frac{1}{xyz} = 2\cos(\theta + \phi + \psi)$$

25.22 ROOTS OF A COMPLEX NUMBER

We know that $\cos\theta + i\sin\theta = \cos(2m\pi + \theta) + i\sin(2m\pi + \theta)$, $m \in I$

$$[\cos\theta + i\sin\theta]^{1/n} = [\cos(2m\pi+\theta) + i\sin(2m\pi+\theta)]^{1/n}$$

$$= \cos\frac{(2m\pi+\theta)}{n} + i\sin\frac{(2m\pi+\theta)}{n}$$

Giving m the values 0, 1, 2, 3, ... $n-1$ successively, we get the following n values of $(\cos\theta + i\sin\theta)^{1/n}$.

when $m = 0$, $\quad \cos\frac{\theta}{n} + i\sin\frac{\theta}{n}$

When $m = 1$, $\quad \cos\left(\frac{2\pi+\theta}{n}\right) + i\sin\left(\frac{2\pi+\theta}{n}\right)$

When $m = 2$, $\quad \cos\left(\frac{4\pi+\theta}{n}\right) + i\sin\left(\frac{4\pi+\theta}{n}\right)$

When $m = n-1$, $\quad \cos\left(\frac{2(n-1)\pi+\theta}{n}\right) + i\sin\left(\frac{2(n-1)\pi+\theta}{n}\right)$

When $m = n$, $\quad \cos\frac{2n\pi+\theta}{n} + i\sin\frac{2n\pi+\theta}{n} = \cos\left(2\pi + \frac{\theta}{n}\right) + i\sin\left(2\pi+\frac{\theta}{n}\right)$

$$= \cos\frac{\theta}{n} + i\sin\frac{\theta}{n}$$

which is the same as the value for $m = 0$. Thus, the values of $(\cos\theta + i\sin\theta)^{1/n}$ for $m = n$, $n+1$, $n+2$ etc., are the mere repetition of the first n values as obtained above.

Example 33. *Solve $x^4 + i = 0$.* *(M.U. 2008)*

Solution. Here, we have

$$x^4 = -i = \cos\frac{\pi}{2} - i\sin\frac{\pi}{2}$$

$$x^4 = \cos\left(2n\pi+\frac{\pi}{2}\right) - i\sin\left(2n\pi+\frac{\pi}{2}\right)$$

$$\Rightarrow \quad x = \left[\cos\left(2n\pi+\frac{\pi}{2}\right) - i\sin\left(2n\pi+\frac{\pi}{2}\right)\right]^{\frac{1}{4}}$$

$$= \cos(4n+1)\frac{\pi}{8} - i\sin(4n+1)\frac{\pi}{8}$$

Putting $n = 0, 1, 2, 3$ we get the roots as

$$x_1 = \cos\frac{\pi}{8} - i\sin\frac{\pi}{8}, \quad x_2 = \cos\frac{5\pi}{8} - i\sin\frac{5\pi}{8}$$

$$x_3 = \cos\frac{9\pi}{8} - i\sin\frac{9\pi}{8}, \quad x_4 = \cos\frac{13\pi}{8} - i\sin\frac{13\pi}{8}$$ **Ans.**

Example 34. *Solve $x^5 = 1 + i$ and find the continued product of the roots.* *(M.U. 2005, 2004)*

Solution. $\quad x^5 = 1 + i$

$$= \sqrt{2}\left(\cos\frac{\pi}{4} + i\sin\frac{\pi}{4}\right) \Rightarrow x = 2^{\frac{1}{10}}\left(\cos\frac{\pi}{4} + i\sin\frac{\pi}{4}\right)^{\frac{1}{5}}$$

$$\Rightarrow \quad x = 2^{\frac{1}{10}}\left[\cos\left(2k\pi+\frac{\pi}{4}\right).\frac{1}{5} + i\sin\left(2k\pi+\frac{\pi}{4}\right).\frac{1}{5}\right]$$

$$= 2^{\frac{1}{10}}\left[\cos\,(8\,k+1)\,\frac{\pi}{20}+i\sin\,(8\,k+1)\,\frac{\pi}{20}\right]$$

The roots are obtained by putting $k = 0, 1, 2, 3, 4...$

$$x_1 = 2^{\frac{1}{10}}\left[\cos\frac{\pi}{20}+i\sin\frac{\pi}{20}\right],\qquad x_2 = 2^{\frac{1}{10}}\left[\cos\frac{9\pi}{20}+i\sin\frac{9\pi}{20}\right]$$

$$x_3 = 2^{\frac{1}{10}}\left[\cos\frac{17\pi}{20}+i\sin\frac{17\pi}{20}\right]\qquad x_4 = 2^{\frac{1}{10}}\left[\cos\frac{5\pi}{4}+i\sin\frac{5\pi}{4}\right]$$

$$x_5 = 2^{\frac{1}{10}}\left[\cos\frac{33\pi}{20}+i\sin\frac{33\pi}{20}\right]$$

$$x_1\,.\,x_2\,.\,x_3.\,x_4.\,x_5 = \left(2^{\frac{1}{10}}\right)^5\left(\cos\frac{\pi}{20}+i\sin\frac{\pi}{20}\right)\left(\cos\frac{9\pi}{20}+i\sin\frac{9\pi}{20}\right)\left(\cos\frac{17\pi}{20}+i\sin\frac{17\pi}{20}\right)$$

$$\left(\cos\frac{5\pi}{4}+i\sin\frac{5\pi}{4}\right)\left(\cos\frac{33\pi}{20}+i\sin\frac{33\pi}{20}\right)$$

$$= 2^{\frac{1}{2}}\left[\cos\left(\frac{\pi}{20}+\frac{9\pi}{20}+\frac{17\pi}{20}+\frac{5\pi}{4}+\frac{33\pi}{20}\right)+i\sin\left(\frac{\pi}{20}+\frac{9\pi}{20}+\frac{17\pi}{20}+\frac{5\pi}{4}+\frac{33\pi}{20}\right)\right]$$

$$= \sqrt{2}\left[\cos\frac{17\pi}{4}+i\sin\frac{17\pi}{4}\right] = \sqrt{2}\left[\cos\left(4\,\pi+\frac{\pi}{4}\right)+i\sin\left(4\,\pi+\frac{\pi}{4}\right)\right]$$

$$= \sqrt{2}\left[\cos\frac{\pi}{4}+i\sin\frac{\pi}{4}\right] = \sqrt{2}\left[\frac{1}{\sqrt{2}}+i\,\frac{1}{\sqrt{2}}\right] = 1+i$$

Ans.

Example 35. *If $\alpha, \alpha^2, \alpha^3, \alpha^4$, are the roots of $x^5 - 1 = 0$ find them and show that $(1-\alpha)(1-\alpha^2)(1-\alpha^3)(1-\alpha^4) = 5$.* *(M.U. 2007)*

Solution. Here, we have

$$x^5 - 1 = 0$$

$$\Rightarrow \quad x^5 = 1 = \cos 0 + i\sin 0$$

$$\Rightarrow \quad x^5 = \cos\,(2k\pi) + i\sin\,(2k\pi)$$

$$\Rightarrow \quad x = (\cos 2k\pi + i\sin 2k\pi)^{1/5} = \cos\frac{2k\pi}{5} + i\sin\frac{2k\pi}{5}$$

Putting $k = 0, 1, 2, 3, 4$, we get the five roots as below

$$x_0 = \cos 0 + i\sin 0,\quad x_1 = \cos\frac{2\pi}{5}+i\sin\frac{2\pi}{5}$$

$$x_2 = \cos\frac{4\pi}{5}+i\sin\frac{4\pi}{5},\quad x_3 = \cos\frac{6\pi}{5}+i\sin\frac{6\pi}{5}$$

$$x_4 = \cos\frac{8\pi}{5}+i\sin\frac{8\pi}{5}$$

Putting $\bar{x}_1 = \cos\frac{2\pi}{5}+i\sin\frac{2\pi}{5} = \alpha$, we see that

$$x_2 = \left(\cos\frac{4\pi}{3}+i\sin\frac{4\pi}{3}\right) = \left(\cos\frac{2\pi}{3}+i\sin\frac{2\pi}{3}\right)^2 = \alpha^2$$

Similarly, $\quad x_3 = \alpha^3$ and $x_4 = \alpha^4$

$\therefore$ The roots are $1, \alpha, \alpha^2, \alpha^3, \alpha^4$

Hence $\quad x^5 - 1 = (x-1)(x-\alpha)(x-\alpha^2)(x-\alpha^3)(x-\alpha^4)$

$$\Rightarrow \quad \frac{x^5-1}{x-1} = (x-\alpha)(x-\alpha^2)(x-\alpha^3)(x-\alpha^4)$$

On dividing $x^5 - 1$ by $x - 1$, we get

$$x^4 + x^3 + x^2 + x + 1 = (x - \alpha)(x - \alpha^2)(x - \alpha^3)(x - \alpha^4)$$

$$\therefore (x - \alpha)(x - \alpha^2)(x - \alpha^3)(x - \alpha^4) = x^4 + x^3 + x^2 + x + 1$$

Putting $x = 1$, we get

$$(1 - \alpha)(1 - \alpha^2)(1 - \alpha^3)(1 - \alpha^4) = 1 + 1 + 1 + 1 + 1 = 5.$$ **Proved.**

Example 36. *If ω is a cube root of unity, prove that*

$$(1 - \omega)^6 = -27$$ *(M.U. 2003)*

Solution. Let $x^3 = 1$

$$\Rightarrow \quad x = (1)^{1/3} = (\cos 0 + i \sin 0)^{1/3} = (\cos 2n\pi + i \sin 2n\pi)^{1/3}$$

$$= \cos\left(\frac{2n\pi}{3}\right) + i \sin\left(\frac{2n\pi}{3}\right)$$

Putting $n = 0, 1, 2$ the roots of unity are

$$x_0 = 1$$

$$x_1 = \cos\frac{2\pi}{3} + i \sin\frac{2\pi}{3} = \omega \text{ (say)}$$

$$x_2 = \cos\frac{4\pi}{3} + i \sin\frac{4\pi}{3} = \left[\cos\frac{2\pi}{3} + i \sin\frac{2\pi}{3}\right]^2 = \omega^2$$

Now,

$$1 + \omega + \omega^2 = 1 + \cos\frac{2\pi}{3} + i \sin\frac{2\pi}{3} + \cos\frac{4\pi}{3} + i \sin\frac{4\pi}{3}$$

$$= 1 + \cos\left(\pi - \frac{\pi}{3}\right) + i \sin\left(\pi - \frac{\pi}{3}\right) + \cos\left(\pi + \frac{\pi}{3}\right) + i \sin\left(\pi + \frac{\pi}{3}\right)$$

$$= 1 - \cos\frac{\pi}{3} + i \sin\frac{\pi}{3} - \cos\frac{\pi}{3} - i \sin\frac{\pi}{3}$$

$$= 1 - 2\cos\frac{\pi}{3} = 1 - 2\left(\frac{1}{2}\right) = 0$$

$$\Rightarrow \quad 1 + \omega + \omega^2 = 0$$

$$\Rightarrow \quad 1 + \omega^2 = -\omega \qquad \ldots(1)$$

Now,

$$(1 - \omega)^6 = [(1 - \omega)^2]^3 = [1 - 2\omega + \omega^2]^3 = [-\omega - 2\omega]^3$$

$$= (-3\omega)^3 = -27\omega^3 = -27$$ [Using (1)] **Proved.**

Example 37. *Use De Moivre's theorem to solve the equation $x^4 - x^3 + x^2 - x + 1 = 0$.*

Solution. $x^4 - x^3 + x^2 - x + 1 = 0$

$$(x + 1)(x^4 - x^3 + x^2 - x + 1) = 0$$

$$x^5 + 1 = 0$$

$$x^5 = -1 = (\cos \pi + i \sin \pi) = \cos(2n\pi + \pi) + i \sin(2n\pi + \pi)$$

$$x = [\cos(2n + 1)\pi + i \sin(2n + 1)\pi]^{1/5}$$

$$= \cos\frac{(2n + 1)\pi}{5} + i \sin\frac{(2n + 1)\pi}{5}$$

When $n = 0, 1, 2, 3, 4$, the values are

$$\cos\frac{\pi}{5} + i \sin\frac{\pi}{5}, \cos\frac{3\pi}{5} + i \sin\frac{3\pi}{5} \ \cos\pi + i \sin\pi, \cos\frac{7\pi}{5} + i \sin\frac{7\pi}{5},$$

$$\cos\frac{9\pi}{5} + i \sin\frac{9\pi}{5}.$$

$\cos \pi + i \sin \pi = -1$, which is rejected as it is corresponding to $x + 1 = 0$.

Hence, the required roots are

$$\cos \frac{\pi}{5} + i \sin \frac{\pi}{5}, \cos \frac{3\pi}{5} + i \sin \frac{3\pi}{5}, \cos \frac{7\pi}{5} + i \sin \frac{7\pi}{5}, \cos \frac{9\pi}{5} + i \sin \frac{9\pi}{5}.$$ **Ans.**

EXERCISE 25.7

Find the values of:

1. $(1 + i)^{1/5}$. **Ans.** $2^{1/10}\left[\cos \frac{1}{5}\left(2n\pi + \frac{\pi}{4}\right) + i \sin \frac{1}{5}\left(2n\pi + \frac{\pi}{4}\right)\right]$, where $n = 0, 1, 2, 3, 4$

2. $(1 + \sqrt{-3})^{3/4}$ **Ans.** $(2)^{3/4}\left[\cos \frac{3}{4}\left(2n\pi + \frac{\pi}{3}\right) + i \sin \frac{3}{4}\left(2n\pi + \frac{\pi}{3}\right)\right]$, where $n = 0, 1, 2, 3$.

3. $(-i)^{1/6}$ **Ans.** $\cos (4n + 1) \frac{\pi}{12} - i \sin (4n + 1) \frac{\pi}{12}$, where $n = 0, 1, 2, 3, 4, 5$.

4. $(1 + i)^{2/3}$ **Ans.** $2^{1/3}\left[\cos\left(\frac{4n\pi}{3} + \frac{\pi}{6}\right) + i \sin\left(\frac{4n\pi}{3} + \frac{\pi}{6}\right)\right]$, where $n = 0, 1, 2$

5. Solve the equation with the help of De Moivre's theore $x^7 - 1 = 0$

Ans. $\cos \frac{2n\pi}{7} + i \sin \frac{2n\pi}{7}$ where $n = 0, 1, 2, 3, 4, 5, 6$.

6. Find the roots of the equation $x^3 + 8 = 0$.

Ans. $2\left[\cos\left(\frac{2n\pi + \pi}{3}\right) + i \sin\left(\frac{2n\pi + \pi}{3}\right)\right]$, where $n = 0, 1, 2$.

7. Use De-Moivre's theorem to solve $x^9 - x^5 + x^4 - 1 = 0$

Ans. $\left[\cos (2n + 1) \frac{\pi}{5} + i \sin (2n + 1) \frac{\pi}{5}\right]$, where $n = 0, 1, 2, 3, 4$,

and $\left[\cos \frac{n\pi}{2} + i \sin \frac{n\pi}{2}\right]$, where $n = 0, 1, 2, 3$.

8. Show that the roots of $(x + 1)^6 + (x - 1)^6 = 0$ are given by

$i \cot \frac{(2n + 1)}{12}$, $n = 0, 1, 2, 3, 4, 5$. Deduce $\tan^2 \frac{\pi}{12} + \tan^2 \frac{3\pi}{12} + \tan^2 \frac{5\pi}{12} = 15$.

9. Show that all the roots of $(x + 1)^7 = (x - 1)^7$ are given by $\pm\, i \cot\left(\frac{n\pi}{7}\right)$, where $n = 1, 2, 3$. Why $n \neq 0$.

25.23 CIRCULAR FUNCTIONS OF COMPLEX NUMBERS

We have already discussed circular functions in terms of exponential functions i.e, Euler's exponential form of circular functions:

$$\boxed{\cos \theta = \frac{e^{i\theta} + e^{-i\theta}}{2}}, \quad \boxed{\sin \theta = \frac{e^{i\theta} - e^{-i\theta}}{2i}}$$

If $\theta = z$, then $\boxed{\cos z = \frac{e^{iz} + e^{-iz}}{2}}$ and $\boxed{\sin z = \frac{e^{iz} - e^{-iz}}{2i}}$

25.24 HYPERBOLIC FUNCTIONS

(*i*) $\sinh x = \frac{e^x - e^{-x}}{2}$ (*ii*) $\cosh x = \frac{e^x + e^{-x}}{2}$ (*iii*) $\tanh x = \frac{e^x - e^{-x}}{e^x + e^{-x}}$

(iv) $\coth x = \frac{e^x + e^{-x}}{e^x - e^{-x}}$ (v) $\text{sech}\, x = \frac{2}{e^x + e^{-x}}$ (*vi*) $\text{cosech}\, x = \frac{2}{e^x - e^{-x}}$

(*vii*) $\cosh x + \sinh x = \frac{e^x + e^{-x}}{2} + \frac{e^x - e^{-x}}{2} = e^x$

(*viii*) $\cosh x - \sinh x = \dfrac{e^x + e^{-x}}{2} - \dfrac{e^x - e^{-x}}{2} = e^{-x}$

(*ix*) $(\cosh x + \sinh x)^n = \cosh nx + \sinh nx.$

25.25 RELATION BETWEEN CIRCULAR AND HYPERBOLIC FUNCTIONS

$\sin ix = i \sinh x$ $\qquad$ $\sinh ix = i \sin x$

$\cos ix = \cosh x$ $\qquad$ $\cosh ix = \cos x$

$\tan ix = i \tanh x$ $\qquad$ $\tanh ix = i \tan x$

25.26 FORMULAE OF HYPERBOLIC FUNCTIONS

A. (1) $\cosh^2 x - \sinh^2 x = 1,$ $\qquad$ (2) $\text{sech}^2 x = 1 - \tanh^2 x,$

(3) $\text{cosech}^2 x = \coth^2 x - 1$

B. (1) $\sinh (x \pm y) = \sinh x \cosh y \pm \cosh x \sinh y$

(2) $\cosh (x \pm y) = \cosh x \cosh y \pm \sinh x \sinh y$

(3) $\tanh (x \pm y) = \dfrac{\tanh x \pm \tanh y}{1 \pm \tanh x \tanh y}$

C. (1) $\sinh 2x = 2 \sinh x \cosh x$ $\qquad$ (2) $\cosh 2x = \cosh^2 x + \sinh^2 x$

(3) $\cosh 2x = 2 \cosh^2 x - 1$ $\qquad$ (4) $\cosh 2x = 1 + 2 \sinh^2 x$

(5) $\tanh 2x = \dfrac{2 \tanh x}{1 + \tanh^2 x}$

D. (1) $\sinh x + \sinh y = 2 \sinh \dfrac{x + y}{2} \cosh \dfrac{x - y}{2}$

(2) $\sinh x - \sinh y = 2 \cosh \dfrac{x + y}{2} \sinh \dfrac{x - y}{2}$

(3) $\cosh x + \cosh y = 2 \cosh \dfrac{x + y}{2} \cosh \dfrac{x - y}{2}$

(4) $\cosh x - \cosh y = 2 \sinh \dfrac{x + y}{2} \sinh \dfrac{x - y}{2}$

Note: For proof, put $\sinh x = \dfrac{e^x - e^{-x}}{2}$ and $\cosh x = \dfrac{e^x + e^{-x}}{2}$

Example 38. *Prove that*

$$(\cosh x - \sinh x)^n = \cosh nx - \sinh nx. \qquad \text{(M.U. 2001, 2002)}$$

Solution. L.H.S. $= (\cosh x - \sinh x)^n$

$$= \left[\frac{e^x + e^{-x}}{2} - \frac{e^x - e^{-x}}{2}\right]^n = \left[\frac{2e^{-x}}{2}\right]^n = (e^{-x})^n = e^{-nx} \qquad \text{...(1)}$$

R.H.S. $= \cosh nx - \sinh nx$

$$= \left(\frac{e^{nx} + e^{-nx}}{2} - \frac{e^{nx} + e^{-nx}}{2}\right) = \frac{2e^{-nx}}{2} = e^{-nx} \qquad \text{.(2)}$$

From (1) and (2), we have

L.H.S. = R.H.S. **Proved.**

Example 39. *If* $x = 2 \sin \alpha \cosh \beta$, $y = 2 \cosh \alpha \sinh \beta$, *show that*

$$\text{cosec} (\alpha - i\beta) + \text{cosec} (\alpha + i\beta) = \frac{4x}{x^2 + y^2}$$

Solution. We know that $\text{cosec} (\alpha + i\beta) = \dfrac{1}{\sin (\alpha + i\beta)} = \dfrac{1}{\sin \alpha \cos i\beta + \cos \alpha \sin i\beta}$

$$= \frac{1}{\sin\alpha\cosh\beta + i\cos\alpha\sinh\beta} = \frac{1}{\frac{x}{2} + i\frac{y}{2}} = \frac{2}{x+iy} \qquad \text{... (1) (Given)}$$

$$\operatorname{cosec}(\alpha - i\beta) = \frac{2}{x-iy} \qquad \text{... (2)}$$

Adding (1) and (2), we get

$$\operatorname{cosec}(\alpha - i\beta) + \operatorname{cosec}(\alpha + i\beta) = \frac{2}{x-iy} + \frac{2}{x+iy} = \frac{4x}{x^2+y^2}$$ **Proved.**

Example 40. *If* $\tan(x+iy) = i$, *where* x *and* y *are real, prove that* x *is indeterminate and* y *is infinite.*

Solution. $\tan(x+iy) = i \Rightarrow \tan(x-iy) = -i$

$$\tan 2x = \tan(\overline{x+iy} + \overline{x-iy}) = \frac{\tan(x+iy) + \tan(x+iy)}{1 - \tan(x+iy)\tan(x-iy)}$$

$$= \frac{i-i}{1-i(-i)} = \frac{i-i}{1-1} = \frac{0}{0}, \text{ which is indeterminate.}$$

Also $$\tan 2iy = \tan[(x+iy)-(x-iy)] = \frac{\tan(x+iy) - \tan(x-iy)}{1+\tan(x+iy)\tan(x-iy)}$$

$$= \frac{i-(-i)}{1+i(-i)} = \frac{2i}{1+1} = i$$

$$i\tanh 2y = i \quad \Rightarrow \quad \tanh 2y = 1 \quad \Rightarrow \quad 2y = \tanh^{-1}(1) = \frac{1}{2}\log\frac{1+1}{1-1} = \infty$$

$\therefore$ y is infinite. **Proved**

Example 41. *If* $\tan(\theta + i\phi) = \cos\alpha + i\sin\alpha$, *prove that:*

$$\theta = \frac{n\pi}{2} + \frac{\pi}{4} \text{ and } \phi = \frac{1}{2}\log\tan\left(\frac{\pi}{4} + \frac{\alpha}{2}\right)$$ *(Nagpur University, Summer 2002, Winter 2001*

Solution. We have, $\tan(\theta + i\phi) = \cos\alpha + i\sin\alpha$

$\therefore$ $$\tan(\theta - i\phi) = \cos\alpha - i\sin\alpha$$

But $$\tan 2\theta = \tan[(\theta + i\phi) + (\theta - i\phi)]$$

$$= \frac{\tan(\theta+i\phi) + \tan(\theta - i\phi)}{1 - \tan(\theta + i\phi)\tan(\theta - i\phi)} = \frac{\cos\alpha + i\sin\alpha + \cos\alpha - i\sin\alpha}{1 - (\cos\alpha + i\sin\alpha)(\cos\alpha - i\sin\alpha)}$$

$$= \frac{2\cos\alpha}{1-(\cos^2\alpha + \sin^2\alpha)} = \frac{2\cos\alpha}{1-1} = \infty = \tan\frac{\pi}{2}$$

$\therefore$ $$2\theta = \frac{\pi}{2} \text{ or for general values,}$$

$$2\theta = n\pi + \frac{\pi}{2} \Rightarrow \theta = \frac{n\pi}{2} + \frac{\pi}{4}$$

Again, $$\tan(2i\phi) = \tan[(\theta + i\phi) - (\theta - i\phi)] = \frac{\tan(\theta + i\phi) - \tan(\theta - i\phi)}{1 + \tan(\theta + i\phi)\tan(\theta - i\phi)}$$

$$= \frac{\cos\alpha + i\sin\alpha - (\cos\alpha - i\sin\alpha)}{1 + (\cos\alpha + i\sin\alpha)(\cos\alpha - i\sin\alpha)}$$

$$= \frac{2i\sin\alpha}{1 + \cos^2\alpha + \sin^2\alpha} = \frac{2i\sin\alpha}{1+1} = \frac{2i\sin\alpha}{2} = i\sin\alpha$$

$\Rightarrow$ $$i\tanh 2\phi = i\sin\alpha \qquad (\because \tan ix = i\tanh x)$$

$\Rightarrow$ $$\tanh 2\phi = \sin\alpha$$

i.e., $$\frac{e^{2\phi} - e^{-2\phi}}{e^{2\phi} + e^{-2\phi}} = \frac{\sin\alpha}{1}$$

$$\therefore \frac{e^{2\phi} - e^{-2\phi} + e^{2\phi} + e^{-2\phi}}{(e^{2\phi} + e^{-2\phi}) - (e^{2\phi} - e^{-2\phi})} = \frac{1 + \sin\alpha}{1 - \sin\alpha} \qquad \text{(Componendo and dividendo)}$$

i.e.
$$\frac{2e^{2\phi}}{2e^{-2\phi}} = \frac{1 - \cos\left(\frac{\pi}{2} + \alpha\right)}{1 + \cos\left(\frac{\pi}{2} + \alpha\right)} \qquad \because \cos\left(\frac{\pi}{2} + \theta\right) = -\sin\theta$$

$$\Rightarrow \quad e^{4\phi} = \frac{2\sin^2\left(\frac{\pi}{4} + \frac{\alpha}{2}\right)}{2\cos^2\left(\frac{\pi}{4} + \frac{\alpha}{2}\right)}$$

$$\Rightarrow \quad e^{4\phi} = \tan^2\left(\frac{\pi}{4} + \frac{\alpha}{2}\right) \quad \Rightarrow \quad e^{2\phi} = \tan\left(\frac{\pi}{4} + \frac{\alpha}{2}\right)$$

Hence,
$$2\phi = \log_e \tan\left(\frac{\pi}{4} + \frac{\alpha}{2}\right) \quad \Rightarrow \quad \phi = \frac{1}{2}\log_e \tan\left(\frac{\pi}{4} + \frac{\alpha}{2}\right)$$
Proved.

Example 42. *If $u = \log_e \tan\left(\frac{\pi}{4} + \frac{\theta}{2}\right)$, prove that*

$$\tanh\frac{u}{2} = \tan\frac{\theta}{2}$$

Solution. Here, we have,

$$u = \log_e \tan\left(\frac{\pi}{4} + \frac{\theta}{2}\right) \quad \Rightarrow \quad e^u = \tan\left(\frac{\pi}{4} + \frac{\theta}{2}\right)$$

$$\Rightarrow \quad e^u = \frac{\tan\frac{\pi}{4} + \tan\frac{\theta}{2}}{1 - \tan\frac{\pi}{4}\cdot\tan\frac{\theta}{2}} = \frac{1 + \tan\frac{\theta}{2}}{1 - \tan\frac{\theta}{2}} \qquad \text{...(1)}$$

$$\Rightarrow \quad e^{-u} = \frac{1 - \tan\frac{\theta}{2}}{1 + \tan\frac{\theta}{2}} \qquad \text{... (2)}$$

By componendo and dividendo on (1), we have

$$\frac{e^u + 1}{e^u - 1} = \frac{2}{2\tan\frac{\theta}{2}} \quad \therefore \quad \frac{e^u - 1}{e^u + 1} = \tan\frac{\theta}{2} \qquad \text{...(3)}$$

Now,
$$\tanh\frac{u}{2} = \frac{e^{\frac{u}{2}} - e^{-\frac{u}{2}}}{e^{\frac{u}{2}} + e^{-\frac{u}{2}}} = \frac{e^{\frac{u}{2}} - e^{-\frac{u}{2}}}{e^{\frac{u}{2}} + e^{-\frac{u}{2}}} \cdot \frac{e^{\frac{u}{2}}}{e^{\frac{u}{2}}}$$

$$\Rightarrow \quad \tanh\frac{u}{2} = \frac{e^u - 1}{e^u + 1}$$

$$\Rightarrow \quad \tanh\frac{u}{2} = \tan\frac{\theta}{2} \qquad \text{[Using (3) and (4)]}$$
Proved.

Example 43. *If $\cosh x = \sec\theta$, prove that:*

(i) $$\theta = \frac{\pi}{2} - 2\tan^{-1}(e^{-x})$$

(ii) $$\tanh\frac{\pi}{2} = \tan\frac{\theta}{2}$$
(*M.U.* 2003, 2005)

Solution. (*i*) Let $\tan^{-1} e^{-x} = \alpha$

$\Rightarrow \quad e^{-x} = \tan\alpha \quad \text{and} \quad \alpha = \tan^{-1}(e^{-x}) \quad \text{...(1)}$

$\Rightarrow \quad e^{x} = \cot\alpha \quad \text{...(2)}$

Now, $$\sec\theta = \cosh x = \frac{e^x + e^{-x}}{2} \quad \text{...(3) (Given)}$$

Putting the values of e^{-x} and e^x from (1) and (2) in (3), we get

$$\sec\theta = \frac{\cot\alpha + \tan\alpha}{2}$$

$$\therefore \quad 2\sec\theta = \cot\alpha + \tan\alpha = \frac{\cos\alpha}{\sin\alpha} + \frac{\sin\alpha}{\cos\alpha} = \frac{\cos^2\alpha + \sin^2\alpha}{\sin\alpha\cos\alpha}$$

$$= \frac{2}{2\sin\alpha\cos\alpha} \quad [\because \cos^2\alpha + \sin^2\alpha = 1]$$

$$= \frac{2}{\sin 2\alpha}$$

$$\therefore \quad \cos\theta = \sin 2\alpha$$

$$\Rightarrow \quad \cos\theta = \cos\left(\frac{\pi}{2} - 2\alpha\right)$$

$$\therefore \quad \theta = \frac{\pi}{2} - 2\alpha = \frac{\pi}{2} - 2\tan^{-1}(e^{-x}) \quad \text{[From (1)]}$$ **Proved.**

(*ii*) We have,

$$\cosh x = \sec\theta \quad \text{(Given)}$$

$$\Rightarrow \quad \frac{e^x + e^{-x}}{2} = \sec\theta \quad \left[\because \cosh x = \frac{e^x + e^{-x}}{2}\right]$$

$$\therefore \quad e^x - 2\sec\theta + e^{-x} = 0$$

$$\therefore \quad (e^x)^2 - 2e^x\sec\theta + 1 = 0$$

Solving the quadratic equation in e^x.

$$e^x = \frac{2\sec\theta \pm \sqrt{4\sec^2\theta - 4}}{2}$$

$$\Rightarrow \quad e^x = \sec\theta \pm \sqrt{\sec^2\theta - 1}$$

$$\Rightarrow \quad e^x = \sec\theta \pm \tan\theta \quad \text{...(4)}$$

Now, $$\tanh\frac{x}{2} = \frac{e^{\frac{x}{2}} - e^{-\frac{x}{2}}}{e^{\frac{x}{2}} + e^{-\frac{x}{2}}} = \frac{e^x - 1}{e^x + 1} \quad \text{...(5)}$$

Putting the value of e^x from (4) in (5), we get

$$\tanh\frac{x}{2} = \frac{\sec\theta + \tan\theta - 1}{\sec\theta + \tan\theta + 1} \quad \text{[Using (1)]}$$

$$= \frac{1 + \sin\theta - \cos\theta}{1 + \sin\theta + \cos\theta} = \frac{(1 - \cos\theta) + \sin\theta}{(1 + \cos\theta) + \sin\theta}$$

$$= \frac{2\sin^2\frac{\theta}{2} + 2\sin\frac{\theta}{2}\cos\frac{\theta}{2}}{2\cos^2\frac{\theta}{2} + 2\sin\frac{\theta}{2}\cos\frac{\theta}{2}} = \frac{\sin\frac{\theta}{2}}{\cos\frac{\theta}{2}} = \tan\frac{\theta}{2}$$ **Proved.**

EXERCISE 25.8

1. If $\tan\left(\frac{\pi}{8} + i\alpha\right) = x + iy$, prove that $x^2 + y^2 + 2x = 1$.
2. If $\cot\left(\frac{\pi}{8} + i\alpha\right) = x + iy$, prove that $x^2 + y^2 - 2x = 1$.
3. Prove that if $(1 + i\tan\alpha)^{1 + i\tan\beta}$ can have real values, one of them is $(\sec\alpha)^{\sec^2\beta}$
4. If $\frac{(1+i)^{x+iy}}{(1-i)^{x-iy}} = \alpha + i\beta$, prove that the value of $\tan^{-1}\frac{\beta}{\alpha}$ *is* $\frac{\pi x}{2} + y\log 2$.
5. If $\tanh x = \frac{1}{2}$, find the value of $\sinh 2x$. **Ans.** $\frac{4}{3}$
6. If $\sin\alpha\cosh\beta = \frac{x}{2}$, $\cos\alpha\sinh\beta = \frac{y}{2}$, show that

   (*i*) $\operatorname{cosec}(\alpha - i\beta) + \operatorname{cosec}(\alpha + i\beta) = \frac{4x}{x^2 + y^2}$

   (*ii*) $\operatorname{cosec}(\alpha - i\beta) - \operatorname{cosec}(\alpha + i\beta) = \frac{4iy}{x^2 + y^2}$
7. Show that $\tan\left(\frac{u + iv}{2}\right) = \frac{\sin u + i\sinh v}{\cos u + \cosh v}$
8. If $\cot(\alpha + i\beta) = x + iy$, prove that

   (*i*) $x^2 + y^2 - 2x\cot 2\alpha = 1$

   (*ii*) $x^2 + y^2 + 2y\coth 2\beta + 1 = 0$
9. If $\tan\frac{x}{2} = \tanh\frac{u}{2}$ prove that

   (*i*) $\sinh u = \tan x$

   (*ii*) $\cosh u = \sec x$.
10. Solve the following equation for real values of x.

    $17\cosh x + 18\sinh x = 1$ **Ans.** $-\log 5$

25.27 SEPARATION OF REAL AND IMAGINARY PARTS OF CIRCULAR FUNCTIONS

Example 44. *Separate the following into real and imaginary parts:*

(*i*) *sin* $(x + iy)$ (*ii*) *cos* $(x + iy)$ (*iii*) *tan* $(x + iy)$

(*iv*) *cot* $(x + iy)$ (*v*) *sec* $(x + iy)$ (*vi*) *cosec* $(x + iy)$.

Solution. (*i*) $\sin(x + iy) = \sin x\cos iy + \cos x\sin(iy) = \sin x\cosh y + i\cos x\sinh y$.

(*ii*) $\cos(x + iy) = \cos x\cos(iy) - \sin x\sin(iy) = \cos x\cosh y - i\sin x\sinh y$.

(*iii*)
$$\tan(x + iy) = \frac{\sin(x+iy)}{\cos(x+iy)} = \frac{2\sin(x+iy)\cos(x-iy)}{2\cos(x+iy)\cos(x-iy)}$$
$$= \frac{\sin 2x + \sin(2iy)}{\cos 2x + \cos 2iy} = \frac{\sin 2x + i\sinh 2y}{\cos 2x + \cosh 2y}$$
$$\left\{\begin{array}{l}\because \quad 2\sin A.\cos B = \sin(A+B) + \sin(A-B)\\ \text{and } 2\cos A.\cos B = \cos(A+B) + \cos(A-B)\end{array}\right\}$$

(*iv*)
$$\cot(x + iy) = \frac{\cos(x+iy)}{\sin(x+iy)} = \frac{2\cos(x+iy)\sin(x-iy)}{2\sin(x+iy)\sin(x-iy)}$$
$$= \frac{\sin 2x - \sin(2iy)}{\cos(2iy) - \cos 2x} = \frac{\sin 2x - i\sinh 2y}{\cosh 2y - \cos 2x}$$

(*v*)
$$\sec(x + iy) = \frac{1}{\cos(x+iy)} = \frac{2\cos(x-iy)}{2\cos(x+iy)\cos(x-iy)}$$
$$= \frac{2[\cos x\cos(iy) + \sin x\sin(iy)]}{\cos 2x + \cos(2iy)} = \frac{2[\cos x\cosh y + i\sin x\sinh y]}{\cos 2x + \cosh 2y}$$

(*vi*) cosec $(x + iy) = \frac{1}{\sin(x+iy)} = \frac{2\sin(x-iy)}{2\sin(x+iy)\sin(x-iy)} = \frac{2[\sin x\cos(iy) - \cos x\sin(iy)]}{\cos(2iy) - \cos 2x}$

$$= \frac{2[\sin x\cosh y - i\cos x\sinh y]}{\cosh 2y - \cos 2x}$$ **Ans.**

Example 45. *If* $\tan(A + iB) = x + iy$, *prove that*

$\tan 2A = \frac{2x}{1-x^2-y^2}$ *and* $\tanh 2B = \frac{2x}{1+x^2+y^2}$ (*Nagpur University, Summer 2000*)

Solution. $\tan(A + iB) = x + iy$; $\tan(A - iB) = x - iy$

$$\tan 2A = \tan(A + iB + A - iB)$$

$$= \frac{\tan(A+iB) + \tan(A-iB)}{1 - \tan(A+iB)\tan(A-iB)}$$

$$\tan 2A = \frac{(x+iy)+(x-iy)}{1-(x+iy)(x-iy)} = \frac{2x}{1-\left(x^2+y^2\right)} = \frac{2x}{1-x^2-y^2}$$

Again $\tan 2iB = \tan(A + iB - A + iB) = \frac{\tan(A+iB) - \tan(A-iB)}{1+\tan(A+iB)\tan(A-iB)}$

$$\tan 2iB = \frac{(x+iy)-(x-iy)}{1+(x+iy)(x-iy)} = \frac{(2y)i}{1+x^2+y^2}$$

$$\tanh 2B = \frac{2y}{1+x^2+y^2}$$ $\tan ix = i\tanh x$ **Proved.**

Example 46. *If* $\sin(\alpha + i\beta) = x + iy$, *prove that*

(*a*) $\frac{x^2}{\cosh^2\beta} + \frac{y^2}{\sinh^2\beta} = 1$ (*b*) $\frac{x^2}{\sin^2\alpha} - \frac{y^2}{\cos^2\alpha} = 1$

Solution. (*a*) $x + iy = \sin(\alpha + i\beta) = \sin\alpha\cosh\beta + i\cos\alpha\sinh\beta$

Equating real and imaginary parts, we get

$$x = \sin\alpha\cosh\beta,\ y = \cos\alpha\sinh\beta$$

$$\sin\alpha = \frac{x}{\cosh\beta} \text{ and } \cos\alpha = \frac{y}{\sinh\beta}$$

Squaring and adding, $\sin^2\alpha + \cos^2\alpha = \frac{x^2}{\cosh^2\beta} + \frac{y^2}{\sinh^2\beta}$

$$\Rightarrow \quad 1 = \frac{x^2}{\cosh^2\beta} + \frac{y^2}{\sinh^2\beta}$$ **Proved.**

(*b*) Again $\cosh\beta = \frac{x}{\sin\alpha}$ and $\sinh\beta = \frac{y}{\cos\alpha}$

$$\cosh^2\beta - \sinh^2\beta = \frac{x^2}{\sin^2\alpha} - \frac{y^2}{\cos^2\alpha}$$

$$1 = \frac{x^2}{\sin^2\alpha} - \frac{y^2}{\cos^2\alpha}$$ **Proved.**

25.28 SEPARATION OF REAL AND IMAGINARY PARTS OF HYPERBOLIC FUNCTIONS

Example 47. *Separate the following into real and imaginary parts of hyperbolic functions.*

(*a*) *sinh* $(x + iy)$ (*b*) *cosh* $(x + iy)$ (*c*) *tanh* $(x + iy)$

Solution. (*a*) $\sinh(x + iy) = \sinh x\cosh(iy) + \cosh x\sinh(iy)$

$$= \sinh x\cos y + i\sin y\cosh x.$$ **Ans.**

(*b*) $\cosh(x + iy) = \cosh x \cosh(iy) - \sinh x \sinh iy = \cosh x \cos y - i \sinh x \sin y.$

Ans.

(*c*) $$\tanh(x + iy) = \frac{\sinh(x+iy)}{\cosh(x+iy)} = \frac{-i \sin i(x+iy)}{\cos i(x+iy)}$$

$$= \frac{-i \sin(ix - y)}{\cos(ix - y)} = \frac{-i\, 2\sin(ix-y)\cos(ix+y)}{2\cos(ix-y)\cos(ix+y)} \quad \text{(Note this step)}$$

$$= -i\frac{\sin 2ix - \sin 2y}{\cos 2ix + \cos 2y} = -i\frac{i\sinh 2x - \sin 2y}{\cosh 2x + \cos 2y} = \frac{\sinh 2x + i\sin 2y}{\cosh 2x + \cos 2y}$$

$$= \frac{\sinh 2x}{\cosh 2x + \cos 2y} + i\frac{\sin 2y}{\cosh 2x + \cos 2y}$$

Ans.

Example 48. *If* $\tan(x + iy) = \sin(u + iv)$, *prove that*

$$\frac{\sin 2x}{\sinh 2y} = \frac{\tan u}{\tanh v}$$

Solution. Now $\tan(x + iy) = \sin(u + iv)$ separating the real and imaginary parts of both sides, we have

$$\frac{\sin 2x}{\cos 2x + \cosh 2y} + \frac{i\sinh 2y}{\cos 2x + \cosh 2y} = \sin u \cosh v + i \cos u \sinh v$$

Equating real and imaginary parts, we get

$$\frac{\sin 2x}{\cos 2x + \cosh 2y} = \sin u \cosh v \quad \text{...(1)}$$

and $$\frac{\sinh 2y}{\cos 2x + \cosh 2y} = \cos u \sinh v \quad \text{...(2)}$$

Dividing (1) by (2), we obtain

$$\frac{\sin 2x}{\sinh 2y} = \frac{\sin u \cosh v}{\cos u \sinh v}$$

$$\Rightarrow \quad \frac{\sin 2x}{\sinh 2y} = \frac{\tan u}{\tanh v}$$

Proved.

Example 49. *If* $\sin(\theta + i\phi) = \tan\alpha + i\sec\alpha$, *show that* $\cos 2\theta \cosh 2\phi = 3$

Solution. $\sin(\theta + i\phi) = \tan\alpha + i\sec\alpha$

$\sin\theta\cosh\phi + i\cos\theta\sinh\phi = \tan\alpha + i\sec\alpha$

Equating real and imaginary parts, we get

$$\sin\theta\cosh\phi = \tan\alpha \quad \text{...(1)}$$

$$\cos\theta\sinh\phi = \sec\alpha \quad \text{...(2)}$$

We know that

$$\sec^2\alpha - \tan^2\alpha = 1$$

$$\cos^2\theta\sinh^2\phi - \sin^2\theta\cosh^2\phi = 1 \quad \text{[From (1) and (2)]}$$

$$\left(\frac{1+\cos 2\theta}{2}\right)\left(\frac{\cosh 2\phi - 1}{2}\right) - \left(\frac{1-\cos 2\theta}{2}\right)\left(\frac{\cosh 2\phi + 1}{2}\right) = 1$$

$[-1 + \cosh 2\phi - \cos 2\theta + \cos 2\theta\cosh 2\phi] - [\cosh 2\phi + 1 - \cos 2\theta\cosh 2\phi - \cos 2\theta] = 4$

$\Rightarrow \quad -2 + 2\cos 2\theta\cosh 2\phi = 4$

$\Rightarrow \quad 2\cos 2\theta\cosh 2\phi = 6 \quad \Rightarrow \quad \cos 2\theta\cosh 2\phi = 3$ **Proved.**

Example 50. *If* $\sinh(\theta + i\phi) = \cos\alpha + i\sin\alpha$, *prove that* $\sinh^4\theta = \cos^2\alpha = \cos^4\phi$.

Solution. $\sinh(\theta + i\phi) = \cos\alpha + i\sin\alpha$

$\sinh\theta\cos\phi + i\sin\phi\cosh\theta = \cos\alpha + i\sin\alpha$

Equating real and imaginary parts, we have

$$\sinh\theta\cos\phi = \cos\alpha \text{ and} \qquad ...(1)$$

$$\sin\phi\cosh\theta = \sin\alpha \qquad ...(2)$$

Let us eliminate ϕ from (1) and (2).

$$\cos\phi = \frac{\cos\alpha}{\sinh\theta} \text{ and } \sin\phi = \frac{\sin\alpha}{\cosh\theta}$$

Squaring and adding, we get

$$1 = \frac{\cos^2\alpha}{\sinh^2\theta} + \frac{\sin^2\alpha}{\cosh^2\theta} \Rightarrow \frac{\cos^2\alpha}{\sinh^2\theta} = 1 - \frac{\sin^2\alpha}{\cosh^2\theta}$$

$$\Rightarrow \frac{\cos^2\alpha}{\sinh^2\theta} = 1 - \frac{1-\cos^2\alpha}{1+\sinh^2\theta} = \frac{1+\sinh^2\theta - 1 + \cos^2\alpha}{1+\sinh^2\theta}$$

$$\frac{\cos^2\alpha}{\sinh^2\theta} = \frac{\sinh^2\theta + \cos^2\alpha}{1+\sinh^2\theta}$$

$\sinh^4\theta + \sinh^2\theta\cos^2\alpha = \cos^2\alpha + \cos^2\alpha\sinh^2\theta$

$\Rightarrow \quad \sinh^4\theta = \cos^2\alpha$ **Proved.**

For second result, let us eliminate θ.

$$\sinh\theta = \frac{\cos\alpha}{\cos\phi} \quad \text{and} \quad \cosh\theta = \frac{\sin\alpha}{\sin\phi}$$

$$\cosh^2\theta - \sinh^2\theta = \frac{\sin^2\alpha}{\sin^2\phi} - \frac{\cos^2\alpha}{\cos^2\phi} \Rightarrow 1 = \frac{1-\cos^2\alpha}{1-\cos^2\phi} - \frac{\cos^2\alpha}{\cos^2\phi}$$

$$\Rightarrow \frac{\cos^2\alpha}{\cos^2\phi} = \frac{1-\cos^2\alpha - 1 + \cos^2\phi}{1-\cos^2\phi}$$

$$\frac{\cos^2\alpha}{\cos^2\phi} = \frac{\cos^2\phi - \cos^2\alpha}{1-\cos^2\phi}$$

$\Rightarrow \cos^4\phi - \cos^2\phi\cos^2\alpha = \cos^2\alpha - \cos^2\alpha\cos^2\phi$

$\Rightarrow \quad \cos^4\phi = \cos^2\alpha.$ **Proved.**

Example 51. *If $e^z = \sin(u+iv)$ and $z = x+iy$, prove that*

$$2e^{2x} = \cosh 2v - \cos 2u \qquad (M.U.\ 2006)$$

Solution. We have, $e^z = \sin(u+iv)$

$\Rightarrow \quad e^{x+iy} = \sin(u+iv)$

$\Rightarrow \quad e^x . e^{iy} = \sin u \cos iv + \cos u \sin iv$

$\Rightarrow \quad e^x(\cos y + i\sin y) = \sin u\cosh v + i\cos u\sinh v$

Equating real and imaginary parts, we get

$$e^x\cos y = \sin u\cosh v$$

and $\quad e^x\sin y = \cos u\sinh v$

Squaring and adding, we get

$$e^{2x}(\cos^2 y + \sin^2 y) = \sin^2 u\cosh^2 v + \cos^2 u\sinh^2 v$$

$$\Rightarrow e^{2x} = (1-\cos^2 u)\cosh^2 v + \cos^2 u(\cosh^2 v - 1)$$

$$\Rightarrow e^{2x} = \cosh^2 v - \cos^2 u$$

$$\Rightarrow e^{2x} = \frac{1}{2}(1+\cosh 2v) - \frac{1}{2}(1+\cos 2u)$$

$$\Rightarrow e^{2x} = \frac{1}{2}(\cosh 2v - \cos 2u)$$

$\Rightarrow \quad 2e^{2x} = \cosh 2v - \cos 2u$ **Proved.**

Example 52. *If* $\sin(\theta + i\phi) = \cos\alpha + i\sin\alpha$, *prove that*

$$\cos^4\theta = \sin^2\alpha = \sinh^4\phi. \qquad (M.U.\ 2003,\ 2004)$$

Solution. Here, we have

$$\sin(\theta + i\phi) = \cos\alpha + i\sin\alpha$$

$$\Rightarrow \quad \sin\theta\cosh\phi + i\cos\theta\sinh\phi = \cos\alpha + i\sin\alpha$$

Equating real and imaginary parts, we get

$$\sin\theta\cosh\phi = \cos\alpha \quad \Rightarrow \quad \cosh\phi = \frac{\cos\alpha}{\sin\theta} \qquad ...(1)$$

and
$$\cos\theta\sinh\phi = \sin\alpha \quad \Rightarrow \quad \sinh\phi = \frac{\sin\alpha}{\cos\theta} \qquad ...(2)$$

But
$$\cosh^2\phi - \sinh^2\phi = 1$$

$$\Rightarrow \quad \frac{\cos^2\alpha}{\sin^2\theta} - \frac{\sin^2\alpha}{\cos^2\theta} = 1 \qquad \text{[Using (1) and (2)]}$$

$$\Rightarrow \quad \cos^2\alpha.\cos^2\theta - \sin^2\alpha.\sin^2\theta = \sin^2\theta\cos^2\theta$$

$$\Rightarrow \quad (1 - \sin^2\alpha)\cos^2\theta - \sin^2\alpha.\sin^2\theta = (1 - \cos^2\theta)\cos^2\theta$$

$$\Rightarrow \quad \cos^2\theta - \sin^2\alpha(\cos^2\theta + \sin^2\theta) = \cos^2\theta - \cos^4\theta$$

$$\Rightarrow \quad \sin^2\alpha = \cos^4\theta \qquad ...(3)$$

Again
$$\sin^2\theta + \cos^2\theta = 1$$

$$\therefore \quad \frac{\cos^2\alpha}{\cosh^2\phi} + \frac{\sin^2\alpha}{\sinh^2\phi} = 1 \qquad \text{[Using (1) and (2)]}$$

$$\Rightarrow \quad \cos^2\alpha.\sinh^2\phi + \sin^2\alpha\cosh^2\phi = \sinh^2\phi\cosh^2\phi$$

$$\Rightarrow \quad (1 - \sin^2\alpha)\sinh^2\phi + \sin^2\alpha(1 + \sinh^2\phi) = \sinh^2\phi(1 + \sinh^2\phi)$$

$$\Rightarrow \quad \sinh^2\phi - \sin^2\alpha\sinh^2\phi + \sin^2\alpha + \sin^2\alpha\sinh^2\phi = \sinh^2\phi + \sinh^4\phi$$

$$\Rightarrow \quad \sin^2\alpha = \sinh^4\phi. \qquad ...(4)$$

From (3) and (4), we have $\cos^4\theta = \sin^2\alpha = \sinh^4\phi$ **Proved.**

Example 53. *If* $\operatorname{cosec}\left(\frac{\pi}{4} + ix\right) = u + iv$, *prove that*

$$(u^2 + v^2)^2 = 2(u^2 - v^2) \qquad (M.U.\ 2009)$$

Solution. Here, we have

$$u + iv = \operatorname{cosec}\left(\frac{\pi}{4} + ix\right)$$

$$= \frac{1}{\sin\left(\frac{\pi}{4} + ix\right)} \Rightarrow = \frac{1}{\sin\frac{\pi}{4}\cos ix + \cos\frac{\pi}{4}\sin ix}$$

$$= \frac{1}{\frac{1}{\sqrt{2}}\cosh x + \frac{1}{\sqrt{2}}i\sinh x} = \frac{\sqrt{2}}{\cosh x + i\sinh x}$$

$$= \frac{\sqrt{2}(\cosh x - i\sinh x)}{\cosh^2 x + \sinh^2 x} = \frac{\sqrt{2}(\cosh x - i\sinh x)}{\cosh 2x}$$

Equating real and imaginary parts, we get $u = \dfrac{\sqrt{2}\cosh x}{\cos h\, 2x}$, $v = -\dfrac{\sqrt{2}\sinh x}{\cosh 2x}$

Squaring and adding, we get

$$u^2 + v^2 = \frac{2(\cosh^2 x + \sinh^2 x)}{\cosh^2 2x} = \frac{2\cosh 2x}{\cosh^2 2x}$$

$$\Rightarrow \qquad (u^2 + v^2)^2 = \left(\frac{2}{\cosh 2x}\right)^2 = \frac{4}{\cosh^2 2x} \qquad ...(1)$$

Also, $$u^2 - v^2 = \frac{2}{\cosh^2 2x}(\cosh^2 x - \sinh^2 x) = \frac{2}{\cosh^2 2x} \qquad ...(2)$$

From (1) and (2), we have

$$(u^2 + v^2)^2 = 2(u^2 - v^2)$$ **Proved.**

Example 54. *Separate into real and imaginary parts* $\sqrt{i}^{\sqrt{i}}$. *(M.U. 2008)*

Solution. We have,

$$\sqrt{i} = i^{\frac{1}{2}} = \left(\cos\frac{\pi}{2} + i\sin\frac{\pi}{2}\right)^{\frac{1}{2}}$$

$$= \cos\frac{\pi}{4} + i\sin\frac{\pi}{4}$$

$$= \frac{1}{\sqrt{2}} + i\frac{1}{\sqrt{2}}$$

Also, $$\sqrt{i} = \left(\cos\frac{\pi}{2} + i\sin\frac{\pi}{2}\right)^{\frac{1}{2}} = \left(e^{i\frac{\pi}{2}}\right)^{\frac{1}{2}} = e^{i\frac{\pi}{4}}$$

$$\therefore \quad \left(\sqrt{i}\right)^{\sqrt{i}} = \left(e^{i\frac{\pi}{4}}\right)^{\left(\frac{1}{\sqrt{2}} + i.\frac{1}{\sqrt{2}}\right)} = e^{i\frac{\pi}{4\sqrt{2}} - \frac{\pi}{4\sqrt{2}}}$$

$$= e^{-\frac{\pi}{4\sqrt{2}}}.e^{i\frac{\pi}{4\sqrt{2}}} = e^{-\frac{\pi}{4\sqrt{2}}}\left(\cos\frac{\pi}{4\sqrt{2}} + i\sin\frac{\pi}{4\sqrt{2}}\right)$$

$$\therefore \quad \text{Real part} = e^{-\frac{\pi}{4\sqrt{2}}}\cos\left(\frac{\pi}{4\sqrt{2}}\right)$$

$$\text{Imaginary part} = e^{-\frac{\pi}{4\sqrt{2}}}\sin\left(\frac{\pi}{4\sqrt{2}}\right)$$ **Ans.**

EXERCISE 25.9

Separate into real and imaginary parts.

1. sech $(x + iy)$ **Ans.** $\dfrac{2\cosh x\cos y - 2i\sinh x\sin y}{\cosh 2x + \cos 2y}$

2. coth $i(x + iy,$ **Ans.** $\dfrac{-\sinh 2y - i\sin 2x}{\cosh 2x - \cos 2y}$

3. coth $(x + iy)$ **Ans.** $\dfrac{\sinh 2x - i\sin 2y}{\cosh 2x - \cos 2y}$

4. If $\sin(\theta + i\phi) = p(\cos\alpha + i\sin\alpha)$, prove that
$$p^2 = \frac{1}{2}[\cosh 2\phi - \cos 2\theta], \ \tan\alpha = \tanh\phi\cot\theta$$

5. If $\sin(\alpha + i\beta) = x + iy$, prove that $x^2\operatorname{sech}^2\beta + y^2\operatorname{cosech}^2\beta = 1$ and $x^2\operatorname{cosec}^2\alpha - y^2\sec^2\alpha = 1$

6. If $\cos(\theta + i\phi) = r(\cos\alpha + i\sin\alpha)$, prove that $\theta = \dfrac{1}{2}\log\left[\dfrac{\sin(\theta - \alpha)}{\sin(\theta + \alpha)}\right]$

7. If $\tan\left(\dfrac{\pi}{6} + i\alpha\right) = x + iy$, prove that $x^2 + y^2 + \dfrac{2x}{\sqrt{3}} = 1$

8. If $\tan(A+B) = \alpha + i\beta$, show that $\dfrac{1-(\alpha^2+\beta^2)}{1+(\alpha^2+\beta^2)} = \dfrac{\cos 2A}{\cosh 2B}$

9. If $\dfrac{x+iy-c}{x+iy+c} = e^{u+iv}$, prove that

$$x = -\frac{c\sinh u}{\cosh u - \cos v}, \qquad y = \frac{c\sinh v}{\cosh u - \cos v}$$

Further, if $v = (2n+1)\dfrac{\pi}{2}$, prove that $x^2 + y^2 = c^2$ where n is an integer.

10. If $\dfrac{u-1}{u+1} = \sin(x+iy)$, where $u = \alpha + i\beta$ show that the argument of u is $\theta + \phi$ where

$$\tan\theta = \frac{\cos x \sinh y}{1+\sin x\cosh y} \text{ and , } \tan\phi = \frac{\cos x\sinh y}{1-\sin x\sinh y}$$

11. If $A + iB = C\tan(x+iy)$, prove that $\tan 2x = \dfrac{2CA}{C^2 - A^2 - B^2}$

12. If $\cosh(\alpha + i\beta) = x + iy$, prove that

$$(a)\ \frac{x^2}{\cosh^2\alpha} + \frac{y^2}{\sinh^2\alpha} = 1 \quad (b)\ \frac{x^2}{\cos^2\beta} - \frac{y^2}{\sin^2\beta} = 1$$

13. If $\cos(\theta + i\phi) = R(\cos\alpha + i\sin\alpha)$, prove that $\phi = \dfrac{1}{2}\log_e\left[\dfrac{\sin(\theta-\alpha)}{\sin(\theta+\alpha)}\right]$

14. If $\cos(\alpha+i\beta)\cos(\gamma+i\delta) = 1$, prove that $\tanh^2\delta\cosh^2\beta = \sin^2\alpha$

15. If $\dfrac{u-1}{u+1} = \sin(x+iy)$, find u. **Ans.** $\tan^{-1}\dfrac{2\cos x\sinh y}{\cos^2 x - \sinh^2 y}$

25.29 LOGARITHMIC FUNCTION OF A COMPLEX VARIABLE

Example 55. *Define logarithm of a complex number.*

Solution. If z and w are two complex numbers and $z = e^w$ then $w = \log z$, and if $w = \log z$; then $z = e^w$

Here $\log z$ is a many valued function. General value of $\log z$ is defined by Log z, where Log $z = \log z + 2n\pi i$.

Example 56. *Separate $\log(x+iy)$ into its real and imaginary parts.*

Solution. Let $x = r\cos\theta$ ...(1)

and $y = r\sin\theta$ ...(2)

Squaring and adding (1) and (2) we have $x^2 + y^2 = r^2$

$$\therefore \quad r = \sqrt{x^2+y^2},$$

We have, $\tan\theta = \dfrac{y}{x} \Rightarrow \theta = \tan^{-1}\left(\dfrac{y}{x}\right)$ [Dividing (2) by (1)]

$$\begin{aligned}\therefore \quad \log(x+iy) &= \log[r(\cos\theta + i\sin\theta)]\\ &= [\log r + \log(\cos\theta + i\sin\theta)]\\ \log(x+iy) &= \log r + \log[\cos(2n\pi+\theta) + i\sin(2n\pi+\theta)]\\ &= \log r + \log e^{i(2n\pi+\theta)} = \log r + i(2n\pi+\theta)\\ \text{Log}(x+iy) &= \log\sqrt{x^2+y^2} + i\left(2n\pi + \tan^{-1}\frac{y}{x}\right)\end{aligned}$$

and $\log(x+iy) = \log\sqrt{x^2+y^2} + i\tan^{-1}\dfrac{y}{x}$ **Ans.**

Example 57. *Find the general value of Log $(1+i)$ + Log $(1-i)$.*

Solution. $1 + i = \sqrt{2}\left(\cos\dfrac{\pi}{4} + i\sin\dfrac{\pi}{4}\right) = \sqrt{2}\,e^{i\frac{\pi}{4}}$

$$\log(1+i) = \log \sqrt{2}\,.\,e^{i\frac{\pi}{4}} = \log\sqrt{2} + i\frac{\pi}{4}$$

$$\text{Log}(1+i) = \log\sqrt{2} + i\frac{\pi}{4} + 2\,n\,\pi\,i = \log\sqrt{2} + \left(2n\pi + \frac{\pi}{4}\right)i$$

$$\text{Log}(1-i) = \log\sqrt{2} + \left(2n\pi - \frac{\pi}{4}\right)i$$

Hence, $\text{Log}(1+i) + \text{Log}(1-i) = \left[\log\sqrt{2} + \left(2n\pi + \frac{\pi}{4}\right)i\right] + \left[\log\sqrt{2} + \left(2n\pi - \frac{\pi}{4}\right)i\right]$

$$= 2\log\sqrt{2} + 4n\pi i = \log 2 + 4n\pi i$$ **Ans.**

Example 58. *Show that* $\log \dfrac{x+iy}{x-iy} = 2\,i\,\tan^{-1}\dfrac{y}{x}$. *(Nagpur University, Winter 2003)*

Solution. Let $\log(x+iy) = \log(r\cos\theta + ir\sin\theta) = \log r\,e^{i\theta}$

$$= \log r + i\,\theta \qquad \begin{bmatrix} x = r\cos\theta \\ y = r\sin\theta \end{bmatrix}$$

Similarly, $\log(x-iy) = \log r - i\,\theta$

$$\log\frac{x+iy}{x-iy} = \log(x+iy) - \log(x-iy) = (\log r + i\theta) - (\log r - i\theta) = 2i\,\theta$$

$$= 2\,i\tan^{-1}\frac{y}{x}.$$ **Proved.**

Example 59. *Show that for real values of a and b*

$$e^{2ai\cot^{-1}b}\left[\frac{b\,i-1}{b\,i+1}\right]^{-a} = 1$$ *(M.U. 2008)*

Solution. Consider $\dfrac{b\,i-1}{b\,i+1} = \dfrac{b\,i+i^2}{b\,i-i^2} = \dfrac{b+i}{b-i}$

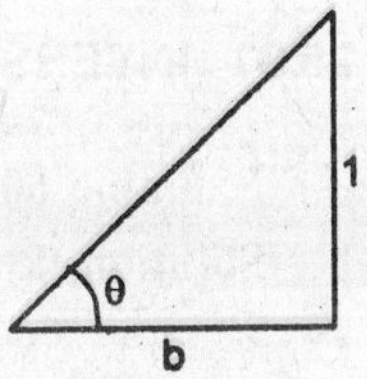

$$\Rightarrow \left(\frac{b\,i-1}{b\,i+1}\right)^{-a} = \left(\frac{b+i}{b-i}\right)^{-a}$$

$$\log\left[\frac{b\,i-1}{b\,i+1}\right]^{-a} = \log\left(\frac{b+i}{b-i}\right)^{-a} = -a\,[\log(b+i) - \log(b-i)]$$

$$= -a\left[\log\sqrt{b^2+1} + i\tan^{-1}\frac{1}{b} - \log\sqrt{b^2+1} + i\tan^{-1}\frac{1}{b}\right]$$

$$= -2ai\tan^{-1}\frac{1}{b}$$

$$\left(\frac{b\,i-1}{b\,i+1}\right)^{-a} = e^{-2ai\tan^{-1}\left(\frac{1}{b}\right)} \qquad \begin{bmatrix} \text{If } \cot\theta = b,\ \tan\theta = \dfrac{1}{b} \\ \text{Since } \cot^{-1}b = \tan^{-1}\left(\dfrac{1}{b}\right) \end{bmatrix}$$

$$e^{2ai\cot^{-1}b}\left(\frac{b\,i-1}{b\,i+1}\right)^{-a} = \left[e^{2ai\tan^{-1}\left(\frac{1}{b}\right)}\right].\left[e^{-2ai\tan^{-1}\left(\frac{1}{b}\right)}\right] = 1$$ **Proved.**

EXERCISE 25.10

1. Find the general value of Log i. **Ans.** $(4n+1)\dfrac{\pi i}{2}$
2. Express Log (-5) in terms of $a + ib$. **Ans.** $\log 5 + i(2n+1)\pi$
3. Find the value of z if

   (*a*) $\cos z = 2$. **Ans.** $z = 2n\pi \pm i\log\left(2+\sqrt{3}\right)$

   (*b*) $\cosh z = -1$. **Ans.** $z = (2n+1)\pi i$

4. Find the general and principal values of i^i **Ans.** $e^{-\left(2n\pi+\frac{\pi}{2}\right)}$, $e^{-\frac{\pi}{2}}$
5. If $i^{(\alpha+i\beta)} = x + iy$, prove that $x^2 + y^2 = e^{-(4m+1)\pi\theta}$.
6. Prove that $\log \dfrac{1}{1-e^{i\theta}} = \log\left(\dfrac{1}{2}\text{cosec}\,\theta\right) + i\left(\dfrac{\pi}{2}-\dfrac{\theta}{2}\right)$.
7. Show that $\log\sin(x+iy) = \dfrac{1}{2}\log\dfrac{\cosh 2y - \cos 2x}{2} + i\tan^{-1}(\cot x\tanh y)$.
8. Prove that $\tan\left[i\log\dfrac{a-ib}{a+ib}\right] = \dfrac{2ab}{a^2-b^2}$.
9. $\log\dfrac{\cos(x-iy)}{\cos(x+iy)} = 2i\tan^{-1}(\tan x\tanh y)$.
10. Separate $i^{(1+i)}$ into real and imaginary parts. **Ans.** $ie^{-\frac{\pi}{2}}$

25.30 INVERSE FUNCTIONS

If $\sin\theta = \dfrac{1}{2}$ then $\theta = \sin^{-1}\left(\dfrac{1}{2}\right)$, so here θ is called inverse sine of $\dfrac{1}{2}$.

Similarly, we can define inverse hyperbolic function sinh, cosh, tanh, etc. If $\cosh\theta = z$ then $\theta = \cosh^{-1} z$.

25.31 INVERSE HYPERBOLIC FUNCTIONS

Example 60. *Prove that* $\sinh^{-1} x = \log\left(x+\sqrt{x^2+1}\right)$ (*M.U. 2009*)

Solution. Let $\sinh^{-1} x = y \quad\Rightarrow\quad x = \sinh y$

$$x = \frac{e^y - e^{-y}}{2}$$

$$\Rightarrow \qquad e^y - e^{-y} = 2x$$

$$\Rightarrow \qquad e^{2y} - 2x\,e^y - 1 = 0$$

This is quadratic in e^y.

$$e^y = \frac{2x \pm \sqrt{4x^2+4}}{2} = x \pm \sqrt{x^2+1}$$

$$y = \log\left(x+\sqrt{x^2+1}\right) \qquad \text{(Taking positive sign only)}$$

$$\sinh^{-1} x = \log\left(x+\sqrt{x^2+1}\right)$$ **Proved.**

Example 61. *Prove that* $\cosh^{-1} x = \log\left(x+\sqrt{x^2-1}\right)$ (*M.U. 2009*)

Solution. Let $y = \cosh^{-1} x \;\Rightarrow\; x = \cosh y$

$$x = \frac{e^y + e^{-y}}{2} \quad\Rightarrow\quad 2x = e^y + e^{-y}$$

$\Rightarrow \quad e^{2y} - 2x\, e^y + 1 = 0$ (This is quadratic in e^y)

$$\Rightarrow \quad e^y = \frac{2x \pm \sqrt{4x^2-4}}{2} = x \pm \sqrt{x^2-1}$$

$$\Rightarrow \quad y = \log\,(x + \sqrt{x^2-1}) \qquad \text{(Taking positive sign only)}$$

$$\Rightarrow \quad \cosh^{-1} x = \log\,(x + \sqrt{x^2-1})$$ **Proved.**

Example 62. *Prove that* $\tanh^{-1} x = \frac{1}{2} \log \frac{1+x}{1-x}$

Solution. Let $\tanh^{-1} x = y \Rightarrow x = \tanh y$

$$x = \frac{e^y - e^{-y}}{e^y + e^{-y}}$$

Applying componendo and dividendo, we obtain

$$\frac{1+x}{1-x} = \frac{e^y}{e^{-y}} = e^{2y}, \quad 2y = \log \frac{1+x}{1-x}$$

$$\Rightarrow \quad \tanh^{-1} x = \frac{1}{2} \log \frac{1+x}{1-x}.$$

Similarly, $\coth^{-1} x = \frac{1}{2} \log \frac{x+1}{x-1}$ **Proved.**

Example 63. *Prove that* $\operatorname{sech}^{-1} x = \log \frac{1+\sqrt{1-x^2}}{x}$

Solution. Let $y = \operatorname{sech}^{-1} x \Rightarrow x = \operatorname{sech} y$

$$x = \frac{2}{e^y + e^{-y}} \Rightarrow x = \frac{2e^y}{e^{2y}+1}$$

$$\Rightarrow \quad xe^{2y} - 2e^y + x = 0 \Rightarrow e^y = \frac{2 \pm \sqrt{4-4x^2}}{2x} = \frac{1 \pm \sqrt{1-x^2}}{x}$$

We take only positive sign

$$e^y = \frac{1+\sqrt{1-x^2}}{x} \Rightarrow y = \log \frac{1+\sqrt{1-x^2}}{x}$$

$$\operatorname{sech}^{-1} x = \log \frac{1+\sqrt{1-x^2}}{x}$$

Similarly, $\operatorname{cosech}^{-1} x = \log \frac{1+\sqrt{1+x^2}}{x}$ **Proved.**

Example 64. *If $x + iy = \cos(\alpha + i\beta)$ or if $\cos^{-1}(x + iy) = \alpha + i\beta$ express x and y in terms of α and β. Hence show that $\cos^2 \alpha$ and $\cosh^2 \beta$ are the roots of the equation $\lambda^2 - (x^2 + y^2 + 1)\lambda + x^2 = 0$.* *(M.U. 2002, 2004)*

Solution. Here, we have

$$\cos(\alpha + i\beta) = x + iy$$

$$\Rightarrow \cos\alpha \cos i\beta - \sin\alpha \sin i\beta = x + iy$$

$$\Rightarrow \cos\alpha \cosh\beta - i \sin\alpha \sinh\beta = x + iy$$

Equating real and imaginary parts, we get

$$\cos\alpha \cosh\beta = x \text{ and } \sin\alpha \sinh\beta = -y$$

We want to find the equation whose roots are $\cos^2\alpha$ and $\cosh^2\beta$.

Now, $x^2 + y^2 + 1 = \cos^2\alpha \cos h^2 \beta + \sin^2\alpha \sinh^2\beta + 1$

$$= \cos^2\alpha \cosh^2\beta + (1 - \cos^2\alpha)(\cosh^2\beta - 1) + 1$$

$$= \cos^2\alpha\cosh^2\beta + \cosh^2\beta - 1 - \cos^2\alpha\cosh^2\beta + \cos^2\alpha + 1$$
$$= \cos^2\alpha + \cosh^2\beta$$

Sum of the roots $= \cos^2\alpha + \cosh^2\beta$

$$= x^2 + y^2 + 1$$

And product of the roots $= \cos^2\alpha\cosh^2\beta$

$$= x^2$$

Hence, the equation whose roots are $\cos^2\alpha$, $\cosh^2\beta$ is

$\lambda^2 - (x^2 + y^2 + 1)\lambda + x^2 = 0$ **Proved.**

Example 65. *Separate into real and imaginary part* $\cos^{-1}\left(\frac{3i}{4}\right)$ *(M.U. 2003)*

Solution. Let $\cos^{-1}\left(\frac{3i}{4}\right) = x + iy$

$$\Rightarrow \quad \frac{3i}{4} = \cos(x + iy)$$

$$\Rightarrow \quad \frac{3i}{4} = \cos x\cosh y - i\sin x\sinh y$$

Equating real and imaginary parts, we get

$$\therefore \quad \cos x\cosh y = 0 \Rightarrow \cos x = 0 \Rightarrow x = \frac{\pi}{2}$$

and $\quad \sin x\sinh y = \frac{3}{4}$

$$-1\sinh y = \frac{3}{4} \qquad\qquad \sin x = \sin\left(\frac{\pi}{2}\right) = 1$$

$$\therefore \quad \sinh y = -\frac{3}{4}$$

$$\Rightarrow \quad y = \log\left(\frac{-3}{4} + \sqrt{1 + \frac{9}{16}}\right) \Rightarrow y = \log\left(\frac{-3}{4} + \frac{5}{4}\right) = -\log 2 = \log\left(\frac{1}{2}\right)$$

$\therefore$ Real part $= \frac{\pi}{2}$ and imaginary Part $= -\log 2$ **Proved.**

25.32 SOME OTHER INVERSE FUNCTIONS

Example 66. *Separate* $\tan^{-1}(\cos\theta + i\sin\theta)$ *into real and imaginary parts.* *(M.U. 2009)*

Solution. Let $\tan^{-1}(\cos\theta + i\sin\theta) = x + iy$

$$\Rightarrow \quad \cos\theta + i\sin\theta = \tan(x + iy)$$

Similarly, $\quad \cos\theta - i\sin\theta = \tan(x - iy)$

$$\tan 2x = \tan[(x + iy) + (x - iy)] = \frac{\tan(x+iy) + \tan(x-iy)}{1 - \tan(x+iy)\tan(x-iy)}$$

$$= \frac{(\cos\theta + i\sin\theta) + (\cos\theta - i\sin\theta)}{1 - (\cos\theta + i\sin\theta)(\cos\theta - i\sin\theta)} = \frac{2\cos\theta}{1 - (\cos^2\theta + \sin^2\theta)}$$

$$= \frac{2\cos\theta}{1-1} = \frac{2\cos\theta}{0} = \infty = \tan\frac{\pi}{2}$$

$$\tan 2x = \tan\left(n\pi + \frac{\pi}{2}\right) \Rightarrow 2x = n\pi + \frac{\pi}{2} \Rightarrow x = \frac{n\pi}{2} + \frac{\pi}{4}$$

Now, $\tan 2iy = \tan[(x + iy) - (x - iy)] = \frac{\tan(x+iy) - \tan(x-iy)}{1 + \tan(x+iy)\tan(x-iy)}$

$$= \frac{(\cos\theta + i\sin\theta) - (\cos\theta - i\sin\theta)}{1 + (\cos\theta + i\sin\theta)(\cos\theta - i\sin\theta)} = \frac{2i\sin\theta}{1 + (\cos^2\theta + \sin^2\theta)} = \frac{2i\sin\theta}{1+1} = i\sin\theta$$

$$i\tanh 2y = i\sin\theta \quad \Rightarrow \quad \frac{e^{2y} - e^{-2y}}{e^{2y} + e^{-2y}} = \frac{\sin\theta}{1}$$

By componendo and dividendo, we have

$$\frac{2e^{2y}}{2e^{-2y}} = \frac{1+\sin\theta}{1-\sin\theta} \Rightarrow e^{4y} = \frac{1+\cos\left(\frac{\pi}{2}-\theta\right)}{1-\cos\left(\frac{\pi}{2}-\theta\right)} = \frac{1+2\cos^2\left(\frac{\pi}{4}-\frac{\theta}{2}\right)-1}{1-\left[1-2\sin^2\left(\frac{\pi}{4}-\frac{\theta}{2}\right)\right]}$$

$$= \frac{\cos^2\left(\frac{\pi}{4}-\frac{\theta}{2}\right)}{\sin^2\left(\frac{\pi}{4}-\frac{\theta}{2}\right)} = \cot^2\left(\frac{\pi}{4}-\frac{\theta}{2}\right) \Rightarrow e^{2y} = \cot\left(\frac{\pi}{4}-\frac{\theta}{2}\right)$$

$$\Rightarrow \quad 2y = \log\cot\left(\frac{\pi}{4}-\frac{\theta}{2}\right) \quad \Rightarrow \quad y = \frac{1}{2}\log\cot\left(\frac{\pi}{4}-\frac{\theta}{2}\right)$$

$$\text{Imaginary part} = \frac{1}{2}\log\cot\left(\frac{\pi}{4}-\frac{\theta}{2}\right)$$

$$\text{Real part} = \frac{n\pi}{2}+\frac{\pi}{4}$$

$$\tan^{-1}(\cos\theta + i\sin\theta) = \frac{n\pi}{2}+\frac{\pi}{4}+\frac{i}{2}\log\cot\left(\frac{\pi}{4}-\frac{\theta}{2}\right)$$ **Ans.**

Example 67. *Separate* $\sin^{-1}(\alpha + i\beta)$ *into real and imaginary parts.*

Solution. Let $\sin^{-1}(\alpha + i\beta) = x + iy$

$$\alpha + i\beta = \sin(x+iy)$$

$$\Rightarrow \quad \alpha + i\beta = \sin x\cos iy + \cos x\sin iy$$

$$= \sin x\cosh y + i\cos x\sinh y$$

Equating real and imaginary parts, we have

$$\alpha = \sin x\cosh y \qquad \text{...(1)}$$

and $$\beta = \cos x\sinh y \qquad \text{....(2)}$$

We know that $\cosh^2 y - \sinh^2 y = 1$

$$\left(\frac{\alpha}{\sin x}\right)^2 - \left(\frac{\beta}{\cos x}\right)^2 = 1 \qquad \left[\begin{aligned}\cosh y &= \frac{\alpha}{\sin x}\\ \sinh y &= \frac{\beta}{\cos x}\end{aligned}\right]$$

$$\Rightarrow \quad \alpha^2\cos^2 x - \beta^2\sin^2 x = \sin^2 x\cos^2 x$$

$$\Rightarrow \quad \alpha^2(1-\sin^2 x) - \beta^2\sin^2 x = \sin^2 x(1-\sin^2 x)$$

$$\Rightarrow \quad \sin^4 x - (\alpha^2+\beta^2+1)\sin^2 x + \alpha^2 = 0$$

This is quadratic equation in $\sin^2 x$.

$$\sin^2 x = \frac{(\alpha^2+\beta^2+1) \pm \sqrt{(\alpha^2+\beta^2+1)^2 - 4\alpha^2}}{2}$$

$$\Rightarrow \quad \sin x = \sqrt{\frac{1}{2}\left[(\alpha^2+\beta^2+1) \pm \sqrt{(\alpha^2+\beta^2+1)^2 - 4\alpha^2}\right]}$$

$$\Rightarrow \quad x = \sin^{-1}\sqrt{\frac{1}{2}\left[(\alpha^2+\beta^2+1) \pm \sqrt{(\alpha^2+\beta^2+1)^2 - 4\alpha^2}\right]}$$

We know that $\sin^2 x + \cos^2 x = 1$

$$\Rightarrow \quad \left(\frac{\alpha}{\cosh y}\right)^2 + \left(\frac{\beta}{\sinh y}\right)^2 = 1 \qquad \left[\begin{array}{l}\text{From (1) and (2)} \\ \sin x = \dfrac{\alpha}{\cosh y} \\ \cos x = \dfrac{\beta}{\sinh y}\end{array}\right]$$

$$\Rightarrow \quad \alpha^2 \sinh^2 y + \beta^2 \cosh^2 y = \sinh^2 y \cosh^2 y$$

$$\Rightarrow \quad \alpha^2 \sinh^2 y + \beta^2 (1 + \sinh^2 y) = \sinh^2 y (1 + \sinh^2 y)$$

$$\Rightarrow \quad \sinh^4 y - (\alpha^2 + \beta^2 - 1) \sinh^2 y - \beta^2 = 0$$

This is quadratic equation in $\sinh^2 y$.

$$\sinh^2 y = \frac{(\alpha^2+\beta^2-1) \pm \sqrt{(\alpha^2+\beta^2-1)^2 + 4\beta^2}}{2}$$

$$\Rightarrow \quad \sinh y = \sqrt{\frac{1}{2}\left[(\alpha^2+\beta^2-1) \pm \sqrt{(\alpha^2+\beta^2-1)^2+4\beta^2}\right]}$$

$$\Rightarrow \quad y = \sinh^{-1} \sqrt{\frac{1}{2}\left[(\alpha^2+\beta^2-1) \pm \sqrt{(\alpha^2+\beta^2-1)^2+4\beta^2}\right]}$$

$$\text{Real part} = \sin^{-1} \sqrt{\frac{1}{2}\left[(\alpha^2+\beta^2+1) \pm \sqrt{(\alpha^2+\beta^2+1)^2-4\alpha^2}\right]}$$

$$\text{Imaginary part} = \sinh^{-1} \sqrt{\frac{1}{2}\left[(\alpha^2+\beta^2-1) \pm \sqrt{(\alpha^2+\beta^2-1)^2+4\beta^2}\right]}$$

Ans.

Example 68. *Separate $\tan^{-1}(a + i\,b)$ into real and imaginary parts.*

(Nagpur University, Summer 2008, 2004)

Solution. Let $\tan^{-1}(a + i\,b) = x + i\,y$

$$\therefore \quad \tan(x + i\,y) = a + i\,b$$

On both sides for i write $-i$ we get,

$$\therefore \quad \tan(x - i\,y) = a - i\,b$$

Now,

$$\tan 2x = \tan[(x + i\,y) + (x - i\,y)]$$

$$= \frac{\tan(x + i\,y) + \tan(x - i\,y)}{1 - \tan(x + i\,y)\tan(x - i\,y)} = \frac{a + i\,b + a - i\,b}{1 - (a + i\,b)(a - i\,b)} = \frac{2a}{1 - a^2 - b^2}$$

$$2x = \tan^{-1}\left[\frac{2a}{1-a^2-b^2}\right] \Rightarrow x = \frac{1}{2}\tan^{-1}\left[\frac{2a}{1-a^2-b^2}\right] \qquad \text{...(2)}$$

and

$$\tan(2\,y\,i) = \tan[(x + i\,y) - (x - i\,y)]$$

$$= \frac{\tan(x + i\,y) - \tan(x - i\,y)}{1 + \tan(x + i\,y)\tan(x - i\,y)} = \frac{a + bi - a + bi}{1 + (a + bi)(a - bi)}$$

$$i \tanh 2y = \frac{2bi}{1+a^2+b^2} \quad \text{so, } \tanh 2y = \frac{2b}{1+a^2+b^2}$$

$$2y = \tanh^{-1}\left[\frac{2b}{1+a^2+b^2}\right]$$

$$\text{so } y = \frac{1}{2}\tanh^{-1}\left[\frac{2b}{1+a^2+b^2}\right] \qquad \text{...(3)}$$

From (1), (2) and (3), we have

$$\tan^{-1}(a+ib) = \frac{1}{2}\tan^{-1}\left[\frac{2a}{1-a^2-b^2}\right] + \frac{i}{2}\tanh^{-1}\left[\frac{2b}{1+a^2+b^2}\right] \qquad \textbf{Ans.}$$

Example 69. *Show that* $\tan^{-1} i\left(\frac{x-a}{x+a}\right) = \frac{i}{2}\log\left(\frac{x}{a}\right)$. (*M.U. 2006, 2002*)

Solution. Let $\tan^{-1} i\left(\frac{x-a}{x+a}\right) = u + iv$ ...(1)

$$\Rightarrow \quad \tan(u+iv) = i\left(\frac{x-a}{x+a}\right) \text{ and } \tan(u-iv) = -i\left(\frac{x-a}{x+a}\right)$$

$$\tan 2u = \tan[(u+iv)+(u-iv)] = \frac{\tan(u+iv)+\tan(u-iv)}{1-\tan(u+iv)\tan(u-iv)}$$

$$= \frac{ix-ia-ix+ia}{x+a} = 0$$

$$\therefore \quad \tan 2u = 0 \Rightarrow 2u = 0 \Rightarrow u = 0$$

Putting the value of u in (1), we get

$$\therefore \quad \tan^{-1} i\left(\frac{x-a}{x+a}\right) = iv \qquad \therefore \quad i\left(\frac{x-a}{x+a}\right) = \tan iv = i\tanh v$$

$$\therefore \quad \frac{x-a}{x+a} = \tanh v = \frac{e^v - e^{-v}}{e^v + e^{-v}}$$

By Componendo and dividendo, we get

$$\frac{2x}{2a} = \frac{2e^v}{2e^{-v}} \Rightarrow \frac{x}{a} = e^{2v} \Rightarrow v = \frac{1}{2}\log\left(\frac{x}{a}\right)$$

$$\therefore \tan^{-1} i\left(\frac{x-a}{x+a}\right) = u + iv = 0 + \frac{i}{2}\log\frac{x}{a} = \frac{i}{2}\log\left(\frac{x}{a}\right) \qquad \textbf{Proved.}$$

Example 70. *Prove that*

(*i*) $\cosh^{-1}\sqrt{1+x^2} = \sinh^{-1} x$ (*M.U. 2007*)

(*ii*) $\cosh^{-1}\sqrt{1+x^2} = \tanh^{-1}\left(\frac{x}{\sqrt{1+x^2}}\right)$ (*M.U. 2002*)

Solution. (*i*) Let $\cosh^{-1}\sqrt{1+x^2} = y$ ...(1)

$$\Rightarrow \quad \sqrt{1+x^2} = \cosh y \qquad ...(2)$$

On squaring both sides, we get

$$1 + x^2 = \cosh^2 y$$

$$\therefore \quad x^2 = \cosh^2 y - 1$$

$$\Rightarrow \quad x^2 = \sinh^2 y$$

$$\Rightarrow \quad x = \sinh y \qquad ...(3)$$

$$\Rightarrow \quad y = \sinh^{-1} x$$

$$\Rightarrow \quad \cosh^{-1}\sqrt{1+x^2} = \sinh^{-1} x \qquad \text{[Using (1)]} \ \textbf{Proved}$$

(*ii*) Dividing (3) by (2), we get

$$\frac{\sinh y}{\cosh y} = \frac{x}{\sqrt{1+x^2}}$$

$$\Rightarrow \qquad \tanh y = \frac{x}{\sqrt{1+x^2}} \Rightarrow y = \tanh^{-1}\left(\frac{x}{\sqrt{1+x^2}}\right)$$

$$\Rightarrow \qquad \cosh^{-1}\sqrt{1+x^2} = \tanh^{-1}\left(\frac{x}{\sqrt{1+x^2}}\right) \quad \text{[Using (1)]}$$ **Proved.**

EXERCISE 25.11

1. Prove that $\sin^{-1}(\operatorname{cosec}\theta) = \frac{\pi}{2} + i\log\cot\frac{\theta}{2}$.
2. If $\tan(\alpha + i\beta) = x + iy$, prove that
   (*a*) $x^2 + y^2 + 2x\cot 2\alpha = 1$
   (*b*) $x^2 + y^2 - 2y\coth 2\beta = -1$.
3. If $\tan(\theta + i\phi) = \sin(x + iy)$, then prove that
   $\coth y \sinh 2\phi = \cot x \sin 2\theta$.
4. If $\sin^{-1}(\cos\theta + i\sin\theta) = x + iy$, show that
   (*a*) $x = \cos^{-1}\sqrt{\sin\theta}$
   (*b*) $y = \log\left[\sqrt{\sin\theta} + \sqrt{1+\sin\theta}\right]$.
5. Separate into real and imaginary parts $\sin^{-1}(e^{i\theta})$

   Ans. $\cos^{-1}\sqrt{\sin\theta} + i\log\left[\sqrt{\sin\theta} + \sqrt{1+\sin\theta}\right]$
6. Prove that

$$\tan^{-1}\left(\frac{\tan 2\theta + \tan 2\phi}{\tan 2\theta - \tan 2\phi}\right) + \tan^{-1}\left(\frac{\tan\theta - \tan\phi}{\tan\theta + \tan\phi}\right) = \tan^{-1}(\cot\theta \coth\phi)$$

7. Prove that $\tanh^{-1} x = \sinh^{-1}\frac{x}{\sqrt{1-x^2}}$.
8. Prove that $\tanh^{-1}(\sin\theta) = \cosh^{-1}(\sec\theta)$
9. Prove that

$$\cosh^{-1}\left(\frac{b + a\cos x}{a + b\cos x}\right) = \log\left[\frac{\sqrt{b+a} + \sqrt{b-a}\tan\frac{x}{2}}{\sqrt{b+a} - \sqrt{b-a}\tan\frac{x}{a}}\right]$$

10. Prove that $\tan^{-1}(e^{i\theta}) = \frac{n\pi}{2} + \frac{\pi}{4} = \log\tan\left(\frac{\pi}{4} - \frac{\theta}{2}\right)$
11. If $\cosh^{-1}(x + iy) + \cosh^{-1}(x - iy) = \cosh^{-1} a$, prove that
    $2(a-1)x^2 + 2(a+1)y^2 = a^2 - 1$.
12. Prove that : $\tanh^{-1}\cos\theta = \cosh^{-1}\operatorname{cosec}\theta$
13. Prove that : $\sinh^{-1}\tan\theta = \log(\sec\theta + \tan\theta)$
14. Prove that : $\sinh^{-1}\tan\theta = \log\tan\left(\frac{\theta}{2} + \frac{\pi}{4}\right)$

Separate into real and imaginary parts

15. $\cos^{-1} e^{i\theta}$ or $\cos^{-1}(\cos\theta + i\sin\theta)$ **Ans.** $\sin^{-1}\sqrt{\sin\theta} + i\log\left(\sqrt{1+\sin\theta} - \sqrt{\sin\theta}\right)$
16. If $\sinh^{-1}(x + iy) + \sinh^{-1}(x - iy) = \sinh^{-1} a$, prove that
    $2(x^2 + y^2)\sqrt{a^2+1} = a^2 - 2x^2 - 2y^2$.

EXPANSION OF TRIGONOMETRIC FUNCTIONS

26.1 EXPANSION OF sin $n\theta$, cos $n\theta$ IN POWERS OF sin θ, cos θ

By De-Moivre's theorem, we know that

$$\cos n\,\theta + i\sin n\theta = (\cos + i\sin\theta)^n. \qquad \text{(Binomial Theorem)}$$

$$= {}^nC_0(\cos\theta)^n + {}^nC_1(\cos\theta)^{n-1}(i\sin\theta) + {}^nC_2(\cos\theta)^{n-2}(i\sin\theta)^2$$
$$+ {}^nC_3(\cos\theta)^{n-3}(i\sin\theta)^3 + {}^nC_4(\cos\theta)^{n-4}(i\sin\theta)^4$$
$$+ {}^nC_5(\cos\theta)^{n-5}(i\sin\theta)^5 + {}^nC_6(\cos\theta)^{n-6}(i\sin\theta)^6$$
$$+ {}^nC_7(\cos\theta)^{n-7}(i\sin\theta)^7 + \ldots + {}^nC_n(\cos\theta)^{n-n}(i\sin\theta)^n$$
$$= \cos^n\theta + i\,{}^nC_1\cos^{n-1}\theta\sin\theta - {}^nC_2\cos^{n-2}\theta\sin^2\theta$$
$$- i\,{}^nC_3\cos^{n-3}\theta\sin^3\theta + {}^nC_4\cos^{n-4}\theta\sin^4\theta + i\,{}^nC_5\cos^{n-5}\theta\sin^5\theta$$
$$- {}^nC_6\cos^{n-6}\theta\sin^6\theta - i\,{}^nC_7\cos^{n-7}\theta\sin^7\theta + \ldots + (i\sin\theta)^n$$
$$= \cos^n\theta - {}^nC_2\cos^{n-2}\theta\sin^2\theta + {}^nC_4\cos^{n-4}\theta\sin^4\theta$$
$$- {}^nC_6\cos^{n-6}\theta\sin^6\theta + \ldots + i\,[{}^nC_1\cos^{n-1}\theta\sin\theta$$
$$- {}^nC_3\cos^{n-3}\theta\sin^3\theta + {}^nC_5\cos^{n-5}\theta\sin^5\theta - {}^nC_7\cos^{n-7}\theta\sin^7\theta \ldots]$$

Equating real and imaginary parts, we get

$$\boldsymbol{\cos n\,\theta = \cos^n\theta - {}^nC_2\cos^{n-2}\theta\sin^2\theta + {}^nC_4\cos^{n-4}\theta\sin^4\theta + \ldots} \qquad \ldots(1)$$

$$\boldsymbol{\sin n\,\theta = {}^nC_1\cos^{n-1}\theta\sin\theta - {}^nC_3\cos^{n-3}\theta\sin^3\theta + {}^nC_5\cos^{n-5}\theta\sin^5\theta + \ldots} \qquad \ldots(2)$$

Replacing every $\sin^2\theta$ by $(1-\cos^2\theta)$ in (1) and every $\cos^2\theta$ by $(1-\sin^2\theta)$ in (2), we get the expansions of $\cos n\,\theta$ in powers of $\cos\theta$ and $\sin n\,\theta$ in powers of $\sin\theta$.

Dividing (2) by (1), we get

$$\frac{\sin n\,\theta}{\cos n\,\theta} = \frac{{}^nC_1\cos^{n-1}\theta\sin\theta - {}^nC_3\cos^{n-3}\theta\sin^3\theta + {}^nC_5\cos^{n-5}\theta\sin^5\theta\ldots}{\cos^n\theta - {}^nC_2\cos^{n-2}\theta\sin^2\theta + {}^nC_4\cos^{n-4}\theta\sin^4\theta - \ldots}$$

and dividing numerator and denominator by $\cos^n\theta$, we get

$$\boldsymbol{\tan n\,\theta = \frac{{}^nC_1\tan\theta - {}^nC_3\tan^3\theta + {}^nC_5\tan^5\theta\ldots}{1 - {}^nC_2\tan^2\theta + {}^nC_4\tan^4\theta\ldots}}$$

Example 1. *Expand* $\cos 6\,\theta$ *and* $\sin 6\,\theta$ *in terms of* $\cos\theta$ *and* $\sin\theta$.

Solution. $\cos 6\,\theta + i\sin 6\,\theta = (\cos\theta + i\sin\theta)^6$

Expansion by Binomial theorem

$$\cos 6\,\theta + i \sin 6\,\theta = \cos^6\theta + {}^6C_1 \cos^5\theta\,(i\sin\theta) + {}^6C_2\cos^4\theta\,(i\sin\theta)^2$$
$$+ {}^6C_3\cos^3\theta\,(i\sin\theta)^3 + {}^6C_4\cos^2\theta\,(i\sin\theta)^4 + {}^6C_5\cos\theta\,(i\sin\theta)^5 + {}^6C_6\,(i\sin\theta)^6$$
$$= \cos^6\theta + i\,6\cos^5\theta\sin\theta - 15\cos^4\theta\sin^2\theta - i.20\cos^3\theta\sin^3\theta + 15\cos^2\theta\sin^4\theta$$
$$+ i\,6\cos\theta\sin^5\theta - \sin^6\theta$$

Equating real and imaginary parts, we have

$$\cos 6\,\theta = \cos^6\theta - 15\cos^4\theta\sin^2\theta + 15\cos^2\theta\sin^4\theta - \sin^6\theta$$
$$\sin 6\,\theta = 6\cos^5\theta\sin\theta - 20\cos^3\theta\sin^3\theta + 6\cos\theta\sin^5\theta$$

Ans.

Example 2. *Prove that* $\sin 7\,\theta = 7\sin\theta - 56\sin^3\theta + 112\sin^5\theta - 64\sin^7\theta$.

Solution. $\cos 7\,\theta + i\sin 7\,\theta = (\cos\theta + i\sin\theta)^7$

$$= \cos^7\theta + {}^7C_1\cos^6\theta\,(i\sin\theta) + {}^7C_2\cos^5\theta\,(i\sin\theta)^2 + {}^7C_3\cos^4\theta\,(i\sin\theta)^3 +$$
$${}^7C_4\cos^3\theta\,(i\sin\theta)^4 + {}^7C_5\cos^2\theta\,(i\sin\theta)^5 + {}^7C_6\cos\theta\,(i\sin\theta)^6 + {}^7C_7\,(i\sin\theta)^7$$

Equating imaginary parts, we get

$$\sin 7\,\theta = 7\cos^6\theta\sin\theta - 35\cos^4\theta\sin^3\theta + 21\cos^2\theta\sin^5\theta - \sin^7\theta$$
$$= 7\,(1-\sin^2\theta)^3\sin\theta - 35\,(1-\sin^2\theta)^2\sin^3\theta + 21\,(1-\sin^2\theta)\sin^5\theta - \sin^7\theta$$
$$= 7\,(1 - 3\sin^2\theta + 3\sin^4\theta - \sin^6\theta)\sin\theta - 35\,(1 - 2\sin^2\theta + \sin^4\theta)\sin^3\theta$$
$$+ 21\sin^5\theta - 21\sin^7\theta - \sin^7\theta$$
$$= 7\sin\theta - 21\sin^3\theta + 21\sin^5\theta - 7\sin^7\theta - 35\sin^3\theta$$
$$+ 70\sin^5\theta - 35\sin^7\theta + 21\sin^5\theta - 22\sin^7\theta$$
$$= 7\sin\theta - 56\sin^3\theta + 112\sin^5\theta - 64\sin^7\theta.$$

Proved.

Example 3. *Expand* $\tan 9\,\theta$ *in powers of* $\tan\theta$.

Solution. We know,

$$\tan n\theta = \frac{{}^nC_1\tan\theta - {}^nC_3\tan^3\theta + {}^nC_5\tan^5\theta + \ldots}{1 - {}^nC_2\tan^2\theta + {}^nC_4\tan^4\theta - {}^nC_6\tan^6\theta + \ldots}$$

$$\tan 9\theta = \frac{{}^9C_1\tan\theta - {}^9C_3\tan^3\theta + {}^9C_5\tan^5\theta - {}^9C_7\tan^7\theta + \tan^9\theta}{1 - {}^9C_2\tan^2\theta + {}^9C_4\tan^4\theta - {}^9C_6\tan^6\theta + {}^9C_8\tan^8\theta}$$

$$= \frac{9\tan\theta - 84\tan^3\theta + 126\tan^5\theta - 36\tan^7\theta + \tan^9\theta}{1 - 36\tan^2\theta + 126\tan^4\theta - 84\tan^6\theta + 9\tan^8\theta}$$

Ans.

EXERCISE 26.1

Prove that

1. $\cos 4\theta = \cos^4\theta - 6\cos^2\theta\sin^2\theta + \sin^4\theta$.
2. $\sin 4\theta = 4\cos^3\theta\sin\theta - 4\cos\theta\sin^3\theta$.
3. $\dfrac{\sin 6\theta}{\cos\theta} = 32\sin^5\theta - 32\sin^3\theta + 6\sin\theta$.
4. $\sin 7\theta = 7\cos^6\theta\sin\theta - 35\cos^4\theta\sin^3\theta + 21\cos^2\theta\sin^5\theta - \sin^7\theta$.
5. $\dfrac{\sin 7\theta}{\sin\theta} = 7 - 56\sin^2\theta + 112\sin^4\theta - 64\sin^6\theta$.
6. $\cos 8\,\theta = \cos^8\theta - 28\cos^6\theta\sin^2\theta + 70\cos^4\theta\sin^4\theta - 28\cos^2\theta\sin^6\theta + \sin^8\theta$.

7. $\sin 10\theta = 10\cos^9\theta \sin\theta - 120\cos^7\theta\sin^3\theta + 210\cos^5\theta\sin^5\theta - 120\cos^3\theta\sin^7\theta + 10\cos\theta\sin^9\theta$

8. $1 + \cos 10\theta = 2(16\cos^5\theta - 20\cos^3\theta + 5\cos\theta)^2$.

9. $1 - \cos 10\theta = 2(16\sin^5\theta - 20\sin^3\theta + 5\sin\theta)^2$

10. $\tan 5\theta = \dfrac{5\tan\theta - 10\tan^3\theta + \tan^5\theta}{1 - 10\tan^2\theta + 5\tan^4\theta}$

26.2 EXPANSION OF $\cos^n\theta \sin^n\theta$ IN TERMS OF SINES AND COSINES OF MULTIPLES OF θ

Method :

Let $x = \cos\theta + i\sin\theta$, then $\dfrac{1}{x} = \cos\theta - i\sin\theta$

$$x + \frac{1}{x} = 2\cos\theta \text{ and } x - \frac{1}{x} = 2i\sin\theta$$

Again
$$x^n = \cos n\theta + i\sin n\theta, \text{ and}$$
$$\frac{1}{x^n} = \cos n\theta - i\sin n\theta$$
$$x^n + \frac{1}{x^n} = 2\cos n\theta \text{ and } x^n - \frac{1}{x^n} = 2i\sin n\theta$$

To expand $\cos^n\theta$: Start from $(2\cos\theta)^n = \left(x+\dfrac{1}{x}\right)^n$

Expand R.H.S. and substitute the values of $\left(x+\dfrac{1}{x}\right), \left(x^2+\dfrac{1}{x^2}\right)$ etc.

To expand $\sin^n\theta$: Start from $(2i\sin\theta)^n = \left(x-\dfrac{1}{x}\right)^n$

Expand R.H.S. and substitute the values of $\left(x-\dfrac{1}{x}\right), \left(x^2-\dfrac{1}{x^2}\right)$ etc.

Example 4. *Express $\sin^5\theta$ in terms of sines of multiples of θ.*

Solution. $(2i\sin\theta)^5 = \left(x - \dfrac{1}{x}\right)^5$ $\quad \left[\begin{array}{l} x = \cos\theta + i\sin\theta \\ \dfrac{1}{x} = \cos\theta - i\sin\theta \end{array}\right]$

$$i\,32\sin^5\theta = x^5 + 5x^4\left(-\frac{1}{x}\right) + 10x^3\left(-\frac{1}{x}\right)^2 + 10x^2\left(-\frac{1}{x}\right)^3 + 5x\left(-\frac{1}{x}\right)^4 + \left(-\frac{1}{x}\right)^5$$
$$= \left(x^5 - \frac{1}{x^5}\right) - 5\left(x^3 - \frac{1}{x^3}\right) + 10\left(x - \frac{1}{x}\right)$$
$$= 2i\sin 5\theta - 5(2i\sin 3\theta) + 10(2i\sin\theta)$$
$$16\sin^5\theta = \sin 5\theta - 5\sin 3\theta + 10\sin\theta$$
$$\sin^5\theta = \frac{1}{16}(\sin 5\theta - 5\sin 3\theta + 10\sin\theta)$$ **Ans.**

Example 5. *Prove that*
$-2^{12}\cos^6\theta\sin^7\theta = \sin 13\theta - \sin 11\theta - 6\sin 9\theta + 6\sin 7\theta + 15\sin 5\theta - 15\sin 3\theta - 20\sin\theta$

Solution. $x^n = (\cos\theta + i\sin\theta)^n = \cos n\theta + i\sin n\theta$ $\quad \left[\begin{array}{l} x = \cos\theta + i\sin\theta \\ \dfrac{1}{x} = \cos\theta - i\sin\theta \end{array}\right]$

$$\frac{1}{x^n} = (\cos\theta + i\sin\theta)^n = \cos n\theta - i\sin n\theta$$

$$x^n + \frac{1}{x^n} = 2\cos n\theta \text{ and } x^n - \frac{1}{x^n} = 2i\sin n\theta \qquad \left[\begin{array}{l} x+\dfrac{1}{x}=2\cos\theta \\ x-\dfrac{1}{x}=2i\sin\theta \end{array}\right]$$

$$(2\cos\theta)^6(2i\sin\theta)^7 = \left(x+\frac{1}{x}\right)^6\left(x-\frac{1}{x}\right)^7 = \left(x^2-\frac{1}{x^2}\right)^6\left(x-\frac{1}{x}\right)$$

$$= \left[x^{12}+6x^{10}\left(-\frac{1}{x^2}\right)+15x^8\left(-\frac{1}{x^2}\right)^2+20x^6\left(-\frac{1}{x^2}\right)^3+ 15x^4\left(-\frac{1}{x^2}\right)^4+6x^2\left(-\frac{1}{x^2}\right)^5+\left(-\frac{1}{x^2}\right)^6\right]\left(x-\frac{1}{x}\right)$$

$$= \left[x^{12}-6x^8+15x^4-20+\frac{15}{x^4}-\frac{6}{x^8}+\frac{1}{x^{12}}\right]\left[x-\frac{1}{x}\right]$$

$$= x^{13}-6x^9+15x^5-20x+\frac{15}{x^3}-\frac{6}{x^7}+\frac{1}{x^{11}}-x^{11}+6x^7-15x^3+\frac{20}{x}-\frac{15}{x^5}+\frac{6}{x^9}-\frac{1}{x^{13}}$$

$$= \left(x^{13}-\frac{1}{x^{13}}\right)-\left(x^{11}-\frac{1}{x^{11}}\right)-6\left(x^9-\frac{1}{x^9}\right)+6\left(x^7-\frac{1}{x^7}\right)+ 15\left(x^5-\frac{1}{x^5}\right)-15\left(x^3-\frac{1}{x^3}\right)-20\left(x-\frac{1}{x}\right)$$

$$= 2i\sin 13\theta - 2i\sin 11\theta - 6(2i\sin 9\theta) + 6(2i\sin 7\theta) + 15(2i\sin 5\theta) - 15(2i\sin 3\theta) - 20(2i\sin\theta)$$

$$-2^{12}\cos^6\theta\sin^7\theta = \sin 13\theta - \sin 11\theta - 6\sin 9\theta + 6\sin 7\theta + 15\sin 5\theta - 15\sin 3\theta - 20\sin\theta$$ **Proved.**

EXERCISE 26.2

1. Express $\sin^7\theta$ as a sum of sines of multiples of θ.

   Ans. $-\dfrac{1}{64}[\sin 7\theta - 7\sin 5\theta + 12\sin 3\theta - 35\sin\theta]$

2. Express $\cos^8\theta$ as a sum of cosines of multiples of θ.

   Ans. $\dfrac{1}{128}[\cos 8\theta + 8\cos 6\theta + 28\cos 4\theta + 56\cos 2\theta + 35]$

3. **Prove that $2^7\cos^3\theta\sin^5\theta = \sin 8\theta - 2\sin 6\theta - 2\sin 4\theta + 6\sin 2\theta$.**
4. Prove that $32\cos^6\theta = \cos 6\theta + 6\cos 4\theta + 15\cos 2\theta + 10$
5. Prove that $\sin^8\theta = 2^{-7}(\cos 8\theta - 8\cos 6\theta + 28\cos 4\theta - 56\cos 2\theta + 35)$
6. $32\sin^4\theta\cos^2\theta = \cos 6\theta - 2\cos 4\theta - \cos 2\theta + 2$.
7. $\sin^5\theta\cos^2\theta = \dfrac{1}{64}(\sin 7\theta - 3\sin 5\theta + \sin 3\theta + 5\sin\theta)$
8. Expand $\cos^5\theta\sin^7\theta$ in a series of sines and of multiples of θ.

   Ans. $-2^{-11}(\sin 12\theta - 2\sin 10\theta - 4\sin 8\theta + 10\sin 6\theta + 5\sin 4\theta - 20\sin 2\theta)$

26.3 SUMMATION OF SINES AND COSINES SERIES: 'C + *i* S' METHOD

Sum of cosines series is denoted by C and sum of sines series is denoted by S.

(*i*) Suppose we have to find C, the sum of cosines series. Then write a similar series of sines, S.

(*ii*) Multiply the sines series by i and add to series of cosines, we will get $C + iS$ as an exponential series. The sum of exponential series is calculated by using any one of th following series.

(*a*) **Geometric Series :** $a + ar + ar^2 + \ldots + ar^{n-1} = \dfrac{a(1-r^n)}{1-r}$

$$a + ar + ar^2 + \ldots \infty = \frac{a}{1-r}$$

(*b*) **Binomial series :** $1 + nx + \dfrac{n(n-1)}{2!}x^2 + \ldots \infty = (1+x)^n$

(*c*) **Exponential series :** $1 + x + \dfrac{x^2}{2!} + \dfrac{x^3}{3!} + \ldots \infty = e^x$

(*d*) **sin, cos, sinh or cosh series :**

$$x - \frac{x^3}{3!} + \frac{x^5}{5!} - \ldots \infty = \sin x \quad \text{and}$$

$$1 - \frac{x^2}{2!} + \frac{x^4}{4!} + \ldots \infty = \cos x$$

$$x + \frac{x^3}{3!} + \frac{x^5}{5!} + \ldots \infty = \sinh x \quad \text{and}$$

$$1 + \frac{x^2}{2!} + \frac{x^4}{4!} + \ldots \infty = \cosh x$$

(*e*) **Logarithmic series :** $x - \dfrac{x^2}{2} + \dfrac{x^3}{3} - \ldots \infty = \log(1+x)$

$$-x - \frac{x^2}{2} - \frac{x^3}{3} \ldots \infty = \log(1-x)$$

(*f*) **Gregory's series :** $x - \dfrac{x^3}{3} + \dfrac{x^5}{5} - \ldots \infty = \tan^{-1} x$

$$x + \frac{x^3}{3} + \frac{x^5}{5} + \ldots \infty = \tanh^{-1} x = \frac{1}{2}\log\frac{1+x}{1-x}$$

Example 6. *Sum the series*

$\cos\alpha + x\cos(\alpha+\beta) + \dfrac{x^2}{2!}\cos(\alpha+2\beta) + \ldots \infty$ (*Nagpur University Summer 2005*)

Solution. Let $C = \cos\alpha + x\cos(\alpha+\beta) + \dfrac{x^2}{2!}\cos(\alpha+2\beta) + \ldots \infty$

and $S = \sin\alpha + x\sin(\alpha+\beta) + \dfrac{x^2}{2!}\sin(\alpha+2\beta) + \ldots \infty$

$\therefore$

$$C + iS = (\cos\alpha + i\sin\alpha) + x[\cos(\alpha+\beta) + i\sin(\alpha+\beta)] + \frac{x^2}{2!}[\cos(\alpha+2\beta) + i\sin(\alpha+2\beta)] + \ldots \infty$$

$$= e^{i\alpha} + xe^{i(\alpha+\beta)} + \frac{x^2}{2!}e^{i(\alpha+2\beta)} + \ldots \infty$$

$$= e^{i\alpha}\left[1 + x\,e^{i\beta} + \frac{x^2 e^{i2\beta}}{2!} + \ldots \infty\right]$$

This is an exponential series.

$$= e^{i\alpha}\left(e^{x.e^{i\beta}}\right) = e^{i\alpha}\,.\,e^{x(\cos\beta + i\sin\beta)} = e^{x\cos\beta}\,.\,e^{i(\alpha + x\sin\beta)}$$

$$= e^{x\cos\beta}[\cos(\alpha + x\sin\beta) + i\sin(\alpha + x\sin\beta)]$$

Equating real parts, we have

$$C = e^{x\cos\beta}\,.\cos(\alpha + x\sin\beta)$$ **Ans.**

Example 7. *Sum the series*

$$\sin\alpha + \sin(\alpha+\beta) + \sin(\alpha+2\beta) \ldots + \sin(\alpha + \overline{n-1}\,\beta)$$

Solution. Let $S = \sin\alpha + \sin(\alpha+\beta) + \sin(\alpha+2\beta) + \ldots + \sin(\alpha + \overline{n-1}\,\beta)$

and $C = \cos\alpha + \cos(\alpha+\beta) + \cos(\alpha+2\beta) + \ldots + \cos(\alpha + \overline{n-1}\,\beta)$

$$\therefore \quad C + iS = [\cos\alpha + i\sin\alpha] + [\cos(\alpha+\beta) + i\sin(\alpha+\beta)] + [\cos(\alpha+2\beta) + i\sin(\alpha+2\beta)] + \ldots + [\cos(\alpha+\overline{n-1}\,\beta) + i\sin(\alpha+\overline{n-1}\,\beta)]$$

$$= e^{i\alpha} + e^{i(\alpha+\beta)} + e^{i(\alpha+2\beta)} + \ldots + e^{i(\alpha+\overline{n-1}\,\beta)}$$

$$= e^{i\alpha}\,[1 + e^{i\beta} + e^{i2\beta} + \ldots + e^{i\overline{n-1}\,\beta}]$$

This is a geometric series.

$$= e^{i\alpha}\,\frac{1-(e^{i\beta})^n}{1-e^{i\beta}} = e^{i\alpha}\cdot\frac{1-e^{in\beta}}{1-e^{i\beta}} = e^{i\alpha}\,\frac{1-\cos n\beta - i\sin n\beta}{1-\cos\beta - i\sin\beta}$$

$$= e^{i\alpha}\,\frac{2\sin^2\frac{n\beta}{2} - 2i\sin\frac{n\beta}{2}\cos\frac{n\beta}{2}}{2\sin^2\frac{\beta}{2} - 2i\sin\frac{\beta}{2}\cos\frac{\beta}{2}}$$

$$= e^{i\alpha}\,\frac{\sin\frac{n\beta}{2}}{\sin\frac{\beta}{2}} \times \frac{\sin\frac{n\beta}{2} - i\cos\frac{n\beta}{2}}{\sin\frac{\beta}{2} - i\cos\frac{\beta}{2}} = e^{i\alpha}\,\frac{\cos\frac{n\beta}{2} + i\sin\frac{n\beta}{2}}{\cos\frac{\beta}{2} + i\sin\frac{\beta}{2}} \times \frac{\sin\frac{n\beta}{2}}{\sin\frac{\beta}{2}}$$

$$= \frac{e^{i\alpha}\,.\,e^{i\frac{n\beta}{2}}}{e^{i\frac{\beta}{2}}}\sin\frac{n\beta}{2}\,\text{cosec}\,\frac{\beta}{2} = e^{i\left[\alpha+(n-1)\frac{\beta}{2}\right]}\sin\frac{n\beta}{2}\,\text{cosec}\,\frac{\beta}{2}$$

$$= \left[\cos\left\{\alpha+(n-1)\frac{\beta}{2}\right\} + i\sin\left\{\alpha+(n-1)\frac{\beta}{2}\right\}\right]\sin\frac{n\beta}{2}\,\text{cosec}\,\frac{\beta}{2}$$

Equating imaginary parts, we get $S = \sin\left\{\alpha+(n-1)\frac{\beta}{2}\right\}\sin\frac{n\beta}{2}\,\text{cosec}\,\frac{\beta}{2}$ **Ans.**

Example 8. *Sum the series*

$$n\sin\alpha + \frac{n(n+1)}{1.2}\sin 2\alpha + \frac{n(n+1)(n+2)}{1.2.3}\sin 3\alpha + \ldots \infty$$

(Nagpur University, Winter 2001)

Solution.

Let $S = n\sin\alpha + \frac{n(n+1)}{1.2}\sin 2\alpha + \frac{n(n+1)(n+2)}{1.2.3}\sin 3\alpha + \ldots \infty$

and $C = n\cos\alpha + \frac{n(n+1)}{1.2}\cos 2\alpha + \frac{n(n+1)(n+2)}{1.2.3}\cos 3\alpha + \ldots \infty$

$$\therefore \quad C + iS = n(\cos\alpha + i\sin\alpha) + \frac{n(n+1)}{1.2}(\cos 2\alpha + i\sin 2\alpha) + \frac{n(n+1)(n+2)}{1.2.3}(\cos 3\alpha + i\sin 3\alpha) + \ldots \infty$$

$$= n\,e^{i\alpha} + \frac{n(n+1)}{1.2}e^{i2\alpha} + \frac{n(n+1)(n+2)}{1.2.3}e^{i3\alpha} + \ldots \infty$$

$$= -1 + 1 + (-n)(-e^{i\alpha}) + \frac{-n(-n-1)}{1.2}(-e^{i\alpha})^2 + \frac{-n(-n-1)(-n-2)}{1.2.3}(-e^{i\alpha})^3 + \ldots \infty$$

This is a binomial series.

$$= -1 + [1 - e^{i\alpha}]^{-n} = -1 + (1 - \cos\alpha - i\sin\alpha)^{-n}$$

$$= -1 + \left[2\sin^2\frac{\alpha}{2} - i2\sin\frac{\alpha}{2}\cos\frac{\alpha}{2}\right]^{-n} = -1 + \left(2\sin\frac{\alpha}{2}\right)^{-n}\left[\sin\frac{\alpha}{2} - i\cos\frac{\alpha}{2}\right]^{-n}$$

$$= -1 + \left(2\sin\frac{\alpha}{2}\right)^{-n}\left[\cos\left(\frac{\pi}{2} - \frac{\alpha}{2}\right) - i\sin\left(\frac{\pi}{2} - \frac{\alpha}{2}\right)\right]^{-n}$$

$$= -1 + \frac{\operatorname{cosec}^n\frac{\alpha}{2}}{2^n}\left[\cos n\left(\frac{\pi}{2} - \frac{\alpha}{2}\right) + i\sin n\left(\frac{\pi}{2} - \frac{\alpha}{2}\right)\right]$$

Equating imaginary parts, we get $S = \dfrac{\operatorname{cosec}^n\frac{\alpha}{2}}{2^n}\sin\dfrac{n}{2}(\pi - \alpha)$ **Ans.**

Example 9. *Sum the series*

$$\frac{\sin\theta}{1!} - \frac{\sin 2\theta}{2!} + \frac{\sin 3\theta}{3!} - \frac{\sin 4\theta}{4!} + \ldots\infty$$ *(Nagpur University, Summer 2001)*

Solution. Let $S = \dfrac{\sin\theta}{1!} - \dfrac{\sin 2\theta}{2!} + \dfrac{\sin 3\theta}{3!} - \dfrac{\sin 4\theta}{4!} + \ldots\infty$

and $C = \dfrac{\cos\theta}{1!} - \dfrac{\cos 2\theta}{2!} + \dfrac{\cos 3\theta}{3!} - \dfrac{\cos 4\theta}{4!} + \ldots\infty$

$$\therefore\quad C + iS = \frac{1}{1!}(\cos\theta + i\sin\theta) - \frac{1}{2!}(\cos 2\theta + i\sin 2\theta) + \frac{1}{3!}(\cos 3\theta + i\sin 3\theta) - \ldots\infty$$

$$= \frac{1}{1!}e^{i\theta} - \frac{1}{2!}e^{2i\theta} + \frac{1}{3!}e^{3i\theta} - \ldots\infty = 1 - \left[1 - \frac{e^{i\theta}}{1!} + \frac{e^{2i\theta}}{2!} - \frac{e^{3i\theta}}{3!} + \ldots\infty\right]$$

$$= 1 - e^{-e^{i\theta}} \qquad \left(\because\ 1 - \frac{x}{1!} + \frac{x^2}{2!} - \frac{x^3}{3!} + \ldots\infty = e^{-x}\right)$$

$$= 1 - e^{-(\cos\theta + i\sin\theta)} = 1 - e^{-\cos\theta}\,.(e^{-i\sin\theta})$$

$$= 1 - e^{-\cos\theta}\,.[\cos(\sin\theta) - i\sin(\sin\theta)]$$

$$= [1 - e^{-\cos\theta}\,.\cos(\sin\theta)] + i[e^{-\cos\theta}\,.\sin(\sin\theta)]$$

Equating imaginary parts on both sides, we get

$S = e^{-\cos\theta}\,.\sin(\sin\theta)$ **Ans.**

Example 10. *Sum the series*

$$1 + \frac{1}{2}\cos 2\theta + \frac{1.3}{2.4}\cos 4\theta + \frac{1.3.5}{2.4.6}\cos 6\theta + \ldots\infty$$

(Nagpur University, Winter 2003, 2000)

Solution. Let $C = 1 + \dfrac{1}{2}\cos 2\theta + \dfrac{1.3}{2.4}\cos 4\theta + \dfrac{1.3.5}{2.4.6}\cos 6\theta + \ldots\infty$

and $S = \dfrac{1}{2}\sin 2\theta + \dfrac{1.3}{2.4}\sin 4\theta + \dfrac{1.3.5}{2.4.6}\sin 6\theta + \ldots\infty$

$$\therefore\quad C + iS = 1 + \frac{1}{2}(\cos 2\theta + i\sin 2\theta) + \frac{1.3}{2.4}(\cos 4\theta + i\sin 4\theta)$$

$$+ \frac{1.3.5}{2.4.6}(\cos 6\theta + i\sin 6\theta) + \ldots\infty$$

$$= 1 + \frac{1}{2}e^{2i\theta} + \frac{1.3}{2.4}e^{4i\theta} + \frac{1.3.5}{2.4.6}e^{6i\theta} + \ldots\infty$$

$$= 1-\left(-\frac{1}{2}\right)e^{2i\theta}+\frac{-\frac{1}{2}\left(-\frac{1}{2}-1\right)}{1.2}(e^{2i\theta})^2-\frac{-\frac{1}{2}\left(-\frac{1}{2}-1\right)\left(-\frac{1}{2}-2\right)}{1.2.3}(e^{2i\theta})^3+\ldots\infty$$

(Binomial Theorem)

$$= (1-e^{2i\theta})^{-1/2} = (1-\cos 2\theta - i\sin 2\theta)^{-1/2}$$

$$= [2\sin^2\theta - i\,.\,2\sin\theta\cos\theta]^{-1/2} = (2\sin\theta)^{-1/2}(\sin\theta - i\cos\theta)^{-1/2}$$

$$= (2\sin\theta)^{-1/2}\left[\cos\left(\frac{\pi}{2}-\theta\right) - i\sin\left(\frac{\pi}{2}-\theta\right)\right]^{-1/2}$$

$$= (2\sin\theta)^{-1/2}\left[\cos\left(\frac{\pi}{4}-\frac{\theta}{2}\right) - i\sin\left(\frac{\pi}{4}-\frac{\theta}{2}\right)\right]$$

(De-Moivre's Theorem)

Equating real parts on both sides, we get

$$C = (2\sin\theta)^{-1/2}\cos\left(\frac{\pi}{4}-\frac{\theta}{2}\right)$$ **Ans.**

Example 11. *Sum the series*

$$\sin\alpha - \frac{\sin(\alpha+2\beta)}{2!} + \frac{\sin(\alpha+4\beta)}{4!} - \ldots\infty$$

Solution. Let $S = \sin\alpha - \frac{\sin(\alpha+2\beta)}{2!} + \frac{\sin(\alpha+4\beta)}{4!} - \ldots\infty$

and $C = \cos\alpha - \frac{\cos(\alpha+2\beta)}{2!} + \frac{\cos(\alpha+4\beta)}{4!} - \ldots\infty$

$$\therefore \quad C+iS = (\cos\alpha + i\sin\alpha) - \frac{1}{2!}[\cos(\alpha+2\beta) + i\sin(\alpha+2\beta)] + \frac{1}{4!}[\cos(\alpha+4\beta) + i\sin(\alpha+4\beta)] - \ldots\infty$$

$$= e^{i\alpha} - \frac{e^{i(\alpha+2\beta)}}{2!} + \frac{e^{i(\alpha+4\beta)}}{4!} - \ldots\infty = e^{i\alpha}\left[1 - \frac{e^{i2\beta}}{2!} + \frac{e^{i4\beta}}{4!} - \ldots\infty\right]$$

This series of R.H.S. is a cosine series.

$$= e^{i\alpha}\cos(e^{i\beta}) = (\cos\alpha + i\sin\alpha)\cos(\cos\beta + i\sin\beta)$$

$$= (\cos\alpha + i\sin\alpha)[\cos(\cos\beta)\cos(i\sin\beta) - \sin(\cos\beta)\sin(i\sin\beta)]$$

$$= (\cos\alpha + i\sin\alpha)[\cos(\cos\beta)\cosh(\sin\beta) - i\sin(\cos\beta)\sinh(\sin\beta)]$$

Equating imaginary parts, we get

$$S = \sin\alpha\cos(\cos\beta)\cosh(\sin\beta) - \cos\alpha\sin(\cos\beta)\sinh(\sin\beta)$$ **Ans.**

Example 12. *Sum the series*

$$x\sin\theta - \frac{1}{2}\,.\,x^2\sin 2\theta + \frac{1}{3}\,.\,x^3\sin 3\theta - \ldots\infty$$

(*Nagpur University, Winter 2004*)

Solution. Let $S = x\sin\theta - \frac{1}{2}\,.\,x^2\sin 2\theta + \frac{1}{3}\,.\,x^3\sin 3\theta - \ldots\infty$

and $C = x\cos\theta - \frac{1}{2}\,.\,x^2\cos 2\theta + \frac{1}{3}\,.\,x^3\cos 3\theta - \ldots\infty$

$$\therefore \quad C+iS = x(\cos\theta + i\sin\theta) - \frac{x^2}{2}(\cos 2\theta + i\sin 2\theta) + \frac{x^3}{3}(\cos 3\theta + i\sin 3\theta) - \ldots\infty$$

$$= x\,e^{i\theta} - \frac{x^2}{2}e^{2i\theta} + \frac{x^3}{3}e^{i3\theta} - \ldots\infty$$

The series of R.H.S. is a logarithmic series.

$$= \log(1 + x\,e^{i\theta}) = \log(1 + x\cos\theta + i\,x\sin\theta)$$

Separating real and imaginary parts,

$$= \log\sqrt{(1 + x\cos\theta)^2 + (x\sin\theta)^2} + i\tan^{-1}\frac{x\sin\theta}{1 + x\cos\theta}$$

Equating imaginary parts, we have

$$S = \tan^{-1}\frac{x\sin\theta}{1 + x\cos\theta} \qquad [\text{except when } x\cos\theta = -1] \quad \textbf{Ans.}$$

Example 13. *Find sum of the series*

$$c\sin\alpha + c^2\sin 2\alpha + c^3\sin 3\alpha + \ldots \qquad \textit{(Nagpur University Summer 2002)}$$

Solution. Let $S = c\sin\alpha + c^2\sin 2\alpha + c^3\sin 3\alpha + \ldots$

and $C = c\cos\alpha + c^2\cos 2\alpha + c^3\cos 3\alpha + \ldots$

$$C + iS = c(\cos\alpha + i\sin\alpha) + c^2(\cos 2\alpha + i\sin 2\alpha) + c^3(\cos 3\alpha + i\sin 3\alpha) + \ldots$$

$$C + iS = c\,e^{i\alpha} + c^2 e^{i2\alpha} + c^3 e^{i3\alpha} + \ldots$$

$$= \frac{ce^{i\alpha}}{1 - ce^{i\alpha}} = \frac{ce^{i\alpha}(1 - ce^{-i\alpha})}{(1 - ce^{i\alpha})(1 - ce^{-i\alpha})} \quad \textit{(Infinite Geometric series)} \quad \left(S = \frac{a}{1 - r}\right)$$

$$= \frac{ce^{i\alpha} - c^2}{1 - c(e^{i\alpha} + e^{-i\alpha}) + c^2} = \frac{c(\cos\alpha + i\sin\alpha) - c^2}{1 - 2c\cos\alpha + c^2}$$

Equating imaginary part, we get

$$\therefore \quad S = \frac{c\sin\alpha}{1 + c^2 - 2c\cos\alpha} \qquad \textbf{Ans.}$$

Example 14. *Find the sum of the series*

$$c\sin\theta - \frac{1}{2}c^2\sin 2\theta + \frac{1}{3}c^3\sin 3\theta - \ldots\infty \qquad \textit{(Nagpur University, Summer 2003)}$$

Solution. Let $S = c\sin\theta - \frac{1}{2}c^2\sin 2\theta + \frac{1}{3}c^3\sin 3\theta - \ldots\infty$

and $C = c\cos\theta - \frac{1}{2}c^2\cos 2\theta + \frac{1}{3}c^3\cos 3\theta - \ldots\infty$

$$\therefore \quad C + iS = c(\cos\theta + i\sin\theta) - \frac{1}{2}c^2(\cos 2\theta + i\sin 2\theta) + \frac{1}{3}c^3(\cos 3\theta + i\sin 3\theta) - \ldots\infty$$

$$= c\,e^{i\theta} - \frac{1}{2}c^2 e^{2i\theta} + \frac{1}{3}c^3 e^{3i\theta} - \ldots \text{ to } \infty$$

$$= c\,e^{i\theta} - \frac{1}{2}(ce^{i\theta})^2 + \frac{1}{3}(ce^{i\theta})^3 - \ldots \text{ to } \infty$$

$$= \log_e(1 + ce^{i\theta}) \qquad \left[\log(1 + x) = x - \frac{x^2}{2} + \frac{x^3}{3} + \ldots\right]$$

$$= \log_e(1 + c\cos\theta + i\,c\sin\theta) \qquad \ldots(1)$$

Now, put $1 + c\cos\theta + i\,c\sin\theta = r(\cos\alpha + i\sin\alpha)$

$$\therefore \quad 1 + c\cos\theta = r\cos\theta, \quad c\sin\theta = r\sin\theta$$

$$r^2 = (1 + c\cos\theta)^2 + c^2\sin^2\theta = 1 + 2c\cos\theta + c^2\cos^2\theta + c^2\sin^2\theta$$

$$= 1 + 2\,c\cos\theta + c^2(\cos^2\theta + \sin^2\theta)$$

$$r^2 = 1 + 2\,c\cos\theta + c^2$$

$$\therefore \quad r^2 = \sqrt{1 + 2c \cos\theta + c^2} \quad \text{and} \quad \alpha = \tan^{-1} \frac{c \sin\theta}{1 + c\cos\theta}$$

From equation (1), we have

$$C + iS = \log_e \{r (\cos\alpha + i \sin\alpha)\} = \log_e re^{i\alpha} = \log_e r + \log_e e^{i\alpha}$$

$$\therefore \quad C + iS = \log_e r + i\alpha$$

Putting the values of r and α, we get

$$C + iS = \log_e \sqrt{1 + 2c \cos\theta + c^2} + \tan^{-1} \frac{c \sin\theta}{1 + c\cos\theta}$$

Equating imaginary part, we get

$$S = \tan^{-1} \frac{c \sin\theta}{1 + c\cos\theta}$$ **Ans.**

Example 15. *Sum the series*

$$c \cos\alpha - \frac{c^2}{2} \cos 2\alpha + \frac{c^3}{3} \cos 3\alpha - \ldots \infty$$

where, $0 < c < 1$. *(Nagpur University, Winter 2000)*

[**Hint.** From example 12

$$c \cos\alpha - \frac{c^2}{2} \cos 2\alpha + \frac{c^3}{3} \cos 3\alpha \ldots = \log_e \sqrt{1 + 2c \cos\alpha + c^2}\Bigg]$$ **Ans.**

Example 16. *Sum the series* $\cos\theta + \sin\theta \cos 2\theta + \dfrac{\sin^2\theta}{1.2} \cos 3\theta + \ldots \infty$

(Nagpur University, Summer 2004)

Solution. Let $\quad C = \cos\theta + \sin\theta . \cos 2\theta + \dfrac{\sin^2\theta}{1.2} \cos 3\theta + \ldots \infty$

and $\quad S = \sin\theta + \sin\theta . \sin 2\theta + \dfrac{\sin^2\theta}{1.2} \sin 3\theta + \ldots \infty$

$$\therefore \quad C + iS = (\cos\theta + i\sin\theta) + \sin\theta(\cos 2\theta + i \sin 2\theta) + \frac{\sin^2\theta}{2!}(\cos 3\theta + i \sin 3\theta) + \ldots \infty$$

$$= e^{i\theta} + \sin\theta . e^{2i\theta} + \frac{\sin^2\theta}{2!} e^{3i\theta} + \ldots \infty$$

$$= e^{i\theta} \left[1 + \frac{\sin\theta}{1!} e^{i\theta} + \frac{\sin^2\theta}{2!} e^{2i\theta} + \ldots \infty\right]$$

$$= e^{i\theta} . e^{\sin\theta . e^{i\theta}} \qquad \left(\because e^x = 1 + x + \frac{x^2}{2!} + \ldots \infty\right)$$

$$= e^{i\theta} . e^{\sin\theta(\cos\theta + i\sin\theta)} = e^{\sin\theta . \cos\theta} . e^{i(\theta + \sin^2\theta)}$$

$$= e^{\sin\theta \ \cos\theta} . [\cos(\theta + \sin^2\theta) + i \sin(\theta + \sin^2\theta)]$$

Equating real parts on both sides, we get

$$C = e^{\sin\theta \ \cos\theta} . \cos(\theta + \sin^2\theta)$$ **Ans.**

EXERCISE 26.3

Find the sum of the following series :

1. $\sin\alpha + x \sin(\alpha + \beta) + \dfrac{x^2}{2!} \sin(\alpha + 2\beta) + \ldots \infty$ **Ans.** $e^{x \cos\beta} \sin(\alpha + x\beta)$

2. $\cos\alpha + \cos(\alpha + \beta) + \cos(\alpha + 2\beta) + \ldots + \cos(\alpha + \overline{n-1}\beta)$

Ans. $\cos\left\{\alpha + (n-1)\dfrac{\beta}{2}\right\} \sin\dfrac{n\beta}{2} \operatorname{cosec}\dfrac{\beta}{2}$

3. $n \cos \alpha + \dfrac{n(n+1)}{1.2} \cos 2\alpha + \dfrac{n(n+1)(n+2)}{1.2.3} \cos 3\alpha + \ldots \infty$

Ans. $-1 + \dfrac{\operatorname{cosec}^n \dfrac{\alpha}{2}}{2^n} \cos \dfrac{n}{2}(\pi - \alpha)$

4. $e^{\alpha} \sin \beta - \dfrac{e^{3\alpha}}{3} \sin 3\beta + \dfrac{e^{5\alpha}}{5} \sin 5\beta + \ldots \infty$ **Ans.** $\dfrac{1}{2} \tanh^{-1}(\operatorname{sech} \alpha \sin \beta)$

5. $\sin \alpha . \cos \alpha - \dfrac{1}{2} \sin^2 \alpha \cos 2\alpha + \dfrac{1}{3} \sin^3 \alpha \cos 3\alpha + \ldots \infty$ **Ans.** $\log (1 + \sin \alpha)$

6. $-\dfrac{1}{2} \sin \alpha + \dfrac{1.3}{2.4} \sin 2\alpha - \dfrac{1.3.5}{2.4.6} \sin 3\alpha + \ldots \infty$ **Ans.** $-\dfrac{\sin \dfrac{\alpha}{4}}{\sqrt{2 \cos \dfrac{\alpha}{2}}}$

7. $1 + \dfrac{1}{2} \cos 2\alpha - \dfrac{1.3}{2.4} \cos 4\alpha + \dfrac{1.3.5}{2.4.6} \cos 6\alpha + \ldots \infty$ **Ans.** $\sqrt{[\cos \alpha(1 + \cos \alpha)]}$

8. $\sin \alpha + \dfrac{1}{3} \sin 3\alpha + \dfrac{1}{3^2} \sin 5\alpha + \ldots \infty$ **Ans.** $\dfrac{6 \sin \alpha}{5 - 3 \cos 2\alpha}$

9. $1 + x \cos \theta + x^2 \cos 2\theta + x^3 \cos 3\alpha + \ldots$ **Ans.** $\dfrac{1 - x \cos \theta - x^n \cos n\theta + x^{n+1} \cos (n-1)\theta}{1 - 2x \cos \theta + x^2}$

10. $1 - \dfrac{1}{2} \cos \theta + \dfrac{1.3}{2.4} \cos 2\theta - \dfrac{1.3.5}{2.4.6} \cos 3\theta + \ldots$ **Ans.** $\left(2 \cos \dfrac{\theta}{2}\right)^{-1/2} \cos \dfrac{\theta}{4}$

11. $\sin \theta - \dfrac{\sin 2\theta}{2!} + \dfrac{\sin 3\theta}{3!} - \ldots \infty$ **Ans.** $e^{-\cos \theta} \sin (\sin \theta)$

12. $1 + \dfrac{1}{3} \cos x + \dfrac{1}{9} \cos 2x + \dfrac{1}{27} \cos 3x + \ldots \infty$ **Ans.** $\dfrac{9 - 3 \cos x}{10 - 6 \cos x}$

13. $x \cos \theta - \dfrac{x^2}{2} \cos 2\theta + \dfrac{x^3}{3} \cos 3\theta - \ldots \infty$ **Ans.** $\dfrac{1}{2} \log (1 + 2x \cos \theta + x^2)$

26.4 APPROXIMATION

Example 17. *If* $\dfrac{\sin x}{x} = \dfrac{2399}{2400}$, *find an approximate value of x in radians.*

Solution. $\dfrac{\sin x}{x} = \dfrac{2399}{2400}$

We know that $\sin x = x - \dfrac{x^3}{3!} + \dfrac{x^5}{5!} - \ldots$ or $\dfrac{\sin x}{x} = 1 - \dfrac{x^2}{6} + \dfrac{x^4}{120}$

$$\frac{2399}{2400} = 1 - \frac{x^2}{6} \quad \text{(Ignoring } x^4 \text{ and higher powers)}$$

$$\frac{x^2}{6} = 1 - \frac{2399}{2400} \Rightarrow \frac{x^2}{6} = \frac{1}{2400}$$

$$x^2 = \frac{1}{400} \quad \text{or} \quad x = \frac{1}{20} \text{ radian.}$$ **Ans.**

Example 18. *If* $\cos x = \dfrac{1681}{1682}$, *find x approximately.*

Solution. $\cos x = \dfrac{1681}{1682}$

$$1 - 2\sin^2\frac{x}{2} = \frac{1681}{1682} \quad \text{or} \quad 2\sin^2\frac{x}{2} = \frac{1}{1682} \quad \text{or} \quad \sin^2\frac{x}{2} = \frac{1}{3364}$$

$$\sin\frac{x}{2} = \frac{1}{58}$$

We know that $\lim\limits_{x \to 0} \dfrac{\sin\frac{x}{2}}{\frac{x}{2}} = 1 \Rightarrow \sin\frac{x}{2} = \frac{x}{2} \Rightarrow \frac{x}{2} = \frac{1}{58} \Rightarrow x = \frac{1}{29}$ radian

$$x = 1° \; 58.5'.$$

Ans.

Example 19. *Solve* $\cos\left(\frac{\pi}{3} + \theta\right) = 0.49.$

Solution.

$$\cos\left(\frac{\pi}{3} + \theta\right) = 0.49$$

$$\cos\frac{\pi}{3}\cos\theta - \sin\frac{\pi}{3}\sin\theta = 0.49$$

$$\frac{1}{2}\cos\theta - \frac{\sqrt{3}}{2}\sin\theta = 0.49$$

$$\frac{1}{2}\left[1 - \frac{\theta^2}{2!} + \ldots\right] - \frac{\sqrt{3}}{2}\left[\theta - \frac{\theta^3}{3!} + \ldots\right] = 0.49$$

$$\frac{1}{2} - \frac{\theta^2}{4} - \frac{\sqrt{3}}{2}\theta + \frac{\sqrt{3}}{12}\theta^3 + \ldots = 0.49$$

$$\frac{1}{2} - \frac{\sqrt{3}}{2}\theta + \ldots = 0.49 \quad \text{(Ignoring higher powers of } \theta)$$

$$\Rightarrow \quad \frac{\sqrt{3}}{2}\theta = 0.5 - 0.49 \Rightarrow \theta = \frac{0.02}{\sqrt{3}} = 0.011547$$

$$\theta = 39.696'.$$

Ans.

Example 20. *Evaluate* $\lim\limits_{\theta \to 0} \dfrac{\tan\theta - \sin\theta}{\theta^3}$

Solution.

$$\lim_{\theta \to 0} \frac{\tan\theta - \sin\theta}{\theta^3} = \lim_{\theta \to 0} \frac{\left[\theta + \frac{\theta^3}{3} + \frac{2}{15}\theta^5 + \ldots\right] - \left[\theta - \frac{\theta^3}{3!} + \frac{\theta^5}{5!} - \ldots\right]}{\theta^3}$$

$$= \lim_{\theta \to 0} \frac{1 + \frac{\theta^2}{3} + \frac{2}{15}\theta^4 + \ldots - 1 + \frac{\theta^2}{6} - \frac{\theta^4}{120} + \ldots}{\theta^2} = \lim_{\theta \to 0} \frac{\frac{\theta^2}{2} + \frac{\theta^4}{8} + \ldots}{\theta^2}$$

$$= \lim_{\theta \to 0}\left(\frac{1}{2} + \frac{\theta^2}{8}\right) = \frac{1}{2}.$$

Ans.

EXERCISE 26.4

1. If $\dfrac{\sin x}{x} = \dfrac{559}{600}$, find an approximate value of x. **Ans.** $x = \dfrac{1}{10}$ radian

2. If $\dfrac{\sin x}{x} = \dfrac{5045}{5046}$, find an approximate value of x. **Ans.** $x = 1°, 58'$

3. Solve $\sin\left(\dfrac{x}{6} + x\right) = 0.51$ approximately. **Ans.** $x = 39.7'$

4. Evaluate $\lim\limits_{x \to 0} \dfrac{\sin x - x\cos x}{x^3}$. **Ans.** $\dfrac{1}{3}$

CHAPTER

27

FUNCTIONS OF COMPLEX VARIABLE, ANALYTIC FUNCTION

27.1 INTRODUCTION

The theory of functions of a complex variable is of atmost importance in solving a large number of problems in the field of engineering and science. Many complicated integrals of real functions are solved with the help of functions of a complex variable.

27.2 COMPLEX VARIABLE

$x + iy$ is a complex variable and it is denoted by z.

(1) $z = x + iy$. where $i = \sqrt{-1}$ (Cartesian form)

(2) $z = r(\cos\theta + i\sin\theta)$(Polar form)

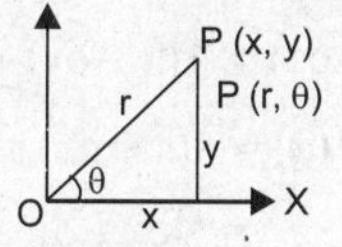

(3) $z = re^{i\theta}$ (Exponential form)

27.3 FUNCTIONS OF A COMPLEX VARIABLE

$f(z)$ is a function of a complex variable z and is denoted by w.

$$w = f(z)$$
$$w = u + iv$$

where u and v are the real and imaginary parts of $f(z)$.

27.4 LIMIT OF A FUNCTION OF A COMPLEX VARIABLE

Let $f(z)$ be a single valued function defined at all points in some neighbourhood of point z_0. Then the limit of $f(z)$ as z approaches z_0 is w_0.

$$\lim_{z \to z_0} f(z) = w_0$$

Example. Prove that $\lim_{z \to 1-i} \frac{(z^2 + 4z + 3)}{z + 1} = 4 - i$

Solution. $\lim_{z \to 1-i} \frac{z^2 + 4z + 3}{z + 1} = \lim_{z \to 1-i} \frac{(z+1)(z+3)}{(z+1)} = \lim_{z \to 1-i} (z+3) = (1-i) + 3 = 4 - i$ **Proved.**

27.5 CONTINUITY

The function $f(z)$ of a complex variable z is said to be continuous at the point z_0 if for any given positive number $\in$, we can find a number δ such that $|f(z) - f(z_0)| < \in$

for all points z of the domain satisfying

$$|z - z_0| < \delta$$

$f(z)$ is said to be continuous at $z = z_0$ if

$$\lim_{z \to z_0} f(z) = f(z_0)$$

27.6 DIFFERENTIABILITY

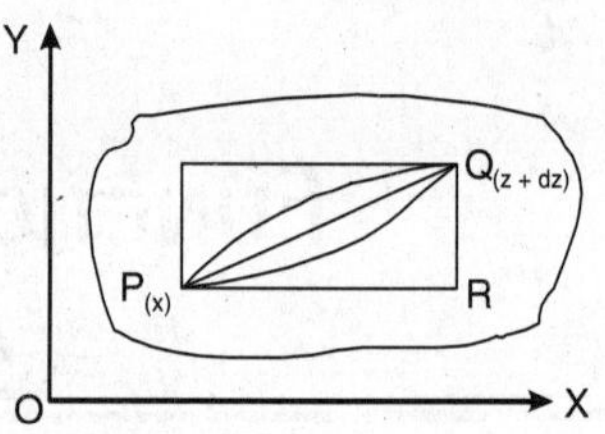

Let $f(z)$ be a single valued function of the variable z, then

$$f'(z) = \lim_{\delta z \to 0} \frac{f(z + \delta z) - f(z)}{\delta z}$$

provided that the limit exists and is independent of the path along which $\delta z \to 0$.
Let P be a fixed point and Q be a neighbouring point. The point Q may approach P along any straight line or curved path.

Example 1. *Consider the function*

$$f(z) = 4x + y + i(-x + 4y)$$

and discuss $\dfrac{df}{dz}$.

Solution. Here, $f(z) = 4x + y + i(-x + 4y)$

$$= u + iv$$

so $u = 4x + y$

and $v = -x + 4y$

$$f(z + \delta z) = 4(x + \delta x) + (y + \delta y) - i(x + \delta x) + 4i(y + \delta y)$$

$$f(z + \delta z) - f(z) = 4(x + \delta x) + (y + \delta y) - i(x + \delta x) + 4i(y + \delta y) - 4x - y + ix - 4iy$$

$$= 4\delta x + \delta y - i\delta x + 4i\delta y$$

$$\frac{f(z + \delta z) - f(z)}{\delta z} = \frac{4\delta x + \delta y - i\delta x + 4i\delta y}{\delta x + i\delta y}$$

$$\Rightarrow \quad \frac{\delta f}{\delta z} = \frac{4\delta x + \delta y - i\delta x + 4i\delta y}{\delta x + i\delta y} \quad \ldots (1)$$

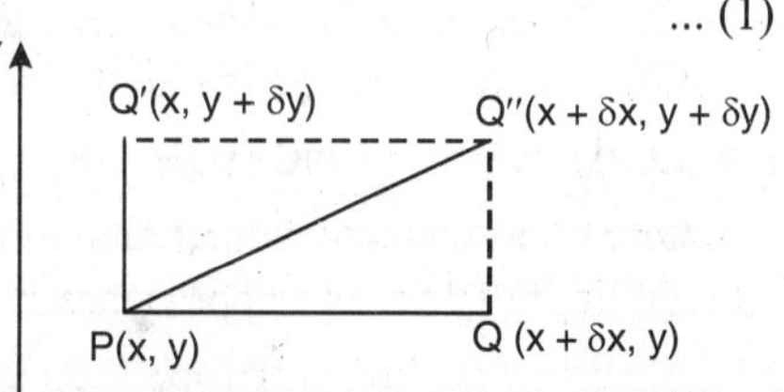

(*a*) **Along real axis:** If Q is taken on the horizontal line through $P(x, y)$ and Q then approaches P along this line, we shall have $\delta y = 0$ and $\delta z = \delta x$.

$$\frac{\delta f}{\delta z} = \frac{4\delta x - i\delta x}{\delta x} = 4 - i$$

(*b*) **Along imaginary axis:** If Q is taken on the vertical line through P and then Q approaches P along this line, we have

$$z = x + iy = 0 + iy, \ \delta z = i\delta y, \ \delta x = 0.$$

Putting these values in (1), we have

$$\frac{\delta f}{\delta z} = \frac{\delta y + 4i\delta y}{i\delta y} = \frac{1}{i}(1 + 4i) = 4 - i$$

(*c*) **Along a line $y = x$:** If Q is taken on a line $y = x$.

$$z = x + iy = x + ix = (1 + i)x$$

$$\delta z = (1 + i)\delta x \quad \text{and} \quad \delta y = \delta x$$

On putting these values in (1), we have

$$\frac{\delta f}{\delta z}=\frac{4\delta x+\delta x-i\delta x+4i\delta x}{\delta x+i\delta x}=\frac{4+1-i+4i}{1+i}=\frac{5+3i}{1+i}=\frac{(5+3i)(1-i)}{(1+i)(1-i)}=4-i$$

In all the three different paths approaching Q from P, we get the same values of $\frac{\delta f}{\delta z}=4-i$. In such a case, the function is said to be differentiable at the point z in the given region.

Example 2. *If* $f(z)=\begin{cases}\dfrac{x^3y(y-ix)}{x^6+y^2}, & z\neq 0,\\ 0, & z=0\end{cases}$ *then discuss* $\frac{df}{dz}$ *at* $z=0$.

Solution. If $z\to 0$ along radius vector $y=mx$

$$\lim_{z\to 0}\frac{f(z)-f(0)}{z}=\lim_{z\to 0}\left[\frac{\dfrac{x^3y(y-ix)}{x^6+y^2}-0}{x+iy}\right]=\lim_{z\to 0}\left[\frac{-ix^3y(x+iy)}{(x^6+y^2)(x+iy)}\right]$$

$$=\lim_{z\to 0}\left[\frac{-ix^3y}{x^6+y^2}\right]=\lim_{z\to 0}\left[\frac{-ix^3(mx)}{x^6+m^2x^2}\right] \qquad [\because\ y=mx]$$

$$=\lim_{x\to 0}\left[\frac{-imx^2}{x^4+m^2}\right]=0$$

But along $y=x^3$

$$\lim_{z\to 0}\frac{f(z)-f(0)}{z}=\lim_{z\to 0}\left[\frac{-ix^3y}{x^6+y^2}\right]=\lim_{x\to 0}\frac{-ix^3(x^3)}{x^6+(x^3)^2}=-\frac{i}{2}$$

In different paths we get different values of $\frac{df}{dz}$ *i.e.* 0 and $\frac{-i}{2}$. In such a case, the function is not differentiable at $z=0$.

Theorem: Continuity is a necessary condition but not sufficient condition for the existence of a finite derivative.

Proof. We have, $f(z_0+\delta z)-f(z_0)=\delta z\left\{\dfrac{f(z_0+\delta z)-f(z_0)}{\delta z}\right\}$ ... (1)

Taking limit of both sides of (1), as $\delta z\to 0$, we ge

$$\lim_{\delta z\to 0}[f(z_0+\delta z)-f(z_0)]=0.f'(z_0)\ \Rightarrow\ \lim_{\delta z\to 0}[f(z_0+\delta z)-f(z_0)]=0$$

$$\Rightarrow\ \lim_{z\to z_0}[f(z)-f(z_0)]=0\ \Rightarrow\ \lim_{z\to z_0}f(z)=f(z_0)$$

$\Rightarrow$ $f(z)$ is continuous at $z=z_0$. **Proved.**

The converse of the above theorem is not true.

This can be shown by the following example.

Example 3. *Prove that the function* $f(z)=|z|^2$ *is continuous everywhere but no where differentiable except at the origin.*

Solution. Here, $f(z)=|z|^2$.

$\therefore$ But $|z|=\sqrt{(x^2+y^2)}\ \Rightarrow\ |z|^2=x^2+y^2$

Since x^2 and y^2 are polynomial so x^2+y^2 is continuous everywhere, therefore, $|z|^2$ is continuous everywhere.

Now, we have $f'(z) = \lim\limits_{\delta z \to 0} \dfrac{f(z+\delta z) - f(z)}{\delta z}$

$$f'(z) = \lim_{\delta z \to 0} \frac{|z+\delta z|^2 - |z|^2}{\delta z} \qquad (z\bar{z} = |z|^2)$$

$$= \lim_{\delta z \to 0} \frac{(z+\delta z)(\bar{z}+\delta \bar{z}) - z\bar{z}}{\delta z} = \lim_{\delta z \to 0} \frac{z\bar{z} + z\delta\bar{z} + \delta z.\bar{z} + \delta z.\delta\bar{z} - z\bar{z}}{\delta z}$$

$$= \lim_{\delta z \to 0} \frac{z.\delta\bar{z} + \delta z.\bar{z} + \delta z \delta\bar{z}}{\delta z} = \lim_{\delta z \to 0} \left\{ \bar{z} + \delta\bar{z} + z\frac{\delta\bar{z}}{\delta z} \right\} = \lim_{\delta z \to 0} \left\{ \bar{z} + z\frac{\delta\bar{z}}{\delta z} \right\} \qquad ...(1)$$

[Since, $\delta z \to 0$ so $\delta\bar{z} \to 0$]

Let $\delta z = r(\cos\theta + i\sin\theta)$ and $\delta\bar{z} = r(\cos\theta - i\sin\theta)$

$$\Rightarrow \frac{\delta\bar{z}}{\delta z} = \frac{\cos\theta - i\sin\theta}{\cos\theta + i\sin\theta} \qquad \Rightarrow \frac{\delta\bar{z}}{\delta z} = (\cos\theta - i\sin\theta)(\cos\theta + i\sin\theta)^{-1}$$

$$\Rightarrow \frac{\delta\bar{z}}{\delta z} = (\cos\theta - i\sin\theta)(\cos\theta - i\sin\theta) \Rightarrow \frac{\delta\bar{z}}{\delta z} = (\cos\theta - i\sin\theta)^2$$

$$\Rightarrow \frac{\delta\bar{z}}{\delta z} = \cos 2\theta - i\sin 2\theta$$

Since $\dfrac{\delta\bar{z}}{\delta z}$ depends on θ. It means for different values of θ, $\dfrac{\delta\bar{z}}{\delta z}$ has different values.

It means $\dfrac{\delta\bar{z}}{\delta z}$ has different values for different z.

Therefore $\lim\limits_{\delta z \to 0} \dfrac{\delta\bar{z}}{\delta z}$ does not tend to a unique limit when $z \neq 0$.

Thus, from (1), it follows that $f'(z)$ is not unique and hence $f(z)$ is not differentiable when $z \neq 0$.
But when $z = 0$ then $f'(z) = 0$ *i.e.*, $f'(0) = 0$ and is unique.
Hence, the function is differentiable at $z = 0$. **Ans.**

27.7 ANALYTIC FUNCTION

A function $f(z)$ is said to be **analytic** at a point z_0, if f is differentiable not only at z_0 but at every point of some neighbourhood of z_0.
A function $f(z)$ is analytic in a domain if it is **analytic** at every point of the domain.
The point at which the function is not differentiable is called a **singular point** of the function.
An analytic function is also known as "holomorphic", "regular", "monogenic".
Entire Function. A function which is analytic everywhere (for all z in the complex plane) is known as an entire function.
For Example 1. Polynomials rational functions are entire.

2. $|\bar{z}|^2$ is differentiable only at $z = 0$. So it is no where analytic.

Note: (*i*) An entire is always analytic, differentiable and continuous function. But converse is not true.

(*ii*) Analytic function is always differentiable and continuous. But converse is not true.

(*iii*) A differentiable function is always continuous. But converse is not true

27.8 THE NECESSARY CONDITION FOR $f(z)$ TO BE ANALYTIC

Theorem. *The necessary conditions for a function $f(z) = u + iv$ to be analytic at all the points in a region R are*

(*i*) $\dfrac{\partial u}{\partial x} = \dfrac{\partial v}{\partial y}$ (*ii*) $\dfrac{\partial u}{\partial y} = -\dfrac{\partial v}{\partial x}$ provided $\dfrac{\partial u}{\partial x}, \dfrac{\partial u}{\partial y}, \dfrac{\partial v}{\partial x}, \dfrac{\partial v}{\partial y}$ exist.

Proof: Let $f(z)$ be an analytic function in a region R,

$$f(z) = u + iv,$$

where u and v are the functions of x and y.

Let δu and δv be the increments of u and v respectively corresponding to increments δx and δy of x and y.

$$\therefore \quad f(z+\delta z) = (u+\delta u) + i(v+\delta v)$$

Now $\dfrac{f(z+\delta z)-f(z)}{\delta z} = \dfrac{(u+\delta u)+i(v+\delta v)-(u+iv)}{\delta z} = \dfrac{\delta u + i\delta v}{\delta z} = \dfrac{\delta u}{\delta z} + i\dfrac{\delta v}{\delta z}$

$$\lim_{\delta z\to 0} \frac{f(z+\delta z)-f(z)}{\delta z} = \lim_{\delta z\to 0}\left(\frac{\delta u}{\delta z} + i\frac{\delta v}{\delta z}\right) \text{ or } f'(z) = \lim_{\delta z\to 0}\left(\frac{\delta u}{\delta z} + i\frac{\delta v}{\delta z}\right) \quad \ldots (1)$$

since δz can approach zero along any path.

(*a*) **Along real axis (*x*-axis)**

$z = x + iy$ but on x-axis, $y = 0$

$\therefore$ $z = x$, $\delta z = \delta x$, $\delta y = 0$

Putting these values in (1), we have

$$f'(z) = \lim_{\delta x\to 0}\left(\frac{\delta u}{\delta x} + i\frac{\delta v}{\delta x}\right) = \frac{\partial u}{\partial x} + i\frac{\partial v}{\partial x} \quad \ldots (2)$$

(*b*) **Along imaginary axis (*y*-axis)**

$z = x + iy$ but on y-axis, $x = 0$

$z = 0 + iy$ $\delta x = 0$, $\delta z = i\delta y$.

Putting these values in (1), we get

$$f'(z) = \lim_{\delta y\to 0}\left(\frac{\delta u}{i\delta y} + \frac{i\delta v}{i\delta y}\right) = \lim_{\delta y\to 0}\left(-i\frac{\delta u}{\delta y} + \frac{\delta v}{\delta y}\right) = -i\frac{\partial u}{\partial y} + \frac{\partial v}{\partial y} \quad \ldots (3)$$

If $f(z)$ is differentiable, then two values of $f'(z)$ must be the same.

Equating (2) and (3), we get

$$\frac{\partial u}{\partial x} + i\frac{\partial v}{\partial x} = -i\frac{\partial u}{\partial y} + \frac{\partial v}{\partial y}$$

Equating real and imaginary parts, we have

$$\frac{\partial u}{\partial x} = \frac{\partial v}{\partial y}, \qquad \frac{\partial v}{\partial x} = -\frac{\partial u}{\partial v}$$

$$\boxed{\frac{\partial u}{\partial x} = \frac{\partial v}{\partial y}}$$

$$\boxed{\frac{\partial u}{\partial y} = -\frac{\partial v}{\partial x}}$$ **are known as Cauchy Riemann equations.**

27.9 SUFFICIENT CONDITION FOR $f(z)$ TO BE ANALYTIC

Theorem. *The sufficient condition for a function $f(z) = u + iv$ to be analytic at all the points in a region R are* (*i*) $\dfrac{\partial u}{\partial x}=\dfrac{\partial v}{\partial y},\quad \dfrac{\partial u}{\partial y}=-\dfrac{\partial v}{\partial x}$

(*ii*) $\dfrac{\partial u}{\partial x},\ \dfrac{\partial u}{\partial y},\ \dfrac{\partial v}{\partial x},\ \dfrac{\partial v}{\partial y}$ *are continuous functions of x and y in region R.*

Proof. Let $f(z)$ be a single-valued function having

$$\frac{\partial u}{\partial x},\ \frac{\partial u}{\partial y},\ \frac{\partial v}{\partial x},\ \frac{\partial v}{\partial y}$$

at each point in the region R. Then the $C-R$ equations are satisfied.

By Taylor's Theorem:

$$f(z+\delta z)=u(x+\delta x,\ y+\delta y)+iv(x+\delta x,\ y+\delta y)$$

$$=u(x,y)+\left(\frac{\partial u}{\partial x}\delta x+\frac{\partial u}{\partial y}\delta y\right)+\ldots+i\left[v(x,y)+\left(\frac{\partial v}{\partial x}\delta x+\frac{\partial v}{\partial y}\delta y\right)+\ldots\right]$$

$$=[u(x,y)+iv(x,y)]+\left[\frac{\partial u}{\partial x}\cdot\delta x+i\frac{\partial v}{\partial x}\cdot\delta x\right]+\left[\frac{\partial u}{\partial y}\delta y+i\frac{\partial v}{\partial y}\cdot\delta y\right]+\ldots$$

$$=f(z)+\left(\frac{\partial u}{\partial x}+i\frac{\partial v}{\partial x}\right)\delta x+\left(\frac{\partial u}{\partial y}+i\frac{\partial v}{\partial y}\right)\delta y+\ldots$$

(Ignoring the terms of second power and higher powers)

$$\Rightarrow\quad f(z+\delta z)-f(z)=\left(\frac{\partial u}{\partial x}+i\frac{\partial v}{\partial x}\right).\delta x+\left(\frac{\partial u}{\partial y}+i\frac{\partial v}{\partial y}\right)\delta y \qquad \ldots(1)$$

We know $C-R$ equations *i.e.*,

$$\frac{\partial u}{\partial x}=\frac{\partial v}{\partial y} \text{ and } \frac{\partial u}{\partial y}=-\frac{\partial v}{\partial x}$$

Replacing $\dfrac{\partial u}{\partial y}$ and $\dfrac{\partial v}{\partial y}$ by $-\dfrac{\partial v}{\partial x}$ and $\dfrac{\partial u}{\partial x}$ respectively in (1), we get

$$f(z+\delta z)-f(z)=\left(\frac{\partial u}{\partial x}+i\frac{\partial v}{\partial x}\right).\delta x+\left(-\frac{\partial v}{\partial x}+i\frac{\partial u}{\partial x}\right).\delta y \qquad \text{(taking } i \text{ common)}$$

$$=\left(\frac{\partial u}{\partial x}+i\frac{\partial v}{\partial x}\right).\delta x+\left(i\frac{\partial v}{\partial x}+\frac{\partial u}{\partial x}\right).i\delta y=\left(\frac{\partial u}{\partial x}+i\frac{\partial v}{\partial x}\right).(\delta x+i\delta y)=\left(\frac{\partial u}{\partial x}+i\frac{\partial v}{\partial x}\right).\delta z$$

$$\Rightarrow\quad \frac{f(z+\delta z)-f(z)}{\delta z}=\frac{\partial u}{\partial x}+i\frac{\partial v}{\partial x}$$

$$\Rightarrow\quad \lim_{\delta z\to 0}\frac{f(z+\delta z)-f(z)}{\delta z}=\frac{\partial u}{\partial x}+i\frac{\partial v}{\partial x}$$

$$\Rightarrow\quad \boxed{f'(z)=\frac{\partial u}{\partial x}+i\frac{\partial v}{\partial x}}$$

$$\Rightarrow\quad \boxed{f'(z)=\frac{\partial v}{\partial y}-i\frac{\partial u}{\partial y}}$$

Proved.

Remember: **1.** If a function is analytic in a domain D, then u, v satisfy $C-R$ conditions at all points in D.

2. $C-R$ conditions are necessary but not sufficient for analytic function.

3. $C-R$ conditions are sufficient if the partial derivative are continuous.

Example 4. *Determine whether $\frac{1}{z}$ is analytic or not ?* *(R.G.P.V. Bhopal, III Sem., June 2003)*

Solution. Let $w = f(z) = u + iv = \frac{1}{z} \Rightarrow u + iv = \frac{1}{x+iy} = \frac{x-iy}{x^2+y^2}$

Equating real and imaginary parts, we get

$$u = \frac{x}{x^2+y^2}, \qquad v = \frac{-y}{x^2+y^2}$$

$$\frac{\partial u}{\partial x} = \frac{(x^2+y^2).1 - x.2x}{(x^2+y^2)^2} = \frac{y^2-x^2}{(x^2+y^2)^2}, \qquad \frac{\partial u}{\partial y} = \frac{-2xy}{(x^2+y^2)^2}.$$

$$\frac{\partial v}{\partial x} = \frac{2xy}{(x^2+y^2)^2}, \qquad \frac{\partial v}{\partial y} = \frac{y^2-x^2}{(x^2+y^2)^2}$$

Thus, $\frac{\partial u}{\partial x} = \frac{\partial v}{\partial y}$ and $\frac{\partial u}{\partial y} = -\frac{\partial v}{\partial x}$.

Thus C – R equations are satisfied. Also partial derivatives are continuous except at (0, 0).

Therefore $\frac{1}{z}$ is analytic everywhere except at $z = 0$.

Also $$\frac{dw}{dz} = -\frac{1}{z^2}$$

This again shows that $\frac{dw}{dz}$ exists everywhere except at $z = 0$. Hence $\frac{1}{z}$ is analytic everywhere except at $z = 0$. **Ans.**

Example 5. *Show that the function e^x (cos y + i sin y) is an analytic function, find its derivative.* *(R.G.P.V., Bhopal, III Semester, June 2008)*

Solution. Let $e^x (\cos y + i \sin y) = u + iv$

So, $e^x \cos y = u$ and $e^x \sin y = v$ then $\frac{\partial u}{\partial x} = e^x \cos y, \quad \frac{\partial v}{\partial y} = e^x \cos y$

$$\frac{\partial u}{\partial y} = -e^x \sin y, \qquad \frac{\partial v}{\partial x} = e^x \sin y$$

Here we see that $\frac{\partial u}{\partial x} = \frac{\partial v}{\partial y}, \quad \frac{\partial u}{\partial y} = -\frac{\partial v}{\partial x}$

These are $C - R$ equations and are satisfied and the partial derivatives are continuous.

Hence, $e^x(\cos y + i \sin y)$ is analytic.

$$f(z) = u + iv = e^x(\cos y + i \sin y) \text{ and } \frac{\partial u}{\partial x} = e^x \cos y, \quad \frac{\partial v}{\partial x} = e^x \sin y$$

$$f'(z) = \frac{\partial u}{\partial x} + i\frac{\partial v}{\partial x} = e^x \cos y + ie^x \sin y = e^x(\cos y + i \sin y) = e^x.e^{iy} = e^{x+iy} = e^z.$$

Which is the required derivative. **Ans.**

Example 6. *Test the analyticity of the function w = sin z and hence derive that:*

$$\frac{d}{dz}(\sin z) = \cos z$$

Solution. $w = \sin z = \sin(x + iy) = \sin x \cos iy + \cos x \sin iy$

$$= \sin x \cosh y + i \cos x \sinh y$$

$$u = \sin x \cosh y, \quad v = \cos x \sinh y$$

$$\frac{\partial u}{\partial x} = \cos x \cosh y, \quad \frac{\partial u}{\partial y} = \sin x \sinh y \quad \begin{bmatrix} \cos iy = \cosh y \\ \sin iy = i\sinh y \end{bmatrix}$$

$$\frac{\partial v}{\partial x} = -\sin x \sinh y, \quad \frac{\partial v}{\partial y} = \cos x \cosh y$$

Thus $\dfrac{\partial u}{\partial x} = \dfrac{\partial v}{\partial y}$ and $\dfrac{\partial u}{\partial y} = -\dfrac{\partial v}{\partial x}$

So $C - R$ equations are satisfied and partial derivatives are continuous.

Hence, sin z is an analytic function.

$$\frac{d}{dz}(\sin z) = \frac{d}{dz}[\sin x \cosh y + i \cos x \sinh y]$$

$$= \frac{\partial}{\partial x}(\sin x \cosh y + i \cos x \sinh y)$$

$$\cosh x = \frac{e^x + e^{-x}}{2} \quad \ldots (1)$$

$$\sinh x = \frac{e^x - e^{-x}}{2} \quad \ldots (2)$$

$$\cos x = \frac{e^{ix} + e^{-ix}}{2} \quad \ldots (3)$$

$$\sin x = \frac{e^{ix} - e^{-ix}}{2i} \quad \ldots (4)$$

From (1) $\cosh ix = \dfrac{e^{ix} + e^{-ix}}{2} = \cos x$

From (3) $\cos ix = \dfrac{e^{i(ix)} + e^{-i(ix)}}{2} = \dfrac{e^x + e^{-x}}{2} = \cosh x$

From (4) $\sin ix = \dfrac{e^{i(ix)} - e^{-i(ix)}}{2i} = i\dfrac{e^x - e^{-x}}{2} = i \sinh x$

From (2) $\sinh ix = \dfrac{e^{ix} - e^{-ix}}{2} = i \sin x$

$$= \cos x \cosh y - i \sin x \sinh y = \cos x \cos iy - \sin x \sin iy$$

$$= \cos(x + iy) = \cos z$$

Ans.

Example 7. *Show that the real and imaginary parts of the function w = log z satisfy the Cauchy-Riemann equations when z is not zero. Find its derivative.*

Solution. To separate the real and imaginary parts of log z, we put $x = r\cos\theta$; $y = r\sin\theta$

$$w = \log z = \log(x + iy)$$

$$\Rightarrow \quad u + iv = \log(r\cos\theta + ir\sin\theta) = \log r(\cos\theta + i\sin\theta) = \log_e r.e^{i\theta}$$

$$= \log_e r + \log_e e^{i\theta} = \log r + i\theta = \log\sqrt{x^2 + y^2} + i\tan^{-1}\frac{y}{x} \quad \begin{bmatrix} r = \sqrt{x^2 + y^2} \\ \theta = \tan^{-1}\dfrac{y}{x} \end{bmatrix}$$

So $$u = \log\sqrt{x^2 + y^2} = \frac{1}{2}\log(x^2 + y^2), \quad v = \tan^{-1}\frac{y}{x}$$

On differentiating u, v, we get

$$\frac{\partial u}{\partial x} = \frac{1}{2}\frac{1}{x^2 + y^2}.(2x) = \frac{x}{x^2 + y^2} \quad \ldots (1)$$

$$\frac{\partial v}{\partial y} = \frac{1}{1 + \frac{y^2}{x^2}}\left(\frac{1}{x}\right) = \frac{x}{x^2 + y^2} \quad \ldots (2)$$

From (1) and (2), $\dfrac{\partial u}{\partial x} = \dfrac{\partial v}{\partial y}$ ... (A)

Again differentiating u, v, we have

$$\frac{\partial u}{\partial y} = \frac{1}{2}\frac{1}{x^2 + y^2}(2y) = \frac{y}{x^2 + y^2} \quad \ldots (3)$$

$$\frac{\partial v}{\partial x} = \frac{1}{1 + \frac{y^2}{x^2}}\left(-\frac{y}{x^2}\right) = -\frac{y}{x^2 + y^2} \quad \ldots (4)$$

From (3) and (4), we have

$$\frac{\partial u}{\partial y} = -\frac{\partial v}{\partial x} \quad \text{... (B)}$$

Equations (A) and (B) are $C-R$ equations and partial derivatives are continuous. Hence, $w = \log z$ is an analytic function except

when $\quad x^2 + y^2 = 0 \Rightarrow x = y = 0 \Rightarrow x + iy = 0 \Rightarrow z = 0$

Now $\quad w = u + iv$

$$\frac{dw}{dz} = \frac{\partial u}{\partial x} + i\frac{\partial v}{\partial x} = \frac{x}{x^2+y^2} - i\frac{y}{x^2+y^2} = \frac{x-iy}{x^2+y^2}$$

$$= \frac{x-iy}{(x+iy)(x-iy)} = \frac{1}{x+iy} = \frac{1}{z}$$

Which is the required derivative. **Ans.**

Example 8. *Find the values of* C_1 and C_2 *such that the function*

$$f(z) = x^2 + C_1 y^2 - 2xy + i(C_2 x^2 - y^2 + 2xy) \text{ is analytic. Also find } f'(z).$$

Solution. Let $\quad f(z) = u + iv = x^2 + C_1 y^2 - 2xy + i(C_2 x^2 - y^2 + 2xy)$

Equating real and imaginary parts, we get

$$u = x^2 + C_1 y^2 - 2xy \text{ and } v = C_2 x^2 - y^2 + 2xy$$

$$\frac{\partial u}{\partial x} = 2x - 2y \text{ and } \frac{\partial v}{\partial x} = 2C_2 x + 2y$$

$$\frac{\partial u}{\partial y} = 2C_1 y - 2x \text{ and } \frac{\partial v}{\partial y} = -2y + 2x$$

$C-R$ equations are

$$\left.\begin{aligned} \frac{\partial u}{\partial x} &= \frac{\partial v}{\partial y} \\ \frac{\partial u}{\partial y} &= -\frac{\partial v}{\partial x} \end{aligned}\right] \Rightarrow \begin{aligned} 2x - 2y &= -2y + 2x \quad &\text{...(1)} \\ 2C_1 y - 2x &= -2C_2 x - 2y \quad &\text{...(2)} \end{aligned}$$

From (2) equating the coefficient of x and y.

$$2C_1 = -2 \quad \Rightarrow C_1 = -1$$

$$-2 = -2C_2 \Rightarrow C_2 = 1$$

Hence, $\quad C_1 = -1$ and $C_2 = 1$ **Ans.**

On putting the value of C_2, we get

$$\frac{\partial u}{\partial x} = 2x - 2y, \qquad \frac{\partial v}{\partial x} = 2x + 2y$$

$$f'(z) = \frac{\partial u}{\partial x} + i\frac{\partial v}{\partial x} = (2x-2y) + i(2x+2y) = 2[(x+ix) + (-y+iy)]$$

$$= 2[(1+i)x + i(1+i)y] = 2(1+i)(x+iy) = 2(1+i)z$$

This is the required derivative. **Ans.**

Example 9. *Discuss the analyticity of the function* $f(z) = z\bar{z}$.

Solution. $\quad f(z) = z\bar{z} = (x+iy)(x-iy) = x^2 - i^2 y^2 = x^2 + y^2$

$$f(z) = x^2 + y^2 = u + iv.$$

$$u = x^2 + y^2,\ v = 0$$

At origin, $\dfrac{\partial u}{\partial x} = \lim\limits_{h \to 0} \dfrac{u(0+h,\ 0) - u(0,\ 0)}{h} = \lim\limits_{h \to 0} \dfrac{h^2}{h} = 0$

$$\frac{\partial u}{\partial y} = \lim_{k \to 0} \frac{u(0, 0+k) - u(0,0)}{k} = \lim_{k \to 0} \frac{k^2}{k} = 0$$

Also, $\dfrac{\partial v}{\partial x} = \lim\limits_{h \to 0} \dfrac{v(0+h, 0) - v(0,0)}{h} = 0$

$$\frac{\partial v}{\partial y} = \lim_{k \to 0} \frac{v(0, 0+k) - v(0,0)}{k} = 0$$

Thus, $\dfrac{\partial u}{\partial x} = \dfrac{\partial v}{\partial y}$ and $\dfrac{\partial u}{\partial y} = -\dfrac{\partial v}{\partial x}$.

Hence, C – R equations are satisfied at the origin.

$$f'(0) = \lim_{z \to 0} \frac{f(z) - f(0)}{z} = \lim_{z \to 0} \frac{(x^2 + y^2) - 0}{x + iy}$$

Let $z \to 0$ along the line $y = mx$, then

$$f'(0) = \lim_{x \to 0} \frac{(x^2 + m^2 x^2)}{(x + imx)} = \lim_{x \to 0} \frac{(1 + m^2)x}{1 + im} = 0$$

Therefore, $f'(0)$ is unique. Hence the function $f(z)$ is analytic at $z = 0$. **Ans.**

Example 10. *Show that the function $f(z) = u + iv$, where*

$$f(z) = \frac{x^3(1+i) - y^3(1-i)}{x^2 + y^2}, \quad z \neq 0$$
$$= 0, \quad z = 0$$

satisfies the Cauchy-Riemann equations at $z = 0$. Is the function analytic at $z = 0$? Justify your answer.

Solution. $f(z) = \dfrac{x^3(1+i) - y^3(1-i)}{x^2 + y^2} = u + iv$

$$u = \frac{x^3 - y^3}{x^2 + y^2}, \qquad v = \frac{x^3 + y^3}{x^2 + y^2}$$

[By differentiation the value of $\dfrac{\partial u}{\partial x}, \dfrac{\partial u}{\partial y}, \dfrac{\partial v}{\partial x}, \dfrac{\partial v}{\partial y}$ at (0, 0) we get $\dfrac{0}{0}$, so we apply first principle method]

At the origin

$$\frac{\partial u}{\partial x} = \lim_{h \to 0} \frac{u(0+h,\ 0) - u(0,\ 0)}{h} = \lim_{h \to 0} \frac{\frac{h^3}{h^2}}{h} = 1 \qquad \text{(Along } x\text{- axis)}$$

$$\frac{\partial u}{\partial y} = \lim_{k \to 0} \frac{u(0,\ 0+k) - u(0,\ 0)}{k} = \lim_{k \to 0} \frac{\frac{-k^3}{k^2}}{k} = -1 \qquad \text{(Along } y\text{- axis)}$$

$$\frac{\partial v}{\partial x} = \lim_{h \to 0} \frac{v(0+h,\ 0) - v(0,\ 0)}{h} = \lim_{h \to 0} \frac{\frac{h^3}{h^2}}{h} = 1 \qquad \text{(Along } x\text{- axis)}$$

$$\frac{\partial v}{\partial y} = \lim_{k \to 0} \frac{v(0,\ 0+k) - v(0,\ 0)}{k} = \lim_{k \to 0} \frac{\frac{k^3}{k^2}}{k} = 1 \qquad \text{(Along } y\text{-axis)}$$

Thus we see that

$$\frac{\partial u}{\partial x}=\frac{\partial v}{\partial y} \qquad \text{and} \qquad \frac{\partial u}{\partial y}=-\frac{\partial v}{\partial x}$$

Hence, Cauchy-Riemann equations are satisfied at $z = 0$.

Again $$f'(0)=\lim_{z\to 0}\frac{f(0+z)-f(0)}{z}=\lim_{z\to 0}\left[\frac{\dfrac{x^3-y^3+i(x^3+y^3)}{x^2+y^2}-(0)}{x+iy}\right]$$

$$=\lim_{z\to 0}\left[\frac{x^3-y^3+i(x^3+y^3)}{x^2+y^2}\cdot\frac{1}{x+iy}\right]$$

Now let $z\to 0$ along $y = x$, then

$$f'(0)=\lim_{x\to 0}\frac{x^3-x^3+i(x^3+x^3)}{x^2+x^2}\left(\frac{1}{x+ix}\right)$$

$$=\frac{2i}{2(1+i)}=\frac{i}{1+i}=\frac{i(1-i)}{(1+i)(1-i)}=\frac{i+1}{1+1}=\frac{1}{2}(1+i) \qquad \text{... (1)}$$

Again let $z\to 0$ along $y = 0$, then

$$f'(0)=\lim_{x\to 0}\frac{x^3+ix^3}{x^2}\cdot\frac{1}{x}=(1+i) \qquad [\text{Increment} = z] \qquad \text{... (2)}$$

From (1) and (2), we see that $f'(0)$ is not unique. Hence the function $f(z)$ is not analytic at $z = 0$. **Ans.**

Example 11. *Show that the function defined by* $f(z)=\sqrt{|xy|}$ *Satisfies Cauchy-Riemann equation at the origin but is not analytic at that point.*

Solution. Let $f(z)=u+iv=\sqrt{|xy|}$

Equating real and imaginary parts, we get $u=\sqrt{|xy|}$, $v = 0$

At origin

$$\frac{\partial u}{\partial x}=\lim_{h\to 0}\frac{u(0+h,0)-u(0,0)}{h}=\lim_{h\to 0}\frac{0-0}{h}=0$$

$$\frac{\partial u}{\partial y}=\lim_{k\to 0}\frac{u(0,0+k)-u(0,0)}{k}=\lim_{k\to 0}\frac{0-0}{k}=0$$

Also $$\frac{\partial v}{\partial x}=\lim_{h\to 0}\frac{v(0+h,0)-v(0,0)}{h}=\lim_{h\to 0}\frac{0-0}{h}=0$$

$$\frac{\partial v}{\partial y}=\lim_{k\to 0}\frac{v(0,0+k)-v(0,0)}{k}=\lim_{k\to 0}\frac{0-0}{k}=0$$

From the above results, it is clear that

$$\frac{\partial u}{\partial x}=\frac{\partial v}{\partial y} \text{ and } \frac{\partial u}{\partial y}=-\frac{\partial v}{\partial x}$$

Hence, C– R equations are satisfied at the origin

$$f'(0)=\lim_{z\to 0}\frac{f(z)-f(0)}{z}=\lim_{z\to 0}\frac{\sqrt{|xy|}-0}{x+iy}$$

Let $z \to 0$ along the line $y = mx$, then

$$f'(0) = \lim_{x\to 0} \frac{\sqrt{|mx^2|} - 0}{x(1+im)} = \lim_{x\to 0} \frac{\sqrt{|m|}}{1+im}$$

Thus, the limit on R.H.S. depends upon m and hence will have different values for different values of m.

Therefore, $f'(0)$ is not unique.

Hence the function $f(z)$ is not analytic at $z = 0$. **Ans.**

Example 12. *Show that the function*

$$f(z) = e^{-z^{-4}}, \qquad (z \neq 0) \quad \text{and}$$
$$f(0) = 0$$

is not analytic at $z = 0$,
although, Cauchy-Riemann equations are satisfied at the point. How would you explain this.

Solution. $$f(z) = u + iv = e^{-z^{-4}} = e^{-(x+iy)^{-4}} = e^{-\frac{1}{(x+iy)^4}}$$

$$\Rightarrow \qquad u + iv = e^{-\frac{(x-iy)^4}{(x^2+y^2)^4}} = e^{-\frac{1}{(x^2+y^2)^4}[(x^4+y^4-6x^2y^2)-i4xy(x^2-y^2)]}$$

$$\Rightarrow \qquad u + iv = e^{-\frac{x^4+y^4-6x^2y^2}{(x^2+y^2)^4}} . e^{\frac{-i4xy(x^2-y^2)}{(x^2+y^2)^4}}$$

$$\Rightarrow \qquad u + iv = e^{-\frac{x^4+y^4-6x^2y^2}{(x^2+y^2)^4}}\left[\cos\frac{4xy(x^2-y^2)}{(x^2+y^2)^4} - i\sin\frac{4xy(x^2-y^2)}{(x^2+y^2)^4}\right]$$

Equating real and imaginary parts, we get

$$u = e^{-\frac{x^4+y^4-6x^2y^2}{(x^2+y^2)^4}}\cos\frac{4xy(x^2-y^2)}{(x^2+y^2)^4}, \quad v = e^{-\frac{x^4+y^4-6x^2y^2}{(x^2+y^2)^4}}\sin\frac{4xy(x^2-y^2)}{(x^2+y^2)^4}$$

At $z = 0$ $$\frac{\partial u}{\partial x} = \lim_{h\to 0} \frac{u(0+h,\ 0) - u(0,\ 0)}{h} = \lim_{h\to 0} \frac{e^{-h^{-4}}}{h} = \lim_{h\to 0} \frac{1}{he^{\frac{1}{h^4}}}$$

$$= \lim_{h\to 0}\left[\frac{1}{h\left[1 + \frac{1}{h^4} + \frac{1}{2!h^8} + \frac{1}{3!h^{12}} + \ldots\right]}\right], \quad \left(e^x = 1 + x + \frac{x^2}{2!} + \ldots\right)$$

$$= \lim_{h\to 0}\left[\frac{1}{\left[h + \frac{1}{h^3} + \frac{1}{2\,h^7} + \frac{1}{6h^{11}} \ldots\right]}\right] = \frac{1}{0 + \infty} = \frac{1}{\infty} = 0$$

$$\frac{\partial u}{\partial y} = \lim_{k\to 0} \frac{u(0,\ 0+k) - u(0,\ 0)}{k} = \lim_{k\to 0} \frac{e^{-k^{-4}}}{k} = \lim_{k\to 0} \frac{1}{k\,e^{\frac{1}{k^4}}} = 0$$

$$\frac{\partial v}{\partial x} = \lim_{h\to 0} \frac{v(0+h,\ 0) - v(0,\ 0)}{h} = \lim_{h\to 0} \frac{e^{-h^{-4}}}{h} = \lim_{h\to 0} \frac{1}{h.e^{\frac{1}{h^4}}} = 0$$

$$\frac{\partial v}{\partial y} = \lim_{k\to 0} \frac{v(0, 0+k) - v(0, 0)}{k} = \lim_{k\to 0} \frac{e^{-k^{-4}}}{k} = \lim_{k\to 0} \frac{1}{k.e^{\frac{1}{k^4}}} = 0$$

Hence $\frac{\partial u}{\partial x} = \frac{\partial v}{\partial y}$ and $\frac{\partial u}{\partial y} = -\frac{\partial v}{\partial x}$ ($C - R$ equations are satisfied at $z = 0$)

But $f'(0) = \lim_{z\to 0} \frac{f(z) - f(0)}{z} = \lim_{z\to 0} \frac{e^{-z^{-4}}}{z}$

Along $z = re^{i\frac{\pi}{4}}$

$$f'(0) = \lim_{r\to 0} \frac{e^{-r^{-4}} \cdot e^{-\left(e^{i\frac{\pi}{4}}\right)^{-4}}}{re^{i\frac{\pi}{4}}} = \lim_{r\to 0} \frac{e^{-r^{-4}} \cdot e^{-\left(\cos\frac{\pi}{4} + i\sin\frac{\pi}{4}\right)^{-4}}}{re^{i\frac{\pi}{4}}}$$

$$= \lim_{r\to 0} \frac{e^{-r^{-4}}\, e^{-\cos\pi}}{re^{i\frac{\pi}{4}}} = \lim_{r\to 0} \frac{e^{-r^{-4}} \cdot e}{re^{i\frac{\pi}{4}}} = \infty$$

Showing that $f'(z)$ does not exist at $z = 0$. Hence $f(z)$ is not analytic at $z = 0$. **Proved.**

Example 13. *Examine the nature of the function*

$$f(z) = \frac{x^2 y^5 (x + iy)}{x^4 + y^{10}} ; z \neq 0$$

$$f(0) = 0$$

in the region including the origin.

Solution. Here $f(z) = u + iv = \frac{x^2 y^5 (x + iy)}{x^4 + y^{10}} ; z \neq 0$

Equating real and imaginary parts, we get

$$u = \frac{x^3 y^5}{x^4 + y^{10}}, \quad v = \frac{x^2 y^6}{x^4 + y^{10}}$$

$$\frac{\partial u}{\partial x} = \lim_{h\to 0} \frac{u(0 + h, 0) - u(0, 0)}{h} = \lim_{h\to 0} \frac{\frac{0}{h^4}}{h} = \lim_{h\to 0} \frac{0}{h} = 0$$

$$\frac{\partial u}{\partial y} = \lim_{k\to 0} \frac{u(0, 0 + k) - u(0, 0)}{k} = \lim_{k\to 0} \frac{\frac{0}{k^{10}}}{k} = \lim_{k\to 0} \frac{0}{k} = 0$$

$$\frac{\partial v}{\partial x} = \lim_{h\to 0} \frac{v(0 + h, 0) - v(0, 0)}{h} = \lim_{h\to 0} \frac{\frac{0}{h^4}}{h} = \lim_{h\to 0} \frac{0}{h} = 0$$

$$\frac{\partial v}{\partial y} = \lim_{k\to 0} \frac{v(0, 0 + k) - v(0, 0)}{k} = \lim_{k\to 0} \frac{\frac{0}{k^{10}}}{k} = \lim_{k\to 0} \frac{0}{k} = 0$$

From the above results, it is clear that

$$\frac{\partial u}{\partial x} = \frac{\partial v}{\partial y} \quad \text{and} \quad \frac{\partial u}{\partial y} = -\frac{\partial v}{\partial x}$$

Hence, C-R equations are satisfied at the origin.

But $f'(0) = \lim_{z\to 0} \dfrac{f(0+z)-f(0)}{z} = \lim_{\substack{x\to 0\\ y\to 0}} \left[\dfrac{x^2y^5(x+iy)}{x^4+y^{10}} - 0\right]\cdot\dfrac{1}{x+iy}$ (Increment = z)

$$= \lim_{\substack{x\to 0\\ y\to 0}} \frac{x^2y^5}{x^4+y^{10}}$$

Let $z \to 0$ along the radius vector $y = mx$, then

$$f'(0) = \lim_{x\to 0} \frac{m^5x^7}{x^4+m^{10}x^{10}} = \lim_{x\to 0} \frac{m^5x^3}{1+m^{10}x^6} = \frac{0}{1} = 0 \qquad \text{... (1)}$$

Again let $z \to 0$ along the curve $y^5 = x^2$

$$f'(0) = \lim_{x\to 0} \frac{x^4}{x^4+x^4} = \frac{1}{2} \qquad \text{... (2)}$$

(1) and (2) shows that $f'(0)$ does not exist. Hence, $f(z)$ is not analytic at origin although Cauchy-Riemann equations are satisfied there. **Ans.**

27.10 C–R EQUATIONS IN POLAR FORM

$$\frac{\partial u}{\partial r} = \frac{1}{r}\frac{\partial v}{\partial \theta}$$

$$\frac{\partial u}{\partial \theta} = -r\frac{\partial v}{\partial r}$$

(RGPV., Bhopal, III Sem. Dec. 2007)

Proof. We know $x = r\cos\theta$, and u is a function of x and y.

$$z = x+iy = r(\cos\theta + i\sin\theta) = re^{i\theta}$$

$$u+iv = f(z) = f(re^{i\theta}) \qquad \text{... (1)}$$

Differentiating (1) partially w.r.t., "r", we get

$$\frac{\partial u}{\partial r} + i\frac{\partial v}{\partial r} = f'(re^{i\theta}).e^{i\theta} \qquad \text{... (2)}$$

Differentiating (1) w.r.t. "θ", we get

$$\frac{\partial u}{\partial \theta} + i\frac{\partial v}{\partial \theta} = f'(re^{i\theta})re^{i\theta}i \qquad \text{... (3)}$$

Substituting the value of $f'(re^{i\theta})e^{i\theta}$ from (2) in (3), we obtain

$$\frac{\partial u}{\partial \theta} + i\frac{\partial v}{\partial \theta} = r\left(\frac{\partial u}{\partial r} + i\frac{\partial v}{\partial r}\right)i \quad \text{or} \quad \frac{\partial u}{\partial \theta} + i\frac{\partial v}{\partial \theta} = ir\frac{\partial u}{\partial r} - r\frac{\partial v}{\partial r}$$

Equating real and imaginary parts, we get

$$\boxed{\frac{\partial u}{\partial \theta} = -r\frac{\partial v}{\partial r}} \quad \Rightarrow \quad \frac{\partial v}{\partial r} = \frac{-1}{r}\frac{\partial u}{\partial \theta}$$

And $\boxed{\dfrac{\partial u}{\partial r} = \dfrac{1}{r}\dfrac{\partial v}{\partial \theta}}$ **Proved.**

27.11 DERIVATIVE OF W IN POLAR FORM

We know that $w = u + iv$, $\dfrac{\partial w}{\partial x} = \dfrac{\partial u}{\partial x} + i\dfrac{\partial v}{\partial x}$

But $$\frac{dw}{dz} = \frac{\partial w}{\partial r}\frac{\partial r}{\partial x} + \frac{\partial w}{\partial \theta}\frac{\partial \theta}{\partial x} = \frac{\partial w}{\partial r}\cos\theta - \left(\frac{\partial u}{\partial \theta} + i\frac{\partial v}{\partial \theta}\right)\frac{\sin\theta}{r}$$

$$= \frac{\partial w}{\partial r}\cos\theta - \left(-r\frac{\partial v}{\partial r} + i\cdot r\frac{\partial u}{\partial r}\right)\frac{\sin\theta}{r} \qquad \frac{\partial u}{\partial \theta} = -r\frac{\partial v}{\partial r}$$

$$= \frac{\partial w}{\partial r}\cos\theta - i\left(\frac{\partial u}{\partial r} + i\frac{\partial v}{\partial r}\right)\sin\theta \qquad \frac{\partial v}{\partial \theta} = r\frac{\partial u}{\partial r}$$

$$= \frac{\partial w}{\partial r}\cos\theta - i\frac{\partial}{\partial r}(u + iv)\sin\theta = \frac{\partial w}{\partial r}\cos\theta - i\frac{\partial w}{\partial r}\sin\theta \qquad [\because w = u + iv]$$

$$= (\cos\theta - i\sin\theta)\frac{\partial w}{\partial r} \qquad \text{... (1)}$$

Second form of $\dfrac{dw}{dz}$

$$\frac{dw}{dz} = \frac{\partial w}{\partial r}\frac{\partial r}{\partial x} + \frac{\partial w}{\partial \theta}\frac{\partial \theta}{\partial x} = \frac{\partial(u + iv)}{\partial r}\cos\theta - \frac{\partial w}{\partial \theta}\frac{\sin\theta}{r} \qquad [w = u + iv]$$

$$= \left(\frac{\partial u}{\partial r} + i\frac{\partial v}{\partial r}\right)\cos\theta - \frac{\partial w}{\partial \theta}\frac{\sin\theta}{r}$$

$$= \left(\frac{1}{r}\frac{\partial v}{\partial \theta} - i\frac{1}{r}\cdot\frac{\partial u}{\partial \theta}\right)\cos\theta - \frac{\partial w}{\partial \theta}\cdot\frac{\sin\theta}{r} \qquad \left[\begin{array}{c}\dfrac{\partial u}{\partial \theta} = -r\dfrac{\partial v}{\partial r}\\ \dfrac{\partial v}{\partial \theta} = r\dfrac{\partial u}{\partial r}\end{array}\right]$$

$$= -\frac{i}{r}\left(\frac{\partial u}{\partial \theta} + i\frac{\partial v}{\partial \theta}\right)\cos\theta - \frac{\partial w}{\partial \theta}\left(\frac{\sin\theta}{r}\right)$$

$$= -\frac{i}{r}\frac{\partial}{\partial \theta}(u + iv)\cos\theta - \frac{\partial w}{\partial \theta}\left(\frac{\sin\theta}{r}\right) \qquad [w = 0]$$

$$= -\frac{i}{r}\frac{\partial w}{\partial \theta}\cos\theta - \frac{\partial w}{\partial \theta}\frac{\sin\theta}{r} = -\frac{i}{r}(\cos\theta - i\sin\theta)\frac{\partial w}{\partial \theta} \qquad \text{... (2)}$$

$$\boxed{\frac{dw}{dz} = (\cos\theta - i\sin\theta)\frac{\partial w}{\partial r}} \qquad \left[-\frac{i}{r}\frac{\partial w}{\partial \theta} = \frac{\partial w}{\partial r}\right]$$

$$\boxed{\frac{dw}{dz} = -\frac{i}{r}(\cos\theta - i\sin\theta)\frac{\partial w}{\partial \theta}}$$

These are the two forms for $\dfrac{dw}{dz}$.

Example 14. *If n is real, show that $r^n(\cos n\theta + i\sin n\theta)$ is analytic except possible when $r = 0$ and that its derivative is*

$$nr^{n-1}[\cos(n-1)\theta + i\sin(n-1)\theta].$$

Solution. Let $w = f(z) = u + iv = r^n(\cos n\theta + i\sin n\theta)$

Here, $u = r^n\cos n\theta, \quad v = r^n\sin n\theta$

then, $$\frac{\partial u}{\partial r} = nr^{n-1}\cos n\theta, \quad \frac{\partial v}{\partial r} = nr^{n-1}\sin n\theta$$

$$\frac{\partial u}{\partial \theta} = -nr^n\sin n\theta, \quad \frac{\partial v}{\partial \theta} = nr^n\cos n\theta$$

Here, $$\frac{\partial u}{\partial r} = nr^{n-1} \cos n\theta = \frac{1}{r}\left(nr^n \cos n\theta\right)$$

$$\frac{\partial u}{\partial r} = \frac{1}{r}\frac{\partial v}{\partial \theta} \qquad ...(1)$$

and $$\frac{\partial v}{\partial r} = nr^{n-1} \sin n\theta \qquad \frac{\partial v}{\partial r} = -\frac{1}{r}(-nr^n \sin n\theta)$$

$$\frac{\partial v}{\partial r} = -\frac{1}{r}\frac{\partial u}{\partial \theta} \qquad ...(2)$$

Equations (1) and (2) satisfied C-R equations.

We have, $$\frac{dw}{dz} = (\cos\theta - i\sin\theta)\frac{\partial w}{\partial r} = (\cos\theta - i\sin\theta)\left(\frac{\partial u}{\partial r} + i\frac{\partial v}{\partial r}\right)$$

$$= (\cos\theta - i\sin\theta)\,(nr^{n-1} \cos n\theta + inr^{n-1} \sin r\theta)$$

$$= (\cos\theta - i\sin\theta)\, nr^{n-1} (\cos n\theta + i\sin n\theta)$$

$$= nr^{n-1} \{(\cos n\theta \cos\theta + \sin n\theta \sin\theta) + i\,(\sin n\theta \cos\theta - \cos n\theta \sin\theta)\}$$

$$= nr^{n-1} \{\cos(n-1)\,\theta + i\sin(n-1)\theta\}$$

This exists for all finite values of r including zero, except when $r = 0$ and $n \le 1$. **Proved.**

EXERCISE 27.1

Determine which of the following functions are analytic:

1. $x^2 + iy^2$ **Ans.** Analytic at all points $y = x$ **2.** $2xy + i(x^2 - y^2)$ **Ans.** Not analytic

3. $\dfrac{x-iy}{x^2+y^2}$ **Ans.** Not analytic **4.** $\dfrac{1}{(z-1)(z+1)}$ **Ans.** Analytic at all points, except $z = \pm 1$

5. $\dfrac{x-iy}{x-iy+a}$ **Ans.** Not analytic **6.** $\sin x \cosh y + i\cos x \sinh y$ **Ans.** Yes, analytic

7. $xy + iy^2$ **Ans.** Yes, analytic at origin

8. Discuss the analyticity of the function $f(z) = z\bar{z} + \bar{z}^2$ in the complex plane, where z is the complex conjugate of z. Also find the points where it is differentiable but not analytic.

Ans. Differentiable only at $z = 0$, No where analytic.

9. Show the function of $\bar{z}$ is not analytic any where.

10. If $\begin{cases} \dfrac{x^2 y\,(y - ix)}{x^4 + y^2}, & \text{when } z \ne 0 \\ 0, & \text{when } z = 0 \end{cases}$

prove that $\dfrac{f(z) - f(0)}{z} \to 0$, as $z \to 0$, along any radius vector but not as $z \to 0$ in any manner.

(AMIETE, Dec. 2010)

11. If $f(z)$ is an analytic function with constant modulus, show that $f(z)$ is constant. *(AMIETE, Dec. 2009)*

Choose the correct answer :

12. The Cauchy-Riemann equations for $f(z) = u(x, y) + iv(x, y)$ to be analytic are :

(*a*) $\dfrac{\partial^2 u}{\partial x^2} + \dfrac{\partial^2 u}{\partial y^2} = 0,\ \dfrac{\partial^2 v}{\partial x^2} + \dfrac{\partial^2 v}{\partial y^2} = 0$ (*b*) $\dfrac{\partial u}{\partial x} = -\dfrac{\partial v}{\partial y},\ \dfrac{\partial u}{\partial y} = -\dfrac{\partial v}{\partial x}$

(*c*) $\dfrac{\partial u}{\partial x} = \dfrac{\partial v}{\partial y},\ \dfrac{\partial u}{\partial y} = -\dfrac{\partial v}{\partial x}$ (*d*) $\dfrac{\partial u}{\partial x} = -\dfrac{\partial v}{\partial y},\ \dfrac{\partial u}{\partial y} = \dfrac{\partial v}{\partial x}$ **Ans.** (*c*)

(R.G.P.V., Bhopal, III Semester, Dec. 2006)

13. Polar form of *C-R* equations are :

(*a*) $\frac{\partial u}{\partial \theta}=\frac{1}{r}\frac{\partial v}{\partial r},\quad \frac{\partial u}{\partial r}=r\frac{\partial v}{\partial \theta}$ (*b*) $\frac{\partial u}{\partial \theta}=r\frac{\partial v}{\partial r},\quad \frac{\partial u}{\partial r}=\frac{1}{r}\frac{\partial v}{\partial \theta}$

(*c*) $\frac{\partial u}{\partial r}=\frac{1}{r}\frac{\partial v}{\partial \theta},\quad \frac{\partial u}{\partial \theta}=-r\frac{\partial v}{\partial r}$ (*d*) $\frac{\partial u}{\partial r}=r\frac{\partial v}{\partial \theta},\quad \frac{\partial u}{\partial \theta}=-r\frac{\partial v}{\partial r}$ **Ans.** (*c*)

(R.G.P.V., Bhopal, III Semester, June, 2007)

14. The curve $u(x, y) = C$ and $v(x, y) = C^1$ are orthogonal if

(*a*) *u* and *v* are complex functions (*b*) $u + iv$ is an analytic function.

(*c*) $u - v$ is an analytic function. (*d*) $u + v$ is an analytic function **Ans.** (*b*)

15. If $f(z)=\frac{1}{2}\log_e(x^2+y^2)+i\tan^{-1}\left(\frac{\alpha x}{y}\right)$ **be an analytic function if α is equal to**

(*a*) + 1 (*b*) − 1 (*c*) + 2 (*d*) − 2 *(AMIETE, Dec. 2009)* **Ans.** (*a*)

27.12 ORTHOGONAL CURVES

(U.P. III Semester, June 2009)

Two curves are said to be orthogonal to each other, when they intersect at right angle at each of their points of intersection.

The analytic function $f(z)=u+iv$ consists of two families of curves $u(x, y) = c_1$ and $v(x, y) = c_2$ which form an orthogonal system.

$$u(x, y) = c_1 \qquad \ldots(1)$$

$$v(x, y) = c_2 \qquad \ldots(2)$$

Differentiating (1), $\frac{\partial u}{\partial x}dx+\frac{\partial u}{\partial y}dy=0$

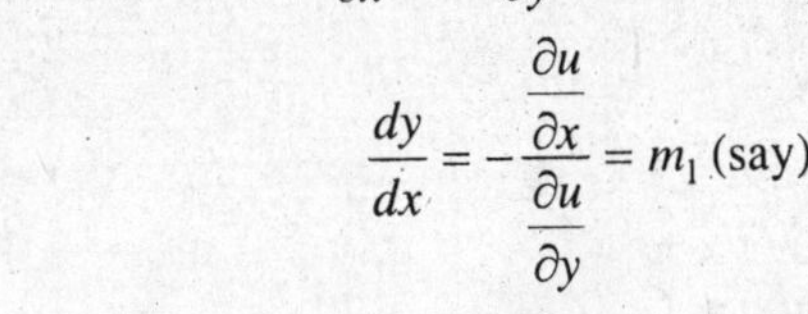

$$\Rightarrow \qquad \frac{dy}{dx}=-\frac{\frac{\partial u}{\partial x}}{\frac{\partial u}{\partial y}}=m_1 \text{ (say)}$$

Similarly from (2), $\frac{dy}{dx}=-\frac{\frac{\partial v}{\partial x}}{\frac{\partial v}{\partial y}}=m_2$ *(say)*

The product of two slopes

$$m_1m_2=\left(-\frac{\frac{\partial u}{\partial x}}{\frac{\partial u}{\partial y}}\right)\left(-\frac{\frac{\partial v}{\partial x}}{\frac{\partial v}{\partial y}}\right)=\left(-\frac{\frac{\partial u}{\partial x}}{\frac{\partial u}{\partial y}}\right)\left(-\frac{-\frac{\partial u}{\partial y}}{\frac{\partial u}{\partial x}}\right) \qquad (C-R \text{ equations})$$

$$= -1$$

Since $m_1 m_2 = -1$, two curves $u = c_1$ and $v = c_2$ are orthogonal, and c_1, c_2 are parameters, $u = c_1$ and $v = c_2$ form an orthogonal system.

27.13 HARMONIC FUNCTION

(U.P., III Semester 2009-2010)

Any function which satisfies the Laplace's equation is known as a harmonic function.

Theorem. If $f(z)=u+iv$ is an analytic function, then *u* and *v* are both harmonic functions.

Proof. Let $f(z) = u + iv$, be an analytic function, then we have

$$\left.\begin{aligned}\frac{\partial u}{\partial x}&=\frac{\partial v}{\partial y} \qquad \ldots(1)\\ \frac{\partial u}{\partial y}&=-\frac{\partial v}{\partial x} \qquad \ldots(2)\end{aligned}\right] \quad C-R \text{ equations.}$$

Differentiating (1) with respect to x, we get $\dfrac{\partial^2 u}{\partial x^2}=\dfrac{\partial^2 v}{\partial x\,\partial y}$ ... (3)

Differentiating (2) w.r.t. 'y' we have $\dfrac{\partial^2 u}{\partial y^2}=-\dfrac{\partial^2 v}{\partial y\,\partial x}$ ... (4)

Adding (3) and (4) we have $\dfrac{\partial^2 u}{\partial x^2}+\dfrac{\partial^2 u}{\partial y^2}=\dfrac{\partial^2 v}{\partial x\,\partial y}-\dfrac{\partial^2 v}{\partial y\,\partial x}$

$$\frac{\partial^2 u}{\partial x^2}+\frac{\partial^2 u}{\partial y^2}=0 \qquad \left(\frac{\partial^2 v}{\partial x\,\partial y}=\frac{\partial^2 v}{\partial y\,\partial x}\right)$$

Similarly $\dfrac{\partial^2 v}{\partial x^2}+\dfrac{\partial^2 v}{\partial y^2}=0$

Therefore both u and v are harmonic functions.

Such functions u, v are called **Conjugate harmonic functions if $u + iv$ is also analytic function.**

Example 15. *Define a harmonic function and conjugate harmonic function. Find the harmonic conjugate function of the function $U(x, y) = 2x(1 - y)$. (U.P., III Semester Dec. 2009)*

Solution. See Art. 27.13.

Here, we have $U(x, y) = 2x(1 - y)$ Let V be the harmonic conjugate of U.

By total differentiation

$$\begin{aligned} dV &= \frac{\partial V}{\partial x}dx+\frac{\partial V}{\partial y}dy \\ &= -\frac{\partial U}{\partial y}dx+\frac{\partial U}{\partial x}dy \\ &= -(-2x)\,dx+(2-2y)\,dy + \\ &= 2x\,dx+(2\,dy-2y\,dy)+C \\ V &= x^2+2y-y^2+C \end{aligned}$$

Hence, the harmonic conjugate of U is $x^2 + 2y - y^2 + C$ **Ans.**

Example 16. *Show that the function $u=\dfrac{1}{2}\log(x^2+y^2)$ is harmonic. Find its harmonic conjugate.*

Solution. $u=\dfrac{1}{2}\log(x^2+y^2)$

$$\frac{\partial u}{\partial x}=\frac{1}{2}\,\frac{1}{x^2+y^2}\,.\,(2x)=\frac{x}{x^2+y^2}, \qquad \frac{\partial u}{\partial y}=\frac{y}{x^2+y^2}$$

$$\frac{\partial^2 u}{\partial x^2}=\frac{(x^2+y^2)\,.\,1-x(2x)}{(x^2+y^2)^2}=\frac{y^2-x^2}{(x^2+y^2)^2}$$

$$\frac{\partial^2 u}{\partial y^2}=\frac{(x^2+y^2)\,.\,1-y(2y)}{(x^2+y^2)^2}=\frac{x^2-y^2}{(x^2+y^2)^2} \quad\Rightarrow\quad \frac{\partial^2 u}{\partial x^2}+\frac{\partial^2 u}{\partial y^2}=0$$

Hence u is a harmonic function.

Let v be the harmonic conjugate of u.

$$dv=\frac{\partial v}{\partial x}dx+\frac{\partial v}{\partial y}dy=-\frac{\partial u}{\partial y}dx+\frac{\partial u}{\partial x}dy \qquad \text{(By C – R equations)}$$

$$dv=-\frac{y}{x^2+y^2}dx+\frac{x}{x^2+y^2}dy$$

$$dv = \frac{xdy - ydx}{x^2 + y^2} = d\left(\tan^{-1}\frac{y}{x}\right)$$

Integrating, we get $v = \tan^{-1}\frac{y}{x} + C.$, where C is a real constant.

This is the required harmonic conjugate. **Ans.**

Example 17. *Prove that $u = x^2 - y^2$ and $v = \frac{y}{x^2 + y^2}$ are harmonic functions of (x, y), but are not harmonic conjugates.*

Solution. We have, $u = x^2 - y^2$

$$\frac{\partial u}{\partial x} = 2x, \qquad \frac{\partial^2 u}{\partial x^2} = 2, \qquad \frac{\partial u}{\partial y} = -2y, \qquad \frac{\partial^2 u}{\partial y^2} = -2$$

$$\frac{\partial^2 u}{\partial x^2} + \frac{\partial^2 u}{\partial y^2} = 2 - 2 = 0$$

u (x, y) satisfies Laplace equation, hence u (x, y) is harmonic

$$v = \frac{y}{x^2 + y^2}, \qquad \frac{\partial v}{\partial x} = -\frac{2xy}{(x^2 + y^2)^2}$$

$$\frac{\partial^2 v}{\partial x^2} = \frac{(x^2 + y^2)^2(-2y) - (-2xy)2(x^2 + y^2)2x}{(x^2 + y^2)^4}$$

$$= \frac{(x^2 + y^2)(-2y) - (-2xy)4x}{(x^2 + y^2)^3} = \frac{6x^2y - 2y^3}{(x^2 + y^2)^3}$$

$$\frac{\partial v}{\partial y} = \frac{(x^2 + y^2)\cdot 1 - y(2y)}{(x^2 + y^2)^2} = \frac{x^2 - y^2}{(x^2 + y^2)^2}$$

$$\frac{\partial^2 v}{\partial y^2} = \frac{(x^2 + y^2)^2(-2y) - (x^2 - y^2)2(x^2 + y^2)(2y)}{(x^2 + y^2)^4} = \frac{(x^2 + y^2)(-2y) - (x^2 - y^2)(4y)}{(x^2 + y^2)^3}$$

$$= \frac{-2x^2y - 2y^3 - 4x^2y + 4y^3}{(x^2 + y^2)^3} = \frac{-6x^2y + 2y^3}{(x^2 + y^2)^3} \qquad \text{... (2)}$$

On adding (1) and (2), we get $\frac{\partial^2 v}{\partial x^2} + \frac{\partial^2 v}{\partial y^2} = 0$

v (x, y) also satisfies Laplace equations, hence v (x, y) is also harmonic function.

But $\frac{\partial u}{\partial x} \neq \frac{\partial v}{\partial y}$ and $\frac{\partial u}{\partial y} \neq -\frac{\partial v}{\partial x}$

Therefore u and v are not harmonic conjugates. **Proved.**

Example 18. *Show that the function $x^2 - y^2 + 2y$ which is harmonic remains harmonic under the transformation $z = w^3$*

Solution. $u = x^2 - y^2 + 2y$

$$\frac{\partial u}{\partial x} = 2x, \qquad \frac{\partial^2 u}{\partial x^2} = 2 \qquad \text{... (1)}$$

$$\Rightarrow \quad \frac{\partial u}{\partial y} = -2y + 2 \qquad \frac{\partial^2 u}{\partial y^2} = -2 \qquad \text{... (2)}$$

Adding (1) and (2), we get

$$\frac{\partial^2 u}{\partial x^2} + \frac{\partial^2 u}{\partial y^2} = 2 - 2 = 0 \Rightarrow \frac{\partial^2 u}{\partial x^2} + \frac{\partial^2 u}{\partial y^2} = 0$$

Hence function is harmonic.

Transformation: $z = w^3$, $z = re^{i\theta}$ and $w = Re^{i\phi}$

$$\Rightarrow \quad re^{i\theta} = \left(Re^{i\phi}\right)^3 \quad \Rightarrow \quad re^{i\theta} = R^3 e^{3i\phi}$$

By comparing both side $r = R^3$, $\theta = 3\phi$

Given function, $f(x, y) = x^2 - y^2 + 2y$ where $x = r\cos\theta$ and $y = r\sin\theta$

$$f(r\cos\theta,\ r\sin\theta) = (r\cos\theta)^2 - (r\sin\theta)^2 + 2\times r\sin\theta$$

$$= r^2\cos^2\theta - r^2\sin^2\theta + 2r\sin\theta$$

$$= r^2(\cos^2\theta - \sin^2\theta) + 2r\sin\theta = r^2\cos 2\theta + 2r\sin\theta$$

$$f(R^3\cos 3\phi,\ R^3\sin 3\phi) = R^6\cos 6\phi + 2R^3\sin 3\phi$$

This is a function in cosine and sine. Hence it will be harmonic function. **Proved.**

Example 19. *If $u(x, y)$ and $v(x, y)$ are harmonic functions in a region R, prove that the function*

$$\left[\left(\frac{\partial u}{\partial y} - \frac{\partial v}{\partial x}\right) + i\left(\frac{\partial u}{\partial x} + \frac{\partial v}{\partial y}\right)\right]$$

is an analytic function of $z = x + iy$. (*R.G.P.V., Bhopal, III Semester, Dec. 2004*)

Solution. Since $u(x, y)$ and $v(x, y)$ are harmonic functions in a region R, therefore

$$\frac{\partial^2 u}{\partial x^2} + \frac{\partial^2 u}{\partial y^2} = 0 \quad \text{... (1)}$$

and
$$\frac{\partial^2 v}{\partial x^2} + \frac{\partial^2 v}{\partial y^2} = 0 \quad \text{... (2)}$$

Let
$$F(z) = R + iS = \left(\frac{\partial u}{\partial y} - \frac{\partial v}{\partial x}\right) + i\left(\frac{\partial u}{\partial x} + \frac{\partial v}{\partial y}\right)$$

Equating real and imaginary parts, we get

$$R = \frac{\partial u}{\partial y} - \frac{\partial v}{\partial x},$$

$$\frac{\partial R}{\partial x} = \frac{\partial^2 u}{\partial x\,\partial y} - \frac{\partial^2 v}{\partial x^2} \quad \text{... (3)}$$

$$\frac{\partial R}{\partial y} = \frac{\partial^2 u}{\partial y^2} - \frac{\partial^2 v}{\partial x\,\partial y} \quad \text{... (4)}$$

$$S = \frac{\partial u}{\partial x} + \frac{\partial v}{\partial y}$$

$$\frac{\partial S}{\partial x} = \frac{\partial^2 u}{\partial x^2} + \frac{\partial^2 v}{\partial x\,\partial y} \quad \text{... (5)}$$

$$\frac{\partial S}{\partial y} = \frac{\partial^2 u}{\partial x\,\partial y} + \frac{\partial^2 v}{\partial y^2} \quad \text{... (6)}$$

Putting the value of $\frac{\partial^2 u}{\partial x^2}$ from (1) in (5), we get

$$\frac{\partial S}{\partial x} = -\frac{\partial^2 u}{\partial y^2} + \frac{\partial^2 v}{\partial x\, \partial y} \qquad \text{... (7)}$$

Putting the value of $\frac{\partial^2 v}{\partial y^2}$ from (2) in (6), we get

$$\frac{\partial S}{\partial y} = \frac{\partial^2 u}{\partial x\, \partial y} - \frac{\partial^2 v}{\partial x^2} \qquad \text{... (8)}$$

From (3) and (8), $\quad \frac{\partial R}{\partial x} = \frac{\partial S}{\partial y}$

From (4) and (7), $\quad \frac{\partial R}{\partial y} = -\frac{\partial S}{\partial x}$

Therefore, C-R equations are satisfied and hence the given function is analytic. **Proved.**

27.14 APPLICATIONS TO FLOW PROBLEMS

As the real part u and imaginary part v of an analytic function $f(z)$ are the solution of Laplace equation.

$$\frac{\partial^2 u}{\partial x^2} + \frac{\partial^2 u}{\partial y^2} = 0$$

and
$$\frac{\partial^2 v}{\partial x^2} + \frac{\partial^2 v}{\partial y^2} = 0$$

So, we get a solution to a number of field and flow problems.

27.15 VELOCITY POTENTIAL

Consider the two dimensional irrotational motion of an incompressible fluid, in planes parallel to xy-plane.

Let $\vec{v}$ be the velocity of a fluid particle, then it can be expressed as

$$\vec{V} = v_x\, \hat{i} + v_y\, \hat{j} \qquad \text{...(1)}$$

Since the motion is irrotational, there exists a scalar function $\phi(x, y)$, such that

$$\vec{V} = \nabla\phi(x, y) = \frac{\partial \phi}{\partial x}\hat{i} + \frac{\partial \phi}{\partial y}\hat{j} \qquad \text{...(2)}$$

From (1) and (2), we have $\quad v_x = \frac{\partial \phi}{\partial x} \quad$ and $\quad v_y = \frac{\partial \phi}{\partial y} \qquad$ ...(3)

The scalar function $\phi(x, y)$, which gives the velocity components, is called the velocity potential function or simply the velocity potential.

Also the fluid being incompressible, div $\vec{V} = 0$

$$\Rightarrow \quad \left(\hat{i}\frac{\partial}{\partial x} + \hat{j}\frac{\partial}{\partial y}\right).(v_x\hat{i} + v_y\hat{j}) = 0$$

$$\Rightarrow \quad \frac{\partial v_x}{\partial x} + \frac{\partial v_y}{\partial y} = 0 \qquad \text{...(4)}$$

Substituting the values of v_x and v_y from (3) in (4), we get

$$\frac{\partial}{\partial x}\left(\frac{\partial \phi}{\partial x}\right)+\frac{\partial}{\partial y}\left(\frac{\partial \phi}{\partial y}\right) = 0 \quad \Rightarrow \quad \left(\frac{\partial^2 \phi}{\partial x^2}\right)+\left(\frac{\partial^2 \phi}{\partial y^2}\right)=0$$

Thus the function ϕ is harmonic.

This is the real part of an analytic function.

$$f(z) = \phi(x, y) + i\,\psi(x, y)$$

$$\frac{df}{dz} = \frac{\partial \phi}{\partial x}+i\frac{\partial \psi}{\partial x} = \frac{\partial \phi}{\partial x}-i\frac{\partial \phi}{\partial y} \qquad \left(\frac{\partial \psi}{\partial x}=-\frac{\partial \phi}{\partial y}\right)$$

$$= v_x - iv_y$$

The magnitude of the resultant velocity

$$= \left|\frac{df}{dz}\right| = \sqrt{v_x^2+v_y^2}.$$

$\phi(x, y) = C_1$ and $\psi(x, y) = C_2$ are called equipotential lines and lines of force respectively.

In heat flow problem the curves $\phi(x, y) = C_1$ and $\psi(x, y) = C_2$ are known as isothermals and heat flow lines respectively.

27.16 METHOD TO FIND THE CONJUGATE FUNCTION

Case I. Given. If $f(z)=u+iv$, and u is known.

To find. v, conjugate function.

Method. We know that
$$dv=\frac{\partial v}{\partial x}\cdot dx+\frac{\partial v}{\partial y}\cdot dy \qquad \text{... (1)}$$

Replacing $\frac{\partial v}{\partial x}$ by $-\frac{\partial u}{\partial y}$ and $\frac{\partial v}{\partial y}$ by $\frac{\partial u}{\partial x}$ in (1), we get [C-R equations]

$$dv=-\frac{\partial u}{\partial y}\cdot dx+\frac{\partial u}{\partial x}\cdot dy$$

$$v=-\int \frac{\partial u}{\partial y}dx+\int \frac{\partial u}{\partial x}\cdot dy$$

$$\Rightarrow \qquad v=\int M\,dx+\int N\,dy \qquad \text{... (2)}$$

where
$$M=-\frac{\partial u}{\partial y} \text{ and } N=\frac{\partial u}{\partial x}$$

so that
$$\frac{\partial M}{\partial y}=\frac{\partial}{\partial y}\left(-\frac{\partial u}{\partial y}\right)=-\frac{\partial^2 u}{\partial y^2} \text{ and } \frac{\partial N}{\partial x}=\frac{\partial}{\partial x}\left(\frac{\partial u}{\partial x}\right)=\frac{\partial^2 u}{\partial x^2}$$

since u is a conjugate function, so $\frac{\partial^2 u}{\partial x^2}+\frac{\partial^2 u}{\partial y^2}=0$

$$\Rightarrow \qquad -\frac{\partial^2 u}{\partial y^2}=\frac{\partial^2 u}{\partial x^2} \qquad \Rightarrow \qquad \frac{\partial M}{\partial y}=\frac{\partial N}{\partial x} \qquad \text{... (3)}$$

Equation (3) satisfies the condition of an exact differential equation

So equation (2) can be integrated and thus v is determined.

Case II. Similarly, if $v = v(x, y)$ is given

To find out u.

We know that
$$du = \frac{\partial u}{\partial x}dx + i\frac{\partial u}{\partial y}dy \qquad \text{... (4)}$$

On substituting the values of $\frac{\partial u}{\partial x}$ and $\frac{\partial u}{\partial y}$ in (4) , we get

$$du = \frac{\partial v}{\partial y} dx - \frac{\partial v}{\partial x} dy$$

On integrating, we get

$$u = \int \frac{\partial v}{\partial y} dx - \int \frac{\partial v}{\partial x} dy \qquad ...(5)$$

(since v is already known so $\frac{\partial v}{\partial y}, \frac{\partial v}{\partial x}$ on R.H.S. are also known)

Equation (5) is an exact differential equation. On solving (5), u can be determined. Consequently $f(z) = u + iv$ can also be determined.

Example 20. *Let $f(z) = u(x, y) + iv(x, y)$ be an analytic function. If $u = 3x - 2xy$, then find v and express $f(z)$ in terms of z.*

Solution. Here, we have $\quad u = 3x - 2xy$

$$\frac{\partial u}{\partial x} = 3 - 2y, \qquad \frac{\partial u}{\partial y} = -2x$$

We know that $\quad dv = \frac{\partial v}{\partial x} dx + \frac{\partial v}{\partial y} dy \qquad$ (Total differentiation)

$$= \left(-\frac{\partial u}{\partial y}\right) dx + \left(\frac{\partial u}{\partial x}\right) dy \qquad \left(\frac{\partial v}{\partial x} = -\frac{\partial u}{\partial y}, \frac{\partial v}{\partial y} = \frac{\partial u}{\partial x}\right)$$

$$= 2x\, dx + (3 - 2y)\, dy$$

$$v = \int 2x\, dx + \int (3 - 2y)\, dy = x^2 + 3y - y^2 + c$$

$$f(z) = u(x, y) + iv(x, y) = (3x - 2xy) + i(x^2 + 3y - y^2 + c)$$

$$= (ix^2 - iy^2 - 2xy) + (3x + 3yi) + ic = i(x^2 - y^2 + 2ixy) + 3(x + iy) + ic$$

$$= i(x + iy)^2 + 3(x + iy) + ic = iz^2 + 3z + ic \qquad \textbf{Ans.}$$

Which is the required expression of $f(z)$ in terms of z.

Example 21. *Prove that $u = x^2 - y^2 - 2xy - 2x + 3y$ is harmonic. Find a function v such that $f(z) = u + iv$ is analytic. Also express $f(z)$ in terms of z.*

(R.G.P.V., Bhopal, III Semester, June 2005)

Solution. We have,

$$u = x^2 - y^2 - 2xy - 2x + 3y$$

$$\frac{\partial u}{\partial x} = 2x - 2y - 2 \quad \Rightarrow \quad \frac{\partial^2 u}{\partial x^2} = 2$$

$$\frac{\partial u}{\partial y} = -2y - 2x + 3 \quad \Rightarrow \quad \frac{\partial^2 u}{\partial y^2} = -2$$

$$\frac{\partial^2 u}{\partial x^2} + \frac{\partial^2 u}{\partial y^2} = 2 - 2 \quad \Rightarrow \quad \frac{\partial^2 u}{\partial x^2} + \frac{\partial^2 u}{\partial y^2} = 0$$

Since Laplace equation is satisfied, therefore u is harmonic. **Proved.**

We know that $\quad dv = \frac{\partial v}{\partial x} dx + \frac{\partial v}{\partial y} dy$

$$\Rightarrow \quad dv = -\frac{\partial u}{\partial y} dx + \frac{\partial u}{\partial x} dy \qquad ...(1) \qquad \left[\because \frac{\partial v}{\partial x} = -\frac{\partial u}{\partial y} \text{ and } \frac{\partial v}{\partial y} = \frac{\partial u}{\partial x}\right]$$

Putting the values of $\frac{\partial u}{\partial y}$ and $\frac{\partial u}{\partial x}$ in (1), we get

$$dv = -(-2y - 2x + 3)\,dx + (2x - 2y - 2)dy$$

$$\Rightarrow \quad v = \int (2y + 2x - 3)dx + \int (-2y - 2)dy + C \qquad \text{(Ignoring } 2x\text{)}$$

Hence, $\quad v = 2xy + x^2 - 3x - y^2 - 2y + C$ **Ans.**

Now,

$$\begin{aligned} f(z) &= u + iv \\ &= (x^2 - y^2 - 2xy - 2x + 3y) + i(2xy + x^2 - 3x - y^2 - 2y) + iC \\ &= (x^2 - y^2 + 2ixy) + (ix^2 - iy^2 - 2xy) - (2 + 3i)x - i(2 + 3i)y + iC \\ &= (x^2 - y^2 + 2ixy) + i(x^2 - y^2 + 2ixy) - (2 + 3i)x - i(2 + 3i)y + iC \\ &= (x + iy)^2 + i(x + iy)^2 - (2 + 3i)(x + iy) + iC \\ &= z^2 + iz^2 - (2 + 3i)z + iC \\ &= (1 + i)z^2 - (2 + 3i)z + iC \end{aligned}$$

Which is the required expression of $f(z)$ in terms of z. **Ans.**

Example 22. *Define a harmonic function. Show that the function* $u(x, y) = x^4 - 6x^2y^2 + y^4$ *is harmonic. Also find the analytic function* $f(z) = u(x, y) + iv(x, y)$.

Solution. See Art. 27.13 on page 685 for definition of harmonic function.

We have,

$$u(x, y) = x^4 - 6x^2y^2 + y^4 \qquad \frac{\partial u}{\partial x} = 4x^3 - 12xy^2$$

$$\frac{\partial u}{\partial y} = -12x^2y + 4y^3, \qquad \frac{\partial^2 u}{\partial x^2} = 12x^2 - 12y^2 \qquad \ldots (1)$$

$$\frac{\partial^2 u}{\partial y^2} = -12x^2 + 12y^2 \qquad \ldots (2)$$

Adding (1), and (2), we get

$$\frac{\partial^2 u}{\partial x^2} + \frac{\partial^2 u}{\partial y^2} = 12x^2 - 12y^2 - 12x^2 + 12y^2$$

$$\Rightarrow \quad \frac{\partial^2 u}{\partial x^2} + \frac{\partial^2 u}{\partial y^2} = 0$$

Hence, u is a harmonic function. **Proved.**

Let us find out v:

We know that $dv = \frac{\partial v}{\partial x}dx + \frac{\partial v}{\partial y}dy$

$$\Rightarrow \quad dv = -\frac{\partial u}{\partial y}dx + \frac{\partial u}{\partial x}dy \qquad \left[\because \frac{\partial v}{\partial x} = -\frac{\partial u}{\partial y} \text{ and } \frac{\partial v}{\partial y} = \frac{\partial u}{\partial x}\right]$$

$$\Rightarrow \quad dv = (12x^2y - 4y^3)\,dx + (4x^3 - 12xy^2)\,dy$$

$$v = \int (12x^2y - 4y^3)\,dx + \int (4x^3 - 12xy^2)\,dy$$

(y is constant) (Integrate only those terms which do not contain x)

$$v = 4x^3y - 4xy^3 + C$$

$$f(z) = u + iv = x^4 - 6x^2y^2 + y^4 + i4x^3y - 4ixy^3 + iC$$

$$f(z) = x^4 + 4x^3(iy) + 6x^2(iy)^2 + 4x(iy)^3 + (iy)^4 + iC$$

$$= (x+iy)^4 + iC$$

$$= z^4 + iC \qquad [\because z = x+iy]$$

This is the required analytic function. **Ans.**

Example 23. *If $w = \phi + i\psi$ represents the complex potential for an electric field and*

$$\psi = x^2 - y^2 + \frac{x}{x^2+y^2},$$

determine the function ϕ.

Solution. $w = \phi + i\psi$ and $\psi = x^2 - y^2 + \dfrac{x}{x^2+y^2}$

$$\frac{\partial \psi}{\partial x} = 2x + \frac{(x^2+y^2)\cdot 1 - x \cdot 2x}{(x^2+y^2)^2} = 2x + \frac{y^2 - x^2}{(x^2+y^2)^2}$$

$$\frac{\partial \psi}{\partial y} = -2y - \frac{x(2y)}{(x^2+y^2)^2} = -2y - \frac{2xy}{(x^2+y^2)^2}$$

We know that, $d\phi = \dfrac{\partial \phi}{\partial x} dx + \dfrac{\partial \phi}{\partial y} dy = \dfrac{\partial \psi}{\partial y} dx - \dfrac{\partial \psi}{\partial x} dy$

$$= \left(-2y - \frac{2xy}{(x^2+y^2)^2}\right) dx - \left(2x + \frac{y^2 - x^2}{(x^2+y^2)^2}\right) dy$$

$$\phi = \int \left[-2y - \frac{2xy}{(x^2+y^2)^2}\right] dx + c$$

This is an exact differential equation.

$$\phi = -2xy + \frac{y}{x^2+y^2} + C$$ **Ans.**

Which is the required function.

Example 24. *An electrostatic field in the xy-plane is given by the potential function $\phi = 3x^2y - y^3$, find the stream function.* *(R.G.P.V., Bhopal, III Semester, Dec. 2001)*

Solution. Let $\psi(x, y)$ be a stream function

We know that $d\psi = \dfrac{\partial \psi}{\partial x} dx + \dfrac{\partial \psi}{\partial y} dy = \left(-\dfrac{\partial \phi}{\partial y}\right) dx + \left(\dfrac{\partial \phi}{\partial x}\right) dy$ [C-R equations]

$$= \{-(3x^2 - 3y^2)\}\, dx + 6xy\, dy$$

$$= -3x^2\, dx + (3y^2\, dx + 6xy\, dy)$$

$$= -d(x^3) + 3d(xy^2)$$

$$\psi = \int -d(x^3) + 3d(xy^2) + c$$

$$\psi = -x^3 + 3xy^2 + c$$

ψ is the required stream function. **Ans.**

Example 25. *Find the imaginary part of the analytic function whose real part is* $x^3 - 3xy^2 + 3x^2 - 3y^2$. *(R.G.P.V., Bhopal, III Semester, Dec. 2008, 2005)*

Solution.

Let $u(x, y) = x^3 - 3xy^2 + 3x^2 - 3y^2$

$$\frac{\partial u}{\partial x} = 3x^2 - 3y^2 + 6x$$

$$\frac{\partial u}{\partial y} = -6xy - 6y$$

We know that

$$dv = \frac{\partial v}{\partial x} dx + \frac{\partial v}{\partial y} dy \quad \Rightarrow \quad dv = -\frac{\partial u}{\partial y} dx + \frac{\partial u}{\partial x} dy$$

$$\Rightarrow \qquad dv = (6xy + 6y)\, dx + (3x^2 - 3y^2 + 6x)\, dy$$

This is an exact differential equation.

$$v = \int (6xy + 6y)\, dx + \int -3y^2 dy + C$$

$$= 3x^2 y + 6xy - y^3 + C$$

Which is the required imaginary part. **Ans.**

Example 26. *If $u - v = (x - y)(x^2 + 4xy + y^2)$ and $f(z) = u + iv$ is an analytic function of $z = x + iy$, find $f(z)$ in terms of z.*

Solution. $u + iv = f(z) \quad \Rightarrow \quad iu - v = i f(z)$

Adding these, $(u - v) + i(u + v) = (1 + i) f(z)$

Let $U + iV = (1 + i) f(z)$ where $U = u - v$ and $V = u + v$

$$F(z) = (1 + i) f(z)$$

$$U = u - v = (x - y)(x^2 + 4xy + y^2)$$

$$= x^3 + 3x^2y - 3xy^2 - y^3$$

$$\frac{\partial U}{\partial x} = 3x^2 + 6xy - 3y^2,$$

$$\frac{\partial U}{\partial y} = 3x^2 - 6xy - 3y^2$$

We know that $dV = \frac{\partial V}{\partial x} \cdot dx + \frac{\partial V}{\partial y} dy = -\frac{\partial U}{\partial v} \cdot dx + \frac{\partial U}{\partial x} \cdot dy$ [C-R equations]

On putting the values of $\frac{\partial U}{\partial x}$ and $\frac{\partial U}{\partial y}$, we get

$$= (-3x^2 + 6xy + 3y^2)\, dx + (3x^2 + 6xy - 3y^2)\, .\, dy$$

Integrating, we get

$$V = \underset{(y \text{ as constant})}{\int (-3x^2 + 6xy + 3y^2) dx} + \underset{(\text{Ignoring terms of } x)}{\int (-3y^2) dy}$$

$$= -x^3 + 3x^2y + 3xy^2 - y^3 + c$$

$$F(z) = U + iV$$

$$= (x^3 + 3x^2y - 3xy^2 - y^3) + i(-x^3 + 3x^2y + 3xy^2 - y^3) + ic$$

$$= (1 - i)x^3 + (1 + i)3x^2y - (1 - i)3xy^2 - (1 + i)y^3 + ic$$

$$= (1 - i)x^3 + i(1 - i)3x^2y - (1 - i)3xy^2 - i(1 - i)y^3 + ic$$

$$= (1 - i)[x^3 + 3ix^2y - 3xy^2 - iy^3] + ic$$

$$= (1 - i)(x + iy)^3 + iC = (1 - i)z^3 + ic$$

$$(1 + i) f(z) = (1 - i) z^3 + ic, \qquad [F(z) = (1 + i) f(z)]$$

$$f(z) = \frac{1-i}{1+i} z^3 + \frac{ic}{1+i} = -\frac{i(1+i)}{(1+i)} z^3 + \frac{i(1-i)}{(1+i)(1-i)} c = -iz^3 + \frac{1+i}{2} c$$ **Ans.**

Example 27. *If $f(z) = u + iv$, is any analytic function of the complex variable z and $u - v = e^x(\cos y - \sin y)$, find $f(z)$ in terms of z*

Solution. $u + iv = f(z) \quad \Rightarrow \quad iu - v = i f(z)$

Adding, we have

$$u + iv + iu - v = f(z) + i f(z)$$

$$(u - v) + i(u + v) = (1 + i) f(z) = F(z) \text{ say}$$

Put $u - v = U$ and $u + v = V$, then $F(z) = U + iV$ is an analytic function.

Now $U = e^x (\cos y - \sin y)$

$$\therefore \quad \frac{\partial U}{\partial x} = e^x(\cos y - \sin y) \text{ and } \frac{\partial U}{\partial y} = e^x(-\sin y - \cos y)$$

We know that $dV = \frac{\partial V}{\partial x}dx + \frac{\partial V}{\partial y}dy = -\frac{\partial U}{\partial y}dx + \frac{\partial U}{\partial x}dy$

$$= e^x (\sin y + \cos y)\, dx + e^x (\cos y - \sin y) dy.$$

Integrating, we have

$$V = e^x (\sin y + \cos y) + c$$

$$F(z) = U + iV$$

$$= e^x (\cos y - \sin y) + ie^x (\sin y + \cos y) + ic$$

$$= e^x (\cos y + i \sin y) + ie^x (\cos y + i \sin y) + ic$$

$$= e^x . e^{iy} + ie^x e^{iy} + ic = e^{x+iy} + ie^{x+iy} + ic = e^z + ie^z + ic$$

$$(1 + i) f(z) = (1 + i) e^z + ic$$

$$f(z) = e^z + \frac{ic}{1+i}, \quad f(z) = e^z + c_1$$ **Ans.**

This is the required result.

Example 28. *If $f(z) = u + iv$ is an analytic function of $z = x + iy$ and*
$u - v = e^{-x} [(x - y) \sin y - (x + y) \cos y]$ (*U.P. III Semester, 2009-2010*)

Solution. We know that,

$$f(z) = u + iv \quad \ldots (1)$$

$$i f(z) = i u - v \quad \ldots (2)$$

$$F(z) = U + iV$$

$$U = u - v = e^{-x} [(x - y) \sin y - (x + y) \cos y]$$

$$\frac{\partial U}{\partial x} = -e^{-x} [(x - y) \sin y - (x + y) \cos y] + e^{-x} [\sin y - \cos y]$$

$$\frac{\partial U}{\partial y} = e^{-x} [(x - y) \cos y - \sin y - (x + y)(-\sin y) - \cos y)]$$

We know that,

$$dV = \frac{\partial V}{\partial x}dx + \frac{\partial V}{\partial y}dy = -\frac{\partial U}{\partial y}dx + \frac{\partial U}{\partial x}dy \quad \text{[C – R equations]}$$

$$= -e^{-x} [(x - y) \cos y - \sin y + (x + y) \sin y - \cos y]\, dx$$
$$- e^{-x} [(x - y) \sin y - (x + y) \cos y - \sin y + \cos y]\, dy$$

$$= -e^{-x} x \{(\cos y + \sin y)\, dx - e^{-x} (-y \cos y - \sin y + y \sin y - \cos y)\, dx$$
$$- e^{-x} [(x - y) \sin y - (x + y) \cos y - \sin y + \cos y]\, dy$$

$$V = (\cos y + \sin y)(x e^{-x} + e^{-x}) + e^{-x} (-y \cos y - \sin y + y \sin y - \cos y) + C$$

$$F(z) = U + iV$$

$$F(z) = e^{-x} [(x - y) \sin y - (x + y) \cos y] + i e^{-x} [x \cos y + \cos y + x \sin y + \sin y$$
$$- y \cos y - \sin y + y \sin y - \cos y] + iC$$

$$= e^{-x} [\{x \sin y - y \sin y - x \cos y - y \cos y\} + i \{x \cos y + x \sin y - y \cos y + y \sin y\}] + iC$$

$$= e^{-x} [(x + i y) \sin y - (x + i y) \cos y + (-y + i x) \sin y + (-y + i x) \cos y] + iC$$

$$= e^x [(x + iy) \sin y - (x + iy) \cos y + i(x + iy) \sin y + i(x + iy) \cos y] + iC$$
$$= e^x (x + iy) [\sin y - \cos y + i \sin y + i \cos y] + iC$$
$$= e^x (x + iy) [(1 + i) \sin y + i(1 + i) \cos y] + iC$$
$$(1 + i) f(z) = e^x (x + iy)(1 + i)(\sin y + i \cos y) + iC$$
$$f(z) = e^x (x + iy)(\sin y + i \cos y) + \frac{iC}{1+i}$$
$$= i z e^x (\cos y - i \sin y) + \frac{iC}{1+i}$$
$$= i z e^x e^{iy} = i z e^{(x+iy)} = i z e^z + \frac{iC}{1+i}$$ **Ans.**

Let $\phi_1 (x, y) = -e^x [(x - y) \sin y - (x + y) \cos y] + e^x [\sin y - \cos y]$

$\phi_1 (z, 0) = -e^z [z \sin 0 - z \cos 0] + e^z [\sin 0 + \cos 0]$

$= -e^z [z - 1]$

Let $\phi_2 (x, y) = e^x [(x - y) \cos y - \sin y + (x + y) \sin y - \cos y]$

$\phi_2 (z, 0) = e^z [(z) \cos 0 - \sin 0 + z \sin 0 - \cos 0]$

$= e^z [z - 1]$

$F(z) = U + iV$

$$F'(z) = \frac{\partial U}{\partial x} + i \frac{\partial V}{\partial x} = \frac{\partial U}{\partial x} - i \frac{\partial U}{\partial y} = f_1 (z, 0) - i f_2 (z, 0)$$
$$= e^z (z - 1) - i e^z (z - 1) = (1 - i) e^z (z - 1) = (1 - i) e^z (z - 1)$$
$$F(z) = (1 - i) \left[z \frac{e^{-z}}{-1} - \int \frac{e^{-z}}{-1} dz \right] + C = (1 - i)[-z e^{-z} - e^z] + C$$
$$(1 + i) f(z) = (-1 + i)(z + 1) e^z + C$$
$$f(z) = \frac{(-1+i)}{1+i} (z+1) e^{-z} + C = \frac{(-1+i)(1-i)}{(1+i)(1-i)} (z+1) e^{-z} + C$$
$$= i(z + 1) e^{-z} + C$$ **Ans.**

Example 29. *Let* $f(z) = u(r, \theta) + iv(r, \theta)$ *be an analytic function and* $u = -r^3 \sin 3\theta$, *then construct the corresponding analytic function* $f(z)$ *in terms of* z.

Solution. $u = -r^3 \sin 3\theta$

$$\frac{\partial u}{\partial r} = -3r^2 \sin 3\theta, \qquad \frac{\partial u}{\partial \theta} = -3r^3 \cos 3\theta$$

We know that $dv = \frac{\partial v}{\partial r} dr + \frac{\partial v}{\partial \theta} d\theta$

$$= \left(-\frac{1}{r} \frac{\partial u}{\partial \theta} \right) dr + \left(r \frac{\partial u}{\partial r} \right) d\theta$$

$\left(C - R \text{ equations} \quad \frac{\partial u}{\partial r} = \frac{1}{r} \frac{\partial v}{\partial \theta} \quad \frac{\partial u}{\partial \theta} = -r \frac{\partial v}{\partial r} \right)$

$$= -\frac{1}{r} (-3r^3 \cos 3\theta) dr + r(-3r^2 \sin 3\theta) d\theta$$
$$= 3r^2 \cos 3\theta \cdot dr - 3r^3 \sin 3\theta \, d\theta$$
$$v = \int \overline{(3r^2 \cos 3\theta) dr} - c = r^3 \cos 3\theta + c$$
$$f(z) = u + iv = -r^3 \sin 3\theta + ir^3 \cos 3\theta + ic = ir^3 (\cos 3\theta + i \sin 3\theta) + ic$$
$$= ir^3 e^{i3\theta} + ic = i(re^{i\theta})^3 + ic = iz^3 + ic$$ **Ans.**

This is the required analytic function.

Example 30. *Find analytic function $f(z) = u(r,\theta) + iv(r,\theta)$ such that*

$$v(r,\theta) = r^2 \cos 2\theta - r\cos\theta + 2.$$

Solution. We have, $\quad v = r^2 \cos 2\theta - r\cos\theta + 2 \quad \text{... (1)}$

Differentiating (1), we get

$$\frac{\partial v}{\partial \theta} = -2r^2 \sin 2\theta + r\sin\theta \quad \text{... (2)}$$

$$\frac{\partial v}{\partial r} = 2r\cos 2\theta - \cos\theta \quad \text{... (3)}$$

Using $C - R$ equations in polar coordinates, we get

$$r\frac{\partial u}{\partial r} = \frac{\partial v}{\partial \theta} = -2r^2 \sin 2\theta + r\sin\theta \quad \text{[From (2)]}$$

$$\Rightarrow \quad \frac{\partial u}{\partial r} = -2r\sin 2\theta + \sin\theta \quad \text{... (4)}$$

$$-\frac{1}{r}\frac{\partial u}{\partial \theta} = \frac{\partial v}{\partial r} = 2r\cos 2\theta - \cos\theta \quad \text{[From (3)]}$$

$$\Rightarrow \quad \frac{\partial u}{\partial \theta} = -2r^2 \cos 2\theta + r\cos\theta \quad \text{... (5)}$$

By total differentiation formula

$$du = \frac{\partial u}{\partial r}dr + \frac{\partial u}{\partial \theta}d\theta = (-2r\sin 2\theta + \sin\theta)dr + (-2r^2\cos 2\theta + r\cos\theta)d\theta$$

$$= -[(2r\,dr)\sin 2\theta + r^2(2\cos 2\theta\, d\theta)] + [\sin\theta \cdot dr + r(\cos\theta\, d\theta)]$$

$$= -[(2r\,dr)\sin 2\theta - \sin\theta dr] + [-r^2 2\cos 2\theta\, d\theta + r\cos\theta\, d\theta]$$

$$= -d(r^2 \sin 2\theta) + d(r\sin\theta) \quad \text{(Exact differential equation)}$$

Integrating, we get

$$u = -r^2 \sin 2\theta + r\sin\theta + c$$

Hence, $\quad f(z) = u + iv$

$$= (-r^2 \sin 2\theta + r\sin\theta + c) + i(r^2\cos 2\theta - r\cos\theta + 2)$$

$$= ir^2(\cos 2\theta + i\sin 2\theta) - ir(\cos\theta + i\sin\theta) + 2i + c$$

$$= ir^2 e^{2i\theta} - ir e^{i\theta} + 2i + c \;= i(r^2 e^{2i\theta} - r e^{i\theta}) + 2i + c. \qquad \textbf{Ans.}$$

This is the required analytic function.

Example 31. *Deduce the following with the polar form of Cauchy-Riemann equations :*

(*a*) $\dfrac{\partial^2 u}{\partial r^2} + \dfrac{1}{r}\dfrac{\partial u}{\partial r} + \dfrac{1}{r^2}\dfrac{\partial^2 u}{\partial \theta^2} = 0 \qquad$ (*b*) $f'(z) = \dfrac{r}{z}\left[\dfrac{\partial u}{\partial r} + i\dfrac{\partial v}{\partial r}\right]$

Solution. We know that polar form of C-R equations are

$$\frac{\partial u}{\partial r} = \frac{1}{r}\frac{\partial v}{\partial \theta} \quad \text{... (1)}$$

$$\frac{\partial u}{\partial \theta} = -r\frac{\partial v}{\partial r} \quad \text{... (2)}$$

(*a*) Differentiating (1) partially w.r.t. r., we get

$$\frac{\partial^2 u}{\partial r^2} = -\frac{1}{r^2}\frac{\partial v}{\partial \theta} + \frac{1}{r}\frac{\partial^2 v}{\partial r\partial \theta} \quad \text{... (3)}$$

Differentiating (2) partially w.r.t. θ, we have

$$\frac{\partial^2 u}{\partial \theta^2} = -r\frac{\partial^2 v}{\partial \theta\, \partial r} \qquad ...(4)$$

Thus using (1), (3) and (4), we get

$$\frac{\partial^2 u}{\partial r^2} + \frac{1}{r}\frac{\partial u}{\partial r} + \frac{1}{r^2}\frac{\partial^2 u}{\partial \theta^2} = -\frac{1}{r^2}\frac{\partial v}{\partial \theta} + \frac{1}{r}\frac{\partial^2 v}{\partial r\, \partial \theta} + \frac{1}{r}\left(\frac{1}{r}\frac{\partial v}{\partial \theta}\right) + \frac{1}{r^2}\left(-r\frac{\partial^2 v}{\partial \theta\, \partial r}\right) = 0 \qquad \left[\frac{\partial^2 v}{\partial \theta \partial r} = \frac{\partial^2 v}{\partial \theta\, \partial r}\right]$$

Proved.

(*b*) Now,

$$r\left(\frac{\partial u}{\partial r} + i\frac{\partial v}{\partial r}\right) = r\left[\left(\frac{\partial u}{\partial x}\frac{\partial x}{\partial r} + \frac{\partial u}{\partial y}\frac{\partial y}{\partial r}\right) + i\left(\frac{\partial v}{\partial x}\frac{\partial x}{\partial r} + \frac{\partial v}{\partial y}\frac{\partial y}{\partial r}\right)\right]$$

$$= r\left[\left(\frac{\partial u}{\partial x}\cos\theta + \frac{\partial u}{\partial y}\sin\theta\right) + i\left(\frac{\partial v}{\partial x}\cos\theta + \frac{\partial v}{\partial y}\sin\theta\right)\right]$$

$$= r\cos\theta\left(\frac{\partial u}{\partial x} + i\frac{\partial v}{\partial x}\right) + r\sin\theta\left(\frac{\partial u}{\partial y} + i\frac{\partial v}{\partial y}\right)$$

$$= x\left(\frac{\partial u}{\partial x} + i\frac{\partial v}{\partial x}\right) + iy\left(\frac{\partial v}{\partial y} - i\frac{\partial u}{\partial y}\right) \qquad \text{(By C-R equations)}$$

$$= x f'(z) + iy f'(z) = (x + iy) f'(z) = z f'(z).$$

$$\therefore \qquad f'(z) = \frac{r}{z}\left(\frac{\partial u}{\partial r} + i\frac{\partial v}{\partial r}\right)$$

Proved.

27.17 MILNE THOMSON METHOD (To construct an Analytic function)

By this method $f(z)$ is directly constructed without finding v and the method is given below:

Since $\quad z = x + iy$ and $\bar{z} = x - iy$

$$\therefore \qquad x = \frac{z + \bar{z}}{2}, \; y = \frac{z - \bar{z}}{2i}$$

$$f(z) \equiv u(x, y) + iv(x, y) \qquad ... (1)$$

$$f(z) \equiv u\left(\frac{z+\bar{z}}{2}, \frac{z-\bar{z}}{2i}\right) + iv\left(\frac{z+\bar{z}}{2}, \frac{z-\bar{z}}{2i}\right)$$

This relation can be regarded as a formal identity in two independent variables z and $\bar{z}$. Replacing $\bar{z}$ by z, we get

$$f(z) \equiv u(z, 0) + iv(z, 0)$$

Which can be obtained by replacing x by z and y by 0 in (1)

Case I. If *u* is given

We have $\quad f(z) = u + iv$

$$\therefore \qquad f'(z) = \frac{\partial u}{\partial x} + i\frac{\partial v}{\partial x}, \qquad f'(z) = \frac{\partial u}{\partial x} - i\frac{\partial u}{\partial y} \qquad (C - R \text{ equations})$$

If we write $\quad \dfrac{\partial u}{\partial x} = \phi_1(x, y), \qquad \dfrac{\partial u}{\partial y} = \phi_2(x, y)$

$$f'(z) = \phi_1(x, y) - i\phi_2(x, y) \;\text{ or }\; f'(z) = \phi_1(z, 0) - i\phi_2(z, 0)$$

On integrating $\quad f(z) = \displaystyle\int \phi_1(z, 0)\, dz - i\int \phi_2(z, 0) dz + c$

Case II. If *v* is given

$$f(z) = u + iv$$

$$f'(z) = \frac{\partial u}{\partial x} + i\frac{\partial v}{\partial x} = \frac{\partial v}{\partial y} + i\frac{\partial v}{\partial x} = \psi_1(x, y) + i\,\psi_2(x, y)$$

when $\qquad \psi_1(x, y) = \dfrac{\partial v}{\partial y}, \qquad \psi_2(x, y) = \dfrac{\partial v}{\partial x}.$

$$f(z) = \int \psi_1(z, 0)\, dz + i \int \psi_2(z, 0)\, dz + c$$

27.18 WORKING RULE: TO CONSTRUCT AN ANALYTIC FUNCTION BY MILNE THOMSON METHOD

Case I. When u is given

Step 1. Find $\dfrac{\partial u}{\partial x}$ and equate it to $\phi_1(x, y)$.

Step 2. Find $\dfrac{\partial u}{\partial y}$ and equate it to $\phi_2(x, y)$.

Step 3. Replace x by z and y by 0 in $\phi_1(x, y)$ to get $\phi_1(z, 0)$.

Step 4. Replace x by z and y by 0 in $\phi_2(x, y)$ to get $\phi_2(z, 0)$.

Step 5. Find $f(z)$ by the formula $f(z) = \int \{\phi_1(z, 0) - i\phi_2(z, 0)\}\, dz + c$

Case II. When v is given

Step 1. Find $\dfrac{\partial v}{\partial x}$ and equate it to $\psi_2(x, y)$.

Step 2. Find $\dfrac{\partial v}{\partial y}$ and equate it to $\psi_1(x, y)$.

Step 3. Replace x by z and y by 0 in $\psi_1(x, y)$ to get $\psi_1(z, 0)$.

Step 4. Replace x by z and y by 0 in $\psi_2(x, y)$ to get $\psi_2(z, 0)$.

Step 5. Find $f(z)$ by the formula

$$f(z) = \int \{\psi_1(z, 0) + i\psi_2(z, 0)\}\, dz + c$$

Case III. When $u - v$ *is given.*

We know that $\qquad f(z) = u + iv \qquad$ ...(1)

$\qquad if(z) = iu - v \qquad$ ...(2) [Multiplying by i]

Adding (1) and (2), we get

$$(1 + i) f(z) = (u - v) + i(u + v)$$

$\Rightarrow \qquad F(z) = U + iV$

where $\qquad F(z) = (1 + i) f(z) \qquad$ ...(3) $\begin{bmatrix} U = u - v \\ V = u + v \end{bmatrix}$

Here, $\qquad U = (u - v)$ is given

Find out F(z) by the method described in case I, then substitute the value of F(z) in (3), we get

$$f(z) = \frac{F(z)}{1 + i}$$

Case IV. When $u + v$ is given.

We know that $\qquad f(z) = u + iv \qquad$ ...(1)

$\qquad if(z) = iu - v \qquad$ [Multiplying by i]...(2)

Adding (1) and (2), we get

$$(1 + i) f(z) = (u - v) + i(u + v)$$

$\Rightarrow \qquad F(z) = U + iV$

where $\qquad F(z) = (1 + i) f(z) \qquad$ ...(3) $\begin{bmatrix} U = u - v \\ V = u + v \end{bmatrix}$

Here, $\qquad V = (u + v)$ is given

Find out F(z) by the method described in case II, then substitute the value of F(z) in (3), we get

$$f(z) = \frac{F(z)}{1+i}$$

Example 32. *If $u = x^2 - y^2$, find a corresponding analytic function.*

Solution. $\frac{\partial u}{\partial x} = 2x = \phi_1(x, y), \quad \frac{\partial u}{\partial y} = -2y = \phi_2(x, y)$

On replacing x by z and y by 0, we have

$\phi_1(z, 0) = 2z$ and $\phi_2(z, 0) = 0 \quad f(z) = \int [\phi_1(z,0) - i\phi_2(z,0)]dz + C$

$$= \int [2z - i(0)]dz + c = \int 2z\,dz + c = z^2 + C$$ **Ans.**

This is the required analytic function.

Example 33. *Show that the function $u = e^{-2xy}\sin(x^2 - y^2)$ is harmonic. Find the conjugate function v and express u + iv as an analytic function of z.*

(R.G.P.V. Bhopal, III Semester, June, 2007, Dec. 2006)

Solution. We have,

$$u = e^{-2xy}\sin(x^2 - y^2) \qquad \dots (1)$$

Differentiating (1), w.r.t. x, we get

$$\frac{\partial u}{\partial x} = 2x\,e^{-2xy}\cos(x^2 - y^2) - 2ye^{-2xy}\sin(x^2 - y^2)$$

$$\Rightarrow \quad \frac{\partial u}{\partial x} = e^{-2xy}[2x\cos(x^2 - y^2) - 2y\sin(x^2 - y^2)] = \phi_1(x, y) \qquad \dots (2)$$

$$\phi_1(z, 0) = 2z\cos z^2$$

Differentiating (1), w.r.t. y, we get

$$\frac{\partial u}{\partial y} = -2y\,e^{-2xy}\cos(x^2 - y^2) - 2x\,e^{-2xy}\sin(x^2 - y^2)$$

$$\Rightarrow \quad \frac{\partial u}{\partial y} = e^{-2xy}[-2y\cos(x^2 - y^2) - 2x\sin(x^2 - y^2)] = \phi_2(x, y) \qquad \dots (3)$$

$$\phi_2(z, 0) = -2z\sin z^2$$

Differentiating (2), w.r.t, 'x', we get

$$\frac{\partial^2 u}{\partial x^2} = -2ye^{-2xy}[2x\cos(x^2 - y^2) - 2y\sin(x^2 - y^2)]$$
$$+ e^{-2xy}[2\cos(x^2 - y^2) + 2x(2x)\{-\sin(x^2 - y^2)\} - 2y(2x)\cos(x^2 - y^2)]$$

$$\Rightarrow \quad \frac{\partial^2 u}{\partial x^2} = e^{-2xy}[-4xy\cos(x^2 - y^2) + 4y^2\sin(x^2 - y^2) + 2\cos(x^2 - y^2)$$
$$-4x^2\sin(x^2 - y^2) - 4xy\cos(x^2 - y^2)]$$

$$= e^{-2xy}[-8xy\cos(x^2 - y^2) + 4y^2\sin(x^2 - y^2) + 2\cos(x^2 - y^2) - 4x^2\sin(x^2 - y^2)] \qquad \dots (4)$$

Differentiating (3), w.r.t. 'y', we get

$$\frac{\partial^2 u}{\partial y^2} = -2xe^{-2xy}[-2y\cos(x^2 - y^2) - 2x\sin(x^2 - y^2)]$$
$$+ e^{-2xy}\left[-2\cos(x^2 - y^2) + 2y(-2y)\sin(x^2 - y^2) - 2x(-2y)\cos(x^2 - y^2)\right]$$

$$\Rightarrow \quad \frac{\partial^2 u}{\partial y^2} = e^{-2xy}[4xy\cos(x^2 - y^2) + 4x^2\sin(x^2 - y^2) - 2\cos(x^2 - y^2)$$
$$-4y^2\sin(x^2 - y^2) + 4xy\cos(x^2 - y^2)]$$

$$\frac{\partial^2 u}{\partial y^2} = e^{-2xy}\,[8xy\cos(x^2-y^2) + 4x^2\sin(x^2-y^2) - 2\cos(x^2-y^2) - 4y^2\sin(x^2-y^2)] \quad \ldots (5)$$

Adding (4) and (5), we get $\frac{\partial^2 u}{\partial x^2} + \frac{\partial^2 u}{\partial y^2} = 0$

Which proves that u is harmonic.

Now we have to express $u + iv$ as a function of z

$$f(z) = \int [\phi_1(z, 0) - i\,\phi_2(z, 0)]\, dz = \int [2z\cos z^2 - i(-2z\sin z^2)]\, dz$$

$$= \sin z^2 - i\cos z^2 + C = -i(\cos z^2 + i\sin z^2) + C = -i\,e^{iz^2} + C$$ **Ans.**

Example 34. *If* $u = \frac{\sin 2x}{\cosh 2y + \cos 2x}$, *find* $f(z)$.

(R.G.P.V., Bhopal, III Semester, Dec. 2003)

Solution. $\frac{\partial u}{\partial x} = \frac{(\cosh 2y + \cos 2x)2\cos 2x - \sin 2x(-2\sin 2x)}{(\cosh 2y + \cos 2x)^2}$

$$= \frac{2\cosh 2y\cos 2x + 2(\cos^2 2x + \sin^2 2x)}{(\cosh 2y + \cos 2x)^2} = \frac{2\cosh 2y\cos 2x + 2}{(\cosh 2y + \cos 2x)^2} = \phi_1(x, y) \quad \ldots (1)$$

On putting $x = z$ and $y = 0$ in (1), we get

$$\phi_1(z, 0) = \frac{2\cos 2z + 2}{(1 + \cos 2z)^2}$$

$$\frac{\partial u}{\partial y} = \frac{-\sin 2x(2\sinh 2y)}{(\cosh 2y + \cos 2x)^2} = \frac{-2\sin 2x\sinh 2y}{(\cosh 2y + \cos 2x)^2} = \phi_2(x, y) \quad \ldots (2)$$

On putting $x = z$, $y = 0$ in (2), we get

$\phi_2(z, 0) = 0$

$$f(z) = \int [\phi_1(z,0) - i\phi_2(z,0)]\, dz + C = \int \frac{(2\cos 2z + 2)}{(1 + \cos 2z)^2}\, dz + C = 2\int \frac{1}{1 + \cos 2z}\, dz + C$$

$$= 2\int \frac{1}{2\cos^2 z}\, dz + C = \int \sec^2 z\, dz + C = \tan z + C$$ **Ans.**

which is the required function.

Example 35. *Find the analytic function* $f(z) = u + iv$, *given that* $v = e^x(x\sin y + y\cos y)$.

Solution. $\frac{\partial v}{\partial x} = e^x(x\sin y + y\cos y) + e^x\sin y = \psi_2(x, y) \quad \Rightarrow \quad \psi_2(z, 0) = 0$

$$\frac{\partial v}{\partial y} = e^x(x\cos y + \cos y - y\sin y) = \phi_1(x, y) \quad \Rightarrow \quad \psi_1(z, 0) = ze^z + e^z$$

$$f(z) = \int [\psi_1(z,0) + i\psi_2(z,0)]\, dz + C$$

$$= \int [e^z(z+1) + i(0)]\, dz + c = (z+1)e^z - \int e^z\, dz + C$$

$$= (z+1)e^z - e^z + c = z\,e^z + C$$ **Ans.**

Which is the required function.

Example 36. *Show that* $e^x(x\cos y - y\sin y)$ *is a harmonic function. Find the analytic function for which* $e^x(x\cos y - y\sin y)$ *is imaginary part.*

(U.P., III Semester, June 2009, R.G.P.V., Bhopal, III Semester, June2004)

Solution. Here $v = e^x(x\cos y - y\sin y)$

Differentiating partially w.r.t. x and y, we have

$$\frac{\partial v}{\partial x} = e^x\ (x\cos y - y\sin y) + e^x\cos y = \psi_2(x, y), \quad \text{(say)} \quad \ldots (1)$$

$$\frac{\partial v}{\partial y} = e^x\ (-x\sin y - y\cos y - \sin y) = \psi_1\ (x,\ y) \quad \text{(say)} \quad \ldots (2)$$

$$\frac{\partial^2 v}{\partial x^2} = e^x\ (x\cos y - y\sin y) + e^x\cos y + e^x\cos y$$

$$= e^x\ (x\cos y - y\sin y + 2\cos y) \quad \ldots (3)$$

and $$\frac{\partial^2 v}{\partial y^2} = e^x\ (-x\cos y + y\sin y - 2\cos y) \quad \ldots (4)$$

Adding equations (3) and (4), we have

$$\frac{\partial^2 v}{\partial x^2} + \frac{\partial^2 v}{\partial y^2} = 0 \quad \Rightarrow v \text{ is a harmonic function.}$$

Now putting x = z, y = 0 in (1) and (2), we get

$$\psi_2\ (z, 0) = ze^z + e^z \qquad \psi_1\ (z, 0) = 0$$

Hence by Milne-Thomson method, we have

$$f\ (z) = \int [\psi_1\ (z,\ 0) + i\,\psi_2\ (z,\ 0)]\ dz + C$$

$$= \int [0 + i\,(ze^z + e^z)]\ dz + C = i\,(ze^z - e^z + e^z) + C = i\,z\,e^z + C.$$

This is the required analytic function. **Ans.**

Example 37. *If $u - v = (x - y)\ (x^2 + 4\ xy + y^2)$ and $f\ (z) = u + iv$ is an analytic function of $z = x + iy$, find $f\ (z)$ in terms of z by Milne Thomson method.*

Solution. We know that

$$f\ (z) = u + iv \quad \ldots (1)$$

$$i\,f\ (z) = i\ u - v \quad \ldots (2)$$

Adding (1) and (2), we get

$$(1 + i)\,f\ (z) = (u - v) + i\ (u + v)$$

Let $$F\ (z) = U + i\ V$$

$$U = u - v = (x - y)\ (x^2 + 4xy + y^2)$$

$$\frac{\partial U}{\partial x} = (x^2 + 4xy + y^2) + (x - y)\ (2x + 4y)$$

$$= x^2 + 4xy + y^2 + 2x^2 + 4xy - 2xy - 4y^2 = 3x^2 + 6xy - 3y^2$$

$$\phi_1\ (x,\ y) = 3x^2 + 6xy - 3y^2$$

$$\phi_1\ (z,\ 0) = 3z^2$$

$$\frac{\partial U}{\partial y} = -\ (x^2 + 4xy + y^2) + (x - y)\ (4x + 2y)$$

$$= -\ x^2 - 4xy - y^2 + 4x^2 + 2xy - 4xy - 2y^2 = 3x^2 - 6xy - 3y^2$$

$$\phi_2\ (x,\ y) = 3x^2 - 6xy - 3y^2$$

$$\phi_2\ (z,\ 0) = 3z^2$$

$$F\ (z) = \text{U} + \text{iV}$$

$$F'\ (z) = \frac{\partial U}{\partial x} + i\,\frac{\partial V}{\partial x} = \frac{\partial U}{\partial x} - i\,\frac{\partial U}{\partial y} = \phi,\ (z,\ 0) - i\ \phi_2\ (z,\ 0) = 3z^2 - i\ 3z^2$$

$$= 3(1-i)z^2$$

$$F(z) = (1-i)z^3 + C$$

$$(1+i)f(z) = (1-i)z^3 + C$$

$$f(z) = \frac{1-i}{1+i}z^3 + \frac{C}{1+i} = \frac{(1-i)(1-i)}{(1+i)(1-i)}z^3 + C_1$$

$$= \frac{1-2i+(-i)^2}{1+1}z^3 + C_1 = \frac{1-2i-1}{2}z^3 + C_1 = -iz^3 + C_1 \qquad \textbf{Ans.}$$

Note: This example has already been solved on page 694 as Example 26.

Example 38. *If* $f(z) = u + iv$ *is an analytic function of* z *and* $u - v = \dfrac{\cos x + \sin x - e^{-y}}{2\cos x - 2\cosh y}$, *prove that*

$$f(z) = \frac{1}{2}\left[1 - \cot\frac{z}{2}\right] \text{ when } f\left(\frac{\pi}{2}\right) = 0. \quad \textit{(R.G.P.V. Bhopal, III Semester, Dec. 2007)}$$

Solution. We know that $\quad f(z) = u + iv$

$\therefore \quad i f(z) = iu - v \qquad$ [Multiplying by i]

On adding, we get $\quad (1+i)f(z) = (u-v) + i(u+v)$

$\Rightarrow \quad F(z) = U + iV \qquad \begin{bmatrix} U = u - v \\ V = u + v \end{bmatrix}$

$\Rightarrow \qquad (1+i)f(z) = F(z)$

We have, $\quad U = u - v = \dfrac{\cos x + \sin x - e^{-y}}{2\cos x - 2\cosh y}$

$$\Rightarrow \quad U = \frac{\cos x + \sin x - \cosh y + \sinh y}{2\cos x - 2\cosh y} \qquad [\because e^{-y} = \cosh y - \sinh y]$$

$$= \frac{\cos x - \cosh y}{2(\cos x - \cosh y)} + \frac{\sin x + \sinh y}{2(\cos x - \cosh y)} = \frac{1}{2} + \frac{\sin x + \sinh y}{2(\cos x - \cosh y)} \qquad ...(1)$$

Differentiating (1) w.r.t. x partially, we get

$$\frac{\partial U}{\partial x} = \frac{1}{2}\left[\frac{(\cos x - \cosh y)\cos x - (\sin x + \sinh y)(-\sin x)}{(\cos x - \cosh y)^2}\right]$$

$$= \frac{1}{2}\left[\frac{(\cos^2 x + \sin^2 x - \cosh y \cos x + \sinh y \sin x)}{(\cos x - \cosh y)^2}\right] \qquad \begin{bmatrix} (1+i)f(z) = F(z) \\ u - v = U \\ u + v = V \end{bmatrix}$$

$$f(x, y) = \frac{1}{2}\left[\frac{1 - \cosh y \cos x + \sinh y \sin x}{(\cos x - \cosh y)^2}\right] = \frac{1}{2}\,\frac{1 - \cos iy \cos x - i \sin iy \sin x}{(\cos x - \cosh y)^2}$$

Replacing x by z and y by 0 in (2), we get

$$\phi_1(z, 0) = \frac{1}{2}\left[\frac{1 - \cos z}{(\cos z - 1)^2}\right] = \frac{1}{2}\,\frac{-(1 - \cos z)}{(1 - \cos z)^2} = \frac{1}{2(1 - \cos z)}$$

Differentiating (1) partially w.r.t. y, we get

$$\frac{\partial U}{\partial y} = \frac{1}{2}\left[\frac{(\cos x - \cosh y).\cosh y - (\sin x + \sinh y)(-\sinh y)}{(\cos x - \cosh y)^2}\right]$$

$$= \frac{1}{2}\left[\frac{(\cos x \cosh y) + \sin x \sinh y - (\cosh^2 y - \sinh^2 y)}{(\cos x - \cosh y)^2}\right]$$

$$\phi_2(x, y) = \frac{1}{2}\left[\frac{\cos x \cosh y + \sin x \sinh y - 1}{(\cos x - \cosh y)^2}\right] \quad ...(3)$$

Replacing x by z and y by 0 in (3), we have

$$\phi_2(z, 0) = \frac{1}{2}\left[\frac{\cos z - 1}{(\cos z - 1)^2}\right] = \frac{1}{2}\cdot\left(\frac{-1}{1-\cos z}\right)$$

$$F'(z) = \frac{\partial U}{\partial x} + i\frac{\partial V}{\partial x} = \frac{\partial U}{\partial x} - i\frac{\partial U}{\partial y} \quad \text{[C–R equations]}$$

$$= \phi_1(z, 0) - i\,\phi_2(z, 0)$$

By Milne Thomson Method,

$$F(z) = \int[\phi_1(z, 0) - i\phi_2(z, 0)]\,dz$$

$$= \int\left[\frac{1}{2}\cdot\frac{1}{(1-\cos z)} + \frac{i}{2}\cdot\frac{1}{1-\cos z}\right]dz$$

$$= \frac{1+i}{2}\int\frac{1}{2\sin^2 z/2}\,dz = \frac{1+i}{4}\int \text{cosec}^2(z/2)\,dz$$

$$= \left(\frac{1+i}{4}\right)\cdot\frac{(-\cot z/2)}{\left(\frac{1}{2}\right)} + C = -\left(\frac{1+i}{2}\right)\cot\frac{z}{2} + C \qquad F(z) = (1+i)f(z)$$

$$\Rightarrow \quad (1+i)f(z) = -\left(\frac{1+i}{2}\right)\cot\frac{z}{2} + C \quad \Rightarrow \quad f(z) = -\frac{1}{2}\cot\frac{z}{2} + \frac{C}{1+i} \quad ...(4)$$

On putting $z = \frac{\pi}{2}$ in (4), we get

$$f\left(\frac{\pi}{2}\right) = -\frac{1}{2}\cot\frac{\pi}{4} + \frac{C}{1+i}$$

$$0 = -\frac{1}{2} + \frac{C}{1+i} \Rightarrow \frac{C}{1+i} = \frac{1}{2} \qquad \left[f\left(\frac{\pi}{2}\right) = 0, \text{ given}\right]$$

On putting the value of $\frac{C}{1+i}$ in (4), we get

$$f(z) = -\frac{1}{2}\cot\frac{z}{2} + \frac{1}{2}$$

Hence, $\quad f(z) = \frac{1}{2}\left(1 - \cot\frac{z}{2}\right)$, when $f\left(\frac{\pi}{2}\right) = 0$. **Proved.**

EXERCISE 27.2

Show that the following functions are harmonic and determine the conjugate functions.

1. $u = 2x(1-y)$ **Ans.** $v = x^2 - y^2 + 2y + C$ **2.** $u = 2x - x^3 + 3xy + 3xy^2$

Ans. $v = -\frac{3}{2}(x^2 - y^2) + \frac{3}{2}y^2 + y^3 + 2y + C$

Determine the analytic function, whose real part is

3. $\log\sqrt{x^2+y^2}$ **Ans.** $\log z + C$ **4.** $\cos x \cosh y$ **Ans.** $\cos z + c$

5. $e^{-x}(\cos y + \sin y)$ *(AMIETE, June 2010)* **Ans.** $e^{-z}(1 + i + C$

6. $e^{2x}(x\cos 2y - y\sin 2y)$ **Ans.** $ze^{2z} + iC$ **7.** $e^{-x}(x\cos y + y\sin y)$ and $f(0) = i$. **Ans.** $ze^{-z} + i$

Determine the analytic function, whose imaginary part is

8. $v = \log(x^2+y^2) + x - 2y$ *(GBTU, III Sem. April 2012)* **Ans.** $2i\log z - (2-i)z + C$

9. $v = \sinh x \cos y$ **Ans.** $\sin iz + C$

10. $v = \left(r - \frac{1}{r}\right)\sin\theta$ **Ans.** $z + \frac{1}{z} + C$

11. If $f(z) = u + iv$ is an analytic function of $z = x + iy$ and $u - v = \frac{e^y - \cos x + \sin x}{\cosh y - \cos x}$, find $f(z)$ subject to the condition that $f\left(\frac{\pi}{2}\right) = \frac{3-i}{2}$. **Ans.** $f(z) = \cot\frac{z}{2} + \frac{1-i}{2}$

12. Find an analytic function $f(z) = u(r, \theta) + iv(r, \theta)$ such that $V(r,\theta) = r^2\cos 2\theta - r\cos\theta + 2$.
Ans. $i\,[z^2 - z + 2]$

13. Show that the function $u = x^2 - y^2 - 2xy - 2x - y - 1$ is harmonic. Find the conjugate harmonic function v and express $u + iv$ as a function of z where $z = x + iy$.
Ans. $(1 + i)\,z^2 + (-2 + i)\,z - 1$

14. Construct an analytic function of the form $f(z) = u + iv$, where v is $\tan^{-1}(y/x)$, $x \neq 0$, $y \neq 0$.
Ans. $\log cz$

15. Show that the function $u = e^{-2xy}\sin(x^2 - y^2)$ is harmonic. Find the conjugate function v and express $u + iv$ as an analytic function of z. **Ans.** $v = e^{-2xy}\cos(x^2 - y^2) + C$
$f(z) = -\,ie^{iz^2} + C_1$

16. Show that the function $v(x, y) = e^x \sin y$ is harmonic. Find its conjugate harmonic function $u(x, y)$ and the corresponding analytic function $f(z)$. *(AMIETE, June 2009)*
Ans. $u = e^x\cos y$, $f(z) = e^z$

Choose the correct answer:

21. The harmonic conjugate of $u = x^3 - 3xy^2$ is
(*a*) $y^3 - 3xy^2$ (*b*) $3x^2 y - y^3$ (*c*) $3xy^2 - y^3$ (*d*) $3xy^2 - x^3$ *(AMIETE, June 2010)* **Ans.** (*b*)

27.19 PARTIAL DIFFERENTIATION OF FUNCTION OF COMPLEX VARIABLE

Example 39. *Prove that*

$$4\frac{\partial^2}{\partial z\,\partial \bar{z}} = \left[\frac{\partial^2}{\partial x^2} + \frac{\partial^2}{\partial y^2}\right]$$

Solution. We know that

$$x + iy = z \quad \ldots (1), \qquad x - iy = \bar{z} \quad \ldots (2)$$

From (1) and (2), we get

$$x = \frac{1}{2}(z + \bar{z}), \qquad y = \frac{-i}{2}(z - \bar{z})$$

$$\Rightarrow \quad \frac{\partial x}{\partial z} = \frac{1}{2}, \qquad \frac{\partial y}{\partial z} = -\frac{i}{2}$$

$$\text{and} \quad \frac{\partial x}{\partial \bar{z}} = \frac{1}{2}, \qquad \frac{\partial y}{\partial \bar{z}} = \frac{i}{2}$$

We know that,

$$\frac{\partial}{\partial z} = \frac{\partial}{\partial x}\left(\frac{\partial x}{\partial z}\right) + \frac{\partial}{\partial y}\left(\frac{\partial y}{\partial z}\right) = \frac{\partial}{\partial x}\left(\frac{1}{2}\right) + \frac{\partial}{\partial y}\left(\frac{-i}{2}\right) = \frac{1}{2}\left(\frac{\partial}{\partial x} - i\frac{\partial}{\partial y}\right) \quad \ldots(3)$$

$$\frac{\partial}{\partial \bar{z}} = \frac{\partial}{\partial x}\left(\frac{\partial x}{\partial \bar{z}}\right) + \frac{\partial}{\partial y}\left(\frac{\partial y}{\partial \bar{z}}\right) = \frac{\partial}{\partial x}\left(\frac{1}{2}\right) + \frac{\partial}{\partial y}\left(\frac{i}{2}\right) = \frac{1}{2}\left(\frac{\partial}{\partial x} + \frac{i\partial}{\partial y}\right) \quad \ldots (4)$$

From (3) and (4), we get

$$\frac{\partial^2}{\partial z\,\partial \bar{z}} = \frac{\partial}{\partial z}\left(\frac{\partial}{\partial \bar{z}}\right) = \frac{1}{4}\left(\frac{\partial}{\partial x} - i\frac{\partial}{\partial y}\right)\left(\frac{\partial}{\partial x} + i\frac{\partial}{\partial y}\right)$$

$$= \frac{1}{4}\left(\frac{\partial^2}{\partial x^2} + \frac{\partial^2}{\partial y^2} + i\frac{\partial^2}{\partial x \partial y} - i\frac{\partial^2}{\partial x \partial y}\right) = \frac{1}{4}\left(\frac{\partial^2}{\partial x^2} + \frac{\partial^2}{\partial y^2}\right)$$

$$\Rightarrow \qquad 4\frac{\partial^2}{\partial z\, \partial \bar{z}} = \left[\frac{\partial^2}{\partial x^2} + \frac{\partial^2}{\partial y^2}\right]$$ **Proved.**

Example 40. *If $f(z)$ is a harmonic function of z, show that*

$$\left\{\frac{\partial}{\partial x}|f(z)|\right\}^2 + \left\{\frac{\partial}{\partial y}|f(z)|\right\}^2 = |f'(z)|^2$$ (*U.P. III Semester, June 2009*)

Solution. Since $f(z) = u(x, y) + i\,v(x, y)$

so $$|f(z)| = \sqrt{u^2 + v^2} \qquad \text{... (1)}$$

Differentiating (1) partially w.r.t. 'x', we get

$$\frac{\partial}{\partial x}|f(z)| = \frac{\partial}{\partial x}(\sqrt{u^2 + v^2})$$

$$= \frac{1}{2}(u^2 + v^2)^{\frac{-1}{2}}\left(2u\frac{\partial u}{\partial x} + 2v\frac{\partial v}{\partial x}\right) = \frac{u\frac{\partial u}{\partial x} + v\frac{\partial v}{\partial x}}{\sqrt{u^2 + v^2}} = \frac{u\frac{\partial u}{\partial x} + v\frac{\partial v}{\partial x}}{|f(z)|} \qquad \text{... (2)}$$

Similarly $$\frac{\partial}{\partial y}|f(z)| = \frac{u\frac{\partial u}{\partial y} + v\frac{\partial v}{\partial y}}{|f(z)|} \qquad \text{... (3)}$$

Squaring (2) and (3) adding, we get

$$\left[\frac{\partial}{\partial x}|f(z)|\right]^2 + \left[\frac{\partial}{\partial y}|f(z)|\right]^2 = \frac{\left(u\frac{\partial u}{\partial x} + v\frac{\partial v}{\partial x}\right)^2 + \left(u\frac{\partial u}{\partial y} + v\frac{\partial v}{\partial y}\right)^2}{|f(z)|^2}$$

$$= \frac{\left(u\frac{\partial u}{\partial x}\right)^2 + 2uv\frac{\partial u}{\partial x}\frac{\partial v}{\partial x} + \left(v\frac{\partial v}{\partial x}\right)^2 + \left(u\frac{\partial u}{\partial y}\right)^2 + 2u\frac{\partial u}{\partial y}.v\frac{\partial v}{\partial y} + \left(v\frac{\partial v}{\partial y}\right)^2}{|f(z)|^2}$$

By C-R equation $\frac{\partial u}{\partial x} = \frac{\partial v}{\partial y}$ and $\frac{\partial u}{\partial y} = -\frac{\partial v}{\partial x}$

Now, $$2uv\frac{\partial u}{\partial x}\frac{\partial v}{\partial x} = 2uv\left(\frac{\partial v}{\partial y}\right)\left(-\frac{\partial u}{\partial y}\right)$$

Putting the value of $2uv\frac{\partial u}{\partial x}.\frac{\partial v}{\partial x} = -2uv\frac{\partial v}{\partial y}.\frac{\partial u}{\partial y}$ in (4), we get

$$\left[\frac{\partial}{\partial x}|f(z)|\right]^2 + \left[\frac{\partial}{\partial y}|f(z)|\right]^2 = \frac{\left(u\frac{\partial u}{\partial x}\right)^2 - 2uv\frac{\partial v}{\partial y}.\frac{\partial u}{\partial y} + \left(v\frac{\partial v}{\partial x}\right)^2 + \left(u\frac{\partial u}{\partial y}\right)^2 + 2uv\frac{\partial u}{\partial y}.\frac{\partial v}{\partial y} + \left(v\frac{\partial v}{\partial y}\right)^2}{|f(z)|^2}$$

$$= \frac{u^2\left(\frac{\partial u}{\partial x}\right)^2 + u^2\left(\frac{\partial u}{\partial y}\right)^2 + v^2\left(\frac{\partial v}{\partial x}\right)^2 + v^2\left(\frac{\partial v}{\partial y}\right)^2}{|f(z)|^2}$$

$$= \frac{u^2\left[\left(\frac{\partial u}{\partial x}\right)^2 + \left(\frac{\partial u}{\partial y}\right)^2\right] + v^2\left[\left(\frac{\partial v}{\partial x}\right)^2 + \left(\frac{\partial v}{\partial y}\right)^2\right]}{|f(z)|^2}$$

$$= \frac{u^2\left[\left(\frac{\partial u}{\partial x}\right)^2 + \left(\frac{-\partial v}{\partial x}\right)^2\right] + v^2\left[\left(\frac{\partial v}{\partial x}\right)^2 + \left(\frac{\partial u}{\partial x}\right)^2\right]}{|f(z)|^2} \qquad \text{[C – R equations]}$$

$$= \frac{(u^2+v^2)\left[\left(\frac{\partial u}{\partial x}\right)^2 + \left(\frac{\partial v}{\partial x}\right)^2\right]}{|f(z)|^2} = \frac{|f(z)|^2\left[\left(\frac{\partial u}{\partial x}\right)^2 + \left(\frac{\partial v}{\partial x}\right)^2\right]}{|f(z)|^2} = \left(\frac{\partial u}{\partial x}\right)^2 + \left(\frac{\partial v}{\partial x}\right)^2$$

$$[\,|f(x)|^2 = u^2 + v^2\,]$$

$$= |f'(z)|^2$$ **Proved.**

Example 41. *Prove that* $\left(\frac{\partial^2}{\partial x^2} + \frac{\partial^2}{\partial y^2}\right)|u|^P = P(P-1)|u|^{P-2}|f'(z)|^2$

Solution. We know that $\frac{\partial^2}{\partial x^2} + \frac{\partial^2}{\partial y^2} = 4\frac{\partial^2}{\partial z\,\partial \bar{z}}$ [Example 39, page 705]

$$\left(\frac{\partial^2}{\partial x^2} + \frac{\partial^2}{\partial y^2}\right)|u|^P = \frac{1}{2^P}\frac{4\partial^2}{\partial z \partial \bar{z}}[f(z) + f(\bar{z})]^P \qquad \left[\because\ u = \frac{1}{2}[f(z) + f(\bar{z})]\right]$$

$$= \frac{4}{2^P}\frac{\partial}{\partial \bar{z}}P[f(z)+f(\bar{z})]^{P-1}f'(z) = \frac{1}{2^{P-2}}P(P-1)[f(z)+f(\bar{z})]^{P-2}f'(z)f'(\bar{z})$$

$$= P(P-1)\left[\frac{1}{2}\{f(z)+f(\bar{z})\}\right]^{P-2}[f'(z)f'(\bar{z})] = P(P-1)|u|^{P-2}|f'(z)|^2$$

Proved.

Example 42. *Prove that* $\left(\frac{\partial^2}{\partial x^2} + \frac{\partial^2}{\partial y^2}\right)\log|f'(z)| = 0$

Solution. We have, $\left(\frac{\partial^2}{\partial x^2} + \frac{\partial^2}{\partial y^2}\right) = 4\frac{\partial^2}{\partial z\,\partial \bar{z}}$ (*Example 39 on page 705*)

Hence $\left(\frac{\partial^2}{\partial x^2} + \frac{\partial^2}{\partial y^2}\right)\{\log|f'(z)|\} = 4\frac{\partial^2}{\partial z\,\partial \bar{z}}\{\log|f'(z)|\} = 4\frac{\partial^2}{\partial z \partial \bar{z}}\frac{1}{2}\log|f'(z)|^2$

$$= 2\frac{\partial^2}{\partial z \partial \bar{z}}\log\{f'(z)f'(\bar{z})\}$$

$$= 2\frac{\partial^2}{\partial z \partial \bar{z}}[\log f'(z) + \log f'(\bar{z})] = 2\frac{\partial}{\partial z}\left(0 + \frac{1}{f'(\bar{z})}f''(\bar{z})\right)$$

$$= 2\frac{\partial}{\partial z}\frac{f''(\bar{z})}{f'(\bar{z})} = 2 \times 0 = 0$$ **Proved.** [$\bar{z}$ is constant in regards to differentiation *w.r.t.* z]

Example 43. *Prove that* $\left(\frac{\partial^2}{\partial x^2} + \frac{\partial^2}{\partial y^2}\right)|Rf(z)|^2 = 2|f'(z)|^2$

Solution. $f(z) = u + iv$ or $R\,f(z) = u \Rightarrow$ Real part of $f(z) = u$

$$\frac{\partial}{\partial x}u^2 = 2u\frac{\partial u}{\partial x}$$

$$\frac{\partial^2}{\partial x^2}u^2 = 2\left(\frac{\partial u}{\partial x}\right)^2 + 2u\frac{\partial^2 u}{\partial x^2} \qquad \text{... (1)}$$

Similarly, $$\frac{\partial^2}{\partial y^2}u^2 = 2\left(\frac{\partial u}{\partial y}\right)^2 + 2u\frac{\partial^2 u}{\partial y^2} \qquad \text{... (2)}$$

Adding (1) and (2), we get

$$\left(\frac{\partial^2}{\partial x^2}+\frac{\partial^2}{\partial y^2}\right)u^2 = 2\left[\left(\frac{\partial u}{\partial x}\right)^2+\left(\frac{\partial u}{\partial y}\right)^2\right]+2u\left(\frac{\partial^2 u}{\partial x^2}+\frac{\partial^2 u}{\partial y^2}\right)$$

$$= 2\left[\left(\frac{\partial u}{\partial x}\right)^2+\left(\frac{\partial u}{\partial y}\right)^2\right]+0 = 2\left[\left(\frac{\partial u}{\partial x}\right)^2+\left(-\frac{\partial v}{\partial x}\right)^2\right] = 2|f'(z)|^2 \qquad \left(\frac{\partial u}{\partial y}=-\frac{\partial v}{\partial x}\right)$$

$$\Rightarrow \qquad \left(\frac{\partial^2}{\partial x^2}+\frac{\partial^2}{\partial y^2}\right)|R\, f(z)|^2 = 2\,|f'(z)|^2$$

Proved.

Example 44. *If $f(z)$ is regular function of z, show that*

$$\left(\frac{\partial^2}{\partial x^2}+\frac{\partial^2}{\partial y^2}\right)|f(z)|^2 = 4\,|f'(z)|^2$$

(R.G.P.V., Bhopal, III Semester, June 2004)

Solution. $f(z) = u + iv$

$$|f(z)|^2 = u^2+v^2 \qquad \text{... (1)}$$

Let $\phi = u^2 + v^2$

Differentiating (1) *w.r.t.* x, we get

$$\frac{\partial \phi}{\partial x} = 2u\frac{\partial u}{\partial x}+2v\frac{\partial v}{\partial x}$$

$$\frac{\partial^2 \phi}{\partial x^2} = 2\left[u\frac{\partial^2 u}{\partial x^2}+\left(\frac{\partial u}{\partial x}\right)^2+v\frac{\partial^2 v}{\partial x^2}+\left(\frac{\partial v}{\partial x}\right)^2\right] \qquad \text{...(2)}$$

Similarly,

$$\frac{\partial^2 \phi}{\partial y^2} = 2\left[u\frac{\partial^2 u}{\partial y^2}+\left(\frac{\partial u}{\partial y}\right)^2+v\frac{\partial^2 v}{\partial y^2}+\left(\frac{\partial v}{\partial y}\right)^2\right] \qquad \text{... (3)}$$

Adding (2) and (3), we get

$$\frac{\partial^2 \phi}{\partial x^2}+\frac{\partial^2 \phi}{\partial y^2} = 2\left[u\left(\frac{\partial^2 u}{\partial x^2}+\frac{\partial^2 u}{\partial y^2}\right)+v\left(\frac{\partial^2 v}{\partial x^2}+\frac{\partial^2 v}{\partial y^2}\right)+\left\{\left(\frac{\partial u}{\partial x}\right)^2+\left(\frac{\partial u}{\partial y}\right)^2+\left(\frac{\partial v}{\partial x}\right)^2+\left(\frac{\partial v}{\partial y}\right)^2\right\}\right] \qquad \text{... (4)}$$

By C – R equations

$$\left(\frac{\partial u}{\partial x}\right)^2 = \left(\frac{\partial v}{\partial y}\right)^2$$

$$\left(\frac{\partial u}{\partial y}\right)^2 = \left(-\frac{\partial v}{\partial x}\right)^2 = \left(\frac{\partial v}{\partial x}\right)^2 \qquad \text{... (5)}$$

By Laplace equations

$$\frac{\partial^2 u}{\partial x^2}+\frac{\partial^2 u}{\partial y^2} = 0$$

$$\frac{\partial^2 v}{\partial x^2}+\frac{\partial^2 v}{\partial y^2} = 0$$

On putting the values of $\left(\frac{\partial^2 u}{\partial x^2}+\frac{\partial^2 u}{\partial y^2}\right)$, $\left(\frac{\partial^2 v}{\partial x^2}+\frac{\partial^2 v}{\partial y^2}\right)$, $\frac{\partial v}{\partial x}, \frac{\partial v}{\partial y}$ from (5) in (4), we get

$$\frac{\partial^2 \phi}{\partial x^2}+\frac{\partial^2 \phi}{\partial y^2} = 4\left[\left(\frac{\partial u}{\partial x}\right)^2+\left(\frac{\partial v}{\partial y}\right)^2\right], \quad \left(\frac{\partial^2}{\partial x^2}+\frac{\partial^2}{\partial y^2}\right)\phi = 4\left|\frac{\partial u}{\partial x}+i\frac{\partial v}{\partial x}\right|^2$$

$$\left(\frac{\partial^2}{\partial x^2}+\frac{\partial^2}{\partial y^2}\right)|f(z)|^2 = 4\,|f'(z)|^2$$

Proved.

Example 45. *If $|f(z)|$ is constant, prove that $f(z)$ is also constant.*

Solution.
$$f(z) = u + iv$$
$$|f(z)|^2 = u^2 + v^2$$
$$|f(z)| = \text{constant} = c \text{ (given)}$$
$$u^2 + v^2 = c^2 \qquad \text{... (1)}$$

Differentiating (1) w.r.t. x, we get $2u\frac{\partial u}{\partial x} + 2v\frac{\partial v}{\partial x} = 0 \quad \Rightarrow \quad u\frac{\partial u}{\partial x} + v\frac{\partial v}{\partial x} = 0$ ... (2)

Differentiating (1) w.r.t. 'y', we get $2u\frac{\partial u}{\partial y} + 2v\frac{\partial v}{\partial y} = 0$
$$-u\frac{\partial v}{\partial x} + v\frac{\partial u}{\partial x} = 0 \qquad \text{... (3)}$$

Squaring (2) and (3) and then adding, we get
$$u^2\left(\frac{\partial u}{\partial x}\right)^2 + v^2\left(\frac{\partial v}{\partial x}\right)^2 + u^2\left(\frac{\partial v}{\partial x}\right)^2 + v^2\left(\frac{\partial u}{\partial x}\right)^2 = 0$$
$$(u^2 + v^2)\left[\left(\frac{\partial u}{\partial x}\right)^2 + \left(\frac{\partial v}{\partial x}\right)^2\right] = 0$$
$$\left(\frac{\partial u}{\partial x}\right)^2 + \left(\frac{\partial v}{\partial x}\right)^2 = 0$$

As
$$f(z) = u + iv \Rightarrow f'(z) = \frac{\partial u}{\partial x} + i\frac{\partial v}{\partial x}$$
$$\overline{f'(z)} = \frac{\partial u}{\partial x} - i\frac{\partial v}{\partial x}$$
$$|f'(z)|^2 = \left(\frac{\partial u}{\partial x}\right)^2 + \left(\frac{\partial v}{\partial x}\right)^2 = 0$$
$$|f'(z)|^2 = 0 \quad \Rightarrow \quad f(z) \text{ is constant.}$$
Proved.

EXERCISE 27.3

1. If $f(z) = u + iv$ is an analytic function of $z = x + iy$, and ψ is any function of x and y with differential coefficients of the first two orders, then show that
$$\left(\frac{\partial \psi}{\partial x}\right)^2 + \left(\frac{\partial \psi}{\partial y}\right)^2 = \left[\left(\frac{\partial \psi}{\partial u}\right)^2 + \left(\frac{\partial \psi}{\partial v}\right)^2\right]|f'(z)|^2$$
and
$$\frac{\partial^2 \psi}{\partial x^2} + \frac{\partial^2 \psi}{\partial y^2} = \left(\frac{\partial^2 \psi}{\partial u^2} + \frac{\partial^2 \psi}{\partial v^2}\right)|f'(z)|^2.$$

2. If $|f'(z)|$ is the product of a function of x and a function of y, show that
$$f'(z) = \exp(\alpha z^2 + \beta z + \gamma)$$
where α is a real and β and γ are complex constants.

Choose the correct alternative:

3. If $|f(z)|$ is constant then $f(z)$ is
(*a*) Variable (*b*) Partially variable and constant (*c*) Constant (*d*) None of these **Ans.** (*c*)

4. If $f(z) = u + iv$ then $|f(z)|$ is equal to
(*a*) $\sqrt{u^2 + v^2}$ (*b*) $u^2 + v^2$ (*c*) $u + v$ (*d*) $\sqrt{u^2 - v^2}$ **Ans.** (*a*)

5. If $z = r(\cos\theta + i\sin\theta)$ then $|z|^3$ is equal to
(*a*) $(\cos\theta + i\sin\theta)^3$ (*b*) $r^3(\cos\theta + i\sin\theta)^3$ (*c*) $r^3/2$ (*d*) r^3 **Ans.** (*d*)

CONFORMAL TRANSFORMATION

28.1 GEOMETRICAL REPRESENTATION

To draw a curve of complex variable (x, y) on z-plane we take two axes *i.e.*, one real axis and the other imaginary axis. A number of points (x, y) are plotted on z-plane, by taking different value of z (different values of x and y). The curve C is drawn by joining the plotted points. The diagram obtained is called *Argand diagram* in z-plane.

But a complex function $w = f(z)$ *i.e.*, $(u + iv) = f(x + iy)$ involves four variables x, y and u, v.

A figure of only three dimensions (x, y, z) is possible in a plane. A figure of four dimensional region for x, y, u, v is not possible.

So, we choose two complex planes z-plane and w-plane. In the w-plane we plot the corresponding points $w = u + iv$. By joining these points we have a corresponding curve C' in w-plane.

28.2 TRANSFORMATION

For every point (x, y) in the z-plane, the relation $w = f(z)$ defines a corresponding point (u, v) in the w-plane. We call this "transformation or mapping of z-plane into w-plane". If a point z_0 maps into the point w_0, w_0 is also known as the image of z_0.

If the point $P(x, y)$ moves along a curve C in z-plane, the point $P'(u, v)$ will move along a corresponding curve C' in w-plane, then we say that a curve C in the z-plane is mapped into the corresponding curve C' in the w-plane by the relation $w = f(z)$.

Example 1. *Transform the rectangular region ABCD in z-plane bounded by $x = 1$, $x = 3$; $y = 0$ and $y = 3$. Under the transformation $w = z + (2 + i)$.*

Solution. Here,
$$w = z + (2 + i)$$
$$\Rightarrow \quad u + iv = x + iy + (2 + i)$$
$$= (x + 2) + i(y + 1)$$

By equating real and imaginary quantities, we have $u = x + 2$ and $v = y + 1$.

z-plane	w-plane	z-plane	w-plane
x	$u = x + 2$	y	$v = y + 1$
1	$= 1 + 2 = 3$	0	$= 0 + 1 = 1$
3	$= 3 + 2 = 5$	3	$= 3 + 1 = 4$

Here the lines $x = 1$, $x = 3$; $y = 0$ and $y = 1$ in the z-plane are transformed onto the line $u = 3$, $u = 5$; $v = 1$ and $v = 4$ in the w-plane. The region $ABCD$ in z-plane is transformed into the region $EFGH$ in w-plane.

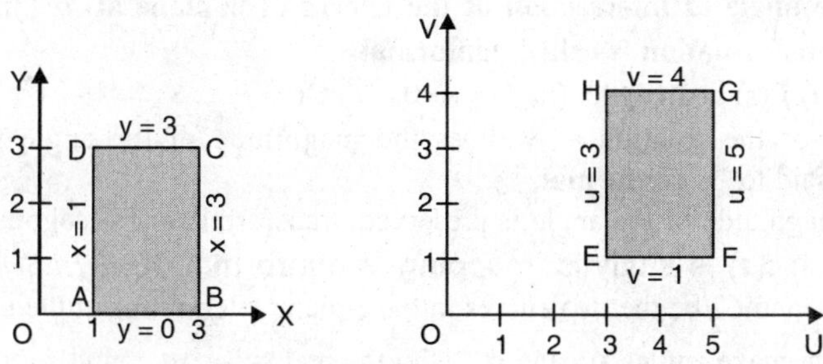

Example 2. *Transform the curve* $x^2 - y^2 = 4$ *under the mapping* $w = z^2$.

Solution. $w = z^2 = (x+iy)^2, \; u + iv = x^2 - y^2 + 2ixy$

This gives $u = x^2 - y^2$ and $v = 2xy$

Table of (x, y) and (u, v)

x	2	2.5	3	3.5	4	4.5	5
$y = \pm\sqrt{x^2 - 4}$	0	± 1.5	± 2.2	± 2.9	± 3.5	± 4.1	± 4.6
$u = x^2 - y^2$	4	4	4	4	4	4	4
$v = 2xy$	0	± 7.5	± 13.2	± 20.3	± 28	± 36.9	± 46

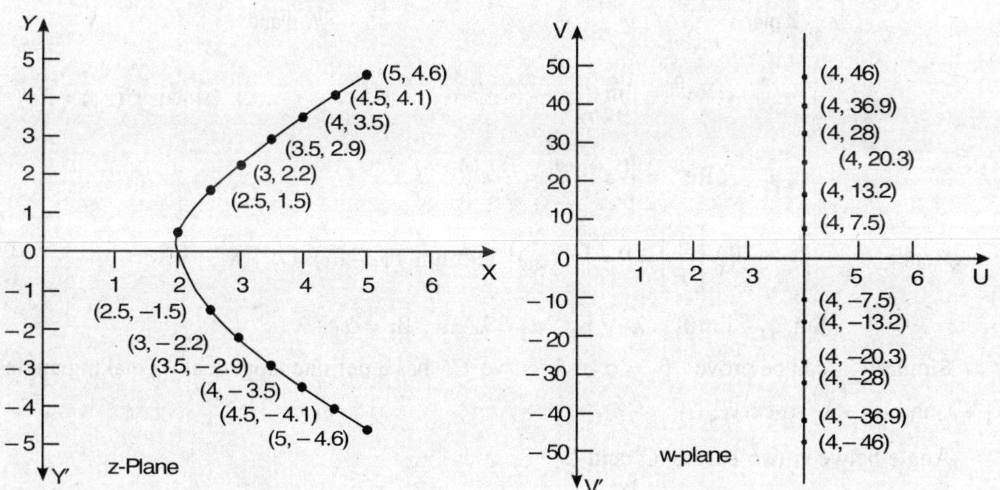

Image of the curve $x^2 - y^2 = 4$ is a straight line, $u = 4$ parallel to the v-axis in w-plane. **Ans.**

28.3 CONFORMAL TRANSFORMATION

(*U.P. III Semester Dec., 2006, 2005*)

Let two curves C, C_1 in the z-plane intersect at the point P and the corresponding curve C', C_1' in the w-plane intersect at P'. **If the angle of intersection of the curves at P in z-plane**

is the same as the angle of intersection of the curves of w-plane at P' in magnitude and sense, then the transformation is called conformal:

conditions: (*i*) $f(z)$ is analytic. (*ii*) $f(z) \neq 0$ Or

If the sense of the rotation as well as the magnitude of the angle is preserved, the transformation is said to be **conformal**.

If only the magnitude of the angle is preserved, transformation is **Isogonal**.

28.4 THEOREM. If $f(z)$ is analytic, mapping is conformal (*U.P. III Semester Dec. 2005*)

Proof. Let C_1 and C_2 be the two curves in the z-plane intersecting at the point z_0 and let the tangents at this point make angles α_1 and α_2 with the real axis. Let z_1 and z_2 be the points on the curves C_1 and C_2 near to z_0 at the same distance r from z_0, so that we have

$$z_1 - z_0 = r e^{i\theta_1}, \quad z_2 - z_0 = r e^{i\theta_2}$$

As $\quad r \to 0, \theta_1 \to \alpha_1$ and $\theta_2 \to \alpha_2$.

Let the image of the curves C_1, C_2 be C_1' and C_2' in w-plane and images of z_0, z_1 and z_2 be w_0, w_1 and w_2.

Let $$w_1 - w_0 = r.e^{i\phi_1}, \quad w_2 - w_0 = r.e^{i\phi_2}$$

$$f'(z_0) = \lim_{z_1 \to z_0} \frac{w_1 - w_0}{z_1 - z_0}$$

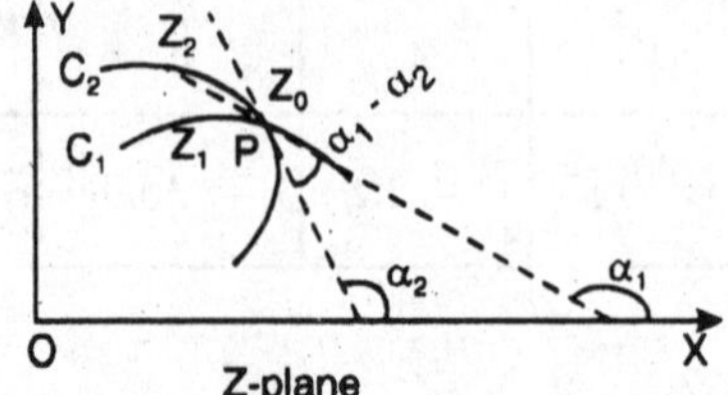

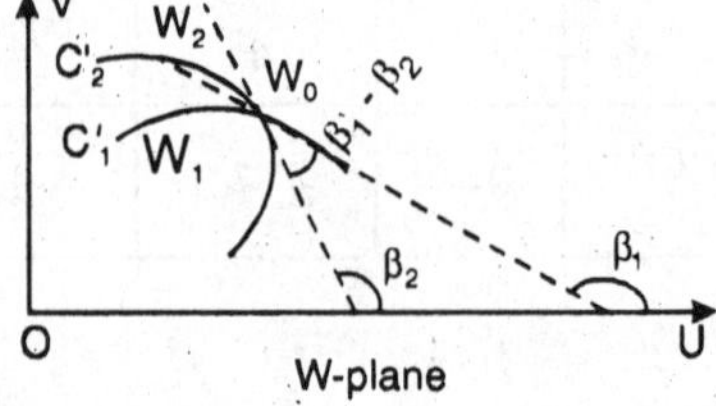

$$\mathrm{R}e^{i\lambda} = \lim_{r \to 0} \frac{r_1 e^{i\phi_1}}{r e^{i\theta_1}} \qquad \text{(since } f'(z_0) = \mathrm{R}e^{i\lambda}\text{)}$$

$$\mathrm{R}e^{i\lambda} = \frac{r_1}{r} e^{i\phi_1 - i\theta_1} = \frac{r_1}{r} e^{i(\phi_1 - \theta_1)}$$

Hence $$\lim_{r \to 0} \left[\frac{r_1}{r}\right] = R = |f'(z_0)| \text{ and } \lim (\phi_1 - \theta_1) = \lambda$$

$\Rightarrow \quad \lim \phi_1 - \lim \theta_1 = \lambda$ or $\beta_1 - \alpha_1 = \lambda$ *i.e.*, $\beta_1 = \alpha_1 + \lambda$

Similarly it can be proved $\beta_2 = \alpha_2 + \lambda$ curve C_1' has a definite tangent at w_0 making angles $\alpha_1 + \lambda$ and $\alpha_2 + \lambda$ so curve C_2'.

Angle between two curves C_1' and C_2'

$$= \beta_1 - \beta_2 = (\alpha_1 + \lambda) - (\alpha_2 + \lambda) = (\alpha_1 - \alpha_2)$$

so the transformation is conformal at each point where $f'(z) = 0$

Note 1. The point at which $f'(z) = 0$ is called a **critical point** of the transformation. Also the points where $\frac{dw}{dz} \neq 0$ are called **ordinary points**.

2. Let $\phi = \alpha_1 - \alpha_2$ or $\alpha_1 = \alpha_2 + \phi$ shows that the tangent at P to the curve is rotated through an $\angle\phi$ = amp $\{f'(z)\}$ under the given transformation.

Angle of rotation $= \tan^{-1}\dfrac{v}{u}$.

3. In formal transformation, element of arc passing through P is magnified by the factor $|f'(z)|$.

The area element is also magnified by the factor $|f'(z)|$ or $J = \dfrac{\partial(u,v)}{\partial(x,y)}$ in a conformal transformation.

$$J = \frac{\partial(u,v)}{\partial(x,y)} = \begin{vmatrix} \frac{\partial u}{\partial x} & \frac{\partial u}{\partial y} \\ \frac{\partial v}{\partial x} & \frac{\partial v}{\partial y} \end{vmatrix} = \begin{vmatrix} \frac{\partial u}{\partial x} & -\frac{\partial v}{\partial x} \\ \frac{\partial v}{\partial x} & \frac{\partial u}{\partial x} \end{vmatrix} = \left(\frac{\partial u}{\partial x}\right)^2 + \left(\frac{\partial v}{\partial x}\right)^2$$

$$= \left|\frac{\partial u}{\partial x} + i\frac{\partial v}{\partial x}\right|^2 = |f'(z)|^2 = |f'(x+iy)|^2$$

$|f'(z)|$ is called the **coefficient of magnification.**

4. Conjugate functions remain conjugate functions after conformal transformation. A function which is the solution of Laplace's equation, its transformed function again remains the solution of Laplace's equation after conformal transformation.

28.5 THEOREM

Prove that an analytic function $f(z)$ ceases to be conformal at the points where $f'(z) = 0$. *(U.P. III Semester, Dec. 2006)*

Proof. Let $f'(z) = 0$ and $f'(z_0) = 0$ at $z = z_0$

Suppose that $f'(z_0)$ has a zero of order $(n-1)$ at the point z_0, then first $(n-1)$ derivatives of $f(z)$ vanish at z_0 but $f^n(z_0) \neq 0$, hence at any point z in the neighbourhood of z_0, we have by Taylor's Theorem.

$$f(z) = f(z_0) + a_n(z - z_0)^n + \ldots\ldots$$

where $\quad a_n = \dfrac{f^n(z_0)}{n!}$, so that $a_n \neq 0$.

Hence, $\quad f(z_1) - f(z_0) = a_n(z_1 - z_0)^n + \ldots\ldots$

i.e. $\quad w_1 - w_0 = a_n(z_1 - z_0)^n + \ldots.$

or $\quad \rho_1 e^{i\phi_1} = |a_n| \cdot r^n e^{i(n\theta_1 + \lambda)} + \ldots\ldots \quad$ **where λ = amp a_n**

Hence, $\quad \text{Lim } \phi_1 = \text{Lim } (n\theta_1 + \lambda) = n\alpha_1 + \lambda$

Similary, $\quad \text{Lim } \phi_2 = n\alpha_2 + \lambda.$

Thus the curves γ_1 and γ_2 still have definite tangents at w_0.

But the angle between the tangents

$$= \text{Lim}\,\phi_2 - \text{Lim}\,\phi_1 = n(\alpha_2 - \alpha_1).$$

So magnitude of the angle is not preserved.

Also the linear magnification $R = \text{Lim}\,(\rho_1 / r) = 0$.

Hence, the conformal property does not hold good at a point where $f'(z) = 0$.

Example 3. *If* $u = 2x^2 + y^2$ *and* $v = \dfrac{y^2}{x}$, *show that the curves u = constant and v= constant cut orthogonally at all intersections but that the transformation w = u + iv is not conformal.*

(*Q. Bank U.P. III Semester 2002*)

Solution. For the curve, $2x^2 + y^2 = u$

$$2x^2 + y^2 = \text{constant} = k_1 \text{ (say)} \qquad ...(1)$$

Differentiating (1), we get

$$4x + 2y\frac{dy}{dx} = 0 \quad \Rightarrow \quad \frac{dy}{dx} = -\frac{2x}{y} \qquad ...(2)$$

$$\frac{y^2}{x} = v$$

For the curve, $\dfrac{y^2}{x}$ = constant = k_2 (say),

$$\Rightarrow \qquad y^2 = k_2 x. \qquad ...(3)$$

Differentiating (3), we get

$$2y\frac{dy}{dx} = k_2 \quad \Rightarrow \quad \frac{dy}{dx} = \frac{k_2}{2y} = \frac{y^2}{x} \times \frac{1}{2y} = \frac{y}{2x} \qquad ...(4)$$

From (2) and (4), we see that

$$m_1 m_2 = \left(\frac{-2x}{y}\right)\left(\frac{y}{2x}\right) = -1$$

Hence, two curves cut orthogonally.

However, since $\dfrac{\partial u}{\partial x} = 4x, \qquad \dfrac{\partial u}{\partial y} = 2y$

$$\frac{\partial v}{\partial x} = -\frac{y^2}{x^2}, \qquad \frac{\partial v}{\partial y} = \frac{2y}{x}$$

The Cauchy-Riemann equations are not satisfied by u and v.

Hence, the function $u + iv$ is not analytic. So, the transformation is not conformal. **Proved**

Example 4. (*i*) *For the conformal transformation* $w = z^2$, *show that*

(*a*) *The coefficient of magnification at* $z = 2 + i$ *is* $2\sqrt{5}$.

(*b*) *The angle of rotation at* $z = 2 + i$ *is* $\tan^{-1} 0.5$.

(*Q. Bank U.P. III Semester 2002*)

(*ii*) *For the conformal transformation* $w = z^2$, *show that*

(*a*) *The co-efficient of magnification at* $z = 1 + i$ *is* $2\sqrt{2}$.

(*b*) *The angle of rotation at* $z = 1 + i$ *is* $\dfrac{\pi}{4}$.

Solution. (*i*) Let $\qquad w = f(z) = z^2$

$$\therefore \qquad f'(z) = 2z$$

$$f'(2 + i) = 2(2 + i) = 4 + 2i.$$

(*a*) Coefficient of magnification at $z = 2 + i$ is $= |f'(2 + i)| = |4 + 2i| = \sqrt{16 + 4} = 2\sqrt{5}$.

(*b*) Angle of rotation at $z = 2 + i$ is = amp. $f'(2 + i) = (4 + 2i) = \tan^{-1}\left(\dfrac{2}{4}\right) = \tan^{-1}(0.5)$.

(*ii*) Here $\qquad f(z) = w = z^2$

$$\therefore \qquad f'(z) = 2z$$

and $\qquad f'(1 + i) = 2(1 + i) = 2 + 2i$

$\therefore$ (*a*) The co-efficient of magnification at $z = 1 + i$ is $|f'(1+i)| = |2+2i| = \sqrt{4+4} = 2\sqrt{2}$

(*b*) The angle of rotation at $z = 1 + i$ is amp. $|f'(1+i)| = 2\,(1+i) = 2 + 2i = \tan^{-1}\dfrac{2}{2} = \dfrac{\pi}{4}$

Ans.

Standard transformations

28.6 TRANSLATION $\quad w = z + C,$

where
$$C = a + ib$$
$$u + iv = x + iy + a + ib$$
$$u = x + a \text{ and } v = y + b$$
$$x = u - a \text{ and } y = v - b$$

On substituting the values of x and y in the equation of the curve to be transformed, we get the equation of the image in the w-plane.

The point $P\,(x, y)$ in the z-plane is mapped onto the point $P'(x+a,\ y+b)$ in the w-plane. Similarly other points of z-plane are mapped onto w-plane. Thus if w-plane is superimposed on the z-plane, the figure of w-plane is shifted through a vector C.

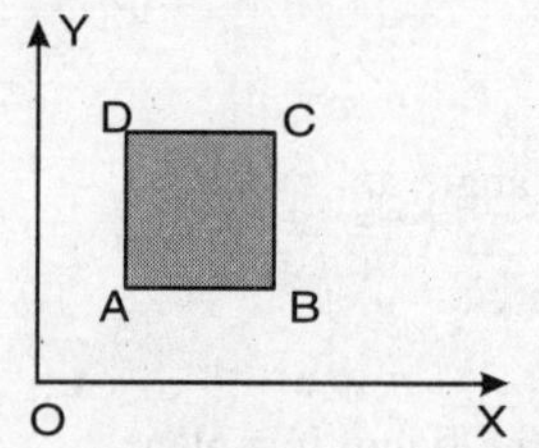

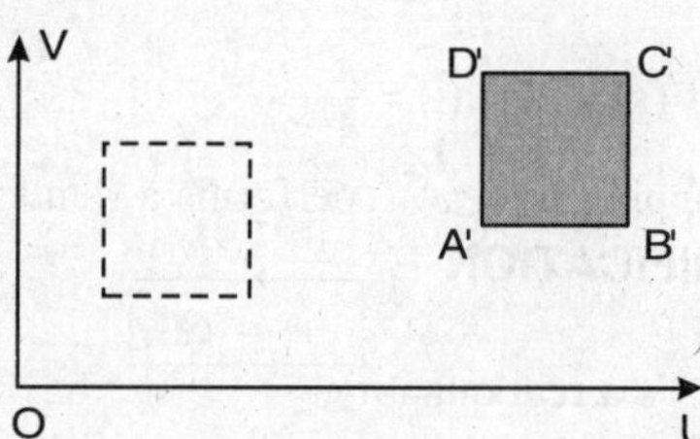

In other words the transformation is mere translation of the axes.

28.7 ROTATION $\quad w = ze^{i\theta}$

The figure in z-plane rotates through an angle θ in anticlockwise in w-plane.

Example 5. *Consider the transformation $w = ze^{i\pi/4}$ and determine the region R' in w-plane corresponding to the triangular region R bounded by the lines $x = 0$, $y = 0$ and $x + y = 1$ in z-plane.*

Solution.
$$w = ze^{i\pi/4}$$
$$w = (x+iy)\left(\cos\frac{\pi}{4} + i\sin\frac{\pi}{4}\right)$$
$$\Rightarrow \qquad u + iv = (x+iy)\left(\frac{1+i}{\sqrt{2}}\right) = \frac{1}{\sqrt{2}}[x - y + i(x+y)]$$

Equating real and imaginay parts, we get
$$\Rightarrow \qquad u = \frac{1}{\sqrt{2}}(x-y), \qquad v = \frac{1}{\sqrt{2}}(x+y)$$

(*i*) Put $x = 0$, $\quad u = -\dfrac{1}{\sqrt{2}}y, \quad v = \dfrac{1}{\sqrt{2}}y$ or $v = -u$

(*ii*) Put $y = 0$, $\quad u = \dfrac{1}{\sqrt{2}}x, \quad v = \dfrac{1}{\sqrt{2}}x$ or $v = u$

(*iii*) Putting $x + y = 1$ in (1), we get $v = \dfrac{1}{\sqrt{2}}$

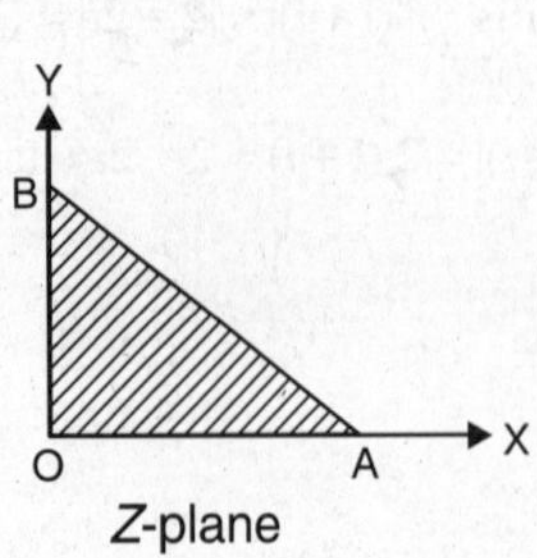

Z-plane

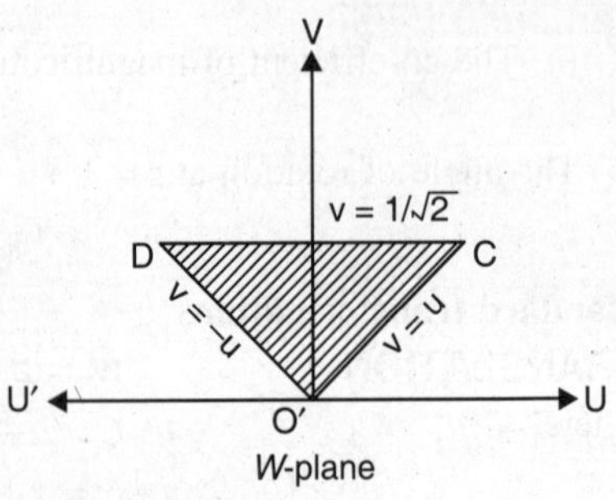

W-plane

Hence the triangular region OAB in z-plane is mapped on a triangular region O′CD of w-plane bounded by the lines $v = u$, $v = -u$, $v = \frac{1}{\sqrt{2}}$.

$$f'(z) = \frac{1}{\sqrt{2}}(1+i)$$

$$f(z) = \frac{1}{\sqrt{2}}[(x-y)+i(x+y)]$$

Amp. $f'(z) = \tan^{-1}(1) = \frac{\pi}{4}$

The mapping $w = ze^{i\pi/4}$ performs a rotation of R through an angle $\pi/4$. **Ans.**

28.8 MAGNIFICATION

$$w = cz$$

where c is a real quantity.

(*i*) The figure in w-plane is magnified c-times the size of the figure in z-plane.

(*ii*) Both figures in z-plane and w-plane are similar.

Example 6. *Determine the region in w-plane on the transformation of rectangular region enclosed by x = 1, y = 1, x = 2 and y = 2 in the z-plane. The transformation is w = 3z.*

Solution. We have, $w = 3z$

$$u + iv = 3(x + iy)$$

Equating the real and imaginary parts, we get

$$u = 3x \quad \text{and} \quad v = 3y$$

z-plane		w-plane	
x	y	$u = 3x$	$v = 3y$
1	1	3	3
2	2	6	6

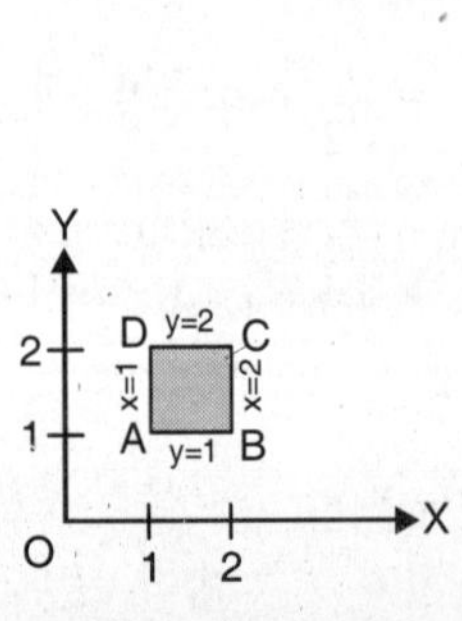

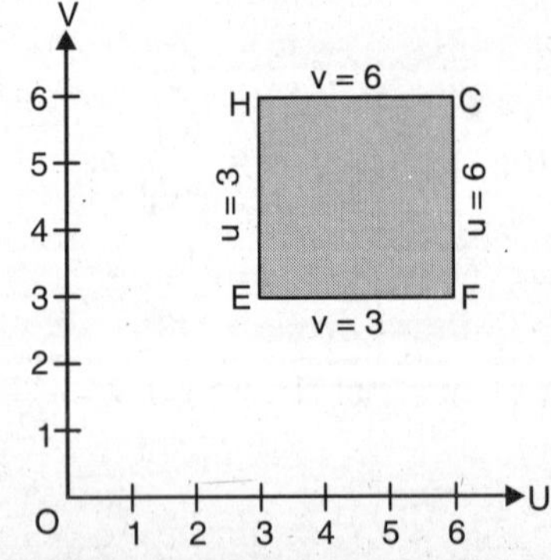

28.9 MAGNIFICATION AND ROTATION

$$w = c\,z \quad \ldots (1)$$

where c, z, w all are complex numbers.

$$c = ae^{i\alpha}, \quad z = re^{i\theta}, \quad w = Re^{i\phi}$$

Putting these values in (1), we have

$$Re^{i\phi} = (ae^{i\alpha})(re^{i\theta}) = are^{i(\theta+\alpha)}$$

i.e. $$R = ar \text{ and } \phi = \theta + \alpha$$

Thus we see that the transform $w = c\,z$ corresponding to a rotation, together with magnification.

Algebraically $\quad w = c\,z$ or $u + iv = (a + ib)(x + iy)$

$$\Rightarrow \quad u + iv = ax - by + i(ay + bx)$$

$$u = ax - by \text{ and } v = ay + bx$$

On solving these equations, we can get the values of x and y.

$$x = \frac{au + bv}{a^2 + b^2}, \; y = \frac{-bu + av}{a^2 + b^2}$$

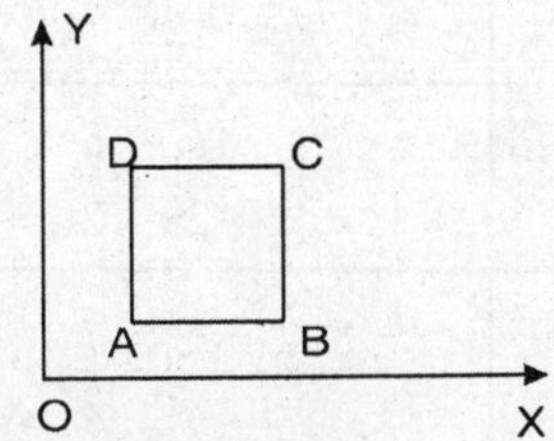

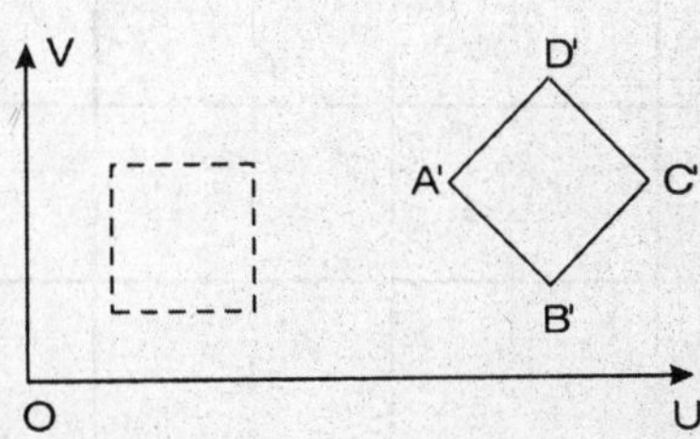

On putting values of x and y in the equation of the curve to be transformed we get the equation of the image.

Example 7. *Find the image of the triangle with vertices at i, 1 + i, 1 − i in the z-plane, under the transformation*

(*i*) $w = 3z + 4 - 2i$, $\qquad$ (*ii*) $w = e^{\frac{5\pi i}{3}}.z - 2 + 4i$

Solution. (*i*) $\qquad w = 3z + 4 - 2i$

$$\Rightarrow \quad u + iv = 3(x + iy) + 4 - 2i \quad \Rightarrow \quad u = 3x + 4, \; v = 3y - 2$$

S. No.	x	y	$u = 3x + 4$	$v = 3y - 2$
1.	0	1	4	1
2.	1	1	7	1
3.	1	−1	7	− 5

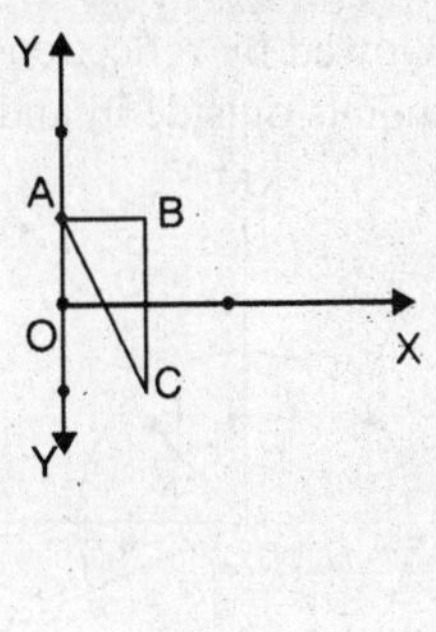

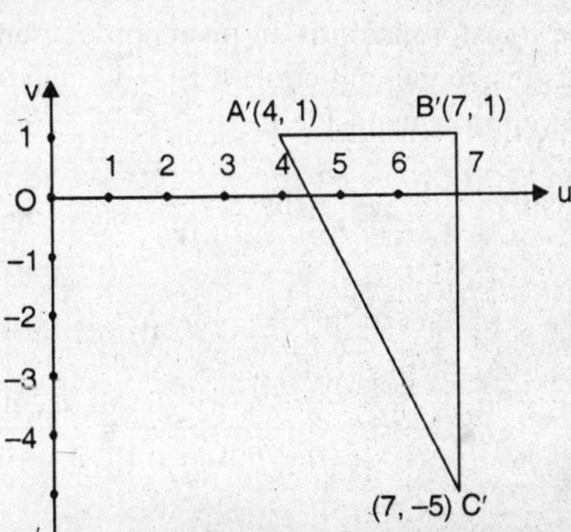

(ii) $$w = e^{\frac{5\pi i}{3}} . z - 2 + 4i$$

$$\Rightarrow \quad u + iv = \left(\cos\frac{5\pi}{3} + i\sin\frac{5\pi}{3}\right)(x+iy) - 2 + 4i$$

$$= \left(\frac{1}{2} - \frac{\sqrt{3}}{2}i\right)(x+iy) - 2 + 4i$$

$$= \frac{x}{2} - 2 + \frac{\sqrt{3}}{2}y + i\left(-\frac{\sqrt{3}}{2}x + \frac{y}{2} + 4\right)$$

$$\Rightarrow \quad u = \frac{x}{2} - 2 + \frac{\sqrt{3}}{2}y \quad \text{and} \quad v = -\frac{\sqrt{3}}{2}x + \frac{y}{2} + 4$$

S.No.	z-Plane		w-plane	
	x	y	$u = \frac{x}{2} - 2 + \frac{\sqrt{3}}{2}y$	$v = -\frac{\sqrt{3}}{2}x + \frac{y}{2} + 4$
1.	0	1	$-2 + \frac{\sqrt{3}}{2}$	$\frac{9}{2}$
2.	1	1	$-\frac{3}{2} + \frac{\sqrt{3}}{2}$	$-\frac{\sqrt{3}}{2} + \frac{9}{2}$
3.	1	–1	$-\frac{3}{2} - \frac{\sqrt{3}}{2}$	$-\frac{\sqrt{3}}{2} + \frac{7}{2}$

28.10 INVERSION AND REFLECTION

$$w = \frac{1}{z} \qquad \ldots (1)$$

If $z = re^{i\theta}$ and $w = Re^{i\phi}$

Putting these values in (1), we get

$$Re^{i\phi} = \frac{1}{re^{i\theta}} = \frac{1}{r}e^{-i\theta}$$

On equating, $$R = \frac{1}{r} \text{ and } \phi = -\theta$$

Thus the point $P(r, \theta)$ in the z-plane is mapped onto the point $P'\left(\frac{1}{r}, -\theta\right)$ in the w-plane. Hence the transformation is an inversion of z and followed by reflection into the real axis. The points inside the unit circle ($|z| = 1$) map onto points outside it, and points outside the unit circle into points inside it.

Algebraically $$w = \frac{1}{z} \text{ or } z = \frac{1}{w}$$

$$x + iy = \frac{1}{u + iv}$$

$$\Rightarrow \quad x + iy = \frac{u - iv}{(u + iv)(u - iv)} = \frac{u - iv}{u^2 + v^2}$$

$$x = \frac{u}{u^2+v^2}, \; y = -\frac{v}{u^2+v^2}$$

Let the circle $x^2+y^2+2gx+2fy+c = 0$ ... (1) be in z-plane.

On substituting the values of x and y in (1), we get

$$\frac{u^2}{(u^2+v^2)^2}+\frac{v^2}{(u^2+v^2)^2}+2g\frac{u}{u^2+v^2}+2f\frac{(-v)}{u^2+v^2}+c=0$$

This is the equation of circle in w-plane. This shows that a circle in z-plane transforms to another circle in w-plane.

But a circle through origin transforms into a straight line.

Example 8. *Find the image of* $|\, z - 3i \,| = 3$ *under the mapping* $w = \frac{1}{z}$.

(*Uttarakhand, III Semester 2008*)

Solution. $w = \frac{1}{z} \quad \Rightarrow \quad z = \frac{1}{w}$

$$\Rightarrow \quad x+iy = \frac{1}{u+iv} = \frac{u-iv}{(u+iv)(u-iv)} = \frac{u-iv}{u^2+v^2}$$

$$\Rightarrow \quad x = \frac{u}{u^2+v^2} \quad \text{and} \quad y = -\frac{v}{u^2+v^2} \quad \text{... (1)}$$

The given curve is $|\, z - 3i \,| = 3$

$\Rightarrow \quad |\, x + iy - 3i \,| = 3 \quad \Rightarrow \quad x^2 + (y-3)^2 = 9$ ... (2)

On substituting the values of x and y from (1) into (2), we get

$$\frac{u^2}{(u^2+v^2)^2}+\left(-\frac{v}{u^2+v^2}-3\right)^2 = 9$$

$$\frac{u^2}{(u^2+v^2)^2}+\frac{(-v-3u^2-3v^2)^2}{(u^2+v^2)^2} = 9$$

$$\Rightarrow \quad u^2+(-v-3u^2-3v^2)^2 = 9(u^2+v^2)^2$$

$$\Rightarrow \quad u^2+v^2+9u^4+9v^4+6u^2v+6v^3+18u^2v^2 = 9u^4+18u^2v^2+9v^4$$

$$\Rightarrow \quad u^2+v^2+6u^2v+6v^3 = 0$$

$$\Rightarrow \quad u^2+v^2+6v(u^2+v^2) = 0$$

$$\Rightarrow \quad (u^2+v^2)(6v+1) = 0$$

$\Rightarrow \quad 6v+1=0$ is the equation of the image. **Ans.**

Aliter. $|z-3i| = 3, \; z = \frac{1}{w}$

$$\left|\frac{1}{w}-3i\right| = 3 \quad \Rightarrow \quad |\,1-3iw\,| = 3\,|\,w\,|$$

$$\Rightarrow \quad |1-3i(u+iv)| = 3\,|u+iv| \quad \Rightarrow \quad |1+3v-3iu| = 3\,|u+iv|$$

$$\Rightarrow \quad (1+3v)^2+9u^2+9(u^2+v^2) \quad \Rightarrow \quad 1+6v+9v^2+9u^2 = 9(u^2+v^2)$$

$\Rightarrow \quad 1 + 6v = 0$ **Ans.**

Aliter. $|z-3i| = 3 \quad \Rightarrow \quad z-3i = 3e^{i\theta} \quad \Rightarrow \quad z = 3i+3e^{i\theta}$

$$w = \frac{1}{z} = \frac{1}{3i + 3e^{i\theta}} \qquad \Rightarrow \qquad 3w = \frac{1}{i + e^{i\theta}}$$

$$\Rightarrow \qquad 3(u + iv) = \frac{1}{i + \cos + i\sin\theta}$$

$$(3u + 3iv) = \frac{\cos\theta - i(1 + \sin\theta)}{\cos^2\theta + (1 + \sin\theta)^2} \qquad \Rightarrow \qquad 3v = -\frac{1 + \sin\theta}{2 + 2\sin\theta} = -\frac{1}{2}$$

$$6v + 1 = 0$$

Ans.

Example 9. *Image of* $| z + 1 | = 1$ *under the mapping* $\omega = \frac{1}{z}$ is

(*a*) $2v + 1 = 1$ (*b*) $2v - 1 = 0$ (*c*) $2u + 1 = 0$ (*d*) $2u - 1 = 0$

(*AMIETE, June 2009*)

Solution. $w = \frac{1}{z} \Rightarrow u + iv = \frac{1}{x + iy} = \frac{x - iy}{x^2 + y^2}$

$$\Rightarrow \quad u = \frac{x}{x^2 + y^2}, \qquad v = -\frac{y}{x^2 + y^2}$$

Given $| z + 1 | = 1 \Rightarrow | x + iy + 1 | = 1 \Rightarrow (x + 1)^2 + y^2 = 1$

$\Rightarrow x^2 + y^2 + 2x = 0 \Rightarrow x^2 + y^2 = -2x \Rightarrow \frac{1}{2} = \frac{-x}{x^2 + y^2} = -u$

$\Rightarrow \frac{1}{2} = -u \qquad \Rightarrow 2u + 1 = 0$

Hence (c) is correct answer. **Ans.**

Example 10. *Show that under the transformation* $w = \frac{1}{z}$, *the image of the hyperbola* $x^2 - y^2 = 1$ *is the lemniscate* $R^2 = \cos 2\phi$.

Solution. $x^2 - y^2 = 1$

Putting $x = r\cos\theta$

and $y = r\sin\theta$

$$\Rightarrow \quad r^2\cos^2\theta - r^2\sin^2\theta = 1 \quad \Rightarrow \quad r^2(\cos^2\theta - \sin^2\theta) = 1$$

$$\Rightarrow \quad r^2\cos 2\theta = 1 \qquad \ldots (1)$$

And $w = \frac{1}{z} \Rightarrow z = \frac{1}{w} \Rightarrow re^{i\theta} = \frac{1}{Re^{i\phi}} \Rightarrow re^{i\theta} = \frac{1}{R}e^{-i\phi}$

$\therefore \qquad r = \frac{1}{R}$ and $\theta = -\phi$

Putting the values of r and θ in (1), we get

$$\frac{1}{R^2}\cos 2(-\phi) = 1 \quad \Rightarrow \quad R^2 = \cos 2\phi$$

Proved.

EXERCISE 28.1

1. Find the image of the semi infinite, strip $x > 0$, $0 < y < 2$ under the transformation $w = iz + 1$.

Ans. Strip $-1 < u < 1$, $v > 0$

2. Determine the region in the w-plane in which the rectangle bounded by the lines $x = 0$, $y = 0$, $x = 2$ and $y = 1$ is mapped under the transformation $w = \sqrt{2}\, e^{i\pi/4} z$. (*Q. Bank U.P. III Semester 2002*)

Ans. Region bounded by the lines $v = -u$, $v = u$, $u + v = 4$ and $v - u = 2$.

3. Show that the condition for transformation $w = a^2 + b/cz + d$ to make the circle $|w| = 1$ correspond to a straight line in the z-plane is $(a) = (c)$.

4. What is the region of the w-plane in two ways the rectangular region in the z-plane bounded by the lines $x = 0$, $y = 0$, $x = 1$ and $y = 2$ is mapped under the transformation $w = z + (2 - i)$?

Ans. Region bounded by $u = 2$, $v = -1$, $u = 3$ and $v = 1$.

5. For the mapping $w(z) = 1/z$, find the image of the family of circles $x^2 + y^2 = ax$, where a is real.

Ans. $u = \frac{1}{a}$ is a straight line || to v-axis.

6. Show that the function $w = \frac{4}{z}$ transforms the straight line $x = c$ in the z-plane into a circle in the w-plane.

7. If $(w+1)^2 = \frac{4}{z}$, then prove that the unit circle in the w-plane corresponds to a parabola in the z-plane, and the inside of the circle to the outside of the parabola.

8. Find the image of $|z - 2i| = 2$ under the mapping $w = \frac{1}{z}$ *(Q. Bank U.P. 2002)* **Ans.** $4v + 1 = 0$

28.11 BILINEAR TRANSFORMATION (Mobius Transformation)

$$\boxed{w = \frac{az+b}{cz+d}} \qquad ad - bc \neq 0 \qquad \text{... (1)}$$

(1) is known as bilinear transformation.

If $\quad ad - bc \neq 0$ then $\frac{dw}{dz} \neq 0$ *i.e.* transformation is conformal.

From (1), $$z = \frac{-dw+b}{cw-a} \qquad \text{...(2)}$$

This is also bilinear except $w = \frac{a}{c}$.

Note. From (1), every point of z-plane is mapped into unique point in w-plane except $z = -\frac{d}{c}$.

From (2), every point of w-plane is mapped into unique point in z-plane except $w = \frac{a}{c}$.

28.12 INVARIANT POINTS OF BILINEAR TRANSFORMATION.

We know that $$w = \frac{az+b}{cz+d} \qquad \text{... (1)}$$

If z maps into itself, then $w = z$

(1) becomes $$z = \frac{az+b}{cz+d} \qquad \text{... (2)}$$

Roots of (2) are the invariants or fixed points of the bilinear transformation.
If the roots are equal, the bilinear transformation is said to be parabolic.

28.13 CROSS-RATIO

If there are four points z_1, z_2, z_3, z_4 taken in order, then the ratio $\frac{(z_1 - z_2)(z_3 - z_4)}{(z_2 - z_3)(z_4 - z_1)}$ is called the cross-ratio of z_1, z_2, z_3, z_4.

28.14 THEOREM

A bilinear transformation preserves cross-ratio of four points

Proof. We know that $\quad w = \dfrac{az+b}{cz+d}.$

As w_1, w_2, w_3, w_4 are images of z_1, z_2, z_3, z_4 respectively, so

$$w_1 = \frac{az_1+b}{cz_1+d}, \quad w_2 = \frac{az_2+b}{cz_2+d}$$

$$\therefore \qquad w_1 - w_2 = \frac{(ad-bc)}{(cz_1+d)(cz_2+d)}(z_1 - z_2) \qquad ...(1)$$

Similarly

$$w_2 - w_3 = \frac{ad-bc}{(cz_2+d)(cz_3+d)}(z_2 - z_3) \qquad ...(2)$$

$$w_3 - w_4 = \frac{ad-bc}{(cz_3+d)(cz_4+d)}(z_3 - z_4) \qquad ...(3)$$

$$w_4 - w_1 = \frac{ad-bc}{(cz_4+d)(cz_1+d)}(z_4 - z_1) \qquad ...(4)$$

From (1), (2), (3) and (4), we have

$$\frac{(w_1-w_2)(w_3-w_4)}{(w_1-w_4)(w_3-w_2)} = \frac{(z_1-z_2)(z_3-z_4)}{(z_1-z_4)(z_3-z_2)}$$

$$\Rightarrow \qquad (w_1, w_2, w_3, w_4) = (z_1, z_2, z_3, z_4).$$

28.15 PROPERTIES OF BILINEAR TRANSFORMATION

1. A bilinear transformation maps circles into circles.

2. A bilinear transformation preserves cross ratio of four points.

If four points z_1, z_2, z_3, z_4 of the z-plane map onto the points w_1, w_2, w_3, w_4 of the w-plane respectively.

$$\Rightarrow \qquad \frac{(w_1-w_2)(w_3-w_4)}{(w_1-w_4)(w_3-w_2)} = \frac{(z_1-z_2)(z_3-z_4)}{(z_1-z_4)(z_3-z_2)}$$

Hence, under the **bilinear** transform of four points cross-ratio is preserved.

28.16 METHODS TO FIND BILINEAR TRANSFORMATION

1. By finding a, b, c, d for $w = \dfrac{az+b}{cz+d}$ with the given conditions.

2. With the help of cross-ratio

$$\boxed{\frac{(w-w_1)(w_2-w_3)}{(w-w_3)(w_2-w_1)} = \frac{(z-z_1)(z_2-z_3)}{(z-z_3)(z_2-z_1)}}$$

Example 11. *Find the bilinear transformation which maps the points $z = 1, i, -1$ into the points $w = i, 0, -i$.*

Hence find the image of $|z| < 1$.

(U.P., III Semester, 2008, Summer 2002, U.P. (Agri. Engg.) 2002)

Solution. Let the required transformation be $w = \dfrac{az+b}{cz+d}$

$$\text{or} \qquad w = \frac{\frac{a}{d}z + \frac{b}{d}}{\frac{c}{d}z+1} = \frac{pz+q}{rz+1} \qquad ...(1) \qquad \left[p = \frac{a}{d}, q = \frac{b}{d}, r = \frac{c}{d}\right]$$

On substituting the values of z and corresponding values of w in (1), we get

z	w
1	i
i	0
-1	$-i$

$$i = \frac{p+q}{r+1} \Rightarrow p+q = ir+i \quad \text{... (2)}$$

$$0 = \frac{pi+q}{ri+1} \Rightarrow pi+q = 0 \quad \text{... (3)}$$

$$-i = \frac{-p+q}{-r+1} \Rightarrow -p+q = ir-i \quad \text{... (4)}$$

On subtracting (4) from (2), we get $2p = 2i$ or $p = i$

On putting the value of p in (3), we have i (i) + $q = 0$ or $q = 1$

On substituting the values of p and q in (2), we obtain

$$i+1 = i\,r + i \quad \text{or} \quad 1 = i\,r \quad \text{or} \quad r = -i$$

By using the values of p, q, r and (1), we have

$$w = \frac{iz+1}{-iz+1}$$

$$u+iv = \frac{i(x+iy)+1}{-i(x+iy)+1} = \frac{(ix-y+1)(ix+y+1)}{(-ix+y+1)(ix+y+1)} = \frac{-x^2-y^2+1+2ix}{x^2+(y+1)^2}$$

Equating real and imaginary parts, we get

$$u = \frac{-x^2-y^2+1}{x^2+(y+1)^2} \quad \text{... (5)}$$

But $|z| < 1 \Rightarrow x^2+y^2 < 1 \quad \Rightarrow \quad 1-x^2-y^2 > 0$

From (5) $\quad u > 0$ As denominator is positive **Ans.**

Example 12. *Find a bilinear transformation which maps the points i, –i, 1 of the z-plane into 0, 1, ∞ of the w-plane respectively.* (Q. Bank U.P. III Semester 2002)

Solution. By Cross ratio

$$\frac{(w-w_1)(w_2-w_3)}{(w-w_3)(w_2-w_1)} = \frac{(z-z_1)(z_2-z_3)}{(z-z_3)(z_2-z_1)} \quad \text{...(1)}$$

z	w
i	0
–i	1
1	∞

Here $w_1 = 0$, $w_2 = 1$, $w_3 = \infty$ and $z_1 = i$, $z_2 = -i$, $z_3 = 1$

From (1),

$$\frac{(w-w_1)\left(\frac{w_2}{w_3}-1\right)}{\left(\frac{w}{w_3}-1\right)(w_2-w_1)} = \frac{(z-z_1)(z_2-z_3)}{(z-z_3)(z_2-z_1)}$$

$$\Rightarrow \quad \frac{(w-0)\left(\frac{1}{\infty}-1\right)}{\left(\frac{w}{\infty}-1\right)(1-0)} = \frac{(z-i)(-i-1)}{(z-1)(-i-i)}$$

$$\Rightarrow \quad w = \frac{(z-i)(i+1)}{(z-1)(2i)} = \frac{(z-i)(-1+i)}{(z-1)(-2)} = \frac{(i-1)z+(i+1)}{-2z+2} \quad \textbf{Ans.}$$

Example 13. *Find the bilinear transformation which maps the points z = 1, – i, – 1 to the points w = i, 0, – i respectively. Show also that transformation maps the region outside the circle | z | = 1 into the half-plane R(w) ≥ 0.* (Q. Bank U.P. III Semester 2002)

Solution. On putting $z = 1, -i, -1$ and $w = i, 0, -i$ in

$$\frac{(w-w_1)(w_2-w_3)}{(w-w_3)(w_2-w_1)} = \frac{(z-z_1)(z_2-z_3)}{(z-z_3)(z_2-z_1)}$$

z	w
1	i
–i	0
–1	–i

$$\Rightarrow \quad \frac{(w-i)(0+i)}{(w+i)(-i)} = \frac{(z-1)(-i+1)}{(z+1)(-i-1)}$$

$$\frac{i-w}{i+w} = \frac{(z-1)(i-1)}{(z+1)(i+1)}$$

$$\frac{i-w}{i+w} = \frac{(i-1)\,z+1-i}{(i+1)\,z+1+i}$$

$$\Rightarrow \quad \frac{2i}{-2w} = \frac{2iz+2}{-2z-2i} \qquad \text{(Applying Componendo and Dividendo)}$$

$$\Rightarrow \quad \frac{i}{-w} = \frac{iz+1}{-(z+i)} \quad \Rightarrow \quad w = \frac{i(z+i)}{iz+1}$$

$$\Rightarrow \quad w = \frac{iz-1}{iz+1} \qquad \text{... (1)} \qquad \textbf{Ans.}$$

From (1), $\qquad z = i\left(\dfrac{w+1}{w-1}\right)$

$|z| \geq 1$ is transformed into $\left|\dfrac{w+1}{w-1}\right| |i| \geq 1$

$$\Rightarrow \quad |w+1|^2 \geq |w-1|^2 \quad \Rightarrow \quad |u+iv+1|^2 \geq |u+iv-1|^2$$

$$\Rightarrow \quad (u+1)^2+v^2 \geq (u-1)^2+v^2 \quad \Rightarrow \quad u \geq 0$$

$$\Rightarrow \quad \text{R}(w) \geq 0.$$

Thus exterior of the circle $|z| = 1$ is transformed into the half-plane $\text{R}(w) \geq 0$. **Proved.**

Example 14. *Find the bilinear transformation which maps the points $z = 0, -1, i$ onto $w = i, 0, \infty$. Also find the image of the unit circle $|z| = 1$.*

[*Uttarakhand, III Semester 2008, U.P. III semester (C.O.) 2003*]

Solution. On putting $z = 0, -1, i$ into $w = i, 0, \infty$ respectively in

$$\frac{(w-w_1)(w_2-w_3)}{(w-w_3)(w_2-w_1)} = \frac{(z-z_1)(z_2-z_3)}{(z-z_3)(z_2-z_1)} \qquad \text{...(1)}$$

z	w
0	i
–1	0
i	∞

$$\Rightarrow \quad \frac{(w-w_1)\left(\dfrac{w_2}{w_3}-1\right)}{\left(\dfrac{w}{w_3}-1\right)(w_2-w_1)} = \frac{(z-z_1)(z_2-z_3)}{(z-z_3)(z_2-z_1)}$$

$$\Rightarrow \quad \frac{(w-i)(-1)}{(-1)(0-i)} = \frac{(z-0)(-1-i)}{(z-i)(-1-0)} \quad \Rightarrow \quad \left(\frac{w-i}{-i}\right) = \frac{z(1+i)}{z-i}$$

$$\Rightarrow \quad w-i = \frac{(-i+1)z}{z-i} \quad \Rightarrow \quad w = \frac{(1-i)z}{z-i}+i = \frac{(1-i)z+iz+1}{z-i}$$

$$\Rightarrow \quad w = \frac{z+1}{z-i} \qquad \text{...(2)} \qquad \textbf{Ans.}$$

From (2) $$z = \frac{iw+1}{w-1} \qquad ...(3) \qquad \left[\text{Inverse transformation is } z = \frac{dw+b}{cw-a}\right]$$

And $$|z| = 1$$

$$\Rightarrow \quad \left|\frac{iw+1}{w-1}\right| = 1 \qquad \Rightarrow \quad |1 + iw| = |w - 1|$$

$$\Rightarrow \quad |1 + i(u + iv)| = |u + iv - 1| \quad \Rightarrow \quad |1 - v + iu| = |u - 1 + iv|$$

$$\Rightarrow \quad (1 - v)^2 + u^2 = (u - 1)^2 + v^2 \quad \Rightarrow \quad 1 + v^2 - 2v + u^2 = u^2 + 1 - 2u + v^2$$

$$\Rightarrow \quad u - v = 0 \quad \Rightarrow \quad v = u$$ **Ans.**

Example 15. *Find the fixed points and the normal form of the following bilinear transformations.*

(*a*) $w = \frac{3z-4}{z-1}$ and (*b*) $w = \frac{z-1}{z+1}$

Discuss the nature of these transformations.

Solution. (*a*) The fixed points are obtained by

$$z = \frac{3z-4}{z-1} \quad \text{or} \quad z^2 - 4z + 4 = 0 \quad \text{or} \quad (z-2)^2 = 0 \Rightarrow z = 2$$

$z = 2$ in the only fixed point. This transformation is parabolic.

Normal Form

$$w = \frac{3z-4}{z-1} \quad \Rightarrow \quad \frac{1}{w-2} = \frac{1}{\frac{3z-4}{z-1} - 2} = \frac{z-1}{3z-4-2z+2} = \frac{z-1}{z-2}$$

$$\Rightarrow \quad \frac{1}{w-2} = \frac{1}{z-2} + 1$$

(*b*) The fixed points are obtained by

$$z = \frac{z-1}{z+1} \quad \Rightarrow \quad z^2 + z = z - 1 \Rightarrow z^2 = -1 \Rightarrow z = \pm i$$

Hence $\pm\, i$ are the two fixed points.

Normal Form

$$w = \frac{z-1}{z+1}$$

$$w - i = \frac{z-1}{z+1} - i = \frac{z-1-i(z+1)}{z+1} \qquad ...(1)$$

and $$w + i = \frac{z-1}{z+1} + i = \frac{z-1+i(z+1)}{z+1} \qquad ...(2)$$

On dividing (1) by (2), we get

$$\frac{w-i}{w+i} = \frac{z-1-iz-i}{z-1+iz+i} = \frac{(1-i)(z-i)}{(1+i)(z+i)} = \frac{(-i^2-i)(z-i)}{(1+i)(z+i)}$$

$$\frac{w-1}{w+1} = -i\left(\frac{z-i}{z+i}\right) = k\left(\frac{z-i}{z+i}\right) \quad \text{where } k = -i$$

The transformation is elliptic. **Ans.**

Example 16. *The fixed points of the transformation* $w = \dfrac{2z-5}{z+4}$ *are given by*

(*a*) $\left(\dfrac{5}{2}\ \ 0\right)$ (*b*) (–4, 0) (*c*) (–1 + 2*i*, – 1 – 2*i*) (*d*) $(-1+\sqrt{6}, -1-\sqrt{6})$

(*AMIETE, Dec. 2010*)

Solution. Here $f(z) = \dfrac{2z-5}{z+4}$

In the case of fixed point $z = \dfrac{2z-5}{z+4}$

$\Rightarrow z^2 + 4z = 2z - 5 \qquad \Rightarrow \quad z^2 + 2z + 5 = 0$

$\Rightarrow z = \dfrac{-2 \pm \sqrt{4-20}}{2} = \dfrac{-2 \pm 4i}{2} = -1 \pm 2i$

$z = -1 \pm 2i$ are the only fixed points

Hence (c) is correct answer. **Ans.**

Example 17. *Show that the transformation* $w = i\,\dfrac{1-z}{1+z}$ *transforms the circle* $|z|=1$ *onto the real axis of the w-plane and the interior of the circle into the upper half of the w-plane.*

(*U.P., III Semester, Dec. 2003*)

Solution. $w = i\left(\dfrac{1-z}{1+z}\right)$

$$(u+iv) = i\left(\frac{1-(x+iy)}{1+(x+iy)}\right) = \frac{(i-ix+y)}{[1+(x+iy)]}\frac{[(1+x)-iy]}{[(1+x)-iy]}$$

$$= \frac{i+ix+y-ix-ix^2-xy+y+xy-iy^2}{(1+x)^2+y^2} = \frac{y-xy+y+xy+i+ix-ix-ix^2-iy^2}{(1+x)^2+y^2}$$

$$= \frac{2y + i(1-x^2-y^2)}{(1+x)^2+y^2}$$

Equating the real and imaginary parts, we get

$$u = \frac{2y}{(1+x)^2+y} \qquad \text{... (1)}$$

and

$$v = \frac{1-(x^2+y^2)}{(1+x)^2+y^2} \qquad \text{... (2)}$$

when $x^2+y^2=1$, then $v = \dfrac{1-1}{(1+x)^2+y^2} = 0$

$v=0$ is the equation of the real axis in the w – plane. **Proved.**

(*b*) Now the equation of the interior of the circle is $x^2+y^2<1$.

Dividing (1) by (2), we get

$$\frac{u}{v} = \frac{2y}{1-(x^2+y^2)}, \qquad u-u\,(x^2+y^2) = 2vy, \qquad u(x^2+y^2) = u$$

$$x^2+y^2 = 1-\frac{2vy}{u}, \quad 1-\frac{2vy}{u} < 1 \qquad [\text{as } x^2+y^2<1]$$

$$-\frac{2vy}{u} < 0, \quad 2vy>0$$

$$v>0$$

$v>0$ is the equation of the upper half of w-plane. **Proved.**

Example 18. *Show that* $\omega = \dfrac{i-z}{i+z}$ *maps the real axis of the z-plane into the circle* $|w| = 1$ *and (ii) the half-plane* $y > 0$ *into the interior of the unit circle* $|\omega| < 1$ *in the w-plane.*

(U.P., III Semester, Dec. 2005, 2002)

Solution. We have $\omega = \dfrac{i-z}{i+z}$

$$|\omega| = \left|\frac{i-z}{i+z}\right| = \frac{|i-z|}{|i+z|} = \frac{|i-x-iy|}{|i+x+iy|}$$

$$|\omega| = \left|\frac{-x+i(1-y)}{x+i(1+y)}\right|, \quad |\omega| = \frac{\sqrt{x^2+(1-y)^2}}{\sqrt{x^2+(1+y)^2}}$$

Now the real axis in z-plane *i.e.* $y = 0$, transform into

$$|\omega| = \frac{\sqrt{x^2+1}}{\sqrt{x^2+1}} = 1, \quad |\omega| = 1 \qquad\qquad |z| = 1$$

Hence the real axis in the z-plane is mapped into the circle $|\omega|=1$.

(*ii*) The interior of the circle *i.e.* $|w|<1$ gives.

$$\frac{\sqrt{x^2+(1-y)^2}}{\sqrt{x^2+(1+y)^2}} < 1$$

$$\Rightarrow \quad \frac{x^2+(1-y)^2}{x^2+(1+y)^2} < 1 \quad \Rightarrow \quad x^2+(1-y)^2 < x^2+(1+y)^2$$

$$\Rightarrow \quad 1+y^2-2y < 1+y^2+2y \Rightarrow -4y < 0, \Rightarrow \quad y>0.$$

Thus the upper half of the z-plane corresponds to the interior of the circle $|w| = 1$. **Proved.**

Example 19. *Show that the transformation* $w = \dfrac{3-z}{z-2}$ *transforms the circle with centre* $\left(\dfrac{5}{2}, 0\right)$ *and radius* $\dfrac{1}{2}$ *in the z-plane into the imaginary axis in the w-plane and the interior of the circle into the right half of the plane.* *(A.M.I.E.T.E. Summer 2000)*

Solution. $w = \dfrac{3-z}{z-2} \Rightarrow u+iv = \dfrac{3-x-iy}{x+iy-2} \Rightarrow (u+iv)(x+iy-2) = 3-x-iy$

$$\Rightarrow \quad ux + iuy - 2u + ivx - vy - 2iv = 3 - x - iy$$

$$\Rightarrow \quad ux - 2u - vy + i(uy + vx - 2v) = 3 - x - iy$$

Equating real and imaginary quantities, we have

$$ux - vy - 2u = 3 - x \text{ and } vx - 2v + uy = -y$$

$$\Rightarrow \quad (u+1)x - vy = 2u+3 \text{ and } vx + (u+1)y = 2v$$

On solving the equations for x and y, we have

$$x = \frac{2u^2+2v^2+5u+3}{u^2+v^2+2u+1}, \quad y = \frac{-v}{u^2+v^2+2u+1}$$

Here, the equation of the given circle is $\left(x-\frac{5}{2}\right)^2+y^2=\frac{1}{4}$ ... (1)

Putting the values of x and y in (1), we have

$$\left(\frac{2u^2+2v^2+5u+3}{u^2+v^2+2u+1}-\frac{5}{2}\right)^2+\left(\frac{-v}{u^2+v^2+2u+1}\right)^2=\frac{1}{4}$$

$$\Rightarrow \quad \left(\frac{-u^2-v^2+1}{2(u^2+v^2+2u+1)}\right)^2+\left(\frac{-v}{u^2+v^2+2u+1}\right)^2=\frac{1}{4}$$

$$\Rightarrow \quad (-u^2-v^2+1)^2+4v^2=(u^2+v^2+2u+1)^2$$

$$\Rightarrow \quad (u^2+v^2-1)^2+4v^2=[(u^2+v^2-1)+(2u+2)]^2$$

$$\Rightarrow \quad (u^2+v^2-1)^2+4v^2=(u^2+v^2-1)^2+(2u+2)^2+2(u^2+v^2-1)(2u+2)$$

$$\Rightarrow \quad v^2=(u+1)^2+(u^2+v^2-1)(u+1)$$

$$\Rightarrow \quad v^2=u^2+2u+1+u^3+uv^2-u+u^2+v^2-1$$

$$\Rightarrow \quad 0=u^3+2u^2+u+uv^2$$

$$\Rightarrow \quad u(u^2+2u+1+v^2)=0 \Rightarrow u=0 \text{ i.e., equation of imaginary axis.}$$

Equation of the interior of the circle is $\left(x-\frac{5}{2}\right)^2+y^2<\frac{1}{4}$

Then corresponding equation in u, v is

$$u(u^2+2u+1+v^2)>0 \text{ or } u[(u+1)^2+v^2]>0$$

As $(u+1)^2+v^2>0$ so $u>0$ *i.e.*, equation of the right half plane. **Ans.**

28.17 INVERSE POINT WITH RESPECT TO A CIRCLE

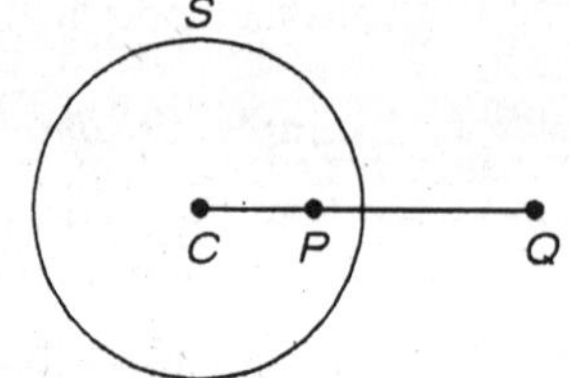

Two points P and Q are said to be the inverse points with respect to a circle S if they are collinear with the centre C on the same side of it, and if the product of their distances from the centre is equal to r^2 where r is the radius of the circle.

Thus when P and Q are the inverse points of the circle, then the three points C, P, Q are collinear, and also $CP.CQ = r^2$

Example 20. *Show that the inverse of a point a, with respect to the circle* $|z-c| = R$ *is the point* $c+\frac{R^2}{\bar{a}-\bar{c}}$

Solution. Let b be the inverse point of the point a' with respect to the circle $|z-c| = R$.

Condition I. The points a, b, c are collinear. Hence

$$\arg(\bar{b}-\bar{c}) = \arg(\bar{a}-\bar{c}) = -\arg(\bar{a}-\bar{c}) \qquad (\text{since } \arg z = -\arg \bar{z})$$

$$\Rightarrow \quad \arg(\bar{b}-\bar{c})+\arg(\bar{a}-\bar{c})=0 \text{ or } \arg(\bar{b}-\bar{c})(\bar{a}-\bar{c})=0$$

$\therefore$ $(\bar{b}-\bar{c})(\bar{a}-\bar{c})$ is real, so that

$$(\bar{b}-\bar{c})(\bar{a}-\bar{c})=|(\bar{b}-\bar{c})(\bar{a}-\bar{c})|$$

Condition II. $|\bar{b}-\bar{c}||\bar{a}-\bar{c}|=R^2 \Rightarrow |\bar{b}-\bar{c}||\bar{a}-\bar{c}|=R^2$ $\{|z|=|\bar{z}|\}$

$$|(\bar{b}-\bar{c})(\bar{a}-\bar{c})|=R^2 \Rightarrow (\bar{b}-\bar{c})(\bar{a}-\bar{c})=R^2 \Rightarrow \bar{b}-\bar{c}=\frac{R^2}{\bar{a}-\bar{c}}$$

$$\Rightarrow \qquad b=c+\frac{R^2}{\bar{a}-\bar{c}}.$$ **Proved.**

Example 21. *Find a Mobius transformation which maps the circle $|w| \leq 1$ into the circle $|z-1|<1$ and maps $w=0$, $w=1$ respectively into $z=\frac{1}{2}$, $z=0$.*

Solution. Let the transformation be,

$$w=\frac{az+b}{cz+d} \qquad \ldots (1)$$

z	w
$\frac{1}{2}$	0
0	1

Since, $w=0$ maps into $z=\frac{1}{2}$,
From (1), we get

$$0=\frac{\frac{a}{2}+b}{\frac{c}{2}+d} \Rightarrow \frac{a}{2}+b=0 \qquad \ldots (2)$$

Since $w=1$ maps into $z=0$, from (1), we get

$$1=\frac{0+b}{0+d} \Rightarrow b=d \qquad \ldots (3)$$

Here $|w|=1$ corresponding to $|z-1|=1$

Therefore points $w, \frac{1}{\bar{w}}$ inverse with respect to the circle $|w|=1$ correspond to the points $z, 1+\frac{1}{\bar{z}-1}$ inverse with respect to the circle $|z-1|=1$

[z and $a+\frac{R^2}{\bar{z}-\bar{a}}$ are inverse points on the circle $|z-a|=R$]

Particular $w=0$ and ∞ correspond to

$$z=\frac{1}{2}, 1+\frac{1}{\frac{1}{2}-1} \Rightarrow z=\frac{1}{2}, -1$$

Since $w=0$ maps into $z=-1$, from (1), we get

$$\infty=\frac{-a+b}{-c+d} \Rightarrow -c+d=0 \Rightarrow c=d \qquad \ldots (4)$$

From (2), (3) and (4), $\quad b=-\frac{a}{2}, \quad b=c=d$

From (1) $$w=\frac{az+b}{cz+d}=\frac{-2bz+b}{bz+b}=\frac{-2z+1}{z+1}$$ **Ans.**

Example 22. *Find two bilinear transformations whose fixed points are 1 and 2.*

(Q. Bank U.P.T.U. 2002)

Solution. We have, $$w=\frac{az+b}{cz+d} \qquad \ldots (1)$$

Fixed points are given by

$$z=\frac{az+b}{cz+d}$$

$$\Rightarrow \qquad cz^2 - (a-d)z - b = 0 \qquad \Rightarrow \qquad z^2 - \frac{(a-d)}{c} - \frac{b}{c} = 0 \qquad \text{... (2)}$$

Fixed points are 1 and 2, so

$$(z-1)(z-2) = 0$$

$$\Rightarrow \qquad z^2 - 3z + 2 = 0$$

Equating the coefficeints of z and constants in (2) and (3), we get

$$\therefore \qquad \frac{a-d}{c} = 3 \quad \text{and} \quad -\frac{b}{c} = 2$$

$$\Rightarrow \qquad b = -2c \quad \text{and} \quad d = a - 3c$$

Putting the values of b and d in (1), we get

$$w = \frac{az - 2c}{cz + a - 3c} \text{ has its fixed points at } z = 1 \text{ and } z = 2.$$

Taking $a = 1$, $c = -1$ and $a = 2$, $c = -1$, we have

$$w = \frac{z+2}{4-z} \quad \text{and} \quad w = \frac{2(z+1)}{5-z}$$

Ans.

Example 23. *Show that the transformation* $w = \dfrac{2z+3}{z-4}$ *maps the circle* $x^2 + y^2 - 4x = 0$ *onto the straight line* $4u + 3 = 0$.

Solution. We have, $\quad w = \dfrac{2z+3}{z-4}$

The inverse transformation is $z = \dfrac{4w+3}{w-2}$ ... (1)

Now the circle $x^2 + y^2 - 4x = 0$ can be written as $z\bar{z} - 2(z + \bar{z}) = 0$ $\quad \begin{bmatrix} z = x + iy \\ \bar{z} = x - iy \end{bmatrix}$

Substituting for z and $\bar{z}$ from (1), we get

$$\frac{4w+3}{w-2} \cdot \frac{4\bar{w}+3}{\bar{w}-2} - 2\left(\frac{4w+3}{w-2} + \frac{4\bar{w}+3}{\bar{w}-2}\right) = 0$$

$$\Rightarrow \quad 16w\bar{w} + 12w + 12\bar{w} + 9 - 2(4w\bar{w} + 3\bar{w} - 8w - 6 + 4w\bar{w} + 3w - 8\bar{w} - 6) = 0$$

$$\Rightarrow \quad 22(w + \bar{w}) + 33 = 0 \quad \Rightarrow \quad 22(2u) + 33 = 0 \Rightarrow 4u + 3 = 0 \qquad \begin{bmatrix} w = u + iv \\ \bar{w} = u - iv \end{bmatrix}$$

Thus, circle is transformed into a straight line. **Ans.**

Example 24. *Show that the transformation* $w = \dfrac{5-4z}{4z-2}$ *transform the circle* $|z| = 1$ *into a circle of radius unity in w-plane and find the centre of the circle.*

(Q. Bank U.P. III Semester 2002)

Solution. Here, $\quad w = \dfrac{5-4z}{4z-2}$

$$\Rightarrow \qquad z = \frac{2w+5}{4w+4} \quad \Rightarrow \quad |z| = \left|\frac{2w+5}{4w+4}\right|$$

$$|z| = 1 \quad \Rightarrow \quad \left|\frac{2w+5}{4w+4}\right| = 1$$

$$\Rightarrow \qquad |2w+5| = |4w+4|$$

$\Rightarrow \quad |2u + 5 + 2iv| = |4u + 4 + 4iv|$ [$\because w = u + iv$]

$\Rightarrow \quad (2u + 5)^2 + 4v^2 = (4u + 4)^2 + (4v)^2$ ...(1)

$\Rightarrow \quad 4u^2 + 25 + 20u + 4v^2 = 16u^2 + 16 + 32u + 16v^2$

$\Rightarrow \quad 12u^2 + 12v^2 + 12u - 9 = 0$

$\Rightarrow \quad u^2 + v^2 + u - \frac{3}{4} = 0.$

This is the equation of circle in w plane. ...(2)

Now we have to find its centre.

$u^2 + v^2 + 2gu + 2fv + c = 0$ ...(3)

From (2) and (3), $\quad g = \frac{1}{2},\ f = 0,\ c = -\frac{3}{4}$

Centre is $(-g, -f)$ *i.e.*, $\left(-\frac{1}{2}, 0\right)$

$$\text{Radius} = \sqrt{g^2 + f^2 - c} = \sqrt{\frac{1}{4} + 0 + \frac{3}{4}} = 1$$

Thus (2) is circle with its centre at $\left(-\frac{1}{2}, 0\right)$ and of radius unity in w-plane. **Proved.**

Example 25. *Find the image of* $x^2 + y^2 - 4y + 2 = 0$ *under the mapping* $w = \frac{z - i}{iz - 1}$.

(*Q. Bank U.P. III Semester 2002*)

Solution. $\quad w = \frac{z - i}{iz - 1} \Rightarrow w(iz - 1) = z - i$

$x^2 + y^2 - 4y + 2 = 0$ ... (1)

$\Rightarrow \quad z = \frac{w - i}{iw - 1}$

$\Rightarrow \quad x + iy = \frac{u + i(v - 1)}{iu - (v + 1)}$ ...(2)

$\therefore \Rightarrow \quad x - iy = \frac{u - i(v - 1)}{-iu - (v + 1)}$ ...(3)

Multiplying (2) and (3) we get

$\Rightarrow \quad x^2 + y^2 = \frac{u^2 + (v - 1)^2}{u^2 + (v + 1)^2}$ ... (4)

Subtracting (3) from (2), we get

$2iy = \frac{-2iu^2 - 2i(v^2 - 1)}{u^2 + (v + 1)^2}$ ... (5)

Putting the values of $(x^2 + y^2)$ and y in (1), we get

$$\frac{u^2 + (v - 1)^2}{u^2 + (v + 1)^2} + 4 \cdot \frac{u^2 + (v^2 - 1)}{u^2 + (v + 1)^2} + 2 = 0$$

$\Rightarrow \quad u^2 + (v - 1)^2 + 4[u^2 + (v^2 - 1)] + 2[u^2 + (v + 1)^2] = 0$

$\Rightarrow \quad 7(u^2 + v^2) + 2v - 1 = 0$

This is the image. **Ans.**

EXERCISE 28.2

1. Find the bilinear transformation that maps the points $z_1 = 2$, $z_2 = i$, $z_3 = -2$ into the points $w_1 = 1$, $w_2 = i$ and $w_3 = -1$ respectively. **Ans.** $w = \dfrac{3z + 2i}{iz + 6}$

2. Determine the bilinear transformation which maps $z_1 = 0$, $z_2 = 1$, $z_3 = \infty$ onto $w_1 = i$, $w_2 = -1$, $w_3 = -i$ respectively. **Ans.** $w = \dfrac{z - i}{iz - 1}$

3. Verify that the equation $w = \dfrac{1 + iz}{1 + z}$ maps the exterior of the circle $|z| = 1$ into the upper half plane $v > 0$.

4. Find the bilinear transformation which maps 1, i, -1 to 2, i, -2 respectively. Find the fixed and critical points of the transformation. **Ans.** i, $2i$

5. Show that the transformation $w = \dfrac{i(1 - z)}{1 + z}$ maps the circle $|z| = 1$ into the real axis of the w-plane and the interior of the circle $|z| < 1$ into the upper half of the w-plane.

6. Show that the transformation $w = \dfrac{iz + 2}{4z + i}$ transforms the real axis in the z-plane into circle in the w-plane. Find the centre and the radius of this circle. *(A.M.I.E.T.E., Winter 2000)* **Ans.** $\left(0, \dfrac{7}{8}\right), \dfrac{9}{8}$

7. If z_0 is the upper half of the z-plane show that the bilinear transformation

$$w = e^{i\alpha}\left(\frac{z - z_0}{z - \overline{z_0}}\right)$$

maps the upper half of the z-plane into the interior of the unit circle at the origin in the w-plane.

8. Find the condition that the transformation $w = \dfrac{az + b}{cz + d}$ transforms the unit circle in the w-plane into straight line in the z-plane. **Ans.** If $\left|\dfrac{c}{a}\right| = 1$ or $|a| = |c|$

9. Prove that $w = \dfrac{z}{1 - z}$ maps the upper half of the z-plane onto the upper half of the w-plane. What is the image of the circle $|z| = 1$ under this transformation ? **Ans.** Straight line $2u + 1 = 0$

10. Show that the map of the real axis of the z-plane on the w-plane by the transformation $w = \dfrac{1}{z + i}$ is a circle and find its centre and radius. **Ans.** Centre $\left(0, -\dfrac{1}{2}\right)$, Radius $= \dfrac{1}{2}$

11. Find the invariant points of the transformation $w = -\left(\dfrac{2z + 4i}{iz + 1}\right)$. Prove also that these two points together with any point z and its image w, form a set of four points having a constant cross ratio. **Ans.** $4i$ and $-i$

12. Show that under the transformation $w = \dfrac{z - i}{z + i}$, the real axis in z-plane is mapped into the circle $|w| = 1$. What portion of the z-plane corresponds to the interior of the circle ?
Ans. The half z-plane above the real axis corresponds to the interior of the circle $|w| = 1$.

13. Discuss the application of the transformation $w = \dfrac{iz + 1}{z + i}$ to the areas in the z-plane which are respectively inside and outside the unit circle with its centre at the origin.

14. What is the form of a bilinear transformation which has one fixed point α and the other fixed point ∞?

Choose the correct alternative:

15. The fixed points of the mapping w = (5z + 4)/(z + 5) are

(*i*) – 4/5, –5 (*ii*) 2, 2 (*iii*) – 2, – 2 (*iv*) 2, –2 **Ans.** (*iv*)

16. The invariant points of the bilinear transformation are

(*i*) 1 ± 2*i* (*ii*) –1 ± 2i (*iii*) ± 2i (*iv*) invariant point does not exit

(*AMIETE, June 2010*) **Ans.** (*iv*)

28.18 TRANSFORMATION: $w = z^2$

(*U.P., III Semester, Summer 2002*)

$$w = z^2$$

$$u + iv = (x + iy)^2 = x^2 - y^2 + 2ixy$$

Equating real and imaginary parts, we get $u = x^2 - y^2,\ v = 2xy$

(*i*) (*a*) Any line parallel to *x*-axis, *i.e.*, *y* = *c*, maps into

$$u = x^2 - c^2, \quad v = 2cx$$

Eliminating *x*, we get $v^2 = 4c^2(u + c^2)$ ... (1) which is a parabola.

(*b*) Any line parallel to *y*-axis, *i.e.*, *x* = *b*, maps into a curve

$$u = b^2 - y^2, \quad v = 2by$$

Eliminating *y*, we get $v^2 = -4b^2(u - b^2)$, ... (2) which is a parabola.

(*c*) The rectangular region bounded by the lines *x* = 1, *x* = 2, and *y* = 1, *y* = 2 maps into the region bounded by the parabolas.

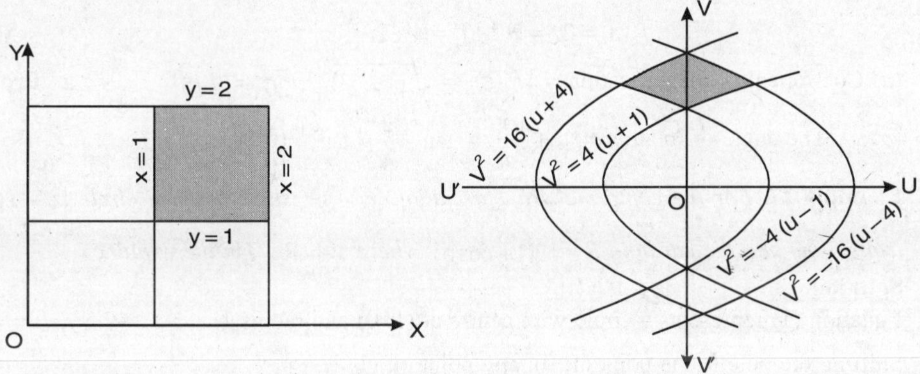

By putting *x* = 1 = *b* in (2) we get $v^2 = -4(u - 1)$,

By putting *x* = 2 = *b* in (2) we get $v^2 = -16(u - 4)$ and

By putting *y* = 1 = *c* in (1) we get $v^2 = 4(u + 1)$,

By putting *y* = 2 = *c* in (1) we get $v^2 = 16(u + 4)$

(*ii*) (*a*) In polar co-ordinates: $z = re^{i\theta}, w = Re^{i\phi}$

$$w = z^2$$

$$Re^{i\phi} = r^2 e^{2i\theta}$$

Then $R = r^2,\ \phi = 2\theta$

In *z*-plane, a circle *r* = *a* maps into $R = a^2$ in *w*-plane.

Thus, circles with centre at the origin map into circles with centre at the origin.

(*b*) If $\theta = 0, \phi = 0$, *i.e.*, real axis in *z*-plane maps into real axis in *w*-plane

If $\theta = \frac{\pi}{2}$, $\phi = \pi$, *i.e.*, the positive imaginary axis in z-plane maps into negative real axis in w-plane.

Thus, the first quadrant in z-plane $0 \le \theta \le \frac{\pi}{2}$, maps into upper half of w-plane $0 \le \phi \le \pi$.

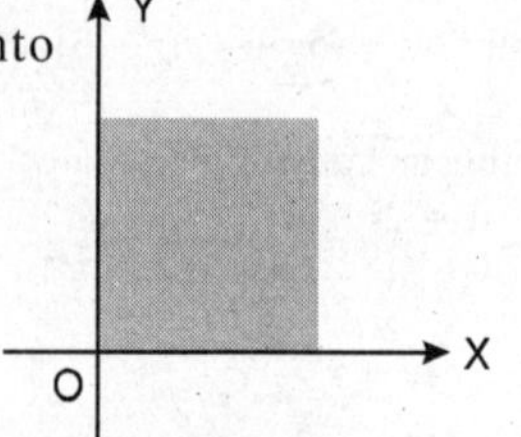

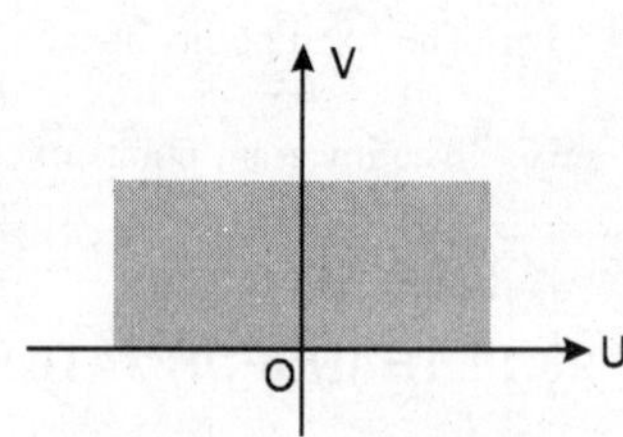

The angles in z-plane at origin maps into double angle in w-plane at origin.

Hence, the mapping $w = z^2$ is not conformal at the origin.

It is conformal in the entire z-plane except origin. Since $\frac{dw}{dz} = 2z = 0$ for $z = 0$, therefore, it is critical point of mapping.

Example 26. *For the conformal transformation* $w = z^2$, *show that*

(*a*) *the coefficient of magnification at* $z = 2 + i$ *is* $2\sqrt{5}$

(*b*) *the angle of rotation at* $z = 2 + i$ *is* $\tan^{-1}$ (0.5).

Solution.

$$z = 2 + i$$

$$f(z) = w = z^2$$

$$= (2+i)^2 = 4 - 1 + 4i = 3 + 4i$$

$$f'(z) = 2z = 2(2+i) = 4 + 2i$$

(*a*) Coefficient of magnification $= |f'(z)| = \sqrt{4^2 + 2^2} = \sqrt{20} = 2\sqrt{5}$ **Proved.**

(*b*) The angle of rotation $= \tan^{-1} \frac{v}{u} = \tan^{-1} \frac{2}{4} = \tan^{-1}$ (0.5) **Proved.**

Example 27. *For the conformal transformation* $w = z^2$, *show that the circle* $|z - 1| = 1$ *transforms into the cardioid* $R = 2(1 + \cos\phi)$ *where* $w = \text{R}e^{i\phi}$ *in the w-plane.*

Solution. $|z - 1| = 1$ ... (1)

Equation (1) represents a circle with centre at (1, 0) and radius 1.

Shifting the pole to the point (1, 0), any point on (1) is $1 + e^{i\theta}$

Transformation is under $w = z^2$.

$$\text{R}e^{i\phi} = (1 + e^{i\theta})^2$$

$$= e^{i\theta}\left(e^{\frac{i\theta}{2}} + e^{-\frac{i\theta}{2}}\right)^2$$

$$= e^{i\theta}\left(2\cos\frac{\theta}{2}\right)^2 = 4e^{i\theta}\cos^2\frac{\theta}{2}$$

This gives $R = 4\cos^2\frac{\theta}{2}$,

$\Rightarrow$ $R = 2\left(2\cos^2\frac{\phi}{2}\right)$ $[\phi = \theta]$

$\Rightarrow$ $R = 2(\cos\phi + 1)$ **Proved**

28.19 TRANSFORMATION: $w = Z^n$ $(n \in N)$

$$Re^{i\phi} = (re^{i\theta})^n = r^n e^{in\theta}$$

Hence, $R = r^n,\ \phi = n\theta$

Mapping of simple figures

z-plane	w-plane
Circle, $r = a$	Circle, $R = a^n$
The initial line, $\theta = 0$	The initial line, $\phi = 0$
The straight line, $\theta = \theta_0$	The straight line, $\phi = n\,\theta_0$

28.20 TRANSFORMATION: $w = z + \frac{1}{z}$

$$\frac{dw}{dz} = 1 - \frac{1}{z^2}$$

At $z = \pm 1$, $\frac{dw}{dz} = 0$, so transformation is not conformal at $z = \pm 1$.

$$w = z + \frac{1}{z} = r(\cos\theta + i\sin\theta) + \frac{1}{r(\cos\theta + i\sin\theta)}$$

$$= r(\cos\theta + i\sin\theta) + \frac{1}{r}(\cos\theta - i\sin\theta)$$

$$u + iv = \left(r + \frac{1}{r}\right)\cos\theta + i\left(r - \frac{1}{r}\right)\sin\theta$$

$$u = \left(r + \frac{1}{r}\right)\cos\theta \quad \text{and} \quad v = \left(r - \frac{1}{r}\right)\sin\theta$$

$$\frac{u}{r + \frac{1}{r}} = \cos\theta \quad \text{and} \quad \frac{v}{r - \frac{1}{r}} = \sin\theta$$

$$\sin^2\theta + \cos^2\theta = \frac{u^2}{\left(r + \frac{1}{r}\right)^2} + \frac{v^2}{\left(r - \frac{1}{r}\right)^2} \quad \Rightarrow \quad 1 = \frac{u^2}{\left(r + \frac{1}{r}\right)^2} + \frac{v^2}{\left(r - \frac{1}{r}\right)^2}$$

z-plane	w-plane
Circle, $r = r$ Circle, $r = 1$	Ellipses Lines $u = 2$
Lines, $\theta = \theta_0$	Hyperbola : $\frac{u^2}{4\cos^2\theta} - \frac{v^2}{4\sin^2\theta} = 1$

28.21 TRANSFORMATION: $w = e^z$

$$u + iv = e^{x+iy} = e^x(\cos y + i\sin y)$$

Equating real and imaginary parts, we have

$$u = e^x \cos y, \qquad v = e^x \sin y$$

Again $w = e^z$

$$Re^{i\phi} = e^{x+iy} = e^x . e^{iy}$$

Hence $\quad R = e^x$ or $x = \log_e R$ and $y = \phi$

Mapping of simple figures

z-plane	w-plane
The straight line $x = c$	Circle $R = e^c$
y-axis ($x = 0$)	Unit Circle $R = e^0 = 1$
Region between $y = 0$, $y = \pi$	Upper half plane
Region between $y = 0$, $y = -\pi$	Lower half plane
Region between the lines $y = c$ and $y = c + 2\pi$	Whole plane

Example 28. *Find the image and draw a rough sketch of the mapping of the region* $1 \le x \le 2$ and $2 \le y \le 3$ *under the mapping* $w = e^z$.

Solution. $\quad z = x + iy$

Let $\quad w = Re^{i\phi} \quad$ ... (1)

But $\quad w = e^z = e^{x+iy} \quad$ ... (2)

From (1) and (2); $\quad Re^{i\phi} = e^{x+iy} = e^x . e^{iy}$

Equating real and imaginary parts, we get $R = e^x$ ... (3) and $\phi = y$

(*i*) Here $1 \le x$, then $R = e^1$ is circle of radius $e^1 = 2.7$

$x = 2$, then $R = e^2$ represents a circle of radius $e^2 = 7.4$

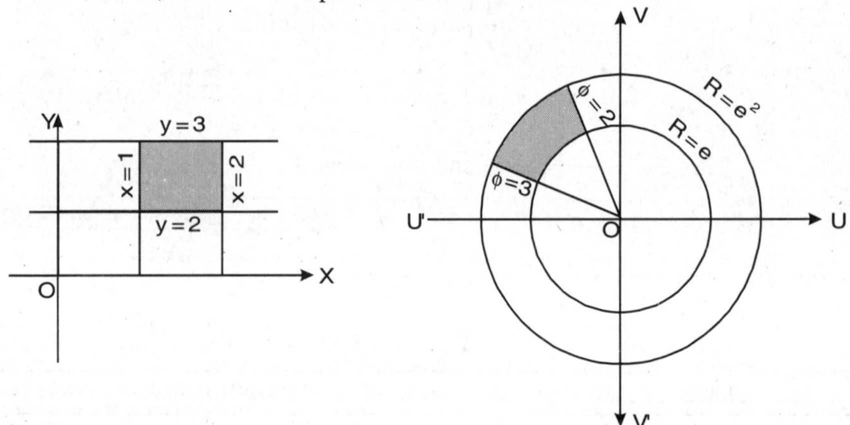

(*ii*) $y = 2$, then $\phi = 2$ represents radial line making an angle of 2 radians with the x-axis. $y = 3$, then $\phi = 3$ represents radial line making an angle 3 radians with x-axis.

Hence, the mapping of the region $1 \le x \le 2$ and $2 \le y \le 3$ maps the shaded sectors in the figure. **Ans.**

Example 29. *Find the image of the strip* $-\frac{\pi}{2} < x < \frac{\pi}{2}, 1 < y < 2$ *under the mapping* $w(z) = \sin z$.

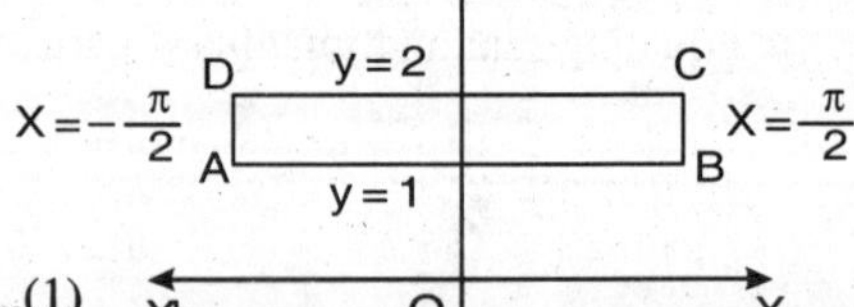

Solution. $w(z) = \sin z = \sin (x + iy)$

$$= \sin x \cos iy + \cos x \sin iy$$

$$u + iv = \sin x \cosh y + i \cos x \sinh y$$

$$u = \sin x \cosh y \Rightarrow \sin x = \frac{u}{\cosh y} \quad ...(1)$$

$$v = \cos x \sinh y \Rightarrow \cos x = \frac{v}{\sinh y} \qquad \text{...(2)}$$

Eliminating x from (1) and (2), we get

$$\sin^2 x + \cos^2 x = \frac{u^2}{\cosh^2 y} + \frac{v^2}{\sinh^2 y} \Rightarrow 1 = \frac{u^2}{\cosh^2 y} + \frac{v^2}{\sinh^2 y}$$

Hence $y = 2$, maps into the ellipse

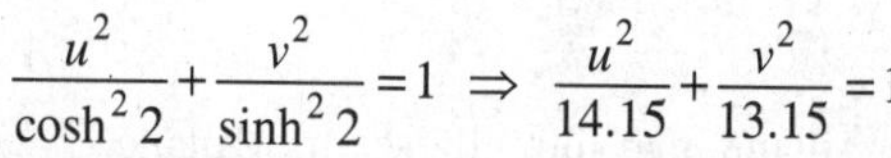

$$\frac{u^2}{\cosh^2 2} + \frac{v^2}{\sinh^2 2} = 1 \Rightarrow \frac{u^2}{14.15} + \frac{v^2}{13.15} = 1$$

Also $y = 1$, maps into the ellipse.

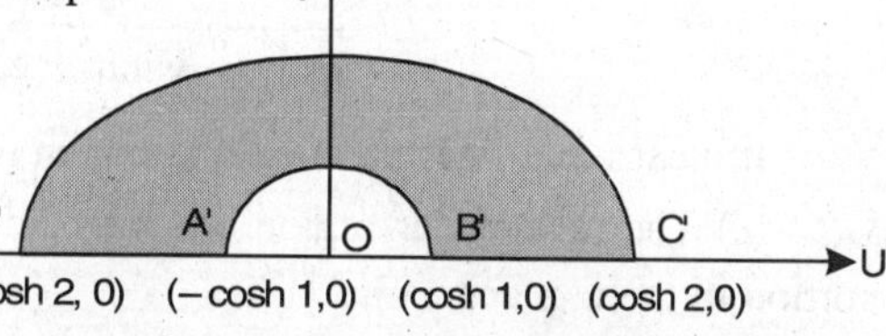

The image of $A\left(\frac{-\pi}{2}, 1\right)$ in z-plane is $(-\cosh 1, 0)$ *i.e.* $(-1.543, 0)$ in w-plane

The image of the point $D\left(-\frac{\pi}{2}, 2\right)$ in z-plane is $(-\cosh 2, 0)$ *i.e.*, $(-7.524, 0)$.

Hence, AD line in z-plane maps into $A'D'$ line in w-plane.

The image of $B\left(\frac{\pi}{2}, 1\right)$ is $(\cosh 1, 0)$ *i.e.*, $(1.543, 0)$ in w-plane.

The image of $C\left(\frac{\pi}{2}, 2\right)$ is $(\cosh 2, 0)$ *i.e.*, $(7.524, 0)$ in w-plane.

Hence, BC line maps into $B'C'$ line in w-plane.

Hence, the strip $\frac{-\pi}{2} < x < \frac{\pi}{2}$, $1 < y < 2$ maps into the shaded region of w-plane bounded by the ellipses and u-axis. **Ans.**

28.22 TRANSFORMATION:

$$\mathbf{w = \cosh z}$$

$$u + iv = \cosh(x + iy) = \cos i(x + iy) = \cos(ix - y)$$

$$= \cos ix \cos y + \sin ix \sin y = \cosh x \cos y + i \sinh x \sin y$$

So $\quad u = \cosh x \cos y, \quad v = \sinh x \sin y$

On eliminating x, we get $(\cosh^2 x - \sinh^2 x = 1)$ $\quad \frac{u^2}{\cos^2 y} - \frac{v^2}{\sin^2 y} = 1 \qquad \text{... (1)}$

If y is eliminated $(\cos^2 y + \sin^2 y = 1)$ $\quad \frac{u^2}{\cosh^2 x} + \frac{v^2}{\sinh^2 x} = 1 \qquad \text{... (2)}$

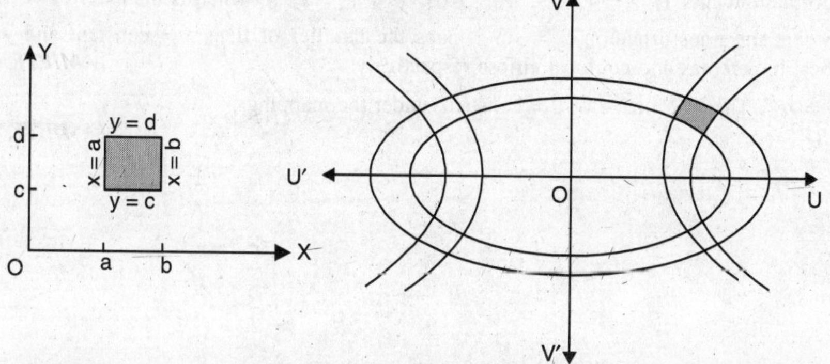

(*a*) On putting $y = a$ (constant) in (1), we get

$$\frac{u^2}{\cos^2 a} - \frac{v^2}{\sin^2 a} = 1 \text{ i.e., Hyperbola.}$$

It shows that the lines parallel to x-axis in the z-plane map into hyperbola in the w-plane.

(*b*) On substituting $x = b$ (constant) in (2), we obtain

$$\frac{u^2}{\cosh^2 b} + \frac{v^2}{\sinh^2 b} = 1$$

It means that lines parallel to y-axis in the z-plane map into ellipses in w-plane.

(*c*) The rectangular region $a \le x \le b,\ c \le y \le d$ in the z-plane transforms into the shaded portion in the w-plane.

EXERCISE 28.3

1. Determine the region of the w-plane into which the region bounded by $x = 1$, $y = 1$, $x + y = 1$ is mapped by the transformation $w = z^2$. *(U.P. III Semester Dec. 2004)*

Ans. $4u + v^2 = 4,\ 4u - v^2 = -4,\ u^2 - 2v^2 = 1$

2. By the transformation $w = z^2$, show that the circle $| z - a | = c$ in the z-plane correspond to the limacon in the w-plane. **Ans.** $R = 2c(a + c\cos\phi)$

3. Under the mapping $w = z^2$, show that the family of circles $| w - 1 | = c$ is transformed into the family of leminscates $|z - 1|\,|z + 1| = c$, where c is the parameter.

4. Discuss the conformal transformation $w = \sqrt{z}$.

5. Show that the transformation $w(z + i)^2 = 1$ maps the inside of the circle $|z| = 1$ in the z-plane on the exterior of the parabola.

6. Show that the transformation $w = z + \frac{1}{z}$ maps the circle $| r | = c$ into the ellipse in w-plane. Discuss the case $c = 1$. $u = \left(c + \frac{1}{c}\right)\cos\theta,\quad v = \left(c - \frac{1}{c}\right)\sin\theta$

Discuss the case when $c = 1$.

7. Show that the transformation $w = z + 1/z$ converts the straight line $\arg z = \alpha\ (|\alpha| < \pi/2)$ into a branch of hyperbola of eccentricity $\sec\alpha$.

8. If $w = z + \frac{a^2}{z}$, prove that when z describes the circle $x^2 + y^2 = a^2$, w describes a st. line of length $4a$.

Also prove that if z describes the circle $x^2 + y^2 = b^2,\ (b > a)$, w describes an ellipse whose foci are the extermities of the above line.

9. Find the region of the z-plane which corresponds to the interior of the circle $| w | = 1$ by means of the transformation $(w + 1)^2 z = 4$ **Ans.** $y^2 + 4x - 4 > 0$ which is the exterior of the parabola.

10. Show that the transformation $\omega = \sin z$ maps the families of lines x = constant and y = cos into confocal hyperbolas and confocal ellipse respectively. *(AMIETE, Dec. 2009)*

11. Let $z = re^{i\theta}$. Then the image of θ = constant, under the mapping $w = Re^{i\theta} = iz^2$ is *(AMIETE, Dec. 2009)*

28.23 SCHWARZ-CHRISTOFFEL TRANSFORMATION

The interior of a polygon of w-plane is mapped into the upper half of the z-plane and the sides of the polygon into the real axis. This transformation is called Schwarz-Christoffel transformation.

The formula of the mapping is

$$\frac{dw}{dz} = A(z-x_1)^{\alpha_1/\pi-1}(z-x_2)^{\alpha_2/\pi-1}....(z-x_n)^{\alpha_n/\pi-1} \quad ... (1)$$

where A is a complex constant.

$\alpha_1, \alpha_2, ...\alpha_n$ are the interior angles of the polygon of the w-plane with vertices $w_1, w_2, ...w_n$ are mapped into the points $x_1, x_2, ...x_n$ on the real axis of the z-plane.

$$w = A\int (z-x_1)^{\alpha_1/\pi-1}(z-x_2)^{\alpha_2/\pi-1}....(z-x_n)^{\alpha_n/\pi-1}dz + B \quad .. (2)$$

Proof. From (1), we have

$$\text{Arg. } dw = \arg dz + \arg A + \left(\frac{\alpha_1}{\pi}-1\right)\arg(z-x_1)+...+\left(\frac{\alpha_n}{\pi}-1\right)\arg(z-x_n) \quad ... (3)$$

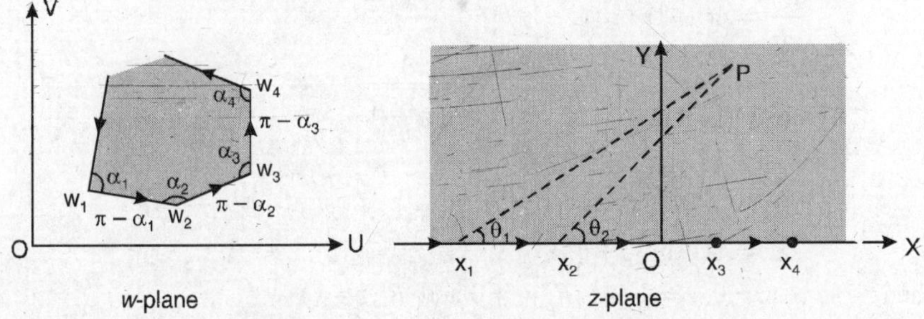

w-plane *z*-plane

As z moves from left towards x_1, let w move along a side of a polygon towards the vertex w_1. So long as z remains to the left of x_1, the arg of w remains unchanged.

When z crosses x_1 from left to right, $\theta_1 = \arg(z-x_1)$ changes from π to zero, while all other terms in (3) remain uchanged. Hence, the arg dw decreases by $\left(\frac{\alpha_1}{\pi}-1\right)\pi = \alpha_1 - \pi$, *i.e.*, increases by $\pi - \alpha$, in the anticlockwise direction. It means amp $\frac{dw}{dz}$ increases by $\pi - \alpha_1$. Thus, the direction of w_1 turns through the angle $\pi - \alpha_1$ and w now moves along the side $w_1 w_2$ of the polygon.

Similarly, as the point z moves from left to right of x_2 the point w moves along the side $w_1 w_2$ of the polygon. Now as the point z crosses x_2 from left to right, the amptitude $(z-x_2)$ changes from π to 0, while the amplitude of all the other terms unaltered. Thus, amp dw increasingly $\left(\frac{\alpha_2}{\pi}-1\right)(-\pi)$ by $\pi - \alpha_2$. Hence, the direction at the point w_2 rotates through an angle $\pi - \alpha_2$. Consequently w begins moving along the third side of the polygon towards w_3.

By continuing the process we find that as the point w moves along the sides of the polygon in the w-plane, the point z moves along the x-axis in the z-plane.

It can also be proved that interior of polygon (if it is closed) maps into upper half of the z-plane

Example 30. *Find the transformation which transforms, the semi infinite strip bounded by $v = 0$, $v = \pi$ and $u = 0$ onto the upper half z-plane.*

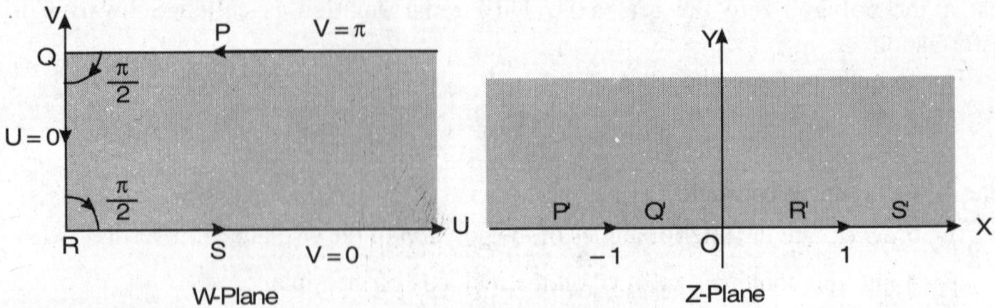

Solution.

Let the points, P, Q, R, S map into P', Q', R', S' respectively. Let us consider $PQRS$ as a limiting case of polygon with vertices Q and R and the third vertex P or S at infinity.
By Schwarz-Christoffel transformation

$$\frac{dw}{dz} = A(z+1)^{\frac{\pi/2}{\pi}-1}(z-1)^{\frac{\pi/2}{\pi}-1} \qquad \left\{\angle Q = \frac{\pi}{2}, \angle R = \frac{\pi}{2}\right\}$$

$$= A(z+1)^{-1/2}(z-1)^{-1/2} = \frac{A}{\sqrt{z^2-1}}$$

$$w = A\int \frac{1}{\sqrt{z^2-1}}\, dz + B$$

$$w = A\cosh^{-1} z + B \qquad \dots (1)$$

when $\quad w = 0,\ z = 1$ (At R, $w = 0$ and $R', z = 1$)

$$0 = A\cosh^{-1}(1) + B \qquad [\cosh^{-1} x = \log(x + \sqrt{x^2-1})]$$

$$0 = A\log(1+\sqrt{1-1}) + B$$

$$0 = 0 + B \text{ or } B = 0$$

Equation (1) is reduced to $w = A\cosh^{-1} z$ ... (2)

At Q, $w = \pi i$; at Q', $z = -1$,

Putting these values of w, z in (2), we get $\pi i = A\cosh^{-1}(-1) = A(\pi i) \Rightarrow A = 1$

Equation (2) is reduced to $w = \cosh^{-1} z$ or $\cosh z = w$ is the required transformation. **Ans.**

Example 31. *Find a transformation which maps the interior of a triangle ABC in w-plane onto the upper half of the z-plane.*

Solution. Let us consider a triangle ABC in w-plane. Suppose vertices A, B, of the triangle ABC map into A', B' and C be mapped into C' at infinity.

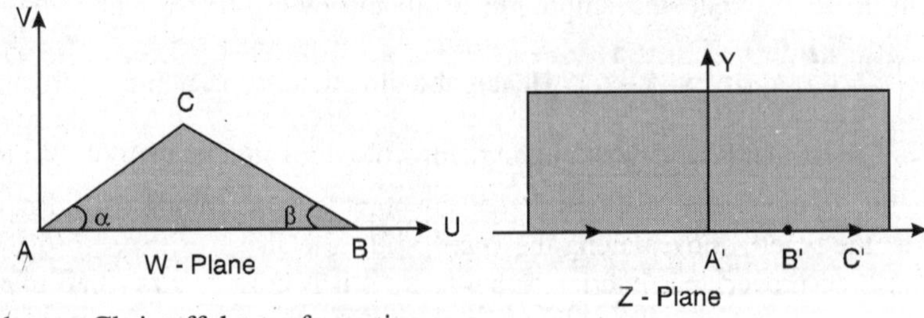

By Schwarz-Christoffel transformation

$$[\angle A = \alpha,\ \angle B = \beta;\ \text{At } A', z = 0;\ \text{At } B', z = 1]$$

$$\frac{dw}{dz} = K(z-0)^{\theta/\pi-1}(z-1)^{\beta/\pi-1} = Kz^{\alpha/\pi-1}(z-1)^{\beta/\pi-1}$$

$$w = \int Kz^{\alpha/\pi-1}(z-1)^{\beta/\pi-1}dz + k_1 \qquad \text{... (1)}$$

(At A, w = 0; at A', z = 0, so $k_1 = 0$)

At B, $w = 1$, at B', $z = 1$, we get $1 = K\int_0^1 z^{\alpha/\pi-1}(1-z)^{\beta/\pi-1}dz$

$$1 = K\frac{\Gamma\left(\frac{\alpha}{\pi}\right)\Gamma\left(\frac{\beta}{\pi}\right)}{\Gamma\left(\frac{\alpha+\beta}{\pi}\right)} \quad \text{or} \quad K = \frac{\Gamma\left(\frac{\alpha+\beta}{\pi}\right)}{\Gamma\left(\frac{\alpha}{\pi}\right)\Gamma\left(\frac{\beta}{\pi}\right)}$$

The required transformation, from (1)

$$w = \frac{\Gamma\left(\frac{\alpha+\beta}{\pi}\right)}{\Gamma\left(\frac{\alpha}{\pi}\right)\Gamma\left(\frac{\beta}{\pi}\right)}\int_0^z z^{\alpha/\pi-1}(1-z)^{\beta/\pi-1}dz$$ **Ans.**

EXERCISE 28.4

1. Find the transformation which will map the interior of the infinite strip bounded by the lines $v = 0$, $v = \pi$ onto the upper half of the z-plane. **Ans.** $w = \log z$
2. Determine the function which maps the semi-infinite strip bounded by $v = -b$, $u = 0$ and $v = b$ into upper half of the z-plane. **Ans.** $w = \frac{2b}{\pi}\cosh^{-1} z - bi$
3. Find the transformation that maps the semi-infinite strip $u = b$, $u = -b$, $v = 0$ in w-plane into the upper half of z-plane. **Ans.** $z = \sin\frac{\pi w}{2b}$

COMPLEX INTEGRATION

(Cauchy's Integral Theorem, Cauchy's Integral Formula for Derivatives of analytic function)

29.1 INTRODUCTION (LINE INTEGRAL)

In case of real variable, the path of integration of $\int_a^b f(x)\,dx$ is always along the x-axis from $x = a$ to $x = b$. But in case of a complex function $f(z)$ the path of the definite integral $\int_a^b f(z)\,dz$ can be along any curve from $z = a$ to $z = b$.

$z = x + iy \quad \Rightarrow \quad dz = dx + idy$ ... (1) $dz = dx$ if $y = 0$ ... (2) $dz = idy$ if $x = 0$ ... (3)

In (1), (2), (3) the direction of dz are different. Its value depends upon the path (curve) of integration. But the value of integral from a to b remains the same along any regular curve from a to b.

In case the initial point and final point coincide so that c is a closed curve, then this integral is called *contour integral* and is denoted by $\oint_C f(z)\,dz$.

If $f(z) = u(x, y) + iv(x, y)$, then since $dz = dx + idy$,
we have

$$\oint_C f(z)\,dz = \int_C (u + iv)(dx + idy) = \int_C (u\,dx - v\,dy) + i\int_C (v\,dx + u\,dy)$$

which shows that the evaluation of the line integral of a complex function can be reduced to the evaluation of two line integrals of real functions.

Let us consider a few examples:

Real integral

Example 1. *Find the value of the integral* $\int_C (x+y)\,dx + x^2y\,dy$

(a) along $y = x^2$, having (0, 0), (3, 9) end points.

(b) along $y = 3x$ between the same points.

Do the values depend upon path.

Solution. $\int_C (x+y)\,dx + x^2y\,dy$ ... (1)

Let $\qquad P = x + y,\ Q = x^2y$

$$\frac{\partial P}{\partial y} = 1, \qquad \frac{\partial Q}{\partial x} = 2xy \qquad \text{or} \qquad \frac{\partial P}{\partial y} \neq \frac{\partial Q}{\partial x}$$

The integrals are not independent of path.

(*a*) Along $y = x^2 \Rightarrow dy = 2x\,dx$

Putting the values of *y* and *dy* in (1), we get

$$\int_0^3 (x + x^2)\,dx + x^2x^2(2x\,dx) = \int_0^3 (x + x^2 + 2x^5)\,dx = \left[\frac{x^2}{2} + \frac{x^3}{3} + \frac{x^6}{3}\right]_0^3$$

$$= \frac{9}{2} + 9 + 243 = 256\frac{1}{2}$$

Which is the required value of given integral.

(*b*) Along $y = 3x$, $dy = 3dx$

Substituting these values in (1), we get

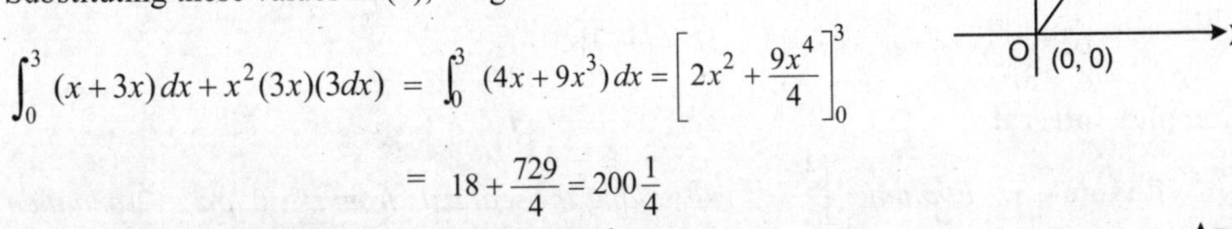

$$\int_0^3 (x + 3x)\,dx + x^2(3x)(3dx) = \int_0^3 (4x + 9x^3)\,dx = \left[2x^2 + \frac{9x^4}{4}\right]_0^3$$

$$= 18 + \frac{729}{4} = 200\frac{1}{4}$$

Which is the required value of given integral. **Ans.**

Example 2. *Evaluate* $\int_{1-i}^{2+i} (2x + iy + 1)\,dz$ *along the two paths:*

(*i*) $x = t + 1,\ y = 2t^2 - 1$

(*ii*) *The straight line joining 1 – i and 2 + i.* (*RGPV, Bhopal, Dec. 2008*)

Solution. (i) Here we have

$$x = t + 1 \Rightarrow dx = dt$$
$$y = 2t^2 - 1 \Rightarrow dy = 4t\,dt$$
$$dz = dx + i\,dy$$
$$= dt + i\,4t\,dt = (1 + i4t)\,dt$$

The limits of the integral are (1, – 1) and (2 , 1).

Corresponding to these points, the limits of *t* are 0 and 1. $\left[\begin{array}{lll}\text{If } t = 0, & x = 1, & y = -1\\ \text{If } t = 1, & x = 2, & y = 1\end{array}\right]$

$$\int_{1-i}^{2+i} (2x + iy + 1)\,dz$$

$$= \int_0^1 [2(t + 1) + i(2t^2 - 1) + 1]\,(1 + 4it)\,dt$$

$$= \int_0^1 [\{2(t + 1) + i(2t^2 - 1) + 1\} + \{8it(t + 1) - 4t(2t^2 - 1) + 4it\}]\,dt$$

$$= \int_0^1 [2(t + 1) + 1 - 4t(2t^2 - 1) + i\{2t^2 - 1 + 8t(t + 1) + 4t\}]\,dt$$

$$= \int_0^1 \{(2t + 2 + 1 - 8t^3 + 4t) + i(2t^2 - 1 + 8t^2 + 8t + 4t)\}\,dt$$

$$= \int_0^1 \{(-8t^3 + 6t + 3) + i(10t^2 + 12t - 1)\}\,dt$$

$$= \left[-2t^4 + 3t^2 + 3t + i\left(\frac{10}{3}t^3 + 6t^2 - t\right)\right]_0^1$$

$$= -2 + 3 + 3 + i\left(\frac{10}{3} + 6 - 1\right) - 0 = 4 + \frac{25}{3}i$$

(ii) $\int_{1-i}^{2+i} (2x + iy + 1)\, dz$

The equation of the straight line joining (1, –1) and (2, 1) is

$$y + 1 = \frac{1+1}{2-1}(x-1) \quad \Rightarrow y + 1 = 2x - 2 \quad \Rightarrow y = 2x - 3 \quad \Rightarrow dy = 2dx$$

$$\int_{(1,-1)}^{(2,1)} (2x + iy + 1)(dx + i\,dy) = \int_1^2 (2x + 2ix - 3i + 1)(dx + 2idx)$$

$$= \int_1^2 (2x + 2ix - 3i + 1)(1 + 2i)\, dx = \int_1^2 (2x + 2ix - 3i + 1 + 4ix - 4x + 6 + 2i)\, dx$$

$$= \int_1^2 (-2x + 6ix - i + 7)\, dx = (-x^2 + 3ix^2 - ix + 7x)_1^2$$

$$= [-4 + 12i - 2i + 14 + 1 - 3i + i - 7]$$

$$= 4 + 8i$$

Ans.

Complex Integral

Example 3. *Evaluate* $\int_0^{2+i} (\bar{z})^2 dz$ *along the real axis from z = 0 to z = 2 and then along a line parallel to y-axis from z = 2 to z = 2 + i.*

(R.G.P.V., Bhopal, III Semester, June 2005)

Solution. $\int_0^{2+i} (\bar{z})^2 dz = \int_0^{2+i} (x - iy)^2 (dx + idy)$

$$= \int_{OA} (x^2)\, dx + \int_{AB} (2 - iy)^2\, i\,dy$$

[Along OA, $y = 0$, $dy = 0$, x varies 0 to 2.
Along AB, $x = 2$, $dx = 0$ and y varies 0 to 1

Y, B, 1, A, X, O, 1, 2

$$= \int_0^2 x^2 dx + \int_0^1 (2 - iy)^2\, i\,dy$$

$$= \left[\frac{x^3}{3}\right]_0^2 + i\int_0^1 (4 - 4iy - y^2)\, dy = \frac{8}{3} + i\left[4y - 2iy^2 - \frac{y^3}{3}\right]_0^1$$

$$= \frac{8}{3} + i\left[4 - 2i - \frac{1}{3}\right] = \frac{8}{3} + \frac{i}{3}(11 - 6i) = \frac{1}{3}(8 + 11i + 6) = \frac{1}{3}(14 + 11i)$$

Which is the required value of the given integral. **Ans.**

Example 4. *Find the value of the integral* $\int_0^{1+i} (x - y + ix^2)\, dz$

(a) *Along the straight line from z = 0 to z = 1 + i;*

(b) *Along real axis from z = 0 to z = 1 and then along a line parallel to the imaginary axis from z = 1 to z = 1 + i.*

Solution. (a) Along OA line: Equation of a straight line OA passing through (0, 0) and (1, 1) is

$$y - 0 = \frac{1-0}{1-0}(x - 0) \quad \Rightarrow \quad y = x.$$

Y, A(1, i), O, X

$$z = x + iy = x + ix = (1+i)x$$

$$dz = (1+i)\, dx$$

$$\int_{OA} (x - y + ix^2)\, dz = \int_0^1 (x - x + ix^2)(1+i)dx$$

$$= (1+i)\int_0^1 ix^2 dx = (1+i)i\left(\frac{x^3}{3}\right)_0^1 = (1+i)\frac{i}{3} = \frac{1}{3}(i-1)$$

Ans.

Required integral $= \int_{OB} (x - y + ix^2)\,dz + \int_{BA} (x - y + ix^2)\,dz$ ... (1)

Now first integral $= \int_{OB} (x - y + ix^2)\,dz = \int_0^1 (x - 0 + ix^2)\,dx$

(Along OB, $y = 0$, $dy = 0$, x varies 0 to 1)

$$= \left(\frac{x^2}{2} + i\frac{x^3}{3}\right)_0^1 = \frac{1}{2} + \frac{i}{3} \qquad ...(2)$$

Y, A (1 + i), dz = idy, O, X = 1, B, X

Now second integral $= \int_{BA} (x - y + ix^2)\,dz$

$$= \int_0^1 (1 - y + i)\,i\,dy \qquad [z = x + iy = 1 + iy \Rightarrow dz = idy]$$

(Along BA, $x = 1$, $dx = 0$, and y varies 0 to 1.)

$$= i\left(y - \frac{y^2}{2} + iy\right)_0^1 = i\left(1 - \frac{1}{2} + i\right) = \frac{i}{2} - 1 \qquad ...(3)$$

Substituting the values of the first and second integral from (2) and (3) in (1), we get

Required integral $= \left(\frac{1}{2} + \frac{i}{3}\right) + \left(\frac{i}{2} - 1\right) = -\frac{1}{2} + \frac{5}{6}i$ **Ans.**

Example 5. *Evaluate* $\int_0^{1+i} (x^2 - iy)\,dz$, *along the path*

(*a*) $y = x$ *(R.G.P.V., Bhopal, III Semester, Dec. 2007)* (*b*) $y = x^2$.

Solution. (*a*) Along the line $y = x$,

$dy = dx$ so that $dz = dx + idy \Rightarrow dz = dx + idx = (1 + i)\,dx$

$\therefore \int_0^{1+i} (x^2 - iy)\,dz$

[On putting $y = x$ and $dz = (1 + i)dx$]

$$= \int_0^1 (x^2 - ix)(1 + i)\,dx$$

$$= (1 + i)\left[\frac{x^3}{3} - i\frac{x^2}{2}\right]_0^1 = (1 + i)\left(\frac{1}{3} - \frac{1}{2}i\right)$$

$$= \frac{(1 + i)(2 - 3i)}{6} = \frac{5}{6} - \frac{1}{6}i.$$

Y, P(1, 1), y = x, y = x², O, X

Which is the required value of the given integral. **Ans.**

(*b*) Along the parabola $y = x^2$, $dy = 2x\,dx$ so that

$dz = dx + idy \Rightarrow dz = dx + 2ix\,dx = (1 + 2ix)\,dx$

and x varies from 0 to 1.

$$\therefore \int_0^{1+i} (x^2 - iy)\,dx = \int_0^1 (x^2 - ix^2)(1 + 2ix)dx = \int_0^1 x^2(1 - i)(1 + 2ix)\,dx$$

$$= (1 - i)\int_0^1 x^2(1 + 2ix)\,dx = (1 - i)\left[\frac{x^3}{3} + i\frac{x^4}{2}\right]_0^1$$

$$= (1 - i)\left[\frac{1}{3} + \frac{1}{2}i\right] = \frac{(1 - i)(2 + 3i)}{6} = \frac{1}{6}(2 + 3i - 2i + 3) = \frac{5}{6} + \frac{1}{6}i$$

Which is the required value of the given integral. **Ans.**

Example 6. *Integrate z^2 along the straight line OM and also along the path OLM consisting of two line segments OL and OM where O is the origin, L is the point z = 3 and M the point z = 3 + i.* (R.G.P.V., Bhopal, III Semester, Dec. 2004)

Solution. We have, $z = x + iy$, $dz = dx + idy$. curve C is line OM.

$$\int_C z^2\, dz = \int_C (x+iy)^2\,(dx+idy) = \int_C (x^2-y^2+2ixy)\,(dx+idy) \quad \text{... (1)}$$

The point M is $z = 3 + i$, i.e, M is (3, 1).

The equation of the line OM is $y-0 = \dfrac{1-0}{3-0}(x-0)$ i.e., $x = 3y$

Now on the line OM, $x = 3y$, $\therefore\ dx = 3dy$ and y varies from 0 to 1. Therefore, from (1), we have

$$\int_{OM} z^2\, dz = \int_0^1 (9y^2 - y^2 + 2i3y.y)\,.(3\,dy+idy) = \int_0^1 (8+6i)\,(3+i)\,y^2\,dy$$

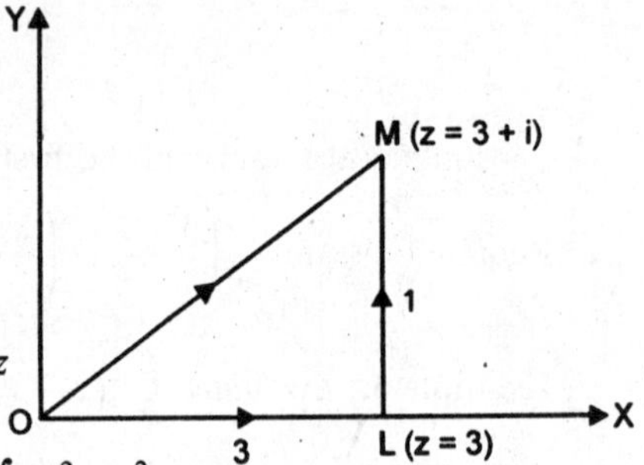

$$= (18+26i)\left[\frac{y^3}{3}\right]_0^1 = \frac{1}{3}(18+26i)$$

$$= 6+i\,\frac{26}{3} \quad \text{... (2)}$$

Now, we have to integrate along OL and LM.

Again, $$\int_{OLM} z^2 dz = \int_{OL+LM} z^2\, dz = \int_{OL} z^2\, dz + \int_{LM} z^2 dz$$

$$= \int_{OL} (x^2-y^2+2ixy)\,(dx+idy) + \int_{LM} (x^2-y^2+2ixy)\,(dx+idy) \quad \text{... (3)}$$

Y, M (z = 3 + i), 1, O, 3, L (z = 3), X

On the line OL, $y = 0$ $\therefore\ dy = 0$ and x varies from 0 to 3.
On the line LM, $x = 3$ $\therefore\ dx = 0$ and y varies from 0 to 1.
$\therefore$ From (3), we obtain

$$\int_{OLM} z^2 dz = \int_0^3 x^2\, dx + \int_0^1 (9-y^2+6iy)\,idy = \left[\frac{x^3}{3}\right]_0^3 + i\left[9\,y - \frac{y^3}{3} + 3iy^2\right]_0^1$$

$$= 9+i\left[9-\frac{1}{3}+3i\right] = 6+i\,\frac{26}{3} \quad \text{... (4)}$$

Which is the required result. **Ans.**

Example 7. *Evaluate $\int_C (12z^2 - 4iz)\, dz$ along the curve C joining the points (1, 1) and (2, 3)* (U.P., III Semester, Dec. 2009)

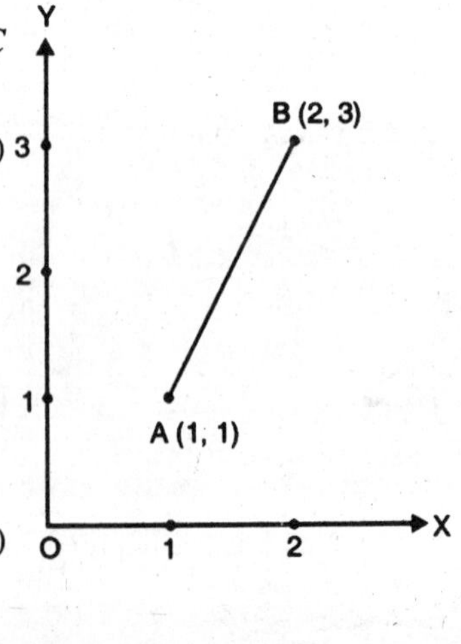

Y, B (2, 3), 3, 2, 1, A (1, 1), O, 1, 2, X

Solution. Here, we hae

$$\int_C (12z^2 - 4iz)\, dz$$

$$= \int_C [12\,(x+iy)^2 - 4i\,(x+iy)]\,(dx + i\,dy)$$

$$= \int_C [12\,(x^2 - y^2 + 2i\,xy) - 4ix + 4y]\,(dx + i\,dy)$$

$$= \int_C (12\,x^2 - 12y^2 + 24\,ixy - 4ix + 4y)\,(dx + i\,dy) \quad \text{... (1)}$$

Equation of the line AB passing through (1, 1) and (2, 3) is

$$y-1 = \frac{3-1}{2-1}(x-1)$$

$$y-1 = 2(x-1) \Rightarrow y = 2x-1 \Rightarrow dy = 2\,dx$$

Putting the values of y and dy in (1), we get

$$= \int_1^2 (12x^2 - 12(2x-1)^2 + 24\,ix(2x-1) - 4ix + 4(2x-1))\,[dx + i2dx)$$

$$= \int_1^2 [12x^2 - 48x^2 + 48x - 12 + 48\,ix^2 - 24\,ix - 4ix + 8x - 4]\,(1+2i)\,dx$$

$$= (1+2i)\int_1^2 [-36+48i)\,x^2 + (56-28i)\,x - 16]\,dx$$

$$= (1+2i)\left[(-36+48i)\frac{x^3}{3} + (56-28i)\frac{x^2}{2} - 16x\right]_1^2$$

$$= (1+2\text{i})\left[(-36+48i)\frac{8}{3} + (56-28i)\,2 - 16\times 2 - (36+48i)\frac{1}{3} - (56-28i)\frac{1}{2} + 16\right]$$

$$= (1+2\text{i})\,(-96 + 128\,i + 112 - 56i - 32 + 12 - 16i - 28 + 14\,i + 16$$

$$= (1+2i)\,(-16+70i) = -16 + 70i - 32i - 140 = -156 + 38i \qquad \textbf{Ans.}$$

Example 8. *Evaluate the line integral* $\int_C z^2\,dz$ *where C is the boundary of a triangle with vertices 0, 1 + i, – 1 + i clockwise.*

Solution. Here, $I = \int_c z^2\,dz$

where C is the boundary of a triangle with vertices 0, $1+i$, $-1+i$ clockwise.

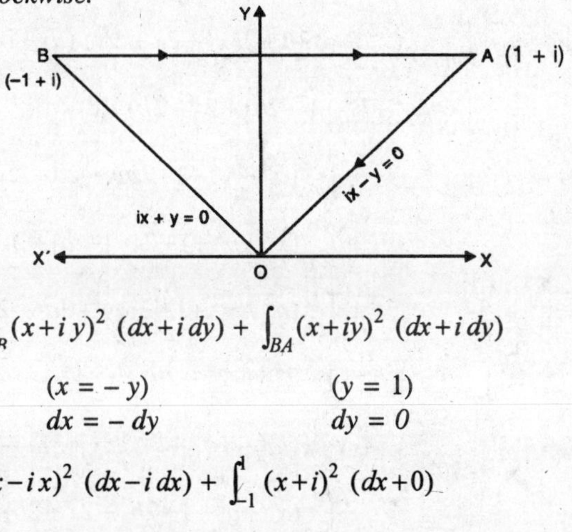

$$I = \int_{AO}(x+iy)^2\,(dx+i\,dy) + \int_{OB}(x+i\,y)^2\,(dx+i\,dy) + \int_{BA}(x+iy)^2\,(dx+i\,dy)$$

$$(x = y) \qquad\qquad (x = -y) \qquad\qquad (y = 1)$$

$$dx = dy \qquad\qquad dx = -dy \qquad\qquad dy = 0$$

$$= \int_1^0 (x+ix)^2\,(dx+i\,dx) + \int_0^{-1}(x-ix)^2\,(dx-i\,dx) + \int_{-1}^1 (x+i)^2\,(dx+0)$$

$$= \int_1^0 x^2\,(1+i)^2\,(1+i)\,dx + \int_0^{-1} x^2\,(1-i)^2\,(1-i)\,dx + \int_{-1}^1 (x+i)^2\,dx$$

$$= (1+i)^3\int_1^0 x^2\,dx + (1-i)^3\int_0^{-1} x^2\,dx + \int_{-1}^1 (x+i)^2\,dx$$

$$= (1+i)^3\left[\frac{x^3}{3}\right]_1^0 + (1-i)^3\left[\frac{x^3}{3}\right]_0^{-1} + \left[\frac{(x+i)^3}{3}\right]_{-1}^1$$

$$= (1+i^3+3i+3i^2)\left[0-\frac{1}{3}\right] + (1-i^3-3i+3i^2)\left[\frac{(-1)^3}{3} - 0\right] + \left[\frac{(1+i)^3}{3} - \frac{(-1+i)^3}{3}\right]$$

$$= (1-i+3i-3)\left(-\frac{1}{3}\right) + (1+i-3i-3)\left(-\frac{1}{3}\right) + \frac{1}{3}(1+i^3+3i+3i^2+1-i^3-3i+3i^2)$$

$$= (2i-2)\left(-\frac{1}{3}\right)+(-2-2i)\left(-\frac{1}{3}\right)+\frac{1}{3}[1-i+3i-3+1+i-3i-3]$$

$$= \frac{1}{3}[2-2i+2i+2-4]=0$$

Which is the required value of the given integral. **Ans.**

By Green's Theorem

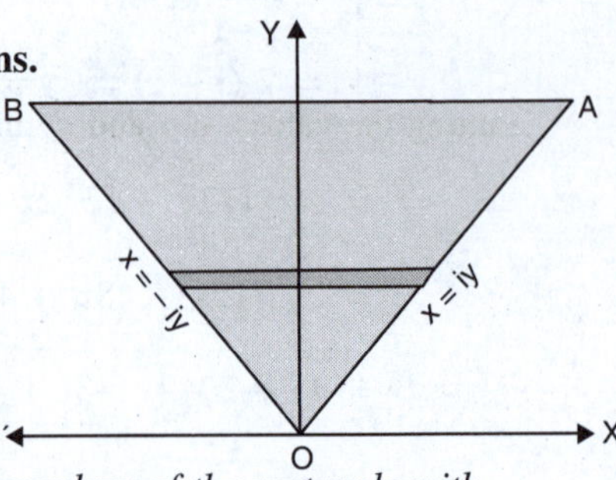

$$I = \int (x+iy)^2\,dx+(x+iy)^2\,i\,dy$$

$$= \iint [2i(x+iy)-2(x+iy)i]\,dx\,dy$$

$$= 0 \qquad \textbf{Ans.}$$

Which is the required value of the given integral.

Example 9. *Evaluate* $\int_C (z+1)^2\,dz$ *where C is the boundary of the rectangle with vertices at the points a + ib, – a + ib, – a – ib, a – ib.*

Solution. Here, $I = \int_C (z+1)^2\,dz$

Where C is the boundary of the rectangle with vertices at the points $a+ib,\ -a+ib,$ $-a-ib$ and $a-ib$.

$$I = \int_{ABCD} (x+iy+1)^2\,(dx+i\,dy)$$

$$= \int_{ABCD} (x^2-y^2+1+2ixy+2x+2iy)\,(dx+i\,dy)$$

$$= \int_{AB} (x^2+b^2+1-2bx+2x-2b)\,dx$$

$$+\int_{BC} (a^2-y^2+1-2iay-2a+2iy)\,(i\,dy)$$

$$+\int_{CD} [x^2+b^2+1+2ix(-ib)+2x+2i(-ib)]\,dx$$

$$+\int_{DA} (a^2-y^2+1+2iay+2a+2iy)\,(i\,dy)$$

Along AB	$y = ib$	$dy = 0$
Along BC	$x = -a$	$dx = 0$
Along CD	$y = -ib$	$dy = 0$
Along DA	$x = a$	$dx = 0$

$$= \int_a^{-a} (x^2+b^2+1-2bx+2x-2b)\,dx$$

$$+\int_{ib}^{-ib} (ia^2-iy^2+i+2ay-2ia-2y)\,dy$$

$$+\int_{-a}^{a} (x^2+b^2+1+2bx+2x+2b)\,dx+\int_{-ib}^{ib} (a^2i-iy^2+i-2ay+2ia-2y)\,dy$$

$$\left[\int_{-a}^{a} f(x)\,dx = 2\int_0^a f(x)\,dx, \text{ if } f(x) \text{ is even.} \quad = 0, \text{ if } f(x) \text{ is odd}\right]$$

$$= -2\int_0^a (x^2+b^2+1-2b)\,dx-2\int_0^{ib} (ia^2-iy^2+i-2ia)\,dy+2\int_0^a (x^2+b^2+1+2b)\,dx$$
$$+2\int_0^{ib} (a^2i-iy^2+i+2ia)\,dy$$

$$= -2\left[\frac{x^3}{3}+b^2x+x-2bx\right]_0^a-2\left[ia^2y-i\frac{y^3}{3}+iy-2iay\right]_0^{ib}+2\left[\frac{x^3}{3}+b^2x+x+2bx\right]_0^a$$
$$+2\left[a^2iy-\frac{iy^3}{3}+iy+2iay\right]_0^{ib}$$

$$= -2\left[\frac{a^3}{3}+ab^2+a-2ab\right]-2\left[-a^2b-\frac{b^3}{3}-b+2ab\right]+2\left[\frac{a^3}{3}+ab^2+a+2ab\right]$$
$$+2\left[-a^2b-\frac{b^3}{3}-b-2ab\right]$$

$$= 2\left[-\frac{a^3}{3}-ab^2-a+2ab+a^2b+\frac{b^3}{3}+b-2ab+\frac{a^3}{3}+ab^2+a+2ab-a^2b-\frac{b^3}{3}-b-2ab\right]$$

$= 0$

Which is the required value of the given integral. **Ans.**

Example 10. *Evaluate* $\int (z-a)^n\,dz$ *where c is the circle with centre a and r. Discuss the case when n = –1.*

Solution. The equation of circle C is $|z-a| = r$ or $z-a = re^{i\theta}$ where θ varies from 0 to 2π

$$dz = ire^{i\theta}d\theta$$

$$\oint_C (z-a)^n\,dz = \int_0^{2\pi} r^n e^{in\theta}.ire^{i\theta}\,d\theta$$

$$= ir^{n+1}\int_0^{2\pi} e^{i(n+1)\theta}\,d\theta = ir^{n+1}\left[\frac{e^{i(n+1)\theta}}{i(n+1)}\right]_0^{2\pi} \qquad [\because n \neq -1]$$

$$= \frac{r^{n+1}}{n+1}[e^{i2(n+1)\pi}-1]$$

$$= \frac{r^{n+1}}{n+1}[\cos 2(n+1)\pi + i\sin 2(n+1)\pi - 1] = \frac{r^{n+1}}{n+1}[1+0i-1]$$

$= 0.$ [When $n \neq -1$]

z = re^{iθ}, r, θ, C

Which is the required value of the given integral.

When $n=-1$,

$$\oint_C \frac{dz}{z-a} = \int_0^{2\pi}\frac{ire^{i\theta}d\theta}{re^{i\theta}} = i\int_0^{2\pi} d\theta = 2\pi i.$$ **Ans.**

Example 11. *Evaluate the integral* $\int_c |z|\,dz$, *where c is the straight line from z = – i to z = i.*

Solution. Equation of the straight line AB from $z = -i$ to $z = i$ is $x = 0$

$\Rightarrow$ $z = x + iy = 0 + iy = iy$ so that $dz = idy$

$$\therefore \quad \int_c |z|\,dz = \int_{-1}^{1} |iy|\,i\,dy = i\int_{-1}^{0}(-y)\,dy + i\int_0^1 y\,dy$$

$$= -i\left(\frac{y^2}{2}\right)_{-1}^{0} + i\left(\frac{y^2}{2}\right)_0^1 = -i\left(-\frac{1}{2}\right)+i\left(\frac{1}{2}\right) = i$$

Y, A, i, X', O, X, B, –i, Y'

Which is the required value of the given integral. **Ans.**

Example 12. *Evaluate the integral* $\int_c |z|\,dz$, *where c is the left half of the unit circle |z| = 1 from z = –i to z = i.*

Solution. For a point on the unit circle $|z| = 1$,

$$z = e^{i\theta}$$

$$\therefore \quad dz = ie^{i\theta}\,d\theta.$$

The points $z = -i$ and i correspond to $\theta = \frac{3\pi}{2}$ and $\theta = \frac{\pi}{2}$ respectively.

$$\therefore \quad \int_c |z|\,dz = \int_{3\pi/2}^{\pi/2} 1 \cdot e^{i\theta}\, i\,d\theta$$

$$= \left(e^{i\theta}\right)_{3\pi/2}^{\pi/2} = e^{i\pi/2} - e^{3i\pi/2}$$

$$= \cos\frac{\pi}{2} + i\sin\frac{\pi}{2} - \cos\frac{3\pi}{2} - i\sin\frac{3\pi}{2}$$

$$= 0 + i - 0 - i(-1) = 2i.$$

Y, i, $\theta = \frac{\pi}{2}$, X', X, O, $-i$, $\theta = \frac{3\pi}{2}$, Y'

Which is the required value of the given integral. **Ans.**

Example 13. *Evaluate the integral* $\int_c \log z\, dz$, *where c is the unit circle* $|z| = 1$.

Solution. Here, $\quad c \equiv |z| = 1$ ...(1)

$$\int_c \log z\, dz = \int_c \log(x+iy)\, dz = \int_c \left[\frac{1}{2}\log(x^2+y^2) + i\tan^{-1}\frac{y}{x}\right] dz$$

$$= \int_c \left[\frac{1}{2}\log 1 + i\tan^{-1}\frac{y}{x}\right] dz \qquad [\because x^2+y^2 = 1]$$

$$= \int_c \left[0 + i\tan^{-1}\frac{y}{x}\right] dz = i\int_c \tan^{-1}\left(\frac{y}{x}\right) dz \qquad ...(2)$$

On the unit circle, $\quad z = e^{i\theta}$

$\therefore \quad dz = ie^{i\theta}\, d\theta$

Now (2) becomes,

$$\int_c \log z\, dz = i\int_0^{2\pi} \tan^{-1}(\tan\theta)\, ie^{i\theta}\, d\theta = -\int_0^{2\pi} \theta e^{i\theta}\, d\theta$$

$$= -\left[\left(\theta\frac{e^{i\theta}}{i}\right)_0^{2\pi} - \int_0^{2\pi} 1 \cdot \frac{e^{i\theta}}{i}\, d\theta\right] = -\left[\frac{2\pi}{i}e^{2\pi i} - \frac{1}{i}\left(\frac{e^{i\theta}}{i}\right)_0^{2\pi}\right]$$

$$= -\left[\frac{2\pi}{i}e^{2\pi i} + e^{2\pi i} - 1\right] = 2\pi i e^{2\pi i} + 1 - e^{2\pi i}$$

$$= (2\pi i - 1)\, e^{2\pi i} + 1 = (2\pi i - 1)(\cos 2\pi + i\sin 2\pi) + 1$$

$$= 2\pi i - 1 + 1 = 2\pi i$$

Which is the required value of the given integral. **Ans.**

EXERCISE 29.1

1. Integrate $f(z) = x^2 + ixy$ from $A(1, 1)$ to $B(2, 8)$ along

(*i*) the straight line AB; (*ii*) the curve C, $x = t$, $y = t^3$. **Ans.** (*i*) $-\frac{1}{3}(147 - 71)i$ (*ii*) $-\left(\frac{1094}{21} - \frac{124i}{5}\right)$

2. Evaluate $\int_{1-i}^{2+i} (2x + iy + 1)\, dz$ along

(*i*) $x = t + 1$, $y = 2t^2 - 1$; (*ii*) the straight line joining $1 - i$ and $2 + i$. **Ans.** (*i*) $4 + \frac{25}{3}i$ (*ii*) $4 + 8i$

(*R.G.P.V., Bhopal, Dec. 2008*)

3. Evaluate the line integral $\int_c (3y^2 dx + 2y\, dy)$, where c is the circle $x^2 + y^2 = 1$, counter clockwise from (1, 0) to (0, 1). **Ans.** -1.

4. Integrate xz along the straight line from A (1, 1) to B (2, 4) in the complex plane. Is the value the same if the path of integration from A to B is along the curve $x = t,\ y = t^2$? **Ans.** $-\dfrac{151}{15}+\dfrac{45i}{4}$

5. Evaluate $\displaystyle\int_0^{2+i} (\bar{z})^2 dz$, along the line $y = x/2$.

(U.P., III Semester, June 2009) **Ans.** $\dfrac{5}{3}(2-i)$

Choose the correct answer:

6. The value of the integral $\displaystyle\int_C (4x^3 dx + 3y^2 z^2 dy + 2y^3 z\, dz)$ where C is any path joining A (– 1, 1, 0) to B (1, 2, 1) is

(*i*) 1 (*ii*) 0 (*iii*) – 8 (*iv*) 8 **Ans.** (*iv*)

7. The value of $\displaystyle\int_C \frac{4z^2+z+5}{z-4}\, dz$, where C : $9x^2 + 4y^2 = 36$

(*i*) – 1 (*ii*) 1 (*iii*) 2 (*iv*) 0 *(AMIETE, June 2009)* **Ans.** (*iv*)

29.2 IMPORTANT DEFINITIONS

(*i*) **Simply connected Region.** A connected region is said to be a simply connected if all the interior points of a closed curve C drawn in the region D are the points of the region D.

(*ii*) **Multi-Connected Region.** Multi-connected region is bounded by more than one curve. We can convert a multi-connected region into a simply connected one, by giving it one or more cuts.

Note. A function $f(z)$ is said to be **meromorphic** in a region R if it is analytic in the region R except at a finite number of poles.

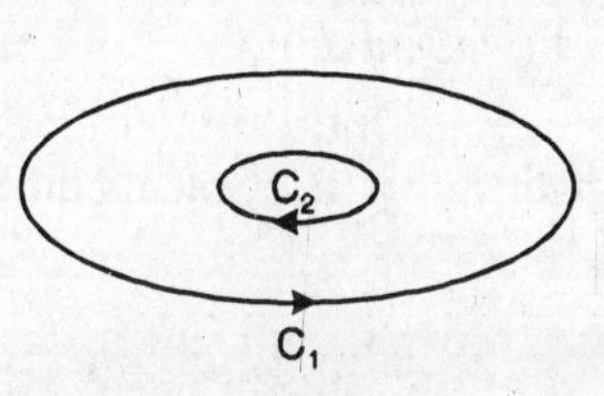

Multi-Connected Region

Simply Connected Region

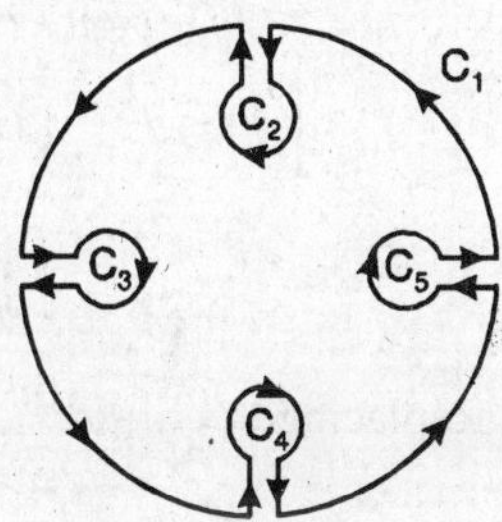

Simply Connected Region

(*iii*) **Single-valued and Multi-valued function**

If a function has only one value for a given value of z, then it is a single valued function.

For example $f(z) = z^2$

If a function has more than one value, it is known as multi-valued function,

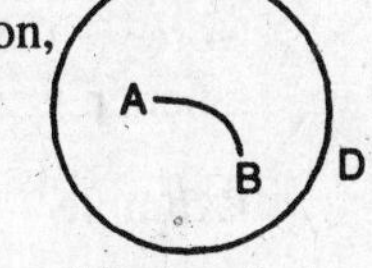

For example $f(z) = z^{\frac{1}{2}}$

(*iv*) **Limit of a function**

A function $f(z)$ is said to have a limit l at a point $z = z_0$, if for a given an arbitrary chosen positive number ε, there exists a positive number δ , such that

$$|f(z)-l| < \varepsilon \text{ for } |z-z_0| < \delta$$

It may be written as $\lim\limits_{z\to z_0} f(z) = l$

(*v*) **Continuity**

A function $f(z)$ is said to be continuous at a point $z = z_0$ if for a given an arbitrary positive number ε , there exists a positive number δ , such that

$$|f(z)-f(z_0)|<\varepsilon \text{ for } |z-z_0|<\delta$$

In other words, a function $f(z)$ is continuous at a point $z = z_0$ if

(a) $f(z_0)$ exists (b) $\lim_{z\to z_0} f(z) = f(z)_{z=0}$

(vi) **Multiple point.** If an equation is satisfied by more than one value of the variable in the given range, then the point is called a multiple point of the arc.

(vii) **Jordan arc.** A continuous arc without multiple points is called a Jordan arc.

Regular arc. If the derivatives of the given function are also continuous in the given range, then the arc is called a regular arc.

(viii) **Contour.** A contour is a Jordan curve consisting of continuous chain of a finite number of regular arcs.

The contour is said to be closed if the starting point A of the arc coincides with the end point B of the last arc.

(ix) **Zeros of an Analytic function.**

The value of z for which the analytic function $f(z)$ becomes zero is said to be the zero of f(z). **For example,** (1) Zeros of z^2-3z+2 are $z=1$ and $z=2$.

(2) Zeros of $\cos z$ is $\pm(2n-1)\dfrac{\pi}{2}$, where $n=1, 2, 3$.........

29.3 CAUCHY'S INTEGRAL THEOREM

(*AMIETE, Dec. 2009, U.P. III Semester, 2009-2010, R.G.P.V., Bhopal, III Semester, Dec. 2002*)

If a function $f(z)$ is analytic and its derivative $f'(z)$ continuous at all points inside and on a simple closed curve c, then $\int_c f(z)\,dz = 0$.

Proof. Let the region enclosed by the curve c be R and let

$$f(z)=u+iv,\quad z=x+iy,\quad dz=dx+idy$$

$$\int_c f(z)\,dz=\int_c (u+iv)(dx+idy)=\int_c (u\,dx-v\,dy)+i\int_c (v\,dx+u\,dy)$$

$$=\iint_R\left(-\frac{\partial v}{\partial x}-\frac{\partial u}{\partial y}\right)dx\,dy+i\iint_c\left(\frac{\partial u}{\partial x}-\frac{\partial v}{\partial y}\right)dx\,dy \qquad \text{(By Green's theorem)}$$

Replacing $-\dfrac{\partial v}{\partial x}$ by $\dfrac{\partial u}{\partial y}$ and $\dfrac{\partial v}{\partial y}$ by $\dfrac{\partial u}{\partial x}$, we get

$$\int_c f(z)\,dz=\iint_R\left(\frac{\partial u}{\partial y}-\frac{\partial u}{\partial y}\right)dx\,dy+i\iint_c\left(\frac{\partial u}{\partial x}-\frac{\partial u}{\partial x}\right)dx\,dy=0+i0=0$$

$$\Rightarrow \int_c f(z)\,dz=0$$ **Proved.**

Note. If there is no pole inside and on the contour then the value of the integral of the function is zero.

Example 14. *Find the integral* $\displaystyle\int_c \frac{3z^2+7z+1}{z+1}\,dz$, where C is the circle $|z| = \dfrac{1}{2}$.

Solution. Poles of the integrand are given by putting the denominator equal to zero.

$z+1=0 \Rightarrow z=-1$

The given circle $|z| = \dfrac{1}{2}$ with centre at $z = 0$ *and radius* $\dfrac{1}{2}$ does not enclose any singularity of the given function.

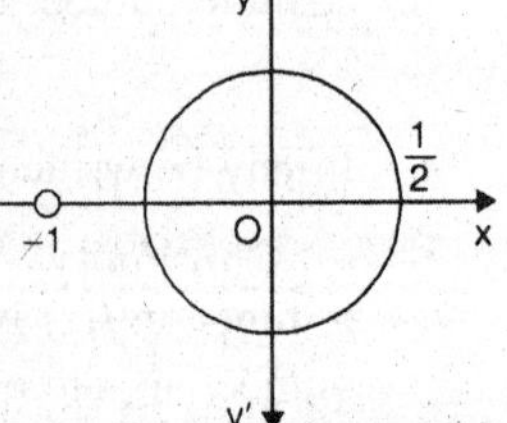

$$\int_C \frac{3z^2+7z+1}{z+1}\,dz=0$$ (*By Cauchy Integral theorem*). **Ans.**

Example 15. *Find the value of* $\int_c \frac{z+4}{z^2+2z+5}\, dz$, *if C is the circle* $|z+1| = 1$.

Solution. Poles of integrand are given by putting the denominator equal to zero.

$$z^2+2z+5=0$$

$$z = \frac{-2 \pm \sqrt{4-20}}{2} = \frac{-2 \pm 4i}{2} = -1 \pm 2i$$

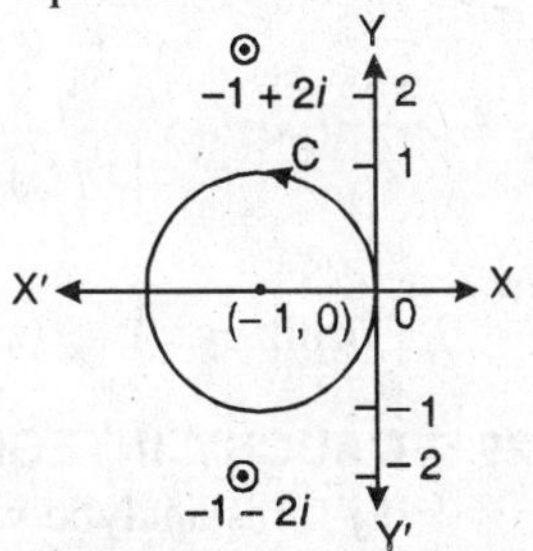

The given circle $|z+1| = 1$ with centre at $z = -1$ and radius unity does not enclose any singularity of the function $\frac{z+4}{z^2+2z+5}$.

$$\therefore \quad \int_c \frac{z+4}{z^2+2z+5}\, dz = 0 \text{ (By Cauchy Integral Theorem)} \qquad \textbf{Ans.}$$

Example 16. *Evaluate* $\oint_c \frac{e^{-z}}{z+1}\, dz$, *where c is the circle* $|z| = \frac{1}{2}$.

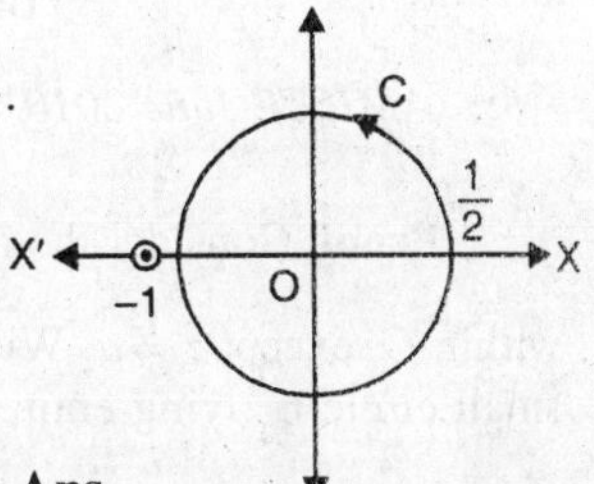

Solution. The point $z = -1$ lies outside the circle $|z| = \frac{1}{2}$.

$\therefore$ The function $\frac{e^{-z}}{z+1}$ is analytic within and on C.

By Cauchy's integral theorem, we have $\oint_c \frac{e^{-z}}{z+1}\, dz = 0.$ **Ans.**

Example 17. *Evaluate:* $\oint_C \frac{2z^2+5}{(z+2)^3(z^2+4)}\, dz.$

where C is the square with the vertices at $1+i,\ 2+i,\ 2+2i,\ 1+2i$.

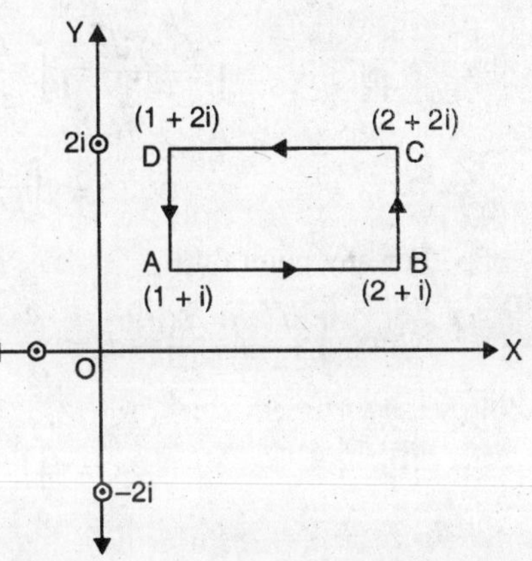

Solution. Here, $\quad f(z) = \frac{2z^2+5}{(z+2)^3(z^2+4)}$

Poles are given by

$z = -2$ (pole of order 3)

$z = \pm 2i$ (simple poles).

Since the obtained poles do not lie inside the contour C with vertices at $1+i$, $2+i$, $2+2i$ and $1+2i$, hence by Cauchy Integral theorem.

$$\oint_C \frac{2z^2+5}{(z+2)^3(z^2+4)}\, dz = 0. \qquad \textbf{Ans.}$$

29.4 EXTENSION OF CAUCHY'S THEOREM TO MULTIPLE CONNECTED REGION

If $f(z)$ is analytic in the region R between two simple closed curves c_1 and c_2 then

$$\int_{c_1} f(z)dz = \int_{c_2} f(z)dz$$

Proof. $\int f(z)\,dz = 0$

where the path of integration is along AB, and curves c_2 in clockwise direction and along BA and along c_1 in anticlockwise direction.

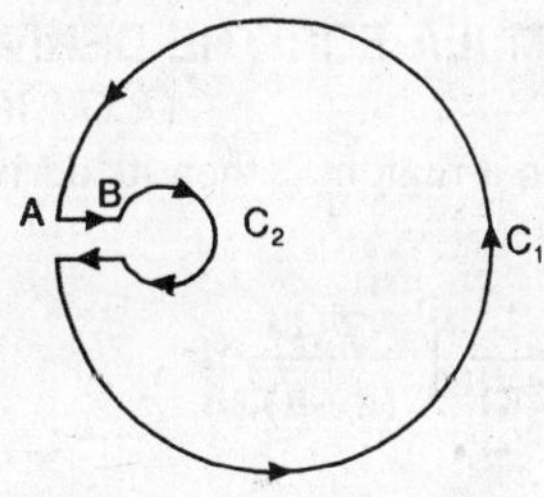

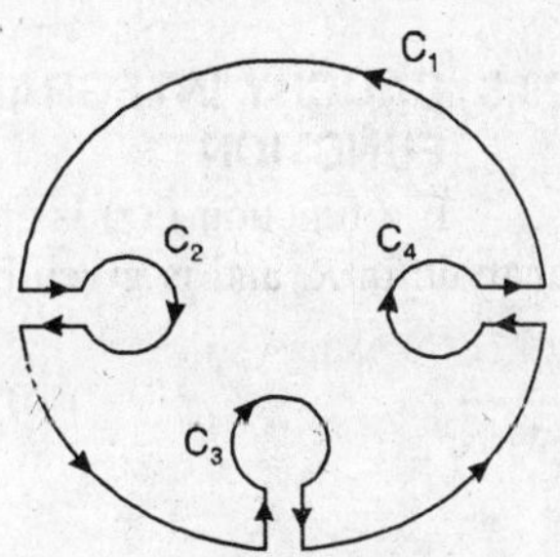

$$\int_{AB} f(z)dz - \int_{C_2} f(z)dz + \int_{BA} f(z)dz + \int_{C_1} f(z)dz = 0$$

$$\Rightarrow \qquad -\int_{C_2} f(z)\,dz + \int_{C_1} f(z)\,dz = 0 \qquad \text{as } \int_{AB} f(z)\,dz = -\int_{BA} f(z)\,dz$$

$$\int_{C_1} f(z)\,dz = \int_{C_2} f(z)\,dz$$

Proved.

Corollary. $\int_{C_1} f(z)dz = \int_{C_2} f(z)dz + \int_{C_3} f(z)dz + \int_{C_4} f(z)dz$

29.5 CAUCHY INTEGRAL FORMULA

If $f(z)$ is analytic within and on a closed curve C, and if a is any point within C, then

$$\boldsymbol{f(a) = \frac{1}{2\pi i}\int_C \frac{f(z)}{z-a}\,dz}$$

(AMIETE June 2010, U.P., III Semester Dec. 2009 R.G.P.V., Bhopal, III Semester, June 2008)

Proof. Consider the function $\frac{f(z)}{z-a}$, which is analytic at all points within C, except $z = a$. With the point a as centre and radius r, draw a small circle C_1 lying entirely within C.

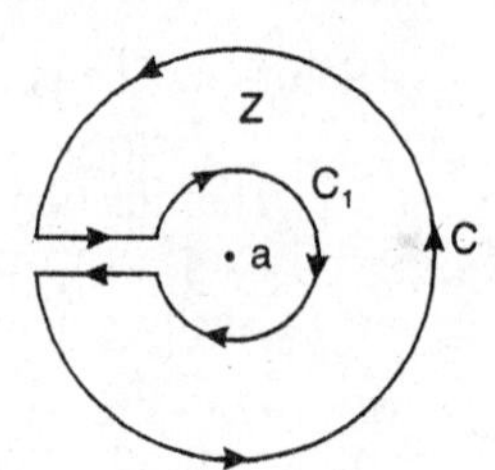

Now $\frac{f(z)}{z-a}$ is analytic in the region between C and C_1; hence by Cauchy's Integral Theorem for multiple connected region, we have

$$\int_C \frac{f(z).dz}{z-a} = \int_{C_1} \frac{f(z)}{z-a}\,dz = \int_{C_1} \frac{f(z)-f(a)+f(a)}{z-a}\,dz$$

$$= \int_{C_1} \frac{f(z)-f(a)}{z-a}\,dz + f(a)\int_{C_1} \frac{dz}{z-a} \qquad \ldots (1)$$

For any point on C_1.

Now, $\int_{C_1} \frac{f(z)-f(a)}{z-a}\,dz = \int_0^{2\pi} \frac{f(a+re^{i\theta})-f(a)}{re^{i\theta}}\, ire^{i\theta}\,d\theta \qquad [\,z-a = re^{i\theta} \text{ and } dz = ire^{i\theta}d\theta\,]$

$$= \int_0^{2\pi} [f(a+re^{i\theta}) - f(a)]\; id\theta = 0 \qquad \text{(where } r \text{ tends to zero).}$$

$$\int_{C_1} \frac{dz}{z-a} = \int_0^{2\pi} \frac{ire^{i\theta}d\theta}{re^{i\theta}} = \int_0^{2\pi} id\theta = i[\theta]_0^{2\pi} = 2\pi i$$

Putting the values of the integrals in R.H.S. of (1), we have

$$\int_C \frac{f(z)\,dz}{z-a} = 0 + f(a)\,(2\pi i) \Rightarrow f(a) = \frac{1}{2\pi i}\int_C \frac{f(z)}{z-a}\,dz$$

Proved.

29.6 CAUCHY INTEGRAL FORMULA FOR THE DERIVATIVE OF AN ANALYTIC FUNCTION

(R.G.P.V., Bhopal, III Semester, Dec. 2007)

If a function $f(z)$ is analytic in a region R, then its derivative at any point $z = a$ of R is also analytic in R, and is given by

$$\boldsymbol{f'(a) = \frac{1}{2\pi i}\int_C \frac{f(z)}{(z-a)^2}\,dz}$$

where c is any closed curve in R surrounding the point $z = a$.

Proof. We know Cauchy's Integral formula

$$f(a) = \frac{1}{2\pi i}\int_c \frac{f(z)}{(z-a)}dz \qquad \dots (1)$$

Differentiating (1) w.r.t. 'a', we get

$$f'(a) = \frac{1}{2\pi i}\int_c f(z)\frac{\partial}{\partial a}\left(\frac{1}{z-a}\right)dz$$

$$\boxed{f'(a) = \frac{1}{2\pi i}\int_c \frac{f(z)}{(z-a)^2}dz}$$

Similarly, $$\boxed{f''(a) = \frac{2!}{2\pi i}\int_c \frac{f(z)\,dz}{(z-a)^3}}$$

$$\boxed{f^n(a) = \frac{n!}{2\pi i}\int_c \frac{f(z)\,dz}{(z-a)^{n+1}}}$$

29.7 MORERA THEOREM (Converse of Cauchy's Theorem)

If a function $f(z)$ is continuous in region D and if the integral $\int f(z)\,dz$, taken around any simple closed contour in D, is zero then $f(z)$ is an analytic function inside D.

Proof. $\int_{z_0}^{z} f(z)\,dz$ is independent of path from z_0 fixed point to a variable point z and hence must be function of z only. Thus $\int_{z_0}^{z} f(z)\,dz = \phi(z)$

$$\int (u+iv)(dx+idy) = U+iV \quad \text{and} \quad f(z) = u+iv$$

$$\Rightarrow \qquad \int_{(x_0,y_0)}^{(x,y)} (udx - vdy) = U \text{ and } \int_{(x_0,y_0)}^{(x,y)} vdx + udy = V$$

Differentiating under the sign of integral, we get

$$\frac{\partial U}{\partial x} = u, \quad \frac{\partial V}{\partial x} = v, \qquad \frac{\partial U}{\partial y} = -v, \quad \frac{\partial V}{\partial y} = u$$

$$\therefore \qquad \frac{\partial U}{\partial x} = \frac{\partial V}{\partial y} \text{ and } \frac{\partial U}{\partial y} = \frac{-\partial V}{\partial x}$$

Thus, U and V satisfy $C - R$ equations.

$\therefore$ $\qquad \phi(z) = U + iV$ is an analytic function.

$$\phi'(z) = \frac{\partial U}{\partial x} + i\frac{\partial V}{\partial x} = u + iv = f(z)$$

$f(z)$ is the derivative of an analytic function $\phi(z)$. **Proved.**

29.8 CAUCHY'S INEQUALITY

If $f(z)$ is analytic within a circle C *i.e.*, $|z-a| = R$ and if $|f(z)| \le M$ on C, then

$$|f^n(a)| \le \frac{Mn!}{R^n}$$

Proof. We know that $f^n(a) = \frac{n!}{2\pi i}\int_c \frac{f(z)dz}{(z-a)^{n+1}} \le \frac{n!}{|2\pi i|}\int_c \frac{|f(z)|\,|dz|}{|z-a|^{n+1}}$

$$\le \frac{n!}{2\pi}\frac{M}{R^{n+1}}\int_0^{2\pi} R\,d\theta \qquad [\text{since } z = Re^{i\theta},\ |dz| = |iRe^{i\theta}d\theta| = Rd\theta]$$

$$\le \frac{n!}{2\pi}\frac{M}{R^{n+1}}2\pi R \le \frac{Mn!}{R^n}$$ **Proved.**

Second Proof. Let a and b be any two points of z plane. Draw a large circle c_n with centre at the origin, of radius R, enclosing the points a and b. So that

$R > |a|$, and also $R > |b|$.

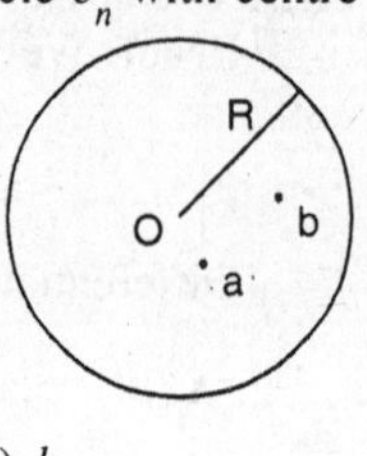

By Cauchy's integral formula

$$\int_c \frac{f(z)\,dz}{z-a} = 2\pi i f(a) \text{ and } \int_c \frac{f(z)}{z-b}dz = 2\pi i f(b)$$

$$2\pi i[f(a)-f(b)] = \int_c \frac{f(z)dz}{z-a} - \int_c \frac{f(z)}{z-b}dz = \int_c \frac{a-b}{(z-a)(z-b)}f(z)dz$$

$$|2\pi i\,[f(a)-f(b)]| = \left|\int_c \frac{a-b}{(z-a)(z-b)} f(z)\,dz\right|$$

$$|f(a)-f(b)| \le \left|\frac{1}{2\pi i}\right| \int_c \frac{|a-b||f(z)||dz|}{|z-a||z-b|}$$

$$\le \frac{1}{2\pi}|a-b|M \int_c \frac{|dz|}{(|z|-|a|)(|z|-|b|)} \quad \text{[Since } f(z) \le M]$$

$$= \frac{1}{2\pi} \frac{|a-b|M}{(R-|a|)(R-|b|)} \int_c |dz| \quad \text{[Since } |z| = R]$$

$$= \frac{1}{2\pi} \frac{(a-b)M}{(R-|a|)(R-|b|)} \int_0^{2\pi} R\,d\theta \quad [z = Re^{i\theta},\ |dz| = R\,d\theta]$$

$$= \frac{|a-b|M\,2\pi R}{2\pi(R-|a|)(R-|b|)} = \frac{|a-b|M}{R\left(1-\frac{a}{R}\right)\left(1-\frac{b}{R}\right)}$$

$$= 0 \text{ as } R \to \infty$$

$\therefore\ f(a) = f(b)$. Since this holds for all values of a and b, therefore $f(x)$ is constant.

29.9 LIOUVILLE THEOREM

(U.P., III Semester, June 2009)

If a function $f(z)$ is analytic for all finite values of z, and is bounded, is a constant.

Proof. Since $f(z)$ is bounded so $|f(z)| \le M$, where M is positive constant.

Let z_1, z_2 be any two points of the z-plane.

Let us draw a circle with centre at origin and large radius R enclosing the points z_1 and z_2. So that $R > |z_1|$ and $R > |z_2|$

$$f(z_1) - f_2(z_2) = \frac{1}{2\pi i}\int_c \frac{f(z)}{z-z_1}dz - \frac{1}{2\pi i}\int_c \frac{f(z)}{z-z_2}dz$$

(Cauchy's Integral formula)

$$= \frac{1}{2\pi i}\int_c \frac{z_1-z_2}{(z-z_1)(z-z_2)}\cdot f(z)\,dz$$

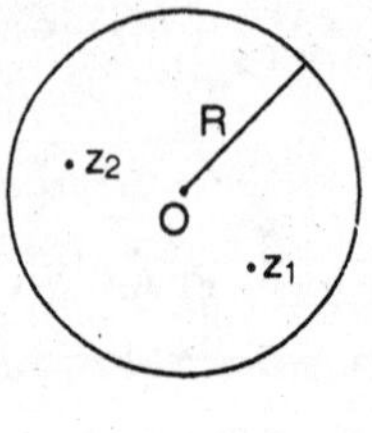

$$|f(z_1)-f(z_2)| = \left|\frac{1}{2\pi i}\int_c \frac{(z_1-z_2)f(z)}{(z-z_1)(z-z_2)}dz\right|$$

$$\le \left|\frac{1}{2\pi i}\right| \int_c \frac{|(z_1-z_2)||f(z)|}{|z-z_1||z-z_2|}|dz|$$

$$\le \frac{1}{2\pi}|z_1-z_2|M \cdot \int_c \frac{|dz|}{[|z|-|z_1|][|z|-|z_2|]} \quad [|f(z)| < M]$$

$$\leq \frac{1}{2\pi} \cdot \frac{|z_1 - z_2| M}{[R - |z_1|][R - |z_2|]} \int_c |dz| \qquad \text{(Since } |z| = R\text{)}$$

$$= \frac{1}{2\pi} \frac{|z_1 - z_2| M}{[R - |z_1|[R - |z_2|]} \int_0^{2\pi} R\, d\theta \qquad (\because z = Re^{i\theta}), \quad |dz| = R\, d\theta$$

$$= \frac{(z_1 - z_2) M . 2\pi R}{2\pi [R - |z_1|] [R - |z_2|]} = 0 \qquad \text{Since } R \to \infty$$

Hence, $\quad f(z_1) = f(z_2)$

$f(z)$ is constant. **Proved.**

Alternative. On putting $n = 1$ in Cauchy's inequality

$$|f'(z)| \leq \frac{M}{R}$$

As $R \to \infty$, $f'(z) = 0$, *i.e.*, $f(z)$ is constant for all finite values of z. **Proved.**

29.10 FUNDAMENTAL THEOREM OF ALGEBRA

Every polynomial of degree ≥ 1 has alteast one zero (root) in C.

Proof. Let $f(z)$ be a polynomial of degree ≥ 1. Suppose, $f(z)$ has no zero in C, then

$f(z) \neq 0$ for all z.

Further $f(z)$ is an entire function in the complex plane.

$\therefore \quad \dfrac{1}{f(z)}$ is also an entire function. Also as $z \to \infty$, $f(z) \to \infty$

$\therefore \quad \dfrac{1}{f(z)} \to 0$ as $z \to \infty$ $\qquad \therefore \quad \dfrac{1}{f(z)}$ is a bounded function.

Thus, by Liouville's theorem $\dfrac{1}{f(z)}$ is a constant function.

$\therefore$ $f(z)$ is a constant function and hence it is a polynomial of degree zero which is a contradiction.

Hence, $f(z)$ has at least one root in C. **Proved.**

29.11 POISSON INTEGRAL FORMULA FOR A CIRCLE

If $f(z)$ is analytic within and on the circle C given by $|z| = R$ and $z = re^{i\theta}$ is any point within C, then

$$f(re^{i\theta}) = \frac{1}{2\pi} \int_0^{2\pi} \frac{(R^2 - r^2) f(Re^{i\phi})}{R^2 - 2rR\cos(\theta - \phi) + r^2} d\phi$$

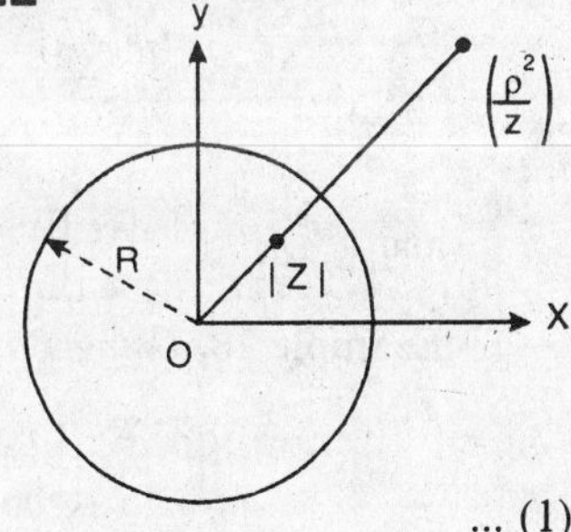

Proof. Since $z = re^{i\theta}$ is any point within C, by Cauchy's integral formula, we have

$$f(z) = \frac{1}{2\pi i} \oint_c \frac{f(w)}{w - z} dw \qquad \ldots (1)$$

The inverse of the point z w.r.t. C is given by $\dfrac{R^2}{\bar{z}}$ and lies outside C.

$\therefore$ By Cauchy's theorem, we have

$$0 = \oint_c \frac{f(w)}{w - \dfrac{R^2}{\bar{z}}} dw \quad \text{or} \quad 0 = \frac{1}{2\pi i} \oint_c \frac{f(w)}{w - \dfrac{R^2}{\bar{z}}} dw \qquad \ldots (2)$$

Subtracting (2) from (1), we get

$$f(z)=\frac{1}{2\pi i}\oint_C\left(\frac{1}{w-z}-\frac{1}{w-\frac{R^2}{\bar{z}}}\right)f(w)dw=\frac{1}{2\pi i}\oint_C\frac{z\bar{z}-R^2}{\bar{z}w^2-(z\bar{z}+R^2)w+R^2z}f(w)\,dw$$

$$=\frac{1}{2\pi i}\oint_C\frac{r^2-R^2}{\bar{z}w^2-(r^2+R^2)w+R^2z}f(w)dw\ \ldots(3)\qquad [\text{since } z\bar{z}=|z|^2=r^2]$$

Taking $w=\text{Re}^{i\phi}$, we have from (3),

$$f(re^{i\theta})=\frac{1}{2\pi i}\int_0^{2\pi}\frac{(r^2-R^2)f(Re^{i\phi}).Rie^{i\phi}d\phi}{re^{-i\theta}.R^2e^{2i\phi}-(r^2+R^2)Re^{i\phi}+R^2re^{i\theta}}$$

$$=\frac{1}{2\pi}\int_0^{2\pi}\frac{(r^2-R^2)f(Re^{i\phi}).e^{i\phi}d\phi}{rRe^{i(2\phi-\theta)}-(r^2+R^2)e^{i\phi}+Rre^{i\theta}}$$

Dividing the numerator and denominator by $e^{i\phi}$, we get

$$=\frac{1}{2\pi}\int_0^{2\pi}\frac{(r^2-R^2)f(Re^{i\phi})d\phi}{rRe^{i(\phi-\theta)}-(r^2+R^2)+Rre^{i(\theta-\phi)}}=\frac{1}{2\pi}\int_0^{2\pi}\frac{(R^2-r^2)f(Re^{i\phi})}{r^2+R^2-rR[e^{i(\theta-\phi)}+e^{-i(\theta-\phi)}]}d\phi$$

$$=\frac{1}{2\pi}\int_0^{2\pi}\frac{(R^2-r^2)f(Re^{i\phi})}{R^2-2rR\cos(\theta-\phi)+r^2}d\phi\qquad [\text{Since } e^{ix}+e^{-ix}=2\cos x.]$$

This is called *Poisson's integral formula for a circle*. It expresses the values of a harmonic function within a circle in terms of its values on the boundary.

If $f(re^{i\theta})=u(r,\theta)+iv(r,\theta)$ and $f(Re^{i\phi})=u(R,\phi)+iv(R,\phi)$, then we have

$$u(r,\theta)+iv(r,\theta)=\frac{1}{2\pi}\int_0^{2\pi}\frac{(R^2-r^2)[u(R,\phi)+iv(R,\phi)]}{R^2-2rR\cos(\theta-\phi)+r^2}d\phi$$

Equating real and imaginary parts, we get

$$u(r,\theta)=\frac{1}{2\pi}\int_0^{2\pi}\frac{(R^2-r^2)u(R,\phi)}{R^2-2rR\cos(\theta-\phi)+r^2}d\phi$$

and
$$v(r,\theta)=\frac{1}{2\pi}\int_0^{2\pi}\frac{(R^2-r^2)v(R,\phi)}{R^2-2rR\cos(\theta-\phi)+r^2}d\phi$$

Example 18. *Using Poisson's integral formula for a circle, show that*

$$\int_0^{2\pi}\frac{e^{\cos\phi}\cos(\sin\phi)}{5-4\cos(\theta-\phi)}d\phi=\frac{2\pi}{3}e^{\cos\theta}\cos(\sin\theta).$$

Solution. Poisson's integral formula for the circle $|z|=R$ is

$$f(re^{i\theta})=\frac{1}{2\pi}\int_0^{2\pi}\frac{(R^2-r^2)f(Re^{i\phi})}{R^2-2rR\cos(\theta-\phi)+r^2}d\phi\qquad \ldots(1)$$

Comparing the given integral with the integral on R.H.S. of (1), we get

$$R^2+r^2=5,\ rR=2,\ f(Re^{i\phi})=e^{\cos\phi}\cos(\sin\phi)$$

whence $R=2, r=1, f(re^{i\theta})=e^{\cos\theta}\cos(\sin\theta)$

Substituting these values in (1), we get

$$e^{\cos\theta}\cos(\sin\theta)=\frac{1}{2\pi}\int_0^{2\pi}\frac{(2^2-1^2)\,e^{\cos\phi}\cos(\sin\phi)}{5-4\cos(\theta-\phi)}\,d\phi$$

$$\therefore \quad \int_0^{2\pi}\frac{e^{\cos\phi}\cos(\sin\phi)}{5-4\cos(\theta-\phi)}\,d\phi=\frac{2\pi}{3}e^{\cos\theta}\cos(\sin\theta).$$ **Proved.**

29.12 POISSON INTEGRAL FORMULA FOR A HALF PLANE

If $f(z)$ is an analytic function in the upper half of the z-plane and $a=\alpha+i\beta$ is any point in this upper half plane, then

$$f(a)=\frac{1}{\pi}\int_{-\infty}^{\infty}\frac{\beta f(x)}{(x-\alpha)^2+\beta^2}\,dx \qquad \ldots (1)$$

Consider a semi-circle of radius R with centre at the origin. Let C denote the boundary of the semi-circle along with its bounding diameter. Since $a=\alpha+i\beta$ is an interior point, $\bar{a}=\alpha-i\beta$ lies outside C.

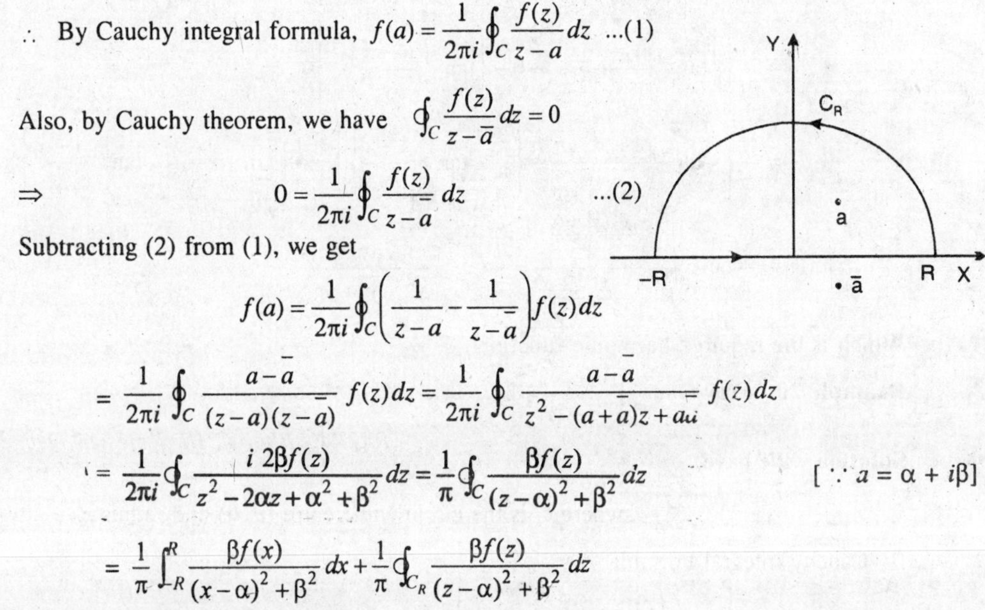

$\therefore$ By Cauchy integral formula, $f(a)=\dfrac{1}{2\pi i}\oint_C\dfrac{f(z)}{z-a}\,dz$ ...(1)

Also, by Cauchy theorem, we have $\oint_C\dfrac{f(z)}{z-\bar{a}}\,dz=0$

$$\Rightarrow \qquad 0=\frac{1}{2\pi i}\oint_C\frac{f(z)}{z-\bar{a}}\,dz \qquad \ldots(2)$$

Subtracting (2) from (1), we get

$$f(a)=\frac{1}{2\pi i}\oint_C\left(\frac{1}{z-a}-\frac{1}{z-\bar{a}}\right)f(z)\,dz$$

$$=\frac{1}{2\pi i}\oint_C\frac{a-\bar{a}}{(z-a)(z-\bar{a})}f(z)\,dz=\frac{1}{2\pi i}\oint_C\frac{a-\bar{a}}{z^2-(a+\bar{a})z+a\bar{a}}f(z)\,dz$$

$$=\frac{1}{2\pi i}\oint_C\frac{i\,2\beta f(z)}{z^2-2\alpha z+\alpha^2+\beta^2}\,dz=\frac{1}{\pi}\oint_C\frac{\beta f(z)}{(z-\alpha)^2+\beta^2}\,dz \qquad [\because a=\alpha+i\beta]$$

$$=\frac{1}{\pi}\int_{-R}^{R}\frac{\beta f(x)}{(x-\alpha)^2+\beta^2}\,dx+\frac{1}{\pi}\oint_{C_R}\frac{\beta f(z)}{(z-\alpha)^2+\beta^2}\,dz$$

where C_R is the semi-circular arc of C.

As $R\to\infty$, the second integral becomes zero.

$$\therefore \qquad f(a)=\frac{1}{\pi}\int_{-\infty}^{\infty}\frac{\beta f(x)}{(x-\alpha)^2+\beta^2}\,dx \qquad \ldots (3)$$

This is called *Poisson integral formula for a half-plane*. It expresses the values of a harmonic function in the upper half-plane in terms of the values on the real axis.

If $\qquad f(a)=f(\alpha+i\beta)=u(\alpha,\beta)+iv(\alpha,\beta)$

and $\qquad f(x)=u(x,0)+iv(x,0)$, then we have

$$u(\alpha,\beta)+iv(\alpha,\beta)=\frac{1}{\pi}\int_{-\infty}^{\infty}\frac{\beta[u(x,0)+iv(x,0)]}{(x-\alpha)^2+\beta^2}\,dx$$

Equating real and imaginary parts, we get

$$u(\alpha,\beta) = \frac{1}{\pi}\int_{-\infty}^{\infty} \frac{\beta u(x,0)}{(x-\alpha)^2+\beta^2}\,dx \text{ and } v(\alpha,\beta) = \frac{1}{\pi}\int_{-\infty}^{\infty} \frac{\beta v(x,0)}{(x-\alpha)^2+\beta^2}\,dx.$$

Example 19. *Find a function harmonic in the upper half of the z-plane which takes the following values on the x-axis.*

$$G(x) = \begin{cases} 1, & x < -1 \\ 0, & -1 < x < 1 \\ -1, & x > 1 \end{cases}$$

Solution. Let $\phi(x, y)$ be the function harmonic in the upper half of the z-plane, then

$$\phi(x,y) = \frac{1}{\pi}\int_{-\infty}^{\infty} \frac{yG(w)}{(w-x)^2+y^2}\,dw \qquad \text{(Poisson's Integral Formula)}$$

[Obtained from (Article 35.12) on page 949, on replacing α, β by x, y and the variable of integration x by w]

$$= \frac{1}{\pi}\left[\int_{-\infty}^{-1} \frac{y.1}{(w-x)^2+y^2}\,dw + \int_{-1}^{1} \frac{y.0}{(w-x)^2+y^2}\,dw + \int_{1}^{\infty} \frac{y.(-1)}{(w-x)^2+y^2}\,dw\right]$$

$$= \frac{1}{\pi}\left[\left\{\tan^{-1}\frac{w-x}{y}\right\}_{-\infty}^{-1} + 0 - \left\{\tan^{-1}\frac{w-x}{y}\right\}_{1}^{\infty}\right]$$

$$= \frac{1}{\pi}\left[-\tan^{-1}\frac{1+x}{y} + \frac{\pi}{2} - \left\{\frac{\pi}{2} - \tan^{-1}\frac{1-x}{y}\right\}\right] = \frac{1}{\pi}\left[\tan^{-1}\frac{1-x}{y} - \tan^{-1}\frac{1+x}{y}\right]$$

$$= \frac{1}{\pi}\tan^{-1}\frac{\frac{1-x}{y} - \frac{1+x}{y}}{1+\frac{1-x}{y}.\frac{1+x}{y}} = \frac{1}{\pi}\tan^{-1}\frac{2xy}{x^2-y^2-1}.$$

Which is the required harmonic function. **Ans.**

Example 20. *Prove that* $\int_C \frac{dz}{z-a} = 2\pi i$, *where C is the circle* $|z-a| = r$

(*R.G.P.V., Bhopal, III Semester, Dec. 2006*)

Solution. We have,

$$\int_C \frac{dz}{z-a}, \text{ where } C \text{ is the circle with centre } (a, 0) \text{ and radius } r.$$

By Cauchy Integral Formula

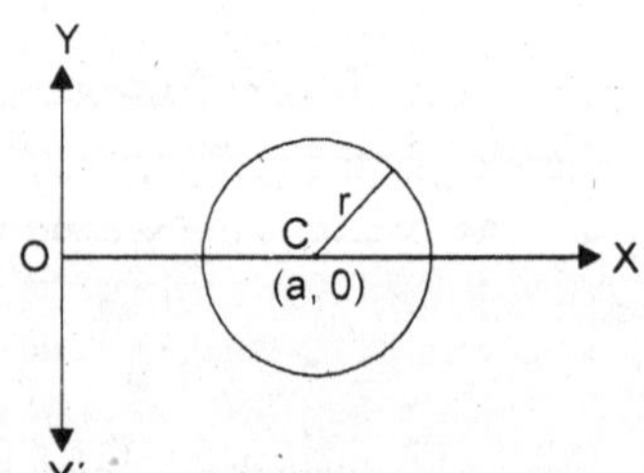

$$\left[\int_C \frac{f(z)}{z-a}\,dz = 2\pi i\, f(a)\right]$$

$$\int_C \frac{dz}{z-a} = 2\pi i\ (1)$$

$$\Rightarrow \quad \int_C \frac{dz}{z-a} = 2\pi i \qquad \textbf{Proved.}$$

Example 21. *Evaluate the following integral*

$$\int_C \frac{1}{z}\cos z dz.$$

where C is the ellipse $9x^2 + 4y^2 = 1$.

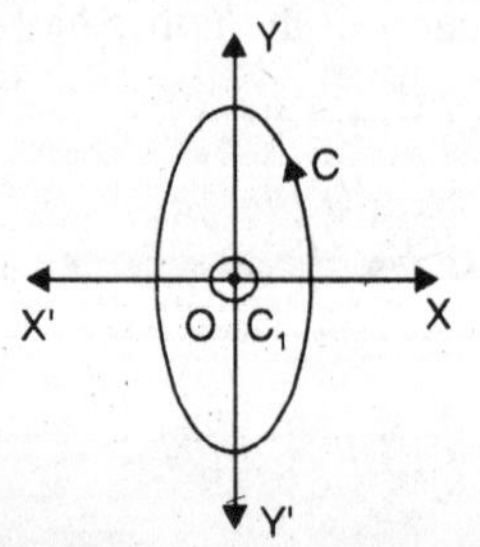

Solution. The given ellipse $9x^2 + 4y^2 = 1$ encloses a simple pole $z = 0$.

By Cauchy Integral formula

$$\int_c \frac{\cos z}{z}\,dz = 2\pi i(\cos z)_{z=0} = 2\pi i.$$

Which is the required value of the given integral. **Ans.**

Example 22. *Use Cauchy's integral formula to evaluate* $\int_c \frac{z}{(z^2-3z+2)}\,dz$

where c is the circle $|z-2| = \frac{1}{2}$ *(U.P. III Semester, June 2009)*

Solution. Here, we have

$$\int_c \frac{z}{(z^2-3z+2)}\,dz$$

The poles are determined by putting the denominator equal to zero

i.e.; $z^2 - 3z + 2 = 0 \quad \Rightarrow \quad (z-1)\quad(z-2) = 0$

$\Rightarrow \quad z = 1, 2$

So, there are two poles $z = 1$ and $z = 2$.

There is only one pole at $z = 2$ inside the given circle.

$$\int_c \frac{z}{(z^2-3z+2)}\,dz = \int_c \frac{z}{(z-1)(z-2)}\,dz$$

$$= \int_c \frac{\frac{z}{z-1}}{z-2}\,dz \qquad \left[\int_c \frac{f(z)}{z-a}\,dz = 2\pi i\, f(a)\right]$$

$$= 2\pi i\left[\frac{z}{z-1}\right]_{z=2} = 2\pi i\left(\frac{2}{2-1}\right) = 4\pi i$$ **Ans.**

Example 23. *Use Cauchy's integral formula to calculate*

$\int_C \frac{2z+1}{z^2+z}\,dz$ *where C is* $|z| = \frac{1}{2}$. *(AMIETE, Dec. 2009)*

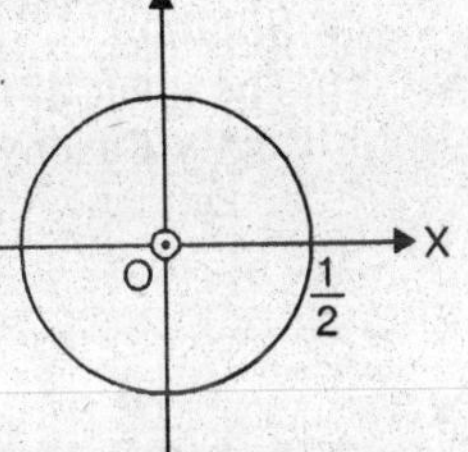

Solution. Poles are given by

$$z^2 + z = 0$$

$\Rightarrow \quad z(z+1) = 0 \quad \Rightarrow \quad z = 0, -1$

$|z| = \frac{1}{2}$ is a circle with centre at origin and radius $\frac{1}{2}$.

Therefore it encloses only one pole $z = 0$.

$$\therefore \quad \int_C \frac{2z+1}{z(z+1)}\,dz = \int_C \frac{\frac{2z+1}{z+1}}{z}\,dz = 2\pi i\left[\frac{2z+1}{z+1}\right]_{z=0} = 2\pi i$$ **Ans.**

Example 24. *Evaluate by Cauchy's integral formula*

$$\int_C \frac{dz}{z(z+\pi i)}, \text{ where } C \text{ is } |z+3i| = 1$$

Solution. Poles of the integrand are

$z = 0, -\pi i$ (simple poles)

The given curve C is a circle with centre at $z = -3i$ *i.e.* at $(0, -3)$ and radius 1.

Clearly, only the pole $z = -\pi i$ lies inside the circle.

$$\therefore \quad \int_C \frac{dz}{z(z+\pi i)} = \int_C \frac{\frac{1}{z}}{z+\pi i}\,dz$$

$$= 2\pi i\left(\frac{1}{z}\right)_{z=-\pi i}$$

[By Cauchy's Integral formula]

$$= \frac{2\pi i}{-\pi i} = -2$$

Which is the required value of the given integral. **Ans.**

Example 25. *Evaluate the complex integral*

$$\int_c \tan z . dz \text{ where } c \text{ is } |z| = 2.$$

Solution. $\int_C \tan z . dz = \int_C \frac{\sin z}{\cos z} . dz$

$|z| = 2$, is a circle with centre at origin and radius = 2.

Poles are given by putting the denominator equal to zero.

$$\cos z = 0,\ z = -\frac{\pi}{2}, \frac{\pi}{2}, \frac{3\pi}{2}, \ldots$$

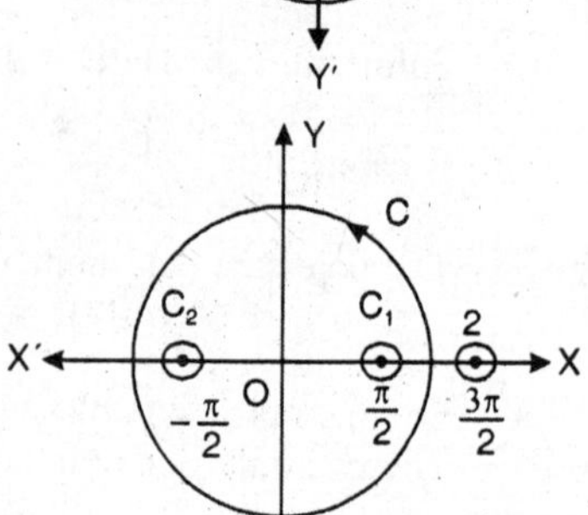

The integrand has two poles at $z = \frac{\pi}{2}$ and $z = -\frac{\pi}{2}$ inside the given circle $|z| = 2$.

On applying Cauchy integral formula

$$\int_C \frac{\sin z}{\cos z} dz = \int_{c_1} \frac{\sin z}{\cos z} dz + \int_{c_2} \frac{\sin z}{\cos z} dz = 2\pi i\ [\sin z]_{z=\frac{\pi}{2}} + 2\pi i\ [\sin z]_{z=\frac{-\pi}{2}}$$

$$= 2\pi i(1) + 2\pi i(-1) = 0$$

Which is the required value of the given integral. **Ans.**

Example 26. *Evaluate* $\oint_C \frac{e^{-z}}{z+1} dz$, *where C is the circle* $|z| = 2$

Solution. $f(z) = e^{-z}$ is an analytic function

The point $z = -1$ lies inside the circle $|z| = 2$.

$\therefore$ By Cauchy's integral formula,

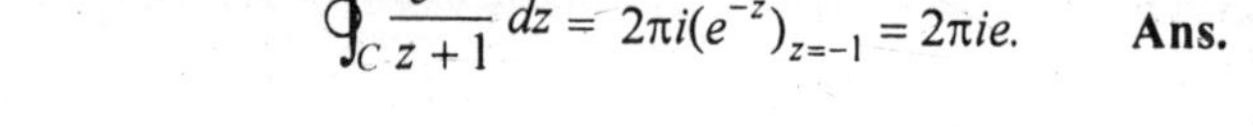

$$\oint_C \frac{e^{-z}}{z+1} dz = 2\pi i (e^{-z})_{z=-1} = 2\pi i e. \qquad \textbf{Ans.}$$

Example 27. *Evaluate:* $\int_C \frac{e^z}{(z-1)(z-4)} dz$ *where C is the circle* $|z| = 2$ *by using Cauchy's Integral Formula.* *(R.G.P.V., Bhopal, III Semester, June 2006)*

Solution. We have,

$$\int_C \frac{e^z}{(z-1)(z-4)} dz \quad \text{where C is the circle with centre at origin and radius 2.}$$

Poles are given by putting the denominator equal to zero.

$$(z-1)(z-4) = 0$$

$$\Rightarrow \qquad z = 1, 4$$

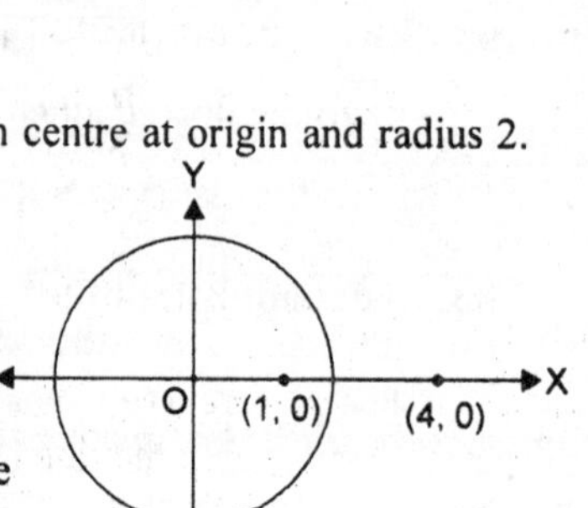

Here there are two simple poles at $z = 1$ and $z = 4$.

There is only one pole at $z = 1$ inside the contour. Therefore

$$\int_C \frac{e^z}{(z-1)(z-4)}\,dz = \int \frac{\frac{e^z}{(z-4)}}{(z-1)}\,dz = 2\pi i\left[\frac{e^z}{z-4}\right]_{z=1} \quad \text{(By Cauchy Integral Theorem)}$$

$$= 2\pi i\left(\frac{e}{1-4}\right) = -\frac{2\pi i e}{3}$$

Which is the required value of the given integral. **Ans.**

Example 28. *If* $f(z_1) = \int_C \frac{3z^2+7z+1}{z-z_1}dz$, *where C is the circle* $x^2 + y^2 = 4$, *find the values of*

(*i*) $f(3)$, (*ii*) $f'(1-i)$, (*iii*) $f''(1-i)$.

Solution. The given circle C is $x^2 + y^2 = 4$ or $|z| = 2$.

The point $z = 3$ lies outside the circle $|z| = 2$.

(*i*) $f(3) = \oint_C \frac{3z^2+7z+1}{z-3}\,dz$ and $\frac{3z^2+7z+1}{z-3}$ is analytic within and on C.

∴ By Cauchy's integral theorem, we have

$$\oint_C \frac{3z^2+7z+1}{z-3}\,dz = 0 \Rightarrow f(3) = 0.$$ **Ans.**

(*ii*) $z_1 = 1 - i$ lies inside the circle *C*.

By Cauchy's Integral formula, we have

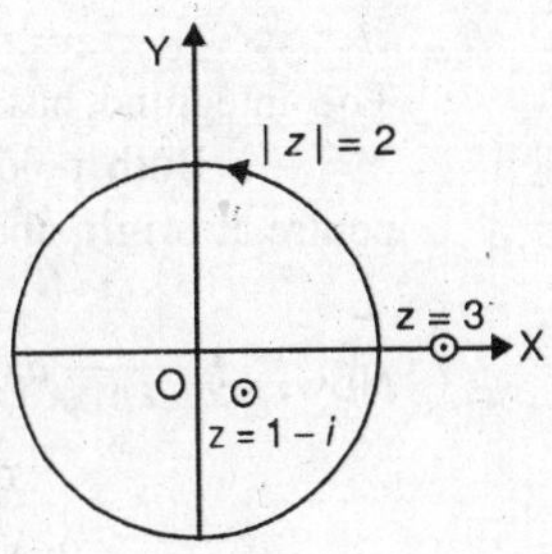

$$\int_C \frac{3z^2+7z+1}{z-z_1}\,dz = 2\pi i\,(3z^2+7z+1)_{z=z_1}$$

$$f(z) = 2\pi i\,(3z^2 + 7z + 1)$$

$$f'(z) = 2\pi i\,(6z + 7)$$

$$f'(1-i) = 2\pi i\,[6(1-i) + 7]$$

$$\Rightarrow \quad f'(1-i) = 2\pi i\,[13 - 6i]$$

$$\Rightarrow \quad f'(1-i) = 2\pi\,[6 + 13i]$$ **Ans.**

(*iii*)

$$f''(z) = 2\pi i\; 6$$

$$f''(1-i) = 12\pi i$$ **Ans.**

Example 29. *Evaluate*

$$\int_C \frac{e^z}{z^2+1}dz \text{ over the circular path } |z| = 2.$$

Solution. Poles of the integrand are given by putting the denominator equal to zero.

$$z^2+1=0 \Rightarrow z^2=-1 \Rightarrow z=\pm i$$

The integrand has two simple poles at $z = i$ and $z = -i$.

Both poles are inside the given circle with centre at origin and radius 2.

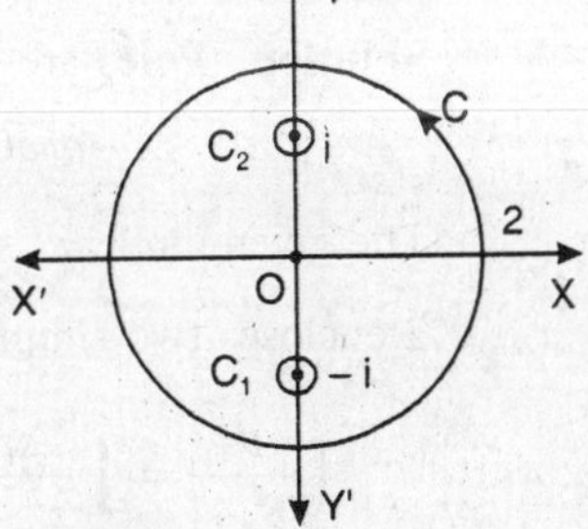

$$\int_C \frac{1}{2i}\left(\frac{e^z}{z-i} - \frac{e^z}{z+i}\right)dz = \int_C \frac{1}{2i}\frac{e^z}{z-i}dz - \frac{1}{2i}\int_C \frac{e^z}{z+i}dz = \frac{1}{2i}\left[2\pi i(e^z)_{z=i} - 2\pi i(e^z)_{z=-i}\right]$$

$$= \frac{2\pi i}{2i}[e^i - e^{-i}] = \pi\; 2\; i\sin 1 = 2\pi i \sin 1$$

Which is the required value of the given integral. **Ans.**

Second Method. $\int_C \frac{e^z}{z^2+1}dz = \int_C \frac{e^z dz}{(z+i)(z-i)} = \int_{C_1} \frac{\frac{e^z}{z-i}}{z+i}dz + \int_{C_2} \frac{\frac{e^z}{z+i}}{z-i}dz$

$$= 2\pi i \left(\frac{e^z}{z-i}\right)_{z=-i} + 2\pi i \left(\frac{e^z}{z+i}\right)_{z=i} = \left[2\pi i \frac{e^{-i}}{-i-i} + 2\pi i \frac{e^i}{i+i}\right] = \pi[-e^{-i} + e^i]$$

$$= \pi(2i \sin 1) = 2\pi i \sin 1$$

Which is the required value of the given integral. **Ans.**

Example 30. *State the Cauchy's integral formula. Show that* $\int_C \frac{e^{zt}}{z^2+1} dz = 2\pi i \sin t$

if $t > 0$ *and C is the circle* $|z| = 3$ *(U.P., III Semester, Dec. 2009)*

Solution. See Art. 29.5 on page 754

Here, we have $\int_C \frac{e^{zt}}{z^2+1} dz$

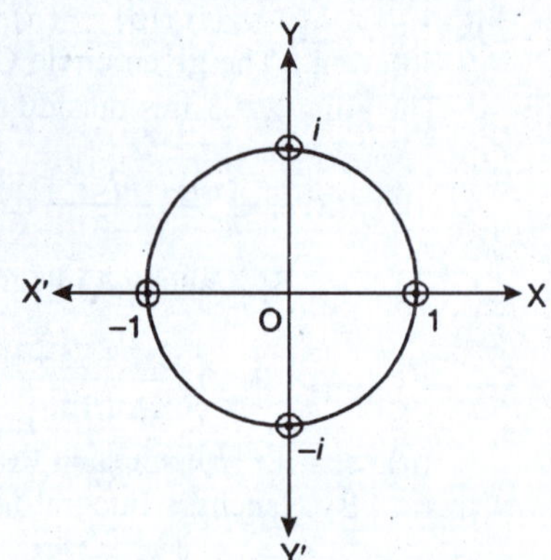

The poles are determined by putting the denominator equal to zero.

i.e., $z^2 + 1 = 0$

$\Rightarrow \quad z^2 = -1$

$\Rightarrow \quad z = \pm\sqrt{-1} = \pm i$

$\Rightarrow \quad z = i, -i$

The integrand has two simple poles at $z = i$ and at $z = -i$. Both poles are inside the given circle with centre at origin and radius 3.

Now, $\int_C \frac{e^{zt}}{z^2+1} dz = \frac{1}{2i} \int_C \left(\frac{e^{zt}}{z-i} - \frac{e^{xt}}{z+i}\right) dz$ [By partial fraction]

$$= \frac{1}{2i}\left[\int_{C_1} \frac{e^{zt}}{z-i} dz - \int_{C_2} \frac{e^{zt}}{z+i} dz\right] = \frac{1}{2i}\left[2\pi i\,(e^{zt})_{z=i} - 2\pi i\,(e^{zt})_{z=-i}\right]$$

$$= \frac{2\pi i}{2i}\left[e^{ti} - e^{-ti}\right] = \pi . 2i \sin t$$

Example 31. *Evaluate* $\int_C \frac{dz}{z^2-1}$, *where C is the circle* $x^2 + y^2 = 4$.

Solution. Poles are given by putting the denominator equal to zero.

$$z^2 - 1 = 0, z^2 = 1, \ z = \pm 1$$

The given circle $x^2 + y^2 = 4$ with centre at $z = 0$ and radius 2 encloses two simple poles at $z = 1$ and $z = -1$.

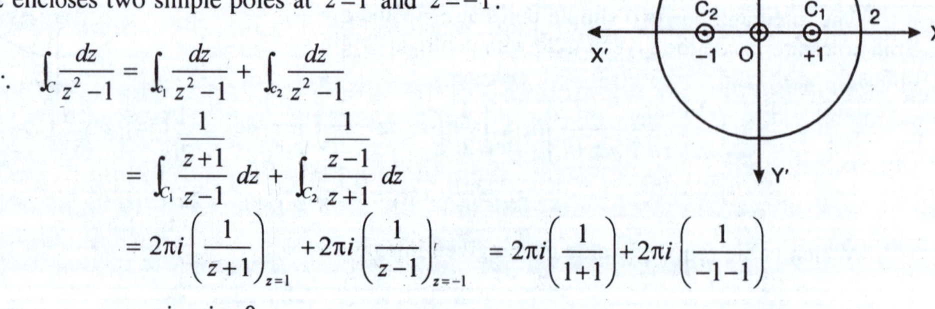

$$\therefore \quad \int_C \frac{dz}{z^2-1} = \int_{C_1} \frac{dz}{z^2-1} + \int_{C_2} \frac{dz}{z^2-1}$$

$$= \int_{C_1} \frac{\frac{1}{z+1}}{z-1} dz + \int_{C_2} \frac{\frac{1}{z-1}}{z+1} dz$$

$$= 2\pi i \left(\frac{1}{z+1}\right)_{z=1} + 2\pi i \left(\frac{1}{z-1}\right)_{z=-1} = 2\pi i \left(\frac{1}{1+1}\right) + 2\pi i \left(\frac{1}{-1-1}\right)$$

$$= \pi i - \pi i = 0$$

Which is the required value of the given integral. **Ans.**

Example 32. *Evaluate the following integral using Cauchy integral formula*

$$\int_c \frac{4-3z}{z(z-1)(z-2)}\,dz \quad \textit{where c is the circle } |z| = \frac{3}{2}.$$

(AMIETE, Dec. 2009, R.G.P.V., Bhopal, III Semester, June 2008)

Solution. Poles of the integrand are given by putting the denominator equal to zero.

$$z(z-1)(z-2) = 0 \text{ or } z = 0,\ 1,\ 2$$

The integrand has three simple poles at $z = 0, 1, 2$.

The given circle $|z| = \frac{3}{2}$ with centre at $z = 0$ and radius $= \frac{3}{2}$ encloses two poles $z = 0$, and $z = 1$.

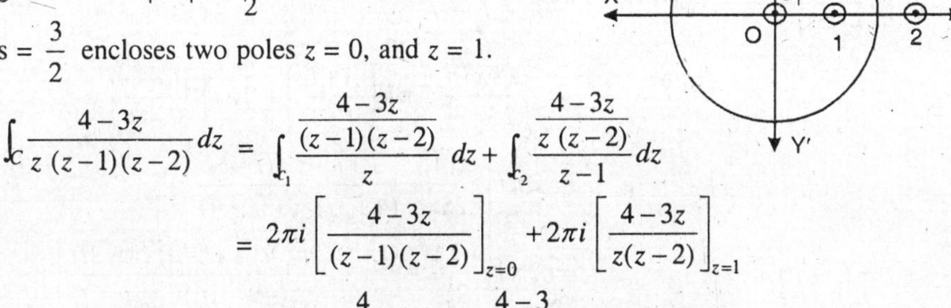

$$\int_C \frac{4-3z}{z(z-1)(z-2)}\,dz = \int_{c_1} \frac{\frac{4-3z}{(z-1)(z-2)}}{z}\,dz + \int_{c_2} \frac{\frac{4-3z}{z(z-2)}}{z-1}\,dz$$

$$= 2\pi i\left[\frac{4-3z}{(z-1)(z-2)}\right]_{z=0} + 2\pi i\left[\frac{4-3z}{z(z-2)}\right]_{z=1}$$

$$= 2\pi i.\frac{4}{(-1)(-2)} + 2\pi i\frac{4-3}{1(1-2)} = 2\pi i(2-1)$$

$$= 2\pi i$$

Which is the required value of the given integral. **Ans.**

Example 33. *Evaluate* $\int_c \frac{z^2-2z}{(z+1)^2(z^2+4)}\,dz$

where c is the circle $|z| = 10$. *(U.P. III Semester, June 2009)*

Solution. Here, we have

$$\int_c \frac{z^2-2z}{(z+1)^2(z^2+4)}\,dz$$

The poles are determined by putting the denominator equal to zero.

i.e.; $(z+1)^2(z^2+4) = 0$

$\Rightarrow$ $z = -1, -1$ and $z = \pm 2i$

The circle $|z| = 10$ with centre at origin and radius $= 10$.
encloses a pole at $z = -1$ of second order and simple poles $z = \pm 2i$

The given integral $= I_1 + I_2 + I_3$

$$I_1 = \int_{c_1} \frac{z^2-2z}{(z+1)^2(z^2+4)}\,dz = \int_{c_1} \frac{\frac{z^2-2z}{z^2+4}}{(z+1)^2}\,dz = 2\pi i\left[\frac{d}{dz}\frac{z^2-2z}{z^2+4}\right]_{z=-1}$$

$$= 2\pi i\left[\frac{(z^4+4)(2z-2)-(z^2-2z)\,2z}{(z^2+4)^2}\right]_{z=-1} = 2\pi i\left[\frac{(1+4)(-2-2)-(1+2)\,2(-1)}{(1+4)^2}\right]$$

$$= 2\pi i\left(-\frac{14}{25}\right) = \frac{-28\pi i}{25}$$

$$I_2 = \int_{c_2} \frac{\dfrac{z^2-2z}{(z+1)^2(z+2i)}}{(z-2i)}\,dz = 2\pi i\left[\frac{z^2-2z}{(z+1)^2(z+2i)}\right]_{z=2i} = 2\pi i\left[\frac{-4-4i}{(2i+1)^2(2i+2i)}\right] = 2\pi i\,\frac{(1+i)}{4+3i}$$

$$I_3 = \int_{c_3} \frac{\dfrac{z^2-2z}{(z+1)^2(z-2i)}}{(z+2i)}\,dz = 2\pi i\left[\frac{z^2-2z}{(z+1)^2(z-2i)}\right]_{z=-2i}$$

$$= 2\pi i\left[\frac{-4+4i}{(-2i+1)^2(-2i-2i)}\right] = 2\pi i\,\frac{(i-1)}{(3i-4)}$$

$$\int_c \frac{z^2-2z}{(z+1)^2(z^2+4)}\,dz = I_1 + I_2 + I_3$$

$$= \frac{-28\pi i}{25} + 2\pi i\left(\frac{1+i}{4+3i}\right) + 2\pi i\left(\frac{i-1}{3i-4}\right)$$

$$= 2\pi i\left[\frac{-14}{25} + \frac{1+i}{(4+3i)} + \frac{(i-1)}{(3i-4)}\right]$$

$$= 2\pi i\left[\frac{-14}{25} + \frac{(1+i)(3i-4)+(i-1)(4+3i)}{(-9-16)}\right]$$

$$= \frac{2\pi i}{-25}\,[14 + (3i-4-3-4i) + (4i-3-4-3i)]$$

$$= 0$$ **Ans.**

Example 34. *Integrate* $\dfrac{1}{(z^3-1)^2}$ *the counter clock-wise sense around the circle* $|z-1| = 1$.

Solution. Poles of the given function are found by putting denominator equal to zero.

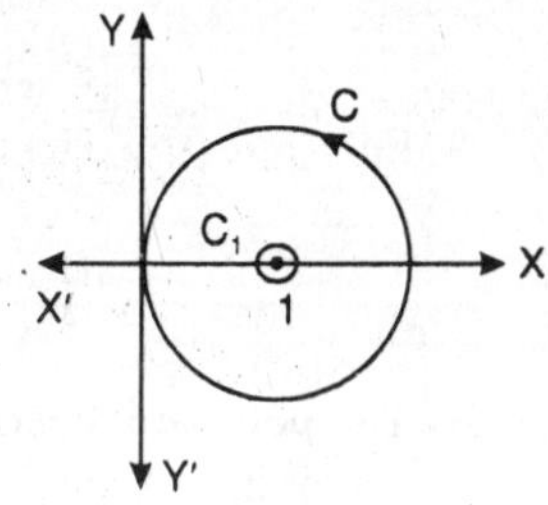

$$(z^3-1)^2 = 0,$$

$$(z-1)^2(z^2+z+1)^2 = 0$$

$$z = 1, 1, \qquad z = \frac{-1\pm\sqrt{1-4}}{2} = \frac{-1\pm\sqrt{-3}}{2} = -\frac{1}{2}\pm\frac{\sqrt{3}}{2}i$$

The circle $|z-1| = 1$ with centre at $z = 1$ and unit radius encloses a pole of order two at $z = 1$. By Cauchy Integral formula

$$\int_C \frac{1}{(z^3-1)^2}dz = \int_{C_1} \frac{1}{(z-1)^2(z^2+z+1)^2}dz = \int_{C_1} \frac{\dfrac{1}{(z^2+z+1)^2}}{(z-1)^2}dz$$

$$= 2\pi i\left[\frac{d}{dz}\frac{1}{(z^2+z+1)^2}\right]_{z=1} = 2\pi i\left[\frac{-2(2z+1)}{(z^2+z+1)^3}\right]_{z=1} = 2\pi i\left[\frac{-2(2+1)}{(1+1+1)^3}\right] = -\frac{4\pi i}{9}$$ **Ans.**

Example 35. *Find the value of* $\displaystyle\int_C \frac{3z^2+z}{z^2-1}dz$.

If C is circle $|z-1| = 1$ *(R.G.P.V., Bhopal, III Semester, June 2007)*

Solution. Poles of the integrand are given by putting the denominator equal to zero.

$$z^2-1 = 0,\ z^2 = 1,\ z = \pm 1$$

The circle with centre $z = 1$ and radius unity encloses a simple pole at $z = 1$.
By Cauchy Integral formula

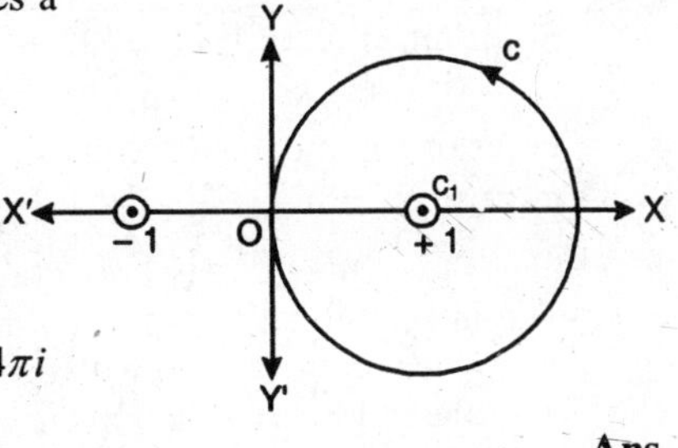

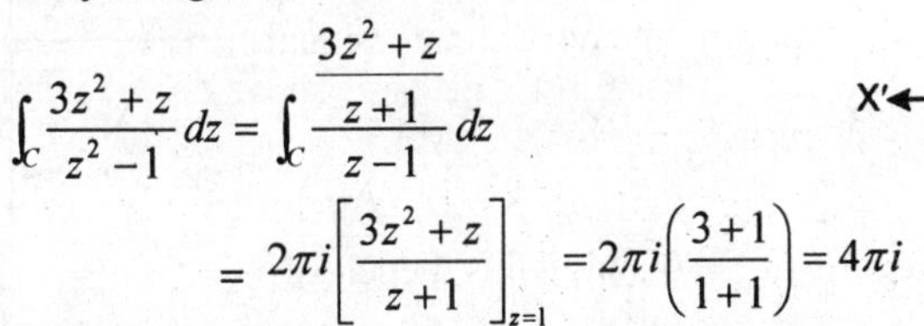

$$\int_C \frac{3z^2+z}{z^2-1}\,dz = \int_C \frac{\frac{3z^2+z}{z+1}}{z-1}\,dz$$

$$= 2\pi i\left[\frac{3z^2+z}{z+1}\right]_{z=1} = 2\pi i\left(\frac{3+1}{1+1}\right) = 4\pi i$$

Which is the required value of the given integral. **Ans.**

Example 36. *Evaluate* $\oint_C \frac{z^2+1}{z^2-1}dz$ *where C is circle,*

(*i*) $|z| = \frac{3}{2}$ (*ii*) $|z-1| = 1,$ (*iii*) $|z| = \frac{1}{2}.$

Solution. Poles of the integrand are given by putting the denominator equal to zero.

i.e.; $z^2 - 1 = 0 \Rightarrow z = 1, -1$

(*i*) $|z| = \frac{3}{2}$ is equation of circle C with centre O and radius $\frac{3}{2}$

Both poles $z = 1, -1$ lie inside C.

$$\oint_C \frac{z^2+1}{z^2-1}\,dz = \oint_{C_1} \frac{\left(\frac{z^2+1}{z-1}\right)}{z+1}\,dz + \oint_{C_2} \frac{\left(\frac{z^2+1}{z+1}\right)}{z-1}\,dz$$

$$= 2\pi i\left[\frac{z^2+1}{z-1}\right]_{z=-1} + 2\pi i\left[\frac{z^2+1}{z+1}\right]_{z=1} = 2\pi i\left[\frac{1+1}{-1-1}\right] + 2\pi i\left[\frac{1+1}{1+1}\right] = -2\pi i + 2\pi i = 0.$$

Which is the required value of the given integral. **Ans.**

(*ii*) $|z-1| = 1$ is equation of circle C with centre 1 and radius 1 encloses only pole $z = 1$.

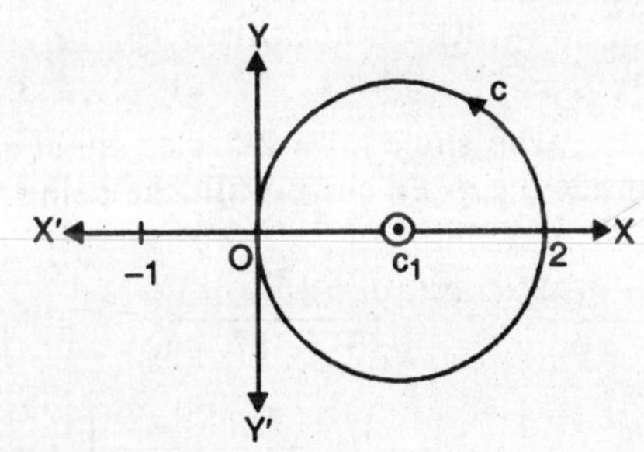

$$\oint_C \frac{z^2+1}{z^2-1}\,dz = \oint_{C_1} \frac{\left(\frac{z^2+1}{z+1}\right)}{z-1}\,dz.$$

$$= 2\pi i\left[\frac{z^2+1}{z+1}\right]_{z=1}$$

$$= 2\pi i\left[\frac{1+1}{1+1}\right] = 2\pi i$$

(*iii*) $|z| = \frac{1}{2}$ is equation of circle C with centre O and radius $\frac{1}{2}$.

There is no pole inside C.

Hence, $\oint_C \frac{z^2+1}{z^2-1}\,dz = 0$. **Ans.**

Example 37. *Use Cauchy integral formula to evaluate.*

$$\int_C \frac{\sin \pi z^2 + \cos \pi z^2}{(z-1)(z-2)}\,dz$$

where C is the circle $|z| = 3$.

(AMIETE, Dec. 2010, R.G.P.V., Bhopal, III Semester, June 2003)

Solution. $\oint \frac{\sin \pi z^2 + \cos \pi z^2}{(z-1)(z-2)} dz$

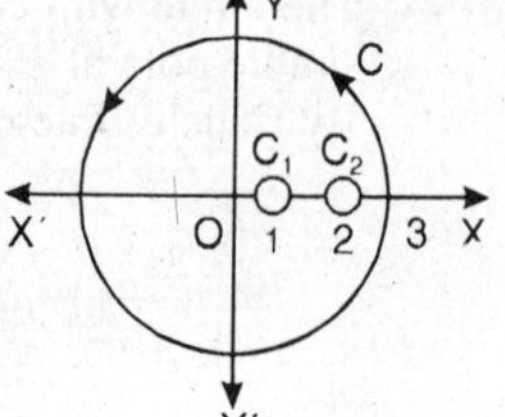

Poles of the integrand are given by putting the denominator equal to zero.

$$(z-1)(z-2) = 0 \Rightarrow z = 1, 2$$

The integrand has two poles at $z = 1, 2$.
The given circle $|z| = 3$ with centre at $z = 0$ and radius 3 encloses both the poles $z = 1$, and $z = 2$.

$$\oint_C \frac{\sin \pi z^2 + \cos \pi z^2}{(z-1)(z-2)} dz = \int_{C_1} \frac{\frac{\sin \pi z^2 + \cos \pi z^2}{(z-2)}}{(z-1)} dz + \int_{C_2} \frac{\frac{\sin \pi z^2 + \cos \pi z^2}{(z-1)}}{(z-2)} dz$$

$$= 2\pi i \left[\frac{\sin \pi z^2 + \cos \pi z^2}{z-2}\right]_{z=1} + 2\pi i \left[\frac{\sin \pi z^2 + \cos \pi z^2}{z-1}\right]_{z=2}$$

$$= 2\pi i \left(\frac{\sin \pi + \cos \pi}{1-2}\right) + 2\pi i \left(\frac{\sin 4\pi + \cos 4\pi}{2-1}\right)$$

$$= 2\pi i \left(\frac{-1}{-1}\right) + 2\pi i \left(\frac{1}{1}\right) = 4\pi i$$

Which is the required value of the given integral. **Ans.**

Example 38. *Evaluate the following complex integration using Cauchy's integral formula*

$$\int_C \frac{3z^2 + z + 1}{(z^2 - 1)(z+3)} dz$$

where C is the circle $|z| = 2$.

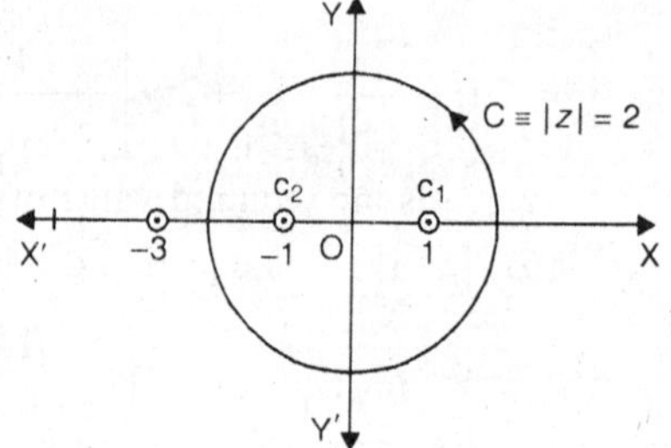

Solution. Poles of the integrand are given by putting the denominator equal to zero.

i.e., $(z^2 - 1)(z + 3) = 0$

$\Rightarrow \quad z = 1, -1, -3$ (Simple poles)

The circle $|z| = 2$ has centre at $z = 0$ and radius 2. Clearly the poles $z = 1$ and $z = -1$ lie inside the given circle while the pole $z = -3$ lies outside it.

$$\therefore \quad \int_C \frac{3z^2 + z + 1}{(z^2 - 1)(z+3)} dz = \int_{C_1} \frac{\frac{3z^2 + z + 1}{(z+1)(z+3)}}{z-1} dz + \int_{C_2} \frac{\frac{3z^2 + z + 1}{(z-1)(z+3)}}{z+1} dz$$

$$= 2\pi i \left[\frac{3z^2 + z + 1}{(z+1)(z+3)}\right]_{z=1} + 2\pi i \left[\frac{3z^2 + z + 1}{(z-1)(z+3)}\right]_{z=-1} \quad \text{(Using Cauchy's Integral formula)}$$

$$= 2\pi i \left(\frac{5}{8}\right) + 2\pi i \left(-\frac{3}{4}\right) = 2\pi i \left(\frac{-1}{8}\right) = -\frac{\pi i}{4}$$

Which is the required value of the given integral. **Ans.**

Example 39. *Using Cauchy's integral formula, evaluate* $\frac{1}{2\pi i} \int_C \frac{ze^z}{(z-a)^3} dz$, *where the point a lies within the closed curve C.*

Solution. $$\int_C \frac{ze^z}{(z-a)^3} dz = \frac{2\pi i}{2!} \left[\frac{d^2}{dz^2}(ze^z)\right]_{z=a} = \frac{2\pi i}{2} \left[\frac{d}{dz}\{(z+1)e^z\}\right]_{\text{at } z=a}$$

$$= \frac{2\pi i}{2}\left[(z+1)e^z + e^z \cdot 1\right]_{\text{at } z=a} = \frac{2\pi i}{2}\left[(z+2)e^z\right]_{\text{at } z=a} = 2\pi i\,\frac{(a+2)e^a}{2} = \pi i(a+2)e^a$$

$$\frac{1}{2\pi i}\oint_C \frac{ze^z}{(z-a)^3}\,dz = \frac{\pi i}{2\pi i}(a+2)e^a = \frac{1}{2}(a+2)e^a$$

Which is the required value of the given integral. **Ans.**

Example 40. *Derive Cauchy Integral Formula.*

Evaluate $\displaystyle\int_C \frac{e^{3iz}}{(z+\pi)^3}\,dz$

where C is the circle $|z-\pi| = 3.2$

Solution. Here, $I = \displaystyle\oint_C \frac{e^{3iz}}{(z+\pi)^3}\,dz$

Where C is a circle $\{|z-\pi| = 3.2\}$ with centre (π, 0) and radius 3.2.

Poles are determined by putting the denominator equal to zero.

$$(z+\pi)^3 = 0 \quad \Rightarrow \quad z = -\pi, -\pi, -\pi$$

There is a pole at $z = -\pi$ of order 3. But there is no pole within C.

By Cauchy Integral Formula $\displaystyle\oint_C \frac{e^{3iz}}{(z+\pi)^3}\,dz = 0$ **Ans.**

Example 41. *Verify, Cauchy theorem by integrating* e^{iz} *along the boundary of the triangle with the vertices at the points* $1+i$, $-1+i$ *and* $-1-i$.

Solution.

$$\int_{AB} e^{iz}dz = \left(\frac{e^{iz}}{i}\right)_{1+i}^{-1+i}$$

$$= \frac{1}{i}\left[e^{i(-1+i)} - e^{i(1+i)}\right]$$

$$= \frac{1}{i}\left[e^{-i-1} - e^{i-1}\right] \quad \text{...(1)}$$

$$\int_{BC} e^{iz}dz = \left[\frac{e^{iz}}{i}\right]_{-1+i}^{-1-i} = \frac{1}{i}\left[e^{i(-1-i)} - e^{i(-1+i)}\right]$$

$$= \frac{1}{i}\left[e^{-i+1} - e^{-i-1}\right] \quad \text{...(2)}$$

$$\int_{CA} e^{iz}dz = \left[\frac{e^{iz}}{i}\right]_{-1-i}^{1+i} = \frac{1}{i}\left[e^{i(1+i)} - e^{i(-1-i)}\right] = \frac{1}{i}\left[e^{i-1} - e^{-i+1}\right] \quad \text{...(3)}$$

On adding (1), (2) and (3), we get

$$\int_{AB} e^{iz}dz + \int_{BC} e^{iz}dz + \int_{CA} e^{iz}dz = \frac{1}{i}\left[\left(e^{-i-1} - e^{i-1}\right) + \left(e^{-i+1} - e^{-i-1}\right) + \left(e^{i-1} - e^{-i+1}\right)\right]$$

$$\Rightarrow \quad \int_{\Delta ABC} e^{iz}\,dz = 0 \quad \text{...(4)}$$

The given function has no pole. So there is no pole in Δ ABC.

The given function e^{iz} is analytic inside and on the triangle ABC.

By Cauchy Theorem, we have $\displaystyle\oint_C e^{iz}dz = 0$ ...(5)

From (4) and (5) theorem is verified.

EXERCISE 29.2

Evaluate the following

1. $\int_C \frac{1}{z-a}\,dz$, where c is a simple closed curve and the point $z = a$ is
(i) outside c; (ii) inside c. **Ans.** (i) 0 (ii) $2\pi i$

2. $\int_c \frac{e^z}{z-1}\,dz$, where c is the circle $|z| = 2$. **Ans.** $2\pi i e$

3. $\int_C \frac{\cos \pi z}{z-1}\,dz$, where c is the circle $|z| = 3$. **Ans.** $-2\pi i$

4. $\int_C \frac{\cos \pi z^2}{(z-1)(z-2)}\,dz$, where c is the circle $|z| = 3$. **Ans.** $4\pi i$

5. $\int_C \frac{e^{-z}}{(z+2)^5}\,dz$, where c is the circle $|z| = 3$. **Ans.** $\frac{\pi i e^2}{12}$

6. $\int_C \frac{e^{2z}}{(z+1)^4}\,dz$, where c is the circle $|z| = 2$. **Ans.** $\frac{8\pi}{3} i e^{-2}$

7. $\int_c \frac{2z^2+z}{z^2-1}\,dz$ where c is the circle $|z-1| = 1$ **Ans.** $3\pi i$

8. $\int_C \frac{e^z}{z^2(z+1)^3}\,dz$, C : $|z| = 2$. (*AMIETE, June 2009*) **Ans.** $2\pi i\left(\frac{11}{e}-2\right)$

Choose the correct alternative:

9. The value of the integral $\int_C \frac{z^2+1}{(z+1)(z+2)}\,dz$, where C is $|z| = \frac{3}{2}$ is
(i) πi (ii) 0 (iii) $2\pi i$ (iv) $4\pi i$ **Ans.** (iv)
(*AMIETE, June 2010*)

10. Cauchy's Integral formula states that if $f(z)$ is analytic within a and on a closed curve C and if a is any point within C then $f(a) =$: (*R.G.P.V., Bhopal, III Semester, June 2007*)
(i) $\frac{1}{2\pi i}\oint \frac{f(z)dz}{z-a}$ (ii) $\frac{1}{2\pi i}\oint f(z)dz$ (iii) $\frac{1}{2\pi i}\oint \frac{dz}{z-a}$ (iv) none of these. **Ans.** (i)

11. The value of $\int_C \frac{z^2-z+1}{z-1}\,dz$, C being $|z| = \frac{1}{2}$ is :
(i) $2\pi i$ (ii) $\frac{1}{2\pi i}$ (iii) 0 (iv) πi (*R.G.P.V., Bhôpal, III Sem., Dec. 2006*) **Ans.** (iii)

12. If $f(z) = \frac{z^2}{(z-1)^2(z+2)}$, then Res. $f(-2)$ is :
(i) $\frac{5}{9}$ (ii) $\frac{4}{9}$ (iii) $\frac{1}{9}$ (iv) $\frac{3}{9}$ (*RGPV, Bhopal, III Sems, Dec. 2006*) **Ans.** (ii)

13. Let $f(z) = \frac{1}{(z-2)^4(z+3)^6}$, then $z = 2$ and $z = -3$ are the poles of order :
(i) 6 and 4 (ii) 2 and 3 (iii) 3 and 4 (iv) 4 and 6 (*RGPV, Bhopal, III Sem., June 2006*) **Ans.** (iv)

14. The value of the integral $\int_C \frac{z+1}{z^3-2z^2}\,dz$, where C is the circle $|z| = 1$ is equal to.
(i) $2\pi i$ (ii) $-\frac{2}{3}\pi i$ (iii) zero (iv) $-\frac{3}{2}\pi i$. *AMIETE, Dec. 2010*) **Ans.** (iv)

CHAPTER 30

TAYLOR'S AND LAURENT'S SERIES

30.1 INTRODUCTION

An analytic function within a circle can be expanded by Taylor's series.

If a function which is not analytic within a circle is expanded by Laurent's series.

30.2 CONVERGENCE OF A SERIES OF COMPLEX TERMS

Let $$(u_1 + iv_1) + (u_2 + iv_2) + (u_3 + iv_3) + \ldots + (u_n + iv_n) + \ldots \qquad \ldots (1)$$

be an infinite series of complex terms: $u_1, u_2, \ldots v_1, v_2 \ldots$ being real numbers.

(*a*) If the series Σu_n and Σv_n converge to the sums U and V then series (1) is said to converge to the sum $U + iV$.

(*b*) If (1) is a convergent series, then

$$\lim_{n\to\infty}(u_n + iv_n) = 0$$

(*c*) The series (1) is said to be **absolutely convergent** if the series

$$|u_1 + iv_1| + |u_2 + iv_2| + |u_3 + iv_3| + \ldots + |u_n + iv_n| + \ldots$$

is convergent. Since $|u_n|$ and $|v_n|$ are both less than $|u_n + iv_n|$.

(*d*) Let the series

$$a_1(z) + a_2(z) + a_3(z) + \ldots + a_n(z) + \ldots \qquad \ldots (2)$$

converge to the sum $S(z)$ and $S_n(z)$ be the sum of its first n terms.

The series (2) is said to be absolutely convergent in region R, if corresponding to any positive number ε, there exists a positive number N.

$$|S(z) - S_n(z)| < \varepsilon \text{ for } n > N$$

(*e*) Weirstras's, M-test holds good for series of complex terms also.

Series (2) is uniformly convergent in a region R if there is a convergent series ΣM_n.

Such that $\quad |a_n(z)| \le M_n$

A uniformly convergent series can be integrated term by term.

30.3 POWER SERIES

A series in powers of $(z - z_0)$ is called power series.

$$\sum_{n=0}^{\infty} a_n (z - z_0)^n = a_0 + a_1(z - z_0) + a_2(z - z_0)^2 + \ldots + a_n(z - z_0)^n + \ldots \qquad \ldots (1)$$

Here $a_0, a_1, a_2, \ldots, a_n, \ldots$ are known as the coefficient of the series.

Here z is a complex variable and z_0 is called the centre of the series.
(1) is also called the power series about the point z_0

$$\sum_{n=0}^{\infty} a_n z^n = a_0 + a_1 z + a_2 z^2 + ... + a_n z^n + ...$$

Here the centre of the series is zero.

30.4 REGION OF CONVERGENCE

The region of convergence is the set of all points z for which the series converges.
There are three distinct possibilities for a convergent series.

(1) The series converges only at the point $z = z_0$

(2) The series converges for all the points in the whole plane.

(3) The series converges everywhere inside a circular plane $|z - z_0| < R$, where R is the radius of convergence and diverges everywhere outside the circle/circular ring.

30.5 RADIUS OF CONVERGENCE OF POWER SERIES

Consider the power series $\Sigma a_n z^n$.
By Cauchy theorem on limits, radius of convergence R is given by

(i) $\dfrac{1}{R} = \lim\limits_{n \to \infty} |a_n|^{1/n}$ (ii) $\dfrac{1}{R} = \lim\limits_{n \to \infty} \left|\dfrac{a_{n+1}}{a_n}\right|$

Example 1. *Find the radius of convergence of the power series*

$$f(z) = \sum_{n=0}^{\infty} \frac{z^n}{n!}$$

Solution. Here, $a_n = \dfrac{1}{n!} \Rightarrow a_{n+1} = \dfrac{1}{(n+1)!}$

Radius of convergence is given by

$$\frac{1}{R} = \lim_{n \to \infty} \left|\frac{a_{n+1}}{a_n}\right| = \lim_{n \to \infty} \frac{1}{n+1} = 0$$

$$R = \infty$$

Hence, the radius of convergence of the given power series is ∞. **Ans.**

Example 2. *Find the radius of convergence of the power series*

$$f(z) = \sum_{n=0}^{\infty} \frac{z^n}{2^n + 3}$$

Solution. Here,

$$a_n = \frac{1}{2^n + 3} \Rightarrow a_{n+1} = \frac{1}{2^{n+1} + 3}$$

Radius of convergence is given by

$$\frac{1}{R} = \lim_{n \to \infty} \left|\frac{a_{n+1}}{a_n}\right| = \lim_{n \to \infty} \frac{2^n + 3}{2^{n+1} + 3} = \lim_{n \to \infty} \frac{1 + \frac{3}{2^n}}{2 + \frac{3}{2^n}} = \frac{1}{2}$$

$$R = 2$$

Hence, the radius of convergence of the given power series is 2. **Ans.**

Example 3. *Find the radius of convergence of the power series:*

$$f(z) = \sum_{n=0}^{\infty} \frac{z^n}{n^n}$$

Solution. Here,

$$a_n = \frac{1}{n^n}, \qquad a_{n+1} = \frac{1}{(n+1)^{n+1}}$$

Radius of convergence is given by

$$\frac{1}{R} = \lim_{n\to\infty}\left|\frac{a_{n+1}}{a_n}\right| = \lim_{n\to\infty}\frac{n^n}{(n+1)^{n+1}}$$

$$= \lim_{n\to\infty}\frac{n^n}{(n+1)^n\,(n+1)} = \lim_{n\to\infty}\frac{1}{\left(1+\frac{1}{n}\right)^n(n+1)} = 0$$

$\Rightarrow$ $R = \infty$

Hence, the radius of convergence of the given power series is ∞. **Ans.**

Example 4. *Find the radius of convergence of the power series:*

$$f(z) = \sum_{n=0}^{\infty}\frac{n!}{n^n}z^n$$

Solution. Here, $a_n = \dfrac{n!}{n^n}$

$\therefore$ $a_{n+1} = \dfrac{(n+1)!}{(n+1)^{n+1}}$

Now,

$$\frac{a_{n+1}}{a_n} = \frac{(n+1)!}{(n+1)^{n+1}}\cdot\frac{n^n}{n!} = \frac{n!}{(n+1)^n}\cdot\frac{n^n}{n!} = \frac{n^n}{(n+1)^n} = \left(\frac{n}{n+1}\right)^n = \frac{1}{\left(1+\frac{1}{n}\right)^n}$$

Radius of convergence is given by

$$\frac{1}{R} = \lim_{n\to\infty}\left|\frac{a_{n+1}}{a_n}\right| = \lim_{n\to\infty}\frac{1}{\left(1+\frac{1}{n}\right)^n} = \frac{1}{e}$$

$\Rightarrow$ $R = e.$

Hence, the radius of convergence of the given power series is e. **Ans.**

EXERCISE 30.1

Find the radius of convergence of following power series:

1. $\displaystyle\sum_{n=1}^{\infty}\frac{z^n}{n^p}$ **Ans.** 1
2. $\displaystyle\sum_{n=0}^{\infty}\frac{z^n}{3^{n+1}}$ **Ans.** 3
3. $\displaystyle\sum_{n=0}^{\infty}\left(1+\frac{1}{n}\right)^{n^2} z^n$ **Ans.** $\frac{1}{e}$
4. $\displaystyle\sum_{m=0}^{\infty}(5+12i)^m z^m$ **Ans.** 13
5. $\displaystyle\sum_{n=0}^{\infty}\frac{2n+3}{(2n+5)(n+5)}z^n$ **Ans.** 1

30.6 METHOD OF EXPANSION OF A FUNCTION

(1) Taylor's series (2) Binomial series (3) Exponential series

30.7 TAYLOR'S THEOREM

If a function $f(z)$ is analytic at all points inside a circle C, with its centre at the point a and radius R, then at each point z inside C.

$$f(z) = f(a) + f'(a)(z-a) + \frac{f''(a)}{2!}(z-a)^2 + \ldots + \frac{f^n(a)}{n!}(z-a)^n + \ldots$$

Proof. Take any point z inside C. Draw a circle C_1 with centre a, enclosing the point z. Let w be a point on circle C_1.

$$\frac{1}{w-z}=\frac{1}{w-a+a-z}=\frac{1}{w-a-(z-a)}$$

$$=\frac{1}{(w-a)}\frac{1}{\left(1-\frac{z-a}{w-a}\right)}=\frac{1}{w-a}\left(1-\frac{z-a}{w-a}\right)^{-1}$$

C

C_1

a z w

Apply Binomial theorem

$$\frac{1}{w-z}=\frac{1}{w-a}\left[1+\frac{z-a}{w-a}+\left(\frac{z-a}{w-a}\right)^2+\ldots+\left(\frac{z-a}{w-a}\right)^n+\ldots\right]$$

$$\Rightarrow \quad \frac{1}{w-z}=\frac{1}{w-a}+\frac{z-a}{(w-a)^2}+\frac{(z-a)^2}{(w-a)^3}+\ldots+\frac{(z-a)^n}{(w-a)^{n+1}}+\ldots \qquad \ldots (1)$$

As $\quad |z-a|<|w-a| \quad \Rightarrow \quad \frac{|z-a|}{|w-a|}<1$,

So the series converges uniformly. Hence the series is integrable.
Multiply (1) by $f(w)$.

$$\frac{f(w)}{w-z}=\frac{f(w)}{w-a}+(z-a)\frac{f(w)}{(w-a)^2}+(z-a)^2\frac{f(w)}{(w-a)^3}+\ldots+(z-a)^n\frac{f(w)}{(w-a)^{n+1}}+,\ldots$$

On integrating w.r.t. 'w', we get

$$\int_{C_1}\frac{f(w)}{w-z}dw=\int_{C_1}\frac{f(w)}{w-a}dw+(z-a)\int_{C_1}\frac{f(w)}{(w-a)^2}dw+(z-a)^2\int_{C_1}\frac{f(w)}{(w-a)^3}dw+$$

$$\ldots+(z-a)^n\int_{C_1}\frac{f(w)}{(w-a)^{n+1}}dw+\ldots \qquad \ldots (2)$$

We know that

$$\int_{C_1}\frac{f(w)}{w-z}dz=2\pi i\,f(z) \text{ and } \int_{C_1}\frac{f(w)}{w-a}dw=2\pi i\,f(a)$$

$$\int_{C_1}\frac{f(w)}{(w-a)^2}dw=2\pi i\,f'(a) \text{ and so on.}$$

Substituting these values in (2), we get
Taylor's series as given below

$$\boldsymbol{f(z)=f(a)+f'(a)\,(z-a)+\frac{f''(a)}{2!}(z-a)^2+\ldots+\frac{f^n(a)(z-a)^n}{n!}+\ldots} \qquad \ldots (3)$$ **Proved.**

Corollary 1. Putting $z = a + h$ in (3), we get

$$f(a+h)=f(a)+hf'(a)+\frac{h^2}{2!}f''(a)+\ldots\ldots+\frac{h^n}{n!}f^n(a)+\ldots$$

Corollary 2. If $a = 0$, the series (3) becomes

$$f(z)=f(0)+z\,f'(0)+\frac{z^2}{2!}f''(0)+\ldots\ldots+\frac{z^n}{n!}f^n(0)+\ldots$$

This series is called **Maclaurin series.**

Example 5. *Expand e^z about a.*

Solution. Here, we have

$$f(z)=e^z \qquad \Rightarrow \qquad f(a)=e^a$$

$$\Rightarrow \quad f'(z) = e^z \quad \Rightarrow \quad f'(a) = e^a$$
$$f''(z) = e^z \quad \Rightarrow \quad f''(a) = e^a$$
$$\ldots\ldots\ldots\ldots\ldots\ldots\ldots\ldots\ldots$$
$$\ldots\ldots\ldots\ldots\ldots\ldots\ldots\ldots\ldots$$
$$f^n(z) = e^z \quad \Rightarrow \quad f^n(a) = e^a$$

By Taylor's series of $f(z)$ is

$$f(z) = f(a) + (z-a) f'(a) + \frac{(z-a)^2}{2!} f''(a) + \ldots\ldots\ldots\ldots$$

$$e^z = e^a + \frac{(z-a)}{1!} e^a + \frac{(z-a)^2}{2!} e^a + \ldots\ldots\ldots \qquad \textbf{Ans.}$$

Example 6. *Expand the function*

$$f(z) = \frac{1}{z}$$

about $z = 2$ *in Taylor's series. Obtain its radius of convergence.*

Solution. Here, we have,

$$f(z) = \frac{1}{z}$$

$$\Rightarrow \quad f'(z) = -\frac{1}{z^2}$$

$$\Rightarrow \quad f''(z) = \frac{2}{z^3}$$

$$\ldots\ldots\ldots\ldots\ldots\ldots\ldots\ldots\ldots$$

$$f^n(z) = (-1)^n \frac{n!}{z^{n+1}}$$

By Taylor's series

$$f(z) = f(2) + (z-2)\ f'(2) + \frac{(z-2)^2}{2!} f''(2) + \ldots\ldots\ldots + \frac{(z-2)^n}{n!} f^n(2) + \ldots\ldots$$

$$= \frac{1}{2} + (z-2)\left(-\frac{1}{2^2}\right) + \frac{(z-2)^2}{2!}\left(\frac{2}{2^3}\right) + \ldots\ldots \frac{(z-2)^n}{n!} (-1)^n \frac{n!}{2^{n+1}} + \ldots\ldots$$

$$= \frac{1}{2} - \frac{1}{4}(z-2) + \frac{1}{8}(z-2)^2 - \ldots\ldots\ldots + \frac{(-1)^n}{2^{n+1}} (z-2)^n + \ldots\ldots\ldots \qquad \textbf{Ans.}$$

Alternative.

We can expand the given function by Binomial expansion.

$$\frac{1}{z} = \frac{1}{2+z-2} = \frac{1}{2}\left[\frac{1}{1+\frac{z-2}{2}}\right] = \frac{1}{2}\left[\left(1+\frac{z-2}{2}\right)^{-1}\right] \qquad \left|\frac{z-2}{2}\right| < 1$$

$$= \frac{1}{2}\left[1 - \frac{z-2}{2} + \frac{(-1)(-2)}{2!}\left(\frac{z-2}{2}\right)^2 + \frac{(-1)(-2)(-3)}{3!}\left(\frac{z-2}{2}\right)^3 + \ldots\ldots\right]$$

$$= \frac{1}{2} - \frac{z-2}{4} + \frac{1}{8}(z-2)^2 - \frac{1}{16}(z-2)^3 + \ldots\ldots \qquad \textbf{Ans.}$$

Radius of convergence $\dfrac{1}{R} = \lim\limits_{n\to\infty}\left(\dfrac{a^{n+1}}{a^n}\right) = \lim\limits_{n\to\infty}\left|\dfrac{(-1)^{n+1}}{2^{n+2}} \cdot \dfrac{2^{n+1}}{(-1)^n}\right| = \dfrac{1}{2}$

$$\Rightarrow \quad R = 2 \qquad \textbf{Ans.}$$

Example 7. *Expand $f(z) = \cosh z$ about $\pi\, i$.*

Solution. Here, we have

$$f(z) = \cosh z = \frac{e^z + e^{-z}}{2} \Rightarrow f(\pi\, i) = \cosh(\pi\, i)$$

$$f'(z) = \sinh z \quad \Rightarrow \quad f'(\pi i) = \sinh(\pi i)$$

$$f''(z) = \cosh z \quad \Rightarrow \quad f''(\pi i) = \cosh(\pi i)$$

$$f'''(z) = \sinh z \quad \Rightarrow \quad f'''(\pi i) = \sinh(\pi i)$$

By Taylor's series

$$f(z) = f(\pi i) + (z - \pi i) f'(\pi i) + \frac{(z-\pi i)^2}{2!} f''(\pi i) + \frac{(z-\pi i)^3}{3!} f'''(\pi i) + ...$$

$$= \cosh \pi i + (z - \pi i) \sinh \pi i + \frac{(z-\pi i)^2}{2!} \cosh(\pi i) + \frac{(z-\pi i)^3}{3!} \sinh(\pi i) + ...$$ **Ans.**

Example 8. *Expand* $f(z) = \dfrac{a}{bz+c}$ *about* z_0.

Solution. Here, we have

$$f(z) = \frac{a}{bz+c} = \frac{a}{bz - bz_0 + bz_0 + c} = \frac{a}{b(z-z_0) + bz_0 + c}$$

$$= \frac{a}{(bz_0+c)\left[1 + \left(\dfrac{b(z-z_0)}{bz_0+c}\right)\right]} = \frac{a}{bz_0+c}\left[1 + \frac{b(z-z_0)}{bz_0+c}\right]^{-1} \quad \text{(Binomial series)}$$

$$= \frac{a}{bz_0+c}\left[1 - \frac{b(z-z_0)}{bz_0+c} + \frac{(-1)(-2)}{2!}\left(\frac{b(z-z_0)}{bz_0+c}\right)^2 + \frac{(-1)(-2)(-3)}{3!}\left(\frac{b(z-z_0)}{bz_0+c}\right)^3 + ...\right]$$

$$= \frac{a}{bz_0+c}\left[1 - \frac{b(z-z_0)}{bz_0+c} + \left(\frac{b(z-z_0)}{bz_0+c}\right)^2 - \left(\frac{b(z-z_0)}{bz_0+c}\right)^3 + ...\right]$$

$$= \frac{a}{bz_0+c}\left[1 - \frac{b}{bz_0+c}(z-z_0) + \left(\frac{b}{bz_0+c}\right)^2 (z-z_0)^2 - \left(\frac{b}{bz_0+c}\right)^3 (z-z_0)^3 + ...\right]$$

Radius of curvature

$$\frac{1}{R} = \lim_{n\to\infty}\left(\frac{a^{n+1}}{a^n}\right) = \lim_{n\to\infty}\frac{\left(\dfrac{b}{bz_0+c}\right)^{n+1}}{\left(\dfrac{b}{bz_0+c}\right)^n} = \lim_{n\to\infty}\frac{b}{bz_0+c} = \frac{b}{bz_0+c}$$

$$\Rightarrow \quad R = \frac{bz_0+c}{b} = z_0 + \frac{c}{b}$$ **Ans.**

Example 9. *Show that :*

$$\log z = (z-1) - \frac{(z-1)^2}{2} + \frac{(z-1)^3}{3} + ...$$

Solution. Let $f(z) = \log(z) \quad \Rightarrow \quad f(1) = \log 1 = 0$

$$f'(z) = \frac{1}{z}, \qquad f'(1) = \frac{1}{1} = 1$$

$$f''(z) = -\frac{1}{z^2}, \qquad f''(1) = -1$$

$$f'''(z) = \frac{2 \times 1}{z^3}, \qquad f'''(1) = 2$$

$$f^{iv}(z) = -3 \times 2 \times 1 \times \frac{1}{z^4}, \quad f^{iv}(1) = -3!$$

By Taylor series

$$f(z) = f(a) + f'(a)\,.\,(z-a) + \frac{f''(a)(z-a)^2}{2!} + \frac{f'''(a)}{3!}(z-a)^3 + ...$$

$$f(z) = \log z = \log\,(1 + \overline{z-1})$$

On substituting the values of $f(1), f'(1), f''(1)$ etc., we get

$$\log z = 0 + 1(z-1) - \frac{1}{2!}(z-1)^2 + \frac{2}{3!}(z-1)^3 - \frac{3!}{4!}(z-1)^4 + ...$$

$$\Rightarrow \qquad \log z = (z-1) - \frac{1}{2}(z-1)^2 + \frac{1}{3}(z-1)^3 - \frac{1}{4}(z-1)^4 + ...$$ **Proved.**

Example 10. *Expand* $\dfrac{1}{z^2 - 3z + 2}$ *in the region*

(a) $|z| < 1$ *(b)* $|z| > 2.$ *(R.G.P.V., Bhopal, III Semester, Dec. 2005)*

Solution. Here, $f(z) = \dfrac{1}{z^2 - 3z + 2} = \dfrac{1}{(z-1)(z-2)} = \dfrac{1}{z-2} - \dfrac{1}{z-1}$

(a) If $|z| < 1$

Taking common, bigger term out of $|z|$ and 2, here 2 is bigger than $|z|$. So we take 2 common.

$$f(z) = \frac{1}{-2\left(1 - \frac{z}{2}\right)} + \frac{1}{1-z}$$

$$= -\frac{1}{2}\left(1 - \frac{z}{2}\right)^{-1} + (1-z)^{-1}$$

$$= -\frac{1}{2}\left(1 + \frac{z}{2} + \frac{z^2}{4} + \frac{z^3}{8} + ...\right) + (1 + z + z^2 + z^3 + ...)$$

[By Binomial theorem]

$$= \frac{1}{2} + \frac{3z}{4} + \frac{7}{8}z^2 + \frac{15}{16}z^3 + ...$$

Which is the required expansion.

(b) If $|z| > 2$

We have, $\quad f(z) = \dfrac{1}{z-2} - \dfrac{1}{z-1}$

Taking common, bigger term out of $|z|$ and 2, here z is bigger than 2. So we take $|z|$ common.

$$f(z) = \frac{1}{z\left(1 - \frac{2}{z}\right)} - \frac{1}{z\left(1 - \frac{1}{z}\right)} = \frac{1}{z}\left(1 - \frac{2}{z}\right)^{-1} - \frac{1}{z}\left(1 - \frac{1}{z}\right)^{-1}$$

$$= \frac{1}{z}\left(1 + \frac{2}{z} + \frac{4}{z^2} + \frac{8}{z^3} + ...\right) - \frac{1}{z}\left(1 + \frac{1}{z} + \frac{1}{z^2} + \frac{1}{z^3} + ...\right)$$ [By Binomial theorem]

$$= z^{-2} + 3z^{-3} + 7z^{-4} + ...$$

Which is the required expansion. **Ans.**

Example 11. *Show that when* $|z+1|<1$,

$$z^{-2} = 1 + \sum_{n=1}^{\infty} (n+1)(z+1)^n .$$

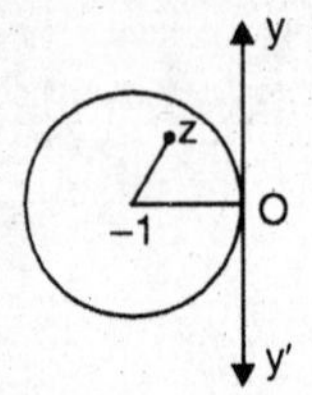

Solution. $f(z) = z^{-2} = \dfrac{1}{z^2} = \dfrac{1}{[(z+1)-1]^2}$

Taking common, bigger term out of 1 and $|z+1|$, here, $1 > |z+1|$.
So, we take 1 common.

$$f(z) = \frac{1}{[1-(z+1)]^2} = [1-(z+1)]^{-2} = 1 + 2(z+1) + 3(z+1)^2 + 4(z+1)^3 + \ldots\ldots$$

$$= 1 + \sum_{n=1}^{\infty} (n+1)(z+1)^n \qquad \textbf{Proved.}$$

Example 12. *Find the first four terms of the Taylor's series expansion of the complex variable function*

$$f(z) = \frac{z+1}{(z-3)(z-4)}$$

about $z = 2$. *Find the region of convergence.*

Solution. $f(z) = \dfrac{z+1}{(z-3)(z-4)}$

If centre of a circle is at $z = 2$, then the distances of the singularities $z = 3$ and $z = 4$ from the centre are 1 and 2.
Hence if a circle is drawn with centre $z = 2$ and radius 1, then within the circle $|z-2| = 1$, the given function $f(z)$ is analytic, hence it can be expanded in a Taylor's series within the circle $|z-2| = 1$, which is therefore the circle of convergence.

$$f(z) = \frac{z+1}{(z-3)(z-4)} = \frac{-4}{z-3} + \frac{5}{z-4} \qquad \text{(By Partial fraction method)}$$

$$= \frac{-4}{(z-2)-1} + \frac{5}{(z-2)-2} = 4[1-(z-2)]^{-1} - \frac{5}{2}\left[1 - \frac{z-2}{2}\right]^{-1} \qquad [|z-2|<1]$$

$$= 4[1 + (z-2) + (z-2)^2 + (z-2)^3 + \ldots] - \frac{5}{2}\left[1 + \frac{z-2}{2} + \frac{(z-2)^2}{4} + \frac{(z-2)^3}{8} + \ldots\right]$$

$$= \left(4 - \frac{5}{2}\right) + \left(4 - \frac{5}{4}\right)(z-2) + \left(4 - \frac{5}{8}\right)(z-2)^2 + \left(4 - \frac{5}{16}\right)(z-2)^3 \ldots$$

$$= \frac{3}{2} + \frac{11}{4}(z-2) + \frac{27}{8}(z-2)^2 + \frac{59}{16}(z-2)^3 + \ldots \qquad \textbf{Ans.}$$

Alternative method. In obtaining the Taylor series we evaluate the coefficients by contour integration.

$$f(z) = \frac{z+1}{(z-3)(z-4)}, \qquad f(2) = \frac{2+1}{(2-3)(2-4)} = \frac{3}{2}.$$

To make the differentiation easier let us convert the given fraction into partial fractions

$$f(z) = \frac{-4}{z-3} + \frac{5}{z-4}$$

$$f'(z) = \frac{4}{(z-3)^2} - \frac{5}{(z-4)^2}, \qquad f'(2) = \frac{4}{(2-3)^2} - \frac{5}{(2-4)^2} = \frac{11}{4}$$

$$f''(z)=\frac{-8}{(z-3)^3}+\frac{10}{(z-4)^3}, \qquad f''(2)=\frac{-8}{(2-3)^3}+\frac{10}{(2-4)^3}=\frac{27}{4}$$

$$f'''(z)=\frac{24}{(z-3)^4}-\frac{30}{(z-4)^4}. \qquad f'''(2)=\frac{24}{(2-3)^4}-\frac{30}{(2-4)^4}=\frac{177}{8}$$

Taylor series is $f(z)=f(a)+(z-a)f'(a)+\frac{(z-a)^2}{2!}f''(a)+\frac{(z-a)^3}{3!}f'''(a)+\ldots$

$$\frac{z+1}{(z-3)(z-4)}=\frac{3}{2}+(z-2)\frac{11}{4}+\frac{(z-2)^2}{2!}\left(\frac{27}{4}\right)+\frac{(z-2)^3}{3!}\frac{177}{8}+\ldots$$

$$=\frac{3}{2}+(z-2)\frac{11}{4}+(z-2)^2\cdot\frac{27}{8}+(z-2)^3\cdot\frac{59}{16}+\ldots$$ **Ans.**

Example 13. *Find the first three terms of the Taylor series expansion of* $f(z)=\frac{1}{z^2+4}$ *about* $z=-i$. *Find the region of convergence.*

Solution. $f(z)=\frac{1}{z^2+4}$

Poles are given by $z^2+4=0$, $\Rightarrow$ $z^2=-4$, $\Rightarrow$ $z=\pm 2i$

If the centre of a circle is $z=-i$, then the distances of the singularities $z=2i$ and $z=-2i$ from the centre are 3 and 1. Hence if a circle of radius 1 is drawn with centre at $z=-i$, then within the circle $|z+i|=1$, the given function $f(z)$ is analytic. Thus the function can be expanded in Taylor series within the circle $|z+i|=1$, which is therefore the circle of convergence.

$$f(z)=\frac{1}{z^2+4}=\frac{1}{(z+2i)(z-2i)}=\frac{1}{4i}\left[\frac{1}{z-2i}-\frac{1}{z+2i}\right]$$

$$=\frac{1}{4i}\left[\frac{1}{(z+i)-3i}-\frac{1}{(z+i)+i}\right]$$

$$=\frac{1}{4i}\left[-\frac{1}{3i}\frac{1}{1-\frac{z+i}{3i}}-\frac{1}{i}\frac{1}{1+\frac{z+i}{i}}\right]=\frac{1}{4}\left[\frac{1}{3}\frac{1}{1-\frac{z+i}{3i}}+\frac{1}{1+\frac{z+i}{i}}\right] \quad \left[\begin{array}{l}3>|z+i|\\1>|z+i|\end{array}\right]$$

$$=\frac{1}{4}\left[\frac{1}{3}\frac{1}{1+\frac{i}{3}(z+i)}+\frac{1}{1-i(z+i)}\right]=\frac{1}{4}\left[\frac{1}{3}\left\{1+\frac{i}{3}(z+i)\right\}^{-1}+\{1-i(z+i)\}^{-1}\right]$$

$$=\frac{1}{4}\left[\frac{1}{3}\left\{1-\frac{i}{3}(z+i)+\frac{(-1)(-2)}{2!}\left(\frac{i}{3}\right)^2(z+i)^2+\frac{(-1)(-2)(-3)}{3!}\left(\frac{i}{3}\right)^3(z+i)^3+\ldots\right\}\right.$$
$$\left.+\left\{1+i(z+i)+\frac{(-1)(-2)}{2!}\cdot(-i)^2(z+i)^2+\frac{(-1)(-2)(-3)}{3!}(-i)^3(z+i)^3+\ldots\right\}\right]$$

$$=\frac{1}{4}\left[\frac{1}{3}-\frac{i}{9}(z+i)-\frac{1}{27}(z+i)^2+\frac{i}{81}(z+i)^3+\ldots+1+i(z+i)-(z+i)^2-i(z+i)^3+\ldots\right]$$

$$=\frac{1}{4}\left[\frac{4}{3}+\frac{8i}{9}(z+i)-\frac{28}{27}(z+i)^2-\frac{80}{81}i(z+i)^3+\ldots\right]$$

$$=\frac{1}{3}+\frac{2i}{9}(z+i)-\frac{7}{27}(z+i)^2-\frac{20}{81}i(z+i)^3+\ldots$$ **Ans.**

Alternative method

$$f(z) = \frac{1}{z^2+4}$$

By Taylor expansion $f(z) = f(a) + (z-a)f'(a) + \frac{(z-a)^2}{2!}f''(a) + \frac{(z-a)^3}{3!}f'''(a) + \ldots$

Putting $a = -i$ in above, we get

$$f(z) = f(-i) + (z+i)f'(-i) + \frac{(z+i)^2}{2!}f''(-i) + \ldots$$

$$f(z) = \frac{1}{z^2+4} \quad \Rightarrow \quad f(-i) = \frac{1}{(-i)^2+4} = \frac{1}{-1+4} = \frac{1}{3}$$

$$f'(z) = \frac{-2z}{(z^2+4)^2} \quad \Rightarrow \quad f'(-i) = \frac{2i}{(-1+4)^2} = \frac{2i}{9}$$

$$f''(z) = -\frac{(z^2+4)^2(2) - 2z \cdot 2(z^2+4)2z}{(z^2+4)^4}$$

$$f''(z) = -\frac{(z^2+4)2 - 8z^2}{(z^2+4)^3} \quad \Rightarrow \quad f''(-i) = -\frac{(-1+4)2 - 8(-1)}{(-1+4)^3} = -\frac{14}{27}$$

On substituting the value of $f(-i), f'(-i), f''(-i)$, we get

$$f(z) = \frac{1}{3} + (z+i)\left(\frac{2i}{9}\right) + \frac{(z+i)^2}{2!}\left(-\frac{14}{27}\right) + \ldots$$

$$\Rightarrow \quad f(z) = \frac{1}{3} + \frac{2i}{9}(z+i) - \frac{7}{27}(z+i)^2 + \ldots$$

Region of convergence is $|z+i| < 1$. **Ans.**

Example 14. *For the function* $f(z) = \frac{4z-1}{z^4-1}$, *find all Taylor series about the centre zero.* (*U.P., III Semester, Dec. 2006*)

Solution. $f(z) = \frac{4z-1}{z^4-1}$,

Poles are determined by $z^4 - 1 = 0$

$$\Rightarrow \quad (z-1)(z+1)(z^2+1) = 0$$

$$\Rightarrow \quad z = 1, -1, \pm i$$

By Partial fractions

$$f(z) = \frac{\frac{3}{4}}{z-1} + \frac{\frac{5}{4}}{z+1} + \frac{-2z+\frac{1}{2}}{z^2+1}$$

$$= -\frac{3}{4}\frac{1}{1-z} + \frac{5}{4}\frac{1}{1+z} + \left(-2z+\frac{1}{2}\right)\frac{1}{1+z^2}$$

$$= -\frac{3}{4}(1-z)^{-1} + \frac{5}{4}(1+z)^{-1} + \left(-2z+\frac{1}{2}\right)(1+z^2)^{-1}$$

$$= -\frac{3}{4}[1+z+z^2+z^3+z^4+\ldots] + \frac{5}{4}[1-z+z^2-z^3+z^4\ldots] + \left(-2z+\frac{1}{2}\right)[1-z^2+z^4+\ldots]$$

$$= -\frac{3}{4} - \frac{3}{4}z - \frac{3}{4}z^2 - \frac{3}{4}z^3 - \frac{3}{4}z^4 + \ldots\ldots\ldots + \frac{5}{4} - \frac{5}{4}z + \frac{5}{4}z^2 - \frac{5}{4}z^3 + \frac{5}{4}z^4 + \ldots$$

$$-2z + 2z^3 - 2z^5 + \ldots + \frac{1}{2} - \frac{z^2}{2} + \frac{z^4}{2} + \ldots$$

$$= 1 - 4z + z^4 + \ldots\ldots$$ **Ans.**

Which is the required series.

Example 15. *Find Taylor expansion of* $f(z) = \dfrac{2z^3+1}{z^2+z}$ *about the point* $z = 1$.

Solution. $f(z) = \dfrac{2z^3+1}{z(z+1)}$, singularities are given by $z.(z+1) = 0 \Rightarrow z=0,\ z=-1$

If centre of the circle is at $z = 1$, then the distance of the singularities $z = 0$ and $z = -1$ from the centre are 1 and 2. Hence, if a circle is drawn with centre $z=1$ and radius 1, then within the circle $|z-1| = 1$, the given function $f(z)$ is analytic & therefore, it can be expanded in a Taylor series within the circle $|z-1|=1$, which is thus the circle of convergence.

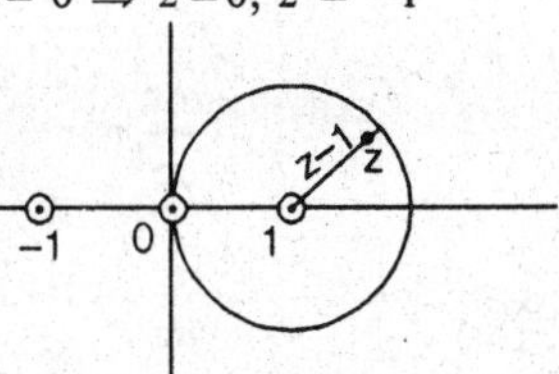

$$\frac{2z^3+1}{z(z+1)} = 2z-2+\frac{1}{z+1}+\frac{1}{z} = 2z-2+\frac{1}{z-1+2}+\frac{1}{z-1+1} \quad [|z-1|<1]$$

$$= 2z-2+\frac{1}{2}\left(1+\frac{z-1}{2}\right)^{-1}+[1+(z-1)]^{-1}$$

$$= 2z-2+\frac{1}{2}\left[1-\left(\frac{z-1}{2}\right)+\left(\frac{z-1}{2}\right)^2-\left(\frac{z-1}{2}\right)^3+....\right]+[1-(z-1)+(z-1)^2-(z-1)^3+...]$$

$$= 2z-2+\frac{3}{2}-\frac{3}{2}\left(\frac{z-1}{2}\right)+\frac{9}{8}(z-1)^2-\frac{17}{16}(z-1)^3+....$$

Which is the required expansion. **Ans.**

Example 16. *Expand* $\cos z$ *in a Taylor series about* $z = \dfrac{\pi}{4}$.

Solution. Here $f(z) = \cos z$, $f'(z) = -\sin z$, $f''(z) = -\cos z$, $f'''(z) = \sin z$, ...

$$\therefore \quad f\left(\frac{\pi}{4}\right) = \frac{1}{\sqrt{2}},\ f'\left(\frac{\pi}{4}\right) = -\frac{1}{\sqrt{2}},\ f''\left(\frac{\pi}{4}\right) = -\frac{1}{\sqrt{2}},\ f'''\left(\frac{\pi}{4}\right) = \frac{1}{\sqrt{2}}, ...$$

Hence $\cos z = f(z)$

$$= f\left(\frac{\pi}{4}\right)+\left(z-\frac{\pi}{4}\right)f'\left(\frac{\pi}{4}\right)+\frac{\left(z-\frac{\pi}{4}\right)^2}{2!}f''\left(\frac{\pi}{4}\right)+\frac{\left(z-\frac{\pi}{4}\right)^3}{3!}f'''\left(\frac{\pi}{4}\right)+......$$

$$= \frac{1}{\sqrt{2}}\left[1-\left(z-\frac{\pi}{4}\right)-\frac{1}{2!}\left(z-\frac{\pi}{4}\right)^2+\frac{1}{3!}\left(z-\frac{\pi}{4}\right)^3+......\right]$$ **Ans.**

Which is the required expansion.

Example 17. *Expand the function* $\dfrac{\sin z}{z-\pi}$ *about* $z = \pi$.

Solution. Putting $z - \pi = t$, we have

$$\frac{\sin z}{z-\pi} = \frac{\sin(\pi+t)}{t} = \frac{-\sin t}{t}$$

$$= -\frac{1}{t}\left(t-\frac{t^3}{3!}+\frac{t^5}{5!}......\right) = -1+\frac{t^2}{3!}-\frac{t^4}{5!}+...... = -1+\frac{(z-\pi)^2}{3!}-\frac{(z-\pi)^4}{5!}+......$$

Which is the required expansion. **Ans.**

Example 18. *Expand the function* $\sin^{-1} z$ *in powers of* z. *(U.P. III Semester, Dec. 2006)*

Solution. Let $w = \sin^{-1} z$

$$\frac{dw}{dz} = \frac{1}{\sqrt{1-z^2}} = (1-z^2)^{-\frac{1}{2}} \qquad \text{... (1)}$$

On expanding the R.H.S. of (1) by Binomial Theorem, we get

$$\frac{dw}{dz} = 1 - \frac{1}{2}(-z^2) + \frac{\frac{-1}{2}\left(-\frac{3}{2}\right)}{2!}(-z^2)^2 + \frac{\frac{-1}{2}\left(-\frac{3}{2}\right)\left(\frac{-5}{2}\right)}{3!}(-z^2)^3 + \ldots$$

$$\frac{dw}{dz} = 1 + \frac{z^2}{2} + \frac{3}{8}z^4 + \frac{5}{16}z^6 + \ldots.$$

On integrating, we get

$$w = z + \frac{z^3}{6} + \frac{3z^5}{40} + \frac{5z^7}{112} + \ldots\ldots + C \qquad \text{... (2)}$$

On putting $z = 0$ and $w = \sin^{-1} z = 0$ in (2), we get

$$0 = 0 + C \quad \Rightarrow \quad C = 0$$

On putting the value of C in (2), we get

$$\sin^{-1} z = z + \frac{z^3}{6} + \frac{3z^5}{40} + \frac{5z^7}{112} + \ldots\ldots\ldots$$

Which is the required expansion. **Ans.**

Example 19. *Expand the function $f(z) = \tan^{-1} z$ in powers of z. (U.P. III Sem. 2009-2010)*

Solution. We have, $f(z) = \tan^{-1} z$

$$\Rightarrow \quad \frac{d\,f(z)}{dz} = \frac{1}{1+z^2} = (1+z^2)^{-1} \qquad \text{... (1)}$$

On expanding the R.H.S of (1) by Binomial Theorem, we get

$$\frac{d\,f(z)}{dz} = 1 - z^2 + \frac{-1(-1-1)}{2!}(z^2)^2 + \frac{-1(-1-1)(-1-2)}{3!}(z^2)^3 + \ldots..$$

$$\Rightarrow \quad \frac{d\,f(z)}{dz} = 1 - z^2 + z^4 - z^6 + \ldots\ldots.$$

On integration, we get

$$f(z) = z - \frac{z^3}{3} + \frac{z^5}{5} - \frac{z^7}{7} + \ldots\ldots\ldots\ldots + C \qquad \text{... (2)}$$

On putting $z = 0$ and $f(z) = \tan^{-1} z = 0$ in (2), we get

$$0 = 0 + C \quad \Rightarrow \quad C = 0$$

On putting the value of C in (2), we get

$$\tan^{-1} z = z - \frac{z^3}{3} + \frac{z^5}{5} - \frac{z^7}{7} + \ldots\ldots\ldots$$ **Ans.**

EXERCISE 30.2

Expand the following functions in Taylor's series

1. $\frac{1}{z+1}$, about $z = 1$. **Ans.** $-\frac{1}{2}\left(1 + \frac{z}{2} + \frac{z^2}{4} + \frac{z^3}{8} + \ldots\right) - (z^{-1} + z^{-2} + z^{-3} + \ldots)$

2. $\frac{z-1}{z+1}$ (*a*) about $z = 0$ (*b*) about $z = 1$.

Ans. (*a*) $-1 + 2(z - z^2 + z^3 - z^4 + \ldots)$ (*b*) $\frac{1}{2}(z-1) - \frac{1}{2^2}(z-1)^2 + \frac{1}{2^3}(z-1)^3 - \ldots.$

3. $\dfrac{1}{4-3z}$ about $(1+i)$ **Ans.** $\dfrac{1}{1-3i}\sum_{n=0}^{\infty}(-1)^n\left(\dfrac{-3}{1-3i}\right)^n(z-(1+i))^n$

Region of convergence $|z-(1+i)|<\sqrt{\dfrac{10}{3}}$

4. $\dfrac{1}{z^2-z-6}$ about (*a*) $z=-1$, (*b*) $z=1$ **Ans.** (*a*) $\dfrac{1}{20}\sum_{n=0}^{\infty}\dfrac{(-4)^{n+1}-1}{4^n}(z+1)^n$

Region of convergence $|z+1|\le 1$.

Ans. (*b*) $-\dfrac{1}{10}\sum_{n=0}^{\infty}\left(\dfrac{z-1}{2}\right)^n-\dfrac{1}{15}\sum_{n=0}^{\infty}(-1)^n\left(\dfrac{z-1}{3}\right)^n$

Region of convergence $\left|\dfrac{z-1}{2}\right|<1$ and $\left|\dfrac{z-1}{3}\right|<1$ common region $|z-1|<2$.

5. $\dfrac{2z^2+9z+5}{z^3+z^2-8z-12}$ about $z=1$ **Ans.** $\sum_{n=0}^{\infty}\left[(-1)^n\dfrac{n+1}{3^{n+2}}-\dfrac{1}{2^n}\right](z-1)^n$

Region of convergence $|z-1|<2$.

6. $\log\left(\dfrac{1+z}{1-z}\right)$ about $z=0$ **Ans.** $\sum_{n=0}^{\infty}\dfrac{2z^{2n+1}}{2n+1}$, $|z|<1$

7. $\dfrac{1}{z^2+(1+2i)z+2i}$ about $z=0$. **Ans.** $\dfrac{1}{(1-2i)}\left[\sum_{n=0}^{\infty}\left\{\left(\dfrac{1}{2i}\right)^{n+1}-1\right\}(-1)^n z^n\right]$

8. $(z+i)^{\frac{1}{2}}$, about $z=0$ **Ans.** $\dfrac{1+i}{\sqrt{2}}\left[\sum_{n=0}^{\infty}{}^{\frac{1}{2}}C_r\left(\dfrac{z}{i}\right)^r\right]$, $R=1$

9. $\dfrac{1}{(z^2-1)(z^2-2)}$ about $z=0$. **Ans.** $\sum_{n=0}^{\infty}\left[1-\dfrac{1}{2^{n+1}}\right]z^{2n}$, $R=1$.

10. $\tan z$ about $z=0$ **Ans.** $z+\dfrac{z^3}{3}+\dfrac{2z^5}{15}+\ldots$

11. $z\cot z$ about $z=0$ **Ans.** $1-\dfrac{z^2}{3}-\dfrac{z^4}{45}+\ldots\ldots$ **12.** $\dfrac{e^z}{1+e^z}$ about $z=0$ **Ans.** $\dfrac{1}{2}+\dfrac{z}{4}-\dfrac{z^3}{48}$

30.8 LAURENT'S THEOREM

(*U.P., III Semester, Dec. 2009*)

If we are required to expand $f(z)$ about a point where $f(z)$ is not analytic, then it is expanded by Laurent's Series and not by Taylor's Series.

Statement. If $f(z)$ is analytic on c_1 and c_2, and the annular region R bounded by the two concentric circles c_1 and c_2 of radii r_1 and r_2 $(r_2<r_1)$ and with centre at a, then for all z in R

$$f(z)=a_0+a_1(z-a)+a_2(z-a)^2+\ldots+\frac{b_1}{z-a}+\frac{b_2}{(z-a)^2}+\ldots$$

where

$$a_n=\frac{1}{2\pi i}\int_{c_1}\frac{f(w)}{(w-a)^{n+1}}dw,$$

$$b_n=\frac{1}{2\pi i}\int_{c_2}\frac{f(w)}{(w-a)^{-n+1}}dw$$

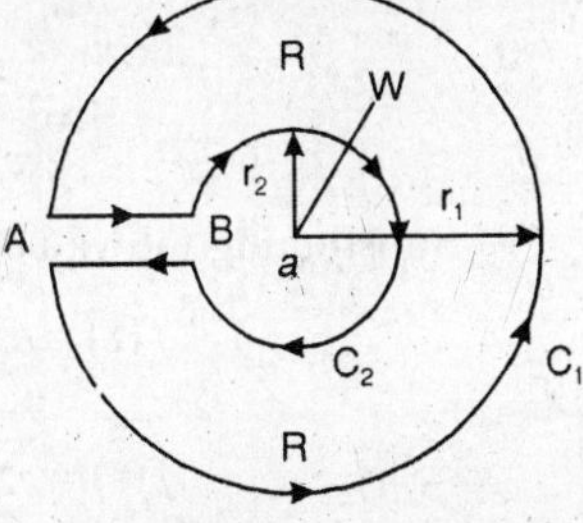

Proof. By introducing a cross cut AB, multi connected region R is converted to a simply connected region. Now $f(z)$ is analytic in this region.

Now by Cauchy's integral formula

$$f(z)=\frac{1}{2\pi i}\int_{c_1}\frac{f(w)\,dw}{w-z}+\frac{1}{2\pi i}\int_{AB}\frac{f(w)}{w-z}\,dw-\frac{1}{2\pi i}\int_{c_2}\frac{f(w)\,dw}{w-z}+\frac{1}{2\pi i}\int_{BA}\frac{f(w)}{w-z}\,dw$$

Integral along c_2 is clockwise, so it is negative. Integrals along AB and BA cancel.

$$\therefore \qquad f(z)=\frac{1}{2\pi i}\int_{c_1}\frac{f(w)\,dw}{w-z}-\frac{1}{2\pi i}\int_{c_2}\frac{f(w)\,dw}{w-z} \qquad \text{... (1)}$$

For the first integral, $\frac{f(w)}{w-z}$ can be expanded exactly as in Taylor's series as z lies on c_1.

$$\left(\frac{|z-a|}{|w-a|}\leq 1 \qquad \because w \text{ lies on } c_1\right)$$

$$\frac{1}{2\pi i}\int_{c_1}\frac{f(w)\,dw}{w-z}=\frac{1}{2\pi i}\int_{c_1}\frac{f(w)}{w-a}\,dw+\frac{(z-a)}{2\pi i}\int_{c_1}\frac{f(w)}{(w-a)^2}\,dw+\frac{(z-a)^2}{2\pi i}\int_{c_1}\frac{f(w)}{(w-a)^3}\,dw+\ldots$$

$$=a_0+a_1(z-a)+a_2(z-a)^2+ \qquad \text{... (2)} \qquad \left[a_n=\frac{1}{2\pi i}\int_{c_1}\frac{f(w)}{(w-a)^{n+1}}\,dw\right]$$

In the second integral, z lies on c_2. Therefore

$$|w-a|<|z-a| \quad \text{or} \quad \frac{|w-a|}{|z-a|}<1$$

so here
$$\frac{1}{w-z}=\frac{1}{w-a+a-z}=\frac{1}{(w-a)-(z-a)}$$
$$=-\frac{1}{(z-a)}\frac{1}{\left(1-\dfrac{w-a}{z-a}\right)}=-\frac{1}{z-a}\left(1-\frac{w-a}{z-a}\right)^{-1}$$
$$=-\frac{1}{z-a}\left[1+\frac{w-a}{z-a}+\left(\frac{w-a}{z-a}\right)^2+\ldots+\left(\frac{w-a}{z-a}\right)^{n+1}+\ldots\right]$$

Multiplying by $-\frac{f(w)}{2\pi i}$, we get

$$-\frac{1}{2\pi i}\frac{f(w)}{w-z}=\frac{1}{2\pi i}\frac{f(w)}{z-a}+\frac{1}{2\pi i}\frac{(w-a)}{(z-a)^2}f(w)+\frac{1}{2\pi i}\frac{(w-a)^2}{(z-a)^3}f(w)+\ldots$$
$$=\left(\frac{1}{z-a}\right)\frac{1}{2\pi i}f(w)+\frac{1}{(z-a)^2}\frac{1}{2\pi i}\frac{f(w)}{(w-a)^{-1}}+\frac{1}{(z-a)^3}\frac{1}{2\pi i}\frac{f(w)}{(w-a)^{-2}}+\ldots$$

Integrating, we have

$$-\frac{1}{2\pi i}\int_{c_2}\frac{f(w)}{w-z}\,dw=\left(\frac{1}{z-a}\right)\frac{1}{2\pi i}\int_{c_2}f(w)\,dw+\frac{1}{(z-a)^2}\frac{1}{2\pi i}\int_{c_2}\frac{f(w)\,dw}{(w-a)^{-1}}$$
$$+\frac{1}{(z-a)^3}\frac{1}{2\pi i}\int_{c_2}\frac{f(w)\,dw}{(w-a)^{-2}}+\ldots,$$

$$=\frac{b_1}{z-a}+\frac{b_2}{(z-a)^2}+\frac{b_3}{(z-a)^3}+ \qquad \text{... (3)} \qquad \left[b_n=\frac{1}{2\pi i}\int_{c_2}\frac{f(w)\,dw}{(w-a)^{-n+1}}\right]$$

Substituting the values of both integrals from (2) and (3) in (1), we get

$$f(z)=a_0+a_1(z-a)+a_2(z-a)^2+b_1(z-a)^{-1}+b_2(z-a)^{-2}+\ldots$$

$$\Rightarrow \qquad f(z)=\sum_{n=0}^{n=\infty}a_n(z-a)^n+\sum_{n=1}^{n=\infty}\frac{b_n}{(z-a)^n}$$ **Proved.**

Note. To expand a function by Laurent Theorem is cumbersome. By Binomial theorem, the expansion of a function can be done easily.

Example 20. *Expand* $f(z)=\dfrac{1}{(z-1)(z-2)}$ *for* $1<|z|<2$

Solution. $f(z)=\dfrac{1}{(z-1)(z-2)}=\dfrac{1}{z-2}-\dfrac{1}{z-1}$

In first bracket $|z|<2$, we take out 2 as common and from second

bracket z is taken out common as $1<|z|$.

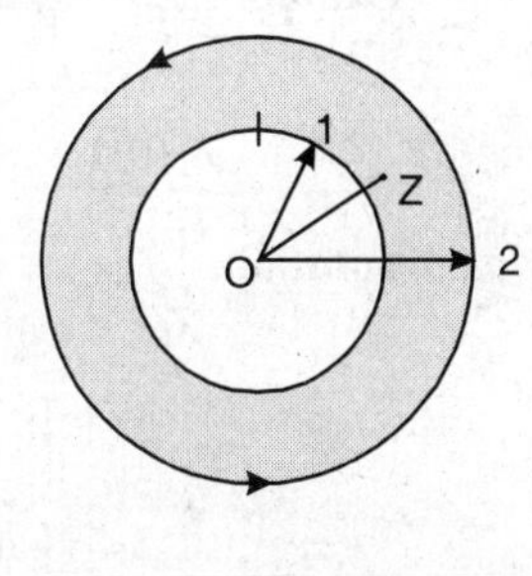

$$f(z)=-\frac{1}{2}\left(\frac{1}{1-\frac{z}{2}}\right)-\frac{1}{z}\left(\frac{1}{1-\frac{1}{z}}\right)=-\frac{1}{2}\left(1-\frac{z}{2}\right)^{-1}-\frac{1}{z}\left(1-\frac{1}{z}\right)^{-1}$$

$$=-\frac{1}{2}\left[1+\frac{z}{2}+\frac{z^2}{4}+\frac{z^3}{8}+\ldots\right]-\frac{1}{z}\left[1+\frac{1}{z}+\frac{1}{z^2}+\frac{1}{z^3}+\ldots\right]$$

$$=-\frac{1}{2}-\frac{z}{4}-\frac{z^2}{8}-\frac{z^3}{16}\ldots-\frac{1}{z}-\frac{1}{z^2}-\frac{1}{z^3}-\frac{1}{z^4}\ldots$$

Which is the required expansion. **Ans.**

Example 21. *Obtain the Taylor or Laurent series which represents the function*

$$f(z)=\frac{1}{(1+z^2)(z+2)} \text{ when}$$

(*i*) $1<|z|<2$; (*ii*) $|z|>2$

Solution. $f(z)=\dfrac{1}{(1+z^2)(z+2)}=\dfrac{-\frac{z}{5}+\frac{2}{5}}{1+z^2}+\dfrac{\frac{1}{5}}{z+2}=-\dfrac{1}{5}\dfrac{z-2}{1+z^2}+\dfrac{1}{5}\dfrac{1}{z+2},$ $\quad 1<|z|<2$

(*i*) In first expression $1<|z|$ and in second expression $|z|<2$

$$f(z)=-\frac{1}{5}\frac{1}{z^2}\frac{z-2}{1+\frac{1}{z^2}}+\frac{1}{5}\frac{1}{2}\frac{1}{1+\frac{z}{2}}$$

$$=-\frac{1}{5z^2}(z-2)\left(1+\frac{1}{z^2}\right)^{-1}+\frac{1}{10}\left(1+\frac{z}{2}\right)^{-1}$$

$$=-\frac{1}{5}\left(\frac{1}{z}-\frac{2}{z^2}\right)\left(1-\frac{1}{z^2}+\frac{1}{z^4}-\frac{1}{z^6}+\ldots\right)+\frac{1}{10}\left[1-\frac{z}{2}+\frac{z^2}{4}-\frac{z^3}{8}+\ldots\right]$$

$$=\frac{1}{5}\left[-\frac{1}{z}+\frac{1}{z^3}-\frac{1}{z^5}+\frac{1}{z^7}+\ldots+\frac{2}{z^2}-\frac{2}{z^4}+\frac{2}{z^6}-\frac{2}{z^8}+\ldots+\frac{1}{2}-\frac{z}{4}+\frac{z^2}{8}-\frac{z^3}{16}\ldots\right]$$

$$=\frac{1}{5}\left[\ldots-2z^{-8}+z^{-7}+2z^{-6}-z^{-5}-2z^{-4}+z^{-3}+2z^{-2}-z^{-1}+\frac{1}{2}-\frac{z}{4}+\frac{z^2}{8}-\frac{z^3}{16}\ldots\right]$$

Which is the required series. **Ans.**

(*ii*) Here $|z|>2$

$$f(z)=-\frac{1}{5}\frac{z-2}{1+z^2}+\frac{1}{5}\frac{1}{z+2}=-\frac{1}{5}\frac{1}{z^2}\frac{z-2}{1+\frac{1}{z^2}}+\frac{1}{5}\frac{1}{z}\frac{1}{1+\frac{2}{z}}$$

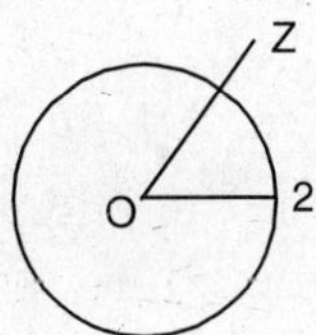

$$=-\frac{1}{5z^2}(z-2)\left[1+\frac{1}{z^2}\right]^{-1}+\frac{1}{5z}\left[1+\frac{2}{z}\right]^{-1}$$

$$=\frac{1}{5}\left[-\frac{1}{z}+\frac{2}{z^2}\right]\left[1-\frac{1}{z^2}+\frac{1}{z^4}-\frac{1}{z^6}+\ldots\right]+\frac{1}{5z}\left[1-\frac{2}{z}+\frac{4}{z^2}-\frac{8}{z^3}+\frac{16}{z^4}+\ldots\right]$$

$$=\frac{1}{5}\left[-\frac{1}{z}+\frac{1}{z^3}-\frac{1}{z^5}+\frac{1}{z^7}+\ldots+\frac{2}{z^2}-\frac{2}{z^4}+\frac{2}{z^6}-\frac{2}{z^8}+\ldots+\frac{1}{z}-\frac{2}{z^2}+\frac{4}{z^3}-\frac{8}{z^4}+\ldots\right]$$

Which is the required series. **Ans.**

Example 22. *Expand* $f(z)=\dfrac{1}{(z+1)(z+3)}$ *in Laurent series valid for*

(*i*) $1<|z|<3$ (*ii*) $|z|>3$

(*iii*) $0<|z+1|<2$ (*iv*) $|z|<1$

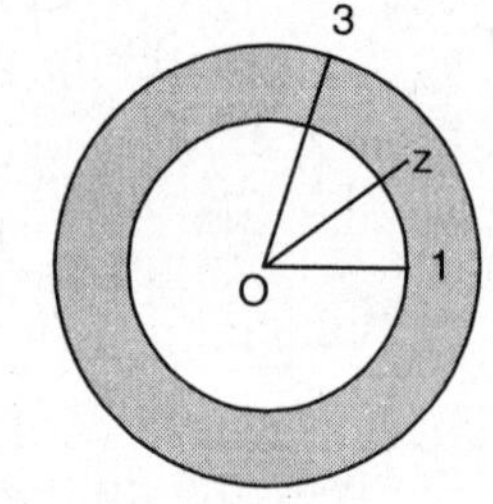

Solution. $f(z)=\dfrac{1}{(z+1)(z+3)}=\dfrac{1}{2}\left(\dfrac{1}{z+1}-\dfrac{1}{z+3}\right)$

(*i*) $1<|z|<3. \quad\Rightarrow\quad \dfrac{1}{|z|}<1$ and $\dfrac{|z|}{3}<1.$

$\Rightarrow \quad 1<|z|$ and $|z|<3.$

$$f(z)=\frac{1}{2}\left[\frac{1}{z\left(1+\frac{1}{z}\right)}-\frac{1}{3\left(1+\frac{z}{3}\right)}\right]=\frac{1}{2}\left[\frac{1}{z}\left(1+\frac{1}{z}\right)^{-1}-\frac{1}{3}\left(1+\frac{z}{3}\right)^{-1}\right]$$

$$=\frac{1}{2}\left[\frac{1}{z}\left(1-\frac{1}{z}+\frac{1}{z^2}-\frac{1}{z^3}\ldots\right)-\frac{1}{3}\left(1-\frac{z}{3}+\frac{z^2}{3^2}-\frac{z^3}{3^3}\ldots\right)\right]$$

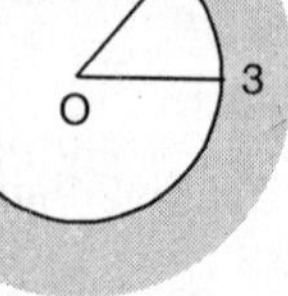

$$=\left(\frac{1}{2z}-\frac{1}{2z^2}+\frac{1}{2z^3}-\frac{1}{2z^4}\ldots\right)-\frac{1}{6}+\frac{z}{18}-\frac{z^2}{54}+\frac{z^3}{162}\ldots$$

$$=-\frac{1}{2z^4}+\frac{1}{2z^3}-\frac{1}{2z^2}+\frac{1}{2z}-\frac{1}{6}+\frac{z}{18}-\frac{z^2}{54}+\frac{z^3}{162}\ldots$$

Which is the required series. **Ans.**

(*ii*) $|z|>3 \quad\Rightarrow\quad \dfrac{3}{|z|}<1$

$$f(z)=\frac{1}{2}\left[\frac{1}{z}\left(1+\frac{1}{z}\right)^{-1}-\frac{1}{z}\left(1+\frac{3}{z}\right)^{-1}\right]$$

$$=\frac{1}{2z}\left(1-\frac{1}{z}+\frac{1}{z^2}-\frac{1}{z^3}\ldots\right)-\frac{1}{2z}\left(1-\frac{3}{z}+\frac{3^2}{z^2}\ldots\right)$$

$$=\left(\frac{1}{2z}-\frac{1}{2z}\right)+\left(\frac{-1}{2z^2}+\frac{3}{2z^2}\right)+\left(\frac{1}{2z^3}-\frac{9}{2z^3}\right)+\left(-\frac{1}{2z^4}+\frac{27}{2z^4}\right)$$

$$=\frac{1}{z^2}-\frac{4}{z^3}+\frac{13}{z^4}-\frac{50}{z^5}+\ldots.$$

Which is required expansion. **Ans.**

(*iii*) $0<|z+1|<2.$

$|z+1|>0$ and $|z+1|<2.$

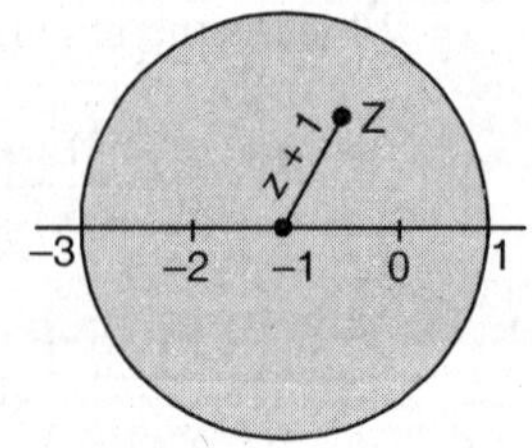

$\dfrac{|z+1|}{2}<1.$

$$f(z)=\frac{1}{2}\left[\frac{1}{1+z}-\frac{1}{z+1+2}\right]=\frac{1}{2}\left[\frac{1}{1+z}-\left(2+\overline{z+1}\right)^{-1}\right]$$

$$= \frac{1}{2(1+z)} - \frac{1}{2.2}\left(1+\frac{z+1}{2}\right)^{-1}$$

$$= \frac{1}{2(1+z)} - \frac{1}{4}\left[1-\frac{(z+1)}{2}+\frac{(z+1)^2}{4}-...\right]$$

$$= \frac{1}{2(1+z)} - \frac{1}{4}+\frac{(z+1)}{8}-\frac{(z+1)^2}{16}+....$$

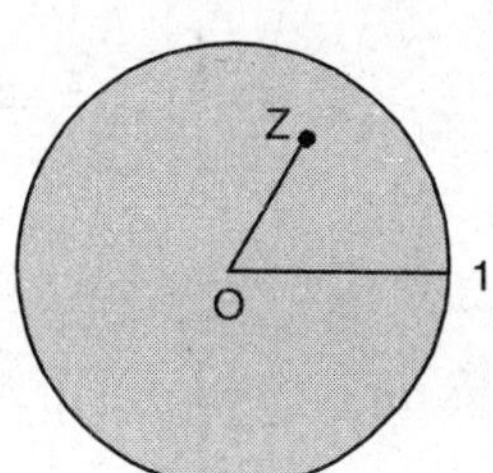

Which is required expansion. **Ans.**

(*iv*) $|z|<1$

$$f(z) = \frac{1}{2}\left[\frac{1}{(1+z)}-\frac{1}{3+z}\right]$$

$$= \frac{1}{2}(1+z)^{-1} - \frac{1}{2}(3+z)^{-1}$$

$$= \frac{1}{2}(1+z)^{-1} - \frac{1}{6}\left(i+\frac{z}{3}\right)^{-1} \qquad \left[|z|<1 \text{ and } \frac{|z|}{3}<1.\right]$$

$$= \frac{1}{2}(1-z+z^2-z^3+....)-\frac{1}{6}\left(1-\frac{z}{3}+\frac{z^2}{9}-\frac{z^3}{27}+...\right)$$

$$= \left(\frac{1}{2}-\frac{1}{6}\right)+\left(\frac{-z}{2}+\frac{z}{18}\right)+\left(\frac{z^2}{2}-\frac{z^2}{54}\right)+\left(-\frac{z^3}{2}+\frac{z^3}{162}\right)+...$$

$$= \frac{1}{3}-\frac{4}{9}z+\frac{13}{27}z^2-\frac{40}{81}z^3+...$$

Which is required expansion. **Ans.**

Example 23. *Expand the following function in a Laurent series about the point z = 0.*

$$f(z) = \frac{1-\cos z}{z^3}$$

Solution. $f(z) = \dfrac{1-\cos z}{z^3} = \dfrac{1}{z^3}\left[1-\left\{1-\dfrac{z^2}{2!}+\dfrac{z^4}{4!}-......\right\}\right]$

$$= \frac{1}{z^3}\left(\frac{z^2}{2!}-\frac{z^4}{4!}+\frac{z^6}{6!}-......\right) = \frac{1}{2!z}-\frac{1}{4!}z+\frac{1}{6!}z^3-......$$

$$= \sum_{n=2}^{\infty}\frac{(-1)^n z^{2n-5}}{(2n-2)!}$$

which is a Laurent's series. **Ans.**

Example 24. *Find the terms in the Laurent expansion of* $\dfrac{1}{z(e^z-1)}$ *for the region* $0<|z|<2\pi$.

(*AMIETE, June 2010, 2009*)

Solution. $f(z) = \dfrac{1}{z(e^z-1)} = \dfrac{1}{z\left[\left(1+z+\dfrac{z^2}{2!}+\dfrac{z^3}{3!}+......\right)-1\right]}$

$$= z^{-1}\left(z+\frac{z^2}{2!}+\frac{z^3}{3!}+......\right)^{-1} = z^{-2}\left(1+\frac{z}{2!}+\frac{z^2}{3!}+......\right)^{-1}$$

$$= z^{-2}\left[1-\left(\frac{z}{2}+\frac{z^2}{6}+\frac{z^3}{24}+...\right)+\left(\frac{z}{2}+\frac{z^2}{6}+\frac{z^3}{24}+...\right)^2-\left(\frac{z}{2}+\frac{z^2}{6}+\frac{z^3}{24}+...\right)^3+...\right]$$

$$= z^{-2}\left[1-\left(\frac{z}{2}+\frac{z^2}{6}+\frac{z^3}{24}+\frac{z^4}{120}+......\right)+\frac{1}{4}z^2\left(1+\frac{z}{3}+\frac{z^2}{12}+......\right)^2 - \frac{1}{8}z^3\left(1+\frac{z}{3}+......\right)^3+\frac{z^4}{16}(1+......)^4+......\right]$$

$$= z^{-2}\left[1-\frac{z}{2}+z^2\left(\frac{1}{4}-\frac{1}{6}\right)-z^3\left(\frac{1}{8}-\frac{1}{6}+\frac{1}{24}\right)+z^4\left(\frac{1}{16}-\frac{1}{8}+\frac{1}{24}+\frac{1}{36}-\frac{1}{120}\right)+......\right]$$

$$= z^{-2}\left[1-\frac{z}{2}+\frac{z^2}{12}-\frac{z^4}{720}+......\right] = z^{-2}-\frac{1}{2}z^{-1}+\frac{1}{12}-\frac{z^2}{720}+......$$ **Ans.**

30.9 IF $f(z)$ HAS A POLE AT $z = a$ THEN $|f(z)| \to \infty$ AS $z \to a$

Solution. Suppose the pole is of order m, then $f(z) = \sum_{n=1}^{\infty} a_n(z-a)^n + \sum_{n=0}^{m} b_n(z-a)^{-n}$

Its principal part is $\sum_{n=0}^{m} b_n(z-a)^{-n}$

$$\sum_{n=1}^{m} b_n(z-a)^{-n} = \frac{b_1}{z-a}+\frac{b_2}{(z-a)^2}+\frac{b_3}{(z-a)^3}+...+\frac{b_m}{(z-a)^m}$$

$$= \frac{1}{(z-a)^m}[b_m+b_{m-1}(z-a)+...b_1(z-a)^{m-1}]$$

$$= \frac{1}{(z-a)^m}\left[b_m+\sum_{n=1}^{m-1} b_n(z-a)^{m-n}\right]$$

$$\left|\sum_{n=1}^{m} b_n(z-a)^{-n}\right| = \left|\frac{1}{(z-a)^m}\left[b_m+\sum_{n=1}^{m-1} b_n(z-a)^{m-n}\right]\right|$$

$$\geq \left|\frac{1}{(z-a)^m}\right|\left\{|b_m|-\sum_{n=1}^{m-1}|b_n||z-a|^{m-n}\right\}$$

This tends to b_m $\quad |a_1+a_2| \geq |a_1|-|a_2|$

As $z \to a$ $\quad$ R.H.S $= \infty$.

Example 25. *Write all possible Laurent series for the function*

$$f(z) = \frac{1}{z(z+2)^3}$$

about the pole $z = -2$. Using appropriate Laurent series

Solution. To expand $\dfrac{1}{z(z+2)^3}$ about $z = -2$, *i.e.*, in powers of $(z + 2)$, we put $z + 2 = t$.

Then $\quad f(z) = \dfrac{1}{z(z+2)^3} = \dfrac{1}{(t-2)t^3} = \dfrac{1}{t^3}\cdot\dfrac{1}{t-2}$

$$= \frac{1}{t^3}\cdot\frac{1}{-2}\cdot\left(\frac{1}{1-\frac{t}{2}}\right) = -\frac{1}{2t^3}\left(1-\frac{t}{2}\right)^{-1}$$

$[0 < |z-2| < 1$ or $0 < |t| < 1]$

Y
X' X
−2 −1 O

$$= -\frac{1}{2t^3}\left[1+\frac{t}{2}+\frac{t^2}{4}+\frac{t^3}{8}+\frac{t^4}{16}+\frac{t^5}{32}+...\right]$$

$$= -\frac{1}{2t^3} - \frac{1}{4t^2} - \frac{1}{8t} - \frac{1}{16} - \frac{t}{32} - \frac{t^2}{64} + ...$$

$$= -\frac{1}{2(z+2)^3} - \frac{1}{4(z+2)^2} - \frac{1}{8(z+2)} - \frac{1}{16} - \frac{z+2}{32} - \frac{(z+2)^2}{64} + ... \qquad \textbf{Ans.}$$

Example 26. *Expand* $\dfrac{e^z}{(z-1)^2}$ *about* $z = 1$

Solution. Let. $f(z) = \dfrac{e^z}{(z-1)^2} = \dfrac{e^{1+t}}{t^2}$ [Put $z - 1 = t \Rightarrow z = 1 + t$]

$$= \frac{e \cdot e^t}{t^2} = \frac{e}{t^2}\left[1 + t + \frac{t^2}{2!} + \frac{t^3}{3!} + ...\right] = e\left[\frac{1}{t^2} + \frac{1}{t} + \frac{1}{2!} + \frac{t}{3!} ...\right]$$

$$= e\left[\frac{1}{(z-1)^2} + \frac{1}{z-1} + \frac{1}{2!} + \frac{z-1}{3!} + ...\right]$$

Which is required expansion. **Ans.**

Example 27. *Expand* $f(z) = \sin\left\{c\left(z + \dfrac{1}{z}\right)\right\}$

Solution. $f(z)$ is not analytic at $z = 0$.
Therefore $f(z)$ can be expanded by Laurent theorem.

$$\therefore \qquad f(z) = \sum_{n=0}^{\infty} a_n z^n + \sum_{n=1}^{\infty} \frac{b_n}{z^n}$$

where $\qquad a_n = \dfrac{1}{2\pi i}\displaystyle\int_c \frac{f(z)dz}{(z-0)^{n+1}}$ and $b_n = \dfrac{1}{2\pi i}\displaystyle\int_c f(z) z^{n-1} dz$

Now, $\qquad a_n = \dfrac{1}{2\pi i}\displaystyle\int_c \frac{\sin c\left(z + \frac{1}{z}\right) dz}{z^{n+1}}$ $\qquad \left[\begin{array}{l} z = e^{i\theta} \\ dz = i e^{i\theta} d\theta \end{array}\right]$

$$= \frac{1}{2\pi i}\int_0^{2\pi} \frac{\sin[c(2\cos\theta)]\, i e^{i\theta}\, d\theta}{e^{(n+1)i\theta}} = \frac{1}{2\pi}\int_0^{2\pi} \sin(2c\cos\theta)\, e^{-ni\theta}\, d\theta$$

$$= \frac{1}{2\pi}\int_0^{2\pi} \sin(2c\cos\theta)\cos n\theta\, d\theta - \frac{i}{2\pi}\int_0^{2\pi} \sin(2c\cos\theta)\sin n\theta d\theta$$

$$= \frac{1}{2\pi}\int_0^{2\pi} \sin(2c\cos\theta)\cos n\theta\, d\theta + 0 \qquad \left[\text{If } f(2a-x) = -f(x) \text{ then } \int_0^{2a} f(x)dx = 0\right]$$

Similarly, $\qquad b_n = \dfrac{1}{2\pi}\displaystyle\int_0^{2\pi} \sin(2c\cos\theta)\cos n\theta d\theta$

Since the function remains unaltered by putting z for $\dfrac{1}{z}$.

$$f(z) = a_0 + \sum_{n=1}^{\infty} a_n z^n + \sum_{n=1}^{\infty} \frac{b_n}{z^n} \qquad [a_n = b_n]$$

$$= a_0 + \Sigma a_n\left(z^n + \frac{1}{z^n}\right) \qquad \textbf{Ans.}$$

Example 28. *Expand* $f(z) = e^{\frac{c}{2}\left(z-\frac{1}{z}\right)} = \sum_{n=0}^{\infty} a_n z^n$

Solution. $f(z)$ is also analytic function at $z = 0$ so $f(z)$ can be expanded by Taylor's theorem.

$$f(z) = \sum_{n=0}^{\infty} a^n z^n$$

where

$$a_n = \frac{1}{2\pi i}\int_c \frac{f(z)dz}{z^{n+1}}$$

$$= \frac{1}{2\pi i}\int_c \frac{e^{\frac{c}{2}\left(z-\frac{1}{z}\right)}dz}{z^{n+1}} \qquad [z = e^{i\theta},\ dz = i\,e^{i\theta}\,d\theta]$$

$$= \frac{\frac{1}{2\pi i}\int_0^{2\pi} e^{\frac{c}{2}(2i\sin\theta)}\,ie^{i\theta}\,d\theta}{e^{(n+1)i\theta}} = \frac{1}{2\pi}\int_0^{2\pi} e^{ci\sin\theta}e^{-ni\theta}d\theta$$

$$= \frac{1}{2\pi}\int_0^{2\pi} e^{i(c\sin\theta-n\theta)}\,d\theta$$

$$= \frac{1}{2\pi}\int_0^{2\pi}[\cos(c\sin\theta-n\theta)-i\sin(c\sin\theta-n\theta)]d\theta$$

$$= \frac{1}{2\pi}\int_0^{2\pi}\cos(c\sin\theta-n\theta)\,d\theta \qquad \left\{\begin{array}{l}\text{If } f(2a-x)=-f(x)\\ \text{then } \int_0^{2a} f(x)dx = 0\end{array}\right.$$

$b_n = (-1)^n a^n$ since $f(z)$ remains unaltered if $\frac{-1}{z}$ is written for z

So,

$$f(z) = \Sigma a_n z^n + \Sigma \frac{b_n}{z^n}$$

$$= \sum_{n=0}^{\infty} a_n(z^n) + (-1)^n \sum_{n=1}^{n} \frac{a_n}{z^n} = \sum_{-\infty}^{\infty} a_n z^n$$

Ans.

Example 29. *Find the Laurent expansion for* $f(z) = \frac{7z-2}{z^3-z^2-2z}$ *in the regions given by*

(*i*) $0 < |z+1| < 1$ (*ii*) $1 < |z+1| < 3$ (*AMIETE, June 2010*) (*iii*) $|z+1| > 3$.

Solution. We have,

$$f(z) = \frac{7z-2}{z^3-z^2-2z} = \frac{1}{z} - \frac{3}{z+1} + \frac{2}{z-2} \qquad \text{[By partial fractions]}$$

$$= \frac{1}{(z+1)-1} - \frac{3}{z+1} + \frac{2}{(z+1)-3}$$

(*i*) $0 < |z+1| < 1$

$$f(z) = -\{1-(z+1)\}^{-1} - \frac{3}{z+1} - \frac{2}{3}\left\{1-\left(\frac{z+1}{3}\right)\right\}^{-1}$$

$$= -\left[1+(z-1)+(z-1)^2+(z-1)^3+\ldots\ldots\right] - \frac{3}{z+1}$$

$$-\frac{2}{3}\left[1+\frac{(z+1)}{3}+\frac{(z+1)^2}{9}+\frac{(z+1)^3}{27}+\ldots\ldots\right]$$

Which is the required Laurent expansion.

(*ii*) $1 < |z+1| < 3$

$$f(z) = \frac{1}{(z+1)-1} - \frac{3}{z+1} + \frac{2}{(z+1)-3}$$

Taking common, bigger term out of $|z+1|$ and 1in first fraction, here $|z+1|$ is bigger than 1, so we take $|z+1|$ common from the first fraction.

Similarly, we take 3 common from the third fraction as 3 is bigger than $|z+3|$.

$$\Rightarrow \quad f(z) = \frac{1}{z+1}\left[\frac{1}{1-\left(\frac{1}{z+1}\right)}\right] - \frac{3}{z+1} - \frac{2}{3}\left[\frac{1}{1-\left(\frac{z+1}{3}\right)}\right]$$

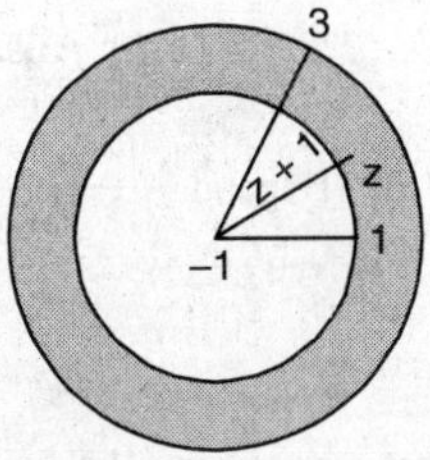

$$= \frac{1}{(z+1)}\left(1-\frac{1}{z+1}\right)^{-1} - \frac{3}{z+1} - \frac{2}{3}\left\{1-\left(\frac{z+1}{3}\right)\right\}^{-1}$$

$$= \frac{1}{z+1}\left[1+\frac{1}{z+1}+\frac{1}{(z+1)^2}+\frac{1}{(z+1)^3}+\ldots\right]$$

$$-\frac{3}{z+1} - \frac{2}{3}\left[1+\frac{z+1}{3}+\frac{(z+1)^2}{9}+\frac{(z+1)^3}{27}+\ldots\ldots\right]$$

Which is the required Laurent expansion.

(*iii*) $|z+1| > 3$.

$$f(z) = \frac{1}{(z+1)-1} - \frac{3}{(z+1)} + \frac{2}{(z+1)-3}$$

Taking common bigger term out of $|z+1|$ and 3 here $|z+1|$ is greater than 3, so we take $|z+1|$ common from third fraction.

Similarly, $|z+1|$ is also greater than 1, so we take $|z+1|$ common from the first fraction.

$$f(z) = \frac{1}{z+1}\left(\frac{1}{1-\frac{1}{z+1}}\right) - \frac{3}{z+1} + \frac{2}{z+1}\left(\frac{1}{1-\frac{3}{z+1}}\right)$$

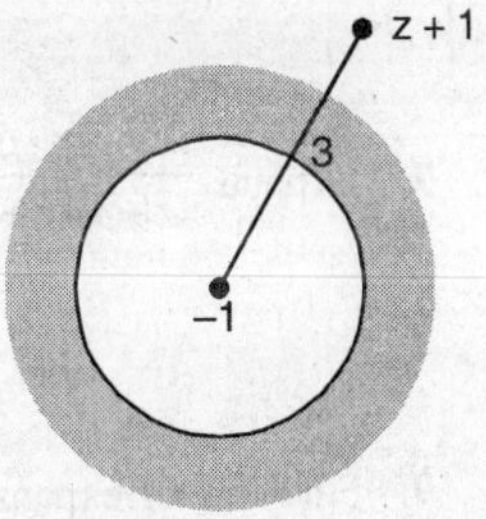

$$\Rightarrow \quad f(z) = \frac{1}{z+1}\left(1-\frac{1}{z+1}\right)^{-1} - \frac{3}{z+1} + \frac{2}{z+1}\left(1-\frac{3}{z+1}\right)^{-1}$$

$$= \frac{1}{z+1}\left[1+\frac{1}{z+1}+\frac{1}{(z+1)^2}+\frac{1}{(z+1)^3}+\ldots\ldots\right]$$

$$-\frac{3}{z+1} + \frac{2}{z+1}\left[1+\frac{3}{z+1}+\frac{9}{(z+1)^2}+\frac{27}{(z+1)^3}+\ldots\ldots\right]$$

Which is the required Laurent expansion. **Ans.**

EXERCISE 30.3

1. Find the Taylor's and Laurent series which represents the function $\dfrac{z^2-1}{(z+2)(z+3)}$ when

(*i*) $|z|<2$ (*ii*) $2<|z|<3$.

Ans. (*i*) $-\dfrac{1}{6}+\dfrac{5}{36}z+\dfrac{17}{216}z^2-\dfrac{115}{1296}z^3+\ldots$ (*ii*) $-\dfrac{5}{3}+\dfrac{3}{z}-\dfrac{6}{z^2}+\dfrac{12}{z^3}-\dfrac{24}{z^4}+\ldots\dfrac{8z}{9}-\dfrac{8z^2}{27}+\dfrac{8z^3}{81}+\ldots$

2. Find four terms of the Laurent series expansion valid in the region $0<|z-1|<1$ for the function $f(z)=\dfrac{2z+1}{z^3+z^2-2z}$

Ans. $-\dfrac{1}{2}[1-(z-1)+(z-1)^2-(z-1)^3+\ldots]-\dfrac{1}{6}\left[1-\dfrac{z-1}{3}+\dfrac{(z-1)^2}{9}-\dfrac{(z-1)^3}{27}+\ldots\right]+\dfrac{1}{z-1}$

3. Expand $\dfrac{z}{(z^2-1)(z^2+4)}$ in $1<|z|<2$

Ans. $\dfrac{1}{10}\left[\left(\dfrac{2}{z}+\dfrac{2}{z^3}+\dfrac{2}{z^5}+\ldots\right)-\left(\dfrac{z}{2}+\dfrac{z^3}{8}+\ldots\right)\right]$

4. Represent the function $f(z)=\dfrac{4z+3}{z(z-3)(z+2)}$ in Laurent series

(*i*) within $|z|=1$ (*ii*) in the annular region between $|z|=2$ and $|z|=3$.

Ans. (*i*) $-\dfrac{1}{2z}+\displaystyle\sum_{n=0}^{\infty}\left[(-1)^{n+1}\dfrac{1}{2^{n+2}}-\dfrac{1}{3^{n+1}}\right]z^n$. (*ii*) $-\dfrac{1}{2z}-\dfrac{1}{2z}\displaystyle\sum_{n=0}^{\infty}(-1)^n\left(\dfrac{2}{z}\right)^n-\dfrac{1}{3}\sum_{n=0}^{\infty}\left(\dfrac{z}{3}\right)^n$

5. Write all possible Laurent Series for the function $f(z)=\dfrac{z^2}{(z-1)^2(z+3)}$ about the singularity $z=1$, stating the region of convergence in each case.

Ans. When $|z-1|>4$, $\dfrac{1}{z-1}-\dfrac{2}{(z-1)^2}+\dfrac{9}{(z-1)^3}-\dfrac{36}{(z-1)^4}+\ldots$

When $0<|z-1|<4$, $\dfrac{1}{4}\left[1-\dfrac{1}{(z-1)^2}+\dfrac{7}{4}\dfrac{1}{z-1}+\dfrac{9}{16}-\dfrac{9}{64}(z-1)-\ldots\right]$

6. Obtain the expansion

$$f(z)=f(a)+2\left\{\frac{z-a}{2}f'\left(\frac{z+a}{2}\right)+\frac{(z-a)^3}{2^3.3!}f'''\left(\frac{z+a}{2}\right)+\frac{(z-a)^5}{2^5.5!}f^{v}\left(\frac{z+a}{2}\right)+\ldots\right\}$$

7. Expand $\dfrac{z^2-6z-1}{(z-1)(z+2)(z-3)}$ in $3<|z+2|<5$.

Ans. $\dfrac{2}{z+2}+\dfrac{3}{(z+2)^2}+\dfrac{3^2}{(z+2)^3}+\ldots+\dfrac{1}{5}\left[1+\dfrac{z+2}{5}+\dfrac{(z+2)^2}{5}+\dfrac{(z+2)^3}{5^3}+\ldots\right]$

9. Find Taylor expansion of $f(z)=\dfrac{2z^3+1}{z^2+z}$ about the point $z=i$. $\left[\textbf{Hint. } f(z)=2z-2+\dfrac{1}{z}+\dfrac{1}{z+1}\right]$

Ans. $\left(\dfrac{i}{2}-\dfrac{3}{2}\right)+\left(3+\dfrac{i}{2}\right)(z-i)+\displaystyle\sum_{n=2}^{\infty}(-1)^n\left\{\dfrac{1}{(1+i)^{n+1}}+\dfrac{1}{(i)^{n+1}}\right\}(z-i)^n$

10. Find the Laurent's series of $f(z)=\dfrac{1}{z^2(1-z^2)}$ and determine the precise region of its convergence.

(*AMIETE, Dec. 2010*)

CHAPTER

31

THE CALCULUS OF RESIDUES (INTEGRATION)

31.1 ZERO OF ANALYTIC FUNCTION

A zero of analytic function $f(z)$ is the value of z for which $f(z) = 0$.

Example 1. *Find out the zeros and discuss the nature of the singularities of*

$$f(z) = \frac{(z-2)}{z^2} \sin\left(\frac{1}{z-1}\right)$$ *(R.G.P.V. Bhopal, III Semester, Dec. 2004)*

Solution. Poles of $f(z)$ are given by equating to zero the denominator of $f(z)$ i.e. $z = 0$ is a pole of order two.

zeros of $f(z)$ are given by equating to zero the numerator of $f(z)$ *i.e.*, $(z-2) \sin\left(\frac{1}{z-1}\right) = 0$

$\Rightarrow$ Either $z - 2 = 0$ or $\sin\left(\frac{1}{z-1}\right) = 0$

$\Rightarrow$ $z = 2$ and $\frac{1}{z-1} = n\pi$

$\Rightarrow$ $z = 2,$ $z = \frac{1}{n\pi} + 1, \; n = \pm 1, \pm 2, \ldots\ldots\ldots$

Thus, $z = 2$ is a simple zero. The limit point of the zeros are given by

$$z = \frac{1}{n\pi} + 1 \quad (n = \pm 1, \pm 2, \ldots\ldots\ldots) \text{ is } z = 1.$$

Hence $z = 1$ is an isolated essential singularity. **Ans.**

31.2 SINGULAR POINT

A point at which a function $f(z)$ is not analytic is known as a singular point or **singularity** of the function.

For example, the function $\frac{1}{z-2}$ has a singular point at $z - 2 = 0$ or $z = 2$.

Isolated singular point. If $z = a$ is a singularity of $f(z)$ and if there is no other singularity within a small circle surrounding the point $z = a$, then $z = a$ is said to be an isolated singularity of the function $f(z)$; otherwise it is called non-isolated.

For example, the function $\frac{1}{(z-1)(z-3)}$ has two isolated singlar points, namely $z = 1$ and $z = 3$. $[(z-1)(z-3) = 0$ or $z = 1, 3]$.

Example of non-isolated singularity. Function $\dfrac{1}{\sin\frac{\pi}{z}}$ is not analytic at the points where $\sin\frac{\pi}{z}=0$, *i.e.*, at the points $\frac{\pi}{z}=n\pi$ *i.e.*, the points $z=\frac{1}{n}$ $(n=1,2,3,\ldots)$. Thus $z=1,\frac{1}{2},\frac{1}{3},\ldots,z=0$ are the points of singularity. $z=0$ is the **non-isolated singularity** of the function $\dfrac{1}{\sin\frac{\pi}{z}}$ because in the neighbourhood of $z=0$, there are infinite number of other singularities $z=\frac{1}{n}$, where n is very large.

Pole of order *m*. Let a function $f(z)$ have an isolated singular point $z=a$, $f(z)$ can be expanded in a Laurent's series around $z=a$, giving

$$f(z)=a_0+a_1(z-a)+a_2(z-a)^2+\ldots$$
$$+\frac{b_1}{z-a}+\frac{b_2}{(z-a)^2}+\ldots+\frac{b_m}{(z-a)^m}+\frac{b_{m+1}}{(z-a)^{m+1}}+\frac{b_{m+2}}{(z-a)^{m+2}}+\ldots \quad \ldots (1)$$

In some cases it may happen that the coefficients $b_{m+1}=b_{m+2}=b_{m+3}=0$, then (1) reduces to

$$f(z)=a_0+a_1(z-a)+a_2(z-a)^2+\ldots+\frac{b_1}{(z-a)}+\frac{b_2}{(z-a)^2}+\ldots+\frac{b_m}{(z-a)^m}$$

$$f(z)=a_0+a_1(z-a)+a_2(z-a)^2+\ldots..+\frac{1}{(z-a)^m}\{b_1(z-a)^{m-1}+b_2(z-a)^{m-2}$$
$$+b_3(z-a)^{m-3}+\ldots.+b_m\}$$

then $z=a$ is said to be a **pole of order *m*** of the function $f(z)$, when $m=1$, the pole is said to be **simple pole**. In this case

$$f(z)=a_0+a_1(z-a)+a_2(z-a)^2+\ldots+\frac{b_1}{z-a}$$

If the number of the terms of negative powers in expansion (1) is infinite, then $z=a$ is called an essential singular point of $f(z)$.

Example 2. *Define the singularity of a function. Find the singularity (ties) of the functions*

(*i*) $f(z)=\sin\frac{1}{z}$ (*ii*) $g(z)=\dfrac{e^{\frac{1}{z}}}{z^2}$ (*U.P. III Semester, 2009-2010*)

Solution. **See Art. 31.2 on page 793 for definition.**

(*i*) We know that

$$\sin\frac{1}{z}=\frac{1}{z}-\frac{1}{3!\,z^3}+\frac{1}{5!\,z^5}+\ldots\ldots\ldots+(-1)^n\frac{1}{(2n+1)!\,z^{2n+1}}$$

Obviously, there is a number of singularity.

$\sin\frac{1}{z}$ is not analytic at $z=0$. $\left(\frac{1}{z}=\infty \text{ at } z=0\right)$

Hence, $\sin\frac{1}{z}$ has a singularity at $z=0$.

(ii) Here, we have $g(z)=\dfrac{e^{\frac{1}{z}}}{z^2}$

We know that, $\left(\dfrac{1}{z^2}\right)\left(e^{\frac{1}{z}}\right)=\dfrac{1}{z^2}\left(1+\dfrac{1}{z}+\dfrac{1}{2!\,z^2}+\dfrac{1}{3!\,z^3}+\ldots.+\dfrac{1}{n!\,z^n}\ldots.+\right)$

$$= \frac{1}{z^2} + \frac{1}{z^3} + \frac{1}{2!\,z^4} + \frac{1}{3!\,z^5} + \dots\dots \frac{1}{n!\,z^{n+2}} + \dots\dots$$

Here, $f(z)$ has infinite number of terms in negative powers of z.
Hence, $f(z)$ has essential singularity at $z = 0$. **Ans.**

Example 3. *Find the pole of the function* $\dfrac{e^{z-a}}{(z-a)^2}$

Solution. $$\frac{e^{z-a}}{(z-a)^2} = \frac{1}{(z-a)^2}\left[1 + (z-a) + \frac{(z-a)^2}{2!} + \dots\right]$$

The given function has negative power 2 of $(z - a)$.
So, the given function has a pole at $z = a$ of order 2. **Ans.**

Example 4. *Find the poles of* $f(z) = \sin\left(\dfrac{1}{z-a}\right)$

Solution. $$\sin\left(\frac{1}{z-a}\right) = \frac{1}{z-a} - \frac{1}{3!}\frac{1}{(z-a)^3} + \frac{1}{5!}\frac{1}{(z-a)^5} - \dots\dots$$

The given function $f(z)$ has infinite number of terms in the negative powers of $z - a$.
So, $f(z)$ has essential singularity at $z = a$. **Ans.**

Example 5. *Find the pole of* $f(z) = \dfrac{\sin(z-a)}{(z-a)^4}$

Solution. $$\frac{\sin(z-a)}{(z-a)^4} = \frac{1}{(z-a)^4}\left[(z-a) - \frac{(z-a)^3}{3!} + \frac{(z-a)^5}{5!} - \frac{(z-a)^7}{7!} + \dots\dots\right]$$

$$= \frac{1}{(z-a)^3}\left[1 - \frac{(z-a)^2}{3!} + \frac{(z-a)^4}{5!} - \frac{(z-a)^6}{7!} + \dots\right]$$

The given function has a negative power 3 of $(z - a)$.
So, $f(z)$ has a pole at $z = a$ of order 3. **Ans.**

Example 6. *Prove that* $f(z) = \lim\limits_{z\to a} e^{\frac{1}{z-a}}$ *does not exist.*

Solution. $$\lim_{z\to a} e^{\frac{1}{z-a}} = \lim_{z\to a}\left(1 + \frac{1}{z-a} + \frac{1}{2!(z-a)^2} + \frac{1}{3!(z-a)^3} + \dots + \frac{1}{n!(z-a)^n} + \dots.\ \infty\right)$$

Here $n \to \infty$, $f(z)$ has infinite number of terms in negative power of $(z - a)$.
Thus, $f(z)$ has essential singularity at $z = a$.

Hence, $f(z) = \lim\limits_{z\to a} e^{\frac{1}{z-a}}$ does not exit. **Ans.**

Example 7. *Discuss singularity of* $\dfrac{1}{1-e^z}$ *at* $z = 2\pi i$.

Solution. We have, $$f(z) = \frac{1}{1-e^z}$$

The poles are determined by putting the denominator equal to zero.

i.e., $$1 - e^z = 0$$

$\Rightarrow$ $$e^z = 1 = (\cos 2n\pi + i \sin 2n\pi) = e^{2n\pi i}$$

$\Rightarrow$ $$z = 2n\pi i \qquad (n = 0, \pm 1, \pm 2, \dots\dots)$$

Clearly $z = 2\pi i$ is a simple pole. **Ans.**

Example 8. *Discuss singularity of* $\dfrac{\cot \pi z}{(z-a)^2}$ *at* $z = a$ *and* $z = \infty$.

(R.G.P.V., Bhopal, III Semester, Dec. 2002)

Solution. Let $$f(z) = \frac{\cot \pi z}{(z-a)^2} = \frac{\cos \pi z}{\sin \pi z\,(z-a)^2}$$

The poles are given by putting the denominator equal to zero.

i.e., $\sin \pi z\,(z-a)^2 = 0 \quad \Rightarrow \quad (z-a)^2 = 0 \quad \text{or} \quad \sin \pi z = 0 = \sin n\pi$

$\Rightarrow \qquad z = a, \ \pi z = n\pi, \qquad (n \in \text{I})$

$\Rightarrow \qquad z = a, n$

$f(z)$ has essential singularity at $z = \infty$.

Also, $z = a$ being repeated twice gives the double pole. **Ans.**

Example 9. *Show that* $e^{-\left(\frac{1}{z^2}\right)}$ *has no singularities.*

Solution. $$f(z) = e^{-\left(\frac{1}{z^2}\right)} = \frac{1}{e^{(1/z^2)}}$$

The poles are determined by putting the denominator

$$e^{\left(\frac{1}{z^2}\right)} = 0 \qquad \ldots(1)$$

It is not possible to find the value of z which can satisfy equation (1).

Hence, there is no pole or singularity of the given function. **Proved.**

Example 10. *Find the nature of singularities of*

$$f(z) = \frac{z - \sin z}{z^3} \quad at \ z = 0.$$

Solution. $$f(z) = \frac{1}{z^3}(z - \sin z) = \frac{1}{z^3}\left[z - \left(z - \frac{z^3}{3!} + \frac{z^5}{5!} - \frac{z^7}{7!} + \ldots\right)\right]$$

$$= \frac{1}{z^3}\left(\frac{z^3}{3!} - \frac{z^5}{5!} + \frac{z^7}{7!} - \ldots\right) = \frac{1}{3!} - \frac{z^2}{5!} + \frac{z^4}{7!} - \ldots$$

There is no negative power of z

Hence, there is no pole. **Ans.**

Example 11. *Determine the poles of the function z*

$$f(z) = \frac{1}{z^4 + 1}$$ *(R.G.P.V., Bhopal, III Semester, June 2003)*

Solution. $$f(z) = \frac{1}{z^4 + 1}$$

The poles of $f(z)$ are determined by putting the denominator equal to zero.

i.e., $$z^4 + 1 = 0 \quad \Rightarrow \quad z^4 = -1$$

$$z = (-1)^{\frac{1}{4}} = (\cos \pi + i \sin \pi)^{\frac{1}{4}}$$

$$= [\cos(2n+1)\pi + i \sin(2n+1)\pi]^{\frac{1}{4}} \quad \text{[By De Moiver's theorem]}$$

$$= \left[\cos \frac{(2n+1)\pi}{4} + i \sin \frac{(2n+1)\pi}{4}\right]$$

If $n = 0$, Pole at $\qquad z = \left[\cos\frac{\pi}{4} + i\sin\frac{\pi}{4}\right] = \left(\frac{1}{\sqrt{2}} + i\frac{1}{\sqrt{2}}\right)$

If $n = 1$, Pole at $\qquad z = \left[\cos\frac{3\pi}{4} + i\sin\frac{3\pi}{4}\right] = \left(-\frac{1}{\sqrt{2}} + i\frac{1}{\sqrt{2}}\right)$

If $n = 2$, Pole at $\qquad z = \left[\cos\frac{5\pi}{4} + i\sin\frac{5\pi}{4}\right] = \left(-\frac{1}{\sqrt{2}} - i\frac{1}{\sqrt{2}}\right)$

If $n = 3$, Pole at $\qquad z = \left[\cos\frac{7\pi}{4} + i\sin\frac{7\pi}{4}\right] = \left(\frac{1}{\sqrt{2}} - i\frac{1}{\sqrt{2}}\right)$ **Ans.**

Example 12. *Show that the function e^z has an isolated essential singularity at $z = \infty$.*
(R.G.P.V., Bhopal, III Semester, Dec. 2003)

Solution. Let $f(z) = e^z$

Putting $z = \frac{1}{t}$, we get $f\left(\frac{1}{t}\right) = e^{\frac{1}{t}} = 1 + \frac{1}{t} + \frac{1}{2!.t^2} + \frac{1}{3!\,t^3} + \ldots..$

Here, the principal part of $f\left(\frac{1}{t}\right)$;

$$\frac{1}{t} + \frac{1}{2!\,t^2} + \frac{1}{3!\,t^3} + \ldots$$

Contains infinite number of terms.

Hence $t = 0$ is an isolated essential singularity of $e^{\frac{1}{t}}$ and $z = \infty$ is an isolated essential singularity of e^z. **Ans.**

EXERCISE 31.1

Find the poles or singularity of the following functions:

1. $\dfrac{1}{(z-2)(z-3)}$ **Ans.** 2 simple poles at $z = 2$ and $z = 3$.

2. $\dfrac{e^z}{(z-2)^3}$ **Ans.** Pole at $z = 2$ of order 3.

3. $\dfrac{1}{\sin z - \cos z}$ **Ans.** Simple pole at $z = \frac{\pi}{4}$

4. $\cot\dfrac{1}{z}$ **Ans.** Essential singularity at $z = 0$

5. $z \operatorname{cosec} z$ **Ans.** Non-isolated essential singularity

6. $\sin\dfrac{1}{z}$ **Ans.** Essential singularity

Choose the correct alternative :

7. Let $f(z) = \dfrac{1}{(z-2)^4 (z+3)^6}$, then $z = 2$ and $z = -3$ are the poles of order :

(*a*) 6 and 4 (*b*) 2 and 3 (*c*) 3 and 4 (*d*) 4 and 6 **Ans.** (*d*)

(R.G.P.V., Bhopal III Semester, June 2007)

31.3 THEOREM

If $f(z)$ has a pole at $z = a$, then $|f(z)| \to \infty$ as $z \to a$.

Proof. Let $z = a$ be a pole of order m of $f(z)$. Then by Laurent's theorem

$$f(z) = \sum_0^\infty a_n (z-a)^n + \sum_1^m b_n (z-a)^{-n}$$

$$= \sum_0^\infty a_n (z-a)^n + \frac{b_1}{z-a} + \frac{b_2}{(z-a)^2} + \ldots.. + \frac{b_m}{(z-a)^m}$$

$$= \sum_0^\infty a_n (z-a)^n + \frac{1}{(z-a)^m}[b_1(z-a)^{m-1} + b_2(z-a)^{m-2} + \ldots\ldots + b_{m-1}(z-a) + b_m]$$

$$= \sum_0^\infty a_n (z-a)^n + \frac{\varphi(z)}{(z-a)^m}$$

Now $\varphi(z) \to b_m$ as $z \to a$.

Hence $|f(z)| \to \infty$ as $z \to a$. **Proved.**

Example 13. *If an analytic function f(z) has a pole of order m at z = a, then $\frac{1}{f(z)}$ has a zero of order m at z = a.*

Solution. If $f(z)$ has a pole of order m at $z = a$, then

$$f(z) = \frac{\varphi(z)}{(z-a)^m} \quad \text{where } \varphi(z) \text{ is analytic and non-zero at } z = a.$$

$$\therefore \quad \frac{1}{f(z)} = \frac{(z-a)^m}{\varphi(z)}$$

Clearly, $\frac{1}{f(z)}$ has a zero of order m at $z = a$, since $\phi(a) \neq 0$.

31.4 DEFINITION OF THE RESIDUE AT A POLE

Let $z = a$ be a pole of order m of a function $f(z)$ and C_1 circle of radius r with centre at $z = a$ which does not contain any other singularities except at $z = a$ then $f(z)$ is analytic within the annulus $r < |z-a| < R$ can be expanded within the annulus. Laurent's series:

$$f(z) = \sum_{n=0}^{\infty} a_n (z-a)^n + \sum_{n=1}^{\infty} b_n (z-a)^{-n} \quad \ldots(1)$$

where $$a_n = \frac{1}{2\pi i}\int_C \frac{f(z)\,dz}{(z-a)^{n+1}} \quad \ldots(2)$$

and $$b_n = \frac{1}{2\pi i}\int_{C_1} \frac{f(z)}{(z-a)^{-n+1}}\,dz \quad \ldots(3)$$

$|z-a| = r$ being the circle C_1.

Particularly, $$b_1 = \frac{1}{2\pi i}\int_{C_1} f(z)\,dz$$

The coefficient b_1 is called residue of $f(z)$ at the pole $z = a$. It is denoted by symbol Res. $(z = a) = b_1$.

31.5 RESIDUE AT INFINITY

Residue of $f(z)$ at $z = \infty$ is defined as $-\frac{1}{2\pi i}\int_C f(z)\,dz$ where the integration is taken round C in anti-clockwise direction.

where C is a large circle containing all finite singularities of $f(z)$.

31.6 METHOD OF FINDING RESIDUES

(*a*) **Residue at simple pole**

(*i*) If $f(z)$ has a simple pole at $z = a$, then

$$\text{Res } f(a) = \lim_{z\to a}(z-a)f(z)$$

Proof. $$f(z) = a_0 + a_1(z-a) + a_2(z-a)^2 + \ldots + \frac{b_1}{z-a}$$

$$\Rightarrow \quad (z-a)f(z) = a_0(z-a) + a_1(z-a)^2 + a_2(z-a)^3 + \ldots + b_1$$

$$\Rightarrow \quad b_1 = (z-a)f(z) - [a_0(z-a) + a_1(z-a)^2 + a_2(z-a)^3 + \ldots]$$

Taking limit as $z \to a$, we have $b_1 = \lim_{z \to a}(z-a)f(z)$

$$\textbf{Res (at } z = a\textbf{)} = \lim_{z \to a} (z-a)f(z)$$ **Proved.**

(*ii*) If $f(z)$ is of the form $f(z) = \dfrac{\phi(z)}{\psi(z)}$ where $\psi(a) = 0$, but $\phi(a) \neq 0$

$$\textbf{Res (at } z = a\textbf{)} = \frac{\phi(a)}{\psi'(a)}$$

Proof. $f(z) = \dfrac{\phi(z)}{\psi(z)}$

$$\text{Res (at } z = a) = \lim_{z \to a}(z-a)f(z) = \lim_{z \to a}(z-a)\frac{\phi(z)}{\psi(z)}$$

$$= \lim_{z \to a} \frac{(z-a)[\phi(a) + (z-a)\,\phi'(a) + \ldots]}{\psi(a) + (z-a)\psi'(a) + \dfrac{(z-a)^2}{2!}\psi''(a) + \ldots} \quad \text{(By Taylor's Theorem)}$$

$$= \lim_{z \to a} \frac{(z-a)\,[\phi(a) + (z-a)\phi'(a) + \ldots]}{(z-a)\psi'(a) + \dfrac{(z-a)^2}{2!}\psi''(a) + \ldots} \quad [\text{since } \psi(a) = 0]$$

$$= \lim_{z \to a} \frac{\phi(a) + (z-a)\,\phi'(a) + \ldots}{\psi'(a) + \dfrac{z-a}{2!}\psi''(a) + \ldots}$$

$$\textbf{Res (at } z = a\textbf{)} = \frac{\phi(a)}{\psi'(a)}$$ **Proved.**

(*b*) **Residue at a pole of order *n*.** If $f(z)$ has a pole of order n at $z = a$, then

$$\textbf{Res (at } z = a\textbf{)} = \frac{1}{(n-1)!}\left\{\frac{d^{n-1}}{dz^{n-1}}[(z-a)^n f(z)]\right\}_{z=a}$$

Proof. If $z = a$ is a pole of order n of function $f(z)$, then by Laurent's theorem

$$f(z) = a_0 + a_1(z-a) + a_2(z-a)^2 + \ldots + \frac{b_1}{z-a} + \frac{b_2}{(z-a)^2} + \ldots + \frac{b_n}{(z-a)^n}$$

Multiplying by $(z-a)^n$, we get

$$(z-a)^n f(z) = a_0(z-a)^n + a_1(z-a)^{n+1} + a_2(z-a)^{n+2} + \ldots$$
$$+ \ b_1(z-a)^{n-1} + b_2(z-a)^{n-2} + b_3(z-a)^{n-3} + \ldots + b_n$$

Differentiating both sides w.r.t. 'z' $n-1$ times and putting $z = a$, we get

$$\left\{\frac{d^{n-1}}{dz^{n-1}}[(z-a)^n f(z)]\right\}_{z=a} = (n-1)!\,b_1$$

$$\Rightarrow \quad b_1 = \frac{1}{(n-1)!}\left\{\frac{d^{n-1}}{dz^{n-1}}[(z-a)^n f(z)]\right\}_{z=a}$$

$$\textbf{Residue } f(a) = \frac{1}{(n-1)!}\left\{\frac{d^{n-1}}{dz^{n-1}}[(z-a)^n f(z)]\right\}_{z=a}$$

(*c*) **Residue at a pole $z = a$ of any order (simple or of order m)**

$$\textbf{Res } f(a) = \textbf{coefficient of } \frac{1}{t}$$

Proof. If $f(z)$ has a pole of order m, then by Laurent's theorem

$$f(z) = a_0 + a_1(z-a) + a_2(z-a)^2 ... + \frac{b_1}{z-a} + \frac{b_2}{(z-a)^2} + ... + \frac{b_m}{(z-a)^m}$$

If we put $z - a = t$ or $z = a + t$, then

$$f(a+t) = a_0 + a_1 t + a_2 t^2 + ... + \frac{b_1}{t} + \frac{b_2}{t^2} + ... + \frac{b_m}{t^m}$$

Res $f(a) = b_1$, **Res $f(a)$ = coefficient of $\frac{1}{t}$**

Rule. Put $z = a + t$ in the function $f(z)$, expand it in powers of t. Coefficient of $\frac{1}{t}$ is the residue of $f(z)$ at $z = a$.

(*d*) **Residue of $f(z)$ at $z = \infty$** $= \lim_{z\to\infty} \{-z f(z)\}$

or The residue of $f(z)$ at infinity $= -\frac{1}{2\pi i}\int_C f(z)\, dz$

31.7 RESIDUE BY DEFINITION

Example 14. *Fina the residue at $z = 0$ of $z \cos \frac{1}{z}$.*

Solution. Expanding the function in powers of $\frac{1}{z}$, we have

$$z \cos\frac{1}{z} = z\left[1 - \frac{1}{2z^2} + \frac{1}{4!\, z^4} - .....\right] = z - \frac{1}{2z} + \frac{1}{24z^3} - ......$$

This is the Laurent's expansion about $z = 0$.

The coefficient of $\frac{1}{z}$ in it is $-\frac{1}{2}$. So the residue of $z \cos\frac{1}{z}$ at $z = 0$ is $-\frac{1}{2}$. **Ans.**

Example 15. *Find the residue of $f(z) = \frac{z^3}{z^2 - 1}$ at $z = \infty$.*

Solution. We have, $f(z) = \frac{z^3}{z^2 - 1}$

$$f(z) = \frac{z^3}{z^2\left(1 - \frac{1}{z^2}\right)} = z\left(1 - \frac{1}{z^2}\right)^{-1} = z\left(1 + \frac{1}{z^2} + \frac{1}{z^4} + ...\right) = z + \frac{1}{z} + \frac{1}{z^3} + ...$$

$$\text{Residue at infinity} = -\left(\text{coeff. of } \frac{1}{z}\right) = -1.$$ **Ans.**

31.8 FORMULA: RESIDUE = $\lim_{z\to a} (z - a) f(z)$

Example 16. *Determine the pole and residue at the pole of the function* $f(z) = \frac{z}{z-1}$

Solution. The poles of $f(z)$ are given by putting the denominator equal to zero.

$\therefore$ $z - 1 = 0 \Rightarrow z = 1$

The function $f(z)$ has a simple pole at $z = 1$.

Residue is calculated by the formula

$$\text{Residue} = \lim_{z\to a} (z - a) f(z)$$

Residue of $f(z)$ at (at $z = 1$) $= \lim_{z\to 1} (z-1)\left(\frac{z}{z-1}\right) = \lim_{z\to 1} (z) = 1$

Hence, $f(z)$ has a simple pole at $z = 1$ and residue at the pole is 1. **Ans.**

Example 17. *Determine the poles and the residue at simple pole of the function*

$$f(z) = \frac{z^2}{(z-1)^2(z+2)}$$

Solution. The pole of $f(z)$ are given by putting the denominator equal to zero.

$$(z-1)^2(z+2) = 0 \qquad \Rightarrow \quad z = 1, 1, -2$$

The function $f(z)$ has simple pole at $z = -2$ and at $z = 1$ pole of second order.

Residue of $f(z)$ at $z = -2$ is $= \lim_{z \to -2} (z+2) f(z)$ $\quad$ [Residue $= \lim_{z \to a} (z-a)\, f(z)$]

$$= \lim_{z \to -2} (z+2) \frac{z^2}{(z-1)^2(z+2)} = \lim_{z \to -2} \frac{z^2}{(z-1)^2} = \frac{(-2)^2}{(-2-1)^2} = \frac{4}{9}$$

Hence, residue at simple pole is $\frac{4}{9}$. **Ans.**

Example 18. *Find the order of each pole and residue at it of* $\frac{1-2z}{z(z-1)(z-2)}$.

(R.G.P.V., Bhopal, III Semester, Dec. 2001)

Solution. Let $f(z) = \frac{1-2z}{z(z-1)(z-2)}$

The poles of $f(z)$ are given by $z(z-1)(z-2) = 0$

$\Rightarrow$ $\quad z = 0, 1, 2$ all are simple poles.

Residue of $f(z)$ at $(z = 0)$ $= \lim_{z \to 0} (z-0)\, f(z) = \lim_{z \to 0} \frac{z(1-2z)}{z(z-1)(z-2)}$

$$= \lim_{z \to 0} \frac{1-2z}{(z-1)(z-2)} = \frac{1}{2}$$

Residue of $f(z)$ at $(z = 1)$ $= \lim_{z \to 1} (z-1)\, f(z) = \lim_{z \to 1} \frac{(z-1)(1-2z)}{z(z-1)(z-2)} = \lim_{z \to 1} \frac{1-2z}{z(z-2)} = 1$

Residue of $f(z)$ at $(z = 2)$ $= \lim_{z \to 2} (z-2)\, f(z) = \lim_{z \to 2} \frac{(z-2)(1-2z)}{z(z-1)(z-2)}$

$$= \lim_{z \to 2} \frac{1-2z}{z(z-1)} = -\frac{3}{2}$$

Hence, the residue of $f(z)$ at $z = 0, z = 1$ and $z = 2$ are $\frac{1}{2}$, 1 and $-\frac{3}{2}$ respectively. **Ans.**

Example 19. *Determine the residue of* $f(z) = \frac{z^3}{(z-1)^4(z-2)(z-3)}$ *at its simple poles.*

Solution. The poles of $f(z)$ are determined by putting the denominator equal to zero.

i.e. $\quad (z-1)^4(z-2)(z-3) = 0$

$\Rightarrow$ $\quad z = 1, 1, 1. 1$ and $z = 2$ and $z = 3$

The simple poles of the function $f(z)$ are at $z = 2$ and $z = 3$.

Pole at $z = 2$

Residue, $\quad R(2) = \lim_{z \to 2} (z-2) \frac{z^3}{(z-1)^4(z-2)(z-3)}$

[Residue $R(2) = \lim_{z \to 2} [(z-2)\, f(z)]$]

$$= \lim_{z \to 2} \frac{z^3}{(z-1)^4(z-3)} = \frac{(2)^3}{(1)^4(-1)} = -8$$

Pole at $z = 3$

Residue, $$R(3) = \lim_{z\to 3} (z-3)\frac{z^3}{(z-1)^4 (z-2)(z-3)}$$

$$= \lim_{z\to 3} \frac{z^3}{(z-1)^4 (z-2)} = \frac{(3)^3}{(3-1)^4 (3-2)} = \frac{27}{16}$$

Hence, residue at $z = 2$ and $z = 3$ are -8 and $\frac{27}{16}$ respectively. **Ans.**

Example 20. *Evaluate the residues of* $\frac{z^2}{(z-1)(z-2)(z-3)}$ *at* $z = 1, 2, 3$ *and infinity and show that their sum is zero.* *(R.G.P.V., Bhopal, III Semester Dec. 2002)*

Solution. Let $f(z) = \frac{z^2}{(z-1)(z-2)(z-3)}$

The poles of $f(z)$ are determined by putting the denominator equal to zero.

$$\therefore \quad (z-1)(z-2)(z-3) = 0 \Rightarrow z = 1, 2, 3$$

Residue of $f(z)$ at $(z = 1)$ $= \lim_{z\to 1} (z-1) f(z) = \lim_{z\to 1} (z-1) . \frac{z^2}{(z-1)(z-2)(z-3)}$

$$= \lim_{z\to 1} \frac{z^2}{(z-2)(z-3)} = \frac{1}{2}$$

Residue of $f(z)$ at $(z = 2)$ $= \lim_{z\to 2} (z-2) f(z) = \lim_{z\to 2} (z-2) \frac{z^2}{(z-1)(z-2)(z-3)}$

$$= \lim_{z\to 2} \frac{z^2}{(z-1)(z-3)} = \frac{4}{(1)(-1)} = -4$$

Residue of $f(z)$ at $(z = 3)$ $= \lim_{z\to 3} (z-3) f(z)$

$$= \lim_{z\to 3} (z-3) \frac{z^2}{(z-1)(z-2)(z-3)} = \lim_{z\to 3} \frac{z^2}{(z-1)(z-2)} = \frac{9}{2}$$

Residue of $f(z)$ at $(z=\infty)$ $= \lim_{z\to\infty} -z f(z) = \frac{-z(z^2)}{(z-1)(z-2)(z-3)}$

$$= \lim_{z\to\infty} \frac{-1}{\left(1-\frac{1}{z}\right)\left(1-\frac{2}{z}\right)\left(1-\frac{3}{z}\right)} = -1$$

Sum of the residues at all the poles of $f(z) = \frac{1}{2} - 4 + \frac{9}{2} - 1 = 0$

Hence, the sum of the residues is zero. **Proved.**

31.9 RESIDUE FORMULA: $f(a) = \frac{1}{(n-1)!}\left\{\frac{d^{n-1}}{dz^{n-1}} [(z-a)^n f(z)]\right\}_{z=a}$

Example 21. *Find the residue of a function*

$$f(z) = \frac{z^2}{(z+1)^2 (z-2)} \quad \textit{at its double pole.}$$

Solution. We have, $$f(z) = \frac{z^2}{(z+1)^2 (z-2)}$$

Poles are determined by putting denominator equal to zero.

i.e., $\quad (z+1)^2 (z-2) = 0$

$\Rightarrow \qquad z = -1, -1 \quad \text{and} \quad z = 2$

The function has a double pole at $z = -1$

$$\text{Residue at } (z=-1) = \lim_{z\to -1} \frac{1}{(2-1)!}\left[\frac{d}{dz}\left\{(z+1)^2 \frac{z^2}{(z+1)^2 (z-2)}\right\}\right]$$

$$= \left[\frac{d}{dz}\left(\frac{z^2}{z-2}\right)\right]_{z=-1} = \left(\frac{(z-2)\,2z - z^2\cdot 1}{(z-2)^2}\right)_{z=-1} = \left[\frac{z^2-4z}{(z-2)^2}\right]_{z=-1} = \frac{(-1)^2-4(-1)}{(-1-2)^2}$$

$$\text{Residue at } (z=-1) = \frac{1+4}{9} = \frac{5}{9}$$ **Ans.**

Example 22. *Find the residue of* $\dfrac{1}{(z^2+1)^3}$ *at* $z = i$.

Solution. Let $\qquad f(z) = \dfrac{1}{(z^2+1)^3}$

The poles of $f(z)$ are determined by putting denominator equal to zero.

i.e.; $\qquad (z^2+1)^3 = 0$

$\Rightarrow \qquad (z+i)^3 (z-i)^3 = 0$

$\Rightarrow \qquad z = \pm i$

Here, $z = i$ is a pole of order 3 of $f(z)$.

Residue at $z = i$

$$= \lim_{z\to i} \frac{1}{(3-1)!}\left\{\frac{d^{3-1}}{dz^{3-1}}\left[(z-i)^3 \frac{1}{(z^2+1)^3}\right]\right\} = \lim_{z\to i} \frac{1}{2!}\left\{\frac{d^2}{dz^2}\left(\frac{1}{(z+i)^3}\right)\right\}$$

$$= \lim_{z\to i} \frac{1}{2}\left(\frac{3\times 4}{(z+i)^5}\right) = \frac{1}{2}\times\frac{12}{(i+i)^5} = \frac{6}{32i} = \frac{3}{16i} = -\frac{3i}{16}$$

Hence, the residue of the given function at $z = i$ is $-\dfrac{3i}{16}$. **Ans.**

31.10 FORMULA: RES. (AT $z = a$) = $\dfrac{\phi(a)}{\psi'(a)}$

Example 23. *Determine the poles and residue at each pole of the function* $f(z) = \cot z$.

Solution. $\qquad f(z) = \cot z = \dfrac{\cos z}{\sin z}$

The poles of the function $f(z)$ are given by

$$\sin z = 0, \; z = n\pi, \; \text{ where } n = 0, \pm 1, \pm 2, \pm 3 \ldots$$

$$\text{Residue of } f(z) \text{ at } z = n\pi \text{ is } = \frac{\cos z}{\frac{d}{dz}(\sin z)} = \frac{\cos z}{\cos z} = 1 \quad \left[\text{Res. at } (z=a) = \frac{\phi(a)}{\psi'(a)}\right]$$ **Ans.**

Example 24. *Determine the poles of the function and residue at the poles.*

$$f(z) = \frac{z}{\sin z}$$

Solution. $\qquad f(z) = \dfrac{z}{\sin z}$

Poles are determined by putting $\sin z = 0 = \sin n\pi \quad \Rightarrow z = n\pi$

$$\text{Residue} = \left(\frac{z}{\cos z}\right)_{z=n\pi} \qquad \left[\text{Residue} = \frac{\phi(a)}{\psi'(a)}\right]$$

$$= \frac{n\pi}{\cos n\pi} = \frac{n\pi}{(-1)^n}$$

Hence, the residue of the given function at pole $z = n\pi$ is $\frac{n\pi}{(-1)^n}$. **Ans.**

31.11 FORMULA: RESIDUE = COEFFICIENT OF $\frac{1}{t}$ in $f(1 + t)$ where $z = \frac{1}{t}$

Example 25. *Find the residue of* $\frac{z^3}{(z-1)^4(z-2)(z-3)}$ *at a pole of order 4.*

Solution. The poles of $f(z)$ are determined by $(z-1)^4(z-2)(z-3) = 0 \Rightarrow z = 1, 2, 3$

Here $z = 1$ is a pole of order 4.

$$f(z) = \frac{z^3}{(z-1)^4(z-2)(z-3)} \quad ...(1)$$

Putting $z - 1 = t$ or $z = 1 + t$ in (1), we get

$$f(1+t) = \frac{(1+t)^3}{t^4(t-1)(t-2)} = \frac{1}{t^4}(t^3+3t^2+3t+1)(1-t)^{-1}\frac{1}{2}\left(1-\frac{t}{2}\right)^{-1}$$

$$= \frac{1}{2}\left(\frac{1}{t}+\frac{3}{t^2}+\frac{3}{t^3}+\frac{1}{t^4}\right)(1+t+t^2+t^3+...)\times\left(1+\frac{t}{2}+\frac{t^2}{4}+\frac{t^3}{8}...\right)$$

$$= \frac{1}{2}\left(\frac{1}{t}+\frac{3}{t^2}+\frac{3}{t^3}+\frac{1}{t^4}\right)\left(1+\frac{3}{2}t+\frac{7}{4}t^2+\frac{15}{8}t^3+...\right) = \frac{1}{2}\left(\frac{1}{t}+\frac{9}{2t}+\frac{21}{4}\frac{1}{t}+\frac{15}{8}\frac{1}{t}\right)+...$$

$$= \frac{1}{2}\left(1+\frac{9}{2}+\frac{21}{4}+\frac{15}{8}\right)\frac{1}{t} \qquad \left[\text{Res } f(a) = \text{coeffi. of } \frac{1}{t}\right]$$

Coefficient of $\frac{1}{t} = \frac{1}{2}\left(1+\frac{9}{2}+\frac{21}{4}+\frac{15}{8}\right) = \frac{101}{16}$,

Hence, the residue of the given function at a pole of order 4 is $\frac{101}{16}$. **Ans.**

Example 26. *Find the residue of* $f(z) = \frac{ze^z}{(z-a)^3}$ *at its pole.*

Solution. The pole of $f(z)$ is given by $(z-a)^3 = 0$ *i.e.*, $z = a$

Here $z = a$ is a pole of order 3.

Putting $z - a = t$ where t is small.

$$f(z) = \frac{ze^z}{(z-a)^3} \Rightarrow f(z) = \frac{(a+t)e^{a+t}}{t^3} = \left(\frac{a}{t^3}+\frac{1}{t^2}\right)e^{a+t} = e^a\left(\frac{a}{t^3}+\frac{1}{t^2}\right)e^t \qquad (z = a + t)$$

$$= e^a\left(\frac{a}{t^3}+\frac{1}{t^2}\right)\left(1+\frac{t}{1!}+\frac{t^2}{2!}+...\right) = e^a\left[\frac{a}{t^3}+\frac{a}{t^2}+\frac{a}{2t}+\frac{1}{t^2}+\frac{1}{t}+\frac{1}{2}+...\right]$$

$$= e^a\left[\frac{1}{2}+\left(\frac{a}{2}+1\right)\frac{1}{t}+(a+1)\frac{1}{t^2}+(a)\frac{1}{t^3}+...\right]$$

Coefficient of $\frac{1}{t} = e^a\left(\frac{a}{2}+1\right)$

Hence the residue at $z = a$ is $e^a\left(\frac{a}{2}+1\right)$ **Ans.**

Example 27. *Find the sum of the residues of the function* $f(z)=\dfrac{\sin z}{z\cos z}$ *at its poles inside the circle* $|z|=2$.

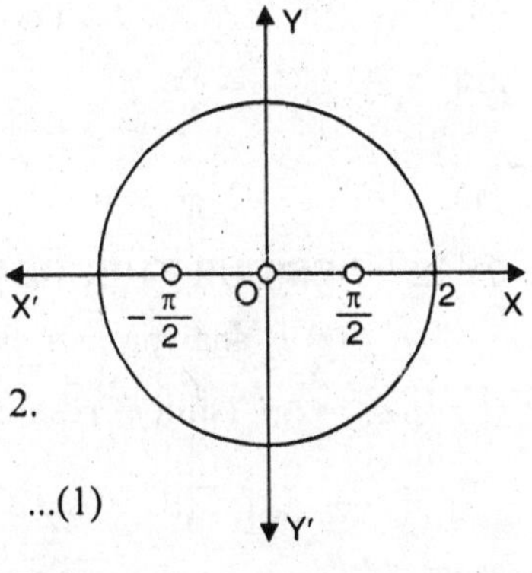

Solution. We have,

$$f(z)=\frac{\sin z}{z\cos z}$$

The pole can be determined by putting denominator

$$z\cos z=0$$

$$\Rightarrow \qquad z=0,\ \pm\frac{\pi}{2},\ \pm\frac{3\pi}{2},\ \ldots\ldots$$

Of these poles only $z=0,\ z=\pm\dfrac{\pi}{2}$ lie inside a circle $|z|=2$.

Residue of $f(z)$ at $z=0$ is $\lim\limits_{z\to 0}|z\cdot f(z)|=\lim\limits_{z\to 0}\dfrac{\sin z}{\cos z}=0$ ...(1)

Residue of $f(z)$ at $z=\dfrac{\pi}{2}$ is

$$\lim_{z\to\frac{\pi}{2}}\left(z-\frac{\pi}{2}\right)f(z)=\lim_{z\to\frac{\pi}{2}}\frac{\left(z-\frac{\pi}{2}\right)\sin z}{z\cos z}\qquad\left[\text{From } \frac{0}{0}\right]$$

$$=\lim_{z\to\frac{\pi}{2}}\frac{\left(z-\frac{\pi}{2}\right)\cos z+\sin z}{\cos z-z\sin z}\qquad\text{[By L' Hopital's Rule]}$$

$$=\frac{1}{-\frac{\pi}{2}}=-\frac{2}{\pi}\qquad\text{...(2)}$$

Similarly, residue of $f(z)$ at $z=-\dfrac{\pi}{2}$ is $\dfrac{2}{\pi}$ ...(3)

$\therefore$ Sum of the residues $=0-\dfrac{2}{\pi}+\dfrac{2}{\pi}=0.$ **Ans.**

EXERCISE 31.2

1. Determine the poles of the following functions. Find the order of each pole.

(*i*) $\dfrac{z^2}{(z-a)(z-b)(z-c)}$ **Ans.** Simple poles at $z=a,\ z=b,\ z=c$

(*ii*) $\dfrac{z-3}{(z-2)^2(z+1)}$ **Ans.** Pole at $z=2$ of second order and $z=-1$of first order.

(*iii*) $\dfrac{ze^{iz}}{z^2+a^2}$ **Ans.** Poles at $z=\pm ia$, order 1.

(*iv*) $\dfrac{1}{(z-1)(z-2)}$ **Ans.** $z=2,\ z=1$

Find the residue of

2. $\dfrac{z^3}{(z-2)(z-3)}$ at its poles. **Ans.** $27,\ -8$

3. $\dfrac{z^2}{z^2+a^2}$ at $z=ia$. **Ans.** $\dfrac{1}{2}ia$

4. $\dfrac{1}{(z^2+a^2)^2}$ at $z=ia$ **Ans.** $-\dfrac{i}{4a^3}$

5. $\tan z$ at its pole. **Ans.** $f\left(n+\dfrac{\pi}{2}\right)=-1$ at its pole

6. $z^2 e^{1/z}$ at the point $z = 0$. **Ans.** $\frac{1}{6}$

7. $z^2 \sin\left(\frac{1}{z}\right)$ at $z = 0$ **Ans.** $-\frac{1}{6}$

8. $\frac{1}{z^2(z-i)}$ at $z = i$ **Ans.** -1

9. $\frac{e^{2z}}{1-e^z}$ at its pole **Ans.** -1

10. $\frac{1+e^z}{\sin z + z\cos z}$ at $z = 0$ **Ans.** 1

11. $\frac{1}{z(e^z-1)}$ at its poles **Ans.** $-\frac{1}{2}$

31.12 RESIDUE THEOREM

If $f(z)$ is analytic in a closed curve C, except at a finite number of poles within C, then $\int_c f(z)\,dz = 2\pi i$ **(sum of residues at the poles within** C**).**

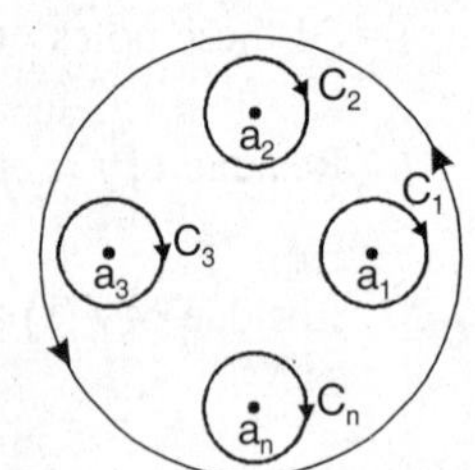

Proof. Let $C_1, C_2, C_3, \ldots, C_n$ be the non-intersecting circles with centres at $a_1, a_2, a_3, \ldots, a_n$ respectively, and radii so small that they lie entirely within the closed curve C. Then $f(z)$ is analytic in the multiple connected region lying between the curves C and $C_1, C_2, \ldots, C_n$.

Applying Cauchy's theorem

$$\int_C f(z)\,dz = \int_{C_1} f(z)\,dz + \int_{C_2} f(z)\,dz + \int_{C_3} f(z)\,dz + \ldots + \int_{C_n} f(z)\,dz.$$

$$= 2\pi i\,[\text{Res } f(a_1) + \text{Res } f(a_2) + \text{Res } f(a_3) + \ldots + \text{Res } f(a_n)]$$ **Proved.**

Example 28. *Evaluate the following integral using residue theorem*

$$\int_c \frac{1+z}{z\,(2-z)}\,dz$$

where c is the circle $|z| = 1$.

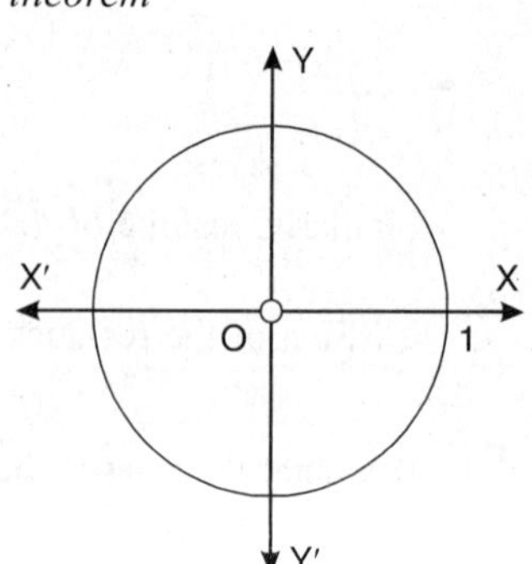

Solution. The poles of the integrand are given by putting the denominator equal to zero.

$$z(2-z) = 0 \quad \text{or} \quad z = 0,\ 2$$

The integrand is analytic on $|z| = 1$ and all points inside except $z = 0$, as a pole at $z = 0$ is inside the circle $|z| = 1$. Hence by residue theorem

$$\int_c \frac{1+z}{z(2-z)}\,dz = 2\pi i \quad [\text{Res } f(0)] \qquad \ldots (1)$$

$$\text{Residue } f(0) = \lim_{z\to 0} z.\frac{1+z}{z\,(2-z)} = \lim_{z\to 0}\frac{1+z}{2-z} = \frac{1}{2}$$

Putting the value of Residue $f(0)$ in (1), we get

$$\int_c \frac{1+z}{z(2-z)}\,dz = 2\pi i\left(\frac{1}{2}\right) = \pi i$$ **Ans.**

Example 29. *Evaluate the following integral using residue theorem*

$$\int_c \frac{4-3z}{z(z-1)(z-2)}\,dz$$

where c is the circle $|z| = \frac{3}{2}$.

Solution. The poles of the function $f(z)$ are given by equating the denominator to zero.

$$z(z-1)(z-2) = 0, \quad z = 0, 1, 2$$

The function has poles at $z = 0$, $z = 1$ and $z = 2$ of which the given circle encloses the pole at $z = 0$ and $z = 1$.

Residue of $f(z)$ at the simple pole $z = 0$ is

$$= \lim_{z \to 0} z \frac{4-3z}{z(z-1)(z-2)} = \lim_{z \to 0} \frac{4-3z}{(z-1)(z-2)}$$

$$= \frac{4-0}{(0-1)(0-2)} = 2$$

Residue of $f(z)$ at the simple pole $z = 1$ is

$$= \lim_{z \to 1} (z-1) \frac{4-3z}{z(z-1)(z-2)} = \lim_{z \to 1} \frac{4-3z}{z(z-2)} = \frac{4-3}{1(1-2)} = -1$$

By Cauchy's integral formula

$$\int_C f(z)dz = 2\pi i \times \text{sum of the residue within } c$$

$$= 2\pi i \times (2-1) = 2\pi i \qquad \textbf{Ans.}$$

Example 30. *Evaluate*

$$\int_C \frac{12z-7}{(z-1)^2\,(2z+3)}\,dz, \text{ where } C \text{ is the circle}$$

(*i*) $|z| = 2$ (*ii*) $|z+i| = \sqrt{3}$

Solution. We have, $f(z) = \dfrac{12z-7}{(z-1)^2(2z+3)}$

Poles are given by

$$z = 1 \text{ (double pole) and } z = -\frac{3}{2} \text{ (simple pole)}$$

Residue at $(z = 1)$ is

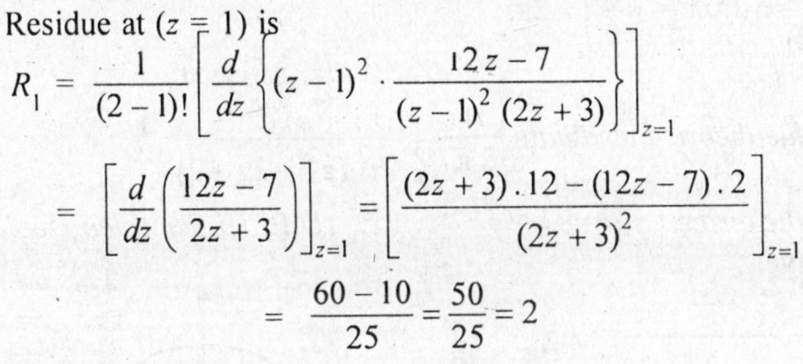

$$R_1 = \frac{1}{(2-1)!}\left[\frac{d}{dz}\left\{(z-1)^2 \cdot \frac{12z-7}{(z-1)^2\,(2z+3)}\right\}\right]_{z=1}$$

$$= \left[\frac{d}{dz}\left(\frac{12z-7}{2z+3}\right)\right]_{z=1} = \left[\frac{(2z+3).12-(12z-7).2}{(2z+3)^2}\right]_{z=1}$$

$$= \frac{60-10}{25} = \frac{50}{25} = 2$$

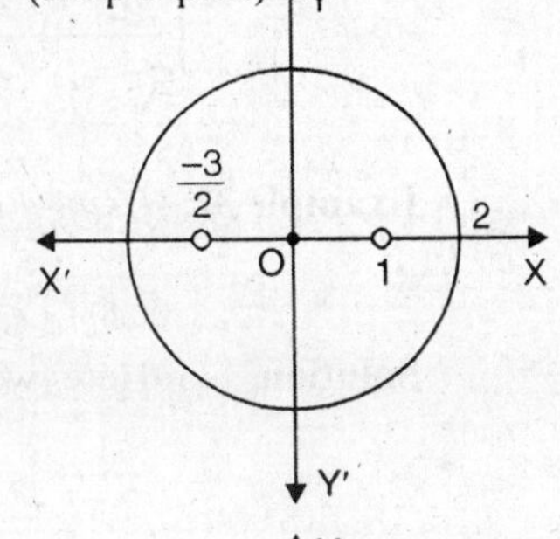

Residue at simple pole $\left(z = -\dfrac{3}{2}\right)$ is

$$R_2 = \lim_{z \to -3/2} \left(z + \frac{3}{2}\right) . \frac{12z-7}{(z-1)^2\,(2z+3)}$$

$$= \lim_{z \to -3/2} \frac{1}{2} . \frac{(12z-7)}{(z-1)^2} = -2.$$

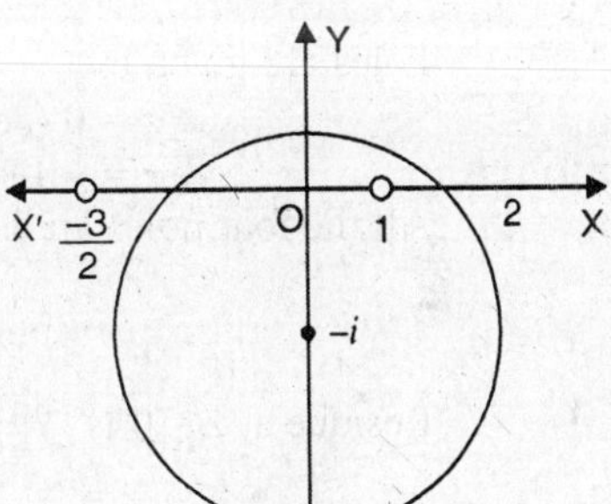

(*i*) The contour $|z| = 2$ encloses both the poles 1 and $-\dfrac{3}{2}$.

$\therefore$ The given integral $= 2\pi i\,(R_1 + R_2) = 2\pi i\,(2-2) = 0$.

(*ii*) The contour $|z+i| = \sqrt{3}$ is a circle of radius $\sqrt{3}$ and centre at $z = -i$.

The distances of the centre from $z = 1$ and $-\dfrac{3}{2}$ are respectively $\sqrt{2}$ and $\sqrt{\dfrac{13}{4}}$. The first of these is $< \sqrt{3}$ and the second is $> \sqrt{3}$.

$\therefore$ The second contour includes only the first singularity $z = 1$.

Hence, the given integral $= 2\pi i\,(R_1) = 2\pi i\,(2) = 4\pi i$. **Ans.**

Example 31. *Determine the poles of the following function and residue at each pole:*

$$f(z)=\frac{z^2}{(z-1)^2(z+2)} \text{ and hence evaluate}$$

$$\int_c \frac{z^2 dz}{(z-1)^2(z+2)} \quad \text{where } c: |z| = 3. \qquad (\textit{R.G.P.V. Bhopal, III Sem. Dec. 2007})$$

Solution. $f(z)=\dfrac{z^2}{(z-1)^2(z+2)}$

Poles of $f(z)$ are given by $(z-1)^2(z+2) = 0$ *i.e.* $z = 1, 1, -2$

The pole at $z = 1$ is of second order and the pole at $z = -2$ is simple.

$$\text{Residue of } f(z) \text{ (at } z=1) = \lim_{z\to 1} \frac{1}{(2-1)!}\frac{d}{dz}\frac{(z-1)^2 z^2}{(z-1)^2(z+2)}$$

$$= \lim_{z\to 1}\frac{d}{dz}\frac{z^2}{z+2} = \lim_{z\to 1}\frac{(z+2)\,2z-1.z^2}{(z+2)^2}$$

$$= \lim_{z\to 1}\frac{z^2+4z}{(z+2)^2} = \frac{1+4}{(1+2)^2} = \frac{5}{9}$$

$$\text{Residue of } f(z) \text{ (at } z=-2) = \lim_{z\to -2}\frac{(z+2)z^2}{(z-1)^2(z+2)} = \lim_{z\to -2}\frac{z^2}{(z-1)^2} = \frac{4}{(-2-1)^2} = \frac{4}{9} \qquad \textbf{Ans.}$$

$$\int_c \frac{z^2 dz}{(z-1)^2(z+2)} = 2\pi i\left(\frac{5}{9}+\frac{4}{9}\right) = 2\pi i \qquad \textbf{Ans.}$$

Example 32. *Using Residue theorem, evaluate* $\dfrac{1}{2\pi i}\displaystyle\int_C \frac{e^{zt}dz}{z^2(z^2+2z+2)}$

where C is the cirlce $|z| = 3$. (*U.P., III Semester, Dec. 2009*)

Solution. Here, we have

$$\frac{1}{2\pi i}\int_C \frac{e^{zt}dz}{z^2(z^2+2z+2)}$$

Poles are given by

$z = 0$ (double ple)

$z = -1 \pm$ i (simple poles)

All the four poles are inside the given circle.

Y, X′, X, Y′, −3, 3, −1 + i, −1 − i

$|z| = 3$

$$\text{Residue at z = 0 is } \lim_{z\to 0}\frac{d}{dz}z^2\frac{e^{zt}}{z^2(z^2+2z+2)}$$

$$= \lim_{z\to 0}\frac{d}{dz}\frac{e^{zt}}{z^2+2z+2}$$

$$= \lim_{z\to 0}\frac{(z^2+2z+2)\,t\,e^{zt}-(2z+2)\,e^{zt}}{(z^2+2z+2)^2}$$

$$= \frac{2t\,e^{\circ}-2e^{\circ}}{4} = \frac{(t-1)}{2}$$

$$z^2+2z+2=0$$
$$\Rightarrow z^2\ 2z+1=-1$$
$$\Rightarrow (z+1)^2=-1$$
$$\Rightarrow z+1=\pm i$$
$$\Rightarrow z=-1\pm i$$

Residue at $z = -1 + i$

$$= \lim_{z \to -1+i} \frac{(z+1-i)\, e^{zt}}{z^2 (z+1-i)(z+1+i)} = \lim_{z \to -1+i} \frac{e^{zt}}{z^2 (z+1+i)}$$

$$= \frac{e^{(-1+i)t}}{(-1+i)^2 (-1+i+1+i)} = \frac{e^{(-1+i)t}}{(1-2i-1)(2i)} = \frac{e^{(-1+i)t}}{4}$$

Similarly Residue at $z = -1 - i = \dfrac{e^{(-1-i)t}}{4}$

$$\int \frac{e^{zt}}{z^2 (z^2+2z+2)}\, dz = 2\pi i \text{ (Sum of the Residues)}$$

$$\Rightarrow \quad \frac{1}{2\pi i} \int \frac{e^{zt}}{z^2 (z^2+2z+2)}\, dz = \frac{t-1}{2} + \frac{e^{(-1+i)t}}{4} + \frac{e^{(-1-i)t}}{4}$$

$$= \frac{t-1}{2} + \frac{e^{-t}}{4} (e^{it} + e^{-it}) = \frac{t-1}{2} + \frac{e^{-t}}{4} (2\cos t)$$

$$= \frac{t-1}{2} + \frac{e^{-t}}{2} \cos t$$

Ans.

Example 33. *Evaluate* $\oint_C \frac{1}{\sinh z}\, dz$, *where C is the circle* $|z| = 4$.

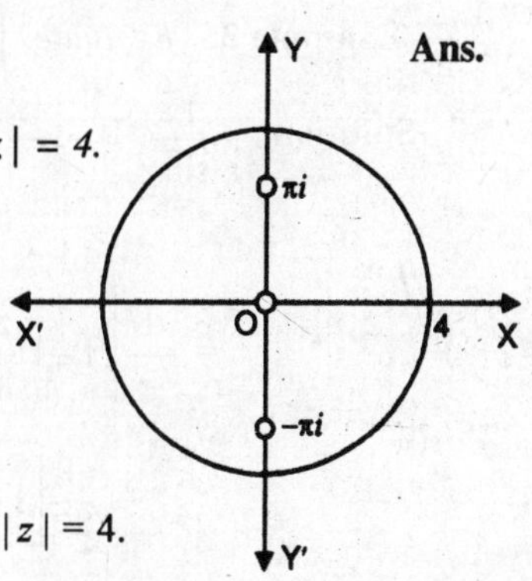

Solution. Here, $f(z) = \dfrac{1}{\sinh z}$.

Poles are given by

$$\sinh z = 0$$

$$\Rightarrow \quad \sin iz = 0$$

$$\Rightarrow \quad z = n\pi i \text{ where } n \text{ is an integer.}$$

Out of these, the poles $z = -\pi i$, 0 and πi lie inside the circle $|z| = 4$.

The given function $\dfrac{1}{\sinh z}$ is of the form $\dfrac{\phi(z)}{\psi(z)}$

Its pole at $z = a$ is $\dfrac{\phi(a)}{\psi'(a)}$.

$$\text{Residue (at } z = -\pi i) = \frac{1}{\cosh(-\pi i)} = \frac{1}{\cos i(-\pi i)} = \frac{1}{\cos \pi} = \frac{1}{-1} = -1$$

$$\text{Residue (at } z = 0) = \frac{1}{\cosh 0} = \frac{1}{1} = 1$$

$$\text{Residue (at } z = \pi i) = \frac{1}{\cosh(\pi i)} = \frac{1}{\cos i(\pi i)} = \frac{1}{\cos(-\pi)} = \frac{1}{\cos \pi} = \frac{1}{-1} = -1$$

Residue at $-\pi i$, 0, πi are respectively -1, 1 and -1.

Hence, the required integral $= 2\pi i(-1 + 1 - 1) = -2\pi i$. **Ans.**

Example 34. *Obtain Laurent's expansion for the function* $f(z) = \dfrac{1}{z^2 \sinh z}$ *at the isolated singularity and hence evaluate* $\oint_C \dfrac{1}{z^2 \sinh z}\, dz$, *where C is the circle* $|z - 1| = 2$.

Solution. Here,

$$f(z) = \frac{1}{z^2 \sinh z} = \frac{2}{z^2 (e^z - e^{-z})} = \frac{2}{z^2 \left[\left(1 + z + \frac{z^2}{2!} + \frac{z^3}{3!} + \ldots\right) - \left(1 - z + \frac{z^2}{2!} - \frac{z^3}{3!} + \ldots\right)\right]}$$

$$= \frac{2}{z^2\left(2z + \frac{2z^3}{3!} + \frac{2z^5}{5!} + \dots\right)} = \frac{1}{z^3\left(1 + \frac{z^2}{3!} + \frac{z^4}{5!} + \dots\right)}$$

$$= z^{-3}\left[1 + \left(\frac{z^2}{6} + \frac{z^4}{120}\right) + \dots\right]^{-1}$$

$$= z^{-3}\left(1 - \frac{z^2}{6} - \frac{z^4}{120} + \dots\right)$$

$$= z^{-3} - \frac{z^{-1}}{6} - \frac{z}{120} + \dots = \frac{1}{z^3} - \frac{1}{6z} - \frac{z}{120} + \dots$$

which is the required Laurent's expansion.

Only pole $z = 0$ of order three lies inside the circle $C \equiv |z - 1| = 2$.

Residue of $f(z)$ at $(z = 0)$ is

$$= \text{coeff. of } \frac{1}{z} \text{ in the Laurent's expansion of } f(z) = -\frac{1}{6}. \quad \textbf{Ans.}$$

Example 35. *Evaluate* $\int_c \frac{dz}{z \sin z}$ *: c is the unit circle about origin.*

Solution. $$\frac{1}{z \sin z} = \frac{1}{z\left[z - \frac{z^3}{3!} + \frac{z^5}{5!} - \dots\right]} = \frac{1}{z^2\left[1 - \frac{z^2}{3!} + \frac{z^4}{5!} - \dots\right]}$$

$$= \frac{1}{z^2}\left[1 - \left(\frac{z^2}{6} - \frac{z^4}{120}\dots\right)\right]^{-1} = \frac{1}{z^2}\left[1 + \left(\frac{z^2}{6} - \frac{z^4}{120}\dots\right) + \left(\frac{z^2}{6} - \frac{z^4}{120}\dots\right)^2 \dots\right]$$

$$= \frac{1}{z^2}\left[1 + \frac{z^2}{6} - \frac{z^4}{120} + \frac{z^4}{36} + \dots\right] = \frac{1}{z^2} + \frac{1}{6} - \frac{z^2}{120} + \frac{z^2}{36} \dots$$

$$= \frac{1}{z^2} + \frac{1}{6} + \frac{7}{360}z^2 \dots$$

This shows that $z = 0$ is a pole of order 2 for the function $\frac{1}{z \sin z}$ and the residue at the pole is zero, (coefficient of $\frac{1}{z}$).

Now the pole at $z = 0$ lies within C.

$$\therefore \quad \int \frac{1}{z \sin z} dz = 2\pi i \text{ (Sum of Residues)} = 0 \qquad \textbf{Ans.}$$

Example 36. *Find the value of* $\oint_C z e^{1/z} dz$*, around the unit circle.*

Solution. The only singularity of $z\, e^{1/z}$ is at the origin. Expanding $e^{1/z}$, we have

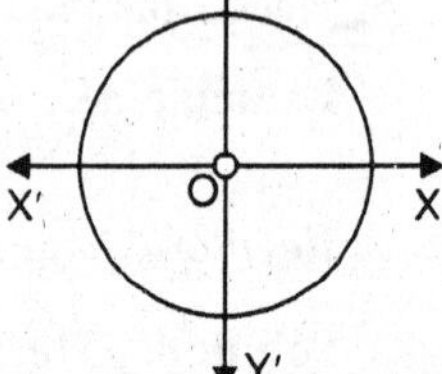

$$z\, e^{1/z} = z\left[1 + \frac{1}{z} + \frac{1}{2z^2} + \frac{1}{6z^3} + \dots\right]$$

$$= z + 1 + \frac{1}{2z} + \frac{1}{6z^2} + \dots$$

Residue at origin = coeff. of $\frac{1}{z} = \frac{1}{2}$.

Hence, the required integral $= 2\pi i\left(\frac{1}{2}\right) = \pi i.$ **Ans.**

Example 37. *Find the value of the complex integral* $\int_c z^4 e^{1/z} dz$, *where c is* $|z| = 1$.

Solution. $\int_c z^4 \cdot e^{\frac{1}{z}} dz$ where c is $|z| = 1$

At $z = 0$, there is a simple pole. Now we have to find out the residue at $z = 0$.

Put $\quad z - 0 = t$

$$z^4 e^{\frac{1}{z}} = t^4 e^{\frac{1}{t}}$$

$$= t^4\left[1+\frac{1}{t}+\frac{1}{2!t^2}+\frac{1}{3!}\frac{1}{t^3}+\frac{1}{4!}\frac{1}{t^4}+\frac{1}{5!}\frac{1}{t^5}+...\right]$$

$$= t^4+t^3+\frac{t^2}{2!}+\frac{t}{3!}+\frac{1}{4!}+\frac{1}{5!}\frac{1}{t}+...$$

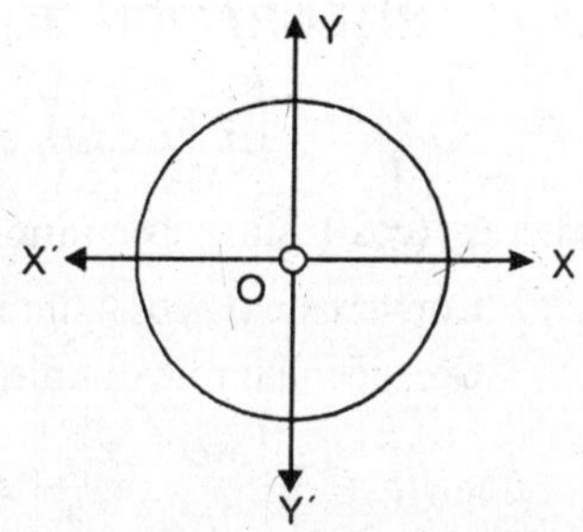

Coefficient of $\frac{1}{t} = \frac{1}{5!} = \frac{1}{120}$

$$\int_c z^4 e^{\frac{1}{z}} dz = 2\pi i \text{ Residue}$$

$$\Rightarrow \quad \int_c z^4 e^{\frac{1}{z}} dz = (2\pi i)\left(\frac{1}{120}\right) = \frac{\pi i}{60}$$ **Ans.**

EXERCISE 31.3

Evaluate the following complex integrals :

1. $\int_c \frac{1-2z}{z(z-1)(z-2)} dz$, where c is the circle $|z| = 1.5$ **Ans.** $3\pi i$

2. $\int_c \frac{z^2 e^{zt}}{z^2+1} dz$, where c is the circle $|z| = 2$ **Ans.** $-2\pi i \sin t$

3. $\int_c \frac{z-1}{(z+1)^2(z-2)} dz$, where c is the circle $|z - i| = 2$. **Ans.** $-\frac{2\pi i}{9}$

4. $\int_c \frac{2z^2+z}{z^2-1} dz$, where c is the circle $|z-1| = 1$. **Ans.** $3\pi i$

5. $\int_c \frac{e^{2z}+z^2}{(z-1)^5} dz$, where c is the circle $|z| = 2$ **Ans.** $\frac{4\pi e^2 i}{3}$

6. $\int_c \frac{dz}{(z^2+1)(z^2-4)}$, where c is the circle $|z| = 1.5$ **Ans.** 0

7. $\int_c \frac{4z^2-4z+1}{(z-2)(z^2+4)} dz$, where c is the circle $|z| = 1$ **Ans.** 0

8. $\int_c \frac{\sin z}{z^6} dz$, where c is the circle $|z| = 2$ **Ans.** $\frac{\pi i}{60}$

9. Let $\left[\frac{P(z)}{Q(z)}\right]$, where both $P(z)$ and $Q(z)$ are complex polynomial of degree two. If $f(0) = f(-1) = 0$ and only singularity of $f(z)$ is of order 2 at $z = 1$ with residue -1, then find $f(z)$. **Ans.** $f(z) = -\frac{1}{3}\frac{z(z+1)}{(z-1)^2}$

31.13 EVALUATION OF REAL DEFINITE INTEGRALS BY CONTOUR INTEGRATION

A large number of real definite integrals, whose evaluation by usual methods become sometimes very tedious, can be easily evaluated by using Cauchy's theorem of residues. For finding we take a closed curve C, find the poles of the function $f(z)$ and calculate residues at those poles only which lie within the curve C.

$$\int_C f(z)\,dz = 2\pi i \text{ (sum of the residues of } f(z) \text{ at the poles within } C)$$

We call the curve, a contour and the process of integration along a contour is called contour integration.

31.14 INTEGRATION ROUND UNIT CIRCLE OF THE TYPE

$$\int_0^{2\pi} f(\cos\theta, \sin\theta)\,d\theta$$

where $f(\cos\theta, \sin\theta)$ is a rational function of $\cos\theta$ and $\sin\theta$.

Convert $\sin\theta$, $\cos\theta$ into z.

Consider a circle of unit radius with centre at origin, as contour.

$$\sin\theta = \frac{e^{i\theta} - e^{-i\theta}}{2i} = \frac{1}{2i}\left[z - \frac{1}{z}\right], \qquad z = re^{i\theta} = 1.\,e^{i\theta} = e^{i\theta}$$

$$\cos\theta = \frac{e^{i\theta} + e^{-i\theta}}{2} = \frac{1}{2}\left[z + \frac{1}{z}\right]$$

As we know

$$z = e^{i\theta} \Rightarrow dz = e^{i\theta}\, i\,d\theta = z\,i\,d\theta \text{ or } d\theta = \frac{dz}{iz}$$

The integrand is converted into a function of z.

Then apply Cauchy's residue theorem to evaluate the integral.

Some examples of these are illustrated below.

Example 38. *Evaluate the integral:*

$$\int_0^{2\pi} \frac{d\theta}{5 - 3\cos\theta}$$

(R.G.P.V., Bhopal, III Semester, June 2007)

Solution.
$$\int_0^{2\pi} \frac{d\theta}{5 - 3\cos\theta} = \int_0^{2\pi} \frac{d\theta}{5 - 3\left(\frac{e^{i\theta} + e^{-i\theta}}{2}\right)} \qquad \left[\begin{array}{l} e^{i\theta} = z \Rightarrow i.e.^{i\theta}\,d\theta = dz \\ d\theta = \dfrac{dz}{iz} \end{array}\right]$$

$$= \int_0^{2\pi} \frac{2\,d\theta}{10 - 3e^{i\theta} - 3e^{-i\theta}}$$

$$= \int_C \frac{2}{10 - 3z - \frac{3}{z}} \frac{dz}{iz} = \frac{2}{i}\int_C \frac{dz}{10z - 3z^2 - 3}$$

[C is the unit circle $|z| = 1$]

$$= -\frac{2}{i}\int_C \frac{dz}{3z^2 - 10z + 3}$$

$$= -\frac{2}{i}\int_C \frac{dz}{(3z-1)(z-3)} = 2i\int_C \frac{dz}{(3z-1)(z-3)}$$

Let $\quad I = 2i\displaystyle\int_C \frac{dz}{(3z-1)(z-3)}$

Poles of the integrand are given by

$$(3z-1)(z-3) = 0 \quad \Rightarrow \quad z = \frac{1}{3}, 3$$

There is only one pole at $z = \frac{1}{3}$ inside the unit circle C.

$$\text{Residue at } z = \frac{1}{3} \quad = \lim_{z \to \frac{1}{3}} \left(z - \frac{1}{3}\right) f(z) = \lim_{z \to \frac{1}{3}} \frac{2i\left(z - \frac{1}{3}\right)}{(3z-1)(z-3)} = \lim_{z \to \frac{1}{3}} \frac{2i}{3(z-3)}$$

$$= \frac{2i}{3\left(\frac{1}{3} - 3\right)} = -\frac{i}{4}$$

Hence, by Cauchy's Residue Theorem

$$I = 2\pi i \text{ (Sum of the residues within Contour)} = 2\pi i\left(-\frac{i}{4}\right) = \frac{\pi}{2}$$

$$\int_0^{2\pi} \frac{d\theta}{5 - 3\cos\theta} = \frac{\pi}{2}$$ **Ans.**

Example 39. *Use residue calculus to evaluate the following integral*

$$\int_0^{2\pi} \frac{1}{5 - 4\sin\theta} d\theta$$

Solution. Let $\quad I = \displaystyle\int_0^{2\pi} \frac{1}{5 - 4\sin\theta}\, d\theta = \int_0^{2\pi} \frac{1}{5 - 4\left(\frac{e^{i\theta} - e^{-i\theta}}{2i}\right)} d\theta$

$$= \int_0^{2\pi} \frac{d\theta}{5 + 2ie^{i\theta} - 2ie^{-i\theta}} \qquad \left[\text{writing } e^{i\theta} = z,\ d\theta = \frac{dz}{iz}\right]$$

$$= \int_c \frac{1}{5 + 2iz - \frac{2i}{z}} \frac{dz}{iz} \qquad \text{where } c \text{ is the unit circle } |z| = 1.$$

$$= \int_c \frac{dz}{5iz - 2z^2 + 2}$$

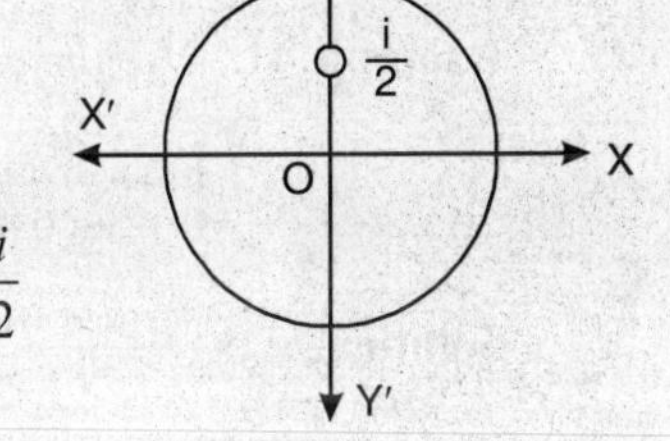

Poles of integrand are given by

$$-2z^2 + 5iz + 2 = 0 \Rightarrow z = \frac{-5i \pm \sqrt{-25+16}}{-4} = \frac{-5i \pm 3i}{-4} = 2i,\ \frac{i}{2}$$

Only $z = \frac{i}{2}$ lies inside c.

Residue at the simple pole at $z = \frac{i}{2}$ is

$$\lim_{z \to \frac{i}{2}} \left(z - \frac{i}{2}\right) \times \left[\frac{1}{(2z-i)(-z+2i)}\right] = \lim_{z \to \frac{i}{2}} \frac{1}{2(-z+2i)} = \frac{1}{2\left(-\frac{i}{2} + 2i\right)} = \frac{1}{3i}$$

Hence, by Cauchy's residue theorem

$$I = 2\pi i \times \text{ Sum of residues within the contour} = 2\pi i \times \frac{1}{3i} = \frac{2\pi}{3}$$

Hence, given integral $= \frac{2\pi}{3}$ **Ans.**

Example 40. *Evaluate* $\displaystyle\int_0^{2\pi} \frac{d\theta}{a + b\sin\theta}$ if $a > |b|$

(GBTU, III Sem. April 2012, U.P. III Semester 2009-2010)

Solution. Let $\quad I = \displaystyle\int_0^{2\pi} \frac{d\theta}{a + b\sin\theta}$

$$= \int_0^{2\pi} \frac{1}{a + b\,\frac{e^{i\theta} - e^{-i\theta}}{2i}}\, d\theta \qquad \left[\text{Writing } e^{i\theta} = z,\ d\theta = \frac{dz}{iz}\right]$$

$$= \int_C \frac{1}{a + \frac{b}{2i}\left(z - \frac{1}{z}\right)} \frac{dz}{iz} \quad \text{(where } C \text{ is the unit circle } |z| = 1$$

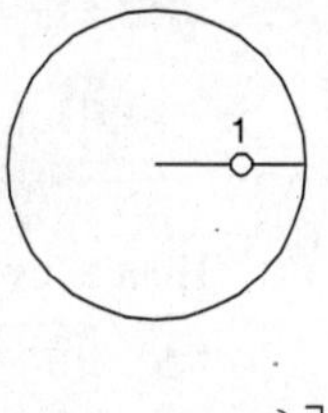

$$= \int_c \frac{2}{bz^2 + 2aiz - b}\, dz$$

$$= \frac{1}{b} \int \frac{2dz}{z^2 + \frac{2aiz}{b} - 1}$$

$$= \frac{1}{b} \int_c \frac{2}{(z-\alpha)(z-\beta)}\, dz \qquad \left[bz^2 + 2aiz - b = b\left\{z^2 + \frac{2aiz}{b} - 1\right\}\right]$$

Where $\alpha + \beta = -\frac{2ai}{b}$

$\alpha\beta = -1 \quad \Rightarrow \quad |\alpha\beta| = 1$

$$\left[\begin{aligned}(\alpha-\beta)^2 &= (\alpha+\beta)^2 - 4\alpha\beta \\ &= -\frac{4a^2}{b^2} + 4 \\ \alpha - \beta &= 2\,\frac{\sqrt{b^2 - a^2}}{b}\end{aligned}\right]$$

$|\alpha| < 1$ then $|\beta| > 1$

i.e.; Pole lies at $z = \alpha$ in the unit circle.

$$\text{Residue at } z = a = \lim_{z \to \alpha} (z-\alpha) \frac{2}{(z-\alpha)(z-\beta)} = \frac{2}{\alpha - \beta} = \frac{b}{\sqrt{b^2 - a^2}} = \frac{b}{i\sqrt{a^2 - b^2}}$$

$$\int_0^{2\pi} \frac{1}{a + b\sin\theta}\, d\theta = \frac{1}{b} \int_c \frac{2}{z^2 + 2\frac{aiz}{b} - 1}\, dz = 2\pi i \frac{b}{bi\sqrt{a^2 - b^2}} = \frac{2\pi}{\sqrt{a^2 - b^2}} \qquad \textbf{Ans.}$$

Example 41. *Use the complex variable technique to find the value of the integral :*

$$\int_0^{2\pi} \frac{d\theta}{2 + \cos\theta}.$$

(R.G.P.V., Bhopal, III Semester, Dec. 2003)

Solution. Let $I = \int_0^{2\pi} \frac{d\theta}{2+\cos\theta} = \int_0^{2\pi} \frac{d\theta}{2 + \frac{e^{i\theta} + e^{-i\theta}}{2}} = \int_0^{2\pi} \frac{2d\theta}{4 + e^{i\theta} + e^{-i\theta}}$

Put $e^{i\theta} = z$ so that $e^{i\theta}(i\,d\theta) = dz \Rightarrow iz\,d\theta = dz \Rightarrow d\theta = \frac{dz}{iz}$

$$I = \int_C \frac{2\frac{dz}{iz}}{4 + z + \frac{1}{z}} \qquad \text{where } c \text{ denotes the unit circle } |z| = 1.$$

$$= \frac{1}{i} \int_c \frac{2\,dz}{z^2 + 4z + 1}$$

The poles are given by putting the denominator equal to zero.

$$z^2 + 4z + 1 = 0 \text{ or } z = \frac{-4 \pm \sqrt{16-4}}{2} = \frac{-4 \pm 2\sqrt{3}}{2} = -2 \pm \sqrt{3}$$

The pole within the unit circle C is a simple pole at $z = -2 + \sqrt{3}$. Now we calculate the residue at this pole.

$$\text{Residue at } (z=-2+\sqrt{3}) = \lim_{z\to(-2+\sqrt{3})} \frac{1}{i}\,\frac{(z+2-\sqrt{3})2}{(z+2-\sqrt{3})(z+2+\sqrt{3})}$$

$$= \lim_{z\to(-2+\sqrt{3})} \frac{2}{i(z+2+\sqrt{3})} = \frac{2}{i(-2+\sqrt{3}+2+\sqrt{3})} = \frac{1}{\sqrt{3}i}$$

Hence by Cauchy's Residue Theorem, we have

$$\int_0^{2\pi} \frac{d\theta}{2+\cos\theta} = 2\pi i \text{ (sum of the residues within the contour)}$$

$$= 2\pi i\,\frac{1}{i\sqrt{3}} = \frac{2\pi}{\sqrt{3}}$$ **Ans.**

Example 42. *Using complex variable techniques evaluate the real integral*

$$\int_0^{2\pi} \frac{\sin^2\theta}{5-4\cos\theta}\,d\theta$$

Solution. If we write $z = e^{i\theta}$

$$\cos\theta = \frac{1}{2}\left(z+\frac{1}{z}\right),\quad d\theta = \frac{dz}{iz}$$

and so $$I = \int_0^{2\pi} \frac{\sin^2\theta}{5-4\cos\theta}\,d\theta = \frac{1}{2}\int_0^{2\pi} \frac{1-\cos 2\theta}{5-4\cos\theta}\,d\theta$$

$$\text{I} = \text{Real part of } \frac{1}{2}\int_0^{2\pi} \frac{1-\cos 2\theta - i\sin 2\theta}{5-4\cos\theta}\,d\theta \qquad \left[\begin{array}{l}\text{where } C \text{ is a circle of unit}\\ \text{radius with centre z} = 0\end{array}\right]$$

$$= \text{Real part of } \frac{1}{2}\int_0^{2\pi} \frac{1-e^{2i\theta}}{5-4\cos\theta}\,d\theta$$

$$= \text{Real part of } \frac{1}{2}\int_c \frac{1-z^2}{5-2(z+\frac{1}{z})}\left(\frac{dz}{iz}\right) = \text{Real part of } \frac{1}{2i}\int_c \frac{1-z^2}{5z-2z^2-2}\,dz$$

$$= \text{Real part of } \frac{1}{2i}\int_c \frac{z^2-1}{2z^2-5z+2}\,dz$$

Poles are determined by $2z^2-5z+2=0$ or $(2z-1)(z-2)=0$ or $z = \frac{1}{2}, 2$

So inside the contour C there is a simple pole at $z = \frac{1}{2}$

$$\text{Residue at the simple pole } \left(z=\frac{1}{2}\right) = \lim_{z\to\frac{1}{2}}\left(z-\frac{1}{2}\right)\frac{z^2-1}{(2z-1)(z-2)}$$

$$= \lim_{z\to\frac{1}{2}} \frac{z^2-1}{2(z-2)} = \frac{\frac{1}{4}-1}{2\left(\frac{1}{2}-2\right)} = \frac{1}{4}$$

$$I = \text{Real part of } \frac{1}{2i}\int_c \frac{(z^2-1)}{2z^2-5z+2}\,dz = \frac{1}{2i}2\pi i \text{ (sum of the residues)}$$

$$\Rightarrow \int_0^{2\pi} \frac{\sin^2\theta}{5-4\cos\theta}\,d\theta = \pi\left(\frac{1}{4}\right) = \frac{\pi}{4}$$ **Ans.**

Example 43. *Using contour integration, evaluate the real integral*

$$\int_0^\pi \frac{1+2\cos\theta}{5+4\cos\theta}\,d\theta$$ *(R.G.P.V., Bhopal, III Semester, Dec. 2004)*

Solution. Let $I = \int_0^\pi \frac{1+2\cos\theta}{5+4\cos\theta}\,d\theta = \frac{1}{2}\int_0^{2\pi} \frac{1+2\cos\theta}{5+4\cos\theta}\,d\theta$ [Even function]

$$= \text{Real part of } \frac{1}{2}\int_0^{2\pi} \frac{1+2e^{i\theta}}{5+4\cos\theta}\,d\theta$$

$$= \text{Real part of } \frac{1}{2}\int_0^{2\pi} \frac{1+2e^{i\theta}}{5+2(e^{i\theta}+e^{-i\theta})}\,d\theta$$

writing $e^{i\theta} = z,\ d\theta = \frac{dz}{iz}$ where C is the unit circle $|z| = 1$.

$$= \text{Real part of } \frac{1}{2}\int_C \frac{1+2z}{5+2\left(z+\frac{1}{z}\right)}\frac{dz}{iz}, = \text{Real part of } \frac{1}{2}\int_C \frac{-i(1+2z)}{2z^2+5z+2}\,dz$$

$$= \text{Real part of } \frac{1}{2}\int_C \frac{-i(2z+1)}{(2z+1)(z+2)}\,dz = \text{Real part of } -\frac{i}{2}\int_C \frac{1}{z+2}\,dz$$

Pole is given by $z + 2 = 0$ *i.e.* $z = -2$.

Thus there is no pole of $f(z)$ inside the unit circle C. Hence $f(z)$ is analytic in C.

By Cauchy's Theorem $\int_C f(z)\,dz = 0$ if $f(z)$ is analytic in C.

$\therefore$ I = Real part of zero = 0

Hence, the given integral = 0 **Ans.**

Example 44. *Using complex variables, evaluate the real integral*

$$\int_0^{2\pi} \frac{d\theta}{1-2p\sin\theta+p^2},\ \text{where } p^2<1.$$

Solution. $\int_0^{2\pi} \frac{d\theta}{1-2p\sin\theta+p^2} = \int_0^{2\pi} \frac{d\theta}{1-2p\frac{(e^{i\theta}-e^{-i\theta})}{2i}+p^2}$

Let $I = \int_0^{2\pi} \frac{d\theta}{1+ip(e^{i\theta}-e^{-i\theta})+p^2}$

Writing $z = e^{i\theta},\ dz = ie^{i\theta}d\theta = i\,z\,d\theta,\ d\theta = \frac{dz}{z\,i}$

$$I = \int_C \frac{1}{1+i\,p\left(z-\frac{1}{z}\right)+p^2}\frac{dz}{z\,i}$$ where c is the unit circle $|z| = 1$.

$$= \int_C \frac{dz}{zi-pz^2+p+p^2zi} = \int_C \frac{dz}{-pz^2+ip^2z+zi+p} = \int_C \frac{dz}{(iz+p)(izp+1)}$$

Poles are given by $(iz+p)(ipz+1) = 0$

$\Rightarrow$ $z = -\frac{p}{i} = ip$ and $z = -\frac{1}{pi} = \frac{i}{p}$ $|ip|<1$ and $\left|\frac{i}{p}\right|>1$ as $p^2<1$

pi is the only pole inside the unit circle.

$$\text{Residue } (z = pi) = \lim_{z\to pi}\frac{(z-pi)}{(iz+p)(i\,z\,p+1)} = \lim_{z\to pi}\left[\frac{1}{i(i\,z\,p+1)}\right] = \frac{1}{i}\frac{1}{(-p^2+1)}$$

Hence by Cauchy's residue theorem

$$\int_0^{2\pi} \frac{d\theta}{1-2p\sin\theta+p^2} = 2\pi i\left(\frac{1}{i}\frac{1}{1-p^2}\right) = \frac{2\pi}{1-p^2}$$ **Ans.**

Example 45. *Apply calculus of residue to prove that :*

$$\int_0^{2\pi} \frac{\cos 2\theta\, d\theta}{1-2a\cos\theta+a^2} = \frac{2\pi a^2}{1-a^2}, \qquad (a^2 < 1)$$

(*R.G.P.V., Bhopal, III Semester, June 2003*)

Solution. Let $I = \int_0^{2\pi} \frac{\cos 2\theta\, d\theta}{1-2a\cos\theta+a^2} = \int_0^{2\pi} \frac{\cos 2\theta\, d\theta}{1-a\,(e^{i\theta}+e^{-i\theta})+a^2}$

$$= \text{Real part of } \int_0^{2\pi} \frac{e^{2i\theta}}{(1-ae^{i\theta})\,(1-ae^{-i\theta})}\,d\theta$$

$$= \text{Real part of } \oint_C \frac{z^2}{(1-az)\left(1-\frac{a}{z}\right)}\frac{dz}{iz} \qquad \left[\text{Put } e^{i\theta} = z \text{ so that } d\theta = \frac{dz}{iz}\right]$$

$$= \text{Real part of } \oint_C \frac{-iz^2}{(1-az)\,(z-a)}\,dz \qquad [C \text{ is the unit circle } |z| = 1]$$

Poles of $\frac{-iz^2}{(1-az)\,(z-a)}$ are given by

$$(1-az)\,(z-a) = 0$$

Thus, $z = \frac{1}{a}$ and $z = a$ are the simple poles. Only $z = a$ lies within the unit circle C as $a < 1$.

The residue of $f(z)$ at $(z = a)$ $= \lim_{z\to a} (z-a)\frac{-iz^2}{(1-az)\,(z-a)} = \lim_{z\to a} \frac{-iz^2}{(1-az)} = -\frac{ia^2}{1-a^2}$

Hence, by Cauchy's Residue Theorem, we have

$\oint_C f(z)\,dz = 2\pi i$ [Sum of residues within the contour]

$$= 2\pi i\left(-\frac{ia^2}{1-a^2}\right) = \frac{2\pi a^2}{1-a^2} \quad \text{which is purely real.}$$

Thus, $I = \text{Real part of } \oint_C f(z)\,dz = \frac{2\pi a^2}{1-a^2}$

Hence, $\int_0^{2\pi} \frac{\cos 2\theta}{1-2a\cos\theta+a^2} = \frac{2\pi a^2}{1-a^2}.$ **Proved.**

Example 46. *Using complex variable techniques, evaluate the integral*

$$\int_0^{2\pi} \frac{\sin^2\theta - 2\cos\theta}{2+\cos\theta}\,d\theta.$$

Solution. $\int_0^{2\pi} \frac{\sin^2\theta - 2\cos\theta}{2+\cos\theta}\,d\theta = \int_0^{2\pi} \frac{\frac{1}{2}-\frac{1}{2}\cos 2\theta - 2\cos\theta}{2+\cos\theta}\,d\theta$

$$= \frac{1}{2}\int_0^{2\pi} \frac{1-\cos 2\theta - 4\cos\theta}{2+\cos\theta}\,d\theta = \text{Real part of } \frac{1}{2}\int_0^{2\pi} \frac{1-e^{2i\theta}-4e^{i\theta}}{2+\cos\theta}\,d\theta$$

write $e^{i\theta} = z$ so that $i e^{i\theta} d\theta = dz$ or $i z d\theta = dz$ or $d\theta = \dfrac{dz}{iz}$

$$= \text{Real part of } \frac{1}{2}\int_c \frac{1 - z^2 - 4z}{2 + \frac{1}{2}\left(z + \frac{1}{z}\right)} \frac{dz}{iz} = \text{Real part of } \frac{1}{i}\int_c \frac{(1 - z^2 - 4z)dz}{4z + z^2 + 1}$$

The poles are given by $z^2 + 4z + 1 = 0$

$$z = \frac{-4 \pm \sqrt{16 - 4}}{2} = -2 \pm \sqrt{3}$$

The pole within the unit circle C is $-2 + \sqrt{3}$

Residue at the simple pole $z = -2 + \sqrt{3}$

$$= \lim_{z \to -2+\sqrt{3}} (z + 2 - \sqrt{3}) \frac{1 - z^2 - 4z}{(z + 2 - \sqrt{3})(z + 2 + \sqrt{3})} = \lim_{z \to -2+\sqrt{3}} \frac{1 - z^2 - 4z}{z + 2 + \sqrt{3}}$$

$$= \frac{1 - (-2 + \sqrt{3})^2 - 4(-2 + \sqrt{3})}{(-2 + \sqrt{3}) + 2 + \sqrt{3}} = \frac{1}{\sqrt{3}}$$

$$\text{Real part of } \frac{1}{i}\int_c \frac{(1 - z^2 - 4z)dz}{4z + z^2 + 1} = \text{Real part of } \left(\frac{1}{i}\right) 2\pi i \text{ (Residue)}$$

$$= \text{Real part of } 2\pi\left(\frac{1}{\sqrt{3}}\right) \quad \text{or} \quad I = \frac{2\pi}{\sqrt{3}}$$

Hence, the given integral $= \dfrac{2\pi}{\sqrt{3}}$. **Ans.**

Example 47. *Evaluate:* $\displaystyle\int_0^{2\pi} \frac{\cos 2\theta}{5 + 4\cos\theta}\, d\theta$ *by using contour integration.*

(R.G.P.V., Bhopal, III Semester, June 2007)

Solution.

$$\left[\begin{array}{l} e^{i\theta} = z \\ \Rightarrow \; i.e.^{i\theta} d\theta = dz \\ \Rightarrow d\theta = \dfrac{dz}{ie^{i\theta}} = \dfrac{dz}{iz} \end{array}\right]$$

[C is the unit circle $|z| = 1$]

Let $$I = \int_0^{2\pi} \frac{\cos 2\theta}{5 + 4\cos\theta}\, d\theta$$

$$= \text{Real part of } \int_0^{2\pi} \frac{\cos 2\theta + i\sin 2\theta}{5 + 4\cos\theta}\, d\theta$$

$$= \text{Real part of } \int_0^{2\pi} \frac{e^{2i\theta}}{5 + 2(e^{i\theta} + e^{-i\theta})}\, d\theta$$

$$= \text{Real part of } \oint_C \frac{z^2}{5 + 2\left(z + \frac{1}{z}\right)} \frac{dz}{iz}$$

$$= \text{Real part of } \oint_C \frac{z^2}{5z + 2z^2 + 2} \frac{dz}{i}$$

$$= \text{Real part of } \oint_C \frac{-iz^2}{2z^2 + 5z + 2}\, dz$$

$$= \text{Real part of } \oint_C \frac{-iz^2}{(2z + 1)(z + 2)}\, dz$$

Poles are determined by putting denominator equal to zero.

$$(2z+1)(z+2)=0 \quad \Rightarrow \quad z=-\frac{1}{2},\,-2$$

The only simple pole at $z=-\frac{1}{2}$ is inside the contour.

$$\text{Residue at } \left(z=-\frac{1}{2}\right) = \lim_{z\to-\frac{1}{2}}\left(z+\frac{1}{2}\right)f(z) = \lim_{z\to-\frac{1}{2}}\left(z+\frac{1}{2}\right)\frac{-iz^2}{(2z+1)(z+2)}$$

$$= \lim_{z\to-\frac{1}{2}}\frac{-iz^2}{2(z+2)} = \frac{-i\left(-\frac{1}{2}\right)^2}{2\left(-\frac{1}{2}+2\right)} = \frac{-i}{12}$$

By Cauchy's Integral Theorem

$$\int_C f(z)\,dz = 2\pi i \text{ (Sum of the residues within } C\text{)}$$

$$= 2\pi i\left(\frac{-i}{12}\right) = \frac{\pi}{6}, \text{ which is real}$$

$$\int_0^{2\pi}\frac{\cos 2\theta}{5+4\cos\theta}\,d\theta = \frac{\pi}{6}$$ **Ans.**

Example 48. *Evaluate contour integration of the real integral*

$$\int_0^{2\pi}\frac{\cos 3\theta}{5-4\cos\theta}\,d\theta.$$ *(U.P., III Sem., 2009, R.G.P.V., Bhopal, III Semester, Dec. 2007)*

Solution. $\displaystyle\int_0^{2\pi}\frac{\cos 3\theta}{5-4\cos\theta}\,d\theta$ = Real part of $\displaystyle\int_0^{2\pi}\frac{e^{3i\theta}}{5-4\cos\theta}\,d\theta$

= Real part of $\displaystyle\int_0^{2\pi}\frac{e^{3i\theta}}{5-2(e^{i\theta}+e^{-i\theta})}\,d\theta$ On writing $z=e^{i\theta}$ and $d\theta=\frac{dz}{iz}$

= Real part of $\displaystyle\int_c \frac{z^3}{5-2\left(z+\frac{1}{z}\right)}\frac{dz}{iz}$ c is the unit circle.

= Real part of $\displaystyle\frac{1}{i}\int_c\frac{z^3}{5z-2z^2-2}\,dz$ = Real part of $\displaystyle -\frac{1}{i}\int\frac{z^3}{2z^2-5z+2}\,dz$

= Real part of $\displaystyle i\int\frac{z^3}{(2z-1)(z-2)}\,dz$

Poles are given by $(2z-1)(z-2)=0$ *i.e.* $z=\frac{1}{2},\ z=2$

$z=\frac{1}{2}$ is the only pole inside the unit circle.

$$\text{Residue }\left(\text{at } z=\frac{1}{2}\right) = \lim_{z\to\frac{1}{2}}\frac{i\left(z-\frac{1}{2}\right)z^3}{(2z-1)(z-2)}$$

$$= \lim_{z\to\frac{1}{2}}\frac{iz^3}{2(z-2)} = \frac{i\frac{1}{8}}{2\left(\frac{1}{2}-2\right)} = -\frac{i}{24}$$

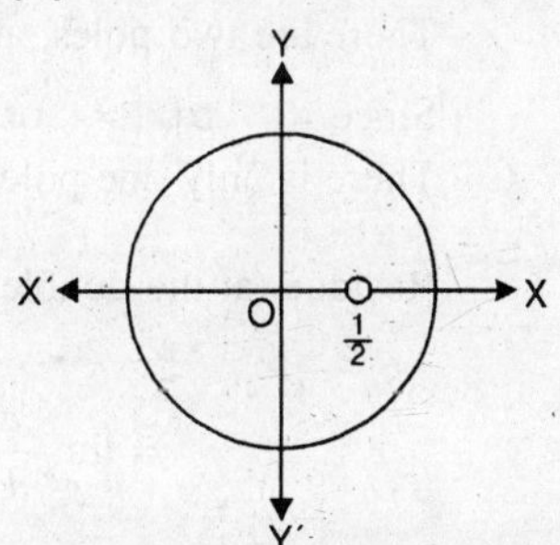

$$\int_0^{2\pi} \frac{\cos 3\theta}{5-4\cos\theta}\,d\theta = \text{Real part of } 2\pi i\left(-\frac{i}{24}\right) = \frac{\pi}{12}$$ **Ans.**

Question. *Evaluate* : $\int_0^{\infty} \frac{\cos 3\theta}{5+4\cos\theta}\,d\theta$ *(U.P. III Semester, Dec. 2008, 2006)*

Example 49. *Use the residue theorem to show that*

$$\int_0^{2\pi} \frac{d\theta}{(a+b\cos\theta)^2} = \frac{2\pi a}{(a^2-b^2)^{3/2}} \qquad \textit{where } a>0,\, b>0,\, a>b.$$

(R.G.P.V., Bhopal, III Semester, June 2004)

Solution. $\int_0^{2\pi} \frac{d\theta}{(a+b\cos\theta)^2} = \int_0^{2\pi} \frac{d\theta}{\left(a+b\cdot\frac{e^{i\theta}+e^{-i\theta}}{2}\right)^2}$

Put $e^{i\theta} = z$, so that $e^{i\theta}(id\theta) = dz \Rightarrow izd\theta = dz \Rightarrow d\theta = \frac{dz}{iz}$

$$= \int_c \frac{1}{\left\{a+\frac{b}{2}\left(z+\frac{1}{z}\right)\right\}^2}\frac{dz}{iz} \qquad \text{where } C \text{ is the unit circle } |z| = 1.$$

$$\int_c \frac{1}{\left(a+\frac{bz}{2}+\frac{b}{2z}\right)^2}\frac{dz}{iz} = \int_c \frac{-4iz}{\left(a+\frac{bz}{2}+\frac{b}{2z}\right)^2}\frac{dz}{(2z)^2}$$

$$= \int_c \frac{-4izdz}{(bz^2+2az+b)^2} = \frac{-4i}{b^2}\int_c \frac{z\,dz}{\left(z^2+\frac{2az}{b}+1\right)^2}$$

The poles are given by putting the denominator equal to zero.

i.e., $$\left(z^2+\frac{2a}{b}z+1\right)^2 = 0$$

$\Rightarrow$ $$(z-\alpha)^2(z-\beta)^2 = 0$$

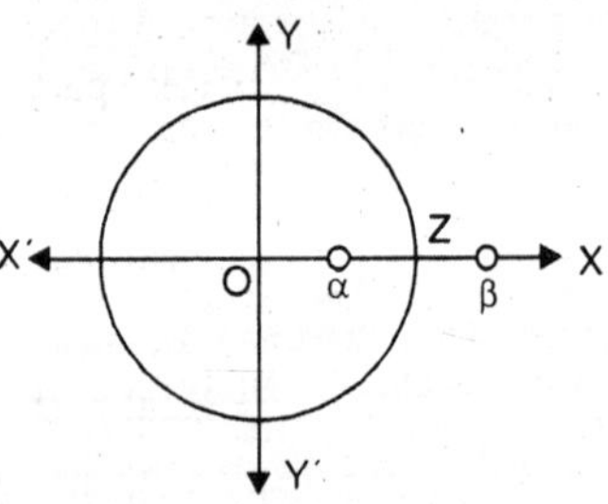

where $$\alpha = \frac{-\frac{2a}{b}+\sqrt{\frac{4a^2}{b^2}-4}}{2} = \frac{-a+\sqrt{a^2-b^2}}{b}$$

$$\beta = \frac{-\frac{2a}{b}-\sqrt{\frac{4a^2}{b^2}-4}}{2} = \frac{-a-\sqrt{a^2-b^2}}{b}$$

There are two poles, at $z = \alpha$ and at $z = \beta$, each of order 2.

Since $|\alpha\beta| = 1$ or $|\alpha|\,|\beta| = 1$ if $|\alpha| < 1$ then $|\beta| > 1$.

There is only one pole $[|\alpha| < 1]$ of order 2 within the unit circle c.

Residue (at the double pole $z = \alpha$) $= \lim_{z\to\alpha} \frac{d}{dz}(z-\alpha)^2 \frac{(-4iz)}{b^2(z-\alpha)^2(z-\beta)^2}$

$$= \lim_{z\to\alpha} \frac{d}{dz}\frac{-4iz}{b^2(z-\beta)^2}$$

$$= -\frac{4i}{b^2}\lim_{z\to\alpha}\frac{(z-\beta)^2.1-2(z-\beta)z}{(z-\beta)^4} = \frac{-4i}{b^2}\lim_{z\to\alpha}\frac{z-\beta-2z}{(z-\beta)^3} = \frac{-4i}{b^2}\lim_{z\to\alpha}\frac{-(z+\beta)}{(z-\beta)^3}$$

$$= \frac{4i}{b^2}\frac{(\alpha+\beta)}{(\alpha-\beta)^3} = \frac{4i}{b^2}\frac{\alpha+\beta}{[(\alpha+\beta)^2-4\alpha\beta]^{\frac{3}{2}}} = \frac{4i}{b^2}\frac{\frac{-2a}{b}}{\left[\left(-\frac{2a}{b}\right)^2-4(1)\right]^{\frac{3}{2}}}$$

$$= \frac{-8ai}{(4a^2-4b^2)^{\frac{3}{2}}} = -\frac{ai}{(a^2-b^2)^{\frac{3}{2}}}$$

Hence, $\displaystyle\int_0^{2\pi}\frac{d\theta}{(a+b\cos\theta)^2} = 2\pi i\times\frac{-ai}{(a^2-b^2)^{3/2}} = \frac{2\pi a}{(a^2-b^2)^{3/2}}$ **Proved.**

Example 50. *Show by the method of residues, that*

$$\int_0^{\pi}\frac{a\,d\theta}{a^2+\sin^2\theta} = \frac{\pi}{\sqrt{1+a^2}}$$

Solution. Let $\displaystyle I = \int_0^{\pi}\frac{a\,d\theta}{a^2+\sin^2\theta} = \int_0^{\pi}\frac{2a\,d\theta}{2a^2+2\sin^2\theta}$ $\quad(\cos 2\theta = 1-2\sin^2\theta)$

$$= \int_0^{\pi}\frac{2a\,d\theta}{2a^2+1-\cos 2\theta} = \int_0^{2\pi}\frac{a\,d\phi}{2a^2+1-\cos\phi} \qquad [\text{Putting } 2\theta=\phi,\ 2d\theta=d\phi]$$

$$= \int_0^{2\pi}\frac{a\,d\phi}{2a^2+1-\frac{1}{2}(e^{i\phi}+e^{-i\phi})} = \int_0^{2\pi}\frac{2a\,d\phi}{4a^2+2-(e^{i\phi}+e^{-i\phi})}$$

Writing $e^{i\phi} = z$, $e^{i\phi}(id\phi) = dz$ or $z(id\phi) = dz$, $d\phi = \dfrac{dz}{iz}$

$$= \int_C \frac{2a}{4a^2+2-\left(z+\frac{1}{z}\right)}\cdot\frac{dz}{iz} \qquad \text{where } C \text{ is unit circle } |z| = 1$$

$$= \frac{2a}{i}\int_C\frac{dz}{(4a^2+2)z-z^2-1} = \frac{2a}{-i}\int_c\frac{dz}{z^2-(4a^2+2)z+1}$$

$$= 2ai\int_C\frac{dz}{z^2-(4a^2+2)z+1}$$

The poles are given by $z^2-(4a^2+2)z+1 = 0$

$$\Rightarrow \quad z = \frac{(4a^2+2)\pm\sqrt{(4a^2+2)^2-4}}{2} = \frac{(4a^2+2)\pm\sqrt{16a^4+16a^2}}{2}$$

$$= 2a^2+1\pm 2a\sqrt{a^2+1}$$

Let $\alpha = 2a^2+1+2a\sqrt{a^2+1}$

$\beta = 2a^2+1-2a\sqrt{a^2+1}$

$$z^2-(4a^2+2)z+1 = (z-\alpha)(z-\beta)$$

$$I = 2ai\int\frac{dz}{(z-\alpha)(z-\beta)}$$

Product of the roots $= \alpha\beta = 1$ or $|\alpha\beta| = 1$

But $\quad |\alpha| > 1 \therefore |\beta| < 1$

Only β lies inside the circle c.

Now we calculate the residue at $z = \beta$.

$$\text{Residue (at } z=\beta\text{) is} = \lim_{z\to\beta}(z-\beta)\frac{2ai}{(z-\alpha)(z-\beta)} = \lim_{z\to\beta}\frac{2ai}{(z-\alpha)}$$

$$= \frac{2ai}{(\beta-\alpha)} = \frac{2ai}{(2a^2+1-2a\sqrt{a^2+1})-(2a^2+1+2a\sqrt{a^2+1})}$$

$$= \frac{2ai}{-4a\sqrt{a^2+1}} = -\frac{i}{2\sqrt{a^2+1}}$$

Hence by Cauchy's residue theorem

$$I = 2\pi i \text{ (sum of the residues within the contour } c)$$

$$= 2\pi i\frac{-i}{2\sqrt{a^2+1}} = \frac{\pi}{\sqrt{a^2+1}}$$

Hence; $\displaystyle\int_0^{\pi}\frac{a\,d\theta}{a^2+\sin^2\theta} = \frac{\pi}{\sqrt{1+a^2}}$ **Proved.**

Example 51. *Evaluate by Contour integration:*

$$\int_0^{2\pi} e^{\cos\theta}\cos(\sin\theta - n\theta)\,d\theta\cdot$$

Solution. Let $\quad I = \displaystyle\int_0^{2\pi} e^{\cos\theta}[\cos(\sin\theta - n\theta) + i\sin(\sin\theta - n\theta)]\,d\theta$

$$= \int_0^{2\pi} e^{\cos\theta}e^{i(\sin\theta - n\theta)}d\theta = \int_0^{2\pi} e^{\cos\theta + i\sin\theta}\cdot e^{-ni\theta}d\theta$$

$$= \int_0^{2\pi} e^{e^{i\theta}}\cdot e^{-in\theta}d\theta \quad \text{...(1)}$$

Put $e^{i\theta} = z$ so that $d\theta = \dfrac{dz}{iz}$ then,

$$\text{I} = \int_c e^z\cdot\frac{1}{z^n}\cdot\frac{dz}{iz} = -i\int_c\frac{e^z}{z^{n+1}}dz$$

Pole is at $z = 0$ of order $(n+1)$.
It lies inside the unit circle.
Residue of $f(z)$ at $z = 0$ is

$$\frac{1}{(n+1-1)!}\left[\frac{d^n}{dz^n}\left\{z^{n+1}\cdot\frac{-ie^z}{z^{n+1}}\right\}\right]_{z=0} = \frac{-i}{n!}\left[\frac{d^n}{dz^n}(e^z)\right]_{z=0} = \frac{-i}{n!}\left(e^z\right)_{z=0} = \frac{-i}{n!}$$

$\therefore$ By Cauchy's Residue theorem,

$$\text{I} = 2\pi i\left(\frac{-i}{n!}\right) = \frac{2\pi}{n!}$$

Comparing real part of $\displaystyle\int_0^{2\pi} e^{\cos\theta}[\cos(\sin\theta - n\theta) + i\sin(\sin\theta - n\theta)]\,d\theta = \frac{2\pi}{n!}$,

we have

$$\int_0^{2\pi} e^{\cos\theta}\cos(\sin\theta - n\theta)\,d\theta = \frac{2\pi}{n!}$$ **Ans.**

EXERCISE 31.4

Evaluate the following integrals:

1. $\int_0^{2\pi} \frac{\sin^2\theta}{a+b\cos\theta}\, d\theta$ (*R.G.P.V., Bhopal, III Semester, June 2008*) **Ans.** $\frac{2\pi}{b^2}\{a-\sqrt{a^2-b^2}\},\ a>b>0$

2. $\int_0^{2\pi} \frac{(1+2\cos\theta)^n \cos n\theta}{3+2\cos\theta}\, d\theta$ **Ans.** $\frac{2\pi}{\sqrt{5}}(3-\sqrt{5})^n,\ n>0$

3. $\int_0^{2\pi} \frac{4}{5+4\sin\theta}\, d\theta$ **Ans.** $\frac{8\pi}{3}$

4. $\int_0^{\pi} \frac{d\theta}{17-8\cos\theta}$ **Ans.** $\frac{\pi}{15}$

5. $\int_0^{\pi} \frac{d\theta}{a+b\cos\theta}$, *where* $a > |b|$. *Hence or otherwise evaluate* $\int_0^{2\pi} \frac{d\theta}{\sqrt{2}-\cos\theta}$. **Ans.** $\frac{\pi}{\sqrt{a^2-b^2}};\pi$

31.15 EVALUATION OF $\int_{-\infty}^{\infty} \frac{f_1(x)}{f_2(x)} dx$ where $f_1(x)$ and $f_2(x)$ are polynomials in x.

Such integrals can be reduced to contour integrals, if

(*i*) $f_2(x)$ has no real roots.

(*ii*) the degree of $f_2(x)$ is greater than that of $f_1(x)$ by at least two.

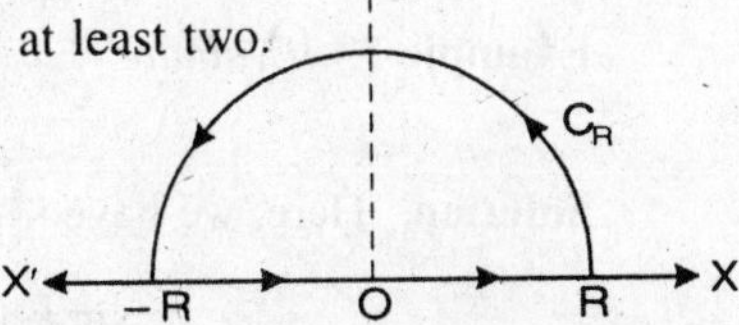

Procedure: Let $f(x) = \frac{f_1(x)}{f_2(x)}$

Consider $\int_C f(z)\,dz$

where C is a curve, consisting of the upper half C_R of the circle $|z| = R$, and part of the real axis from $-R$ to R.

If there are no poles of $f(z)$ on the real axis, the circle $|z| = R$ which is arbitrary can be taken such that there is no singularity on its circumference C_R in the upper half of the plane, but possibly some poles inside the contour C specified above.

Using Cauchy's theorem of residues, we have

$$\int_C f(z)\,dz = 2\pi i \times \text{ (sum of the residues of } f(z) \text{ at the poles within } C)$$

i.e. $$\int_{-R}^{R} f(x)\,dx + \int_{CR} f(z)\,dz = 2\pi i \text{ (sum of residues within } C)$$

$$\Rightarrow \int_{-R}^{R} f(x)\,dx = -\int_{CR} f(z)\,dz + 2\pi i \text{ (sum of residues within } C)$$

$$\therefore \lim_{R\to\infty}\int_{-R}^{R} f(x)\,dx = -\lim_{R\to\infty}\int_{CR} f(z)\,dz + 2\pi i \text{ (sum of residues within } C) \quad \text{... (1)}$$

Now, $$\lim_{R\to\infty}\int_{CR} f(z)\,dz = \int_0^{\pi} f(Re^{i\theta})Ri\,e^{i\theta}\,d\theta$$
$$= 0 \quad \text{when } R\to\infty$$

(1) reduces $$\int_{-\infty}^{\infty} f(x)dx = 2\pi i \text{ (sum of residues within } C)$$

Example 52. *Evaluate* $\int_0^{\infty} \frac{\cos mx}{(x^2+1)}dx$. (*R.G.P.V., Bhopal, III Semester, Dec. 2006*)

Solution. $\int_0^{\infty} \frac{\cos mx}{x^2+1}dx$

Consider the integral $\int_C f(z)dz$, where

$f(z) = \frac{e^{imz}}{z^2+1}$, taken round the closed contour c consisting of the upper half of a large circle $|z| = R$ and the real axis from $-R$ to R.

Poles of $f(z)$ are given by

$$z^2+1=0 \text{ i.e. } z^2=-1 \text{ i.e. } z=\pm i$$

The only pole which lies within the contour is at $z = i$.
The residue of $f(z)$ at $z = i$

$$=\lim_{z\to i}\frac{(z-i)\,e^{imz}}{(z^2+1)}=\lim_{z\to i}\frac{e^{imz}}{z+i}=\frac{e^{-m}}{2i}$$

Hence by Cauchy's residue theorem, we have

$$\int_C f(z)\,dz = 2\pi i\times \text{ sum of the residues}$$

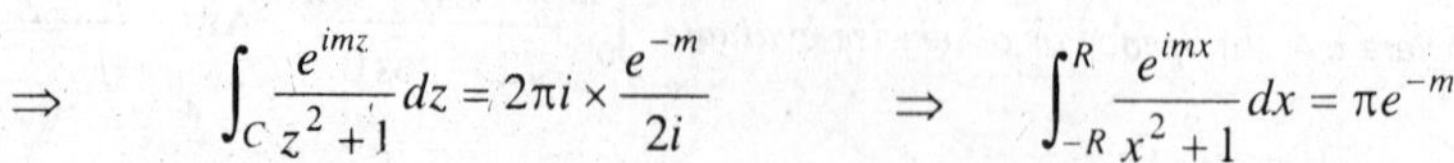

$$\Rightarrow \quad \int_C \frac{e^{imz}}{z^2+1}\,dz = 2\pi i\times\frac{e^{-m}}{2i} \quad\Rightarrow\quad \int_{-R}^{R}\frac{e^{imx}}{x^2+1}\,dx=\pi e^{-m}$$

Equating real parts, we have

$$\int_{-\infty}^{\infty}\frac{\cos mx}{x^2+1}\,dx=\pi e^{-m} \quad\Rightarrow\quad \int_0^{\infty}\frac{\cos mx}{x^2+1}\,dx=\frac{\pi e^{-m}}{2}$$ **Ans.**

Example 53. *Evaluate* $\int_{-\infty}^{\infty}\frac{x\sin\pi x}{x^2+2x+5}\,dx$ (*U.P. III Semester 2009-2010*)

Solution. Here, we have $\int_{-\infty}^{\infty}\frac{x\sin\pi x}{x^2+2x+5}\,dx$

Let us consider $\int \frac{z\sin\pi z}{z^2+2z+5}\,dz$

The pole can be determined by putting the denominator equal to zero.

$$z^2+2z+5=0 \quad\Rightarrow\quad z=\frac{-2\pm\sqrt{4-20}}{2} \quad\Rightarrow\quad z=-1\pm 2i$$

Out of two poles, only $z=-1+2i$ is inside the contour.
Residue at $z = -1+2i$

$$=\lim_{z\to -1+2i}(z+1-2i)\frac{z\sin\pi z}{z^2+2z+5}=\lim_{z\to -1+2i}(z+1-2i)\frac{z\sin\pi z}{(z+1-2i)(z+1+2i)}$$

$$=\lim_{z\to -1+2i}\frac{z\sin\pi z}{(z+1+2i)}=\frac{(-1+2i)\sin\pi(-1+2i)}{(-1+2i+1+2i)}$$

$$=\frac{(-1+2i)\sin\pi(-1+2i)}{4i}$$

$$\int_{-R}^{R}\frac{z\sin\pi z}{z^2+2z+5}\,dz = 2\pi i\text{ (Residue)}$$

$$=2\pi i\,\frac{(-1+2i)\sin\pi(-1+2i)}{4i}=\frac{\pi}{2}(2i-1)\sin(-\pi+2\pi i)$$

$$=\frac{\pi}{2}(2i-1)(-\sin 2\pi i) \qquad \left[\begin{aligned}\sin(-\pi+\theta)&=-\sin(\pi-\theta)\\&=-\sin\theta\end{aligned}\right.$$

$$=\frac{\pi}{2}(1-2i)\sin 2\pi i \qquad =\frac{\pi}{2}(1-2i)\,i\sinh 2\pi$$

$$=\frac{\pi}{2}(i+2)\sinh 2\pi \qquad \text{(Taking real parts)}$$

Hence $\int_{-\infty}^{\infty}\frac{x\sin\pi x}{x^2+2x+5}=\pi\sinh 2\pi$ **Ans.**

Example 54. *Use contour integration to evaluate the real integral* $\int_0^{\infty} \frac{dx}{(1+x^2)^3}$.

Solution. Consider $\int_C f(z)dz$, where $f(z) = \frac{1}{(z^2+1)^3}$ taken round the closed contour C consisting of real axis and upper half C_R of a large semi-circle $|z| = R$.

Poles of $f(z)$ are given by

$$(z^2+1)^3 = 0 \quad i.e. \quad (z-i)^3(z+i)^3 = 0$$

i.e. $z = \pm i$ are the poles each of order 3.

The only pole which lies within C is $z = i$ of order 3.

$\therefore$ Residue of $\frac{1}{(z-i)^3} \cdot \frac{1}{(z+i)^3}$ (at $z = i$)

$$= \frac{1}{2}\left[\frac{d^2}{dz^2}(z-i)^3 \cdot \frac{1}{(z-i)^3(z+i)^3}\right]_{z=i} = \frac{1}{2}\left[\frac{d^2}{dz^2}\frac{1}{(z+i)^3}\right]_{z=i} = \frac{1}{2}\left[\frac{(-3)(-4)}{(z+i)^5}\right]_{z=i} = \frac{3}{16i}$$

Hence by Cauchy's residue theorem, we have

$$\int f(z)dz = 2\pi i \times \text{ sum of residues within } c.$$

$$\int_{-R}^{R} f(x)\,dx + \int_{CR} f(z)\,dz = 2\pi i \times \frac{3}{16i} \Rightarrow \int_{-R}^{R} \frac{1}{(x^2+1)^3}dx + \int_{CR} \frac{1}{(z^2+1)^3}dz = \frac{3\pi}{8} \quad \ldots (1)$$

Now,

$$\left|\int_{CR} \frac{1}{(z^2+1)^3}dz\right| \le \int_{CR} \left|\frac{1}{(z^2+1)^3}\right||dz| \le \int_{CR} \frac{|dz|}{(|z|^2-1)^3}$$

$$= \int_0^{\pi} \frac{R\,d\theta}{(R^2-1)^3} \qquad [\text{since } z = Re^{i\theta}, \quad |dz| = R\,d\theta]$$

$$= \frac{\pi R}{(R^2-1)^3}, \text{ which} \to 0 \text{ as } R \to \infty$$

Hence making $R \to \infty$, relation (1) becomes

$$\int_{-\infty}^{\infty} \frac{1}{(x^2+1)^3}dx = \frac{3\pi}{8} \qquad \text{or} \qquad \int_0^{\infty} \frac{1}{(x^2+1)^3}dx = \frac{3\pi}{16} \qquad \textbf{Ans.}$$

Example 55. *Evaluate by the method of complex variables, the integral*

$$\int_{-\infty}^{\infty} \frac{x^2}{(1+x^2)^3}dx$$

Solution. Consider $\int_C \frac{z^2}{(1+z^2)^3}dz$ where c is a closed contour consisting of the upper half C_R of a large circle $|z| = R$ and the real axis from $-R$ to $+R$.

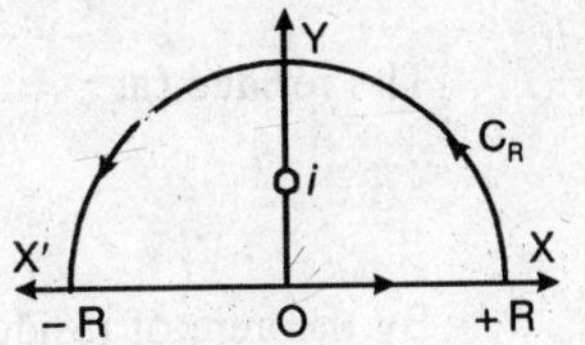

Poles of $\frac{z^2}{(1+z^2)^3}$ are given by

$$(z^2+1)^3 = 0 \quad \text{or} \quad z^2 = -1 \Rightarrow z = \pm i$$

$\therefore$ $z = i$ and $z = -i$ are the two poles each of order 3. But only $z = i$ lies within the contour.

To get residue at $z = i$, put $z = i + t$, then

$$\frac{z^2}{(1+z^2)^3} = \frac{(i+t)^2}{[1+(i+t)^2]^3} = \frac{-1+2it+t^2}{[1-1+2it+t^2]^3}$$

$$= \frac{(-1+2it+t^2)}{(2it)^3\left(1+\frac{1}{2i}t\right)^3} = \frac{(-1+2it+t^2)}{-8it^3}\left(1+\frac{t}{2i}\right)^{-3} = -\frac{1}{8i}\left(-\frac{1}{t^3}+\frac{2i}{t^2}+\frac{1}{t}\right)\left(1-\frac{3t}{2i}+\frac{(-3)(-4)}{2}\frac{t^2}{-4}+\dots\right)$$

$$= -\frac{1}{8i}\left[-\frac{1}{t^3}+\frac{2i}{t^2}+\frac{1}{t}\right]\left[1-\frac{3}{2i}t-\frac{3}{2}t^2+\dots\right]$$

Here coefficient of $\frac{1}{t}$ is $\frac{-1}{8i}\left(\frac{3}{2}-3+1\right)$ or $\frac{i}{8}\left(-\frac{1}{2}\right)$ or $\frac{-i}{16}$ which is therefore the residue at $z = i$.

Hence by Cauchy's residue theorem, we have

$$\int f(z)\,dz = 2\pi i \times \text{sum of the residues within } c$$

i.e.
$$\int_{-R}^{R} f(x)\,dx + \int_{C_R} f(z)\,dz = 2\pi i\left(-\frac{i}{16}\right)$$

$$\int_{-R}^{R} \frac{x^2}{(1+x^2)^3}\,dx + \int_{C_R} \frac{z^2}{(1+z^2)^3}\,dz = \frac{\pi}{8} \qquad \dots (1)$$

Now
$$\left|\int_{C_R} \frac{z^2}{(1+z^2)^3}\,dz\right| \le \int_{C_R} \frac{|z|^2\,|dz|}{|1+z^2|^3} \le \frac{R^2}{(R^2-1)^3}\int_0^{\pi} R\,d\theta$$

since
$$z = Re^{\theta i}, \quad |dz| = R\,d\theta \quad = \frac{R^3\pi}{(R^2-1)^3}, \text{ which } \to 0 \text{ as } R\to\infty$$

Hence by making $R \to \infty$, equation (1) becomes $\displaystyle\int_{-\infty}^{\infty} \frac{x^2}{(1+x^2)^3}\,dx = \frac{\pi}{8}$ **Ans.**

Example 56. *Evaluate* $\displaystyle\int_{-\infty}^{\infty} \frac{x^2\,dx}{(x^2+1)(x^2+4)}$

Solution. We consider $\displaystyle\int_C \frac{z^2\,dz}{(z^2+1)(z^2+4)} = \int_C f(z)\,dz$

where C is the contour consisting of the semi-circle C_R of radius R together with the part of the real axis from $-R$ to $+R$.

The integral has simple poles at

$$z = \pm i,\ z = \pm 2i$$

of which $z = i,\ 2i$ only lie inside C.

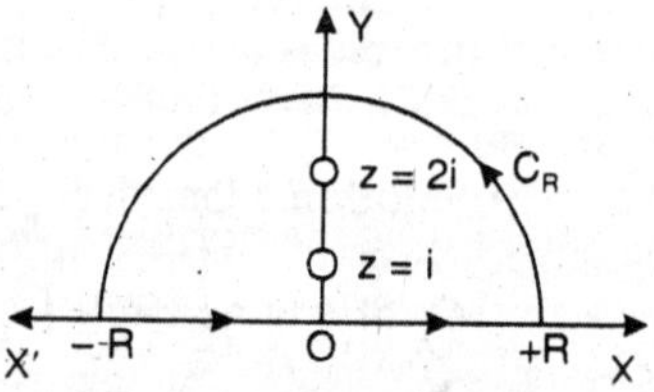

$$\text{The residue (at } z = i) = \lim_{z\to i}(z-i)\frac{z^2}{(z+i)(z-i)(z^2+4)}$$

$$= \lim_{z\to i}\frac{z^2}{(z+i)(z^2+4)} = \frac{-1}{2i(-1+4)} = \frac{-1}{6i}$$

$$\text{The residue (at } z = 2i) = \lim_{z\to 2i}(z-2i).\frac{z^2}{(z^2+1)(z+2i)(z-2i)}$$

$$= \lim_{z\to 2i}\frac{z^2}{(z^2+1)(z+2i)} = \frac{(2i)^2}{(-4+1)(2i+2i)} = \frac{1}{3i}$$

By theorem of residue;

$$\int_C f(z)\,dz = 2\pi i[\text{Res } f(i) + \text{Res } f(2i)] = 2\pi i\left(-\frac{1}{6i}+\frac{1}{3i}\right) = \frac{\pi}{3}$$

i.e.
$$\int_{-R}^{R} f(x)\,dx + \int_{C_R} f(z)\,dz = \frac{\pi}{3} \qquad \dots (1)$$

Hence by making $R \to \infty$, relation (1) becomes

$$\int_{-\infty}^{\infty} f(x)dx + \lim_{z\to\infty} \int_{C_R} f(z)\, dz = \frac{\pi}{3}$$

Now $\quad R \to \infty, \int_{C_R} f(z)dz$ vanishes.

For any point on C_R as $|z| \to \infty, f(z) = 0$

$$\lim_{|z|\to\infty} \int_{C_R} f(z)\, dz = 0, \int_{-\infty}^{\infty} f(x)\, dx = \frac{\pi}{3}$$

$$\Rightarrow \qquad \int_{-\infty}^{\infty} \frac{x^2 dx}{(x^2+1)(x^2+4)} = \frac{\pi}{3}$$ **Ans.**

Example 57. *Using the complex variable techniques, evaluate the integral*

$$\int_{-\infty}^{\infty} \frac{1}{x^4+1}\, dx$$ *(AMIETE, June 2010, U.P. III Semester, Dec. 2006)*

Solution. $\int_{-\infty}^{\infty} \frac{1}{x^4+1}\, dx$

Consider $\int_C f(z)\,dz$, where $f(z) = \frac{1}{z^4+1}$

taken around the closed contour consisting of real axis and upper half C_R, *i.e.* $|z| = R$.
Poles of $f(z)$ are given by

$$z^4+1 = 0 \text{ i.e. } z^4 = -1 = (\cos\pi + i\sin\pi)$$

$$\Rightarrow \qquad z^4 = [\cos(2n+1)\pi + i\sin(2n+1)\pi]$$

$$z = [\cos(2n+1)\pi + i\sin(2n+1)\pi]^{\frac{1}{4}} = \left[\cos(2n+1)\frac{\pi}{4} + i\sin(2n+1)\frac{\pi}{4}\right]$$

If $n = 0, \quad z_1 = \left(\cos\frac{\pi}{4} + i\sin\frac{\pi}{4}\right) = \left(\frac{1}{\sqrt{2}} + i\frac{1}{\sqrt{2}}\right) = e^{i\frac{\pi}{4}}$

$n = 1, \quad z_2 = \left(\cos\frac{3\pi}{4} + i\sin\frac{3\pi}{4}\right) = \left(-\frac{1}{\sqrt{2}} + i\frac{1}{\sqrt{2}}\right) = e^{i\frac{3\pi}{4}}$

$n = 2, \quad z_3 = \left(\cos\frac{5\pi}{4} + i\sin\frac{5\pi}{4}\right) = \left(-\frac{1}{\sqrt{2}} - i\frac{1}{\sqrt{2}}\right)$

$n = 3, \quad z_4 = \left(\cos\frac{7\pi}{4} + i\sin\frac{7\pi}{4}\right) = \left(\frac{1}{\sqrt{2}} - i\frac{1}{\sqrt{2}}\right)$

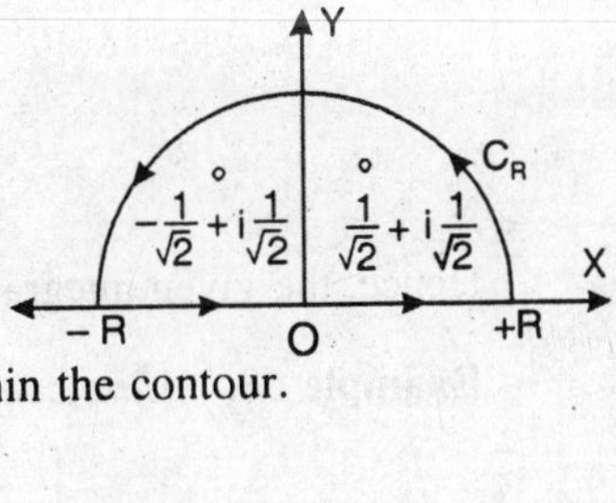

Y
C_R
$-\frac{1}{\sqrt{2}} + i\frac{1}{\sqrt{2}}$
$\frac{1}{\sqrt{2}} + i\frac{1}{\sqrt{2}}$
X
– R
O
+R

There are four poles, but only two poles at z_1 and z_2 lie within the contour.

$$\text{Residue}\left(\text{at } z = e^{\frac{i\pi}{4}}\right) = \left[\frac{1}{\frac{d}{dz}(z^4+1)}\right]_{z=e^{\frac{i\pi}{4}}} = \left[\frac{1}{4z^3}\right]_{z=e^{\frac{i\pi}{4}}} = \frac{1}{4\left(e^{i\frac{\pi}{4}}\right)^3} = \frac{1}{4e^{i\frac{3\pi}{4}}}$$

$$= \frac{1}{4}e^{-i\frac{3\pi}{4}} = \frac{1}{4}\left[\cos\frac{3\pi}{4} - i\sin\frac{3\pi}{4}\right] = \frac{1}{4}\left[-\frac{1}{\sqrt{2}} - i\frac{1}{\sqrt{2}}\right]$$

$$\text{Residue}\left(\text{at } z = e^{\frac{3i\pi}{4}}\right) = \left[\frac{1}{\frac{d}{dz}(z^4+1)}\right]_{z=e^{\frac{3i\pi}{4}}} = \frac{1}{[4z^3]_{z=\frac{3i\pi}{4}}} = \frac{1}{4\left(e^{i\frac{3\pi}{4}}\right)^3} = \frac{1}{4e^{i\frac{9\pi}{4}}}$$

$$= \frac{1}{4}e^{-i\frac{9\pi}{4}} = \frac{1}{4}\left(\cos\frac{9\pi}{4} - i\sin\frac{9\pi}{4}\right) = \frac{1}{4}\left(\frac{1}{\sqrt{2}} - i\frac{1}{\sqrt{2}}\right)$$

$$\int_C f(z)\,dz = 2\pi i \text{ (sum of residues at poles within c)}$$

$$\int_{-R}^{R} f(z)\,dz + \int_{C_R} f(z)\,dz = 2\pi i \text{ (sum of the residues)}$$

$$\int_{-R}^{R} \frac{1}{x^4+1}\,dx + \int_{C_R} \frac{1}{z^4+1}\,dz = 2\pi i, \text{ (sum of the residues)}$$

Now, $\left|\int_{C_R} \frac{1}{z^4+1}\,dz\right| \le \int_{C_R} \frac{1}{|z^4+1|}|dz|$

$$\le \int_{C_R} \frac{1}{(|z^4|-1)}|dz| \qquad [\text{Since } z = Re^{i\theta}, |dz| = |Re^{i\theta}\, i\, d\theta| = R\,d\theta]$$

$$\le \int_0^{\pi} \frac{1}{R^4-1} R\,d\theta \le \frac{R}{R^4-1}\int_0^{\pi} d\theta$$

$$\le \frac{R\pi}{R^4-1} = \frac{\pi/R^3}{1-1/R^4} \qquad \text{which} \to 0 \text{ as } R \to \infty.$$

Hence, $\int_{-R}^{R} \frac{1}{x^4+1}\,dx = 2\pi i$ (Sum of the residues within contour)

As $R \to \infty$

Hence, $\int_{-\infty}^{\infty} \frac{1}{x^4+1}\,dx = 2\pi i$ (Sum of the residues within contour) ... (1)

$$\int_{-\infty}^{\infty} \frac{1}{x^4+1}\,dx = 2\pi i\left[\frac{1}{4}\left(-\frac{1}{\sqrt{2}} - i\frac{1}{\sqrt{2}}\right) + \frac{1}{4}\left(\frac{1}{\sqrt{2}} - i\frac{1}{\sqrt{2}}\right)\right]$$

$$= \frac{\pi}{2}i\left(-\frac{1}{\sqrt{2}} - i\frac{1}{\sqrt{2}} + \frac{1}{\sqrt{2}} - i\frac{1}{\sqrt{2}}\right) = \frac{\pi i}{2}\left(-i\frac{2}{\sqrt{2}}\right) = \frac{\pi}{\sqrt{2}}$$

Hence, the given integral $= \frac{\pi}{\sqrt{2}}$ **Ans.**

Example 58. *Using complex variable techniques, evaluate the real integral*

$$\int_0^{\infty} \frac{dx}{1+x^6}$$

Solution. Let $f(z) = \frac{1}{1+z^6}$

We consider $\int_C \frac{1}{1+z^6}\,dz$

where C is the contour consisting of the semi-circle C_R of radius R together with the part of real axis from $-R$ to R.

Poles are given by $1+z^6=0$

$$z^6=-1=\cos\pi+i\sin\pi=\cos(2n\pi+\pi)+i\sin(2n\pi+\pi)$$
$$=e^{(2n+1)\pi i}$$

$$z=e^{\frac{2n+1}{6}\pi i}=\left[\cos\frac{2n\pi+\pi}{6}+i\sin\frac{2n\pi+\pi}{6}\right]\quad\text{where } n=0,1,2,3,4,5$$

If $n=0$, $\quad z=e^{\frac{\pi i}{6}}=\cos\frac{\pi}{6}+i\sin\frac{\pi}{6}=\frac{\sqrt{3}}{2}+\frac{i}{2}$

If $n=1$, $\quad z=e^{\frac{i\pi}{2}}=\cos\frac{\pi}{2}+i\sin\frac{\pi}{2}=i$

If $n=2$ $\quad z=e^{i\frac{5\pi}{6}}=\cos\frac{5\pi}{6}+i\sin\frac{5\pi}{6}=\frac{-\sqrt{3}}{2}+\frac{i}{2}$

If $n=3$, $\quad z=e^{i\frac{7\pi}{6}}=\cos\frac{7\pi}{6}+i\sin\frac{7\pi}{6}=\frac{-\sqrt{3}}{2}-\frac{i}{2}$

If $n=4$, $\quad z=e^{i\frac{3\pi}{2}}=\cos\frac{3\pi}{2}+i\sin\frac{3\pi}{2}=-i$

If $n=5$, $\quad z=e^{i\frac{11\pi}{6}}=\cos\frac{11\pi}{6}+i\sin\frac{11\pi}{6}=\frac{\sqrt{3}}{2}-\frac{i}{2}$

Only first three poles *i.e.*, $e^{i\frac{\pi}{6}}, e^{i\frac{\pi}{2}}, e^{i\frac{5\pi}{6}}$ are inside the contour.

$$\text{Residue}\left(\text{at } z=e^{\frac{i\pi}{6}}\right)=\lim_{z\to e^{\frac{i\pi}{6}}}\frac{1}{\frac{d}{dz}(1+z^6)}=\lim_{z\to e^{\frac{i\pi}{6}}}\frac{1}{6z^5}=\frac{1}{6}e^{\frac{-i5\pi}{6}}$$

$$\text{Residue}\left(\text{at } z=e^{\frac{i\pi}{2}}\right)=\lim_{z\to e^{\frac{i\pi}{2}}}\frac{1}{\frac{d}{dz}(1+z^6)}=\lim_{z\to e^{\frac{i\pi}{2}}}\frac{1}{6z^5}=\frac{1}{6}e^{\frac{-i5\pi}{2}}$$

$$\text{Residue}\left(\text{at } z=e^{\frac{i5\pi}{6}}\right)=\lim_{z\to e^{\frac{i5\pi}{6}}}\frac{1}{\frac{d}{dz}(1+z^6)}=\lim_{z\to e^{\frac{i5\pi}{6}}}\frac{1}{6z^5}=\frac{1}{6}e^{\frac{-i25\pi}{6}}$$

$$\text{Sum of the residues}=\frac{1}{6}\left[e^{\frac{-5i\pi}{6}}+e^{-\frac{i5\pi}{2}}+e^{-\frac{i25\pi}{6}}\right]=\frac{1}{6}\left(-\frac{\sqrt{3}}{2}-\frac{i}{2}+0-i+\frac{\sqrt{3}}{2}-\frac{i}{2}\right)=\frac{1}{6}(-2i)=-\frac{i}{3}$$

$$\Rightarrow\quad \int_C\frac{dz}{1+z^6}=2\pi i\ (\text{Residue})=2\pi i\left(-\frac{i}{3}\right)=\frac{2\pi}{3}$$

$$\Rightarrow\quad \int_{-\infty}^{\infty}\frac{dx}{1+x^6}=\frac{2\pi}{3}$$

$$\Rightarrow\quad \int_0^{\infty}\frac{dx}{1+x^6}=\frac{\pi}{3}$$

Ans.

Example 59. *Using complex variables, evaluate the real integral*

$$\int_0^{\infty}\frac{\cos 3x\, dx}{(x^2+1)(x^2+4)}$$

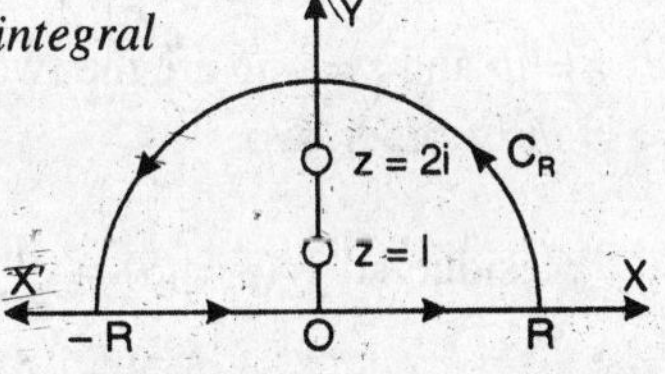

Solution. Let $\quad f(z)=\dfrac{e^{3iz}}{(z^2+1)(z^2+4)}$

Poles are given by

$$(z^2+1)(z^2+4)=0$$

i.e., $z^2+1=0$ or $z=\pm i$

$z^2+4=0$ or $z=\pm 2i$

Let C be a closed contour consisting of the upper half C_R of a large circle $|z|=R$ and the real axis from $-R$ to $+R$. The poles at $z=i$ and $z=2i$ lie within the contour.

$$\text{Residue (at } z=i) = \lim_{z\to i}\frac{(z-i)e^{3iz}}{(z^2+1)(z^2+4)} = \lim_{z\to i}\frac{e^{3iz}}{(z+i)(z^2+4)} = \frac{e^{-3}}{6i}$$

$$\text{Residue (at } z=2i) = \lim_{z\to 2i}\frac{(z-2i)e^{3iz}}{(z^2+1)(z^2+4)} = \lim_{z\to 2i}\frac{e^{3iz}}{(z^2+1)(z+2i)} = \frac{e^{-6}}{-12i}$$

By theorem of Residue $\int_C f(z)dz = 2\pi i$ [Sum of Residues]

$$\int_{-R}^{R}\frac{e^{3iz}dz}{(z^2+1)(z^2+4)} + \int_{C_R}\frac{e^{3iz}dz}{(z^2+1)(z^2+4)} = 2\pi i\left[\frac{e^{-3}}{6i}+\frac{e^{-6}}{-12i}\right]$$

$$\left[\int_{C_R}\frac{e^{3izdz}}{(z^2+1)(z^2+4)} = 0 \text{ as } z=Re^{i\theta} \text{ and } R\to\infty\right]$$

$$\int_{-R}^{R}\frac{e^{3ix}}{(x^2+1)(x^2+4)}dx = \pi\left[\frac{e^{-3}}{3}-\frac{e^{-6}}{6}\right]$$

$$\int_0^{\infty}\frac{\cos 3x\,dx}{(x^2+1)(x^2+4)} = \text{Real part of } \frac{1}{2}\int_{-\infty}^{\infty}\frac{e^{3ix}dx}{(x^2+1)(x^2+4)}$$

$$= \text{Real part of } \frac{\pi}{2}\left(\frac{e^{-3}}{3}-\frac{e^{-6}}{6}\right)$$

Hence, given integral $= \dfrac{\pi}{2}\left[\dfrac{e^{-3}}{3}-\dfrac{e^{-6}}{6}\right]$ **Ans.**

Example 60. *Using the calculus of residues, evaluate the integral given by the following:*

$$\int_0^{\infty}\frac{\cos ax}{(x^2+b^2)^2}dx,\ a>0,\ b>0$$

Solution. Consider the integral $\int_C f(z)\,dz$

where $$f(z)=\frac{e^{iaz}}{(z^2+b^2)^2}$$

taken around the closed contour C consisting of the upper half of a large circle $|z|=R$ and the real axis from $-R$ to R.

Poles of $f(z)$ are given by $(z^2+b^2)^2=0$

i.e., $z=ib$ and $z=-ib$ are the two poles of order two. The only pole which lies within the contour is $z=ib$ of order two.

$$\text{Residue at } (z=ib) = \lim_{z\to ib}\frac{d}{dz}(z-ib)^2\frac{e^{iaz}}{(z^2+b^2)^2} = \lim_{z\to ib}\frac{d}{dz}\frac{e^{iaz}}{(z+ib)^2}$$

$$= \lim_{z \to ib} \frac{(z+ib)^2 iae^{iaz} - e^{iaz} 2(z+ib)}{(z+ib)^4} = \lim_{z \to ib} \frac{[(z+ib)ia-2]e^{iaz}}{(z+ib)^3}$$

$$= \frac{[(2ib)ia-2]e^{ia(ib)}}{(2ib)^3} = \frac{(-2ab-2)e^{-ab}}{-8ib^3} = \frac{(ab+1)e^{-ab}}{4ib^3}$$

Hence, by Cauchy's residue theorem, we have

$$\int_C f(z)dz = 2\pi i \times \text{Sum of the residues within } C$$

$$\int_{-R}^{R} f(x)dx + \int_{C_R} f(z)dz = 2\pi i \frac{(ab+1)e^{-ab}}{4ib^3}$$

$$\int_{-R}^{R} \frac{e^{iax}}{(x^2+b^2)^2} dx + \int_{C_R} \frac{e^{iaz}}{(z^2+b^2)^2} dz = \frac{\pi(ab+1)e^{-ab}}{2b^3} \quad \ldots (1)$$

Now $\left| \int_{C_R} \frac{e^{iaz} dz}{(z^2+b^2)^2} \right| \le \int_{C_R} \frac{|e^{iaz}||dz|}{(z^2+b^2)^2} \le \int_{C_R} \frac{|e^{iaz}||dz|}{[|z|^2-b^2]^2}$

$$\le \int_0^{\pi} \frac{e^{-aR\sin\theta} Rd\theta}{(R^2-b^2)^2} \le \frac{R}{(R^2-b^2)^2} \int_0^{\pi} e^{-aR\sin\theta} d\theta \le \frac{R}{(R^2-b^2)^2} 2\int_0^{\frac{\pi}{2}} e^{-aR\frac{2\theta}{\pi}} d\theta$$

$$\le \frac{R}{a(R^2-b^2)^2}(1-e^{-aR}) \quad \text{which} \to 0, \text{ as } R \to \infty$$

Hence by making $R \to \infty$, (1) becomes $\int_{-\infty}^{\infty} \frac{e^{iax} dx}{(x^2+b^2)^2} = \frac{\pi(ab+1)e^{-ab}}{2b^3}$

Equating real parts, we have $\int_{-\infty}^{\infty} \frac{\cos ax}{(x^2+b^2)^2} dx = \frac{\pi(ab+1)e^{-ab}}{2b^3}$

$$\Rightarrow \int_0^{\infty} \frac{\cos ax dx}{(x^2+b^2)^2} = \frac{\pi(ab+1)e^{-ab}}{4b^3} \qquad \textbf{Ans.}$$

Example 61. *Using complex variable techniques, evaluate the real integral*

$$\int_0^{\infty} \frac{\cos 2x}{(x^2+9)^2(x^2+16)} dx$$

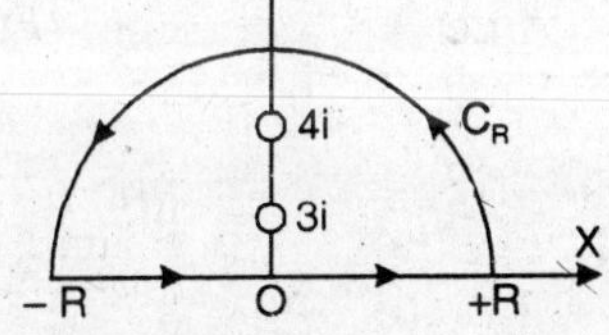

Solution. Consider the integral $\int_C f(z)dz$,

where $\quad f(z) = \frac{e^{2iz}}{(z^2+9)^2(z^2+16)}$,

taken around the closed contour C consisting of the upper half of a large circle $|z| = R$ and the real axis from $-R$ to R.

Poles of $f(z)$ are given by

$$(z^2+9)^2(z^2+16) = 0$$

i.e. $\quad (z+3i)^2(z-3i)^2(z+4i)(z-4i) = 0$

i.e. $\quad z = 3i, -3i, 4i, -4i$

The poles which lie within the contour are $z = 3i$ of the second order and $z = 4i$ simple pole.

Residue of $f(z)$ at $z = 3i$

$$= \frac{1}{1!}\left[\frac{d}{dz}\left\{(z-3i)^2 \frac{e^{2iz}}{(z-3i)^2(z+3i)^2(z^2+16)}\right\}\right]_{z=3i} = \left[\frac{d}{dz}\left\{\frac{e^{2iz}}{(z+3i)^2(z^2+16)}\right\}\right]_{z=3i}$$

$$= \left[\frac{(z+3i)^2(z^2+16)2ie^{2iz} - e^{2iz}[2(z+3i)(z^2+16)+2z(z+3i)^2]}{(z+3i)^4(z^2+16)^2}\right]_{z=3i}$$

$$= \left[\frac{(z+3i)(z^2+16)2i\,e^{2iz} - e^{2iz}[2(z^2+16)+2z(z+3i)]}{(z+3i)^3(z^2+16)^2}\right]_{z=3i}$$

$$= \frac{6i\times 7\times 2i\,e^{-6} - e^{-6}(2\times 7+6i\times 6i)}{(6i)^3(7)^2} = \frac{e^{-6}[-84+22]i}{216\times 49} = \frac{e^{-6}(-62)i}{216\times 49} = -\frac{i31e^{-6}}{108\times 49}$$

$$\text{Residue of } f(z) \text{ at } (z=4i) = \lim_{z\to 4i}(z-4i)\frac{e^{2iz}}{(z^2+9)^2(z-4i)(z+4i)}$$

$$= \frac{e^{-8}}{(-16+9)^2(4i+4i)} = \frac{e^{-8}}{49\times 8i} = \frac{-ie^{-8}}{392}$$

$$\text{Sum of the residues} = -\frac{i31e^{-6}}{108\times 49} - \frac{ie^{-8}}{392}$$

Hence by Cauchy's Residue Theorem, we have

$$\int_C f(z)\,dz = 2\pi i \times \text{Sum of the residues within } C$$

i.e.
$$\int_{-R}^{R} f(x)dx + \int_{C_R} f(z)\,dz = 2\pi i \times \text{sum of residues}$$

or
$$\int_{-R}^{R} \frac{e^{2ix}}{(x^2+9)^2(x^2+16)}dx + \int_{C_R} \frac{e^{2iz}}{(z^2+9)^2(z^2+16)}dz = 2\pi i\times \text{ Sum of residues} \quad \ldots (1)$$

Now let $R\to\infty$, so as to show that the second integral in above relation vanishes. For any point on C_R, as $|z|\to\infty$

Let
$$F(z) = \frac{1}{z^6}\frac{e^{2iz}}{\left(1+\frac{9}{z^2}\right)^2\left(1+\frac{16}{z^2}\right)}$$

$$\lim_{|z|\to\infty} F(z) = 0 \quad \text{or} \quad \int_{C_R}\frac{e^{2iz}}{(z^2+9)^2(z^2+16)}dz = 0 \text{ as } z\to\infty$$

Hence by making $R\to\infty$, relation (1) becomes

$$\therefore \quad \int_{-\infty}^{\infty}\frac{e^{2ix}}{(x^2+9)^2(x^2+16)}dx = 2\pi i\left[\frac{-i31e^{-6}}{108\times 49} - i\frac{e^{-8}}{392}\right] = \frac{2\pi}{196}\left[\frac{31e^{-6}}{27}+\frac{e^{-8}}{2}\right]$$

Equating real parts, we have

$$\int_{-\infty}^{\infty}\frac{\cos 2x dx}{(x^2+9)^2(x^2+16)} = \frac{\pi}{98}\left(\frac{31e^{-6}}{27}+\frac{e^{-8}}{2}\right)$$

$$\int_0^{\infty}\frac{\cos 2x}{(x^2+9)^2(x^2+16)}dx = \frac{\pi}{196}\left(\frac{31e^{-6}}{27}+\frac{e^{-8}}{2}\right) \quad \left[\because \int_{-\infty}^{\infty} f(x)dx = 2\int_0^{\infty} f(x)dx \text{ If } f(x) \text{ is even function.}\right]$$ **Ans.**

EXERCISE 31.5

Evaluate the following :

1. $\int_0^{\infty} \frac{1}{1+x^2}\,dx$ **Ans.** $\frac{\pi}{2}$

2. $\int_{-\infty}^{\infty} \frac{1}{(x^2+1)^2}\,dx$ **Ans.** $\frac{\pi}{2}$

3. $\int_0^{\infty} \frac{x^3 \sin x}{(x^2+a^2)(x^2+b^2)}\,dx$ **Ans.** $\frac{\pi}{2(a^2-b^2)}[a^2e^{-a}-b^2e^{-b}]$

4. $\int_{-\infty}^{\infty} \frac{\cos x}{(x^2+a^2)(x^2+b^2)}\,dx,\quad a>b>0$ **Ans.** $\frac{\pi}{a^2-b^2}\left(\frac{e^{-b}}{b}-\frac{e^{-a}}{a}\right)$

5. Show that $\int_0^{\infty} \frac{\cos x}{x^2+a^2}\,dx = \frac{\pi e^{-a}}{2a}$

6. Show that $\int_0^{\infty} \frac{x^3 \sin x}{(x^2+a^2)}\,dx = -\frac{\pi}{4}(a-2)\,a^{-a},\ a>0$

Evaluate the following :

7. $\int_{-\infty}^{\infty} \frac{\sin mx}{x(x^2+a^2)}\,dx,\quad m>0,\ a>0$ **Ans.** $\frac{\pi}{a^2}(2-e^{-ma})$

8. $\int_0^{\infty} \frac{x^2}{x^6+1}\,dx$ **Ans.** $\frac{\pi}{6}$

9. $\int_0^{\infty} \frac{x\sin ax}{x^4+a^4}\,dx$ **Ans.** $\frac{\pi}{2a^2}e^{-\frac{a^2}{\sqrt{2}}}\sin\frac{a^2}{\sqrt{2}}$

10. $\int_0^{\infty} \frac{x^6}{(a^4+x^4)^2}\,dx$ **Ans.** $\frac{3\pi\sqrt{2}}{16a},\ a>0$

11. $\int_0^{\infty} \frac{\cos x^2+\sin x^2-1}{x^2}\,dx$ **Ans.** 0

12. $\int_0^{\infty} \frac{\cos mx}{x^4+x^2+1}\,dx$ **Ans.** $\frac{\pi}{\sqrt{3}}\sin\frac{1}{2}\left(m+\frac{\pi}{3}\right)e^{-\frac{1}{2}m\sqrt{3}}$

13. $\int_0^{\infty} \frac{\log(1+x^2)}{1+x^2}\,dx$ **Ans.** $\pi \log 2$

31.16 RECTANGULAR CONTOUR

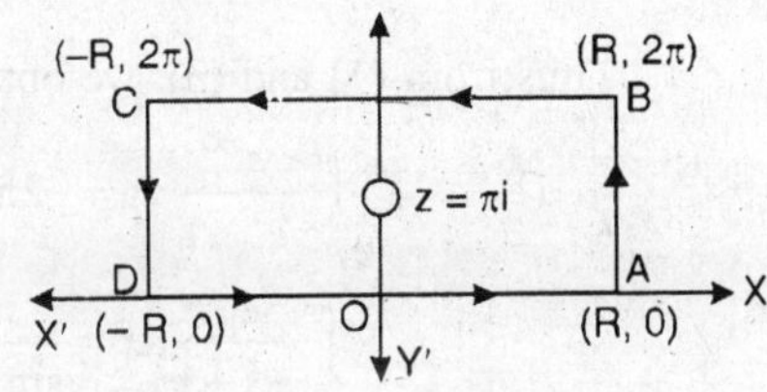

Example 62. *Evaluate* $\int_{-\infty}^{+\infty} \frac{e^{ax}}{e^x+1}\,dx$

Solution. We consider $\int_C \frac{e^{az}}{e^z+1}\,dz = \int_C f(z)\,dz$

where C is the rectangle $ABCD$ with vertices at $(R, 0), (R, 2\pi), (-R, 2\pi)$ and $(-R, 0)$.

$f(z)$ has simple poles, $e^z = -1$

$$= \cos(2n+1)\pi + i\sin(2n+1)\pi = e^{i(2n+1)\pi}$$

$\Rightarrow \qquad z = (2n+1)\pi i, \qquad$ where $n = 0, \pm 1, \pm 2 \ldots$

The only pole inside the rectangle is $z = \pi i$.

$\therefore$ By Residue theorem

$$\int_C f(z)dz = 2\pi i \text{ Residue } f(\pi i) = 2\pi i \left[\frac{e^{az}}{\frac{d}{dz}(e^z+1)}\right]_{z=\pi i} \qquad \left[R(a) = \frac{\phi(a)}{\psi'(a)}\right]$$

$$= 2\pi i \left[\frac{e^{az}}{e^z}\right]_{z=\pi i} = 2\pi i \frac{e^{a\pi i}}{e^{\pi i}} = -2\pi i\, e^{a\pi i} \qquad \left[\begin{array}{l}\because e^{\pi i} = \cos\pi + i\sin\pi \\ = -1+0 \\ = -1\end{array}\right]$$

Also $$\int_C f(z)dz = \int_{AB} f(z)dz + \int_{BC} f(z)dz + \int_{CD} f(z)dz + \int_{DA} f(z)dz \quad \text{... (1)}$$

$$= \int_0^{2\pi} f(R+iy)idy + \int_R^{-R} f(x+2\pi i)dx + \int_{2\pi}^0 (-R+iy)idy + \int_{-R}^R f(x)dx \quad \text{... (2)}$$

[$\because z = R + iy$ along AB, $z = x + 2\pi i$ along BC, $z = -R + iy$ along CD and $z = x$ along DA].

$$\int_C f(z)dz = i\int_0^{2\pi} \frac{e^{a(R+iy)}}{e^{R+iy}+1}dy - \int_{-R}^{+R} \frac{e^{a(x+2\pi i)}}{e^{x+2\pi i}+1}dx - i\int_0^{2\pi} \frac{e^{a(-R+iy)}}{e^{-R+iy}+1}dy + \int_{-R}^{R} \frac{e^{ax}}{e^x+1}dx$$

Now for any two complex numbers z_1, z_2, $|z_1| \geq |z_2|$

we have $$|z_1 + z_2| \geq |z_1| - |z_2|.$$

So that $|e^{R+iy} + 1| \geq e^R - 1$. Also $|e^{a(R+iy)}| = e^{aR}$.

$\therefore$ For the integrand of first integral in (2), we have

$$\left|\frac{e^{a(R+iy)}}{e^{R+iy}+1}\right| \leq \frac{e^{aR}}{e^R - 1} \text{ which} \to 0, \text{ as } R \to \infty \quad [\because a > 1]$$

Similarly, for the integrand of the third integral in (2), we get

$$\left|\frac{e^{a(-R+iy)}}{e^{-R+iy}+1}\right| \leq \frac{e^{-aR}}{1-e^{-r}} \text{ which also} \to 0, \text{ as } R \to \infty \quad [\because a < 0]$$

Hence as $R \to \infty$, since the first and third integrals in (2) approach zero, we get

$$\int_C f(z)dz = -e^{2a\pi i}\int_{-\infty}^{\infty} \frac{e^{ax}}{e^x+1}dx + \int_{-\infty}^{\infty} \frac{e^{ax}}{e^x+1}dx$$

$$= (1 - e^{2a\pi i})\int_{-\infty}^{\infty} \frac{e^{ax}}{e^x+1}dx. \quad \text{... (3)}$$

Thus from (1) and (3), we obtain

$$(1 - e^{2a\pi i})\int_{-\infty}^{\infty} \frac{e^{ax}}{e^x+1}dx = -2\pi i e^{a\pi i} \quad \text{or} \quad \int_{-\infty}^{\infty} \frac{e^{ax}}{e^x+1}dx = \frac{2\pi i}{e^{a\pi i} - e^{-a\pi i}}$$

$$\int_{-\infty}^{\infty} \frac{e^{ax}}{e^x+1}dx = \frac{\pi}{\sin a\pi}$$ **Ans.**

Example 63. *By integrating* e^{-z^2} *round the rectangle whose vertices are 0, R, R + ia, ia, show that*

(*i*) $\int_0^{\infty} e^{-x^2} \cos 2ax\, dx = \frac{e^{-a^2}}{2}\sqrt{\pi}$ and (*ii*) $\int_0^{\infty} e^{-x^2} \sin 2ax\, dx = e^{-a^2}\int_0^a e^{y^2} dy$.

Solution. Let $$f(z) = e^{-z^2}$$

$$\therefore \quad \int_C f(z)dz = \int_C e^{-z^2} dz.$$

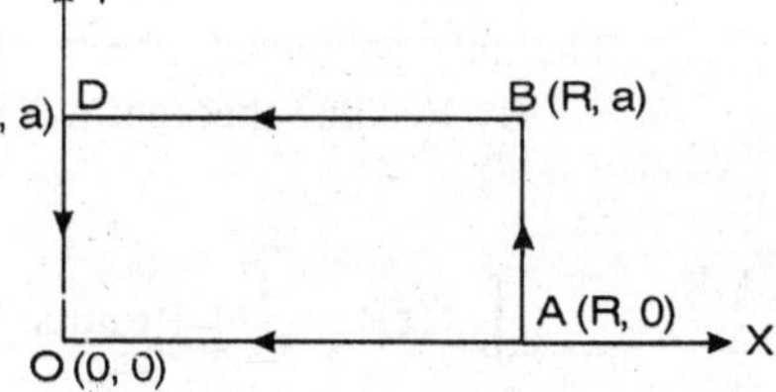

where C is the closed contour, a rectangle $OABD$.

Since $f(z)$ is analytic within and on the contour C (There is no pole within rectangle $OABD$). Hence by Cauchy's residue theorem, we have

$$\int_{OABD} e^{-z^2} dz = 0$$

i.e. $$\int_{OA} e^{-z^2} dz + \int_{AB} e^{-z^2} dz + \int_{BD} e^{-z^2} dz + \int_{DO} e^{-z^2} dz = 0 \quad \text{... (1)}$$

Since on OA, $z = x$, $dz = dx$. On AB, $z = R + iy$, $dz = idy$
On BD, $z = x + ia$, $dz = dx$. On DO, $z = iy$, $dz = idy$

Hence (1) becomes

$$\int_0^R e^{-x^2}\,dx + \int_0^a e^{-(R+iy)^2}.idy + \int_R^0 e^{-(x+ia)^2}.dx + \int_a^0 e^{-(iy)^2}.idy = 0 \qquad \ldots (2)$$

Now $$\left|\int_0^a e^{-(R+iy)^2}.idy\right| \le \left|\int_0^a e^{-(R+iy)^2}\right| |idy| \le \int_0^a e^{-R^2+y^2}\,dy$$

$$\le \int_0^a e^{-R^2+a^2}.dy \qquad [\text{since } y \le a \text{ on } AB]$$

$$\le e^{-R^2+a^2}.a = 0 \text{ as } R \to \infty$$

Hence by making $R \to \infty$, equation (2) becomes

$$\int_0^\infty e^{-(x+ia)^2}\,dx = \int_0^\infty e^{-x^2}\,dx - i\int_0^a e^{y^2}\,dy$$

$$\Rightarrow \quad \int_0^\infty e^{(-x^2+a^2-2aix)}\,dx = \frac{\sqrt{\pi}}{2} - i\int_0^a e^{y^2}\,dy \qquad \left[\text{since } \int_0^\infty e^{-x^2}\,dx = \frac{\sqrt{\pi}}{2}\right]$$

$$\int_0^\infty e^{-x^2+a^2}.e^{-2aix}\,dx = \frac{\sqrt{\pi}}{2} - i\int_0^a e^{y^2}\,dy$$

$$\int_0^\infty e^{-x^2+a^2}(\cos 2ax - i\sin 2ax)\,dx = \frac{\sqrt{\pi}}{2} - i\int_0^a e^{y^2}\,dy$$

$$\int_0^\infty e^{-x^2}(\cos 2ax - i\sin 2ax)\,dx = \frac{\sqrt{\pi}}{2}e^{-a^2} - ie^{-a^2}\int_0^a e^{y^2}\,dy$$

Equating real and imaginary parts, we have

$$\int_0^\infty e^{-x^2}\cos 2ax\,dx = \frac{e^{-a^2}}{2}\sqrt{\pi}$$

$$\int_0^\infty e^{-x^2}\sin 2ax\,dx = e^{-a^2}\int_0^a e^{y^2}\,dy$$ **Proved.**

EXERCISE 31.6

1. Using contour integration, show that

$$\int_0^\infty \frac{x^6\,dx}{(a^4+x^4)^2} = \frac{3\sqrt{2}\pi}{16a}, \qquad (a > 0)$$

2. Using method of contour integration, evaluate

$$\int_0^\infty \frac{x\sin ax\,dx}{x^4+4}$$ **Ans.** $\frac{\pi}{8}e^{-a}\sin a$

3. Integrating $\frac{e^{iz}}{z+a}$ along the boundary of the square defined by $x = 0$, $x = R$, $y = 0$, $y = R$.

Prove that (*i*) $\int_0^\infty \frac{\cos x}{x+a}\,dx = \int_0^\infty \frac{xe^{-ax}}{1+x^2}\,dx$ (*ii*) $\int_0^\infty \frac{\sin x}{x+a}\,dx = \int_0^\infty \frac{e^{-ax}}{1+x^2}\,dx$

4. Evaluate, using Cauchy's integral formula

$\oint_c \frac{\cos \pi z\,dz}{z^2-1}$ around a rectangle

(*i*) $2 \pm i, -2 \pm i$ **Ans.** 0 (*ii*) $-i, 2-i, 2+i$ and i **Ans.** $-\pi i$

5. By integrating $\dfrac{e^{iaz^2}}{\sinh \pi z}$ round the rectangle with vertices $\pm R \pm \dfrac{i}{2}$, show that

$$\int_0^\infty \frac{\cos(ax^2)\cosh(ax)}{\cosh \pi x}\,dx = \frac{1}{2}\cos\left(\frac{a}{4}\right) \text{ and } \int_0^\infty \frac{\sin(ax^2)\cosh(ax)}{\cosh \pi x}\,dx = \frac{1}{2}\sin\left(\frac{a}{4}\right) \qquad (0 < a \le \pi)$$

31.17 INDENTED SEMI-CIRCULAR CONTOUR

When the integrand has a simple pole on real axis, it is deleted from the region by indenting the contour (a small semi-circle having pole is drawn)

Example 64. *By contour integration, prove that*

$$\int_0^\infty \frac{\sin mx}{x}\,dx = \frac{\pi}{2}.$$

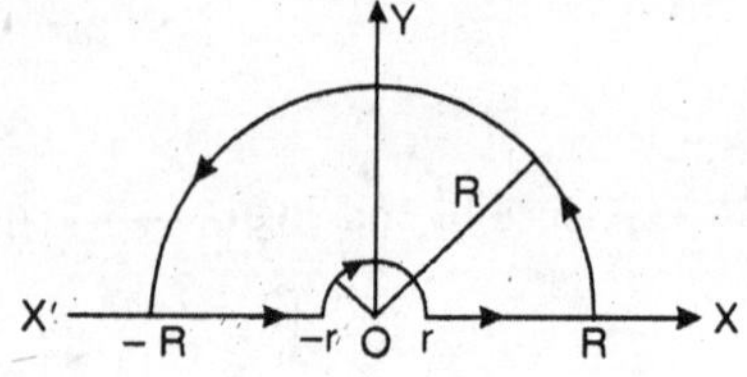

Solution. Consider the integral $\displaystyle\int_C \frac{e^{miz}}{z}\,dz$

where C is a large semi-circle $|z| = R$ indented at $z = 0$ (pole), let r be the radius of indentation. There is no singularity within the given contour.

Hence by Cauchy Theorem.

$$\int_C \frac{e^{miz}}{z}\,dz = 0$$

i.e.,
$$\int_{-R}^{-r} \frac{e^{mix}}{x}\,dx + \int_{C_r} \frac{e^{miz}}{z}\,dz + \int_r^R \frac{e^{mix}}{x}\,dx + \int_{C_R} \frac{e^{miz}}{z}\,dz = 0 \qquad \text{... (1)}$$

Substituting $-x$ for x in the first integral and combining it with the third integral, we get

$$\int_r^R \frac{e^{mix} - e^{-imx}}{x}\,dx + \int_{C_2} \frac{e^{miz}}{z}\,dz + \int_{C_1} \frac{e^{miz}}{z}\,dz = 0 \quad \left[z = Re^{i\theta} \Rightarrow dz = Rie^{i\alpha}\,d\theta\right]$$

$$\Rightarrow \qquad 2i\int_r^R \frac{\sin mx}{x}\,dx + \int_{C_2} \frac{e^{imz}}{z}\,dz + \int_{C_1} \frac{e^{imz}}{z}\,dz = 0 \qquad \text{... (2)}$$

Now
$$\int_{C_2} \frac{e^{imz}}{z}\,dz = \int_{C_2} \frac{1}{z}\,dz + \int_{C_2} \frac{e^{imz} - 1}{z}\,dz \qquad \text{... (3)}$$

On $\qquad C_2\ z = re^{i\theta}$

$$\therefore \qquad \int_{C_2} \frac{1}{z}\,dz = \int_\pi^0 \frac{re^{i\theta}\,i\,d\theta}{re^{i\theta}} = -\int_0^\pi i\,d\theta = -i\pi$$

Also
$$\left|\int_{C_2} \frac{e^{imx} - 1}{z}\,dz\right| \le M\int_{C_2} \frac{|dz|}{|z|} = \pi M$$

where M is the maximum value on C_2 of $|e^{imx} - 1| = |e^{imr(\cos\theta + i\sin\theta)} - 1|$

Clearly, $M \to 0$ as $r \to 0$

$$\therefore \text{ From (3)}, \qquad \int_{C_2} \frac{e^{imz}}{z}\,dz = -i\pi \qquad \text{... (4)}$$

Putting $\qquad z = Re^{i\theta}$ in the integral over C_1, we get

$$\int_{C_1} \frac{e^{imz}}{z}\,dz = \int_0^\pi \frac{e^{imR(\cos\theta + i\sin\theta)}}{Re^{i\theta}}\,Re^{i\theta}i\,d\theta = i\int_0^\pi e^{imR\cos\theta}.e^{-mR\sin\theta}\,d\theta$$

Since $|e^{imR\cos\theta}| \le 1$

$$\therefore \quad \left|\int_{C_1} \frac{e^{imz}}{z} dz\right| \le \int_0^{\pi} e^{-mR\sin\theta} d\theta = 2\int_0^{\pi/2} e^{-mR\sin\theta} d\theta$$

Also $\dfrac{\sin\theta}{\theta}$ continuously decreases from 1 to $\dfrac{2}{\pi}$ as θ increases from 0 to $\dfrac{\pi}{2}$.

$$\therefore \quad \text{For } 0 \le \theta \le \frac{\pi}{2}, \quad \frac{\sin\theta}{\theta} \ge \frac{2}{\pi} \quad \text{or} \quad \sin\theta \ge \frac{2\theta}{\pi}$$

$$\therefore \quad \left|\int \frac{e^{imz}}{z} dz\right| \le 2\int_0^{\pi/2} e^{-2mR\theta/\pi} d\theta = \left[-\frac{\pi}{mR} e^{-2mR\theta/\pi}\right]_0^{\pi/2} = \frac{\pi}{mR}(1-e^{-mR})$$

As $\quad R \to \infty, \dfrac{\pi}{mR}(1-e^{-mR}) \to 0$

$$\therefore \quad \int_{C_1} \frac{e^{imz}}{z} dz = 0$$

Hence from (2), on taking the limit as $r \to 0$ and $R \to \infty$, we get

$$2i\int_0^{\infty} \frac{\sin mx}{x} dx - i\pi = 0 \quad \text{or} \quad \int_0^{\infty} \frac{\sin mx}{x} dx = \frac{\pi}{2}. \qquad \textbf{Ans.}$$

Example 65. *Show that, if* $a \ge b \ge 0$, *then* $\displaystyle\int_0^{\infty} \frac{\cos 2ax - \cos 2bx}{x^2} dx = \pi(b-a)$

Solution. Consider the integral $\displaystyle\int_C f(z)dz$

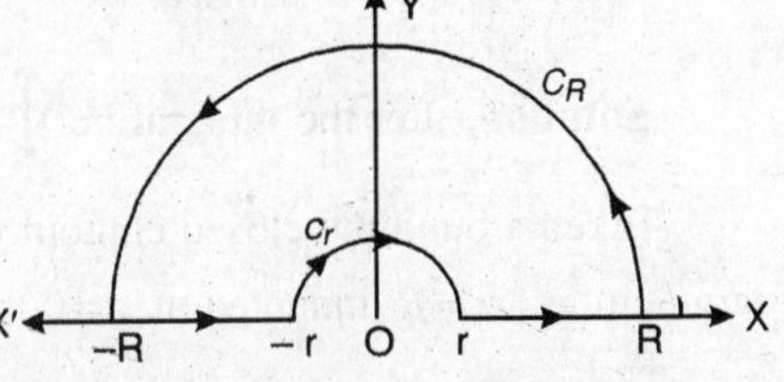

where $$f(z) = \frac{e^{i2az} - e^{i2bz}}{z^2}$$

and C is a large semi-circle $|z| = R$ indented at $z = 0$ (pole), let r be the radius of indentation. Now there is no singularity within the given contour.

$$\int_C f(z)dz = 0 \quad \text{(By Cauchy Integral Theorem)}$$

$$\Rightarrow \quad \int_{-R}^{-r} f(x)\,dx + \int_{C_R} f(z)\,dz + \int_r^R f(x)\,dx + \int_{C_R} f(z)\,dz = 0 \qquad \dots (1)$$

$$\text{Now} \quad \left|\int_{C_R} f(z)\,dz\right| \le \int_{C_R} \frac{|e^{2iaz} - e^{2ibz}|}{|z|^2} |dz|$$

$$\le \int_{C_R} \frac{|e^{2iaz}| + |e^{2ibz}|}{|z|^2} |dz|$$

$$= \int_0^{\pi} \frac{e^{-2aR\sin\theta} + e^{-2bR\sin\theta}}{R^2} R d\theta \qquad (z = Re^{i\theta})$$

$$\le \frac{2}{R}\int_0^{\frac{\pi}{2}} \left[e^{\frac{-4aR\theta}{\pi}} + e^{\frac{-4bR\theta}{\pi}}\right] d\theta \qquad \text{[By Jordan's inequality]}$$

$$= \frac{2}{R}\left[\frac{\pi}{4aR}(1-e^{-2aR}) + \frac{\pi}{4aR}(1-e^{-2bR})\right] = 0 \text{ as } R \to \infty$$

We have, $\lim_{z\to 0}\{z(f(z))\} = \lim_{z\to 0}\left\{z\frac{e^{2iaz}-e^{2ibz}}{z^2}\right\} = \lim_{z\to 0}\{2i(a-b)-2(a^2-b^2)z^2....\} = 2i(a-b)$

$$\lim_{r\to 0}\int_{C_r} f(z)dz = -i(\pi-0)\times 2i(a-b) = -2\pi(b-a)$$

Hence, by making $R\to\infty$ and $r\to 0$, equation (1) reduces to

$$\int_{-\infty}^{0} f(x)dx - 2\pi(b-a) + \int_{0}^{\infty} f(x)dx + 0 = 0 \Rightarrow \int_{-\infty}^{\infty} f(x)dx = 2\pi(b-a)$$

$$\int_{-\infty}^{\infty}\frac{e^{2iax}-e^{i2bx}}{x^2}dx = 2\pi(b-a)$$

$$\int_{-\infty}^{\infty}\frac{(\cos 2ax + i\sin 2ax)-(\cos 2bx + i\sin 2bx)}{x^2}dx = 2\pi(b-a)$$

Equating real parts, we get

$$\int_{-\infty}^{\infty}\frac{\cos 2ax-\cos 2bx}{x^2}dx = 2\pi(b-a) \qquad \left[\int_{-a}^{a} f(x)dx = 2\int_{0}^{a} f(x)dx \text{ If } f(x) \text{ is even function.}\right]$$

Hence, $\displaystyle\int_{0}^{\infty}\frac{\cos 2ax-\cos 2bx}{x^2}dx = \pi(b-a)$ **Proved.**

Example 66. *Using contour integration method, prove the integral*

(*i*) $\displaystyle\int_{0}^{\infty}\frac{x^{a-1}}{1+x}dx = \frac{\pi}{\sin\pi a}$. $(0<a<1)$ (*ii*) $\displaystyle\int_{0}^{\infty}\frac{x^{a-1}}{1-x}dx = \pi\cot\pi a.$

Solution. Let the integral be $\int_C f(z)\,dz$, where $f(z) = \dfrac{z^{a-1}}{1-z}$

Taken around the closed contour C consisting of real axis from $-R$ to R, and upper half of a large circle $|z| = R$ indented at $z=0$, $z=1$, the radii of indentations being r and r' respectively.

The singularities of $f(z)$ are $z=0$, $z=1$ which have been avoided by the identation, so there are no singularities within the contour.

Hence, by Cauchy's residue theorem, we have

$$\int_{-R}^{-r} f(x)\,dx + \int_{Cr} f(z)dz + \int_{r}^{1-r'} f(x)dx + \int_{Cr'} f(z)dz + \int_{1+r'}^{R} f(x)dx + \int_{CR} f(z)dz = 0 \quad ...(1)$$

Since $\displaystyle\lim_{z\to\infty} z\,f(z) = \lim_{z\to\infty} z\frac{z^{a-1}}{1-z} = \lim_{z\to\infty}\frac{z^a}{1-z} = 0,\ 0<a<1.$

$$\therefore \lim_{R\to\infty}\int_{C_R} f(z)dz = i(\pi-0)\,0 = 0$$

Again, $\displaystyle\lim_{z\to 0}\{z\,f(z)\} = \lim_{z\to 0}\left\{\frac{z.z^{a-1}}{1-z}\right\} = \lim_{z\to 0}\left(\frac{z^a}{1-z}\right) = 0,\ a>0.$

$$\therefore \lim_{r\to 0}\int_{Cr} f(z)dz = -i(\pi-0).0 = 0$$

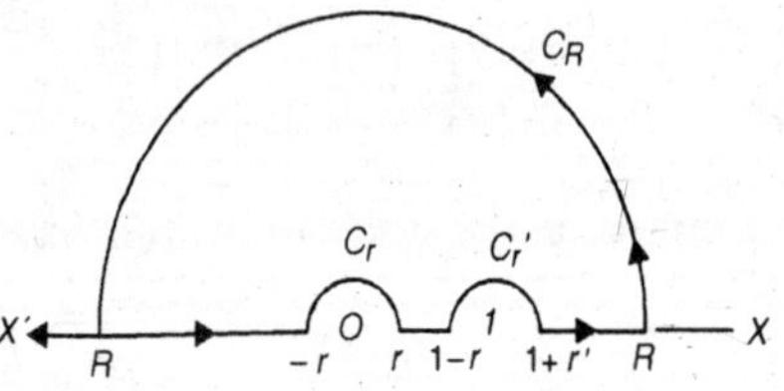

Also, $\displaystyle\lim_{r\to 1}\{(z-1)f(z)\} = \lim_{z\to 1}\left\{(z-1)\frac{z^{a-1}}{1-z}\right\} = -1$

$$\therefore \lim_{r\to 0}\int_{C_{r'}} f(z)dz = -i(\pi-0)(-1) = i\pi$$

Hence making $R\to\infty, r\to 0, r'\to 0$, we have from (1)

$$\int_{-\infty}^{0} f(x)\,dx+\int_{0}^{1} f(x)\,dx+\pi i+\int_{1}^{\infty} f(x)\,dx=0$$

$$\Rightarrow \quad \int_{-\infty}^{\infty} f(x)\,dx+\pi i = 0 \qquad \Rightarrow \quad \int_{-\infty}^{\infty}\frac{x^{a-1}}{1-x}\,dx+\pi i = 0$$

$$\Rightarrow \quad \int_{-\infty}^{0}\frac{x^{a-1}}{1-x}\,dx+\int_{0}^{\infty}\frac{x^{a-1}}{1-x}\,dx = -\pi i \qquad \Rightarrow \quad -\int_{0}^{-\infty}\frac{x^{a-1}}{1-x}\,dx+\int_{0}^{\infty}\frac{x^{a-1}}{1-x}\,dx = -\pi i$$

Putting $-x$ for x in the first integral, we have

$$\int_{0}^{\infty}\frac{(-1)^{a-1}x^{a-1}}{1+x}\,dx+\int_{0}^{\infty}\frac{x^{a-1}}{1-x}\,dx = -\pi i$$

$$\Rightarrow \quad \int_{0}^{\infty}\frac{(e^{i\pi})^{a-1}x^{a-1}}{1+x}\,dx+\int_{0}^{\infty}\frac{x^{a-1}}{1-x}\,dx = -\pi i$$

$$\Rightarrow \quad \int_{0}^{\infty}\frac{e^{-i\pi}e^{ia\pi}x^{a-1}}{1+x}\,dx+\int_{0}^{\infty}\frac{x^{a-1}}{1-x}\,dx = -\pi i$$

$$\therefore \quad -e^{ia\pi}\int_{0}^{\infty}\frac{x^{a-1}}{1+x}\,dx+\int_{0}^{\infty}\frac{x^{a-1}}{1-x}\,dx = -\pi i \qquad \textbf{[since } e^{-i\pi}=-1\textbf{]}$$

$$-(\cos a\pi+i\sin a\pi)\int_{0}^{\infty}\frac{x^{a-1}}{1+x}\,dx+\int_{0}^{\infty}\frac{x^{a-1}}{1-x}\,dx = -\pi i$$

Equating imaginary and real parts, we have

$$-\sin a\pi\int_{0}^{\infty}\frac{x^{a-1}}{1+x}\,dx = -\pi \Rightarrow \int_{0}^{\infty}\frac{x^{a-1}}{1+x}\,dx = \frac{\pi}{\sin a\pi} \qquad \text{...(1)}$$

$$\text{and} \quad -\cos a\pi\int_{0}^{\infty}\frac{x^{a-1}}{1+x}\,dx+\int_{0}^{\infty}\frac{x^{a-1}}{1-x}\,dx = 0$$

$$\Rightarrow \quad -\cos a\pi\int_{0}^{\infty}\frac{x^{a-1}}{1+x}\,dx = -\int_{0}^{\infty}\frac{x^{a-1}}{1-x}\,dx$$

$$\Rightarrow \quad \cos a\pi\times\frac{\pi}{\sin a\pi} = \int_{0}^{\infty}\frac{x^{a-1}}{1-x}\,dx \qquad \text{[From (1)]}$$

Thus $$\int_{0}^{\infty}\frac{x^{a-1}}{1-x}\,dx = \pi\cot a\pi$$ **Ans.**

EXERCISE 31.7

Using the method of contour integration, evaluate the following :

1. $\int_{0}^{\infty}\frac{\cos x}{x}\,dx$ **Ans.** 0 **2.** $\int_{0}^{\infty}\frac{\log(1+x^2)}{x^{1+\alpha}}\,dx,\ 0<\alpha<1$ **Ans.** 0

3. $\int_{0}^{\infty}\frac{\log x}{(1+x)^3}\,dx$ **Ans.** $-\frac{1}{2}$ **4.** $\int_{-\infty}^{\infty}\frac{1}{x^3+1}\,dx$ **Ans.** $\frac{\pi}{\sqrt{3}}$

5. $\int_{-\infty}^{\infty}\frac{x^2}{(x^2+1)^2}\,dx$ **Ans.** $\frac{\pi}{2}$ **6.** $\int_{0}^{\infty}\frac{1}{x+1}\,dx$ **Ans.** $\frac{\pi}{2}$

7. $\int_{0}^{\infty}\frac{\log x}{1+x^2}\,dx$ **Ans.** $\frac{\pi^3}{8}$

SERIES SOLUTIONS OF SECOND ORDER DIFFERENTIAL EQUATIONS

32.1 INTRODUCATION

We have already studied how to find the solution of a differential equation with constant coefficients.

There are many differential equations with variable coefficients like Bessels equation, Legender's equation, Hermite's equations, Laguerrer's differential equation whose solution are not the combination of elementary functions.

The solutions are infinite series. In this chapter we will solve the differential equations by power series method and Frobenius method (extended power series method).

32.2 POWER SERIES SOLUTIONS OF DIFFERENTIAL EQUATIONS

We know that the solution of the differential equation

$$\frac{d^2y}{dx^2} - y = 0$$

are $\qquad y = e^x = 1 + \frac{x}{1!} + \frac{x^2}{2!} + \frac{x^3}{3!} + \ldots$ and $y = e^{-x} = 1 - \frac{x}{1!} + \frac{x^2}{2!} - \frac{x^3}{3!} + \ldots$

These are power series solution of the given differential equation.

Another example of the differential equation $\frac{d^2y}{dx^2} + y = 0$

is satisfied by the power series

$$y = \sin x = x - \frac{x^3}{3!} + \frac{x^5}{5!} - \frac{x^7}{7!} + \ldots$$

This idea leads to the methods of obtaining the solution of a linear differential equation of second order in series form.

The solution of the differential equation will be a series of ascending powers of x, the infinite series solution obtained will have its own region of convergence or validity.

32.3 ANALYTIC FUNCTION

A function $f(x)$ which can be expended in Taylor's series on interval containing the point x_0. The series converges to $f(x)$ for all x in the interval of convergence.

32.4 ORDINARY POINT

Consider the equation

$$\frac{d^2y}{dx^2} + P\frac{dy}{dx} + Qy = 0$$

where P, Q are polynomials in x.

$x = a$ is an ordinary point of the above equation if the denominators of P and Q do not vanish for $x = a$. *i.e.* $(P \neq \infty, Q \neq \infty)$

For example :

(*i*) $$(1+x^2)\frac{d^2y}{dx^2}+x\frac{dy}{dx}-y=0 \Rightarrow \frac{d^2y}{dx^2}+\frac{x}{1+x^2}\frac{dy}{dx}-\frac{1}{1+x^2}y=0$$

Here, denominator of P and Q *i.e.*, $1 + x^2$ is not equal to zero at $x = 0$. Therefore $x = 0$ is an ordinary point for this differential equation.

(*ii*) $$x^2\frac{d^2y}{dx^2}+(2x^2-x)\frac{dy}{dx}+y=0 \Rightarrow \frac{d^2y}{dx^2}+\frac{2x^2-x}{x^2}\frac{dy}{dx}+\frac{1}{x^2}y=0$$

Here, the denominator of P and Q *i.e.*, x^2 is equal to zero for $x = 0$ $(P = \infty, Q = \infty)$. So, $x = 0$ is not ordinary point for this equation.

Note. In this section, we have to solve those differential equation whose ordinary point is at $x = 0$.

32.5 SOLUTION OF THE DIFFERENTIAL EQUATION WHEN *X* = 0 IS AN ORDINARY POINT *i.e.* WHEN *P* DOES NOT VANISH FOR *X* = 0.

(*i*) Let $y = a_0 + a_1x + a_2x^2 + a_3x^3 + ... + a_k x^k + ... = \sum_{k=0}^{\infty} a_k x^k$ be the solution of the given differential equation.

(*ii*) Find $\frac{dy}{dx}, \frac{d^2y}{dx^2}$ etc.

$$\frac{dy}{dx} = a_1 + 2a_2x + 3a_3x^2 + ... + ka_k x^{k-1} + ... + ... = \sum_{k=1}^{\infty} ka_k x^{k-1}$$

$$\frac{d^2y}{dx^2} = 2a_2 + 2.3\, a_3x + ... + a_k\, k(k-1)\, x^{k-2} + ... = \sum_{k=2}^{\infty} a_k \,.\, k\,(k-1)\,.\, x^{k-2}$$

(*iii*) Substitute the values of $y, \frac{dy}{dx}, \frac{d^2y}{dx^2}$ etc. in the given differential equation.

(*iv*) Calculate a_0, a_1, a_2 .... coefficients of various powers of x by equating the coefficients to zero.

(*v*) Substitute the values of a_0, a_1, a_2 ... in the differential equation to get the required series solution.

32.6 WORKING RULE TO SOLVE A DIFFERENTIAL EQUATION IF *X* = 0 IS AN ORDINARY POINT OF THE EQUATION

Step 1. Assume the solution $y = \Sigma a_k x^k$ *i.e.*,

$$y = a_0 + a_1x + a_2x^2 + a_3x^3 + a_4x^4 + ...$$

Step 2. Find $\frac{dy}{dx}$ and $\frac{d^2y}{dx^2}$ of step (1).

Step 3. Substitute the values of $y, \frac{dy}{dx}$ and $\frac{d^2y}{dx^2}$ in the given differential equation.

Step 4. Equate to zero the coefficients of various power of x, to find out a_2, a_3 ..... in terms of a_0 and a_1.

Step 5. Equate to zero the coefficient of x^k to get a recurrence relation of a's.

Step 6. Substitute $k = 0, 1, 2, 3,$ ..... in the recurrence relations to get the values of $a_2, a_3,$ .... $a_n,$ ....

Step 7. Substitute the values of $a_2, a_3,$ ...... etc (obtained in step 6) in the solution (1)

When $x = 0$ is the ordinary point.

Example 1. *Solve in series the equation* $\dfrac{d^2y}{dx^2}+x^2y=0.$

Solution. We have, $\dfrac{d^2y}{dx^2}+x^2y=0$ ... (1)

The denominator of P and Q is not zero so $x = 0$ is the ordinary point.

Let $y = a_0 + a_1x + a_2x^2 + a_3x^3 + a_4x^4 + a_5x^5 + a_6x^6 + a_7x^7 + a_8x^8 + ... + a_nx^n + ...$ ... (2)

Here $P \neq \infty$, and $Q \neq \infty$ for $x = 0$. So, $x = 0$ is the ordinary point of the equation (1).

Then $\dfrac{dy}{dx} = a_1 + 2a_2 x + 3\, a_3\, x^2 + 4\, a_4\, x^3 + 5\, a_5\, x^4 + 6\, a_6\, x^5 + 7\, a_7\, x^6 + 8\, a_8\, x^7 + ...$

$$\frac{d^2y}{dx^2} = 2a_2 + 2\,.\,3\, a_3\, x + 3\,.\,4\, a_4\, x^2 + 4\,.\,5\, a_5\, x^3 + 5\,.\,6\, a_6\, x^4 + 6\,.\,7\, a_7\, x^5 + 7\,.\,8\, a_8\, x^6 + ...$$

Substituting the values of $\dfrac{d^2y}{dx^2}$ and y in (1), we get

$$2\, a_2 + 2.3\, a_3\, x + 3.4\, a_4\, x^2 + 4.5\, a_5\, x^3 + 5.6\, a_6\, x^4 + 6.7\, a_7\, x^5 + 7.8\, a_8\, x^6 + ...$$
$$+ x^2\,(a_0 + a_1\, x + a_2\, x^2 + a_3\, x^3 + a_4\, x^4 + a_5\, x^5 + a_6\, x^6 + a_7\, x^7 + ...) = 0$$
$$2\, a_2 + 6\, a_3\, x + (a_0 + 12\, a_4)\, x^2 + (a_1 + 20a_5)\, x^3 + (a_2 + 30\, a_6)\, x^4 + ...$$
$$+ [a_{n-2} + (n+2)\,(n+1)\, a_{n+2}]\, x^n + ... = 0$$

Equating to zero the coefficients of the various powers of x, we obtain

$$a_2 = 0,\quad a_3 = 0$$

$$a_0 + 12a_4 = 0 \quad i.e.\quad a_4 = -\frac{1}{12}a_0$$

$$a_1 + 20\, a_5 = 0 \quad i.e.\quad a_5 = -\frac{1}{20}a_1$$

$$a_2 + 30\, a_6 = 0 \quad i.e.\quad a_6 = -\frac{1}{30}\, a_2 = 0 \qquad (a_2 = 0)$$

and so on. In general

$$a_{n-2} + (n+2)\,(n+1)\, a_{n+2} = 0 \quad\Rightarrow\quad \boxed{a_{n+2} = -\frac{a_{n-2}}{(n+1)(n+2)}}$$

Putting $n = 5$, $\quad a_7 = -\dfrac{a_3}{6\times 7} = 0 \qquad (a_3 = 0)$

Putting $n = 6$, $\quad a_8 = -\dfrac{a_4}{7\times 8} = \dfrac{a_0}{12\times 7\times 8}$

Putting $n = 7$, $\quad a_9 = -\dfrac{a_5}{8\times 9} = \dfrac{a_1}{20\times 8\times 9}$

Putting $n = 8$, $\quad a_{10} = -\dfrac{a_6}{9\times 10} = 0, \qquad (a_6 = 0)$

Putting $n = 9$, $\quad a_{11} = -\dfrac{a_7}{11\times 10} = 0, \qquad (a_7 = 0)$

Putting $n = 10$, $\quad a_{12} = -\dfrac{a_8}{12\times 11} = -\dfrac{a_0}{12\times 8\times 7\times 11\times 12}$

Substituting these values in (2), we get

$$y = a_0 + a_1 x - \frac{1}{12}a_0 x^4 - \frac{a_1}{20}x^5 + \frac{a_0}{12\times 7\times 8}x^8 + \frac{a_1}{20\times 8\times 9}x^9 - \frac{a_0}{12\times 8\times 7\times 11\times 12}x^{12} + ...$$

$$y = a_0\left(1 - \frac{1}{12}x^4 + \frac{x^8}{12\times 7\times 8} - \frac{x^{12}}{12\times 8\times 7\times 11\times 12} + ...\right) + a_1\left(x - \frac{x^5}{20} + \frac{x^9}{20\times 8\times 9} - ...\right)$$ **Ans.**

Example 2. *Solve the following differential equation in series*

$$(1-x^2)\frac{d^2y}{dx^2}-x\frac{dy}{dx}+4y=0. \qquad \text{(U.P. II Semester summer 2006)}$$

Solution. We have,

$$(1-x^2)\frac{d^2y}{dx^2}-x\frac{dy}{dx}+4y=0 \Rightarrow \frac{d^2y}{dx^2}-\frac{x}{1-x^2}\frac{dy}{dx}+\frac{4}{1-x^2}y=0 \qquad \text{... (1)}$$

Here, $P \neq \infty$ and $Q \neq \infty$ for $x = 0$. So, $x = 0$ is an ordinary point of the given equation.
Assume the solution

$$y = \Sigma a_k x^k$$

$$\Rightarrow \quad y = a_0 + a_1x + a_2x^2 + a_3x^3 + \ldots + a_nx^n + \ldots\ldots \qquad \text{... (2)}$$

$$y' = \Sigma a_k (k) x^{k-1}$$

$$y'' = \Sigma a_k (k)(k-1)x^{k-2}$$

Putting these values in the given equation, we get

$$(1-x^2)\Sigma a_k (k)(k-1)x^{k-2} - x\Sigma a_k (k)x^{k-1} + 4\Sigma a_k x^k = 0$$

$$\Rightarrow \quad \Sigma a_k (k)(k-1)x^{k-2} - \Sigma a_k (k)(k-1)x^k - \Sigma a_k (k)x^k + 4\Sigma a_k x^k = 0$$

$$\Rightarrow \quad \Sigma a_k (k)(k-1)x^{k-2} - \Sigma a_k x^k \{k(k-1)+k-4\} = 0$$

$$\Rightarrow \quad \Sigma a_k k(k-1)x^{k-2} - \Sigma a_k x^k (k^2-4) = 0$$

Now, equating the coefficient of x^k equal to zero. By putting $k + 2$ for k in first summation and $k = k$ in second summation, we have

$$a_{k+2}(k+2)(k+1) - a_k(k^2-4) = 0$$

$$a_{k+2}(k+2)(k+1) = a_k(k^2-4)$$

$$a_{k+2} = \frac{k^2-4}{(k+2)(k+1)}a_k = \frac{k-2}{k+1}a_k \quad \Rightarrow \quad \boxed{a_{k+2} = \frac{k-2}{k+1}a_k}$$

If $k = 0$, $\quad a_2 = -2a_0$

If $k = 1$, $\quad a_3 = -\frac{1}{2}a_1$

If $k = 2$, $\quad a_4 = \frac{0}{3}a_2 = 0$

If $k = 3$, $\quad a_5 = \frac{1}{4}a_3 = \frac{1}{4}\left(-\frac{1}{2}\right)a_1 = -\frac{a_1}{8}$

If $k = 4$, $\quad a_6 = \frac{2}{5}a_4 = \frac{2}{5}\times 0 = 0$

and so on.

Substituting these values is (2), we get

$$y = a_0 + a_1x + (-2a_0)x^2 + \left(-\frac{1}{2}a_1\right)x^3 + 0x^4 + \left(\frac{-a_1}{8}\right)x^5 + 0.x^6 + \ldots\ldots$$

$$= a_0 + a_1x - 2a_0x^2 - \frac{1}{2}a_1x^3 - \frac{a_1}{8}x^5 + \ldots\ldots$$

$$= a_0(1-2x^2) + a_1x\left(1 - \frac{x^2}{2} - \frac{x^4}{8} + \ldots\ldots\right) \qquad \textbf{Ans.}$$

Example 3. *Find the power series solution of $(1 - x^2)\, y'' - 2xy' + 2y = 0$ about $x = 0$.*
(A.M.I.E.T.E., Winter 2000)

Solution. Here, we have

$$(1-x^2)\frac{d^2y}{dx^2} - 2x\frac{dy}{dx} + 2y = 0 \Rightarrow \frac{d^2y}{dx^2} - \frac{2x}{1-x^2}\frac{dy}{dx} + \frac{2}{1-x^2}y = 0$$

Here $P \neq \infty$ and $Q \neq \infty$ for $x = 0$. So $x = 0$ is an ordinary point.

Let $y = a_0 + a_1 x + a_2 x^2 + a_3 x^3 + \dots + a_n x^n + \dots$ be the required solution.

$$y = \sum_{k=0}^{\infty} a_k x^k$$

Then $$\frac{dy}{dx} = \sum_{k=1}^{\infty} a_k \cdot k\, x^{k-1}, \qquad \frac{d^2y}{dx^2} = \sum_{k=2}^{\infty} a_k \cdot k(k-1)x^{k-2}$$

Substituting the values of $y, \frac{dy}{dx}$ and $\frac{d^2y}{dx^2}$ in the given equation, we get

$$(1-x^2)\Sigma a_k\, k\cdot(k-1)x^{k-2} - 2x\Sigma a_k \cdot k\, x^{k-1} + 2\Sigma a_k\, x^k = 0$$

$$\Rightarrow \quad \Sigma a_k \cdot k\cdot(k-1)x^{k-2} - \Sigma a_k k(k-1)x^k - 2\,\Sigma a_k . k x^k + 2\Sigma a_k x^k = 0$$

$$\Rightarrow \quad \Sigma a_k \cdot k\cdot(k-1)x^{k-2} - \Sigma[k(k-1) + 2k - 2]a_k x^k = 0$$

$$\Rightarrow \quad \Sigma a_k \cdot k\cdot(k-1)x^{k-2} - \Sigma(k^2 + k - 2)a_k x^k = 0$$

where the first summation extends over all values of k from 2 to ∞ and the second from $k = 0$ to ∞.

Now equating the coefficient of x^k equal to zero, we have

$$\Rightarrow \quad (k+2)\,(k+1)a_{k+2} - (k^2 + k - 2)a_k = 0$$

$$\Rightarrow \quad a_{k+2} = \frac{k^2 + k - 2}{(k+2)(k+1)}a_k = \frac{(k+2)(k-1)}{(k+2)(k+1)}a_k$$

$$\Rightarrow \quad \boxed{a_{k+2} = \frac{k-1}{k+1}a_k}$$

For $k = 0$ $$a_2 = -a_0,\ a_3 = 0,\ a_4 = \frac{a_2}{3} = -\frac{a_0}{3},\ a_5 = \frac{2}{4}a_3 = 0$$

For $k = 4$ $$a_6 = \frac{3}{5}a_4 = \frac{3}{5}\left(-\frac{a_0}{3}\right) = -\frac{a_0}{5},\ a_7 = \frac{4}{6}a_5 = 0, \text{ etc.}$$

$$y = a_0 + a_1 x + a_2 x^2 + a_3 x^3 + a_4 x^4 + a_5 x^5 + a_6 x^6 + a_7 x^7 + \dots$$

$$= a_0 + a_1 x - a_0 x^2 + 0 - \frac{a_0}{3}x^4 + 0 - \frac{a_0}{5}x^6 + 0 + \dots$$

$$\Rightarrow \quad y = a_0\left[1 - x^2 - \frac{x^4}{3} - \frac{x^6}{5} + \dots\right] + a_1 x$$ **Ans.**

32.7 SINGULAR POINTS ABOUT $x = a$

Definition. Consider the equation

$$y'' + P(x)\,y' + Q(x)y = 0 \qquad \dots (1)$$

and assume that functions P and Q are not analytic ($P = \infty$ or $Q = \infty$) at $x = a$, so that $x = a$ is not *ordinary point* but a *singular point of* (1).

There are two types of singular points. (1) Regular singular point, (2) Irregular singular points.

1. Regular Singular Point:

If $(x-a)P$ and $(x-a)^2 Q$ are not infinite at $x = a$, then $x = a$ is a regular singular point.

2. Irregular Singular Point:

If $(x-a)P$ and $(x-a)^2 Q$ are infinite at $x = a$, then $x = a$ is an irregular singular point.

Example 4. *Solve the differential equation $y'' + (x-1)^2 y' - 4(x-1)y = 0$ in series about the ordinary point $x = 1$.*

Solution. Put $\quad x = t + 1 \qquad$ (or $x - 1 = t$)

$$\frac{dy}{dx} = \frac{dy}{dt} \bullet \frac{dt}{dx} = \frac{dy}{dt} \qquad \left(\because \frac{dt}{dx} = 1\right)$$

$$\Rightarrow \quad \frac{d}{dx} \equiv \frac{d}{dt}$$

$$\frac{d^2y}{dx^2} = \frac{d}{dx}\left(\frac{dy}{dx}\right) = \frac{d}{dt}\left(\frac{dy}{dt}\right) = \frac{d^2y}{dt^2}$$

$\therefore$ The given equation becomes,

$$\frac{d^2y}{dt^2} + t^2 y' - 4t\, y = 0 \qquad \text{...(1)}$$

Now, $t = 0$ is an ordinary point. [given]

Assume the solution to be

$$y = a_0 + a_1 t + a_2 t^2 + a_3 t^3 + \ldots + a_n t^n + \ldots \qquad \text{...(2)}$$

then $\quad y' = a_1 + 2a_2 t + 3a_3 t^2 + \ldots + na_n t^{n-1} + \ldots$

and $\quad y'' = 2a_2 + 3.2.\, a_3 t + \ldots + n(n-1)\, a_n t^{n-2} + \ldots$

Substituting these values is equation (1), we get

$$[2a_2 + 3.2\,.\, a_3 t + 4.3\,.\, a_4 t^2 + \ldots + n(n-1)\, a_n t^{n-2} + \ldots]$$
$$+ t^2 [a_1 + 2a_2 t + 3a_3 t^2 + 4a_4 t^3 + \ldots + n\, a_n t^{n-1} + \ldots]$$
$$- 4t\, [a_0 + a_1 t + a_2 t^2 + a_3 t^3 + \ldots + a_n t^n + \ldots] = 0$$

Coeffieient of $t^0 = 0$

$$\Rightarrow \quad 2a_2 = 0 \quad \Rightarrow \quad \boxed{a_2 = 0}$$

Coefficient of $t = 0$

$$\Rightarrow \quad 3.2.\, a_3 - 4a_0 = 0 \quad \Rightarrow \quad \boxed{a_3 = \frac{2a_0}{3}}$$

Coefficient of $t^2 = 0$

$$\Rightarrow \quad 4.3.\, a_4 + a_1 - 4a_1 = 0$$

$$\Rightarrow \quad 12a_4 = 3a_1 \quad \Rightarrow \quad \boxed{a_4 = \frac{a_1}{4}}$$

Coefficient of $t^3 = 0$

$$\Rightarrow \quad 5.4.\, a_5 + 2a_2 - 4a_2 = 0 \Rightarrow a_5 = \frac{1}{10} a_2 \quad \Rightarrow \quad \boxed{a_5 = 0}$$

Coefficient of $t^4 = 0$

$$\Rightarrow \quad 6.5.\, a_6 + 3a_3 - 4a_3 = 0$$

$$a_6 = \frac{a_3}{6.5} = \frac{2a_0}{6.5.3} \quad \Rightarrow \quad \boxed{a_6 = \frac{a_0}{45}}$$

Now, coefficient of $t^n = 0$

$\Rightarrow (n+2)(n+1)a_{n+2} + (n-1)a_{n-1} - 4a_{n-1} = 0$

$$\Rightarrow \quad a_{n+2} = -\frac{(n-5)}{(n+2)(n+1)} a_{n-1}$$

Putting $n = 5, 6, 7, 8, \ldots$, we get

$$a_7 = 0$$

$$a_8 = \frac{-1}{8.7} a_5 = 0$$

$$a_9 = \frac{-2}{9.8} a_6 = \frac{-2}{9.8}\frac{a_0}{45} = -\frac{a_0}{1620}$$

and so on.

Substituting these values in (2), we get

$$y = a_0 + a_1 t + \frac{2}{3} a_0 t^3 + \frac{a_1}{4} t^4 + \frac{a_0}{45} t^6 - \frac{a_0}{1620} t^9 + \ldots$$

$$= a_0\left(1 + \frac{2}{3}t^3 + \frac{1}{45}t^6 - \frac{1}{1620}t^9 + \ldots\right) + a_1\left(t + \frac{t^4}{4} + \ldots\right)$$

$$\Rightarrow \quad y = a_0\left[1 + \frac{2}{3}(x-1)^3 + \frac{1}{45}(x-1)^6 - \frac{1}{1620}(x-1)^9 + \ldots\right] + a_1\left[(x-1) + \frac{(x-1)^4}{4} + \ldots\right]$$

Where a_0 and a_1 are constants. **Ans.**

Note. Example 4 is solved about the regular singular point $x = 1$. Now we will solve the problems about the regular singular point $x = 0$.

We can also find the solution about a point other than $x = 0$, say about $x = a$. In this case we have to find out the series solution of powers of $(x - a)$, and the series is valid (convergent) around the point $x = a$.

In this method first we shift the origin to the point $x = c$, by putting $x = t + c$. The differential equation so obtained is solved by the method already discussed.

EXERCISE 32.1

Solve the following differential equation by power series method :

1. $\dfrac{d^2y}{dx^2} + xy = 0$ **Ans.** $y = a_0\left(1 - \dfrac{x^3}{3!} + \dfrac{4x^6}{6!} - \dfrac{28x^9}{9!} + \ldots\right) + a_1\left(x - \dfrac{2x^4}{4!} + \dfrac{10x^7}{7!} + \ldots\right)$

(AMIETE, June 2010)

2. $y'' - xy' + x^2 y = 0$ **Ans.** $y = a_0\left(1 - \dfrac{1}{12}x^4 - \ldots\right) + a_1\left(x + \dfrac{1}{6}x^3 - \dfrac{1}{40}x^5 \ldots\right)$

3. $(x^2+1)y'' + xy' - xv = 0$ (U.P. (C.O.) 2008)

Ans. $y = a_0\left(1 + \dfrac{x^3}{6} - \dfrac{3}{40}x^5 + \ldots\right) + a_1\left(x - \dfrac{x^3}{6} + \dfrac{x^4}{12} + \dfrac{3}{40}x^5 \ldots\right)$

4. $y'' - 2x^2 y' + 4xy = x^2 + 2x + 4$

Ans. $y = a_0\left(1 - \dfrac{2}{3}x^2 - \dfrac{2}{45}x^6 - \dfrac{2}{405}x^9 \ldots\right) + a_1\left(x - \dfrac{1}{6}x^4 - \dfrac{1}{63}x^7 - \dfrac{1}{567}x^{10} \ldots\right)$

$+ 2x^2 + \dfrac{1}{3}x^3 + \dfrac{1}{12}x^4 + \dfrac{1}{45}x^6 + \dfrac{1}{126}x^7 + \dfrac{1}{405}x^9 + \dfrac{1}{1134}x^{10} + \ldots$

5. $(x^2+2)y'' + xy' - (1+xy) = 0$ **Ans.** $y = a_0\left(1 + \dfrac{1}{4}x^2 + \dfrac{1}{12}x^3 - \dfrac{1}{32}x^4 \ldots\right) + a_1\left(x + \dfrac{1}{24}x^4 + \ldots\right)$

32.8 SINGULAR POINT ABOUT $x = 0$.

Consider the differential equation

$$\frac{d^2y}{dx^2} + P\frac{dy}{dx} + Qy = 0$$

If $(x - 0)\,P$ and $(x - 0)^2\,Q$ are not infinite at $x = 0$, then $x = 0$ is a regular singular point. Otherwise it is an irregular singular point.

Note: In this section we will solve those differential equation where x_0 is a regular singular point.

Example 5. *Find regular singular points of the differential equation.*

$$2x^2\frac{d^2y}{dx^2} + 3x\frac{dy}{dx} + (x^2 - 4)\,y = 0 \qquad \text{... (1)}$$

Solution. We have, $\dfrac{d^2y}{dx^2} + \dfrac{3}{2x}\dfrac{dy}{dx} + \dfrac{x^2-4}{2x^2}\;y = 0$

$$P = \frac{3}{2x} \quad \text{and } Q = \frac{x^2-4}{2x^2}$$

P and Q are not analytic (infinity) at $x = 0$. So, $x = 0$ is not ordinary point but as $(x - 0)\,P$ and $(x - 0)^2\,Q$ are analytic (not infinite) so $x = 0$ is a regular singular point.

$$x \cdot P = x\left(\frac{3}{2x}\right) = \frac{3}{2} \neq \infty \text{ at } x = 0.$$

$$x^2 Q = x^2 \cdot \frac{x^2-4}{2x^2} = \frac{1}{2}(x^2 - 4) \neq \infty \text{ at } x = 0$$

Ans.

Example 6. *Find regular singular points of the differential equation.*

$$x(x-2)^2 y'' + 2(x-2)y' + (x+3)y = 0 \qquad \text{... (1)}$$

Solution. Here, we have $P = \dfrac{2(x-2)}{x\,(x-2)^2} = \dfrac{2}{x\,(x-2)}$ and $Q = \dfrac{x+3}{x\,(x-2)^2}$

P and Q are not analytic ($P = \infty$, $Q = \infty$) at $x = 0$ and $x = 2$.

Hence both these points are singular points of (1).

(*i*) At $x = 0$

$$xP = x \cdot \frac{2}{x\,(x-2)} = \frac{2}{x-2} \neq \infty \quad \text{at } x = 0$$

$$x^2 Q = x^2 \cdot \frac{x+3}{x\,(x-2)^2} = \frac{x\,(x+3)}{(x-2)^2} \neq \infty \quad \text{at } x = 0$$

Hence, $x\,P$ and $x^2\,Q$ are analytic ($xP \neq \infty$, $x^2P \neq \infty$) at $x = 0$. So $x = 0$ is a regular singular point.

(*ii*) At $x = 2$

$$(x-2)P = (x-2) \cdot \frac{2}{x(x-2)} = \frac{2}{x} \neq \infty \text{ at } x = 2$$

$$(x-2)^2 Q = (x-2)^2 \frac{(x+3)}{x(x-2)^2} = \frac{x+3}{x} \neq \infty \text{ at } x = 2.$$

Since both $(x - 2)\,P$ and $(x - 2)^2 Q$ are analytic ($(x - 2)\,P \neq \infty$, $(x - 2)^2 Q \neq 0$) at $x = 2$, so $x = 2$ is a regular singular point.

The solution of a differential equation about a regular singular point can be obtained.

The cases of irregular singular points are beyond the scope of this book. **Ans.**

32.9 FROBENIUS METHOD

If $x = 0$ is a regular singularity of the equation.

$$\frac{d^2y}{dx^2} + P(x)\frac{dy}{dx} + Q(x)\,y = 0 \qquad [P_0(0) = 0] \quad \ldots (1)$$

Then the series solution is

$$\boxed{y = x^m (a_0 + a_1 x + a_2 x^2 + a_3 x^3 + \ldots) = \sum_{k=0}^{\infty} a_k x^{m+k}}$$

The value of m will be determined by substituting the expressions for $y, \frac{dy}{dx}, \frac{d^2y}{dx^2}$ in (1), we get the identity.

On equating the coefficient of lowest power of x in the identity to zero, a quadratic equation in m (**indicial equation**) is obtained.

Thus, we will get two values of m. The series solution of (1) will depend on the nature of the roots of the indicial equation.

(*i*) **Case 1 : When roots m_1, m_2 are distinct and not differing by an integer i.e.** $m_1 - m_2 \neq 0$ or a positive integer. *e.g.*, $m_1 = \frac{1}{2}, m_2 = 2$.

The complete solution is $\boxed{y = c_1(y)_{m_1} + c_2(y)\; m_2}$

(*ii*) **Case 2 : When roots m_1, m_2 *are equal i.e.* $m_1 = m_2$**

$$\boxed{y = c_1(y)_{m_1} + c_2\left(\frac{\partial y}{\partial m}\right)_{m_1}}$$

(*iii*) **Case 3 : When roots m_1, m_2 are distinct and differ by an integer** ($m_1 < m_2$) *e.g.*, $m_1 = \frac{3}{2}, m_2 = \frac{5}{2}$ or $m_1 = 2, m_2 = 4$.

If some of the coefficients of y series become infinite when $m = m_1$, to overcome this difficulty, replace a_0 by $b_0\,(m - m_1)$. We get a solution which is only a constant multiple of the first solution.

$$\boxed{a_0 = b_0(m - m_1)}$$

Complete solution is $\boxed{y = c_1(y)_{m_1} + c_2\left(\frac{\partial y}{\partial m}\right) m_1}$

(*iv*) **Case 4 : Roots are distinct and differing by an integer, making some coefficient indeterminate**

Complete solution is $y = c_1(y)_{m_1} + c_2(y)_{m_2}$

if the coefficients do not become infinite when $m_1 = m_2$.

Case I : When the roots are distinct and not differing by an integer.

Example 7. *Find solution in generalized series form about $x = 0$ of the differential equation*

$$3x\frac{d^2y}{dx^2} + 2\frac{dy}{dx} + y = 0$$

Solution. We have, $$3x\frac{d^2y}{dx^2} + 2\frac{dy}{dx} + y = 0 \qquad \ldots (1)$$

Here, $xP(x)$ and $x^2Q(x)$ are analytic (not infinite). So, $x = 0$ is a regular singular point, we assume the solution in the form

$$y = \sum_{k=0}^{\infty} a_k x^{m+k}$$

Such that
$$\frac{dy}{dx} = \sum_{k=0}^{\infty} a_k (m+k) x^{m+k-1}$$

$$\frac{d^2y}{dx^2} = \sum_{k=0}^{\infty} a_k (m+k)(m+k-1) x^{m+k-2}$$

Substituting for y, $\frac{dy}{dx}$ and $\frac{d^2y}{dx^2}$ in the given equation (1), we get

$$3\Sigma a_k (m+k)(m+k-1) x^{m+k-1} + 2\Sigma a_k (m+k) x^{m+k-1} + \Sigma a_k x^{m+k} = 0$$

$$\Rightarrow \quad \Sigma a_k [3(m+k)(m+k-1) + 2(m+k)] x^{m+k-1} + \Sigma a_k x^{m+k} = 0 \quad \ldots (2)$$

The coefficient of the lowest degree term x^{m-1} in the identity (2) is obtained by putting $k = 0$ in first summation only and equating it to zero. Then the **indicial equation** is

$$a_0 [3m(m-1) + 2m] = 0 \Rightarrow a_0 [3m^2 - m] = 0 \Rightarrow \boxed{a_0 m(3m-1) = 0}$$

Since $a_0 \neq 0$, $m = 0$, or $1/3$

The coefficient of next lowest degree term x^m in the identity (2) is obtained by putting $k = 1$ in first summation and $k = 0$ in the second summation and equating it to zero.

$$a_1 [3(m+1)m + 2(m+1)] + a_0 = 0$$

$$\Rightarrow \quad a_1 [3m^2 + 5m + 2] + a_0 = 0 \Rightarrow a_1 (3m+2)(m+1) + a_0 = 0$$

$$\Rightarrow \quad a_1 = -\frac{1}{(3m+2)(m+1)} a_0$$

Equating to zero the coefficient of x^{m+k}, the recurrence relation is given by

$$a_{k+1} [3(m+k+1)(m+k) + 2(m+k+1)] + a_k = 0.$$

$$\Rightarrow \quad a_{k+1} (m+k+1)(3m+3k+2) + a_k = 0 \Rightarrow \boxed{a_{k+1} = \frac{-1}{(m+k+1)(3m+3k+2)} a_k}$$

This gives

For $k = 0$, $\quad a_1 = \frac{-1}{(m+1)(3m+2)} a_0$

For $k = 1$, $\quad a_2 = \frac{-1}{(m+2)(3m+5)} a_1 = \frac{1}{(m+1)(m+2)(3m+2)(3m+5)} a_0$

For $k = 2$, $\quad a_3 = \frac{-1}{(m+3)(3m+8)} a_2$

$$= \frac{-1}{(m+1)(m+2)(m+3)(3m+2)(3m+5)(3m+8)} a_0$$

For $m = 0$

$$a_1 = -\frac{1}{2} a_0, \quad a_2 = \frac{1}{20} a_0, \quad a_3 = -\frac{1}{480} a_0$$

Hence, for $m = 0$, $\quad y_1 = a_0 \left(1 - \frac{1}{2}x + \frac{1}{20}x^2 - \frac{1}{480}x^3 + \ldots\right)$

For $m = \frac{1}{3}$

$$a_1 = -\frac{1}{4}a_0, \qquad a_2 = \frac{1}{56}a_0, \qquad a_3 = \frac{-1}{1680}a_0$$

Hence for $m = \frac{1}{3}$, the second solution is

$$y_2 = a_0\left(x^{\frac{1}{3}} - \frac{1}{4}x^{\frac{4}{3}} + \frac{1}{56}x^{\frac{7}{3}} - \frac{1}{1680}x^{\frac{10}{3}} + \ldots\right)$$

Thus the complete solution is

$$y = Ay_1 + By_2$$

$$y = a_0\left(1 - \frac{x}{2} + \frac{x^2}{20} - \frac{x^3}{480} + \ldots\right) + b_0 x^{1/3}\left(1 - \frac{x}{4} + \frac{x^2}{56} - \frac{x^3}{1680} + \ldots\right) \quad \textbf{Ans.}$$

Example 8. *Solve* $9x(1-x)y'' - 12y' + 4y = 0$ ...(1)

Solution. Here, $xP(x)$ and $x^2Q(x)$ are analytic (not infinite) at $x = 0$. So, it is a regular singular point.

Let $y = \sum_{k=0}^{\infty} a_k x^{m+k}$ be the solution of (1)

Differentiating twice in succession, we get

$$y' = \sum_{k=0}^{\infty} a_k (m+k) x^{m+k-1}, \qquad y'' = \sum_{k=0}^{\infty} a_k (m+k)(m+k-1) x^{m+k-2}$$

Substituting the values of y' and y'' in (1), we have

$$9x(1-x)\sum_{k=0}^{\infty} a_k (m+k)(m+k-1)x^{m+k-2} - 12\sum_{k=0}^{\infty} a_k (m+k)x^{m+k-1} + 4\sum_{k=0}^{\infty} a_k x^{m+k} = 0$$

$$\Rightarrow 9\sum_{k=0}^{\infty} a_k (m+k)(m+k-1)x^{m+k-1} - 9\sum_{k=0}^{\infty} a_k (m+k)(m+k-1)x^{m+k}$$

$$-12\sum_{k=0}^{\infty} a_k (m+k)x^{m+k-1} + 4\sum_{k=0}^{\infty} a_k x^{m+k} = 0$$

$$\Rightarrow 3\sum_{=0}^{\infty} a_k (m+k)(3m+3k-7)x^{m+k-1} - \sum_{k=0}^{\infty} a_k [9(m+k)^2 + (9m+9k+4)]\, x^{m+k} = 0$$

$$\Rightarrow 3\sum_{k=0}^{\infty} a_k (m+k)(3m+3k-7)x^{m+k-1} - \sum_{k=0}^{\infty} a_k (3m+3k-4)(3m+3k+1)x^{m+k} = 0 \qquad \ldots (2)$$

The coefficient of the lowest degree term x^{m-1} in the identity (2) is obtained by putting $k = 0$ in first summation only and equating it to zero. Then the **indicial equation** is

$$3a_0 m(3m-7) = 0 \Rightarrow m = 0,\ m = \frac{7}{3}$$

The coefficient of next lowest degree term x^m in the identity (2) is obtained by putting $k = 1$ in first summation and $k = 0$ in the second summation and equating it to zero.

$$3a_1 (m+1)(3m+3-7) - a_0 (3m-4)(3m+1) = 0$$

$$a_1 = \frac{(3m-4)(3m+1)}{3(m+1)(3m-4)}a_0 = \frac{3m+1}{3m+3}a_0$$

Equating to zero the coefficient of x^{m+k}, the recurrence relation is given by

$$3a_{k+1}(m+k+1)(3m+3k-4)-a_k(3m+3k-4)(3m+3k+1)=0$$

$$a_{k+1}=\frac{(m+3k-4)(3m+3k+1)}{3(m+k+1)(3m+3k-4)}a_k$$

$$\boxed{\boldsymbol{a_{k+1}=\frac{3m+3k+1}{3m+3k+3}a_k}}$$

$m=0$

$$a_{k+1}=\frac{3k+1}{3k+3}a_k$$

$k=0,\quad a_1=\frac{1}{3}a_0$

$k=1,\quad a_2=\frac{2}{3}a_1=\frac{2}{9}a_0$

$k=2,\quad a_3=\frac{7}{9}a_2=\frac{7}{9}\cdot\frac{2}{9}a_0=\frac{14}{81}a_0$

and so on

$m=\frac{7}{3}$

$$a_{k+1}=\frac{3k+8}{3k+10}a_k$$

$k=0,\quad a_1=\frac{8}{10}a_0=\frac{4}{5}a_0$

$k=1,\quad a_2=\frac{11}{13}a_1=\frac{11}{13}\times\frac{4}{5}a_0=\frac{44}{65}a_0$

$k=2,\quad a_3=\frac{14}{16}a_2=\frac{7}{8}\times\frac{44}{65}a_0=\frac{77}{130}a_0$

and so on

$$y_1=a_0\left(1+\frac{1}{3}x+\frac{2}{9}x^2+\frac{14}{81}x^3+\dots\right),\quad y_2=a_0x^{\frac{7}{3}}\left(1+\frac{4}{5}x+\frac{44}{65}x^2+\frac{77}{130}x^3+\dots\right)$$

Since two solutions are linearly independent, the general solution of (1) may be written as

$y=Ay_1+By_2$ **Ans.**

Example 9. *Solve the following equation in power series about $x=0$.*

$$2x^2\frac{d^2y}{dx^2}+x\frac{dy}{dx}-(x+1)y=0$$ (*U.P. II Semester Summer 2005*)

Solution. Given equation is $2x^2\frac{d^2y}{dx^2}+x\frac{dy}{dx}-(x+1)y=0$ ... (1)

Here, $x\,P(x)$ and $x^2Q(x)$ are analytic (not infinite) at $x=0$. So, $x=0$ is a regular singular point, we assume the solution in the form

$$y=\sum_{k=0}^{\infty}a_kx^{m+k}$$

$$\frac{dy}{dx}=\Sigma a_k(m+k)x^{m+k-1}$$

$$\frac{d^2y}{dx^2}=\Sigma a_k(m+k)(m+k-1)x^{m+k-2}$$

Putting the values of y, $\frac{dy}{dx}$ and $\frac{d^2y}{dx^2}$ in equation (1), we get

$$2x^2\Sigma a_k(m+k)(m+k-1)x^{m+k-2}+x\Sigma a_k(m+k)x^{m+k-1}-(x+1)\Sigma a_kx^{m+k}=0$$

$$\Rightarrow\quad 2\Sigma a_k(m+k)(m+k-1)x^{m+k}+\Sigma a_k(m+k)x^{m+k}-\Sigma a_kx^{m+k+1}-\Sigma a_kx^{m+k}=0$$

$$\Rightarrow\quad \Sigma a_k\left[2(m+k)(m+k-1)+(m+k)-1\right]x^{m+k}-\Sigma a_kx^{m+k+1}=0 \quad \dots (2)$$

Equating the coefficient of lowest degree term x^m to zero by putting $k=0$ in first summation of (2)

$$\boxed{\boldsymbol{a_0[2m(m-1)+m-1]=0}}$$ [Indicial equation]

Let $\quad a_0 \neq 0$

$$2m(m-1)+m-1=0 \quad \Rightarrow \quad (m-1)(2m+1)=0 \quad \Rightarrow \quad m=1, -\frac{1}{2}$$

Now equating the coefficients of next lowest degree terms to zero by putting $k = 1$ and $k = 0$ in first and second summation of (2) respectively, we get

$$a_1[2(m+1)m+m+1-1]-a_0=0 \quad \Rightarrow \quad a_1(2m^2+2m+m)=a_0$$

$$\Rightarrow \qquad a_1=\frac{a_0}{m(2m+3)}$$

Now equating to zero the coefficient of x^{m+k+1} by putting $k = k + 1$ in first summation and $k = k$ in second summation, we get

$$a_{k+1}\,[2(m+k+1)(m+k)+(m+k+1)-1]-a_k=0$$

$$\Rightarrow \qquad a_{k+1}[(m+k+1)\{2(m+k)+1\}-1]=a_k$$

$$\Rightarrow \qquad a_{k+1}[(m+k+1)(2m+2k+1)-1]=a_k$$

$$\Rightarrow \qquad \boxed{\boldsymbol{a_{k+1}=\frac{a_k}{(m+k+1)(2m+2k+1)-1}}}$$

If $k = 1$, $\qquad a_2=\dfrac{a_1}{(m+2)(2m+3)-1},$

If $k = 2$, $\qquad a_3=\dfrac{a_2}{(m+3)(2m+5)-1}$

If $k = 3$, $\qquad a_4=\dfrac{a_3}{(m+4)(2m+7)-1}$

$m=1$	$m=-\dfrac{1}{2}$
$a_1=\dfrac{a_0}{5},$	$a_1=-a_0,$
$a_2=\dfrac{a_1}{14}=\dfrac{a_0}{70},$	$a_2=-\dfrac{a_0}{2},$
$a_3=\dfrac{a_2}{27}=\dfrac{a_0}{1890},$	$a_3=\dfrac{a_2}{9}=\dfrac{a_0}{9(-2)}=-\dfrac{a_0}{18},$
$a_4=\dfrac{a_3}{44}=\dfrac{a_0}{44\times1890}$	$a_4=\dfrac{a_3}{20}=-\dfrac{a_0}{18\times20}=-\dfrac{a_0}{360}$

We have, $\quad y = x^m\,(a_0 + a_1x + a_2x^2 + a_3x^3 + a_4x^4 + \ldots\ldots)$

For $\boldsymbol{m = 1}$

$$(y)_{m=1}=x\left(a_0+\frac{a_0}{5}x+\frac{a_0}{70}x^2+\frac{a_0}{1890}x^3+\frac{a_0}{44(1890)}x^4+\ldots\right)$$

$$\Rightarrow \qquad (y)_{m=1}=a_0x\left(1+\frac{x}{5}+\frac{x^2}{70}+\frac{x^3}{1890}+\frac{x^4}{44(1890)}+\ldots\right)$$

For $\boldsymbol{m=-\frac{1}{2}}$

$$(y)_{m=-\frac{1}{2}} = x^{-\frac{1}{2}}\left(a_0 - a_0 x - \frac{a_0}{2}x^2 - \frac{a_0}{18}x^3 - \frac{a_0}{360}x^4 + ...\right)$$

$$\Rightarrow \quad (y)_{m=-\frac{1}{2}} = a_0 x^{-\frac{1}{2}}\left(1 - x - \frac{1}{2}x^2 - \frac{1}{18}x^3 - \frac{1}{360}x^4 + ...\right)$$

Thus roots of indicial equation are distinct and not differing by an integer. Its solution is given by

$$y = c_1 (y)_{m=1} + c_2 (y)_{m=-\frac{1}{2}}$$

Thus, the required solution is

$$y = c_1 a_0 x\left(1 + \frac{x}{5} + \frac{x^2}{70} + \frac{x^3}{1890} + \frac{1}{44(1890)}x^4 + ...\right)$$

$$+ c_2 a_0 x^{-1/2}\left(1 - x - \frac{x^2}{2} - \frac{x^3}{18} - \frac{x^4}{360} - ...\right)$$

$$\Rightarrow \quad y = Ax\left(1 + \frac{x}{5} + \frac{x^2}{70} + \frac{x^3}{1890} + \frac{x^4}{44(1890)} + ...\right) + Bx^{-1/2}\left(1 - x - \frac{x^2}{2} - \frac{x^3}{18} - \frac{x^4}{360} .....\right)$$ **Ans.**

Example 10. *Using Frobenius method, obtain a series soultion in powers of x for differential equation :* $2x(1-x)\frac{d^2y}{dx^2} + (1-x)\frac{dy}{dx} + 3y = 0.$ *(U.P. II. Sem 2010, June 2001)*

Solution. Here, we have

$$2x(1-x)\frac{d^2y}{dx^2} + (1-x)\frac{dy}{dx} + 3y = 0 \qquad ...(1)$$

Here $xP_1(x)$ and $x^2P_2(x)$ are analytic (Not infinite) at $x = 0$. So, $x = 0$ is a regular singular point, we assume the solution in the form

$$y = \sum_{k=0}^{\infty} a_k x^{m+k}$$

such that

$$\frac{dy}{dx} = \sum_{k=0}^{\infty} a_k (m+k) x^{m+k-1}$$

$$\frac{d^2y}{dx^2} = \sum_{k=0}^{\infty} a_k (m+k)(m+k-1) x^{m+k-2}$$

Substituting the expressions for $y, \frac{dy}{dx}, \frac{d^2y}{dx^2}$ in (1), we have

$$2x(1-x)\sum_{k=0}^{\infty} a_k (m+k)(m+k-1) x^{m+k-2} + (1-x)\sum_{k=0}^{\infty} a_k (m+k) x^{m+k-1}$$

$$+ 3\sum_{k=0}^{\infty} a_k x^{m+k} = 0$$

$$\Rightarrow \quad \sum_{k=0}^{\infty} 2a_k (m+k)(m+k-1) x^{m+k-1} - 2\sum_{k=0}^{\infty} a_k (m+k)(m+k-1) x^{m+k}$$

$$+ \sum_{k=0}^{\infty} a_k (m+k) x^{m+k-1} - \sum_{k=0}^{\infty} a_k (m+k) x^{m+k} + 3\sum_{x=0}^{\infty} a_k x^{m+k} = 0$$

Collecting the coefficients of like powers of x, we get

$$\sum_{k=0}^{\infty}[-2a_k(m+k)(m+k-1)-a_k(m+k)+3a_k]x^{m+k}+$$
$$\sum_{k=0}^{\infty}[2a_k(m+k)(m+k-1)+a_k(m+k)]x^{m+k-1}=0$$

$$\Rightarrow \sum_{k=0}^{\infty}a_k[-2m^2-2mk+2m-2mk-2k^2+2k-m-k+3]x^{m+k}+$$
$$\sum_{k=0}^{\infty}a_k[2m^2+2mk-2m+2mk+2k^2-2k+m+k]x^{m+k-1}=0$$

$$\Rightarrow \sum_{k=0}^{\infty}a_k[-2m^2-4mk+m-2k^2+k+3]x^{m+k}$$
$$+\sum_{k=0}^{\infty}a_k[2m^2+(4k-1)m+2k^2-k)x^{m+k-1}=0$$

$$\Rightarrow \sum_{k=0}^{\infty}a_k(m+k+1)(-2m-2k+3)x^{m+k}+\sum_{k=0}^{\infty}a_k(m+k)(2m+2k-1)x^{m+k-1}=0 \quad \ldots (2)$$

The coefficient of the lowest degree term x^{m-1} in (2) is obtained by putting $k = 0$ in second summation of (2) only and equating it to zero.

Then the *indicial equation* is

$$a_0 m(2m-1)=0 \Rightarrow m=0,\ m=\frac{1}{2},\ a_0\neq 0$$

The coefficient of the next lowest degree term x^m in (2) is obtained by putting $k = 0$ in first summation and $k = 0$ in the second summation, we get

$$a_0(m+1)(-2m+3)+a_1(m+1)(2m+1)=0$$

$$a_1=-\frac{(m+1)(-2m+3)}{(m+1)(2m+1)} \quad \Rightarrow \quad a_1=\frac{2m-3}{2m+1}$$

Equating the coefficient of x^2.

On putting $k \to k+1$ in second summation of (2), we get the coefficient of x^m and equation to zero, we get

$$a_k(m+k+1)(-2m-2k+3)+a_{k+1}(m+k+1)(2m+2k+1)=0$$

$$\Rightarrow \quad a_{k+1}=-\frac{(m+k+1)(-2m-2k+3)}{(m+k+1)(2m+2k+1)}a_k$$

$$\Rightarrow \quad a_{k+1}=-\frac{-2m-2k+3}{2m+2k+1}a_k=\frac{2m+2k-3}{2m+2k+1}a_k$$

If $k=0$, $\quad a_1=\dfrac{2m-3}{2m+1}a_0$

If $k=1$, $\quad a_2=\dfrac{2m-1}{2m+3}a_1$

If $k=2$, $\quad a_3=\dfrac{2m+1}{2m+5}a_2$

If $k=3$, $\quad a_4=\dfrac{2m+3}{2m+7}a_3$

If $k=4$, $\quad a_5=\dfrac{2m+5}{2m+9}a_4$

$$m = 0$$

$$a_1 = -3a_0$$

$$a_2 = -\frac{1}{3}a_1 = -\frac{1}{3}(-3a_0) = a_0$$

$$a_4 = \frac{3}{7}a_3 = \frac{3}{7}\left(-\frac{1}{5}\right)a_0 = \frac{3}{35}a_0$$

$$a_5 = \frac{5}{9}a_n = \frac{5}{9}\left(\frac{3}{35}\right)a_0 = \frac{1}{21}a_0$$

$$y_1 = a_0\left(1 - 3x + x^2 + \frac{1}{5}x^3 + \frac{3}{35}x^4 + \frac{1}{21}x^5 + ...\right)$$

$$y_1 = a_0\left(1 - 3x + \frac{3x^2}{1.3} + \frac{3x^3}{3.5} + \frac{3x^4}{5.7} + \frac{3x^5}{7.9} .....\right)$$

$$m = \frac{1}{2}$$

$$a_1 = \frac{2\left(\frac{1}{2}\right) - 3}{2\left(\frac{1}{2}\right) + 1}a_0 = -a_0$$

$$a_2 = \frac{2\left(\frac{1}{2}\right) - 1}{2\left(\frac{1}{2}\right) + 3}a_1 = -\frac{1}{4}a_0 = 0$$

$$a_3 = a_4 = a_5 = ... = 0$$

$$y_2 = a_0x^{\frac{1}{2}} - a_0x^{\frac{3}{2}}$$

$$= a_0\sqrt{x}\,(1 - x)$$

General Solution is

$$y = Ay_1 + By_2$$

$$y = A\left(1 - 3x + \frac{3x^2}{1.3} + \frac{3x^3}{3.5} + \frac{3}{5.7}x^4 + \frac{3}{7.9}x^5 + ...\right) + B\sqrt{x}\,(1 - x) \qquad \textbf{Ans.}$$

EXERCISE 32.2

Solve in series the following differential equation :

1. $2x^2\dfrac{d^2y}{dx^2} - x\dfrac{dy}{dx} + (1 - x^2)y = x^2$

Ans. $y = ax\left(1 + \dfrac{x^2}{2.5} + \dfrac{x^4}{2.4.5.9} + ...\right) + bx^{\frac{1}{2}}\left(1 + \dfrac{x^2}{2.3} + \dfrac{x^4}{2.4.3.7} + ...\right)$

2. $2x\,(1 - x)\dfrac{d^2y}{dx^2} + (5 - 7x)\dfrac{dy}{dx} - 3y = 0.$ **Ans.** $y = a\left(1 + \dfrac{3}{5}x + \dfrac{3}{7}x^2 + \dfrac{3}{9}x^3 + ...\right) + bx^{-\frac{3}{2}}$

3. $2x^2\dfrac{d^2y}{dx^2} + (2x^2 - x)\dfrac{dy}{dx} + y = 0$

Ans. $ax\left(1 - \dfrac{2}{3}x + \dfrac{2^2}{3.5}x^2 - \dfrac{2^3}{3.5.7}x^3 + ...\right) + b\sqrt{x}\left(1 - x + \dfrac{1}{2}x^2 - \dfrac{1}{6}x^3 + ...\right)$

4. $x\,(2 + x^2)\dfrac{d^2y}{dx^2} - \dfrac{dy}{dx} - 6xy = 0$

Ans. $y = a\left(1 + 3x^2 + \dfrac{3}{5}x^4 - \dfrac{1}{15}x^6 + ...\right) + bx^{\frac{3}{2}}\left(1 + \dfrac{3}{8}x^2 - \dfrac{3.1}{8.16}x^4 + \dfrac{5.3.1}{8.16.24}x^6 ....\right)$

Case II. When the roots of indicial equation are equal.

Example 11. *Solve* $x\,(x - 1)\,y'' + (3x - 1)\,y' + y = 0$

Solution. $x(x - 1)y'' + (3x - 1)\,y' + y = 0 \qquad ...\,(1)$

Here $xP\,(x)$ and $x^2Q\,(x)$ are analytic (Not infinite) at $x = 0$. So, $x = 0$ is a regular singular point, we assume the solution in the form

$$y = \sum_{k=0}^{\infty} a_k\, x^{m+k}$$

such that $$\frac{dy}{dx} = \sum_{k=0}^{\infty} a_k\,(m+k)\,x^{m+k-1}$$

$$\frac{d^2y}{dx^2} = \sum_{k=0}^{\infty} a_k\,(m+k)(m+k-1)\,x^{m+k-2}$$

Substituting the expressions for $y, \frac{dy}{dx}, \frac{d^2y}{dx^2}$ in (1), we have

$$x(x-1)\,\Sigma a_k\,(m+k)(m+k-1)\,x^{m+k-2} + (3x-1)\,\Sigma a_k\,(m+k)\,x^{m+k-1} + \Sigma a_k\,x^{m+k} = 0$$

$$\Rightarrow \quad \Sigma a_k\,(m+k)(m+k-1)\,x^{m+k} - \Sigma a_k\,(m+k)(m+k-1)\,x^{m+k-1} + 3\,\Sigma a_k\,(m+k)\,x^{m+k} - \Sigma a_k\,(m+k)\,x^{m+k-1} + \Sigma a_k\,x^{m+k} = 0$$

$$\Rightarrow \quad \Sigma a_k\,[(m+k)(m+k-1) + 3(m+k) + 1]\,x^{m+k} - \Sigma a_k\,[(m+k)(m+k-1) + (m+k)]\,x^{m+k-1} = 0$$

$$\Rightarrow \quad \Sigma a_k\,[(m+k)(m+k+2) + 1]\,x^{m+k} - \Sigma a_k\,(m+k)^2\,x^{m+k-1} = 0 \qquad \text{... (2)}$$

The coefficient of lowest degree term x^{m-1} in (2) is obtained by putting $k = 0$ in the second summation only of (2) and equating it to zero. Then the **indicial equation** is

$$\boxed{a_0(m+0)^2 = 0} \quad \Rightarrow \quad m = 0,\ 0 \text{ as } a_0 \neq 0$$

The coefficient of the next lowest degree term x^m in (2) is obtained by putting $k = 0$ in the first summation and $k = 1$ in the second summation only of (2) and equating it to zero, we get

$$a_0\,[(m+0)(m+2) + 1] - a_1\,(m+1)^2 = 0$$

$$\Rightarrow \quad a_0(m^2 + 2m + 1) - a_1(m^2 + 2m + 1) = 0$$

$$a_1 - a_0 = 0 \quad \Rightarrow \quad a_1 = a_0 \quad (\text{as } m = 0)$$

Equating the coefficient of x^{m+k} to zero, the recurrence relation is given by

$$a_k\,[(m+k)(m+k+2) + 1] - a_{k+1}\,(m+k+1)^2 = 0$$

$$a_k\,(m+k+1)^2 - a_{k+1}\,(m+k+1)^2 = 0$$

Hence, $$a_{k+1} = a_k$$

$$y = x^m\,[a_0 + a_1\,x + a_2\,x^2 + \,.........]$$

$$y = a_0\,x^m\,[1 + x + x^2 + x^3 + \,........]$$

For $m = 0$

when $m = 0, 0$, this gives only one solution instead of two.

Second solution is given by

$$\left(\frac{\partial y}{\partial m}\right)_{m=0} \quad \text{and} \quad y_1 = a_0\,(1 + x + x^2 + x^3)$$

$$\frac{\partial y}{\partial m} = a_0\,x^m \log x\,[1 + x + x^2 + x^3 + \,.....]$$

$$y_2 = a_0 \log x\,[1 + x + x^2 + x^3 + \,.......] \qquad m = 0$$

$$y_1 = a_0\,[1 + x + x^2 + x^3 + \,.......] \qquad m = 0$$

$$y = A\,y_1 + B\,y_2$$

$$y = A\,[1 + x + x^2 + x^3 + \,......] + B \log x\,(1 + x + x^2 + x^3 + \,.......)$$ **Ans.**

Example 12. *Using extended power series method find one solution of the differential equation $xy''+y'+x^2y=0$. Indicate the form of a second solution which is linearly independent of the first obtained above.* (*U.P. II Semester, June 2007*)

Solution. Here $xP(x)$ and $x^2Q(x)$ are analytic (not infinity). So, $x=0$ is a regular singular point of the equation, $x\dfrac{d^2y}{dx^2}+\dfrac{dy}{dx}+x^2y=0$

Let $$y=\Sigma a_k x^{m+k} \qquad \text{... (1)}$$

$$\frac{dy}{dx}=\Sigma a_k(m+k)x^{m+k-1}$$

$$\frac{d^2y}{dx^2}=\Sigma a_k(m+k)(m+k-1)x^{m+k-2}$$

Substituting the values of y, $\dfrac{dy}{dx}$ and $\dfrac{d^2y}{dx^2}$ in (1), we get

$$x\,\Sigma a_k(m+k)(m+k-1)x^{m+k-2}+\Sigma a_k(m+k)x^{m+k-1}+x^2\,\Sigma a_k x^{m+k}=0$$

$\Rightarrow$ $$\Sigma a_k(m+k)(m+k-1)x^{m+k-1}+\Sigma a_k(m+k)x^{m+k-1}+\Sigma a_k x^{m+k+2}=0$$

$\Rightarrow$ $$\Sigma a_k[(m+k)(m+k-1)+(m+k)]x^{m+k-1}+\Sigma a_k x^{m+k+2}=0$$

$\Rightarrow$ $$\Sigma a_k(m+k)^2x^{m+k-1}+\Sigma a_k x^{m+k+2}=0 \qquad \text{... (2)}$$

The coefficient of lowest degree term x^{m-1} in (2) is obtained by putting $k=0$ in first summation of (2) only and equating it to zero. Then the **indicial equation** is

$$\boxed{a_0 m^2=0} \Rightarrow m^2=0 \text{ or } m=0,0$$

The coefficient of the next lowest degree term x^m in (2) is obtained by putting $k=1$ in first summation only and equating it to zero.

$$a_1(m+1)^2=0 \qquad \Rightarrow \qquad a_1=0$$

Equating the coefficient of x^{m+1} for $k=2$, we get $a_2(m+2)^2=0 \Rightarrow a_2=0$

Equating the coefficient of x^{m+k+2} to zero, we have $a_{k+3}(m+k+3)^2+a_k=0$

$$\boxed{a_{k+3}=-\frac{a_k}{(m+k+3)^2}}$$

If $k=0$, $$a_3=-\frac{1}{(m+3)^2}a_0$$

If $k=1$, $$a_4=-\frac{1}{(m+4)^2}a_1=0,\ a_7=0,\ a_{10}=0$$

If $k=2$, $$a_5=-\frac{1}{(m+5)^2}a_2=0,\ a_8=0,\ a_{11}=0$$

If $k=3$, $$a_6=-\frac{1}{(m+6)^2}a_3=\frac{1}{(m+3)^2(m+6)^2}a_0$$

$$a_9=-\frac{1}{(m+9)^2}a_6=-\frac{1}{(m+3)^2(m+6)^2(m+9)^2}a_0$$

$$y=x^m a_0\left[1-\frac{x^3}{(m+3)^2}+\frac{x^6}{(m+3)^2(m+6)^2}-\frac{x^9}{(m+3)^2(m+6)^2(m+9)^2}+\dots\right] \qquad \text{... (3)}$$

For $m = 0$

To get the first solution, put $m = 0$ in (3), then

$$y_1 = a_0 \left[1 - \frac{x^3}{3^2} + \frac{x^6}{3^2 \times 6^2} - \frac{x^9}{3^2 \times 6^2 \times 9^2} + \ldots\right] \qquad \ldots (4)$$

To get the second independent solution, differentiate (3) *w.r.t. m*. Then

$$\frac{\partial y}{\partial m} = (x^m \log x)\, a_0 \left[1 - \frac{x^3}{(m+3)^2} + \frac{x^6}{(m+3)^2 (m+6)^2} - \frac{x^9}{(m+3)^2 (m+6)^2 (m+9)^2} + \ldots\right]$$

$$+ x^m a_0 \left[\frac{2x^3}{(m+3)^3} - \frac{2x^6}{(m+3)^3 (m+6)^2} - \frac{2x^6}{(m+3)^2 (m+6)^3} + \frac{2x^9}{(m+3)^3 (m+6)^2 (m+9)^2} + \frac{2x^9}{(m+3)^2 (m+6)^3 (m+9)^2} + \frac{2x^9}{(m+3)^2 (m+6)^2 (m+9)^3} + \ldots\right] \qquad \ldots (5)$$

Putting $m = 0$ in (5), we get

$$y_2 = (\log x) a_0 \left[1 - \frac{x^3}{3^2} + \frac{x^6}{3^2 \times 6^2} - \frac{x^9}{3^2 \times 6^2 \times 9^2} + \ldots\right]$$

$$+ a_0 \left[\frac{2x^3}{3^3} - \frac{2x^6}{3^3 \times 6^2} - \frac{2x^6}{3^2 \times 6^3} + \frac{2x^9}{3^3 \times 6^2 \times 9^2} + \frac{2x^9}{3^2 \times 6^3 \times 9^2} + \frac{2x^9}{3^2 \times 6^2 \times 9^3} + \ldots\right] \qquad \ldots (6)$$

Hence, the general solution is given by (4) and (6)

$$y = c_1 y_1 + c_2 y_2$$

$$y = c_1 a_0 \left[1 - \frac{x^3}{3^2} + \frac{x^6}{3^2 \times 6^2} - \frac{x^9}{3^2 \times 6^2 \times 9^2} + \ldots\right]$$

$$+ c_2 (\log x) a_0 \left[1 - \frac{x^3}{3^2} + \frac{x^6}{3^2 \times 6^2} - \frac{x^9}{3^2 \times 6^2 \times 9^2} + \ldots\right]$$

$$+ c_2 a_0 \left[\frac{2x^3}{3^3} - \frac{2x^6}{3^2 \times 6^6}\left(\frac{1}{3} + \frac{1}{6}\right) + \frac{2x^9}{3^2 \times 6^2 \times 9^2}\left(\frac{1}{3} + \frac{1}{6} + \frac{1}{9}\right) + \ldots\right]$$

$$\Rightarrow \quad = (c_1 + c_2 \log x)\, a_0 \left[1 - \frac{x^3}{3^2} + \frac{x^6}{3^2 \times 6^2} - \frac{x^9}{3^2 \times 6^2 \times 9^2} + \ldots\right]$$

$$+ 2 \cdot c_2 a_0 \left[\frac{x^3}{3^3} - \frac{x^6}{3^5 \times 2^2}\left(1 + \frac{1}{2}\right) + \frac{2x^9}{3^9 \times 2^2}\left(1 + \frac{1}{2} + \frac{1}{3}\right) + \ldots\right] \quad \textbf{Ans.}$$

Example 13. *Solve in series the differential equation:*

$$x \frac{d^2 y}{dx^2} + \frac{dy}{dx} - y = 0 \qquad \textit{[U.P., II Semester, (C.O.) 2003]}$$

Solution. Comparing with the equation

$$\frac{d^2 y}{dx^2} + P(x) \frac{dy}{dx} + Q(x)\, y = 0, \text{ we get}$$

$$P(x) = \frac{1}{x} \quad \text{and} \quad Q(x) = -\frac{1}{x}$$

Since at $x = 0$, both $P(x)$ and $Q(x)$ are not analytic $\therefore$ $x = 0$ is a singular point.

Also, $\quad x\,P(x) = 1$ and $x^2\,Q(x) = -x$

Both $x\,P(x)$ and $x^2\,Q(x)$ are analytic at $x = 0$ $\therefore$ $x = 0$ is a regular singular point.

Let us assume

$$y = a_0 x^m + a_1 x^{m+1} + a_2 x^{m+2} + a_3 x^{m+3} + \ldots \qquad \ldots(1)$$

Then, $\quad y' = m a_0 x^{m-1} + (m+1)\,a_1 x^m + (m+2)\,a_2 x^{m+1} + (m+3)\,a_3 x^{m+2} + \ldots$

and $\quad y'' = m(m-1)\,a_0 x^{m-2} + (m+1)\,m a_1 x^{m-1} + (m+2)(m+1)\,a_2 x^m + (m+3)(m+2)\,a_3 x^{m+1} + \ldots$

Substituting these values in the given equation, we get

$$x\,[m(m-1)\,a_0 x^{m-2} + (m+1)m\,a_1 x^{m-1} + (m+2)(m+1)\,a_2 x^m + (m+3)(m+2)\,a_3 x^{m+1} + \ldots]$$
$$+ [m\,a_0 x^{m-1} + (m+1)\,a_1 x^m + (m+2)\,a_2 x^{m+1} + (m+3)\,a_3 x^{m+2} + \ldots]$$
$$- [a_0 x^m + a_1 x^{m+1} + a_2 x^{m+2} + a_3 x^{m+3} + \ldots] = 0$$

Now, coefficient of $x^{m-1} = 0$

$\Rightarrow \quad m(m-1)\,a_0 + m a_0 = 0$

$\Rightarrow \quad m^2 a_0 = 0 \quad \Rightarrow \quad m^2 = 0 \qquad (\because a_0 \neq 0)$

Which is Indicial equation.

It roots are $\boxed{m = 0, 0}$ *which are equal.*

Coefficient of $x^m = 0$

$\Rightarrow \quad (m+1)\,m a_1 + (m+1)a_1 - a_0 = 0 \quad \Rightarrow \quad (m+1)^2\,a_1 = a_0$

$\Rightarrow \quad \boxed{a_1 = \dfrac{a_0}{(m+1)^2}}$

Coefficient of $x^{m+1} = 0$

$\Rightarrow \quad (m+2)(m+1)\,a_2 + (m+2)\,a_2 - a_1 = 0 \Rightarrow (m+2)^2\,a_2 = a_1$

$\Rightarrow \quad a_2 = \dfrac{a_1}{(m+2)^2} \Rightarrow \boxed{a_2 = \dfrac{a_0}{(m+1)^2\,(m+2)^2}}$

Similarly, $\quad a_3 = \dfrac{a_0}{(m+1)^2\,(m+2)^2\,(m+3)^2}$ and so on.

From (1), $y = a_0 x^m \left[1 + \dfrac{x}{(m+1)^2} + \dfrac{x^2}{(m+1)^2\,(m+2)^2} + \dfrac{x^3}{(m+1)^2\,(m+2)^2\,(m+3)^2} + \ldots\right] \qquad \ldots(2)$

Now, $\quad y_1 = (y)_{m=0} = a_0 \left[1 + x + \dfrac{x^2}{(2!)^2} + \dfrac{x^3}{(3!)^2} + \ldots\right] \qquad \ldots(3)$

To get the second independent solution, differentiate (1) partially w.r.t. m, we get

$$\frac{\partial y}{\partial m} = a_0 x^m \log x \left[1 + \frac{x}{(m+1)^2} + \frac{x^2}{(m+1)^2\,(m+2)^2} + \frac{x^3}{(m+1)^2\,(m+2)^2\,(m+3)^2} + \ldots\right]$$
$$+ a_0 x^m \left[-\frac{2x}{(m+1)^3} - \frac{2}{(m+1)^2\,(m+2)^2}\left\{\frac{1}{m+1} + \frac{1}{m+2}\right\}x^2\right.$$
$$\left. - \frac{2}{(m+1)^2\,(m+2)^2\,(m+3)^2}\left\{\frac{1}{m+1} + \frac{1}{m+2} + \frac{1}{m+3}\right\}x^3 - \ldots\right]$$

The second solution is $y_2 = \left(\dfrac{\partial y}{\partial m}\right)_{m=0} = a_0 \log x\left[1 + x + \dfrac{x^2}{(2!)^2} + \dfrac{x^3}{(3!)^2} + \ldots\right]$

$$- 2a_0\left[x + \frac{1}{(2!)^2}\left(1+\frac{1}{2}\right)x^2 + \frac{1}{(3!)^2}\left(1+\frac{1}{2}+\frac{1}{3}\right)x^3 + \ldots\right]$$

$$= y_1 \log x - 2a_0\left[x + \frac{1}{(2!)^2}\left(1+\frac{1}{2}\right)x^2 + \frac{1}{(3!)^2}\left(1+\frac{1}{2}+\frac{1}{3}\right)x^3 + \ldots\right]$$

Hence the complete solution is

$$y = c_1y_1 + c_2y_2 = (c_1a_0 + c_2a_0 \log x)\left[1 + x + \frac{x^2}{(2!)^2} + \frac{x^3}{(3!)^2} + \ldots\right]$$

$$- 2c_2a_0\left[x + \frac{1}{(2!)^2}\left(1+\frac{1}{2}\right)x^2 + \frac{1}{(3!)^2}\left(1+\frac{1}{2}+\frac{1}{3}\right)x^3 + \ldots\right]$$

$$\Rightarrow \quad y = (A + B\log x)\left[1 + x + \frac{x^2}{(2!)^2} + \frac{x^3}{(3!)^2} + \ldots\right]$$

$$-2B\left[x + \frac{1}{(2!)^2}\left(1+\frac{1}{2}\right)x^2 + \frac{1}{(3!)^2}\left(1+\frac{1}{2}+\frac{1}{3}\right)x^3 + \ldots\right]$$

Where $c_1a_0 = A,\ c_2a_0 = B.$ **Ans.**

EXERCISE 32.3

Solve in series the following differential equations : (First part of the solution is denoted by y_1).

1. $x\dfrac{d^2y}{dx^2} + \dfrac{dy}{dx} + xy = 0$

Ans. $y = a\left(1 - \dfrac{x^2}{2^2} + \dfrac{x^4}{2^2.4^2} - \dfrac{x^6}{2^2.4^2.6^2} + \ldots.\right) + b\left[y_1 \log x + a_0\left\{\dfrac{x^2}{2^2} - \dfrac{1}{2^2.4^2}\left(1+\dfrac{1}{2}\right)x^4 + \ldots\right\}\right]$

2. $(x - x^2)\dfrac{d^2y}{dx^2} + (1 - 5x)\dfrac{dy}{dx} - 4y = 0$

Ans. $y = a\,(1^2 + 2^2x + 3^2x^2 + 4^2x^3 + \ldots) + b(y_1 \log x - 2a_0(1.2x + 2.3x^2 + 3.4x^3 + \ldots)$

3. $x\dfrac{d^2y}{dx^2} + (1 + x)\dfrac{dy}{dx} + 2y = 0$

Ans. $y = a\left(1 - 2x + \dfrac{3}{2!}x^2 - \dfrac{4}{3!}x^3 + \ldots\right) + b\left[y_1 \log x + a_0\left(3x - \dfrac{13}{4}x^2 + \ldots.\right)\right]$

4. $x^2\dfrac{d^2y}{dx^2} - x(1 + x)\dfrac{dy}{dx} + y = 0$

Ans. $y = ax\left(1 + x + \dfrac{1}{2}x^2 + \dfrac{1}{2.3}x^3 + \ldots.\right) + b\left[y_1 \log x + a_0x^2\left(-1 - \dfrac{3}{4}x + \ldots\right)\right]$

Case III : When m_1 and m_2 are distinct and differing by an integer, then

$$y = c_1(y)_{m_1} + c_2\left(\frac{\partial y}{\partial m}\right)_{m_2} \qquad \left[\begin{array}{r}\text{If coefficient} = \infty \\ \text{when } m = m_2\end{array}\right]$$

Example 14. *Solve*

$$x^2\frac{d^2y}{dx^2}+x\frac{dy}{dx}+(x^2-4)y=0 \qquad \ldots (1)$$

Solution. Here $xP(x)$ and $x^2Q(x)$ are analytic (not infinite) at $x = 0$. So, $x = 0$ is regular singular point of this equation.

Let $y=\Sigma a_k x^{m+k}$

$$\frac{dy}{dx}=\Sigma a_k(m+k)x^{m+k-1}$$

$$\frac{d^2y}{dx^2}=\Sigma a_k(m+k)(m+k-1)x^{m+k-2}$$

Substituting the values of $\frac{d^2y}{dx^2}$, $\frac{dy}{dx}$ and y in (1), we get

$$x^2\,\Sigma a_k(m+k)(m+k-1)x^{m+k-2}+x\Sigma a_k(m+k)x^{m+k-1}+(x^2-4)\,\Sigma a_k x^{m+k}=0$$

$$\Rightarrow \quad \Sigma a_k[(m+k)(m+k-1)+(m+k)-4]x^{m+k}+\Sigma a_k x^{m+k+2}=0$$

$$\Rightarrow \quad \Sigma a_k(m+k+2)(m+k-2)x^{m+k}+\Sigma a_k x^{m+k+2}=0 \qquad \ldots (2)$$

The coefficient of lowest degree term x^m in (2) is obtained by putting $k = 0$ in first summation only and equating it to zero. Then the indicial equation is

$$\boxed{a_0(m+2)\,(m-2)=0} \Rightarrow m=2,-2$$

The coefficient of next lowest term x^{m+1} in (2) is obtained by putting $k = 1$ in first summation only and equating it to zero.

$$a_1(m+3)(m-1)=0 \Rightarrow a_1=0$$

Equating to zero the coefficient of x^{m+k+2}, we get

$$a_{k+2}(m+k+4)(m+k)+a_k=0 \quad \Rightarrow \quad \boxed{a_{k+2}=-\frac{a_k}{(m+k+4)(m+k)}}$$

$$a_1=a_3=a_5=\ldots=0$$

$$a_2=-\frac{a_0}{m(m+4)}$$

$$a_4=-\frac{a_2}{(m+2)(m+6)}=\frac{a_0}{m(m+2)(m+4)(m+6)}$$

$$a_6=-\frac{a_4}{(m+4)(m+8)}=-\frac{a_0}{m(m+2)(m+4)^2(m+6)(m+8)}$$

Hence

$$y=a_0x^m\left[1-\frac{x^2}{m(m+4)}+\frac{x^4}{m(m+2)(m+4)(m+6)}-\frac{x^6}{m(m+2)(m+4)^2(m+6)(m+8)}+\ldots\right] \qquad \ldots (3)$$

Putting $m = 2$ in (3), we get

$$y_1=a_0x^2\left[1-\frac{x^2}{2\times 6}+\frac{x^4}{2\times 4\times 6\times 8}-\frac{x^6}{2\times 4\times 6^2\times 8\times 10}+\ldots\right] \qquad \ldots (4)$$

For $m = -2$

Coefficient of x^4, x^6 etc. in (3) becomes infinite on putting $m = -2$. To overcome this difficulty, we put

$a_0 = b_0(m+2)$ in (1) and we get

$$y = b_0 x^m \left[(m+2) - \frac{(m+2)x^2}{m(m+4)} + \frac{x^4}{m(m+4)(m+6)} - \frac{x^6}{m(m+4)^2(m+6)(m+8)} + \ldots\right] \quad \ldots (5)$$

On differentiating (5) w.r.t. 'm', we get

$$\frac{\partial y}{\partial m} = b_0(x^m \cdot \log x)\left[(m+2) - \frac{(m+2)x^2}{m(m+4)} + \frac{x^4}{m(m+4)(m+6)} - \frac{x^6}{m(m+4)^2(m+6)(m+8)} + \ldots\right]$$

$$+ b_0 x^m \left[1 - \frac{(m+2)x^2}{m(m+4)}\left(\frac{1}{m+2} - \frac{1}{m} - \frac{1}{m+4}\right) + \frac{x^4}{m(m+4)(m+6)}\left(-\frac{1}{m} - \frac{1}{m+4} - \frac{1}{m+6}\right) + \ldots\right]$$

On replacing m by -2, we get

$$\left(\frac{\partial y}{\partial m}\right)_{m=-2} = (b_0 x^{-2} \log x)\left[0 - 0 + \frac{x^4}{(-2)(2)(4)} - \frac{x^6}{(-2)(2)^2(4)(6)} + \ldots\right]$$

$$+ b_0 x^{-2}\left[1 + \frac{x^2}{2^2} + \frac{x^4}{2^2 \times 4}\left(\frac{1}{4}\right) + \ldots\right]$$

$$\Rightarrow \quad y_2 = b_0 x^2 \log x\left(-\frac{1}{2^2 \times 4} + \frac{x^2}{2^3 \times 4 \times 6} - \frac{x^4}{2^3 \times 4^2 \times 6 \times 8} + \ldots\right) + b_0 x^{-2}\left(1 + \frac{x^2}{2^2} + \frac{x^4}{2^2 \times 4^2} + \ldots\right)$$

General solution is $y = c_1 y_1 + c_2 y_2$

$$y = c_1 x^2\left(1 - \frac{x^2}{2 \times 6} + \frac{x^4}{2 \times 4 \times 6 \times 8} - \frac{x^6}{2 \times 4 \times 6^2 \times 8 \times 10} + \ldots\right)$$

$$+ c_2\left[x^2 \log x\left(-\frac{1}{2^2 \times 4} + \frac{x^2}{2^3 \times 4 \times 6} - \frac{x^4}{2^3 \times 4^2 \times 6 \times 8} + \ldots\right) + x^{-2}\left(1 + \frac{x^2}{2^2} + \frac{x^4}{2^2 \times 4^2} + \ldots\right)\right]$$

Ans.

Example 15. *Solve:* $x(1-x)\frac{d^2y}{dx^2} - (1+3x)\frac{dy}{dx} - y = 0$ ...(1)

Solution. Here, $xP(x)$ and $x^2Q(x)$ are analytic (not infinite) at $x = 0$. So, $x = 0$ is regular singular point of the equation (1).

Let $\quad y = \Sigma a_k x^{m+k}$

$$\frac{dy}{dx} = \Sigma a_k (m+k) x^{m+k-1}$$

$$\frac{d^2y}{dx^2} = \Sigma a_k (m+k)(m+k-1) x^{m+k-2}$$

Substituting the values of $\frac{d^2y}{dx^2}, \frac{dy}{dx}$ and y in (1), we get

$$\sum_{k=0}^{\infty} a_k[(x - x^2)(m+k)(m+k-1)x^{m+k-2} - \sum_{k=0}^{\infty} a_k(1+3x)(m+k)x^{m+k-1} - \sum_{k=0}^{\infty} a_k x^{m+k} = 0$$

$$\Rightarrow \quad \sum_{k=0}^{\infty} a_k[(m+k)(m+k-1)x^{m+k-1} - \sum_{k=0}^{\infty} a_k(m+k)(m+k-1)x^{m+k} - \sum_{k=0}^{\infty} a_k(m+k)x^{m+k-1}$$

$$-3\sum_{k=0}^{\infty} a_k(m+k)x^{m+k} - \sum_{k=0}^{\infty} a_k x^{m+k} = 0$$

$$-\sum_{k=0}^{\infty} a_k[(m+k)(m+k-1) + 3(m+k) + 1]\, x^{m+k} + \sum_{k=0}^{\infty} a_k[(m+k)(m+k-1) - (m+k)]\, x^{m+k-1} = 0$$

$$\Rightarrow \qquad -\sum_{k=0}^{\infty} a_k[(m+k)^2+2(m+k)+1]x^{m+k}+\sum_{k=0}^{\infty} a_k[(m+k)^2-2(m+k)]\,x^{m+k-1}=0$$

$$\Rightarrow \qquad -\sum_{k=0}^{\infty} a_k(m+k+1)^2x^{m+k}+\sum_{k=0}^{\infty} a_k(m+k)(m+k-2)x^{m+k-1}=0 \qquad ...(2)$$

Now (2), being an identity, we will equate the coefficients of various powers of x to zero.

The coefficient of lowest degree term x^{m-1} is obtained by putting $k=0$ in the second summation of (2) and equating it to zero. Then the **indicial equation** is

$$\boxed{a_0 m(m-2)=0} \Rightarrow \qquad m=0,\,2$$

The coefficient of x^{m+k-1} is obtained by putting $k=k-1$ in the first summation and $k=k$ in the second summation of (2), and equating it to zero.

$$-a_{k-1}(m+k)^2+a_k(m+k)(m+k-2)=0$$

$$\Rightarrow \qquad -a_{k-1}(m+k)+a_k(m+k-2)=0$$

$$\Rightarrow \qquad a_k=\frac{(m+k)}{(m+k-2)}a_{k-1} \qquad ...(3)$$

If $k=1$, $\quad a_1=\dfrac{m+1}{m-1}a_0$

If $k=2$, $\quad a_2=\dfrac{m+2}{m}a_1=\left(\dfrac{m+2}{m}\right)\left(\dfrac{m+1}{m-1}\right)a_0$

If $k=3$, $\quad a_3=\left(\dfrac{m+3}{m+1}\right)a_2=\left(\dfrac{m+3}{m+1}\right)\left(\dfrac{m+2}{m}\right)\left(\dfrac{m+1}{m-1}\right)a_0$

We know that $\quad y=\sum_{k=0}^{\infty} a_k x^{m+k}=a_0x^m+a_1x^{m+1}+a_2x^{m+2}+\ldots$

$$y=a_0x^m+\left(\frac{m+1}{m-1}\right)a_0x^{m+1}+\frac{(m+1)(m+2)}{(m-1)m}a_0x^{m+2}+\frac{(m+1)(m+2)(m+3)}{(m-1)m(m+1)}a_0x^{m+3}+\ldots..$$

$$\Rightarrow \quad y=a_0x^m\left[1+\left(\frac{m+1}{m-1}\right)x+\frac{(m+1)(m+2)}{(m-1)m}x^2+\frac{(m+1)(m+2)(m+3)}{(m-1)m(m+1)}x^3+\ldots.\right] \qquad ....(4)$$

For $m=2$

$$y=a_0x^2\left[1+3x+6x^2+10x^3+\ldots.\right]=a_0\,u \qquad ..(5)$$

For $m=0$

If we put $m=0$ in the above series (4), the coefficients become infinite. To remove this difficulty, we modify the form (4) of y by putting $a_0=mb,\ b\neq 0$

$$y=bx^m\left[m+\frac{m(m+1)}{(m-1)}x+\frac{m(m+1)(m+2)}{(m-1)m}x^2+\frac{m(m+1)(m+2)(m+3)}{(m-1)m(m+1)}x^3+\ldots..\right] \qquad ...(6)$$

Now equation (6), gives only one solution instead of two solutions. The second solution is given by $\dfrac{\partial y}{\partial m}$.

$$\frac{\partial y}{\partial m}=bx^m\log x\left[m+\frac{m(m+1)}{m-1}x+\frac{(m+1)(m+2)}{m-1}x^2+\ldots...\right]$$

$$+bx^m\left[1+\frac{m^2-2m-1}{(m-1)^2}x+\frac{m^2-m-5}{(m-1)^2}x^2+\frac{m^2-2m-11}{(m-1)^2}x^3+\ldots..\right]$$

Putting $m = 0$, we have

$$\left(\frac{\partial y}{\partial m}\right)_{m=0} = b\log x\left[\frac{2}{-1}x^2+.....\right]+b\left[1+\frac{(-1)}{1}x+(-5)x^2+(-11)x^3+.....\right]$$

$$= -b\log x\,[2x^2+.....]+b\,[1-x-5x^2-11x^3+...]$$

$$= b\,v$$

The complete solution of (1) is given by $y = a_0 u + bv$ **Ans.**

Example 16. *Solve in series the differential equation*

$$x^2\frac{d^2y}{dx^2}+5x\frac{dy}{dx}+x^2y=0 \qquad \textit{(U.P., II Semester, 2002)}$$

Solution. Comparing the given equation with the form

$$\frac{d^2y}{dx^2}+P(x)\frac{dy}{dx}+Q(x)\,y=0, \text{ we get}$$

$$P(x)=\frac{5}{x}, \qquad Q(x)=1$$

At $x = 0$, since $P(x)$ is not analytic therefore $x = 0$ is a singular point.

Also, $x\,P(x) = 5$

$x^2\,Q(x) = x^2$

Since both $x\,P(x)$ and $x^2\,Q(x)$ are analytic at $x = 0$ therefore $x = 0$ is a regular singular point.

Let us assume

$$y = a_0x^m + a_1x^{m+1} + a_2x^{m+2} + a_3x^{m+3} + ... \qquad ...(1)$$

$$\therefore \frac{dy}{dx} = ma_0x^{m-1} + (m+1)\,a_1x^m + (m+2)\,a_2x^{m+1} + ... \qquad ...(2)$$

and $$\frac{d^2y}{dx^2} = m(m-1)\,a_0x^{m-2} + (m+1)\,ma_1x^{m-1} + (m+2)(m+1)\,a_2x^m + ... \qquad ...(3)$$

Substituting the above values in given equation, we get

$$x^2\,[m(m-1)\,a_0x^{m-2} + (m+1)\,ma_1\,x^{m-1} + (m+2)(m+1)\,a_2x^m + ...]$$
$$+ 5x\,[ma_0\,x^{m-1} + (m+1)\,a_1x^m + (m+2)\,a_2\,x^{m+1} + ...]$$
$$+ x^2\,[a_0x^m + a_1\,x^{m+1} + a_2x^{m+2} + a_3x^{m+3} + ...] = 0 \qquad ...(4)$$

Equating the coefficient of lowest power of x to zero, we get

$$m(m-1)\,a_0 + 5ma_0 = 0 \qquad [\text{coeff. of } x^m = 0]$$

$$\Rightarrow \quad (m^2+4m)\,a_0 = 0$$

$$\Rightarrow \quad m(m+4) = 0 \qquad \text{(Indicial equation)} \qquad (\because a_0 \neq 0)$$

$$\Rightarrow \quad \boxed{m = 0,\ -4}$$

Hence the roots are distinct and differing by an integer. Equating to zero, the coefficients of successive powers of x, we get

Coefficient of $x^{m+1} = 0$

$$(m+1)\,ma_1 + 5(m+1)\,a_1 = 0$$

$$\Rightarrow \quad (m+5)(m+1)\,a_1 = 0 \Rightarrow \boxed{a_1 = 0} \qquad ...(5) \qquad [\because m \neq -5, -1]$$

Coefficient of $x^{m+2} = 0$

$$(m+2)(m+1)\,a_2 + 5(m+2)\,a_2 + a_0 = 0$$

$$(m+2)(m+6)\,a_2 + a_0 = 0$$

$$\boxed{a_2 = \frac{-a_0}{(m+2)(m+6)}} \qquad ...(6)$$

Again, Coefficient of $x^{m+3} = 0$

$$(m+3)(m+2)\,a_3 + 5\,(m+3)\,a_3 + a_1 = 0$$

$$(m+3)(m+7)\,a_3 + a_1 = 0$$

$$\Rightarrow \qquad a_3 = \frac{-a_1}{(m+3)(m+7)}$$

$$\Rightarrow \qquad \boxed{a_3 = 0} \qquad \ldots(7)$$

Similarly, $a_5 = a_7 = a_9 = \ldots\ldots = 0$

Now, coefficient of $x^{m+4} = 0$

$$(m+4)(m+3)\,a_4 + 5\,(m+4)\,a_4 + a_2 = 0$$

$$\Rightarrow \qquad (m+4)(m+8)\,a_4 = -a_2$$

$$a_4 = \frac{-a_2}{(m+4)(m+8)} = \frac{a_0}{(m+2)(m+4)(m+6)(m+8)} \text{ etc.} \qquad \ldots(8)$$

These give $y = a_0 x^m \left[1 - \frac{x^2}{(m+2)(m+6)} + \frac{x^4}{(m+2)(m+4)(m+6)(m+8)} - \ldots\ldots\right]$ ...(9)

Putting $m = 0$ in (9), we get

$$y_1 = (y)_{m=0} = a_0\left[1 - \frac{x^2}{2.6} + \frac{x^4}{2.4.6.8} - \ldots\right] \qquad \ldots(10)$$

If we put $m = -4$ in the series given by equation (9), the coefficients become infinite. To avoid this difficulty, we put $a_0 = b_0\,(m+4)$, so that

$$y = b_0 x^m\left[(m+4) - \frac{(m+4)\,x^2}{(m+2)(m+6)} + \frac{x^4}{(m+2)(m+6)(m+8)} - \ldots\right] \qquad \ldots(11)$$

Now, $\dfrac{\partial y}{\partial m} = \log x\, b_0 x^m\left[1 + \dfrac{m^2+8m+20}{(m^2+8m+12)^2}x^2 - \dfrac{(3m^2+32m+76)}{(m^3+16m^2+76m+96)^2}x^4 + \ldots\right]$

Second solution is given by

$$y_2 = \left(\frac{\partial y}{\partial m}\right)_{m=-4} = \log x\, b_0 x^{-4}\left(1 + \frac{x^2}{4} - \frac{x^4}{4} + \ldots\right)$$

$$= b_0\,x^{-4}\log x\left[0 - 0 + \frac{x^4}{(-2)(2)(4)} - \frac{x^6}{16} + \ldots\right] + b_0\,x^{-4}\left(1 + \frac{x^2}{4} - \frac{x^4}{4} + \ldots\right)$$

$$= b_0\,x^{-4}\log x\left(\frac{-x^4}{16} - \frac{x^6}{16} - \ldots\right) + b_0\,x^{-4}\left(1 + \frac{x^2}{4} - \frac{x^4}{4} + \ldots\right)$$

Hence the complete solution is given by

$$y = c_1 y_1 + c_2 y_2$$

$$= c_1 a_0\left(1 - \frac{x^2}{12} + \frac{x^4}{384} - \ldots\right) + c_2 b_0\,x^{-4}\log x\left(-\frac{x^4}{16} - \frac{x^6}{16} - \ldots\right) + c_2 b_0\,x^{-4}\left(1 + \frac{x^2}{4} - \frac{x^4}{4} + \ldots\right)$$

$$\therefore \quad y = A\left(1 - \frac{x^2}{12} + \frac{x^4}{384} - \ldots\right) + Bx^{-4}\left(1 + \frac{x^2}{4} - \frac{x^4}{4} + \ldots\right) - B\log x\left(\frac{1}{16} + \frac{x^2}{16} + \ldots\right)$$

where $A = c_1 a_0$ and $B = c_2 b_0$. **Ans.**

EXERCISE 32.4

Solve in series the following differential equation

1. $x^2\dfrac{d^2y}{dx^2}+x\dfrac{dy}{dx}+(x^2-1)y=0$

$$\textbf{Ans. } y=a_0x\left(1-\frac{x^2}{2.4}+\frac{x^4}{2.4^2.6}-\ldots\right)+b_0x^{-1}\log x\left[-\frac{1}{2}x^2+\frac{1}{2^2.4}x^4-\frac{1}{2^2.4^2.6}x^6+\ldots\right]$$
$$+b_0x^{-1}\left[1+\frac{x^2}{2^2}-\frac{3}{2^2.2^3}x^4+\ldots\right]$$

2. $x(1-x)\dfrac{d^2y}{dx^2}-3x\dfrac{dy}{dx}-y=0$

$$\textbf{Ans. } y=a_0x(1+2x+3x^2+\ldots..)+b_0\log x[x+2x^2+3x^3+\ldots.]+b_0(1+x+x^2+\ldots.)$$

Case IV. If the roots differ by an integer such that one or more coefficients are indeterminate.

Example 17. *Find the extended power series solution of the differential equation*

$$x^2y''+4xy'+(x^2+2)y=0 \qquad \ldots (1)$$

Solution. Here $xP(x)$ and $x^2Q(x)$ are analytic (not infinity) at $x = 0$. So, $x = 0$ is a regular singular point of this equation

Let $y=\Sigma a_kx^{m+k}$ be the required solution of the given equation.

Then $$\frac{dy}{dx}=\Sigma a_k(m+k)x^{m+k-1},\qquad \frac{d^2y}{dx^2}=\Sigma a_k(m+k)(m+k-1)x^{m+k-2}$$

Substituting the values of y, $\dfrac{dy}{dx}$ and $\dfrac{d^2y}{dx^2}$ in the given equation, we get

$$x^2\,\Sigma a_k(m+k)(m+k-1)x^{m+k-2}+4x\,\Sigma a_k(m+k)x^{m+k-1}+(x^2+2)\,\Sigma a_kx^{m+k}=0$$
$$\Sigma a_k(m+k)(m+k-1)x^{m+k}+4\Sigma a_k(m+k)x^{m+k}+\Sigma a_kx^{m+k+2}+\Sigma 2a_kx^{m+k}=0$$
$$\Sigma a_k[(m+k)(m+k-1)+4(m+k)+2]x^{m+k}+\Sigma a_kx^{m+k+2}=0$$
$$\Sigma a_k[(m+k)^2+3(m+k)+2]x^{m+k}+\Sigma a_kx^{m+k+2}=0 \qquad \ldots (2)$$

The coefficient of lowest degree term x^m in (2) is obtained by putting $k = 0$ in first summation only and equating it to zero. Then the indicial equation is

$$\boxed{a_0(m^2+3m+2)=0}$$

$$a_0\neq 0,\ m^2+3m+2=0 \;\Rightarrow\; (m+1)(m+2)=0 \;\Rightarrow\; m=-1,-2$$

The coefficient of next lowest degree term x^{m+1} in (2) is obtained by putting $k = 1$ in first summation only and equating it to zero.

$$a_1[m^2+5m+6]=0 \text{ or } a_1(m+2)(m+3)=0 \Rightarrow a_1=\frac{0}{(m+2)\,(m+3)}$$

when $m = -2$, a_1 becomes indeterminate $\left(\dfrac{0}{0}\right)$. When $m = -2$ we get the identity $a_1(0) = 0$ which is satisfied by every value of a_1. Therefore in this case we can take a_1 as arbitrary constant.

Equating to zero the coefficient of x^{m+k+2}

$$a_{k+2}[(m+2+k)^2+3(m+2+k)+2]+a_k=0$$
$$\Rightarrow\quad a_{k+2}[m^2+(2k+4+3)m+(k+2)^2+3(k+2)+2]+a_k=0$$
$$\Rightarrow\quad a_{k+2}[m^2+(2k+7)m+k^2+7k+12]+a_k=0$$

$$\Rightarrow \qquad \boxed{a_{k+2} = -\frac{1}{m^2 + (2k+7)m + k^2 + 7k + 12} a_k}$$

For $k = 0$, $\quad a_2 = -\dfrac{1}{m^2 + 7m + 12} a_0 = -\dfrac{1}{(m+3)(m+4)} a_0$

$k = 1 \quad a_3 = -\dfrac{1}{m^2 + 9m + 20} a_1 = -\dfrac{1}{(m+4)(m+5)} a_1$

$k = 2 \quad a_4 = -\dfrac{1}{m^2 + 11m + 30} a_2 = \dfrac{1}{(m+3)(m+4)(m+5)(m+6)} a_0$

$k = 3 \quad a_5 = -\dfrac{1}{m^2 + 13m + 42} a_3 = \left\{\dfrac{1}{(m+4)(m+5)(m+6)(m+7)} a_1\right\}$

For $m = -1$

$$a_2 = -\frac{1}{6} a_0, \quad a_3 = \frac{1}{12} a_1, \quad a_4 = \frac{1}{120} a_0, \quad a_5 = \frac{1}{360} a_1$$

First solution is $\quad y_1 = x^{-1}\left[1 - \dfrac{1}{6}x^2 + \dfrac{1}{120}x^4 + \ldots\right] a_0 + \left[1 - \dfrac{1}{12}x^2 + \dfrac{x^4}{360} + \ldots\right] a_1$

For $m = -2$

$$a_2 = -\frac{1}{2} a_0, \quad a_3 = -\frac{1}{6} a_1, \quad a_4 = \frac{1}{24} a_0, \quad a_5 = \frac{1}{120} a_1$$

Second solution is $\quad y_2 = x^{-2}\left[1 - \dfrac{x^2}{2} + \dfrac{x^4}{24} + \ldots\right] a_0 + \left[\dfrac{1}{x} - \dfrac{x}{6} + \dfrac{x^3}{120} + \ldots\right] a_1$

$$y_2 = x^{-2}\left[\left\{1 - \frac{x^2}{2} + \frac{x^4}{24} + \ldots\right\} a_0 + \left\{x - \frac{x^3}{6} + \frac{x^5}{120} + \ldots\right\} a_1\right]$$

$$y_2 = x^{-2}\left[a_0 \cos x + a_1 \sin x\right]$$

Thus the complete solution is $y = Ay_1 + By_2$ **Ans.**

Example 18. *Solve*

$$x^2 y'' + (x^2 + x) y' + (x - 9) y = 0 \qquad \ldots (1)$$

Solution. Here, $xP(x)$ and $x^2Q(x)$ are analytic (not infinity) at $x = 0$. So, $x = 0$ is a regular singular point of (1).

Let $\quad y = \sum_{k=0}^{\infty} a_k x^{m+k}$

Differentiating it, we get

$$y' = \sum_{k=0}^{\infty} (m+k) a_k x^{m+k-1} \text{ and } y'' = \sum_{k=0}^{\infty} (m+k)(m+k-1) a_k x^{m+k-2}$$

Substituting the values of y, y' and y'' in (1), we get

$$x^2 \sum_{k=0}^{\infty} (m+k)(m+k-1) a_k x^{m+k-2} + (x^2 + x) \sum_{k=0}^{\infty} (m+k) a_k x^{m+k-1} + (x-9) \sum_{k=0}^{\infty} a_k x^{m+k} = 0$$

$$\Rightarrow \sum_{k=0}^{\infty} (m+k)(m+k-1) a_k x^{m+k} + \sum_{k=0}^{\infty} (m+k) a_k x^{m+k+1} + \sum_{k=0}^{\infty} (m+k) a_k x^{m+k} + \sum_{k=0}^{\infty} a_k x^{m+k+1} - \sum_{k=0}^{\infty} 9 a_k x^{m+k} = 0$$

$$\Rightarrow \sum_{k=0}^{\infty} (m+k)^2 a_k x^{m+k} + \sum_{k=0}^{\infty} (m+k+1) a_k x^{m+k+1} - 9 \sum_{k=0}^{\infty} a_k x^{m+k} = 0$$

$$\Rightarrow \quad \sum_{k=0}^{\infty} [(m+k)^2 - 9] a_k x^{m+k} + \sum_{k=0}^{\infty} (m+k+1) a_k x^{m+k+1} = 0$$

$$\Rightarrow \quad \sum_{k=0}^{\infty} [(m+k)^2 - 9] a_k x^{m+k} + \sum_{k=1}^{\infty} (m+k) a_{k-1} x^{m+k} = 0 \quad \text{... (2)}$$

The coefficient of lowest degree term x^m in (2) is obtained by putting $k = 0$ in first summation only and equating it to zero.

$$\boxed{(m^2 - 9) a_0 = 0} \quad \Rightarrow \quad m^2 - 9 = 0 \quad \Rightarrow \quad m = \pm 3$$

The coefficient of next lowest degree x^{m+1} in (2) is obtained by putting $k = 1$ in first summation and second summation, we get

$$[(m+1)^2 - 9] a_1 + (m+1) a_0 = 0$$

$$a_1 = \frac{(m+1) a_0}{9 - (m+1)^2}$$

Equating to zero the coefficient of x^{m+k}, we get $[(m+k)^2 - 9] a_k + (m+k) a_{k-1} = 0$

$$\boxed{a_k = \frac{(m+k)\ a_{k-1}}{9 - (m+k)^2}} \qquad k \geq 1$$

For $k = 2$, $$a_2 = \frac{(m+2) a_1}{9 - (m+2)^2} = \frac{(m+1)(m+2) a_0}{[9 - (m+1)^2][9 - (m+2)^2]}$$

For $k = 3$, $$a_3 = \frac{(m+3) a_2}{9 - (m+3)^2} = \frac{(m+1)(m+2)(m+3) a_0}{[9 - (m+1)^2][9 - (m+2)^2][9 - (m+3)^2]}$$

and so on.

$$y = a_0 x^m + a_1 x^{m+1} + a_2 x^{m+2} + a_3 x^{m+3} + \ldots$$

$$y = a_0 x^m \left[1 + \frac{(m+1)}{9 - (m+1)^2} x + \frac{(m+1)(m+2)}{[9 - (m+1)^2][9 - (m+2)^2]} x^2 + \frac{(m+1)(m+2)(m+3)}{[9 - (m+1)^2][9 - (m+2)^2][9 - (m+3)^2]} x^3 + \ldots \right]$$

For $m = 3$

$$y = a_0 x^3 \left[1 + \frac{4}{-7} x + \frac{4 \times 5}{(-7)(-16)} x^2 + \frac{4 \times 5 \times 6}{(-7)(-16)(-27)} x^3 + \ldots \right] \quad \text{... (3)}$$

For $m = -3$

Coefficients of x^3, x^4, x^5 become zero and the coefficients of remaining terms *indeterminate*.

Thus for $m = -3$

$$y = a_0 x^{-3} \left[1 + \frac{-2}{5} x + \frac{(-2)(-1)}{(5)(8)} x^2 \right] + a_6 x^3 \left[1 + \frac{4}{-7} x + \frac{(4)(5)}{(-7)(-16)} x^2 + \frac{(4)(5)(6)}{(-7)(-16)(-27)} x^3 \ldots \right] \quad \text{... (4)}$$

Series (3) is a constant multiple of second series in (4). Solution (4) contains two arbitrary constants, so it may be taken as the required solution.

Note. In general, if $m_1 - m_2 = a$ positive integer and some coefficients become indeterminate when $m = m_2$, the complete solution is given by putting $m = m_2$ is y which then contains two arbitrary constants. The result by putting $m = m_1$ in y nearly gives a numerical multiple of one of the series contained in the first solution.

Example 19. *Solve in series the differential equation:*

$$xy'' + 2y' + xy = 0. \qquad \text{(U.P., II Semester, 2003)}$$

Solution. Comparing the given equation with the form

$$\frac{d^2y}{dx^2} + P(x)\frac{dy}{dx} + Q(x)\,y = 0, \text{ we get}$$

$$P(x) = \frac{2}{x} \quad \text{and} \quad Q(x) = 1$$

At $x = 0$, $P(x)$ is not analytic $\therefore$ $x = 0$ is a *singular point.*

Also, $x\,P(x) = 2$ and $x^2\,Q(x) = x^2$

At $x = 0$, since $x\,P(x)$ and $x^2\,Q(x)$ are analytic $\therefore$ $x = 0$ is a *regular singular point.*

Let us assume

$$y = a_0 x^m + a_1 x^{m+1} + a_2 x^{m+2} + a_3 x^{m+3} + \ldots \qquad \ldots(1)$$

Then, $\dfrac{dy}{dx} = ma_0 x^{m+1} + (m+1)\,a_1 x^m + (m+2)\,a_2 x^{m+1} + (m+3)\,a_3 x^{m+2} + \ldots\ldots$

and $\dfrac{d^2y}{dx^2} = m(m-1)\,a_0 x^{m-2} + (m+1)\,ma_1 x^{m-1} + (m+2)(m+1)\,a_2 x^m + (m+3)(m+2)\,a_3 x^{m+1} + \ldots$

Substituting these values in the given equation, we get

$$x\,[m(m-1)\,a_0 x^{m-2} + (m+1)\,m\,a_1 x^{m-1} + (m+2)(m+1)\,a_2 x^m + (m+3)(m+2)\,a_3 x^{m+1} + \ldots]$$
$$+ 2\,[m\,a_0 x^{m-1} + (m+1)\,a_1 x^m + (m+2)\,a_2 x^{m+1} + \ldots]$$
$$+ x\,[a_0 x^m + a_1 x^{m+1} + a_2 x^{m+2} + a_3 x^{m+3} + \ldots] = 0$$

Now, Coefficient of $x^{m-1} = 0$

$\Rightarrow$ $m(m-1)\,a_0 + 2m\,a_0 = 0$

$(m^2 + m)\,a_0 = 0$

$\Rightarrow$ $m^2 + m = 0$ (Indicial equation) $[\because a_0 \neq 0]$

$\Rightarrow$ $\boxed{m = 0, -1}$

Hence, roots are distinct and differ by an integer.

Coefficient of $x^m = 0$

$\Rightarrow$ $(m+1)\,m\,a_1 + 2(m+1)a_1 = 0$

$\Rightarrow$ $(m+1)(m+2)\,a_1 = 0$

$\Rightarrow$ $(m+1)\,a_1 = 0$ $[\because m + 2 \neq 0]$

Since $m + 1$ may be zero, hence a_1 is arbitrary (or takes the form $\frac{0}{0}$). In other words, a_1 becomes indeterminate.

Hence the solution will contain a_0 and a_1 as arbitrary constants. The complete solution will be given by putting $m = -1$ in y.

Now, Coefficient of $x^{m+1} = 0$

$\Rightarrow$ $(m+2)(m+1)\,a_2 + 2(m+2)\,a_2 + a_0 = 0$

$\Rightarrow$ $(m+2)(m+3)\,a_2 + a_0 = 0$

$$\boxed{a_2 = \frac{-a_0}{(m+2)(m+3)}}$$

Coefficient of $x^{m+2} = 0$

$\Rightarrow$ $(m+3)(m+2)\,a_3 + 2(m+3)\,a_3 + a_1 = 0$

$(m+3)(m+4)\,a_3 + a_1 = 0$

$$\boldsymbol{a_3 = \frac{-a_1}{(m+3)(m+4)}}$$

Coefficient of $x^{m+3} = 0$

$\Rightarrow (m+4)(m+3)\,a_4 + 2(m+4)\,a_4 + a_2 = 0$

$\Rightarrow$ $(m+4)(m+5)\,a_4 = -a_2$

$\Rightarrow$ $a_4 = \dfrac{-a_2}{(m+4)(m+5)}$

$\Rightarrow$ $$\boldsymbol{a_4 = \frac{a_0}{(m+2)(m+3)(m+4)(m+5)}}$$

Coefficient of $x^{m+4} = 0$

$(m+5)(m+4)\,a_5 + 2(m+5)\,a_5 + a_3 = 0$

$(m+5)(m+6)\,a_5 = -a_3$

$$\boldsymbol{a_5 = \frac{a_1}{(m+3)(m+4)(m+5)(m+6)}}$$

and so on.

Substituting these values in equation (1), we get

$$y = x^m\left[a_0 + a_1x - \frac{a_0}{(m+2)(m+3)}x^2 - \frac{a_1}{(m+3)(m+4)}x^3 + \frac{a_0}{(m+2)(m+3)(m+4)(m+5)}x^4 + \frac{a_1}{(m+3)(m+4)(m+5)(m+6)}x^5 + \ldots\right]$$

$$y = x^m\left[a_0\left\{1 - \frac{x^2}{(m+2)(m+3)} + \frac{x^4}{(m+2)(m+3)(m+4)(m+5)} - \ldots\right\} + a_1\left\{x - \frac{x^3}{(m+3)(m+4)} + \frac{x^5}{(m+3)(m+4)(m+5)(m+6)} - \ldots\right\}\right]$$

Now, $$(y)_{m=-1} = x^{-1}\left[a_0\left(1 - \frac{x^2}{1.2} + \frac{x^4}{1.2.3.4} - \ldots\right) + a_1\left(x - \frac{x^3}{2.3} + \frac{x^5}{2.3.4.5} - \ldots\right)\right]$$

$$= x^{-1}\,[a_0 \cos x + a_1 \sin x]$$

Hence complete solution is given by

$$y = (y)_{m=-1} \quad \Rightarrow \quad y = \frac{1}{x}(a_0 \cos x + a_1 \sin x).$$ **Ans.**

Note. All those problems, in which $x = 0$, was an ordinary point of $y'' + P(x)\,y' + Q(x)\,y = 0$, can also be solved by Frobenius method as given in Art. 32.9 and explained in above illustrative example.

EXERCISE 32.5

Solve in series the following differential equations:

1. $(1-x^2)\dfrac{d^2y}{dx^2} + 2x\dfrac{dy}{dx} + y = 0$ **Ans.** $y = a_0\left(1 - \frac{1}{2}x^2 + \frac{1}{8}x^4 + \ldots\ldots\right) + b\left(x - \frac{1}{2}x^3 + \frac{1}{40}x^5 + \ldots\ldots\right)$

(AMIETE, June 2010)

2. $(2+x^2)\dfrac{d^2y}{dx^2} + x\dfrac{dy}{dx} + (1+x)\,y = 0$

Ans. $y = a_0\left(1 - \frac{1}{4}x^2 - \frac{1}{12}x^3 + \frac{5}{56}x^4 + \ldots\right) + b\left(x - \frac{1}{6}x^3 - \frac{1}{24}x^4 + \ldots\ldots\right)$

3. Find the power series solution about the point $x_0 = 2$ of the equation $y'' + (x-1)\,y' + y = 0$ *(AMIETE, Dec. 2009)*

Ans. $y = a_0\,[1 - 1/2\,(x-2)^2 + 1/6\,(x-2)^3 + \ldots\ldots..]$ $+ b\,[(x-2) - 1/2\,(x-2)^2 - 1/6\,(x-2)^3 + \ldots..]$

CHAPTER

33

BESSEL'S FUNCTIONS)

33.1 BESSEL'S EQUATION

The differential equation

$$x^2\frac{d^2y}{dx^2}+x\frac{dy}{dx}+(x^2-n^2)y=0$$

is called the Bessel's differential equation, and particular solutions of this equation are called Bessel's functions of order n.

We find the Bessel's equation while solving Laplace equation in polar coordinates by the method of separation of variables. This equation has a number of applications in engineering.

Bessel's functions are involved in

(*i*) The Oscillatory motion of a hanging chain
(*ii*) Euler's theory of a circular membrane
(*iii*) The studies of planetary motion
(*iv*) The propagation of waves
(*v*) The Elasticity
(*vi*) The fluid motion
(*vii*) The potential theory
(*viii*) Cylindrical and spherical waves
(*ix*) Theory of plane waves

Bessel's functions are also known as cylindrical and spherical function.

33.2 SOLUTION OF BESSEL'S EQUATION

$$x^2\frac{d^2y}{dx^2}+x\frac{dy}{dx}+(x^2-n^2)y=0. \quad \text{... (1)}$$

Let $\quad y=\sum_{r=0}^{\infty}a_r x^{m+r}$ or $y=a_0x^m+a_1x^{m+1}+a_2x^{m+2}+\ldots \quad$ ... (2)

So that $\quad \frac{dy}{dx}=\sum_{r=0}^{\infty}a_r(m+r)x^{m+r-1}$

and $\quad \frac{d^2y}{dx^2}=\sum_{r=0}^{\infty}a_r(m+r)(m+r-1)x^{m+r-2}$

Substituting these values in (1), we get

$$x^2\sum_{r=0}^{\infty}a_r(m+r)(m+r-1)x^{m+r-2}+x\sum_{r=0}^{\infty}a_r(m+r)x^{m+r-1}+(x^2-n^2)\sum_{r=0}^{\infty}a_r x^{m+r}=0$$

$$\Rightarrow \sum_{r=0}^{\infty} a_r (m+r)(m+r-1)x^{m+r} + \sum_{r=0}^{\infty} a_r (m+r)x^{m+r} + \sum_{r=0}^{\infty} a_r x^{m+r+2} - n^2 \sum_{r=0}^{\infty} a_r x^{m+r} = 0$$

$$\Rightarrow \sum_{r=0}^{\infty} a_r [(m+r)(m+r-1)+(m+r)-n^2]x^{m+r} + \sum_{r=0}^{\infty} a_r x^{m+r+2} = 0$$

$$\Rightarrow \sum_{r=0}^{\infty} a_r [(m+r)^2 - n^2]x^{m+r} + \sum_{r=0}^{\infty} a_r x^{m+r+2} = 0. \qquad \text{... (3)}$$

Equating the coefficient of lowest degree term of x^m in the identity (3) to zero, by putting $r = 0$ in the first summation we get the indicial equation.

$$a_0[(m+0)^2 - n^2] = 0. \qquad (r = 0)$$

$$\Rightarrow \qquad m^2 = n^2 \text{ i.e. } m = n,\ m = -n \qquad a_0 \neq 0$$

Equating the coefficient of the next lowest degree term x^{m+1} in the identity (3), we put $r = 1$ in the first summation

$$a_1\ [(m+1)^2 - n^2] = 0 \text{ i.e. } a_1 = 0, \text{ since } (m+1)^2 - n^2 \neq 0$$

Equating the coefficient of x^{m+r+2} in (3) to zero, to find relation in successive coefficients, we get

$$a_{r+2}[(m+r+2)^2 - n^2] + a_r = 0$$

$$\Rightarrow \qquad a_{r+2} = -\frac{1}{(m+r+2)^2 - n^2} . a_r$$

Therefore, $a_3 = a_5 = a_7 = ... = 0$, since $a_1 = 0$

If $r = 0$, $\quad a_2 = -\dfrac{1}{(m+2)^2 - n^2} a_0$

If $r = 2$, $\quad a_4 = -\dfrac{1}{(m+4)^2 - n^2} a_2 = \dfrac{1}{[(m+2)^2 - n^2][(m+4)^2 - n^2]} a_0$ and so on.

On substituting the values of the coefficients a_1, a_2, a_3, a_4 .... in (2), we have

$$y = a_0 x^m - \frac{a_0}{(m+2)^2 - n^2} x^{m+2} + \frac{a_0}{[(m+2)^2 - n^2][(m+4)^2 - n^2]} x^{m+4} + ...$$

$$y = a_0 x^m \left[1 - \frac{1}{(m+2)^2 - n^2} x^2 + \frac{1}{[(m+2)^2 - n^2][(m+4)^2 - n^2]} x^4 - ...\right]$$

For $m = n$

$$y = a_0 x^n \left[1 - \frac{1}{4(n+1)} x^2 + \frac{1}{4^2 .2!(n+1)(n+2)} x^4 - ...\right]$$

where a_0 is an arbitrary constant.

For $m = -n$

$$y = a_0 x^{-n} \left[1 - \frac{1}{4(-n+1)} x^2 + \frac{1}{4^2 .2!(-n+1)(-n+2)} x^4 - ...\right]$$

33.3 BESSEL'S FUNCTIONS, $J_n(x)$

The Bessel's equation is $x^2 \dfrac{d^2y}{dx^2} + x\dfrac{dy}{dx} + (x^2 - n^2)y = 0 \qquad$ ... (1)

Solution of (1) is

$$y = a_0 x^n \left[1 - \frac{x^2}{2.2(n+1)} + \frac{x^4}{2.4.2^2 (n+1)(n+2)} - ... + (-1)^r \frac{x^{2r}}{(2^r r!).2^r (n+1)(n+2)...(n+r)} + ...\right]$$

$$= a_0 x^n \sum_{r=0}^{\infty} (-1)^r \frac{x^{2r}}{2^{2r}.r!(n+1)(n+2)...(n+r)}$$

where a_0 is an arbitrary constant.

If $\quad a_0 = \dfrac{1}{2^n \overline{|(n+1)}}$

he above solution is called Bessel's function denoted by $J_n(x)$

Thus $\quad J_n(x) = \dfrac{1}{2^n \overline{|(n+1)}} \Sigma(-1)^r \dfrac{x^{n+2r}}{2^{2r}.r!(n+1)(n+2)...(n+r)} \qquad (\overline{|n+1} = n!)$

$$\Rightarrow \quad J_n(x) = \left(\frac{x}{2}\right)^n \left\{ \frac{1}{\overline{|(n+1)}} - \frac{1}{1!\overline{|(n+2)}}\left(\frac{x}{2}\right)^2 + \frac{1}{2!\overline{|(n+3)}}\left(\frac{x}{2}\right)^4 - \frac{1}{3!\overline{|(n+4)}}\left(\frac{x}{2}\right)^6 + ... \right\}$$

$$\Rightarrow \quad J_n(x) = \frac{x^n}{2^n \overline{|n+1}} \left[1 - \frac{x^2}{2.(2n+2)} + \frac{x^4}{2.4.(2n+2)(2n+4)} + \cdots \right] \quad ... (2)$$

$$\Rightarrow \quad J_n(x) = \sum_{r=0}^{\infty} \frac{(-1)^r}{r!\overline{|(n+r+1)}} \left(\frac{x}{2}\right)^{n+2r} \quad \Rightarrow J_n(x) = \sum_{r=0}^{\infty} \frac{(-1)^r}{r!(n+r)!} \left(\frac{x}{2}\right)^{n+2r}$$

If $n = 0$, $J_0(x) = \Sigma \dfrac{(-1)^r}{(r!)^2}\left(\dfrac{x}{2}\right)^{2r} \Rightarrow J_0(x) = 1 - \dfrac{x^2}{2^2} + \dfrac{x^4}{2^2.4^2} - \dfrac{x^6}{2^2.4^2.6^2} + ...$

If $n = 1$, $J_1(x) = \dfrac{x}{2} - \dfrac{x^3}{2^2.4} + \dfrac{x^5}{2^2.4^2.6} - ...$

We draw the graph of these two functions. Both the functions are oscillatory with a varying period and a decreasing amplitude.

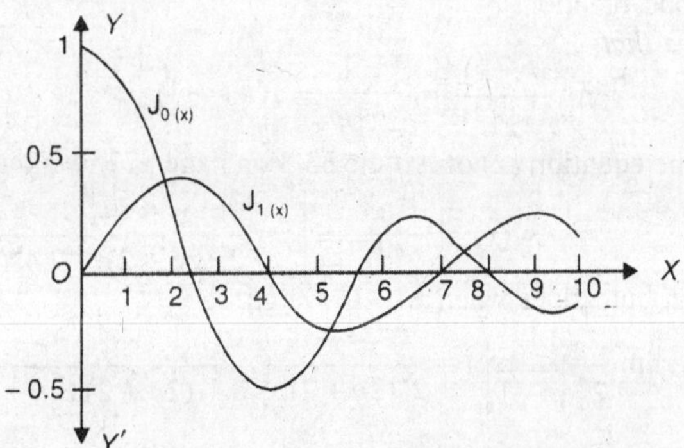

Replacing n by $-n$ in (2), we get $J_{-n}(x) = \displaystyle\sum_{r=0}^{\infty} \frac{(-1)^r}{r!\,\overline{|(-n+r+1)}} \left(\frac{x}{2}\right)^{-n+2r}$

Case I. If n is not integer or zero, then the complete solution of (1) is

$$y = A J_n(x) + B J_{-n}(x)$$

Case II. If $n = 0$, then $y_1 = y_2$ and complete solution of (1) is the Bessel's function of order zero.

Case III. If n is positive integer, then y_2 is not the solution of (1). And y_1 fails to give a solution for negative values of n. Let us find out the general solution when n is an integer.

Example 1. *Show that Bessel's function $J_n(x)$ is an even function when n is even and is odd function when n is odd.* (*U.P., II Semester, 2009*)

Solution. We know that

$$J_n(x) = \sum_{r=0}^{\infty} \frac{(-1)^r}{r!\,\lceil n+r+1} \left(\frac{x}{2}\right)^{n+2r} \qquad \text{...(1)}$$

Replacing x by $-x$ in (1), we get

$$J_n(-x) = \sum_{r=0}^{\infty} \frac{(-1)^r}{r!\,\lceil n+r+1} \left(\frac{-x}{2}\right)^{n+2r} \qquad \text{...(2)}$$

Case I. If n is even, then $n + 2r$ is even $\Rightarrow \left(\frac{-x}{2}\right)^{n+2r} = \left(\frac{x}{2}\right)^{n+2r}$

Thus (2), becomes

$$J_n(-x) = -\sum_{r=0}^{\infty} \frac{(-1)^r}{r!\,\lceil n+r+1} \left(\frac{x}{2}\right)^{n+2r}$$

$$= J_n(x) \qquad \left[\begin{array}{c}\text{For even function} \\ f(-x) = f(x)\end{array}\right]$$

Hence, $J_n(x)$ is even function.

Case II. If n is odd, then $n + 2r$ is odd $\Rightarrow \left(\frac{-x}{2}\right)^{n+2r} = -\left(\frac{x}{2}\right)^{n+2r}$

Thus (2), becomes

$$J_n(-x) = -\sum_{r=0}^{\infty} \frac{(-1)^r}{r!\,\lceil n+r+1} \left(\frac{x}{2}\right)^{n+2r}$$

$$= -J_n(x) \qquad \left[\begin{array}{c}\text{For odd function} \\ f(-x) = -f(x)\end{array}\right]$$

Hence, $J_n(x)$ is odd function. **Proved.**

Example 2. *Prove that:*

$$\lim_{x \to 0} \frac{J_n(x)}{x^n} = \frac{1}{2^n \lceil n+1}; \; (n > -1).$$

Solution. From the equation (2) of Article 33.3 on page 872, we know that

$$J_n(x) = \frac{x^n}{2^n \lceil n+1} \left[1 - \frac{x^2}{2.(2n+2)} + \frac{x^4}{2.4.(2n+2)(2n+4)} - \dots\right]$$

On taking limit on both sides when $x \to 0$, we get

$$\lim_{x \to 0} \frac{J_n(x)}{x^n} = \lim_{x \to 0} \frac{1}{2^n \lceil n+1} \left[1 - \frac{x^2}{2.(2n+2)} + \frac{x^4}{2.4.(2n+2)(2n+4)} - \dots\right]$$

$$= \frac{1}{2^n \lceil n+1}$$

Proved.

33.4. BESSEL'S FUNCTION OF THE SECOND KIND OF ORDER *n*.

$$x^2 \frac{d^2y}{dx^2} + x\frac{dy}{dx} + (x^2 - n^2)y = 0 \qquad \text{... (1)}$$

Let $y = u(x) J_n(x)$ be the second solution of the Bessel's equation when n is integer.

$$\frac{dy}{dx} = u' J_n + u J'_n$$

$$\frac{d^2y}{dx^2} = u'' J_n + 2u' J'_n + u J''_n$$

Substituting these values of y, y', y'' in (1), we get

$$x^2 (u'' J_n + 2u' J'_n + u J''_n) + x (u' J_n + uJ'_n) + (x^2 - n^2) u J_n = 0$$

$$\Rightarrow \quad u [x^2 J''_n + x J'_n + (x^2 - n^2) J_n] + x^2 u'' J_n + 2x^2 u' J'_n + x u' J_n = 0 \quad \text{... (2)}$$

$$\Rightarrow \quad x^2 J''_n + x J'_n + (x^2 - n^2) J_n = 0 \quad \text{[Since } J_n \text{ is a solution of (1)]}$$

(2) becomes $x^2 u'' J_n + 2x^2 u' J_n' + xu' J_n = 0$ ... (3)

Dividing (3) by $x^2 u' J_n$, we have

$$\frac{u''}{u'} + 2\frac{J_n'}{J_n} + \frac{1}{x} = 0 \quad \text{... (4)}$$

(4) Can also be written as

$$\frac{d}{dx}[\log u'] + 2\frac{d}{dx}[\log J_n] + \frac{d}{dx}(\log x) = 0$$

$$\Rightarrow \quad \frac{d}{dx}[\log u' + 2\log J_n + \log x] = 0$$

$$\Rightarrow \quad \frac{d}{dx}[\log (u'.J_n^2.x) = 0 \quad \text{... (5)}$$

Integrating (5), we get

$$\log u'.J_n^2.x = \log C_1$$

$$\Rightarrow \quad u'.J_n^2.x = C_1 \quad \Rightarrow \quad u' = \frac{C_1}{J_n^2.x} \quad \text{... (6)}$$

On integrating (6), we obtain

$$u = \int \frac{C_1}{J_n^2.x} dx + C_2$$

Putting the value of u in the assumed solution $y = u(x). J_n(x)$, we get

$$y = \left[\int \frac{C_1 dx}{J_n^2(x).x} + C_2\right] J_n(x)$$

$$y = C_2 J_n(x) + C_1 J_n(x) \int \frac{dx}{x J_n^2(x)} \quad \Rightarrow \quad = C_2 J_n(x) + C_1 y_n(x)$$

where, $y_n(x) = J_n(x) \int \frac{dx}{x J_n^2(x)}$

The function $y_n(x)$ is known as Bessel's function of second kind of order n. It is also called **Neumann function.**

When n is not an integer.

$$y_n(x) = \frac{1}{\sin n\pi}[J_n(x)\cos n\pi - J_{-n}(x)]$$

When n is an integer

$$y_n(x) = \lim_{m\to n}\left[\frac{1}{\sin m\pi}\{J_m(x)\cos mx - J_{-m}(x)\}\right]$$

General solution of Bessel's Equation is

$$y = AJ_n(x) + BJ_{-n}(x)$$

Example 3. *Prove that,* $J_{-n}(x) = (-1)^n J_n(x)$
where n is a positive integer. (*A.M.I.E.T.E., Winter 2001*)

Solution. $$J_{-n}(x) = \sum_{r=0}^{\infty} (-1)^r \frac{1}{r! \, \lceil(r-n+1)} \left(\frac{x}{2}\right)^{-n+2r}$$

$$\Rightarrow \quad = \sum_{r=0}^{r=n-1} \frac{(-1)^r \left(\frac{x}{2}\right)^{-n+2r}}{r!\,\overline{|(-n+r+1)}} + \sum_{r=n}^{\infty} \frac{(-1)^r \left(\frac{x}{2}\right)^{-n+2r}}{r!\,\overline{|(-n+r+1)}}$$

$$\Rightarrow \quad = 0 + \sum_{r=n}^{\infty} \frac{(-1)^r \left(\frac{x}{2}\right)^{-n+2r}}{r!\,\overline{|(-n+r+1)}} \qquad \text{(since – ve integer = } \infty)$$

On putting $r = n + k$, we get

$$J_{-n}(x) = \sum_{k=0}^{\infty} \frac{(-1)^{n+k} \left(\frac{x}{2}\right)^{n+2k}}{(n+k)!\,\overline{|(k+1)}} = (-1)^n \sum_{k=0}^{\infty} \frac{(-1)^k \left(\frac{x}{2}\right)^{n+2k}}{(n+k)!\,k!}$$

$$= (-1)^n J_n(x)$$

Proved.

Example 4. *Prove that*

(*a*) $J_{1/2}(x) = \sqrt{\left(\frac{2}{\pi x}\right)} \sin x$ (*AMIETE, June 2010, U.P., II Semester, 2009*)

(*b*) $J_{-1/2}(x) = \sqrt{\left(\frac{2}{\pi x}\right)} \cos x$ (*AMIETE, June 2009*)

Solution. We know that,

$$J_n(x) = \frac{x^n}{2^n \overline{|n+1}} \left[1 - \frac{x^2}{2.2(n+1)} + \frac{x^4}{2.4.2^2(n+1)(n+2)} - \ldots\ldots\right] \qquad \ldots (1)$$

(*a*) Substituting $n = \frac{1}{2}$ in (1), we obtain

$$J_{1/2}(x) = \frac{x^{1/2}}{2^{1/2} \overline{\left|\frac{1}{2}+1\right.}} \left[1 - \frac{x^2}{2.2.\left(\frac{1}{2}+1\right)} + \frac{x^4}{2.4.2^2\left(\frac{1}{2}+1\right)\left(\frac{1}{2}+2\right)} - \ldots\ldots\right]$$

$$= \frac{\sqrt{x}}{\sqrt{2}\,\overline{|3/2}} \left[1 - \frac{x^2}{2.3} + \frac{x^4}{2.3.4.5} \ldots\right] = \frac{\sqrt{x}}{\sqrt{2}.\frac{1}{2}\overline{\left|\frac{1}{2}\right.}} \frac{1}{x}\left[x - \frac{x^3}{3!} + \frac{x^5}{5!} - \ldots\right]$$

$$= \frac{1}{\sqrt{2x}.\ \frac{1}{2}\sqrt{\pi}} \sin x = \sqrt{\left(\frac{2}{\pi x}\right)} \sin x \qquad \left(\text{since } \overline{\left|\frac{1}{2}\right.} = \sqrt{\pi}\right)$$

Proved.

(*b*) Again substituting $n = -\frac{1}{2}$ in (1), we have

$$J_{-1/2}(x) = \frac{x^{-1/2}}{2^{-1/2} \overline{\left|-\frac{1}{2}+1\right.}} \left[1 - \frac{x^2}{2.2.\left(-\frac{1}{2}+1\right)} + \frac{x^4}{2.4.2^2\left(-\frac{1}{2}+1\right)\left(-\frac{1}{2}+2\right)} - \ldots\right]$$

$$= \frac{\sqrt{2}}{\sqrt{x}\,\overline{\left|\frac{1}{2}\right.}} \left[1 - \frac{x^2}{2!} + \frac{x^4}{4!} - \ldots\right] = \sqrt{\left(\frac{2}{\pi x}\right)} \cos x \qquad \left(\text{since } \overline{\left|\frac{1}{2}\right.} = \sqrt{\pi}\right)$$

Proved.

Example 5. *Show that*

$$\sqrt{\left(\frac{1}{2}\pi x\right)}\, J_{\frac{3}{2}}(x) = \frac{\sin x}{x} - \cos x$$

Solution. We know that

$$J_n(x) = \frac{x^n}{2^n \overline{|(n+1)}}\left[1 - \frac{x^2}{2.2.(n+1)} + \frac{x^4}{2.4.2^2(n+1)(n+2)} - \ldots\right] \qquad \ldots (1)$$

Putting $n = \frac{3}{2}$ in (1), we get

$$J_{\frac{3}{2}}(x) = \frac{x^{3/2}}{2^{3/2}\overline{\left|\frac{3}{2}+1\right.}}\left[1 - \frac{x^2}{2.2.\left(\frac{3}{2}+1\right)} + \frac{x^4}{2.4.2^2.\left(\frac{3}{2}+1\right)\left(\frac{3}{2}+2\right)} - \ldots\right]$$

$$= \frac{x^{-\frac{1}{2}}}{2\sqrt{2}.\frac{3}{2}.\frac{1}{2}\sqrt{\pi}}\left[x^2 - \frac{x^4}{2.5} + \frac{x^6}{2.4.5.7} - \ldots\right] = \sqrt{\left(\frac{2}{\pi x}\right)}\left[\frac{x^2}{2!} - \frac{x^2}{3!} + \frac{x^4}{5!} - \frac{x^4}{4!} + \frac{x^6}{6!} - \frac{x^6}{7!} + \ldots\right]$$

$$= \sqrt{\frac{2}{\pi x}}\left[\frac{1}{x}\left\{x - \frac{x^3}{3!} + \frac{x^5}{5!} - \ldots\right\} - \left\{1 - \frac{x^2}{2!} + \frac{x^4}{4!} - \ldots\right\}\right] = \sqrt{\frac{2}{\pi x}}\left[\frac{\sin x}{x} - \cos x\right]$$

$$\Rightarrow \qquad \sqrt{\frac{\pi x}{2}}\, J_{\frac{3}{2}}(x) = \frac{\sin x}{x} - \cos x$$

Proved.

Example 6. *Show that* $\sqrt{\left(\frac{1}{2}\pi x\right)}\, J_{\frac{3}{-2}}(x) = -\sin x - \frac{\cos x}{x}$

Solution. We know that

$$J_n(x) = \frac{x^n}{2^n \overline{|n+1}}\left[1 - \frac{x^2}{2.2.(n+1)} + \frac{x^4}{2.4.2^2\,(n+1)(n+2)} - \ldots\right]$$

Multiplying numerator and denominator by $(n + 1)$, we get

$$J_n(x) = \frac{x^n(n+1)}{2^n \overline{|n+2}}\left[1 - \frac{x^2}{2.2.(n+1)} + \frac{x^4}{2.4.2^2(n+1)(n+2)} - \ldots\right]$$

Putting $n = -\frac{3}{2}$, we get

$$J_{\frac{3}{-2}}(x) = \frac{x^{-3/2}\left(-\frac{3}{2}+1\right)}{2^{-3/2}\overline{\left|\frac{-3}{2}+2\right.}}\left[1 + \frac{x^2}{2} - \frac{x^4}{2.4} - \ldots\right] = \frac{x^{-\frac{3}{2}}\left(-\frac{1}{2}\right)}{2^{-3/2}\overline{\left|\frac{1}{2}\right.}}\left[1 + \frac{x^2}{2} - \frac{x^4}{2.4}\ldots\right]$$

$$= -\sqrt{\left(\frac{2}{\pi x}\right)}\frac{1}{x}\left[1 + \frac{x^2}{2} - \frac{x^4}{2.4}\ldots\right] = \sqrt{\left(\frac{2}{\pi x}\right)}\left[-\frac{1}{x}\left\{1 + x^2\left(1 - \frac{1}{2}\right) - x^4\left(\frac{1}{6} - \frac{1}{2.4}\right)\ldots\right\}\right]$$

$$= \sqrt{\left(\frac{2}{\pi x}\right)}\left[-\frac{1}{x}\left(1 - \frac{x^2}{2!} + \frac{x^4}{4!} - \ldots\right) - \left(x - \frac{x^3}{3!} + \ldots\right)\right] = \sqrt{\left(\frac{2}{\pi x}\right)}\left[-\frac{1}{x}\cos x - \sin x\right]$$

Hence, $\sqrt{\left(\frac{\pi x}{2}\right)}\, J_{-\frac{3}{2}}(x) = -\frac{1}{x}\cos x - \sin x$ **Proved.**

33.5 RECURRENCE FORMULAE

These formulae are very useful in solving the questions. So, they are to be committed to memory.

1.	$x J_n' = n J_n - x J_{n+1}$
2.	$x J_n' = -n J_n + x J_{n-1}$
3.	$2 J_n' = J_{n-1} - J_{n+1}$
4.	$2n J_n = x (J_{n-1} + J_{n+1})$ (*AMIETE, June 2010*)
5.	$\frac{d}{dx}(x^{-n} J_n) = -x^{-n} J_{n+1}$
6.	$\frac{d}{dx}(x^n J_n) = x^n J_{n-1}$ (*AMIETE, June 2010*)

Formula I. $x J_n' = n J_n - x J_{n+1}$

Proof. We know that

$$J_n = \sum_{r=0}^{\infty} \frac{(-1)^r}{r! \, \overline{|(n+r+1)}} \left(\frac{x}{2}\right)^{n+2r}$$

Differentiating with respect to x, we get

$$J_n' = \sum \frac{(-1)^r (n+2r)}{r! \, \overline{|(n+r+1)}} \left(\frac{x}{2}\right)^{n+2r-1} \frac{1}{2}$$

$$\Rightarrow \quad x J_n' = n \sum \frac{(-1)^r}{r! \, \overline{|(n+r+1)}} \left(\frac{x}{2}\right)^{n+2r} + x \sum \frac{(-1)^r . 2r}{2 . r! \, \overline{|(n+r+1)}} \left(\frac{x}{2}\right)^{n+2r-1}$$

$$= n J_n + x \sum_{r=1}^{\infty} \frac{(-1)^r}{(r-1)! \, \overline{|(n+r+1)}} \left(\frac{x}{2}\right)^{n+2r-1}$$

$$= n J_n + x \sum_{s=0}^{\infty} \frac{(-1)^{s+1}}{s! \, \overline{|(n+s+2)}} \left(\frac{x}{2}\right)^{n+2s+1} \quad \text{[Putting } r-1=s\text{]}$$

$$= n J_n - x \sum_{s=0}^{\infty} \frac{(-1)^s}{s! \, \overline{|[(n+1)+s+1]}} \left(\frac{x}{2}\right)^{(n+1)+2s}$$

$$\boxed{x J_n' = n J_n - x J_{n+1}}$$

Proved.

Formula II. $x J_n' = -n J_n + x J_{n-1}$ (*U.P., II Semester, summer 2006*)

Proof. We know that

$$J_n = \sum_{r=0}^{\infty} \frac{(-1)^r}{r! \, \overline{|(n+r+1)}} \left(\frac{x}{2}\right)^{n+2r}$$

Differentiating w.r.t. 'x', we get

$$J_n' = \sum_{r=0}^{\infty} \frac{(-1)^r (n+2r)}{r! \, \overline{|(n+r+1)}} \left(\frac{x}{2}\right)^{n+2r-1} \frac{1}{2}$$

$$xJ_n' = \sum_{r=0}^{\infty} \frac{(-1)^r (n+2r)}{r!\,\overline{|(n+r+1)}} \left(\frac{x}{2}\right)^{n+2r} = \sum_{r=0}^{\infty} \frac{(-1)^r [(2n+2r)-n]}{r!\,\overline{|(n+r+1)}} \left(\frac{x}{2}\right)^{n+2r}$$

$$= \sum_{r=0}^{\infty} \frac{(-1)^r (2n+2r)}{r!\,\overline{|(n+r+1)}} \left(\frac{x}{2}\right)^{n+2r} - n \sum_{r=0}^{\infty} \frac{(-1)^r}{r!\,\overline{|(n+r+1)}} \left(\frac{x}{2}\right)^{n+2r}$$

$$= \sum_{r=0}^{\infty} \frac{(-1)^r 2}{r!\,\overline{|(n+r)}} \left(\frac{x}{2}\right)^{n+2r} - nJ_n = x \sum_{r=0}^{\infty} \frac{(-1)^r}{r!\,\overline{|[(n-1)+r+1]}} \left(\frac{x}{2}\right)^{(n-1+2r)} - nJ_n$$

$\Rightarrow$ $\boxed{xJ_n' = xJ_{n-1} - nJ_n}$ **Proved.**

Formula III. $\mathbf{2J'_n = J_{n-1} - J_{n+1}}$

Proof. We know that $xJ_n' = nJ_n - xJ_{n+1}$ ... (1) (Recurrence formula *I*)

$xJ_n' = -nJ_n + xJ_{n-1}$ ... (2) (Recurrence formula *II*)

Adding (1) and (2), we get

$2xJ_n' = -xJ_{n+1} + xJ_{n-1} \Rightarrow$ $\boxed{2J'_n = J_{n-1} - J_{n+1}}$ **Proved.**

Formula IV. $\mathbf{2nJ_n = x(J_{n-1} + J_{n+1})}$ (*U.P. II Semester, June 2007*)

Proof. We know that

$xJ_n' = nJ_n - xJ_{n+1}$ ... (1) (Recurrence formula *I*)

$xJ_n' = -nJ_n + xJ_{n-1}$ ... (2) (Recurrence formula *II*)

Subtracting (2) from (1), we get

$0 = 2nJ_n - xJ_{n+1} - xJ_{n-1} \Rightarrow$ $\boxed{2nJ_n = x(J_{n-1} + J_{n+1})}$ ... (3) **Proved.**

The following examples are solved by using Recurrence formula IV.

Example 7. *Prove that*

$$J_{5/2}(x) = \sqrt{\frac{2}{\pi x}}\left[\left(\frac{3-x^2}{x^2}\right)\sin x - \frac{3\cos x}{x}\right]$$ (*AMIETE, June 2010, Q. Bank U.P.T.U. 2002*)

Solution. From Recurrence relation (4), we have

$$2nJ_n(x) = x[J_{n-1}(x) + J_{n+1}(x)] \quad ...(1)$$

$$J_{n+1}(x) = \frac{2n}{x} J_n(x) - J_{n-1}(x)$$

Putting $n = 1/2$ in (1), we get

$$J_{3/2}(x) = \frac{1}{x} J_{1/2}(x) - J_{-1/2}(x) \quad ...(2)$$

Again putting $n = \frac{3}{2}$ in (1), we get

$$J_{5/2}(x) = \frac{3}{2} J_{3/2}(x) - J_{1/2}(x) \quad ...(3)$$

Putting the value of $J_{\frac{3}{2}}(x)$ from (2) in (3), we get

$$J_{5/2}(x) = \frac{3}{x}\left[\frac{1}{x} J_{1/2}(x) - J_{-1/2}(x)\right] - J_{1/2}(x) \quad ...(4)$$

$$= \left(\frac{3}{x^2} - 1\right) J_{1/2}(x) - \frac{3}{x} J_{-1/2}(x)$$

Putting the values of $J_{\frac{1}{2}}(x)$ and $J_{-\frac{1}{2}}(x)$ in (4), we get

$$J_{5/2}(x) = \left(\frac{3-x^2}{x^2}\right)\sqrt{\frac{2}{\pi x}}\sin x - \frac{3}{x}\cdot\sqrt{\frac{2}{\pi x}}\cos x = \sqrt{\frac{2}{\pi x}}\left[\left(\frac{3-x^2}{x^2}\right)\sin x - \frac{3}{x}\cos x\right] \quad \textbf{Proved.}$$

Example 8. *Express $J_6(x)$ in terms of $J_0(x)$ and $J_1(x)$.* *(U.P., II Semester, 2009)*

Solution. We know that

$$J_{n+1} = \frac{2n}{x}J_n - J_{n-1} \qquad ...(1) \quad \left[\begin{array}{c}\text{From Recurrence} \\ \text{relation (4)}\end{array}\right]$$

Putting $n = 5$ in (1), we get

$$J_6 = \frac{10}{x}J_5 - J_4 \qquad ...(2)$$

Putting $n = 4$ in (1), we get

$$J_5 = \frac{8}{3}J_4 - J_3$$

Putting the value of J_5 in (2), we get

$$J_6 = \frac{10}{x}\left[\frac{8}{x}J_4 - J_3\right] - J_4 = \left[\left(\frac{10}{x}\right)\left(\frac{8}{x}\right) - 1\right]J_4 - \frac{10}{x}J_3 \qquad ...(3)$$

On putting the value of J_4 in (3), we get

$$J_6 = \left[\left(\frac{10}{x}\right)\left(\frac{8}{x}\right) - 1\right]\left(\frac{6}{x}J_3 - J_2\right) - \frac{10}{x}J_3$$

$$= \left[\left(\frac{10}{x}\right)\left(\frac{8}{x}\right)\left(\frac{6}{x}\right) - \frac{16}{x}\right]J_3 - \left[\left(\frac{10}{x}\right)\left(\frac{8}{x}\right) - 1\right]J_2 \qquad ...(4)$$

$$\left[\begin{array}{c}\text{If } n = 2 \text{ in (1), then} \\ J_3 = \frac{4}{x}J_2 - J_1\end{array}\right]$$

On putting the values of J_3 in (4), we get

$$J_6 = \left[\left(\frac{10}{x}\right)\left(\frac{8}{x}\right)\left(\frac{6}{x}\right) - \frac{16}{x}\right]\left[\frac{4}{x}J_2 - J_1\right] - \left[\left(\frac{10}{x}\right)\left(\frac{8}{x}\right) - 1\right]J_2$$

$$= \left[\left(\frac{10}{x}\right)\left(\frac{8}{x}\right)\left(\frac{6}{x}\right)\left(\frac{4}{x}\right) - \left(\frac{16}{x}\right)\left(\frac{4}{x}\right) - \left(\frac{10}{x}\right)\left(\frac{8}{x}\right) + 1\right]J_2$$

$$- \left[\left(\frac{10}{x}\right)\left(\frac{8}{x}\right)\left(\frac{6}{x}\right) - \frac{16}{x}\right]J_1 \qquad ...(5) \qquad \left[\begin{array}{c}\text{If } n = 1 \text{ in (1) then} \\ J_2 = \frac{2}{x}J_1 - J_0\end{array}\right]$$

On putting the value of J_2 in (5), we get

$$J_6 = \left[\left(\frac{10}{x}\right)\left(\frac{8}{x}\right)\left(\frac{6}{x}\right)\left(\frac{4}{x}\right) - \left(\frac{16}{x}\right)\left(\frac{4}{x}\right) - \left(\frac{10}{x}\right)\left(\frac{8}{x}\right) + 1\right]\left[\frac{2}{x}J_1 - J_0\right]$$
$$- \left[\left(\frac{10}{x}\right)\left(\frac{8}{x}\right)\left(\frac{6}{x}\right) - \frac{16}{x}\right]J_1$$

$$\Rightarrow \quad J_6 = \left[\left(\frac{10}{x}\right)\left(\frac{8}{x}\right)\left(\frac{6}{x}\right)\left(\frac{4}{x}\right)\left(\frac{2}{x}\right) - \left(\frac{16}{x}\right)\left(\frac{4}{x}\right)\left(\frac{2}{x}\right) - \left(\frac{10}{x}\right)\left(\frac{8}{x}\right)\left(\frac{2}{x}\right) + \frac{2}{x}\right.$$
$$\left. - \left(\frac{10}{x}\right)\left(\frac{8}{x}\right)\left(\frac{6}{x}\right) + \frac{16}{x}\right]J_1$$
$$- \left[\left(\frac{10}{x}\right)\left(\frac{8}{x}\right)\left(\frac{6}{x}\right)\left(\frac{4}{x}\right) - \left(\frac{16}{x}\right)\left(\frac{4}{x}\right) - \left(\frac{10}{x}\right)\left(\frac{8}{x}\right) + 1\right]J_0$$

$$\Rightarrow \quad J_6 = \left[\frac{3840}{x^5} - \frac{128}{x^3} - \frac{160}{x^3} + \frac{2}{x} - \frac{480}{x^3} + \frac{16}{x}\right] J_1 - \left[\frac{1920}{x^4} - \frac{64}{x^2} - \frac{80}{x^2} + 1\right] J_0$$

$$= \left[\frac{3840}{x^5} - \frac{768}{x^3} + \frac{18}{x}\right] J_1 - \left[\frac{1920}{x^4} - \frac{144}{x^2} + 1\right] J_0 \qquad \textbf{Ans.}$$

Example 9. *Express J_5 in terms of J_1 and J_2.* (*AMIETE, Dec. 2009*)

Solution. Putting $n = 4$ in recurrence formula (4), we get

$$n = 4,\ 8J_4 = x(J_3 + J_5) \qquad \Rightarrow \quad J_5 = \frac{8}{x} J_4 - J_3$$

$$n = 3,\ J_4 = \frac{6}{x} J_3 - J_2, \qquad J_5 = \frac{8}{x}\left(\frac{6}{x} J_3 - J_2\right) - J_3 = \frac{48}{x^2} J_3 - \frac{8}{x} J_2 - J_3$$

$$J_5 = \frac{48}{x^2}\left(\frac{4}{x} J_2 - J_1\right) - \frac{8}{x} J_2 - \left(\frac{4}{x} J_2 - J_1\right) - \frac{4}{x} J_2 + J_1$$

$$\Rightarrow \quad J_5 = \frac{192}{x^3} J_2 - \frac{48}{x^2} J_1 - \frac{8}{x} J_2 - \frac{4}{x} J_2 + J_1$$

$$= \left(\frac{192}{x^3} - \frac{12}{x}\right) J_2 + \left(1 - \frac{48}{x^2}\right) J_1 \qquad \textbf{Ans.}$$

Formula V. $\dfrac{d}{dx}(x^{-n} \cdot J_n) = -x^{-n} J_{n+1}$

Proof. We know that $x J_n' = n J_n - x J_{n+1}$ (Recurrence formula *I*)

Multiplying by x^{-n-1}, we obtain $x^{-n} J_n' = n x^{-n-1} J_n - x^{-n} J_{n+1}$

i.e., $x^{-n} J_n' - n x^{-n-1} J_n = -x^{-n} J_{n+1}$

$$\Rightarrow \quad \boxed{\frac{d}{dx}(x^{-n} J_n) = -x^{-n} J_{n+1}} \qquad \textbf{Proved.}$$

Example 10. *If $n > -1$, show that:*

$$\int_0^x x^{-n} J_{n+1}(x)\, dx = \frac{1}{2^n \lfloor n+1} - x^{-n} J_n(x)$$

Solution. From relation (5), we know that

$$\frac{d}{dx}[x^{-n} J_n(x)] = -x^{-n} J_{n+1}(x)$$

Integrating it between 0 and x, we get

$$\int_0^x x^{-n} J_{n+1}(x)\, dx = -\left[x^{-n} J_n(x)\right]_0^x$$

$$= -x^{-n} J_n(x) + \underset{x \to 0}{Lt}\left[\frac{J_n(x)}{x^n}\right] \qquad \left[\begin{array}{l}\text{If } n = 3 \text{ in (1), then} \\ J_4 = \frac{6}{x} J_3 - J_2\end{array}\right]$$

$$= -x^{-n} J_n(x) + \frac{1}{2^n \lfloor n+1}$$

$$= \frac{1}{2^n \lfloor n+1} - x^{-n} J_n(x) \qquad \textbf{Proved.}$$

Formula VI. $\dfrac{d}{dx}(x^n J_n) = x^n J_{n-1}$ (*U.P., II Semester, 2004, 2005*)

Proof. We know that

$$x J_n' = -n J_n + x J_{n-1} \qquad \text{(Recurrence formula } II\text{)}$$

Multiplying by x^{n-1}, we have

$$x^n J_n' = -nx^{n-1} J_n + x^n J_{n-1} \quad i.e., \quad x^n J_n' + nx^{n-1} J_n = x^n J_{n-1}$$

$$\Rightarrow \quad \boxed{\frac{d}{dx}(x^n J_n) = x^n J_{n-1}}$$ **Proved.**

Example 11. *Prove that* $\frac{d}{dx}[J_0(x)] = -J_1(x)$

Solution. We know that $\frac{d}{dx}\left[x^{-n} J_n(x)\right] = -x^{-n} J_{n+1}(x)$ (Recurrence relation *VI)*

On putting $n = 0$ in the formula VI, we get $\frac{d}{dx}[x^0 J_0(x)] = -x^0 J_1(x)$

$$\frac{d}{dx}[J_0(x)] = -J_1(x)$$ **Proved.**

Example 12. *Prove that*

$$\int x J_0^2(x)\, dx = \frac{1}{2} x^2 [J_0^2(x) + J_1^2(x).] + c$$ *(U.P., II Semester, 2003,2004)*

Solution. L.H.S $= \int x J_0^2(x)\, dx$

On integrating by parts, we get

$$= J_0^2(x).\frac{x^2}{2} - \int 2J_0(x)\, J_0'(x).\frac{x^2}{2}\, dx + c$$

Putting the value of $J_0'(x) = -J_1(x)$, we get

$$= \frac{x^2}{2} J_0^2(x) - \int x^2 J_0(x)\{-J_1(x)\}\, dx + c$$

$$= \frac{x^2}{2} J_0^2(x) + \int x J_1(x) . x J_0(x)\, dx + c$$

Putting the value of $xJ_0(x)$ from recurrence relation VI, we get

$$= \frac{x^2}{2} J_0^2(x) - \int x J_1(x) . \frac{d}{dx}[x J_1(x)]\, dx + c$$

$$= \frac{x^2}{2} J_0^2(x) + \frac{[x J_1(x)]^2}{2} + c \quad \left[\because \int t\,dt = \frac{t^2}{2}, \text{ put } x\, J_1(x) = t\right]$$

$$= \frac{x^2}{2}[J_0^2(x) + J_1^2(x)] + c$$ **Proved.**

Example 13. *Show that* (*a*) $J_{n+3} + J_{n+5} = \frac{2}{x}(n+4) J_{n+4}$

(*b*) *Express* $J_4(x)$ *in terms of* $J_0(x)$ *and* $J_1(x)$ *(A.M.I.E.T.E., Summer, 2007, 2002)*

Solution. (*a*) Recurrence relation IV is $2n J_n = x(J_{n-1} + J_{n+1}) \Rightarrow J_{n-1} + J_{n+1} = \frac{2n}{x} J_n$

Putting $n + 4$ for n, we have $J_{n+3} + J_{n+5} = \frac{2(n+4)}{x} J_{n+4}$ **Proved.**

(*b*) We know that $J_{n+1} = \frac{2n}{x} J_n - J_{n-1}$ [From Recurrence relation (4)]

If $n = 1$, $J_2 = \frac{2}{x} J_1 - J_0$,

If $n=2$, $J_3 = \frac{4}{x} J_2 - J_1$,

If $n=3$, $J_4 = \frac{6}{x} J_3 - J_2 = \frac{6}{x}\left(\frac{4}{x} J_2 - J_1\right) - J_2$

$$= \frac{24}{x^2} J_2 - \frac{6}{x} J_1 - J_2 \quad = \frac{24}{x^2}\left(\frac{2}{x} J_1 - J_0\right) - \frac{6}{x} J_1 - \left(\frac{2}{x} J_1 - J_0\right)$$

$$= \frac{48}{x^3} J_1 - \frac{24}{x^2} J_0 - \frac{6}{x} J_1 - \frac{2}{x} J_1 + J_0 \quad = \left(\frac{48}{x^3} - \frac{8}{x}\right) J_1 + \left(1 - \frac{24}{x^2}\right) J_0 \quad \textbf{Ans.}$$

Example 14. *Prove that* $J_2'(x) = \left(1 - \frac{4}{x^2}\right) J_1(x) + \frac{2}{x} J_0(x)$ *where* $J_n(x)$ *is the Bessel's function of first kind.* (*AMIETE, Dec. 2010, U.P., III Semester, June 2008, Winter 2001*)

Solution. $x J_n' = -n J_n + x J_{n-1}$ (Recurrence formula *II*) ... (1)

On putting $n = 2$ in (1), we have $x J_2' = -2 J_2 + x J_1$

$$\Rightarrow \quad J_2' = -\frac{2}{x} J_2 + J_1 \qquad \text{... (2)}$$

$$x J_n' = n J_n - x J_{n+1} \qquad \text{(Recurrence formula } I) \text{ ... (3)}$$

From (1) and (3), we have $-n J_n + x J_{n-1} = n J_n - x J_{n+1}$

On putting $n = 1$, $-J_1 + x J_0 = J_1 - x J_2$

$$\Rightarrow \quad -\frac{1}{x} J_1 + J_0 = \frac{1}{x} J_1 - J_2 \quad \Rightarrow \quad J_2 = \frac{2}{x} J_1 - J_0 \qquad \text{... (4)}$$

Putting the value of J_2 from (4) in (2), we get

$$J_2' = -\frac{2}{x}\left(\frac{2}{x} J_1 - J_0\right) + J_1 = -\frac{4}{x^2} J_1 + \frac{2}{x} J_0 + J_1 = \left(1 - \frac{4}{x^2}\right) J_1 + \frac{2}{x} J_0 \quad \textbf{Proved.}$$

Example 15. *Using the recurrence relation, show that*

$$4 J_n''(x) = J_{n-2}(x) - 2 J_n(x) + J_{n+2}(x).$$

Solution. $2 J_n' = J_{n-1} - J_{n+1}$ (Recurrence formula *III*) ... (1)

On differentiating (1), we have

$$2 J_n'' = J_{n-1}' - J_{n+1}' \qquad \text{... (2)}$$

Replacing n by $n - 1$ and n by $n + 1$ in (1), we have

$$2 J_{n-1}' = J_{n-2} - J_n \quad \Rightarrow \quad J_{n-1}' = \frac{1}{2} J_{n-2} - \frac{1}{2} J_n \qquad \text{... (3)}$$

and
$$2 J_{n+1}' = J_n - J_{n+2} \quad \Rightarrow \quad J_{n+1}' = \frac{1}{2} J_n - \frac{1}{2} J_{n+2} \qquad \text{....(4)}$$

Putting the values of J_{n-1}' and J_{n+1}' from (3) and (4) in (2), we get

$$2 J_n'' = \frac{1}{2}[J_{n-2} - J_n] - \frac{1}{2}[J_n - J_{n+2}]$$

$$\Rightarrow \quad 4 J_n'' = J_{n-2} - J_n - J_n + J_{n+2}$$

$$\Rightarrow \quad 4 J_n'' = J_{n-2} - 2 J_n + J_{n+2} \qquad \textbf{Proved.}$$

Example 16. *Show that*

$$\frac{d}{dx}(J_n^2 + J_{n+1}^2) = 2\left(\frac{n}{x} J_n^2 - \frac{n+1}{x} J_{n+1}^2\right)$$

(*U.P. II Semester Summer 2005*)

Solution. $x J_n' = n J_n - x J_{n+1}$ (Recurrence formula *I*) ... (1)

and $x J_n' = -n J_n + x J_{n-1}$ (Recurrence formula *II*) ... (2)

Putting $(n + 1)$ for n in (2), we get

$$x J'_{n+1} = -(n+1) J_{n+1} + x J_n \qquad \text{... (3)}$$

Now $\frac{d}{dx}(J_n^2 + J_{n+1}^2) = 2 J_n . J'_n + 2 J_{n+1} . J'_{n+1}$

$$= 2 J_n \cdot \frac{1}{x}(n J_n - x J_{n+1}) + 2 J_{n+1} \cdot \frac{1}{x}\left[-(n+1) J_{n+1} + x J_n)\right] \text{[From (1) \& (3)]}$$

$$= 2\left[\frac{n}{x} J_n^2 - \frac{n+1}{x} J_{n+1}^2\right]$$

Proved.

Example 17. *Prove that following relation:*

$$x^2 J''_n(x) = (n^2 - n - x^2) J_n(x) + x J_{n+1}(x)$$

(A.M.I.E.T.E., Summer 2001, U.P. II Semester Summer, 2007, 2006)

Solution. $x^2 \frac{d^2y}{dx^2} + x\frac{dy}{dx} + (x^2 - n^2) y = 0$ (Bessel's equation) ... (1)

$J_n(x)$ is the solution of (1)

So $x^2 J''_n + x J'_n + (x^2 - n^2) J_n = 0$ ... (2)

We know that

$$x J'_n = n J_n - x J_{n+1} \qquad \text{(Recurrence relation } I\text{) ... (3)}$$

Putting the value of $x J'_n$ from (3) in (2), we get

$$x^2 J_n'' + n J_n - x J_{n+1} + (x^2 - n^2) J_n = 0$$

$$x^2 J_n'' = -n J_n + x J_{n+1} + (n^2 - x^2) J_n$$

$\Rightarrow$ $x^2 J_n'' = (n^2 - n - x^2) J_n + x J_{n+1}$ **Proved.**

Example 18. *Show that* $J_1''(x) = -J_1(x) + \frac{1}{x} J_2(x)$ *(U.P. II Semester 2010)*

Solution. From example 17, we know that:

$$x^2 J_n''(x) = (n^2 - n - x^2) J_n(x) + x J_{n+1}(x), \qquad \text{... (1)}$$

Putting $n = 1$, we get

$$x^2 J_1''(x) = (1^2 - 1 - x^2) J_1(x) + x J_{1+1}(x)$$

$\Rightarrow$ $x^2 J_1''(x) = -x^2 J_1(x) + x J_2(x)$

$\Rightarrow$ $J_1'' = -J_1 + \frac{1}{x} J_2$ **Proved.**

Example 19. *Prove that following*

$$J_3(x) + 3J'_0(x) + 4J'''_0(x) = 0$$

(U.P. III Semester, Winter 2001, A.M.I.E.T.E., Summer 2000)

Solution. We know that

$$2J'_n = J_{n-1} - J_{n+1} \qquad \text{(Recurrence relation } II\text{)}$$

Differentiating and multiplying by 2, we get

$$2^2 J''_n = 2J'_{n-1} - 2J'_{n+1} = (J_{n-2} - J_n) - (J_n - J_{n+2})$$

$$= J_{n-2} - 2J_n + J_{n+2}$$

Differentiating again and multiplying by 2, we get

$$2^3 J'''_n = 2J'_{n-2} - 4J'_n + 2J'_{n+2}$$

$$= (J_{n-3} - J_{n-1}) - 2(J_{n-1} - J_{n+1}) + (J_{n+1} - J_{n+3})$$

$$= J_{n-3} - 3 J_{n-1} + 3 J_{n+1} - J_{n+3}$$

Putting $n = 0$, we get

$$2^3 . J_0''' = J_{-3} - 3J_{-1} + 3J_1 - J_3 = (-1)^3 J_3 - 3(-1)J_1 + 3J_1 - J_3$$
$$= -2J_3 + 6J_1$$
$$4J_0''' = -J_3 + 3J_1$$
$$= -J_3 + 3(-J_0') \qquad \text{[From example 11, } J_1 = -J_0'\text{]}$$
$$J_3 + 3J_0' + 4J_0''' = 0$$
$$\Rightarrow J_3(x) + 3J_0'(x) + 4J_0'''(x) = 0$$

Proved.

Example 20. *Prove that* $\frac{d}{dx}(x\,J_n J_{n+1}) = x(J_n^2 - J_{n+1}^2)$

Solution.
$$\frac{d}{dx}(x\,J_n J_{n+1}) = J_n J_{n+1} + x\frac{d}{dx}(J_n J_{n+1})$$
$$= J_n J_{n+1} + x(J_n J_{n+1}' + J_n' J_{n+1})$$
$$= J_n J_{n+1} + (xJ_n')J_{n+1} + J_n(x\,J_{n+1}') \qquad \text{... (1)}$$

Recurrence formula *I*, $x\,J_n' = n\,J_n - x\,J_{n+1}$ .... (2)

Recurrence formula *II*, $x\,J_n' = -n\,J_n + x\,J_{n-1}$

Putting $n+1$ for n, $x\,J_{n+1}' = -(n+1)\,J_{n+1} + x\,J_n$ ... (3)

Putting the values of $x\,J_n'$ and $x\,J_{n+1}'$ from (2) and (3) in (1), we obtain

$$\frac{d}{dx}(x\,J_n J_{n+1}) = J_n\,J_{n+1} + (n\,J_n - x\,J_{n+1})\,J_{n+1} + J_n[-(n+1)\,J_{n+1} + x\,J_n]$$
$$= (1+n-n-1)\,J_n \,.\, J_{n+1} + x\,(J_n^2 - J_{n+1}^2)$$
$$= x(J_n^2 - J_{n+1}^2)$$

Proved.

Example 21. *Prove that*

$$\int J_3(x)\,dx + J_2(x) + \frac{2}{x}J_1(x) = 0$$

(A.M.I.E.T.E., Summer 2000)

Solution. We know that

$$\frac{d}{dx}[x^{-n}J_n(x)] = -x^{-n}J_{n+1}(x) \qquad \text{(Recurrence Relation } V\text{)}$$

Integrating above relation, we get

$$x^{-n}J_n(x) = -\int x^{-n}J_{n+1}(x)\,dx \qquad \text{... (1)}$$

On taking $n = 2$ in (1), we have

$$\int x^{-2}J_3(x)\,dx = -x^{-2}J_2(x) \qquad \text{... (2)}$$

Again
$$\int J_3(x)\,dx = \int x^2(x^{-2})J_3(x)dx$$
$$= x^2\int (x^{-2})J_3(x)dx - \int\left(2x\int (x^{-2}J_3(x)dx\right)dx \qquad \text{... (3)}$$

Putting the value of $\int x^{-2}J_3(x)dx$ from (2) in (3), we get

$$\int J_3(x)dx = x^2(-x^{-2}J_2(x)) - \int 2x(-x^{-2}J_2(x))dx$$
$$= -J_2(x) + 2\int x^{-1}J_2(x)\,dx = -J_2(x) + 2(-x^{-1}J_1(x))$$

On using (1), again, when $n = 1$

Hence,
$$\int J_3(x)dx + J_2(x) + \frac{2}{x}J_1(x) = 0$$

Proved.

Example 22. *Show that* $\int_0^x x^{n+1} J_n(x)\,dx = x^{n+1} J_{n+1}(x)$ $(n > -1)$

Solution. Recurrence relation *VI* is $\frac{d}{dx}[x^n J_n(x)] = x^n J_{n-1}(x)$

Putting $n+1$ for n, we get $\frac{d}{dx}[x^{n+1} J_{n+1}(x)] = x^{n+1} J_n(x)$

Integrating both sides w.r.t. x between 0 and x, we get

$$x^{n+1} J_{n+1}(x) = \int_0^x x^{n+1} J_n(x)\,dx\,.$$ **Proved.**

Example 23. *Prove that*

$$2^r \cdot J_n^r = J_{n-r} - r J_{n-r+2} + \frac{r(r-1)}{2!} J_{n-r+4} + \ldots + (-1)^r J_{n+r}$$

Solution. We know that

$$2J_n' = J_{n-1} - J_{n+1} \qquad \text{(Recurrence formula } III)$$

Differentiating, we get

$$2J_n'' = J_{n-1}' - J_{n+1}' \qquad \Rightarrow \qquad 4J_n'' = 2J_{n-1}' - 2J_{n+1}'$$

$$\Rightarrow \qquad 2^2 J_n'' = (J_{n-2} - J_n) - (J_n - J_{n+2})$$

$$2^2 \cdot J_n'' = J_{n-2} - 2J_n + J_{n+2}$$

Again differentiating and multiplying by 2, we get

$$2^3 J_n''' = 2J_{n-2}' - 2^2 J_n' + 2J_{n+2}'$$

$$= [J_{n-3} - J_{n-1}] - 2(J_{n-1} - \ldots J_{n+1}) + (J_{n+1} - J_{n+3})$$

$$= J_{n-3} - 3J_{n-1} + 3J_{n+1} - \ldots J_{n+3}$$

And so on

$$2^r J_n^r = J_{n-r} - r J_{n-r+2} + \ldots + (-1)^r J_{n+r}$$

Example 24. *Show that*

$$\frac{x}{2} J_{n-1} = n J_n - (n+2) J_{n+2} + (n+4) J_{n+4} + \ldots$$

Solution. We know that

$$2nJ_n = x[J_{n+1} + J_{n-1}] \qquad \text{(Recurrence Relation } IV)$$

$$\Rightarrow \qquad nJ_n = \frac{x}{2}[J_{n+1} + J_{n-1}]$$

$$\Rightarrow \qquad \frac{x}{2} J_{n-1} = nJ_n - \frac{x}{2} J_{n+1} \qquad \ldots (1)$$

Replacing n by $n+2$, we get $\frac{x}{2} J_{n+1} = (n+2) J_{n+2} - \frac{x}{2} J_{n+3}$ ... (2)

On putting the value of $\frac{x}{2} J_{n+1}$ from (2) in (1), we get

$$\frac{x}{2} J_{n-1} = nJ_n - \left[(n+2) J_{n+2} - \frac{x}{2} J_{n+3}\right] = nJ_n - (n+2) J_{n+2} + \frac{x}{2} J_{n+3}$$

On putting the value of $\frac{x}{2} J_{n+3}$, we get

$$= nJ_n - (n+2) J_{n+2} + (n+4) J_{n-4} - \frac{x}{2} J_{n+5} \quad \text{and so on}$$

$$= nJ_n - (n+2) J_{n+2} + (n+4) J_{n+4} - (n+6) J_{n+6} + \ldots..$$

Example 25. *Prove that*

$$J_n' = \frac{2}{x}\left[\frac{n}{2} J_n - (n+2) J_{n+2} + (n+4) J_{n+1} - \cdots\right]$$

Solution. From Recurrence formula (2), we have

$$J_n' = -\frac{n}{x} J_n + J_{n-1}$$

From example 24, Page 886, putting value of J_{n-1}, we get

$$= -\frac{n}{x} J_n + \frac{2}{x}[nJ_n - (n+2) J_{n+2} + \cdots]$$

$$= \frac{2}{x}\left[\frac{n}{2} J_n - (n+2) J_{n+2} + \ldots\right]$$ **Proved.**

Example 26. *Prove that*

$$\frac{d}{dx}\left[\frac{J_{-n}(x)}{J_n(x)}\right] = \frac{-2\sin n\pi}{\pi x J_n^2}$$

Solution. $x^2 \frac{d^2y}{dx^2} + x\frac{dy}{dx} + (x^2 - n^2)y = 0$ (Bessel's Equation)

$$\frac{d^2y}{dx^2} + \frac{1}{x}\frac{dy}{dx} + \left(1 - \frac{n^2}{x^2}\right)y = 0 \qquad \text{(Dividing by } x^2\text{)}$$

As J_n and J_{-n} are the solutions of the above equation, so

$$\Rightarrow \quad J_n'' + \frac{1}{x} J_n' + \left(1 - \frac{n^2}{x^2}\right) J_n = 0 \qquad \ldots (1)$$

$$\Rightarrow \quad J_{-n}'' + \frac{1}{x} J'_{-n} + \left(1 - \frac{n^2}{x^2}\right) J_{-n} = 0 \qquad \ldots (2)$$

On multiplying (1) by J_{-n} and (2) by J_n, we get

$$J_n'' J_{-n} + \frac{1}{x} J_n' J_{-n} + \left(1 - \frac{n^2}{x^2}\right) J_n . J_{-n} = 0 \qquad \ldots (3)$$

$$J_{-n}'' . J_n + \frac{1}{x} J'_{-n} . J_n + \left(1 - \frac{n^2}{x^2}\right) J_{-n} . J_n = 0 \quad (n \to -n) \qquad \ldots (4)$$

On subtracting (4) from (3), we get

$$J_n'' J_{-n} - J_{-n}'' . J_n + \frac{1}{x}\left(J'_n . J_{-n} - J'_{-n} . J_n\right) = 0$$

$$\Rightarrow \quad \frac{J_n'' J_{-n} - J_{-n}'' J_n}{J'_n J_{-n} - J'_{-n} J_n} = -\frac{1}{x}$$

On integrating, we get

$$\log\,(J_n' . J_{-n} - J_{-n}' . J_n) = -\log x + \log C = \log \frac{C}{x}$$

Therefore, $J_{-n} J_n' - J_n J_{-n}' = \frac{C}{x}$ ...(5)

Using definition of J_n and J_{-n}, (5) becomes

$$\frac{1}{2^{-n}\overline{|(-n+1)}}\left[x^{-n} - \frac{x^{-n+2}}{2\cdot(-2n+2)} + \frac{x^{-n+4}}{2\cdot 4\cdot(-2n+2)(-2n+4)} - \ldots\right]$$

$$\times \frac{1}{2^n \overline{|(n+1)}}\left[nx^{n-1} - \frac{(n+2)x^{n+1}}{2\cdot(2n+2)} + \frac{(n+4)x^{n+3}}{2\cdot4\cdot(2n+2)(2n+4)} - \ldots\right]$$

$$-\frac{1}{2^n \overline{|(n+1)}}\left[x^n - \frac{x^{n+2}}{2\cdot(2n+2)} + \frac{x^{n+4}}{2\cdot4\cdot(2n+2)(2n+4)} - \ldots\right]$$

$$\times \frac{1}{2^{-n} \overline{|(-n+1)}}\left[-nx^{-n-1} - \frac{(-n+2)x^{-n+1}}{2\cdot(-2n+2)} + \frac{(-n+4)x^{-n+3}}{2\cdot4(-2n+2)(-2n+4)}\right] = \frac{C}{x}$$

Now comparing the coefficients of $1/x$ on both sides, we get

$$\frac{n}{\overline{|n+1}\ \overline{|-n+1}} - \frac{-n}{\overline{|n+1}\ \overline{|-n+1}} = C$$

$$\frac{1}{\overline{|(-n+1)}\ \overline{|(n+1)}}[n-(-n)] = C$$

$$\Rightarrow \qquad C = \frac{2\sin n\pi}{\pi} \qquad \left[\text{Since } \overline{|(n)}\ \overline{|(1-n)} = \frac{\pi}{\sin n\pi}\right]$$

Substituting the value of C in (5), we get

$$J_n' \cdot J_{-n} - J_{-n}' J_n = \frac{2\sin n\pi}{\pi x}$$

$$\Rightarrow \qquad \frac{J_n' J_{-n} - J_{-n}' J_n}{J_n^2} = \frac{2\sin n\pi}{\pi x J_n^2}$$

$$\Rightarrow \qquad \frac{d}{dx}\left(\frac{J_{-n}(x)}{J_n(x)}\right) = \frac{-2\sin n\pi}{\pi x J_n^2}$$

Proved.

33.6 EQUATIONS REDUCIBLE TO BESSEL'S EQUATION

There are some differential equations which can be reduced to Bessel's equation and therefore, can be solved.

(*a*) We shall reduce the following differential equation to Bessel's equation.

$$x^2 \frac{d^2y}{dx^2} + x\frac{dy}{dx} + (k^2x^2 - n^2)y = 0 \qquad \ldots (1)$$

Put $t = kx$, $\quad \frac{dt}{dx} = k, \ \frac{dy}{dx} = \frac{dy}{dt}\frac{dt}{dx} = k\frac{dy}{dt}$

$$\frac{d^2y}{dx^2} = \frac{d}{dx}\left(k\frac{dy}{dt}\right) = \frac{d}{dt}\left(k\frac{dy}{dt}\right)\frac{dt}{dx} = k^2\frac{d^2y}{dt^2}$$

Thus (1) becomes

$$\left(\frac{t^2}{k^2}\right)\left(k^2\frac{d^2y}{dt^2}\right) + \left(\frac{t}{k}\right)\left(k\frac{dy}{dt}\right) + (t^2 - n^2)y = 0 \quad \Rightarrow \quad t^2\frac{d^2y}{dt^2} + t\frac{dy}{dt} + (t^2 - n^2)y = 0$$

$\therefore$ Its solution is $y = c_1 J_n(t) + c_2 J_{-n}(t)$,

Hence solution of (1) is $y = c_1 J_n(kx) + c_2 J_{-n}(kx)$. **Ans.**

(*b*) Let us reduce the following differential equation to Bessel's equation.

$$x\frac{d^2y}{dx^2} + a\frac{dy}{dx} + k^2xy = 0 \qquad \ldots (2)$$

Put $y = x^n z$, $\quad \frac{dy}{dx} = x^n\frac{dz}{dx} + nx^{n-1}.z$

$$\frac{d^2y}{dx^2} = x^n \frac{d^2y}{dx^2} + n x^{n-1} \frac{dz}{dx} + n x^{n-1} \frac{dz}{dx} + n(n-1).x^{n-2}.z$$

$$= x^n \frac{d^2y}{dx^2} + 2n x^{n-1} \frac{dz}{dx} + n(n-1)x^{n-2}.z.$$

Then (2) becomes

$$x\left[x^n \frac{d^2z}{dx^2} + 2n x^{n-1} \frac{dz}{dx} + n(n-1)x^{n-2} z\right] + a\left[x^n \frac{dz}{dx} + n x^{n-1} z\right] + k^2.x.x^n z = 0$$

$$\Rightarrow \qquad x^{n+1} \frac{d^2z}{dx^2} + (2n+a)x^n \frac{dz}{dx} + [k^2x^2 + n^2 + (a-1)n]x^{n-1} z = 0 \qquad \text{... (3)}$$

Dividing (3) by x^{n-1}, we get

$$x^2 \frac{d^2z}{dx^2} + (2n+a)x\frac{dz}{dx} + [k^2x^2 + n^2 + (a-1)n]z = 0$$

Let us put $2n + a = 1$, then $\quad x^2 \frac{d^2z}{dx^2} + x\frac{dz}{dx} + (k^2x^2 - n^2)z = 0$

Its solution is $\quad z = c_1 J_n(kx) + c_2 J_{-n}(kx)$

Hence, the solution of (2) is $y = x^n [c_1 J_n(kx) + c_2 J_{-n}(kx)]$, $n \notin 1$ **Ans.**

where $\quad n = \frac{1-a}{2}$

(c) To reduce the following differential equation to Bessel's equation.

$$\boldsymbol{x\frac{d^2y}{dx^2} + c\frac{dy}{dx} + k^2x^r y = 0} \qquad \text{... (4)}$$

Put $\quad x = t^m,\ t = x^{1/m}$

So that $\quad \frac{dy}{dx} = \frac{dy}{dt}\frac{dt}{dx} = \frac{dy}{dt}\left(\frac{1}{m} x^{\frac{1}{m}-1}\right) = \frac{dy}{dt}\left(\frac{1}{m} t^{1-m}\right)$

$$\frac{d^2y}{dx^2} = \frac{d}{dx}\left(\frac{1}{m} t^{1-m} \frac{dy}{dt}\right) = \frac{d}{dt}\left(\frac{1}{m} t^{1-m} \frac{dy}{dt}\right)\frac{dt}{dx}$$

$$= \left(\frac{1}{m} t^{1-m} \frac{d^2y}{dt^2} + \frac{1}{m}(1-m)t^{-m} \frac{dy}{dt}\right)\frac{1}{m} t^{\frac{1}{m}-1} = \frac{1}{m^2} t^{2-2m} \frac{d^2y}{dt^2} + \frac{1-m}{m^2} t^{1-2m} \frac{dy}{dt}$$

Now, (4) becomes

$$t^m\left[\frac{1}{m^2} t^{2-2m} \frac{d^2y}{dt^2} + \frac{1-m}{m^2} t^{1-2m} \frac{dy}{dt}\right] + c\frac{1}{m} t^{1-m} \frac{dy}{dt} + k^2 t^{mr} y = 0$$

$$\Rightarrow \qquad \frac{1}{m^2} t^{2-m} \frac{d^2y}{dt^2} + \frac{1-m+cm}{m^2} t^{1-m} \frac{dy}{dt} + k^2 t^{mr} y = 0$$

On multiplying by $\frac{m^2}{t^{1-m}}$, we get

$$t\frac{d^2y}{dt^2} + (1-m+cm)\frac{dy}{dt} + (km)^2 t^{mr+m-1} y = 0 \qquad \text{... (5)}$$

Let us put $\quad a = 1-m+cm$ and $m = \frac{1}{r+1} \Rightarrow mr + m - 1 = \frac{r}{r+1} + \frac{1}{r+1} - 1 = 0$

Thus (5) becomes $t\frac{d^2y}{dt^2} + a\frac{dy}{dt} + (km)^2 y = 0$

Its solution is $y = t^n [c_1 J_n(kmt) + c_2 J_{-n}(kmt)]$

Solution of (4) is $y = x^{n/m} [c_1 J_n(km x^{1/m}) + c_2 J_{-n}(km x^{1/m})]$ **Ans.**

33.7 ORTHOGONALITY OF BESSEL FUNCTION

(*AMIETE, June 2009*)

$$\int_0^1 x J_n(\alpha x).J_n(\beta x)\,dx = 0$$

where α and β are the roots of $J_n(x) = 0$.

Proof. We know that

$$x^2 \frac{d^2y}{dx^2} + x\frac{dy}{dx} + (\alpha^2 x^2 - n^2)y = 0 \qquad \text{... (1)}$$

$$\Rightarrow \qquad x^2 \frac{d^2z}{dx^2} + x\frac{dz}{dx} + (\beta^2 x^2 - n^2)z = 0 \qquad \text{... (2)}$$

Solution of (1) and (2) are $y = J_n(\alpha x)$, $z = J_n(\beta x)$ respectively.

Multiplying (1) by $\frac{z}{x}$ and (2) by $-\frac{y}{x}$ and adding, we get

$$x\left(z\frac{d^2y}{dx^2} - y\frac{d^2z}{dx^2}\right) + \left(z\frac{dy}{dx} - y\frac{dz}{dx}\right) + (\alpha^2 - \beta^2)\,xyz = 0.$$

$$\Rightarrow \qquad \frac{d}{dx}\left[x\left(z\frac{dy}{dx} - y\frac{dz}{dx}\right)\right] + (\alpha^2 - \beta^2)\,xyz = 0 \qquad \text{... (3)}$$

Integrating (3) w.r.t. 'x' between the limits 0 and 1, we get

$$\left[x\left(z\frac{dy}{dx} - y\frac{dz}{dx}\right)\right]_0^1 + (\alpha^2 - \beta^2)\int_0^1 x\,y\,z\,dx = 0$$

$$\Rightarrow \qquad (\beta^2 - \alpha^2)\int_0^1 x\,y\,z\,dx = \left[x\left(z\frac{dy}{dx} - y\frac{dz}{dx}\right)\right]_0^1 = \left[z\frac{dy}{dx} - y\frac{dz}{dx}\right]_{x=1} \qquad \text{... (4)}$$

Putting the values of $y = J_n(\alpha x)$, $\frac{dy}{dx} = \alpha J_n'(\alpha x)$, $z = J_n(\beta x)$, $\frac{dz}{dx} = \beta$, $J_n'(\beta x)$ in (4), we get

$$(\beta^2 - \alpha^2)\int_0^1 x J_n(\alpha x).J_n(\beta x)\,dx = \left[\alpha J_n'(\alpha x) J_n(\beta x) - \beta J_n'(\beta x) J_n(\alpha x)\right]_{x=1}$$

$$= \alpha J_n'(\alpha) J_n(\beta) - \beta J_n'(\beta) J_n(\alpha) \qquad \text{... (5)}$$

Since α, β are the roots of $J_n(x) = 0$, so $J_n(\alpha) = J_n(\beta) = 0$.

Putting the values of $J_n(\alpha) = J_n(\beta) = 0$ in (5), we get

$$(\beta^2 - \alpha^2)\int_0^1 x J_n(\alpha x).J_n(\beta x)\,dx = 0$$

$$\Rightarrow \qquad \boxed{\int_0^1 x J_n(\alpha x).J_n(\beta x)\,dx = 0}$$ **Proved.**

Example 27. *Prove that*

$$\int_0^1 x[J_n(\alpha x)]^2\,dx = \frac{1}{2}[J_n'(\alpha)]^2$$

Solution. From (5) of article 33.7, we know that

$$(\beta^2 - \alpha^2)\int_0^1 x J_n(\alpha x).J_n(\beta x)\,dx = \alpha J_n'(\alpha).J_n(\beta) - \beta J_n'(\beta).J_n(\alpha)$$

when $\qquad \beta = \alpha$

We also know that $J_n(\alpha) = 0$. Let β be a neighbouring value of α, which tends to α. Then

$$\lim_{\beta\to\alpha} \int_0^1 xJ_n(\alpha x) J_n(\beta x)\, dx = \lim_{\beta\to\alpha} \frac{0+\alpha J_n'(\alpha).J_n(\beta)}{\beta^2-\alpha^2}$$

As the limit is of the form $\frac{0}{0}$, we apply L' Hopital's rule

$$\int_0^1 xJ_n^2(\alpha x)\, dx = \lim_{\beta\to\alpha} \frac{0+\alpha J_n'(\alpha).J_n'(\beta)}{2\beta} = \frac{1}{2}[J_n'(\alpha)]^2 \quad [\because \alpha=\beta] \quad \textbf{Proved.}$$

33.8 A GENERATING FUNCTION FOR $J_n(x)$

Prove that $J_n(x)$ is the coefficient of z^n in the expansion of $e^{\frac{x}{2}\left(z-\frac{1}{z}\right)}$

Proof. We know that $e^t = 1+t+\frac{t^2}{2!}+\frac{t^3}{3!}+...$

$$e^{\frac{xz}{2}} = 1+\left(\frac{xz}{2}\right)+\frac{1}{2!}\left(\frac{xz}{2}\right)^2+\frac{1}{3!}\left(\frac{xz}{2}\right)^3+... \qquad ...(1)$$

$$e^{-\frac{x}{2z}} = 1-\left(\frac{x}{2z}\right)+\frac{1}{2!}\left(\frac{x}{2z}\right)^2-\frac{1}{3!}\left(\frac{x}{2z}\right)^3+... \qquad ...(2)$$

On multiplying (1) and (2), we get

$$e^{\frac{x}{2}\left(z-\frac{1}{z}\right)} = \left[1+\left(\frac{xz}{2}\right)+\frac{1}{2!}\left(\frac{xz}{2}\right)^2+\frac{1}{3!}\left(\frac{xz}{2}\right)^3+...\right]\times\left[1-\frac{x}{2z}+\frac{1}{2!}\left(\frac{x}{2z}\right)^2-\frac{1}{3!}\left(\frac{x}{2z}\right)^3+...\right] \qquad ...(3)$$

The coefficient of z^n in the product of (3), we get

$$= \frac{1}{n!}\left(\frac{x}{2}\right)^n - \frac{1}{(n+1)!}\left(\frac{x}{2}\right)^{n+2} + \frac{1}{2!(n+2)!}\left(\frac{x}{2}\right)^{n+4} - ... = J_n(x)$$

Similarly, coefficient of z^{-n} in the product of (3) $= J_{-n}(x)$

$$\therefore \quad e^{\frac{x}{2}\left(z-\frac{1}{z}\right)} = J_0 + zJ_1 + z^2J_2 + z^3J_3 + ... + z^{-1}J_{-1} + z^{-2}J_{-2} + z^{-3}J_{-3} + ...$$

$$\boxed{e^{\frac{x}{2}\left(z-\frac{1}{z}\right)} = \sum_{n=-\infty}^{\infty} z^n J_n(x)}$$

For this reason $e^{\frac{x}{2}\left(z-\frac{1}{z}\right)}$ is known as the generating function of Bessel's functions. **Proved.**

Cor. In the expansion of (3), coefficient of z^0

$$= 1-\frac{x^2}{2^2}+\frac{x^4}{2^2.4^2}-... \quad \text{or} \quad J_0 = 1-\frac{x^2}{2^2}+\frac{x^4}{2^2.4^2}-...$$

33.9 TRIGONOMETRIC EXPANSION INVOLVING BESSEL FUNCTIONS

We know that

$$e^{\frac{x}{2}\left(z-\frac{1}{z}\right)} = J_0 + zJ_1 + z^2J_2 + z^3J_3 + ... + z^{-1}J_{-1} + z^{-2}J_{-2} + z^{-3}J_{-3} + ... \qquad ...(1)$$

Putting $z = e^{i\theta}$ in (1), we get

$$e^{\frac{x}{2}(e^{i\theta}-e^{-i\theta})} = J_0 + J_1 e^{i\theta} + J_2 e^{2i\theta} + J_3 e^{3i\theta} + ... + J_{-1}e^{-i\theta} + J_{-2}e^{-2i\theta} + J_{-3}e^{-3i\theta} + ...$$

$$\left(\text{since } \frac{e^{i\theta}-e^{-i\theta}}{2i} = \sin\theta\right)$$

$$e^{ix\sin\theta} = J_0 + J_1 e^{i\theta} + J_2 e^{2i\theta} + J_3 e^{3i\theta} + \ldots - J_1 e^{-i\theta} + J_2 e^{-2i\theta} - J_3 e^{-3i\theta} - \ldots$$

(since $J_{-n} = (-1)^n J_n$)

$$\Rightarrow \quad \cos(x\sin\theta) + i\sin(x\sin\theta) = J_0 + J_1(e^{i\theta} - e^{-i\theta}) + J_2(e^{2i\theta} + e^{-2i\theta}) + J_3(e^{3i\theta} - e^{-3i\theta}) + \ldots$$

$$\Rightarrow \quad \cos(x\sin\theta) + i\sin(x\sin\theta) = J_0 + J_1(2i\sin\theta) + J_2(2\cos 2\theta) + J_3(2i\sin 3\theta) + \ldots$$

Now equating real and imaginary parts, we get

$$\cos(x\sin\theta) = J_0 + 2J_2\cos 2\theta + 2J_4\cos 4\theta + \ldots \quad \ldots (2)$$

$$\sin(x\sin\theta) = 2J_1\sin\theta + 2J_3\sin 3\theta + 2J_5\sin 5\theta + \ldots \quad \ldots (3)$$

On putting $\theta = \frac{\pi}{2} - \alpha$ in (2) and (3), we get

$$\cos(x\cos\alpha) = J_0 - 2J_2\cos 2\alpha + 2J_4\cos 4\alpha - \ldots$$

$$\sin(x\cos\alpha) = 2J_1\cos\alpha + 2J_3\cos 3\alpha + 2J_5\cos 5\alpha - \ldots$$

Example 28. *Prove that*

$$\cos x = J_0 - 2J_2 + 2J_4 - \ldots$$

$$\sin x = 2J_1 - 2J_3 + 2J_5 + \ldots$$

Solution. We know that

$$\cos(x\sin\theta) = J_0 + 2J_2\cos 2\theta + 2J_4\cos 4\theta + \ldots \quad \ldots (1)$$

$$\sin(x\sin\theta) = 2J_1\sin\theta + 2J_3\sin 3\theta + 2J_5\sin 5\theta + \ldots \quad \ldots (2)$$

Putting $\theta = \frac{\pi}{2}$ in (1) and (2), we get

$$\cos x = J_0 - 2J_2 + 2J_4 - \ldots$$

and $\sin x = 2J_1 - 2J_3 + 2J_5 - \ldots$ **Proved.**

Example 29. *Prove that*

$$x\sin x = 2[2^2 J_2 - 4^2 J_4 + 6^2 J_6 - \ldots]$$

$$x\cos x = 2[1^2 J_1 - 3^2 J_3 + 5^2 J_5 + \ldots]$$

Solution. We know that

$$\cos(x\sin\theta) = J_0 + 2J_2\cos 2\theta + 2J_4\cos 4\theta + \ldots \quad \ldots (1)$$

and $\sin(x\sin\theta) = 2J_1\sin\theta + 2J_3\sin 3\theta + 2J_5\sin 5\theta + \ldots \quad \ldots (2)$

Differentiating (1) w.r.t. "θ", we get

$$[-\sin(x\sin\theta)]x\cos\theta = 0 - 4J_2\sin 2\theta - 8J_4\sin 4\theta + \ldots \quad \ldots (3)$$

Again differentiating (3) w.r.t., "θ", we get

$$[-\sin(x\sin\theta)](-x\sin\theta) + [-\cos(x\sin\theta)(x\cos\theta)]x\cos\theta$$

$$= -8J_2\cos 2\theta - 32J_4\cos 4\theta + \ldots \quad \ldots (4)$$

Now putting $\theta = \frac{\pi}{2}$ in (4), we get

$$x\sin x = 8J_2 - 32J_4 + \ldots = 2[2^2 J_2 - 4^2 J_4 + 6^2 J_6 - \ldots]$$

Similarly differentiating (2) twice and putting $\theta = \frac{\pi}{2}$, we have

$$x\cos x = 2[1^2 J_1 - 3^2 J_3 + 5^2 J_5 - \ldots]$$ **Proved.**

Example 30. *Prove that* $J_0^2 + 2J_1^2 + 2J_2^2 + ... = 1$

$$\int_0^\pi 2\sin^2 n\theta\, d\theta = \pi$$
$$\int_0^\pi 2\cos^2 n\theta\, d\theta = \pi$$
$$\int_0^\pi 2\sin n\theta \sin m\theta\, d\theta = 0$$
$$\int_0^\pi 2\cos n\theta \cos m\theta\, d\theta = 0$$

Solution. $J_0 + 2J_2 \cos 2\theta + 2J_4 \cos 4\theta + ...$

$$= \cos(x \sin\theta) \quad \text{[From (33.9)]} \quad ...(1)$$

$$2J_1 \sin\theta + 2J_3 \sin 3\theta + 2J_5 \sin 5\theta + ... = \sin(x \sin\theta) \quad ...(2)$$

Now squaring (1) and integrating w.r.t. 'θ' between the limits 0 and π, we get

$$J_0^2 \pi + 2J_2^2 \pi + 2J_4^2 \pi + ... = \int_0^\pi \cos^2(x\sin\theta)\, d\theta \quad ...(3)$$

Also squaring (2) and integrating w.r.t. "θ" between the limits 0 and π, we get

$$2J_1^2\pi + 2J_3^2\pi + 2J_5^2\pi + ... = \int_0^\pi \sin^2(x\sin\theta)\, d\theta \quad ...(4)$$

Adding (3) and (4), we get

$$\pi[J_0^2 + 2J_1^2 + 2J_2^2 + 2J_3^2 + ...] = \int_0^\pi \cos^2(x\sin\theta)\, d\theta + \int_0^\pi \sin^2(x\sin\theta)\, d\theta = \int_0^\pi d\theta = \pi$$

$$\Rightarrow \quad J_0^2 + 2J_1^2 + 2J_2^2 + 2J_3^2 + ... = 1 \qquad \textbf{Proved.}$$

33.10 BESSEL'S INTEGRAL

To prove that

(*a*) $J_0(x) = \frac{1}{\pi}\int_0^\pi \cos(x\sin\theta)\, d\theta$ (*b*) $J_n(x) = \frac{1}{\pi}\int_0^\pi \cos(n\theta - x\sin\theta)\, d\theta$

Proof. We know that

$$\cos(x\sin\theta) = J_0 + 2J_2\cos 2\theta + 2J_4 \cos 4\theta + ... \quad ...(1)$$

$$\sin(x\sin\theta) = 2J_1\sin\theta + 2J_3\sin 3\theta + 2J_5\sin 5\theta + ... \quad ...(2)$$

(*a*) Integrating (1) between the limits 0 and π, we have

$$\int_0^\pi \cos(x\sin\theta)\, d\theta = \int_0^\pi (J_0 + 2J_2\cos 2\theta + 2J_4\cos 4\theta + ...)\, d\theta$$

$$= J_0 \int_0^\pi d\theta + 2J_2 \int_0^\pi \cos 2\theta\, d\theta + 2J_4 \int_0^\pi \cos 4\theta\, d\theta + ...$$

$$= J_0\pi + 0 + 0$$

i.e. $$J_0 = \frac{1}{\pi}\int_0^\pi \cos(x\sin\theta)\, d\theta \qquad \textbf{Proved.}$$

(*b*) Multiplying (1) by cos $n\theta$ and integrating between the limits 0 and π, we have

$$\int_0^\pi \cos(x\sin\theta)\cos n\theta\, d\theta = \int_0^\pi [J_0\cos n\theta + 2J_2\cos 2\theta\cos n\theta + 2J_4\cos 4\theta\cos n\theta + ...]\, d\theta$$

$$= J_0\int_0^\pi \cos n\theta\, d\theta + 2J_2\int_0^\pi \cos 2\theta\cos n\theta\, d\theta + ...$$

$$= 0, \quad \text{if } n \text{ is odd} \quad ...(3)$$

$$= \pi J_n, \quad \text{if } n \text{ is even} \quad ...(4)$$

Again multiplying (2) by sin $n\theta$ and integrating between the limits 0 and π, we have

$$\int_0^\pi \sin(x\sin\theta)\sin n\theta\, d\theta = \int_0^\pi (2J_1\sin\theta\sin n\theta + 2J_3\sin 3\theta\sin n\theta + ...)\, d\theta$$

$$= 2J_1\int_0^\pi \sin\theta\sin n\theta\, d\theta + 2J_3\int_0^\pi \sin 3\theta\sin n\theta\, d\theta + ...$$

$$= 0 \quad \text{if } n \text{ is even} \quad ...(5)$$

$$= \pi J_n \quad \text{if } n \text{ is odd} \quad ...(6)$$

Adding (3) and (6) or (4) and (5), we get

$$\int_0^{\pi} [\cos(x\sin\theta)\cos n\theta + \sin(x\sin\theta)\sin n\theta]\,d\theta = \pi J_n$$

$$\Rightarrow \quad \int_0^{\pi} \cos(n\theta - x\sin\theta)\,d\theta = \pi J_n \text{ or } J_n = \frac{1}{\pi}\int_0^{\pi} \cos(n\theta - x\sin\theta)\,d\theta$$ **Proved.**

EXERCISE 33.1

Prove that

1. $J_1(x) = \frac{x}{2} - \frac{x^3}{2^3.1!.2!} + \frac{x^5}{2^5.2!.3!} - \frac{x^7}{2^7.3!.4!} + \ldots$
2. (a) $J_0(2) = 0.224$ (b) $J_1 = 0.44.$
3. $J_2 = J_0'' - x^{-1}J_0'$
4. $\frac{d}{dx}[xJ_1(x)] = xJ_0(x)$
5. $\frac{d}{dx}[J_n^2(x) + J_{n+1}^2(x)] = 2\left(\frac{n}{x}J_n^2(x) - \frac{n+1}{x}J_{n+1}^2(x)\right)$
6. $\int_0^{\pi} x^{n+1}J_n(x)\,dx = x^{n+1}(x),$ $n > -1$
7. $x^2 J_n''(x) = (n^2 - n - x^2)J_n(x) + xJ_{n+1}(x)$ *(A.M.I.E.T.E., Summer 2001)*
8. $\int x^2 J_0 J_1\,dx = \frac{1}{2}x^2 J_0' + c$
9. $J_{3/2}(x)\sin x - J_{-3/2}(x)\cos(x) = \frac{\sqrt{2\pi}}{x^3}$
10. $J_1'' = \left(\frac{2}{x^2} - 1\right)J_1(x) - \frac{1}{x}J_0(x)$
11. $J_0^2 + 2(J_1^2 + J_2^2 + J_3^2 + \ldots) = 1$
12. If $J_0(2) = a$, $J_1(2) = b$ find $J_2(2)$, $J_1'(2)$, $J'_2(2)$ in terms of a and b where $J_n(x)$ is the Bessel function of first kind.

    Ans. $J_2(2) = b - a$, $J'_1(2) = a - \frac{b}{2}$, $J_2'(2) = a$
13. Prove that $J_n(x) = 0$ has no repeated root except $x = 0$.
14. Integrate $\int x^3 J_0(x)\,dx$, where $J_n(x)$ is the Bessel's function of first kind, in terms of $J_0(x)$, $J_1(x)$ and $J_2(x)$.

33.11 FOURIER-BESSEL EXPANSION

If a function $f(x)$ is continuous and has a finite number of oscillations in the interval $0 \le x \le a$, then $f(x)$ can be expanded in a series.

$$f(x) = C_1 J_n(\alpha_1 x) + C_2 J_n(\alpha_2 x) + C_3 J_n(\alpha_3 x) + \ldots + C_n J_n(\alpha_n x) + \ldots\ldots$$

$$\Rightarrow \quad f(x) = \sum_{i=1}^{\infty} C_i J_n(\alpha_i x).$$

where $\alpha_1, \alpha_2, \alpha_3$ are the roots of the equation $J_n(x) = 0$.

Solution. [The orthogonal property of Bessel's functions enables us to expand a function in terms of Bessel's function].

Let $$f(x) = \sum_{i=1}^{\infty} C_i J_n(\alpha_i x) \qquad \ldots (1)$$

Multiplying both sides of (1) by $x J_n(\alpha_j x)$, we get

$$x\,f(x)J_n(\alpha_j x) = \sum_{i=1}^{\infty} C_i x\,J_n(\alpha_j x).J_n(\alpha_i x) \qquad \text{... (2)}$$

Integrating both sides of (2) from $x = 0$ to $x = a$, we have

$$\int_0^a x\,f(x).J_n(\alpha_j x)dx = \sum_{i=1}^{\infty} C_i \int_0^a x\,J_n(\alpha_j x).J_n(\alpha_i x)\,dx \qquad \text{... (3)}$$

By orthogonal property of Bessel's functions, we know that

$$\int_0^a x\,J_n(\alpha_i x).J_n(\alpha_j x)dx = \begin{cases} 0 & \text{if } i \neq j \\ \dfrac{a^2}{2} J_{(n+1)}^2(\alpha_i a) & \text{if } i = j \end{cases}$$

On applying this property on the right-hand side of (3), it reduces to

$$\int_0^a x\,f(x)J_n(\alpha_i x)\,dx = C_i.\frac{a^2}{2} J_{n+1}^2(\alpha_i a)$$

$$\Rightarrow \qquad C_i = \frac{2\int_0^a x\,f(x)J_n(\alpha_i x)dx}{a^2.J_{n+1}^2(\alpha_i a)}$$

By putting the values of the coefficient $C's$ in (1), we get the Fourier-Bessel Expansions.
Ans.

Example 31. *Show that* $\sum_{i=1}^{\infty} \dfrac{2J_0(a_i x)}{a_i J_1(a_i)} = 1$, *where* a_1, a_2, a_3 .... *are the roots of* J_0 *(x).*

Solution. Let $f(x) = \sum_{i=1}^{\infty} C_i J_n(a_i x)$, ... (1)

then $$C_n = \frac{2}{J_{n+1}^2(a_i)} \int_0^1 x\,J_n(a_i x)\,f(x)\,dx \qquad \text{... (2)}$$

Putting $f(x) = 1$ and $n = 0$ in (1), we get

$$1 = \sum_{i=1}^{\infty} C_i J_0(a_i x) \qquad \text{...(3)}$$

$$C_i = \frac{2}{J_1^2(a_i)} \int_0^1 xJ_0(a_i x)\,dx = \frac{2}{J_1^2(a_i)}\left[\frac{J_1(a_i)}{a_i}\right] = \frac{2}{a_i J_1(a_i)}$$

Substituting the values of $C_i f(x)$ and n in (3), we obtain

$$1 = \sum_{i=1}^{\infty} \frac{2}{a_i J_1(a_i)} J_0(a_i x)$$

$$\Rightarrow \qquad \sum \frac{2J_0(a_i x)}{a_i J_1(a_i)} = 1 \qquad \textbf{Proved.}$$

Example 32. *Expand* $f(x) = x^2$ *in the interval* $0 < x < 2$ *in terms of* $J_2(\alpha_n x)$ *where* α_n *are the roots of* $J_2(2\alpha_n) = 0$.

Solution. $f(x) = x^2$

$$x^2 = \sum_{i=1}^{\infty} C_i J_2(\alpha_i x) \qquad \text{... (1)}$$

Multiplying both sides of (1) by $xJ_2(\alpha_j x)$, we get

$$x^3 J_2(\alpha_j .x) = \sum_{i=1}^{\infty} C_i x J_2(\alpha_i x).J_2(\alpha_j .x) \qquad \text{... (2)}$$

Integrating (2) w.r.t. x from $x = 0$ to $x = 2$, we get

$$\int_0^2 x^3 . J_2(\alpha_i\, x)\, dx = \sum_{i=1}^{\infty} C_i \int_0^2 x J_2(\alpha_i x). J_2(\alpha_j . x)\, dx$$

$$\left[\frac{x^3 . J_3(\alpha_i x)}{\alpha_i}\right]_0^2 = C_i \int_0^2 x J_2^2(\alpha_i x)\, dx \quad (i = j) \text{ (other integrals are zero)}$$

$$\frac{8 J_3(2\alpha_i)}{\alpha_i} = C_i \frac{2^2}{2} J_3^2(2\alpha_i)$$

$$C_i = \frac{8 J_3(2\alpha_i)}{\alpha_i} \frac{2}{4 J_3^2(2\alpha_i)} = \frac{4}{\alpha_i J_3(2\alpha_i)}$$

On putting the values of coefficients C_i in (1), we get

$$x^2 = \sum_{i=1}^{\infty} \frac{4 J_2(\alpha_i x)}{\alpha_i J_3(2\alpha_i)}$$ **Ans.**

EXERCISE 33.2

1. Expand $f(x) = x^3$ in the interval $0 < x < 3$ in terms of Bessel's functions $J_1(\alpha_n x)$ where α_n are the roots of $J_1(3\alpha) = 0$. **Ans.** $x^3 = \sum_{i=1}^{\infty} \frac{6}{\alpha_i^2} \frac{1}{J_2^2(3\alpha_i)} \{3\alpha_i J_2(3\alpha_i) - 2J_3(3\alpha_i)\}$

2. Show that the Fourier-Bessel series in $J_2(\alpha_i x)$ for $f(x) = x^2$, $(0 < x < \alpha)$ where α_i are positive roots of $J_2(x) = 0$ is

$$x^2 = 2a^2 \sum_{i=1}^{\infty} \frac{J_2(\alpha_i x)}{a\alpha_i J_3(\alpha_i a)}$$

33.12 BER AND BEI FUNCTIONS

The following differential equation is useful in certain problems in electrical engineering.

$$x \frac{d^2 y}{dx^2} + \frac{dy}{dx} - i x y = 0 \qquad \text{... (1)}$$

This equation (1) is a particular case of the differential equation

$$x^2 \frac{d^2 y}{dx^2} + x \frac{dy}{dx} + (k^2 x^2 - n^2)\, y = 0 \qquad \text{... (2)}$$

On putting $n = 0$, $k^2 = -i$ in equation (2), we get

$$x \frac{d^2 y}{dx^2} + \frac{dy}{dx} - i x y = 0$$

Its solution is $y = J_0(k x) = J_0[(-1)^{1/2} x] = J_0(i^{3/2} x)$

$$y = J_0(i^{3/2} x) = 1 - \frac{i^3 x^2}{2^2} + \frac{i^6 x^4}{(2!)^2 . 2^4} - \frac{i^9 x^6}{(3!)^2 . 2^6} + \frac{i^{12} x^8}{(4!)^2 . 2^8} \ldots$$

$$= \left[1 - \frac{x^4}{2^2 \cdot 4^2} + \frac{x^8}{2^2 \cdot 4^2 \cdot 6^2 \cdot 8^2} \ldots\right] + i\left[\frac{x^2}{2^2} - \frac{x^6}{2^2 \cdot 4^2 \cdot 6^2} + \frac{x^{10}}{2^2 \cdot 4^2 \cdot 6^2 \cdot 8^2 \cdot 10^2} + \ldots\right]$$

$$= 1 + \sum_{k=1}^{\infty} (-1)^k \frac{x^{4k}}{2^2 \cdot 4^2 \cdot 6^2 \ldots (4k)^2} + i\left[-\sum_{k=1}^{\infty} \frac{(-1)^k x^{4k-2}}{2^2 \cdot 4^2 \cdot 6^2 \cdot 8^2 \ldots (4k-2)^2}\right]$$

$J_0(i^{3/2}x) =$ Ber (Bessel real) + Bei (Bessel imaginary)

$$\text{Ber } x = 1+\sum_{k=1}^{\infty}(-1)^k \frac{x^{4k}}{2^2 \cdot 4^2 \cdot 6^2 \ldots\ldots\ldots (4k)^2}$$

$$\text{Bei } x = -\sum_{k=1}^{\infty} \frac{(-1)^k x^{4k-2}}{2^2 \cdot 4^2 \cdot 6^2 . 8^2 \ldots\ldots (4k-2)^2}$$

Example 33. *Show that* $\frac{d}{dx}$ *(x Ber′ x) = – x Bei x.*

Solution. We know that

$$\text{Ber } x = 1-\frac{x^4}{2^2 \cdot 4^2}+\frac{x^8}{2^2 \cdot 4^2 \cdot 6^2 \cdot 8^2}+\ldots\ldots\infty$$

On differentiating, $$\text{Ber}' x = -\frac{4x^3}{2^2 \cdot 4^2}+\frac{8x^7}{2^2 \cdot 4^2 \cdot 6^2 \cdot 8^2}+\ldots\infty$$

$$(x \text{ Ber}' x) = -\frac{x^4}{2^2 \cdot 4}+\frac{x^8}{2^2 \cdot 4^2 \cdot 6^2 \cdot 8}+\ldots$$

$$\frac{d}{dx}(x \text{ Ber}' x) = -\frac{x^3}{2^2}+\frac{x^7}{2^2 \cdot 4^2 \cdot 6}+\ldots$$

$$= -x\left[\frac{x^2}{2^2}-\frac{x^6}{2^2 \cdot 4^2 \cdot 6^2}+\ldots\ldots\infty\right] = -x \text{ Bei } x$$ **Proved.**

Example 34. *Show that* $\frac{d}{dx}$ *(x* Bei′ *x) = x Ber x.*

Solution. We know that Bei $x = \frac{x^2}{2^2}-\frac{x^6}{2^2 \cdot 4^4 \cdot 6^2}+\frac{x^{10}}{2^2 \cdot 4^2 \cdot 6^2 \cdot 8^2 \cdot 10^2}\ldots$ ... (1)

On differentiating (1) w.r.t. 'x', we get

$$\text{Bei}'x = \frac{x}{2}-\frac{x^5}{2^2 \cdot 4^2 \cdot 6}+\frac{x^9}{2^2 \cdot 4^2 \cdot 6^2 \cdot 8^2 \cdot 10}\ldots\ldots\infty$$

$$(x \text{ Bei}' x) = \frac{x^2}{2}-\frac{x^6}{2^2 \cdot 4^2 \cdot 6}+\frac{x^{10}}{2^2 \cdot 6^2 \cdot 8^2 \cdot 10}\ldots\ldots\infty$$

$$\frac{d}{dx}(x \text{Bei}' x) = x-\frac{x^5}{2^2 \cdot 4^2}+\frac{x^9}{2^2 \cdot 4^2 \cdot 6^2 \cdot 8^2}\ldots\ldots\infty$$

$$= x\left[1-\frac{x^4}{2^2 \cdot 4^2}+\frac{x^8}{2^2 \cdot 4^2 \cdot 6^2 \cdot 8^2}\ldots\ldots\infty\right] = x \text{ Ber } x$$ **Proved.**

Example 35. *Show that*

$$\int_0^a x[Ber^2 x+Bei^2 x]\, dx = a[Ber\ a\ Bei'\ a-Bei\ a\ Ber'\ a]$$

Solution. $\frac{d}{dx}(x \text{ Bei}' x) = x \text{ Ber } x$ ... (1) (See example 34)

and $\frac{d}{dx}(x \text{ Ber}' x) = -x \text{Bei} x$ ... (2) (See example 33)

Multiplying (1) by Ber x and (2) by Bei x and subtracting, we get

$$Ber\, x \frac{d}{dx}(x \text{ Bei}' x) - Bei\, x \frac{d}{dx}(x\ Ber'\ x) = x \text{ Ber}^2 x + x \text{Bei}^2 x \quad \ldots (3)$$

Integrating both sides of (3) from 0 to a, we get

$$\int_0^a \left[\text{Ber } x \frac{d}{dx}(x \text{Bei}' x) - \text{Bei}\, x \frac{d}{dx}(x \text{Ber}' x)\right] dx = \int_0^a (x\, Ber^2 x + x \text{Bei}^2 x) dx$$

$$\Rightarrow \quad \int_0^a x(\text{Ber}^2 x + \text{Bei}^2 x)\, dx = \int_0^a \left[\text{Ber } x \frac{d}{dx}(x \text{Bei}' x) - \text{Bei } x \frac{d}{dx}(x \text{ Ber}' x)\right] dx$$

On adding and subtracting Bei$'$ $x \dfrac{d}{dx}(x \text{ Ber } x)$ on R.H.S.

$$= \int_0^a \left[\text{Ber } x \frac{d}{dx}(x \text{ Ber}' x) + \text{Bei}' x \frac{d}{dx}(x \text{Bei } x) - \text{Bei}' x \frac{d}{dx}(x \text{Ber}' x) - \text{Ber}' x \frac{d}{dx}(x \text{Ber } x)\right] dx$$

$$= \int_0^a \frac{d}{dx}[x(\text{Ber } x \text{ Bei}' x - \text{Bei } x \text{ Ber}' x)]\, dx$$

$$= [x\, (\text{Ber } x \text{ Bei}' x - \text{Bei } x \text{ Ber}' x)]_0^a$$

$$= a\, [\text{Ber } a \text{ Bei}' a - \text{Bei } a \text{ Ber}' a]$$

Proved.

OBJECTIVE TYPE QUESTIONS

Choose the correct or the best of the answers/statements given in the following parts:

1. $J_{\frac{1}{2}}$ is given by

(*i*) $\sqrt{\dfrac{2\pi}{x}} \sin x$ (*ii*) $\sqrt{\dfrac{2\pi}{x}} \cos x$ (*iii*) $\sqrt{\dfrac{\pi}{2x}} \cos x$ (*iv*) $\sqrt{\dfrac{2}{\pi x}} \sin x$

(*UP., II Semester, 2010*) **Ans.** (*iv*)

2. If $J_{n+1}(x) = \dfrac{2}{x} J_n(x) - J_0(x)$, then n is

(*i*) 0 (*ii*) 2 (*iii*) -1 (*iv*) None of these **Ans.** (*iv*)

3. $J_0' =$ (*R.G.P.V. Bhopal, II Semester, Feb 2006*)

(*i*) J_1 (*ii*) $-J_1$ (*iii*) $J_2 - J_0$ (*iv*) J_2 **Ans.** (*ii*)

4. $J_{-\frac{1}{2}}(x) =$ (*R.G.P.V. Bhopal, II Semester, June 2006*)

(*i*) $\sqrt{\dfrac{2}{\pi x}} \sin x$ (*ii*) $\sqrt{\dfrac{2}{\pi}} \sin x$ (*iii*) $\sqrt{\dfrac{2}{\pi x}} \cos x$ (*iv*) $\sqrt{\dfrac{2\pi}{x}} \cos x$ **Ans.** (*iii*)

(*R.G.P.V. Bhopal, II Semester, Feb. 2006*)

5. Bessel's equation of order zero is:

(*i*) $xy + y = 0$ (*ii*) $xy_2 + xy_1 + y = 0$

(*iii*) $xy_2 + y_1 + xy = 0$ (*iv*) $xy_2 + y_1 + xy = 0$ **Ans.** (*iv*)

(*R.G.P.V. Bhopal, II Semester, Dec. 2007, Feb. 2006*)

6. The value of the integral $\int x^2 J_1(x)\, dx$ is

(*i*) $x^2 J_1(x) + c$ (*ii*) $x^2 J_{-1}(x) + c$

(*iii*) $x^2 J_2(x) + c$ (*iv*) $x^2 J_{-2}(x) + c$ **Ans.** (*iii*)

(*AMIETE, June 2010*)

CHAPTER

34

LEGENDRE'S FUNCTIONS

34.1 LEGENDRE'S EQUATION

The differential equation $(1-x^2)\frac{d^2y}{dx^2} - 2x\frac{dy}{dx} + n(n+1)\,y = 0$ ... (1)

is known as Legendre's equation. The above equation can also be written as

$$\frac{d}{dx}\left\{(1-x^2)\frac{dy}{dx}\right\} + n(n+1)\,y = 0 \qquad n \in I$$

This equation can be integrated in series of ascending or descending powers of x. *i.e.*, series in ascending or descending powers of x can be found which satisfy the equation (1).

Let the series in descending powers of x be

$$y = x^m (a_0 + a_1 x^{-1} + a_2 x^{-2} + \ldots) \qquad \ldots (2)$$

$$\Rightarrow \qquad y = \sum_{r=0}^{\infty} a_r\, x^{m-r}$$

so that

$$\frac{dy}{dx} = \sum_{r=0}^{\infty} a_r (m-r)^{m-r-1}$$

and

$$\frac{d^2y}{dx^2} = \sum_{r=0}^{\infty} a_r (m-r)(m-r-1)\, x^{m-r-2}$$

Substituting these values in (1), we have

$$(1-x^2)\sum_{r=0}^{\infty} a_r (m-r)(m-r-1)\, x^{m-r-2} - 2x\sum_{r=0}^{\infty} a_r (m-r)\, x^{m-r-1} + n(n+1)\sum_{r=0}^{\infty} a_r\, x^{m-r} = 0$$

$$\Rightarrow \qquad \sum_{r=0}^{\infty} \left[(m-r)(m-r-1)\, x^{m-r-2} + \{n(n+1) - 2(m-r) - (m-r)(m-r-1)\}\, x^{m-r}\right] a_r = 0$$

$$\Rightarrow \qquad \sum_{r=0}^{\infty} [(m-r)(m-r-1)\, x^{m-r-2} + \{n(n+1) - (m-r)(m-r+1)\}\, x^{m-r}]\, a_r \equiv 0 \qquad \ldots (3)$$

The equation (3) is an identity and therefore coefficients of various powers of x must vanish. Now equating to zero the coefficients of x^m *i.e.* by substituting $r = 0$ in the second summation, we get,

$$a_0\, \{n(n+1) - m(m+1)\} = 0$$

But $a_0 \neq 0$, as it is the coefficient of the very first term in the series

Hence $$n(n+1) - m(m+1) = 0 \quad \ldots (4)$$

i.e., $$n^2 + n - m^2 - m = 0 \quad \Rightarrow \quad (n^2 - m^2) + (n - m) = 0$$

$\Rightarrow$ $$(n - m)(n + m + 1) = 0,$$ This is the indicial equation.

which gives $$m = n \quad \text{or} \quad m = -n - 1 \quad \ldots (5)$$

Next equating to zero the coefficient of x^{m-1} by putting $r = 1$, in the second summation

$$a_1[n(n+1) - (m-1)m] = 0$$

$$\Rightarrow \quad a_1(n^2 + n - m^2 + m) = 0 \quad \Rightarrow \quad a_1[(n^2 - m^2) + n + m] = 0$$

$$\Rightarrow \quad a_1[(m+n)(m-n-1)] = 0$$

which gives $$a_1 = 0 \quad \ldots (6)$$

Since $(m + n)(m - n - 1) \neq 0$, by (5)

Again to find a relation in successive coefficients a_r, etc., equating the coefficient of x^{m-r-2} to zero, we get

$$(m-r)(m-r-1)a_r + [n(n+1) - (m-r-2)(m-r-1)]a_{r+2} = 0 \quad \ldots (7)$$

Now
$$\left\{\begin{aligned} n(n+1) - (m-r-2)(m-r-1) &= n^2 + n - (m-r-1-1)(m-r-1) \\ &= -[(m-r-1)^2 - (m-r-1) - n^2 - n] \\ &= -[(m-r-1+n)(m-r-1-n) - (m-r-1+n)] \\ &= -[(m-r-1+n)(m-r-1-n-1)] \\ &= -(m-r+n-1)(m-r-n-2) \end{aligned}\right\}$$

On simplification (7) becomes

$$\Rightarrow \quad (m-r)(m-r-1)a_r - (m-r+n-1)(m-r-n-2)a_{r+2} = 0$$

$$\Rightarrow \quad a_{r+2} = \frac{(m-r)(m-r-1)}{(m-r+n-1)(m-r-n-2)} a_r \quad \ldots (8)$$

Now since $a_1 = a_3 = a_5 = a_7 = \ldots = 0$

For the two values given by (5) there arises following two cases.

Case I : When $m = n$

$$a_{r+2} = -\frac{(n-r)(n-r-1)}{(2n-r-1)(r+2)} a_r \quad \text{[From (8)]}$$

If $r = 0$ $$a_2 = -\frac{n(n-1)}{(2n-1)2} a_0$$

If $r = 2$, $$a_4 = -\frac{(n-2)(n-3)}{(2n-3)\times 4} a_2 = \frac{n(n-1)(n-2)(n-3)}{(2n-1)(2n-3)2\cdot 4} a_0$$

and so on and $$a_1 = a_3 = a_5 = \ldots = 0$$

Hence the series (2) becomes

$$y = a_0\left[x^n - \frac{n(n-1)}{(2n-1).2}x^{n-2} + \frac{n(n-1)(n-2)(n-3)}{(2n-1)(2n-3)\,2.4}.x^{n-4} - \ldots\right]$$

Which is a solution of (1).

Case II : When $m = -(n+1)$, we have

$$a_{r+2} = \frac{(n+r+1)(n+r+2)}{(r+2)(2n+r+3)} a_r \quad \text{[From (8)]}$$

If $r = 0$, $$a_2 = \frac{(n+1)(n+2)}{2(2n+3)} a_0;$$

If $r = 2$, $\quad a_4 = \dfrac{(n+3)(n+4)}{4.(2n+5)} a_2 = \dfrac{(n+1)(n+2)(n+3)(n+4)}{2.4(2n+3)(2n+5)} a_0$ and so on.

Hence the series (2) in this case becomes

$$y = a_0 \left[x^{-n-1} + \frac{(n+1)(n+2)}{2.(2n+3)} x^{-n-3} + \frac{(n+1)(n+2)(n+3)(n+4)}{2 \cdot 4(2n+3)(2n+5)} \cdot x^{-n-5} + \ldots \right] \quad \ldots (9)$$

This gives another solution of (1) in a series of descending powers of x.

Note. If we want to integrate the Legendre's equation in a series of ascending powers of x. we may proceed by taking

$$y = a_0 x^k + a_1 x^{k+1} + a_2 x^{k+2} + \ldots = \sum_{r=0}^{\infty} a_r x^{k+r}$$

But integration in descending powers of x is more important than that in ascending powers of x.

34.2 LEGENDRE'S POLYNOMIAL $P_n(x)$.

Definition :

The Legendre's Equation is

$$(1-x^2)\frac{d^2y}{dx^2} - 2x\frac{dy}{dx} + n(n+1)\, y = 0 \quad \ldots (1)$$

The solution of the above equation in the series of descending powers of x is

$$y = a_0 \left[x^n - \frac{n(n-1)}{(2n-1)\,2} x^{n-2} + \frac{n(n-1)(n-2)(n-3)}{(2n-1)(2n-3)\,2 \cdot 4} x^{n-4} \ldots \right]$$

where a_0 is an arbitrary constant.

Now if n is a positive integer and $a_0 = \dfrac{1.3.5 \ldots (2n-1)}{n!}$ the above solution is $P_n(x)$, so that

$$P_n(x) = \frac{1.3.5 \ldots (2n-1)}{n!} \left[x^n - \frac{n(n-1)}{(2n-1) \cdot 2} x^{n-2} + \ldots \right]$$

Note 1. This is a terminating series.

When n is even, it contains $\frac{1}{2}n+1$ terms, the last term being

$$(-1)^{\frac{1}{2}} \frac{n(n-1)(n-2)\ldots 1}{(2n-1)(2n-3)\ldots(n+1)\, 2.4.6 \ldots (n-1)}$$

And when n is odd it contains $\frac{1}{2}(n+1)$ terms and the last term in this case is

$$(-1)^{\frac{1}{2}(n-1)} \frac{n(n-1)(n-2)\ldots 3.2}{(2n-1)(2n-3)\ldots(n+2)\, 2.4 \ldots (n-1)} x$$

$P_n(x)$ is called the *Legendre's function of the first kind.*

Note. $P_n(x)$ is that solution of Legendre's equation (1) which is equal to unity when $x = 1$.

34.3 LEGENDRE'S FUNCTION OF THE SECOND KIND *i.e.* $Q_n(x)$

Another solution of Legendre's equation

$$(1-x^2)\frac{d^2y}{dx^2} - 2x\frac{dy}{dx} + n(n+1)\, y = 0$$

when n is a positive integer

$$y = a_0 \left[x^{-n-1} + \frac{(n+1)(n+2)}{2 \cdot (2n+3)} x^{-n-3} + \ldots \right]$$

If we take $$a_0 = \frac{n!}{1.3.5\ldots(2n+1)}$$

the above solution is called $Q_n(x)$, so that

$$Q_n(x) = \frac{n!}{1.3.5\ldots(2n+1)}\left[x^{-n-1} + \frac{(n+1)(n+2)}{2\cdot(2n+3)}x^{-n-3} + \ldots\right]$$

The series for $Q_n(x)$ is a non-terminating series.

34.4 GENERAL SOLUTION OF LEGENDRE'S EQUATION

Since $P_n(x)$ and $Q_n(x)$ are two independent solutions of Legendre's equation, therefore the most general solution of Legendre's equation is

$$y = AP_n(x) + B\,Q_n(x)$$

Where A and B are two arbitrary constants.

34.5 RODRIGUE'S FORMULA

(AMIETE, June 2010, 2009, U.P., II Semester, 2010, 2007, 2004)

$$P_n(x) = \frac{1}{2^n \cdot n!}\frac{d^n}{dx^n}(x^2-1)^n$$

Proof. Let $$v = (x^2-1)^n \qquad \ldots (1)$$

Then $$\frac{dv}{dx} = n(x^2-1)^{n-1}(2x)$$

Multiplying both sides by $(x^2 - 1)$, we get

$$(x^2-1)\frac{dv}{dx} = 2n(x^2-1)^n x.$$

$$\Rightarrow \quad (x^2-1)\frac{dv}{dx} = 2nvx \qquad \text{[Using (1)]} \qquad \ldots (2)$$

Now differentiating (2), $(n + 1)$ times by Leibnitz's theorem, we have

$$(x^2-1)\frac{d^{n+2}v}{dx^{n+2}} + {}^{(n+1)}C_1\,(2x)\frac{d^{n+1}v}{dx^{n+1}} + {}^{(n+1)}C_2(2)\frac{d^n v}{dx^n} = 2n\left[x\frac{d^{n+1}v}{dx^{n+1}} + {}^{(n+1)}C_1(1)\frac{d^n v}{dx^n}\right]$$

$$\Rightarrow \quad (x^2-1)\frac{d^{n+2}v}{dx^{n+2}} + 2x\left[{}^{(n+1)}C_1 - n\right]\frac{d^{n+1}v}{dx^{n+1}} + 2\left[{}^{n+1}C_2 - n.{}^{(\iota+1)}C_1\right]\frac{d^n v}{dx^n} = 0$$

$$\Rightarrow \quad (x^2-1)\frac{d^{n+2}v}{dx^{n+2}} + 2x\frac{d^{n+1}v}{dx^{n+1}} - n(n+1)\frac{d^n v}{dx^n} = 0 \qquad \ldots (3)$$

If we put $\frac{d^n v}{dx^n} = y$, (3) becomes

$$(x^2-1)\frac{d^2y}{dx^2} + 2x\frac{dy}{dx} - n(n+1)\,y = 0$$

$$\Rightarrow \quad (1-x^2)\frac{d^2y}{dx^2} - 2x\frac{dy}{dx} + n(n+1)\,y = 0$$

This shows that $y = \frac{d^n v}{dx^n}$ is a solution of Legendre's equation.

$$\therefore \quad C\frac{d^n v}{dx^n} = P_n(x) \qquad \ldots (4)$$

Where C is a constant.

But $$v = (x^2 - 1)^n = (x + 1)^n (x - 1)^n$$

so that $$\frac{d^n v}{dx^n} = (x+1)^n \frac{d^n}{dx^n}(x-1)^n + {}^nC_1 \cdot n(x+1)^{n-1} \cdot \frac{d^{n-1}}{dx^{n-1}}(x-1)^n + \ldots + (x-1)^n \frac{d^n}{dx^n}(x+1)^n = 0$$

when $$x = 1, \quad \text{then } \frac{d^n v}{dx^n} = 2^n \cdot n!$$

All the other terms disappear as $(x - 1)$ is a factor in every term except first.

Therefore when $x = 1$, (4) gives

$$C \cdot 2^n \cdot n! = P_n(1) = 1 \qquad [P_n(1) = 1]$$

$$C = \frac{1}{2^n \cdot n!}. \qquad \ldots(5)$$

Substituting the value of C from (5) in (4), we have

$$P_n(x) = \frac{1}{2^n \cdot n!} \frac{d^n v}{dx^n}$$

$$P_n(x) = \frac{1}{2^n \cdot n!} \frac{d^n}{dx^n}(x^2 - 1)^n \qquad [\because v = (x^2 - 1)^n]$$

Example 1. *Show that* $\int_{-1}^{+1} P_n(x)\, dx = 0, \quad n \neq 0$

and $\int_{-1}^{+1} P_n(x)\, dx = 2, \quad n = 0$

Solution.

(*i*) We know that $P_n(x) = \frac{1}{2^n \cdot n!} \frac{d^n}{dx^n}(x^2 - 1)^n$

Integrating, we get $$\int_{-1}^{+1} P_n(x)\, dx = \frac{1}{2^n\, n!} \int_{-1}^{+1} \frac{d^n}{dx^n}(x^2 - 1)^n\, dx = \frac{1}{2^n\, n!}\left[\frac{d^{n-1}}{dx^{n-1}}(x^2 - 1)^n\right]_{-1}^{+1}$$

$$= \frac{1}{2^n\, n!}[0 - 0] = 0$$

(*ii*) When $n = 0$

$$\int_{-1}^{1} P_0(x)\, dx = \int_{-1}^{+1} 1 \cdot dx = [x]_{-1}^{+1} = 2$$ **Proved.**

Example 2. *Let $P_n(x)$ be the Legendre polynomial of degree n. Show that for any function, $f(x)$, for which the nth derivative is continuous,*

$$\int_{-1}^{1} f(x) P_n(x)\, dx = \frac{(-1)^n}{2^n\, n!} \int_{-1}^{1} (x^2 - 1)^n f^n(x)\, dx.$$

Solution. $$\int_{-1}^{1} f(x) P_n(x)\, dx = \int_{-1}^{1} f(x) \cdot \frac{1}{2^n\, n!} \frac{d^n}{dx^n}(x^2 - 1)^n\, dx. \qquad \left[P_n(x) = \frac{1}{2^n\, n!} \frac{d^n}{dx^n}(x^2 - 1)^n\right]$$

$$= \frac{1}{2^n\, n!} \int_{-1}^{1} f(x) \cdot \frac{d^n}{dx^n}(x^2 - 1)^n\, dx$$

Integrating by parts, we get

$$= \frac{1}{2^n\, n!}\left[f(x) \cdot \frac{d^{n-1}}{dx^{n-1}}(x^2 - 1)^n - \int f'(x) \cdot \frac{d^{n-1}}{dx^{n-1}}(x^2 - 1)^n\, dx\right]_{-1}^{1}$$

$$= \frac{1}{2^n\, n!}\left[0 - \int_{-1}^{1} f'(x)\cdot\frac{d^{n-1}}{dx^{n-1}}(x^2-1)^n\,dx\right] = \frac{(-1)}{2^n\, n!}\int_{-1}^{1} f'(x)\cdot\frac{d^{n-1}}{dx^{n-1}}(x^2-1)^n\,dx$$

Again integrating by parts, we have

$$= \frac{(-1)}{2^n\, n!}\left[f'(x)\cdot\frac{d^{n-2}}{dx^{n-2}}(x^2-1)^n - \int f''(x)\cdot\frac{d^{n-2}}{dx^{n-2}}(x^2-1)^n\,dx\right]_{-1}^{1}$$

$$= \frac{(-1)^2}{2^n\, n!}\int_{-1}^{1} f''(x)\cdot\frac{d^{n-2}}{dx^{n-2}}(x^2-1)^n\,dx$$

Similarly, integrating $(n-2)$ times by parts, we get

$$\int_{-1}^{1} f(x)\,P_n(x)\,dx = \frac{(-1)^{n-2}}{2^n\, n!}\int_{-1}^{1} f^n(x)\,(x^2-1)^n\,dx = \frac{(-1)^n}{2^n\, n!}\int_{-1}^{+1} f^n(x)\,(x^2-1)^n\,dx$$

Proved.

34.6 LEGENDRE'S POLYNOMIALS

$$P_n(x) = \frac{1}{2^n\cdot n!}\cdot\frac{d^n}{dx^n}(x^2-1)^n \qquad \text{(Rodrigue's formula)}$$

If $n = 0$, $\qquad P_0(x) = \frac{1}{2^0\cdot 0!} = 1$

If $n = 1$, $\qquad P_1(x) = \frac{1}{2^1\cdot 1!}\frac{d}{dx}(x^2-1) = \frac{1}{2}(2x) = x$

If $n = 2$, $\qquad P_2(x) = \frac{1}{2^2\cdot 2!}\frac{d^2}{dx^2}(x^2-1)^2 = \frac{1}{8}\frac{d}{dx}[2(x^2-1)(2x)]$

$$= \frac{1}{2}[(x^2-1)\cdot 1 + 2x\cdot x] = \frac{1}{2}(3x^2-1)$$

Similarly $\quad P_3(x) = \frac{1}{2}(5x^3-3x)$

$$P_4(x) = \frac{1}{8}(35x^4 - 30x^2 + 3)$$

$$P_5(x) = \frac{1}{8}(63x^5 - 70x^3 + 15x)$$

$$P_6(x) = \frac{1}{16}(231x^6 - 315x^4 + 105x^2 - 5)$$

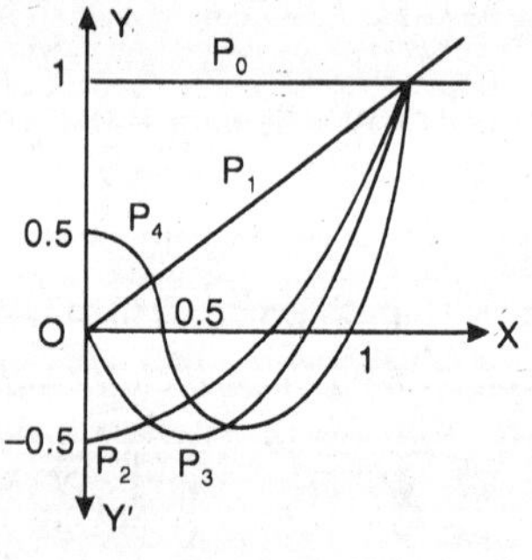

....

$$P_n(x) = \sum_{r=0}^{n} \frac{(-1)^r\,(2n-2r)!}{2^n\cdot r!\,(n-r)!\,(n-2r)!}\,x^{n-2r}$$

where $\qquad N = \frac{n}{2}$ if n is even.

$$N = \frac{1}{2}(n-1) \text{ if } n \text{ is odd}$$

Note. We can evaluate $P_n(x)$ by differentiating $(x^2-1)^n$, n times.

$$\frac{d^n}{dx^n}(x^2-1)^n = \sum_{r=0}^{r=n} {}^nC_r (x^2)^{n-r}(-1)^r = \sum_{r=0}^{r=n}(-1)^r \frac{n!}{r!(n-r)!}x^{2n-2r}$$

$$P_n(x) = \frac{1}{2^n \cdot n!}\frac{d^n}{dx^n}(x^2-1)^n = \frac{1}{2^n \cdot n!}\sum_{r=0}^{r=n}(-1)^r \frac{n!}{r!(n-r)!}(x^{2n-2r})$$

$$= \sum_{r=0}^{N}\frac{(-1)^r\,(2n-2r)!}{2^n \cdot r!(n-r)!(n-2r)!}x^{n-2r}$$

Either x^0 or x^1 is in the last term.

$$\therefore \qquad n-2r=0 \qquad \text{or} \qquad r=\frac{n}{2} \qquad (n \text{ is even})$$

$$n-2r=1 \qquad \text{or} \qquad r=\frac{1}{2}(n-1) \qquad (n \text{ is odd})$$

Example 3. *Express $f(x) = 4x^3 + 6x^2 + 7x + 2$ in terms of Legendre Polynomials.*

Solution. Let $4x^3 + 6x^2 + 7x + 2 \equiv a\,P_3(x) + b\,P_2(x) + c\,P_1(x) + d\,P_0(x)$ ... (1)

$$\equiv a\left(\frac{5x^3}{2}-\frac{3x}{2}\right)+b\left(\frac{3x^2}{2}-\frac{1}{2}\right)+c(x)+d\,(1)$$

$$\equiv \frac{5ax^3}{2}-\frac{3ax}{2}+\frac{3bx^2}{2}-\frac{b}{2}+cx+d$$

$$4x^3+6x+7x+2 \equiv \frac{5ax^3}{2}+\frac{3bx^2}{2}+\left(\frac{-3a}{2}+c\right)x-\frac{b}{2}+d.$$

Equating the coefficients of like powers of x, we have

$$4=\frac{5a}{2}, \qquad \Rightarrow \qquad a=\frac{8}{5}$$

$$6=\frac{3b}{2} \qquad \Rightarrow \qquad b=4$$

$$7=\frac{-3a}{2}+c \qquad \Rightarrow \qquad 7=\frac{-3}{2}\left(\frac{8}{5}\right)+c \qquad \Rightarrow \qquad c=\frac{47}{5}$$

$$2=\frac{b}{-2}+d \qquad \Rightarrow \qquad 2=\frac{4}{-2}+d \qquad \Rightarrow \qquad d=4$$

Putting the values of a, b, c, d in (1), we get

$$4x^3+6x^2+7x+2 = \frac{8}{5}P_3(x)+4P_2(x)+\frac{47}{5}P_1(x)+4P_0(x)$$ **Ans.**

Example 4. *Express the polynomial*

$f(x) = 4x^3 - 2x^2 - 3x + 8$ in terms of Legendre Polynomials.

(U.P., II Semester, 2009)

Solution. Let

$$4x^3-2x^2-3x+8 \equiv a\,P_3(x)+bP_2(x)+cP_1(x)+dP_0(x) \qquad ...(1)$$

$$\equiv a\left(\frac{5x^3}{2}-\frac{3x}{2}\right)+b\left(\frac{3x^2}{2}-\frac{1}{2}\right)+c\,(x)+d\,(1)$$

$$\equiv \frac{5ax^3}{2}-\frac{3ax}{2}+\frac{3bx^2}{2}-\frac{b}{2}+cx+d$$

$$\Rightarrow \qquad 4x^3 - 2x^2 - 3x + 8 \equiv \frac{5ax^3}{2} + \frac{3bx^2}{2} + \left(\frac{-3a}{2} + c\right) x - \frac{b}{2} + d$$

Equating the coefficients of like powers of x, we have

$$4 = \frac{5a}{2} \qquad \Rightarrow \qquad a = \frac{8}{5}$$

$$-2 = \frac{3b}{2} \qquad \Rightarrow \qquad b = -\frac{4}{3}$$

$$-3 = -\frac{3a}{2} + c \qquad \Rightarrow \qquad c = -3 + \frac{3}{2}\left(\frac{8}{5}\right) = -3 + \frac{12}{5} = -\frac{3}{5}$$

$$8 = -\frac{b}{2} + d \qquad \Rightarrow \qquad d = 8 + \frac{b}{2} = 8 + \frac{1}{2}\left(-\frac{4}{3}\right) = 8 - \frac{2}{3} = \frac{22}{3}$$

Putting these values in (1), we get

$$f(x) = \frac{8}{5} P_3(x) - \frac{4}{3} P_2(x) - \frac{3}{5} P_1(x) + \frac{22}{3} P_0(x)$$ **Ans.**

Example 5. *Show that* $\quad x^4 = \frac{1}{35}[8P_4(x) + 20\,P_2(x) + 7P_0(x)]$

Solution. We know that $P_4(x) = \frac{1}{8}(35x^4 - 30x^2 + 3)$

$$P_2(x) = \frac{1}{2}(3x^2 - 1),\ P_0(x) = 1$$

$$\text{R.H.S.} = \frac{1}{35}[8P_4(x) + 20P_2(x) + 7P_0(x)]$$

$$= \frac{1}{35}[(35x^4 - 30x^2 + 3) + 10(3x^2 - 1) + 7]$$

$$= x^4 = \text{L.H.S.}$$ **Proved.**

34.7 A GENERATING FUNCTION OF LEGENDRE'S POLYNOMIAL

Prove that $P_n(x)$ is the coefficient of z^n in the expansion of $(1 - 2xz + z^2)^{-1/2}$ in ascending powers of x. *(U.P. II Semester summer 2005)*

Proof. $\quad (1 - 2xz + z^2)^{-1/2} = [1 - z(2x - z)]^{-1/2}$

Expanding R.H.S. by Binomial Theorem, we have

$$(1 - 2xz + z^2)^{\frac{-1}{2}} = 1 + \frac{1}{2} z(2x - z) + \frac{-\frac{1}{2}\left(-\frac{3}{2}\right)}{2!} z^2 (2x - z)^2 + \ldots$$

$$+ \frac{-\frac{1}{2}\left(-\frac{3}{2}\right)\left(-\frac{5}{2}\right)\ldots\left(-\frac{1}{2} - n + 1\right)}{n!}(-z)^n (2x - z)^n + \ldots \ldots \quad (1)$$

Now coefficient of z^n in $(n+1)$ th term *i.e.* $\frac{\left(-\frac{1}{2}\right)\left(-\frac{3}{2}\right)\left(-\frac{5}{2}\right)\ldots\left(-\frac{1}{2} - n + 1\right)}{n!}(-z)^n (2x - z)^n$

$$= \frac{\left(-\frac{1}{2}\right)\left(-\frac{3}{2}\right)\left(-\frac{5}{2}\right)\ldots\left(-\frac{1}{2} - n + 1\right)}{n!}(-1)^n (2x)^n$$

$$= \frac{1.3.5...(2n-1)}{2^n \cdot n!}(2)^n \cdot x^n = \frac{1.3.5...(2n-1)}{n!}x^n \qquad ...(2)$$

Coefficient of z^n in nth term *i.e.* $\dfrac{\left(-\frac{1}{2}\right)\left(-\frac{3}{2}\right)\left(-\frac{5}{2}\right)...\left(-\frac{1}{2}-n+2\right)}{(n-1)!}(-z)^{n-1}(2x-z)^{n-1}$

$$= \frac{\left(-\frac{1}{2}\right)\left(-\frac{3}{2}\right)\left(-\frac{5}{2}\right)...\left(-\frac{1}{2}-n+2\right)}{(n-1)!}(-1)^{n-1}[-(n-1)(2x)^{n-2}]$$

$$= \frac{1.3.5...(2n-3)}{2^{n-1}\cdot(n-1)!}(2)^{n-2}(n-1)x^{n-2} = \frac{1.3.5...(2n-3)}{2\cdot(n-1)!}(n-1)x^{n-2}$$

$$= \frac{1.3.5...(2n-3)}{2\cdot(n-1)!} \times \frac{(2n-1)}{(2n-1)}(n-1)x^{n-2} = \frac{1.3.5...(2n-3)(2n-1)}{n!} \times \frac{n(n-1)}{2(2n-1)}x^{n-2} \qquad ...(3)$$

Coefficient of x^n in

$$\frac{\left(-\frac{1}{2}\right)\left(-\frac{3}{2}\right)\left(-\frac{5}{2}\right)...\left(-\frac{1}{2}-n+3\right)}{(n-2)!}z^{n-2}(2x-z)^{(n-2)}$$

$$= \frac{\left(-\frac{1}{2}\right)\left(-\frac{3}{2}\right)\left(-\frac{5}{2}\right)...\left(-\frac{1}{2}-n+3\right)}{(n-2)!} \times (-1)^{n-2} \times \frac{(n-2)(n-3)}{2!}(2x)^{n-4}$$

$$= \frac{1.3.5...(2n-5)}{2^{n-2}(n-2)!} \times \frac{(n-2)(n-3)}{2!}(2x)^{n-4}$$

$$= \frac{1.3.5...(2n-5)(2n-3)(2n-1)}{4(n-2)!} \times \frac{(n-2)(n-3)}{2(2n-3)(2n-1)}x^{n-4}$$

$$= \frac{1.3.5...(2n-1)}{4n(n-1)(n-2)!} \times \frac{n(n-1)(n-2)(n-3)}{2(2n-3)(2n-1)}x^{n-4}$$

$$= \frac{1.3.5...(2n-1)}{n!} \times \frac{n(n-1)(n-2)(n-3)}{2.4(2n-1)(2n-3)}x^{n-4} \qquad ...(4)$$

and so on.

Thus coefficient of z^n in the expansion of (1) is sum of (2), (3) and (4) etc.

$$= \frac{1.3.5...(2n-1)}{n!}\left[x^n - \frac{n(n-1)}{2(2n-1)}\cdot x^{n-2} + \frac{n(n-1)(n-2)(n-3)}{2.4(2n-1)(2n-3)}x^{n-4} - ...\right] = P_n(x)$$

Thus coefficients of z, z^2, z^3 ... etc. in (1) are $P_1(x), P_2(x), P_3(x)$ ...

Hence

$$(1-2xz+z^2)^{-1/2} = P_0(x) + zP_1(x) + z^2P_2(x) + z^3P_3(x) + ... + z^n P_n(x) + ..$$

i.e.,
$$(1-2xz+z^2)^{-1/2} = \sum_{n=0}^{n=\infty} P_n(x)\cdot z^n.$$
Proved.

Example 6. *Prove that $P_n(1) = 1$.*

Solution. We know that

$$(1 - 2xz + z^2)^{-1/2} = 1 + zP_1(x) + z^2 P_2(x) + z^3 P_3(x) + \ldots + z^n P_n(x) + \ldots$$

Substituing 1 for x in the above equation, we get

$$(1 - 2z + z^2)^{-1/2} = 1 + z P_1(1) + z^2 P_2(1) + z^3 P_3(1) + \ldots + z^n P_n(1) + \ldots$$

$$[(1 - z)^2]^{-1/2} = \sum_{n=0}^{\infty} z^n P_n(1) \quad \Rightarrow \quad (1 - z)^{-1} = \sum z^n P_n(1)$$

$$\Rightarrow \quad \sum z^n P_n(1) = (1 - z)^{-1} = 1 + z + z^2 + z^3 + \ldots + z^n + \ldots$$

Equating the coefficients of z^n on both sides, we get

$$P_n(1) = 1$$ **Proved.**

Example 7. *Prove that* $\sum_{n=0}^{\infty} P_n(x) = \dfrac{1}{\sqrt{2-2x}}$.

Solution. We know that $(1-2xz+z^2)^{-\frac{1}{2}} = \sum_{n=0}^{\infty} z^n P_n(x$ ... (1)

Putting $z = 1$ in (1), we get

$$(1-2x+1)^{-\frac{1}{2}} = \sum_{n=0}^{\infty} P_n(x)$$

$$\frac{1}{\sqrt{2-2x}} = \sum_{n=0}^{\infty} P_n(x)$$ **Proved.**

Example 8. *Prove that :* $\sum_{n=0}^{\infty} \dfrac{x^{n+1}}{n+1} P_n(x) = \dfrac{1}{2} \log\left(\dfrac{1+x}{1-x}\right)$

Solution. We know that $\sum_{n=0}^{\infty} h^n P_n(x) = (1 - 2xh + h^2)^{-1/2}$

Intergrating both sides w.r.t. h from 0 to h, we get

$$\sum_{n=0}^{\infty} \frac{h^{n+1}}{n+1} P_n(x) = \int_0^h \frac{dh}{\sqrt{1-2hx+h^2}} = \int_0^h \frac{dh}{\sqrt{(h-x)^2+(1-x^2)}} \; ; if\ |x| < 1 \quad \left[\begin{array}{l}\text{Here } x \text{ is constant}\\ h \text{ is variable.}\end{array}\right]$$

$$= \log \frac{(h-x)+\sqrt{h^2-2hx+1}}{1-x} \qquad \left[\int \frac{dh}{\sqrt{h^2+a^2}} = \log \frac{h+\sqrt{h^2+a^2}}{a}\right]$$

Putting $h = x$ in the expression, we get

$$\sum_{n=0}^{\infty} \frac{x^{n+1}}{n+1} P_n(x) = \log\left(\frac{\sqrt{1-x^2}}{1-x}\right) = \frac{1}{2} \log\left(\frac{1+x}{1-x}\right)$$ **Proved.**

Example 9. *Show that*

$P_n(-x) = (-1)^n P_n(x)$ and $P_n(-1) = (-1)^n$. *(AMIETE, June 2010)*

Solution. We know that

$$(1-2xz+z^2)^{-\frac{1}{2}} = \sum_{n=0}^{\infty} z^n P_n(x) \qquad \ldots (1)$$

Putting $-x$ for x in both sides of (1), we get

$$(1+2xz+z^2)^{-\frac{1}{2}} = \sum_{n=0}^{\infty} z^n P_n(-x) \qquad \ldots (2)$$

Again putting $-z$ for z in (1), we obtain

$$(1+2xz+z^2)^{-\frac{1}{2}} = \sum_{n=0}^{\infty}(-1)^n z^n P_n(x) \qquad \text{... (3)}$$

Form (2) and (3), we have $\sum_{n=0}^{\infty} z^n P_n(-x) = \sum_{n=0}^{\infty}(-1)^n z^n P_n(x)$ ... (4)

Comparing the coefficients of z^n from both sides of (4), we obtain

$$P_n(-x) = (-1)^n P_n(x) \qquad \text{.... (5)}$$

Putting $x = 1$ in (5), we get $P_n(-1) = (-1)^n P_n(1) = (1)(-1)^n$ $[P_n(1) = 1]$

(1) If n is even, then from (5)

$$P_n(-x) = P_n(x),$$

So $P_n(x)$, is even function of x.

(*ii*) If n is odd, then from (5)

$P_n(-x) = -P_n(x)$, so $P_n(x)$ is an odd function. **Proved.**

Example 10. *Prove that* (*i*) $P'_n(1) = \frac{1}{2} n(n+1)$

(*ii*) $P'_n(-1) = (-1)^{n-1} \frac{n}{2}(n+1)$

Solution. Legendre's equation is

$$(1-x^2)\frac{d^2y}{dx^2} - 2x\frac{dy}{dx} + n(n+1)y = 0 \qquad \text{... (1)}$$

$$(1-x^2)P_n''(x) - 2xP_n'(x) + n(n+1)P_n(x) = 0 \qquad \text{... (2)}$$

[$P_n(x)$ is the solution of (1)]

(*i*) Putting $x = 1$ in (2), we get

$0 - 2P_n'(1) + n(n+1)P_n(1) = 0 \Rightarrow 2P_n'(1) = n(n+1)(1)$ $[P_n(1) = 1]$

$\Rightarrow$ $P_n'(1) = \frac{1}{2}n(n+1)$ **Proved.**

(*ii*) On putting $x = -1$ in (2), we get

$$0 + 2P_n'(-1) + n(n+1)P_n(-1) = 0$$

$\Rightarrow$ $2P_n'(-1) + n(n+1)(-1)^n P_n(1) = 0$

$\Rightarrow$ $2P_n'(-1) + n(n+1)(-1)^n(1) = 0 \Rightarrow 2P_n'(-1) - n(n+1)(-1)^{n-1} = 0$

$\Rightarrow$ $P_n'(-1) = (-1)^{n-1}\frac{1}{2}n(n+1)$ **Proved.**

Example 11. *Show that*

(*i*) $P_{2n}(0) = (-1)^n \frac{1.3.5...(2n-1)}{2.4.6...2n}$ (*ii*) $P_{2n+1}(0) = 0$.

(*U.P., II Semester, Summer 2008, 2005*)

Solution. We know that $\Sigma z^{2n}P_{2n}(x) = (1-2xz+z^2)^{-1/2}$

On putting $x = 0$

$$\Sigma z^{2n}P_{2n}(0) = (1+z^2)^{-1/2}$$

$$= 1 + \left(-\frac{1}{2}\right)z^2 + \frac{\left(-\frac{1}{2}\right)\left(-\frac{3}{2}\right)}{2!}(z^2)^2 + \frac{\left(-\frac{1}{2}\right)\left(-\frac{3}{2}\right)\left(-\frac{5}{2}\right)}{3!}(z^2)^3$$

$$+ ... + \frac{\left(-\frac{1}{2}\right)\left(-\frac{3}{2}\right)\left(-\frac{5}{2}\right)...\left(-\frac{1}{2}-n+1\right)}{n!}(z^2)^n + ...$$

Equating the coefficient of z^{2n} both sides, we get

$$P_{2n}(0) = \frac{\left(-\frac{1}{2}\right)\left(-\frac{3}{2}\right)\left(-\frac{5}{2}\right)\cdots\left(-\frac{1}{2}-n+1\right)}{n!} = (-1)^n \frac{1.3.5...(2n-1)}{2^n.n!}$$

$$= (-1)^n \frac{1\cdot 3\cdot 5...(2n-1)}{2\cdot 4\cdot 6\cdot 8...2n}$$ **Proved.**

Equating the coefficient of z^{2n+1} of both sides, **we get $P_{2n+1}(0) = 0$**

Example 12. *Show that*

$$\frac{1-z^2}{(1-2xz+z^2)^{3/2}} = \sum_{n=0}^{\infty} (2n+1)\, P_n \cdot z^n$$

Solution. We know that

$$(1-2xz+z^2)^{-\frac{1}{2}} = \sum_{n=0}^{\infty} z^n\ P_n(x) \qquad \dots (1)$$

Differentiating both sides of (1) with respect to z, we get

$$-\frac{1}{2}(1-2xz+z^2)^{-3/2}(-2x+2z) = \sum_{n=0}^{\infty} n\, z^{n-1} \cdot P_n(x)$$

$$\Rightarrow \quad \frac{x-z}{(1-2xz+z^2)^{3/2}} = \sum_{n=0}^{\infty} n\, z^{n-1}\ P_n(x) \qquad \dots (2)$$

Multiplying both sides of (2) by $2z$, we get

$$\frac{2xz-2z^2}{(1-2xz+z^2)^{3/2}} = \sum_{n=0}^{\infty} 2n\, z^n\ P_n(x) \qquad \dots (3)$$

On adding (1) and (3), we get

$$(1-2xz+z^2)^{-\frac{1}{2}} + \frac{2xz-2z^2}{(1-2xz+z^2)^{3/2}} = \sum_{n=0}^{\infty} z^n P_n(x) + \sum_{n=0}^{\infty} 2n z^n\ P_n(x)$$

$$\Rightarrow \quad \frac{1-2xz+z^2+2xz-2z^2}{(1-2xz+z^2)^{3/2}} = \sum_{n=0}^{\infty} (2n+1)\, z^n\ P_n(x)$$

$$\Rightarrow \quad \frac{1-z^2}{(1-2xz+z^2)^{3/2}} = \sum_{n=0}^{\infty} (2n+1)\, z^n P_n$$ **Proved.**

Example 13. *Prove that* $\dfrac{1+z}{z\sqrt{(1-2xz+z^2)}} - \dfrac{1}{z} = \displaystyle\sum_{n=0}^{\infty} (P_n + P_{n+1})\, z^n$

(*A.M.I.E.T.E., Summer 2001*)

Solution.

$$\text{R.H.S.} = \sum_{n=0}^{\infty} (P_n + P_{n+1})\, z^n = \sum_{n=0}^{\infty} z^n P_n + \sum_{n=0}^{\infty} z^n P_{n+1}$$

$$= \sum_{n=0}^{\infty} z^n P_n + \frac{1}{z}\sum_{n=0}^{\infty} z^{n+1} P_{n+1} \qquad \dots (1)$$

But $$\sum_{n=0}^{\infty} z^n P_n = P_0 + zP_1 + z^2P_2 + z^3P_3 + \dots$$

And $$\sum_{n=0}^{\infty} z^{n+1}\ P_{n+1} = zP_1 + z^2P_2 + z^3\ P_3 + \dots.$$

$$= -P_0 + P_0 + zP_1 + z^2P_2 + z^3P_3 + \dots. = -P_0 + \Sigma\ z^n P_n$$

Putting the value of $\sum_{n=0}^{\infty} z^{n+1} P_{n+1}$ in (1), we get

$$\text{R.H.S.} = \sum_{n=0}^{\infty} z^n P_n + \frac{1}{z}\left[\Sigma z^n P_n - P_0\right] = \left(1+\frac{1}{z}\right)\sum_{n=0}^{\infty} z^n P_n - \frac{P_0}{z}$$

$$= \left(1+\frac{1}{z}\right)(1-2xz+z^2)^{-\frac{1}{2}} - \frac{1}{z} = \text{L.H.S.} \qquad (P_0 = 1)$$

$$= \frac{1+z}{z\sqrt{(1-2xz+z^2)}} - \frac{1}{z}$$

Proved.

Example 14. *Prove that*

$$P_n\left(-\frac{1}{2}\right) = P_0\left(-\frac{1}{2}\right)P_{2n}\left(\frac{1}{2}\right) + P_1\left(-\frac{1}{2}\right)P_{2n-1}\left(\frac{1}{2}\right) + \ldots\ldots + P_{2n}\left(-\frac{1}{2}\right)P_0\left(\frac{1}{2}\right)$$

Solution. We know that

$$(1-2xz+z^2)^{-\frac{1}{2}} = \sum_{n=0}^{\infty} z^n P_n(x) \qquad \ldots (1)$$

Substituting $\frac{1}{2}$ for x in (1), we have

$$(1-z+z^2)^{-\frac{1}{2}} = \sum_{n=0}^{\infty} z^n P_n\left(\frac{1}{2}\right) \qquad \ldots (2)$$

Again putting $-\frac{1}{2}$ for x in (1), we have

$$(1+z+z^2)^{-\frac{1}{2}} = \sum_{n=0}^{\infty} z^n P_n\left(-\frac{1}{2}\right) \qquad \ldots (3)$$

Replacing z by z^2 in (3), we obtain

$$(1+z^2+z^4)^{-\frac{1}{2}} = \sum_{n=0}^{\infty} z^{2n} P_n\left(-\frac{1}{2}\right) \qquad \ldots (4)$$

But

$$(1+z^2+z^4)^{-\frac{1}{2}} = [(1+z^2)^2 - z^2]^{-\frac{1}{2}}$$

$$= [(1+z^2+z)(1+z^2-z)]^{-\frac{1}{2}}$$

$$= (1+z+z^2)^{-\frac{1}{2}}(1-z+z^2)^{-\frac{1}{2}} \qquad \ldots (5)$$

Putting the values of $(1+z+z^2)^{-\frac{1}{2}}$ and $(1-z+z^2)^{-\frac{1}{2}}$ from (3) and (2) in (5), we get

$$(1+z^2+z^4)^{-\frac{1}{2}} = \sum_{n=0}^{\infty} z^n P_n\left(-\frac{1}{2}\right) \cdot \sum_{n=0}^{\infty} z^n P_n\left(\frac{1}{2}\right) \qquad \ldots (6)$$

Now substituting the value of $(1+z^2+z^4)^{-\frac{1}{2}}$ from (4) in (6), we get

$$\sum_{n=0}^{\infty} z^{2n} P_n\left(-\frac{1}{2}\right) = \sum_{n=0}^{\infty} z^n P_n\left(-\frac{1}{2}\right) \cdot \sum_{n=0}^{\infty} z^n P_n\left(\frac{1}{2}\right)$$

$$= \left[P_0\left(-\frac{1}{2}\right) + zP_1\left(-\frac{1}{2}\right) + \ldots + z^{2n-1}P_{2n-1}\left(-\frac{1}{2}\right) + z^{2n}P_{2n}\left(-\frac{1}{2}\right) + \ldots\right]$$

$$\left[P_0\left(\frac{1}{2}\right) + zP_1\left(\frac{1}{2}\right) + \ldots\ldots + z^{2n-1}P_{2n-1}\left(\frac{1}{2}\right) + z^{2n}P_{2n}\left(\frac{1}{2}\right) + \ldots\right]$$

On equating the coefficients of z^{2n} on both sides, we have

$$P_n\left(-\frac{1}{2}\right) = P_0\left(-\frac{1}{2}\right) P_{2n}\left(\frac{1}{2}\right) + P_1\left(-\frac{1}{2}\right) P_{2n-1}\left(\frac{1}{2}\right)$$

$$+ \ldots + P_{2n-1}\left(-\frac{1}{2}\right) P_1\left(\frac{1}{2}\right) + P_{2n}\left(-\frac{1}{2}\right) P_0\left(\frac{1}{2}\right) \quad \textbf{Proved.}$$

Example 15. *Prove that*

$$1+\frac{1}{2} P_1(\cos\theta)+\frac{1}{3}P_2(\cos\theta)+\frac{1}{4}P_3(\cos\theta)+\ldots\ldots = \log\frac{1+\sin\frac{\theta}{2}}{\sin\frac{\theta}{2}}$$

Solution. We know that

$$\sum_{n=0}^{\infty} z^n P_n(x) = (1-2xz+z^2)^{-\frac{1}{2}} = [z^2-2xz+x^2+1-x^2]^{-\frac{1}{2}}$$

$$= \left[(z-x)^2+(1-x^2)\right]^{-\frac{1}{2}} = \frac{1}{\sqrt{(z-x)^2+(\sqrt{1-x^2})^2}} \qquad \ldots (1)$$

On integrating both sides of (1) w.r.t., 'z' between the limits 0 to 1, we have

$$\sum_{n=0}^{\infty} P_n(x)\int_0^1 z^n\,dz = \int_0^1 \frac{dz}{\sqrt{(z-x)^2+(\sqrt{1-x^2})^2}}$$

$$\sum_{n=0}^{\infty} P_n(x)\cdot\left[\frac{z^{n+1}}{n+1}\right]_0^1 = \log\left[(z-x)+\sqrt{(z-x)^2+(1-x^2)}\right]_0^1$$

$$\left[\int\frac{dx}{\sqrt{(x^2+a^2)}} = \log\left\{x+\sqrt{(x^2+a^2)}\right\}\right]$$

$$= \log\left[(1-x)+\sqrt{(1-2xz+z^2)}\right]_0^1$$

$$\sum_{n=0}^{\infty} P_n(x)\left(\frac{1}{n+1}\right) = \log\,[(1-x)+\sqrt{(1-2x+1)}] - \log(-x+1)$$

$$= \log\left[1-x+\sqrt{2-2x}\right] - \log(-x+1)$$

Replacing x by $\cos\theta$, we get

$$\sum_{n=0}^{\infty}\frac{P_n(\cos\theta)}{n+1} = \log[1-\cos\theta+\sqrt{2-2\cos\theta}] - \log(-\cos\theta+1) \qquad \left[\cos\theta = 1-2\sin^2\frac{\theta}{2}\right]$$

$$= \log\left[1-\left(1-2\sin^2\frac{\theta}{2}\right)+\sqrt{2-2\left(1-2\sin^2\frac{\theta}{2}\right)}\right] - \log\left\{1-\left(1-2\sin^2\frac{\theta}{2}\right)\right\}$$

$$= \log\left\{2\sin^2\frac{\theta}{2}+2\sin\frac{\theta}{2}\right\} - \log 2\sin^2\frac{\theta}{2} = \log\frac{\left\{2\sin^2\frac{\theta}{2}+2\sin\frac{\theta}{2}\right\}}{2\sin^2\frac{\theta}{2}}$$

$$\Rightarrow \quad \sum_{n=0}^{\infty}\frac{P_n(\cos\theta)}{n+1} = \log\frac{\sin\frac{\theta}{2}+1}{\sin\frac{\theta}{2}}$$

$$\Rightarrow \quad \log\frac{1+\sin\frac{\theta}{2}}{\sin\frac{\theta}{2}} = P_0(\cos\theta)+\frac{1}{2}P_1(\cos\theta)+\frac{1}{3}P_2(\cos\theta)+\frac{1}{4}P_3(\cos\theta)+\ldots$$

$$\Rightarrow \quad \log\frac{1+\sin\frac{\theta}{2}}{\sin\frac{\theta}{2}} = 1+\frac{1}{2}P_1(\cos\theta)+\frac{1}{3}P_2(\cos\theta)+\frac{1}{4}P_3(\cos\theta)+.... \qquad \textbf{Proved.}$$

34.8 ORTHOGONALITY OF LEGENDRE POLYNOMIALS

(AMIETE, June 2010)

$$\int_{-1}^{+1} P_m(x)\cdot P_n(x)\,dx = 0 \qquad n \neq m$$

Proof. $P_n(x)$ is a solution of

$$(1-x^2)\frac{d^2y}{dx^2} - 2x\frac{dy}{dx} + n(n+1)\,y = 0 \qquad ...(1)$$

$P_m(x)$ is the solution of

$$(1-x^2)\frac{d^2z}{dx^2} - 2x\frac{dz}{dx} + m(m+1)\,z = 0 \qquad ...(2)$$

Multiplying (1) by z and (2) by y and subtracting, we get

$$(1-x^2)\left[z\frac{d^2y}{dx^2} - y\frac{d^2z}{dx^2}\right] - 2x\left[z\frac{dy}{dx} - y\frac{dz}{dx}\right] + [n(n+1) - m(m+1)]\,yz = 0$$

$$(1-x^2)\left[\left\{z\frac{d^2y}{dx^2} + \frac{dz}{dx}\frac{dy}{dx}\right\} - \left\{\frac{dy}{dx}\frac{dz}{dx} + y\frac{d^2z}{dx^2}\right\}\right] - 2x\left\{\frac{zdy}{dx} - \frac{ydz}{dx}\right\} + (n-m)(n+m+1)yz = 0$$

$$\Rightarrow \quad \frac{d}{dx}\left[(1-x^2)\left(z\frac{dy}{dx} - y\frac{dz}{dx}\right)\right] + (n-m)(n+m+1)\,yz = 0$$

Now integrating from – 1 to 1, we get

$$\left[(1-x^2)\left(z\frac{dy}{dx} - y\frac{dz}{dx}\right)\right]_{-1}^{+1} + (n-m)(n+m+1)\int_{-1}^{+1} y.z\,dx = 0.$$

$$\Rightarrow \quad 0 + (n-m)(n+m+1)\int_{-1}^{+1} y\cdot z\,dx = 0 \qquad \{y = P_n(x),\ z = P_m(x)\}$$

$$\int_{-1}^{+1} P_n(x)\cdot P_m(x)\,dx = 0 \quad \text{if } n \neq m. \qquad \textbf{Proved.}$$

Example 16. *Prove that*

$$\int_{-1}^{+1} [P_n(x)]^2\,dx = \frac{2}{2n+1}$$ *(U.P., II Semester, June 2008, 2004, 2002)*

and hence show that $\int_{-1}^{+1} P_3^2(x)\,dx = \frac{2}{7}$ *(AMIETE, June 2010)*

Solution. We know that $(1 - 2xz + z^2)^{-1/2} = \Sigma z^n P_n(x)$

Squaring both sides, we get

$$(1 - 2xz + z^2)^{-1} = \Sigma z^{2n} P_n^2(x) + 2\Sigma z^{m+n} P_m(x)\cdot P_n(x)$$

Integrating both sides between – 1 and + 1, we have

$$\int_{-1}^{+1} \Sigma z^{2n}\cdot P_n^2(x)\,dx + \int_{-1}^{+1} 2\Sigma z^{m+n}\cdot P_m(x)\cdot P_n(x)\,dx = \int_{-1}^{+1} (1-2xz+z^2)^{-1}\,dx$$

$$\int_{-1}^{+1} \Sigma z^{2n} P_n^2(x)\ dx + 0 = \int_{-1}^{+1} \frac{1}{1-2xz+z^2}\,dx$$

$$\Rightarrow \qquad \Sigma z^{2n} \int_{-1}^{+1} P_n^2(x)\, dx = -\frac{1}{2z} \{\log(1 - 2xz + z^2)\}_{-1}^{+1}$$

$$= -\frac{1}{2z} \log \frac{1 - 2z + z^2}{1 + 2z + z^2} = -\frac{1}{2z} \log \left(\frac{1-z}{1+z}\right)^2$$

$$= \frac{1}{z} \log \frac{1+z}{1-z} = \frac{1}{z} [\log(1+z) - \log(1-z)]$$

$$= \frac{1}{z}\left[\left(z - \frac{z^2}{2} + \frac{z^3}{3} - \frac{z^4}{4} + \ldots + \frac{z^{2n+1}}{2n+1} + \ldots\right) - \left(-z - \frac{z^2}{2} - \frac{z^3}{3} - \frac{z^4}{4} - \ldots - \frac{z^{2n+1}}{2n+1} - \ldots\right)\right]$$

$$\Sigma z^{2n} \int_{-1}^{+1} P_n^2(x)\, dx = \frac{2}{z}\left[z + \frac{z^3}{3} + \frac{z^5}{5} + \ldots + \frac{z^{2n+1}}{2n+1} + \ldots\right] = 2\left[1 + \frac{z^2}{3} + \frac{z^4}{5} + \ldots + \frac{z^{2n}}{2n+1} + \ldots\right]$$

Equating the coefficient of z^{2n} on both sides, we have

$$\int_{-1}^{+1} [P_n(x)]^2\, dx = \frac{2}{2n+1}.$$ **Proved.**

Hence $$\int_{-1}^{+1} P_3^2(x)\, dx = \frac{2}{2 \times 3 + 1} = \frac{2}{7}.$$ **Proved.**

Example 17. *Prove that*

$$\int_{-1}^{1} P_n(x)\,(1 - 2xt + t^2)^{-1/2}\, dx = \frac{2t^n}{2n+1}$$

where n is a positive integer.

Solution. L.H.S. $= \int_{-1}^{1} P_n(x)\,(1 - 2xt + t^2)^{-1/2}\, dx = \int_{-1}^{1} P_n(x) \left\{\sum t^n P_n(x)\right\} dx$

$$= \int_{-1}^{1} P_n(x)\,[P_0(x) + t\, P_1(x) + t^2\, P_2(x) + t^3\, P_3(x) + \cdots + t^n\, P_n(x) + \cdots]$$

$$= t^n \int_{-1}^{1} P_n^2(x)\, dx$$ $\left[\text{All other terms vanish since } \int_{-1}^{1} P_m(x)\, P_n(x)\, dx = 0;\ m \neq n\right]$

$$= t^n \cdot \frac{2}{2n+1} = \text{R.H.S.}$$ [By II orthogonal property] **Proved.**

Example 18. *Show that*

$$\int_{-1}^{1} (1 - x^2)\, P_m'\, P_n'\, dx = \begin{cases} 0, & \text{when } m \neq n \\ \dfrac{2n(n+1)}{2n+1}, & \text{when } m = n \end{cases}$$

where dashes denote differentiation w.r.t. x. *(U.P. II Semester summer 2006)*

Solution. We have, $\int_{-1}^{1} (1 - x^2)\, P_m'\, P_n'\, dx$

$$= \left[(1 - x^2)\, P_m'\, P_n\right]_{-1}^{1} - \int_{-1}^{1} P_n \left[\frac{d}{dx}(1 - x^2)\, P_m'\right] dx$$ [Integrating by parts]

$$= 0 - \int_{-1}^{1} P_n \frac{d}{dx}\{(1 - x^2)\, P_m'\}\, dx$$

$$= -\int_{-1}^{1} P_n \{-m(m+1) P_m\}\, dx$$

$\left[\text{From Legendre's differential equation } \frac{d}{dx}\{(1 - x^2)\, P_m'\} + m(m+1)\, P_m = 0 \text{ (Orthogonality)}\right]$

$$= m(m+1)\int_{-1}^{1} P_n\, P_m\ dx = m(m+1)\cdot 0 = 0 \qquad \left[\int_{-1}^{+1} P_n\, P_m dx = 0, \text{ when } m \neq n\right]$$

By putting $m = n$ in example 18, we can also prove that

$$\int_{-1}^{1}(1-x^2)\,(P_n')^2\ dx = \cdot\frac{2n(n+1)}{2n+1}$$

(Orthogonality Property) **Proved.**

Example 19. *Assuming that a polynomial f(x) of degree n can be written as*

$$f(x) = \sum_{m=0}^{\infty} C_m P_m(x),$$

show that $$C_m = \frac{2m+1}{2}\int_{-1}^{1} f(x)P_m(x)dx$$

Solution. We have, $f(x) = \sum_{m=0}^{\infty} C_m P_m(x)$

$$= C_0P_0(x) + C_1P_1(x) + C_2P_2(x) + C_3P_3(x) + C_4P_4(x) + \ldots + C_mP_m(x) + \ldots.$$

Multiplying both sides by $P_m(x)$, we get

$$P_m(x)\,f(x) = C_0P_0(x)\,P_m(x) + C_1P_1(x)\,P_m(x) + C_2P_2(x)\,P_m(x) + \ldots + C_m\,P_m^2(x) + \ldots$$

$$\int_{-1}^{+1} f(x)P_m(x)\,dx = \int_{-1}^{+1}\left[C_0P_0(x)P_m(x) + C_1P_1(x)P_m(x) + C_2P_2(x)P_m(x) + \ldots C_mP_m^2(x) + \ldots\right]dx$$

$$= \left[0+0+\ldots+C_m\frac{2}{2m+1}+\ldots\right] = \frac{2C_m}{2m+1}$$

$$C_m = \frac{2m+1}{2}\int_{-1}^{+1} f(x)P_m(x)\,dx$$ **Proved.**

Example 20. *Using the Rodrigue's formula for Legendre function, prove that*

$\int_{-1}^{+1} x^m P_n(x)\,dx = 0,$ *where m, n are positive integers and m < n.*

Solution. $$\int_{-1}^{+1} x^m P_n(x)\,dx = \int_{-1}^{+1} x^m \frac{1}{2^n\ n!}\frac{d^n}{dx^n}(x^2-1)^n\,dx = \frac{1}{2^n\ n!}\int_{-1}^{+1} x^m \frac{d^n}{dx^n}(x^2-1)^n\,dx$$

On integrating by parts, we get

$$= \frac{1}{2^n\ n!}\left[\left\{x^m\frac{d^{n-1}}{dx^{n-1}}(x^2-1)^n\right\}_{-1}^{+1} - \int_{-1}^{+1} m\,x^{m-1}\frac{d^{n-1}}{dx^{n-1}}(x^2-1)^n\,dx\right]$$

$$= 0 - \frac{m}{2^n\ n!}\int_{-1}^{+1} x^{m-1}\frac{d^{n-1}}{dx^{n-1}}(x^2-1)^n\,dx$$

Similarly, $$\frac{m}{2^n n!}\int_{-1}^{+1} x^{m-1}\frac{d^{n-1}}{dx^{n-1}}(x^2-1)\,dx = -(-1)^2\frac{m(m-1)}{2^2\ n!}\int_{-1}^{+1} x^{m-2}\frac{d^{n-2}}{dx^{n-2}}(x^2-1)^n\,dx$$

Integrating R.H.S., $m-2$ times, we get

$$\int_{-1}^{+1} x^m P_n(x)\,dx = (-1)^m\frac{m(m-1)\ldots 1}{2^n\ n!}\int_{-1}^{+1}\frac{d^{n-m}}{dx^{n-m}}(x^2-1)^n\,dx$$

$$= \frac{(-1)^m\ m!}{2^n\ n!}\int_{-1}^{+1}\frac{d^{n-m}}{dx^{n-m}}(x^2-1)^n\,dx$$

$$= \frac{(-1)^m\ m!}{2^n\ n!}\left[\frac{d^{n-m-1}}{dx^{n-m-1}}(x^2-1)^n\right]_{-1}^{+1} = 0$$ **Proved.**

34.9 RECURRENCE FORMULAE

The following recurrence formulae are derived from the generating function. These formulae are very useful in solving the questions. So they are to be committed to memory.

1.	$(n+1)P_{n+1}(x)$	$= (2n+1)\,x\,P_n(x) - n\,P_{n-1}(x)$
2.	$n\,P_n(x)$	$= x\,P_n'(x) - P_{n-1}'(x)$
3.	$(2n+1)\,P_n(x)$	$= P_{n+1}'(x) \;\; P_{n-1}'(x)$
4.	$P_n'(x)$	$= x\,P_{n-1}' + n\,P_{n-1}(x)$
5.	$(x^2-1)\,P_n'(x)$	$= n[xP_n - P_{n-1}]$
6.	$(x^2-1)\,P_n'(x)$	$= (n+1)[P_{n+1}(x) - x\,P_n(x)]$

Formula I. $\boxed{(n+1)\,P_{n+1} = (2n+1)\,x\,P_n - n\,P_{n-1}}$

Solution. We know that, $(1-2xz+z^2)^{-1/2} = \Sigma z^n P_n(x)$

Differentiating w.r.t, 'z', we get

$$-\frac{1}{2}(1-2xz+z^2)^{-3/2}(-2x+2z) = \Sigma\, nz^{n-1}P_n(x)$$

Multiplying both sides by $(1-2xz+z^2)$, we get

$$(1-2xz+z^2)^{-1/2}(x-z) = (1-2xz+z^2)\Sigma\, n z^{n-1}\,P_n(x)$$

$$(x-z)\,\Sigma z^n P_n(x) = (1-2xz+z^2)\Sigma\, nz^{n-1}P_n(x) \qquad \ldots (1)$$

Equating the coefficients of z^n on both sides of (1), we get

$$xP_n(x) - P_{n-1}(x) = (n+1)P_{n+1}(x) - 2x\,(n)\,P_n(x) + (n-1)\,P_{n-1}(x)$$

$$(n+1)\,P_{n+1}(x) = (2n+1)\,x\,P_n - nP_{n-1}$$ **Proved.**

Formula II. $\boxed{nP_n = x\,P_n' - P_{n-1}'}$ *(U.P., II Semester, summer, 2009, 2006)*

Solution. We know that, $(1-2xz+z^2)^{-1/2} = \Sigma z^n\,P_n(x) \qquad \ldots (1)$

Differentiating (1) with respect to z, we get

$$-\frac{1}{2}(1-2xz+z^2)^{-3/2}(-2x+2z) = \Sigma\, nz^{n-1}P_n(x)$$

$$\Rightarrow \qquad (x-z)\,(1-2xz+z^2)^{-3/2} = \Sigma\; nz^{n-1}P_n(x) \qquad \ldots (2)$$

Differentiating (1) with respect to x, we get

$$-\frac{1}{2}(1-2xz+z^2)^{-3/2}\,(-2z) = \Sigma\; z^n\; P_n'(x)$$

$$\Rightarrow \qquad z(1-2xz+z^2)^{-3/2} = \Sigma\, z^n\; P_n'(x) \qquad \ldots (3)$$

Dividing (2) by (3), we get

$$\frac{x-z}{z} = \frac{\Sigma\, nz^{n-1}P_n(x)}{\Sigma\, z^n P_n'(x)}$$

$$\Rightarrow \qquad (x-z)\,\Sigma\, z^n P_n'(x) = \Sigma\; nz^n\,P_n(x)$$

Equating coefficients of z^n on both sides, we get

$$x\,P_n'(x) - P_{n-1}'(x) = nP_n(x)$$ **Proved.**

Formula III. $\boxed{P'_{n+1} - P'_{n-1} = (2n+1)P_n}$

Solution. We know that, $(n+1)P_{n+1} = (2n+1)xP_n - nP_{n-1}$ (Recurrence Formula *I*) ... (1)

Differentiating (1) w.r.t. 'x', we get

$$(n+1)P'_{n+1} = (2n+1)P_n + (2n+1)xP'_n - nP'_{n-1} \quad \text{... (2)}$$

$$xP'_n - P'_{n-1} = nP_n \quad \text{(Recurrence formula } II\text{) ... (3)}$$

Substituting the value of xP'_n from (3) into (2), we get

$$(n+1)P'_{n+1} = (2n+1)P_n + (2n+1)[nP_n + P'_{n-1}] - nP'_{n-1}$$

$$\Rightarrow \quad (n+1)P'_{n+1} - (n+1)P'_{n-1} = (2n+1)(1+n)P_n$$

$$\Rightarrow \quad (2n+1)P_n = P'_{n+1} - P'_{n-1}$$ **Proved.**

Formula IV. $\boxed{P'_n - xP'_{n-1} = nP_{n-1}}$ (*U.P. II Semester, 2010*)

Solution. We know that, $(n+1)P_{n+1} = (2n+1)xP_n - nP_{n-1}$ (Formula *I*)

Replacing n by $n - 1$, we get

$$nP_n = (2n-1)xP_{n-1} - (n-1)P_{n-2}$$

Differentiating the above formula w.r.t. 'x', we get

$$nP'_n = (2n-1)P_{n-1} + (2n-1)xP'_{n-1} - (n-1)P'_{n-2}$$

$$\Rightarrow n[P'_n - xP'_{n-1}] - (n-1)[xP'_{n-1} - P'_{n-2}] = (2n-1)P_{n-1}$$

$$n[P'_n - xP'_{n-1}] - (n-1)[(n-1)P_{n-1}] = (2n-1)P_{n-1} \quad \text{(Form formula } II\text{)}$$

$$\Rightarrow \quad n[P'_n - xP'_{n-1}] = [(n-1)^2 + (2n-1)]P_{n-1} = n^2 P_{n-1}$$

$$\Rightarrow \quad P'_n - xP'_{n-1} = nP_{n-1}.$$ **Proved.**

Formula V. $\boxed{(x^2-1)P'_n = n[xP_n - P_{n-1}]}$

Solution. We know that, $P'_n - xP'_{n-1} = nP_{n-1}$ ... (1) (Recurrence Formula *III*)

$xP'_n - P'_{n-1} = nP_n$ ... (2) (Recurrence Formula *II*)

Multiplying (2) by x and subtracting from (1), we get

$$(1-x^2)P'_n = n(P_{n-1} - xP_n).$$ **Proved.**

Formula VI. $\boxed{(x^2-1)P'_n = (n+1)(P_{n+1} - xP_n).}$ (*U.P. II Semester, June 2007*)

Solution. We know that, $(n+1)P_{n+1} = (2n+1)xP_n - nP_{n-1}$ (Recurrence formula *I*)

Which can be written as

$$(n+1)(P_{n+1} - xP_n) = n(xP_n - P_{n-1}) \quad \text{... (1)}$$

But $(x^2-1)P'_n = n(xP_n - P_{n-1})$ ... (2) (Recurrence formula *V*)

From (1) and (2), we get

$$\Rightarrow \quad (x^2-1)P'_n = (n+1)(P_{n+1} - xP_n).$$ **Proved.**

Example 21. *If $P_n(x)$ is a Legendre polynomial of degree n and α is such that $P_n(\alpha) = 0$. Show that $P_{n-1}(\alpha)$ and $P_{n+1}(\alpha)$ are of opposite signs.*

Solution. From Recurrence relation (1), we have

$$(2n+1)xP_n(x) = (n+1)P_{n+1}(x) + nP_{n-1}(x) \quad \text{...(1)}$$

Putting $x = \alpha$ in (1), we get

$$(2n+1)\,\alpha\,.\,P_n(\alpha) = (n+1)\,P_{n+1}(\alpha) + nP_{n-1}(\alpha) \qquad \text{...(2)}$$

Putting $P_n(\alpha) = 0$ (given) in (2), we get

$$(2n+1)\,\alpha.\,0 = (n+1)\,P_{n+1}(\alpha) + nP_{n-1}(\alpha)$$

$$\Rightarrow \qquad \frac{P_{n+1}(a)}{P_{n-1}(a)} = \frac{n}{n+1} \qquad \text{...(3)}$$

As n is a positive integer so R.H.S. of (3) is negative. Hence (3) shows that $P_{n+1}(\alpha)$ and $P_{n-1}(\alpha)$ are of opposite signs. **Proved.**

Example 22. *Prove that* $\int P_n dx = \frac{1}{2n+1}[P_{n+1} - P_{n-1}] + C$

Solution. We know that $(2n+1)P_n = P'_{n+1} - P'_{n-1}$ (Recurrence relation *III*)

Integrating $(2n+1)\int P_n\, dx = P_{n+1} - P_{n-1} + A$

$$\int P_n dx = \frac{P_{n+1} - P_{n-1}}{2n+1} + C \qquad \textbf{Proved.}$$

Example 23. *Show that* $P'_{n+1} + P'_n = P_0 + 3P_1 + 5P_2 + \ldots + (2n+1)P_n$

Solution. We know that

$$(2n+1)P_n = P'_{n+1} - P'_{n-1} \qquad \text{(Recurrence formula III)}$$

Putting $n = 1, 2, 3, \ldots n$ in succession, we get

$$3P_1 = P'_2 - P'_0 \qquad \text{... (1)}$$

$$5P_2 = P'_3 - P'_1 \qquad \text{... (2)}$$

$$7P_3 = P'_4 - P'_2 \qquad \text{... (3)}$$

....

....

$$(2n-1)P_{n-1} = P'_n - P'_{n-2}$$

$$(2n+1)P_n = P'_{n+1} - P'_{n-1}$$

Adding (1), (2), (3) etc., we get

$$3P_1 + 5P_2 + 7P_3 + \ldots + (2n-1)P_{n-1} + (2n+1)P_n = P'_{n+1} + P'_n - P'_0 - P'_1 = P'_{n+1} + P'_n - P_0$$

$$\Rightarrow P'_{n+1} + P'_n = P_0 + 3P_1 + 5P_2 + 7P_3 + \ldots + (2n+1)P_n \qquad [P'_1 = 1 = P_0 \text{ and } P'_0 = 0] \quad \textbf{Proved.}$$

Example 24. *Prove that*

$$(2n+1)(x^2-1)P'_n(x) = n(n+1)\{P_{n+1}(x) - P_{n-1}(x)\}$$

Solution. We know that

$$(x^2-1)P'_n(x) = (n+1)[P_{n+1}(x) - xP_n(x)] \qquad \text{... (1)}$$

(Recurrence Formula *VI*)

$$(x^2-1)P'_n(x) = n[xP_n(x) - P_{n-1}(x)] \qquad \text{... (2)}$$

(Recurrence Formula *V*)

$P_n(x)$ does not occur in the required result so we have to eliminate $P_n(x)$.

Multiplying (1) by n and (2) by $(n+1)$ and then adding, we get

$$[n+(n+1)](x^2-1)P'_n(x) = n(n+1)P_{n+1}(x) - n(n+1)P_{n-1}(x)$$

$$\Rightarrow \qquad (2n+1)(x^2-1)P'_n(x) = n(n+1)[P_{n+1}(x) - P_{n-1}(x)] \qquad \textbf{Proved.}$$

Example 25. *Prove that:*

$$\int_{-1}^{1}(x^2-1)\,P_{n+1}\,P_n'\,dx=\frac{2n\,(n+1)}{(2n+1)\,(2n+3)}$$

Solution. $$\text{L.H.S.}=\int_{-1}^{1}(x^2-1)\,P_{n+1}\,P_n'\,dx \qquad ...(1)$$

From Recurrence relation (5),

$$n\,(P_{n-1}-xP_n)=(1-x^2)P_n'$$

$$\Rightarrow \qquad (x^2-1)\,P_n'=n(xP_n-P_{n-1}) \qquad ...(2)$$

Now, $$\int_{-1}^{1}(x^2-1)\,P_{n+1}\,P_n'\,dx=\int_{-1}^{1}\left\{(x^2-1)\,P_n'\right\}P_{n+1}\,dx$$

Putting the value of $(x^2-1)\,P_n$ from (2) in (1), we get

$$=\int_{-1}^{1}n\left(xP_n-P_{n-1}\right)P_{n+1}\,dx$$

$$=n\int_{-1}^{1}xP_n\,P_{n+1}\,dx-n\int_{-1}^{1}P_{n-1}\,P_{n+1}\,dx$$

$$=n\int_{-1}^{1}xP_n\,P_{n+1}\,dx+0 \qquad ...(3)\left[\because \int_{-1}^{1}P_{n-1}\,P_{n+1}\,dx=0\right]$$

From Recurrence relation (1), we have

$$(2n+1)\,xP_n=(n+1)\,P_{n+1}+nP_{n-1}$$

$$xP_n=\frac{(n+1)\,P_{n+1}+nP_{n-1}}{(2n+1)} \qquad ...(4)$$

Putting the value of xP_n from (4) in (3), we get

$$\int_{-1}^{1}(x^2-1)\,P_{n+1}\,P_n'\,dx=n\int_{-1}^{1}\left[\frac{(n+1)\,P_{n+1}+n\,P_{n-1}}{2n+1}\right].P_{n+1}\,dx \qquad \text{[From (2)]}$$

$$=\frac{n\,(n+1)}{2n+1}\int_{-1}^{1}P_{n+1}'dx+\frac{n^2}{2n+1}\int_{-1}^{1}P_{n-1}\,P_{n+1}\,dx$$

$$=\frac{n\,(n+1)}{2n+1}\bullet\frac{2}{2\,(n+1)+1}+0=\frac{2n\,(n+1)}{(2n+1)\,(2n+3)} \qquad \textbf{Proved.}$$

34.10 LAPLACE'S FIRST DEFINITE INTEGRAL FOR $P_n\,(x)$

$$P_n\,(x)=\frac{1}{\pi}\int_0^{\pi}\,[x\pm\sqrt{(x^2-1)}\cos\phi]^n\,d\phi$$

where n is a positive integer.

Solution. We know that

$$\int_0^{\pi}\frac{d\phi}{a\pm b\cos\phi}=\frac{\pi}{\sqrt{(a^2-b^2)}} \qquad \text{(If } a>b\text{)} \qquad ...\ (1)$$

Put $$a=1-xz \text{ and } b=z^2\sqrt{(x-1)}\text{, so that}$$

$$a^2-b^2=(1-xz)^2-z^2(x^2-1)=1-2xz+z^2$$

Substituting the values of a, b and a^2-b^2 in (1), we get

$$\int_0^{\pi}\frac{d\phi}{(1-xz)\pm z\sqrt{(x^2-1)}\,\cos\phi}=\frac{\pi}{\sqrt{(1-2xz+z^2)}}$$

$$\Rightarrow \qquad \pi(1-2xz+z^2)^{-\frac{1}{2}} = \int_0^{\pi} \frac{d\phi}{1-z\{x \pm \sqrt{(x^2-1)}\cos\phi\}}$$

$$= \int_0^{\pi} [1-z\{x \pm \sqrt{(x^2-1)}\cos\phi\}]^{-1} d\phi$$

$$\pi \sum_{n=0}^{\infty} z^n P_n(x) = \int_0^{\pi} [1-t]^{-1} \, d\phi$$

where $t = z\,\{x \mp \sqrt{(x^2-1)}\cos\phi\}$ and $z\,\{x \pm (\sqrt{x^2-1})\cos\phi\} < 1$ and z is small quantity.

$$= \int_0^{\pi} [1+t+t^2+....]\, d\phi = \int_0^{\pi} \sum_{n=0}^{\infty} t^n \, d\phi$$

$$= \int_0^{\pi} \sum_{n=0}^{\infty} z^n [x \pm \sqrt{(x^2-1)}\cos\phi]^n \, d\phi$$

$$\pi \sum_{n=0}^{\infty} z^n P_n(x) = \int_{n=0}^{\pi} z^n \{x \pm \sqrt{(x^2-1)}\cos\phi\}^n \cdot d\phi$$

Equating the coefficients of z^n from both the sides, we obtain

$$\pi\, P_n(x) = \int_0^{\pi} \left\{x \mp \sqrt{(x^2-1)}\cos\phi\right\}^n d\phi$$

$$\Rightarrow \qquad P_n(x) = \frac{1}{\pi}\int_0^{\pi} \left\{x \mp \sqrt{(x^2-1)}\cos\phi\right\}^n d\phi \qquad \textbf{Proved.}$$

34.11 LAPLACE'S SECOND DEFINITE INTEGRAL FOR P_n (X)

$$P_n(x) = \frac{1}{\pi}\int_0^{\pi} \frac{d\phi}{\left\{x \pm \sqrt{(x^2-1)}\cos\phi\right\}^{n+1}}$$

where n is a positive integer.

Solution. We know that

$$\int_0^{\pi} \frac{d\pi}{a \pm b\cos\phi} = \frac{\pi}{\sqrt{(a^2-b^2)}} \qquad \text{(If } a > b) \quad ... (1)$$

Putting $a = xz - 1$ and $b = z\sqrt{(x^2-1)}$, we get

$$a^2 - b^2 = (x\,z-1)^2 - z^2(x^2-1) = x^2z^2 - 2\,x\,z + 1 - z^2x^2 + z^2 = 1 - 2\,x\,z + z^2$$

On substituting the values of a, b and $a^2 - b^2$ in (1), we get

$$\int_0^{\pi} \frac{d\pi}{z\{x \pm \sqrt{(x^2-1)}\cos\phi\} - 1} = \pi(1-2xz+z^2)^{-\frac{1}{2}}$$

Put $z\{x \pm \sqrt{(x^2-1)}\cos\phi\} = t$ and $|\,t\,| > 1$

$$\int_0^{\pi} \frac{d\phi}{t-1} = \frac{\pi}{z}\left\{1 - \frac{2x}{z} + \frac{1}{z^2}\right\}^{-\frac{1}{2}}$$

$$\Rightarrow \qquad \int_0^{\pi} \frac{1}{t}\frac{d\phi}{1-\frac{1}{t}} = \frac{\pi}{z}\sum_{n=0}^{\infty} \frac{1}{z^n} P_n(x)$$

$$\Rightarrow \qquad \int_0^{\pi} \frac{1}{t}\left(1 + \frac{1}{t} + \frac{1}{t^2} + ...\right) d\phi = \pi \sum_{n=0}^{\infty} \frac{1}{z^{n+1}} P_n(x)$$

$$\Rightarrow \qquad \int_0^{\pi} \sum_{n=0}^{\infty} \frac{1}{t^{n+1}}\, d\phi = \pi \sum_{n=0}^{\infty} \frac{1}{z^{n+1}} P_n(x)$$

On putting the value of t, we get

$$\Rightarrow \int_0^{\pi} \sum_{n=0}^{\infty} \frac{d\phi}{z^{n+1}\{x \pm \sqrt{(x^2-1)}\cos\phi\}^{n+1}} = \pi \sum_{n=0}^{\infty} \frac{1}{z^{n+1}} P_n(x) \qquad \ldots (2)$$

Equating the coefficient of $\frac{1}{z^{n+1}}$ on both sides of (2), we get

$$\int_0^{\pi} \frac{d\phi}{[x \pm \sqrt{(x^2-1)}\cos\phi]^{n+1}} = \pi P_n(x)$$

$$\Rightarrow \qquad P_n(x) = \frac{1}{\pi}\int_0^{\pi} \frac{d\phi}{[x \pm \sqrt{(x^2-1)}\cos\phi]^{n+1}}$$

Proved.

Example 26. *Prove that*

$$\int_{-1}^{+1} x^2 P_{n+1}(x) \cdot P_{n-1}(x)\, dx = \frac{2n(n+1)}{(2n-1)(2n+1)(2n+3)}$$

Deduce the value of $\int_0^1 x^2 P_{n+1} \cdot P_{n-1}\, dx.$

Solution. The recurrence formula *1* is

$$(n+1) P_{n+1} = (2n+1) x P_n - n P_{n-1}$$

$$\Rightarrow \qquad (2n+1) x P_n = (n+1) P_{n+1} + nP_{n-1}$$

Replacing n by $(n + 1)$ and $(n - 1)$, we have

$$(2n+3) xP_{n+1} = (n+2) P_{n+2} + (n+1) P_n \qquad \ldots (1)$$

$$(2n-1) x P_{n-1} = nP_n + (n-1) P_{n-2} \qquad \text{[Replacing } n \text{ by } (n-2)] \ldots (2)$$

Multiplying (1) and (2) and integrating in the limits – 1 to +1, we have

$$(2n+3)(2n-1)\int_{-1}^{+1} x^2 P_{n+1} \cdot P_{n-1} dx = n(n+1)\int_{-1}^{1} P_n^2\, dx + n(n+2)\int_{-1}^{+1} P_n \cdot P_{n+2}\, dx$$

$$+(n^2-1)\int_{-1}^{+1} P_n P_{n-2} dx + (n-1)(n+2)\int_{-1}^{+1} P_{n+2} \cdot P_{n-2}\, dx$$

$$= n(n+1)\int_{-1}^{1} P_n^2 dx + 0 + 0 + 0 \qquad \text{(Orthogonality property)}$$

$$= n(n+1) \cdot \frac{2}{(2n+1)}$$

$$\int_{-1}^{+1} x^2 \cdot P_{n+1} \cdot P_{n-1} dx = \frac{2n(n+1)}{(2n-1)(2n+1)(2n+3)} \qquad \text{as R.H.S. is even function.}$$

and $$\int_0^1 x^2 \cdot P_{n+1} \cdot P_{n-1}\, dx = \frac{n(n+1)}{(2n-1)(2n+1)(2n+3)}$$

Proved.

Example 27. *Prove that*

$$P_{-(n+1)}(x) = P_n(x)$$

Solution. From Laplace's first integral

$$P_n(x) = \frac{1}{\pi}\int_0^{\pi} [x \pm \sqrt{(x^2-1)}\cos\phi]^n\, d\phi \qquad \ldots (1)$$

On putting – $(n + 1)$ for n in (1), we get

$$P_{-(n+1)}(x) = \frac{1}{\pi}\int_0^{\pi} \frac{d\phi}{[x \pm \sqrt{(x^2-1)}\cos\phi]^{n+1}}$$

$$= P_n(x) \text{ (from Laplace's second integral)}$$

Proved.

EXERCISE 34.1

1. Express in terms of Legendre Polynomials.

(a) $1 + x - x^2$ **Ans.** $-\frac{2}{3}P_2(x) + P_1(x) + \frac{2}{3}P_0(x)$

(b) $x^3 + 1$ **Ans.** $\frac{2}{5}P_3(x) + \frac{3}{5}P_1(x) + P_0(x)$

(c) $1 + 2x - 3x^2 + 4x^3$ **Ans.** $\frac{8}{5}P_3(x) - 2P_2(x) + \frac{22}{5}P_1(x)$

(d) $x^4 + 2x^3 - 6x^2 + 5x - 3$ *(AMIETE, Dec. 2009)*

Ans. $\frac{8}{35}P_4(x) + \frac{4}{5}P_3(x) - \frac{24}{7}P_2(x) + \frac{31}{5}P_1(x) - \frac{24}{5}P_0(x)$

(e) $x^4 + 3x^3 - x^2 + 5x - 2$ *(AMIETE, June 2010)*

Ans. $\frac{8}{35}P_4(x) + \frac{6}{5}P_3(x) - \frac{2}{21}P_2(x) + \frac{34}{5}P_1(x) - \frac{32}{15}$

2. Show that

(a) $x^5 = \frac{8}{63}\left[P_5(x) + \frac{7}{2}P_3(x) + \frac{27}{8}P_1(x)\right]$ (b) $x^3 = \frac{2}{5}P_3(x) + \frac{3}{5}P_1(x)$

Evaluate the following :

3. $\int_{-1}^{+1} x^3 P_4(x)dx$ **Ans.** 0

4. $\int_{-1}^{+1} x^3 P_3(x)dx$ *(A.M.I.E.T.E., Summer 2001)* **Ans.** $\frac{4}{35}$

Prove that

5. $\int_{-1}^{+1} x^n P_n(x)\,dx = \dfrac{\frac{1}{2}(n+1)}{2^n\left(\frac{2n+3}{2}\right)} - \dfrac{2^{n+1}(n!)^2}{(2n+1)}$

6. $\int_{-1}^{+1} x P_n(x) P_{n-1}(x)\,dx = \frac{2n}{4n^2-1}$

7. $\int_0^1 P_{2n}(x)\cdot P_{2n+1}(x)dx = \int_0^1 P_{2n}(x)\cdot P_{2n-1}(x)dx$

8. $\int_{\infty}^1 P_n(x)dx = \frac{1}{2n+1}\left[P_{n-1}(x) - P_{n+1}(x)\right]$

9. $P_n(x) = P'_{n+1}(x)$

10. $\frac{d}{dx}\left[(1-x)\frac{d}{dx}P_n(x)\right] + n(n+1)P_n(x) = 0$

11. $\int_{-1}^{+1} x\cdot P_n\cdot\frac{d}{dx}P_m dx$ is either 0, 2 or $\frac{2n}{2n+1}$

12. $P_n(x) = P'_{n+1}(x) - 2xP'_n(x) + P'_{n-1}(x)$

13. $P'_{2n+1} = (2n+1)P_{2n} + 2nxP_{2n-1} + (2n-1)x^2P_{n-2} + ... + 2x^{2n-1}P_1 + x^{2n}P_0$

14. $(n+1)\{P_nP'_{n+1} - P_{n+1}P'_n\} = P_0^2 + 3P_1^2 + 5P_2^2 + ... + (2n+1)P_n^2$

15. $(n+1)^2P_n^2 - (x^2-1)P_n^{-2} = P_0^2 + 3P_1^2 + 5P_2^2 + ... + (2n+1)P_n^2$

16. $\int_0^{\pi} P_n(\cos\theta)(\cos n\theta\, d\theta) = \beta\left(n+\frac{1}{2}, \frac{1}{2}\right)$ if n is a positive integer.

17. $u = (1-2xz+z^2)^{-\frac{1}{2}}$ is a solution of the equation $z\frac{\partial^2 v}{\partial x^2}\left\{(1-x^2)\frac{\partial\theta}{\partial x}\right\} = 0$?

34.12 FOURIER-LEGENDRE EXPANSION

Let $f(x)$ be a function defined from $x = -1$ to $x = 1$.

The Fourier-Legendre expansion of $f(x)$

$$f(x) = C_1 P_1(x) + C_2 P_2(x) + C_3 P_3(x) + \dots \qquad \dots (1)$$

$$\Rightarrow \qquad C_n = \frac{2n+1}{2}\int_{-1}^{+1} f(x) P_n(x)\, dx$$

Multiplying both sides of (1) by $P_n(x)$, we have

$$f(x)\cdot P_n(x) = C_1 P_1(x)\cdot P_n(x) + C_2 P_2(x) P_n(x) + \dots + C_n P_n^2(x) + \dots \qquad \dots (2)$$

Integrating both sides of (2), we get

$$\int_{-1}^{+1} f(x)\cdot P_n(x)\, dx = C_1\int_{-1}^{+1} P_1(x)\cdot P_n(x) dx + C_2\int_{-1}^{+1} P_2(x)\cdot P_n(x)\, dx + \dots + C_n\int_{-1}^{+1} P_n^2(x) dx + \dots$$

$$\int_{-1}^{+1} f(x)\cdot P_n(x)\, dx = C_n\int_{-1}^{+1} P_n^2(x) dx \qquad \text{(Other integrals are equal to zero)}$$

$$= C_n \frac{2}{2n+1}$$

$$\Rightarrow \qquad C_n = \frac{2n+1}{2}\int_{-1}^{+1} f(x)\cdot P_n(x) dx \quad \text{or} \quad C_n = \left(n+\frac{1}{2}\right)\int_{-1}^{+1} f(x)\cdot P_n(x)\, dx.$$

Example 28. *Evaluate*

$$\int_{-1}^{+1} x^2 P_n^2(x) dx \qquad \textit{(U.P. II Semester 2010)}$$

Solution. We know that

$$(n+1) P_{n+1} = (2n+1)\, x\, P_n(x) - n P_{n-1}(x) \qquad \text{(Recurrence formula I)}$$

$$(2n+1)\, x\, P_n(x) = (n+1)\, P_{n+1}(x) + n P_{n-1}(x)$$

Squaring both sides, we get

$$(2n+1)^2 x^2 P_n^2(x) = (n+1)^2 P_{n+1}^2(x) + n^2 P_{n-1}^2(x) + 2n(n+1) P_{n+1}(x) P_{n-1}(x)$$

Integrating both sides w.r.t. x between -1 and $+1$, we get

$$(2n+1)^2\int_{-1}^{+1} x^2 P_n^2(x)\, dx = (n+1)^2\int_{-1}^{+1} P_{n+1}^2(x)\, dx + n^2\int_{-1}^{+1} P_{n-1}^2(x)\, dx + 2n(n+1)\int_{-1}^{+1} P_{n+1}(x) P_{n-1}(x)\, dx$$

$$= (n+1)^2 \frac{2}{2(n+1)+1} + n^2 \frac{2}{2(n-1)+1} + 0 = \frac{2(n+1)^2}{2n+3} + \frac{2n^2}{2n-1}$$

$$\int_{-1}^{+1} x^2 P_n^2(x)\, dx = \frac{2(n+1)^2}{(2n+1)^2(2n+3)} + \frac{2n^2}{(2n+1)^2(2n-1)} \qquad \textbf{Ans.}$$

Example 29. *Find the sum of the first (n + 1) terms of the series* $\sum_{n=0}^{\infty} (2m+1)\, P_m(x) P_m(y)$

[*Christoffel's Summation Formula*]

Solution. We know that

$$(2m+1)\, x\, P_m(x) = (m+1)\, P_{m+1}(x) + m\, P_{m-1}(x) \qquad \dots (1)$$

and

$$(2m+1)\, y\, P_m(y) = (m+1)\, P_{m+1}(y) + m\, P_{m-1}(y) \qquad \dots (2)$$

Multiplying (1) by $P_m(y)$ and (2) by $P_m(x)$ and subtracting, we get

$$(2m+1)(x-y) P_m(x) P_m(y) = (m+1)\{P_{m+1}(x) P_m(y) - P_m(x) P_{m+1}(y)\}$$
$$- m\{P_m(x) P_{m-1}(y) - P_{m-1}(x) P_m(y)\}$$

Now putting $m = 1, 2, 3, \ldots\ldots n$ in succession, we have

$$3(x-y)P_1(x)P_1(y) = 2\{P_2(x)P_1(y) - P_1(x)P_2(y)\} - 1\{P_1(x)P_0(y) - P_0(x)P_1(y)\}$$

$$5(x-y)P_2(x)P_2(y) = \{P_3(x)P_2(y) - P_2(x)P_3(y)\} - 2\{P_2(x)P_1(y) - P_1(x)P_2(y)\}$$

. . . .

And again

$$(2n+1)(x-y)P_n(x)P_n(y) = (n+1)\{P_{n+1}(x)P_n(y) - P_n(x) + P_{n+1}(y)\}$$
$$- n\{P_n(x)P_{n-1}(y) - P_{n-1}(x)P_n(y)\}$$

Adding these equations, we get

$$(x-y)[3P_1(x)P_1(y) + 5P_2(x)P_2(y) + \ldots\ldots\ldots + (2n+1)P_n(x)\cdot P_n(y)]$$
$$= (n+1)[P_{n+1}(x)P_n(y) - P_n(x)P_{n+1}(y)] - [P_1(x)P_0(y) - P_0(x)P_1(y)]$$

But $\quad P_0(x) = P_0(y) = 1$ and $P_1(x) = x,\ P_1(y) = y$

Therefore $P_1(x)P_2(y) - P_0(x)P_1(y) = x - y = (x-y)P_0(x)P_0(y)$

$$(x-y)[P_0(x)P_0(y) + 3P_1(x), P_1(y) + 5P_2(x)P_2(y) + \ldots + (2n+1)P_n(x)P_n(y)]$$
$$= (n+1)[P_{n+1}(x)P_n(y) - P_n(x)P_{n+1}(y)]$$

$$\sum_{n=0}(2m+1)P_m(x)\cdot P_m(y) = (n+1)\frac{P_{n+1}(x)\cdot P_n(y) - P_n(x)P_{n+1}(y)}{x-y}$$ **Ans.**

Example 30. *Prove that*

$$\int_{-1}^{+1}\left(\frac{dP_n(x)}{dx}\right)^2 dx = n(n+1)$$

Solution. We know that (*Christoffel's series*)

$$\frac{d}{dx}P_n(x) = (2n-1)P_{n-1}(x) + (2n-5)P_{n-3}(x) + (2n-9)P_{n-5}(x) + \ldots$$

the last term being $3P_1$ if n is even and P_0 if n is odd.
Squaring both sides and then integrating , we have

$$\int_{-1}^{+1}\left(\frac{dP_n(x)}{dx}\right)^2 dx = (2n-1)^2\int_{-1}^{+1}P_{n-1}^2(x)\,dx + (2n-5)^2\int_{-1}^{+1}P_{n-3}^2(x)\,dx + \ldots\ldots$$
$$+2(2n-1)(2n-5)\int_{-1}^{+1}P_{n-1}(x)\cdot P_{n-3}(x)\,dx + \ldots$$
$$= (2n-1)^2\frac{2}{2(n-1)+1} + (2n-5)^2\frac{2}{2(n-3)+1} + \ldots$$
$$= \frac{2(2n-1)^2}{(2n-1)} + \frac{2(2n-5)^2}{2n-5} + \ldots\ldots$$
$$= 2(2n-1) + 2(2n-5) + \ldots$$
$$= 2[(2n-1) + (2n-5) + \ldots\ldots] \qquad \ldots (1)$$

(*i*) If n is even, then the last term is $3P_1$

$$(3)^2\int P_1^2(x)\,dx = 9\frac{2}{2\cdot 1+1} = 2\times 3 = 6$$

Equation (1) becomes

$$\int_{-1}^{+1}\left[\frac{d P_n(x)}{dx}\right]^2 dx = 2[(2n-1)+(2n-5)+......+3] \text{ (This is an } A.\ P.\text{, Number of terms} = \frac{n}{2})$$

$$= 2\frac{n}{2\times 2}\left[2\times 3+\left(\frac{n}{2}-1\right)4\right] = \frac{n}{2}(6+2n-4) = n(n+1)$$

(*ii*) If n is odd, then the last term is $P_0(x)$.

$$\int_{-1}^{+1} P_0^2(x)\,dx = \int_{-1}^{+1} dx = 2$$

$$\int_{-1}^{+1}\left(\frac{d P_n(x)}{dx}\right)^2 dx = 2[(2n-1)+(2n-5)+......+1]$$

$$\left[\text{ This is an A.P. Here number of terms} = \frac{n+1}{2}\right]$$

$$= 2\frac{n+1}{2\times 2}\left[2\times 1+\left(\frac{n+1}{2}-1\right)4\right] = \frac{n+1}{2}(2+2n+2-4) = n(n+1) \textbf{ Ans.}$$

Example 31. *Show that*

$$x P_n'(x) = n P_n(x)+(2n-3) P_{n-2}(x)+(2n-7) P_{n-4}(x)+....$$

Solution. We know that

$$n P_n(x) = x P_n'(x) - P_{n-1}'(x) \qquad \text{(Recurrence formula II)}$$

$$\Rightarrow \qquad x P_n'(x) = n P_n(x)+P_{n-1}'(x) \qquad ... (1)$$

$$P_{n+1}'(x) = (2n+1)\ P_n(x)+P_{n-1}'(x) \qquad \text{(Recurrence formula III)} \quad ... (2)$$

Putting $n-2$ for n in (2), we get

$$P_{n-1}'(x) = (2n-3)\ P_{n-2}(x)+P_{n-3}'(x) \qquad ... (3)$$

Putting $n-4$ for n in (2), we get

$$P_{n-3}'(x) = (2n-7)\ P_{n-4}(x)+P_{n-5}'(x) \qquad ... (4)$$

Putting $n-6$ for n in (2), we have

$$P_{n-5}(x) = (2n-11)\ P_{n-6}(x)+P_{n-7}'(x) \qquad ... (5)$$

and so on.

Adding (1), (3), (4), (5) etc., we get

$$x P_n'(x) = nP_n(x)+(2n-3)\ P_{n-2}(x)+(2n-7) P_{n-4}(x)+(2n-11)\ P_{n-6}(x)+... \qquad \textbf{Proved.}$$

Example 32. *Prove that* $\int_{-1}^{+1} x P_n P_n'\, dx = \frac{2n}{2n+1}$

Solution. In the last example we have proved that

$$x P_n'(x) = n P_n(x)+(2n-3)\ P_{n-2}(x)+(2n-7)\ P_{n-4}(x)+(2n-11) P_{n-6}(x)+....$$

Multiplying the above expansion by P_n and integrating between -1 and $+1$, we get

$$\int_{-1}^{+1} x\cdot P_n(x) P_n'(x)dx = n\int_{-1}^{+1} P_n^2(x)\,dx+(2n-3)\int_{-1}^{+1} P_n(x)\cdot P_{n-2}(x)dx+(2n-7)\int_{-1}^{+1} P_n(x)\cdot P_{n-4}(x)$$

$$+(2n-11)\int_{-1}^{+1} P_n(x)\cdot P_{n-6}(x)dx+.....$$

$$= n\cdot\frac{2}{2n+1}+0+0+..... = \frac{2n}{2n+1} \qquad \textbf{Proved.}$$

Example 33. *Expand the function*

$$f(x) = \begin{cases} 0, & -1 < x < 0 \\ 1, & 0 < x < 1 \end{cases}$$

in terms of Legendre polynomials.

Solution. Let $\quad f(x) = \sum_{n=0}^{\infty} C_n P_n(x)\,dx,$ then

$$C_n = \left(n+\frac{1}{2}\right)\int_{-1}^{+1} f(x) P_n(x)\,dx$$

$$= \left(n+\frac{1}{2}\right)\left[\int_{-1}^{0} 0\,P_n(x)\,dx + \int_0^1 1\,P_n(x)\,dx\right] = \left(n+\frac{1}{2}\right)\int_0^1 P_n(x)\,dx$$

$$= \left(n+\frac{1}{2}\right)\left[\int_0^1 P_0(x)\,dx + \int_0^1 P_1(x)\,dx + \int_0^1 P_2(x)dx + \int_0^1 P_3(x)dx + \ldots\right]$$

$$C_0 = \frac{1}{2}\int_0^1 P_0(x)\,dx = \frac{1}{2}\int_0^1 1.\,dx = \frac{1}{2}[x]_0^1 = \frac{1}{2}$$

$$C_1 = \frac{3}{2}\int_0^1 P_1(x)dx = \frac{3}{2}\int_0^1 x\,dx = \frac{3}{2}\left[\frac{x^2}{2}\right]_0^1 = \frac{3}{4}$$

$$C_2 = \frac{5}{2}\int_0^1 P_2(x)dx = \frac{5}{2}\int_0^1 \frac{1}{2}(3x^2-1)dx = \frac{5}{4}(x^3-x)_0^1 = 0$$

$$C_3 = \frac{7}{2}\int_0^1 P_3(x)dx = \frac{7}{2}\int_0^1 \frac{1}{2}(5x^3-3x)dx = \frac{7}{4}\left[\frac{5x^4}{4} - \frac{3x^2}{2}\right]_0^1 = \frac{-7}{16}$$

$$C_4 = \frac{9}{2}\int_0^1 P_4(x)\,dx = \frac{9}{2}\int_0^1 \frac{1}{8}(35x^4 - 30x^2 + 3)dx$$

$$= \frac{9}{16}\left[7x^5 - 10x^3 + 3x\right]_0^1 = 0$$

$$C_5 = \frac{11}{2}\int_0^1 P_5(x)\,dx = \frac{11}{2}\int_0^1 \frac{1}{8}(63x^5 - 70x^3 + 15x)\,dx$$

$$= \frac{11}{16}\left[\frac{21}{2}x^6 - \frac{35}{2}x^4 + \frac{15}{2}x^2\right]_0^1 = \frac{11}{32}$$

Hence $\quad f(x) = \frac{1}{2}P_0(x) + \frac{3}{4}P_1(x) - \frac{7}{16}P_3(x) + \frac{11}{32}P_5(x) + \ldots.$ **Ans.**

Example 34. *Express the function*

$$f(x) = \begin{cases} 0, & -1 < x \le 0 \\ x, & 0 < x < 1 \end{cases}$$

in Fourier-Legendre expansion.

Solution. Let $\quad f(x) = \sum_{n=0}^{\infty} C_n P_n(x),$ then

$$C_n = \left(n+\frac{1}{2}\right)\int_{-1}^{+1} f(x) P_n(x)\,dx$$

$$C_n = \left(n+\frac{1}{2}\right)\int_{-1}^{0} 0\cdot P_n(x)\,dx + \left(n+\frac{1}{2}\right)\int_0^1 x\cdot P_n(x)\,dx = \left(n+\frac{1}{2}\right)\int_0^1 x\cdot P_n(x)\,dx$$

$$C_0 = \frac{1}{2}\int_0^1 x \cdot P_0(x)\,dx = \frac{1}{2}\int_0^1 x \cdot 1\,dx = \frac{1}{2}\int_0^1 x\,dx = \frac{1}{2}\left[\frac{x^2}{2}\right]_0^1 = \frac{1}{4}$$

$$C_1 = \frac{3}{2}\int_0^1 x \cdot P_1(x)\,dx = \frac{3}{2}\int_0^1 x \cdot x\,dx = \frac{3}{2}\int_0^1 x^2\,dx = \frac{3}{2}\left[\frac{x^3}{3}\right]_0^1 = \frac{1}{2}$$

$$C_2 = \frac{5}{2}\int_0^1 x \cdot P_2(x)\,dx = \frac{5}{2}\int_0^1 x \cdot \frac{3x^2-1}{2}\,dx = \frac{5}{4}\int_0^1 (3x^3 - x)dx = \frac{5}{4}\left(\frac{3x^4}{4} - \frac{x^2}{2}\right)_0^1 = \frac{5}{16}$$

$$C_3 = \frac{7}{2}\int_0^1 x \cdot P_3(x)\,dx = \frac{7}{2}\int_0^1 x \cdot \frac{5x^3-3x}{2}\,dx = \frac{7}{4}\int_0^1 (5x^4 - 3x^2)\,dx = \frac{7}{4}(x^5 - x^3)_0^1 = 0$$

$$C_4 = \frac{9}{2}\int_0^1 x \cdot P_4(x)\,dx = \frac{9}{2}\int_0^1 x \cdot \frac{35x^4 - 30x^2 + 3}{8}\,dx$$

$$= \frac{9}{16}\int_0^1 (35x^5 - 30x^3 + 3x)\,dx = \frac{9}{16}\left[\frac{35x^6}{6} - \frac{15}{2}x^4 + \frac{3x^2}{2}\right]_0^1 = \frac{-3}{32}$$

$$C_5 = \frac{11}{2}\int_0^1 x \cdot P_5(x)\,dx = \frac{11}{2}\int_0^1 x \cdot \frac{63x^5 - 70x^3 + 15x}{8}\,dx$$

$$= \frac{11}{16}\int_0^1 (63x^6 - 70x^4 + 15x^2)\,dx = \frac{11}{16}\left[9x^7 - 14x^5 + 5x^3\right]_0^1 = 0$$

Hence $f(x) = \frac{1}{4}P_0(x) + \frac{1}{2}P_1(x) + \frac{5}{16}P_2(x) - \frac{3}{32}P_4(x) + ...$ **Ans.**

Example 35. *Expand the function*

$$f(x) = \begin{cases} 0 \ , & -1 < x \le 0 \\ x^2 \ , & 0 < x < 1 \end{cases}$$

in terms of Legendre Polynomials.

Ans. $f(x) = \frac{1}{6}P_0(x) + \frac{3}{8}P_1(x) + \frac{1}{3}P_2(x) + \frac{7}{48}P_3(x) - \frac{11}{384}P_5(x) + ...$

34.13 STRUM-LIOUVILLE EQUATION

$$\frac{d}{dx}\left[P(x)\cdot\frac{dy}{dx}\right] + [\lambda\, q(x) + r(x)]y = 0$$

Solution. We know that Bessel's equation is

$$X^2\frac{d^2y}{dX^2} + X\frac{dy}{dX} + (X^2 - n^2)\,y = 0 \qquad ... (1)$$

Substituting $X = kx$ in (1), we get

$$\frac{dy}{dX} = \frac{dy}{dx}\frac{dx}{dX} = \frac{dy}{dx}\frac{1}{k}$$

$$k^2x^2\left(\frac{d^2y}{dx^2}\,\frac{1}{k^2}\right) + (k\,x)\left(\frac{dy}{dx}\frac{1}{k}\right) + (k^2x^2 - n^2)y = 0$$

$$\Rightarrow \qquad x^2\,\frac{d^2y}{dx^2} + x\frac{dy}{dx} + (k^2\,x^2 - n^2)\,y = 0$$

$$x\frac{d^2y}{dx^2}+\frac{dy}{dx}+\left(k^2x-\frac{n^2}{x}\right)=0$$

$$\left(x\frac{d^2y}{dx^2}+\frac{dy}{dx}\right)+\left(\lambda x-\frac{n^2}{x}\right)y=0 \qquad \text{(Put } k^2=\lambda\text{)}$$

$$\frac{d}{dx}\left(x\frac{dy}{dx}\right)+\left(\lambda x-\frac{n^2}{x}\right)y=0 \qquad \dots (2)$$

Equations (1) and (2) are of the form.

$$\frac{d}{dx}\left[P(x)\cdot\frac{dy}{dx}\right]+[\lambda\, q(x)+r(x)]\,y=0 \qquad \dots (3)$$

Equation (3) is known as the Strum-Liouville equation.

Equation (3) with the following conditions is known as Strum-Liouville problem.

$$\alpha_1\, y(a)+\alpha_2 y'(a)=0$$

$$\beta_1 y(b)+\beta_2 y'(b)=0$$

Solution of Strum-Liouville problem is called an eigen function where λ is an eigen value.

Particular Case. Putting $p = 1$, $q = 1$, $r = 0$ in (3), we have

$$\frac{d^2y}{dx^2}+\lambda y=0$$

Now taking conditions as $\alpha_1=\beta_1=1$ and $\alpha_2=\beta_2=0$

$$y(a)=0 \quad \text{and} \quad y(b)=0$$

Hence $\left.\begin{array}{r} y''+\lambda y=0 \\ y(a)=0,\ y(b)=0 \end{array}\right]$ simplest form of Strum-Liouville problem.

34.14 ORTHOGONALITY

$$\int_a^b P(x)\,y_m(x)\cdot y_n(x)\,dx=0, \quad m\neq n$$

$$\int_a^b P(x)[y_m(x)]^2\,dx=\|\,y_m\,\|, \quad m=n$$

where $\|\,y_m\,\|$ is the norm of y_m.

Orthonormal : If the function is orthogonal and have norm equal to 1, then the function is known as orthonormal.

34.15 ORTHOGONALITY OF EIGEN FUNCTIONS

If P, q, r and r' are the functions in Strum-Liouville equation and $\lambda_m(x)$, $\lambda_n(x)$ be the eigen functions of Strum-Liouville problem, then

$$(\lambda_m-\lambda_n)\int_a^b q\,y_m y_n\,dx = y_m(P\,y'n)-y_n(P\,y'm)$$

$$=\frac{d}{dx}[(P\,y_n')\,y_m-(P\,y_m')\,y_n]_a^b$$

$$=P(b)\left[y_n'(b)\,y_m(b)-y_m'(b)\,y_n(b)\right]-P(a)\left[y_n'(a)\,y_m(a)-y_n'(a)\,y_n(a)\right] \qquad \dots (1)$$

$$=0 \text{ if}$$

(i) $y(a) = y(b)$ (ii) $y'(a) = y'(b)$

(iii) $\alpha_1 y(a) + \alpha_2 y'(a) = 0$ (iv) $\beta_1 y(b) + \beta_2 y'(b) = 0$

Equation (1) becomes $\int_a^b q\, y_m y_n\, dx = 0 \qquad (m \neq n)$

It means that eigen functions y_m, y_n are orthogonal with the weight $q(x)$.

OBJECTIVE TYPE QUESTIONS

Choose the correct or the best of the answer given in the following parts :

1. Let $P_n(x)$ be Legendre polynomial of degree $n > 1$, then

$\int_{-1}^{+1} (1+x) P_n(x) dx$ is equal to

(i) 0. (ii) $1/(2n+1)$. (iii) $2/(2n+1)$. (iv) $n/(2n+1)$

Ans. (i)

2. The Rodrigue formula for Legendre Polynomial $P_n(x)$ is given by:

(i) $P_n(x) = \dfrac{1}{n!2^n} \dfrac{d^n}{dx^n} (x^2-1)^n$ (ii) $P_n(x) = \dfrac{n!}{2^n} \dfrac{d^n}{dx^n} (x^2-1)^{n-1}$

(iii) $P_n(x) = \dfrac{n!}{2^{n-1}} \dfrac{d^n}{dx^n} (x^2-1)^{n-1}$ (iv) $P_n(x) = \dfrac{1}{n!2^n} (x^2-1)^n$ **Ans.** (i)

(*U.P., II Semester, 2009*)

3. The integral $\int_0^\pi P_n(\cos\theta) \sin 2\theta\, d\theta$, $n > 1$, where $P_n(x)$ is the Legendre polynomial of degree n equals to

(i) 1 (ii) $\dfrac{1}{2}$ (iii) 0 (iv) 2 (*AMIETE, Dec. 2004*) **Ans.** (iii)

CHAPTER

35

HERMITE FUNCTION

35.1 INTRODUCTION

In this chapter we will learn Hermite function in detail.

35.2 HERMITE'S EQUATION

$$\frac{d^2y}{dx^2} - 2x\frac{dy}{dx} + 2ny = 0 \qquad ....(1)$$

The solution of (1) is known as Hermite's polynomial.

Solution Here, we have

$$\frac{d^2y}{dx^2} - 2x\frac{dy}{dx} + 2ny = 0 \qquad ....(1)$$

Suppose its series solution is

$$y = a_0 x^m + a_1 x^{m+1} + a_2 x^{m+2} + a_3 x^{m+3} + \ldots\ldots + a_k x^{m+k}$$

or
$$y = \sum_{k=0}^{\infty} a_k x^{m+k} \qquad ....(2)$$

$$\frac{dy}{dx} = \sum a_k (m+k) x^{m+k-1}$$

$$\frac{d^2y}{dx^2} = \sum_{k=0}^{\infty} a_k (m+k)(m+k-1)\, x^{m+k-2}$$

Putting the values of y, $\frac{dy}{dx}$ and $\frac{d^2y}{dx^2}$ in (1), we get

$$\Rightarrow \quad \sum a_k (m+k)(m+k-1) x^{m+k-2} - 2x\sum a_k (m+k) x^{m+k-1} + 2n\sum a_k x^{m+k} = 0$$

$$\Rightarrow \quad \sum a_k (m+k)(m+k-1) x^{m+k-2} - 2\sum a_k (m+k) x^{m+k} + 2n\sum a_k\, x^{m+k} = 0$$

$$\Rightarrow \quad \sum a_k (m+k)(m+k-1) x^{m+k-2} - 2\sum a_k \left[(m+k) - n\right] x^{m+k} = 0 \qquad ....(3)$$

This equation holds good for $k = 0$ and all positive integer. By our assumption k cannot be negative.

To get the lowest degree term x^{m-2}, we put $k = 0$ in the first summation of (3) and we cannot have x^{m-2} from the second summation. Since $k \neq -2$.

The coefficient of x^{m-2} is

$$a_0\, m(m-1) = 0 \;\Rightarrow\; m = 0,\ m = 1, \text{ since } a_0 \neq 0 \qquad \text{........(4)}$$

This is the **indicial equation.**

Now equating the coefficient of next lowest degree term x^{m-1} to zero in (3), we get (by putting $k = 1$ in the first summation and we cannot have x^{m-1} from the second summation since $k \neq -1$.)

$$a_1 m(m+1) = 0 \;\Rightarrow\; \begin{bmatrix} a_1 \text{ may or may not be zero when } m = 0 \\ a_1 = 0, \text{ when } m = 1 \end{bmatrix} \quad \begin{pmatrix} m+1 \neq 0 \text{ as m is} \\ \text{already equal to zero.} \end{pmatrix}$$

Again equating the coefficient of the general term x^{m+k} to zero, we get

$a_{k+2}\,(m+k+2)\,(m+k+1) - 2a_k\,(m+k-n) = 0$

$$a_{k+2} = \frac{2(m+k-n)}{(m+k+2)(m+k+1)}a_k \qquad \text{.........(5)}$$

If $m = 0$, then, $$a_{k+2} = \frac{2(k-n)}{(k+2)(k+1)}a_k \qquad \text{.........(6)}$$

If $m = 1$, then, $$a_{k+2} = \frac{2(k+1-n)}{(k+3)(k+2)}a_k \qquad \text{.........(7)}$$

Case I. When $m = 0$, $a_{k+2} = \dfrac{2(k-n)}{(k+2)(k+1)}a_k$

If $k = 0$, then, $$a_2 = \frac{-2n}{2}a_0 = -na_0$$

If $k = 1$, then, $$a_3 = \frac{2(1-n)}{6}a_1 = -2\frac{(n-1)}{3!}a_1$$

If $k = 2$, then, $$a_4 = \frac{2(2-n)}{12}a_2 = 2\frac{(2-n)}{12}(-na_0) = (2)^2\frac{n(n-2)}{4!}a_0$$

If $k = 3$, then, $$a_5 = \frac{2(3-n)}{20}a_3 = \frac{2(3-n)}{20}\left(-\frac{2(n-1)}{3!}a_1\right) = (2)^2\frac{(n-1)(n-3)}{5!}a_1$$

$$a_{2r} = \frac{(-2)^r\, n(n-2)(n-4)\ldots\ldots(n-2r+2)}{(2r)!}a_0$$

$$a_{2r+1} = \frac{(-2)^r\,(n-1)(n-3)\ldots\ldots(n-2r+1)}{(2r+1)!}a_1 = 0$$

When $m = 0$, then there are two possibilities

Possibility I. When, $a_1 = 0$, then $a_3 = a_5 = a_7 = a_{2r+1} = \ldots\ldots = 0$.

Possibility II. When $a_1 \neq 0$

$$y = \sum_{k=0}^{\infty} a_k x^k$$

i.e. $y = a_0 + a_1x + a_2x^2 + a_3x^3 + a_4x^4 + a_5x^5 + \ldots\ldots\ldots\ldots$

$\quad = a_0 + a_2x^2 + a_4x^4 + \ldots\ldots + a_1x + a_3x^3 + a_5x^5 \qquad \text{....(8)}$

Putting the values of a_0, a_1, a_2, a_3, a_4 and a_5 in (8), we get

$$= a_0\left[1-\frac{2n}{2!}x^2+\frac{2^2 n(n-2)}{4!}x^4-\ldots+(-1)^r\frac{2}{(2r)!}n(n-2)\ldots(n-2r+2)x^{2r}+\ldots\right]$$

$$+a_1 x\left[1-\frac{2(n-1)}{3!}x^2+\frac{2^2(n-1)(n-3)}{5!}-\ldots\right.$$

$$\left.+(-1)^r\frac{2^r}{(2r+1)!}(n-1)(x-3)\ldots(n-2r+1)x^{2r}+\ldots\right] \quad \ldots\ldots..(9)$$

$$= a_0\left[1+\sum_{r=1}^{\infty}\frac{(-1)^r 2^r}{(2r)!}n(n-2)\ldots(n-2r+2)x^{2r}\right]$$

$$+a_0\left[x+\sum_{r=1}^{\infty}\frac{(-1)^r 2^r}{(2r+1)}(n-1)(n-3)\ldots(n-2r+2)x^{2r+1}\right] \quad (\text{If } a_1 = a_0) \quad \ldots\ldots(10)$$

Case II. When $m = 1$, then $a_1 = 0$ and so by putting $k = 0, 1, 2, 3, \ldots\ldots.$ in (7), we get

$$a_{k+2} = \frac{2(k+1-n)}{(k+3)(k+2)}a_k$$

$$a_2 = -\frac{2(n-1)}{3!}a_0$$

$$a_4 = \frac{2^2(n-1)(n-3)}{5!}a_0$$

....

$$a_{2r} = (-1)^r\frac{2^r(n-1)(n-3)\ldots(n-2r+1)}{(2r+1)!}a_0$$

Hence, the solution is

$$= a_0 x\left[1-\frac{2(n-1)}{3!}x^2+\frac{2^2(n-1)(n-3)}{5!}x^4\ldots+\frac{(-1)^r 2^r(n-1)(n-3)\ldots(n-2r+1)}{(2r+1)!}x^{2r}+\ldots.\right.$$

.....(11)

It is clear that the solution (11) is included in the second part of (9) except that a_0 is replaced by a_1 and hence in order that the Hermite equation may have two independent solutions, a_1 must be zero, even if $m = 0$ and then (9) reduces to

$$y = a_0\left[1-\frac{2n}{2!}x^2+\frac{2^2 n(n-2)}{4!}x^4-\ldots+(-1)^r\frac{2^r}{(2r)!}n(n-2)\ldots(n-2r+2)x^{2r}+\ldots\right]$$

.......(12)

The complete integral of (1) is then given by

$$y = A\left[1-\frac{2n}{2!}x^2+\frac{2^2 n(n-2)}{4!}x^4-\ldots\right]+B\left[1-\frac{2(n-1)}{3!}x^2+\frac{2^2(n-1)(n-3)}{5!}x^4\ldots\right]$$

........(13)

where A and B are arbitrary constants.

35.3 GENERATING FUNCTION OF HERMITE POLYNOMIALS (RODRIGUE FORMULA)

We know that

$$e^{x^2}\frac{\partial^n}{\partial t^n}e^{\{-(t-x)^2\}}=H_n(x)+H_{n+1}(x)t+H_{n+2}(x).t^2+\ldots\ldots \qquad \ldots\ldots\ldots(1)$$

Now differentiating $e^{\{-(t-x)^2\}}$ w.r.t., t, we get

$$\frac{\partial}{\partial t}e^{\{-(t-x)^2\}}=-2(t-x)e^{\{-(t-x)^2\}}$$

Taking limit when $t\to 0$, we get

$$\lim_{t\to 0}\frac{\partial}{\partial t}e^{\{-(t-x)^2\}}=2xe^{-x^2} \qquad \ldots\ldots\ldots\ldots(2)$$

Again differentiating $e^{\{-(t-x)^2\}}$ w.r.t. 'x', we get

$$\frac{\partial}{\partial x}e^{\{-(t-x)^2\}}=(-1)^2(t-x)\,e^{\{-(t-x)^2\}}$$

Taking limit when $t\to 0$, we get

$$\lim_{t\to 0}\frac{\partial}{\partial x}e^{\{-(t-x)^2\}}=-2xe^{-x^2} \qquad \ldots\ldots\ldots(3)$$

From (2) and (3), we have

$$\lim_{t\to 0}\frac{\partial}{\partial t}e^{\{-(t-x)^2\}}=(-1)^1\lim_{t\to 0}\frac{\partial}{\partial x}e^{\{-(t-x)^2\}}$$

Similarly,

$$\lim_{t\to 0}\frac{\partial^2}{\partial t^2}e^{\{-(t-x)^2\}}=(-1)^2\lim_{t\to 0}\frac{\partial^2}{\partial x^2}e^{\{-(t-x)^2\}}$$

– – – – – – – – – – – –

– – – – – – – – – – – –

$$\lim_{t\to 0}\frac{\partial^n}{\partial t^n}e^{\{-(t-x)^2\}}=(-1)^n\lim_{t\to 0}\frac{\partial^n}{\partial x^n}e^{\{-(t-x)^2\}}=(-1)^n\frac{d^n}{dx^n}e^{-x^2}$$

[differentiating n times] ....(4)

Putting $t = 0$ in (1), we get

$$\lim_{t\to 0}e^{x^2}\frac{\partial^n}{\partial t^n}e^{\{-(t-x)^2\}}=H_n(x) \qquad \ldots\ldots\ldots\ldots(5)$$

Putting the value of $\lim_{t\to 0}\frac{\partial^n}{\partial t^n}e^{\{-(t-x)^2\}}$ from (4) in (5), we get

$$(-1)^n e^{x^2}\frac{d^n}{dx^n}e^{-x^2}=H_n(x)$$

$$H_n(x)=(-1)^n e^{x^2}\frac{d^n}{dx^n}e^{-x^2} \qquad \ldots\ldots\ldots(6)$$

***n* = 0**

On putting $n = 0$ in (6), we get

$$H_0(x) = (-1)^0 e^{x^2} e^{-x^2} = 1$$

$$H_0(x) = 1$$

***n* = 1**

On putting $n = 1$ in (6), we get

$$H_1(x) = (-1)^1 e^{x^2} \frac{d}{dx} e^{-x^2} = -e^{x^2}(-2x)e^{-x^2} = 2x$$

$$H_1(x) = 2x$$

***n* = 2**

On putting $n = 2$ in (6), we get

$$H_2(x) = (-1)^2 e^{x^2} \frac{d^2}{dx^2} e^{-x^2} = e^{x^2} \frac{d}{dx}\left(-2xe^{-x^2}\right)$$

$$= e^{x^2}\left[-2e^{x^2} - 2x(-2x)e^{-x^2}\right.$$

$$= -2 + 4x^2$$

$$H_2(x) = 4x^2 - 2$$

***n* = 3**

On putting $n = 3$ in (6), we get

$$H_3(x) = (-1)^3 e^{x^2} \frac{d^3}{dx^3}\left(e^{-x^2}\right) = -e^{x^2} \frac{d^2}{dx^2}\left(-2xe^{-x^2}\right)$$

$$= -e^{x^2} \frac{d}{dx}\left(-2e^{-x^2} + (-2x)(-2x)e^{-x^2}\right)$$

$$= -e^{x^2} \frac{d}{dx}\left(-2 + 4x^2\right)e^{-x^2} = -e^{x^2}\left[8xe^{-x^2} + \left(4x^2 - 2\right)(-2x)e^{-x^2}\right]$$

$$= -\left[8x + \left(4x^2 - 2\right)(-2x)\right] = -8x + 8x^3 - 4x = 8x^3 - 12x$$

$$H_3(x) = 8x^3 - 12x$$

Similarly $H_4(x) = 16x^4 - 48x^2 + 12$

$H_5(x) = 32x^5 - 160x^3 + 120x$

$H_6(x) = 64x^6 - 480x^4 + 720x^2 - 120$

$H_7(x) = 128x^7 - 1344x^5 + 3360x^3 - 1680x$

Example 1. *Convert Hermite polynomial*

$$2H_4(x) + 3H_3(x) - H_2(x) + 5H_1(x) + 6H_0$$

into ordinary polynomial.

Solution. Here, we have

$$2H_4(x) + 3H_3(x) - H_2(x) + 5H_1(x) + 6H_0$$

$$= 2\left[16x^4 - 48x^2 + 12\right] + 3\left\{8x^3 - 12x\right\} - \left(4x^2 - 2\right) + 5(2x) + 6(1)$$

$$= 32x^4 - 96x^2 + 24 + 24x^3 - 36x - 4x^2 + 2 + 10x + 6$$

$$= 32x^4 + 24x^3 - 100x^2 - 26x + 32$$ **Ans.**

Example 2. *Convert ordinary polynomial*

$$64x^4 + 8x^3 - 32x^2 + 40x + 10$$

into Hermite polynomial.

Solution. Here, we have

Let $64x^4 + 8x^3 - 32x^2 + 40x + 10 = AH_4(x) + BH_3(x) + CH_2(x) + DH_1(x) + EH_0(x)$

$$= A(16x^4 - 48x^2 + 12) + B(8x^3 - 12x) + C(4x^2 - 2) + D(2x) + E(1)$$

$$= 16Ax^4 + 8Bx^3 (-48A + 4C)x^2 + (-12B + 2D)x + 12A - 2C + E$$

Equating the coefficients of like powers of x, we get

$$16A = 64 \Rightarrow A = 4$$

$$8B = 8 \Rightarrow B = 1$$

$$-48A + 4C = -32 \Rightarrow 4C = -32 + 192 \Rightarrow C = 40$$

$$-12B + 2D = 40 \Rightarrow -12 + 2D = 40 \Rightarrow 2D = 52 \Rightarrow D = 26$$

$$12A - 2C + E = 10 \Rightarrow 12 \times 4 - 2(40) + E = 10 \Rightarrow E = 42$$

The required Hermite polynomial is

$$4H_4(x) + H_3(x) + 40H_2(x) + 26H_1(x) + 42H_0(x)$$ **Ans.**

35.4 ORTHOGONAL PROPERTY

The *orthogonal property* of Hermites polynomials is

$$\int_{-\infty}^{\infty} e^{-x^2} H_m(x) H_n(x)\,dx = \begin{cases} 0, & m \neq n \\ 2^n n! \sqrt{\pi}, & m = n \end{cases}$$

Solution. We know that,

$$e^{\{x^2-(t_1-x)^2\}} = \sum \frac{H_n(x)}{n!} t_1^n \qquad \text{(generating function)} \;.......(1)$$

and

$$e^{\{x^2-(t_2-x)^2\}} = \sum \frac{H_m(x)}{m!} t_2^m \qquad ....(2)$$

Multiplying (1) and (2), we get

$$e^{\{x^2-(t_1-x)^2\}} . e^{\{x^2-(t_2-x)^2\}} = \left[\sum_{n=0}^{\infty} \frac{H_n(x)}{n!} t_1^n\right]\left[\sum_{m=0}^{\infty} \frac{H_m(x)}{m!} t_2^m\right]$$

$$= \sum_{\substack{n=0\\m=0}}^{\infty} \left[H_n(x) | H_m(x)|\right] \frac{t_1^n . t_2^m}{n!m!}$$

Multiplying both the sides of this equation by e^{-x^2} and then integrating with the limits from $-\infty$ to ∞, we have

$$\sum_{nm}\left[\int_{-\infty}^{\infty} e^{-x^2} H_n(x) H_m(x)\,dx\right]\frac{t_1^n . t_2^m}{n!m!} = e^{-x^2} \int_{-\infty}^{\infty} e^{\{x^2-(t_1-x)^2\}} . e^{\{x^2-(t_2-x)^2\}} dx$$

$$= \int_{-\infty}^{\infty} e^{\{x^2-(t_1-x)^2\}-(t_2-x)^2} dx$$

$$= e^{\{-(t_1^2+t_2^2)\}} \int_{-\infty}^{\infty} e^{\{-x^2+2x(t_1+t_2)\}} dx \qquad ....(3)$$

We have already learnt that $\int_{-\infty}^{\infty} e^{\{-ax^2+2bx\}} dx = \sqrt{\frac{\pi}{2}} e^{\frac{b^2}{a}}$ [standard formula].......(4)

Replacing $2b$ by $(t_1 + t_2)$ and a by 1 in (4), we get

$$\int_{-\infty}^{\infty} e^{\{-x^2+2x(t_1+t_2)\}} dx = \sqrt{\pi}\, e^{(t_1+t_2)^2} \qquad ....(5)$$

Putting the value of $\int_{-\infty}^{\infty} e^{\left\{-x^2+2x(t_1+t_2)\right\}} dx$ from (5) in R.H.S. of (3), we get

$$e^{\left\{-(t_1+t_2)^2\right\}}.\sqrt{\pi}e^{(t_1+t_2)^2} = \sqrt{\pi}\, e^{-t_1^2-t_2^2+t_1^2+t_2^2+2t_1t_2} = \sqrt{\pi}\, e^{2t_1t_2}$$

$$= \sqrt{\pi}\left[1+2t_1t_2+\frac{(2t_1t_2)^2}{2!}+\frac{(2t_1t_2)^3}{3!}+........\right] = \sqrt{\pi}\sum\frac{(2t_1t_2)^n}{n!}$$

$$= \sqrt{\pi}\sum\frac{2^n t_1^n t_2^n}{n!} = \sqrt{\pi}\sum_{\substack{m=0\\n=0}}^{\infty} 2^n t_1^n t_2^m \delta_{m,n} \qquad \left[t_2^n = t_2^m \delta_{n,m}\right]$$

From (3) , we have

$$\sum_{nm}\left[\int_{-\infty}^{\infty} e^{-x^2} H_n(x) H_m(x)\, dx\right]\frac{t_1^n t_2^m}{n!m!} = \sqrt{\pi}\sum_{nm}\frac{2^n}{n!} t_1^n t_2^m \delta_{n,m}$$

On equating the coefficients of t_1^n, t_2^m on both sides., we get

$$\int_{-\infty}^{\infty} e^{-x^2}\frac{H_n(x) H_m(x)}{n!m!} dx = \frac{\sqrt{\pi}2^n}{n!}\delta_{n,m}$$

$$\Rightarrow \qquad \int_{-\infty}^{\infty} e^{-x^2} H_n(x) H_m(x)\, dx = \sqrt{\pi}2^n m!\delta_{n,m}$$

$$\Rightarrow \qquad \int_{-\infty}^{\infty} e^{-x^2} H_m(x) H_n(x)\, dx = \begin{cases} 0 & m \neq n \\ 2^n n!\sqrt{\pi}, & m = n \end{cases} \qquad \left[\begin{aligned}\delta_{n,m} &= 0, \text{if } m \neq n \\ &= 1, \text{if } m = n\end{aligned}\right]$$

Proved.

Example 3. *Find the value of* $\int_{-\infty}^{\infty} e^{-x^2} H_2(x) H_3(x)\, dx$.

Solution. We know that

$$\int_{-\infty}^{\infty} e^{-x^2} H_m(x) H_n(x) = 0 \text{ if } m \neq n$$

Here $m = 2$ and $n = 3$, $m \neq n$

Hence, $\int_{-\infty}^{\infty} e^{-x^2} H_2(x) H_3(x) = 0$ **Ans.**

Example 4. *Find the value of* $\int_{-\infty}^{\infty} e^{-x^2}.[H_2(x)]^2\, dx$

Solution. We know that

$$\int_{-\infty}^{\infty} e^{-x^2}[H_n(x)]^2\, dx = 2^n (n)!\sqrt{\pi}$$

$$\int_{-\infty}^{\infty} e^{-x^2}[H_2(x)]^2\, dx = 2^2 (2!)\sqrt{\pi} = 8\sqrt{\pi}$$ **Ans.**

35.5 RECURRENCE FORMULAE FOR H_n (X) OF HERMITE EQUATION.

Four recurrence Relations

1. $2n\, H_{n-1}(x) = H'_n(x)$
2. $2x\, H_n(x) = 2nH_{n-1}(x) + H_{n+1}(x)$
3. $H'_n(x) = 2x\, H_n(x) - H_{n+1}(x)$
4. $H'_n(x) = x\, H'_n(x) + 2n\, H_n(x) = 0$

Recurrence Relation I

Hermite equation is $y'' - 2xy' + 2ny = 0$ for integral values taking $v = n$.

Also, $e^{2tx-t^2} = \sum_{n=0}^{\infty} \frac{H_n(x)t^n}{n.!}$ ....(1)

I. Differentiating partially *w.r.t.* x, we have

$$2t\,e^{2tx-t^2} = \sum_{n=0}^{\infty} H_n'(x)\frac{t^n}{n!} \quad \text{.........(2)}$$

Substituting the value of $2te^{2tx-t^2}$ from (1) in (2), we get

i.e $$2t\sum_{n=0}^{\infty} \frac{H_n(x)t^n}{n.!} = \sum_{n=0}^{\infty} H_n'(x)\frac{t^n}{n!}$$

$$\Rightarrow \quad 2\sum_{n=0}^{\infty} \frac{H_n(x)t^{n+1}}{n!} = \sum_{n=0}^{\infty} H_n'(x)\frac{t^n}{n!}$$

On replacing n by $n-1$ on L.H.S, we get

$$2\frac{H_{n-1}(x)t^n}{(n-1)!} = H_n'(x)\frac{t^n}{n!}$$

$$\Rightarrow \quad \frac{2nH_{n-1}(x)t^n}{n!} = H_n'(x)\frac{t^n}{n!}$$

On equating the coefficients of $\frac{t^n}{n!}$ on both sides , we get

$$2\frac{n!}{(n-1)!}H_{n-1}(x) = H'_n(x)$$

i.e $$2nH_{n-1}(x) = H_n'(x) \quad \text{.........(3)}$$

$$\mathbf{2nH_{n-1}(x) = H_n'(x)}$$

II. Differentiating partially *w.r.t.* 't', both sides of (1), we get

$$2(x-t)e^{2tx-t^2} = \sum_{n=0}^{\infty} H_n(x)\frac{nt^{n-1}}{(n-1)!}$$

$$2(x-t)e^{2tx-t^2} = \sum_{n=1}^{\infty} H_n(x)\frac{t^{n-1}}{(n-1)!} \quad (n = 0 \text{ vanishes on R.H.S}) \quad \text{........(4)}$$

On putting $n = 0$ R.H.S becomes zero.

Putting the value of e^{2tx-t^2} from (1) in (4) , we get

$$2(x-t)\sum_{n=0}^{\infty} H_n(x)\frac{t^n}{n!} = \sum_{n=1}^{\infty} H_n(x)\frac{t^{n-1}}{(n-1)!}$$

or $$2x\sum_{n=0}^{\infty} H_n(x)\frac{t^n}{n!} - 2\sum_{n=0}^{\infty} H_n(x)\frac{t^{n+1}}{n!} = \sum_{n=1}^{\infty} H_n(x)\frac{t^{n-1}}{(n-1)!}$$

Equating the coefficients of t^n on either side, we get

$$2x\frac{H_n(x)}{n!}-2\frac{H_{n-1}(x)}{(n-1)!}=\frac{H_{n+1}(x)}{n!} \qquad \text{(Replacing } n \text{ by } n+1 \text{ on R.H.S)}$$

i.e. $$2xH_n(x) = 2nH_{n-1}(x) + H_{n+1}(x)$$

$$\mathbf{2xH_n(x) = 2nH_{n-1}(x) + H_{n+1}(x)} \qquad ....(5)$$

III. Eliminating $H_{n-1}(x)$ from recurrence relation (1) and (2), we get

$$2xH_n(x) = H'_n(x) + H_{n+1}(x)$$

$$H'_n(x) = 2xH_n(x) - H_{n+1}(x)$$

$$\mathbf{H'_n(x) = 2xH_n(x) - H_{n+1}(x)}$$

IV. Differentiating recurrence relation (3), *w.r.t. x*, we get

$$H''_n(x) = 2xH'_n(x) + 2H_n(x) - H'_{n+1}(x)$$

Putting $H'_{n+1}(x) = 2(n+1)H_n(x)$ obtained from recurrence relation (1) on replacing n by $n+1$, we get

$$H''_n(x) = 2xH'_n(x) + 2H_n(x) - 2(n+1)H_n(x)$$

$$\Rightarrow \qquad H''_n(x) - 2x\,H'_n(x) + 2n\,H_n(x) = 0$$

which clearly follows that $y = H_n(x)$ is a solution of Hermite eqaution.

$$H''_n(x) - 2x\,H'_n(x) + 2n\,H_n(x) = 0$$

Example 5. *Prove that*

$$H_{2n}(0) = (-1)^n \cdot 2^{2n}\left(\frac{1}{2}\right)^n$$

Solution. We know that

$$H_{2n}(x) = \sum \frac{(-1)^n (2m)!(2x)^{2n+2x}}{x!(2n-2x)!} \qquad ...\ (1) \qquad \begin{bmatrix}\text{Even Hermite}\\ \text{polynomial}\end{bmatrix}$$

Putting $x = 0$ in (1), we get

$$H_{2n}(0) = \frac{(-1)^n (2n)!}{(n)!} = (-1)^n \frac{(2n)(2n-1)(2n-2).....1}{n(n-1)(n-2)........1}$$

$$= (-1)^n \frac{2(2n-1)2(2n-3)2(2n-5)2.....2.1}{n!} n!$$

$$= (-1)^n 2^n . 2^n \frac{(2n-1)}{2}\frac{(2n-3)}{2}\frac{(2n-5)}{2}............\frac{3}{2}.\frac{1}{2}$$

$$= (-1)^n 2^{2n}\left(\frac{1}{2}\right)\left(\frac{3}{2}\right)\left(\frac{5}{2}\right)\left(\frac{7}{2}\right).........\left(\frac{2n-3}{2}\right)\left(\frac{2n-1}{2}\right)$$

$$= (-1)^n 2^{2n}\left(\frac{1}{2}\right)^m$$

Proved.

Exmaple 6. *Prove that*

$$H'_{2n+1}(0) = (-1)^n 2^{2n+1}\left(\frac{3}{2}\right)^n$$

Solution. We know that by Recurrence Relation I

$$H'_n(x) = 2n\, H_{n-1}(x) \qquad ....(1)$$

Replacing n by $2n+1$ in (1), we get

$$H'_{2n+1}(x) = 2(2n+1)\, H_{2n}(x) \qquad ....(2)$$

Putting $x = 0$ in (2), we get

$$H'_{2n+1}(0) = 2(2n+1)\, H_{2n}(0)$$

$$= 2(2n+1)(-1)^n\, 2^{2n}\left(\frac{1}{2}\right)^n$$

$$= (2n+1)(-1)^n\, 2^{2n+1}\left[\frac{(2n-1)(2n-3).......3.1}{2^n}\right]$$

[Using previous exp]

$$= (-1)^n\, 2^{2n+1}\left[\frac{3}{2}\left(\frac{3}{2}+1\right).......\left(\frac{3}{2}+n-1\right)\right]$$

$$= (-1)^n \cdot 2^{2n+1}\left(\frac{3}{2}\right)^n$$

Proved.

Example 7. *Prove that*

$$H_{2n+1}(0) = 0$$

Solution. We know that

$$H_{2n+1}(x) = \sum_{k=0}^{2n+1/2} \frac{(-1)^k\,(2n+1)!(2x)^{2n+1-2k}}{k!(2n+1-2k)} \qquad \left[\begin{matrix}\text{Odd Hermite}\\ \text{Polynomial}\end{matrix}\right]$$

Putting $x = 0$ in above, we get

$\therefore \qquad H_{2n+1}(0) = 0$ **Proved.**

Example 8. *Prove that*

$$H'_{2n}(0) = 0$$

Solution. From recurrence relation I, we know that

$$H'_n(x) = 2n\, H_{n-1}(x)$$

Replacing n by $2n$, we get

$$H'_{2n}(x) = 2(2n)\, H_{2n-1}(x) \qquad ....(1)$$

Putting $x = 0$ in (1), we get

$$H'_{2n}(0) = 4n\, H_{2n-1}(0)$$

$$= 4n\,(0) \qquad \left[\begin{matrix}\text{From example 7}\\ H_{2n-1}(0) = 0\end{matrix}\right]$$

$$= 0$$

Example 9. *Prove that*

$$\frac{d^m}{dx^m}\{H_n(x)\} - \frac{2^n\,(n)!}{(n-m)!} H_{n-m} \qquad m < n$$

Solution. From recurrence relation I, we know that

$$H'_n(x) = 2n\, H_{n-1}(x) \qquad \text{........(1)}$$

$$\Rightarrow \qquad \frac{d}{dx}\{H_n(x)\} = 2nH_{n-1}(x)$$

$$\Rightarrow \qquad \frac{d^2}{dx^2}\{H_n(x)\} = 2n\frac{d}{dx}\left[H_{n-1}(x)\right]$$

$$= 2n\, H'_{n-1}(x)$$

$$= 2n\,[2\,(n-1)\, H_{n-2}(x)] \qquad \text{[From (1)]}$$

$$= 2^2 n\,(n-1)\, H_{n-2}(x)$$

Similarly $\dfrac{d^3}{dx^3}\{H_n(x)\} = 2^3 n(n-1)(n-2)\, H_{n-3}(x)$

Proceeding similarly m times, we get

$$\frac{d^m}{dx^m}\{H_n(x)\} = 2^m\, n(n-1)......(n-m+1)\, H_{n-m}(x), \cdot\ m < n$$

$$= \frac{2^m}{(n-m)!} H_{n-m}(x) \qquad \textbf{Proved.}$$

Example 10. *Prove that* $H_n(-x) = (-1)^n H_n(x)$.

Solution. Here, we have $\displaystyle\sum_{n=0}^{\infty} \frac{H_n(x)\,t^n}{n!} = e^{2tx-t^2} = e^{2tx} e^{-t^2} = \sum_{n=0}^{\infty} \frac{(2x)^n\, t^n}{n!} \times \sum_{n=0}^{\infty} \frac{(-1)\,t^{2n}}{n!}$

$$= \sum_{n=0}^{\infty} \sum_{k=0}^{n/2} \frac{(-1)^k (2x)^{n-2k}}{k!(n-2k)!}$$

Equating coefficient of $\dfrac{t^n}{n!}$ on either side, we get

$$H_n(x) = \sum_{k=0}^{n/2} \frac{(-1)^k\, n!(2x)^{n-2k}}{k!(n-2k)!}$$

Replacing x by $-x$, we get

$$H_n(-x) = \sum_{k=0}^{n/2} \frac{(-1)^k\, n!(-2x)^{n-2k}}{k!(n-2k)!}$$

$$= \sum_{k=0}^{n/2} \frac{(-1)^k\,(-1)^{n-2k}\, n!(2x)^{n-2k}}{k!(n-2k)!}$$

$$= (-1)^n \sum_{k=0}^{n/2} \frac{(-1)^k\, n!(2x)^{n-2k}}{k!(n-2k)!} = (-1)^n\, H_n(x)$$

Example 11. *Prove that*

$$\int_{-\infty}^{\infty} xe^{-x^2} H_n(x)\, H_m(x)\, dx = \sqrt{\pi}\left[2^{n-1} n!\delta_{m,n-1} + 2^n (n+1)\delta_{n+1,m}\right].$$

Solution. Integrating by parts, we have

$$\int_{-\infty}^{\infty} xe^{-x^2} H_n(x) H_m(x)\, dx = \left[-\frac{1}{2}e^{-x^2} H_n(x) H_m(x)\, dx\right]_{-\infty}^{\infty}$$

$$+\frac{1}{2}\int_{-\infty}^{\infty} e^{-x^2} \frac{d}{dx}\{H_n(x) H_m(x)\}\, dx$$

$$= 0 + \frac{1}{2}\int_{-\infty}^{\infty} e^{-x^2} \frac{d}{dx}\{H_n(x) H_m(x)\}\, dx \quad \text{(Orthogonality property)}$$

Differentiating by product rule *w.r.t.* 'x' under the sign of integration, we get

$$= \frac{1}{2}\int_{-\infty}^{\infty} e^{-x^2} \{H'_n(x) H_m(x) + H_n(x) H'_m(x)\}\, dx$$

By putting the values of $H'_n(x) = 2n H_{n-1}(x)$ and $H'_m(x) = 2mH_{m-1}(x)$ by recurrence relation (1), we get

$$= \frac{1}{2}\int_{-\infty}^{\infty} e^{-x^2} [2n H_{n-1}(x) H_m(x) + 2m H_n(x) H_{m-1}(x)]\, dx$$

$$= n\int_{-\infty}^{\infty} e^{-x^2} H_{n-1}(x) H_m(x)\, dx + m\int_{-\infty}^{\infty} e^{-x^2} H_n(x) H_{m-1}(x)\, dx$$

$$= n\sqrt{\pi}\, 2^{n-1} (n-1)!\, \delta_{m,n-1} + m\sqrt{\pi}\, 2^n\, n!\, \delta_{n,m-1} \qquad \text{(by orthogonal properties)}$$

By putting the value of $\delta_{n,m-1} = \delta_{n+1,m}$, we get

$$= \sqrt{\pi}\left[2^{n-1}\, n!\, \delta_{m,n-1} + 2^n (n+1)!\, \delta_{n+1,m}\right]$$

Proved.

EXERCISE 35.1

1. Find the value of $\int_{-\infty}^{\infty} e^{-x^2} H_{20}(x) H_{10}(x)\, dx$ **Ans.** 0

2. Find the value of $\int_{-\infty}^{\infty} e^{-x^2} [H_9(x)]^2\, dx$ **Ans.** $2^9 (9)! \sqrt{\pi}$

3. Find the value of $H_2(x)$ in terms of x. **Ans.** $4x^2 - 2$
4. Convert the Hermite polynomial $2H_3 - 4H_2 + H_1 + H_0$ into ordinary polynomial. **Ans.** $16x^3 - 16x^2 - 22x + 9$
5. Convert $8x^3 + 8x^2 - 6x + 2$ into Hermite polynomial **Ans.** $H_3 + 2H_2 + 3H_1 + 6H_0$

6. Verify, $P_n(x) = \dfrac{2}{\sqrt{\pi} n!}\int_0^{\infty} t^n e^{-t^2} H_n(xt)\, dt.$

7. Show that if m is an integer, $\int_{-\infty}^{\infty} x^m e^{-x^2} H_n(x)\, dx = 0.$

8. Show that if m is an integer, $\int_{-\infty}^{\infty} x^m e^{-x^2} H_n(x)\, dx = 0.$

CHAPTER

36

LAGUERRE'S FUNCTION

36.1 LAGUERRE'S FUNCTION

The Laguerre's differential equation is $x\frac{d^2y}{dx^2}+(1-x)\frac{dy}{dx}+ny=0$

Solution. Here, we have

$$x\frac{d^2y}{dx^2}+(1-x)\frac{dy}{dx}+ny=0 \qquad ....(1)$$

$$\Rightarrow \qquad \frac{d^2y}{dx^2}+\left(\frac{1-x}{x}\right)\frac{dy}{dx}+\frac{n}{x}y=0$$

Here $x = 0$ is a regular singularity of (1). So we will solve it by series solution method. :

Let $$y=\sum_{k=0}^{\infty}a_k x^{m+k}$$

$$\left[y=a_0x^m+a_1x^{m+1}+a_2x^{m+2}+a_3x^{m+3}+.......+a_kx^{m+k}+.......\right] \qquad ....(2)$$

$$\Rightarrow \qquad \frac{dy}{dx}=\sum_{k=0}^{\infty}(m+k)\,a_kx^{m+k-1}$$

$$\Rightarrow \qquad \frac{d^2y}{dx^2}=\sum_{k=0}^{\infty}(m+k)(m+k-1)\,a_kx^{m+k-2}$$

Putting the values of y, $\frac{dy}{dx}$ and $\frac{d^2y}{dx^2}$ in (1), we get

$$x\sum_{k=0}^{\infty}(m+k)(m+k-1)\,a_kx^{m+k-2}+(1-x)\sum_{k=0}^{\infty}(m+k)\,a_kx^{m+k-1}+n\sum_{k=0}^{\infty}a_kx^{m+k}=0$$

$$\Rightarrow \qquad \sum_{k=0}^{\infty}(m+k)(m+k-1)\,a_kx^{m+k-1}+\sum_{k=0}^{\infty}(m+k)\,a_kx^{m+k-1}-\sum_{k=0}^{\infty}(m+k)\,a_kx^{m+k}+n\sum_{k=0}^{\infty}a_kx^{m+k}=0$$

$$\Rightarrow \qquad \sum_{k=0}^{\infty}a_k\left\{(m+k)(m+k-1)+(m+k)\right\}x^{m+k-1}-\sum_{k=0}^{\infty}a_k\left\{m+k-n\right\}x^{m+k}=0$$

$$\Rightarrow \quad \sum_{k=0}^{\infty} a_k \left\{(m+k)^2 - (m+k) + (m+k)\right\} x^{m+k-1} - \sum_{k=0}^{\infty} a_k \{m+k-n\} x^{m+k} = 0$$

$$\Rightarrow \quad \sum_{k=0}^{\infty} a_k (m+k)^2 x^{m+k-1} - \sum_{k=0}^{\infty} a_k (m+k-n) x^{m+k} = 0 \qquad \text{........(3)}$$

Equating to zero the coefficient of lowest degree term x^{m-1}

($x^{m-1}$is obtained by putting $k = 0$ in the first summation of (3), but we cannot put $k = -1$ in second summation to get x^{m-1} since k is always positive.)

$$a_0\, m^2 = 0 \qquad \text{(Indicial equation)}$$

$$\Rightarrow m = 0\,,\, m = 0,\;\; a_0 \neq 0$$

Again equating to zero, the coefficient of x^{m+k} in (2), we get

(To get x^{m+k} we put $k \Rightarrow k + 1$ in first summation and $k \Rightarrow k$ in second summation of (2).

$$(m + k + 1)^2 a_{k+1} - (m + k - n)\, a_k = 0$$

$$\Rightarrow \quad a_{k+1} = \frac{(m+k-n)}{(m+k+1)^2} a_k \qquad \text{........(4)}$$

For $m = 0$ in (4), we have

$$a_{k+1} = \frac{k-n}{(k+1)^2} a_k$$

If $k = 0$, then $\quad a_1 = -n a_0$

If $k = 1$, then $\quad a_2 = \dfrac{1-n}{4} a_1 = \dfrac{(n-1)}{4} n\, a_0$

If $k = 2$, then $\quad a_3 = \dfrac{2-n}{9} a_2 = \left(\dfrac{(2-n)}{9}\right)\left(\dfrac{(n-1)\,n}{4}\right) a_0 = (-1)^3 \dfrac{n(n-1)(n-2)}{(3!)^2} a_0$

\- - - - - - - - - - - - - -

\- - - - - - - - - - - - - -

$$a_k = (-1)^k \frac{n(n-1)(n-2)---(n-k+1)}{(k!)^2} a_0$$

On putting these values of the coefficients a_1, a_2, a_3 .... a_k and $m = 0$ in (2), we get

$y = a_0 + a_1 x + a_2 x^2 + a_3 x^3$ .... $+ a_k x^k +$ ....

$$y = a_0 - n\, a_0\, x + \frac{n(n-1)}{(2!)^2} a_0 x^2 - \frac{n(n-1)(n-2)}{(3!)^2} a_0 x^3 + \text{........}$$

$$+ (-1)^k \frac{n(n-1)(n-2)\text{.........}(n-k+1)}{(k!)^2} a_0 x^k + \text{.......}$$

$$= a_0\left[1 - nx + \frac{n(n-1)}{(2!)^2}x^2 - \frac{n(n-1)(n-2)}{(3!)^2}x^3 + \ldots\ldots\ldots + (-1)^k \frac{n(n-1)(n-2)---(n-k+1)}{(k!)^2}x^k + ---\right]$$

$$= a_0 \sum_{k=0}^{\infty} (-1)^k \frac{(n!)}{(k!)^2 (n-k)!} x^k \text{ ,where } n \text{ is positive.}$$

If we take $a_0 = n!$, then solution of (1) becomes Laguerre's polynomial

$$y = n!\left[1 - nx + \frac{n(n-1)}{(2!)^2}x^2 - \frac{n(n-1)(n-2)}{(3!)^2}x^3 + \ldots\ldots\ldots\ldots + \frac{(-1)^k n(n-1)(x-2)\ldots\ldots(n-k+1)}{(k!)^2}x^k + \ldots\ldots\ldots\ldots\right]$$

$$\Rightarrow \qquad L_n(x) = (-1)^n\left[x^n - \frac{n^2}{1!}x^{n-1} + \frac{n^2(n-1)^2}{2!}x^{n-2} + \ldots\ldots\ldots + (-1)^n n!\right]$$

This is the expression for Laguerre's polynomial.

36.2 LAGUERRE'S FUNCTION FOR DIFFERENT VALUES OF *n*.

$L_n(0) = n!$

$L_0(x) = 1$

$L_1(x) = 1 - x$

$L_2(x) = x^2 - 4x + 2$

$L_3(x) = -x^3 + 9x^2 - 18x + 6$

$L_4(x) = x^4 - 16x^3 + 72x^2 - 96x + 48$

and so on.

36.3 GENERATING FUNCTION OF LAGUERRE POLYNOMIAL

$$(1-t)\sum_{n=0}^{\infty} \frac{L_n(x)}{n!} t^n = e^{\frac{-xt}{1-t}}$$

Solution : Here, we have

$$(1-t)\sum_{n=0}^{\infty} \frac{L_n(x)}{n!} t^n = e^{\frac{-xt}{1-t}}$$

$$\sum_{n=0}^{\infty} \frac{L_n(x)t^n}{n!} = \frac{1}{1-t} e^{\frac{-xt}{1-t}}$$

$$= \frac{1}{1-t}\left[1 - \frac{xt}{(1-t)} + \frac{x^2t^2}{2!(1-t)^2} - \ldots + \frac{(-1)^k x^k t^k}{k!(1-t)^k} + \ldots\right] = \sum_{k=0}^{\infty} - \frac{(-1)^k x^k t^k}{k!(1-t)^{k+1}}$$

$$= \sum_{k=0}^{\infty} - \frac{(-1)^k x^k t^k}{k!}(1-t)^{-(k+1)}$$

$$= \sum_{k=0}^{\infty} \frac{(-1)^k x^k}{k!} t^k \left[1 + (k+1)t + \frac{(k+1)(k+2)}{2!} t^2 + ... + \frac{(k+1)(k+2)...(k+1)}{l!} t^l + ... \right]$$

$$= \sum_{k=0}^{\infty} \sum_{l=0}^{\infty} \frac{(-1)^k (k+1) l}{k! l!} x^k t^{k+1} \text{ ,where } (k+1)_l = \frac{\Gamma(k+1+l)}{\Gamma(k+1)}$$

Equating the coefficients of t^n on both sides, we get
On putting $l = n - k$, we get the coefficient of x^k.

$$\frac{L_n(x)}{n!} = \sum_{k=0}^{\infty} \frac{(-1)^k (k+1)_{n-k}}{r!(n-k)!} x^k$$

Here $$(k+1)_{n-k} = \frac{\Gamma(k+1+n-k)}{\Gamma(k+1)} + \frac{\Gamma(n+1)}{\Gamma(k+1)} = \frac{n!}{k!}$$

$$\frac{(-1)^k}{(n-k)!} = \frac{(-1)^k n(n-1)...(n-k+1)}{n!}$$

$$= \frac{(-n)(-n+1)(-n+2)...(-n+k-1)}{n!} = \frac{(-n)_k}{n!}$$

$$L_n(x) = n! \sum_{k=0}^{\infty} \frac{(-n)_k}{n!} . \frac{n!}{(k!)^2} x^k = n! \sum_{k=0}^{\infty} \frac{(-n)_k}{(k!)^2} x^k$$

$$= n! \left[1 + \frac{(-n)}{1!1!} x + \frac{(-n)(-n+1)}{2!\,2!} x^2 + \frac{(-n)(-n+1)(-n+2)}{3!\,3!} x^3 + ... \right]$$

$$= n! F(-n, 1; x)$$

From which it follows that $L_n(x)$ is a polynomial of degree n in x and that the coefficient of x^n is $(-1)^n$.

36.4 RECURRENCE RELATION

Relation I : We know that

$$e^{\frac{-xt}{1-t}} = (1-t) \sum_{n=0}^{\infty} \frac{L_n(x)}{n!} t^n \qquad ....(1)$$

Differentiating (1) *w.r.t* 't',we get

$$-\frac{x}{(1-t)^2} e^{\frac{-xt}{1-t}} = (1-t) \sum_{n=0}^{\infty} \frac{L_n(x) t^{n-1}}{(n-1)!} - \sum_{n=0}^{\infty} \frac{L_n(x) t^n}{n!}$$

$$\Rightarrow \quad -\frac{x}{(1-t)^2} (1-t) \sum_{n=0}^{\infty} \frac{L_n(x) t^n}{n!} = (1-t) \sum_{n=0}^{\infty} \frac{L_n(x) t^{n-1}}{(n-1)!} - \sum_{n=0}^{\infty} \frac{L_n(x) t^n}{n!}$$

$$\Rightarrow \quad x \sum_{n=0}^{\infty} \frac{L_n(x) t^n}{n!} + (1-t)^2 \sum_{n=0}^{\infty} \frac{L_n(x) t^{n-1}}{(n-1)!} - (1-t) \sum_{n=0}^{\infty} \frac{L_n(x) t^n}{n!} = 0$$

Equating to zero the coefficient of t^n, we get

$$x\frac{L_n(x)}{n!}+\frac{L_{n+1}(x)}{n!}-2\frac{L_n(x)}{(n-1)!}+\frac{L_{n-1}(x)}{(n-2)!}-\frac{L_n(x)}{n!}+\frac{L_{n-1}(x)}{(n-1)!}=0$$

$\Rightarrow \quad L_{n-1}(x) + (x - 2n - 1)\, L_n(x) + n^2 L_{n-1}(x) = 0$

Relation II : We know that

$$e^{\frac{-xt}{1-t}} = (1-t)\sum_{n=0}^{\infty}\frac{L_n(x)}{n!}t^n \qquad \text{.......(1)}$$

On differentiating (1) w.r.t. x, we get

$$-\left(\frac{t}{1-t}\right)e^{\frac{-xt}{1-t}} = (1-t)\sum_{n=0}^{\infty}\frac{L'_n(x)t^n}{n!}$$

$$\Rightarrow \quad -\left(\frac{t}{1-t}\right)(1-t)\sum_{n=0}^{\infty}\frac{L_n(x)}{n!}t^n = (1-t)\sum_{n=0}^{\infty}\frac{L'_n(x)}{n!}t^n$$

$$\Rightarrow \quad t\sum_{n=0}^{\infty}\frac{L_n(x)t^n}{n!} + (1-t)\sum_{n=0}^{\infty}\frac{L'_n(x)t^n}{n!} = 0$$

Equating to zero the coefficient of t^n, we get

$$\frac{L'_n(x)}{n!}-\frac{L'_{n-1}(x)}{(n-1)!}+\frac{L_n(x)}{(n-1)!}=0$$

i.e. $\quad L'_n(x) - n\, L'_{n-1}(x) + n\, L_{n-1}(x) = 0$

Relation III : By recurrence relation (1), we have

$$L_{n+1}(x) + (x - 2n - 1)\, L_n(x) + x^2 L_{n-1}(x) = 0 \qquad \text{.....(1)}$$

On differentiating (1) w.r.t 'x', we get

$$L'_{n+1}(x) - (x - 2n - 1)\ L'_n(x) + \ L_n(x) + n^2 L'_{n-1}(x) = 0. \qquad \text{.......(2)}$$

Differentiating (2) w.r.t 'x', we get

$$L''_{n+1}(x) + (x - 2n - 1)\ L''_n(x) + \ 2L'_n(x) + n^2 L''_{n-1}(x) = 0.$$

Let us replace n by $(n + 1)$, we get

$$L''_{n+2}(x) + (x - 2n - 3)\ L''_{n-1}(x) + (n + 1)^2 L_n''(x) + 2L'_{n+1}(x) = 0. \qquad \text{.......(3)}$$

From recurrence relation (2), we have

$$L_n'(x) = \text{n}\ \{L'_{n-1}(x) - L_{n-1}(x)\}$$

On replacing n by $(n + 1)$, we get

$$L'_{n+1}(x) + (n + 1)\ \{L'_n(x) - L_n(x)\} \qquad \text{.....(4)}$$

On differentiating (4), we get

$$L''_{n+1}(x) = (n + 1)\ \{L_n''(x) - L_n'(x)\}. \qquad \text{.......(5)}$$

Again replacing n by $(n + 1)$, we get

$$L'_{n+2}(x) = (n + 2)\ \{L''_{n+1}(x) - L'_{n+1}(x)\} \qquad \text{.......(6)}$$

From (3),

$$(n+2)\{L''_{n+1}(x) - L'_{n+1}(x)\} + (x-2n-3)L''_{n+1}(x) + (n+1)^2 L_n''(x) + 2L'_{n+1}(x) = 0$$

$$(x-n-1)\, L''_{n+1}(x) - n\, L'_{n+1}(x) + (n+1)^2 L_n''(x) = 0$$

Eliminating $L''_{n+1}(x)$ and $L'_{n+1}(x)$ by (4) and (5), we get

$$x\, L''_n(x) + (1-x)\, L'_n(x) + n\, L_n(x) = 0$$

So, $t\, y = A\, L_n(x)$ is a solution of Laguerre equation.

Relation IV : We know that

$$(1-t)^{-1} e^{\left(1-\frac{1}{1-t}\right)x} = \sum_{n=0}^{\infty} \frac{L_n(x)t^n}{n!} \qquad \text{........(1)}$$

Differentiating (1) w.r.t 't' n times by Leibnitz theorem, we have

$$e^x \frac{d^n}{dt^n}\left[(1-t)^{-1} e^{-\frac{x}{1-t}}\right] = L_n(x) + L_{n+1}(x) + \ldots\ldots$$

Now, $$\frac{d}{dt}\left[(1-t)^{-1} e^{-\frac{x}{1-t}}\right] = \frac{1-x-t}{(1-t)^3} e^{-\frac{x}{1-t}}$$

Taking limit when $t \to 0$, we get

$$\lim_{t\to 0} \frac{d}{dt}\left[(1-t)^{-1} e^{-\frac{x}{1-t}}\right] = (1-x)e^{-x}\frac{d}{dx}\left(xe^{-x}\right)$$

Similarly

$$\lim_{t\to 0} \frac{d^2}{dt^2}\left[(1-t)^{-1} e^{-\frac{x}{1-t}}\right] = \frac{d^2}{dx^2}\left(x^2 e^{-x}\right)$$

$$\Rightarrow \qquad \lim_{t\to 0} \frac{d^n}{dt^n}\left[(1-t)^{-1} e^{-\frac{x}{1-t}}\right] = \frac{d^n}{dx^n}\left(x^n e^{-x}\right)$$

Hence, proceeding to the limit as $t \to 0$, we get

$$e^x \frac{d^n}{dx^n}\left(x^n e^{-x}\right) = L_n(x)$$

36.5 ORTHOGONAL PROPERTY

Let $$f_n(x) = \frac{1}{n!} e^{-x/2} L_n(x) \qquad \text{........(1)}$$

$$\int_0^{\infty} f_m(x)\, f_n(x)\, dx = \int_0^{\infty} e^{-x} \frac{L_m(x)}{m!} \frac{L_n(x)}{n!}\, dx = \delta_{m,n}$$

Over the interval $0 \le x \le \infty$ 1 when $\delta_{m,n} = \begin{cases} 0 \text{ for } m \ne n \\ 1 \text{ for } m = n \end{cases}$

From recurrence relation (4), we know that

$$L_n(x) = e^x \frac{d^n}{dx^n}\left(x^n e^{-x}\right)$$

So, we get

$$\int_0^\infty e^{-x}x^m L_n(x)\,dx = \int_0^\infty e^{-x}x^m e^x \frac{d^n}{dx^n}\left(x^n e^{-x}\right)dx = \int_0^\infty x^m \frac{d^n}{dx^n}\left(x^n e^{-x}\right)dx$$

Integrating the R.H.S. by parts, we get

$$\int_0^\infty e^{-x}x^m L_n(x)\,dx = \left[x^m \frac{d^{n-1}}{dx^{n-1}} x^n e^{-x}\right]_0^\infty - \int_0^\infty mx^{m-1}\frac{d^{n-1}}{dx^{n-1}}\left(x^n e^{-x}\right)dx$$

$$= (-1)m\int_0^\infty e^{m-1}\frac{d^{n-1}}{dx^{n-1}}\left(x^n e^{-x}\right)dx$$

$$= (-1)^2 m(m-1)\int_0^\infty e^{m-2}\frac{d^{n-2}}{dx^{n-2}}\left(x^n e^{-x}\right)dx$$

$$= \ldots\ldots\ldots\ldots\ldots\ldots\ldots\ldots\ldots\ldots$$

$$= (-1)^n m!\int_0^\infty \frac{d^{n-m}}{dx^{n-m}}\left(x^n e^{-x}\right)dx$$

$$= 0 \text{ if } n > m.$$

Replacing n by m, we get $(m < n)$

$$\int_0^\infty e^{-x}x^n L_m\,dx = 0 \text{ for } m < n$$

$$\Rightarrow \quad \int_0^\infty e^{-x}L_m(x)\,L_n(x)\,dx = 0, \text{ if } m \neq n$$

$$\Rightarrow \quad \int_0^\infty e^{-x}\frac{L_m(x)}{m!}\frac{L_n(x)}{n!}\,dx = 0, \quad \text{if } m \neq n$$

Taking $m = n$, then $L_n(x)$ is $(-1)^n x^n$,

$$\therefore \quad \int_0^\infty e^{-x}\{L_n(x)\}^2\,dx = (-1)^n\int_0^\infty e^{-x}x^n L_n(x)\,dx = (-1)^n\int_0^\infty e^{-x}x^n e^x\frac{d^n}{dx^n}\left(x^n e^{-x}\right)dx$$

$$= (-1)^n\int_0^\infty x^n\frac{d^n}{dx^n}\left(x^n e^{-x}\right)dx = (-1)^{2n}\,n!\int_0^\infty x^n e^{-x}\,dx = (n!)^2$$

$$\Rightarrow \quad \int_0^\infty e^{-x/2}\frac{L_n(x)}{n!}e^{-x/2}\frac{L_n(x)}{n!}\,dx = 1 \qquad \ldots\ldots(3)$$

Combining (2) and (3), we get

$$\int_0^\infty f_m(x)\,f_n(x)\,dx = \int_0^\infty e^{-x/2}\frac{L_m(x)}{m!}e^{-x/2}\frac{L_n(x)}{n!}\,dx = \delta_{m,n}$$

EXERCISE 36.1

Prove the following :

1. $L_3^1(x) = -18 + 18x - 3x^2$ **2.** $L_4^2(x) = 144 - 96x + 12x^2$

3. $L_4^4(x) = 24.$

4. Find a series solution of

$xy'' + (1 + x)\,y' + y = 0.$ **Ans.** $y = e^{-x}L_0(x) = e^{-x}$

Chebyshev Polynomials

37.1 INTRODUCTION

In numerical analysis, main problem is approximating a function. For better approximation of a function the error should be minimum. To make the error minimum we use the least squares method. On the other hand, we may choose the approximation such that maximum component of error is minimised. This introduces Chebyshev polynomial. We apply Chebyshev polynomial in the economization of power series.

37.2 CHEBYSHEV POLYNOMIALS (Tchebcheff or Tschebyscheff polynomials)

The Chebyshev polynomials of first kind

$$T_n(x) = \cos(n\cos^{-1} x)$$

and the second kind

$$U_n(x) = \sin(n\cos^{-1} x)$$

where n is non-negative integer

37.3 Chebyshev Equation

$$(1-x^2)\frac{d^2y}{dx^2} - x\frac{dy}{dx} + n^2y = 0 \qquad \dots (1)$$

Prove that $T_n(x)$ and $U_n(x)$ are tne independent solutions of the Chebyshev equation.

Proof.

Let $\quad y = T_n(x) = \cos(n\cos^{-1} x) = \cos n\theta$

$$\frac{dy}{dx} = \frac{dy}{d\theta}\frac{d\theta}{dx} = -\frac{n\sin n\theta}{-\sin\theta}$$

$$\left\{\cos\theta = x \Rightarrow -\sin\theta\frac{d\theta}{dx} = 1 \text{ or } \frac{d\theta}{dx} = -\frac{1}{\sin\theta}\right\}$$

or $$\frac{dy}{dx} = \frac{n\sin n\theta}{\sin\theta}$$

$$\frac{d^2y}{dx^2} = \frac{\sin\theta\,(n^2\cos n\theta) - n\sin n\theta\cdot\cos\theta}{\sin^2\theta}\frac{d\theta}{dx}$$

$$= \frac{n^2\sin\theta\cos n\theta - n\sin n\theta\cdot\cos\theta}{\sin^2\theta}\left(-\frac{1}{\sin\theta}\right)$$

$\Rightarrow$ $$= \frac{-n^2\cos n\theta + \left(\dfrac{n\sin n\theta}{\sin\theta}\right)(\cos\theta)}{\sin^2\theta}$$

$$\Rightarrow \qquad \frac{d^2y}{dx^2} = \frac{-n^2 y + x\dfrac{dy}{dx}}{1-x^2}$$

$$\Rightarrow \qquad (1-x^2)\frac{d^2y}{dx^2} - x\frac{dy}{dx} + n^2 y = 0$$

Which is satisfied by $T_n(x)$.

Similarly we can prove that $U_n = \sin(n\cos^{-1} x)$ is a solution of Chebyshev equation.

Hence $T_n(x)$ and $U_n(x)$ both are the solutions of (1). But $U_n(x)$ can not be expressed as a constant multiple of $T_n(x)$ as shown below :

$$T_n(1) = \cos(n\cos^{-1} 1) = \cos n(0) = \cos 0 = 1$$

$$U_n(1) = \sin(n\cos^{-1} 1) = \sin n0 = 0$$

Thus $T_n(x)$ and $U_n(x)$ are independent solutions of Chebyhev's equation.

37.4 ORTHOGONAL PROPERTIES OF CHEBYSHEV POLYNOMIALS.

$$\int_{-1}^{1} \frac{T_m(x)T_n(x)}{\sqrt{1-x^2}}\,dx = \begin{cases} 0, & m \neq n \neq 0 \\ \dfrac{\pi}{2} & m = n \neq 0 \\ \pi & m = n = 0 \end{cases}$$

Proof. We know that

$$T_m(x) = \cos(m\cos^{-1} x) = \cos m\theta \quad \left\{ \begin{aligned} \theta &= \cos^{-1} x \\ x &= \cos\theta \\ dx &= -\sin\theta\, d\theta \end{aligned} \right.$$

$$T_n(x) = \cos(n\cos^{-1} x) = \cos n\theta$$

$\therefore$ (a) If $m \neq n \neq 0$ $\displaystyle\int_{-1}^{1} \frac{T_m(x).T_n(x)}{\sqrt{1-x^2}}dx = \int_{\pi}^{0} \frac{\cos m\theta \cos n\theta}{\sin\theta}(-\sin\theta\, d\theta)$

$$= \int_0^{\pi} \cos m\theta \cos n\theta\, d\theta$$

$$= \frac{1}{2}\int_0^{\pi} [(\cos(m+n)\theta + \cos(m-n)\theta)]\, d\theta$$

$$= \frac{1}{2}\left[\frac{\sin(m+n)\theta}{m+n} + \frac{\sin(m-n)\theta}{m-n}\right]_0^{\pi}$$

$$= 0 \qquad\qquad m \neq n$$

(b) If $m = n \neq 0$

$$\int_{-1}^{+1} \frac{T_n(x).T_n(x)dx}{\sqrt{1-x^2}} = \int_{\pi}^{0} \frac{\cos n\theta . \cos n\theta}{\sin\theta}(-\sin\theta\, d\theta)$$

$$= \int_0^{\pi} \cos^2 n\theta\, d\theta = \frac{1}{2}\int_0^{\pi} (\cos 2\theta + 1)\, d\theta$$

$$= \frac{1}{2}\left[\frac{\sin 2\theta}{2} + \theta\right]_0^\pi = \frac{\pi}{2}$$

(c) If $m = n = 0$

$$\int_{-1}^{+1} \frac{T_0(x)\, T_0(x)}{\sqrt{1-x^2}}\, dx = \int_\pi^0 \frac{(1)(1)}{\sin\theta}(-\sin\theta\, d\theta) = \int_0^\pi d\theta = (\theta)_0^\pi = \pi$$

Note: (1) Similarly we can prove that $\int_{-1}^{+1} \frac{U_m(x) U_n(x)}{\sqrt{1-x^2}} dx = \begin{cases} 0, & m \neq n \\ \frac{\pi}{2}, & m = n \neq 0 \\ \pi, & m = n = 0 \end{cases}$

(2) The polynomials $T_n(x)$ are orthogonal with the function $\frac{1}{\sqrt{1-x^2}}$.

Example 1. *Prove that (a)* $T_{-n}(x) = T_n(x)$

(b) $T_0(x) = 1$ (c) $T_1(x) = x$

Solution. The Chebyshev polynomial of degree n over the integral $[-1, 1]$ is defined as

$$T_n(x) = \cos(n \cos^{-1} x) \quad \text{... (1)}$$

On putting $-n$ for n in (1) we get

$$T_{-n} = \cos(-n \cos^{-1} x)$$
$$= \cos(n \cos^{-1} x)$$
$$= T_n$$

$\Rightarrow$ $T_n = T_{-n}$

(b) Let $\cos^{-1} x = \theta$ so that $x = \cos\theta$

On putting $\cos^{-1} x = \theta$ in (1), it becomes

$$T_n(x) = \cos n\theta \quad \text{... (2)}$$
$$T_0(x) = \cos 0 = 1$$

On putting $n = 0$ in (1), we get $T(n) = \cos\theta = 1$

(c) If $n = 1$ On putting $n = 1$ in (1) we get, $T_1(x) = \cos\theta = x$

37.5 RECURRENCE RELATION OF CHEBYSHEV POLYNOMIALS

(I) **Formula I** $T_{n+1}(x) = 2x\, T_n(x) - T_{n-1}(x)$

$$\cos(n-1)\theta + \cos(n+1)\theta = 2\cos n\theta . \cos\theta \quad \text{... (1) (Trigonometric identity)}$$

On putting the values of $\cos(n-1)\theta$, $\cos(n+1)\theta$, $\cos n\theta$, $\cos\theta$ in (1) we get

$$T_{n-1}(x) + T_{n+1}(x) = 2x T_n(x)$$

$\Rightarrow$ $$T_{n+1}(x) = 2x\, T_n(x) - T_{n-1}(x) \quad \text{... (2)}$$

This is the required recurrence relation of Chebyshev polynomials

Similarly we can prove that $U_{n+1}(x) = 2xU_n(x) - U_{n-1}(x)$

On substituting $n = 1$, in (2) we have

$$T_2(x) = 2x\, T_1(x) - T_0(x)$$
$$= 2x(x) - 1 = 2x^2 - 1$$

If $n = 2$, $T_3(x) = 2x\, T_2(x) - T_1(x) = 2x(2x^2 - 1) - x$
$$= 4x^3 - 3x$$

If $n = 3$, $\quad T_4(x) = 2x\,T_3(x) - T_2(x)$

$$= 2x(4x^3 - 3x) - (2x^2 - 1)$$

$$= 8x^4 - 8x^2 + 1$$

In this way

$$T_0(x) = 1$$

$$T_1(x) = x$$

$$T_2(x) = 2x^2 - 1$$

$$T_3(x) = 4x^3 - 3x$$

$$T_4(x) = 8x^4 - 8x^2 + 1$$

$$T_5(x) = 16x^5 - 20x^3 + 5x$$

$$T_6(x) = 32\,x^6 - 48\,x^4 + 18\,x^2 - 1$$

Here the coefficient of x^n in $T_n(x)$ is always 2^{n-1}.

Graph of Chebyshev polynomials.

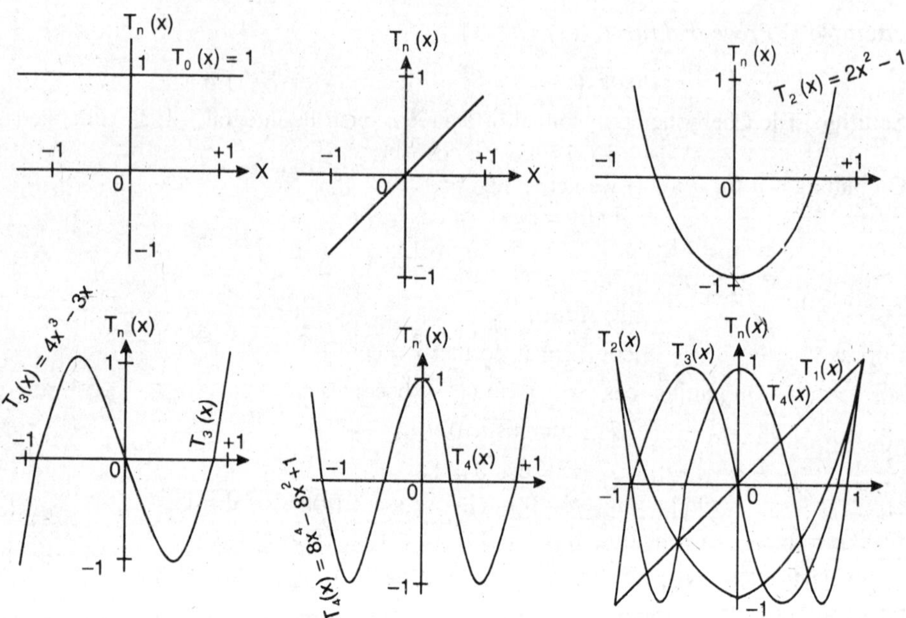

37.6 POWERS OF X IN TERMS OF $T_n(x)$

$$1 = T_0(x)$$

$$x = T_1(x)$$

$$x^2 = \frac{1}{2}[T_0(x) + T_2(x)]$$

$$x^3 = \frac{1}{4}[3T_1(x) + T_3(x)]$$

$$x^4 = \frac{1}{8}[3T_0(x) + 4T_2(x) + T_4(x)]$$

$$x^5 = \frac{1}{16}[10T_1(x) + 5T_3(x) + T_5(x)]$$

$$x^6 = \frac{1}{32}[10T_0(x) + 15T_2(x) + 6T_4(x) + T_6(x)]$$

and so on.

Recurrence Formula II

$$(1-x^2)\,T_n'(x) = -nx\,T_n(x) + nT_{n-1}(x)$$

Solution. We know that

$$T_n(x) = \cos(n\cos^{-1}x) = \cos n\theta \qquad \theta = \cos^{-1}x = x = \cos\theta$$

$$T_n^1(x) = -\sin n\theta\left(n\frac{d\theta}{dx}\right) = \frac{-\sin n\theta}{1}\left(-\frac{n}{\sin\theta}\right) = \frac{n\sin n\theta}{\sin\theta} \qquad 1 = -\sin\theta\frac{d\theta}{dx}$$

$$\frac{d\theta}{dx} = -\frac{1}{\sin\theta}$$

Multiplying both sides by $(1 - x^2)$ we get

$$(1-x^2)T_n'(x) = (1-x^2)\frac{n\sin n\theta}{\sin\theta} = \frac{(l-\cos^2\theta)n\sin n\theta}{\sin\theta} = \frac{(\sin^2\theta)(n\sin n\theta)}{\sin\theta} = n\sin n\theta\sin\theta \qquad \dots (1)$$

$$-nxT_n(x) + nT_{n-1}(x) = -n\cos\theta\cos n\theta + n\cos(n-1)\theta$$
$$= n[-\cos\theta\cos n\theta + \cos(n\theta-\theta)]$$
$$= n[-\cos\theta\cos n\theta + \cos n\theta\cos\theta + \sin n\theta\sin\theta]$$
$$= n\sin n\theta.\sin\theta \qquad \dots (2)$$

From (1) and (2)

$$(1-x^2)T_n'(x) = -nxT_n(x) + nT_{n-1}(x)$$ **Proved.**

Example 2. *Prove that*

$$[T_n(x)]^2 - T_{n+1}(x)T_{n-1}(x) = 1 - x^2$$

Solution. Let $\cos^{-1}x = \theta \Rightarrow x = \cos\theta$

$$[T_n(x)]^2 - T_{n+1}(x)T_{n-1}(x) = [\cos(n\cos^{-1}x)]^2 - [\cos\{(n+1)\cos^{-1}x\}][\cos\{(n-1)\cos^{-1}x\}]$$
$$= \cos^2 n\theta - \cos(n+1)\theta\cos(n-1)\theta$$
$$= \cos^2 n\theta - \frac{1}{2}[\cos 2n\theta + \cos 2\theta]$$
$$= \cos^2 n\theta - \frac{1}{2}[2\cos^2 n\theta - 1 + 2\cos^2 - 1]$$
$$= \cos^2 n\theta - \cos^2 n\theta + \frac{1}{2} - \cos^2\theta + \frac{1}{2}$$
$$= 1 - \cos^2\theta = 1 - x^2$$ **Proved**

Example 3. *Prove that*

$$T_n(x) = \frac{1}{2}\left[\left\{x + i\sqrt{(1-x^2)}\right\}^n + \left\{x - i\sqrt{(1-x^2)}\right\}^n\right]$$

Solution. We know that

$$T_n(x) = \cos(n\cos^{-1}x) = \cos n\theta \qquad \dots (1)$$

$$\theta = \cos^{-1} x \Rightarrow x = \cos\theta$$

$$\text{R.H.S.} = \frac{1}{2}[\{x + i\sqrt{(1-x^2)}\}^n + \{x - i\sqrt{(1-x^2)}\}^n]$$

$$= \frac{1}{2}[\{\cos\theta + i\sqrt{(1-\cos^2\theta)}\}^n + \{\cos\theta - i\sqrt{(1-\cos^2\theta)}\}^n]$$

$$= \frac{1}{2}[\{\cos\theta + i\sin\theta\}^n + \{\cos\theta - i\sin\theta\}^n]$$

$$= \frac{1}{2}[\cos n\theta + i\sin n\theta + \cos n\theta - i\sin n\theta] \quad \textit{(De Moivre's theorem)}$$

$$= \cos n\theta \qquad \ldots (2)$$

From (1) and (2) we have

$$T_n(x) = \frac{1}{2}[\{x + i\sqrt{(1-x^2)}\}^n + \{x - i\sqrt{(1-x^2)}\}^n]$$ **Proved.**

Example 4. *Prove that*

$$U_n(x) = -\frac{1}{2}i\left[\left\{x + i\sqrt{(1-x^2)}\right\}^n - \left\{x - i\sqrt{(1-x^2)}\right\}^n\right]$$

Solution. We know that

$$U_n(x) = \sin(n\cos^{-1} x) = \sin n\theta \qquad \ldots (1) \quad \cos^{-1} x = \theta \Rightarrow x = \cos\theta$$

$$\text{R.H.S.} = -\frac{1}{2}i[\{x + i\sqrt{(1-x^2)}\}^n - \{x - i\sqrt{(1-x^2)}\}^n]$$

$$= \frac{-i}{2}[\{\cos\theta + i\sqrt{(1-\cos^2\theta)}\}^n - \{\cos\theta - i\sqrt{(1-\cos^2\theta)}\}^n]$$

$$= -\frac{i}{2}[\{\cos\theta + i\sin\theta\}^n - \{\cos(-\theta) + i\sin(-\theta)\}^n]$$

$$= -\frac{i}{2}[\{\cos n\theta + i\sin n\theta\} - \{\cos(-n\theta) + i\sin(-n\theta)\}] \quad \text{(De Moivre's Theorem)}$$

$$= \frac{-i}{2}[\cos n\theta + i\sin n\theta - \cos n\theta + i\sin n\theta]$$

$$= \frac{-i}{2}(2i\sin n\theta) = \sin n\theta \qquad \ldots (2)$$

From (1) and (2) we have

$$U_n(x) = -\frac{1}{2}i[\{x + i\sqrt{(1-x^2)}\}^n - \{x - i\sqrt{(1-x^2)}\}^n]$$ **Proved.**

37.7 RECURRENCE FORMULAE FOR $u_n(x)$

(I) $U_{n+1}(x) - 2x\,U_n(x) + U_{n-1}(x) = 0$

Proof $U_n(x) = \sin(n\cos^{-1} x) = \sin n\theta$ $\qquad (x = \cos\theta)$

Now $U_{n+1}(x) - 2x\,U_n(x) + U_{n-1}(x) = \sin(n+1)\theta - 2\cos\theta\sin n\theta + \sin(n-1)\theta,$

$$= \sin(n+1)\theta - [\sin(n+1)\theta + \sin(n-1)\theta] + \sin(n-1)\theta$$

$$= 0$$ **Proved.**

(ii) $(1-x^2)U_n'(x) = -nx\,U_n(x) + nU_{n-1}(x)$ *(U.P., III semestor 2002)*

Proof. We know that

$$U_n(x) = \sin(n\cos^{-1}x) = \sin n\theta$$

$$U_n'(x) = \cos n\theta\left(n\frac{d\theta}{dx}\right) = \cos n\theta\left(\frac{-n}{\sin\theta}\right) = -n\frac{\cos n\theta}{\sin\theta}$$

$$\begin{aligned} &\theta = \cos^{-1}x \\ &x = \cos\theta \\ &1 = -\sin\theta\frac{d\theta}{dx} \\ &\frac{d\theta}{dx} = \frac{-1}{\sin\theta} \end{aligned}$$

Multiplying both sides by $\sqrt{(1-x^2)}$ we get

$$\sqrt{(1-x^2)}\,U_n'(x) = -n\sqrt{(1-x^2)}\frac{\cos n\theta}{\sin\theta} = -n\sqrt{(1-\cos^2\theta)}\frac{\cos n\theta}{\sin\theta}$$

$$= \frac{n\sin^2\theta\cos n\theta}{\sin\theta} = -n\cos n\theta\sin\theta \quad \text{... (1)}$$

$$\begin{aligned} \text{R.H.S.} &= -nxU_n(x) + nU_{n-1}(x) \\ &= -n\cos\theta\sin n\theta + n\sin(n-1)\theta \\ &= -n[\cos\theta\sin n\theta - \sin(n\theta-\theta)] \\ &= -n[\cos\theta\sin n\theta - \sin n\theta\cos\theta + \cos n\theta\sin\theta] \\ &= -n(\cos n\theta\sin\theta) = -n\cos n\theta\sin\theta \quad \text{... (2)} \end{aligned}$$

From (1) and (2) we get

$$(1-x^2)U_n'(x) = -nxU_n(x) + nU_{n-1}(x)$$

Proved.

Example 5. *Show that* $\dfrac{1}{\sqrt{(1-x^2)}}U_n(x)$ *satisfies the differential equation*

$$(1-x^2)\frac{d^2y}{dx^2} - 3x\frac{dy}{dx} + (n^2-1)y = 0$$

Solution. Let $y = \dfrac{1}{\sqrt{(1-x^2)}}U_n(x)$

$$= \frac{\sin(n\cos^{-1}x)}{\sqrt{1-x^2}} = \frac{\sin n\theta}{\sqrt{1-x^2}}$$

$$\left[\begin{aligned} &\cos^{-1}x = \theta \\ &\frac{-1}{\sqrt{1-x^2}} = \frac{d\theta}{dx} \end{aligned}\right]$$

Differentiating both sides we get

$$\frac{dy}{dx} = \frac{\sqrt{1-x^2}\,.\cos n\theta\left(n\frac{d\theta}{dx}\right) - \frac{1}{2}(1-x^2)^{\frac{-1}{2}}(-2x).\sin n\theta}{(1-x^2)}$$

$$\Rightarrow \quad = \frac{n\sqrt{1-x^2}\cos n\theta.\left(\frac{-1}{\sqrt{(1-x^2)}}\right) + \frac{x}{\sqrt{1-x^2}}\sin n\theta}{(1-x^2)}$$

$$\Rightarrow \quad (1-x^2)\frac{dy}{dx} = -n\cos n\theta + \frac{x\sin n\theta}{\sqrt{1-x^2}}$$

$$\Rightarrow \quad (1-x^2)\frac{dy}{dx} = -n\cos n\theta + xy$$

Again differentiating we get

$$(1-x^2)\frac{d^2y}{dx^2}-2x\frac{dy}{dx}=n^2\sin n\theta\frac{d\theta}{dx}+x\frac{dy}{dx}+1.y$$

$$\Rightarrow\quad (1-x^2)\frac{d^2y}{dx^2}-3x\frac{dy}{dx}=-n^2\frac{\sin n\theta}{\sqrt{1-x^2}}+y$$

$$(1-x^2)\frac{d^2y}{dx^2}-3x\frac{dy}{dx}=-n^2y+y$$

$$\Rightarrow\quad (1-x^2)\frac{d^2y}{dx^2}-3\times\frac{dy}{dx}+(n^2-1)\,y=0$$ **Proved.**

Example 6. *Show that*

$$\sqrt{(1-x^2)}\,T_n(x)=U_{n+1}(x)-x\,U_n(x)$$

Solution. Let $\cos^{-1}x=\theta\Rightarrow x=\cos\theta$

$$\text{L.H.S.}=\sqrt{(1-x^2)}\;T_n(x)=\sqrt{(1-\cos^2\theta)}\cos(n\cos^{-1}x)=\sin\theta\cos n\theta$$

$$\begin{aligned}\text{R.H.S.}&=U_{n+1}(x)-xU_n(x)\\&=\sin(n+1)\theta-\cos\theta\sin n\theta\\&=\sin(n\theta+\theta)-\cos\theta\sin n\theta\\&=\sin n\theta\cos\theta+\cos n\theta\sin\theta-\cos\theta\sin n\theta\\&=\cos n\theta.\sin\theta\\&=\text{L.H.S.}\end{aligned}$$ **Proved.**

Example 7. *Show that*

$$\sum_{r=0}^{n}T_{2r}(x)=\frac{1}{2}\left\{1+\frac{1}{\sqrt{(1-x^2)}}U_{2n+1}(x)\right\}$$

Solution. Let $\cos^{-1}x=\theta\quad\Rightarrow\quad x=\cos\theta$

$$\sum_{r=0}^{n}T_{2r}(x)=\sum_{r=0}^{n}\cos(2r\cos^{-1}x)=\sum_{r=0}^{n}\cos(2r\theta)$$

$$=\text{Real part of}\sum_{r=0}^{n}[(\cos(2r\theta)+i\sin(2r\theta)]=\text{Real part of}\sum_{r=0}^{n}e^{i2r\theta}$$

$$=\text{Real part of }[1+e^{i2\theta}+e^{i4\theta}+e^{i6\theta}+\ldots+e^{i2n\theta}]$$

(This is G.P. of $(n+1)$ terms with common ratio $e^{2i\theta}$)

$$=\text{Real part of}\ \frac{1-e^{(i2\theta)(n+1)}}{1-e^{i2\theta}}\qquad\left(\text{sum }=\frac{1-r^{n+1}}{1-r}\right)$$

$$=\text{Real part of}\ \frac{[1-e^{i(2n+2)\theta}][1-e^{-i2\theta}]}{(1-e^{i2\theta})(1-e^{-i2\theta})}$$

$$=\text{Real part of}\ \frac{1-e^{i(2n+2\theta)}-e^{-i2\theta}+e^{i2n\theta}}{1+1-e^{i2\theta}-e^{-i2\theta}}$$

$$= \text{Real part of } \frac{1-\{\cos(2n+2)\theta - i\sin(2n+2)\theta\} - \{\cos 2\theta - i\sin 2\theta\} + \{\cos 2n\theta - i\sin 2n\theta\}}{2-2\cos 2\theta}$$

$$= \frac{1-\cos(2n+2)\theta - \cos 2\theta + \cos 2n\theta}{2(1-\cos 2\theta)} = = \frac{1}{2}\left[\frac{1-\cos 2\theta}{1-\cos\theta} - \frac{\cos(2n+2)}{1-\cos 2\theta}\cos 2n\theta\right]$$

$$= \frac{1}{2}\left[1 - \frac{\cos(2n+2)\theta - \cos 2n\theta}{1-\cos 2\theta}\right] = \frac{1}{2}\left[1 + \frac{2\sin(2n+1)\theta\sin\theta}{2\sin^2\theta}\right]$$

$$= \frac{1}{2}\left[1 + \frac{\sin(2n+1)\theta}{\sin\theta}\right] = \frac{1}{2}\left[1 + \frac{U_{(2n+1)}(x)}{\sqrt{1-x^2}}\right]$$ **Proved.**

Example 8. *Prove that*

$$T_n(x) = \sum_{r=0}^{N} (-1)^r \frac{\lfloor n}{\lfloor 2r \ \lfloor n-2r}(1-x^2)x^{n-2r} \qquad N = \frac{n}{2} \text{ if } n \text{ is even}$$

$$N = \frac{n-1}{2} \text{ if } n \text{ is odd}$$

Solution. We know that $T_n(x) = x^n - {}^nC_2 x^{n-2}(1-x^2) + {}^nC_4 x^{n-4}(1-x^2)^2 + \ldots$

$$T_n(x) = \frac{1}{2}\left[\left\{x + i\sqrt{(1-x^2)}\right\}^n + \left\{x - i\sqrt{(1-x^2)}\right\}^n\right] \qquad \ldots (1)$$

On applying Binomial theorem on (1) we get

$$T_n = \frac{1}{2}\Big[x^n + {}^nC_1 x^{n-1}(i\sqrt{1-x^2}) + \ldots + {}^nC_r x^{n-r}(-i\sqrt{1-x^2})^r \ldots$$

$$+ \{x^n + {}^nC_1 x^{n-1}(-i\sqrt{1-x^2}) + \ldots + {}^nC_r x^{n-r}(-i\sqrt{1-x^2})^r + \ldots\}\Big].$$

$$= \frac{1}{2}\left[\left\{\sum_{r=0}^{n} {}^nC_r x^{n-r}(i\sqrt{1-x^2})^r\right\} + \left\{\sum_{r=0}^{n} {}^nC_r x^{n-r}(-i\sqrt{1-x^2})^r\right\}\right]$$

$$= \frac{1}{2}\sum_{r=0}^{n} {}^nC_r\, x^{n-r}(1-x^2)^{\frac{r}{2}}(i)^r\{1+(-1)^r\} \qquad \ldots (2)$$

(*a*) If r is odd or $r = 2s + 1$ $\qquad r \le n$

$$T_n = \sum_{s=0}^{\frac{n-1}{2}} {}^nC_{2s+1}\, x^{n-2s-1}(i)^{2s+1}[1+(-1)^{2s+1}]$$

$$2s+1 \le n$$

$$T_n = \sum_{s=0}^{\frac{n-1}{2}} {}^nC_{2s+1}\, x^{n-2s-1}(i)^{2s+1}(1-1) = 0 \qquad s \le \frac{n-1}{2}$$

(*b*) If r is even or $r = 2s$

$$T_n = \frac{1}{2}\sum_{s=0}^{\frac{n}{2}} {}^nC_{2s}\, x^{n-2s}(1-x^2)^s(i)^{2s}\{1+(-1)^{2s}\} \qquad \left[\begin{array}{c} r \le n \\ 2s \le n \\ s \le \frac{n}{2} \\ \text{if } n \text{ is even}\end{array}\right.$$

$$= \frac{1}{2}\sum_{s=0}^{\frac{n}{2}} {}^nC_{2s}\, x^{n-2s}(1-x^2)^s(-1)^s\{1+1\}$$

$$= \sum_{s=0}^{\frac{n}{2}} (-1)^s\, {}^nC_{2s}\, x^{n-2s}(1-x^2)^s$$

$$= \sum_{s=0}^{\frac{n}{2}} (-1)^s \frac{n!}{(n-2s)!\,(2s)!}(1-x^2)^s x^{n-2s}$$

(*ii*) If n is odd or $n \equiv 2s + 1$ $\qquad s = \frac{n-1}{2}$

$$T_n = \sum_{s=0}^{\frac{n-1}{2}} (-1)^s \frac{n!}{(n-2s)!\,(2s)!}(1-x^2)^s\, x^{n-2s}$$

Hence $T_n = x^n - {}^nC_2\, x^{n-2}(1-x^2) + {}^nC_4\, x^{n-4}(1-x^2)^2 + \ldots$ **Proved.**

Example 9. *Find the value of the following:*

(*i*) $U_0(x)$ (*ii*) $U_1(x)$ (*iii*) $U_2(x)$ (*iv*) $U_3(x)$

Solution. We know that

$$U_n(x) = \sin(n\cos^{-1}x)$$

1, $\sqrt{1-x^2}$, θ, x

If $n = 0$, $\qquad U_0(x) = \sin(0) = 0$

If $n = 1$, $\qquad U_1(x) = \sin(\cos^{-1}x) = \sin\sin^{-1}\sqrt{(1-x^2)} = \sqrt{(1-x^2)}$

$\left[\begin{array}{l}\theta = \cos^{-1}x \\ \cos\theta = x\end{array}\right.$

If $n = 2$, $U_2(x) = \sin(2\cos^{-1}x) = \sin 2\theta = 2\sin\theta\cos\theta = 2x\sqrt{1-x^2}$

If $n = 3$, $U_3(x) = \sin(3\cos^{-1}x) = \sin 3\theta = 3\sin\theta - 4\sin^3\theta$

$$= 3\sqrt{1-x^2} - 4(\sqrt{1-x^2})^3 = \sqrt{1-x^2}\,[3-4(1-x^2)] = \sqrt{1-x^2}\,[4x^2-1]$$

Here we see that $U_n(x)$ is not polynomial.

But if we define

$$U_n(x) = \frac{\sin\{(n+1)\cos^{-1}x\}}{\sin(\cos^{-1}x)} = \frac{U_{n+1}}{\sqrt{1-x^2}}$$

Then $U_n(x)$ is a polynomial in x of degree n.

37.8 GENERATING FUNCTION FOR $T_n(x)$

$$\frac{1-t^2}{1-2tx+t^2} = T_0(x) + 2\sum_{n=1}^{\infty} T_n(x).t^n$$

Proof L.H.S. $= \frac{1-t^2}{1-2tx+t^2}$ $\qquad$ Put $\quad x = \cos\theta = \frac{e^{i\theta}+e^{-i\theta}}{2}$

$$= \frac{1-t^2}{1-t(e^{i\theta}+e^{-i\theta})+t^2} = \frac{1-t^2}{(1-te^{i\theta})-te^{-i\theta}+t^2} = \frac{1-t^2}{(1-te^{i\theta})-te^{-i\theta}(1-te^{i\theta})}$$

$$= \frac{1-t^2}{(1-te^{i\theta})(1-te^{-i\theta})} = (1-t^2)(1-te^{i\theta})^{-1}(1-te^{-i\theta})^{-1}$$

$$= (1-t^2)(1+te^{i\theta}+t^2e^{2i\theta}+...+t^r e^{ri\theta}+...)[1+te^{-i\theta}+t^2e^{-2i\theta}+...+t^s e^{-si\theta}+...]$$

$$= (1-t^2)\sum_{r=0}^{\infty} t^r e^{ri\theta} \sum_{s=0}^{\infty} t^s e^{-si\theta} = (1-t^2)\sum_{\substack{r=0\\s=0}}^{\infty} t^{r+s} e^{(r-s)i\theta}$$

$$= \sum_{\substack{r=0\\s=0}}^{\infty} i^{i(r-s)\theta} t^{r+s} - \sum_{\substack{r=0\\s=0}}^{\infty} i^{i(r-s)\theta} t^{r+s+2} \qquad ... (1)$$

On putting $r = s = 0$ in the first summation of (1) we get coefficient of $t_0 = e^{i(0-0)0} = e^0 = 1 = T_0(x)$.

On putting $r + s = n$ or $s = n - r$ in the first summation and $r + s + 2 = n$ or $s = n - r - 2$ in the second summation, we get

$$\text{Coefficient of } t^n = \sum_{r=0}^{n} e^{i[r-(n-r)]\theta} - \sum_{r=0}^{n-2} e^{i[r-(n-r-2)]\theta}$$

$$= \sum_{r=0}^{n} e^{i(-n+2r)\theta} - \sum_{r=0}^{n-2} e^{i[-n+2r+2]\theta}$$

$$= e^{-in\theta}\sum_{r=0}^{n} e^{i(2r)\theta} - e^{-i(n-2)\theta}\sum_{r=0}^{n-2} e^{i(2r)\theta}$$

$$= e^{-in\theta}[1+e^{2i\theta}+e^{4i\theta}+...\text{to }(n+1)\text{ terms}] - e^{-i(n-2)\theta}[1+e^{2i\theta}+e^{4i\theta}+...\text{to }(n-1)\text{ terms}]$$

$$= e^{-in\theta}\frac{1-(e^{2i\theta})^{n+1}}{1-e^{2i\theta}} - e^{-i(n-2)\theta}\frac{1-(e^{2i\theta})^{n-1}}{1-e^{2i\theta}} \qquad \text{[Sum of G.P.]}$$

$$= \frac{e^{-in\theta}-e^{i(n+2)\theta}}{1-e^{2i\theta}} - \frac{e^{-i(n-2)\theta}-e^{in\theta}}{1-e^{2i\theta}}$$

$$= \frac{e^{-in\theta}-e^{i(n+2)\theta}-e^{-i(n-2)\theta}+e^{in\theta}}{1-e^{2i\theta}} = \frac{e^{-in\theta}(1-e^{2i\theta})+e^{in\theta}(1-e^{2i\theta})}{1-e^{2i\theta}}$$

$$= \frac{(1-e^{2i\theta})(e^{in\theta}+e^{-in\theta})}{1-e^{2i\theta}} = (e^{in\theta}+e^{-in\theta}) = 2\cos n\theta = 2T_n(x)$$

or $$\frac{1-t^2}{1-2tx+t^2} = T_0(x) + 2\sum_{n=0}^{\infty} T_n(x)\, t^n$$ **Proved.**

Example 10. *Show that*

(*i*) $T_n(-1) = (-1)^n$ (*ii*) $T_{2n}(0) = (-1)^n$ (*iii*) $T_{2n+1}(0) = 0$

Solution. We know that

$$T_n(x) = \cos(n\cos^{-1}x) \qquad ... (1)$$

(*i*) Replacing x by -1 in (1) we get

$$T_n(-1) = \cos[n\cos^{-1}(-1)] = \cos[n\pi] = (-1)^n$$

(*ii*) Replacing x by 0 and n by $2n$ in (1) we get

$$T_{2n}(0) = \cos[2n\cos^{-1}(0)] = \cos\left(2n\frac{\pi}{2}\right) = \cos n\pi = (-1)^n$$

(*iii*) Replacing x by 0 and n by $2n + 1$, we get

$$T_{2n+1}(0) = \cos[(2n+1)\cos^{-1}(0)] = \cos[(2n+1)\frac{\pi}{2}] = 0$$ **Proved.**

Example 11. *Show that*
(*i*) $U_n(1) = 0$ (*ii*) $U_n(-1) = 0$, (*iii*) $U_{2n}(0) = 0$, (*iv*) $U_{2n+1}(0) = (-1)^n$

Solution. We know that

$$U_n(x) = \sin(n\cos^{-1}x) \qquad \text{... (1)}$$

(*i*) Replacing x by 1 in (1), we get

$$U_n(1) = \sin(n\cos^{-1}1) = \sin(n\times 0) = 0$$

(*ii*) Replacing x by (-1) in (1) we get

$$U_n(-1) = \sin(n\cos^{-1}(-1)] = \sin(n\pi) = 0$$

(*iii*) Replacing x by 0 and n by $2n$ in (1) we get

$$U_{2n}(0) = \sin(2n\cos^{-1}0) = \sin\left(2n\frac{\pi}{2}\right) = \sin(n\pi) = 0$$

(*iv*) Replacing x by 0 and n by $2n + 1$, in (1) we get

$$U_{2n+1}(0) = \sin[(2n+1)\cos^{-1}0] = \sin(2n+1)\frac{\pi}{2} = \sin\left(n\pi + \frac{\pi}{2}\right) = (-1)^n$$ **Proved.**

Example 12. *Show that*

$$2\{T_n(x)\}^2 = 1 + T_{2n}(x)$$

Solution.

$$2[T_n(x)]^2 = 2[\cos(n\cos^{-1}x)]^2 \qquad \left[\begin{matrix}\cos^{-1}x = 0\\ \Rightarrow x = \cos\theta\end{matrix}\right]$$

$$= 2[\cos n\theta]^2$$
$$= \cos 2n\theta + 1$$
$$= \cos[2n\cos^{-1}x] + 1$$
$$= T_{2n}(x) + 1$$ **Proved.**

Example 13. *Show that*

$$T_{m+n}(x) + T_{m-n}(x) = 2T_m(x)T_n(x)$$

Solution. $T_{m+n}(x) + T_{m-n}(x) = \cos[(m+n)\cos^{-1}x] + \cos[(m-n)\cos^{-1}x]$

$$= \cos\ [m+n]\theta + \cos(m-n)\theta \qquad (\cos^{-1}x = \theta)$$
$$= 2\cos m\theta\ \cos n\theta$$
$$= 2\cos[m\cos^{-1}x]\cos[n\cos^{-1}x]$$
$$= 2T_m(x).T_n(x)$$ **Proved.**

Example 14. *Show that* $T_n'(x) = \dfrac{n}{\sqrt{1-x^2}}u_n(x)$

Solution. $T_n(x) = \cos(n\cos^{-1} x) = \cos n\theta$ $\qquad \cos^{-1} x = \theta$

$$T_n'(x) = -\sin n\theta . n\frac{d\theta}{dx} \qquad \frac{-1}{\sqrt{1-x^2}} = \frac{d\theta}{dx}$$

$$= -\sin n\theta\left(-\frac{n}{\sqrt{1-x^2}}\right)$$

$$= \frac{n\sin n\theta}{\sqrt{1-x^2}} = \frac{n\sin[n\cos^{-1} x]}{\sqrt{1-x^2}}$$

$$= \frac{n u_n(x)}{\sqrt{1-x^2}}$$ **Ans.**

EXERCISE 37.1

1. Express the following Chebyshev functions :
   $T_2(x) + 2T_1(x) + 2T_0(x)$ into ordinary polynomials. **Ans.** $2x^2 + 2x + 1$
2. Express the polynomial
   $12x^3 + 6x^2 + 4x + 1$
   in Chebyshev polynomical (*a*) first kind $T_n(x)$.
   (*b*) second kind $U_n(x) = \dfrac{\sin\{(n+1)\cos^{-1} x\}}{\sin(\cos^{-1} x)}$

   Ans. (*a*) $3T_3(x) + 3T_2(x) + 13T_1(x) + 4T_0(x)$ (*b*) $\frac{3}{2}U_3(x) + \frac{3}{2}U_2(x) + 5U_1(x) + \frac{5}{2}U_0(x)$
3. Express the polynomial
   $16x^4 + 4x^3 + 2x^2 + 4x + 5$ into the Chebyshev polynomial of first kind.
   Ans. $2T_4(x) + T_3(x) + 9T_2(x) + 7T_1(x) + 12T_0(x)$
4. If $U_n(x) = \dfrac{\sin\{(n+1)\cos^{-1} x\}}{\sin(\cos^{-1} x)}$, show that

   (*a*) $U_n(-x) = (-1)^n U_n(x)$ (*b*) $U_n(1) = n+1$

   (c) $U_n(-1) = (-1)^n(n+1)$
5. Prove that
   $U_{n+1}(x) - 2x\,U_n(x) + U_{n-1}(x) = 0$
6. Prove that
   (*a*) $\int_{-1}^{+1} \sqrt{1-x^2}\, U_m(x)\, U_n(x) dx = 0,\ m \neq n$ (*b*) $\int_{-1}^{+1} \sqrt{1-x^2}\, (U_n(x))^2 dx = \frac{\pi}{2}$
8. Prove that

$$U_n(x) = \sum_{r=0}^{\frac{1}{2}(n-1)} (-1)^r \frac{n!}{(2r+1)!\ (n-2r-1)!} (1-x^2)^{r+\frac{1}{2}} x^{n-2r-1}$$

$$= {}^{n+1}C_1 x^n - {}^{n+1}C_3\, x^{n-2}(1-x^2) + {}^{n+1}C_5 x^{n-4}(1-x^2)^2 .....$$

7. Prove that $T_n(x) = U_n(x) - xU_{n-1}(x)$

GAMMA, BETA FUNCTION

38.1 GAMMA FUNCTION

(*U.P. I Semester Dec. 2007*)

$$\int_0^\infty e^{-x} x^{n-1} dx \qquad (n > 0)$$

is called gamma function of n. It is also written as $\overline{|n} = \int_0^\infty e^{-x} x^{n-1} dx$

$$\boxed{\int_0^\infty e^{-x}\, x^{n-1}\, dx = \overline{|n}}$$

Example 1. *Prove that* $\overline{|1} = 1$

Solution. We know that, $\overline{|n} = \int_0^\infty e^{-x} x^{n-1} dx$

Put $n = 1$, $\overline{|1} = \int_0^\infty e^{-x}\, x^{1-1}\, dx = \int_0^\infty e^{-x} dx = \left[\frac{e^{-x}}{-1}\right]_0^\infty = 1$ **Proved.**

Example 2. *Prove that*

(*i*) $\overline{|n+1} = n\overline{|n}$ (*ii*) $\overline{|n+1} = n!$ (***Reduction formula***)

Solution.

(*i*) We know that, $\overline{|n} = \int_0^\infty x^{n-1} e^{-x} dx$ ...(1)

Integrating by parts, we have

$$\overline{|n} = \left[x^{n-1}\frac{e^{-x}}{-1}\right]_0^\infty - (n-1)\int_0^\infty x^{n-2}\frac{e^{-x}}{-1} dx$$

$$= \lim_{x\to 0}\left\{\left(1 - \frac{x}{1!} + \frac{x^2}{2!} - ... + (-1)^n \frac{x^n}{n!} + ... + \infty\right)x^{n-1}\right\} + (n-1)\int_0^\infty x^{n-2}\, e^{-x}\, dx$$

$$= 0 + (n-1)\int_0^\infty x^{n-2} e^{-x} dx$$

$$\therefore \quad \overline{|n} = (n-1)\overline{|n-1} \qquad ...(2)$$

$$\boxed{\overline{|n+1} = n\overline{|n}}$$ Replacing n by $(n + 1)$ **Proved.**

(*ii*) Replacing n by $n - 1$ in (2), we get

$$\overline{|n-1} = (n-2)\overline{|n-2}$$

Putting the value $\overline{|n-1}$ in (2), we get

$$\overline{|n}=(n-1)(n-2)\overline{|n-2}$$

Similarly, $\overline{|n} = (n-1)(n-2) \ldots\ldots 3.2.1 \overline{|1}$ ...(3)

Putting the value of $\overline{|1}$ in (3), we have

$$\overline{|n} = (n-1)(n-2) \ldots\ldots 3.2.1.1$$

$$\overline{|n} = (n-1)\,!$$

Replacing n by $n+1$, we have $\overline{|n+1} = n\,!$ **Proved.**

Example 3. *Evaluate* $\overline{\left|-\frac{1}{2}\right.}$.

Solution. $\overline{|n+1} = n\overline{|n}$

$$\overline{\left|-\frac{1}{2}+1\right.} = -\frac{1}{2}\overline{\left|-\frac{1}{2}\right.} \Rightarrow \overline{\left|\frac{1}{2}\right.} = -\frac{1}{2}\overline{\left|-\frac{1}{2}\right.} \Rightarrow \sqrt{\pi} = -\frac{1}{2}\overline{\left|-\frac{1}{2}\right.} \Rightarrow \overline{\left|-\frac{1}{2}\right.} = -2\sqrt{\pi}$$ **Ans.**

Example 4. *Evaluate* $\int_0^\infty \sqrt[4]{x}\, e^{-\sqrt{x}}\, dx$

Solution. Let $I = \int_0^\infty x^{1/4}\, e^{-\sqrt{x}}\, dx$ ...(1)

Putting $\sqrt{x} = t \Rightarrow x = t^2$ so that $dx = 2t\, dt$ in (1), we get

$$I = \int_0^\infty t^{1/2} e^{-t}\, 2t\, dt = 2\int_0^\infty t^{3/2} e^{-t} dt = 2\int_0^\infty t^{\frac{5}{2}-1} e^{-t}\, dt$$

$$= 2\overline{\left|\frac{5}{2}\right.}$$ [By definition]

$$= 2\cdot\frac{3}{2}\overline{\left|\frac{3}{2}\right.} = 2\cdot\frac{3}{2}\cdot\frac{1}{2}\overline{\left|\frac{1}{2}\right.} = \frac{3}{2}\sqrt{\pi}$$ **Ans.**

Example 5. *Evaluate* $\int_0^\infty \sqrt{x}\, e^{-\sqrt[3]{x}}\, dx$.

Solution. Let $I = \int_0^\infty \sqrt{x}\, e^{-\sqrt[3]{x}}\, dx$ ...(1)

Putting $\sqrt[3]{x} = t \Rightarrow x = t^3$ so that $dx = 3t^2\, dt$ in (1), we get

$$I = \int_0^\infty t^{3/2} e^{-t}\, 3t^2\, dt = 3\int_0^\infty t^{7/2} e^{-t} dt = 3\int_0^\infty t^{\frac{9}{2}-1} e^{-t}\, dt = 3\overline{\left|\frac{9}{2}\right.} = 3\cdot\frac{7}{2}\cdot\frac{5}{2}\cdot\frac{3}{2}\cdot\frac{1}{2}\overline{\left|\frac{1}{2}\right.} = \frac{315}{16}\sqrt{\pi}$$ **Ans.**

Example 6. *Evaluate* $\int_0^\infty x^{n-1}\, e^{-h^2x^2}\, dx$.

Solution. Let $I = \int_0^\infty x^{n-1}\, e^{-h^2x^2}\, dx$ ...(1)

Putting $t = h^2x^2 \Rightarrow x = \frac{\sqrt{t}}{h}$ so that $dx = \frac{dt}{2h\sqrt{t}}$, we get

$$I = \int_0^\infty \left(\frac{\sqrt{t}}{h}\right)^{n-1} e^{-t}\frac{dt}{2h\sqrt{t}}$$

$$= \frac{1}{2h^n}\int_0^\infty t^{\frac{n-1}{2}} e^{-t}\frac{dt}{\sqrt{t}} = \frac{1}{2h^n}\int_0^\infty t^{\frac{n-2}{2}} e^{-t} dt = \frac{1}{2h^n}\overline{\left|\frac{n}{2}\right.}$$ **Ans.**

Example 7. *Evaluate* $\int_0^\infty \frac{x^a}{a^x}\,dx$, *hence show that* $\int_0^\infty \frac{x^7}{7^x}\,dx = \frac{7!}{(\log 7)^8}$ $(a > 1)$

Solution. Here, we have $\int_0^\infty \frac{x^a}{a^x}\,dx$ ...(1)

Putting $a^x = e^t \Rightarrow x \log a = t \Rightarrow x = \frac{t}{\log a}, \Rightarrow dx = \frac{dt}{\log a}$ in (1), we have

$$\int_0^\infty \frac{x^a}{a^x}\,dx = \int_0^\infty \left(\frac{t}{\log a}\right)^a e^{-t}\,\frac{dt}{\log a} = \frac{1}{(\log a)^{a+1}}\int_0^\infty e^{-t}t^a\,dt = \frac{1}{(\log a)^{a+1}}\int_0^\infty t^{(a+1)-1}\,e^{-t}\,dt$$

$$= \frac{1}{(\log a)^{a+1}}\overline{|a+1}$$

On putting $a = 7$, we get $\int_0^\infty \frac{x^7}{7^x}\,dx = \frac{7!}{(\log 7)^8}$ **Ans.**

38.2 PROVE THAT

$$\int_0^1 x^m (\log x)^n\,dx = \frac{(-1)^n}{(m+1)^{n+1}}\overline{|n+1}$$

Proof : Put $\log x = -t$ so that $x = e^{-t} \Rightarrow dx = -e^{-t}dt$

$\therefore$ $x^m = e^{-mt}$

$(\log x)^n = (-t)^n$

Now, $\int_0^1 x^m (\log x)^n\,dx = \int_\infty^0 e^{-mt}(-t)^n\,(-e^{-t})\,dt = \int_0^\infty (-1)^n e^{-mt-t}t^n\,dt$

Putting $(m+1)\,t = u$ so that $(m+1)\,dt = du$, we get

$$\therefore \quad I = \int_0^\infty (-1)^n e^{-u}.\frac{u^n}{(m+1)^n}\,\frac{du}{(m+1)}$$

$$= \frac{(-1)^n}{(m+1)^{n+1}}\int_0^\infty e^{-u}u^n\,du = \frac{(-1)^n}{(m+1)^{n+1}}\int_0^\infty e^{-u}\,u^{(n+1)-1}\,du = \frac{(-1)^n}{(m+1)^{n+1}}\overline{|n+1}$$ **Proved.**

Example 8. *Prove that* $\int_0^1 (x\log x)^4\,dx = \frac{4!}{5^5}$ *(M.U. II Semester, 2009)*

Solution. We know that

$$\int_0^1 x^m (\log x)^n\,dx = \frac{(-1)^n}{(m+1)^{n+1}}\cdot\overline{|n+1} \qquad \text{...(1) [From Art 38.2]}$$

Now, $\int_0^1 (x\log x)^4\,dx = \int_0^1 x^4 (\log x)^4\,dx$

Putting $m = n = 4$ in (1), we get

$$\int_0^1 x^4 (\log x)^4\,dx = \frac{(-1)^4}{(4+1)^{4+1}}\overline{|4+1} = \frac{\overline{|5}}{5^5} = \frac{4!}{5^5}$$ **Proved.**

Example 9. *Evaluate* $\int_0^1 \frac{dx}{\sqrt{-\log x}}$

Solution. Let $-\log x = y \Rightarrow \log x = -y \Rightarrow e^{-y} = x$ so that $dx = -e^{-y}\,dy$

$$\int_0^1 \frac{dx}{\sqrt{-\log x}} = \int_0^1 \frac{-e^{-y}\,dy}{\sqrt{y}} = \int_0^\infty y^{-\frac{1}{2}}\,e^{-y}\,dy = \overline{\left|\tfrac{1}{2}\right.} = \sqrt{\pi}$$ **Ans.**

Example 10. *Evaluate* $\int_0^1 x^{n-1}\left[\log_e\left(\frac{1}{x}\right)\right]^{m-1} \cdot dx$

Solution: Put $\log_e \frac{1}{x} = t$ or $x = e^{-t}$ $\quad \therefore dx = -e^{-t}\, dt$

$$\int_0^1 x^{n-1}\left[\log_e\left(\frac{1}{x}\right)\right]^{m-1} dx = \int_\infty^0 (e^{-t})^{n-1}[t]^{m-1}(-e^{-t}dt) = \int_0^\infty e^{-nt}t^{m-1}\, dt$$

Putting $\quad nt = u \Rightarrow t = \frac{u}{n}$ so that $dt = \frac{du}{n}$

$$= \int_0^\infty e^{-u}\left(\frac{u}{n}\right)^{m-1}\frac{du}{n} = \frac{1}{n^m}\int_0^\infty e^{-u}u^{m-1}du = \frac{1}{n^m}\overline{|m} \qquad \textbf{Ans.}$$

38.3 TRANSFORMATION OF GAMMA FUNCTION

Prove that $\quad$ (*i*) $\int_0^\infty e^{-ky}y^{n-1}dy = \frac{\overline{|n}}{k^n}$ (*AMIETE, Dec. 2010*) $\qquad$ (*ii*) $\overline{\left|\frac{1}{2}\right.} = \sqrt{\pi}$

(*iii*) $\int_0^1\left(\log\frac{1}{y}\right)^{n-1} dy = \overline{|n}$ $\qquad$ (*iv*) $\Gamma\, n = \frac{1}{n}\int_0^\infty e^{-x^{\frac{1}{n}}}dx$

Solution: We know that $\overline{|n} = \int_0^\infty x^{n-1}e^{-x}dx$ ...(1)

(*i*) Replace x by $k\,y$, so that $dx = k\,dy$; then (1) becomes

$$\overline{|n} = \int_0^\infty (k\,y)^{n-1}e^{-ky}\,k\,dy.$$

$$\overline{|n} = k^n\int_0^\infty e^{-ky}y^{n-1}dy$$

$$\therefore \quad \boxed{\int_0^\infty e^{-ky}y^{n-1}\,dy = \frac{\overline{|n}}{k^n}} \qquad \text{...(2) } \textbf{Proved.}$$

(*ii*) Replace x^n by y, so that $n\,x^{n-1}\, dx = dy$ in (1), then

$$\overline{|n} = \int_0^\infty y^{\frac{n-1}{n}}e^{-y^{1/n}}\frac{dy}{nx^{n-1}} = \int_0^\infty y^{\frac{n-1}{n}}e^{-y^{1/n}}\frac{dy}{ny^{\frac{n-1}{n}}} = \frac{1}{n}\int_0^\infty e^{-y^{1/n}}dy$$

When $n = \frac{1}{2}$, $\qquad \overline{\left|\frac{1}{2}\right.} = \frac{1}{\frac{1}{2}}\int_0^\infty e^{-y^2}dy = 2\left[\frac{1}{2}\sqrt{\pi}\right]$

$$\boxed{\overline{\left|\frac{1}{2}\right.} = \sqrt{\pi}} \qquad \textbf{Proved.}$$

(*iii*) Putting $e^{-x} = y$, so that $-e^{-x}\,dx = dy$ and $-x = \log y$, $x = \log\frac{1}{y}$, (1) becomes

$$\overline{|n} = -\int_1^0\left(\log\frac{1}{y}\right)^{n-1} y\cdot\frac{dy}{e^{-x}} = \int_0^1\left(\log\frac{1}{y}\right)^{n-1}y\cdot\frac{dy}{y} = \int_0^1\left(\log\frac{1}{\mathbf{y}}\right)^{n-1}\boldsymbol{dy} \qquad \textbf{Proved.}$$

(*iv*) We know that, $\qquad \Gamma(n) = \int_0^\infty e^{-x}.\,x^{n-1}\,dx$ ...(1)

Putting $\qquad x^n = y \Rightarrow x = y^{1/n}$ so that $dx = \frac{1}{n}y^{\frac{1}{n}-1}dy$ in (1), we get

$$\Gamma n = \int_0^\infty e^{-y^{1/n}} y^{\frac{n-1}{n}} \cdot \frac{1}{n} y^{\frac{1}{n}-1} dy = \frac{1}{n}\int_0^\infty e^{-y^{\frac{1}{n}}} dy$$

$$\Gamma n = \frac{1}{n}\int_0^\infty e^{-x^{\frac{1}{n}}} dx.$$ **Proved.**

EXERCISE 38.1

Evaluate:

1. (*i*) $\left|\overline{-\frac{3}{2}}\right.$ (*ii*) $\left|\overline{-\frac{15}{2}}\right.$ (*iii*) $\left|\overline{\frac{7}{2}}\right.$ (*iv*) $\left|\overline{0}\right.$

 Ans. (*i*) $\frac{4}{3}\sqrt{\pi}$ (*ii*) $\frac{2^8\sqrt{\pi}}{15\times 13\times 11\times 9\times 7\times 5\times 3}$ (*iii*) $\frac{15\sqrt{\pi}}{8}$ (*iv*) ∞

2. $\int_0^\infty \sqrt{x}\, e^{-x} dx$ **Ans.** $\left|\overline{\frac{3}{2}}\right.$

3. $\int_0^\infty x^4 e^{-x^2} dx$ **Ans.** $\frac{3\sqrt{\pi}}{8}$

4. $\int_0^\infty e^{-h^2x^2} dx$ **Ans.** $\frac{\sqrt{\pi}}{2h}$

5. $\int_0^\infty \int_0^\infty e^{-\left(ax^2+by^2\right)} x^{2m-1} y^{2n-1} dx\, dy, \quad a, b, m, n > 0$ **Ans.** $\frac{\left|\overline{m}\right.\left|\overline{n}\right.}{4\, a^m b^n}$

6. $\int_0^1 (x\log x)^3 dx$ **Ans.** $-\frac{3}{128}$

7. $\int_0^1 \frac{dx}{\sqrt{x\log\frac{1}{x}}}$ **Ans.** $\sqrt{2\pi}$

8. Prove that 1.3.5.... $(2\,n\ -1) = \frac{2^n\left|\overline{n+\frac{1}{2}}\right.}{\sqrt{\pi}}$

9. $\int_0^\infty e^{-y^{1/m}} dy = m\left|\overline{m}\right.$

10. $\int_0^\infty \int_0^\infty e^{-(ax^2+by^2)} x^{2m-1} y^{2n-1} dx\, dy = \frac{\Gamma(m)\,\Gamma(n)}{4a^m b^n}$, where a, b, m, n are positive.

11. $\int_0^{\pi/2} \frac{d\theta}{(a\cos^4\theta + b\sin^4\theta)} = \frac{(\Gamma\, 1/4)^2}{4(ab)^{1/4}\sqrt{\pi}}$ **[Hint.** Put $\tan\theta = t$ then $bt^4 = az$]

38.4 BETA FUNCTION

(U.P. I Semester Dec. 2007)

$$\int_0^1 x^{l-1}(1-x)^{m-1} dx \qquad (l > 0,\ m > 0)$$

is called the Beta function of l, m. It is also written as

$$\beta\,(l,\ m) = \int_0^1 x^{l-1}(1-x)^{m-1} dx$$

38.5 EVALUATION OF BETA FUNCTION

$$\beta\,(l,\ m) = \frac{\left|\overline{l}\right.\left|\overline{m}\right.}{\left|\overline{l+m}\right.}$$

Solution. We have, $\beta\,(l,\ m) = \int_0^1 x^{l-1}(1-x)^{m-1} dx = \int_0^1 (1-x)^{m-1} x^{l-1} dx$

Integrating by parts, we have

$$= \left[(1-x)^{m-1}\frac{x^l}{l}\right]_0^1 + (m-1)\int_0^1 (1-x)^{m-2}\left(\frac{x^l}{l}\right) dx$$

$$= \frac{(m-1)}{l}\int_0^1 (1-x)^{m-2} x^l dx$$

Again integrating by parts, we get

$$= \frac{(m-1)(m-2)}{l(l+1)}\int_0^1 (1-x)^{m-3}x^{l+1}dx = \frac{(m-1)(m-2)...2.1}{l(l+1).....(l+m-2)}\int_0^1 x^{l+m-2}dx$$

$$= \frac{(m-1)(m-2)...2.1}{l(l+1).....(l+m-2)}\left[\frac{x^{l+m-1}}{l+m-1}\right]_0^1 = \frac{(m-1)(m-2)...2.1}{l(l+1).....(l+m-2)(l+m-1)}$$

$$= \frac{(m-1)!}{l(l+1).....(l+m-2)(l+m-1)}\times\frac{(l-1)(l-2).....1}{(l-1)(l-2).....1}$$

$$= \frac{(m-1)!\ (l-1)!}{1.2....(l-2)(l-1).l(l+1)...(l+m-2)(l+m-1)} = \frac{(l-1)!(m-1)!}{(l+m-1)!} = \frac{\Gamma l\ \Gamma m}{\Gamma l+m}$$

And if only l is positive integer and not m then

$$\boxed{\beta(l, m) = \frac{(l-1)!}{m(m+1)...(m+l-1)}}$$ **Ans.**

38.6 A PROPERTY OF BETA FUNCTION

$$\beta(l, m) = \beta(m, l)$$

Solution. We have

$$\beta(l, m) = \int_0^1 x^{l-1}(1-x)^{m-1}dx \qquad \left[\int_0^a f(x)dx = \int_0^a f(a-x)dx\right]$$

$$= \int_0^1 (1-x)^{l-1}[1-(1-x)]^{m-1}dx = \int_0^1 (1-x)^{l-1}x^{m-1}dx$$

$$= \int_0^1 x^{m-1}(1-x)^{l-1}dx$$

l and m are interchanged.

$$\boxed{\beta(l, m) = \beta(m, l)}$$ **Proved.**

Example 11. *Evaluate* $\int_0^1 x^4\left(1-\sqrt{x}\right)^5 dx$

Solution. Let $\sqrt{x} = t \Rightarrow x = t^2$ so that $dx = 2\,t\,dt$

$$\int_0^1 x^4\left(1-\sqrt{x}\right)^5 dx = \int_0^1 (t^2)^4(1-t)^5(2\,t\,dt)$$

$$= 2\int_0^1 t^9(1-t)^5 dt = 2\beta(10,6) = 2\frac{\Gamma 10\ \Gamma 6}{\Gamma 16} = 2.\frac{9!\,5!}{(15)!}$$

$$= 2.\frac{5!}{10\times11\times12\times13\times14\times15} = \frac{2\times1\times2\times3\times4\times5}{10\times11\times12\times13\times14\times15} = \frac{1}{11\times13\times7\times15} = \frac{1}{15015}$$ **Ans.**

Example 12. *Evaluate* $\int_0^1 (l-x^3)^{-\frac{l}{2}}dx$

Solution. Let $x^3 = y \Rightarrow x = y^{1/3}$ so that $dx = \frac{1}{3}y^{-\frac{2}{3}}dy$

$$\int_0^1 (1-x^3)^{-\frac{1}{2}}dx = \int_0^1 (1-y)^{-\frac{1}{2}}\left(\frac{1}{3}y^{-\frac{2}{3}}dy\right)$$

$$= \frac{1}{3}\int_0^1 y^{-\frac{2}{3}}(1-y)^{-\frac{1}{2}}dy = \frac{1}{3}\beta\left(\frac{1}{3},\frac{1}{2}\right) = \frac{1}{3}\frac{\Gamma\frac{1}{3}\ \Gamma\frac{1}{2}}{\Gamma\frac{5}{6}}$$ **Ans.**

38.7 TRANSFORMATION OF BETA FUNCTION

We know that

$$\beta(l, m) = \int_0^1 x^{l-1}(1-x)^{m-1}\, dx,$$

Putting $x = \dfrac{1}{1+y}$ so that $dx = -\dfrac{1}{(1+y)^2}dy$ and $1-x = \dfrac{y}{1+y}$ in (1), we get

$$\beta(l, m) = \int_\infty^0 \left(\frac{1}{1+y}\right)^{l-1}\left(\frac{y}{1+y}\right)^{m-1}\left[-\frac{1}{(1+y)^2}dy\right] = \int_0^\infty \frac{y^{m-1}}{(1+y)^{l+m}}dy$$

Since l, m can be interchanged in $\beta(l, m)$,

$$\beta(l, m) = \int_0^\infty \frac{y^{l-1}}{(1+y)^{m+l}}dy \Rightarrow \beta(l, m) = \int_0^\infty \frac{x^{l-1}}{(1+x)^{m+l}}dx \quad ...(1)$$

Example 13. *Evaluate* $\displaystyle\int_0^1 \frac{x^{m-l}+x^{n-l}}{(1+x)^{m+n}}dx$

Solution. We know that

$$\beta(m, n) = \int_0^\infty \frac{x^{m-1}}{(1+x)^{m+n}}dx \Rightarrow \int_0^\infty \frac{x^{m-1}}{(1+x)^{m+n}}dx = \beta(m,n)$$

$$\Rightarrow \int_0^1 \frac{x^{m-1}}{(1+x)^{m+n}}dx + \int_1^\infty \frac{x^{m-1}}{(1+x)^{m+n}}dx = \beta(m, n) \quad ...(1)$$

Consider $\displaystyle\int_1^\infty \frac{x^{m-1}}{(1+x)^{m+n}}dx = \int_1^0 \frac{\left(\frac{1}{t}\right)^{m-1}}{\left(1+\frac{1}{t}\right)^{m+n}}\left(-\frac{1}{t^2}dt\right) = \int_0^1 \frac{\left(\frac{1}{t}\right)^{m-1}\frac{1}{t^2}}{\left(\frac{1}{t}\right)^{m+n}(t+1)^{m+n}}dt$ $\left(\text{Put } x = \frac{1}{t}\right)$

$$= \int_0^1 \frac{t^{n-1}}{(1+t)^{m+n}}dt = \int_0^1 \frac{x^{n-1}}{(1+x)^{m+n}}dx$$

Putting the value of $\displaystyle\int_1^\infty \frac{x^{m-1}}{(1+x)^{m+n}}dx$ in (1), we get

$$\int_0^1 \frac{x^{m-1}}{(1+x)^{m+n}}dx + \int_0^1 \frac{x^{n-1}}{(1+x)^{m+n}}dx = \beta(m, n)$$

$$\Rightarrow \int_0^1 \frac{x^{m-1}+x^{n-1}}{(1+x)^{m+n}}dx = \beta(m, n)$$ **Ans.**

38.8 RELATION BETWEEN BETA AND GAMMA FUNCTIONS

We know that, $\Gamma l = \int_0^\infty e^{-x}x^{l-1}dx,$ [Put $zx = y$]

$$\frac{\Gamma l}{z^l} = \int_0^\infty e^{-zx}x^{l-1}dx$$

$$\Gamma l = \int_0^\infty z^l e^{-zx}x^{l-1}dx$$

Multiplying both sides by $e^{-z}z^{m-1}$, we have

$$\Gamma l . e^{-z}.z^{m-1} = \int_0^\infty e^{-z}.z^{m-1}.z^l.e^{-zx}.x^{l-1}dx$$

$$\Gamma l . e^{-z} . z^{m-1} = \int_0^\infty e^{-(1+x)z} . z^{l+m-1} . x^{l-1} dx$$

Integrating both sides w.r.t. 'z', we get

$$\int_0^\infty \Gamma l\, e^{-z} z^{m-1}\, dz = \int_0^\infty \int_0^\infty e^{-(1+x)z} z^{l+m-1} x^{l-1}\, dx\, dz$$

$$\Gamma l\, \Gamma m = \int_0^\infty x^{l-1}\, dx \int_0^\infty e^{-(1+x)z} z^{l+m-1} dz$$

$$= \int_0^\infty x^{l-1} dx \cdot \frac{\Gamma l+m}{(1+x)^{l+m}}$$ [From (1), Art 38.7]

$$\Gamma l\, \Gamma m = \Gamma l+m \int_0^\infty \frac{x^{l-1}}{(1+x)^{l+m}} dx = \Gamma l+m \ . \ \beta\,(l, m)$$

$$\therefore \quad \boldsymbol{\beta\,(l, m) = \frac{\Gamma l\, \Gamma m}{\Gamma l+m}}$$

This is the required relation.

38.9. SHOW THAT

$$\int_0^{\frac{\pi}{2}} \sin^p\theta \cos^q\theta d\theta = \frac{\Gamma\left(\frac{p+1}{2}\right) \Gamma\left(\frac{q+1}{2}\right)}{2\Gamma\left(\frac{p+q+2}{2}\right)}$$

Solution. We know that

$$\beta\,(m, n) = \int_0^1 x^{m-1} (1-x)^{n-1}\, dx \qquad ...(1)$$

Putting $x = \sin^2\theta,\ dx = 2\sin\theta\cos\theta\, d\theta$

and $1 - x = 1 - \sin^2\theta = \cos^2\theta$

Then (1) becomes

$$\beta\,(m,n) = \int_0^{\frac{\pi}{2}} \sin^{2m-2}\theta \cos^{2n-2}\theta\ 2\sin\theta\cos\theta\, d\theta$$

or

$$\frac{\Gamma m\, \Gamma n}{\Gamma m+n} = 2\int_0^{\frac{\pi}{2}} \sin^{2m-1}\theta \cos^{2n-1}\theta\, d\theta$$

Putting $2m - 1 = p, \quad i.e., \quad m = \frac{p+1}{2}$

and $2n - 1 = q, \quad i.e., \quad n = \frac{q+1}{2}$

$$\frac{\Gamma \frac{p+1}{2}\, \Gamma \frac{q+1}{2}}{\Gamma \frac{p+q+2}{2}} = 2\int_0^{\frac{\pi}{2}} \sin^p\theta \cdot \cos^q\theta\, d\theta$$

$$\boldsymbol{\int_0^{\frac{\pi}{2}} \sin^p\theta \cos^q\theta\, d\theta = \frac{\Gamma \frac{p+1}{2}\, \Gamma \frac{q+1}{2}}{2\,\Gamma \frac{p+q+2}{2}}}$$

Proved.

Example 14. *Prove that* $\int_0^{\frac{\pi}{2}} \sqrt{\sin\theta}\, d\theta \times \int_0^{\frac{\pi}{2}} \frac{1}{\sqrt{\sin\theta}}\, d\theta = \pi$ (*AMIETE, June 2009*)

Solution. L.H.S. $= \int_0^{\frac{\pi}{2}} \sqrt{\sin\theta}\, d\theta \times \int_0^{\frac{\pi}{2}} \frac{1}{\sqrt{\sin\theta}}\, d\theta$

$$= \int_0^{\frac{\pi}{2}} \sin^{\frac{1}{2}}\theta \cos^0\theta\, d\theta \times \int_0^{\frac{\pi}{2}} \sin^{-\frac{1}{2}}\theta \cos^0\theta\, d\theta$$

$$= \frac{\Gamma\left(\frac{\frac{1}{2}+1}{2}\right)\Gamma\left(\frac{0+1}{2}\right)}{2\,\Gamma\left(\frac{3}{4}+\frac{1}{2}\right)} \times \frac{\Gamma\left(\frac{-\frac{1}{2}+1}{2}\right)\Gamma\left(\frac{0+1}{2}\right)}{2\,\Gamma\left(\frac{1}{4}+\frac{1}{2}\right)} = \frac{\Gamma\frac{3}{4}\,\Gamma\frac{1}{2}}{2\,\Gamma\frac{5}{4}} \times \frac{\Gamma\frac{1}{4}\,\Gamma\frac{1}{2}}{2\,\Gamma\frac{3}{4}}$$

$$= \frac{\Gamma\frac{1}{2}\,\Gamma\frac{1}{2}\,\Gamma\frac{1}{4}}{4\,\Gamma\frac{5}{4}} = \frac{(\sqrt{\pi})(\sqrt{\pi})\,\Gamma\frac{1}{4}}{4\cdot\frac{1}{4}\,\Gamma\frac{1}{4}} = \pi = \text{R.H.S.}$$ **Proved.**

Example 15. *Prove that* $\int_0^1 \frac{x^2}{\sqrt{(1-x^4)}}\, dx \times \int_0^1 \frac{dx}{\sqrt{(1+x^4)}} = \frac{\pi}{4\sqrt{2}}$ (*AMIETE, Dec. 2009*)

Solution. Here, we have $\int_0^1 \frac{x^2}{\sqrt{1-x^4}}\, dx \times \int_0^1 \frac{dx}{\sqrt{1+x^4}}$ ... (1)

Let $I_1 = \int_0^1 \frac{x^2}{\sqrt{1-x^4}}\, dx$ Put $x^2 = \sin\theta \Rightarrow 2x\, dx = \cos\theta\, d\theta$

$x = \sqrt{\sin\theta} \Rightarrow dx = \frac{1}{2}(\sin\theta)^{-\frac{1}{2}} \cos\theta\, d\theta$

$$= \int_0^{\frac{\pi}{2}} \frac{\sin\theta}{\sqrt{1-\sin^2\theta}} \frac{1}{2}(\sin\theta)^{-\frac{1}{2}} \cos\theta\, d\theta.$$

$$= \frac{1}{2}\int_0^{\frac{\pi}{2}} \frac{(\sin\theta)^{1-\frac{1}{2}}}{\cos\theta}.\cos\theta\, d\theta = \frac{1}{2}\int_0^{\frac{\pi}{2}} (\sin\theta)^{\frac{1}{2}} (\cos\theta)^0\, d\theta$$

$$= \frac{1}{2}\left(\frac{\Gamma\left(\frac{\frac{1}{2}+1}{2}\right)\Gamma\left(\frac{0+1}{2}\right)}{2\,\Gamma\left(\frac{3}{4}+\frac{1}{2}\right)}\right) = \frac{1}{4}\,\frac{\Gamma\frac{3}{4}\,\Gamma\frac{1}{2}}{\Gamma\frac{5}{4}} \qquad \text{... (2)}$$

Let $I_2 = \int_0^1 \frac{dx}{\sqrt{1+x^4}}$ $\left[\begin{array}{l} \text{Put } x^2 = \tan\theta \Rightarrow x = \sqrt{\tan\theta} \\ \Rightarrow dx = \frac{1}{2}(\tan\theta)^{-\frac{1}{2}} \sec^2\theta\, d\theta \end{array}\right]$

$$= \frac{1}{2}\int_0^{\frac{\pi}{4}} \frac{(\tan\theta)^{-\frac{1}{2}} \sec^2\theta}{\sqrt{1+\tan^2\theta}}\, d\theta = \frac{1}{2}\int_0^{\frac{\pi}{4}} (\tan\theta)^{-\frac{1}{2}} \sec\theta\, d\theta$$

$$= \frac{1}{2}\int_0^{\frac{\pi}{4}} (\sin\theta)^{-\frac{1}{2}}.(\cos\theta)^{-\frac{1}{2}}\, d\theta = \frac{1}{2}\int_0^{\frac{\pi}{4}} \sqrt{\frac{2}{2\sin\theta\cos\theta}}\, d\theta$$

$$= \frac{1}{\sqrt{2}} \int_0^{\frac{\pi}{2}} \frac{d\theta}{\sqrt{\sin 2\theta}} \qquad \text{Put } 2\theta = t \quad \Rightarrow \quad d\theta = \frac{dt}{2}$$

$$= \frac{1}{2\sqrt{2}} \int_0^{\frac{\pi}{2}} \frac{dt}{\sqrt{\sin t}} = \frac{1}{2\sqrt{2}} \int_0^{\frac{\pi}{2}} (\sin t)^{-\frac{1}{2}} (\cos\theta)^{\circ}\, dt$$

$$= \frac{1}{2\sqrt{2}} \left(\frac{\Gamma\left(\frac{1-\frac{1}{2}}{2}\right) \Gamma\left(\frac{0+1}{2}\right)}{2\, \Gamma\left(\frac{1}{4}+\frac{1}{2}\right)} \right) = \frac{1}{4\sqrt{2}} \frac{\Gamma\left(\frac{1}{4}\right)\Gamma\left(\frac{1}{2}\right)}{\Gamma\left(\frac{3}{4}\right)} \qquad \ldots (3)$$

Putting the value in (1) from equation (1) and (2), we get

$$\int_0^1 \frac{x^2 dx}{\sqrt{1-x^4}} \times \int_0^1 \frac{dx}{\sqrt{1-x^4}} = \frac{1}{4} \frac{\Gamma\left(\frac{3}{4}\right)\Gamma\left(\frac{1}{2}\right)}{\Gamma\left(\frac{5}{4}\right)} \times \frac{1}{4\sqrt{2}} \frac{\Gamma\left(\frac{1}{4}\right)\Gamma\left(\frac{1}{2}\right)}{\Gamma\left(\frac{3}{4}\right)} = \frac{1}{16\sqrt{2}} \frac{\Gamma\left(\frac{1}{2}\right)\Gamma\left(\frac{1}{2}\right)\Gamma\left(\frac{1}{4}\right)}{\Gamma\left(\frac{5}{4}\right)} = \frac{1}{16\sqrt{2}} = \frac{\pi\,\Gamma\left(\frac{1}{4}\right)}{\frac{1}{4}\Gamma\left(\frac{1}{4}\right)} = \frac{\pi}{4\sqrt{2}}$$ **Proved.**

Example 16. *Prove that* $\int_0^1 \frac{dx}{\sqrt{1-x^4}} = \frac{\sqrt{\pi}}{4} \frac{\Gamma\left(\frac{1}{4}\right)}{\Gamma\left(\frac{3}{4}\right)}$ *(AMIETE, June 2010)*

Solution. Here, we have $\int_0^1 \frac{dx}{\sqrt{1-x^4}}$, Put $x^2 = \sin\theta \quad \Rightarrow \quad x = \sqrt{\sin\theta}$

$$\Rightarrow \quad dx = \frac{1}{2} (\sin\theta)^{-\frac{1}{2}} \cos\theta . d\theta$$

$$= \frac{1}{2} \int_0^{\frac{\pi}{2}} \frac{(\sin\theta)^{-\frac{1}{2}} \cos\theta}{\sqrt{1-\sin^2\theta}} = \frac{1}{2} \int_0^{\frac{\pi}{2}} \frac{(\sin\theta)^{-\frac{1}{2}} \cos\theta}{\cos\theta} d\theta$$

$$= \frac{1}{2} \int_0^{\frac{\pi}{2}} (\sin\theta)^{-\frac{1}{2}} (\cos\theta)^0\, d\theta = \frac{1}{2} \left(\frac{\Gamma\left(\frac{-\frac{1}{2}+1}{2}\right) \Gamma\left(\frac{0+1}{2}\right)}{2\,\Gamma\left(\frac{1}{4}+\frac{1}{2}\right)} \right) = \frac{1}{4} \frac{\Gamma\left(\frac{2}{4}\right)\Gamma\left(\frac{1}{2}\right)}{\Gamma\left(\frac{3}{4}\right)} = \frac{\sqrt{\pi}}{4} \frac{\Gamma\left(\frac{1}{4}\right)}{\Gamma\left(\frac{3}{4}\right)}$$ **Proved.**

Example 17. *Find the value of* $\Gamma\left(\frac{1}{2}\right)$.

Solution. We have already solved this problem in Art. 38.3 (ii) Transformation of the Gamma Function.

Now, by **Second method:** We know that,

$$\int_0^{\frac{\pi}{2}} \sin^p\theta \cdot \cos^q\theta\, d\theta = \frac{\Gamma\left(\frac{p+1}{2}\right)\Gamma\left(\frac{q+1}{2}\right)}{2\,\Gamma\left(\frac{p+q+2}{2}\right)}$$

Putting $p = q = 0$, we get $\int_0^{\frac{\pi}{2}} d\theta = \frac{\Gamma\left(\frac{1}{2}\right)\Gamma\left(\frac{1}{2}\right)}{2\,\Gamma 1} \Rightarrow [\theta]_0^{\pi/2} = \frac{1}{2}\left(\Gamma\frac{1}{2}\right)^2 \Rightarrow \frac{\pi}{2} = \frac{1}{2}\left(\Gamma\frac{1}{2}\right)^2$

$$\Rightarrow \quad \left(\Gamma\frac{1}{2}\right)^2 = \pi \quad \Rightarrow \quad \Gamma\frac{1}{2} = \sqrt{\pi}$$ **Ans.**

Example 18. *Show that*

$$\int_0^{\frac{\pi}{2}} \sqrt{\cot\theta}\, d\theta = \frac{1}{2}\left|\frac{1}{4}\right|\frac{3}{4}$$

Solution. We know that

$$\int_0^{\frac{\pi}{2}} \sin^p\theta \cdot \cos^q\theta\, d\theta = \frac{\left|\frac{p+1}{2}\right|\frac{q+1}{2}}{2\left|\frac{p+q+2}{2}\right.} \qquad ...(1)$$

$$\int_0^{\frac{\pi}{2}} \sqrt{\cot\theta}\, d\theta = \int_0^{\frac{\pi}{2}} \frac{\cos^{1/2}\theta}{\sin^{1/2}\theta}\, d\theta = \int_0^{\frac{\pi}{2}} \sin^{-1/2}\theta \cos^{1/2}\theta\, d\theta$$

On applying formula (1), we have

$$\int_0^{\frac{\pi}{2}} \sqrt{\cot\theta}\, d\theta = \frac{\left|\frac{-\frac{1}{2}+1}{2}\right|\frac{\frac{1}{2}+1}{2}}{2\left|\frac{-\frac{1}{2}+\frac{1}{2}+2}{2}\right.} = \frac{\left|\frac{1}{4}\right|\frac{3}{4}}{2\left|1\right.} = \frac{1}{2}\left|\frac{1}{4}\right|\frac{3}{4} \qquad \textbf{Proved.}$$

Example 19. *Using Beta and Gamma functions, evaluate*

$$\int_0^1 \left(\frac{x^3}{1-x^3}\right)^{\frac{1}{2}} dx$$

Solution.

$$\int_0^1 \left(\frac{x^3}{1-x^3}\right)^{\frac{1}{2}} dx \qquad ...(1)$$

Putting $x^3 = \sin^2\theta$, so that $x = \sin^{\frac{2}{3}}\theta$, $dx = \frac{2}{3}\sin^{-\frac{1}{3}}\theta\cos\theta\, d\theta$ in (1), we get

$$\int_0^1 \left(\frac{x^3}{1-x^3}\right)^{\frac{1}{2}} dx = \int_0^{\pi/2} \left(\frac{\sin^2\theta}{1-\sin^2\theta}\right)^{\frac{1}{2}} \frac{2}{3}\sin^{-\frac{1}{3}}\theta\cos\theta\, d\theta$$

$$= \frac{2}{3}\int_0^{\pi/2} \left(\frac{\sin\theta}{\cos\theta}\right) \sin^{-\frac{1}{3}}\theta\cos\theta\, d\theta = \frac{2}{3}\int_0^{\pi/2} \sin^{\frac{2}{3}}\theta\, d\theta$$

$$= \frac{2}{3}\,\frac{\left|\frac{\frac{2}{3}+1}{2}\right|\frac{0+1}{2}}{2\left|\frac{\frac{2}{3}+1+1}{2}\right.} = \frac{2}{3}\,\frac{\left|\frac{5}{6}\right|\frac{1}{2}}{\left|\frac{4}{3}\right.} = \frac{2}{3}\,\frac{\sqrt{\pi}\left|\frac{5}{6}\right.}{\left|\frac{4}{3}\right.} \qquad \textbf{Ans.}$$

Example 20. *Evaluate* $\int_{-1}^{+1} (1+x)^{p-1}(1-x)^{q-1}\, dx$.

Solution. Put $x = \cos 2\theta$, then $dx = -2\sin 2\theta\, d\theta$

$$\int_{-1}^{+1} (1+x)^{p-1}(1-x)^{q-1}\, dx = \int_{\frac{\pi}{2}}^{0} (1+\cos 2\theta)^{p-1}(1-\cos 2\theta)^{q-1}(-2\sin 2\theta\, d\theta)$$

$$= \int_{\frac{\pi}{2}}^{0} \left(1+2\cos^2\theta-1\right)^{p-1}\left(1-1+2\sin^2\theta\right)^{q-1}(-4\sin\theta\cos\theta\, d\theta)$$

$$= 4\int_0^{\frac{\pi}{2}} 2^{p-1}\cos^{2p-2}\theta \,.\, 2^{q-1}\sin^{2q-2}\theta \,.\, \sin\theta\cos\theta\, d\theta = 2^{p+q}\int_0^{\infty} \sin^{2q-1}\theta\cos^{2p-1}\theta\, d\theta$$

$$= 2^{p+q}\frac{\left\lceil\frac{2q}{2}\right.\left\lceil\frac{2p}{2}\right.}{2\left\lceil\frac{2p+2q}{2}\right.} = 2^{p+q-1}\frac{\lceil p\,\lceil q}{\lceil p+q} \qquad \textbf{Ans.}$$

Example 21. *Show that* $\lceil n\,\lceil 1-n = \dfrac{\pi}{\sin n\pi}$ $\quad$ *(0 < n < 1)*

Solution. We know that

$$\beta(m, n) = \int_0^{\infty}\frac{x^{n-1}}{(1+x)^{m+n}}\,dx \qquad \text{[From (1), Art 38.7]}$$

$$\frac{\lceil m\,\lceil n}{\lceil m+n} = \int_0^{\infty}\frac{x^{n-1}}{(1+x)^{m+n}}\,dx$$

Putting $m + n = 1$ or $m = 1 - n$, we get

$$\frac{\lceil 1-n\,\lceil n}{\lceil 1} = \int_0^{\infty}\frac{x^{n-1}}{(1+x)^1}\,dx$$

$$\lceil 1-n\,\lceil n = \int_0^{\infty}\frac{x^{n-1}}{1+x}\,dx \qquad \left[\int_0^{\infty}\frac{x^{n-1}}{1+x}\,dx = \frac{\pi}{\sin n\pi}\right]$$

$$\Rightarrow \quad \lceil n\,\lceil 1-n = \frac{\pi}{\sin n\pi} \qquad \textbf{Proved.}$$

Example 22. *Assuming* $\lceil n\,\lceil 1-n = \pi \operatorname{cosec} n\pi.\ 0 < n < 1,$ *show that*

$$\int_0^{\infty}\frac{x^{p-1}}{1+x}\,dx = \left(\frac{\pi}{\sin p\pi}\right).\ 0 < p < 1 \qquad \text{(U.P., I Semester, Dec 2009)}$$

Solution: Here, we have $\pi \operatorname{cosec} n\pi = \lceil n\,\lceil 1-n$

$$\Rightarrow \quad \frac{\pi}{\sin n\pi} = \lceil n\,\lceil 1-n$$

We know that $\displaystyle\int_0^{\infty}\frac{x^{n-1}}{1+x}\,dx = \frac{\lceil 1-n\,\lceil n}{\lceil 1}$ ...(1)

Setting $m + n = 1$ so that $m = 1 - n$ in (1), we get

$$\int_0^{\infty}\frac{x^{n-1}}{(1+x)^{m+n}}\,dx = \frac{\lceil m\,\lceil n}{\lceil m+n} = \beta(m,n) \qquad \textbf{Proved.}$$

Example 23. *Prove that* $\left\lceil\left(\frac{1}{4}\right)\right.\left\lceil\left(\frac{3}{4}\right)\right. = \pi\sqrt{2}$

Solution. Putting $n = \frac{1}{4}$ in result of example 22, we obtain

$$\left\lceil\left(\frac{1}{4}\right)\right.\left\lceil\left(1-\frac{1}{4}\right)\right. = \frac{\pi}{\sin\frac{\pi}{4}}$$

$$\Rightarrow \quad \left\lceil\left(\frac{1}{4}\right)\right.\left\lceil\left(\frac{3}{4}\right)\right. = \frac{\pi}{\left(\frac{1}{\sqrt{2}}\right)} \Rightarrow \left\lceil\left(\frac{1}{4}\right)\right.\left\lceil\left(\frac{3}{4}\right)\right. = \pi\sqrt{2} \qquad \textbf{Proved.}$$

Example 24. *Evaluate* $\int_0^1 \frac{dx}{(1-x^n)^{\frac{1}{n}}}$

Solution. Let $x^n = \sin^2\theta$ or $x = \sin^{2/n}\theta$

So that $$dx = \frac{2}{n}\sin^{2/n-1}\theta\cos\theta\, d\theta$$

$$\int_0^1 \frac{dx}{(1-x^n)^{\frac{1}{n}}} = \int_0^{\frac{\pi}{2}} \frac{\frac{2}{n}\sin^{2/n-1}\theta\cos\theta}{(1-\sin^2\theta)^{1/n}}\,d\theta = \frac{2}{n}\int_0^{\frac{\pi}{2}} \frac{\sin^{2/n-1}\theta\cos\theta}{(\cos^2\theta)^{1/n}}\,d\theta$$

$$= \frac{2}{n}\int_0^{\frac{\pi}{2}} \sin^{2/n-1}\theta\cos^{1-2/n}\theta\, d\theta$$

$$= \frac{2}{n}\,\frac{\left|\overline{\frac{\frac{2}{n}-1+1}{2}}\right.\left|\overline{\frac{1-\frac{2}{n}+1}{2}}\right.}{2\left|\overline{\frac{\frac{2}{n}-1+1+2-\frac{2}{n}}{2}}\right.} = \frac{1}{n}\,\frac{\left|\overline{\frac{1}{n}}\right.\left|\overline{\frac{n-1}{n}}\right.}{\left|\overline{1}\right.} \qquad \left(\because\ \left|\overline{\frac{1}{n}}\right.\left|\overline{1-\frac{1}{n}}\right. = \frac{\pi}{\sin\frac{\pi}{n}}\right)$$

$$= \frac{\pi}{n\sin\frac{\pi}{n}}$$ **Ans.**

Example 25. *Show that* $\int_0^a \frac{dx}{\sqrt[n]{(a^n-x^n)}} = \frac{\pi}{n}\operatorname{cosec}\left(\frac{\pi}{n}\right)$, *where* $n > 1$.*(M.U. II Semester 2009)*

Solution. Let $x^n = a^n\sin^2\theta \quad\Rightarrow\quad x = a\sin^{\frac{2}{n}}\theta$

So that $$dx = \frac{2a}{n}\sin^{\frac{2}{n}-1}\theta\cos\theta\, d\theta$$

$$\therefore \int_0^a \frac{dx}{\sqrt[n]{a^n-x^n}} = \int_0^{\frac{\pi}{2}} \frac{a\times\frac{2}{n}\sin^{\frac{2}{n}-1}\theta\cos\theta}{(a^n-a^n\sin^2\theta)^{\frac{1}{n}}}\,d\theta = \frac{2}{n}\int_0^{\frac{\pi}{2}} \frac{\sin^{\frac{2}{n}-1}\theta\cos\theta}{\cos^{\frac{2}{n}}\theta}\,d\theta$$

$$= \frac{2}{n}\int_0^{\frac{\pi}{2}} \sin^{\frac{2}{n}-1}\theta\cos^{1-\frac{2}{n}}\theta\, d\theta = \frac{2}{n}\,\frac{\left|\overline{\frac{\frac{2}{n}-1+1}{2}}\right.\left|\overline{\frac{1-\frac{2}{n}+1}{2}}\right.}{2\left|\overline{\frac{\frac{2}{n}-1+1+2-\frac{2}{n}}{2}}\right.}$$

$$= \frac{1}{n}\,\frac{\left|\overline{\frac{1}{n}}\right.\left|\overline{\frac{n-1}{n}}\right.}{\left|\overline{1}\right.} \qquad \left[\left|\overline{\frac{1}{n}}\right.\left|\overline{1-\frac{1}{n}}\right. = \frac{\pi}{\sin\frac{\pi}{n}}\right]$$

$$= \frac{\pi}{n\sin\frac{\pi}{n}} = \frac{\pi}{n}\operatorname{cosec}\left(\frac{\pi}{n}\right)$$ **Proved.**

Example 26. *Show that* $\int_0^{\frac{\pi}{2}} \tan^P \theta \, d\theta = \frac{\pi}{2} \sec \frac{P\pi}{2}$ *and indicate the restriction on the values of P.*

Solution. $\int_0^{\frac{\pi}{2}} \tan^P \theta \, d\theta = \int_0^{\frac{\pi}{2}} \sin^P \theta \cos^{-P} \theta \, d\theta$

$$= \frac{\Gamma\frac{P+1}{2}\,\Gamma\frac{-P+1}{2}}{2\,\Gamma\frac{P+1-P+1}{2}} \quad \begin{bmatrix} 1-P>0 \\ 1>P \end{bmatrix} \qquad = \frac{\Gamma\frac{P+1}{2}\,\Gamma\frac{-P+1}{2}}{2\,\Gamma 1} \quad \begin{bmatrix} 1+P>0 \\ P>-1 \end{bmatrix}$$

$$= \frac{1}{2}\Gamma\frac{P+1}{2}\,\Gamma\frac{-P+1}{2} \;\; \therefore 1 > P > -1 = \frac{1}{2}\frac{\pi}{\sin\frac{P+1}{2}\pi} = \frac{1}{2}\frac{\pi}{\cos\frac{P\pi}{2}} = \frac{\pi}{2}\sec\frac{P\pi}{2}$$ **Proved.**

38.10 DUPLICATION FORMULA

$$\Gamma m\,\Gamma\left(m+\frac{1}{2}\right) = \frac{\sqrt{\pi}}{2^{2m-1}}\Gamma 2m, \text{ where } m \text{ is positive.}$$

Hence show that $\beta(m, m) = 2^{1-2m}\,\beta\left(m, \frac{1}{2}\right)$

(AMIETE, Dec. 2010, U.P., II Semester; Summer 2001)

Proof. We know that $\frac{\Gamma\frac{p+1}{2}\,\Gamma\frac{q+1}{2}}{2\,\Gamma\frac{p+q+2}{2}} = \int_0^{\frac{\pi}{2}} \sin^p\theta \cdot \cos^q \theta \, d\theta$

Putting $q = p$, we get $\frac{\Gamma\frac{p+1}{2}\,\Gamma\frac{p+1}{2}}{2\,\Gamma(p+1)}$

$$= \int_0^{\frac{\pi}{2}} \sin^p\theta \cdot \cos^p\theta \, d\theta = \int_0^{\frac{\pi}{2}} (\sin\theta\cos\theta)^p \, d\theta$$

$$= \int_0^{\frac{\pi}{2}} \frac{1}{2^p}(2\sin\theta\cos\theta)^p \, d\theta = \frac{1}{2^p}\int_0^{\frac{\pi}{2}} (\sin 2\theta)^p \, d\theta$$

[Putting $2\theta = t \Rightarrow d\theta = \frac{dt}{2}$]

$$= \frac{1}{2^p}\int_0^{\pi} \sin^p t \frac{dt}{2} = \frac{1}{2^p}\cdot\frac{1}{2}\cdot 2\int_0^{\frac{\pi}{2}} \sin^p t \, dt = \frac{1}{2^p}\int_0^{\frac{\pi}{2}} \sin^p t \cos^0 t \, dt$$

$$= \frac{1}{2^p}\frac{\Gamma\frac{p+1}{2}\,\Gamma\frac{0+1}{2}}{2\,\Gamma\frac{p+2}{2}} \quad \Rightarrow \quad \frac{\Gamma\frac{p+1}{2}\,\Gamma\frac{p+1}{2}}{2\,\Gamma(p+1)} = \frac{1}{2^p}\frac{\Gamma\frac{p+1}{2}\,\Gamma\frac{1}{2}}{2\,\Gamma\frac{p+2}{2}}$$

$$\Rightarrow \quad \frac{\Gamma\frac{p+1}{2}}{\Gamma(p+1)} = \frac{1}{2^p}\frac{\Gamma\frac{1}{2}}{\Gamma\frac{p+2}{2}} \quad \Rightarrow \quad \frac{\Gamma\frac{p+1}{2}}{\Gamma(p+1)} = \frac{1}{2^p}\frac{\sqrt{\pi}}{\Gamma\frac{p+2}{2}}$$

Take $\frac{p+1}{2} = m \quad \Rightarrow \quad p = 2m - 1$

$$\Rightarrow \quad \frac{\Gamma m}{\Gamma 2m} = \frac{1}{2^{2m-1}} \frac{\sqrt{\pi}}{\Gamma\left(\frac{2m+1}{2}\right)} \qquad \ldots(1)$$

$$\Gamma m \, \Gamma\left(m+\frac{1}{2}\right) = \frac{\sqrt{\pi}}{2^{2m-1}} \Gamma 2m$$ **Proved.**

Multiplying both sides of (1) by Γm, we have

$$\frac{\Gamma m \, \Gamma m}{\Gamma 2m} = 2^{1-2m} \frac{\Gamma\frac{1}{2} \, \Gamma m}{\Gamma\left(m+\frac{1}{2}\right)}$$

$$\boxed{\beta(m, m) = 2^{1-2m} \beta\left(m, \frac{1}{2}\right)}$$ **Proved.**

Example 27. *For a β function, show that*

$\beta(p, q) = \beta(p+1, q) + \beta(p, q+1)$ (*U.P., Ist Semester, Dec 2008*)

Solution. $\beta(p+1, q) + \beta(p, q+1)$

$$= \int_0^1 x^p (1-x)^{q-1} \, dx + \int_0^1 x^{p-1} (1-x)^q \, dx$$

$$= \int_0^1 x^{p-1} (1-x)^{q-1} [x + 1 - x] \, dx \qquad \text{(Taking common)}$$

$$= \int_0^1 x^{p-1} (1-x)^{q-1} dx = \beta(p, q)$$ **Proved.**

Example 28. *Prove that* $\beta(m, m) \times \beta\left(m+\frac{1}{2}, m+\frac{1}{2}\right) = \frac{\pi}{m} 2^{1-4m}$ (*M.U. II, Semester, 2008*)

Solution. L.H.S. $= \beta(m, m) \times \beta\left(m+\frac{1}{2}, m+\frac{1}{2}\right) = \frac{\Gamma m \, \Gamma m}{\Gamma 2m} \times \frac{\Gamma\left(m+\frac{1}{2}\right) \Gamma\left(m+\frac{1}{2}\right)}{\Gamma(2m+1)}$

$$= \frac{(\Gamma m)^2}{\Gamma 2m} \times \frac{\left(\Gamma\left(m+\frac{1}{2}\right)\right)^2}{2m \, \Gamma 2m} \qquad [\because \Gamma(2m+1) = 2m \, \Gamma 2m]$$

$$= \left(\frac{\Gamma m \, \Gamma\left(m+\frac{1}{2}\right)}{\Gamma 2m}\right)^2 \cdot \frac{1}{2m} = \left(\frac{\sqrt{\pi}}{2^{2m-1}}\right)^2 \frac{1}{2m}$$

$$= \frac{\pi}{2^{4m-2}} \cdot \frac{1}{2m}$$

$$\left[\begin{array}{c}\text{By duplication formula} \\ 2^{2m-1} \cdot \Gamma m \cdot \Gamma\left(m+\frac{1}{2}\right) = \sqrt{\pi} \cdot \sqrt{2m}\end{array}\right]$$

$$\frac{\Gamma m \, \Gamma\left(m+\frac{1}{2}\right)}{\Gamma 2m} = \frac{\sqrt{\pi}}{2^{2m-1}}$$

$$= \frac{\pi}{m} \cdot 2^{1-4m}$$

$= $ R.H.S. **Proved.**

38.11 TO SHOW THAT

$$\Gamma\left(\frac{1}{n}\right) \Gamma\left(\frac{2}{n}\right) \Gamma\left(\frac{3}{n}\right) \ldots\ldots \Gamma\left(\frac{n-1}{n}\right) = \frac{(2\pi)^{\left(\frac{n-1}{2}\right)}}{n^{1/2}}$$

where n is a positive integer than one.

Proof. Let $$P = \Gamma\left(\frac{1}{n}\right)\Gamma\left(\frac{2}{n}\right)\Gamma\left(\frac{3}{n}\right)\cdots\cdots\Gamma\left(\frac{n-2}{n}\right)\Gamma\left(\frac{n-1}{n}\right)$$

$$= \Gamma\left(\frac{1}{n}\right)\Gamma\left(\frac{2}{n}\right)\Gamma\left(\frac{3}{n}\right)\cdots\cdots\Gamma\left(1-\frac{2}{n}\right)\Gamma\left(1-\frac{1}{n}\right) \qquad \ldots(1)$$

Writing the value of P in the reverse order, we have

$$P = \Gamma\left(1-\frac{1}{n}\right)\Gamma\left(1-\frac{2}{n}\right)\cdots\cdots\Gamma\left(\frac{3}{n}\right)\Gamma\left(\frac{2}{n}\right)\Gamma\left(\frac{1}{n}\right) \qquad \ldots(2)$$

Multiplying (1) and (2), we get

$$P^2 = \left(\Gamma\left(\frac{1}{n}\right)\Gamma\left(1-\frac{1}{n}\right)\right)\left(\Gamma\left(\frac{2}{n}\right)\Gamma\left(1-\frac{2}{n}\right)\right)\cdots\cdots$$

$$\left(\Gamma\left(1-\frac{2}{n}\right)\Gamma\left(\frac{2}{n}\right)\right)\left(\Gamma\left(1-\frac{1}{n}\right)\Gamma\left(\frac{1}{n}\right)\right)$$

$$P^2 = \frac{\pi}{\sin\left(\frac{\pi}{n}\right)}\cdot\frac{\pi}{\sin\left(\frac{2\pi}{n}\right)}\cdot\frac{\pi}{\sin\left(\frac{3\pi}{n}\right)}\cdots\cdots\frac{\pi}{\sin\frac{(n-1)\pi}{n}} \qquad \left[\because \Gamma(n)\,\Gamma(1-n) = \frac{\pi}{\sin n\pi}\right]$$

$$\Rightarrow \qquad P^2 = \frac{\pi^{n-1}}{\sin\left(\frac{\pi}{n}\right)\sin\left(\frac{2\pi}{n}\right)\sin\left(\frac{3\pi}{n}\right)\cdots\cdots\sin\left\{\frac{(n-1)\pi}{n}\right\}} \qquad \ldots(3)$$

But from Trigonometry, we know that

$$\frac{\sin n\theta}{\sin\theta} = 2^{n-1}\sin\left(\theta+\frac{\pi}{n}\right)\sin\left(\theta+\frac{2\pi}{n}\right)\cdots\cdots\sin\left\{\theta+\frac{(n-1)\pi}{n}\right\} \qquad \ldots(4)$$

Take Limit as $\theta \to 0$,

$$\operatorname*{Lt}_{\theta\to 0}\frac{\sin n\theta}{\sin\theta} = \operatorname*{Lt}_{\theta\to 0}\left(n\cdot\frac{\sin n\theta}{n\theta}\cdot\frac{\theta}{\sin\theta}\right) = n$$

On putting this limitm in (4), we get

$$n = 2^{n-1}\sin\frac{\pi}{n}\sin\frac{2\pi}{n}\sin\frac{3\pi}{n}\ldots\sin\frac{(n-1)\pi}{n} \qquad \text{[From (4)]}$$

$$\Rightarrow \qquad \sin\frac{\pi}{n}\cdot\sin\frac{2\pi}{n}\cdot\sin\frac{3\pi}{n}\ldots\sin\frac{(n-1)\pi}{n} = \frac{n}{2^{n-1}}$$

Substituting this in equation (3), we obtain

$$P^2 = \frac{\pi^{n-1}}{\left(\frac{n}{2^{n-1}}\right)} = \frac{(2\pi)^{n-1}}{n} \qquad \therefore \qquad P = \frac{(2\pi)^{(n-1)/2}}{n^{1/2}}$$

$$\Rightarrow \qquad \boxed{\Gamma\left(\frac{1}{n}\right)\Gamma\left(\frac{2}{n}\right)\Gamma\left(\frac{3}{n}\right)\cdots\cdots\Gamma\left(\frac{n-1}{n}\right) = \frac{(2\pi)^{(n-1)/2}}{n^{1/2}}}$$

38.12 TO SHOW THAT

(*i*) $\int_0^\infty e^{-ax}\,x^{n-1}\cos bx\,dx = \dfrac{\Gamma(n)\cos n\theta}{(a^2+b^2)^{n/2}}$ [*U.P., I Semester, (C.O.) 2004*]

(*ii*) $\int_0^\infty e^{-ax}\,x^{n-1}\sin bx\,dx = \dfrac{\Gamma(n)\sin n\theta}{(a^2+b^2)^{n/2}}$ where $\theta = \tan^{-1}\left(\dfrac{b}{a}\right)$ [*U.P., I Semester, 2003*]

Proof. We know that $\int_0^\infty e^{-ax} . x^{n-1}\, dx = \dfrac{\Gamma(n)}{a^n}$, where a, n are positive.

Put $ax = z$ so that $\qquad dx = \dfrac{dz}{a}$

$$\therefore \qquad \int_0^\infty e^{-ax}\, x^{n-1}\, dx = \int_0^\infty e^{-z}\left(\frac{z}{a}\right)^{n-1} . \frac{dz}{a} = \frac{1}{a^n}\int_0^\infty e^{-z}\, z^{n-1}\, dz = \frac{\Gamma(n)}{a^n}$$

Replacing a by $(a + ib)$, we have

$$\int_0^\infty e^{-(a+ib)x}\, x^{n-1}\, dx = \frac{\Gamma n}{(a+ib)^n} \qquad ...(1)$$

Now $\qquad e^{-(a+ib)x} = e^{-ax} . e^{-ibx} = e^{-ax}(\cos bx - i \sin bx)$

Putting the value of $e^{-(a+ib)x}$ in (1), we get

$$\int_0^\infty e^{-ax}(\cos bx - i \sin bx)\, x^{n-1}\, dx = \frac{\overline{|n}}{(a+ib)^n} \qquad ...(2)$$

Putting $a = r \cos \theta$ and $b = r \sin \theta$ so that $r^2 = a^2 + b^2$ and $\theta = \tan^{-1} \dfrac{b}{a}$

$$(a + ib)^n = (r \cos \theta + ir \sin \theta)^n = r^n (\cos \theta + i \sin \theta)^n$$

$$(a + ib)^n = r^n (\cos n\theta + i \sin n\theta) \qquad \text{[De Moivre's Theorm]}$$

Putting the value of $(a + ib)^n$ in (2), we have

$$\int_0^\infty e^{-ax}(\cos bx - i \sin bx)\, x^{n-1}\, dx = \frac{\overline{|n}}{r^n (\cos n\theta + i \sin n\theta)}$$

$$= \frac{\overline{|n}}{r^n}(\cos n\,\theta + i \sin n\,\theta)^{-1}$$

$$= \frac{\overline{|n}}{r^n}(\cos n\theta - i \sin n\theta)$$

Now, equating real and imaginary parts on the two sides, we get

(*i*) $\qquad \int_0^\infty e^{-ax}\, x^{n-1} \cos bx\, dx = \dfrac{\overline{|n}}{r^n} \cos n\,\theta$ and

(*ii*) $\qquad \int_0^\infty e^{-ax}\, x^{n-1} \sin bx\, dx = \dfrac{\overline{|n}}{r^n} \sin n\,\theta$

where $\qquad r^2 = a^2 + b^2 \qquad$ and $\qquad \theta = \tan^{-1} \dfrac{b}{a}$. **Proved.**

Example 29. *Evaluate:*

(*i*) $\int_0^\infty \cos x^2\, dx \qquad$ (*ii*) $\int_{-\infty}^\infty \cos \dfrac{\pi x^2}{2}\, dx.$

Solution. (*i*) We know that

$$\int_0^\infty e^{-ax} . x^{n-1} \cos bx\, dx = \frac{\Gamma n \cos n\theta}{(a^2 + b^2)^{n/2}}, \text{ where } \theta = \tan^{-1}\left(\frac{b}{a}\right)$$

Put $a = 0$, $\qquad \int_0^\infty x^{n-1} \cos bx\, dx = \dfrac{\Gamma n}{b^n} \cos \dfrac{n\pi}{2}$ $\qquad \left[\begin{aligned}\theta &= \tan^{-1}\left(\frac{b}{0}\right)\\ &= \tan^{-1}(\infty) = \frac{\pi}{2}\end{aligned}\right]$

Put $x^n = z$ so that $\qquad x^{n-1}\, dx = \dfrac{dz}{n}$ and $x = z^{1/n}$

then, $\qquad \int_0^\infty \cos bz^{1/n}\, dz = \dfrac{n\,\Gamma n}{b^n} \cos \dfrac{n\pi}{2}$

$$\Rightarrow \qquad \int_0^\infty \cos(bx^{1/n})\,dx = \frac{\Gamma(n+1)}{b^n}\cos\frac{n\pi}{2} \qquad ...(1)$$

Here $b = 1,\ n = \dfrac{1}{2}$

$$\therefore \qquad \int_0^\infty \cos x^2\,dx = \Gamma\left(\frac{3}{2}\right)\cos\frac{\pi}{4} = \frac{\sqrt{\pi}}{2}\cdot\frac{1}{\sqrt{2}} = \frac{\sqrt{\pi}}{2\sqrt{2}} \qquad \textbf{Ans.}$$

$$(ii)\ I = \int_{-\infty}^\infty \cos\frac{\pi x^2}{2}\,dx = 2\int_0^\infty \cos\frac{\pi x^2}{2}\,dx \qquad [f(-x) = f(x)] \qquad ...(2)$$

Putting $b = \dfrac{\pi}{2}$ and $n = \dfrac{1}{2}$ in equation (1), we get

$$\int_0^\infty \cos\left(\frac{\pi}{2}x^2\right)dx = \frac{\Gamma\left(\frac{3}{2}\right)}{\left(\frac{\pi}{2}\right)^{1/2}}\cos\frac{\pi}{4}$$

$$\therefore \text{ From (2),} \qquad I = 2\,\frac{\Gamma\left(\frac{3}{2}\right)}{\left(\frac{\pi}{2}\right)^{1/2}}\cos\frac{\pi}{4} = 2.\frac{1}{2}\sqrt{\pi}.\frac{\sqrt{2}}{\sqrt{\pi}}.\frac{1}{\sqrt{2}} = 1. \qquad \textbf{Ans.}$$

Example 30. *Evaluate:* $\int_0^1 \log\Gamma(x)\,dx$

Solution. Let $\quad I = \int_0^1 \log\Gamma(x)\,dx$

$$= \int_0^1 \log\Gamma(1-x)\,dx \qquad \left[\int_0^a f(x)\,dx = \int_0^a f(a-x)\,dx\right]$$

Adding (1) and (2), we get

$$2I = \int_0^1 (\log\Gamma(x) + \log\Gamma(1-x))\,dx$$

$$= \int_0^1 \log(\Gamma(x)\,\Gamma(1-x))\,dx = \int_0^1 \log\left(\frac{\pi}{\sin\pi x}\right)dx \qquad \left[\overline{|x}\,\overline{|1-x} = \frac{\pi}{\sin\pi x}\right]$$

$$= \int_0^1 (\log\pi - \log\sin\pi x)\,dx = \int_0^1 \log\pi\,dx - \int_0^1 \log\sin\pi x\,dx$$

$$= I_1 - I_2 \qquad ...(3)$$

where $\quad I_1 = \int_0^1 \log\pi\,dx = \log\pi$

$$I_2 = \int_0^1 \log\sin\pi x\,dx \qquad \left[\text{Put } \pi x = t \Rightarrow dx = \frac{1}{\pi}\,dt\right]$$

$$= \int_0^\pi \log\sin t\left(\frac{dt}{\pi}\right) = \frac{1}{\pi}\cdot 2\int_0^{\pi/2} \log\sin t\,dt = \frac{2}{\pi}\left(-\frac{\pi}{2}\log 2\right) = -\log 2$$

From (3), $\quad 2I = \log\pi + \log 2 = \log 2\pi$

$$I = \frac{1}{2}\log 2\pi. \qquad \textbf{Ans.}$$

Example 31. *Prove that* $\int_0^\infty x^2 e^{-x^4}\,dx \times \int_0^\infty \frac{e^{-x^2}}{\sqrt{x}}\,dx = \frac{\pi}{4\sqrt{2}}$ *(M.U., II Semester, 2008)*

Solution. Here, we have

$$\int_0^\infty x^2 e^{-x^4}\, dx \qquad \ldots(1)$$

Putting $x^4 = t \Rightarrow x = t^{\frac{1}{4}}$, $\quad dx = \frac{1}{4} t^{-\frac{3}{4}}\, dt$ in (1), we get

$$\int_0^\infty t^{\frac{1}{2}} e^{-t}\left(\frac{1}{4} t^{\frac{-3}{4}}\, dt\right) = \frac{1}{4}\int_0^\infty e^{-t} t^{-\frac{1}{4}}\, dt = \frac{1}{4}\int_0^\infty e^{-t}\, t^{\frac{3}{4}-1}\, dt = \frac{1}{4}\Gamma\left(\frac{3}{4}\right) \qquad \ldots(2)$$

Now, we calculate the value of $\int_0^\infty \frac{e^{-x^2}}{\sqrt{x}}\, dx$ $\qquad \ldots(3)$

Putting $x^2 = t$, $\quad x = t^{\frac{1}{2}} \quad \Rightarrow \quad dx = \frac{1}{2} t^{-\frac{1}{2}}\, dt$ in (2), we get

$$\int_0^\infty e^{-t} t^{-\frac{1}{4}}\left(\frac{1}{2} t^{\frac{-1}{2}}\, dt\right) = \frac{1}{2}\int_0^\infty e^{-t} t^{-\frac{3}{4}}\, dt = \frac{1}{2}\int_0^\infty e^{-t}\, t^{\frac{1}{4}-1}\, dt = \frac{1}{2}\Gamma\left(\frac{1}{4}\right) \qquad \ldots(4)$$

From (3) and (4), we get

$$\int_0^\infty x^2 e^{-x^4}\, dx \cdot \int_0^\infty \frac{e^{-x^2}}{\sqrt{x}}\, dx = \left(\frac{1}{4}\Gamma\left(\frac{3}{4}\right)\right)\left(\frac{1}{2}\Gamma\left(\frac{1}{4}\right)\right) = \frac{1}{8}\,\frac{\Gamma\left(\frac{3}{4}\right)\Gamma\left(\frac{1}{4}\right)}{\Gamma\left(\frac{3}{4}+\frac{1}{4}\right)}$$

[Duplication formula

$$2^{2m-1}\Gamma(m)\,\Gamma\left(m+\frac{1}{2}\right) = \sqrt{\pi}\,\Gamma(2m)$$

Putting $m = \frac{1}{4}$, we get

$$2^{-\frac{1}{2}}\Gamma\left(\frac{1}{4}\right)\Gamma\left(\frac{3}{4}\right) = \sqrt{\pi}\,\Gamma\left(\frac{1}{2}\right)$$

$$\Gamma\left(\frac{1}{4}\right)\Gamma\left(\frac{3}{4}\right) = \sqrt{\pi}\sqrt{\pi}\,(2)^{\frac{1}{2}} = \pi\sqrt{2}\text{]}$$

$$= \frac{1}{8}(\pi\sqrt{2}) = \frac{\pi}{4\sqrt{2}}$$

Proved.

Example 32. *Show that*

$$\int_0^1 \frac{x^{m-1}(1-x)^{n-1}}{(a+x)^{m+n}}\, dx = \frac{\beta(m,n)}{a^n(1+a)^m}$$

Solution. Put $\frac{x}{a+x} = \frac{t}{a+1}$

$$(a+1)\,x = t\,(a+x) \;\Rightarrow\; x = \frac{at}{a+1-t}$$

$$dx = \frac{(a+1-t)\,a\,dt - a\,t\,(-dt)}{(a+1-t)^2} = \frac{\left(a^2+a-at+at\right)}{(a+1-t)^2}\,dt = \frac{a(a+1)}{(a+1-t)^2}\,dt$$

$$\text{z}\int_0^1 \frac{x^{m-1}(1-x)^{n-1}}{(a+x)^{m+n}}\, dx = \int_0^1 \frac{\left(\frac{at}{a+1-t}\right)^{m-1}\cdot\left(1-\frac{at}{a+1-t}\right)^{n-1}}{\left(a+\frac{at}{a+1-t}\right)^{m+n}}\,\frac{a(a+1)}{(a+1-t)^2}\,dt$$

$$= \int_0^1 \frac{(at)^{m-1}(a+1-t-at)^{n-1}}{(a^2+a-at+at)^{m+n}}\,a\,(a+1)\,dt$$

$$= \int_0^1 \frac{a^{m-1}\, t^{m-1}(a+1)^{n-1}(1-t)^{n-1}}{a^{m+n}(a+1)^{m+n}}\, a\,(a+1)\, dt$$

$$= \frac{1}{a^n (a+1)^m} \int_0^1 t^{m-1}(1-t)^{n-1}\, dt \;=\; \frac{1}{a^n (a+1)^m}\, \beta\,(m, n)$$ **Proved.**

Example 33. *Prove that* $\int_0^\infty \frac{x^4\left(1+x^5\right)}{(1+x)^{15}}\, dx = \frac{1}{5005}$. *(M.U. II Semester, 2008)*

Solution. Let $I = \int_0^\infty \frac{x^4\left(1+x^5\right)}{(1+x)^{15}}\, dx$

$$\Rightarrow \qquad I = \int_0^\infty \frac{x^4}{(1+x)^{15}}\, dx + \int_0^\infty \frac{x^9}{(1+x)^{15}}\, dx = I_1 + I_2 \qquad \text{...(1)}$$

Now, put $x = \frac{t}{1-t}$, when $x = 0$, $t = 0$; when $x = \infty$, $t = 1$

$$1 + x = 1 + \frac{t}{1-t} = \frac{1}{1-t} \quad \Rightarrow \quad dx = \frac{dt}{(1-t)^2}$$

$$\therefore \qquad I_1 = \int_0^1 \left(\frac{t}{1-t}\right)^4 .(1-t)^{15} . \frac{1}{(1-t)^2}\, dt = \int_0^1 t^4\,(1+t)^9\, dt = \beta\,(5, 10) \qquad \text{...(2)}$$

and

$$I_2 = \int_0^1 \left(\frac{t}{1-t}\right)^9 .(1-t)^{15} . \frac{dt}{(1-t)^2} = \int_0^1 t^9\,(1-t)^4\, dt = \beta\,(10, 5) \qquad \text{...(3)}$$

$\therefore \qquad I = I_1 + I_2$

$= \beta\,(5, 10) + \beta\,(10, 5)$ [Using (2) and (3)]

$= \beta\,(5, 10) + \beta\,(5, 10)$ [$\because\ \beta(m, n) = \beta(n, m)$]

$$= 2\,\beta\,(5, 10) = \frac{2\,\Gamma 5\,\Gamma 10}{\Gamma 15} = \frac{2.4!.9!}{14!}$$

$$= \frac{2\times4\times3\times2\times1\times9!}{14\times13\times12\times11\times10\times9!} = \frac{1}{7\times13\times11\times5} = \frac{1}{5005}$$ **Proved.**

Example 34. *Prove that*

$$\int_0^1 \frac{x^3 - 2x^4 + x^5}{(1+x)^7}\, dx = \frac{1}{960}.$$ *(M.U., II Semester, 2008)*

Solution. Let $I = \int_0^1 \frac{x^3 - 2x^4 + x^5}{(1+x)^7}\, dx$

$$\Rightarrow \qquad I = \int_0^1 \frac{x^3}{(1+x)^7}\, dx - 2\int_0^1 \frac{x^4}{(1+x)^7}\, dx + \int_0^1 \frac{x^5}{(1+x)^7}\, dx$$

$$\Rightarrow \qquad I = I_1 - 2\, I_2 + I_3 \qquad \text{...(1)}$$

Now put $x = \frac{t}{1-t}$ so that $1 + x = 1 + \frac{t}{1-t} \Rightarrow 1 + x = \frac{1}{1-t} \Rightarrow dx = \frac{d\,t}{(1-t)^2}$

$$\therefore \qquad I_1 = \int_0^1 \left(\frac{t}{1-t}\right)^3 .(1-t)^7 . \frac{1}{(1-t)^2}\, d\,t$$

$$= \int_0^1 t^3 (1-t)^2\, dt = \int_0^1 t^{4-1} (1-t)^{3-1}\, dt = \beta\,(4, 3) \qquad ...(2)$$

and $$I_2 = \int_0^1 \left(\frac{t}{1-t}\right)^4 (1-t)^7 \cdot \frac{1}{(1-t)^2}\, dt$$

$$= \int_0^1 t^4 (1-t)\, dt = \int_0^1 t^{5-1} (1-t)^{2-1}\, dt = \beta\,(5, 2) \qquad ...(3)$$

Also, $$I_3 = \int_0^1 \left(\frac{t}{1-t}\right)^5 (1-t)^7 \frac{dt}{(1-t)^2} = \int_0^1 t^5 (1-t)^0\, dt = \int_0^1 t^{6-1} (1-t)^{1-1}\, dt = \beta\,(6, 1) \qquad ...(4)$$

Putting the values of I_1, I_2 and I_3 in (1), we get

$$I = \beta\,(4, 3) - 2\,\beta\,(5, 2) + \beta\,(6, 1) = \frac{\overline{|4}\,\overline{|3}}{\overline{|4+3}} - 2\frac{\overline{|5}\,\overline{|2}}{\overline{|5+2}} + \frac{\overline{|6}\,\overline{|1}}{\overline{|6+1}}$$

$$= \frac{\overline{|4}\,\overline{|3}}{\overline{|7}} - 2\frac{\overline{|5}\,\overline{|2}}{\overline{|7}} + \frac{\overline{|6}\,\overline{|1}}{\overline{|7}} = \frac{\overline{|3}}{4\times5\times6} - 2\frac{1}{5\times6} + \frac{1}{6}$$

$$= \frac{1}{60} - \frac{1}{15} + \frac{1}{6} = \frac{1-4+10}{60} = \frac{7}{60}$$ **Ans.**

EXERCISE 38.2

Prove that

1. $\int_0^{\frac{\pi}{2}} \sin^2\theta \cos^4\theta\, d\theta = \frac{\pi}{32}$

2. $\int_0^{\frac{\pi}{2}} \sin^6\theta\, d\theta = \frac{5\pi}{32}$

3. $\int_0^{\frac{\pi}{2}} \sin^{m-1}(2\theta)\, d\theta = \frac{2^{m-1}}{m!}\left(\overline{\left|\frac{m}{2}\right.}\right)^2$ *(Nagpur University, Winter 2003)*

4. $\int_0^{\pi} \sin^5 x\,(1-\cos x)^3\, dx = \frac{32}{21}$ *(Nagpur University, Winter 2001)*

5. $\int_0^2 x\,(8-x^3)^{\frac{1}{3}}\, dx = \frac{1}{9}\,\overline{\left|\frac{1}{3}\right.}\,\overline{\left|\frac{2}{3}\right.}$ *(Nagpur University, Summer 2005)*

6. $\int_0^{2\pi} x\sqrt{2ax - x^2}\, dx = \frac{\pi a^3}{2}$ *(Nagpur University, Summer 2002)*

7. $\beta\,(m+1, n) = \frac{m}{m+n}\,\beta\,(m, n)$

8. $\beta\,(m, n+1) = \frac{n}{m+n}\,\beta\,(m, n)$

9. $\int_0^1 \sqrt{x}\,\sqrt[3]{1-x^2}\, dx = \frac{\overline{\left|\frac{3}{4}\right.}\,\overline{\left|\frac{4}{3}\right.}}{2\,\overline{\left|\frac{7}{12}\right.}}$

10. $\int_0^1 (1-x^n)^{-\frac{1}{2}}\, dx = \frac{\overline{\left|\frac{1}{n}\right.}\,\overline{\left|\frac{1}{2}\right.}}{n\,\overline{\left|\frac{n+2}{2n}\right.}}$

11. $\int_0^1 (1-x^{1/n})^m\, dx = \frac{\overline{|m}\,\overline{|n}}{\overline{|m+n}}$

12. $\int_1^{\infty} \frac{dx}{x^{P+1}(x-1)^q} = \beta(P+q, 1-q)$ if $-P < q < 1$

13. $\int_0^1 x^m (1-x^n)^P\, dx = \frac{1}{n}\,\frac{\overline{\left|\frac{m+1}{2}\right.}\,\overline{|P+1}}{\overline{\left|\frac{m+1}{n} + P + 1\right.}}$

14. $\int_0^b (x-a)^m (b-x)^n\, dx = (b-a)^{m+n+1} \cdot \beta\,(m+1, n+1)$

15. $\int_a^b (x-a)^{m-1} (b-x)^{n-1}\, dx = (b-a)^{m+n-1} \cdot \beta\,(m, n)$

16. $\int_3^7 \sqrt[4]{(x-3)(7-x)}\, dx = \dfrac{2\left(\sqrt{\frac{1}{4}}\right)^2}{3\sqrt{\pi}}$ **[Hint.** Put $x = 4\,t + 3$]

17. $\displaystyle\int_0^{\frac{\pi}{2}} \frac{d\theta}{\sqrt{1-\frac{1}{2}\sin^2\theta}} = \frac{\left(\sqrt{\frac{1}{4}}\right)^2}{4\sqrt{\pi}}$

18. If $\int_0^\infty e^{-x}\, x^{n-1}\, dx = I_n$ for $n > 0$ find $\dfrac{I_{n+1}}{I_n}$ *(A.M.I.E., Summer 2000)* **Ans.** n

Show that :

19. $\displaystyle\int_0^\infty \sqrt{x}e^{-x^2}\, dx \times \int_0^\infty \frac{e^{-x^2}}{\sqrt{x}}\, dx = \frac{\pi}{2\sqrt{2}}$

20. $\displaystyle\int_0^1 \frac{x^2 dx}{\sqrt{1-x^4}} \int_0^1 \frac{dx}{\sqrt{1+x^4}} = \frac{\pi}{4\sqrt{2}}$

21. $\displaystyle\int_0^\infty x^2 e^{-x^4}\, dx \times \int_0^\infty e^{-x^4}\, dx = \frac{\pi}{8\sqrt{2}}$

22. $\displaystyle\int_0^{\pi/2} \sin^3 x \cos^{5/2} x\, dx = \frac{8}{77}$

23. $\displaystyle\int_0^1 \frac{x^{m-1}\,(1-x)^{n-1}}{(a+bx)^{m+n}}\, dx = \frac{\beta(m, n)}{a^n\,(a+b)^m}$

24. $\displaystyle\frac{\beta\,(p, q+1)}{q} = \frac{\beta\,(p+1, q)}{p} = \frac{\beta\,(p, q)}{p+q}$

38.13 DOUBLE INTEGRATION

Example 35. *Evaluate* $\displaystyle\iint_A \frac{dx\, dy}{\sqrt{xy}}$ *using the substitutions*

$$x = \frac{u}{1+v^2}, \quad y = \frac{uv}{1+v^2}$$

where A is bounded by $x^2 + y^2 - x = 0,\ y = 0,\ y > 0.$

Solution. Here $\displaystyle\sqrt{xy} = \sqrt{\left(\frac{u}{1+v^2}\right)\left(\frac{uv}{1+v^2}\right)} = \frac{u\sqrt{v}}{1+v^2}$

$$dx\ dy = \left|\frac{\partial(x, y)}{\partial(u, v)}\right| du\, dv$$

$$= \begin{vmatrix} \frac{\partial x}{\partial u} & \frac{\partial x}{\partial v} \\ \frac{\partial y}{\partial u} & \frac{\partial y}{\partial v} \end{vmatrix} du\, dv = \begin{vmatrix} \frac{1}{1+v^2} & -\frac{2uv}{(1+v^2)^2} \\ \frac{v}{1+v^2} & \frac{u(1-v^2)}{(1+v^2)^2} \end{vmatrix} du\, dv$$

$$= \left[\frac{u(1-v^2)}{(1+v^2)^3} + \frac{2uv^2}{(1+v^2)^3}\right] du\, dv = \left[\frac{u - uv^2 + 2uv^2}{(1+v^2)^3}\right] du\, dv$$

$$= \frac{u(1+v^2)}{(1+v^2)^3}\, du\, dv = \frac{u}{(1+v^2)^2}\, du\, dv$$

Also the circle $x^2 + y^2 - x = 0$ is transformed into

$$\frac{u^2}{(1+v^2)^2} + \frac{u^2v^2}{(1+v^2)^2} - \frac{u}{1+v^2} = 0 \Rightarrow \frac{u^2(1+v^2)}{(1+v^2)^2} - \frac{u}{1+v^2} = 0$$

$$\Rightarrow \quad \frac{u^2}{1+v^2}-\frac{u}{1+v^2} = 0 \Rightarrow u^2-u=0 \Rightarrow u(u-1)=0 \Rightarrow u=0, u=1$$

Further $\quad y = 0 \Rightarrow \frac{uv}{1+v^2}=0 \Rightarrow u=0, v=0$

and $y>0 \Rightarrow uv>0$ either both u and v are positive or both negative.

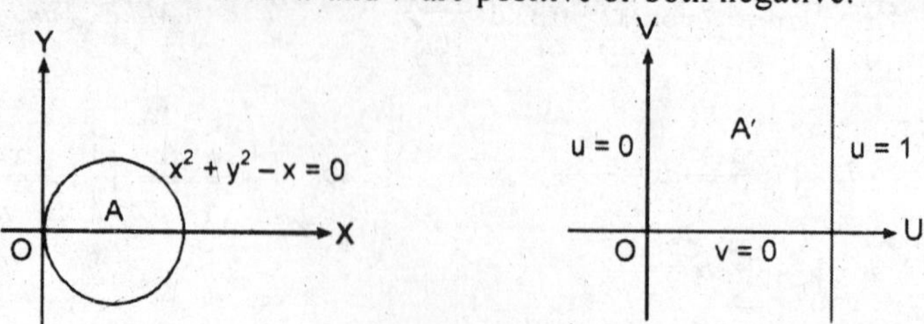

The area A, *i.e.*, $x^2+y^2-x=0$ is transformed into 'A'' bounded by $u=0$, $v=0$ and $u=1$ and $v=\infty$.

$$\iint \frac{dx\,dy}{\sqrt{x}} = \int_0^1\int_0^\infty \frac{\frac{u}{(1+v^2)^2}}{\frac{u\sqrt{v}}{1+v^2}}\,du\,dv = \int_0^1\int_0^\infty \frac{1}{\sqrt{v}(1+v^2)}\,dv\,du$$

On putting $v=\tan\theta$, $dv=\sec^2\theta\,d\theta$

$$= \int_0^1\int_0^{\frac{\pi}{2}} \frac{\sec^2\theta}{\sqrt{\tan\theta}\left(1+\tan^2\theta\right)}\,d\theta\,du = \int_0^1 du\int_0^{\frac{\pi}{2}}\sqrt{\frac{\cos\theta}{\sin\theta}}\,d\theta = \int_0^1 du\int_0^{\frac{\pi}{2}} \sin^{-\frac{1}{2}}\theta\cos^{\frac{1}{2}}\theta\,d\theta$$

Duplication formula $\boxed{m}\;\boxed{m+\frac{1}{2}} = \frac{\sqrt{\pi}}{2^{2m-1}}\boxed{2m}$

$$= \int_0^1 du\,\frac{\boxed{-\frac{1}{2}+1}\;\boxed{\frac{1}{2}+1}}{\boxed{2}\;\;\boxed{2}} = \frac{1}{2}\int_0^1 du\,\boxed{\frac{1}{4}}\;\boxed{\frac{3}{4}} = \frac{1}{2}\int_0^1 du\left[\frac{\sqrt{\pi}}{2^{\frac{-1}{2}}}\boxed{\frac{1}{2}}\right]$$

$$\frac{1}{2}\int_0^1 du\,\sqrt{2}\sqrt{\pi}\cdot\sqrt{\pi} = \frac{\pi}{\sqrt{2}}[u]_0^1 = \frac{\pi}{\sqrt{2}}$$ **Ans.**

Example 36. *Prove that*

$$\iint_D x^{l-1}y^{m-1}dx\,dy = \frac{\boxed{l}\;\boxed{m}}{\boxed{l+m+1}}h^{l+m}$$

where D is the domain $x\ge 0$, $y\ge 0$ and $x+y\le h$. (*U.P., I Semester, Dec. 2004*)

Solution. Putting $x=Xh$ and $y=Yh$, $dx\,dy=h^2\,dX\,dY$

$$\iint_D x^{l-1}y^{m-1}dx\,dy = \iint_{D'}(Xh)^{l-1}(Yh)^{m-1}h^2dX\,dY$$

where D' is the domain

$X\ge 0, \quad Y\ge 0, \quad X+Y\le 1$

$$= h^{l+m}\int_0^1\int_0^{1-X} X^{l-1}Y^{m-1}dX\,dY = h^{l+m}\int_0^1 X^{l-1}dX\int_0^{1-X} Y^{m-1}dY$$

$$= h^{l+m}\int_0^1 X^{l-1}dX\left[\frac{Y^m}{m}\right]_0^{1-X} = \frac{h^{l+m}}{m}\int_0^1 X^{l-1}(1-X)^m\,dX$$

$$= \frac{h^{l+m}}{m}\beta(l,m+1) = \frac{h^{l+m}}{m}\frac{\boxed{l}\;\boxed{m+1}}{\boxed{l+m+1}} = \frac{h^{l+m}}{m}\frac{m\boxed{l}\;\boxed{m}}{\boxed{l+m+1}} = h^{l+m}\frac{\boxed{l}\;\boxed{m}}{\boxed{l+m+1}}$$ **Proved.**

38.14 DIRICHLET'S INTEGRAL (Triple Integration)

If l, m, n are all positive, then the triple integral

$$\iiint_V x^{l-1} y^{m-1} z^{n-1} dx\, dy\, dz = \frac{\lceil l \lceil m \lceil n}{\lceil l+m+n+1}$$

where V is the region $x \geq 0$, $y \geq 0$, $z \geq 0$ and $x + y + z \leq 1$.

Proof. Putting $y + z \leq 1 - x = h$. Then $z \leq h - y$

$$\iiint_V x^{l-1} y^{m-1} z^{n-1} dx\, dy\, dz = \int_0^1 x^{l-1} dx \int_0^{1-x} y^{m-1} dy \int_0^{1-x-y} z^{n-1}\, dz$$

$$= \int_0^1 x^{l-1} dx \left[\int_0^h \int_0^{h-y} y^{m-1} z^{n-1}\, dy\, dz \right] \qquad \text{[Put } x = h\text{]}$$

$$= \int_0^1 x^{l-1}\, dx \left[\frac{\lceil m \lceil n}{\lceil m+n+1} h^{m+n} \right] = \int_0^1 x^{l-1} dx \left[\frac{\lceil m \lceil n}{\lceil m+n+1} h^{m+n} \right]$$

$$= \frac{\lceil m \lceil n}{\lceil m+n+1} \int_0^1 x^{l-1} (1-x)^{m+n} dx = \frac{\lceil m \lceil n}{\lceil m+n+1} \beta(l, m+n+1)$$

$$= \frac{\lceil m \lceil n}{\lceil m+n+1} \frac{\lceil l \lceil m+n+1}{\lceil l+m+n+1}$$

$$\boxed{\iiint_V x^{l-1} y^{m-1} z^{n-1} dx\, dy\, dz = \frac{\lceil l \lceil m \lceil n}{\lceil l+m+n+1}}$$

Note. $\displaystyle\iiint_V x^{l-1} y^{m-1} z^{n-1} dx\, dy\, dz = \frac{\lceil l \lceil m \lceil n}{\lceil l+m+n+1} h^{l+m+n}$

where V is the domain, $x \geq 0$, $y \geq 0$, $z \geq 0$ and $x + y + z \leq h$.

Corollary: Dirichlet's theorem for n variables, the theorem states that

$$\iiint \cdots \int x_1^{l_1-1} x_2^{l_2-1} \ldots x_n^{l_n-1}\, dx_1\, dx_2\, dx_3 \ldots dx_n = \frac{\lceil l_1 \lceil l_2 \lceil l_3 \ldots \lceil l_n}{\lceil 1 + l_1 + l_2 + \ldots + l_n} h^{l_1 + l_2 + \ldots + l_n}$$

38.15 LIOUVILLE'S EXTENSION OF DIRICHLET THEOREM

If the variables x, y, z are all positive such that $h_1 < x + y + z < h_2$, then

$$\iiint f(x+y+z)\, x^{l-1} y^{m-1} z^{n-1}\, dx\, dy\, dz = \frac{\lceil l \lceil m \lceil n}{\lceil l+m+n} \int_{h_1}^{h_2} f(u)\, u^{l+m+n-1}\, du$$

Proof. By Dirichlet Theorem, we have

$$I = \iiint X^{l-1}\, Y^{m-1}\, Z^{n-1}\, dX\, dY\, dZ = \frac{\lceil l \lceil m \lceil n}{\lceil 1+l+m+n} \qquad \ldots(1)$$

Under the condition $x + y + z \leq u \Rightarrow \dfrac{x}{u} + \dfrac{y}{u} + \dfrac{z}{u} \leq 1$

Putting $X = \dfrac{x}{u}$, $\quad Y = \dfrac{y}{u}$ and $Z = \dfrac{z}{u}$ so that

$dX = \dfrac{dx}{u}$, $\quad dY = \dfrac{dy}{u}$, $\quad dZ = \dfrac{dz}{u}$ in (1), we get

$$\iiint \left(\frac{x}{u}\right)^{l-1} \left(\frac{y}{u}\right)^{m-1} \left(\frac{z}{u}\right)^{n-1} \frac{dx}{u} \frac{dy}{u} \frac{dz}{u} = \frac{\lceil l \lceil m \lceil n}{\lceil 1+l+m+n}$$

$$\iiint x^{l-1}\, y^{m-1}\, z^{n-1}\, dx\, dy\, dz = u^{l+m+n} \cdot \frac{\overline{|l}\,\overline{|m}\,\overline{|n}}{\overline{|1+l+m+n}}$$

Similarly, if $x + y + z \leq u + \delta u$ then

$$\iiint x^{l-1}\, y^{m-1}\, z^{n-1}\, dx\, dy\, dz = (u+\delta u)^{l+m+n} \frac{\overline{|l}\,\overline{|m}\,\overline{|n}}{\overline{|1+l+m+n}}$$

Hence, value of the integral I extended to all such values of the variables as make the sum of the variables lie between u and $u + \delta u$ is given by

$$I = (u+\delta u)^{l+m+n} \frac{\overline{|l}\,\overline{|m}\,\overline{|n}}{\overline{|1+l+m+n}} - u^{l+m+n} \frac{\overline{|l}\,\overline{|m}\,\overline{|n}}{\overline{|1+l+m+n}}$$

$$\iiint x^{l-1}\, y^{m-1}\, z^{n-1}\, dx\, dy\, dz = \frac{\overline{|l}\,\overline{|m}\,\overline{|n}}{\overline{|1+l+m+n}} [(u+\delta u)^{l+m+n} - u^{l+m+n}]$$

$$= \frac{\overline{|l}\,\overline{|m}\,\overline{|n}}{\overline{|1+l+m+n}} u^{l+m+n} \left[\left(1+\frac{\delta u}{u}\right)^{l+m+n} - 1\right]$$

$$= \frac{\overline{|l}\,\overline{|m}\,\overline{|n}}{\overline{|1+l+m+n}} u^{l+m+n} \left[1+(l+m+n)\frac{\delta u}{u}+\ldots\ldots-1\right]$$

$$= \frac{\overline{|l}\,\overline{|m}\,\overline{|n}}{\overline{|1+l+m+n}} u^{l+m+n} (l+m+n)\frac{\delta u}{u} = \frac{\overline{|l}\,\overline{|m}\,\overline{|n}}{\overline{|l+m+n}} u^{l+m+n-1}\, \delta u$$

Let us consider $\iiint f(x+y+z)\, x^{l-1}\, y^{m-1}\, z^{n-1}\, dx\, dy\, dz$

Under the condition $h_1 \leq x + y + z \leq h_2$

When $x + y + z$ lies between u and $u + \delta u$, the value of $f(x + y + z)$ can only differ from $f(u)$ by a small quantity of the same order as δu. Hence, neglecting square of δu, the part of the integral

$$\iiint f(x+y+z)\, x^{l-1}\, y^{m-1}\, z^{n-1}\, dx\, dy\, dz = \frac{\overline{|l}\,\overline{|m}\,\overline{|n}}{\overline{|l+m+n}} f(u) u^{l+m+n-1}\, \delta u$$

(supposing the sum of variables to be between u and $u + \delta u$)

So $$\boldsymbol{\iiint f(x+y+z)\, x^{l-1}\, y^{m-1}\, z^{n-1}\, dx\, dy\, dz = \frac{\overline{|l}\,\overline{|m}\,\overline{|n}}{\overline{|(l+m+n)}} \int_{h_1}^{h_2} f(u)\, u^{l+m+n-1} du}$$

Example 37. *Show that* $\iiint \frac{dx\, dy\, dz}{(x+y+z+1)^3} = \frac{1}{2}\log 2 - \frac{5}{16}$, *the integral being taken throughout the volume bounded by planes* $x = 0,\ y = 0,\ z = 0,\ x + y + z = 1$.

Solution. By Liouville's theorem when $0 < x + y + z < 1$

$$\iiint \frac{dx\, dy\, dz}{(x+y+z+1)^3} = \iiint \frac{x^{1-1}\, y^{1-1}\, z^{1-1}\, dx\, dy\, dz}{(x+y+z+1)^3} \qquad (0 \leq x + y + z \leq 1)$$

$$= \frac{\overline{|1}\,\overline{|1}\,\overline{|1}}{\overline{|1+1+1}} \int_0^1 \frac{1}{(u+1)^3} u^{3-1}\, du = \frac{1}{2}\int_0^1 \frac{u^2}{(u+1)^3}\, du$$

$$= \frac{1}{2}\int_0^1 \left[\frac{1}{u+1} - \frac{2}{(u+1)^2} + \frac{1}{(u+1)^3}\right] du \qquad \text{(Partial fractions)}$$

$$= \frac{1}{2}\left[\log(u+1)+\frac{2}{u+1}-\frac{1}{2(u+1)^2}\right]_0^1$$

$$= \frac{1}{2}\left[\log 2+2\left(\frac{1}{2}-1\right)-\left(\frac{1}{8}-\frac{1}{2}\right)\right]=\frac{1}{2}\log 2-\frac{5}{16}$$ **Proved.**

Example 38. *Find the value of* $\iiint \log(x+y+z)\,dx\,dy\,dz$ *the integral extending over all positive and zero values of x, y, z subject to the condition* $x + y + z < 1$.
(*U.P., I Sem. 2001*)

Solution. By Liouville's theorem when $0 < x + y + z < 1$

$\iiint \log(x+y+z)\,dx\,dy\,dz$

$$= \iiint \log(x+y+z)\,x^{1-1}\,y^{1-1}\,z^{1-1}\,dx\,dy\,dz = \frac{\overline{|1}\,\overline{|1}\,\overline{|1}}{\overline{|1+1+1}}\int_0^1 (\log u)u^{1+1+1-1}\,du$$

$$= \frac{1}{\overline{|3}}\int_0^1 u^2\log u\,du=\frac{1}{2}\left[\log u\left(\frac{u^3}{3}\right)-\frac{1}{3}\frac{u^3}{3}\right]_0^1=\frac{1}{2}\left(-\frac{1}{9}\right)=-\frac{1}{18}$$ **Ans.**

Example 39. *Evaluate* $\iiint\frac{dx_1\,dx_2\ldots dx_n}{\sqrt{1-x_1^2-x_2^2\ldots-x_n^2}}$, *integral being extended to all positive values of the variables for which the expression is real.* (*U.P., II Semester, Summer 2001*)

Solution. $\sqrt{1-x_1^2-x_2^2\ldots x_n^2}$ is real only when $x_1^2+x_2^2\ldots x_n^2<1$

Hence, the given integral is extended for all positive values of the variables $x_1, x_2 \ldots$ and x_n such that $0<x_1^2+x_2^2+\ldots+x_n^2<1$

Let us now put $x_1^2=u_1$ *i.e.* $x_1=u_1^{\frac{1}{2}}$ so that $dx_1=\frac{1}{2}u_1^{-\frac{1}{2}}\,du_1$

$x_2^2=u_2$ *i.e.* $x_2=u_2^{\frac{1}{2}}$ so that $dx_2=\frac{1}{2}u_2^{-\frac{1}{2}}\,du_2$

$x_n^2=u_n$ *i.e.* $x_n=u_n^{\frac{1}{2}}$ so that $dx_n=\frac{1}{2}u_n^{-\frac{1}{2}}\,du_n$

Making these substitutions, the given condition becomes $0<u_1+u_2+\ldots+u_n<1$.
Hence, the required integral becomes

$$= \frac{1}{2^n}\iiint\frac{u_1^{-\frac{1}{2}}\,.\,u_2^{-\frac{1}{2}}\ldots u_n^{-\frac{1}{2}}\,.\,du_1\,.\,du_2\ldots du_n}{\sqrt{1-u_1-u_2\ldots-u_n}}$$

$$= \frac{1}{2^n}\iiint\frac{u_1^{\frac{1}{2}-1}\,.\,u_2^{\frac{1}{2}-1}\ldots u_n^{\frac{1}{2}-1}\,.\,du_1\,.\,du_2\ldots du_n}{\sqrt{1-u_1-u_2\ldots-u_n}}$$

$$= \frac{1}{2^n}\frac{\overline{\left|\frac{1}{2}\right.}\;\overline{\left|\frac{1}{2}\right.}\ldots\overline{\left|\frac{1}{2}\right.}}{\overline{\left|\frac{1}{2}+\frac{1}{2}+\ldots+\frac{1}{2}\right.}}\int_0^1\frac{1}{\sqrt{1-u}}\,.\,u^{\left(\frac{1}{2}+\frac{1}{2}+\ldots+\frac{1}{2}\right)-1}\,du$$

By Liouville's Extension of Dirichlet's Therorem

$$= \frac{1}{2^n} \frac{\left(\Gamma\left(\frac{1}{2}\right)\right)^n}{\Gamma\left(\frac{n}{2}\right)} \int_0^1 \frac{1}{\sqrt{1-u}} \,.\, u^{\frac{n}{2}-1} \, du$$

$$= \frac{1}{2^n} \frac{(\sqrt{\pi})^n}{\Gamma\left(\frac{n}{2}\right)} \int_0^{\frac{\pi}{2}} \frac{1}{\sqrt{1-\sin^2\theta}} \,.\, (\sin^2\theta)^{\frac{n}{2}-1} \; 2\sin\theta\cos\theta \, d\theta \qquad \text{(Put } u = \sin^2\theta\text{)}$$

$$= \frac{1}{2^{n-1}} \frac{(\sqrt{\pi})^n}{\Gamma\left(\frac{n}{2}\right)} \int_0^{\frac{\pi}{2}} \sin^{n-1}\theta \, (\cos\theta)^{\circ} \, d\theta = \frac{1}{2^{n-1}} \frac{\pi^{\frac{n}{2}}}{\Gamma\left(\frac{n}{2}\right)} \frac{\Gamma\left(\frac{n}{2}\right)\Gamma\left(\frac{1}{2}\right)}{2\,\Gamma\left(\frac{n+1}{2}\right)} = \frac{1}{2^n} \frac{\pi^{\frac{n+1}{2}}}{\Gamma\left(\frac{n+1}{2}\right)} \qquad \textbf{Ans.}$$

Example 40. *Evaluate* $\iiint \frac{\sqrt{1-x^2-y^2-z^2}}{1+x^2+y^2+z^2}\, dx\, dy\, dz$, *integral being taken over all positive values of* x, y, z *such that* $x^2+y^2+z^2 \le 1$.

Solution. Putting $x^2 = u$, $y^2 = v$, $z^2 = w$ so that $u + v + w \le 1$

Also, $x = \sqrt{u} \Rightarrow dx = \frac{1}{2\sqrt{u}}\, du$

$y = \sqrt{v} \Rightarrow dy = \frac{1}{2\sqrt{v}}\, dv$

$z = \sqrt{w} \Rightarrow dz = \frac{1}{2\sqrt{w}}\, dw$

$\therefore$ The given integral

$$= \iiint \sqrt{\frac{1-(u+v+w)}{1+(u+v+w)}} \frac{du\, dv\, dw}{8\sqrt{uv\, w}}$$

$$= \frac{1}{8} \iiint u^{1/2-1}\, v^{1/2-1}\, w^{1/2-1} \sqrt{\frac{1-(u+v+w)}{1+(u+v+w)}}\, du\, dv\, dw$$

$$= \frac{1}{8} \frac{\Gamma\left(\frac{1}{2}\right)\Gamma\left(\frac{1}{2}\right)\Gamma\left(\frac{1}{2}\right)}{\Gamma\left(\frac{1}{2}+\frac{1}{2}+\frac{1}{2}\right)} \int_0^1 \sqrt{\frac{1-u}{1+u}}\; u^{\frac{1}{2}+\frac{1}{2}+\frac{1}{2}-1}\, du \quad \text{[Using Liouville's extension]}$$

$$= \frac{1}{8} \frac{\left(\Gamma\left(\frac{1}{2}\right)\right)^3}{\frac{1}{2}\Gamma\left(\frac{1}{2}\right)} \int_0^1 \frac{(1-u)}{\sqrt{1-u^2}}\, u^{1/2}\, du = \frac{\pi}{4} \int_0^1 \frac{(1-\sqrt{t})}{\sqrt{1-t}}\, t^{1/4}\, \frac{dt}{2\sqrt{t}} \qquad \text{where } u^2 = t$$

$$= \frac{\pi}{8} \int_0^1 \frac{(1-\sqrt{t})\, t^{-1/4}}{\sqrt{1-t}}\, dt = \frac{\pi}{8}\left[\int_0^1 t^{\frac{3}{4}-1} (1-t)^{1/2-1}\, dt - \int_0^1 t^{5/4-1} (1-t)^{1/2-1}\, dt\right]$$

$$= \frac{\pi}{8}\left[\beta\left(\frac{3}{4}, \frac{1}{2}\right) - \beta\left(\frac{5}{4}, \frac{1}{2}\right)\right] \qquad \textbf{Ans.}$$

Example 41. *Find the area and the mass contained m the first quadrant enclosed by the curve* $\left(\frac{x}{a}\right)^{\alpha}+\left(\frac{y}{b}\right)^{\beta}=1$ *where* $\alpha>0$, $\beta>0$ *given that density at any point p (xy) is k* $\sqrt{xy}$.

(U.P. 1 Semester 2008)

Solution. Here, we have $\left(\frac{x}{a}\right)^{\alpha}+\left(\frac{y}{b}\right)^{\beta}=1$

Put $\left(\frac{x}{a}\right)^{\alpha}=\cos^2 t$	and $\left(\frac{y}{b}\right)^{\beta}=\sin^2 t$
$\Rightarrow \quad x=a\cos^{\frac{2}{\alpha}} t$	$\Rightarrow y=b\sin^{\frac{2}{\beta}} t$

Y, X′, O, X, Y′

$$dx=\frac{2}{\alpha}a\cos^{\frac{2}{\alpha}-1} t\,(-\sin t)dt$$

$$\text{Area}=\int y\,dx=\int_0^{\frac{\pi}{2}}(b\sin^{\frac{2}{\beta}} t)\left(-\frac{2a}{\alpha}\cos^{\frac{2}{\alpha}-1} t\sin t\right)dt$$

$$=\frac{-2ab}{\alpha}\int_0^{\frac{\pi}{2}}\sin^{\frac{2}{\beta}+1} t\cos^{\frac{2}{\alpha}-1} t\,dt=\frac{-2ab}{\alpha}\,\frac{\Gamma\left(\frac{\frac{2}{\beta}+1+1}{2}\right)\Gamma\left(\frac{\frac{2}{\alpha}-1+1}{2}\right)}{2\,\Gamma\left(\frac{\beta+1}{\beta}+\frac{1}{\alpha}\right)}$$

$$=\frac{-ab}{\alpha}\,\frac{\Gamma\left(1+\frac{1}{\beta}\right)\Gamma\left(\frac{1}{\alpha}\right)}{\Gamma\left(\frac{\alpha+\beta+\alpha\beta}{\alpha\beta}\right)}=-ab\,\frac{\frac{1}{\alpha}\Gamma\left(\frac{1}{\alpha}\right)\Gamma\left(1+\frac{1}{\beta}\right)}{\Gamma\left(\frac{\alpha+\beta+\alpha\beta}{\alpha\beta}\right)}=-ab\,\frac{\Gamma\left(\frac{1}{\alpha}+1\right)\Gamma\left(1+\frac{1}{\beta}\right)}{\Gamma\left(\frac{\alpha+\beta+\alpha\beta}{\alpha\beta}\right)}$$

$$\text{Required area}=ab\,\frac{\Gamma\left(\frac{\alpha+1}{\alpha}\right)\Gamma\left(\frac{\beta+1}{\beta}\right)}{\Gamma\left(\frac{\alpha+\beta+\alpha\beta}{\alpha\beta}\right)}$$ **Ans.**

$$\text{Density}=k\sqrt{xy}$$

$$\text{Mass}=\text{Area}\times\text{Density}=\int\left[y\,dx\,k\sqrt{xy}\right]=4k\int_0^{\frac{\pi}{2}}x^{\frac{1}{2}}y^{\frac{3}{2}}\,dx$$

$$=4k\int_0^{\frac{\pi}{2}}(a\cos^{\frac{2}{\alpha}} t)^{\frac{1}{2}}(b\sin^{\frac{2}{\beta}} t)^{\frac{3}{2}}\frac{2}{\alpha}a\cos^{\frac{2}{\alpha}-1} t\,(-\sin t)\,dt$$

$$=4k\int_0^{\frac{\pi}{2}}a^{\frac{1}{2}}\cos^{\frac{1}{\alpha}} t\,b^{\frac{3}{2}}\sin^{\frac{3}{\beta}} t\,\frac{2}{\alpha}a\cos^{\frac{2}{\alpha}-1} t\,(-\sin t\,dt)$$

$$=4k\,a^{\frac{1}{2}+1}b^{\frac{3}{2}}\cdot\frac{2}{\alpha}\int_0^{\frac{\pi}{2}}\sin^{\frac{3}{\beta}+1} t\cos^{\frac{1}{\alpha}+\frac{2}{\alpha}-1} t\,dt$$

$$=\frac{8k}{\alpha}a^{\frac{3}{2}}b^{\frac{3}{2}}\int_0^{\frac{\pi}{2}}\sin^{\frac{3}{\beta}+1} t\cos^{\frac{3}{\alpha}-1} t\,dt$$ (–ve sign to be neglected)

$$= \frac{8k}{\alpha} a^{\frac{3}{2}} b^{\frac{3}{2}} \frac{\Gamma\left(\frac{\frac{3}{\beta}+1+1}{2}\right) \Gamma\left(\frac{\frac{3}{\alpha}-1+1}{2}\right)}{2\,\Gamma\left(\frac{\frac{3}{\beta}+1+1+\frac{3}{\alpha}-1+1}{2}\right)} = \frac{8k}{\alpha} a^{\frac{3}{2}} b^{\frac{3}{2}} \frac{\Gamma\left(\frac{3+2\beta}{2\beta}\right)\Gamma\left(\frac{3}{2\alpha}\right)}{2\,\Gamma\left(\frac{3}{2\alpha}+\frac{3}{2\beta}+1\right)} \qquad \textbf{Ans.}$$

Example 42. *Find the mass of an octant of the ellipsoid* $\frac{x^2}{a^2}+\frac{y^2}{b^2}+\frac{z^2}{c^2}=1$, *the density at any point being* $\rho = k\,x\,y\,z$. (*U.P., I Semester, 2009*)

Solution.

$$\text{Mass} = \iiint \rho\, dv = \iiint (k\,x\,y\,z)\, dx\, dy\, dz$$

$$= k \iiint (x\,dx)(y\,dy)(z\,dz) \qquad \ldots(1)$$

Putting $\frac{x^2}{a^2} = u,\ \frac{y^2}{b^2} = v,\ \frac{z^2}{c^2} = w$ and $u+v+w=1$

so that $\frac{2x\,dx}{a^2} = du,\ \frac{2y\,dy}{b^2} = dv,\ \frac{2z\,dz}{c^2} = dw$

$$\text{Mass} = k\iiint \left(\frac{a^2\,du}{2}\right)\left(\frac{b^2\,dv}{2}\right)\left(\frac{c^2\,dw}{2}\right)$$

$$= \frac{ka^2b^2c^2}{8}\iiint du\,dv\,dw, \qquad \text{where } u+v+w \le 1.$$

$$= \frac{ka^2b^2c^2}{8}\iiint u^{1-1}\,v^{1-1}\,w^{1-1}\,du\,dv\,dw$$

$$= \frac{ka^2b^2c^2}{8}\,\frac{\Gamma(1)\Gamma(1)\Gamma(1)}{\Gamma(3+1)} = \frac{ka^2b^2c^2}{8\times 6} = \frac{ka^2b^2c^2}{48} \qquad \textbf{Ans.}$$

Example 43. *Find the mass of a solid* $\left(\frac{x}{a}\right)^p + \left(\frac{y}{b}\right)^q + \left(\frac{z}{c}\right)^r = 1$, *the density at any point being* $\rho = kx^{l-1}y^{m-1}z^{n-1}$, *where x, y, z are all positive.*

Solution. Here, we have

$$\left(\frac{x}{a}\right)^p + \left(\frac{y}{b}\right)^q + \left(\frac{z}{c}\right)^r = 1$$

Put $\frac{x}{a} = u \Rightarrow x = au \Rightarrow dx = a\,du$

$\frac{y}{b} = v \Rightarrow y = bv \Rightarrow dy = b\,dv$

$\frac{z}{c} = w \Rightarrow z = cw \Rightarrow dz = c\,dw$

$$\text{Density} = k\,x^{l-1}\,y^{m-1}\,z^{n-1}$$

$$\text{Mass} = \text{Volume} \times \text{Density}$$

$$= \iiint dx\,dy\,dz.\,kx^{l-1}y^{m-1}z^{n-1}$$

$$= k\iiint (au)^{l-1}\,(bv)^{m-1}\,(cw)^{n-1}\,a\,du.b\,dv.c\,dw$$

$$= k\,a^l\,b^m\,c^n \iiint u^{l-1}\,v^{m-1}\,w^{n-1}\,du\,dv\,dw$$

$$= k.\,a^l.\,b^m\,c^n\,\frac{\Gamma(l)\,\Gamma(m)\,\Gamma(n)}{\sqrt{l+m+n+1}} \qquad \textbf{Ans.}$$

Example 44. *Evaluate* $I = \iiint_V x^{\alpha-1}\,y^{\beta-1}\,z^{\gamma-1}\,dx\,dy\,dz$, *where V is the region in the first octant bonded by sphere* $x^2+y^2+z^2=1$ *and the coordinate planes.*

[*U.P., I Semester (C.O.) 2003*]

Solution. Let $x^2 = u \Rightarrow x = \sqrt{u} \quad \therefore \; dx = \frac{1}{2\sqrt{u}}\,du$

$y^2 = v \Rightarrow y = \sqrt{v} \quad \therefore \; dy = \frac{1}{2\sqrt{v}}\,dv$

$z^2 = w \Rightarrow z = \sqrt{w} \quad \therefore \; dz = \frac{1}{2\sqrt{w}}\,dw$

Then, $u + v + w = 1$. Also, $u \geq 0, v \geq 0, w \geq 0$.

$$I = \iiint_V (\sqrt{u})^{\alpha-1} (\sqrt{v})^{\beta-1} (\sqrt{w})^{\gamma-1} \frac{du}{2\sqrt{u}}.\frac{dv}{2\sqrt{v}}.\frac{dw}{2\sqrt{w}}$$

$$= \frac{1}{8}\iiint u^{(\alpha/2)-1}\, v^{(\beta/2)-1}\, w^{(\gamma/2)-1}\, du\, dv\, dw$$

$$= \frac{1}{8}\frac{\Gamma(\alpha/2)\,\Gamma(\beta/2)\,\Gamma(\gamma/2)}{\Gamma\left((\alpha/2)+(\beta/2)+(\gamma/2)+1\right)}$$ **Ans.**

EXERCISE 38.3

Evaluate:

1. $\iiint e^{x+y+z}\, dx\, dy\, dz$ taken over the positive octant such that $x + y + z \leq 1$. **Ans.** $\frac{e-2}{2}$

2. $\iiint \frac{dx\, dy\, dz}{(a^2 - x^2 - y^2 - z^2)}$ for all positive values of the variables for which the expression is real.

   [**Hint.** $a^2 - x^2 - y^2 - z^2 > 0 \Rightarrow 0 < x^2 + y^2 + z^2 < a^2$] **Ans.** $\frac{\pi^2 a^2}{8}$

3. $\iiint_R (x+y+z+1)^2\, dx\, dy\, dz$ where R is defined by $x \geq 0, y \geq 0, z \geq 0, x + y + z \leq 1$.

   Ans. $\frac{31}{60}$

4. $\iiint x^{-\frac{1}{2}}\, y^{-\frac{1}{2}}\, z^{-\frac{1}{2}} (1-x-y-z)^{\frac{1}{2}}\, dx\, dy\, dz,\; x + y + z \leq 1, x > 0, y > 0, z > 0$ **Ans.** $\frac{\pi^2}{4}$

5. Show that $\iiint \frac{dx\, dy\, dz}{(x+y+z+1)^2} = \frac{3}{4} - \log 2$, the integral being taken throughout the volume bounded by the planes $x = 0, y = 0, z = 0$ and $x + y + z = 1$.

Choose the correct alternative:

6. $\Gamma\left(\frac{1}{2}\right)$ is equal to

   (*i*) π (*ii*) $\frac{1}{2!}$ (*iii*) $\sqrt{\pi}$ (*iv*) $\frac{\pi}{2}$ **Ans.** (*iii*)

   (*R.G.P.V. Bhopal, I Semester, June 2006*)

7. $\beta\,(m, n) =$

   (*i*) $\int_0^{\pi} \sin^{2m-1}\theta \cos^{2n-1}\theta\, d\theta$ (*ii*) $\int_0^{\frac{\pi}{2}} \sin^{2m-1}\theta \cos^{2n-1}\theta\, d\theta$

   (*iii*) $2\int_0^{\frac{\pi}{2}} \sin^{2m-1}\theta \cos^{2n-1}\theta\, d\theta$ (*iv*) $\int_0^{\frac{\pi}{2}} \sin^{m}\theta \cos^{n}\theta\, d\theta$ **Ans.** (*iii*)

   (*R.G.P.V. Bhopal, I Semester Dec. 2006*)

8. What is the relation between Beta and Gamma functions?

   (*i*) $\beta\,(m, l) = \frac{\Gamma m + \Gamma l}{\Gamma(l+m)}$ (*ii*) $\beta\,(l, m) = \frac{\Gamma l\, \Gamma m}{\Gamma(l+m)}$

(*iii*) $\beta(l, m) = \frac{\overline{|l} + \overline{|m}}{\overline{|m+l}}$ (*iv*) $\beta(l, m) = \frac{\overline{|l}\,\overline{|m}}{\overline{|m}\,\overline{|l}}$ **Ans.** (*ii*)

(*R.G.P.V. Bhopal, I Semester Dec. 2006*)

9. The value of $\overline{|3.5}$

(*i*) $\frac{15}{4}\sqrt{\pi}$ (*ii*) $\frac{15}{8}\sqrt{\pi}$ (*iii*) $\frac{15}{2}\sqrt{\pi}$ (*iv*) $\frac{15}{16}\sqrt{\pi}$ (*AMIETE, June 2010*) **Ans.** (*ii*)

10. The value of $\int_{-\infty}^{\infty} e^{-x^2}\, dx$ is

(*i*) $\frac{\sqrt{\pi}}{2}$ (*ii*) $\overline{\left|\frac{\pi}{2}\right.}$ (*iii*) $\sqrt{\pi}$ (*iv*) $\frac{\pi}{\sqrt{2}}$

(*AMIETE, June 2010*) **Ans.** (*iii*)

11. In terms of Beta function $\int_0^{\frac{\pi}{2}} \sin^7\theta \sqrt{\cos\theta}\, d\theta$ is

(*i*) $\frac{1}{2}\beta\left(4, \frac{3}{4}\right)$ (*ii*) $\frac{1}{2}\beta\left(2, \frac{1}{4}\right)$ (*iii*) $\frac{1}{2}\beta\left(3, \frac{3}{4}\right)$ (*iv*) None of these

(*AMIETE, June 2010*) **Ans.** (*i*)

12. If l, m, n are all positive, then the triple integral

$$\iiint_V x^{l-1} y^{m-1} z^{n-1}\, dx\, dy\, dz = \frac{\overline{|l}\,\overline{|m}\,\overline{|n}}{\overline{|l+m+n}},$$

where V is the region $x \geq 0$, $y \geq 0$, $z \geq 0$ and $x + y + z \leq 1$. (*U.P., Ist Semester, 2005*) **Ans.** True

38.16 ELLIPTIC INTEGRALS

Draw a circle with AA' (diameter) the major axis of the ellipse $\frac{x^2}{a^2} + \frac{y^2}{b^2} = 1$. This circle is called the auxiliary circle $x^2 + y^2 = a^2$. The co-ordinates of a point P on the ellipse are ($a \sin\phi$, $b \cos\phi$)).

$x = a \sin\phi$, $y = b\cos\phi$ is the parametric equation of the ellipse.

Now the length of the arc BP of the ellipse

$$= \int_0^{\phi} \sqrt{\left\{\left(\frac{dx}{d\phi}\right)^2 + \left(\frac{dy}{d\phi}\right)^2\right\}}\, d\phi = \int_0^{\phi} \sqrt{(a^2\cos^2\phi + b^2\sin^2\phi)}\, d\phi \qquad \{b^2 = a^2(1 - e^2)\}$$

$$= \int_0^{\phi} \sqrt{\{a^2\cos^2\phi + (a^2 - a^2e^2)\sin^2\phi\}}\, d\phi = \boldsymbol{a\int_0^{\phi} \sqrt{(1 - e^2\sin^2\phi)}\, d\phi}$$

where e is the eccentricity of the ellipse.

This integral cannot be evaluated in the form of the elementary function. It defines a new function, called *elliptic function.* This integral is called the elliptic integrals as it is derived from the determination of the Perimeter of the ellipse. This integral cannot be evaluated by standard methods of integration. First the integrand $\sqrt{1 - e^2\sin^2\phi}$ is expanded as power series and then is integrated term by term.

38.17 DEFINITION AND PROPERTY

Elliptic integral of first kind $= F(k, \phi) = \int_0^{\phi} \frac{1}{\sqrt{(1 - k^2\sin^2\phi)}}\, d\phi \qquad k^2 < 1$

Elliptic integral of second kind $= E(k, \phi) = \int_0^{\phi} \sqrt{(1-k^2 \sin^2 \phi)}\, d\phi$ $\qquad k^2 < 1$

Here k is known as modulus and ϕ amplitude.

The following results are easy to prove

$$F(0, \phi) = E(0, \phi) = \phi$$

$$F(1, \phi) = \log(\tan\phi + \sec\phi)$$

$$E(1, \phi) = \sin\phi$$

If $\phi = \frac{\pi}{2}$ is the upper limit of the integral then the integral is called *complete elliptic integral* as under:

$$F(k) = \int_0^{\frac{\pi}{2}} \frac{d\phi}{\sqrt{1-k^2 \sin^2 \phi}} \qquad \ldots (1)$$

and

$$E(k) = \int_0^{\frac{\pi}{2}} \sqrt{1-k^2 \sin^2 \phi}\; d\phi \qquad \ldots (2)$$

These integrals can be evaluated by expanding the integrand in binomial series and integrating term by term.

$$(1-k^2 \sin^2 \phi)^{-1/2} = 1 + \frac{k^2}{2} \sin^2 \phi + \frac{1.3}{2.4} k^4 \sin^4 \phi + \ldots\ldots$$

$$F(k, \phi) = \int_0^{\phi} \frac{d\phi}{\sqrt{(1-k^2 \sin^2 \phi)}} = \phi + \frac{k^2}{2} \int_0^{\phi} \sin^2 \phi\, d\phi + \frac{1.3}{2.4} k^4 \int_0^{\phi} \sin^4 \phi\, d\phi + \ldots \qquad \ldots (3)$$

which can be evaluated by the *Reduction Formula*

$$\int_0^{\phi} \sin^n \phi\, d\phi = -\frac{\sin^{n-1} \phi \cos \phi}{n} + \frac{n-1}{n} \int_0^{\phi} \sin^{n-2} \phi\, d\phi$$

From (3), we get

$$K(k) = \int_0^{\frac{\pi}{2}} \frac{d\phi}{\sqrt{(1-k^2 \sin^2 \phi)}} = \frac{\pi}{2} + \frac{k^2}{2}\left(\frac{1}{2}\,\frac{\pi}{2}\right) + \frac{1.3}{2.4} k^4 \left(\frac{3.1}{4.2}\,\frac{\pi}{2}\right) + \ldots$$

or

$$K(k) = \frac{\pi}{2}\left[1 + \frac{k^2}{4} + \frac{9k^4}{64} + \ldots\ldots\right]$$

If $k = \sin 10'$

$$K = \frac{\pi}{2}[1 + 0.00754 + 0.00012 + \ldots\ldots] = 1.5828$$

The elliptic integrals are periodic functios with a period π.

$$F(k, \phi + P\pi) = PF(k, \pi) + F(k, \phi), P = 0, 1, 2\ldots\ldots$$

$$E(k, \phi + P\pi) = PE(k, \pi) + E(k, \phi), P = 0, 1, 2\ldots\ldots$$

$$F(k, \phi + P\pi) = 2\,PF(k) + F(k, \phi)$$

$$E(k, \phi + P\pi) = 2\,PE(k) + E(k, \phi)$$

If we substitute $\sin\phi = x$, $d\phi = \frac{dx}{\sqrt{1-x^2}}$ in (1) and (2), we have

$$F_1(k, x) = \int_0^x \frac{dx}{\sqrt{\{(1-x^2)(1-k^2x^2)\}}}$$

$$E_1(k, x) = \int_0^x \sqrt{\left(\frac{1-k^2x^2}{1-x^2}\right)}\, dx$$

These are known as Jacobi's form of elliptic integrals.

Example 45. *Express* $\int_0^{\frac{\pi}{2}} \sqrt{\cos x}\, dx$ *in terms of elliptic integrals.*

Solution. Substitute $\cos x = \cos^2 \phi$

so that $\quad x = \cos^{-1} \cos^2\phi,\ dx = \dfrac{2\cos\phi \sin\phi\, d\phi}{\sqrt{1-\cos^4\phi}} = \dfrac{2\cos\phi\, d\phi}{\sqrt{1+\cos^2\phi}}$

$$\int_0^{\frac{\pi}{2}} \sqrt{\cos x}\, dx = \int_0^{\frac{\pi}{2}} \frac{2\cos^2\phi\, d\phi}{\sqrt{1+\cos^2\phi}} = 2\int_0^{\frac{\pi}{2}} \frac{(1+\cos^2\phi)-1}{\sqrt{1+\cos^2\phi}}\, d\phi$$

$$= 2\left\{\int_0^{\frac{\pi}{2}} \sqrt{1+\cos^2\phi}\, d\phi - \int_0^{\pi/2} \frac{1}{\sqrt{1+\cos^2\phi}}\, d\phi\right\}$$

$$= 2\left\{\int_0^{\frac{\pi}{2}} \sqrt{(2-\sin^2\phi}\, d\phi - \int_0^{\frac{\pi}{2}} \frac{1}{\sqrt{(2-\sin^2\phi)}}\, d\phi\right\}$$

$$= 2\sqrt{2}\int_0^{\frac{\pi}{2}} \sqrt{\left(1-\frac{1}{2}\sin^2\phi\right)}\ d\phi - \frac{2}{\sqrt{2}}\int_0^{\frac{\pi}{2}} \frac{1}{\sqrt{\left(1-\frac{1}{2}\sin^2\phi\right)}}\, d\phi$$

$$= 2\sqrt{2}\, E\left(\frac{1}{\sqrt{2}}\right) - \sqrt{2}K\left(\frac{1}{\sqrt{2}}\right)$$ **Ans.**

Example 46. *Express* $\int_0^{\frac{\pi}{2}} \frac{dx}{\sqrt{(2-\cos x)}}$ *in terms of elliptic integrals.*

Solution. $\int_0^{\frac{\pi}{2}} \frac{dx}{\sqrt{(2-\cos x)}} = \int_0^{\frac{\pi}{2}} \frac{dx}{\sqrt{\left\{2-\left(2\cos^2\frac{x}{2}-1\right)\right\}}}$

$$= \int_0^{\frac{\pi}{2}} \frac{dx}{\sqrt{\left(3-2\cos^2\frac{x}{2}\right)}} = \frac{1}{\sqrt{3}}\int_0^{\frac{\pi}{2}} \frac{dx}{\sqrt{\left(1-\frac{2}{3}\cos^2\frac{x}{2}\right)}}$$

On Putting $x = \pi - 2\phi$, so that $dx = -2\, d\phi$

$$= \frac{1}{\sqrt{3}}\int_{\frac{\pi}{2}}^{\frac{\pi}{4}} \frac{-2d\phi}{\sqrt{1-\frac{2}{3}\cos^2\left(\frac{\pi}{2}-\phi\right)}} = \frac{-2}{\sqrt{3}}\int_{\frac{\pi}{2}}^{\frac{\pi}{4}} \frac{d\phi}{\sqrt{\left(1-\frac{2}{3}\sin^2\phi\right)}}$$

$$= \frac{2}{\sqrt{3}}\left[\int_0^{\frac{\pi}{2}} \frac{d\phi}{\sqrt{\left(1-\frac{2}{3}\sin^2\phi\right)}} - \int_0^{\frac{\pi}{4}} \frac{d\phi}{\sqrt{\left(1-\frac{2}{3}\sin^2\phi\right)}}\right] = \frac{2}{\sqrt{3}}\left[K\left(\frac{\sqrt{2}}{3}\right) - F\left(\frac{\sqrt{2}}{3}, \frac{\pi}{4}\right)\right]$$ **Ans.**

Example 47. *Show that* $\int_0^{\frac{a}{2}} \frac{dx}{\sqrt{(2ax-x^2)(a^2-x^2)}} = \frac{2}{3a} K\left(\frac{1}{3}\right)$

Solution. On substituting $x = \frac{a}{2}(1-\sin\theta)$ so that $dx = -\frac{a}{2} 2\cos\theta\, d\theta$

Upper limit, $x = \frac{a}{2}$, $\frac{a}{2} = \frac{a}{2}(1-\sin\theta) \Rightarrow \quad \theta = 0$

Lower limit, $x = 0$, $0 = \frac{a}{2}(1-\sin\theta) \Rightarrow \quad \theta = \frac{\pi}{2}$

$$2ax - x^2 = (2a)\frac{a}{2}(1-\sin\theta) - \frac{a^2}{4}(1-\sin\theta)^2 = \frac{a^2}{4}[4 - 4\sin\theta - 1 + 2\sin\theta - \sin^2\theta]$$

$$= \frac{a^2}{4}(3 - 2\sin\theta - \sin^2\theta) = \frac{a^2}{4}(1-\sin\theta)(3+\sin\theta)$$

$$a^2 - x^2 = a^2 - \frac{a^2}{4}(1-\sin\theta)^2 = \frac{a^2}{4}[4 - 1 - \sin^2\theta + 2\sin\theta] = \frac{a^2}{4}[3 + 2\sin\theta - \sin^2\theta]$$

$$= \frac{a^2}{4}(1+\sin\theta)(3-\sin\theta)$$

$$\int_0^{\frac{a}{2}} \frac{dx}{\sqrt{(2ax-x^2)(a^2-x^2)}} = \int_{\frac{\pi}{2}}^{0} \frac{-\frac{a}{2}\cos\theta\, d\theta}{\sqrt{\frac{a^2}{4}(1-\sin\theta)(3+\sin\theta)\frac{a^2}{4}(1+\sin\theta)(3-\sin\theta)}}$$

$$= \int_0^{\frac{\pi}{2}} \frac{\cos\theta\, d\theta}{\frac{a}{2}\sqrt{(1-\sin^2\theta)(9-\sin^2\theta)}} = \frac{2}{a}\int_0^{\frac{\pi}{2}} \frac{d\theta}{(9-\sin^2\theta)}$$

$$= \frac{2}{3a}\int_0^{\frac{\pi}{2}} \frac{d\theta}{\sqrt{\left(1-\left(\frac{1}{3}\right)^2\sin^2\theta\right)}} = \frac{2}{3a} K\left(\frac{1}{3}\right)$$

Proved.

EXERCISE 38.4

Show that

1. $\int_0^{\pi} \frac{dx}{\sqrt{(1-k^2\sin^2\phi)}} = \frac{1}{k} F\left(\frac{1}{k}, x\right)$

2. $\int_0^{\frac{\pi}{2}} \frac{dx}{\sqrt{(1+3\sin^2 x)}} = \frac{1}{2} K\left(\frac{\sqrt{3}}{2}\right)$

3. $\int_0^{\frac{\pi}{6}} \frac{dx}{\sqrt{\sin x}} = \sqrt{2}\left[K\left(\frac{1}{\sqrt{2}}\right) - F\left(\frac{1}{\sqrt{2}}, \frac{\pi}{4}\right)\right]$

4. $\int_0^{\phi} \frac{\sin\phi}{\sqrt{(1-k^2\sin^2\phi)}} d\phi = \frac{1}{k^2}[F(k,\phi) - E(k,\phi)]$

5. $\int_0^{1} \frac{dx}{\sqrt{(1-x^4)}} = \frac{1}{\sqrt{2}} K\left(\frac{1}{\sqrt{2}}\right)$

6. $\int_0^{\phi} \sqrt{(1-k^2\sin^2\phi)}\, d\phi = \left(\frac{1}{k} - k\right) F\left(\frac{1}{k}, x\right) + KE\left(\frac{1}{k}, x\right)$ and $k\sin\phi < 1$ **[Hint. Put $\sin x = k\sin\phi$]**

38.18 ERROR FUNCTION

1. $\frac{2}{\sqrt{\pi}}\int_0^x e^{-t^2}\,dt$ is called error function of x and is also written as $erf(x)$.

2. $\frac{2}{\sqrt{\pi}}\int_x^\infty e^{-t^2}\,dt$ is called complementary error function of x and is also written as $erf_c(x)$.

3. Important formula.

$$\int_0^\infty e^{-t^2}\,dt = \frac{\sqrt{\pi}}{2}$$

Example 48. *Prove that* $erf(0) = 0$

Solution. $erf(x) = \frac{2}{\sqrt{\pi}}\int_0^x e^{-t^2}\,dt$

$$erf(0) = \frac{2}{\sqrt{\pi}}\int_0^0 e^{-t^2}\,dt = 0$$ **Proved.**

Example 49. *Prove that* $erf(\infty) = 1$

Solution. $erf(x) = \frac{2}{\sqrt{\pi}}\int_0^x e^{-t^2}\,dt$

$$erf(\infty) = \frac{2}{\sqrt{\pi}}\int_0^\infty e^{-t^2}\,dt = \frac{2}{\sqrt{\pi}}\frac{\sqrt{\pi}}{2} = 1$$ **Proved.**

Example 50. *Prove that* $erf(x) + erf_c(x) = 1$

Solution. $erf(x) + erf_c(x) = \frac{2}{\sqrt{\pi}}\int_0^x e^{-t^2}\,dt + \frac{2}{\sqrt{\pi}}\int_x^\infty e^{-t^2}\,dt = \frac{2}{\sqrt{\pi}}\left[\int_0^x e^{-t^2}\,dt + \int_x^\infty e^{-t^2}\,dt\right]$

$$= \frac{2}{\sqrt{\pi}}\int_0^\infty e^{-t^2}\,dt = \frac{2}{\sqrt{\pi}}\frac{\sqrt{\pi}}{2} = 1$$ **Proved.**

Example 51. *Prove that* $erf(-x) = -erf(x)$

Solution. $erf(x) = \frac{2}{\sqrt{\pi}}\int_0^x e^{-t^2}\,dt$

$$erf(-x) = \frac{2}{\sqrt{\pi}}\int_0^{-x} e^{-t^2}\,dt$$ [Put $t = -\mu$]

$$= \frac{2}{\sqrt{\pi}}\int_0^x e^{-u^2}(-du) = -\frac{2}{\sqrt{\pi}}\int_0^x e^{-u^2}(du) = -erf(x)$$ **Proved.**

Example 52. *Show that* $\int_a^b e^{-x^2}\,dx = \frac{\sqrt{\pi}}{2}[erf(b) - erf(a)]$

Solution. $\frac{\sqrt{\pi}}{2}[erf(b) - erf(a)]$

$$= \frac{\sqrt{\pi}}{2}\left[\frac{2}{\sqrt{\pi}}\int_0^b e^{-t^2}\,dt - \frac{2}{\sqrt{\pi}}\int_0^a e^{-t^2}\,dt\right] = \int_0^b e^{-t^2}\,dt - \int_0^a e^{-t^2}\,dt$$

$$= \int_0^b e^{-t^2}\,dt + \int_a^0 e^{-t^2}\,dt = \int_a^b e^{-t}\,dt = \int_a^b e^{-x^2}\,dx$$ **Proved.**

Example 53. *Show that*

$$\int_0^{\infty} e^{-x^2-2bx}\,dx = \frac{\sqrt{\pi}}{2} e^{b^2}\,[1-erf(b)]$$

Solution. $\displaystyle\int_0^{\infty} e^{-x^2-2bx}\,dx = \int_0^{\infty} e^{-x^2-2bx-b^2+b^2}\,dx = \int_0^{\infty} e^{-(x+b)^2}\cdot e^{b^2}\,dx$

$$= e^{b^2}\left[\int_b^{\infty} e^{-t^2}\,dt + \int_0^{b} e^{-t^2}\,dt\right] \qquad [\text{Put } x+b=t]$$

$$= e^{b^2}\left[-\int_b^{0} e^{-t^2}\,dt + erf(\infty)\right]$$

$$= e^{b^2}\left[\frac{\sqrt{\pi}}{2} - \frac{\sqrt{\pi}}{2}erf(b)\right] = e^{b^2}\frac{\sqrt{\pi}}{2}[1-erf(b)]$$ **Proved.**

Example 54. *Prove that*

$$\frac{d}{dx}[erf_c(\alpha x)] = \frac{-2a}{\sqrt{\pi}}e^{-\alpha^2 x^2}$$

Solution. $\displaystyle\frac{d}{dx}[erf_c(\alpha\, x)] = \frac{d}{dx}\left[\frac{2}{\sqrt{\pi}}\int_{\alpha x}^{\infty} e^{-t^2}\,dt\right]$

On applying the rule of differentiation under integral sign, we get

$$= \frac{2}{\sqrt{\pi}}\left[\int_{\alpha x}^{\infty}\left(\frac{\partial}{\partial x}e^{-t^2}\right)dt + \frac{d}{dx}(\infty)e^{-\infty} - \frac{d}{dx}(\alpha x)e^{-\alpha^2 x^2}\right]$$

$$= \frac{2}{\sqrt{\pi}}\left[0+0-\alpha.e^{-\alpha^2 x^2}\right] = -\frac{2\alpha}{\sqrt{\pi}}e^{-\alpha^2 x^2}$$ **Proved.**

EXERCISE 38.5

Prove that

1. $erf_c(x) + erf_c(-x) = 2$
2. $erf_c(-x) = 1 + erf_c(x)$
3. $erf(x) = \dfrac{2}{\sqrt{\pi}}\left[x - \dfrac{x^3}{3} + \dfrac{1}{2!}\cdot\dfrac{x^5}{5} - \dfrac{1}{3!}\cdot\dfrac{x^7}{7} + \ldots\ldots\right]$
4. $\displaystyle\int_0^{\infty} e^{-(x+a)^2}\,dx = \frac{\sqrt{\pi}}{2}[1-erf(a)]$
5. $\displaystyle\int_0^{t} erf_c(ax)dx = t\,erf_c(at) - \frac{e^{-a^2t^2}}{a\sqrt{\pi}} + \frac{1}{a\sqrt{\pi}}$
6. $\dfrac{d}{dx}[erf(ax)] = \dfrac{2a}{\sqrt{\pi}}e^{-a^2x^2}$
7. $\dfrac{d}{dx}[erf(\sqrt{x})] = \dfrac{e^{-x}}{\sqrt{\pi x}}$
8. $\dfrac{d}{dx}[erf(\sqrt{x})] = \dfrac{2}{\sqrt{\pi}}e^{-x^2}$

38.19 DIFFERENTIATION UNDER THE INTEGRAL SIGN

The value of a definite integral $\int_a^b f(x,\alpha)dx$ is a function of α (parameter), $F(\alpha)$ say. To find $F'(\alpha)$, first we have to evaluate the integral $\int_a^b f(x,\alpha)dx$ and then differentiate $F(\alpha)$ w.r.t. α. However, it is not always possible to evaluate the integral and then to find its derivative. Such problems are solved by reversing the order of the integration and differentiation *i.e.*, first differentiate $f(x,\alpha)$ partially w.r.t. "α" and then integrate it.

38.20 LEIBNITZ'S RULE

If $f(x,\alpha)$ and $\dfrac{\partial f(x,\alpha)}{\partial \alpha}$ be continuous functions of x and α, then

$$\frac{d}{d\alpha}\left[\int_a^b f(x,\alpha)dx\right] = \int_a^b \frac{\partial f(x,\alpha)}{\partial \alpha}dx$$

Proof. Let $$\int_a^b f(x,\alpha)\,dx = F(\alpha)$$

then $$F(\alpha+\delta\alpha) = \int_a^b f(x,\alpha+\delta\alpha)dx$$

Hence $$F(\alpha+\delta) - F(\alpha) = \int_a^b f(x,\alpha+\delta\alpha)dx - \int_a^b f(x,\alpha)dx$$

$$= \int_a^b [f(x,\alpha+\delta\alpha) - f(x,\alpha)]dx$$

$$\frac{F(\alpha+\delta\alpha)-F(\alpha)}{\delta\alpha} = \int_a^b \frac{f(x,\alpha+\delta\alpha)-f(x,\alpha)}{\delta\alpha}dx$$

Taking limits of both sides as $\delta\alpha \to 0$, we have

$$\frac{\partial F}{d\alpha} = \int_a^b \frac{\partial f(x,\alpha)}{\partial\alpha}dx$$

The above formula is useful for evaluating definite integrals which are otherwise impossible to evaluate.

Example 55. *Evaluate* $$\int_0^\infty \frac{\tan^{-1}(ax)}{x(1+x^2)}dx$$

Solution. Let $$I = \int_0^\infty \frac{\tan^{-1}(ax)}{x(1+x^2)}dx$$

$$\therefore \quad \frac{dI}{da} = \int_0^\infty \frac{\partial}{\partial a}\frac{\tan^{-1}(ax)}{x(1+x^2)}dx \qquad ...(1)$$

$$= \int_0^\infty \frac{1}{x(1+x^2)}\cdot\frac{x}{1+a^2x^2}dx = \int_0^\infty \frac{1}{(1+x^2)(1+a^2x^2)}dx$$

Breaking the integrand into partial fractions,

$$= \int_0^\infty \frac{1}{1-a^2}\left[\frac{1}{1+x^2} - \frac{a^2}{1+a^2x^2}\right]dx = \frac{1}{1-a^2}\left[\tan^{-1}x - a\tan^{-1}ax\right]_0^\infty$$

$$= \frac{1}{1-a^2}\left[\frac{\pi}{2} - a\frac{\pi}{2}\right] = \frac{\pi}{2}\frac{1-a}{1-a^2} = \frac{\pi}{2}\cdot\frac{1}{1+a}$$

Now, integrating with respect to "a", we get $I = \dfrac{\pi}{2}\log(1+a) + c$ ....(2)

From (1), when $a = 0$, then $I = 0$

Putting $a = 0$ and $I = 0$ in (2), we get $c = 0$

Hence (2) gives $\quad I = \dfrac{\pi}{2}\log(1+a)$ **Ans.**

Example 56. *Evaluate* $\int_0^{\infty} \frac{e^{-x}}{x}\left(a-\frac{1}{x}+\frac{1}{x}e^{-ax}\right)dx$

using the rule of differentiation under the sign of integration.

Solution. Let $$I=\int_0^{\infty} \frac{e^{-x}}{x}\left(a-\frac{1}{x}+\frac{1}{x}e^{-ax}\right)dx \quad ...(1)$$

$$\frac{dI}{da}=\int_0^{\infty}\frac{\partial}{\partial a}\left[\frac{e^{-x}}{x}\left(a-\frac{1}{x}+\frac{1}{x}e^{-2ax}\right)\right]dx=\int_0^{\infty}\frac{e^{-x}}{x}\left(1-0-\frac{x}{x}e^{-ax}\right)dx$$

$$=\int_0^{\infty}\frac{e^{-x}}{x}\left(1-e^{-ax}\right)dx \quad ...(2)$$

$$\frac{d^2I}{da^2}=\int_0^{\infty}\frac{\partial}{\partial a}\left[\frac{e^{-x}}{x}\left(1-e^{-ax}\right)\right]dx=\int_0^{\infty}\frac{e^{-x}}{x}\left(xe^{-ax}\right)dx=\int_0^{\infty}e^{-(a+1)x}dx$$

$$=\left[\frac{e^{-(a+1)x}}{-(a+1)}\right]_0^{\infty}=\left[0+\frac{1}{a+1}\right]=\frac{1}{a+1} \quad ...(3)$$

Integrating w.r.t. a, we have $\frac{dI}{da}=\log(a+1)+c_1$ ...(4)

Putting $a=0$ in (2), we get $\frac{dI}{da}=0$

Putting $a=0$ and $\frac{dI}{da}=0$ in (4), we get $c_1=0$

From (4), $\frac{dI}{da}=\log(a+1)$

$$I=\int\log(a+1)\,da=\log(a+1)\cdot a-\int\frac{a}{a+1}da=a\log(a+1)-\int\left(1-\frac{1}{a+1}\right)da$$

$$I=a\log(a+1)-a+\log(a+1)+c_2$$

$$I=(a+1)\log(a+1)-a+c_2 \quad ...(5)$$

Putting $a=0$ in (1), we get $I=0$

Putting $a=0$, $I=0$ in (5), we get

$$0=c_2$$

Hence (5) gives

$$I=(a+1)\log(a+1)-a$$ **Ans.**

Example 57. *Evaluate the integral*

$$\int_0^{\infty}\frac{e^{-x}\sin bx}{x}dx$$

Solution. Let $$I=\int_0^{\infty}\frac{e^{-x}\sin bx}{x}dx \quad ...(1)$$

$$\frac{dI}{db}=\int_0^{\infty}\frac{\partial}{\partial b}\left(\frac{e^{-x}\sin bx}{x}\right)dx=\int_0^{\infty}\frac{e^{-x}\cdot x\cos bx}{x}dx=\int_0^{\infty}e^{-x}\cos bx\,dx$$

[We know that $\int e^{ax}\cos bx\,dx=\frac{e^{ax}}{a^2+b^2}(a\cos bx+b\sin bx)$]

$$=\left[\frac{e^{-x}}{1+b^2}(-\cos bx+b\sin bx)\right]_0^{\infty}$$

$$\frac{dI}{db}=\frac{1}{1+b^2} \qquad ....(2)$$

Integrating both sides of (2) w.r.t. 'b', we have $I=\tan^{-1}b+c$ ...(3)

On putting $b=0$ in (1), we have $I=0$

On putting $b=0$, $I=0$ in (3), we get $c=0$

Hence (3) gives

$$I=\tan^{-1}b$$

or $$\int_0^\infty \frac{e^{-x}\sin bx}{x}\,dx=\tan^{-1}b$$ **Ans.**

Example 58. *Find the value of* $\int_0^\pi \frac{dx}{a+b\cos x}$ *(when* $a>0,\ |b|<a$*)*

and deduce that $\int_0^\pi \frac{dx}{(a+b\cos x)^2}=\frac{\pi a}{(a^2-b^2)^{3/2}}$

Solution. Let $I=\int_0^\pi \frac{dx}{a+b\cos x}=\int_0^\pi \frac{dx}{a\left(\cos^2\frac{x}{2}+\sin^2\frac{\pi}{2}\right)+b\left(\cos^2\frac{x}{2}-\sin^2\frac{x}{2}\right)}$

$$=\int_0^\pi \frac{dx}{(a+b)\cos^2\frac{x}{2}+(a-b)\sin^2\frac{x}{2}}=\frac{1}{a-b}\int_0^\infty \frac{\sec^2\frac{x}{2}\,dx}{\frac{a+b}{a-b}+\tan^2\frac{x}{2}}$$

$$=\frac{2}{a-b}\,\frac{\sqrt{a-b}}{a+b}\left[\tan^{-1}\left\{\tan\frac{x}{2}\cdot\frac{\sqrt{a-b}}{a+b}\right\}\right]_0^\pi=\frac{2}{a-b}\,\frac{\sqrt{a-b}}{a+b}\left[\tan^{-1}\infty-\tan^{-1}0\right]$$

Now differentiating both sides w.r.t. 'a' we get

$$\frac{dI}{da}=-\frac{1}{2}\frac{2\pi a}{(a^2-b^2)^{3/2}} \quad \text{or} \quad \int_0^\pi \frac{\partial}{\partial a}\left(\frac{1}{a+b\cos x}\right)dx=-\frac{1}{2}\frac{2\pi a}{(a^2-b^2)^{3/2}}$$

$$\Rightarrow \qquad \int_0^\pi \frac{-1}{(a+b\cos x)^2}\,dx=\frac{-\pi a}{(a^2-b^2)^{3/2}}$$

$$\Rightarrow \qquad \int_0^\pi \frac{1}{(a+b\cos x)^2}\,dx=\frac{\pi a}{(a^2-b^2)^{3/2}}$$ **Ans.**

EXERCISE 38.6

Prove that:

1. $\int_0^\infty \frac{1-e^{-ax}}{x}e^{-x}\,dx=\log(1+a)$ 2. $\int_0^\infty \frac{e^{-ax}-e^{-bx}}{x}\,dx=\log\frac{b}{a}$

3. $\int_0^\infty \frac{e^{-ax}\sin x}{x}\,dx=\cot^{-1}\alpha$ and hence deduce that $\int_0^\infty \frac{\sin x}{x}\,dx=\frac{\pi}{2}$

4. $\int_0^1 \frac{x^a-x^b}{\log x}\,dx=\log\frac{a+1}{b+1}$

5. $\int_0^\infty \frac{\cos \lambda x}{x}\left(e^{-ax} - e^{-bx}\right)dx = \frac{1}{2}\log\frac{b^2+\lambda^2}{a^2+\lambda^2}, (a > 0,\ b > 0)$

6. $\int_0^\infty e^{-bx^2}\cos 2ax\, dx = \frac{1}{2}\frac{\sqrt{\pi}}{b}e^{-a^2/b}$ $(b > 0)$ Assume $\int_0^\infty e^{-x^2}dx = \frac{\sqrt{\pi}}{2}$

Evaluate the following.

7. $\int_0^{\frac{\pi}{2}} \log(\alpha^2\cos^2\theta + \beta^2\sin^2\theta)\, d\theta \quad (\alpha > 0, \beta > 0)$ **Ans.** $\pi\log\frac{\alpha+\beta}{2}$

8. $\int_0^\infty \frac{\log(1+a^2x^2)}{1+b^2x^2}dx$ **Ans.** $\frac{\pi}{l}\log\frac{a+b}{b}$

9. $\int_0^{\frac{\pi}{2}} \log\left(\frac{a+b\sin\theta}{a-b\sin\theta}\right)\cdot\frac{d\theta}{\sin\theta}$ **Ans.** $\pi\sin^{-1}\frac{b}{a}$

10. $\int_0^{\frac{\pi}{2}} \frac{\log(1+\cos\alpha\cdot\cos x)}{\cos x}dx$ **Ans.** $\frac{1}{2}\left(\frac{\pi^2}{4}-\alpha^2\right)$

Prove that

11. $\int_0^{\frac{\pi}{2}} \frac{\log(1+y\sin^2 x)}{\sin^2 x}dx = \pi[\sqrt{(1+y)}-1]$ When $y > 1$.

12. $\int_0^{\frac{\pi}{2}} \frac{dx}{(a^2+\sin^2 x + b^2\cos^2 x)^2} = \frac{\pi(a^2+b^2)}{4a^3b^3}$

38.21 RULE OF DIFFERENTIATION UNDER THE INTEGRAL SIGN WHEN THE LIMITS OF INTEGRATION ARE FUNCTIONS OF THE PARAMETER

If $f(x, \alpha), \frac{\partial f(x,\alpha)}{\partial \alpha}$ be continuous functions of x and α, then

$$\frac{d}{d\alpha}\left\{\int_{\phi(\alpha)}^{\psi(\alpha)} f(x,\alpha)dx\right\} = \int_{\phi(\alpha)}^{\psi(\alpha)}\frac{\partial f(x,\alpha)}{\partial\alpha}dx - \frac{d\phi}{d\alpha}f[\phi(\alpha),\alpha] + \frac{d\psi}{d\alpha}f[\psi(\alpha),\alpha]$$

Example 59. *Verify the rule of differentiation under the sign of integration for* $\int_0^{a^2}\tan^{-1}\frac{x}{a}dx$

Solution. Let $I = \int_0^{a^2}\tan^{-1}\frac{x}{a}dx$

$$\frac{d}{d\alpha}\left\{\int_{\phi(\alpha)}^{\psi(\alpha)} f(x,\alpha)dx\right\} = \int_{\phi(\alpha)}^{\psi(\alpha)}\frac{\partial f(x,\alpha)}{\partial\alpha}dx - \frac{d\phi}{d\alpha}f[\phi(\alpha),\alpha] + \frac{d\psi}{d\alpha}f[\psi(\alpha),\alpha]$$

$$\frac{dI}{da} = \int_0^{a^2}\left[\frac{\partial}{\partial\alpha}\left(\tan^{-1}\frac{x}{a}\right)\right]dx - 0 + 2a\left[\tan^{-1}\frac{a^2}{a}\right]$$

$$= \int_0^{a^2}\frac{1}{1+\frac{x^2}{a^2}}\left(-\frac{x^2}{a}\right)dx + 2a\tan^{-1}a = \int_0^{a^2} -\frac{x}{a^2+x^2}dx + 2a\tan^{-1}a$$

$$= \left[-\frac{1}{2}\log(a^2+x^2)\right]_0^{a^2} + 2a\tan^{-1}a = -\frac{1}{2}\log(a^2+a^4) + \frac{1}{2}\log a^2 + 2a\tan^{-1}a$$

$$= -\frac{1}{2}\log\frac{a^2+a^4}{a^2} + 2a\tan^{-1}a = -\frac{1}{2}\log(a^2+1) + 2a\tan^{-1}a \quad ...(1)$$

Now integration by parts

$$I = \int_0^{a^2} \tan^{-1}\frac{x}{a}\cdot 1\, dx = \left[\left(\tan^{-1}\frac{x}{a}\right)\cdot x\right]_0^{a^2} - \int_0^{a^2} \frac{1}{1+\frac{x^2}{a^2}}\cdot\frac{1}{a}\cdot x\, dx$$

$$= \left[a^2\tan^{-1}\frac{a^2}{a}\right] - \int_0^{a^2}\frac{ax}{a^2+x^2}dx = a^2\tan^{-1}a - \frac{a}{2}[\log(x^2+a^2)]_0^{a^2}$$

$$= a^2\tan^{-1}a - \frac{a}{2}[\log(a^4+a^2) - \log a^2] = a^2\tan^{-1}a - \frac{a}{2}\log\frac{a^4+a^2}{a^2}$$

$$= a^2\tan^{-1}a - \frac{a}{2}\log(a^2+1)$$

$$\frac{dI}{da} = \left[a^2\frac{1}{1+a^2} + 2a\tan^{-1}a\right] - \left[\frac{a}{2}\frac{2a}{a^2+1} + \frac{1}{2}\log(a^2+1)\right]$$

$$= \frac{a^2}{1+a^2} + 2a\tan^{-1}a - \frac{a^2}{a^2+1} - \frac{1}{2}\log(a^2+1) = 2a\tan^{-1}a - \frac{1}{2}\log(a^2+1) \qquad ...(2)$$

From (1) and (2), the rule is verified.

Example 60. *Evaluate* $\int_0^{\alpha}\frac{\log(1+\alpha x)}{1+x^2}dx$ *and hence show that*

$$\int_0^1 \frac{\log(1+x)}{1+x^2}dx = \frac{\pi}{8}\log_e 2$$

Solution. Let $\quad I = \int_0^{\alpha}\frac{\log(1+\alpha x)}{1+x^2}dx \qquad ...(1)$

$$\left[\frac{d}{d\alpha}\left\{\int_{\phi(\alpha)}^{\psi(\alpha)} f(x,\alpha)dx\right\} = \int_{\phi(\alpha)}^{\psi(\alpha)}\frac{\partial f(x,\alpha)}{\partial\alpha}dx - \frac{d\phi}{d\alpha}f[\phi(\alpha),\alpha] + \frac{d\psi}{d\alpha}f[\psi(\alpha),\alpha]\right]$$

$$\frac{dI}{d\alpha} = \int_0^{\alpha}\frac{\partial}{\partial\alpha}\left\{\frac{\log(1+\alpha x)}{1+x^2}\right\}dx + \frac{d\alpha}{d\alpha}f(\alpha,\alpha) = \int_0^{\alpha}\frac{x}{(1+x^2)(1+\alpha x)}dx + \frac{\log(1+\alpha^2)}{1+\alpha^2}$$

Converting into partial fractions,

$$= -\frac{\alpha}{1+\alpha^2}\int_0^{\alpha}\frac{dx}{1+\alpha x} + \frac{1}{2(1+\alpha^2)}\int_0^{\alpha}\frac{2x}{1+x^2}dx + \frac{\alpha}{1+\alpha^2}\int_0^{\alpha}\frac{dx}{1+x^2} + \frac{\log(1+\alpha^2)}{1+\alpha^2}$$

$$= -\frac{\alpha}{1+\alpha^2}\left[\frac{1}{\alpha}\log(1+\alpha x)\right]_0^{\alpha} + \frac{1}{2(1+\alpha^2)}[\log(1+x^2)]_0^{\alpha} + \frac{\alpha}{1+\alpha^2}[\tan^{-1}x]_0^{\alpha} + \frac{\log(1+\alpha^2)}{1+\alpha^2}$$

$$= -\frac{1}{1+\alpha^2}\log(1+\alpha^2) + \frac{\log(1+\alpha^2)}{2(1+\alpha^2)} + \frac{\alpha}{1+\alpha^2}\tan^{-1}\alpha + \frac{\log(1+\alpha^2)}{1+\alpha^2}$$

$$= \frac{\log(1+\alpha^2)}{2(1+\alpha^2)} + \frac{\alpha}{1+\alpha^2}\tan^{-1}\alpha$$

On integrating both sides w.r.t. α, we have

$$I = \frac{1}{2}\int\log(1+\alpha^2)\cdot\frac{1}{1+\alpha^2}d\alpha + \int\frac{\alpha\tan^{-1}\alpha}{1+\alpha^2}d\alpha + c$$

$$I = \frac{1}{2}\log(1+\alpha^2)\cdot\tan^{-1}\alpha - \frac{1}{2}\int\frac{2\alpha}{1+\alpha^2}\cdot\tan^{-1}\alpha\, d\alpha + \int\frac{\alpha\tan^{-1}\alpha}{1+\alpha^2}d\alpha$$

$$I = \frac{1}{2}\log(1+\alpha^2)\cdot\tan^{-1}\alpha + c \qquad \ldots (2)$$

From (1), when $\alpha = 0$, then $I = 0$. From (2), when $\alpha = 0$, $I = 0$, then $c = 0$

Hence (2) gives

$$I = \frac{1}{2}\log(1+\alpha^2)\cdot\tan^{-1}\alpha \qquad \ldots (3)$$

On putting $\alpha = 1$ in (3), we have

$$\int_0^1 \frac{\log(1+x)}{1+x^2}dx = \frac{1}{2}\log(1+1)\cdot\tan^{-1}(1) = \frac{1}{2}(\log_e 2)\frac{\pi}{4} = \frac{\pi}{8}\log_e 2$$ **Ans.**

Example 61. *Evaluate* $\int_{\frac{\pi}{6a}}^{\frac{\pi}{2a}} \frac{\sin ax}{x}dx$

Solution. Let $$I = \int_{\frac{\pi}{6a}}^{\frac{\pi}{2a}} \frac{\sin ax}{x}dx \qquad \ldots (1)$$

$$\left[\frac{d}{d\alpha}\left\{\int_{\phi(\alpha)}^{\psi(\alpha)} f(x,\alpha)dx\right\} = \int_{\phi(\alpha)}^{\psi(\alpha)} \frac{\partial f(x,\alpha)}{\partial\alpha}dx - \frac{d\phi}{d\alpha}f[\phi(\alpha),\alpha] + \frac{d\psi}{d\alpha}f[\psi(\alpha),\alpha]\right]$$

$$\frac{dI}{da} = \int_{\frac{\pi}{6a}}^{\frac{\pi}{2a}} \frac{\partial}{\partial a}\left(\frac{\sin ax}{x}\right)dx - \frac{d}{da}\left(\frac{\pi}{6a}\right)\cdot\frac{\left[\sin a\cdot\frac{\pi}{6a}\right]}{\frac{\pi}{6a}} + \frac{d}{da}\left(\frac{\pi}{2a}\right)\cdot\frac{\sin a\cdot\frac{\pi}{2a}}{\frac{\pi}{2a}}$$

$$= \int_{\frac{\pi}{6a}}^{\frac{\pi}{2a}} \frac{x\cos ax}{x}dx + \frac{\pi}{6a^2}\cdot\frac{6a}{\pi}\sin\frac{\pi}{6} - \frac{\pi}{2a^2}\cdot\frac{2a}{\pi}\sin\frac{\pi}{2}$$

$$= \int_{\frac{\pi}{6a}}^{\frac{\pi}{2a}} \cos ax dx + \frac{1}{2a} - \frac{1}{a} = \left[\frac{\sin ax}{a}\right]_{\frac{\pi}{6a}}^{\frac{\pi}{2a}} - \frac{1}{2a}$$

$$= \frac{1}{a}\left[\sin a\cdot\frac{\pi}{2a} - \sin a\cdot\frac{\pi}{6a}\right] - \frac{1}{2a}$$

$$= \frac{1}{a}\left[\sin\frac{\pi}{2} - \sin\frac{\pi}{6}\right] - \frac{1}{2a} = \frac{1}{a}\left[1 - \frac{1}{2}\right] - \frac{1}{2a} = \frac{1}{2a} - \frac{1}{2a} = 0$$

Integrating we have $I = \text{Constant}$ **Ans.**

Example 62. *If* $y = \int_0^x f(t)\sin[k(x-t)]\,dt$, *prove that y satisfies the differential equation*

$$\frac{d^2y}{dx^2} + k^2y = kf(x)$$

Solution. $$y = \int_0^x f(t)\sin[k(x-t)]\,dt$$

$$\frac{dy}{dx} = \int_0^x \frac{\partial}{\partial x}[f(t)\sin\{k(x-t\}]dt - 0\ \frac{d}{dx}(x)\cdot f(x)\sin\{k(x-x)\}$$

$$-\int_0^x f(t)\,k\cos\{k(x-t)\}\cdot dt = k\int_0^x f(t)\cos\{k(x-t)\}\,dt$$

Again applying the same rule

$$\frac{d^2y}{dx^2} = k\left[\int_0^x \frac{\partial}{\partial x}\{f(t)\cos k(x-t)\}\,dt - 0 + \frac{d}{dx}(x)\cdot f(x)\cos k(x-x)\right]$$

$$= -k^2\int_0^x f(t)\sin[k(x-t)]\,dt + k f(x) = -k^2 y + k f(x)$$

$$\Rightarrow \qquad \frac{d^2y}{dx^2} + k^2 y = k f(x)$$ **Proved.**

Example 63. *Using differentition under integral sign, evaluate* $\int_0^1 \frac{x^\alpha - 1}{\log x}\,dx,\ \alpha \geq 0$

(*AMIETE, June 2010*)

Solution. Here, we have

$$I = \int_0^1 \frac{x^\alpha - 1}{\log x}\,dx \qquad \ldots (1)$$

$$\left[\frac{d}{d\alpha}\left\{\int_{\phi(\alpha)}^{\psi(\alpha)} f(x,\alpha)\,dx\right\} = \int_{\phi(\alpha)}^{\psi(\alpha)} \frac{\partial f(x,\alpha)}{\partial \alpha}\,dx - \frac{d\phi}{d\alpha} f[\phi(\alpha),\alpha] + \frac{d\psi}{d\alpha} f[\psi(\alpha),\alpha]\right]$$

$$\frac{dI}{d\alpha} = \int_0^1 \frac{\partial}{\partial\alpha}\left(\frac{x^\alpha - 1}{\log x}\right)dx - \frac{d(0)}{d(\alpha)} f(\theta,\alpha) + \frac{d(1)}{d\alpha} f(1,\alpha)$$

$$= \int_0^1 \frac{x^\alpha \log x}{\log x}\,dx - 0 + 0 = \int_0^1 x^\alpha\,dx$$

$$= \left[\frac{x^{\alpha+1}}{\alpha+1}\right]_0^1 = \frac{1}{\alpha+1}$$

$$\frac{dI}{d\alpha} = \frac{1}{\alpha+1}$$

Now, integrating both sides *w.r.t.* α, we get

$$I = \int \frac{d\alpha}{\alpha+1} = \log(\alpha+1) + C \qquad \ldots (2)$$

From (1), when $\alpha = 0$, then $I = 0$

Putting $\alpha = 0$ and $I = 0$ in (2), we get $C = 0$

Hence (2) becomes $I = \log(\alpha + 1)$ **Ans.**

INFINITE SERIES

39.1 SEQUENCE

A *sequence* is a succession of numbers or terms formed according to some definite rule. The nth term in a sequence is denoted by u_n.

For example, if $u_n = 2n + 1$.

By giving different values of n in u_n, we get different terms of the sequence.

Thus, $u_1 = 3$, $u_2 = 5$, $u_3 = 7$, ...

A sequence having unlimited number of terms is known as an *infinite sequence.*

39.2 LIMIT

If a sequence tends to a limit l, then we write $\lim_{n \to \infty} (u_n) = l$

39.3 CONVERGENT SEQUENCE

If the limit of a sequence is finite, the sequence is *convergent*. If the limit of a sequence does not tend to a finite number, the sequence is said to be *divergent*.

e.g., $1, \frac{1}{4}, \frac{1}{9}, \frac{1}{16}, \ldots, \frac{1}{n^2} + \ldots$. is a convergent sequence.

$3, 5, 7, \ldots, (2n + 1), \ldots$ is a divergent sequence.

39.4 BOUNDED SEQUENCE

$u_1, u_2, u_3 \ldots, u_n \ldots$ is a bounded sequence if $u_n < k$ for every n.

39.5 MONOTONIC SEQUENCE

The sequence is either increasing or decreasing, such sequences are called *monotonic.*

e.g., $1, 4, 7, 10, \ldots$ is a monotonic sequence.

$1, \frac{1}{2}, \frac{1}{3}, \frac{1}{4}, \ldots$ is also a monotonic sequenece.

$1, -1, 1, -1, 1, \ldots$ is not a monotonic sequence.

A sequence which is monotonic and bounded is a convergent sequence.

EXERCISE 39.1

Determine the general term of each of the following sequence. Prove that the following sequences are convergent.

1. $\frac{1}{2}, \frac{1}{4}, \frac{1}{8}, \frac{1}{16}, \ldots$ **Ans.** $\frac{1}{2^n}$

2. $\frac{1}{2}, \frac{2}{3}, \frac{3}{4}, \frac{4}{5}, \ldots$ **Ans.** $\frac{n}{n+1}$

3. $1, -1, 1, -1, \ldots$ **Ans.** $(-1)^{n-1}$

4. $\frac{1^2}{1!}, \frac{2^2}{2!}, \frac{3^2}{3!}, \frac{4^2}{4!}, \frac{5^2}{5!}, \ldots$ **Ans.** $\frac{n^2}{n!}$

Which of the following sequences are convergent ?

5. $u_n = \dfrac{n+1}{n}$ **Ans.** Convergent

6. $u_n = 3n$ **Ans.** Divergent

7. $u_n = n^2$ **Ans.** Divergent

8. $u_n = \dfrac{1}{n}$ **Ans.** Convergent

39.6 REMEMBER THE FOLLOWING LIMITS

(*i*) $\lim\limits_{n \to \infty} x^n = 0$ if $x < 1$ and $\lim\limits_{n \to \infty} x^n = \infty$ if $x > 1$

(*ii*) $\lim\limits_{n \to \infty} \dfrac{x^n}{n!} = 0$ for all values of x

(*iii*) $\lim\limits_{n \to \infty} \dfrac{\log n}{n} = 0$

(*iv*) $\lim\limits_{n \to \infty} \left(1 + \dfrac{1}{n}\right)^n = e$

(*v*) $\lim\limits_{n \to \infty} (n)^{1/n} = 1$

(*vi*) $\lim\limits_{n \to \infty} [n!]^{1/n} = \infty$

(*vii*) $\lim\limits_{n \to \infty} \left[\dfrac{(n!)}{n}\right]^{1/n} = \dfrac{1}{e}$

(*viii*) $\lim\limits_{n \to \infty} n\, x^n = 0$ if $x < 1$

(*ix*) $\lim\limits_{n \to \infty} n^h = \infty$

(*x*) $\lim\limits_{n \to \infty} \dfrac{1}{n^h} = 0$

(*xi*) $\lim\limits_{x \to \infty} \left[\dfrac{a^x - 1}{x}\right] = \log a$ or $\lim\limits_{n \to \infty} \dfrac{a^{1/n} - 1}{1/n} = \log a$

(*xii*) $\lim\limits_{x \to 0} \dfrac{\sin x}{x} = 1$

(*xiii*) $\lim\limits_{x \to 0} \dfrac{\tan x}{x} = 1$

39.7 SERIES

A *series* is the sum of a sequence.

Let $u_1, u_2, u_3, \ldots\ldots\ldots, u_n, \ldots\ldots\ldots$ be a given sequence. Then, the expression $u_1 + u_2 + u_3 + \ldots\ldots\ldots + u_n + \ldots\ldots\ldots$ is called the series associated with the given sequence.

For example, $1 + 3 + 5 + 7 + \ldots$ is a series.

If the number of terms of a series is limited, the series is called *finite*. When the number of terms of a series are unlimited, it is called an *infinite series*.

$$u_1 + u_2 + u_3 + u_4 + \ldots + u_n + \ldots \infty$$

is called an infinite series and it is denoted by $\sum\limits_{n=1}^{\infty} u_n$ or $\Sigma\, u_n$. The sum of the first n terms of a series is denoted by S_n.

39.8 CONVERGENT, DIVERGENT AND OSCILLATORY SERIES

Consider the infinite series $\Sigma\, u_n = u_1 + u_2 + u_3 + \ldots + u_n + \ldots \infty$

$$S_n = u_1 + u_2 + u_3 + \ldots + u_n$$

Three cases arise:

(*i*) If S_n tends to a finite number as $n \to \infty$, the series $\Sigma\, u_n$ is said to be *convergent.*

(*ii*) If S_n tends to infinity as $n \to \infty$, the series $\Sigma\, u_n$ is said to be *divergent.*

(*iii*) If S_n does not tend to a unique limit, finite or infinite, the series $\Sigma\, u_n$ is called *oscillatory.*

39.9 PROPERTIES OF INFINITE SERIES

1. The nature of an infinite series does not change:

(*i*) by multiplication of all terms by a constant k.

(*ii*) by addition or deletion of a finite number of terms.

2. If two series Σu_n and Σv_n are convergent, then $\Sigma (u_n + v_n)$ is also convergent.

Example 1. *Examine the nature of the series* $1 + 2 + 3 + 4 + \ldots + n + \ldots \infty$.

Solution. Let $$S_n = 1 + 2 + 3 + 4 + \ldots + n = \frac{n(n+1)}{2}$$ [Series in A.P.]

Since $$\lim_{n \to \infty} S_n = \lim_{n \to \infty} \frac{n(n+1)}{2} \Rightarrow \infty$$

Hence, this series is divergent. **Ans.**

Example 2. *Test the convergence of the series* $1 + \frac{1}{2} + \frac{1}{4} + \frac{1}{8} + \ldots \infty$

Solution. Let $$S_n = 1 + \frac{1}{2} + \frac{1}{4} + \frac{1}{8} + \ldots \infty$$ [Series in G.P.]

$$= \frac{1}{1 - \frac{1}{2}} = 2 \qquad \left(S_n = \frac{a}{1-r}\right)$$

$$\lim_{n \to \infty} S_n = 2$$

Hence, the series is convergent. **Ans.**

Example 3. *Prove that the following series:*

$\frac{2}{3!} + \frac{3}{4!} + \frac{4}{5!} + \ldots$ *is convergent and find its sum.* (*M.U. 2008*)

Solution. Here, $$u_n = \frac{n+1}{(n+2)!} = \frac{n+2-1}{(n+2)!} = \frac{n+2}{(n+2)!} - \frac{1}{(n+2)!}$$

$$= \frac{1}{(n+1)!} - \frac{1}{(n+2)!}$$

$$S_n = \left(\frac{1}{2!} - \frac{1}{3!}\right) + \left(\frac{1}{3!} - \frac{1}{4!}\right) + \left(\frac{1}{4!} - \frac{1}{5!}\right) + \ldots$$

$$+ \left(\frac{1}{(n+1)!} - \frac{1}{(n+2)!}\right) = \frac{1}{2!} - \frac{1}{(n+2)!}$$

$$\lim_{n \to \infty} S_n = \lim_{n \to \infty} \left[\frac{1}{2!} - \frac{1}{(n+2)!}\right] = \frac{1}{2}$$

$\therefore \quad \Sigma u_n$ converges and its limit is $\frac{1}{2}$. **Ans.**

Example 4. *Discuss the nature of the series* $2 - 2 + 2 - 2 + 2 - \ldots \infty$.

Solution. Let $$S_n = 2 - 2 + 2 - 2 + 2 - \ldots \infty$$
$= 0$ if n is even
$= 2$ if n is odd.

Hence, S_n does not tend to a unique limit, and, therefore, the given series is oscillatory. **Ans.**

EXERCISE 39.2

Discuss the nature of the following series:

1. $1 + 4 + 7 + 10 + \ldots \infty$ **Ans.** Divergent
2. $1 + \frac{5}{4} + \frac{6}{4} + \frac{7}{4} + \ldots \infty$ **Ans.** Divergent
3. $6 - 5 - 1 + 6 - 5 - 1 + 6 - 5 - 1 + \ldots \infty$ **Ans.** Oscillatory
4. $3 + \frac{3}{2} + \frac{3}{2^2} + \ldots \infty$ **Ans.** Convergent
5. $1^2 + 2^2 + 3^2 + 4^2 + \ldots \infty$ **Ans.** Divergent
6. $1 - \frac{1}{2} + \frac{1}{4} - \frac{1}{8} + \ldots \infty$ **Ans.** Convergent
7. $\frac{1}{1.3} + \frac{1}{3.5} + \frac{1}{5.7} + \ldots \infty$ **Ans.** Convergent
8. $\frac{1}{2!} + \frac{2}{3!} + \frac{3}{4!} + \ldots \infty$ **Ans.** Convergent
9. $\log 3 + \log \frac{4}{3} + \log \frac{5}{4} + \ldots \infty$ **Ans.** Divergent
10. $\sum \log \frac{n}{n+1}$ **Ans.** Divergent
11. $\Sigma (\sqrt{n+1} - \sqrt{n})$ **Ans.** Divergent
12. $\sum \frac{1}{n(n+2)}$ **Ans.** Convergent
13. $\sum \frac{1}{n(n+1)(n+2)(n+3)}$ **Ans.** Convergent
14. $\sum \frac{n}{(n+1)(n+2)(n+3)}$ **Ans.** Convergent
15. $\sum \frac{2n+1}{n^2(n+1)^2}$ **Ans.** Convergent

39.10 PROPERTIES OF GEOMETRIC SERIES

The series $1 + r + r^2 + r^3 + \ldots \infty$ is

(*i*) convergent if $|r| < 1$ (*ii*) divergent if $r \geq 1$ (*iii*) oscillatory if $r \leq -1$.

Proof.
$$S_n = 1 + r + r^2 + \ldots + r^{n-1} = \frac{1 - r^n}{1 - r}$$

(*i*) When $|r| < 1$,
$$\lim_{n \to \infty} r^n = 0$$
$$\lim_{n \to \infty} S_n = \lim_{n \to \infty} \frac{1 - r^n}{1 - r} = \frac{1 - 0}{1 - r} = \frac{1}{1 - r}$$

Hence, the series is convergent.

(*ii*) (*a*) When $r > 1$,
$$\lim_{n \to \infty} r^n = \infty \quad \therefore \quad \lim_{n \to \infty} S_n = \lim_{n \to \infty} \frac{r^n - 1}{r - 1} \Rightarrow \infty$$

Hence, the series is divergent.

(*b*) When $r = 1$, the series becomes $1 + 1 + 1 + 1 + \ldots \infty$
$$S_n = 1 + 1 + 1 + 1 + \ldots = n$$
$$\lim_{n \to \infty} S_n = \lim_{n \to \infty} n = \infty$$

Hence, the series is divergent.

(*iii*) (*a*) When $r = -1$, the series becomes $1 - 1 + 1 - 1 + 1 - \ldots \infty$
$$S_n = 0 \text{ if } n \text{ is even}$$
$$= 1 \text{ if } n \text{ is odd}$$

Hence, the series is oscillatory.

(*b*) When $r < -1$, let $r = -k$ where $k > 1$.

$$r^n = (-k)^n = (-1)^n k^n$$

$$\lim_{n \to \infty} S_n = \lim_{n \to \infty} \frac{1 - r^n}{1 - r} = \lim_{n \to \infty} \frac{1 - (-1)^n k^n}{1 - (-k)}$$

$$= +\infty \text{ if } n \text{ is odd}$$

$$= -\infty \text{ if } n \text{ is even}$$

Hence, the series is oscillatory. **Proved.**

EXERCISE 39.3

Test the nature of the following series :

1. $1 + \frac{1}{2} + \frac{1}{2^2} + \frac{1}{2^3} + \ldots \infty$ **Ans.** Convergent **2.** $1 + \frac{3}{4} + \frac{9}{16} + \frac{27}{64} + \ldots \infty$ **Ans.** Convergent

3. $1 - \frac{1}{3} + \frac{1}{9} - \frac{1}{27} + \ldots \infty$ **Ans.** Convergent **4.** $1 - 2 + 4 - 8 + \ldots \infty$ **Ans.** Oscillatory

5. $2 + 3 + \frac{9}{2} + \frac{27}{4} + \ldots \infty$ **Ans.** Divergent **6.** $1 + \frac{4}{3} + \left(\frac{4}{3}\right)^2 + \left(\frac{4}{3}\right)^3 + \ldots \infty$ **Ans.** Divergent

7. State, which one of the alternatives in the following is correct:

The series $1 - 1 + 1 - 1 + \ldots$ is

(*i*) Convergent with its sum equal to 0. (*ii*) Convergent with its sum equal to 1.

(*iii*) Divergent. (*iv*) Oscillatory. **Ans.** Oscillatory series

39.11 POSITIVE TERM SERIES

If all terms after few negative terms in an infinite series are positive, such a series is a positive term series.

e.g., $-10 - 6 - 1 + 5 + 12 + 20 + \ldots$ is a positive term series.

By omitting the negative terms, the nature of a positive term series remains unchanged.

39.12 NECESSARY CONDITIONS FOR CONVERGENT SERIES

For every convergent series Σu_n.

$$\lim_{n \to \infty} u_n = 0$$

Solution. Let $S_n = u_1 + u_2 + u_3 + \ldots + u_n$

$$\lim_{n \to \infty} S_n = k \quad \text{(a finite quantity)}$$

Also $\lim_{n \to \infty} S_{n-1} = k$ (a finite quantity)

$$S_n = S_{n-1} + u_n$$

$$u_n = S_n - S_{n-1}$$

$$\lim_{n \to \infty} u_n = \lim_{n \to \infty} [S_n - S_{n-1}] = 0$$

$$\lim_{n \to \infty} u_n = 0$$

Corollary. Converse of the above theorem is not true.

e.g., $1 + \frac{1}{\sqrt{2}} + \frac{1}{\sqrt{3}} + \frac{1}{\sqrt{4}} + \ldots + \frac{1}{\sqrt{n}} + \ldots \infty$ is divergent.

$$S_n = 1 + \frac{1}{\sqrt{2}} + \frac{1}{\sqrt{3}} + \frac{1}{\sqrt{4}} + \ldots + \frac{1}{\sqrt{n}}$$

$$> \frac{1}{\sqrt{n}} + \frac{1}{\sqrt{n}} + \frac{1}{\sqrt{n}} + \frac{1}{\sqrt{n}} + \ldots + \frac{1}{\sqrt{n}}$$

$$> \frac{n}{\sqrt{n}} > \sqrt{n}$$

$$\lim_{n \to \infty} S_n = \lim_{n \to \infty} \sqrt{n} = \infty$$

Thus, the series is divergent, although $\lim_{n \to \infty} u_n = \lim_{n \to \infty} \frac{1}{\sqrt{n}} = 0$

So $\lim_{n \to \infty} u_n = 0$ is a necessary condition but not a sufficient condition for convergence.

Note: 1. Test for divergence

If $\lim_{n \to \infty} u_n \neq 0$, *the series* $\Sigma\, u_n$ *must be divergent.*

2. ***To determine the nature of a series we have to find*** S_n. ***Since it is not possible to find*** S_n ***for every series, we have to device tests for convergence without involving*** S_n.

39.13 CAUCHY'S FUNDAMENTAL TEST FOR DIVERGENCE

If $\lim_{n \to \infty} u_n \neq 0$, the series is divergent.

Example 5. *Test for convergence of the series* $1 + \frac{2}{3} + \frac{3}{4} + \frac{4}{5} + \ldots + \frac{n}{n+1} + \ldots \infty$

Solution. Here, $\lim_{n \to \infty} u_n = \lim_{n \to \infty} \frac{n}{n+1} = \lim_{n \to \infty} \frac{1}{1 + \frac{1}{n}} = 1 \neq 0$

Hence, by **Cauchy's Fundamental Test** for divergence, the series is divergent. **Ans.**

Example 6. *Test for convergence the series* $1 + \frac{3}{5} + \frac{8}{10} + \frac{15}{17} + \ldots + \frac{2^n - 1}{2^n + 1} + \ldots \infty$

Solution. Here, $\lim_{n \to \infty} u_n = \lim_{n \to \infty} \frac{2^n - 1}{2^n + 1} = \lim_{n \to \infty} \frac{1 - \frac{1}{2^n}}{1 + \frac{1}{2^n}} = 1 \neq 0$

Hence, by **Cauchy's Fundamental Test** for divergence the series is divergent. **Ans.**

Example 7. Test the convergence of the following series:

$$\sqrt{\frac{1}{4}} + \sqrt{\frac{2}{6}} + \sqrt{\frac{3}{8}} + \cdots + \sqrt{\frac{n}{2(n+1)}} + \cdots$$ *(M.D.U., 2000)*

Solution. Here, we have

$$\sqrt{\frac{1}{4}} + \sqrt{\frac{2}{6}} + \sqrt{\frac{3}{8}} + \cdots + \sqrt{\frac{n}{2(n+1)}} + \cdots$$

$$u_n = \sqrt{\frac{n}{2(n+1)}} = \frac{1}{\sqrt{2\left(1 + \frac{1}{n}\right)}}$$

$$\lim_{n \to \infty} u_n = \lim_{n \to \infty} \frac{1}{\sqrt{2\left(1 + \frac{1}{n}\right)}} = \frac{1}{\sqrt{2}} \neq 0$$

$\Rightarrow \Sigma\, u_n$ does not converge.

The given series is a series of + ve terms,

Hence by Cauchy fundamental test for divergence, the series is divergent. **Ans.**

EXERCISE 39.4

Examine for convergence:

1. $\frac{1}{\sqrt{2}}+\frac{2}{\sqrt{5}}+\frac{4}{\sqrt{17}}+\ldots+\frac{2^n}{\sqrt{4^n+1}}+\ldots\infty$ **Ans.** Divergent

2. $\sum_{n=1}^{\infty}\frac{n}{n+1}$ **Ans.** Divergent

3. $\sum_{n=1}^{\infty}\sqrt{\frac{n}{n+1}}$ **Ans.** Divergent

4. $\sum \cos\frac{1}{n}$ **Ans.** Divergent

5. $1+\frac{1}{2}+2+\frac{1}{3}+3+\frac{1}{4}+4+\ldots$ **Ans.** Divergent

6. $\Sigma\,(6-n^2)$ **Ans.** Divergent

7. $\Sigma\,(-2^n)$ **Ans.** Divergent

8. $\Sigma\,3^{n+1}$ **Ans.** Divergent

39.14 *p*-SERIES

The series $\frac{1}{1^p}+\frac{1}{2^p}+\frac{1}{3^p}+\ldots\infty$ *is* (*i*) *convergent if* $p>1$ (*ii*) *Divergent if* $p\le 1$.

(*MDU, Dec. 2010*)

Solution. Case 1: ($p > 1$)
The given series can be grouped as

$$\frac{1}{1^p}+\left(\frac{1}{2^p}+\frac{1}{3^p}\right)+\left(\frac{1}{4^p}+\frac{1}{5^p}+\frac{1}{6^p}+\frac{1}{7^p}\right)+\left(\frac{1}{8^p}+\frac{1}{9^p}+\frac{1}{10^p}+\frac{1}{11^p}+\frac{1}{12^p}+\frac{1}{13^p}+\frac{1}{14^p}+\frac{1}{15^p}\right)+\ldots$$

Now

$$\frac{1}{1^p}=1 \qquad \ldots(1)$$

$$\frac{1}{2^p}+\frac{1}{3^p}<\frac{1}{2^p}+\frac{1}{2^p}=\frac{2}{2^p} \qquad \ldots(2)$$

$$\frac{1}{4^p}+\frac{1}{5^p}+\frac{1}{6^p}+\frac{1}{7^p}<\frac{1}{4^p}+\frac{1}{4^p}+\frac{1}{4^p}+\frac{1}{4^p}=\frac{4}{4^p} \qquad \ldots(3)$$

$$\frac{1}{8^p}+\frac{1}{9^p}+\ldots+\frac{1}{15^p}<\frac{1}{8^p}+\frac{1}{8^p}+\ldots+\frac{1}{8^p}=\frac{8}{8^p} \qquad \ldots(4)$$

On adding (1), (2), (3) and (4), we get:

$$\frac{1}{1^p}+\left(\frac{1}{2^p}+\frac{1}{3^p}\right)+\left(\frac{1}{4^p}+\frac{1}{5^p}+\frac{1}{6^p}+\frac{1}{7^p}\right)+\left(\frac{1}{8^p}+\frac{1}{9^p}+\ldots+\frac{1}{15^p}\right)+\ldots$$

$$<\frac{1}{1^p}+\frac{2}{2^p}+\frac{4}{4^p}+\frac{8}{8^p}+\ldots$$

$$<1+\left(\frac{1}{2}\right)^{p-1}+\left(\frac{1}{2}\right)^{2p-2}+\left(\frac{1}{2}\right)^{3p-3}+\ldots$$

$$<\frac{1}{1-\left(\frac{1}{2}\right)^{p-1}} \qquad \left[\text{G.P., } r=\left(\frac{1}{2}\right)^{p-1},\ S=\frac{1}{1-r}\right]$$

$<$ Finite number if $p>1$.

Hence, the given series is convergent when $p > 1$.

Case 2: $p = 1$

When $p = 1$, the given series becomes

$$1+\frac{1}{2}+\left(\frac{1}{3}+\frac{1}{4}\right)+\left(\frac{1}{5}+\frac{1}{6}+\frac{1}{7}+\frac{1}{8}\right)+\left(\frac{1}{9}+\frac{1}{10}+\ldots+\frac{1}{16}\right)+\ldots$$

$$1+\frac{1}{2} = 1+\frac{1}{2} \qquad \ldots(1)$$

$$\frac{1}{3}+\frac{1}{4} > \frac{1}{4}+\frac{1}{4}=\frac{1}{2} \qquad \ldots(2)$$

$$\frac{1}{5}+\frac{1}{6}+\frac{1}{7}+\frac{1}{8} > \frac{1}{8}+\frac{1}{8}+\frac{1}{8}+\frac{1}{8}=\frac{4}{8}=\frac{1}{2} \qquad \ldots(3)$$

$$\frac{1}{9}+\frac{1}{10}+\ldots+\frac{1}{16} > \frac{1}{16}+\frac{1}{16}+\ldots+\frac{1}{16}=\frac{8}{16}=\frac{1}{2} \qquad \ldots(4)$$

On adding (1), (2), (3) and (4), we get

$$1+\frac{1}{2}+\left(\frac{1}{3}+\frac{1}{4}\right)+\left(\frac{1}{5}+\frac{1}{6}+\frac{1}{7}+\frac{1}{8}\right)+\left(\frac{1}{9}+\frac{1}{10}+\ldots+\frac{1}{16}\right)+\ldots$$

$$> 1+\frac{1}{2}+\frac{1}{2}+\frac{1}{2}+\frac{1}{2}+\ldots$$

$$> 1+\frac{n}{2} \quad (n \to \infty)$$

$$> \infty$$

Hence, the given series is divergent when $p = 1$.

Case 3: $p < 1$

$$\frac{1}{2^p} > \frac{1}{2}, \qquad \frac{1}{3^p} > \frac{1}{3}, \qquad \frac{1}{4^p} > \frac{1}{4} \text{ and so on}$$

Therefore, $\frac{1}{1^p}+\frac{1}{2^p}+\frac{1}{3^p}+\frac{1}{4^p}+\ldots > 1+\frac{1}{2}+\frac{1}{3}+\frac{1}{4}+\ldots$

$>$ divergent series $(p = 1)$ [From Case 2]

$$\left[\text{As the series on R.H.S. } \left(1+\frac{1}{2}+\frac{1}{3}+\frac{1}{4}+\ldots\right) \text{ is divergent}\right]$$

Hence, the given series is divergent when $p < 1$.

39.15 COMPARISON TEST

If two positive terms Σu_n and Σv_n be such that

$\lim_{n \to \infty} \frac{u_n}{v_n} = k$ **(finite number), then both series converge or diverge together.**

Proof. By definition of limit there exists a positive number ε, however small, such that

$$\left|\frac{u_n}{v_n}-k\right| < \varepsilon \text{ for } n > m \qquad i.e., \; -\varepsilon < \frac{u_n}{v_n}-k < +\varepsilon$$

$$k-\varepsilon < \frac{u_n}{v_n} < k+\varepsilon \text{ for } n > m$$

Ignoring the first m terms of both series, we have

$$k - \varepsilon < \frac{u_n}{v_n} < k + \varepsilon \text{ for all } n. \qquad ...(1)$$

Case 1. $\Sigma\, v_n$ is convergent, then

$$\lim_{n \to \infty} (v_1 + v_2 + ...\, v_n) = h \text{ (say)} \qquad \text{where } h \text{ is a finite number.}$$

From (1), $u_n < (k + \varepsilon)\, v_n$ for all n.

$$\lim_{n \to \infty} (u_1 + u_2 + ... + u_n) < (k + \varepsilon) \lim_{n \to \infty} (v_1 + v_2 + ... + v_n) = (k + \varepsilon)h$$

Hence, $\Sigma\, u_n$ is also convergent.

Case 2. $\Sigma\, v_n$ is divergent, then

$$\lim_{n \to \infty} (v_1 + v_2 + ... + v_n) \to \infty \qquad ...(2)$$

Now from (1) $\qquad k - \varepsilon < \dfrac{u_n}{v_n}$

$$u_n > (k - \varepsilon)v_n \text{ for all } n$$

$$\lim_{n \to \infty} (u_1 + u_2 + ... + u_n) > (k - \varepsilon) \lim_{n \to \infty} (v_1 + v_2 + ... + v_n)$$

From (2), $\displaystyle\lim_{n \to \infty} (u_1 + u_2 + ... + u_n) \to \infty$

Hence, $\Sigma\, u_n$ is also divergent.

Note. For testing the convergence of a series, this Comparison Test is very useful. We choose $\Sigma\, v_n$ (p-series) in such a way that

$$\lim_{n \to \infty} \frac{u_n}{v_n} = \text{finite number.}$$

Then the nature of both the series is the same. The nature of $\Sigma\, v_n$ (p-series) is already known, so the nature of $\Sigma\, u_n$ is also known.

Example 8. *Test the series* $\displaystyle\sum_{n=1}^{\infty} \frac{1}{n + 10}$ *for convergence or divergence.*

Solution. Here, $\qquad u_n = \dfrac{1}{n + 10}$

Let $\qquad v_n = \dfrac{1}{n}$

$$\lim_{n \to \infty} \frac{u_n}{v_n} = \lim_{n \to \infty} \frac{n}{n + 10} = \lim_{n \to \infty} \frac{1}{1 + \frac{10}{n}} = 1 = \text{finite number.}$$

According to Comparison Test both series converge or diverge together, but $\Sigma\, v_n$ is divergent as $p = 1$.

$\therefore \quad \Sigma\, u_n$ is also divergent. **Ans.**

Example 9. *Test the convergence of the following series:*

$$\frac{1}{\sqrt{1} + \sqrt{2}} + \frac{1}{\sqrt{2} + \sqrt{3}} + \frac{1}{\sqrt{3} + \sqrt{4}} + \cdots \qquad \textit{(M.D.U., 2000)}$$

Solution. Here, we have

$$\frac{1}{\sqrt{1} + \sqrt{2}} + \frac{1}{\sqrt{2} + \sqrt{3}} + \frac{1}{\sqrt{3} + \sqrt{4}} + \cdots$$

$$u_n = \frac{1}{\sqrt{n}+\sqrt{n+1}} = \frac{1}{\sqrt{n}\left[1+\sqrt{1+\frac{1}{n}}\right]}$$

Let us compare Σu_n with Σv_n, where

$$v_n = \frac{1}{\sqrt{n}}$$

$$\frac{u_n}{v_n} = \frac{1}{\sqrt{n}\left[1+\sqrt{1+\frac{1}{n}}\right]}\frac{\sqrt{n}}{1} = \frac{1}{1+\sqrt{1+\frac{1}{n}}}$$

$$\lim_{n\to\infty}\frac{u_n}{v_n} = \lim_{n\to\infty}\frac{1}{1+\sqrt{1+\frac{1}{n}}} = \frac{1}{1+1} = \frac{1}{2}$$

Which is finite and non-zero.

$\therefore \Sigma u_n$ and Σv_n, converge or diverge together since $\Sigma v_n = \sum \frac{1}{n^{\frac{1}{2}}}$ is of the form $\sum \frac{1}{n^p}$.

$$P = \frac{1}{2} < 1$$

$\therefore$ Σv_n is divergent $\Rightarrow$ Σu_n is also divergent. **Ans.**

Example 10. *Examine the convergence of the series:*

$$\sum\left(\sqrt[3]{n^3+1}-n\right)$$ *(M.D.U. 2003)*

Solution. Here, we have $\sum\left(\sqrt[3]{n^3+1}-n\right)$

$$u_n = (n^3+1)^{\frac{1}{3}} - n = \left[n^3\left(1+\frac{1}{n^3}\right)\right]^{\frac{1}{3}} - n$$

$$= n\left[\left(1+\frac{1}{n^3}\right)^{\frac{1}{3}} - 1\right] = n\left[1+\frac{1}{3}.\frac{1}{n^3}+\frac{\frac{1}{3}\left(\frac{1}{3}-1\right)}{2!}\left(\frac{1}{n^3}\right)^2+\ldots-1\right]$$

$$= \frac{n}{n^3}\left[\frac{1}{3}-\frac{1}{9}.\frac{1}{n^3}\ldots\right] = \frac{1}{n^2}\left[\frac{1}{3}-\frac{1}{9}.\frac{1}{n^3}\ldots\right]$$

Let $v_n = \frac{1}{n^2}$

$$\frac{u_n}{v_n} = \frac{1}{n^2}\left[\frac{1}{3}-\frac{1}{9}.\frac{1}{n^3}\ldots\right].\frac{n^2}{1} = \left[\frac{1}{3}-\frac{1}{9}.\frac{1}{n^3}\ldots\right]$$

$$\Rightarrow \lim_{n\to\infty}\frac{u_n}{v_n} = \lim_{n\to\infty}\left(\frac{1}{3}-\frac{1}{9}.\frac{1}{n^3}\ldots\right) = \frac{1}{3}$$

which is finite and non-zero.

$\therefore$ $\sum u_n$ and $\sum v_n$ converge or diverge together.

Since $\sum v_n = \sum \frac{1}{n^2}$ is of the form $\sum \frac{1}{n^p}$ with $p = 2 > 1$

$\therefore \quad \sum v_n$ is convergent $\Rightarrow \sum u_n$ is convergent. **Ans.**

Example 11. *Test the convergence of the following series*

$$\frac{1}{1+2^{-1}}+\frac{2}{1+2^{-2}}+\frac{3}{1+2^{-3}}+\cdots \qquad (M.D.U.,\ 2000)$$

Solution. Here, we have

$$\frac{1}{1+2^{-1}}+\frac{2}{1+2^{-2}}+\frac{3}{1+2^{-3}}+\cdots$$

Here $$u_n = \frac{n}{1+2^{-n}} = \frac{n}{1+\frac{1}{2^n}}$$

Let $$u_n = n$$

Let us compare $\sum u_n$ with $\sum v_n$,

$$\frac{u_n}{v_n} = \frac{n}{1+\frac{1}{2^n}} \cdot \frac{1}{n} = \frac{1}{1+\frac{1}{2^n}}$$

$$\lim_{n\to\infty}\frac{u_n}{v_n} = \lim_{n\to\infty}\frac{1}{1+\frac{1}{2^n}} = \frac{1}{1+0} = 1$$

Which is finite and non-zero.

$\therefore \sum u_n$ and $\sum v_n$, converge or diverge together since $\sum v_n = \sum \frac{1}{n}$ is of the form $\sum \frac{1}{n^p}$ with $p = 1$.

$\therefore \sum v_n$ is divergent $\Rightarrow \sum u_n$ is also divergent. **Ans.**

Example 12. *Examine the convergence of the series* $\frac{\sqrt{2}-1}{3^3-1}+\frac{\sqrt{3}-1}{4^3-1}+\frac{\sqrt{4}-1}{5^3-1}+\cdots$

(M.D.U., 2000)

Solution. Here, we have

$$\frac{\sqrt{2}-1}{3^3-1}+\frac{\sqrt{3}-1}{4^3-1}+\frac{\sqrt{4}-1}{5^3-1}+\cdots$$

Here $$u_n = \frac{\sqrt{n+1}-1}{(n+2)^3-1} = \frac{\sqrt{n}\left(\sqrt{1+\frac{1}{n}}-\frac{1}{\sqrt{n}}\right)}{n^3\left[\left(1+\frac{2}{n}\right)^3-\frac{1}{n^3}\right]} = \frac{\sqrt{1+\frac{1}{n}}-\frac{1}{\sqrt{n}}}{n^{\frac{5}{2}}\left[\left(1+\frac{2}{n}\right)^3-\frac{1}{n^3}\right]}$$

Let $$v_n = \frac{1}{n^{\frac{5}{2}}}$$

Let us compare $\sum u_n$ with $\sum v_n$

$$\frac{u_n}{v_n} = \frac{\sqrt{1+\frac{1}{n}}-\frac{1}{\sqrt{n}}}{n^{\frac{5}{2}}\left[\left(1+\frac{2}{n}\right)^3-\frac{1}{n^3}\right]} \times \frac{n^{\frac{5}{2}}}{1} = \frac{\sqrt{1+\frac{1}{n}}-\frac{1}{\sqrt{n}}}{\left[\left(1+\frac{2}{n}\right)^3-\frac{1}{n^3}\right]}$$

$$\lim_{n\to\infty}\frac{u_n}{v_n} = \lim_{n\to\infty}\frac{\sqrt{1+\frac{1}{n}-\frac{1}{\sqrt{n}}}}{\left[\left(1+\frac{2}{n}\right)^3-\frac{1}{n^3}\right]} = \frac{\sqrt{1+0-0}}{(1-0)^3-0} = 1$$

Which is finite and non-zero.

$\therefore \sum u_n$ and $\sum v_n$, converge or diverge together since $\sum v_n$, $= \sum \frac{1}{n^{\frac{5}{2}}}$ is of the form

$$\sum\frac{1}{n^p}. \qquad \text{where } P = \frac{5}{2} > 1.$$

$\therefore \sum v_n$ is convergent $\Rightarrow \sum u_n$ is convergent. **Ans.**

Example 13. *Test the convergence and divergence of the following series.*

$$\sum_{n=1}^{\infty}\frac{2n^2+3n}{5+n^5}$$

(*Gujarat, I Semester, Jan. 2009*)

Solution. Here,

$$u_n = \frac{2n^2+3n}{5+n^5} = \frac{n^2\left(2+\frac{3}{n}\right)}{n^5\left(\frac{5}{n^5}+1\right)} = \frac{1}{n^3}\,\frac{2+\frac{3}{n}}{\frac{5}{n^5}+1}$$

Let $\qquad v_n = \dfrac{1}{n^3}$

By Comparison Test

$$\lim_{n\to\infty}\frac{u_n}{v_n} = \lim_{n\to\infty}\frac{n^3\left(2+\frac{3}{n}\right)}{n^3\left(\frac{5}{n^5}+1\right)} = \lim_{n\to\infty}\frac{2+\frac{3}{n}}{\frac{5}{n^5}+1} = 2 = \text{Finite number.}$$

According to comparison test both series converge or diverge together but $\Sigma\, v_n$ is convergent as $p = 2$.

Hence, the given series is convergent. **Ans.**

Example 14. *Test the following series for convergence* $\dfrac{2}{1^p}+\dfrac{3}{2^p}+\dfrac{4}{3^p}+\dfrac{5}{4^p}+\ldots$

Solution. Given series is $\dfrac{2}{1^p}+\dfrac{3}{2^p}+\dfrac{4}{3^p}+\dfrac{5}{4^p}+\ldots$

Here $\qquad u_n = \dfrac{n+1}{n^p} = \dfrac{1+\frac{1}{n}}{n^{p-1}}$

Let $\qquad v_n = \dfrac{1}{n^{p-1}} \qquad \therefore \qquad \dfrac{u_n}{v_n} = \dfrac{1+\frac{1}{n}}{n^{p-1}}\times\dfrac{n^{p-1}}{1} = 1+\dfrac{1}{n}$

$$\lim_{n\to\infty}\frac{u_n}{v_n} = 1$$

Therefore, both the series are either convergent or divergent.

But Σv_n is convergent if $p-1>1$, *i.e.*, if $p>2$ (P series)

and is divergent if $p-1\le 1$, *i.e.*, if $p\le 2$

$\therefore$ The given series is convergent if $p>2$ and divergent if $p\le 2$. **Ans.**

EXERCISE 39.5

Examine the convergence or divergence of the following series:

1. $2 + \frac{3}{2}.\frac{1}{4} + \frac{4}{3}.\frac{1}{4^2} + \frac{5}{4}.\frac{1}{4^3} + \dots \infty$ **Ans.** Convergent

2. $1 + \frac{1.2}{1.3} + \frac{1.2.3}{1.3.5} + \frac{1.2.3.4}{1.3.5.7} + \dots \infty$ **Ans.** Convergent

3. $\frac{1}{1.2} + \frac{2}{3.4} + \frac{3}{5.6} + \dots \infty$ **Ans.** Divergent

4. $\frac{1}{1.2.3} + \frac{1}{2.3.4} + \frac{1}{3.4.5} + \dots\dots \infty$ **Ans.** Convergent *(M.D. University, Dec. 2004)*

5. $1 + \frac{2^2}{2!} + \frac{3^2}{3!} + \frac{4^2}{4!} + \dots \infty$ **Ans.** Convergent

6. $\frac{1}{1+2} + \frac{2}{1+2^2} + \frac{3}{1+2^3} + \dots$ **Ans.** Convergent *(M.D. University, 2001)*

7. $\frac{1}{3} + \frac{2!}{3^2} + \frac{3!}{3^3} + \dots \infty$ **Ans.** Convergent

8. $\sum_{n=1}^{\infty} \frac{1}{\sqrt{n} + \sqrt{n+1}}$ **Ans.** Divergent

9. $\sum_{n=1}^{\infty} \frac{2n^3 + 5}{4n^5 + 1}$ **Ans.** Convergent

10. $\sum_{n=1}^{\infty} \frac{a^n}{x^n + n^a}$ **Ans.** If $x > a$, convergent; if $x \le a$, Divergent

11. $\sum_{n=1}^{\infty} \frac{\sqrt{n}}{n^2 + 1}$ **Ans.** Convergent

12. $\sum_{n=1}^{\infty} \sqrt{(n^2+1)} - n$ **Ans.** Divergent

13. $\sum_{n=1}^{\infty} \left[\sqrt{(n^4+1)} - \sqrt{(n^4-1)}\right]$ **Ans.** Convergent

14. $\sum_{n=1}^{\infty} \frac{2^n + 1}{3^n + n}$ **Ans.** Convergent

15. $\sum_{n=1}^{\infty} \frac{n^n}{n!}$ **Ans.** Convergent

16. $\sum_{n=1}^{\infty} \frac{n^2}{e^n}$ **Ans.** Convergent

39.16 D'ALEMBERT'S RATIO TEST

Statement. ***If Σu_n is a positive term series such that $\lim_{n \to \infty} \frac{u_{n+1}}{u_n} = k$ then***

(i) the series is convergent if $k < 1$ (ii) the series is divergent if $k > 1$

Solution.

Case I. When $\lim_{n \to \infty} \frac{u_{n+1}}{u_n} = k < 1$

By definition of a limit, we can find a number r (< 1) such that

$$\frac{u_{n+1}}{u_n} < r \text{ for all } n \ge m \qquad \left[\frac{u_2}{u_1} < r, \frac{u_3}{u_2} < r, \frac{u_4}{u_3} < r \dots\right]$$

Omitting the first m terms, let the series be

$$u_1 + u_2 + u_3 + u_4 + \dots \infty$$

$$= u_1\left(1 + \frac{u_2}{u_1} + \frac{u_3}{u_1} + \frac{u_4}{u_1} + \dots\right) = u_1\left(1 + \frac{u_2}{u_1} + \frac{u_3}{u_2}.\frac{u_2}{u_1} + \frac{u_4}{u_3}.\frac{u_3}{u_2}.\frac{u_2}{u_1} + \dots \infty\right)$$

$$< u_1(1 + r + r^2 + r^3 + \dots \infty) \qquad (r < 1)$$

$$= \frac{u_1}{1-r}, \text{ which is a finite quantity.}$$

Hence, $\Sigma\, u_n$ is convergent.

Case 2. When $\lim\limits_{n\to\infty} \dfrac{u_{n+1}}{u_n} = k > 1$

By definition of limit, we can find a number m such that $\dfrac{u_{n+1}}{u_n} \geq 1$ for all $n \geq m$

$$\frac{u_2}{u_1} > 1, \qquad \frac{u_3}{u_2} > 1, \qquad \frac{u_4}{u_3} > 1$$

Ignoring the first m terms, let the series be

$$u_1 + u_2 + u_3 + u_4 + \ldots \infty$$

$$= u_1\left(1 + \frac{u_2}{u_1} + \frac{u_3}{u_1} + \frac{u_4}{u_1} + \ldots\right) = u_1\left(1 + \frac{u_2}{u_1} + \frac{u_3}{u_2}.\frac{u_2}{u_1} + \frac{u_4}{u_3}.\frac{u_3}{u_2}.\frac{u_2}{u_1} + \ldots\infty\right)$$

$$\geq u_1\,(1 + 1 + 1 + 1 \ldots \text{ to } n \text{ terms}) = nu_1$$

$$\left[\because \lim_{n\to\infty} (u_1 + u_2 + \ldots + u_n) = nu_1\right]$$

$$\lim_{n\to\infty} S_n \geq \lim_{n\to\infty} nu_1 = \infty$$

Hence, $\Sigma\, u_n$ is divergent.

Note. When $\dfrac{u_{n+1}}{u_n} = 1 \qquad (k = 1)$

The ratio test fails.

For Example. Consider the series whose n^{th} term $= \dfrac{1}{n}$

$$\lim_{n\to\infty} \frac{u_{n+1}}{u_n} = \lim_{n\to\infty} \frac{\frac{1}{n+1}}{\frac{1}{n}} = \lim_{n\to\infty} \frac{n}{n+1} = \lim_{n\to\infty} \frac{1}{1+\frac{1}{n}} = 1 \qquad \ldots(1)$$

Consider the second series whose n^{th} term is $\dfrac{1}{n^2}$.

$$\lim_{n\to\infty} \frac{u_{n+1}}{u_n} = \lim_{n\to\infty} \frac{\frac{1}{(n+1)^2}}{\frac{1}{n^2}} = \lim_{n\to\infty} \left(\frac{n}{n+1}\right)^2 = 1 \qquad \ldots(2)$$

Thus, from (1) and (2) in both cases $\lim\limits_{n\to\infty} \dfrac{u_{n+1}}{u_n} = 1$

But we know that the first series is divergent as $p = 1$.
The second series is convergent as $p = 2$.

Hence, when $\lim\limits_{n\to\infty} \dfrac{u_{n+1}}{u_n} = 1$, the series may be convergent or divergent.

Thus, ratio test fails when $k = 1$.

Example 15. *Test for convergence of the series whose n^{th} term is* $\dfrac{n^2}{2^n}$

Solution. Here, we have $u_n = \dfrac{n^2}{2^n}, \qquad u_{n+1} = \dfrac{(n+1)^2}{2^{n+1}}$

By D'Alembert's Test

$$\lim_{n\to\infty} \frac{u_{n+1}}{u_n} = \lim_{n\to\infty} \frac{(n+1)^2}{2^{n+1}}.\frac{2^n}{n^2} = \lim_{n\to\infty} \frac{1}{2}\left(1+\frac{1}{n}\right)^2 = \frac{1}{2} < 1$$

Hence, the series is convergent by D'Alembert's Ratio Test. **Ans.**

Example 16. *Test for convergence the series whose n^{th} term is* $\dfrac{2^n}{n^3}$

Solution. Here, we have $u_n = \dfrac{2^n}{n^3}$, $\quad u_{n+1} = \dfrac{2^{n+1}}{(n+1)^3}$

By D'Alembert's Ratio Test

$$\frac{u_{n+1}}{u_n} = \frac{2^{n+1}}{(n+1)^3} \cdot \frac{n^3}{2^n} = \frac{2}{\left(1+\frac{1}{n}\right)^3} \Rightarrow \lim_{n \to \infty} \frac{u_{n+1}}{u_n} = \lim_{n \to \infty} \frac{2}{\left(1+\frac{1}{n}\right)^3} = 2 > 1$$

Hence, the series is divergent. **Ans.**

Example 17. *Discuss the convergence of the series*:

$$\sum \frac{\sqrt{n}}{\sqrt{n^2+1}} x^n \qquad (x > 0)$$

(M.D.University, Dec., 2001)

Solution. Here, we have

$$u_n = \sqrt{\frac{n}{n^2+1}}\, x^n$$

$$\therefore \qquad u_{n+1} = \sqrt{\frac{n+1}{(n+1)^2+1}} . x^{n+1}$$

$$\frac{u_n}{u_{n+1}} = \sqrt{\frac{n}{n+1}} . \sqrt{\frac{n^2+2n+2}{n^2+1}} \cdot \frac{1}{x} = \sqrt{\frac{1}{1+\frac{1}{n}} \cdot \frac{\left(1+\frac{2}{n}+\frac{2}{n^2}\right)}{\left(1+\frac{1}{n^2}\right)}} \cdot \frac{1}{x}$$

$$\lim_{n \to \infty} \frac{u_n}{u_{n+1}} = \lim_{n \to \infty} \sqrt{\frac{1}{1+\frac{1}{n}} \cdot \frac{\left(1+\frac{2}{n}+\frac{2}{n^2}\right)}{\left(1+\frac{1}{n^2}\right)}} \cdot \frac{1}{x} = \frac{1}{x}$$

$\therefore$ By D' Alembert's Ratio Test, $\sum u_n$ converges if $\frac{1}{x} > 1$, *i.e.* $x < 1$ and diverges if $\frac{1}{x} < 1$ *i.e.*, $x > 1$.

When $x = 1$, the Ratio Test fails.

When $x = 1$, $u_n = \sqrt{\dfrac{n}{n^2+1}} = \sqrt{\dfrac{n}{n^2\left(1+\frac{1}{n^2}\right)}} = \dfrac{1}{\sqrt{n}} \cdot \dfrac{1}{\sqrt{1+\frac{1}{n^2}}}$

$$v_n = \frac{1}{\sqrt{n}},$$

$$\frac{u_n}{v_n} = \frac{1}{\sqrt{n}} \cdot \frac{1}{\sqrt{1+\frac{1}{n^2}}} \cdot \frac{\sqrt{n}}{1} = \frac{1}{\sqrt{1+\frac{1}{n^2}}}$$

$$\lim_{n \to \infty} \frac{u_n}{v_n} = \lim_{n \to \infty} \frac{1}{\sqrt{1+\frac{1}{n^2}}} = 1$$

Which is finite and non-zero.

$\therefore$ By comparison test, $\sum u_n$ and $\sum v_n$ converge or diverge together.

Since $\sum v_n = \sum \frac{1}{\sqrt{n}}$ is of the form $\sum \frac{1}{n^p}$ with $p = \frac{1}{2} < 1$.

$\sum v_n$ diverges $\Rightarrow \sum u_n$ diverges.

Hence, the given series $\sum u_n$ converges as $x < 1$ and diverges *as* $x \geq 1$. **Ans.**

EXERCISE 39.6

Test the convergence for series:

1. $\sum_{n=1}^{n} \frac{n^2}{3^n}$ **Ans.** Convergent

2. $\sum_{n=1}^{n} \frac{n!}{n^n}$ **Ans.** Convergent

3. $\left(\frac{1}{3}\right)^2 + \left(\frac{1.2}{3.5}\right)^2 + \left(\frac{1.2.3}{3.5.7}\right)^2 + \ldots \infty$ **Ans.** Convergent

4. $\frac{2}{1} + \frac{2.5.8}{1.5.9} + \frac{2.5.8.11}{1.5.9.13} + \ldots \infty$ **Ans.** Convergent

5. $\sum_{n=1}^{n} \frac{n!.2^n}{n^n}$ **Ans.** Convergent

6. $\sum_{n=1}^{\infty} \frac{x^{n-1}}{n.3^n}$ **Ans.** Convergent if $x > 3$, Divergent if $x < 3$

7. Prove that, if $u_{n+1} = \frac{k}{1+u_n}$, where $k > 0$, $u_1 > 0$, then the series $\Sigma\, u_n$ converges to the positive root of the equation $x^2 + x = k$.

39.17 RAABE'S TEST (HIGHER RATIO TEST)

> ***If*** $\Sigma\, u_n$ ***is a positive term series such that*** $\lim_{n \to \infty} n\left(\frac{u_n}{u_{n+1}} - 1\right) = k$, ***then***
>
> ***(i) the series is convergent if*** $k > 1$ ***(ii) the series is divergent if*** $k < 1$.

Proof. Case I. $k > 1$

Let p be such that $k > p > 1$ and compare the given series $\Sigma\, u_n$ with $\sum \frac{1}{n^p}$ which is convergent as $p > 1$.

$$\frac{u_n}{u_{n+1}} > \frac{(n+1)^p}{n^p} \quad \text{or} \quad \left(\frac{u_n}{u_{n+1}}\right) > \left(1 + \frac{1}{n}\right)^p > 1 + \frac{p}{n} + \frac{p(p-1)}{2!}\frac{1}{n^2} + \ldots$$

(Binomial Theorem)

$$n\left(\frac{u_n}{u_{n+1}} - 1\right) > p + \frac{p\,(p-1)}{2!}\frac{1}{n} + \ldots.$$

If $$\lim_{n \to \infty} n\left(\frac{u_n}{u_{n+1}} - 1\right) > p$$

and $k > p$ which is true as $k > p > 1$; Σu_n is convergent when $k > 1$.

Case II. $k < 1$ Same steps as in Case I.

Notes:

1. Raabe's Test fails if $k = 1$
2. Raabe's Test is applied only when D'Alembert's Ratio Test fails.

Example 18. *Test the convergence for the series* $\frac{x}{1.2}+\frac{x^2}{3.4}+\frac{x^3}{5.6}+\frac{x^4}{7.8}+\ldots$ *(M.U. 2009)*

Solution. Here, $u_n = \frac{x^n}{(2n-1)\,2n}$ and $u_{n+1} = \frac{x^{n+1}}{(2n+1)(2n+2)}$

By D'Alembert's Test

$$\frac{u_{n+1}}{u_n} = \frac{x^{n+1}}{(2n+1)(2n+2)} \times \frac{(2n-1)2n}{x^n} = \frac{x\left(1-\frac{1}{2n}\right)}{\left(1+\frac{1}{2n}\right)\left(1+\frac{2}{2n}\right)}$$

$$\lim_{n\to\infty} \frac{u_{n+1}}{u_n} = x$$

(*i*) If $x < 1$, $\Sigma\, u_n$ is convergent (*ii*) If $x > 1$, $\Sigma\, u_n$ is divergent (*iii*) If $x = 1$, Test fails.

Let us apply **Raabe's Test** when $x = 1$

$$\lim_{n\to\infty} n\left(\frac{u_n}{u_{n+1}} - 1\right) = \lim_{n\to\infty} n\left[\frac{(2n+1)(2n+2)}{2n(2n-1)} - 1\right]$$

$$= \lim_{n\to\infty} n\left[\frac{(2n+1)(2n+2)-2n(2n-1)}{2n\,(2n-1)}\right]$$

$$= \lim_{n\to\infty} n\left[\frac{(8n+2)}{(2n)\,(2n-1)}\right] = \lim_{n\to\infty} \frac{\left(2+\frac{2}{4n}\right)}{1\left(1-\frac{1}{2n}\right)} = 2$$

So the series is convergent.

Hence we can say that the given series is convergent if $x \le 1$ and divergent, if $x > 1$. **Ans.**

Example 19. *Test the following series for convergence* $\sum \frac{1}{\sqrt{n+1}-1}$.

Solution. Here, $u_n = \frac{1}{\sqrt{n+1}-1}$, $u_{n+1} = \frac{1}{\sqrt{n+2}-1}$

$$\frac{u_{n+1}}{u_n} = \frac{\sqrt{n+1}-1}{\sqrt{n+2}-1}$$

$$\lim_{n\to\infty} \frac{u_{n+1}}{u_n} = \lim_{n\to\infty} \frac{\sqrt{1+\frac{1}{n}} - \frac{1}{n^2}}{\sqrt{1+\frac{2}{n}} - \frac{1}{n^2}} = 1$$

D'Alembert's test fails.

By Raabe's test.

$$\lim_{n\to\infty} n\left(\frac{u_n}{u_{n+1}} - 1\right) = \lim_{n\to\infty} n\left(\frac{\sqrt{n+2}-1}{\sqrt{n+1}-1} - 1\right)$$

$$= \lim_{n\to\infty} n\left[\frac{\sqrt{n+2}-1-\sqrt{n+1}+1}{\sqrt{n+1}-1}\right] = \lim_{n\to\infty} n\left[\frac{\sqrt{n+2}-\sqrt{n+1}}{\sqrt{n+1}-1}\right]$$

$$= \lim_{n\to\infty} n\left[\frac{\sqrt{1+\frac{2}{n}}-\sqrt{1+\frac{1}{n}}}{\sqrt{1+\frac{1}{n}-\frac{1}{n^2}}}\right] = 0 < 1$$

Hence, $\Sigma\, u_n$ is divergent. **Ans.**

Example 20. *Discuss the convergence of the series:*

$$\frac{x}{1}+\frac{1}{2}\cdot\frac{x^3}{3}+\frac{1\cdot3}{2\cdot4}\frac{x^5}{5}+\frac{1\cdot3\cdot5}{2\cdot4\cdot6}\cdot\frac{x^7}{7}+\ldots\ (x>0)$$ (*M.D.U. Dec., 2001*)

Solution. Here, we have

$$\frac{x}{1}+\frac{1}{2}\cdot\frac{x^3}{3}+\frac{1\cdot3}{2\cdot4}\frac{x^5}{5}+\frac{1\cdot3\cdot5}{2\cdot4\cdot6}\cdot\frac{x^7}{7}+\ldots$$

Neglecting the first term, we have

$$u_n = \frac{1\cdot3\cdot5\ldots(2n-1)}{2\cdot4\cdot5\ldots(2n)}\cdot\frac{x^{2n+1}}{2n+1}$$

and $$u_{n+1} = \frac{1\cdot3\cdot5\ldots(2n-1)\,(2n+1)}{2\cdot4\cdot5\ldots(2n)\,(2n+2)}\cdot\frac{x^{2n+3}}{2n+3}$$

$$\frac{u_n}{u_{n+1}} = \frac{(2n-1)\,x^{2n+1}}{2n\,(2n+1)}\times\frac{2n\,(2n+2)\,(2n+3)}{(2n-1)\,(2n+1)\,x^{2n+3}} = \frac{(2n+2)\,(2n+3)}{(2n+1)\,(2n+1)}\,\frac{1}{x^2}$$

$$\therefore \qquad = \frac{2n+2}{2n+1}\cdot\frac{2n+3}{2n+1}\cdot\frac{1}{x^2} = \frac{2n\left(1+\frac{1}{n}\right).2n\left(1+\frac{3}{2n}\right)}{2n\left(1+\frac{1}{2n}\right).2n\left(1+\frac{1}{2n}\right)}\cdot\frac{1}{x^2} = \frac{\left(1+\frac{1}{n}\right)\left(1+\frac{3}{2n}\right)}{\left(1+\frac{1}{2n}\right)^2}\cdot\frac{1}{x^2}$$

$$\lim_{n\to\infty}\frac{u_n}{u_{n+1}} = \lim_{n\to\infty}\frac{\left(1+\frac{1}{n}\right)\left(1+\frac{3}{2n}\right)}{\left(1+\frac{1}{2n}\right)^2}\cdot\frac{1}{x^2} = \frac{1}{x^2}$$

$\therefore$ Ratio Test, $\sum u_n$ is convergent if $\frac{1}{x^2} > 1$.

i.e.; $x^2 < 1$ and divergent if $\frac{1}{x^2} < 1$. *i.e.*, $x^2 > 1$.

If $x^2 = 1$, then Ratio Test fails.

Now Raabe's test

When $x^2 = 1$, we have $$\frac{u_n}{u_{n+1}} = \frac{(2n+2)(2n+3)}{(2n+1)^2} = \frac{4n^2+10n+6}{4n^2+4n+1}$$

$$\lim_{n\to\infty} n\left(\frac{u_n}{u_{n+1}}-1\right) = \lim_{n\to\infty} n\left(\frac{4n^2+10n+6}{4n^2+4n+1}-1\right)$$

$$= \lim_{n\to\infty}\frac{6n^2+5n}{4n^2+4n+1} = \lim_{n\to\infty}\frac{6+\frac{5}{n}}{4+\frac{4}{n}+\frac{1}{n^2}} = \frac{6}{4} = \frac{3}{2} > 1$$

$\therefore$ By Raabe's Test, the series converges.

Hence, $\sum u_n$ is convergent if $x^2 \le 1$ and divergent if $x^2 > 1$. **Ans.**

Example 21. *Test the following series for convergence*

$$\frac{1}{2}x + x^2 + \frac{9}{8}x^3 + x^4 + \frac{25}{32}x^5 + \dots \infty$$

Solution. Here, $u = \dfrac{n^2.x^n}{2^n}$ $\quad u_{n+1} = \dfrac{(n+1)^2.x^{n+1}}{2^{n+1}}$

By D'Alembert's Test

$$\frac{u_{n+1}}{u_n} = \frac{(n+1)^2 x^{n+1}}{2^{n+1}} \cdot \frac{2^n}{n^2 x^n} = \left(\frac{n+1}{n}\right)^2 \frac{x}{2}$$

$$\lim_{n \to \infty} \frac{u_{n+1}}{u_n} = \lim_{n \to \infty} \left(1 + \frac{1}{n}\right)^2 \frac{x}{2} = \frac{x}{2}$$

(*i*) If $\dfrac{x}{2} < 1$ or $x < 2$, then $\Sigma\, u_n$ is convergent. (*ii*) If $\dfrac{x}{2} > 1$ or $x > 2$, then $\Sigma\, u_n$ is divergent.

(*iii*) If $\dfrac{x}{2} = 1$ or $x = 2$, then the test fails.

Let us apply **Raabe's test**

$$n\left(\frac{u_n}{u_{n+1}} - 1\right) = n\left[\frac{n^2}{(n+1)^2}\frac{2}{2} - 1\right] = n\left[\frac{n^2 - n^2 - 2n - 1}{(n+1)^2}\right] = \frac{-2n^2 - n}{(n+1)^2}$$

$$\lim_{n \to \infty} n\left(\frac{u_n}{u_{n+1}} - 1\right) = \lim_{n \to \infty} \frac{-2 - \frac{1}{n}}{\left(1 + \frac{1}{n}\right)^2} = -2 < 1$$

Hence, $\Sigma\, u_n$ is divergent if $x \geq 2$, and convergent if $x < 2$. **Ans.**

Example 22. *Show that the series* $\dfrac{1}{x} + \dfrac{2!}{x(x+1)} + \dfrac{3!}{x(x+1)(x+2)} + \dots$ *converges if $x > 2$ and diverges if $x \leq 2$.*

Solution. Here, $u_n = \dfrac{n!}{x(x+1)\,(x+2) \dots (x+n-1)}$

$$u_{n+1} = \frac{(n+1)!}{x(x+1)\,(x+2) \dots (x+n-1)\,(x+n)}$$

By D'Alembert's test

$$\frac{u_{n+1}}{u_n} = \frac{n+1}{(x+n)}, \quad \lim_{n \to \infty} \frac{u_{n+1}}{u_n} = \frac{1 + \frac{1}{n}}{1 + \frac{x}{n}} = 1$$

Test fails. Let us apply **Raabe's Test.**

$$\lim_{n \to \infty} n\left(\frac{u_n}{u_{n+1}} - 1\right) = \lim_{n \to \infty} n\left(\frac{x+n}{n+1} - 1\right) = \lim_{n \to \infty} n\left(\frac{x-1}{n+1}\right) = \lim_{n \to \infty} \frac{x-1}{1 + \frac{1}{n}} = x - 1$$

If $x - 1 > 1$ or $x > 2$, then $\Sigma\, u_n$ is convergent.
If $x - 1 < 1$ or $x < 2$, then $\Sigma\, u_n$ is divergent. **Ans.**

Example 23. *Discuss the convergence of the series* $\dfrac{x^2}{2 \log 2} + \dfrac{x^3}{3 \log 3} + \dfrac{x^4}{4 \log 4} + \dots$

Solution. Here, we have $\dfrac{x^2}{2\log 2}+\dfrac{x^3}{3\log 3}+\dfrac{x^4}{4\log 4}+\ldots$

$$u_n=\frac{x^{n+1}}{(n+1)\log(n+1)},\qquad u_{n+1}=\frac{x^{n+2}}{(n+2)\log(n+2)}$$

By D'Alembert's Test

$$\lim_{n\to\infty}\frac{u_{n+1}}{u_n}=\lim_{n\to\infty}\frac{x^{n+2}}{(n+2)\log(n+2)}\times\frac{(n+1)\log(n+1)}{x^{n+1}}$$

$$=\lim_{n\to\infty}x\left(\frac{n+1}{n+2}\right)\frac{\log(n+1)}{\log(n+2)}$$

$$=\lim_{n\to\infty}x\left(\frac{1+\frac{1}{n}}{1+\frac{2}{n}}\right)\frac{\log n+\log\left(1+\frac{1}{n}\right)}{\log n+\log\left(1+\frac{2}{n}\right)}$$

$$=\lim_{n\to\infty}x\left(\frac{1+\frac{1}{n}}{1+\frac{2}{n}}\right)\left[\frac{\log n+\frac{1}{n}-\frac{1}{2}\frac{1}{n^2}+\ldots}{\log n+\frac{2}{n}-\frac{1}{2}.\frac{4}{n^2}+\ldots}\right]$$

$$=\lim_{n\to\infty}x\left(\frac{1+\frac{1}{n}}{1+\frac{2}{n}}\right)\left[\frac{1+\frac{1}{n\log n}+\ldots}{1+\frac{2}{n\log n}+\ldots}\right]=x$$

(*i*) When $x<1$, the series is convergent (*ii*) When $x>1$, the series is divergent.
(*iii*) When $x=1$, the test fails.

Let us apply Raabe's Test

$$\frac{u_n}{u_{n+1}}=\left(\frac{n+2}{n+1}\right)\frac{\log(n+2)}{\log(n+1)}=\left(\frac{n+2}{n+1}\right)\frac{\log n+\log\left(1+\frac{2}{n}\right)}{\log n+\log\left(1+\frac{1}{n}\right)}$$

By D'Alembert's Test

$$=\left(\frac{n+2}{n+1}\right)\frac{\log n+\frac{2}{n}-\frac{1}{2}.\frac{4}{n^2}+\ldots}{\log n+\frac{1}{n}-\frac{1}{2}.\frac{1}{n^2}+\ldots}=\left(\frac{n+2}{n+1}\right)\frac{1+\frac{2}{n\log n}+\ldots}{1+\frac{1}{n\log n}+\ldots}$$

$$=\frac{n+2}{n+1}\left(1+\frac{2}{n\log n}\right)\left(1+\frac{1}{n\log n}\right)^{-1}=\frac{n+2}{n+1}\left(1+\frac{2}{n\log n}\right)\left(1-\frac{1}{n\log n}\right)$$

$$=\frac{n+2}{n+1}\left(1+\frac{2}{n\log n}-\frac{1}{n\log n}+\ldots\right)=\left(\frac{n+2}{n+1}\right)\left[1+\frac{1}{n\log n}\right]$$

$$\lim_{n\to\infty}\frac{u_n}{u_{n+1}}=\lim_{n\to\infty}\left[\frac{1+\frac{2}{n}}{1+\frac{1}{n}}\right]\left[1+\frac{1}{n\log n}\right]=1+\frac{1}{n\log n}$$

$$n\left[\frac{u_n}{u_{n+1}}-1\right]=n\left[1+\frac{1}{n\log n}-1\right]=\frac{1}{\log n}=0<1$$

Thus the series is divergent when $x=1$.

Hence, the series converges if $x < 1$ and diverges if $x \geq 1$. **Ans.**

Example 24. *Test the series for convergence*

$$1+\frac{\alpha.\beta}{1.\gamma}x+\frac{\alpha(\alpha+1).\beta(\beta+1)}{1.2.\gamma(\gamma+1)}x^2+\frac{\alpha(\alpha+1)(\alpha+2).\beta(\beta+1)(\beta+2)}{1.2.3.\gamma(\gamma+1)(\gamma+2)}x^3+...$$

Solution. $u_n = \dfrac{\alpha(\alpha+1)(\alpha+2)\,...\,[\alpha+(n-1)]\,.\,\beta(\beta+1)\,...\,[\beta+(n-1)]}{n!\,\gamma(\gamma+1)\,...\,[\gamma+(n-1)]}x^n$

$$u_{n+1} = \frac{\alpha(\alpha+1)(\alpha+2)...[\alpha+(n-1)](\alpha+n).\beta(\beta+1)...[\beta+(n-1)](\beta+n)}{(n+1)!\,\gamma(\gamma+1)\,...\,[\gamma+(n-1)]\,(\gamma+n)}x^{n+1}$$

By D'Alembert's Test

$$\lim_{n\to\infty}\frac{u_{n+1}}{u_n} = \lim_{n\to\infty}\frac{(\alpha+n)(\beta+n)}{(n+1)(\gamma+n)}x = \lim_{n\to\infty}\frac{\left(1+\frac{\alpha}{n}\right)\left(1+\frac{\beta}{n}\right)}{\left(1+\frac{1}{n}\right)\left(1+\frac{\gamma}{n}\right)}.x$$

$$\lim_{n\to\infty}\frac{u_{n+1}}{u_n} = x$$

(*i*) If $x < 1$, the series is convergent. (*ii*) If $x > 1$, the series is divergent.

(*iii*) If $x = 1$, the test fails.

Let us apply Raabe's Test

$$n\left(\frac{u_n}{u_{n+1}}-1\right) = n\left[\frac{(n+1)(\gamma+n)}{(\alpha+n)(\beta+n)}-1\right] = n\left[\frac{n\gamma+n^2+\gamma+n-\alpha\beta-n\alpha-n\beta-n^2}{(\alpha+n)(\beta+n)}\right]$$

$$\lim_{n\to\infty} n\left(\frac{u_n}{u_{n+1}}-1\right) = \lim_{n\to\infty}\frac{\gamma+\frac{\gamma}{n}+1-\frac{\alpha\beta}{n}-\alpha-\beta}{\left(\frac{\alpha}{n}+1\right)\left(\frac{\beta}{n}+1\right)} = \gamma+1-\alpha-\beta$$

(*i*) If $\gamma+1-\alpha-\beta > 1$ or $\gamma > \alpha+\beta$, then $\Sigma\, u_n$ is convergent.

(*ii*) If $\gamma+1-\alpha-\beta < 1$ or $\gamma < \alpha+\beta$, then $\Sigma\, u_n$ is divergent. **Ans.**\

EXERCISE 39.7

Determine the nature of the following series:

1. $1+\frac{2!}{2^2}+\frac{3!}{3^2}+\frac{4!}{4^2}+...\infty$ **Ans.** Divergent

2. $\frac{1}{1}+\frac{1.3}{1.4}+\frac{1.3.5}{1.4.7}+\frac{1.3.5.7}{1.4.7.10}+...\infty$ **Ans.** Convergent

3. $1+\frac{1+\alpha}{1+\beta}+\frac{(1+\alpha)(2+\alpha)}{(1+\beta)(2+\beta)}+...\infty$ **Ans.** If $\beta-\alpha > 1$, convergent. If $\beta-\alpha \leq 1$, Divergent.

4. $\sum_{n=1}^{\infty}\frac{n^3}{e^n}$ **Ans.** Convergent

5. $x+\frac{2x^2}{2!}+\frac{3x^3}{3!}+\frac{4x^4}{4!}+...\infty$ **Ans.** Convergent

6. $1+\frac{x}{2}+\frac{x^2}{3}+\frac{x^3}{4}+...$ **Ans.** Divergent

7. $1+\frac{1}{2}x+\frac{1}{5}x^2+\frac{1}{10}x^3+...$ **Ans.** Convergent if $-1 \leq x < 1$ and divergent if $|x| > 1$

9. $1+\frac{(1!)^2}{2!}x^2+\frac{(2!)^2}{4!}x^4+\frac{(3!)^2}{6!}x^6+\ldots\infty\ (x>0)$

Ans. If $x^2<4$, convergent; and divergent if $x^2\geq 4$

Find the values of x for which the following series converges:

10. $x^2(\log 2)^q+x^3(\log 3)^q+x^4(\log 4)^q+\ldots$

Ans. If $x<1$, convergent; and divergent if $x\geq 1$

11. $\sum_{n=0}^{\infty}\frac{(-1)^n\, n!\, x^n}{10^n}$ 12. $\sum\frac{x^n}{2n\,(2n+1)}$ **Ans.** If $x\leq 1$, convergent; and if $x>1$, divergent

13. $\sum\frac{1.2\ldots n}{4.7\ldots(3n+1)}x^n$ **Ans.** If $0<x<3$, convergent and divergent if $x\geq 3$.

14. $1+\frac{(1!)^2}{2!}x+\frac{(2!)^2}{4!}x^2+\frac{(3!)^2}{6!}x^3+\ldots\ldots$ *(M.D.U., Dec. 2010)*

Ans. convergent if $x<4$; divergent if $x\geq 4$.

39.18. GAUSS'S TEST

Statement. ***If Σu_n is a positive term series such that***

$$\frac{u_n}{u_{n+1}}=\alpha+\frac{\beta}{n}+\frac{\gamma}{n^2}\qquad \textit{where } \alpha>0$$

(i) if $\alpha>1$, convergent if $\alpha<1$, divergent, whatever β may be

(ii) if $\alpha=1$ and $\begin{cases}\beta>1, \text{ convergent}\\ \beta\leq 1, \text{ divergent}\end{cases}$

Example 25. *Test for convergence the series* $\frac{2^2.4^2}{3^2.5^2}+\frac{2^2.4^2.6^2}{3^2.5^2.7^2}+\frac{2^2.4^2.6^2.8^2}{3^2.5^2.7^2.9^2}+\ldots$

Solution. The given series is $\frac{2^2.4^2}{3^2.5^2}+\frac{2^2.4^2.6^2}{3^2.5^2.7^2}+\frac{2^2.4^2.6^2.8^2}{3^2.5^2.7^2.9^2}+\ldots$

$$+\frac{2^2.4^2.6^2.8^2\ldots(2n+2)^2}{3^2.5^2.7^2.9^2\ldots(2n+3)^2}+\ldots\infty$$

$$u_{n+1}=\frac{2^2.4^2.6^2.8^2\ldots(2n+2)^2\,(2n+4)^2}{3^2.5^2.7^2.9^2\ldots(2n+3)^2\,(2n+5)^2}$$

By D'Alembert's Test

$$\frac{u_{n+1}}{u_n}=\frac{(2n+4)^2}{(2n+5)^2}=\frac{4n^2+16n+16}{4n^2+20n+25}=\frac{4+\frac{16}{n}+\frac{16}{n^2}}{4+\frac{20}{n}+\frac{25}{n^2}}$$

$$\lim_{n\to\infty}\frac{u_{n+1}}{u_n}=\lim_{n\to\infty}\frac{4+\frac{16}{n}+\frac{16}{n^2}}{4+\frac{20}{n}+\frac{25}{n^2}}=1$$

D'Alembert's Test fails. Let us apply **Raabe's Test.**

$$\lim_{n\to\infty}n\left(\frac{u_n}{u_{n+1}}-1\right)=\lim_{n\to\infty}n\left(\frac{4n^2+20n+25}{4n^2+16n+16}-1\right)$$

$$=\lim_{n\to\infty}\left(\frac{4n^2+9n}{4n^2+16n+16}\right)=\lim_{n\to\infty}\left[\frac{4+\frac{9}{n}}{4+\frac{16}{n}+\frac{16}{n^2}}\right]=1,\ \text{Raabe's Test fails}$$

Let us apply **Gauss's Test**

$$\frac{u_n}{u_{n+1}} = \frac{(2n+5)^2}{(2n+4)^2} = \frac{\left(1+\frac{5}{2n}\right)^2}{\left(1+\frac{2}{n}\right)^2} = \left(1+\frac{5}{n}+\frac{25}{4n^2}\right)\left(1+\frac{2}{n}\right)^{-2}$$

$$= \left(1+\frac{5}{n}+\frac{25}{4n^2}\right)\left(1-\frac{4}{n}+\frac{(-2)\times(-3)}{2!}\frac{4}{n^2}+\ldots\right) = \left(1+\frac{5}{n}+\frac{25}{4n^2}\right)\left(1-\frac{4}{n}+\frac{12}{n^2}+\ldots\right)$$

$$= 1-\frac{4}{n}+\frac{12}{n^2}+\frac{5}{n}-\frac{20}{n^2}+\frac{25}{4n^2}+\ldots = 1+\frac{1}{n}-\frac{7}{n^2} \qquad \left(\frac{u_n}{u_{n+1}} = \alpha+\frac{\beta}{n}+\frac{\gamma}{n^2}\right)$$

Hence, $\alpha = 1$, $\beta = 1$. Thus, the series is divergent. **Ans.**

39.19 CAUCHY'S INTEGRAL TEST

Statement. *A positive term series $f(1) + f(2) + f(3) + \ldots + f(n) + \ldots$ where $f(n)$ decreases as n increases, converges or diverges according to the integral*

$$\int_1^\infty f(x)\,dx$$

is finite or infinite.

Proof. In the figure, the area under the curve from $x = 1$ to $x = n + 1$ lies between the sum of the areas of small rectangles (small height) and sum of the areas of large rectangles (large height).

[$f(1), f(2)$ ... represent the height of the rectangles]

$$\Rightarrow \quad f(1)+f(2)+\ldots+f(n) \geq \int_1^{n+1} f(x)\,dx \geq f(2)+f(3)+\ldots+f(n+1)$$

$$S_n \geq \int_1^{n+1} f(x)\,dx \geq S_{n+1} - f(1)$$

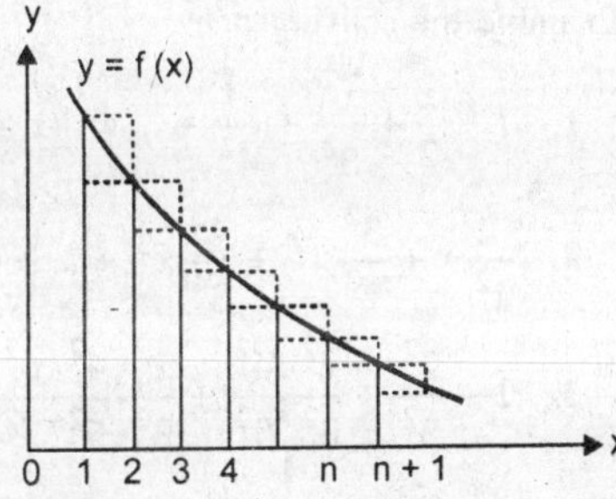

As $n \to \infty$, from the second inequality that if the integral has a finite value then $\lim_{n\to\infty} S_{n+1}$ is also finite, so $\Sigma f(n)$ is convergent.

Similarly, if the integral is infinite, then from the first inequality that $\lim_{n\to\infty} S_n \to \infty$, so the series is divergent.

Example 26. *Apply the integral test to determine the convergence of the p-series*

$$\frac{1}{1^p}+\frac{1}{2^p}+\frac{1}{3^p}+\ldots+\frac{1}{n^p}+\ldots\infty$$

Solution. (*i*) When $p > 1$, $\quad f(x) = \frac{1}{x^p}$

$$\int_1^\infty f(x)\,dx = \lim_{m\to\infty}\int_1^m \frac{1}{x^p}\,dx = \lim_{m\to\infty}\left[\frac{x^{1-p}}{1-p}\right]_1^m = \lim_{m\to\infty}\frac{1}{1-p}(m^{1-p}-1)$$

$$= \lim_{m\to\infty}\frac{1}{1-p}\left[\frac{1}{m^{p-1}}-1\right] = \frac{1}{p-1}, \text{ which is finite.}$$

By Cauchy's Integral Test, the series is convergent for $p > 1$.

(*ii*) When $p < 1$,

$$\int_1^\infty f(x)\,dx = \frac{1}{1-p}\left[\lim_{m\to\infty}(m^{1-p}-1)\right] \to \infty$$

Thus, the series is divergent, if $p < 1$.
(*iii*) When $p = 1$,

$$\int_1^\infty \frac{1}{x}\,dx = [\log x]_1^\infty \to \infty$$

Thus, the series is divergent.
Hence, $\Sigma \frac{1}{n^p}$ is convergent if $p > 1$ and divergent if $p \le 1$. **Ans.**

Example 27. *Examine the convergence of* $\sum_{n=2}^{\infty} \frac{1}{n \log n}$.

Solution. Here $f(x) = \frac{1}{x \log x}$

$$\int_2^\infty \frac{1}{x \log x}\,dx = \lim_{m\to\infty}\left[\log \log x\right]_2^m = \lim_{m\to\infty}[\log \log m - \log \log 2] \to \infty$$

By Cauchy's Integral Test the series is divergent. **Ans.**

Example 28. *Examine the convergence of* $\sum_{x=1}^{\infty} x\, e^{-x^2}$

Solution. Here $\quad f(x) = x\, e^{-x^2}$

Now, $\int_1^\infty x\, e^{-x^2}\,dx = \lim_{m\to\infty}\left[\frac{e^{-x^2}}{-2}\right]_1^m = \lim_{m\to\infty}\left[\frac{e^{-m^2}}{-2} + \frac{e^{-1}}{2}\right] = \frac{e^{-1}}{2} = \frac{1}{2e}$, which is finite.

Hence, the given series is convergent. **Ans.**

EXERCISE 39.8

Examine the convergence:

1. $1 + \frac{x}{2} + \frac{x^2}{3^2} + \frac{x^2}{4^3} + \ldots \infty \quad (x > 0)$ **Ans.** Convergent

2. $\frac{2}{1^2}x + \frac{3^2}{2^3}x^2 + \frac{4^3}{3^4}x^3 + \ldots + \frac{(n+1)^n}{n^{n+1}}x^n + \ldots$ **Ans.** $x < 1$, convergent; $x \ge 1$, divergent

3. $1 + \frac{2^2}{3^2} + \frac{2^2.4^2}{3^2.5^2} + \frac{2^2.4^2.6^2}{3^2.5^2.7^2} + \ldots \infty$ **Ans.** Divergent

4. $\sum_{n=1}^{\infty} \frac{1}{\sqrt{n}}$ **Ans.** Divergent

5. $\sum_{n=1}^{\infty} \frac{1}{n^2+1}$ **Ans.** Convergent

6. $\sum_{n=1}^{\infty} \frac{1}{n^n}$, **Ans.** Convergent

7. $\sum_{n=1}^{\infty} n^2 e^{-n^3}$ **Ans.** Convergent

8. $\sum_{n=1}^{\infty} \frac{1}{n(\log n)^2}$ **Ans.** Convergent

39.20 CAUCHY'S ROOT TEST

Statement. *If* $\Sigma\, u_n$ *is positive term series such that* $\lim_{n\to\infty}(u_n)^{1/n} = k$, *then*

(*i*) *if* $k < 1$, *the series converges.* (*ii*) *if* $k > 1$, *the series diverges.*

Proof. By definition of limit

$$|(u_n)^{1/n} - k| < \varepsilon \text{ for } n > m$$

$$k - \varepsilon < (u_n)^{1/n} < k + \varepsilon \text{ for } n > m$$

(*i*) $\quad k < 1$

$$k + \varepsilon < r < 1$$

$$(u_n)^{1/n} < k \quad \Rightarrow u_n < k^n$$

$$u_1 + u_2 + \ldots \infty < k + k^2 + \ldots + k^n + \ldots \infty$$

$$< \frac{1}{1-k} \quad \text{(a finite quantity)}$$

$\therefore$ The series is convergent.

(*ii*) $\quad k > 1$

$$k - \varepsilon > 1$$

$$(u_n)^{1/n} > k - \varepsilon > 1$$

$$u_n > 1$$

$$S_n = u_1 + u_2 + \ldots u_n > n$$

$$\lim_{n \to \infty} S_n \to \infty$$

$\therefore$ The series is divergent.

(*iii*) $\quad k = 1$

If $\lim_{n \to \infty} (u_n)^{1/n} = 1$, the test fails.

For example, $\quad \Sigma u_n = \sum \frac{1}{n^p}$

$$\lim_{n \to \infty} (u_n)^{1/n} = \lim_{n \to \infty} \left(\frac{1}{n^p}\right)^{1/n} = \lim_{n \to \infty} \left(\frac{1}{n^{1/n}}\right)^{-p} = 1 \text{ for all } p,\ k = 1$$

But $\sum \frac{1}{n^p}$ is convergent for $p > 1$ and divergent for $p \leq 1$.

Thus, we cannot say whether Σu_n is convergent or divergent for $k = 1$.

Example 29. *Examine the convergence of the series* $\sum \frac{1}{\left(1 + \frac{1}{n}\right)^{n^2}}$ *(MDU, Dec. 2010)*

Solution. Here, $\quad u_n = \frac{1}{\left(1 + \frac{1}{n}\right)^{n^2}} \Rightarrow (u_n)^{1/n} = \left[\frac{1}{\left(1 + \frac{1}{n}\right)^{n^2}}\right]^{\frac{1}{n}} = \frac{1}{\left(1 + \frac{1}{n}\right)^n}$

$$\lim_{n \to \infty} (u_n)^{1/n} = \lim_{n \to \infty} \frac{1}{\left(1 + \frac{1}{n}\right)^n} = \frac{1}{e} < 1$$

Hence, the given series is convergent. **Ans.**

Example 30. *Test the following series for convergence*

$$\sum_{n=1}^{\infty} \frac{(n+1)^n x^n}{n^{n+1}}$$

(M.D.U. Dec., 2001)

Solution. Here, we have

$$\sum_{n=1}^{\infty} \frac{(n+1)^n x^n}{n^{n+1}}$$

Here, $\quad u_n = \frac{(n+1)^n x^n}{n^{n+1}} = \left[\frac{(n+1)x}{n}\right]^n \cdot \frac{1}{n}$

$$\Rightarrow \qquad (u_n)^{\frac{1}{n}} = \frac{(n+1)x}{n}\cdot\frac{1}{\frac{1}{n^n}} = \left(1+\frac{1}{n}\right)x\cdot\frac{1}{\frac{1}{n^n}}$$

$$\Rightarrow \qquad \lim_{n\to\infty}(u_n)^{\frac{1}{n}} = \left[\lim_{n\to\infty}\left(1+\frac{1}{n}\right)x\right]\left[\lim_{n\to\infty}\frac{1}{\frac{1}{n^n}}\right]$$

$$= (1+0)\,x\cdot\frac{1}{1} = x \qquad \left[\because \lim_{n\to\infty} n^{\frac{1}{n}} = 1\right]$$

$\therefore$ By Cauchy's root test, $\sum u_n$ is convergent if $x < 1$ and divergent if $x > 1$. The test fails when $x = 1$.

When $x = 1$, $\qquad u_n = \dfrac{(n+1)^n}{n^{n+1}} = \dfrac{1}{n}\cdot\dfrac{(n+1)^n}{n^n} = \dfrac{1}{n}\left(1+\dfrac{1}{n}\right)^n$

Let $\qquad v_n = \dfrac{1}{n}$,

$$\frac{u_n}{v_n} = \left(1+\frac{1}{n}\right)^n$$

$\displaystyle\lim_{n\to\infty}\frac{u_n}{v_n} = \lim_{n\to\infty}\left(1+\frac{1}{n}\right)^n = e$, which is finite and non-zero.

$\therefore$ By camparison test, $\sum u_n$ and $\sum v_n$ converge or diverge together.

Since $\sum v_n = \sum \frac{1}{n}$ is of the form $\sum \frac{1}{n^p}$ with $p = 1$,

$\sum v_n$ divergnt $\Rightarrow \sum u_n$ also divergent

Hence, $\sum u_n$ is convergent is $x < 1$ and divergent if $x \geq 1$. **Ans.**

Example 31. *Discuss the convergence of the following series:*

$$\left(\frac{2^2}{1^2}-\frac{2}{1}\right)^{-1} + \left(\frac{3^3}{2^3}-\frac{3}{2}\right)^{-2} + \left(\frac{4^4}{3^4}-\frac{4}{3}\right)^{-3} + \ldots \infty$$

Solution. Here, $u_n = \left[\dfrac{(n+1)^{n+1}}{n^{n+1}} - \dfrac{(n+1)}{n}\right]^{-n}$

$$[u_n]^{1/n} = \left[\left[\frac{(n+1)^{n+1}}{n^{n+1}} - \frac{n+1}{n}\right]^{-n}\right]^{\frac{1}{n}} = \left[\frac{(n+1)^{n+1}}{n^{n+1}} - \frac{n+1}{n}\right]^{-1}$$

$$\lim_{n\to\infty}(u_n)^{1/n} = \lim_{n\to\infty}\left[\left(1+\frac{1}{n}\right)^{n+1} - \left(1+\frac{1}{n}\right)\right]^{-1} = (e-1)^{-1} = \frac{1}{e-1} < 1$$

Hence, the given series is convergent. **Ans.**

EXERCISE 39.9

Discuss the convergence of the following series:

1. $\displaystyle\sum_{n=1}^{\infty}\frac{1}{n^n}$ **Ans.** Convergent

2. $\displaystyle\sum_{n=1}^{\infty}\left(1+\frac{1}{n}\right)^{n^2}$ **Ans.** Divergent

3. $\sum_{n=1}^{\infty}\left(1+\frac{1}{\sqrt{n}}\right)^{-n^{3/2}}$ **Ans.** Convergent 4. $\sum_{n=1}^{\infty}\frac{1}{(\log n)^n}$ **Ans.** Convergent

5. $\Sigma\, n^{-k}$ **Ans.** If $k > 1$, convergent

6. $\sum_{n=2}^{\infty}\frac{1}{n(\log n)^k}$ **Ans.** If $k > 1$, convergent; and divergent if $k \le 1$.

7. $\Sigma\,(n\log n)^{-1}(\log\log n)^{-k}$ **Ans.** If $k > 1$, convergent; and divergent if $k \le 1$.

8. $\sum\left(1-\frac{1}{n}\right)^{n^2}$ **Ans.** Convergent 9. $\sum\frac{x^n}{n^n}$ **Ans.** Convergent

10. $(a+b)+(a^2+b^2)+(a^3+b^3)+\ldots$ **Ans.** Convergent if $a < 1, b < 1$; divergent if $a \ge 1, b \ge 1$

39.21 LOGARITHMIC TEST

If $\Sigma\, u_n$ is a positive term series such that $\lim_{n\to\infty}\left(n\log\frac{u_n}{u_{n+1}}\right) = k$

(*i*) If $k > 1$, then the series is convergent. (*ii*) If $k < 1$, then the series is divergent.

Proof. (*i*) If $k > 1$

Compare $\Sigma\, u_n$ with $\sum\frac{1}{n^p}$, if $k > p > 1$, then $\Sigma\, u_n$ converges.

$$\frac{u_n}{u_{n+1}} = \frac{(n+1)^p}{n^p} = \left(1+\frac{1}{n}\right)^p \quad \ldots(1)$$

Taking logarithm of both sides of (1), we have:

$$\log\frac{u_n}{u_{n+1}} > p\log\left(1+\frac{1}{n}\right) \quad \left[\log(1+x) = x - \frac{x^2}{2}+\frac{x^3}{3}-\frac{x^4}{4}+\ldots\right]$$

if $$\log\frac{u_n}{u_{n+1}} > p\left(\frac{1}{n}-\frac{1}{2n^2}+\frac{1}{3n^3}-\frac{1}{4n^4}+\ldots\right)$$

if $$n\log\frac{u_n}{u_{n+1}} > p\left(1-\frac{1}{2n}+\frac{1}{3n^2}-\frac{1}{4n^3}+\ldots\right)$$

$$\lim_{n\to\infty} n\log\frac{u_n}{u_{n+1}} > p$$

i.e., $k > p$ which is true as $k > p > 1$. $\left[\lim_{n\to\infty} n\log\frac{u_n}{u_{n+1}} = k\right]$

Hence, $\Sigma\, u_n$ is convergent.
When $p < 1$
Similarly, when $p < 1$, $\Sigma\, u_n$ is divergent.
When $p = 1$, the test fails.

Example 32. *Test the convergence of the series* $x+\frac{2^2.x^2}{2!}+\frac{3^3.x^3}{3!}+\frac{4^4.x^4}{4!}+\ldots\infty$

(*MDU, Dec. 2010*)

Solution. Here the series is $x+\frac{2^2.x^2}{2!}+\frac{3^3.x^3}{3!}+\frac{4^4.x^4}{4!}+\ldots+\frac{n^n.x^n}{n!}+\ldots\infty$

$$u_n = \frac{n^n.x^n}{n!} \quad\text{and}\quad u_{n+1} = \frac{(n+1)^{n+1}.x^{n+1}}{(n+1)!}$$

$$\frac{u_n}{u_{n+1}} = \frac{n^n . x^n}{n!} \frac{(n+1)!}{(n+1)^{n+1} . x^{n+1}} = \frac{n^n}{(n+1)^n} \frac{1}{x} = \frac{1}{\left(1+\frac{1}{n}\right)^n} \frac{1}{x}$$

$$\lim_{n \to \infty} \frac{u_n}{u_{n+1}} = \lim_{n \to \infty} \frac{1}{\left(1+\frac{1}{n}\right)^n} \frac{1}{x} = \frac{1}{e} . \frac{1}{x}$$

If $\frac{1}{e\,x} > 1$ or $x < \frac{1}{e}$, the series is convergent.

If $\frac{1}{e\,x} < 1$ or $\frac{1}{e} < x$, the series is divergent. If $\frac{1}{e\,x} = 1$ or $x = \frac{1}{e}$, the test fails.

$$\log \frac{u_n}{u_{n+1}} = \log \frac{1}{\left(1+\frac{1}{n}\right)^n} . e = \log e - \log \left(1+\frac{1}{n}\right)^n$$

$$= 1 - n \log \left(1+\frac{1}{n}\right) = 1 - n\left[\frac{1}{n} - \frac{1}{2n^2} + \frac{1}{3n^3} + \ldots\right]$$

$$= 1 - 1 + \frac{1}{2n} - \frac{1}{3n^2} + \ldots = \frac{1}{2n} - \frac{1}{3n^2}$$

$$\lim_{n \to \infty} n \log \frac{u_n}{u_{n+1}} = \lim_{n \to \infty} \left[\frac{1}{2} - \frac{1}{3n}\right] = \frac{1}{2} < 1.$$

Thus, the series is divergent. **Ans.**

Example 33. *Discuss the convergence of the series:*

$$1 + \frac{x}{2} + \frac{2!}{3^2} x^2 + \frac{3!}{4^3} x^3 + \frac{4!}{5^4} x^4 + \ldots\ldots\ldots \infty \; (x > 0)$$

(M.D.University, I Semester, 2009)

Solution. Here, we have

$$1 + \frac{x}{2} + \frac{2!}{3^2} x^2 + \frac{3!}{4^3} x^3 + \frac{4!}{5^4} x^4 + \ldots\ldots\ldots \infty$$

Neglecting the first term, we get

$$u_n = \frac{n!}{(n+1)^n} x^n \quad \text{and} \quad u_{n+1} = \frac{(n+1)!}{(n+2)^{n+1}} x^{n+1}$$

$$\frac{u_n}{u_{n+1}} = \frac{n!}{(n+1)^n} . x^n . \frac{(n+2)^{n+1}}{(n+1)! . x^{n+1}} = \frac{(n+2)^{n+1}}{(n+1)^n . (n+1)} . \frac{1}{x}$$

$$\lim_{n\to\infty} \frac{u_n}{u_{n+1}} = \lim_{n\to\infty} \frac{(n+2)^{n+1}}{(n+1)^n . (n+1)} . \frac{1}{x}$$

$$= \lim_{n\to\infty} \frac{n^{n+1}\left(1+\frac{2}{n}\right)^{n+1}}{n^n\left(1+\frac{1}{n}\right)^n . n\left(1+\frac{1}{n}\right)} . \frac{1}{x} = \lim_{n\to\infty} \frac{\left(1+\frac{2}{n}\right)^n\left(1+\frac{2}{n}\right)}{\left(1+\frac{1}{n}\right)^n\left(1+\frac{1}{n}\right)} . \frac{1}{x} \quad \ldots(1)$$

$$= \frac{e^2}{e} . \frac{1}{x} = \frac{e}{x} . \quad \left[\because \lim_{n\to\infty}\left(1+\frac{a}{n}\right)^n = \lim_{n\to\infty}\left\{\left(1+\frac{a}{n}\right)^{\frac{n}{a}}\right\} = e\right.$$

$$\left|\; \lim_{n\to\infty}\left(1+\frac{2}{n}\right)^n = \lim_{n\to\infty}\left[\left(1+\frac{2}{n}\right)^{\frac{n}{2}}\right]^2 = e^2\right.$$

$\therefore$ By D' Alembert's ratio test, the series converges if $1 < \frac{e}{x}$ or if $x < e$ and diverges if $\frac{e}{x} < 1$ or if $e < x$.

If $x = e$, the ratio test fails, $\because \lim_{n\to\infty} \frac{u_n}{u_{n+1}} = 1$.

Now when $x = e$

Putting the value of x in (1), we get

$$\frac{u_n}{u_{n+1}} = \frac{\left(1+\frac{2}{n}\right)^{n+1}}{\left(1+\frac{1}{n}\right)^{n+1}} \cdot \frac{1}{e}$$

Since the expression $\frac{u_n}{u_{n+1}}$ involves the number e, so we do not apply Raabe's test but apply logarithmic test.

$$\therefore \quad \log \frac{u_n}{u_{n+1}} = (n+1)\log\left(1+\frac{2}{n}\right) - (n+1)\log\left(1+\frac{1}{n}\right) - \log e$$

$$= (n+1)\left[\log\left(1+\frac{2}{n}\right) - \log\left(1+\frac{1}{n}\right)\right] - 1$$

$$= (n+1)\left[\left(\frac{2}{n} - \frac{1}{2}\cdot\frac{4}{n^2} + \frac{1}{3}\cdot\frac{8}{n^3} \ldots\right) - \left(\frac{1}{n} - \frac{1}{2n^2} + \frac{1}{3n^3} + \ldots\right)\right] - 1$$

$$= (n+1)\left[\frac{1}{n} - \frac{3}{2n^2} + \frac{7}{3n^3} - \ldots\right] - 1$$

$$= \left(1 - \frac{3}{2n}\right) + \left(\frac{1}{n} - \frac{3}{n^2} + \ldots\right) - 1 = \frac{1}{n} - \frac{3}{n^2} - \frac{3}{2n^2}$$

$$= 1 - \frac{1}{2n} - \frac{3}{2n^2} + \ldots - 1 = -\frac{1}{2n} - \frac{3}{2n^2}$$

$$\therefore \quad \lim_{n\to\infty} n \log \frac{u_n}{u_{n+1}} = \lim_{n\to\infty} n\left[-\frac{1}{2n} - \frac{3}{2n^2} + \ldots\right] = \lim_{n\to\infty}\left(-\frac{1}{2} - \frac{3}{2n} + \ldots\right) = -\frac{1}{2} < 1$$

$\therefore$ By log test, the series diverges.

Hence, the given series $\sum u_n$ converges if $x < e$ and diverges if $x \geq e$. **Ans.**

EXERCISE 39.10

Examine the convergence for the following series :

1. $\frac{1^2}{4^2} + \frac{5^2}{8^2} + \frac{9^2}{12^2} + \frac{13^2}{16^2} + \ldots\infty$ **Ans.** Convergent

2. $1 + \frac{1!}{2}x + \frac{2!}{3^2}x^2 + \frac{3!}{4^3}x^3 + \ldots\infty$ **Ans.** If $x < e$, convergent and divergent if $x \geq e$

3. $\frac{1^2}{2^2} + \frac{1^2.3^2}{2^2.4^2}x + \frac{1^2.3^2.5^2}{2^2.4^2.6^2}x^2 + \ldots\infty$ **Ans.** Convergent if $x < 1$, and divergent if $x \geq 1$

4. $\frac{a+x}{1!} + \frac{(a+2x)^2}{2!} + \frac{(a+3x)^3}{3!} + \ldots$ **Ans.** Convergent if $x < \frac{1}{e}$, divergent if $x \geq \frac{1}{e}$

39.22 DE MORGAN'S AND BERTRAND'S TEST

If Σu_n is a positive term series such that

$$\lim_{n \to \infty} \left[\left\{ n\left(\frac{u_n}{u_{n+1}} - 1 \right) - 1 \right\} \log n \right] = k$$

then the series is convergent if $k > 1$ and divergent if $k < 1$.

Example 34. *Test for convergence the series* $1^p + \left(\frac{1}{2}\right)^p + \left(\frac{1 \times 3}{2 \times 4}\right)^p + \left(\frac{1 \times 3 \times 5}{2 \times 4 \times 6}\right)^p + \ldots$

Solution. The given series is :

$$1^p + \left(\frac{1}{2}\right)^p + \left(\frac{1\times3}{2\times4}\right)^p + \left(\frac{1\times3\times5}{2\times4\times6}\right)^p + \ldots$$

Here
$$u_n = \left[\frac{1\times3\times5\times\ldots\times(2n-3)}{2\times4\times6\times\ldots\times(2n-2)}\right]^p$$

$\therefore$
$$u_{n+1} = \left[\frac{1\times3\times5\times\ldots(2n-3)(2n-1)}{2\times4\times6\times\ldots\times(2n-2)(2n)}\right]^p$$

$\therefore$
$$\frac{u_{n+1}}{u_n} = \left(\frac{2n-1}{2n}\right)^p = \left(1 - \frac{1}{2n}\right)^p$$

$$\lim_{n\to\infty} \frac{u_{n+1}}{u_n} = 1$$

$\therefore$ **D'Alembert's Test fails.**

Now let us apply Raabe's Test.

Here $n\left(\frac{u_n}{u_{n+1}} - 1\right) = n\left[\left(1 - \frac{1}{2n}\right)^{-p} - 1\right] = n\left[1 + \frac{p}{2n} + \frac{p(p+1)}{8n^2} + \ldots - 1\right] = \frac{p}{2} + \frac{p(p+1)}{8n} + \ldots$

$\therefore$
$$\lim_{n\to\infty} n\left[\frac{u_n}{u_{n+1}} - 1\right] = \frac{p}{2}$$

If $\frac{p}{2} > 1$, *i.e.*, $p > 2$, the series is convergent and divergent if $\frac{p}{2} < 1$, *i.e.*, $p < 2$.

This test fails if $\frac{p}{2} = 1$, *i.e.*, $p = 2$.

Now let us apply **De Morgan's Test**. When $p = 2$

$$n\left[\frac{u_n}{u_{n+1}} - 1\right] = 1 + \frac{3}{4n} + \ldots$$

Now, $\lim_{n\to\infty} \left[n\left(\frac{u_n}{u_{n+1}} - 1\right) - 1\right] \log n = \lim_{n\to\infty} \left[1 + \frac{3}{4n} + \ldots - 1\right] \log n$

39.23 CAUCHY'S CONDENSATION TEST

If $\phi(n)$ is positive for all positive integral values of n and continually diminishes as n increases and if a be a positive integer greater than 1, then the two series $\Sigma\,\phi(n)$ and $\Sigma\, a^n \phi(a^n)$ are either both convergent or both divergent.

Example 35. *Show that the series*

$$1+\frac{1}{2(\log 2)^p}+\frac{1}{3(\log 3)^p}+\ldots+\frac{1}{n(\log n)^p}+\ldots$$

is convergent if $p > 1$ and divergent if $p = 1$ or $p < 1$.

Solution. We apply Cauchy's Condensation Test.

Here $$\phi(n) = \frac{1}{n(\log n)^p}$$

$\therefore$ nth term of the second series $\Sigma\, a^n \phi(a^n)$ is :

$$a^n\left[\frac{1}{a^n(\log a^n)^p}\right] \text{ i.e., } \frac{1}{(\log a^n)^p} \text{ i.e., } \frac{1}{(n\log a)^p} \text{ i.e., } \frac{1}{(\log a)^p}\times\frac{1}{n^p}$$

$\therefore$ The given series will be convergent or divergent if $\sum\left[\frac{1}{(\log a)^p}\times\frac{1}{n^p}\right]$ is convergent or divergent, i.e., if $\sum\frac{1}{n^p}$ is convergent or divergent.

But we know the $\sum\frac{1}{n^p}$ is convergent when $p > 1$ and divergent if $p = 1$ or < 1.

Hence, the given series is convergent if $p > 1$ and divergent if $p = 1$ or < 1. **Proved.**

39.24 ALTERNATING SERIES

A series in which the terms are alternately negative is called the alternating series.

e.g., $$u_1 - u_2 + u_3 - u_4 + \ldots\infty$$

39.25 LEIBNITZ'S RULE FOR CONVERGENCE OF AN ALTERNATING SERIES

(*i*) Each term is numerically less than its preceeding term.

(*ii*) $\lim\limits_{n\to\infty} u_n = 0$

Exmaple 36. *Discuss the convergence of the series* $\sum\limits_{n=1}^{\infty}(-1)^n\frac{n}{n^2+1}$.

Solution. The terms of the given series are alternately positive and negative;

(*i*) $$|u_n| = \frac{n}{n^2+1} \quad\text{and}\quad |u_{n+1}| = \frac{(n+1)}{(n+1)^2+1}$$

$$|u_n| - |u_{n+1}| = \frac{n}{n^2+1} - \frac{(n+1)}{(n+1)^2+1} = \frac{n(n+1)^2 + n - (n+1)(n^2+1)}{(n^2+1)[(n+1)^2+1]}$$

$$= \frac{n^2+n-1}{(n^2+1)[(n+1)^2+1]} = +\text{ve}$$

and each term is numerically less than its preceeding term.

(ii) $$\lim_{n\to\infty} |u_n| = \lim_{n\to\infty} \frac{n}{n^2+1} = \lim_{n\to\infty} \frac{1}{n+\frac{1}{n}} = 0$$

Both conditions are satisfied.

Hence, by Leibnitz's rule, the given series is convergent. **Ans.**

Example 37. *Test the convergence of the series* $\frac{1}{6}-\frac{2}{11}+\frac{3}{16}-\frac{4}{21}+\frac{5}{26}-\ldots\infty$

Solution. The terms of the given series are alternately positive and negative.

$$u_n = (-1)^{n-1}\frac{n}{5n+1}$$

$$|u_n| = \frac{n}{5n+1} \text{ and } |u_{n+1}| = \frac{n+1}{5(n+1)+1}$$

(i) $$|u_n| - |u_{n+1}| = \frac{n}{5n+1} - \frac{n+1}{5(n+1)+1} = \frac{5n^2+6n-5n^2-5n-n-1}{(5n+1)(5n+6)}$$

$$= \frac{-1}{(5n+1)(5n+6)}$$

$\therefore$ $$|u_n| > |u_{n+1}|$$

Thus each term is not numerically less than its preceeding terms.

(ii) $$\lim_{n\to\infty} u_n = \lim_{n\to\infty}\frac{n}{5n+1} = \lim_{n\to\infty}\frac{1}{5+\frac{1}{n}} = \frac{1}{5} \neq 0$$

$$\lim_{n\to\infty} u_n \neq 0$$

Both conditions for convergence are not satisfied.

Hence, the series is not convergent. It is oscillatory. **Ans.**

Example 38. *Test the following series for convergence and absolute convergence*:

$$1-\frac{1}{2\sqrt{2}}+\frac{1}{3\sqrt{3}}-\frac{1}{4\sqrt{4}}+\ldots$$ (*M.D.U. Dec., 2002*)

Solution. The given series is

$$\sum_{n=1}^{\infty} u_n = \sum_{n=1}^{\infty}(-1)^{n-1}.\frac{1}{n\sqrt{n}} = \sum_{n=1}^{\infty}(-1)^{n-1}\frac{1}{n^{\frac{3}{2}}} = \sum_{n=1}^{\infty}(-1)^{n-1}a_n$$

It is an alternating series.

Here, $$a_n = \frac{1}{n^{\frac{3}{2}}}$$

$$a_{n+1} = \frac{1}{(n+1)^{\frac{3}{2}}}$$

Since, $$\frac{1}{n^{\frac{3}{2}}} > \frac{1}{(n+1)^{\frac{3}{2}}}\ \forall\ n \qquad (\therefore a_n > a_{n+1}\ \forall\ n)$$

Also, $$\lim_{n\to\infty} a_n = \lim_{n\to\infty}\frac{1}{n^{\frac{3}{2}}} = 0$$

$\therefore$ By Leibnitz's test, the series $\sum u_n$ is convergent.

$$|u_n| = \frac{1}{n^{\frac{3}{2}}}$$

Now $\sum |u_n| = \sum \frac{1}{n^{\frac{3}{2}}}$ is convergent $\left(\because p = \frac{3}{2} > 1\right)$

Hence, the series $\sum u_n$ is absolutely convergent. **Ans.**

Example 39. *Test the convergence of the series:*

$$\sum_{n=1}^{\infty} \frac{(-1)^n (x+1)^n}{2^n n^2}$$

(M.D. University, I Semester, 2009)

Solution. Here, we have

$$\sum_{n=1}^{\infty} \frac{(-1)^n (x+1)^n}{2^n n^2} \qquad ...(1)$$

Here, $u_n = \frac{(-1)^n (x+1)^n}{2^n . n^2}$ and $u_{n+1} = \frac{(-1)^{n+1} (x+1)^{n+1}}{2^{n+1} . (n+1)^2}$

$$\frac{|u_n|}{|u_{n+1}|} = \frac{|x+1|^n}{2^n . n^2} . \frac{2^{n+1} . (n+1)^2}{|x+1|^{n+1}}$$

$$= 2\left(\frac{n+1}{n}\right)^2 . \frac{1}{|x+1|} = 2\left(1+\frac{1}{n}\right)^2 . \frac{1}{|x+1|}$$

$$\therefore \lim_{n\to\infty} \frac{|u_n|}{|u_{n+1}|} = \lim_{n\to\infty} 2\left(1+\frac{1}{n}\right)^2 . \frac{1}{|x+1|} = \frac{2}{|x+1|}$$

$\therefore$ By ratio test, the series $\sum |u_n|$ is convergent if

$1 < \frac{2}{|x+1|}$ *i.e.,* if $|x + 1| < 2$

i.e., if $-2 < x + 1 < 2$ *i.e.,* if $-3 < x < 1$

Also $\sum |u_n|$ is divergent if $\frac{2}{|x+1|} < 1$

i.e., if $|x + 1| > 2$ *i.e.,* if $x + 1 > 2$ or $x + 1 < -2$ *i.e.,* if $x > 1$ or $x < -3$.

Ratio test fails when $x = 1$ or -3.

When $x = 1$, $\sum u_n = \sum \frac{(-1)^n . 2^n}{2^n . n^2} = \sum \frac{(-1)^n}{n^2} = \sum (-1)^n . v_n$ From (1)

It is an alternating series.

Here, $v_n = \frac{1}{n^2}$, $v_{n+1} = \frac{1}{(n+1)^2}$

Clearly $v_n > v_{n+1}$ $\forall n$

Also $\lim_{n\to\infty} v_n = \lim_{n\to\infty} \frac{1}{n^2} = 0$

$\therefore$ By Leibnitz's test, $\sum u_n$ is convergent.

When $x = -3$, $\sum u_n = \sum \frac{(-1)^n . (-2)^n}{2^n . n^2} = \sum \frac{(-1)^{2n} 2^n}{2^n . n^2} = \sum \frac{1}{n^2}$

Which is convergent.

Hence, the given series is convergent if $-3 \le x \le 1$ and divergent if $x > 1$ or $x < -3$

EXERCISE 39.11

Discuss the convergence of the following series :

1. $1-\frac{1}{\sqrt{2}}+\frac{1}{\sqrt{3}}-\frac{1}{\sqrt{4}}+...$ **Ans.** Convergent
2. $1-2x+3x^2-4x^3+...\infty\ (x<1)$ **Ans.** Convergent
3. $\frac{x}{1+x}-\frac{x^2}{1+x^2}+\frac{x^3}{1+x^3}-\frac{x^4}{1+x^4}+...\infty\ (0<x<1)$ **Ans.** Convergent
4. $\frac{1}{1^p}-\frac{1}{2^p}+\frac{1}{3^p}-\frac{1}{4^p}+...$ **Ans.** If $p>0$, convergent; oscillatory if $p<0$.
5. $\sum_{n=1}^{\infty}(-1)^{n-1}\frac{n}{2n-1}$ **Ans.** Oscillatory
6. Show that the series $\frac{1}{x+1}-\frac{1}{x+2}+\frac{1}{x+3}-\frac{1}{x+4}+...$ is convergent for all real values of x other than negative integers.
7. Prove that the series $x-\frac{x^2}{2}+\frac{x^3}{3}-...$ converges if $-1\ x\le 1$.

39.26 ALTERNATING CONVERGENT SERIES

There are two types of alternating convergent series :

(1) Absolutely convergent series. (2) Conditionally convergent series.
Absolutely convergent series. If $u_1+u_2+u_3+...$ be such that $|u_1|+|u_2|+|u_3|+...$ be convergent then $u_1+u_2+u_3+...\infty$ is called absolutely convergent.

Example 40. *Show that the series $1-\frac{1}{2}+\frac{1}{3}-\frac{1}{4}+\frac{1}{5}-\frac{1}{6}+...$ is convergent but not absolutely convergent.*

Solution. $1-\frac{1}{2}+\frac{1}{3}-\frac{1}{4}+...$

The terms of the series are alternately positive and negative.

(*i*) $|u_{n+1}|<|u_n|$ as $\frac{1}{n+1}<\frac{1}{n}$ (*ii*) $\lim_{n\to\infty} u_n=\lim_{n\to\infty}\frac{1}{n}=0$

Both conditions are satisfied. Hence, the given series is convergent.

But $1+\frac{1}{2}+\frac{1}{3}+\frac{1}{4}+...\infty$ is divergent since in *p-series*, $p=1$.

Hence, the given series is conditionally convergent. **Ans.**

Example 41. *What can you say about the series $1-\frac{1}{2^2}+\frac{1}{3^2}-\frac{1}{4^2}+...\infty$?*

Solution. $1-\frac{1}{2^2}+\frac{1}{3^2}-\frac{1}{4^2}+...$

$$|u_n|=\frac{1}{n^2},\quad \text{and}\quad |u_{n+1}|=\frac{1}{(n+1)^2}$$

(*i*) $|u_{n+1}|<|u_n|$ (*ii*) $\lim_{n\to\infty} u_n=\lim_{n\to\infty}\frac{1}{n^2}=0$

Thus, the given series is convergent by Leibnitz's rule.

And $1+\frac{1}{2^2}+\frac{1}{3^2}+\frac{1}{4^2}+....$ is also convergent since in p-series, $p=2>1$.

Both the conditions are satisfied.
Hence, the given series is absolutely convergent. **Ans.**

Example 42. *Discuss the series for convergence* $1-\frac{1}{3}+\frac{1}{2}-\frac{1}{3^3}+\frac{1}{2^2}-\frac{1}{3^5}+\frac{1}{2^3}-\frac{1}{3^7}+\ldots$

Solution. The given series is rewitten as

$$1+\frac{1}{2}+\frac{1}{2^2}+\frac{1}{2^3}+\ldots-\left(\frac{1}{3}+\frac{1}{3^3}+\frac{1}{3^5}+\frac{1}{3^7}+\ldots\right)$$

$$\lim_{n\to\infty} S_n = \frac{1}{1-\frac{1}{2}}-\frac{\frac{1}{3}}{1-\frac{1}{3^2}} = 2-\frac{3}{8}=1\frac{5}{8}$$

The given series is convergent.

Again $1+\frac{1}{2}+\frac{1}{2^2}+\frac{1}{2^3}+\ldots+\frac{1}{3}+\frac{1}{3^3}+\frac{1}{3^5}+\ldots$

$$\lim_{n\to\infty} S_n = \frac{1}{1-\frac{1}{2}}+\frac{\frac{1}{3}}{1-\frac{1}{3}} \qquad \left(\lim_{n\to\infty} S_n = \frac{a}{1-r}\right)$$

$$\lim_{n\to\infty} S_n = 2+\frac{3}{8}=\frac{19}{8}$$

Both the conditions are satisfied.
This series is also convergent.
Hence, the given series is absolutely convergent. **Ans.**

Example 43. *Test the convergence and divergence of the series*

$$5-\frac{10}{3}+\frac{20}{9}-\frac{40}{27}+\ldots\ldots\ldots$$ (*Gujarat, I Semester, Jan. 2009*)

Solution. The terms of the given series are alternately positive and negative and the given series is geometric infinite series.

(*i*) $$S = 5-\frac{10}{3}+\frac{20}{9}-\frac{40}{27}+\frac{80}{81}+\ldots\ldots\ldots$$

Here $a = 5$ and $r = -\frac{2}{3}$, $S = \frac{a}{1-r}$

$$S = \frac{5}{1-\left(-\frac{2}{3}\right)}=\frac{5}{1+\frac{2}{3}}=\frac{5}{\frac{5}{3}}=3$$

Sum of the series is finite.
Hence, the given series is convergent.

(*ii*) Again $5+\frac{10}{3}+\frac{20}{9}+\frac{40}{27}+\frac{80}{81}+\ldots$

This is also G.P.

Here, $a = 5$ and $r = \frac{2}{3}$

$$S = \frac{a}{1-r}, \qquad S = \frac{5}{1-\frac{2}{3}}=\frac{5}{\frac{1}{3}}=15$$

Again sum of the positive terms is finite.
Thus the series is also convergent.
Both of the conditions are satisfied.
Hence, the given series is absolutely convergent. **Ans.**

Example 44. *Test the following series for convergence and divergence.*

$$\sum_{n=1}^{\infty} \tan^{-1}\left(\frac{1}{n^2+n+1}\right) \qquad \textit{(Gujarat, I Semester, Jan. 2009)}$$

Solution. Let $u_n = \tan^{-1} \dfrac{1}{1+n(n+1)}$

$$u_n = \tan^{-1} \frac{(n+1)-n}{1+n(n+1)}$$

$$u_n = \tan^{-1}(n+1) - \tan^{-1}(n)$$

$$u_{n+1} = \tan^{-1}(n+2) - \tan^{-1}(n+1)$$

By D' Alembert's test

$$\lim_{n\to\infty} \frac{u_{n+1}}{u_n} = \lim_{n\to\infty} \frac{\tan^{-1}(n+2) - \tan^{-1}(n+1)}{\tan^{-1}(n+1) - \tan^{-1} n} = \lim_{n\to\infty} \frac{\dfrac{1}{1+(n+2)^2} - \dfrac{1}{1+(n+1)^2}}{\dfrac{1}{1+(n+1)^2} - \dfrac{1}{1+n^2}}$$

[L'Hopital Rule]

$$= \lim_{n\to\infty} \frac{\dfrac{1+(n+1)^2-1-(n+2)^2}{[1+(n+2)^2][1+(n+1)^2]}}{\dfrac{1+n^2-1-(n+1)^2}{[1+(n+1)^2][1+n^2]}} = \lim_{n\to\infty} \frac{(2n+3)(-1)}{(2n+1)(-1)} \times \frac{1+n^2}{1+(n+2)^2}$$

$$= \lim_{n\to\infty} \frac{(2n+3)(1+n^2)}{(2n+1)[n^2+4n+5]} = \lim_{n\to\infty} \frac{\left(2+\dfrac{3}{n}\right)\left(\dfrac{1}{n^2}+1\right)}{\left(2+\dfrac{1}{n}\right)\left(1+\dfrac{4}{n}+\dfrac{5}{n^2}\right)} = \frac{2}{2} = 1$$

Test fails.

Let us apply **Raabe's Test.**

$$\lim_{n\to\infty} n\left(\frac{u_n}{u_{n+1}} - 1\right) = \lim_{n\to\infty} n\left[\frac{(2n+1)(n^2+4n+5)}{(2n+3)(n^2+1)} - 1\right]$$

$$= \lim_{n\to\infty} n\left[\frac{(2n+1)(n^2+4n+5)-(2n+3)(n^2+1)}{(2n+3)(n^2+1)}\right]$$

$$= \lim_{n\to\infty} n\left[\frac{2n^3+8n^2+10n+n^2+4n+5-2n^3-2n-3n^2-3}{(2n+3)(n^2+1)}\right]$$

$$= \lim_{n\to\infty} n\left[\frac{6n^2+12n+2}{(2n+3)(n^2+1)}\right] = \lim_{n\to\infty} \frac{6+\dfrac{12}{n}+\dfrac{2}{n^2}}{\left(2+\dfrac{3}{n}\right)\left(1+\dfrac{1}{n^2}\right)} = \frac{6}{2} = 3 > 1$$

Hence, the given series is convergent. **Ans.**

EXERCISE 39.12

Discuss the convergence of the following series :

1. $1-\frac{1}{2}+\frac{1}{4}-\frac{1}{8}+...\infty$ **Ans.** Absolutely convergent

2. $1-\frac{2}{5}+\frac{3}{10}-\frac{4}{17}+...+\frac{(-1)^{n-1}n}{n^2+1}+...\infty$ **Ans.** Conditionaliy convergent

3. $1-\frac{2}{3}+\frac{3}{3^2}-\frac{4}{3^3}+...$ **Ans.** Absolutely convergent

4. $1+\frac{1}{2^2}-\frac{1}{3^2}-\frac{1}{4^2}+\frac{1}{5^2}+\frac{1}{6^2}-\frac{1}{7^2}-\frac{1}{8^2}+...$ **Ans.** Absolutely convergent

5. $1-\frac{1}{\sqrt{2}}+\frac{1}{\sqrt{3}}-\frac{1}{\sqrt{4}}+...\infty$ **Ans.** Conditionally convergent

6. $\frac{\sin x}{1^3}-\frac{\sin 2x}{2^3}+\frac{\sin 3x}{3^3}+...$ **Ans.** Absolutely convergent

39.27 POWER SERIES IN *x*

$$a_0+a_1x+a_2x^2+...+a_nx^n+...$$

is power series in x, here a's are independent of x.

Proof. $u_n = a_nx^n$ and $u_{n+1} = a_{n+1}x^{n+1}$

D'Alembert's Ratio Test

$$\lim_{n\to\infty}\frac{u_{n+1}}{u_n} = \lim_{n\to\infty}\frac{a_{n+1}x^{n+1}}{a_nx^n} = \lim_{n\to\infty}\frac{a_{n+1}}{a_n}x$$

If $\frac{a_{n+1}}{a_n}=k,$

If $|kx|<1 \Rightarrow |x|<\frac{1}{k}$, then the series is convergent.

Thus, the power series is convergent if $-\frac{1}{k}<x<\frac{1}{k}$.

Thus, the interval of the power series is $-\frac{1}{k}$ to $\frac{1}{k}$ for convergence. Outside this interval the series is divergent. **Ans.**

Example 45. *Find the values of x for which the series* $x-\frac{x^2}{2^2}+\frac{x^3}{3^2}-\frac{x^4}{4^2}+...\infty$ *converges.*

Solution. Here $u_n = (-1)^{n-1}\frac{x^n}{n^2}$, and $u_{n+1} = (-1)^n\frac{x^{n+1}}{(n+1)^2}$

$$\frac{u_{n+1}}{u_n} = -\frac{n^2}{(n+1)^2}x \quad\Rightarrow\quad \lim_{n\to\infty}\frac{u_{n+1}}{u_n} = -\lim_{n\to\infty}\frac{1}{\left(1+\frac{1}{n}\right)^2}x=-x$$

By **D'Alembert's Test** the given series is convergent for $|x|<1$ and divergent if $|x|>1$.

At $x = 1$. The series becomes $1-\frac{1}{2^2}+\frac{1}{3^2}-\frac{1}{4^2}+...$

This is an alternately convergent series.

At x = – 1. The series becomes $-1-\frac{1}{2^2}-\frac{1}{3^2}-\frac{1}{4^2}+...$

This is also convergent series, $p = 2$.
Hence, the interval of convergence is $-1 \le x \le 1$. **Ans.**

39.28 EXPONENTIAL SERIES

$$e^x = 1 + x + \frac{x^2}{2!} + \frac{x^3}{3!} + \dots + \frac{x^n}{n!} + \dots$$

is convergent for all values of x.

Proof. Here, we have $u_n = \dfrac{x^{n-1}}{(n-1)!}$, and $u_{n+1} = \dfrac{x^n}{n!}$

$$\lim_{n \to \infty} \frac{u_{n+1}}{u_n} = \lim_{n \to \infty} \frac{x^n}{n!} \frac{(n-1)!}{x^{n-1}} = \lim_{n \to \infty} \frac{x}{n} = 0 < 1$$

Hence, by D'Alembert's Test the exponential series is convergent for all values of x.

39.29 LOGARITHMIC SERIES

$$\log(1+x) = x - \frac{x^2}{2} + \frac{x^3}{3} - \frac{x^4}{4} + \dots + (-1)^{n-1} \frac{x^n}{n} + \dots$$

is convergent for $-1 < x \le 1$.

Proof. Here, $u_n = (-1)^{n-1} \dfrac{x^n}{n}$, and $u_{n+1} = (-1)^n \dfrac{x^{n+1}}{n+1}$

$$\lim_{n \to \infty} \frac{u_{n+1}}{u_n} = \lim_{n \to \infty} \frac{(-1)^n x^{n+1}}{n+1} \cdot \frac{n}{(-1)^{n-1} x^n} = -\lim_{n \to \infty} \left(\frac{n}{n+1}\right) x = -\lim_{n \to \infty} \left(\frac{1}{1+\frac{1}{n}}\right) x = -x$$

Thus, the series is convergent for $|x| < 1$ and divergent for $|x| > 1$.

At $x = 1$. The series becomes

$$1 - \frac{1}{2} + \frac{1}{3} - \frac{1}{4} + \dots \infty \text{ which is convergent.}$$

At $x = -1$. The series becomes

$$-1 - \frac{1}{2} - \frac{1}{3} - \frac{1}{4} + \dots \infty \text{ which is divergent.}$$

39.30 BINOMIAL SERIES

$$(1+x)^n = 1 + nx + \frac{n(n-1)}{2!} x^2 + \frac{n(n-1)(n-2)}{3!} x^3 + \dots$$

is convergent for $|x| < 1$.

Proof.
$$u_r = \frac{n(n-1)\dots(n-r+2)}{(r-1)!} x^{r-1}$$

$$u_{r+1} = \frac{n(n-1)\dots(n-r+1)}{r!} x^r$$

$$\lim_{r \to \infty} \frac{u_{r+1}}{u_r} = \lim_{r \to \infty} \frac{n-r+1}{r} x = \lim_{r \to \infty} \left(\frac{n+1}{r} - 1\right) x = -x \text{ for } r > n+1$$

If $|x| < 1$, the series is convergent by D'Alembert's Test.

EXERCISE 39.13

Test the convergence of the following series

1. $1 + x + 2x^2 + 3x^3 + \dots + nx^n + \dots \infty$ **Ans.** Convergent if $-1 < x > 1$

2. $\dfrac{1}{2} + \dfrac{x}{3} + \dfrac{x^2}{4} + \dfrac{x^3}{5} + \dots + \dfrac{x^n}{n+2} + \dots \infty$ **Ans.** Convergent for $-1 < x < 1$

3. $1 + \dfrac{1}{1!} + \dfrac{1}{2!} + \dfrac{1}{3!} + \dots \infty$ **Ans.** Convergent

4. $1 + \dfrac{4}{1!} + \dfrac{4^2}{2!} + \dfrac{4^3}{3!} + \dots \infty$ **Ans.** Convergent

5. $3-\frac{3^2}{2}+\frac{3^3}{3}-\frac{3^4}{4}+...\infty$ **Ans.** Convergent 6. $\sqrt{2}-\frac{2}{2}+\frac{2^{\frac{3}{2}}}{3}-\frac{2^2}{4}+...\infty$ **Ans.** Convergent

7. $\frac{1}{1-x}+\frac{1}{2(1-x)^2}+\frac{1}{3(1-x)^3}+...\infty$ **Ans.** Convergent, if $x \neq 1$

39.31 UNIFORM CONVERGENCE

If for a given $\varepsilon > 0$, a number N can be found *independent of* x, such that for every x in the interval (a, b), the series is said to be uniformly convergent in the interval (a, b).

Example 46. *Discuss the uniform convergence of the series*

$$1 + x + x^2 + ...\infty$$

Solution. $S_n = 1 + x + x^2 + ... + x^n = \frac{1-x^n}{1-x}$

$$S_\infty(x) = \lim_{n\to\infty}\frac{1-x^n}{1-x}=\frac{1}{1-x} \qquad \text{for } |x| < 1$$

$$|S_\infty(x) - S_n(x)| = \left|\frac{1}{1-x}-\frac{1-x^n}{1-x}\right| = \left|\frac{x^n}{1-x}\right| = \frac{|x|^n}{1-x} < \varepsilon$$

if $|x|^n < \varepsilon(1-x)$

Let $|x|^N = \varepsilon(1-x) \Rightarrow N = \frac{\log \varepsilon(1-x)}{\log|x|}$

In the interval $\left(-\frac{1}{2},\frac{1}{2}\right)$, N has maximum value. $N = \dfrac{\log\frac{\varepsilon}{2}}{\log\frac{1}{2}}$

Hence, the given series is uniformly convergent in the interval $\left(-\frac{1}{2},\frac{1}{2}\right)$.

Note. The series is convergent in $(-1, 1)$ but not uniformly convergent.

39.32 ABEL'S TEST

If $v_n(x)$ be either monotonic decreasing in n for each fixed x in (a, b) or monotonic increasing in n for each fixed x in (a, b), $\Sigma\, a_n(x)\, v_n(x)$ is uniformly convergent in (a, b) if

(*i*) $\Sigma\, a_n(x)$ is uniformly convergent in (a, b).

(*ii*) There exists k such that $|v_n(x)| < k$ for all n when $a \le x < b$.

Example 47. *Prove that* $\frac{x^n}{n^3}$ *is uniformly convergent in* $(-1, 1)$.

Solution. (*i*) $\sum\frac{1}{n^3}$ is uniformly convergent. (*ii*) $|x^n| < k$ for all n when $-1 < k < 1$

Hence, $\sum\frac{x^n}{n^3}$ is uniformly convergent by Abel's Test. **Ans.**

EXERCISE 39.14

Prove that the following series are uniformly convergent in $(-1, 1)$.

1. $\sum\frac{x^n}{n^2}$ 2. $\sum\frac{x^n}{n(n+1)}$ 3. $\sum\frac{x^{2n}}{x^{2n}+n^2}$

4. Prove that the series $1+\frac{e^{-2x}}{2^2-1}-\frac{e^{-4x}}{4^2-1}+\frac{e^{-6x}}{6^2-1}+...\infty$ is uniformly convergent with regard to x in $x \ge 0$.

CHAPTER

40

FOURIER SERIES

40.1 PERIODIC FUNCTIONS

If the value of each ordinate $f(t)$ repeats itself at equal intervals in the abscissa, then $f(t)$ is said to be a periodic function.

If $f(t) = f(t + T) = f(t + 2T) = \ldots$ then T is called the period of the function $f(t)$.

For example:

The period of $\sin x$, $\cos x$, $\sec x$, and $\operatorname{cosec} x$ is 2π.

The period of $\tan x$ and $\cot x$ is π.

$\sin x = \sin(x + 2\pi) = \sin(x + 4\pi) = \ldots$ so $\sin x$ is a periodic function with the period 2π.

$$\sin 5x = \sin(5x + 2\pi) = \sin 5\left(x + \frac{2\pi}{5}\right), \text{ Period} = \frac{2\pi}{5}$$

$$\cos 3x = \cos(3x + 2\pi) = \cos 3\left(x + \frac{2\pi}{3}\right), \text{Period} = \frac{2\pi}{3}$$

$$\cos\frac{2n\pi x}{k} = \cos\left(\frac{2n\pi x}{k} + 2\pi\right) = \cos\frac{2n\pi}{k}\left(x + \frac{2\pi k}{2n\pi}\right)$$

$$= \cos\frac{2n\pi}{k}\left(x + \frac{k}{n}\right), \text{ Period} = \frac{k}{n}$$

$$\tan 2x = \tan(2x + \pi) = \tan 2\left(x + \frac{\pi}{2}\right), \text{ Period} = \frac{\pi}{2}$$

This is also called sinusoidal periodic function.

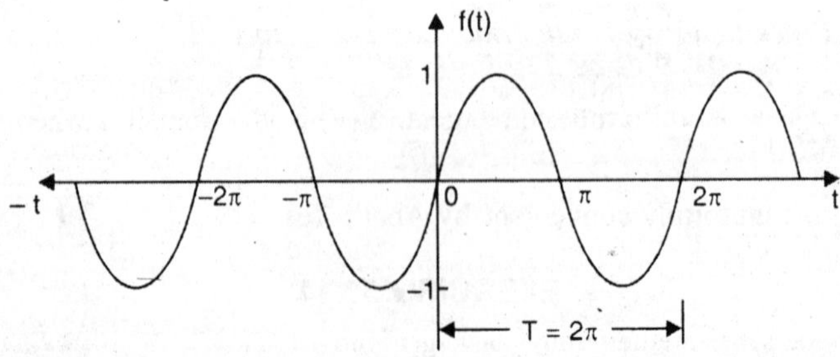

40.2 FOURIER SERIES

Here we will express a non-sinusoidal periodic function into a fundamental and its harmonics. A series of sines and cosines of an angle and its multiples of the form.

$$\frac{a_0}{2} + a_1 \cos x + a_2 \cos 2x + a_3 \cos 3x + ... + a_n \cos nx + ...$$
$$+ b_1 \sin x + b_2 \sin 2x + b_3 \sin 3x + ... + b_n \sin nx + ...$$
$$= \frac{a_0}{2} + \sum_{n=1}^{\infty} a_n \cos nx + \sum_{n=1}^{\infty} b_n \sin nx.$$

is called the Fourier series, where $a_0, a_1, a_2, ... a_n, ... b_1, b_2, b_3 ... b_n ...$ are constants.

A periodic function $f(x)$ can be expanded in a Fourier Series. The series consists of the following:

(*i*) A constant term a_0 (called d.c. component in electrical work).

(*ii*) A component at the fundamental frequency determined by the values of a_1, b_1.

(*iii*) Components of the harmonics (multiples of the fundamental frequency) determined by $a_2, a_3 ... b_2, b_3 ....$ And $a_0, a_1, a_2 ... ,b_1, b_2 ...$ are known as *Fourier coefficients* or Fourier constants.

Note. (1) When the function and its derivatives are continuous then the function can be expanded in powers of x by Maclaurin's theorem.

(2) But by Fourier series we can expand continuous and discontinuous both types of functions under certain conditions.

40.3 DIRICHLET'S CONDITIONS FOR A FOURIER SERIES

If the function $f(x)$ for the interval $(-\pi, \pi)$

(1) is single-valued **(2)** is bounded

(3) has at most a finite number of maxima and minima.

(4) has only a finite number of discontinuous

(5) is $f(x + 2\pi) = f(x)$ for values of x outside $[-\pi, \pi]$, then

$$S_P(x) = \frac{a_0}{2} + \sum_{n=1}^{P} a_n \cos nx + \sum_{n=1}^{P} b_n \sin nx$$

converges to $f(x)$ as $P \to \infty$ at values of x for which $f(x)$ is continuous and the sum of the series is equal to $\frac{1}{2}[f(x+0) + f(x-0)]$ at points of discontinuity.

40.4 ADVANTAGES OF FOURIER SERIES

1. Discontinuous function can be represented by Fourier series. Although derivatives of the discontinuous functions do not exist. (This is not true for Taylor's series).

2. The Fourier series is useful in expanding the periodic functions since outside the closed interval, there exists a periodic extension of the function.

3. Expansion of an oscillating function by Fourier series gives all modes of oscillation (fundamental and all overtones) which is extremely useful in physics.

4. Fourier series of a discontinuous function is not uniformly convergent at all points.

5. Term by term integration of a convergent Fourier series is always valid, and it may be valid if the series is not convergent. However, term by term, differentiation of a Fourier series is not valid in most cases.

40.5 USEFUL INTEGRALS

The following integrals are useful in Fourier Series.

(*i*) $\int_0^{2\pi} \sin nx\, dx = 0$ (*ii*) $\int_0^{2\pi} \cos nx\, dx = 0$

(*iii*) $\int_0^{2\pi} \sin^2 nx\, dx = \pi$ (*iv*) $\int_0^{2\pi} \cos^2 nx\, dx = \pi$

(v) $\int_0^{2\pi} \sin nx \cdot \sin mx \, dx = 0$ (vi) $\int_0^{2\pi} \cos nx \cos mx \, dx = 0$

(vii) $\int_0^{2\pi} \sin nx \cdot \cos mx \, dx = 0$ (viii) $\int_0^{2\pi} \sin nx \cdot \cos nx \, dx = 0$

(ix) $[uv]_1 = uv_1 - u'v_2 + u''v_3 - u'''v_4 + \ldots$

where $v_1 = \int v \, dx, v_2 = \int v_1 \, dx$ and so on $u' = \frac{du}{dx}$, $u'' = \frac{d^2u}{dx^2}$ and so on and

(x) $\sin n\pi = 0$, $\cos n\pi = (-1)^n$ where $n \in I$

40.6 DETERMINATION OF FOURIER COEFFICIENTS (EULER'S FORMULAE)

$$f(x) = \frac{a_0}{2} + a_1 \cos x + a_2 \cos 2x + \ldots + a_n \cos nx + \ldots$$
$$+ b_1 \sin x + b_2 \sin 2x + \ldots + b_n \sin nx + \ldots \quad \ldots(1)$$

(*i*) **To find a_0 :** Integrate both sides of (1) from $x = 0$ to $x = 2\pi$.

$$\int_0^{2\pi} f(x)\, dx = \frac{a_0}{2}\int_0^{2\pi} dx + a_1 \int_0^{2\pi} \cos x \, dx + a_2 \int_0^{2\pi} \cos 2x \, dx + \ldots + a_n \int_0^{2\pi} \cos nx \, dx + \ldots$$
$$+ b_1 \int_0^{2\pi} \sin x \, dx + b_2 \int_0^{2\pi} \sin 2x \, dx + \ldots + b_n \int_0^{2\pi} \sin nx \, dx + \ldots$$

$$= \frac{a_0}{2} \int_0^{2\pi} dx$$ (other integrals = 0 by formulae (*i*) and (*ii*) of Art 40.5)

$$\int_0^{2\pi} f(x)\, dx = \frac{a_0}{2} 2\pi \quad \Rightarrow \quad \boxed{a_0 = \frac{1}{\pi}\int_0^{2\pi} f(x)\, dx} \quad \ldots(2)$$

(*ii*) **To find a_n :** Multiply each side of (1) by $\cos nx$ and integrate from $x = 0$ to $x = 2\pi$.

$$\int_0^{2\pi} f(x) \cos nx \, dx = \frac{a_0}{2}\int_0^{2\pi} \cos nx \, dx + a_1 \int_0^{2\pi} \cos x \cos nx \, dx + \ldots + a_n \int_0^{2\pi} \cos^2 nx \, dx \ldots$$
$$+ b_1 \int_0^{2\pi} \sin x \cos nx \, dx + b_2 \int_0^{2\pi} \sin 2x \cos nx \, dx + \ldots$$

$$= a_n \int_0^{2\pi} \cos^2 nx \, dx = a_n \pi$$ (Other integrals = 0, by formulae Art. 40.5)

$$\therefore \quad \boxed{a_n = \frac{1}{\pi}\int_0^{2\pi} f(x) \cos nx \, dx} \quad \ldots(3)$$

By taking $n = 1, 2 \ldots$ we can find the values of $a_1, a_2 \ldots$.

(*iii*) **To find b_n :** Multiply each side of (1) by $\sin nx$ and integrate from $x = 0$ to $x = 2\pi$.

$$\int_0^{2\pi} f(x) \sin nx \, dx = \frac{a_0}{2}\int_0^{2\pi} \sin nx \, dx + a_1 \int_0^{2\pi} \cos x \sin nx \, dx + \ldots + a_n \int_0^{2\pi} \cos nx \sin nx \, ax + \ldots$$
$$\ldots + b_1 \int_0^{2\pi} \sin x \sin nx \, dx + \ldots + b_n \int_0^{2\pi} \sin^2 nx \, dx + \ldots$$

$$= b_n \int_0^{2\pi} \sin^2 nx \, dx$$

$$= b_n \pi$$ (All other integrals = 0, Article No. 40.5)

$$\therefore \quad \boxed{b_n = \frac{1}{\pi}\int_0^{2\pi} f(x) \sin nx \, dx} \quad \ldots(4)$$

Note : To get similar formula of a_0, $\frac{1}{2}$ has been written with a_0 in Fourier series.

Example 1. *Find the Fourier series representing*

$$f(x) = x, \quad 0 < x < 2\pi$$

and sketch its graph from $x = -4\pi$ to $x = 4\pi$.

Solution. Let $f(x) = \frac{a_0}{2} + a_1 \cos x + a_2 \cos 2x + \ldots + b_1 \sin x + b_2 \sin 2x + \ldots$ ...(1)

Hence $a_0 = \frac{1}{\pi}\int_0^{2\pi} f(x)\,dx = \frac{1}{\pi}\int_0^{2\pi} x\,dx = \frac{1}{\pi}\left[\frac{x^2}{2}\right]_0^{2\pi} = 2\pi$

$$a_n = \frac{1}{\pi}\int_0^{2\pi} f(x)\cos nx\,dx = \frac{1}{\pi}\int_0^{2\pi} x\cos nx\,dx$$

$$= \frac{1}{\pi}\left[x\frac{\sin nx}{n} - 1.\left(-\frac{\cos nx}{n^2}\right)\right]_0^{2\pi} = \frac{1}{\pi}\left[\frac{\cos 2n\pi}{n^2} - \frac{1}{n^2}\right] = \frac{1}{n^2\pi}(1-1) = 0$$

$$b_n = \frac{1}{\pi}\int_0^{2\pi} f(x)\sin nx\,dx = \frac{1}{\pi}\int_0^{2\pi} x\sin nx\,dx$$

$$= \frac{1}{\pi}\left[x\left(-\frac{\cos nx}{n}\right) - 1.\left(\frac{-\sin nx}{n^2}\right)\right]_0^{2\pi} = \frac{1}{\pi}\left[\frac{-2\pi\cos 2n\pi}{n}\right] = -\frac{2}{n}$$

Substituting the values of $a_0, a_1, a_2 \ldots, b_1, b_2 \ldots$ in (1), we get

$$x = \pi - 2\left[\sin x + \frac{1}{2}\sin 2x + \frac{1}{3}\sin 3x + \ldots\right]$$ **Ans.**

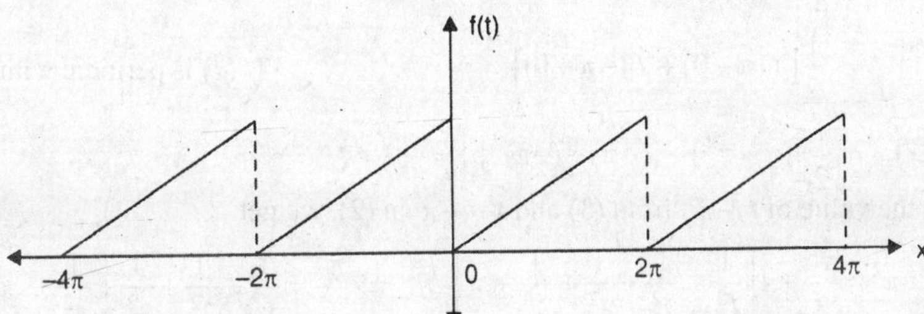

Example 2. *Given that* $f(x) = x + x^2$ *for* $-\pi < x < \pi$, *find the Fourier expression of* $f(x)$.

Deduce that $\frac{\pi^2}{6} = 1 + \frac{1}{2^2} + \frac{1}{3^2} + \frac{1}{4^2} + \ldots$

(U.P., II Semester, Summer 2003, Uttarakhand, June 2009)

Solution. Let $x + x^2 = \frac{a_0}{2} + a_1\cos x + a_2\cos 2x + \ldots + b_1\sin x + b_2\sin 2x + \ldots$ ...(1)

$$a_0 = \frac{1}{\pi}\int_{-\pi}^{\pi} f(x)\,dx = \frac{1}{\pi}\int_{-\pi}^{\pi}(x + x^2)\,dx$$

$$= \frac{1}{\pi}\left[\frac{x^2}{2} + \frac{x^3}{3}\right]_{-\pi}^{\pi} = \frac{1}{\pi}\left[\frac{\pi^2}{2} + \frac{\pi^3}{3} - \frac{\pi^2}{2} + \frac{\pi^3}{3}\right] = \frac{2\pi^2}{3}$$

$$a_n = \frac{1}{\pi}\int_{-\pi}^{\pi} f(x)\cos nx\,dx = \int_{-\pi}^{\pi}(x + x^2)\cos nx\,dx$$

$$= \frac{1}{\pi}\left[(x+x^2)\frac{\sin nx}{n} - (2x+1)\frac{(-\cos nx)}{n^2} + (2)\left(-\frac{\sin nx}{n^3}\right)\right]_{-\pi}^{\pi}$$

$$= \frac{1}{\pi}\left[(2\pi+1)\frac{\cos n\pi}{n^2} - (-2\pi+1)\frac{\cos(-n\pi)}{n^2}\right] = \frac{1}{\pi}\left[4\pi\frac{\cos n\pi}{n^2}\right] = \frac{4(-1)^n}{n^2}$$

$$b_n = \frac{1}{\pi}\int_{-\pi}^{\pi} f(x)\sin nx\,dx = \frac{1}{\pi}\int_{-\pi}^{\pi}(x+x^2)\sin nx\,dx$$

$$= \frac{1}{\pi}\left[(x+x^2)\left(-\frac{\cos nx}{n}\right) - (2x+1)\left(\frac{-\sin nx}{n^2}\right) + 2\frac{\cos nx}{n^3}\right]_{-\pi}^{\pi}$$

$$= \frac{1}{\pi}\left[-(\pi+\pi^2)\frac{\cos n\pi}{n} + 2\frac{\cos n\pi}{n^3} + (-\pi+\pi^2)\frac{\cos n\pi}{n} - 2\frac{\cos n\pi}{n^3}\right]$$

$$= \frac{1}{\pi}\left[-\frac{2\pi}{n}\cos n\pi\right] = -\frac{2}{n}(-1)^n$$

Substituting the values of a_0, a_n, b_n, in (1), we get

$$x + x^2 = \frac{\pi^2}{3} + 4\left[-\cos x + \frac{1}{2^2}\cos 2x - \frac{1}{3^2}\cos 3x + \ldots\right]$$
$$-2\left[-\sin x + \frac{1}{2}\sin 2x - \frac{1}{3}\sin 3x + \ldots\right] \quad \ldots(2)$$

Here, $f(x) = x + x^2$ is valid for all values of x *between* $-\pi$ and π but not at the end points $-\pi$ and π due to open interval.

$$f(-\pi) = \frac{1}{2}[f(-\pi-0) + f(-\pi+0)]$$

$$= \frac{1}{2}[f(\pi-0) + f(-\pi+0)] \qquad [f(x) \text{ is periodic with period } 2\pi]$$

$$= \frac{1}{2}\left[(\pi+\pi^2) + \{(-\pi)+(-\pi)^2\}\right] = \pi^2 \quad \ldots(3)$$

Putting the value of $f(-\pi)$ from (3) and $x = -\pi$ in (2), we get

$$\pi^2 = \frac{\pi^2}{3} + 4\left[\frac{1}{1^2} + \frac{1}{2^2} + \frac{1}{3^2} + \frac{1}{4^2} + \ldots\right] \quad \Rightarrow \frac{\pi^2}{6} = \frac{1}{1^2} + \frac{1}{2^2} + \frac{1}{3^2} + \frac{1}{4^2} + \ldots \qquad \textbf{Ans.}$$

Example 3. *Find the Fourier series expansion for* $f(x) = x + \frac{x^2}{4}$, $-\pi \le x \le \pi$

(U.P. II Semester, 2009)

Solution. Let $x + \frac{x^2}{4} = \frac{a_0}{2} + a_1\cos x + a_2\cos 2x + \ldots\ldots + b_1\sin x + b_2\sin 2x + \ldots\ldots$

where $$a_0 = \frac{1}{\pi}\int_{-\pi}^{\pi} f(x)\,dx = \frac{1}{\pi}\int_{-\pi}^{\pi}\left(x + \frac{x^2}{4}\right)dx$$

$$= \frac{1}{\pi}\left[\frac{x^2}{2} + \frac{x^3}{12}\right]_{-\pi}^{\pi} = \frac{1}{\pi}\left[\frac{\pi^2}{2} + \frac{\pi^3}{12} - \frac{\pi^2}{2} + \frac{\pi^3}{12}\right] = \frac{1}{\pi}\left[\frac{2\pi^3}{12}\right] = \frac{\pi^2}{6}$$

$$a_n = \frac{1}{\pi}\int_{-\pi}^{\pi} f(x)\cos nx\,dx = \frac{1}{\pi}\int_{-\pi}^{\pi}\left(x + \frac{x^2}{4}\right)\cos nx\,dx$$

$$= \frac{1}{\pi}\left[\left(x + \frac{x^2}{4}\right)\left(\frac{\sin nx}{n}\right) - \left(1 + \frac{2x}{4}\right)\left(\frac{-\cos nx}{n^2}\right) + \frac{1}{2}\left(-\frac{\sin nx}{n^3}\right)\right]_{-\pi}^{\pi}$$

$$= \frac{1}{\pi}\left[\left(\pi+\frac{\pi^2}{4}\right)\left(\frac{\sin n\pi}{n}\right)+\left(1+\frac{2\pi}{4}\right)\left(\frac{\cos n\pi}{n^2}\right)-\frac{1}{2}\left(\frac{\sin n\pi}{n^3}\right)-\left(-\pi+\frac{\pi^2}{4}\right)\left(\frac{\sin(-n\pi)}{n}\right)\right.$$
$$\left.-\left(1-\frac{2\pi}{4}\right)\frac{\cos(-n\pi)}{n^2}+\frac{1}{2}\frac{\sin(-n\pi)}{n^3}\right]$$

$$= \frac{1}{\pi}\left[\left(1+\frac{\pi}{2}\right)\frac{(-1)^n}{n^2}-\left(1-\frac{\pi}{2}\right)\frac{(-1)^n}{n^2}\right] = \frac{(-1)^n}{n^2\pi}\left[1+\frac{\pi}{2}-1+\frac{\pi}{2}\right] = \frac{(-1)^n}{n^2\pi}(\pi) = \frac{(-1)^n}{n^2}$$

$$a_1 = -1, \quad a_2 = \frac{1}{4}, \quad a_3 = -\frac{1}{9}, \quad a_4 = \frac{1}{16} \quad \ldots\ldots\ldots$$

$$b_n = \frac{1}{\pi}\int_{-\pi}^{\pi} f(x)\sin nx\,dx = \frac{1}{\pi}\int_{-\pi}^{\pi}\left(x+\frac{x^2}{4}\right)\sin nx\,dx$$

$$= \frac{1}{\pi}\int_{-\pi}^{\pi} x\sin nx\,dn+\frac{1}{\pi}\int_{-\pi}^{\pi}\frac{x^2}{4}\sin nx\,dn$$

Even function   Odd function

$$= \frac{2}{\pi}\int_0^{\pi} x\sin nx\,dx+0$$

$$= \frac{2}{\pi}\left[x\left(-\frac{\cos nx}{n}\right)-(1)\left(\frac{-\sin nx}{n^2}\right)\right]_0^{\pi}$$

$$= \frac{2}{\pi}\left[-\pi\frac{\cos n\pi}{n}+\frac{\sin n\pi}{n^2}\right]$$

$$= \frac{2}{\pi}\left[-\pi\frac{(-1)^n}{n}\right] = -\frac{2(-1)^n}{n}$$

$$b_1 = \frac{2}{1}, \quad b_2 = -1, \quad b_3 = \frac{2}{3}, \quad b_4 = -\frac{1}{2} \ldots.$$

Hence, Fourier series of the given function is

$$f(x) = \frac{a_0}{2}+\sum_{n=1}^{\infty} a_n\cos nx+\sum_{n=1}^{\infty} b_n\sin nx = \frac{\pi^2}{12}+\sum\frac{(-1)^n}{n^2}\cos nx-\sum_{n=1}^{\infty}\frac{2}{n}(-1)^n\sin nx$$

$$f(x) = \frac{\pi^2}{12}-\cos x+\frac{1}{4}\cos 2x-\frac{1}{9}\cos 3x+\frac{1}{16}\cos 4x+\ldots.+2\sin x-\sin 3x+\frac{2}{3}\sin 4x-\frac{1}{2}\sin 4x+\ldots.$$

Ans.

EXERCISE 40.1

1. Find a Fourier series to represent, $f(x) = \pi - x$ for $0 < x < 2\pi$.

Ans. $2\left[\sin x+\frac{1}{2}\sin 2x+\frac{1}{3}\sin 3x+\ldots+\frac{1}{n}\sin nx+\ldots\right]$

2. Find a Fourier series to represent $x - x^2$ from $x = -\pi$ to π and show that

$$\frac{\pi^2}{12} = \frac{1}{1^2}-\frac{1}{2^2}+\frac{1}{3^2}-\frac{1}{4^2}+\ldots$$

Ans. $-\frac{\pi^2}{3}+4\left[\frac{\cos x}{1^2}-\frac{\cos 2x}{2^2}+\frac{\cos 3x}{3^2}-\frac{\cos 4x}{4^2}+\ldots\right]+2\left[\frac{\sin x}{1}-\frac{\sin 2x}{2}+\frac{\sin 3x}{3}-\frac{\sin 4x}{4}+\ldots\right]$

3. Find a Fourier series to represent the function $f(x) = e^x$, for $-\pi < x < \pi$ and hence derive a series for $\dfrac{\pi}{\sinh \pi}$.

Ans. $\dfrac{2\sinh\pi}{\pi}\left[\dfrac{1}{2} - \dfrac{1}{1^2+1}\cos x + \dfrac{1}{2^2+1}\cos 2x - \dfrac{1}{3^2+1}\cos 3x + \dots + \dfrac{1}{1^2+1}\sin x - \dfrac{2}{2^2+1}\sin 2x - \dfrac{3}{3^2+1}\sin 3x \dots\right]$, $\dfrac{\pi}{\sinh\pi} = 1 + 2\left[-\dfrac{1}{2} + \dfrac{1}{5} - \dfrac{1}{10} + \dots\right]$

4. Obtain the Fourier series for $f(x) = e^{x}$ in the interval $0 \le x < 2\pi$.

Ans. $\dfrac{1-e^{-2\pi}}{\pi}\left[\dfrac{1}{2} + \dfrac{1}{2}\cos x + \dfrac{1}{5}\cos 2x + \dfrac{1}{10}\cos 3x + \dfrac{1}{2}\sin x + \dfrac{2}{5}\sin 2x + \dfrac{3}{10}\sin 3x + \dots\right]$

5. If $f(x) = \left(\dfrac{\pi - x}{2}\right)^2$, $0 < x < 2\pi$, show that $f(x) = \dfrac{\pi^2}{12} + \sum_{n=1}^{\infty}\dfrac{\cos nx}{n^2}$

6. Prove that $x^2 = \dfrac{\pi^2}{3} + 4\sum_{n=1}^{\infty}(-1)^n\dfrac{\cos nx}{n^2}$, $-\pi < x < \pi$. Hence show that

(*i*) $\Sigma\dfrac{1}{(2n-1)^2} = \dfrac{\pi^2}{8}$ (*ii*) $\Sigma\dfrac{1}{n^4} = \dfrac{\pi^4}{90}$

7. If $f(x)$ is a periodic function defined over a period $(0, 2\pi)$ by $f(x) = \dfrac{(3x^2 - 6x\pi + 2\pi^2)}{12}$.

Prove that $f(x) = \sum_{n=1}^{\infty}\dfrac{\cos nx}{n^2}$ and hence show that $\dfrac{\pi^2}{6} = 1 + \dfrac{1}{2^2} + \dfrac{1}{3^2} + \dots$

40.7 FOURIER SERIES FOR DISCONTINUOUS FUNCTIONS

Let the function $f(x)$ be defined by

$$f(x) = f_1(x), \quad c < x < x_0$$
$$= f_2(x), \quad x_0 < x < c + 2\pi,$$

where x_0 is the point of discontinuity in the interval $(c, c + 2\pi)$.

In such cases also, we obtain the Fourier series for $f(x)$ in the usual way. The values of a_0, a_n, b_n are evaluated by

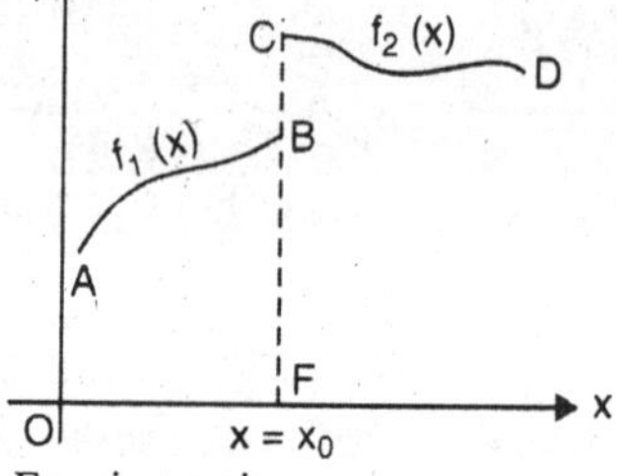

$$a_0 = \frac{1}{\pi}\left[\int_c^{x_0} f_1(x)\,dx + \int_{x_0}^{2+2\pi} f_2(x)\,dx\right];$$

$$a_n = \frac{1}{\pi}\left[\int_c^{x_0} f_1(x)\cos nx\,dx + \int_{x_0}^{2+2\pi} f_2(x)\cos nx\,dx\right]$$

$$b_n = \frac{1}{\pi}\left[\int_c^{x_0} f_1(x)\sin nx\,dx + \int_{x_0}^{c+2\pi} f_2(x)\cos nx\,dx\right]$$

If $x = x_0$ is the point of finite discontinuity, then the sum of the Fourier series

$$= \frac{1}{2}\left[\lim_{h\to 0} f(x_0 - h) + \lim_{h\to 0} f(x_0 + h)\right]$$

$$= \frac{1}{2}[f(x_0 - 0) + f(x_0 + 0)] = \frac{1}{2}(FB + FC)$$

Remarks.

1. It may be seen from the graph, that at a point of finite discontinuity $x = x_0$, there is a finite jump equal to BC in the value of the function $f(x)$ at $x = x_0$.
2. A given function $f(x)$ may be defined by different formulae in different regions. Such types of functions are quite common in Fourier Series.
3. At a point of discontinuity the sum of the series is equal to the mean of the limits on the right and left.

40.8 FUNCTION DEFINED IN TWO OR MORE SUB-RANGES

Example 4. *Find the fourier series to represent the function f (x) given by :*

$$f(x)=\begin{bmatrix} -k & \text{for } -\pi < x < 0 \\ k & \text{for } 0 < x < \pi \end{bmatrix}$$ *Hence show that :*

$$1-\frac{1}{3}+\frac{1}{5}-\frac{1}{7}+\ldots.=\frac{\pi}{4}.$$ (*U.P. II Semester 2010*)

Solution. $f(x)=\begin{cases} -k, & -\pi < x < 0 \\ k, & 0 < x < \pi \end{cases}$

$$a_0=\frac{1}{\pi}\int_{-\pi}^{\pi} f(x)\,dx=\frac{1}{\pi}\left[\int_{-\pi}^{0}-k\,dx+\int_0^{\pi}k\,dx\right]=\frac{1}{\pi}\left[[-kx]_{-\pi}^{0}+[kx]_0^{\pi}\right]$$

$$=\frac{1}{\pi}k[0-\pi+\pi-0]=0$$

$$a_n=\frac{1}{\pi}\int_{-\pi}^{\pi} f(x)\cos nx\,dx=\frac{1}{\pi}\left[\int_{-\pi}^{0}-k\cos nx\,dx+\int_0^{\pi}k\cos nx\,dx\right]$$

$$=\frac{1}{\pi}k\left[-\left\{\frac{\sin nx}{n}\right\}_{-\pi}^{0}+\left\{\frac{\sin nx}{n}\right\}_0^{\pi}\right]=\frac{1}{\pi}k[-0+0]=0$$

$$b_n=\frac{1}{\pi}\int_{-\pi}^{\pi} f(x)\sin nx\,dx=\frac{1}{\pi}\left[\int_{-\pi}^{0}-k\sin nx\,dx+\int_0^{\pi}k\sin nx\,dx\right]$$

$$=\frac{1}{\pi}k\left[\left\{\frac{\cos nx}{n}\right\}_{-\pi}^{0}-\left\{\frac{\cos nx}{n}\right\}_0^{\pi}\right]$$

$$=\frac{1}{\pi}k\left[\frac{1}{n}-\frac{(-1)^n}{n}-\frac{(-1)^n}{n}+\frac{1}{n}\right]=\frac{1}{\pi}k\left[\frac{2}{n}-\frac{2(-1)^n}{n}\right]$$

If n is even $b_n=0$

If n is odd $b_n=\frac{4k}{n\pi}$

$$f(x)=a_0+\sum_{n=1}^{\infty}a_n\cos nx+\sum_{n=1}^{\infty}b_n\sin nx \qquad \ldots(1)$$

$$f(x)=\sum_{n=1}^{\infty}b_n\sin nx=b_1\sin x+b_2\sin 2x+b_3\sin 3x+b_4\sin 4x+\ldots$$

Thus required Fourier sine series is

$$f(x)=\frac{4k}{\pi}\sin x+\frac{4k}{3\pi}\sin 3x+\frac{4k}{5\pi}\sin 5x+\ldots$$

$$\Rightarrow \quad f(x)=\frac{4k}{\pi}\left[\sin x+\frac{1}{3}\sin 3x+\frac{1}{5}\sin 5x+\ldots\right] \qquad \ldots(2)$$

Putting $x=\frac{\pi}{2}$ in (2), we get

$$k=\frac{4k}{\pi}\left[\sin\frac{\pi}{2}+\frac{1}{3}\sin\frac{3\pi}{2}+\frac{1}{5}\sin\frac{5\pi}{2}+\ldots.\right]$$

$$\Rightarrow \quad 1=\frac{4}{\pi}\left[1+\frac{1}{3}(-1)+\frac{1}{5}(1)+\frac{1}{7}(-1)+\ldots\right]$$

$$=\frac{4}{\pi}\left[1-\frac{1}{3}+\frac{1}{5}-\frac{1}{7}+\ldots\right] \Rightarrow \frac{\pi}{4}=1-\frac{1}{3}+\frac{1}{5}-\frac{1}{7}+\ldots$$ **Proved.**

Example 5. *Find the Fourier series of the function*

$$f(x) = \begin{cases} -1 & \text{for } -\pi < x < -\dfrac{\pi}{2} \\ 0 & \text{for } -\dfrac{\pi}{2} < x < \dfrac{\pi}{2} \\ +1 & \text{for } \dfrac{\pi}{2} < x < \pi. \end{cases}$$

Solution. Let $f(x) = \dfrac{a_0}{2} + a_1 \cos x + a_2 \cos 2x + \dots + b_1 \sin x + b_2 \sin 2x + \dots$ ...(1)

$$a_0 = \frac{1}{\pi}\int_{-\pi}^{\pi} f(x)\,dx = \frac{1}{\pi}\int_{-\pi}^{-\pi/2}(-1)\,dx + \frac{1}{\pi}\int_{-\pi/2}^{\pi/2} 0\,dx + \frac{1}{\pi}\int_{\pi/2}^{\pi} 1\,dx$$

$$= \frac{1}{\pi}[-x]_{-\pi}^{-\pi/2} + \frac{1}{\pi}[x]_{\pi/2}^{\pi} = \frac{1}{\pi}\left[\frac{\pi}{2} - \pi + \pi - \frac{\pi}{2}\right] = 0$$

$$a_n = \frac{1}{\pi}\int_{-\pi}^{\pi} f(x)\cos nx\,dx$$

$$= \frac{1}{\pi}\int_{-\pi}^{-\pi/2}(-1)\cos nx\,dx + \frac{1}{\pi}\int_{-\pi/2}^{\pi/2}(0)\cos nx\,dx + \frac{1}{\pi}\int_{\pi/2}^{\pi}(1)\cos nx\,dx$$

$$= -\frac{1}{\pi}\left[\frac{\sin nx}{n}\right]_{-\pi}^{-\pi/2} + \frac{1}{\pi}\left[\frac{\sin nx}{n}\right]_{\pi/2}^{\pi}$$

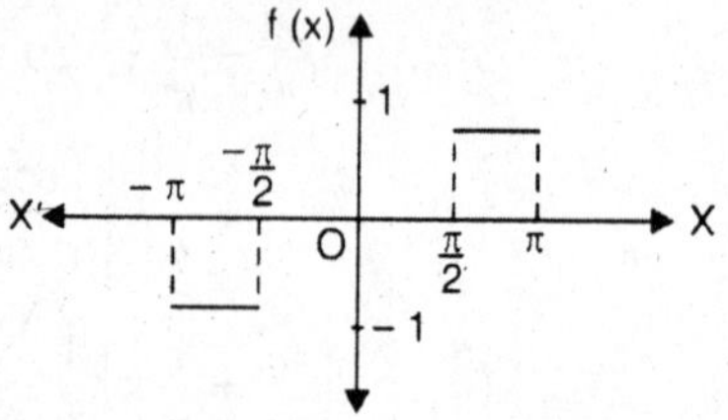

$$= -\frac{1}{\pi}\left[-\frac{\sin\frac{n\pi}{2}}{n}\right] + \frac{1}{\pi}\left[-\frac{\sin\frac{n\pi}{2}}{n}\right] = 0$$

$$b_n = \frac{1}{\pi}\int_{-\pi}^{\pi} f(x)\sin nx\,dx$$

$$= \frac{1}{\pi}\int_{-\pi}^{-\pi/2}(-1)\sin nx\,dx + \frac{1}{\pi}\int_{-\pi/2}^{\pi/2}(0)\sin nx\,dx$$

$$+ \frac{1}{\pi}\int_{\pi/2}^{\pi}(1)\sin nx\,dx = \frac{1}{\pi}\left[\frac{\cos nx}{n}\right]_{-\pi}^{-\pi/2} - \frac{1}{\pi}\left[\frac{\cos nx}{n}\right]_{\pi/2}^{\pi}$$

$$= \frac{1}{n\pi}\left[\cos\frac{n\pi}{2} - \cos n\pi\right] - \frac{1}{n\pi}\left(\cos n\pi - \cos\frac{n\pi}{2}\right) = \frac{2}{n\pi}\left[\cos\frac{n\pi}{2} - \cos n\pi\right]$$

$$b_1 = \frac{2}{\pi}, \qquad b_2 = -\frac{2}{\pi}, \qquad b_3 = \frac{2}{3\pi}$$

Putting the values of a_0, a_n, b_n in (1), we get

$$f(x) = \frac{1}{\pi}\left[2\sin x - 2\sin 2x + \frac{2}{3}\sin 3x + \dots\right]$$ **Ans.**

40.9 DISCONTINUOUS FUNCTIONS

At a point of discontinuity, Fourier series gives the value of $f(x)$ as the arithmetic mean of left and right limits.

At the point of discontinuity, $x = c$

At $x = c$, $f(x) = \dfrac{1}{2}[f(c-0) + f(c+0)]$

Example 6. *Find the Fourier series for* $f(x)$, *if* $f(x) = \begin{cases} -\pi, & -\pi < x < 0 \\ x, & 0 < x < \pi \end{cases}$

Deduce that $\frac{1}{1^2} + \frac{1}{3^2} + \frac{1}{5^2} + \ldots = \frac{\pi^2}{8}$

Solution. Let $f(x) = \frac{a_0}{2} + a_1 \cos x + a_2 \cos 2x + \ldots + a_n \cos nx + \ldots$

$$+ b_1 \sin x + b_2 \sin 2x + \ldots + b_n \sin nx + \ldots \qquad \ldots(1)$$

$$a_0 = \frac{1}{\pi} \int_{-\pi}^{\pi} f(x)\, dx$$

Then $a_0 = \frac{1}{\pi}\left[\int_{-\pi}^{0} (-\pi)\, dx + \int_0^{\pi} x\, dx\right] = \frac{1}{\pi}\left[-\pi\, (x)_{-\pi}^{0} + (x^2/2)_0^{\pi}\right] = \frac{1}{\pi}(-\pi^2 + \pi^2/2) = -\frac{\pi}{2};$

$$a_n = \frac{1}{\pi} \int_{-\pi}^{\pi} f(x) \cos nx\, dx$$

$$a_n = \frac{1}{\pi}\left[\int_{-\pi}^{0} (-\pi) \cos nx\, dx + \int_0^{\pi} x \cos nx\, dx\right] = \frac{1}{\pi}\left[-\pi\left(\frac{\sin nx}{n}\right)_{-\pi}^{0} + \left(\frac{x \sin nx}{n} + \frac{\cos nx}{n^2}\right)_0^{\pi}\right].$$

$$= \frac{1}{\pi}\left[0 + \frac{1}{n^2}\cos n\pi - \frac{1}{n^2}\right] = \frac{1}{\pi n^2}(\cos n\pi - 1)$$

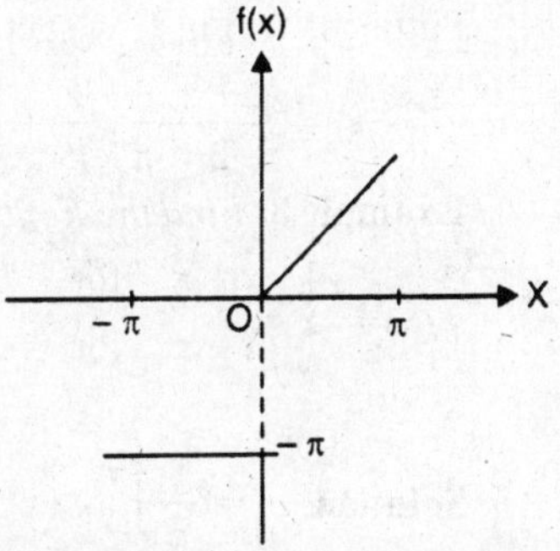

$$b_n = \frac{1}{\pi} \int_{-\pi}^{\pi} f(x) \sin nx\, dx$$

$$= \frac{1}{\pi}\left[\int_{-\pi}^{0} (-\pi) \sin nx\, dx + \int_0^{\pi} x \sin nx\, dx\right]$$

$$= \frac{1}{\pi}\left[\left(\frac{\pi \cos nx}{n}\right)_{-\pi}^{0} + \left(-x\frac{\cos nx}{n} + \frac{\sin nx}{n^2}\right)_0^{\pi}\right]$$

$$= \frac{1}{\pi}\left[\frac{\pi}{n}(1 - \cos n\pi) - \frac{\pi}{n}\cos n\pi\right] = \frac{1}{n}(1 - 2 \cos n\pi)$$

$$f(x) = -\frac{\pi}{4} - \frac{2}{\pi}\left(\cos x + \frac{\cos 3x}{3^2} + \frac{\cos 5x}{5^2} + \ldots\right) + 3 \sin x - \frac{\sin 2x}{2} + \frac{3 \sin 3x}{3} - \frac{\sin 4x}{4} \qquad \ldots(2)$$

Putting $x = 0$ in (2), we get $f(0) = -\frac{\pi}{4} - \frac{2}{\pi}\left(1 + \frac{1}{3^2} + \frac{1}{5^2} + \ldots \infty\right)$ $\qquad \ldots(3)$

Now, $f(x)$ is discontinuous at $x = 0$.
But $f(0 - 0) = -\pi$ and $f(0 + 0) = 0$

$$\therefore \quad f(0) = \frac{1}{2}\,[f(0 - 0) + f(0 + 0)] = -\pi/2 \qquad \ldots(4)$$

Form (3) and (4) $-\frac{\pi}{2} = -\frac{\pi}{4} - \frac{2}{\pi}\left[\frac{1}{1^2} + \frac{1}{3^2} + \frac{1}{5^2} + \ldots\right] \Rightarrow \frac{\pi^2}{8} = \frac{1}{1^2} + \frac{1}{3^2} + \frac{1}{5^2} + \ldots$ **Proved.**

Example 7. *Obtain Fourier Series of the function*

$$f(x) = \begin{cases} x, & -\pi < x < 0 \\ -x, & 0 < x < \pi \end{cases}$$

and hence show that

$$\frac{1}{1^2} + \frac{1}{3^2} + \frac{1}{5^2} + \ldots\ldots\ldots\ldots = \frac{\pi^2}{8}$$

(U.P., II Semester, June 2008, 2002)

Solution. We have, $f(x) = \begin{cases} x, & -\pi < x < 0 \\ -x, & 0 < x < \pi \end{cases}$

Here $f(x)$ is an even function so $b_n = 0$

$$a_0 = \frac{2}{\pi}\int_0^{\pi} f(x)\,dx = \frac{2}{\pi}\int_0^{\pi} -x\,dx = -\frac{2}{\pi}\left[\frac{x^2}{2}\right]_0^{\pi} = -\frac{2}{\pi}\left[\frac{\pi^2}{2}\right] = -\pi$$

$$a_n = \frac{2}{\pi}\int_0^{\pi} f(x)\cos nx\,dx = \frac{2}{\pi}\int_0^{\pi} -x\cos nx\,dx = -\frac{2}{\pi}\left[x\frac{\sin nx}{n} + \frac{\cos nx}{n^2}\right]_0^{\pi}$$

$$= -\frac{2}{\pi}\left[\frac{(-1)^n}{n^2} - \frac{1}{n^2}\right] = \frac{2}{\pi n^2}[1-(-1)^n] = \begin{cases} 0, & n \text{ is even} \\ \dfrac{4}{\pi n^2}, & n \text{ is odd} \end{cases}$$

Fourier series

$$f(x) = -\frac{\pi}{2} + \frac{4}{\pi}\left[\frac{\cos x}{1^2} + \frac{\cos 3x}{3^2} + \frac{\cos 5x}{5^2} + \ldots\right] \qquad \ldots (1)$$ **Ans.**

Now $f(x)$ is discontinuous at $x = 0$.

At $x = 0$ the point of discontinuity $f[0-0) = 0$ and $f(0+0) = 0]$

$$f(0) = \frac{1}{2}\,[f(0-0) + f(0+0) = \frac{1}{2}\,(0+0) = 0]$$

Putting $x = 0$ in 1, we get

$$0 = -\frac{\pi}{2} + \frac{4}{\pi}\left(\frac{1}{1^2} + \frac{1}{3^2} + \frac{1}{5^2} + \ldots\ldots\right) \Rightarrow \frac{1}{1^2} + \frac{1}{3^2} + \frac{1}{5^2} + \ldots\ldots = \frac{\pi^2}{8}$$ **Ans.**

Example 8. *Find the Fourier series of the function defined as*

$$f(x) = \begin{cases} x + \pi, & \text{for } 0 \le x \le \pi, \\ -x - \pi, & \text{for } -\pi \le x < 0 \end{cases} \qquad \textit{and} \qquad f(x + 2\pi) = f(x).$$

(U.P., II Semester Summer 2006)

Solution. $a_0 = \frac{1}{\pi}\int_{-\pi}^{\pi} f(x)\,dx = \frac{1}{\pi}\int_{-\pi}^{0} f(x)\,dx + \frac{1}{\pi}\int_0^{\pi} f(x)\,dx$

$$= \frac{1}{\pi}\int_{-\pi}^{0}(-x-\pi)\,dx + \frac{1}{\pi}\int_0^{\pi}(x+\pi)\,dx = \frac{1}{\pi}\left(-\frac{x^2}{2} - \pi x\right)_{-\pi}^{0} + \frac{1}{\pi}\left(\frac{x^2}{2} + \pi x\right)_0^{\pi}$$

$$= \frac{1}{\pi}\left(\frac{\pi^2}{2} - \pi^2\right) + \frac{1}{\pi}\left(\frac{\pi^2}{2} + \pi^2\right) = \pi\left(\frac{1}{2} - 1\right) + \pi\left(\frac{1}{2} + 1\right) = \pi$$

$$a_n = \frac{1}{\pi}\int_{-\pi}^{0} f(x)\cos nx\,dx = \frac{1}{\pi}\int_{-\pi}^{0} f(x)\cos nx\,dx + \frac{1}{\pi}\int_0^{\pi} f(x)\cos nx\,dx$$

$$= \frac{1}{\pi}\int_{-\pi}^{0}(-x-\pi)\cos nx\,dx + \frac{1}{\pi}\int_0^{\pi}(x+\pi)\cos nx\,dx$$

$$= \frac{1}{\pi}\left[(-x-\pi)\frac{\sin nx}{n} - (-1)\left\{-\frac{\cos nx}{n^2}\right\}\right]_{-\pi}^{0} + \frac{1}{\pi}\left[(x+\pi)\frac{\sin nx}{n} - (-1)\left\{-\frac{\cos nx}{n^2}\right\}\right]_0^{\pi}$$

$$= \frac{1}{\pi}\left[-\frac{1}{n^2} + \frac{(-1)^n}{n^2}\right] + \frac{1}{\pi}\left[\frac{(-1)^n}{n^2} - \frac{1}{n^2}\right] = \frac{2}{n^2\pi}[(-1)^n - 1] = \frac{-4}{n^2\pi}, \text{ if } n \text{ is odd.}$$

$= 0$, if n is even.

$$b_n = \frac{1}{\pi}\int_{-\pi}^{\pi} f(x)\sin nx\,dx = \frac{1}{\pi}\int_{-\pi}^{0} f(x)\sin nx\,dx + \frac{1}{\pi}\int_0^{\pi} f(x)\sin nx\,dx$$

$$= \frac{1}{\pi}\int_{-\pi}^{0}(-x-\pi)\sin nx\,dx + \frac{1}{\pi}\int_0^{\pi}(x+\pi)\sin nx\,dx$$

$$= \frac{1}{\pi}\left[(-x-\pi)\left(-\frac{\cos nx}{n}\right)-(-1)\left(-\frac{\sin nx}{n^2}\right)\right]_{-\pi}^{0}$$

$$+\frac{1}{\pi}\left[(x+\pi)\left(-\frac{\cos nx}{n}\right)-(-1)\left(-\frac{\sin nx}{n^2}\right)\right]_{0}^{\pi}$$

$$= \frac{1}{\pi}\left[\frac{\pi}{n}\right]+\frac{1}{\pi}\left[-\frac{2\pi}{n}(-1)^n+\frac{\pi}{n}\right]=\frac{1}{n}[(1)-2(-1)^n+(1)]=\frac{2}{n}[1-(-1)^n]$$

$$= \frac{4}{n}, \quad \text{if } n \text{ is odd.}$$

$$= 0, \quad \text{if } n \text{ is even.}$$

Fourier series is

$$f(x) = \frac{a_0}{2}+a_1\cos x+a_2\cos 2x+\ldots+b_1\sin x+b_2\sin 2x+\ldots$$

$$f(x) = \frac{\pi}{2}-\frac{4}{\pi}\left(\frac{\cos x}{1^2}+\frac{\cos 3x}{3^2}+\ldots\right)+4\left(\frac{\sin x}{1}+\frac{\sin 3x}{3}+\ldots\right)$$ **Ans.**

EXERCISE 40.2

1. Find the Fourier series of the function

$$f(x) = \begin{cases} -1 & \text{for } -\pi < x < 0 \\ 1 & \text{for } 0 < x < \pi \end{cases}$$ *(U.P. II Semester 2005)*

where $f(x+2\pi) = f(x)$. **Ans.** $\frac{4}{\pi}\left[\frac{1}{1}\sin x+\frac{1}{3}\sin 3x+\frac{1}{5}\sin 5x+\frac{1}{7}\sin 7x+\ldots\right]$

2. Find the Fourier series of the function

$$f(x) = \begin{cases} 0 & \text{for } -\pi \le x \le 0 \\ 1 & \text{for } 0 < x < \frac{\pi}{2} \\ 0 & \text{for } \frac{\pi}{2} \le x \le \pi \end{cases}$$

Ans. $\frac{1}{4}+\frac{1}{\pi}\left[\frac{\cos x}{1}-\frac{\cos 3x}{3}+\frac{\cos 5x}{5}+\ldots+\frac{\sin x}{1}+\frac{\sin 2x}{2}+\frac{\sin 3x}{3}+\ldots\right]$

3. Obtain a Fourier series to represent the following periodic function

$f(x) = 0$ when $0 < x < \pi$

$f(x) = 1$ when $\pi < x < 2\pi$ **Ans.** $\frac{1}{2}-\frac{2}{\pi}\left(\sin x+\frac{1}{3}\sin 3x+\frac{1}{5}\sin 5x+\ldots\right)$

4. Find the Fourier expansion of the function defined in a single period by the relations.

$$f(x) = \begin{cases} 1 & \text{for } 0 < x < \pi \\ 2 & \text{for } \pi < x < 2\pi \end{cases}$$

and from it deduce that $\frac{\pi}{4} = 1-\frac{1}{3}+\frac{1}{5}-\frac{1}{7}+\ldots$ **Ans.** $\frac{3}{2}-\frac{2}{\pi}\left(\sin x+\frac{1}{3}\sin 3x+\frac{1}{5}\sin 5x+\ldots\right)$

5. Find a Fourier series to represent the function

$$f(x) = \begin{cases} 0 & \text{for } -\pi < x \le 0 \\ \frac{1}{4}\pi x & \text{for } 0 < x < \pi \end{cases}$$ and hence deduce that $\frac{\pi^2}{8} = \frac{1}{1^2}+\frac{1}{3^2}+\frac{1}{5^2}+\ldots$

Ans. $\frac{\pi^2}{16}+\sum_{n=1}^{\infty}\left(\frac{[(-1)^n-1]}{4n^2}\cos nx-\frac{(-1)^n\pi}{4n}\sin nx+\ldots\right)$

6. Find the Fourier series for $f(x)$, if

$$f(x) = \begin{cases} -\pi & \text{for } -\pi < x \le 0 \\ x & \text{for } 0 < x < \pi \\ \dfrac{-\pi}{2} & \text{for } x = 0 \end{cases}$$

Deduce that $\dfrac{1}{1^2}+\dfrac{1}{3^2}+\dfrac{1}{5^2}+\ldots=\dfrac{\pi^2}{8}$

Ans. $-\dfrac{\pi}{4}-\dfrac{2}{\pi}\left(\cos x+\dfrac{\cos 3x}{3^2}+\dfrac{\cos 5x}{5^2}+\ldots\right)+3\sin x-\dfrac{1}{2}\sin 2x+\dfrac{3}{3}\sin 3x-\dfrac{1}{4}\sin 4x+\ldots$

7. Obtain a Fourier series to represent the function

$f(x) = |x|$ for $-\pi < x < \pi$

and hence deduce $\dfrac{\pi^2}{8} = \dfrac{1}{1^2}+\dfrac{1}{3^2}+\dfrac{1}{5^2}+\ldots$ **Ans.** $\dfrac{\pi}{2}-\dfrac{4}{\pi}\left[\cos x+\dfrac{1}{3^2}\cos 3x+\dfrac{1}{5^2}\cos 5x+\ldots\right]$

8. Expand as a Fourier series, the function $f(x)$ defined as

$$f(x)\begin{cases} \pi + x & \text{for } -\pi < x < -\dfrac{\pi}{2} \\ \dfrac{\pi}{2} & \text{for } -\dfrac{\pi}{2} < x < \dfrac{\pi}{2} \\ \pi - x & \text{for } \dfrac{\pi}{2} < x < \pi \end{cases}$$

Ans. $\dfrac{3\pi}{8}+\dfrac{2}{\pi}\left[\dfrac{1}{1^2}\cos x-\dfrac{2}{2^2}\cos 2x+\dfrac{1}{3^2}\cos 3x+\ldots\right]$

9. Obtain a Fourier series to represent the function

$f(x) = |\sin x|$ for $-\pi < x < \pi$

$\left\{\textbf{Hint}\ \begin{array}{lll} f(x) = -\sin x & \text{for} & -\pi < x < 0 \\ \quad\ \ = \sin x & \text{for} & 0 < x < \pi \end{array}\right\}$

Ans. $\dfrac{2}{\pi}-\dfrac{4}{\pi}\left[\dfrac{1}{3}\cos 2x+\dfrac{1}{15}\cos 4x+\dfrac{1}{35}\cos 6x+\ldots\right]$

10. An alternating current after passing through a rectifier has the form

$i = I\sin\theta$ for $0 < \theta < \pi$

$\ \ = 0$ for $\pi < \theta < 2\pi$

Find the Fourier series of the function.

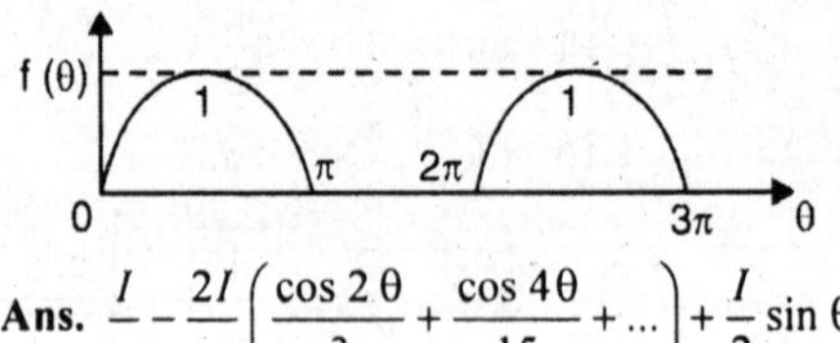

Ans. $\dfrac{I}{\pi}-\dfrac{2I}{\pi}\left(\dfrac{\cos 2\theta}{3}+\dfrac{\cos 4\theta}{15}+\ldots\right)+\dfrac{I}{2}\sin\theta$

11. If $f(x) = 0$ for $-\pi < x < 0$

$\quad = \sin x$ for $0 < x < \pi$

Prove that $f(x) = \dfrac{1}{\pi}+\dfrac{\sin x}{2}-\dfrac{2}{\pi}\displaystyle\sum_{m=1}^{\infty}\dfrac{\cos 2mx}{4m^2-1}$. Hence show that $\dfrac{1}{1.3}-\dfrac{1}{3.5}+\dfrac{1}{5.7}\ldots\infty=\dfrac{1}{4}(\pi-2)$

40.10 EVEN FUNCTION AND ODD FUNCTION

(*a*) Even Function

A function $f(x)$ is said to be even (or symmetric) function if, $f(-x) = f(x)$

The graph of such a function is symmetrical with respect to y-axis [$f(x)$ axis]. Here y-axis is a mirror for the reflection of the curve.

The area under such a curve from $-\pi$ to π is double the area from 0 to π.

$$\therefore \qquad \int_{-\pi}^{\pi} f(x)\,dx = 2\int_0^{\pi} f(x)\,dx$$

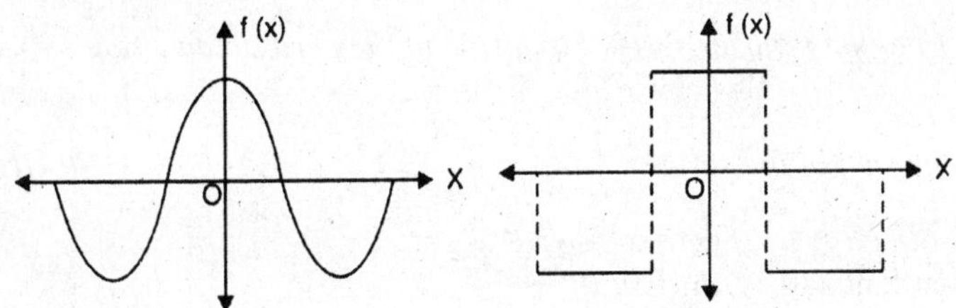

(*b*) **Odd Function**

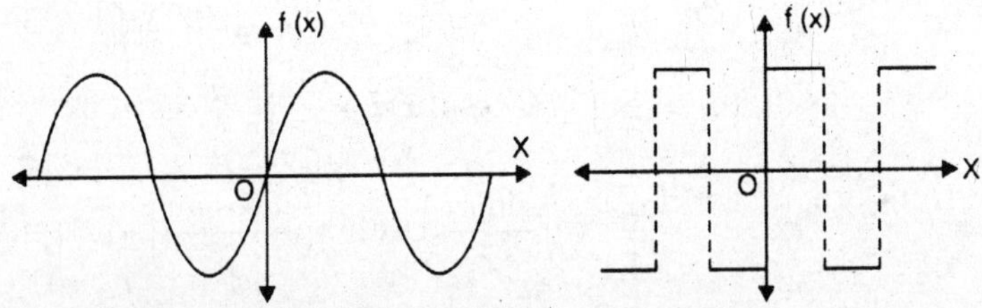

A function $f(x)$ is called odd (or skew symmetric) function if

$$f(-x) = -f(x)$$

Here the area under the curve from $-\pi$ to π is zero.

$$\int_{-\pi}^{\pi} f(x)\, dx = 0$$

Expansion of an even function:

$$a_0 = \frac{1}{\pi}\int_{-\pi}^{\pi} f(x)\, dx = \frac{2}{\pi}\int_0^{\pi} f(x)\, dx$$

$$a_n = \frac{1}{\pi}\int_{-\pi}^{\pi} f(x)\cos nx\, dx = \frac{2}{\pi}\int_0^{\pi} f(x)\cos nx\, dx$$

As $f(x)$ and $\cos nx$ are both even functions, therefore, the product of $f(x)$. $\cos nx$ is also an even function.

$$b_n = \frac{1}{\pi}\int_{-\pi}^{\pi} f(x)\sin nx\, dx = 0$$

As $\sin nx$ is an odd function so $f(x)$. $\sin nx$ is also an odd function. We need not to calculate b_n. It saves our labour a lot.

The series of the even function will contain only cosine terms. (*U.P. II Semester 2010*)

Expansion of an odd function :

$$a_0 = \frac{1}{\pi}\int_{-\pi}^{\pi} f(x)\, dx = 0$$

$$a_n = \frac{1}{\pi}\int_{-\pi}^{\pi} f(x)\cos nx\, dx = 0 \qquad [f(x).\ \cos nx \text{ is odd function.}]$$

$$b_n = \frac{1}{\pi}\int_{-\pi}^{\pi} f(x)\sin nx\, dx = \frac{2}{\pi}\int_0^{\pi} f(x)\sin nx\, dx$$

[$f(x)$. $\sin nx$ is even function.]

The series of the odd function will contain only sine terms.

The function shown below is neither odd nor even so it contains both sine and cosine terms

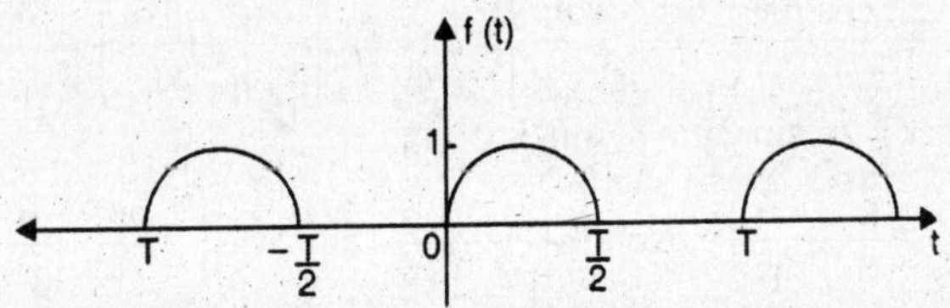

Example 9. *Find the Fourier series expansion of the periodic function of period* 2π

$$f(x) = x^2, \ -\pi \le x \le \pi.$$

Hence, find the sum of the series $\frac{1}{1^2} - \frac{1}{2^2} + \frac{1}{3^2} - \frac{1}{4^2} + \ldots$ (*U.P., II Semester 2004*)

Solution. $f(x) = x^2, \ -\pi \le x \le \pi$

This is an even function. $\therefore \ b_n = 0$ $[f(-x) = f(x)]$

$$a_0 = \frac{2}{\pi}\int_0^{\pi} f(x)\,dx = \frac{2}{\pi}\int_0^{\pi} x^2\,dx = \frac{2}{\pi}\left[\frac{x^3}{3}\right]_0^{\pi} = \frac{2\pi^2}{3}$$

$$a_n = \frac{2}{\pi}\int_0^{\pi} f(x)\cos nx\,dx = \frac{2}{\pi}\int_0^{\pi} x^2 \cos nx\,dx$$

$$= \frac{2}{\pi}\left[x^2\left(\frac{\sin nx}{n}\right) - (2x)\left(-\frac{\cos nx}{n^2}\right) + (2)\left(-\frac{\sin nx}{n^3}\right)\right]_0^{\pi}$$

$$= \frac{2}{\pi}\left[\frac{\pi^2 \sin n\pi}{n} + \frac{2\pi\cos n\pi}{n^2} - \frac{2\sin n\pi}{n^3}\right] = \frac{4(-1)^n}{n^2}$$

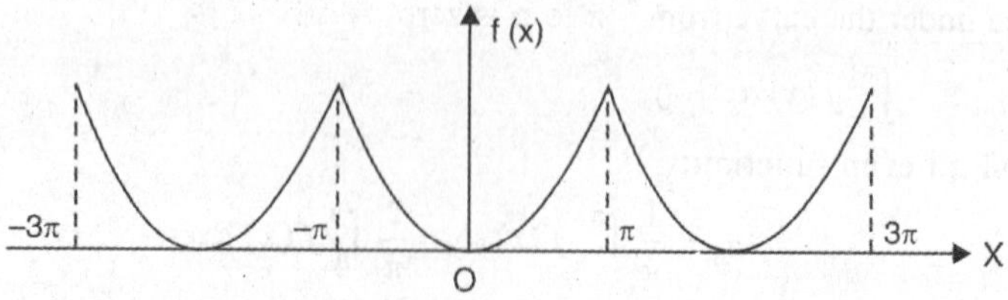

Fourier series is $f(x) = \frac{a_0}{2} + a_1 \cos x + a_2 \cos 2x + a_3 \cos 3x + \ldots + a_n \cos nx + \ldots$

$$x^2 = \frac{\pi^2}{3} - 4\left[\frac{\cos x}{1^2} - \frac{\cos 2x}{2^2} + \frac{\cos 3x}{3^2} - \frac{\cos 4x}{4^2} + \ldots\right]$$

On putting $x = 0$, we have

$$0 = \frac{\pi^2}{3} - 4\left[\frac{1}{1^2} - \frac{1}{2^2} + \frac{1}{3^2} - \frac{1}{4^2} \ldots\right]$$

$$\Rightarrow \quad \frac{1}{1^2} - \frac{1}{2^2} + \frac{1}{3^2} - \frac{1}{4^2} \ldots = \frac{\pi^2}{12}$$ **Ans.**

Example 10. *Obtain a Fourier expression for* $f(x) = x^3$ *for* $-\pi < x < \pi$.

Solution. $f(x) = x^3$ is an odd function.

$\therefore \quad a_0 = 0$ and $a_n = 0$ $[f(-x) = -f(x)]$

$$b_n = \frac{2}{\pi}\int_0^{\pi} f(x)\sin nx\,dx = \frac{2}{\pi}\int_0^{\pi} x^3 \sin nx\,dx \quad \left[\int uv = uv_1 - u'v_2 + u''v_3 - u'''v_4 + \ldots\right]$$

$$= \frac{2}{\pi}\left[x^3\left(-\frac{\cos nx}{n}\right) - 3x^2\left(-\frac{\sin nx}{n^2}\right) + 6x\left(\frac{\cos nx}{n^3}\right) - 6\left(\frac{\sin nx}{n^4}\right)\right]_0^{\pi}$$

$$= \frac{2}{\pi}\left[-\frac{\pi^3\cos n\pi}{n} + \frac{6\pi\cos n\pi}{n^3}\right] = 2.(-1)^n\left[-\frac{\pi^2}{n} + \frac{6}{n^3}\right]$$

$$f(x) = b_1 \sin x + b_2 \sin 2x + b_3 \sin 3x + \ldots\ldots$$

$$\therefore \quad x^3 = 2\left[-\left(-\frac{\pi^2}{1} + \frac{6}{1^3}\right)\sin x + \left(-\frac{\pi^2}{2} + \frac{6}{2^3}\right)\sin 2x - \left(-\frac{\pi^2}{3} + \frac{6}{3^3}\right)\sin 3x + \ldots\right]$$ **Ans.**

Example 11. *Expand the function $f(x) = x \sin x$, as a Fourier series in the interval $-\pi \le x \le \pi$.*

Hence deduce that $\dfrac{1}{1.3} - \dfrac{1}{3.5} + \dfrac{1}{5.7} - \dfrac{1}{7.9} + \ldots = \dfrac{\pi - 2}{4}$

(*MDU, Dec. 2010, U.P., II Sem., Summer 2008, 2001, Uttarakhand, II Sem., June 2007*)

Solution. $f(x) = x \sin x$

$$a_0 = \frac{2}{\pi}\int_0^{\pi} f(x)\,dx = \frac{2}{\pi}\int_0^{\pi} x \sin x\,dx \qquad \text{(Here } x \sin x \text{ is an even function)}$$

$$= \frac{2}{\pi}[x(-\cos x) - (1)(-\sin x)]_0^{\pi} = \frac{2}{\pi}(\pi) = 2$$

$$a_n = \frac{2}{\pi}\int_0^{\pi} f(x)\cos nx\,dx = \frac{2}{\pi}\int_0^{\pi} x \sin x \cos nx\,dx$$

$$= \frac{1}{\pi}\int_0^{\pi} x\{\sin(n+1)x - \sin(n-1)x\}\,dx$$

$$= \frac{1}{\pi}\int_0^{\pi} x \sin(n+1)x\,dx - \frac{1}{\pi}\int_0^{\pi} x \sin(n-1)x\,dx$$

$$= \frac{1}{\pi}\left[x\left(-\frac{\cos(n+1)x}{n+1}\right)_0^{\pi} - (1)\left\{-\frac{\sin(n+1)x}{(n+1)^2}\right\}\right]_0^{\pi}$$

$$- \frac{1}{\pi}\left[x\left(-\frac{\cos(n-1)x}{(n-1)}\right) - (1)\left\{-\frac{\sin(n-1)x}{(n-1)^2}\right\}\right]_0^{\pi}$$

$$= \frac{1}{\pi}\left[-\pi\frac{(-1)^{(n+1)}}{n+1} + 0\right] - \frac{1}{\pi}\left[-\pi\frac{(-1)^{(n-1)}}{n-1} - 0\right]$$

$$= -\frac{(-1)^{n+1}}{n+1} + \frac{(-1)^{n-1}}{n-1} = (-1)^{n+1}\left[-\frac{1}{n+1} + \frac{1}{n-1}\right] = \frac{2(-1)^{n+1}}{n^2-1}$$

$$a_1 = \frac{2}{\pi}\int_0^{\pi} x \sin x \cos x\,dx = \frac{1}{\pi}\int_0^{\pi} x \sin 2x\,dx$$

$$= \frac{1}{\pi}\left[x\left(-\frac{\cos 2x}{2}\right) - (1)\left(\frac{-\sin 2x}{4}\right)\right]_0^{\pi} = \frac{1}{\pi}\left[\frac{-\pi}{2}\right] = -\frac{1}{2}$$

$$b_n = 0 \qquad \text{[As } x \sin x \sin nx \text{ is an odd function]}$$

Hence $f(x) = 1 - \dfrac{1}{2}\cos x + 2\displaystyle\sum_{n=2}^{\infty}\frac{(-1)^{n+1}}{n^2-1}\cos nx = 1 - \frac{1}{2}\cos x + 2\sum_{n=2}^{\infty}\frac{(-1)^{n+1}}{(n-1)(n+1)}\cos nx$

$$x \sin x = 1 + 2\left[-\frac{1}{4}\cos x - \frac{1}{1.3}\cos 2x + \frac{1}{2.4}\cos 3x - \frac{1}{3.5}\cos 4x + \ldots\right] \qquad \ldots (1)$$

Putting $x = \dfrac{\pi}{2}$ in (1), we get $\qquad \dfrac{\pi}{2} = 1 + 2\left\{\dfrac{1}{1.3} - \dfrac{1}{3.5} + \dfrac{1}{5.7} - \ldots\right\}$

or $\qquad \dfrac{\pi}{4} = \dfrac{1}{2} + \dfrac{1}{1.3} - \dfrac{1}{3.5} + \dfrac{1}{5.7} - \ldots \Rightarrow \dfrac{\pi}{4} - \dfrac{1}{2} = \dfrac{1}{1.3} - \dfrac{1}{3.5} + \dfrac{1}{5.7} - \ldots$

$\Rightarrow \qquad \dfrac{\pi - 2}{4} = \dfrac{1}{1.3} - \dfrac{1}{3.5} + \dfrac{1}{5.7} - \ldots$ **Proved.**

Example 12. *Find the Fourier Series expansion for the function*

$f(x) = x \cos x, -\pi < x < \pi.$ (*U.P., II Semester, Summer 2002*)

Solution. Since $x \cos x$ is an odd function therefore, $a_0 = a_n = 0$.

Let $x \cos x = \Sigma\, b_n \sin bx$, where

$$b_n = \frac{2}{\pi}\int_0^{\pi} x \cos x . \sin nx \, dx = \frac{1}{\pi}\int_0^{\pi} x \{\sin (n+1)x + \sin (n-1)x\}\, dx$$

$$= \frac{1}{\pi}\int_0^{\pi} x \sin (n+1)x \, dx + \frac{1}{\pi}\int_0^{\pi} x \sin (n-1)x \, dx$$

$$= \frac{1}{\pi}\left[x\left(\frac{-\cos (n+1)x}{n+1}\right) + \frac{\sin (n+1)x}{(n+1)^2}\right]_0^{\pi} + \frac{1}{\pi}\left[-x\frac{\cos (n-1)x}{n-1} + \frac{\sin (n-1)x}{(n-1)^2}\right]_0^{\pi}$$

$$= \frac{1}{\pi}\left[x.\left\{-\frac{\cos (n+1)x}{(n+1)} - \frac{\cos (n-1)x}{(n-1)}\right\} + 1.\left\{\frac{\sin (n+1)x}{(n+1)^2} + \frac{\sin (n-1)x}{(n-1)^2}\right\}\right]_0^{\pi}$$

$$= \frac{1}{\pi}\left[\pi.\left\{-\frac{\cos (n+1)\pi}{(n+1)} - \frac{\cos (n-1)\pi}{(n-1)}\right\}\right]$$

$$\Rightarrow \quad b_n = \left\{-\frac{(-1)^{n+1}}{n+1} - \frac{(-1)^{n-1}}{n-1}\right\}, \quad n \neq 1$$

$$b_n = -(-1)^{n+1}\left[\frac{1}{n+1} + \frac{1}{n-1}\right]$$

$$= -\left\{\frac{1}{(n+1)} + \frac{1}{(n-1)}\right\} = \frac{-2n}{n^2-1}, \qquad \text{If } n \text{ is odd; } n \neq 1.$$

$$\text{But } b_n = \left\{\frac{1}{n+1} + \frac{1}{n-1}\right\} = \frac{2n}{n^2-1}, \qquad \text{If } n \text{ is even; } n \neq 1$$

$$\text{If} \quad n = 1, \text{ then } b_1 = \frac{2}{\pi}\int_0^{\pi} x \cos x . \sin x \, dx = \frac{1}{\pi}\int_0^{\pi} x . \sin 2x \, dx$$

$$= \frac{1}{\pi}\left[x.\left(-\frac{\cos 2x}{2}\right) - 1\left(-\frac{\sin 2x}{4}\right)\right]_0^{\pi} = \frac{1}{\pi}\left[\pi\left(-\frac{1}{2}\right)\right] = -\frac{1}{2}$$

$$\therefore\ x \cos x = b_1 \sin x + b_2 \sin 2x + b_3 \sin 3x + \ldots$$

$$= -\frac{1}{2}\sin x + \frac{4 \sin 2x}{2^2-1} - \frac{6 \sin 3x}{3^2-1} + \ldots$$

Ans.

40.11 HALF-RANGE SERIES, PERIOD 0 TO π

The given function is defined in the interval $(0, \pi)$ and it is immaterial whatever the function may be outside the interval $(0, \pi)$. To get the series of cosines only we assume that $f(x)$ is an even function in the interval $(-\pi, \pi)$.

$$\boldsymbol{a_n = \frac{2}{\pi}\int_0^{\pi} f(x) \cos nx \, dx \ \text{ and } b_n = 0}$$

To expand $f(x)$ as a sine series we extend the function in the interval $(-\pi, \pi)$ as an odd function.

$$\boldsymbol{b_n = \frac{2}{\pi}\int_0^{\pi} f(x) \sin nx \, dx \ \text{ and } a_n = 0}$$

Example 13. *Represent the following function by a Fourier sine series :*

$$f(t) = \begin{cases} t, & 0 < t \le \dfrac{\pi}{2} \\ \dfrac{\pi}{2}, & \dfrac{\pi}{2} < t \le \pi \end{cases}$$

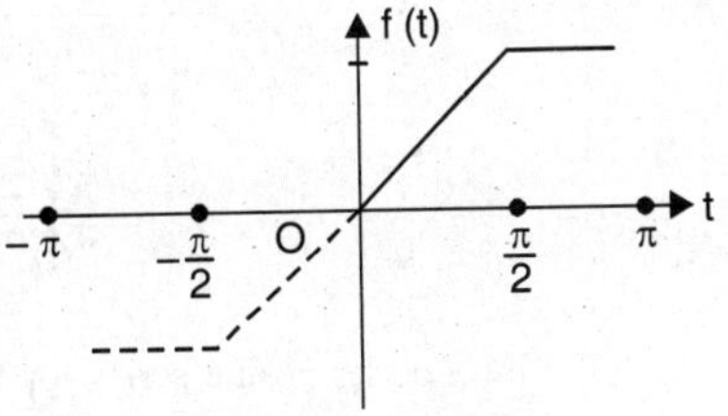

Solution. $b_n = \dfrac{2}{\pi} \int_0^{\pi} f(t) \sin nt \, dt$

$$= \frac{2}{\pi} \int_0^{\pi/2} t \sin nt \, dt + \frac{2}{\pi} \int_{\pi/2}^{\pi} \frac{\pi}{2} \sin nt \, dt$$

$$= \frac{2}{\pi} \left[t \left(-\frac{\cos nt}{n} \right) - (1) \left(-\frac{\sin nt}{n^2} \right) \right]_0^{\pi/2} + \frac{2}{\pi} \frac{\pi}{2} \left[-\frac{\cos nt}{n} \right]_{\pi/2}^{\pi}$$

$$= \frac{2}{\pi} \left[-\frac{\pi}{2} \frac{\cos \frac{n\pi}{2}}{n} + \frac{\sin \frac{n\pi}{2}}{n^2} \right] + \left[-\frac{\cos n\pi}{n} + \frac{\cos \frac{n\pi}{2}}{n} \right]$$

$$b_1 = \frac{2}{\pi} \left[-\frac{\pi}{2} \cos \frac{\pi}{2} + \sin \frac{\pi}{2} \right] + \left[-\cos \pi + \cos \frac{\pi}{2} \right] = \frac{2}{\pi} [0 + 1] + [1] = \frac{2}{\pi} + 1$$

$$b_2 = \frac{2}{\pi} \left[-\frac{\pi}{2} \frac{\cos \pi}{2} + \frac{\sin \pi}{2^2} \right] + \left[-\frac{\cos 2\pi}{2} + \frac{\cos \pi}{2} \right] = \frac{2}{\pi} \left[-\frac{\pi}{2} \frac{(-1)}{2} + 0 \right] + \left[-\frac{1}{2} - \frac{1}{2} \right]$$

$$= \frac{2}{\pi} \left[\frac{\pi}{4} \right] - 1 = \frac{1}{2} - 1 = -\frac{1}{2}$$

$$b_3 = \frac{2}{\pi} \left[-\frac{\pi}{2} \frac{\cos \frac{3\pi}{2}}{3} + \frac{\sin \frac{3\pi}{2}}{3^2} \right] + \left[-\frac{\cos 3\pi}{3} + \frac{\cos \frac{3\pi}{2}}{3} \right]$$

$$= \frac{2}{\pi} \left[-\frac{\pi}{2} (0) - \frac{1}{9} \right] + \left[\frac{1}{3} + 0 \right] = -\frac{2}{9\pi} + \frac{1}{3}$$

$$f(t) = \left(\frac{2}{\pi} + 1 \right) \sin t - \frac{1}{2} \sin 2t + \left(-\frac{2}{9\pi} + \frac{1}{3} \right) \sin 3t + \ldots$$ **Ans.**

Example 14. *Find the Fourier sine series for the function*

$$f(x) = e^{ax} \text{ for } 0 < x < \pi$$

where a is constant.

Solution. $b_n = \dfrac{2}{\pi} \int_0^{\pi} e^{ax} \sin nx \, dx \qquad \left[\int e^{ax} \sin bx \, dx = \dfrac{e^{ax}}{a^2 + b^2} [a \sin bx - b \cos bx] \right]$

$$= \frac{2}{\pi} \left[\frac{e^{ax}}{a^2 + n^2} (a \sin nx - n \cos nx) \right]_0^{\pi}$$

$$= \frac{2}{\pi} \left[\frac{e^{a\pi}}{a^2 + n^2} (a \sin n\pi - n \cos n\pi) + \frac{n}{a^2 + n^2} \right]$$

$$= \frac{2}{\pi} \frac{n}{a^2 + n^2} \left[-(-1)^n e^{a\pi} + 1 \right] = \frac{2n}{(a^2 + n^2)\pi} [1 - (-1)^n e^{a\pi}]$$

$$b_1 = \frac{2(1+e^{a\pi})}{(a^2+1^2)\pi}, \qquad b_2 = \frac{2.2.(1-e^{a\pi})}{.(a^2+2^2)\pi}$$

$$e^{ax} = \frac{2}{\pi}\left[\frac{1+e^{a\pi}}{a^2+1^2}\sin x + \frac{2(1-e^{a\pi})}{a^2+2^2}\sin 2x + \ldots\right]$$ **Ans.**

EXERCISE 40.3

1. Find the Fourier cosine series for the function

$$f(x) = \begin{cases} 1 & \text{for } 0 < x < \dfrac{\pi}{2} \\ 0 & \text{for } \dfrac{\pi}{2} < x < \pi. \end{cases}$$

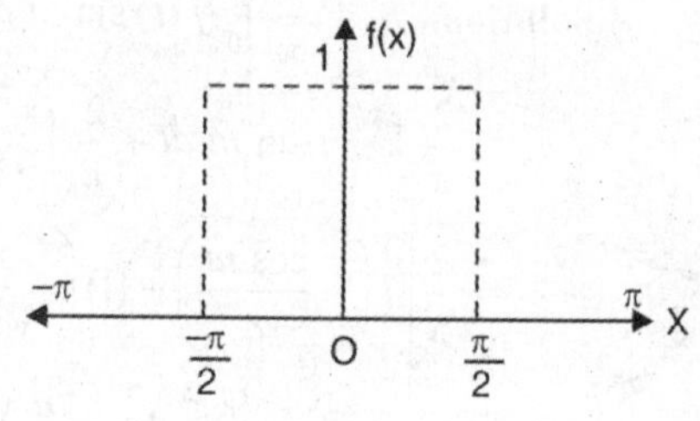

Ans. $\frac{1}{2} + \frac{2}{\pi}\left[\cos x - \frac{1}{3}\cos 3x + \frac{1}{5}\cos 5x - \ldots\right]$

2. Find a series of cosine of multiples of x which will represent $f(x)$ in $(0, \pi)$ where

$$f(x) = \begin{cases} 0 & \text{for } 0 < x < \dfrac{\pi}{2} \\ \dfrac{\pi}{2} & \text{for } \dfrac{\pi}{2} < x < \pi \end{cases}$$

Deduce that $1 - \frac{1}{3} + \frac{1}{5} - \frac{1}{7} + \ldots \infty = \frac{\pi}{4}$ **Ans.** $\frac{\pi}{4} - \cos x + \frac{1}{3}\cos 3x - \frac{1}{5}\cos 5x + \ldots$

3. Express $f(x) = x$ as a sine series in $0 < x < \pi$. **Ans.** $2\left[\sin x - \frac{1}{2}\sin 2x + \frac{1}{3}\sin 3x - \ldots\right]$

4. Find the cosine series for $f(x) = \pi - x$ in the interval $0 < x < \pi$.

Ans. $\frac{\pi}{2} + \frac{4}{\pi}\left(\cos x + \frac{\cos 3x}{3^2} + \frac{\cos 5x}{5^2} + \ldots\right)$

5. If $f(x) = \begin{cases} 0 & \text{for } 0 < x < \dfrac{\pi}{2} \\ \pi - x, & \text{for } \dfrac{\pi}{2} < x < \pi \end{cases}$

Show that: (*i*) $f(x) = \frac{4}{\pi}\left(\sin x - \frac{1}{3^2}\sin 3x + \frac{1}{5^2}\sin 5x - \ldots\right)$

(*ii*) $f(x) = \frac{\pi}{4} - \frac{2}{\pi}\left(\frac{1}{1^2}\cos 2x + \frac{1}{3^2}\cos 6x + \frac{1}{5^2}\cos 10x + \ldots\right)$

6. Obtain the half-range cosine series for $f(x) = x^2$ in $0 < x < \pi$.

Ans. $\frac{\pi^2}{3} - \frac{4}{\pi}\left(\cos x - \frac{1}{2^2}\cos 2x + \frac{1}{3^2}\cos 3x - \ldots\right)$

7. Find (*i*) sine series and (*ii*) cosine series for the function $f(x) = e^x$ for $0 < x < \pi$.

Ans. (*i*) $\frac{2}{\pi}\sum_1^\infty n\left[\frac{1-(-1)^n e^\pi}{n^2+1}\right]\sin nx$ (*ii*) $\frac{e^\pi - 1}{\pi} - \frac{2}{\pi}\sum_1^\infty \frac{1-(-1)^n e^\pi}{n^2+1}\cos nx$

8. If $f(x) = x + 1$, for $0 < x < \pi$, find its Fourier (*i*) sine series (*ii*) cosine series. Hence deduce that

(*i*) $1 - \frac{1}{3} + \frac{1}{5} - \frac{1}{7} + \ldots = \frac{\pi}{4}$ (*ii*) $1 + \frac{1}{3^2} + \frac{1}{5^2} + \frac{1}{7^2} + \ldots = \frac{\pi^2}{8}$

Ans. (*i*) $\frac{2}{\pi}\left[(\pi+2)\sin x - \frac{\pi}{2}\sin 2x + \frac{1}{3}(\pi+2)\sin 3x - \frac{\pi}{4}\sin 4x + \ldots\right]$

(*ii*) $\frac{\pi}{2} + 1 - 4\left(\cos x + \frac{\cos 3x}{3^2} + \frac{\cos 5x}{5^2} + \ldots\right)$

9. Find the Fourier series expansion of the function
$f(x) = \cos(sx),\ -\pi \le x \le \pi$

where s is a fraction. Hence, show that $\cot \theta = \frac{1}{\theta} + \frac{2\theta}{\theta^2 - \pi^2} + \frac{2\theta}{\theta^2 - 4\pi^2} + \ldots$

Ans. $\frac{\sin \pi x}{\pi s} + \frac{1}{\pi} \Sigma \left(\frac{\sin(s\pi + n\pi)}{s+n} + \frac{\sin(s\pi - n\pi)}{s-n} \right) \cos nx$

40.12 CHANGE OF INTERVAL AND FUNCTIONS HAVING ARBITRARY PERIOD

In electrical engineering problems, the period of the function is not always 2π but T or $2c$. This period must be converted to the length 2π. The independent variable x is also to be changed proportionally.

Let the function $f(x)$ be defined in the interval $(-c, c)$. Now we want to change the function to the period of 2π so that we can use the formulae of a_n, b_n as discussed in Article 40.6.

$\therefore$ $2c$ is the interval for the variable x.

$\therefore$ 1 is the interval for the variable $= \frac{x}{2c}$

$\therefore$ 2π is the interval for the variable $= \frac{x \cdot 2\pi}{2c} = \frac{\pi x}{c}$

so put $z = \frac{\pi x}{c}$ or $x = \frac{zc}{\pi}$

Thus the function $f(x)$ of period $2c$ is transformed to the function

$f\left(\frac{cz}{\pi}\right)$ or $F(z)$ of period 2π.

$F(z)$ can be expanded in the Fourier series.

$$F(z) = f\left(\frac{cz}{\pi}\right) = \frac{a_0}{2} + a_1 \cos z + a_2 \cos 2z + \ldots + b_1 \sin z + b_2 \sin 2z + \ldots$$

where
$$a_0 = \frac{1}{\pi}\int_0^{2\pi} F(z)\,dz = \frac{1}{\pi}\int_0^{2\pi} f\left(\frac{cz}{\pi}\right) dz$$

$$= \frac{1}{\pi}\int_0^{2c} f(x)\, d\left(\frac{\pi x}{c}\right) = \frac{1}{c}\int_0^{2c} f(x)\,dx \qquad \left[\text{put } z = \frac{\pi x}{c}\right]$$

$$\boxed{a_0 = \frac{1}{c}\int_0^{2c} f(x)\,dx}$$

$$a_n = \frac{1}{\pi}\int_0^{2\pi} F(z)\cos nz\,dz = \frac{1}{\pi}\int_0^{2\pi} f\left(\frac{cz}{\pi}\right)\cos nz\,dz$$

$$= \frac{1}{\pi}\int_0^{2c} f(x)\cos\frac{n\pi x}{c}\, d\left(\frac{\pi x}{c}\right) = \frac{1}{c}\int_0^{2c} f(x)\cos\frac{n\pi x}{c}\,dx. \qquad \left[\text{Put } z = \frac{\pi x}{c}\right]$$

$$\boxed{a_n = \frac{1}{c}\int_0^{2c} f(x)\cos\frac{n\pi x}{c}\,dx}$$

Similarly,
$$\boxed{b_n = \frac{1}{c}\int_0^{2c} f(x)\sin\frac{n\pi x}{c}\,dx}$$

Example 15. *Find the Fourier series corresponding to the function $f(x)$ defined in $(-2, 2)$ as follows*

$$f(x) = \begin{cases} 2 & in \quad -2 \le x \le 0 \\ x & in \quad 0 < x < 2 \end{cases}$$

Solution. Here the interval is $(-2, 2)$ and $c = 2$

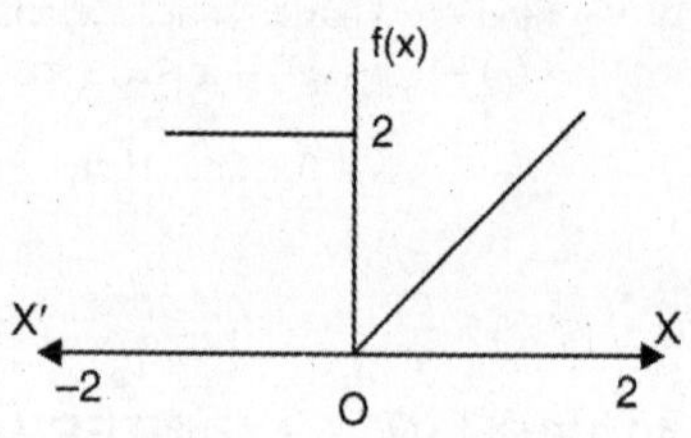

$$a_0 = \frac{1}{c}\int_{-c}^{c} f(x)\,dx = \frac{1}{2}\left[\int_{-2}^{0} 2.dx + \int_{0}^{2} x.dx\right]$$

$$= \frac{1}{2}\left[[2x]_{-2}^{0} + \left(\frac{x^2}{2}\right)_0^2\right] = \frac{1}{2}[4+2] = 3$$

$$a_n = \frac{1}{c}\int_{-c}^{c} f(x)\cos\left(\frac{n\pi x}{c}\right)dx = \frac{1}{2}\left[\int_{-2}^{0} 2\cos\frac{n\pi x}{2}\,dx + \int_{0}^{2} x\cos\frac{n\pi x}{2}\,dx\right]$$

$$= \frac{1}{2}\left[\frac{4}{n\pi}\left(\sin\frac{n\pi x}{2}\right)_{-2}^{0} + \left(x\frac{2}{n\pi}\sin\frac{n\pi x}{2} + \frac{4}{n^2\pi^2}\cos\frac{n\pi x}{2}\right)_0^2\right]$$

$$= \frac{1}{2}\left[\frac{4}{n^2\pi^2}\cos n\pi - \frac{4}{n^2\pi^2}\right] = \frac{2}{n^2\pi^2}[(-1)^n - 1]$$

$$= -\frac{4}{n^2\pi^2}, \qquad \text{when } n \text{ is odd}$$

$$= 0, \qquad \text{when } n \text{ is even.}$$

$$b_n = \frac{1}{c}\int_{-c}^{c} f(x)\sin\frac{n\pi x}{c}\,dx = \frac{1}{2}\int_{-2}^{0} 2\sin\frac{n\pi x}{2}\,dx + \frac{1}{2}\int_{0}^{2} x\sin\frac{n\pi x}{2}\,dx$$

$$= \frac{1}{2}\left[2\left(-\frac{2}{n\pi}\cos\frac{n\pi x}{2}\right)\right]_{-2}^{0} + \frac{1}{2}\left[x\left(-\frac{2}{n\pi}\cos\frac{n\pi x}{2}\right) + (1)\frac{4}{n^2\pi^2}\sin\frac{n\pi x}{2}\right]_0^2$$

$$= \frac{1}{2}\left[-\frac{4}{n\pi} + \frac{4}{n\pi}\cos n\pi\right] + \frac{1}{2}\left[-\frac{4}{n\pi}\cos n\pi + \frac{4}{n^2\pi^2}\sin n\pi\right] = \frac{1}{2}\left[-\frac{4}{n\pi}\right] = -\frac{2}{n\pi}$$

$$f(x) = \frac{a_0}{2} + a_1\cos\frac{\pi x}{c} + a_2\cos\frac{2\pi x}{c} + a_3\cos\frac{3\pi x}{c} + \ldots$$

$$+ b_1\sin\frac{\pi x}{c} + b_2\sin\frac{2\pi x}{c} + b_3\sin\frac{3\pi x}{c} + \ldots$$

$$= \frac{3}{2} - \frac{4}{\pi^2}\left\{\frac{1}{1^2}\cos\frac{\pi x}{2} + \frac{1}{3^2}\cos\frac{3\pi x}{2} + \ldots\right\}$$

$$- \frac{2}{\pi}\left\{\frac{1}{1}\sin\frac{\pi x}{2} + \frac{1}{2}\sin\frac{2\pi x}{2} + \frac{1}{3}\sin\frac{3\pi x}{2} + \ldots\right\} \qquad \textbf{Ans.}$$

Example 16. *A periodic function of period 4 is defined as*

$$f(x) = |x|, \; -2 < x < 2.$$

Find its Fourier series expansion.

Solution. $f(x) = |x| \qquad -2 < x < 2$

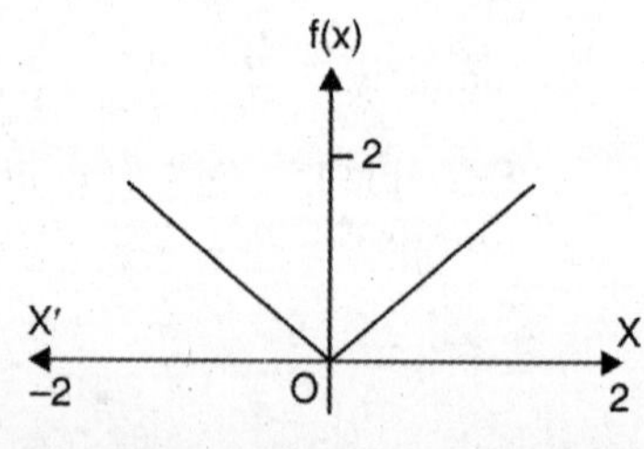

$$\Rightarrow \qquad f(x) = \begin{cases} x, & 0 < x < 2 \\ -x, & -2 < x < 0 \end{cases}$$

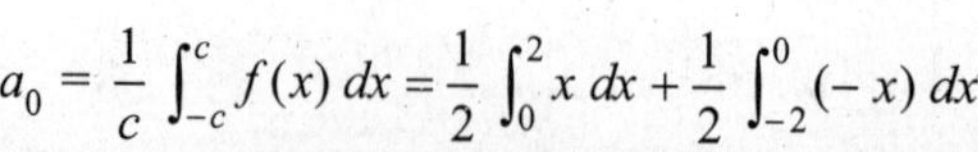

$$a_0 = \frac{1}{c}\int_{-c}^{c} f(x)\,dx = \frac{1}{2}\int_{0}^{2} x\,dx + \frac{1}{2}\int_{-2}^{0}(-x)\,dx$$

$$= \frac{1}{2}\left[\frac{x^2}{2}\right]_0^2 + \frac{1}{2}\left[\frac{-x^2}{2}\right]_{-2}^0 = \frac{1}{4}(4-0) + \frac{1}{4}(0+4) = 2$$

$$a_n = \frac{1}{c}\int_{-c}^{c} f(x)\cos\frac{n\pi x}{c}\,dx = \frac{1}{2}\int_0^2 x\cos\frac{n\pi x}{2}\,dx + \frac{1}{2}\int_{-2}^{0}(-x)\cos\frac{n\pi x}{2}\,dx$$

$$= \frac{1}{2}\left[x\left(\frac{2}{n\pi}\sin\frac{n\pi x}{2}\right) - (1)\left(-\frac{4}{n^2\pi^2}\cos\frac{n\pi x}{2}\right)\right]_0^2$$

$$+\frac{1}{2}\left[(-x)\left(\frac{2}{n\pi}\sin\frac{n\pi x}{2}\right) - (-1)\left(-\frac{4}{n^2\pi^2}\right)\cos\frac{n\pi x}{2}\right]_{-2}^0$$

$$= \frac{1}{2}\left[0 + \frac{4}{n^2\pi^2}(-1)^n - \frac{4}{n^2\pi^2}\right] + \frac{1}{2}\left[0 - \frac{4}{n^2\pi^2} + \frac{4}{n^2\pi^2}(-1)^n\right]$$

$$= \frac{1}{2}\,\frac{4}{n^2\pi^2}[(-1)^n - 1 - 1 + (-1)^n] = \frac{4}{n^2\pi^2}[(-1)^n - 1]$$

$$= -\frac{8}{n^2\pi^2} \qquad \text{(If } n \text{ is odd.)}$$

$$= 0 \qquad \text{(If } n \text{ is even)}$$

$b_n = 0$ as $f(x)$ is even function.

Fourier series is

$$f(x) = \frac{a_0}{2} + a_1\cos\frac{\pi x}{c} + a_2\cos\frac{2\pi x}{c} + \ldots + b_1\sin\frac{\pi x}{c} + b_2\sin\frac{2\pi x}{c} + \ldots$$

$$f(x) = 1 - \frac{8}{\pi^2}\left[\frac{\cos\frac{\pi x}{2}}{1^2} + \frac{\cos\frac{3\pi x}{2}}{3^2} + \frac{\cos\frac{5\pi x}{2}}{5^2} + \ldots\right]$$ **Ans.**

Example 17. *Prove that*

$$\frac{l}{2} - x = \frac{l}{\pi}\sum_{1}^{\infty}\frac{1}{n}\sin\frac{2n\pi x}{l},\ 0 < x$$

Solution. $f(x) = \frac{1}{2} - x$

$$a_0 = \frac{1}{l/2}\int_0^l f(x)\,dx = \frac{2}{l}\int_0^l\left(\frac{1}{2} - x\right)dx = \frac{2}{l}\left[\frac{lx}{2} - \frac{x^2}{2}\right]_0^l = 0$$

$$a_n = \frac{1}{l/2}\int_0^l f(x)\cos\frac{n\pi x}{l/2}\,dx = \frac{2}{l}\int_0^l\left(\frac{l}{2} - x\right)\cos\frac{2n\pi x}{l}\,dx$$

$$= \frac{2}{l}\left[\left(\frac{l}{2} - x\right)\frac{l}{2n\pi}\sin\frac{2n\pi x}{l} + (-1)\frac{l^2}{4n^2\pi^2}\cos\frac{2n\pi x}{l}\right]_0^l$$

$$= \frac{2}{l}\left[0 - \frac{l^2}{4n^2\pi^2}\cos 2n\pi + \frac{l^2}{4n^2\pi^2}\right]$$

$$= \frac{2}{l}\,\frac{l^2}{4n^2\pi^2}(-\cos 2\,n\,\pi + 1) = \frac{l}{2\,n^2\pi^2}(-1+1) = 0$$

$$b_n = \frac{1}{l/2}\int_0^l f(x)\sin\frac{n\pi x}{l/2}\,dx = \frac{2}{l}\int_0^l \left(\frac{1}{2} - x\right)\sin\frac{2n\pi x}{l}\,dx$$

$$= \frac{2}{l}\left[\left(\frac{1}{2} - x\right)\left(-\frac{1}{2n\pi}\cos\frac{2n\pi x}{l}\right) - (-1)\left(-\frac{l^2}{4n^2\pi^2}\sin\frac{2n\pi x}{l}\right)\right]_0^l$$

$$= \frac{2}{l}\left[\frac{l}{2}\,\frac{l}{2n\pi}\cos 2\,n\,\pi - 0 + \frac{l}{2}\cdot\frac{l}{2n\pi}(1)\right] = \frac{2}{l}\left[\frac{l^2}{2n\pi}\right] = \frac{l}{n\pi}$$

Fourier series is

$$f(x) = \frac{a_0}{2} + a_1\cos\frac{\pi x}{l/2} + a_2\cos\frac{2\pi x}{l/2} + a_3\cos\frac{3\pi x}{l/2} + \ldots$$

$$+ b_1\sin\frac{\pi x}{l/2} + b_2\sin\frac{2\pi x}{l/2} + b_3\sin\frac{3\pi x}{l/2} + \ldots$$

$$\frac{l}{2} - x = \frac{l}{\pi}\sin\frac{2\pi x}{l} + \frac{l}{2\pi}\sin\frac{4\pi x}{l} + \frac{l}{3\pi}\sin\frac{6\pi x}{l} + \ldots$$

$$= \frac{l}{\pi}\sum_1^\infty \frac{1}{n}\sin\frac{2n\pi x}{l}$$ **Proved.**

40.13 HALF PERIOD SERIES

Cosine series: $f(x) = \frac{a_0}{2} + a_1\cos\frac{\pi x}{c} + a_2\cos\frac{2\pi x}{c} + \ldots + a_n\cos\frac{n\pi x}{c} + \ldots$

where $a_0 = \frac{2}{c}\int_0^c f(x)\,dx,\ a_n = \frac{2}{c}\int_0^c f(x)\cos\frac{n\pi x}{c}\,dx$

Sine series: $f(x) = b_1\sin\frac{\pi x}{c} + b_2\sin\frac{2\pi x}{c} + \ldots + b_n\sin\frac{n\pi x}{c} + \ldots$

where $b_n = \frac{2}{c}\int_0^c f(x)\sin\frac{n\pi x}{c}\,dx.$

Example 18. *Expand for $f(x) = k$ for $0 < x < 2$ in a half range sine series.*

(*U.P., II Semester, June 2007*)

Solution. $f(x) = k$

$$b_n = \frac{2}{c}\int_0^c f(x)\cdot\sin\frac{n\pi x}{c}\,dx \text{ in half range } (0, c) = \frac{2}{2}\int_0^2 k\sin\frac{n\pi x}{2}\,dx$$

$$= k\,\frac{2}{n\pi}\left(-\cos\frac{n\pi x}{2}\right)_0^2 = \frac{2k}{n\pi}[-\cos n\pi + 1]$$

Half range sine series is

$$f(x) = \sum_{n=1}^\infty b_n\sin\frac{n\pi x}{2}$$

$$k = \sum_{n=1}^\infty \frac{2k}{n\pi}[1 - \cos n\,\pi]\sin\frac{n\pi x}{2}$$ **Ans.**

Example 19. *Obtain the half-range sine series for the function $f(x) = x^2$ in the interval $0 < x < 3$.* (*U.P., II Semester, Summer 2002*)

Solution. We know that half range sine series is given by $f(x) = \Sigma b_n \sin nx$

Where $\quad b_n = \frac{2}{c}\int_0^c f(x)\sin\frac{n\pi x}{c}\,dx$ in the half-range $(0, c)$.

Here, we have half range $0 < x < 3$ and $f(x) = x^2$

$$b_n = \frac{2}{3}\int_0^3 x^2 \sin\frac{n\pi x}{3}\,dx$$

$$= \frac{2}{3}\left[x^2\left(\frac{3}{n\pi}\right)\left(-\cos\frac{n\pi x}{3}\right) + 2x\times\left(\frac{3}{n\pi}\right)\left(\frac{3}{n\pi}\right)\sin\frac{n\pi x}{3} - 2\left(\frac{3}{n\pi}\right)\left(\frac{3}{n\pi}\right)\left(\frac{3}{n\pi}\right)\left(-\cos\frac{n\pi x}{3}\right)\right]_0^3$$

$$\Rightarrow \quad b_n = \frac{2}{3}\left[\left\{-\frac{27}{n\pi}(-1)^n - \frac{54}{n^3\pi^3}(-1)^n\right\} + \frac{54}{n^3\pi^3}\right]$$

$$\Rightarrow \quad b_n = \frac{2}{3}\left[\frac{54}{n^3\pi^3}\{1-(-1)^n\} - \frac{27}{n\pi}(-1)^n\right] \Rightarrow b_n = \frac{2}{3}\left[\frac{108}{n^3\pi^3} + \frac{27}{n\pi}\right] \text{ when } n \text{ is odd}$$

And $b_n = -\frac{18}{n\pi}$ when n is even

$\therefore$ Half range sine series

$$f(x) = b_1\sin x + b_2\sin 2x + b_3 \sin 3x + \ldots$$

$$= \frac{2}{3}\left[\frac{108}{\pi^3}\left(\frac{\sin x}{1^3} + \frac{\sin 3x}{3^3} + \frac{\sin 5x}{5^3} + \ldots\right) + \frac{27}{\pi}\left(\frac{\sin x}{1} + \frac{\sin 3x}{3} + \frac{\sin 5x}{5} + \ldots\right)\right]$$

$$- \frac{18}{\pi}\left(\frac{\sin 2x}{2} + \frac{\sin 4x}{4} + \ldots\right) \quad \textbf{Ans.}$$

Example 20. *Expand $f(x) = e^x$ in a cosine series over $(0, 1)$.*

Solution. Here, we have $f(x) = e^x$ and $c = 1$

$$a_0 = \frac{2}{c}\int_0^c f(x)\,dx = \frac{2}{1}\int_0^1 e^x\,dx = 2(e-1)$$

$$a_n = \frac{2}{c}\int_0^c f(x)\cos\frac{n\pi x}{c}\,dx = \frac{2}{1}\int_0^1 e^x \cos\frac{n\pi x}{1}\,dx$$

$$= 2\left[\frac{e^x}{n^2\pi^2+1}(n\pi\sin n\pi x + \cos n\pi x)\right]_0^1$$

$$= 2\left[\frac{e^1}{n^2\pi^2+1}(n\pi\sin n\pi + \cos n\pi) - \frac{1}{n^2\pi^2+1}\right]$$

$$= \frac{2}{n^2\pi^2+1}[(-1)^n e - 1]$$

$$f(x) = \frac{a_0}{2} + a_1\cos\pi x + a_2\cos 2\pi x + a_3\cos 3\pi x + \ldots$$

$$e^x = e - 1 + 2\left[\frac{-e-1}{\pi^2+1}\cos\pi x + \frac{e-1}{4\pi^2+1}\cos 2\pi x + \frac{-e-1}{9\pi^2+1}\cos 3\pi x + \ldots\right] \quad \textbf{Ans.}$$

Example 21. *Find the Fourier half-range cosine series of the function*

$$f(t) = \begin{cases} 2t, & 0 < t < 1 \\ 2(2-t), & 1 < t < 2 \end{cases}$$ *(U.P., II Semester, Summer 2007, 2006, 2001)*

Solution. $f(t) = \begin{cases} 2t, & 0 < t < 1 \\ 2(2-t), & 1 < t < 2 \end{cases}$

Let $$f(t) = \frac{a_0}{2} + a_1 \cos \frac{\pi t}{c} + a_2 \cos \frac{2\pi t}{c} + a_3 \cos \frac{3\pi t}{c} + \ldots$$

$$+ b_1 \sin \frac{\pi t}{c} + b_2 \sin \frac{2\pi t}{c} + b_3 \sin \frac{3\pi t}{c} + \ldots \quad \ldots(1)$$

Here, $c = 2$, because it is half range series.

Hence, $a_0 = \frac{2}{c} \int_0^c f(t)\, dt = \frac{2}{2} \int_0^1 2t\, dt + \frac{2}{2} \int_1^2 2(2-t)\, dt$

$$= [t^2]_0^1 + \left[2\left(2t - \frac{t^2}{2} \right) \right]_1^2 = 1 + [4t - t^2]_1^2 = 1 + (8 - 4 - 4 + 1) = 2$$

$$a_n = \frac{2}{c} \int_0^c f(t) \cos \frac{n\pi t}{c}\, dt = \frac{2}{2} \int_0^1 2t \cos \frac{n\pi t}{2}\, dt + \frac{2}{2} \int_1^2 2(2-t) \cos \frac{n\pi t}{2}\, dt$$

$$= \left[2t \left(\frac{2}{n\pi} \sin \frac{n\pi t}{2} \right) - (2)\left(-\frac{4}{n^2\pi^2} \cos \frac{n\pi t}{2} \right) \right]_0^1$$

$$+ \left[(4 - 2t) \left(\frac{2}{n\pi} \sin \frac{n\pi t}{2} \right) - (-2)\left(-\frac{4}{n^2\pi^2} \cos \frac{n\pi t}{2} \right) \right]_1^2$$

$$= \left[\frac{4}{n\pi} \sin \frac{n\pi}{2} + \frac{8}{n^2\pi^2} \cos \frac{n\pi}{2} - \frac{8}{n^2\pi^2} \right] + \left[0 - \frac{8}{n^2\pi^2} \cos n\pi - \frac{4}{n\pi} \sin \frac{n\pi}{2} + \frac{8}{n^2\pi^2} \cos \frac{n\pi}{2} \right]$$

$$= \frac{16}{n^2\pi^2} \cos \frac{n\pi}{2} - \frac{8}{n^2\pi^2} - \frac{8}{n^2\pi^2} \cos n\pi = \frac{8}{n^2\pi^2} \left[2\cos \frac{n\pi}{2} - 1 - \cos n\pi \right]$$

$$f(t) = 1 + \frac{8}{\pi^2} \sum_{n=1}^{\infty} \left[2\cos \frac{n\pi}{2} - 1 - \cos n\pi \right] \cos \frac{n\pi t}{2}$$ **Ans.**

Example 22. *Obtain the Fourier cosine series expansion of the periodic function defined by*

$f(t) = \sin\left(\frac{\pi t}{l} \right),\ 0 < t < l$ *(U.P., II Semester, Summer 2001)*

Solution. We have, $f(t) = \sin\left(\frac{\pi t}{l} \right),\ 0 < t < l$

$$a_0 = \frac{2}{l}\int_0^l \sin\left(\frac{\pi t}{l}\right) dt = \frac{2}{l}\left(-\frac{l}{\pi}\cos\frac{\pi t}{l}\right)_0^l = -\frac{2}{\pi}(\cos\pi - \cos 0) = -\frac{2}{\pi}(-1-1) = \frac{4}{\pi}$$

$$a_n = \frac{2}{l}\int_0^l \sin\left(\frac{\pi t}{l}\right)\cos\frac{n\pi t}{l}\, dt = \frac{1}{l}\int_0^l \left[\sin\left(\frac{\pi t}{l}+\frac{n\pi t}{l}\right) - \sin\left(\frac{n\pi t}{l}-\frac{\pi t}{l}\right)\right] dt$$

$$= \frac{1}{l}\int_0^l \sin(n+1)\frac{\pi t}{l}\, dt - \frac{1}{l}\int_0^l \sin(n-1)\frac{\pi t}{l}\, dt$$

$$= \frac{1}{l}\left[-\frac{l}{(n+1)\pi}\cos\frac{(n+1)\pi t}{l}\right]_0^l - \frac{1}{l}\left[\frac{l}{(n-1)\pi}\cos\frac{(n-1)\pi t}{l}\right]_0^l$$

$$= \frac{-1}{(n+1)\pi}[\cos(n+1)\pi - \cos 0] + \frac{1}{(n-1)\pi}[\cos(n-1)\pi - \cos 0]$$

$$= \frac{1}{(n+1)\pi}[(-1)^{n+1} - 1] + \frac{1}{(n-1)\pi}[(-1)^{n+1} - 1]$$

$$= (-1)^{n+1}\left[\frac{1}{(n+1)\pi} + \frac{1}{(n-1)\pi}\right] + \frac{1}{(n+1)\pi} - \frac{1}{(n-1)\pi}$$

$$= (-1)^{n+1}\frac{2}{(n^2-1)\pi} - \frac{2}{(n^2-1)\pi} = \frac{2}{(n^2-1)\pi}[(-1)^{n+1} - 1]$$

$$= \frac{-4}{(n^2-1)\pi}, \quad \text{when } n \text{ is even}$$

$$= 0, \quad \text{when } n \text{ is odd.}$$

The above formula for finding the value of a_1 is not applicable.

$$a_1 = \frac{2}{l}\int_0^l \sin\frac{\pi t}{l}\cos\frac{\pi t}{l}\, dt = \frac{1}{l}\int_0^l \sin\frac{2\pi t}{l}\, dt$$

$$= \frac{1}{l}\left(-\frac{l}{2\pi}\cos\frac{2\pi t}{l}\right)_0^l = -\frac{l}{2\pi l}(\cos 2\pi - \cos 0) = -\frac{1}{2\pi}(1-1) = 0$$

$$f(t) = \frac{a_0}{2} + a_1\cos\frac{\pi t}{l} + a_2\cos\frac{2\pi t}{l} + a_3\cos\frac{3\pi t}{l} + a_4\cos\frac{4\pi t}{l} + \ldots$$

$$= \frac{2}{\pi} - \frac{4}{\pi}\left[\frac{1}{3}\cos\frac{2\pi t}{l} + \frac{1}{15}\cos\frac{4\pi t}{l} + \frac{1}{35}\cos\frac{6\pi t}{l} + \ldots\right]$$ **Ans.**

Example 23. *Find the Fourier cosine series expansion of the periodic function of period 1*

$$f(x) = \begin{cases} \frac{1}{2} + x, & -\frac{1}{2} < x \le 0 \\ \frac{1}{2} - x, & 0 < x < \frac{1}{2} \end{cases}$$

Solution. Let $f(x) = \frac{a_0}{2} + a_1\cos\frac{\pi x}{c} + a_2\cos\frac{2\pi x}{c} + \ldots$ ...(1)

as $f(x)$ is a cosine series.

Here $2c = 1 \Rightarrow c = \dfrac{1}{2}$

$$a_0 = \frac{1}{c}\int_{-c}^{c} f(x)\,dx = \frac{1}{1/2}\int_{-1/2}^{0}\left(\frac{1}{2}+x\right)dx + \frac{1}{1/2}\int_{0}^{1/2}\left(\frac{1}{2}-x\right)dx$$

$$= 2\left[\frac{x}{2}+\frac{x^2}{2}\right]_{-1/2}^{0} + 2\left[\frac{x}{2}-\frac{x^2}{2}\right]_{0}^{1/2} = 2\left[\frac{1}{4}-\frac{1}{8}\right]+2\left[\frac{1}{4}-\frac{1}{8}\right]=\frac{1}{2}$$

$$a_n = \frac{1}{c}\int_{-c}^{c} f(x)\cos\frac{n\pi x}{c}\,dx$$

$$= \frac{1}{1/2}\int_{-1/2}^{0}\left(\frac{1}{2}+x\right)\cos\frac{n\pi x}{1/2}\,dx + \frac{1}{1/2}\int_{0}^{1/2}\left(\frac{1}{2}-x\right)\cos\frac{n\pi x}{1/2}\,dx$$

$$= 2\int_{-1/2}^{0}\left(\frac{1}{2}+x\right)\cos 2n\pi x\,dx + 2\int_{0}^{1/2}\left(\frac{1}{2}-x\right)\cos 2n\pi x\,dx$$

$$= 2\left[\left(\frac{1}{2}+x\right)\frac{\sin 2n\pi x}{2n\pi} - (1)\left(-\frac{\cos 2n\pi x}{4n^2\pi^2}\right)\right]_{-1/2}^{0}$$

$$+ 2\left[\left(\frac{1}{2}-x\right)\frac{\sin 2n\pi x}{2n\pi} - (-1)\left(\frac{-\cos 2n\pi x}{4n^2\pi^2}\right)\right]_{0}^{1/2}$$

$$= 2\left[0+\frac{1}{4n^2\pi^2}-\frac{(-1)^n}{4n^2\pi^2}\right]+2\left[0-\frac{(-1)^n}{4n^2\pi^2}+\frac{1}{4n^2\pi^2}\right]=\frac{1}{\pi^2}\left[\frac{1}{n^2}-\frac{(-1)^n}{n^2}\right]$$

$$= \frac{2}{n^2\pi^2} \qquad \text{(if } n \text{ is odd)}$$

$$= 0 \qquad \text{(if } n \text{ is even)}$$

Substituting the values of $a_0, a_1, a_2, a_3, \ldots$ in (1), we have

$$f(x) = \frac{1}{4}+\frac{2}{\pi^2}\left[\frac{\cos 2\pi x}{1^2}+\frac{\cos 6\pi x}{3^2}+\frac{\cos 10\pi x}{5^2}+\ldots\right]$$ **Ans.**

Example 24. *Let* $f(x) = \begin{cases} wx, & \text{where } 0 \le x \le \dfrac{l}{2} \\ w(l-x), & \text{where } \dfrac{l}{2} \le x \le l \end{cases}$

Show that $$f(x) = \frac{4wl}{\pi^2}\sum_{n=0}^{\infty}\frac{(-1)^n}{(2n+1)^2}\sin\frac{(2n+1)\pi x}{l}$$

Hence, obtain the sum of the series

$$1+\frac{1}{3^2}+\frac{1}{5^2}+\ldots$$ *(U.P., Second Semester 2003)*

Solution. Half range sine series is

$$f(x) = \sum_{n=1}^{\infty} b_n \sin\frac{n\pi x}{l} \qquad \ldots(1)$$

where, $$b_n = \frac{2}{l}\int_0^l f(x)\sin\frac{n\pi x}{l}\,dx$$

$$b_n = \frac{2}{l}\left[\int_0^{\frac{l}{2}} f(x)\sin\frac{n\pi x}{l}\,dx + \int_{\frac{l}{2}}^{l} f(x)\sin\frac{n\pi x}{l}\,dx\right]$$

$$= \frac{2}{l}\left[\int_0^{\frac{l}{2}} wx \sin\frac{n\pi x}{l}\,dx + \int_{\frac{l}{2}}^{l} w(l-x)\sin\frac{n\pi x}{l}\,dx\right]$$

$$= \frac{2}{l}\left[\left\{wx\frac{\left(-\cos\frac{n\pi x}{l}\right)}{\left(\frac{n\pi}{l}\right)} - w\frac{\left(-\sin\frac{n\pi x}{l}\right)}{\left(\frac{n\pi}{l}\right)^2}\right\}_0^{\frac{l}{2}} + \left\{w(l-x)\frac{\left(-\cos\frac{n\pi x}{l}\right)}{\left(\frac{n\pi}{l}\right)} - w(-1)\frac{\left(-\sin\frac{n\pi x}{l}\right)}{\left(\frac{n\pi}{l}\right)^2}\right\}_{\frac{l}{2}}^{l}\right]$$

$$= \frac{2}{l}\left[\left\{\frac{wl}{2}\frac{\left(-\cos\frac{n\pi}{2}\right)}{\left(\frac{n\pi}{l}\right)} + \frac{w\sin\frac{n\pi}{2}}{\left(\frac{n\pi}{l}\right)^2} - 0 - 0\right\} - \left\{\frac{wl}{2}\frac{\left(-\cos\frac{n\pi}{2}\right)}{\left(\frac{n\pi}{l}\right)} - \frac{w\sin\frac{n\pi}{2}}{\left(\frac{n\pi}{l}\right)^2}\right\}\right]$$

$$= \frac{2w}{l}\left[-\frac{l^2}{2n\pi}\cos\frac{n\pi}{2} + \frac{l^2}{n^2\pi^2}\sin\frac{n\pi}{2} + \frac{l^2}{2n\pi}\cos\frac{n\pi}{2} + \frac{l^2}{n^2\pi^2}\sin\frac{n\pi}{2}\right]$$

$$= \frac{2w}{l}\left[\frac{l^2}{n^2\pi^2}\sin\frac{n\pi}{2} + \frac{l^2}{n^2\pi^2}\sin\frac{n\pi}{2}\right] = \frac{4wl^2}{ln^2\pi^2}\sin\frac{n\pi}{2}$$

$$b_n = \frac{4wl}{n^2\pi^2}\sin\frac{n\pi}{2}, \text{ when } n \text{ is odd.}$$

$$b_n = 0, \quad \text{when } n \text{ is even.}$$

Now, putting the value of b_n in (1), we get

$$f(x) = \sum_{n=1}^{\infty}\frac{4\,wl}{n^2\pi^2}\sin\frac{n\pi}{2}\sin\frac{n\pi x}{l}, \text{ when } n \text{ is odd}$$

$$f(x) = \sum_{n=1}^{\infty}\frac{4\,wl}{(2n+1)^2\pi^2}\sin\frac{(2n+1)\pi}{2}\sin\frac{(2n+1)\pi x}{l}$$

$n \to 2n+1$

$\sin(2n+1)\frac{\pi}{2}$

$= \sin\left(n\pi + \frac{\pi}{2}\right)$

$= \cos n\pi = (-1)^n$

$$\boxed{f(x) = \frac{4wl}{\pi^2}\sum_{n=0}^{\infty}\frac{(-1)^n}{(2n+1)^2}\sin\frac{(2n+1)\pi x}{l}}$$

$$f(x) = \frac{4wl}{\pi^2}\left[\frac{1}{1^2}\sin\frac{\pi x}{l} - \frac{1}{3^2}\sin\frac{3\pi x}{l} + \frac{1}{5^2}\sin\frac{5\pi x}{l} + \ldots\right] \qquad \ldots(2)$$

Putting $x = \frac{l}{2}$, $f(x) = wx$ and $f\left(\frac{l}{2}\right) = \frac{wl}{2}$ in (2), we get

$$\frac{wl}{2} = \frac{4wl}{\pi^2}\left[\frac{1}{1^2}\sin\frac{\pi}{2} - \frac{1}{3^2}\sin\frac{3\pi}{2} + \frac{1}{5^2}\sin\frac{5\pi}{2} + \ldots\right]$$

$$\Rightarrow \frac{\pi^2}{8} = \frac{1}{1^2} + \frac{1}{3^2} + \frac{1}{5^2} + \ldots$$ **Ans.**

Example 25. *Find the half period sine series for f (x) given in the range (0, l) by the graph OPQ as shown in figure.* *(U.P. II semester, 2009)*

Solution. The equation of line *OP* is $y - 0 = \dfrac{0-k}{l-\dfrac{l}{2}}(x-l) \Rightarrow y = \dfrac{2kx}{l} + 2k$

and the equation of the line *PQ* is $y = -\dfrac{kx}{\dfrac{l}{2}} \Rightarrow y = -\dfrac{2kx}{l}$

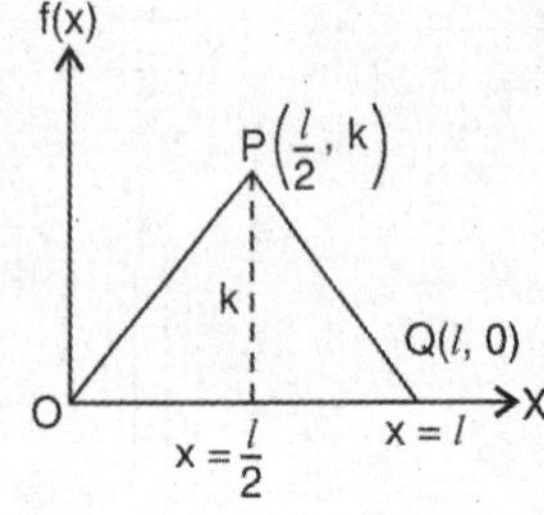

$f(x)$ is the half period

$$f(x) = \begin{cases} \dfrac{2kx}{l}, & 0 < x < \dfrac{l}{2} \\ -\dfrac{2kx}{l} + 2k, & \dfrac{l}{2} < x < l \end{cases}$$

$f(x)$ is half period series. It is to be expanded as sine series.
Here, $a_0 = 0$ and $a_n = 0$

$$b_n = \frac{2}{l}\int_0^{\frac{l}{2}} f(x)\sin\frac{n\pi x}{l}\,dx + \frac{2}{l}\int_{\frac{l}{2}}^{l} f(x).\sin\frac{n\pi x}{l}\,dx$$

$$= \frac{2}{l}\int_0^{\frac{l}{2}} \frac{2kx}{l}\sin\frac{n\pi x}{l}\,dx + \frac{2}{l}\int_{\frac{l}{2}}^{l}\left(\frac{-2kx}{l} + 2k\right)\sin\frac{n\pi x}{l}\,dx$$

$$= \frac{4k}{l^2}\int_0^{\frac{l}{2}} x\sin\frac{n\pi x}{l}\,dx + \frac{4k}{l^2}\int_{\frac{l}{2}}^{l}(-x+l)\sin\frac{n\pi x}{l}\,dx$$

$$= \frac{4k}{l^2}\left[x\left(\frac{-l}{n\pi}\cos\frac{n\pi x}{l}\right) - \left(\frac{-l^2}{n^2\pi^2}\sin\frac{n\pi x}{l}\right)\right]_0^{\frac{l}{2}} + \frac{4k}{l^2}\left[(-x+l)\left(\frac{-l}{n\pi}\cos\frac{n\pi x}{l}\right) - (-1)\left(\frac{-l^2}{n^2\pi^2}\sin\frac{n\pi x}{l}\right)\right]_{\frac{l}{2}}^{l}$$

$$= \frac{4k}{l^2}\left[-\frac{l}{2}\left(\frac{l}{n\pi}\right)\cos\frac{n\pi\frac{l}{2}}{l} + \frac{l^2}{n^2\pi^2}\sin\frac{n\pi\frac{l}{2}}{l}\right]$$

$$+ \frac{4k}{l^2}\left[(-l+l)\left(-\frac{l}{n\pi}\cos n\pi\right) - \frac{l^2}{n^2\pi^2}\sin n\pi - \left(-\frac{l}{2}+l\right)\left(-\frac{l}{n\pi}\cos\frac{n\pi\frac{l}{2}}{l}\right) + \frac{l^2}{n^2\pi^2}\sin\frac{n\pi\frac{l}{2}}{l}\right]$$

$$= \frac{4k}{l^2}\left[-\frac{l^2}{2n\pi}\cos\frac{n\pi}{2} + \frac{l^2}{n^2\pi^2}\sin\frac{n\pi}{2}\right] + \frac{4k}{l^2}\left[0 - 0 + \frac{l^2}{2n\pi}\cos\frac{n\pi}{2} + \frac{l^2}{n^2\pi^2}\sin\frac{n\pi}{2}\right]$$

$$= \frac{4k}{l^2}\left(\frac{l^2}{2n\pi}\right)\left[-\cos\frac{n\pi}{2} + \frac{2}{n\pi}\sin\frac{n\pi}{2}\right] + \frac{4k}{l^2}\left[\frac{l^2}{2n\pi}\cos\frac{n\pi}{2} + \frac{l^2}{n^2\pi^2}\sin\frac{n\pi}{2}\right]$$

$$= \frac{4k}{l^2}\left(\frac{l^2}{2n\pi}\right)\left[-\cos\frac{n\pi}{2} + \frac{2}{n\pi}\sin\frac{n\pi}{2} + \cos\frac{n\pi}{2} + \frac{2}{n\pi}\sin\frac{n\pi}{2}\right]$$

$$= \frac{2k}{n\pi}\left[\frac{4}{n\pi}\sin\frac{n\pi}{2}\right] = \frac{8k}{n^2\pi^2}\sin\frac{n\pi}{2}$$

Hence, Fourier series of $f(x)$ is

$$f(x) = \frac{8k}{\pi^2}\sum_{n=1}^{\infty}\frac{1}{n^2}\sin\frac{n\pi}{2}\sin\frac{n\pi x}{l}$$ **Ans.**

EXERCISE 40.4

1. Find the Fourier series to represent $f(x)$, where

$$f(x) = \begin{cases} -a, & -c < x < 0 \\ a, & 0 < x < c \end{cases}$$ **Ans.** $\frac{4a}{\pi}\left[\sin\frac{\pi x}{c} + \frac{1}{3}\sin\frac{3\pi x}{c} + \frac{1}{5}\sin\frac{5\pi x}{c} + \ldots\right]$

2. Find the half-range sine series for the function

$f(x) = 2x - 1 \quad 0 < x < 1$ **Ans.** $-\frac{2}{\pi}\left[\sin 2\pi x + \frac{1}{2}\sin 4\pi x + \frac{1}{3}\sin 6\pi x + \ldots\right]$

3. Express $f(x) = x$ as a cosine, half range series in $0 < x < 2$.

Ans. $1 - \frac{8}{\pi^2}\left[\frac{1}{1^2}\cos\frac{\pi x}{2} + \frac{1}{3^2}\cos\frac{3\pi x}{2} + \frac{1}{5^2}\cos\frac{5\pi x}{2} + \ldots\right]$

4. Find the Fourier series of the function

$$f(x) = \begin{cases} -2 & \text{for } -4 < x < -2 \\ x & \text{for } -2 < x < 2 \\ 2 & \text{for } 2 < x < 4 \end{cases}$$

Ans. $\frac{4}{\pi} + \frac{8}{\pi^2}\sin\frac{\pi x}{4} - \frac{2}{\pi}\sin\frac{2\pi x}{4} + \left(\frac{4}{3\pi} - \frac{8}{3^2\pi}\right)\sin\frac{3\pi x}{4} - \frac{2}{2\pi}\sin\frac{4\pi x}{4} + \ldots$

5. Find the Fourier series to represent
$f(x) = x^2 - 2$ from $-2 < x < 2$.

Ans. $-\frac{2}{3} - \frac{16}{\pi^2}\left[\cos\frac{\pi x}{2^2} - \frac{1}{4}\cos\pi x + \frac{1}{9}\cos\frac{3\pi x}{2} + \ldots\right]$

6. If $f(x) = e^{-x}$, $-c < x < c$, show that

$$f(x) = (e^c - e^{-c})\left\{\frac{1}{2c} - c\left(\frac{1}{c^2+\pi^2}\cos\frac{\pi x}{c} - \frac{1}{c^2+4\pi^2}\cos\frac{2\pi x}{c} + \ldots\right)\right.$$
$$\left.- \pi\left(\frac{1}{c^2+\pi^2}\sin\frac{\pi x}{c} - \frac{2}{c^2+4\pi^2}\sin\frac{2\pi x}{c}\ldots\right)\right\}$$ *(MDU, Dec. 2010)*

7. A sinusoidal voltage E sin ω t is passed through a half wave rectifier which clips the negative portion of the wave. Develop the resulting portion of the function

$$u(t) = \begin{cases} 0, & \text{when } -\frac{T}{2} < t < 0 \\ E\sin\omega t, & \text{when } 0 < t < \frac{T}{2} \end{cases} \qquad \left(T = \frac{2\pi}{\omega}\right)$$

Ans. $\frac{E}{\pi} + \frac{E}{2}\sin\omega t - \frac{2E}{\pi}\left[\frac{1}{1.3}\cos 2\omega t + \frac{1}{3.5}\cos 4\omega t + \frac{1}{5.7}\cos 6\omega t + \ldots\right]$

8. A periodic square wave has a period 4. The function generating the square is

$$f(t) = \begin{cases} 0 & \text{for } -2 < t < -1 \\ k & \text{for } -1 < t < 1 \\ 0 & \text{for } 1 < t < 2 \end{cases}$$

Find the Fourier series of the function. **Ans.** $f(t) = \frac{k}{2} + \frac{2k}{\pi}\left[\cos\frac{\pi t}{2} - \frac{1}{3}\cos\frac{3\pi t}{2} + \ldots\right]$

9. Find a Fourier series to represent x^2 in the interval $(-l, l)$.

Ans. $\frac{l^2}{3} - \frac{4l^2}{\pi^2}\left[\cos\pi x - \frac{\cos\pi x}{2^2} + \frac{\cos 3\pi x}{3^2}\right]$

40.14. PARSEVAL'S FORMULA

$$\int_{-c}^{c}[f(x)]^2\,dx = c\left\{\frac{1}{2}a_0^2+\sum_{n=1}^{\infty}(a_n^2+b_n^2)\right\}$$

We know that $f(x) = \dfrac{a_0}{2}+\sum_{n=1}^{\infty}\left(a_n\cos\dfrac{n\pi x}{c}+b_n\sin\dfrac{n\pi x}{c}\right)$ ...(1)

Multiplying (1) by $f(x)$, we get

$$[f(x)]^2 = \frac{a_0}{2}f(x)+\sum_{n=1}^{\infty}a_n\,f(x)\cos\frac{n\pi x}{c}+\sum_{n=1}^{\infty}b_n\,f(x)\sin\frac{n\pi x}{c} \quad ..(2)$$

Intergrating term by term from $-c$ to c, we have

$$\int_{-c}^{c}[f(x)]^2\,dx = \frac{a_0}{2}\int_{-c}^{c}f(x)\,dx+\sum_{n=1}^{\infty}a_n\int_{-c}^{c}f(x)\cos\frac{n\pi x}{c}\,dx$$
$$+\sum_{n=1}^{\infty}b_n\int_{-c}^{c}f(x)\sin\frac{n\pi x}{c}\,dx \quad ...(3)$$

In article 40.12, we have the following results

$$\int_{-c}^{c}f(x)\,dx = c\,a_0$$

$$\int_{-c}^{c}f(x)\cos\frac{n\pi x}{c}\,dx = c\,a_n$$

$$\int_{-c}^{c}f(x)\sin\frac{n\pi x}{c}\,dx = c\,b_n$$

On putting these integrals in (3), we get

$$\int_{-c}^{c}[f(x)]^2\,dx = c\frac{a_0^2}{2}+\sum_{n=1}^{\infty}c\,a_n^2+\sum_{n=1}^{\infty}c\,b_n^2 = c\left[\frac{a_0^2}{2}+\sum_{n=1}^{\infty}(a_n^2+b_n^2)\right]$$

This is the Parseval's formula.

Note. 1. If $0<x<2c$, then $\int_0^{2c}[f(x)]^2\,dx = c\left[\dfrac{a_0^2}{2}+\sum_{n=1}^{\infty}(a_n^2+b_n^2)\right]$

2. If $0<x<c$ (Half range consine series), $\int_0^{c}[f(x)]^2\,dx = \dfrac{c}{2}\left[\dfrac{a_0^2}{2}+\sum_{n=1}^{\infty}a_n^2\right]$

3. If $0<x<c$ (Half range sine series), $\int_0^{c}[f(x)]^2\,dx = \dfrac{c}{2}\left[\sum_{n=1}^{\infty}b_n^2\right]$

4. R.M.S. $= \left\{\dfrac{\int_a^b[f(x)]^2\,dx}{b-a}\right\}^{\frac{1}{2}}$

Example 26. *By using the series for $f(x)=1$ in $0<x<\pi$ show that*

$$\frac{\pi^2}{8} = 1+\frac{1}{3^2}+\frac{1}{5^2}+\frac{1}{7^2}+\ldots$$

Solution. Sine series is $f(x)=\Sigma\,b_n\sin nx$

$$b_n = \frac{2}{\pi}\int_0^{\pi}f(x)\sin nx\,dx$$

$$= \frac{2}{\pi}\int_0^{\pi}(1)\sin nx\,dx = \frac{2}{\pi}\left(\frac{-\cos nx}{n}\right)_0^{\pi} = \frac{-2}{n\pi}[\cos n\pi-1] = \frac{-2}{n\pi}[(-1)^n-1]$$

$$= \frac{4}{n\pi} \text{ if } n \text{ is odd} \quad = 0 \quad \text{if } n \text{ is even}$$

Then the sine series is

$$1 = \frac{4}{\pi}\sin x + \frac{4}{3\pi}\sin 3x + \frac{4}{5\pi}\sin 5x + \frac{4}{7\pi}\sin 7x + ...$$

$$\int_0^c [f(x)]^2\, dx = \frac{c}{2}\left[b_1^2 + b_2^2 + b_3^2 + b_4^2 + b_5^2 + ...\right]$$

$$\int_0^\pi (1)^2\, dx = \frac{\pi}{2}\left[\left(\frac{4}{\pi}\right)^2 + \left(\frac{4}{3\pi}\right)^2 + \left(\frac{4}{5\pi}\right)^2 + \left(\frac{4}{7\pi}\right)^2 + ...\right]$$

$$[x]_0^\pi = \left(\frac{\pi}{2}\right)\left(\frac{16}{\pi^2}\right)\left[1 + \frac{1}{3^2} + \frac{1}{5^2} + \frac{1}{7^2} + ...\right]$$

$$\pi = \frac{\pi}{2}\left(\frac{16}{\pi^2}\right)\left[1 + \frac{1}{3^2} + \frac{1}{5^2} + \frac{1}{7^2} + ...\right]$$

$$\frac{\pi^2}{8} = 1 + \frac{1}{3^2} + \frac{1}{5^2} + \frac{1}{7^2} + ...$$ **Proved.**

Example 27. *If* $f(x) = \begin{cases} \pi x & 0 < x < 1 \\ \pi(2 - x), & 1 < x < 2 \end{cases}$

using half range cosine series, show that

$$\frac{1}{1^4} + \frac{1}{3^4} + \frac{1}{5^4} + ... = \frac{\pi^4}{96}$$

Solution. Half range consine series is

$$f(x) = \frac{a_0}{2} + \Sigma\, a_n \cos\frac{n\pi x}{c}$$

where $$a_0 = \frac{2}{c}\int_0^c f(x)\, dx = \frac{2}{2}\left[\int_0^1 \pi x\, dx + \int_1^2 \pi(2 - x)\, dx\right]$$

$$= \pi\left(\frac{x^2}{2}\right)_0^1 + \pi\left(2x - \frac{x^2}{2}\right)_1^2 = \frac{\pi}{2} + \pi\left[(4-2) - \left(2 - \frac{1}{2}\right)\right] = \pi$$

$$a_n = \frac{2}{c}\int_0^c f(x)\cos\frac{n\pi x}{c}\, dx$$

$$= \frac{2}{2}\left[\int_0^1 \pi x\cos\frac{n\pi x}{2}\, dx + \int_1^2 \pi(2 - x)\cos\frac{n\pi x}{2}\, dx\right]$$

$$= \pi\left[x\frac{\sin\frac{n\pi x}{2}}{\frac{n\pi}{2}} - \left(\frac{-\cos\frac{n\pi x}{2}}{\frac{n^2\pi^2}{4}}\right)\right]_0^1 + \pi\left[(2 - x)\frac{\sin\frac{n\pi x}{2}}{\frac{n\pi}{2}} - (-1)\left(\frac{-\cos\frac{n\pi x}{2}}{\frac{n^2\pi^2}{4}}\right)\right]_1^2$$

$$= \pi\left[\frac{2}{n\pi}\sin\frac{n\pi}{2} + \frac{4}{n^2\pi^2}\cos\frac{n\pi}{2} - \frac{4}{n^2\pi^2}\right] + \pi\left[0 - \frac{4}{n^2\pi^2}\cos n\pi - \frac{2}{n\pi}\sin\frac{n\pi}{2} + \frac{4}{n^2\pi^2}\cos\frac{n\pi}{2}\right]$$

$$= \pi\left[\frac{8}{n^2\pi^2}\cos\frac{n\pi}{2} - \frac{4}{n^2\pi^2} - \frac{4}{n^2\pi^2}\cos n\pi\right] = \frac{4}{n^2\pi}\left[2\cos\frac{n\pi}{2} - 1 - \cos n\pi\right]$$

$$a_1 = 0,\ a_2 = \frac{-4}{\pi},\ a_3 = 0,\ a_4 = 0,\ a_5 = 0,\ a_6 = \frac{-4}{9\pi} \ldots$$

$$\int_0^c [f(x)]^2\, dx = \frac{c}{2}\left[\frac{a_0^2}{2} + a_1^2 + a_2^2 + a_3^2 + \ldots\right]$$

$$\int_0^1 (\pi x)^2\, dx + \int_1^2 \pi^2 (2-x)^2\, dx = \frac{2}{2}\left[\frac{\pi^2}{2} + \frac{16}{\pi^2} + \frac{16}{81\pi^2} + \ldots\right]$$

$$\pi^2\left[\frac{x^3}{3}\right]_0^1 - \pi^2\left[\frac{(2-x)^3}{3}\right]_1^2 = \frac{\pi^2}{2} + \frac{16}{\pi^2} + \frac{16}{81\pi^2} + \ldots$$

$$\frac{\pi^2}{3} - \pi^2\left(0 - \frac{1}{3}\right) = \frac{\pi^2}{2} + \frac{16}{\pi^2}\left[1 + \frac{1}{81} + \ldots\right]$$

$$\frac{2\pi^2}{3} - \frac{\pi^2}{2} = \frac{16}{\pi^2}\left[1 + \frac{1}{3^4} + \frac{1}{5^4} + \ldots\right]$$

$$\frac{\pi^2}{6} = \frac{16}{\pi^2}\left[1 + \frac{1}{3^4} + \frac{1}{5^4} + \ldots\right]$$

$$\frac{\pi^4}{96} = 1 + \frac{1}{3^4} + \frac{1}{5^4} + \ldots$$ **Ans.**

Example 28. *Prove that for $0 < x < \pi$*

(*a*) $x(\pi - x) = \dfrac{\pi^2}{6} - \left[\dfrac{\cos 2x}{1^2} + \dfrac{\cos 4x}{2^2} + \dfrac{\cos 6x}{3^2} + \ldots\right]$

(*b*) $x(\pi - x) = \dfrac{8}{\pi}\left[\dfrac{\sin x}{1^2} + \dfrac{\sin 3x}{3^2} + \dfrac{\sin 5x}{5^2} + \ldots\right]$

Deduce from (a) and (b) respectively that

(*c*) $\displaystyle\sum_{n=1}^{\infty} \frac{1}{n^4} = \frac{\pi^4}{90}$ (*d*) $\displaystyle\sum_{n=1}^{\infty} \frac{1}{n^6} = \frac{\pi^4}{945}$

Solution. (a) Half range cosine series

$$a_0 = \frac{2}{\pi}\int_0^\pi x(\pi - x) = \frac{2}{\pi}\left[\frac{\pi x^2}{2} - \frac{x^3}{3}\right]_0^\pi = \frac{2}{\pi}\left[\frac{\pi^3}{2} - \frac{\pi^3}{3}\right] = \frac{\pi^2}{3}$$

$$a_n = \frac{2}{\pi}\int_0^\pi x(\pi - x)\cos nx\, dx$$

$$= \frac{2}{\pi}\left[(\pi x - x^2)\frac{\sin nx}{n} - (\pi - 2x)\left(\frac{-\cos nx}{n^2}\right) + (-2)\left(\frac{-\sin nx}{n^3}\right)\right]_0^\pi$$

$$= \frac{2}{\pi}\left[0 - \frac{\pi(-1)^n}{n^2} + 0 - \frac{\pi}{n^2}\right] = \frac{2}{\pi}\left(\frac{\pi}{n^2}\right)[-(-1)^n - 1]$$

$$= -\frac{4}{n^2}$$ (when n is even)

$$= 0$$ (when n is odd)

Hence, $x(\pi - x) = \dfrac{\pi^2}{6} - 4\left[\dfrac{\cos 2x}{2^2} + \dfrac{\cos 4x}{4^2} + \ldots\right]$

By Parseval's formula $\frac{2}{\pi}\int_0^{\pi} x^2 (\pi - x)^2 \, dx = \frac{a_0^2}{2} + \Sigma a_n^2$

$$\frac{2}{\pi}\int_0^{\pi} (\pi^2 x^2 - 2\pi x^3 + x^4)\, dx = \frac{1}{2}\left(\frac{\pi^4}{9}\right) + 16\left[\frac{1}{2^4} + \frac{1}{4^4} + \frac{1}{6^4} + \ldots\right]$$

$$\frac{2}{\pi}\left[\frac{\pi^2 x^3}{3} - \frac{2\pi x^4}{4} + \frac{x^5}{5}\right]_0^{\pi} = \frac{\pi^4}{18} + \left[\frac{1}{1^4} + \frac{1}{2^4} + \frac{1}{3^4} + \ldots\right]$$

$$\frac{2}{\pi}\left[\frac{\pi^5}{3} - \frac{2\pi^5}{4} + \frac{\pi^5}{5}\right] = \frac{\pi^4}{18} + \left[\frac{1}{1^4} + \frac{1}{2^4} + \frac{1}{3^4} + \ldots\right]$$

$$\frac{\pi^4}{15} = \frac{\pi^4}{18} + \sum_{n=1}^{\infty}\frac{1}{n^4} \qquad \text{or} \qquad \sum_{n=1}^{\infty}\frac{1}{n^4} = \frac{\pi^4}{90}$$

(*b*) Half range sine series

$$b_n = \frac{2}{\pi}\int_0^{\pi} x(\pi - x)\sin nx \, dx$$

$$= \frac{2}{\pi}\left[(\pi x - x^2)\left(\frac{-\cos nx}{n}\right) - (\pi - 2x)\left(\frac{-\sin nx}{n^2}\right) + (-2)\frac{\cos nx}{n^3}\right]_0^{\pi}$$

$$= \frac{2}{\pi}\left[-2\frac{(-1)^n}{n^3} + \frac{2}{n^3}\right] = \frac{4}{\pi n^3}[-(-1)^n + 1]$$

$$= \frac{8}{n^3 \pi} \qquad \text{(when } n \text{ is odd)}$$

$$= 0 \qquad \text{(when } n \text{ is even)}$$

$$\therefore \quad x(\pi - x) = \frac{8}{\pi}\left[\frac{\sin x}{1^3} + \frac{\sin 3x}{3^3} + \frac{\sin 5x}{5^3} + \ldots\right]$$

By Parseval's formula

$$\frac{2}{\pi}\int_0^{\pi} x^2 (\pi - x^2)\, dx = \Sigma b_n^2$$

$$\frac{\pi^4}{15} = \frac{64}{\pi^2}\left[\frac{1}{1^6} + \frac{1}{3^6} + \frac{1}{5^6} + \ldots\right]$$

$$\frac{\pi^6}{960} = \frac{1}{1^6} + \frac{1}{3^6} + \frac{1}{5^6} + \ldots.$$

Let $S = \frac{1}{1^6} + \frac{1}{2^6} + \frac{1}{3^6} + \frac{1}{4^6} + \ldots = \left(\frac{1}{1^6} + \frac{1}{3^6} + \frac{1}{5^6}\ldots\right) + \left(\frac{1}{2^6} + \frac{1}{4^6} + \frac{1}{6^6} + \ldots\right)$

$$S = \frac{\pi^6}{960} + \left(\frac{1}{2^6} + \frac{1}{4^6} + \frac{1}{6^6} + \ldots\right) = \frac{\pi^6}{960} + \frac{1}{2^6}\left[\frac{1}{1^6} + \frac{1}{2^6} + \frac{1}{3^6} + \ldots\right]$$

$$S = \frac{\pi^6}{960} + \frac{S}{64}$$

$$S - \frac{S}{64} = \frac{\pi^6}{960} \quad \Rightarrow \quad \frac{63}{64}S = \frac{\pi^6}{960}$$

$$S = \frac{\pi^6}{960} \times \frac{64}{63} = \frac{\pi^6}{945}$$

$$\sum_{n=1}^{\infty}\frac{1}{n^6} = \frac{\pi^6}{945}$$

Proved.

EXERCISE 40.5

1. Prove that in $0 < x < c$,

$$x = \frac{c}{2} - \frac{4c}{\pi^2}\left(\cos\frac{\pi x}{c} + \frac{1}{3^2}\cos\frac{3\pi x}{c} + \frac{1}{5^2}\cos\frac{5\pi x}{c} + \ldots\right) \text{ and deduce that}$$

(i) $\frac{1}{1^4} + \frac{1}{3^4} + \frac{1}{5^4} + \ldots = \frac{\pi^4}{96}$ (ii) $\frac{1}{1^4} + \frac{1}{2^4} + \frac{1}{3^4} + \frac{1}{4^4} + \ldots = \frac{\pi^4}{90}$

40.15 FOURIER SERIES IN COMPLEX FORM

Fourier series of a function $f(x)$ of period $2l$ is

$$f(x) = \frac{a_0}{2} + a_1 \cos\frac{\pi x}{l} + a_2 \cos\frac{2\pi x}{l} + \ldots + a_n \cos\frac{n\pi x}{l} + \ldots$$

$$+ b_1 \sin\frac{\pi x}{l} + b_2 \sin\frac{2\pi x}{l} + \ldots + b_n \sin\frac{n\pi x}{l} + \ldots \quad \ldots(1)$$

We know that $\cos x = \frac{e^{ix} + e^{-ix}}{2}$ and $\sin x = \frac{e^{ix} - e^{-ix}}{2i}$

On putting the values of $\cos x$ and $\sin x$ in (1), we get

$$f(x) = \frac{a_0}{2} + a_1 \frac{e^{\frac{i\pi x}{l}} + e^{\frac{-i\pi x}{l}}}{2} + a_2 \frac{e^{\frac{2i\pi x}{l}} + e^{\frac{-2i\pi x}{l}}}{2} + \ldots + b_1 \frac{e^{\frac{i\pi x}{l}} - e^{\frac{i\pi x}{l}}}{2i} + b_2 \frac{e^{\frac{2i\pi x}{l}} - e^{\frac{-2i\pi x}{l}}}{2i} + \ldots$$

$$= \frac{a_0}{2} + (a_1 - ib_1)\, e^{\frac{i\pi x}{l}} + (a_2 - ib_2)\, e^{\frac{2i\pi x}{l}} + \ldots + (a_1 + ib_1)\, e^{\frac{-i\pi x}{l}} + (a_2 + ib_2)\, e^{\frac{-2i\pi x}{l}} + \ldots$$

$$= c_0 + c_1 e^{\frac{i\pi x}{l}} + c_2 e^{\frac{2i\pi x}{l}} + \ldots + c_{-1} e^{\frac{-i\pi x}{l}} + c_{-2}\, e^{-\frac{2i\pi x}{l}} + \ldots$$

$$= c_0 + \sum_{n=1}^{\infty} c_n e^{\frac{in\pi x}{l}} + \sum_{n=1}^{\infty} c_{-n} e^{\frac{-in\pi x}{l}}$$

$$c_n = \frac{1}{2}(a_n - ib_n),\ c_{-n} = \frac{1}{2}(a_n + ib_n)$$

here $c_0 = \frac{a_0}{2} = \frac{1}{2}\frac{1}{l}\int_0^{2l} f(x)\, dx$

$$c_n = \frac{1}{2}\left[\frac{1}{l}\int_0^{2l} f(x) \cos\frac{n\pi x}{l}\, dx - \frac{i}{l}\int_0^{2l} f(x) \sin\frac{n\pi x}{l}\, dx\right]$$

$$\Rightarrow \quad c_n = \frac{1}{2}\frac{1}{l}\int_0^{2l} f(x)\left\{\cos\frac{n\pi x}{l} - i \sin\frac{n\pi x}{l}\right\} dx$$

$$c_n = \frac{1}{2l}\int_0^{2l} f(x)\, e^{\frac{-in\pi x}{l}}\, dx,$$

$$c_{-n} = \frac{1}{2l}\int_0^{2l} f(x)\, e^{\frac{in\pi x}{l}}\, dx$$

Example 29. *Obtain the complex form of the Fourier series of the function*

$$f(x) = \begin{cases} 0, & -\pi \le x \le 0 \\ 1, & 0 \le x \le \pi \end{cases}$$

Solution. $c_0 = \frac{1}{2\pi}\int_0^{\pi} dx = \frac{1}{2}$

$$c_n = \frac{1}{2\pi}\int_{-\pi}^{\pi} f(x)\, e^{-inx}\, dx$$

$$= \frac{1}{2\pi}\left[\int_{-\pi}^{0} 0\cdot e^{-inx}\, dx + \int_{0}^{\pi} 1\cdot e^{-inx}\, dx\right] = \frac{1}{2\pi}\int_0^{\pi} e^{-inx}\, dx = \frac{1}{2\pi}\left[\frac{e^{-inx}}{-in}\right]_0^{\pi}$$

$$= -\frac{1}{2n\pi i}[e^{-inx} - 1] = -\frac{1}{2n\pi i}[\cos n\pi - i\sin n\pi - 1] = -\frac{1}{2n\pi i}[(-1)^n - 0 - 1]$$

$$= \begin{cases} \dfrac{1}{in\pi}, & n \text{ is odd} \\ 0\;, & n \text{ is even} \end{cases}$$

$$f(x) = \frac{1}{2} + \frac{1}{i\pi}\left[\frac{e^{ix}}{1} + \frac{e^{3ix}}{3} + \frac{e^{5ix}}{5} + \ldots\right] + \frac{1}{i\pi}\left[\frac{e^{-ix}}{-1} + \frac{e^{-3ix}}{-3} + \frac{e^{-5ix}}{-5} + \ldots\right]$$

$$= \frac{1}{2} - \frac{1}{i\pi}\left[\left(e^{ix} - e^{-ix}\right) + \frac{1}{3}\left(e^{3ix} - e^{-3ix}\right) + \frac{1}{5}\left(e^{5ix} - e^{-5ix}\right) + \ldots\right]$$

$$= \frac{1}{2} - \frac{2}{\pi}\left[\sin x + \frac{1}{3}\sin 3x + \frac{1}{5}\sin 5x + \ldots\ldots\right]$$ **Ans.**

EXERCISE 40.6

Find the complex form of the Fourier series of

1. $f(x) = e^{-x},\ -1 \le x \le 1.$ **Ans.** $\sum_{n=-\infty}^{\infty} \frac{(-1)^n (1 - in\pi)}{1 + n^2\pi^2} \sinh 1 . e^{in\pi x}$

2. $f(x) = e^{ax},\ -l < x < l$ **Ans.** $\frac{2}{\pi} - \frac{2}{\pi}\left[\frac{e^{2it} + e^{-2it}}{1\cdot 3} + \frac{e^{4it} + e^{-4it}}{3\cdot 5} + \frac{e^{6it} + e^{-6it}}{5 - 7} + \ldots\right]$

3. $f(x) = \cos ax,\ -\pi < x < \pi$ **Ans.** $\frac{a}{\pi}\sin a\pi \sum_{-\infty}^{\infty} \frac{(-1)^n e^{inx}}{a^2 - n^2}$

40.16 PRACTICAL HARMONIC ANALYSIS

Some times the function is not given by a formula, but by a graph or by a table of corresponding values. The process of finding the Fourier series for a function given by such values of the function and independent variable is known as Harmonic Analysis. The Fourier constants are evaluated by the following formulae :

(1) $$a_0 = \frac{1}{\pi}\int_0^{2\pi} f(x)\, dx$$

$$= 2\frac{1}{2\pi - 0}\int_0^{2\pi} f(x)\, dx \qquad \left[\text{Mean} = \frac{1}{b-a}\int_a^b f(x)\, dx\right]$$

$\Rightarrow$ $$\boldsymbol{a_0 = 2\ [\text{mean value of } f(x) \text{ in } (0, 2\pi)]}$$

(2) $$a_n = \frac{1}{\pi}\int_0^{2\pi} f(x)\cos nx\, dx = 2\frac{1}{2\pi - 0}\int_0^{2\pi} f(x)\cos nx\, dx$$

$$\boldsymbol{= 2\ [\text{mean value of } f(x) \cos nx \text{ in } (0, 2\pi)]}$$

(3) $$b_n = \frac{1}{\pi}\int_0^{2\pi} f(x)\sin nx\, dx = 2\frac{1}{2\pi - 0}\int_0^{2\pi} f(x)\sin nx\, dx$$

$$\boldsymbol{= 2\ [\text{mean value of } f(x) \sin nx \text{ in } (0, 2\pi)]}$$

Fundamental or first harmonic. The term $(a_1 \cos x + b_1 \sin x)$ in Fourier series is called the fundamental or first harmonic.

Second harmonic. The term $(a_2 \cos 2x + b_2 \sin 2x)$ in Fourier series is called the second harmonic and so on.

Example 30. *Find the Fourier series as far as the second harmonic to represent the function given by table below :*

x	0°	30°	60°	90°	120°	150°	180°	210°	240°	270°	300°	330°
$f(x)$	2.34	3.01	3.69	4.15	3.69	2.20	0.83	0.51	0.88	1.09	1.19	1.64

Solution.

$x°$	$\sin x$	$\sin 2x$	$\cos x$	$\cos 2x$	$f(x)$	$f(x)\sin x$	$f(x)\sin 2x$	$f(x)\cos x$	$f(x)\cos 2x$
0°	0	0	1	1	2.34	0	0	2.340	2.340
30°	0.50	0.87	0.87	0.50	3.01	1.505	2.619	2.619	1.505
60°	0.87	0.87	0.50	– 0.50	3.69	3.210	3.210	1.845	– 1.845
90°	1.00	0	0	– 1.00	4.15	4.150	0	0	– 4.150
120°	0.87	– 0.87	– 0.50	– 0.50	3.69	3.210	– 3.210	– 1.845	– 1.845
150°	0.50	– 0.87	– 0.87	0.50	2.20	1.100	– 1.914	– 1.914	1.100
180°	0	0	– 1	1.00	0.83	0	0	– 0.830	0.830
210°	– 0.50	0.87	– 0.87	0.50	0.51	– 0.255	0.444	– 0.444	0.255
240°	– 0.87	0.87	– 0.50	– 0.50	0.88	– 0.766	0.766	–0.440	– 0.440
270°	– 1.00	0	0	– 1.00	1.09	– 1.090	0	0	– 1.090
300°	– 0.87	– 0.87	0.50	–0.50	1.19	– 1.035	– 1.035	0.595	– 0.595
330°	– 0.50	– 0.87	0.87	0.50	1.64	– 0.820	– 1.427	1.427	0.820
					25.22	9.209	– 0.547	3.353	– 3.115

$$a_0 = 2 \times \text{Mean of } f(x) = 2 \times \frac{25.22}{12} = 4.203$$

$$a_1 = 2 \times \text{Mean of } f(x) \cos x = 2 \times \frac{3.353}{12} = 0.559$$

$$a_2 = 2 \times \text{Mean of } f(x) \cos 2x = 2 \times \frac{-3.115}{12} = -0.519$$

$$b_1 = 2 \times \text{Mean of } f(x) \sin x = 2 \times \frac{9.209}{12} = 1.535$$

$$b_2 = 2 \times \text{Mean of } f(x) \sin 2x = 2 \times \frac{-0.547}{12} = -0.091$$

Fourier series is

$$f(x) = \frac{a_0}{2} + a_1 \cos x + a_2 \cos 2x + \ldots + b_1 \sin x + b_2 \sin 2x + \ldots$$

$$= 2.1015 + 0.559 \cos x - 0.519 \cos 2x + \ldots + 1.535 \sin x - 0.091 \sin 2x + \ldots$$ **Ans.**

Example 31. *A machine completes its cycle of operations every time as certain pulley completes a revolution. The displacement $f(x)$ of a point on a certain portion of the machine is given in the table given below for twelve positions of the pulley, x being the angle in degree turned through by the pulley. Find a Fourier series to represent $f(x)$ for all values of x.*

x	30°	60°	90°	120°	150°	180°	210°	240°	270°	300°	330°	360°
$f(x)$	7.976	8.026	7.204	5.676	3.674	1.764	0.552	0.262	0.904	2.492	4.736	6.824

Solution.

x	$\sin x$	$\sin 2x$	$\sin 3x$	$\cos x$	$\cos 2x$	$\cos 3x$	$f(x)$	$f(x) \times \sin x$	$f(x) \times \sin 2x$	$f(x) \times \sin 3x$	$f(x) \times \cos x$	$f(x) \times \cos 2x$	$f(x) \times \cos 3x$
30°	0.50	0.87	1	0.87	0.50	0	7.976	3.988	6.939	7.976	6.939	3.988	0
60°	0.87	0.87	0	0.50	– 0.50	– 1	8.026	6.983	6.983	0	4.013	4.013	– 8.026
90°	1.00	0	– 1	0	– 1	0	7.204	7.204	0	– 7.204	0	– 7.204	0
120°	0.87	– 0.87	0	– 0.50	– 0.50	1	5.676	4.938	– 4.939	0	– 2.838	– 2.838	5.676
150°	0.50	– 0.87	1	– 0.87	0.50	0	3.674	1.837	– 3.196	– 3.196	– 3.196	1.837	0
180°	0	0	0	– 1	1	– 1	1.764	0	0	– 1.764	– 1.764	1.764	– 1.764
210°	– 0.50	0.87	– 1	– 0.87	0.50	0	0.552	– 0.276	0.480	0.480	–0.480	0.276	0
240°	– 0.87	0.87	0	– 0.50	– 0.50	1	0.262	– 0.228	0.228	– 0.131	– 0.131	0.131	0.262
270°	– 1.00	0	1	0	– 1.00	0	0.904	– 0.904	0	0	0	– 0.904	0
300°	– 0.87	– 0.87	0	0.50	– 0.50	– 1	2.492	– 2.168	– 2.168	1.246	1.246	–1.296	– 2.492
330°	– 0.50	– 0.87	– 1	0.87	0.50	0	4.736	–2.368	– 4.120	4.120	4.120	2.368	0
360°	0	0	0	1	1	1	6.824	0	0	0	6.824	6.824	6.824
						Σ	50.09	19.206	0.207	0.062	14.733	0.721	0.460

$$a_0 = 2 \times \text{Mean value of } f(x) \quad = 2 \times \frac{50.09}{12} = 8.34$$

$$a_1 = 2 \times \text{Mean value of } f(x) \cos x = 2 \times \frac{14.733}{12} = 2.45$$

$$a_2 = 2 \times \text{Mean value of } f(x) \cos 2x = 2 \times \frac{0.721}{12} = 0.12$$

$$a_3 = 2 \times \text{Mean value of } f(x) \cos 3x = 2 \times \frac{0.460}{12} = 0.08$$

$$b_1 = 2 \times \text{Mean value of } f(x) \sin x = 2 \times \frac{19.206}{12} = 3.16$$

$$b_2 = 2 \times \text{Mean value of } f(x) \sin 2x = 2 \times \frac{0.207}{12} = 0.03$$

$$b_3 = 2 \times \text{Mean value of } f(x) \sin 3x = 2 \times \frac{0.062}{12} = 0.01$$

Fourier series is

$$f(x) = \frac{a_0}{2} + a_1 \cos x + a_2 \cos 2x + a_3 \cos 3x + \ldots + b_1 \sin x + b_2 \sin 2x + b_3 \sin 3x + \ldots$$

$$= 4.17 + 2.45 \cos x + 0.12 \cos 2x + 0.08 \cos 3x + \ldots$$
$$+ 3.16 \sin x + 0.03 \sin 2x + 0.01 \sin 3x + \ldots$$ **Ans.**

Example 32. *Obtain the constant term and the coefficient of the first sine and cosine terms in the Fourier series of $f(x)$ as given in the following table.*

x	0	1	2	3	4	5
$f(x)$	9	18	24	28	26	20

Solution.

x	$\frac{x\pi}{3}$	$sin\frac{\pi x}{3}$	$cos\frac{\pi x}{3}$	$f(x)$	$f(x)\,sin\frac{\pi x}{3}$	$f(x)\,cos\frac{\pi x}{3}$
0	0	0	1.0	9	0	9
1	$\frac{\pi}{3}$	0.87	0.5	18	15.66	9
2	$\frac{2\pi}{3}$	0.87	– 0.5	24	20.88	– 12
3	$\frac{3\pi}{3}$	0	– 1.0	28	0	– 28
4	$\frac{4\pi}{3}$	– 0.87	– 0.5	26	– 22.62	– 13
5	$\frac{5\pi}{3}$	– 0.87	0.5	20	– 17.4	10
				$\Sigma = 125$	$\Sigma = -3.468$	$\Sigma = 25$

$$a_0 = 2 \text{ Mean value of } f(x) = 2 \times \frac{125}{6} = 41.67$$

$$a_1 = 2 \text{ Mean value of } f(x) \cos\frac{\pi x}{3} = 2 \times \frac{-25}{6} = -8.33$$

$$b_1 = 2 \text{ Mean value of } f(x) \sin\frac{\pi x}{3} = 2 \times \frac{-3.48}{6} = -1.16$$

Fourier series is
$$f(x) = \frac{a_0}{2} + a_1 \cos\frac{\pi x}{3} + \ldots + b_1 \sin\frac{\pi x}{3} + \ldots$$

$$= 20.84 - 8.33 \cos\frac{\pi x}{3} + \ldots - 1.16 \sin\frac{\pi x}{3} + \ldots$$ **Ans.**

EXERCISE 40.7

1. In a machine the displacement $f(x)$ of a given point is given for a certain angle $x°$ as follows:

$x°$	0°	30°	60°	90°	120°	150°	180°	210°	240°	270°	300°	330°
$f(x)$	7.9	8.0	7.2	5.6	3.6	1.7	0.5	0.2	0.9	2.5	4.7	6.8

Find the coefficient of sin 2 x in the Fourier series representing the above variations. **Ans.** – 0.072

2. The displacement $f(x)$ of a part of a machine is tabulated with corresponding angular moment 'x' of the crank. Express $f(x)$ as a Fourier series upto third harmonic.

$x°$	0°	30°	60°	90°	120°	150°	180°	210°	240°	270°	300°	330°
$f(x)$	1.80	1.10	0.30	0.16	0.50	1.30	2.16	1.25	1.30	1.52	1.76	2.00

Ans. $f(x) = 1.26 + 0.04 \cos x + 0.53 \cos 2x - 0.1 \cos 3x + \ldots$
$- 0.63 \sin x - 0.23 \sin 2x + 0.085 \sin 3x + \ldots$

3. Fourier coefficient 'a_0' in Fourier series expansion of a function represents the:
(*i*) maximum value of the function (*ii*) 2 mean value of the function
(*iii*) minimum value of the function (*iv*) None of these (*U.P. II Semester 2010*) **Ans.** (*ii*)

4. If the Fourier series of $f(x)$ has only cosine terms then $f(x)$ must be :
(*i*) odd function (*ii*) even function **Ans.** (*ii*) (*U.P. II Semester 2010*)

CHAPTER

41

INTEGRAL TRANSFORMS

41.1 INTRODUCTION

Integral transforms are used in the solution of partial differential equations. The choice of a particular transform to be used for the solution of a differential equations depends upon the nature of the boundary conditions of the equation and the facility with which the transform $F(s)$ can be converted to give $f(x)$.

41.2 INTEGRAL TRANSFORMS

The integral transform $F(s)$ of a function $f(x)$ with the kernel $k(s, x)$ is defined as

$$I[f(x)] = F(s) = \int_a^b f(x)\, k(s, x)\, dx.$$

For example ·

1. Laplace transform with the kernel $k(s, x) = e^{-sx}$

$$L[f(x)] = F(s) = \int_0^\infty f(x).e^{-sx}\, dx$$

2. Fourier Complex transform with the kernel $k(s, k) = e^{isx}$

$$F[f(x)] = F(s) = \frac{1}{\sqrt{2\pi}} \int_{-\infty}^{\infty} f(x)\, e^{isx}\, dx.$$

$$f(x) = \frac{1}{\sqrt{2\pi}} \int_{-\infty}^{\infty} F(s)\, e^{-isx}\, ds \qquad \text{(Inversion formula)}$$

3. Fourier Sine transform with the kernel $k(s, x) = \sin sx$

$$F_s[f(x)] = F_s(s) = \sqrt{\frac{2}{\pi}} \int_0^\infty f(x) \sin sx.dx$$

$$f(x) = \sqrt{\frac{2}{\pi}} \int_0^\infty F_s(s) \sin sx\, ds \qquad \text{(Inversion formula)}$$

4. Fourier Cosine transform with the kernel $k(s, x) = \cos sx$

$$F_c[f(x)] = F_c(s) = \sqrt{\frac{2}{\pi}} \int_0^\infty f(x) \cos sx.dx$$

$$f(x) = \sqrt{\frac{2}{\pi}} \int_0^\infty F_c(s) \cos sx\, ds \qquad \text{(Inversion formula)}$$

5. Hankel Transform with the kernel $(k, s) = x J_n(sx)$

$$H[f(x)] = F(s) = \int_0^\infty f(x).x J_n(sx)\, dx$$

$$f(x)=\frac{1}{\sqrt{2\pi}}\int_{-\infty}^{\infty}F(s)\,e^{-isx}\,ds \qquad \text{(Inversion formula)}$$

6. Hilbert Transform with the kernel $k(s,x)=\dfrac{1}{s-x}$

$$F(s)=\frac{1}{\pi}\int_{-\infty}^{\infty}\frac{f(x)}{s-x}\,dx$$

$$f(x)=\frac{-1}{\pi}\int_{-\infty}^{\infty}\frac{F(s)}{s-x}\,ds \qquad \text{(Inversion formula)}$$

7. Mellin transform with the kernel $k(s,x)=x^{s-1}$

$$M[f(x)]=F(s)=\int_0^{\infty}f(x).x^{s-1}dx.$$

The students have already done "Laplace transform" and also learnt to solve the ordinary differential equations by using Laplace transforms.

Integral transforms are used in solving the partial differential equation with boundary conditions.

List of Formulae of Fourier Integrals

1. Fourier Integral for $f(x)$ is $\quad f(x)=\dfrac{1}{\pi}\displaystyle\int_0^{\infty}\int_{-\infty}^{\infty}f(t)\cos u(t-x)\,du\,dt$

2. Fourier Sine Integral for $f(x)$ is $\quad f(x)=\dfrac{2}{\pi}\displaystyle\int_0^{\infty}\int_0^{\infty}f(t)\sin ut\sin ux\,du\,dt$

3. Fourier Cosine Integral for $f(x)$ is $f(x)=\dfrac{2}{\pi}\displaystyle\int_0^{\infty}\int_0^{\infty}f(t)\cos ut\cos ux\,du\,dt$

41.3 FOURIER INTEGRAL THEOREM

It states that $f(x)=\dfrac{1}{\pi}\displaystyle\int_0^{\infty}\int_{-\infty}^{\infty}f(t)\cos u(t-x)\,dt\,du$

Proof. We know that Fourier series of a function $f(x)$ in $(-c, c)$ is given by

$$f(x)=\frac{a_0}{2}+\sum_{n=1}^{\infty}a_n\cos\frac{n\pi x}{c}+\sum_{n=1}^{\infty}b_n\sin\frac{n\pi x}{c} \qquad \text{... (1)}$$

where a_0, a_n and b_n are given by

$$a_0=\frac{1}{c}\int_{-c}^{c}f(t)dt,$$

$$a_n=\frac{1}{c}\int_{-c}^{c}f(t)\cos\frac{n\pi t}{c}\,dt$$

$$b_n=\frac{1}{c}\int_{-c}^{c}f(t)\sin\frac{n\pi t}{c}\,dt$$

Substituting the values of a_0, a_n and b_n in (1), we get

$$f(x)=\frac{1}{2c}\int_{-c}^{c}f(t)\,dt+\sum_{n=1}^{\infty}\frac{1}{c}\int_{-c}^{c}f(t)\cos\frac{n\pi t}{c}\cos\frac{n\pi x}{c}\,dt+\sum_{n=1}^{\infty}\frac{1}{c}\int_{-c}^{c}f(t)\sin\frac{n\pi t}{c}\sin\frac{n\pi x}{c}\,dt$$

$$=\frac{1}{2c}\int_{-c}^{c}f(t)\,dt+\sum_{n=1}^{\infty}\frac{1}{c}\int_{-c}^{c}f(t)\left[\cos\frac{n\pi t}{c}\cos\frac{n\pi x}{c}+\sin\frac{n\pi t}{c}\sin\frac{n\pi x}{c}\right]dt$$

$$= \frac{1}{2c}\int_{-c}^{c} f(t)\,dt + \sum_{n=1}^{\infty}\frac{1}{c}\int_{-c}^{c} f(t)\cos\left(\frac{n\pi t}{c} - \frac{n\pi x}{c}\right)dt$$

$$= \frac{1}{2c}\int_{-c}^{c} f(t)\,dt + \sum_{n=1}^{\infty}\frac{1}{c}\int_{-c}^{c} f(t)\cos\frac{n\pi}{c}(t-x)\,dt$$

$$= \frac{1}{2c}\int_{-c}^{c} f(t)\left\{1 + 2\sum_{n=1}^{\infty}\cos\frac{n\pi}{c}(t-x)\right\}dt \qquad \ldots (2)$$

Since cosine functions are even functions *i.e.*, $\cos(-\theta) = \cos\theta$ the expression

$$1 + 2\sum_{n=1}^{\infty}\cos\frac{n\pi}{c}(t-x) = \sum_{n=-\infty}^{\infty}\cos\frac{n\pi}{c}(t-x)$$

Therefore, (2) becomes

$$f(x) = \frac{1}{2c}\int_{-c}^{c} f(t)\left\{\sum_{-\infty}^{\infty}\cos\frac{n\pi}{c}(t-x)\right\}dt$$

$$= \frac{1}{2\pi}\int_{-c}^{c} f(t)\left\{\frac{\pi}{c}\sum_{-\infty}^{\infty}\cos\frac{n\pi}{c}(t-x)\right\}dt \qquad \ldots (3)$$

Let us now assume that c increases indefinitely, so that we may write $\frac{n\pi}{c} = u$ and $\frac{\pi}{c} = du$.

This assumption gives

$$\lim_{c\to\infty}\left\{\frac{\pi}{c}\sum_{-\infty}^{\infty}\cos\frac{n\pi}{c}(t-x)\right\} = \int_{-\infty}^{\infty}\cos u(t-x)\,du$$

$$= 2\int_{0}^{\infty}\cos u(t-x)\,du \qquad \ldots (4)$$

Substituting in (3) from (4), we obtain

$$f(x) = \frac{1}{2\pi}\int_{-\infty}^{\infty} f(t)\left\{2\int_{0}^{\infty}\cos u(t-x)\,du\right\}dt \qquad \ldots (5)$$

Thus $$\boxed{f(x) = \frac{1}{\pi}\int_{0}^{\infty}\int_{-\infty}^{\infty} f(t)\cos u(t-x)\,du\,dt}$$ **Proved.**

Note. We have assumed the following conditions on $f(x)$.

(*i*) $f(x)$ is defined as single-valued except at finite points in $(-c, c)$.

(*ii*) $f(x)$ is periodic outside $(-c, c)$ with period $2c$.

(*iii*) $f(x)$ and $f'(x)$ are sectionally continuous in $(-c, c)$.

(*iv*) $\int_{-\infty}^{\infty}|f(x)|\,dx$ converges, *i.e.*, $f(x)$ is absolutely integrable in $(-\infty, \infty)$.

41.4 FOURIER SINE AND COSINE INTEGRALS

$$f(x) = \frac{2}{\pi}\int_{0}^{\infty}\sin ux\,du\int_{0}^{\infty} f(t)\sin ut\,dt \qquad \textbf{(Fourier Sine Integral)}$$

$$f(x) = \frac{2}{\pi}\int_{0}^{\infty}\cos ux\,du\int_{0}^{\infty} f(t)\cos ut\,dt \qquad \textbf{(Fourier Cosine Integral)}$$

Proof. We know that,

$$\cos u(t-x) = \cos(ut - ux)$$

$$\Rightarrow \quad \cos u(t-x) = \cos ut \cos ux + \sin ut \sin ux$$

Then equation (5) of article 1.3, can be written as

$$f(x) = \frac{1}{\pi}\int_0^{\infty}\int_{-\infty}^{\infty} f(t)\,(\cos ut \cos ux + \sin ut \sin ux)\,du\,dt$$

$$\Rightarrow \quad f(x) = \frac{1}{\pi}\int_0^{\infty}\int_{-\infty}^{\infty} f(t)\cos ut\,\cos ux\,du\,dt + \frac{1}{\pi}\int_0^{\infty}\int_{-\infty}^{\infty} f(t)\sin ut\,\sin ux\,du\,dt \quad \text{... (6)}$$

Case 1. When $f(t)$ is odd.

$\therefore$ $f(t)\cos ut$ is odd hence $\int_0^{\infty}\int_{-\infty}^{\infty} f(t)\cos ut\,\cos ux\,du\,dt = 0$

$$\left[\begin{array}{ll}\text{For odd function,} & \int_{-a}^{a} f(x)dx = 0 \\ \text{For even function,} & \int_{-a}^{a} f(x)\,dx = 2\int_0^{a} f(x)\,dx\end{array}\right]$$

From (6), we have

$$\therefore \quad \boxed{f(x) = \frac{2}{\pi}\int_0^{\infty}\sin ux\,du\int_0^{\infty} f(t)\sin ut\,dt} \quad \text{... (7)}$$

The relation (7) is called **Fourier sine integral.**

Case 2. When $f(t)$ is even.

$\therefore$ $f(t)\sin ut$ is odd hence $\int_0^{\infty}\int_{-\infty}^{\infty} f(t)\sin ut\,\sin ux\,du\,dt = 0$

$\therefore$ $f(t)\cos ut$ is even.

From (6), we have

$$\boxed{f(x) = \frac{2}{\pi}\int_0^{\infty}\cos ux\,du\int_0^{\infty} f(t)\cos ut\,dt} \quad \text{... (8)}$$

The relation (8) is known as **Fourier cosine integral.**

41.5 FOURIER'S COMPLEX INTEGRAL

$$f(x) = \frac{1}{2\pi}\int_{-\infty}^{\infty} e^{-iux}\,du\int_{-\infty}^{\infty} f(t)e^{iut}\,dt$$

Proof. We know that $\int_{-a}^{a} f(x)dx = 0$ if $f(x)$ is odd function.

$$\therefore \quad \int_{-\infty}^{\infty}\sin u(t-x)\,du = 0 \qquad \text{[since } \sin u(t-x) \text{ is odd.]}$$

Obviously we have

$$\frac{1}{2\pi}\int_{-\infty}^{\infty} f(t)\,dt\int_{-\infty}^{\infty}\sin u(t-x)\,du = 0$$

$$\Rightarrow \quad \frac{i}{2\pi}\int_{-\infty}^{\infty} f(t)\,dt\int_{-\infty}^{\infty}\sin u(t-x)\,du = 0 \quad \text{(Multiplying by } i\text{)} \quad \text{... (9)}$$

On adding (5) and (9), we have

$$f(x) = \frac{1}{2\pi}\int_{-\infty}^{\infty}\int_{-\infty}^{\infty} f(t)\cos u(t-x)\,du\,dt + \frac{i}{2\pi}\int_{-\infty}^{\infty} f(t)\,dt\int_{-\infty}^{\infty}\sin u(t-x)\,du.$$

$$= \frac{1}{2\pi}\int_{-\infty}^{\infty} f(t)\,dt \int_{-\infty}^{\infty} [\cos u(t-x) + i \sin u(t-x)]\,du$$

$$= \frac{1}{2\pi}\int_{-\infty}^{\infty} f(t)\,dt \int_{-\infty}^{\infty} e^{iu(t-x)}\,du$$

$$\Rightarrow \qquad \boxed{f(x) = \frac{1}{2\pi}\int_{-\infty}^{\infty} e^{-iux}\,du \int_{-\infty}^{\infty} f(t)e^{iut}\,dt} \qquad \ldots (10)$$

Relation (10) is called **Fourier's Complex Integral.**

Some Useful Results

1. $\boxed{\int_0^{\infty} \frac{\sin ax}{x}\,dx = \frac{\pi}{2}}$ $\qquad (a > 0)$

2. $\boxed{\int_0^{\infty} e^{-x^2}\,dx = \frac{\sqrt{\pi}}{2}}$

3. $\boxed{\int_{-\infty}^{\infty} \frac{\sin mx}{(x-b)^2 + a^2}\,dx = \frac{\pi}{a} e^{-am} \sin bm,}$ $\qquad (m > 0)$

Examples based on Fourier integral

Example 1. *Express the function*

$$f(x) = \begin{cases} 1 & \text{when} \quad |x| \le 1 \\ 0 & \text{when} \quad |x| > 1 \end{cases}$$

as a Fourier integral. Hence evaluate

$$\int_0^{\infty} \frac{\sin\lambda \cos\lambda x}{\lambda}\,d\lambda$$

(U.P. & Uttarakhand, III Semester 2008, Dec. 2004)

Solution. The Fourier Integral for $f(x)$ is $\frac{1}{\pi}\int_0^{\infty}\int_{-\infty}^{\infty} f(t) \cos u(t-x)\,du\,dt$

On replacing u by λ, we have

$$f(x) = \frac{1}{\pi}\int_0^{\infty}\int_{-\infty}^{\infty} f(t) \cos\lambda(t-x)\,dt\,d\lambda$$

$$= \frac{1}{\pi}\int_0^{\infty}\int_{-1}^{1} \cos\lambda(t-x)\,dt\,d\lambda \qquad \text{(Since } f(t) = 1\text{)}$$

$$= \frac{1}{\pi}\int_0^{\infty}\left[\frac{\sin\lambda(t-x)}{\lambda}\right]_{-1}^{1} d\lambda = \frac{1}{\pi}\int_0^{\infty}\left[\frac{\sin\lambda(1-x)}{\lambda} - \frac{\sin\lambda(-1-x)}{\lambda}\right] d\lambda$$

$$= \frac{1}{\pi}\int_0^{\infty} \frac{\sin\lambda(1-x) + \sin\lambda(1+x)}{\lambda}\,d\lambda \qquad \left[\sin C + \sin D = 2 \sin\frac{C+D}{2}\cos\frac{C-D}{2}\right]$$

$$= \frac{1}{\pi}\int_0^{\infty} \frac{2\sin\frac{[\lambda(1-x)+\lambda(1+x)]}{2}\cos\frac{[\lambda(1-x)-\lambda(1+x)]}{2}}{\lambda}$$

Thus, $\qquad f(x) = \frac{2}{\pi}\int_0^{\infty} \frac{\sin\lambda\cos\lambda x}{\lambda}\,d\lambda$ **Ans.**

$$\Rightarrow \quad \int_0^\infty \frac{\sin\lambda\cos\lambda x}{\lambda}\,d\lambda = \frac{\pi}{2}f(x)$$

$$\Rightarrow \quad \int_0^\infty \frac{\sin\lambda\cos\lambda x}{\lambda}\,d\lambda = \begin{cases} \frac{\pi}{2}\times 1 = \frac{\pi}{2} & \text{for } |x|<1 \\ \frac{\pi}{2}\times 0 = 0 & \text{for } |x|>1 \end{cases} \qquad \left[\begin{array}{l} f(x)=1 \text{ for } |x|\le 1 \\ f(x)=0 \text{ for } |x|>1 \end{array}\right.$$

For $|x| = 1$, which is a point of discontinuity of $f(x)$, value of integral $= \dfrac{\pi/2+0}{2} = \dfrac{\pi}{4}$ **Ans.**

Example 2. *Find the Fourier Sine integral for*

$$f(x) = e^{-\beta x} \qquad (\beta > 0) \qquad \textit{(U.P. III Semester Comp., 2004)}$$

hence show that $\dfrac{\pi}{2}e^{-\beta x} = \displaystyle\int_0^\infty \frac{\lambda\sin\lambda x}{\beta^2+\lambda^2}\,d\lambda$

Solution. The Fourier Sine Integral of $f(x)$ is

$$f(x) = \frac{2}{\pi}\int_0^\infty \sin ux\,du \int_0^\infty f(t)\sin ut\,dt$$

On replacing u by λ, we get

$$f(x) = \frac{2}{\pi}\int_0^\infty \sin\lambda x\,d\lambda \int_0^\infty f(t)\sin\lambda t\,dt \qquad \ldots (1)$$

Putting the value of $f(x)$ in (1), we get

$$e^{-\beta x} = \frac{2}{\pi}\int_0^\infty \sin\lambda x\,d\lambda \int_0^\infty e^{-\beta t}\sin\lambda t\,dt \qquad \left[\int e^{ax}\sin bx\,dx = \frac{e^{ax}}{a^2+b^2}(a\sin bx - b\cos bx)\right]$$

$$= \frac{2}{\pi}\int_0^\infty \sin\lambda x\,d\lambda\left[\frac{e^{-\beta t}}{\beta^2+\lambda^2}(-\beta\sin\lambda t - \lambda\cos\lambda t)\right]_0^\infty = \frac{2}{\pi}\int_0^\infty \sin\lambda x\,d\lambda\left[0+\frac{\lambda}{\beta^2+\lambda^2}\right]$$

$$e^{-\beta x} = \frac{2}{\pi}\int_0^\infty \frac{\lambda\sin\lambda x}{\beta^2+\lambda^2}\,d\lambda \quad \text{or} \quad \frac{\pi}{2}e^{-\beta x} = \int_0^\infty \frac{\lambda\sin\lambda x}{\beta^2+\lambda^2}\,d\lambda. \qquad \textbf{Proved.}$$

Example 3. *Using Fourier Cosine Integral representation of an appropriate function, show that*

$$\int_0^\infty \frac{\cos wx}{k^2+w^2}\,dw = \frac{\pi e^{-kx}}{2k}, \quad x>0, \quad k>0.$$

Solution. We know that Fourier Cosine Integral is $f(x) = \dfrac{2}{\pi}\displaystyle\int_0^\infty \cos ux\,du \int_0^\infty f(t)\cos ut\,dt$

Putting the value of $f(t)$ and replacing u by w, we get

$$e^{-kx} = \frac{2}{\pi}\int_0^\infty \cos wx\,dw \int_0^\infty e^{-kt}\cos wt\,dt \qquad \left[f(t) = e^{-kt}\right]$$

$$= \frac{2}{\pi}\int_0^\infty \cos wx\,dw\left[\frac{e^{-kt}}{k^2+w^2}\{-k\cos wt + w\sin wt\}\right]_0^\infty \qquad \left\{\int e^{ax}\cos bx\,dx = \frac{e^{ax}}{a^2+b^2}[a\cos bx + b\sin bx]\right\}$$

$$e^{-kx} = \frac{2}{\pi}\int_0^\infty \cos wx\,dw\left[0+\frac{k}{k^2+w^2}\right]$$

$$e^{-kx} = \frac{2k}{\pi}\int_0^\infty \frac{\cos wx\,dw}{k^2+w^2} \quad \Rightarrow \quad \int_0^\infty \frac{\cos wx}{k^2+w^2}\,dw = \frac{\pi e^{-kx}}{2k} \qquad \textbf{Proved.}$$

Example 4. *Using Fourier integral representation, show that*

$$\int_0^{\infty} \frac{\cos(\lambda x)}{(1+\lambda^2)} d\lambda = \frac{\pi}{2} e^{-x}, (x>0)$$ (*U.P. & Uttarakhand, III Semester 2008*)

Solution. Fourier cosine integral is

$$F(x) = \frac{2}{\pi} \int_0^{\infty} \cos\lambda x \int_0^{\infty} F(t) \cos\lambda t \, dt \, d\lambda \Rightarrow e^{-x} = \frac{2}{\pi} \int_0^{\infty} \cos\lambda x \int_0^{\infty} e^{-t} \cos\lambda t \, dt \, d\lambda$$

[Put $F(x) = e^{-x}$]

$$e^{-x} = \frac{2}{\pi} \int_0^{\infty} \cos\lambda x \left[\frac{e^{-t}}{1+\lambda^2}(-\cos\lambda t + \lambda \sin\lambda t)\right]_0^{\infty} d\lambda$$

$$e^{-x} = \frac{2}{\pi} \int_0^{\infty} \cos\lambda x . \left(\frac{1}{1+\lambda^2}\right) d\lambda \Rightarrow \int_0^{\infty} \frac{\cos\lambda x}{1+\lambda^2} d\lambda = \frac{\pi}{2} e^{-x}, x \geq 0$$ **Proved**

Example 5. *Using Fourier integral representation, show that*

$$\int_0^{\infty} \frac{\cos x\omega + \omega \sin x\omega}{1+\omega^2} d\omega = \begin{cases} 0, & if\ x<0 \\ \pi/2, & if\ x=0 \\ \pi e^{-x}, & if\ x>0 \end{cases}$$

Solution. Putting $k = 1$ in example 4, we have

$$\int_0^{\infty} \frac{\cos x\omega}{1+\omega^2} d\omega = \frac{\pi}{2} e^{-x}, x > 0 \qquad \text{... (1)}$$

Putting $\beta = 1$ and $\lambda = \omega$ in example 2, we have

$$\int_0^{\infty} \frac{\omega \sin x\omega}{1+\omega^2} d\omega = \frac{\pi}{2} e^{-x}, x > 0 \qquad \text{... (2)}$$

Adding (1) and (2), we get

$$\int_0^{\infty} \frac{\cos x\omega + \omega \sin x\omega}{1+\omega^2} d\omega = \frac{\pi}{2} e^{-x} + \frac{\pi}{2} e^{-x} \qquad \text{... (3)}$$

Case I. When $x < 0 \Rightarrow x$ is replaced by $-x$ in (3), we have

$$\int_0^{\infty} \frac{\cos(-x\omega) + \omega \sin(-x\omega)}{1+\omega^2} d\omega = \int_0^{\infty} \frac{\cos(x\omega) - \omega \sin(x\omega)}{1+\omega^2} d\omega = \frac{\pi}{2} e^{-x} - \frac{\pi}{2} e^{-x} = 0$$

Case II. When $x = 0$, L.H.S. of (3) becomes

$$\int_0^{\infty} \frac{\cos 0\omega + \omega \sin 0\omega}{1+\omega^2} d\omega = \int_0^{\infty} \frac{d\omega}{1+\omega^2} = \left(\tan^{-1}\omega\right)_0^{\infty} = \frac{\pi}{2}$$

Case III. When $x > 0$, (3) is

$$\int_0^{\infty} \frac{\cos x\omega + \omega \sin x\omega}{1+\omega^2} d\omega = \frac{\pi}{2} e^{-x} + \frac{\pi}{2} e^{-x}$$

$$= \pi e^{-x}.$$

From case I, case II and case III, we get

$$\int_0^{\infty} \frac{\cos x\omega + \omega \sin x\omega}{1+\omega^2} d\omega = \begin{cases} 0, & \text{if } x<0 \\ \frac{\pi}{2}, & \text{if } x=0 \\ \pi e^{-x}, & \text{if } x>0 \end{cases}$$ **Proved.**

Example 6. *Find the complex form of the Fourier integral representation of*

$$f(x) = \begin{cases} e^{-kx}, & x > 0 \text{ and } k > 0 \\ 0, & \text{otherwise} \end{cases}$$

(*U.P. III semester Dec. 2005; U.P. III Semester Dec. 2004*)

Solution. We have, $f(x) = \begin{cases} e^{-kx}, & x > 0 \text{ and } k > 0 \\ 0, & \text{otherwise} \end{cases}$

We know that the complex form of Fourier integral representation of $f(x)$ is given

$$f(x) = \frac{1}{2\pi} \int_{-\infty}^{\infty} e^{-iux}\, du \int_{-\infty}^{\infty} f(t)\, e^{iut}\, dt$$

On replacing u by λ, we have

$$f(x) = \frac{1}{2\pi} \int_{-\infty}^{\infty} e^{-i\lambda x}\, d\lambda \int_{-\infty}^{\infty} f(t)\, e^{i\lambda t}\, dt \qquad \dots (1)$$

On putting the value of $f(t)$ in (1), we get

$$f(x) = \frac{1}{2\pi} \int_{-\infty}^{\infty} e^{-i\lambda x}\, d\lambda \left[\int_0^{\infty} e^{-kt}\, e^{i\lambda t}\, dt \right] = \frac{1}{2\pi} \int_{-\infty}^{\infty} e^{-i\lambda x}\, d\lambda \left[\int_0^{\infty} e^{-(k - i\lambda)t}\, dt \right]$$

Subtracting (2) from (1)

$$= \frac{1}{2\pi} \int_{-\infty}^{\infty} e^{-i\lambda x}\, d\lambda \left[\frac{e^{-(k-i\lambda)t}}{-(k-i\lambda)} \right]_0^{\infty} = \frac{1}{2\pi} \int_{-\infty}^{\infty} e^{-i\lambda x}\, d\lambda \left[0 - \frac{1}{-k + i\lambda} \right] = \frac{1}{2\pi} \int_{-\infty}^{\infty} \frac{e^{-i\lambda x}}{k - i\lambda}\, d\lambda.$$

EXERCISE 41.1

1. Find the Fourier Sine Integral representation of

$$f(x) = \begin{cases} 0, & 0 < x < 1 \\ k, & 1 < x < 2 \\ 0, & x > 2 \end{cases}$$ where k is constant. **Ans.** $f(x) = \frac{2k}{\pi} \int_0^{\infty} \left(\frac{\cos\lambda - \cos 2\lambda}{\lambda} \right) \sin \lambda x\, d\lambda$

2. Find Fourier Sine Integral representation of

$f(x) = x^2,\ 0 \le x \le 1.$ **Ans.** $f(x) = \frac{2}{\pi} \int_0^{\infty} \left[\left(\frac{-1}{\lambda} + \frac{2}{\lambda^3} \right) \cos\lambda + \frac{2\sin\lambda}{\lambda^2} - \frac{2}{\lambda^3} \right] \sin \lambda x\, d\lambda$

3. Find Fourier Sine Integral of

$$f(x) = \begin{cases} x & \text{for } 0 < x < 1 \\ 2 - x & \text{for } 1 < x < 2. \\ 0 & \text{for } x > 2 \end{cases}$$ **Ans.** $f(x) = \frac{2}{\pi} \int_0^{\infty} \left[\frac{4\sin u}{u} - \frac{4\sin 2u}{u^2} + \frac{2\cos u}{u} - \frac{\cos 2u}{u^2} - \frac{1}{u^2} \right] \sin 4x\, d\lambda$

4. Find Fourier Cosine Integral representation of

$$f(x) = \begin{cases} \sin x\ , & 0 \le x \le \pi \\ 0\ , & x > \pi \end{cases}$$ **Ans.** $f(x) = -\frac{2}{\pi} \int_0^{\infty} \left(\frac{1 + \cos\lambda\pi}{\lambda^2 - 1} \right) \cos \lambda x\, d\lambda$

5. Find Fourier Cosine Integral representation of

$$f(x) = \begin{cases} x\ , & 0 \le x \le 2 \\ 0\ , & \text{otherwise} \end{cases}$$ **Ans.** $f(x) = \frac{2}{\pi} \int_0^{\infty} \left(\frac{2\sin 2\lambda}{\lambda} + \frac{\cos 2\lambda - 1}{\lambda^2} \right) \cos \lambda x\, d\lambda$

6. Find the Fourier Cosine Integral of the function e^{-ax}. Hence show that

$$\int_0^{\infty} \frac{\cos 2x}{\lambda^2 + 1^2}\, d\lambda = \frac{\pi}{2} e^{-x}, \quad x \ge 0$$ **Ans.** $\frac{2a}{\pi} \int_0^{\infty} \frac{\cos x}{\lambda^2 + a^2}\, d\lambda.$

7. Express $f(x) = \begin{cases} 1 & \text{for } 0 \le x \le \pi \\ 0 & \text{for } x > \pi \end{cases}$ as a Fourier Sine Integral and hence evaluate

$$\int_0^{\infty} \frac{1 - \cos\pi\lambda}{\lambda} \sin \lambda x\, d\lambda$$ **Ans.** $\frac{\pi}{4}$

8. Find the Fourier Integral representation of the function

$$f(x) = \begin{cases} 1 & \text{for } |x| < 1 \\ 0 & \text{for } |x| > 1 \end{cases}$$

Evaluate $\int_0^\infty \frac{\cos ux \sin u}{u} du$ at $x = 1$ and $\int_0^\infty \frac{\sin u}{u} du$ **Ans.** $\frac{2}{u} \sin u \cos ux, \ \frac{\pi}{2}, \ \frac{\pi}{2}$

(U.P. III Semester, 2008)

9. Using Fourier Integral representation, show that

$$\int_0^\infty \frac{\lambda \sin \lambda x}{(\lambda^2+\alpha^2)(\lambda^2+\beta^2)} d\lambda = \frac{\pi}{2} \frac{(e^{-\alpha x} - e^{-\beta x})}{(\beta^2 - \alpha^2)}$$

Ans. $\frac{6}{\pi} \int_0^\infty \frac{\lambda \sin \lambda x}{(\lambda^2+1)(\lambda^2+4)} d\lambda$

Hence find the Fourier Sine integral representation of $e^{-x} - e^{-2x}$.

10. If $f(x) = \begin{Bmatrix} 0, & x < 0 \\ x, & 0 \le x \le \pi \\ 0, & x > \pi \end{Bmatrix}$ then show that

$$f(x) = \frac{1}{\pi \lambda^2} \int_0^\infty [\lambda \pi \sin \lambda(\pi - x) + \cos \lambda(\pi - x) - \cos \lambda x] \, d\lambda$$

11. Using Fourier Integral formula, prove that $\int_0^\infty \left(\frac{\lambda^2+2}{\lambda^4+4}\right) \cos \lambda x \, d\lambda = \frac{\pi}{2} e^{-x} \cos x$, if $x > 0$.

12. Using Fourier Integral method, prove that $\int_0^\infty \left(\frac{\sin \pi\lambda}{1-\lambda^2}\right) \sin \lambda x \, d\lambda = \begin{cases} \frac{\pi}{2} \sin x, & \text{if } 0 \le x \le \pi \\ 0, & \text{if } x > \pi \end{cases}$

13. Solve the integral equation

$$\int_0^\infty f(x) \cos \lambda x \, dx = e^{-\lambda}$$

Ans. $f(x) = \frac{2}{\pi(1+x^2)}$

(U.P. III Semester, Dec. 2005)

41.6 FOURIER TRANSFORMS

We have done in Article 41.5 that

$$f(x) = \frac{1}{2\pi} \int_{-\infty}^\infty e^{-iux} du \int_{-\infty}^\infty f(t) e^{iut} dt$$

$$= \frac{1}{2\pi} \int_{-\infty}^\infty e^{-isx} ds \int_{-\infty}^\infty f(t) e^{ist} dt = \frac{1}{\sqrt{2\pi}} \int_{-\infty}^\infty e^{-isx} ds \left[\frac{1}{\sqrt{2\pi}} \int_{-\infty}^\infty f(t) e^{ist} dt \right] \quad (u = s) \quad \ldots (1)$$

Putting $\frac{1}{\sqrt{2\pi}} \int_{-\infty}^\infty f(t) e^{ist} dt = F(s)$ in (1), we get ... (2)

$$\Rightarrow \quad f(x) = \frac{1}{\sqrt{2\pi}} \int_{-\infty}^\infty e^{-isx} . F(s) \, ds \quad \ldots (3)$$

In (2), $F(s)$ is called the **Fourier transform** of $f(x)$.

In (3), $f(x)$ is called the **inverse Fourier transform** of $F(s)$.

Note: For reasons of symmetry, we multiply both $f(x)$ and $F(s)$ by $\sqrt{\frac{1}{2\pi}}$ instead of having the factor $\frac{1}{2\pi}$ in only one function. Thus, we obtain the definition of Fourier transform as

$$F(s) = F[f(x)] = \frac{1}{\sqrt{2\pi}} \int_{-\infty}^\infty f(t) e^{ist} dt$$

$$f(x) = \frac{1}{\sqrt{2\pi}} \int_{-\infty}^\infty F(s) e^{-isx} ds$$

41.7 FOURIER SINE AND COSINE TRANSFORMS

Fourier Sine Transform

From equation (7) of Article 47.4 we know that

$$f(x) = \frac{2}{\pi}\int_0^\infty \sin sx\, ds \int_0^\infty f(t) \sin st\, dt = \sqrt{\frac{2}{\pi}}\int_0^\infty \sin sx\, ds\left[\sqrt{\frac{2}{\pi}}\int_0^\infty f(t)\sin st\, dt\right] \quad (s = u)$$

$$= \sqrt{\frac{2}{\pi}}\int_0^\infty \sin sx\, ds\, F(s)$$

$$F(s) = F_s[f(x)] = \sqrt{\frac{2}{\pi}}\int_0^\infty f(t)\sin st\, dt \quad \text{... (1)}$$

$$f(x) = \sqrt{\frac{2}{\pi}}\int_0^\infty F(s)\sin sx\, ds \quad \text{... (2)}$$

In equation (1), $F(s)$ is called **Fourier Sine transform** of $f(x)$.
In equation (2), $f(x)$ is called the **Inverse Fourier Sine transform** of $F(s)$.

Fourier Cosine Transform

From equation (8) of Article 41.4 on page 1085 we have

$$f(x) = \sqrt{\frac{2}{\pi}}\int_0^\infty \cos sx\, ds\left[\sqrt{\frac{2}{\pi}}\int_0^\infty f(t)\cos st\, dt\right] = \sqrt{\frac{2}{\pi}}\int_0^\infty \cos sx\, ds\, F(s)$$

$$F(s) = F_c[f(x)] = \sqrt{\frac{2}{\pi}}\int_0^\infty f(t)\cos st\, dt \quad \text{... (3)}$$

$$f(x) = \sqrt{\frac{2}{\pi}}\int_0^\infty \cos sx\, F(s)\, ds \quad \text{... (4)}$$

In equation (3), $F(s)$ is called **Fourier Cosine transform** of $f(x)$.
In equation (4), $f(x)$ is called the **Inverse Fourier Cosine transform of** $F(s)$.

Examples based on Fourier Transform.

Example 7. *Find the Fourier transform of*

$$f(x) = \begin{cases} 1 & \text{for } |x| < a \\ 0 & \text{for } |x| > a \end{cases}$$

Solution. The Fourier transform of a function $f(x)$ is given by

$$F[f(x)] = \frac{1}{\sqrt{2\pi}}\int_{-\infty}^{\infty} f(x).e^{isx}\, dx \quad \text{... (1)}$$

Substituting the value of $f(x)$ in (1), we get

$$F[f(x)] = \frac{1}{\sqrt{2\pi}}\int_{-a}^{a} 1.e^{isx}\, dx = \left[\frac{e^{isx}}{is}\right]_{-a}^{a} = \frac{1}{\sqrt{2\pi}}\frac{1}{(is)}\left[e^{ias} - e^{-ias}\right]$$

$$= \frac{1}{\sqrt{2\pi}}\frac{2}{s}.\frac{e^{ias} - e^{-ias}}{2i} = \frac{1}{\sqrt{2\pi}}\frac{2\sin sa}{s} = \sqrt{\frac{2}{\pi}}\frac{\sin sa}{s}$$ **Ans.**

Example 8. *Find the Fourier transform of* *(Uttarakhand, III Semester, 2009)*

$$f(x) = \begin{cases} 1 - x^2 & \text{if } |x| \le 1. \\ 0 & \text{if } |x| > 1. \end{cases}$$

and use it to evaluate $\int_0^\infty \left(\frac{x\cos x - \sin x}{x^3}\right)\cos\frac{x}{2}\, dx.$

Solution. We have $f(x)=\begin{cases}1-x^2, & -1<x<1\\ 0, & |x|>1\end{cases}$

The Fourier transform of a function $f(x)$ is given by

$$F[f(x)]=\frac{1}{\sqrt{2\pi}}\int_{-\infty}^{\infty} f(x).e^{isx}.dx \quad \text{... (1)}$$

Substituting the values of $f(x)$ in (1), we get

$$F[f(x)]=\frac{1}{\sqrt{2\pi}}\int_{-1}^{1}(1-x^2)e^{isx}dx$$

Integrating by parts, we get $\left[\int[uv]=uv_1-u'v_2+u''v_3...\right]$

$$F[f(x)]=\frac{1}{\sqrt{2\pi}}\left[(1-x^2)\frac{e^{isx}}{is}-(-2x)\frac{e^{isx}}{(is)^2}+(-2)\frac{e^{isx}}{(is)^3}\right]_{-1}^{1}$$

$$\Rightarrow \quad F(s)=\frac{1}{\sqrt{2\pi}}\left[-2\frac{e^{is}}{s^2}+2\frac{e^{is}}{is^3}-2\frac{e^{-is}}{s^2}-\frac{2e^{-is}}{is^3}\right]$$

$$=\frac{1}{\sqrt{2\pi}}\left[-\frac{2}{s^2}(e^{is}+e^{-is})+\frac{2}{is^3}(e^{is}-e^{-is})\right]$$

$$=\frac{1}{\sqrt{2\pi}}\left[-\frac{2}{s^2}(2\cos s)+\frac{2}{is^3}(2i\sin s)\right]=\frac{1}{\sqrt{2\pi}}\frac{4}{s^3}[-s\cos s+\sin s]$$

Ans.

By inversion formula for Fourier transform

$$f(x)=\frac{1}{\sqrt{2\pi}}\int_{-\infty}^{\infty}F(s)e^{-isx}dx \quad \text{... (2)}$$

Putting the value of $F(s)$ in (2), we get

$$f(x)=\frac{1}{\sqrt{2\pi}}\int_{-\infty}^{\infty}\frac{1}{\sqrt{2\pi}}\frac{4}{s^3}(\sin s-s\cos s)\,e^{-isx}\,ds$$

$$=\frac{2}{\pi}\int_{-\infty}^{\infty}\frac{1}{s^3}(\sin s-s\cos s)(\cos sx-i\sin sx)\,ds$$

$$=\frac{2}{\pi}\left[\int_{-\infty}^{\infty}\frac{1}{s^3}\sin s\cos sx\,ds-\int_{-\infty}^{\infty}\frac{i}{s^3}\sin s\sin sx\,ds-\int_{-\infty}^{\infty}\frac{1}{s^3}s\cos s\cos sx\,ds\right.$$
$$\left.+i\int_{-\infty}^{\infty}\frac{s}{s^3}\cos s\sin sx\,ds\right]$$

$$=\frac{4}{\pi}\left[\int_0^{\infty}\frac{1}{s^3}\sin s\cos sx\,ds-0-\int_0^{\infty}\frac{s}{s^3}\cos s\cos sx\,ds+0\right]$$

[I, III, functions are even and II, IV functions are odd]

$$=\frac{4}{\pi}\int_0^{\infty}\left(\frac{\sin s-s\cos s}{s^3}\right)\cos sx\,ds$$

Putting $x=\frac{1}{2}$, we get

$$f\left(\frac{1}{2}\right)=\frac{4}{\pi}\int_0^{\infty}\left(\frac{\sin s-s\cos s}{s^3}\right)\cos\frac{s}{2}ds \qquad \left[f\left(\frac{1}{2}\right)=1-\left(\frac{1}{2}\right)^2=\frac{3}{4}\right]$$

$$\frac{3}{4}=-\frac{4}{\pi}\int_0^{\infty}\left(\frac{s\cos s-\sin s}{s^3}\right)\cos\frac{s}{2}ds$$

$$-\frac{3\pi}{16} = \int_0^\infty \left(\frac{s\cos s - \sin s}{s^3}\right)\cos\frac{s}{2}\,ds$$

Hence, $\int_0^\infty \left(\frac{s\cos s - \sin s}{s^3}\right)\cos\frac{s}{2}\,ds = -\frac{3\pi}{16}$ **Ans.**

Example 9. *Find the Fourier transform of* e^{-ax^2}, where $a > 0$.

(*U.P. III Semester, Dec. 2002*)

Solution. The Fourier transform of $f(x)$:

$$F\{f(x)\} = \frac{1}{\sqrt{2\pi}}\int_{-\infty}^{\infty} f(x)e^{isx}\,dx \qquad \text{... (1)}$$

Putting the value of $f(x)$ in (1), we get

$$F\left\{e^{-ax^2}\right\} = \frac{1}{\sqrt{2\pi}}\int_{-\infty}^{\infty} e^{-ax^2}e^{isx}\,dx = \frac{1}{\sqrt{2\pi}}\int_{-\infty}^{\infty} e^{-x^2a-\frac{s^2}{4a}+isx+\frac{s^2}{4a}}\,dx = \frac{1}{\sqrt{2\pi}}\int_{-\infty}^{\infty} e^{-\left(x\sqrt{a}-\frac{is}{2\sqrt{a}}\right)^2} e^{-\frac{s^2}{4a}}\,dx$$

Putting $x\sqrt{a}-\frac{is}{2\sqrt{a}} = y,\quad dx = \frac{dy}{\sqrt{a}}$, we get $= \frac{e^{-s^2/4a}}{\sqrt{2\pi}}\int_{-\infty}^{\infty} e^{-y^2}\frac{dy}{\sqrt{a}}$.

$$= \frac{e^{-s^2/4a}}{\sqrt{2\pi}} \times \frac{\sqrt{\pi}}{\sqrt{a}} \qquad \left(\text{Since } \int_{-\infty}^{\infty} e^{-y^2}dy = \sqrt{\pi}\right)$$

$$F\left\{e^{-ax^2}\right\} = \frac{e^{-s^2/4a}}{\sqrt{2a}}$$ **Ans.**

Example 10. *Find the Fourier transform of the function*

$$f(x) = \begin{cases} \frac{1}{2\in}, & |x| \le \in \\ 0, & x > \in \end{cases}$$

Solution. We have, $f(x) = \begin{cases} \frac{1}{2\in}, & |x| \le \in \\ 0, & x > \in \end{cases}$

The Fourier transform of the function $f(x)$ is given by

$$F(s) = \frac{1}{\sqrt{2\pi}}\int_{-\infty}^{\infty} f(x)e^{isx}\,dx \qquad \text{..(1)}$$

Substituting the value of $f(x)$ in (1), we get

$$F[f(x)] = \frac{1}{\sqrt{2\pi}}\int_{-\in}^{\in}\frac{1}{2\in}e^{isx}\,dx = \frac{1}{\sqrt{2\pi}}\cdot\frac{1}{2\in}\int_{-\in}^{\in}(\cos sx + i\sin sx)\,dx$$

$$= \frac{1}{\sqrt{2\pi}}\cdot\frac{1}{2\in}\left[\int_{-\in}^{\in}\cos sx\,dx + i\int_{-\in}^{\in}\sin sx\,dx\right] = \frac{1}{\sqrt{2\pi}}\frac{2}{2\in}\left[\int_0^{\in}\cos sx\,dx + 0\right]$$

[First function is even and second is odd]

$$= \frac{1}{\sqrt{2\pi}}\cdot\frac{1}{\in}\left[\frac{\sin sx}{s}\right]_0^{\in} = \frac{1}{\sqrt{2\pi}}\cdot\frac{1}{\in}\left[\frac{\sin s\in - \sin 0}{s}\right] = \frac{1}{\sqrt{2\pi}}\frac{\sin s\in}{s\in}$$ **Ans.**

Example 11. *Find the Fourier transform of function*

$$f(t) = \begin{cases} t, & \text{for } |t| < a \\ 0, & \text{for } |t| > a \end{cases}$$

Solution. We have,

$$f(t) = \begin{cases} t, & for\,|t| < a \\ 0, & for\,|t| > a \end{cases}$$

The Fourier Transform of the function $f(t)$ is given by

$$F\,[f(t)] = \frac{1}{\sqrt{2\pi}} \int_{-\infty}^{\infty} f(t) e^{ist}\, dt \qquad ...(1)$$

Substituting the value of $f(t)$ in (1), we get

$$F\,[f(t)] = \frac{1}{\sqrt{2\pi}} \int_{-a}^{a} t\, e^{ist}\, dt = \frac{1}{\sqrt{2\pi}} \int_{-a}^{a} t\,(\cos st + i \sin st)\, dt$$

$$= \frac{1}{\sqrt{2\pi}} \left[0 + 2 \int_0^a it \,\sin st\, dt \right] \qquad \text{[First is odd and second is even]}$$

$$= \frac{2i}{\sqrt{2\pi}} \left[\left\{ t\left(\frac{-\cos st}{s} \right) \right\}_0^a - \int_0^a 1 \left(\frac{-\cos st}{s} \right) dt \right] = \frac{2i}{\sqrt{2\pi}} \left[-\frac{a}{s} \cos as + \frac{1}{s} \left[\frac{\sin st}{s} \right]_0^a \right]$$

$$= \frac{2i}{\sqrt{2\pi}} \left[-\frac{a}{s} \cos as + \frac{1}{s^2} \sin as \right] = \frac{2i}{\sqrt{2\pi}} \frac{1}{s^2} [\sin sa - as \cos as] \qquad \textbf{Ans.}$$

Example 12. *Find the Fourier transform of the function shown in the adjoining figure.*

Solution. Here,
$$f(t) = \begin{cases} 2, & \text{for } -1 < t < 1 \\ 1, & \text{for } -2 < t < -1 \\ 1, & \text{for } 1 < t < 2 \end{cases}$$

The Fourier transform of the function $f(t)$ is given by

$$F\,[f(t)] = \frac{1}{\sqrt{2\pi}} \int_{-\infty}^{\infty} f(t)\, e^{ist}\, dt \qquad ...\ (1)$$

Putting the value of $f(t)$ in (1), we get

$$= \frac{1}{\sqrt{2\pi}} \left[\int_{-1}^{1} 2e^{ist}\, dt + \int_{-2}^{-1} 1 \cdot e^{ist}\, dt + \int_1^2 1 \cdot e^{ist}\, dt \right] = \frac{1}{\sqrt{2\pi}} \left[2\left(\frac{e^{ist}}{is} \right)_{-1}^{1} + \left(\frac{e^{ist}}{is} \right)_{-2}^{-1} + \left(\frac{e^{ist}}{is} \right)_1^2 \right]$$

$$= \frac{1}{\sqrt{2\pi}} \left[2\left(\frac{e^{is} - e^{-is}}{is} \right) + \left(\frac{e^{-is} - e^{-2is}}{is} \right) + \left(\frac{e^{2is} - e^{is}}{is} \right) \right]$$

$$= \frac{1}{\sqrt{2\pi s}} \left[2\left(\frac{e^{is} - e^{-is}}{i} \right) + \left(\frac{e^{2is} - e^{-2is}}{i} \right) - \left(\frac{e^{is} - e^{-is}}{i} \right) \right] = \frac{1}{\sqrt{2\pi s}} \left[\left(\frac{e^{is} - e^{-is}}{i} \right) + \left(\frac{e^{2is} - e^{-2is}}{i} \right) \right]$$

$$= \frac{1}{\sqrt{2\pi s}} [2 \sin s + 2 \sin 2s] = \frac{1}{\sqrt{2\pi s}} [2 \sin s + 4 \sin s \cos s]$$

$$= \frac{2}{\sqrt{2\pi s}} \sin s\,(1 + 2\cos s) = = \sqrt{\frac{2}{\pi}} \cdot \frac{\sin s}{s} (1 + 2\cos s) \qquad \textbf{Ans.}$$

Example 13. *Find the Fourier transform of the function*

$$f(x) = \begin{cases} 1 + \dfrac{x}{a}, & (-a < x < 0) \\ 1 - \dfrac{x}{a}, & 0 < x < a \\ 0, & otherwise \end{cases}$$

(U.P., III Semester, Summer 2002)

Solution. Fourier transform of $f(x)$ is given by

$$F[f(x)] = \frac{1}{\sqrt{2\pi}}\int_{-\infty}^{\infty} f(x)\cdot e^{isx}dx \qquad \ldots (1)$$

Putting the value of $f(x)$ in (1), we get

$$F[f(x)] = \frac{1}{\sqrt{2\pi}}\int_{-a}^{0}\left(1+\frac{x}{a}\right)e^{isx}\,dx + \frac{1}{\sqrt{2\pi}}\int_{0}^{a}\left(1-\frac{x}{a}\right)e^{isx}dx$$

$$= \frac{1}{\sqrt{2\pi}}\left[\left(1+\frac{x}{a}\right)\times\frac{e^{isx}}{is}-\left(\frac{1}{a}\right)\frac{e^{isx}}{-s^2}\right]_{-a}^{0} + \frac{1}{\sqrt{2\pi}}\left[\left(1-\frac{x}{a}\right)\frac{e^{isx}}{is}-\left(-\frac{1}{a}\right)\frac{e^{isx}}{-s^2}\right]_{0}^{a}$$

$$= \frac{1}{\sqrt{2\pi}}\left[\frac{1}{is}+\frac{1}{a}\cdot\frac{1}{s^2}+\frac{1}{a}\frac{e^{-isa}}{-s^2}\right] + \frac{1}{\sqrt{2\pi}}\left[\frac{1}{a}\cdot\frac{e^{isa}}{-s^2}-\frac{1}{is}+\frac{1}{a}\frac{1}{s^2}\right]$$

$$= \frac{1}{\sqrt{2\pi}}\left[\frac{2}{as^2}+\frac{1}{-as^2}\left(e^{isa}+e^{-isa}\right)\right]$$

$$= \frac{1}{\sqrt{2\pi}}\left[\frac{2}{as^2}-\frac{2}{as^2}\cos sa\right] = \frac{1}{\sqrt{2\pi}}\,\frac{2}{as^2}[1-\cos as]$$

$$= \frac{2}{\sqrt{2\pi}\,as^2}\,2\sin^2\frac{as}{2} = \frac{2\sqrt{2}\,\sin^2\frac{as}{2}}{\sqrt{\pi}\,as^2} \qquad \textbf{Ans.}$$

Examples based on Fourier Sine Transform:

Example 14. *Find Fourier Sine transform of* $\frac{1}{x}$.

Solution. Here, $f(x) = \frac{1}{x}$

$$F_s[f(x)] = \sqrt{\frac{2}{\pi}}\int_0^{\infty} f(x)\sin sx\,dx \qquad \ldots (1)$$

Putting the value of $f(x)$ in (1), we get

$$F_s\left(\frac{1}{x}\right) = \sqrt{\frac{2}{\pi}}\int_0^{\infty}\frac{\sin sx}{x}dx = \sqrt{\frac{2}{\pi}}\int_0^{\infty}\frac{\sin\theta}{\frac{\theta}{s}}\frac{d\theta}{s} \qquad \text{Putting } s\,x = \theta \text{ so that } s\,dx = d\theta$$

$$= \sqrt{\frac{2}{\pi}}\int_0^{\infty}\frac{\sin\theta}{\theta}d\theta = \sqrt{\frac{2}{\pi}}\left(\frac{\pi}{2}\right) \qquad \text{[Some useful result I on page 1087]} \ \textbf{Ans.}$$

$$= \sqrt{\frac{\pi}{2}}$$

Example 15. *Find the Fourier Sine Transform of* e^{-ax}

Solution. Here, $f(x) = e^{-ax}$

The Fourier sine transform of $f(x)$:

$$F_s[f(x)] = \sqrt{\frac{2}{\pi}}\int_0^{\infty} f(x)\sin sx\,dx \qquad \ldots (1)$$

On putting the value of $f(x)$ in (1), we get

$$F_s[e^{-ax}] = \sqrt{\frac{2}{\pi}}\int_0^{\infty} e^{-ax}\sin sx\,dx$$

...rts, we get

$$= \sqrt{\frac{2}{\pi}} \left[\frac{e^{-ax}}{a^2+s^2} [-a \sin sx - s \cos sx] \right]_0^\infty \qquad \left[\int e^{ax} \sin bx \, dx = \frac{e^{ax}}{a^2+b^2} (a \sin bx - b \cos bx) \right]$$

$$= \sqrt{\frac{2}{\pi}} \left[0 - \frac{1}{a^2+s^2} (-s) \right] = \sqrt{\frac{2}{\pi}} \left(\frac{s}{a^2+s^2} \right) \qquad \textbf{Ans.}$$

Examples based on Fourier Cosine transform

Example 16. *Find the Fourier Cosine Transform of* $f(x) = e^{-ax}$.

Solution. The Fourier Cosine Transform is

$$F_c[f(x)] = \sqrt{\frac{2}{\pi}} \int_0^\infty f(x) \cos sx \, dx \qquad \text{... (1)}$$

Putting the value of $f(x)$ in (1), we get

$$F_c[e^{-ax}] = \sqrt{\frac{2}{\pi}} \int_0^\infty e^{-ax} \cos sx \, dx \qquad \left[\int e^{ax} \cos bx \, dx = \frac{e^{ax}}{a^2+b^2} (a \cos bx + b \sin bx) \right]$$

$$= \sqrt{\frac{2}{\pi}} \left[\frac{e^{-ax}}{a^2+s^2} \{-a \cos sx + s \sin sx\} \right]_0^\infty = \sqrt{\frac{2}{\pi}} \left[0 + \frac{a}{a^2+s^2} \right] = \sqrt{\frac{2}{\pi}} \frac{a}{a^2+s^2} \qquad \textbf{Ans.}$$

Example 17. *Find the Fourier Cosine Transform of* $f(x) = 5e^{-2x} + 2e^{-5x}$

Solution. The Fourier Cosine Transform of $f(x)$ is given by

$$F_c\{f(x)\} = \sqrt{\frac{2}{\pi}} \int_0^\infty f(x) \cos sx \, dx \qquad \text{... (1)}$$

Putting the value of $f(x)$ in (1), we get $F_c\{f(x)\} = \int_0^\infty (5e^{-2x} + 2e^{-5x}) \cos sx \, dx$

$$= 5 \int_0^\infty e^{-2x} \cos sx \, dx + 2 \int_0^\infty e^{-5x} \cos sx \, dx$$

$$= 5 \left[\frac{e^{-2x}}{(-2)^2 + s^2} (-2 \cos sx + s \sin sx) \right]_0^\infty + 2 \left[\frac{e^{-5x}}{(-5)^2 + s^2} (-5 \cos sx + s \sin sx) \right]_0^\infty$$

$$= 5 \left[0 - \frac{1}{4+s^2} (-2) \right] + 2 \left[0 - \frac{1}{25+s^2} (-5) \right] = 5 \left(\frac{2}{s^2+4} \right) + 2 \left(\frac{5}{s^2+25} \right)$$

$$= 10 \left(\frac{1}{s^2+4} + \frac{1}{s^2+25} \right) \qquad \textbf{Ans.}$$

Example 18. *Obtain Fourier Cosine Transfrom of*

$$f(x) = \begin{cases} x, & \text{for } 0 < x < 1 \\ 2 - x, & \text{for } 1 < x < 2 \\ 0, & \text{for } x > 2. \end{cases} \qquad \textit{(U.P., III Semester Dec. 2002)}$$

Solution. Fourier Cosine Transform

$$F_c\{f(x)\} = \sqrt{\frac{2}{\pi}} \int_0^\infty f(x) \cos sx \, dx \qquad \text{... (1)}$$

Putting the value of $f(x)$ in (1), we get

$$F_c[f(x)] = \sqrt{\frac{2}{\pi}}\left[\int_0^1 x\cos sx\,dx + \int_1^2 (2-x)\cos sx\,dx + \int_2^\infty 0.\cos sx\,dx\right]$$

$$= \sqrt{\frac{2}{\pi}}\left[\left\{x\frac{\sin sx}{s} - \left(-\frac{\cos sx}{s^2}\right)\right\}_0^1 + \left\{(2-x)\frac{\sin sx}{s} - (-1)\left(-\frac{\cos sx}{s^2}\right)\right\}_1^2\right] + 0$$

$$= \sqrt{\frac{2}{\pi}}\left[\left\{\left(\frac{\sin s}{s} + \frac{\cos s}{s^2}\right) - \frac{1}{s^2}\right\} + \left\{\left(-\frac{\cos 2s}{s^2}\right) - \left(\frac{\sin s}{s} - \frac{\cos s}{s^2}\right)\right\}\right]$$

$$= \sqrt{\frac{2}{\pi}}\left[\frac{2\cos s}{s^2} - \frac{1}{s^2} - \frac{\cos 2s}{s^2}\right] = \sqrt{\frac{2}{\pi}}\left[\frac{2\cos s - 1 - (2\cos^2 s - 1)}{s^2}\right]$$

$$= \sqrt{\frac{2}{\pi}}\left[\frac{2\cos s - 1 - 2\cos^2 s + 1}{s^2}\right] = \sqrt{\frac{2}{\pi}}\,\frac{2\cos s\,(1-\cos s)}{s^2}$$ **Ans.**

Examples based on Fourier Sine and Fourier Cosine Transform

Example 19. *Find Fourier Sine and Cosine Transform of (a)* x^{n-1}. *(b)* $\frac{1}{\sqrt{x}}$.

(U.P. III Semester (Comp.) 2004)

Solution. Here, $f(x) = x^{n-1}$

The Fourier Sine Transform of $f(x)$:

$$F_s[f(x)] = \sqrt{\frac{2}{\pi}}\int_0^\infty f(x)\sin sx\,dx \qquad \text{... (1)}$$

Putting the value of $f(x)$ in (1), we get

$$(a) \quad F_s(x^{n-1}) = \sqrt{\frac{2}{\pi}}\int_0^\infty \sin sx\cdot x^{n-1}\,dx \qquad \text{... (2)}$$

The Fourier Cosine Transform of $f(x)$:

$$F_c[f(x)] = \sqrt{\frac{2}{\pi}}\int_0^\infty f(x)\cos sx\,dx \qquad \text{... (3)}$$

Putting the value of $f(x)$ in (3), we get

$$F_c(x^{n-1}) = \sqrt{\frac{2}{\pi}}\int_0^\infty \cos sx\cdot x^{n-1}\,dx \qquad \text{... (4)}$$

Multiplying (2) by i and adding to (4), we have

$$F_c(x^{n-1}) + iF_s(x^{n-1}) = \sqrt{\frac{2}{\pi}}\int_0^\infty (\cos sx + i\sin sx)\,x^{n-1}\,dx = \sqrt{\frac{2}{\pi}}\int_0^\infty e^{isx}x^{n-1}\,dx$$

On putting $i\,s\,x = -t$ so that $isdx = -dt$, we get

$$= \sqrt{\frac{2}{\pi}}\int_0^\infty e^{-t}\left(-\frac{t}{is}\right)^{n-1}\left(-\frac{dt}{is}\right) = \sqrt{\frac{2}{\pi}}\,\frac{1}{(is)^n}(-1)^n\int_0^\infty e^{-t}t^{n-1}dt$$

$$= \sqrt{\frac{2}{\pi}}\,\frac{(i)^{2n}}{(i)^n s^n}\,\Gamma n = \sqrt{\frac{2}{\pi}}\,\frac{(i)^n}{s^n}\,\Gamma n \qquad \left[\int_0^\infty e^{-t}t^{n-1}dt = \Gamma n\right]$$

$$= \sqrt{\frac{2}{\pi}}\,\frac{\left(\cos\frac{\pi}{2} + i\sin\frac{\pi}{2}\right)^n \Gamma n}{s^n} = \sqrt{\frac{2}{\pi}}\,\frac{\left(\cos\frac{n\pi}{2} + i\sin\frac{n\pi}{2}\right)\Gamma n}{s^n}$$

Equating real and imaginary parts, we get

$$F_c(x^{n-1}) = \sqrt{\frac{2}{\pi}}\,\frac{\Gamma n}{s^n}\cos\frac{n\pi}{2} \qquad \text{... (5)}$$

$$F_s(x^{n-1}) = \sqrt{\frac{2}{\pi}}\,\frac{\Gamma n}{s^n}\sin\frac{n\pi}{2} \qquad \text{... (6)}\ \textbf{Ans.}$$

(*b*) Putting $n = \frac{1}{2}$ in (5), we get $F_c\left(\frac{1}{\sqrt{x}}\right) = \sqrt{\frac{2}{\pi}}\,\frac{\Gamma\frac{1}{2}}{s^{\frac{1}{2}}}\cos\frac{\pi}{4} = \sqrt{\frac{2}{\pi}}\,\frac{\sqrt{\pi}}{\sqrt{s}}\,\frac{1}{\sqrt{2}} = \frac{1}{\sqrt{s}}$

Putting $n = \frac{1}{2}$ in (6), we get $F_s\left(\frac{1}{\sqrt{x}}\right) = \sqrt{\frac{2}{\pi}}\,\frac{\Gamma\frac{1}{2}}{s^{\frac{1}{2}}}\sin\frac{\pi}{4} = \sqrt{\frac{2}{\pi}}\,\frac{\sqrt{\pi}}{\sqrt{s}}\,\frac{1}{\sqrt{2}} = \frac{1}{\sqrt{s}}$ **Ans.**

Second Method.

Example 20. *Find the Fourier Sine Transform of*

$$f(x) = \frac{e^{-ax}}{x}. \qquad \textit{(U.P. III Seamester 2008)}$$

Solution. The Sine Transform of the function $f(x)$ is given by

$$F(s) = \sqrt{\frac{2}{\pi}}\int_0^\infty f(x)\,\sin sx\,dx \qquad \text{... (1)}$$

Substituting the value of $f(x)$ in (1), we get

$$F(s) = \sqrt{\frac{2}{\pi}}\int_0^\infty \frac{e^{-ax}}{x}\sin sx\,dx$$

Differentiating both sides w.r.t. 's', we get

$$\frac{d}{ds}[F(s)] = \sqrt{\frac{2}{\pi}}\int_0^\infty \frac{e^{-ax}}{x}(x\cos sx)\,dx$$

$$= \sqrt{\frac{2}{\pi}}\int_0^\infty e^{-ax}\cos sx\,dx = \sqrt{\frac{2}{\pi}}\left[\frac{e^{-ax}}{a^2+s^2}(-a\cos + \sin sx)\right]_0^\infty$$

$$= \sqrt{\frac{2}{\pi}}\,\frac{a}{s^2+a^2} \qquad \text{... (2)}$$

Integrating (2) w.r.t. 's', we get

$$F(s) = \sqrt{\frac{2}{\pi}}\int\frac{a}{s^2+a^2}\,ds = \sqrt{\frac{2}{\pi}}\tan^{-1}\frac{s}{a} + C \qquad \text{... (3)}$$

For $s = 0$, $F(s) = 0$

Putting s = 0, $F(s) = 0$ in (3), we get

$$0 = 0 + C \quad \text{or} \quad C = 0$$

On putting the value of C in (3), we get

$$F(s) = \sqrt{\frac{2}{\pi}}\tan^{-1}\frac{s}{a} \qquad \textbf{Ans.}$$

Example 21. *Find Fourier Cosine Transform of* $e^{-a^2x^2}$ *and hence evaluate Fourier Sine Transform of* $xe^{-a^2x^2}$

Solution. *Here,* $f(x) = e^{-a^2x^2}$

The Fourier Cosine Transform of $f(x)$:

$$F_c[f(x)] = \sqrt{\frac{2}{\pi}}\int_0^\infty f(x)\cos sx\,dx \qquad \text{... (1)}$$

Putting the value of $f(x)$ in (1), we get

$$F_c(e^{-a^2x^2}) = \sqrt{\frac{2}{\pi}} \int_0^\infty e^{-a^2x^2} \cdot \cos sx \, dx$$

$$= \text{Real part of } \sqrt{\frac{2}{\pi}} \int_0^\infty e^{-a^2x^2} \cdot e^{isx} \, dx \qquad [\cos sx + i \sin sx = e^{isx}]$$

$$= \text{Real part of } \sqrt{\frac{2}{\pi}} \int_0^\infty e^{-a^2x^2+isx} \, dx = \frac{1}{a\sqrt{2}} e^{-\frac{s^2}{4a^2}}$$ (See Example 9 on page 1294)

We know that

$$F_s[(x f(x)] = -\frac{d}{ds} F_c f(x)$$ (See Example 22 page 1300)

$$F_s(xe^{-a^2x^2}) = -\frac{d}{ds} F_c(e^{-a^2x^2}) \qquad \left[\because f(x) = e^{-a^2x^2}\right]$$

$$= -\frac{d}{ds} \frac{1}{a\sqrt{2}} e^{-\frac{s^2}{4a^2}} = \frac{1}{a\sqrt{2}} e^{-\frac{s^2}{4a^2}} \frac{s}{2a^2} = \frac{s}{2\sqrt{2}a^3} e^{-\frac{s^2}{4a^2}}$$ **Ans.**

Example 22. *Find Fourier Cosine Transform of* $\frac{1}{1+x^2}$ *and hence find Fourier Sine Transform of* $\frac{x}{1+x^2}$. *(Uttrakhand, III Semester, June 2009; U.P. III Semester Dec. 2004)*

Solution. Here, $f(x) = \frac{1}{1+x^2}$

The Fourier Cosine Transform of $f(x)$:

$$F_c[f(x)] = \sqrt{\frac{2}{\pi}} \int_0^\infty f(x) \cos sx \, dx \qquad \text{... (1)}$$

Putting the value of $f(x)$ in (1), we get

$$F_c(s) = \sqrt{\frac{2}{\pi}} \int_0^\infty \frac{1}{1+x^2} \cos sx \, dx = I \text{ (say)} \qquad \text{...(2)}$$

Differentiating w.r.t. 's', we get

$$\frac{dI}{ds} = \sqrt{\frac{2}{\pi}} \int_0^\infty \frac{1}{1+x^2} \left(\frac{d}{ds} \cos sx\right) dx$$

$$\Rightarrow \quad \frac{dI}{ds} = \sqrt{\frac{2}{\pi}} \int_0^\infty \frac{-x \sin sx}{1+x^2} dx = -\sqrt{\frac{2}{\pi}} \int_0^\infty \frac{x^2 \sin sx}{x(1+x^2)} dx$$

$$= -\sqrt{\frac{2}{\pi}} \int_0^\infty \frac{(1+x^2-1) \sin sx}{x(1+x^2)} dx$$

$$= -\sqrt{\frac{2}{\pi}} \int_0^\infty \frac{(1+x^2) \sin sx}{x(1+x^2)} dx + \sqrt{\frac{2}{\pi}} \int_0^\infty \frac{\sin sx}{x(1+x^2)} dx$$

$$= -\sqrt{\frac{2}{\pi}} \int_0^\infty \frac{\sin sx}{x} dx + \sqrt{\frac{2}{\pi}} \int_0^\infty \frac{\sin sx}{x(1+x^2)} dx \qquad \left[\int_0^\infty \frac{\sin sx}{x} dx = \frac{\pi}{2}\right]$$

$$\Rightarrow \quad \frac{dI}{ds} = -\sqrt{\frac{2}{\pi}} \cdot \frac{\pi}{2} + \sqrt{\frac{2}{\pi}} \int_0^\infty \frac{\sin sx}{x(1+x^2)} dx \qquad \text{...(3)]}$$

Again differentiating, we get

$$\Rightarrow \quad \frac{d^2I}{ds^2} = 0 + \sqrt{\frac{2}{\pi}} \int_0^\infty \frac{\frac{d}{ds}\sin sx}{x(1+x^2)}\, dx$$

$$\Rightarrow \quad \frac{d^2I}{ds^2} = \sqrt{\frac{2}{\pi}} \int_0^\infty \frac{x\cos sx}{x(1+x^2)}\, dx = \sqrt{\frac{2}{\pi}} \int_0^\infty \frac{\cos sx}{1+x^2}\, dx \qquad \text{[From (2)]}$$

$$\Rightarrow \quad \frac{d^2I}{ds^2} - I = 0 \Rightarrow D^2 I - I = 0 \Rightarrow (D^2 - 1)\, I = 0$$

A.E. is $m^2 - 1 = 0 \Rightarrow m = \pm 1$

C.F. $= C_1 e^s + C_2 e^{-s}$ and P.I. $= 0$

$$\Rightarrow \quad I = C_1 e^s + C_2 e^{-s} \qquad \text{...(4)}$$

$$\therefore \quad \frac{dI}{ds} = C_1 e^s - C_2 e^{-s} \qquad \text{...(5)}$$

Putting $s = 0$ in (2), we get

$$I = \sqrt{\frac{2}{\pi}} \int_0^\infty \frac{1}{1+x^2}\, dx = \sqrt{\frac{2}{\pi}} \left[\tan^{-1} x\right]_0^\infty = \sqrt{\frac{2}{\pi}} \left[\tan^{-1}\infty - 0\right] = \sqrt{\frac{2}{\pi}}\left(\frac{\pi}{2}\right) = \sqrt{\frac{\pi}{2}} \qquad \text{...(6)}$$

Again putting s = 0 in (3), we get $\frac{dI}{ds} = -\sqrt{\frac{\pi}{2}}$ ...(7)

Putting $s = 0$ and equating (4) and (6) ; (5) and (7), we get

$$C_1 + C_2 = \sqrt{\frac{\pi}{2}} \text{ and } C_1 - C_2 = -\sqrt{\frac{\pi}{2}}$$

On solving, we get $C_1 = 0, \quad C_2 = \sqrt{\frac{\pi}{2}}$

Putting the values of C_1 and C_2 in(4), we get

$$I = \sqrt{\frac{\pi}{2}}\, e^{-s}$$

$$\Rightarrow \quad \sqrt{\frac{2}{\pi}} \int_0^\infty \frac{\cos sx}{1+x^2}\, dx = \sqrt{\frac{\pi}{2}}\, e^{-s} \qquad \left[I = \sqrt{\frac{2}{\pi}} \int_0^\infty \frac{\cos sx}{1+x^2}\, dx\right]$$

Differentiating w.r.t. s, we get

$$\sqrt{\frac{2}{\pi}} \int_0^\infty \frac{-x\sin sx}{1+x^2}\, dx = -\sqrt{\frac{\pi}{2}}\, e^{-s}$$

$$\Rightarrow \quad \sqrt{\frac{2}{\pi}} \int_0^\infty \frac{x\sin sx}{1+x^2}\, dx = \sqrt{\frac{\pi}{2}}\, e^{-s}$$

$$\Rightarrow \quad \int_0^\infty \frac{x\sin sx}{1+x^2}\, dx = \frac{\pi}{2}\, e^{-s} \qquad \textbf{Ans.}$$

Example 23. *Taking the function* $f(x) = \begin{cases} 1, & 0 < x < \pi \\ 0, & x > \pi \end{cases}$

show that $\int_0^\infty \left(\frac{1-\cos s\pi}{s}\right) . \sin sx\, ds = \begin{cases} \frac{\pi}{2}, & 0 < x < \pi \\ 0, & x > \pi \end{cases}$

Solution. We have, $f(x) = \begin{cases} 1, & 0 < x < \pi \\ 0, & x > \pi \end{cases}$

The Fourier Sine Transform of $f(x)$:

$$F_s[f(x)] = \sqrt{\frac{2}{\pi}} \int_0^\infty f(x) \sin sx \, dx \qquad \dots (1)$$

Putting the value of $f(x)$ in (1), we get

$$= \sqrt{\frac{2}{\pi}} \int_0^\pi 1 \cdot \sin sx \, dx + \sqrt{\frac{2}{\pi}} \int_\pi^\infty 0. \sin sx \, dx = \sqrt{\frac{2}{\pi}} \left[-\frac{\cos sx}{s} \right]_0^\pi = \sqrt{\frac{2}{\pi}} \left(\frac{1 - \cos s\pi}{s} \right)$$

By inverse formula for Fourier sine transform

$$\sqrt{\frac{2}{\pi}} \int_0^\infty \sqrt{\frac{2}{\pi}} \left(\frac{1 - \cos s\pi}{s} \right) \sin sx \, ds = \begin{cases} 1, & 0 < x < \pi \\ 0, & x > \pi \end{cases}$$

$$\Rightarrow \quad \frac{2}{\pi} \int_0^\infty \left(\frac{1 - \cos s\pi}{s} \right) \sin sx \, ds = \begin{cases} 1, & 0 < x < \pi \\ 0, & x > \pi \end{cases}$$

$$\Rightarrow \quad \int_0^\infty \left(\frac{1 - \cos s\pi}{s} \right) \sin sx \, ds = \begin{cases} \frac{\pi}{2}, & 0 < x < \pi \\ 0, & x > \pi \end{cases}$$

Ans.

Example 24. *Find the Fourier sine transform of* $e^{-|x|}$.

Hence evaluate $\int_0^\infty \frac{x \sin mx}{1+x^2} dx$ *(U.P. III Semester, Dec. 2003)*

Solution. In the interval (0, ∞), x is always positive therefore $e^{-|x|} = e^{-x}$

Now, Fourier sine transform of e^{-x} is given by

$$F_s\{e^{-x}\} = \sqrt{\frac{2}{\pi}} \int_0^\infty e^{-x} \sin sx \, dx = \sqrt{\frac{2}{\pi}} \left[\frac{e^{-x}}{(-1)^2 + s^2} (-\sin sx - s \cos sx) \right]_0^\infty$$

$$= \sqrt{\frac{2}{\pi}} \frac{1}{1+s^2} (0 + s) = \sqrt{\frac{2}{\pi}} \left(\frac{s}{1+s^2} \right) = F(s) \qquad \dots (1)$$

Now, the inverse sine transform of F (s), is e^{-x}. Using inverse formula for the Sine Transform, we get

$$e^{-x} = \sqrt{\frac{2}{\pi}} \int_0^\infty F(s) \sin sx \, ds = \sqrt{\frac{2}{\pi}} \int_0^\infty \sqrt{\frac{2}{\pi}} \left(\frac{s}{1+s^2} \right) \sin sx \, ds \qquad \text{[Using (1)]}$$

Replacing x by m, we get

$$e^{-m} = \frac{2}{\pi} \int_0^\infty \frac{s}{1+s^2} \sin ms \, ds$$

Replacing s by x, we get

$$e^{-m} = \frac{2}{\pi} \int_0^\infty \frac{x \sin mx}{1+x^2} dx \qquad \left[\int_a^b F(s) \, ds = \int_a^b f(x) \, dx \right]$$

Hence, we get $\int_0^\infty \frac{x \sin mx}{1+x^2} dx = \frac{\pi}{2} e^{-m}$. **Ans.**

Example 25. *Find the function whose Sine Transform is* $\frac{e^{-as}}{s}$

Solution. Here, $F_s[f(x)] = \frac{e^{-as}}{s}$

The inverse Fourier Sine Transform of $F_s[f(x)]$ or $F_s(s)$:

$$f(x)=\sqrt{\frac{2}{\pi}}\int_0^{\infty}F(s)\,.\,\sin sx\,ds \qquad \ldots (1)$$

Putting the value of $F(s)$ in (1), we get $f(x)=\sqrt{\frac{2}{\pi}}\int_0^{\infty}\frac{e^{-as}}{s}\sin sx\,dx$

Differentiating w.r.t 'x' we get

$$\frac{df}{dx}=\sqrt{\frac{2}{\pi}}\int_0^{\infty}\frac{e^{-as}}{s}\frac{d}{dx}(\sin sx)\,ds=\sqrt{\frac{2}{\pi}}\int_0^{\infty}\frac{e^{-as}}{s}(s\cos sx)\,ds$$

$$=\sqrt{\frac{2}{\pi}}\int_0^{\infty}e^{-as}\cos sx\,ds=\sqrt{\frac{2}{\pi}}\left[\frac{e^{-as}}{a^2+x^2}(-a\cos sx+x\sin sx)\right]_0^{\infty}$$

$$=\sqrt{\frac{2}{\pi}}\left[0-\frac{1}{a^2+x^2}(-a)\right]=\sqrt{\frac{2}{\pi}}\frac{a}{a^2+x^2} \qquad \ldots (2)$$

Integrating both sides, of (2) w.r.t. 'x', we get

$$f(x)=\left(\sqrt{\frac{2}{\pi}}\,a\right)\frac{1}{a}\tan^{-1}\frac{x}{a}+C \qquad =\sqrt{\frac{2}{\pi}}\,.\,\tan^{-1}\frac{x}{a}+C \qquad \ldots (3)$$

On putting $x=0$ and $f(0)=0$ in (2), we get

$$f(0)=0+C \qquad \Rightarrow 0=0+C \qquad \Rightarrow C=0$$

On putting the value of C in (3), we get

$$f(x)=\sqrt{\frac{2}{\pi}}\tan^{-1}\frac{x}{a}$$

$$F_s^{-1}\left(\frac{e^{-as}}{s}\right)=\sqrt{\frac{2}{\pi}}\tan^{-1}\frac{x}{a} \qquad \ldots (3)$$

On substituting $a=0$ in (3), we have

$$F^{-1}\left(\frac{1}{s}\right)=\sqrt{\frac{2}{\pi}}\,.\,\frac{\pi}{2}=\sqrt{\frac{\pi}{2}} \qquad \textbf{Ans.}$$

Example 26. *Find the Fourier Cosine transform of*

$$f(x)=\begin{cases} x & \text{for } 0<x<\frac{1}{2} \\ 1-x & \text{for } \frac{1}{2}<x<1 \\ 0 & \text{for } x>1. \end{cases}$$

Write the inverse transform.

Soluiton. The Fourier Cosine Transform of $f(x)$:

$$F_c[f(x)]=\sqrt{\frac{2}{\pi}}\int_0^{\infty}f(x)\cdot\cos sx\,dx \qquad \ldots (1)$$

Putting the value of $f(x)$ in (1), we get

$$=\sqrt{\frac{2}{\pi}}\int_0^{1/2}x\cos sx\,dx+\sqrt{\frac{2}{\pi}}\int_{1/2}^{1}(1-x)\cos sx\,dx$$

$$=\sqrt{\frac{2}{\pi}}\left[x\frac{\sin sx}{s}-\left(\frac{-\cos sx}{s^2}\right)\right]_0^{1/2}+\sqrt{\frac{2}{\pi}}\left[(1-x)\frac{\sin sx}{s}-(-1)\frac{(-\cos sx)}{s^2}\right]_{1/2}^{1}$$

$$= \sqrt{\frac{2}{\pi}}\left[\frac{1}{2}\frac{\sin\frac{s}{2}}{s}+\frac{\cos\frac{s}{2}}{s^2}-\frac{1}{s^2}\right]+\sqrt{\frac{2}{\pi}}\left[-\frac{\cos s}{s^2}-\frac{1}{2}\frac{\sin\frac{s}{2}}{s}+\frac{\cos\frac{s}{2}}{s^2}\right]$$

$$= \sqrt{\frac{2}{\pi}}\left[-\frac{\cos s}{s^2}+\frac{2\cos s/2}{s^2}-\frac{1}{s^2}\right] \qquad \textbf{Ans.}$$

Example 27. *Solve the integral equation*

$$\int_0^\infty f(x)\cos sx\,dx = \begin{cases}1-s, & 0\le s\le 1\\ 0, & s>1\end{cases}$$

Hence prove that $\int_0^\infty \frac{\sin^2 t}{t^2}dt = \frac{\pi}{2}$. *(U.P. III Semester (SUM) 2004)*

Solution. $F_c(s) = \sqrt{\frac{2}{\pi}}\int_0^\infty f(x)\cos sx\,dx$ and $F_c(s) = \begin{cases}1-s, & 0\le s\le 1\\ 0, & s>1\end{cases}$

$\therefore$ By inversion formula for Fourier cosine transform, we have

$$f_c(x) = \sqrt{\frac{2}{\pi}}\int_{-\infty}^{\infty} F_c(s)\cos sx\,ds \qquad \text{...(1)}$$

Putting the value of $F_c(s)$ in (1), we get

$$f(x) = \sqrt{\frac{2}{\pi}}\left[\int_0^1 (1-s)\cos sx\,ds + \int_1^\infty 0.\cos sx\,ds\right] = \sqrt{\frac{2}{\pi}}\int_0^1 (1-s)\cos sx\,ds$$

$$= \sqrt{\frac{2}{\pi}}\left[(1-s)\cdot\frac{\sin sx}{x}-(-1)\cdot\frac{-\cos sx}{x^2}\right]_0^1 = \sqrt{\frac{2}{\pi}}\left[0-\frac{\cos x}{x^2}+\frac{1}{x^2}\right] = \sqrt{\frac{2}{\pi}}\left[\frac{1-\cos x}{x^2}\right] \qquad \text{...(2)}$$

Deduction. Since $\sqrt{\frac{2}{\pi}}\int_0^\infty f(x)\cos sx\,dx = \begin{cases}1-s, & 0\le s\le 1\\ 0, & s>1\end{cases}$ where $f(x) = \sqrt{\frac{2}{\pi}}\left[\frac{1-\cos x}{x^2}\right]$

From (1) and (2), we have

$$\sqrt{\frac{2}{\pi}}\int_0^\infty \sqrt{\frac{2}{\pi}}\left(\frac{1-\cos x}{x^2}\right)\cos sx\,dx = \begin{cases}1-s, & 0\le s\le 1\\ 0, & s>1\end{cases} \qquad \text{... (3)}$$

On putting $s = 0$ in (3), we have

$$\frac{2}{\pi}\int_0^\infty \frac{1-\cos x}{x^2}dx = 1 \qquad \Rightarrow \int_0^\infty \frac{2\sin^2\frac{x}{2}}{x^2}dx = \frac{\pi}{2}$$

Putting $x = 2t$ so that $dx = 2dt$, we get $\int_0^\infty \frac{\sin^2 t}{t^2}dt = \frac{\pi}{2}$. **Proved.**

Example 28. *Show that*

(*a*) $F_s[x\,f(x)] = -\frac{d}{ds}F_c(s)$ (*b*) $F_c[x\,f(x)] = \frac{d}{ds}F_s(s)$

and hence find Fourier Cosine and Sine Transform of $x\,e^{-ax}$.

Solution. (*a*) The Fourier Cosine Transform of $f(x)$

$$F_c[f(x)] = \sqrt{\frac{2}{\pi}}\int_0^\infty f(x)\cos sx\,dx \qquad \text{... (1)}$$

Differentiating (1), w.r.t. '*s*', we get $\frac{d}{ds}F_c[f(x)] = -\sqrt{\frac{2}{\pi}}\int_0^\infty x\,f(x)\sin sx\,dx$

$$\frac{d}{ds}F_c(s) = -F_s\{x\,f(x)\} \qquad \text{... (2)}$$

(*b*) The Fourier Sine Transform of $f(x)$

$$F_s\,[f(x)] = \sqrt{\frac{2}{\pi}} \int_0^\infty f(x) \sin sx\, dx$$

Differentiating w.r.t 's' we get

$$\frac{d}{ds} F_s\,[f(x)] = \sqrt{\frac{2}{\pi}} \int_0^\infty f(x) \frac{d}{ds} (\sin sx)\, dx = \sqrt{\frac{2}{\pi}} \int_0^\infty x f(x) \cos sx\, dx$$

$$\frac{d}{ds} F_s(s) = F_c\,\{x f(x)\} \qquad \ldots (3)$$

(c) $F_c\,\{x f(x)\} = \dfrac{d}{ds} F_s\,[f(x)]$ [From (3)] $\Rightarrow$ $F_c\,(xe^{-ax}) = \dfrac{d}{ds} F_s(e^{-ax})$

$$= \frac{d}{ds}\left[\sqrt{\frac{2}{\pi}}\, \frac{s}{a^2+s^2}\right] \qquad \text{(Using example ???on page ???)}$$

$$= \sqrt{\frac{2}{\pi}}\, \frac{(a^2+s^2)-s(2s)}{(a^2+s^2)^2} = \sqrt{\frac{2}{\pi}}\, \frac{a^2-s^2}{(a^2+s^2)^2}$$

(d) $F_s\{x f(x)\} = \dfrac{-d}{ds} F_c\,\{f(x)\}$ [From (2)]

$$F_s\,(xe^{-ax}) = -\frac{d}{ds} F_c\,(e^{-ax}) = -\frac{d}{ds}\left(\sqrt{\frac{2}{\pi}}\, \frac{u}{a^2+s^2}\right)$$

$$= \sqrt{\frac{2}{\pi}}\, \frac{2as}{(a^2+s^2)^2}$$

Ans.

Example 29. *If* $F_c(s) = \dfrac{1}{2} \tan^{-1}\left(\dfrac{2}{s^2}\right)$, *then find* $f(x)$.

Solution. $\tan^{-1}\left(\dfrac{2}{s^2}\right) = \tan^{-1}\left[\dfrac{2}{(s^2-1)+1}\right] = \tan^{-1}\left[\dfrac{(s+1)-(s-1)}{1+(s+1)\,(s-1)}\right]$

$$= \tan^{-1}\left(\frac{\dfrac{1}{s-1} - \dfrac{1}{s+1}}{1+\left(\dfrac{1}{s+1}\right)\left(\dfrac{1}{s-1}\right)}\right) = \tan^{-1}\left(\frac{1}{s-1}\right) - \tan^{-1}\left(\frac{1}{s+1}\right)$$

The inversion Fourier Cosine formula is

$$f(x) = \sqrt{\frac{2}{\pi}} \int_0^\infty F(s) \cos sx\, dx \qquad \ldots (1)$$

Putting the value of F (s) in (1), we get

$$f(x) = \sqrt{\frac{2}{\pi}} \int_0^\infty \frac{1}{2} \tan^{-1}\left(\frac{2}{s^2}\right) \cos sx\, ds = \sqrt{\frac{1}{2\pi}} \int_0^\infty \left\{\tan^{-1}\left(\frac{1}{s-1}\right) - \tan^{-1}\left(\frac{1}{s+1}\right)\right\} \cos sx\, ds$$

$$= \sqrt{\frac{1}{2\pi}}\left[\int_0^\infty \tan^{-1}\left(\frac{1}{s-1}\right) \cos sx\, ds - \int_0^\infty \tan^{-1}\left(\frac{1}{s+1}\right) \cos sx\, ds\right]$$

$$= I_1 - I_2 = I \text{ (say)} \qquad \ldots (1)$$

Where $$I_1 = \sqrt{\frac{1}{2\pi}}\left[\left\{\tan^{-1}\left(\frac{1}{s-1}\right)\cdot\frac{\sin sx}{x}\right\}_0^\infty - \int_0^\infty \frac{-1}{(s-1)^2}\cdot\frac{1}{\left\{1+\frac{1}{(s-1)^2}\right\}}\cdot\frac{\sin sx}{x}\,ds\right]$$

$$I_1 = 0 + \frac{1}{x}\sqrt{\frac{1}{2\pi}}\int_0^\infty \frac{\sin sx}{(s-1)^2+1}\,ds$$

Similarly, $$I_2 = \frac{1}{x}\sqrt{\frac{1}{2\pi}}\int_0^\infty \frac{\sin sx}{(s+1)^2+1}\,ds$$

On putting the value of I_1 and I_2 in (1), we get

$$I = \frac{1}{x}\sqrt{\frac{1}{2\pi}}\int_0^\infty\left\{\frac{\sin sx}{(s-1)^2+1}-\frac{\sin sx}{(s+1)^2+1}\right\}ds$$

$$= \frac{1}{x}\sqrt{\frac{1}{2\pi}}\,\pi\left[-\pi e^{-x}\sin x - e^{-x}\sin(-x)\right]$$ [See formula 3 on page 5]

$$\Rightarrow \quad f(x) = \frac{1}{x}\sqrt{\frac{1}{2\pi}}\,\pi\left[e^{-x}\sin x + e^{-x}\sin x\right] = \frac{\pi}{x}\sqrt{\frac{2}{\pi}}e^{-x}\sin x$$

$$= \sqrt{\frac{\pi}{2}}\left[\frac{e^{-x}\sin x}{x}\right]$$ **Ans.**

Example 30. *By finding the Fourier Transform of* $f(x) = e^{-a^2x^2}$, $a > 0$, *show that the transform of* $e^{-\frac{x^2}{2}}$ is $e^{-\frac{s^2}{2}}$.

Solution. The Fourier Transform of $f(x)$ is

$$F_c[f(x)] = \sqrt{\frac{2}{\pi}}\int_0^\infty f(x)\cos sx\,dx \qquad \dots(1)$$

Putting the value of $f(x)$ in (1), we get

$$F\{e^{-a^2x^2}\} = \sqrt{\frac{2}{\pi}}\int_{-\infty}^{\infty} e^{-a^2x^2}e^{isx}\,dx = \sqrt{\frac{2}{\pi}}\int_{-\infty}^{\infty} e^{-a^2x^2+isx+\frac{s^2}{4a^2}}\,e^{\frac{-s^2}{4a^2}}\,dx$$

$$= \sqrt{\frac{2}{\pi}}\int_{-\infty}^{\infty} e^{-\left[(ax)^2-\left(\frac{is}{2a}\right)^2+2(ax)\left(\frac{is}{2a}\right)\right]}\,e^{-\frac{s^2}{4a^2}}\,dx = \sqrt{\frac{2}{\pi}}\int_{-\infty}^{\infty} e^{-\left(ax-\frac{is}{2a}\right)^2}\,e^{-\frac{s^2}{4a^2}}\,dx$$

$\left[\text{Putting } ax-\frac{is}{2a} = t \text{ and } a>0 \text{ so that } a\,dx = dt \Rightarrow dx = \frac{dt}{a}\right]$

$$= e^{-\frac{s^2}{4a^2}}\cdot\sqrt{\frac{2}{\pi}}\int_{-\infty}^{\infty} e^{-t^2}\frac{dt}{a}$$

$$= e^{-\frac{s^2}{4a^2}}\cdot\frac{1}{a}\sqrt{\frac{2}{\pi}}\times\frac{\sqrt{\pi}}{2} \qquad \left[\text{since } \int_0^\infty e^{-t^2}dt = \frac{\sqrt{\pi}}{2}\right]$$

$$= \frac{1}{\sqrt{2}a}e^{-\frac{s^2}{4a^2}}$$

Putting $a = \frac{1}{\sqrt{2}}$, we get

$$F\left\{e^{-\frac{x^2}{2}}\right\} = e^{-\frac{s^2}{2}}$$

Thus, $e^{-\frac{x^2}{2}}$ is self-reciprocal. **Proved.**

EXERCISE 41.2

1. Find the Fourier Transform of $f(x)$ if

$$f(x)=\begin{cases}x, & |x|\le a\\ 0, & |x|>a\end{cases}$$

Ans. $\frac{1}{\sqrt{2\pi}}\frac{2i}{s^2}(as\cos as-\sin as)$

2. Show that the Fourier Transform of

$$f(x)=\begin{cases}a-|x| & \text{for } |x|<a\\ 0 & \text{for } |x|>a>0\end{cases}$$

is $\sqrt{\frac{2}{\pi}}\left(\frac{1-\cos as}{s^2}\right)$.

Hence show that $\int_0^\infty\left(\frac{\sin t}{t}\right)^2 dt=\frac{\pi}{2}$

3. Show that the Fourier Transform of

$$f(x)=\begin{cases}\frac{\sqrt{2\pi}}{2a} & \text{for } |x|\le a\\ 0 & \text{for } |x|>a\end{cases}$$

is $\frac{\sin sa}{sa}$

4. Find Fourier Transform of $e^{-a|x|}$ if $a>0$ and $x>0$. **Ans.** $\frac{2a}{a^2+s^2}$

5. Find Fourier Transform of $\frac{1}{\sqrt{|x|}}$.

6. Find the Fourier Transform of $f(x)=\begin{cases}e^{ikx}, & a<x<b\\ 0, & x<a \text{ and } x>b\end{cases}$ **Ans.** $\frac{i}{\sqrt{2\pi}(k+s)}\left[e^{i(k+s)a}-e^{i(k+s)b}\right]$

7. Show that the Fourier Transform of

$$f(x)=\begin{cases}0 & \text{for } x<\alpha\\ 1 & \text{for } \alpha<x<\beta\\ 0 & \text{for } x>\beta\end{cases}$$

is $\frac{1}{\sqrt{2\pi}}\frac{2i}{s^2}(as\cos as-\sin as)$

8. If $F(s)$ is the Fourier Transform of $f(x)$, prove that

$$f[e^{iax}f(x)]=F(s+a)$$

9. Find Fourier transform of

$$F(x)=\begin{cases}x^2, & |x|<a\\ 0, & |x|>a\end{cases}.$$

Ans. $\left(\frac{2a^2}{s}-\frac{4}{s^3}\right)\sin as+\frac{4a}{s^2}\cos as$

10. Show that the Fourier Sine Transform of $f(x)=\begin{cases}x & \text{for } 0<x<1\\ 2-x & \text{for } 1<x<2\\ 0 & \text{for } x>2\end{cases}$ is $\frac{2\sin s(1-\cos s)}{s^2}$.

11. Show that the Fourier Sine Transform of $\frac{x}{1+x^2}$ is $\sqrt{\frac{\pi}{2}}\,as\,e^{-as}$

12. Find Fourier Sine Transform of

$$f(x)=\frac{1}{x(x^2+a^2)}$$ (*U.P.T.U. 2001*)

Ans. $\frac{\pi}{2a^2}(1-e^{-as})$

13. Find Fourier Sine Transform of $xe^{-x^2/2}$. **Ans.** $\frac{1}{2} s e^{-\frac{s^2}{2}}$

14. Find Fourier Sine Transform of $\frac{e^{-ax}-e^{-bx}}{x}$ **Ans.** $\tan^{-1}\frac{s}{a}-\tan^{-1}\frac{s}{b}$

15. Find Fourier Sine transform of $\frac{\cosh ax}{\sinh \pi x}$ **Ans.** $\frac{\sinh s}{2(\cos a+\cosh s)}$

16. Find the Fourier Cosine Transform of $\frac{e^{ax}+e^{-ax}}{e^{\pi x}-e^{-\pi x}}$ **Ans.** $\cos\frac{a}{2}\left(\frac{e^{\frac{s}{2}}+e^{-\frac{s}{2}}}{2\cos sa+e^{s}+e^{-s}}\right)$

17. Find the Fourier Sine and Cosine Transform of $ae^{-\alpha x}+be^{-\beta x}$ $\alpha, \beta > 0$.

Ans. $\frac{as}{s^2+\alpha^2}+\frac{bs}{s^2+\beta^2}, \frac{a\alpha}{s^2+\alpha^2}+\frac{b\beta}{s^2+\beta^2}$

18. If $F(s)$ is the Fourier Transform of $f(x)$, prove that $F[x^n f(x)]=(-i)^n \frac{d^n}{ds^n}\{F(s)\}$

[**Hint:** See Art. 1.8 Properties of Fourier Transform on page 35]

19. Find the function $f(x)$ if its Cosine Transform is

$$F_c(s)=\begin{cases}\frac{1}{2\pi}\left(a-\frac{s}{2}\right), & s<2a\\ 0, & s\geq 2a\end{cases}$$ **Ans.** $\frac{2\sin^2 ax}{\pi^2 x^2}$

20. Find $f(x)$ if its Fourier Sine Transform is $\frac{s}{1+s^2}$ **Ans.** e^{-x}

21. Find $f(x)$ if its Fourier Sine Transform is $e^{-\pi s}$ **Ans.** $\frac{2}{\pi}\left(\frac{s}{1+s^2}\right)$

22. Find $f(x)$ if its Fourier Sine Transform is $\frac{\pi}{2}$. **Ans.** $\frac{1}{x}$

23. Find $f(x)$ if its Fourier Sine Transform is $(2\pi s)^{\frac{1}{2}}$. **Ans.** $\frac{1}{x\sqrt{x}}$

24. Find $f(x)$ if its Fourier Sine Transform is $\begin{cases}\sin s, & 0<s<\pi\\ 0, & s\geq\pi\end{cases}$ **Ans.** $\frac{2}{\pi}\cdot\frac{\sin \pi x}{(1-x^2)}$

25. Find $f(x)$ whose Fourier Sine Transform is se^{-as} **Ans.** $\frac{2}{\pi}\frac{\sin 2\theta}{a^2+x^2}$, where $\tan\theta=\frac{x}{a}$

41.8 PROPERTIES OF FOURIER TRANSFORMS

(1) **LINEAR PROPERTY.** If $F_1(s)$ and $F_2(s)$ are Fourier transforms of $f_1(x)$ and $f_2(x)$ respectively, then

$$F[af_1(x)+bf_2(x)]=aF_1(s)+bF_2(s)$$

where a and b are constants.

We know that $$F_1(s)=\frac{1}{\sqrt{2\pi}}\int_{-\infty}^{\infty} e^{isx}\cdot f_1(x)\,dx$$

and $$F_2(s)=\frac{1}{\sqrt{2\pi}}\int_{-\infty}^{\infty} e^{isx} f_2(x)\,dx$$

$$F[af_1(x)+bf_2(x)]=\frac{1}{\sqrt{2\pi}}\int_{-\infty}^{\infty} e^{isx}\left[af_1(x)+bf_2(x)\right]dx$$

$$= a\frac{1}{\sqrt{2\pi}}\int_{-\infty}^{\infty} e^{isx} f_1(x)dx + b\int_{-\infty}^{\infty} e^{isx} f_2(x)dx = aF_1(s) + bF_2(s)$$ **Proved.**

(2) CHANGE OF SCALE PROPERTY

If $F(s)$ is the complex Fourier transform of $f(x)$, then $F\{f(ax)\} = \frac{1}{a}F\left(\frac{s}{a}\right)$

Proof. We know that $F(s) = \frac{1}{\sqrt{2\pi}}\int_{-\infty}^{\infty} e^{isx} f(x)\,dx$

$$F\{f(ax)\} = \frac{1}{\sqrt{2\pi}}\int_{-\infty}^{\infty} e^{isx} f(ax)\,dx \qquad \left[\text{Put } ax = t \Rightarrow dx = \frac{dt}{a}\right]$$

$$= \frac{1}{\sqrt{2\pi}}\int_{-\infty}^{\infty} e^{is\frac{t}{a}} f(t)\frac{dt}{a} = \frac{1}{a}\frac{1}{\sqrt{2\pi}}\int_{-\infty}^{\infty} e^{i\left(\frac{s}{a}\right)t} f(t)\,dt = \frac{1}{a}F\left(\frac{s}{a}\right)$$ **Proved.**

(3) SHIFTING PROPERTY

If $F(s)$ is the complex Fourier transform of $f(x)$, then

$$F\{f(x-a)\} = e^{isa}F(s)$$

Proof. $F(s) = \int_{-\infty}^{\infty} e^{isx} f(x)\,dx$

$$F\{f(x-a)\} = \frac{1}{\sqrt{2\pi}}\int_{-\infty}^{\infty} e^{isx} f(x-a)\,dx \qquad [\text{Put } x - a = t, \text{ so that } dx = dt]$$

$$= \frac{1}{\sqrt{2\pi}}\int_{-\infty}^{\infty} e^{is(t+a)} f(t)\,dt = \frac{e^{isa}}{\sqrt{2\pi}}\int_{-\infty}^{\infty} e^{ist} f(t)\,dt = e^{isa}F(s)$$ **Proved.**

(4) $F\{e^{iax}f(x)\} = F(s+a)$

Proof. $F\{e^{iax}f(x)\} = \frac{1}{\sqrt{2\pi}}\int_{-\infty}^{\infty} e^{iax} f(x)e^{isx}dx = \frac{1}{\sqrt{2\pi}}\int_{-\infty}^{\infty} e^{i(s+a)x} f(x)\,dx$

$= F(s + a)$ **Proved.**

(5) MODULATION THEOREM [*U.P. III semester Dec. 2005*]

If $F(s)$ is the complex Fourier transform of $f(x)$, then

$$F\{f(x)\cos ax\} = \frac{1}{2}[F(s+a) + F(s-a)]$$

Proof. We know that $F(s) = \frac{1}{\sqrt{2\pi}}\int_{-\infty}^{\infty} e^{isx} f(x)\,dx$

$$F\{f(x)\cos ax\} = \frac{1}{\sqrt{2\pi}}\int_{-\infty}^{\infty} e^{isx} f(x)\cos ax\,dx = \frac{1}{\sqrt{2\pi}}\int_{-\infty}^{\infty} e^{isx} f(x)\frac{e^{iax}+e^{-iax}}{2}dx$$

$$= \frac{1}{2}\frac{1}{\sqrt{2\pi}}\int_{-\infty}^{\infty} e^{isx} f(x)e^{iax}dx + \frac{1}{2}\frac{1}{\sqrt{2\pi}}\int_{-\infty}^{\infty} e^{isx} f(x)\,e^{-iax}dx$$

$$= \frac{1}{2}\frac{1}{\sqrt{2\pi}}\int_{-\infty}^{\infty} e^{i(s+a)x} f(x)\,dx + \frac{1}{2}\frac{1}{\sqrt{2\pi}}\int_{-\infty}^{\infty} e^{i(s-a)x} f(x)dx$$

$$= \frac{1}{2}F(s+a) + \frac{1}{2}F(s-a) = \frac{1}{2}[F(s+a) + F(s-a)]$$ **Proved.**

(6) If $F\{f(x)\} = F(s)$, then $F\{x^n f(x)\} = (-i)^n \frac{d^n}{ds^n}F(s)$.

Proof. We know that $F(s)=\frac{1}{\sqrt{2\pi}}\int_{-\infty}^{\infty} e^{isx} f(x)\,dx$ ... (1)

Differentiating (1) w.r.t. s both sides, n times, we get

$$\frac{d^n F(s)}{ds^n}=\frac{1}{\sqrt{2\pi}}\int_{-\infty}^{\infty}(ix)^n e^{isx} f(x)\,dx=(i)^n\frac{1}{\sqrt{2\pi}}\int_{-\infty}^{\infty} x^n e^{isx}\cdot f(x)\cdot dx=(i)^n F(x^n f(x))$$

$$F(x^n f(x))=(-i)^n\frac{d^n\{F(s)\}}{ds^n}$$

(7) $\boldsymbol{F\{f'(x)\}=is\,F(s)}$ **if** $\boldsymbol{f(x)\to 0}$ **as** $\boldsymbol{x\to\pm\infty}$

Proof. $F\{f'(x)\}=\frac{1}{\sqrt{2\pi}}\int_{-\infty}^{\infty} e^{isx} f'(x)\,dx=\frac{1}{\sqrt{2\pi}}\int_{-\infty}^{\infty} e^{isx} d\{f(x)\}dx$

$$=\frac{1}{\sqrt{2\pi}}\left[\left\{e^{isx}f(x)\right\}_{-\infty}^{\infty}-is\int_{-\infty}^{\infty} e^{isx} f(x)\,dx\right]=\frac{1}{\sqrt{2\pi}}\left[0-is\int_{-\infty}^{\infty} e^{isx} f(x)\,dx\right]$$

$=-is\,F(s).$ **Proved.**

(8) $\boldsymbol{F\left\{\int_a^x f(x)\,dx\right\}=\frac{F(s)}{(-is)}}$

Proof. Let $f_1(x)=\int_a^x f(x)\,dx \Rightarrow f_1'(x)=f(x)$

$$F\{f'(x)\}=(-is)F_1(s)=(-is)F\{f_1(x)\}=-is\,F\left\{\int_a^x f(x)\,dx\right\}$$

$$F\left\{\int_a^x f(x)\,dx\right\}=\frac{1}{(-is)}F\{f_1'(x)\}=\frac{1}{(-is)}F\{f(x)\}=\frac{F(s)}{(-is)}$$ **Proved.**

Note. $F_s(s)$ and $F_c(s)$ are Fourier Sine and Cosine transforms of $f(x)$ respectively.

Properties.

1. $F_s\{af(x)+bg(x)\}=aF_s\{f(x)\}+b\,F_s\{g(x)\}$

2. $F_c\{af(x)+bg(x)\}=aF_c\{f\{x\}\}+b\,F_c\{g(x)\}$

3. $F_s\{f(ax)\}=\frac{1}{a}F_s\left(\frac{s}{a}\right)$ **4.** $F_c\{f(ax)\}=\frac{1}{a}F_c\left(\frac{s}{a}\right)$

5. $F_s[f(x)\sin ax]=\frac{1}{2}[F_c(s-a)-F_c(s+a)]$

6. $F_c\{f(x)\sin ax\}=\frac{1}{2}[F_s(s+a)-F_s(s-a)]$

7. $F_s\{f(x)\cos ax\}=\frac{1}{2}[F_s(s+a)+F_s(s-a)]$

Proof of (5) : $F_s\{f(x)\sin ax\}=\int_0^{\infty} f(x)\sin ax\,.\,\sin sx\,dx$

$$=\frac{1}{2}\int_0^{\infty} f(x)\{\cos(s-a)x-\cos(s+a)x\}dx$$

$$=\frac{1}{2}\left[\int_0^{\infty} f(x)\cos(s-a)x\,dx-\int_0^{\infty} f(x)\cos(s+a)x\,dx\right]$$

$$=\frac{1}{2}[F_c(s-a)-F_c(s+a)]$$ **Proved**

41.9 CONVOLUTION

The Convolution of two functions $f(x)$ and $g(x)$ is defined as

$$f(x)*g(x)=\int_{-\infty}^{\infty} f(u)\,g(x-u)\,du$$

Convolution Theorem on Fourier Transform *(U.P., III Semester, Dec. 2006)*

The Fourier transform of the convolution of $f(x)$ and $g(x)$ is the product of their Fourier transforms, *i.e.*,

$$F[f(x)*g(x)]=F[f(x)]\cdot F[g(x)]$$

Proof. We know that $f(x)*g(x)=\dfrac{1}{\sqrt{2\pi}}\displaystyle\int_{-\infty}^{\infty} f(u)\cdot g(x-u)\,du$ ... (1)

Taking Fourier transform of both sides of (1), we have

$$F[f(x)*g(x)]=F\left[\int_{-\infty}^{\infty} f(u)\cdot g(x-u)\,du\right]$$

$$=\frac{1}{\sqrt{2\pi}}\int_{-\infty}^{\infty}\left[\frac{1}{\sqrt{2\pi}}\int_{-\infty}^{\infty} f(u)\cdot g(x-u)\,du\right]e^{isx}dx$$

$$=\frac{1}{\sqrt{2\pi}}\int_{-\infty}^{\infty} f(u)\,du\,\frac{1}{\sqrt{2\pi}}\left\{\int_{-\infty}^{\infty} g(x-u)\,e^{isx}dx\right\}=\frac{1}{\sqrt{2\pi}}\int_{-\infty}^{\infty} f(u)\cdot du\cdot F\{g(x-u)\}$$

$$=\frac{1}{\sqrt{2\pi}}\int_{-\infty}^{\infty} f(u)du\cdot e^{ius}G(s) \qquad \text{(Using shifting property)}$$

$$=G(s)\cdot\frac{1}{\sqrt{2\pi}}\int_{-\infty}^{\infty} f(u)\,e^{ius}du=G(s)\cdot F(s)=F(s)\cdot G(s)$$ **Proved.**

By inversion

$$F^{-1}\{F(s)\cdot G(s)\}=f*g=F^{-1}\{F(s)\}*F^{-1}\{G(s)\}$$

Question: Verify the above statement to find the Fourier Inverse Transform of e^{-as} sin bs. *(U.P., III Semester, Dec. 2006)*

41.10 PARSEVAL'S IDENTITY FOR FOURIER TRANSFORMS

(U.P. III Semester Dec. 2005)

If the Fourier transforfm of $f(x)$ and $g(x)$ be $F(s)$ and $G(s)$ respectively, then

(*i*) $\displaystyle\int_{-\infty}^{\infty} F(x)\,\overline{G}(s)\,ds=\int_{-\infty}^{\infty} f(x)\,\overline{g}(x)\,dx$

where $\overline{G}(s)$ is the complex conjugate of $G(s)$ and $\overline{g}(x)$ is the complex conjugate of $g(x)$

(*ii*) $\displaystyle\int_{-\infty}^{\infty}[F(s)]^2\,ds=\int_{-\infty}^{\infty}|f(x)|^2\,dx$

Proof. (*i*) $\displaystyle\int_{-\infty}^{\infty}[f(x)\overline{g}(x)dx=\int_{-\infty}^{\infty} f(x)\left[\frac{1}{\sqrt{2\pi}}\int_{-\infty}^{\infty}\overline{G}(s)\,e^{isx}ds\right]dx$

Since $$\overline{g}(x)=\frac{1}{\sqrt{2\pi}}\int_{-\infty}^{\infty}\overline{G}(s)\,e^{isx}ds$$

$$\int_{-\infty}^{\infty} f(x)\overline{g}(x)dx=\frac{1}{\sqrt{2\pi}}\int_{-\infty}^{\infty}\overline{G}(s)ds\cdot\int_{-\infty}^{\infty} f(x)\,\underline{e}^{isx}dx$$

$$\left[\text{since } \frac{1}{\sqrt{2\pi}}\int_{-\infty}^{\infty} f(x)\,e^{isx}dx=F(s)\right] \text{ Fourier Transform}$$

$$= \int_{-\infty}^{\infty} \overline{G}(s)\, F(s)\, ds \qquad \text{... (1)}$$

Putting $g(x) = f(x)$ in (1), we get

$$\int_{-\infty}^{\infty} F(s) \cdot \overline{F}(s) = \int_{-\infty}^{\infty} f(x) \cdot \bar{f}(x)\, dx$$

$$\Rightarrow \quad \int_{-\infty}^{\infty} [F(s)]^2\, ds = \int_{-\infty}^{\infty} [f(x)]^2\, dx$$

Proved.

41.11 PARSEVAL'S IDENTITY FOR COSINE TRANSFORM

(i) $\dfrac{2}{\pi} \displaystyle\int_0^{\infty} F_C(s) \cdot G_C(s)\, ds = \int_0^{\infty} f(x) \cdot g(x)\, dx$ (ii) $\dfrac{2}{\pi} \displaystyle\int_0^{\infty} |F_c(s)|^2\, dx = \int_0^{\infty} |f(x)|^2\, dx$

41.12 PARSEVAL'S IDENTITY FOR SINE TRANSFORM

(i) $\dfrac{2}{\pi} \displaystyle\int_0^{\infty} F_s(s) \cdot G_s(s)\, ds = \int_0^{\infty} f(x) \cdot g(x)\, dx$ (ii) $\dfrac{2}{\pi} \displaystyle\int_0^{\infty} |F_s(s)|^2\, ds = \int_0^{\infty} |f(x)|^2\, dx$

Example 31. *Using Parseval's identity, show that*

$$\int_0^{\infty} \frac{x^2 dx}{(x^2+1)^2} = \frac{\pi}{4}$$

Solution. Let $f(x) = \dfrac{x}{x^2+1}$ so that $F_s(s) = \dfrac{\pi}{2} e^{(-s)}$

By Parseval's identity for sine transformation

$$\frac{2}{\pi} \int_0^{\infty} |F_s(s)|^2 ds = \int_0^{\infty} |f(x)|^2 dx$$

$$\int_0^{\infty} \left| \frac{x}{x^2+1} \right|^2 dx = \frac{2}{\pi} \int_0^{\infty} \left| \frac{\pi}{2} e^{-s} \right|^2 ds$$

$$= \left(\frac{2}{\pi}\right)\left(\frac{\pi^2}{4}\right) \int_0^{\infty} \left| e^{-2s} \right| ds = \frac{\pi}{2}\left[\frac{e^{-2s}}{-2}\right]_0^{\infty} = \frac{\pi}{2}\left[0 + \frac{1}{2}\right] = \frac{\pi}{4}$$

Proved.

Example 32. *Using Parseval's identity, prove that*

$$\int_0^{\infty} \frac{dt}{(a^2+t^2)(b^2+t^2)} = \frac{\pi}{2ab(a+b)}$$

Solution. Let $f(x) = e^{-ax},\ g(x) = e^{-bx}$

Then $\quad F_c(s) = \dfrac{a}{a^2+s^2},\ G_c(s) = \dfrac{b}{b^2+s^2}$

By Parseval's identity for Fourier cosine transformation

$$\frac{2}{\pi} \int_0^{\infty} F_c(s) G_c(s)\, ds = \int_0^{\infty} f(x) \cdot g(x)\, dx \qquad \text{... (1)}$$

On substituting the values of $F_c(s)$, $G_c(s)$, $f(x)$ and $g(x)$ in (1), we get

$$\frac{2}{\pi} \int_0^{\infty} \left(\frac{a}{a^2+s^2}\right)\left(\frac{b}{b^2+s^2}\right) ds = \int_0^{\infty} e^{-ax} \cdot e^{-bx} dx$$

$$\Rightarrow \quad \frac{2}{\pi} \int_0^{\infty} \frac{ab}{(a^2+s^2)} \frac{ds}{(b^2+s^2)} = \int_0^{\infty} e^{-(a+b)x} dx = \left[\frac{e^{-(a+b)x}}{-(a+b)}\right]_0^{\infty} = \left[0 + \frac{1}{a+b}\right]$$

$$\Rightarrow \quad \int_0^{\infty} \frac{ds}{(a^2+s^2)(b^2+s^2)} = \frac{\pi}{2ab} \frac{1}{a+b}$$

$$\int_0^{\infty} \frac{dt}{(a^2+t^2)(b^2+t^2)} = \frac{\pi}{2ab(a+b)}$$

Proved

Example 33. *Using Parseval's identity, prove* $\int_0^{\infty}\left(\frac{\sin t}{t}\right)^2 dt = \frac{\pi}{2}$.

Solution. By example 24, we know that

if $$f(x) = \begin{cases} 1 & \text{for } |x| < a \\ 0 & \text{for } |x| > a > 0 \end{cases}$$

then $$F(s) = \sqrt{\frac{2}{\pi}}\frac{\sin as}{s}$$

Using Parseval's identity

$$\int_{-\infty}^{\infty} |f(t)|^2 dt = \int_{-\infty}^{\infty} |F(s)|^2 ds$$

$$\therefore \quad \int_{-a}^{a} (1)^2 dt = \int_{-\infty}^{\infty} \frac{2}{\pi}\left(\frac{\sin as}{s}\right)^2 ds$$

$$2a = \frac{2}{\pi}\int_{-\infty}^{\infty}\left(\frac{\sin as}{s}\right)^2 ds$$

Putting $as = t$, we get $$2a = \frac{2}{\pi}\int_{-\infty}^{\infty}\left(\frac{\sin t}{\frac{t}{a}}\right)^2 \frac{dt}{a} \qquad \left(\because a\,ds = dt \Rightarrow ds = \frac{dt}{a}\right)$$

$$a\pi = a\int_{-\infty}^{\infty}\left(\frac{\sin t}{t}\right)^2 dt \quad \Rightarrow \quad \pi = \int_{-\infty}^{\infty}\left(\frac{\sin t}{t}\right)^2 dt$$

$$\Rightarrow \quad \int_0^{\infty}\left(\frac{\sin t}{t}\right)^2 dt = \frac{\pi}{2}$$ **Proved.**

Exampe. 34. *Find the Fourier transform of*

$$f(x) = \begin{cases} 1 - |x| & \text{if } |x| < 1 \\ 0 & \text{if } |x| > 1 \end{cases}$$

and hence find the value of $\int_0^{\infty}\frac{\sin t}{t}dt$.

Solution. $$F\{f(x)\} = \frac{1}{\sqrt{2\pi}}\int_{-\infty}^{\infty}(1-|x|)e^{isx}dx$$

$$= \frac{1}{\sqrt{2\pi}}\int_{-\infty}^{\infty}(1-|x|)(\cos sx + i\sin sx)\,dx$$

$$= \frac{1}{\sqrt{2\pi}}\int_{-1}^{1}(1-|x|)\cos sx\,dx + \frac{i}{\sqrt{2\pi}}\int_{-1}^{1}(1-|x|)\sin sx\,dx$$

(Even function) (Odd function)

$$= \frac{2}{\sqrt{2\pi}}\int_0^1 (1-x)\cos sx\,dx + 0 = \sqrt{\frac{2}{\pi}}\left[\left\{(1-x)\frac{\sin sx}{s}\right\}_0^1 + \int_0^1 \frac{\sin sx}{s}dx\right]$$

$$= \sqrt{\frac{2}{\pi}}\left[0 + \left\{\frac{-\cos sx}{s^2}\right\}_0^1\right] = \sqrt{\frac{2}{\pi}}\left(\frac{1-\cos s}{s^2}\right)$$

Using Parseval's identity, we get $\int_{-\infty}^{\infty}|F(s)|^2 ds = \int_{-\infty}^{\infty}|f(t)|^2 dt$

$$\frac{2}{\pi}\int_{-\infty}^{\infty}\frac{(1-\cos s)^2}{s^4}ds = \int_{-1}^{+1}(1-|x|)^2 dx$$

$$\frac{4}{\pi}\int_0^{\infty}\frac{\left(1-1+2\sin^2\frac{s}{2}\right)^2}{s^4}\,ds=\int_{-1}^{+1}(1+x^2-2x)\,dx$$

(L.H.S. is even function, $2x$ on R.H.S. is odd function)

$$\frac{16}{\pi}\int_0^{\infty}\frac{\sin^4\frac{s}{2}}{s^4}\,ds=2\int_0^1(1+x^2)\,dx=2\left(x+\frac{x^3}{3}\right)_0^1=\frac{8}{3}$$

Putting $\frac{s}{2}=x$, we get

$$\frac{16}{\pi}\int_0^{\infty}\frac{\sin^4 x}{16x^4}2\,dx=\frac{8}{3},\qquad \int_0^{\infty}\frac{\sin^4 x}{x^4}\,dx=\frac{4\pi}{3}$$ **Ans.**

Example 35. *Solve for f(x) from the integral equation*

$$\int_0^{\infty}f(x)\cos sx\,dx=e^{-s}$$

Solution. $$\int_0^{\infty}f(x)\cos sx\,dx=e^{-s} \qquad \ldots (1)$$

Multiplying (1) by $\sqrt{\frac{2}{\pi}}$, we get $\sqrt{\frac{2}{\pi}}\int_0^{\infty}f(x)\cos sx\,dx=\sqrt{\frac{2}{\pi}}\,e^{-s}$

$$F_c\{f(x)\}=\sqrt{\frac{2}{\pi}}\,e^{-s}$$

$$f(x)=F_c^{-1}\left[\sqrt{\frac{2}{\pi}}e^{-s}\right]=\sqrt{\frac{2}{\pi}}\left[\sqrt{\frac{2}{\pi}}\int_0^{\infty}e^{-s}\cos sx\,ds\right]$$

$$=\frac{2}{\pi}\int_0^{\infty}e^{-s}\cos sx\,ds=\frac{2}{\pi}\left[\frac{e^{-s}}{1+x^2}\{-\cos sx+x\sin sx\}\right]_0^{\infty}=\frac{2}{\pi}\frac{1}{1+x^2}$$ **Ans.**

Example. 36. *Solve for f (x) from the integral equation*

$$\int_0^{\infty}f(x)\sin sx\,dx=\begin{cases}1 \text{ for } 0\le s<1\\ 2 \text{ for } 1\le s<2\\ 0 \text{ for } s\ge 2\end{cases}$$

Solution. Multiplying by $\sqrt{\frac{2}{\pi}}$ both sides of the given equation, we get

$$\sqrt{\frac{2}{\pi}}\int_0^{\infty}f(x)\sin sx\,dx=\begin{cases}\sqrt{\frac{2}{\pi}} & \text{for } 0\le s<1\\ 2\sqrt{\frac{2}{\pi}} & \text{for } 1\le s<2\\ 0 & \text{for } s\ge 2\end{cases}$$

$$F_s[f(x)]=\begin{cases}\sqrt{\frac{2}{\pi}} & \text{for } 0\le s<1\\ 2\sqrt{\frac{2}{\pi}} & \text{for } 1\le s<2\\ 0 & \text{for } s\ge 2\end{cases}$$

$$f(x)=F_s^{-1}\text{ (R.H.S.) }=\sqrt{\frac{2}{\pi}}\int_0^1\sqrt{\frac{2}{\pi}}\sin sx\,ds+\sqrt{\frac{2}{\pi}}\int_1^2 2\sqrt{\frac{2}{\pi}}\sin sx\,ds$$

$$= \frac{2}{\pi}\left[\frac{-\cos sx}{x}\right]_0^1 + \frac{4}{\pi}\left[\frac{-\cos sx}{x}\right]_1^2 = \frac{2}{\pi}\left(\frac{1-\cos x}{x}\right) + \frac{4}{\pi}\left(\frac{\cos x - \cos 2x}{x}\right)$$

$$= \frac{2}{\pi x}[1-\cos x + 2\cos x - 2\cos 2x]$$

$$f(x) = \frac{2}{\pi x}(1+\cos x - 2\cos 2x)$$ **Ans.**

Example 37. *Prove (i)* $F\{f^n(x)\} = (-is)^n F(s)$

(ii) Hence solve for f (x) if $\int_{-\infty}^{\infty} f(t) e^{|x-t|} dt = \phi(x)$ *is known.*

Proof. (*i*) See property (7) on page 1110, $F\{f'(x)\} = -is\, F(s)$

Similarly $\quad F\{f''(x)\} = (-is)^2 F(s)$

$$F\{f'''(x)\} = (-is)^3 F(s)$$

Using integration by parts successively and making assumptions that $f, f', \ldots f^{(n-1)} \to 0$ as $f(x) \to \pm\infty$.

$$F\{f^n(x)\} = \frac{1}{\sqrt{2\pi}}\int_{-\infty}^{\infty} e^{isx}\frac{d^n}{dx^n}f(x)\cdot dx = (-is)^n F(s).$$

(*ii*) $$\frac{1}{\sqrt{2\pi}}\phi(x) = \frac{1}{\sqrt{2\pi}}\int_{-\infty}^{\infty} f(t)e^{|x-t|}dt, \text{ from the given equation } = f(x)*e^{-|x|}$$

By convolution theorem,

$$\frac{1}{\sqrt{2\pi}}\bar{\phi}(s) = F(s)\cdot\sqrt{\frac{2}{\pi}}\frac{1}{1+s^2}$$

$$F(s) = \frac{1}{2}(1+s^2)\bar{\phi}(s) = \frac{1}{2}[\bar{\phi}(s) - (-is)^2\bar{\phi}(s)]$$

$$\therefore \quad f(x) = \frac{1}{2}\phi(x) - \frac{1}{2}\phi''(x) \text{ using the result derived in } (i)$$

EXERCISE 41.3

Using Parseval's identity

1. Evaluate $\int_0^{\infty}\left(\frac{1-\cos x}{x}\right)^2 dx$ **Ans.** $\frac{\pi}{2}$

2. Prove that $\int_0^{\infty}\frac{\sin at}{t(a^2+t^2)}dt = \frac{\pi}{2}\frac{1-e^{-a^2}}{a^2}$

41.13 FOURIER TRANSFORM OF DERIVATIVES

We have already seen that,

$$F\{f^n(x)\} = (-is)^n F(s)$$

(*i*) $\therefore F\left(\frac{\partial^2 u}{\partial x^2}\right) = (-is)^2 F\{u(x)\} = -s^2\bar{u}$ where $\bar{u}$ is Fourier transform of u w.r.t. x.

(*ii*) $$F_c\{f'(x)\} = -\sqrt{\frac{2}{\pi}}f(o) + s\,F_s(s)$$

$$\text{L.H.S.} = \sqrt{\frac{2}{\pi}}\int_0^{\infty} f'(x)\cdot\cos sx\, dx = \sqrt{\frac{2}{\pi}}\int_0^{\infty}\cos sx\, d\{f(x)\}$$

$$= \sqrt{\frac{2}{\pi}}\left[\{f(x)\cos sx\}_0^\infty + s\int_0^\infty f(x)\sin sx\,dx\right]$$

$$= s\,F_s(s) - \sqrt{\frac{2}{\pi}}f(0) \text{ assuming } f(x) \to 0 \text{ as } x \to \infty$$

(*iii*) $$F_s\{f'(x)\} = \sqrt{\frac{2}{\pi}}\int_0^\infty \sin sx\,d\,[f(x)] = \sqrt{\frac{2}{\pi}}\left[\{f(x)\sin sx\}_0^\infty - s\int_0^\infty f(x)\cos sx\,dx\right]$$

$$= -s\,F_c(s)$$

(*iv*) $$F_c\{f''(x)\} = \sqrt{\frac{2}{\pi}}\int_0^\infty \cos sx\,d\,[f'(x)] = \sqrt{\frac{2}{\pi}}\left[\{f'(x)\cos sx\}_0^\infty + s\int_0^\infty f'(x)\sin sx\,dx\right]$$

$$= -\sqrt{\frac{2}{\pi}}f'(0) + s\,F_s\{f'(x)\} = -s^2\,F_c(s) - \sqrt{\frac{2}{\pi}}f'(0) \text{ assuming } f(x),\ f'(x) \to 0 \text{ as } x \to \infty$$

(*v*) $$F_s\{f''\}(x) = \sqrt{\frac{2}{\pi}}\left[\int_0^\infty \sin sx\,d[f'(x)]\right] = \sqrt{\frac{2}{\pi}}\left[(f'(x)\sin sx)_0^\infty - s\int_0^\infty f'(x)\cos sx\,dx\right]$$

$$= -s\,F_c\{f'(x)\} = -s\left[s\,F_s(s) - \sqrt{\frac{2}{\pi}}f(0)\right]$$

$$= -s^2\,F_s(s) + \sqrt{\frac{2}{\pi}}\,s\,f(0) \text{ assuming } f(x),\ f'(x) \to 0 \text{ as } x \to \infty.$$

41.14 RELATIONSHIP BETWEEN FOURIER AND LAPLACE TRANSFORMS

Consider

$$f(t) = \begin{cases} e^{-st}g(t) \text{ for } t > 0 \\ 0 \quad \text{ for } t < 0 \end{cases} \qquad \dots (1)$$

Then the Fourier transform of $f(t)$ is given by

$$F\{f(t)\} = \frac{1}{\sqrt{2\pi}}\int_{-\infty}^{\infty} e^{ist}f(t)\,dt = \frac{1}{\sqrt{2\pi}}\int_0^\infty e^{ist}f(t)\,dt$$

$$= \frac{1}{\sqrt{2\pi}}\int_0^\infty e^{(is-x)t}g(t)\,dt = \frac{1}{\sqrt{2\pi}}\int_0^\infty e^{-pt}g(t)\,dt \text{ where } p = x - is = \frac{1}{\sqrt{2\pi}}L\{g(t)\}$$

$\therefore$ Fourier transform of $f(t) = \frac{1}{\sqrt{2\pi}} \times$ Laplace transform of $g(t)$ defined by (1).

41.15 SOLUTION OF BOUNDARY VALUE PROBLEMS BY USING INTEGRAL TRANSFORM

Solution of heat conduction problems by Laplace transform.

Example 38. *Use Fourier sine transform to solve the equation*

$$\frac{\partial u}{\partial t} = k\frac{\partial^2 u}{\partial x^2}$$

under the conditions

(*i*) $u(0, t) = 0$ (*ii*) $u(x, 0) = e^{-x}$

(*iii*) *$u(x, t)$ is bounded.* *(U.P. III Semester, Dec. 2006)*

Solution. The given equation is

$$\frac{\partial u}{\partial t} = k\frac{\partial^2 u}{\partial x^2} \qquad \dots (1)$$

Taking Fourier sine transform of both the sides, we get

$$\int_0^\infty \frac{\partial u}{\partial t}\sin sx\,dx = k\int_0^\infty \frac{\partial^2 u}{\partial x^2}\sin sx\,dx$$

$$\frac{d\bar{u}}{dt} = k\left[s(u)_{x=0} - s^2\,\bar{u}\right]$$

$$\Rightarrow \quad \frac{d\bar{u}}{dt} + ks^2\,\bar{u} = 0 \qquad \left[\text{where } \bar{u} = \int_0^\infty u\sin sx\,dx\right]$$

$$\Rightarrow \quad D\bar{u} + ks^2\,\bar{u} = 0$$

$$\Rightarrow \quad (D + ks^2)\,\bar{u} = 0$$

A.E. is $\quad m + ks^2 = 0 \;\Rightarrow\; m = -ks^2$

$$\text{C.F.} = c_1\,e^{-ks^2t}$$

$$\bar{u} = c_1\,e^{-ks^2t} \qquad \ldots(2)$$

At $t = 0$, $\quad (\bar{u})_{t=0} = \int_0^\infty (u)_{t=0}\sin sx\,dx = \int_0^\infty e^{-x}\sin sx\,dx$

$$= \frac{e^{-s}}{1+s^2}\left[-\sin sx - s\cos sx\right]_0^\infty \qquad \left[\int e^{ax}\sin bx\,dx = \frac{e^{ax}}{a^2+b^2}(a\sin bx - b\cos bx)\right].$$

$$= \frac{s}{1+s^2} \qquad \ldots(3)$$

From (2), $\quad (\bar{u})_{t=0} = c_1 \qquad \ldots(4)$

From (3) and (4), we have

$$c_1 = \frac{s}{1+s^2}$$

From (2), $\quad \bar{u} = \dfrac{s}{1+s^2}\,e^{-ks^2t}$

Taking inverse Fourier sine transform, we get

$$u(x, t) = \sqrt{\frac{2}{\pi}}\int_0^\infty \sqrt{\frac{2}{\pi}}\,\frac{s}{1+s^2}\,e^{-ks^2t}\sin sx\,ds.$$

$$\Rightarrow \quad u(x, t) = \frac{2}{\pi}\int_0^\infty \frac{s}{1+s^2}\,e^{-ks^2t}\sin sx\,ds \qquad \textbf{Ans.}$$

Example 39. *Solve the equation* $\dfrac{\partial u}{\partial t} = \dfrac{\partial^2 u}{\partial x^2}, \quad x>0,\ t>0$

subject to the conditions:

(i) $u = 0$, *when* $x = 0,\ t > 0$ $\qquad$ *(ii)* $u = \begin{cases}1, & 0 \le x < 1\\ 0 & x > 1\end{cases}$, *when* $t = 0$

(iii) $u\,(x, t)$ *is bounded.* (*U.P., III Semester Dec. 2003*)

Solution. In view of the initial condition (*i*), we apply Fourier Sine Transform to both sides of the given equation.

$$\int_0^\infty \frac{\partial u}{\partial t}\sin sx\,dx = \int_0^\infty \frac{\partial^2 u}{\partial x^2}\sin sx\,dx$$

$$\Rightarrow \quad \frac{\partial}{\partial t}\int_0^\infty u\sin sx\,dx = -s^2\,\bar{u}(s) + su(0)$$ (See Article 41.13 on page 1115)

On putting $u = 0$ and $x = 0$, we get

$$\Rightarrow \quad \frac{\partial \bar{u}}{\partial t} + s^2 \bar{u} = 0, \quad \Rightarrow \quad \left(\frac{\partial}{\partial t} + s^2\right) \bar{u} = 0$$

Its solution *i.e.* complementary function is $\bar{u}(s,t) = A e^{-s^2 t}$ ... (1)

Since we have assumed

$$\bar{u} = \bar{u}(s,t) = \int_0^\infty u(x,t) \sin sx \, dx, \quad \Rightarrow \quad \bar{u}(s,0) = \int_0^\infty u(x,0) \sin sx \, dx$$

$$\Rightarrow \quad \bar{u}(s,0) = \int_0^1 1 . \sin sx \, dx = \left[\frac{-\cos sx}{s}\right]_0^1 = \frac{1-\cos s}{s} \qquad [u(x, 0) = 1] \text{ ... (2)}$$

From (1) and (2), we get $A = \bar{u}(s,0) = \dfrac{1-\cos s}{s}$

On putting the value of A in (1), we get

Hence, $\bar{u}(s,t) = \dfrac{1-\cos s}{s} e^{-s^2 t}$

Now applying the Inverse Fourier Sine Transform, we have

$$u(x,t) = \frac{2}{\pi} \int_0^\infty \left(\frac{1-\cos s}{s}\right) . e^{-s^2 t} \, ds,$$

which is the required solution. **Ans.**

Solution of wave equation by Laplace transform

Example 40. *An infinitely long string having one end at $x = 0$ is initialy at rest along x-axis. The end $x = 0$ is given a transverse displacement $f(t)$, when $t > 0$. Find the displacement of any point of the string at any time.*

Solution. Let $y(x, t)$ be the displacement, then wave equation is

$$\frac{\partial^2 y}{\partial x^2} = c^2 \frac{\partial^2 y}{\partial t^2} \qquad \text{... (1)}$$

subject to the conditions

$y(x, 0) = 0$ ... (2) $\quad \dfrac{\partial y}{\partial t}(x,0) = 0$ ... (3)

$y(0, t) = f(t)$ ... (4) $\quad y(x, t)$ is bounded ... (5)

On taking Laplace transform of (1), we have

$$L\left(\frac{\partial^2 y}{\partial x^2}\right) = c^2 L \frac{\partial^2 y}{\partial t^2}$$

$$s^2 \bar{y} - s y(x,0) - \frac{\partial y}{\partial t}(x,0) = c^2 \frac{d^2 \bar{y}}{dx^2} \qquad \text{... (6)}$$

On putting $y(x, 0) = 0$, $\dfrac{\partial y}{\partial t}(x, 0) = 0$ in (6) get

$$s^2 \bar{y} = c^2 \frac{d^2 \bar{y}}{dx^2} \quad \text{or} \quad \frac{d^2 \bar{y}}{dx^2} = \left(\frac{s}{c}\right)^2 \bar{y} \qquad \text{... (7)}$$

Laplace transform of (4), $\bar{y}(0, s) = \bar{f}(s)$ at $x = 0$ ... (8)

On solving (7), we get $\bar{y} = A e^{\frac{sx}{c}} + B e^{-\frac{sx}{c}}$ ... (9)

According to condition (5), y is finite at $x \to \infty$, this gives $A = 0$.

Equation (9) becomes

$$\bar{y} = B e^{-\frac{sx}{c}} \qquad \text{... (10)}$$

Putting the value of $\bar{y}(0, s) = \bar{f}(s)$ at $x = 0$ in (10), we get $\bar{f}(s) = B$

Thus (10) becomes $$y = \bar{f}(s) \cdot e^{-\frac{sx}{c}}$$

To get y from $\bar{y}$, we use complex inversion formula

$$y = \frac{1}{2\pi i} \int_{a-i\infty}^{a+i\infty} e^{\left(t-\frac{x}{c}\right)s} - f(s)\, ds$$

Hence $$y = f\left(t - \frac{x}{c}\right)$$ **Ans.**

Solution of Transmission Lines equations by Laplace Transformations.

Example 41. *A semi-infinite transmission line, of negligible inductance and leakage per unit length has its voltage and current equal to zero. A constant voltage v_0 is applied at the sending end ($x = 0$) at $t = 0$. Find the voltage and current at any point ($x > 0$) at any instant.*

Solution. Let v and i be the voltage and current at any point x and at any time t.

$$\left.\begin{aligned} -\frac{\partial v}{\partial x} &= Ri + L\frac{\partial i}{\partial t} \\ -\frac{\partial i}{\partial x} &= c\frac{\partial v}{\partial t} + GV \end{aligned}\right]$$

On putting $L = 0$, $G = 0$ in above equations, we get

$$\frac{\partial v}{\partial x} = -Ri \qquad \text{... (1)}$$

$$\frac{\partial i}{\partial x} = -c\frac{\partial v}{\partial t} \qquad \text{... (2)}$$

Conditions are $v(x, 0) = 0$ ... (3)

$i(x, 0) = 0$ ... (4)

$v(0, t) = v_0$ ... (5)

$v(x, t)$ finite for all x and t. ... (6)

Applying Laplace transform of (1) and (2), we get

$$\frac{d\bar{v}}{dx} = -R\bar{i} \qquad \text{... (7)}$$

$$\frac{di}{dx} = -c(s\bar{v} - v) \text{ and } v(x, 0) = 0 \text{ or } \frac{di}{dx} = -cs\bar{v} \qquad \text{... (8)}$$

Differentiating (7) w.r.t. 'x', we get $\dfrac{d^2\bar{v}}{dx^2} = -R\dfrac{di}{dx}$

$$\Rightarrow \quad \frac{d^2\bar{v}}{dx^2} = -R(-cs\bar{v}) \qquad \left(\frac{d\bar{v}}{dx} = -cs\bar{v}\right)$$

$$\Rightarrow \quad \frac{d^2\bar{v}}{dx^2} = Rcs\bar{v} \Rightarrow \frac{d^2\bar{v}}{dx^2} - Rcs\bar{v} = 0$$

$$\Rightarrow \quad (D^2 - Rcs)\bar{v} = 0 \Rightarrow D = \pm\sqrt{Rcs}) \qquad \text{... (9)}$$

Laplace transform of (5) is $\bar{v}(0, s) = \dfrac{v_0}{s}$ ... (10)

And Laplace transform of (6) is $\bar{v}(x, s)$ remains finite as $x \to \infty$. ... (11)

Equation (9) is an ordinary differential equation and its solution is

$$\bar{v} = Ae^{\sqrt{Rcs}\,x} + Be^{-\sqrt{Rcs}\,x} \qquad \text{... (12)}$$

As $x \to \infty$, $\bar{v}$ remains finite only when $A = 0$.

So (12) becomes $$\bar{v} = B e^{-\sqrt{Rcs}\,x} \qquad \dots (13)$$

Putting $\bar{v} = \frac{v_0}{s}$ in (13), we get $\frac{v_0}{s} = B e^{-\sqrt{Rcs}\,x}$ ...(14)

Putting $x = 0$ in (14), we get $\frac{v_0}{s} = B$

Substituting the value of B in (13) we have

$$\bar{v} = \frac{v_0}{s} e^{-\sqrt{Rcs}\,x}$$

On applying inversion transform, we get

$$v = v_0\, L^{-1}\left[\frac{e^{-\sqrt{Rcs}\,x}}{s}\right] = v_0\, erf\, c\left[\frac{x\sqrt{Rc}}{2\sqrt{t}}\right]$$

$$v = v_0 \frac{x\sqrt{Rc}}{2\sqrt{\pi}} \int_0^t u^{-\frac{3}{2}} e^{-\left(\frac{Rcx^2}{4u}\right)} \cdot du \qquad \dots (15)$$

From (1) $$i = \frac{-1}{r}\frac{\partial v}{\partial x} \qquad \dots (16)$$

On differentiating (15), we get $\frac{\partial v}{\partial x} = \frac{v_0\, x\sqrt{Rc}}{2\sqrt{\pi}} t^{-\frac{3}{2}} e^{-\frac{Rcx^2}{4t}}$

Putting the value of $\frac{\partial v}{\partial x}$ in (16), we obtain $i = \frac{v_0 x}{2\sqrt{\pi}}\sqrt{\frac{c}{R}} \cdot t^{-\frac{3}{2}} e^{-\frac{Rcx^2}{4t}}$ **Ans.**

Solution of partial differential Equations by Fourier Transform

Example 42. *Solve* $\frac{\partial^2 u}{\partial t^2} = \alpha^2 \frac{\partial^2 u}{\partial x^2}$, $\quad -\infty < x < \infty, \quad t \geq 0$

with conditions $u(x, 0) = f(x)$,

$\frac{\partial u}{\partial t}(x, 0) = g(x)$ *and assuming* u, $\frac{\partial u}{\partial x} \to 0$ *as* $x \to \pm\infty$.

Solution. Taking Fourier transform on both sides of the differential equation,

$\frac{d^2\bar{u}}{dt^2} = \alpha^2(-s^2\bar{u})$ where $\bar{u}$ is Fourier transform of u with respect to x.

$$\frac{d^2\bar{u}}{dt^2} + \alpha^2 s^2 \bar{u} = 0$$

Auxiliary equation is $D^2 + \alpha^2 s^2 = 0 \Rightarrow D = \pm\, i\alpha s$

$$\therefore \quad \bar{u}(s,t) = Ae^{i\alpha st} + Be^{-i\alpha st} \qquad \dots (1)$$

Since $$u(x, 0) = f(x) \text{ and } \frac{\partial u}{\partial t}(x,0) = g(x),$$

$\bar{u}(s,0) = F(s)$ and $\frac{d\bar{u}}{dt}(s,0) = G(s)$ on taking transform.

Using these condition in (1), we get

$$\bar{u}(s,0) = A + B = F(s) \qquad \dots (2)$$

$$\frac{d\bar{u}}{dt}(s, 0) = i\alpha s(A - B) = G(s) \qquad \dots (3)$$

Solving (2) and (3), we get $A=\frac{1}{2}\left[F(s)+\frac{G(s)}{i\alpha s}\right] \Rightarrow B=\frac{1}{2}\left[F(s)-\frac{G(s)}{i\alpha s}\right]$

Using these values in (1), we get

$$\bar{u}(s,t)=\frac{1}{2}\left[F(s)+\frac{G(s)}{i\alpha s}\right]e^{i\alpha st}+\frac{1}{2}\left[F(s)-\frac{G(s)}{i\alpha s}\right]e^{-i\alpha st} \quad \ldots (4)$$

By inversion theorem, (4) reduces to

$$u(x,\,t)=\frac{1}{2}\left[f(x-\alpha t)-\frac{1}{\alpha}\int_{\alpha}^{x-\alpha t}g(\theta)\,d\theta\right]+\frac{1}{2}\left[f(x+\alpha t)+\frac{1}{\alpha}\int_{\alpha}^{x+\alpha t}g(\theta)\,d\theta\right]$$

Using the result $F\left(\int_{\alpha}^{x}f(t)\,dt\right)=\frac{F(s)}{(-is)}$ **Ans.**

41.16 FOURIER TRANSFORMS OF PARTIAL DERIVATIVE OF A FUNCTION

$$F_f\left[\frac{\partial^2 u}{\partial x^2}\right]=-s^2\,F(u) \text{ where } F(u) \text{ is Fourier transform of } u \text{ w.r.t. } x.$$

$$F_s\left[\frac{\partial^2 u}{\partial x^2}\right]=s\,(u)_{x=0}-s^2\,F_s(u) \qquad \text{(sine transform)}$$

$$F_c\left[\frac{\partial^2 u}{\partial x^2}\right]=-\left[\frac{\partial u}{\partial x}\right]_{x=0}-s^2\,F_c(u) \qquad \text{(cosine transform)}$$

Proof. Let $F\,[u\,(x,\,t)]$ be the Fourier transform of the function $u\,(x,\,t)$, *i.e.*

$$F[u(x,t)]=\int_{-\infty}^{\infty}e^{isx}u(x,t)\,dx$$

The Fourier transform of $\frac{\partial^2 u}{\partial x^2}$ is given by $F\left[\frac{\partial^2 u}{\partial x^2}\right]=\int_{-\infty}^{\infty}e^{isx}\,\frac{\partial^2 u}{\partial x^2}\,dx.$

Integrating by parts, we have

$$F\left[\frac{\partial^2 u}{\partial x^2}\right]=\left[e^{isx}\,\frac{\partial u}{\partial x}-\int ise^{isx}\,\frac{\partial u}{\partial x}\,dx\right]_{-\infty}^{\infty}$$

$$=\left[e^{isx}\,\frac{\partial u}{\partial x}-is\,e^{isx}u+\int (is)^2\,e^{isx}u\,dx\right]_{-\infty}^{\infty} \qquad \text{Again integrating}$$

$$=\left[0-0-s^2\int_{-\infty}^{\infty}e^{isx}.u\,dx\right] \qquad \left[\begin{array}{l}u=0,\ \frac{\partial u}{\partial x}=0\\ \text{when } x\to\infty\end{array}\right]$$

Thus $$F\left[\frac{\partial^2 u}{\partial x^2}\right]=-s^2\,F[u(x,t)]$$

Similarly, the Fourier sine transform of $\frac{\partial^2 u}{\partial x^2}$ is given by

$$F_s\left[\frac{\partial^2 u}{\partial x^2}\right]=\int_0^{\infty}\frac{\partial^2 u}{\partial x^2}\sin sx\,dx$$

$$\Rightarrow \qquad F_s\left[\frac{\partial^2 u}{\partial x^2}\right]=s\,[u(x,t)]_{x=0}-s^2\,F_s\,[u(s,t)] \qquad \text{(sine transform)}$$

and $$F_c\left[\frac{\partial^2 u}{\partial x^2}\right]=-\left[\frac{\partial u}{\partial x}\right]_{x=0}-s^2\,F_c\,[u(s,t)] \qquad \text{(cosine transform)}$$

41.17 APPLICATIONS TO SIMPLE HEAT TRANSFER EQUATIONS

Solution of heat conduction problems by Fourier sine Transforms

Example 43. *Solve* $\frac{\partial u}{\partial t}=\frac{\partial^2 u}{\partial x^2}$, $0 < x\, 4,\ t > 0$, *subject to conditions* $u\,(0, t) = 0,\ u\,(4, 0) = 0,$ $u\,(x, 0) = 2x.$

Solution. Here $\frac{\partial u}{\partial t}=\frac{\partial^2 u}{\partial x^2} \qquad 0<x<4,\ t>0$ ... (1)

subject to boundary conditions; $u\,(0, t) = 0, \quad u\,(4, t) = 0$ ... (2)

and initial condition: $u\,(x, 0) = 2x$ ... (3)

On applying Fourier sine transform on (1), we get (initial conditions are given)

$$\int_0^4 \frac{\partial u}{\partial t}\sin\frac{s\pi x}{4}\,dx=\int_0^4\frac{\partial^2 u}{\partial x^2}\sin\frac{s\pi x}{4}\,dx$$

Integrating by parts, we get

$$\frac{d}{dt}\int_0^4 u\,(x,t)\sin\frac{s\pi x}{4}\,dx=\left[\frac{\partial u}{\partial x}\sin\frac{s\pi x}{4}\right]_0^4-\frac{s\pi}{4}\int_0^4\frac{\partial u}{\partial x}\cos\frac{s\pi x}{4}\,dx$$

$$\Rightarrow \qquad \frac{d}{dt}(\bar{u}_s)=0-\frac{s\pi}{4}\left\{\left[u(x,t)\cos\frac{s\pi x}{4}\right]_0^4+\frac{s\pi}{4}\int_0^4 u\,(x,t\sin\frac{s\pi x}{4}\,dx\right\}$$

$$\Rightarrow \qquad =-\frac{s\pi}{4}[u\,(4,t)\cos s\pi-u\,(0,t)]-\frac{s^2\pi^2}{16}\bar{u}_s \qquad \text{... (4)}$$

Putting the values of u (0, t) and u (4, t) from (2) in (4), we get

$$\frac{d}{dt}\bar{u}_s=0-\frac{s^2\pi^2}{16}\bar{u}_s \quad\Rightarrow\quad \left(\frac{1}{\bar{u}_s}\right)d\bar{u}_s=-\left(\frac{s^2\pi^2}{16}\right)dt, \quad \text{whose solution is}$$

$$\bar{u}_s\,(s,t)=ce^{-s^2\pi^2 t/16}$$

On putting $t = 0$ in (5), we have $c=\bar{u}_s\,(s,0)$ ... (5)

Now $\bar{u}_s\,(s,t)=\int_0^4 u\,(x,t)\sin\frac{s\pi x}{4}\,dx$ ... (6) (By definition)

On putting $t = 0$ in (6), we get $\bar{u}_s(s,0)=\int_0^4 u\,(x,0)\sin\frac{s\pi x}{4}\,dx$

$$c=\int_0^4 u\,(x,0)\sin\frac{s\pi x}{4}\,du \quad [\because c=\bar{u}_s\,(s,0)] \qquad [\because u\,(x,0)=2x]$$

$$=\int_0^4 2x\sin\frac{s\pi x}{4}\,dx=\left[(2x)\times\left(-\frac{4}{s\pi}\cos\frac{s\pi x}{4}\right)-(2)\times\left(\frac{-16}{s^2\pi^2}\sin\frac{s\pi x}{4}\right)\right]_0^4 \quad \text{(Integrating by parts)}$$

$$=2\times 4\left(-\frac{4}{s\pi}\cos\pi\right)-0=\frac{-32}{s\pi}(-1)^5=\frac{32}{s\pi}(-1)^{s+1}$$

Putting the value of c in (5), we get $\bar{u}_s(s,t)=(32/s\pi)\,(-1)^{s+1}\ e^{-\frac{s^2\pi^2 t}{.16}}$ ... (7)

Now, taking the inverse finite Fourier sine transform, we get

$$u\,(x,t)=\frac{2}{4}\sum_{s=1}^{\infty}\bar{u}_s\,(s,t)\sin\frac{s\pi x}{4} \qquad \text{... (8)}$$

Putting the value of $\bar{u}_s(s,t)$ from (7) in (8), we get

$$u\,(x,t)=\frac{16}{\pi}\sum_{s=1}^{\infty}\frac{(-1)^{s+1}}{s}e^{-s^2\pi^2 t/16}\sin\frac{s\pi x}{4}$$ **Ans.**

Example 44. *Solve* $\frac{\partial u}{dt} = k\frac{\partial^2 u}{\partial x^2}$ *for* $x \geq 0, t \geq 0$ *under the given conditions* $u = u_0$ *at* $x = 0$, $t > 0$ *with initial condition* $u(x, 0) = 0$, $x \geq 0$

Solution. Taking Fourier sine transforms

$$F_s\left(\frac{\partial u}{\partial t}\right) = F_s\left(k\frac{\partial^2 u}{\partial x^2}\right) \qquad \frac{d\bar{u}}{dt} = k\left[-s^2\bar{u} + \sqrt{\frac{2}{\pi}}\, s\, u(0,t)\right]$$

$$= -ks^2\bar{u} + \sqrt{\frac{2}{\pi}}\, k s u_0 \text{ where } \bar{u} \text{ is the Fourier sine transform of } u.$$

$$\frac{d\bar{u}}{dt} + ks^2\bar{u} = \sqrt{\frac{2}{\pi}}\, s k u_0$$

This is linear in $\bar{u}$.

$$\therefore \qquad \bar{u}\, e^{ks^2 t} = \sqrt{\frac{2}{\pi}} k u_0 \int s\, e^{ks^2 t} dt = \sqrt{\frac{2}{\pi}} \frac{u_0}{s} e^{ks^2 t} + c \qquad \text{... (1)}$$

Since, $u(x, 0) = 0$, $\bar{u}(s, 0) = 0$. Using this in (1), we have

$$0 = \sqrt{\frac{2}{\pi}} \frac{u_0}{s} + c \quad \Rightarrow \quad c = -\sqrt{\frac{2}{\pi}} \frac{u_0}{s}$$

$$e^{ks^2 t} = \sqrt{\frac{2}{\pi}} \frac{u_0}{s}\left(e^{ks^2 t} - 1\right) \quad \Rightarrow \quad \bar{u} = \sqrt{\frac{2}{\pi}} \frac{u_0}{s}\left(1 - e^{ks^2 t}\right)$$

By inversion theorem, we have

$$u(x,t) = \frac{2u_0}{\pi} \int_0^\infty \left(\frac{1 - e^{ks^2 t}}{s}\right) \sin sx\, ds. \qquad \textbf{Ans.}$$

Example 45. *Solve* $\frac{\partial u}{\partial t} = k\frac{\partial^2 u}{\partial x^2}$ *for* $0 \leq x < \infty, t > 0$ *given the conditions*

(i) $u(x, 0) = 0$ *for* $x \geq 0$ *(ii)* $\frac{\partial u}{\partial x}(0,t) = -a$ *(constant)*

(iii) $u(x, t)$ *is bounded.*

Solution. In this problem, $\frac{\partial u}{\partial x}$ at $x = 0$ is given. Hence, take Fourier cosine transform on both sides of the given equation.

$$F_c\left(\frac{\partial u}{\partial t}\right) = F_c\left(k\frac{\partial^2 u}{\partial x^2}\right)$$

$$\frac{d\bar{u}}{dt} = k\left(-s^2\bar{u} - \sqrt{\frac{2}{\pi}} \cdot \frac{\partial u}{\partial x}(0, t)\right) = -ks^2\bar{u} + \sqrt{\frac{2}{\pi}}\, ka \qquad \text{[Using condition } (ii)]$$

$$\frac{d\bar{u}}{dt} + ks^2\bar{u} = \sqrt{\frac{2}{\pi}}\, ka$$

This is linear in $\bar{u}$. Therefore, solving

$$\bar{u} e^{ks^2 t} = \int \sqrt{\frac{2}{\pi}}\, ka\, e^{ks^2 t} dt = \sqrt{\frac{2}{\pi}} ka \frac{e^{ks^2 t}}{ks^2} + c$$

$$\bar{u}(s,t) = \sqrt{\frac{2}{\pi}} \frac{a}{s^2} + c e^{-ks^2 t} \qquad \text{... (1)}$$

Since $u(x, 0) = 0$ for $x \geq 0$.

$$\bar{u}(s, 0) = 0.$$

Using this in (1), we get

$$\bar{u}(s,0) = c + \sqrt{\frac{2}{\pi}} \frac{a}{s^2} = 0$$

$$\therefore \qquad c = -\sqrt{\frac{2}{\pi}}\frac{a}{s^2}$$

Substituting this in (1), we have $\bar{u}(s, t) = \sqrt{\frac{2}{\pi}}\frac{a}{s^2}\left(1 - e^{-ks^2 t}\right)$

By inversion theorem, we have $u(x,t) = \frac{2}{\pi}\cdot a\int_0^\infty \frac{1 - e^{-ks^2 t}}{s^2}\cos sx\, ds.$ **Ans.**

EXERCISE 41.4

1. Use Fourier transform to solve the equation $\frac{\partial u}{\partial t} = \frac{\partial^2 u}{\partial x^2}$

under the conditions $u = 0$ at $x = 0$

$$u = \begin{cases} 1, & 0 < x < 1 \\ 0, & x \geq 1 \end{cases} \quad \text{when } t = 0$$

and u is bounded. **Ans.** $u = \frac{2}{\pi}\int_{\pi}^{\infty} \frac{1 - \cos s}{s} e^{-s^2 t} \sin sx\, ds$

2. Use Fourier sine transform to solve the equation $\frac{\partial u}{\partial t} = 2\frac{\partial^2 u}{\partial x^2}$

Under the conditions $u(0, t) = 0$, $u(x, 0) = e^{-x}$, $u(x, t)$ is bounded.

Ans. $u(x,t) = \frac{2}{\pi}\int_0^\infty \frac{s}{s^2+1} e^{-2s^2 t} \sin sx\, ds$

3. A tightly stretched string with fixed end points $x = b$ and $x = c$ is initially in a position given by $y = b \sin\left(\frac{\pi x}{c}\right)$. It is released from rest in this position. Show by the method of Laplace transform that the displacement y at any distance x from one end and at any time t is given by

$$y = b\sin\frac{\pi x}{c}\cos\frac{\pi q}{c}t.$$

and y satisfies the equation $\frac{\partial^2 y}{\partial x^2} = a^2\frac{\partial^2 y}{\partial t^2}$

4. A string is stretched tightly between $x = 0$ and $x = l$ and both ends are given displacement $y = a \sin pt$ perpendicular to the string. If the string satisfies the differential equation

$$\frac{\partial^2 y}{\partial x^2} = \frac{1}{c^2}\frac{\partial^2 y}{\partial t^2}$$

Show that the oscillations of the string are given by

$$y = a\sec\frac{Pl}{2c}\cos\left(\frac{Px}{c} - \frac{Pl}{2c}\right)\sin pt.$$

5. An infinite cable with resistance R ohms/km, capacitance C Farads/km, and negligible inductance and leakage is subjected to constant E.M.F. E_0 at the home end at time $t = 0$. Using the operational method show that the entering current at any subsequent time t is

$$I(t) = E_0\left(\frac{C}{\pi Rt}\right)^{1/2}$$

6. Solve the equation for high voltage semi-infinite line with the following initial and boundary conditions $v(x, t) = 0$ and $i(x, 0) = 0$, $v(0, t) = v_0\, u(t)$, $v(x, t)$ is finite as $x \to \infty$.

Ans. $v = v_0\, u[t - x\sqrt{LC}]$, for $x \leq \frac{t}{\sqrt{LC}}$ and

$v = 0$ for $x > \frac{t}{\sqrt{LC}}$

7. Solve $\frac{\partial u}{\partial t} = \frac{\partial^2 u}{\partial x^2}$ if

(i) $\frac{\partial u}{\partial x}(0,t) = 0$ for $t > 0$. (ii) $u(x, 0) = \begin{cases} x & 0 \le x \le 0 \\ 0 & x > 1 \end{cases}$

(iii) and $u(x, t)$ is bounded for $x > 0$, $t > 0$ **Ans.** $\left[u(x,t) = \frac{2}{\pi}\int_0^\infty \left(\frac{\sin s}{s} + \frac{\cos s - 1}{s^2}\right) e^{-s^2 t} \cos sx\, ds\right]$

41.18 FINITE FOURIER TRANSFORMS

Let $f(x)$ denote a function which is sectionally continuous over the range $(0, l)$. Then the **Finite Fourier sine transform** of $f(x)$ on this interval is defined as

$$F_s(p) = \bar{f}_s(P) = \int_0^l f(x) \sin \frac{p\pi x}{l} dx$$

where p is an integer (Instead of s, we take p as a parameter)

Inversion formula for sine transform

If $\bar{f}_s(p) = F_s(p)$ is the finite Fourier sine transform of $f(x)$ in $(0, l)$ then the inversion formula for sine transform is

$$f(x) = \frac{2}{l}\sum_{p=l}^{\infty} \bar{f}_s(P) \sin \frac{p\pi x}{l}$$

Proof. For the given function $f(x)$ in $(0, l)$, if we find the half range Fourier sine series, we get.

$$f(x) = \sum_{n=1}^{\infty} b_n \sin \frac{n\pi x}{l} \quad \ldots (1)$$

where $$b_n = \frac{2}{l}\int_0^l f(x) \sin \frac{n\pi x}{l} dx$$

$\therefore$ $$b_p = \frac{2}{l}\int_0^l f(x) \sin \frac{p\pi x}{l} dx = \frac{2}{l}\bar{f}_s(p) \text{ by definition}$$

Substituting in (1), we get $f(x) = \frac{2}{l}\sum_{p=1}^{\infty} f_s(p) \sin \frac{p\pi x}{l}$

Finite Fourier Cosine Transform

Let $f(x)$ denote a sectionally continuous function in $(0, l)$.

Then the Finite Fourier cosine transform of $f(x)$ over $(0, l)$ is defined as

$$F_c(p) = \bar{f}_c(p) = \int_0^l f(x) \cos \frac{p\pi x}{l} dx \text{ where } p \text{ is an integer.}$$

Inversion formula for cosine transform

If $\bar{f}_c(P)$ is the finite Fourier cosine transform of $f(x)$ in $(0, l)$. then the inversion formula for cosine transform is

$$f(x) = \frac{1}{l}\bar{f}_c(0) + \frac{2}{l}\sum_{p=l}^{\infty} \bar{f}_c(p) \cos \frac{p\pi x}{l}$$

where $$\bar{f}_c(0) = \int_0^l f(x)\, dx.$$

Proof. If we find half range Fourier cosine series for $f(x)$ in $(0, l)$, we obtain.

$$f(x) = \frac{a_0}{2} + \sum_{n=1}^{\infty} a_n \cos \frac{n\pi x}{l} \qquad \ldots (2)$$

where
$$a_n = \frac{2}{l} \int_0^l f(x) \cos \frac{n\pi x}{l} \, dx$$

$\therefore$
$$a_p = \frac{2}{l} \bar{f}_c(p)$$

$$a_0 = \frac{2}{l} \int_0^l f(x)\, dx = \frac{2}{l} \bar{f}_c(0).$$

Substituting in (2), we get

$$f(x) = \frac{1}{l} f_c(0) + \frac{2}{l} \sum_{p=1}^{\infty} f_c(p) \cos \frac{p\pi x}{l}$$

Example. 46. *Find the finite Fourier sine and cosine transforms of*

(i) $f(x) = 1$ in $(0, \pi)$
(ii) $f(x) = x$ in $(0, l)$
(iii) $f(x) = x^2$ in $(0, l)$
(iv) $f(x) = 1$ in $0 < x < \pi/2$
$= -1$ in $\pi/2 < x < \pi$
(v) $f(x) = x^3$ in $(0, l)$
(vi) $f(x) = e^{ax}$ in $(0, l)$

Solution.

(i) $$\bar{f}_s(p) = F_s(1) = \int_0^{\pi} 1.\sin \frac{p\pi x}{\pi} dx = \left(-\frac{\cos px}{p}\right)_0^{\pi} = \frac{1 - \cos p\pi}{p} \qquad [\text{If } p \neq 0]$$

$$\bar{f}_c(p) = \int_0^{\pi} 1 \cdot \cos px \, dx = \left(\frac{\sin px}{p}\right)_0^{\pi} = \frac{1}{p}(0 - 0) = 0$$

(ii) $$\bar{f}_s(p) = F_s(p) = \int_0^l x \sin \frac{p\pi x}{l} dx$$

$$= \left[(x) \left(\frac{-\cos \frac{p\pi x}{l}}{\frac{p\pi}{l}} \right) - (1) \left(-\frac{\sin \frac{p\pi x}{l}}{\frac{p^2\pi^2}{l^2}} \right) \right]_0^l = \frac{-l}{p\pi}(l \cos p\pi)$$

$$= \frac{-l^2}{p\pi}(-1)^p \qquad [\text{If } p \neq 0]$$

$$\bar{f}_c(p) = F_c(x) = \int_0^l x \cos \frac{p\pi x}{l} dx$$

$$= \left[(x) \left(\frac{\sin \frac{p\pi x}{l}}{\frac{p\pi}{l}} \right) - (1) \left(-\frac{\cos \frac{p\pi x}{l}}{\frac{p^2\pi^2}{l^2}} \right) \right]_0^l = \frac{l^2}{p^2\pi^2}[(-1)^p - 1] \qquad [\text{If } p \neq 0]$$

(iii) $$\bar{f}_s(p) = F_s(x^2) = \int_0^l x^2 \sin \frac{p\pi x}{l} dx$$

$$= \left[(x^2)\left(-\frac{\cos\frac{p\pi x}{l}}{\frac{p\pi}{l}} \right) - (2x)\left(-\frac{\sin\frac{p\pi x}{l}}{\frac{p^2\pi^2}{l^2}} \right) + (2)\left(\frac{\cos\frac{p\pi x}{l}}{\frac{p^3\pi^3}{l^3}} \right) \right]_0^l$$

$$= \frac{-l^3}{p\pi}(-1)^p + \frac{2l^3}{p^3\pi^3}\left[(-1)^p - 1\right] \qquad \text{[If } p \neq 0 \text{]}$$

$$\bar{f}_c(p) = \int_0^l (x^2)\cos\frac{p\pi x}{l}\,dx$$

$$= \left[(x^2)\left(\frac{\sin\frac{p\pi x}{l}}{\frac{p\pi}{l}} \right) - (2x)\left(-\frac{\cos\frac{p\pi x}{l}}{\frac{p^2\pi^2}{l^2}} \right) + (2)\left(-\frac{\sin\frac{p\pi x}{l}}{\frac{p^3\pi^3}{l^3}} \right) \right]_0^l = \frac{2l^3}{p^2\pi^2}(-1)^p \qquad \text{[If } p \neq 0 \text{]}$$

(iv) $$F_s\{f(x)\} = \int_0^{\pi/2} \sin px\,dx + \int_{\pi/2}^{\pi} (-1)\sin px\,dx$$

$$= \left(-\frac{\cos px}{p} \right)_0^{\pi/2} + \left(\frac{\cos px}{p} \right)_{\pi/2}^{\pi} = -\frac{1}{p}\left(\cos\frac{p\pi}{2} - 1 \right) + \frac{1}{p}\left(\cos p\pi - \cos\frac{p\pi}{2} \right)$$

$$= -\frac{1}{p}\left(\cos p\pi - 2\cos\frac{p\pi}{2} - 1 \right) \qquad \text{[If } p \neq 0 \text{]}$$

$$F_c(f(x)) = \int_0^{\pi/2} \cos px\,dx - \int_{\pi/2}^{\pi} \cos px\,dx$$

$$= \left(\frac{\sin px}{p} \right)_0^{\pi/2} - \left(\frac{\sin px}{p} \right)_{\pi/2}^{\pi} = \frac{2}{p}\sin\frac{p\pi}{2} \qquad \text{[If } p \neq 0 \text{]}$$

(v) $$F_s(x^3) = \int_0^l x^3 \sin\frac{p\pi x}{l}\,dx$$

$$= \left[(x^3)\left(-\frac{\cos\frac{p\pi x}{l}}{\frac{p\pi}{l}} \right) - (3x^2)\left(\frac{\sin\frac{p\pi x}{l}}{\frac{p^2\pi^2}{l^2}} \right) + (6x)\left(\frac{\cos\frac{p\pi x}{l}}{\frac{p^3\pi^3}{l^3}} \right) - (6)\left(\frac{\sin\frac{p\pi x}{l}}{\frac{p^4\pi^4}{l^4}} \right) \right]_0^l$$

$$= -\frac{l^4}{p\pi}(-1)^p + \frac{6\,l^4}{p^3\pi^3}(-1)^p \qquad \text{[If } p \neq 0 \text{]}$$

$$F_c(x^3) = \int_0^l x^3 \cos\frac{p\pi x}{l}\,dx$$

$$= \left[(x^3)\left(\frac{\sin\frac{p\pi x}{l}}{\frac{p\pi}{l}} \right) - (3x^2)\left(-\frac{\cos\frac{p\pi x}{l}}{\frac{p^2\pi^2}{l^2}} \right) + (6x)\left(\frac{-\sin\frac{p\pi x}{l}}{\frac{p^3\pi^3}{l^3}} \right) - (6)\left(\frac{\cos\frac{p\pi x}{l}}{\frac{p^4\pi^4}{l^4}} \right) \right]_0^l$$

$$= \frac{3l^4}{\pi^2 p^2}(-1)^p - \frac{6l^4}{p^4\pi^4}[(-1)^p - 1] \qquad \text{[If } p \neq 0 \text{]}$$

(vi) $F_s(e^{ax}) = \int_0^l e^{ax} \sin \frac{p\pi x}{l} dx$

$$= \left\{ \frac{e^{ax}}{a^2 + \frac{p^2\pi^2}{l^2}} \left[a \sin \frac{p\pi x}{l} - \frac{p\pi}{l} \cos \frac{p\pi x}{l} \right] \right\}_0^l = \frac{e^{al}}{a^2 + \frac{p^2\pi^2}{l^2}} \cdot \left(-\frac{p\pi}{l} (-1)^p \right) + \frac{1}{a^2 + \frac{p^2\pi^2}{l^2}} \left(\frac{p\pi}{l} \right)$$

$$F_c(e^{ax}) = \int_0^l e^{ax} \cos \frac{p\pi x}{l} dx$$

$$F_c(e^{ax}) = \left\{ \frac{e^{ax}}{a^2 + \frac{p^2\pi^2}{l^2}} \left[a \cos \frac{p\pi x}{l} + \frac{p\pi}{l} \sin \frac{p\pi x}{l} \right] \right\}_0^l = \frac{e^{al}}{a^2 + \frac{p^2\pi^2}{l^2}} a(-1)^p - \frac{1}{a^2 + \frac{p^2\pi^2}{l^2}} (a) \quad \textbf{Ans.}$$

Example 47. *Find f (x) if its finite Fourier sine transform is* $\frac{2\pi}{p^3}(-1)^{p-1}$ *for* $p = 1, 2, \ldots,$ $0 < x < \pi$.

Solution. By inversion Theorem, we have

$$f(x) = \frac{2}{\pi} \sum_{p=1}^{\infty} \frac{2\pi}{p^3} (-1)^{p-1} \sin px \qquad = 4 \sum_{p=1}^{\infty} \frac{(-1)^{p-1}}{p^3} \sin px \qquad \textbf{Ans.}$$

Example 48. *Find f (x) if its finite Fourier sine transform is given by*

(i) $F_s(p) = \frac{1 - \cos p\pi}{p^2 \pi^2}$ *for* $p = 1, 2, 3, \ldots$ *and* $0 < x < \pi$

(ii) $F_s(p) = \frac{16(-1)^{p-1}}{p^3}$ *for* $p = 1, 2, 3, \ldots.$ *and* $0 < x < 8$

(iii) $F_s(p) = \frac{\cos \frac{2\pi p}{3}}{(2p+1)^2}$ *for* $p = 1, 2, 3, \ldots$ *and* $0 < x < 1$.

Solution. By inversion theorem, we have

(i) $f(x) = \frac{2}{\pi} \sum_{p=1}^{\infty} \left(\frac{1 - \cos p\pi}{p^2\pi^2} \right) . \sin px = \frac{2}{\pi^3} \sum_{p=1}^{\infty} \left(\frac{1 - \cos p\pi}{p^2} \right) . \sin px$

(ii) $f(x) = \frac{2}{l} \sum_{p=1}^{\infty} F_s(p) \sin\left(\frac{p\pi x}{l} \right) = \frac{2}{8} \sum_{p=1}^{\infty} \frac{16(-1)^{p-1}}{p^3} \sin\left(\frac{p\pi x}{8} \right)$ [Since $l = 8$]

$$= 4 \sum_{p=1}^{\infty} \frac{(-1)^{p-1}}{p^3} \sin\left(\frac{p\pi x}{8} \right)$$

(iii) $f(x) = \frac{2}{l} \sum_{p=1}^{\infty} F_s(p) \sin\left(\frac{p\pi x}{l} \right) = 2 \sum_{p=1}^{\infty} \frac{\cos\left(\frac{2\pi p}{3} \right)}{(2p+1)^2} \sin(p\pi x)$ [Since $l = 1$] **Ans.**

Example 49. *Find f (x) if its finite Fourier cosine transform is*

(i) $F_c(p) = \frac{1}{2p} \sin\left(\frac{p\pi}{2} \right)$ *for* $p = 1, 2, 3, \ldots.$

$$= \frac{\pi}{4} \qquad \textit{for } p = 0 \textit{ given } 0 < x < 2\pi$$

$$(ii)\ F_c(p) = \frac{6\sin\frac{p\pi}{2} - \cos p\pi}{(2p+1)\pi} \qquad \textit{for } p = 1, 2, 3, \ldots$$

$$= \frac{2}{\pi} \qquad \textit{for } p = 0 \textit{ given } 0 < x < 4$$

$$(iii)\ F_c(p) = \frac{\cos\left(\frac{2p\pi}{3}\right)}{(2p+1)^2} \qquad \textit{for } p = 1, 2, 3, \ldots$$

$$= 1 \qquad \textit{for } p = 0 \textit{ given } 0 < x < 1$$

Solution. By inversion theorem, we have

$$f(x) = \frac{1}{l}F_c(0) + \frac{2}{l}\sum_{p=1}^{\infty} F_c(p).\cos\frac{p\pi x}{l}$$

(*i*) Here $F_c(0) = \pi/4$ and $l = 2\pi$

$$\therefore \quad f(x) = \frac{1}{2\pi}\left(\frac{\pi}{4}\right) + \frac{2}{2\pi}\sum_{p=1}^{\infty}\frac{1}{2p}\sin\left(\frac{p\pi}{2}\right)\cos\left(\frac{p\pi x}{2\pi}\right)$$

$$= \frac{1}{8} + \frac{1}{2\pi}\sum_{p=1}^{\infty}\frac{1}{p}\sin\left(\frac{p\pi}{2}\right)\cos\left(\frac{px}{2}\right)$$

(*ii*) Here $F_c(0) = \frac{2}{\pi}$ and $l = 4$

$$f(x) = \frac{1}{4}\left(\frac{2}{\pi}\right) + \frac{2}{4}\sum_{p=1}^{\infty}\frac{\left(6\sin\frac{p\pi}{2} - \cos p\pi\right)}{(2p+1)\pi}\cos\left(\frac{p\pi x}{4}\right)$$

$$= \frac{1}{2\pi} + \frac{1}{2\pi}\sum_{p=1}^{\infty}\frac{\left(6\sin\frac{p\pi}{2} - \cos p\pi\right)}{(2p+1)}.\cos\left(\frac{p\pi x}{4}\right)$$

(*iii*) Here $F_c(0) = 1,\ l = 1$

$$\therefore \quad f(x) = \frac{1}{1} + \frac{2}{1}\sum_{p=1}^{\infty}\frac{1}{(2p+1)^2}\cos\left(\frac{2p\pi}{3}\right).\cos(p\pi x)$$

$$= 1 + 2\sum_{p=1}^{\infty}\frac{\cos\left(\frac{2p\pi}{3}\right)}{(2p+1)^2}\cos(p\pi x)$$ **Ans.**

Example 50. *Find the finite Fourier sine transform of $f(x) = 1$ in $(0, \pi)$. Use the inversion theorem and find Fourier series for $f(x) = 1$ in $(0, \pi)$. Hence prove*

$$(i)\ 1 - \frac{1}{3} + \frac{1}{5} - \frac{1}{7} + \ldots = \pi/4 \qquad (ii)\ \frac{1}{1^2} + \frac{1}{3^2} + \frac{1}{5^2} + \ldots = \pi^2/8$$

Solution. $$F_s(1) = \int_0^{\pi} 1.\sin\left(\frac{p\pi x}{\pi}\right)dx = \left[-\frac{\cos px}{p}\right]_0^{\pi}$$

$$\bar{f}_s(p) = \frac{1 - \cos p\pi}{p} \quad \text{if } p \neq 0$$

By inversion theorem,

$$f(x)=\frac{2}{l}\sum_{p=1}^{\infty} F_s\,(p)\sin\frac{p\pi x}{l}$$

$$1=\frac{2}{\pi}\sum_{p=1}^{\infty} \frac{1-(-1)^p}{p}.\sin px \qquad \text{[Since } l=\pi\text{]}$$

$$1=\frac{4}{\pi}\left[\sin x+\frac{1}{3}\sin 3x+\frac{1}{5}\sin 5x+...\right] \qquad [p \text{ is odd}] \quad .... (1)$$

This is the half range Fourier sine series for $f(x)=1$ in $(0,\pi)$ getting $x=\pi/2$.

On putting $x=\frac{\pi}{2}$ in (1), we get $\frac{4}{\pi}\left[1-\frac{1}{3}+\frac{1}{5}-\frac{1}{7}+....\right]=1$

$$\therefore \qquad 1-\frac{1}{3}+\frac{1}{5}-\frac{1}{7}+....=\frac{\pi}{4}$$

In the half Fourier sine series $l_n=\frac{4}{\pi}.\frac{1}{n}$ for n odd

By using Parseval's Theorem

$$\text{(Range)}\left[\frac{1}{2}\sum b_n^2\right]=\int_0^{\pi}(1)^2\,dx$$

$$\pi\left[\frac{1}{2}.\frac{16}{\pi^2}\sum_{n=1,3,5....}^{\pi}\frac{1}{n^2}\right]=\pi \qquad i.e., \quad \sum_{n=1,3,5}^{\infty}\frac{1}{n^2}=\frac{\pi^2}{8} \qquad \textbf{Ans.}$$

(*ii*) $\frac{1}{1^2}+\frac{1}{3^2}+\frac{1}{5^2}+.........=\frac{\pi^2}{8}$

EXERCISE 41.5

Find the finite Fourier sine and cosine transforms of

1. $f(x)=2x$ in (0, 4) **Ans.** $F_s(s)=\begin{cases}\frac{32}{s\pi}(1-\cos s\pi), & s\neq 0\\ 0, & s=0\end{cases}$, $F_c(s)=\begin{cases}\frac{32}{s^2\pi^2}(\cos s\pi-1), & s\neq 0\\ 16, & s=0\end{cases}$

2. $f(x)=x$ in $(0,\pi)$ **Ans.** $F_s(s)=\begin{cases}\frac{\pi}{s}(-1)^{s+1}, & s\neq 0\\ 0, & s=0\end{cases}$, $F_c(s)=\begin{cases}\frac{(-1)^s-1}{s^2}, & s\neq 0\\ \frac{\pi^2}{2}, & s=0\end{cases}$

3. $f(x)=\cos ax$ in $(0,\pi)$ Ans. $F_s(s)=\frac{s}{s^2-a^2}[1-(-1)^s\cos a\pi]$

4. $f(x)=1-\frac{x}{\pi}$ in $(0,\pi)$ Ans. $F_s(s)=\frac{1}{s}$

5. $f(x)=\begin{cases}x \text{ in } (0,\pi/2)\\ \pi-x \text{ in } (\pi/2,\pi)\end{cases}$ **Ans.** $F_s(s)=\frac{2}{s^2}\sin\frac{s\pi}{2}$

6. $f(x)=e^{-ax}$ in $(0,l)$ **Ans.** $F_s(e^{-ax})=\frac{e^{-al}}{a^2+\frac{s^2a^2}{l^2}}\left(-\frac{s\pi}{l}(-1)^s\right)+\frac{1}{a^2+\frac{s^2\pi^2}{l^2}}\left(\frac{s\pi}{l}\right)$

$$F_c(e^{-ax})=\frac{ae^{-al}}{a^2+\frac{s^2\pi^2}{l^2}}(-1)^{s+1}+\frac{a}{a^2+\frac{s^2\pi^2}{l^2}}$$

7. $(x)=\frac{\pi}{3}-x+\frac{x^2}{2\pi}$ in $(0,\pi)$ **Ans.** $F_s(s)=\frac{\pi(-1)^p}{6p}+\frac{1(-1)^p}{\pi p^3}+\frac{\pi}{3p}-\frac{1}{\pi p^3}$

8. Find finite Fourier cosine transform of $\left(1-\frac{x}{\pi}\right)^2, 0 < x < \pi$ **Ans.** $F_c(s) = \begin{cases} \frac{2}{\pi s^2}, & s \neq 0 \\ \frac{\pi}{3}, & s = 0 \end{cases}$

9. Find $f(x)$ if $\bar{f}_c(s) = \begin{cases} \frac{\sin\left(\frac{s\pi}{2}\right)}{2s}, & s = 1, 2, 3 \ldots \\ \frac{\pi}{4} & s = 0 \text{ given } 0 < x < 2x. \end{cases}$ **Ans.** $\frac{1}{8} + \frac{1}{\pi}\sum_{s=1}^{\infty} \frac{\sin\frac{s\pi}{2}}{2s} \cos\frac{sx}{2}$

41.19 FINITE FOURIER SINE AND COSINE TRANSFORMS OF DERIVATIVES

Using the definition and the integration by parts, we can easily prove the following results. For $0 \leq x \leq l$,

(*i*) $$F_s(f'(x)) = -\frac{p\pi}{l}\bar{f}_c(p)$$

(*ii*) $$F_c\{f'(x)\} = f(l)(-1)^p - f(0) + \frac{p\pi}{l}\bar{f}_s(p)$$

(*iii*) $$F_s\{f''(x)\} = -\frac{p^2\pi^2}{l^2}\bar{f}_s(p) + \frac{p\pi}{l}\left[f(0) - (-1)^n f(l)\right]$$

(*iv*) $$F_c\{f''(x)\} = -\frac{p^2\pi^2}{l^2}\bar{f}_c(p) + f'(l)(-1)^p - f'(0)$$

Proof: (*i*) $$F_s(f'(x)) = \int_0^l f'(x)\sin\frac{p\pi x}{l}dx = \int_0^l \sin\frac{p\pi x}{l}\cdot d\{f(x)\}$$

$$= \left(f(x)\sin\frac{p\pi x}{l}\right)_0^l - \int_0^l f(x)\frac{p\pi}{l}.\cos\frac{p\pi x}{l}dx$$

$$= -\frac{p\pi}{l}\bar{f}_c(p) \quad \ldots (1)$$

(*ii*) $$F_c\{f'(x)\} = \int_0^l f'(x)\cos\frac{p\pi x}{l}dx = \left(f(x)\cos\frac{p\pi x}{l}\right)_0^l - \int_0^l f(x)\cdot\left(-\frac{p\pi}{l}\sin\frac{p\pi x}{l}\right)dx$$

$$= (-1)^P \bar{f}(l) - f(0) + \frac{p\pi}{l}\bar{f}_s(p) \quad \ldots (2)$$

(*iii*) $$F_s[f''(x)] = \int_0^l \sin\frac{p\pi x}{l}d[f'(x)]$$

$$= \left(f'(x)\sin\frac{p\pi x}{l}\right)_0^l - \frac{p\pi}{l}\int_0^l f'(x)\cos\frac{p\pi x}{l}ds$$

$$= -\frac{p\pi}{l}\left[(-1)^p f(l) - f(0) + \frac{p\pi}{l}\bar{f}_s(p)\right] \quad \text{[Using (2)]}$$

$$= -\frac{p^2\pi^2}{l^2}\bar{f}_s(p) + \frac{p\pi}{l}\left[f(0) - (-1)^p f(l)\right] \quad \ldots (3)$$

(iv) $$F_c\{f''(x)\} = \int_0^l \cos\frac{p\pi x}{l}\, d[f'(x)]$$

$$= \left[f'(x)\cos\frac{p\pi x}{l}\right]_0^l + \frac{p\pi}{l}\int_0^l f'(x)\sin\frac{p\pi x}{l}\,dx \qquad \text{[Using (1)]}$$

$$= (-1)^p f'(l) - f'(0) + \frac{p\pi}{l}\left[-\frac{p\pi}{l}\bar{f}_c(p)\right] \qquad \ldots (4)$$

$$= -\frac{p^2\pi^2}{l^2}\bar{f}_c(p) + f'(l)(-1)^p - f'(0)$$

Note. if $u = u(x, t)$, then

$$F_s\left[\frac{\partial u}{\partial x}\right] = \frac{-p\pi}{l}F_c(u)$$

$$F_c\left[\frac{\partial u}{\partial x}\right] = \frac{p\pi}{l}F_s(u) - u(0, t) + (-1)^p u(l, t)$$

$$F_s\left[\frac{\partial^2 u}{\partial x^2}\right] = \frac{p^2\pi^2}{l^2}F_s(u) + \frac{p\pi}{l}[u(0, t) - (-1)^p u(l, t)]$$

$$F_c\left[\frac{\partial^2 u}{\partial x^2}\right] = -\frac{p^2\pi^2}{l^2}F_c(u) + \frac{\partial u}{\partial x}(l, t)\cos p\pi - \frac{\partial u}{\partial x}(0, t)$$

Example 51. *Using finite Fourier transform, solve*

$$\frac{\partial u}{\partial t} = \frac{\partial^2 u}{\partial x^2}, \text{ given } u(0, t) = 0 \text{ and } u(4, t) = 0$$

and $u(x, 0) = 2x$ *where* $0 < x < 4,\ t > 0$.

Solution. Since $u(0, t)$ given, take finite Fourier sine transform

$$\int_0^4 \frac{\partial u}{\partial t}\sin\frac{p\pi x}{4}\,dx = \int_0^4 \frac{\partial^2 u}{\partial x^2}\sin\frac{p\pi x}{4}\,dx$$

$$\frac{d\bar{u}_s}{dt} = F_s\left(\frac{\partial^2 u}{\partial x^2}\right) = -\frac{p^2\pi^2}{16}\bar{u}_s + \frac{p\pi}{4}[u(0, t) - (-1)^n u(4, t)]$$

$$= -\frac{p^2\pi^2}{16}\bar{u}_s \text{ using } u(0, t) = 0,\ u(4, t) = 0$$

$$\frac{d\bar{u}_s}{\bar{u}_s} = -\frac{p^2\pi^2}{16}dt$$

Integrating $$\log \bar{u}_s = -\frac{p^2\pi^2}{16}t + c$$

$$\bar{u}_s = A e^{-\frac{\pi^2 p^2}{16}t} \qquad \ldots (1)$$

Since $u(x, 0) = 2x$

$$\bar{u}_s(p, 0) = \int_0^4 (2x)\sin\left(\frac{p\pi x}{4}\right)dx = \left[(2x)\left(-\frac{4}{p\pi}\cos\left(\frac{p\pi x}{4}\right) + (2)\frac{16}{p^2\pi^2}\sin\left(\frac{p\pi x}{4}\right)\right)\right]_0^4$$

$$= -\frac{32}{p\pi}\cos p\pi \qquad \ldots (2)$$

Using (2) in (1), we get $\bar{u}_s\ (p, 0) = A = -\frac{32}{p\pi}\cos p\pi = -\frac{32}{p\pi}(-1)^p$

Substituting in (1), we get

$$\therefore \qquad \bar{u}_s = -\frac{32}{p\pi}(-1)^p\ e^{-\frac{p^2\pi^2}{16}t}$$

By inversion Theorem, we get

$$u(x, t) = \frac{2}{4}\sum_{p=1}^{\infty}\frac{32}{p\pi}(-1)^{p+1}e^{-\frac{p^2\pi^2}{16}t}\sin\left(\frac{p\pi x}{4}\right) \qquad \textbf{Ans.}$$

Example 52. *Solve* $\frac{\partial u}{\partial t} = \frac{\partial^2 u}{\partial x^2}, \qquad 0 < x < 6, t > 0$

given $\frac{\partial u}{\partial x}(0, t) = 0, \frac{\partial u}{\partial x}(6, t) = 0$ *and* $u\ (x, 0) = 2x.$

Solution. Since $\frac{\partial u}{\partial x}(0, t)$ is given, use finite Fourier cosine transform of

$$\frac{\partial u}{\partial t} = \frac{\partial^2 u}{\partial x^2}$$

$$\int_0^6 \frac{\partial u}{\partial t}\cos\frac{p\pi x}{6}dx = \int_0^6 \frac{\partial^2 u}{\partial x^2}\cos\frac{p\pi x}{6}dx$$

$$\frac{d\bar{u}_c}{dt} = -\frac{p^2\pi^2}{36}\bar{u}_c + \frac{\partial u}{\partial x}(6, t)\cos p\pi - \frac{\partial u}{\partial x}(0, t) = -\frac{p^2\pi^2}{36}\bar{u}_c$$

$$\Rightarrow \qquad \frac{d\bar{u}_c}{\bar{u}_c} = -\frac{p^2\pi^2}{36}dt \qquad \Rightarrow \qquad \log\ \bar{u}_c = -\frac{p^2\pi^2}{36}t + c$$

$$\bar{u}_c = A\ e^{-\frac{p^2\pi^2}{36}t} \qquad \text{... (1)}$$

$$u\ (x,\ 0) = 2x\,.$$

$\therefore$ At $t = 0$

$$\bar{u}_c\ (p,\ 0) - \int_0^6 (2x)\cos\frac{p\pi x}{6} = \frac{72}{p^2\pi^2}(\cos\ p\pi - 1) \qquad \text{... (2)}$$

Using this in (1), we get

$$\bar{u}_c(p, 0) = A = \frac{72}{p^2\pi^2}(\cos\ p\pi - 1)$$

Substituting in (1), we get

$$\bar{u}_c\ (p,\ t) = \frac{72}{p^2\pi^2}(\cos p\pi)$$

By inversion theorem, we get

$$u(x, t) = \frac{1}{l}\bar{f}_c(0) + \frac{2}{l}\sum_{p=1}^{\infty}\bar{f}_c(p)\cos\left(\frac{p\pi x}{l}\right)$$

$$= \frac{1}{6}\int_0^6 (2x)\ dx + \frac{2}{6}\sum_{p=1}^{\infty}\frac{72}{p^2\pi^2}(\cos p\pi - 1)\ e^{-\frac{p^2\pi^2}{36}t}.\cos\left(\frac{p\pi x}{6}\right)$$

$$= 6 + \frac{24}{\pi^2}\sum_{p=1}^{\infty}\frac{(\cos p\pi - 1)}{p^2}e^{-\frac{p^2\pi^2}{36}t}\cos\left(\frac{p\pi x}{6}\right). \qquad \textbf{Ans.}$$

Example 53. *Solve* $\frac{\partial u}{\partial t} = 2\frac{\partial^2 u}{\partial x^2}$, $\quad 0 < x < 4,\ t > 0$

given $u(0, t) = 0;\quad u(4, t) = 0;\quad u(x, 0) = 3\sin \pi x - 2\sin 5\pi x$.

Solution. Since $u(0, t)$ is given, take finite Fourier sine transform. The equation becomes (as in example 62 on page 68).

$$\frac{d\bar{u}_c}{dt} = 2\left[-\frac{p^2\pi^2}{16}\bar{u}_s + \frac{p\pi}{4}\{u(0, t) - (-1)^p u(4, t)\}\right] = -\frac{p^2\pi^2}{8}\bar{u}_s$$

Solving, we get $\quad \bar{u}_s = A e^{-\frac{p^2\pi^2}{8}t}$ ... (1)

$$u(x, 0) = 3\sin \pi x - 2\sin 5\pi x$$

Taking sine Transform, $\bar{u}_s(p, 0) = \int_0^4 (3\sin \pi x - 2\sin 5\pi x)\sin\frac{p\pi x}{4}\,dx$

$$= 0,\ \text{if } p \neq 4 \quad \text{or} \quad p \neq 20.$$

If $p = 4$, $\quad \bar{u}_s(4, 0) = 6$

If $p = 20$, $\quad \bar{u}_s(20, 0) = -4$

$$u(x, t) = \frac{2}{4}\sum_{p=1}^{\infty} \bar{u}_s(p, t)\sin\left(\frac{p\pi x}{4}\right)$$

$$= \frac{1}{2}[6\,e^{-\frac{p^2\pi^2}{8}t}\sin \pi x - 4e^{-\frac{p^2\pi^4}{8}t}\sin 5\pi x]$$

where p in the first term is 4 and p in the second term is 20

$$= 3e^{-2\pi^2 t}\sin \pi x - 2e^{-50\pi^2 t}\sin 5\pi x.$$ **Ans.**

EXERCISE 41.6

1. Solve $\frac{\partial u}{\partial t} = \frac{\partial^2 u}{\partial x^2}$, $0 < x < 6,\ t > 0$ given that $u(0, t) = 0 = u(6, t)$ and $u(x, 0) = \begin{cases} 1 \text{ for } 0 < x < 3 \\ 0 \text{ for } 3 < x < 6 \end{cases}$

 Ans. $u(x,t) = \frac{2}{\pi}\sum_{p=1}^{\pi}\left(\frac{1 - \cos\frac{p\pi}{2}}{p}\right) e^{-\frac{p^2\pi^2 t}{36}}\sin\left(\frac{p\pi x}{6}\right)$

2. Solve $\frac{\partial v}{\partial t} = \frac{\partial^2 v}{\partial x^2}$ subject to conditions $v(0, t) = 1$, $v(\pi, t) = 3$

 $v(x, 0) = 1$ for $0 < x < \pi,\ t > 0$ **Ans.** $v(x, t) = \frac{4}{\pi}\sum_{p=1}^{\infty}\frac{\cos p\pi}{p}e^{-p^2 t}\sin px + 1 + \frac{2x}{\pi}$

3. Solve $\frac{\partial \theta}{\partial t} = \frac{\partial^2 \theta}{\partial x^2}$

 given $\theta(0, t) = 0$, $\quad \theta(\pi, t) = 0$, $\quad \theta(x, 0) = 2x$ for $0 < x < \pi,\ t > 0$

 Ans. $\theta(x, t) = 4\sum_{n=1}^{\infty}\frac{1}{n}(-1)^{n+1} e^{-n^2 t}\sin nx$

TABLE

S.No.	Function $f(x)$	Fourier Sine Transform $F_s(s)$
1.	$\begin{cases} 1, & 0 < x < b \\ 0, & x > b \end{cases}$	$\dfrac{1-\cos bs}{s}$
2.	x^{-1}	$\dfrac{x}{2}$
3.	$\dfrac{x}{x^2+b^2}$	$\dfrac{\pi}{2}e^{-bs}$
4.	e^{-bx}	$\dfrac{s}{s^2+b^2}$
5.	$x^{n-1}e^{-bx}$	$\dfrac{\Gamma(n)\sin(n\tan^{-1}s/b)}{(s^2+b^2)^{n/2}}$
6.	xe^{-bx^2}	$\dfrac{\sqrt{\pi}}{4b^{3/2}}se^{-s^2/4b}$
7.	$x^{-1/2}$	$\sqrt{\dfrac{\pi}{2s}}$
8.	x^{-n}	$\dfrac{\pi s^{n-1}\csc(n\pi/2)}{2\Gamma(n)}, \quad 0 < n < 2$
9.	$\dfrac{\sin bx}{x}$	$\dfrac{1}{2}\text{In}\left(\dfrac{s+b}{s-b}\right)$
10.	$\dfrac{\sin bx}{x^2}$	$\begin{cases} \pi s/2, & s < b \\ \pi b/2, & s > b \end{cases}$
11.	$\dfrac{\cos bx}{x}$	$\begin{cases} 0\ , & s < b \\ \pi/4, & s = b \\ \pi/2, & s > b \end{cases}$
12.	$\tan^{-1}(x/b)$	$\dfrac{\pi}{2s}e^{-bs}$
13.	$\csc bx$	$\dfrac{\pi}{2b}\tanh\dfrac{\pi s}{2b}$
14.	$\dfrac{1}{e^{2x}-1}$	$\dfrac{\pi}{4}\cot h\left(\dfrac{\pi s}{2}\right)-\dfrac{1}{2s}$

		Fourier cos Transform
15.	$\begin{cases}1, & 0<x<b \\ 0, & x>b\end{cases}$	$\dfrac{\sin bs}{s}$
16.	$\dfrac{1}{x^2+b^2}$	$\dfrac{\pi e^{-bs}}{2b}$
17.	e^{-bx}	$\dfrac{b}{s^2+b^2}$
18.	$x^{n-1}e^{-bx}$	$\dfrac{\Gamma(n)\cos(n\tan^{-1}s/b)}{(s^2+b^2)^{n/2}}$
19.	xe^{-bx^2}	$\dfrac{1}{2}\sqrt{\dfrac{\pi}{b}}\,e^{-s^{2/4b}}$
20.	$x^{-1/2}$	$\sqrt{\dfrac{\pi}{2s}}$
21.	x^{-n}	$\dfrac{\pi s^{n-1}\sec(n\pi/2)}{2\Gamma(n)}, \; 0<n<1$
22.	$\ln\left(\dfrac{x^2+b^2}{x^2+c^2}\right)$	$\dfrac{e^{-cs}-e^{-bs}}{\pi s}$
23.	$\dfrac{\sin bx}{x^2}$	$\begin{cases}\pi/2, & s<b \\ \pi/4, & s=b \\ 0, & s>b\end{cases}$
24.	$\sin bx^2$	$\dfrac{\sqrt{\pi}}{8b}\left(\cos\dfrac{s^2}{4b}-\sin\dfrac{s^2}{4b}\right)$
25.	$\cos bx^2$	$\sqrt{\dfrac{\pi}{8b}}\left(\cos\dfrac{s^2}{4b}+\sin\dfrac{s^2}{4b}\right)$
26.	sech bx	$\dfrac{\pi}{2b}\,\text{sech}\,\dfrac{s\pi}{2b}$
27.	$\dfrac{\cosh(\sqrt{\pi}x/2)}{\cosh(\sqrt{\pi}\,x)}$	$\dfrac{\sqrt{\pi}}{2}\dfrac{\cosh(\sqrt{\pi}s/2)}{\cosh(\sqrt{\pi}\,s)}$
28.	$\dfrac{e^{-b\sqrt{x}}}{\sqrt{x}}$	$\sqrt{\dfrac{\pi}{2s}}\{\cos(2b\sqrt{s})-\sin(2b\sqrt{s})\}$

LAPLACE TRANSFORM

42.1 INTRODUCTION

Laplace transforms help in solving the differential equations with boundary values without finding the general solution and the values of the arbitrary constants.

42.2 LAPLACE TRANSFORM

Definition. Let $f(t)$ be function defined for all positive values of t, then

$$F(s)=\int_0^{\infty} e^{-st} f(t)\,dt$$

provided the integral exists, is called the **Laplace Transform** of $f(t)$. It is denoted as

$$\boxed{L[f(t)]=F(s)=\int_0^{\infty} e^{-st} f(t)\,dt}$$

42.3 IMPORTANT FORMULAE

1. $L(1)=\frac{1}{s}$
2. $L(t^n)=\frac{n!}{s^{n+1}}$, when $n=0, 1, 2, 3...$
3. $L(e^{at})=\frac{1}{s-a}$ $(s>a)$
4. $L(\cosh at)=\frac{s}{s^2-a^2}$ $(s^2>a^2)$
5. $L(\sinh at)=\frac{a}{s^2-a^2}$ $(s^2>a^2)$
6. $L(\sin at)=\frac{a}{s^2+a^2}$ $(s>0)$
7. $L(\cos at)=\frac{s}{s^2+a^2}$ $(s>0)$

1. $\boxed{L(1)=\frac{1}{s}}$

Proof. $L(1) = \int_0^{\infty} 1\cdot e^{-st}\,dt=\left[\frac{e^{-st}}{-s}\right]_0^{\infty}=-\frac{1}{s}\left[\frac{1}{e^{st}}\right]_0^{\infty}=-\frac{1}{s}[0-1]=\frac{1}{s}$

Hence $L(1) = \frac{1}{s}$ **Proved.**

2. $\boxed{L(t^n)=\frac{n!}{s^{n+1}}}$ where n and s are positive.

Proof. $L(t^n) = \int_0^{\infty} e^{-st} t^n\,dt$

Putting $st=x$ or $t=\frac{x}{s}$ or $dt=\frac{dx}{s}$

Thus, we have $\quad L(t^n) = \int_0^\infty e^{-x}\left(\frac{x}{s}\right)^n \frac{dx}{s} \quad \Rightarrow \quad L(t^n) = \frac{1}{s^{n+1}}\int_0^\infty e^{-x}.x^n\,dx$

$$\Rightarrow L(t^n) = \frac{\overline{|n+1}}{s^{n+1}} \quad \Rightarrow L(t^n) = \frac{n!}{s^{n+1}} \quad \left[\begin{array}{l}\overline{|n+1} = \int_0^\infty e^{-x}.x^n\,dx \\ \text{and} \quad \overline{|n+1} = n!\end{array}\right]$$

Proved.

3. $\boxed{\mathbf{L(e^{at}) = \frac{1}{s-a}}}$, where $s > a$

Proof. $L(e^{at}) = \int_0^\infty e^{-st}.e^{at}\,dt = \int_0^\infty e^{-st+at}.dt$

$$= \int_0^\infty e^{(-s+a)t}.dt = \int_0^\infty e^{-(s-a)t}.dt = \left[\frac{e^{-(s-a)t}}{-(s-a)}\right]_0^\infty = -\frac{1}{s-a}\left[\frac{1}{e^{(s-a)t}}\right]_0^\infty$$

$$= \frac{-1}{(s-a)}(0-1) = \frac{1}{s-a}$$

Proved.

4. $\boxed{\mathbf{L(\cosh at) = \frac{s}{s^2-a^2}}}$

Proof. $L(\cosh at) = L\left[\frac{e^{at}+e^{-at}}{2}\right] \qquad \left(\therefore \cosh at = \frac{e^{at}+e^{-at}}{2}\right)$

$$= \frac{1}{2}L(e^{at}) + \frac{1}{2}L(e^{-at}) = \frac{1}{2}\left[\frac{1}{s-a}+\frac{1}{s+a}\right] \qquad \left[L(e^{at}) = \frac{1}{s-a}\right]$$

$$= \frac{1}{2}\left[\frac{s+a+s-a}{s^2-a^2}\right] = \frac{s}{s^2-a^2}$$

Proved.

5. $\boxed{\mathbf{L(\sinh at) = \frac{a}{s^2-a^2}}}$

Proof. $L(\sinh at) = L\left[\frac{1}{2}\left(e^{at}-e^{-at}\right)\right]$

$$= \frac{1}{2}[L(e^{at}) - L(e^{-at})] = \frac{1}{2}\left[\frac{1}{s-a}-\frac{1}{s+a}\right] = \frac{1}{2}\left[\frac{s+a-s+a}{s^2-a^2}\right]$$

$$= \frac{a}{s^2-a^2}$$

Proved.

6. $\boxed{\mathbf{L(\sin at) = \frac{a}{s^2+a^2}}}$

Proof. $L(\sin at) = L\left[\frac{e^{iat}-e^{-iat}}{2i}\right] \qquad \left[\because \sin at = \frac{e^{iat}-e^{-iat}}{2i}\right]$

$$= \frac{1}{2i}\left[L\left(e^{iat}-e^{-iat}\right)\right] = \frac{1}{2i}\left[L\left(e^{iat}\right) - L\left(e^{-iat}\right)\right]$$

$$= \frac{1}{2i}\left[\frac{1}{s-ia}-\frac{1}{s+ia}\right] = \frac{1}{2i}\frac{s+ia-s+ia}{s^2+a^2} = \frac{1}{2i}\frac{2ia}{s^2+a^2} = \frac{a}{s^2+a^2}$$

Proved.

7. $\boxed{\mathbf{L(\cos at) = \frac{s}{s^2+a^2}}}$

Proof. $\quad L(\cos at) = L\left(\frac{e^{iat}+e^{-iat}}{2}\right) \qquad \left[\because \cos at = \frac{e^{iat}+e^{-iat}}{2}\right]$

$$= \frac{1}{2}[L(e^{iat} + e^{-iat})] = \frac{1}{2}[L(e^{iat}) + L(e^{-iat})] = \frac{1}{2}\left[\frac{1}{s-ia} + \frac{1}{s+ia}\right] = \frac{1}{2}\frac{s+ia+s-ia}{s^2+a^2}$$

$$= \frac{s}{s^2+a^2}$$ **Proved.**

Example 1. *Find the Laplace transform of f(t) defined as*

$$f(t) = \begin{cases} \dfrac{t}{k}, & when \ \ 0 < t < k \\ 1, & when \ \ t > k \end{cases}$$

Solution. $L[f(t)] = \int_0^k \frac{t}{k} e^{-st} dt + \int_k^\infty 1.e^{-st} dt = \frac{1}{k}\left[\left(t\frac{e^{-st}}{-s}\right)_0^k - \int_0^k \frac{e^{-st}}{-s} dt\right] + \left[\frac{e^{-st}}{-s}\right]_k^\infty$

$$= \frac{1}{k}\left[\frac{ke^{-ks}}{-s} - \left(\frac{e^{-st}}{s^2}\right)_0^k\right] + \frac{e^{-ks}}{s} = \frac{1}{k}\left[\frac{ke^{-ks}}{-s} - \frac{e^{-sk}}{s^2} + \frac{1}{s^2}\right] + \frac{e^{-ks}}{s}$$

$$= -\frac{e^{-sk}}{s} - \frac{1}{k}\frac{e^{-ks}}{s^2} + \frac{1}{k}\frac{1}{s^2} + \frac{e^{-ks}}{s} = \frac{1}{ks^2}[-e^{-ks} + 1]$$ **Ans.**

Example 2. *Find the Laplace transform of the function f(t)* $= \begin{cases} t-1, & 1 < t < 2 \\ 3-t, & 2 < t < 3 \end{cases}$.

(U.P., II Semester, 2009)

Solution. The given function is periodic with period 3.

$$L[f(t)] = \int_1^3 f(t)\, e^{-st} dt$$

$$= \left[\int_1^2 f(t)\, e^{-st} dt + \int_2^3 f(t)\, e^{-st} dt\right]$$

$$= \left[\int_1^2 (t-1)\, e^{-st} dt + \int_2^3 (3-t)\, e^{-st} dt\right]$$

$$= \left[\left\{(t-1)\frac{e^{-st}}{-s} - \frac{e^{-st}}{(-s)^2}\right\}_1^2 + \left\{(3-t)\frac{e^{-st}}{-s} + \frac{e^{-st}}{(-s)^2}\right\}_2^3\right]$$

$$= \left[\left\{\frac{e^{-2s}}{-s} - \frac{e^{-2s}}{s^2} + \frac{e^{-s}}{s^2}\right\} + \left\{\frac{e^{-3s}}{s^2} - \frac{e^{-2s}}{-s} - \frac{e^{-2s}}{s^2}\right\}\right]$$

$$= \left[-\frac{e^{-2s}}{s} - \frac{e^{-2s}}{s^2} + \frac{e^{-s}}{s^2} + \frac{e^{-3s}}{s^2} + \frac{e^{-2s}}{s} - \frac{e^{-2s}}{s^2}\right]$$

$$= \left[\frac{1}{s^2}(-e^{-2s} + e^{-s} + e^{-3s} - e^{-2s})\right] = \frac{1}{s^2}[e^{-s} - 2e^{-2s} + e^{-3s}]$$ **Ans.**

Example 3. *Find the Laplace transform of* $F(t) = \begin{cases} 1, & 0 \le t < 1 \\ t, & 1 \le t < 2 \\ t^2, & 2 \le t < \infty \end{cases}$ *(Q. Bank U.P. 2001)*

Solution. Here, we have $F(t) = \begin{cases} 1, & 0 \le t < 1 \\ t, & 1 \le t < 2 \\ t^2, & 2 \le t < \infty \end{cases}$

$$L[F(t)] = \int_0^\infty e^{-st} \cdot F(t) dt = \int_0^1 e^{-st} dt + \int_1^2 t\, e^{-st} dt + \int_2^\infty t^2 e^{-st} dt$$

$$= \left(\frac{e^{-st}}{-s}\right)_0^1 + \left(t\frac{e^{-st}}{-s} - \frac{e^{-st}}{s^2}\right)_1^2 + \left(t^2\frac{e^{-st}}{-s}\right)_2^\infty - \int_2^\infty 2t \cdot \frac{e^{-st}}{-s}\,dt$$

$$= \left(\frac{1-e^{-s}}{s}\right) + \left(\frac{-2}{s}e^{-2s} - \frac{e^{-2s}}{s^2}\right) - \left(\frac{e^{-s}}{-s} - \frac{e^{-s}}{s^2}\right) + \frac{4}{s}e^{-2s} + \frac{2}{s}\int_2^\infty t\,e^{-st}\,dt$$

$$= \frac{1}{s} + \frac{2}{s}e^{-2s} + \frac{e^{-s}}{s^2} - \frac{e^{-2s}}{s^2} + \frac{2}{s}\left[\left(t\frac{e^{-st}}{-s}\right)_2^\infty - \int_2^\infty 1 \cdot \frac{e^{-st}}{-s}\,dt\right]$$

$$= \frac{1}{s} + \frac{2}{s}e^{-2s} + \frac{e^{-s}}{s^2} - \frac{e^{-2s}}{s^2} + \frac{2}{s}\left[\frac{2}{s}e^{-2s} + \frac{1}{s}\left(\frac{e^{-st}}{-s}\right)_2^\infty\right]$$

$$= \frac{1}{s} + \frac{2}{s}e^{-2s} + \frac{e^{-s}}{s^2} + \frac{3}{s^2}e^{-2s} + \frac{2}{s^3}e^{-2s}.$$ **Ans.**

Example 4. *Find the Laplace transform of* $f(t) = \begin{cases} t^2, & 0 < t < 2 \\ t-1, & 2 < t < 3 \\ 7, & t > 3 \end{cases}$

(U.P, II Semester, June 2007)

Solution. $L[f(t)] = \int_0^\infty e^{-st} f(t)\,dt = \int_0^2 t^2 e^{-st}\,dt + \int_2^3 (t-1)e^{-st}\,dt + \int_3^\infty 7e^{-st}\,dt$

$$\left[\int I\,II = I\,II_1 - I'\,II_{11} + I''\,II_{111} - \ldots\right]$$

$$= \left[t^2\left(\frac{e^{-st}}{(-s)}\right) - 2t\frac{e^{-st}}{(-s)^2} + 2\frac{e^{-st}}{(-s)^3}\right]_0^2 + \left[(t-1)\left(\frac{e^{-st}}{(-s)}\right) - \frac{e^{-st}}{(-s)^2}\right]_2^3 + 7\left[\frac{e^{-st}}{-s}\right]_3^\infty$$

$$= \left[-4\left(\frac{e^{-2s}}{s}\right) - 4\left(\frac{e^{-2s}}{s^2}\right) - 2\left(\frac{e^{-2s}}{s^3}\right) + \frac{2}{s^3}\right] + \left[2\left(\frac{e^{-3s}}{-s}\right) - \frac{e^{-3s}}{s^2} + \frac{e^{-2s}}{s} + \frac{e^{-2s}}{s^2}\right] + 7\left(0 + \frac{e^{-3s}}{s}\right)$$

$$= \frac{2}{s^3} + e^{-2s}\left[-\frac{4}{s} - \frac{4}{s^2} - \frac{2}{s^3}\right] + e^{-3s}\left[-\frac{2}{s} - \frac{1}{s^2}\right] + e^{-2s}\left[\frac{1}{s} + \frac{1}{s^2}\right] + e^{-3s}\left[\frac{7}{s}\right]$$

$$= \frac{2}{s^3} + e^{-2s}\left[-\frac{4}{s} - \frac{4}{s^2} - \frac{2}{s^3} + \frac{1}{s} + \frac{1}{s^2}\right] + e^{-3s}\left[-\frac{2}{s} - \frac{1}{s^2} + \frac{7}{s}\right]$$

$$= \frac{2}{s^3} + e^{-2s}\left[-\frac{3}{s} - \frac{3}{s^2} - \frac{2}{s^3}\right] + e^{-3s}\left[\frac{5}{s} - \frac{1}{s^2}\right] = \frac{2}{s^3} - \frac{e^{-2s}}{s^3}\left(2 + 3s + 3s^2\right) + \frac{e^{-3s}}{s^2}(5s-1)$$ **Ans.**

Example 5. *Find the Laplace transform of (1 + sin 2t).*

Solution. Laplace transform of $(1 + \sin 2t)$

$$= \int_0^\infty e^{-st}(1 + \sin 2t)\,dt = \int_0^\infty e^{-st}\left(1 + \frac{e^{2it} - e^{-2it}}{2i}\right)dt$$

$$= \frac{1}{2i}\int_0^\infty [2ie^{-st} + e^{(-s+2i)t} - e^{(-s-2i)t}]\,dt = \frac{1}{2i}\left[\frac{2ie^{-st}}{-s} + \frac{e^{(-s+2i)t}}{-s+2i} - \frac{e^{(-s-2i)t}}{-s-2i}\right]_0^\infty$$

$$= \frac{1}{2i}\left[\left(0 + \frac{2i}{s}\right) + \frac{1}{-s+2i}(0-1) - \frac{1}{-s-2i}(0-1)\right]$$

$$= \frac{1}{2i}\left[\frac{2i}{s} + \frac{1}{s-2i} - \frac{1}{s+2i}\right] = \frac{1}{2}\left[\frac{2}{s} + \frac{4}{s^2+4}\right] = \frac{1}{s} + \frac{2}{s^2+4}$$ **Ans.**

Alternate Method

$$L(1+\sin 2t) = L(1) + L\sin 2t = \frac{1}{s} + \frac{2}{s^2+4}$$ **Ans.**

42.4 PROPERTIES OF LAPLACE TRANSFORM

(1) $L[af_1(t)+bf_2(t)] = a\,L[f_1(t)]+b\,L[f_2(t)]$

Proof. $L[af_1(t)+bf_2(t)] = \int_0^\infty e^{-st}[af_1(t)+bf_2(t)]\,dt$

$$= a\int_0^\infty e^{-st} f_1(t)\,dt + b\int_0^\infty e^{-st} f_2(t)\,dt$$

$$= a\,L[f_1(t)] + b\,L[f_2(t)]$$ **Proved.**

42.5 CHANGE OF SCALE PROPERTY

If $L\{f(t)\} = F(s)$ then $\boxed{L\{f(at)\} = \frac{1}{a}F\left(\frac{s}{a}\right)}$

Proof. $L\{f(at)\} = \int_0^\infty e^{-st} f(at)dt = \int_0^\infty e^{-\left(\frac{s}{a}\right)u} f(u)\frac{du}{a}$ $\left[\text{Put } at = u \Rightarrow dt = \frac{du}{a}\right]$

$$= \frac{1}{a}\int_0^\infty e^{-\left(\frac{s}{a}\right)u} f(u)\,du = \frac{1}{a}\int_0^\infty e^{-\left(\frac{s}{a}\right)t} f(t)\,dt$$

$$= \frac{1}{a}\int_0^\infty e^{-St} f(t)\,dt = \frac{1}{a}L\{f(t)\} = \frac{1}{a}F(S)$$ $\left[\text{Put } S = \frac{s}{a}\right]$

$$= \frac{1}{a}F\left(\frac{s}{a}\right)$$ **Proved.**

Example 6. *If* $L\{J_0(\sqrt{t})\} = \frac{e^{-\frac{1}{4s}}}{s}$, *find* $L\{J_0(2\sqrt{t})\}$.

Solution. Here, we have

$$L\{J_0(\sqrt{t})\} = \frac{e^{-\frac{1}{4s}}}{s}$$

By change of scale property,

$$L\{J_0(\sqrt{4t})\} = \frac{1}{4}\cdot\left\{\frac{e^{-\frac{1}{4(s/4)}}}{(s/4)}\right\}$$

$$\Rightarrow \quad L\{J_0(2\sqrt{t})\} = \frac{1}{s}e^{-1/s}$$ **Ans.**

(2) First Shifting Theorem. If $L\{f(t)\} = F(s)$, then

$$\boxed{L[e^{at} f(t)] = F(s-a)}$$

Proof. $L[e^{at} f(t)] = \int_0^\infty e^{-st}.e^{at} f(t)\,dt = \int_0^\infty e^{-(s-a)t} f(t)\,dt$

$$= \int_0^\infty e^{-rt} f(t)\,dt \quad \text{where } r = s-a$$

$$= F(r) = F(s-a)$$ **Proved.**

With the help of this property, we can have the following important results :

1. $L(e^{at}t^n) = \frac{n!}{(s-a)^{n+1}}$

2. $L(e^{at}\cosh bt) = \dfrac{s-a}{(s-a)^2-b^2}$ **3.** $L(e^{at}\sinh bt) = \dfrac{b}{(s-a)^2-b^2}$

4. $L(e^{at}\sin bt) = \dfrac{b}{(s-a)^2+b^2}$ **5.** $L(e^{at}\cos bt) = \dfrac{s-a}{(s-a)^2+b^2}$

42.6 HEAVISIDE'S SHIFTING THEOREM (Second Translation Property)

If $L\{f(t)\} = F(s)$ and $g(t) = \begin{cases} f(t-a), & t>a \\ 0, & 0<t<a \end{cases}$ then prove that

$$\boxed{L\{g(t)\} = e^{-as}F(s)}$$

(U.P. II Semester, Summer 2006)

Proof.
$$L\{g(t)\} = \int_0^\infty e^{-st}g(t)dt$$
$$= \int_0^a e^{-st}g(t)dt + \int_a^\infty e^{-st}g(t)dt \qquad [g(t)=0, \text{ when } 0<t<a]$$
$$= 0 + \int_a^\infty e^{-st}.f(t-a)dt = \int_a^\infty e^{-st}f(t-a)dt \qquad [\text{Put } t-a=u \Rightarrow dt=du]$$
$$= \int_0^\infty e^{-s(u+a)}f(u)\,du = e^{-sa}\int_0^\infty e^{-su}f(u)du = e^{-as}\int_0^\infty e^{-st}f(t)\,dt \quad \textbf{Proved.}$$

$$\boxed{\mathbf{L\{g(t)\} = e^{-as}F(s)}}$$

Example 7. *Find the Laplace transform of* $\cos^2 t$.

Solution. We know that $\cos 2t = 2\cos^2 t - 1$

$$\cos^2 t = \frac{1}{2}[\cos 2t + 1]$$
$$L(\cos^2 t) = L\left[\frac{1}{2}(\cos 2t+1)\right] = \frac{1}{2}[L(\cos 2t) + L(1)]$$
$$= \frac{1}{2}\left[\frac{s}{s^2+(2)^2}+\frac{1}{s}\right] = \frac{1}{2}\left[\frac{s}{s^2+4}+\frac{1}{s}\right] \qquad \textbf{Ans.}$$

Example 8. *If* $L(\cos^2 t) = \dfrac{s^2+2}{s(s^2+4)}$, *find* $L(\cos^2 at)$. *(U.P., II Semester, Summer 2006)*

Solution. We have, $L(\cos^2 t) = \dfrac{s^2+2}{s(s^2+4)}$

By change of scale property, we have

$$L(\cos^2 at) = \frac{1}{a}\cdot\frac{\left(\frac{s}{a}\right)^2+2}{\frac{s}{a}\left[\left(\frac{s}{a}\right)^2+4\right]} = \frac{1}{a}\left[\frac{s^2+2a^2}{\frac{s}{a}(s^2+4a^2)}\right] = \frac{s^2+2a^2}{s(s^2+4a^2)} \qquad \textbf{Ans.}$$

Example 9. *Find the Laplace transform of* $t^{-\frac{1}{2}}$

Solution. We know that $L(t^n) = \dfrac{\overline{|n+1}}{s^{n+1}}$

Put $n = -\dfrac{1}{2}$, $L\left(t^{-1/2}\right) = \dfrac{\overline{\left|-\frac{1}{2}+1\right.}}{s^{-1/2+1}} = \dfrac{\overline{\left|\frac{1}{2}\right.}}{\sqrt{s}} = \dfrac{\sqrt{\pi}}{\sqrt{s}}$, where $\overline{\left|\frac{1}{2}\right.} = \sqrt{\pi}$ **Ans.**

Example 10. *Find the Laplace transform of 2 sin 2t cos 4t.*

Solution. We have

$$f(t) = 2 \sin 2t \cos 4\,t = \sin \frac{2t+4t}{2} + \sin \frac{2t-4t}{2} = \sin 3t - \sin t$$

$$\mathrm{L}\, f(t) = L\,(\sin 3\,t) - L\,(\sin t) = \frac{3}{s^2+9} - \frac{1}{s^2+1} \qquad \textbf{Ans.}$$

Example 11. *Find the Laplace transform of 4 sin³ t.*

Solution. We have

$$f(t) = 4 \sin^3 t = 3 \sin t - \sin 3t \qquad [\sin 3t = 3 \sin t - 4 \sin^3 t]$$

$$\mathrm{L}\, f(t) = 3\, L \sin t - \mathrm{L} \sin 3t = \frac{3}{s^2+1} - \frac{3}{s^2+9} \qquad \textbf{Ans.}$$

Example 12. *Find the Laplace transform of 4 cosh 2t sin 4t*

Solution. We have

$$f(t) = 4 \cosh 2t \sin 4t = 4\left(\frac{e^{2t}+e^{-2t}}{2}\right)\left(\frac{e^{4it}-e^{-4it}}{2i}\right)$$

$$= \frac{1}{i}\left[e^{(2+4i)t} - e^{(2-4i)t} + e^{(-2+4i)t} - e^{(-2-4i)t}\right]$$

$$L\,[f(t)] = -i\left[L(e^{(2+4i)t}) - L\,(e^{(2-4i)t}) + L\,(e^{(-2+4i)t}) - L\,(e^{(-2-4i)t})\right]$$

$$= -i\left[\frac{1}{s-2-4i} - \frac{1}{s-2+4i} + \frac{1}{s+2-4i} - \frac{1}{s+2+4i}\right]$$

$$= -i\left[\left(\frac{1}{s-2-4i} - \frac{1}{s+2+4i}\right) - \left(\frac{1}{s-2+4i} - \frac{1}{s+2-4i}\right)\right]$$

$$= -i\left[\frac{4+8i}{s^2-(2+4i)^2} - \frac{4-8i}{s^2-(2-4i)^2}\right] \qquad \textbf{Ans.}$$

EXERCISE 42.1

Find the Laplace transforms of the following:

1. $t + t^2 + t^3$ **Ans.** $\frac{1}{s^2}+\frac{2}{s^3}+\frac{6}{s^4}$ **2.** $\sin t \cos t$ **Ans.** $\frac{1}{s^2+4}$

3. $t^{7/2} e^{5t}$ *(M.D.U. Dec. 2009)* **Ans.** $\frac{105\sqrt{\pi}}{16\,(s-5)^{9/2}}$

4. $\sin^3 2\,t$ **Ans.** $\frac{48}{(s^2+4)(s^2+36)}$

5. $e^{-t}\cos^2 t$ **Ans.** $\frac{1}{2s+2}+\frac{s+1}{2s^2+4s+10}$ **6.** $\sin 2t \cos 3t$ **Ans.** $\frac{2(s^2-5)}{(s^2+1)(s^2+25)}$

7. $\sin 2\,t \sin 3\,t$ **Ans.** $\frac{12s}{(s^2+1)(s^2+25)}$

8. $\cos at \sinh at$ **Ans.** $\frac{1}{2}\left[\frac{s-a}{(s-a)^2+a^2} - \frac{s+a}{(s+a)^2+a^2}\right]$

9. $\sinh^3 t$ **Ans.** $\frac{6}{(s^2-1)(s^2-9)}$ **10.** $\cos t \cos 2\,t$ **Ans.** $\frac{s(s^2+5)}{(s^2+1)(s^2+9)}$

11. $\cosh at \sin at$ **Ans.** $\frac{a(s^2+2a^2)}{s^4+4a^4}$

12. $f(t)=\begin{cases}\cos\left(t-\dfrac{2\pi}{3}\right), & t>\dfrac{2\pi}{3}\\ 0, & t<\dfrac{2\pi}{3}\end{cases}$ **Ans.** $e^{\frac{-2\pi s}{3}}\cdot\dfrac{s}{s^2+1}$

42.7 EXISTENCE THEOREM

According to this theorem $\int_0^\infty e^{-st} f(t)\,dt$ exists if $\int_0^\lambda e^{-st} f(t)dt$ can actually be evaluated and its limit as $\lambda \to \infty$ exists.

Otherwise we may use the following theorem:

If $f(t)$ is continuous and $\lim\limits_{t\to\infty}\left[e^{-at} f(t)\right]$ is finite, then Laplace transform of $f(t)$ *i.e.* $\int_0^\infty e^{-st} f(t)\,dt$ exists for $s > a$.

It should however, be kept in mind that the above foresaid conditions are sufficient but not necessary.

For example; $L\left(\dfrac{1}{\sqrt{t}}\right)$ exists though $\dfrac{1}{\sqrt{t}}$ is infinite at $t = 0$. Similarly a function $f(t)$ for which $\lim\limits_{t\to\infty}\left[e^{-at} f(t)\right]$ is finite and having a finite discontinuity will have a Laplace transform of $s > a$.

42.8 LAPLACE TRANSFORM OF THE DERIVATIVE OF *f(t)*

$$L[f'(t)] = sL[f(t)] - f(0) \qquad \text{where} \qquad L[f(t)] = F(s).$$

Proof. $L[f'(t)] = \int_0^\infty e^{-st} f'(t)dt$

Integrating by parts, we get

$$L[f'(t)] = \left[e^{-st} f(t)\right]_0^\infty - \int_0^\infty (-se^{-st}) f(t)\,dt$$

$$= -f(0) + s\int_0^\infty e^{-st} f(t)dt \qquad (e^{-st} f(t) = 0, \text{ when } t = \infty)$$

$$= -f(0) + s\,L[f(t)]$$

$$\boxed{L[f'(t)] = s\,L[f(t)] - f(0).}$$ **Proved.**

Note. Roughly, Laplace transform of **derivative** of $f(t)$ corresponds to **multiplication** of the Laplace transform of $f(t)$ by s.

42.9 LAPLACE TRANSFORM OF DERIVATIVE OF ORDER *n* *(M.D.U. Dec. 2009)*

$$L[f^n(t)] = s^n L[f(t)] - s^{n-1} f(0) - s^{n-2} f'(0) - s^{n-3} f''(0) - \ldots - f^{n-1}(0)$$

Proof. We have already proved in Article 42.8 that

$$L[f'(t)] = s\,L[f(t)] - f(0) \qquad \ldots(1)$$

Replacing $f(t)$ by $f'(t)$ and $f'(t)$ by $f''(t)$ in (1), we get

$$L[f''(t)] = s\,L[f'(t)] - f'(0) \qquad \ldots(2)$$

Putting the value of $L[f'(t)]$ from (1) in (2), we have

$$L[f''(t)] = s\,[s\,L[f(t)] - f(0)] - f'(0)$$

$$\Rightarrow \quad L[f''(t)] = s^2 L[f(t)] - sf(0) - f'(0)$$

Similarly, $$L[f'''(t)] = s^3 L[f(t)] - s^2 f(0) - s f'(0) - f''(0)$$

$$L[f^{iv}(t)] = s^4 L[f(t)] - s^3 f(0) - s^2 f'(0) - s f''(0) - f'''(0)$$

$$\boxed{L[f^n(t)] = s^n L[f(t)] - s^{n-1} f(0) - s^{n-2} f'(0) - s^{n-3} f''(0) + \ldots\ldots - f^{n-1}(0)}$$

Example 13. *Given* $L\left(2\sqrt{\frac{t}{\pi}}\right) = \frac{1}{s^{3/2}}$, *show that* $L\left(\frac{1}{\sqrt{\pi t}}\right) = \frac{1}{\sqrt{s}}$. *(U.P., II Semester, 2005)*

Solution. Let $F(t) = 2\sqrt{\frac{t}{\pi}} \Rightarrow F'(t) = \frac{1}{\sqrt{\pi t}}$. Also $F(0) = 0$

$$\therefore \quad L\{F'(t)\} = s\,L\{F(t)\} - F(0) = s.\ \frac{1}{s^{3/2}} - 0$$

$$\therefore \quad L\left(\frac{1}{\sqrt{\pi t}}\right) = \frac{1}{\sqrt{s}}.$$ **Proved.**

Example 14. *Find the Laplace transform of* $\sin\sqrt{t}$ *; Hence find* $L\left(\frac{\cos\sqrt{t}}{2\sqrt{t}}\right)$

Solution. $$\sin\sqrt{t} = \sqrt{t} - \frac{(\sqrt{t})^3}{3!} + \frac{(\sqrt{t})^5}{5!} - \ldots\ldots$$

$$= t^{1/2} - \frac{t^{3/2}}{3!} + \frac{t^{5/2}}{5!} - \ldots\ldots \qquad \left[\sin x = x - \frac{x^3}{3!} + \frac{x^5}{5!} - \ldots\ldots\right]$$

$$\therefore \quad L(\sin\sqrt{t}) = L\left(t^{1/2} - \frac{t^{3/2}}{3!} + \frac{t^{5/2}}{5!} - \ldots\ldots\right) = \frac{\Gamma 3/2}{s^{3/2}} - \frac{\Gamma 5/2}{3!s^{5/2}} + \frac{\Gamma 7/2}{5!s^{7/2}} - \ldots\ldots$$

$$= \frac{\sqrt{\pi}}{2s^{3/2}}\left\{1 - \left(\frac{1}{2^2 s}\right) + \frac{1}{2!}\left(\frac{1}{2^2 s}\right)^2 - \ldots\ldots\right\},$$

$$= \frac{\sqrt{\pi}}{2s^{3/2}}\left\{1 - \left(\frac{1}{2^2 s}\right) + \frac{1}{2!}\left(\frac{1}{2^2 s}\right)^2 - \ldots\ldots\right\} \qquad \left[e^x = 1 + \frac{x}{1!} + \frac{x^2}{2!} + \frac{x^3}{3!} + \ldots\ldots\right]$$

$$\Rightarrow \quad L(\sin\sqrt{t}) = \frac{\sqrt{\pi}}{2s^{3/2}} e^{-(1/4s)}$$

Now, $$L\left[\frac{d}{dt}(\sin\sqrt{t})\right] = s\,L(\sin\sqrt{t}) - 0 \qquad \left[\because F(0) = 0 \text{ and } L\left[\frac{d}{dt}[F(t)]\right] = sF(s)\right]$$

$$L\left(\frac{\cos\sqrt{t}}{2\sqrt{t}}\right) = \frac{\sqrt{\pi}}{2\sqrt{s}} e^{-\left(\frac{1}{4s}\right)} \Rightarrow L\left(\frac{\cos\sqrt{t}}{\sqrt{t}}\right) = \frac{\sqrt{\pi}}{\sqrt{s}} e^{-1/4s}$$ **Ans.**

42.10 LAPLACE TRANSFORM OF INTEGRAL OF *f(t)*

$$\boxed{L\left[\int_0^t f(t)\,dt\right] = \frac{1}{s} F(s)} \qquad \text{where} \quad L[f(t)] = F(s)$$

Proof. Let $\phi(t) = \int_0^t f(t)\,dt$ and $\phi(0) = 0$ then $\phi'(t) = f(t)$

We know the formula of Laplace transforms of $\phi'(t)$ *i.e.*

$$L[\phi'(t)] = s\,L[\phi(t)] - \phi(0)$$

$$\Rightarrow \quad L[\phi'(t)] = s\,L[\phi(t)] \qquad [\phi(0) = 0]$$

$$\Rightarrow \qquad L[\phi(t)] = \frac{1}{s} L[\phi'(t)]$$

Putting the values of $\phi(t)$ and $\phi'(t)$, we get

$$L\left[\int_0^t f(t)\,dt\right] = \frac{1}{s} L[f(t)] \Rightarrow \boxed{\mathbf{L}\left[\int_0^t f(t)\,dt\right] = \frac{1}{s} F(s)}$$ **Proved.**

Note. (1) Laplace transform of **Integral** of $f(t)$ corresponds to the division of the Laplace transform of $f(t)$ by s.

(2) $$\int_0^t f(t)\,dt = L^{-1}\left[\frac{1}{s} F(s)\right]$$

42.11 LAPLACE TRANSFORM OF *t.f(t)* (Multiplication by *t*)

If $\qquad L[f(t)] = F(s), \qquad$ then

$$L[t^n f(t)] = (-1)^n \frac{d^n}{ds^n}[F(s)].$$ *(U.P., II Semester, Summer 2005)*

Proof. $$L[f(t)] = F(s) = \int_0^\infty e^{-st} f(t)\,dt \qquad ...(1)$$

Differentiating (1) w.r.t. 's', we get

$$\frac{d}{ds}[F(s)] = \frac{d}{ds}\left[\int_0^\infty e^{-st} f(t)\,dt\right] = \int_0^\infty \frac{\partial}{\partial s}\left(e^{-st}\right) f(t)\,dt$$

$$= \int_0^\infty \left(-t e^{-st}\right).f(t)\,dt = \int_0^\infty e^{-st}\left[-t.f(t)\right] dt$$

$$= L[-t f(t)] \Rightarrow L[t f(t)] = (-1)^1 \frac{d}{ds}[F(s)]$$

Similarly, $$L\left[t^2 f(t)\right] = (-1)^2 \frac{d^2}{ds^2}[F(s)]$$

$$L\left[t^3 f(t)\right] = (-1)^3 \frac{d^3}{ds^3}[F(s)]$$

$$\boxed{L\left[t^n f(t)\right] = (-1)^n \frac{d^n}{ds^n}[F(s)]}$$ **Proved.**

42.12 INITIAL AND FINAL VALUE THEOREMS

(*a*) Initial Value Theorem. $L\{f(t)\} = F(s)$

$$\Rightarrow \qquad \lim_{t\to 0} f(t) = \operatorname{Lim}_{s\to\infty}[sF(s)],$$ provided the limit exists.

Proof. $$L\{f'(t)\} = s\,L\{f(t)\} - f(0)$$

$$\Rightarrow \qquad \int_0^\infty e^{-st} f'(t)\,dt = s\,F(s) - f(0)$$

$$\Rightarrow \qquad \operatorname{Lim}_{s\to\infty} \int_0^\infty e^{-st} f'(t)\,dt = \operatorname{Lim}_{s\to\infty}[sF(s) - f(0)]$$

$$\Rightarrow \qquad \operatorname{Lim}_{s\to\infty}[s\,F(s)] = f(0) + \int_0^\infty \left(\operatorname{Lim}_{s\to\infty} e^{-st}\right) f'(t)\,dt$$

$$= f(0)+\int_0^\infty (0)\, f'(t)\, dt \qquad (\because \lim_{s\to\infty} e^{-st}=0)$$

$$= f(0)+0 = f(0) = \lim_{t\to 0} f(t)$$

(b) Final Value Theorem. $L\{f(t)\} = F(s)$

$\Rightarrow$ $\lim_{t\to\infty} f(t) = \lim_{s\to 0}[sF(s)]$, provided the limits exist.

Proof. $L\{f'(t)\} = sL\{f(t)\} - f(0) \Rightarrow \int_0^\infty e^{-st} f'(t)\, dt = sF(s) - f(0)$

$$\Rightarrow \lim_{s\to 0}\int_0^\infty e^{-st} f'(t)\, dt = \lim_{s\to 0}[sF(s) - f(0)]$$

$$\Rightarrow \lim_{s\to 0}[s\,F(s) - f(0)] = \lim_{s\to 0}\int_0^\infty e^{-st} f'(t)\, dt$$

$$\Rightarrow \lim_{s\to 0}[s\,F(s)] - f(0) = \lim_{s\to 0}\int_0^\infty e^{-st} f'(t)\, dt$$

$$\Rightarrow \lim_{s\to 0}[sF(s)] - f(0) = \int_0^\infty \lim_{s\to 0} e^{-st} f'(t)\, dt = \int_0^\infty (1) f'(t)\, dt \qquad \left[\because \lim_{s\to 0} e^{-st} = 1\right]$$

$$\Rightarrow \lim_{s\to 0}[sF(s)] = \lim_{t\to\infty} f(t)$$

Example 15. *If* $L\{F(t)\} = \dfrac{1}{s(s+\beta)}$ *then, find* $\lim_{t\to\infty} F(t)$

Solution. By final-value theorem,

$$\lim_{t\to\infty} F(t) = \lim_{s\to 0} sL\{F(t)\} = \lim_{s\to 0}\frac{s}{s(s+\beta)} = \lim_{s\to 0}\frac{1}{(s+\beta)} = \frac{1}{\beta} \quad \textbf{Ans.}$$

42.13. EXPONENTIAL INTEGRAL FUNCTION $\int_t^\infty \left(\frac{e^{-x}}{x}\right) dx$

Let $\quad f(t) = \int_t^\infty \frac{e^{-x}}{x}\, dx$

$\Rightarrow \quad f'(t) = -\dfrac{e^{-t}}{t} \quad \Rightarrow \quad tf'(t) = -e^{-t}$ [Here – ve sign appears due to lower limit]

Taking Laplace Transform of $tf'(t)$, we get $L\{t f'(t)\} = L\{-e^{-t}\} = -L\{e^{-t}\}$

$$\Rightarrow \quad -\frac{d}{ds}[sF(s) - f(0)] = -\frac{1}{s+1}$$

$$\Rightarrow \quad \frac{d}{ds}[s\,F(s)] = \frac{1}{s+1} \qquad \left[\because f(0) = \text{constant} \therefore \frac{d}{ds} f(0) = 0\right]$$

Integrating both the sides, we get

$$s\,F(s) = \log(s+1) + C \qquad \text{...(1)}$$

Now, by final value theorem, we have

$$\lim_{s\to 0} sF(s) = \lim_{t\to\infty} f(t) \qquad \text{...(2)}$$

Hence, $\lim_{s\to 0}[sF(s)] = \lim_{s\to 0}[\log(s+1) + C] = 0 + C = C$ ...(3)

Also, $\quad \lim_{t\to\infty}[f(t)] = \lim_{t\to\infty}\int_t^\infty \left(\frac{e^{-x}}{x}\right) dx = 0$ ...(4)

Putting the values of $\lim_{s\to 0}[s\,F(s)]$ and $\lim_{t\to\infty}[f(t)]$ from (3) and (4) in (2), we get

$$C = 0.$$

Hence from (1), $sF(s) = \log(s+1) \Rightarrow F(s) = \left\{\frac{\log(s+1)}{s}\right\}$

$$\Rightarrow \quad \boxed{L\int_t^\infty \left(\frac{e^{-x}}{x}\right)dx = \left[\frac{\log(s+1)}{s}\right]}$$

Example 16. *Find the Laplace Transform of t sin at.*

Solution. $L(t\sin at) = L\left(t\frac{e^{iat}-e^{-iat}}{2i}\right) = \frac{1}{2i}[L(t.e^{iat}) - L(te^{-iat})]$

$$= \frac{1}{2i}\left[-\frac{d}{ds}\frac{1}{s-ia} + \frac{d}{ds}\frac{1}{s+ia}\right] = \frac{1}{2i}\left[\frac{1}{(s-ia)^2} - \frac{1}{(s+ia)^2}\right] = \frac{1}{2i}\left[\frac{(s+ia)^2-(s-ia)^2}{(s-ia)^2(s+ia)^2}\right]$$

$$= \frac{1}{2i}\frac{(s^2+2ias-a^2)-(s^2-2ias-a^2)}{(s^2+a^2)^2} = \frac{1}{2i}\frac{4ias}{(s^2+a^2)^2} = \frac{2as}{(s^2+a^2)^2} \quad \textbf{Ans.}$$

Example 17. *Find the Laplace transform of t sinh at.*

Solution. $L(\sinh at) = \frac{a}{s^2-a^2}$

$$L[t\sinh at] = -\frac{d}{ds}\left(\frac{a}{s^2-a^2}\right) = \frac{2as}{(s^2-a^2)^2} \quad \textbf{Ans.}$$

Example 18. *Find the Laplace transform of t^2 cos at*

Solution. $L(\cos at) = \frac{s}{s^2+a^2}$

$$L(t^2\cos at) = (-1)^2\frac{d^2}{ds^2}\left[\frac{s}{s^2+a^2}\right] = \frac{d}{ds}\frac{(s^2+a^2).1-s(2s)}{(s^2+a^2)^2} = \frac{d}{ds}\frac{a^2-s^2}{(s^2+a^2)^2}$$

$$= \frac{(s^2+a^2)^2(-2s)-(a^2-s^2).2(s^2+a^2)(2s)}{(s^2+a^2)^4} = \frac{(s^2+a^2)(-2s)-(a^2-s^2)4s}{(s^2+a^2)^3}$$

$$= \frac{-2s^3-2a^2s-4a^2s+4s^3}{(s^2+a^2)^3} = \frac{2s(s^2-3a^2)}{(s^2+a^2)^3} \quad \textbf{Ans.}$$

Example 19. *Obtain the Laplace transform of $t^2e^t.\sin 4t$.*

(Uttarakhand II Sem., Summer 2010, U.P. II Semester, Summer 2002)

Solution. $L(\sin 4t) = \frac{4}{s^2+16}$,

$$L(e^t\sin 4t) = \frac{4}{(s-1)^2+16}$$

$$L(te^t\sin 4t) = -\frac{d}{ds}\left(\frac{4}{s^2-2s+17}\right) = \frac{4(2s-2)}{(s^2-2s+17)^2}$$

$$L\left(t^2 e^t \sin 4t\right) = -\frac{d}{ds}\left(\frac{4(2s-2)}{\left(s^2-2s+17\right)^2}\right) = -4\frac{\left(s^2-2s+17\right)^2 2-(2s-2)2\left(s^2-2s+17\right)(2s-2)}{\left(s^2-2s+17\right)^4}$$

$$= -4\frac{(s^2-2s+17)\,2-2(2s-2)^2}{(s^2-2s+17)^3} = \frac{-4\left(2s^2-4s+34-8s^2+16s-8\right)}{\left(s^2-2s+17\right)^3}$$

$$= \frac{-4\left(-6s^2+12s+26\right)}{\left(s^2-2s+17\right)^3} = \frac{8\left[3s^2-6s-13\right]}{(s^2-2s+17)^3}$$ **Ans.**

Example 20. *Find the Laplace transform of the function*

$$f(t) = te^{-t}\sin 2t$$ *(U.P. II Semester, Summer 2002)*

Solution. $L[\sin 2t] = \dfrac{2}{s^2+4}$

$$L\left[e^{-t}\sin 2t\right] = \frac{2}{(s+1)^2+4} = F(s) \quad \text{(say)}$$

$$L\left(te^{-t}\sin 2t\right) = -F'(s) = -\frac{d}{ds}\left[\frac{2}{(s+1)^2+4}\right] = \frac{2\cdot 2(s+1)}{\left[(s+1)^2+4\right]^2} = \frac{4(s+1)}{\left[(s+1)^2+4\right]^2}$$ **Ans.**

EXERCISE 42.2

Find the Laplace transforms of the following :

1. $t\,e^{at}$ **Ans.** $\dfrac{1}{(s-a)^2}$

2. $t\cosh at$ **Ans.** $\dfrac{s^2+a^2}{\left(s^2-a^2\right)^2}$

3. $t\cos t$ **Ans.** $\dfrac{s^2-1}{\left(s^2+1\right)^2}$

4. $t\cosh t$ **Ans.** $\dfrac{s^2+1}{\left(s^2-1\right)^2}$

5. $t^2\sin t$ **Ans.** $\dfrac{2\left(3s^2-1\right)}{\left(s^2+1\right)^3}$

6. t^3e^{-3t} **Ans.** $\dfrac{6}{(s+3)^4}$

7. $t\sin^2 3t$ **Ans.** $\dfrac{1}{2}\left[\dfrac{1}{s^2}-\dfrac{s^2-36}{\left(s^2+36\right)^2}\right]$

8. $te^{at}\sin at$ **Ans.** $\dfrac{2a(s-a)}{\left(s^2-2as+2a^2\right)^2}$

9. $te^{-t}\cosh t$ **Ans.** $\dfrac{s^2+2s+2}{\left(s^2+2s\right)^2}$

10. $t^2e^{-2t}\cos t$ **Ans.** $\dfrac{2(s^3+6s^2+9s+2)}{(s^2+4s+5)^3}$

11. $\int_0^t e^{-2t}\,t\sin^3 t\,dt$ **Ans.** $\dfrac{3(s+2)}{2s}\left[\dfrac{1}{\left[(s+2)^2+9\right]^2}-\dfrac{1}{\left[(s+2)^2+1\right]^2}\right]$

12. If $f(t)$ is continuous, except for an ordinary discontinuity at $t = a$, $(a > 0)$ as given in the figure, then show that

$$L\left[f'(t)\right] = s\left[f(t)\right] - f(0) - e^{-as}\left[f(a+0)-f(a-0)\right]$$

(U.P. II Semester 2003)

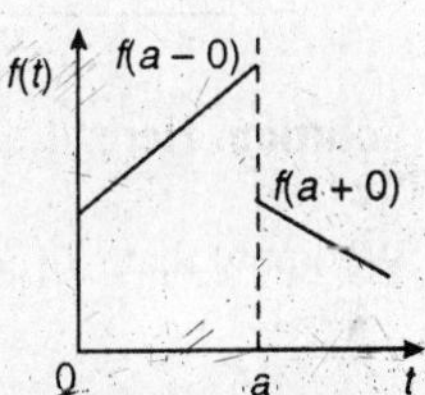

13. Pick the correct statement for final value theorem of Laplace transform:

(*i*) $\underset{t\to 0}{\text{Lt}}\ f(t) = \underset{s\to\infty}{\text{Lt}}\ s\,F(s)$ (*ii*) $\underset{t\to\infty}{\text{Lt}}\ f(t) = \underset{s\to 0}{\text{Lt}}\ s\,F(s)$

(*U.P. II Semester 2010*) **Ans.** (*ii*)

42.14 LAPLACE TRANSFORM OF $\frac{1}{t}f(t)$ (Division by *t*)

If $L[f(t)] = F(s)$, then $L\left[\frac{1}{t}f(t)\right] = \int_s^\infty f(s)\,ds$ (*U.P. II Semester Summer, 2007, 2005*)

Proof. We know that $L[f(t)] = F(s)$ or $F(s) = \int_0^\infty e^{-st} f(t)\,dt$ ...(1)

Integrating (1) w.r.t. '*s*', we have

$$\int_s^\infty F(s)\,ds = \int_s^\infty\left[\int_0^\infty e^{-st} f(t)\,dt\right]ds = \int_0^\infty f(t)\left[\int_s^\infty e^{-st}\,ds\right]dt = \int_0^\infty f(t)\left[\frac{e^{-st}}{-t}\right]_s^\infty dt$$

$$= \int_0^\infty \frac{-f(t)}{t}\left[e^{-st}\right]_s^\infty dt = \int_0^\infty \frac{-f(t)}{t}\left[0 - e^{-st}\right]dt = \int_0^\infty e^{-st}\left\{\frac{1}{t}f(t)\right\}dt = L\left[\frac{1}{t}f(t)\right]$$

$$\Rightarrow \boxed{L\left[\frac{1}{t}f(t)\right] = \int_0^\infty F(s)\,ds}$$ **Proved.**

Cor. $L^{-1}\int_s^\infty F(s)\,ds = \frac{1}{t}f(t)$

Example 21. *Find the Laplace transform of* $\frac{\sin 2t}{t}$

Solution. $L(\sin 2t) = \frac{2}{s^2+4}$

$$L\left(\frac{\sin 2t}{t}\right) = \int_s^\infty \frac{2}{s^2+4}\,ds = 2.\frac{1}{2}\left[\tan^{-1}\frac{s}{2}\right]_s^\infty = \left[\tan^{-1}\infty - \tan^{-1}\frac{s}{2}\right] = \frac{\pi}{2} - \tan^{-1}\frac{s}{2}$$

$$= \cot^{-1}\frac{s}{2}$$ **Ans.**

Example 22. *Find the Laplace transform of* $f(t) = \int_0^t \frac{\sin at}{t}\,dt$

(*M.D.U., Dec. 2009, U.P., II Semester, Summer 2005*)

Solution. $L(\sin at) = \frac{a}{s^2+a^2}$

$$L\left(\frac{\sin at}{t}\right) = \int_s^\infty \frac{a}{s^2+a^2}\,ds = \left[\tan^{-1}\frac{s}{a}\right]_s^\infty = \frac{\pi}{2} - \tan^{-1}\frac{s}{a} = \cot^{-1}\frac{s}{a}$$

Hence, $L\left[\int_0^t \frac{\sin at}{t}\,dt\right] = \frac{1}{s}\cot^{-1}\frac{s}{a}$ **Ans.**

Example 23. *Find the Laplace transform of :*

$\frac{\cos at - \cos bt}{t}$ (*Uttarakhand, II Semester, June 2007, U.P., II Semester, 2004*)

Solution. Here, $f(t) = \frac{\cos at - \cos bt}{t}$

We know that, $L(\cos at - \cos bt) = L(\cos at) - L(\cos bt) = \frac{s}{s^2+a^2} - \frac{s}{s^2+b^2}$

$$L\left(\frac{\cos at-\cos bt}{t}\right)=\int_s^\infty\left(\frac{s}{s^2+a^2}-\frac{s}{s^2+b^2}\right)ds=\left[\frac{1}{2}\log\left(s^2+a^2\right)-\frac{1}{2}\log\left(s^2+b^2\right)\right]_s^\infty$$

$$=\frac{1}{2}\left[\log\frac{s^2+a^2}{s^2+b^2}\right]_s^\infty=\frac{1}{2}\left[\log\frac{1+\frac{a^2}{s^2}}{1+\frac{b^2}{s^2}}\right]_s^\infty=\frac{1}{2}\log 1-\frac{1}{2}\log\frac{1+\frac{a^2}{s^2}}{1+\frac{b^2}{s^2}}=0-\frac{1}{2}\log\frac{s^2+a^2}{s^2+b^2}\quad[\log 1=0]$$

$$=\frac{1}{2}\log\frac{s^2+b^2}{s^2+a^2}$$ **Ans.**

Example 24. *If* $f(t)=\dfrac{e^{at}-\cos bt}{t}$, *find the Laplace transform of* $f(t)$.

(U.P. II Semester, Summer 2003)

Solution. $f(t)=\dfrac{e^{at}-\cos bt}{t}=\dfrac{e^{at}}{t}-\dfrac{\cos bt}{t}$

We know that, $L\left(e^{at}-\cos bt\right)=\left(\dfrac{1}{s-a}-\dfrac{s}{s^2+b^2}\right)$

$$\therefore\quad L\left(\frac{e^{at}-\cos bt}{t}\right)=\int_s^\infty\left(\frac{1}{s-a}-\frac{s}{s^2+b^2}\right)ds=\left[\log(s-a)-\frac{1}{2}\log\left(s^2+b^2\right)\right]_s^\infty$$

$$=\left[\frac{2\log(s-a)-\log\left(s^2+b^2\right)}{2}\right]_s^\infty=\frac{1}{2}\left[\log(s-a)^2-\log\left(s^2+b^2\right)\right]_s^\infty$$

$$=\frac{1}{2}\left[\log\frac{(s-a)^2}{s^2+b^2}\right]_s^\infty=\frac{1}{2}\left[\log\left\{\frac{\left(1-\frac{a}{s}\right)^2}{1+\frac{b^2}{s^2}}\right\}\right]_s^\infty$$

$$=\frac{1}{2}\left[0-\log\frac{\left(1-\frac{a}{s}\right)^2}{\left(1+\frac{b^2}{s^2}\right)}\right]=\frac{1}{2}\left[\log\frac{s^2+b^2}{(s-a)^2}\right]$$ **Ans.**

Example 25. *Find the Laplace transform of* $\dfrac{1-\cos t}{t^2}$.

Solution. L $(1-\cos t)=L(1)-$ L $(\cos t)=\dfrac{1}{s}-\dfrac{s}{s^2+1}$

$$L\left[\frac{1-\cos t}{t}\right]=\int_s^\infty\left(\frac{1}{s}-\frac{s}{s^2+1}\right)ds=\left[\log s-\frac{1}{2}\log\left(s^2+1\right)\right]_s^\infty$$

$$=\frac{1}{2}\left[\log s^2-\log\left(s^2+1\right)\right]_s^\infty=\frac{1}{2}\left[\log\frac{s^2}{s^2+1}\right]_s^\infty$$

$$=\frac{1}{2}\left[\log\frac{1}{\left(1+\frac{1}{s^2}\right)}\right]_s^\infty=\frac{1}{2}\left[0-\log\frac{s^2}{s^2+1}\right]=-\frac{1}{2}\log\frac{s^2}{s^2+1}$$

Again, $L\left[\frac{1-\cos t}{t^2}\right] = -\frac{1}{2}\int_s^\infty \log\frac{s^2}{s^2+1}\,ds = -\frac{1}{2}\int_s^\infty\left(\log\frac{s^2}{s^2+1}.1\right)ds$

Integrating by parts, we have,

$$= -\frac{1}{2}\left[\log\frac{s^2}{s^2+1}.s - \int \frac{s^2+1}{s^2}\frac{(s^2+1)2s - s^2(2s)}{(s^2+1)^2}.sds\right]_s^\infty$$

$$= -\frac{1}{2}\left[s\log\frac{s^2}{s^2+1} - 2\int\frac{1}{s^2+1}ds\right]_s^\infty = -\frac{1}{2}\left[s\log\frac{s^2}{s^2+1} - 2\tan^{-1}s\right]_s^\infty$$

$$= -\frac{1}{2}\left[0 - 2\left(\frac{\pi}{2}\right) - s\log\frac{s^2}{s^2+1} + 2\tan^{-1}s\right] = -\frac{1}{2}\left[-\pi - s\log\frac{s^2}{s^2+1} + 2\tan^{-1}s\right]$$

$$= \frac{\pi}{2} + \frac{s}{2}\log\frac{s^2}{s^2+1} - \tan^{-1}s = \left(\frac{\pi}{2} - \tan^{-1}s\right) + \frac{s}{2}\log\frac{s^2}{s^2+1} = \cot^{-1}s + \frac{s}{2}\log\frac{s^2}{s^2+1}$$ **Ans.**

Example 26. *Evaluate* $L\left[e^{-4t}\frac{\sin 3t}{t}\right]$

Solution. $L\,[\sin 3t] = \frac{3}{s^2+3^2}$

$$\Rightarrow \quad L\left[\frac{\sin 3t}{t}\right] = \int_s^\infty \frac{3}{s^2+9}ds = \left[\frac{3}{3}\tan^{-1}\frac{s}{3}\right]_s^\infty = \left[\tan^{-1}\frac{s}{3}\right]_s^\infty = \frac{\pi}{2} - \tan^{-1}\frac{s}{3} = \cot^{-1}\frac{s}{3}$$

$$L\left[e^{-4t}\frac{\sin 3t}{t}\right] = \cot^{-1}\frac{s+4}{3} = \tan^{-1}\frac{3}{s+4}$$ **Ans.**

EXERCISE 42.3

Find Laplace transform of the following:

1. $\frac{1}{t}(1-e^t)$ **Ans.** $\log\frac{s-1}{s}$
2. $\frac{1}{t}\left(e^{-at} - e^{-bt}\right)$ **Ans.** $\log\frac{s+b}{s+a}$
3. $\frac{1}{t}(1-\cos at)$ **Ans.** $-\frac{1}{2}\log\frac{s^2}{s^2+a^2}$
4. $\frac{1}{t}\sin^2 t$ **Ans.** $\frac{1}{4}\log\frac{s^2+4}{s^2}$
5. $\frac{1}{t}\sinh t$ **Ans.** $-\frac{1}{2}\log\frac{s-1}{s+1}$
6. $\frac{1}{t}\left(e^{-t}\sin t\right)$ **Ans.** $\cot^{-1}(s+1)$
7. $\frac{1}{t}(1-\cos t)$ **Ans.** $\frac{1}{2}[\log(s^2+1) - \log s^2]$
8. $\int_0^\infty \frac{1}{t}e^{-2t}\sin t\,dt$ **Ans.** $\frac{1}{s}\cot^{-1}(s+2)$
9. $\int_0^\infty \frac{e^{-t}-e^{-3t}}{t}dt$ **Ans.** $\log 3$

42.15 LAPLACE TRANSFORM OF ERROR FUNCTION

Example 27. *Find* $L\left\{erf\sqrt{t}\right\}$ *and hence prove that*

$$L\left\{t.erf\,2\sqrt{t}\right\} = \frac{3s+8}{s^2(s+4)^{3/2}}$$ *(U.P. II Semester, Summer 2001)*

Solution. We know that $erf\sqrt{t} = \frac{2}{\sqrt{\pi}}\int_0^{\sqrt{t}} e^{-x^2}dx$

$$= \frac{2}{\sqrt{\pi}} \int_0^{\sqrt{t}} \left(1 - x^2 + \frac{x^4}{2!} - \frac{x^6}{3!} + \ldots\right) dx = \frac{2}{\sqrt{\pi}} \left[x - \frac{x^3}{3} + \frac{x^5}{10} - \frac{x^7}{42} \ldots\right]_0^{\sqrt{t}}$$

$$= \frac{2}{\sqrt{\pi}} \left[\sqrt{t} - \frac{t^{3/2}}{3} + \frac{t^{5/2}}{10} - \frac{t^{7/2}}{42} + \ldots\right]$$

$$L\{erf\sqrt{t}\} = \frac{2}{\sqrt{\pi}} \left[\frac{\left|\frac{3}{2}\right.}{s^{3/2}} - \frac{\left|\frac{5}{2}\right.}{3s^{5/2}} + \frac{\left|\frac{7}{2}\right.}{10s^{7/2}} - \frac{\left|\frac{9}{2}\right.}{42s^{9/2}} + \ldots\right]$$

$$= \frac{2}{\sqrt{\pi}} \left[\frac{\frac{1}{2}\left|\frac{1}{2}\right.}{s^{3/2}} - \frac{\frac{3}{2}\cdot\frac{1}{2}\left|\frac{1}{2}\right.}{3s^{5/2}} + \frac{\frac{5}{2}\cdot\frac{3}{2}\cdot\frac{1}{2}\left|\frac{1}{2}\right.}{10s^{7/2}} - \frac{\frac{7}{2}\cdot\frac{5}{2}\cdot\frac{3}{2}\cdot\frac{1}{2}\left|\frac{1}{2}\right.}{42s^{9/2}} + \ldots\right] \qquad \left[\therefore \left|\frac{1}{2}\right. = \sqrt{\pi}\right]$$

$$= \frac{1}{s^{3/2}} - \frac{1}{2}\frac{1}{s^{5/2}} + \frac{1.3}{2.4}\frac{1}{s^{7/2}} - \frac{1}{2}\cdot\frac{3}{4}\cdot\frac{5}{6}\frac{1}{s^{9/2}} + \ldots$$

$$= \frac{1}{s^{3/2}} \left[1 - \frac{1}{2}\cdot\frac{1}{s} + \frac{1}{2}\cdot\frac{3}{4}\cdot\frac{1}{s^2} - \frac{1}{2}\cdot\frac{3}{4}\cdot\frac{5}{6}\cdot\frac{1}{s^3} + \ldots\right]$$

$$= \frac{1}{s^{3/2}} \left[1 - \frac{1}{2}\cdot\frac{1}{s} + \frac{\left(-\frac{1}{2}\right)\left\{-\frac{3}{2}\right\}}{2!}\frac{1}{s^2} + \frac{\left(-\frac{1}{2}\right)\left(-\frac{3}{2}\right)\left(-\frac{5}{2}\right)}{3!}\frac{1}{s^3} + \ldots\right]$$

$$= \frac{1}{s^{3/2}} \left[1 + \frac{1}{s}\right]^{-\frac{1}{2}} = \frac{1}{s^{3/2}} \left[\frac{s}{s+1}\right]^{\frac{1}{2}} = \frac{1}{s\sqrt{(s+1)}} \qquad \textbf{Ans.}$$

Now, $L\{erf(2\sqrt{t})\} = L\{erf\sqrt{4t}\} = \frac{1}{4}\frac{1}{\frac{s}{4}\sqrt{\frac{s}{4}+1}} = \frac{2}{s\sqrt{s+4}}$

$$L\{t.erf(2\sqrt{t})\} = -\frac{d}{ds}\frac{2}{\sqrt{s^3 + 4s^2}} = -2\left(-\frac{1}{2}\right)\left[s^3 + 4s^2\right]^{-\frac{3}{2}}\left(3s^2 + 8s\right)$$

$$= \frac{3s^2 + 8s}{\left(s^3 + 4s^2\right)^{3/2}} = \frac{s(3s+8)}{s^3(s+4)^{3/2}} = \frac{3s+8}{s^2(s+4)^{3/2}} \qquad \textbf{Proved.}$$

42.16 COMPLEMENTARY ERROR FUNCTION

This function is defined by

$$erf_c\left(\sqrt{t}\right) = 1 - erf\left(\sqrt{t}\right) = 1 - \frac{2}{\sqrt{\pi}} \int_0^{\sqrt{t}} e^{-x^2} dx$$

Now, $L\{erf_c(\sqrt{t})\} = L\left\{1 - \frac{2}{\sqrt{\pi}}\int_0^{\sqrt{t}} e^{-x^2} dx\right\} = L(1) - \frac{2}{\sqrt{\pi}} L\left\{\int_0^{\sqrt{t}} e^{-x^2} dx\right\} = \frac{1}{s} - \frac{1}{s\sqrt{s+1}}$

$$= \frac{\sqrt{(s+1)}-1}{s\sqrt{(s+1)}} = \frac{\left\{\sqrt{(s+1)}-1\right\}\left\{\sqrt{(s+1)}+1\right\}}{s\sqrt{(1+s)}\left\{\sqrt{(s+1)}+1\right\}}$$

$$= \frac{s+1-1}{s\sqrt{s+1}\,(\sqrt{s+1}+1)} = \frac{1}{\sqrt{(s+1)}\left\{\sqrt{(s+1)}+1\right\}}$$

$$\therefore L\left[erf_c\,(\sqrt{t})\right] = \frac{1}{\sqrt{s+1}\left\{\sqrt{s+1}+1\right\}}$$ **Ans.**

42.17 UNIT STEP FUNCTION

With the help of unit step functions, we can find the inverse transform of functions, which cannot be determined with previous methods.

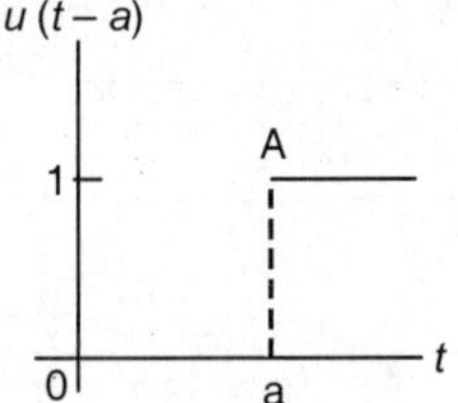

The unit step function $u(t - a)$ is defined as follows:

$$u(t-a) = \begin{cases} 0, \text{ when } t < a \\ 1, \text{ when } t \ge a \end{cases} \quad \text{where } a \ge 0$$

42.18 LAPLACE TRANSFORM OF UNIT FUNCTION

$$L[u(t-a)] = \frac{e^{-as}}{s}$$

Proof. $L\left[u(t-a)\right] = \int_0^\infty e^{-st} u(t-a)\,dt = \int_0^a e^{-st}.0dt + \int_a^\infty e^{-st}.1\,dt = 0 + \left[\frac{e^{-st}}{-s}\right]_a^\infty$

$$L[u(t-a)] = \frac{e^{-as}}{s}$$ **Proved.**

Example 28. *Express the following function in terms of unit step functions and find its Laplace transform:*

$$f(t) = \begin{cases} 8, & t < 2 \\ 6, & t \ge 2 \end{cases}$$

Solution. $f(t) = \begin{cases} 8+0, & t<2 \\ 8-2, & t \ge 2. \end{cases} = 8 + \begin{cases} 0, & t<2 \\ -2, & t \ge 2 \end{cases} = 8 + (-2)\begin{cases} 0, & t<2 \\ 1, & t\ge 2 \end{cases} = 8 - 2u(t-2)$

$$L\{f(t)\} = 8L(1) - 2\,Lu(t-2) = \frac{8}{s} - 2\frac{e^{-2s}}{s}$$ **Ans.**

Example 29. *Draw the graph of u(t – a) – u (t – b).*

Solution. As in Art 42.17 the graph of $u(t-a)$ is a straight line parallel to t-axis from A to ∞.

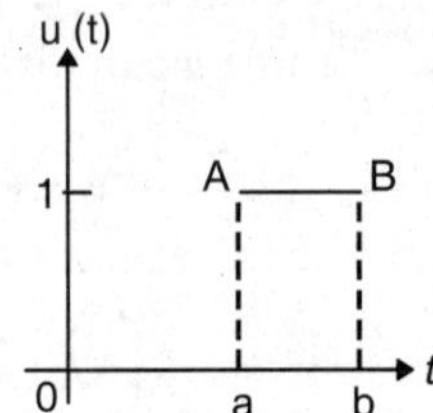

Similarly, the graph of $u(t-b)$ is a straight line parallel to t-axis from B to ∞.

Hence, the graph of $u(t - a) - u(t-b)$ is AB.

42.19 SECOND SHIFTING THEOREM

If $L[f(t)] = F(s)$, then $\boxed{L[f(t-a).u(t-a)] = e^{-as}F(s)}$

Proof. $L[f(t-a).u(t-a)] = \int_0^\infty e^{-st}[f(t-a).u(t-a)]\,dt$

$$= \int_0^a e^{-st} f(t-a).0\,dt + \int_a^\infty e^{-st} f(t-a)(1)\,dt = \int_a^\infty e^{-st} f(t-a)\,dt$$

$$= \int_0^\infty e^{-s(u+a)} f(u)\,du, \qquad \text{where } u = t-a$$

$$= e^{-sa}\int_0^\infty e^{-su}.f(u)\,du = e^{-sa}F(s)$$ **Proved.**

Example 30. *Express the following function in terms of unit step function and find its Laplace transform:*

$$f(t) = \begin{cases} E, & a < t < b \\ 0, & t \geq b \end{cases}$$

Solution. $$f(t) = E\begin{cases} 1, & a < t < b \\ 0, & t \geq b \end{cases}$$ $[L[f(t-a).u(t-a)] = e^{-as}F(s)]$

$$L\{f(t)\} = E\left[\frac{e^{-as}}{s} - \frac{e^{-bs}}{s}\right]$$ **Ans.**

Example 31. *Express the following function in terms of unit step function :*

$$f(t) = \begin{cases} t-1, & 1 < t < 2 \\ 3-t, & 2 < t < 3 \end{cases}$$

and find its Laplace transform. (*U.P.; II Semester, 2009*)

Solution. $$f(t) = \begin{cases} t-1, & 1 < t < 2 \\ 3-t, & 2 < t < 3 \end{cases}$$

$$= (t-1)[u(t-1)-u(t-2)] + (3-t)[u(t-2)-u(t-3)]$$

$$= (t-1)u(t-1) - (t-1)u(t-2) + (3-t)u(t-2) + (t-3)u(t-3)$$

$$= (t-1)u(t-1) - 2(t-2)u(t-2) + (t-3)u(t-3)$$

$$= e^{-s}L(t) - 2e^{-2s}L(t) - e^{-3s}L(t)$$

$[L[f(t-a).u(t-a)] = e^{-as}F(s)]$

$$L[f(t)] = \frac{e^{-s}}{s^2} - 2\frac{e^{-2s}}{s^2} + \frac{e^{-3s}}{s^2}$$ **Ans.**

Example 32. *Find L{F(t)} if*

$$F(t) = \begin{cases} \sin\left(t - \frac{\pi}{3}\right), & t > \frac{\pi}{3} \\ 0, & t < \frac{\pi}{3} \end{cases}$$

Solution. $$L\{F(t)\} = e^{-s\frac{\pi}{3}} L(\sin t)$$ $\left[\because a = \frac{\pi}{3}\right]$

$$= e^{-s\frac{\pi}{3}} \cdot \frac{1}{s^2+1}$$ (Using second shifting property) **Ans.**

42.20 THEOREM. $L[f(t)u(t-a)] = e^{-as}L[f(t+a)]$

Proof. $L[f(t).u(t-a)] = \int_0^\infty e^{-st}[f(t).u(t-a)]dt$

$$= \int_0^a e^{-st}[f(t).u(t-a)]dt + \int_a^\infty e^{-st}.[f(t).u(t-a)]dt = 0 + \int_a^\infty e^{-st}.f(t)(1)dt$$

$$= \int_a^\infty e^{-s(y+a)}.f(y+a)dy = e^{-as}\int_a^\infty e^{-sy}.f(y+a)\,dy \qquad (t-a=y)$$

$$= e^{-as}\int_a^\infty e^{-st}f(t+a)dt = e^{-as}L[f(t+a)]$$ **Proved.**

Example 33. *Find the Laplace transform of* $t^2\,u\,(t-3)$.

Solution. $t^2.u(t-3) = \left[(t-3)^2+6(t-3)+9\right]u(t-3)$

$$= (t-3)^2.u(t-3)+6(t-3).u(t-3)+9u(t-3)$$

$$L[t^2.u(t-3)] = L[(t-3)^2.u(t-3)]+6L[(t-3).u(t-3)]+9L[u(t-3)]$$

$$= e^{-3s}\left[\frac{2}{s^3}+\frac{6}{s^2}+\frac{9}{s}\right]$$

Aliter. $L[t^2u(t-3)] = e^{-3s}L(t+3)^2 = e^{-3s}L\left[t^2+6t+9\right] = e^{-3s}\left[\frac{2}{s^3}+\frac{6}{s^2}+\frac{9}{s}\right]$ **Ans.**

Example 34. *Find the Laplace transform of* $e^{-2t}u_\pi(t)$ *where*

$$u_\pi(t) = \begin{cases} 0; & t<\pi \\ 1; & t>\pi \end{cases}$$

Solution. $u_\pi(t) = \begin{cases} 0; & t<\pi \\ 1; & t>\pi \end{cases}$

$$u_\pi(t) = u\,(t-\pi)$$

$$L\,[u_\pi(t)] = L\,[u\,(t-\pi)] = \frac{e^{-\pi s}}{s}$$

$$L\left[e^{-2t}u_\pi(t)\right] = \frac{e^{-\pi(s+2)}}{s+2}$$ **Ans.**

Example 35. *Express the following function in terms of unit step function and find its Laplace transform* $f(t) = \begin{cases} 0, & 0<t<1 \\ t-1, & 1<t<2 \\ 1, & 2<t \end{cases}$ *(U.P. II Semester, Summer 2002)*

Solution. The above function shown in the figure is expressed in algebraic form

$$f(t) = \begin{cases} 0, & 0<t<1 \\ t-1, & 1<t<2 \\ 1, & 2<t \end{cases} \qquad \dots (1)$$

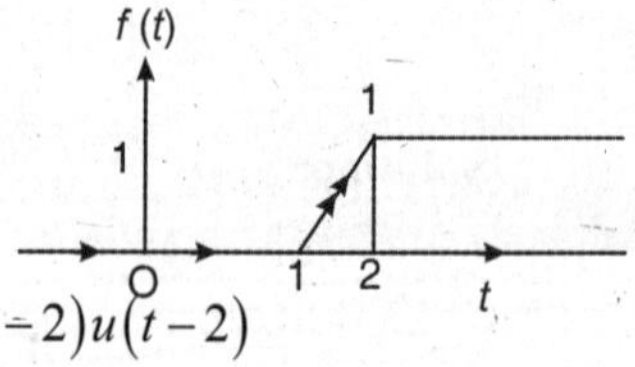

$$f(t) = (t-1)\left[u(t-1)-u(t-2)\right]+u(t-2)$$

$$= (t-1)u(t-1)-u(t-2)\{t-1-1\} = (t-1)u(t-1)-(t-2)u(t-2)$$

$$Lf(t) = L(t-1)u(t-1)-L(t-2)u(t-2) = \frac{e^{-s}}{s^2}-\frac{e^{-2s}}{s^2}$$ **Ans.**

Example 36. *Represent f(t) = sin 2t, 2π < t < 4π and f(t) = 0 otherwise, in terms of unit step function and then find its Laplace transform.*

Solution. $f(t)=\begin{cases}\sin 2t, & 2\pi<t<4\pi\\ 0, & \text{otherwise}\end{cases}$

$$f(t) = \sin 2t\left[u(t-2\pi)-u(t-4\pi)\right]$$

$$L[f(t)] = L[\sin 2t.\,u(t-2\pi)] - L[\sin 2t.\,u(t-4\pi)]$$

$$= e^{-2\pi s} L\sin 2(t+2\pi) - e^{-4\pi s} L\sin 2(t+4\pi)$$

$$= e^{-2\pi s} L\sin 2t - e^{-4\pi s} L\sin(2t)$$

$$= e^{-2\pi s}\frac{2}{s^2+4} - e^{-4\pi s}\frac{2}{s^2+4} = \left(e^{-2\pi s}-e^{-4\pi s}\right)\frac{2}{s^2+4}$$ **Ans.**

Example 37. *A function f (t) obeys the equation f (t) + 2 $\int_0^t f(t)\,dt$ = cosh 2t*

Find the Laplace transform of f (t). (*U.P. II Semester Summer 2006*)

Solution. We have, $f(t) + 2\int_0^t f(t)\,dt = \cosh 2t$

Taking Laplace transformation of both the sides, we get

$$L\{f(t)\} + 2L\int_0^t f(t)\,dt = L(\cosh 2t) \Rightarrow F(s) + 2.\frac{1}{s}F(s) = \frac{s}{s^2-4}$$

$$\Rightarrow F(s)\left\{1+\frac{2}{s}\right\} = \frac{s}{s^2-4} \Rightarrow F(s)\left\{\frac{s+2}{s}\right\} = \frac{s}{s^2-4}$$

$$\Rightarrow F(s) = \left(\frac{s}{s^2-4}\right)\left(\frac{s}{s+2}\right) \Rightarrow F(s) = \frac{s^2}{(s^2-4)(s+2)}$$ **Ans.**

EXERCISE 42.4

Find the Laplace transform of the following:

1. $f(t)=\begin{cases}t-1, & 1<t<2\\ 0, & \text{otherwise}\end{cases}$ **Ans.** $\dfrac{e^{-s}-e^{-2s}}{s^2}-\dfrac{e^{-2s}}{s}$

2. $e^t u(t-1)$ **Ans.** $\dfrac{e^{-(s-1)}}{s-1}$

3. $\dfrac{1-e^{2t}}{t}+tu(t)+\cosh t.\cos t$ **Ans.** $\log\dfrac{s-2}{s}+\dfrac{1}{s^2}+\dfrac{s^3}{s^4+4}$

4. $t^2u(t-2)$ **Ans.** $\dfrac{e^{-2s}}{s^3}\left(4s^2+4s+2\right)$

5. $\sin t\, u(t-4)$ **Ans.** $\dfrac{e^{-4s}}{s^2+1}[\cos 4 + s\sin 4]$

6. $f(t)=K(t-2)[u(t-2)-u(t-3)]$ **Ans.** $\dfrac{K}{s^2}\left[e^{-2s}-(s+1)e^{-3s}\right]$

7. $f(t)=K\dfrac{\sin \pi t}{T}[u(t-2T)-u(t-3T)]$ **Ans.** $\dfrac{K\pi T}{s^2T^2+\pi^2}\left(e^{-2sT}-e^{-3sT}\right)$

Express the following in terms of unit step functions and obtain Laplace transforms.

8. $f(t)=\begin{cases}t, & 0<t<2\\ 0, & 2<t\end{cases}$ **Ans.** $u(t)-u(t-2), \dfrac{1-(2s+1)e^{-2s}}{s^2}$

9. $f(t)=\begin{cases}\sin t, & 0<t<\pi\\ t, & t>\pi\end{cases}$ **Ans.** $\dfrac{1+e^{-\pi s}}{s^2+1}+\dfrac{e^{-\pi s}(\pi s+1)}{s^2}$

10. $f(t)=\begin{cases} 4, & 0<t<1 \\ -2, & 0<t<3 \\ 5, & t>3 \end{cases}$ **Ans.** $\dfrac{4-6e^{-s}+7e^{-3s}}{s}$

42.21. PERIODIC FUNCTIONS

Let $f(t)$ be a periodic function with period T, then

$$\boxed{L[f(t)]=\frac{\int_0^T e^{-st} f(t)\,dt}{1-e^{-sT}}}$$

Proof. $L[f(t)]=\int_0^\infty e^{-st} f(t)\,dt = \int_0^T e^{-st} f(t)\,dt+\int_T^{2T} e^{-st} f(t)\,dt+\int_{2T}^{3T} e^{-st} f(t)\,dt+\ldots$

Substituting $t = u + T$ in second integral and $t = u + 2\,T$ in third integral, and so on.

$$L[f(t)]= \int_0^T e^{-st} f(t)\,dt+\int_0^T e^{-s(u+T)} f(u+T)\,du+\int_0^T e^{-s(u+2T)} f(u+2T)\,du+\ldots$$

$$= \int_0^T e^{-st} f(t)\,dt+e^{-sT}\int_0^T e^{-su} f(u)\,du+e^{-2sT}\int_0^T e^{-su} f(u)\,du+\ldots$$

$$\left[f(u)=f(u+T)=f(u+2T)=f(u+3T)=\ldots.\right]$$

$$= \int_0^T e^{-st} f(t)\,dt+e^{-sT}\int_0^T e^{-st} f(t)\,dt+e^{-2sT}\int_0^T e^{-st} f(t)\,dt+\ldots$$

$$= \left[1+e^{-sT}+e^{-2sT}+e^{-3sT}+\ldots\right]\int_0^T e^{-st} f(t)\,dt \qquad \left[1+a+a^2+a^3+\ldots=\frac{1}{1-a}\right]$$

$$= \frac{1}{1-e^{-sT}}\int_0^T e^{-st} f(t)\,dt.$$ **Proved.**

Example 38. *Find the Laplace transform of the waveform*

$$f(t)=\left(\frac{2t}{3}\right), 0\le t\le 3.$$

Solution. $$L[f(t)]=\frac{1}{1-e^{-sT}}\int_0^T e^{-st} f(t)\,dt$$

$$L\left[\frac{2t}{3}\right]=\frac{1}{1-e^{-3s}}\int_0^3 e^{-st}\left(\frac{2}{3}t\right)dt=\frac{1}{1-e^{-3s}}\frac{2}{3}\left[\frac{te^{-st}}{-s}-(1)\frac{e^{-st}}{s^2}\right]_0^3$$

$$=\frac{2}{3}\frac{1}{1-e^{-3s}}\left[\frac{3e^{-3s}}{-s}-\frac{e^{-3s}}{s^2}+\frac{1}{s^2}\right]=\frac{2}{3}\cdot\frac{1}{1-e^{-3s}}\left[\frac{3e^{-3s}}{-s}+\frac{1-e^{-3s}}{s^2}\right]$$

$$=\frac{2e^{-3s}}{-s\left(1-e^{-3s}\right)}+\frac{2}{3s^2}$$ **Ans.**

Example 39. *Draw the graph and find the Laplace transform of the triangular wave function of period 2 C given by*

$$f(t)=\begin{cases} t, & 0<t\le C \\ 2C-t, & C<t<2C \end{cases}$$ *(Uttarakhand, II Semester, June 2007)*

Solution. Period = $2C = T$

Laplace transform of periodic function $f(t)$

$$L\{f(t)\}=\frac{\int_0^T e^{-st} f(t)\,dt}{1-e^{-sT}}$$

$$L\{f(t)\}=\frac{1}{1-e^{-2Cs}}\int_0^{2C} e^{-st} f(t)\,dt \qquad (T=2c)$$

On putting the values of $f(t)$, we get

$$L\left[f(t)\right]=\frac{1}{1-e^{-2Cs}}\left[\int_0^C e^{-st}t\,dt+\int_C^{2C}e^{-st}(2C-t)\,dt\right]$$

$$=\frac{1}{1-e^{-2Cs}}\left[\left\{\frac{te^{-st}}{-s}-1.\frac{e^{-st}}{(-s)^2}\right\}_0^C+\left\{(2C-t)\frac{e^{-st}}{(-s)}-(-1)\frac{e^{-st}}{(-s)^2}\right\}_C^{2C}\right]$$

$$=\frac{1}{1-e^{-2Cs}}\left[\left\{\frac{C.e^{-Cs}}{-s}-\frac{e^{-Cs}}{(-s)^2}-0+\frac{1}{s^2}\right\}+\left\{(2C-2C)\frac{e^{-2Cs}}{(-s)}+\frac{e^{-2Cs}}{s^2}-\left((2C-C)\frac{e^{-Cs}}{-s}+\frac{e^{-Cs}}{s^2}\right)\right\}\right]$$

$$=\frac{1}{1-e^{-2Cs}}\left\{-\frac{Ce^{-Cs}}{s}-\frac{e^{-Cs}}{s^2}+\frac{1}{s^2}+\frac{e^{-2Cs}}{s^2}+\frac{Ce^{-Cs}}{s}-\frac{e^{-Cs}}{s^2}\right\}$$

$$=\frac{1}{1-e^{-2Cs}}\left\{\frac{1}{s^2}(1-2e^{-Cs}+e^{-2Cs})\right\}=\frac{\left(1-e^{-Cs}\right)^2}{s^2\left(1+e^{-Cs}\right)\left(1-e^{-Cs}\right)}=\frac{1-e^{-Cs}}{s^2\left(1+e^{-Cs}\right)}$$ **Ans.**

Example 40. *Draw the graph of the periodic function*

$$f(t)=\begin{cases} t, & 0<t<\pi \\ \pi-t, & \pi<t<2\pi \end{cases}$$

and find its Laplace transform. (*U.P. Second Semester, 2003*)

Solution. Period $=2\pi=T$

Laplace transform of Periodic functions

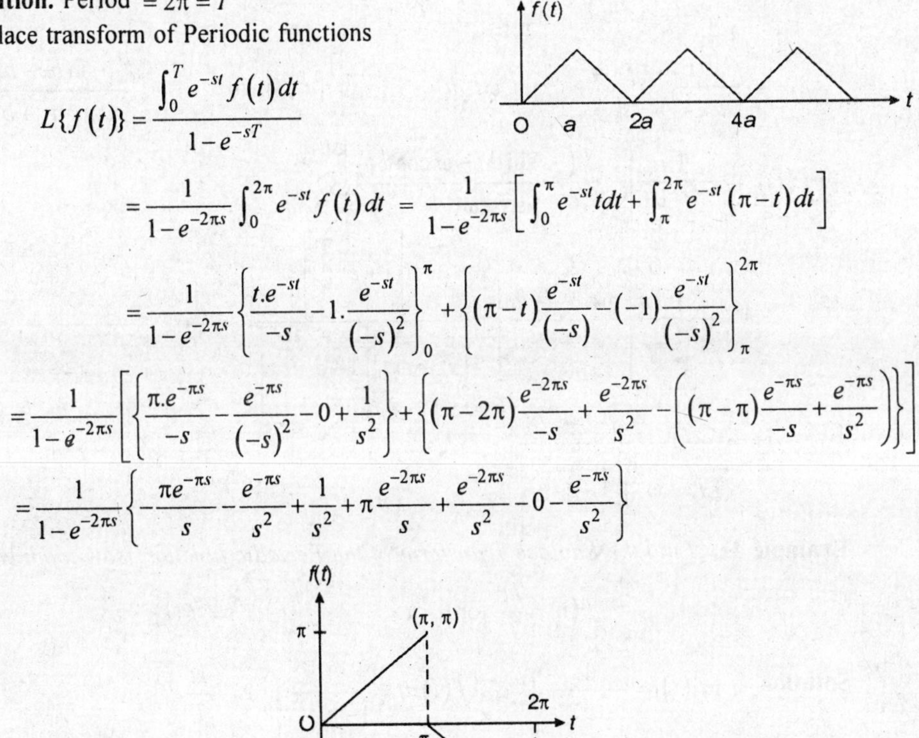

$$L\{f(t)\}=\frac{\int_0^T e^{-st}f(t)\,dt}{1-e^{-sT}}$$

$$=\frac{1}{1-e^{-2\pi s}}\int_0^{2\pi}e^{-st}f(t)\,dt=\frac{1}{1-e^{-2\pi s}}\left[\int_0^{\pi}e^{-st}t\,dt+\int_{\pi}^{2\pi}e^{-st}(\pi-t)\,dt\right]$$

$$=\frac{1}{1-e^{-2\pi s}}\left\{\frac{t.e^{-st}}{-s}-1.\frac{e^{-st}}{(-s)^2}\right\}_0^{\pi}+\left\{(\pi-t)\frac{e^{-st}}{(-s)}-(-1)\frac{e^{-st}}{(-s)^2}\right\}_{\pi}^{2\pi}$$

$$=\frac{1}{1-e^{-2\pi s}}\left[\left\{\frac{\pi.e^{-\pi s}}{-s}-\frac{e^{-\pi s}}{(-s)^2}-0+\frac{1}{s^2}\right\}+\left\{(\pi-2\pi)\frac{e^{-2\pi s}}{-s}+\frac{e^{-2\pi s}}{s^2}-\left((\pi-\pi)\frac{e^{-\pi s}}{-s}+\frac{e^{-\pi s}}{s^2}\right)\right\}\right]$$

$$=\frac{1}{1-e^{-2\pi s}}\left\{-\frac{\pi e^{-\pi s}}{s}-\frac{e^{-\pi s}}{s^2}+\frac{1}{s^2}+\pi\frac{e^{-2\pi s}}{s}+\frac{e^{-2\pi s}}{s^2}-0-\frac{e^{-\pi s}}{s^2}\right\}$$

$$=\frac{1}{1-e^{-2\pi s}}\left\{-\frac{\pi}{s}e^{-\pi s}+\frac{\pi}{s}e^{-2\pi s}+\frac{1}{s^2}-\frac{1}{s^2}.e^{-\pi s}+\frac{1}{s^2}e^{-2\pi s}-\frac{e^{-\pi s}}{s^2}\right\}$$

$$= \frac{1}{1-e^{-2\pi s}}\left[\frac{\pi}{s}\left(e^{-2\pi s}-e^{-\pi s}\right)+\frac{1}{s^2}\left(1+e^{-2\pi s}-2e^{-\pi s}\right)\right] = \frac{-\pi s e^{-\pi s}\left(1-e^{-\pi s}\right)+\left(1-e^{-\pi s}\right)^2}{s^2\left(1+e^{-\pi s}\right)\left(1-e^{-\pi s}\right)}$$

$$= \frac{-\pi s e^{-\pi s}+1-e^{-\pi s}}{s^2\left(1+e^{-\pi s}\right)}$$

Ans.

Example 41. *Find the Laplace transform of the function (Half wave rectifier)*

$$f(t) = \begin{cases} \sin \omega t & \text{for } 0 < t < \dfrac{\pi}{\omega} \\ 0 & \text{for } \dfrac{\pi}{\omega} < t < \dfrac{2\pi}{\omega}. \end{cases}$$

(U.P. II Semester, 2010, Summer 2002)

Solution. $L[f(t)] = \dfrac{1}{1-e^{-sT}}\displaystyle\int_0^T e^{-st} f(t)\,dt$

$$= \frac{1}{1-e^{-\frac{2\pi s}{\omega}}}\int_0^{2\pi/\omega} e^{-st} f(t)\,dt \qquad \left[\begin{array}{l} f(t) \text{ is a periodic function} \\ T = \dfrac{2\pi}{\omega}\end{array}\right]$$

$$= \frac{1}{1-e^{-\frac{2\pi s}{\omega}}}\left[\int_0^{\pi/\omega} e^{-st}\sin\omega t\,dt + \int_{\pi/\omega}^{2\pi/\omega} e^{-st}\times 0\times dt\right]$$

$$= \frac{1}{1-e^{-\frac{2\pi s}{\omega}}}\int_0^{\pi/\omega} e^{-st}\sin\omega t\,dt \qquad \left[\int e^{ax}\sin bx\,dx = e^{ax}\frac{(a\sin bx - b\cos bx)}{a^2+b^2}\right]$$

$$L[f(t)] = \frac{1}{1-e^{-\frac{2\pi s}{\omega}}}\left[\frac{e^{-st}(-s\sin\omega t - \omega\cos\omega t)}{s^2+\omega^2}\right]_0^{\pi/\omega}$$

$$= \frac{1}{1-e^{-\frac{2\pi s}{\omega}}}\left[\frac{\omega e^{-\frac{\pi}{\omega}s}+\omega}{s^2+\omega^2}\right] = \frac{\omega\left[1+e^{-\frac{\pi s}{\omega}}\right]}{\left(s^2+\omega^2\right)\left[1-e^{-\frac{2\pi s}{\omega}}\right]} = \frac{\omega\left[1+e^{-\frac{\pi}{\omega}s}\right]}{(s^2+\omega^2)\left(1-e^{-\frac{\pi}{\omega}s}\right)\left(1+e^{-\frac{\pi}{\omega}s}\right)}$$

$$= \frac{\omega}{\left(s^2+\omega^2\right)\left[1-e^{-\frac{\pi s}{\omega}}\right]}$$

Ans.

Example 42. *Find the Laplace Transform of the Periodic function (saw tooth wave)*

$$f(t) = \frac{kt}{T} \text{ for } 0 < t < T, \quad f(t+T) = f(t)$$

Solution. $L[f(t)] = \dfrac{1}{1-e^{-sT}}\displaystyle\int_0^T e^{-st} f(t)\,dt = \dfrac{1}{1-e^{-sT}}\displaystyle\int_0^T e^{-st}\frac{kt}{T}dt$

$$= \frac{1}{1-e^{-sT}}\frac{k}{T}\int_0^T e^{-st}.t\,dt = \frac{k}{T\left(1-e^{-sT}\right)}\left[t\frac{e^{-st}}{-s} - \int 1.\frac{e^{-st}}{-s}dt\right]_0^T \quad \text{Integrating by parts}$$

$$= \frac{k}{T\left(1-e^{-sT}\right)}\left[\frac{te^{-st}}{-s} - \frac{e^{-st}}{s^2}\right]_0^T = \frac{k}{T\left(1-e^{-sT}\right)}\left[\frac{Te^{-sT}}{-s} - \frac{e^{-sT}}{s^2} + \frac{1}{s^2}\right]$$

$$= \frac{k}{T\left(1-e^{-sT}\right)}\left[\frac{Te^{-sT}}{-s}+\frac{1}{s^2}\left(1-e^{-sT}\right)\right] = -\frac{ke^{-sT}}{s\left(1-e^{-sT}\right)}+\frac{k}{Ts^2}$$ **Ans.**

Example 43. *Obtain Laplace transform of rectangular wave given by*

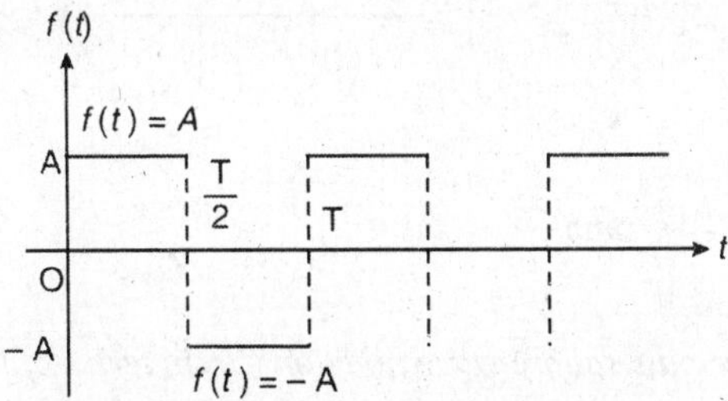

Solution. We know that Laplace transform of a periodic function *i.e.*,

$$L f(t) = \frac{\int_0^T e^{-st} f(t)\,dt}{1-e^{-sT}} = \frac{\int_0^{\frac{T}{2}} e^{-st} A\,dt + \int_{\frac{T}{2}}^{T} e^{-st}(-A)\,dt}{1-e^{-sT}}$$

$$= A\frac{\left[\frac{e^{-st}}{-s}\right]_0^{\frac{T}{2}} - \left[\frac{e^{-st}}{-s}\right]_{\frac{T}{2}}^{T}}{1-e^{-sT}} = \frac{A}{1-e^{-sT}}\left[-\frac{e^{-\frac{sT}{2}}}{s}+\frac{1}{s}+\frac{e^{-sT}}{s}-\frac{e^{-\frac{sT}{2}}}{s}\right]$$

$$= \frac{A}{s\left(1-e^{-sT}\right)}\left[1-2e^{-\frac{sT}{2}}+e^{-sT}\right] = \frac{A}{s\left(1-e^{-sT}\right)}\left[1-e^{-\frac{sT}{2}}\right]^2$$

$$= \frac{A\left[1-e^{-\frac{sT}{2}}\right]^2}{s\left(1+e^{\frac{-sT}{2}}\right)\left(1-e^{-\frac{sT}{2}}\right)} = \frac{A}{s}\frac{\left(1-e^{-\frac{sT}{2}}\right)}{\left(1+e^{-\frac{sT}{2}}\right)} = \frac{A\left(e^{\frac{sT}{4}}-e^{-\frac{sT}{4}}\right)}{s\left(e^{\frac{sT}{4}}+e^{-\frac{sT}{4}}\right)} = \frac{A}{s}\tanh\frac{sT}{4}$$ **Ans.**

Example 44. *Draw the graph of the following periodic function and find its Laplace transform:*

$$f(t) = \begin{cases} t & \text{for } 0 < t \le a \\ 2a - t & \text{for } a < t < 2a \end{cases}$$ *(U.P. II Semester, Summer 2002)*

Solution. The given function is periodic with period $2a$.

$$\therefore \quad L[f(t)] = \frac{1}{1-e^{-2as}}\int_0^{2a} f(t)e^{-st}\,dt$$

$$= \frac{1}{1-e^{-2as}}\left[\int_0^a f(t)e^{-st}\,dt + \int_a^{2a} f(t)e^{-st}\,dt\right]$$

$$= \frac{1}{1-e^{-2as}}\left[\int_0^a t\,e^{-st}\,dt + \int_a^{2a}(2a-t)\,e^{-st}\,dt\right]$$

$$= \frac{1}{1-e^{-2as}}\left[\left\{t\frac{e^{-st}}{-s}-\frac{e^{-st}}{(-s)^2}\right\}_0^a + \left\{(2a-t)\frac{e^{-st}}{-s}+\frac{e^{-st}}{(-s)^2}\right\}_a^{2a}\right]$$

$$= \frac{1}{1-e^{-2as}}\left[-\frac{ae^{-as}}{s}-\frac{e^{-as}}{s^2}+\frac{1}{s^2}+\frac{e^{-2as}}{s^2}+\frac{ae^{-as}}{s}-\frac{e^{-as}}{s^2}\right] = \frac{1}{1-e^{-2as}}\left[\frac{1}{s^2}+\frac{e^{-2as}}{s^2}-2\frac{e^{-as}}{s^2}\right]$$

$$= \frac{1}{s^2}\frac{1}{\left(1-e^{-2as}\right)}\left(1+e^{-2as}-2e^{-as}\right) = \frac{1}{s^2}\frac{\left(1-e^{-as}\right)^2}{\left(1+e^{-as}\right)\left(1-e^{-as}\right)} = \frac{1}{s^2}\left[\frac{1-e^{-as}}{1+e^{-as}}\right]$$

$$= \frac{1}{s^2}\left[\frac{e^{\frac{as}{2}}-e^{-\frac{as}{2}}}{e^{\frac{as}{2}}+e^{-\frac{as}{2}}}\right] = \frac{1}{s^2}\tanh\frac{as}{2}$$ **Ans.**

Example 45. *A periodic square wave function f(t), in terms of unit step functions, is written as*

$$f(t) = k\left[u_0(t) - 2u_a(t) + 2u_{2a}(t) - 2u_{3a}(t) + \ldots\right]$$

Show that the Laplace transform of f (t) is given by

$$L\left[f(t)\right] = \frac{k}{s}\tanh\left(\frac{as}{2}\right)$$

Solution. $f(t) = k\left[u_0(t) - 2u_a(t) + 2u_{2a}(t) - 2u_{3a}(t) + \ldots\right]$

$$f(t) = k\,[u(t-0) - 2u(t-a) + 2u(t-2a) - 2u(t-3a) + \ldots\ldots]$$

$$L\,[f(t)] = k\,[L\,u(t-0) - 2\,L\,u(t-a) + 2\,L\,u(t-2a) - 2\,L\,u(t-3a) + \ldots\ldots]$$

$$= k\left[\frac{1}{s} - 2\frac{e^{-as}}{s} + 2\frac{e^{-2as}}{s} - 2\frac{e^{-3as}}{s} + \ldots\right] = \frac{k}{s}\left[1 - 2e^{-as} + 2e^{-2as} - 2e^{-3as} + \ldots\right]$$

$$= \frac{k}{s}\left[1 - 2(e^{-as} - e^{-2as} + e^{-3as} - \ldots)\right] = \frac{k}{s}\left[1 - 2\frac{e^{-as}}{1+e^{-as}}\right] = \frac{k}{s}\left[\frac{1+e^{-as}-2e^{-as}}{1+e^{-as}}\right]$$

$$= \frac{k}{s}\left[\frac{1-e^{-as}}{1+e^{-as}}\right] = \frac{k}{s}\left[\frac{e^{\frac{as}{2}}-e^{\frac{-as}{2}}}{e^{\frac{as}{2}}+e^{-\frac{as}{2}}}\right] = \frac{k}{s}\tanh\frac{as}{2}$$ **Ans.**

EXERCISE 42.5

1. Find the Laplace transform of the periodic function

   $f(t) = e^t$ for $0 < t < 2\pi$ **Ans.** $\dfrac{e^{2(1-s)\pi} - 1}{(1-s)\left(1-e^{-2\pi s}\right)}$

2. Obtain Laplace transform of full wave rectified sine wave given by

   $f(t) = \sin\omega t,\quad 0 < t < \dfrac{\pi}{\omega}$ **Ans.** $\dfrac{\omega}{\left(s^2+\omega^2\right)}\coth\dfrac{\pi s}{2\omega}$

3. Find the Laplace transform of the staircase function

   $f(t) = kn,\ np < t < (n+1)p,\ n = 0, 1, 2, 3$ **Ans.** $\dfrac{ke^{ps}}{s\left(1-e^{-ps}\right)}$

Find Laplace transform of the following:

4. $f(t) = t^2,\ 0 < t < 2,\ f(t+2) = f(t)$ **Ans.** $\dfrac{2 - e^{-2s} - 4se^{-2s} - 4s^2e^{-2s}}{s^3\left(1-e^{-2s}\right)}$

5. $f(t) = \begin{cases} 1, & 0 \le t \le \dfrac{a}{2} \\ -1, & \dfrac{a}{2} \le t < a \end{cases}$ *(U.P. II Semester, 2004)* **Ans.** $\dfrac{1}{s}\tanh\dfrac{as}{4}$

6. $f(t)=\begin{cases}\cos\omega t, & 0<t<\dfrac{\pi}{\omega}\\[2mm] 0, & \dfrac{\pi}{\omega}<t<\dfrac{2\pi}{\omega}\end{cases}$ **Ans.** $\dfrac{s}{\left(s^2+\omega^2\right)\left(1-e^{-\frac{\pi s}{\omega}}\right)}$

7. $f(t)=\begin{cases}t, & 0<t<1\\ 0, & 1<t<2\end{cases}\quad f(t+2)=f(t)$ **Ans.** $\dfrac{1-e^{-s}(s+1)}{s^2\left(1-e^{-2s}\right)}$

8. $f(t)=\begin{cases}\dfrac{2t}{T}, & 0\le t\le\dfrac{T}{2}\\[2mm] \dfrac{2}{T}(T-t), & \dfrac{T}{2}\le t\le T\end{cases}\quad f(t+T)=f(t)$ **Ans.** $\dfrac{2}{Ts^2}\tanh\dfrac{sT}{4}-\dfrac{1}{s\left(e^{\frac{sT}{2}}+1\right)}$

42.22 IMPULSE FUNCTION

When a large force acts for a short time, then the product of the force and the time is called impulse in applied mechanics. The unit impulse function is the limiting function.

$$\delta(t-1)=\begin{cases}\dfrac{1}{\varepsilon}, & a<t<a+\varepsilon\\ 0, & \text{otherwise}\end{cases}$$

The value of the function (height of the strip in the figure) becomes infinite as $\varepsilon\to 0$ and the area of the rectangle is unity.

(1) The Unit Impulse function is defined as follows:

$$\delta(t-a)=\begin{cases}\infty & \text{for } t=a\\ 0 & \text{for } t\ne a\end{cases}$$

and $\displaystyle\int_0^\infty \delta(t-a).dt=1$ [Area of strip = 1]

(2) Laplace Transform of Unit Impulse function

$$\int_0^\infty f(t)\ \delta(t-a)dt=\int_a^{a+\varepsilon} f(t).\frac{1}{\varepsilon}dt$$

$\left\{\begin{array}{l}\text{Mean value Theorem}\\ \displaystyle\int_a^b f(t)dt=(b-a)f(\eta)\end{array}\right.$

$$=(a+\varepsilon-a)f(\eta).\frac{1}{\varepsilon}$$ where $a<\eta<a+\varepsilon$

$$=f(\eta)$$

Property I. $\boxed{\displaystyle\int_0^\infty f(t)\delta(t-a)dt=f(a)}$ as $\varepsilon\to 0$

Note. If $f(t)=e^{-st}$ and $L[\delta(t-a)]=e^{-as}$

Example 46. *Evaluate* $\displaystyle\int_{-\infty}^{\infty} e^{-5t}\delta(t-2)\,dt.$

Solution. $\displaystyle\int_{-\infty}^{\infty} e^{-5t}\delta(t-2)\,dt=e^{-5\times 2}=e^{-10}$ **Ans.**

Property II: $\boxed{\displaystyle\int_{-\infty}^{\infty} f(t)\delta'(t-a)dt=-f'(a)}$

Proof. $\displaystyle\int_{-\infty}^{\infty} f(t)\delta'(t-a)dt=\left[f(t).\delta(t-a)\right]_{-\infty}^{\infty}-\int_{-\infty}^{\infty} f'(t)\delta(t-a)\,dt$

$$=0-0-f'(a)=-f'(a)$$

Example 47. *Find the Laplace transform of* $t^3\delta(t-4)$

Solution. $Lt^3\delta(t-4)=\int_0^{\infty} e^{-st}t^3\delta(t-4)\,dt = 4^3 e^{-4s}$ **Ans.**

EXERCISE 42.6

Evaluate the following:

1. $\int_0^{\infty} e^{-3t}\delta(t-4)\,dt$ **Ans.** e^{-12}

2. $\int_{-\infty}^{\infty} \sin 2t\, \delta\left(t-\frac{\pi}{4}\right)dt$ **Ans.** 1

3. $\int_{-\infty}^{\infty} e^{-3t}\delta'(t-2)\,dt$ **Ans.** $3e^{-6}$

Find Laplace transform of

4. $\frac{\delta(t-4)}{t}$ **Ans.** $\frac{e^{-4s}}{4}$

5. $\cos t \log t\; \delta\,(t-\pi)$ **Ans.** $-e^{-\pi s}\log \pi$

6. $e^{-4t}\delta(t-3)$ **Ans.** $e^{-3(s+4)}$

42.23 CONVOLUTION THEOREM

If $\quad L[f_1(t)]=F_1(s)$ and $L[f_2(t)]=F_2(s)$

then $\quad L\left\{\int_0^t f_1(x)\,f_2(t-x)\,dx\right\}=F_1(s)\cdot F_2(s)$

or

$$\boldsymbol{L^{-1}(F_1(s).F_2(s))=\int_0^t f_1(x)\,f_2(t-x)\,dx}$$

Proof. We have

$$L\left\{\int_0^{\infty} f_1(x)\,f_2(t-x)\,dx\right\}=\int_0^{\infty} e^{-st}\left[\int_0^t f_1(x)\,f_2(t-x)\,dx\right]dt \qquad \text{(By Definition)}$$

where the double integral is taken over the infinite region in the first quadrant lying between the lines $x = 0$ and $x = t$.

Here first we are integrating w.r.t. "x", within limits $x = 0$ and $x = t$, and then we will integrate w.r.t. "t" with limits $t = 0$ and $t = \infty$.

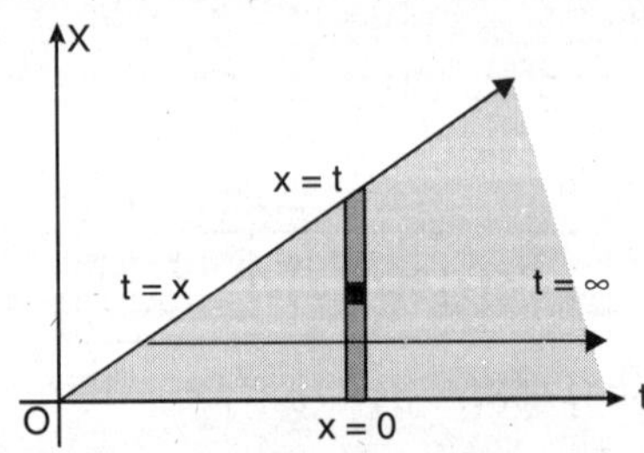

On changing the order of integration first we integrate w.r.t. "t" with limits $t = x$ and $t = \infty$ and then w.r.t. "x" with limits $x = 0$ and $x = \infty$.

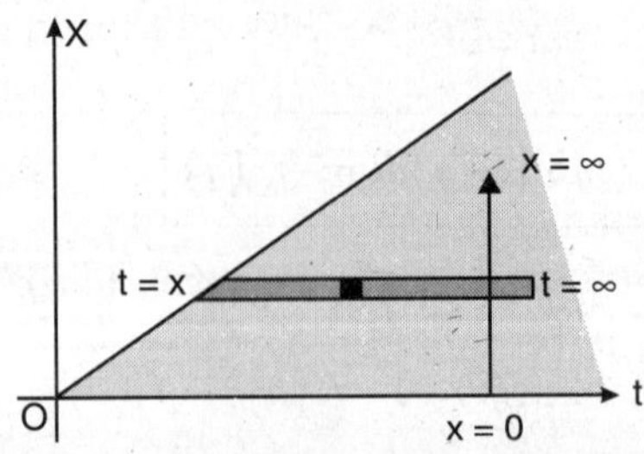

On changing the order of integration, the integral becomes

$$\int_0^\infty dx\left[\int_x^\infty e^{-st} f_1(x).f_2(t-x)\,dt\right]$$

$$= \int_0^\infty dx\left[\int_x^\infty e^{-s(t-x+x)} f_1.(x).f_2(t-x)\,dt\right] = \int_0^\infty dx\left[\int_x^\infty e^{-s(t-x)}.e^{-sx} f_1(x).f_2(t-x)\,dt\right]$$

$$= \int_0^\infty e^{-sx} f_1(x)\,dx\left[\int_x^\infty e^{-s(t-x)} f_2(t-x)\,dt\right] = \int_0^\infty e^{-sx} f_1(x)\,dx\left[\int_x^\infty e^{-sz} f_2(z)\,dz\right]$$

[Put $t-x=z \Rightarrow dt = dz$]

$$= \int_0^\infty e^{-sx} f_1(x)\,dx \int_0^\infty e^{-sz} f_2(z)\,dz,$$

Lower limit $x-x=z \Rightarrow z=0$]

$$= \int_0^\infty e^{-sx} f_1(x) F_2(s)\,dx = \left[\int_0^\infty e^{-sx} f_1(x)\,dx\right] F_2(s) = F_1(s)F_2(s)$$ **Proved.**

Example 48. *Find the Laplace transform of* $\int_0^t e^x.\sin(t-x)\,dx$

Solution. By Convolution Theorem

$$L\int_0^t f_1(x)\,f_2(t-x)\,dx = F_1(s).F_2(s)$$

$$\Rightarrow L\int_0^t e^x.\sin(t-x)\,dx = L(e^x).L\sin t = \frac{1}{s-1}\,\frac{1}{s^2+1} = \frac{1}{(s-1)(s^2+1)}$$ **Ans.**

Note. Convolution Theorem is generally used to find Inverse Laplace transform of the product of two functions, discussed in the next chapter.

42.24 LAPLACE TRANSFORM OF BESSEL FUNCTIONS $J_0(x)$ and $J_1(x)$

We know that

$$J_n(x) = \frac{x^2}{2^n\,\Gamma(n+1)}\left[1-\frac{x^2}{2.(2n+2)}+\frac{x^4}{2.4(2n+2)(2n+4)}-\ldots..\right]$$

$$J_0(t) = \left[1-\frac{t^2}{2^2}+\frac{t^4}{2^2.4^2}-\frac{t^6}{2^2.4^2.6^2}+\ldots\right]$$

Taking Laplace transforms of both sides, we have

$$LJ_0(t) = \frac{1}{s}-\frac{1}{2^2}.\frac{2!}{s^3}+\frac{1}{2^2.4^2}\frac{4!}{s^5}-\frac{1}{2^2.4^2.6^2}\frac{6!}{s^7}+\ldots$$

$$= \frac{1}{s}\left[1-\frac{1}{2}\left(\frac{1}{s^2}\right)+\frac{1.3}{2.4}\left(\frac{1}{s^4}\right)-\frac{1.3.5}{2.4.6}\left(\frac{1}{s^6}\right)+\ldots\right]$$

$$= \frac{1}{s}\left[1+\left(-\frac{1}{2}\right)\left(\frac{1}{s^2}\right)+\frac{\left(-\frac{1}{2}\right)\left(-\frac{3}{2}\right)}{2!}\left(\frac{1}{s^2}\right)^2+\frac{\left(-\frac{1}{2}\right)\left(-\frac{3}{2}\right)\left(-\frac{5}{2}\right)}{3!}\left(\frac{1}{s^2}\right)^3+\ldots\right]$$

$$= \frac{1}{s}\left[1+\frac{1}{s^2}\right]^{-\frac{1}{2}}$$ (By Binomial theorem)

$$= \frac{1}{s}\left[\frac{s^2+1}{s^2}\right]^{-\frac{1}{2}} = \frac{1}{s}\left[\frac{s^2}{s^2+1}\right]^{\frac{1}{2}} = \frac{1}{\sqrt{s^2+1}}$$ ... (1) **Ans.**

We know that $Lf(at) = \frac{1}{a}F\left(\frac{s}{a}\right)$

$$LJ_0(at) = \frac{1}{a}\frac{1}{\sqrt{\frac{s^2}{a^2}+1}} = \frac{1}{\sqrt{s^2+a^2}}$$ [From (1)]

$$LJ_1(x) = -LJ_0'(x) = -[s\,LJ_0(x) - J_0(0)] = -\left[s.\frac{1}{\sqrt{s^2+1}} - 1\right] = 1 - \frac{s}{\sqrt{s^2+1}}$$ **Ans.**

EXERCISE 42.7

Find the Laplace transform of the following:

1. $e^{ax} J_0(bx)$ **Ans.** $\dfrac{1}{\sqrt{s^2+2as+a^2+b^2}}$

2. $x J_0(ax)$ **Ans.** $\dfrac{s}{(s^2+a^2)^{3/2}}$

3. $x J_1(x)$ **Ans.** $\dfrac{1}{(s^2+1)^{3/2}}$

42.25 EVALUATION OF INTEGRALS

We can evaluate number of integrals having lower limit 0 and upper limit ∞ by the help of Laplace transform.

Example 49. *Evaluate* $\int_0^\infty te^{-3t}\sin t\,dt$

Solution. $\int_0^\infty te^{-3t}\sin t\,dt = \int_0^\infty te^{-st}\sin t\,dt$ $(s = 3)$

$$= L(t\sin t) = -\frac{d}{ds}\left(\frac{1}{s^2+1}\right) = \frac{2s}{(s^2+1)^2}$$

Putting $s = 3$, we get $= \dfrac{2\times 3}{(3^2+1)^2} = \dfrac{6}{100} = \dfrac{3}{50}$ **Ans.**

Example 50. *Evaluate* $\int_0^\infty \frac{e^{-t}\sin t}{t}dt$ and $\int_0^\infty \frac{\sin t}{t}dt$ *(U.P., II Semester, 2009)*

Solution. $\int_0^\infty \frac{e^{-t}\sin t}{t}dt = \int_0^\infty e^{-st}\frac{\sin t}{t}dt$ $(s = 1)$

$$= L\left[\frac{\sin t}{t}\right] = \int_s^\infty \frac{1}{s^2+1}ds = \left[\tan^{-1}s\right]_s^\infty = \frac{\pi}{2} - \tan^{-1}s \quad \text{...(1)}$$

$$= \frac{\pi}{2} - \tan^{-1}(1) = \frac{\pi}{2} - \frac{\pi}{4} = \frac{\pi}{4}$$ $(s = 1)$ **Ans.**

On putting $s = 0$ in (1), we get $\int_0^\infty \frac{\sin t}{t}dt = \frac{\pi}{2} - \tan^{-1}(0) = \frac{\pi}{2}$ **Ans.**

Example 51. *Evaluate* $\int_0^\infty \frac{e^{-at} - e^{-bt}}{t}dt$

Solution. $\int_0^\infty e^{-st}(e^{-at} - e^{-bt})dt = L(e^{-at} - e^{-bt}) = L(e^{-at}) - L(e^{-bt}) = \left(\frac{1}{s+a} - \frac{1}{s+b}\right)$

$$\int_0^\infty e^{-st}\left(\frac{e^{-at} - e^{-bt}}{t}\right)dt = L\frac{1}{t}(e^{-at} - e^{-bt}) = \int_s^\infty \left(\frac{1}{s+a} - \frac{1}{s+b}\right)ds$$

$$= [\log(s+a) - \log(s+b)]_s^\infty$$

$$= \left[\log\left(\frac{s+a}{s+b}\right)\right]_s^\infty = \left[\log \frac{1+\frac{a}{s}}{1+\frac{b}{s}}\right]_s^\infty = \left[\log 1 - \log\frac{s+a}{s+b}\right] = \log\frac{s+b}{s+a}$$

Putting s = 0 in above, we get $\int_0^\infty \left(\frac{e^{-at}-e^{-bt}}{t}\right) dt = \log\left(\frac{b}{a}\right)$ **Ans.**

Example 52. *Show that* $\int_0^\infty t^3 e^{-t} \sin t \, dt = 0$

Solution. $L\{t^3 \sin t\} = (-1)^3 \frac{d^3}{ds^3} L\{\sin t\}$

$$\Rightarrow \quad \int_0^\infty e^{-st} t^3 \sin t \, dt = \frac{-d^3}{ds^3} \frac{1}{s^2+1}$$

$\left[\begin{array}{l}\text{This is G.P.}\\ \text{Sum} = \frac{a}{1-r}\end{array}\right]$

$$= -\frac{d^2}{ds^2}\left[-\frac{2s}{(s^2+1)^2}\right] = \frac{d}{ds}\left[\frac{(s^2+1)^2(2) - 2s\,[2(s^2+1)](2s)}{(s^2+1)^4}\right]$$

$$= \frac{d}{ds}\left[\frac{2(s^2+1)-8s^2}{(s^2+1)^3}\right] = \frac{d}{ds}\left[\frac{-6s^2+2}{(s^2+1)^3}\right] = \frac{(s^2+1)^3(-12s) - (-6s^2+2)3(s^2+1)^2(2s)}{(s^2+1)^6}$$

$$= \frac{(s^2+1)(-12s) - (-6s^2+2)6s}{(s^2+1)^4} = \frac{-12s^3-12s+36s^3-12s}{(s^2+1)^4}$$

$$\int_0^\infty e^{-st} t^3 \sin t \, dt = \frac{24s^3-24s}{(s^2+1)^4} = \frac{24s(s^2-1)}{(s^2+1)^4} \quad \ldots (1)$$

Putting $s = 1$ in (1), we get $\int_0^\infty e^{-t} t^3 \sin t \, dt = 0$ **Ans.**

Example 53. *Evaluate* $\int_0^\infty t^2 e^{3t} \sin^2 t \, dt.$

Solution. We have, $\sin^2 t = \frac{1}{2}(1-\cos 2t)$

$$\Rightarrow \quad L\{\sin^2 t\} = \frac{1}{2}[L\{1\} - L\{\cos 2t\}] = \frac{1}{2}\left[\frac{1}{s} - \frac{s}{s^2+4}\right]$$

$$\Rightarrow \quad L\left[t^2 \sin^2 t\right] = (-1)^2 \cdot \frac{d^2}{ds^2}\left[\frac{1}{2}\left\{\frac{1}{s} - \frac{s}{s^2+4}\right\}\right]$$

$$\Rightarrow \quad \int_0^\infty e^{-st} . t^2 \sin^2 t \, dt = \frac{1}{2}\frac{d}{ds}\left[\frac{d}{ds}\left\{\frac{1}{s} - \frac{s}{s^2+4}\right\}\right] = \frac{1}{2}\frac{d}{ds}\left[-\frac{1}{s^2} - \frac{(s^2+4)(1) - s(2s)}{(s^2+4)^2}\right]$$

$$= \frac{1}{2}\frac{d}{ds}\left[-\frac{1}{s^2} - \frac{-s^2+4}{(s^2+4)^2}\right] = \frac{1}{2}\left[\frac{2}{s^3} - \frac{(s^2+4)^2(-2s) - (-s^2+4)2(s^2+4)(2s)}{(s^2+4)^4}\right]$$

$$= \frac{1}{2}\left[\frac{2}{s^3} - \frac{(s^2+4)(-2s) - (-s^2+4)4s}{(s^2+4)^3}\right] \quad \ldots (1)$$

Putting the value of $s = -3$ in (1), we get

$$\int_0^\infty e^{3t} t^2 \sin^2 t \, dt = \frac{1}{2}\left[\frac{2}{-27} - \frac{(13)6 - (-5)(-12)}{(9+4)^3}\right]$$

$$= -\frac{1}{27} - \frac{9}{(13)^3} = \frac{-2197-243}{59319} = \frac{-2440}{59319}$$ **Ans.**

EXERCISE 42.8

Evaluate the following by using Laplace Transform:

1. $\int_0^\infty te^{-4t}\sin t\,dt$ **Ans.** $\frac{8}{289}$
2. $\int_0^\infty \frac{e^{-2t}\sinh t\sin t}{t}dt$ **Ans.** $\frac{1}{2}\tan^{-1}\frac{1}{2}$
3. $\int_0^\infty \frac{\sin^2 t}{t^2}dt$ **Ans.** $i\frac{5}{2}$
4. $\int_0^\infty \frac{e^{-t}-e^{-4t}}{t}dt$ **Ans.** log 4

42.26 FORMULATION OF LAPLACE TRANSFORM

S.No.	$f(t)$	$F(s)$
1.	e^{at}	$\frac{1}{s-a}$
2.	t^n	$\frac{\lceil n+1}{s^{n+1}}$ or $\frac{n!}{s^{n+1}}$
3.	sin at	$\frac{a}{s^2+a^2}$
4.	cos at	$\frac{s}{s^2+a^2}$
5.	sinh at	$\frac{a}{s^2-a^2}$
6.	cosh at	$\frac{s}{s^2-a^2}$
7.	$u(t-a)$	$\frac{e^{-as}}{s}$
8.	$\delta(t-a)$	e^{-as}
9.	$e^{bt}\sin at$	$\frac{a}{(s-b)^2+a^2}$
10.	$e^{bt}\cos at$	$\frac{s-b}{(s-b)^2+a^2}$
11.	$\frac{t}{2a}\sin at$	$\frac{s}{(s^2+a^2)^2}$
12.	$t\cos at$	$\frac{s^2-a^2}{(s^2+a^2)^2}$
13.	$\frac{1}{2a^3}(\sin at-at\cos at)$	$\frac{1}{(s^2+a^2)^2}$
14.	$\frac{1}{2a}(\sin at+at\cos at)$	$\frac{s^2}{(s^2+a^2)^2}$

42.27 PROPERTIES OF LAPLACE TRANSFORM

S.No.	Property	f(t)	F (s)
1.	Scaling	$f(at)$	$\frac{1}{a}F\left(\frac{s}{a}\right)$, $a > 0$
2.	Derivative	$\frac{df(t)}{dt}$ $\frac{d^2 f(t)}{dt^2}$ $\frac{d^3 f(t)}{dt^3}$	$s\,F(s) - f(0)$, $s > 0$ $s^2 F(s) - sf(0) - f'(0)$, $s > 0$ $s^3 F(s) - s^2 f(0) - s\,f'(0) - f''(0)$, $s > 0$
3.	Integral	$\int_0^t f(t)\,dt$	$\frac{1}{s}F(s)$, $s > 0$
4.	Initial Value	$\lim_{t\to 0} f(t)$	$\lim_{s\to\infty} sF(s)$
5.	Final Value	$\lim_{t\to\infty} f(t)$	$\lim_{s\to\infty} sF(s)$
6.	First shifting	$e^{-at} f(t)$	$F(s + a)$
7.	Second shifting	$f(t)\,u(t - a)$	$e^{-a}\,L\,f(t + a)$
8.	Multiplication by t	$t\,f(t)$	$-\frac{d}{ds}F(s)$
9.	Multiplication by t^n	$t^n f(t)$	$(-1)^n \frac{d^n}{ds^n}F(s)$
10.	Division by t	$\frac{1}{t}f(t)$	$\int_s^\infty F(s)\,ds$
11.	Periodic function	$f(t)$	$\frac{\int_0^T e^{-st} f(t)}{1 - e^{-st}}$ $f(t + T) = f(t)$
12.	Convolution	$f(t) * g(t)$	$F(s)\,G(s)$

OBJECTIVE TYPE QUESTIONS

Choose the correct alternative :

1. Laplace transform of $t^3 e^{-3t}$ is :

(i) $\frac{7}{(s+4)^3}$ (ii) $\frac{s}{(s+3)^3}$ (iii) $\frac{6}{(s+3)^4}$ (iv) $\frac{2}{(s+6)^3}$ **Ans.** (iii)

(R.G.P.V., Bhopal, II Semester, Feb. 2006)

2. Laplace transform of $e^{-2t}\sin 4t$ is :

(i) $\dfrac{2}{s^2+4s+20}$ (ii) $\dfrac{s-2}{s^2+4s+20}$ (iii) $\dfrac{s-4}{s^2+4s+20}$ (iv) $\dfrac{4}{s^2+4s+20}$ **Ans.** (iv)

(R.G.P.V., Bhopal, II Semester, June 2007)

3. If $\{F(t)\} = \bar{f}(s)$, then $L\left\{\int_0^t F(x)\,dx\right\}$ is :

(i) $\int_0^s \bar{f}(s)\,ds$ (ii) $\int_0^s \frac{1}{s}\bar{f}(s)\,ds$ (iii) $\frac{1}{s}\bar{f}(s)$ (iv) $s\,f(s)$ **Ans.** (iii)

(R.G.P.V., Bhopal, II Semester, June 2006)

4. If $L\{F(t)\} = \bar{f}(s)$, then $L\{t\,F(t)\}$ is :

(i) $\bar{f}'(s)$ (ii) $-\bar{f}'(s)$ (iii) $\bar{f}'(s)+\bar{f}(s)$ (iv) $s\bar{f}'(s)+\bar{f}(s)$ **Ans.** (ii)

(R.G.P.V., Bhopal, II Semester, June 2006)

5. Laplace transform of $\dfrac{\cos at - \cos bt}{t}$ is

(i) $\log\dfrac{s^2+b^2}{s^2+a^2}$ (ii) $\dfrac{1}{2}\log\dfrac{s^2+a^2}{s^2+b^2}$ (iii) $\dfrac{1}{2}\log\dfrac{s^2+b^2}{s^2+a^2}$ (iv) $\log\dfrac{s+b}{s+a}$ **Ans.** (iii)

(R.G.P.V., Bhopal, II Semester, Feb. 2006, 2005)

6. The Laplace transform of the function

$$F(t) = \begin{cases} 1, & 0 \le t < 2 \\ -1, & 2 \le t < 4 \end{cases},\ f(t+4) = f(t) \text{ is given as,}$$

(i) $\dfrac{1-e^{-2s}}{s\left(1+e^{-2s}\right)}$ (ii) $\dfrac{1+e^{-2s}}{s\left(1+e^{-2s}\right)}$ (iii) 0 (iv) $\dfrac{s+1}{s-1}$ **Ans.** (i)

(U.P., II Semester, 2009)

Fill in the blank for each of the following question:

7. The Laplace transform of

$\int_0^t\int_0^t\int_0^t \cos au\,du\,du\,du$ is given as ...... *[U.P.T.U. (SUM) 2009]* **Ans.** $\dfrac{1}{s^2(s^2+a^2)}$

CHAPTER

43

INVERSE LAPLACE TRANSFORMS

(SOLUTION OF DIFFERENTIAL EQUATIONS)

43.1 INVERSE LAPLACE TRANSFORMS

If $F(s)$ is the Laplace Transform of a function $f(t)$, then $f(t)$ is known as Inverse Laplace Transform. Now we will discuss how to find $f(t)$ when $F(s)$ is given.

If $L[f(t)] = F(s)$, then $L^{-1}[F(s)] = f(t)$, where L^{-1} is called the Inverse Laplace Transform operator.

From the application point of view, the Inverse Laplace Transform is very useful.

Inverse Laplace Transform is used in solving differential equations without finding the general solution and arbitrary constants.

43.2 IMPORTANT FORMULAE

1. $L^{-1}\left(\frac{1}{s}\right) = 1$

2. $L^{-1}\frac{1}{s^n} = \frac{t^{n-1}}{(n-1)!}$

3. $L^{-1}\frac{1}{s-a} = e^{at}$

4. $L^{-1}\frac{s}{s^2-a^2} = \cosh at$ *(R.G.P.V., Bhopal, Dec. 2007)*

5. $L^{-1}\frac{1}{s^2-a^2} = \frac{1}{a}\sinh at$

6. $L^{-1}\frac{1}{s^2+a^2} = \frac{1}{a}\sin at$

7. $L^{-1}\frac{s}{s^2+a^2} = \cos at$

8. $L^{-1}F(s-a) = e^{at}f(t)$

9. $L^{-1}\frac{1}{(s-a)^2+b^2} = \frac{1}{b}e^{at}\sin bt$

10. $L^{-1}\frac{s-a}{(s-a)^2+b^2} = e^{at}\cos bt$

11. $L^{-1}\frac{1}{(s-a)^2-b^2} = \frac{1}{b}e^{at}\sinh bt$

12. $L^{-1}\frac{s-a}{(s-a)^2-b^2} = e^{at}\cosh bt$

13. $L^{-1}\frac{1}{(s^2+a^2)^2} = \frac{1}{2a^3}(\sin at - at\cos at)$

14. $L^{-1}\frac{s}{(s^2+a^2)^2} = \frac{1}{2a}t\sin at$

15. $L^{-1}\frac{s^2-a^2}{(s^2+a^2)^2} = t\cos at$

16. $L^{-1}(1) = s(t)$

17. $L^{-1}\frac{s^2}{(s^2+a^2)^2} = \frac{1}{2a}[\sin at + at\cos at]$

18. $L^{-1}\left\{\frac{1}{s}F(s)\right\} = \int_0^t f(t)\,dt$

Example 1. *Show that* $\frac{1}{s^{1/2}} = L\left[\frac{1}{\sqrt{\pi t}}\right]$. *(U.P., II Semester, Summer 2005)*

Solution. We have to show that $\frac{1}{s^{1/2}} = L\left[\frac{1}{\sqrt{\pi t}}\right]$.

Now, $$L^{-1}\left\{\frac{1}{s^n}\right\} = \frac{t^{n-1}}{(n-1)!} = \frac{t^{n-1}}{\overline{|n}}$$

So $$L^{-1}\left\{\frac{1}{s^{1/2}}\right\} = \frac{t^{\frac{1}{2}-1}}{\overline{|1/2}} = \frac{t^{-1/2}}{\overline{|1/2}} = \frac{t^{-1/2}}{\sqrt{\pi}}$$

$$\Rightarrow \quad L^{-1}\left\{\frac{1}{s^{1/2}}\right\} = \frac{1}{\sqrt{\pi t}} \quad \Rightarrow \quad \frac{1}{s^{1/2}} = L\left[\frac{1}{\sqrt{\pi t}}\right]$$ **Proved.**

Example 2. *Find the inverse Laplace Transform of the following:*

(*i*) $\frac{1}{s-2}$ (*ii*) $\frac{1}{s^2-9}$ (*iii*) $\frac{1}{s^2+25}$

(*iv*) $\frac{s}{s^2+9}$ (*v*) $\frac{1}{(s-2)^2+1}$ (*vi*) $\frac{s-1}{(s-1)^2+4}$

(*vii*) $\frac{1}{(s+3)^2-4}$ (*viii*) $\frac{s+2}{(s+2)^2-25}$

Solution.

(*i*) $L^{-1}\frac{1}{s-2} = e^{2t}$ (*ii*) $L^{-1}\frac{1}{s^2-9} = L^{-1}\frac{1}{3}\cdot\frac{3}{s^2-(3)^2} = \frac{1}{3}\sinh 3t$

(*iii*) $L^{-1}\frac{s}{s^2-16} = L^{-1}\frac{s}{s^2-(4)^2} = \cosh 4t$ (*iv*) $L^{-1}\frac{1}{s^2+25} = \frac{1}{5}\frac{5}{s^2+(5)^2} = \frac{1}{5}\sin 5t$

(*v*) $L^{-1}\frac{s}{s^2+9} = \frac{s}{s^2+(3)^2} = \cos 3t$ (*vi*) $L^{-1}\frac{1}{(s-2)^2+1} = e^{2t}\sin t$

(*vii*) $L^{-1}\frac{1}{(s+3)^2-4} = \frac{1}{2}\frac{2}{(s+3)^2-(2)^2} = \frac{1}{2}e^{-3t}\sinh 2t$

(*viii*) $L^{-1}\frac{s+2}{(s+2)^2-25} = L^{-1}\frac{(s+2)}{(s+2)^2-(5)^2} = e^{-2t}\cosh 5t$

Example 3. *Find* $L^{-1}\frac{s^2+2s+6}{s^3}$ *(M.D.U. 2010)*

Solution. Here, we have

$$L^{-1}\frac{s^2+2s+6}{s^3} = L^{-1}\left[\frac{1}{s}+\frac{2}{s^2}+\frac{6}{s^3}\right] = 1+\frac{2t}{1!}+\frac{6}{2!}t^2$$
$$= 1+2t+3t^2$$ **Ans.**

Example 4. *Find Inverse Laplace Transform of*

(*a*) $\left\{\frac{6}{2s-3}-\frac{3+4s}{9s^2-16}+\frac{8-6s}{16s^2+9}\right\}$ (*b*) $\frac{2s-5}{9s^2-25}$

(*c*) $\frac{s-2}{6s^2+20}$ *(U.P. II. Semester Summer* 2001)

Solution.

(*a*) $L^{-1}\left\{\frac{6}{2s-3}-\frac{3}{9s^2-16}-\frac{4s}{9s^2-16}+\frac{8}{16s^2+9}-\frac{6s}{16s^2+9}\right\}$

$$= L^{-1}\left\{\frac{3}{s-\frac{3}{2}} - \frac{\frac{1}{3}}{s^2-\left(\frac{4}{3}\right)^2} - \frac{\frac{4}{9}s}{s^2-\left(\frac{4}{3}\right)^2} + \frac{\frac{1}{2}}{s^2+\left(\frac{3}{4}\right)^2} - \frac{\frac{3}{8}s}{s^2+\left(\frac{3}{4}\right)^2}\right\}$$

$$= L^{-1}\left\{\frac{3}{s-\frac{3}{2}} - \frac{1}{4}\frac{\frac{4}{3}}{s^2-\left(\frac{4}{3}\right)^2} - \frac{4}{9}\frac{s}{s^2-\left(\frac{4}{3}\right)^2} + \frac{2}{3}\frac{\frac{3}{4}}{s^2+\left(\frac{3}{4}\right)^2} - \frac{3}{8}\frac{s}{s^2+\left(\frac{3}{4}\right)^2}\right\}$$

$$= 3e^{\frac{3}{2}t} - \frac{1}{4}\sinh\frac{4}{3}t - \frac{4}{9}\cosh\frac{4}{3}t + \frac{2}{3}\sin\frac{3}{4}t - \frac{3}{8}\cos\frac{3}{4}t \qquad \textbf{Ans.}$$

(b) $$L^{-1}\frac{2s-5}{9s^2-25} = L^{-1}\left[\frac{2s}{9s^2-25} - \frac{5}{9s^2-25}\right] = L^{-1}\left[\frac{2s}{9\left[s^2-\left(\frac{5}{3}\right)^2\right]} - \frac{5}{9\left[s^2-\left(\frac{5}{3}\right)^2\right]}\right]$$

$$= \frac{2}{9}\cosh\frac{5}{3}t - \frac{1}{3}L^{-1}\left(\frac{\frac{5}{3}}{s^2-\left(\frac{5}{3}\right)^2}\right) = \frac{2}{9}\cosh\frac{5t}{3} - \frac{1}{3}\sin\frac{5t}{3} \qquad \textbf{Ans.}$$

(c) $$L^{-1}\frac{s-2}{6s^2+20} = L^{-1}\frac{s}{6s^2+20} - L^{-1}\frac{2}{6s^2+20} = \frac{1}{6}L^{-1}\frac{s}{s^2+\frac{10}{3}} - \frac{1}{3}L^{-1}\frac{1}{s^2+\frac{10}{3}}$$

$$= \frac{1}{6}L^{-1}\frac{s}{s^2+\frac{10}{3}} - \frac{1}{3}\times\sqrt{\frac{3}{10}}\,L^{-1}\frac{\sqrt{\frac{10}{3}}}{s^2+\frac{10}{3}} = \frac{1}{6}\cos\sqrt{\frac{10}{3}}\,t - \frac{1}{\sqrt{30}}\sin\sqrt{\frac{10}{3}}\,t \qquad \textbf{Ans.}$$

Example 5. *Find the inverse Laplace transform of following function:*

$$\frac{14s+10}{49s^2+28s+13}.$$ [U.P., II Semester, 2007]

Solution. The given function can be written as

$$\frac{14s+10}{49s^2+28s+13} = \frac{14s+10}{(7s+2)^2+9} = \frac{14\left(s+\frac{2}{7}\right)+6}{49\left(s+\frac{2}{7}\right)^2+9}$$

$$\therefore \quad L^{-1}\left(\frac{14\,s+10}{49\,s^2+28\,s+13}\right) = L^{-1}\left[\frac{14\left(s+\frac{2}{7}\right)+6}{49\left(s+\frac{2}{7}\right)^2+9}\right] = e^{-\frac{2t}{7}}L^{-1}\left(\frac{14\,s+6}{49\,s^2+9}\right)$$

$$= e^{-\frac{2}{7}t}L^{-1}\frac{14}{49}\left(\frac{s+\frac{6}{14}}{s^2+\frac{9}{49}}\right) = e^{-\frac{2t}{7}}\left[\frac{14}{49}L^{-1}\left(\frac{s}{s^2+\frac{9}{49}}\right) + \left(\frac{14}{49}\right)\left(\frac{6}{14}\right)L^{-1}\left(\frac{1}{s^2+\frac{9}{49}}\right)\right]$$

$$= e^{-\frac{2t}{7}}\left[\frac{2}{7}\cos\frac{3}{7}t + \frac{6}{49}.\frac{7}{3}\sin\frac{3}{7}t\right]$$

$$= \frac{2}{7}e^{-\frac{2t}{7}}\left(\cos\frac{3}{7}t + \sin\frac{3}{7}t\right) \qquad \textbf{Ans.}$$

EXERCISE 43.1

Find the Inverse Laplace Transform of the following:

1. $\dfrac{3s-8}{4s^2+25}$ **Ans.** $\dfrac{3}{4}\cos\dfrac{5t}{2}-\dfrac{4}{5}\sin\dfrac{5t}{2}$

2. $\dfrac{3\left(s^2-2\right)^2}{2s^5}$ **Ans.** $\dfrac{3}{2}-3t^2+\dfrac{1}{2}t^4$

3. $\dfrac{2s-5}{4s^2+25}+\dfrac{4s-18}{9-s^2}$ **Ans.** $\dfrac{1}{2}\left(\cos\dfrac{5t}{2}-\sin\dfrac{5t}{2}\right)-4\cosh 3t+6\sinh 3t$

4. $\dfrac{5s-10}{9s^2-16}$ **Ans.** $\dfrac{5}{9}\cosh\dfrac{4}{3}t-\dfrac{5}{6}\sinh\dfrac{4}{3}t$

5. $\dfrac{1}{4s}+\dfrac{16}{1-s^2}$ **Ans.** $\dfrac{1}{4}-16\sinh t$

6. $L^{-1}\left\{\dfrac{1}{s^n}\right\}$ exist only when the value of n is :

   (*i*) Negative integer (*ii*) Positive integer
   (*iii*) Zero (*iv*) None of these **Ans.** (*ii*) (*U.P. II Semester, 2010*)

43.3 MULTIPLICATION BY *S*

$$L^{-1}[s\,F(s)]=\frac{d}{dt}f(t)+f(0)\,\delta(t)$$

Example 6. *Find the Inverse Laplace Transform of* (*i*) $\dfrac{s}{s^2+1}$ (*ii*) $\dfrac{s}{4s^2-25}$ (*iii*) $\dfrac{3s}{2s+9}$

Solution.

(*i*) $L^{-1}\dfrac{1}{s^2+1}=\sin t$

$$L^{-1}\frac{s}{s^2+1}=\frac{d}{dt}(\sin t)+\sin(0)\,\delta(t)=\cos t \qquad \textbf{Ans.}$$

(*ii*) $$L^{-1}\frac{1}{4s^2-25}=\frac{1}{4}L^{-1}\frac{1}{s^2-\frac{25}{4}}=\frac{1}{4}\cdot\frac{2}{5}L^{-1}\frac{\frac{5}{2}}{s^2-\left(\frac{5}{2}\right)^2}=\frac{1}{10}\sinh\frac{5}{2}t$$

$$L^{-1}\frac{s}{4s^2-25}=\frac{1}{10}\frac{d}{dt}\sinh\frac{5}{2}t+\frac{1}{10}\sinh\frac{5}{2}(0)\,\delta(t)$$

$$=\frac{1}{10}\left(\frac{5}{2}\right)\cosh\frac{5}{2}t=\frac{1}{4}\cosh\frac{5}{2}t \qquad \textbf{Ans.}$$

(*iii*) $$L^{-1}\frac{3}{2s+9}=\frac{3}{2}L^{-1}\frac{1}{s+\frac{9}{2}}=\frac{3}{2}e^{-\frac{9}{2}t}$$

$$L^{-1}\frac{3s}{2s+9}=\frac{3}{2}\frac{d}{dt}\left(e^{-\frac{9}{2}t}\right)+\frac{3}{2}e^{-\frac{9}{2}(0)\delta(t)}=\frac{3}{2}\left(-\frac{9}{2}\right)e^{-\frac{9}{2}t}+\frac{3}{2}=-\frac{27}{4}e^{-\frac{9}{2}t}+\frac{3}{2} \qquad \textbf{Ans.}$$

EXERCISE 43.2

Find the Inverse Laplace Transform of the following:

1. $\dfrac{s}{s+5}$ **Ans.** $-5e^{-5t}$ 2. $\dfrac{2s}{3s+6}$ **Ans.** $-\dfrac{4}{3}e^{-2t}$

3. $\dfrac{s}{2s^2-1}$ **Ans.** $\dfrac{1}{2}\cosh\dfrac{t}{2}$ 4. $\dfrac{s^2}{s^2+a^2}$ **Ans.** $-a\sin at+1$ 5. $\dfrac{s^2+4}{s^2+9}$ **Ans.** $-\dfrac{5}{3}\sin 3t+1$

43.4 DIVISION BY *s* (MULTIPLICATION BY $\frac{1}{s}$)

$$L^{-1}\left[\frac{F(s)}{s}\right] = \int_0^t \left[L^{-1}[F(s)]\right] dt = \int_0^t f(t)\, dt$$

Example 7. *Find the Inverse Laplace Transform of*

(i) $\dfrac{1}{s(s+a)}$ (ii) $\dfrac{1}{s(s^2+1)}$ (iii) $\dfrac{s^2+3}{s(s^2+9)}$

Solution.

(i) $L^{-1}\left(\dfrac{1}{s+a}\right) = e^{-at}$

$$L^{-1}\left[\frac{1}{s(s+a)}\right] = \int_0^t L^{-1}\left(\frac{1}{s+a}\right) dt = \int_0^t e^{-at}\, dt = \left[\frac{e^{-at}}{-a}\right]_0^t$$

$$= \frac{e^{-at}}{-a} + \frac{1}{a} = \frac{1}{a}\left[1 - e^{-at}\right] \quad \textbf{Ans.}$$

(ii) $L^{-1}\dfrac{1}{s^2+1} = \sin t$

$$L^{-1}\frac{1}{s}\left(\frac{1}{s^2+1}\right) = \int_0^t L^{-1}\left(\frac{1}{s^2+1}\right) dt = \int_0^t \sin t\, dt = [-\cos t]_0^t = -\cos t + 1 \quad \textbf{Ans.}$$

(iii) $L^{-1}\dfrac{s^2+3}{s(s^2+9)} = L^{-1}\left[\dfrac{s^2+9-6}{s(s^2+9)}\right] = L^{-1}\left[\dfrac{1}{s} - \dfrac{6}{s(s^2+9)}\right]$

$$= 1 - 2\int_0^t \sin 3t\, dt = 1 + 2\times\frac{1}{3}[\cos 3t\,]_0^t = 1 + \frac{2}{3}\cos 3t - \frac{2}{3}$$

$$= \frac{2}{3}\cos 3t + \frac{1}{3} = \frac{1}{3}[\,2\cos 3t + 1\,] \quad \textbf{Ans.}$$

EXERCISE 43.3

Find the Inverse Laplace Transform of the following:

1. $\dfrac{1}{2s(s-3)}$ **Ans.** $\dfrac{1}{2}\left[\dfrac{e^{3t}}{3} - 1\right]$
2. $\dfrac{1}{s(s+2)}$ **Ans.** $\dfrac{1-e^{-2t}}{2}$
3. $\dfrac{1}{s(s^2-16)}$ **Ans.** $\dfrac{1}{16}[\cosh 4t - 1]$
4. $\dfrac{1}{s(s^2+a^2)}$ **Ans.** $\dfrac{1-\cos at}{a^2}$
5. $\dfrac{s^2+2}{s(s^2+4)}$ **Ans.** $\cos^2 t$
6. $\dfrac{1}{s^2(s+1)}$ **Ans.** $t - 1 + e^{-t}$
7. $\dfrac{1}{s^3(s^2\ \ 1)}$ **Ans.** $\dfrac{t^2}{2} + \cos t - 1$

43.5 FIRST SHIFTING PROPERTY

If $L^{-1} F(s) = f(t)$, then $\boxed{L^{-1} F(s+a) = e^{-at} L^{-1}[F(s)]}$

Example 8. *Find the Inverse Laplace Transform of*

(i) $\dfrac{1}{(s+2)^5}$ (ii) $\dfrac{s}{s^2+4s+13}$ (iii) $\dfrac{1}{9s^2+6s+1}$

Solution.

(*i*) $L^{-1}\frac{1}{s^5}=\frac{t^4}{4!}$

then $L^{-1}\frac{1}{(s+2)^5}=e^{-2t}.\frac{t^4}{4!}$ **Ans.**

(*ii*) $L^{-1}\left(\frac{s}{s^2+4s+13}\right)=L^{-1}\frac{s+2-2}{(s+2)^2+(3)^2}=L^{-1}\frac{s+2}{(s+2)^2+(3)^2}-L^{-1}\frac{2}{(s+2)^2+(3)^2}$

$=e^{-2t}L^{-1}\frac{s}{s^2+3^2}-e^{-2t}L^{-1}\frac{2}{3}\left(\frac{3}{s^2+3^2}\right)=e^{-2t}\cos 3t-\frac{2}{3}e^{-2t}\sin 3t$ **Ans.**

(*iii*) $L^{-1}\frac{1}{9s^2+6s+1}=L^{-1}\frac{1}{(3s+1)^2}=\frac{1}{9}L^{-1}\frac{1}{\left(s+\frac{1}{3}\right)^2}=\frac{1}{9}e^{-t/3}L^{-1}\frac{1}{s^2}$

$=\frac{1}{9}e^{-t/3}t=\frac{t\,e^{-t/3}}{9}$ **Ans.**

Example 9. *Find the Inverse Laplace Transform of* $\frac{s+1}{s^2-6s+25}$

(*U.P., II Semester 2010*)

Solution. $L^{-1}\left(\frac{s+1}{s^2-6s+25}\right)=L^{-1}\left[\frac{s+1}{(s-3)^2+(4)^2}\right]=L^{-1}\left[\frac{s-3+4}{(s-3)^2+(4)^2}\right]$

$=L^{-1}\left[\frac{s-3}{(s-3)^2+(4)^2}\right]+L^{-1}\left[\frac{4}{(s-3)^2+(4)^2}\right]$

$=e^{3t}\cos 4t+e^{3t}\sin 4t.$ **Ans.**

EXERCISE 43.4

Obtain the Inverse Laplace Transform of the following:

1. $\frac{s+8}{s^2+4s+5}$ **Ans.** $e^{-2t}(\cos t+6\sin t)$
2. $\frac{s}{(s+3)^2+4}$ **Ans.** $e^{-3t}(\cos 2t-1.5\sin 2t)$
3. $\frac{s}{(s+7)^4}$ **Ans.** $e^{-7t}\frac{t^2}{6}(3-7t)$
4. $\frac{s+2}{s^2-2s-8}$ **Ans.** $e^{-t}(\cosh 3t+\sinh 3t)$
5. $\frac{s}{s^2+6s+25}$ **Ans.** $e^{-3t}\left[\cos 4t-\frac{3}{4}\sin 4t\right]$
6. $\frac{1}{2(s-1)^2+32}$ **Ans.** $\frac{e^t}{8}\sin 4t$
7. $\frac{s-4}{4(s-3)^2+16}$ **Ans.** $\frac{1}{4}e^{3t}\cos 2t-\frac{1}{8}e^{3t}\sin 2t$

43.6 SECOND SHIFTING PROPERTY

$$\boxed{L^{-1}\left[e^{-as}F(s)\right]=f(t-a)\,u(t-a)}$$

Example 10. *Obtain Inverse Laplace Transform of*

(*i*) $\frac{e^{-\pi s}}{(s+3)}$ (*ii*) $\frac{e^{-s}}{(s+1)^3}$

Solution.

(*i*) $L^{-1}\frac{1}{s+3}=e^{-3t},\quad L^{-1}\frac{e^{-\pi s}}{s+3}=e^{-3(t-\pi)}u(t-\pi)$ **Ans.**

(*ii*) $L^{-1}\dfrac{1}{s^3}=\dfrac{t^2}{2!} \quad\Rightarrow\quad L^{-1}\dfrac{1}{(s+1)^3}=e^{-t}\dfrac{t^2}{2!}$

$$L^{-1}\frac{e^{-s}}{(s+1)^3}=e^{-(t-1)}\cdot\frac{(t-1)^2}{2!}\cdot u(t-1)$$ **Ans.**

Example 11. *Evaluate*

$$L^{-1}\left[\frac{e^{-s}-3e^{-3s}}{s^2}\right]$$

(*U.P. II Semester, Summer 2002*)

Solution. $L^{-1}\left[\dfrac{e^{-s}-3e^{-3s}}{s^2}\right]=L^{-1}\left[\dfrac{e^{-s}}{s^2}-\dfrac{3e^{-3s}}{s^2}\right]$ ...(1)

We know that $L\,[u\,(t-a)]=\dfrac{e^{-as}}{s}$

and $L\,[(t-a)\,u(t-a)]=\dfrac{e^{-as}}{s^2}$

Using these results in (1), we get

$$\therefore\quad L^{-1}\left[\frac{e^{-s}-3e^{-3s}}{s^2}\right]=(t-1)\,u\,(t-1)-3(t-3)\,u\,(t-3)$$ **Ans.**

Example 12. *Find the Inverse Laplace Transform of*

$$\frac{se^{-s/2}+\pi e^{-s}}{s^2+\pi^2}$$

in terms of unit step functions.

Solution. $L^{-1}\dfrac{\pi}{s^2+\pi^2}=\sin\pi t$

$$L^{-1}\left[e^{-s}\frac{\pi}{s^2+\pi^2}\right]=\sin\pi(t-1).u(t-1)=-\sin(\pi t).u(t-1) \quad ...(1)$$

and $L^{-1}\dfrac{s}{s^2+\pi^2}=\cos\pi t$

$$L^{-1}\left[e^{-s/2}\frac{s}{s^2+\pi^2}\right]=\cos\pi\left(t-\frac{1}{2}\right).u\left(t-\frac{1}{2}\right)=\sin\pi t\,.\,u\left(t-\frac{1}{2}\right) \quad ...(2)$$

On adding (1) and (2), we get

$$L^{-1}\left[\frac{e^{-s/2}s+e^{-s}.\pi}{s^2+\pi^2}\right]=\sin(\pi t).u\left(t-\frac{1}{2}\right)-\sin(\pi t).u(t-1)$$

$$=\sin\pi t\left[u\left(t-\frac{1}{2}\right)-u\,(t-1)\right]$$ **Ans.**

Example 13. *Find the value of*

Solution $\dfrac{1}{(s^2+a^2)^2}=\dfrac{1}{s}\dfrac{s}{(s^2+a^2)^2}=-\dfrac{1}{2s}\dfrac{d}{ds}\left(\dfrac{1}{s^2+a^2}\right)$

$$\Rightarrow L^{-1}\left\{\frac{1}{(s^2+a^2)^2}\right\}=L^{-1}\left\{-\frac{1}{2s}\frac{d}{ds}\left(\frac{1}{s^2+a^2}\right)\right\}$$

$$=-\frac{1}{2s}\left\{-t\frac{1}{a}\sin at\right\}=-\frac{1}{2a}\frac{1}{s}\{t-\sin at\}=\frac{1}{2a}\int_0^t \sin at\,dt$$

$$= \frac{1}{2a}\left[t\left(\frac{-\cos at}{a}\right) - \int_0 \frac{\cos at}{a}\,dt\right]_0^t = \frac{1}{2a}\left[-\frac{t}{a}\cos at + \frac{\sin at}{a^2}\right]_0^t$$

$$= \frac{1}{2a^3}\left[-at\cos at + \sin at\right]$$ **Ans.**

EXERCISE 43.5

Obtain Inverse Laplace Transform of the following:

1. $\dfrac{e^{-s}}{(s+2)^3}$ **Ans.** $e^{-(t-2)}\dfrac{(t-2)^2}{2}u(t-2)$

2. $\dfrac{e^{-2s}}{(s+1)(s^2+2s+2)}$ **Ans.** $e^{-(t-2)}\,|\,1-\cos(t-2)\,|\,u(t-2)$

3. $\dfrac{e^{-s}}{\sqrt{s+1}}$ **Ans.** $\dfrac{e^{-(t-1)}}{\sqrt{\pi(t-1)}}.u(t-1)$

4. $\dfrac{e^{-\frac{\pi}{2}s} + e^{-\frac{3\pi}{2}s}}{s^2+1}$ **Ans.** $\cot t\left[u\left(t-\frac{3\pi}{2}\right) - u\left(t-\frac{\pi}{2}\right)\right]$

5. $\dfrac{e^{-4s}(s+2)}{s^2+4s+5}$ **Ans.** $e^{-2(t-u)}\cos(t-u)\,u(t-4)$

6. $\dfrac{e^{-as}}{s^2}$ **Ans.** $f(t) = t - a$, when $t > a$

$= 0$, when $t < a$

7. $\dfrac{e^{-\pi s}}{s^2+1}$ **Ans.** $-\sin t.\, u(t-\pi)$

43.7 INVERSE LAPLACE TRANSFORMS OF DERIVATIVES

$$L^{-1}\left[\frac{d}{ds}F(s)\right] = -t\,L^{-1}[F(s)] = -t\,f(t) \quad \Rightarrow \quad \boxed{L^{-1}[F(s)] = -\frac{1}{t}L^{-1}\left[\frac{d}{ds}F(s)\right]}$$

Example 14. *Find* $L^{-1}\left\{\log\dfrac{s+1}{s-1}\right\}$. (*Uttarakhand, II Semester, June 2010, 2009,* 2007)

Solution. $L^{-1}\left\{\log\left(\dfrac{s+1}{s-1}\right)\right\} = -\dfrac{1}{t}L^{-1}\left[\dfrac{d}{ds}\log\left(\dfrac{s+1}{s-1}\right)\right]$

$$= -\frac{1}{t}L^{-1}\left[\frac{d}{ds}\log(s+1) - \frac{d}{ds}\log(s-1)\right] = -\frac{1}{t}L^{-1}\left[\frac{1}{s+1} - \frac{1}{s-1}\right]$$

$$= -\frac{1}{t}\left[e^{-t} - e^{t}\right] = \frac{1}{t}\left[e^{t} - e^{-t}\right]$$ **Ans.**

Example 15. *Find the Inverse Laplace Transform of F(s)* $= \log\dfrac{s+a}{s+b}$

(*U.P., II Semester, Summer* 2003)

Solution. $L^{-1}\log\left(\dfrac{s+a}{s+b}\right) = -\dfrac{1}{t}L^{-1}\left[\dfrac{d}{ds}\log\dfrac{s+a}{s+b}\right]$

$$= -\frac{1}{t}L^{-1}\left[\frac{d}{ds}\log(s+a) - \frac{d}{ds}\log(s+b)\right] = -\frac{1}{t}L^{-1}\left[\frac{1}{s+a} - \frac{1}{s+b}\right]$$

$$= -\frac{1}{t}\left[e^{-at} - e^{-bt}\right] = \frac{1}{t}\left(e^{-bt} - e^{-at}\right)$$ **Ans.**

Example 16. *Obtain the Inverse Laplace Transform of log* $\dfrac{s^2-1}{s^2}$.

Solution. $L^{-1}\left[\log\dfrac{s^2-1}{s^2}\right]=-\dfrac{1}{t}L^{-1}\left[\dfrac{d}{ds}\log\dfrac{s^2-1}{s^2}\right]$

$=-\dfrac{1}{t}L^{-1}\left[\dfrac{d}{ds}\left\{\log\left(s^2-1\right)-2\log s\right\}\right]=-\dfrac{1}{t}L^{-1}\left[\dfrac{2s}{s^2-1}-\dfrac{2}{s}\right]=-\dfrac{1}{t}[\,2\cosh t-2\,]$

$=\dfrac{2}{t}[\,1-\cosh t\,]$ **Ans.**

Example 17. *Find the function whose Laplace transform is*

log $\left(1+\dfrac{1}{s}\right)$ *(U.P., II Semester, June 2007)*

Solution. $L^{-1}\left[\log\left(1+\dfrac{1}{s}\right)\right]=\dfrac{1}{t}L^{-1}\left[\dfrac{d}{ds}\log\left(\dfrac{s+1}{s}\right)\right]$

$=-\dfrac{1}{t}L^{-1}\left[\left(\dfrac{s}{s+1}\right)\left(-\dfrac{1}{s^2}\right)\right]=-\dfrac{1}{t}L^{-1}\left[-\dfrac{1}{s\,(s+1)}\right]$

$=-\dfrac{1}{t}L^{-1}\left[\dfrac{1}{s+1}-\dfrac{1}{s}\right]$ (Partial fraction)

$=-\dfrac{1}{t}[e^{-t}-1]=\dfrac{1}{t}[1-e^{-t}]$ **Ans.**

Example 18. *Find the inverse Laplace transform of* $\tan^{-1}\left(\dfrac{2}{s^2}\right)$

Solution. Here, we have $L^{-1}\left[\tan^{-1}\left(\dfrac{2}{s^2}\right)\right]=-\dfrac{1}{t}L^{-1}\left[\dfrac{d}{ds}\tan^{-1}\dfrac{2}{s^2}\right]$

$=-\dfrac{1}{t}L^{-1}\left[\dfrac{1}{1+\dfrac{4}{s^4}}\left(-\dfrac{4}{s^3}\right)\right]=-\dfrac{1}{t}L^{-1}\left[\dfrac{s^4}{s^4+4}\left(-\dfrac{4}{s^3}\right)\right]=\dfrac{1}{t}L^{-1}\left[\dfrac{4s}{s^4+4}\right]$

$=\dfrac{4}{t}L^{-1}\left[\dfrac{s}{s^4+4}\right]=\dfrac{4}{t}L^{-1}\left[\dfrac{s}{(s^2+2s+2)\,(s^2-2s+2)}\right]$

$=\dfrac{4}{t}L^{-1}\left[-\dfrac{1}{4}\dfrac{1}{(s^2+2s+2)}+\dfrac{1}{4}\dfrac{1}{(s^2-2s+2)}\right]$ (By partial fraction)

$=\dfrac{1}{t}L^{-1}\left[-\dfrac{1}{(s^2+2s+2)}+\dfrac{1}{(s^2-2s+2)}\right]=\dfrac{1}{t}L^{-1}\left[-\dfrac{1}{(s+1)^2+1}+\dfrac{1}{(s-1)^2+1}\right]$

$=\dfrac{1}{t}\left[-e^{-t}\sin t+e^{t}\sin t\right]=\dfrac{\sin t}{t}\left[e^{t}-e^{-t}\right]$ **Ans.**

Example 19. *Find Inverse Laplace Transform of tan*$^{-1}\dfrac{1}{s}$.

Solution. $L^{-1}\left(\tan^{-1}\dfrac{1}{s}\right)=-\dfrac{1}{t}L^{-1}\left[\dfrac{d}{ds}\tan^{-1}\dfrac{1}{s}\right]$ *(M.D.U., 2010)*

$=-\dfrac{1}{t}L^{-1}\left[\dfrac{1}{1+\dfrac{1}{s^2}}\left(-\dfrac{1}{s^2}\right)\right]=\dfrac{1}{t}L^{-1}\left[\dfrac{1}{1+s^2}\right]=\dfrac{\sin t}{t}$ **Ans.**

Example 20. *Find* $L^{-1}\left[\tan^{-1}(1+s)\right]$ *(M.D.U. 2010)*

Solution. $L^{-1}\left[\tan^{-1}(1+s)\right] = -\frac{1}{t}L^{-1}\left[\frac{d}{ds}\tan^{-1}(1+s)\right]$

$$= -\frac{1}{t}L^{-1}\left[\frac{1}{1+(s+1)^2}\right] = -\frac{1}{t}L^{-1}\left[\frac{1}{(s+1)^2+1}\right]$$

$$= -\frac{1}{t}e^{-t}\sin t$$ **Ans.**

Example 21. *Find the inverse Laplace transform of*

$$\cot^{-1}\left(\frac{s}{2}\right)$$ *(Q.Bank U.P. 2001)*

Solution.

Let $L^{-1}\left[\cot^{-1}\left(\frac{s}{2}\right)\right] = f(t) \Rightarrow L^{-1}\left[\frac{d}{ds}\cot^{-1}\left(\frac{s}{2}\right)\right] = -t\,f(t)$

$$\Rightarrow \quad L^{-1}\left[\frac{-1}{1+\frac{s^2}{4}}\,\frac{1}{2}\right] = -t\,f(t) \Rightarrow L^{-1}\left[\frac{2}{s^2+2}\right] = t\,f(t)$$

$$\Rightarrow \quad \sin 2t = t\,f(t) \quad \Rightarrow \quad f(t) = \frac{1}{t}\sin 2t$$ **Ans.**

Example 22. *Obtain the Inverse Laplace Transform of* $\cot^{-1}\left(\frac{s+3}{2}\right)$

(U. P., II Semester, Summer 2002)

Solution. We know that $L^{-1}\left[F(s)\right] = -\frac{1}{t}L^{-1}\left[\frac{d}{ds}F(s)\right]$

$$\therefore \quad L^{-1}\left[\cot^{-1}\left(\frac{s+3}{2}\right)\right] = -\frac{1}{t}L^{-1}\left[\frac{d}{ds}\cot^{-1}\left(\frac{s+3}{2}\right)\right]$$

$$= -\frac{1}{t}L^{-1}\left\{\frac{-\frac{1}{2}}{1+\left(\frac{s+3}{2}\right)^2}\right\} = \frac{1}{2t}L^{-1}\left\{\frac{4}{4+(s+3)^2}\right\}$$

$$= \frac{1}{t}L^{-1}\left\{\frac{2}{2^2+(s+3)^2}\right\} = \frac{1}{t}e^{-3t}L^{-1}\left(\frac{2}{2^2+s^2}\right)$$

$$= \frac{e^{-3t}}{t}\sin 2t$$ **Ans.**

Example 23. *Find the inverse Laplace transform of* $\frac{2as}{(s^2+a^2)^2}$

Solution. $L^{-1}\left(\frac{a}{s^2+a^2}\right) = \sin at$

$$L^{-1}\left[\frac{d}{ds}\left\{\frac{a}{s^2+a^2}\right\}\right] = -t\sin at \Rightarrow L^{-1}\left\{\frac{-2as}{(s^2+a^2)^2}\right\} = -t\sin at$$

$$\Rightarrow \qquad L^{-1}\left\{\frac{2as}{(s^2+a^2)^2}\right\} = t \sin at \qquad \textbf{Ans.}$$

Example 24. *Find the inverse Laplace transform of* $\dfrac{s^2-a^2}{(s^2+a^2)^2}$

Solution. We know that

$$L^{-1}\left\{\frac{s}{s^2+a^2}\right\} = \cos at$$

$$\therefore \qquad L^{-1}\left[\frac{d}{ds}\left\{\frac{a}{s^2+a^2}\right\}\right] = -t\cos at$$

$$\Rightarrow \quad L^{-1}\left\{\frac{(s^2+a^2).1 - s\,(2s)}{(s^2+a^2)}\right\} = -t\cos at \quad \Rightarrow \quad L^{-1}\left[\frac{a^2-s^2}{(s^2+a^2)^{\,2}}\right] = -t\cos at$$

$$\therefore \qquad L^{-1}\left[\frac{s^2-a^2}{(s^2+a^2)^{\,2}}\right] = t\cos at \qquad \textbf{Ans.}$$

EXERCISE 43.6

Obtain Inverse Laplace Transform of the following:

1. $\log\left(1+\dfrac{\omega^2}{s^2}\right)$ **Ans.** $-\dfrac{2}{t}\cos\omega t + 2$
2. $\log\left(1+\dfrac{1}{s^2}\right)$ **Ans.** $\dfrac{2}{t}[1-\cos\omega t]$
3. $\dfrac{s}{1+s^2+s^4}$ **Ans.** $\dfrac{2}{\sqrt{3}}\sin\dfrac{\sqrt{3}}{2}t\sinh\dfrac{t}{2}$
4. $\dfrac{s}{\left(s^2+a^2\right)^2}$ **Ans.** $\dfrac{t\sin at}{2a}$
5. $s\log\dfrac{s}{\sqrt{s^2+1}}+\cot^{-1}s$ **Ans.** $\dfrac{1-\cos t}{t^2}$
6. $\dfrac{1}{2}\log\left\{\dfrac{s^2+b^2}{(s-a)^2}\right\}$ **Ans.** $\dfrac{e^{-at}-\cos bt}{t}$

43.8 INVERSE LAPLACE TRANSFORM OF INTEGRALS

$$L^{-1}\left[\int_s^\infty F(s)\,ds\right] = \frac{f(t)}{t} = \frac{1}{t}L^{-1}[F(s)] \quad \text{or} \quad \boxed{L^{-1}[F(s)] = t\,L^{-1}\left[\int_s^\infty F(s)\,ds\right]}$$

Example 25. *Obtain* $L^{-1}\dfrac{2s}{\left(s^2+1\right)^2}$

Solution. $$L^{-1}\frac{2s}{\left(s^2+1\right)^2} = t\,L^{-1}\int_s^\infty \frac{2s\,ds}{\left(s^2+1\right)^2} = t\,L^{-1}\left[-\frac{1}{s^2+1}\right]_s^\infty = t\,L^{-1}\left[-0+\frac{1}{s^2+1}\right]$$

$$= t\sin t \qquad \textbf{Ans.}$$

43.9 PARTIAL FRACTIONS METHOD

Example 26. *Find the Inverse Laplace Transform of* $\dfrac{1}{s^2-5s+6}$.

Solution. Let us convert the given function into partial fractions.

$$L^{-1}\left[\frac{1}{s^2-5s+6}\right] = L^{-1}\left[\frac{1}{s-3}-\frac{1}{s-2}\right]$$

$$= L^{-1}\left(\frac{1}{s-3}\right) - L^{-1}\left(\frac{1}{s-2}\right) = e^{3t}-e^{2t} \qquad \textbf{Ans.}$$

Example 27. *Find the inverse Laplace transform of*

$$\frac{s^3}{s^4 - a^4}$$ *(Q. Bank U.P. 2001)*

Solution. Here, we have

$$L^{-1}\left(\frac{s^3}{s^4 - a^4}\right) = L^{-1}\left[s\left\{\frac{s^2}{(s^2 - a^2)(s^2 + a^2)}\right\}\right] = L^{-1}\left[\frac{s}{2}\left(\frac{1}{s^2 - a^2} + \frac{1}{s^2 + a^2}\right)\right]$$

$$= \frac{1}{2} L^{-1}\left(\frac{s}{s^2 - a^2} + \frac{s}{s^2 + a^2}\right) \quad \text{(By partial fractions)}$$

$$= \frac{1}{2}(\cosh at + \cos at)$$ **Ans.**

Example 28. *Find the Inverse Laplace Transforms of* $\dfrac{s+4}{s(s-1)(s^2+4)}$.

Solution. Let us first resolve $\dfrac{s+4}{s(s-1)(s^2+4)}$ into partial fractions.

$$\frac{s+4}{s(s-1)(s^2+4)} \equiv \frac{A}{s} + \frac{B}{s-1} + \frac{Cs+D}{s^2+4} \quad \text{... (1)}$$

$$s + 4 \equiv A(s-1)(s^2+4) + Bs(s^2+4) + (Cs+D)s(s-1)$$

Putting $s = 0$, we get $4 = -4A \quad \Rightarrow \quad A = -1$

Putting $s = 1$, we get $5 = B \,.\, 1 \,.\, (1 + 4) \quad \Rightarrow B = 1$

Equating the coefficients of s^3 on both sides of (1), we have

$$0 = A + B + C \quad \Rightarrow \quad 0 = -1 + 1 + C \quad \Rightarrow \quad C = 0.$$

Equating the coefficients of s on both sides of (1), we get

$$1 = 4A + 4B - D \quad \Rightarrow \quad 1 = -4 + 4 - D \quad \Rightarrow \quad D = -1.$$

On putting the values of A, B, C, D in (1), we get

$$\frac{s+4}{s(s-1)(s^2+4)} = -\frac{1}{s} + \frac{1}{s-1} - \frac{1}{s^2+4}$$

$$\therefore \quad L^{-1}\left[\frac{s+4}{s(s-1)(s^2+4)}\right] = L^{-1}\left[-\frac{1}{s} + \frac{1}{s-1} - \frac{1}{s^2+4}\right]$$

$$= -L^{-1}\left(\frac{1}{s}\right) + L^{-1}\left(\frac{1}{s-1}\right) - \frac{1}{2}L^{-1}\left(\frac{2}{s^2+2^2}\right) = -1 + e^t - \frac{1}{2}\sin 2t.$$ **Ans.**

Example 29. *Find the inverse Laplace transform of*

$$\frac{1}{s^4 + 4}$$ *[U.P., II Semester, (SUM) 2007]*

Solution. Here, we have

$$s^4 + 4 = (s^2 + 2)^2 - (2s)^2 = (s^2 - 2s + 2)(s^2 + 2s + 2)$$

$$\frac{1}{s^4 + 4} = \frac{1}{(s^2 - 2s + 2)(s^2 + 2s + 2)}$$

$$= \frac{1}{4s}\left[\frac{1}{s^2 - 2s + 2} - \frac{1}{s^2 + 2s + 2}\right] \quad \left(\text{By partial fractions}\right) \quad \text{...(1)}$$

Now, $$L^{-1}\left(\frac{1}{s^2-2s+2}\right) = L^{-1}\left[\frac{1}{(s-1)^2+1}\right] = e^t \sin t$$

and $$L^{-1}\left(\frac{1}{s^2+2s+2}\right) = L^{-1}\left[\frac{1}{(s+1)^2+1}\right] = e^{-t} \sin t$$

$$\therefore \quad \frac{1}{4}L^{-1}\left(\frac{1}{s^2-2s+2}-\frac{1}{s^2+2s+2}\right) = \frac{1}{4}(e^t - e^{-t}) \sin t$$

Hence, $$L^{-1}\left[\frac{1}{4s}\left(\frac{1}{s^2-2s+2}-\frac{1}{s^2+2s+2}\right)\right] = \frac{1}{4}\int_0^t (e^t - e^{-t}) \sin t \, dt$$

$$\Rightarrow \quad L^{-1}\left(\frac{1}{s^2+4}\right) = \frac{1}{4}\left[\frac{e^t}{2}(\sin t - \cos t) - \frac{e^{-t}}{2}(-\sin t - \cos t)\right]$$

$$= \frac{1}{4}\left[\sin t\left(\frac{e^t+e^{-t}}{2}\right) - \cos t\left(\frac{e^t-e^{-t}}{2}\right)\right]$$

$$\Rightarrow \quad L^{-1}\left(\frac{1}{s^2+4}\right) = \frac{1}{4}[\sin t \cosh t - \cos t \sinh t]$$ **Ans.**

Example 30. *Find the inverse Laplace transform of*

$$\frac{s}{s^4+4a^4}$$

Solution.
$$s^4 + 4a^4 = (s^2 + 2a^2)^2 - (2as)^2 = (s^2 - 2as + 2a^2)(s^2 + 2as + 2a^2)$$
$$= \{(s-a)^2 + a^2\}\{(s+a)^2 + a^2\}$$

$$\frac{s}{s^4+4a^4} = \frac{s}{\{(s-a)^2+a^2\}\{(s+a)^2+a^2\}}$$

$$= \frac{1}{4a}\left[\frac{1}{(s-a)^2+a^2} - \frac{1}{(s+a)^2+a^2}\right] \quad \text{(By partial fraction)}$$

$$\therefore \quad L^{-1}\left(\frac{s}{s^4+4a^4}\right) = \frac{1}{4a}\left[L^{-1}\left\{\frac{1}{(s-a)^2+a^2}\right\} - L^{-1}\left\{\frac{1}{(s+a)^2+a^2}\right\}\right]$$

$$= \frac{1}{4a}\left[\frac{1}{a}e^{at}\sin at - e^{-at}\frac{1}{a}\sin at\right]$$

$$= \frac{1}{2a^2}\sin at\left(\frac{e^{at}-e^{-at}}{2}\right) = \frac{1}{2a^2}\sin at \sinh at.$$ **Ans.**

Example 31. *Find the Inverse Laplace Transform of* $\frac{e^{-cs}}{s^2(s+a)}$, $c > 0$.

(U.P. II Semester, Summer 2002)

Solution. We have,

$$L^{-1}\left[\frac{e^{-cs}}{s^2(s+a)}\right] = L^{-1}\left[-\frac{e^{-cs}}{a^2 s} + \frac{e^{-cs}}{as^2} + \frac{e^{-cs}}{a^2(s+a)}\right] \quad \text{(By Partial fractions)}$$

$$= L^{-1}\left[\left(\frac{-1}{a^2}\frac{e^{-cs}}{s}\right) + \left(\frac{1}{a}\right)\frac{e^{-cs}}{s^2} + \left(\frac{1}{a^2}\right)\frac{e^{-c(s+a)}}{e^{-ca}(s+a)}\right]$$

$$= -\frac{1}{a^2}u(t-c) + \frac{1}{a}(t-c)\,u(t-c) + \frac{1}{a^2e^{-ca}}e^{at}\,u(t-c)$$

$$= u(t-c)\left[\frac{-1}{a^2}+\frac{1}{a}(t-c)+\frac{1}{a^2}e^{a(c+t)}\right]$$, where $u(t-c)$ is unit step function. **Ans.**

Example 32. *Find the Inverse Laplace Transform of*

$$\frac{5s+3}{(s-1)(s^2+2s+5)}$$ *(U.P. II Semester Summer 2005)*

Solution. $L^{-1}\left\{\frac{5s+3}{(s-1)(s^2+2s+5)}\right\}$

Let $$\frac{5s+3}{(s-1)(s^2+2s+5)}=\frac{A}{s-1}+\frac{Bs+C}{s^2+2s+5}$$

$$5s+3=A(s^2+2s+5)+(Bs+C)(s-1)$$
$$5s+3=s^2(A+B)+s(2A-B+C)+(5A-C)$$

Comparing the coefficients of s^2, s and constant, we get

$$A+B=0 \quad \ldots(1)$$
$$2A-B+C=5 \quad \ldots(2)$$
$$5A-C=3 \quad \ldots(3)$$

On adding equations (1) and (2), we have $3A+C=5$ ... (4)

Adding equations (3) and (4), we get $8A=8 \quad \Rightarrow \quad A=1$

Putting A = 1 in (3), we get $C=2$

Putting $A=1$, $C=2$ in (2), we get

$$B=-1$$

Thus $$\frac{5s+3}{(s-1)(s^2+2s+5)}=\frac{1}{s-1}+\frac{-s+2}{s^2+2s+5}=\frac{1}{s-1}-\frac{s-2}{(s+1)^2+2^2}$$

$$=\frac{1}{s-1}-\frac{s+1}{(s+1)^2+2^2}+\frac{3}{(s+1)^2+2^2}$$

$$L^{-1}\left\{\frac{5s+3}{(s-1)(s^2+2s+5)}\right\}=L^{-1}\left\{\frac{1}{s-1}\right\}+L^{-1}\left\{\frac{3}{(s+1)^2+2^2}\right\}-L^{-1}\left\{\frac{s+1}{(s+1)^2+2^2}\right\}$$

$$=e^t+3e^{-t}L^{-1}\left\{\frac{1}{s^2+2^2}\right\}-e^{-t}L^{-1}\left\{\frac{s}{s^2+2^2}\right\}$$

$$=e^t+3e^{-t}\cdot\frac{1}{2}\sin 2t-e^{-t}\cos 2t$$ **Ans.**

Example 33. *Find the Inverse Laplace Transform of* $\frac{s^2}{(s^2+a^2)(s^2+b^2)}$.

Solution. Let us convert the given function into partial fractions.

$$L^{-1}\left[\frac{s^2}{(s^2+a^2)(s^2+b^2)}\right]=L^{-1}\left[\frac{a^2}{a^2-b^2}\cdot\frac{1}{s^2+a^2}-\frac{b^2}{a^2-b^2}\cdot\frac{1}{s^2+b^2}\right]$$

$$=\frac{1}{a^2-b^2}L^{-1}\left[\frac{a^2}{s^2+a^2}-\frac{b^2}{s^2+b^2}\right]=\frac{1}{a^2-b^2}\left[a^2\left(\frac{1}{a}\sin at\right)-b^2\left(\frac{1}{b}\sin bt\right)\right]$$

$$=\frac{1}{a^2-b^2}[a\sin at-b\sin bt]$$ **Ans.**

Note: This question is also solved by using the Convolution Theorem as an example 37.

EXERCISE 43.7

Find the Inverse Laplace Transforms of the following by partial fractions method:

1. $\dfrac{1}{s^2-7s+12}$ **Ans.** $e^{4t}-e^{3t}$

2. $\dfrac{s+2}{s^2-4s+13}$ **Ans.** $e^{2t}\cos 3t+\dfrac{4}{3}e^{2t}\sin 3t$

3. $\dfrac{3s+1}{(s-1)(s^2+1)}$ **Ans.** $e^t-2\cos t+\sin t$

4. $\dfrac{11s^2-2s+5}{2s^3-3s^2-3s+2}$ **Ans.** $2e^{-t}+5e^{2t}-\dfrac{3}{2}e^{t/2}$

5. $\dfrac{2s^2-6s+5}{(s-1)(s-2)(s-3)}$ **Ans.** $\dfrac{1}{2}e^t-e^{2t}+\dfrac{5}{2}e^{3t}$

6. $\dfrac{s-4}{(s-4)^2+9}$ **Ans.** $e^{4t}\cos 3t$

7. $\dfrac{16}{(s^2+2s+5)^2}$ **Ans.** $e^{-t}(\sin 2t-2t\cos 2t)$

8. $\dfrac{1}{(s+1)(s^2+2s+2)}$ **Ans.** $e^{-t}(1-\cos t)$

9. $\dfrac{1}{(s-2)(s^2+1)}$ **Ans.** $\dfrac{1}{5}e^{2t}-\dfrac{1}{5}\cos t-\dfrac{2}{5}\sin t$

10. $\dfrac{s^2-6s+7}{(s^2-4s+5)^2}$ **Ans.** $te^{2t}\{\cos t-\sin t\}$

43.10 INVERSE LAPLACE TRANSFORM BY CONVOLUTION

$$L\left\{\int_0^t f_1(x)*f_2(t-x)\,dx\right\}=F_1(s).F_2(s) \text{ or } \boxed{\int_0^t f_1(x).f_2(t-x)\,dx=L^{-1}[F_1(s).F_2(s)]}$$

Example 34. *Use convolution theorem to evaluate:*

$$L^{-1}\left\{\frac{s}{(s^2+4)^2}\right\}$$ (*U.P., II Semester, 2010*)

Solution. $$\frac{s}{(s^2+4)^2}=\frac{1}{s^2+4}\cdot\frac{s}{s^2+4}$$

Let $$F_1(s)=\frac{1}{s^2+4} \text{ and } F_2(s)=\frac{s}{s^2+4}$$

and $$L^{-1}[F_1(s)]=L^{-1}\left(\frac{1}{s^2+4}\right)=\frac{1}{2}\sin 2t$$

and $$L^{-1}[F_2(s)]=L^{-1}\left(\frac{s}{s^2+4}\right)=\cos 2t$$

According to Convolution Theorem

$$L^{-1}[F_1(s).F_2(s)]=\int_0^t f_1(x).f_2(t-x)dx=\int_0^t \frac{1}{2}\sin 2x\cos 2(t-x)dx$$

$$=\frac{1}{4}\int_0^t[\sin(2x+2t-2x)+\sin(2x-2t+2x)]\,dx=\frac{1}{4}\int_0^t[\sin 2t+\sin(4x-2t)]dx$$

$$=\frac{1}{4}[x\sin 2t-\frac{1}{4}\cos(4x-2t)]_0^t=\frac{1}{4}\left[t\sin 2t-\frac{1}{4}\cos(4t-2t)+\frac{1}{4}\cos(-2t)\right]$$

$$=\frac{1}{4}\left[t\sin 2t-\frac{1}{4}\cos 2t+\frac{1}{4}\cos 2t\right]=\frac{1}{4}\sin 2t$$ **Ans.**

Example 35. *Use convolution theorem to find the inverse of the function* $\dfrac{1}{(s^2+a^2)^2}$.

Solution. We know that (*U.P., II Semester, 2009*)

$$L^{-1}\left[\frac{1}{s^2+a^2}\right]=\frac{1}{a}\sin at$$

Hence by convolution theorem

$$L^{-1}\frac{1}{(s^2+a^2)}\frac{1}{(s^2+a^2)}=\int_0^t\frac{1}{a}\sin ax\frac{1}{a}\sin a(t-x)\,dx$$

$$= \frac{1}{a^2}\int_0^t \frac{1}{2}[\cos(ax - at + ax) - \cos(ax + at - ax)]\, dx \quad \left\{\sin A \sin B = \frac{1}{2}[\cos(A-B) - \cos(A+B)]\right\}$$

$$= \frac{1}{2a^2}\int_0^t [\cos(2ax - at) - \cos at]\, dx = \frac{1}{2a^2}\left[\frac{1}{2a}\sin(2ax - at) - x\cos at\right]_0^t$$

$$= \frac{1}{2a^2}\left[\frac{1}{2a}\sin(2at - at) - t\cos at - \frac{1}{2a}\sin(-at)\right] = \frac{1}{2a^2}\left[\frac{1}{2a}\sin at - t\cos at + \frac{1}{2a}\sin at\right]$$

$$= \frac{1}{2a^2}\left[\frac{2}{2a}\sin at - t\cos at\right] = \frac{1}{2a^3}[\sin at - at\cos at]$$ **Ans.**

Example 36. *State convolution theorem and hence find*

$$L^{-1}\left\{\frac{1}{(s+2)^2(s-2)}\right\}$$

(Uttarakhand, II Semester, June 2007)

Solution. Convolution Theorem (See Art 43.10 on page 1185)

Let $L\{f_1(t)\} = F_1(s)$ and Let $L\{f_2(t)\} = F_2(s)$

$$F_1(s) = \frac{1}{(s+2)^2} \quad \text{and} \quad F_2(s) = \frac{1}{s-2}$$

$$f_1(t) = L^{-1}\left[\frac{1}{(s+2)^2}\right] = t\, e^{-2t}$$

$$f_2(t) = L^{-1}\left[\frac{1}{(s-2)}\right] = e^{2t}$$

According to Convolution Theorem

$$L^{-1}[F_1(s).F_2(s)] = \int_0^t f_1(x) f_2(t-x)\, dx$$

$$L^{-1}\left[\frac{1}{(s+2)^2(s-2)}\right] = \int_0^t x\, e^{-2x}\cdot e^{2(t-x)}\, dx = \int_0^t x e^{2t-4x}\, dx$$

$$= \left[x\frac{e^{2t-4x}}{-4} - \int 1.\frac{e^{(2t-4x)}}{-4}\, dx\right]_0^t = \left[-\frac{x}{4}e^{2t-4x} + \frac{1}{4}\left\{\frac{e^{(2t-4x)}}{-4}\right\}\right]_0^t$$

$$= \frac{-t}{4}e^{2t-4t} - \frac{1}{16}e^{2t-4t} + \frac{1}{16}e^{2t} = \frac{-t}{4}e^{-2t} - \frac{1}{16}e^{-2t} + \frac{1}{16}e^{2t}$$

$$= \frac{e^{2t}}{16} - \frac{1}{16}e^{-2t}[4t+1]$$ **Ans.**

Example 37. *Using the Convolution Theorem find*

$$L^{-1}\left\{\frac{s^2}{(s^2+a^2)(s^2+b^2)}\right\} \quad a \neq b.$$

(M.D.U., 2009, U.P. II Semester Summer 2006, 2004)

Solution. We have, $L(\cos at) = \frac{s}{s^2+a^2}$ and $L(\cos bt) = \frac{s}{s^2+b^2}$

Hence, by the convolution theorem

$$L\left\{\int_0^t \cos ax \cos b(t-x)\, dx\right\} = \frac{s^2}{(s^2+a^2)(s^2+b^2)}$$

Therefore,

$$L^{-1}\left\{\frac{s^2}{(s^2+a^2)(s^2+b^2)}\right\} = \int_0^t \cos ax \cos b(t-x)\, dx$$

$$= \frac{1}{2}\int_0^t \{\cos(ax+bt-bx)+\cos(ax-bt+bx)\}\,dx$$

$$= \frac{1}{2}\int_0^t \cos\left[(a-b)x+bt\right]dx + \frac{1}{2}\int_0^t \cos\left[(a+b)x-bt\right]dx$$

$$= \left[\frac{\sin[(a-b)x+bt]}{2(a-b)}\right]_0^t + \left[\frac{\sin[(a+b)x-bt]}{2(a+b)}\right]_0^t = \frac{\sin at - \sin bt}{2(a-b)} + \frac{\sin at + \sin bt}{2(a+b)}$$

$$= \frac{a\sin at - b\sin bt}{a^2-b^2}$$ **Ans.**

Example 38. *Evaluate* $L^{-1}\left\{\frac{s}{(s^2+1)(s^2+4)}\right\}$ (*U.P., II Semester, Summer* 2002)

Solution. We know that $L^{-1}\frac{s}{s^2+1} = \cos x$ and $L^{-1}\frac{2}{s^2+2^2} = \sin 2x$

$$L^{-1}\left(\frac{s}{(s^2+1)(s^2+4)}\right) = \frac{1}{2}L^{-1}\left[\left(\frac{s}{s^2+1}\right)\left(\frac{2}{s^2+4}\right)\right]$$

$$= \frac{1}{2}\int_0^t \sin 2x\cos(t-x)\,dx$$ [By Convolution Th.]

$$= \int_0^t \sin x\cos x\,\{\cos t\cos x + \sin t\sin x\}\,dx = \int_0^t\left[\sin x\cos^2 x\cos t + \sin^2 x\cos x\sin t\right]dx$$

$$= \left[-\frac{\cos^3 x}{3}\cos t + \frac{\sin^3 x}{3}\sin t\right]_0^t = -\frac{\cos^4 t}{3} + \frac{\sin^4 t}{3} + \frac{\cos t}{3} = \frac{1}{3}\left[\sin^4 t - \cos^4 t\right] + \frac{\cos t}{3}$$

$$= \frac{1}{3}(\sin^2 t + \cos^2 t)(\sin^2 t - \cos^2 t) + \frac{\cos t}{3} = \frac{1}{3}(\sin^2 t - \cos^2 t) + \frac{\cos t}{3} = -\frac{1}{3}\cos 2t + \frac{\cos t}{3}$$

$$= \frac{1}{3}(\cos t - \cos 2t)$$ **Ans.**

Example 39. *Obtain* $L^{-1}\frac{1}{s(s^2+a^2)}$.

Solution. $L^{-1}\frac{1}{s} = 1$ and $L^{-1}\frac{1}{s^2+a^2} = \frac{\sin at}{a}$.

$$L^{-1}\{F_1(s).F_2(s)\} = \int_0^t f_1(t)\,f_2(t-x)\,dx$$ (Convolution Theorem)

Hence by the Convolution Theorem

$$L^{-1}\left[\frac{1}{s}.\frac{1}{s^2+a^2}\right] = \int_0^t \frac{\sin a(t-x)}{a}\,dx = \left[\frac{-\cos(at-ax)}{-a^2}\right]_0^t$$

$$= \frac{1}{a^2}[1-\cos at]$$ **Ans.**

Example 40. *Using Convolution Theorem, prove that*

$$L^{-1}\left[\frac{1}{s^3(s^2+1)}\right] = \frac{t^2}{2} + \cos t - 1$$ (*U.P., II Semester, Summer* 2005)

Solution. We know that,

$$L^{-1}\left\{\frac{1}{s^3}\right\} = \frac{t^2}{2!}$$

$$L^{-1}\left\{\frac{1}{s^2+1}\right\} = \sin t$$

Using Convolution Theorem,

$$L^{-1}\left\{\frac{1}{s^3\left(s^2+1\right)}\right\} = \int_0^t \frac{(t-x)^2}{2!}\sin x\,dx$$

$$= \frac{1}{2}\int_0^t \left(t^2+x^2-2tx\right)\sin x\,dx = \frac{1}{2}\left[\left(t^2+x^2-2tx\right)(-\cos x) - \int(2x-2t)(-\cos x)\,dx\right]_0^t$$

$$= \frac{1}{2}\left[(t^2+x^2-2tx)(-\cos x) + 2\int(x-t)\cos x\,dx\right]_0^t$$

$$= \frac{1}{2}\left[(t^2+x^2-2tx)(-\cos x) + 2(x-t)\sin x + 2\cos x\right]_0^t$$

$$= \frac{1}{2}\left[(t^2+t^2-2t^2)(-\cos t) + 0 + 2\cos t + t^2\cos 0 - 2\cos 0\right]$$

$$= \frac{1}{2}[2\cos t + t^2 - 2] = \cos t + \frac{t^2}{2} - 1 = \frac{t^2}{2} + \cos t - 1$$ **Ans.**

EXERCISE 43.8

Obtain the Inverse Laplace Transform by convolution:

1. $\dfrac{s^2}{\left(s^2+a^2\right)^2}$ **Ans.** $\dfrac{1}{2}t\cos at + \dfrac{1}{2a}\sin at$

2. $\dfrac{1}{\left(s^2+1\right)^3}$ **Ans.** $\dfrac{1}{8}\left[\left(3-t^2\right)\sin t - 3t\cos t\right]$

3. $\dfrac{s}{\left(s^2+a^2\right)^2}$ **Ans.** $\dfrac{t\sin at}{2a}$

4. $\dfrac{1}{s^2\left(s^2-a^2\right)}$ **Ans.** $\dfrac{1}{a^3}[-at + \sinh at]$

5. $\dfrac{1}{(s+1)\left(s^2+1\right)}$ **Ans.** $\dfrac{1}{2}\left(\cos t - \sin t - e^{-t}\right)$

43.11 HEAVISIDE INVERSE FORMULA OF $\dfrac{F(s)}{G(s)}$

If $F(s)$ and G (s) be two polynomials in S. The degree of F(s) is less than that of G (s). Let $\alpha_1, \alpha_2, \alpha_3, \ldots \alpha_n$ be n roots of the equation $G(s) = 0$

Inverse Laplace formula of $\dfrac{F(s)}{G(s)}$ is given by

$$L^{-1}\left\{\frac{F(s)}{G(s)}\right\} = \sum_{i=1}^{n}\frac{F(\alpha_i)}{G'(\alpha_i)}e^{\alpha_i t}$$

Example 41. *Find* $L^{-1}\left\{\dfrac{2s^2+5s-4}{s^3+s^2-2s}\right\}$

Solution. Let $F(s) = 2s^2 + 5s - 4$

and $G(s) = s^3 + s^2 - 2s = s(s^2+s-2) = s(s+2)(s-1)$

$G'(s) = 3s^2 + 2s - 2$

$G(s) = 0$ has three roots, 0, 1, – 2

$\Rightarrow$ $\alpha_1 = 0, \quad \alpha_2 = 1, \quad \alpha_3 = -2$

By Heaviside Inverse formula

$$L^{-1}\left\{\frac{F(s)}{G(s)}\right\} = \sum_{i=1}^{n}\frac{F(\alpha_i)}{G'(\alpha_i)}e^{\alpha_i t}$$

$$= \left\{\frac{F(\alpha_1)}{G'(\alpha_1)}\right\}e^{t\alpha_1} + \frac{F(\alpha_2)}{G'(\alpha_2)}e^{t\alpha_2} + \frac{F(\alpha_3)}{G'(\alpha_3)}e^{t\alpha_3} = \frac{F(0)}{G'(0)}e^0 + \frac{F(1)}{G'(1)}e^t + \frac{F(-2)}{G'(-2)}e^{-2t}$$

$$= \frac{-4}{-2}e^0 + \frac{3}{3}e^t + \frac{(-6)}{(6)}e^{-2t} = 2 + e^t - e^{-2t} \qquad \textbf{Ans.}$$

Example 42. *Find* $L^{-1}\left[\dfrac{2s^2 - 6s + 5}{s^3 - 6s^2 + 11s - 6}\right]$ *(U.P. II Semester,* 2004)

Solution. Let
$$F(s) = 2s^2 - 6s + 5$$
$$G(s) = s^3 - 6s^2 + 11s - 6 = (s-1)(s-2)(s-3)$$
$G(s) = 0$ has three roots, 1, 2, 3.
$$\Rightarrow \quad \alpha_1 = 1, \quad \alpha_2 = 2, \quad \alpha_3 = 3$$
$$G'(s) = 3s^2 - 12s + 11$$

By Heaviside Inverse formula, we have $L^{-1}\left\{\dfrac{F(s)}{G(s)}\right\} = \sum_{i=1}^{n} \dfrac{F(\alpha_i)}{G'(\alpha_i)} e^{t\alpha_i}$

$$L^{-1}\left\{\frac{2s^2 - 6s + 5}{s^3 - 6s^2 + 11s - 6}\right\} = \frac{F(\alpha_1)}{G'(\alpha_1)}e^{t\alpha_1} + \frac{F(\alpha_2)}{G'(\alpha_2)}e^{t\alpha_2} + \frac{F(\alpha_3)}{G'(\alpha_3)}e^{t\alpha_3}$$

$$= \frac{F(1)}{G'(1)}e^t + \frac{F(2)}{G'(2)}e^{2t} + \frac{F(3)}{G'(3)}e^{3t} = \frac{(1)}{(2)}e^t + \frac{(1)}{(-1)}e^{2t} + \frac{(5)}{(2)}e^{3t} = \frac{1}{2}e^t - e^{2t} + \frac{5}{2}e^{3t} \qquad \textbf{Ans.}$$

EXERCISE 43.9

Using Heaviside expansion formula, find the Inverse Laplace Transform of the following :

1. $\dfrac{s-1}{s^2+3s+2}$ **Ans.** $-2e^{-t} + 3e^{-2t}$
2. $\dfrac{s}{(s-1)(s-2)(s-3)}$ **Ans.** $\dfrac{1}{2}e^t - 2e^{2t} + \dfrac{3}{2}e^{3t}$
3. $\dfrac{2s+3}{(s-2)(s-3)(s-4)}$ **Ans.** $\dfrac{7}{2}e^{2t} - 9e^{3t} + \dfrac{11}{2}e^{4t}$
4. $\dfrac{11s^2 - 2s + 5}{2s^3 - 3s^2 - 3s + 2}$ **Ans.** $2e^{-2t} + 5\cdot e^{2t} - \dfrac{3}{2}e^{\frac{t}{2}}$

43.12 SOLUTION OF DIFFERENTIAL EQUATIONS BY LAPLACE TRANSFORMS

Ordinary linear differential equations with constant coefficients can be easily solved by the Laplace Transform method, without finding the general solution and the arbitrary constants. The method will be clear from the following examples:

Example 43. *Solve the following equation by Laplace transform*

$$y''' - 2y'' + 5y' = 0\ ;\ y = 0, \quad y' = 1 \text{ at } t = 0 \text{ and } y = 1 \text{ at } t = \frac{\pi}{8}.$$

(Q. Bank U.P., II Semester 2001)

Solution. Here, we have
$$y''' - 2y'' + 5y' = 0 \qquad \text{...(1)}$$
Taking Laplace transform on both sides of (1), we get
$$L(y''') - 2L(y'') + 5L(y') = L(0)$$
$$\Rightarrow s^3\bar{y} - s^2y(0) - sy'(0) - y''(0) - 2[s^2\bar{y} - sy(0) - y'(0)] + 5[s\bar{y} - y(0)] = 0$$
$$\Rightarrow (s^3 - 2s^2 + 5s)\bar{y} - s - k + 2 = 0 \qquad \text{[Let } y''(0) = k \text{ (say)]}$$
$$\Rightarrow \quad \bar{y} = \frac{(k-2)+s}{s(s^2-2s+5)}$$
$$= \left(\frac{k-2}{5}\right)\left(\frac{1}{s} - \frac{s-2}{s^2-2s+5}\right) + \frac{1}{s^2-2s+5}$$

$$= \left(\frac{k-2}{5}\right)\frac{1}{s} - \left(\frac{k-2}{5}\right)\left\{\frac{(s-1)-1}{(s-1)^2+4}\right\} + \frac{1}{(s-1)^2+4}$$

$$\Rightarrow \quad \bar{y} = \left(\frac{k-2}{5}\right)\frac{1}{s} - \left(\frac{k-2}{5}\right)\left\{\frac{(s-1)}{(s-1)^2+4}\right\} + \left(\frac{k+3}{10}\right).\left\{\frac{2}{(s-1)^2+4}\right\} \quad \ldots(3)$$

Taking Inverse Laplace transform on both sides of (3), we get

$$y = \left(\frac{k-2}{5}\right) - \left(\frac{k-2}{5}\right)e^t \cos 2t + \left(\frac{k+3}{10}\right)e^t \sin 2t \quad \ldots (4)$$

Puttting $y\left(\frac{\pi}{8}\right) = 1$, we get $1 = \left(\frac{k-2}{5}\right) - \left(\frac{k-2}{5}\right)e^{\pi/8}.\frac{1}{\sqrt{2}} + \left(\frac{k+3}{10}\right)e^{\pi/8}.\frac{1}{\sqrt{2}}$ ...(5)

$\Rightarrow \qquad k = 7$

On putting the value of k in (4), we get (on simplification)

Hence required solution is

$$y = 1 + e^t (\sin 2t - \cos 2t)$$ **Ans.**

Example 44. *Using Laplace transforms, find the solution of the initial value problem*

$$y'' - 4y' + 4y = 64 \sin 2t$$
$$y(0) = 0, \; y'(0) = 1.$$

Solution. Here, we have $y'' - 4y' + 4y = 64 \sin 2t$ ... (1)

$$y(0) = 0, \; y'(0) = 1.$$

Taking Laplace transform of both sides of (1), we have

$$[s^2\bar{y} - sy(0) - y'(0)] - 4[s\bar{y} - y(0)] + 4\bar{y} = \frac{64 \times 2}{s^2+4} \quad \ldots (2)$$

On putting the values of $y(0)$ and $y'(0)$ in (2), we get

$$s^2\bar{y} - 1 - 4s\bar{y} + 4\bar{y} = \frac{128}{s^2+4}$$

$$\Rightarrow \quad (s^2 - 4s + 4)\bar{y} = 1 + \frac{128}{s^2+4}, \quad \Rightarrow \quad (s-2)^2\bar{y} = 1 + \frac{128}{s^2+4}$$

$$\Rightarrow \quad \bar{y} = \frac{1}{(s-2)^2} + \frac{128}{(s-2)^2(s^2+4)} = \frac{1}{(s-2)^2} - \frac{8}{s-2} + \frac{16}{(s-2)^2} + \frac{8s}{s^2+4}$$

$$y = L^{-1}\left[-\frac{8}{s-2} + \frac{17}{(s-2)^2} + \frac{8s}{s^2+4}\right]$$

$$y = -8e^{2t} + 17t\,e^{2t} + 8 \cos 2t$$ **Ans.**

Example 45. *Using Laplace transforms, find the solution of the initial value problem*

$$y'' + 9y = 6 \cos 3t$$
$$y(0) = 2, \; y'(0) = 0$$

(U. P. II Semester Summer 2006)

Solution. We have, $y'' + 9y = 6 \cos 3t$ ... (1)

$$y(0) = 2, \; y'(0) = 0$$

Taking Laplace transform of (1), we get

$$[s^2\bar{y} - sy(0) - y'(0)] + 9\bar{y} = 6\frac{s}{s^2+9} \quad \ldots (2)$$

Putting the values of $y(0)$ and $y'(0)$ in (2), we have

$$s^2\bar{y} - 2s + 9\bar{y} = \frac{6s}{s^2+9}$$

$$\Rightarrow \quad (s^2+9)\bar{y} = 2s + \frac{6s}{s^2+9} \quad \Rightarrow \quad \bar{y} = \frac{2s}{s^2+9} + \frac{6s}{(s^2+9)^2}$$

$$y = L^{-1}\frac{2s}{s^2+9} + L^{-1}\frac{6s}{(s^2+9)^2} = 2\cos 3t + L^{-1}\frac{d}{ds}\left[\frac{-3}{(s^2+9)}\right]$$

$$= 2 \cos 3t - t \sin 3t$$ **Ans.**

Example 46. *Using Laplace transformation solve the following differential equation:*

$$\frac{d^2x}{dt^2}+9x=\cos 2t, \quad \textit{if } x(0)=1, \quad x\left(\frac{\pi}{2}\right)=-1$$ (U. P. II Semester, Summer 2002)

Solution. $\frac{d^2x}{dt^2}+9x=\cos 2t$ ... (1)

Taking Laplace transform of both the sides of (1), we get

$$L\frac{d^2x}{dt^2}+9\,L\,x=L\cos 2t$$

$$\Rightarrow \quad s^2\bar{x}-sx(0)-x'(0)+9\bar{x}=\frac{s}{s^2+4} \quad \text{... (2)}$$

On putting $x(0)=1$ in (2), we get

$$s^2\bar{x}-s+9\bar{x}-x'(0)=\frac{s}{s^2+4}$$

$$(s^2+9)\,\bar{x}=s+\frac{s}{s^2+4}+x'(0)=\frac{s(s^2+4)+s}{s^2+4}+x'(0)=\frac{s^3+5s}{s^2+4}+x'(0)$$

$$\Rightarrow \quad \bar{x}=\frac{(s^3+5s)}{(s^2+4)(s^2+9)}+\frac{x'(0)}{s^2+9}=\frac{1}{5}\frac{s}{s^2+4}+\frac{4}{5}\frac{s}{s^2+9}+\frac{x'(0)}{s^2+9}$$

Taking the Inverse Laplace Transform, we get

$$x(t)=\frac{1}{5}L^{-1}\frac{s}{s^2+4}+\frac{4}{5}L^{-1}\frac{s}{s^2+9}+L^{-1}\frac{x'(0)}{s^2+9}$$

$$x(t)=\frac{1}{5}\cos 2t+\frac{4}{5}\cos 3t+\frac{x'(0)\sin 3t}{3} \quad \text{... (3)}$$

On putting $x\left(\frac{\pi}{2}\right)=-1$ in (3), we get

$$-1=-\frac{1}{5}+0-\frac{x'(0)}{3} \quad \Rightarrow \quad x'(0)=\frac{12}{5}$$

On putting the value of $x'(0)$ in (3), we get

$$x=\frac{1}{5}\cos 2t+\frac{4}{5}\cos 3t+\frac{12}{5}\frac{\sin 3t}{3}=\frac{1}{5}[\cos 2t+4\cos 3t+4\sin 3t]$$ **Ans.**

Example 47. *Solve, using Laplace transform method*

$$y(0)=-2,\ y'(0)=8;\quad y''+4y'+4y=6e^{-t}$$ (*U.P., II Semester, 2007*)

Solution. Here, we have

$$y''+4y'+4y=6e^{-t} \quad \text{...(1)}$$

Taking Laplace transform on both sides of (1), we get

$$L(y'')+4L(y')+4L(y)=6L(e^{-t})$$

$$\Rightarrow [s^2\bar{y}-sy(0)-y'(0)]+4[s\,\bar{y}-y(0)]+4\bar{y}=\frac{6}{s+1} \quad \text{... (2) [Here } \bar{y}=L(y)]$$

Putting the values of $y(0)=-2$ and $y'(0)=8$ in (2), we get

$$(s^2+4s+4)\,\bar{y}+2s-8+8=\frac{6}{s+1}$$

$$\Rightarrow \quad (s^2+4s+4)\,\bar{y}+2s=\frac{6}{s+1}$$

$$(s+2)^2\,\bar{y}+2s=\frac{6}{s+1}$$

$$\bar{y} = \frac{6}{(s+1)(s+2)^2} - \frac{2s}{(s+2)^2} = 6\left[\frac{1}{s+1} - \frac{1}{s+2} - \frac{1}{(s+2)^2}\right] - \frac{2\{(s+2)-2\}}{(s+2)^2}$$

$$= \frac{6}{s+1} - \frac{6}{s+2} - \frac{6}{(s+2)^2} - \frac{2}{s+2} + \frac{4}{(s+2)^2} = \frac{6}{s+1} - \frac{8}{s+2} - \frac{2}{(s+2)^2} \quad ...(3)$$

Taking inverse Laplace transform on both sides of (3), we get

$$y = 6e^{-t} - 8e^{-2t} - 2t\,e^{-2t} \qquad \textbf{Ans.}$$

Example 48. *Solve the following differential equation using Laplace transform*

$$\frac{d^3y}{dt^3} - 3\frac{d^2y}{dt^2} + 3\frac{dy}{dt} - y = t^2 e^t$$

where $y(0) = 1$, $\left(\frac{dy}{dt}\right)_{t=0} = 0$, $\left(\frac{d^2y}{dt^2}\right)_{t=0} = -2$ *(U. P., II Semester, (SUM) 2008)*

Solution. Here we have equation

$$y''' - 3y'' + 3y' - y = t^2 e^t \quad ... (1)$$

Taking Laplace transform on both side of equation (1), we get

$$L(y''') - 3L(y'') + 3L(y') - L(y) = L(t^2 e^t)$$

$$\Rightarrow [s^3\bar{y} - s^2y(0) - sy'(0) - y''(0)] - 3[s^2\bar{y} - sy(0) - y'(0)] + 3[s\bar{y} - y(0)] - \bar{y} = \frac{2}{(s-1)^3}$$

... (2) [where $\bar{y} = L(y)$]

Putting the values of $y(0)$, $y'(0)$ and y'' at $x = 0$ in (2), we get

$$\Rightarrow (s^3\bar{y} - s^2 + 2) - 3(s^2\bar{y} - s) + 3(s\bar{y} - 1) - \bar{y} = \frac{2}{(s-1)^3}$$

$$\Rightarrow \quad (s^3 - 3s^2 + 3s - 1)\,\bar{y} - s^2 + 3s - 1 = \frac{2}{(s-1)^3}$$

$$\Rightarrow \quad (s-1)^3\,\bar{y} = s^2 - 3s + 1 + \frac{2}{(s-1)^3}$$

$$\Rightarrow \quad \bar{y} = \frac{(s-1)^2}{(s-1)^3} - \frac{s}{(s-1)^3} + \frac{2}{(s-1)^6} = \frac{1}{s-1} - \frac{(s-1)+1}{(s-1)^3} + \frac{2}{(s-1)^6}$$

$$= \frac{1}{s-1} - \frac{1}{(s-1)^2} - \frac{1}{(s-1)^3} + \frac{2}{(s-1)^6} \quad ...(3)$$

Taking inverse Laplace transform on both sides of equation (3), we get

$$y = e^t - te^t - \frac{t^2}{2}e^t + \frac{t^5 e^t}{60} = \left(1 - t - \frac{t^2}{2} + \frac{t^5}{60}\right)e^t \qquad \textbf{Ans.}$$

Example 49. *Solve* $\frac{d^2y}{dx^2} + 2\frac{dy}{dx} + 5y = e^{-x}\sin x$ *where* $y(0) = 0$, $y'(0) = 1$.

(U.P., II Semester, 2004)

Solution. Here, we have $\frac{d^2y}{dx^2} + 2\frac{dy}{dx} + 5y = e^{-x}\sin x$

Taking the Laplace Transform of both the sides, we get

$$[s^2\bar{y} - sy(0) - \bar{y}(0)] + 2[s\bar{y} - y(0)] + 5\bar{y} = L(e^{-x}\sin x)$$

$$[s^2\bar{y} - sy(0) - y'(0)] + 2[s\bar{y} - y(0)] + 5\bar{y} = \frac{1}{(s+1)^2 + 1} \quad ... (1)$$

On substituting the values of $y(0)$ and $y'(0)$ in (1), we get

$$(s^2\,\bar{y}-1)+2(s\,\bar{y})+5\bar{y}=\frac{1}{s^2+2s+2}$$

$$(s^2+2s+5)\,\bar{y}=1+\frac{1}{s^2+2s+2}=\frac{s^2+2s+3}{s^2+2s+2}$$

$$\bar{y}=\frac{s^2+2s+3}{\left(s^2+2s+5\right)\left(s^2+2s+2\right)}$$

On resolving the R.H.S. into partial fractions, we get

$$\bar{y}=\frac{2}{3}\,\frac{1}{s^2+2s+5}+\frac{1}{3}\,\frac{1}{s^2+2s+2}$$

On inversion, we obtain

$$y=\frac{2}{3}L^{-1}\,\frac{1}{s^2+2s+5}+\frac{1}{3}L^{-1}\,\frac{1}{s^2+2s+2}$$

$$\Rightarrow\quad y=\frac{1}{3}L^{-1}\,\frac{2}{(s+1)^2+(2)^2}+\frac{1}{3}L^{-1}\,\frac{1}{(s+1)^2+(1)^2}$$

$$\Rightarrow\quad y=\frac{1}{3}e^{-x}\sin 2x+\frac{1}{3}e^{-x}\sin x\quad\Rightarrow\quad y=\frac{1}{3}e^{-x}(\sin x+\sin 2x)$$ **Ans.**

Example 50. *Solve the equation by the transform method:*

$$\frac{dy}{dt}+2y+\int_0^t y\,dt=\sin t,\qquad y(0)=1$$

(*R. G. P. V. Bhopal, June* 2003)

Solution. Taking Laplace transform of the given equation, we get

$$[s\,\bar{y}-y(0)]+2\bar{y}+\frac{\bar{y}}{s}=\frac{1}{s^2+1}\quad\ldots(1)\qquad\left[\because L\left\{\int_0^t y\,dt=\frac{\bar{y}}{s}\right\}\right]$$

Putting the values of $y(0)=1$ in (1), we get

$$[s\,\bar{y}-1]+2\,\bar{y}+\frac{\bar{y}}{s}=\frac{1}{s^2+1}$$

$$\bar{y}\left(s+2+\frac{1}{s}\right)=1+\frac{1}{s^2+1}\qquad[\because y(0)=1]$$

$$\frac{1}{s}\,\bar{y}(s^2+2s+1)=\frac{s^2+1+1}{s^2+1}$$

$$\frac{1}{s}\,\bar{y}\,(s+1)^2=\frac{s^2+2}{s^2+1}\quad\Rightarrow\quad\bar{y}\,(s+1)^2=\frac{s^3+2s}{s^2+1}$$

$$\Rightarrow\quad\bar{y}=\frac{s^3+2s}{(s+1)^2\,(s^2+1)}=\frac{1}{s+1}-\frac{3}{2\,(s+1)^2}+\frac{1}{2\,(s^2+1)}\qquad\text{[By partial fractions]}$$

Taking Inverse Laplace Transform, we have

$$y=e^{-t}-\frac{3}{2}t\,e^{-t}+\frac{1}{2}\sin t$$ **Ans.**

Example 51. *Solve the following differential equation using Laplace transform:*

$$x\frac{d^2y}{dx^2}+\frac{dy}{dx}+xy=0,\qquad\text{given } y(0)=2,\ y'(0)=0.$$

(*Uttarakhand, II Semester, June* 2007)

Solution. Taking Laplace transform on both sides of the given equation, we have

$$L[xy'']+L[y']+L[xy]=L(0)$$

$$\Rightarrow\quad-\frac{d}{ds}L(y'')+L(y')-\frac{d}{ds}L(y)=0$$

$$\Rightarrow \quad -\frac{d}{ds}\{s^2\bar{y} - sy(0) - y'(0)\} + \{s\bar{y} - y(0)\} - \frac{d\bar{y}}{ds} = 0 \quad \text{... (1)}$$

Putting $y(0) = 2$ and $y'(0) = 0$ in (1), we get

$$-\frac{d}{ds}\{s^2\bar{y} - 2s - 0\} + \{s\bar{y} - 2\} - \frac{d\bar{y}}{ds} = 0$$

$$\Rightarrow \quad -s^2\frac{d\bar{y}}{ds} - (2s)\bar{y} + 2 + s\bar{y} - 2 - \frac{d\bar{y}}{ds} = 0$$

$$\Rightarrow \quad (s^2+1)\frac{d\bar{y}}{ds} + s\bar{y} = 0 \quad \text{... (2)}$$

Separating the variables, we have

$$\frac{d\bar{y}}{\bar{y}} + \frac{s\,ds}{s^2+1} = 0 \quad \text{... (3)}$$

On integrating, we have

$$\log\bar{y} + \frac{1}{2}\log(s^2+1) = \log C$$

$$\Rightarrow \quad \log\bar{y} = \log C - \log\sqrt{s^2+1} \quad \Rightarrow \quad \log\bar{y} = \log\frac{C}{\sqrt{s^2+1}}$$

$$\Rightarrow \quad \bar{y} = \frac{C}{\sqrt{s^2+1}} \quad \text{... (4)}$$

Taking Inverse Laplace Transform, we get

$$y = L^{-1}\left[\frac{C}{\sqrt{s^2+1}}\right]$$

$$\Rightarrow \quad y = C\,J_0(x) \quad \text{... (5)}$$ [See Art. 42.24]

At $x = 0$ $\quad y(0) = C\,J_0(0)$ ... (6)

Putting $y(0) = 2$ and $J_0(0) = 1$ in (6), we get

$$2 = C(1) \quad \Rightarrow \quad C = 2$$

On putting the value of C in (5), we get

$$y = 2\,J_0(x)$$ **Ans.**

Example 52. *Using Laplace transforms, find the solution of the initial value problem*

$$y'' + 9y = 9\,u(t-3), \qquad y(0) = y'(0) = 0$$

where $u(t-3)$ is the unit step functions.

Solution. We have, $\quad y'' + 9y = 9\,u(t-3)$ ... (1)

Taking Laplace transform of (1), we have

$$s^2\bar{y} - sy(0) - y'(0) + 9\bar{y} = 9\frac{e^{-3s}}{s} \quad \text{... (2)}$$

Putting the values of $y(0) = 0$ and $y'(0) = 0$ in (2), we get

$$s^2\bar{y} + 9\bar{y} = 9\frac{e^{-3s}}{s}$$

$$(s^2+9)\bar{y} = 9\frac{e^{-3s}}{s}$$

$$\bar{y} = \frac{9e^{-3s}}{s(s^2+9)} \quad \Rightarrow \quad y = L^{-1}\frac{9e^{-3s}}{s(s^2+9)} \quad \text{... (3)}$$

We know that $\quad L^{-1}\dfrac{3}{s^2+9} = \sin 3t$

$$\Rightarrow \qquad 3\,L^{-1}\frac{3}{s\,(s^2+9)}=3\int_0^t \sin 3t\,dt=-[\cos 3t]_0^t=1-\cos 3t \qquad \ldots (4)$$

[Using second shifting theorem]

Using (4), we get the inverse of (3)

$$y=[\,1-\cos 3\,(t-3)\,]\,u\,(t-3) \qquad \textbf{Ans.}$$

Example 53. *Solve, by the method of Laplace transform, the differential equation*

$$(D^2+n^2)\,x=a\sin(nt+\alpha),$$

$$x=Dx=0 \text{ at } t=0. \qquad (U.P., II\ Semester,\ Summer,\ 2010,\ 2002)$$

Solution. Taking Laplace transform of the given differential equation, we get

$$[\,s^2\,\bar{x}-sx\,(0)-x'\,(0)\,]+n^2\,\bar{x}=a\,L\sin(nt+\alpha)$$

$$=a\,L\,[\,\sin nt\cos\alpha+\cos nt\sin\alpha\,]=a\left[\frac{n}{s^2+n^2}\cos\alpha+\frac{s}{s^2+n^2}\sin\alpha\right]$$

$$=\frac{a\,[n\cos\alpha+s\sin\alpha\,]}{s^2+n^2} \qquad \ldots (1)$$

Putting $x\,(0)=x'\,(0)=0$ in (1), we get

$$s^2\,\bar{x}+n^2\,\bar{x}=a\left[\frac{n\cos\alpha+s\sin\alpha}{s^2+n^2}\right] \quad\Rightarrow\quad (s^2+n^2)\,\bar{x}=\frac{a\,[\,n\cos\alpha+s\sin\alpha\,]}{s^2+n^2}$$

$$\Rightarrow \qquad \bar{x}=\frac{a\,n\cos\alpha+as\sin\alpha}{(s^2+n^2)^2}$$

$$\Rightarrow \qquad \bar{x}=a\cos\alpha\,.\,\frac{n}{(s^2+n^2)^2}+a\sin\alpha\,.\,\frac{s}{(s^2+n^2)^2} \qquad \ldots (2)$$

Taking the Inverse Laplace Transform of (2), we get

$$x=a\cos\alpha\,L^{-1}\left[\frac{n}{(s^2+n^2)^2}\right]+(\,a\sin\alpha\,)\,L^{-1}\left[\frac{s}{(s^2+n^2)^2}\right] \qquad \ldots (3)$$

Let us find out the inverse of the term on R.H.S.

$$\text{But} \quad L^{-1}\left\{\frac{2s}{(s^2+n^2)^2}\right\}=L^{-1}\frac{d}{ds}\frac{1}{s^2+n^2}=-\frac{t}{n}\sin nt \qquad \left[-\frac{d}{ds}[\,F\,(s)\,]=t\,f(t)\right]$$

$$\Rightarrow \qquad L^{-1}\left\{\frac{s}{(s^2+n^2)^2}\right\}=\frac{t}{2n}\sin nt \qquad \ldots (4)$$

$$\text{Again} \quad L^{-1}\frac{1}{s}\left\{\frac{s}{(s^2+n^2)^2}\right\}=\frac{1}{2n}\int_0^t t\sin nt\,dt \qquad \left[\frac{F(s)}{s}=\int_0^t f\,(t)\,dt\right]$$

$$L^{-1}\left[\frac{1}{(s^2+n^2)^2}\right]=\frac{1}{2n}\left[t\left(\frac{-\cos nt}{n}\right)+\frac{1}{n^2}\sin nt\right]_0^t$$

$$L^{-1}\left[\frac{n}{(s^2+n^2)^2}\right]=\frac{1}{2n^2}[\,-nt\cos nt+\sin nt\,] \qquad \ldots (5)$$

Putting the values of $L^{-1}\left[\dfrac{s}{(s^2+n^2)^2}\right]$ from (4) and $L^{-1}\left[\dfrac{n}{(s^2+n^2)^2}\right]$ from (5) in (3), we get

$$x=(\,a\cos\alpha\,)\,\frac{1}{2n^2}[\,\sin nt-nt\cos nt\,]+(\,a\sin\alpha\,)\,\frac{t}{2n}\sin nt$$

$$= \frac{a}{2n^2}[\cos\alpha \sin nt - nt\cos\alpha\cos nt + nt\sin\alpha\sin nt]$$

$$= \frac{a}{2n^2}[\cos\alpha\sin nt - nt(\cos nt\cos\alpha - \sin nt\sin\alpha)] = \frac{a}{2n^2}[\sin nt\cos\alpha - nt\cos(nt+\alpha)]$$

Which is the required solution. **Ans.**

Example 54. *Solve* $t\,y'' + 2\,y' + t\,y = \cos t$, *if* $y(0) = 1,\ y'(0) = 0.$

Solution. We have, $t\,y'' + 2\,y' + t\,y = \cos t$

$$\Rightarrow \quad L\{t y''\} + L\{2y'\} + L\{t y\} = L\{\cos t\}$$

$$\Rightarrow \quad -\frac{d}{ds}L\{y''\} + 2L\{y'\} - \frac{d}{ds}L\{y\} = L\{\cos t\}$$

$$\Rightarrow \quad -\frac{d}{ds}[s^2\bar{y} - sy(0) - y'(0)] + 2[s\bar{y} - y(0)] - \frac{d}{ds}\bar{y} = \frac{s}{s^2+1}$$

$$\Rightarrow \quad -\frac{d}{ds}[s^2\bar{y} - s] + 2s\bar{y} - 2 - \frac{d}{ds}\bar{y} = \frac{s}{s^2+1}$$

$$\Rightarrow \quad -2s\bar{y} - s^2\frac{d\bar{y}}{ds} + 1 + 2s\bar{y} - 2 - \frac{d\bar{y}}{ds} = \frac{s}{s^2+1}$$

$$\Rightarrow \quad (s^2+1)\frac{d\bar{y}}{ds} + 1 = \frac{-s}{s^2+1} \qquad \Rightarrow \qquad \frac{d\bar{y}}{ds} = \frac{-s}{(s^2+1)^2} - \frac{1}{s^2+1}$$

Taking Inverse Laplace Transform, we get

$$L^{-1}\frac{d}{ds}(\bar{y}) = L^{-1}\left[\frac{-s}{(s^2+1)^2} - \frac{1}{s^2+1}\right] \qquad \left[L^{-1}\frac{d}{ds}r \quad = (-1)^1 t^1 f(t)\right]$$

$$\Rightarrow \quad (-1)^1 t^1\, y = -\frac{1}{2}L^{-1}\left\{\frac{2s}{(s^2+1)^2}\right\} - L^{-1}\left\{\frac{1}{s^2+1}\right\} = -\frac{1}{2}t\sin t - \sin t$$

$$\Rightarrow \quad y = \frac{1}{2}\left(1 + \frac{2}{t}\right)\sin t$$ **Ans.**

Example 55. *Solve* $[t D^2 + (1 - 2t) D - 2]\,y = 0$, *where* $y(0) = 1, y'(0) = 2$

(*M.D.U., 2010, U. P. II Semester, June* 2002)

Solution. Here, $t D^2 y + (1 - 2t)\,Dy - 2y = 0 \quad \Rightarrow \quad ty'' + y' - 2\,t\,y' - 2y = 0$

Taking Laplace transform of given differential equation, we get

$$L(t\,y'') + L(y') - 2L(t\,y') - 2L(y) = 0$$

$$\Rightarrow \quad -\frac{d}{ds}L\{y''\} + L\{y'\} + 2\frac{d}{ds}L(y') - 2L(y) = 0$$

$$\Rightarrow \quad -\frac{d}{ds}[s^2\bar{y} - sy(0) - y'(0)] + [s\bar{y} - y(0)] + 2\frac{d}{ds}[s\bar{y} - y(0)] - 2\bar{y} = 0$$

Putting the values of $y(0)$ and $y'(0)$, we get

$$-\frac{d}{ds}(s^2\bar{y} - s - 2) + (s\bar{y} - 1) + 2\frac{d}{ds}(s\bar{y} - 1) - 2\bar{y} = 0 \qquad [\because y(0) = 1, y'(0) = 2]$$

$$\Rightarrow \quad -s^2\frac{d\bar{y}}{ds} - 2s\bar{y} + 1 + s\bar{y} - 1 + 2\left(s\frac{d\bar{y}}{ds} + \bar{y}\right) - 2\bar{y} = 0$$

$$\Rightarrow \quad -(s^2 - 2s)\frac{d\bar{y}}{ds} - s\bar{y} = 0$$

$$\Rightarrow \quad -\frac{d\bar{y}}{\bar{y}} - \frac{1}{s-2}ds = 0 \quad (\textit{Separating the variables})$$

$$\Rightarrow \quad \int \frac{d\,\bar{y}}{\bar{y}} + \int \frac{ds}{s-2} = 0 \quad \Rightarrow \quad \log \bar{y} + \log (s-2) = \log C \quad \Rightarrow \quad \log \bar{y}\,(s-2) = \log C$$

$$\Rightarrow \quad \bar{y}\,(s-2) = C \Rightarrow \bar{y} = \frac{C}{s-2} \quad \Rightarrow \quad y = C\,L^{-1}\left\{\frac{1}{s-2}\right\} \Rightarrow y = C\,e^{2t} \quad \ldots (1)$$

$$y\,(0) = C\,e^0 \quad \ldots (2)$$

Putting $y\,(0) = 1$ in (2), we get $1 = C\,e^0 \quad \Rightarrow \quad C = 1.$

Putting $C = 1$ in (1), we get $y = e^{2t}$

This is the required solution. **Ans.**

Example 56. *Using Laplace transform, solve the following differential equation: -*

$$y'' + 2t\,y' - y = t$$

when $y\,(0) = 0$ *and* $y'\,(0) = 1$ (*U.P., II Semester, Summer* 2003)

Solution. We have, $y'' + 2t\,y' - y = t$ ... (1)

Taking Laplace transform of (1), we get

$$[\,s^2\,\bar{y} - sy\,(0) - y'\,(0)\,] - 2\frac{d}{ds}[\,s\,\bar{y} - y\,(0)\,] - \bar{y} = \frac{1}{s^2} \quad \ldots (2)$$

On putting $y\,(0) = 0$ and $y'\,(0) = 1$ in (2), we get

$$(\,s^2\,\bar{y} - 1\,) - 2\frac{d}{ds}(\,s\,\bar{y} - 0\,) - \bar{y} = \frac{1}{s^2}$$

$$\Rightarrow \quad (\,s^2\,\bar{y} - 1\,) - 2\,\bar{y} - 2s\frac{d\,\bar{y}}{ds} - \bar{y} = \frac{1}{s^2} \Rightarrow -2s\frac{d\,\bar{y}}{ds} + (\,s^2 - 3\,)\,\bar{y} = \frac{1}{s^2} + 1 = \frac{1+s^2}{s^2}$$

$$\Rightarrow \quad \frac{d\,\bar{y}}{ds} - \frac{s^2-3}{2s}\,\bar{y} = \frac{1+s^2}{-2s^3} \quad \Rightarrow \quad \frac{d\,\bar{y}}{ds} - \left(\frac{s}{2} - \frac{3}{2s}\right)\bar{y} = -\frac{1}{2s^3} - \frac{1}{2s} \quad \ldots (3)$$

Thus (3) is a linear differential equation.

$$\text{I.F.} = e^{\frac{1}{2}\int\left(\frac{3}{s} - s\right)ds} = e^{\frac{1}{2}\left(3\log s - \frac{s^2}{2}\right)} = e^{-\frac{s^2}{4}}\,.\,s^{\frac{3}{2}}$$

Solution of differential equation (3) is

$$\bar{y}\,e^{-\frac{s^2}{4}}\,.\,s^{\frac{3}{2}} = -\frac{1}{2}\int\left(\frac{1}{s^3} + \frac{1}{s}\right)s^{\frac{3}{2}}\,.\,e^{-\frac{s^2}{4}}\,ds = -\frac{1}{2}\int\left(\sqrt{s} + \frac{1}{s^{\frac{3}{2}}}\right)e^{-\frac{s^2}{4}}\,ds$$

Put $s^2 = 4\,z \quad \Rightarrow \quad s = 2\sqrt{z}$ so that $ds = \dfrac{dz}{\sqrt{z}}$

$$\bar{y}\,s^{\frac{3}{2}}\,.\,e^{-\frac{s^2}{4}} = -\frac{1}{2}\int\left(\sqrt{2}\,z^{\frac{1}{4}} + \frac{1}{2\sqrt{2}}\,z^{-\frac{3}{4}}\right)e^{-z}\,\frac{dz}{\sqrt{z}}$$

$$= -\frac{1}{\sqrt{2}}\int\left(z^{-\frac{1}{4}} + \frac{1}{4}\,z^{-\frac{5}{4}}\right)e^{-z}\,dz = -\frac{1}{\sqrt{2}}\int z^{-\frac{1}{4}}\,e^{-z}\,dz - \frac{1}{4\sqrt{2}}\int z^{-\frac{5}{4}}\,e^{-z}\,dz$$

$$= -\frac{1}{\sqrt{2}}\left[z^{-\frac{1}{4}}\,\frac{e^{-z}}{-1} + \int\left(-\frac{1}{4}\right)z^{-\frac{5}{4}}\,e^{-z}\,dz\right] - \frac{1}{4\sqrt{2}}\int z^{-\frac{5}{4}}\,e^{-z}\,dz + C$$

$$= \frac{1}{\sqrt{2}}\,e^{-z}\,z^{-\frac{1}{4}} + C = \frac{1}{\sqrt{2}}\,e^{-\frac{s^2}{4}}\left(\frac{s^2}{4}\right)^{-\frac{1}{4}} + C = \frac{1}{\sqrt{s}}\,e^{-\frac{s^2}{4}} + C \Rightarrow \bar{y} = \frac{1}{s^2} + C$$

(Particular case)

$$\Rightarrow \quad \bar{y} = \frac{1}{s^2} + c \quad \Rightarrow \quad y = L^{-1}\left(\frac{1}{s^2} + c\right) = t + c$$

$y = t$ (c must vanish if $\bar{y}$ is a transform since $\bar{y} \to 0$ as $s \to \infty$) **Ans.**

Example 57. *A particle moves in a line so that its displacement x from a fixed point O at any time t, is given by*

$$\frac{d^2x}{dt^2} + 4\frac{dx}{dt} + 5x = 80 \sin 5t$$

Using Laplace transform, find its displacement at any time t if initially particle is at rest at x = 0 *(U.P., II Semester 2009)*

Solution. Here, we have

$$\frac{d^2x}{dt^2} + 4\frac{dx}{dt} + 5x = 80 \sin 5t, \; x(0) = 0, x'(0) = 0 \qquad ...(1)$$

Taking Laplace transform of both sides of (1), we get

$$[s^2\bar{x} - sx(0) - x'(0)] + 4[s\bar{x} - x(0)] + 5\bar{x} = L[80 \sin 5t]$$

$$[s^2\bar{x} - 0 - 0] + 4s\bar{x} + 5\bar{x} = 80\left(\frac{5}{s^2+25}\right), \qquad (s^2+4s+5)\bar{x} = \frac{400}{s^2+25}$$

$$\bar{x} = \left(\frac{1}{s^2+4s+5}\right)\left(\frac{400}{s^2+25}\right)$$

$$\bar{x} = \frac{2s+18}{s^2+4s+5} - \frac{2s+10}{s^2+25} \qquad \text{[By Partial fraction]}$$

$$= \frac{2(s+2)+14}{(s+2)^2+1} - \frac{2s}{s^2+(5)^2} - \frac{10}{s^2+(5)^2}$$

$$= \frac{2(s+2)}{(s+2)^2+1} + \frac{14}{(s+2)^2+1} - \frac{2s}{s^2+(5)^2} - \frac{10}{s^2+(5)^2}$$

$$\Rightarrow \quad x = 2L^{-1}\frac{(s+2)}{(s+2)^2+1} + 14L^{-1}\frac{1}{(s+2)^2+1} - 2L^{-1}\frac{s}{s^2+(5)^2} - 2L^{-1}\frac{5}{s^2+(5)^2}$$

$$= 2e^{-2t}\cos t + 14e^{-2t}\sin t - 2\cos 5t - 2\sin 5t$$

$$= 2e^{-2t}(\cos t + 7\sin t) - 2(\cos 5t + \sin 5t)$$ **Ans.**

43.13 ELECTRIC CIRCUIT

Consider an electric circuit consisting of a resistance R, inductance L, a condensor of capacity C and electromotive power of voltage E in a series. A switch is also connected in the circuit. Here,

$$i = \frac{dq}{dt}$$

Voltage developed by Ri, $L\frac{di}{dt}$ and $\frac{q}{C}$

By Kirchhoff low $\boxed{L\frac{di}{dt} + Ri + \frac{q}{C} = E}$

R
L
C
E
S

Example 58. *A resistance R in series with inductance L is connected with e.m.f. E (t). The current i is given by*

$$L\frac{di}{dt} + Ri = E$$

If the switch is connected at t = 0 and disconnected at t = a, find the current i in terms of t.

(U.P. II Semester, Summer 2001)

Solution. Conditions under which current i flows are

$$E(t) = \begin{cases} E, & 0 < t < a \\ 0, & t > a \end{cases} \qquad [\, i = 0 \text{ at } t = 0]$$

Given equation is $L\dfrac{di}{dt} + Ri = E$ ... (1)

Taking Laplace transform of (1), we get

$$L[\, s\,\bar{i} - i(0)\,] + R\,\bar{i} = \int_0^\infty e^{-st}\, E\, dt$$

$$L\, s\,\bar{i} + R\,\bar{i} = \int_0^\infty e^{-st}\, E\, dt \qquad [\, i(0) = 0]$$

$$(Ls + R)\,\bar{i} = \int_0^\infty e^{-st} \,.\, E\, dt = \int_0^a e^{-st}\, E\, dt + \int_a^\infty e^{-st}\, E\, dt$$

$$= E\left[\frac{e^{-st}}{-s}\right]_0^a + 0 = \frac{E}{s}[\, 1 - e^{-as}\,] = \frac{E}{s} - \frac{E}{s} e^{-as}$$

$$\Rightarrow \qquad \bar{i} = \frac{E}{s(Ls + R)} - \frac{Ee^{-as}}{s(Ls + R)}$$

Taking Inverse Laplace Transform, we obtain

$$i = \text{Inverse Lap.}\left[\frac{E}{s(Ls + R)}\right] - \text{Inverse Lap.}\left[\frac{E\, e^{-as}}{s(Ls + R)}\right] \qquad \text{... (2)}$$

Now we have to find the value of Inverse Lap. $\left[\dfrac{E}{s(Ls + R)}\right]$

$$\text{Inverse Lap.}\left[\frac{E}{s(Ls + R)}\right] = \frac{E}{L}\ \text{Inverse Lap.}\left[\frac{1}{s\left(s + \frac{R}{L}\right)}\right]$$

$$= \frac{E}{L}\frac{L}{R}\ \text{Inverse Lap.}\left[\frac{1}{s} - \frac{1}{s + \frac{R}{L}}\right] = \frac{E}{R}\left[1 - e^{-\frac{R}{L}t}\right]$$

(Resolving into partial fractions)

and $\quad \text{Inverse Lap.}\left[\dfrac{E\, e^{-as}}{s(Ls + R)}\right] = \dfrac{E}{R}\left[1 - e^{-\frac{R}{L}(t-a)}\right] u(t - a)$

(By the Second Shifting Theorem)

On substituting the values of the inverse transforms in (2), we get

$$i = \frac{E}{R}\left[1 - e^{-\frac{R}{L}t}\right] - \frac{E}{R}\left[1 - e^{-\frac{R}{L}(t-a)}\right] u(t - a)$$

Hence $\qquad i = \dfrac{E}{R}\left[1 - e^{-\frac{R}{L}t}\right]$ for $0 < t < a$, $\qquad [\, u(t - a) = 0]$

$$i = \frac{E}{R}\left[1 - e^{-\frac{R}{L}t}\right] - \frac{E}{R}\left[1 - e^{-\frac{R}{L}(t-a)}\right]$$

for $t > a$ $\quad [u(t - a) = 1]$

$$= \frac{E}{R}\left[e^{-\frac{R}{L}(t-a)} - e^{-\frac{R}{L}t}\right] = \frac{E}{R} e^{-\frac{R}{L}t}\left[e^{\frac{Ra}{L}} - 1\right] \qquad \textbf{Ans.}$$

Example 59. *Voltage Ee^{-at} is applied at $t = 0$ to a circuit of inductance L and resistance R. Show that the current at time t is* $\frac{E}{R-aL}(e^{-at} - e^{-Rt/L})$. [U.P., II Semester, (SUM) 2007]

Solution. We know that

$$L\frac{dI}{dt} + RI = Ee^{-at}, \quad \text{...(1)}$$

where $I(0) = 0$

Taking Laplace transform of both sides of (1), we get

$$L[s\bar{I} - I(0)] + R\bar{I} = \frac{E}{s+a} \quad \text{...(2)}$$

Putting I (0) = 0 in (2), we get

$$(Ls + R)\bar{I} = \frac{E}{s+a}$$

$$\Rightarrow \quad \bar{I} = \frac{E}{(s+a)(Ls+R)} = \frac{E}{R-aL}\left(\frac{1}{s+a} - \frac{1}{Ls+R}\right)$$

$$= \frac{E}{R-aL}\left(\frac{1}{s+a} - \frac{1}{s+R/L}\right) \quad \text{...(3)}$$

Taking the Inverse Laplace transform of both sides of (3), we get

$$I = \frac{E}{R-aL}L^{-1}\left\{\frac{1}{s+a} - \frac{1}{s+R/L}\right\} = \frac{E}{R-aL}[e^{-at} - e^{-Rt/L}] \quad \textbf{Ans.}$$

Example 60. *Using the Laplace transform, find the current i (t) in the LC - circuit. Assuming L = 1 henry, C = 1 farad, zero initial current and charge on the capacitor, and*

$v(t) = t$, *when* $0 < t < 1$

$= 0$ *otherwise.*

C L V(t)

Solution. The differential equation for L and C circuit is given by

$$L\frac{d^2q}{dt^2} + \frac{q}{C} = E \quad \text{... (1)}$$

Putting $L = 1$, $C = 1$, $E = v(t)$ in (1), we get

$$\frac{d^2q}{dt^2} + q = v(t) \quad \text{... (2)}$$

Taking Laplace Transform of (2), we have

$$s^2\bar{q} - sq(0) - q'(0) + \bar{q} = \int_0^\infty v(t)\,e^{-st}\,dt$$

Substituting $q(0) = 0$, and $q'(0) = 0$, we get

$$s^2\bar{q} + \bar{q} = \int_0^1 te^{-st}\,dt + \int_1^\infty 0\,e^{-st}\,dt$$

$$(s^2+1)\,\bar{q} = \left[t\frac{e^{-st}}{-s}\right]_0^1 - \int_0^1 \frac{e^{-st}}{-s}\,dt = \frac{e^{-s}}{-s} - \left[\frac{e^{-st}}{s^2}\right]_0^1 = -\frac{e^{-s}}{s} - \frac{e^{-s}}{s^2} + \frac{1}{s^2}$$

$$\bar{q} = \frac{1}{s^2+1}\left[-\frac{e^{-s}}{s} - \frac{e^{-s}}{s^2} + \frac{1}{s^2}\right]$$

$$\bar{q} = \frac{-e^{-s}}{s(s^2+1)} - \frac{e^{-s}}{s^2(s^2+1)} + \frac{1}{s^2(s^2+1)}$$

Taking Inverse Laplace Transform, we get

$$q = \text{Inverse Lap.}\ \frac{-e^{-s}}{s\,(s^2+1)} - \text{Inverse Lap.}\ \frac{e^{-s}}{s^2\,(s^2+1)} + \text{Inverse Lap.}\ \frac{1}{s^2\,(s^2+1)} \qquad \text{... (3)}$$

We know that

$$\text{Inverse Lap.}\ [\,e^{-as}\,F(s)\,] = f(t-a)\,u(t-a)$$

$$\text{Inverse Lap.}\ \frac{1}{s(s^2+1)} = \int_0^t \sin t\,dt = [\,-\cos t\,]_0^t = 1-\cos t \qquad \text{... (4)}$$

$$\text{Inverse Lap.}\ \frac{1}{s^2\,(s^2+1)} = \int_0^t (1-\cos t)\,dt = t - \sin t \qquad \text{... (5)}$$

In view of this, we have

$$\text{Inverse Lap.}\ \left[\frac{-e^{-s}}{s\,(s^2+1)}\right] = -\,[\,1-\cos(t-1)\,]\,u(t-1) \qquad \text{[From (4)]}$$

$$\text{Inverse Lap.}\ \frac{e^{-s}}{s^2\,(s^2+1)} = [\,(t-1) - \sin(t-1)\,]\,u(t-1) \qquad \text{[From (5)]}$$

Putting the above values in (3), we get

$$q = -\,[1-\cos(t-1)]\,u(t-1) - [(t-1) - \sin(t-1)]\,u(t-1) + t - \sin t \qquad \textbf{Ans.}$$

EXERCISE 43.10

Solve the following differential equations:

1. $\dfrac{d^2y}{dx^2} + y = 0$ where $y = 1$ and $\dfrac{dy}{dx} = -1$ at $x = 0$. **Ans.** $y = \cos x - \sin x$

2. $\dfrac{d^2y}{dx^2} + 2\dfrac{dy}{dx} + 5y = 0$, where $y = 2$, $\dfrac{dy}{dx} = -4$ at $x = 0$. **Ans.** $y = e^{-x}\,(2\cos 2x - \sin 2x)$

3. $\dfrac{d^3y}{dx^3} + 2\dfrac{d^2y}{dx^2} - \dfrac{dy}{dx} - 2y = 0$, given $y = \dfrac{dy}{dx} = 0$, $\dfrac{d^2y}{dx^2} = 6$ at $x = 0$.

Ans. $y = e^x - 3e^{-x} + 2e^{-2x}$

4. $\dfrac{d^2y}{dx^2} + y = 3\cos 2x$, where $y = \dfrac{dy}{dx} = 0$ at $x = 0$. **Ans.** $y = \cos x - \cos 2x$.

5. $\dfrac{d^2y}{dx^2} + \dfrac{dy}{dx} - 2y = 1 - 2x$, given $y = 0$, $\dfrac{dy}{dx} = 4$ at $x = 0$. **Ans.** $y = e^x - e^{-2x} + x$

6. $\dfrac{d^2y}{dx^2} - 3\dfrac{dy}{dx} + 2y = 4e^{2x}$, given $y = -3$, and $\dfrac{dy}{dx} = 5$ at $x = 0$

Ans. $y = -7e^x + 4e^{2x} + 4x\,e^{2x}$

7. $\dfrac{d^2y}{dx^2} - 3\dfrac{dy}{dx} + 2y = 4x + e^{2x}$, where $y = 1$, $\dfrac{dy}{dx} = -1$ at $x = 0$.

Ans. $y = 3 + 2x + \dfrac{1}{2}e^{3x} -- 2e^{2x} - \dfrac{1}{2}e^x$.

8. $\dfrac{d^3y}{dx^3} + 2\dfrac{d^2y}{dx^2} - \dfrac{dy}{dx} - 2y = 0$, where $y = 1$, $\dfrac{dy}{dx} = 2$, $\dfrac{d^2y}{dx^2} = 2$ at $x = 0$.

Ans. $y = \dfrac{5}{3}e^x - e^{-x} + \dfrac{1}{2}e^{-2x}$

9. $(D^2 - D - 2)\,x = 20\sin 2t,\ x_0 = -1,\ x_1 = 2$ **Ans.** $x = 2e^{2t} - 4e^{-t} + \cos 2t - 3\sin 2t$

10. $(D^3 + D^2)\,x = 6t^2 + 4,\ x(0) = 0,\ x'(0) = 2,\ x''(0) = 0$

Ans. $x = \dfrac{1}{2}t^4 - 2t^3 + 8t^2 - 16t + 16 - 16e^{-t}$

11. $\dfrac{d^2x}{dt^2} - 2\dfrac{dx}{dt} + x = e^t$, where $x(0) = 2$, $\dfrac{dx}{dt} = -1$ at $t = 0$

Ans. $x = 2e^t - 3\,t\,e^t + \dfrac{1}{2}t^2e^t$

12. $y'' + 2y' + y = t\,e^{-t}$ if $y(0) = 1$, $y'(0) = -2$. **Ans.** $y = \left(1 - t + \dfrac{t^3}{6}\right)e^{-t}$

13. $\dfrac{d^2y}{dx^2} + y = x\cos 2x$, where $y = \dfrac{dy}{dx} = 0$ at $x = 0$. **Ans.** $y = \dfrac{4}{9}\sin 2x - \dfrac{5}{9}\sin x - \dfrac{x}{3}\cos 2x$

14. $\dfrac{d^3y}{dx^3} - 3\dfrac{d^2y}{dx^2} + 3\dfrac{dy}{dx} - y = x^2e^{2x}$, where $y = 1$, $\dfrac{dy}{dx} = 0$, $\dfrac{d^2y}{dx^2} = -2$ at $x = 0$.

Ans. $y = e^{2x}(x^2 - 6x + 12) - e^x(15x^2 + 7x + 11)$.

15. $y'' + 4y' + 3y = t$, $t > 0$; given that $y(0) = 0$ and $y'(0) = 1$.

Ans. $y = -\dfrac{4}{9} + \dfrac{t}{6} + e^{-t} - \dfrac{5}{9}e^{-3t}$

43.14 SOLUTION OF SIMULTANEOUS DIFFERENTIAL EQUATIONS BY LAPLACE TRANSFORMS

Simultaneous differential equations can also be solved by Laplace Transform method.

Example 61. *Solve* $\dfrac{dx}{dt} + y = 0$ *and* $\dfrac{dy}{dt} - x = 0$ *under the condition* $x(0) = 1$, $y(0) = 0$.

Solution. We have,
$$x' + y = 0 \quad \text{... (1)}$$
and
$$y' - x = 0 \quad \text{... (2)}$$
Taking the Laplace transform of (1) and (2), we get
$$[\,s\,\bar{x} - x(0)\,] + \bar{y} = 0 \quad \text{... (3)}$$
$$[\,s\,\bar{y} - y(0)\,] - \bar{x} = 0 \quad \text{... (4)}$$
On substituting the values of $x(0)$ and $y(0)$ in (3) and (4), we get
$$s\,\bar{x} - 1 + \bar{y} = 0 \quad \text{... (5)}$$
$$s\,\bar{y} - \bar{x} = 0 \quad \text{... (6)}$$
Solving (5) and (6) for $\bar{x}$ and $\bar{y}$, we get
$$\bar{x} = \frac{s}{s^2+1}, \qquad \bar{y} = \frac{1}{s^2+1},$$
On inversion, we obtain
$$x = L^{-1}\left(\frac{s}{s^2+1}\right), \qquad y = L^{-1}\left(\frac{1}{s^2+1}\right),$$
$$x = \cos t, \qquad y = \sin t \qquad \textbf{Ans.}$$

Example 62. *Solve the following simultaneous differential equations by Laplace transform*
$$3\frac{dx}{dt} - y = 2t, \quad \frac{dx}{dt} + \frac{dy}{dt} - y = 0$$
with the conditions $x(0) = y(0) = 0$. [*U.P., II Semester, (SUM) 2008*]

Solution. Here, we have
$$3\frac{dx}{dt} - y = 2t, \quad \text{...(1)}$$
and
$$\frac{dx}{dt} + \frac{dy}{dt} - y = 0 \quad \text{...(2)}$$
Taking Laplace transform on both sides of equation (1), we get
$$3L(x') - L(y) = L(2t)$$
$$3[s\bar{x} - x(0)] - \bar{y} = \frac{2}{s^2} \qquad [\text{where } L(x) = \bar{x} \text{ and } L(y) = \bar{y}]$$

$$3s\bar{x} - \bar{y} = \frac{2}{s^2} \quad \text{...(3)}$$

Again taking Laplace transform on both sides of equation (2), we get

$$L(x') + L(y') - L(y) = L(0)$$

$$\Rightarrow [s\bar{x} - x(0)] + [s\bar{y} - y(0)] - \bar{y} = 0 \Rightarrow s\bar{x} + (s-1)\bar{y} = 0 \quad \text{...(4)}$$

Multiplying equation (4) by 3, we get

$$3s\bar{x} + 3(s-1)\bar{y} = 0 \quad \text{...(5)}$$

Subtracting equation (3) from (5), we get

$$(3s-2)\bar{y} = -\frac{2}{s^2}$$

$$\Rightarrow \quad \bar{y} = -\frac{2}{s^2(3s-2)} = \frac{1}{s^2} + \frac{3}{2s} - \frac{3}{2\left(s - \frac{2}{3}\right)}$$

Taking inverse Laplace transform on both sides, we get

$$y = t + \frac{3}{2} - \frac{3}{2} e^{\frac{2t}{3}} \quad \text{...(6)}$$

Substituting $\bar{y}$ in (3), we get

$$3s\bar{x} - \frac{1}{s^2} - \frac{3}{2s} + \frac{3}{2\left(s - \frac{2}{3}\right)} = \frac{2}{s^2} \Rightarrow 3s\bar{x} = \frac{3}{s^2} + \frac{3}{2s} - \frac{3}{2\left(s - \frac{2}{3}\right)}$$

$$\Rightarrow \quad \bar{x} = \frac{1}{s^3} + \frac{1}{2s^2} - \frac{1}{2s\left(s - \frac{2}{3}\right)} \Rightarrow \bar{x} = \frac{1}{s^3} + \frac{1}{2s^2} - \frac{3}{4}\left(\frac{1}{s - \frac{2}{3}} - \frac{1}{s}\right)$$

Taking inverse Laplace transform on both sides, we get

$$x = \frac{t^2}{2} + \frac{t}{2} - \frac{3}{4} e^{\frac{2t}{3}} + \frac{3}{4} \quad \text{...(7)}$$

Equation (6) and (7) when taken together, give the complete solution. **Ans.**

Example 63. *Solve the simultaneous equations:*

$$\frac{dx}{dt} - y = e^t,$$

$$\frac{dy}{dt} + x = \sin t, \text{ given } x(0) = 1,\ y(0) = 0,$$ *(U.P., II Semester, Summer 2006)*

Solution. $\frac{dx}{dt} - y = e^t$ ... (1) $\quad \frac{dy}{dt} + x = \sin t$ ... (2)

Taking Laplace transform of (1), we get

$$[s\bar{x} - x(0)] - \bar{y} = \frac{1}{s-1}$$

i.e.,

$$s\bar{x} - 1 - \bar{y} = \frac{1}{s-1} \quad [\because x(0) = 1]$$

$$s\bar{x} - \bar{y} = 1 + \frac{1}{s-1}$$

$$s\bar{x} - \bar{y} = \frac{s}{s-1} \quad \text{... (3)}$$

Taking Laplace Transform of (2), we get

$$[s\bar{y} - y(0)] + \bar{x} = \frac{1}{s^2+1} \quad [y(0) = 0]$$

$$\bar{x} + s\,\bar{y} = \frac{1}{s^2+1} \quad \text{... (4)}$$

Solving (3) and (4) for $\bar{x}$ and $\bar{y}$, we have

$$\bar{x} = \frac{s^2}{(s-1)(s^2+1)} + \frac{1}{(s^2+1)^2} = \frac{1}{2}\left[\frac{1}{s-1} + \frac{s}{s^2+1} + \frac{1}{s^2+1}\right] + \frac{1}{(s^2+1)^2} \quad \text{... (5)}$$

and $$\bar{y} = \frac{s}{(s^2+1)^2} - \frac{s}{(s-1)(s^2+1)} = \frac{s}{(s^2+1)^2} - \frac{1}{2}\left[\frac{1}{s-1} - \frac{s}{s^2+1} + \frac{1}{s^2+1}\right] \quad \text{... (6)}$$

Taking Inverse Laplace Transform of (5), we get

$$x = \frac{1}{2} L^{-1}\left[\frac{1}{s-1} + \frac{s}{s^2+1} + \frac{1}{s^2+1}\right] + L^{-1}\left[\frac{1}{(s^2+1)^2}\right]$$

$$= \frac{1}{2}[e^t + \cos t + \sin t] + \frac{1}{2}(\sin t - t\cos t) \quad \left[\because L^{-1}\left[\frac{1}{(s^2+a^2)^2}\right] = \frac{1}{2a^2}(\sin at - at\cos at)\right]$$

$$= \frac{1}{2}[e^t + \cos t + 2\sin t - t\cos t]$$

Taking Inverse Laplace Transform of (6), we get

$$y = L^{-1}\left[\frac{s}{(s^2+1)^2}\right] - \frac{1}{2} L^{-1}\left[\frac{1}{s-1} - \frac{s}{s^2+1} + \frac{1}{s^2+1}\right]$$

$$= \frac{1}{2} t\sin t - \frac{1}{2}[e^t - \cos t + \sin t] \quad \left[\because L^{-1}\left[\frac{s}{(s^2+a^2)^2}\right] = \frac{1}{2a^2} t\sin at\right]$$

$$= \frac{1}{2}[t\sin t - e^t + \cos t - \sin t]$$

Hence, $x = \dfrac{1}{2}(e^t + \cos t + 2\sin t - t\cos t)$

$y = \dfrac{1}{2}(t\sin t - e^t + \cos t - \sin t)$ **Ans.**

[$i = 0$ at $t = 0$]

Example 64. *Use Laplace transform to solve:*

$$\frac{dx}{dt} + y = \sin t, \quad \frac{dy}{dt} + x = \cos t$$

given that $x = 2$, $y = 0$ at $t = 0$. [*U.P., II Semester, 2004*]

Solution. Here, we have

$$\frac{dx}{dt} + y = \sin t \quad \text{...(1)}$$

$$\frac{dy}{dt} + x = \cos t \quad \text{...(2)}$$

Taking Laplace transform of (1) and (2), we get

$$s\bar{x} - x(0) + \bar{y} = \frac{1}{s^2+1} \Rightarrow s\bar{x} + \bar{y} = \frac{1}{s^2+1} + 2 \quad \text{...(3)}$$

and $$s\bar{y} - y(0) + \bar{x} = \frac{s}{s^2+1} \Rightarrow \bar{x} + s\bar{y} = \frac{s}{s^2+1} \quad \text{...(4)}$$

Solving (3) and (4) for $\bar{x}$ and $\bar{y}$, we get

$$\bar{x} = \frac{2s}{s^2-1} \text{ and } \bar{y} = \frac{1}{1+s^2} + \frac{2}{1-s^2}$$

$$\bar{x} = \frac{1}{s+1} + \frac{1}{s-1} \quad \text{...(5)}$$

and $$\bar{y} = \frac{1}{1+s^2} + \frac{1}{s+1} - \frac{1}{s-1} \quad \text{... (6) (By partial fractions)}$$

Taking Inverse Laplace transform on both sides of (5) and (6), we get

$$x = e^{-t} + e^{t}$$

and $$y = \sin t + e^{-t} - e^{t}$$ **Ans.**

Example 65. *The co-ordinates (x, y) of a particle moving along a plane curve at any time t are given by*

$$\frac{dy}{dt} + 2x = \sin 2t, \quad \frac{dx}{dt} - 2y = \cos 2t; \; (t > 0)$$

It is given that at t = 0, x = 1 and y = 0. Show using transforms that the particle moves along the curve $4x^2 + 4xy + 5y^2 = 4$. [*U.P. II Semester 2003*]

Solution. Here, we have

$$\begin{bmatrix} \dfrac{dy}{dt} + 2x = \sin 2t \\ \dfrac{dx}{dt} - 2y = \cos 2t \end{bmatrix} \Rightarrow \begin{bmatrix} 2x + Dy = \sin 2t & \text{...(1)} \\ Dx - 2y = \cos 2t, & \text{...(2)} \end{bmatrix}$$

Taking Laplace transform of (1) on both sides, we get

$$2\bar{x} + s\bar{y} - y(0) = \frac{2}{s^2+4} \qquad \text{where } \bar{x} = L(x) \text{ and } \bar{y} = L(y)$$

$$2\bar{x} + s\bar{y} = \frac{2}{s^2+4} \qquad \text{...(3) } [\because y(0) = 0]$$

Again, taking Laplace transform of equation, (2) on both sides, we get

$$s\bar{x} - x(0) - 2\bar{y} = \frac{s}{s^2+4}, \text{ where } \bar{x} = L(x) \text{ and } \bar{y} = L(y)$$

$$\Rightarrow \quad s\bar{x} - 2\bar{y} = \frac{s}{s^2+4} + 1 \qquad \text{...(4) } [\because x(0) = 0]$$

Multiplying equation (3) by 2 and equation (4) by s and then adding, we get

$$s\bar{x} + s^2\bar{x} = \frac{4}{s^2+4} + \frac{s^2}{s^2+4} + s$$

$$(4 + s^2)\,\bar{x} = 1 + s$$

$$\bar{x} = \frac{1+s}{4+s^2} = \frac{1}{s^2+4} + \frac{s}{s^2+4} \qquad \text{...(5)}$$

Taking Inverse Laplace transform of (5), we get

$$x = \frac{1}{2}\sin 2t + \cos 2t \qquad \text{...(6)}$$

Again, multiplying (3) by s and (4) by 2 then subtracting equation (6) from (3), we get

$$s^2\bar{y} + 4\bar{y} = \frac{2s}{s^2+4} - \frac{2s}{s^2+4} - 2$$

$$\Rightarrow \quad \bar{y} = \frac{-2}{s^2+4} \qquad \text{...(7)}$$

Taking Inverse Laplace transform of (7), we get $y = -\sin 2t$

Now, $$4x^2 = 4\left[\frac{1}{4}\sin^2 2t + \cos^2 2t + \sin 2t \cos 2t\right]$$

$$5y^2 = 5\sin^2 2t$$

$$4xy = 4\left[\left(\frac{1}{2}\sin 2t + \cos 2t\right).(-\sin 2t)\right]$$

$$= -(2\sin^2 2t + 4\sin 2t\cos 2t)$$

$$\therefore \qquad 4x^2 + 5y^2 + 4xy = 4\sin^2 2t + 4\cos^2 2t = 4$$ **Ans.**

Example 66. *Solve the following simultaneous differential equations by Laplace transform*

$$\frac{dx}{dt} + 4\frac{dy}{dt} - y = 0; \quad \frac{dx}{dt} + 2y = e^{-t}$$

with conditions $x(0) = y(0) = 0.$ [*U.P., II Semester, 2008*]

Solution. Here, we have

$$\frac{dx}{dt} + 4\frac{dy}{dt} - y = 0 \qquad \text{...(1)}$$

and $$\frac{dx}{dt} + 2y = e^{-t} \qquad \text{...(2)}$$

Taking Laplace transform on both sides of equation (1), we get

$$L(x') + 4L(y') - L(y) = L(0)$$

$$\Rightarrow \quad s\bar{x} - x(0) + 4[s\bar{y} - y(0)] - \bar{y} = 0$$

$$\Rightarrow \quad s\bar{x} + (4s - 1)\bar{y} = 0 \qquad \text{...(3)}$$

Again, taking Laplace transform on both sides of equation (2), we get

$$L(x') + 2L(y) = L(e^{-t})$$

$$\Rightarrow \quad s\bar{x} - x(0) + 2\bar{y} = \frac{1}{s+1} \quad \Rightarrow \quad s\bar{x} + 2\bar{y} = \frac{1}{s+1} \qquad \text{...(4)}$$

Subtracting (4) from (3), we get

$$(4s - 3)\bar{y} = -\frac{1}{s+1}$$

$$\bar{y} = -\frac{1}{(s+1)(4s-3)} = -\frac{1}{7}\left(\frac{-1}{s+1} + \frac{1}{s-3/4}\right) = \frac{1}{7}\left(\frac{1}{s+1} - \frac{1}{s-3/4}\right) \qquad \text{...(5)}$$

Taking inverse Laplace transform on both sides of (5), we get

$$y = \frac{1}{7}\left(e^{-t} - e^{\frac{3t}{4}}\right) \qquad \text{...(6)}$$

Substituting $\bar{y}$ in (4), we get

$$s\bar{x} + \frac{2}{7}\left(\frac{1}{s+1} - \frac{1}{s-3/4}\right) = \frac{1}{s+1}$$

$$\Rightarrow \quad s\bar{x} = \frac{5}{7(s+1)} + \frac{2}{7(s-3/4)}$$

$$\Rightarrow \quad \bar{x} = \frac{5}{7s(s+1)} + \frac{2}{7s(s-3/4)} = \frac{5}{7}\left(\frac{1}{s} - \frac{1}{s+1}\right) + \frac{8}{21}\left(\frac{1}{s-3/4} - \frac{1}{s}\right)$$

$$= \frac{1}{3s} - \frac{5}{7(s+1)} + \frac{8}{21(s-3/4)} \qquad \text{...(7)}$$

Taking Inverse Laplace transform on both sides of (7), we get

$$x = \frac{1}{3} - \frac{5}{7}e^{-t} + \frac{8}{21}e^{\frac{3t}{4}}$$ **Ans.**

Example 67. *Using Laplace Transformation, solve*

$$(D-2)\,x-(D+1)\,y=6\,e^{3t} \quad \dots (1)$$

$$(2D-3)\,x+(D-3)\,y=6e^{3t} \quad \dots (1)$$

Given $x=3,\ y=0$ *when* $t=0$. (*U.P., II Semester Summer* 2001)

Solution. Taking Laplace transformation of the given equations, we get

$$\begin{bmatrix} L\,Dx-2\,Lx-LDy-Ly=6\,Le^{3t} \\ 2L\,Dx-3Lx+LDy-3Ly=6\,Le^{3t} \end{bmatrix}$$

$$\Rightarrow \begin{bmatrix} s\,\bar{x}-x(0)-2\,\bar{x}-s\bar{y}+y(0)-\bar{y}=6\dfrac{1}{s-3} \\ 2s\,\bar{x}-2x(0)-3\,\bar{x}+s\,\bar{y}-y(0)-3\,\bar{y}=\dfrac{6}{s-3} \end{bmatrix}$$

$$\Rightarrow \begin{bmatrix} (s-2)\,\bar{x}-(s+1)\,\bar{y}-3=\dfrac{6}{s-3} \\ (2s-3)\,\bar{x}+(s-3)\,\bar{y}-6=\dfrac{6}{s-3} \end{bmatrix} \Rightarrow \begin{bmatrix} (s-2)\,\bar{x}-(s+1)\,\bar{y}=\dfrac{3s-3}{s-3} \\ (2s-3)\,\bar{x}+(s-3)\,\bar{y}=\dfrac{6s-12}{s-3} \end{bmatrix}$$

$$\Rightarrow \begin{bmatrix} (s-3)\,(s-2)\,\bar{x}-(s-3)\,(s+1)\,\bar{y}=3s-3 \\ (s+1)\,(2s-3)\,\bar{x}+(s+1)\,(s-3)\,\bar{y}=\dfrac{(s+1)\,(6s-12)}{s-3} \end{bmatrix} \text{ on adding, we get:}$$

$$(3s^2-6s+3)\,x=3\,(s-1)+\frac{6\,(s^2-s-2)}{s-3} \Rightarrow \bar{x}=\frac{3\,(s-1)}{3\,(s-1)^2}+\frac{6\,(s^2-s-2)}{3\,(s-1)^2\,(s-3)}$$

$$x=L^{-1}\left[\frac{1}{s-1}+\frac{2}{(s-1)^2}+\frac{2}{s-3}\right]=e^t+2t\,e^t+2e^{3t}$$

Putting the value of x in (1), we get

$$(D-2)\,(e^t+2te^t+2e^{3t})-(D+1)\,y=6e^{3t}$$

$$\Rightarrow \quad e^t+2te^t+2e^t+6e^{3t}-2e^t-4te^t-4e^{3t}-(D+1)\,y=6e^{3t}$$

$$\Rightarrow \quad (D+1)\,y=e^t-2te^t-4e^{3t} \quad \dots (2)$$

Taking Laplace transform of (2), we get

$$s\,\bar{y}-y(0)+\bar{y}=\frac{1}{s-1}-\frac{2}{(s-1)^2}-\frac{4}{s-3}$$

$$\Rightarrow \quad (s+1)\,\bar{y}=\frac{1}{s-1}-\frac{2}{(s-1)^2}-\frac{4}{s-3} \qquad [\because y(0)=0]$$

$$\Rightarrow \quad \bar{y}=\frac{1}{s^2-1}-\frac{2}{(s+1)\,(s-1)^2}-\frac{4}{(s+1)\,(s-3)}$$

$$=\frac{1}{s^2-1}-\frac{\frac{1}{2}}{s+1}+\frac{\frac{1}{2}}{s-1}-\frac{1}{(s-1)^2}+\frac{1}{s+1}-\frac{1}{s-3}$$

$$\Rightarrow \quad \bar{y}=\frac{1}{s^2-1}+\frac{\frac{1}{2}}{s+1}+\frac{\frac{1}{2}}{s-1}-\frac{1}{(s-1)^2}-\frac{1}{s-3}$$

$$\Rightarrow \quad y=L^{-1}\left[\frac{1}{s^2-1}+\frac{1}{2}\frac{1}{s+1}+\frac{1}{2}\frac{1}{s-1}-\frac{1}{s-3}-\frac{1}{(s-1)^2}\right]$$

$$\Rightarrow \qquad y = \sinh t + \frac{1}{2}e^{-t} + \frac{1}{2}e^{t} - e^{3t} - te^{t}$$

$$y = \sinh t + \cosh t - e^{3t} - te^{t}$$ **Ans.**

Example 68. *Solve the simultaneous equations*

$$(D^2 - 3)x - 4y = 0$$
$$x + (D^2 + 1)y = 0$$

for $t > 0$, given that $x = y = \frac{dy}{dt} = 0$ and $\frac{dx}{dt} = 2$ at $t = 0$. [U.P., II Semester, 2004]

Solution. Here, we have

$$(D^2 - 3)x - 4y = 0 \qquad ...(1)$$
$$x + (D^2 + 1)y = 0 \qquad ...(2)$$

Taking Laplace transform of (1) and (2), we get

$$s^2\bar{x} - sx(0) - x'(0) - 3\bar{x} - 4\bar{y} = 0$$

i.e., $$(s^2 - 3)\bar{x} - 4\bar{y} = 2 \qquad ...(3)[\because x(0) = 0, x'(0) = 2]$$

and $$\bar{x} + s^2\bar{y} - sy(0) - y'(0) + \bar{y} = 0$$

i.e., $$\bar{x} + (s^2 + 1)\bar{y} = 0 \qquad ...(4)[\because y(0) = 0, y'(0) = 0]$$

Solving (3) and (4) for $\bar{x}$ and $\bar{y}$, we get

$$\bar{x} = \frac{2(s^2+1)}{(s^2-1)^2} = \frac{1}{(s-1)^2} + \frac{1}{(s+1)^2} \qquad ...(5)$$

and $$\bar{y} = -\frac{2}{(s^2-1)^2} = -\frac{1}{2}\left[\frac{1}{s+1} - \frac{1}{s-1} - \frac{1}{(s+1)^2} + \frac{1}{(s-1)^2}\right] \qquad ...(6)$$

Taking Inverse Laplace transform of both sides of (5) and (6), we get

$$x = L^{-1}\left[\frac{1}{(s-1)^2} + \frac{1}{(s+1)^2}\right] = te^{t} + te^{-t} = 2t\left(\frac{e^{t}+e^{-t}}{2}\right) = 2t\cosh t$$

and $$y = -\frac{1}{2}L^{-1}\left(\frac{1}{s+1} - \frac{1}{s-1} - \frac{1}{(s+1)^2} + \frac{1}{(s-1)^2}\right)$$

$$= -\frac{1}{2}(e^{-t} - e^{t} - te^{-t} + te^{t}) = \frac{e^{t}-e^{-t}}{2} - t\left(\frac{e^{t}-e^{-t}}{2}\right) = (1-t)\sinh t$$

Hence, $$x = 2t\cosh t,\ y = (1-t)\sinh t.$$ **Ans.**

EXERCISE 43.11

Solve the following:

1. $\frac{dx}{dt} + 4y = 0$, $\frac{dy}{dt} - 9x = 0$. Given $x = 2$ and $y = 1$ at $t = 0$.

Ans. $x = -\frac{2}{3}\sin 6t + 2\cos 6t,\ y = \cos 6t + 3\sin 6t$

2. $4\frac{dy}{dt} + \frac{dx}{dt} + 3y = 0,\ 3\frac{dx}{dt} + 2x + \frac{dy}{dt} = 1$ under the condition $x = y = 0$ at $t = 0$.

Ans. $x = \frac{1}{2} - \frac{1}{5}e^{-t} - \frac{3}{10}e^{-\frac{6}{11}t},\ y = \frac{1}{5}e^{-t} - \frac{1}{5}e^{-\frac{6}{11}t}$

3. $\frac{dx}{dt} + 5x - 2y = t, \frac{dy}{dt} + 2x + y = 0$ being given $x = y = 0$ when $t = 0$.

Ans. $x = -\frac{1}{27}(1 + 6t)\, e^{-3t} + \frac{1}{27}(1 + 3t),\ y = -\frac{2}{27}(2 + 3t)\, e^{-3t} - \frac{2t}{9} + \frac{4}{27}$

43.15 SOLUTION OF PARTIAL DIFFERENTIAL EQUATION BY LAPLACE TRANSFORM

Example 69. *Solve the differential equation using Laplace transform method:*

$\frac{\partial y}{\partial t} = 3\frac{\partial^2 y}{\partial t^2}$ *where* $y\left(\frac{\pi}{2}, t\right) = 0, \left(\frac{\partial y}{\partial x}\right)_{x=0} = 0$ *and* $y(x, 0) = 30 \cos 5x$.

(U.P., II Semester Summer 2005)

Solution. Given equation is

$$\frac{\partial y}{\partial t} = 3\frac{\partial^2 y}{\partial t^2}$$

Taking Laplace transform of both sides, we get

$$sL\{y\} - y(x, 0) = 3\frac{d^2}{dx^2} L\{y\}, \qquad [\text{Let } L\{y\} = \bar{y}]$$

$$s\,\bar{y} - y(x, 0) = 3\frac{d^2\,\bar{y}}{dx^2}$$

$$3\frac{d^2\bar{y}}{dx^2} - s\,\bar{y} = -30 \cos 5x$$

$$\left(D^2 - \frac{s}{3}\right)\bar{y} = -10 \cos 5x$$

A.E. is $\quad m^2 - \frac{s}{3} = 0 \quad \Rightarrow \quad m = \pm\sqrt{\frac{s}{3}}$

$$C.F. = C_1\, e^{x\sqrt{s/3}} + C_2\, e^{-x\sqrt{s/3}}$$

$$P.I. = \frac{1}{\left(D^2 - \frac{s}{3}\right)}(-10 \cos 5x)$$

$$P.I. = \frac{30 \cos 5x}{75 + s}$$

Thus $\quad \bar{y} = C_1\, e^{-x\sqrt{s/3}} + C_2\, e^{x\sqrt{s/3}} + \frac{30 \cos 5x}{75 + s} \quad \ldots (1)$

$$\frac{\partial y}{\partial x} = 0 \quad \text{when} \quad x = 0$$

$$\Rightarrow \quad L\left\{\frac{\partial y}{\partial x}\right\} = 0 \quad \text{at} \quad x = 0$$

$$\Rightarrow \quad \frac{d\bar{y}}{dx} = 0 \quad \text{at} \quad x = 0$$

Again, $\quad y\left(\frac{\pi}{2}, t\right) = 0 \quad \Rightarrow \quad L\left\{y\left(\frac{\pi}{2}, t\right)\right\} = 0 \quad \Rightarrow \quad \bar{y}\left(\frac{\pi}{2}, s\right) = 0$

$$\frac{d\,\bar{y}}{dx} = \sqrt{\frac{s}{3}}\left[Ae^{x\sqrt{s/3}} - Be^{-x\sqrt{s/3}} - \frac{150 \sin 5x}{75 + s}\right]$$

Putting $\quad \frac{d\,\bar{y}}{dx} = 0 \quad \text{at} \quad x = 0$

$$0 = \sqrt{\frac{s}{3}}\,[A - B] \qquad \Rightarrow \qquad A = B$$

Equation (1) becomes,

$$\bar{y} = A\left[e^{x\sqrt{s/3}} + e^{-x\sqrt{s/3}}\right] + \frac{30\cos 5x}{75+s} \qquad ..(2)$$

Subjecting this to the condition

$$\bar{y}\left(\frac{\pi}{2}, s\right) = 0$$

$$0 = A\left[e^{\frac{\pi}{2}\sqrt{s/3}} + e^{-\frac{\pi}{2}\sqrt{s/3}}\right] + \frac{30\cos(5\pi/2)}{75+s}$$

$$24\cosh\left[\frac{\pi}{2}\sqrt{\frac{s}{3}}\right] = 0 \qquad \Rightarrow \qquad A = 0$$

From equation (2), $\bar{y} = \dfrac{30\cos 5x}{75+s}$, taking inverse Laplace, we get

$$y = 30e^{-75t}\cos 5x \qquad \textbf{Ans.}$$

EXERCISE 43.12

Solve the following differential equation using Laplace transform method:

1. $\dfrac{\partial u}{\partial t} = \dfrac{\partial^2 u}{\partial x^2}$ if $u(x, 0) = \sin \pi x$ **Ans.** $u = \sin \pi x.\, e^{-p^2 t}$

2. $\dfrac{\partial u}{\partial t} = 4\dfrac{\partial^2 u}{\partial x^2}$ if $u(x, 0) = 4x - \dfrac{1}{2}x^2$ **Ans.** $u = \left(4x - \dfrac{x^2}{2}\right)e^{-p^2 t}$

3. $\dfrac{\partial^2 u}{\partial t^2} = \dfrac{\partial^2 u}{\partial x^2}$ if $u(x, 0) = \dfrac{1}{2}x(1-x)$

Ans. $u = \dfrac{x}{2}(1-x)\cos pt + C_2 \sin pt\,(C_3 \cos px + C_4 \sin px)$

4. $16\dfrac{\partial^2 u}{\partial t^2} = \dfrac{\partial^2 u}{\partial x^2}$ if $u(x, 0) = x^2(5-x)$

Ans. $u = x^2(5-x)\cos pt + C_4 \sin pt\left(C_1 \cos\dfrac{px}{4} + C_2 \sin\dfrac{px}{4}\right)$

5. $\dfrac{\partial u}{\partial t} = \dfrac{\partial^2 u}{\partial x^2}$ if $u(x, 0) = \begin{cases} 2x, & \text{when } 0 \le x \le \frac{1}{2} \\ 2(1-x), & \text{when } \frac{1}{2} \le x \le 1 \end{cases}$

Z - TRANSFORMS

44.1 INTRODUCTION

Z-transform plays an important role in discrete analysis. Its role in discrete analysis is the same as that of Laplace and Fourier transforms in continuous system. Communication is one of the field whose development is based on discrete analysis. Difference equations are also based on discrete system and their solutions and analysis are carried out by Z- transform.

44.2 SEQUENCE

Sequence $\{f(k)\}$ is an ordered list of real or complex numbers.

44.3 REPRESENTATION OF A SEQUENCE

First Method.

The elementary way is to list all the members of the sequence such as

$$\{f(k)\} = \{15, 10, 7, \underset{\uparrow}{4}, 1, -1, 0, 3, 6\}$$

The symbol $\uparrow$ is used to denote the term in zero position *i.e.*, $k = 0$, k is an index of position of a term in the sequence.

$$\{g(k)\} = \{15, 10, 7, 4, 1, \underset{\uparrow}{-1}, 0, 3, 6\}$$

Two sequences $\{f(k)\}$ and $\{g(k)\}$ have the same terms but these sequences are not identically the same as the zeroth term of those sequences are different.

In case the symbol $\uparrow$ is not given then left hand end term is considered as the term corresponding to $k = 0$.

In sequence $\{8, 6, 3, -1, 0, 1, 4, 5\}$

the zeroth term is 8, the left hand end term.

Second Method.

The second way of specifying the sequence is to define the general term of the sequence $\{f(k)\}$ as function of k.

For example, $$f(k) = \frac{1}{3^k}$$

This sequence represents $$\left\{\ldots\frac{1}{3^{-3}}, \frac{1}{3^{-2}}, \frac{1}{3^{-1}}, \underset{\substack{\uparrow \\ k=0}}{1}, \frac{1}{3}, \frac{1}{3^2}, \frac{1}{3^3}\ldots\right\}$$

If $f(k)=\frac{1}{3^k}, -4 \le k \le 3$ $\left\{\frac{1}{3^{-4}}, \frac{1}{3^{-3}}, \frac{1}{3^{-2}}, \frac{1}{3^{-1}}, 1, \frac{1}{3^1}, \frac{1}{3^2}, \frac{1}{3^3}\right\}$

44.4 BASIC OPERATIONS ON SEQUENCES

Let $\{f(k)\}$ and $\{g(k)\}$ be two sequences having same number of terms.

Addition. $\{f(k)\} + \{g(k)\} = \{f(k) + g(k)\}$

Multiplication. Let a be a scalar, then $a\{f(k)\} = \{a f(k)\}$

Linearity. $a\{f(k)\} + b\{g(k)\} = \{a f(k) + b g(k)\}$

EXERCISE 44.1

1. Write down the term corresponding to $k = 2$

$$(6, 7, 5, 1, 0, \underset{\uparrow}{4}, 6, 8, 10)$$

Ans. 8

2. Write down the term corresponding to $k = -3$.

$$(20, 16, 14, 13, 12, \underset{\uparrow}{10}, 5, 1, 0)$$

Ans. 14

3. Write down the sequence $\{f(k)\}$ where $f(k)=\frac{1}{2^k}$ **Ans.** $\left\{\ldots\frac{1}{2^{-3}}, \frac{1}{2^{-2}}, \frac{1}{2^{-1}}, 1, \frac{1}{2^1}, \frac{1}{2^2}, \frac{1}{2^3}\ldots\right\}$

4. Write down the sequence $\{f(k)\}$ where $f(k)=\frac{1}{4^k}, \quad -3 \le k \le 4$

Ans. $\left\{\frac{1}{4^{-3}}, \frac{1}{4^{-2}}, \frac{1}{4^{-1}}, 1, \frac{1}{4^1}, \frac{1}{4^2}, \frac{1}{4^3}, \frac{1}{4^4}\right\}$

5. Write down the sequence $\frac{1}{2}\{f(k)\}$, where $f(k)=\frac{1}{3^k}$

Ans. $\left\{\frac{27}{2}, \frac{9}{2}, \frac{3}{2}, \frac{1}{2}, \frac{1}{6}, \frac{1}{18}, \frac{1}{54}\ldots\right\}$

6. What sequence is generated when $f(k) = \begin{cases} 0, & k < 0 \\ \cos\frac{k}{2}, & k \ge 0 \end{cases}$

Ans. $\left\{\ldots 0, 0, 1, \cos\frac{1}{2}, \cos 1, \cos\frac{3}{2}\ldots\right\}$

7. Write the sequence $\frac{1}{3}\{f(k)\} + \frac{1}{4}\{g(k)\} = \{F(k)\}$. Where $f(k) = \frac{1}{3^k}$, $g(k)$

$= \begin{cases} 0, & k < 0 \\ 4, & k \ge 0 \end{cases}$ **Ans.** $\{F(k)\} = \begin{cases} \frac{1}{3^{k+1}}, & k < 0 \\ \frac{1}{3^{k+1}} + 1, & k \ge 0 \end{cases}$

44.5 Z-TRANSFORM

Definition. The Z- transform of a sequence $\{f(k)\}$ is denoted as $Z[\{f(k)\}]$.

It is defined as $Z[\{f(k)\}] = F(z) = \sum_{k=-\infty}^{\infty} f(k) z^{-k} = \sum_{k=-\infty}^{\infty} \frac{f(k)}{z^k}$

where **1.** Z is a complex number.

2. Z is an operator of Z-transform.

3. $F(z)$ is the Z transform of $\{f(k)\}$.

Example 1. *If* $f(k) = \{15, 10, 7, \underset{\uparrow}{4}, 1, -1, 0, 3, 6\}$, *then*

$$Z[\{f(k)\}] = F(z) = 15z^3 + 10z^2 + 7z + 4 + \frac{1}{z} - \frac{1}{z^2} + 0 + \frac{3}{z^4} + \frac{6}{z^5}$$

Example 2. *If* $g(k) = \{15, 10, 7, 4, 1, -1, 0, \underset{\uparrow}{3}, 6\}$

$$Z[\{g(k)\}] = F(z) = 15z^7 + 10z^6 + 7z^5 + 4z^4 + z^3 - z^2 + 0 + 3 + \frac{6}{z}$$

Example 3. *The Z-transform of the sequence {8, 6, 3 –1, 0, 14, 5} is*

$$8 + \frac{6}{z} + \frac{3}{z^2} - \frac{1}{z^3} + 0 + \frac{14}{z^5} + \frac{5}{z^6}$$

Example 4. *If* $f(k) = \frac{1}{3^k}$ *then* $Z[\{f(k)\}] = \ldots + 27z^3 + 9z^2 + 3z + 1 + \frac{1}{3z} + \frac{1}{9z^2} + \frac{1}{27z^3} + \ldots$

Example 5. *If* $f(k) = \frac{1}{3^k}, \ -4 \le k \le 3$, *then*

$$Z[\{f(k)\} = 81z^4 + 27z^3 + 9z^2 + 3z + 1 + \frac{1}{3z} + \frac{1}{9z^2} + \frac{1}{27z^3}$$

Example 6. *Find Z-transform of the sequence* $\left\{\frac{1}{2^k}\right\} -4 \le k \le 4$

Solution. $F(z) = \sum_{k=-4}^{4} \frac{1}{2^k} z^{-k} = 16z^4 + 8z^3 + 4z^2 + 2z + 1 + \frac{1}{2z} + \frac{1}{4z^2} + \frac{1}{8z^3} + \frac{1}{16z^4}$ **Ans.**

Example 7. *Find Z-transform of the sequence* $\{a^k\}$, $k \ge 0$.

Solution. $F(z) = \sum_{k=0}^{\infty} a^k z^{-k} = 1 + \frac{a}{z} + \frac{a^2}{z^2} + \frac{a^3}{z^3} + \ldots$

This is a Geometrical series whose sum $= \frac{a}{1-r} = \frac{1}{1-\frac{a}{z}} = \frac{z}{z-a}$ **Ans.**

44.6 PROPERTIES OF Z-TRANSFORMS

Linearity.

Theorem 1: If $\{f(k)\}$ and $\{g(k)\}$ are such that they can be added and a and b are constants, then

$$Z\{af(k) + bg(k)\} = a\,Z[\{f(k)\}] + b\,Z[\{g(k)\}]$$

Proof. $Z[\{af(k) + bg(k)\}] = \sum_{k=-\infty}^{\infty} [af(k) + bg(k)]\, z^{-k}$ [By definition]

$$= \sum_{k=-\infty}^{\infty} [af(k)z^{-k} + bg(k)z^{-k}] = a\sum_{k=-\infty}^{\infty} f(k)z^{-k} + b\sum_{k=-\infty}^{\infty} g(k)z^{-k}$$

$$= aZ[\{f(k)\}] + bZ[\{g(k)\}]$$ **Proved.**

Example 8. *Find the Z transform of* $\{f(k)\}$, *where*

$$f(k) = \begin{cases} 5^k, & k < 0 \\ 3^k, & k \ge 0 \end{cases}$$

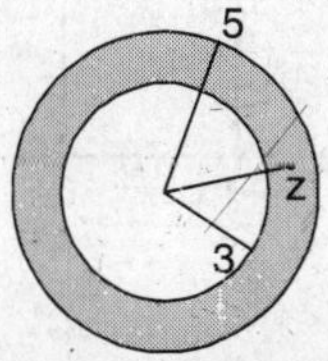

Solution. $Z[\{f(k)\}] = \sum_{k=-\infty}^{-1} 5^k z^{-k} + \sum_{k=0}^{\infty} 3^k z^{-k}$

$$= [\ldots + 5^{-3} z^3 + 5^{-2} z^2 + 5^{-1} z^1] + \left[1 + \frac{3}{z^{-1}} + \frac{9}{z^{-2}} + \frac{27}{z^{-3}} + \ldots\right] \qquad \text{[G. P.]}$$

$$= \frac{5^{-1} z}{1 - 5^{-1} z} + \frac{1}{1 - \frac{3}{z^{-1}}} = \frac{z}{5 - z} + \frac{z}{z - 3} \qquad \left[S = \frac{a}{1 - r}\right]$$

$$= \frac{z^2 - 3z + 5z - z^2}{(5 - z)(z - 3)} = \frac{-2z}{z^2 - 8z + 15}, \qquad \left|\frac{z}{5}\right| < 1, \ \left|\frac{3}{z}\right| < 1$$

Two series are convergent in annulus. Here $3 < |z|$ and $|z| < 5$. **Ans.**

Example 9. *Find the Z-transform of* $\{a^{|k|}\}$

Solution. $Z\left[\{a^{|k|}\}\right] = \sum_{k=-\infty}^{\infty} a^{|k|} z^{-k} = \sum_{k=-\infty}^{-1} a^{-k} z^{-k} + \sum_{k=0}^{\infty} a^k z^{-k}$

$$= [\ldots + a^3 z^3 + a^2 z^2 + az] + [1 + az^{-1} + a^2 z^{-2} + a^3 z^{-3} + \ldots]$$

These are two G.P. and sum of G.P. $= \dfrac{a}{1 - r}$

$$= \frac{az}{1 - az} + \frac{1}{1 - az^{-1}}, \qquad |az| < 1 \text{ and } |az^{-1}| < 1|$$

$$= \frac{az}{1 - az} + \frac{z}{z - a} = \frac{az(z - a) + z(1 - az)}{(1 - az)(z - a)} = \frac{az^2 - a^2 z + z - az^2}{(1 - az)(z - a)} = \frac{z - a^2 z}{(1 - az)(z - a)} \qquad \textbf{Ans.}$$

Example 10. *Find the Z-transform of* $\left\{\left(\frac{1}{2}\right)^{|k|}\right\}$

Solution. $Z\left[\left\{\left(\frac{1}{2}\right)^{|k|}\right\}\right] = \sum_{-\infty}^{\infty} \left(\frac{1}{2}\right)^{|k|} z^{-k} = \sum_{k=-\infty}^{-1} \left(\frac{1}{2}\right)^{-k} z^{-k} + \sum_{k=0}^{\infty} \left(\frac{1}{2}\right)^{-k} z^{-k}$

$$= \left(\ldots + \frac{z^4}{16} + \frac{z^3}{8} + \frac{z^2}{4} + \frac{z}{2}\right) + \left(1 + \frac{1}{2z} + \frac{1}{4z^2} + \frac{1}{8z^3} + \ldots\right)$$

These infinite series are G.P., and sum of a G.P. $= \dfrac{a}{1 - r}$

$$= \frac{\frac{z}{2}}{1 - \frac{z}{2}} + \frac{1}{1 - \frac{1}{2z}} = \frac{z}{2 - z} + \frac{2z}{2z - 1} = \frac{2z^2 - z + 4z - 2z^2}{(2 - z)(2z - 1)} = \frac{3z}{(2 - z)(2z - 1)} \qquad \textbf{Ans.}$$

Example 11. *Express the signals shown below in terms of unit impulse functions and hence find the Z-transform.*

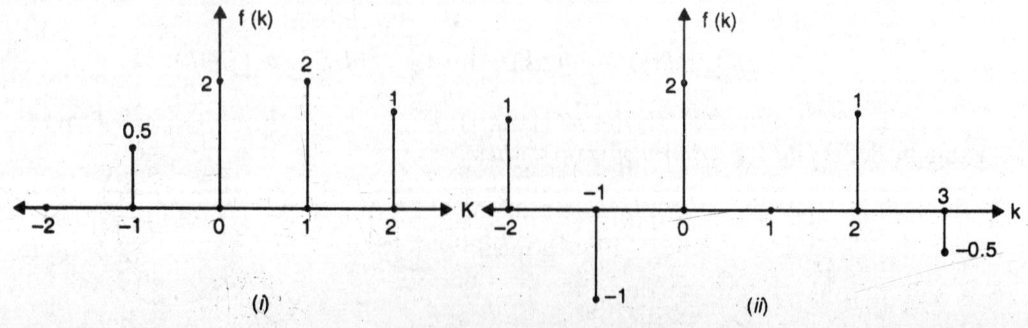

Solution. (*i*) $f(k) = 0.5\,\delta(k+1) + 2\,\delta(k) + 2\,\delta(k-1) + \delta(k-2)$

(*ii*) $f(k) = \delta(k+2) - \delta(k+1) + 2\,\delta(k) + \delta(k-2) - 0.5\,\delta(k-3)$

Example 12. *Find the Z-transform of unit impulse*

$$\delta(k) = \begin{cases} 1, & k = 0 \\ 0, & k \neq 0 \end{cases}$$

Solution. $Z[\{f(k)\}] = \sum_{k=-\infty}^{\infty} \delta(k) z^{-k} = [.....0 + 0 + 1 + 0 + 0....] = 1$ **Ans.**

Example 13. *Find the Z-transform of discrete unit step*

$$U(k) = \begin{cases} 0, & k < 0 \\ 1, & k \geq 0 \end{cases}$$

Solution. $Z[\{U(k)\}] = \sum_{k=0}^{\infty} U(k) z^{-k} = [1 + z^{-1} + z^{-2} + z^{-3} + ...]$

$$= \frac{1}{1-z^{-1}} = \frac{1}{1-\frac{1}{z}} = \frac{z}{z-1} \qquad \left[\text{This is G.P., its sum is } \frac{a}{1-r}\right] \quad \textbf{Ans.}$$

Example 14. *Find the Z-transform of sin α k,* $k \geq 0$

Solution. $Z[\{\sin \alpha k\}] = \sum_{k=0}^{\infty} \sin \alpha k\, z^{-k} = \sum_{k=0}^{\infty} \frac{e^{i\alpha k} - e^{-i\alpha k}}{2i} z^{-k}$

$$= \frac{1}{2i}\sum_{k=0}^{\infty} e^{i\alpha k} z^{-k} - \frac{1}{2i}\sum_{k=0}^{\infty} e^{-i\alpha k} z^{-k} = \frac{1}{2i}\sum_{k=0}^{\infty} (e^{i\alpha} z^{-1})^k - \frac{1}{2i}\sum_{k=0}^{\infty} (e^{-i\alpha} z^{-1})^k$$

$$= \frac{1}{2i}\left[1 + (e^{i\alpha} z^{-1}) + (e^{i\alpha} z^{-1})^2 + ...\right] - \frac{1}{2i}\left[1 + (e^{-i\alpha} z^{-1}) + (e^{-i\alpha} z^{-1})^2 + ...\right]$$

These infinite series are G.P. and sum of a G.P. $= \frac{a}{1-r}$

$$= \frac{1}{2i}\frac{1}{1-e^{i\alpha} z^{-1}} - \frac{1}{2i}\frac{1}{1-e^{-i\alpha} z^{-1}} = \frac{1}{2i}\frac{z}{z-e^{i\alpha}} - \frac{1}{2i}\frac{z}{z-e^{-i\alpha}}$$

$$= \frac{1}{2i}\left[\frac{z}{z-e^{i\alpha}} - \frac{z}{z-e^{-i\alpha}}\right] = \frac{1}{2i}\frac{z(z-e^{-i\alpha}) - z(z-e^{i\alpha})}{(z-e^{i\alpha})(z-e^{-i\alpha})}$$

$$= \frac{1}{2i}\frac{z(e^{i\alpha} - e^{-i\alpha})}{z^2 - z(e^{i\alpha} + e^{-i\alpha}) + 1} = \frac{z \sin \alpha}{z^2 - 2z\cos\alpha + 1} \quad \textbf{Ans.}$$

Example 15. *Find the Z-transform of* $c^k \cosh \alpha k$. $(k \geq 0)$.

Solution. $Z[\{c^k \cosh \alpha k\}] = \sum_{k=0}^{\infty} (c^k \cosh \alpha k) z^{-k} = \sum_{k=0}^{\infty} \frac{c^k}{2}(e^{\alpha k} + e^{-\alpha k}) z^{-k}$

$$= \frac{1}{2}\sum_{k=0}^{\infty} (ce^{\alpha} z^{-1})^k + \frac{1}{2}\sum_{k=0}^{\infty} (ce^{-\alpha} z^{-1})^k$$

$$= \frac{1}{2}\left[1 + (ce^{\alpha} z^{-1}) + (ce^{\alpha} z^{-1})^2 + ...\right] + \frac{1}{2}\left[1 + (ce^{-\alpha} z^{-1}) + (ce^{-\alpha} z^{-1})^2 + ...\right] \quad \left[|\,z\,| > |\,ce^{\alpha}\,|, |\,z\,| > |\,ce^{-\alpha}\,|\right]$$

$$=\frac{1}{2}\frac{1}{1-ce^{\alpha}z^{-1}}+\frac{1}{2}\frac{1}{1-ce^{-\alpha}z^{-1}}=\frac{1}{2}\frac{1-ce^{-\alpha}z^{-1}+1-ce^{\alpha}z^{-1}}{(1-ce^{\alpha}z^{-1})(1-ce^{-\alpha}z^{-1})}$$

$$=\frac{1-c(\frac{e^{\alpha}+e^{-\alpha}}{2})z^{-1}}{1+c^2z^{-2}-ce^{\alpha}z^{-1}-ce^{-\alpha}z^{-1}}=\frac{1-(c\cosh\alpha)z^{-1}}{1+c^2z^{-2}-2cz^{-1}\cosh\alpha}\qquad \text{(Multiply by } z^2\text{)}$$

$$=\frac{z(z-c\cosh\alpha)}{z^2+c^2-2cz\cosh\alpha}$$

Ans.

Corollary. If $c = 1$, then

$$Z[\{\cosh\alpha k\}]=\frac{z(z-\cosh\alpha)}{z^2-2z\cosh\alpha+1}$$

Example 16. *Find the Z-transform of* $\cosh\left(\frac{k\pi}{2}+\alpha\right)$ *(U.P., III Semester, 2008)*

Solution. $F(z)=\sum_{k=0}^{\infty}\cosh\left(\frac{k\pi}{2}+\alpha\right)z^{-k}=\sum_{k=0}^{\infty}\frac{e^{\frac{k\pi}{2}+\alpha}+e^{-(\frac{k\pi}{2}+\alpha)}}{2}z^{-k}$

$$=\frac{1}{2}\sum_{k=0}^{\infty}e^{\frac{k\pi}{2}+\alpha}z^{-k}+\frac{1}{2}\sum_{k=0}^{\infty}e^{-\frac{k\pi}{2}-\alpha}z^{-k}=\frac{1}{2}e^{\alpha}\sum_{k=0}^{\infty}\left(e^{\frac{\pi}{2}}z^{-1}\right)^k+\frac{1}{2}e^{-\alpha}\sum_{k=0}^{\infty}\left(e^{\frac{-\pi}{2}}z^{-1}\right)^k$$

$$=\frac{1}{2}e^{\alpha}\left[1+\left(e^{\frac{\pi}{2}}z^{-1}\right)+\left(e^{\frac{\pi}{2}}z^{-1}\right)^2+\ldots\right]+\frac{1}{2}e^{-\alpha}\left[1+\left(e^{-\frac{\pi}{2}}z^{-1}\right)+\left(e^{-\frac{\pi}{2}}z^{-1}\right)^2+\ldots\right]$$

(Sum of geometrical series $=\frac{a}{1-r}$)

$$=\frac{1}{2}e^{\alpha}\frac{1}{1-e^{\frac{\pi}{2}}z^{-1}}+\frac{1}{2}e^{-\alpha}\frac{1}{1-e^{-\frac{\pi}{2}}z^{-1}}=\frac{1}{2}\frac{e^{\alpha}\left(1-e^{-\frac{\pi}{2}}z^{-1}\right)+e^{-\alpha}\left(1-e^{\frac{\pi}{2}}z^{-1}\right)}{\left(1-e^{\frac{\pi}{2}}z^{-1}\right)\left(1-e^{-\frac{\pi}{2}}z^{-1}\right)}$$

$$=\frac{\frac{e^{\alpha}+e^{-\alpha}}{2}-\frac{e^{\alpha-\frac{\pi}{2}}+e^{-\alpha+\frac{\pi}{2}}}{2}.z^{-1}}{1-e^{\frac{\pi}{2}}z^{-1}-e^{-\frac{\pi}{2}}z^{-1}+z^{-2}}=\frac{\cosh\alpha-\cosh\left(\alpha-\frac{\pi}{2}\right).z^{-1}}{1-\left(2\cosh\frac{\pi}{2}\right)z^{-1}+z^{-2}}=\frac{z^2\cosh\alpha-z\cosh\left(\frac{\pi}{2}-\alpha\right)}{z^2-2z\cosh\frac{\pi}{2}+1}$$

Ans.

Example 17. *Find Z-transform of sin* $(3k+5)$.

Solution. $F(z)=\sum_{k=0}^{\infty}\sin(3k+5)z^{-k}=\sum_{k=0}^{\infty}\frac{e^{i(3k+5)}-e^{-i(3k+5)}}{2i}z^{-k}$

$$=\frac{1}{2i}\sum_{k=0}^{\infty}e^{i(3k+5)}z^{-k}-\frac{1}{2i}\sum_{k=0}^{\infty}e^{-i(3k+5)}z^{-k}=\frac{1}{2i}e^{i5}\sum_{k=0}^{\infty}(e^{3i}z^{-1})^k-\frac{1}{2i}e^{-5i}\sum_{k=0}^{\infty}(e^{-3i}z^{-1})^k$$

$$=\frac{1}{2i}e^{5i}[1+(e^{3i}z^{-1})+(e^{3i}z^{-1})^2+\ldots]-\frac{1}{2i}e^{-5i}[1+(e^{-3i}z^{-1})+(e^{-3i}z^{-1})^2+\ldots]$$

$$= \frac{e^{i5}}{2i}\frac{1}{1-e^{3i}z^{-1}} - \frac{1}{2i}e^{-5i}\frac{1}{1-e^{-3i}z^{-1}} \qquad \left[S = \frac{a}{1-r}\right]$$

$$= \frac{1}{2i}\frac{e^{i5}(1-e^{-3i}z^{-1}) - e^{-5i}(1-e^{3i}z^{-1})}{(1-e^{3i}z^{-1})(1-e^{-3i}z^{-1})} = \frac{1}{2i}\frac{(e^{i5}-e^{-5i}) - e^{2i}z^{-1} + e^{-2i}z^{-1}}{1-e^{3i}z^{-1} - e^{-3i}z^{-1} + z^{-2}}$$

$$= \frac{\dfrac{e^{i5}-e^{-5i}}{2i} - z^{-1}\dfrac{e^{2i}-e^{-2i}}{2i}}{1-(e^{3i}+e^{-3i})z^{-1}+z^{-2}} = \frac{\sin 5 - z^{-1}\sin 2}{1-(2\cos 3)z^{-1}+z^{-2}}$$

$$= \frac{z^2 \sin 5 - z\sin 2}{z^2 - 2z\cos 3 + 1} \qquad |z| > 1$$ **Ans.**

Example 18. *Find the Z-transform of $c^k \cos \alpha k$, $k \geq 0$.* (*U.P., III Semester, Dec. 2004*)

Solution. $F(z) = \sum_{k=0}^{\infty} [c^k \cos \alpha k] z^{-k}$

$$= \sum_{k=0}^{\infty} c^k \left[\frac{e^{i\alpha k} + e^{-i\alpha k}}{2}\right] z^{-k} = \sum_{k=0}^{\infty}\left(\frac{1}{2}c^k e^{i\alpha k}\right)z^{-k} + \sum_{k=0}^{\infty}\left(\frac{1}{2}c^k e^{-i\alpha k}\right)z^{-k}$$

$$= \frac{1}{2}\sum_{k=0}^{\infty}(ce^{i\alpha}z^{-1})^k + \frac{1}{2}\sum_{k=0}^{\infty}(ce^{-i\alpha}z^{-1})^k \qquad ({}^nC_r = {}^nC_{n-r})$$

$$= \frac{1}{2}\left[1 + (ce^{i\alpha}z^{-1}) + (ce^{i\alpha}z^{-1})^2 + \ldots\right] + \frac{1}{2}\left[1 + (ce^{-i\alpha}z^{-1}) + (ce^{-i\alpha}z^{-1})^2 + \ldots\right]$$

This is a Geometric series whose sum is $\frac{a}{1-r}$

$$= \frac{1}{2}\left[\frac{1}{1-ce^{i\alpha}z^{-1}}\right] + \frac{1}{2}\left[\frac{1}{1-ce^{-i\alpha}z^{-1}}\right], \qquad [|z| > |c|]$$

$$= \frac{1}{2}\frac{1-ce^{-i\alpha}z^{-1} + 1 - ce^{i\alpha}z^{-1}}{(1-ce^{i\alpha}z^{-1})(1-ce^{-i\alpha}z^{-1})} = \frac{1}{2}\frac{2-c(e^{i\alpha}+e^{-i\alpha})z^{-1}}{\left[1-ce^{-i\alpha}z^{-1} - ce^{i\alpha}z^{-1} + c^2z^{-2}\right]}$$

$$= \frac{1 - c\cos\alpha . z^{-1}}{1 - 2cz^{-1}\cos\alpha + c^2 z^{-2}} \qquad [|z| > |c|]$$

$$= \frac{z^2 - cz\cos\alpha}{z^2 - 2cz\cos\alpha + c^2}$$ **Ans.**

Corollary. If $c = 1$, then

$$Z[\{\cos \alpha k\}] = \frac{z^2 - z\cos\alpha}{z^2 - 2z\cos\alpha + 1}$$

Example 19. *Find the Z-transform of* $\left\{\cos\left(\frac{k\pi}{8} + \alpha\right)\right\}$.

Solution. $Z\left[\left\{\cos\left(\frac{k\pi}{8}+\alpha\right)\right\}\right] = \sum \cos(\frac{k\pi}{8}+\alpha)z^{-k} = \sum\left\{\cos\frac{k\pi}{8}\cos\alpha - \sin\frac{k\pi}{8}\sin\alpha\right\}z^{-k}$

$$= \sum \cos\frac{k\pi}{8}\cos\alpha z^{-k} - \sum \sin\frac{k\pi}{8}\sin\alpha z^{-k} = \cos\alpha\sum\cos\frac{k\pi}{8}z^{-k} - \sin\alpha\sum\sin\frac{k\pi}{8}z^{-k}$$

$$= \cos\alpha \frac{z^2 - z\cos\frac{\pi}{8}}{z^2 - 2z\cos\frac{\pi}{8} + 1} - \sin\alpha \frac{z\sin\frac{\pi}{8}}{z^2 - 2z\cos\frac{\pi}{8} + 1}$$ [See Example 14, 18]

$$= \frac{(z^2 - z\cos\frac{\pi}{8})\cos\alpha - z\sin\frac{\pi}{8}\sin\alpha}{z^2 - 2z\cos\frac{\pi}{8} + 1} = \frac{z^2\cos\alpha - z[\cos\frac{\pi}{8}\cos\alpha + \sin\frac{\pi}{8}\sin\alpha]}{z^2 - 2z\cos\frac{\pi}{8} + 1}$$

$$= \frac{z^2\cos\alpha - z\cos(\frac{\pi}{8} - \alpha)}{z^2 - 2z\cos\frac{\pi}{8} + 1}$$ **Ans.**

Example 20. *Find the Z-transform of* $\{^nC_k\}$ $(0 \le k \le n)$.

Solution. $Z\left[\{^nC_k\}\right] = \sum_{k=0}^{n} {^nC_k} z^{-k} = 1 + {^nC_1}z^{-1} + {^nC_2}z^{-2} + {^nC_3}z^{-3} + \ldots + {^nC_n}z^{-n}$

$= (1 + z^{-1})^n$ (This is the expansion of Binomial theorem.) **Ans.**

Example 21. *Find Z-transform of* $\{ ^{k+n}C_n \}$

Solution. $Z\left[\{^{k+n}C_n\}\right] = \sum_{k=0}^{\infty} {^{k+n}C_n} z^{-k}$ $\begin{pmatrix} k+n > n \\ k > 0 \end{pmatrix}$

$= \sum_{k=0}^{\infty} {^{k+n}C_k} z^{-k}$ $({^nC_r} = {^nC_{n-r}})$

$= 1 + {^{n+1}C_1}z^{-1} + {^{n+2}C_2}z^{-2} + {^{n+3}C_3}z^{-3} + \ldots.$

$= 1 + (n+1)z^{-1} + \frac{(n+2)(n+1)}{2!}z^{-2} + \frac{(n+3)(n+2)(n+1)}{3!}(z^{-3}) + \ldots$

$= 1 + (-n-1)(-z^{-1}) + \frac{(-n-1)(-n-2)}{2!}(-z^{-1})^2 + \frac{(-n-1)(-n-2)(-n-3)}{3!}(-z^{-1})^3 + \ldots$

This is the expansion of Binomial theorem.

$= (1 - z^{-1})^{-n-1} = (1 - z^{-1})^{-(n+1)}$ **Ans.**

Example 22. *Find the Z-transform of* $\{^{k+n}C_n a^k\}$.

Solution. $Z\left[\{^{k+n}C_n a^k\}\right] = \sum_{k=0}^{\infty} {^{k+n}C_n} a^k z^{-k} = \sum_{k=0}^{\infty} {^{k+n}C_k} a^k z^{-k}$ $({^nC_r} = {^nC_{n-r}})$

$= \sum_{k=0}^{\infty} {^{k+n}C_k} (az^{-1})^k$ $|z| > |a|$

$= (1 - a z^{-1})^{-(n+1)}$ (See Example 21)

Corollary. If $n = 1$

$$Z\left[\{^{k+1}C_1 a^k\}\right] = (1 - az^{-1})^{-2} = \frac{z^2}{(z-a)^2}$$

If $n = 2$

$$Z\left[\{^{k+2}C_2 a^k\}\right] = (1 - az^{-1})^{-3} = \frac{z^3}{(z-a)^3}$$

If $n \to n-1$

$$Z\left[\left\{{}^{k+n-1}C_{n-1}a^k\right\}\right] = (1-az^{-1})^{-(n-1+1)} = (1-az^{-1})^{-n} = \frac{z^n}{(z-a)^n}$$ **Ans.**

Example 23. *Find the Z-transform of* $\left\{\dfrac{a^k}{k!}\right\}$. $(k \geq 0)$

Solution. $$Z\left[\left\{\frac{a^k}{k!}\right\}\right] = \sum_{k=0}^{\infty}\frac{a^k}{k!}z^{-k} = \sum_{k=0}^{\infty}\frac{(az^{-1})^k}{k!} = 1+\frac{az^{-1}}{1!}+\frac{(az^{-2})^2}{2!}+\frac{(az^{-1})^3}{3!}+\ldots$$

This is exponential series.

$$= e^{az^{-1}} = e^{\frac{a}{z}}$$ **Ans.**

Example 24. *Find the Z-transform of* $f(k) = \dfrac{1}{(k+1)!}$, $k \geq 0$. *(Q. Bank U.P., III Sem. 2002)*

Solution. $$Z\left[\frac{1}{(k+1)!}\right] = \sum_{k=0}^{\infty}\frac{1}{(k+1)!}z^{-k} = 1+\frac{1}{2!}z^{-1}+\frac{1}{3!}z^{-2}+\ldots$$

$$= z\left[z^{-1}+\frac{1}{2!}z^{-2}+\frac{1}{3!}z^{-3}+\ldots\right] = z\left[1+z^{-1}+\frac{1}{2!}z^{-2}+\frac{1}{3!}+z^{-3}\ldots\ldots-1\right]$$

$$= z\left(e^{z^{-1}}-1\right) = z\left(e^{\frac{1}{z}}-1\right)$$ **Ans.**

Example 25. *Find the Z-transform of*

(*i*) $u(k-1)$ (*ii*) $4k\,\delta(k-1)\,;\, k \geq 0$ (*iii*) $\delta(k-n)\,;\, k \geq 0$

Solution.

(*i*) $$Z\{f(k)\} = Zu(k-1) = \sum_{k=1}^{\infty}1.z^{-k} = \frac{1}{z}+\frac{1}{z^2}+\frac{1}{z^3}+\ldots = \frac{1}{z}\left(1+\frac{1}{z}+\frac{1}{z^2}+\ldots\right)[u(k-1)=1]$$

This infinite series is a G.P., whose sum is $\dfrac{a}{1-r}$

$$= \frac{1}{z}\frac{1}{\left(1-\frac{1}{z}\right)} \qquad \left[\text{if}\left|\frac{1}{z}\right|<1\right]$$

$$= \frac{1}{z-1} \text{ if } |z|>1$$

(*ii*) $$Z\{f(k)\} = Z\{4^k\,\delta(k-1)] = \sum_{k=0}^{\infty}4^k\,\delta(k-1)\,z^{-k} = \frac{4}{z}. \qquad [\delta(k-1) \Rightarrow k=1]$$

(*iii*) $$Z\{f(k)\} = Z[\delta(k-n)] = \sum_{k=0}^{\infty}\delta(k-n)\,z^{-k} = \frac{1}{z^n},$$ n is (+)ve integer. $[\delta(k-n) \Rightarrow k=n]$

Example 26. *Determine the Z-transform of*

$f(k) = \delta(k+1)\} + 3\delta(k) + 6\delta(k-3) - \delta(k-4)$

Solution. By linearity property, we have

$F(z) = Z\{f(k)\} = Z\{\delta(k+1)\} + 3\,Z\{\delta(k)\} + 6\,Z\{\delta(k-3)] - Z\{\delta(k-4)\}$

$[\delta(k+1) \Rightarrow k=-1,\ \delta(k) \Rightarrow k=0,\ \delta(k-3) \Rightarrow k=3,\ \delta(k-4) \Rightarrow k=4]$

$= z+3+6z^{-3}-z^{-4}$ **Ans.**

EXERCISE 44.2

Find the Z-transform of the following for $(k \geq 0)$:

1. 2^k **Ans.** $\dfrac{z}{z-2}, \quad |z|>2$

2. $\sin 2k$ **Ans.** $\dfrac{z \sin 2}{z^2 - 2z\cos 2 + 1}$

3. $\sin \alpha k$, **Ans.** $\dfrac{z \sin \alpha}{z^2 - 2z \cos\alpha + 1}$

4. $\sinh \dfrac{k\pi}{2}$ **Ans.** $\dfrac{z \sinh \frac{\pi}{2}}{z^2 - 2z\cosh\frac{\pi}{2} + 1}$

5. $\sin(3k+5)$ **Ans.** $\dfrac{z^2 \sin 5 - z \sin 2}{z^2 - 2z\cos 3 + 1}, \quad |z|>1$

6. $\sin\left(\dfrac{k\pi}{2} + \alpha\right)$ **Ans.** $\dfrac{z^2 \sin\alpha + z\cos\alpha}{z^2+1}, \quad |z|>1$

7. $c^k \sinh \alpha k$ **Ans.** $\dfrac{cz \sinh \alpha}{z^2 - 2cz\cosh\alpha + 1}$

8. $\cos \alpha k$ **Ans.** $\dfrac{z(z-\cos\alpha)}{z^2 - 2z\cos\alpha + 1}$

9. $\cos\left(\dfrac{k\pi}{2} + \dfrac{\pi}{4}\right)$ **Ans.** $\dfrac{z^2 - z}{\sqrt{2}(z^2+1)}$

10. $\cos\left(\dfrac{k\pi}{8} + \alpha\right)$ **Ans.** $\dfrac{z^2\cos\alpha - z\cos\left(\frac{\pi}{8} - \alpha\right)}{z^2 - 2z\cos\frac{\pi}{8} + 1}$

11. $\cosh \alpha k$ **Ans.** $\dfrac{z(z-\cosh\alpha)}{z^2 - 2z\cosh\alpha + 1}$

12. $a\cos\alpha k + b\sin\alpha k$ **Ans.** $\dfrac{a - z^{-1}(a\cos\alpha - b\sin\alpha)}{1 - 2z^{-1}\cos\alpha + z^{-2}}, \quad |z|>1$

13. $\dfrac{a^k}{k!}$ **Ans.** $e^{\frac{a}{z}}$

14. $\dfrac{1}{k}, k>0$ **Ans.** $-\log(1-z^{-1})$

15. $a^k, \quad k<0$ **Ans.** $\dfrac{-1}{(1-az^{-1})}$

16. $2^{|k|}$ **Ans.** $\dfrac{-3z}{(1-2z)(z-2)}$

17. $f(k) = \begin{cases} 1, & \text{for } k \geq 0 \\ 0, & \text{for } k < 0 \end{cases}$ **Ans.** $\dfrac{z}{z-1}, \quad |z|>1$

44.7 CHANGE OF SCALE

Theorem. If $Z[\{f(k)\}] = F(z)$ then $Z[\{a^k f(k)\}] = F\left(\dfrac{z}{a}\right)$

Proof. $$F(z) = Z[\{f(k)\}] = \sum_{k=-\infty}^{\infty} f(k) z^{-k}$$

Substituting $\dfrac{z}{a}$ for z, we get $$F\left(\frac{z}{a}\right) = \sum_{k=-\infty}^{\infty} f(k)\left(\frac{z}{a}\right)^{-k} \qquad ...(1)$$

But $$Z\left[\{a^k f(k)\}\right] = \sum_{k=-\infty}^{\infty} a^k f(k) z^{-k} = \sum_{k=-\infty}^{\infty} f(k)\left(\frac{z}{a}\right)^{-k} \qquad ...(2)$$

From (1) and (2), we get $Z\left[\{a^k f(k)\}\right] = F\left(\dfrac{z}{a}\right)$ **Proved.**

Example 27. *Find the Z-transform of a^k, $k \geq 0$.*

Solution. We know that

$$Z\left[\{1\}\right]=\frac{z}{z-1}$$ (See example 13)

For the given sequence, by the scale change formula the Z-transform

$$Z\left[\left\{a^{k}.1\right\}\right]=\frac{\frac{z}{a}}{\frac{z}{a}-1}=\frac{z}{z-a}$$ **Ans.**

Example 28. *Find the Z-transform of c^k sin α k, k $\geq$ 0.*

Solution. We know that

$$Z\left[\{\sin \alpha k\}\right]=\frac{z \sin \alpha}{z^{2}-2z \cos \alpha+1}$$ (See example 14)

By applying the formula of change of scale, we get

$$Z\left[\left\{c^{k} \sin \alpha k\right\}\right]=\frac{\frac{z}{c}\sin \alpha}{\left(\frac{z}{c}\right)^{2}-2\left(\frac{z}{c}\right)\cos \alpha+1}=\frac{cz \sin \alpha}{z^{2}-2cz \cos \alpha+c^{2}}$$ **Ans.**

44.8 SHIFTING PROPERTY

Theorem. If $Z[\{f(k)\}]=F(z)$,

then $Z[\{f(k \pm n)\}]=z^{\pm n} F(z)$

Proof. $Z[\{f(k \pm n)\}]=\sum_{k=-\infty}^{\infty} f(k \pm n) z^{-k}=z^{\pm n} \sum_{k=-\infty}^{\infty} f(k \pm n) z^{-(k \pm n}$ $(r=k \pm n)$

$$=z^{\pm n} \sum_{k=-\infty}^{\infty} f(r) z^{-r}=z^{\pm n} F(z)$$ **Proved.**

Case I. $Z[\{f(k+n)\}]=f(k+n) z^{-k},\ k \geq 0=z^{n} \sum_{k=0}^{\infty} f(k+n) z^{-(k+n)}$

Put $r=k+n$

$$=z^{n} \sum_{r=n}^{\infty} f(r) z^{-r}=z^{n} \sum_{r=0}^{\infty} f(r) z^{-r}-z^{n} \sum_{r=0}^{n-1} f(r) z^{-r}=z^{n} F(z)-z^{n} \sum_{r=0}^{n-1} f(r) z^{-r}$$

$$=z^{n} F(z)-\sum_{r=0}^{n-1} f(r) z^{n-r}=z^{n} F(z)-\sum_{r=0}^{n-1} f(r) z^{n-r}$$ **Ans.**

Case II. $Z[\{f(k-n)\}]=\sum_{k=0}^{\infty} f(k-n) z^{-k}=z^{-n} \sum_{k=0}^{\infty} f(k-n) z^{-(k-n)}$, $k \geq 0$

Put $r=k-n$

$$=z^{-n} \sum_{r=-n}^{\infty} f(r) z^{-r}$$

$$=z^{-n} \sum_{r=0}^{\infty} f(r) z^{-r}+z^{n} \sum_{r=-n}^{-1} f(r) z^{-r}=2^{-n} F(z)+\sum_{r=-n}^{-1} f(r) z^{-n-r}$$ Put $r=-m$

$$=z^{-n} F(z)+\sum_{m=1}^{n} f(-m) z^{-n+m}$$ **Ans.**

Corollary 1. If $\{f(k)\}$ is casual sequence, then

$$Z[\{f(k-n)\}]=z^{-n} F(z)$$

Since $f(-1) = f(-2) = f(-3) = \ldots = f(-n) = 0$

Corollary 2. For casual sequence

$$Z[\{f(k-1)\}] = z^{-1} F(z) \text{ as } f(-1) = 0.$$
$$Z[\{f(k+1)\}] = z F(z) - z f(0)$$
$$Z[\{f(k+2)\}] = z^2 F(z) - z^2 f(0) - z f(1)$$

44.9 INVERSE Z-TRANSFORM

Finding the sequence $\{f(k)\}$ from $F(z)$ is defined as inverse Z-transform. It is denoted as

$$Z^{-1} F(z) = \{f(k)\} \qquad Z^{-1} \text{ is the inverse Z-transform.}$$

44.10 THEOREM

If $\{f(k)\} = F(z)$, $\{g(k)\} = G(z)$, a and b are constants,

then $Z^{-1}[a F(z) + b G(z)] = aZ^{-1}[F(z)] + bZ^{-1}[G(z)]$

Proof. We know that

$$Z[\{af(k) + bg(k)\}] = a Z[\{f(k)\}] + b Z[\{g(k)\}] = aF(z) + b G(z)$$
$$\therefore \quad Z^{-1}[a F(z) + bG(z)] = \{af(k) + bg(k)\} = a\{f(k)\} + b\{g(k)\}$$
$$= a Z^{-1}[F(z)] + bZ^{-1}[G(z)]$$

Proved.

Example 29. *Find the inverse Z-transform of* $\dfrac{1}{z-2}$

Solution. $F(z) = \dfrac{1}{z-2}$

Case I. If $\left|\dfrac{2}{z}\right| < 1$, $F(z) = \dfrac{1}{z}\dfrac{1}{1-2z^{-1}} = z^{-1}(1-2z^{-1})^{-1} = z^{-1}[1 + 2z^{-1} + 2^2 z^{-2} + \ldots]$

$$= z^{-1} + 2z^{-2} + 2^2 z^{-3} + \ldots$$
$$\{f(k)\} = \{2^{k-1}\}, \qquad k \geq 1$$

Case II. If $\left|\dfrac{z}{2}\right| < 1$

$$F(z) = \frac{1}{z-2} = -\frac{1}{2}\frac{1}{1-\dfrac{z}{2}} = \frac{-1}{2}\left(1-\frac{z}{2}\right)^{-1} = -\frac{1}{2}\left[1 + \frac{z}{2} + \frac{z^2}{2^2} + \ldots\right]$$
$$\{f(k)\} = \{-2^{k-1}\}, \qquad k \leq 0$$

Note. The inverse Z-transform can only be settled when region of convergence (ROC) is given.

44.11 SOLUTION OF DIFFERENCE EQUATIONS

Example 30. *Solve the difference equation*

$$y_{k+1} - 2y_{k-1} = 0, \quad k \geq 1, \quad y_{(0)} = 1$$

Solution. $y_{k+1} - 2y_{k-1} = 0$ ...(1)

Taking the Z-transform of both sides of (1), we get

$$Z[y_{k+1} - 2y_{k-1}] = 0$$
$$Z[y_{k+1}] - 2Z[y_{k-1}] = 0$$
$$zY(z) - y_0 z - 2Y(z) = 0 \qquad (y_0 = 1)$$
$$(z-2)\,Y(z) - z = 0$$

$$Y(z)=\frac{z}{z-2}$$

$$\{y_{(k)}\}=Z^{-1}\left[\frac{z}{z-2}\right]=Z^{-1}\left[\frac{1}{1-2z^{-1}}\right]$$

$$= Z^{-1}\,[1-2z^{-1}]^{-1} = 1 + 2z^{-1} + (2z^{-1})^2 + \ldots \quad = \{2^k\},\quad k \ge 0 \quad \textbf{Ans.}$$

44.12 MULTIPLICATION BY *K*

Theorem. If $Z\,[\{f(k)\}] = F(z)$, then $Z\,[\{k\,f(k)\}] = -z\dfrac{d}{dz}F(z)$

Proof. $Z[\{kf(k)\}] = \sum\limits_{k=-\infty}^{\infty} kf(k)z^{-k} = -z\sum\limits_{k=-\infty}^{\infty} -kf(k)z^{-k-1} = -z\sum\limits_{k=-\infty}^{\infty} f(k)(-kz^{-k-1})$

$$= -z\sum_{k=-\infty}^{\infty} f(k)\frac{d}{dz}(z^{-k}) = -z\frac{d}{dz}\sum_{k=-\infty}^{\infty} f(k)z^{-k} = -z\frac{d}{dz}F(z) \qquad \textbf{Proved.}$$

In general $\quad Z\,[\{k^n f(k)\}] = \left(-z\dfrac{d}{dz}\right)^n F(z)$

Corollary 1. If $f(k) = 1$ then $Z\,[\{1\}] = (1-z^{-1})^{-1}$, $k \ge 0$

Putting these values in the above theorem, we get

$$Z[\{k\}] = -z\frac{d}{dz}(1-z^{-1})^{-1} = z(1-z^{-1})^{-2}\left(\frac{1}{z^2}\right) = z^{-1}(1-z^{-1})^{-2} \qquad \textbf{Ans.}$$

Corollary 2. If $f(k) = 1$, then $Z\,[\{1\}] = (1-z^{-1})^{-1} = \dfrac{z}{z-1}$ and $n = 2$

$$= Z[\{k^2\}] = \left(-z\frac{d}{dz}\right)^2\frac{z}{z-1} = \left(-z\frac{d}{dz}\right)\left(-z\frac{d}{dz}\right)\frac{z}{z-1}$$

$$= -z\frac{d}{dz}\left[-z\frac{(z-1).1-z.1}{(z-1)^2}\right] = -z\frac{d}{dz}\left[\frac{z}{(z-1)^2}\right] = -z\frac{(z-1)^2.1-z.2(z-1)}{(z-1)^4} = -z\frac{z-1-2z}{(z-1)^3}$$

$$= -z\frac{-z-1}{(z-1)^3} = \frac{z^2+z}{(z-1)^3} = \frac{z(z+1)}{(z-1)^3} \qquad \textbf{Ans.}$$

44.13 DIVISION BY *K*

Theorem. If $\quad Z\,[\{f(k)\}] = F(z)$, $\quad$ then $Z\left[\left\{\dfrac{f(k)}{k}\right\}\right] = -\int^z z^{-1}F(z)dz$

Proof. $\quad Z\left[\left\{\dfrac{f(k)}{k}\right\}\right] = \sum\limits_{k=-\infty}^{\infty}\dfrac{f(k)}{k}z^{-k}$

$$= \sum_{k=-\infty}^{\infty} f(k)\left(\frac{1}{k}z^{-k}\right) = -\sum_{k=-\infty}^{\infty} f(k)\int^z z^{-k-1}dz$$

$$= -\int^z \sum_{k=-\infty}^{\infty} f(k)z^{-k-1}dz = -\int^z z^{-1}\sum_{k=-\infty}^{\infty} f(k)z^{-k}dz = -\int^z z^{-1}F(z)dz$$

$$Z\left[\left\{\frac{f(k)}{k}\right\}\right] = -\int^z z^{-1}F(z)dz$$

44.14 INITIAL VALUE

Theorem. If $Z[\{f(k)\}] = F(z)$, $k \geq 0$

then $$f(0) = \lim_{z\to\infty} F(z).$$

Proof. $$Z[\{f(k)\}] = \sum_{k=0}^{\infty} f(k) z^{-k} = F(z)$$

$$f(0) + f(1)\,z^{-1} + f(2)\,z^{-2} + \ldots = F(z)$$

Taking the limit, $z \to \infty$, we get

$$f(0) = \lim_{z\to\infty} F(z)$$ **Proved.**

44.15 FINAL VALUE

Theorem. $\lim_{k\to\infty} f(k) = \lim_{z\to 1} (z-1) F(z)$

Proof. $$Z[\{f(k+1) - f(k)\}] = \sum_{k=0}^{\infty} [f(k+1) - f(k)] z^{-k}$$

$$zF(z) - f(0) - F(z) = \lim_{n\to\infty} \sum_{k=0}^{n} [f(k+1) - f(k)] z^{-k}$$

$$\lim_{z\to 1} (z-1) F(z) = f(0) + \lim_{z\to 1} \lim_{n\to\infty} \sum [f(k+1) - f(k)] z^{-k}$$

By changing the order of limits, we get

$$\lim_{z\to 1} (z-1) F(z) = f(0) + \lim_{n\to\infty} \sum_{k=0}^{n} \lim_{z\to 1} [f(k+1) - f(k)] z^{-k}$$

$$= \lim_{n\to\infty} \left[f(0) + \sum_{k=0}^{n} \{f(k+1) - f(k)\} \right]$$

$$= \lim [f(0) - f(0) + f(1) - f(1) + f(2) - f(2) + \ldots + f(n+1) - f(n)]$$

$$= \lim_{n\to\infty} f(n+1) = \lim_{n\to\infty} f(n) = \lim_{k\to\infty} f(k)$$

44.16 PARTIAL SUM

Theorem. If $Z[\{f(k)\}] = F(z)$ then $Z\left[\left\{ \sum_{n\to-\infty}^{k} f(n) \right\}\right] = \dfrac{F(z)}{1 - z^{-1}}$

Proof. Let $\{g(k)\}$ be a sequence such that $g(k) = \sum_{n=-\infty}^{k} f(n)$

We are required to find $Z[\{g(k)\}]$,

We know that $g(k) - g(k-1) = \sum_{n=-\infty}^{k} f(n) - \sum_{n=-\infty}^{k-1} f(n) = -f(k)$

$$Z[\{g(k)\} - \{g(k-1)\}] = Z[\{f(k)\}]$$

$$Z[\{g(k)\}] - Z[\{g(k-1)\}] = F(z)$$

$$G(z) - z^{-1} G(z) = F(z)$$

$$\sum_{n=-\infty}^{k} f(k) = G(z) = \frac{F(z)}{1 - z^{-1}}$$ **Proved.**

44.17 CONVOLUTION

Let two sequences be $\{f(k)\}$ and $\{g(k)\}$ and the convolution of $\{f(k)\}$ and $\{g(k)\}$ be $\{h(k)\}$ and denoted as

$$\{h(k)\} = \{f(k)\} * \{g(k)\}$$

where $$\{h(k)\} = \sum_{n=-\infty}^{\infty} f(n)g(k-n) \qquad ...(1)$$

$$= \sum_{n=-\infty}^{\infty} g(n)f(k-n) = \{g(k)\} * \{f(k)\}$$

Proof. Z-transform of (1) is

$$Z\{h(k)\} = \sum_{k=-\infty}^{\infty}\left(\sum_{n=-\infty}^{\infty} f(n)g(k-n)\right)z^{-k} = \sum_{n=-\infty}^{\infty}\sum_{k=-\infty}^{\infty} f(n)g(k-n)z^{-k}$$

$$= \sum_{n=-\infty}^{\infty} f(n)z^{-n} \sum_{k=-\infty}^{\infty} g(k-n)z^{-(k-n)} = \left[\sum_{n=-\infty}^{\infty} f(n)z^{-n}\right] G(z) = F(z)G(z)$$

Note. Region of convergence of $H(z)$ is the common region of convergence of $F(z)$ and $G(z)$.

44.18 CONVOLUTION PROPERTY OF CASUAL SEQUENCE

$$F(z) = \{f(0) + f(1)z^{-1} + f(2)z^{-2} + f(3)z^{-3} + ...\}$$

and $$G(z) = \{g(0) + g(1)z^{-1} + g(2)z^{-2} + g(3)z^{-3} + ...\}$$

Now

$$F(z)G(z) = \{f(0) + f(1)z^{-1} + f(2)z^{-2} + f(3)z^{-3} + \ldots$$
$$\{g(0) + g(1)z^{-1} + g(2)z^{-2} + g(3)z^{-3} + ...\}$$
$$= f(0)\,g(0) + \{f(1)\,g(0) + f(0)\,g(1)\}\,z^{-1} +$$
$$\{f(0)\,g(2) + f(1)\,g(1) + f(2)\,g(0)\}\,z^{-2} + \ldots$$
$$= h(0) + h(1)z^{-1} + h(2)z^{-2} + \ldots = Z\{h(k)\}$$

$$Z\{h(k)\} = Z\{f(k)\} * \{g(k)\}$$ **Proved.**

Example 31. *Evaluate the Z-transform of the sequence*

$$\{f(k)\} = \sum_{k=0}^{\infty} 2^k \sum_{k=0}^{\infty} 3^k$$

Solution. $Z\left[\{2^k\}\right] = 1 + 2z^{-1} + 2^2z^{-2} + 2^3z^{-3} + ... = \dfrac{1}{1-2z^{-1}}$

Similarly, $$Z\left[\{3^k\}\right] = \frac{1}{1-3z^{-1}}$$

$$Z\left[\{f(k)\}\right] = Z\{2^k\}\{3^k\} = \frac{1}{1-2z^{-1}} \cdot \frac{1}{1-3z^{-1}} = \frac{1}{(1-2z^{-1})(1-3z^{-1})}$$ **Ans.**

EXERCISE 44.3

Find the Z-transform of the following for $(k > 0)$:

1. $e^{\alpha k}$ **Ans.** $\dfrac{1}{1-z^{-1}e^{\alpha k}}$ **2.** $\sin 5k$ **Ans.** $\dfrac{z\sin 5}{z^2 - 2z\cos 5 + 1}$

3. $\cos 3k$ **Ans.** $\dfrac{z(z-\cos 3)}{z^2 - 2z\cos 3 + 1}$ **4.** $\sinh 7k$ **Ans.** $\dfrac{z\sinh 7}{z^2 - 2z\cosh 7 + 1}$

5. $\cosh 9k$ **Ans.** $\dfrac{z(z-\cosh 9)}{z^2 - 2z\cosh 9 + 1}$ **6.** $\sin(5k+3)$ **Ans.** $\dfrac{z^2\sin 3 + z\sin 2}{z^2 - 2z\cos 5 + 1}$

7. $\cos\left(\frac{k\pi}{3}+5\right)$ **Ans.** $\dfrac{z^2\cos 5 - z\cos(\frac{\pi}{3}-5)}{z^2-2z\cos\frac{\pi}{3}+1}$

8. $\cos\left(\frac{k\pi}{5}+6\right)$ **Ans.** $\dfrac{z^2\cosh 6 - z\cosh(\frac{\pi}{5}-6)}{z^2-2z\cos\frac{\pi}{5}+1}$ 9. $3^k\cosh 5k$ **Ans.** $\dfrac{z(z-3\cosh 5)}{z^2-2z\cosh 5+9}$

44.19 TRANSFORM OF IMPORTANT SEQUENCES

S. No.	*Sequence*	*Z-transform*	
1	$\{f(k)\}$	$F(z)$	
2	$\delta(k)$	1	
3	$U(k)$ or 1	$(1-z^{-1})^{-1}$	
4	k	$-z\frac{d}{dz}(1-z^{-1})^{-1}$	$\|z\|>1$
5	k^n	$\left(-z\frac{d}{dz}\right)^n(1-z^{-1})^{-1}$	$\|z\|>1$
6	${}^kC_n$	$z^{-n}(1-z^{-1})^{-(n+1)}$	$\|z\|>1$
7	${}^{k+n}C_n.a^k$	$(1-az^{-1})^{-(n+1)}$	$\|z\|>\|a\|$
8	${}^nC_k$	$(1-z^{-1})^n$	$0\le k\le n, \|z\|>0$
9	a^k	$(1-az^{-1})^{-1}$	$\|z\|>\|a\|$
10	k^na^k	$\left(-z\frac{d}{dz}\right)^n(1-az^{-1})^{-1}$	$\|z\|>\|a\|$
11	a^k	$-(1-az^{-1})^{-1}$	$\|z\|>\|a\|$
12	k^na^k	$\left(-z\frac{d}{dz}\right)^n(1-az^{-1})^{-1}$	$\|z\|<\|a\|$
13	$a^{\|k\|}$	$\dfrac{(1-a^2)(1-az)}{(1-az^{-1})^{-1}}$	$\|a\|<\|z\|<\frac{1}{\|a\|}$
14	$\sin\alpha k$	$\dfrac{z\sin\alpha}{z^2-2z\cos\alpha+1}$	$k\ge 0$
15	$c^k\sin\alpha k$	$\dfrac{cz\sin\alpha}{z^2-2cz\cos\alpha+c^2}$	$k\ge 0$
16	$\cos\alpha k$	$\dfrac{z^2-z\cos\alpha}{z^2-2z\cos\alpha+1}$	$k\ge 0$
17	$c^k\cos\alpha k$	$\dfrac{z^2-cz\cos\alpha}{z^2-2cz\cos\alpha+c^2}$	$k\ge 0$
18	$\cosh\alpha k$	$\dfrac{z^2-z\cosh\alpha}{z^2-2z\cosh\alpha+1}$	$k\ge 0$

19	$c^k \cosh \alpha k$	$\dfrac{z^2 - cz\cosh\alpha}{z^2 - 2cz\cosh\alpha + c^2}$	$k \geq 0$
20	$\sinh \alpha k$	$\dfrac{z\sinh\alpha}{z^2 - 2z\cosh\alpha + 1}$	$k \geq 0$
21	$c^k \sinh \alpha k$	$\dfrac{cz\sinh\alpha}{z^2 - 2cz\cosh\alpha + c^2}$	$k \geq 0$
22	Change of scale	$a^k f(k)$	$F\left(\dfrac{z}{a}\right)$
23	Shifting	$f(k \pm n)$	$z^{\pm n} F(z)$
24		$f(k - n)$	$z^{-n}F(z) + \sum_{r=1}^{n} f(-r)z^{-n+r}$, $k < 0$
25	Casual sequence	$f(k + n)$	$z^n F(z) - \sum_{r=0}^{n-1} f(r)z^{n-r}$
26		$f(k - n)$	$z^{-n} F(z)$
27	Multiplication by k	$k f(k)$	$-z\dfrac{d}{dz}F(z)$
28	Multiplication by k	$k^n f(k)$	$\left(-z\dfrac{d}{dz}\right)^n F(z)$
29	Division by k	$\dfrac{1}{k} f(k)$	$-\int^{z} z^{-1}F(z)dz$
30	Initial value theorem	$f(0)$	$\lim_{z\to\infty} F(z)$
31	Final value theorem	$\lim_{k\to\infty} f(k)$	$\lim_{z\to 1} (z-1)F(z)$
32	Partial sum	$\sum_{n\to-\infty}^{k} f(n)$	$\dfrac{F(z)}{z-1}$
33		$\sum_{n\to-\infty}^{\infty} f(n)$	$F(1)$
34	Convolution	$f(k)^* g(k)$	$F(z) . G(z)$

Inverse Z-transforms

S. No		*Inverse Z-transform*	
		$\lvert z \rvert > \lvert a \rvert$, $k > 0$	$\lvert z \rvert < \lvert a \rvert$, $k < 0$
1	$\dfrac{z}{z-a}$	$a^k U(k)$	$-a^k$
2	$\dfrac{z^2}{(z-a)^2}$	$(k+1)a^k$	$-(k+1)a^k$

3	$\dfrac{z^3}{(z-a)^3}$	$\dfrac{1}{2!}(k+1)(k+2)a^k U(k)$	$-\dfrac{1}{2!}(k+1)(k+2)a^k U(-k+2)$
4	$\dfrac{z^n}{(z-a)^n}$	$\dfrac{1}{(n-1)!}(k+1)...(k+n-1)$ $a^k U(k)$	$-\dfrac{1}{(n-1)!}(k+1)...(k+n-1)a^k$
5	$\dfrac{1}{z-a}$	$a^{k-1}U(k-1)$	$-a^{k-1}U(-k)$
6	$\dfrac{1}{(z-a)^2}$	$(k-1)a^{k-2}U(k-2)$	$-(k-1)a^{k-2}U(-k+1)^{\hat{u}}$
7	$\dfrac{1}{(z-a)^3}$	$\dfrac{1}{2}(k-2)(k-1)a^{k-3}U(k-3)$	$-\dfrac{1}{2}(k-2)\,(k-)\,a^{k-3}\,U\,(-k+2)$

44.20 INVERSE OF Z-TRANSFORM BY DIVISION

From Z-transform $F(z)$, we find the sequence $\{f(k)\}$ if $F(z)$ is a rational function of z. Region of convergence must be given.

1. By Direct Division

Example 32. *Find* $Z^{-1}\left[\dfrac{1}{z-2}\right]$

Solution. Case I. $|z|>2$

$$z-2\ \overline{\big)\ 1\ }\ \left(\frac{1}{z}+\frac{2}{z^2}+\frac{4}{z^3}\right.$$

$$\begin{array}{l} 1 \\ 1-\dfrac{2}{z} \\ \hline \dfrac{2}{z} \\ \dfrac{2}{z}-\dfrac{4}{z^2} \\ \hline \dfrac{4}{z^2} \\ \dfrac{4}{z^2}-\dfrac{8}{z^3} \\ \hline \dfrac{8}{z^3} \end{array}$$

$$\frac{1}{z-2}=\frac{1}{z}+\frac{2}{z^2}+\frac{4}{z^3}+\frac{8}{z^4}+...$$

$$= z^{-1}+2z^{-2}+4z^{-3}+8\,z^{-4}+\ldots+2^{k-1}z^{-k}+\ldots$$

$$= \{2^{k-1}\}\ z^{-k}$$

$$Z^{-1}\left[\frac{1}{z-2}\right]=\{2^{k-1}\}$$ **Ans.**

Case II. $|z|<2$

$$-2+z\ \overline{\big)\ 1\ }\ \left(-\frac{1}{2}-\frac{z}{4}-\frac{z^2}{8}-\frac{z^3}{16}\right.$$

$$\begin{array}{l} 1 \\ 1-\dfrac{z}{2} \\ \hline \dfrac{z}{2} \\ \dfrac{z}{2}-\dfrac{z^2}{4} \\ \hline \dfrac{z^2}{4} \\ \dfrac{z^2}{4}-\dfrac{z^3}{8} \\ \hline \dfrac{z^3}{8} \\ \dfrac{z^3}{8}-\dfrac{z^4}{16} \\ \hline \dfrac{z^4}{16} \end{array}$$

$$\frac{1}{z-2}=-\frac{1}{2}-\frac{z}{4}-\frac{z^2}{8}-\frac{z^3}{16}-...$$

$$=-\frac{1}{2}\left[1+\frac{z}{2}+\frac{z^2}{2^2}+\frac{z^3}{2^3}+...+\frac{z^k}{2^k}+...\right]$$

$$=-\frac{1}{2}-\frac{z}{2^2}-\frac{z^2}{2^3}-\frac{z^3}{2^4}-...-\frac{z^k}{2^{k+1}}-...=\{2^{-k-1}\}z^k$$

$$Z^{-1}\left[\frac{1}{z-2}\right]=\{2^{-k-1}\}$$ **Ans.**

44.21 BY BINOMIAL EXPANSION AND PARTIAL FRACTION

Example 33. *Find the inverse Z-transform of* $\dfrac{4z}{z-a}$

(*i*) $|z|>|a|$ (*ii*) $|z|<|a|$

Solution. Case I. $|z|>|a|$

$$\frac{4z}{z-a}=\frac{4z}{z}\frac{1}{1-\frac{a}{z}}=4\left(1-\frac{a}{z}\right)^{-1}=4\left[1+\frac{a}{z}+\frac{a^2}{z^2}+\frac{a^3}{z^3}+\dots\right]$$

$$=4+4a\,z^{-1}+4a^2z^{-2}+4a^3z^{-3}+\dots+4a^kz^{-k}+\dots=\{4a^k\}z^{-k}$$

$$Z^{-1}\left(\frac{4z}{z-a}\right)=\{4a^k\}$$ **Ans.**

Case II. $|z|<|a|$

$$\frac{4z}{z-a}=-\frac{1}{a}\frac{z}{\left(1-\frac{z}{a}\right)}=\frac{-z}{a}\left(1-\frac{z}{a}\right)^{-1}$$

$$=-\frac{z}{a}\left[1+\frac{z}{a}+\frac{z^2}{a^2}+\frac{z^3}{a^3}+\dots\right]=-\frac{z}{a}-\frac{z^2}{a^2}-\frac{z^3}{a^3}-\frac{z^4}{a^4}-\dots$$

$$\{f(k)\}=\left\{\dots-\frac{1}{a^4},-\frac{1}{a^3},-\frac{1}{a^2},-\frac{1}{a}\right\}$$ **Ans.**

EXERCISE 44.4

Find the inverse of the *Z*-transform of the following :

1. $\dfrac{z^2}{\left(z-\frac{1}{4}\right)\left(z-\frac{1}{5}\right)}$, $|z|<\dfrac{1}{5}$ **Ans.** $4\left(\dfrac{1}{5}\right)^k-5\left(\dfrac{1}{4}\right)^k$

2. $\dfrac{z}{\left(z-\frac{1}{4}\right)\left(z-\frac{1}{5}\right)}$, $|z|>\dfrac{1}{4}$ **Ans.** $20\left[\left(\dfrac{1}{4}\right)^k-\left(\dfrac{1}{5}\right)^k\right]U(k)$

3. $\dfrac{z^3}{\left(z-\frac{1}{2}\right)^2(z-1)}$, $|z|<\dfrac{1}{2}$ **Ans.** $-4(1)^k+(k+3)\left(\dfrac{1}{2}\right)^k$, $(k<0)$

4. $\dfrac{z+1}{z^2-2z+1}$, $|z|>1$ **Ans.** $(2k+1)\,U(k)$

44.22 INVERSE OF Z-TRANSFORM BY PARTIAL FRACTIONS

Let $f(z)=\dfrac{R(z)}{D(z)}$ [If the degree of Numerator < the degree of Denominator.]

and $F(z)=Q(z)+\dfrac{R(z)}{D(z)}$ [If the degree of Numerator > the degree of Denominator.]

$\dfrac{R(z)}{D(z)}$ is then expressed into partial fractions.

We convert $\dfrac{F(z)}{z}$ into partial fractions and not that of $F(z)$.

Let $\dfrac{F(z)}{z}=\dfrac{A}{z-a}+\dfrac{B}{z-b}+\dfrac{C}{z-c}+\dfrac{D}{(z-c)^2}+\dfrac{E}{(z-c)^3}+\dfrac{Mz+N}{z^2+pz+q}$

then $\quad F(z)=A\dfrac{z}{z-a}+B\dfrac{z}{z-b}+C\dfrac{z}{z-c}+D\dfrac{z}{(z-c)^2}+E\dfrac{z}{(z-c)^3}+\dfrac{z(Mz+N)}{z^2+pz+q}$

$$Z^{-1}F(z)=AZ^{-1}\frac{z}{z-a}+BZ^{-1}\frac{z}{z-b}+CZ^{-1}\frac{z}{z-c}+DZ^{-1}\frac{z}{(z-c)^2}$$
$$+EZ^{-1}\frac{z}{(z-c)^3}+Z^{-1}\frac{z(Mz+N)}{z^2+pz+q}$$

(*i*) Linear non-repeated factor

Let the linear non-repeated factor be $\dfrac{z}{z-a}$.

$$Z^{-1}\left(\frac{z}{z-a}\right)=Z^{-1}\frac{1}{1-az^{-1}}=\{a^k\} \qquad \text{If } |z|>|a|$$

$$Z^{-1}\left(\frac{z}{z-a}\right)=-Z^{-1}\frac{\frac{z}{a}}{1-\frac{z}{a}}=\{-a^k\},\ k<0, \qquad \text{If } |z|<|a|$$

(*ii*) Linear repeated factor

Let the linear repeated factor be $\dfrac{z}{(z-b)^r}$. $\qquad r\ge 2, |z|>|b|$

$$Z^{-1}\frac{z}{(z-b)^r}=Z^{-1}\left[z^{-(r-1)}\frac{z^r}{(z-b)^r}\right]$$
$$=\frac{(k+2-r)(k+3-r)...kU(k-r+1)b^{(k-r+1)}}{(r-1)!}$$

$$Z^{-1}\frac{z}{(z-b)^r}=-\frac{(k+2-r)(k+3-r)......kb^{k-r+1}}{(r-1)!}U(-k-r-1),\ \text{If } |z|<|b|$$

(*iii*) Quadratic non-repeated factor

Let the quadratic non-repeated factor be

$$\frac{Mz^2+Nz}{z^2+pz+q}... \qquad ...(1) \qquad \text{If } |z|>0$$

Compare (1) with $Z[\{c^k\cos\alpha k\}]=\dfrac{z^2-cz\cos\alpha}{z^2-2cz\cos\alpha+c^2}$

$$p=-2c\cos\alpha,\ q=c^2$$

or with $Z[\{c^k\cosh\alpha k\}]=\dfrac{z^2-cz\cosh\alpha}{z^2-2cz\cosh\alpha+c^2}$

or $\qquad p=2c\cosh\alpha, \qquad q=c^2$

$\dfrac{p}{-2c}=\cos\alpha\Rightarrow\left|\dfrac{p}{2c}\right|<1\text{ or }>1$, c is given by $(-2c\cos\alpha)$ and α by $(-2c\cosh\alpha)$

Case I. If $\left|\dfrac{p}{2c}\right|<1$

$$\frac{Mz^2+Nz}{z^2+pz+q}=\frac{Mz(z-c\cos\alpha)+\dfrac{Mc\cos\alpha+N}{c\sin\alpha}(cz\sin\alpha)}{z^2-2cz\cos\alpha+c^2}$$

$$=\frac{M(z^2-cz\cos\alpha)}{z^2-2cz\cos\alpha+c^2}+\frac{Mc\cos\alpha+N}{c\sin\alpha}.\frac{cz\sin\alpha}{z^2-2cz\cos\alpha+c^2}$$

$$Z^{-1}\frac{Mz^2+Nz}{z^2+pz+q}=M\{c^k\cos\alpha k\}+\left(\frac{Mc\cos\alpha+N}{c\sin\alpha}\right)[c^k\sin\alpha k]$$

Case II. If $\left|\dfrac{p}{2c}\right|>1$

$$\frac{Mz^2+Nz}{z^2+pz+q}=\frac{Mz(z-c\cosh\alpha)+\dfrac{Mc\cosh\alpha+N}{c\sinh\alpha}(cz\sinh\alpha)}{z^2-2cz\cosh\alpha+c^2}$$

$$=\frac{Mz(z-c\cosh\alpha)}{z^2-2cz\cosh\alpha+c^2}+\frac{Mc\cosh\alpha+N}{c\sinh\alpha}.\frac{cz\sinh\alpha}{z^2-2cz\cosh\alpha+c^2}$$

$$=\frac{M(z^2-cz\cosh\alpha)}{z^2-2cz\cosh\alpha+c^2}+\frac{Mc\cosh\alpha+N}{c\sinh\alpha}\ \frac{cz\sinh\alpha}{z^2-2cz\cosh\alpha+c^2}$$

Similarly, $Z^{-1}\left[\dfrac{Mz^2+Nz}{z^2+pz+q}\right]=M\{c^k\cosh\alpha k\}+\dfrac{Mc\cosh\alpha+N}{c\sinh\alpha}[c^k\sinh\alpha]$

Example 34. *Find the inverse Z-transform of* $\dfrac{1}{(z-3)(z-2)}$

(*i*) $|z|<2$ (*ii*) $2<|z|<3$ (*iii*) $|z|>3$

Solution. $F(z)=\dfrac{1}{(z-3)(z-2)}=\dfrac{1}{z-3}-\dfrac{1}{z-2}$

Case (*i*). $|z|<2$

$$F(z)=-\frac{1}{3}\frac{1}{1-\frac{z}{3}}+\frac{1}{2}\frac{1}{1-\frac{z}{2}}=-\frac{1}{3}\left(1-\frac{z}{3}\right)^{-1}+\frac{1}{2}\left(1-\frac{z}{2}\right)^{-1}$$

$$=-\frac{1}{3}\left(1+\frac{z}{3}+\frac{z^2}{3^2}+\frac{z^3}{3^3}+\ldots\right)+\frac{1}{2}\left[1+\frac{z}{2}+\frac{z^2}{2^2}+\frac{z^3}{2^3}+\ldots\right]$$

$$=-\frac{1}{3}-\frac{z}{3^2}-\frac{z^2}{3^3}-\frac{z^3}{3^4}-\ldots+\frac{1}{2}+\frac{z}{2^2}+\frac{z^2}{2^3}+\frac{z^3}{2^4}+\ldots\ldots\ldots$$ **Ans.**

Case (*ii*). $2<|z|<3$

$$F(z)=-\frac{1}{3}\frac{1}{1-\frac{z}{3}}-\frac{1}{z}\frac{1}{1-\frac{2}{z}}=-\frac{1}{3}\left[1-\frac{z}{3}\right]^{-1}-\frac{1}{z}\left[1-\frac{2}{z}\right]^{-1}$$

$$=-\frac{1}{3}\left(1+\frac{z}{3}+\frac{z^2}{3^2}+\frac{z^3}{3^3}+\ldots\right)-\frac{1}{z}\left[1+\frac{2}{z}+\frac{2^2}{z^2}+\frac{2^3}{z^3}+\ldots\right]$$

$$=-\frac{1}{3}-\frac{z}{3^2}-\frac{z^2}{3^3}-\frac{z^3}{3^4}-\ldots-\frac{1}{z}-\frac{2}{z^2}-\frac{2^2}{z^3}-\frac{2^3}{z^4}-\ldots$$

$$=\ldots\ldots-\frac{z^3}{3^4}-\frac{z^2}{3^3}-\frac{z}{3^2}-\frac{1}{3}-\frac{1}{z}-\frac{2}{z^2}-\frac{2^2}{z^3}-\frac{2^3}{z^4}-\ldots\ldots$$

$$=\ldots\ldots-\frac{z^3}{3^4}-\frac{z^2}{3^3}-\frac{z}{3^2}-\frac{1}{3}-z^{-1}-2z^{-2}-2^2z^{-3}-2^3z^{-4}-\ldots\ldots$$

$$\Rightarrow f(k) = \{-2^{k-1}\}, \quad k > 0$$

$$\Rightarrow f(k) = \{-3^{k-1}\}, \quad k \le 0$$ **Ans.**

Case (*iii*). $|z| > 3$

$$F(z) = \frac{1}{z}\frac{1}{1-\frac{3}{z}} - \frac{1}{z}\frac{1}{1-\frac{2}{z}} = \frac{1}{z}\left(1-\frac{3}{z}\right)^{-1} - \frac{1}{z}\left(1-\frac{2}{z}\right)^{-1}$$

$$= \frac{1}{z}\left[1+\frac{3}{z}+\frac{3^2}{z^2}+\frac{3^3}{z^3}+\ldots\right] - \frac{1}{z}\left(1+\frac{2}{z}+\frac{2^2}{z^2}+\frac{2^3}{z^3}+\ldots\right)$$

$$= \frac{1}{z}+\frac{3}{z^2}+\frac{3^2}{z^3}+\frac{3^3}{z^4}+\ldots-\frac{1}{z}-\frac{2}{z^2}-\frac{2^2}{z^3}-\frac{2^3}{z^4}+\ldots$$

$$= \{3^{k-1} - 2^{k-1}\}\, z^{-k}, \qquad k \ge 1$$

$$= 0 \qquad , \qquad k \le 1$$ **Ans.**

Example 35. *Find the inverse of Z-transform of* $\frac{1}{(z-5)^3}$, $|z| > 5$

Solution. $F(z) = \frac{1}{(z-5)^3} = \frac{1}{z^3}\frac{1}{\left(1-\frac{5}{z}\right)^3} = \frac{1}{z^3}(1-5z^{-1})^{-3}$

$$= z^{-3}\left[1 + 15z^{-1} + 6\,(5z^{-1})^2 + 10(5\,z^{-1})^3 + \ldots\ldots + \frac{(n+1)(n+2)}{2}(5z^{-1})^n + \ldots\ldots\right]$$

$$= z^{-3}\left\{\frac{(k+1)(k+2)}{2}5^k\right\}z^{-k} = \left\{\frac{(k+1)(k+2)}{2}5^k\right\}z^{-k-3}$$

Replacing k by $k-3$ we get

$$= \left\{\frac{(k-3+1)(k-3+2)}{2}5^{k-3}\right\}z^{-k} = \left\{\frac{(k-2)(k-1)}{2}5^{k-3}\right\}z^{-k}$$

$$Z^{-1}F(z) = f(k) = \begin{cases} \frac{(k-2)(k-1)}{2}5^{k-3} & , \quad k \ge 3 \\ 0 & , \quad k < 3 \end{cases}$$ **Ans.**

Example 36. *Obtain* $Z^{-1}\frac{1}{\left(z-\frac{1}{2}\right)\left(z-\frac{1}{3}\right)}$. *When* (*i*) $\frac{1}{3} < |z| < \frac{1}{2}$ (*ii*) $\frac{1}{2} < |z|$

Solution. $F(z) = \frac{1}{\left(z-\frac{1}{2}\right)\left(z-\frac{1}{3}\right)} = 6\left[\frac{1}{\left(z-\frac{1}{2}\right)} - \frac{1}{\left(z-\frac{1}{3}\right)}\right] = \frac{6}{z-\frac{1}{2}} - \frac{6}{z-\frac{1}{3}}$

(*i*) $\frac{1}{3} < |z| < \frac{1}{2}$, $F(z) = \frac{6}{-\frac{1}{2}(1-2z)} - \frac{1}{z}\frac{6}{(1-\frac{1}{3}z^{-1})} = -12(1-2z)^{-1} - \frac{6}{z}(1-\frac{1}{3}z^{-1})^{-1}$

$$= -12[1+(2z)+(2z)^2+(2z)^3+\ldots] - \frac{6}{z}\left(1+\frac{1}{3z}+\frac{1}{(3z)^2}+\frac{1}{(3z)^3}+\ldots\right)$$

$$= -12(2z)^k - \frac{6}{z}\left(\frac{1}{3z}\right)^k = -12(2z)^k - 6\left(\frac{1}{3^k z^{k+1}}\right)$$

$f(k) = -\frac{6}{3^{k-1}}$, if $k > 0$

$f(k) = -12\,.\,2^{-k}$, if $k < 0$

(*ii*) $\frac{1}{2}<|z|$

$$F(z)=\frac{6}{z-\frac{1}{2}}-\frac{6}{z-\frac{1}{3}}=\frac{1}{z}\frac{6}{(1-\frac{1}{2z})}-\frac{1}{z}\frac{6}{(1-\frac{1}{3z})}=6z^{-1}\left(1-\frac{1}{2}z^{-1}\right)^{-1}-6z^{-1}\left(1-\frac{1}{3}z^{-1}\right)^{-1}$$

$$=6z^{-1}\left[1+\frac{1}{2}z^{-1}+\left(\frac{1}{2}\right)^2z^{-2}+\left(\frac{1}{2}\right)^3z^{-3}+...\right]-6z^{-1}\left[1+\frac{1}{3}z^{-1}+\left(\frac{1}{3}\right)^2z^{-2}+\left(\frac{1}{3}\right)^3z^{-3}+...\right]$$

$$=6\left[z^{-1}+\frac{1}{2}z^{-2}+\left(\frac{1}{2}\right)^2z^{-3}+\left(\frac{1}{2}\right)^3z^{-4}+...\right]-6\left[z^{-1}+\frac{1}{3}z^{-2}+\left(\frac{1}{3}\right)^2z^{-3}+\left(\frac{1}{3}\right)^3z^{-4}+...\right]$$

$$=6\left\{\frac{1}{2^{k+1}}\right\}z^{-k}-6\left\{\frac{1}{3^{k+1}}\right\}z^{-k}\Rightarrow Z^{-1}F(z)=f(k)=6\left[\frac{1}{2^{k-1}}-\frac{1}{3^{k-1}}\right],\qquad k\geq 1$$

Ans.

Example 37. *Obtain* $Z^{-1}\frac{2z^2-10z+13}{(z-3)^2(z-2)}$, *when* $2<|z|<3$

Solution. Let $\frac{2z^2-10z+13}{(z-3)^2(z-2)}=\frac{A}{(z-3)^2}+\frac{B}{z-3}+\frac{C}{z-2}$

Converting into partial fractions, we get

$$=\frac{1}{(z-3)^2}+\frac{1}{z-3}+\frac{1}{z-2}$$

$$=\frac{1}{9}\frac{1}{\left(1-\frac{z}{3}\right)^2}-\frac{1}{3}\frac{1}{1-\frac{z}{3}}+\frac{1}{z}\frac{1}{1-\frac{2}{z}}=\frac{1}{9}\left(1-\frac{z}{3}\right)^{-2}-\frac{1}{3}\left(1-\frac{z}{3}\right)^{-1}+\frac{1}{z}\left(1-\frac{2}{z}\right)^{-1}$$

$$=\frac{1}{9}\left[1+\frac{2z}{3}+\frac{3z^2}{9}+\frac{4z^3}{27}+...\right]-\frac{1}{3}\left[1+\frac{z}{3}+\frac{z^2}{9}+\frac{z^3}{27}+....\right]+\frac{1}{z}\left[1+\frac{2}{z}+\frac{4}{z^2}+\frac{8}{z^3}+...\right]$$

$$=\frac{1}{3^2}+\frac{2z}{3^3}+\frac{3z^2}{3^4}+\frac{4z^3}{3^5}...-\frac{1}{3}-\frac{z}{3^2}-\frac{z^2}{3^3}-\frac{z^3}{3^4}-....+\frac{1}{z}+\frac{2}{z^2}+\frac{4}{z^3}+\frac{8}{z^4}+...$$

$$=\{2^{k-1}\}z^{-k},\qquad k\geq 1 \qquad ...(1)$$

$$=\left[\frac{k+1}{3^{k+2}}-\frac{1}{3^{k+1}}\right]z^k,\ k<0\quad\text{or}=\frac{-k-2}{3^{-k+2}},\ k\leq 0 \qquad ...(2)$$

Hence, $Z^{-1}[F(z)]=2^{k-1}$ if $k\geq 1$, and $Z^{-1}[F(z)]=-(k+2)\,3^{k-2},\ k\leq 0$ **Ans.**

Example 38. *Find* $Z^{-1}\frac{3z^2+4z}{z^2-z+1}$ $\qquad |z|>1.$

Solution. Let $c^2=1\ \Rightarrow\ c=\pm 1$

If $\qquad c=1,\quad \left|\frac{p}{2c}\right|=\left|\frac{-1}{2\times 1}\right|=\frac{1}{2}<1$

So $\qquad -1=-2c\cos\alpha$ or $-1=-2\times 1\times\cos\alpha\ \Rightarrow\ \cos\alpha=\frac{1}{2}$

$$\alpha=\frac{\pi}{3},\quad \sin\alpha=\sin\frac{\pi}{3}=\frac{\sqrt{3}}{2}$$

$$\frac{3z^2+4z}{z^2-z+1}=\frac{3z(z-c\cos\alpha)+\frac{(3c\cos\alpha+4)}{c\sin\alpha}cz\sin\alpha}{z^2-2cz\cos\alpha+c^2}$$

$$\frac{3z^2+4z}{z^2-z+1}=\frac{3z(z-c\cos\alpha)}{z^2-2cz\cos\alpha+c^2}+\frac{\dfrac{3c\cos\alpha+4}{c\sin\alpha}cz\sin\alpha}{z^2-2cz\cos\alpha+c^2}$$

Putting the values of c and α, in the coefficient of $cz \sin \alpha$, we get

$$=\frac{3z(z-c\cos\alpha)}{z^2-2cz\cos\alpha+c^2}+\frac{\dfrac{3\times1\times\frac{1}{2}+4}{1\times\frac{\sqrt{3}}{2}}cz\sin\alpha}{z^2-2cz\cos\alpha+c^2}$$

$$Z^{-1}\left[\frac{3z^2+4z}{z^2-z+1}\right]=\left[\left\{3c^k\cos\alpha k\right\}+\left\{\frac{11}{\sqrt{3}}c^k\sin\alpha k\right\}\right]U_k$$

Putting the values of c and α, we get

$$Z^{-1}\left[\frac{3z^2+4z}{z^2-z+1}\right]=\left[3(1)^k\left\{\cos\frac{\pi}{3}k\right\}+\frac{11}{3}(1)^k\left\{\sin\frac{\pi}{3}k\right\}\right]U_k$$

$$=\left[3\left\{\cos\frac{\pi k}{3}\right\}+\frac{11}{\sqrt{3}}\left\{\sin\frac{\pi}{3}k\right\}\right]U_k$$

Ans.

Example 39. *Obtain* $Z^{-1}\dfrac{2z^2+3z}{z^2+z+\frac{1}{9}}$ $\left(|z|>\dfrac{1}{3}\right)$

Solution. $c^2=\dfrac{1}{9}\Rightarrow c=\pm\dfrac{1}{3},\ p=1$

If $\quad c=-\dfrac{1}{3},\ \left|\dfrac{p}{2c}\right|=\left|\dfrac{1}{2\times-\frac{1}{3}}\right|=\dfrac{3}{2}>1$

Hence, $\quad 1=-2c\cosh\alpha \quad$ or $\quad 1=-2\left(-\dfrac{1}{3}\right)\cosh\alpha$

or $\quad \cosh\alpha=\dfrac{3}{2},\ \sinh\alpha=\dfrac{\sqrt{5}}{2} \qquad (\cosh^2\alpha-\sinh^2\alpha=1)$

$$\frac{2z^2+3z}{z^2+z+\frac{1}{9}}=\frac{2z(z-c\cosh\alpha)+\dfrac{(2c\cosh\alpha+3)}{c\sinh\alpha}cz\sinh\alpha}{z^2-2cz\cosh\alpha+c^2}$$

Putting the values of c and α in the coefficient of $cz \sin \alpha$, we get

$$=2\frac{z(z-c\cosh\alpha)}{z^2-2cz\cosh\alpha+c^2}+\frac{\dfrac{\left[2\left(-\frac{1}{3}\right)\left(\frac{3}{2}\right)+3\right]}{-\frac{1}{3}\times\frac{\sqrt{5}}{2}}cz\sinh\alpha}{z^2-2cz\cosh\alpha+c^2}$$

$$=2\frac{z(z-c\cosh\alpha)}{z^2-2cz\cosh\alpha+c^2}+\frac{\left(\dfrac{-1+3}{\frac{-\sqrt{5}}{6}}\right)cz\sinh\alpha}{z^2-2cz\cosh\alpha+c^2}=2\frac{z(z-c\cosh\alpha)}{z^2-2cz\cosh\alpha+c^2}-\frac{12}{\sqrt{5}}\frac{cz\sinh\alpha}{z^2-2cz\cosh\alpha+c^2}$$

$$Z^{-1}\frac{2z^2+3z}{z^2+z+\frac{1}{9}}=2\left(-\frac{1}{3}\right)^k\cosh\alpha k-\frac{12}{\sqrt{5}}\left(-\frac{1}{3}\right)^k\sinh\alpha,\ k\geq 0$$

where $\cosh\alpha = 3/2$. **Ans.**

EXERCISE 44.5

Evaluate inverse Z-transform of the following:

1. $\frac{z}{3-z},\quad |z|<3$ **Ans.** $\{3^k\},\ k<0$
2. $\frac{2z^2-5z}{(z-2)(z-3)},\quad |z|>3$ **Ans.** $\{2^k+3^k\},\ k\geq 0$
3. $\frac{z}{z-e^{\alpha}}$ **Ans.** $\{e^{k\alpha}\}$
4. $\frac{z^3}{z^3-27},\quad |z|>3$ **Ans.** $\{f(k)\}=\{3^k\}=0$, for $k=0, 3, 6, 9$
5. $\frac{2z}{(z-2)^2},\quad |z|>2$ **Ans.** $\{k\,2^k\},\ k\geq 0$
6. $\frac{z^2+z\cos\alpha}{z^2-2z\cos\alpha+1},\quad |z|>1$ **Ans.** $\{\cos\alpha\, k\},\ k\geq 0$
7. $-\log(1-2z^{-1}),\ |z|>2$ **Ans.** $\left\{\frac{2^k}{k}\right\},\ k\geq 1$
8. $\frac{4z}{4z+1}+\frac{5z}{5z+1},\quad \frac{1}{5}<|z|<\frac{1}{4}$ **Ans.** $f(k)=\begin{cases}-\left(-\frac{1}{4}\right)^k, & k<0\\ \left(\frac{1}{5}\right)^k, & k\geq 0\end{cases}$
9. $\frac{z(z^2+4z+1)}{(z-1)^4},\quad |z|>1$ **Ans.** $\{k^3\}$
10. $\frac{ze^{-a}}{(z-e^{-a})^2},\quad |z|>|e^{-a}|$ **Ans.** $\{k\,e^{-ak}\}$
11. $\frac{ze^{-a}\sin b}{z^2-2e^{-a}z\cos b+e^{-2a}},\quad |z|>|e^{-a}|$ **Ans.** $\{e^{-ak}\sin bk\}$
12. $\frac{z(z-e^{-a}\cos b)}{z^2-2e^{-a}z\cos b+e^{-2a}}$ **Ans.** $\{e^{-ak}\cos bk\}$
13. $\frac{z^2\sin\beta+\frac{z}{\sqrt{2}}(\cos\beta-\sin\beta)}{z^2-\sqrt{2}z+1},\quad |z|>1$ **Ans.** $\left\{\sin(\frac{k\pi}{4}+\beta)\right\},\quad k\geq 0$
14. $\frac{z(z+a)}{(z-a)^3},\quad |z|>a$ **Ans.** $\{k^2a^{k-1}U(k-1)\},\ k\geq 0$
15. $\frac{15z}{(4-z)(4z-1)},\quad \frac{1}{4}<|z|<4$ **Ans.** $\left\{\left(\frac{1}{4}\right)^{|k|}\right\}$

44.23 INVERSION OF Z-TRANSFORM BY RESIDUE METHOD

Take the contour c such that all the poles of the function z lie within the contour. Then by Residue Method

$f(k) = \Sigma$ Residue of $z^{k-1} F(z)$ at its poles.

where Residue for simple pole $z = z_i$ is $= \left[(z - z_i) z^{k-1} F(z)\right]_{z = z_i}$

Residue of order r at the pole $z = z_i$

$$\left[\frac{1}{(r-1)!}\frac{d^{r-1}}{dz^{r-1}}(z - z_i)^r z^{k-1} F(z)\right]_{z = z_i}$$

Example 40. *Obtain* $Z^{-1}\dfrac{z}{(z-2)(z-3)}$

Solution. The poles are determined by

$(z-2)(z-3) = 0 \quad \Rightarrow \quad z = 2, 3$

There are two poles. Let us consider the contour $|z| > 3$.

$$\text{Residue at } (z=2) = \left[(z-2)z^{k-1}\frac{z}{(z-2)(z-3)}\right]_{z=2} = \left[\frac{z^k}{z-3}\right]_{z=2} = \frac{2^k}{-1}$$

$$\text{Residue at } (z=3) = \left[(z-3)z^{k-1}\frac{z}{(z-2)(z-3)}\right]_{z=3} = \left[\frac{z^k}{z-2}\right]_{z=3} = \frac{3^k}{1}$$

Hence, $f(k) =$ Sum of the residues $= 3^k - 2^k$. **Ans.**

Example 41. *Obtain* $Z^{-1}\dfrac{3z^2 - 18z + 26}{(z-2)(z-3)(z-4)}$

Solution. The poles are determined by $(z-2)(z-3)(z-4) = 0 \Rightarrow z = 2, 3, 4$.

There are three poles. Let us consider the contour $|z| > 4$.

$$\text{Residue at } (z=2) = \left[\frac{(z-2)z^{k-1}[3z^2 - 18z + 26]}{(z-2)(z-3)(z-4)}\right]_{z=2}$$

$$= \left[\frac{3z^{k+1} - 18z^k + 26z^{k-1}}{(z-3)(z-4)}\right]_{z=2} = \frac{3.2^{k+1} - 18.2^k + 26.2^{k-1}}{(-1)(-2)}$$

$$= 3.2^k - 9.2^k + 13.2^{k-1} = -6.2^k + 13.2^{k-1} = -12.2^{k-1} + 13.2^{k-1} = 2^{k-1}$$

$$\text{Residue at } (z=3) = \left[\frac{(z-3)z^{k-1}[3z^2 - 18z + 26]}{(z-2)(z-3)(z-4)}\right]_{z=3} = \left[\frac{3z^{k+1} - 18z^k + 26z^{k-1}}{(z-2)(z-4)}\right]_{z=3}$$

$$= \frac{3.3^{k+1} - 18.3^k + 26.3^{k-1}}{1(-1)} = -3.3^{k+1} + 18.3^k - 26.3^{k-1}$$

$$= -27.3^{k-1} + 54.3^{k-1} - 26.3^{k-1} = 3^{k-1}$$

$$\text{Residue at } (z=4) = \left[(z-4)\frac{z^{k-1}[3z^2 - 18z + 26]}{(z-2)(z-3)(z-4)}\right]_{z=4} = \left[\frac{3z^{k+1} - 18z^k + 26z^{k-1}}{(z-2)(z-3)}\right]_{z=4}$$

$$= \frac{3.4^{k+1} - 18.4^k + 26.4^{k-1}}{(2)(1)} = \frac{3}{2}.4^{k+1} - 9.4^k + 13.4^{k-1}$$

$$= 24.4^{k-1} - 36.4^{k-1} + 13.4^{k-1} = 4^{k-1}$$

Hence $f(k) =$ Sum of the residues $= 2^{k-1} + 3^{k-1} - 4^{k-1},\ k > 0$ **Ans.**

Example 42. *Obtain* $Z^{-1}\dfrac{z(3z^2-6z+4)}{(z-1)^2(z-2)}$

Solution. The poles are determined by $(z-1)^2(z-2)=0 \quad \Rightarrow \quad z=1,1,2$

There are two poles, simple pole at $z=2$ and pole of order 2 at $z=1$. Let us consider the contour $|z|>2$.

Residue at $(z=2)$ is $\left[\dfrac{(z-2)z^{k-1}.z[3z^2-6z+4]}{(z-1)^2(z-2)}\right]_{z=2} = \left[\dfrac{3z^{k+2}-6z^{k+1}+4z^k}{(z-1)^2}\right]_{z=2}$

$$=3.2^{k+2}-6.2^{k+1}+4.2^k = 12.2^k-12.2^k+4.2^k=4.2^k=2^{k+2}$$

Residue at $(z=1)=\dfrac{d}{dz}\left[(z-1)^2\dfrac{z^{k-1}.z(3z^2-6z+4)}{(z-1)^2(z-2)}\right]_{z=1} = \dfrac{d}{dz}\left[\dfrac{3z^{k+2}-6z^{k+1}+4z^k}{z-2}\right]_{z=1}$

$$=\left[\frac{(z-2)[3(k+2)z^{k+1}-6(k+1)z^k+4kz^{k-1}]-(3z^{k+2}-6z^{k+1}+4z^k).1}{(z-2)^2}\right]_{z=1}$$

$$=-\{(3k+6)-6k-6+4k\}-\{3-6+4\}$$

$$=-3k-6+6k+6-4k-3+6-4=-k-1$$

$f(k)=$ Sum of the residues $=[2^{k+2}-k-1]U(k)$ **Ans.**

Example 43. *Find* $Z^{-1}\dfrac{9z^3}{(3z-1)^2(z-2)}$

Solution. The poles are determined by putting the denominator equal to zero, *i.e.*,

$$(3z-1)^2(z-2)=0 \quad \Rightarrow z=\frac{1}{3},\frac{1}{3},2$$

There are two poles *i.e.*, one simple pole at $z=2$ and second pole of order 2 at $z=\dfrac{1}{3}$.

Let us consider the contour $|z|>2$.

Residue at $(z=2)=\left[(z-2).z^{k-1}.\dfrac{9z^3}{(3z-1)^2(z-2)}\right]_{z=2} = \left[\dfrac{9z^{k+2}}{(3z-1)^2}\right]_{z=2} = \dfrac{9.2^{k+2}}{25}$

Residue at $\left(z=\dfrac{1}{3}\right)=\dfrac{d}{dz}\left[\dfrac{\left(z-\frac{1}{3}\right)^2.z^{k-1}.9z^3}{(3z-1)^2.(z-2)}\right]_{z=\frac{1}{3}} = \dfrac{d}{dz}\left[\dfrac{9z^{k+2}}{9(z-2)}\right]_{z=\frac{1}{3}} = \dfrac{d}{dz}\left(\dfrac{z^{k+2}}{z-2}\right)_{z=\frac{1}{3}}$

$$=\left[\frac{(z-2)(k+2)z^{k+1}-z^{k+2}}{(z-2)^2}\right]_{z=\frac{1}{3}} = \left[\frac{(k+2)z^{k+2}-2(k+2)z^{k+1}-z^{k+2}}{(z-2)^2}\right]_{z=\frac{1}{3}}$$

$$=\left[\frac{z^{k+2}(k+2-1)-2(k+2)z^{k+1}}{(z-2)^2}\right]_{z=\frac{1}{3}}$$

$$=\left[\frac{(k+1)z^{k+2}-2(k+2)z^{k+1}}{(z-2)^2}\right]_{z=\frac{1}{3}} = \left[\frac{z^{k+1}\{(k+1)z-2(k+2)\}}{(z-2)^2}\right]_{z=\frac{1}{3}}$$

$$= \frac{\left(\frac{1}{3}\right)^{k+1}\left\{(k+1)\frac{1}{3}-2(k+2)\right\}}{\frac{25}{9}} = \frac{\left(\frac{1}{3}\right)^{k+1}.9.\frac{1}{3}[k+1-6(k+2)]}{25}$$

$$= \left(\frac{1}{3}\right)^{k}\left(\frac{k+1-6k-12}{25}\right) = \frac{\left(\frac{1}{3}\right)^{k}(-5k-11)}{25}$$

Hence, $f(k)$ = Sum of residues $= \frac{9}{25}.2^{k+2} - \left(\frac{1}{3}\right)^{k}\frac{(5k+11)}{25}$. **Ans.**

Example 44. *Find* $Z^{-1}\frac{z^2}{z^2+4}$, $\qquad |z| > 2$

Solution. The poles are determined by $z^2 + 4 = 0 \quad \Rightarrow \quad z = \pm\, 2i$

There are two poles at $z = 2i$ and $z = -\,2i$.

Let us consider a contour $|z| > 2$

Residue at $(z = 2i)$ $= \left[\frac{(z-2i)z^{k-1}.z^2}{z^2+4}\right]_{z=2i} = \left[\frac{z^{k+1}}{z+2i}\right]_{z=2i} = \frac{(2i)^{k+1}}{4i} = 2^{k-1}(i)^k$

$$= 2^{k-1}\left[\cos\frac{\pi}{2} + i\sin\frac{\pi}{2}\right]^k = 2^{k-1}\left[\cos\frac{k\pi}{2} + i\sin\frac{k\pi}{2}\right]$$

Residue at $(z = -\,2i)$ $= \left[(z+2i)z^{k-1}\frac{z^2}{z^2+4}\right]_{z=-2i} = \left[\frac{z^{k+1}}{z-2i}\right]_{z=-2i} = \frac{(-2i)^{k+1}}{-4i}$

$$= 2^{k-1}(-i)^k = 2^{k-1}\left[\cos\frac{\pi}{2} - i\sin\frac{\pi}{2}\right]^k = 2^{k-1}\left[\cos\frac{k\pi}{2} - i\sin\frac{k\pi}{2}\right]$$

$f(k)$ = Sum of the residues $= 2^{k-1}\left[\cos\frac{k\pi}{2} + i\sin\frac{k\pi}{2}\right] + 2^{k-1}\left[\cos\frac{k\pi}{2} - i\sin\frac{k\pi}{2}\right]$

$$f(k) = 2^{k-1}\left[2\cos\frac{k\pi}{2}\right] = 2^k\cos\frac{k\pi}{2}$$ **Ans.**

44.24 CONVOLUTION METHOD

We know that $\qquad Z\{f * g\} = F(z)\, G(z)$

$$Z^{-1}\{F(z).\, G(z)\} = f * g = \sum_{n=0}^{k} f(n)\, g(k-n).$$

44.25 INVERSE OF Z-TRANSFORM BY CONVOLUTION METHOD

Example 45. *Using Convolution theorem, evaluate*

$$Z^{-1}\left\{\frac{z^2}{(z-1)(z-3)}\right\}.$$ *(Q. Bank U.P., III Semester 2002)*

Solution. We know that

$$Z^{-1}\{F(z).\, G(z)\} = f * g$$

Let $\qquad F(z) = \frac{z}{z-1} \qquad \therefore \quad f(k) = (1)^k$

$$G(z) = \frac{z}{z-3} \qquad \therefore\ g(k) = (3)^k$$

Now, $\quad Z^{-1}\{F(z) \,.\, G(z)\} = (1)^k * (3)^k$

$$= \sum_{n=0}^{k} 1^n\, 3^{k-n} = 3^k \sum_{n=0}^{k} \left(\frac{1}{3}\right)^n \qquad (a \text{ G.P.})$$

$$= 3^k \left\{ \frac{\left(\frac{1}{3}\right)^{k+1} - 1}{\frac{1}{3} - 1} \right\} = \frac{(1)^{k+1} - (3)^{k+1}}{1-3} = \left\{ \frac{1}{2}(3^{k+1} - 1) \right\} \qquad \textbf{Ans.}$$

44.26 INVERSE OF Z-TRANSFORM BY POWER SERIES METHOD

Example 46. *Find* $Z^{-1}\left[\log \frac{z}{z+1}\right]$ *by power series method.*

Solution. Let $F(z) = \log\left(\frac{z}{z+1}\right) = \log\left(\frac{1}{1+\frac{1}{z}}\right) = -\log\left(1+\frac{1}{z}\right)$

$$= -\frac{1}{z} + \frac{1}{2}\left(\frac{1}{z}\right)^2 - \frac{1}{3}\left(\frac{1}{z}\right)^3 + \frac{1}{4}\left(\frac{1}{z}\right)^4 + \ldots..$$

$$= -\frac{1}{z} + \frac{1}{2z^2} - \frac{1}{3z^3} + \frac{1}{4z^4} + \ldots\ldots + \frac{(-1)^k}{k z^k} + \ldots..$$

$$f(k) = Z^{-1}[F(z)] = \begin{cases} 0\ , & \text{for } k = 0 \\ \frac{(-1)^k}{k}, & \text{othersie} \end{cases} \qquad \textbf{Ans.}$$

EXERCISE 44.6

Find the inverse Z-transform of the following functions by residue method:

1. $\frac{z}{(z-1)(z-2)}$ **Ans.** $2^k - 1,\ k \geq 0$
2. $\frac{3z^2 + 2z}{z^2 - 3z + 2},\ 1 |<| z | < 2$ **Ans.** $\{-5\}\ (k \geq 0),\ -8\,(2)^k,\ k < 0$
3. $\frac{2z^2 + 3z}{z^2 + z + 1},\ | z | > 1$ **Ans.** $\left[2\{\cos\frac{2\pi k}{3}\} + \frac{4}{\sqrt{3}}\{\sin\frac{2\pi k}{3}\}\right] U(k)$
4. $\frac{z(z+1)}{(z-1)(z^2 + z + 1)},\ | z | > 1$ **Ans.** $\frac{2}{3}\left[1 - \cos\frac{2\pi k}{3}\right] U(k)$
5. $\frac{16z^3}{(4z-1)^2\,(z-1)}$ **Ans.** $\left[\frac{16}{9} - \frac{1}{9}(3k+7)\left(\frac{1}{4}\right)^k\right] U(k)$
6. $\frac{z^2}{z^2 + 1}$ **Ans.** $\cos\frac{k\pi}{2}$

44.27 DIFFERENCE EQUATION

Difference equation is the equation between the differences of an unknown function.

For example, $\Delta y_n + 2y_n = 0 \qquad (\Delta y_n = y_{n+1} - y_n)$ ... (1)

$\Delta^2 y_n + 5\Delta y_n + 6y_n = 0$ ... (2)

Second way. To express the difference equation.

Putting the value of $\Delta = E - 1$

(1) becomes, $(E-1)y_n + 2y_n = 0$

$Ey_n + y_n = 0$ ... (3)

(2) becomes, $(E-1)^2 y_n + 5(E-1)y_n + 6y_n = 0$

$\Rightarrow (E^2 - 2E + 1)y_n + 5(E-1)y_n + 6y_n = 0$

$\Rightarrow E^2 y_n + 3Ey_n + 2y_n = 0$ ... (4)

Third way

(3) can be written as $y_{n+1} + y_n = 0 \qquad [Ey_n = y_{n+1}]$ ... (5)

(4) can be written as $y_{n+2} + 3y_{n+1} + 2y_n = 0 \qquad [E^2 y_n = y_{n+2}]$ ... (6)

44.28 ORDER OF A DIFFERENCE EQUATION = HIGHEST POWER OF E

Order of a difference equation is the difference between the largest and the smallest arguments involved in the difference equation, divided by the unit of interval.

$$\text{Thus the order of (5)} = \frac{\text{Largest argument} - \text{Smallest argument}}{\text{Unit of interval}} = \frac{(n+1)-n}{1} = 1$$

$$\text{Similarly, order of (6)} = \frac{n+2-n}{1} = 2$$

44.29 DEGREE OF A DIFFERENCE EQUATION

The degree of a difference equation is defined to be the highest power of $f(x)$.

44.30 SOLUTION OF A DIFFERENCE EQUATION

A solution of a difference equation is any function which satisfies the given equation. The general solution of a difference equation is defined as the solution which involves as many arbitrary constants as the order of the difference equation.

The particular solution is a solution obtained from the general solution by assigning particular values to periodic constants.

44.31 THEOREM

Prove that $Z(y_{k+n}) = z^n\left(\bar{y} - y_0 - \frac{y_1}{z} - \ldots - \frac{y_{n-1}}{z^{n-1}}\right)$, where $Z(y_k) = \bar{y}$

Proof. L.H. S. $= Z(y_{k+n}) = \sum_{k=0}^{\infty} y_{k+n} z^{-k} = z^n \sum_{k=0}^{\infty} y_{k+n} z^{-(n+k)}$ ...(1)

On putting $m = n + k$ in (1), we get

$$Z(y_{k+n}) = z^n \sum_{m=n}^{\infty} y_m z^{-m} = z^n \left[\sum_{m=0}^{\infty} y_m z^{-m} - \sum_{m=0}^{n-1} y_m z^{-m}\right]$$

$$= z^n \left[\bar{y} - y_0 - \frac{y_1}{z} - \frac{y_2}{z^2} - \ldots - \frac{y_{n-1}}{z^{n-1}}\right]$$

Remember

For $n = 1$, $Z(y_{k+1}) = (z\bar{y} - z\,y_0)$; **For $n = 2$,** $Z(y_{k+2}) = (z^2\bar{y} - z^2 y_0 - zy_1)$

For $n = 3$, $Z(y_{k+3}) = (z^3\bar{y} - z^3 y_0 - z^2 y_1 - z\,y_2)$

Note. If $Z(y_k) = \bar{y}$, then $Z(y_{k-n}) = Z^{-n}\bar{y}$

Example 47. *Solve the difference equation*

$$6y_{k+2} - y_{k-1} - y_k = 0, \quad y(0) = 0,\ y(1) = 1 \text{ by Z-transform.}$$

Solution. $6y_{k+2} - y_{k+1} - y_k = 0$ ...(1)

Taking the Z-transform of both sides of (1), we get

$$Z[6y_{k+2} - y_{k+1} - y_k] = 0$$

$$Z(6y_{k+2}) - Z(y_{k+1}) - Z(y_k) = 0$$

$$6[z^2\bar{y} - z^2 y(0) - z\,y(1)] - [z\bar{y} - zy(0)] - \bar{y} = 0 \qquad \text{... (2)}$$

On putting the values of $y(0)$ and $y(1)$ in (2), we get

$$6z^2\bar{y} - 6z - z\bar{y} - \bar{y} = 0 \quad \Rightarrow \quad (6z^2 - z - 1)\,\bar{y} = 6z$$

$$\bar{y} = \frac{6z}{6z^2 - z - 1} = \frac{6z}{(3z+1)(2z-1)} = \frac{z^{-1}}{\left(1+\frac{z^{-1}}{3}\right)\left(1-\frac{z^{-1}}{2}\right)} = \frac{\frac{6}{5}}{1-\frac{z^{-1}}{2}} - \frac{\frac{6}{5}}{1+\frac{z^{-1}}{3}}$$

$$y_k = Z^{-1}\left[\frac{\frac{6}{5}}{1-\frac{z^{-1}}{2}}\right] - Z^{-1}\left[\frac{\frac{6}{5}}{1+\frac{z^{-1}}{3}}\right] = \frac{6}{5}\left(\frac{1}{2}\right)^k - \frac{6}{5}\left(-\frac{1}{3}\right)^k = \frac{6}{5}\left[\left(\frac{1}{2}\right)^k - \left(-\frac{1}{3}\right)^k\right] \qquad \textbf{Ans.}$$

Example 48. *Solve the difference equation*

$$y_{k+3} - 3y_{k+2} + 3y_{k+1} - y_k = U(k)$$

$$y(0) = y(1) = y(2) = 0 \quad \text{by Z-transforms.}$$

Solution. $y_{k+3} - 3y_{k+2} + 3y_{k+1} - y_k = U(k)$ ...(1)

Taking the Z-transform of both sides of (1), we get

$$Z[y_{k+3} - 3y_{k+2} + 3y_{k+1} - y_k] = Z\,U(k),$$

$$\Rightarrow Z[y_{k+3}] - 3Z[y_{k+2}] + 3Z[y_{k+1}] - Z[y_k] = Z\,U(k)$$

$$\Rightarrow [z^3\bar{y} - z^3 y(0) - z^2 y(1) - zy(2)] - 3[z^2\bar{y} - z^2 y(0) - z\,y(1)]$$

$$+ 3[z\bar{y} - z\,y(0)] - \bar{y} = Z\,U(k)$$

Putting the values of $y(0) = y(1) = y(2) = 0$ in the above equation, we get

$$z^3\bar{y} - 3z^2\bar{y} + 3z\bar{y} - \bar{y} = \frac{1}{1-z^{-1}}$$

$$[z^3 - 3z^2 + 3z - 1]\,\bar{y} = \frac{1}{1-z^{-1}} \Rightarrow (z-1)^3\,\bar{y} = \frac{1}{1-z^{-1}}$$

$$\bar{y} = \frac{1}{(z-1)^3(1-z^{-1})} = \frac{1}{z^3(1-z^{-1})^3(1-z^{-1})} = z^{-3}(1-z^{-1})^{-4}$$

$$y_k = \text{coeff. of } z^{-k} \text{ in } z^{-3}(1-z^{-1})^{-4} = \text{coeff. of } z^{-k-3} \text{ in } (1-z^{-1})^{-4}$$

$$= \frac{(k-2)(k-1)k}{6},\ k \geq 3$$

Ans.

Example 49. *Solve by Z-transform.* $y_{k+1} + \frac{1}{4} y_k = \left(\frac{1}{4}\right)^k$, $(k \geq 0), y(0) = 0$

Solution. $y_{k+1} + \frac{1}{4} y_k = \left(\frac{1}{4}\right)^k$ ... (1)

Taking Z-transform of both sides of (1), we get

$$Z[y_{k+1} + \frac{1}{4} y_k] = Z\left[\left(\frac{1}{4}\right)^k\right]$$

$$Z[y_{k+1}] + Z\left[\frac{1}{4} y_k\right] = Z\left[\left(\frac{1}{4}\right)^k\right]$$

$$z\bar{y} - zy(0) + \frac{1}{4}\bar{y} = \frac{1}{1-\frac{1}{4}z^{-1}}, \qquad |z| > \frac{1}{4}$$

$$z\bar{y} - 0 + \frac{1}{4}\bar{y} = \frac{1}{1-\frac{1}{4}z^{-1}} \quad \Rightarrow \quad (z+\frac{1}{4})\,\bar{y} = \frac{1}{1-\frac{1}{4}z^{-1}}$$

$$\bar{y} = \frac{1}{z+\frac{1}{4}} \times \frac{1}{1-\frac{1}{4}z^{-1}} = \frac{z^{-1}}{1+\frac{1}{4}z^{-1}} \times \frac{1}{1-\frac{1}{4}z^{-1}} = \frac{-2}{1+\frac{1}{4}z^{-1}} + \frac{2}{1-\frac{1}{4}z^{-1}}$$

$$y_{(k)} = Z^{-1}\left[\frac{-2}{1+\frac{1}{4}z^{-1}}\right] + Z^{-1}\left[\frac{2}{1-\frac{1}{4}z^{-1}}\right] = Z^{-1}\left[-2(1+\frac{1}{4}z^{-1})^{-1}\right] + Z^{-1}\left[2(1-\frac{1}{4}z^{-1})^{-1}\right]$$

$$= -2\left(-\frac{1}{4}\right)^k + 2\left(\frac{1}{4}\right)^k$$

Ans.

Example 50. *Solve* $y_k + \frac{1}{4} y_{k-1} = U_{(k)} + \frac{1}{3} U_{(k-1)}$

Solution. $y_k + \frac{1}{4} y_{k-1} = U_{(k)} + \frac{1}{3} U_{(k-1)}$ ...(1)

Taking the *Z*-transform of both sides of (1), we get

$$Z[\{y_k\}] + \frac{1}{4} Z[\{y_{k-1}\}] = Z[\{U_k\}] + \frac{1}{3}[\{U_{k-1}\}]$$

$$\bar{y} + \frac{1}{4}z^{-1}\bar{y} = \left[1+\frac{1}{3}z^{-1}\right]U(k), \qquad \left[1+\frac{1}{4}z^{-1}\right]\bar{y} = (1+\frac{1}{3}z^{-1})\,U(k)$$

$$\bar{y} = \frac{1+\frac{1}{3}z^{-1}}{1+\frac{1}{4}z^{-1}} = \frac{z+\frac{1}{3}}{z+\frac{1}{4}}$$

There is only one simple pole at $z=-\frac{1}{4}$

Let us consider the contour $|z|>\frac{1}{4}$.

$$\text{Residue at }\left(z=-\frac{1}{4}\right)=\left[(z+\frac{1}{4})z^{k-1}\frac{z+\frac{1}{3}}{z+\frac{1}{4}}\right]_{z=-\frac{1}{4}}=\left[z^k+\frac{z^{k-1}}{3}\right]_{z=-\frac{1}{4}}$$

$$=\left(-\frac{1}{4}\right)^k+\frac{1}{3}\left(-\frac{1}{4}\right)^{k-1}=-\frac{1}{4}\left(-\frac{1}{4}\right)^{k-1}+\frac{1}{3}\left(-\frac{1}{4}\right)^{k-1}=\frac{1}{12}\left(-\frac{1}{4}\right)^{k-1}$$

Hence, $y_k=\text{Residue}=\frac{1}{12}\left(-\frac{1}{4}\right)^{k-1}$

Example 51. *Solve* $y_k+\frac{1}{25}y_{k-2}=\left(\frac{1}{5}\right)^k\cos\frac{k\pi}{2}$ $(k\geq 0)$ *by residue method.*

Solution. $$y_k+\frac{1}{25}y_{k-2}=\left(\frac{1}{5}\right)^k\cos\frac{k\pi}{2} \qquad \text{... (1)}$$

Taking Z-transform of both sides of (1), we obtain

$$Z\left[y_k+\frac{1}{25}y_{k-2}\right]=Z\left(\frac{1}{5}\right)^k\cos\frac{k\pi}{2}\quad\Rightarrow\quad \bar{y}+\frac{1}{25}z^{-2}\bar{y}=\frac{z^2}{z^2+\frac{1}{25}}$$

$$\Rightarrow\quad\left[1+\frac{1}{25}z^{-2}\right]\bar{y}=\frac{z^2}{z^2+\frac{1}{25}}\quad\Rightarrow\bar{y}=\frac{z^2}{\left(1+\frac{1}{25}z^{-2}\right)\left(z^2+\frac{1}{25}\right)}=\frac{z^4}{\left(z^2+\frac{1}{25}\right)^2}$$

There are two poles of second order at $z=\frac{i}{5}$ and $z=-\frac{i}{5}$

Let us consider a contour $|z|>\left|\frac{1}{5}\right|$

$$\text{Residue at }\left(z=\frac{i}{5}\right)=\left[\frac{d}{dz}(z-\frac{i}{5})^2\frac{z^{k-1}z^4}{(z^2+\frac{1}{25})^2}\right]_{z=\frac{i}{5}}=\left[\frac{d}{dz}\frac{z^{k+3}}{(z+\frac{i}{5})^2}\right]_{z=\frac{i}{5}}$$

$$=\left[\frac{(z+\frac{i}{5})^2(k+3)z^{k+2}-z^{k+3}2(z+\frac{i}{5})}{(z+\frac{i}{5})^4}\right]_{z=\frac{i}{5}}=\left[\frac{(z+\frac{i}{5})(k+3)z^{k+2}-2z^{k+3}}{(z+\frac{i}{5})^3}\right]_{z=\frac{i}{5}}$$

$$=\frac{\left(\frac{2i}{5}\right)(k+3)\left(\frac{i}{5}\right)^{k+2}-2\left(\frac{i}{5}\right)^{k+3}}{\left(\frac{2i}{5}\right)^3}=\left(\frac{5}{2i}\right)^3\left[(2k+6)\left(\frac{i}{5}\right)^{k+3}-2\left(\frac{i}{5}\right)^{k+3}\right]$$

$$= \left(\frac{5}{2i}\right)^3 \left[(2k+4)\left(\frac{i}{5}\right)^{k+3}\right] = \left(\frac{1}{8}\right)(2k+4)\left(\frac{i}{5}\right)^k = \frac{1}{4}(k+2)\left(\frac{1}{5}\right)^k [\cos\frac{\pi}{2} + i\sin\frac{\pi}{2}]^k$$

$$= \frac{1}{4}(k+2)\left(\frac{1}{5}\right)^k \left[\cos\frac{k\pi}{2} + i\sin\frac{k\pi}{2}\right]$$

$$\text{Residue at } \left(z = -\frac{i}{5}\right) = \frac{1}{4}(k+2)\left(\frac{1}{5}\right)^k \left[\cos\frac{k\pi}{2} - i\sin\frac{k\pi}{2}\right]_{(i \to -i)}$$

$$y_k = \text{Sum of the residues}$$

$$\Rightarrow \qquad y_k = \frac{1}{2}(k+2)\frac{1}{5^k}\cos\frac{k\pi}{2} \qquad \textbf{Ans.}$$

EXERCISE 44.7

Solve the following difference equations by Z-transform:

1. $y_k - \frac{5}{6} y_{k-1} + \frac{1}{6} y_{k-2} = U_{(k)}$ **Ans.** $y_k = \left[3 - 3\left(\frac{1}{2}\right)^k + \left(\frac{1}{3}\right)^k\right] U(k)$

2. $6\, y_{k+2} + 5 y_{k+1} - y_k = 6\, U(k)$ **Ans.** $y_k = \left[\frac{6}{7}(k+1) - \frac{78}{49} + \frac{36}{49}\left(-\frac{1}{6}\right)^k\right] U(k)$

3. $y_{k+1} - y_{k-1} = U(k),\ y(0) = 0$ **Ans.** $\left[\frac{k}{2} + \frac{1}{4}\right] U(k)$

4. $y_{k+1} - 2y_k + y_{k-1} = a^k,\ a \neq 1$

 Ans. $y_k = \frac{1}{a}(k+1)U(k) - \frac{a}{(a-1)^2}U(k) + \frac{a}{(a-1)^2}a^k U(k) + \frac{1}{1-a} kU(k-1)$

5. $y_k + \frac{1}{9} y_{k-2} = \left(\frac{1}{3}\right)^k \cos\frac{k\pi}{2},\ (k \geq 0)$ **Ans.** $y_k = \left(\frac{k+2}{2}\right)\left(\frac{1}{3}\right)^k \cos\frac{k\pi}{2} U(k)$

6. $y_{k+2} - 3y_{k+1} - 4\, y_k = 0; y_0 = 3, y_1 = -2$ **Ans.** $y_k = \frac{1}{5}(4)^k + \frac{14}{5}(-1)^k$

7. $y_{k+2} + y_{k+1} - 2\, y_k = 0; y_0 = 4, y_1 = 0$ **Ans.** $y_k = \frac{8}{3} + \frac{4}{3}(-2)$

8. $y_{k+2} - 2y_{k+1} + y_k = k; y_{(0)} = 0, y_{(1)} = 0$ **Ans.** $y_k = \frac{k-1}{4}\{1 - (-1)^k\}$

9. $y_{k+2} - 4\, y_k = 0;\ y_{(0)} = 0, y_{(1)} = 2$ **Ans.** $y_k = 2^{k-1} + (-2)^{k-1}$

 (Uttarakhand III Semester, 2008)

10. $y_{k+2} - 2y_{k+1} + y_k = 2^k; y_{(0)} = 2, y_{(1)} = 1$ **Ans.** $y_k = 1 - 2k + 2^k$

11. $8\, y_{k+2} - 6\, y_{k+1} + y_k = 5\sin\left(\frac{k\pi}{2}\right)$ *(U.P. III Semester, Dec. 2006)*

HANKEL TRANSFORM

45.1 HANKEL TRANSFORM

If $J_n(sx)$ be the Bessel function of the first kind of order n, then the Hankel transform of a function $f(x)$, $(0 < x < \infty)$ denoted by $F(s)$ is defined as $H(s) = \int_0^\infty f(x).\ xJ_n(sx)\,dx$

Here $x\,J_n(sx)$ is the *Kernel of the transformation,*

45.2 THE FORMULAE USED IN FINDING THE HANKEL TRANSFORMS.

Recurrence relations for Bessel's functions

1. $x\,J'_n = nJ_n - x\,J_{n+1}$

2. $xJ'_n = x\,J_{n-1} - nJ_n$

3. $2J'_n = J_{n-1} - J_{n+1}$

4. $2n\,J_n = x[J_{n-1} + J_{n+1}]$

5. $\frac{d}{dx}(x^{-n}J_n) = -x^{-n}\,J_{n+1}$

6. $\frac{d}{dx}(x^n\,J_n) = x^n\,J_{n-1}$ or $x^n\,J_n = \int x^n\,J_{n-1}\,dx$

From (6) Recurrence relation we can find the definite Integrals

7. $\int_0^\infty x^n J_{n-1}(x)\,dx = [x^n J_n(x)]_0^\infty$ ... (7)

In (7) we put $n = 1$ and substitute $J(x)$ by $J(sx)$, we get (8),

8. $\int_0^\infty x\,J_0(xs)\,dx = \left[\frac{x}{s}J_1(xs)\right]_0^\infty$ ... (8)

Example 1. *Evaluate* $\int_0^a x^2\,J_1(sx)\,dx$.

Solution. $\int_0^a x^2\,J_1(sx)\,dx = \left[\frac{x^2}{s}J_2(sx)_0^a\right] = \frac{a^2}{s}J_2(as)$ **Ans.**

Example 2. *Evaluate* $\int x^3.J_0(sx)\,dx$

Solution. $\int_0^a x^3\,J_0(sx)\,dx = \int_0^a x^2.\{x\,J_0(sx)\}\,dx$

Integrating by parts we get

$$= \left[x^2.\left(\frac{x}{s}J_1(ax)\right)\right]_0^a - \int_0^a (2x)\left\{\frac{x}{s}J_1(sx)\right\}dx \qquad \left[\int x^n J_{n-1}(x)\,dx = x^n J_n(x)\right]$$

$$= \left[\frac{x^2}{s} J_1(sx)\right]_0^a - \left[2x.\left\{\frac{x}{s^2} J_2(sx)\right\}\right]_0^a = \frac{a^3}{s} J_1(as) - \frac{2}{s}\int_0^a x^2 J_1(sx)dx$$

$$= \frac{a^3}{s} J_1(as) - \frac{2}{s^2}\left[x^2 J_2(sx)\right]_0^a = \frac{a^3}{s} J_1(as) - \frac{2a^2}{s^2} J_2(as) \qquad \textbf{Ans.}$$

Example 3. *Evaluate* $\int_0^a x(a^2 - x^2) J_0(sx)\, dx$

Solution. $\int_0^a x(a^2 - x^2) J_0(sx)\, dx = \int_0^a (a^2 - x^2).\{x J_0(sx)\}\, dx$

Integrating by parts, we get

$$= \left[(a^2 - x^2).\frac{x}{s} J_1(sx)\right]_0^a - \int_0^a (-2x)\frac{x}{s} J_1(sx)\, dx \qquad \left[\int x^n J_{n-1}(x)dx = x^n J_n(x)\right]$$

$$= 0 + 2\int_0^a x.\frac{x}{s} J_1(sx)\, dx = \frac{2}{s}\int_0^a x^2 J_1(sx) dx$$

$$= \left[\frac{2}{s}\frac{x^2}{s} J_2(sx)\right]_0^a = \left[\frac{2x^2}{s^2} J_2(sx)\right]_0^a = \frac{2a^2}{s^2} J_2(as)$$

$$= \frac{2a^2}{s^2}\left[\frac{2}{as} J_1(as) - J_0(as)\right] = \frac{4a}{s^3} J_1(as) - \frac{2a^2}{s^2} J_0(as) \qquad \textbf{Ans.}$$

45.3 SOME MORE INTEGRALS INVOLVING EXPONENTIAL FUNCTIONS AND BESSEL'S FUNCTION

1. $\int_0^\infty e^{-ax} J_0(sx)\, dx = (a^2 + s^2)^{\frac{-1}{2}}$

2. $\int_0^\infty e^{-ax} J_1(sx)\, dx = \frac{1}{s} - \frac{a}{s\sqrt{a^2 + s^2}}$

3. $\int_0^\infty x e^{-ax} J_0(sx)\, dx = a(a^2 + s^2)^{-3/2}$

4. $\int_0^\infty x e^{-ax} J_1(sx)\, dx = s(a^2 + s^2)^{-3/2}$

5. $\int_0^\infty \frac{e^{-ax}}{x} J_1(sx)\, dx = \frac{(a^2 + s^2)^{\frac{1}{2}} - a}{s}$

6. $J_n(x) = \frac{x^n}{2^n\,(n+1)!}\left[1 - \frac{x^2}{1!.2^2(n+1)} + \frac{x^4}{2!.2^2.4^2(n+1)(n+2)} \cdots\right]$

7. $J_{\frac{1}{2}}(x) = \sqrt{\frac{2}{\pi x}} \sin x$

8. $J_{\frac{-1}{2}}(x) = \sqrt{\frac{2}{\pi x}} \cos x.$

LINEARITY PROPERTY

Theorem 1. $H\{f(x) + g(x)\} = H\{f(x)\} + H\{g(x)\}$

Proof. $H\{f(x) + g(x)\} = \int_0^\infty x\{f(x) + g(x)\} J_n(sx)\, dx$

$$= \int_0^\infty x f(x). J_n(sx)\, dx + \int_0^\infty x g(x) J_n(sx)\, dx$$

$$= H\{f(x)\} + H\{g(x)\} \qquad \textbf{Proved.}$$

Theorem 2. $H\{f(ax)\} = a^{-2} H\left(\frac{s}{a}\right)$ (Similarity Theorem)

Proof. We know that

$$H\{f(ax)\} = \int_0^\infty x\, f(ax)\, J_n(sx)\, dx$$

$$= \int_0^\infty (ax)\, f(ax)\, J_n\left(\frac{s}{a} ax\right) d(ax).\frac{1}{a^2} = \frac{1}{a^2}\int_0^\infty t\, f(t)\, J_n\left(\frac{s}{a} t\right) dt \quad \text{(Putting } t = ax\text{)}$$

$$= a^{-2} H\left(\frac{s}{a}\right)$$

Proved.

Example 4. *Find the Hankel transform of the function*

$$f(x) = \begin{cases} 1 & 0 < x < a,\ n = 0 \\ 0 & x > a,\ n = 0 \end{cases}$$

Solution. Let $H(s)$ be the Hankel Transform of $f(x)$.

$$H(s) = \int_0^\infty f(x)\, x J_0(sx)\, dx = \int_0^a 1.x\, J_0(sx) dx + \int_a^\infty a\, x\, J_0(sx) dx$$

$$= \int_0^a x\, J_0(sx)\, dx + 0 = \int_0^a x\, J_0(sx)\, dx = \left[\frac{x}{s} J_1(sx)\right]_0^a = \frac{a}{s} J_1(as) \quad \textbf{Ans.}$$

Example 5. *Find the Hankel Transform of the function*

$$f(x) = \begin{cases} x^n, 0 < x < a, & n > -1 \\ 0, \quad x > a, & n > -1 \end{cases}$$

Solution. Let $H(s)$ be the Hankel Transform of $f(x)$.

$$H(s) = \int_0^\infty f(x).x\, J_n(sx)\, dx = \int_0^a x^n.x\, J_n(sx)\, dx + \int_a^\infty 0.x\, J_n(sx)\, dx$$

$$= \int_0^a x^{n+1} J_n(sx)\, dx = \left[\frac{x^{n+1}}{s} J_{n+1}(sx)\right]_0^a = \frac{a^{n+1}}{s} J_{n+1}(as) \quad \textbf{Ans.}$$

Example 6. *Find the Hankel Transform*

$$f(x) = \begin{cases} a^2 - x^2,\ 0 < x < a & n = 0 \\ 0, \quad x > a & n = 0 \end{cases}$$

Solution. Let $H(s)$ be the Hankel transform of $f(x)$.

$$H(s) = \int_0^\infty f(x).x\, J_n(sx)\, dx = \int_0^a (a^2 - x^2).x\, J_0(sx)\, dx + \int_a^\infty 0.x\ J_0(sx)\, dx$$

$$= \int_0^a (a^2 - x^2)\, x\, J_0(sx)\, dx = a^2 \int_0^a x\, J_0(sx) dx - \int_0^a x^3\, J_0(dx) dx \quad \dots (1)$$

Let us find out the above two integrals

$$a^2 \int_0^a x J_0(sx) dx = a^2 \left[\frac{x}{s} J_1(sx)\right]_0^a = a^2 \frac{a}{s} J_1(as) = \frac{a^3}{s} J_1(as) \quad \dots (2)$$

$$\int_0^a x^3 J_0(sx) dx = \frac{4a^3}{s} J_1(sx) - \frac{2a^2}{s^2} J_2(as) \quad \dots (3)$$

(See Example 2 on page 1245)

On putting these values from (2) and (3) in (1), we get

$$\text{H}(s) = \frac{a^3}{s} J_1(as) - \frac{a^3}{s} J_1(sx) + \frac{2a^2}{s^2} J_2(as) = \frac{2a^2}{s^2} J_2(as)$$

$$= \frac{2a^2}{s^2}\left[\frac{2}{as}J_1(as) - J_0(as)\right] = \frac{4a^2}{s^3}J_1(as) - \frac{2a^2}{s^2}J_0(as)$$

Example 7. *Find the Hankel transform of* $\frac{e^{-ax}}{x^2}$, $n = 1$.

Solution. Let $H(s)$ be the Hankel Transform of $f(x)$

i.e.,
$$H(s) = \int_0^\infty f(x).x.J_n(sx)\,dx = \int_0^\infty \frac{e^{-ax}}{x^2}.x.J_1(sx)\,dx$$

$$= \int_0^\infty \frac{e^{-ax}}{x}.J_1(sx)\,dx = \frac{(a^2+s^2)^{\frac{1}{2}} - a}{s} \qquad \textbf{Ans.}$$

Example 8. *Find the Hankel Transform of the function*

$$\frac{e^{-ax}}{x}, \quad n = 0$$

Solution. Let $H(s)$ be the Hankel Transform of this function $f(x)$.

i.e.
$$H(s) = \int_0^\infty f(x).xJ_n(sx)\,dx = \int_0^\infty \frac{e^{-ax}}{x}.x\,J_0(sx)\,dx$$

$$= \int_0^\infty e^{-ax}.J_0(sx)\,dx = (a^2+s^2)^{\frac{-1}{2}} \qquad \textbf{Ans.}$$

Example 9. *Show that if* $n = 0$, *the Hankel transform*

$$H\left\{\frac{\sin ax}{x}\right\} = \begin{cases} 0 & \text{if } s > a \\ \dfrac{1}{\sqrt{a^2-s^2}} & \text{if } 0 < s < a \end{cases}$$ *(U.P. III Semester, Summer 2002)*

Solution.
$$H(s) = \int_0^\infty f(x).dJ_n(sx)dx$$

$$H\left\{\frac{\sin ax}{x}\right\} = \int_0^\infty \frac{\sin ax}{x}.x\,J_0(sx)\,dx = \int_0^\infty \sin ax\,J_0(sx)dx$$

$$= \text{Imaginary part of } \int_0^a -e^{-iax}J_0(sx)\,dx$$

$$= \text{Imaginary part of } \left\{-(i^2a^2+s^2)^{\frac{-1}{2}}\right\}$$

$$H\left(\frac{\sin ax}{x}\right) = \text{Imaginary part of } \frac{-1}{\sqrt{s^2-a^2}}$$

Case 1. $s > a$, $\quad H\left(\frac{\sin ax}{x}\right) = 0$

Case 2. $0 < s < a$

$$H\left(\frac{\sin ax}{x}\right) = \text{Imaginary part of } \frac{-1}{i\sqrt{a^2-s^2}} \qquad 0 < s < a$$

$$= \text{Imaginary part of } \frac{i}{\sqrt{a^2-s^2}} = \frac{1}{\sqrt{a^2-s^2}} \qquad \textbf{Proved.}$$

Example 10. *Find the Hankel transform of the function* $\frac{e^{-ax}}{x}$, $n = 1$.

Solution. Let $H(s)$ be the Hankel Transform of the function $f(x)$.

i.e. $$H(s) = \int_0^\infty f(x).x\, J_n(sx)\,dx = \int_0^\infty \frac{e^{-ax}}{x}.x\; J_1(sx)(dx)$$

$$= \int_0^\infty e^{-ax}.J_1(sx)\,dx = \frac{1}{s} - \frac{a}{s\left(s^2+a^2\right)^{1/2}} \qquad \textbf{Ans.}$$

Example 11. *Find the Hankel Transform of* e^{-ax}. $n = 0$.

Solution. Let $H(s)$ be the Hankel transform of $f(x)$,

$$H(s) = \int_0^\infty f(x)\,x\,J_n(x)\,dx.$$

$$H(s) = \int_0^\infty e^{-ax}x\,J_0(sx)\,dx = \frac{a}{(a^2+s^2)^{\frac{3}{2}}} \qquad \textbf{Ans.}$$

Example 12. *Find the Hankel transform of* e^{-ax}, $n = 1$.

Solution. Let H (s) be the Hankel Transform of $f(x)$.

$$H(s) = \int_0^\infty f(x).x\,J_n(x)\,dx$$

i.e. $$H(s) = \int_0^\infty e^{-ax}.x.J_1(sx)\,dx = s\,(a^2+s^2)^{\frac{-3}{2}} \qquad \textbf{Ans.}$$

45.4 INVERSION FORMULA FOR HANKEL TRANSFORM

If $H(s)$ be the Hankel transform of the function $f(x)$ for $-\infty < x < \infty$

i.e. $$H(s) = \int_{-\infty}^\infty f(x)\,x\,J_n(sx)\,dx$$

Then $$f(x) = \int_{-\infty}^\infty H(s)\,s\,J_n(sx)\,ds$$

is said to be the inversion formula for the Hankel transform $H(s)$ and we may write

$$f(x) = H^{-1}[H(s)]$$

We know that in Fourier transform

$$F(s.t) = \int_{-\infty}^\infty \int_{-\infty}^\infty f(s,y)\,e^{i(sx+ty)}\,dx\,dv \qquad ...(1)$$

then $$f(x,y) = \frac{1}{4\pi^2}\int_{-\infty}^\infty \int_{-\infty}^\infty F(s,t)\,e^{i(sx+ty)}\,ds\,dt$$

On putting $x = r\cos\theta$, $y = r\sin\theta$, $(s = p\cos\alpha)$, $t = p\sin\alpha$ in (1), we get

$$F(p,\alpha) = \frac{1}{4\pi^2}\int_0^\infty r\,dr \int_0^{2\pi} f(r,\theta)\,e^{irp\cos(\theta-\alpha)}\,d\theta \qquad ...(2)$$

and $$f(r,\theta) = \frac{1}{4\pi^2}\int_0^\infty p\,dp \int_0^{2\pi} F(\rho,\alpha)\,e^{-irp\cos(\theta-\alpha)}\,d\alpha \qquad ...(3)$$

On putting $f(r)e^{-in\theta}$ for $f(r,\theta)$ in (2), we get

$$F(p,\alpha) = \int_0^\infty f(r)\,r\,dr \int_0^{2\pi} e^{i\{-n\theta+pr\cos(0-\alpha)\}}\,d\theta \qquad ...(4)$$

In (4), we put $\phi = \alpha - \theta - \frac{\pi}{2}$, we get

$$F(p,\alpha) = \int_0^\infty f(r) r\, dr \int_0^{2\pi} e^{i\{n(\phi+\frac{\pi}{2}-\alpha)+pr\cos(\phi+\frac{\pi}{2})\}} d\phi$$

$$= \int_0^\infty f(r) r\, dr . e^{in(\frac{\pi}{2}-\alpha)} \int_0^{2\pi} e^{i(n\phi - pr\sin\phi)} d\phi = \int_0^\infty f(r) r\, dr . 2\pi e^{in(\frac{\pi}{2}-\alpha)} J_n(pr)\, dr$$

$$\because \quad J_n(pr) = \frac{1}{2\pi}\int_0^{2\pi} e^{i(n\phi - pr\sin\phi)} d\phi$$

$$= 2\pi e^{in(\frac{\pi}{2}-\alpha)} \int_0^\infty f(r).r\, J_n(pr)\, dr = 2\pi e^{in(\frac{\pi}{2}-\alpha)} F(p) \quad \ldots (5)$$

Putting $f(r,\theta) = f(r) e^{-in\theta}$ and using (3) and (5), we have

$$f(r) e^{-in\theta} = \frac{1}{4\pi^2}\int_0^\infty p.dp \int_0^{2\pi} 2\pi e^{in(\frac{\pi}{2}-\alpha)} F(p)\, e^{-ipr\cos(\theta-\alpha)} d\alpha$$

$$= \frac{1}{2\pi}\int_0^\infty pF(p)\, dp \int_0^{2\pi} e^{i\{n(\frac{\pi}{2}-\alpha) - pr\cos(\theta-\alpha)\}} d\alpha$$

Substituting $\Psi = \theta - \alpha + \frac{\pi}{2}$,

$$f(r) e^{-in\theta} = \frac{1}{2\pi}\int_0^\infty pF(p)\, dp \int_0^{2\pi} e^{i\{n(\Psi-\theta - pr\cos\left(\frac{\pi}{2}-\Psi\right)} d\Psi$$

$$= \frac{1}{2\pi}\int_0^\infty p\, F(p)\, dp e^{-in\theta} \int_0^{2\pi} e^{i(n\Psi - pr\sin\Psi)} d\psi$$

$$f(r) = \frac{1}{2\pi}\int_0^\infty p\, F(p)\, dp . 2\pi\, J_n(pr)\, dp$$

$$f(r) = \int_0^\infty F(s) s\, J_n(sr)\, ds$$

or

$$f(x) = \int_0^\infty F(s).s\, J_n(sx)\, ds$$

This is the required inversion formula.

Example 13. *Find $H^{-1}[e^{-as}]$, when $n = 0$.*

Solution. $f(x) = \int_0^\infty s\, H(s)\, J_n(sx)\, ds = \int_0^\infty s e^{-as} J_0(sx)\, ds = \dfrac{a}{(a^2+x^2)^{3/2}}$ **Ans.**

Example 14. *Find $H^{-1}[s^{-2} e^{-as}]$ when $n = 1$.*

Solution.

$$H^{-1}[s^2 e^{-as}] = \int_0^\infty sH(s) J_n(sx)\, ds = \int_0^\infty s^{-2} e^{-as} s\, J_1(sx)\, ds$$

$$= \int_0^\infty \frac{1}{s} e^{-as} J_1(sx)\, \mathrm{ds}$$

$$= \frac{(a^2+x^2)^{\frac{1}{2}} - a}{x}$$ **Ans.**

45.5 PARSIVAL'S THEOREM FOR HANKEL TRANSFORM

Let $F(s)$ and $G(s)$ be the Hankel Transforms of the functions $f(x)$ and $g(x)$. Then

$$\int_0^\infty x.f(x).g(x)\,dx = \int_0^\infty s\,F(s).G(s)\,ds$$

Proof: On putting the value of $G(s)$ in $\int_0^\infty s\,F(s).G.(s)\,ds$, we get

$$\int_0^\infty s.F(s).G(s)ds = \int_0^\infty s.F(s)ds \int_0^\infty g(x).x\,J_n(sx)\,dx$$

$$= \int_0^\infty x\,g(x)\,dx \int_0^\infty s\,F(s).J_n(sx)\,ds \quad \text{(On changing the order of integration)}$$

$$= \int_0^\infty x\,g(x)\,dx.f(x).$$

Proved.

45.6 HANKEL TRANSFORMATION OF THE DERIVATIVE OF A FUNCTION

$$H\left\{\frac{df}{dx}\right\}_n = -s\left[\frac{n+1}{2n}H\{f(x)\}_{n-1} - \frac{n-1}{2n}H_{n+1}\{f(x)\}\right]$$

Proof. If $H(s)$ be the Hankel transformation of order n of $f(x)$

i.e. $\quad H(s) = \int_0^\infty x\,f(x)\,J_n(sx)\,dx$, then the Hankel transformation of $\frac{df}{dx}$ is

$$H\left\{\frac{df}{dx}\right\}_n = \int_0^\infty x\,\frac{df}{dx}\,J_n(sx)\,dx$$

On integrating by parts, we get

$$= [x\,f(x).J_n(sx)]_0^\infty - \int_0^\infty f(x)\frac{d}{dx}[x\,J_n(sx)]\,dx = 0 - \int_0^\infty f(x)[1.J_n(sx) + xsJ'(sx)]\,dx \quad \text{...(1)}$$

Assuming that $x\,f(x) \to 0$ and $x \to 0, x \to \infty$

Putting $\quad sx\,J_n'(sx) = sx\,J_{n-1}(sx) - n\,J_n(sx)$ in (1), we get

$$= -\int_0^\infty f(x)\,J_n(sx)\,dx - \int_0^\infty f(x)\{xs\,J_{n-1}(sx) - nJ_n(sx)\}dx$$

$$= (n-1)\int_0^\infty f(x)\,J_n(sx)\,dx - s\int_0^\infty x\,f(x)\,J_{n-1}(sx)\,dx$$

$$= (n-1)\int_0^\infty f(x)\,J_n(sx)\,dx - s\,H_{n-1}(s)$$

The recurrence relation (4) is

$$2n\,J_n(x) = x\,J_{n-1}(x) + x\,J_{n+1}(x)$$

On replacing x by sx, we get

$$2n\,J_n(x) = sx\,J_{n-1}(x) + sx\,J_{n+1}(x)$$

$$2n\int_0^\infty f(x)\,J_n(sx)\,dx = s\left[\int_0^\infty x\,f(x)\,J_{n-1}(sx)dx + \int_0^\infty sx\,f(x)\,J_{n+1}(sx)\,dx\right]$$

$$= s\,H_{n-1}(s) + s\,H_{n+1}(s)$$

$$\Rightarrow \quad \int_0^\infty f(x)\,J_n(sx)\,dx = \frac{s}{2n}H_{n-1}(s) + \frac{s}{2n}H_{n+1}(s) \quad \text{... (2)}$$

On putting the value of $\int_0^\infty f(x)\,J_n(sx)\,dx$ from (2) in (1), we get

$$H_n\left\{\frac{df}{dx}\right\} = s\left[\frac{n-1}{2n}H_{n-1}(s) + \frac{n-1}{2n}H_{n+1}(s)\right] - sH_{n-1}(s)$$

$$H_n\left(\frac{df}{dx}\right) = -s\left[\frac{n+1}{2n}H_{n-1}(s) - \frac{n-1}{2n}H_{n+1}(s)\right] \quad \text{... (3)}$$

This is the required formula for the Hankel transform of $\frac{df}{dx}$. **Proved.**

On replacing n by $n-1$ in (3), we get

$$H_{n-1}\left(\frac{df}{dx}\right) = -s\left[\frac{n}{2(n-1)}H_{n-2}(s) - \frac{n-2}{2(n-1)}H_n(s)\right] \quad \text{...(4)}$$

Putting $(n+1)$ for n in (3) we get

$$H_{n+1}\left(\frac{df}{dx}\right) = -s\left[\frac{n+2}{2(n+1)}H_n(s) - \frac{n}{2(n+1)}H_{n+2}(s)\right] \quad \text{... (5)}$$

From (3), (4), (5) and replacing $\frac{df}{dx}$ by $\frac{d^2f}{dx^2}$, we have

$$H_n\left(\frac{d^2f}{dx^2}\right) = -s\left[\frac{n+1}{2n}H_{n-1}\left(\frac{df}{dx}\right) - \frac{n-1}{2n}H_{n+2}\left(\frac{df}{dx}\right)\right] \quad \text{... (6)}$$

$$= \frac{s^2}{4}\left[\frac{n+1}{n-1}H_{n-2}(s) - 2\frac{n^2-3}{n^2-1}H_n(s) + \frac{n-1}{n+1}H_{n+2}(s)\right] \quad \text{... (7)}$$

Corollary. Putting $n = 1, 2, 3$ in (3), we get

$$H_1\left(\frac{df}{dx}\right) = -s\,H_0(s)$$

$$H_2\left(\frac{df}{dx}\right) = -s\left(\frac{3}{4}H_1(s) - \frac{1}{4}H_3(s)\right)$$

$$H_3\left(\frac{df}{dx}\right) = -s\left[\frac{2}{3}H_2(s) - \frac{1}{3}H_4(s)\right]$$

Example 15. *Find Hankel Transforms of the following*

(a) $\frac{d^2f}{dx^2}$ *(b)* $\frac{d^2f}{dx^2} + \frac{1}{x}\frac{df}{dx}$ *(c)* $\frac{d^2f}{dx^2} + \frac{1}{x}\frac{df}{dx} - \frac{n^2}{x^2}f$

Solution.

(a) $H\left\{\frac{d^2f}{dx^2}\right\} = \int_0^\infty \frac{d^2f}{dx^2}.x\,J_n(sx)\,dx$ Integrating by parts, we get

$$= \left[\frac{df}{dx}x.J_n(sx)\right]_0^\infty - \int_0^\infty \frac{df}{dx}.\frac{d}{dx}[x\,J_n(sx)]\,dx$$

Putting $x\,f(x) \to 0$ where $x \to 0$ or $x \to \infty$

$$H\left\{\frac{d^2f}{dx^2}\right\} = 0 - \int_0^\infty \frac{df}{dx}[J_n(sx) + sx\,J'(sx)]\,dx.$$

$$\int_0^\infty x\frac{d^2f}{dx^2}.J_n(sx)\,dx = \int_0^\infty x\frac{d^2f}{dx^2}.J_n(sx)\,dx$$

(b) $\int_0^\infty x\left[\frac{d^2f}{dx^2} + \frac{1}{x}\frac{df}{dx}\right]J_n(sx)\,dx = -s\int_0^\infty \frac{df}{dx}.x\,J_n'(sx)\,dx$

$$= -s\left[f(x).xJ_n'(sx)\right]_0^\infty - \int_0^\infty f(x)\frac{d}{dx}\{xJ_n'(sx)\}dx = s\int_0^\infty f(x)\frac{d}{dx}\{x\,J_n'(sx)\}dx$$

$\because\ x\,f(x) \to 0$ as $x \to 0$ or $x \to \infty$

But $J_n(sx)$ is the solution of Bessel's differential equation, so it satisfies Bessel's equation

$$\frac{d}{dx}\left(x\frac{dy}{dx}\right)+\left(1-\frac{n^2}{x^2}\right)xy=0$$

$$\frac{d}{dx}\left[x\frac{dy}{dx}J_n(x)\right]+\left(1-\frac{n^2}{x^2}\right)x.J_n(x)=0$$

On replacing x by sx we get

$$\frac{1}{s}\frac{d}{dx}[sx\,J_n'(sx)]=-\left(s^2-\frac{n^2}{x^2}\right)\frac{x}{s^2}$$

$$\frac{d}{dx}[x\,J_n'(sx)]=-\left(s^2-\frac{n^2}{x^2}\right)\frac{x}{s}J_n(sx)$$

On putting the value of $\frac{d}{dx}[x\,J_n'(sx)]$ in (1) we get

$$\int_0^\infty\left[\frac{d^2f}{dx^2}+\frac{1}{x}\frac{df}{dx}\right].x\,J_n(sx)\,dx=-s\int_0^\infty f(x)\left(s^2-\frac{n^2}{x^2}\right)\frac{x}{s}J_n(sx)\,dx$$

(c) $$\int_0^\infty\left[\frac{d^2f}{dx^2}+\frac{1}{x}\frac{df}{dx}-\frac{n^2}{x^2}f\right].x\,J_n(sx)\,dx=-s^2\int_0^\infty f(x).x\,J_n(sx)\,dx=-s^2H(s) \qquad \text{... (2)}$$

Deduction I. On putting $n = 0$ in (2), we get

$$\int_0^\infty x\left(\frac{d^2f}{dx^2}+\frac{1}{x}\frac{df}{dx}\right)J_0(sx)\,dx=-s^2H_0(s)$$

Deduction II. On putting $n = 1$ in (2), we get

$$\int_0^\infty x\left(\frac{d^2f}{dx^2}+\frac{1}{x}\frac{df}{dx}-\frac{f}{x^2}\right)J_1(sx)\,dx=-s^2H_1(s)$$

or $$\int_0^\infty x\frac{df}{dx}J_1(sx)=-s\,H_0(s)\ \text{ where }\ H_0(s)=\int_0^\infty x\,f(x)J_0(sx)\,dx$$

Example 16. Find $H\left\{\frac{\partial}{\partial x}\left(\frac{e^{-ax}}{x}\right)\right\}$ when $n = 1$

Solution. $$H\left\{\frac{\partial f}{\partial x}\right\}=\int_0^\infty x\frac{df}{dx}J_1(sx)dx=-sH_0(s)=-s\int_0^\infty x\,f(x)\,J_0(sx)\,dx$$

$$=-s\int_0^\infty x\left(\frac{e^{-ax}}{x}\right)J_0(sx)\,dx=-s\int_0^\infty e^{-ax}J_0(sx)\,dx=\frac{-s}{(a^2+s^2)^{\frac{1}{2}}} \qquad \textbf{Ans.}$$

Example 17. *Evaluate* $H\left\{\frac{\partial^2}{\partial t^2}f(x,t)\right\}$

Solution. $$1+\left\{\frac{\partial^2}{\partial t^2}f(x,t)\right\}=\int_0^\infty x\frac{\partial^2 f}{\partial t^2}J_n(sx)\,dx=\frac{\partial^2}{\partial t^2}\int_0^\infty x\,f(x,t)\,J_n(sx)\,dx$$

$$=\frac{\partial^2}{\partial t^2}H\{f(p,t)\} \qquad \textbf{Ans.}$$

Example 18. *Find* $H\left\{\frac{d^2(e^{-ax})}{dx^2}+\frac{1}{x}\frac{d(e^{-ax})}{dx}\right\}$, *when n = 0.*

Solution. $H\left\{\frac{d^2(e^{-ax})}{dx^2}+\frac{1}{x}\frac{d}{dx}(e^{-ax})\right\} = \int_0^\infty \left\{\frac{d^2(e^{-ax})}{dx^2}+\frac{1}{x}\frac{d(e^{-ax})}{dx}\right\} x.\, J_0(sx)\, dx$

$$= -s^2 H_0(s)$$

$$= -s^2\int_0^\infty e^{-ax}.J_0(sx)\,dx = \frac{-s^2}{(a^2+s^2)^{\frac{1}{2}}}$$ **Ans.**

Application of Hankel Transform to Boundary problems

Example 19. *The magnetic potential V for a circular disc of radius a and strength w, magnetised parallel to its axis, satisfying Laplace's equation is equal to* $2\pi w$ *on the disc itself and vanishes at its exterior points in the plane of the disc. Show that at the points (r, z), z > 0.*

$$V = 2\pi w\int_0^\infty e^{-sz} J_0(sr)J_1(sa)ds.$$

Solution. The magnetic potential V satisfies the Laplace's equation.

$$\frac{\partial^2 V}{\partial r^2}+\frac{1}{r}\frac{\partial V}{\partial r}+\frac{\partial^2 V}{\partial z^2} = 0, \qquad \text{..... (1) } 0 < r < \infty$$

Boundary conditions are

$$V = 2\pi w,\ 0 \le r < a,\ z = 0 \text{ and } V = 0,\ r > a,\ z = 0.$$

Taking Hankel transform of (1), we have

$$\int_0^\infty \left(\frac{\partial^2 V}{\partial r^2}+\frac{1}{r}\frac{\partial V}{\partial r}\right) r\, J_0(sr)dr + \int_0^\infty \frac{\partial^2 V}{\partial z^2}.r\, J_0(sr)dr = 0$$

$$-s^2 H_0(V) + \frac{d^2}{dz^2}H_0(V) = 0 \text{ where } H_0(V) = \int_0^\infty V.r\,J_0(sr)dr$$

$$D^2H_0(V) - s^2 H_0(V) = 0 \Rightarrow (D^2-s^2)H_0(V) = 0$$

Its solution $\quad H_0(V) = Ae^{sz} + Be^{-sz}$ ... (2)

And $\quad H_0(V)_{z=0} = \int_0^a (V)_{z=0}.r\,J_0(sr)dr + \int_a^\infty (V)_{z=0}\, r\, J_0(sr)dr$

$$= \int_0^a 2\pi w r\, J_0(sr)dr + 0 = 2\pi w\int_0^a \frac{1}{s}\frac{d}{dr}(r\,J_1(sr))\,dr$$

$$= \frac{2\pi\, wa}{s} J_1(sa) \qquad \text{... (3)}$$

Putting the values of $H_0(V) = 0$ and $z = \infty$ in (2), we get

$$0 = Ae^{s\infty} \Rightarrow A = 0$$

So (2) reduces to $\quad H_0(V) = Be^{-sz}$ ... (4)

On putting the value of $H(V)_z$ from (4) and z = 0 in (3), we get $\frac{2\pi wa}{f}J_1(sa) = B$

On substituting the value of B in (4), we have $H_0(V) = \frac{2\pi wa}{s}J_1(sa)e^{-sz}$

By inversion formula, we get $V(r, z) = 2\pi wa\int_0^\infty e^{-sz}J_0(sr)\,J_1(as)\,ds$ **Ans.**

Example 20. *Find the potential V (r, z) of a field due to a flat circular disc of unit radius with centre at origin and axis along z-axis, satisfying the differential equation*

$$\frac{\partial^2 V}{\partial r^2}+\frac{1}{r}\frac{\partial V}{\partial r}+\frac{\partial V}{\partial z^2}=0,$$

$$0 \le r \le \infty, \quad z \ge 0 \text{ and}$$

(*i*) $V = V_0$ when $z = 0$, $\qquad 0 \le r < 1$

(*ii*) $\frac{\partial V}{\partial z} = 0$ when $z = 0$, $\qquad r > 1$.

Solution. $$\frac{\partial^2 V}{\partial r^2}+\frac{1}{r}\frac{\partial V}{\partial r}+\frac{\partial^2 V}{\partial z^2}=0 \qquad \text{... (1)}$$

Taking Hankel Transform of (1), we get

$$\int_0^\infty \left(\frac{\partial^2 V}{\partial r^2}+\frac{1}{r}\frac{\partial V}{\partial r}\right) r J_0(sr)\,dr + \int_0^\infty \frac{\partial^2 V}{\partial z^2} r J_0(sr)\,dr = 0$$

$$\Rightarrow \quad -s^2 H_0(V) + \frac{\partial^2}{\partial z^2}\int_0^\infty Vr\, J_0(sr)\,dr = 0 \quad \Rightarrow \quad -s^2 H_0(V) + \frac{\partial^2}{\partial z^2} H_0(V) = 0$$

$$\Rightarrow \quad \frac{d^2 H_0(V)}{dz^2} - s^2 H_0(V) = 0 \quad \Rightarrow \quad (D^2 - s^2) H_0(v) = 0$$

A.t. $\qquad D^2 - s^2 = 0 \quad \Rightarrow \quad D = \pm s$

$$H_0(V) = A e^{sz} + B e^{-sz} \qquad \text{... (2)}$$

On putting $z = \infty$, $H_0(v) = 0$ (as $V = 0$) in (2), we get $0 = Ae^{\infty s} + 0 \Rightarrow A = 0$

On putting $A = 0$ in (2), we have $H_0(v) = B e^{-sz}$ ... (3)

Applying inversion formula to get V

$$V(r, z) = \int_0^\infty B(s) e^{-sz} s\, J_0(sr)\, ds \qquad \text{... (4)}$$

On putting $z = 0$ in (4), we have

$$V(r,0) = \int_0^\infty s\, B(s) J_0(sr)\, ds = V_0 \qquad 0 \le r \le 1$$

On differentiating (4), w.r.t. 'z', we obtain

$$\left(\frac{\partial V}{\partial z}\right) = \int_0^\infty B(s)\,(-s e^{-sz}) s\, J_0(sr)\, ds \qquad \text{... (5)}$$

On putting z = 0 in (5), we get

$$\left(\frac{\partial V}{\partial z}\right)_{z=0} = \int_0^\infty -s^2 B(s)\, J_0(sr)\, ds = 0, \; r > 1 \qquad \text{... (6)}$$

On comparing (4) and (5), we get

$$\int_0^\infty J_0(sr)\frac{\sin s}{s}\, ds = \frac{\pi}{2}, \qquad 0 \le r \le 1 \text{ and}$$

$$\int_0^\infty J_0(sr) \sin s\, ds = 0 \qquad r > 0$$

$$B(s)=\frac{2}{\pi}V_0\frac{\sin s}{s}$$

Hence, the required solution is

$$V(r,\,z)=\frac{2V_0}{\pi}\int_0^{\infty}e^{-sz}\,\frac{\sin s}{s}J_0(sr)\,ds$$

45.7 FINITE HANKEL TRANSMISSION FORMATION

If $f(x)$ be a function satisfying Dirichlet conditions in the interval $(0, a)$ then the

$$f(x)=\frac{2}{a^2}\sum_{i=0}^{\infty}H(si)\frac{J_n(s_ix)}{[J_n'(sia)^2]},\text{ where } H_{(si)}=\int_0^a x\,f(x)\,J_n(s_ix)\,dx$$

Where si is a root of the equation $J_n\,(S_i\;a)=0$

The upper limit a is generally converted to 1 by suitable transformation. All the roots of $J_n(s_i)$ are real and distinct.

Particular case

If $\quad n = 0$ and $a = 1$, then $J_0'(x)=-J_1(x)$

The inversion formula reduces to

$$f(x)=2\sum H(si)\frac{J_0(s_ix)}{\{J_1(s_i)\}^2}$$

Where si are the roots of $J_0(si)\;=0$

General Form

$$f(x)=\sum_{i=0}^{\infty}C_i\;J_n(s_ix),\qquad 0\le x\le a$$

where
$$C_i=\frac{2}{a^2J_{n+1}^2(s_ia)}\int_0^a x\,f(x)\,J_n(s_ix)\,dx=\frac{2H(si)}{a^2\,[J_{n+1}(sia)]^2}=\frac{2H(si)}{a^2\,[J_n'(sia)]^2}$$

so
$$f(x)=\frac{2}{a^2}\sum_{i=0}^{\infty}H(si)\frac{J_n(s_ix)}{[J_n'(sia)]^2}$$

If $\quad a = 1$, then

$$f(x)=2\sum_{i=0}^{\infty}H(si)\frac{J_n(s_ix)}{J_n'^2(s_i)}$$

and $\quad \Sigma\,J_n(s_i)=0$

45.8 ANOTHER FORM OF HANKEL TRANSFORM

If origin is not inluded in the interval and $f(x)$ satisfies the Dirichlets condition $0<b\le x\le a$,

then $\quad [H_n(x)=J_n(x)+iY_n(x)]$

$$H(six)=\int_a^b x\,f(x)[J_n(s_ix)Y_n(s_ia)-Y_n(s_ia)\,J_n(s_ia)]\,dx$$

Where Y_n is the Bessel function of order n of second kind and si is the root of the equation

$$J_n(s_ia)\,Y_n(s_i\,b)-J_n(s_i\,b)Y_n(s_ia)=0$$

Inversion formula

$$f(x)=\sum\frac{2s_i^{\,2}J_n^2(s_ia)\,H(s_i)}{J_n^2(s_i\,b)-J_n^2(s_ia)}[J_n(s_ix)\,Y_n(s_i\,b)-J_n(s_ib)Y_n(s_ia)]$$

Example 21. *Find f (x) if H {f (x)}* $= \frac{c}{s} J_1(sa)$, *s being the root of* $J_0(s\alpha)$.

Solution. We know the inversion formula.

$$f(x) = \frac{2}{a^2} \sum_{i=0}^{\infty} H_n(s) \frac{J_n(sx)}{[J_n'(sa)]^2} = \frac{2}{a} \sum \frac{c}{s} J_1(sa) \frac{J_0(sx)}{[J_0'(sa)]^2}$$

$$= \frac{2}{a} \sum \frac{c}{s} \frac{J_1(sa) J_0(sx)}{[J_1(sa)]^2} \quad [J_0'(x) = -J_1(x)] \quad = \frac{2}{a^2} \sum \frac{c}{s} \frac{J_n(sx)}{J_1(sa)} \qquad \textbf{Ans.}$$

Example 22. *Show that* $\int_0^a x\, J_0(sx)\, dx = J_1(as).\left(\frac{a}{s}\right)$

Solution. We know that

$$\frac{d}{dx}[x^n J_n(x)] = x^n J_{n-1}(x) \qquad \text{....(1) [Recurrence Relation]}$$

Replacing x by sx and putting $n = 1$ in (1), we get

$$\frac{1}{s}\frac{d}{dx}[sx\, J_1(sx)] = x\, J_0(sx), \qquad \frac{1}{s}\frac{d}{dx}[x\, J_1(sx)] = x\, J_0(sx)$$

or
$$\frac{1}{s}[x J_1(sx)]_0^a = \int_0^a x\, J_0(sx)\, dx$$

$$\therefore \quad \frac{a}{s} J_1(sa) = \int_0^a x\, J_0(sx)\, dx \qquad \textbf{Proved.}$$

Example 23. *Find* $H(x^{n-1})$, $n > 0$ *and* $x\, J_{n-1}(sx)$ *is the kernel of the transform.*

Solution. $$H(x^{n-1}) = \int_0^a x^{n-1}.x\, J_{n-1}(s_i x)\, dx = \int_0^a x^n\, J_{n-1}(s_i x)\, dx$$

$$= \frac{1}{s_i}\int_0^a \frac{d}{dx}[x^n\, J_n(s_i x)]\, dx = \frac{1}{s_i}[x^n J_n(s_i x)]_0^a = \frac{1}{s_i} a^n\, J_n(s_i a) \qquad \textbf{Ans.}$$

Example 24. *Find* $H[x^n]$, $n > -1$ *and* $x J_n(si\, x)$ *is the kernel of the transform.*

Solution. $$H(x^n) = \int_0^a x^n x J_n(s_i\, x) dx = \int_0^a x^{n+1} J_n(s_i\, x) dx$$

$$= \int_0^a \left[\frac{1}{\text{si}}\frac{d}{dx} x^{n+1} J_{n+1}(s_i\, x)\right] dx = \left[\frac{1}{s_i} x^{n+1} J_{n+1}(s_i\, x)\right]_0^a = \frac{a^{n+1}}{\text{si}} J_{n+1}(si\, a) \qquad \textbf{Ans.}$$

Example 25. *Show that*

$$\int_0^a x^3 J_0(sx)\, dx = \frac{a^2}{s^2}[2\, J_0(sa) + \left(as - \frac{4}{as}\right) J_1(as)$$

Solution. We know that

$$\frac{d}{dx}[x^n J_n(x)] = x^n J_{n-1}(x) \qquad \text{.... (1) (Recurrence Relation)}$$

Replacing x, by sx and putting $n = 1$ in (1), we get

$$\frac{1}{s}\frac{d}{dx}[x\, J_1(sx)] = x\, J_0(sx)$$

Now
$$\int_0^a x^3\, J_0(sx)\, dx = \int_0^a x^2.x\, J_0(sx)\, dx = \int_0^a x^2\left[\frac{1}{s}\frac{d}{dx}[x\, J_1(sx)]\right] dx$$

Integrating by parts, we get

$$= \left[x^2.\frac{1}{s}x\,J_1(sx)\right]_0^a - \int_0^a \frac{2x}{s}.x\,J_1(sx)\,dx = \frac{a^3}{s}J_1(sa) - \frac{2}{s}\int_0^a x^2 J_1(sx)\,dx$$

$$= \frac{a^3}{s}J_1(sa) - \frac{2}{s}\int_0^a \frac{1}{s}\left[\frac{d}{dx}(x^2 J_2(sx)\right]dx = \frac{a^3}{s}J_1(sa) - \frac{2}{s^2}\left[x^2 J_2(sx)\right]_0^a$$

$$= \frac{a^3}{s}J_1(sa) - \frac{2a^2}{s^2}J_2(as) \qquad \text{.. (2)}$$

We also know that $\frac{2n}{x}J_n(x) = J_{n-1}(x) + J_{n+1}(x)$ (Recurrence Relation) ... (3)

Replacing x by sa and putting $n = 1$ in (3), we get

$$\frac{2}{sa}J_1(sa) = J_0(sa) + J_2(sa)$$

$$\Rightarrow \qquad J_2(sa) = \frac{2}{sa}J_1(sa) - J_0(sa)$$

Substituting the value of $J_2(sa)$ in (2), we get

$$\int_0^a x^3 J_0(sx)\,dx = \frac{a^3}{s}J_1(sa) - \frac{2a^2}{s^2}\left[\frac{2}{sa}J_1(sa) - J_0(sa)\right]$$

$$= \frac{a^3}{s}J_1(sa) - \frac{4a}{s^3}J_1(sa) + \frac{2a^2}{s^2}J_0(sa) = \left[\frac{a^3}{s} - \frac{4a}{s^3}\right]J_1(sa) + \frac{2a^2}{s^2}J_0(sa)$$

$$= \frac{a^2}{s^2}\left[2J_0(as) + \left(as - \frac{4}{as}\right)J_1(as)\right]$$

Proved.

Example 26. *Find $H[1-x^2]$, $x\,J_0(sx)$ being the kernel.*

Solution. $H_0(1-x^2) = \int_0^a (1-x^2).x\,J_0(sx)\,dx = \int_0^a x\,J_0(sx)\,dx - \int_0^a x^3 J_0(sx)\,dx$

$$= \left[\frac{x}{s}J_1(sx)\right]_0^a - \frac{a^2}{s^2}[2J_0(sa) + \left(as - \frac{4}{as}\right)J_1(a)] = \frac{a}{s}J_1(as) - \frac{a^2}{s^2}[2J_0(sa) + (as - \frac{4}{as})J_1(a)]$$

Ans.

Example 27. *Prove that the finite Hankel transform of $\frac{2^{1+n-m}}{m-n}x^n(1-x^2)^{m-n-1}$ is $s^{n-m}.J_m(s)$, for $0 \le x < 1$.*

Solution. We know that $H[f(x)] = \int_0^a f(x).x\,J_n(sx)\,dx$.

$$H\left[\frac{2^{1+n-m}}{m-n}x^n(1-x^2)^{m-n-1}\right] = \int_0^1 \frac{2^{1+n-m}}{m-n}x^n(1-x^2)^{m-n-1}.x\,J_n(sx)\,dx$$

$$= \int_0^1 \frac{2^{1+n-m}}{m-n}x^n(1-x^2)^{m-n-1}.x\sum_{r=0}^{\infty}\frac{(-1)^r}{\lfloor r\,(n+r+1)}\left(\frac{sx}{2}\right)^{n+2r}dx$$

$$= \frac{1}{m-n}\sum_{r=0}^{\infty}\frac{(-1)^r s^{n+2r}}{\lfloor r\; n+r+1\; 2^{-1+m+2r}}\int_0^1 x^{2(n+r)}.(1-x^2)^{m-n-1}.x\,dx$$

[Put $x^2 = t$, $2x\,dx = dt$]

$$= \frac{1}{m-n}\sum_{r=0}^{\infty} \frac{(-1)^s s^{n+2r}}{\lfloor r \, n+r+1 \, 2^{m+2r}} \int_0^1 t^{(n+r+1)-1}(1-t)^{(m-n)-1} dt$$

$$= \frac{1}{m-n}\sum_{r=0}^{\infty} \frac{(-1)^r s^{n+2r}}{\lfloor r \, n+r+1 \; 2^{m+2r}} \frac{n+r+1 \; m-n}{n+r+1+m-n} \qquad \int_0^a t^{m-1}(1-t)^{n-1} dt = \frac{m \; n}{m+n}$$

$$= s^{n-m}\sum_{r=0}^{\infty} \frac{(-1)^r}{m+r+1}\left(\frac{s}{2}\right)^{m+2r} = (s)^{n-m} J_m(s) \qquad \textbf{Proved.}$$

Example 28. *Find* $H_n\left[\frac{J_n(\alpha x)}{J_n(\alpha)}\right]$.

Solution. We know that $J_n(\alpha x)$ and $J_n(s\alpha)$ are the solutions of Bessel's equation.

$$\therefore \qquad x^2 \frac{d^2}{dx^2} J_n(\alpha x) + x\frac{d}{dx} J_n(\alpha x) + (\alpha^2 x^2 - n^2) J_n(\alpha x) = 0 \qquad \ldots (1)$$

$$x^2 \frac{d^2}{dx^2} J_n(sx) + x\frac{d}{dx} J_n(sx) + (s^2 x^2 - n^2) J_n(sx) = 0 \qquad \ldots (2)$$

Multiplying (1) by $J_n(sx)$, (2) by $J_n(\alpha x)$ and subtracting, we get

$$(\alpha^2 - s^2)\, x\, J_n(\alpha x)\, J_n(sx) = \frac{d}{dx}\left[x\left\{J_n(\alpha x)\frac{d}{dx} J_n(sx) - J_n(sx)\frac{d}{dx} J_n(\alpha x)\right\}\right]$$

Integrating with respect to x from 0 to 1 and using $J_n(s) = 0$, we get

$$(\alpha^2 - s^2)\int_0^1 x\, J_n(\alpha x)\, J_n(sx)\, dx = s\left[x\left\{J_n(\alpha x) J_n'(sx) - J_n(sx) J_n'(\alpha x)\right\}\right]_0^1$$

$$= s[J_n(\alpha) J_n'(s) - J_n(s) J_n'(\alpha)]$$

$$= s\, J_n(\alpha)\, J_n'(s) \qquad (J_n(s) = 0)$$

$$\Rightarrow \qquad \int_0^1 \frac{J_n(\alpha x)}{J_n(\alpha)} . x\, J_n(sx)\, dx = \frac{s}{\alpha^2 - s^2} J_n'(s)$$

$$H_n\left\{\frac{J_n(\alpha x)}{J_n(\alpha)}\right\} = \frac{s}{\alpha^2 - s^2} J_n'(s) \qquad \textbf{Proved.}$$

If $n = 0$,

$$H_n\left\{\frac{J_n(\alpha x)}{J_n(\alpha)}\right\} = \frac{s}{\alpha^2 - s^2} J_0'(s) \qquad [J_0'(s) = -J_1(s)]$$

$$= \frac{-s}{\alpha^2 - s^2} J_1(s). \qquad \textbf{Ans.}$$

EXERCISE 45.1

1. Prove that $H_0\left(\frac{a}{(a^2+x^2)^{3/2}}\right] = e^{-as}$

2. Find the Hankel transform of $1 - x^2$, taking $x\, J_0(sx)$ as the kernel, where $0 \le x \le 1$. **Ans.** $\frac{4}{s^3} J_1(s)$

3. Find the Hankel transform of $\frac{\cos ax}{x}$ taking $x\, J_0(sx)$ as the kernel

Ans. (*i*) $(s^2 - a^2)^{-\frac{1}{2}}$ if $s > a$. (*ii*) 0 if $s < a$.

4. Prove that $H\left\{\frac{\sin ax}{a}\right\}_{n=1} = \frac{a}{s(s^2-a^2)^{1/2}}$, $s > a$

5. If $H\{f(x)\} = \frac{c}{s}J_1(sa)$, Prove that

$f(x) = \frac{2}{a^2}\Sigma\frac{cJ_0(sx)}{sJ_1(as)}$ where sum is taken over all the positive roots of the equation $J_0(s_i a) = 0$

6. Prove that $H_n\left(\frac{df}{dx}\right) = \frac{s}{2n}[(n-1)H_{n+1}(f) - (n+1)H_{n-1}(f)]$

and if $n = 1$, then $H_1\left(\frac{df}{dx}\right) = -sH_0(f(x))$

7. Prove that $H_n\left\{\frac{d^2f}{dx^2} + \frac{1}{x}\frac{df}{dx}\right\} = \frac{s}{2}\left[-H_{n-1}\left(\frac{df}{dx}\right) + H_{n+1}\left(\frac{df}{dx}\right)\right]$, When s is a root of $J_n(s) = 0$

8. Prove that $H_0\left\{\frac{d^2f}{dx^2} + \frac{1}{x}\frac{df}{dx}\right\} = -s^2H_0\{f(x)\}$

When s is a root of $sJ_n'(s) + hJ_n(s) = 0$

9. Prove that

$$H_n\left\{\frac{d^2f}{dx^2} + \frac{1}{x}\frac{df}{dx}\left(\frac{n^2}{x^2}f(x)\right)\right\} = -sf(1)J_n'(s) - s^2H_n\{f(x)\}.$$

10. Find the Hankel Transform of $\frac{d^2f}{dx^2} + \frac{1}{x}\frac{df}{dx}$ if s is the root of the equation

$J_n(sa)Y_n(sb) - J_n(sb)Y_n(sa) = 0$.

If $f(a) = 0 = f(b)$, then deduce that $H_0\left[\frac{d^2f}{dx^2} + \frac{1}{x}\frac{df}{dx}\right] = -s^2H_0\{f(x)\}$

11. Find the Hankel transform of $\frac{d^2f}{dx^2} + \frac{1}{x}\frac{df}{dx} - \frac{n^2}{x^2}f(x)$ and s being a root of $J_n(sa) = 0$

Ans. $-saf(a)J_n'(as) - P^2H_n\{f(x)\}$

12. Viscous fluid is contained between two infinitely long concentric circular cylinders of radii a and b. The inner cylinder is kept at rest and outer cylinder suddenly starts rotating with uniform angular velocity ω. Find the velocity v of the fluid if the equation of motion is

$$\frac{\partial^2 v}{\partial r^2} + \frac{1}{r}\frac{\partial v}{\partial r} - \frac{v}{r^2} = \frac{1}{\nu}\frac{\partial v}{\partial t}, \qquad a < r < b, t > 0$$

ν being Kinematic viscosity.

Hint: Take $f^2(s) = \int_a^b f(r).rB_1(sr)dr$, $b > a$

where $B_1(sr) = J_1(sr)Y_1(sa) - Y_1(sr)J_1(sa), Y_1(sr)$ being Bessel's function of second kind of order one, and s is a positive root of

$J_1(sb)Y_1(sa) = Y_1(sb)J_1(sa)$.

Multiplying the given equation by $B_1(sr)$ and integrating w.r.t. 'r' from a to b with boundary conditions $v = b\omega$ when $r = b$

$v = 0$ when $r = a$, $v = 0$ when $t = 0$. **Ans.** $v = \pi b\omega\sum_p \frac{1-e^{-\nu p^2 t}}{J_1^2(sa) - J_1^2(sb)}J_1(sa)J_1(sb)B_1(sr)$

HANKEL TRANSFORM

S. No.	Function f(x)	n	Hankel Transform F (s)
1.	$f(x)=\begin{cases} x^n, & 0<x<a \\ 0, & x>a \end{cases}$	$n>-1$	$\frac{a^{n+1}}{s} J_{n+1}(sx)$
2.	$f(x)=\begin{cases} 1, & 0<x<a \\ 0, & x>a \end{cases}$	$n=0$	$\frac{a}{s} J_1(sx)$
3.	$f(x)=\begin{cases} a^2-x^2, & 0<x<a \\ 0, & x>a \end{cases}$	$n=0$	$\frac{4a}{s^3} J_1(sx)-\frac{2a^2}{s^2} J_0(sx)$
4.	$f(x)=x^{m-1}$	$n>-1$	$\frac{2m\,\Gamma\left(\frac{1}{2}+\frac{1}{2}m+\frac{n}{2}\right)}{s^{m+1}\,\Gamma\left(\frac{1}{2}-\frac{m}{2}+\frac{n}{2}\right)}$
5.	$x^2 e^{-ax}$	$n=1$	$\frac{\sqrt{s^2+a^2}-a}{s}$
6.	e^{-ax}	$n=0$	$a(s^2+a^2)^{-3/2}$
7.	e^{-ax}	$n=1$	$s(s^2+a^2)^{-3/2}$
8.	$x^n e^{-qx^2}$	$n>-1$	$\frac{s}{(2q)^{n+1}} e^{\frac{-s^2}{4q}}$
9.	$\frac{e^{-ax}}{x}$	$n=0$	$(s^2+a^2)^{\frac{-1}{2}}$
10.	$\frac{e^{-ax}}{x}$	$n=1$	$\frac{1}{s}-\frac{a}{s(s^2+a^2)^{\frac{1}{2}}}$
11.	$\frac{\sin ax}{a}$	$n=0$	$\begin{cases} 0, & s>a \\ (a^2-s^2)^{-\frac{1}{2}}, & 0<s<a \end{cases}$
12.	$\frac{\sin ax}{a}$	$n=1$	$\begin{cases} \frac{a}{(s^2-a^2)^{1/2}}, & s>a \\ 0, & s<a \end{cases}$
13.	$\frac{\sin x^2}{x^2}$	$n=0$	$\begin{cases} \sin^{-1}\frac{1}{s}, & s>1 \\ \frac{\pi}{2}, & s<1 \end{cases}$
14.	$\frac{a}{(a^2+x^2)^{3/2}}$	$n=0$	e^{-as}
15.	$x^{m-2}e^{-ax^2}$	$n>-1$	$\frac{s^n\,\Gamma\left(\frac{n}{2}+\frac{m}{2}\right)}{2^{n+1} s^{\frac{n}{2}+\frac{m}{2}}\, 1+n} \times F\left\{\frac{n}{2}+\frac{m}{2},\ n+1,\ -\frac{s^2}{4a}\right\}$

CHAPTER

46

HILBERT TRANSFORM

46.1 INTRODUCTION

Method of separating signals is based on Phase selecting which use phase shifts between the pertinent signals to achieve the desired separation.

On shifting the phase angle of all components of a given signal through ± 90° degrees, the resulting function of time is called as the Hilbert transform of the signal.

The Hilbert transform of $f(t)$ is denoted by $H_i\{f(t)\}$ and is defined as

$$H_i\{f(t)\} = \frac{1}{\pi}\int_{-\infty}^{\infty} \frac{f(s)\,ds}{t-s}$$

LINEARITY PROPERTY

The Hilbert transform of $f(t)$ is a linear operation.

Inverse of Hilbert Transform

It is defined as

$$f(t) = -\frac{1}{\pi}\int_{-\infty}^{\infty} \frac{H_i\{f(t)\}}{t-s}\,ds$$

$f(t)$ and $H_i\{f(t)\}$ make a Hilbert-transform pair.

46.2 ELEMENTARY FUNCTIONS AND THEIR HILBERT TRANSFORM

S.No.	*Functions*	*Hilbert Transforms*
1.	$\cos t$	$H_i\{\cos t\} = \frac{1}{\pi}\int_{-\infty}^{\infty} \frac{\sin s}{s-t}\,ds = -\sin t$
2.	$\sin t$	$H_i\{\sin t\}\frac{1}{\pi}\int_{-\infty}^{\infty} \frac{\sin s}{s-t}\,ds = \cos t$
3.	$\frac{\sin t}{t}$	$H_i\left\{\frac{\sin t}{t}\right\} = \frac{1}{\pi}\int_{-\infty}^{\infty} \frac{\frac{\sin s}{s}}{s-t}\,ds = \frac{\cos t - 1}{t}$
4.	$\frac{1}{1+t^2}$	$H_i\left\{\frac{1}{1+t^2}\right\} = \frac{1}{\pi}\int_{-\infty}^{\infty} \frac{\frac{1}{1+t^2}}{s-t}\,ds = \frac{-t}{1+t^2}$
5.	$\delta\ (t)$	$H_i\{st\} = \frac{1}{\pi}\int_{-\infty}^{\infty} \frac{\delta(s)}{s-t}\,ds = -\frac{1}{\pi t}$
6.	$\frac{1}{t}$	$H_i\left\{\frac{1}{t}\right\} = \frac{1}{\pi}\int_{-\infty}^{\infty} \frac{\frac{1}{s}}{s-t}\,ds = -\pi\delta(t)$

46.3 PROPERTIES

This Hilbert transform differs from the Fourier transform in a way that it operates exclusively in the time domain. Some important properties are listed below :

1. The *amplitude spectrum* of a signal $f(t)$ and its Hilbert transform $H_i\{f(t)\}$ is the same.

$$H_i\{f(t)\} = \frac{1}{\pi t} * f(t) = \frac{1}{\pi}\int_{-\infty}^{\infty}\frac{f(s)}{s-t}ds$$

2. *Inverse Transform* :
   If $H_i\{f(t)\}$ is the Hilbert transform of $f(t)$, then the inverse Hilbert transform is given below.

   If $H_i\{f(t)\} = \frac{1}{\pi}\int_{-\infty}^{\infty}\frac{f(s)}{s-t}ds$, then

$$f(t) = -\frac{1}{\pi}\int_{-\infty}^{\infty}\frac{H_i\{f(s)\}}{s-t}ds$$

3. *Orthogonality.* A signal $f(t)$ and its Hilbert transform $H_i\{f(t)\}$ are orthogonal.

$$\int_{-\infty}^{\infty} f(t)\cdot H_i\{f(t)\}dt = 0$$

46.4 APPLICATIONS

1. **Phase selectively :** Hilbert transform is used to realise phase selectivity in the generation of a special kind of modulation (single side band modulation)
2. It gives a mathematical basis to represent band-past signals.

First Order Lagrange's Linear Partial Differential Equations

47.1 PARTIAL DIFFERENTIAL EQUATIONS

Partial differential equations are those equations which contain partial derivatives, independent variables and dependent variables.

The independent variables will be denoted by x and y and the dependent variable by z. The partial differential coefficients are denoted as follows :

$$\frac{\partial z}{\partial x} = p, \qquad \frac{\partial z}{\partial y} = q. \qquad \frac{\partial^2 z}{\partial x^2} = r, \qquad \frac{\partial^2 z}{\partial x \partial y} = s, \qquad \frac{\partial^2 z}{\partial y^2} = t$$

47.2 ORDER

Order of a partial differential equation is the same as that of the order of the highest differential coefficient in it.

47.3 METHOD OF FORMING PARTIAL DIFFERENTIAL EQUATIONS

A partial differential equation is formed by two methods.

(*i*) By eliminating arbitrary constants.

(*ii*) By eliminating arbitrary functions.

(*i*) **Method of elimination of arbitrary constants**

Example 1. *Find the PDE of all sphere whose centre lie on z-axis and given by equations $x^2 + y^2 + (z - a)^2 = b^2$, a and b being constants.* (*U.P., II Semester, 2009*)

Solution. We have, $x^2 + y^2 + (z - a)^2 = b^2$ ... (1)

(1) contains two arbitrary constants a and b.

Differentiating (1) partially w.r.t. x, we get

$$2x + 2(z - a)\frac{\partial z}{\partial x} = 0$$

$$\Rightarrow \qquad x + (z - a)\,p = 0 \qquad \text{... (2)}$$

Differentiating (1) partially w.r.t. y, we get

$$2y + 2(z - a)\frac{\partial z}{\partial y} = 0$$

$$y + (z - a)\,q = 0 \qquad \text{... (3)}$$

Let us eliminate a from (2) and (3).

From (2) $\qquad (z - a) = -\dfrac{x}{p}$

Putting this value of $z - a$ in (3), we get

$$y - \frac{x}{p} q = 0$$

$$\Rightarrow \qquad y\,p - x\,q = 0 \qquad \textbf{Ans.}$$

(*ii*) **Method of elimination of arbitrary functions**

Example 2. *Form the partial differential equation from*

$$z = f(x^2 - y^2)$$

Solution. We have, $z = f(x^2 - y^2)$ ... (1)

Differentiating (1) w.r.t. x and y, we get

$$p = \frac{\partial z}{\partial x} = f'(x^2 - y^2)\, 2x \qquad \text{... (2)}$$

$$q = \frac{\partial z}{\partial y} = f'(x^2 - y^2)\,(-2y) \qquad \text{... (3)}$$

Dividing (2) by (3), we get $\dfrac{p}{q} = \dfrac{-x}{y} \quad \Rightarrow \quad py = -qx$

$$\Rightarrow \qquad yp + xq = 0 \qquad \textbf{Ans.}$$

EXERCISE 47.1

Form the partial differential equations from:

1. $z = (x + a)(y + b)$ **Ans.** $pq = z$
2. $(x - h)^2 + (y - k)^2 + z^2 = a^2$ **Ans.** $z^2 (p^2 + q^2 + 1) = a^2$
3. $2z = (a\,x + y)^2 + b$ **Ans.** $p\,x + q\,y = q^2$
4. $a\,x^2 + b\,y^2 + z^2 = 1$ **Ans.** $z\,(p\,x + q\,y) = z^2 - 1$
5. $x^2 + y^2 = (z - c)^2 \tan^2 \alpha$ **Ans.** $y\,p - x\,q = 0$
6. $z = f(x^2 + y^2)$ **Ans.** $yp - xq = 0$
7. $2z = \dfrac{x^2}{a^2} + \dfrac{y^2}{b^2}$ (*A.M.I.E., Winter 2001*) **Ans.** $2z = xp + yq$
8. $f(x + y + z, x^2 + y^2 + z^2) = 0$ **Ans.** $(y - z)\,p + (z - x)\,q = x - y$

47.4 SOLUTION OF EQUATION BY DIRECT INTEGRATION

Example 3. *Solve* $\dfrac{\partial^3 z}{\partial x^2 \partial y} = \cos(2x + 3y)$

Solution. We have, $\dfrac{\partial^3 z}{\partial x^2 \partial y} = \cos(2x + 3y)$

Integrating w.r.t. 'x'; we get $\dfrac{\partial^2 z}{\partial x \partial y} = \dfrac{1}{2} \sin(2x + 3y) + f(y)$

Again, integrating w.r.t. x, we get $\dfrac{\partial z}{\partial y} = -\dfrac{1}{4} \cos(2x + 3y) + x \int f(y)\, dx + g(y)$

$$= -\frac{1}{4} \cos(2x + 3y) + x\,\phi(y) + g(y)$$

Integrating w.r.t. 'y', we get $z = -\frac{1}{12}\sin(2x+3y) + x\int \phi(y)\,dy + \int g(y)\,dy$

$$z = -\frac{1}{12}\sin(2x+3y) + x\phi_1(y) + \phi_2(y)$$ **Ans.**

Example 4. *Solve* $\frac{\partial^2 z}{\partial x\,\partial y} = x^2 y$

subject to the conditions $z(x, 0) = x^2$ *and* $z(1, y) = \cos y$.

Solution. We have, $\frac{\partial}{\partial x}\left(\frac{\partial z}{\partial y}\right) = x^2 y$

On integrating w.r.t. x, we obtain

$$\frac{\partial z}{\partial y} = \frac{x^3}{3}y + f(y)$$

Integrating w.r.t. y, we obtain

$$z = \frac{x^3}{3}\cdot\frac{y^2}{2} + \int f(y)\,dy + g(x) \qquad \left[F(y) = \int f(y)\,dy\right]$$

$$\Rightarrow \qquad z = \frac{x^3 y^2}{6} + F(y) + g(x) \qquad \ldots (1)$$

Condition 1 : Putting $z = x^2$ and $y = 0$ in (1), we get

$$x^2 = 0 + F(0) + g(x)$$

Putting the value of $g(x)$ in (1), we get

$$z = \frac{x^3 y^2}{6} + F(y) + x^2 - F(0) \qquad \ldots (2)$$

Condition 2 : $z(1, y) = \cos y$

Putting $x = 1$ and $z = \cos y$ in (2), we get

$$\cos y = \frac{y^2}{6} + F(y) + 1 - F(0) \Rightarrow F(y) = \cos y - \frac{1}{6}y_2 - 1 + F(0)$$

Putting the value of $F(y)$ in (2), we get

$$z = \frac{1}{6}x^3 y^2 + \cos y - \frac{1}{6}y^2 - 1 + F(0) + x^2 - F(0)$$

$$\Rightarrow \qquad z = \frac{1}{6}x^3 y^2 + \cos y - \frac{1}{6}y^2 - 1 + x^2$$ **Ans.**

EXERCISE 47.2

Solve the following:

1. $\frac{\partial^2 z}{\partial x \partial y} = xy^2$ **Ans.** $z = \frac{x^2 y^3}{6} + f(y) + \phi(x)$

2. $\frac{\partial^2 z}{\partial x \partial y} = e^y \cos x$ **Ans.** $z = e^y \sin x + f(y) + \phi(x)$

3. $\frac{\partial^2 z}{\partial x \partial y} = \frac{y}{x} + 2$ **Ans.** $z = \frac{y^2}{2}\log x + 2xy + f(y) + \phi(x)$

4. $\dfrac{\partial^2 z}{\partial x^2} = a^2 z$, when $x = 0$, $\dfrac{\partial z}{\partial x} = a \sin y$ and $\dfrac{\partial z}{\partial y} = 0$ **Ans.** $z = \sin x + e^y \cos x$

5. $\dfrac{\partial^2 z}{\partial x \partial y} = \sin x \sin y$ if $\dfrac{\partial z}{\partial y} = -2 \sin y$ and $z = 0$, when $x = 0$.

Choose the correct answer:

6. The solution of the partial differential equation $\dfrac{\partial^2 z}{\partial y^2} = \sin(xy)$ is

(a) $z = -x^{-2} \sin(xy) + yf(x) + g(x)$ (b) $z = -x^2 \sin(xy) - xf(x) + g(x)$
(c) $z = -y^2 \sin(xy) + yf(x) + g(x)$ (d) $z = -x - 2 \sin(xy) + xf(x) + g(x)$
(AMIETE, June 2009) **Ans.** (*d*)

47.5 LAGRANGE'S LINEAR EQUATION IS AN EQUATION OF THE TYPE

$$Pp + Qq = R$$

where P, Q, R are the functions of x, y, z and $p = \dfrac{\partial z}{\partial x}$, $q = \dfrac{\partial z}{\partial y}$

$$Pp + Qq = R \qquad \dots (1)$$

This form of the equation is obtained by eliminating an arbitrary function f from

$$f(u, v) = 0 \qquad \dots (2)$$

where u, v are functions of x, y, z.

Differentiating (2) partially w.r.t. to x and y.

$$\frac{\partial f}{\partial u}\left(\frac{\partial u}{\partial x} + \frac{\partial u}{\partial z}\frac{\partial z}{\partial x}\right) + \frac{\partial f}{\partial v}\left(\frac{\partial v}{\partial x} + \frac{\partial v}{\partial z}\frac{\partial z}{\partial x}\right) = 0 \qquad \dots (3)$$

and
$$\frac{\partial f}{\partial u}\left(\frac{\partial u}{\partial y} + \frac{\partial u}{\partial z}\frac{\partial z}{\partial y}\right) + \frac{\partial f}{\partial v}\left(\frac{\partial v}{\partial y} + \frac{\partial v}{\partial z}\frac{\partial z}{\partial y}\right) = 0 \qquad \dots (4)$$

Let us eliminate $\dfrac{\partial f}{\partial u}$ and $\dfrac{\partial f}{\partial v}$ from (3) and (4).

From (3), $$\frac{\partial f}{\partial u}\left[\frac{\partial u}{\partial x} + \frac{\partial u}{\partial z} p\right] = -\frac{\partial f}{\partial v}\left[\frac{\partial v}{\partial x} + \frac{\partial v}{\partial z} p\right] \qquad \dots (5)$$

From (4), $$\frac{\partial f}{\partial u}\left[\frac{\partial u}{\partial y} + \frac{\partial u}{\partial z} q\right] = -\frac{\partial f}{\partial v}\left[\frac{\partial v}{\partial y} + \frac{\partial v}{\partial z} q\right] \qquad \dots (6)$$

Dividing (5) by (6), we get

$$\frac{\dfrac{\partial u}{\partial x} + \dfrac{\partial u}{\partial z} \cdot p}{\dfrac{\partial u}{\partial y} + \dfrac{\partial u}{\partial z} \cdot q} = \frac{\dfrac{\partial v}{\partial x} + \dfrac{\partial v}{\partial z} \cdot p}{\dfrac{\partial v}{\partial y} + \dfrac{\partial v}{\partial z} \cdot q}$$

$$\Rightarrow \quad \left[\frac{\partial u}{\partial x} + \frac{\partial u}{\partial z} \cdot p\right]\left[\frac{\partial v}{\partial y} + \frac{\partial v}{\partial z} \cdot q\right] = \left[\frac{\partial u}{\partial y} + \frac{\partial u}{\partial z} \cdot q\right]\left[\frac{\partial v}{\partial x} + \frac{\partial v}{\partial z} \cdot p\right]$$

$$\frac{\partial u}{\partial x} \times \frac{\partial v}{\partial y} + \frac{\partial u}{\partial x} \times \frac{\partial v}{\partial z} \cdot q + \frac{\partial u}{\partial z} \times p \times \frac{\partial v}{\partial y} + \frac{\partial u}{\partial z} \times \frac{\partial v}{\partial z} \cdot pq$$

$$= \frac{\partial u}{\partial y} \times \frac{\partial v}{\partial x} + \frac{\partial u}{\partial y} \times \frac{\partial v}{\partial z} p + \frac{\partial u}{\partial z} \cdot q \times \frac{\partial v}{\partial x} + \frac{\partial u}{\partial z} \times \frac{\partial v}{\partial z} \cdot pq$$

$$\Rightarrow \quad \left[\frac{\partial u}{\partial y}\times\frac{\partial v}{\partial z}-\frac{\partial u}{\partial z}\times\frac{\partial v}{\partial y}\right]p+\left[\frac{\partial u}{\partial z}\times\frac{\partial v}{\partial x}-\frac{\partial u}{\partial x}\times\frac{\partial v}{\partial z}\right]q = \frac{\partial u}{\partial x}\times\frac{\partial v}{\partial y}-\frac{\partial u}{\partial y}\times\frac{\partial v}{\partial x} \quad \text{... (7)}$$

If (1) and (7) are the same, then the coefficients of p, q are equal

$$P=\frac{\partial u}{\partial y}\times\frac{\partial v}{\partial z}-\frac{\partial u}{\partial z}\times\frac{\partial v}{\partial y}, \qquad Q=\frac{\partial u}{\partial z}\times\frac{\partial v}{\partial x}-\frac{\partial u}{\partial x}\times\frac{\partial v}{\partial z} \quad \text{... (8)}$$

$$R=\frac{\partial u}{\partial x}\times\frac{\partial v}{\partial y}-\frac{\partial u}{\partial y}\times\frac{\partial v}{\partial x}$$

Now suppose $u = c_1$ and $v = c_2$ are two solutions, where c_1, c_2 are constants.
Differentiating $u = c_1$ and $v = c_2$

$$\frac{\partial u}{\partial x}dx+\frac{\partial u}{\partial y}dy+\frac{\partial u}{\partial z}dz=0 \quad \text{... (9)}$$

and

$$\frac{\partial v}{\partial x}dx+\frac{\partial v}{\partial y}dy+\frac{\partial v}{\partial z}dz=0 \quad \text{... (10)}$$

Solving (9) and (10), we get

$$\frac{dx}{\frac{\partial u}{\partial y}\times\frac{\partial v}{\partial z}-\frac{\partial u}{\partial z}\times\frac{\partial v}{\partial y}}=\frac{dy}{\frac{\partial u}{\partial z}\times\frac{\partial v}{\partial x}-\frac{\partial u}{\partial x}\times\frac{\partial v}{\partial z}}=\frac{dz}{\frac{\partial u}{\partial x}\times\frac{\partial v}{\partial y}-\frac{\partial u}{\partial y}\times\frac{\partial v}{\partial x}} \quad \text{... (11)}$$

From (8) and (11), we have $\frac{dx}{P}=\frac{dy}{Q}=\frac{dz}{R}$

Solutions of these equations are $u = c_1$ and $v = c_2$

$\therefore$ $f(u, v) = 0$ is the required solution of (1).

47.6 WORKING RULE TO SOLVE Pp + Qq = R

Step 1. Write down the auxiliary equations $\frac{dx}{P}=\frac{dy}{Q}=\frac{dz}{R}$

Step 2. Solve the above auxiliary equations.
Let the two solutions be $u = c_1$ and $v = c_2$.

Step 3. Then $f(u, v) = 0$ or $u = \phi(v)$ is the required solution of

$$Pp + Qq = R.$$

Example 5. *Solve the following partial differential equation*

$$yq - xp = z. \qquad \textit{where } p=\frac{\partial z}{\partial x},\ q=\frac{\partial z}{\partial y}.$$

Solution. We have, $y\,q - x\,p = z$
Here the auxiliary equations are

$$\frac{dx}{-x}=\frac{dy}{y}=\frac{dz}{z}$$

$$-\log x = \log y - \log a \quad \text{(From first two equations)}$$

$$x\,y = a \quad \text{... (1)}$$

$$\log y = \log z + \log b \quad \text{(From last two equations)}$$

$$\frac{y}{z}=b \quad \text{... (2)}$$

From (1) and (2) we get the solution

$$f\left(xy, \frac{y}{z}\right)=0.$$ **Ans.**

Example 6. *Solve* $y^2p - xyq = x(z - 2y)$. *(A.M.I.E., Summer 2001)*

Solution. We have, $y^2p - xyq = x(z - 2y)$

The auxiliary equations are $\dfrac{dx}{y^2} = \dfrac{dy}{-xy} = \dfrac{dz}{x(z-2y)}$ ... (1)

Considering first two members of the equations

$$\frac{dx}{y} = \frac{dy}{-x} \quad \Rightarrow \quad x\,dx = -y\,dy$$

Integrating $\dfrac{x^2}{2} = -\dfrac{y^2}{2} + \dfrac{C_1}{2} \quad \Rightarrow \quad x^2 + y^2 = C_1$ ... (2)

From last two equations of (1), we have $-\dfrac{dy}{y} = \dfrac{dz}{z-2y}$

$$\Rightarrow \quad -z\,dy + 2y\,dy = y\,dz \quad \Rightarrow \quad 2y\,dy = y\,dz + z\,dy$$

On integration, we get $y^2 = yz + C_2$

$y^2 - yz = C_2$ ... (3)

From (2) and (3), we have $x^2 + y^2 = f(y^2 - yz)$ **Ans.**

Example 7. *Solve* $(x^2 - yz)\,p + (y^2 - zx)\,q = z^2 - xy$

(U.P., II Semester, 2010, A.M.I.E., Summer 2001)

Solution. $(x^2 - y\,z)\,p + (y^2 - z\,x)\,q = z^2 - x\,y$ ... (1)

The auxiliary equations are

$$\frac{dx}{x^2 - yz} = \frac{dy}{y^2 - zx} = \frac{dz}{z^2 - xy}$$

$$\Rightarrow \quad \frac{dx - dy}{x^2 - yz - y^2 + zx} = \frac{dy - dz}{y^2 - zx - z^2 + xy} = \frac{dz - dx}{z^2 - xy - x^2 + yz}$$

$$\Rightarrow \quad \frac{dx - dy}{(x-y)(x+y+z)} = \frac{dy - dz}{(y-z)(x+y+z)} = \frac{dz - dx}{(z-x)(x+y+z)}$$

$$\Rightarrow \quad \frac{dx - dy}{x-y} = \frac{dy - dz}{y-z} = \frac{dz - dx}{z-x} \qquad \text{... (2)}$$

Integrating first two members of (2), we have

$$\log(x-y) = \log(y-z) + \log c_1$$

$$\log\frac{x-y}{y-z} = \log c_1 \quad \Rightarrow \quad \frac{x-y}{y-z} = c_1 \qquad \text{... (3)}$$

Similarly from last two members of (2), we have $\dfrac{y-z}{z-x} = c_2$ .... (4)

From (3) and (4), the required solution is

$$f\left[\frac{x-y}{y-z}, \frac{y-z}{z-x}\right] = 0$$ **Ans.**

47.7 METHOD OF MULTIPLIERS

Let the auxiliary equations be $\dfrac{dx}{P} = \dfrac{dy}{Q} = \dfrac{dz}{R}$

l, m, n may be constants or functions of x, y, z then, we have

$$\frac{dx}{P}=\frac{dy}{Q}=\frac{dz}{R}=\frac{l\,dx+m\,dy+n\,dz}{lP+mQ+nR}$$

l, m, n are chosen in such a way that

$$lP + mQ + nR = 0$$

Thus $\qquad l\,dx + m\,dy + n\,dz = 0$

Solve this differential equation, if the solution is $u = c_1$.

Similarly, choose another set of multipliers (l_1, m_1, n_1) and if the second solution is $v = c_2$.

$\therefore$ Required solution is $f(u, v) = 0$.

Example 8. *Solve*

$$(mz-ny)\frac{\partial z}{\partial x}+(nx-lz)\frac{\partial z}{\partial y}=l\,y-mx. \qquad \textit{(A.M.I.E. Winter 2001)}$$

Solution. We have, $(mz-ny)\dfrac{\partial z}{\partial x}+(nx-lz)\dfrac{\partial z}{\partial y}=l\,y-mx$

Here, the auxiliary equations, are $\dfrac{dx}{mz-ny}=\dfrac{dy}{nx-lz}=\dfrac{dz}{ly-mx}$

Using multipliers x, y, z, we get

$$\text{each fraction}=\frac{x\,dx+y\,dy+z\,dz}{x(mz-ny)+y(nx-lz)+z(ly-mx)}=\frac{x\,dx+y\,dy+z\,dz}{0}$$

$\therefore \qquad x\,dx + y\,dy + z\,dz = 0$

which on integration gives

$$x^2 + y^2 + z^2 = c_1 \qquad \text{... (1)}$$

Again using multipliers, l, m, n, we get

$$\text{each fraction}=\frac{l\,dx+m\,dy+n\,dz}{l(mz-ny)+m(nx-lz)+n(ly-mx)}=\frac{l\,dx+m\,dy+n\,dz}{0}$$

$\therefore \qquad l\,dx + m\,dy + n\,dz = 0$

which on integration gives

$$l\,x + m\,y + n\,z = c_2 \qquad \text{... (2)}$$

Hence from (1) and (2), the required solution is

$$x^2+y^2+z^2=f(lx+my+nz)$$ **Ans.**

Example 9. *Solve the partial differential equation* $x(y^2+z)\,p - y(x^2+z)\,q = z(x^2-y^2)$

where, $\qquad p=\dfrac{\partial z}{\partial x}$ *and* $q=\dfrac{\partial z}{\partial y}$. *(U.P., II Semester, 2008)*

Solution. Lagrange's subsidiary equations are

$$\frac{dx}{x(y^2+z)}=\frac{dy}{-y(x^2+z)}=\frac{dz}{z(x^2-y^2)} \qquad \text{...(1)}$$

Using $x, y, -1$ as multipliers, we get

$$\text{each fraction}=\frac{x\,dx+y\,dy-dz}{x^2(y^2+z)-y^2(x^2+z)-z(x^2-y^2)}=\frac{x\,dx+y\,dy-dz}{0}$$

$\therefore \qquad x\,dx + y\,dy - dz = 0$

Integrating, we get

$$\frac{x^2}{2}+\frac{y^2}{2}-z=\frac{C_1}{2}$$

$$\Rightarrow \qquad x^2 + y^2 - 2z = C_1 \qquad \text{...(2)}$$

Again, using $\frac{1}{x}$, $\frac{1}{y}$ and $\frac{1}{z}$ as multipliers, we get

$$\text{each fraction} = \frac{\frac{1}{x}dx + \frac{1}{y}dy + \frac{1}{z}dz}{y^2 + z - x^2 - z + x^2 - y^2} = \frac{\frac{1}{x}dx + \frac{1}{y}dy + \frac{1}{z}dz}{0}$$

$$\therefore \quad \frac{1}{x}dx + \frac{1}{y}dy + \frac{1}{z}dz = 0$$

Integrating, we get

$\log x + \log y + \log z = \log C_2 \quad \Rightarrow \quad xyz = C_2$ ...(3)

Hence the general solution is

$\phi(x^2 + y^2 - 2z, xyz) = 0$ **Ans.**

Example 10. *Find the general solution of*

$$x(z^2 - y^2)\frac{\partial z}{\partial x} + y(x^2 - z^2)\frac{\partial z}{\partial y} = z(y^2 - x^2)$$ (*AMIETE, Dec. 2009*)

Solution. $x(z^2 - y^2)\frac{\partial z}{\partial x} + y(x^2 - z^2)\frac{\partial z}{\partial y} = z(y^2 - x^2)$ ... (1)

$\therefore$ The auxiliary simultaneous equations are

$$\frac{dx}{x(z^2 - y^2)} = \frac{dy}{y(x^2 - z^2)} = \frac{dz}{z(y^2 - x^2)} \quad \text{... (2)}$$

Using multipliers x, y, z, we get

$$\text{Each term of (2)} = \frac{x\,dx + y\,dy + z\,dz}{x^2(z^2 - y^2) + y^2(x^2 - z^2) + z^2(y^2 - x^2)} = \frac{x\,dx + y\,dy + z\,dz}{0}$$

$\therefore \quad x\,dx + y\,dy + z\,dz = 0$ ... (3)

On integration $x^2 + y^2 + z^2 = C_1$

Again (2) can be written as

$$\frac{\frac{dx}{x}}{z^2 - y^2} = \frac{\frac{dy}{y}}{x^2 - z^2} = \frac{\frac{dz}{z}}{y^2 - x^2} = \frac{\frac{dx}{x} + \frac{dy}{y} + \frac{dz}{z}}{(z^2 - y^2) + (x^2 - z^2) + (y^2 - x^2)} \quad \text{... (4)}$$

$$= \frac{\frac{dx}{x} + \frac{dy}{y} + \frac{dz}{z}}{0} \quad \Rightarrow \quad \frac{dx}{x} + \frac{dy}{y} + \frac{dz}{z} = 0$$

On integration, we get

$\Rightarrow \quad \log x + \log y + \log z = \log C_2$

$\log x\,y\,z = \log C_2 \quad \Rightarrow \quad x\,y\,z = C_2$ .... (5)

From (3) and (5), the general solution is

$$xyz = f(x^2 + y^2 + z^2)$$ **Ans.**

Example 11. *Solve the partial differential equation*

$$\frac{y - z}{yz}p + \frac{z - x}{zx}q = \frac{x - y}{xy}$$ (*AMIETE, June 2010, 2009*)

Solution. We have, $\frac{y - z}{yz}p + \frac{z - x}{zx}q = \frac{x - y}{xy}$

Multiplying by xyz, we get

$$x(y-z)p + y(z-x)q = z(x-y)$$

$$\frac{dx}{x(y-z)} = \frac{dy}{y(z-x)} = \frac{dz}{z(x-y)} = \frac{dx+dy+dz}{x(y-z)+y(z-x)+z(x-y)} \quad \dots (1)$$

$$= \frac{dx+dy+dz}{0}$$

$$\therefore \quad dx + dy + dz = 0$$

Which on integration gives

$$x + y + z = a \quad \dots (2)$$

Again (1) can be written as

$$\frac{\frac{dx}{x}}{y-z} = \frac{\frac{dy}{y}}{z-x} = \frac{\frac{dz}{z}}{x-y} = \frac{\frac{dx}{x}+\frac{dy}{y}+\frac{dz}{z}}{(y-z)+(z-x)+(x-y)} = \frac{\frac{dx}{x}+\frac{dy}{y}+\frac{dz}{z}}{0} \Rightarrow \frac{dx}{x}+\frac{dy}{y}+\frac{dz}{z} = 0$$

On integration, we get

$$\log x + \log y + \log z = \log b \Rightarrow \log xyz = \log b \Rightarrow xyz = b \quad \dots (3)$$

From (2) and (3) the general solution is

$$xyz = f(x+y+z)$$ **Ans.**

Example 12. *Solve* $(x^2 - y^2 - z^2)p + 2xyq = 2xz.$ *(A.M.I.E., Summer 2004, 2000)*

Solution. We have, $(x^2 - y^2 - z^2)p + 2xyq = 2xz$ ... (1)

Here the auxiliary equations are $\dfrac{dx}{x^2-y^2-z^2} = \dfrac{dy}{2xy} = \dfrac{dz}{2xz}$ ... (2)

From the last two members of (2), we have $\dfrac{dy}{y} = \dfrac{dz}{z}$

which on integration gives

$$\log y = \log z + \log a \Rightarrow \log \frac{y}{z} = \log a$$

$$\Rightarrow \quad \frac{y}{z} = a \quad \dots (3)$$

Using multipliers x, y, z, we have

$$\frac{dx}{x^2-y^2-z^2} = \frac{dy}{2xy} = \frac{dz}{2xz} = \frac{x\,dx + y\,dy + z\,dz}{x(x^2+y^2+z^2)}$$

$$\frac{2x\,dx + 2y\,dy + 2z\,dz}{(x^2+y^2+z^2)} = \frac{dz}{z}$$

which on integration gives

$$\log(x^2+y^2+z^2) = \log z + \log b$$

$$\frac{x^2+y^2+z^2}{z} = b \quad \dots (4)$$

Hence from (3) and (4), the required solution is $x^2+y^2+z^2 = z\,f\left(\dfrac{y}{z}\right)$ **Ans.**

Example 13. *Solve the differential equation*

$$x^2\frac{\partial z}{\partial x} + y^2\frac{\partial z}{\partial y} = (x+y)z$$

Solution. We have, $x^2\dfrac{\partial z}{\partial x} + y^2\dfrac{\partial z}{\partial y} = (x+y)z$ ... (1)

The auxiliary equations of (1) are

$$\frac{dx}{x^2}=\frac{dy}{y^2}=\frac{dz}{(x+y)z} \qquad \text{... (2)}$$

Take first two members of (2) and integrate them

$$-\frac{1}{x}=-\frac{1}{y}+c$$

$$\frac{1}{x}-\frac{1}{y}=c_1 \qquad \text{... (3)}$$

(2) can be written as

$$\frac{\frac{dx}{x}}{x}=\frac{\frac{dy}{y}}{y}=\frac{\frac{dz}{z}}{x+y}=\frac{\frac{dx}{x}+\frac{dy}{y}-\frac{dz}{z}}{(x+y)-(x+y)}$$

$$\Rightarrow \qquad \frac{dx}{x}+\frac{dy}{y}-\frac{dz}{z}=0$$

On integration, we get

$$\Rightarrow \qquad \log x+\log y-\log z=\log c_2$$

$$\Rightarrow \qquad \log\frac{xy}{z}=\log c_2 \quad \Rightarrow \quad \frac{xy}{z}=c_2 \qquad \text{... (4)}$$

From (3) and (4), we have $f\left[\frac{1}{x}-\frac{1}{y},\ \frac{xy}{z}\right]=0.$ **Ans.**

Example 14. *Find the general solution of* $x\frac{\partial z}{\partial x}+y\frac{\partial z}{\partial y}+t\frac{\partial z}{\partial t}=xyt$

Solution. The auxiliary equations are $\frac{dx}{x}=\frac{dy}{y}=\frac{dt}{t}=\frac{dz}{xyt}$ ... (1)

Taking the first two members and integrating, we get

$$\log x=\log y+\log a=\log ay$$

$$\Rightarrow \qquad x=ay, \quad i.e. \quad x/y=a \qquad \text{... (2)}$$

Similarly, from the 2nd and 3rd members $\frac{t}{y}=b$ ... (3)

Multiplying the equation (1) by xyt, we get

$$dz=\frac{ty\,dx}{1}=\frac{tx\,dy}{1}=\frac{xy\,dt}{1}=\frac{ty\,dx+tx\,dy+xy\,dt}{3}$$

Integrating, we get

$$z=\frac{1}{3}xyt+c \ \Rightarrow \ z-\frac{1}{3}xyt=c \qquad \text{... (4)}$$

From (2), (3) and (4) the solution is $z-\frac{1}{3}xyt=f\left(\frac{y}{x}\right)+\phi\left(\frac{t}{y}\right)$ **Ans.**

Example 15. *Solve* $(y+z)\,p-(x+z)\,q=x-y$ *(AMIETE, June 2010)*

Solution. $(y+z)\,p-(x+z)\,q=x-y$ ... (1)

$\therefore$ The auxiliary equations are

$$\frac{dx}{y+z}=\frac{dy}{-(x+z)}=\frac{dz}{x-y} \qquad \text{... (2)}$$

$$\frac{dx}{y+z}=\frac{dy}{-(x+z)}=\frac{dz}{x-y}=\frac{dx+dy+dz}{y+z-(x+z)+x-y}$$

$$\Rightarrow \qquad \frac{dx}{y+z} = \frac{dy}{-(x+z)} = \frac{dz}{x-y} = \frac{dx+dy+dz}{0}$$

Thus, we have $\qquad dx + dy + dz = 0$

Which on integration gives $\quad x + y + z = c_1$ ... (3)

Using multipliers $x, y, -z$ for (2), we get

$$\frac{dx}{y+z} = \frac{dy}{-(x+z)} = \frac{dz}{x-y} = \frac{x\,dx + y\,dy - z\,dz}{x(y+z) - y(x+z) - z(x-y)}$$

$$\Rightarrow \qquad \frac{dx}{y+z} = \frac{dy}{-(x+z)} = \frac{dz}{x-y} = \frac{x\,dx + y\,dy - z\,dz}{0}$$

Integrating $x\,dx + y\,dy - z\,dz = 0$, we get

$$\frac{x^2}{2} + \frac{y^2}{2} - \frac{z^2}{2} = c_2 \quad \Rightarrow \quad x^2 + y^2 - z^2 = 2c_2 \qquad \text{... (4)}$$

From (3) and (4), we get the required solution $f(x + y + z, x^2 + y^2 - z^2) = 0$ **Ans.**

Example 16. *Solve* $z\,p + yq = x$.

Solution. The auxiliary equations are $\dfrac{dx}{z} = \dfrac{dy}{y} = \dfrac{dz}{x}$ ...(1)

(*i*) (*ii*) (*iii*)

From (*i*) and (*iii*)

$$\frac{dx}{z} = \frac{dz}{x} \qquad \Rightarrow \qquad x\,.\,dx = z\,.\,dz$$

$$\Rightarrow \qquad \frac{x^2}{2} = \frac{z^2}{2} - \frac{c_1}{2} \qquad \Rightarrow \qquad x^2 = z^2 - c_1 \qquad \text{... (2)}$$

$$\Rightarrow \qquad z = \sqrt{x^2 + c_1}$$

Putting the value of z in (1), we get

$$\frac{dx}{\sqrt{x^2 + c_1}} = \frac{dy}{y}, \qquad \sinh^{-1}\frac{x}{\sqrt{c_1}} = \log y + c_2 \qquad \text{... (3)}$$

From (2) and (3), the required solution is

$$f(z^2 - x^2) = \sinh^{-1}\frac{x}{\sqrt{c_1}} - \log y$$ **Ans.**

Example 17. *Solve* $px\,(z - 2y^2) = (z - qy)\,(z - y^2 - 2x^3)$. (*A.M.I.E., Summer 2000*)

Solution. $\quad px\,(z - 2y^2) = (z - qy)\,(z - y^2 - 2x^3)$

$$\Rightarrow \qquad px\,(z - 2y^2) + qy\,(z - y^2 - 2x^3) = z\,(z - y^2 - 2x^3) \qquad \text{... (1)}$$

Here the auxiliary equations are

$$\frac{dx}{x(z-2y^2)} = \frac{dy}{y(z-y^2-2x^3)} = \frac{dz}{z(z-y^2-2x^3)} \qquad \text{... (2)}$$

From the last two members of (2), we have $\dfrac{dy}{y} = \dfrac{dz}{z}$

which gives on integration

$$\log y = \log z + \log a \;\Rightarrow\; y = a\,z \qquad \text{... (3)}$$

From the first and third members of (2), we have

$$\frac{dx}{x(z-2y^2)} = \frac{dz}{z(z-y^2-2x^3)}$$

$$\Rightarrow \quad \frac{dx}{x(z-2a^2z^2)} = \frac{dz}{z(z-a^2z^2-2x^3)} \qquad \text{[Using (3), } y = az]$$

$$\frac{dx}{x(1-2a^2z)} = \frac{dz}{z-a^2z^2-2x^3}$$

$$\Rightarrow \quad z\,dx - a^2z^2dx - 2x^3dx = x\,dz - 2a^2\,x\,z\,dz$$

$$\Rightarrow \quad (x\,dz - z\,dx) - a^2\,(2\,x\,z\,dz - z^2dx) + 2x^3\,dx = 0$$

$$\Rightarrow \quad \frac{x\,dz - z\,dx}{x^2} - a^2\frac{(2\,x\,z\,dz - z^2\,dx)}{x^2} + 2\,x\,dx = 0$$

On integrating, we have $\frac{z}{x} - a^2\frac{z^2}{x} + x^2 = b$ ... (4)

From (3) and (4), we have the required solution

$$\frac{y}{z} = f\left(\frac{z}{x} - \frac{a^2z^2}{x} + x^2\right)$$ **Ans**

EXERCISE 47.3

Solve the following partial differential equations:

1. $p\tan x + q\tan y = \tan z$ **Ans.** $f\left(\frac{\sin x}{\sin y}, \frac{\sin y}{\sin z}\right) = 0$

2. $(y-z)\,p + (x-y)\,q = z - x$ **Ans.** $f(x+y+z,\ x^2+2yz) = 0$

3. $(y+zx)\,p - (x+yz)\,q = x^2 - y^2$ **Ans.** $f(x^2+y^2-z^2) = (x-y)^2 - (z+1)^2$

4. $zx\frac{\partial z}{\partial x} - zy\frac{\partial z}{\partial y} = y^2 - x^2$ **Ans.** $f(x^2+y^2+z^2,\ xy) = 0$

5. $pz - qz = z^2 + (x+y)^2$ **Ans.** $[z^2 + (x+y)^2]\,e^{2x} = f(x+y)$

6. $p + q + 2xz = 0$ **Ans.** $f(x-y) = x^2 + \log z$

7. $x^2p + y^2q + z^2 = 0$ **Ans.** $f\left(\frac{1}{y} - \frac{1}{x}, \frac{1}{y} + \frac{1}{z}\right) = 0$

8. $(x^2+y^2)\,p + 2\,x\,y\,q = (x+y)\,z$ *(A.M.I.E., Summer 2000)* **Ans.** $f\left(\frac{x+y}{z}, \frac{2y}{x^2-y^2}\right) = 0$

9. $\frac{\partial z}{\partial x} - 2\frac{\partial z}{\partial y} = 2x - e^y + 1$ **Ans.** $f(2x+y) = z - \frac{(2x+1)^2}{4} - \frac{e^y}{2}$

10. $p + 3q = 5\,z + \tan(y-3x)$ **Ans.** $f(y-3x) = \frac{e^{5x}}{5z + \tan(y-3x)}$

11. $xp - yq + x^2 - y^2 = 0$ **Ans.** $f(xy) = \frac{x^2}{2} + \frac{y^2}{2} + z$

12. $(x+y)\left(\frac{\partial z}{\partial x} + \frac{\partial z}{\partial y}\right) = z - 1$ **Ans.** $f(x-y) = \frac{x+y}{(z-1)^2}$.

3. $(x^3+3xy^2)\frac{\partial z}{\partial x} + (y^3+3x^2y)\frac{\partial z}{\partial y} = 2(x^2+y^2)z$ *(A.M.I.E.T.E., Summer 2000)* **Ans.** $f\left(\frac{xy}{z^2}\right) = \frac{4xy}{(x^2-y^2)^2}$

14. $(z^2 - 2yz - y^2)p + (xy + zx)q = xy - zx$ *(A.M.I.E.T.E., Dec. 2010)*

Ans. $f(z^2 - 2yz - y^2, x^2 + y^2 + z^2) = 0$

15. Find the solution of the equation $\frac{x\partial z}{\partial y} - \frac{y\partial z}{\partial x} = 0$, which passes through the curve $z = 1, x^2 + y^2 = 4$

Ans. $f(x^2 + y^2 - 4, z - 1) = 0$

Indicate True or False for the following statements

16. With usual symbols, the *P.D.E.* $u_{xx} + u^2 u_{yy} = f(xy)$ is non-linear in 'u' and is of second order.

(U.P., II Semester, 2009) **Ans.** True

17. Solution of the *P.D.E.* $\frac{\partial^2 z}{\partial x\, \partial y} = xy^2$ is $z = \frac{x^2 y^3}{6} + f(y) + \phi(x)$ **Ans.** True

47.8 PARTIAL DIFFERENTIAL EQUATIONS NON-LINEAR IN *p* AND *q*.

We give below the methods of solving non-linear partial differntial equations in certain standard form only.

Type I. Equation of the type $f(p, q) = 0$ i.e., equations containing p and q only.

Method. Let the required solution be

$$z = ax + by + c \qquad \text{... (1)}$$

$$\frac{\partial z}{\partial x} = a, \frac{\partial z}{\partial y} = b.$$

On putting these values in $f(p, q) = 0$

we get $f(a, b) = 0,$

From this, find the value of b in terms of a and substitude the value of b in (1), that will be the required solution.

Example 18. *Solve* $p^2 + q^2 = 1$ ... (1)

Solution. Let $z = ax + by + c$ ... (2)

$$p = \frac{\partial z}{\partial x} = a \quad \text{and} \quad q = \frac{\partial z}{\partial y} = b$$

On substituting the values of p and q in (1), we have

$$\therefore \quad a^2 + b^2 = 1 \quad \Rightarrow \quad b = \sqrt{1 - a^2}$$

Putting the value of b in (2), we get

$$z = ax + \sqrt{1 - a^2}\, y + c \qquad \textbf{Ans.}$$

Example 19. *Solve* $x^2p^2 + y^2q^2 = z^2$ *(MDU, May 2010, R.G.P.V. Bhopal, Feb. 2008)*

Solution. This equation can be transformed in the above type.

$$\frac{x^2}{z^2}p^2 + \frac{y^2}{z^2}q^2 = 1 \quad \Rightarrow \quad \left(\frac{x}{z}\frac{\partial z}{\partial x}\right)^2 + \left(\frac{y}{z}\frac{\partial z}{\partial y}\right)^2 = 1$$

$$\left(\frac{\frac{\partial z}{z}}{\frac{\partial x}{x}}\right)^2 + \left(\frac{\frac{\partial z}{z}}{\frac{\partial y}{y}}\right)^2 = 1 \qquad \text{... (1)}$$

Let $\frac{\partial z}{z} = \partial Z, \quad \frac{\partial x}{x} = \partial X, \quad \frac{\partial y}{y} = \partial Y$

$\therefore \quad \log z = Z, \quad \log x = X, \quad \log y = Y$

$\therefore$ (1) can be written as

$$\left(\frac{\partial Z}{\partial X}\right)^2 + \left(\frac{\partial Z}{\partial Y}\right)^2 = 1$$

or $$P^2 + Q^2 = 1 \quad \dots (2)$$

Let the required solution be

$$Z = aX + bY + c$$

$$P = \frac{\partial Z}{\partial X} = a, \quad Q = \frac{\partial Z}{\partial Y} = b$$

From (2) we have

$$a^2 + b^2 = 1 \Rightarrow b = \sqrt{1-a^2}$$

$\therefore$ $$Z = aX + \sqrt{1-a^2}\, Y + c$$

$\Rightarrow$ $$\log z = a \log x + \sqrt{1-a^2} \log y + c$$ **Ans.**

Example 20. *Solve* $(x+y)(p+q)^2 + (x-y)(p-q)^2 = 1$

(R.G.P.V. Bhopal, II Semester, June 2006)

Solution. We have,

$$(x+y)(p+q)^2 + (x-y)(p-q)^2 = 1 \quad \dots (1)$$

Put $x + y = U^2 \Rightarrow 2U\frac{\partial U}{\partial x} = 1$ and $2U\frac{\partial U}{\partial y} = 1$

$\Rightarrow \frac{\partial U}{\partial x} = \frac{1}{2U}$ and $\frac{\partial U}{\partial y} = \frac{1}{2U}$

And $x - y = V^2 \Rightarrow 2V\frac{\partial V}{\partial x} = 1$ and $2V\frac{\partial V}{\partial y} = -1 \Rightarrow \frac{\partial V}{\partial x} = \frac{1}{2V}$ and $\frac{\partial V}{\partial y} = -\frac{1}{2V}$

Also, $$p = \frac{\partial z}{\partial x} = \frac{\partial z}{\partial U}\cdot\frac{\partial U}{\partial x} + \frac{\partial z}{\partial V}\cdot\frac{\partial V}{\partial x} \Rightarrow p = \frac{1}{2U}\cdot\frac{\partial z}{\partial U} + \frac{1}{2V}\cdot\frac{\partial z}{\partial V} \quad \dots (2)$$

and $$q = \frac{\partial z}{\partial y} = \frac{\partial z}{\partial U}\cdot\frac{\partial U}{\partial y} + \frac{\partial z}{\partial V}\cdot\frac{\partial V}{\partial y} \Rightarrow q = \frac{1}{2U}\cdot\frac{\partial z}{\partial U} - \frac{1}{2V}\cdot\frac{\partial z}{\partial V} \quad \dots (3)$$

(2) + (3), $p + q = \frac{1}{U}\frac{\partial z}{\partial U}$, (2) – (3), $p - q = \frac{1}{V}\frac{\partial z}{\partial V}$

On putting the values of $(x+y)$, $(x-y)$, $(p+q)$, $(p-q)$ in (1), we get

$$U^2\left(\frac{1}{U}\frac{\partial z}{\partial U}\right)^2 + V^2\left(\frac{1}{V}\frac{\partial z}{\partial V}\right)^2 = 1, \quad \left(\frac{\partial z}{\partial U}\right)^2 + \left(\frac{\partial z}{\partial V}\right)^2 = 1$$

The complete integral is

$$z = aU + \sqrt{1-a^2}\, V + C \qquad [z = ax + by + c]$$

Hence, $$z = a\sqrt{x+y} + \sqrt{1-a^2}\,\sqrt{x-y} + C$$ **Ans.**

EXERCISE 47.4

Solve the following parital differential equations

1. $pq = 1$ **Ans.** $z = ax + \frac{1}{a}y + c$

2. $\sqrt{p} + \sqrt{q} = 1$ **Ans.** $z = ax + (1-\sqrt{a})^2 y + c$

3. $p^2 - q^2 = 1$ **Ans.** $z = ax + \sqrt{(a^2-1)}\, y + c$

4. $pq + p + q = 0$ **Ans.** $z = ax - \frac{a}{1+a}y + c$

Type II. Equation of the type

$$z = px + qy + f(p, q)$$

Its solution is $z = ax + by + f(a, b)$

Example 21. *Solve* $z = px + qy + p^2 + q^2$

Solution. $z = px + qy + p^2 + q^2$, $[p = a, q = b]$

Its solution is $z = ax + by + a^2 + b^2$ **Ans.**

Example 22. *Solve* $z = px + qy + 2\sqrt{pq}$

Solution. $z = px + qy + 2\sqrt{pq}$

Its solution is $z = ax + by + 2\sqrt{ab}$ **Ans.**

Type III. Equation of the type $f(z, p, q) = 0$ *i.e.* equations not containing x and y.

Let z be a function of u where

$$u = x + ay.$$

$$\frac{\partial u}{\partial x} = 1 \quad \text{and} \quad \frac{\partial u}{\partial y} = a$$

Then $p = \dfrac{\partial z}{\partial x} = \dfrac{\partial z}{\partial u} \cdot \dfrac{\partial u}{\partial x} = \dfrac{\partial z}{\partial u}$; $\quad q = \dfrac{\partial z}{\partial y} = \dfrac{\partial z}{\partial u}\dfrac{\partial u}{\partial y} = \dfrac{\partial z}{\partial u}(a)$

On putting the values of p and q in the given equation $f(z, p, q) = 0$, it becomes $f\left(z, \dfrac{\partial z}{\partial u}, a\dfrac{\partial z}{\partial u}\right) = 0$ which is an ordinary differential equation of the first order.

Example 23. *Solve*

$$p(1 + q) = qz$$ *(MDU, May 2005)*

Solution. $p(1 + q) = qz$ ... (1)

Let $u = x + ay \Rightarrow \dfrac{\partial u}{\partial x} = 1, \dfrac{\partial u}{\partial y} = a$

$$p = \frac{\partial z}{\partial x} = \frac{\partial z}{\partial u}\frac{\partial u}{\partial x} = \frac{dz}{du} \quad \text{and} \quad q = \frac{\partial z}{\partial y} = \frac{\partial z}{\partial u}\frac{\partial u}{\partial y} = a\frac{dz}{du}$$

(1) becomes

$$\frac{dz}{du}\left(1 + a\frac{dz}{du}\right) = a\frac{dz}{du}z \quad \Rightarrow \quad 1 + a\frac{dz}{du} = az$$

$$a\frac{dz}{du} = az - 1 \quad \Rightarrow \quad \frac{dz}{du} = \frac{az - 1}{a}$$

$$\Rightarrow \quad \frac{du}{dz} = \frac{a}{az - 1} \quad \Rightarrow \quad du = \frac{adz}{az - 1}$$

$$u = \log(az - 1) + \log c$$

$$x + ay = \log c(az - 1)$$ **Ans.**

Example 24. *Solve* $z^2(p^2 + q^2) = x^2 + y^2$, *where* $p = \dfrac{\partial z}{\partial x}$ *and* $q = \dfrac{\partial z}{\partial y}$

(MDU, May, 2006, R.G.P.V, Bhopal June 2009, 2008, 2003)

Solution. $z^2(p^2 + q^2) = x^2 + y^2$

The equation can be written as:

$$\left(z\frac{\partial z}{\partial x}\right)^2 + \left(z\frac{\partial z}{\partial y}\right)^2 = x^2 + y^2 \quad ...(1)$$

Let $zdz = dZ$ so that $Z = \dfrac{1}{2}z^2$

Now $\dfrac{\partial Z}{\partial x} = \dfrac{\partial Z}{\partial z}\cdot\dfrac{\partial z}{\partial x} = z\dfrac{\partial z}{\partial x}$ and $\dfrac{\partial Z}{\partial y} = \dfrac{\partial Z}{\partial z}\cdot\dfrac{\partial z}{\partial y} = z\dfrac{\partial z}{\partial y}$

Therefore, equation (1) becomes

$$\left(\frac{\partial Z}{\partial x}\right)^2 + \left(\frac{\partial Z}{\partial y}\right)^2 = x^2 + y^2$$

$$P^2 + Q^2 = x^2 + y^2, \text{ where } P = \frac{\partial Z}{\partial x} \text{ and } \frac{\partial Z}{\partial y} = Q$$

Or $P^2 - x^2 = y^2 - Q^2$ which is of the form $f_1(x, P) = f_2(y, Q)$

Let $P^2 - x^2 = y^2 - Q^2 = a$ then $P = \sqrt{x^2 + a}$ and $Q = \sqrt{y^2 - a}$

Substituting these values of P and Q in

$$dZ = P dx + Q dy, \text{ we get}$$

$$dZ = \sqrt{x^2 + a}\, dx + \sqrt{y^2 - a}\, dy$$

Integrating $Z = \frac{1}{2} x\sqrt{x^2 + a} + \frac{a}{2}\log\left(x + \sqrt{x^2 + a}\right) + \frac{1}{2} y\sqrt{y^2 - a} - \frac{a}{2}\log\left(y + \sqrt{y^2 - a}\right) + b$

$$\Rightarrow \quad z^2 = x\sqrt{x^2 + a} + y\sqrt{y^2 - a} + a\log\frac{x + \sqrt{x^2 + a}}{y + \sqrt{y^2 - a}} + c \text{ (where } c = 2b \;\; Z = \frac{1}{2} Z^2)$$ **Ans.**

Example 25. *Solve $p(1 + q^2) = q(z - a)$.*

Solution. Let $u = x + by$

so that $$p = \frac{dz}{du} \quad \text{and} \quad q = b\frac{dz}{du}$$

Substituting these values of p and q in the given equation, we have

$$\frac{dz}{du}\left[1 + b^2\left(\frac{dz}{du}\right)^2\right] = b\frac{dz}{du}(z - a)$$

$$1 + b^2\left(\frac{dz}{du}\right)^2 = b(z - a) \Rightarrow b^2\left(\frac{dz}{du}\right)^2 = bz - ab - 1$$

$$\frac{dz}{du} = \frac{1}{b}\sqrt{bz - ab - 1}, \qquad \int\frac{b\,dz}{\sqrt{bz - ab - 1}} = \int du + c$$

$$2\sqrt{bz - ab - 1} = u + c$$

$$4(bz - ab - 1) = (u + c)^2$$

$$4(bz - ab - 1) = (x + by + cy)^2$$ **Ans.**

Example 26. *Solve $z^2(p^2x^2 + q^2) = 1$*

(MDU, May 2007, R.G.P.V., Bhopal, II Semester, June 2005)

Solution. $$z^2(p^2x^2 + q^2) = 1 \qquad ...(1)$$

$$\Rightarrow \quad z^2\left[\left(x\frac{\partial z}{\partial x}\right)^2 + \left(\frac{\partial z}{\partial y}\right)^2\right] = 1 \quad \Rightarrow \quad z^2\left[\left(\frac{\frac{\partial z}{\partial x}}{x}\right)^2 + \left(\frac{\partial z}{\partial y}\right)^2\right] = 1$$

$$\Rightarrow \quad z^2\left[\left(\frac{\partial z}{\partial X}\right)^2 + \left(\frac{\partial z}{\partial y}\right)^2\right] = 1 \qquad ...(2)$$

where $$\frac{\partial x}{x} = \partial X \Rightarrow \log x = X, \quad \text{Let } u = X + ay$$

$$\frac{\partial z}{\partial X} = \frac{dz}{du} \quad \text{and} \quad \frac{\partial z}{\partial y} = a\frac{dz}{du}$$

Then (2) becomes

$$z^2\left[\left(\frac{dz}{du}\right)^2+\left(a\frac{dz}{du}\right)^2\right]=1 \Rightarrow \left(\frac{dz}{du}\right)^2+a^2\left(\frac{dz}{du}\right)^2=\frac{1}{z^2}$$

$$\Rightarrow \quad \left(\frac{dz}{du}\right)^2=\frac{1}{z^2(1+a^2)} \Rightarrow \frac{dz}{du}=\frac{1}{z\sqrt{1+a^2}} \Rightarrow z\,dz=\frac{du}{\sqrt{1+a^2}}$$

$$\Rightarrow \quad \int z\,dz=\int\frac{du}{\sqrt{1+a^2}}+c \Rightarrow \frac{z^2}{2}=\frac{u}{\sqrt{1+a^2}}+c$$

$$\sqrt{1+a^2}\,\frac{z^2}{2}=u+c\sqrt{1+a^2}=X+ay+c\sqrt{1+a^2}$$

$$\Rightarrow \quad \sqrt{1+a^2}\,\frac{z^2}{2}=\log x+ay+c\sqrt{1+a^2}$$ **Ans.**

EXERCISE 47.5

Solve the following partial differential equations:

1. $z^2(p^2+q^2+1)=1$ **Ans.** $(1-z^2)^{\frac{1}{2}}=-\dfrac{x+ay}{\sqrt{1+a^2}}+c$

2. $1+q^2=q(z-a)$ **Ans.** $\dfrac{x+by}{b}+\dfrac{1}{4}(z-a)^2=\dfrac{1}{4}(z-a)\sqrt{(z-a)^2-2^2}+y\cosh^{-1}\left(\dfrac{z-a}{2}\right)$

3. $x^2p^2+y^2q^2=z$ **Ans.** $2\sqrt{z}=\dfrac{\log x+a\log y}{\sqrt{1+a^2}}+c$

Type IV. Equation of the type

$$f_1(x,p)=f_2(y,q)$$

In these equations, z is absent and the terms containing x and p can be written on one side and the terms containing y and q can be written on the other side.

Method. Let $f_1(x,p)=f_2(y,q)=a$

$f_1(x,p)=a$, solve it for p.

$f_2(y,q)=a$, solve it for q.

Since $\quad dz=\dfrac{\partial z}{\partial x}dx+\dfrac{\partial z}{\partial y}dy \quad \Rightarrow \quad dz=p\,dx+q\,dy$

$\Rightarrow \quad dz=F_1(x)\,dx+F_2(y)\,dy \quad \Rightarrow \quad z=\int F_1(x)dx+\int F_2(y)dy+c$

Example 27. *Solve* $p-x^2=q+y^2$

Solution. $\quad p-x^2=q+y^2=c \quad$ (say)

i.e. $\quad p=x^2+c$ and $q=c-y^2$

Putting these values of p and q in

$$dz=p\,dx+q\,dy=(x^2+c)\,dx+(c-y^2)\,dy$$

$$z=\left(\frac{x^3}{3}+cx\right)+\left(cy-\frac{y^3}{3}\right)+c_1$$ **Ans.**

Example 28. *Solve* $p^2+q^2=z^2(x+y)$

Solution $\quad p^2+q^2=z^2(x+y) \quad \Rightarrow \quad \left(\dfrac{p}{z}\right)^2+\left(\dfrac{q}{z}\right)^2=(x+y)$

$$\Rightarrow \quad \left(\frac{1}{z}\frac{\partial z}{\partial x}\right)^2+\left(\frac{1}{z}\frac{\partial z}{\partial y}\right)^2=x+y \quad \Rightarrow \quad \left(\frac{\partial z}{\frac{z}{\partial x}}\right)^2+\left(\frac{\partial z}{\frac{z}{\partial y}}\right)^2=x+y$$

$$\Rightarrow \quad \left(\frac{\partial Z}{\partial x}\right)^2 + \left(\frac{\partial Z}{\partial y}\right)^2 = x + y \qquad \Rightarrow \quad \frac{\partial z}{z} = \partial Z \text{ or } \log z = Z$$

$$\Rightarrow \quad P^2 + Q^2 = x + y \qquad \Rightarrow \quad P^2 - x = y - Q^2 = a$$

$$P^2 - x = a \qquad \Rightarrow \quad P = \sqrt{a + x}$$

$$y - Q^2 = a \qquad \Rightarrow \quad Q = \sqrt{y - a}$$

Therefore, the equation $\quad dZ = \dfrac{\partial Z}{\partial x} dx + \dfrac{\partial Z}{\partial y} dy$

$$dZ = P\,dx + Q\,dy$$

gives $\quad dZ = \sqrt{a + x}\, dx + \sqrt{y - a}\, dy$

$$Z = \int \sqrt{a + x}\, dx + \int \sqrt{y - a}\, dy + c$$

$$\Rightarrow \quad \log z = \frac{2}{3}(a + x)^{\frac{3}{2}} + \frac{2}{3}(y - a)^{\frac{3}{2}} + c \qquad \textbf{Ans.}$$

EXERCISE 47.6

Solve the following partial differential equations:

1. $q - p + x - y = 0$ (*MDU, May 2010*) **Ans.** $2z = (x + a)^2 + (y + a)^2 + b$

2. $\sqrt{p} + \sqrt{q} = 2x$ **Ans.** $z = \dfrac{1}{6}(2x - a)^3 + a^2 y + b$

3. $q = xp + p^2$ **Ans.** $z = -\dfrac{x^2}{4} + \left\{\dfrac{x\sqrt{x^2 + 4a}}{4} + a \log\left(x + \sqrt{x^2 + 4a}\right)\right\} + ay + b$

4. $z(p^2 - q^2) = x - y$ **Ans.** $z^{3/2} = (x + a)^{3/2} + (y + a)^{3/2} + b$

5. $p^2 - q^2 = x - y$ (*MDU, May 2010*) **Ans.** $z = \dfrac{2}{3}(x + c)^{3/2} + \dfrac{2}{3}(y + c)^{3/2} + b$

Tick ✓ the correct answer.

6. The partial differential equation from $z = (a + x)^2 + y$ is

(*i*) $z = \dfrac{1}{4}\left(\dfrac{\partial z}{\partial x}\right)^2 + y$ (*ii*) $z = \dfrac{1}{4}\left(\dfrac{\partial z}{\partial y}\right)^2 + y$ (*iii*) $z = \left(\dfrac{\partial z}{\partial x}\right)^2 + y$ (*iv*) $z = \left(\dfrac{\partial z}{\partial y}\right)^2 + y$ **Ans.** (*i*)

7. The solution of $xp + yq = z$ is

(*i*) $f(x, y) = 0$ (*ii*) $f\left(\dfrac{x}{y}, \dfrac{y}{z}\right) = 0$ (*iii*) $f(xy, yz) = 0$ (*iv*) $f(x^2, y^2) = 0$ **Ans.** (*ii*)

8. The solution of $p + q = z$ is

(*i*) $f(x + y, y + \log z) = 0$ (*ii*) $f(xy, y \log z) = 0$

(*iii*) $f(x - y, y - \log z) = 0$ (*iv*) None of these **Ans.** (*iii*)

9. The solution of $(y - z)p + (z - x)q = x - y$ is

(*i*) $f(x + y + z) = xyz$ (*ii*) $f(x^2 + y^2 + z^2) = xyz$

(*iii*) $f(x^2 + y^2 + z^2, x^2y^2z^2) = 0$ (*iv*) $f(x + y + z) = x^2 + y^2 z^2$ **Ans.** (*iv*)

47.9 CHARPIT'S METHOD

General method for solving partial differential equation with two independent variables.

Solution. Let the general partial differential equation be

$$f(x, y, z, p, q) = 0 \qquad \text{... (1)}$$

Since z depends on x, y, we have

$$dz = \frac{\partial z}{\partial x} dx + \frac{\partial z}{\partial y} dy$$

$$dz = p dx + q dy \qquad \text{... (2)}$$

The main thing in Charpit's method is to find another relation between the variables x, y, z and p, q. Let the relation be

$$\phi(x, y, z, p, q) = 0 \qquad \text{... (3)}$$

On solving (1) and (3), we get the values of p and q.

These values of p and q when substituted in (2), it becomes integrable.

To determine ϕ, (1) and (3) are differentiated w.r.t. x and y giving

$$\left.\begin{aligned} &\frac{\partial f}{\partial x} + \frac{\partial f}{\partial z} p + \frac{\partial f}{\partial p}\frac{\partial p}{\partial x} + \frac{\partial f}{\partial q}\frac{\partial q}{\partial x} = 0 \\ &\frac{\partial \phi}{\partial x} + \frac{\partial \phi}{\partial z} p + \frac{\partial \phi}{\partial p}\frac{\partial p}{\partial x} + \frac{\partial \phi}{\partial q}\frac{\partial q}{\partial x} = 0 \end{aligned}\right\} \text{ w.r.t. } x, \text{ (First pair)}$$

$$\left.\begin{aligned} &\frac{\partial f}{\partial y} + \frac{\partial f}{\partial z} p + \frac{\partial f}{\partial p}\frac{\partial p}{\partial y} + \frac{\partial f}{\partial q}\frac{\partial q}{\partial y} = 0 \\ &\frac{\partial \phi}{\partial y} + \frac{\partial \phi}{\partial z} p + \frac{\partial \phi}{\partial p}\frac{\partial p}{\partial y} + \frac{\partial \phi}{\partial q}\frac{\partial q}{\partial y} = 0 \end{aligned}\right\} \text{ w.r.t. } y, \text{ (Second pair)}$$

Eliminating $\frac{\partial p}{\partial x}$ between the equation of first pair, we have

$$-\frac{\partial p}{\partial y} = \frac{\frac{\partial f}{\partial x} + \frac{\partial f}{\partial z} p + \frac{\partial f}{\partial q}\frac{\partial q}{\partial x}}{\frac{\partial f}{\partial p}} = \frac{\frac{\partial \phi}{\partial x} + \frac{\partial \phi}{\partial z} p + \frac{\partial \phi}{\partial q}\frac{\partial q}{\partial x}}{\frac{\partial \phi}{\partial p}}$$

$$\Rightarrow \left(\frac{\partial f}{\partial x}\frac{\partial \phi}{\partial p} - \frac{\partial \phi}{\partial x}\frac{\partial f}{\partial p}\right) + p\left(\frac{\partial f}{\partial z}\frac{\partial \phi}{\partial p} - \frac{\partial \phi}{\partial z}\frac{\partial f}{\partial p}\right) + \frac{\partial q}{\partial x}\left(\frac{\partial f}{\partial q}\frac{\partial \phi}{\partial p} - \frac{\partial \phi}{\partial q}\frac{\partial f}{\partial p}\right) = 0 \qquad \text{... (4)}$$

On eliminating $\frac{\partial q}{\partial y}$ between the equations of second pair, we have

$$\left(\frac{\partial f}{\partial y}\frac{\partial \phi}{\partial q} - \frac{\partial \phi}{\partial y}\frac{\partial f}{\partial q}\right) + q\left(\frac{\partial f}{\partial z}\frac{\partial \phi}{\partial q} - \frac{\partial \phi}{\partial z}\frac{\partial f}{\partial q}\right) + \frac{\partial p}{\partial y}\left(\frac{\partial f}{\partial p}\frac{\partial \phi}{\partial q} - \frac{\partial \phi}{\partial p}\frac{\partial f}{\partial q}\right) = 0 \qquad \text{... (5)}$$

Adding (4) and (5) and keeping in view the relation $\frac{\partial p}{\partial y} = \frac{\partial^2 z}{\partial x \partial y} = \frac{\partial q}{\partial x}$, the terms of the last brackets of (4) and (5) cancel. On rearranging, we get

$$\frac{\partial \phi}{\partial f}\left(\frac{\partial f}{\partial x} + p\frac{\partial f}{\partial z}\right) + \frac{\partial \phi}{\partial q}\left(\frac{\partial f}{\partial y} + q\frac{\partial f}{\partial z}\right) + \frac{\partial \phi}{\partial z}\left(-p\frac{\partial f}{\partial p} - q\frac{\partial f}{\partial q}\right) + \left(-\frac{\partial f}{\partial p}\right)\frac{\partial \phi}{\partial x} + \left(-\frac{\partial f}{\partial q}\right)\frac{\partial \phi}{\partial y} = 0$$

$$\Rightarrow \left(-\frac{\partial f}{\partial q}\right)\left(\frac{\partial \phi}{\partial x}\right) + \left(-\frac{\partial f}{\partial q}\right)\frac{\partial \phi}{\partial y} + \left(-p\frac{\partial f}{\partial p} - q\frac{\partial f}{\partial q}\right)\frac{\partial \phi}{\partial z} + \left(\frac{\partial f}{\partial x} + p\frac{\partial f}{\partial z}\right)\frac{\partial \phi}{\partial p} + \left(\frac{\partial f}{\partial y} + q\frac{\partial f}{\partial z}\right)\frac{\partial \phi}{\partial q} = 0$$

... (6)

Equation (6) is a Lagrange's linear equation of the first order with x, y, z, p, q as independent variables and ϕ as dependent variable. Its subsidiary equations are

$$\frac{dx}{-\frac{\partial f}{\partial p}} = \frac{dy}{-\frac{\partial f}{\partial q}} = \frac{dz}{-p\frac{\partial f}{\partial p} - q\frac{\partial f}{\partial q}} = \frac{dp}{\frac{\partial f}{\partial x} + p\frac{\partial f}{\partial z}} = \frac{dq}{\frac{\partial f}{\partial y} + q\frac{\partial f}{\partial z}} = \frac{\partial \phi}{0} \quad \ldots (7)$$

(Commit to memory)

Any of the integrals of (7) satisfies (6). Such an integral involving p or q or both may be taken as assumed relation (3). However, we should choose the simplest integral involving p and q derived from (7). This relation and equation (1) gives the values p and q. The values of p and q are substituted in (2). On integration new equation (2) gives the solution of (1).

Example 29. *Solve $px + qy = pq$*

Solution. $f(x, y, z, p, q) = 0$ is $px + qy - pq = 0$ ... (1)

$$\frac{\partial f}{\partial x} = p, \quad \frac{\partial f}{\partial y} = q, \quad \frac{\partial f}{\partial z} = 0, \quad \frac{\partial f}{\partial p} = x - q, \quad \frac{\partial f}{\partial q} = y - p$$

Charpit's equations are

$$\frac{dx}{-\frac{\partial f}{\partial p}} = \frac{dy}{-\frac{\partial f}{\partial q}} = \frac{dz}{-p\frac{\partial f}{\partial p} - q\frac{\partial f}{\partial q}} = \frac{dp}{\frac{\partial f}{\partial x} + p\frac{\partial f}{\partial z}} = \frac{dq}{\frac{\partial f}{\partial y} + q\frac{\partial f}{\partial z}} = \frac{d\phi}{0}$$

$$\frac{dx}{-(x-q)} = \frac{dy}{-(y-p)} = \frac{dz}{-p(x-q) - q(y-p)} = \frac{dp}{p} = \frac{dq}{q} = \frac{d\phi}{0}$$

We have to choose the simplest integral involving p and q

$$\Rightarrow \quad \frac{dp}{p} = \frac{dq}{q} \Rightarrow \log p = \log q + \log a \Rightarrow p = aq.$$

Putting for p in the given equation (1), we get

$$q(ax + y) = aq^2 \quad \therefore \quad q = \frac{y + ax}{a}$$

$$\therefore \quad p = aq = y + ax$$

Now $$dz = pdx + qdy \quad \ldots (2)$$

Putting for p and q in (2), we get

$$dz = (y + ax)\, dx + \frac{y + ax}{a} dy$$

$$adz = (y + ax)\, dx + (y + ax)\, dy$$

$$adz = (y + ax)(adx + dy)$$

Integrating $$az = \frac{(y + ax)^2}{2} + b$$ **Ans.**

Example 30. *Solve $(p^2 + q^2)y = qz$* ... (1)

(MDU, Dec. 2007, R.G.P.V. Bhopal, II Semester, June 2007)

Solution. $f(x, y, z, p, q) = 0$ is $(p^2 + q^2)y - qz = 0$

$$\frac{\partial f}{\partial x} = 0, \frac{\partial f}{\partial y} = p^2 + q^2, \frac{\partial f}{\partial z} = -q, \frac{\partial f}{\partial p} = 2py, \frac{\partial f}{\partial q} = 2qy - z$$

Now, Charpit's equations are

$$\frac{dx}{-\frac{\partial f}{\partial p}} = \frac{dy}{-\frac{\partial f}{\partial q}} = \frac{dz}{-p\frac{\partial f}{\partial p} - q\frac{\partial f}{\partial q}} = \frac{dp}{\frac{\partial f}{\partial x} + p\frac{\partial f}{\partial z}} = \frac{dq}{\frac{\partial f}{\partial y} + q\frac{\partial f}{\partial z}}$$

$$\Rightarrow \qquad \frac{dx}{-2py} = \frac{dy}{-2q+z} = \frac{dz}{-2p^2y-2q^2y+qz} = \frac{dp}{-pq} = \frac{dq}{p^2+q^2-q^2} = \frac{d\phi}{0}$$

We have to choose the simplest integral involving p and q

$$\frac{dp}{-pq} = \frac{dq}{p^2} \Rightarrow -\frac{dp}{q} = \frac{dq}{p} \Rightarrow pdp + qdq = 0$$

Integrating $\qquad p^2 + q^2 = a^2 \quad$ (say)

Putting for $p^2 + q^2$ in the equation (1), we get

$$a^2y = qz \Rightarrow q = \frac{a^2y}{z} \text{ so } p = \sqrt{a^2 - q^2} = \sqrt{a^2 - \frac{a^4y^2}{z^2}}$$

$$\Rightarrow \qquad p = \frac{a}{z}\sqrt{z^2 - a^2y^2}$$

Now $\qquad dz = pdx + qdy$

Putting for p and q in (2), we get

$$dz = \frac{a}{z}\sqrt{z^2 - a^2y^2}\, dx + \frac{a^2y}{z} dy \Rightarrow zdz = a\sqrt{z^2 - a^2y^2}\, dx + a^2ydy$$

$$\frac{zdz - a^2ydy}{\sqrt{z^2 - a^2y^2}} = a.dx$$

Integrating, we get

$$\frac{1}{2}\frac{2}{1}\sqrt{z^2 - a^2y^2} = ax + b \qquad \text{(Put } z^2 - a^2y^2 = t\text{)}$$

On squaring, $\qquad z^2 - a^2y = (ax + b)^2$ **Ans.**

Example 31. *Solve* $2z + p^2 + qy + 2y^2 = 0$

(*MDU, Dec. 2006, R.G.P.V. Bhopal, II Semester, June 2006*)

Solution. Here, we have

Let $\qquad f = 2z + p^2 + qy + 2y^2 = 0 \qquad$ ... (1)

$$\Rightarrow \qquad \frac{\partial f}{\partial x} = 0, \frac{\partial f}{\partial z} = 2, \text{ and } \frac{\partial f}{\partial p} = 2p \qquad \text{... (2)}$$

Charpit's equations are

$$\frac{dx}{-\frac{\partial f}{\partial p}} = \frac{dy}{-\frac{\partial f}{\partial q}} = \frac{dz}{-p\frac{\partial f}{\partial p} - q\frac{\partial f}{\partial q}} = \frac{dp}{\frac{\partial f}{\partial x} + p\frac{\partial f}{\partial z}} = \frac{dq}{\frac{\partial f}{\partial y} + q\frac{\partial f}{\partial z}} = \frac{d\phi}{0}$$

Here, we take $$\frac{dp}{\frac{\partial f}{\partial x} + p\frac{\partial f}{\partial z}} = \frac{dx}{-\frac{\partial f}{\partial p}} \qquad \text{... (3)}$$

Putting the values of $\frac{\partial f}{\partial x}, \frac{\partial f}{\partial z}$ and $\frac{\partial f}{\partial p}$ from (2) in (3), we get

$$\frac{dp}{0 + 2p} = \frac{dx}{-2p}$$

$$dp + dx = 0 \Rightarrow p + x = a \Rightarrow p = a - x \qquad \text{... (4)}$$

On putting $p = a - x$ in (1), we get

$$2z + (a - x)^2 + qy + 2y^2 = 0$$

$$q = -\frac{2z + (a - x)^2 + 2y^2}{y} \qquad \text{... (5)}$$

Putting for p from (4) and for q from (5) in $dz = pdx + qdy$, we get

$$dz = (a-x)\,dx - \frac{2z+(a-x)^2+2y^2}{y}\,dy$$

$$\Rightarrow \quad dz-(a-x)dx+y\,dy = -\frac{[2z+(a-x)^2+y^2]}{y}\,dy$$

$$\frac{2[dz-(a-x)dx+ydy]}{2z+(a-x)^2+y^2} = -\frac{2}{y}dy$$

In the L.H.S. the numerator is the differential coefficient of denominator.
Hence integrating, we get

$$\log\{2z+(a-x)^2+y^2\} = \log y^{-2} + \log b \quad \Rightarrow \quad 2z+(a-x)^2+y^2 = \frac{b}{y^2}$$

$$\therefore \quad 2z = \frac{b}{y^2}-(a-x)^2-y^2$$ **Ans.**

Example 32. *Solve by using the charpits method*

$$p^2+qy=z$$ *(MDU, Dec., 2005)*

Solution. $f(x, y, z, p, q) = 0$ is $p^2+qy-z=0$

$$\frac{\partial f}{\partial x}=0, \quad \frac{\partial f}{\partial y}=q, \quad \frac{\partial f}{\partial z}=-1, \quad \frac{\partial f}{\partial p}=2p, \quad \frac{\partial f}{\partial q}=y$$

The subsidiary equations are

$$\frac{dx}{2p}=\frac{dy}{y}=\frac{dz}{2p^2+qy}=\frac{-dp}{-p}=\frac{-dp}{q+q(-1)}$$

$$\frac{dx}{2p}=\frac{dy}{y}=\frac{dz}{2p^2+qy}=\frac{dp}{p}=\frac{dq}{0}$$

$$dq=0 \quad \Rightarrow \quad q=a \quad \text{and} \quad p=\sqrt{z-ay}$$

Substitute the value of p and q in $dz = pdx + qdy$, we get

$$dz = \sqrt{z-ay}\,dx + ady$$

$$\frac{dz-ady}{\sqrt{z-ay}} = dx$$

On integration $$\frac{\sqrt{z-ay}}{\frac{1}{2}} = x+b, \quad 2\sqrt{z-ay} = x+b$$

$$4(z-ay) = (x+b)^2 \quad \Rightarrow \quad z-ay = \frac{(x+b)^2}{4}$$

$$\boxed{z = ay + \frac{x^2+b^2+2xb}{4}}$$ **Ans.**

Example 33. *Solve* $2(z+xp+yq) = yp^2$ *(MDU, Dec., 2008, RGPV, II Sem., Feb., 2006)*

Solution. Here, we have

$$f = 2(z+xp+yq)-yp^2$$

Forming the auxiliary equations

$$\frac{dx}{2x-2yp}=\frac{dy}{2y}=\frac{dz}{2xp-2yp^2+2qy}=\frac{dp}{-(2p+2p)}=\frac{dq}{-(2q-p^2+2q)}$$

$$\Rightarrow \quad \frac{dx}{x-yp}=\frac{dy}{y}=\frac{dz}{xp-yp^2+yq}=\frac{dp}{-2p}=\frac{dq}{-\left(2q-\frac{p^2}{2}\right)}$$

Using second & fourth

$$\frac{dy}{y}=\frac{dp}{-2p} \Rightarrow \log y=-\frac{1}{2}\log p \Rightarrow -2\log y=\log p \Rightarrow p=ay^{-2}=\frac{a}{y^2}$$

Substituting the value of p in given PDE, we get

$$2z+2x\left(\frac{a}{y^2}\right)+2yq = y\left(\frac{a}{y^2}\right)^2$$

$$\Rightarrow \qquad 2yq = y\left(\frac{a}{y^2}\right)^2-2z-2x\left(\frac{a}{y^2}\right) \Rightarrow q=\frac{a^2}{2y^4}-\frac{z}{y}-\frac{ax}{y^3}$$

Now, $\qquad dz = pdx+qdy$

$$\Rightarrow \qquad dz=\frac{a}{y^2}dx+\left(\frac{a^2}{2y^4}-\frac{z}{y}-\frac{ax}{y^3}\right)dy$$

Regrouping the forms, we get

$$\left(\frac{ydz+zdy}{y}\right)=\left(\frac{ay\,dx-ax\,dy}{y^3}\right)+\frac{a^2}{2y^4}dy$$

Multiplying throughout by y, we get

$$y\,dz+z\,dy=a\frac{y\,dx-x\,dy}{y^2}+\frac{a^2}{2y^3}dy$$

$$d(yz)=ad\left(\frac{x}{y}\right)+\frac{a^2}{2}\frac{dy}{y^3}$$

On integration, we get

$$yz=a\frac{x}{y}+\frac{a^2}{2}\left(\frac{1}{-2y^2}\right)+b \Rightarrow z=\frac{ax}{y^2}-\frac{a^2}{4y^3}+\frac{b}{y} \qquad \textbf{Ans.}$$

EXERCISE 47.7

Solve the following partial differential equations:

1. $z=p.q$ **Ans.** $2\sqrt{az}=ax+y+\sqrt{ab}$

2. $(p+q)(px+qy)-1=0$ **Ans.** $z\sqrt{(1+a)}=2\sqrt{(ax+y)}+b$

3. $z=px+qy+p^2+q^2$ **Ans.** $z=ax+by+a^2+b^2$

4. $z=p^2x+q^2y$ *(MDU, May, 2005)* **Ans.** $(1+a)z=[\sqrt{ax}+\sqrt{(b+y)}]^2$

5. $z=pq\,xy$ *(MDU, May, 2008)* **Ans.** $z=ax^by^{1/b}$

6. $pxy+pq+qy=yz$ **Ans.** $\log(z-ax)=y-a\log(a+y)+b$

7. $q+xp=p^2$ *(R.G.P.V. Bhopal, Dec., 2001)* **Ans.** $z=axe^{-y}-\frac{1}{2}a^2e^{-2y}+b$

8. $q=px^2+q^2$ **Ans.** $z=a\log x+\left(\frac{1\pm\sqrt{1-4a}}{2}\right)y+b$

9. $2xz-px^2-2qxy+pq=0$ *(MDU, Dec., 2009, Rajasthan, 2006)* **Ans.** $z=ay+b(x^2-a)$

LINEAR PARTIAL DIFFERENTIAL EQUATIONS WITH CONSTANT COEFFICIENTS OF 2ND ORDER

48.1 LINEAR HOMOGENEOUS PARTIAL DIFFERENTIAL EQUATIONS OF *n*TH ORDER WITH CONSTANT COEFFICIENTS

An equation of the type

$$a_0 \frac{\partial^n z}{\partial x^n} + a_1 \frac{\partial^n z}{\partial x^{n-1} \partial y} + ... + a_n \frac{\partial^n z}{\partial y^n} = F(x, y) \quad ... (1)$$

is called a homogeneous linear partial differential equation of *n*th order with constant coefficients. It is called homogeneous because all the terms contain derivatives of the same order.

Putting $\frac{\partial}{\partial x} = D$ and $\frac{\partial}{\partial y} = D'$ in (1), we get

$$(a_0 D^n + a_1 D^{n-1} D' + ... + a_n D'^n) z = F(x, y)$$

$$\Rightarrow \qquad f(D, D') z = F(x, y)$$

48.2 RULES FOR FINDING THE COMPLEMENTARY FUNCTION

Consider the equation

$$a_0 \frac{\partial^2 z}{\partial x^2} + a_1 \frac{\partial^2 z}{\partial x \partial y} + a_2 \frac{\partial^2 z}{\partial y^2} = 0 \Rightarrow \qquad (a_0 D^2 + a_1 DD' + a_2 D'^2) z = 0$$

Step 1 : Put $D = m$ and $D' = 1$

$$a_0 m^2 + a_1 m + a_2 = 0$$

This is the auxiliary equation.

Step 2 : Solve the auxiliary equation.

Case 1. If the roots of the auxiliary equation are real and different; say m_1, m_2.

Then $\qquad$ C.F. $= f_1(y + m_1 x) + f_2(y + m_2 x)$

Theory: $(D - m_1 D')(D - m_2 D') z = 0 \qquad$ ...(1)

(1) will be satisfied by the solution of

$$(D - m_2 D') z = 0 \quad \Rightarrow \quad p - m_2 q = 0 \qquad ...(2)$$

This is a Lagrange's linear equation. Its subsidiary equations are

$$\frac{dx}{1} = \frac{dy}{-m_2} = \frac{dz}{0} \Rightarrow y + m_2 x = C_1 \text{ and } z = C_2$$

$\therefore$ Solution of (2) is $\quad z = f_2(y + m_2 x)$

Similarly the solution of $(D - m_1 D')z = 0$ is

$$z = f_1(y + m_1 x)$$

Hence the complete solution of (1) is

$$z = f_1(y + m_1 x) + f_2(y + m_2 x)$$

Case 2. If the roots are equal; say m

Then $\quad C.F. = f_1(y + mx) + x f_2(y + mx)$

Theory:

$$(D - m D')(D - m D')z = 0 \quad \text{...(1)}$$

Let $\quad (D - mD')z = u \quad$ ... (2)

(1) becomes $\quad (D - m D')u = 0 \quad$ ... (3)

Solution of (3) is $\quad u = f(y + mx)$

(2) becomes $\quad (D - mD')z = f(y + mx) \quad \Rightarrow p - mq = f(y + mx)$

This is Lagrange's equation and its subsidiary equations are

$$\underset{(i)}{\frac{dx}{1}} = \underset{(ii)}{\frac{dy}{-m}} = \underset{(iii)}{\frac{dz}{f(y+mx)}}$$

From (*i*) and (*ii*), $y + mx = C_1$ and $dz = f(y + mx)\,dx$

$$dz = f(C_1)\,dx \quad \Rightarrow \quad z = f(y + mx).x + f_1(y + mx)$$

S. No.	*Roots of A.E.*	*C.F.*
1.	m_1, m_2, m_3 (different)	$f_1(y + m_1 x) + f_2(y + m_2 x) + f_3(y + m_3 x)$
2.	m_1, m_2, m_3 $\begin{bmatrix} m_2 = m_1 \\ m_3 \neq m_1 \end{bmatrix}$	$f_1(y + m_1 x) + x\, f_2(y + m_1 x) + f_3(y + m_3 x)$
3.	m_1, m_2, m_3 $m_1 = m_2 = m_3$	$f_1(y + m_1 x) + x\, f_2(y + m_1 x) + x^2 f_3(y + m_1 x)$

Example 1. *Solve* $(D^3 - 4D^2 D' + 3DD'^2)z = 0.$

Solution. $\quad (D^3 - 4D^2 D' + 3DD'^2)z = 0. \qquad [D = m,\ D' = 1]$

Its auxiliary equation is $m^3 - 4m^2 + 3m = 0$

$$m(m^2 - 4m + 3) = 0$$

$$m(m - 1)(m - 3) = 0 \quad \Rightarrow \quad m = 0, 1, 3$$

The required solution is

$$z = f_1(y) + f_2(y + x) + f_3(y + 3x)$$ **Ans.**

Example 2. *Solve* $\dfrac{\partial^2 z}{\partial x^2} - 4\dfrac{\partial^2 z}{\partial x\,\partial y} + 4\dfrac{\partial^2 z}{\partial y^2} = 0.$

Solution. $(D^2 - 4DD' + 4D'^2)z = 0$

Its auxiliary equation is $m^2 - 4m + 4 = 0$ $[D = m,\ D' = 1]$

$$\Rightarrow \qquad (m-2)^2 = 0 \Rightarrow \qquad m = 2, 2$$

The required solution is

$$z = f_1(y+2x) + x\, f_2(y+2x)$$ **Ans.**

Example 3. *Solve the linear partial differential equation*

$$\frac{\partial^4 z}{\partial x^4} - 2\frac{\partial^4 z}{\partial x^3 \partial y} + 2\frac{\partial^4 z}{\partial x \partial y^3} - \frac{\partial^4 z}{\partial y^4} = 0.$$ *(Q.Bank U.P. II semester 2002)*

Solution. Here, we have

$$(D^4 - 2D^3D' + 2DD'^3 - D'^4)\, z = 0, \text{ where } D \equiv \frac{\partial}{\partial x} \text{ and } D' \equiv \frac{\partial}{\partial y}$$

Auxiliary equation is

$$m^4 - 2m^3 + 2m - 1 = 0$$

$$m^3 (m-1) - m^2 (m-1) - m\,(m-1) + 1\,(m-1) = 0$$

$$\Rightarrow \qquad (m^3 - m^2 - m + 1)(m-1) = 0$$

$$\Rightarrow \qquad (m+1)(m-1)^3 = 0$$

$$\Rightarrow \qquad m = -1, 1, 1, 1$$

$$\therefore \qquad z = f_1(y-x) + f_2(y+x) + x f_3(y+x) + x^2 f_4(y+x)$$ **Ans.**

Example 4. *Solve the linear partial differential equation* $\dfrac{\partial^4 z}{\partial x^4} + \dfrac{\partial^4 z}{\partial y^4} = 0.$

Solution. $(D^4 + D'^4)\, z = 0$

Auxiliary equation is $m^4 + 1 = 0$

$$\Rightarrow \qquad m^4 + 1 + 2m^2 = 2m^2$$

$$\Rightarrow \qquad (m^2+1)^2 - \left(m\sqrt{2}\right)^2 = 0$$

$$\Rightarrow \qquad \left(m^2 + \sqrt{2}m + 1\right)\left(m^2 - \sqrt{2}m + 1\right) = 0$$

so that $m^2 + \sqrt{2}m + 1 = 0$ or $m^2 - \sqrt{2}m + 1 = 0$

$$\Rightarrow \qquad m = \frac{-1+i}{\sqrt{2}}, \frac{1+i}{\sqrt{2}}, \frac{-1-i}{\sqrt{2}}, \frac{1-i}{\sqrt{2}},$$

Hence
$$z = f_1\left\{y + \left(\frac{-1+i}{\sqrt{2}}\right)x\right\} + f_2\left\{y + \left(\frac{1+i}{\sqrt{2}}\right)x\right\}$$

$$+ f_3\left\{y + \left(\frac{-1-i}{\sqrt{2}}\right)x\right\} + f_4\left\{y + \left(\frac{1-i}{\sqrt{2}}\right)x\right\}$$ **Ans.**

EXERCISE 48.1

Solve the following equations:

1. $\dfrac{\partial^2 z}{\partial x^2} + \dfrac{4\partial^2 z}{\partial x \partial y} - 5\dfrac{\partial^2 z}{\partial y^2} = 0$ **Ans.** $z = f_1(y+x) + f_2(y-5x)$

2. $2\dfrac{\partial^2 z}{\partial x^2} + 5\dfrac{\partial^2 z}{\partial x \partial y} + 2\dfrac{\partial^2 z}{\partial y^2} = 0$ **Ans.** $z = f_1(2y-x) + f_2(y-2x)$

3. $(D^3 - 6D^2D' + 11DD'^2 - 6D'^3)z = 0$ **Ans.** $z = f_1(y+x) + f_2(y+2x) + f_3(y+3x)$

4. $\frac{\partial^2 z}{\partial x^2} - 2\frac{\partial^2 z}{\partial x \partial y} + \frac{\partial^2 z}{\partial y^2} = 0$ **Ans.** $z = f_1(y+x) + x f_2(y+x)$

5. $(D^3 - 6D^2D' + 12DD'^2 - 8D'^3)z = 0$ **Ans.** $z = f_1(y+2x) + x f_2(y+2x) + x^2 f_3(y+2x)$

6. $\frac{\partial^4 z}{\partial x^4} - \frac{\partial^4 z}{\partial y^4} = 0$ **Ans.** $z = f_1(y+x) + f_2(y-x) + f_3(y+ix) + f_4(y-ix)$

7. $\frac{\partial^3 z}{\partial x^3} - 4\frac{\partial^3 z}{\partial x^2 \partial y} + 4\frac{\partial^3 z}{\partial x \partial y^2} = 0$ **Ans.** $z = f_1(y) + f_2(y + 2x) + xf_3(y + 2x)$

8. $\frac{\partial^3 z}{\partial x^3} - 7\frac{\partial^3 z}{\partial x \partial y^2} + 6\frac{\partial^3 z}{\partial y^3} = 0$ **Ans.** $z = f_1(y + x) + f_2(y + 2x) + f_3(y - 3x)$

9. $\frac{\partial^4 z}{\partial x^4} + \frac{\partial^4 z}{\partial y^4} = 2\frac{\partial^4 z}{\partial x^2 \partial y^2}$ **Ans.** $z = f_1(y - x) + x f_2(y - 2x) + f_3(y + x) + xf_4(y + x)$

48.3 GENERAL RULES FOR FINDING THE PARTICULAR INTEGRAL

Given partial differential equation is

$$f(D, D')\, z = F(x, y)$$

$$\text{P.I.} = \frac{1}{f(D, D')} F(x, y)$$

If f (D, D′) is a homogeneous function of D and D′ of degree n and R.H.S. function $\phi(ax + by)$, $e^{(ax + by)}$, $ax + by$, $\sin(ax + by)$. Then the particular integral

$$\text{P.I.} = \frac{1}{F(D, D')} \phi(ax + by)$$

$$\text{P.I. of } \frac{1}{F(D, D')} F(x, y) = \frac{1}{F(a, b)} \int\int\int \cdots \int \phi(u)\, du\, du \ldots du \ (n \text{ times}), \text{ where } u = ax + by$$

GENERAL RULE

Integrate $\phi(u)$ w.r.t. u, n times and after integration replace u by $ax + by$.

Case of failure:

To find P.I.

The given equation is

$$F(D, D')\, z = \phi(ax + by) \text{ and } F(a, b) = 0$$

Procedure: Let $F(D, D')$ is a homogeneous function of degree n.

Differentiating $F(D, D')$ partially w.r.t D and multiply L.H.S. by x, we get

$$x \frac{1}{\frac{\partial}{\partial D} F(D, D')} \phi(ax + by)$$

If F (a, b) is again zero.

Differentiate it second time and multiply by x to get $x^2 \frac{1}{\frac{\partial^2}{\partial D^2} F(D, D')} \phi(ax + by)$

If $\qquad F(a, b) \neq 0$ then stop.

If $F(a, b) = 0$, then repeat the above procedure. After m times differentiating and multiplying x by m times, we get

$$x^m \frac{1}{\frac{\partial^m}{\partial D^m} F(D, D')} \phi(ax+by)$$

Let $\qquad F(a, b) \neq 0$, then $\qquad \left[\text{Put } \frac{\partial^m}{\partial D^m} F(D, D') = \psi(D, D')\right]$

$$\text{P.I.} = x^m \frac{1}{(a, b)} \phi(ax+by)$$

SHORT FORMULAE:

We can find P.I. of function on the R.H.S. of the form e^{ax+by}, $\sin(ax+by)$, $\cos(ax+by)$ either by general formula or by the short formula given below:

(*i*) When $\qquad F(x, y) = e^{ax+by}$

$$\text{P.I.} = \frac{\mathbf{1}}{\boldsymbol{f(D, D')}} \boldsymbol{e^{ax+by}} = \frac{\boldsymbol{e^{ax+by}}}{\boldsymbol{f(a, b)}} \qquad [\text{Put } D = a,\ D' = b]$$

Theory, $\qquad De^{ax+by} = ae^{ax+by},\ D'e^{ax+by} = be^{ax+by}$

$$D^2 e^{ax+by} = a^2 e^{ax+by},\ DD'e^{ax+by} = ab\,e^{ax+by},\ D'^2 e^{ax+by} = b^2 e^{ax+by}$$

$$\therefore \qquad (D^2 + k_1 DD' + k_2 D'^2) e^{ax+by} = (a^2 + k_1 ab + k_2 b^2) e^{ax+by}$$

i.e., $\qquad \boxed{f(D, D') e^{ax+by} = f(a,b) e^{ax+by}}$

$$\Rightarrow \qquad \frac{1}{f(D, D')} f(D, D') e^{ax+by} = \frac{1}{f(D, D')} f(a,b) e^{ax+by}$$

$$\Rightarrow \qquad \boxed{\frac{1}{f(D, D')} e^{ax+by} = \frac{1}{f(a,b)} e^{ax+by}}$$

(*ii*) When $\qquad F(x, y) = \sin(ax+by) \quad$ or $\quad F(x, y) = \cos(ax+by)$

$$\text{P.I.} = \frac{\mathbf{1}}{\boldsymbol{f(D^2, DD', D'^2)}} \mathbf{\sin(ax+by)} = \frac{\mathbf{1}}{\boldsymbol{f(-a^2, -ab, -b^2)}} \mathbf{\sin(ax+by)}$$

Put $\qquad D^2 = -a^2, \quad DD' = -ab, \quad D'^2 = -b^2$

$$\text{P.I.} = \frac{1}{f(D^2, DD', D'^2)} \cos(ax+by) = \frac{1}{f(-a^2, -ab, -b^2)} \cos(ax+by)$$

Theory. $D^2 \sin(ax+by) = -a^2 \sin(ax+by)$

$$DD' \sin(ax+by) = -ab \sin(ax+by)$$

$$D'^2 \sin(ax+by) = -b^2 \sin(ax+by)$$

$$f(D^2, DD', D'^2) \sin(ax+by) = f(-a^2, -ab, -b^2) \sin(ax+by)$$

$$\Rightarrow \qquad \frac{1}{f(D^2, DD', D'^2)} f(D^2, DD', D'^2) \sin(ax+by)$$

$$= \frac{1}{f(D^2, DD', D'^2)} f(-a^2, -ab, -b^2) \sin(ax+by)$$

$$\boxed{\frac{1}{f(-a^2, -ab, -b^2)} \sin(ax+by) = \frac{1}{f(D^2, DD', D'^2)} \sin(ax+by)}$$

Case 1. When R.H.S. $= e^{ax+by}$

Example 5. *Solve:* $\dfrac{\partial^3 z}{\partial x^3} - 3\dfrac{\partial^3 z}{\partial x^2 \partial y} + 4\dfrac{\partial^3 z}{\partial y^3} = e^{x+2y}$ *(U.P. II Semester, June 2007)*

Solution. $\dfrac{\partial^3 z}{\partial x^3} - 3\dfrac{\partial^3 z}{\partial x^2 \partial y} + 4\dfrac{\partial^3 z}{\partial y^3} = e^{x+2y}$

Given equation in symbolic form is

$$(D^3 - 3D^2D' + 4D'^3)z = e^{x+2y}$$

Its A.E. is $m^3 - 3m^2 + 4 = 0 \Rightarrow m = -1, 2, 2.$

$\therefore$ C.F. $= f_1(y-x) + f_2(y+2x) + x f_3(y+2x)$

$$\text{P.I.} = \frac{1}{D^3 - 3D^2D' + 4D'^3} e^{x+2y}$$

Put $D = 1$, $D' = 2$ $\qquad = \dfrac{1}{1-6+32} e^{x+2y} = \dfrac{e^{x+2y}}{27}$

Hence, complete solution is

$$z = f_1(y-x) + f_2(y+2x) + x f_3(y+2x) + \frac{e^{x+2y}}{27}$$ **Ans.**

EXERCISE 48.2

Solve the following equations:

1. $\dfrac{\partial^2 z}{\partial x^2} - \dfrac{\partial^2 z}{\partial y^2} = e^{x+2y}$ **Ans.** $z = f_1(y+x) + f_2(y-x) - \dfrac{e^{x+2y}}{3}$

2. $\dfrac{\partial^2 z}{\partial x^2} - 5\dfrac{\partial^2 z}{\partial x \partial y} + 6\dfrac{\partial^2 z}{\partial y^2} = e^{x+y}$ **Ans.** $z = f_1(y+2x) + f_2(y+3x) + \dfrac{1}{2}e^{x+y}$

3. $\dfrac{\partial^2 z}{\partial x^2} - 4\dfrac{\partial^2 z}{\partial x \partial y} + 4\dfrac{\partial^2 z}{\partial y^2} = e^{2x+y}$ **Ans.** $z = f_1(y+2x) + x f_2(y+2x) + \dfrac{x^2}{2}e^{2x+y}$

4. $\dfrac{\partial^2 z}{\partial x^2} - 7\dfrac{\partial^2 z}{\partial x \partial y} + 12\dfrac{\partial^2 z}{\partial y^2} = e^{x-y}$ **Ans.** $z = f_1(y+3x) + f_2(y+4x) + \dfrac{1}{20}e^{x-y}$

5. $\dfrac{\partial^3 z}{\partial x^3} - 2\dfrac{\partial^3 z}{\partial x^2 \partial y} = 2e^{2x-y}$ **Ans.** $z = f_1(y) + x f_2(y) + f_3(y+2x) + \dfrac{1}{8}e^{2x-y}$

6. $(D^2 - 2DD' + D'^2)z = e^{x+2y}$ **Ans.** $z = f_1(y+x) + x f_2(y+x) + e^{x+2y}$

7. $\dfrac{\partial^2 z}{\partial x^2} - \dfrac{\partial^2 z}{\partial y^2} - 2\dfrac{\partial z}{\partial x} + 2\dfrac{\partial z}{\partial y} = e^{2x+3y}$ **Ans.** $z = f_1(y+x) + e^{2x} f_2(y-x) - \dfrac{1}{3}e^{2x+3y}$

8. $\dfrac{\partial^2 z}{\partial x^2} - 5\dfrac{\partial^2 z}{\partial x \partial y} + 6\dfrac{\partial^2 z}{\partial y^2} = \exp(3x-2y)$ **Ans.** $z = f_1(y+2x) + f_2(y+3x) + \dfrac{1}{63}e^{3x-2y}$

Case II. When R.H.S. = sin $(ax + by)$ or cos $(ax + by)$

Example 6. *Solve the linear partial differential equation*

$$\frac{\partial^2 z}{\partial x^2}+2\frac{\partial^2 z}{\partial x\,\partial y}+\frac{\partial^2 z}{\partial y^2}=\sin(2x+3y) \qquad \text{(U.P. II Semester Summer 2006)}$$

Solution. We have,

$$\frac{\partial^2 z}{\partial x^2}+2\frac{\partial^2 z}{\partial x\,\partial y}+\frac{\partial^2 z}{\partial y^2}=\sin(2x+3y)$$

$$(D^2+2DD'+D'^2)z=\sin(2x+3y) \quad \text{where } D=\frac{\partial}{\partial x} \text{ and } D'=\frac{\partial}{\partial y}$$

Put $D = m, \quad D' = 1$

The auxiliary equation is

$$m^2+2m+1=0$$

$$\Rightarrow \quad (m+1)^2=0 \Rightarrow m=-1,-1 \quad \Rightarrow \quad \text{C.F.} = f_1(y-x)+x\,f_2(y-x)$$

$$\text{P.I.} = \frac{1}{D^2+2DD'+D'^2}\sin(2x+3y) = \frac{1}{-4+2(-6)-9}\sin(2x+3y)$$

$$= \frac{1}{-25}\sin(2x+3y) \qquad \begin{bmatrix} D^2=-2^2=-4 \\ D'^2=-3^2=-9 \\ DD'=-2\times 3=-6 \end{bmatrix}$$

Hence, the complete solution is

$$z = C.F. + P.I.$$

$$\Rightarrow \quad z = f_1(y-x)+x\,f_2(y-x)+\frac{1}{-25}\sin(2x+3y)$$

$$\Rightarrow \quad z = f_1(y-x)+x\,f_2(y-x)-\frac{1}{25}\sin(2x+3y) \qquad \textbf{Ans.}$$

Example 7. *Solve* $(D+1)(D+D'-1)z=\sin(x+2y)$ *(U.P. II Semester, 2010)*

Solution. $\text{C.F.} = e^{-x}\phi(y)+e^{x}\phi_2(y-x)$

$$\text{P.I.} = \frac{1}{(D+1)(D+D'-1)}\sin(x+2y) = \frac{1}{D^2+DD'+D'-1}\sin(x+2y)$$

$$= \frac{1}{-1+(-2)+D'-1}\sin(x+2y) = \frac{1}{D'-4}\sin(x+2y)$$

$$= \frac{D'+4}{(D'^2-16)}\sin(x+2y) = \frac{D'+4}{(-4-16)}\sin(x+2y)$$

$$= -\frac{1}{20}(D'+4)\sin(x+2y) = -\frac{1}{20}[D'\sin(x+2y)+4\sin(x+2y)]$$

$$= -\frac{1}{20}[2\cos(x+2y)+4\sin(x+2y)]$$

Hence, the solution is

$$z = e^{-x}\phi(y)+e^{x}\phi_2(y-x)-\frac{1}{10}[\cos(x+2y)+2\sin(x+2y)] \qquad \textbf{Ans.}$$

Example 8. *Solve* $\frac{\partial^2 z}{\partial x^2} - \frac{\partial^2 z}{\partial x \partial y} = \sin x \cos 2y$ *(U.P., II Semester, June, 2010, 2008)*

Solution. We have, $\frac{\partial^2 z}{\partial x^2} - \frac{\partial^2 z}{\partial x \partial y} = \sin x \cos 2y$

The given equation can be written in the form

$$(D^2 - DD')z = \sin x \cos 2y \text{ where } D = \frac{\partial}{\partial x},\ D' = \frac{\partial}{\partial y}$$

Writing $D = m$ and $D' = 1$, the auxiliary equation is

$$m^2 - m = 0 \quad \Rightarrow \quad m(m-1) = 0 \Rightarrow m = 0, 1$$

$$\therefore \quad \text{C.F.} = f_1(y) + f_2(y+x)$$

$$\text{P.I.} = \frac{1}{D^2 - DD'} \sin x \cos 2y = \frac{1}{D^2 - DD'} \frac{1}{2}[\sin(x+2y) + \sin(x-2y)]$$

$$= \frac{1}{2} \frac{1}{D^2 - DD'} \sin(x+2y) + \frac{1}{2} \frac{1}{D^2 - DD'} \sin(x-2y)$$

Put $D^2 = -1$, $DD' = -2$ in the first integral and $D^2 = -1$, $DD' = 2$ in the second integral.

$$= \frac{1}{2}\left[\frac{\sin(x+2y)}{-1-(-2)}\right] + \frac{1}{2}\left[\frac{\sin(x-2y)}{-1-(2)}\right] = \frac{1}{2}\sin(x+2y) - \frac{1}{6}\sin(x-2y)$$

Hence the complete solution is z = C.F. + P.I.

i.e., $$z = f_1(y) + f_2(y+x) + \frac{1}{2}\sin(x+2y) - \frac{1}{6}\sin(x-2y)$$ **Ans.**

Example 9. *Solve the partial differential equation :*

$$\frac{\partial^2 z}{\partial x^2} - 3\frac{\partial^2 z}{\partial x \partial y} + 2\frac{\partial^2 z}{\partial y^2} = e^{2x-y} + e^{x+y} + \cos(x+2y)$$ *(U.P. II Semester Summer 2006)*

Solution. Given equation is

$$\frac{\partial^2 z}{\partial x^2} - 3\frac{\partial^2 z}{\partial x \partial y} + 2\frac{\partial^2 z}{\partial y^2} = e^{2x-y} + e^{x+y} + \cos(x+2y)$$

Given equation can be written as :

$$(D^2 - 3DD' + 2D'^2)z = e^{2x-y} + e^{x+y} + \cos(x+2y)$$

The auxiliary equation is

$$m^2 - 3m + 2 = 0, \quad \Rightarrow \quad m^2 - 2m - m + 2 = 0 \Rightarrow m(m-2) - 1(m-2) = 0$$

$$(m-1)(m-2) = 0 \qquad m = 1, 2$$

Hence, $$\text{C.F.} = \phi_1(y+x) + \phi_2(y+2x)$$

Now $$\text{P.I.} = \frac{1}{(D-D')(D-2D')}\{e^{2x-y} + e^{x+y} + \cos(x+2y)\}$$

$$= \frac{1}{(D-D')(D-2D')} e^{2x-y} + \frac{1}{(D-D')(D-2D')} e^{x+y} + \frac{1}{(D-D')(D-2D')} \cos(x+2y)$$

$$= I_1 + I_2 + I_3$$

Let $I_1 = \dfrac{1}{(D-D')(D-2D')} e^{2x-y}$ (Replacing D by 2 and D' by – 1.)

$= \dfrac{1}{(2+1)(2+2)} e^{2x-y} = \dfrac{1}{12} e^{2x-y}$

Now, $I_2 = \dfrac{1}{(D-D')(D-2D')} e^{x+y}$, (Replacing D by 1 and D' by 1)

$= \dfrac{1}{(D-D')(-1)} e^{x+y} = -\dfrac{1}{(D-D')} e^{x+y} = -x\dfrac{1}{1} e^{x+y} = -x\, e^{x+y}$

Now, $I_3 = \dfrac{1}{(D-D')(D-2D')} \cos(x+2y) = \dfrac{1}{D^2-3DD'+2D'^2} \cos(x+2y)$

$= \dfrac{1}{-1-3(-2)+2(-4)} \cos(x+2y)$ (Replacing D^2 by –1; DD' by –2; D'^2 by –4)

$= \dfrac{1}{-1+6-8} \cos(x+2y) = -\dfrac{1}{3} \cos(x+2y)$

P.I. $= I_1 + I_2 + I_3$

Thus required $P.I. = \dfrac{1}{12} e^{2x-y} - xe^{x+y} - \dfrac{1}{3} \cos(x+2y)$

Hence, the complete solution is

$z = C.F. + P.I.$

$= \phi_1(y+x) + \phi_2(y+2x) + \dfrac{1}{12} e^{2x-y} - xe^{x+y} - \dfrac{1}{3} \cos(x+2y)$ **Ans.**

Example 10. *Solve the P.D.E.* $\dfrac{\partial^2 z}{\partial x^2} - 2\dfrac{\partial^2 z}{\partial x \partial y} + \dfrac{\partial^2 z}{\partial y^2} = \sin x.$ (*U.P II Semester 2009, 2004*)

Solution. Here, we have

$$\frac{\partial^2 z}{\partial x^2} - 2\frac{\partial^2 z}{\partial x \partial y} + \frac{\partial^2 z}{\partial y^2} = \sin x$$

$\Rightarrow$ $(D^2 - 2DD' + D'^2)z = \sin x$

Its auxiliary equation is $(m^2 - 2m + 1) = 0 \Rightarrow (m-1)^2 = 0 \Rightarrow m = 1, 1$

C.F. $= f_1(x+y) + x f_2(x+y)$

P.I. $= \dfrac{1}{D^2 - 2DD' + D'^2} \sin x = \left[\dfrac{1}{-1-0+0}\right] \sin x = -\sin x$ $\begin{bmatrix} D=1 \\ D'=0 \end{bmatrix}$

Hence, the complete solution is

$z = $ C.F. + P.I. $= f_1(x+y) + x f_2(x+y) - \sin x$ **Ans.**

EXERCISE 48.3

Solve the following equations :

1. $[2D^2 - 5DD' + 2D'^2]z = 5\sin(2x+y)$ **Ans.** $z = f_1(y+2x) + f_2(2y+x) - \dfrac{5}{3} x\cos(2x+y)$

2. $\dfrac{\partial^2 z}{\partial x^2} - \dfrac{\partial^2 z}{\partial x \partial y} = \cos(x+2y)$ **Ans.** $z = f_1(y) + f_2(y+x) + \cos(x+2y)$

3. $(D^2 - DD')z = \cos x \cos 2y$ **Ans.** $z = f_1(y) + f_2(y+x) + \dfrac{1}{2}\cos(x+2y) - \dfrac{1}{6}\cos(x-2y)$

4. $r - 2s = \sin x \cos 2y$ **Ans.** $z = f_1(y) + f_2(y+2x) + \frac{1}{15}(\sin x \cos 2y) + 4 \sin 2y \cos x$

5. $(D^2 + D'^2)\, z = \cos mx. \cos ny$ **Ans.** $z = f_1(y+ix) + f_2(y-ix) - \frac{\cos mx. \cos xy}{(m^2+n^2)}$

6. $(D^3 - 7DD'^2 - 6D'^3)\, z = \sin(x+2y) + e^{3x+y}$

Ans. $z = f_1(y-x) + f_2(y-2x) + f_3(y+3x) - \frac{1}{75}\cos(x+2y) + \frac{x}{20}e^{3x+y}$

7. $(D^2 - DD')\, z = \cos 2y\,(\sin x + \cos x)$ *(U.P; II Semester, 2003)*

Ans. $z = f_1(y) + f_2(y+x) + \frac{1}{2}[\sin(x+2y) + \cos(x+2y)] - \frac{1}{6}[\sin(x-2y) + \cos(x-2y)]$

Case III. When R.H.S. = $\phi\,(ax + by)$ polynomial

Example 11. *Solve* $\frac{\partial^2 z}{\partial x^2} + \frac{\partial^2 z}{\partial x \partial y} - 6\frac{\partial^2 z}{\partial y^2} = x + y.$

Solution. With $D = \frac{\partial}{\partial x}$, $D' = \frac{\partial}{\partial y}$, the given equation can be written in the form

$$(D^2 + DD' - 6D'^2)z = x + y$$

Writing $D = m$ and $D' = 1$, the auxiliary equation is $m^2 + m - 6 = 0$

$$\Rightarrow \quad (m+3)(m-2) = 0 \Rightarrow m = -3, 2$$

$$\therefore \quad \text{C.F.} = f_1(y-3x) + f_2(y+2x)$$

$$\therefore \quad \text{P.I.} = \frac{1}{D^2 + DD' - 6D'^2}(x+y) = \frac{1}{(1)^2 + (1)(1) - 6(1)^2}\iint u\, du\, du \quad [\text{where } u = x+y]$$

$$= \frac{1}{-4}\frac{u^3}{6} = -\frac{u^3}{24} = \frac{(x+y)^3}{-24}$$

The complete solution is $z = f_1(y-3x) + f_2(y+2x) - \frac{(x+y)^3}{24}$ **Ans.**

Example 12. *Solve :* $(D^2 + 2DD' - 8D'^2)z = \sqrt{2x+3y}$

Solution. Here, we have

$$(D^2 + 2DD' - 8D'^2)z = \sqrt{2x+3y}$$

A.E. is $m^2 + 2m - 8 = 0 \Rightarrow (m+4)(m-2) = 0 \Rightarrow m = 2,\ m = -4$

$$C.F. = f_1(y+2x) + f_2(y-4x)$$

$$P.I. = \frac{1}{D^2 + 2DD' - 8D'^2}\sqrt{2x+3y} = \frac{1}{D^2 + 2DD' - 8D'^2}(2x+3y)^{\frac{1}{2}}$$

$$= \frac{1}{(2)^2 + 2(2)(3) - 8(3)^2}\iint u^{\frac{1}{2}}\, du\, du, \quad \text{where } u = 2x + 3y$$

$$= \frac{1}{-56}\frac{u^{5/2}}{\frac{3}{2}\cdot\frac{5}{2}} = -\frac{1}{56}\left[\frac{4}{15}(2x+3y)^{\frac{5}{2}}\right] = -\frac{1}{210}(2x+3y)^{\frac{5}{2}}$$

Hence, the complete solution = *C.F.* + *P.I.*

$$= f_1(y+2x) + f_2(y-4x) - \frac{1}{210}(2x+3y)^{\frac{5}{2}}$$ **Ans.**

Example 13. *Solve* $(D^3 - 3D^2D' - 4DD'^2 + 12D'^3)z = \sin(y + 2x)$.

Solution. Here, we have

$(D^3 - 3D^2D' - 4DD'^2 + 12D'^3)z = 0$ ...(1)

Putting $D = m$ and $D' = 1$ in (1) ; we get

A.E. is $m^3 - 3m^2 - 4m + 12 = 0$

$\Rightarrow \quad m^2(m - 3) - 4(m - 3) = 0$

$\Rightarrow \quad (m^2 - 4)(m - 3) = 0 \Rightarrow m = \pm 2, 3$

$\therefore \quad \text{C.F.} = f_1(y + 2x) + f_2(y - 2x) + f_3(y + 3x)$

$$P.I. = \frac{1}{D^3 - 3D^2D' - 4DD'^2 + 12D'^3} \sin(y + 2x)$$

$$= \frac{1}{2^3 - 3(2)^2(1) - 4(2)(1)^2 + 12(1)^3} \iiint \sin u \, du \, du \, du \quad \left(\begin{array}{l}\text{where } u = y + 2x \\ \text{case of failure}\end{array}\right)$$

$$= x \frac{1}{3D^2 - 6DD' - 4D'^2} \iint \sin u \, du \, du = \frac{x}{3(2)^2 - 6(2)(1) - 4(1)^2}(-\sin u)$$

$$= \frac{x}{4} \sin(y + 2x)$$

Complete solution is $z = \text{C.F.} + \text{P.I.} = f_1(y = 2x) + f_2(y - 2x) + f_3(y + 3x) + \frac{x}{4}\sin(y + 2x)$

Example 14. *Solve :* $(4D^2 - 4DD' + D'^2)z = 16 \log(x + 2y)$

Solution. Here, we have $(4D^2 - 4DD' + D'^2)z = 16 \log(x + 2y)$

Auxiliary equation is

$$4m^2 - 4m + 1 = 0 \quad \Rightarrow \quad (2m - 1)^2 = 0 \quad \Rightarrow m = \frac{1}{2}, \frac{1}{2}$$

$$C.F. = f_1\left(y + \frac{x}{2}\right) + x\, f_2\left(y + \frac{x}{2}\right)$$

$$\text{P.I.} = \frac{1}{4D^2 - 4DD' + D'^2} 16 \log(x + 2y)$$

$$= \frac{1}{4(1)^2 - 4(1)(2) + (2)^2} 16 \iint \log u \, du \, du, \text{ where } u = x + 2y \quad \text{(case of failure)}$$

$$= x \frac{1}{8D - 4D'} 16 \int \log u \, du = x \frac{1}{8(1) - 4(2)} 16 \log u \quad \text{(case of failure)}$$

$$= 16x^2\left(\frac{1}{8}\right) \log u = 16\frac{x^2}{8} \log(x + 2y) = 2x^2 \log(x + 2y)$$

The complete solution $= C.F. + P.I.$

$$= f_1\left(y + \frac{x}{2}\right) + xf_2\left(y + \frac{x}{2}\right) + 2x^2 \log(x + 2y)$$ **Ans.**

EXERCISE 48.4

Solve the following equations

1. $\dfrac{\partial^2 z}{\partial x^2} - \dfrac{\partial^2 z}{\partial y^2} = x - y$ **Ans.** $z = f_1(y - x) + f_2(y + x) + \dfrac{x^3}{6} - \dfrac{x^2 y}{2}$

2. $\dfrac{\partial^2 z}{\partial x^2} + 3\dfrac{\partial^2 z}{\partial x \partial y} - 4\dfrac{\partial^2 z}{\partial y^2} = x + \sin y$ **Ans.** $z = f_1(y + x) + f_2(y - 4x) + \dfrac{x^3}{6} + \dfrac{1}{4}\sin y$

3. $\dfrac{\partial^3 z}{\partial x^2 \partial y} - 2\dfrac{\partial^3 z}{\partial x \partial y^2} + \dfrac{\partial^3 z}{\partial y^3} = \dfrac{1}{x^2}$ **Ans.** $z = f_1(x) + f_2(y + x) + xf_3(y + x) - y \log x$

4. $(D^3 - 4D^2D' + 5DD'^2 - 2D'^3)z = e^{y+2x} + (y + x)^{\frac{1}{2}}$

Ans. $z = f_1(y + x) + xf_2(y + x) + f_3(y + 2x) + xe^{y+x} - \dfrac{x^2}{3}(y + x)^{\frac{3}{2}}$

5. $\dfrac{\partial^3 z}{\partial x^3} - 4\dfrac{\partial^3 z}{\partial x^2 \partial y} + 4\dfrac{\partial^3 z}{\partial x \partial y^2} = 2 \sin (3x + 2y)$ **Ans.** $z = f_1(y + 2x) + f_2(2y + x) - \dfrac{5}{3} x \cos (2x + y)$

6. $\dfrac{\partial^2 z}{\partial x^2} - 4\dfrac{\partial^2 z}{\partial x \partial y} + 3\dfrac{\partial^2 z}{\partial y^2} = \sqrt{x + 3y}$ **Ans.** $z = f_1(y + x) + f_2(y + 3x) + \dfrac{1}{60}(x + 3y)^{\frac{5}{2}}$

Case IV. *When* $F(x, y) = x^m y^n$

$$P.I. = \frac{1}{f(D, D')} x^m y^n = [f(D, D')]^{-1} x^m y^n$$

(*a*) If $m > n$, then $\dfrac{1}{f(D, D')}$ is expanded in the powers of $\dfrac{D'}{D}$.

(*b*) If $m < n$, then $\dfrac{1}{f(D, D')}$ is expanded in the powers of $\dfrac{D}{D'}$.

Example 15. *Solve :* $(D^2 + D'^2)z = x^2y^2$

Solution. Here, we have

$$(D^2 + D'^2)z = x^2y^2$$

Putting $D = m$ and $D' = 1$, we get the A.E. as

$$m^2 + 1 = 0 \quad \Rightarrow \quad m^2 = -1 \quad \Rightarrow \quad m = \pm i.$$

$\therefore$ $C.F. = f_1(y + ix) + f_2(y - ix)$

$$P.I. = \frac{1}{D^2 + D'^2}(x^2y^2) = \frac{1}{D^2} \cdot \frac{1}{\left(1 + \frac{D'^2}{D^2}\right)}(x^2y^2) = \frac{1}{D^2} \cdot \frac{1}{D^2}\left(1 + \frac{D'^2}{D^2}\right)^{-1}(x^2y^2)$$

$$= \frac{1}{D^2}\left(1 - \frac{D'^2}{D^2}\right)(x^2y^2) = \frac{1}{D^2}(x^2y^2) - \frac{D'^2}{D^4}(x^2y^2) = \frac{x^4}{12}y^2 - \frac{1}{D^4}(2x^2)$$

$$= \frac{x^4}{12}y^2 - 2 \cdot \frac{x^6}{3 \cdot 4 \cdot 5 \cdot 6} = \frac{1}{180}(15x^4y^2 - x^6)$$

Thus, the complete solution is

$$z = f_1(y + ix) + f_2(y - ix) + \frac{1}{180}(15x^4y^2 - x^6)$$ **Ans.**

Example 16. *Solve :* $\dfrac{\partial^3 z}{\partial x^3} - \dfrac{\partial^3 z}{\partial y^3} = x^3y^3$ (*Q. Bank U.P. 2002*)

Solution. Here, we have $\dfrac{\partial^3 z}{\partial x^3} - \dfrac{\partial^3 z}{\partial y^3} = x^3y^3 \Rightarrow (D^3 - D'^3)\, z = x^3y^3$

Putting $D = m$ and D' = 1 in above, we have

A.E. is $m^3 - 1 = 0 \Rightarrow m = 1, w, w^2$

Where w is one of the cube roots of unity.

$\therefore$ $C.F. = f_1(y + x) + f_2(y + wx) + f_3(y + w^2x)$

$$P.I. = \frac{1}{D^3 - D'^3}(x^3y^3) = \frac{1}{D^3\left(1 - \frac{D'^3}{D^3}\right)} x^3y^3$$

$$= \frac{1}{D^3}\cdot\left(1 - \frac{D'^3}{D^3}\right)^{-1}(x^3y^3) = \frac{1}{D^3}\left(1 + \frac{D'^3}{D^3}\right)(x^3y^3)$$

$$= \frac{1}{D^3}\left[x^3y^3 + \frac{1}{D^3}D'^3(x^3y^3)\right] = \frac{1}{D^3}\left[x^3y^3 + \frac{1}{D^3}(6x^3)\right]$$

$$= \frac{1}{D^3}(x^3y^3) + \frac{1}{D^6}(6x^3) = \frac{x^6y^3}{6\cdot5\cdot4} + \frac{6x^9}{4\cdot5\cdot6\cdot7\cdot8\cdot9} = \frac{x^6y^3}{120} + \frac{x^9}{10080}$$

Hence, the complete solution is

$$z = C.F. + P.I. = f_1(y + x) + f_2(y + wx) + f_3(y + w^2x) + \frac{x^6y^3}{120} + \frac{x^9}{10080} \qquad \textbf{Ans.}$$

Example 17. *Solve* $\frac{\partial^2 z}{\partial x^2} + 2\frac{\partial^2 z}{\partial x\,\partial y} + \frac{\partial^2 z}{\partial y^2} = x^2 + xy + y^2$

Solution. Here, we have $\frac{\partial^2 z}{\partial x^2} + 2\frac{\partial^2 z}{\partial x\,\partial y} + \frac{\partial^2 z}{\partial y^2} = x^2 + xy + y^2$

A.E. is $m^2 + 2m + 1 = 0 \Rightarrow (m + 1)^2 = 0 \quad \Rightarrow \quad m = -1, -1$

$$C.F. = f_1(y - x) + xf_2(y - x)$$

$$P.I. = \frac{1}{D^2 + 2DD' + D'^2}(x^2 + xy + y^2) = \frac{1}{D^2}\frac{1}{\left(1 + \frac{2D'}{D} + \frac{D'^2}{D^2}\right)}(x^2 + xy + y^2)$$

$$= \frac{1}{D^2}\left(1 + \frac{2D'}{D} + \frac{D'^2}{D^2}\right)^{-1}(x^2 + xy + y^2)$$

$$= \frac{1}{D^2}\left(1 - \frac{2D'}{D} - \frac{D'^2}{D^2} + \frac{4D'^2}{D^2} + ...\right)(x^2 + xy + y^2)$$

$$= \left(\frac{1}{D^2} - \frac{2D'}{D^3} + \frac{3D'^2}{D^4}\right)(x^2 + xy + y^2)$$

$$= \frac{1}{D^2}(x^2 + xy + y^2) - \frac{2D'}{D^3}(x^2 + xy + y^2) + \frac{3D'^2}{D^4}(x^2 + xy + y^2)$$

$$= \left(\frac{x^4}{12} + \frac{x^3y}{6} + \frac{x^2y^2}{2}\right) - \left(\frac{2}{D^3}\right)(x + 2y) + \frac{3}{D^4}(2)$$

$$= \frac{x^4}{12} + \frac{x^3y}{6} + \frac{x^2y^2}{2} - \frac{x^4}{12} - \frac{2x^3y}{3} + \frac{6x^4}{2\cdot3\cdot4}$$

$$= \frac{x^4}{12} + \frac{x^3y}{6} + \frac{x^2y^2}{2} - \frac{x^4}{12} - \frac{2x^3y}{3} + \frac{x^4}{4} = \frac{x^4}{4} - \frac{1}{2}x^3y + \frac{x^2y^2}{2}$$

Hence, the complete solution is

$$z = f_1(y - x) + xf_2(y - x) + \frac{x^4}{4} - \frac{x^3y}{2} + \frac{x^2y^2}{2} \qquad \textbf{Ans.}$$

EXERCISE 48.5

Solve the following equations :

1. $\dfrac{\partial^2 z}{\partial x^2}+3\dfrac{\partial^2 z}{\partial x\,\partial y}+2\dfrac{\partial^2 z}{\partial y^2}=12xy$ *(A.M.I.E., Winter 2001)* **Ans.** $z=f_1(y-x)+f_2(y-2x)+2x^3y-\dfrac{3x^4}{2}$

2. $\dfrac{\partial^2 z}{\partial x^2}-\dfrac{\partial^2 z}{\partial x\partial y}-6\dfrac{\partial^2 z}{\partial y^2}=xy$ **Ans.** $z=f_1(y-2x)+f_2(y+3x)+\dfrac{x^3y}{6}+\dfrac{x^4}{24}$

$\dfrac{\partial^2 z}{\partial x^2}-\dfrac{\partial^2 z}{\partial y^2}-3\dfrac{\partial z}{\partial x}+3\dfrac{\partial z}{\partial y}=xy+e^{x+2y}$ *(Uttarakhand, June 2009, U.P. III Semester, Summer 2002)*

Ans. $z=f_1(y+x)+e^{3x}f_2(y-x)-\dfrac{1}{3}\left(\dfrac{x^2y}{2}+\dfrac{x^3}{6}+\dfrac{x^2}{3}+\dfrac{xy}{3}+\dfrac{2x}{9}\right)-xe^{x+2y}$

4. $(D^3-3D^2D')z=x^2y$ **Ans.** $z=\phi_1(y+x)+\phi_2(y-x)+\dfrac{1}{12}e^{2x-y}-xe^{x+y}-\dfrac{1}{3}\cos(x+2y)$

5. $\dfrac{\partial^3 z}{\partial x^3}-2\dfrac{\partial^3 z}{\partial x^2\partial y}=2e^{2x}+3x^2y$ **Ans.** $z=f_1(y)+xf_2(y)+f_3(y+2x)+\dfrac{1}{60}(15e^{2x}+3x^5y+x^6)$

6. $(D^2-6DD'+9D'^2)z=12x^2+36xy$ **Ans.** $z=f_1(y+3x)+xf_2(y+3x)+6x^3y+10x^4$

7. $(D^2-D'^2+D+3D'-2)z=e^{x-y}-x^2y$

Ans. $z=e^{-2x}f_1(y+x)+e^xf_2(y-x)-\dfrac{1}{4}e^{x-y}+\dfrac{1}{2}\left(x^2y+xy+\dfrac{3}{2}x^2+\dfrac{3y}{2}+3x+\dfrac{21}{4}\right)$

48.4 P.I. OF ANY FUNCTION

If the function on the R.H.S. of the P.D.E. is not of the form given previous cases

$$\text{P.I.}=\frac{1}{F(D,D')}\phi(x,y)$$

$F(D, D')$ is factorized to get

$$F(D,D')=(D-m_1D')(D-m_2D')\ldots\ldots(D-m_nD')$$

$$\text{P.I.}=\frac{1}{(D-m_1D)(D-m_2D')\ldots(D-m_nD')}\phi(x,y)$$

Let us consider

$$\text{P.I.}=\frac{1}{D-m_1D'}\phi(x,y) \qquad \text{(Taking only one term)}$$

$$\Rightarrow \qquad p-m_1q=\phi(x,y)$$

Subsidiary equations are (Lagrange's equations)

$$\frac{dx}{1}=\frac{dy}{-m_1}=\frac{dz}{\phi(x,y)}$$

From the first two

$$dy+m_1dx=0 \qquad\Rightarrow\qquad y+m_1x=c$$

From the first and last equations, we get

$$dz=\phi(x,y)dx=\phi(x,c-m_1x)\,dx$$

$$\Rightarrow \qquad z=\int\phi(x,c-m_1x)dx$$

$$\boxed{\text{P.I.}=\frac{1}{D-mD'}F(x,y)=\int\phi(x,c-mx)dx}$$

where c is replaced by $y+mx$ after integration.

Similarly we repeat the above method to get *P.I.*

Case V. When R.H.S. = Any function

Example 18. *Solve* $(D^2 - DD' - 2D'^2)z = (y-1)e^x$

Solution. $(D^2 - DD' - 2D'^2)z = (y-1)e^x$

A.E. is $m^2 - m - 2 = 0$

$$(m-2)(m+1) = 0 \Rightarrow m = 2, -1$$

$$\text{C.F.} = f_1(y+2x) + f_2(y-x)$$

$$\text{P.I.} = \frac{1}{D^2 - DD' - 2D'^2}(y-1)e^x$$

$$= \frac{1}{(D+D')(D-2D')}(y-1)e^x = \frac{1}{D+D'}\int[(c-2x-1)e^x dx] \qquad \text{[Put } y = c - 2x\text{]}$$

$$= \frac{1}{D+D'}[(c-2x-1)e^x + 2e^x] = \frac{1}{D+D'}[ce^x - 2xe^x + e^x] \qquad \text{[Put c} = y + 2x\text{]}$$

$$= \frac{1}{D+D'}[(y+2x)e^x - 2xe^x + e^x] = \frac{1}{D+D'}[ye^x + e^x]$$

$$= \int[(c+x)e^x + e^x]dx \qquad \text{[Put } y = c + x\text{]}$$

$$= (c+x)e^x - e^x + e^x \qquad \text{[Put } c = y - x\text{]}$$

$$= ce^x + xe^x = (y-x)e^x + xe^x = ye^x$$

Hence, complete solution is

$$\therefore \quad z = f_1(y+2x) + f_2(y-x) + ye^x \qquad \textbf{Ans.}$$

Example 19. *Solve* $\frac{\partial^2 z}{\partial x^2} + \frac{\partial^2 z}{\partial x \partial y} - 6\frac{\partial^2 z}{\partial y^2} = y \cos x$.

(R.G.P.V., Bhopal, June 2009, Feb. 2008, June 2006, 2004, Dec. 2002)

Solution. We have, $\frac{\partial^2 z}{\partial x^2} + \frac{\partial^2 z}{\partial x \partial y} - 6\frac{\partial^2 z}{\partial y^2} = y \cos x$

$$\Rightarrow \quad (D^2 + DD' - 6D'^2)\ z = y \cos x$$

Its auxiliary equation is

$$m^2 + m - 6 = 0 \quad \Rightarrow \quad (m+3)(m-2) = 0 \quad \Rightarrow \quad m = 2, -3$$

$$\therefore \quad \text{C.F.} = f_1(y+2x) + f_2(y-3x)$$

$$\text{P.I.} = \frac{1}{D^2 + DD' - 6D'^2} y \cos x = \frac{1}{(D-2D')(D+3D')} y \cos x$$

$$= \frac{1}{D-2D'}\int(c+3x)\cos x\, dx \qquad \text{[Put } y = c + 3x\text{]}$$

$$= \frac{1}{D-2D'}[(c+3x)\sin x + 3\cos x] = \frac{1}{D-2D'}[y \sin x + 3\cos x] \qquad \text{[Put } c + 3x = y\text{]}$$

$$= \int[(c-2x)\sin x + 3\cos x]dx \qquad \text{[Put } y = c - 2x\text{]}$$

$$= (c-2x)(-\cos x) - 2\sin x + 3\sin x = -y\cos x + \sin x \qquad \text{[Put } c - 2x = y\text{]}$$

Hence, the complete solution is

$$z = f_1(y+2x) + f_2(y-3x) + \sin x - y\cos x \qquad \textbf{Ans.}$$

Example. 20. *Solve:* $(D^2 + D D' - 6 D'^2) z = x^2 \sin (x + y)$

Solution. Here. we have

$$(D^2 + D D' - 6 D'^2) z = x^2 \sin (x + y)$$

Putting $D = m$ and $D' = 1$, we have

A.E. is $m^2 + m - 6 = 0 \Rightarrow (m + 3)(m - 2) = 0 \Rightarrow m = 2, -3$

C.F. $= f_1 (y + 2x) + f_2 (y - 3x)$

$$\text{P.I.} = \frac{1}{D^2 + D D' - 6 D'^2}\left[x^2 \sin (x+y)\right] = \frac{1}{(D-2D')(D+3D')}\left[x^2 \sin (x+y)\right].$$

Let $\frac{1}{D+3D'}[x^2 \sin (x+y)] = u \Rightarrow (D + 3D') u = x^2 \sin (x + y)$

$$u = \int x^2 \sin (x+c+3x)\, dx = \int x^2 \sin (4x+c)\, dx \qquad [y = c + 3x]$$

$$= x^2\left(\frac{-\cos (4x+c)}{4}\right) - (2x)\left(\frac{-\sin (4x+c)}{16}\right) + 2\,\frac{\cos (4x+c)}{64}$$

$$= \left[\frac{-x^2}{4} + \frac{1}{32}\right]\cos (4x+c) + \frac{x}{8}\sin (4x+c) \qquad \ldots(1)$$

On eliminating c, we put $c = y - 3x$ in (1) and get

$$u = \left(\frac{-x^2}{4} + \frac{1}{32}\right)\cos (4x+y-3x) + \frac{x}{8}\sin (4x+y-3x)$$

$$= \left(\frac{-x^2}{4} + \frac{1}{32}\right)\cos (x+y) + \frac{x}{8}\sin (x+y)$$

$$\text{P.I.} = \frac{1}{(D-2\,D')(D+3\,D')}x^2 \sin (x+y) = \frac{1}{(D-2D')}u$$

$$= \frac{1}{D-2\,D'}\left[\left(\frac{-x^2}{4} + \frac{1}{32}\right)\cos (x+y) + \frac{x}{8}\sin (x+y)\right]$$

$$= \int\left[\left(\frac{-x^2}{4} + \frac{1}{32}\right)\cos (x+c-2x) + \frac{x}{8}\sin (x+c-2x)\right]dx \qquad (y = c - 2x)$$

$$= \int\left[\left(\frac{-x^2}{4} + \frac{1}{32}\right)\cos (c-x) + \frac{x}{8}\sin (c-x)\right]dx$$

$$= \left(\frac{-x^2}{4} + \frac{1}{32}\right)\{-\sin (c-x)\} - \left(\frac{-x}{2}\right)[(-\cos (c-x)]$$

$$+ \left(\frac{-1}{2}\right)\sin (c-x) + \frac{x}{8}\cos (c-x) - \frac{1}{8}[-\sin (c-x)]$$

$$= \left(\frac{x^2}{4} - \frac{1}{32} - \frac{1}{2} + \frac{1}{8}\right)\sin (c-x) + \left(\frac{-x}{2} + \frac{x}{8}\right)\cos (c-x)$$

$$= \left(\frac{x^2}{4} - \frac{13}{32}\right)\sin (c-x) - \frac{3x}{8}\cos (c-x) \qquad \ldots(2)(c = 2x + y)$$

On eliminating c, we put $c = 2x + y$ in (2) and get

$$\text{P.I.} = \left(\frac{x^2}{4} - \frac{13}{32}\right) \sin(2x + y - x) - \frac{3x}{8} \cos(2x + y - x)$$

$$= \left(\frac{x^2}{4} - \frac{13}{32}\right) \sin(x + y) - \frac{3x}{8} \cos(x + y)$$

The complete solution = *C.F.* + *P.I.*

$$z = f_1(y + 2x) + f_2(y - 3x) + \left(\frac{x^2}{4} - \frac{13}{32}\right) \sin(x + y) - \frac{3x}{8} \cos(x + y) \textbf{ Ans.}$$

Example 21. *Solve* : $(r - 4t) = \frac{4x}{y^2} - \frac{y}{x^2}$

Solution. Here, we have

$$(D^2 - 4D'^2)\, z = \frac{4x}{y^2} - \frac{y}{x^2}$$

A.E. is $m^2 - 4 = 0 \Rightarrow (m + 2)(m - 2) = 0 \Rightarrow m = 2, -2$

$$\text{C.F.} = f_1(y + 2x) + f_2(y - 2x)$$

$$P.I. = \frac{1}{D^2 - 4D'^2}\left(\frac{4x}{y^2} - \frac{y}{x^2}\right) = \frac{1}{(D + 2D')(D - 2D')}\left(\frac{4x}{y^2} - \frac{y}{x^2}\right)$$

Let $u = \frac{1}{D - 2D'}\left(\frac{4x}{y^2} - \frac{y}{x^2}\right)$

$$\Rightarrow (D - 2D')\, u = \left(\frac{4x}{y^2} - \frac{y}{x^2}\right) \quad \Rightarrow \quad u = \int \left[\frac{4x}{(c - 2x)^2} - \frac{c - 2x}{x^2}\right] dx \quad [y = c - 2x]$$

$$= \int \left[\frac{-2(c - 2x) + 2c}{(c - 2x)^2} - \frac{c}{x^2} + \frac{2}{x}\right] dx = \int \left[\frac{-2}{(c - 2x)} + \frac{2c}{(c - 2x)^2} - \frac{c}{x^2} + \frac{2}{x}\right] dx$$

$$= \log(c - 2x) + \frac{c}{c - 2x} + \frac{c}{x} + 2 \log x$$

On eliminating c, replace c *by* $2x + y$ and have

$$u = \log(2x + y - 2x) + \frac{2x + y}{2x + y - 2x} + \frac{2x + y}{x} + 2 \log x$$

$$= \log y + \frac{2x + y}{y} + 2 + \frac{y}{x} + 2 \log x$$

$$\text{Now P.I.} = \frac{1}{(D + 2D')(D - 2D')}\left(\frac{4x}{y^2} - \frac{y}{x^2}\right) = \frac{1}{(D + 2D')}\, u$$

$$= \frac{1}{D + 2D'}\left[\log y + \frac{2x + y}{y} + 2 + \frac{y}{x} + 2 \log x\right]$$

$$= \int \left[\log(c + 2x) + \frac{2x + c + 2x}{c + 2x} + 2 + \frac{c + 2x}{x} + 2 \log x\right] dx \quad [y = c + 2x]$$

$$= \int \left[\log(c + 2x) + \frac{2x}{c + 2x} + 1 + 2 + \frac{c}{x} + 2 + 2 \log x\right] dx$$

$$= \int \left[\log (c+2x) + \frac{2x+c-c}{2x+c} + 5 + \frac{c}{x} + 2 \log x \right] dx$$

$$= \int \left[\log (c+2x) + 1 - \frac{c}{2x+c} + 5 + \frac{c}{x} + 1 \cdot \log x^2 \, dx \right]$$

$$= \left[x \log (c+2x) - \int x \frac{1}{c+2x} 2\, dx + 6x - \frac{c}{2} \log (c+2x) \right.$$

$$\left. + c \log x + x \log x^2 - \int \frac{2x}{x^2} . x\, dx \right]$$

$$= x \log (c+2x) - \int \frac{c+2x-c}{c+2x} dx + 6x - \frac{c}{2} \log (c+2x)$$

$$+ c \log x + x \log x^2 - \int 2\, dx$$

$$= x \log (c+2x) - x + \frac{c}{2} \log (c+2x) + 6x - \frac{c}{2} \log (c+2x)$$

$$+ c \log x + x \log x^2 - 2x$$

$$= x \log (c+2x) + 3x + c \log x + x \log x^2$$

On eliminating c, replacing c by $y - 2x$ and have

$$= x \log (y - 2x + 2x) + 3x + (y - 2x) \log x + x \log x^2$$

$$= x \log y + 3x + y \log x - 2x \log x + x \log x^2$$

$$= x \log y + 3\, x + y \log x - x \log x^2 + x \log x^2$$

$$= x \log y + 3x + y \log x$$

Hence, the complete solution is

$y = C.F + P.I. = f_1 (y + 2x) + f_2 (y - 2x) + x \log y + 3x + y \log x$ **Ans.**

Example 22. *Solve* : $[D^3 + D^2 D' - D D'^2 - D'^3]\, z = e^x \cos 2y$

Solution. We have

$$[D^3 + D^2 D' - D D'^2 - D'^3]\, z = e^x \cos 2y$$

A.E. is $\quad m^3 + m^2 - m - 1 = 0 \Rightarrow (m + 1)^2 (m - 1) = 0 \Rightarrow m = 1, -1, -1$

$$\text{C.F.} = f_1 (y + x) + f_2 (y - x) + x f_3 (y - x)$$

$$\text{P. I.} = \frac{1}{D^3 + D^2 D' - D D'^2 - D'^3} e^x \cos 2y = \frac{1}{(D + D')^2 (D - D')} e^x \cos 2y$$

Let $\quad u = \dfrac{1}{D - D'} e^x \cos 2y \qquad (y = c - x)$

$$= \int e^x \cos 2 (c - x)\, dx = \frac{e^x}{1+4} [\cos 2 (c - x) - 2 \sin (c - x)] \qquad \dots (1)$$

On eliminating c, replace c by $x + y$ in (1), and have

$$u = \frac{e^x}{5} [\cos 2 (x + y - x) - 2 \sin (x + y - x)] = \frac{e^x}{5} [\cos 2y - 2 \sin 2y]$$

Now $\left[\dfrac{1}{(D + D')(D - D')} \right] e^x \cos 2y = \dfrac{1}{(D + D')} u$

$$= \frac{1}{D + D'} \left[\frac{e^x}{5} (\cos 2y - 2 \sin 2y) \right] \qquad (y = c + x)$$

$$= \int \left[\frac{e^x}{5} \{\cos (2\, c + 2x) - 2 \sin (2c + 2x)\} \right] dx$$

$$= \int \frac{e^x}{5} \cos(2c+2x)\,dx - 2\int \frac{e^x}{5} \sin(2c+2x)\,dx$$

$$= \frac{e^x}{5(1+4)}[\cos(2c+2x) + 2\sin(2c+2x)] - \frac{2e^x}{5(1+4)}[\sin(2c+2x) - 2\cos(2c+2x)]$$

$$= \frac{e^x}{25}\left[\cos(2c+2x) + 2\sin(2c+2x) - 2\sin(2c+2x) + 4\cos(2c+2x)\right]$$

$$= \frac{e^x}{25}\left[5\cos(2c+2x)\right] = \frac{e^x}{5}\cos(2c+2x)$$

On eliminating c, replace c by $y - x$ and have

$$= \frac{e^x}{5}\cos(2y - 2x + 2x) = \frac{e^x}{5}\cos 2y$$

$$\text{P.I.} = \frac{1}{(D+D')}\left(\frac{1}{D+D'}\,\frac{1}{D-D'}\right)(e^x \cos 2y) = \frac{1}{D+D'}\,\frac{e^x}{5}\cos 2y \qquad (y = c + x)$$

$$= \int \frac{e^x}{5}\cos 2(c+x)\,dx = \frac{e^x}{5(1+4)}\left[\cos 2(c+x) + 2\sin 2(c+x)\right]$$

On eliminating c, replace c by $(y - x)$ and get

$$= \frac{e^x}{25}\left[\cos 2(y - x + x) + 2\sin 2(y - x + x)\right] = \frac{e^x}{25}\left[\cos 2y + 2\sin 2y\right]$$

The complete solution *is* z = *C.F.* + *P.I.*

$$= f_1(y+x) + f_2(y-x) + x\,f_3(y-x) + \frac{e^x}{25}(\cos 2y + 2\sin 2y)] \textbf{ Ans.}$$

EXERCISE 48.6

Solve the following equations:

1. $(D - D')(D + 2D')z = (y+1)e^x$ **Ans.** $z = f_1(y+x) + f_2(y-2x) + ye^x$

2. $\frac{\partial^2 z}{\partial x^2} - \frac{\partial^2 z}{\partial y^2} = \tan^3 x \tan y - \tan x \tan^3 y$ **Ans.** $z = f_1(y+x) + f_2(x-y) + \frac{1}{2}\tan x \tan y$

3. $(D^2 - DD' - 2D'^2)z = (2x^2 + xy - y^2)\sin xy - \cos xy$ **Ans.** $z = f_1(y+2x) + f_2(y-x) + \sin xy$

4. $\frac{\partial^2 z}{\partial x^2} + \frac{\partial^2 z}{\partial x\,\partial y} - 2\frac{\partial^2 z}{\partial y^2} = (y-1)\,e^x$ (*Q. Bank, U.P. 2002*)

Ans. $z = f_1(y+x) + f_2(y-2x) + (y-2)\,e^x$

5. $(D^2 - 6DD' + 9D'^2)\,z = 12x^2 + 36xy$ **Ans.** $z = f_1(y+3x) + x f_2(y+3x) + 10x^4 + 6x^3 y$

48.5 NON-HOMOGENEOUS LINEAR EQUATIONS

The linear differential equations which are not homogeneous are called Non-homogeneous Linear Equations.

For example, $3\frac{\partial^2 z}{\partial x^2} + 2\frac{\partial^2 z}{\partial x \partial y} + 4\frac{\partial^2 z}{\partial y^2} + 5\frac{\partial z}{\partial x} + \frac{\partial z}{\partial y} + z = 0$

$$f(D, D') = f_1(x, y)$$

Its solution, z = C.F. + P.I.

Complementary Function: Let the non-homogeneous equation be

$$(D - mD' - a)z = 0 \qquad \text{or} \qquad \frac{\partial z}{\partial x} - m\frac{\partial z}{\partial y} - az = 0$$

$$p - mq = az$$

The Lagrange's subsidiary equations are $\frac{dx}{1} = \frac{dy}{-m} = \frac{dz}{az}$

From first two relations, we have $-mdx = dy$

$$dy + mdx = 0 \Rightarrow \qquad y + mx = c_1 \qquad \text{... (1)}$$

and from first and third relation, we have

$$dx = \frac{dz}{az} \Rightarrow x = \frac{1}{a}\log z + c_2 \Rightarrow z = c_3 e^{ax} \qquad \text{... (2)}$$

From (1) and (2), we have $z = e^{ax}\phi(y + mx)$.

Similarly the solution of $(D - m D' - a)^2 z = 0$ is

$$z = e^{ax}\phi_1(y + mx) + x e^{ax}\phi_2(y + mx)$$

48.6 IF THE EQUATION IS OF THE FORM

$$(\alpha D + \beta D' + \gamma) z = 0 \quad \Rightarrow \quad \alpha p + \beta q = -\gamma z$$

It is of Lagrange's form

Lagrange's subsidiary equations are $\frac{dx}{\alpha} = \frac{dy}{\beta} = \frac{dz}{-\gamma z}$

From first two, we have $\alpha\gamma - \beta x = C_1$

From first and last, we have $\frac{dz}{z} = -\frac{\gamma}{\alpha}dx$

$$\Rightarrow \quad \log z = -\frac{\gamma}{\alpha}x + \log c_2 \quad \Rightarrow \quad z = e^{-\frac{\gamma}{\alpha}x} = \phi(C_1)e^{-\frac{\gamma}{a}x} \quad \Rightarrow \quad z = e^{-\frac{\gamma}{\alpha}x} = \varphi(\alpha y - \beta x)$$

where φ is an arbitrary function.

Example 23. *Solve* $(D + D' - 2)(D + 4D' - 3)z = 0$

Solution. The equation can be rewritten as

$$\{D - (-D)' - 2\}\ \{D - (-4D') - 3\} z = 0$$

Hence the solution is $z = e^{2x}\phi_1(y - mx) + e^{3x}\phi_2(y - 4mx)$ **Ans.**

Example 24. *Solve* $(D + 3D' + 4)^2 z = 0$

Solution. The equation is rewritten as $[D - (-3D') - (-4^2)]z = 0$

Hence the solution is $z = e^{-4x}\phi_1(y - 3x) + x e^{-4x}\phi_2(y - 3x)$ **Ans.**

Example 25. *Solve* $r + 2s + t + 2p + 2q + z = 0$

Solution. The equation is rewritten as

$$(D^2 + 2DD' + D'^2 + 2D + 2D' + 1) z = 0$$

$$\Rightarrow \qquad [(D + D')^2 + 2(D + D') + 1]z = 0$$

$$\Rightarrow \qquad (D + D' + 1)^2 z = 0$$

$$\Rightarrow \qquad [D - (-D') - (-1)]^2 z = 0$$

Hence the solution is

$$z = e^{-x}\phi_1 y(y - x) + xe^{-x}\phi_2(y - x)$$ **Ans.**

Example 26. *Solve* $r-t+p-q=0$

Solution. The equation is rewritten as

$$(D^2-D'^2+D-D')z=0$$

$$\Rightarrow \quad [(D-D')(D+D')+1(D-D')]z=0$$

$$\Rightarrow \quad (D-D')(D+D'+1)z=0$$

Hence the solution is

$$z=\phi_1(y+x)+e^{-x}\phi_2(y-x)$$ **Ans.**

EXERCISE 48.7

Solve the following equations:

1. $(D-D')(D+D'-3)z=0$ **Ans.** $z=\phi(y+x)+e^{3x}\phi_2(y-x)$
2. $(D-D'-1)(D-D'-2)z=0$ **Ans.** $z=e^x\phi_1(y+x)+e^{2x}\phi_2(y+x)$
3. $(D+D'-1)(D+2D'-2)z=0$ **Ans.** $z=e^x\phi_1(y-x)+e^{2x}\phi_2(y-2x)$
4. $(D^2+DD'+D'-1)z=0$ **Ans.** $z=e^{-x}\phi_1(y)+e^x\phi_2(y-x)$
5. $(D^2-DD'-2D'^2+2D+2D')z=0$ **Ans.** $z=\phi_1(y-x)+e^{-2x}\phi_2(y+2x)$
6. $[D^2-D'^2+D+3D'-2]z=0$ **Ans.** $z=e^{-2x}\phi_1(y+x)+e^x\phi_2(y-x)$
7. $(D^2-a^2D'^2+2ab\,D+2ab\,D')z=0$ **Ans.** $z=\phi_1(y-ax)+e^{-2abx}\phi_2(y+ax)$
8. $t+s+q=0$ **Ans.** $z=\phi_1(x)+e^{-x}\phi_2(y-x)$
9. $(D+D'-1)(D+2D'-3)z=0$ **Ans.** $z=e^x\phi_1(y-x)+e^{3x}\phi_2(y-2x)$
10. $(D-2D'+5)^2z=0$ **Ans.** $z=e^{-5x}\phi_1(y+2x)+xe^{-5x}\phi_2(y+2x)$

Particular Integral

Case I. $$\frac{1}{F(D,D')}e^{ax+by}=\frac{1}{F(a,b)}e^{ax+by}$$

Example 27. *Solve* $(D-D'-2)(D-D'-3)z=e^{3x-2y}$

Solution. The complementary function is

$$\text{C.F.}=e^{2x}\phi_1(y+x)+e^{3x}\phi_2(y+x).$$

$$\text{P.I.}=\frac{1}{(D-D'-2)(D-D'-3)}e^{3x-2y}=\frac{1}{[3-(-2)-2][3-(-2)-3]}e^{3x-2y}=\frac{1}{6}e^{3x-2y}$$

Hence, the complete solution is

$$z=e^{2x}\phi_1(y+x)+e^{3x}\phi_2(y+x)+\frac{1}{6}e^{3x-2y}$$ **Ans.**

Case II. $$\frac{1}{F(D^2,DD',D'^2)}\sin(ax+by)=\frac{1}{F(-a^2,-ab,-b^2)}\sin(ax+by)$$

Case III. $$\frac{1}{F(D,D')}x^my^n=[F(D,D')^{-1}]x^my^n$$

Example 28. *Solve* $[D^2-D'^2+D+3D'-2]z=x^2y$

Solution. $(D-D'+2)(D+D'-1)z=0$

$$\text{C.F.} = e^{-2x}\phi_1(y+x)+e^{x}\phi_2(y-x)$$

$$\text{P.I.} = \frac{1}{(D-D'+2)(D+D'-1)}x^2y$$

$$= \frac{1}{D^2-D'^2+D+3D'-2}x^2y = -\frac{1}{2}\left[\frac{1}{1-\frac{3D'}{2}-\frac{D}{2}+\frac{D'^2}{2}-\frac{D^2}{2}}\right]x^2y$$

$$= -\frac{1}{2}[1-\frac{1}{2}(3D'+D-D'^2+D^2)]^{-1}x^2y$$

$$= -\frac{1}{2}[1+\frac{1}{2}(3D'+D-D'^2+D^2)+\frac{1}{4}(3D'+D-D'^2+D^2)^2$$

$$+\frac{1}{8}(3D'+D-D'^2+D^2)^3+.........]x^2y$$

$$= -\frac{1}{2}[1+\frac{1}{2}(3D'+D-D'^2+D^2)+\frac{1}{4}(9D'^2+D^2+6DD'+6D^2D')+\frac{1}{8}(9D^2D')+....]x^2y$$

$$= -\frac{1}{2}[x^2y+\frac{1}{2}(3x^2+2xy-0+2y)+\frac{1}{4}(0+2y+12x+12)+\frac{1}{8}(18)]$$

$$= -\frac{1}{2}\left[x^2y+\frac{3x^2}{2}+xy+y+\frac{y}{2}+3x+3+\frac{9}{4}\right] = -\frac{1}{2}\left[x^2y+\frac{3x^2}{2}+xy+\frac{3y}{2}+3x+\frac{21}{4}\right]$$

Hence, the complete solution is

$$z = e^{-2x}\phi_1(y+x)+e^{x}\phi_2(y-x)-\frac{1}{2}\left(x^2y+\frac{3x^2}{2}+xy+\frac{3y}{2}+3x+\frac{21}{4}\right)$$ **Ans.**

Case IV. $\frac{1}{F(D,D')}[e^{ax+by}\phi(x,y)] = e^{ax+by}\frac{1}{F(D+a,D'+b)}\phi(x,y)$

Example 29. *Solve* $(D-3D'-2)^2z = 2e^{2x}\sin(y+3x)$

Solution. A.E. is $(D-3D'-2)^2 = 0$

$$\text{C.F.} = e^{2x}\phi_1(y+3x)+xe^{2x}\phi_2(y+3x)$$

$$\text{P.I.} = \frac{1}{(D-3D'-2)^2}2e^{2x}.\sin(y+3x)$$

$$= 2e^{2x}\frac{1}{(D+2-3D'^2-2)^2}\sin(y+3x) = 2e^{2x}\frac{1}{(D-3D')^2}\sin(y+3x)$$

$$= 2e^{2x}\frac{1}{D^2+9D'^2-60D'}\sin(y+3x)$$

$$[\,D^2+9D'^2-60D' = -9-9-6(-3) = 0\,]$$

$$= 2e^{2x}.x\frac{1}{2(D-3D')}\sin(y+3x)$$ (As denominator becomes zero)

$$= 2x^2e^{2x}\frac{1}{2}\sin(y+3x)$$ (Again differentiate)

$$= x^2e^{2x}\sin(y+3x)$$

Hence, the complete solution is

$$z = e^{2x}\phi_1(y+3x)+xe^{2x}\phi_2(y+3x)+x^2e^{2x}\sin(y+3x)$$ **Ans.**

Example 30. *Solve* $(D^2 + DD' - 6D'^2)z = x^2 \sin(x+y)$

Solution. $(D^2 + DD' - 6D'^2)z = x^2 \sin(x+y)$

For complementary function

$$(D^2 + DD' - 6D'^2) = 0 \text{ or } (D-2D')(D+3D') = 0$$

$$\text{C.F.} = \phi_1(y+2x) + \phi_2(y-3x)$$

$$\text{P.I.} = \frac{1}{D^2 + DD' - 6D'^2} x^2 \sin(x+y)$$

$$= \text{Imaginary part of } \frac{1}{D^2 + DD' - 6D'^2} x^2 [\cos(x+y) + i\sin(x+y)]$$

$$= \text{Imaginary part of } \frac{1}{D^2 + DD' - 6D'^2} x^2 e^{i(x+y)}$$

$$= \text{Imaginary part of } e^{iy} \frac{1}{D^2 + Di - 6(i)^2} x^2 e^{ix}$$

$$= \text{Imaginary part of } e^{i(x+y)} \frac{1}{(D+i)^2 + (D+i)i + 6} x^2$$

$$= \text{Imaginary part of } e^{i(x+y)} \frac{1}{D^2 + 3iD + 4} x^2$$

$$= \text{Imaginary part of } \frac{e^{i(x+y)}}{4} \frac{1}{1 + \frac{3iD}{4} + \frac{D^2}{4}} x^2$$

$$= \text{Imaginary part of } \frac{e^{i(x+y)}}{4} \left[1 + \frac{3iD}{4} + \frac{D^2}{4}\right]^{-1} x^2$$

$$= \text{Imaginary part of } \frac{e^{i(x+y)}}{4} \left[1 - \frac{3iD}{4} - \frac{D^2}{4} - \frac{9D^2}{16} \ldots\right] x^2$$

$$= \text{Imaginary part of } \frac{e^{i(x+y)}}{4} \left[x^2 - \frac{3ix}{2} - \frac{2}{4} - \frac{9}{16}(2)\right]$$

$$= \text{Imaginary part of } \frac{1}{4}[\cos(x+y) + i\sin(x+y)] \left[x^2 - \frac{3ix}{2} - \frac{13}{8}\right]$$

$$= \frac{1}{4}\left[\sin(x+y)\left(x^2 - \frac{13}{8}\right) - \frac{3}{2} x\cos(x+y)\right]$$

$$= \frac{1}{4}\sin(x+y)\left(x^2 - \frac{13}{8}\right) - \frac{3x}{8}\cos(x+y)$$

Hence, the complete solution is

$$z = \phi_1(y+2x) + \phi_2(y-3x) + \frac{1}{4}\sin(x+y)\left(x^2 - \frac{13}{8}\right) - \frac{3x}{8}\cos(x+y) \quad \textbf{Ans.}$$

EXERCISE 48.8

Solve the following equations:

1. $[D^2+2DD'+D'^2-2D-2D']z=0$ **Ans.** $z=\phi_1(x-y)+e^{2x}\phi_2(x-y)$

2. $(D^2-D'^2-3D+3D')z=e^{x-2y}$ **Ans.** $z=\phi_1(y+x)+e^{3x}\phi_2(y-x)-\frac{1}{12}e^{x-2y}$

3. $(D-D'-1)(D+D'-2)z=e^{2x-y}$ **Ans.** $z=e^x\phi_1(x+y)+e^{2x}\phi_2(y-x)-\frac{1}{2}e^{2x-y}$

4. $(D^2-D'^2-3D+3D')z=e^{x+2y}$ **Ans.** $z=\phi_1(y+x)+e^{3x}\phi_1(x-y)-xe^{x+2y}$

5. $(D+D')(D+D'-2)z=\sin(x+2y)$

Ans. $z=\phi_1(y-x)+e^{2x}(y-x)+\frac{1}{117}[6\cos(x+2y)-9\sin(x+2y)]$

6. $(D^2-DD'-2D)z=\cos(3x+4y)$ **Ans.** $z=\phi_1(y)+e^{2x}\phi_1(y+x)+\frac{1}{15}[\cos(3x+4y)-2\sin(3x+4y)]$

7. $(DD'+D-D'-1)z=xy$ **Ans.** $z=e^{-y}\phi_1(x)+e^x\phi_2(y)-(xy+y-x-1)$

8. $(D+D'-1)(D+2D'-3)z=4+3x+6y$ **Ans.** $z=e^x\phi_1(x-y)+e^{3x}\phi_2(2x-y)+6+x+2y$

9. $D(D+D'-1)(D+3D'-2)z=x^2-4xy+2y^2$

Ans. $z=\phi_1(y)+e^x\phi_2(x-y)+e^{2x}\phi_3(3x-y)+\frac{1}{2}\left[\frac{x^3}{3}-2x^2y+2xy^2-\frac{7}{2}x^2+4xy+\frac{x}{2}\right]$

10. $(D-D'+2)(D+D'-1)z=e^{x-y}-x^2y$

Ans. $z=e^{2y}\phi_1(x+y)+e^x\phi_2(x-y)-\frac{e^{x-y}}{4}+\frac{1}{2}\left[x^2y+xy+\frac{3x^2}{2}+\frac{3}{2}y+3x+\frac{21}{4}\right]$

11. $(D^2-DD'-2D'^2+2D'+2D)z=e^{2x+3y}+\sin(2x+y)+xy$

Ans. $z=\phi_1(x-y)+e^y\phi_2(2x+y)-\frac{1}{10}e^{2x+3y}-\frac{1}{6}\cos(2x+y)+\frac{x}{24}(6xy-6y+9x-2x^2-12)$

48.7 MONGE'S METHOD

Let the equation be

$$Rr+Ss+Tt=V \qquad \text{... (1)}$$

where R, S, T, V are functions of x, y, z, p and q.

$$r=\frac{\partial^2 f}{\partial x^2},\quad s=\frac{\partial^2 f}{\partial x\partial y},\quad t=\frac{\partial^2 f}{\partial y^2}$$

We have
$$dp=\frac{\partial p}{\partial x}dx+\frac{\partial p}{\partial y}dy=rdx+sdy \qquad \text{... (2)}$$

and
$$dq=\frac{\partial q}{\partial x}dx+\frac{\partial q}{\partial y}dy=sdx+tdy \qquad \text{... (3)}$$

From (2) and (3), $r=\dfrac{dp-sdy}{dx}$ and $t=\dfrac{dq-sdx}{dy}$

Putting for r and t in (1), we get

$$R\left(\frac{dp-sdy}{dx}\right)+Ss+T\left(\frac{dq-sdx}{dy}\right)=V$$

$$\Rightarrow \quad Rdpdy+Tdqdx-Vdxdy-s(Rdy^2-Sdxdy+Tdx^2)=0 \quad \text{... (4)}$$

Equation (4) is satisfied if

$$Rdpdy+Tdqdx-Vdxdy=0 \quad \text{... (5)}$$

$$Rdy^2-Sdxdy+Tdx^2=0 \quad \text{.... (6)}$$

Equations (5) and (6) are called **Monge's equations.**
Since (6) can be factorised into two equations.

$$dy-m_1\,dx=0 \text{ and } dy-m_2dx=0$$

Now combine $dy-m_1dx=0$ and equation (5). If need be, we may also use the relation $dz=p.dx+q.dy$ while solving (5) and (6). The solution leads to two integrals

$$u(x,y,z,p,q)=\text{a and } V(x,y,z,p,q)=b$$

Then we get a relation between u and v.

$$V=f_1(u) \quad \text{... (7)}$$

Equation (7) is further integrated by methods of first order equations.

Note. If the intermediate solution is of the form $Pr+Qq=R,$ then we use lagrange's equation.

Example 31. *Solve* $r=a^2t$.

Solution. We have $dp=rdx+sdy$ and $dq=sdx+tdy$ which gives

$$r=\frac{dp-sdy}{dx} \text{ and } t=\frac{dq-sdx}{dy}$$

Putting these values of r and t in $r=a^2t$, we get

$$\frac{dp-sdy}{dx}=a^2\frac{dq-sdx}{dy} \quad \Rightarrow \quad dp\,dy-a^2\,dx\,dq-s(dy^2-a^2\,dx^2)=0$$

Thus, the Monges' equations are $dpdy-a^2dxdq=0$ ... (1)

$$dy^2-a^2dx^2=0 \quad \text{... (2)}$$

(2) can be resolved into factors

$$dy-adx=0 \quad \text{... (3)}$$

and $$dy+adx=0 \quad \text{... (4)}$$

Combining (3) with (1), we get

$$dp(adx)-a^2\,dx\,dq=0 \text{ or } dp-adq=0 \quad \text{... (5)}$$

(3) and (5) on integration give respectively

$$\left.\begin{array}{l} y-ax=A \\ p-aq=B \end{array}\right] \Rightarrow p-aq=f_1(y-ax) \quad \text{... (6)}$$

Similarly combining (4) and (1)

$$p+aq=f_2(y+ax) \quad \text{... (7)}$$

Adding and subtracting (6) and (7), we get

$$p=\frac{1}{2}[f_1(y-ax)+f_2(y+ax)],\ q=\frac{1}{2a}[f_2(y+ax)-f_1(y-ax)]$$

Substituting these values in $dz=pdx+qdy$

$$dz = \frac{1}{2}[f_1(y-ax)+f_2(y+ax)]dx + \frac{1}{2a}[f_2(y+ax)-f_1(y-ax)]dy$$

$$dz = \frac{1}{2a}(dy+adx)f_2(y+ax) - \frac{1}{2a}(dy-adx)f_1(y-ax)$$

Integrating, $z = \frac{1}{2a}\phi_1(y+ax) - \frac{1}{2a}\phi_2(y-ax)$

$\Rightarrow \quad z = F_1(y+ax)+F_2(y-ax)$ **Ans.**

Example 32. *Solve* $r - t\cos^2 x + p\tan x = 0$

Solution. $r = \frac{dp - sdy}{dx}$ and $t = \frac{dq - sdx}{dy}$

Putting for r and t in the given equation, we get

$$\frac{dp-sdy}{dx} - \frac{dq-sdx}{dy}\cos^2 x + p\tan x = 0$$

$\Rightarrow \quad dp\,dy - sdy^2 - dx\,dp\cos^2 x + sdx^2\cos^2 x + p\,dx\,dy\tan x = 0$

$\Rightarrow \quad dp\,dy - dx\,dq\cos^2 x + p\,dx\,dy\tan x - s(dy^2 - dx^2\cos^2 x) = 0$

Monge's equations are

$$dp\,dy - dx\,dq\cos^2 x + p\,dx\,dy\tan x = 0 \qquad ...(1)$$

$$dy^2 - dx^2\cos^2 x = 0 \qquad ...(2)$$

Eq. (2) is factorised $(dy + dx\cos x)(dy - dx\cos x) = 0$

$$dy - dx\cos x = 0 \qquad ...(3)$$

$$dy + dx\cos x = 0 \qquad ...(4)$$

Integrating (3) and (4), we get

$$y - \sin x = A \qquad ...(5)$$

$$y + \sin x = B. \qquad ...(6)$$

Combining (3) and (1), we get

$$dp - dq.\cos x + p\tan x dx = 0$$

$\Rightarrow \quad (dp\sec x + p\sec x\tan x\,dx) - dq = 0$

Integrating $\quad p\sec x - q = B \qquad ...(7)$

Combining (5) and (7), we have

$$p\sec x - q = f_1(y - \sin x) \qquad ...(8)$$

In combining (6) and (7), we get

$$p\sec x + q = f_2(y + \sin x) \qquad ...(9)$$

From (5) and (9)

$$p = \frac{1}{2}\cos x[f_1(y-\sin x)+f_2(y+\sin x)] \text{ and } q = \frac{1}{2}[f_2(y+\sin x) - f_1(y-\sin x)]$$

Putting for p and q in $dz = pdx + qdy$, we get

$$dz = \frac{1}{2}\cos x[f_1(y-\sin x)+f_2(y+\sin x)]dx + \frac{1}{2}[f_2(y+\sin x) - f_1(y-\sin x)]dy$$

$\Rightarrow \quad dz = \frac{1}{2}f_2(y+\sin x)[dy + \cos x dx] - \frac{1}{2}f_1(y-\sin x)[dy - \cos x\,dx]$

Integrating we get $z = \frac{1}{2}F_2(y+\sin x) + F_1(y - \sin x)$ **Ans.**

48.8 CLASSIFICATION OF PARTIAL DIFFERENTIAL EQUATIONS

(*U.P., II Semester, June 2007*)

In practical problems, the following types of equations are generally used :

(*i*) *Wave equation*: $\dfrac{\partial^2 u}{\partial t^2} = c^2 \dfrac{\partial^2 u}{\partial x^2}$

(*ii*) *One-dimensional heat flow*: $\dfrac{\partial u}{\partial t} = c^2 \dfrac{\partial^2 u}{\partial y^2}$

(*iii*) *Two-dimensional heat flow*: $\dfrac{\partial^2 u}{\partial x^2} + \dfrac{\partial^2 u}{\partial y^2} = 0$

(*iv*) *Radio equations*: $-\dfrac{\partial V}{\partial x} = L\dfrac{\partial I}{\partial t}, -\dfrac{\partial I}{\partial x} = C\dfrac{\partial V}{\partial t}$

Consider the Equation. $A\dfrac{\partial^2 u}{\partial x^2} + B\dfrac{\partial^2 u}{\partial x \partial y} + C\dfrac{\partial^2 u}{\partial y^2} + F(x, y, u, p, q) = 0$ ...(1)

Where A, B, C may be constants or functions of x and y. Now the equation (1) is

1. Parabolic; if $B^2 - 4AC = 0$

2. Elliptic; if $B^2 - 4AC < 0$

3. Hyperbolic; if $B^2 - 4AC > 0$

1. Parabolic Equation. The one-dimensional heat conduction equation

$\dfrac{\partial u}{\partial t} = C^2 \dfrac{\partial^2 u}{\partial x^2}$ is an example of parabolic partial differential equation e.g.

One dimensional heat flow equation $\dfrac{\partial u}{\partial t} = \dfrac{\partial^2 u}{\partial x^2}$ is parabolic type

2. Elliptic Equations

The following are the examples of elliptic equations.

Two dimensional heat flow equation in steady state given by $\dfrac{\partial^2 z}{\partial x^2} + \dfrac{\partial^2 z}{\partial y^2} = 0$ is elliptic in nature.

(*i*) Laplace equation: $\dfrac{\partial^2 u}{\partial x^2} + \dfrac{\partial^2 u}{\partial y^2} = 0$

(*ii*) Poisson equation : $\dfrac{\partial^2 u}{\partial x^2} + \dfrac{\partial^2 u}{\partial y^2} = f(x, y)$

Laplace equation arrises in study-state flow and potential problems.

Poisson equations arrises in fluid mechanics electricity and magnetism and torsion problems.

3. Hyperbolic Equations

The wave equation $\dfrac{\partial^2 u}{\partial t^2} = c^2 \dfrac{\partial^2 u}{\partial x^2}$

is the simplest example of hyperbolic partial differential equation.

Remark 1. If A, B, C in (1) are constants, then nature of equation (1) will be the same for all values of x and y.

Remark 2. If A, B, C are functions of x and y in (1), then nature of equation (1) will not for all values of x and y.

Example 33. *Classify the following equations.*

(*a*) $2\dfrac{\partial^2 u}{\partial x^2}+4\dfrac{\partial^2 u}{\partial x\partial y}+3\dfrac{\partial^2 u}{\partial y^2}=0$ (*Q. Bank U.P. II Semester 2002*)

(*b*) $\dfrac{\partial^2 u}{\partial x^2}+4\dfrac{\partial^2 u}{\partial x\partial y}+4\dfrac{\partial^2 u}{\partial y^2}=0$ (*c*) $2\dfrac{\partial^2 u}{\partial x^2}+6\dfrac{\partial^2 u}{\partial x\partial y}+3\dfrac{\partial^2 u}{\partial y^2}=0$

Solution. (*a*) $A = 2,\ B = 4,\ C = 3$

$B^2 - 4AC = (4)^2 - 4(2)(3) < 0$ **Ans.** Elliptic

(*b*) $A = 1,\ B = 4,\ C = 4$

$B^2 - 4AC = (4)^2 - 4(1)(4) = 0$ **Ans.** Parabolic

(*c*) $A = 2,\ B = 6,\ C = 3$

$B^2 - 4AC = (6)^2 - 4(2)(3) = +12 > 0$ **Ans.** Hyperbolic

Example 34. *Determine whether the following equations are hyperbolic, parabolic and elliptic ?*

(*a*) $x^2\dfrac{\partial^2 u}{\partial t^2}-\dfrac{\partial^2 u}{\partial x^2}=u$ (*b*) $t\dfrac{\partial^2 u}{\partial t^2}+2\dfrac{\partial^2 u}{\partial x\partial t}+x\dfrac{\partial^2 u}{\partial x^2}+\dfrac{\partial u}{\partial x}=0$

(*c*) $x\dfrac{\partial^2 u}{\partial t^2}+t\dfrac{\partial^2 u}{\partial x\partial t}+\dfrac{\partial^2 u}{\partial t^2}=0$

Solution. (*a*) Here $A = x^2,\ B = 0,\ C = -1$

Now, $B^2 - 4AC = (0)^2 - 4x^2(-1) = 4x^2$

$\therefore$ It is hyperbolic if $4x^2 > 0$ *i.e.,* $x > 0$

parabolic if $4x^2 = 0$ *i.e.,* $x = 0$

Since $4x^2$ being a square, cannot be negative hence it cannot be elliptic.

(*b*) Here $A = t,\ B = 2,\ C = x$

Now, $B^2 - 4AC = 4 - 4tx$

It is hyperbolic if $4 - 4tx > 0$ *i.e.,* $tx < 1$

elliptic if $4 - 4tx < 0$ *i.e.,* $tx > 1$

and parabolic if $4 - 4tx = 0$ *i.e.,* $tx = 1$.

(*c*) Here $A = x,\ B = t,\ C = 1$

Now, $B^2 - 4AC = (t)^2 - 4(x)(1) = t^2 - 4x$

$\therefore$ It is hyperbolic if $t^2 - 4x > 0$ *i.e.,* $t^2 > 4x$

elliptic if $t^2 - 4x < 0$ *i.e.,* $t^2 < 4x$

and parabolic if $t^2 - 4x = 0$ *i.e.,* $t^2 = 4x$. **Ans.**

Example 35. *Classify the partial differential equation*

$$\frac{\partial^2 u}{\partial t^2}+t\frac{\partial^2 u}{\partial x\partial t}+x\frac{\partial^2 u}{\partial x^2}+2\frac{\partial u}{\partial t}+\frac{\partial u}{\partial x}+6u=0$$ (*Q. Bank U.P. II Semester 2002*)

Solution. We have,

$$\frac{\partial^2 u}{\partial t^2}+t\frac{\partial^2 u}{\partial x\partial t}+x\frac{\partial^2 u}{\partial x^2}+2\frac{\partial u}{\partial t}+\frac{\partial u}{\partial x}+6u=0$$

Here $A = 1,\ B = t,\ C = x$

Now, $B^2 - 4AC = t^2 - 4(1)(x) = t^2 - 4x$

The equation is elliptic if $t^2 - 4x < 0$.

The equation is parabolic if $t^2 - 4x = 0$.

The equation is hyperbolic if $t^2 - 4x > 0$. **Ans.**

Example 36. *Classify the partial differential equation*

$$x^2 \frac{\partial^2 u}{\partial t^2} + 3\frac{\partial^2 u}{\partial x \partial t} + x\frac{\partial^2 u}{\partial x^2} + 17\frac{\partial u}{\partial t} = 100u$$ *(Q. Bank U.P. II Semester 2002)*

Solution. We have,

$$x^2 \frac{\partial^2 u}{\partial t^2} + 3\frac{\partial^2 u}{\partial x \partial t} + x\frac{\partial^2 u}{\partial x^2} + 17\frac{\partial u}{\partial t} = 100u$$

Here $A = x^2, B = 3, C = x$

Now, $B^2 - 4AC = (3)^2 - 4x^2.x = 9 - 4x^3$

The equation is elliptic if $9 - 4x^3 < 0$.

The equation is parabolic if $9 - 4x^3 = 0$.

The equation is hyperbolic if $9 - 4x^3 > 0$. **Ans.**

Example 37. *Show that the equation.*

$$\frac{\partial^2 u}{\partial t^2} = c^2 \frac{\partial^2 u}{\partial x^2} \text{ is hyperbolic.}$$

Solution. We have,

$$\frac{\partial^2 u}{\partial t^2} - c^2 \frac{\partial^2 u}{\partial x^2} = 0.$$

Here $A = 1, B = 0, C = -c^2$

Now, $B^2 - 4AC = (0)^2 - 4(1)(-c^2) = 4c^2$

Hence, $B^2 - 4AC > 0$

Thus, the given equation is hyperbolic. **Proved.**

Example 38. *Classify the following differential equation as to type in the second quadrant of xy-plane* $\sqrt{y^2 + x^2}\, u_{xx} + 2(x - y)\, u_{xy} + \sqrt{y^2 + x^2}\, u_{yy} = 0$

Solution. Here $A = \sqrt{y^2 + x^2}$, $B = 2(x - y)$, $C = \sqrt{y^2 + x^2}$

Now, $B^2 - 4AC = 4(x - y)^2 - 4(y^2 + x^2) = -8xy$

In second quadrant, y is positive while x is – ve.

$\Rightarrow \quad B^2 - 4AC = +\text{ ve} > 0$

Hence, differential equation is hyperbolic. **Ans.**

Example 39. *Match the column for the items of the Left side to that of right side :*

A second order P.D.E. in the function 'u' of two independent variables x, y given with usual symbols $Au_{xx} + Bu_{xy} + Cu_{yy} + F(u) = 0$ *then*

(i) Hyperbolic *(a)* $B^2 - 4AC = 0$

(ii) Parabolic *(b)* $B^2 - 4AC < 0$

(iii) Elliptic *(c)* $B^2 - 4AC > 0$

(iv) Not classifies *(d)* $A = B = C = 0$ *(U.P., II Semester, June 2009)*

Solution.

(*i*) Hyperbolic (*c*) $B^2 - 4AC > 0$

(*ii*) Parabolic (*a*) $B^2 - 4AC = 0$

(*iii*) Elliptic (*b*) $B^2 - 4AC < 0$

(*iv*) Not classifies (*d*) $A = B = C = 0$ Ans.

EXERCISE 48.9

Classify the following partial differential equations:

1. $9\dfrac{\partial^2 u}{\partial x^2} - 6\dfrac{\partial^2 u}{\partial x \partial t} + \dfrac{\partial^2 u}{\partial t^2} = 0$ **Ans.** Parabolic **2.** $3\dfrac{\partial^2 z}{\partial x^2} + 2\dfrac{\partial^2 z}{\partial x \partial y} + 4\dfrac{\partial^2 z}{\partial y^2} = 0$ **Ans.** Elliptic

3. $2\dfrac{\partial^2 z}{\partial x^2} - 6\dfrac{\partial^2 z}{\partial x \partial y} + 3\dfrac{\partial^2 z}{\partial y^2} = 0$ **Ans.** Hyperbolic **4.** $t\dfrac{\partial^2 u}{\partial t^2} + 3\dfrac{\partial^2 u}{\partial x \partial t} + x\dfrac{\partial^2 u}{\partial x^2} + 17\dfrac{\partial u}{\partial x} = 0$

Ans. Hyperbolic if $xt < \dfrac{9}{4}$, parabolic if $xt = \dfrac{9}{4}$, elliptic if $xt > \dfrac{9}{4}$

5. $\dfrac{\partial^2 z}{\partial x^2} = \dfrac{\partial x}{\partial y}$ (*U.P., II Semester, Summer 2003*) **Ans.** Parabolic **6.** $\dfrac{\partial^2 z}{\partial x^2} = \dfrac{\partial^2 z}{\partial y^2}$ **Ans.** Hyperbolic

7. $\dfrac{\partial^2 z}{\partial x^2} + \dfrac{\partial^2 z}{\partial y^2} = 0$ (*U.P., II Semester, 2010*) **Ans.** Elliptic

Choose the correct answer:

8. The complementary function of $r - 7s + 6t = e^{x+y}$ is :

(*i*) $f_1(y - x) + f_2(y - 6x)$ (*ii*) $f_1(y + x) + f_2(y + 6x)$

(*iii*) $f_1(y + 2x) + f_2(y - 2x)$ (*iv*) $f_1(y + 3x) + f_2(y - 4x)$ **Ans.** (*ii*)

(*R.G.P.V Bhopal, II Semester Feb. 2006*)

CHAPTER

49

APPLICATIONS OF PARTIAL DIFFERENTIAL EQUATIONS

49.1 INTRODUCTION

In applied mathematics, the partial differential equations generally arise from the mathematical formulation of physical problems. Subject to certain given conditions, called boundary conditions solving such an equation is known as solving a boundary value problem.

The method of solution of such equations differs from that used in the case of ordinary differential equations. We first find out the general solution of the ordinary differential equation and determine the particular solution with the help of given conditions. Here, from the start, we try to find particular solutions of the partial differential equations which satisfy all the boundary conditions. Method of separation of variables is employed to solve the applied partial differential equation.

49.2 METHOD OF SEPARATION OF VARIABLES (*U.P., II Semester, June 2007*)

In this method, we assume that the dependent variable is the product of two functions, each of which involves only one of the independent variables. So two ordinary differential equations are formed.

Example 1. *Applying the method of separation of variables techniques, find the solution to the P.D.E.*

$$3u_x + 2u_y = 0 \text{ ..., where } u_x = \frac{\partial u}{\partial x}, u_y = \frac{\partial u}{\partial y}.$$

Solution : Here we have

$$\frac{3\,\partial u}{\partial x} + \frac{2\,\partial u}{\partial y} = 0 \qquad ...(1)$$

Let $$u = X(x)\,Y(y) \qquad ...(2)$$

Where X is a function of x only and Y is a function of y only.
On differentiating (2) partially w.r.t. x, we get

$$\frac{\partial u}{\partial x} = \frac{\partial X}{\partial x}.Y \qquad ...(3)$$

On differentiating (2) partially w.r.t. y, we get

$$\frac{\partial u}{\partial y} = X.\frac{\partial Y}{\partial y} \qquad ...(4)$$

Putting the values of $\frac{\partial u}{\partial x}$ and $\frac{\partial u}{\partial y}$ from (3) and (4) in (1), we get

$$3\frac{\partial X}{\partial x}.Y + 2X\frac{\partial Y}{\partial y} = 0 \qquad ...(5)$$

Dividing (5) by XY, we get

$$\frac{3}{X}\frac{\partial X}{\partial x}+\frac{2}{Y}\frac{\partial Y}{\partial y}=0$$

[R.H.S is constant for L.H.S, So we take both equations are equal to k (constant)

$$\Rightarrow \quad \frac{3}{X}\frac{\partial X}{\partial x}=-\frac{2}{Y}\frac{\partial Y}{\partial y}=k \quad\Rightarrow\quad \frac{3}{X}\frac{\partial X}{\partial x}=k \text{ and } -\frac{2}{Y}\frac{\partial Y}{\partial y}=k$$

$$\Rightarrow \quad \frac{\partial X}{X}=\frac{k}{3}\,\partial x \text{ and } \frac{\partial Y}{Y}=-\frac{k}{2}\,\partial y \quad\Rightarrow\quad \log X=\frac{k}{3}x+c_1 \text{ and } \log Y=\frac{-k}{2}y+c_2$$

$$\Rightarrow \quad X=e^{\frac{k}{3}x+c_1} \text{ and } Y=e^{-\frac{k}{2}y+c_2}$$

Putting the values of X and Y in (2), we get

$$u=e^{\frac{k}{3}x+c_1}\,e^{\frac{-k}{2}y+c_2}=e^{k\left(\frac{x}{3}-\frac{y}{2}\right)+c_1+c_2}=e^{k\left(\frac{x}{3}-\frac{y}{2}\right)}\,.\,e^{c_1+c_2}$$

Hence $\quad u=A\,e^{k\left(\frac{x}{3}-\frac{y}{2}\right)}$ [where $A=e^{c_1+c_2}$] **Ans.**

Example 2. *Solve the following equation* $\frac{\partial^2 z}{\partial x^2}-2\frac{\partial z}{\partial x}+\frac{\partial z}{\partial y}=0$ *by the method of separation of variables.* *(AMIETE, June 2009, U.P., II Semester, Summer 2009, 2005)*

Solution. Given equation is

$$\frac{\partial^2 z}{\partial x^2}-2\frac{\partial z}{\partial x}+\frac{\partial z}{\partial y}=0 \qquad \text{... (1)}$$

Let $\quad z=X(x)\,Y(y) \qquad$ ... (2)

where X is a function of x only and Y is a function of y only.

$$\frac{\partial z}{\partial x}=Y\frac{dX}{dx}, \qquad \frac{\partial^2 z}{\partial x^2}=Y\frac{d^2X}{dx^2}$$

$$\frac{\partial z}{\partial y}=X\frac{dY}{dy}$$

Putting all values in equation (1), we get $Y\frac{d^2X}{dx^2}-2Y\frac{dX}{dx}+X\frac{dY}{dy}=0$

Dividing by XY, we have $\frac{1}{X}\frac{d^2X}{dx^2}-\frac{2}{X}\frac{dX}{dx}+\frac{1}{Y}\frac{dY}{dy}=0$

Separating the variables, we have $\frac{1}{X}\frac{d^2X}{dx^2}-\frac{2}{X}\frac{dX}{dx}=-\frac{1}{Y}\frac{dY}{dy}=K\ (let)$

where K is a constant.

$$\frac{1}{X}\frac{d^2X}{dx^2}-\frac{2}{X}\frac{dX}{dx}=K \qquad\Bigg|\qquad -\frac{1}{Y}\frac{dY}{dy}=K$$

$$\frac{d^2X}{dx^2}-2\frac{dX}{dx}=KX \qquad\Bigg|\qquad \frac{dY}{dy}+KY=0$$

$$\Rightarrow \quad (D^2-2D-K)X=0 \qquad\Bigg|\qquad (D+K)Y=0$$

$$\frac{d^2X}{dx^2}-2\frac{dX}{dx}=KX \qquad\qquad \frac{dY}{dy}+KY=0$$

$\Rightarrow \quad (D^2-2D-K)X=0 \qquad\qquad (D+K)Y=0$

A. E. is $\quad m^2-2m-K=0 \qquad\qquad$ A.E. is $m+K=0 \Rightarrow m=-K$

$\Rightarrow \quad m=\frac{2\pm\sqrt{4+4K}}{2} \qquad\qquad \Rightarrow Y=C_3e^{-Ky}$ ...(3)

$\Rightarrow \quad m=1\pm\sqrt{1+K}$

Thus $\quad X=C_1e^{(1+\sqrt{1+K})x}+C_2e^{(1-\sqrt{1+K})x}$ ...(4)

Putting the values of X and Y from (3) and (4) in (2), we get

$$z=\left\{C_1e^{(1+\sqrt{1+K})x}+C_2e^{(1-\sqrt{1+K})x}\right\}C_3e^{-Ky}$$ **Ans.**

Example 3. *Using the method of separation of variables, solve*

$$\frac{\partial u}{\partial x}=2\frac{\partial u}{\partial t}+u$$

where $\quad u(x,0)=6e^{-3x}$ *(U.P. II Semester summer 2006, A.M.I.E.T.E., Summer 2002)*

Solution. $\quad \frac{\partial u}{\partial x}=2\frac{\partial u}{\partial t}+u$ ... (1)

Let $\quad u=X(x).T(t)$ ... (2)

where X is a function of x only and T is a function of t only.

Putting the value of u in (1), we get

$$\frac{\partial(X.T)}{\partial x}=2\frac{\partial}{\partial t}(X.T)+X.T, \qquad T\frac{dX}{dx}=2X\frac{dT}{dt}+X.T$$

On separating the variables, we get

$$\frac{1}{X}\frac{dX}{dx}=\frac{2}{T}\frac{dT}{dt}+1=C$$ [On dividing by XT]

$\frac{1}{X}\frac{dX}{dx}=C \qquad\qquad \frac{2}{T}\frac{dT}{dt}+1=C$

$\Rightarrow \quad \frac{dX}{dx}=CX \qquad\qquad \Rightarrow \quad \frac{aT}{dt}+\frac{T}{2}=\frac{CT}{2}$

$\Rightarrow \quad DX-CX=0 \qquad\qquad \Rightarrow \quad DT-\left(\frac{C}{2}-\frac{1}{2}\right)T=0$

$\Rightarrow \quad (D-C)X=0 \qquad\qquad$ A.E. is $m-\left(\frac{C}{2}-\frac{1}{2}\right)=0 \Rightarrow m=\frac{1}{2}(C-1)$

A.E. is $m-C=0 \Rightarrow m=C \qquad \Rightarrow \quad T=be^{\frac{1}{2}(c-1)t}$

$\Rightarrow \quad X=ae^{cx}$

Putting the values of X and T in (2). we have

$$u=ae^{c^x}.be^{\frac{1}{2}(c--1)t}$$

$$6e^{-3x} = abe^{cx} \quad \Rightarrow \quad ab = 6 \text{ and } c = -3$$

Putting the values of ab and c in (3), we have

$$u = 6e^{-3x+\frac{1}{2}(-3-1)t}$$

$$u = 6e^{-3x-2t}$$ **Ans.**

which is the required solution.

Example 4. *Solve the following equation by the method of separation of variables*

$$\frac{\partial^2 u}{\partial x\, \partial t} = e^{-t} \cos x$$

given that $u = 0$ *when* $t = 0$ *and* $\frac{\partial u}{\partial t} = 0$ *when* $x = 0.$ [*U.P. II Semester, (SUM) 2008*]

Solution. Let $$u = XT \qquad \text{...(1)}$$

where X is a function of x only and T is a function of t only.

Then, $$\frac{\partial u}{\partial t} = \frac{\partial}{\partial t}(XT) = X\frac{dT}{dt}$$

$\therefore$ $$\frac{\partial^2 u}{\partial x \partial t} = \frac{\partial}{\partial x}\left(X\frac{dT}{dt}\right) = \frac{dT}{dt}\cdot\frac{dX}{dx} \qquad \text{...(2)}$$

Substituting the value $\frac{\partial^2 u}{\partial x \partial t}$ from (2) in the given equation, we get

$$\frac{dT}{dt}\frac{dX}{dx} = e^{-t}\cos x$$

Separating the variables, we get

$$e^t\frac{dT}{dt} = \frac{\cos x}{\left(\frac{dX}{dx}\right)} = -p^2 \text{ (say)} \qquad \text{...(3)}$$

Now, $$e^t\frac{dT}{dt} = -p^2$$

$\Rightarrow$ $$dT = -p^2 e^{-t}\, dt$$

On integration, we get

$$T = p^2e^{-t} + c_1 \qquad \text{...(4)}$$

Also, $$\frac{dX}{dx} = -\frac{1}{p^2}\cos x$$

$$dX = -\frac{1}{p^2}\cos x\, dx$$

On integration, we get

$$X = -\frac{1}{p^2}\sin x + c_2 \qquad \text{...(5)}$$

Putting the values of X and T from (4) and (5) in (1), we get

$$u = XT = \left(-\frac{1}{p^2}\sin x + c_2\right)(p^2e^{-t} + c_1) \qquad \text{...(6)}$$

On putting $u = 0$ and $t = 0$ in (6), we get

$$0 = \left(-\frac{1}{p^2}\sin x + c_2\right)(p^2 + c_1)$$

$\Rightarrow$ $$p^2 + c_1 = 0 \Rightarrow c_1 = -p^2$$

Differentiating (6) w.r.t. "t", we get

$$\frac{\partial u}{\partial t} = \left(-\frac{1}{p^2}\sin x + c_2\right)(-p^2e^{-t}) \qquad \text{...(7)}$$

Putting $\frac{\partial u}{\partial t} = 0$ when $x = 0$ in (7), we get

$$0 = c_2 (-p^2 e^{-t})$$

$\Rightarrow$ $c_2 = 0$

Substituting the values of $c_1 = -p^2$ and $c_2 = 0$ in (6), we get

$$u = -\frac{1}{p^2} \sin x\,(p^2 e^{-t} - p^2)$$

$$= (1 - e^{-1}) \sin x$$ **Ans.**

Example 5. *Solve the P.D.E. by separation of variables method,*

$$u_{xx} = u_y + 2u,\ u(0, y) = 0$$

$$\frac{\partial}{\partial x} u(0, y) = 1 + e^{-3y}.$$ *(U.P. II Semester, 2010, 2009)*

Solution. Let $u = XY$ ...(1)

where X is a function of x only and Y is a function of y only.

On differentiating, we get

$$\frac{\partial u}{\partial y} = \frac{\partial}{\partial y}(XY) = X\frac{dY}{dy} = XY' \text{ and}$$

$$\frac{\partial u}{\partial x} = \frac{\partial}{\partial x}(XY) = Y\frac{dX}{dx} \text{ and } \frac{\partial^2 u}{\partial x^2} = \frac{\partial}{\partial x}\left[Y\frac{dX}{dx}\right] = Y\frac{d^2X}{dx^2}$$

Putting the values of u_{xx} and u_y and u in the given equation, we get

$$YX'' = XY' + 2XY$$

On separating the variables, we get

$$\frac{X''}{X} = \frac{Y' + 2Y}{Y}$$

$$\Rightarrow \frac{X''}{X} = \frac{Y'}{Y} + 2 = k \text{ (say)} \qquad ...(2)$$

(*i*) $\frac{X''}{X} = k$ | $\frac{Y'}{Y} + 2 = k$

$\Rightarrow$ $X'' - kX = 0$ | $\Rightarrow$ $\frac{Y'}{Y} = k - 2$

A.E. is $m^2 - k = 0 \Rightarrow m = \pm\sqrt{k}$ | $\Rightarrow$ $\frac{dY}{Y} = (k-2)\,dy$

$\therefore$ $\Rightarrow X = C_1 e^{\sqrt{k}x} + C_2 e^{-\sqrt{k}x}$ | $\Rightarrow$ $\log Y = (k-2)y + \log C_3$

$\therefore$ | $\Rightarrow$ $Y = C_3 e^{(k-2)y}$

On putting the values of X and Y in (1), we get

$$u = (C_1 e^{\sqrt{k}x} + C_2 e^{-\sqrt{k}x})\, C_3 e^{(k-2)y} \qquad ...(3)$$

On putting $x = 0$ and $u = 0$ in (3), we get

$$0 = (C_1 + C_2)\, C_3 e^{(k-2)y} \qquad (C_3 e^{(k-2)y} \neq 0) \qquad ...(4)$$

$\Rightarrow$ $C_1 + C_2 = 0 \Rightarrow C_2 = -C_1$

On putting $C_2 = -C_1$ in (4), we get

$$u = \Sigma C_1 C_3 (e^{\sqrt{k}x} - e^{-\sqrt{k}x})\, e^{(k-2)y} \qquad ...(5)$$

On differentiating (5), w.r.t. x, we get

$$\frac{\partial u}{\partial x} = \Sigma C_1\, C_3\, \sqrt{k}\, (e^{\sqrt{k}x} + e^{-\sqrt{k}x})\, e^{(k-2)y} \qquad \ldots(6)$$

On putting $x = 0$ and $\frac{\partial u}{\partial x} = 1 + e^{-3y}$ in (6), we get

$$1 + e^{-3y} = \Sigma C_1\, C_3\, \sqrt{k}\, (2)\, e^{(k-2)y} = \sum_{n=1}^{\infty} b_n\, e^{(k-2)y}$$

$$1 + e^{-3y} = \Sigma C_1\, C_3\, \sqrt{k}\, (2)\, e^{(k-2)y} = b_1\, e^{(k-2)y} + b_2\, e^{(k-2)y} + b_3 e^{(k-2)y} + \ldots$$

Comparing the coefficients, we get

$$b_1 = 1,\ k - 2 = 0 \Rightarrow k = 2$$

Again comparing, we get

$$2C_1\, C_3\, \sqrt{k} = 1, \;\Rightarrow\; C_1\, C_3 = \frac{1}{2\sqrt{k}} = \frac{1}{2\sqrt{2}} \qquad (\because k = 2)$$

Again comparing b_3 on both the sides, we get

$$b_3 = -1,\ k - 2 = -3 \Rightarrow k = -1$$

Again comparing b_3, we get

$$2C_1\, C_3\, \sqrt{k} = 1 \;\Rightarrow\; C_1\, C_3 = \frac{1}{2\sqrt{k}} = \frac{1}{2\sqrt{-1}} = \frac{1}{2\,i} \qquad (\because k = -1)$$

Hence, $u\,(x,y) = \frac{1}{2\sqrt{2}}\,(e^{\sqrt{2}x} - e^{-\sqrt{2}x}) + \frac{1}{2i}\,(e^{ix} - e^{-ix})\, e^{-3y}$

$$u\,(x,y) = \frac{1}{\sqrt{2}} \sinh \sqrt{2}\, x + e^{-3y} \sin x.$$ **Ans.**

EXERCISE 49.1

Using the method of separation of variables, find the solution of the following equations

1. $2x\frac{\partial z}{\partial x} - 3y\frac{\partial z}{\partial y} = 0$ **Ans.** $z = cx^{\frac{k}{2}}y^{\frac{k}{3}}$

2. $\frac{\partial u}{\partial x} + u = \frac{\partial u}{\partial t}$ if $u = 4e^{-3x}$ when $t = 0$ **Ans.** $u = 4e^{-3x-2t}$

3. $4\frac{\partial u}{\partial x} + \frac{\partial u}{\partial y} = 3u$ and $u = e^{-5y}$ when $x = 0$ (*AMIETE, June 2010*) **Ans.** $u = e^{2x-5y}$

4. $4\frac{\partial u}{\partial t} + \frac{\partial u}{\partial x} = 3u,\ u = 3e^{-x} - e^{-5x}$ at $t = 0$ (*A.M.I.E.T.E., Winter 2000*) **Ans.** $u = 3e^{t-x} - e^{2t-5x}$

5. $3\frac{\partial u}{\partial x} + 2\frac{\partial u}{\partial y} = 0;\ u(x, 0) = 4e^{-x}$ (*A.M.I.E.T.E., Summer 2000*) **Ans.** $u = 4e^{-x+3/2y}$

6. $\frac{\partial u}{\partial x} + \frac{\partial u}{\partial y} = 2(x+y)u$ (*AMIETE, June 2010*) **Ans.** $u = ce^{x^2+y^2+k(x-y)}$

7. $\frac{\partial u}{\partial t} = 4\frac{\partial^2 u}{\partial x^2}$ if $u(x, 0) = 4x - \frac{1}{2}x^2$ **Ans.** $u = \left(4x - \frac{x^2}{2}\right)e^{-p^2t}$

8. $\frac{\partial u}{\partial t} = \frac{\partial^2 u}{\partial x^2}$ if $u(x, 0) = \sin \pi x$ **Ans.** $u = \sin \pi x . e^{-pl}$

9. $\frac{\partial^2 u}{\partial x^2} = \frac{\partial u}{\partial t}$ if $u(x,0) = x^2(25 - x^2)$ **Ans.** $u = x^2(25 - x^2)e^{-p^2t}$

10. $\dfrac{\partial^2 z}{\partial x^2} - 2\dfrac{\partial z}{\partial x} - \dfrac{\partial z}{\partial y} = 0$ **Ans.** $z = c_1 e^{[1+\sqrt{(1+p)}]x+p^2 y} + c_2 e^{[1-\sqrt{1+P}]x+p^2 y}$

11. $\dfrac{\partial^2 u}{\partial t^2} = \dfrac{\partial^2 u}{\partial x^2}$ If $u(x,0) = \dfrac{1}{2}x(1-x)$ **Ans.** $u = \dfrac{x}{2}(1-x)\cos pt + c_2 \sin pt(c_3 \cos px + c_4 \sin px)$

12. $16\dfrac{\partial^2 u}{\partial x^2} = \dfrac{\partial^2 u}{\partial t^2}$ if $u(x,0) = x^2(5-x)$ **Ans.** $u = x^2(5-x)\cos pt + c_4 \sin pt\left(c_1 \cos\dfrac{px}{4} + c_2 \sin\dfrac{px}{4}\right)$

13. $\dfrac{\partial^2 u}{\partial x^2} = \dfrac{\partial u}{\partial y} + 2u$ if $u = 0$ and $\dfrac{\partial u}{\partial x} = 1 + e^{-3y}$ when $x = 0$. **Ans.** $u = \dfrac{1}{2\sqrt{2}}(e^{\sqrt{2}x} - e^{-\sqrt{2}x}) + \dfrac{1}{2i}(e^{ix} - e^{-ix})e^{-3y}$

14. $\dfrac{\partial u}{\partial x} = 2\dfrac{\partial u}{\partial y} + u,\ u(x,0) = 4e^{-3x}$ (*A.M.I.E.T.E., Summer 2001*) **Ans.** $u = 4e^{-(3x+2y)}$

15. $\dfrac{\partial u}{\partial x} - 2\dfrac{\partial u}{\partial y} = u,$ given that $u(x,0) = 3e^{-5x} + 2e^{-3x}$ (*A.M.I.E.T.E., Summer 2001*)

Ans. $u = 3e^{-5x-3y} + 2e^{-3x-2y}$

49.3 EQUATION OF VIBRATING STRING

(*AMIETE, June 2010, U.P. II Semester, June 2007*)

Let us consider small transverse vibrations of an elastic string of length l, which is stretched and then fixed at its two ends. Now we will study the transverse vibration of the string when no external forces act on it. Take an end of the string as the origin and the string in the equilibrium position as the x-axis and the line through the origin and perpendicular to the x-axis as the y-axis. We make the following assumptions:

1. The motion takes place entirely in one plane. This plane is chosen as the *xy plane.*
2. In this plane, each particle of the string moves in a direction perpendicular to the equilibrium position of the string.
3. The tension T caused by stretching the string before fixing it at the end points is constant at all times at all points of the deflected string.
4. The tension T is very large compared with the weight of the string and hence the gravitational force may be neglected.
5. The effect of friction is negligible.
6. The string is perfectly flexible. It can transmit only tension but not bending or shearing forces.
7. The slope of the deflection curve is small at all points and at all times.

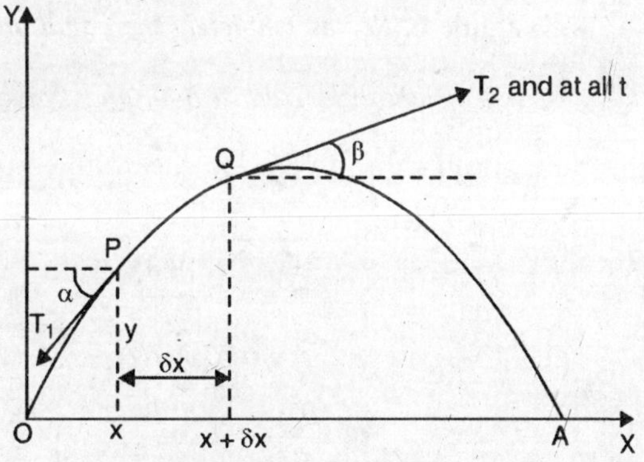

When the string is in motion in the xy-plane, the displacement y of any point of the string is a function of x and time t. Let $P(x, y)$ and $Q(x+\delta x, y+\delta y)$ be two neighbouring points on the string. Let α and β be the inclinations made by the tangents at P and Q respectively with the x-axis. The tension T_1 at P and tension T_2 at Q are tangential. Let m be the mass per unit length of the string which is homogeneous. Consider the motion of the infinitcsimal element PQ of the string. The vertical component of the force to which this element is subjected to is

$$m\delta s\,\frac{\partial^2 y}{\partial t^2} = T_2 \sin\beta - T_1 \sin\alpha \qquad ...(1)$$

Since, there is no motion in the horizontal direction,

$$T_1 \cos\alpha = T_2 \cos\beta = T \text{ (constant)} \qquad \ldots(2)$$

Dividing (1) by T, we get

$$\frac{m\delta s \dfrac{\partial^2 y}{\partial t^2}}{T} = \frac{T_2 \sin\beta}{T} - \frac{T_1 \sin\alpha}{T}$$

$$\frac{m\delta s}{T}\frac{\partial^2 y}{\partial t^2} = \frac{T_2 \sin\beta}{T_2 \cos\beta} - \frac{T_1 \sin\alpha}{T_1 \cos\alpha} \qquad \begin{bmatrix} T = T_1 \cos\alpha \\ T = T_2 \cos\beta \end{bmatrix}$$

$$\Rightarrow \quad \frac{m\delta s}{T}\frac{\partial^2 y}{\partial t^2} = \tan\beta - \tan\alpha \quad \Rightarrow \quad \frac{\partial^2 y}{\partial t^2} = \frac{T}{m\delta s}(\tan\beta - \tan\alpha)$$

Slope $\tan\alpha = \left(\dfrac{\partial y}{\partial x}\right)_x$ at P and slope $\tan\beta = \left(\dfrac{\partial y}{\partial x}\right)_{x+\delta x}$ at Q.

$$\frac{\partial^2 y}{\partial t^2} = \frac{T}{m\delta x}\left[\left(\frac{\partial y}{\partial x}\right)_{x+\delta x} - \left(\frac{\partial y}{\partial x}\right)_x\right] \quad \Rightarrow \quad \frac{\partial^2 y}{\partial t^2} = \frac{T}{m}\left[\frac{\left(\dfrac{\partial y}{\partial x}\right)_{x+\delta x} - \left(\dfrac{\partial y}{\partial x}\right)_x}{\delta x}\right]$$

On taking limit as $\delta x \to 0$, we have

$$\frac{\partial^2 y}{\partial t^2} = \frac{T}{m}\frac{\partial^2 y}{\partial x^2}, \qquad \boxed{\frac{\partial^2 y}{\partial t^2} = a^2 \frac{\partial^2 y}{\partial x^2}} \qquad \left(\text{Take } a^2 = \frac{T}{m}\right)$$

Note. The partial differential equation is knwon as one-dimensional wave equation, because the motion is only in vertical direction (Transverse vibration) not in horizontal direction.

Boundary conditions.

At O, $x = 0$ and $y = 0$, $\dfrac{\partial y}{\partial t} = 0$ as $t = 0$ *At A*, $x = l$ and $y = 0$,

Example 6. *Solve completely the equation* $\dfrac{\partial^2 y}{\partial t^2} = c^2 \dfrac{\partial^2 y}{\partial x^2}$, *representing the vibrations of a string of length l, fixed at both ends, given that* $y(0, t) = 0$, $y(l, t) = 0$; $y(x, 0) = f(x)$ *and* $\dfrac{\partial}{\partial t} y(x,0) = 0,\ 0 < x < l.$ *(U.P. II Semester summer 2005)*

Solution. Here,

$$\frac{\partial^2 y}{\partial t^2} = c^2 \frac{\partial^2 y}{\partial x^2} \qquad \ldots(1)$$

Let
$$y = X(x)\, T(t) \qquad \ldots(2)$$

$$\frac{\partial^2 y}{\partial t^2} = X \frac{d^2 T}{dt^2}$$

$$\frac{\partial^2 y}{\partial x^2} = T \frac{d^2 X}{dx^2}$$

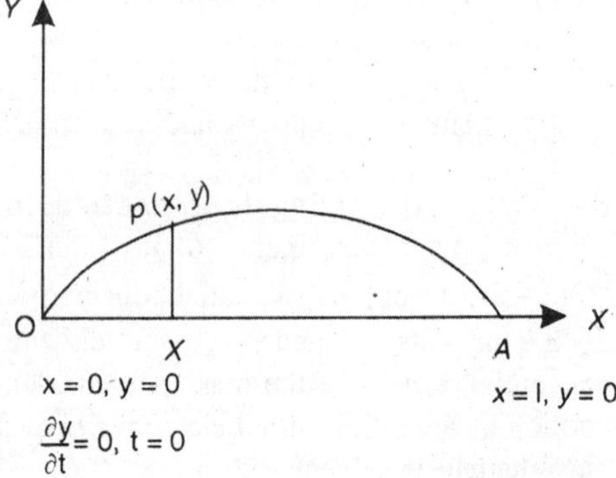

Equation (1) becomes $X \dfrac{d^2 T}{dt^2} = c^2 T \dfrac{d^2 X}{dx^2}$

Separating the variables,

$$\frac{1}{X}\frac{d^2 X}{dx^2} = \frac{1}{c^2 T}\frac{d^2 T}{dt^2} = -p^2 \text{ (let)}$$

If $\frac{1}{X}\frac{d^2X}{dx^2}=-p^2$

$\Rightarrow \quad \frac{d^2X}{dx^2}=-Xp^2$

$\Rightarrow \quad \frac{d^2X}{dx^2}+Xp^2=0$

$\Rightarrow \quad (D^2+p^2)X=0$

A.E. is $m^2+p^2=0 \quad \Rightarrow \quad m=\pm pi$

$X=(C_1\cos px+C_2\sin px)$

If $\frac{1}{c^2T}\frac{\partial^2T}{\partial t^2}=-p^2$

$\Rightarrow \quad (D^2+p^2c^2)T=0$

A.E. is $m^2+p^2c^2=0$

$\Rightarrow \quad m^2=-p^2c^2 \quad \Rightarrow \quad m=\pm pci$

$T=(C_3\cos pct+C_4\sin pct)$

Putting the values of X and T in (2), we get

$$y=(C_1\cos px+C_2\sin px)(C_3\cos pct+C_4\sin pct) \quad ..(3)$$

Now applying the boundary condition

$x=0$ and $y=0$

Putting these values in (3), we get

$0=C_1(C_3\cos pct+C_4\sin pct) \quad \Rightarrow \quad C_1=0$

Equation (3) becomes, $\quad y=C_2\sin px\,(C_3\cos pct+C_4\sin pct) \quad ..(4)$

Putting $x=l$ and $y=0$ in (4), we get

$$0=C_2\sin pl\,(C_3\cos pct+C_4\sin pct) \Rightarrow \sin pl=0=\sin n\pi$$

$$\Rightarrow \quad pl=n\pi \quad \Rightarrow \quad p=\frac{n\pi}{l}$$

On putting $p=\frac{n\pi}{l}$, (4) becomes

$$y=C_2\sin\frac{n\pi x}{l}\left(C_3\cos\frac{n\pi ct}{l}+C_4\sin\frac{n\pi ct}{l}\right) \quad ...(5)$$

On differentiating (5) w.r.t. t, we get

$$\frac{\partial y}{\partial t}=C_2\sin\frac{n\pi x}{l}\left(-\frac{n\pi c}{l}C_3\sin\frac{n\pi ct}{l}+\frac{n\pi c}{l}C_4\cos\frac{n\pi c}{l}t\right) \quad ...(6)$$

On putting $\frac{\partial y}{\partial t}=0$ and $t=0$ in (6), we get

$$0=C_2\sin\frac{n\pi x}{l}.\frac{n\pi c}{l}.C_4 \quad \Rightarrow \quad C_4=0$$

On putting $C_4=0$, (5) becomes

$$y=C_2C_3\sin\frac{n\pi x}{l}\cos\frac{n\pi ct}{l} \quad [\text{let } C_2C_3=b_n] \quad ...(7)$$

Now applying $y=f(x)$ and $t=0$, (7) becomes $f(x)=b_n\sin\frac{n\pi x}{l}$

C_2C_3 can be calculated using Fourier sine series as $b_n=\frac{2}{l}\int_0^l f(x)\sin\frac{n\pi x}{l}dx \quad ... (8)$

Thus, required solution for the given equation is ...(8)

$$y=b_n\sin\frac{n\pi x}{l}\cos\frac{n\pi ct}{l}$$

where b_n is given by equation (8) **Ans.**

Example. 7. *A tightly stretched string with fixed end points $x=0$ and $x=\pi$ is initially at rest in its equilibrium position. If it is set vibrating by giving to each of its points an initial velocity.*

$$\left(\frac{\partial y}{\partial t}\right)_{t=0} = 0.03 \sin x - 0.04 \sin 3x$$

then find the displacement y (x,t) at any point of string at any time t.

Solution. Here we have equation for vibration of a string $\frac{\partial^2 y}{\partial t^2} = c^2 \frac{\partial^2 y}{\partial x^2}$

Its solution is

$$y(x,t) = (c_1 \cos pct + c_2 \sin pct)(c_3 \cos px + c_4 \sin px) \qquad ...(1)$$

[See Example 6 on previous page 1324]

Putting $x = 0$ and $y = 0$ in (1), we get

$$0 = (c_1 \cos pct + c_2 \sin pct)(c_3)$$

$$\Rightarrow \qquad c_3 = 0$$

Putting $c_3 = 0$ in (1), we have

$$y = (c_1 \cos pct + c_2 \sin pct)(c_4 \sin px) \qquad ...(2)$$

Putting $x = \pi$ and $y = 0$ in (2), we get

$$0 = (c_1 \cos pct + c_2 \sin pct)\, c_4 \sin p\pi$$

$$\Rightarrow \qquad 0 = \sin p\pi$$

and $\sin n\pi = \sin p\pi \quad \Rightarrow \quad p = n$

Putting $p = n$ in (2), we get

$$y = (c_1 \cos nct + c_2 \sin nct)(c_4 \sin nx) \qquad ...(3)$$

Putting $t = 0$, $y = 0$ in (3), we get

$$0 = (c_1 + 0)\, c_4 \sin nx \quad (\sin nx \neq 0 \text{ since } \sin nx \text{ is a part of } X)$$

$$\Rightarrow \qquad c_1 = 0$$

On putting $c_1 = 0$ in (3), we get

$$y = (c_2 \sin nct)(c_4 \sin nx)$$

General equation is $\qquad (c_2 c_4 = b_n)$

$$y = \sum_{n=1}^{\infty} b_n \sin nct \sin nx \qquad ...(4)$$

Differentiating (4) w.r.t. 't', we get

$$\frac{\partial y}{\partial t} = \sum_{n}^{\infty} b_n nc \cos nct \sin nx \qquad ...(5)$$

On putting $\frac{\partial y}{\partial t} = 0.03 \sin x - 0.04 \sin 3x$ and $t = 0$ in (5) we get

$$0.03 \sin x - 0.04 \sin 3x = \sum_{n=1}^{\infty} b_n nc \sin nx$$

$$\Rightarrow \quad 0.03 \sin x - 0.04 \sin 3x = cb_1 \sin x + 2cb_2 \sin 2x + 3c_3 b_3 \sin 3x + ...$$

Comparing the coefficient, we get

$$0.03 = cb_1 \Rightarrow b_1 = \frac{0.03}{c}$$

$$0 = 2cb_2 \Rightarrow b_2 = 0$$

$$0.04 = 3c\, b_3 \Rightarrow b_3 = \frac{-0.04}{3c} = \frac{-0.0133}{c}$$

Putting the values of b_1, b_2, b_3 ... in (4), we get

$$y = \frac{0.03}{c} \sin ct \sin x - \frac{0.0133}{c} \sin 3ct \sin 3x$$

Hence $\qquad y = \frac{1}{c}[0.03 \sin ct \sin x - 0.0133 \sin 3x]$ **Ans.**

Example 8. *A string is stretched and fastened to two points l apart. Motion is started by displacing the string in the form* $y = a\sin\frac{\pi x}{l}$ *from which it is released at a time t = 0. Show that the displacement of any point at a distance x from one end at time t is given by*

$$y(x, t) = a\sin\left(\frac{\pi x}{l}\right)\cos\left(\frac{\pi ct}{l}\right)$$

(*A.M.I.E.T.E., Winter 2003, U.P., II Semester, 2004, 2009*)

Solution. Consider an elastic string tightly stretched between two points O and A. Let O be the origin and OA as x-axis. On giving a small displacement to the string, perpendicular to its length (parallel to the y-axis). Let y be the displacement at the point P (x, y) at any time. The wave equation.

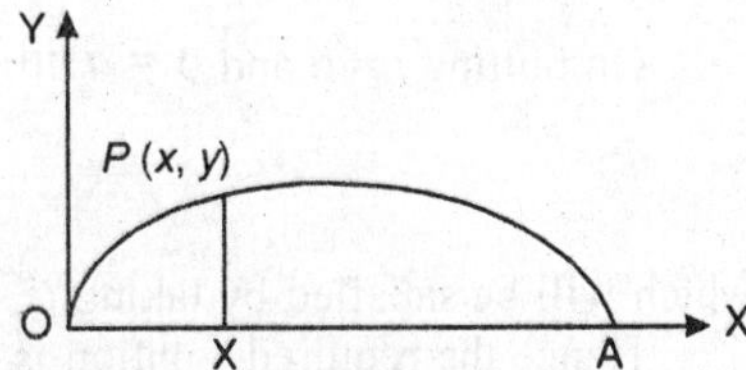

$$\frac{\partial^2 y}{\partial t^2} = c^2\frac{\partial^2 y}{\partial x^2}$$

where c is a constant. The vibration of the string is given by:

$$\frac{\partial^2 y}{\partial t^2} = c^2\frac{\partial^2 y}{\partial x^2} \qquad \text{... (1)}$$

As the end points of the string are fixed, for all time,

$$y(0, t) = 0 \qquad \text{... (2)}$$

and

$$y(l, t) = 0 \qquad \text{... (3)}$$

Since the initial transverse velocity of any point of the string is zero, therefore,

$$\left(\frac{\partial y}{\partial t}\right)_{t=0} = 0 \qquad \text{... (4)}$$

Also

$$y(x,0) = a\sin\frac{\pi x}{l} \qquad \text{... (5)}$$

Now we have to solve (1), subject to the above boundary conditions. Since the vibration of the string is periodic, therefore, the solution of (1) is of the form

$$y(x, t) = (C_1\cos px + C_2\sin px)(C_3\cos cpt + C_4\sin cpt) \qquad \text{... (6)}$$

On putting $x = 0$ and $y = 0$ in (6), we get

$$0 = C_1(C_3\cos c\,pt + C_4\sin c\,pt) \quad \Rightarrow C_1 = 0$$

On putting $C_1 = 0$ in (6), we get

$$y(x, t) = C_2\sin px\,(C_3\cos c\,pt + C_4\sin c\,pt) \qquad \text{... (7)}$$

On differentiating (7) w.r.t. t, we get

$$\frac{\partial y}{\partial t} = C_2\sin px[C_3(-cp\sin c\,pt) + C_4(cp\cos c\,pt)] \qquad \text{...(8)}$$

On putting $\frac{dy}{dt} = 0$ and $t = 0$ in (8), we get

$$0 = C_2\sin px(C_4cp) \quad \Rightarrow \quad C_2C_4cp = 0$$

If $C_2 = 0$, (7) will lead to the trivial solution $y(x,t) = 0$.

Thus, the only possibility is that $C_4 = 0$

On putting $C_4 = 0$ in (7), we get

$$y(x, t) = C_2C_3\sin px\cos c\,pt \qquad \text{...(9)}$$

On putting $x = l$ and $y = 0$ in (9), we get

$$0 = C_2\, C_3 \sin pl \cos cpt, \text{ for all } t.$$

Since C_2 and $C_3 \neq 0$, we have $\sin pl = 0 \therefore pl = n\pi \Rightarrow p = \frac{n\pi}{l}$, where n is an integer.

On putting the value of p in (9), we get

$$y(x, t) = C_2 C_3 \sin \frac{n\pi x}{l} \cos \frac{n\pi ct}{l} \quad \dots (10)$$

On putting $t = 0$ and $y = a \sin \frac{\pi x}{l}$ in (10), we get

$$a \sin \frac{\pi x}{l} = C_2 C_3 \sin \frac{n\pi x}{l}$$

which will be satisfied by taking $C_2\, C_3 = a$ and $n = 1$

Hence the required solution is

$$y(x, t) = a \sin \frac{\pi x}{l} \cos \frac{\pi ct}{l}$$ **Proved.**

Example 9. *The vibrations of an elastic string is governed by the partial differential equation*

$$\frac{\partial^2 u}{\partial t^2} = \frac{\partial^2 u}{\partial x^2}$$

The length of the string is π and the ends are fixed. The initial velocity is zero and the initial deflection is $u(x, 0) = 2\ (\sin x + \sin 3x)$. Find the deflection $u(x, t)$ of the vibrating string for $t > 0$.

Solution. $$\frac{\partial^2 u}{\partial t^2} = \frac{\partial^2 u}{\partial x^2}$$

Solution of the given differential equation is

$$\Rightarrow \quad u = (c_1 \cos pt + c_2 \sin pt)(c_3 \cos px + c_4 \sin px) \quad \dots (1)$$

On putting $x = 0$, $u = 0$ in (1), we get

$$0 = (c_1 \cos pt + c_2 \sin pt) c_3 \quad \Rightarrow c_3 = 0$$

On putting $c_3 = 0$ in (1), it reduces

$$u = (c_1 \cos pt + c_2 \sin pt) c_4 \sin px \quad \dots (2)$$

On putting $x = \pi$ and $u = 0$ in (2), we have

$$0 = (c_1 \cos pt + c_2 \sin pt)\, c_4 \sin p\pi$$

$$\sin\ p\pi = 0 = \sin n\pi \qquad n = 1, 2, 3, 4$$

$$\therefore \quad p\pi = n\pi \quad \text{or} \quad p = n$$

On substituting the value of p in (2), we get

$$u = (c_1 \cos nt + c_2 \sin nt)\, c_4 \sin nx \quad \dots (3)$$

On differentiating (3) w.r.t. "t", we get

$$\frac{\partial u}{\partial t} = (-\,c_1\, n \sin nt + c_2\, n \cos nt)\; c_4 \sin nx \quad \dots (4)$$

On putting $\frac{\partial u}{\partial t} = 0$, $t = 0$ in (4), we have

$$0 = (c_2 n)\ (c_4 \sin nx) \Rightarrow c_2 = 0$$

On putting $c_2 = 0$, (3) becomes

$$u = (c_1 \cos nt)(c_4 \sin nx)$$

$$u = c_1c_4 \cos nt \sin nx = \Sigma b_n \cos nt \sin nx \quad [\because b_n = c_1c_4] \quad ...(5)$$

On putting $u = 2(\sin x + \sin 3x)$ and $t = 0$ in (5), we have

$$2(\sin x + \sin 3x) = b_1 \sin x + b_2 \sin 2x + b_3 \sin 3x + ...$$

Comparing the coefficient of sin on both side, we have

$$b_1 = 2, \; b_2 = 0 \text{ and } b_3 = 2$$

On Substituting the values of b_1, b_2, b_3 in (5), we get

$$u = 2\,[\cos t \sin x + \cos 3t \sin 3x]$$

Ans.

Example 10. *A tightly stretched string with fixed end points $x = 0$ and $x = l$ is initially in a position given by $y = y_0 \sin^3\left(\frac{\pi x}{l}\right)$. If it is released from rest from this position, find the displacement $y(x, t)$.*

Solution. Let the equation to the vibrating string be

$$\frac{\partial^2 y}{\partial t^2} = c^2 \frac{\partial^2 y}{\partial x^2} \quad ... (1)$$

Here the initial conditions are

$$y(0, t) = 0, \; y(l, t) = 0$$

$$\frac{\partial y}{\partial t} = 0 \text{ at } t = 0, \; y(x, 0) = y_0 \sin^3 \frac{\pi x}{l} = \frac{y_0}{4}\left[3\sin\frac{\pi x}{l} - \sin\frac{3\pi x}{l}\right]$$

The solution of (1) is of the form

$$y = (c_1 \cos px + c_2 \sin px)(c_3 \cos pct + c_4 \sin pct) \quad ... (2)$$

On putting $x = 0$ and $y = 0$ in (2), we get

$$0 = c_1\,(c_3 \cos pct + c_4 \sin pct) \qquad \Rightarrow c_1 = 0$$

On putting $c_1 = 0$ in (2), we get

$$y = c_2 \sin px(c_3 \cos pct + c_4 \sin pct) \quad ... (3)$$

On putting $x = l$ and $y = 0$ in (3), we get

$$0 = c_2 \sin pl(c_3 \cos pct + c_4 \sin pct)$$

$$\therefore \; \sin pl = 0 = \sin n\pi \text{ or } pl = n\pi, \text{ or } p = \frac{n\pi}{l} \text{ where } n = 0, 1, 2, 3.....$$

On putting the value of p in (3) we get

$$y = c_2 \sin\frac{n\pi x}{l}\left(c_3 \cos\frac{n\pi ct}{l} + c_4 \sin\frac{n\pi ct}{l}\right) \quad ... (4)$$

On differentiating (4), w.r.t. t, we get

$$\frac{\partial y}{\partial t} = c_2 \sin\frac{n\pi x}{l}\left(-\frac{n\pi c}{l} c_3 \sin\frac{n\pi c}{l} t + c_4 \frac{n\pi c}{l} \cos\frac{n\pi ct}{l}\right)$$

On putting $\frac{\partial y}{\partial t} = 0$ and $t = 0$ in (4), we have

$$0 = c_2 \sin\frac{n\pi x}{l} c_4 \frac{n\pi c}{l} \qquad \Rightarrow c_4 = 0$$

On putting $c_4 = 0$ in (4), we get

$$y = c_2c_3 \sin\frac{n\pi x}{l} \cos\frac{n\pi ct}{l}$$

$$y = b_n \sin\frac{n\pi x}{l} \cos\frac{n\pi c}{l} t \qquad (b_n = c_2c_3)$$

General solution is

$$y = \sum_{n=1}^{\infty} b_n \sin\frac{n\pi x}{l} \cos\frac{n\pi c}{l} t \quad ... (5)$$

On putting $t = 0$ and $y = \frac{y_0}{4}\left[3\sin\frac{\pi x}{l} - \sin\frac{3\pi x}{l}\right]$ in (5), we have

$$\frac{y_0}{4}\left(3\sin\frac{\pi x}{l} - \sin\frac{3\pi x}{l}\right) = \sum_{n=1}^{\infty} b_n \sin\frac{n\pi x}{l}$$

$$\Rightarrow \quad \frac{y_0}{4}\left(3\sin\frac{\pi x}{l} - \sin\frac{3\pi x}{l}\right) = b_1 \sin\frac{\pi x}{l} + b_2 \sin\frac{2\pi x}{l} + b_3 \sin\frac{3\pi x}{l} + \ldots$$

On equating the coefficients of sin on both the sides, we get

$$b_1 = \frac{3y_0}{4}, \quad b_3 = -\frac{y_0}{4}$$

and all others b's are zero.
Hence (5) becomes

$$y = \frac{y_0}{4}\left(3\sin\frac{\pi x}{l}\cos\frac{c\pi t}{l} - \sin\frac{3\pi x}{l}\cos\frac{3c\pi t}{l}\right) \quad \textbf{Ans.}$$

Example 11. *A string is stretched and fastened to two points l apart. Motion is started by displacing the string into the form $y = k(lx - x^2)$ from which it is released at time $t = 0$. Find the displacement of any point on the string at a distance of x from one end at time t.*

(U.P., III Semester, Summer 2002; A.M.I.E.T.E., Summer 2000)

Solution. The vibration of the string is given by

$$\frac{\partial^2 y}{\partial t^2} = c^2 \frac{\partial^2 y}{\partial x^2} \quad \ldots (1)$$

As the end points of the string are fixed for all time,

$$y(0, t) = 0 \quad \ldots (2)$$

and

$$y(l, t) = 0 \quad \ldots (3)$$

since the initial transverse velocity of any point of the string is zero, therefore,

$$\left(\frac{\partial y}{\partial t}\right)_{t=0} = 0 \quad \ldots (4)$$

and

$$y(x, 0) = k(lx - x^2) \quad \ldots (5)$$

Solution of (1) is $\quad y = (c_1 \cos px + c_2 \sin px)(c_3 \cos c\,pt + c_4 \sin cpt) \quad \ldots (6)$

On putting $x = 0$ and $y = 0$ in (6), we get

$$0 = c_1 (c_3 \cos c\,pt + c_4 \sin cpt) \quad \Rightarrow \quad c_1 = 0$$

On putting $c_1 = 0$ in (6), we get

$$y = c_2 \sin px(c_3 \cos cpt + c_4 \sin cpt) \quad \ldots (7)$$

On differentiating (7) w.r.t. t, we get

$$\frac{\partial y}{\partial t} = c_2 \sin px(-c_3 cp \sin cpt + c_4 cp \cos cpt) \quad \ldots (8)$$

On putting $\left(\frac{\partial y}{\partial t}\right) = 0$ and $t = 0$ in (8), we get

$$0 = c_2 \sin px(c_4 cp) \quad \Rightarrow \quad c_4 = 0 \text{ since } c_2 \neq 0$$

On putting $c_4 = 0$ in (7), we get

$$y = c_2 \sin px(c_3 \cos cpt)$$

$$y = c_2 c_3 \sin px \cos cpt \quad \ldots (9)$$

On putting $x = l$ and $y = 0$ in equation (9), we get

$$0 = c_2 c_3 \sin pl \cos cpt \quad \text{or} \quad 0 = \sin pl$$

$$\Rightarrow \qquad \sin n\pi = \sin pl \quad \text{or} \quad pl = n\pi, \quad p = \frac{n\pi}{l} \qquad \text{where } n = 1, 2, 3\ldots\ldots$$

On putting $p = \frac{n\pi}{l}$, equation (9) becomes

$$\Rightarrow \qquad y = b_n \sin\frac{n\pi x}{l}\cos\frac{n\pi c}{l}t \qquad [c_2 c_3 = b_n]$$

We can have any number of solutions by taking different integral values of n and the complete solution will be the sum of these solutions. Thus,

$$y(x, t) = \sum_{n=1}^{\infty} b_n \sin\frac{n\pi x}{l}\cos\frac{n\pi c}{l}t \qquad \ldots (10)$$

On putting $t = 0$ and $y = k(lx - x^2)$ in (10), we get

$$k(l\,x - x^2) = \sum_{n=1}^{\infty} b_n \sin\frac{n\pi x}{l} \qquad \ldots (11)$$

Now it is clear that (11) represents the expansion of $f(x)$ in the form of a Fourier sine series and consequently

$$b_n = \frac{2}{l}\int_0^l f(x)\sin\frac{n\pi x}{l}\,dx \qquad \ldots (12)$$

$$b_n = \frac{2}{l}\int_0^l k\,(lx - x^2)\sin\frac{n\pi x}{l}\,dx$$

$$= \frac{2k}{l}\left[(lx - x^2)\left(-\cos\frac{n\pi x}{l}\right)\frac{l}{n\pi} - (l - 2x)\left(-\sin\frac{n\pi x}{l}\right)\frac{l^2}{n^2\pi^2} + (-2)\left(\cos\frac{n\pi x}{l}\right)\frac{l^3}{n^3\pi^3}\right]_0^l$$

$$= \frac{2k}{l}\left[(-1)^{n+1}\frac{2l^3}{n^3\pi^3} + \frac{2l^3}{n^3\pi^3}\right] = \frac{8l^2 k}{n^3\pi^3} \quad \text{when } n \text{ is odd.}$$

$$= 0, \quad \text{when } n \text{ is even}$$

Putting the value of b_n in (10), we get

$$y = \sum_{n=1}^{\infty} \frac{8l^2 k}{n^3\pi^3}\sin\frac{n\pi x}{l}\cos\frac{n\pi c}{l}t, \quad \text{when } n \text{ is odd.}$$ **Ans.**

Example 12. *A string is stretched between the fixed points (0, 0) and (l, 0) and released at rest from the initial deflection given by*

$$f(x) = \begin{cases} \dfrac{2kx}{l} & \text{when } 0 < x < \dfrac{l}{2} \\ \dfrac{2k}{l}(l - x) & \text{when } \dfrac{l}{2} < x < l \end{cases}$$

Find the deflection of the string at any time t.

Solution. As we have done in the example 11 (see equation 10)

$$y(x, t) = \sum_{n=1}^{\infty} b_n \cos\frac{n\pi ct}{l}\sin\frac{n\pi x}{l} \qquad \ldots (1)$$

$$b_n = \frac{2}{l}\int_0^l f(x)\sin\frac{n\pi x}{l}\,dx$$

$$b_n = \frac{2}{l}\left[\int_0^{l/2} \frac{2kx}{l}\sin\frac{n\pi x}{l}\,dx + \int_{\frac{l}{2}}^{l} \frac{2k}{l}(l-x)\sin\frac{n\pi x}{l}\,dx\right]$$

$$= \frac{4k}{l^2}\left[x\left(-\frac{l}{n\pi}\cos\frac{n\pi x}{l}\right) - 1\left(-\frac{l^2}{n\pi^2}\sin\frac{n\pi x}{l}\right)\right]_0^{\frac{l}{2}}$$

$$+ \frac{4k}{l^2}\left[(l-x)\left(-\frac{l}{n\pi}\cos\frac{n\pi x}{l}\right) - (-1)\left(\frac{-l^2}{n^2\pi^2}\sin\frac{n\pi x}{l}\right)\right]_{\frac{l}{2}}^{l}$$

$$= \frac{4k}{l^2}\left[-\frac{l}{2}\frac{l}{n\pi}\cos\frac{n\pi}{2} + \frac{l^2}{n^2\pi^2}\sin\frac{n\pi}{2}\right] + \frac{4k}{l^2}\left[\frac{l}{2}\frac{l}{n\pi}\cos\frac{n\pi}{2} + \frac{l^2}{n^2\pi^2}\sin\frac{n\pi}{2}\right]$$

$$= \frac{8k}{l^2}\frac{l^2}{n^2\pi^2}\sin\frac{n\pi}{2} = \frac{8k}{n^2\pi^2}\sin\frac{n\pi}{2}$$

On putting the value of b_n in (1) we get

$$y = \sum_{n=1}^{\infty} \frac{8k}{n^2\pi^2}\sin\frac{n\pi}{2}\cos\frac{n\pi ct}{l}\sin\frac{n\pi x}{l}.$$ **Ans.**

Example 13. *A taut string of length 2l is fastened at both ends. The mid point of the string is taken to a height b and then released from the rest in that position. Find the displacement of the string.*

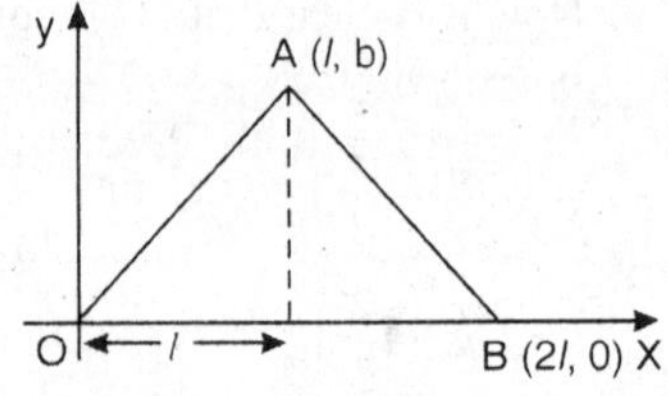

Solution. Taking an end as origin, the boundary conditions are

$$y(0, t) = 0, \quad t \geq 0 \qquad \text{... (1)}$$

$$y(2l, t) = 0, \; t \geq 0 \qquad \text{... (2)}$$

$$\left(\frac{\partial y}{\partial t}\right)_{t=0} = 0,$$

OA passes through (0, 0) and (*l*, *b*)

$$\frac{o}{l}x, \; 0 \leq x \leq l, \quad \Rightarrow \quad y = \frac{bx}{l}$$

Equation of *OA* is $y\,(x, 0) = -\frac{b}{l}(x - 2l),\; l \leq x \leq 2l$ ...(3) $\left[y - y_1 = \frac{y_2 - y_1}{x_2 - x_1}(x - x_1)\right]$

$$\left(\text{Equation of } AB,\; y - 0 = \frac{b-0}{l-2l}(x-2l)\right)$$

Equation of the vibrating string is $\dfrac{\partial^2 y}{\partial t^2} = c^2\,\dfrac{\partial^2 y}{\partial x^2}$ ... (4)

Starting with the solution of (4), we get

$$y(x, t) = (c_1\cos px + c_2\sin px)(c_3\cos pct + c_4\sin pct) \qquad \text{... (5)}$$

On putting $x = 0$, $y = 0$ in (5), we get

$$0 = c_1\,(c_3\cos pct + c_4\sin pct) \Rightarrow c_1 = 0$$

On putting $c_1 = 0$ in (5), we get

$$y(x, t) = c_2\sin px\,(c_3\cos pct + c_4\sin pct) \qquad \text{... (6)}$$

On putting $y = 0, x = 2l$ in (6), we get

$$0 = c_2 \sin p(2l)\,(c_3 \cos pct + c_4 \sin pct) \Rightarrow \sin 2pl = 0 = \sin n\pi \Rightarrow 2pl = n\pi,\ p = \frac{n\pi}{2l}$$

Substituting the value of p in (6), we have

$$y(x, t) = c_2 \sin \frac{n\pi}{2l} x \left(c_3 \cos \frac{n\pi ct}{2l} + c_4 \sin \frac{n\pi ct}{2l} \right) \qquad \ldots (7)$$

On differentiating (7) w.r.t., t, we get

$$\frac{\partial y}{\partial t} - c_2 \sin \frac{n\pi x}{2l} \left[-c_3 \frac{n\pi c}{2l} \sin \frac{n\pi ct}{2l} + c_4 \frac{n\pi c}{2l} \cos \frac{n\pi ct}{2l} \right] \qquad \ldots (8)$$

Putting $\left(\frac{\partial y}{\partial t}\right) = 0,\ t = 0$ in (8), we get

$$0 = c_2 \sin \frac{n\pi x}{2l} \left[0 + c_4 \frac{n\pi c}{2l} \right] \Rightarrow c_4 = 0$$

Putting $c_4 = 0$ in (7), we have

$$y(x, t) = \left(c_2 \sin \frac{n\pi x}{2l} \right) \left(c_3 \cos \frac{n\pi ct}{2l} \right)$$

General solution is

$$y(x,t) = \sum_{n=1}^{\infty} b_n \sin \frac{n\pi x}{2l} \cos \frac{n\pi ct}{2l} \qquad (c_2 c_3 = b_n) \qquad \ldots (9)$$

$$\Rightarrow \quad f(x) = \begin{cases} \dfrac{bx}{l}, & 0 \le x \le l \\ \dfrac{-b}{l}(x - 2l), & l \le x \le 2l \end{cases}$$

$$b_n = \frac{2}{2l} \int_0^{2l} f(x) \sin \frac{n\pi x}{2l}\, dx = \frac{1}{l} \int_0^l f(x) \sin \frac{n\pi x}{2l}\, dx + \frac{1}{l} \int_l^{2l} f(x) \sin \frac{n\pi x}{2l}\, dx$$

$$= \frac{1}{l} \int_0^l \frac{bx}{l} \sin \frac{n\pi x}{2l}\, dx + \frac{1}{l} \int_l^{2l} \frac{-b}{l}(x - 2l) \sin \frac{n\pi x}{2l}\, dx$$

$$= \frac{1}{l} \left[\frac{bx}{l} \frac{2l}{n\pi} \left(-\cos \frac{n\pi x}{2l} \right) - \frac{b}{l} \left(-\frac{4l^2}{n^2\pi^2} \sin \frac{n\pi x}{2l} \right) \right]_0^l$$

$$+ \frac{1}{l} \left[\frac{-b}{l}(x - 2l) \left(-\frac{2l}{n\pi} \cos \frac{n\pi x}{2l} \right) - \left(-\frac{b}{l} \right) - \frac{4l^2}{n^2\pi^2} \sin \frac{n\pi x}{2l} \right]_l^{2l}$$

$$= \frac{b}{l^2} \left[\frac{-2l^2}{n\pi} \cos \frac{n\pi}{2} + \frac{4l^2}{n^2\pi^2} \sin \frac{n\pi}{2} + \frac{2l^2}{n\pi} \cos \frac{n\pi}{2} + \frac{4l^2}{n^2\pi^2} \sin \frac{n\pi}{2} \right]$$

$$= \frac{b}{l^2} \left[\frac{8l^2}{n^2\pi^2} \sin \frac{n\pi}{2} \right] = \frac{8b}{n^2\pi^2} \sin \frac{n\pi}{2} = 0 \text{ for } n \text{ even} = \frac{8b}{n^2\pi^2} \sin \frac{n\pi}{2} \text{ for } n \text{ odd.}$$

Substituting the value of b_n in (9), we get

$$y(x, t) = \frac{8b}{\pi^2} \sum_{n=1}^{\infty} \frac{1}{(2n-1)^2} \sin (2n-1) \frac{\pi}{2} \sin \frac{(2n-1)\pi x}{2l} \cos \frac{(2n-1)\pi ct}{2l} \qquad \textbf{Ans.}$$

Example 14. *The points of trisection of a string are pulled aside through the same distance on opposite sides of the position of equilibrium and the string is released from rest. Derive an expression for the displacement of the string at subsequent time and show that the mid-point of the string always remains at rest.*

Solution. Let the string OA (l) be trisected at B and C. Initially the string is held in the form $OB'C'A$, where $BB' = CC' = a$

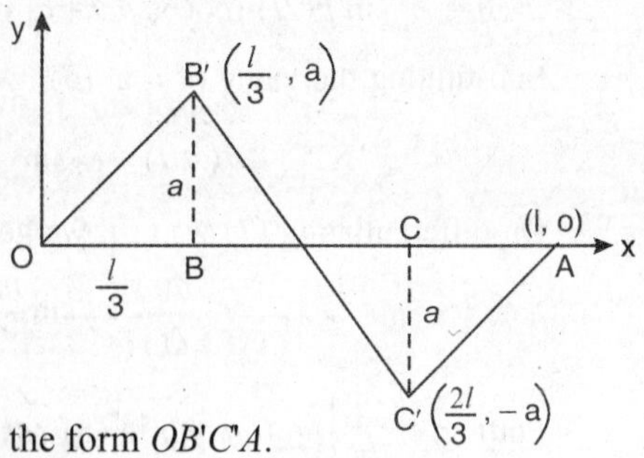

The equation of vibrating string is

$$\frac{\partial^2 y}{\partial t^2} = c^2 \frac{\partial^2 y}{\partial x^2} \quad \text{... (1)}$$

Boundary conditions are

$$y(0, t) = 0,\ y(l, t) = 0,\ \left(\frac{\partial y}{\partial t}\right)_{t=0} = 0$$

Fourth condition is that at $t = 0$, the string rests in the form $OB'C'A$.

Equation of OB' is $y = \dfrac{a}{l/3}x \Rightarrow y = \dfrac{3a}{l}x$ $\qquad (y = mx)$

Equation of $B'C'$ is $y - a = \dfrac{a + a}{\dfrac{l}{3} - \dfrac{2l}{3}}\left(x - \dfrac{l}{3}\right)$ $\qquad \left[y - y_1 = \dfrac{y_2 - y_1}{x_2 - x_1}(x - x_1)\right]$

i.e., $y - a = \dfrac{2a}{\dfrac{-l}{3}}\left(x - \dfrac{l}{3}\right) \Rightarrow y - a = -\dfrac{6a}{l}\left(x - \dfrac{l}{3}\right)$

$$\Rightarrow y = a - \frac{6ax}{l} + 2a \quad \Rightarrow \quad y = 3a - \frac{6ax}{l} \quad \Rightarrow \quad y = \frac{3a}{l}(l - 2x)$$

Equation of $C'A$ is $y - 0 = \dfrac{-a - 0}{\dfrac{2l}{3} - l}(x - l) \quad \Rightarrow \quad y = \dfrac{3a}{l}(x - l)$

Hence fourth condition is

$$f(x) = \begin{cases} \dfrac{3a}{l}x, & 0 \le x \le \dfrac{l}{3} \\ \dfrac{3a}{l}(l - 2x), & \dfrac{l}{3} \le x \le \dfrac{2l}{3} \\ \dfrac{3a}{l}(x - l), & \dfrac{2l}{3} \le x \le l \end{cases}$$

General solution of (1) is

$$y(x, t) = \sum_{n=1}^{\infty} b_n \sin\frac{n\pi x}{l}\cos\frac{n\pi ct}{l} \quad \text{... (2)}$$

On putting $t = 0$ in (2), we get

$$y(x, 0) = \sum_{n=1}^{\infty} b_n \sin\frac{n\pi x}{l}$$

We have to find the value of b_n by Fourier half range formula.

$$b_n = \frac{2}{l}\int_0^l f(x)\sin\frac{n\pi x}{l}\,dx$$

$$= \frac{2}{l}\left[\int_0^{\frac{l}{3}} \frac{3ax}{l}\sin\frac{n\pi x}{l}\,dx + \int_{\frac{l}{3}}^{\frac{2l}{3}} \frac{3a}{l}(l - 2x)\sin\frac{n\pi x}{l}\,dx + \int_{\frac{2l}{3}}^{l} \frac{3a}{l}(x - l)\sin\frac{n\pi x}{l}\,dx\right]$$

$$= \frac{6a}{l^2}\left[x\left(-\frac{l}{n\pi}\cos\frac{n\pi x}{l}\right) - (1)\left(-\frac{l^2}{n^2\pi^2}\sin\frac{n\pi x}{l}\right)\right]_0^{\frac{l}{3}} + \frac{6a}{l^2}\left[(l - 2x)\left(\frac{-l}{n\pi}\cos\frac{n\pi x}{l}\right)\right.$$

$$\left. - (-2)\left(\frac{-l^2}{n^2\pi^2}\sin\frac{n\pi x}{l}\right)\right]_{\frac{l}{3}}^{\frac{2l}{3}} + \frac{6a}{l^2}\left[(x - l)\left(\frac{-l}{n\pi}\cos\frac{n\pi x}{l}\right) - (1)\left(\frac{-l^2}{n^2\pi^2}\sin\frac{n\pi x}{l}\right)\right]_{\frac{2l}{3}}^{l}$$

$$= \frac{6a}{l^2}\left[\left(-\frac{l^2}{3n\pi}\cos\frac{n\pi}{3}+\frac{l^2}{n^2\pi^2}\sin\frac{n\pi}{3}\right)+\frac{l^2}{3n\pi}\cos\frac{2n\pi}{3}-\frac{2l^2}{n^2\pi^2}\sin\frac{2n\pi}{3}\right.$$

$$\left.+\frac{l^2}{3n\pi}\cos\frac{n\pi}{3}+\frac{2l^2}{n^2\pi^2}\sin\frac{n\pi}{3}-\left(\frac{l^2}{3n\pi}\cos\frac{2n\pi}{3}+\frac{l^2}{n^2\pi^2}\sin\frac{2n\pi}{3}\right)\right]$$

$$= \frac{6a}{l^2}.\frac{3l^2}{n^2\pi^2}\left(\sin\frac{n\pi}{3}-\sin\frac{2n\pi}{3}\right)$$

$$= \frac{18a}{n^2\pi^2}\sin\frac{n\pi}{3}[1+(-1)^n] \qquad \left[\sin\frac{2n\pi}{3}=\sin\left(n\pi-\frac{n\pi}{3}\right)=-(-1)^n\sin\frac{n\pi}{3}\right]$$

$= 0$, when n is odd

$= \dfrac{36a}{n^2\pi^2}\sin\dfrac{n\pi}{3}$, when n is even

On substituting the value of b_n in (2), we get

$$y(x,t) = \sum_{n=2}^{\infty}\frac{36a}{n^2\pi^2}\sin\frac{n\pi}{3}\sin\frac{n\pi x}{l}\cos\frac{n\pi ct}{l} \qquad ...(3)\ (n \text{ is even})$$ **Ans.**

Putting $x=\dfrac{l}{2}$ in equation (3), we get

$$y\left(\frac{l}{2},t\right)=\frac{36a}{n^2\pi^2}\sin\frac{n\pi}{3}\sin\frac{n\pi}{2}\cos\frac{n\pi ct}{l}=0 \qquad \left[\sin\frac{n\pi}{2}=0 \text{ as } n \text{ is even}\right]$$

Hence, mid-point of the string is always at rest. **Proved.**

49.4 SOLUTION OF WAVE EQUATION BY D'ALEMBERT'S METHOD

$$\frac{\partial^2 y}{\partial t^2}=c^2\frac{\partial^2 y}{\partial x^2} \qquad ...(1)$$

Let us introduce the two new independent variables $u = x + ct$, $v = x - ct$
So that y becomes a function of u and v. Then,

$$\frac{\partial y}{\partial x}=\frac{\partial y}{\partial u}\frac{\partial u}{\partial x}+\frac{\partial y}{\partial v}\frac{\partial v}{\partial x}=\frac{\partial y}{\partial u}(1)+\frac{\partial y}{\partial v}(1)=\frac{\partial y}{\partial u}+\frac{\partial y}{\partial v}$$

$$\frac{\partial}{\partial x}\equiv\frac{\partial}{\partial u}+\frac{\partial}{\partial v}$$

$$\frac{\partial^2 y}{\partial x^2}=\frac{\partial}{\partial x}\left(\frac{\partial y}{\partial x}\right)=\left(\frac{\partial}{\partial u}+\frac{\partial}{\partial v}\right)\left(\frac{\partial y}{\partial u}+\frac{\partial y}{\partial v}\right)$$

$$=\frac{\partial}{\partial u}\left(\frac{\partial y}{\partial u}+\frac{\partial y}{\partial v}\right)+\frac{\partial}{\partial v}\left(\frac{\partial y}{\partial u}+\frac{\partial y}{\partial v}\right)=\frac{\partial^2 y}{\partial u^2}+2\frac{\partial^2 y}{\partial u\partial v}+\frac{\partial^2 y}{\partial v^2} \qquad ...(2)$$

$$\frac{\partial y}{\partial t}=\frac{\partial y}{\partial u}\frac{\partial u}{\partial t}+\frac{\partial y}{\partial v}\frac{\partial v}{\partial t}=\frac{\partial y}{\partial u}c+\frac{\partial y}{\partial v}(-c)=c\left(\frac{\partial y}{\partial u}-\frac{\partial y}{\partial v}\right) \qquad \left[\because \frac{\partial u}{\partial t}=c,\ \frac{\partial v}{\partial t}=-c\right]$$

$$\frac{\partial}{\partial t}\equiv c\left(\frac{\partial}{\partial u}-\frac{\partial}{\partial v}\right)$$

$$\frac{\partial^2 y}{\partial t^2}=\frac{\partial}{\partial t}\left(\frac{\partial y}{\partial t}\right)=c\left(\frac{\partial}{\partial u}-\frac{\partial}{\partial v}\right)c\left(\frac{\partial y}{\partial u}-\frac{\partial y}{\partial v}\right)=c^2\left(\frac{\partial^2 y}{\partial u^2}-2\frac{\partial^2 y}{\partial u\partial v}+\frac{\partial^2 y}{\partial v^2}\right) \qquad ...(3)$$

Substituting the values of $\dfrac{\partial^2 y}{\partial x^2}$ and $\dfrac{\partial^2 y}{\partial t^2}$ from (2) and (3) in (1), we get

$$c^2\left(\frac{\partial^2 y}{\partial u^2}-2\frac{\partial^2 y}{\partial u\partial y}+\frac{\partial^2 y}{\partial v^2}\right)=c^2\left(\frac{\partial^2 y}{\partial u^2}+2\frac{\partial^2 y}{\partial u\partial v}+\frac{\partial^2 y}{\partial v^2}\right)\Rightarrow\frac{\partial^2 y}{\partial u\partial v}=0 \quad \text{... (4)}$$

Integrating (4) *w.r.t.* v, we get $\frac{\partial y}{\partial u}=f(u)$ ... (5)

where $f(u)$ is constant in respect of v. Again integrating (5) *w.r.t.* 'u' we get

$$y=\int f(u)\,du+\psi(v)$$

where $\psi(v)$ is constant in respect of u

$$y=\phi(u)+\psi(v) \qquad \text{where } \phi(u)=\int f(u)\,du$$

$$\Rightarrow \quad y(x,t)=\phi(x+ct)+\psi(x-ct) \quad \text{... (6)}$$

This is D'Alemberts solution of wave equation (1)

To determine ϕ, ψ, let us apply initial conditions, $y(x,0)=f(x)$ and $\frac{\partial y}{\partial t}=0$ when $t=0$.

Differentiating (6) *w.r.t.* "t", we get

$$\frac{\partial y}{\partial t}=c\phi'(x+ct)-c\psi'(x-ct) \quad \text{... (7)}$$

Putting $\frac{\partial y}{\partial t}=0$, and $t=0$ in (7) we get $0=c\phi'(x)-c\psi'(x)$

$$\Rightarrow \quad \phi'(x)=\psi'(x) \text{ or } \phi(x)=\psi(x)+b \quad \text{... (8)}$$

Again substituting $y=f(x)$ and $t=0$ in (6), we get

$$f(x)=\phi(x)+\psi(x) \text{ or } f(x)=[\psi(x)+b]+\psi(x) \qquad \text{[Using (8)]}$$

$$\Rightarrow \quad f(x)=2\psi(x)+b$$

So that $\psi(x)=\frac{1}{2}[f(x)-b]$ and $\phi(x)=\frac{1}{2}[f(x)+b]$

On putting the values of $\phi(x+ct)$ and $\psi(x-ct)$ in (6), we get

$$y(x,t)=\frac{1}{2}[f(x+ct)+b]+\frac{1}{2}[f(x-ct)-b]$$

$$\Rightarrow \quad y(x,t)=\frac{1}{2}[f(x+ct)+f(x-ct)] \qquad \textbf{Ans.}$$

EXERCISE 49.2

1. A tightly stretched string with fixed end points $x=0$ and $x=l$ is initially at rest in its equilibrium position. If it is set vibrating by giving to each of its points a velocity $\lambda x(l-x)$, find the displacement of the string at any distance x from one end at any time t.

   Ans. $y=\frac{8\lambda l^3}{c\pi^4}\sum_{(n=1)}^{n}\frac{1}{(2n-1)^4}\sin\frac{(2n-1)\pi x}{l}\sin\frac{(2n-1)\pi ct}{l}$

2. A tightly stretched string of length l fastened at both ends, is disturbed from the position of equilibrium imparting to each of its points an initial velocity of magnitude $f(x)$. Show that the solution of the problem.

   $$u(x,t)=\frac{2}{l}\sum_{n=1}^{\infty}\left[\int_0^l f(x)\sin\frac{n\pi}{l}x\,dx\right]\sin\frac{n\pi x}{l}\cos\frac{n\pi\, at}{l}$$

3. Find the solution of the equation $\frac{\partial^2 y}{\partial x^2}=\frac{1}{c^2}\frac{\partial^2 y}{\partial t^2}$ subject to the boundary conditions

   $y(0,t)=0$, $y(l,t)=0$, $y(x,0)=\phi(x)$, $\frac{\partial y}{\partial t}(x,0)=\psi(x)$ **Ans.** $y=\phi(x).\cos\frac{n\pi ct}{l}+\frac{l\psi(x)}{n\pi c}.\frac{\sin\frac{n\pi ct}{l}}{\sin\frac{n\pi x}{l}}$

4. The vibration of an elastic string of length l are governed by the one-dimensional wave equation $\frac{\partial^2 u}{\partial t^2} = c^2 \frac{\partial^2 u}{\partial x^2}$. The string is fixed at the ends.

$u(0,t) = 0 = u(l,t)$ for all t. The initial deflection is $u(x,0) = \begin{cases} x, & 0 < x < \frac{l}{2} \\ l - x, & \frac{l}{2} \le x \le l \end{cases}$

and the initial velocity is zero. Find the deflection of the string at any instant of time.

(*A.M.I.E.T.E., Summer 2001*) **Ans.** $\frac{4l}{\pi^2} \sum_{n=1}^{\infty} \frac{\sin \frac{n\pi}{2}}{n^2} \sin \frac{n\pi x}{l} \cos \frac{n\pi ct}{l}$

5. A tightly stretched violin string of length l is fixed at both ends and is plucked at $x = \frac{l}{3}$ and assumes initially the shape of a triangle of height a. Find the displacement y at any distance x and any time t after the string is released from rest.

Ans. $y(x, t) = \frac{9a}{\pi^2} \sum_{1}^{\infty} \frac{1}{n^2} \sin \frac{n\pi}{3} \cos \frac{n\pi ct}{l} \sin \frac{n\pi x}{l}$

49.5 ONE DIMENSIONAL HEAT FLOW

In this article, we shall consider the flow of heat and the accompanying variation of temperature with position and with time in conducting solids the following empirical laws are taken as the basis of investigation.

1. Heat flows from a higher to lower temperature.
2. The amount of heat required to produce a given temperature change in a body is proportional to the mass of the body and to the temperature change. This constant of proportionality is known as the specific heat (c) of the conducting material.
3. The rate at which heat flows through an area is proportional to the area and to the temperature gradient normal to the area. This constant of proportionality is known as the thermal conductivity (k) of the material.

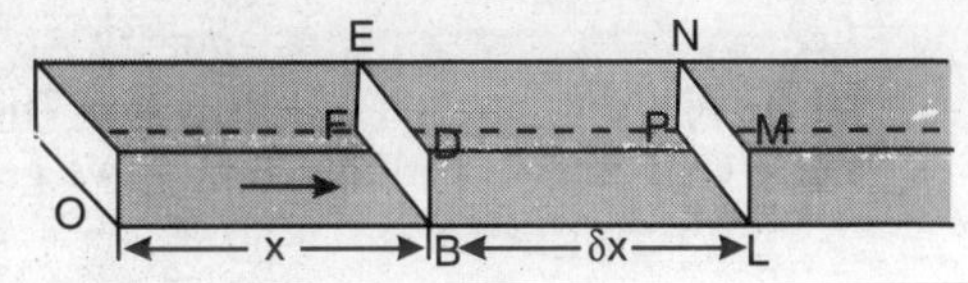

Consider a bar or rod of homogeneous material of density ρ(gr/cm^3) and having a constant cross-sectional area A (cm^2). We suppose that the sides of the bar are insulated and the loss of heat from the sides by conduction or radiations is negligible.

Take an end of the bar as the origin and the direction of heat flow as the positive x-axis. Let c be the specific heat and k the thermal conductivity of the material.

Consider an element between two parallel sections $BDEF$ and $LMNP$ at distances x and $x + \delta x$ from the origin O, the sections being perpendicular to the x-axis. The mass of the element $= A \rho \delta x$.

Let $u(x, t)$ be the temperature at a distance x at time t. By the second law stated above, the rate of increase of heat in the element $= A \rho \delta x\, c \frac{\partial u}{\partial t}$. If R_1 and R_2 are respectively the rates (cal/sec) of inflow and outflow, for the sections $x = x$ and $x = x + \delta x$, then

$$R_1 = -kA \left(\frac{\partial u}{\partial x} \right)_x$$

and $R_2 = -kA \left(\frac{\partial u}{\partial x} \right)_{x + \delta x}$ the negative sign being due to the fact that heat flows from higher to lower temperature. $\left(i.e., \frac{\partial u}{\partial x} \text{ is negative}\right)$

Equating the rates of increase of heat from the two empirical laws.

$$A\rho c\delta x.\frac{\partial u}{\partial t} = R_1 - R_2 = kA\left[\left(\frac{\partial u}{\partial x}\right)_{x+\delta x} - \left(\frac{\partial u}{\partial x}\right)_x\right]$$

$$\therefore \qquad \frac{\partial u}{\partial t} = \frac{k}{\rho c}\left[\frac{\left(\frac{\partial u}{\partial x}\right)_{x+\delta x} - \left(\frac{\partial u}{\partial x}\right)_x}{\delta x}\right]$$

Taking the limit as $\delta x \to 0$ *i.e.* when $x + \delta x \to x$.

$$\frac{\partial u}{\partial t} = \frac{k}{\rho c} \underset{\delta x \to 0}{\text{Lt}} \left[\frac{\left(\frac{\partial u}{\partial x}\right)_{x+\delta x} - \left(\frac{\partial u}{\partial x}\right)_x}{\delta x}\right] = \frac{k}{\rho c}\frac{\partial}{\partial x}\left(\frac{\partial u}{\partial x}\right)$$

i.e.,

$$\frac{\partial u}{\partial t} = \frac{k}{\rho c}\frac{\partial^2 u}{\partial x^2}$$

$\frac{k}{\rho c}$ is called the *diffusivity* (cm²/sec.) of the substance. If we denote it by c^2, the above equation takes the form

$$\frac{\partial u}{\partial t} = c^2\frac{\partial^2 u}{\partial x^2}$$

Example 15. *Solve by the method of separation of variables the P.D.E.*

$$\frac{\partial u}{\partial t} = c^2\frac{\partial^2 u}{\partial x^2}$$

Solution. Here, we have

$$\frac{\partial u}{\partial t} = c^2\frac{\partial^2 u}{\partial x^2} \qquad ...(1)$$

Let $\qquad u = XT \qquad ...(2)$

where X is the function of x only and T is the function of t only.
Differentiating (2) partially w.r.t. t, we get

$$\frac{\partial u}{\partial t} = \frac{\partial}{\partial t}(XT) = X\frac{dT}{dt}$$

and

$$\frac{\partial^2 u}{\partial x^2} = c^2\frac{\partial^2}{\partial x^2}(XT) = \frac{\partial}{\partial x}\left(\frac{\partial}{\partial x}XT\right) = \frac{\partial}{\partial x}\left(T\frac{dX}{dx}\right) = T\frac{d^2X}{dx^2}$$

Putting the values of $\frac{\partial u}{\partial t}$ and $\frac{\partial^2 u}{\partial x^2}$ in (1), we get

$$X\frac{dT}{dt} = c^2 T\frac{d^2X}{dx^2}$$

Separating the variables, we get

$$\frac{1}{c^2}\frac{\frac{dT}{dt}}{T} = \frac{\frac{d^2X}{dx^2}}{X}$$

L.H.S. is constant for x so we take $\frac{1}{c^2}\frac{\frac{dT}{dt}}{T} = -p^2$

R.H.S is constant for t so we take $\frac{\frac{d^2X}{dx^2}}{X} = -p^2$

$$\therefore \quad \frac{1}{c^2}\frac{\frac{dT}{dt}}{T} = \frac{\frac{d^2X}{dx^2}}{X} = -p^2$$

$$\frac{1}{c^2}\frac{\frac{dT}{dt}}{T} = -p^2 \qquad\qquad \frac{\frac{d^2X}{dx^2}}{X} = -p^2$$

$$\frac{dT}{dt} = -p^2c^2\,T \qquad\qquad D^2X = -p^2X$$

$$DT = -p^2c^2\,T$$

A.E. is $m = -p^2c^2$

$$\Rightarrow \quad T = c_1e^{-c^2p^2t} \qquad\qquad \text{A.E. is} \quad m^2 = -p^2$$

$$\Rightarrow \quad m = ip$$

$$X = c_2 \cos px + c_3 \sin px$$

Putting the values of T and X in (1), we get

$$u = c_1e^{-c^2p^2t}\ (c_2\cos px + c_3 \sin px)$$ **Ans.**

Example 16. *Find the temperature in a bar of length 2 whose ends kept at zero and lateral surface insulated if the initial temperature is* $\sin\frac{\pi x}{2} + 3\sin\frac{5\pi x}{2}$.

[*U.P. II Sem., 2007; U.P. II Semester, 2009*]

Solution. Here, we have

One-dimensional heat flow equation is

$$\frac{\partial u}{\partial t} = c^2\frac{\partial^2 u}{\partial x^2}$$

Its solution is

$$u(x, t) = (c_1 \cos px + c_2 \sin px)c_3\,e^{-c^2p^2t} \qquad ...(1)$$

Putting $x = 0$, $u = 0$ in (1), we get

$$0 = c_1c_3e^{-c^2p^2t} \qquad \text{(given)}$$

$$\Rightarrow \quad c_1 = 0$$

Putting $c_1 = 0$ in (1), we get

$$u = c_2c_3 \sin px\, e^{-c^2p^2t} \qquad ...(2)$$

Putting $x = 2$, $u = 0$ in (2), we get (given)

$$0 = c_2c_3 \sin 2p\, e^{-c^2p^2t}$$

$$\Rightarrow \quad \sin 2p = 0 = \sin n\pi$$

$$\Rightarrow \quad p = \frac{n\pi}{2},\ n \in \mathrm{I}$$

Putting the value of P in (2), we get

$$u = b_n \sin\frac{n\pi x}{2}\, e^{-\frac{n^2\pi^2c^2t}{4}} \qquad (c_2c_3 = b_n)$$

Putting $t = 0$ and $u = \sin\frac{\pi x}{2} + 3\sin\frac{5\pi x}{2}$ in (3), we get

$$\sin\left(\frac{\pi x}{2}\right) + 3\sin\left(\frac{5\pi x}{2}\right) = \sum_{n=1}^{\infty} b_n \sin\frac{n\pi x}{2}$$

$$= b_1 \sin\left(\frac{\pi x}{2}\right) + b_2 \sin\left(\frac{2\pi x}{2}\right) + \ldots + b_5 \sin\left(\frac{5\pi x}{2}\right) + \ldots \quad \ldots (4)$$

On equating the coefficients of both sides, we get

$$b_1 = 1 \text{ and } b_5 = 3$$

On putting the values of b_1 and b_5 in (4), we get

Hence, $$u = \sin\left(\frac{\pi x}{2}\right) e^{-\pi^2 c^2 t/4} + 3\sin\left(\frac{5\pi x}{2}\right) e^{-25\pi^2 c^2 t/4}.$$ **Ans.**

Example 17. *Solve the equation* $\frac{\partial u}{\partial t} = \frac{\partial^2 u}{\partial x^2}$ *with boundary condition* $u(x, 0) = 3 \sin n\pi x$, $u(0, t)$, $u(l, t) = 0$ *where* $0 < x < l$. (*Q.Bank U.P. 2002*)

Solution. Here, we have $\frac{\partial u}{\partial t} = \frac{\partial^2 u}{\partial x^2}$

Its solution is $u = = c_1 e^{-p^2 t}(c_2 \cos px + c_3 \sin px)$ ... (1)

Putting $x = 0$, and $u = 0$ in (1), we get (given)

$$0 = c_1 c_2 e^{-p^2 t} \Rightarrow c_2 = 0$$

Putting the value of c_2 in (1), we get $u = c_1 c_3 e^{-p^2 t} \sin px$ ... (2)

Putting $x = l$ and $u = 0$ in (2), we get (given)

$$0 = c_1 c_3 e^{-p^2 t} \sin pl$$

$$\Rightarrow \sin pl = 0 \Rightarrow pl = n\pi \Rightarrow p = \frac{n\pi}{l}$$

Putting the value of p in (2), we get

$$u(x, t) = c_1 c_3 e^{-\frac{n^2\pi^2}{l} t} \sin\frac{n\pi x}{l} = b_n e^{-\frac{n^2\pi^2}{l^2} t} \sin\frac{n\pi x}{l} \qquad (b_n = c_1 c_3)$$

$$u = \sum_{n=1}^{\infty} b_n e^{-(n^2\pi^2 t/l^2)} \sin\frac{n\pi x}{l} \quad \ldots (3) \text{ (general solution)}$$

On putting $t = 0$ and $u = 3\sin n\pi x$ in (3), we get

$$3 \sin n\pi x = \sum_{n=1}^{\infty} b_n \sin\frac{n\pi x}{l} \quad \ldots (4)$$

Equating the coefficients, we get $b_n = 3$, $l = 1$.

On putting the value of b_n and l in (3), we get $u = 3\sum_{n=1}^{\infty} e^{-n^2\pi^2 t} \sin n\pi x$ **Ans.**

Example 18. *A rod of length l with insulated sides is initially at a uniform temperature u. Its ends are suddenly cooled to 0°C and are kept at that temperature. Prove that the temperature function* $u(x,t)$ *is given by*

$$u(x, t) = \sum_{n=1}^{\infty} b_n \sin\frac{n\pi x}{l} . e^{-\frac{c^2\pi^2 n^2 t}{l^2}}$$

where b_n *is determined from the equation.*

$$b_n = \frac{2}{l}\int_0^l u_0(x) \sin\frac{n\pi x}{l} dx$$

Solution. Let heat flow along a bar of uniform cross-section, in the direction perpendicular to the cross-section. Take one end of the bar as origin and the direction of heat flow is along x-axis.

Let the temperature of the bar at any time t at a point x distance from the origin be $u(x, t)$.

Then the equation of one dimensional heat flow is $$\frac{\partial u}{\partial t} = c^2 \frac{\partial^2 u}{\partial x^2} \qquad ...(1)$$

Let us assume that $u = XT$, where X is a function of x alone and T that of t alone.

$$\therefore \quad \frac{\partial u}{\partial t} = X\frac{dT}{dt} \quad \text{and} \quad \frac{\partial^2 u}{\partial x^2} = T\frac{d^2X}{dx^2}$$

Substituting these values in (1), we get $X\dfrac{dT}{dt} = c^2 T\dfrac{d^2X}{dx^2}$

$$\Rightarrow \quad \frac{1}{c^2T}\frac{dT}{dt} = \frac{1}{X}\frac{d^2X}{dx^2} = -p^2 \text{ (constant)} \qquad ...(2)$$

$\dfrac{1}{c^2T}\dfrac{dT}{dt} = -p^2$	$\dfrac{1}{X}\dfrac{d^2X}{dx^2} = -p^2$
$\dfrac{dT}{dt} + p^2c^2T = 0$	$\dfrac{d^2X}{dx^2} + p^2X = 0$
$\Rightarrow \quad DT + p^2c^2T = 0$	$\Rightarrow \quad D^2X + p^2X = 0$
$\Rightarrow \quad (D + p^2c^2)T = 0$	$\Rightarrow \quad (D^2 + p^2)X = 0$
A.E. is $\quad m + p^2c^2 = 0$	A.E. is $\quad m^2 + p^2 = 0$
$\Rightarrow \quad m = -p^2c^2$	$\Rightarrow \quad m = \pm ip$
$\Rightarrow \quad T = c_1 e^{-p^2c^2t}$	$\Rightarrow \quad X = c_2 \cos px + c_3 \sin px$

$$\therefore u = XT \quad \Rightarrow \quad u = c_1e^{-p^2c^2t}(c_2 \cos px + c_3 \sin px) \qquad ...(3)$$

Putting $x = 0$, $u = 0$ in (3), we get

$$0 = c_1e^{-p^2c^2t}(c_2) \Rightarrow c_2 = 0 \qquad [\text{since } c_1 \neq 0]$$

(3) becomes $$u = c_1e^{-p^2c^2t}c_3 \sin px \qquad ...(4)$$

Again putting $x = l$, $u = 0$ in (4), we get

$$0 = c_1e^{-p^2c^2t}c_3 \sin pl \Rightarrow \sin pl = 0 = \sin n\pi \;\Rightarrow\; pl = n\pi$$

$$\Rightarrow \quad p = \frac{n\pi}{l}, \; n \text{ is any integer}$$

Hence (4) becomes $$u = c_1c_3e^{\frac{-n^2c^2\pi^2}{l^2}t} \sin\frac{n\pi x}{l} = b_ne^{-\frac{n^2c^2\pi^2t}{l^2}} \sin\frac{n\pi}{l}x, \quad b_n = c_1c_3$$

This equation satisfies the given conditions for all integral values of n. Hence taking $n = 1, 2, 3, \ldots$, *the most general solution is*

$$u = \sum_{n=1}^{\infty} b_n e^{\frac{-n^2c^2\pi^2t}{l^2}} \sin\frac{n\pi}{l}x$$

$$b_n = \frac{2}{l}\int_0^l u_0 \sin\frac{n\pi x}{l}dx = \frac{2}{l}u_0\left[-\frac{l}{n\pi}\cos\frac{n\pi x}{l}\right]_0^l$$

$$= \frac{2}{l}u_0\left[-\frac{l}{n\pi}(\cos n\pi - \cos 0)\right] = -\frac{2u_0}{n\pi}\left[(-1)^n - 1\right]$$

$$\Rightarrow \qquad b_n = \begin{cases} 0, & \text{if } n \text{ is even} \\ \dfrac{4u_0}{n\pi}, & \text{if } n \text{ is odd} \end{cases}$$

Hence the temperature function

$$u(x, t) = \frac{4u_0}{\pi} \sum_{n=1,3,5....}^{\infty} \frac{1}{n} \sin \frac{n\pi x}{l} e^{-\frac{c^2 n^2 \pi^2 t}{l^2}}$$

$$= \frac{4u_0}{\pi} \sum_{n=1}^{\infty} \frac{1}{2n-1} \sin \frac{(2n-1)\pi x}{l} e^{-\frac{c^2 (2n-1)^2 \pi^2 t}{l^2}}$$

Example 19. *Determine the solution of one dimensional heat equation*

$$\frac{\partial u}{\partial t} = c^2 \frac{\partial^2 u}{\partial x^2}$$

Subject to the boundary conditions $u(0, t) = 0$, $u(l, t) = 0$ $(t > 0)$ and the initial condition $u(x, 0) = x$, l being the length of the bar. (*U.P. II Semester Summer 2006*)

Solution.

$$\frac{\partial u}{\partial t} = c^2 \frac{\partial^2 u}{\partial x^2} \qquad \text{...(1)}$$

Boundary conditions are

$$u(0, t) = 0$$
$$u(l, t) = 0 \qquad (t > 0)$$
$$u(x, 0) = x$$

On solving (1), we get

$$u = c_1 e^{-p^2 c^2 t} (c_2 \cos px + c_3 \sin px) \qquad \text{...(2)}$$

Putting $x = 0$ and $u = 0$ in (2), we get

$$0 = c_1 e^{-p^2 c^2 t} (c_2) \quad \Rightarrow \quad c_2 = 0$$

Putting $c_2 = 0$ in (2), we get

$$u = c_1 e^{-p^2 c^2 t} c_3 \sin px \qquad \text{...(3)}$$

Again putting $x = l$, $u = 0$ in (3), we get

$$0 = c_1 e^{-p^2 c^2 t} c_3 \sin pl$$

$$\Rightarrow \qquad \sin pl = 0 = \sin n\pi \qquad \Rightarrow \qquad pl = n\pi$$

$$\Rightarrow \qquad p = \frac{n\pi}{l}, \; n \text{ is any integer}$$

Hence, (3) becomes $u = c_1 c_3 e^{-\frac{n^2 c^2 \pi^2}{l^2} t} \sin \frac{n\pi x}{l} = b_n e^{-\frac{n^2 c^2 \pi^2 t}{l^2}} \sin \frac{n\pi}{l} x$ ...(4)

On putting $t = 0$ and $u = x$ in (4), we get

$$x = b_n \sin \frac{n\pi}{l} x$$

General solution is

$$x = \sum_{n=1}^{\infty} b_n \sin \frac{n\pi}{l} x$$

Now, $b_n = \frac{2}{l} \int_0^l x \sin \frac{n\pi x}{l} dx = \frac{2}{l} \left[x \cdot \frac{l}{n\pi} \left(-\cos \frac{n\pi x}{l} \right) - (1) \left(\frac{-l^2}{n^2 \pi^2} \sin \frac{n\pi x}{l} \right) \right]_0^l$

$$= \frac{2}{l} \left[\left(l \cdot \frac{l}{n\pi} (-\cos n\pi) + \frac{l^2}{n^2 \pi^2} \sin n\pi \right) - 0 \right] = \frac{2}{l} \left[-\frac{l^2}{n\pi} (-1)^n \right] = (-1)^{n+1} \frac{2l}{n\pi}$$

Putting the value of b_n in (4), we get

$$u = \frac{2l}{\pi}\sum_{n=1}^{\infty}\frac{(-1)^{n+1}}{n}\sin\frac{n\pi x}{l}e^{-\frac{n^2c^2\pi^2 t}{l^2}}$$ **Ans.**

Example 20. *An insulated rod of length l has its ends A and B maintained at 0°C and 100°C respectively until steady state conditions prevail. If B is suddenly reduced to 0°C and maintained at 0°C, find the temperature at a distance x from A at time t.*

(U.P. II Semester Summer, 2004, 2005, AMIE, Summer 2004)

Solution. The initial temperature of the rod can be written as :

$$u(x,t) = 0 + \frac{100}{l}x = \frac{100}{l}x$$

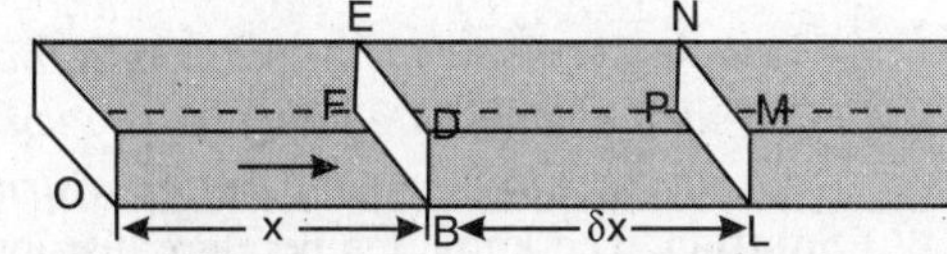

While in steady state, the temperature distribution can be written as

$$u(x,t) = 0 + \frac{0}{l}x = 0$$

To find u in the intermediate period, calculating time from the instant when the end temperature were changed.

$$u = u_1(x) + u_2(x)$$

where $u_2(x)$ is temperature after a sufficient long time and $u_1(x, t)$ is the transient temperature distribution tending to zero as $t \to \infty$. Hence $u_2(x) = 0$.

Also $u_1(x, t)$ satisfies one dimensional heat flow

$$C^2\frac{\partial^2 u}{\partial x^2} = \frac{\partial u}{\partial t}$$

Thus $$u = (C_1\cos px + C_2\sin px)e^{-c^2p^2t} \qquad \dots (1)$$

On putting $x = 0$, $u = 0$ in (1), we get

$$0 = C_1 e^{-p^2c^2t} \quad \Rightarrow \quad C_1 = 0$$

On putting $C_1 = 0$ in (1), we get

$$u = C_2\sin px\, e^{-c^2p^2t} \qquad \dots (2)$$

On putting $x = l$, $u = 0$ in (2), we get

$$0 = C_2\sin pl e^{-p^2c^2t} \quad \Rightarrow \quad \sin pl = 0 = \sin n\pi$$

$$\Rightarrow \quad pl = n\pi \quad \Rightarrow \quad p = \frac{n\pi}{l}$$

On putting the value of p in (2), we get

$$u = C_2\sin\frac{n\pi x}{l}e^{-\frac{n^2\pi^2c^2}{l^2}t} \qquad \dots (3)$$

On putting $t = 0$, $u = \frac{100}{l}x$ in (3), we get

$$\frac{100}{l}x = C_2\sin\frac{n\pi x}{l}$$

$$C_2 = \frac{2}{l}\int_0^l \frac{100}{l}x\cdot\sin\frac{n\pi x}{l}dx$$

$$C_2 = \frac{200}{l^2}\int_0^l x\sin\frac{n\pi x}{l}dx$$

$$C_2 = \frac{200}{l^2}\left[-\frac{xl}{n\pi}\cos\frac{n\pi x}{l} - (-1)\frac{l^2}{n^2\pi^2}\sin\frac{n\pi x}{l}\right]_0^l$$

$$C_2 = \frac{200}{l^2}\left[\frac{-l^2}{n\pi}\cos n\pi\right]$$

$$\Rightarrow \qquad C_2 = -\frac{200}{n\pi}(-1)^n$$

On putting the value of C_2 in (3), we get

$$u = -\frac{200}{n\pi}(-1)^n \sin\frac{n\pi x}{l} e^{-\frac{n^2\pi^2c^2}{l^2}t}$$

$$\Rightarrow \qquad u = (-1)^{n+1}\cdot\frac{200}{n\pi}\sin\frac{n\pi x}{l} e^{-\frac{n^2\pi^2c^2}{l^2}t}$$ **Ans.**

Example 21. *The ends A and B of a rod 20 cm long have the temperatures at 30° C and at 80°C until steady state prevails. The temperature of the ends are changed to 40°C and 60°C respectively. Find the temperature distribution in the rod at time t.*

Solution. The initial temperature distribution in the rod is

$$u_1(x,t) = 30 + \frac{80-30}{20}x = 30 + \frac{50}{20}x \quad i.e., \quad u_1(x,t) = 30 + \frac{5}{2}x$$

and the final distribution (*i.e.* in steady state) is

$$u_2(x) = 40 + \frac{60-40}{2}x = 40 + \frac{20}{20}x = 40 + x$$

To get u in the intermediate period, reckoning time from the instant when the end temperature were changed, we assumed

$$u = u_1(x,t) + u_2(x)$$

where $u_2(x)$ is the steady state temperature distribution in the rod (*i.e.*, temperature after a sufficiently long time) and $u_1(x,t)$ is the transient temperature distribution which tends to zero as t increases.

Thus $$u_2(x) = 40 + x$$

Now $u_1(x, t)$ satisfies the one-dimensional heat-flow equation $c^2\frac{\partial^2 u}{\partial x^2} = \frac{\partial u}{\partial t}$

Hence its solution is $(a_k \cos kx + b_k \sin kx)\, e^{-c^2k^2t}$

Hence u is of the form

$$u = 40 + x + \Sigma(a_k \cos kx + b_k \sin kx)e^{-c^2k^2t}$$

Since $u = 40°$ when $x = 0$ and $u = 60°$ when $x = 20$, we get

$$a_k = 0, \quad \text{and} \quad k = \frac{n\pi}{20}$$

Hence $$u = 40 + x + \sum_{n=1}^{\infty} b_n \sin\frac{n\pi x}{20}\, e^{-c^2\left(\frac{n\pi}{20}\right)^2 t} \qquad ...(1)$$

Using the initial conditions *i.e.*,

$$u = 30 + \frac{5}{2}x \text{ when } t = 0, \text{ we get}$$

$$30 + \frac{5}{2}x = 40 + x + \Sigma b_n \sin\frac{n\pi}{20}x \quad \Rightarrow \quad \frac{3}{2}x - 10 = \Sigma b_n \sin\frac{n\pi x}{20}$$

Hence $$b_n = \frac{2}{20}\int_0^{20}\left(\frac{3}{2}x - 10\right)\sin\frac{n\pi x}{20}dx$$

$$\Rightarrow \qquad = \frac{1}{10}\left[\left(\frac{3x}{2} - 10\right)\left(-\frac{20}{n\pi}\cos\frac{n\pi x}{20}\right) - \frac{3}{2}\left(-\frac{400}{n^2\pi^2}\sin\frac{n\pi x}{20}\right)\right]_0^{20}$$

$$= \frac{1}{10}\left[-20\left(\frac{20}{n\pi}\right)(-1)^n - (-10)\left(-\frac{20}{n\pi}\right)\right] = -\frac{20}{n\pi}[2(-1)^n + 1]$$

Putting this value of b_n in (1), we get

$$\therefore \quad u = 40 + x - \frac{20}{\pi}\sum\left[\left(\frac{2(-1)^n + 1}{n}\right)\sin\frac{n\pi x}{20}.e^{-\left(\frac{cn\pi}{20}\right)^2 t}\right] \quad \textbf{Ans.}$$

EXERCISE 49.3

1. The equation $\frac{\partial u}{\partial t} = \mu\frac{\partial^2 u}{\partial x^2}$ refers to the conduction of heat along a bar, without radiation, show that if $u = Ae^{-gx}\sin(nt - gx)$, where A, g and n are positive constants, then $g = \sqrt{\frac{n}{2\mu}}$

2. Solve $\frac{\partial u}{\partial t} = a^2\frac{\partial^2 u}{\partial x^2}$ under the conditions

   $u'(0,t) = 0 \quad t > 0, \quad u'(\pi, t) = 0, \quad u(x,0) = x^2, 0 < x < \pi$

   Ans. $u(x,t) = \frac{\pi^2}{3} + 4\sum_{n=1}^{\infty}\frac{(-1)^n}{n^2}\cos nx e^{-a^2n^2t}$

3. A uniform rod of length 'a' whose surface is thermally insulated is initially at temperature $\theta = \theta_0$. At time $t = 0$, one end is suddenly cooled to $\theta = 0$ and subsequently maintained at this temperature, the other end remains thermally insulated. Find the temperature $\theta(x,t)$.

   Ans. $\theta(x,t) = \frac{4\theta_0}{\pi}\sum_{n=1}^{\infty}\frac{\sin\frac{(2n+1)\pi x}{2a}}{2n+1}e^{-\frac{(2n+1)^2\pi^2c^2t}{4a^2}}$

4. Solve $\frac{\partial U}{\partial t} = a^2\frac{\partial^2 U}{\partial x^2}$ under the conditions

   (*i*) $U \neq \infty$ if $t \to \infty$; (*ii*) $U(0,t) = U(\pi,t) = 0$; (*iii*) $U(x, 0) = \pi x - x^2$

   Ans. $u = \frac{8}{\pi}\sum_{n=1}^{\infty}\frac{\sin(2n-1)x}{(2n-1)^3}e^{-a^2(2n-1)^2t}$

5. The temperature distribution in a bar of length π, which is perfectly insulated at the ends $x = 0$ and $x = \pi$ is governed by the partial differential equation $\frac{\partial u}{\partial t} = \frac{\partial^2 u}{\partial x^2}$. Assuming the initial temperature as $u(x, 0) = f(x) = \cos 2x$, find the temperature distribution at any instant of time. **Ans.** $u = e^{-4t}\cos 2x$

6. The heat flow in a bar of length 10 cm of homogeneous material is governed by the partial differential equation $u_t = c^2u_{xx}$. The ends of the bar are kept at temperature 0°C, and the initial temperature is $f(x) = x(10 - x)$. Find the temperature in the bar at any instant of time.

   Ans. $u(x,t) = \frac{800}{\pi^3}\sum_{n=1}^{\infty}\frac{1}{(2n-1)^3}\sin\frac{(2n-1)\pi x}{10}e^{-\left[\frac{(2n-1)^2\pi^2c^2t}{100}\right]}$

7. Find the temperature $u(x, t)$ in a bar of length π which is perfectly insulated everywhere including the ends $x = 0$ and $x = \pi$. This leads to the conditions $\frac{\partial u}{\partial x}(0, t) = 0, \frac{\partial u}{\partial x}(\pi, t) = 0$. Further the initial conditions are as given below:

   $u(x, 0) = \begin{cases} x, & 0 < x < \pi/2 \\ \pi x, & \pi/2 \le x < \pi \end{cases}$ Find the solution by the separation of variables.

49.6 TWO DIMENSIONAL HEAT FLOW

When the heat-flow is along curves instead of along straight lines, all the curves lying in parallel planes, then the flow is called two-dimensional. Let us consider now the flow of heat in a metal-plate in the xoy plane. Let the plate be of uniform thickness h, density ρ, thermal conductivity k and the specific heat C. Since the flow is two dimensional, the temperature at any point of the plate is independent of the z-coordinate. The heat-flow lies in the xoy plane and is zero along the direction normal to the xoy plane.

Now, consider a rectangular element $ABCD$ of the plate with sides δx and δy, the edges being parallel to the coordinates axes, as shown in the figure, Then the quantity of heat *entering* the element $ABCD$ per sec. through the surface AB is

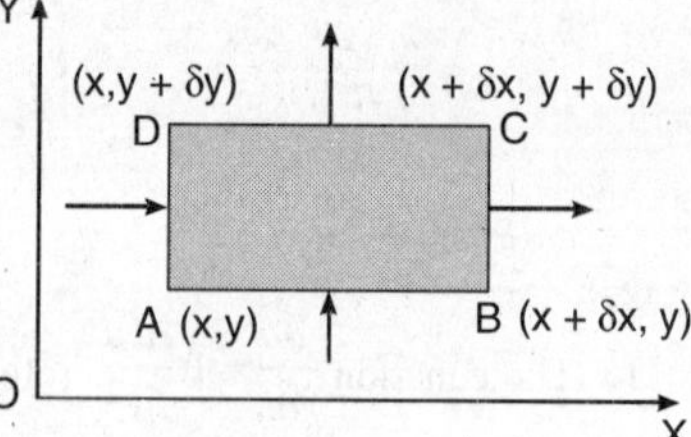

$$= -k\left(\frac{\partial u}{\partial y}\right)_y \delta x.h.$$

Similarly the quantity of heat entering the element $ABCD$ per sec. through the surface AD is

$$= -k\left(\frac{\partial u}{\partial x}\right)_x \delta y.h.$$

The amount of heat which flows out through the surfaces BC and CD are

$$-k\left(\frac{\partial u}{\partial x}\right)_{x+\delta x} .\delta y.h \text{ and } -k\left(\frac{\partial u}{\partial y}\right)_{y+\delta y} .\delta x.h \text{ respectively.}$$

Therefore the total gain of heat by the rectangular element $ABCD$ per sec.

= inflow – outflow.

$$= kh\left[\left\{\left(\frac{\partial u}{\partial x}\right)_{x+\delta x} - \left(\frac{\partial u}{\partial x}\right)_x\right\}\delta y + \left\{\left(\frac{\partial u}{\partial y}\right)_{y+\delta y} - \left(\frac{\partial u}{\partial y}\right)_y\right\}\delta x\right]$$

$$= kh\delta x.\delta y\left[\frac{\left(\frac{\partial u}{\partial x}\right)_{x+\delta x} - \left(\frac{\partial u}{\partial x}\right)_x}{\delta x} + \frac{\left(\frac{\partial u}{\partial y}\right)_{y+\delta y} - \left(\frac{\partial u}{\partial y}\right)_y}{\delta y}\right] \quad ...(1)$$

The rate of gain of heat by the element $ABCD$ is also given by

$$\rho\delta x.\delta y.h.c.\frac{\partial u}{\partial t} \quad ...(2)$$

Equating the two-expressions for gain of heat per sec. from (1) and (2), we have

$$\rho\delta x.\delta y.h.c.\frac{\partial u}{\partial t} = hk\delta x\delta y\left[\frac{\left(\frac{\partial u}{\partial x}\right)_{x+\delta x} - \left(\frac{\partial u}{\partial x}\right)_x}{\delta x} + \frac{\left(\frac{\partial u}{\partial y}\right)_{y+\delta y} - \left(\frac{\partial u}{\partial y}\right)_y}{\delta y}\right]$$

i.e.,

$$\frac{\partial u}{\partial t} = \frac{k}{\rho c}\left[\frac{\left(\frac{\partial u}{\partial x}\right)_{x+\delta x} - \left(\frac{\partial u}{\partial x}\right)_x}{\delta x} + \frac{\left(\frac{\partial u}{\partial y}\right)_{y+\delta y} - \left(\frac{\partial u}{\partial y}\right)_y}{\delta y}\right]$$

Taking the limit as $\delta x \to 0$, $\delta y \to 0$, the above equation reduces to $\frac{\partial u}{\partial t} = \frac{k}{\rho c}\left(\frac{\partial^2 u}{\partial x^2} + \frac{\partial^2 u}{\partial y^2}\right)$

Putting $\alpha^2 = \frac{k}{\rho c}$ as before, the equation becomes,

$$\frac{\partial u}{\partial t} = \alpha^2\left(\frac{\partial^2 u}{\partial x^2} + \frac{\partial^2 u}{\partial y^2}\right) \quad ...(3)$$

The equation (3) gives the temperature distribution of the plate in the transit state.

In the steady-state, u is independent of t, so that $\frac{\partial u}{\partial t} = 0$. Hence the temperature distribution of the plate in the steady-state is $\frac{\partial^2 u}{\partial x^2} + \frac{\partial^2 u}{\partial y^2} = 0.$

Example 22. *Solve* $\frac{\partial^2 u}{\partial x^2} + \frac{\partial^2 u}{\partial y^2} = 0$ *which satisfies the conditions*

$$u(0, y) = u(l, y) = u(x, 0) = 0$$

and $u(x, a) = \sin \frac{n\pi x}{l}$ *(U.P., II Semester, 2004)*

Solution. Consider the heat flow in a metal plate of uniform thickness, in the directions parallel to length and breadth of the plate. There is no heat flow along the normal to the plane of the rectangle.

Let u (x, y) be the temperature at any point (x, y) of the plate at time t is given by

$$\frac{\partial u}{\partial t} = c^2 \left(\frac{\partial^2 u}{\partial x^2} + \frac{\partial^2 u}{\partial y^2} \right) \quad \ldots (1)$$

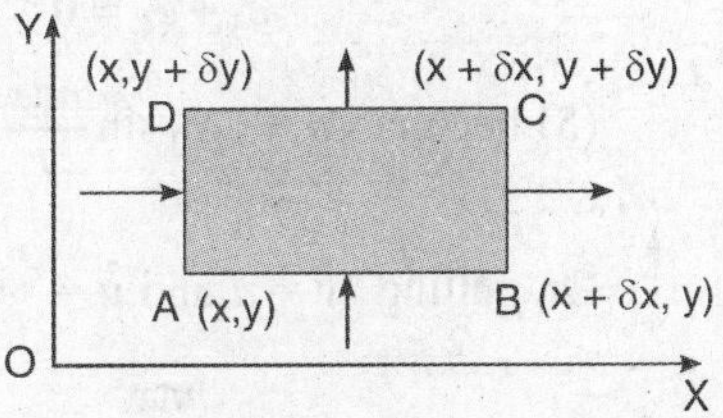

In the steady state, u does not change with t.

$$\therefore \quad \frac{\partial u}{\partial t} = 0$$

(1) becomes $\frac{\partial^2 u}{\partial x^2} + \frac{\partial^2 u}{\partial y^2} = 0$

This is called Laplace equation in two dimensions.

Let $u = X(x).Y(y)$ ...(2)

Putting the values of $\frac{\partial^2 u}{\partial x^2}$ and $\frac{\partial^2 u}{\partial y^2}$ in (1), we have

$$X''Y + XY'' = 0 \quad \ldots (3)$$

$$\Rightarrow \quad \frac{X''}{X} = -\frac{Y''}{Y} = -p^2 \text{ (say)}$$

$D^2X = -p^2X$	$D^2Y = p^2Y$
$\Rightarrow \quad D^2X + p^2X = 0$	$\Rightarrow \quad D^2Y - p^2Y = 0$
$\Rightarrow \quad (D^2 + p^2)X = 0$	$\Rightarrow \quad (D^2 - p^2)Y = 0$
A.E. is $m^2 + p^2 = 0$	*A.E.* is $m^2 - p^2 = 0$
$\Rightarrow \quad m^2 = -p^2$	$\Rightarrow \quad m^2 = p^2$
$\Rightarrow \quad m = \pm ip$	$\Rightarrow \quad m = \pm p$
$X = c_1 \cos px + c_2 \sin px$	$Y = c_3 e^{py} + c_4 e^{-py}$

Putting the values of X and Y in (2), we have

$$u = (c_1 \cos px + c_2 \sin px)(c_3 e^{py} + c_4 e^{-py}) \qquad \text{... (4)}$$

Putting $x = 0$, $u = 0$ in (4), we have

$$0 = c_1 (c_3 e^{py} + c_4 e^{-py})$$

$$\therefore \qquad c_1 = 0$$

(4) is reduced to $u = c_2 \sin px (c_3 e^{py} + c_4 e^{-py})$ ... (5)

On putting $x = l$, $u = 0$ in (5), we have

$$0 = c_2 \sin pl (c_3 e^{py} + c_4 e^{-py})$$

$$c_2 \neq 0 \qquad \therefore \sin pl = 0 = \sin n\pi \quad \Rightarrow \quad pl = n\pi \Rightarrow p = \frac{n\pi}{l}$$

Now (5) becomes $u = c_2 \sin \dfrac{n\pi x}{l} \left(c_3 e^{\frac{n\pi y}{l}} + c_4 e^{\frac{-n\pi y}{l}} \right)$ ... (6)

On putting $u = 0$ and $y = 0$ in (6), we have

$$0 = c_2 \sin \frac{n\pi x}{l} (c_3 + c_4)$$

$$\Rightarrow \qquad c_3 + c_4 = 0 \qquad \Rightarrow \qquad c_3 = -c_4$$

(5) becomes $u = c_2 c_3 \sin \dfrac{n\pi x}{l} \left(e^{\frac{n\pi y}{l}} - e^{\frac{-n\pi y}{l}} \right)$ ... (7)

On putting $y = a$ and $u = \sin \dfrac{n\pi x}{l}$ in (7), we have

$$\sin \frac{n\pi x}{l} = c_2 c_3 \sin \frac{n\pi x}{l} \left(e^{\frac{n\pi a}{l}} - e^{\frac{-n\pi a}{l}} \right) \quad i.e. \quad c_2 c_3 = \frac{1}{e^{\frac{n\pi a}{l}} - e^{\frac{-n\pi a}{l}}}$$

Putting this value in (7), we have

$$u = \sin \frac{n\pi x}{l} \frac{e^{\frac{n\pi y}{l}} - e^{\frac{-n\pi y}{l}}}{e^{\frac{n\pi a}{l}} - e^{\frac{-n\pi a}{l}}} \quad \text{or} \quad u = \sin \frac{n\pi x}{l} \frac{\sinh \frac{n\pi y}{l}}{\sinh \frac{n\pi a}{l}} \qquad \textbf{Ans.}$$

Example 23. *A rectangular plate with insulated surfaces is 10 cm wide and so long compared to its width that it may be considered infinite in length without introducing an appreciable error. If the temperature along the short edge y = 0 is given by*

$$u(x, 0) = 20x, \qquad 0 < x \leq 5$$

$$= 20(10 - x), \qquad 5 < x < 10$$

while the two long edges x = 0 and x = 10 as well as the other short edges are kept at 0°C. Find the steady state temperature at any point (x, y) of the plate.

Solution. In the steady state, the temperature $u(x, y)$ at any point $p(x, y)$ satisfy the equation

$$\frac{\partial^2 u}{\partial x^2} + \frac{\partial^2 u}{\partial y^2} = 0 \qquad \text{... (1)}$$

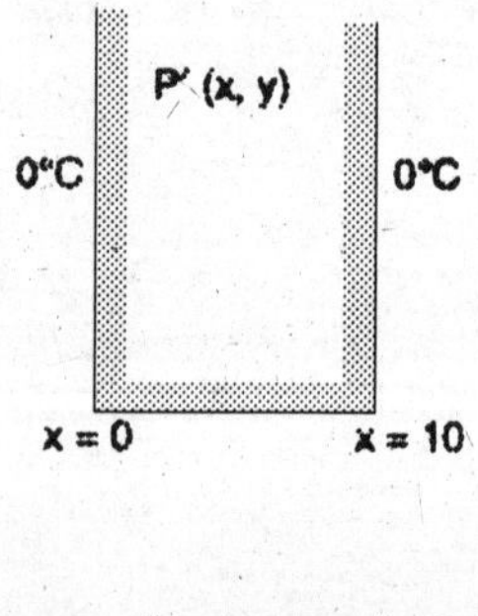

The boundary conditions are

$u(0, y) = 0$ for all values of y ... (2)

$u(10, y) = 0$ for all values of y ... (3)

$u(x, \infty) = 0$ for all values of x ... (4)

$u(x, 0) = 20x, \quad 0 < x \leq 5$

$\quad = 20(10 - x), \quad 5 < x < 10$ ... (5)

P (x, y)
0°C
0°C
x = 0
x = 10

Now three possible solutions of (1) are

$$u = (C_1 e^{px} + C_2 e^{-px})(C_3 \cos py + C_4 \sin py) \quad \text{... (6)}$$

$$u = (C_5 \cos px + C_6 \sin px)(C_7 e^{py} + C_8 e^{-py}) \quad \text{... (7)}$$

$$u = (C_9 x + C_{10})(C_{11} y + C_{12}) \quad \text{... (8)}$$

Of these, we have to choose that solution which is consistent with the physical nature of the problem. The solution (6) and (8) cannot satisfy the condition (2), (3) and (4). Thus, only possible solution is (7) *i.e.,* of the form.

$$u(x, y) = (C_1 \cos px + C_2 \sin px)(C_3 e^{py} + C_4 e^{-py}) \quad \text{... (9)}$$

By (2), $u(0, y) = C_1(C_3 e^{py} + C_4 e^{-py}) = 0$ for all values of y

$\therefore \quad C_1 = 0$

$\therefore$ (9) reduces to $u(x, y) = C_2 \sin px (C_3 e^{py} + C_4 e^{-py})$ ... (10)

By (3), $u(10, y) = C_2 \sin 10p\, (C_3 e^{py} + C_4 e^{-py}) = 0, \; C_2 \neq 0$

$\therefore \quad \sin 10p = 0 = \sin n\pi \Rightarrow 10p = n\pi \Rightarrow p = \dfrac{n\pi}{10}$

Also to satisfy the condition (4) *i.e.* $u = 0$ as $y \to \infty$

$C_3 = 0$

Hence (10) takes the form $u(x, y) = C_2 C_4 \sin px\, e^{-py}$

or $\quad u(x, y) = b_n \sin px . e^{-py}$ where $b_n = C_2 C_4$

$\therefore$ The most general solution that satisfies (2), (3) and (4) is of the form

$$u(x, y) = \sum_{n=1}^{\infty} b_n \sin px\, e^{-py} \quad \text{... (11)}$$

Putting $y = 0$, $u(x, 0) = \sum_{n=1}^{\infty} b_n \sin px$, where $p = \dfrac{n\pi}{10}$

This requires the expansion of u in Fourier series in the interval $x = 0$ and $x = 5$ and from $x = 5$ to $x = 10$.

$$b_n = \frac{2}{10}\int_0^5 20x \sin px\, dx + \frac{2}{10}\int_5^{10} 20(10 - x)\sin px\, dx$$

$$b_n = 4\int_0^5 x \sin px\, dx + 4\int_5^{10} (10 - x)\sin px\, dx$$

$$= 4\left[x\left(\frac{-\cos px}{p}\right) - (1)\left(\frac{-\sin px}{p^2}\right)\right]_0^5 + 4\left[(10 - x)\left(\frac{-\cos px}{p}\right) - (-1)\left(\frac{-\sin px}{p^2}\right)\right]_5^{10}$$

$$= 4\left[\frac{-5\cos 5p}{p} + \frac{\sin 5p}{p^2}\right] + 4\left[0 - \frac{\sin 10p}{p^2} + \frac{5\cos 5p}{p} + \frac{\sin 5p}{p^2}\right]$$

$$= 4\left[\frac{2\sin 5p}{p^2} - \frac{\sin 10p}{p^2}\right] \qquad \left(p = \frac{n\pi}{10}\right)$$

$$= 4\left[\frac{2\sin 5 . \frac{n\pi}{10}}{\frac{n^2\pi^2}{100}} - \frac{\sin 10 . \frac{n\pi}{10}}{\frac{n^2\pi^2}{100}}\right] = \frac{800}{n^2\pi^2}\sin\frac{n\pi}{2} - \frac{400}{n^2\pi^2}\sin n\pi$$

$$= \frac{800}{n^2\pi^2}\sin\frac{n\pi}{2} = 0, \text{ if } n \text{ is even.}$$

$$= \pm\frac{800}{n^2\pi^2}, \qquad \text{if } n \text{ is odd.}$$

$$\Rightarrow \quad b_n = \frac{(-1)^{n+1}800}{(2n-1)^2\pi^2}$$

On putting the value of b_n in (11) the temperature at any point (x, y) is given by

$$u(x, y) = \frac{800}{\pi^2}\sum_{n=1}^{\infty}\frac{(-1)^{n+1}}{(2n-1)^2}\sin\frac{(2n-1)\pi x}{10}\, e^{-\frac{(2n-1)\pi y}{10}} \qquad \textbf{Ans.}$$

Example 24. *A thin rectangular plate whose surface is impervious to heat flow has t = 0 an arbitrary distribution of temperature f (x, y). Its four edges x = 0, x = a, y = 0, y = b are kept at zero temperature. Determine the temperature at a point of the plate as t increases.*

(U.P. III Semester, summer 2002)

Solution. The partial differential equation of two dimensional heat conduction problem is

$$\frac{\partial^2 u}{\partial x^2} + \frac{\partial^2 u}{\partial y^2} = \frac{1}{c^2}\frac{\partial u}{\partial t} \qquad \ldots (1)$$

The boundary conditions are $u(0, y) = u(a, y) = u(x, 0) = u(x, b)$
and the initial condition is $u(x, y) = f(x, y)$

Let the solution be $\qquad u = X.Y.T.$

On putting the values of the derivatives in (1), we get

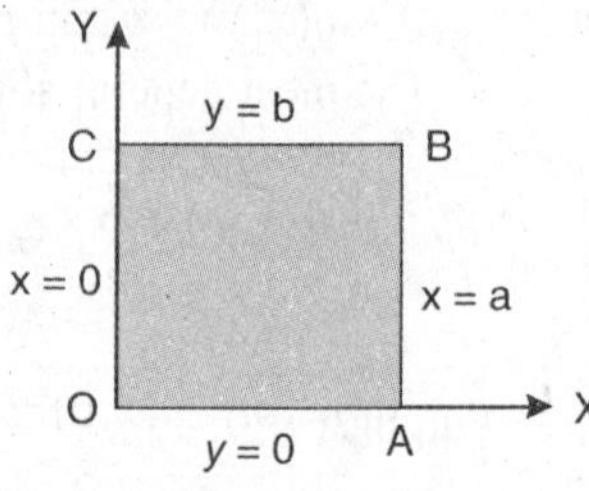

$$YT\frac{d^2X}{dx^2} + XT\frac{d^2Y}{dy^2} = \frac{XY}{c^2}\frac{dT}{dt}$$

Separating the variables, we get, (dividing by XYT)

$$\frac{1}{X}\frac{d^2X}{dx^2} + \frac{1}{Y}\frac{d^2Y}{dy^2} = \frac{1}{c^2T}\frac{dT}{dt}$$

Since X is function of independent variable x alone, Y of y alone and T of t alone, there are three possibilities.

1. $\frac{1}{X}\frac{d^2X}{dx^2} = 0, \quad \frac{1}{Y}\frac{d^2Y}{dy^2} = 0, \quad \frac{1}{c^2T}\frac{dT}{dt} = 0$

2. $\frac{1}{X}\frac{d^2X}{dx^2} = K_1^2, \quad \frac{1}{Y}\frac{d^2Y}{dy^2} = K_2^2, \quad \frac{1}{c^2T}\frac{dT}{dt} = k^2$

3. $\frac{1}{X}\frac{d^2X}{dx^2} = -K_1^2, \quad \frac{1}{Y}\frac{d^2Y}{dy^2} = -K_2^2, \quad \frac{1}{c^2T}\frac{dT}{dt} = -K^2$

$$K^2 = K_1^2 + K_2^2 \qquad \ldots (A)$$

Out of these three solutions, we have to choose that solution which is consistent with the physical nature of the problem. Accordingly third solution is accepted here.

Then $\qquad Y = A\cos K_1x + B\sin K_1x$

$$T = E\,e^{-C^2kt}$$

$$u = XYT$$

$$u = (A\cos K_1x + B\sin K_1x)(C\cos K_2y + D\sin K_2y).E\,e^{-C^2kt} \qquad \ldots (1)$$

Now we apply boundary conditions

On putting $u = 0$ and $x = 0$ in (1), we get

$$0 = A(C\cos K_2 y + D\sin K_2 y)\,E e^{-C^2 kt} \Rightarrow A = 0$$

On putting the value of A in (1), it reduces to

$$u = (B\sin K_1 x)(C\cos K_2 y + D\sin K_2 y).E e^{-C^2 kt}$$

$$\Rightarrow \quad u = (B_1 \sin K_1 x)(C\cos K_2 y + D\sin K_2 y).e^{-C^2 kt} \quad \ldots (2)$$

$$(B_1 = B.E)$$

On putting $u = 0$ and $x = a$ in (2), we get

$$0 = (B_1 \sin K_1 a)(C\cos K_2 y + D\sin K_2 y).e^{-C^2 kt}$$

$$\Rightarrow \quad 0 = B_1 \sin K_1 a \Rightarrow \sin K_1 a = 0 = \sin n\pi \Rightarrow K_1 a = n\pi \Rightarrow K_1 = \frac{n\pi}{a}$$

On putting the value of K_1 in (2), we have

$$u = \left(B_1 \sin\frac{n\pi}{a}x\right)(C\cos K_2 y + D\sin K_2 y).e^{-C^2 kt} \quad \ldots (3)$$

On putting $u = 0$, $y = 0$ in (3), we obtain

$$0 = \left(B_1 \sin\frac{n\pi}{a}x\right)(C)e^{-C^2 kt} \Rightarrow C = 0$$

On substituting the value of C in (3), we have

$$u = \left(B_1 \sin\frac{n\pi x}{a}\right)(D\sin K_2 y)e^{-C^2 kt} \quad \ldots (4)$$

On substituting $u = 0$ and $y = b$ in (4), we have

$$0 = \left(B_1 \sin\frac{n\pi x}{a}\right)(D\sin K_2 b).e^{-C^2 kt} \Rightarrow \sin K_2 b = 0 = \sin m\pi \Rightarrow K_2 = \frac{m\pi}{b}$$

On putting the value of K_2 in (4), we have

$$u = \left(B_1 \sin\frac{n\pi x}{a}\right)\left(D\sin\frac{m\pi y}{b}\right).e^{-C^2 kt} \quad \ldots (5)$$

$$u = A_{mn} \sin\frac{n\pi x}{a}\sin\frac{m\pi y}{b}.e^{-C^2 kt} \quad (B_1 D = A_{mn})$$

But from (A), $$K^2 = K_1^2 + K_2^2 = \pi^2\left(\frac{n^2}{a^2} + \frac{m^2}{b^2}\right)$$

$$\Rightarrow \quad K_{mn}^2 = \pi^2\left(\frac{n^2}{a^2} + \frac{m^2}{b^2}\right)$$

By using K_{mn}, (5) becomes

$$u = \sum_{n=1}^{\infty}\sum_{m=1}^{\infty} A_{mn} \sin\frac{n\pi x}{a}\sin\frac{m\pi y}{b}\, e^{-C^2 k_{mn} t}$$

On applying the initial condition $u = f(x, y)$, $t = 0$

$$f(x, y) = \sum_{n=1}^{\infty}\sum_{m=1}^{\infty} A_{mn} \sin\frac{n\pi x}{a}\sin\frac{m\pi y}{b}$$

Which is the double Fourier sine series of $f(x, y)$

Where $$A_{mn} = \frac{2}{a}\frac{2}{b}\int_{x=0}^{a}\int_{y=0}^{b} \sin\frac{n\pi x}{a}\sin\frac{m\pi y}{b}.f(x, y)\,dx\,dy$$ **Ans.**

Example 25. *A rectangular plate with insulated surface is 8 cm wide and so long compared to its width that it may be considered infinite in length without introducing an appreciable. If the tempearature along one short edge y = 0 is given by*

$$u(x, 0) = 100\sin\frac{\pi x}{8},\ 0 < x < 8$$

While the two long edges x = 0 and x = 8 as well as the other short edge are kept at 0°C, show that the steady state temperature at any plate is given by

$$u(x, y) = 100e^{-\frac{\pi y}{8}} \sin \frac{\pi x}{8}.$$

Soution. Here, we have

Two dimensional heat flow equation in steady state is given by

$$\frac{\partial^2 u}{\partial x^2} + \frac{\partial^2 u}{\partial y^2} = 0$$

Its solution is $u(x, y) = (c_1 \cos px + c_2 \sin px)(c_3 e^{py} + c_4 e^{-py})$ ...(1)

Putting $x = 0$ and $u = 0$ in (1), we get (given)

$$0 = c_1 (c_3 e^{py} + c_4 e^{-py})$$

$$\Rightarrow \quad c_1 = 0$$

Putting the value of c_1 in (1), we get

$$u = c_2 \sin px (c_3 e^{py} + c_4 e^{-py}) \quad ...(2)$$

Again putting $x = 8$ and $u = 0$ in (2), we get (given)

$$0 = c_2 \sin 8p (c_3 e^{py} + c_4 e^{-py})$$

$$\Rightarrow \quad \sin 8p = 0 = \sin n\pi \Rightarrow p = \frac{n\pi}{8} \ (n \in I)$$

Putting the value of p in (2), we get

$$u(x, y) = c_2 \sin \frac{n\pi x}{8} \left(c_3 e^{\frac{n\pi y}{8}} + c_4 e^{-\frac{n\pi y}{8}}\right) \quad ... (3)$$

Putting $y = \infty$ and $u = 0$ in (3), we get (given)

$$0 = c_2 \sin \frac{n\pi x}{8} c_3 e^{\frac{n\pi y}{8}}$$

$$\Rightarrow c_3 = 0$$

Putting $c_3 = 0$ in (3), we get

$$u = c_2 c_4 \sin \frac{n\pi x}{8} e^{-\frac{n\pi y}{8}}$$

$$u = b_n \sin \frac{n\pi x}{8} e^{-\frac{n\pi y}{8}} \quad ... (4)$$

Putting $y = 0$ and $u = 100 \sin \frac{\pi x}{8}$ in (4), we get (given)

$$100 \sin \frac{\pi x}{8} = b_n \sin \frac{n\pi x}{8}$$

$$\Rightarrow b_n = 100,$$

Putting the value of b_n in (4), we get

$$u = 100 \sin \left(\frac{\pi x}{8}\right) e^{-\frac{\pi y}{8}}$$ **Ans.**

Example 26. *Solve* $\frac{\partial^2 u}{\partial x^2} + \frac{\partial^2 u}{\partial y^2} = 0$, $0 < x < \pi$, $0 < y < \pi$, *which satisfies the conditions :*

$u(0, y) = u(\pi, y) = u(x, \pi) = 0$ *and* $u(x, 0) = \sin^2 x$. *(Q.Bank U.P. II Semester 2002)*

Solution. Here, we have

$$\frac{\partial^2 u}{\partial x^2} + \frac{\partial^2 u}{\partial y^2} = 0$$

Its solution is $u(x, y) = (c_1 \cos px + c_2 \sin px)(c_3 e^{py} + c_4 e^{-py})$ ... (1)

Putting $x = 0$ and $u = 0$ in (1), we get (given)

$$0 = c_1(c_3 e^{py} + c_4 e^{-py})$$

$$\Rightarrow c_1 = 0$$

Putting the value of c_1 in (1), we get

$$u(x, y) = c_2 \sin px(c_3 e^{py} + c_4 e^{-py}) \quad \text{... (2)}$$

Putting $x = \pi$ and $u = 0$ in (2), we get (given)

$$0 = c_2 \sin p\pi(c_3 e^{py} + c_4 e^{-py})$$

$$\Rightarrow \quad \sin p\pi = 0 = \sin n\pi$$

$$\Rightarrow \quad p = n$$

Putting $p = n$ in (2), we get $u = c_2 \sin nx(c_3 e^{ny} + c_4 e^{-ny})$

$$= \sin nx(Ae^{ny} + Be^{-ny}) \quad \text{... (3)} \ (c_2 c_3 = A \text{ and } c_2 c_4 = B.)$$

Putting $u = 0$ and $y = \pi$ in (3), we get (given)

$$0 = \sin nx(Ae^{n\pi} + Be^{-n\pi})$$

$$\Rightarrow \quad 0 = Ae^{n\pi} + Be^{-n\pi}$$

$$\Rightarrow \quad Ae^{n\pi} = -Be^{-n\pi} = -\frac{1}{2}b \Rightarrow A = -\frac{1}{2}be^{-n\pi} \text{ and } B = \frac{1}{2}be^{n\pi}$$

On putting the values of A and B in (3), we get

$$u = \sin nx\left[-\frac{1}{2}be^{-n\pi}e^{ny} + \frac{1}{2}be^{n\pi}e^{-ny}\right]$$

$$= \frac{1}{2}b\left[e^{n(\pi - y)} - e^{-n(\pi - y)}\right]\sin nx$$

$$= b \sinh n\,(\pi - y) \sin nx.$$

$$u = \sum_{n=1}^{\infty} b_n \sinh n(\pi - y) \sin nx \quad \text{... (4) (General solution)}$$

Putting $y = 0$ and $u = \sin^2 x$ in (4), we get (given)

$$\sin^2 x = \sum_{n=1}^{\infty} b_n \sinh n\pi \sin nx$$

Here $$b_n \sinh n\pi = \frac{2}{\pi}\int_0^{\pi} \sin^2 x \sin nx\, dx = \frac{1}{\pi}\int_0^{\pi} (1 - \cos 2x) \sin nx\, dx$$

$$= \frac{1}{\pi}\int_0^{\pi}\left[\sin nx - \frac{1}{2}\{\sin (n+2)x + \sin(n-2)x\}\right] dx$$

$$= \frac{1}{\pi}\left[\frac{-\cos nx}{n} + \frac{\cos (n+2)x}{2(n+2)} + \frac{\cos (n-2)x}{2(n-2)}\right]_0^{\pi}$$

$$= \frac{1}{\pi}\left[\frac{-\cos n\pi}{n} + \frac{\cos (n+2)\pi}{2(n+2)} + \frac{\cos (n-2)\pi}{2(n-2)} + \frac{1}{n} - \frac{1}{2(n+2)} - \frac{1}{2(n-2)}\right]$$

$$= \frac{1}{\pi}\left[\left(\frac{-(-1)^n}{n} + \frac{(-1)^{n+2}}{2(n+2)} + \frac{(-1)^{n-2}}{2(n-2)}\right) + \left(\frac{1}{n} - \frac{1}{2(n+2)} - \frac{1}{2(n-2)}\right)\right]$$

$$= \frac{1}{2\pi}\left[\left(\frac{1}{n+2} + \frac{1}{n-2} - \frac{2}{n}\right)\{(-1)^n - 1\}\right], \text{ where } n \neq 2$$

$$b_n \sinh n\pi = \frac{-8}{\pi n\,(n^2-4)}, \text{ when } n \text{ is odd} \qquad \ldots (5)$$

$$b_n \sinh \pi = 0, \text{ when } n \text{ is even and } \neq 2$$

Now, we have to find out $b^2 \sinh 2\pi$.

$$b^2 \sinh 2\pi = \frac{2}{\pi}\int_0^\pi \sin^2 x \sin 2x\, dx$$

$$= \frac{1}{\pi}\int_0^\pi (1-\cos 2x)\sin 2x\, dx = \frac{1}{\pi}\int_0^\pi \left(\sin 2x - \frac{1}{2}\sin 4x\right) dx = \left(\frac{-\cos 2x}{2} + \frac{1}{8}\cos 4x\right)_0^\pi = 0$$

$$\Rightarrow \qquad b^2 = 0 \qquad \ldots (6)$$

On putting the value of b_n from (5) in (4), we get

$$u = \frac{-8}{\pi}\sum_{n=1,3,5,\ldots}^{\infty} \frac{\sin nx \sinh n(\pi - y)}{n(n^2-4)\sin hn\pi} \qquad \textbf{Ans.}$$

EXERCISE 49.4

1. Find by the method of separation of variables, a particular solution of the equation

$$\frac{\partial^2 u}{\partial x^2} + \frac{\partial^2 u}{\partial y^2} = 0$$

that tends to zero as x tends to infinity and is equal to 2 cos y when $x = 0$. **Ans.** $u = 2e^{-x}\cos y$

2. Solve the equation: $u_{xx} + u_{yy} = 0$

$$u(0,y) = u(\pi, y) = 0 \text{ for all } y,$$

$$u(x,0) = k, \qquad 0 < x < \pi$$

$$\lim_{y\to\infty} u(x,y) = 0 \quad 0 < x < \pi$$

Ans. $u(x,y) = \sum_{n=1}^{\infty} b_n \sin nx e^{-ny},\ k = \sum_{n=1}^{\infty} b_n \sin nx$

3. A infinitely long uniform plate is bounded by two parallel edges and an end at right angles to them. The breadth is π. This end is maintained at a temperature u_0 at all points and other edges are at zero temperature. Determine the temperature at any point of the plate in the steady state.

Ans. $u(x,y) = \frac{4u_0}{\pi}\left[e^{-y}\sin x + \frac{1}{3}e^{-3y}\sin 3x + \frac{1}{5}e^{-5y}\sin 5x + \ldots\right]$

4. Solve $\frac{\partial^2 V}{\partial x^2} + \frac{\partial^2 V}{\partial y^2} = 0$, given that

(*i*) $V = 0$ when $x = 0$ and $x = c$; (*ii*) $V \to 0$ as $y \to \infty$; (*iii*) $V = V_0$ when $y = 0$.

Ans. $V(x,y) = \sum_{n=1}^{\infty} b_n \sin\frac{n\pi x}{c} e^{-\frac{n\pi y}{c}},\ V_0 = \sum_{n=1}^{\infty} b_n \sin\frac{n\pi x}{c}$

5. The steady state temperature distribution in a thin plate bounded by the lines $x = 0$, $x = a$, $y = 0$ and $y = \infty$, is governed by the partial differential equation

$$\frac{\partial^2 u}{\partial x^2} + \frac{\partial^2 u}{\partial y^2} = 0.$$

Obtain the steady state temperature distribution under the conditions

$u(0, y) = 0, \qquad u(a, y) = 0, \qquad u(x, \infty) = 0$

$u(x, 0) = x, \qquad 0 \le x \le a/2$

$\qquad = a - x, \qquad a/2 \le x \le a$

Ans. $u(x,y) = \sum_{n=1}^{\infty} \frac{4a}{n^2\pi^2}\sin\frac{n\pi}{2}\sin\frac{n\pi x}{a} e^{-\frac{\pi y}{a}}$

6. An infinitely long plane uniform plate is bounded by two parallel edges and an end at right angles to them. The breadth is π. This end is maintained at temperature u_0 at all points and the other edges are at zero temperature. Determine the temperature at any point of the plate in the steady state.

Ans. $u(x,y)=\dfrac{4u_0}{\pi}\sum_{n=1}^{\infty}\dfrac{1}{n}\sin x e^{-ny}$.

49.7 EQUATION OF HEAT FLOW IN TWO DIMENSIONS IN POLAR COORDINATES

Equation of heat flow in two dimension in polar coordinates.

Consider a sheet of conducting material of uniform density ρ, uniform thickness h, thermal conductivity k and specific heat c. Let O, the pole and OX, the initial line, be taken on the plane of sheet. As we are dealing with two dimensional heat-flow, the temperature function $u(r,\theta,\tau)$ at point (r,θ) of the plate is a function of r, θ and time t.

Consider an element of the sheet included between the circles $r = r$, $r = r + \delta r$ and the straight lines $\theta = \theta$ and $\theta = \theta + \delta\theta$ through the pole. Heat-flow directions are assumed to be positive in the positive directions associated with r and θ. The mass of the element $ABCD = \rho h(r\delta r\rho\theta)$

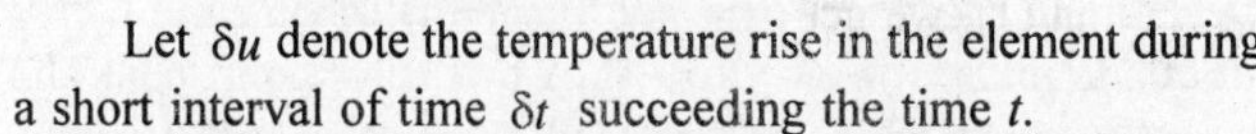

Let δu denote the temperature rise in the element during a short interval of time δt succeeding the time t.

$\therefore$ Rate of increase of heat content in the element

$$= \underset{\delta t\to 0}{\text{Lt}}\ \rho h(r\delta r\delta\theta)c\frac{\delta u}{\delta t} = \rho hcr\delta r\delta\theta\frac{\partial u}{\partial t} \qquad \dots(1)$$

If R_1, R_2, R_3, R_4 are the roots of flow of heat across the sides of the element through the edges AB, CD, AD and BC respectively, then

$$R_1=-k\left(\frac{1}{r}\frac{\partial u}{\partial\theta}\right)_{\theta}h\delta r, \qquad R_2=-k\left(\frac{1}{r}\frac{\partial u}{\partial\theta}\right)_{\theta+\delta\theta}h\delta r$$

$$R_3=-k\left(\frac{\partial u}{\partial r}\right)_{r}hr\delta\theta, \qquad R_4=-k\left(\frac{\partial u}{\partial r}\right)_{r+\delta r}h(r+\delta r)\delta\theta$$

The rate of increase of heat in the element $= R_1 - R_2 + R_3 - R_4$ which is equal to the expression in (1)

$$\therefore\ \rho hcr\,\delta r\,\delta\theta\frac{\partial u}{\partial t}=kh\left[\left\{\left(\frac{1}{r}\frac{\partial u}{\partial\theta}\right)_{\theta+\delta\theta}-\left(\frac{1}{r}\frac{\partial u}{\partial\theta}\right)_{\theta}\right\}\delta r+\left\{\left(\frac{\partial u}{\partial r}\right)_{r+\delta r}-\left(\frac{\partial u}{\partial r}\right)_{r}\right\}r\delta\theta+\delta r\delta\theta\left(\frac{\partial u}{\partial r}\right)_{r+\delta r}\right]$$

Dividing by $\delta r.\delta\theta.hr\rho c$,

$$\frac{\partial u}{\partial t}=\frac{k}{\rho c}\left[\frac{1}{r^2}\frac{\left\{\left(\frac{\partial u}{\partial\theta}\right)_{\theta+\delta\theta}-\left(\frac{\partial u}{\partial\theta}\right)_{\theta}\right\}}{\delta\theta}+\frac{\left(\frac{\partial u}{\partial r}\right)_{r+\delta r}-\left(\frac{\partial u}{\partial r}\right)_{r}}{\delta r}+\frac{1}{r}\left(\frac{\partial u}{\partial r}r+\delta r\right)\right]$$

Taking the limit as $\delta\theta,\ \delta r\to 0$,

$$\frac{\partial u}{\partial t}=\frac{k}{\rho c}\left[\frac{1}{r^2}\frac{\partial^2 u}{\partial\theta^2}+\frac{\partial^2 u}{\partial r^2}+\frac{1}{r}\frac{\partial u}{\partial r}\right]$$

Therefore the equation of heat-flow in polar coordinates is

$$\frac{\partial^2 u}{\partial r^2}+\frac{1}{r}\frac{\partial u}{\partial r}+\frac{1}{r^2}\frac{\partial^2 u}{\partial\theta^2}=\frac{1}{\alpha^2}\frac{\partial u}{\partial t} \qquad \text{where } \alpha^2=\frac{k}{\rho c}$$

In steady-state, $\frac{\partial u}{\partial t}=0$. Hence the equation of heat-flow in steady-state is

$$\nabla^2 u \equiv \frac{\partial^2 u}{\partial r^2}+\frac{1}{r}\frac{\partial u}{\partial r}+\frac{1}{r^2}\frac{\partial^2 u}{\partial \theta^2}=0$$

Example 27. *Solve* $\frac{\partial^2 u}{\partial r^2}+\frac{1}{r}\frac{\partial u}{\partial r}+\frac{1}{r^2}\frac{\partial^2 u}{\partial \theta^2}=0$ *by the method of separation of variables.*

Solution. $\frac{\partial^2 u}{\partial r^2}+\frac{1}{r}\frac{\partial u}{\partial r}+\frac{1}{r^2}\frac{\partial^2 u}{\partial \theta^2}=0$ or $r^2\frac{\partial^2 u}{\partial r^2}+r\frac{\partial u}{\partial r}+\frac{\partial^2 u}{\partial \theta^2}=0$ ... (1)

Let $u = R(r).T(\theta)$

$$\frac{\partial u}{\partial r}=\frac{dR}{dr}.T(\theta) \text{ and } \frac{\partial^2 u}{\partial r^2}=\frac{d^2R}{dr^2}.T(\theta)$$

$$\frac{\partial u}{\partial \theta}=R(r).\frac{dT}{d\theta} \text{ and } \frac{\partial^2 u}{\partial \theta^2}=R(r).\frac{d^2T}{d\theta^2}$$

Putting the values of $\frac{\partial^2 u}{\partial r^2}$, $\frac{\partial u}{\partial r}$ and $\frac{\partial^2 u}{\partial \theta^2}$ in (1), we get

$$r^2.\frac{d^2R}{dr^2}.T(\theta)+r\frac{dR}{dr}.T(\theta)+R(r)\frac{d^2T}{d\theta^2}=0$$

$$\left(r^2\frac{d^2R}{dr^2}+r.\frac{dR}{dr}\right).T+R\frac{d^2T}{d\theta^2}=0$$

$$\frac{r^2\frac{d^2R}{dr^2}+r.\frac{dR}{dr}}{R}=-\frac{1}{T}\frac{d^2T}{d\theta^2}=h \text{ (say)} \qquad ...(2)$$

$$r^2\frac{d^2R}{dr^2}+r.\frac{dR}{dr}-hR=0$$

Put $r=e^z$

$$[D(D-1)+D-h]R=0$$

A.E. is $m^2-h=0$ ...(3)

$\Rightarrow \quad m=\pm\sqrt{h}$

$$R=c_1e^{\sqrt{h}z}+c_2e^{-\sqrt{h}z}$$

$$R=c_1r^{\sqrt{h}}+c_2r^{-\sqrt{h}}$$

$$\frac{d^2T}{d\theta^2}+hT=0$$

$$(D^2+h)T=0$$

A.E. is $m^2+h=0$ ...(4)

$\Rightarrow \quad m=\pm i\sqrt{h}$

$$T=c_3\cos(\sqrt{h}\theta)+c_4\sin(\sqrt{h}\theta)$$

$$u=(c_1r^{\sqrt{h}}+c_2r^{-\sqrt{h}})[c_3\cos(\sqrt{h}\theta)+c_4\sin(\sqrt{h}\theta)] \qquad ...(5)$$

Case 1. If $h=k^2$

(2) becomes $u=(c_1r^k+c_2r^{-k})\,[c_3\cos(k\theta)+c_4\sin(k\theta)]$ ...(6)

Case 2. If $h=0$

On putting $h=0$ in (3), we get

$$D^2=0 \Rightarrow D=0,0$$

$$R=(c_5+Zc_6) \Rightarrow R=[c_5+(\log r)\,c_6]$$

Again on putting $h = 0$ in (4), we get

$$D^2 = 0 \Rightarrow D = 0, 0$$

$$T = c_7 + \theta c_8$$

But $u = \text{R.T.} = (c_5 + c_6 \log r)(c_7 + \theta c_8)$

Case 3. If $h = -p^2$

On putting $h = -p^2$ in (3), we get

$$D^2 + p^2 = 0 \Rightarrow D = \pm ip$$

$$R = c_9 \cos pz + c_{10} \sin(p \log r)$$

Again on putting $h = -p^2$ in (4), we get

$$D^2 - p^2 = 0 \Rightarrow D = \pm p$$

$$T = c_{11}e^{p\theta} + c_{12}e^{-p\theta}$$

$$u = RT = c_9\cos(p \log r) + c_{10} \sin(p \log r)(c_{11}e^{p\theta} + c_{12}e^{-p\theta})$$

Then there are three possible solutions

$$u = (c_1r^k + c_2r^{-k})[c_3\cos(k\theta) + c_4\sin(k\theta)] = (c_5 + c_6\log r)(c_7 + c_8\theta)$$

$$= [c_9\cos(p\log r) + c_{10}\sin(p\log r)][c_{11}e^{p\theta} + c_{12}e^{-p\theta}]$$

On putting $c_3 = a\cos\alpha$ and $c_4 = -a\sin\alpha$ in (6), we get

$$u = (c_1r^k + c_2r^{-k})(a\cos\alpha\cos k\theta - a\sin\alpha\sin k\theta)$$

$$= [(c_1r^k + c_2r^{-k})(a\cos(k\theta + \alpha))]$$

$$= (Ar^k + Br^{-k})\cos(k\theta + \alpha) \quad \text{...(7)} \quad \textbf{Ans.}$$

Example 28. *Solve* $\dfrac{\partial^2 V}{\partial r^2} + \dfrac{1}{r}\dfrac{\partial V}{\partial r} + \dfrac{1}{r^2}\dfrac{\partial^2 V}{\partial \theta^2} = 0$ *with boundary conditions,*

(*i*) **V is finite when $r \to 0$**

(*ii*) $V = \Sigma c_n \cos n\theta$ on $r = a$ *(U.P. II Sem 2010)*

Solution. Solution of the given equation is as equation (7) in above example, we have

$$V = \Sigma(A_n r^n + B_n r^{-n})\cos(\theta + \alpha) \quad \text{...(1)}$$

Case I. On putting $r = a$, $V = \Sigma c_n \cos n\theta$ in (1), we get

$$\Sigma c_n \cos n\theta = \Sigma(A_n a^n + B_n a^{-n})\cos(n\theta + \alpha) \quad \text{...(2)}$$

On equating the constant terms both the sides of (2), we get

$$\Rightarrow \quad c_n = A_n a^n + B_n a^{-n} \quad \text{...(3)}$$

Case II. When $r = 0$ and V is finite

$$\Rightarrow \quad B_n = 0$$

On putting $B_n = 0$ in (2), we get

$$\Sigma c_n \cos n\theta = \Sigma A_n a^n \cos(\theta + \alpha) \quad \text{...(4)}$$

Comparing the constant terms of (4) on both sides, we get

$$c_n = A_n a^n \Rightarrow A_n = \frac{c_n}{a^n} \quad (\alpha = 0)$$

On putting $A_n = \dfrac{c_n}{a^n}$, $B_n = 0$ and $\alpha = 0$ in (1), we get

$$V = \sum\left(\frac{c_n}{a^n}r^n + 0\right)\cos(n\theta + 0) \Rightarrow V = \sum c_n\left(\frac{r}{a}\right)^n \cos n\theta \quad \textbf{Ans.}$$

Example 29. *The diameter of a semicircular plate of radius a is kept at 0°C and the temperature at the semicircular boundary is T°C. Find the steady state temperature in the plate.*

Solution. Let the centre O of the semicircular plate be the pole and the bounding diameter be as the initial line. Let $u(r,\theta)$ be the steady state temperature at any point $P(r,\theta)$ and u satisfies the equation

$$r^2\frac{\partial^2 u}{\partial r^2}+r\frac{\partial u}{\partial r}+\frac{\partial^2 u}{\partial \theta^2}=0 \qquad ...(1)$$

The boundary conditions are

(*i*) $u(r,0)=0 \qquad 0 \le r \le a$

(*ii*) $u(r,\pi)=0 \qquad 0 \le r \le a$

(*iii*) $u(a,\theta)=T$.

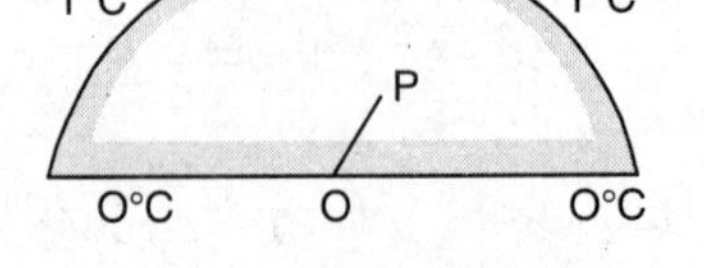

From conditions (*ii*) and (*iii*), we have $u \to 0$ as $r \to 0$. Hence the appropriate solution of (1) is as solved in example 27.

$$u=(c_1 r^p + c_2 r^{-p})(c_3 \cos p\theta + c_4 \sin p\theta) \qquad ...(2)$$

Putting $u\ (r,\ 0) = 0$ in (2), we get

$$0=(c_1 r^p + c_2 r^{-p})\ c_3 \Rightarrow c_3 = 0$$

On putting $c_3 = 0$ in (2), we get

$$u=(c_1 r^p + c_2 r^{-p}) c_4 \sin p\theta \qquad ...(3)$$

Putting $u(r,\pi)=0$ in (3), we get

$$0=(c_1 r^p + c_2 r^{-p}) c_4 \sin p\pi \Rightarrow \sin p\pi = 0 = \sin n\pi \quad \Rightarrow \quad p\pi = n\pi \Rightarrow p = n$$

On putting $p = n$ (3) becomes,

$$u=(c_1 r^n + c_2 r^{-n}) c_4 \sin n\theta \qquad ...(4)$$

Since, $\quad u = 0$ when $r = 0$

$$0 = c_2$$

(4) becomes, $u = c_1 c_4 r^n \sin n\theta$

The most general solution of (1) is

$$u(r,\theta)=\sum_{n=1}^{\infty} b_n r^n \sin n\theta \qquad (c_1 c_2 = b_n) \qquad ...(5)$$

Putting $r = a$ and $u = T$ in (5), we have

$$T=\sum_{n=1}^{\infty} b_n\ a^n \sin n\theta$$

By Fourier half range series, we get

$$b_n a^n = \frac{2}{\pi}\int_0^{\pi} T \sin n\theta\ d\theta = \frac{2}{\pi}T\left(\frac{-\cos n\theta}{n}\right)_0^{\pi} = \frac{2T}{n\pi}[-(-1)^n+1]$$

$$b_n a^n = 0, \qquad \text{when } n \text{ is even,}$$

$$b_n a^n = \frac{4T}{n\pi}, \qquad \text{when } n \text{ is odd.}$$

$$\Rightarrow \qquad b_n = \frac{4T}{n\pi a^n}$$

Hence, (5) becomes

$$u(r,\ \theta)=\frac{4T}{\pi}\left[\frac{r/a}{1}\sin\theta+\frac{(r/a)^3}{3}\sin 3\theta+\frac{(r/a)^5}{5}\sin 5\theta+...\right] \qquad \textbf{Ans.}$$

EXERCISE 49.5

1. Solve the steady-state temperature equation

$$\frac{\partial^2 T}{\partial r^2}+\frac{1}{r}\frac{\partial T}{\partial r}+\frac{1}{r^2}\frac{\partial^2 T}{\partial \theta^2}=0;\quad 10\le r\le 20,\quad 0\le\theta\le 2\pi$$

subject to the following conditions:

$T(10,\theta)=15\cos\theta$ and $T(20,\theta)=30\sin\theta$ **Ans.** $T(r,\theta)=\frac{4T}{\pi}\left[\frac{r}{a}\sin\theta+\frac{1}{3}\left(\frac{r}{a}\right)^3\sin 3\theta+....\right]$

2. A semi-circular plate of radius a has its circumference kept at temperature $u(a,\theta)=k\theta(\pi-\theta)$ while the boundary diameter is kept at zero temperature. Find the steady state temperature distribution $u(r,\theta)$ of the plate assuming the lateral surfaces of the plate to be insulated.

Ans. $u(r,\theta)=\frac{8k}{\pi}\sum_1^\infty\left(\frac{r}{a}\right)^{2n-1}\frac{\sin(2n-1)\theta}{(2n-1)^3}$

3. Find the steady state temperature in a circular plate of radius a which has one half of its circumference at 0°C and the other half at 60°C.

Ans. $u(r,\theta)=50-\frac{200}{n}\sum_{n=1}^\infty\frac{1}{2n-1}\left(\frac{r}{a}\right)^{2n-1}\sin(2n-1)\theta.$

49.8 TRANSMISSION LINE EQUATIONS

$$\frac{\partial^2 V}{\partial x^2}=RC\frac{\partial V}{\partial t}$$

$$\frac{\partial^2 i}{\partial x^2}=RC\frac{\partial i}{\partial t}$$ are called telegraph equations,

where V = potential, i = current, C = capacitance, L = inductance

$$\frac{\partial^2 V}{\partial x^2}=LC\frac{\partial^2 V}{\partial t^2}$$

$$\frac{\partial^2 i}{\partial x^2}=LC\frac{\partial^2 i}{\partial t^2}$$

are called radio equations.

Example 30. *Find the current i and voltage v in a transmission line of length l, t seconds after the ends are suddenly grounded, given that i (x, 0) =* i_0*, v (x, 0) =* $v_0\sin\left(\frac{\pi x}{l}\right)$ *and that R and G are negligible.*

Solution. $$\frac{\partial^2 v}{\partial x^2}=LC\frac{\partial^2 v}{\partial t^2}$$

Let $v=XT$ where X and T are the functions of x and t respectively.

$$\frac{\partial^2 v}{\partial x^2}=T\frac{d^2X}{dx^2}\quad\text{and}\quad\frac{\partial^2 v}{\partial t^2}=X\frac{d^2T}{dt^2}$$

$$T\frac{d^2X}{dx^2}=LCX\frac{d^2T}{dt^2}$$

$$\frac{\frac{d^2X}{dx^2}}{X}=LC\frac{\frac{d^2T}{dt^2}}{T}=-p^2\text{ say}$$

Since the initial conditions suggest the values of v and i are periodic functions,

$$\therefore \quad X = c_1 \cos px + c_2 \sin px$$

$$T = c_3 \cos \frac{pt}{\sqrt{LC}} + c_4 \sin \frac{pt}{\sqrt{LC}}$$

$$\Rightarrow \quad v = (c_1 \cos px + c_2 \sin px)\left(c_3 \cos \frac{pt}{\sqrt{LC}} + c_4 \sin \frac{pt}{\sqrt{LC}} \right) \quad ...(1)$$

When $\quad t = 0, \; v = v_0 \sin \frac{\pi x}{l}$

$$v_0 \sin \frac{\pi x}{l} = (c_1 \cos px + c_2 \sin px) c_3 \quad ...(2)$$

On equating the coefficients, we get

$$c_1 c_3 = 0 \Rightarrow c_1 = 0 \quad \text{and} \quad c_2 c_3 = v_0, \; p = \frac{\pi}{l}$$

(1) becomes

$$v = \sin \frac{\pi x}{l} \left[v_0 \cos \frac{\pi t}{l\sqrt{LC}} + c_2 c_4 \sin \frac{\pi t}{l\sqrt{LC}} \right] \quad ...(3)$$

Now when $\quad t = 0, \; i = i_0$ (constant)

Hence, $\quad \frac{\partial i}{\partial x} = 0$

$$\frac{\partial i}{\partial x} = \frac{-c\partial v}{\partial t} \qquad \therefore \quad \frac{\partial v}{\partial t} = 0 \text{ when } t = 0$$

Now $\quad \frac{\partial v}{\partial t} = \sin \frac{\pi x}{l} \left(\frac{\pi}{l\sqrt{LC}} \right) \left[-v_0 \sin \frac{\pi t}{l\sqrt{LC}} + c_2 c_4 \cos \frac{\pi t}{l\sqrt{LC}} \right] \quad ...(4)$

On putting $\frac{\partial v}{\partial t} = 0$ and $t = 0$ in (4), we get $\quad c_2 c_4 = 0 \Rightarrow c_4 = 0$

(3) is reduced to $\quad v = v_0 \sin \frac{\pi x}{l} \cos \frac{\pi t}{l\sqrt{LC}}$

$$\frac{\partial v}{\partial x} = \frac{\pi}{l} v_0 \cos \frac{\pi x}{l} \cos \frac{\pi t}{l\sqrt{LC}} = -L \frac{\partial i}{\partial t} \quad ... (5)$$

$$\frac{\partial v}{\partial t} = -\frac{v_0 \pi}{l\sqrt{LC}} \sin \frac{\pi x}{l} \sin \frac{\pi t}{l\sqrt{LC}} = \frac{-1}{C} \frac{\partial t}{\partial x} \quad ... (6)$$

Integrating (5) and (6), we get

$$i = -v_0 \sqrt{\frac{C}{L}} \cos \frac{\pi x}{l} \sin \frac{\pi t}{l\sqrt{LC}} + f(x)$$

$$i = -v_0 \sqrt{\frac{C}{L}} \cos \frac{\pi x}{l} \sin \frac{\pi t}{l\sqrt{LC}} + F(t)$$

$\therefore \;$ $f(x)$ and $F(t)$ must be constant only, since $i = i_0$ when $t = 0$

$\therefore \quad$ constant $= i_0 = f(x)$

Hence, $\quad i = i_0 - v_0 \sqrt{\frac{C}{L}} \cos \frac{\pi x}{l} \sin \frac{\pi t}{l\sqrt{LC}}$. **Ans.**

EXERCISE 49.6

1. A transmission line 1,000 miles long is initially under steady state condition with potential 1,300 volts at the sending end ($x = 0$) and 1,200 volts at the receiving end ($x = 1000$). The terminal end of the line is suddenly grounded but the potential at the source is kept at 1,300 volts. Assuming the inductance and leakage to be negligible, find the potential $v(x, t)$, if it satisfies the equation $v_t = (1/RC)v_{xx}$.

Ans. $v(x,y) = \sum_{1}^{\infty} b_n \sin nx\, e^{-ny}$ and $k = \sum_{1}^{\infty} b_n \sin nx$

2. Obtain a solution of the telegraph equation

$$\frac{\partial^2 e}{\partial x^2} = RC\frac{\partial e}{\partial t}$$

suitable for the case when e decays with the time and when there is a steady fall of potential from e_0 to 0 along the line of length l initially and the sending end is suddenly earthed.

Ans. $e(x,t) = \frac{2e_0}{\pi}\sum_{n=1}^{\infty}\frac{1}{n}\sin\frac{n\pi x}{l}e^{\frac{-n^2\pi^2 t}{CRl^2}}$

49.9 LAPLACE EQUATION

Laplace equation is used to solve engineering problems and its theory is called potential theory. Its solutions are known as harmonic functions.

Example 31. *Solve the Laplace equation* $\frac{\partial^2 u}{\partial x^2} + \frac{\partial^2 u}{\partial y^2} = 0$ *in a rectangle in the xy-plane with* $u(x, 0) = 0$, $u(x, b) = 0$, $u(0, y)$ *and* $u(a, y) = f(y)$ *parallel to y-axis.* [*U.P.T.U. 2010, 2008*]

Solution. Here, the Laplace equation is

$$\frac{\partial^2 u}{\partial x^2} + \frac{\partial^2 u}{\partial y^2} = 0 \quad ...(1)$$

Let $u = XY$

Where X is a function of x only Y is a function of y only.

$$\frac{\partial u}{\partial x} = \frac{\partial (XY)}{\partial x} = Y\frac{\partial X}{\partial x}$$

$$\frac{\partial^2 u}{\partial x^2} = \frac{\partial}{\partial x}\left(\frac{\partial u}{\partial x}\right) = \frac{\partial}{\partial x}\left(Y\frac{\partial X}{\partial x}\right) = Y\frac{\partial^2 X}{\partial x^2}$$

$$\frac{\partial u}{\partial y} = \frac{\partial (XY)}{\partial y} = X\frac{\partial Y}{\partial y}$$

$$\frac{\partial^2 u}{\partial y^2} = \frac{\partial}{\partial y}\left(\frac{\partial}{\partial y}(XY)\right) = \frac{\partial}{\partial y}\left(X\frac{\partial Y}{\partial y}\right) = X\frac{\partial^2 Y}{\partial^2 y}$$

Putting the values of $\frac{\partial^2 u}{\partial x^2}$ and $\frac{\partial^2 u}{\partial y^2}$ in (1), we get

$$Y\frac{\partial^2 u}{\partial x^2} + X\frac{\partial^2 u}{\partial y^2} = 0 \text{ or } YX'' + XY'' = 0$$

On dividing by XY and separating the variables, we get

$$\frac{Y''}{Y} = -\frac{X''}{X}$$

Case I. If $\dfrac{Y''}{Y} = -\dfrac{X''}{X} = p^2$(say)

$$\frac{Y''}{Y} = p^2$$

$$Y'' - p^2 Y = 0$$

A. E. is $m^2 - p^2 = 0 \Rightarrow m = \pm p$

$$Y = c_1 e^{py} + c_2 e^{py} \quad ...(2)$$

$$-\frac{X''}{X} = p^2$$

$$\Rightarrow \quad X'' + p^2 X = 0$$

A.E. is $m^2 + p^2 = 0 \Rightarrow m = \pm ip$

$$X = c_3 \cos px + c_4 \sin px \quad ...(3)$$

Putting $y = 0$ and $Y = 0$ in (2), we get

$$c_1 + c_2 = 0 \Rightarrow c_2 = -c_1$$

On putting $y = b$ and Y = 0, we get

$$c_1 e^{pb} + c_2 e^{pb} = 0$$

$$c_1(e^{pb} - e^{pb}) = 0$$

$$c_1 = 0 \qquad [c_1 = c_2 = 0]$$

$$\Rightarrow \quad Y = 0$$

But $\quad u = XY \qquad [Y = 0]$

$$= X(0) = 0, \text{ which is not possible}$$

Case II. If $\dfrac{Y''}{Y} = -\dfrac{X''}{X} = 0$ (say)

$$\frac{Y''}{Y} = 0$$

$$Y'' = 0 \Rightarrow Y = c_5 + c_6 y \quad ...(4)$$

$$-\frac{X''}{X} = 0$$

$$X'' = 0 \Rightarrow X = c_7 + c_8 x$$

Putting $y = 0$, $Y = 0$ in (4), we get

$$0 = c_5$$

Putting $y = b$, $Y = 0$ in (4), we get

$$0 = c_6 b \Rightarrow c_6 = 0$$

$$\Rightarrow \quad Y = 0$$

$$u = XY \qquad [Y = 0]$$

$$= X(0) = 0, \text{ which is not possible.}$$

Case III. If $\dfrac{Y''}{Y} = -\dfrac{X''}{X} = -p^2$

(*i*) $\quad \dfrac{Y''}{Y} = -p^2$

$$\Rightarrow \quad Y'' + p^2 Y = 0$$

A.E. is $m^2 + p^2 = 0 \Rightarrow m = \pm pi$

$$Y = c_9 \cos py + c_{10} \sin py \quad ...(5)$$

$$-\frac{X''}{X} = -p^2$$

$$\Rightarrow \quad X'' - p^2 X = 0$$

A.E. is $m^2 - p^2 = 0 \Rightarrow m = \pm p$

$$X = c_{11} e^{px} + c_{12} e^{-px}$$

On putting $y = 0$, $Y = 0$ in (5), we get

$$\Rightarrow \quad c_9 = 0$$

Putting $y = b$ and $Y = 0$ in (5), we get

$$0 = c_9 \cos pb + c_{10} \sin pb.$$

$$c_{10} \sin pb = 0 \qquad (\because c_9 = 0)$$

$$\Rightarrow \quad \sin pb = 0$$

$$\text{and } \sin n\pi = 0$$

$$\Rightarrow \quad pb = n\pi \Rightarrow p = \frac{n\pi}{b}$$

Putting $c_9 = 0,\ p = \frac{n\pi}{b}$ in (5), we get

$$Y = c_{10} \sin \frac{n\pi y}{b}$$

$$u = XY$$

$$u = c_{10} \sin \frac{n\pi y}{b}\left(c_{11} e^{\frac{n\pi x}{b}} + c_{12} e^{-\frac{n\pi x}{b}}\right) \qquad \text{... (6)}$$

Putting $x = 0$ and $u = 0$ in (6), we get

$$0 = c_{10} \sin \frac{n\pi y}{b}(c_{11} + c_{12})$$

$$c_{11} + c_{12} = 0 \Rightarrow c_{12} = -c_{11}$$

Putting the value of $c_{12} = -c_{11}$ in (6), we get

$$u = c_{10} c_{11} \sin \frac{n\pi y}{b}\left(e^{\frac{n\pi x}{b}} - e^{-\frac{n\pi x}{b}}\right)$$

$$u = 2c_{10} c_{11} \sin \frac{n\pi y}{b} \sinh \frac{n\pi x}{b}$$

$$u = b_n \sin \frac{n\pi y}{b} \sinh \frac{n\pi x}{b} \qquad (b_n = 2\, c_{10}\, c_{11})$$

$$u = \sum_{n=1}^{\infty} b_n \sin \frac{n\pi y}{b} \sinh \frac{n\pi x}{b} \qquad \text{... (7) (General solution)}$$

Putting $x = a$ and $u = f(y)$ in (7), we get

$$u = f(y) = \sum_{n=1}^{\infty} b_n \sin \frac{n\pi y}{b} \sinh \frac{n\pi x}{b}$$ **Ans.**

We also know that

$$\left(\sin \frac{n\pi a}{b}\right) b_n = \frac{2}{b}\int_0^b f(y) \sin \frac{n\pi y}{b}\, dy$$

$$b_n = \frac{2}{b \sinh\left(\frac{n\pi a}{b}\right)} \int_0^b f(y) \sin \frac{n\pi y}{b}\, dy.$$

Example 32. *Find the deflection u(x, y, t) of the square membrane sin with a = b = c = 1, if the initial velocity is zero and the initial deflection f(x, y) = A sin πx sin 2πy.*

(Q.Bank U.P.T.U. 2001)

Solution. Here, the equation of the vibrations of the square membrane

$$\frac{\partial^2 u}{\partial t^2} = c^2\left(\frac{\partial^2 u}{\partial x^2} + \frac{\partial^2 u}{\partial y^2}\right) \qquad \text{...(1)}$$

Let $u = XYT$ ... (2)

Where X is function of x only, Y is a function of y only, and T is a function of t only.

$$\frac{\partial^2 u}{\partial t^2} = \frac{\partial^2}{\partial t^2}(XYT) = XY\frac{d^2 T}{dt^2}$$

$$\frac{\partial^2 u}{\partial x^2} = \frac{\partial^2}{\partial x^2}(XYT) = YT\frac{d^2 X}{dx^2}$$

$$\frac{\partial^2 u}{\partial y^2} = \frac{\partial^2}{\partial y^2}(XYT) = XT\frac{d^2 Y}{dy^2}$$

On putting the values of $\frac{\partial^2 u}{\partial t^2}$, $\frac{\partial^2 u}{\partial x^2}$ and $\frac{\partial^2 u}{\partial y^2}$ in (1), we get

$$XY\frac{d^2T}{dy^2} = c^2\left(YT\frac{d^2X}{dx^2} + XT\frac{d^2Y}{dy^2}\right)$$

On dividing by $c^2\ XYT$, we get

$$\frac{\frac{d^2T}{dt^2}}{c^2T} = \frac{\frac{d^2X}{dx^2}}{X} + \frac{\frac{d^2Y}{dy^2}}{Y}$$

This will be true only when each member is a constant. Choosing suitably, we have

$$\frac{\frac{d^2X}{dx^2}}{X} = -k^2, \quad \frac{\frac{d^2Y}{dy^2}}{Y} = -l^2 \quad \Rightarrow \quad \frac{\frac{d^2T}{dt^2}}{c^2T} = -(k^2+l^2)$$

$\frac{d^2X}{dx^2} + k^2X = 0$	$\frac{d^2Y}{dy^2} + l^2Y = 0$	$\frac{d^2T}{dt^2} + (k^2+l^2)c^2T = 0$
$A.E.\ is\ m^2 + k^2 = 0$	$A.E.\ is\ m^2 + l^2 = 0$	$A.E.\ is\ m^2 + (k^2+l^2)c^2 = 0$
$\Rightarrow m = ik,\ m = -ik$	$\Rightarrow m = il,\ m = -il$	$m = \sqrt{k^2+l^2}\ c,\ m = -i\sqrt{k^2+l^2}\ c$

$X = c_1 \cos kx + c_2 \sin kx \quad Y = c_3 \cos ly + c_4 \sin ly \quad T = c_5 \cos\sqrt{k^2+l^2}\,ct + c_6 \sin\sqrt{k^2+l^2}\,ct.$

Putting the values of X, Y and T in (2), we get

$$u = (c_1 \cos kx + c_2 \sin kx)(c_3 \cos ly + c_4 \sin ly)[c_5 \cos\sqrt{k^2+l^2}\ ct + c_6 \sin\sqrt{k^2+l^2}\ ct] \quad ...(3)$$

On putting $x = 0$ and $u = 0$ in (3), we get (given)

$$0 = c_1(c_3 \cos ly + c_4 \sin ly)[c_5 \cos\sqrt{k^2+l^2}\ ct + c_6 \sin\sqrt{k^2+l^2}\ ct] \quad ...(4)$$

$\Rightarrow c_1 = 0$

On putting $c_1 = 0$ in (3), we get (given)

$$u = c_2 \sin kx(c_3 \cos ly + c_4 \sin ly)(c_5 \cos\sqrt{k^2+l^2}\ ct + c_6 \sin\sqrt{k^2+l^2}\ ct) \quad ...(5)$$

Putting $x = 1$ and $u = 0$ in (5), we get (given)

$$u = c_2 \sin m\pi x(c_3 \cos ly + c_4 \sin ly)(c_5 \cos\sqrt{m^2\pi^2+l^2}\ ct + c_6 \sin\sqrt{m^2\pi^2+l^2}\ ct) \quad ...(7)$$

Putting $y = 0$, $u = 0$ in (7), we get

$$0 = c_2 \sin m\pi x . c_3(c_5 \cos\sqrt{m^2\pi^2+l^2}\ ct + c_6 \sin\sqrt{m^2\pi^2+l^2}\ ct) \quad ...(8)$$

$\Rightarrow c_3 = 0$

Putting the value of c_3 in (7), we get

$$u = c_2c_4 \sin m\pi x \sin ly\ (c_5 \cos\sqrt{m^2\pi^2+l^2}\ ct + c_6 \sin\sqrt{m^2\pi^2+l^2}\ ct) \quad ...(9)$$

Putting $y = 1$ and $u = 0$ in (9), we get (given)

$$0 = c_2c_4 \sin m\pi x \sin l\ (c_5 \cos\sqrt{m^2\pi^2+l^2}\ ct + c_6 \sin\sqrt{m^2\pi^2+l^2}\ ct) \quad ...(6)$$

$\Rightarrow \sin l = 0 = \sin n\pi \quad \Rightarrow \quad l = n\pi$

Putting the value $l = n\pi$ in (9), we get

$$u = c_2c_4 \sin m\pi x \sin n\pi y\ (c_5 \cos\sqrt{m^2\pi^2+n^2\pi^2}\ ct + c_6 \sin\sqrt{m^2\pi^2+n^2\pi^2}\ ct) \quad ...(10)$$

Putting $c_2c_4c_5 = A_{mn}$, $c_2c_4c_6 = B_{mn}$ and $p = \pi c\sqrt{m^2+n^2}$ in (10), we get the general solution as

$$u(x, y, t) = \sum_{m=1}^{\infty}\sum_{n=1}^{\infty} \sin m\pi x \sin n\pi y\,(A_{mn}\cos pt + B_{mn}\sin pt) \qquad \dots (11)$$ **Ans**

EXERCISE 49.7

Indicate True or False for the following :

1. The small transverse vibrations of a string are governed by one dimensional heat equation $y_t = a^2 y_{xx}$. (True/False) (*U.P. II Semester, 2009*) **Ans.** False
2. Two dimensional steady state heat flow is given by Laplace's equation $u_t = a^2(u_{xx} + u_{tt})$. (True/False) (*U.P. II Semester, 2009*) **Ans.** False
3. $\dfrac{\partial^2 y}{\partial t^2} = c^2\dfrac{\partial^2 y}{\partial x^2}$ is a two-dimensional wave equation. **Ans.** True
4. Radio equations are $V_{xx} = LCV_n$ and $I_{xx} = LCI_{tt}$ **Ans.** True
5. The small transverse vibrations of a string are $y_t^2 = a^2 y_{xx}$. **Ans.** False

49.10 VIBRATING MEMBRANE

Now we will discuss the equation of the vibrations of a tightly stretched membrane. Such as membrane of drum. Let T be the tension per unit length and m be the mass per unit area.

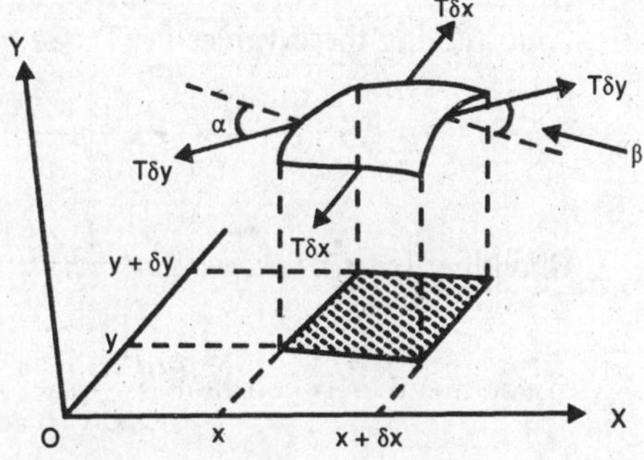

Now we want to discuss the forces acting on an element $\delta x\delta y$ of the membrance. Forces $T\delta x$ and $T\delta y$ act on the edges along the tangent to the membrane. Let the small displacement perpendicular to xy-plane be μ.

$T\delta y$ on the opposite edges of length δy make angles α and β to the horizontal.

Thus Vertical component

$= T\delta y \sin\beta - T\delta y \sin\alpha$

$= T\delta y (\tan\beta - \tan\alpha)$. [tan α = sin α , if α is very small]

$$= T\,\delta y\left[\left(\frac{\partial u}{\partial x}\right)_{x+\delta x} - \left(\frac{\partial u}{\partial x}\right)_x\right] = T\delta y\,\delta x\left[\frac{\left(\frac{\partial u}{\partial x}\right)_{x+\delta x} - \left(\frac{\partial u}{\partial x}\right)_x}{\delta x}\right] \qquad \dots\dots(1)$$

$$= T\,\delta x\,\delta y\frac{\partial^2 u}{\partial x^2} \text{ upto a first order of approximation.}$$

Another vertical component of the force $T\delta x$ acting on the adges of the length δx.

$$= \mathrm{T}\,\delta x\frac{\partial^2 u}{\partial y^2}\delta y \qquad \dots\dots(2)$$

Equation of the motion of the element is

$$\frac{d^2\mathrm{X}}{dx^2}k^2\mathrm{X} = 0, \quad \frac{d^2\mathrm{Y}}{dy^2} + l^2\mathrm{Y} = 0$$

$$m\,\delta x\,\delta y\frac{\partial^2 u}{\partial t^2} = \mathrm{T}\left[\frac{\partial^2 u}{\partial x^2} + \frac{\partial^2 u}{\partial y}\right]\delta x\,\delta y$$

$$\frac{\partial^2 u}{\partial t^2} = C^2\left[\frac{\partial^2 u}{\partial x^2} + \frac{\partial^2 u}{\partial y^2}\right] \quad \text{where } C^2 = \frac{T}{m} \qquad \text{...........(3)}$$

Where m is the mass per unit area of the membrane.

49.11 SOLUTION OF THE EQUATION OF THE VIBRATING MEMBRANE (RECTANGULAR MEMBRANE):

Let u be the solution of equation (3) and we assume that

$$u = X(x) \,.\, Y(y) \,.\, T(t) \qquad \text{........(4)}$$

On differentiating (4), we have

$$\frac{\partial^2 u}{\partial x^2} = Y \cdot T \frac{d^2 X}{dx^2}$$

$$\frac{\partial^2 u}{\partial y^2} = XT \frac{d^2 Y}{dy^2}$$

$$\frac{\partial^2 u}{\partial t^2} = XY \frac{d^2 T}{dt^2}$$

Substituting these values in (3), we get

$$XY\frac{d^2 T}{dt^2} = C^2\left[\frac{d^2 X}{dx^2} YT + XT\frac{d^2 Y}{dy^2}\right]$$

Dividing by $XYTc^2$, we get $\dfrac{1}{C^2}\dfrac{d^2T}{dt^2} = \left[\dfrac{1}{X}\dfrac{d^2X}{dx^2} + \dfrac{1}{Y}\dfrac{d^2Y}{dy^2}\right]$ ....(5)

Each member is constant $\dfrac{d^2X}{dx^2} + k^2 X = 0, \ \dfrac{d^2Y}{dy^2} + l^2 Y = 0$

Putting these values in (5), we get $\dfrac{d^2T}{dt^2} + \left(k^2 + l^2\right)c^2 T = 0$

$\dfrac{d^2X}{dx^2} + k^2X = 0$	$\dfrac{d^2Y}{dy^2} + l^2Y = 0$	$\dfrac{d^2T}{dt^2} + \left(k^2 + l^2\right)c^2 T = 0$
$(D^2 + k^2)\, X = 0$	$(D^2 + l^2)\, Y = 0$	$[D^2 + (k^2 + l^2)\, c^2]T = 0$
$m^2 + k^2 = 0$	$m^2 + l^2 = 0$	$m^2 + [k^2 + l^2]c^2 = 0$
$m = \pm ik$	$m = \pm il$	$m = iC\sqrt{k^2 + l^2}$
$X = c_1 \cos kx + c_2 \sin kx$	$Y = c_3 \cos ly + c_4 \sin ly$	$T = c_5 \cos c\sqrt{k^2 + l^2}$ $+ c_6 \sin c\sqrt{k^2 + l^2}$

Putting the values of X,Y,T in (4), we get

$$u(x, y, t) = (c_1 \cos kx + c_2 \sin kx)\, (c_3 \cos ly + c_4 \sin ly)$$

$$\left[c_5 \cos \sqrt{\left(k^2 + l^2\right)}ct + c_6 \sin \sqrt{\left(k^2 + l^2\right)}ct\right] \qquad \text{........(6)}$$

Suppose if the membrane is rectangular and stretched between the lines

$$x = 0, x = a, y = 0, y = b.$$

Then the boundary conditions are

(i) $u = 0$, when $x = 0$

(ii) $u = 0$, when $x = a$

(iii) $u = 0$, when $y = 0$

(iv) $u = 0$, when $y = b$ for all t.

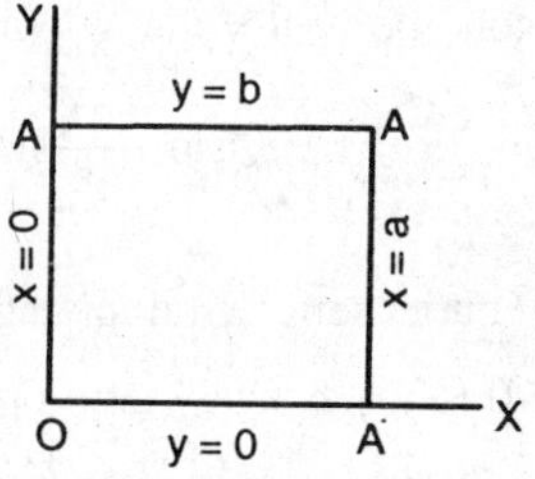

On putting $x = 0$; $u = 0$ in (6), we get

$$0 = c_1\left(c_3 \cos ly + c_4 \sin ly\right)\left[c_5 \cos\sqrt{\left(k^2+l^2\right)}\,ct + c_6 \sin\sqrt{\left(k^2+l^2\right)}\,ct\right]$$

$$c_1 = 0$$

On putting $c_1 = 0$, (6) becomes

$$u = c_2 \sin kx\left(c_3 \cos ly + c_4 \sin ly\right)\left[c_5 \cos x\sqrt{\left(k^2+l^2\right)}ct + c_6 \sin\sqrt{\left(k^2+l^2\right)}ct\right] \quad .....(7)$$

On putting $u = 0$ and $x = a$ in (7), we get

$$0 = c_2 \sin ka\left(c_3 \cos ly + c_4 \sin ly\right)\left[c_5 \cos\sqrt{\left(k^2+l^2\right)}\,ct + c_6 \sin\sqrt{\left(k^2+l^2\right)}\,ct\right]$$

$$\Rightarrow \quad \sin ka = 0 \quad \Rightarrow \sin ka = \sin m\pi \quad \Rightarrow k = \frac{m\pi}{a}$$

On putting the value of k in (7), we get

$$u = c_2 \sin\frac{m\pi x}{a}\left[c_3 \cos ly + c_4 \sin ly\right]\left[c_5 \cos x\sqrt{\frac{m^2\pi^2}{a^2}+l^2}\cdot ct + c_6 \sin\sqrt{\frac{m^2\pi^2}{a^2}+l^2}\cdot ct\right] \quad ....(8)$$

Similarly, applying the conditions (*iii*) and (*iv*) in (8) , we get

$$c_3 = 0 \text{ or } l = \frac{n\pi}{b}, \text{ where } n \text{ is an integer.}$$

(8) becomes,

$$u = c_2 c_4 \sin\frac{m\pi x}{a}\cdot\sin\frac{n\pi y}{b}\left(c_5 \cos x\sqrt{\frac{m^2\pi^2}{a^2}+\frac{n^2\pi^2}{b^2}}\;ct + c_6 \sin\sqrt{\frac{m^2\pi^2}{a^2}+\frac{n^2\pi^2}{b^2}}\;ct\right) \quad ......(9)$$

Let $$p = c\pi\sqrt{\frac{m^2}{a^2}+\frac{n^2}{b^2}}$$

Now, $u = c_2 c_4 \sin\dfrac{m\pi x}{a}\sin\dfrac{n\pi y}{b}\left(c_5 \cos pt + c_6 \sin pt\right)$ [From (9)]

These are the solution of wave equation (1) and are called eigen functions. Choosing the constants c_2 and c_4 so that $c_2 c_4 = 1$, we can write the general solution of (1) as

$$u = \sum_{m=1}^{\infty}\sum_{n=1}^{\infty} \sin\frac{m\pi x}{a}\sin\frac{n\pi y}{b}\left(A_{mn} \cos pt + B_{mn} \sin pt\right) \quad ....(10)$$

Differentiating (10) with respect to 't', we get

$$\frac{\partial u}{\partial t} = \sum_{m=1}^{\infty}\sum_{n=1}^{\infty} \sin\frac{m\pi x}{a}\sin\frac{n\pi y}{b}\left(-pA_{mn} \sin pt + pB_{mn} \cos pt\right) \quad ....(11)$$

On putting $\dfrac{\partial u}{\partial t} = 0$ when $t = 0$ in (11), we get

$$0 = \sum_{m-1}^{\infty}\sum_{n-1}^{\infty} \sin\frac{m\pi x}{a}\sin\frac{n\pi y}{b}\left(pB_{mn} \cos pt\right) \quad \Rightarrow B_{mn} = 0$$

Also using the condition : $u = f(x, y)$ when $t = 0$, we get

$$f(x,y) = \sum_{m=1}^{\infty}\sum_{n=1}^{\infty} A_{mn} \sin\frac{m\pi x}{a}\sin\frac{n\pi y}{b}$$

This is a double Fourier series. Multiplying both sides by $\sin\frac{m\pi x}{a}\sin\frac{n\pi y}{b}$ and integrating from $x = 0$ to $x = a$ and $y = 0$ to $y = b$, every term on the right except one,all become zero. Thus, we get

$$\int_0^a\int_0^b f(x,y) A_{mn} \sin\frac{m\pi x}{a}\sin\frac{n\pi y}{b}\, dydx = \frac{ab}{4}A_{mn}$$

It is the generalised Euler's formula and gives the coefficients in the solution.

On putting the value of A_{mn}, we get the solution of (1),

Example 33. ***Find the deflection u (x, y, t) of the square membrane with a = b = c = 1, if the initial velocity is zero and the initial deflection f(x, y) = A sin πx sin 2πy.***

(U.P .2001)

Solution : The equation of the vibration of the square membrane is

$$\frac{\partial^2 u}{\partial t^2} = c^2\left(\frac{\partial^2 u}{\partial x^2} + \frac{\partial^2 u}{\partial y^2}\right) \qquad \text{........(1)}$$

Boundary conditions are,

$$u(0, y, t) = 0,\ u(1, y, t) = 0$$
$$u(x, 0, t) = 0,\ u(x, 1, t) = 0$$

the initial conditions are

$$u(x, y, 0) = f(x, y) = A \sin px \sin 2py \text{ and } \left(\frac{\partial u}{\partial t}\right)_{t=0} = 0$$

Let, u be the solution of (1) where u = XYT and X is function of x only, Y is a function of y only, and T is a function of t only.

On differentiating u partially with respect to t, x, y, we get

$$\frac{\partial^2 u}{\partial t^2} = \frac{\partial^2}{\partial t^2}(XYT) = XY\frac{d^2T}{dt^2}$$

$$\frac{\partial^2 u}{\partial x^2} = \frac{\partial^2}{\partial x^2}(XYT) = YT\frac{d^2X}{dx^2}$$

$$\frac{\partial^2 u}{\partial y^2} = \frac{\partial^2}{\partial y^2}(XYT) = XT\frac{d^2Y}{dy^2}$$

On putting the derivatives in (1), we get $XY\frac{d^2T}{dt^2} = c^2\left(YT\frac{d^2X}{dx^2} + XT\frac{d^2Y}{dy^2}\right)$

On dividing by XYTc2, we get $\frac{1}{C^2}\frac{1}{T}\frac{d^2T}{dt^2} = \left(\frac{1}{X}\frac{d^2X}{dx^2} + \frac{1}{Y}\frac{d^2Y}{dy^2}\right)$ ....(3)

Equation (3) holds good when each member is a constant. *i.e.* l, and k are constants.

$\frac{1}{X}\frac{d^2X}{dx^2} = l^2$	$\frac{1}{Y}\frac{d^2Y}{dy^2} = -k^2$	$\frac{1}{T}\frac{d^2T}{dt^2} = \left(l^2 + k^2\right)$
$\frac{d^2X}{dx^2} + l^2X = 0,$	$\frac{d^2Y}{dy^2} + k^2Y = 0$ and	$\frac{d^2T}{dt^2} + \left(l^2 + k^2\right)c^2T = 0$
$\frac{d^2X}{dx^2} + l^2X = 0$	$\frac{d^2Y}{dy^2} + k^2Y = 0$	$\frac{d^2T}{dt^2} + \left(l^2 + k^2\right)c^2T = 0$

$(D^2 + l^2) X = 0$	$(D^2 + k^2) Y = 0$	$(D^2 + (l^2 + k^2) t^2 = 0$
$\Rightarrow m^2 + l^2 = 0$	$m^2 + k^2 = 0$	$D = \pm iC\sqrt{l^2 + k^2}$
$m = \pm il$	$\Rightarrow m = \pm ik$	$T = c_5 \cos\sqrt{l^2 + k^2}ct + c_6 \sin\sqrt{l^2 + k^2}ct$

$X = c_1 \cos kx + c_2 \sin kx$ $Y = x_3 \cos ly + c_4 \sin ly$

Putting the values of X, Y and T in (2), we get

$$u = (c_1 \cos kx + c_2 \sin kx)(c_3 \cos ly + c_4 \sin ly)\left[c_5 \cos\sqrt{l^2 + k^2}ct + c_6 \sin\sqrt{l^2 + k^2}ct\right] \quad \text{.......(4)}$$

On putting $x = 0, u = 0$ in (4), we get

$$0 = c_1 (c_3 \cos ly + c_4 \sin ly)\left(c_5 \cos\sqrt{l^2 + k^2}ct + c_6 \sin\sqrt{l^2 + k^2}ct\right) c_1 = 0 \quad \text{........(5)}$$

On putting $c_1 = 0$ in (5), we get

$$u = c_2 \sin kx (c_3 \cos ly + c_4 \sin ly)\left(c_5 \cos\sqrt{l^2 + k^2}ct + c_6 \sin\sqrt{l^2 + k^2}ct\right) \quad \text{.......(6)}$$

On putting $x = 1$ and $u = 0$ in (6), we have

$$0 = c_2 \sin k (c_3 \cos ly + c_4 \sin ly)\left(c_5 \cos\sqrt{l^2 + k^2}ct + c_6 \sin\sqrt{l^2 + k^2}ct\right)$$

$\Rightarrow \quad \sin l = 0 \Rightarrow \quad \sin l = \sin m\pi \quad \Rightarrow \quad l = m\pi$

Putting the value of $l = m\pi$ in (6), we get

$$u = c_2 \sin m\pi x (c_3 \cos ly + c_4 \sin ly)\left(c_5 \cos\sqrt{k^2 + l^2}ct + c_6 \sin\sqrt{k^2 + l^2}ct\right) \quad \text{.......(7)}$$

Putting $y = 0$ and $u = 0$ in (7), we get

$$0 = c_2 \sin m\pi x \cdot c_3 \left(c_5 \cos\sqrt{k^2 + l^2}ct + c_6 \sin\sqrt{k^2 + l^2} \cdot ct\right)$$

$\Rightarrow \quad c_3 = 0$

On putting $c_3 = 0$ in (7), we get

$$u = c_2 c_4 \sin m\pi x \sin ly\left(c_5 \cos\sqrt{k^2 + l^2} \cdot ct + c_6 \sin\sqrt{k^2 + l^2} \cdot ct\right) \quad \text{......(8)}$$

Now on putting $y = 1$ and $u = 0$ in (8), we get

$$u = c_2 c_4 \sin m\pi x \sin l\left(c_5 \cos\sqrt{k^2 + l^2} \cdot ct + c_6 \sin\sqrt{k^2 + l^2} \cdot ct\right)$$

$\Rightarrow \quad \sin l = 0 = \sin n\pi \quad \Rightarrow \quad l = n\pi$

Putting the value of $l = n\pi$ in (8), we get

$$u = c_2 c_4 \sin m\pi x \sin n\pi y\left(c_5 \cos\sqrt{m^2\pi^2 + n^2\pi^2} \cdot ct + c_6 \sin\sqrt{m^2\pi^2 + n^2\pi^2} \cdot ct\right) \quad \text{.....(9)}$$

If we put $p = \pi c\sqrt{m^2 + n^2}$ in (9), we get

$$u = \sin m\pi x \sin n\pi y (A_{mn} \cos pt + B_{mn} \sin pt)$$

The general equation is

$$u = \sum_{m=1}^{\infty}\sum_{n=1}^{\infty} \sin m\pi x \sin n\pi y (A_{mn} \cos pt + B_{mn} \sin pt) \quad \text{......(10)}$$

Differentiating (10) w.r.t 't', we get

$$\frac{\partial u}{\partial t} = \sum_{m=1}^{\infty}\sum_{n=1}^{\infty} \sin m\pi x \sin n\pi y (-p A_{mn} \sin pt + p B_{mn} \cos pt) \quad \text{....(11)}$$

On putting $\frac{\partial u}{\partial t} = 0$ and $t = 0$ in (11), we get

$$0 = \sum_{m=1}^{\infty}\sum_{n=1}^{\infty} \sin m\pi x \sin n\pi y[pB_{mn}] \quad \text{.......(12)}$$

$\Rightarrow$ $\quad B_{mn} = 0$

On putting $u = a \sin \pi x \sin 2\pi y$ and $t = 0$ in (12), we get

$$A \sin \pi x \sin 2\pi y = \sum_{n=1}^{\infty}\sum_{m=1}^{\infty} A_{mn} \sin m\pi x \sin n\pi y$$

$$A_{mn} = \frac{2}{1} \times \frac{2}{1} \int_0^1\int_0^1 A_{mn} \sin \pi x \sin 2\pi y \sin m\pi x \sin n\pi y dx dy$$

$\Rightarrow$ $\quad Am_1 = Am_3 = Am_5 = \ldots\ldots\ldots\ldots = 0$

But, $\quad Am_2 = 4A \int_0^1\int_0^1 \sin \pi x \sin m\pi x \sin^2 2\pi y dx dy \quad$ (of n is odd)

$$= 2A \int_0^1\int_0^1 \sin \pi x \sin m\pi x (1 - \cos 4\pi y)\, dx dy$$

$$= 2A \int_0^1 \sin \pi x \sin m\pi x \left(y - \frac{\sin 4\pi y}{4\pi}\right)_0^1 dx = 2A \int_0^1 \sin \pi x \sin m\pi x\, dx$$

$\Rightarrow$ $\quad A_{22} = A_{32} = A_{42} = \ldots\ldots\ldots\ldots = 0$

$$A_{12} = 2A \int_0^1 \sin^2 n\pi\, dx = A\int_0^1 (1 - \cos 2\pi x) dx = A\left(x - \frac{\sin 2\pi x}{2\pi}\right)_0^1 = A$$

$$u = A \sin \pi x \sin 2\pi y \cos pt$$

where $\quad p = \pi c\sqrt{m^2 + n^2} = \pi(1)\sqrt{1+4} = \pi\sqrt{5}.$

$$u = A \cos \pi\sqrt{5}t \sin \pi x \sin 2\pi y. \quad \textbf{Ans.}$$

Example 34. *Find the deflection u(x, y, t) of a rectangular membrane,* $0 \le x \le a, 0 \le y \le b$, *given that is entire its entire boundary is fixed, initial velocity is zero (starts from rest) and initial deflection f (x, y) = kxy (a – x) (b – y).*

Solution. We know from the Article 23.3, equation (10) that

$$u = \sum_{m=1}^{\infty}\sum_{n=1}^{\infty} \sin\frac{m\pi x}{a} \sin\frac{n\pi y}{b} (A_{mn} \cos pt + B_{mn} \sin pt) \quad \text{......(1)}$$

On differentiating (1), we get

$$\frac{\partial u}{\partial t} = \sum\sum \sin\frac{m\pi x}{a} \sin\frac{n\pi y}{b} (-PA_{mn} \sin pt + B_{mn}P \cos pt) \quad \text{......(2)}$$

On putting initial velocity $\frac{\partial u}{\partial t} = 0$, $t = 0$ in (2), we have

$$0 = \sum\sum \sin\frac{m\pi x}{a} \sin\frac{n\pi y}{b} (PB_{mn} \sin pt)$$

$\Rightarrow$ $\quad B_{mn} = 0$

Putting $B_{mn} = 0$ in (1) we haves

$$u = A_{mn} \sum\sum \sin\frac{m\pi x}{a} \sin\frac{n\pi y}{b} \cos pt \quad \text{......(3)}$$

The membrane starts from rest with initial condition is

$u = kxy\,(a - x)\,(b - y)$, when $t = 0$

$$kxy(a-x)(b-y) = A_{mn}\sum\sum \sin\frac{m\pi x}{a}\sin\frac{n\pi y}{b}$$

$$A_{mn} = \frac{4}{ab}\int_0^a\int_0^b f(x,y)\sin\frac{m\pi x}{a}\sin\frac{n\pi y}{b}\,dy\,dx$$

$$= \frac{4}{ab}\int_0^a\int_0^b kxy(a-x)(b-y)\sin\frac{m\pi x}{a}\sin\frac{n\pi y}{b}\,dy\,dx$$

$$\frac{a\,b\,A_{mn}}{4k} = \int_0^a\left(x(a-x)\sin\frac{m\pi x}{a}\,dx\right)\times\left[\int_0^a y(b-y)\sin\frac{n\pi y}{b}\,dy\right]$$

$$= I_1 \times I_2 \qquad ......(4)$$

$$I_1 = \int_0^a x(a-x)\sin\frac{m\pi x}{a}\,dx$$

$$= \int_0^a x(a-x)\frac{a}{mn}\left(-\cos\frac{m\pi n}{a}\right) - (a-2x)\frac{a^2}{m^2\pi^2}\left(-\sin\frac{m\pi x}{a}\right) + (-2)\frac{a^3}{m^3\pi^3}\cos\left(\frac{m\pi x}{a}\right)\Bigg]_0^a$$

$$= 0 - 0 - \frac{2a^3}{m^3\pi^3}(\cos m\pi - 1)$$

Similarly, $$I_2 = -\frac{2b^3}{n^3\pi^3}(\cos n\pi - 1)$$

Putting the values of I_1 and I_2, we get

$$\frac{ab}{4k}A_{mn} = \left[-\frac{2a^3}{m^3\pi^3}(\cos m\pi - 1)\right]\left[-\frac{2b^3}{n^3\pi^3}(\cos n\pi - 1)\right]$$

$$A_{mn} = \frac{4k}{ab}\left[-\frac{2a^3}{m^3\pi^3}\{(-1)^m - 1\}\right]\left[-\frac{2b^3}{n^3\pi^3}((-1)^n - 1)\right] = \frac{16a^3b^3k}{m^3n^3\pi^6}\left[(-1)^m - 1\right]\left[(-1)^n - 1\right]$$

$$= \begin{bmatrix} \dfrac{64a^2b^2k}{\pi^3 n^3 m^3}, \text{when } m \text{ and } n \text{ are odd.} \\ 0 \quad , \quad \text{otherwise} \end{bmatrix}$$

The required deflection $= u\,(x,\,y,\,t)$

$$= \left[\frac{64a^2bk}{\pi}\sum_{m=1}^{\infty}\sum_{n=1}^{\infty}\frac{1}{m^3}\sin\frac{m\pi x}{a}\right)\times\frac{1}{n^3}\sin\frac{n\pi y}{b}, \text{ where } k^2\pi^2\left[\frac{m^2}{a^2}+\frac{n^2}{b^2}\right]$$

Ans.

ALGEBRAIC AND TRANSCENDENTAL EQUATIONS

(Bisection Method, Regula falsi, Newton Raphson Method)

50.1 INTRODUCTION

In this chapter we will discuss methods of solving polynomial and transcendental equations.

50.2 POLYNOMIAL

An expression of the form

$P_n(x) = a_0 x^n + a_1 x^{n-1} + a_2 x^{n-2} + \ldots.. a_{n-1} x + a_n$

where $a_0, a_1 \ldots.. a_n$ are constants, n is positive integer, is called a polynomial in x of degree n provided $a_0 \neq 0$.

Zero of a Polynomial

Any value of x which makes $P_n(x) = a$, is called the zero of a polynomial $P_n(x)$.

Number of zeroes

Every polynomial $P_n(x)$ of degree n has exactly n zeroes.

Geometrically

The zero of the polynomial $P_n(x)$ is the value of x, where the graph of the $P_n(x)$ crosses the axis of x.

50.3 POLYNOMIAL EQUATION

Let $P_n(x)$ be a polynomial in x then

$P_n(x) = 0$ is known as polynomial equation.

50.4 TRANSCENDENTAL EQUATION

If a function $f(x)$ contains some other functions such as trigonometric, logarithmic, exponential etc. then $f(x) = 0$ is called transcendental equation.

Methods of Solving Equations

The polynomial equations and the transcendental equation can be solved by the following method:

(*i*) Graphical method
(*ii*) Bisection method
(*iii*) Newton-Raphson method
(*iv*) False position method
(*v*) Secant method
(*vi*) Iterative method

50.5 INITIAL APPROXIMATION

If $f(x)$ is continuous in the interval (a, b) and $f(a)$ and $f(b)$ have different signs then the equation $f(x) = 0$ has atleast one root between $x = a$ and $x = b$.

i.e. $a <$ Root $< b$.

Any value between a and b is the first approximate root of the given equation $f(x) = 0$.

If we have to choose an approximate root out of a and b then we have to see whether $f(b)/f(c)$ is nearer to zero. If $f(b)$ is nearer to zero then b is an initial approximate root of the given equation.

After finding out the initial approximate root then for better approximation we have to apply any one of the above methods.

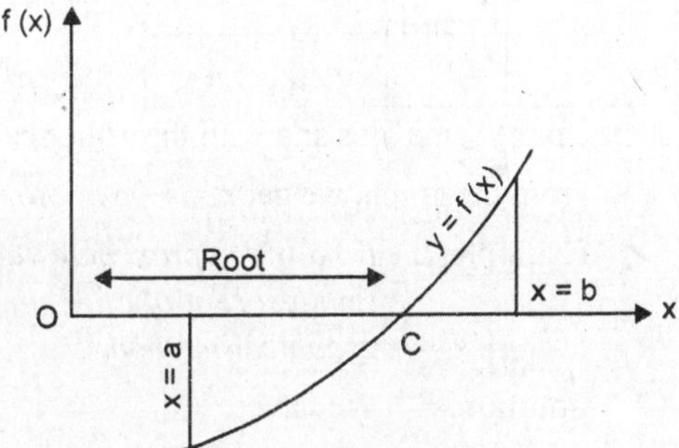

50.6 GRAPHICAL METHOD

There are two ways to solve an equation by graphical method.

Working Rule

Step 1. Find out the interval (a, b) in which a root of $f(x) = 0$ lies.

Step 2. Prepare a table of the values of x in the interval (a, b) and corresponding values of $f(x)$.

Step 3. Plot these points and pass a smooth curve through them by joining the points.

Step 4. Read the abscissa of the point of intersection of the curve $y = f(x)$ with x-axis.

This is a way to find a rough approximation to the root $f(x) = 0$.

The other way

Step 1. As above.

Step 2. Write the equation $f(x) = 0$ as $f_1(x) = f_2(x)$ where $f_2(x)$ contains only terms in x and constant.

Step 3. Draw the graphs of $y = f_1(x)$ and $y = f_2(x)$ on the same scale and with respect to the same axis.

Step 4. Read the abscissa of the point of the intersection of these two curves $y = f_1(x)$ and $y = f_2(x)$.

This is the initial approximation to root of $f(x) = 0$.

Example 1. *Find graphically the positive root of the equation* $x - 1 = \sin x$

Solution. Here, we have

$$x - 1 = \sin x \qquad \text{... (1)}$$

We break the equation (1) into two equations as below

$$y = \sin x \qquad \text{and } y = x - 1$$

To draw the graphs of $y = f_1(x)$ and $y = f_2(x)$.

We prepare the following tables

$y = \sin x$

x	0	$\frac{\pi}{4}$	$\frac{\pi}{2}$	$\frac{3\pi}{4}$	π
$y = \sin x$	0	0.71	1	0.71	0

$y = x - 1$

x	0	1	$\frac{3\pi}{4}$
$y = x - 1$	-1	0	1.4

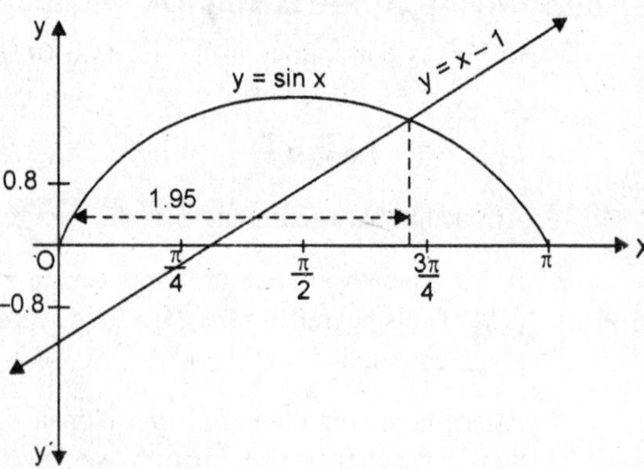

On the same axis and with the same scale construct the graphs of $y = \sin x$ and $y = x - 1$.

From the graph, we get $x = 1.95$ approximately. **Ans.**

Example 2. *Find the approximate value of the smallest root of $e^{-x} - \sin x = 0$, by graphical method.*

Solution. We have,

$$e^{-x} - \sin x = 0 \quad \dots (1)$$

Equation (1) is written in two equations as

$$y = e^{-x} \text{ and } y = \sin x.$$

On the same axis and same scale we draw the graphs of two curves $y = e^{-x}$ and $y = \sin x$. The two curves intersect at $x = 0.6$.

Therefore, the approximate smallest positive root is 0.6. **Ans.**

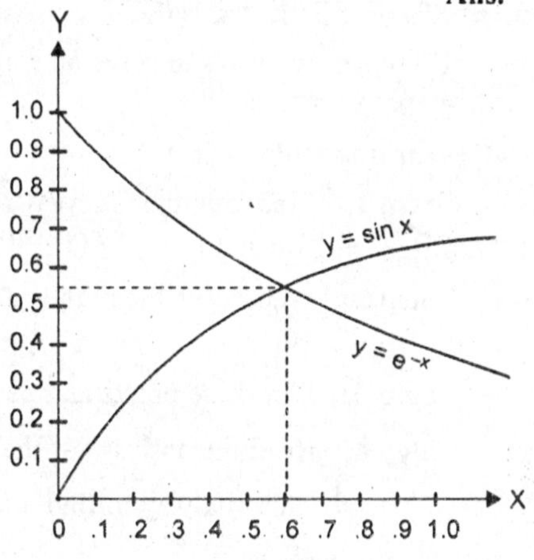

50.7 BISECTION METHOD (BOLZANO METHOD)

If $f(x)$ is continuous in the interval (a, b) such that $f(a)$ and $f(b)$ are of opposite signs, then $f(a) \cdot f(b) < 0$

The curve crosses the x-axis between a and b then the first approximation to the root is

$$x_1 = \frac{1}{2}(a+b)$$

f (a) = (–ve) f (b) = (+ve)

a b

Now there are three cases, if

1. $f(x_1) = 0$, then x_1 is the root of $f(x) = 0$
2. $f(x_1) < 0$, then root lies between x_1 and b. [If $f(a) = -\text{ve}$, $f(b) = +\text{ve}$]
3. $f(x_1) > 0$, then root lies between a and x_1.

Suppose $f(x_1) > 0$, then $a < \text{Root} < x_1$

f (a) = (–ve) $\frac{1}{2}(a+b)$ f (b) = (+ve)

a x_1 b

Second approximation to the root is $x_2 = \frac{1}{2}(a + x_1)$

Again there are three cases, if

1. $f(x_2) = 0$, then x_2 is the root of $f(x) = 0$.
2. $f(x_2) < 0$, then root lies between x_2 and x_1.

f (a) = – ve f (x_2) f (x_1) = +ve

a x_2 x_1 b

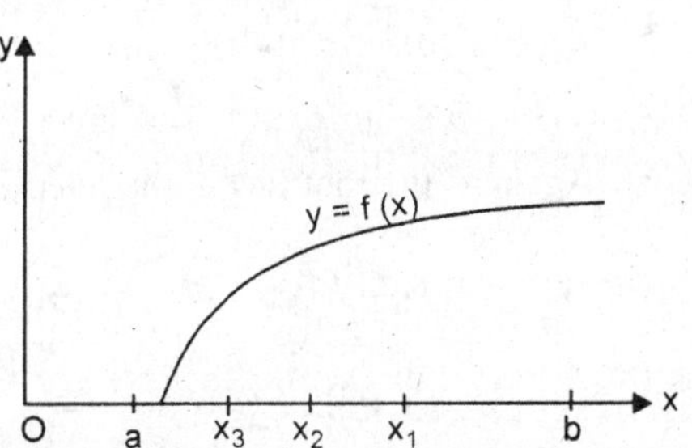

3. $f(x_2) > 0$, then root lies between a and x_2.

Suppose $f(x_2) > 0$.

Third approximation $x_3 = \frac{1}{2}(a + x_2)$

If **1.** $f(x_3) = 0$, then x_3 is the root of $f(x) = 0$

2. $f(x_3) > 0$, then root lies between a and x_3

and so on.

50.8 CONVERGENCE OF BISECTION METHOD

The successive approximations x_n of a root $x = \alpha$ of the equation $f(x) = 0$ is said to converge to $x = \alpha$ with order $q \geq 1$.

If $|x_{n+1} - \alpha| \leq C|x_n - \alpha|$

Here, $q, n > 0$, n and C is some constant greater than 0.

when $q = 1$ and $0 < C < 1$, then the convergence is said to be of first order and C is called the rate of convergence.

$$|x_{n+1} - \alpha| \leq C|x_n - \alpha|$$

50.9 TO SHOW THAT BISECTION METHOD IS ALWAYS CONVERGES

If (p_n, q_n) be the interval at nth step of Bisection and α being the exact root of the equation $f(x) = 0$.

Let x_n be the nth approximation of the root of the equation $f(x) = 0$. Then initially $p_1 = a$, and $q_1 = b$.

x_1 is the first approximation $= \frac{p_1 + q_1}{2}$

i.e.; $p_1 < x_1 < q_1$

Now either the root lies in (a, x_1), or (x_1, b).

Therefore either $(p_2, q_2) = [p_1, x_1]$ or $(p_2, q_2) = [x_1, q_1]$

Either $p_2 = p_1, q_2 = x_1$ or $p_2 = x_1, q_2 = q_1$

$\Rightarrow \quad p_1 \leq p_2, \; q_2 \leq q_1$

Further we have $x_2 = \left(\frac{p_2 + q_2}{2}\right)$ so that $p_2 \leq x_2 \leq q_2$ continuing this process we see that

$$x_n = \left(\frac{p_n + q_n}{2}\right)$$

$p_n \leq x_n \leq q_n$

$p_1 < p_2 < \ldots\ldots\ldots\ldots < p_n$

and $q_1 > q_2 > \ldots\ldots\ldots > q_n$.

Thus $(p_1, p_2, \ldots\ldots\ldots p_n, \ldots\ldots\ldots)$ is a bounded by b non-decreasing sequence.

$q_1, q_2 \ldots\ldots\ldots q_n$ is a bounded non increasing converge.

Let $\lim_{n\to\infty} p_n = p$ and $\lim_{n\to\infty} q_n = q$

But the length of the interval is decreasing at every step so we say that

$$\lim_{n\to\infty} (q_n - p_n) = 0 \quad \Rightarrow q = p.$$

Also, $p_n < x_n < q_n \quad \Rightarrow \lim p_n \le \lim x_n \le \lim q_n$

$\Rightarrow \quad p \le \lim x_n \le q \;\Rightarrow \lim x_n = p = q$ .... (1)

Again as the root lies in the interval $[p_n, q_n]$, we have

$$f(p_n) \cdot f(q_n) < 0 \quad \Rightarrow \quad 0 \ge \lim_{n\to\infty} [f(p_n) \cdot f(q_n)]$$

$$0 \ge f(p) \cdot f(q) \quad \Rightarrow \quad 0 \ge [f(p)]^2 \qquad \text{... (2)}$$

But $[f(p)]^2 \ge 0$, being square. So, we get $f(p) = 0$

$\Rightarrow \quad p$ is a root of the sequence $f(x) = 0$ ... (3)

Thus (1) and (2) imply that the sequence $<x_n>$ converges necessarily to a root of the equation $f(x) = 0$

But $|x_n - \alpha| \le \left(\frac{1}{2}\right)^n (b-a)$, hence, the method shows a linear convergence with a rate $\left(\frac{1}{2}\right)$.

So the method is not rapidly converging. But it is useful in the sense that it converges surely.

Example 3. *Perform five iterations of the bisection method to obtain the smallest positive root of the equation*

$$f(x) = x^3 - 5x + 1 = 0$$

Solution. We have,

$$f(x) = x^3 - 5x + 1 = 0$$

$$f(0) = +1$$

$$f(1) = 1 - 5 + 1 = -3$$

As $f(0) \cdot f(1) < 0$,

The first approximate root lies between 0 and 1.

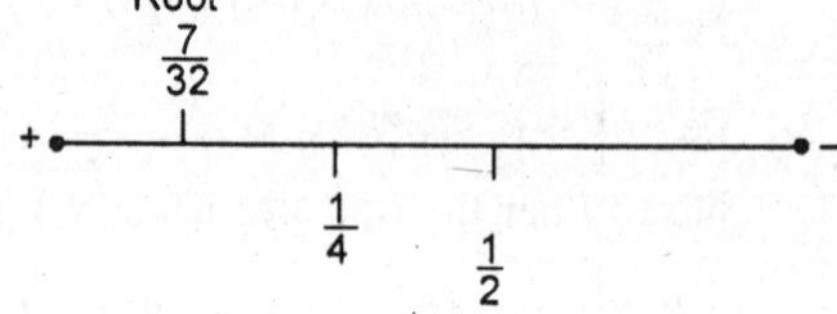

$$x_1 = \frac{0+1}{2} = \frac{1}{2}$$

$$f\left(\frac{1}{2}\right) = \frac{1}{8} - \frac{5}{2} + 1 = -1.375$$

As $f(0)\, f\left(\frac{1}{2}\right) < 0$

The second approximate lies between 0 and $\frac{1}{2}$.

$$x_2 = \frac{0+\frac{1}{2}}{2} = \frac{1}{4}$$

$$f\left(\frac{1}{4}\right) = \frac{1}{64} - \frac{5}{4} + 1 = -0.234375$$

As $f(0) \cdot f\left(\frac{1}{4}\right) < 0$

The third approximate root lies between 0 and $\frac{1}{4}$.

$$x_3 = \frac{0+\frac{1}{4}}{2} = \frac{1}{8}$$

$$f\left(\frac{1}{8}\right) = \frac{1}{512} - \frac{5}{8} + 1 = +0.37695$$

As $$f\left(\frac{1}{8}\right).f\left(\frac{1}{4}\right) < 0$$

The fourth approximate root lies between $\frac{1}{8}$ and $\frac{1}{4}$.

$$x_4 = \frac{\frac{1}{8}+\frac{1}{4}}{2} = \frac{3}{16}$$

$$f\left(\frac{3}{16}\right) = \left(\frac{3}{16}\right)^3 - \frac{15}{16} + 1 = +0.06909$$

As $$f\left(\frac{3}{16}\right).f\left(\frac{1}{4}\right) < 0$$

Then $\frac{3}{16} < \text{root} < \frac{1}{4}$.

$$x_5 = \frac{\frac{3}{16}+\frac{1}{4}}{2} = \frac{7}{32}, \qquad \text{Root} = \frac{7}{32}$$ **Ans.**

Example 4. *Find the real root of the equation*

$$x \sin x + \cos x = 0$$

between (2, 3), using Bisection method.

Solution. Let
$$f(x) = x \sin x + \cos x$$
$$f(2) = 2 \sin 2 + \cos 2 = 1.4024$$
$$f(3) = 3 \sin 3 + \cos 3 = -0.5666$$

As $$f(2).f(3) < 0$$
$$2 < \text{Root} < 3$$

f (2) = +ve   f (3) = –ve
2   2.5   2.75   2.875   3

First approximate root $= \frac{2+3}{2} = 2.5$

$$f(2.5) = 2.5 \sin 2.5 + \cos 2.5 = 0.6950$$

As $$f(2.5).f(3) < 0$$

The second approximate root $= \frac{2.5+3}{2} = 2.75$

$$f(2.75) = 2.75 \sin 2.75 + \cos 2.75 = 0.1253$$

As $$f(2.75).f(3) < 0$$

The third approximate root $= \frac{2.75+3}{2} = 2.875$

$$f(2.875) = 2.875 \sin 2.875 + \cos 2.875 = -0.2073$$

As $$f(2.875)f(2.75) < 0$$

The fourth approximate root $= \frac{2.875 + 2.75}{2} = 2.8125$

$$f(2.8125) = 2.8125 \sin 2.8125 + \cos 2.8125 = -0.0374$$

As $$f(2.75).f(2.8125) < 0$$

The fifth approximate root $= \frac{2.8125+2.75}{2} = 2.78125$

$$f(2.78125) = 2.78125 \sin 2.78125 + \cos 2.78125 = 0.0449$$

As $f(2.78125) \,.\, f(2.8125) < 0$

The sixth approximate root $= \dfrac{2.78125+2.8125}{2} = 2.796875$

$$f(2.796875) = 0.0040 \text{ which is nearly zero.}$$

Hence, the root of the given equation is 2.796875 **Ans.**

EXERCISE 50.1

1. Solve $x^3 - 9x + 1 = 0$ for the root between $x = 2$ and $x = 4$ by the method of Bolzano Method. **Ans.** 2.9375

2. Draw the graph $y = x^5$ and find the solution graphically of the equation $x^5 - x - 0.20 = 0$. **Ans.** -0.2

3. Find the positive root of $x^3 - x = 1$ correct to four decimal places by Bisection method. **Ans.** 1.3248

4. Draw the graph $y = x^3$ and $y = -2x + 20$ and find the appròximate soution of the equation $x^3 + 2x - 20 = 0$ **Ans.** 2.47

5. Find the approximate value of the root $\tan x = 1.2x$ **Ans.** $x = 0.71$ radian

6. Approximate the value of π by solving $\tan \dfrac{x}{4} - 1 = 0$, using Bisection method. **Ans.** 3.146

7. Find the positive root of $x - \cos x = 0$ by Bisection method. **Ans.** 0.7388

8. Find a positive root of the equation $2x = 3 + \cos x$, by Bisection method. **Ans.** 1.524

9. Find a positive root of the equation $e^x = 3x$, by Bisection method. **Ans.** 0.6190

10. Solve $e^x + x^4 + x = 2$ by Bisection method. **Ans.** 0.429494

50.10 NEWTON – RAPHSON METHOD (*U.P., III Semester Dec. 2009*)

Let x_0 be an approximate root of $f(x) = 0$ and let $x_1 = x_0 + h$ be the correct root so that $f(x_0 + h) = 0$

To find *h*, we expand $f(x_0 + h)$ by Taylor's Series

$$f(x_0 + h) = f(x_0) + h f'(x_0) + \frac{h^2}{2!} f''(x_0) + \ldots\ldots \qquad [f(x_0 + h) = 0]$$

$$0 = f(x_0) + h f'(x_0) \quad \text{[Neglecting the second and higher order derivative]}$$

$$h = -\frac{f(x_0)}{f'(x_0)}$$

But $x_1 = x_0 + h$

Putting the value of h, we get $\Rightarrow x_1 = x_0 - \dfrac{f(x_0)}{f'(x_0)}$

x_1 is better approximation than x_0. $\quad x_2 = x_1 - \dfrac{f(x_1)}{f'(x_1)}$

x_2 is better approximation than x_1.

Successive approximations are $x_3, x_4, \ldots\ldots\ldots\ldots x_{n+1}$.

$$\boxed{x_{n+1} = x_n - \frac{f(x_n)}{f'(x_n)}}$$

Which is the Newton - Raphson formula.

Note 1. Newton method is the best known procedure for finding the roots of an equation. It is applicable to the solution of all types of equations *i.e.*, algebraic and transcendental and also useful for calculating complex roots.

2. This method is useful in case of large value of $f'(x)$. For large $f'(x)$, h will be small.

3. This formula converges rapidly. If the initial approximation x_0 is taken very close to the root α.

Thus proper choice of x_0 is very important for the success of this method.

50.11 GEOMETRICAL INTERPRETATION

(AMIETE, June 2010)

Let $P_0 P$ be a curve $y = f(x)$.

Slope of the tangent $P_0 B$ to the curve at the point $P_0(x_0, y_0) = f'(x_0)$.

Tangent $P_0 B$ cuts the x-axis at B *i.e.* $(x_1, 0)$.

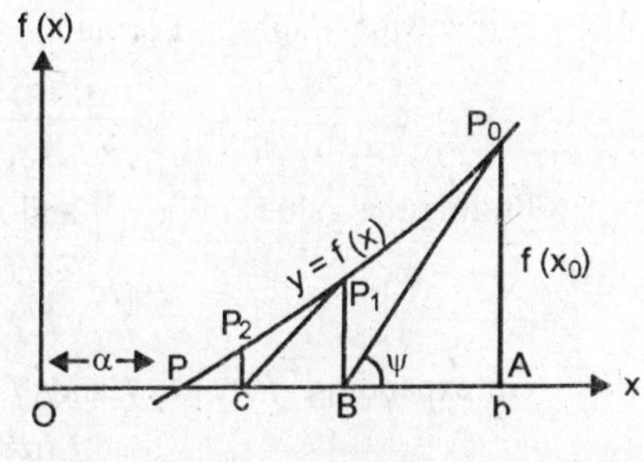

$$\begin{aligned} x_1 &= OB \\ &= OA - AB \\ &= x_0 - P_0 A \cot \Psi \\ &= x_0 - \frac{P_0 A}{\tan \psi} \quad \left[AN = f(x_0), \frac{BA}{AN} = \cot \psi \right] \\ &= x_0 - \frac{f(x_0)}{f'(x_0)} \quad \text{(First approximation)} \end{aligned}$$

The tangent to the curve at P_1 (corresponding to x_1) cuts the axis at $C(x_2, 0)$.

Using x_1 as the starting point, then

Similarly $\quad x_2 = x_1 - \dfrac{f(x_1)}{f'(x_1)}$

Now x_2 is nearer to α than x_1 (second approximation).

The process can be repeated and the root α is approached very fast.

$$x_{n+1} = x_n - \frac{f(x_n)}{f'(x_n)}$$

50.12 CONVERGENCE OF NEWTON – RAPHSON FORMULA

By Newton-Raphson formula

$$x_{n+1} = x_n - \frac{f(x_n)}{f'(x_n)}$$

Let $\quad \phi(x) = x - \dfrac{f(x)}{f'(x)} = \dfrac{x f'(x) - f(x)}{f'(x)}$

On differentiating both sides w.r.t 'x', we get

$$\phi'(x) = \frac{f'(x)\left[f'(x) + x f''(x) - f'(x)\right] - \left[x f'(x) - f(x)\right] f''(x)}{[f'(x)]^2}$$

$$\Rightarrow \quad \phi'(x) = \frac{[f'(x)]^2 + x f'(x) f''(x) - [f'(x)]^2 - x f'(x) f''(x) + f(x) f''(x)}{[f'(x)]^2}$$

$$\Rightarrow \qquad \phi'(x) = \frac{f(x) f''(x)}{[f'(x)]^2}$$

For convergence, $|\phi'(x)|<1$

$$\frac{f(x).f''(x)}{[f'(x)]^2} < 1$$

$$f(x) f''(x) < [f'(x)]^2$$ (*GBTU, Dec.2012*)

50.13 RATE OF CONVERGENCE OF NEWTON - RAPHSON FORMULA

Let x_n (approximate root) differs from the actual root α by a small quantity h_n.

So
$$x_n = \alpha + h_n \quad \text{... (1)}$$
$$x_{n+1} = \alpha + h_{n+1} \quad \text{... (2)}$$

By Newton-Raphson Formula

$$x_{n+1} = x_n - \frac{f(x_n)}{f'(x_n)} \quad \text{... (3)}$$

Putting the values of x_{n+1} and x_n from (1) and (2) in (3), we get

$$\alpha + h_{n+1} = \alpha + h_n - \frac{f(\alpha+h_n)}{f'(\alpha+h_n)} \Rightarrow h_{n+1} = h_n - \frac{f(\alpha+h_n)}{f'(\alpha+h_n)}$$

On expanding $f(\alpha+h_n)$ and $f'(\alpha+h_n)$ by Taylor's Series, we get

$$h_{n+1} = h_n - \frac{f(\alpha)+h_n f'(\alpha)+\frac{1}{2!} h_n^2 f''(\alpha)+.....}{f'(\alpha)+h_n f''(\alpha)+.....}$$

We know that $f(\alpha)=0$, so

$$h_{n+1} = h_n - \frac{h_n f'(\alpha) + \frac{1}{2} h_n^2 f''(\alpha)+...}{f'(\alpha)+h_n f''(\alpha)+.....}$$

$$= \frac{h_n f'(\alpha)+h_n^2 f''(\alpha) - h_n f'(\alpha) - \frac{1}{2} h_n^2 f''(\alpha)+....}{f'(\alpha)+h_n f''(\alpha)+.....} = \frac{\frac{1}{2} h_n^2 f''(\alpha)}{f'(\alpha)+h_n f''(\alpha)}$$

$$h_{n+1} = h_n^2 \left(\frac{f''(\alpha)}{2f'(\alpha)}\right) \text{ approximately} \qquad [f''(\alpha) \text{ neglected}]$$

$$h_{n+1} \propto h_n^2 \qquad \left(\frac{f''(\alpha)}{2f'(\alpha)} \text{ constant}\right)$$

1. It means that subsequent error h_{n+1} at each step is proportional to the square of the previous error h_n. So, the number of correct decimal is approximately doubled at each iteration if $\frac{f''(\alpha)}{2f'(\alpha)}$ is not too large.
2. Convergence is of quadratic order *i.e.* P = 2.

50.14 ORDER OF CONVERGENCE

$$x_n = \alpha + h_n$$
$$x_{n+1} = \alpha + h_{n+1}$$
$$h_{n+1} = h_n^k$$

k is called the order of convergence.
If $k = 1$, the convergence is linear.
If $k = 2$, it is quadratic.

50.15 WORKING RULE TO SOLVE $f(x) = 0$ BY NEWTON - RAPHSON METHOD

Step 1. Choose two close numbers b and c such that $f(b)$ and $f(c)$ are of opposite signs. then the root α lies between b and c.

Step 2. Out of $f(b)$ and $f(c)$ choose which is nearer to zero. If $f(b)$ is nearer to zero then b is an initial approximate root (x_0) of the given equation.

Step 3. Apply Newton Raphson formula to find out better approximate root x

$$x_1 = x_0 - \frac{f(x_0)}{f'(x_0)}$$

Repeat the process to get successive approximation.

Step 4. Stop when two approximate roots are equal.

Example 5. *Find the real root of the following equation, correct to three decimal places, using Newton-Raphson Method.* (*R.G.P.V., III Semester, June 2007*)

$$x^3 - 2x - 5 = 0$$

Solution. $$x^3 - 2x - 5 = 0 \qquad \ldots (1)$$

Let
$$f(x) = x^3 - 2x - 5$$
$$f(2) = 8 - 4 - 5 = -1$$
$$f(2.5) = (2.5)^3 - 2(2.5) - 5 = +5.625$$

Since $f(2)$ and $f(2.5)$ are, of opposite signs, the root of (1) lies between 2 and 2.5; $f(2)$ is near to zero than $f(2.5)$, So 2 is better appropriate root than 2.5.

$$f'(x) = 3x^2 - 2 \Rightarrow f'(2) = 12 - 2 = 10$$

Let 2 be an approximate root of (1). By Newton-Raphson method

$$x_1 = x_0 - \frac{f(x_0)}{f'(x_0)} = 2 - \frac{f(2)}{f'(2)} = 2 - \frac{-1}{10} = 2.1$$

$$f(2.1) = (2.1)^3 - 2(2.1) - 5 = 9.261 - 4.2 - 5 = 0.061$$
$$f'(2.1) = 3(2.1)^2 - 2 = 11.23$$

$$x_2 = 2.1 - \frac{f(2.1)}{f'(2.1)} = 2.1 - \frac{0.061}{11.23} = 2.1 - 0.00543 = 2.09457$$

$$f(2.09457) = (2.09457)^3 - 2(2.09457) - 5$$
$$= 9.1893 - 4.18914 - 5 = 0.00016$$
$$f'(2.09457) = 3(2.09457)^2 - 2 = 13.16167 - 2 = 11.16167$$

$$x_3 = 2.09457 - \frac{f(2.09457)}{f'(2.09457)} = 2.09457 - \frac{0.00016}{11.16167} = 2.09457 - 0.000014 = 2.09456$$

As $x_3 = x_2$ correct upto four places of decimal.

Hence, the root of the given equation is 2.0945 correct upto four places of decimal. **Ans.**

Example 6. *Derive the Newton-Raphson formula for finding a root of a non-linear equation. Find a root of* $f(x) = x^3 + 2x^2 + 10x - 20 = 0$ *up to 10 iterations.* (*U.P., III Semester, Dec. 2009*)

Solution. **For derivation of the formula** **see Art. 50.10 on page 1378.**

$$\text{Newton-Raphson Formula } x_{n+1} = x_n - \frac{f(x_n)}{f'(x_n)}$$

$$f(x) = x^3 + 2x^2 + 10x - 20 \quad \ldots (1) \qquad \Rightarrow f'(x) = 3x^2 + 4x + 10$$

$$f(1) = 1^3 + 2(1)^2 + 10(1) - 20 = 1 + 2 + 10 - 20 = -7$$

$$f(2) = 2^3 + 2(2)^2 + 10(2) - 20 = 8 + 8 + 20 - 20 = 16$$

Since $f(1)$ and $f(2)$ are of opposite signs, so the root of (1) lies between 1 and 2.

As $f(1)$ is near to zero than $f(2)$. So, 1 is better approximate root than 2.

$f'(1) = 3(1)^2 + 4(1) + 10 = 3 + 4 + 10 = 17$

$$x_1 = x_0 - \frac{f(x_0)}{f'(x_0)} \qquad \text{(Newton-Raphson formula)}$$

First Iteration

$$x_1 = x_0 - \frac{f(1)}{f'(1)} = 1 - \frac{-7}{17} = \frac{24}{17} \qquad f\left(\frac{24}{17}\right) = 0.9175656$$

Second Iteration

$$x_2 = x_1 - \frac{f(x_1)}{f'(x_1)} = \frac{24}{17} - \frac{f\left(\frac{24}{17}\right)}{f'\left(\frac{24}{17}\right)} = 1.4117647 - \frac{0.9175656}{21.6262976} \qquad f'\left(\frac{24}{17}\right) = 21.6262976$$

$$= 1.4117647 - 0.04242823 = 1.36933647$$

Third Iteration

$$x_3 = x_2 - \frac{f(x_2)}{f'(x_2)} = 1.36933647 - \frac{f(1.36933647)}{f'(1.36933647)}$$

$$= 1.36933647 - \frac{0.01114811}{21.1025930}$$

$$= 1.36933647 - 0.00052828 \quad = 1.36880819$$

Fourth Iteration

$$x_4 = x_3 - \frac{f(x_3)}{f'(x_3)} = 1.36880819 - \frac{f(1.36880819)}{f'(1.36880819)}$$

$$= 1.36880819 - \frac{0.00000173}{21.09614034} \quad = 1.36880819 - 0.00000008$$

$$= 1.36880811$$

Fifth Iteration

$$x_5 = x_4 - \frac{f(x_4)}{f'(x_4)} = 1.36880811 - \frac{f(1.36880811)}{f'(1.36880811)}$$

$$= 1.36880811 - \frac{0.00000005}{21.09618937} \quad = 1.36880811 - 0.00000000$$

$$= 1.36880811$$

The root of the given equation is 1.36880811 correct upto eigth decimal place after fifth iteration. For the accuracy more than 8 decimal places, we can iterate further. **Ans.**

Example 7. *Find the real root of the equation $x^4 - x - 9 = 0$ by Newton-Raphson Method, correct to three places of decimal.*

(R.G.P.V., Bhopal, III Semester, June 2006)

Solution. $f(x) = x^4 - x - 9 = 0, \qquad f'(x) = 4x^3 - 1$

$f(0) = -9$

$f(1) = 1^4 - 1 - 9 = 1 - 1 - 9 = -9, f(2) = 2^4 - 2 - 9 = 16 - 2 - 9 = 5$

As $f(2)$ is nearer to zero we take 2 as an approximate root of $f(x)$.

$f'(2) = 4(2)^3 - 1 = 4 \times 8 - 1 = 32 - 1 = 31$

By Newton-Raphson method

$$x_1 = x_0 - \frac{f(x_0)}{f'(x_0)}$$

$$x_1 = 2 - \frac{f(2)}{f'(2)} = 2 - \frac{5}{31} = 1.8387$$

$$f(1.8387) = (1.8387)^4 - 1.8387 - 9 = 11.4299 - 1.8387 - 9 = 0.5912$$

$$f'(1.8387) = 4(1.8387)^3 - 1 = 23.8652$$

$$x_2 = 1.8387 - \frac{f(1.8387)}{f'(1.8387)} = 1.8387 - \frac{0.5912}{23.8652} = 1.8387 - 0.02477$$

$$= 1.8139$$

$$f(1.8139) = (1.8139)^4 - 1.8139 - 9 = 0.01173$$

$$f'(1.8139) = 4(1.8139)^3 - 1 = 22.8726$$

$$x_3 = 1.8139 - \frac{f(1.8139)}{f'(1.8139)} = 1.8139 - \frac{0.01173}{22.8726}$$

$$= 1.8139 - 0.0005 = 1.8134$$

$$f(1.8134) = (1.8134)^4 - 1.8134 - 9 = 10.8137 - 1.8134 - 9 = 0.0003$$

$$f'(1.8134) = 4(1.8134)^3 - 1 = 22.8529$$

$$x_4 = 1.8134 - \frac{f(1.8134)}{f'(1.8134)} = 1.8134 - \frac{0.0003}{22.8529} = 1.8134 - 0.000013$$

$$x_4 = 1.8134$$

As x_3 and x_4 are equal, so root = 1.8134

Hence, real root of the given equation is 1.8134 **Ans.**

Example 8. *By using Newton-Raphson's Method, find the root of $x^4 - x - 10 = 0$, which is near to $x = 2$ correct to three places of decimal.*

(GBTU. Dec. 2012, R.G.P.V., Bhopal, III Semester, June 2008, Dec. 2003

Solution. $f(x) = x^4 - x - 10, \qquad f'(x) = 4x^3 - 1$

$$f(2) = 16 - 2 - 10 = 4$$

$$f'(2) = 32 - 1 = 31$$

By Newton-Raphson's method

$$x_1 = x_0 - \frac{f(x_0)}{f'(x_0)} = 2 - \frac{f(2)}{f'(2)} = 2 - \frac{4}{31} = 2 - 0.129 = 1.871$$

$$f(1.871) = (1.871)^4 - 1.871 - 10 = 12.2545 - 1.871 - 10 = 0.3835$$

$$f'(1.871) = 4(1.871)^3 - 1 = 4 \times 6.5497 - 1 = 25.1988$$

$$x_2 = 1.871 - \frac{f(1.871)}{f'(1.871)} = 1.871 - \frac{0.3835}{25.1988} = 1.871 - 0.0152 = 1.8558$$

$$f(1.8558) = (1.8558)^4 - (1.8558) - 10 = 11.8611 - 11.8558 = 0.0053$$

$$f'(1.8558) = 4(1.8558)^3 - 1 = 4 \times 6.3914 - 1 = 24.5656$$

$$x_3 = 1.8558 - \frac{f(1.8558)}{f'(1.8558)} = 1.8558 - \frac{0.0053}{24.5656} = 1.8558 - 0.00022$$

$$= 1.85558$$

As $x_2 = x_3$ correct upto three places of decimals, so the correct root of the given equation is 1.856 **Ans.**

Example 9. *Using Newton – Raphson Method find an iterative scheme to compute the reciprocal of a positive number.* *(AMIETE, Dec. 2010)*

Solution. Let x be the reciprocal of a given positive number N.

$$x = \frac{1}{N} \text{ or } N - \frac{1}{x} = 0$$

$$f(x) = N - \frac{1}{x} = 0, \; f'(x) = \frac{1}{x^2}$$

By Newton - Raphson Method

$$x_{n+1} = x_n - \frac{f(x_n)}{f'(x_n)} = x_n - \frac{N - \frac{1}{x_n}}{\frac{1}{x_n^2}} = x_n - x_n^2\left(N - \frac{1}{x_n}\right)$$

$$= x_n - N x_n^2 + x_n = 2x_n - Nx_n^2 = x_n[2 - N x_n] \qquad \textbf{Ans.}$$

Example 10. *Find the value of* $\frac{1}{18}$ *by Newton - Raphson Method.* *(AMIETE, Dec. 2010)*

Solution. Let $x = \frac{1}{18}$

Let $f(x) = \frac{1}{x} - 18 = 0$ ...(1)

$\Rightarrow f'(x) = -\frac{1}{x^2}$

Let the approximate root of equation (1) be 0.05

By Newton - Raphson Method

$x_1 = x_0(2 - 18x_0) = 0.05(2 - 18 \times 0.05) = 0.05(2 - 0.90) = 0.05 \times 1.10 = 0.055$

$x_2 = x_1(2 - 18x_1) = 0.055(2 - 18 \times 0.055) = 0.055(2 - 0.99) = 0.055 \times 1.01 = 0.05555$

$x_3 = x_2(2 - 18x_2) = 0.05555(2 - 18 \times 0.05555) = 0.05555(1.0001) = 0.05556$

Hence, $\frac{1}{18} = 0.0556$ approximately. **Ans.**

Example 11. *Using Newton-Raphson method find an iterative scheme to compute the cube root of a positive number.* *(AMIETE, Dec. 2010)*

Solution. Let x be the cube root of a given positive number N.

$$x = (N)^{\frac{1}{3}} \text{ or } x^3 = N \text{ or } x^3 - N = 0$$

Let $f(x) = x^3 - N = 0 \Rightarrow f'(x) = 3x^2$

By Newton-Raphson Method

$$x_{n+1} = x_n - \frac{f(x_n)}{f'(x_n)} = x_n - \frac{x_n^3 - N}{3x_n^2} = \frac{3x_n^3 - x_n^3 + N}{3x_n^2}$$

$$= \frac{2x_n^3 + N}{3x_n^2} \text{ for } n = 0, 1, 2.... \qquad \textbf{Ans.}$$

Example 12. *Write the Newton-Raphson procedure for finding* $\sqrt[3]{N}$, *where N is a real number. Use it to find* $\sqrt[3]{18}$ *correct to 2 decimals, assuming 2.5 as the initial approximation.* *(AMIETE, Dec. 2010)*

Solution. Let $x = \sqrt[3]{18} \Rightarrow x^3 = N \Rightarrow x^3 - N = 0$

Let $f(x) = x^3 - N = 0 \Rightarrow f'(x) = 3x^2$

By Newton-Raphson Method,

$$x_{n+1} = x_n - \frac{f(x_n)}{f'(x_n)}$$

$$x_{n+1} = x_n - \frac{x_n^3 - N}{3x_n^2} = \frac{2x_n^3 + N}{3x_n^2}, \quad n = 0, 1, 2, \ldots\ldots \qquad \ldots (1)$$

Putting $N = 18$, x_0 = app. cube root of 18 = 2.5 in (1), we get

$$x_1 = \frac{2(2.5)^3 + 18}{3 \times (2.5)^2} = 2.62667$$

Repeat this method. **Ans.**

Example 13. *Find an iterative formula to find* $\sqrt{N}$ *(where N is a positive number) and hence find*

(a) $\sqrt{5}$, *correct to 5 decimal places.*

(b) Find $\sqrt{24}$. *(A.M.I.E., Summer 2001, 2000)*

Solution. Let x be the square root of a given positive number N.

$$x = (N)^{\frac{1}{2}} \quad \Rightarrow \quad x^2 = N \quad \Rightarrow \quad x^2 - N = 0$$

Let $f(x) = x^2 - N = 0 \quad \Rightarrow \quad f'(x) = 2x$

By Newton-Raphson Method

$$x_{n+1} = x_n - \frac{f(x_n)}{f'(x_n)}$$

$$x_{n+1} = x_n - \frac{x_n^2 - N}{2x_n} = \frac{2x_n^2 - x_n^2 + N}{2x_n} = \frac{x_n^2 + N}{2x_n}$$

$$x_{n+1} = \frac{x_n^2 + N}{2x_n}; \; n = 0, \ldots\ldots$$ This is the iterative formula to find $\sqrt{N}$.

Let $N = 5$, x_0 = app. root of 5 = 2.

$$x_1 = \frac{(2)^2 + 5}{2 \times 2} = 2.25,$$

$$x_2 = \frac{(2.25)^2 + 5}{2 \times 2.25} = 2.23611$$ **Ans.**

(*b*) $N = 24$, x_0 = app. root of 24 = 5

$$x_1 = \frac{(5)^2 + 24}{2 \times 5} = 4.9, \quad x_2 = \frac{(4.9)^2 + 24}{2 \times 4.9} = 4.899$$

$$x_3 = \frac{(4.899)^2 + 24}{2 \times 4.899} = 4.89898$$ **Ans.**

Example 14. *By iteration method find the value of* $(48)^{1/3}$, *correct to three decimal places.* *(R.G.P.V., Bhopal, III Semester Dec. 2003)*

Solution. Let $x = \sqrt[3]{N} \Rightarrow x^3 = N \Rightarrow x^3 - N = 0$

Let $f(x) = x^3 - N = 0 \quad \Rightarrow f'(x) = 3x^2$

By Newton-Raphson Method

$$x_{n+1} = x_n - \frac{f(x_n)}{f'(x_n)}$$

$$x_{n+1} = x_n - \frac{x_n^3 - N}{3x_n^2}, \; n = 0, 1, 2\ldots\ldots \quad \Rightarrow \quad x_{n+1} = \frac{3x_n^3 - x_n^3 + N}{3x_n^2}$$

$$\Rightarrow \quad x_{n+1} = \frac{2x_n^3 + N}{3x_n^2}, \; n = 0, 1, 2, \ldots\ldots \qquad \ldots (1)$$

Putting $N = 48$, x_0 = Approximate root of 48 = 3.5 in (1), we get

$$x_1 = \frac{2(3.5)^3 + 48}{3(3.5)^2} = \frac{133.75}{36.75} = 3.6395$$

$$x_2 = \frac{2(3.6395)^3 + 48}{3(3.6395)^2} = \frac{144.4173}{39.7379} = 3.6342$$

$$x_3 = \frac{2(3.6342)^3 + 48}{3(3.6342)^2} = \frac{143.9967}{39.6222} = 3.6342$$

As $x_2 = x_3$ so $(48)^{\frac{1}{3}} = 3.6342$ is correct upto 4 decimal places.

Hence, $(48)^{\frac{1}{3}} = 3.6342$ **Ans.**

Example 15. *An iterative method for finding $\sqrt{N}$, where N is a real number can be written as $x_{n+1} = \frac{1}{2} x_n \left[\left(3 - \frac{x_n^2}{N}\right) + \frac{3}{4}\left(1 - \frac{x_n^2}{N}\right)^2\right]$. Find the rate of convergence of the method.*

Solution. Let $x = \sqrt{N}$, $x^2 - N = 0$, $f(x) = x^2 - N$

Let the correct root be α.

$$\alpha^2 = N$$

Let x_n be the approximate value of α and the error in x_n be h_n, then

$x_n = \alpha + h_n$ and $x_{n+1} = \alpha + h_{n+1}$

Here the given formula is

$$x_{n+1} = \frac{1}{2} x_n \left[\left(3 - \frac{x_n^2}{N}\right) + \frac{3}{4}\left(1 - \frac{x_n^2}{N}\right)^2\right] \quad \ldots (1)$$

Putting the values of x_n, x_{n+1}, N in (1), we get

$$\alpha + h_{n+1} = \frac{1}{2}(\alpha + h_n)\left[\left(3 - \frac{(\alpha + h_n)^2}{\alpha^2}\right) + \frac{3}{4}\left(1 - \frac{(\alpha + h_n)^2}{\alpha^2}\right)^2\right]$$

$$= \frac{1}{2}(\alpha + h_n)\left[3 - \frac{\alpha^2}{\alpha^2}\left(1 + \frac{h_n}{\alpha}\right)^2\right] + \frac{3}{4}\left[1 - \frac{\alpha^2}{\alpha^2}\left(1 + \frac{h_n}{\alpha}\right)^2\right]^2$$

$$= \frac{1}{2}(\alpha + h_n)\left[\left(3 - 1 - \frac{2h_n}{\alpha} - \frac{h_n^2}{\alpha^2}\right) + \frac{3}{4}\left(1 - 1 - \frac{2h_n}{\alpha} - \frac{h_n^2}{\alpha^2}\right)^2\right]$$

$$= \frac{1}{2}(\alpha + h_n)\left[2 - \frac{2h_n}{\alpha} - \frac{h_n^2}{\alpha^2} + \frac{3}{4}\left(\frac{-2h_n}{\alpha} - \frac{h_n^2}{\alpha^2}\right)^2\right]$$

$$= \frac{1}{2}(\alpha + h_n)\left[2 - \frac{2h_n}{\alpha} - \frac{h_n^2}{\alpha^2} + \frac{3}{4}\left(\frac{4h_n^2}{\alpha^2} + \frac{h_n^4}{\alpha^4} + \frac{4h_n^3}{\alpha^3}\right)\right]$$

$$= \frac{1}{2}(\alpha + h_n)\left[2 - \frac{2h_n}{\alpha} - \frac{h_n^2}{\alpha^2} + \frac{3h_n^2}{\alpha^2} + \frac{3}{4}\frac{h_n^4}{\alpha^4} + \frac{3h_n^3}{\alpha^3}\right]$$

$$= \frac{1}{2}(\alpha + h_n)\left[2 - \frac{2h_n}{\alpha} + \frac{2h_n^2}{\alpha^2} + \frac{3h_n^3}{\alpha^3} + \frac{3}{4}\frac{h_n^4}{\alpha^4}\right]$$

$$= \alpha - h_n + \frac{h_n^2}{\alpha} + \frac{3}{2}\frac{h_n^3}{\alpha^2} + \frac{3}{8}\frac{h_n^4}{\alpha^3} + h_n - \frac{h_n^2}{\alpha} + \frac{h_n^3}{\alpha^2} + \frac{3}{2}\frac{h_n^4}{\alpha^3} + \frac{3}{8}\frac{h_n^5}{\alpha^4}$$

$$h_{n+1} = \frac{5}{2}\frac{h_n^3}{\alpha^2} + \frac{15}{8}\frac{h_n^4}{\alpha^3} + \frac{3}{8}\frac{h_n^5}{\alpha^4}$$

$$h_{n+1} = h_n^3\left(\frac{5}{2\alpha^2}\right) \text{ approximately (Neglecting } h_n^4 \text{ and its higher powers)}$$

$$h_{n+1} \propto h_n^3 \qquad \left(\frac{5}{2\alpha^2} = \text{constant}\right)$$

The rate of convergence is cubic. Here convergence is of order 3. **Ans.**

Example 16. *Apply Newton - Raphson Method to solve*

$$3x - \cos x - 1 = 0$$

(*R.G.P.V., Bhopal, III Semester, Dec. 2006, June 2004, Dec. 2002*)

Solution. Let $f(x) = 3x - \cos x - 1 = 0, \quad f'(x) = 3 + \sin x$

$$f(0) = 0 - 1 - 1 = -2$$

$$f(0.6) = 1.8 - \cos 0.6 - 1 = -0.0253$$

$$f(1) = 3 - \cos 1 - 1 = 1.4597$$

As $f(0.6)$ is nearer to zero, than $f(1)$, so we take first approximate root as 0.6.

By Newton - Raphson Method

$$x_{n+1} = x_n - \frac{f(x_n)}{f'(x_n)}$$

$$x_1 = x - \frac{3x - \cos x - 1}{3 + \sin x} = 0.6 - \frac{3(0.6) - \cos 0.6 - 1}{3 + \sin 0.6} = 0.6071$$

$$x_2 = 0.6071 - \frac{3(0.6071) - \cos 0.6071 - 1}{3 + \sin 0.6071} = 0.6071$$

Since $x_2 = x_1$, therefore the real root of the given equation is 0.6071. **Ans.**

Example 17. *Using Newton's iterative method, find the real root of $x \log_{10} x = 1.2$ correct to five decimal places.*

(*UP III Semester, June 2011, R.G.P.V., Bhopal, III Semester, June 2005, Dec. 2004*)

Solution. Let $f(x) = x \log_{10} x - 1.2$

$$f(1) = -1.2 = -\text{ve},$$

$$f(2) = 2 \log_{10} 2 - 1.2 = -0.59794 = -\text{ve}$$

and $f(3) = 3 \log_{10} 3 - 1.2 = 1.4314 - 1.2 = 0.23136 = +\text{ve}$

$$f(2) \,.\, f(3) < 0$$

So, a root of $f(x) = 0$ lies between 2 and 3.

Let us take $x_0 = 2.$

Also, $f'(x) = \log_{10} x + x.\frac{1}{x}\log_{10} e = \log_{10} x + 0.43429$

$\therefore$ Newton's iteration formula gives

$$x_{n+1} = x_n - \frac{f(x_n)}{f'(x_n)} = x_n - \frac{x_n \log_{10} x_n - 1.2}{\log_{10} x_n + 0.43429} = \frac{x_n \log_{10} x_n + 0.43429 x_n - x_n \log_{10} x_n + 1.2}{\log_{10} x_n + 0.43429}$$

$$= \frac{0.43429x_n + 1.2}{\log_{10} x_n + 0.43429} \quad \text{... (1)}$$

Putting $x_0 = 2$, the first approximation is

$$x_1 = \frac{0.43429 \times x_0 + 1.2}{\log_{10} x_0 + 0.43429} = \frac{0.43429 \times 2 + 1.2}{\log_{10} 2 + 0.43429}$$

$$= \frac{0.86858 + 1.2}{0.30103 + 0.43429} = 2.81$$

Similarly putting $n = 1, 2, 3, 4$ in (1), we get

$$x_2 = \frac{0.43429 \times 2.81 + 1.2}{\log_{10} 2.81 + 0.43429} = 2.741$$

$$x_3 = \frac{0.43429 \times 2.741 + 1.2}{\log_{10} 2.741 + 0.43429} = 2.74065$$

$$x_4 = \frac{0.43429 \times 2.74065 + 1.2}{\log_{10} 2.74065 + 0.43429} = 2.74065$$

Clearly, $x_3 = x_4$

Hence, the required root is 2.74065 correct to five decimal places. **Ans.**

EXERCISE 50.2

1. Using Newton-Raphson Method, find the root that lies in (0,1) of the equation $x^3 - 6x + 4 = 0$, correct to 4 decimal places.
   [Hint: Take $\alpha_0 = 1$] **Ans.** 0.73205
2. Using Newton-Raphson Method, find one negative root of $3x^3 + 8x^2 + 8x + 5 = 0$
   [Hint: Root of $f(-x)$] **Ans.** – 1.67
3. Solve: $x^3 - 3x + 1 = 0$ *(R.G.P.V., Bhopal, III Semester, Dec. 2007)* **Ans.** 1.532
4. Find the real root to four decimals of the equation $x^6 - x^4 - x^3 - 1 = 0$ lying between (1, 2).
   Ans. 1.4036
5. Evaluate $\sqrt{12}$ to four decimal places by Newton's Iterative method. **Ans.** 3.4641
6. Derive the formula $x_{n+1} = \frac{1}{3}\left[2x_n + \frac{a}{x_n^2}\right]$ used for computing successive approximation for cube root of the number a.
   Use it to obtain $\sqrt[3]{25.7}$ correct to three places of decimal. **Ans.** 2.951
7. The value of $N^{1/5}$, N > 0 is to determined numerically. Construct the Newton-Raphson scheme to find the required root. Apply it to find $(29)^{1/5}$, correct to three decimal places, starting with the initial approximation as 2.0. **Ans.** 1.961
8. What do you mean by convergence of the method and its importance in numerical analysis ?
   (R.G.P.V., Bhopal, III Semester, June 2003)
9. Find by the Newton's method, correct to 6 places of decimal the root of the equation
   $x \log_{10} x = 4.772393$ *(R.G.P.V., Bhopal, III Semester, June 2005, Dec. 2004)* **Ans.** 6.089114

10. Find by Newton-Raphson method a root of the equation

$x \sin x + \cos x = 0$

which is near $x = \pi$. (*AMIETE, June 2009*) **Ans.** 2.7985

11. Find by Newton-Raphson method a root of the equation

$e^x = x^3 + \cos 25x$ which is near $x = 4.5$ **Ans.** 4.545

12. Show that the modified Newton-Raphson method

$x_{n+1} = x_n - \dfrac{2f(x_n)}{f'(x_n)}$ gives a quadratic convergence when the equation $f(x) = 0$ has a pair of double root in the neighbourhood $x = x_n$.

13. Solve the equation $\log x = \cos x$ to five decimal places by Newton Raphson method. **Ans.** 1.30295

14. Solve $\tan x = 1.2\, x$. **Ans.** 0.71

15. Find a positive root of $2(x - 3) = \log_{10} x$. **Ans.** 3.256

Choose the correct alternative :

16. If $f(x) = 0$ is an algebraic equation then Newton-Raphson method is given by :

$x_{n+1} = x_n - \dfrac{f(x_n)}{?}$ (*R.G.P.V., Bhopal, III Semester, June 2007*)

(*a*) $f(x_{n-1})$ (*b*) $f'(x_{n-1})$ (*c*) $f'(x_n)$ (*d*) $f''(x_n)$ **Ans.** (*c*)

17. The order of convergence in Newton-Raphson method is :

(*a*) 2 (*b*) 3 (*c*) 0 (*d*) 1 **Ans.** (*a*)

(*R.G.P.V., Bhopal, III Semester, Dec. 2006*)

18. Newton's iterative formula to find the value of $\sqrt{N}$ is

(*a*) $x_{n+1} = \dfrac{1}{2}\left(x_n + \dfrac{1}{N x_n}\right)$ (*b*) $x_{n+1} = x_n(2 - N x_n)$ **Ans.** (*c*)

(*c*) $x_{n+1} = \dfrac{1}{2}\left(x_n + \dfrac{N}{x_n}\right)$ (*d*) $x_{n+1} = \dfrac{x_n^2 + N}{x_n}$ (*R.G.P.V., Bhopal, III Semester, Dec. 2007*)

50.16 REGULA - FALSI METHOD OR FALSE POSITION METHOD

The oldest method for computing the real root of a numerical equation is the method of false position, or "(Regula falsi)".

Let the root lie between *a* and *b*. These numbers *a* and *b* should be as close - together as possible. Since the root lies between *a* and *b*, the graph of $y = f(x)$ must cross the *x*- axis between $x = a$ and $x = b$, and $f(a)$ and $f(b)$ must have opposite signs.

Now since any portion of a smooth curve is practically straight line for a short distance, it is legitimate to assume that change in $f(x)$ is proportional to change in *x* over a short interval. The method of False position is based on this principle, for it assumes that the graph of $y = f(x)$ is a straight line between the points (x_1, y_1) and (x_2, y_2), these points being on opposite sides of *x*-axis.

50.17 RULE OF FALSE POSITION (REGULA FALSI)

Let $f(x) = 0$

Let $y = f(x)$ be represented by the curve *AB*.

The curve *AB* cuts the *x*-axis at *P*.

The real root of (1) is *OP*.

The false position of the curve *AB* is taken as the chord *AB*. The chord *AB* cuts the *x*-axis at *Q*. The approximate root of $f(x) = 0$ is *OQ*.

By this method, we find *OQ*.

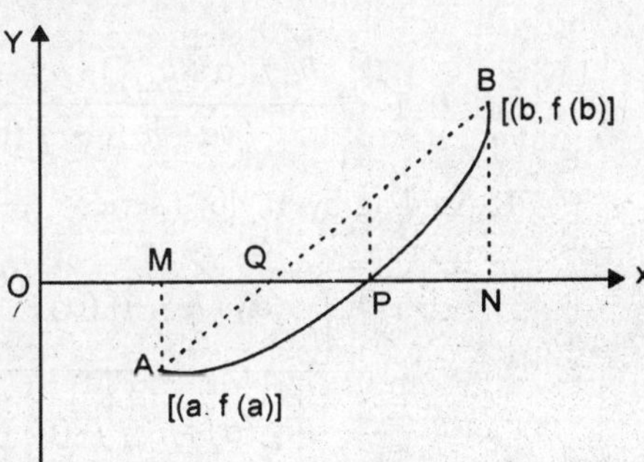

Let $A\ [a, f(a)]$, $B\ [b, f(b)]$ be the extremities of the chord AB

The equation of the chord AB is

$$y - f(a) = \frac{f(b)-f(a)}{b-a}(x-a) \qquad \text{(Two points form)}$$

To find OQ, put $y = 0$

$$0 - f(a) = \frac{f(b)-f(a)}{b-a}(x-a)$$

$$(x-a) = \frac{-(b-a)f(a)}{f(b)-f(a)}$$

$$\Rightarrow \quad x = a + \frac{(a-b)\,f(a)}{f(b)-f(a)} = \frac{a\,f(b)-a\,f(a)+a\,f(a)-b\,f(a)}{f(b)-f(a)}$$

$$\boxed{\boldsymbol{x = \frac{a\,f(b)-b\,f(a)}{f(b)-f(a)}}}$$

Repeat the above rule.

50.18 RATE OF CONVERGENCE OF FALSE POSITION METHOD

Let $f(x) = 0$ ... (1)

and α is the root of (1).

If x_n be the approximate value of α and h_n be error of x_n at the nth stage.

$$x_n = \alpha + h_n \qquad [h_{n+1} = \text{Error at } (n+1) \text{ th stage}]$$

$$x_{n+1} = \alpha + h_{n+1} \text{ and } x_{n+2} = \alpha + h_{n+2}$$

By False position method

$$x_{n+2} = \frac{x_n\,f(x_{n+1}) - x_{n+1}\,f(x_n)}{f(x_{n+1}) - f(x_n)} \qquad \text{... (2)}$$

Putting the values of x_n, x_{n+1} and x_{n+2}, we get

$$\alpha + h_{n+2} = \frac{(\alpha+h_n)\,f(\alpha+h_{n+1}) - (\alpha+h_{n+1})\,f(\alpha+h_n)}{f(\alpha+h_{n+1}) - f(\alpha+h_n)}$$

On removing the brackets and arranging the terms on R.H.S., we get

$$= \frac{\alpha\left[f(\alpha+h_{n+1}) - f(\alpha+h_n)\right] + h_n\,f(\alpha+h_{n+1}) - h_{n+1}\,f(\alpha+h_n)}{f(\alpha+h_{n+1}) - f(\alpha+h_n)}$$

$$\alpha + h_{n+2} = \alpha + \frac{h_n\,f(\alpha+h_{n+1}) - h_{n+1}\,f(\alpha+h_n)}{f(\alpha+h_{n+1}) - f(\alpha+h_n)}$$

$$h_{n+2} = \frac{h_n\,f(\alpha+h_{n+1}) - h_{n+1}\,f(\alpha+h_n)}{f(\alpha+h_{n+1}) - f(\alpha+h_n)}$$

Expanding by Taylor's series

$$= \frac{h_n\left[f(\alpha)+h_{n+1}f'(\alpha)+\frac{h^2_{n+1}}{2!}f''(\alpha)+\ldots\right] - h_{n+1}\left[f(\alpha)+h_n f'(\alpha)+\frac{h_n^2}{2!}f''(\alpha)+\ldots\right]}{\left[f(\alpha)+h_{n+1}f'(\alpha)+\frac{h^2_{n+1}}{2!}f''(\alpha)+\ldots\right] - \left[f(\alpha)+h_n f'(\alpha)+\frac{h_n^2}{2!}f''(\alpha)+\ldots\right]}$$

$$h_{n+2} = \frac{h_n\left[0+h_{n+1}f'(\alpha)+\dfrac{h^2{}_{n+1}}{2!}f''(\alpha)+.....\right]-h_{n+1}\left[0+h_nf'(\alpha)+\dfrac{h_n^2}{2!}f''(\alpha)+....\right]}{\left[0+h_{n+1}f'(\alpha)+\dfrac{h^2{}_{n+1}}{2!}f''(\alpha)+....\right]-\left[0+h_nf'(\alpha)+\dfrac{h_n^2}{2!}f''(\alpha)+....\right]}$$

$$= \frac{h_n h_{n+1} f'(\alpha)+h_n\dfrac{h^2{}_{n+1}}{2!}f''(\alpha)-h_{n+1}h_n f'(\alpha)-h_{n+1}\dfrac{h_n^2}{2!}f''(\alpha)+....}{h_{n+1}f'(\alpha)+\dfrac{h^2{}_{n+1}}{2!}f''(\alpha)-h_n f'(\alpha)-\dfrac{h_n^2}{2!}f''(\alpha)+....}$$

$$= \frac{\dfrac{h_n h_{n+1}^2}{2}f''(\alpha)-\dfrac{h_{n+1}h_n^2}{2}f''(\alpha)}{h_{n+1}f'(\alpha)-h_n f'(\alpha)} \qquad \textit{(Ignoring terms of higher powers of h)}$$

$$= \frac{\dfrac{h_n h_{n+1}}{2}\left[h_{n+1}-h_n\right]f''(\alpha)}{\left[h_{n+1}-h_n\right]f'(\alpha)} = h_n h_{n+1}\frac{f''(\alpha)}{2f'(\alpha)}$$

$$h_{n+2} = h_n h_{n+1}(c) \qquad ...(1) \qquad \left[\text{Put } \frac{f''(\alpha)}{2f'(\alpha)} = \text{constant} = c\right]$$

To find the rate of convergence k such that

$$h_{n+1} = Ah_n^k \qquad [\text{A = constant of proportionality}] \qquad ...(2)$$

$$\Rightarrow h_{n+1}^{\frac{1}{k}} = A^{\frac{1}{k}}h_n$$

$$\Rightarrow h_n = A^{-\frac{1}{k}}h_{n+1}^{\frac{1}{k}} \qquad ...(3)$$

Again, $h_{n+2} = A h_{n+1}{}^k$ ... (4)

Putting the values of h_n and h_{n+2} from (3) and (4) in (1), we get

$$Ah_{n+1}^k = A^{-\frac{1}{k}}h_{n+1}^{\frac{1}{k}}\left[h_{n+1}(c)\right]$$

$$A\,h_{n+1}{}^{k} = A^{-\frac{1}{k}}h_{n+1}^{\frac{1}{k}+1}(c)$$

$$h_{n+1}{}^{k} = A^{-\left(\frac{1}{k}+1\right)}h_{n+1}^{\frac{1}{k}+1}\,c$$

Comparing the powers of h_{n+1} on both sides, we get

$$k = \frac{1}{k}+1$$

$$k^2-k-1 = 0$$

$$k = \frac{1\pm\sqrt{1+4}}{2} \Rightarrow k = \frac{1\pm\sqrt{5}}{2} = \frac{1+2.2361}{2} \qquad \text{(Ignoring - ve sign)}$$

$$\Rightarrow k = \frac{3.2361}{2} = 1.618$$

Putting the value of k in (2), we get $h_{n+1} = Ah_n^{1.618}$

The rate of convergence in False position method is 1.618. **Ans.**

Example 18. *Find an approximate value of the root of the equation $x^3 + x - 1 = 0$ near $x = 1$, using the method of false position (regula falsi) two times.*

Solution. $f(x) = x^3 + x - 1 = 0$

$$f(1) = 1 + 1 - 1 = +1$$

$$f(0.5) = (0.5)^3 + (0.5) - 1 = -0.375, \qquad f(1) . f(0.5) < 0$$

The root lies between 0.5 and 1.

Let $a = 0.5$ and $b = 1$

$$x_1 = \frac{a\,f(b) - b\,f(a)}{f(b) - f(a)} \Rightarrow x_1 = \frac{0.5\,f(1) - 1\,f(0.5)}{f(1) - f(0.5)}$$

$$= \frac{0.5\,(1) - 1\,(-0.375)}{1 + 0.375} = 0.6364$$

Now $f(0.6364) = (0.6364)^3 + 0.6364 - 1 = -0.1059$

and $f(1) = 1$

$\therefore$ Root lies between 0.6364 and 1. $f(0.6364) . f(1) < 0$

$a = 0.6364, \quad b = 1$

$$x_2 = \frac{0.6364\,f(1) - 1\,f(0.6364)}{f(1) - f(0.6364)} = \frac{0.6364 - 1(-0.1059)}{1 + 0.1059}$$

$$= 0.6712$$

Now, $f(0.6712) = -0.0264$ and $f(1) = 1$

$a = 0.6712$ and $b = 1$ $[f(0.6712) . f(1) < 0]$

$$x_3 = \frac{0.6712\,f(1) - 1\,f(0.6712)}{f(1) - f(0.6712)} = \frac{0.6712 - (-0.0264)}{1 - (-0.0264)}$$

$$= 0.6797$$ **Ans.**

Example 19. *find a positive root of $x^3 - 4x + 1 = 0$ by the method of false position.*

(R.G.P.V., Bhopal, III Semester, June 2007)

Solution. Let $f(x) = x^3 - 4x + 1 = 0$, $f'(x) = 3x^2 - 4$

$f(0) = 1$, $f'(0) = -4$

$f(1) = 1 - 4 + 1 = -2$

Since $f(0)$ and $f(1)$ are of opposite signs, so the root lies between 0 and 1.

By False position method

$$x = \frac{a\,f(b) - b\,f(a)}{f(b) - f(a)}$$

$$x_1 = \frac{0\,f(1) - 1\,f(0)}{f(1) - f(0)} = \frac{-1}{-2 - (1)} = \frac{1}{3}$$

$$f\left(\frac{1}{3}\right) = \left(\frac{1}{3}\right)^3 - 4\left(\frac{1}{3}\right) + 1 = \frac{1}{27} - \frac{4}{3} + 1 = \frac{-8}{27}$$

Since $f\left(\frac{1}{3}\right)$ and $f(0)$ are of opposite signs, so the root lies between $\frac{1}{3}$ and 0.

$$x_2 = \frac{\frac{1}{3}f(0) - 0\,f\left(\frac{1}{3}\right)}{f(0) - f\left(\frac{1}{3}\right)} = \frac{\frac{1}{3}(1) - 0}{1 - \left(\frac{-8}{27}\right)} = \frac{9}{35}$$

$$f\left(\frac{9}{35}\right) = \left(\frac{9}{35}\right)^3 - 4\left(\frac{9}{35}\right) + 1 = \frac{729}{42875} - \frac{36}{35} + 1 = -\frac{496}{42875}.$$

Since, $f\left(\frac{9}{35}\right)$ and $f(0)$ are of opposite signs, so the root lies between $\frac{9}{35}$ and 0.

$$x_3 = \frac{\frac{9}{35} f(0) - 0\, f\left(\frac{9}{35}\right)}{f(0) - f\left(\frac{9}{35}\right)} = \frac{\frac{9}{35}(1) - 0}{1 - \left(-\frac{496}{42875}\right)} = \frac{1225}{4819}$$

Since $\left(x_2 = \frac{9}{35} = 0.2571\right) = \left(x_3 = \frac{1225}{4819} = 0.2542\right)$ correct upto two decimal places,

so the root of the given equation is 0.25 **Ans.**

Example 20. *Find the root of the equation $x^3 - 5x - 7 = 0$ which lies between 2 and 3 by the method of false position.* (*R.G.P.V., Bhopal, III Semester, June 2005*)

Solution. Here, we have

Let
$$f(x) = x^3 - 5x - 7 = 0$$
$$f(2) = 8 - 10 - 7 = -9$$
$$f(3) = 27 - 15 - 7 = +5$$

As $f(2)$ and $f(3)$ are of opposite signs, so the root lies between 2 and 3.

$$x_1 = \frac{a\, f(b) - b\, f(a)}{f(b) - f(a)}$$

$$x_1 = \frac{2\, f(3) - 3\, f(2)}{f(3) - f(2)} = \frac{2(5) - 3(-9)}{5 - (-9)} = \frac{37}{14} = 2.6429$$

$$f(2.6429) = (2.6429)^3 - 5\,(2.6429) - 7 = -1.7541$$

Now, $f(2.6429) = -1.7541$ and $f(3) = 5$

∴ Root lies between 2.6429 and 3.

$$a = 2.6429,\ b - 3$$

$$x_2 = \frac{2.6429\, f(3) - 3\, f(2.6429)}{f(3) - f(2.6429)} = \frac{2.6429(5) - 3(-1.7541)}{5 - (-1.7541)} = \frac{18.4768}{6.7541} = 2.7356$$

Now, $f(2.7356) = (2.7356)^3 - 5\,(2.7356) - 7 = -0.2061$

and $f(3) = 5$

∴ Root lies between 2.7356 and 3.

$$a = 2.7356 \quad \text{and } b = 3$$

$$x_3 = \frac{2.7356\, f(3) - 3\, f(2.7356)}{f(3) - f(2.7356)} = \frac{2.7356(5) - 3(-0.2061)}{5 - (-0.2061)} = \frac{14.2963}{5.2061} = 2.7461$$

$$f(2.7461) = (2.7461)^3 - 5\,(2.7461) - 7 = -0.02198$$

Now, $f(2.7461) = -0.02198$ and $f(3) = +5$

∴ Root lies between 2.7461 and 3.

$$a = 2.7461 \text{ and } b = 3$$

$$x_4 = \frac{2.7461\, f(3) - 3\, f(2.7461)}{f(3) - f(2.7461)} = \frac{2.7461(5) \quad 3(-0.02198)}{5 - (-0.02198)} = \frac{13.79644}{5.02198} = 2.7472$$

Hence, the root of the given equation is 2.7472. **Ans.**

Example 21. *Find by the method of regula falsi a root of the equation $x^3 + x^2 - 3x - 3 = 0$ lying between 1 and 2.* (*R.G.P.V., Bhopal, III Semester, June 2005*)

Solution.
$$f(x) = x^3 + x^2 - 3x - 3 = 0$$
$$f(1) = 1 + 1 - 3 - 3 = -4 = -ve$$
$$f(2) = 8 + 4 - 6 - 3 = +3 = +ve$$

The root lies between 1 and 2 as $f(1)$ is –ve and $f(2)$ is +ve.

By Regula Falsi method

$$x_1 = \frac{1f(2) - 2f(1)}{f(2) - f(1)} = \frac{1 \times 3 - 2 \times (-4)}{3 - (-4)} = \frac{11}{7} = 1.571$$

$$f(1.571) = (1.571)^3 + (1.571)^2 - 3(1.571) - 3$$
$$= 3.877 + 2.468 - 4.713 - 3 = -1.368 = -\text{ve}$$

The root lies between 1.571 and 2 as $f(1.571)$ is –ve and $f(2)$ is + *ve*.

$$x_2 = \frac{1.571 f(2) - 2f(1.571)}{f(2) - f(1.571)} = \frac{1.571 \times 3 - 2 \times (-1.368)}{3 - (-1.368)} = \frac{4.713 + 2.736}{4.368} = 1.705$$

$$f(1.705) = (1.705)^3 + (1.705)^2 - 3(1.705) - 3$$
$$= 4.956 + 2.907 - 5.115 - 3 = -0.252 = -ve.$$

The root lies between 1.705 and 2 as $f(1.705)$ is –ve and $f(2)$ is +*ve*.

$$x_3 = \frac{1.705 f(2) - 2f(1.705)}{f(2) - f(1.705)} = \frac{1.705 \times 3 - 2 \times (-0.252)}{3 - (-0.252)} = 1.728$$ **Ans.**

Example 22. *Find the real root of the equation $x \log_{10} x = 1.2$ by the method of false position (i.e. Regula falsi method) correct to four decimal places.*

(*R.G.P.V., Bhopal, III Semester, Dec. 2007, 2002*)

Solution. Let $f(x) = x \log_{10} x - 1.2$

Here $f(2) = 2 \log_{10} 2 - 1.2 = -0.59794;$

$f(3) = 3 \log_{10} 3 - 1.2 = 0.23136$

∴ One root lies between 2 and 3.

Taking $a = 2,\ b = 3,\ f(2) = -0.59794, f(3) = 0.23136$

By method of false position, we have

$$x_1 = \frac{a f(b) - b f(a)}{f(b) - f(a)}$$

$$x_1 = \frac{2f(3) - 3f(2)}{f(3) - f(2)} = \frac{2(0.23136) - 3(-0.59794)}{(0.23136) - (-0.59794)} = 2.72102$$

$$f(2.72102) = (2.72102) \log_{10} 2.72102 - 1.2 = -0.01709$$

Since, $f(2.72102)$ and $f(3)$ are of opposite sign, so the root lies between 2.72102 and 3.

$$x_2 = \frac{2.72102 f(3) - 3f(2.72102)}{f(3) - f(2.72102)} = \frac{2.72102(0.23136) - 3(-0.01709)}{0.23136 - (-0.01709)} = 2.74021$$

Now, $f(2.74021) = -0.00038$.

Since $f(2.74021) f(3) < 0$, so the root lies between 2.74021 and 3.

$$x_3 = \frac{2.74021 f(3) - 3f(2.74021)}{f(3) - f(2.74021)} = \frac{2.74021(0.23136) - 3(-0.00038)}{(0.23136) - (-0.00038)} = 2.74064$$

Again $f(2.74064) = -0.00001$ and $f(3) = 0.23136$

Since $f(2.7406)\,f(3) < 0$, so the root lies between 2.74064 and 3.

$$x_4 = \frac{2.74064\,f(3) - 3\,f(2.74064)}{f(3) - f(2.74064)} = \frac{2.74064(0.23136) - 3(-0.00001)}{0.23136 - (-0.00001)} = 2.74065$$

Hence, the root correct to four decimal places is 2.7407. **Ans.**

EXERCISE 50.3

Solve the following equations by Regula falsi method :

1. $x^3 - 2x - 5 = 0$ **Ans.** 2.0946
2. $x^3 - 10x^2 + 40x - 35 = 0$ **Ans.** 1.1875
3. $x^3 - 9x + 1 = 0$ (Root between 2 &3) **Ans.** 2.9428
4. $x^3 - 3x + 1 = 0$ **Ans.** 1.532 (*R.G.P.V., Bhopal, III Semester, Dec. 2001*)
5. Use the method of false position to find the root of the equation $x^3 - 18 = 0$, given that it lies between 2 and 3. Write down three steps of the procedure. **Ans.** 2.621
6. Find the root of the equation $\tan x + \tanh x = 0$ which lies in the interval (1.6, 3.0) correct to four significant digits using any one of the numerical methods. **Ans.** 2.365 app.
7. $2x - \log x = 6$ **Ans.** 3.257
8. $xe^x = \cos x$ **Ans.** 0.5177
9. $xe^x = \sin x$ **Ans.** – 0.134
10. $x\,e^x = 2$ **Ans.** 0.853
11. $x \sin x + \cos x = 0$ (near $x = \pi$) **Ans.** 2.7985
12. $x = \tan x$ near $x = 4.5$ **Ans.** 4.43464
13. Regular-Falsi method requires initial approximation to the root.

    (*a*) 1 (*b*) 2 (*c*) 3 (*d*) None of above (*AMIETE, Dec. 2010*) **Ans.** (*b*)

50.19 SECANT METHOD (CHORD METHOD)

This method is quite similar to that of False position method and it is improved method over Regula Falsi method. Here it is not necessary to fulfill the condition $f(x_1)\,f(x_2) < 0$.

Here the graph of $f(x)$ is approximated by a secant line (Chord). The root may or may not lie in the interval $[a, b]$

$$\frac{a\,f(b) - b\,f(a)}{f(b) - f(a)} = x_1$$

$$\frac{b\,f(x_1) - x_1\,f(b)}{f(x_1) - f(b)} = x_2$$

$$\frac{x_1\,f(x_2) - x_2\,f(x_1)}{f(x_2) - f(x_1)} = x_3$$

Apply the formula in the following way

I Time: a, b

II Time: b, x_1

III Time: x_1, x_2

IV Time: x_2, x_3

V Time: x_3, x_4

and so on

If $f(x_{n-1}) = f(x_n)$ then the method fails. But the rate of convergence in secant method is greater than that of Regula Falsi.

Example 23. *Given the equation $x^4 - x - 10 = 0$, determine the initial approximations for finding its smallest positive root. Use these to find the root correct to three decimal places with Secant method.*

Solution. Let $f(x) = x^4 - x - 10 = 0$

Here $f(1) = 1 - 1 - 10 = -10 \quad , a = 1$

$f(2) = 16 - 2 - 10 = +4 \quad , b = 2$

1
2
1.71429
1.83853
1.85778
1.85500 1.85563

By Secant Method

$$x_1 = \frac{a\,f(b) - b\,f(a)}{f(b) - f(a)}$$

$$x_1 = \frac{1\,f(2) - 2\,f(1)}{f(2) - f(1)} = \frac{1(4) - 2(-10)}{4 - (-10)} = \frac{24}{14} = \frac{12}{7} = 1.71429$$

$$f(1.71429) = (1.71429)^4 - (1.71429) - 10 = 8.63649 - 1.71429 - 10 = -3.0778$$

Taking $a = 2$, $b = 1.71429$,

$$x_2 = \frac{1.71429\,f(2) - 2\,f(1.71429)}{f(2) - f(1.71429)} = \frac{1.71429(4) - 2(-3.0778)}{(4) - (-3.0778)} = \frac{13.01276}{7.0778} = 1.83853$$

$$f(1.83853) = (1.83853)^4 - (1.83853) - 10 = -0.41283$$

Taking $a = 1.71429$, $b = 1.83853$,

$$x_3 = \frac{1.83853\,f(1.71429) - 1.71429\,f(1.83853)}{f(1.71429) - f(1.83853)}$$

$$= \frac{1.83853(-3.0778) - 1.71429(-0.41283)}{-3.0778 - (-0.41283)} = \frac{-4.95092}{-2.66497} = 1.85778$$

$$f(1.85778) = (1.85778)^4 - (1.85778) - 10 = 0.05401$$

Taking $a = 1.83853$, $b = 1.85778$

$$x_5 = \frac{1.85778\,f(1.83853) - 1.83853\,f(1.85778)}{f(1.83853) - f(1.85778)}$$

$$= \frac{1.85778(-0.41283) - 1.83853(0.05401)}{(-0.41283) - 0.05401} = \frac{-0.86625}{-0.46684} = 1.85561$$

$$f(1.85561) = (1.85561)^4 - (1.85561) - 10 = 11.85623 - 1.85561 - 10 = 0.00062$$

Taking $a = 1.85778$ and $b = 1.85561$,

$$x_6 = \frac{1.85561\,f(1.85778) - 1.85778\,f(1.85561)}{f(1.85778) - f(1.85561)}$$

$$= \frac{1.85561\,(0.05401) - 1.85778(0.00062)}{0.05401 - 0.00062} = \frac{0.099064}{0.05339} = 1.85558$$

$$x_5 = x_6$$

Hence, the root of the given equation is 1.8556 **Ans.**

Example 24. *Two approximations to a real root of the equation* $\cos x - x^2 - x = 0$ *are* -1.5 *and* -1.4*. Use two iterations of the Secant method to find the root (use four decimals arithmetic).*

Solution. Here, $a = -1.5, \quad b = -1.4$

$$f(a) = f(-1.5) = \cos(-1.5) - (-1.5)^2 + 1.5 = -0.6793$$

$$f(b) = f(-1.4) = \cos(-1.4) - (-1.4)^2 + 1.4 = -0.3900$$

By Secant method

$$x_1 = \frac{a\,f(b) - b\,f(a)}{f(b) - f(a)}$$

$$x_1 = \frac{-1.5\,f(-1.4) - (-1.4)\,f(-1.5)}{f(-1.4) - f(-1.5)}$$

$$= \frac{-1.5(-0.3900) - (-1.4)(-0.6793)}{(-0.3900) - (-0.6793)}$$

$$= \frac{0.5850 - 0.9510}{0.2893} = \frac{-0.3660}{0.2893} = -1.2651$$

$$f(x_2) = f(-1.2651) = \cos(-1.2651) - (-1.2651)^2 + 1.2651 = -0.0344$$

$$\begin{matrix} -1.5 \\ -1.4 \\ -1.2651 \end{matrix} \rightarrow -1.2520$$

Taking $a = -1.4,\ b = -1.2651$

$$x_2 = \frac{-1.4\,f(-1.2651) - (-1.2651)\,f(-1.4)}{f(-1.2651) - f(-1.4)} = \frac{-1.4(-0.0344) - (-1.2651)(-0.3900)}{(-0.0344) - (-0.3900)}$$

$$= \frac{0.04816 - 0.4934}{0.3556} = \frac{-0.44524}{0.3556} = -1.2520$$

Hence, the root of the given equation is -1.2520 **Ans.**

Example 25. *Find the root of the equation $x\,e^x = \cos x$ using Secant method to four decimal places.*

Solution. Let $f(x) = \cos x - x\,e^x = 0$

Taking the initial approximate root $a = 0,\ b = 1$

$$f(0) = 1 - 0 = +1 \quad , \qquad a = 0,\ b = 1$$

$$f(1) = \cos 1 - e = 0.5403 - 2.7183 = -2.178$$

$$\begin{matrix} 0 \\ 1 \\ 0.3147 \\ 0.4467 \\ 0.53169 \\ 0.5169 \\ 0.51775 \end{matrix} \rightarrow 0.51776$$

By Secant method

$$x_1 = \frac{a\,f(b) - b\,f(a)}{f(b) - f(a)} = \frac{(0)\,f(1) - 1\,f(0)}{f(1) - f(0)}$$

$$x_1 = \frac{0(-2.178) - 1(1)}{-2.178 - 1} = \frac{-1}{-3.178} = 0.3147$$

Taking $a = 1,\ b = 0.3147$,

$$x_2 = \frac{1\,f(0.3147) - (0.3147)\,f(1)}{f(0.3147) - f(1)} = \frac{0.5198 - (0.3147)(-2.1780)}{0.5198 - (-2.1780)} = 0.4467$$

Taking $a = 0.3147,\ b = 0.4467$

$$x_3 = \frac{0.3147\,f(0.4467) - 0.4467\,f(0.3147)}{f(0.4467) - f(0.3147)}$$

$$= \frac{0.3147(0.2036) - 0.4467(0.5198)}{0.2036 - (0.5198)} = 0.53169$$

Similarly, $x_4 = 0.5169,\ x_5 = 0.51775,\ x_6 = 0.51776$

Hence, the correct root is 0.5178 correct to four places of decimal. **Ans.**

EXERCISE 50.4

Solve the following equations by secant method correct to three decimal places:

1. $x^3 + x^2 + x + 7 = 0$ **Ans.** -2.063
2. $x^3 - 4x - 9 = 0$ **Ans.** 2.707
3. $x^5 - x^4 - x^3 - 1 = 0$ **Ans.** 1.737
4. $x e^x = 1$ **Ans.** 0.567
5. The rate of convergence of the Secant method is

   (*a*) 1.84 (*b*) 2 (*c*) 1.5 (*d*) 1.62 **Ans.** (*a*)

50.20 ITERATION METHOD

Consider the equation $f(x) = 0$ ... (1)

We rewrite the equation in the form

$$x = \phi(x) \qquad \text{.... (2)}$$

Let us draw two curves

$$y = x \text{ and } y = \phi(x)$$

The point of intersection of two curves is the root of (1).

Let $x = x_0$ be an initial approximate root, then first approximation x_1 is found by

$$x_1 = \phi(x_0)$$

Now taking x_1 as initial value, x_2 second approximation is given by

$$x_2 = \phi(x_1) \quad \text{and so on.}$$

$$x_{n+1} = \phi(x_n)$$

This is also known as successive approximation method.

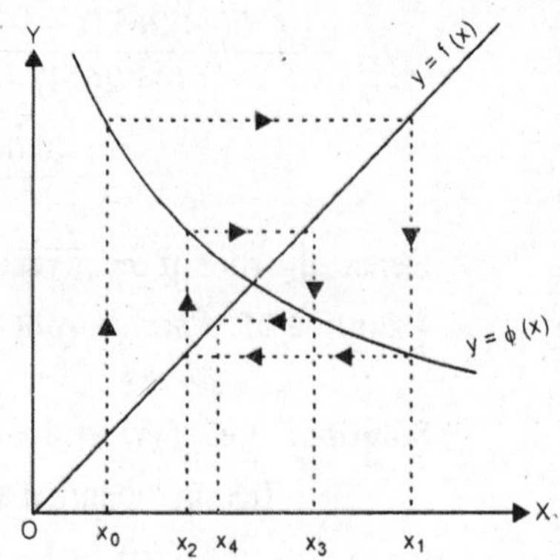

50.21 TEST FOR CONVERGENCE

For convergence it is convenient to identify an interval that contains the root and for which $\phi'(x)$ has small magnitude.

$$x = \phi(x)$$

$$\alpha = \phi(\alpha) \qquad \text{... (1)}$$

$$x_n = f(x_{n-1}) \qquad \text{... (2)}$$

Subtracting (1) from (2), we have

$$x_n - \alpha = \phi(x_{n-1}) - \phi(\alpha) \qquad \text{... (3)}$$

By Mean value theorem

$$\frac{\phi(x_{n-1}) - \phi(\alpha)}{x_{n-1} - \alpha} = \phi'(\xi), \text{ where } x_{n-1} < \xi < \alpha, \qquad \text{... (4)}$$

Substituting the value of $\phi(x_{n-1}) - \phi(\alpha)$ from (4) in (3), we get

$$x_n - \alpha = (x_{n-i} - \alpha)\, \phi'(\xi)$$

$$|x_n - \alpha| \le k\, |x_{n-1} - \alpha| \qquad [\text{If } \phi'(x_i) \le k < 1 \text{ for all } i] \quad \text{... (5)}$$

Similarly, $|x_{n-1} - \alpha| \le k\, |x_{n-2} - \alpha|$ ... (6)

Putting the value of $|x_{n-1} - \alpha|$ from (6) in (5), we have

$$|x_n - \alpha| \le k^2\, |x_{n-2} - \alpha|$$

....

$$|x_n - \alpha| \le k^n \, |x_0 - \alpha|$$

$$|x_n - \alpha| = 0 \qquad \left[\lim_{n\to\infty} k^n = 0\right]$$

So, the approximation converges by this method.

Note. 1. The rate of convergence is more if the value of $\phi'(x)$ is smaller.

2. For real roots, the method is very useful.

Remember. The equation $f(x) = 0$ is written as $x = \phi(x)$.

This form $x = \phi(x)$ can be choosen in many ways. We have to choose $\phi(x)$ in such a way that initial approximation x_0 should satisfy the condition $|\phi'(x_0)| < 1$.

Then $x_0, x_1, x_2 \ldots\ldots\ldots x_n$ converge to the root α of the equation $f(x) = 0$.

Example 26. *Apply the iterative method to find the real roots of $x^3 + x^2 - 1 = 0$, assuming the initial approximation is as $x_0 = 0.8$.*

Solution. Let $f(x) = x^3 + x^2 - 1 = 0$ ... (1)

Equation (1) can be written as $x = \phi(x)$ in many ways

i.e. (i) $x^3 = 1 - x^2 \Rightarrow x = (1-x^2)^{\frac{1}{3}}$, Here $\phi(x) = (1-x^2)^{\frac{1}{3}}$

$$\Rightarrow \quad \phi'(x) = \frac{1}{3}(1-x^2)^{-\frac{2}{3}}(-2x) \quad \Rightarrow \quad |\phi'(0.8)| = 1.05 > 1,$$

Which does not satisfy the condition.

(ii) $x^2 = 1 - x^3 \Rightarrow x = (1-x^3)^{\frac{1}{2}}$.

Here $\phi(x) = (1-x^3)^{\frac{1}{2}}$

$$\Rightarrow \phi'(x) = \frac{1}{2}(1-x^3)^{-\frac{1}{2}}(-3x^2) \qquad \Rightarrow |\phi'(0.8)| = 1.37 > 1$$

Again it also does not satisfy the condition.

(*iii*) $x^3 + x^2 - 1 = 0 \Rightarrow x^2(x+1) = 1 \Rightarrow x^2 = \dfrac{1}{1+x} \quad \Rightarrow x = \dfrac{1}{\sqrt{1+x}}$

Here, $\phi(x) = \dfrac{1}{\sqrt{1+x}}$

$$\Rightarrow \phi'(x) = -\frac{1}{2}(1+x)^{-\frac{3}{2}} \qquad \Rightarrow |\phi'(0.8)| = 0.2 < 1$$

Which satisfies the condition.

By Iterative method

$$x = \phi(x) \qquad \Rightarrow x = \frac{1}{\sqrt{1+x}} \qquad \ldots (2)$$

On putting the initial approximation $x_0 = 0.8$ in (2), we get

$$x_1 = \frac{1}{\sqrt{1+0.8}} = 0.7454$$

Again putting $x_1 = 0.7454$ in (2), we have

$$x_2 = \frac{1}{\sqrt{1+0.7454}} = 0.7569$$

Similarly, putting the successive values of x (approximate root) in (2), we obtain

$$x_3 = \frac{1}{\sqrt{1+0.7569}} = 0.7544$$

$$x_4 = \frac{1}{\sqrt{1+0.7544}} = 0.7550, \qquad x_5 = \frac{1}{\sqrt{1+0.7550}} = 0.7549$$

$$x_6 = \frac{1}{\sqrt{1+0.7549}} = 0.7549 \qquad x_7 = \frac{1}{\sqrt{1+0.7549}} = 0.7549$$

Since $x_6 = x_7$, so the correct root of the given equation is 0.7549 **Ans.**

Example 27. *Find a real root of $2x - \log_{10} x = 7$ correct to three decimal places using iteration method.* (*R.G.P.V., Bhopal, III Semester, Dec. 2006*)

Solution. The given equation is

$$2x - \log_{10} x = 7$$

which can be written as

$$x = \frac{1}{2}(\log_{10} x + 7)$$

Here, $\phi(x) = \frac{1}{2}(\log_{10} x + 7)$ ... (1)

Let $x_0 = 3.8$

Putting the value of $x = 3.8$ in (1), we get

$$x_1 = \frac{1}{2}(\log_{10} 3.8 + 7) = 3.79$$

On putting $x = 3.79$ in (1), we get

$$x_2 = \frac{1}{2}(\log_{10} 3.79 + 7) = 3.7893$$

Again putting $x = 3.7893$ in (1), we get

$$x_3 = \frac{1}{2}(\log_{10} 3.7893 + 7) = 3.7893$$

Since $x_2 = x_3$, the root of the given equation is 3.7893. **Ans.**

Example 28. *Apply iterative scheme method to find the real root of $x\,e^x = 1$, correct to three decimals, assuming the initial approximation as $x_0 = 0.5$.*

Solution.

The condition for the convergence of the iterative scheme is

$$|\phi'(x_k)| < 1.$$

Here $x e^x = 1 \Rightarrow x = e^{-x}$ ... (1)

$\Rightarrow \phi(x) = e^{-x}$

Putting $x = 0.5$ in (1), we get

$$x_1 = e^{-0.5} = 0.6065$$

Again putting $x = 0.6065$ in (1), we have

$$x_2 = e^{-0.6065} = 0.5453$$

Similarly putting the successive values of x in (1), we get

$$x_3 = e^{-.5453} = 0.5797$$
$$x_4 = e^{-.5797} = 0.5601$$
$$x_5 = e^{-.5601} = 0.5712$$
$$x_6 = e^{-.5712} = 0.5648$$
$$x_7 = e^{-.5648} = 0.5685$$
$$x_8 = e^{-.5685} = 0.5664$$

$$\left\{\begin{aligned} \phi'(x) &= -e^{-x} \\ \phi'(0.5) &= -e^{-0.5} \\ &= -0.6065 \\ |\phi'(0.5)| &< 1 \end{aligned}\right.$$

$$x_9 = e^{-.5664} = 0.5676$$
$$x_{10} = e^{-.5676} = 0.5669$$
$$x_{11} = e^{-.5669} = 0.5673$$
$$x_{12} = e^{-0.5673} = 0.5671$$
$$x_{13} = e^{-0.5671} = 0.5672$$

The above iterates are written to show the convergence of the iterates.

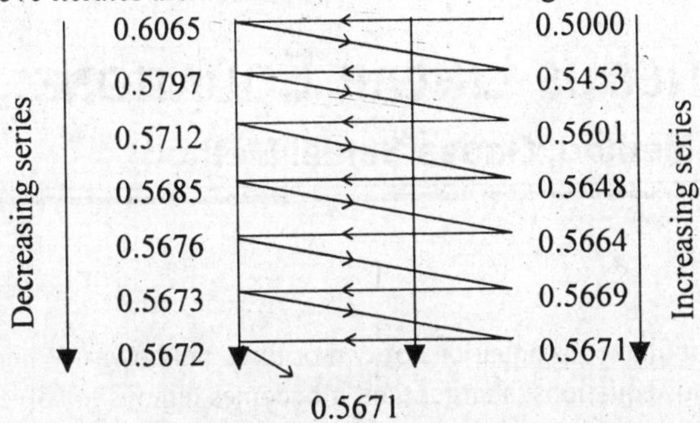

Ans.

Example 29. *The number of positive real roots of the equation* $e^x = 2 \sin x$

(a) zero *(b) 1* *(c) 2* *(d) infinitely many*

Solution. $e^x - 2 \sin x = 0$ ... (1)

(+ ve) (– ve)

On putting positive value of x in (1) we get one change in sign from +ve to -ve, so there is only one real positive root **Ans.** (*b*)

EXERCISE 50.5

Solve the following equations by Iteration method :

1. $1 + \log x = \frac{x}{2}$ **Ans.** 5.36

2. $\sin x = \frac{x+1}{x-1}$ [**Hint.** Approximate root = – 5.5] **Ans.** – 5.5174

3. $e^{-x} - \sin x = 0$ **Ans.** 0.61413

4. Use the method of iteration to find a root, near 2 of the equation $x^3 = x^2 + x - 1$. Carry out 5 iterations. **Ans.** 0.8408

5. $x^3 + x^2 - 100 = 0$ **Ans.** 4.3311 **6.** $x = \frac{1}{2} + \sin x$ **Ans.** 1.4973

7. $\tan x = x$ **Ans.** 4.4346 **8.** Find $\sqrt{30}$ by using iterative process. **Ans.** 5.47722

SIMULTANEOUS LINEAR EQUATIONS

(Crout Method, Gauss Seidel Method)

51.1 INTRODUCTION

We have already solved simultaneous equations of two or three unknowns. When the number of unknowns in simultaneous equations is large, then it becomes tedious to solve them by the known methods. Simultaneous equations of large number of unknowns are very important in the field of science and engineering. Now, we will use the following methods to solve such simultaneous equations

1. Direct method

(*a*) Gauss elimination method (*b*) Gauss-Jordan method

(*c*) Crouts method (Factorisation method)

2. Iterative method

(*a*) Jacobi method

(*b*) Gauss-seidel method

51.2 GAUSS ELIMINATION METHOD

In this method the unknowns of equations below are eliminated and the system is reduced to an upper triangular system. The unknowns are obtained by back substitution.

Let a system of simultaneous equations in n unknowns $x_1, x_2 \ldots\ldots x_n$ be

$$a_{11} x_1 + a_{12} x_2 + \ldots\ldots + a_{1n} x_n = b_1 \qquad \ldots (1)$$

$$a_{21} x_1 + a_{22} x_2 + \ldots\ldots + a_{2n} x_n = b_2 \qquad \ldots (2)$$

....

....

$$a_{n1} x_1 + a_{n2} x_2 + \ldots\ldots + a_{nn} x_n = b_n \qquad \ldots (n)$$

Method to solve the above equations

Step 1. We eliminate x_1 from 2nd, 3rd ....*n*th equation with the help of the first equation

$$a_{11} x_1 + a_{12} x_2 + \ldots\ldots + a_{1n} x_n = b_1$$

$$a'_{22} x_2 + \ldots\ldots + a'_{2n} x_n = b_2'$$

....

....

$$a'_{n2} x_2 + \ldots\ldots + a'_{nn} x_n = b_n'$$

Step 2. We again eliminate x_2 from 3rd, 4th..... *n*th equation with the help of second equation.

$$a_{11} x_1 + a_{12} x_2 + a_{13} x_3 + \ldots\ldots a_{1n} x_n = b_1$$

$$a'_{22} x_2 + a'_{23} x_3 \ldots\ldots + a'_{2n} x_n = b_2'$$

$$a''_{33} x_3 + \ldots\ldots + a''_{3n} x_n = b_3''$$

....

....

$$a''_{n3}x_3 + \ldots\ldots\ldots\ldots + a''_{nn}x_n = b''_n$$

In the third step we will eliminate x_3 and in fourth step x_4 and so on.

Finally the system of equations will be of the following form.

$$a_{11}x_1 + a_{12}x_2 + \ldots\ldots\ldots\ldots a_{1n}x_n = b_1$$
$$a_{22}x_2 + \ldots\ldots\ldots\ldots + a'_{2n}x_n = b'_2$$

....

....

$$c_{nn}x_n = d_n$$

The given system is reduced to the above form *i.e.* triangular form.

Backward Substitution

We first find out the value of x_n from the last equation, then substitute the value of x_n in the $(n-1)$th equation to get the value of x_{n-1}. Again substitute the value of x_{n-1} in $(n-2)$th equation to get the value of x_{n-2}. By this backward substitution we can find the values of all the unknowns.

Example 1. *Solve the following equations by using Gauss-elimination method* :

$$2x_1 + 4x_2 + x_3 = 3$$
$$3x_1 + 2x_2 - 2x_3 = -2$$
$$x_1 - x_2 + x_3 = 6$$

(*R.G.P.V. Bhopal, III Semester, June 2006*)

Solution. Third equation is written as first equation, the system becomes as

$$x_1 - x_2 + x_3 = 6 \quad \ldots(1)$$
$$2x_1 + 4x_2 + x_3 = 3 \quad \ldots(2)$$
$$3x_1 + 2x_2 - 2x_3 = -2 \quad \ldots(3)$$

Step 1. Subtracting 2 (1) from (2), and 3 (1) from (3), we get

$$x_1 - x_2 + x_3 = 6$$
$$6x_2 - x_3 = -9 \quad \ldots(4)$$
$$5x_2 - 5x_3 = -20 \quad \ldots(5)$$

Step 2. Operate $\frac{6}{5}(5) - (4)$

$$x_1 - x_2 + x_3 = 6 \quad \ldots(1)$$
$$6x_2 - x_3 = -9 \quad \ldots(6)$$
$$-5x_3 = -15 \quad \ldots(7)$$

Step 3. Backward substitution

From (7), $x_3 = \dfrac{-15}{-5} = 3$

From (6), $6x_2 - 3 = -9 \quad \Rightarrow 6x_2 = -6 \quad \Rightarrow \quad x_2 = -1$

From (1), $x_1 - (-1) + 3 = 6 \quad \Rightarrow x_1 = 6 - 3 - 1 = 2$

Hence, $x_1 = 2,\ x_2 = -1,\ x_3 = 3$ **Ans.**

EXERCISE 51.1

Solve the following system by Gauss elimination method.

1. $x - y + z = 1,\ -3x + 2y - 3z = -6,\ 2x - 5y + 4z = 5$ **Ans.** $x = -2,\ y = 3,\ z = 6$
2. $x + 10y + z = 12,\ x + y + 10z = 12,\ 10x + y + z = 12$ **Ans.** $x = 1,\ y = 1,\ z = 1$

3. $x - 2y + 9z = 8, \quad 2x - 8y + z = -5, \quad 3x + y - z = 3$ **Ans.** $x = 1, \ y = 1, \ z = 1$

4. $x + 3y + 10z = 23.89, \ 2x + 17y + 4z = 34.84, \ 28x + 4y - z = 31.88$

Ans. $x = 0.99, \ y = 1.50, \ z = 1.84$

5. $2x + 6y - z = -11.98, \ 5x - y + z = 11.01, \ 4x - y + 3z = 10.01$ **Ans.** $x = 1.64, \ y = -2.49, \ z = 0.32$

(R.G.P.V. Bhopal, III Semester, Dec. 2007)

6. $x_1 + 2x_2 + 3x_3 + 4x_4 = 32, \quad 2x_1 - x_2 + 2x_3 - x_4 = 3, \quad 3x_1 + 2x_2 + 4x_3 - x_4 = 17.7$
$8x_1 + 7x_2 + 8x_3 - 5x_4 = 30$

51.3 GAUSS- JORDAN METHOD

This is modification of the Gauss elimination method.

By this method we eliminate unknowns not only from the equations below but also from the equations above. In this way the system is reduced to a diagonal matrix.

Finally each equation consists of only one unknown and thus, we get the solution. Here, the labour of backward substitution for finding the unknowns is saved.

Gauss-Jordan method is modification of Gauss elimination method.

Example 2. *Apply Gauss-Jordan method to solve the equations :*

$$\begin{aligned} x + y + z &= 9 \\ 2x - 3y + 4z &= 13 \\ 3x + 4y + 5z &= 40 \end{aligned}$$

(R.G.P.V. Bhopal, III Semester, Dec. 2007)

Solution. The following system of linear equations can be written in matrix form:

By using Gauss Jordan method we have

$$\begin{bmatrix} 1 & 1 & 1 \\ 2 & -3 & 4 \\ 3 & 4 & 5 \end{bmatrix} \begin{bmatrix} x \\ y \\ z \end{bmatrix} = \begin{bmatrix} 9 \\ 13 \\ 40 \end{bmatrix} \Rightarrow \begin{bmatrix} 1 & 1 & 1 \\ 0 & -5 & 2 \\ 0 & 1 & 2 \end{bmatrix} \begin{bmatrix} x \\ y \\ z \end{bmatrix} = \begin{bmatrix} 9 \\ -5 \\ 13 \end{bmatrix} \begin{matrix} R_2 \to R_2 - 2R_1 \\ R_3 \to R_3 - 3R_1 \end{matrix}$$

$$\Rightarrow \begin{bmatrix} 1 & 1 & 1 \\ 0 & -5 & 2 \\ 0 & 0 & \frac{12}{5} \end{bmatrix} \begin{bmatrix} x \\ y \\ z \end{bmatrix} = \begin{bmatrix} 9 \\ -5 \\ 12 \end{bmatrix} R_3 \to R_3 + \frac{1}{5} R_2 \Rightarrow \begin{bmatrix} 1 & 0 & \frac{7}{5} \\ 0 & -5 & 2 \\ 0 & 0 & \frac{12}{5} \end{bmatrix} \begin{bmatrix} x \\ y \\ z \end{bmatrix} = \begin{bmatrix} 8 \\ -5 \\ 12 \end{bmatrix} R_1 \to R_1 + \frac{1}{5} R_2$$

$$\Rightarrow \begin{bmatrix} 1 & 0 & 0 \\ 0 & -5 & 0 \\ 0 & 0 & \frac{12}{5} \end{bmatrix} \begin{bmatrix} x \\ y \\ z \end{bmatrix} = \begin{bmatrix} 1 \\ -15 \\ 12 \end{bmatrix} \begin{matrix} R_1 \to R_1 - \frac{7}{12} R_3 \\ R_2 \to R_2 - \frac{5}{6} R_3 \end{matrix} \Rightarrow \begin{bmatrix} 1 & 0 & 0 \\ 0 & 1 & 0 \\ 0 & 0 & 1 \end{bmatrix} \begin{bmatrix} x \\ y \\ z \end{bmatrix} = \begin{bmatrix} 1 \\ 3 \\ 5 \end{bmatrix} \begin{matrix} R_2 \to -\frac{1}{5} R_2 \\ R_3 \to \frac{5}{12} R_3 \end{matrix}$$

Hence, $x = 1, \ y = 3, \ z = 5$

EXERCISE 51.2

Solve the following system by Gauss Jordan method.

1. $2x - 6y + 8z = 24, \ 5x + 4y - 3z = 2, \ 3x + y + 2z = 16$ **Ans.** $x = 1, \ y = 3, \ z = 5$

2. $x + 2y + z = 8, \ 2x + 3y + 4z = 20, \ 4x + 3y + 2z = 16$ **Ans.** $x = 1, \ y = 2, \ z = 3$

3. $3x + 4y + 5z = 18, \ 2x - y + 8z = 13, \ 5x - 2y + 7z = 20$ **Ans.** $x = 3, \ y = 1, \ z = 1$

4. $2x - y + 3z = 9,\ x + y + z = 6,\ x - y + z = 2$ **Ans.** $x = 1,\ y = 2,\ z = 3$

5. $10x + y + 2z = 13,\ 3x + 10y + z = 14,\ 2x + 3y + 10z = 15$ **Ans.** $x = 1,\ y = 1,\ z = 1$
(R.G.P.V., Bhopal, III Semester, June 2002)

6. $2x_1 + 2x_2 + x_3 = 6,\ 4x_1 + 2x_2 + 3x_3 = 4,\ x_1 + x_2 + x_3 = 0$
(R.G.P.V., Bhopal, M.C.A. June 2001) **Ans.** $x_1 = 5, x_2 = 1, x_3 = -6$

51.4 FACTORISATION OF A MATRIX

A square matrix A can be factorized into the lower triangular and Upper triangular matrices

$$A = LU$$

If all the principal minors of A are non singular

$$a_{11} \neq 0, \begin{vmatrix} a_{11} & a_{12} \\ a_{21} & a_{22} \end{vmatrix} \neq 0, \begin{vmatrix} a_{11} & a_{12} & a_{13} \\ a_{21} & a_{22} & a_{23} \\ a_{31} & a_{32} & a_{33} \end{vmatrix} \neq 0$$

$$\begin{bmatrix} l_{11} & 0 & 0 \\ l_{21} & l_{22} & 0 \\ l_{31} & l_{32} & l_{33} \end{bmatrix} \begin{bmatrix} 1 & u_{12} & u_{13} \\ 0 & 1 & u_{23} \\ 0 & 0 & 1 \end{bmatrix} = \begin{bmatrix} a_{11} & a_{12} & a_{13} \\ a_{21} & a_{22} & a_{23} \\ a_{31} & a_{32} & a_{33} \end{bmatrix}$$

or

$$\begin{bmatrix} 1 & 0 & 0 \\ l_{21} & 1 & 0 \\ l_{31} & l_{32} & 1 \end{bmatrix} \begin{bmatrix} u_{11} & u_{12} & u_{13} \\ 0 & u_{22} & u_{23} \\ 0 & 0 & u_{33} \end{bmatrix} = \begin{bmatrix} a_{11} & a_{12} & a_{13} \\ a_{21} & a_{22} & a_{23} \\ a_{31} & a_{32} & a_{33} \end{bmatrix}$$

Example 3. *Obtain the LU decomposition of the matrix* $\begin{bmatrix} 1 & \frac{1}{2} & \frac{1}{3} \\ \frac{1}{2} & \frac{1}{3} & \frac{1}{4} \\ \frac{1}{3} & \frac{1}{4} & \frac{1}{5} \end{bmatrix}$

Solution. Let $\begin{bmatrix} l_{11} & 0 & 0 \\ l_{21} & l_{22} & 0 \\ l_{31} & l_{32} & l_{33} \end{bmatrix} \begin{bmatrix} 1 & u_{12} & u_{13} \\ 0 & 1 & u_{23} \\ 0 & 0 & 1 \end{bmatrix} = \begin{bmatrix} 1 & \frac{1}{2} & \frac{1}{3} \\ \frac{1}{2} & \frac{1}{3} & \frac{1}{4} \\ \frac{1}{3} & \frac{1}{4} & \frac{1}{5} \end{bmatrix}$

$$\begin{bmatrix} l_{11} & l_{11}u_{12} & l_{11}u_{13} \\ l_{21} & l_{21}u_{12} + l_{22} & l_{21}u_{13} + l_{22}u_{23} \\ l_{31} & l_{31}u_{12} + l_{32} & l_{31}u_{13} + l_{32}u_{23} + l_{33} \end{bmatrix} = \begin{bmatrix} 1 & \frac{1}{2} & \frac{1}{3} \\ \frac{1}{2} & \frac{1}{3} & \frac{1}{4} \\ \frac{1}{3} & \frac{1}{4} & \frac{1}{5} \end{bmatrix}$$

Equating the corresponding elements of equal matrices, we get

$$l_{11} = 1 \qquad l_{11}u_{12} = \frac{1}{2} \qquad l_{11}u_{13} = \frac{1}{3}$$

$$l_{21} = \frac{1}{2} \qquad l_{21}u_{12} + l_{22} = \frac{1}{3} \qquad l_{21}u_{13} + l_{22}u_{23} = \frac{1}{4}$$

$$l_{31} = \frac{1}{3} \qquad l_{31}\,u_{12} + l_{32} = \frac{1}{4} \qquad l_{31}u_{13} + l_{32}u_{23} + l_{33} = \frac{1}{5}$$

We solve the above equations in the following order

First Column / Row

$$l_{11} = 1, \qquad l_{21} = \frac{1}{2}, \qquad l_{31} = \frac{1}{3}$$

First Row / Column $l_{11}\,u_{12} = \frac{1}{2} \Rightarrow (1)\ u_{12} = \frac{1}{2} \Rightarrow u_{12} = \frac{1}{2}$

$$l_{11}\,u_{13} = \frac{1}{3} \Rightarrow (1)\ u_{13} = \frac{1}{3} \Rightarrow u_{13} = \frac{1}{3}$$

Second Column/Row $l_{21}\,u_{12} + l_{22} = \frac{1}{3} \Rightarrow \left(\frac{1}{2}\right)\left(\frac{1}{2}\right) + l_{22} = \frac{1}{3} \Rightarrow l_{22} = \frac{1}{12}$

$$l_{31}\,u_{12} + l_{32} = \frac{1}{4} \Rightarrow \left(\frac{1}{3}\right)\left(\frac{1}{2}\right) + l_{32} = \frac{1}{4} \Rightarrow l_{32} = \frac{1}{12}$$

Second Row/Column $l_{21}\,u_{13} + l_{22}\,u_{23} = \frac{1}{4} \Rightarrow \left(\frac{1}{2}\right)\left(\frac{1}{3}\right) + \left(\frac{1}{12}\right)u_{23} = \frac{1}{4} \Rightarrow u_{23} = 1$

Third column /Row $l_{31}\,u_{13} + l_{32}\,u_{23} + l_{33} = \frac{1}{5}$

$$\Rightarrow \quad \left(\frac{1}{3}\right)\left(\frac{1}{3}\right) + \left(\frac{1}{12}\right)(1) + l_{33} = \frac{1}{5} \Rightarrow l_{33} = \frac{1}{180}$$

On substituting the values of l_{rs} and u_{rs} in (1), we get

$$\begin{bmatrix} 1 & 0 & 0 \\ \frac{1}{2} & \frac{1}{12} & 0 \\ \frac{1}{3} & \frac{1}{12} & \frac{1}{180} \end{bmatrix} \begin{bmatrix} 1 & \frac{1}{2} & \frac{1}{3} \\ 0 & 1 & 1 \\ 0 & 0 & 1 \end{bmatrix} = \begin{bmatrix} 1 & \frac{1}{2} & \frac{1}{3} \\ \frac{1}{2} & \frac{1}{3} & \frac{1}{4} \\ \frac{1}{3} & \frac{1}{4} & \frac{1}{5} \end{bmatrix}$$

51.5 CROUTS – TRIANGULARISATION METHOD

This method is also called as decomposition method. The coefficients matrix A of the systen $AX = B$ is decomposed or factorised as the product of lower triangular matrix L and an uppe triangular matrix U. This method is based on the fact that every square matrix A is the product of a lower triangular matrix and an upper triangular matrix.

Method

Consider the following equations :

$$\begin{aligned} a_{11}\,x_1 + a_{12}\,x_2 + a_{13}\,x_3 &= b_1 \\ a_{21}\,x_1 + a_{22}\,x_2 + a_{23}\,x_3 &= b_2 \\ a_{31}\,x_1 + a_{32}\,x_2 + a_{33}\,x_3 &= b_3 \end{aligned}$$

These equations are written in the matrix form as

$$\begin{bmatrix} a_{11} & a_{12} & a_{13} \\ a_{21} & a_{22} & a_{23} \\ a_{31} & a_{32} & a_{33} \end{bmatrix} \begin{bmatrix} x_1 \\ x_2 \\ x_3 \end{bmatrix} = \begin{bmatrix} b_1 \\ b_2 \\ b_3 \end{bmatrix} \qquad \text{... (1)}$$

$$AX = B$$

Now, let $A = LU$

$$\begin{bmatrix} a_{11} & a_{12} & a_{13} \\ a_{21} & a_{22} & a_{23} \\ a_{31} & a_{32} & a_{33} \end{bmatrix} = \begin{bmatrix} l_{11} & 0 & 0 \\ l_{21} & l_{22} & 0 \\ l_{31} & l_{32} & l_{33} \end{bmatrix} \begin{bmatrix} 1 & u_{12} & u_{13} \\ 0 & 1 & u_{23} \\ 0 & 0 & 1 \end{bmatrix} \quad \ldots (2)$$

Multiplying the matrices on R.H.S., we get

$$\begin{bmatrix} a_{11} & a_{12} & a_{13} \\ a_{21} & a_{22} & a_{23} \\ a_{31} & a_{32} & a_{33} \end{bmatrix} = \begin{bmatrix} l_{11} & l_{11} u_{12} & l_{11} u_{13} \\ l_{21} & l_{21} u_{12} + l_{22} & l_{21} u_{13} + l_{22} u_{23} \\ l_{31} & l_{31} u_{12} + l_{32} & l_{31} u_{13} + l_{32} u_{23} + l_{33} \end{bmatrix}$$

Equating corresponding elements from both sides, we get

$$\begin{array}{lll} l_{11} = a_{11} & l_{11} u_{12} = a_{12} & l_{11} u_{13} = a_{13} \\ l_{21} = a_{21} & l_{21} u_{12} + l_{22} = a_{22} & l_{21} u_{13} + l_{22} u_{23} = a_{23} \\ l_{31} = a_{31} & l_{31} u_{12} + l_{32} = a_{32} & l_{31} u_{13} + l_{32} u_{23} + l_{33} = a_{33} \end{array}$$

We solve these equations in the following order

Step 1. Solve equations in I column.

Step 2. Solve equations in I row.

Step 3. Solve equations in II column.

Step 4. Solve equations in II row.

Step 5. Solve equations in III column.

Step 6. Solve equations in III row.

Putting LU for A in (1), we have

$$\begin{bmatrix} l_{11} & 0 & 0 \\ l_{21} & l_{22} & 0 \\ l_{31} & l_{32} & l_{33} \end{bmatrix} \begin{bmatrix} 1 & u_{12} & u_{13} \\ 0 & 1 & u_{23} \\ 0 & 0 & 1 \end{bmatrix} \begin{bmatrix} x_1 \\ x_2 \\ x_3 \end{bmatrix} = \begin{bmatrix} b_1 \\ b_2 \\ b_3 \end{bmatrix} \quad \ldots (3)$$

[We will substitute the values of elements which we get on solving the above equations in (3)]

Putting $\begin{bmatrix} 1 & u_{12} & u_{13} \\ 0 & 1 & u_{23} \\ 0 & 0 & 1 \end{bmatrix} \begin{bmatrix} x_1 \\ x_2 \\ x_3 \end{bmatrix} = \begin{bmatrix} p \\ q \\ r \end{bmatrix}$ ... (4)

in (3), we get

$$\begin{bmatrix} l_{11} & 0 & 0 \\ l_{21} & l_{22} & 0 \\ l_{31} & l_{32} & l_{33} \end{bmatrix} \begin{bmatrix} p \\ q \\ r \end{bmatrix} = \begin{bmatrix} b_1 \\ b_2 \\ b_3 \end{bmatrix} \quad \ldots (5)$$

On solving equation (5), we get the value of p, q, r. We substitute the value of p, q, r in (4). On solving (4), we get the values of x_1, x_2, x_3.

Example 4. *Apply Crout's method (Factorization Method) to solve the equations*

$$3x + 2y + 7z = 4;\ 2x + 3y + z = 5;\ 3x + 4y + z = 7$$

(R.G.P.V., Bhopal, III Semester, Dec. 2007, June 2005, Dec. 2004)

Solution. We have,

$$\begin{aligned} 3x + 2y + 7z &= 4 \\ 2x + 3y + z &= 5 \\ 3x + 4y + z &= 7 \end{aligned}$$

These equations are written in the matrix form.

$$\begin{bmatrix} 3 & 2 & 7 \\ 2 & 3 & 1 \\ 3 & 4 & 1 \end{bmatrix} \begin{bmatrix} x \\ y \\ z \end{bmatrix} = \begin{bmatrix} 4 \\ 5 \\ 7 \end{bmatrix}$$

$$\begin{bmatrix} 1 & 0 & 0 \\ l_{21} & 1 & 0 \\ l_{31} & l_{32} & 1 \end{bmatrix} \begin{bmatrix} u_{11} & u_{12} & u_{13} \\ 0 & u_{22} & u_{23} \\ 0 & 0 & u_{33} \end{bmatrix} \begin{bmatrix} x \\ y \\ z \end{bmatrix} = \begin{bmatrix} 4 \\ 5 \\ 7 \end{bmatrix}$$

where
$$\begin{bmatrix} 1 & 0 & 0 \\ l_{21} & 1 & 0 \\ l_{31} & l_{32} & 1 \end{bmatrix} \begin{bmatrix} u_{11} & u_{12} & u_{13} \\ 0 & u_{22} & u_{23} \\ 0 & 0 & u_{33} \end{bmatrix} = \begin{bmatrix} 3 & 2 & 7 \\ 2 & 3 & 1 \\ 3 & 4 & 1 \end{bmatrix} \quad \ldots (1)$$

$$\begin{bmatrix} u_{11} & u_{12} & u_{13} \\ l_{21}u_{11} & l_{21}u_{12} + u_{22} & l_{21}u_{13} + u_{23} \\ l_{31}u_{11} & l_{31}u_{12} + l_{32}u_{22} & l_{31}u_{13} + l_{32}\,u_{23} + u_{33} \end{bmatrix} = \begin{bmatrix} 3 & 2 & 7 \\ 2 & 3 & 1 \\ 3 & 4 & 1 \end{bmatrix} \quad \ldots (2)$$

Equating corresponding elements on both sides of (2), we get

(i) First row

$$u_{11} = 3 \qquad u_{12} = 2, \qquad u_{13} = 7$$

(ii) First column

$$l_{21}\,u_{11} = 2 \quad \Rightarrow \quad l_{21} = \frac{2}{3}$$

$$l_{31}\,u_{11} = 3 \quad \Rightarrow \quad l_{31} = \frac{3}{3} = 1$$

(iii) Second row

$$l_{21}\,u_{12} + u_{22} = 3 \quad \Rightarrow \quad u_{22} = 3 - \left(\frac{2}{3}\right)(2) = 3 - \frac{4}{3} = \frac{5}{3}$$

$$l_{21}\,u_{13} + u_{23} = 1 \quad \Rightarrow \quad u_{23} = 1 - \left(\frac{2}{3}\right)(7) = 1 - \frac{14}{3} = -\frac{11}{3}$$

(iv) Second column

$$l_{31}\,u_{12} + l_{32}\,u_{22} = 4 \quad \Rightarrow \quad (1)(2) + l_{32}\left(\frac{5}{3}\right) = 4 \Rightarrow l_{32} = \frac{3}{5}(4-2)$$

$$\Rightarrow \quad l_{32} = \frac{6}{5}$$

(v) Third row

$$l_{31}\,u_{13} + l_{32}\,u_{23} + u_{33} = 1 \quad \Rightarrow \quad (1)(7) + \left(\frac{6}{5}\right)\left(\frac{-11}{3}\right) + u_{33} = 1$$

$$\Rightarrow \quad u_{33} = 1 - 7 + \frac{66}{15} = -\frac{8}{5}$$

Putting the values of the elements in (1), we get

$$\begin{bmatrix} 1 & 0 & 0 \\ \frac{2}{3} & 1 & 0 \\ 1 & \frac{6}{5} & 1 \end{bmatrix} \begin{bmatrix} 3 & 2 & 7 \\ 0 & \frac{5}{3} & -\frac{11}{3} \\ 0 & 0 & \frac{-8}{5} \end{bmatrix} \begin{bmatrix} x \\ y \\ z \end{bmatrix} = \begin{bmatrix} 4 \\ 5 \\ 7 \end{bmatrix} \quad \ldots (3)$$

Writing $$\begin{bmatrix} 3 & 2 & 7 \\ 0 & \frac{5}{3} & \frac{-11}{3} \\ 0 & 0 & \frac{-8}{5} \end{bmatrix} \begin{bmatrix} x \\ y \\ z \end{bmatrix} = \begin{bmatrix} p \\ q \\ r \end{bmatrix} \quad \ldots (4)$$

in (3), we get

$$\begin{bmatrix} 1 & 0 & 0 \\ \frac{2}{3} & 1 & 0 \\ 1 & \frac{6}{5} & 1 \end{bmatrix} \begin{bmatrix} p \\ q \\ r \end{bmatrix} = \begin{bmatrix} 4 \\ 5 \\ 7 \end{bmatrix} \Rightarrow \begin{bmatrix} p \\ \frac{2}{3}p+q \\ p+\frac{6}{5}q+r \end{bmatrix} = \begin{bmatrix} 4 \\ 5 \\ 7 \end{bmatrix}$$

Solving this system, we have

$$p = 4$$

$$\frac{2}{3}p+q = 5 \Rightarrow q = 5-\frac{8}{3} = \frac{7}{3}$$

$$p+\frac{6}{5}q+r = 7 \Rightarrow r = 7-4-\frac{42}{15} = \frac{1}{5}$$

Putting the values of p, q, r in (4), we get

$$\begin{bmatrix} 3 & 2 & 7 \\ 0 & \frac{5}{3} & \frac{-11}{3} \\ 0 & 0 & -\frac{8}{5} \end{bmatrix} \begin{bmatrix} x \\ y \\ z \end{bmatrix} = \begin{bmatrix} 4 \\ \frac{7}{3} \\ \frac{1}{5} \end{bmatrix}$$

i.e.

$$3x+2y+7z = 4 \quad \ldots (5)$$

$$\frac{5}{3}y-\frac{11}{3}z = \frac{7}{3} \quad \ldots (6)$$

$$-\frac{8}{5}z = \frac{1}{5} \Rightarrow z = -\frac{1}{8} \quad \ldots (7)$$

Putting $z = -\frac{1}{8}$ in (6), we get

$$y = \frac{9}{8}$$

Putting the values of z and y in (5), we get

$$x = \frac{7}{8}$$

Hence, the solution of the given system is

$$x = \frac{7}{8},\ y = \frac{9}{8} \text{ and } z = -\frac{1}{8}.$$ **Ans.**

Example 5. *Solve the following system of equations using the Decomposition method*

$$x_1 + x_2 - x_3 = 2$$
$$2x_1 + 3x_2 + 5x_3 = -3$$
$$3x_1 + 2x_2 - 3x_3 = 6$$

Solution. $$\begin{bmatrix} 1 & 1 & -1 \\ 2 & 3 & 5 \\ 3 & 2 & -3 \end{bmatrix}\begin{bmatrix} x_1 \\ x_2 \\ x_3 \end{bmatrix} = \begin{bmatrix} 2 \\ -3 \\ 6 \end{bmatrix}$$

$$\begin{bmatrix} l_{11} & 0 & 0 \\ l_{21} & l_{22} & 0 \\ l_{31} & l_{32} & l_{33} \end{bmatrix}\begin{bmatrix} 1 & u_{12} & u_{13} \\ 0 & 1 & u_{23} \\ 0 & 0 & 1 \end{bmatrix}\begin{bmatrix} x_1 \\ x_2 \\ x_3 \end{bmatrix} = \begin{bmatrix} 2 \\ -3 \\ 6 \end{bmatrix}$$

where, $$\begin{bmatrix} l_{11} & 0 & 0 \\ l_{21} & l_{22} & 0 \\ l_{31} & l_{32} & l_{33} \end{bmatrix}\begin{bmatrix} 1 & u_{12} & u_{13} \\ 0 & 1 & u_{23} \\ 0 & 0 & 1 \end{bmatrix} = \begin{bmatrix} 1 & 1 & -1 \\ 2 & 3 & 5 \\ 3 & 2 & -3 \end{bmatrix} \qquad \text{... (1)}$$

On multiplication the matrices of L.H.S., we get

$$\begin{bmatrix} l_{11} & l_{11}u_{12} & l_{11}u_{13} \\ l_{21} & l_{21}u_{12}+l_{22} & l_{21}u_{13}+l_{22}u_{23} \\ l_{31} & l_{31}u_{12}+l_{32} & l_{31}u_{13}+l_{32}u_{23}+l_{33} \end{bmatrix} = \begin{bmatrix} 1 & 1 & -1 \\ 2 & 3 & 5 \\ 3 & 2 & -3 \end{bmatrix}$$

Equating the corresponding elements of both the matrices, we get

$$l_{11} = 1 \qquad l_{11}u_{12} = 1 \qquad l_{11}u_{13} = -1$$
$$l_{21} = 2 \qquad l_{21}u_{12}+l_{22} = 3 \qquad l_{21}u_{13}+l_{22}u_{23} = 5$$
$$l_{31} = 3 \qquad l_{31}u_{12}+l_{32} = 2 \qquad l_{31}u_{13}+l_{32}u_{23}+l_{33} = -3$$

On solving these equations, we get

$$l_{11} = 1 \qquad u_{12} = 1 \qquad u_{13} = -1$$
$$l_{21} = 2 \qquad l_{22} = 1 \qquad u_{23} = 7$$
$$l_{31} = 3 \qquad l_{32} = -1 \qquad l_{33} = 7$$

On putting the values of the elements which we have got on solving in (1), we get

$$\begin{bmatrix} 1 & 0 & 0 \\ 2 & 1 & 0 \\ 3 & -1 & 7 \end{bmatrix}\begin{bmatrix} 1 & 1 & -1 \\ 0 & 1 & 7 \\ 0 & 0 & 1 \end{bmatrix}\begin{bmatrix} x_1 \\ x_2 \\ x_3 \end{bmatrix} = \begin{bmatrix} 2 \\ -3 \\ 6 \end{bmatrix} \qquad \text{... (2)}$$

Put $$\begin{bmatrix} 1 & 1 & -1 \\ 0 & 1 & 7 \\ 0 & 0 & 1 \end{bmatrix}\begin{bmatrix} x_1 \\ x_2 \\ x_3 \end{bmatrix} = \begin{bmatrix} p \\ q \\ r \end{bmatrix} \text{ in (2), we have} \qquad \text{... (3)}$$

$$\Rightarrow \begin{bmatrix} 1 & 0 & 0 \\ 2 & 1 & 0 \\ 3 & -1 & 7 \end{bmatrix}\begin{bmatrix} p \\ q \\ r \end{bmatrix} = \begin{bmatrix} 2 \\ -3 \\ 6 \end{bmatrix}$$

$p = 2$

$2p + q = -3 \qquad \Rightarrow \qquad q = -3 - 4 = -7$

$3p - q + 7r = 6 \qquad \Rightarrow 7r = 6 - 6 - 7 \qquad \Rightarrow r = \dfrac{-7}{7} = -1$

Substituting the values of p, q, r in (3), we get

$$\begin{bmatrix} 1 & 1 & -1 \\ 0 & 1 & 7 \\ 0 & 0 & 1 \end{bmatrix}\begin{bmatrix} x_1 \\ x_2 \\ x_3 \end{bmatrix} = \begin{bmatrix} 2 \\ -7 \\ -1 \end{bmatrix}$$

$$x_1 + x_2 - x_3 = 2$$
$$x_2 + 7x_3 = -7$$
$$x_3 = -1$$

By backward substitution, we have

$$x_2 = -7 + 7 = 0$$
$$x_1 + 0 + 1 = 2 \Rightarrow x_1 = 1$$

Hence, solution of the system of given equations is

$$x_1 = 1, \quad x_2 = 0, \quad x_3 = -1$$ **Ans.**

EXERCISE 51.3

Solve the following system by Crout's triangularisation method :

1. $x_1 + 2x_2 + 3x_3 = 14,\ 2x_1 + 5x_2 + 2x_3 = 18,\ 3x_1 + 2x_2 + 5x_3 = 22$ **Ans.** $x_1 = 1,\ x_2 = 2,\ x_3 = 3$
2. $10x + y + z = 12,\ 2x + 10y + z = 13,\ 2x + 2y + 10z = 14$ **Ans.** $x = 1,\ y = 1,\ z = 1$
3. $x + y + z = 1,\ 2x + y - z = 0,\ 3x + 4y + 5z = 4,$ **Ans.** $x = 1, y = -1, z = 1$
4. $10x + y + 2z = 13,\ 3x + 10y + z = 14,\ 2x + 3y + 10z = 15$ **Ans.** $x = 1,\ y = 1,\ z = 1$
5. $2x - 6y + 8z = 24,\ 5x + 4y - 3z = 2,\ 3x + y + 2z = 16$ **Ans.** $x = 1,\ y = 3,\ z = 5$

51.6 CROUT'S METHOD

It requires less computation than that of Gauss method. This method is explained below :
Consider the system :

$$AX = B$$

$$\begin{bmatrix} a_{11} & a_{12} & a_{13} \\ a_{21} & a_{22} & a_{23} \\ a_{31} & a_{32} & a_{33} \end{bmatrix} \begin{bmatrix} x_1 \\ x_2 \\ x_3 \end{bmatrix} = \begin{bmatrix} b_1 \\ b_2 \\ b_3 \end{bmatrix} \quad \text{... (1)}$$

$\Rightarrow$
$$a_{11}x_1 + a_{12}x_2 + a_{13}x_3 = b_1$$
$$a_{21}x_1 + a_{22}x_2 + a_{23}x_3 = b_2$$
$$a_{31}x_1 + a_{32}x_2 + a_{33}x_3 = b_3$$

The augmented matrix of (1) is $\begin{bmatrix} a_{11} & a_{12} & a_{13} & b_1 \\ a_{21} & a_{22} & a_{23} & b_2 \\ a_{31} & a_{32} & a_{33} & b_3 \end{bmatrix}$ ... (2)

Then we consider derived matrix $\begin{bmatrix} a'_{11} & a'_{12} & a'_{13} & b'_1 \\ a'_{21} & a'_{22} & a'_{23} & b'_2 \\ a'_{31} & a'_{32} & a'_{33} & b'_3 \end{bmatrix}$ ... (3)

Which is determined below :

Step 1. To determine the first column :

$$a_{11}' = a_{11} \qquad a_{21}' = a_{21} \qquad a_{31}' = a_{31}$$

Step 2. To determine the first row to the right of the first column.

$$a_{12}' = \frac{a_{12}}{a_{11}}, \quad a_{13}' = \frac{a_{13}}{a_{11}},$$

$$b_1' = \frac{b_1}{a_{11}}$$

except the first element of the first row, other elements of the first row are obtained by dividing the corresponding elements of first row of the matrix (2) by the first element of the first row.

Step 3. To determine the elements of second column except a'_{21}.

$$a'_{22} = a_{22} - a'_{12} \times a'_{21}$$
$$a'_{32} = a_{32} - a'_{12} \times a'_{31}$$

Step 4. To determine the elements of second row except a'_{21}; a'_{22}.

$$a'_{23} = \frac{a_{23} - a'_{13} . a'_{21}}{a'_{22}}$$

$$b'_2 = \frac{b_2 - b'_1 . a'_{21}}{a'_{22}}$$

Step 5. To determine the elements of third column except a'_{13} and a'_{33}.

$$a'_{33} = a_{33} - a'_{23} . a'_{32} - a'_{13} . a'_{31}$$

Step 6. To determine the remaining elements of third row.

$$b'_3 = \frac{b_3 - b'_2 . a'_{32} - b'_1 a'_{31}}{a'_{33}}$$

Step 7. Solution is
$$x_3 = b'_3$$
$$x_2 = b'_2 - a'_{23} . x_3$$
$$x_1 = b'_1 - a'_{13} . x_3 - a'_{12} . x_2$$

Example 6. *Solve the system*

$$x_1 + x_2 + x_3 = 1$$
$$3x_1 + x_2 - 3x_3 = 5$$
$$x_1 - 2x_2 - 5x_3 = 10 \quad \textit{by Crout's method.}$$

Solution. The coefficient matrix is given by

$$\begin{bmatrix} a_{11} & a_{12} & a_{13} \\ a_{21} & a_{22} & a_{23} \\ a_{31} & a_{32} & a_{33} \end{bmatrix} = \begin{bmatrix} 1 & 1 & 1 \\ 3 & 1 & -3 \\ 1 & -2 & -5 \end{bmatrix}$$

The augmented matrix of the coefficient matrix is given by

$$\begin{bmatrix} 1 & 1 & 1 & 1 \\ 3 & 1 & -3 & 5 \\ 1 & -2 & -5 & 10 \end{bmatrix}$$

$$\begin{bmatrix} a_{11}' & a_{12}' & a_{13}' & b_1' \\ a_{21}' & a_{22}' & a_{23}' & b_2' \\ a_{31}' & a_{32}' & a_{33}' & b_3' \end{bmatrix}$$

where, $a'_{11} = a_{11}$, $a'_{21} = a_{21}$, $a'_{31} = a_{31}$

$\Rightarrow$ $a'_{11} = 1$, $a'_{21} = 3$, $a'_{31} = 1$ ... (1)

$$a'_{12} = \frac{a_{12}}{a_{11}} = \frac{1}{1} = 1,\ a'_{13} = \frac{a_{13}}{a_{11}} = \frac{1}{1} = 1,\ b'_1 = \frac{b_1}{a_{11}} = \frac{1}{1} = 1 \quad \text{... (2)}$$

$$a'_{22} = a_{22} - a'_{12} . a'_{21} = 1 - 1 \times 3 = -2,$$
$$a'_{32} = a_{32} - a'_{12} . a'_{31} = -2 - 1.1 = -3 \quad \text{... (3)}$$

$$a'_{23} = \frac{a_{23} - a'_{13} a_{21}}{a'_{22}} = \frac{-3-1\times3}{-2} = 3,\ b'_2 = \frac{b_2 + b'_1 . a'_{21}}{a'_{22}} = \frac{5-1\times3}{-2} = -1 \quad \text{... (4)}$$

$$a'_{33} = a_{33} - a'_{23} . a'_{32} - a'_{13} . a'_{31} = -5-3(-3)-(1)(1) = 3 \quad \text{... (5)}$$

$$b'_3 = \frac{b_3 - b'_2 a'_{32} - b'_1 . a'_{31}}{a'_{33}} = \frac{10-(-1)(-3)-1(1)}{3} = 2 \quad \text{... (6)}$$

$$\therefore \text{ Derived matrix} = \begin{bmatrix} 1 & 1 & 1 & 1 \\ 3 & -2 & 3 & -1 \\ 1 & -3 & 3 & 2 \end{bmatrix}$$

Thus, $x_3 = b'_3 = 2,\ x_2 = b'_2 - a'_{23}x_3 = -1 - 3 \times 2 = -7$
and $x_1 = b'_1 - a'_{13}x_3 - a'_{12}x_2 = 1 - x_3 - x_2 = 1 - 2 + 7 = 6$

Hence, the solution of given system of equations is

$x_1 = 6,\ x_2 = -7,\ x_3 = 2.$ **Ans.**

51.7 ITERATIVE METHODS OR INDIRECT METHODS

We start with an approximation to the true solution and by applying the method repeatedly we get better and better approximation till accurated solution is achieved.

There are two iterative methods for solving simultaneous equations.

(1) Jacobi's method (method of successive correction).

(2) Gauss-Seidel method (Method of successive correction).

51.8 JACOBI'S METHOD

The method is illustrated by taking an example.

Let
$$\left.\begin{aligned} a_{11}x + a_{12}y + a_{13}z &= b_1 \\ a_{21}x + a_{22}y + a_{23}z &= b_2 \\ a_{31}x + a_{32}y + a_{33}z &= b_3 \end{aligned}\right\} \quad \ldots (1)$$

After division by suitable constants and transposition, the equations can be written as

$$\left.\begin{aligned} x &= c_1 - k_{12}y - k_{13}z \\ y &= c_2 - k_{21}x - k_{23}z \\ z &= c_3 - k_{31}x - k_{32}y \end{aligned}\right\} \quad \ldots (2)$$

Let us assume $x = 0$, $y = 0$ and $z = 0$ as first approximation, substituting the values of x, y, z on the right hand side of (2), we get $x = c_1$, $y = c_2$, $z = c_3$. This is the second approximation to the solution of the equations.

Again substituting these values of x, y, z in (2) we get a third approximation.

The process is repeated till two successive approximations are equal or nearly equal.

Note. Condition for using the iterative methods is that the coefficients in the leading diagonal are large compared to the other. If these are not so, then on interchanging the equation we can make the leading diagonal dominant diagonal.

Example 7. *Solve by Jacobi's method*

$$\begin{aligned} 4x + y + 3z &= 17 \\ x + 5y + z &= 14 \\ 2x - y + 8z &= 12 \end{aligned}$$

Solution. The above equations can be written as

$$\left.\begin{aligned} x &= \frac{17}{4} - \frac{y}{4} - \frac{3z}{4} \\ y &= \frac{14}{5} - \frac{x}{5} - \frac{z}{5} \\ z &= \frac{3}{2} - \frac{x}{4} + \frac{y}{8} \end{aligned}\right\} \quad \ldots (1)$$

On substituting $x = y = z = 0$ on the right hand side of (1), we get

$$x = \frac{17}{4},\ y = \frac{14}{5},\ z = \frac{3}{2}$$

Again substituting these values of x, y, z on R.H.S. of (1), we obtain

$$x = \frac{17}{4} - \frac{7}{10} - \frac{9}{8} = \frac{97}{40}$$

$$y = \frac{14}{5} - \frac{17}{20} - \frac{3}{10} = \frac{33}{20}$$

$$z = \frac{3}{2} - \frac{17}{16} + \frac{7}{20} = \frac{63}{80}$$

Again putting these values on R.H.S. of (1) we get next approximations.

$$x = \frac{17}{4} - \frac{33}{80} - \frac{189}{320} = \frac{1039}{320} = 3.25$$

$$y = \frac{14}{5} - \frac{97}{200} - \frac{63}{400} = \frac{863}{400} = 2.16$$

$$z = \frac{3}{2} - \frac{97}{160} + \frac{33}{160} = \frac{176}{160} = 1.1$$

Substituting, again, the values of x, y, z on R.H.S. of (1), we get

$$x = \frac{17}{4} - \frac{2.16}{4} - \frac{3(1.1)}{4} = 2.885$$

$$y = \frac{14}{5} - \frac{3.25}{5} - \frac{1.1}{5} = 1.93$$

$$z = \frac{3}{2} - \frac{3.25}{4} + \frac{2.16}{8} = 0.96$$

Repeating the process for $x = 2.885$, $y = 1.93$, $z = 0.96$, we have

$$x = \frac{17}{4} - \frac{1.93}{4} - \frac{3}{4} \times 0.96 = 4.25 - 0.48 - 0.72 = 3.05$$

$$y = \frac{14}{5} - \frac{2.885}{5} - \frac{0.96}{5} = 2.8 - 0.577 - 0.192 = 2.03$$

$$z = \frac{3}{2} - \frac{2.885}{4} + \frac{1.93}{8} = 1.5 - 0.721 + 0.241 = 1.02$$

This can be written in a table

Iterations	1	2	3	4	5	6
$x = \frac{17}{4} - \frac{y}{4} - \frac{3z}{4}$	0	$\frac{17}{4} = 4.25$	$\frac{97}{40} = 2.425$	$\frac{1039}{320} = 3.25$	2.885	3.05
$y = \frac{14}{5} - \frac{x}{5} - \frac{z}{5}$	0	$\frac{14}{5} = 2.8$	$\frac{33}{20} = 1.65$	$\frac{863}{400} = 2.16$	1.93	2.03
$z = \frac{3}{2} - \frac{x}{4} + \frac{y}{8}$	0	$\frac{3}{2} = 1.5$	$\frac{63}{80} = 0.7875$	$\frac{176}{160} = 1.1$	0.96	1.02

After 6th iteration $x = 3.05$, $y = 2.03$, $z = 1.02$

The actual values are $x = 3$, $y = 2$, $z = 1$ **Ans.**

51.9 GAUSS-SEIDEL METHOD

Gauss-Seidel method is a modification of Jacobi's method. In place of substituting the same se of values in all the three equations (2) of Article 51.8, we use in each step the value obtained in the earlier step.

Step 1. First we put $y = z = 0$ in first of the equation (2) of Article 51.8 and $x = c_1$. Then in second equation we put this value of x *i.e.*, c_1 and $z = 0$ and obtain y. In the third equation we use the values of x and y obtained earlier to get z.

Step 2. We repeat the above procedure. In the first equation we put the values of y and z obtained in step 1 and redetermine x. By using the new value of x and value of z obtained in step 1 we redetermine y and so on.

In other words, the latest values of the unknowns are used in each step.

Consider the following equations

$$a_1 x + b_1 y + c_1 z = d_1$$
$$a_2 x + b_2 y + c_2 z = d_2$$
$$a_3 x + b_3 y + c_3 z = d_3$$

The above equations can be rewritten as

$$x = \frac{1}{a_1}[d_1 - b_1 y - c_1 z], \quad y = \frac{1}{b_2}[d_2 - a_2 x - c_2 z], \quad z = \frac{1}{c_3}[d_3 - a_3 x - b_3 y]$$

Initial approximations

$$x = x_0, \quad y = y_0, \quad z = z_0$$

To find $x = x_1$

$$x_1 = \frac{1}{a_1}[d_1 - b_1 y_0 - c_1 z_0]$$

To find $y = y_1$; put $x = x_1$, $z = z_0$

$$y_1 = \frac{1}{b_2}[d_2 - a_2 x_1 - c_2 z_0]$$

To find $z = z_1$, put $x = x_1$, $y = y_1$

$$z_1 = \frac{1}{c_3}[d_3 - a_3 x_1 - b_3 y_1] \quad \text{and so on.}$$

Note 1. The convergence of Gauss Seidel method is twice as fast as in Jocobi's method.

2. If the absolute value of largest coefficient is greater than the sum of the absolute value of all the remaining coefficient than the method converses for any initial approximation.

Example 8. *Describe a method for solving a system of linear equations. Solve the following system of linear equations using Gauss-Seidel method*

$$23x_1 + 13x_2 + 3x_3 = 29$$
$$5x_1 + 23x_2 + 7x_3 = 37$$
$$11x_1 + x_2 + 23x_3 = 43$$

(U.P., III Semester, Dec. 2009)

Solution. See Art. 51.9 for method.

Here, we have

$$23x_1 + 13x_2 + 3x_3 = 29$$
$$5x_1 + 23x_2 + 7x_3 = 37$$
$$11x_1 + x_2 + 23x_3 = 43$$

Solving each equation of the given system for the unknowns with largest coefficient in terms of the remaining unknowns, we have

$$x_1 = \frac{1}{23}(29 - 13x_2 - 3x_3) \quad \text{... (1)}$$

$$x_2 = \frac{1}{23}(37 - 5x_1 - 7x_3) \quad \text{... (2)}$$

$$x_3 = \frac{1}{23}(43 - 11x_1 - x_2) \quad \text{... (3)}$$

For first iteration

Putting $x_2 = 0,\ x_3 = 0$ in (1), we get

$$x_1 = \frac{1}{23}[29] = 1.26087$$

Putting $x_1 = 1.26087,\ \ x_3 = 0$ in (2), we get

$$x_2 = \frac{1}{23}[37 - 5(1.26087) - 0] = 1.33459$$

Putting $x_1 = 1.26087$ and $x_2 = 1.33459$ in (3), we get

$$x_3 = \frac{1}{23}[43 - 11 \times (1.26087) - 1.33459]$$

$$= \frac{1}{23}[43 - 13.86957 - 1.33459] = 1.20851$$

For the second iteration

Putting $x_2 = 1.33459$ and $x_3 = 1.20851$ in (1), we get

$$x_1 = \frac{1}{23}[29 - 13 \times 1.33459 - 3 \times 1.20851]$$

$$= \frac{1}{23}[29 - 17.34967 - 3.62553] = 0.34890$$

Putting $x_1 = 0.34890$ and $x_3 = 1.20851$ in (2), we get

$$x_2 = \frac{1}{23}[37 - 5 \times 0.34890 - 7 \times 1.20851]$$

$$= \frac{1}{23}[37 - 1.74450 - 8.45957] = 1.16504$$

Putting $x_1 = 0.34890$ and $x_2 = 1.16504$ in (3), we get

$$x_3 = \frac{1}{23}[43 - 11 \times 0.34890 - 1.16504]$$

$$= \frac{1}{23}[43 - 3.8379 - 1.16504] = 1.65205$$

For the third iteration

Putting $x_2 = 1.16504$ and $x_3 = 1.65205$ in (1), we get

$$x_1 = \frac{1}{23}[29 - 13 \times 1.16504 - 3 \times 1.65205]$$

$$= \frac{1}{23}[29 - 15.1502 - 4.95615] = 0.38668$$

Putting $x_1 = 0.38668$ and $x_3 = 1.65205$ in (2), we get

$$x_2 = \frac{1}{23}[37 - 5 \times 0.38668 - 7 \times 1.65205]$$

$$= \frac{1}{23}[37 - 1.9334 - 11.56435] = 1.02184$$

Putting $x_1 = 0.38668$ and $x_2 = 1.02184$ in (3), we get

$$x_3 = \frac{1}{23}[43 - 11 \times 0.38668 - 1.02184]$$

$$= \frac{1}{23}[43 - 4.25348 - 1.02184]$$

$$= 1.640203$$

For the fourth iteration

Putting $x_2 = 1.02184$ and $x_3 = 1.640203$ in (1), we get

$$x_1 = \frac{1}{23}[29 - 13 \times 1.02184 - 3 \times 1.640203]$$

$$x_1 = \frac{1}{23}[29 - 13.28392 - 4.89498] = 0.46937$$

Putting $x_1 = 0.46937$ and $x_3 = 1.640203$ in (2), we get

$$x_2 = \frac{1}{23}[37 - 5 \times 0.46937 - 7 \times 1.640203]$$

$$= \frac{1}{23}[37 - 2.34685 - 11.481421] = 1.007466$$

Putting $x_1 = 0.46937$ and $x_2 = 1.007466$ in (3), we get

$$x_3 = \frac{1}{23}[43 - 11 \times 0.46937 - 1.007466]$$

$$= \frac{1}{23}[43 - 5.16307 - 1.007466] = 1.601281$$

For the fifth iteration

$$x_1 = \frac{1}{23}[29 - 13 \times 1.007466 - 3 \times 1.601281] = 0.48257$$

$$x_2 = \frac{1}{23}[37 - 5 \times 0.48257 - 7 \times 1.601281] = 1.016443$$

$$x_3 = \frac{1}{23}[43 - 11 \times 0.48257 - 1.016443] = 1.594578$$

The following table shows all the iterations

x_1	1.26087	0.34890	0.38668	0.46977	0.48257
x_2	1.33459	1.16504	1.02184	1.007466	1.016443
x_3	1.20851	1.65205	1.640203	1.601167	1.594578

$x_1 = 0.48257,\quad x_2 = 1.016443,\quad x_3 = 1.594578$ **Ans.**

Example 9. *Solve the following system by Gauss-Seidel method :*

$$27x + 6y - z = 85$$
$$6x + 15y + 2z = 72$$
$$x + y + 54z = 110$$

(R.G.P.V., Bhopal, III Semester, June 2008, Dec. 2002)

Solution. We have,

$$27x + 6y - z = 85$$
$$6x + 15y + 2z = 72$$
$$x + y + 54z = 110$$

Solving each equation of the given system for the unknown with the largest coefficient in terms of the remaining unknowns, we have

$$x = \frac{1}{27}(85 - 6y + z) \qquad \text{... (1)}$$

$$y = \frac{1}{15}(72 - 6x - 2z) \qquad \text{... (2)}$$

$$z = \frac{1}{54}(110 - x - y) \qquad \text{... (3)}$$

Starting with $y = 0$, $z = 0$ and putting $y = 0$, $z = 0$ in R.H.S. of equation (1), we get

$$x_1 = \frac{85}{27} = 3.148 = \text{first approximation}$$

Now putting $x = 3.148$, $z = 0$ in equation (2), we get

$$y_1 = \frac{1}{15}(72 - 6 \times 3.148) = 3.541 = \text{first approximation}$$

Again putting $x = 3.148$, $y = 3.541$ in equation (3), we get

$$z_1 = \frac{1}{54}(110 - 3.148 - 3.541) = 1.913 = \text{first approximation}$$

For the second approximation (iteration), we have

$$x_2 = \frac{1}{27}[85 - 6y^{(1)} + z^{(1)}] = \frac{1}{27}(85 - 6 \times 3.541 + 1.913) = 2.432$$

$$y_2 = \frac{1}{15}[72 - 6x^{(2)} - 2z^{(1)}] = \frac{1}{15}[72 - 6 \times 2.432 - 2 \times 1.913]$$

$$= 3.572$$

$$z_2 = \frac{1}{54}[110 - x^{(2)} - y^{(2)}] = \frac{1}{54}[110 - 2.432 - 3.572] = 1.926$$

From the third approximation (iteration), we have

$$x_3 = \frac{1}{27}[85 - 6y^{(2)} + z^{(2)}] = \frac{1}{27}[85 - 6 \times 3.572 + 1.926] = 2.426$$

$$y_3 = \frac{1}{15}[72 - 6x^{(3)} - 2z^{(2)}] = \frac{1}{15}[72 - 6 \times 2.426 - 2 \times 1.926] = 3.573$$

$$z_3 = \frac{1}{54}[110 - x^{(3)} - y^{(3)}] = \frac{1}{54}[110 - 2.426 - 3.573] = 1.926$$

Second and third iterations give practically the same values therefore we can stop.
Hence the solution is $x = 2.426$, $y = 3.573$ and $z = 1.926$
Calculated values of x, y, z are given in the following table.

Iterations	1	2	3
$x = \frac{1}{27}(85 - 6y + z)$	3.148	2.432	2.426
$y = \frac{1}{15}(72 - 6x - 2z)$	3.541	3.572	3.573
$z = \frac{1}{54}(110 - x - y)$	1.913	1.926	1.926

Ans.

Example 10. *Solve the following system :*

$$10x + 2y + z = 9$$
$$2x + 20y - 2z = -44$$
$$-2x + 3y + 10z = 22$$

by Gauss-Seidel method correct to two places of decimal.

(*R.G.P.V., Bhopal, III Semester Dec. 2003*)

Solution. The given equations can be written as

$$x = 0.9 - 0.2y - 0.1z \quad \text{... (1)}$$
$$y = -2.2 - 0.1x + 0.1z \quad \text{... (2)}$$
$$z = 2.2 + 0.2x - 0.3y \quad \text{... (3)}$$

For first iteration

Putting $y = 0, z = 0$ in (1), we get

$$x_1 = 0.9 - 0 - 0 = 0.9$$

Putting $x = 0.9, z = 0$ in (2), we get

$$y_1 = -2.2 - 0.1\ (0.9) + 0.1(0) = -2.29$$

Putting $x = 0.9, y = -2.29$ in (3), we get

$$z_1 = 2.2 + 0.2\ (0.9) - 0.3\ (-2.29) = 3.067$$

For the second iteration

Putting $y = -2.29, z = 3.067$ in (1), we get

$$x_2 = 0.9 - 0.2\ (-2.29) - 0.1\ (3.067) = 1.0513$$

Putting $x = 1.0513,\ z = 3.067$ in (2), we get

$$y_2 = -2.2 - 0.1\ (1.0513) + 0.1(3.067) = -1.99843$$

Putting $x = 1.0513, y = -1.99843$ in (3), we get

$$z_2 = 2.2 + 0.2\ (1.0513) - 0.3\ (-1.99843) = 3.00979$$

For Third iteration

Putting $y = -1.99843$ and $z = 3.00979$ in (1), we get

$$x_3 = 0.9 - 0.2\ (-1.99843) - 0.1\ (3.00979) = 0.99871$$

Putting $x = 0.99871, z = 3.00979$ in (2), we get

$$y_3 = -2.2 - 0.1\ (0.99871) + 0.1\ (3.00979) = -1.99889$$

Putting $x = 0.99871$ and $y = -1.99889$ in (3), we get

$$z_3 = 2.2 + 0.2\ (0.99871) - 0.3\ (-1.99889) = 2.99941$$

For fourth iteration

Proceeding as usual, we get

$$x_4 = 0.9 - (0.2)\ (-1.99889) - (0.1)\ (2.99941)$$
$$= 0.99984$$
$$y_4 = -2.2 - (0.1)\ (0.99984) + (0.1)\ (2.99941)$$
$$= -2.00004$$
$$z_4 = 2.2 + (0.2)\ (0.99984) - (0.3)\ (-2.00004)$$
$$= 2.99998$$

Third and fourth iterations are practically same.

Hence, the solution is $x = 1, y = -2, z = 3$. **Ans.**

The following table shows all the iterations

Iterations	1	2	3	4
$x = 0.9 - 0.2y - 0.1z$	0.9	1.0513	0.99871	0.99984
$y = -2.2 - 0.1x + 0.1z$	–2.29	–1.99843	–1.99889	–2.00004
$z = 2.2 + 0.2x - 0.3y$	3.067	3.00979	2.99941	2.99998

Example 11. *Solve the following equations by Gauss-Seidel method :*

$$83x + 11y - 4z = 95$$

$$7x + 52y + 13z = 104$$

$$3x + 8y + 29z = 71$$

(R.G.P.V., Bhopal, III Semester, Dec. 2003)

Solution. The given equations can be rewritten as

$$x = \frac{1}{83}[95 - 11y + 4z] \quad \text{... (1)}$$

$$y = \frac{1}{52}[104 - 7x - 13z] \quad \text{... (2)}$$

$$z = \frac{1}{29}[71 - 3x - 8y] \quad \text{... (3)}$$

For first approximation

Putting $y = 0$, $z = 0$ in (1), we get

$$x_1 = \frac{95}{83} = 1.145$$

Putting $x = 1.145$, $z = 0$ in (2), we get

$$y_1 = \frac{1}{52}[104 - 7 \times 1.145 - 13 \times 0] = 1.846$$

Putting $x = 1.145$, $y = 1.846$ in (3), we get

$$z_1 = \frac{1}{29}[71 - 3 \times 1.145 - 8 \times 1.846] = 1.821$$

For second approximation

Putting $y = 1.846$, $z = 1.821$ in (1), we get

$$x_2 = \frac{1}{83}[95 - 11 \times 1.846 + 4 \times 1.821] = 0.988$$

Putting $x = 0.988$, $z = 1.821$ in (2), we get

$$y_2 = \frac{1}{52}[104 - 7 \times 0.988 - 13 \times 1.821] = 1.412$$

Putting $x = 0.988$, $y = 1.412$ in (3), we get

$$z_2 = \frac{1}{29}[71 - 3 \times 0.988 - 8 \times 1.412] = 1.957$$

For third approximation

Putting $y = 1.412$, $z = 1.957$ in (1), we get

$$x_3 = \frac{1}{83}[95 - 11 \times 1.412 + 4 \times 1.957] = 1.052$$

Putting $x = 1.052$, $z = 1.957$ in (2), we get

$$y_3 = \frac{1}{52}[104 - 7 \times 1.052 - 13 \times 1.957] = 1.369$$

Putting $x = 1.052$, $y = 1.369$ in (3), we get

$$z_3 = \frac{1}{29}[71 - 3 \times 1.052 - 8 \times 1.369] = 1.962$$

For fourth approximation

Putting $y = 1.369$, $z = 1.962$ in (1), we get

$$x_4 = \frac{1}{83}[95 - 11 \times 1.369 + 4 \times 1.962] = 1.058$$

Putting $x = 1.058$, $z = 1.962$ in (2), we get

$$y_4 = \frac{1}{52}[104 - 7 \times 1.058 - 13 \times 1.962] = 1.367$$

Putting $x = 1.058$, $y = 1.367$ in (3), we get

$$z_4 = \frac{1}{29}[71 - 3 \times 1.058 - 8 \times 1.367] = 1.962$$

For fifth approximation

Putting $y = 1.367$, $z = 1.962$ in (1), we get

$$x_5 = \frac{1}{83}[95 - 11 \times 1.367 + 4 \times 1.962] = 1.058$$

Putting $x = 1.058$, $z = 1.962$ in (2), we get

$$y_5 = \frac{1}{52}[104 - 7 \times 1.058 - 13 \times 1.962] = 1.367$$

Putting $x = 1.058$, $y = 1.367$ in (3), we get

$$z_5 = \frac{1}{29}[71 - 3 \times 1.058 - 8 \times 1.367] = 1.962$$

Thus, fourth and fifth approximations are practically same.
This can be shown in the following table :

Iterations	1	2	3	4	5
$y = \frac{1}{83}(95 - 11y + 4z)$	1.145	0.988	1.052	1.058	1.058
$y = \frac{1}{52}(104 - 7x - 13z)$	1.846	1.412	1.369	1.367	1.367
$z = \frac{1}{29}(71 - 3x - 8y)$	1.821	1.957	1.962	1.962	1.962

Hence, the solution is $x = 1.058$, $y = 1.367$, $z = 1.962$ **Ans.**

Example 12. *Solve the system of equations by Gauss-Seidel iterative method :*

$$\begin{aligned} 54x + y + z &= 110 \\ 2x + 15y + 6z &= 72 \\ -x + 6y + 27z &= 85 \end{aligned}$$

(AMIETE, June 2009, R.G.P.V., Bhopal, III Semester, June 2007)

Solution. The given equations can be rewritten as

$$x = \frac{1}{54}[110 - y - z] \quad \ldots (1)$$

$$y = \frac{1}{15}[72 - 2x - 6z] \quad \ldots (2)$$

$$z = \frac{1}{27}[85 + x - 6y] \quad \ldots (3)$$

For first approximation

Putting $y = 0$, $z = 0$ in (1), we get

$$x_1 = \frac{110}{54} = 2.037$$

Putting $x = 2.037$, $z = 0$ in (2), we get

$$y_1 = \frac{1}{15}[72 - 2(2.037) - 0] = 4.528$$

Putting $x = 2.037$, $y = 4.528$ in (3), we get

$$z_1 = \frac{1}{27}[85 + 2.037 - 6 \times 4.528] = 2.217$$

For second approximation

Putting $y = 4.528$, $z = 2.217$ in (1), we get

$$x_2 = \frac{1}{54}[110 - 4.528 - 2.217] = 1.912$$

Putting $x = 1.912$, $z = 2.217$ in (2), we get

$$y_2 = \frac{1}{15}[72 - 2 \times 1.912 - 6 \times 2.217] = 3.658$$

Putting $x = 1.912$, $y = 3.658$ in (3), we get

$$z_2 = \frac{1}{27}[85 + 1.912 - 6 \times 3.658] = 2.406$$

For third approximation

Putting $y = 3.658$, $z = 2.406$ in (1), we get

$$x_3 = \frac{1}{54}[110 - 3.658 - 2.406] = 1.925$$

Putting $x = 1.925$, $z = 2.406$ in (2), we get

$$y_3 = \frac{1}{15}[72 - 2 \times 1.925 - 6 \times 2.406] = 3.581$$

Putting $x = 1.925$, $y = 3.581$ in (3), we get

$$z_3 = \frac{1}{27}[85 + 1.925 - 6 \times 3.581] = 2.424$$

For fourth approximation

Putting $y = 3.581$, $z = 2.424$ in (1), we get

$$x_4 = \frac{1}{54}[110 - 3.581 - 2.424] = 1.926$$

Putting $x = 1.926$, $z = 2.424$ in (2), we get

$$y_4 = \frac{1}{15}[72 - 2 \times 1.926 - 6 \times 2.424] = 3.574$$

Putting $x = 1.926$, $y = 3.574$ in (3), we get

$$z_4 = \frac{1}{27}[85 + 1.926 - 6 \times 3.574] = 2.425$$

For fifth approximation

Putting $y = 3.574$, $z = 2.425$ in (1), we get

$$x_5 = \frac{1}{54}[110 - 3.574 - 2.425] = 1.926$$

Putting $x = 1.926$, $z = 2.425$ in (2), we get

$$y_5 = \frac{1}{15}[72 - 2 \times 1.926 - 6 \times 2.425] = 3.573$$

Putting $x = 1.926$, $y = 3.573$ in (3), we get

$$z_5 = \frac{1}{27}[85 + 1.926 - 6 \times 3.573] = 2.425$$

Thus, fourth and fifth iterations are practically same.

The calculated values of x, y, z are shown in the table :

Iterations	1	2	3	4	5
$x = \frac{1}{54}(110 - y - z)$	2.037	1.912	1.925	1.926	1.926
$y = \frac{1}{15}(72 - 2x - 6z)$	4.528	3.658	3.581	3.574	3.573
$z = \frac{1}{27}(85 + x - 6y)$	2.217	2.406	2.424	2.425	2.425

Hence, the solution is $x = 1.926,\ y = 3.573$ and $z = 2.425$. **Ans.**

Example 13. *Solve the equations :*

$$10x_1 - 2x_2 - x_3 - x_4 = 3$$
$$-2x_1 + 10x_2 - x_3 - x_4 = 15$$
$$-x_1 - x_2 + 10x_3 - 2x_4 = 27$$
$$-x_1 - x_2 - 2x_3 + 10x_4 = -9$$

By Gauss-Seidel iteration method. *(R.G.P.V., Bhopal, III Semester, Dec. 2006, June 2004)*

Solution. The numbers, 10, 10, 10, 10 in the leading diagonal are the largest, so we can apply Gauss-Seidel method to solve the given equations.

The above equations are rewritten as

$$x_1 = \frac{1}{10}(3 + 2x_2 + x_3 + x_4) \quad \ldots (1)$$

$$x_2 = \frac{1}{10}(15 + 2x_1 + x_3 + x_4) \quad \ldots (2)$$

$$x_3 = \frac{1}{10}(27 + x_1 + x_2 + 2x_4) \quad \ldots (3)$$

$$x_4 = \frac{1}{10}(-9 + x_1 + x_2 + 2x_3) \quad \ldots (4)$$

For first Approximation

Putting $x_2 = 0,\ x_3 = 0,\ x_4 = 0$ in (1), we get $x_1 = \frac{3}{10} = 0.3$

Putting $x_1 = 0.3,\ x_3 = 0,\ x_4 = 0$ in (2), we get $x_2 = \frac{1}{10}[15 + 2(0.3)] = 1.56$

Putting $x_1 = 0.3,\ x_2 = 1.56,\ x_4 = 0$ in (3), we get $x_3 = \frac{1}{10}(27 + 0.3 + 1.56 + 0) = 2.886$

Putting $x_1 = 0.3,\ x_2 = 1.56,\ x_3 = 2.886$ in (4), we get

$$x_4 = \frac{1}{10}[-9 + 0.3 + 1.56 + 2(2.886)] = -0.137$$

For Second Approximation

Putting $x_2 = 1.56,\ x_3 = 2.886,\ x_4 = -0.137$ in (1), we get

$$x_1 = \frac{1}{10}[3 + 2(1.56) + 2.886 - 0.137] = 0.887$$

Putting $x_1 = 0.887,\ x_3 = 2.886,\ x_4 = -0.137$ in (2), we get

$$x_2 = \frac{1}{10}[15 + 2(0.887) + 2.886 - 0.137] = 1.952$$

Putting $x_1 = 0.887,\ x_2 = 1.952,\ x_4 = -0.137$ in (3), we get

$$x_3 = \frac{1}{10}[27 + 0.887 + 1.952 + 2(-0.137)] = 2.957$$

Putting $x_1 = 0.887,\ x_2 = 1.952,\ x_3 = 2.957$ in (4), we get

$$x_4 = \frac{1}{10}[-9 + 0.887 + 1.952 + 2(2.957)] = -0.025$$

For Third Approximation

Putting $x_2 = 1.952,\ x_3 = 2.957,\ x_4 = -0.025$ in (1), we get

$$x_1 = \frac{1}{10}[3 + 2(1.952) + 2.957 - 0.025] = 0.984$$

Putting $x_1 = 0.984,\ x_3 = 2.957,\ x_4 = -0.025$ in (2), we get

$$x_2 = \frac{1}{10}[15 + 2(0.984) + 2.957 - 0.025] = 1.990$$

Putting $x_1 = 0.984,\ x_2 = 1.990,\ x_4 = -0.025$ in (3), we get

$$x_3 = \frac{1}{10}[27 + 0.984 + 1.990 + 2(-0.025)] = 2.992$$

Putting $x_1 = 0.984,\ x_2 = 1.990,\ x_3 = 2.992$ in (4), we get

$$x_4 = \frac{1}{10}[-9 + 0.984 + 1.990 + 2(2.992)] = -0.004$$

For Fourth Approximation

Putting $x_2 = 1.990,\ x_3 = 2.992,\ x_4 = -0.004$ in (1), we get

$$x_1 = \frac{1}{10}[3 + 2(1.990) + 2.992 - 0.004] = 0.997$$

Putting $x_1 = 0.997,\ x_3 = 2.992,\ x_4 = -0.004$ in (2), we get

$$x_2 = \frac{1}{10}[15 + 2(0.997) + 2.992 - 0.004] = 1.998$$

Putting $x_1 = 0.997,\ x_2 = 1.998,\ x_4 = -0.004$ in (3), we get

$$x_3 = \frac{1}{10}[27 + 0.997 + 1.998 + 2(-0.004)] = 2.999$$

Putting $x_1 = 0.997,\ x_2 = 1.998,\ x_3 = 2.999$ in (4), we get

$$x_4 = \frac{1}{10}[-9 + 0.997 + 1.998 + 2(2.999)] = -0.0007$$

Iterations	1	2	3	4
$x_1 = \frac{1}{10}(3 + 2x_2 + x_3 + x_4)$	0.3	0.887	0.984	0.997
$x_2 = \frac{1}{10}(15 + 2x_1 + x_3 + x_4)$	1.56	1.952	1.990	1.998
$x_3 = \frac{1}{10}(27 + x_1 + x_2 + 2x_4)$	2.886	2.957	2.992	2.999
$x_4 = \frac{1}{10}(-9 + x_1 + x_2 + 2x_3)$	– 0.137	– 0.025	– 0.004	– 0.0007

Hence, the solution of the given system of equations is

$x_1 = 0.997,\ x_2 = 1.998,\ x_3 = 2.999,\ x_4 = -0.0007$

Exact value of $x_1 = 1,\ x_2 = 2,\ x_3 = 3$ and $x_4 = 0$. **Ans.**

EXERCISE 51.4

1. Solve by Gauss-Seidel method

$$\begin{aligned} 6x - y - z &= 19, \\ 3x + 4y + z &= 26, \\ x + 2y + 6z &= 22 \end{aligned}$$

Ans. $x = 4,\ y = 3,\ z = 2$

2. Use Gauss-Seidel method to solve the system of equations

$$\begin{aligned} 3x + y + z &= 1 \\ x + 3y - z &= 11 \\ x - 2y + 4z &= 21 \end{aligned}$$

Ans. $x = -7,\ y = 10,\ z = 12$

3. Solve by Gauss-Seidel method to solve

$$\begin{bmatrix} 3 & 1 & 1 \\ 1 & 2 & 2 \\ 2 & 1 & 3 \end{bmatrix} \begin{bmatrix} x_1 \\ x_2 \\ x_3 \end{bmatrix} = \begin{bmatrix} 4 \\ 3 \\ 4 \end{bmatrix}$$

Ans. $x_1 = 1,\ x_2 = \frac{1}{2},\ x_3 = \frac{1}{2}$

4. Use Gauss-Seidel method o solve

$$\begin{aligned} 4x - y + 2z - 2w &= 7 \\ 3x + 5y - z + 6w &= 5 \\ 2x + 3y - 4z + 2w &= -4 \\ x + 2y + 3z - 4w &= 7 \end{aligned}$$

Ans. $x = 1,\ y = \frac{1}{2},\ z = 2,\ w = \frac{1}{4}$

5. Use Jacobi's method to solve

$$\begin{aligned} 10x - 2y - 3z &= 205 \\ 2x - 10y + 2z &= -154 \\ 2x + y - 10z &= -120 \end{aligned}$$

Ans. $x = 32,\ y = 26,\ z = 21$

upto the end of sixth iteration.

6. Use Jacobi's method to solve

$$\begin{aligned} 5x + 2y + z &= 12 \\ x + 4y + 2z &= 15 \\ x + 2y + 5z &= 20 \end{aligned}$$

Ans. $x = 1.08,\ y = 1.95,\ z = 3.16$

upto the end of eighth iteration.

7. Solve Question (6) by Gauss-Seidel method upto fifth iteration **Ans.** $x = 1,\ y = 2,\ z = 3$

8. Solve the following equations by Gauss Seidel method.

$$\begin{aligned} 6x + y + z &= 7 \\ x + 8y + 2z &= 6 \\ 3x + 2y + 10z &= 9 \end{aligned}$$

Initial values are $x = 0.8,\ y = 0.4,\ z = -0.45$

Ans. $x = 1,\ y = \frac{1}{2},\ z = \frac{1}{2}$

9. Solve by Gauss-Seidel method, the following system of equtions:

$$\begin{aligned} 28x + 4y - z &= 32 \\ x + 3y + 10z &= 24 \\ 2x + 17y + 4z &= 35 \end{aligned}$$

(AMIETE, Dec. 2009)

Ans. $x = 0.9935,\ y = 1.5069,\ z = 1.8485$

NUMERICAL TECHNIQUE FOR SOLUTION OF ORDINARY DIFFERENTIAL EQUATIONS

52.1 INTRODUCTION

Differential equations represent a number of problems in the field of engineering and science. The analytical methods of solving differential equations are applicable to a limited type of differential equations. The numerical solutions of the differential equations have become easy for manupulation. These methods are of even greater importance, as the computing machines are now readly available

52.2 SOLUTION OF ORDINARY DIFFERENTIAL EQUATIONS

A number of differential equations cannot be solved by analytical methods. It is, therefore, imperative to solve them by numerical methods. We will discuss the following methods :

(1) Taylor's series method

(2) Picard's method

(3) Runge-Kutta method.

52.3 TAYLOR'S SERIES METHOD

Let us consider the first order differential equation

$$\frac{dy}{dx} = f(x, y) \qquad \ldots (1)$$

under the condition $y = 0$ for $x = x_0$.

Method.

On differentiating (1) again and again, we get $\frac{d^2y}{dx^2}, \frac{d^3y}{dx^3}, \frac{d^4y}{dx^4}$ *etc* .

On putting $x = x_0$ and $y = 0$ in the above equations we get the values of

$$\frac{dy}{dx}, \frac{d^2y}{dx^2}, \frac{d^3y}{dx^3}, \frac{d^4y}{dx^4} \ldots\ldots$$

substuting the values of y', y'', y''', y'''' ... in Taylor's series

$$y = y_0 + (x - x_0)[y'(x_0)] + \frac{(x-x_0)^2}{2!}[y''(x_0)] + \frac{(x-x_0)^3}{3!}[y'''(x_0)] + \ldots$$

Thus we can obtain a power series for $y(x)$ in powers of $(x - x_0)$.

The method is illustrated by the example.

Example 1. *Using Taylor's series method obtain the solution of* $\frac{dy}{dx} = 3x + y^2$ *and* $y = 1$, *when* $x = 0$ *Find the value of y for* $x = 0.1$, *correct to four places of decimals.*

Solution. $$\frac{dy}{dx} = 3x + y^2 \qquad ...(1)$$

$$y(0) = 1 \qquad ...(2)$$

Differentiating (1) w.r.t 'x', we get $\dfrac{d^2y}{dx^2} = 3 + 2y\dfrac{dy}{dx}$ ...(3)

$$\frac{d^3y}{dx^3} = 2y\frac{d^2y}{dx^2} + 2\left(\frac{dy}{dx}\right)^2 \qquad ...(4)$$

$$\frac{d^4y}{dx^4} = 2y\frac{d^2y}{dx^2} + 2\left(\frac{dy}{dx}\right)\left(\frac{d^2y}{dx^2}\right) + 4\left(\frac{dy}{dx}\right)\frac{d^2y}{dx^2} \qquad ...(5)$$

and so on From (1) $\dfrac{dy}{dx} = 0 + (1)^2 = 1$

From (3) $\dfrac{d^2y}{dx^2} = 3+2(1)(1) = 5$

From (4) $\dfrac{d^3y}{dx^3} = 2(1)(5) + 2(1)^2 = 12$

From (5) $\dfrac{d^4y}{dx^4} = 2(1)(12) + 2(1)(5) + 4(1)(5) = 54$

We know by Taylor's series expansion

$$y = y_0 + (x - x_0)(y')_0 + \frac{(x-x_0)^2}{2!}(y'')_0 + \frac{(x-x_0)^3}{3!}(y''')_0 + \frac{(x-x_0)^4}{4!}(y^{iv})_0 + \ldots \qquad ...(6)$$

On substituting the value of $y(0)$, $v'(0)$, $v''(0)$ $v'''(0)$, $y^{iv}(0)$ etc.in (6) we get

$$y = 1 + x + \frac{x^2}{2!}(5) + \frac{x^3}{3!}(12) + \frac{x^4}{4!}(54) + \ldots$$

or $$y(x) = 1 + x + \frac{5}{2}x^2 + 2x^3 + \frac{9}{4}x^4 + \ldots$$

$$y(0.1) = 1 + 0.1 + \frac{5}{2}(0.01) + 2(0.001) + \frac{9}{4}(0.0001) + \ldots$$

$$= 1 + 0.1 + 0.025 + 0.002 + 0.000225 = 1.127225$$ **Ans.**

Example 2. *Use Taylor's series method to solve the equation*

$$\frac{dy}{dx} = -xy, \quad y(0) = 1$$

Solution. $$y' = xy$$

$$y'(0) = 0.$$

Differentiating (1) repeatedly, we find

$$y'' = -xy' - y, \qquad y''(0) = -1$$

$$y''' = -xy'' - 2y', \qquad y'''(0) = 0$$

$$y^{iv} = -xy''' - 3y'', \qquad y^{iv}(0) = 3$$

$$y^{v} = -xy^{iv} - 4y''', \qquad y^{v}(0) = 0$$

$$y^{vi} = -xy^{v} - 5y^{iv}, \qquad y^{vi}(0) = -15$$

By Taylor's series expansion

$$y(x) = y(0) + \frac{x^2}{1!}y'(0) + \frac{x^2}{2!}y''(0) + \frac{x^3}{3!}y'''(0) + \frac{x^4}{4!}y^{iv}(0) + \ldots$$

$$= 1 + 0 + \frac{x^2}{2!}(-1) + 0 + \frac{x^4}{4!}(3) + 0 + \frac{x^6}{6!}(-15) + \ldots = 1 - \frac{x^2}{2} + \frac{x^4}{8} - \frac{x^6}{48} + \ldots \quad \textbf{Ans.}$$

Example 3. *Employ Taylor's series method to obtain approximate value of y at x = 0.2 for the differential equation* $\frac{dy}{dx} = 2y + 3e^x, y(0) = 0$. *Compare the numerical solution with the exact solution.* *(R.G.P.V. Bhopal, III Semester, June 2004)*

Solution.
$$\frac{dy}{dx} = 2y + 3e^x \quad \ldots(1)$$
$$y(0) = 0$$

Putting $x = 0, y = 0$ in (1), we get $\frac{dy}{dx} = 2(0) + 3e^0 = 3$

Differentiating (1) w.r.t. 'x' and putting $x = 0, y = 0$ and earlier obtained values of the derivatives, we get

	$x = 0, y = 0$
$\frac{d^2y}{dx^2} = 2\frac{dy}{dx} + 3e^x$	$\frac{d^2y}{dx^2} = 2(3) + 3 = 9$
$\frac{d^3y}{dx^3} = 2\frac{d^2y}{dx^2} + 3e^x$	$\frac{d^3y}{dx^3} = 2(9) + 3 = 21$
$\frac{d^4y}{dx^4} = 2\frac{d^3y}{dx^3} + 3e^x$	$\frac{d^4y}{dx^4} = 2(21) + 3 = 45$
$\frac{d^5y}{dx^5} = 2\frac{d^4y}{dx^4} + 3e^x$	$\frac{d^5y}{dx^5} = 2(45) + 3 = 93$

By Taylor's series expansion

$$y = y_0 + (x - x_0)\left(\frac{dy}{dx}\right)_{x=x_0} + \frac{(x-x_0)^2}{2!}\left(\frac{d^2y}{dx^2}\right)_{x=x_0} + \frac{(x-x_0)^3}{3!}\left(\frac{d^3y}{dx^3}\right)_{x=x_0} + \ldots \quad \ldots(2)$$

On substituting the values of $y(0), y'(0), y''(0)$ etc. in (2), we get

$$y = 0 + (x-0)(3) + \frac{(x-0)^2}{2}(9) + \frac{(x-0)^3}{6}(21) + \frac{(x-0)^4}{24}(45) + \frac{(x-0)^5}{120}(93) + \ldots$$

$$y = 3x + \frac{9}{2}x^2 + \frac{7}{2}x^3 + \frac{15}{8}x^4 + \frac{31}{40}x^5 + \ldots \quad \ldots(3)$$

To find the value of y when $x = 0.2$

Putting $x = 0.2$ in (3), we get

$$y(0.2) = 3(0.2) + \frac{9}{2}(0.2)^2 + \frac{7}{2}(0.2)^3 + \frac{15}{8}(0.2)^4 + \frac{31}{40}(0.2)^5 + \ldots$$

$$= 0.6 + \frac{9}{2}(0.04) + \frac{7}{2}(0.008) + \frac{15}{8}(0.0016) + \frac{31}{40}(0.00032)$$

$$= 0.6 + 0.18 + 0.028 + 0.003 + 0.000248 = 0.811248.$$

To find exact value of y : $\frac{dy}{dx} - 2y = 3e^x$

$$\text{I.F.} = e^{-2\int dx} = e^{-2x}$$

Solution is $y \,.\, e^{-2x} = \int 3e^{x} \,.\, e^{-2x}\, dx + c = 3\int e^{-x}\, dx + c = -3e^{-x} + c$ ...(4)

Putting $x = 0, y = 0$ in (4), we get $0 = -3 + c \Rightarrow c = 3$

On putting $c = 3$ in (4), we get

$$y.e^{-2x} = -3e^{-x} + 3 \quad \Rightarrow \quad y = -3e^{x} + 3e^{2x} \qquad \text{... (5)}$$

Putting $x = 0.2$ in (5), we get

$$y\,(0.2) = -3e^{0.2} + 3e^{2(0.2)} = -3e^{0.2} + 3e^{0.4} = -3\,(1.221403) + 3\,(1.491825)$$

$$= -3.664209 + 4.475475 = 0.811266$$

Difference between exact value of y and its approximate value

$$= 0.811266 - 0.811248 = 0.000018$$ **Ans.**

EXERCISE 52.1

Using Taylor's method, solve the following differential equations :

1. $\dfrac{dy}{dx} = x + y^2$, given $y(0) = 0$. **Ans.** $y = \dfrac{1}{2}x^2 + \dfrac{1}{20}x^5 + ...$

2. $\dfrac{d^2y}{dx^2} + xy = 0$, subject to $x = 0,\ y = c$ and $\dfrac{dy}{dx} = 0$. **Ans.** $y = c\left(1 - \dfrac{x^3}{3!} + \dfrac{1\times 4}{6!}x^6 \dfrac{1\times 4\times 7}{9!}x^9 + ...\right)$

3. $\dfrac{dy}{dx} = x^2y - 1$, given $y\,(0) = 1$, and find $y\,(0.03)$. **Ans.** $y = 1 - x + \dfrac{x^2}{3} - \dfrac{x^4}{4} + ...$, 0.97001

4. $\dfrac{dy}{dx} - y^2 - x = 0$, for $y\,(0) = 0$, findy when $x = 0.2$ **Ans.** $y = 0.020016$

5. Obtain the linearized form $T\,(x, y)$ of the function $f\,(x, y) = x^2 - xy + \dfrac{1}{2}\,y^2 + 3$ at the function $f\,(x, y) = x^2 - xy + \dfrac{1}{2}\,y^2 + 3$ at the point (3, 2) using the Taylor's series expansion. Find the maximtun error in magnitude in the approximation $T\,(x,y) \approx T\,(x, y)$ over the rectangle $R : |x^{-3}| < 0.1,\ |y - 2| < 0\,.\,1$.

52.4 PICARD'S METHOD OF SUCCESSIVE APPROXIMATIONS

Let us consider the first order differential equation

$$\frac{dy}{dx} = f(x, y) \qquad \text{...(1)}$$

and $y = y_0$ for $x = x_0$

Method. Integrating (1) between the limits x_0 and x, we ge

$$\int_{y_o}^{y} dy = \int_{x_o}^{x} f(x, y)dx \quad \text{or} \quad y = y_o + \int_{x_o}^{x} f(x, y)dx$$

$$y = y_o + \int_{x_o}^{x} f(x, y)dx \qquad \text{... (2)}$$

Equation (2) is the solution of (1). But (2) contains the unknown y under the integral sign on right hand side.

On putting y_0 for y on R.H.S. of (2), we get a first approximation y_1

$$y_1 = y_0 + \int_{x_o}^{x} f(x, y_0)dx$$

From(3) we get the value of y_1 and we put y_1 for y on R.H.S.of (2) to get second approximation y_2.

Thus $$y_2 = y_0 + \int_{x_0}^{x} f(x, y_1)dx$$

Similarly third approximation is $y_3 = y_0 + \int_{x_0}^{x} f(x, y_2)\, dx$ and so on

In this way we get a better approximation each time than the preceding one.

Note. This method is used to solve the differential equation if the succession integration can be performed easily.

The method is now illustrated by an example.

Example 4. *Using Picard's method find a solution of* $\frac{dy}{dx} = 1 + xy$ *upto the third approximation, when* $y(0) = 0$.

Solution. $$\frac{dy}{dx} = 1 + xy \qquad ...(1)$$

Integrating (1) w.r.t. 'x' between the limits 0 and x, we get

$$\int_0^y dy = \int_0^x (1 + xy)dx \text{ or } y = \int_0^x (1 + xy)dx \qquad ...(2)$$

On putting $y(0) = 0$ for y on R.H.S. of (2) we have

$$y_1 = \int_0^x (1 + 0)\, dx \qquad \text{or } y_1 = x$$

On substituting $y_1 = x$ for y on R.H.S. of (2), we obtain

$$y_2 = \int_0^x (1 + x^2)\, dx = x + \frac{x^3}{3}$$

$$y_3 = \int_0^x \left[1 + x\left(x + \frac{x^3}{3}\right)\right] dx = \int_0^x \left[1 + x^2 + \frac{x^3}{3}\right] dx$$

$$y_3 = x + \frac{x^3}{3} + \frac{x^5}{15}.$$ **Ans.**

Example 5. *Use Picard's method to approximate the value of y when x = 0.1, given that y = 1 when x = 0 and* $\frac{dy}{dx} = 3x + y^2$ *(two approximations). (R.G.P.V. Bhopal, III Semester, June 2003)*

Solution. We have,

$$\frac{dy}{dx} = 3x + y^2 \qquad ...(1)$$

Integrating (1) w.r.t 'x', between the limit 0 and x, we get

$$\int_{y_0}^{y} dy = \int_{x_0}^{x} \left(3x + y_0^2\right) dx \Rightarrow y = y_0 + \int_0^x \left(3x + y_0^2\right) dx \qquad ...(2)$$

On putting $y(0) = 1$ for y_0 on R.H.S. of (2), we get

$$y_1 = 1 + \int_0^x (3x + 1)\, dx \Rightarrow y_1 = 1 + x + \frac{3x^2}{2}$$

$$y_2 = 1 + \int_0^x \left(3x + y_1^2\right) dx \qquad ...(3)$$

On substituting the value of y_1 on R.H.S. of (3), we get

$$y_2 = 1 + \int_0^x \left[3x + \left(1 + x + \frac{3x^2}{2}\right)^2\right] dx = 1 + \int_0^x \left[3x + 1 + x^2 + \frac{9x^4}{4} + 2x + 3x^2 + 3x^3\right] dx$$

$$= 1 + \int_0^x \left(1 + 5x + 4x^2 + 3x^3 + \frac{9x^4}{4}\right) dx = 1 + x + \frac{5x^2}{2} + \frac{4x^3}{3} + \frac{3x^4}{4} + \frac{9x^5}{20} \qquad ...(4)$$

Putting the value of $x = 0.1$ in (4), we get

$$= 1 + 0.1 + \frac{5(0.1)^2}{2} + \frac{4(0.1)^3}{3} + \frac{3(0.1)^4}{4} + \frac{9(0.1)^5}{20}$$

$$= 1 + 0.1 + 0.025 + 0.00133 + 0.000075 + 0.0000045 = 1.1264095$$ **Ans.**

Example 6. *Use Picard's method to solve* $\frac{dy}{dx} = -xy,\ y(0) = 1$

Solution. $$y(x) = y_0 - \int_0^x xy\, dx$$

$$= 1 - \int_0^x x(1) dx = 1 - \int_0^x x dx = 1 - \frac{x^2}{2}$$

Now using this value of y, we have

$$y = 1 - \int_0^x x\left(1 - \frac{x^2}{2}\right) dx = 1 - \int_0^x \left(x - \frac{x^3}{2}\right) dx = 1 - \frac{x^2}{2} + \frac{x^4}{8}$$

$$= 1 - \int_0^x x\left(1 - \frac{x^2}{2} + \frac{x^4}{8}\right) dx = 1 - \int_0^x \left(x - \frac{x^3}{2} + \frac{x^5}{2}\right) dx = 1 - \frac{x^2}{2} + \frac{x^4}{8} - \frac{x^6}{48}$$

Repeating once again we shall obtain

$$y = 1 - \frac{x^2}{2} + \frac{x^4}{8} - \frac{x^6}{48} + \frac{x^8}{384}$$ **Ans.**

Example 7. *Use Picard's method to solve the equations*

$$\frac{dx}{dt} = -y, \quad \frac{dy}{dt} = x$$

given that $x = 1$, $y = 0$ when $t = 0$.

Solution. $$\frac{dx}{dt} = -y \qquad ...(1)$$

$$\frac{dy}{dt} = x \qquad ...(2)$$

Integrating (1) w.r.t 't' from $t = 0$ to t we get

$$[x]_1^x = -\int_0^t y\, dt \quad \text{or} \quad x - 1 = -\int_0^t y\, dt$$

$$x = 1 - \int_0^t y\, dt \qquad ...(3)$$

Integrating (2), w.r.t., 't' from $t = 0$ to t, we get

$$[y]_0^y = \int_0^x x\, dt \quad \text{or} \quad y - 0 = \int_0^t x\, dt$$

$$y = \int_0^t x\, dt \qquad ...(4)$$

Replacing y by 0 in (3) and x by 1 in (4), we have

$$x = 1 - \int_0^t 0\, dt = 1 \quad \text{and} \quad y = \int_0^t 1.\, dt = t$$

$$x = 1 - \int_0^t t\,dt = 1 - \frac{t^2}{2}, \quad y = \int_0^t \left(1 - \frac{t^2}{2}\right) dt = t - \frac{t^3}{6}$$

$$x = 1 - \int_0^t \left(t - \frac{t^3}{6}\right) dt = 1 - \frac{t^2}{2} + \frac{t^4}{24}$$

$$y = \int_0^t \left(t - \frac{t^2}{2} + \frac{t^4}{24}\right) dt = t - \frac{t^3}{6} + \frac{t^5}{120},$$ **Ans.**

EXERCISE 52.2

Using Picard's method, solve the following:

1. $\frac{dy}{dx} = x + y^2$, given $y(0) = 0$. *(RGPV., Bhopal, June 2008)*

Ans. $y = \frac{1}{2}x^2 + \frac{1}{20}x^5 + \frac{1}{160}x^8 + \frac{1}{4400}x^{11}$

2. Apply Picard's iteration method to find approximate solutions to the initial value problem

$$y' = 1 + y^2, \; y(0) = 0$$

3. $\frac{dy}{dx} = x - y$, given $y(0) = 1$ and find $y(0.2)$ to five places of decimals.

(RGPV., Bhopal, June 2001, 2000) **Ans.** $y = 1 - x + x^2 - \frac{x^3}{3} + \frac{x^4}{12} - \frac{x^5}{60} + \frac{x^6}{720}$, 0.83746

4. $\frac{dy}{dx}$ $y + x$, given $y(0) = 1$, find $y(1)$, **Ans.** $y = 1 + x + x^2 + \frac{x^3}{3} + \frac{x^4}{12} - \frac{x^5}{60} + \frac{x^6}{120}$ +3.434.

5. $\frac{dy}{dx} = x^2 + y^2$ ' for $y(0) = 0$, find $y(0.4)$. **Ans.** 0.0214.

6. $\frac{dy}{dx}$ $2y + z$, $\frac{dz}{dx} = y + 2z$ given $y(0) = 0$, $z(0) = 1$

Ans. $y = x + 2x^2 + \frac{13}{6}x^3 + \frac{5}{3}x^4 + \ldots, \; z = 1 + 2x + \frac{5}{2}x^2 + \frac{7}{3}x^3 + \frac{41}{40}x^4 + \ldots$

7. Use Picard's method to approximate y when x = 0.1, given that y = 1, when x = 0 and $\frac{dy}{dx} = \frac{y - x}{y + x}$.

(RGPV., Bhopal, III Sem. June 2003) **Ans.** y = 1.0906.

52.5 EULER'S METHOD

This is purely numerical method for solving the first order differential equations. This is an elementary method and which will demonstrate the procedure underlying these methods. This method should not be used for practical solution.

Consider the differential equation

$$\frac{dy}{dx} = f(x, y) \quad \ldots(1)$$

Let $y = \phi(x)$ be the solution of (1). ...(2)

Let $(x_0, y_0), (x_1, y_1) \ldots (x_n, y_n), (x_n + 1, y_{n+1})$ be the points on the curve of (2).

$x_0, x_1, \ldots x_n, x_{n+1} \ldots$ are equispaced at equal interval h.

$$y_{n+1} = \phi(x_{n+1}) \quad [(x_{n+1}, y_{n+1}) \text{ lies on (2).}]$$

$$= \phi(x_n + h) \quad (x_{n+1} = x_n + h)$$

$$= \phi(x_n) + h f'(x_n) + \frac{1}{2}h^2 \phi''(x_n) + \ldots \quad \ldots(3)$$

$$= \phi(x_n) + h\phi'(x_n) \quad (h \text{ is very small})$$

$$= \phi(x_n) + hf(x_n, y_n) \qquad \left[\text{since } \frac{dy}{dx} = f(x, y)\right]$$

$$y_{n+1} = y_n + hf(x_n, y_n) \qquad [\text{since } y_n = \phi(x_n) \text{ from (2)}] \quad ...(4)$$

This formula (4) can be used to find y_{n+1}, where y_n is known.

On substituting the value of y_0, $(n = 0)$ in (4) we get y_1,

Similarly putting the value of $(n = 1)$ in (4), we obtain y_2 and so on.

Note. Since we have neglected $1/2\, h^2 \phi''(x_n)$ and higher powers of h from formula (4) there will be a larger error in y_{n+1}. Therefore it is not used in practical problems.

Geometrically

Let $y = \phi(x)$ be a solution curve PQ. The ordinate of P *i.e.* y_n is known.

Now we have to find the ordinate y_{n+1} of any point Q.

$$y_{n+1} = MQ = MR + RQ = PL + RT + TQ \qquad (TQ = \text{Error})$$

$$= y_n + h \tan\theta = y_n + h\left(\frac{dy}{dx}\right) = y_n + hf(x_n, y_n)$$

Example 8. *Using Euler's method find an approximate value of y corresponding to $x = 2$, given that $\frac{dy}{dx} = x + 2y$ and $y = 1$ when $x = 1$.*

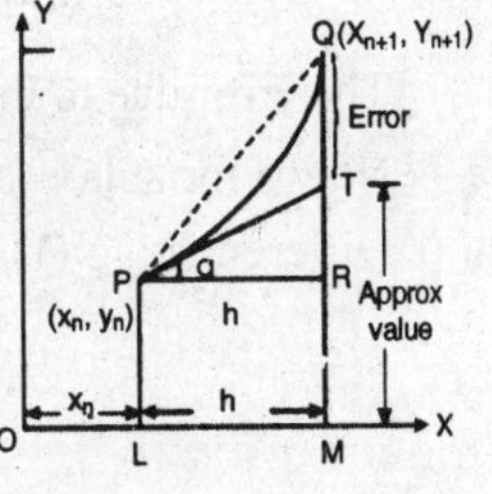

Solution. $f(x, y) = x + 2y$

$$y_{n+1} = y_n + hf(x_n, y_n) = y_n + 0.1(x + 2y)$$

Method: In column 3 we record the value of $x + 2y$ and in column 4 we enter the sum of the value of y and the product of 0.1 with the value of $x + 2y$. This value entered in 4th column is transfered to second column for the next calculation.

x	y	$x + 2y = \frac{dy}{dx}$ old	$y + 0.1\left(\frac{dy}{dx}\right)$ = new y
1.0	1.00	3.00	1.0 + 0.1 (3) = 1.30
1.1	1.3	3.70	1.3 + 0.1 (3.7)= 1.67
1.2	1.67	4.54	1.67 + 0.1 (4.54) = 2.12
1.3	2.12	5.54	2.12 + 0.1 (5.54) = 2.67
1.4	2.67	6.74	2.67 + 0.1 (6.74) = 3.34
1.5	3.34	8.18	3.34 + 0.1 (8.18) = 4.16
1.6	4.16	9.92	4.16 + 0.1 (9.92) =5.15
1.7	5.15	12.00	5.15 + 0.1 (12.0) =6.35
1.8	6.35	14.50	6.35 + 0.1 (14.50) =7.80
1.9	7.80	17.50	7.80 + 0.1 (17.50) = 9.55
2.0	9.55		

Thus the required approximate value of $y = 9.55$ **Ans.**

EXERCISE 52.3

1. Using Euler's method, find an approximate value of corresponding to $x = 1$, given that $\frac{dy}{dx} = x + y$ and $y = 1$ when $x = 0$. **Ans.** 3.18

2. Using Euler's method, find an approximate value of y corresponding to $x = 1.4$, given $\frac{dy}{dx} = xy^{1/2}$ and $y = 1$ when $x = 1$. **Ans.** 1.49857.

3. Using Euler's method, find an approximate value of y corresponding to $x = 1.6$, given $\frac{dy}{dx} = y^2 - \frac{y}{x}$ and $y = 1$ when $x = 1$. **Ans.** 1.1351

4. Using Euler's method to solve the differential equation in six steps $\frac{dy}{dx} = x + y$; $y(0) = 0$ choosing h = 0.2. (*RGPV., Bhopal, III Sem. Dec. 2003*) **Ans.** $y = 0.785984$

52.6 EULER'S MODIFIED FORMULA

In equation (3) of Art 52.14 the expansion of y_{n+1} is

$$y_{n+1} = y_n + hf(x_n, y_n) + \frac{1}{2}h^2\phi''(x_n, y_n) + \frac{1}{\underline{|3}}h^3\,\phi''' + (x_n, y_n) + \ldots \quad \ldots(1)$$

In Euler's formula we omit $\frac{1}{2}h^2\phi''(x_n, y_n)$ and higher powers of h.

The error due to this omission is called **Truncation error.**

Now a formula is derived with small error.

Differentiating (1) w.r.t. x we get

$$\left(\frac{dy}{dx}\right)_{n+1} = \left(\frac{dy}{dx}\right)_n + hf'(x_n, y_n) + \frac{1}{2}h^2\phi'''(x_n, y_n) + \ldots$$

$$f(x_{n+1}, y_{n+1}) = f(x_n, y_n)\ hf'(x_n, y_n)' + \frac{1}{2}h^2\phi'''(x_n, y_n) + \ldots$$

$$= f(x_n, y_n) + h\phi'''(x_n, y_n) + \frac{1}{2}h^2\phi'''(x_n, y_n) + \ldots \quad \ldots(2)$$

Multiplying (2) by $\frac{h}{2}$ and subtracting from (1) we get

$$y_{n+1} - \frac{1}{2}hf(x_{n+1}, y_{n+1}) = y_n + \frac{h}{2}f(x_n, y_n) - \frac{h^3}{12}\phi'''(x_n, y_n)$$

Neglecting terms containing h^3 and higher powers, we obtain

$$y_{n+1} = y_n + h\left[\frac{f(x_n, y_n) + f(x_{n+1}, y_{n+1})}{2}\right] \quad \ldots(3)$$

Equation (3) is the Euler's modified formula.

But $f(x_{n+1}, y_{n+1})$ which occurs on the right hand side of equation (3), cannot be calculated since y_{n+1} is unknown. So first we calculate y_{n+1} from Euler's first formula.

$$y_{n+1} = y_n + hf(x_n, y_n)$$

Thus for each stage we use the following two formulae.

$$y_{n+1} = y_n + hf(x_n, y_n)$$

$$y_{n+1} = y_n + \frac{h}{2}f(x_n, y_n)\ (y_{n+1}, y_{n+1})$$

Example 9. *Apply Euler's modified method to solve* $\frac{dy}{dx} = x + 3y$ *subject to* $y(0) = 1$ *and hence find an approximate value of y when* $x = 1$.

Solution. $f(x,y) = x + 3y$

$$y_{n+1} = y_n + hf(x_{n+1}, y_n)$$

$$y_{n+1} = y_n + \frac{h}{2}[f(x_n, y_n) + f(x_{n+1}, y_{n+1})]$$

This gives $y_{n+1} = y_n + 0.1(x_n + 3y_n)$

$$y_{n+1} = y_n + 0.05\,[x_n + 3y_n) + (x_{n+1} + 3y_{n+1})].$$

Hence the required value of y at $x = 1$ is 21.081.

The exact solution gives $y = 21.873$ for $x = 1$. The error is 0.792 *i.e.*, 3.6%. **Ans.**

Procedure. We calculated y_{n+1} by Euler's formula *i.e.*, $y_{n+1} = y_n + 0.1\,(x_n + 3\,y_n)$ and entered in 5th column. In 7th column we record the sum *i.e.* $x_{n+1} + 3_{n+1}$. Then we computed the value of y_{n+1} by Euler's modified formula *i.e.*,

$$y_{n+1} = y_n + \frac{0.1}{2}\,[(x_n + 3y_n) + (x_{n+1} + 3y_{n+1})].$$ and entered in 8th column.

The following table shows the computation work.

n	x_n	y_n	$x_n + 3\,y_n$	*Eulers formula* y_{n+1}	$x_n + 1$	xn+1+ $3y_{n+1}$	*Eulers* modified y_{n+1}
0	0.0	1	3	1.3	0.1	4	1.35
1	0.1	1.35	4.15	1.765	0.2	5.495	1.832
2	0.2	1.832	5.695	2.402	0.3	7.506	2.492
3	0.3	2.492	7.776	3.270	0.4	10.21	3.391
4	0.4	3.391	10.573	4.448	0.5	13.844	4.612
5	0.5	4.612	14.336	6.046	0.6	18.738	6.266
6	0.6	6.266	19.398	8.206	0.7	25.318	8.502
7	0.7	8.502	26.206	11.123	0.8	34.169	11.521
8	0.8	11.521	35.363	15.057	0.9	46.071	15.593
9	0.9	15.593	47.679	20.361	1.0	62.083	21.081
10	1.0	21.081					

EXERCISE 52.4

1. Using Euler's modified formula, find an approximate value of y when x 0.3, given that $\frac{dy}{dx} = x + y$ and $y = 1$ when $x = 0$. *(RGPV., Bhopal, III Sem. Dec. 2007)* **Ans.** 1.3997

2. Using Euler's modified formula, find an approximate value of y when $x = 0.06$, given that $\frac{dy}{dx} = x^2 + y$ and $y(0) = 1$, taking the interval 0.02. **Ans.** 1.0619

3. Using Euler's modified formula, solve $\frac{dy}{dx} = 1 - 2\,xy$ given $y = 0$ at $x = 0$ from $x = 0$ to 0.6 taking the interval $h = 0.2$. **Ans.** 0.4748

4. Using Euler's modified method to compute y for $x = 0.05$. Given that $\frac{dy}{dx} = x + y$ with initial conditions $x_0 = 0, y_0 = 1$ results correct upto three defimal places. (*RGPV., Bhopal, III Sem. Dec. 2002*) **Ans.** y = 1.0525

52.7 RUNGE'S FORMULA

Euler's modified formula is $y_{n+1} = y_n + \frac{h}{2}[f(x_n, y_n) + f(x_{n+1}, y_{n+1})]$

or
$$y_{n+1} = y_n + \frac{h}{2}[f(x_n, y_n) + f(x_n + n, y_n + hf_n)] \quad \text{... (1)}$$

Let
$$k_1 = hf(x_n, y_n)$$

and
$$k_2 = hf[x_n + h, y_n + hf(x_n, y_n)] \quad \text{or} \quad k_2 = hf(x_n + h, y_n + k_1)$$

Putting the values of k_1 and k_2 in (1) we get

$$y_{n+1} = y_n + \frac{1}{2}(k_1 + k_2) \quad \text{...(2)}$$

This is known as Runge's formula of order 2.

Example 10. *Apply Runge 's formula of order 2 to find approximate value of y when x = 1.1, given* $\frac{dy}{dx} = 3x + y^2$ *and y = 1.2 when x = 1.* (*RGPV., Bhopal, III Sem. June 2005*)

Solution. Here we have $x_0 = 1, y_0 = 1.2, \; h = 0.1$

$$f(x,y) = 3x + y^2, f(x_0, y_0) = 3(1) + (1.2)^2 = 4.44$$

$$k_1 = hf(x_0, y_0) = 0.1 \times 4.44 = 0.444$$

$$k_2 = hf(x_0 + h, y_0 + k_1) = 0.1f(1.1, 1.2 + 0.444) = 0.1f(1.1, 1.644)$$

$$= 0.1[3 \times 1.1 + (1.644)^2] = 0.600$$

$$y_{n+1} = y_n + \frac{1}{2}(k_1 + k_2)$$

$$y_1 = 1.2 + \frac{1}{2}(0.444 + 0.600) = 1.722 \quad \textbf{Ans.}$$

EXERCISE 52.5

1. Apply Runge's formula of second order to find approximate value of y when $x = 1.1$, given that $\frac{dy}{dx} = x - y$ and $y = 1$ when $x = 1$. **Ans.** 1.005

2. Apply Runge's formula of second order to find approximate value of y when $x = 0.02$, given that $\frac{dy}{dx} = x^2 + y$ and $y(0) = 1$. **Ans.** 1.0202.

52.8 RUNGE'S FORMULA (THIRD ORDER)

$$y_1 = y_0 + \frac{1}{6}(k_1 + 4k_4 + k_3)$$

where
$$k_1 = hf(x_0, y_0), k_2 = hf(x_0 + h, y_0 + k_1)$$

$$k_3 = hf(x_0 + h, y_0 + k_2), \; k_4 = hf\left(x_0 + \frac{h}{2}, y_0 + \frac{k_1}{2}\right)$$

This is the Runge's Formula (third order) with an error of the order h^4.

Example 11. *Using Runge 's Formula (third order), solve the differential equation* $\frac{dy}{dx} = x - y$ *subject to y = 1 when x = 1.*

Solution. $f(x, y) = x - y$

Here $h = 0.1,\ x_0 = 1, y_0 = 1$

$k_1 = hf(x_0, y_0) = 0.1\ (x - y) = 0.1\ (1 - 1) = 0$

$k_2 = hf(x_0 + h, y_0 + k_1) = 0.1f\ (1.1., 1 + 0) = 0.1(1.1-1) = 0.01$

$k_3 = hf(x_0 + h, y_0 + k_2) = 0.1f(1.1, 1.01) = 0.1\ (1.1-1.01) = 0.009$

$$k_4 = hf\left(x_0 + \frac{h}{2}, y_0 + \frac{k_1}{2}\right) = 0.1f\left(1.05, 1 + \frac{0}{2}\right) = 0.1(1.05 - 1) = 0.005$$

$$y_1 = y + \frac{1}{6}(k_1 + 4k_4 + k_3)$$

$$y(0.1) = 1 + \frac{1}{6}(0 + 0.02 + 0.009) = 1 + 0.004833 = 1.004833$$ **Ans.**

52.9 RUNGE-KUTTA FORMULA (FOURTH ORDER)

A fourth order Runge's-Kutta Formula for solving the differential equation is

$$\boldsymbol{y = y_0 + \frac{1}{6}(k_1 + 2k_2 + 2k_3 + k_4)}$$

where $$k_1 = hf(x_0, y_0),\quad k_2 = hf\left(x_0 + \frac{h}{2}, y_0 + \frac{k_1}{2}\right)$$

$$k_3 = hf(x_0 + \frac{h}{2}, y_0 + \frac{k_2}{2}),\quad k_4 = hf\left(x_0 + h, y_0 + k_3\right)$$

$$y = y_0 + \frac{1}{6}[k_1 + 2k_2 + 2k_3 + k_4]$$

This is known as Runge-Kutta Formula. The error in this formula is of the order h^5. This method have greater accuracy. No deviatives are required to be tabulated.

It requires only functional values at some selected points on the sub interval.

Example 12. *Apply Runge-Kutta method to find an approximate value of y when x = 0.2, given that*

$$\frac{dy}{dx} = x + y,\ y = 1 \text{ when } x = 0$$

Solution. Let $h = 0.1$

Here $x_0 = 0,\ y_0 = 1, f(x, y) = x + y$

Now $k_1 = hf(x_0, y_0) = 0.1\ (0 + 1) = 0.1$

$$k_2 = hf\left(x_0 + \frac{h}{2}, y_0 + \frac{k_1}{2}\right) = 01\,f(0 + 0.05, 1 + 0.05) = 0.1[0.05 + 1.05] = 0.11$$

$$k_3 = hf\left(x_0 + \frac{h}{2}, y_0 + \frac{k_2}{2}\right) = 01\,f(0 + 0.05, 1 + 0.055) = 0.1(0.05 + 1.055] = 0.1105$$

$$k_4 = hf(x_0 + h, y_0 + k_3) = 0.1f(0 + 0.1,\ 1 + 0.1105)$$

$$= 0.1f(0.1, 1.1105) = 0.1(0.1 + 1.1105) = 0.12105$$

According to Runge-Kutta (Fourth order) formula

$$y = y_0 + \frac{1}{6}[k_1 + 2k_2 + 2k_3 + k_4]$$

$$y_{0.1} = 1 + \frac{1}{6}(0.1 + 0.22 + 0.221 + 0.12105) = 1 + \frac{1}{6}(0.66205) = 1.11034$$

For the second step

$$x_0 = 0.1,\ y_0 = 1.11034,\ h = 0.1$$

$$k_1 = hf(x_0, y_0) = 0.1(0.1 + 1.11034) = 0.121034$$

$$k_2 = hf\left(x_0 + \frac{h}{2},\ y_0 + \frac{k_1}{2}\right) = 0.1 f(0.1 + 0.05,\ 1.11034 + 0.060517)$$

$$= 0.1\ (0.15 + 1.170857) = 0.1320857$$

$$k_3 = hf\left(x_0 + \frac{h}{2},\ y_0 + \frac{k_2}{2}\right) = 0.1 f(0.1 + 0.05,\ 1.11034 + 0.0660428)$$

$$= 0.1\ (0.15 + 1.1763828) = 0.13263828$$

$$k_4 = hf(x_0 + h,\ y_0 + k_3) = 0.1\ (0.1 + 0.1,\ 1.11034 + 0.13263828)$$

$$= 0.1\ (0.2 + 1.24297828) = 0.144297828$$

$$y_{\mathrm{r}} = y_0 + \frac{1}{6}[k_1 + 2k_2 + 2k_3 + k_4]$$

$$= 1.11034 + \frac{1}{6}\ [0.121034 + 2 \times 0.1320857 + 2 \times 0.13263828 + 0.144297828]$$

$$= 1.11034 + \frac{1}{6}\ [0.121034 + 0.2641714 + 0.26527656 + 0.144297828]$$

$$= 1.11034 + \frac{1}{6} \times 0.794779788 = 1.11034 + 0.132463298 = 1.242803298$$ **Ans.**

Example 13. *Apply Range-Kutta method of fourth order to solve :*

$$10\frac{dy}{dx} = x^2 + y^2, \quad y\,(0) = 1 \text{ for } x = 1.$$ *(R. G.P.V. Bhopal, III Semester, Dec. 2002)*

Solution. We have,

$$10\frac{dy}{dx} = x^2 + y^2 \quad \Rightarrow \quad \frac{dy}{dx} = \frac{x^2 + y^2}{10}$$

$$\Rightarrow \quad f(x, y) = \frac{x^2 + y^2}{10}$$

Here, let $h = 0.1$, $x_0 = 0$, $y_0 = 1$.

Now, $k_1 = hf(x_0, y_0) = (0.1) f(0,1) = (0.1)\left(\frac{0+1}{10}\right) = 0.01$

$$k_2 = hf\left(x_0 + \frac{h}{2},\ y_0 + \frac{1}{2}k_1\right) = (0.1) f\left(0 + \frac{0.1}{2},\ 1 + \frac{0.01}{2}\right)$$

$$= (0.1) f(0.05,\ 1.005) = (0.1)\left[\frac{(0.05)^2 + (1.005)^2}{10}\right] = 0.01012525$$

$$k_3 = hf\left(x_0 + \frac{h}{2},\ y_0 + \frac{k_2}{2}\right) = (0.1) f\left(0.05,\ 1 + \frac{0.01012525}{2}\right)$$

$$= (0.1) f(0.05,\ 1.00506263) = (0.1)\left[\frac{(0.05)^2 + (1.00506263)^2}{10}\right] = 0.01012651$$

$$k_4 = hf(x_0 + h,\ y_0 + k_3) = (0.1)f(0.1,\ 1 + 0.01012651)$$

$$= (0.1)f(0.1,\ 1.01012651) = (0.1)\left[\frac{(0.1)^2 + (1.01012651)^2}{10}\right] = 0.010303556$$

$$y_{0.1} = y_0 + \frac{1}{6}(k_1 + 2k_2 + 2k_3 + k_4)$$

$$= 1 + \frac{1}{6}[0.01 + 2(0.01012525) + 2(0.01012651) + 0.010303556]$$

$$= 1 + 0.01013451 = 1.01013451$$

Hence, y at $x = 0.1$ is 1.01013451. **Ans.**

Example 14. *Apply Runge-Kutta method (fourth order), to find an approximate value of y when x = 0.2, given that* $\frac{dy}{dx} = x + y^2$ *and y = 1 when x = 0.*

(*RGPV., Bhopal, III Sem. Dec. 2004, AMIETE, Dec. 2010*)

Solution. Let $h = 0.1$,

Here $x_0 = 0, y_0 = 1,\ f(x, y) = x + y^2$

Now $k_1 = hf(x_0, y_0) = 0.1\ (0 + 1) = 0.1$

$$k_2 = hf\left(x_0 + \frac{h}{2},\ y_0 + \frac{k_1}{2}\right) = 0.1f(0 + 0.05,\ 1 + 0.05) = 0.1[0.05 + (1.05)^2] = 0.11525$$

$$k_3 = hf\left(x_0 + \frac{h}{2},\ y_0 + \frac{k_2}{2}\right) = 0.1f(0 + 0.05,\ 1 + 0.057625)$$

$$= 0.1[0.05 + (1.057625)^2] = 0.11685$$

$$k_4 = hf(x_0 + h,\ y_0 + k_3) = 0.1f(0 + 0.1,\ 1 + 0.11685) = 0.1\ [0.1 + (0.11685)^2] = 0.13474$$

According to Runge-Kutta (fourth order) formula

$$y_1 = y_0 + \frac{1}{6}(k_1 + 2k_2 + 2k_3 + k_4)$$

$$y_{0.1} = 1 + \frac{1}{6}[0.1 + 2(.11525) + 2(0.11685) + 0.13474]$$

$$y_{0.1} = 1 + 0.1165 = 1.1165$$

For the second step $x_0 = 0.1, y_0 = 1.1165$

$$k_1 = 0.1\ (0.1 + 1.2466) = 0.1347$$

$$k_2 = 0.1\ (0.15 + 1.4014) = 0.1551$$

$$k_3 = 0.1\ (0.15 + 1.4259) = 0.1576$$

$$k_4 = 0.1\ (0.2 + 1.6233) = 0.1823$$

$$y_{0.2} = y_{0.1} + \frac{1}{6}(k_1 + 2k_2 + 2k_3 + k_4)$$

$$= 1.1165 + \frac{1}{6}[0.1347 + 2(0.1551) + 2(0.1576) + 0.1823]$$

$$= 1.1165 + 0.1571 = 1.2736$$ **Ans.**

Example 15. *Use the fourth order Runge-Kutta method to find u (0, 2), of the initial value problem* $u' = -2\,t\,u^2$, $u\,(0) = 1$, *using h = 0.2.* (*U.P. III Sem., Dec. 2009*)

Solution. $h = 0.2$

Here $t = 0,\ u = 1, f(t, u) = -2\,tu^2$

$$k_1 = hf(t_0, u_0) = 0.2\,(-2\,tu^2) = 0.2\,(0) = 0$$

$$k_2 = hf\left(t_0 + \frac{h}{2},\ u_0 + \frac{k_1}{2}\right)$$

$$= 0.2f(0.1,\ 1 + 0) = 0.2\,f(0.1,\ 1) = 0.2\,(-2 \times 0.1 \times 1^2) = -0.04$$

$$k_3 = hf\left(t_0 + \frac{h}{2},\ u_0 + \frac{k_2}{2}\right) = 0.2\,f(0 + 0.1,\ 1 - 0.02) = 0.2\,f(0.1,\ 0.98)$$

$$= 0.2f\,[-2 \times 0.1 \times (0.98)^2] = -0.2[0.2 \times 0.9604] = -\ 0.038416$$

$$k_4 = hf(t_0 + h,\ u_0 + k_3) = 0.2\,f(0.2,\ 1 - 0.038416) = 0.2(-2) \times (0.2) \times (0.961584)^2$$

$$= -\ 0.08 \times 0.9246 = -\ 0.073971503$$

$$u = u_0 + \frac{1}{6}[k_1 + 2k_2 + 2k_3 + k_4] = 1 + \frac{1}{6}[0 + 2(-0.04) + 2(-0.038416) + (-0.073968)]$$

$$= 1 - \frac{1}{6}[0.08 + 0.076832 + 0.07391503]$$

$$= 1 - \frac{1}{6}(0.2308) = 1 - 0.03847 = 0.961532749 \qquad \textbf{Ans.}$$

EXERCISE 52.6

1. The initial value problem $y' = x\,(y + x) - 2, y\,(1) = 2$ is given. Find the value of $y\,(1.2)$ with $h = 0.2$ using the Runge-Kutta method of fourth order. **Ans.** $y\,(1.2) = 2.3\ 1\ 38$
2. Use the Runge-Kutta method of fourth order to find $y\,(0.\ 8)$ with $h = 0.2$ for the initial value problem.
$$\frac{dy}{dx} = \sqrt{x + y},\ y(0, 4) = 0.41 \qquad \textbf{Ans. } 0.8489912$$
3. Find $y\,(0.2)$ for the equation
$$\frac{dy}{dx} = -xy, \quad y(0) = 1, \text{ using Runge-Kutta method.}$$
4. Apply the Runge-Kutta method to obtain $y\,(1.1)$ from the differential equation
$$\frac{dy}{dx} = xy^{1/3}, \quad y(1) = 1, \text{ taking } h = 0.1.$$
5. Apply Runge-Kutta (fourth order) formula to find an approximate value of y when $x = 1.1$, given that
$$\frac{dy}{dx} = x - y \text{ and } y = 1 \text{ at } x = 1. \qquad \textbf{Ans. } 1.004837$$
6. Using Ringe-Kutta method of fourth order solve $\dfrac{dy}{dx} = \dfrac{y^2 - x^2}{y^2 + x^2}$ with y (0) = 1 at x = 0.2 and 0.4.

   (*RGPV., Bhopal III Sem. June 2008, 2004*) **Ans.** $y_{1.2} = 1.19600,\ \ y_{1.4} = 1.37527$

52.10 HIGHER ORDER DIFFERENTIAL EQUATIONS

Let $\dfrac{dy}{dx} = f(x, y, z), \dfrac{dz}{dx} = g(x, y, z),\ y(x_0) = y_0,\ z(x_0) = z_0$

Formulae for the application of Runge-Kutta method are as follows :

$$k_1 = hf(x_n, y_n, z_n),\ m_1 = hg(x_n, y_n, z_n)$$

$$k_2 = \left(x_n + \frac{h}{2}, y_n + \frac{k_1}{2}, z_n + \frac{m_1}{2}\right)$$

$$m_2 = hg\left(x_n + \frac{h}{2}, y_n + \frac{k_1}{2}, z_n + \frac{m_1}{2}\right)$$

$$k_3 = hf\left(x_n + \frac{h}{2}, y_n + \frac{k_2}{2}, z_n + \frac{m_2}{2}\right)$$

$$m_3 = hg\left(x_n + \frac{h}{2}, y_n + \frac{k_2}{2}, z_n + \frac{m_2}{2}\right)$$

$$k_4 = hf(x_n + h, y_n + k_3, z_n + m_3)$$

$$m_4 = hg(x_n + h, y_n + k_3, z_n + m_3)$$

$$x_{n+1} = x_n + h$$

$$y_{n+1} = y_n + \frac{1}{6}(k_1 + 2k_2 + 2k_3 + k_4)$$

$$z_{n+1} = z_n + \frac{1}{6}(m_1 + 2m_2 + 2m_3 + m_4)$$

Higher order differential equations are best treated by transforming the given equation into a system of first order simultaneous equations which can be solved by one of the aforesaid methods.

Consider, for example the second order differential equation :

$$\frac{d^2y}{dx^2} = f\left(x, y, \frac{dy}{dx}\right);\ y(x_0) = y_0 \left(\frac{dy}{dx}\right)_{x=x_0} = y_0{}'$$

Substituting $\frac{dy}{dx} = z$, we get

$$\frac{dz}{dx} = f(x, y, z)$$

$$y(x_0) = y_0,\ z(x_0) = y_0{}'$$

These constitute the equivalent system of simultaneous equations.

Example 16. *Use Runge-Kutta method to find y (0.2) for the equation*

$$\frac{d^2y}{dx^2} = x\frac{dy}{dx} - y$$

given that y = 1, $\frac{dy}{dx} = 0$ *when x = 0.*

Solution. Substituting $\frac{dy}{dx} = z = f(x, y, z)$

The given equation reduces to $\frac{d^2y}{dx^2} = xz - y = g(x, y, z)$

The initial conditions are given by $x = 0, y = 1, z = 0$.

Also $h = 0.2$

$$k_1 = hf(x, y, z) = hz = 0.2 \times 0 = 0$$

$$m_1 = hg(x, y, z) = h(xz - y) = 0.2(0 \times 0 - 1) = -0.2$$

$$k_2 = hf\left(x + \frac{h}{2}, y + \frac{k_1}{2}, z + \frac{m_1}{2}\right) = \left(z + \frac{m_1}{2}\right) = 0.2\left(0 - \frac{0.2}{2}\right) = -0.02$$

$$m_2 = hg\left(x+\frac{h}{2},\, y+\frac{k_1}{2},\, z+\frac{m_1}{2}\right) = h\left[\left(x+\frac{h}{2}\right)\left(z+\frac{m_1}{2}\right)-\left(y+\frac{k_1}{2}\right)\right]$$

$$= 0.2\left[\left(0+\frac{0.2}{2}\right)\left(0-\frac{0.02}{2}\right)\left(1+\frac{0}{2}\right)\right] = 0.2[-0.01-1] = -0.202$$

$$k_3 = hf\left(x+\frac{h}{2},\, y+\frac{k_2}{2},\, z+\frac{m_2}{2}\right) = h\left(z+\frac{m_2}{2}\right) = 0.2\left(0-\frac{0.202}{2}\right) = -0.0202$$

$$m_3 = hg\left[x+\frac{h}{2},\, y+\frac{k_2}{2},\, z+\frac{m_2}{2}\right] = h\left[\left(x+\frac{h}{2}\right)\left(z+\frac{m_2}{2}\right)-\left(y+\frac{k_2}{2}\right)\right]$$

$$= 0.2\left[\left(0+\frac{0.2}{2}\right)\left(0-\frac{0.202}{2}\right)-\left(1-\frac{0.02}{2}\right)\right] = 0.2[-0.0101-0.99] = -0.20002$$

$$k_4 = hf(x+h,\, y+k_3,\, z+m_3) = h(z+m_3) = 0.2\,(0-0.20002) = -0.040004$$

$$m_4 = hg(x+h,\, y+k_3,\, z+m_3) = h[(x+h)(z+m_3)-(y+k_3)]$$

$$= 0.2\,[(0.2)(-0.20002)-(1-0.0202)] = 0.2\,[-0.040004-0.9798] = -0.2039608$$

This gives, at $x = 0.2$

$$y(0.2) = y(0)+\frac{1}{6}\,[k_1+2k_2+2k_3+k_4] = 1+\frac{1}{6}\,[0+2(-0.02)+2(-0.0202)+(-0.040004)]$$

$$= 1+\frac{1}{6}\,(-0.04-0.0404-0.040004) = 1+\frac{1}{6}\,(-0.120404) = 0.97993266$$

$$z(0.2) = z(0)+\frac{1}{6}\,(m_1+2m_2+2m_3+m_4)$$

$$= 0+\frac{1}{6}\,[-0.2+2(-0.202)+2(-0.20002)-0.2039608]$$

$$= \frac{1}{6}\,[-0.2-0.404-0.40004-0.2039608]$$

$$= \frac{1}{6}\,(-1.2080008) = -0.201333466$$ **Ans.**

EXERCISE 52.7

1. Find $y(0.4)$ for the equation $\dfrac{dy}{dx} = x^2+y^2$, $y(0) = 0$ by Picard's method. **Ans.** 0.0214.

2. Use Picard's method to solve $\dfrac{dy}{dx} = 2y+z$, $\dfrac{dz}{dx} = y+2z$; given that $y(0) = 0$, $z(0) = 1$.

   Ans. $y = x+2x^2+\frac{13}{6}x^3+\frac{5}{3}x^4+\ldots z = 1+2x+\frac{5}{2}x^2+\frac{7}{3}x^3+\frac{41}{24}x^4+\ldots$

3. Employ Runge-Kuria method to find y for $x = 0.2$ from $\dfrac{d^2y}{dx^2} = x\left(\dfrac{dy}{dx}\right)^2 - y^2$

   given that $y = 1$, $\dfrac{dy}{dx} = 0$ for $x = 0$. **Ans.** $y(0.2) = 0.9801$; $y'(0.2) = -0.1970$

4. Describe Runge-Kutta method (4th order) for obtaining solution of initial value problem :

$$y'' = f(x, y, y'),\; y(x_0) = y_0,\; y'(x_0) = y_0'$$

5. State clearly the conditions under which the method is applicable.

NUMERICAL TECHNIQUES FOR SOLUTION OF PARTIAL DIFFERENTIAL EQUATIONS

53.1 GENERAL LINEAR PARTIAL DIFFERENTIAL EQUATIONS

General partial differential equation is of the form

$$A(x,y)\frac{\partial^2 u}{\partial x^2}+B(x,y)\frac{\partial^2 u}{\partial x\partial y}+C(x,y)\frac{\partial^2 u}{\partial y^2}+D(x,y)\frac{\partial u}{\partial x}+E(x,y)\frac{\partial u}{\partial y}+F(x,y)u+G(x,y)=0$$

This equation is called

(*i*) *Elliptic*, if $B^2-4AC<0$

e.g. $\frac{\partial^2 u}{\partial x^2}+\frac{\partial^2 u}{\partial y^2}=0$ Laplace Equation

$\frac{\partial^2 u}{\partial x^2}+\frac{\partial^2 u}{\partial y^2}=f(x,y)$ Poisson's Equation

(*ii*) *Parabolic*, if $B^2-4AC=0$ *e.g.* $\frac{\partial u}{\partial t}=C^2\frac{\partial^2 u}{\partial x^2}$

One dimensional heat conduction equation.

(*iii*) *Hyperbolic*, if $B^2-4AC>0$ *e.g.* $\frac{\partial^2 u}{\partial t^2}=C^2\frac{\partial^2 u}{\partial x^2}$

Example 1. *Determine the type of* $x^2\frac{\partial^2 u}{\partial y^2}+2xy\frac{\partial^2 u}{\partial x\partial y}+y^2\frac{\partial^2 u}{\partial x^2}=0$

(*A.M.I.E.T.E., Dec. 2006*)

Solution. Here, $A=x^2$, $B=2xy$, $c=y^2$.

$B^2-4AC=4x^2y^2-4x^2y^2=0$. Hence, it is a parabolic equation. **Ans.**

53.2 FINITE-DIFFERENCE APPROXIMATION TO DERIVATIVES

By Taylor formula

$$u(x+h,y)=u(x,y)+\frac{h\partial u}{\partial x}+\frac{1}{2!}h^2\frac{\partial^2 u}{\partial x^2}+\frac{1}{3!}h^3\frac{\partial^3 u}{\partial x^3}+ \quad ...(1)$$

$$u(x-h,y)=u(x,y)-h\frac{\partial u}{\partial x}+\frac{1}{2!}h^2\frac{\partial^2 u}{\partial x^2}-\frac{1}{3!}h^3\frac{\partial^3 u}{\partial x^3} \quad ...(2)$$

From (1) ,neglecting h^2 and higher powers of h, we get

$$\frac{\partial u}{\partial x}\approx\frac{u(x+h,y)-u(x,y)}{h} \quad \text{(Forward difference formula)} \quad ...(3)$$

Form (2),neglecting $h^2$and higher powers of h,we have

$$\frac{\partial u}{\partial x}\approx\frac{u(x,y)-u(x-h,y)}{h} \quad \text{(Backward differencc formula)} \quad ...(4)$$

Subtracting (2) from (1)and neglecting h^3 and higher power of h we get

$$u(x+h,y)-u(x-h,y) \approx 2h\frac{\partial u}{\partial x}$$

or $$\frac{\partial u}{\partial x} \approx \frac{1}{2h}[u(x+h,y)-u(x-h,y)] \quad \text{(Central differemce formula)} \qquad ...(5)$$

Similarly,

$$\frac{\partial u}{\partial y} = \frac{u(x,y+k)-u(x,y)}{k} = \frac{u(x,y)-u(x,y-k)}{k} = \frac{u(x,y+k)-u(x,y-k)}{2k} \qquad ...(6)$$

Adding (1) amd (2). neglecting h^5 and higher powers of h, we get

$$u(x+h, y) + u(x-h,y) = 2u(x,y) + h^2\frac{\partial^2 u}{\partial x^2}$$

$$\frac{\partial^2 u}{\partial x^2} \approx \frac{1}{h^2}[u(x+h,y) - 2u(x,y) + u(x-h,y)] \qquad ...(7)$$

Similarly $$\frac{\partial^2 u}{\partial y^2} \approx \frac{1}{k^2}[u(x,y+k) - 2u(x,y) + u(x,y-k)] \qquad ...(8)$$

and $$\frac{\partial^2 u}{\partial x \partial y} \approx \frac{1}{4hk}[u(x+h,y+k) - u(x-h,y+k) - u(x+h,y-k) + u(x-h,y-k)] \qquad ...(9)$$

The given region (rectangle $ABCD$) is divided into smaller rectangles of sides $\delta x = h$ and $\delta y = k$ The origin is taken at the centre of the rectangle and the coordinates axes are drawn. The rectangle is divided into 36 small rectangles. Here there are 49 mesh-points or lattices or nodal points or grid points.The values of the function u are u_{ij}, $u_{i+2,j...}$ $u_{i,j+1}$, $u_{i,j+2...}$ at the mesh-points.

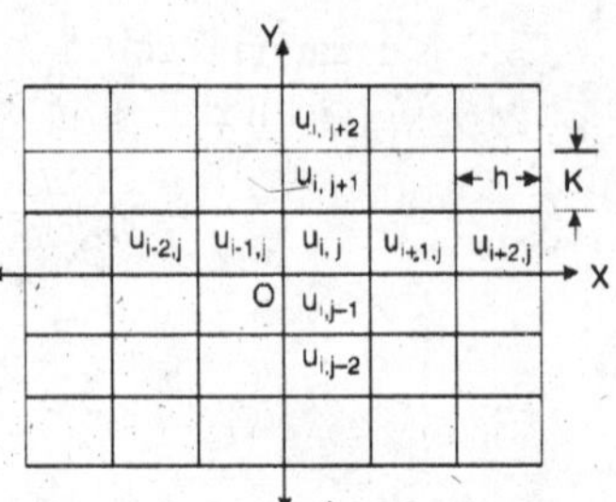

Let these values satisfy the given partial differentiaequation.

At the centre of the rectangle:

Equations (5), (6), (7), (8) and (9) are rewritten on the nodal points as below:

$$\frac{\partial u}{\partial x} = \frac{1}{2h}(u_{i+1,j} - u_{i-1,j}), \qquad \frac{\partial u}{\partial y} = \frac{1}{2k}(u_{i,j+1} - u_{ij-1})$$

$$\frac{\partial^2 u}{\partial x^2} = \frac{1}{h^2}(u_{i+1,j} - 2u_{i,j} + u_{i-1,j})$$

$$\frac{\partial^2 u}{\partial y^2} = \frac{1}{k^2}(u_{i,j+1} - 2u_{i,j} + u_{i,j-1})$$

$$\frac{\partial^2 u}{\partial x \partial y} = \frac{1}{4hk}(u_{i+1,j+1} - u_{i+1,j-1} - u_{i-1,j+1} + u_{i-1,j-1})$$

53.3 SOLUTION OF PARTIAL DIFFERENTIAL EQUATION (LAPLACE EQUATION)

$$\frac{\partial^2 u}{\partial x^2} + \frac{\partial^2 u}{\partial y^2} = 0$$

On substituing the values of $\frac{\partial^2 u}{\partial x^2}$ and $\frac{\partial^2 u}{\partial y^2}$, we get

$$\frac{1}{h^2}[u(x+h,y) - 2u(x,y) + u(x-h,y)] + \frac{1}{k^2}[u(x,y+k) - 2u(x,y) + u(x,y-k)] = 0$$

For values of $h = k$ *i.e.* for square grid of the mesh size h, the above equation can be written as

$$u(x+h,y) - 2u(x,y) + u(x-h,y) + u(x,y+h) - 2u(x,y) + u(x,y-h) = 0$$

$$u(x, y) = \frac{1}{4}[u(x+h, y) + u(x-h, y) + u(x, y+h) + u(x, y-h)]$$

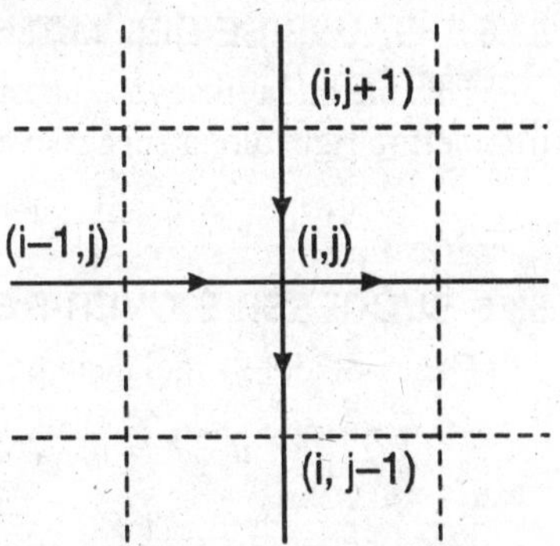

Denoting any mesh point $(x, y) = (ih, jh)$ as simply i, j, the above difference equation can be written as

$$u_i = \frac{1}{4}(u_{i+1,j} + u_{i-1,j} + u_{i,j+1} + u_{i,j-1}) \qquad ...(2)$$

Equation (2) shows that the value of $u(x, y)$ is the average of its four neighbours to the East, West, North and South. The formula (2) is called the Standard five points formula and is written as

$$u_{i+1,j} + u_{i-1,j} + u_{i,j+1} + u_{i,j-1} - 4u_{i,j} = 0$$

This formula is also known as Liebman's averaging procedure.

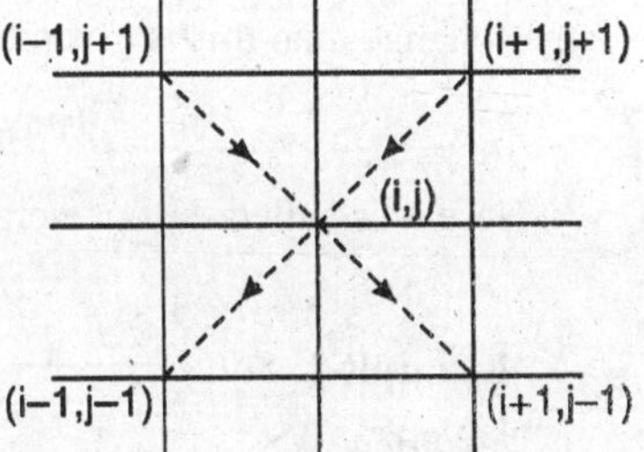

A formula similar to the formula (2) is sometimes used with convenience. It is given as

$$u_{ij} = \frac{1}{4}(u_{i+1,j+1} + u_{i+1,j-1} + u_{i-1,j+1} + u_{i-1,j-1}) \qquad ...(3)$$

This is known as *diagonal five-point formula* as these points lie on the diagonals. Although formula (3) is less accurate than formula (2), still it is a good approximation for obtaining as starting values in the iteration procedure.

Whenever possible, Standard five-point formula is preferred in all commutations.

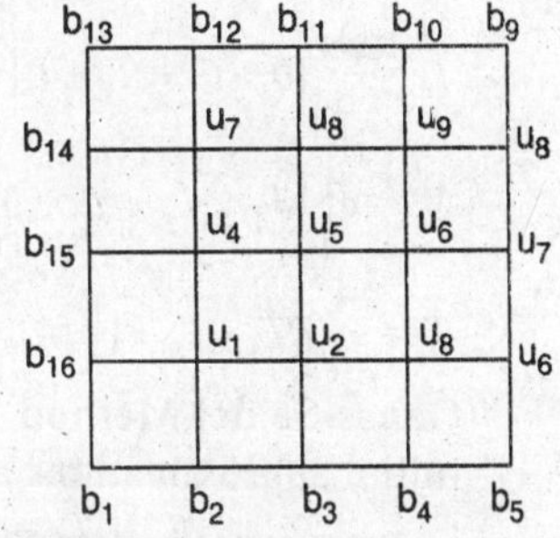

Procedure. We use the following diagonal five point formula to get the initial value of u at the centre.

$$U_5 = \frac{1}{4}[b_1 + b_5 + b_9 + b_{13}]$$

Then the approximate values of diagonal interior points u_1, u_3, u_7, u_9 are calculated by the diagonal five-point formula

$$u_1 = \frac{1}{4}[b_1 + b_3 + b_5 + b_{15}], \qquad u_3 = \frac{1}{4}[b_3 + b_5 + b_7 + u_5]$$

$$u_7 = \frac{1}{4}[b_{15} + u_5 + b_{11} + b_{13}] \qquad u_9 = \frac{1}{4}[u_5 + b_7 + b_9 + b_{11}]$$

The values of the remaining interior points *i.e.* u_2, u_4, u_6 and u_8 are obtained by the standard five point formula.

$$u_2 = \frac{1}{4}[b_3 + u_3 + u_5 + u_1], \qquad u_3 = \frac{1}{4}[u_1 + u_5 + u_7 + u_{15}]$$

$$u_6 = \frac{1}{4}[u_3 + b_7 + u_9 + u_5] \qquad u_8 = \frac{1}{4}[u_5 + u_9 + b_{11} + u_7]$$

Having obtained all values u_1, u_2, $...u_9$ once, their accuracy can be improved by the repeated application of either Jacobi's iteration formula or Gauss-Seidel iteration formula.

53.4 JACOBI'S ITERATION FORMULA

Let $u^n_{i,j}$ be the nth iterative value of u_{ij}. Then Jacobi's iterative procedure is given below.

$$u^{(n+1)_{ij}} = \frac{1}{4}[u^{(n)_{i-1,j}} + u^{(n)_{i+1,}} j + u^{(n)_{i,j-1}} + u^{(n)_{i,j+1}}]$$

53.5 GAUSS-SEIDEL METHOD

This method utilises the latest iterative value available and scans the mesh points symmetrically from left to right along successive rows. The formula is given below.

$$u^{(n+1)}_{ij} = \frac{1}{4}[u^{(n+1)}_{i-1,j} + u^{(n)}_{i+1,j} + u^{(n+1)}_{i,j-1} + u^{(n)}_{i,j+1}]$$

53.6 SUCCESSIVE OVER-RELAXATION OR S.O.R. METHOD

Gauss-Seidel formula can be written as

$$u^{(n+1)}_{ij} = u^{(n)}_{ij} + \frac{1}{4}[u^{(n+1)}_{i-1,j} + u^{(n)}_{i+1,j} + u^{(n+1)}_{i,j-1} + u^{(n)}_{i,j+1} - 4u^{(n)}_{ij}] = u^{(n)}_{ij} + \frac{1}{4}R_{i,j}$$

It gives the change $\frac{1}{4}R_{i,j}$ in the value of $u_{i,j}$ for one Gauss-Seidel iteration. In the *S.O.R.* method, larger change than this is given to $u^{(n)}_{ij}$ and the iteration formula is given below:

$$u^{(n+1)}_{ij} = u^{(n)}_{i,j} + \frac{1}{4}wR_{i,j} = \frac{1}{4}w[u^{(n+1)}_{i-1,j} + u^{(n)}_{i+1,j} + u^{(n+1)}_{i,j-1} + u^{(n)}_{i,j+1}] + (1-w)u^{(n)}_{i,j}$$

Here w is called the accelerating factor and lies between 1 and 2.

Example 1. *Solve* $\frac{\partial^2 T}{\partial x^2} + \frac{\partial^2 T}{\partial y^2} = 0$ *in the domain of the figure given below by Gauss-Seidel method.*

Solution.

$T_1 = \frac{1}{4}[0 + T_2 + T_4 + 0]$ $\quad T_2 = \frac{1}{4}[0 + T_3 + T_5 + T_1]$ $\quad T_3 = \frac{1}{4}[0 + 0 + T_2 + T_6]$

$T_4 = \frac{1}{4}[T_1 + T_5 + T_7 + 0]$ $\quad T_5 = \frac{1}{4}[T_2 + T_6 + T_8 + T_4]$ $\quad T_6 = \frac{1}{4}[T_3 + 0 + T_9 + T_5]$,

$T_7 = \frac{1}{4}[T_4 + T_8 + 1 + 0]$ $\quad T_8 = \frac{1}{4}[T_5 + T_9 + 1 + T_7]$ $\quad T_9 = \frac{1}{4}[T_6 + 0 + 1 + T_8]$

Gauss-Seidel Method

Initial approximations are

$$T_1 = T_2 = T_3 = T_4 = T_5 = T_6 = T_7 = T_8 = T_9 = 0$$

Ten successive iterates are given below:

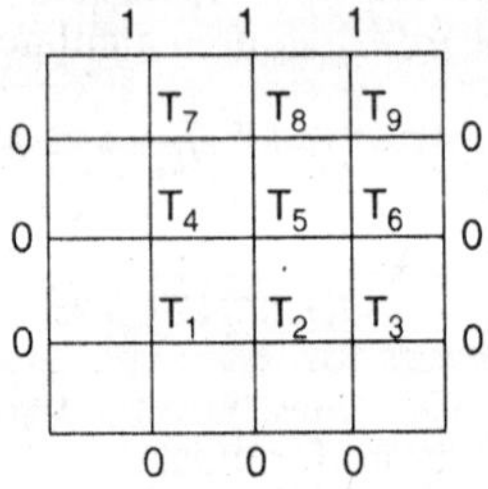

		1	1	1	
0		T_7	T_8	T_9	0
0		T_4	T_5	T_6	0
0		T_1	T_2	T_3	0
		0	0	0	

First Iteration

$T_1^{(n+1)} = \frac{1}{4}[0 + T_4^{(n)} + T_2^{(n)} + 0]$, $\qquad T_1^{(1)} = \frac{1}{4}[0+0+0+0] = 0$

$T_2^{(n+1)} = \frac{1}{4}[T_1^{(n+1)} + T_5^{(n)} + T_3^{(n)} + 0]$, $\qquad T_2^{(1)} = \frac{1}{4}[0+0+0+0] = 0$

$T_3^{(n+1)} = \frac{1}{4}[T_2^{(n+1)} + T_6^{(n)} + 0 + 0]$, $\qquad T_3^{(1)} = \frac{1}{4}[0+0+0+0] = 0$

$T_4^{(n+1)} = \frac{1}{4}[0 + T_7^{(n)} + T_5^{(n)} + T_1^{(n+1)}]$, $\qquad T_4^{(1)} = \frac{1}{4}[0+0+0+0] = 0$

$T_5^{(n+1)} = \frac{1}{4}[T_4^{(n+1)} + T_8^{(n)} + T_6^{(n)} + T_2^{(n+1)}]$, $\qquad T_5^{(1)} = \frac{1}{4}[0+0+0+0] = 0$

$T_6^{(n+1)} = \frac{1}{4}[T_5^{(n+1)} + T_9^{(n)} + 0 + T_3^{(n+1)}]$, $\qquad T_6^{(1)} = \frac{1}{4}[0+0+0+0] = 0$

$T_7^{(n+1)} = \frac{1}{4}[0 + 1 + T_8^{(n)} + T_4^{(n+1)}]$, $\qquad T_7^{(1)} = \frac{1}{4}[0+1+0+0] = 0.25$

$T_8^{(n+1)} = \frac{1}{4}[T_7^{(n+1)} + 1 + T_9^{(n)} + T_5^{(n+1)}]$, $\qquad T_8^{(1)} = \frac{1}{4}[0.25+1+0+0] = 0.312$

$T_9^{(n+1)} = \frac{1}{4}[T_8^{(n+1)} + 1 + 0 + T_6^{(n+1)}]$ $\qquad T_9^{(1)} = \frac{1}{4}[0.312 + 1 + 0 + 0] = 0.328$

Second Iteration

$T_1^{(2)} = \frac{1}{4}[0+0+0+0] = 0,$ $\qquad T_2^{(2)} = \frac{1}{4}[0+0+0+0] = 0$

$T_3^{(2)} = \frac{1}{4}[0+0+0+0] = 0,$ $\qquad T_4^{(2)} = \frac{1}{4}[0+0.25+0+0] = 0.062$

$T_5^{(2)} = \frac{1}{4}[0+0.312+0+0] = 0.078,$ $\qquad T_6^{(2)} = \frac{1}{4}[0+0.328+0+0] = 0.082$

$T_7^{(2)} = \frac{1}{4}[0+1+0.312+0] = 0.328,$ $\qquad T_8^{(2)} = \frac{1}{4}[0.25+1+0.328+0] = 0.394$

$T_9^{(2)} = \frac{1}{4}[0.312+1+0+0] = 0.328$

Third Iteration

$T_1^{(3)} = \frac{1}{4}[0+0.062+0+0] = 0.016,$ $\qquad T_2^{(3)} = \frac{1}{4}[0.016+0.078+0+0] = 0.024$

$T_3^{(3)} = \frac{1}{4}[0.024+0.082+0+0] = 0.027,$ $\qquad T_4^{(3)} = \frac{1}{4}[0+0.328+0.078+0.016] = 0.106$

$T_5^{(3)} = \frac{1}{4}[0.106+0.394+0.082+0.024] = 0.152,$

$T_6^{(3)} = \frac{1}{4}[0.152+0.328+0+0.027] = 0.127$

$T_7^{(3)} = \frac{1}{4}[0+1+0.394+0.106] = 0.375$ $\qquad T_8^{(3)} = \frac{1}{4}[0.375+1+0.382+0.152] = 0.464$

$T_9^{(3)} = \frac{1}{4}[0.464+1+0+0.127] = 0.398$

and so on. **Ans.**

Iteration	T_1	T_2	T_3	T_4	T_5	T_6	T_7	T_8	T_9
4	0.032	0.053	0.045	0.140	0.196	0.160	0.401	0.499	0.415
5	0.048	0.072	0.058	0.161	0.223	0.174	0.415	0.513	0.422
6	0.058	0.085	0.065	0.174	0.236	0.181	0.422	0.520	0.425
7	0.065	0.092	0.068	0.181	0.244	0.184	0.425	0.524	0.427
8	0.068	0.095	0.070	0.184	0.247	0.186	0.427	0.525	0.428
9	0.070	0.097	0.071	0.186	0.249	0.187	0.428	0.526	0.428
10	0.071	0.098	0.071	0.187	0.250	0.187	0.428	0.526	0.428

Example 2. *Solve* $\frac{\partial^2 u}{\partial x^2} + \frac{\partial^2 u}{\partial y^2} = 0$ *by Leibman's iteration process for the domain of the figure given below:*

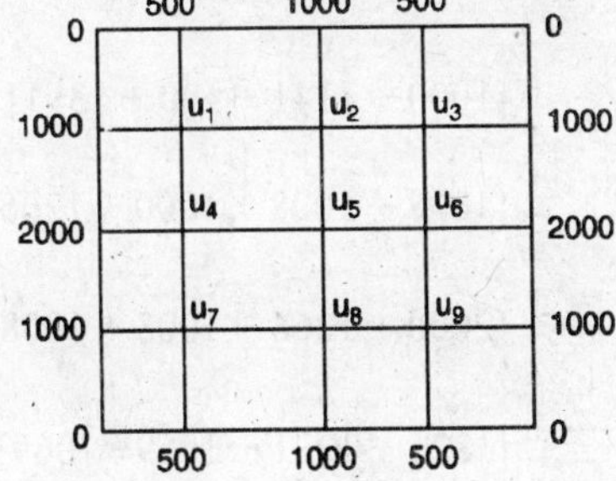

Solution. Values given on the figure are symmetrical about middle line.

$\therefore$ $u_1 = u_3 = u_9 = u_7$

$u_2 = u_8,\ u_4 = u_6$

$$u_5 = \frac{1}{4}(2000 + 2000 + 1000 + 1000) = 1500 \quad \text{(Standard formula)}$$

$$u_1 = \frac{1}{4}[0 + 1000 + 1500 + 2000] = 1125 \quad \text{(Diag. formula)}$$

Similarly $u_1 = u_3 = u_9 = u_7 = 1125$

$$u_2 = \frac{1}{4}(1000 + 1125 + 1500 + 1125) \approx 1188 \quad \text{(Standard formula)}$$

Similarly $u_8 = u_2 = 1188$

$$u_4 = \frac{1}{4}[1125 + 2000 + 1125 + 1500] \approx 1438 \quad \text{(Standard formula)}$$

Similarly $u_4 = u_6 = 1438$

So $u_1 = 1125,\ u_2 = 1188,\ u_3 = 1125,\ u_4 = 1438,\ u_5 = 1500,\ u_6 = 1438,\ u_7 = 1125,\ u_8 = 1188,$ $u_9 = 1125$

Gauss-Seidel Method:

$$u^{(n+1)_{i,j}} = \frac{1}{4}[u_{i-1,j}^{(n+1)} + u_{i+1,j}^{(n)} + u^{(n+1)_{i,j-1}} + u^{(n)_{i,j+1}}]$$

$$u_1^{(n+1)} = \frac{1}{4}[1000 + u_2^{(n)} + 500 + u_4^{(n)}] = u_3^{(n+1)} = u_5^{(n+1)} = u_7^{(n+1)}$$

$$u_2^{(n+1)} = \frac{1}{4}[u_1^{(n+1)} + u_3^{(n+1)} + 1000 + u_5^{(n+1)}] = u_8^{(n+1)}$$

$$u_4^{(n+1)} = \frac{1}{4}[2000 + u_5^{(n)} + u_1^{(n+1)} + u_7^{(n+1)}] = u_6^{(n+1)}$$

$$u_5^{(n+1)} = \frac{1}{4}[u_4^{(n+1)} + u_6^{(n+1)} + u_2^{(n+1)} + u_8^{(n+1)}]$$

First Iteration

$$u_1^{(1)} = \frac{1}{4}[1000 + 1188 + 500 + 1438] \approx 1032 = u_1^{(3)} = u_9^{(1)} = u_7^{(1)}$$

$$u_2^{(1)} = \frac{1}{4}[1032 + 1032 + 1000 + 1500] = 1141 = u_8^{(1)}$$

$$u_4^{(n+1)} = \frac{1}{4}[2000 + 1500 + 1032 + 1032] = 1391 = u_6^{(1)}$$

$$u_5^{(n+1)} = \frac{1}{4}[1091 + 1391 + 1141 + 1141] = 1266$$

Second Iteration

$$u_1^{(2)} = \frac{1}{4}[1000 + 1141 + 500 + 1391] = 1008 = u_3^{(2)} = u_9^{(2)} = u_7^{(2)}$$

$$u_2^{(2)} = \frac{1}{4}[1008 + 1008 + 1000 + 1266] = 1069 = u_8^{(2)}$$

$$u_4^{(2)} = \frac{1}{4}[2000 + 1266 + 1008 + 1008] = 1321 = u_6^{(2)}$$

$$u_5^{(2)} = \frac{1}{4}[1321 + 1321 + 1069 + 1069] = 1195$$

Similarly

Iteration	$u_1 = u_3 = u_9 = u_7$	$u_2 = u_8$	$u_4 = u_6$	u_5
Third	973	1035	1288	1162
Fourth	956	1019	1269	1144
Fifth	947	1010	1260	1135
Sixth	942	1005	1255	1130
Seventh	940	1003	1253	1128
Eighth	939	1002	1252	1127
Ninth	939	1001	1251	1126

Very small difference is in the eighth and ninth iteration

Thus,

$$u_1 = u_3 = u_7 = u_9 = 939$$

$$u_2 = u_8 = 1001,$$

$$u_4 = u_6 = 1251,$$

$$u_5 = 1126$$

Ans.

Example 3. *Solve* $\frac{\partial^2 u}{\partial x^2} + \frac{\partial^2 u}{\partial y^2} = 0$ *in the domain of the figure given below by*

(a) Jacobi's method, (b) Gauss-Seidel method and

(c) Successive Over-Relaxation method

Solution. **(*a*) Jacobi's Method**

$$u_1^{(1)} = \frac{1}{4}[0+0+0+1] = 0.25$$

$$u_2^{(2)} = \frac{1}{4}[0+0+0+1] = 0.25$$

u_1	u_2	u_3	u_4
0.25	0.3125	0.5625	0.46875
0.21875	0.17187	0.42187	0.39844
0.14844	0.13672	0.38672	0.38086
0.13086	0.12793	0.37793	0.37646
0.12646	0.12573	0.37573	0.37537

$$u_3^{(1)} = \frac{1}{4}[1+1+0+0] = 0.5$$

$$u_4^{(1)} = \frac{1}{4}[1+1+0+0] = 0.5$$

Seven successive iterates are given below:

u_1	u_2	u_3	u_4
0.1875	0.1875	0.4375	0.4375
0.15625	0.15625	0.40625	0.40625
0.14062	0.14062	0.39062	0.39062
0.13281	0.13281	0.38281	0.38281
0.12891	0.12891	0.37891	0.37891
0.12695	0.12695	0.37695	0.37695
0.12598	0.12598	0.37598	0.37518

(b) Gauss-Seidel Method

Five successive iterates are given below:

(c) Successive Over-Relaxation method Three successive iterates are given below:

u_1	u_2	u_3	u_4
0.275	0.35062	0.35062	0.35062
0.16534	0.10683	0.38183	0.37432
0.11785	0.12181	0.37216	0.37341

53.7 POISSON EQUATION

$$\nabla^2 u = \frac{\partial^2 u}{\partial x^2} + \frac{\partial^2 u}{\partial y^2} = f(x, y)$$

In this case the standard five-point formula is of the form

$$u_{i-1,j} + u_{i+1,j} + u_{i,j-1} + u_{i,j+1} - 4u_{i,j} = h^2 f(ih, jh)$$

On applying the above formula we get equations. These equations can be solved by Gauss-Seidel iteration method.

Example 4. *Solve the Poisson's equation* $\frac{\partial^2 u}{\partial x^2} + \frac{\partial^2 u}{\partial y^2} = 8x^2 y^2$ *for the square mesh of the figure given below with u (x, y) = 0 on the boundary and mesh length = 1.*

Solution. Here $h = 1$

The standard five-point formula for the given equation is

$$u_{i-1,j} + u_{i+1,j} + u_{i,j+1} + u_{i,j-1} - 4u_{i,j} = 8i^2 j^2 \quad \text{...(1)}$$

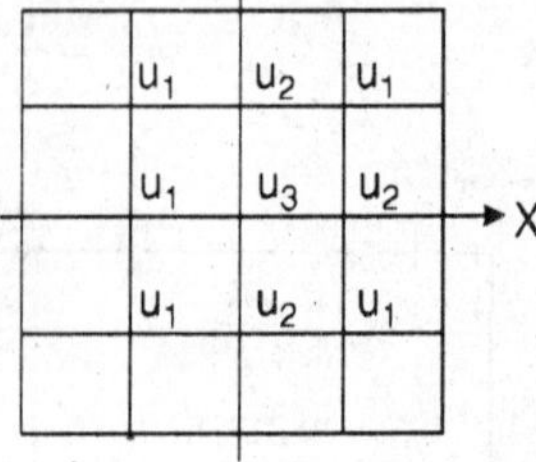

For u_1 ($i = -1, j = +1$), equation (1) becomes

$$0 + u_2 + 0 + u_2 - 4u_1 = 8(-1)^2 (1)^2 \text{ or } 4u_1 = 2u_2 - 8 \quad \text{...(2)}$$

For u_2 ($i = 0, j = 1$), equation (1) becomes

$$u_1 + u_1 + 0 + u_3 - 4u_2 = 0 \text{ or } 4u_2 = 2u_1 + u_3 \quad \text{...(3)}$$

For u_3 ($i = 0, j = 0$), equation (1) becomes

$$u_2 + u_2 + u_2 + u_2 - 4u_3 = 0 \text{ or } 4u_3 = 4u_2 \text{ or } u_3 = u_2 \quad \text{...(4)}$$

Putting u_2 for u_3 in (3), we get $4u_2 = 2u_1 + u_2$ or $3u_2 = 2u_1$

Putting $\frac{2u_1}{3}$ for u_2 in (2), we get

$$4u_1 = \frac{4u_1}{3} - 8 \quad \text{or} \quad 12u_1 = 4u_1 - 24$$

$$8u_1 = -24 \quad \text{or} \quad u_1 = -3$$

$$u_2 = \frac{2u_1}{3} = \left(\frac{2}{3} \times -3\right) = -2, \qquad u_3 = u_2 = -2$$

$$u_1 = -3, \qquad u_2 = -2, \; u_3 = -2$$

Ans.

EXERCISE 53.1

1. Given the values of u (x, y) on the boundary of the square in the figure given below, evaluate the function u (x, y) satisfying the Laplace equation

$$\frac{\partial^2 u}{\partial x^2} + \frac{\partial^2 u}{\partial y^2} = 0$$

at pivotal points of this figure.

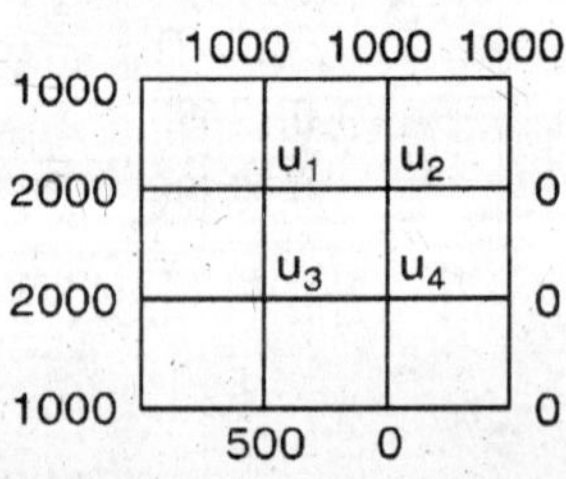

Ans. $u_1 = 1208$, $u_2 = 792$, $u_3 = 1042$, $u_4 = 458$, $u_5 = 0.625$, $u_6 = 1.25$

2. *Solve* $\dfrac{\partial^2 u}{\partial x^2}+\dfrac{\partial^2 u}{\partial y^2}=0$

for the square mesh with boundary values as shown in the figure given below. Iterate until the maximum difference between successive values at any point is less than 0.005.

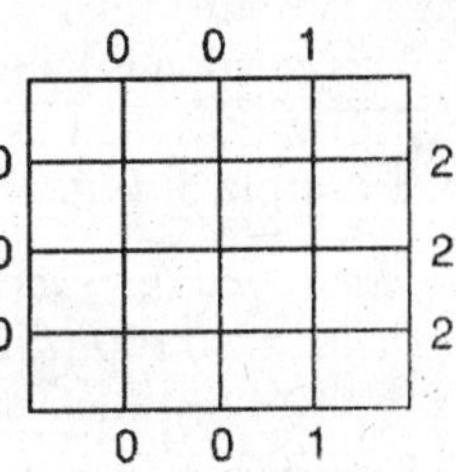

Ans. $u_1 = 10.188,\ u_2 = 0.5,\ u_3 = 1.188,\ u_4 = 0.25,$
$u_5 = 0.625,\ u_6 = 1.25.$

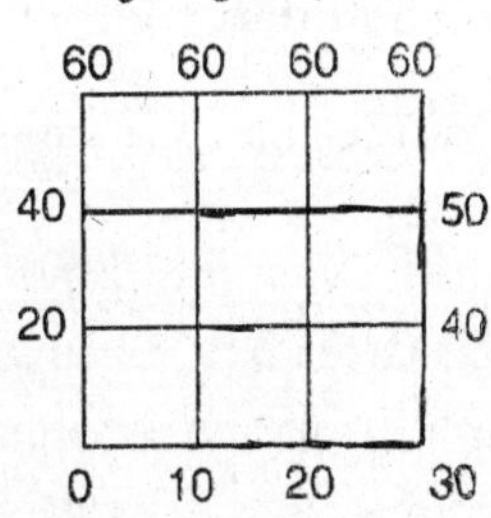

3. Solve $\dfrac{\partial^2 u}{\partial x^2}+\dfrac{\partial^2 u}{\partial t^2}=0$ within the square given in the figure below.

Ans. $u_1 = 26.66,\ u_2 = 33.33,\ u_3 = 43.33,\ u_4 = 46.66.$

4. Solve $\dfrac{\partial^2 u}{\partial x^2}+\dfrac{\partial^2 u}{\partial y^2}=-10(x^2+y^2+10)$

over the square with $x = 0 = y,\ x = 3 = y$ with

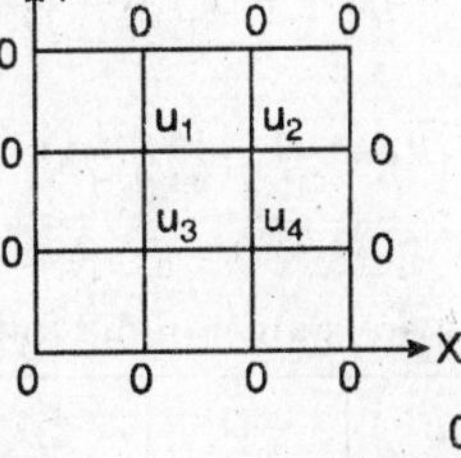

$u = 0$ on the boundary and mess length = 1. **Ans.**
$u_1 = 75,\ u_2 = 82.5,\ u_3 = 67.05,\ u_4 = 75.$

5. Solve $\dfrac{\partial^2 u}{\partial x^2}+\dfrac{\partial^2 u}{\partial y^2}=0$ with boundary values as shown in the figure given below.

0 0 1
0 2
0 2
0 2
0 0 1

Ans. $u_1 = 10.188,\ u_2 = 0.5,\ u_3 = 1.188,$
$u_4 = 0.25,\ u_s = 0.625,\ u_6 = 1.25$

53.8 HEAT EQUATION (PARABOLIC EQUATION)

$$\frac{\partial u}{\partial t}=c^2\frac{\partial^2 u}{\partial x^2}$$

We know that

$$\frac{\partial u}{\partial t}=\frac{u(x,t+k)-u(x,t)}{k}$$

$$\frac{\partial^2 u}{\partial x^2}=\frac{u(x+h,t)-2u(x-h,t)}{h^2}$$

On putting the values of $\dfrac{\partial u}{\partial t}$ and $\dfrac{\partial^2 u}{\partial x^2}$ in (1), we get

$$\frac{u(x,t+k)-u(x,t)}{k}=c^2\,\frac{u(x+h,t)-2u(x,t)+u(x-h,t)}{h^2}$$

$$u(x,\ t+k)=\frac{c^2k}{h^2}u(x+h,t)-\frac{2c^2k}{h^2}u(x,t)+u(x,t)+\frac{c^2k}{h^2}u(x-h,t)$$

or

$$u(x,\ t+k)=au(x+h,\ t)+(1-2a)u(x,\ t)+au(x-h,t)$$

If $a=\dfrac{1}{2}$

$$u(x,\ t+k)=\frac{1}{2}u(x+h,\ t)+\frac{1}{2}u(x-h,t)\ \text{ or }\ u_{i,\ j+1}=\frac{1}{2}[u_{i+1,j}+u_{i-1,\ j}]$$

It means that the value of u at x_i, at time t is the mean of the values of u at x_{i-1} and x_{i+1} at the previous time t_j.

This relation is known as Bendre-Schimidt recurrence relation.

Example 5. *Find the values of u (x, t) satisfying the parabolic equation* $\frac{\partial u}{\partial t} = 4\frac{\partial^2 u}{\partial x^2}$ *with boundary conditions u (0, t) = 0 = u (8, t) and u (x, 0) =* $4x - \frac{1}{2}x^2$ *at the points.*

$x = i : i = 0, 1, 2, 3, \ldots, 7$ *and* $t = \frac{1}{8} j : j = 0, 1, 2, 3, \ldots, 5.$

Solution. $c^2 = 4,\ h = 1,\ k = -\frac{1}{8} \qquad a = \frac{c^2 k}{h^2} = \frac{4 \times 1/8}{(1)} = \frac{1}{2}$

Then the equation is

$$u_{i,j+1} = \frac{1}{2}(u_{i-1,j} + u_{i+1,j})$$

Given $u(0, t) = 0 = u(8, t)$

or $u(0, j) = 0 = u(8, j)$ for all values of $j = 1\ 2, 3, 4, 5$

and $u(x, 0) = 4x - \frac{1}{2}x^2 \qquad u_{i,0} = 4i - \frac{1}{2}i^2$

$u_{0,0} = 0,\ u_{1,0} = 4(1) - \frac{1}{2}(1)^2 = 3.5, \qquad u_{2,0} = 4(2) - \frac{1}{2}(2)^2 = 6$

$u_{3,0} = 7.5, \qquad u_{4,0} = 8, \qquad u_{5,0} = 7.5,\ u_{6,0} = 6,\ u_{7,0} = 3.5.$

These entries are shown in the following table :

j \ i	0	1	2	3	4	5	6	7	8
0	0	3.5	6	7.5	8	7.5	6	3.5	0
1	0	3	5.5	7	7.5	7	5.5	3	0
2	0	2.75	5	6.5	7	6.5	5	2.75	0
3	0	2.5	4.625	6	6.5	6	4.625	2.5	0
4	0	2.3125	4.25	5.5625	6	5.5625	4.25	2.3125	0
5	0	2.125	3.9375	5.125	5.5625	5.125	3.9375	2.125	0

Putting $j = 0$ in (1) we have

$u_{i,1} = \frac{1}{2}(u_{i-1,0} + u_{i+1,0}) \qquad u_{1,1} = \frac{1}{2}(u_{0,0} + u_{2,0}) = \frac{1}{2}(0 + 6) = 3$

$u_{2,1} = \frac{1}{2}(u_{1,0} + u_{3,0}) = \frac{1}{2}(3.5 + 7.5) = 5.5 \qquad u_{3,1} = \frac{1}{2}(u_{2,0} + u_{4,0}) = \frac{1}{2}(6 + 8) = 7$

$u_{4,1} = \frac{1}{2}(u_{3,0} + u_{5,0}) = \frac{1}{2}(7.5 + 7.5) = 7.5 \qquad u_{5,1} = \frac{1}{2}(u_{4,0} + u_{6,0}) = \frac{1}{2}(8 + 6) = 7$

$u_{6,1} = \frac{1}{2}(u_{5,0} + u_{7,0}) = \frac{1}{2}(7.5 + 3.5) = 5.5 \qquad u_{7,1} = \frac{1}{2}(u_{6,0} + u_{8,0}) = \frac{1}{2}(6 + 0) = 3$

Putting $j = 1$ in (1), we get

$u_{i,2} = \frac{1}{2}(u_{i-1,1} + u_{i+1,1}) \qquad u_{1,2} = \frac{1}{2}(u_{0,1} + u_{2,0}) = \frac{1}{2}(0 + 5.5) = 2.75$

$u_{2,2} = \frac{1}{2}(u_{1,1} + u_{3,1}) = \frac{1}{2}(3 + 7) = 5 \qquad u_{3,2} = 6.5,\ u_{4,2} = 7,\ u_{5,2} = 6.5,\ u_{6,2} = 5,\ u_{7,2} = 2.75$

Putting $j = 2$ in (1), we get

$u_{i,3} = \frac{1}{2}(u_{i-1,2} + u_{i+1,2}) \qquad u_{1,3} = \frac{1}{2}(u_{0,2} + u_{2,2}) = \frac{1}{2}(0 + 5) = 2.5$

$u_{2,3}=\frac{1}{2}(u_{1,2}+u_{3,2})=\frac{1}{2}(2.75+6.5)=4.625$ $u_{3,3}=6, u_{4,3}=6.5, u_{5,3}=6, u_{6,3}=4.625, u_{7,3}=2.5$

Putting $j = 3$ in (1), we get

$$u_{i,4}=\frac{1}{2}(u_{i-1,3}+u_{i+1,3}) \qquad u_{1,4}=\frac{1}{2}(u_{0,3}+u_{2,3})=\frac{1}{2}(0+4\,625)=2.3125$$

$$u_{2,4}=\frac{1}{2}(u_{1,3}+u_{3,2})=\frac{1}{2}(2.5+6)=4.25$$

$u_{3,4}=0.5625, u_{4,4}=6, u_{5,4}=5.5625, u_{6,4}=4.25, u_{7,4}=2.3125$

Putting $j = 4$ in (1), we get

$$u_{i,5}=\frac{1}{2}(u_{i-1,4}+u_{i+1,4}) \qquad u_{1,5}=\frac{1}{2}(u_{0,4}+u_{2,4})=\frac{1}{2}(0+4.25)=2.125$$

$$u_{2,5}=\frac{1}{2}(u_{1,4}+u_{3,4})=\frac{1}{2}(2.125+5.5625)=3.9375$$

$u_{3,5}=5.125, u_{4,5}=5.625, u_{5,5}=5.125, u_{6,5}=3.9375, u_{7,5}=2.125$

53.9 WAVE EQUATION (HYPERBOLIC EQUATION)

$$\frac{\partial^2 u}{\partial t^2}=c^2\frac{\partial^2 u}{\partial x^2} \qquad \text{... (1)}$$

We know that

$$\frac{\partial^2 u}{\partial t^2}=\frac{u(x,t+k)-2u(x,t)+u(x,t-k)}{k^2} \quad \text{and} \quad \frac{\partial^2 u}{\partial x^2}=\frac{u(x+h,t)-2u(x,t)+u(x-h,t)}{h^2}$$

Putting the values of $\frac{\partial^2 u}{\partial t^2}$ and $\frac{\partial^2 u}{\partial x^2}$ in (1) we have

$$\frac{u(x,t+k)-2u(x,t)+u(x,t-k)}{k^2}=c^2\frac{u(x+h,t)-2u(x,t)+u(x-h,t)}{h^2}$$

or $u(x,t+k)-2u(x,t)+u(x,t-k)=\frac{c^2k^2}{h^2}[u(x+h,t)-2u(x,t)+u(x-h,t)]$

or $u_{i,j+1}-2u_{i,j}+u_{i,j-1}=a^2c^2[u_{i+1,j}-2u_{i,j}+u_{i-1,j}]$ $\left(a=\frac{k}{h}\right)$

or $u_{i,j+1}=2(1-a^2c^2)u_{i,j}+a^2c^2(u_{i-1,j}+u_{i+1,j})-u_{i,j-1}$ ...(2)

If $a^2c^2=1$, Equation (2) reduces to

$$u_{i,j+1}=u_{i-1,j}+u_{i+1,j}-u_{i,j-1} \qquad \text{...(3)}$$

Example 6. *Solve* $\frac{\partial^2 u}{\partial t^2}=\frac{\partial^2 u}{\partial x^2}$ *with conditions* $u(0,t)=u(1,t)=0$, $u(x,0)=1/2\,x(1-x)$ *and* $u(x,0)=0$, *taking* $h=k=0.1$ *for* $0\le t\le 0.4$. *Compare your solution with the exact solution* $x=0.5$ *and* $t=0.3$.

Solution. $c^2=1$. The difference equation for the given equation is

$$u_{i,j+1}=2(1-\alpha^2)u_{i,j}+\alpha^2(u_{i-1,j}+u_{i+1,j})-u_{i,j-1} \qquad \text{...(1)}$$

where $\alpha=\frac{k}{h}$ But $\alpha=\frac{0.1}{0.1}=1$

Equation (1) reduces to $u_{i,j+1}=u_{i-1,j}+u_{i+1,j}-u_{i,j-1}$ ...(2)

$$u(0,t) = u(1,t) = 0, u_{0j} = 0 \text{ and } u_{10j} = 0$$

i.e., the entries in the first column are zero.

since $u(x, 0) = \frac{1}{2}x(1-x)$ $\quad u(i, 0) = \frac{1}{2} i(1-i)$

= 0.045, 0.08, 0.105, 0.120, 0.125, 0.120, 0.105 for

i = 0.1, 0.2, 0.3, 0.4, 0.5, 0.6, 0.7 at $t = 0$

These are the entries of the fist row. Finally since $u_x(x, 0) = 0$

$$\therefore \quad \frac{u_{i,j+1} - u_{i,j}}{k} = 0 \text{ for } j = 0 \ (t = 0), \quad u_{i,1} = u_{i,9}$$

Putting $j = 0$ in equation (2), we get

$$u_{i,1} = u_{i-1,0} + u_{i+1,0} - u_{i,-1} = u_{i-1,0} + u_{i+1,0} - u_{i,1} \quad (u_{i,1} = u_{i,-1})$$

$$2u_{i,1} = u_{i-1,0} + u_{i+1,0}, \quad u_{i,1} = \frac{1}{2}[u_{i-1,0} + u_{i+1,0}]$$

For $i = 1$ $\quad u_{1,1} = \frac{1}{2}[u_{0,0} + u_{2,0}] = \frac{1}{2}[0 + .080] = 0.040$

For $i = 2$ $\quad u_{2,1} = \frac{1}{2}[u_{1,0} + u_{3,0}] = \frac{1}{2}[0.045 + 0.105] = 0.075$

For $i = 3$ $\quad u_{3,1} = [u_{2,0} + u_{4,0}] = \frac{1}{2}[0.08 + 0.120] = 0.100$

For $i = 4$ $\quad u_{4,1} = \frac{1}{2}[u_{3,0} + u_{5,0}] = \frac{1}{2}[0.105 \ 0.125] = 0.115$

For $i = 5$ $\quad u_{5,1} = \frac{1}{2}[u_{4,0} + u_{6,0}] = \frac{1}{2}[0.120 + 0.120] = 0.120$

For $i = 6$ $\quad u_{6,1} = \frac{1}{2}[u_{5,0} + u_{7,0}] = \frac{1}{2}[0.125 + 0.105] = 0.115$

Putting $j = 1$ in equation (2), we get

$$u_{i,2} = u_{i-1,1} + u_{i+1,} - u_{i,0}$$

For $i = 1$ $\quad u_{1,2} = u_{0,1} + u_{2,1} - u_{1,0} = 0 + 0.075 - 0.045 = 0.03$

For $i = 2$ $\quad u_{2,2} = u_{1,1} + u_{3,1} - u_{2,0} = 0.040 + 0.100 - 0.08 = 0.060$

For $i = 3$ $\quad u_{3,2} = u_{2,1} + u_{4,1} - u_{3,0} = 0.075 + 0.115 - 0.105 = 0.085$

For $i = 4$ $\quad u_{4,2} = u_{3,1} + u_{5,1} - u_{4,0} = 0.100 + 0.120 - 0.120 = 0.100$

For $i = 5$ $\quad u_{5,2} = u_{4,1} + u_{6,1} - u_{5,0} = 0.115 + 0.115 - 0.125 = 0.105$

Similarly for $j = 2$

$$u_{i,3} = u_{i-1,2} + u_{i+1,2} - u_{i,1}$$

$$u_{1,3} = 0.020, \ u_{2,3} = 0.040, u_{3,3} = 0.060, u_{4,3} = 0.075, u_{5,3} = 0.80$$

For $j = 3$ $\quad u_{i,4} = u_{i-1,3} + u_{i+1,3} - u_{i,2}$

$$u_{1,4} = 0.010, \quad u_{2,4} = 0.02,$$

$$u_{3,4} = 0.030, \quad u_{4,4} = 0.040, \ u_{5,4} = 0.048$$

		0	*0.1*	*0.2*	*0.3*	*0.4*	*0.5*	*0.6*
	i / *j*	*0*	*1*	*2*	*3*	*4*	*5*	*6*
0	*0*	0	0.045	0.080	0.105	0.120	0.125	0.120
0.1	1	0	0.040	0.075	0.100	0.115	0.120	0.115
0.2	2	0	0.030	0.060	0.085	0.100	0.105	
0.3	3	0	0.020	0.040	0.060	0.075	0.080	
0.4	4	0	0.010	0.020	0.030	0.040	0.048	

The analytical (exact) solution of the given equation is

$$u = \frac{2}{\pi^3}\sum_{n=1}^{\infty}\frac{1}{n^3}(1-\cos n\pi)\sin n\pi x\cos n\pi t$$

Comparison of two solutions is given below:

$t = 0.3$	$x=$	0.1	0.2	0.3	0.4	0.5
Numercial solution	$u =$	0.02	0.04	0.06	0.075	0.08
Exact solution	$u=$	0.02	0.04	0.06	0.075	0.08

Ans.

EXERCISE 53.2

1. Solve $\dfrac{\partial^2 u}{\partial x^2} = 2\dfrac{\partial u}{\partial t}$

under conditions $u(0, t) = u(4,t)=0$ and $u(x,0)= x(4-x)$,taking $h=1$,find the values upto $t = 5$.

Ans. $u_{1,0} = 3, u_{2,0} = 4, u_{3,0} = 3;\ u_{1,1} = 2, u_{2,1} = 3, u_{3,1} = 2$

$u_{1,2} = 1.5, u_{2,2} = 2, u_{3,2} = 1.5;\ u_{1,3} = 1, u_{2,3} = 1.5, u_{3,3} = 1$

$u_{1,4} = 0.75, u_{2,4} = 1, u_{3,4} = 0.75;\ u_{1,5} = 0.5, u_{2,5} = 0.75, u_{3,5} = 0.50$

2. Solve $\dfrac{\partial u}{\partial t} = \dfrac{\partial^2 u}{\partial x^2};\ 0 \le x \le 1, t \ge 0$ under the conditions that

$u(0, t) = u(1, t) = 0$ and $u(x, 0) = 2x$ for $0 \le x \le \dfrac{1}{2} = (1-x)$ for $\dfrac{1}{2} \le x \le 1$

Ans. $u_1 = 0.1989, u_2 = 0.3956, u_3 = 0.5834, u_4 = 0.7381, u_6 = 0.7591$

3. Solve $\dfrac{\partial u}{\partial t} = \dfrac{\partial^2 u}{\partial x^2};\ 0 \le x \le 1,\ t \ge 0$ under the conditions that

$$\left.\begin{aligned} u &= 0, \text{at } x = 0 \\ u &= 0, \text{at } x = 1 \end{aligned}\right] t \ge 0$$

$$u = \sin \pi x \text{ at } t = 0,\ 0 \le x \le 1$$

find u for $x = 0.8$ at $t = 1$. **Ans.** 0.4853

4. Solve $\dfrac{\partial^2 u}{\partial x^2} = \dfrac{\partial u}{\partial t}$ under the conditions that $u(0, t) = u(5, t) = 0$, $u(x, 0) = x^2(25 - x^2)$ *at* taking $h = 1$ and $k = \dfrac{1}{2}$

Ans. $u_{1,0} = 24, u_{2,0} = 84, u_{3,0} = 144, u_{4,0} = 144$

$u_{1,1} = 42, u_{2,1} = 78, u_{3,1} = 78, u_{4,1} = 57$

$u_{1,2} = 39, u_{2,2} = 60, u_{3,2} = 67.5, u_{4,2} = 39$

$u_{1,3} = 30, u_{2,3} = 53.25, u_{3,3} = 49.5, u_{4,3} = 33.75$

$u_{1,4} = 26.625, u_{2,4} = 39.75, u_{3,4} = 43.5, u_{4,4} = 24.75$

$u_{1,5} = 19.875, u_{2,5} = 35.06, u_{3,5} = 32.25, u_{4,5} = 21.75$

5. Solve $16\dfrac{\partial^2 u}{\partial x^2} = \dfrac{\partial^2 u}{\partial t^2}$ taking $h = 1$, upto $t = 1.25$, under the conditions

$u(0, t) = u(5, t) = 0$, $u_t(x, 0) = 0$ and $u(x, 0) = x^2(5 - x)$.

Ans. $u_{1,0} = 4,\ u_{2,0} = 12,\ u_{3,0} = 18,\ u_{4,0} = 16$

$u_{1,1} = 4,\ u_{2,1} = 12,\ u_{3,1} = 18,\ u_{4,1} = 16$

$u_{1,2} = 8,\ u_{2,2} = 10,\ u_{3,2} = 10,\ u_{4,2} = 2$

$u_{1,3} = 6,\ u_{2,3} = 6,\ u_{3,3} = -6,\ u_{4,3} = -6$

$u_{1,4} = -2,\ u_{2,4} = -10,\ u_{3,4} = -10,\ u_{4,4} = -8$

CHAPTER

54

Calculus Of Variation

54.1 INTRODUCTION

The calculus of variations primarily deals with finding maximum or minimum value of a definite integral involving a certain function.

54.2 FUNCTIONALS

A simple example of functional is the shortest length of a curve through two points $A(x_1, y_1)$ and $B(x_2, y_2)$. In other words, the determination of the curve $y = y(x)$ for which $y = (x_1) = y_1,\ y(x_2) = y_2$ such that

$$\int_{x_1}^{x_2} \sqrt{1+\left(\frac{dy}{dx}\right)} dx \qquad \ldots(1)$$

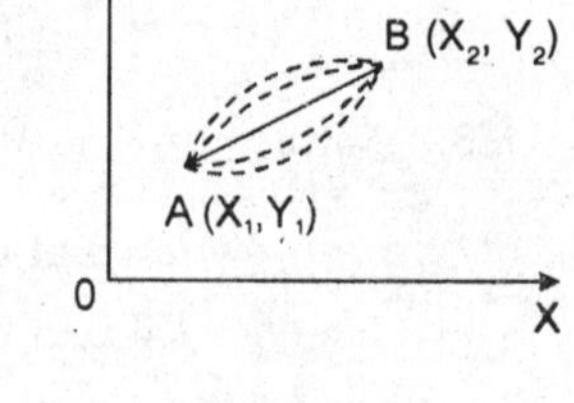

is a minimum.

An integral such as (1) is called a *Functional.*

In general, it is required to find the curve $y = y(x)$ where $y(x_1) = y_1$ and $y(x_2) = y_2$ such that for a given function $f\left(x, y, \frac{dy}{dx}\right)$,

$$\int_{x_1}^{x_2} f\left(x, y, \frac{dy}{dx}\right) dx \qquad \ldots(2)$$

is maximum or minimum.

Integral (2) is known as the functional.

In differential calculus, we find the maximum or minimum value of functions. But the calculus of variations deals with the problems of maxima or minima of functionals.

A functional $I[y(x)]$ is said to be linear if it satisfies.

(*i*) $I[cy(x)] = c\, I[y(x)]$, where c is an arbitrary constant.

(*ii*) $I[y_1(x) + y_2(x)] = I[y_1(x)] + I[y_2(x)]$, where $y_1(x) \in M$ and $y_2(x) \in M$.

54.3 DEFINITION

A functional $I[y(x)]$ is maximum on a curve $y = y(x)$, if the values of $I[y(x)]$ on any curve close to $y = y_1(x)$ do not exceed $I[y_1(x)]$. It means $\Delta I = I[y(x)] - I[y_1(x)] \le 0$ and $\Delta I = 0$ *on* $y = y_1(x)$.

In case of minimum of $I[y(x)]$, $\Delta I = 0$.

Extremal: A function $y = y(x)$ which extremizes a functional is called extremal or extremizing function.

54.4 EULER'S EQUATION IS

$$\frac{\partial f}{\partial y} - \frac{d}{dx}\left(\frac{\partial f}{\partial y}\right) = 0$$

This is the necessary condition for $I=\int_{x_1}^{x_2} f(x,y,y')\,dx$ to be maximum or minimum.

Proof: Let $y = y(x)$ be the curve AB which makes the given function I an extremum.
Consider a family of neighbouring curves

$$Y = y(x) + \alpha\eta(x) \qquad \ldots (1)$$

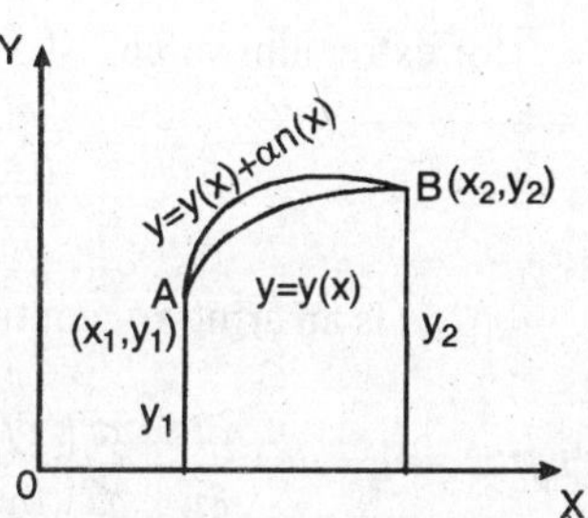

where a is a parameter, $-r\alpha r$ and $\eta(x)$ is an arbitrary differentiable function.

At the end points A and B,

$$\eta(x_1) = \eta(x_2) = 0$$

when $\alpha = 0$, neighbouring curves become $y = y(x)$, which is extremal.

The family of neighbouring curves is called the family of *comparison functions.*

If in the functional $\int_{x_1}^{x_2} f(x,y,y')\,dx$ We replace y by Y, we get

$$\int_{x_1}^{x^2} f(x,Y,Y')\,dx = \int_{x_1}^{x_2} f[x, y(x)+\alpha\eta(x), y'(x)+\alpha\eta'(x)]\,dx.$$

which is a function of α, say $I(\alpha)$.

$$\therefore \qquad I(\alpha) = \int_{x_1}^{x_2} f(x,Y,Y')\,dx$$

For $\alpha = 0$, the neighbouring curves become the extremal, an extremum for $\alpha = 0$.
The necessary condition for this is $I'(\alpha) = 0$ ...(2)
Differentiating I under the integral sign by Leibnitz's rule, we have

$$I'(\alpha) = \int_{x_1}^{x_2} \left(\frac{\partial f}{\partial x}\frac{\partial x}{\partial \alpha} + \frac{\partial f}{\partial Y}\frac{\partial Y}{\partial \alpha} + \frac{\partial f}{\partial Y'}\frac{\partial Y'}{\partial \alpha} \right) dx$$

$$I'(\alpha) = \int_{x_1}^{x_2} \left(\frac{\partial f}{\partial Y}\frac{\partial Y}{\partial \alpha} + \frac{\partial f}{\partial Y'}\frac{\partial Y'}{\partial \alpha} \right) dx \qquad \left(\frac{\partial x}{\partial \alpha} = 0 \text{ as } \alpha \text{ is independent of } x \right) \qquad \ldots (3)$$

On differentiating (1), *w. r. t.* 'x', we get, $Y' = y'(x) + \alpha\eta'(x)$

Again differentiating *w.r. t.* 'α', we get $\dfrac{\partial Y'}{\partial \alpha} = \eta'(x)$

Differentiating (1), *w. r. t.*, we get $\qquad \alpha\dfrac{\partial Y}{\partial \alpha} = \eta(x)$

Now (3) becomes $I'(\alpha) = \displaystyle\int_{x_1}^{x_2} \left[\frac{\partial f}{\partial Y}\eta(x) + \frac{\partial f}{\partial Y'}\eta'(x) \right] dx$

Integrating the second term on the right by parts, we get

$$= \int_{x_1}^{x_2} \frac{\partial f}{\partial Y}\eta(x)\,dx + \left[\left\{ \frac{\partial f}{\partial Y'}\eta(x) \right\}_{x_1}^{x_2} - \int_{x_1}^{x_2} \frac{d}{dx}\left(\frac{\partial f}{\partial Y'} \right)\eta(x)\,dx \right]$$

$$= \int_{x_1}^{x_2} \frac{\partial f}{\partial Y} \eta(x)\,dx + \left[\frac{\partial f}{\partial Y'}\eta(x_2) - \frac{\partial f}{\partial Y'}\eta(x_1)\right] - \int_{x_1}^{x_2} \frac{d}{dx}\left[\frac{\partial f}{\partial Y'}\right]\eta(x)\,dx$$

$$= \int_{x_1}^{x_2} \frac{\partial f}{\partial Y}\eta(x)\,dx + 0 - \int_{x_1}^{x_2} \frac{d}{dx}\left(\frac{\partial f}{\partial Y'}\right)\eta(x)\,dx = \int_{x_1}^{x_2}\left[\frac{\partial f}{\partial Y} - \frac{d}{dx}\left(\frac{\partial f}{\partial Y'}\right)\right]\eta(x)\,dx \left[\eta(x_1) = \eta(x_2) = 0\right]$$

for extremum value, $I'(\alpha) = 0$

$$0 = \int_{x_1}^{x_2}\left[\frac{\partial f}{\partial Y} - \frac{d}{dx}\left(\frac{\partial f}{\partial Y'}\right)\right]\eta(x)\,dx$$

$\eta(x)$ is an arbitrary continuous function.

$\therefore$ $\dfrac{\partial f}{\partial y} - \dfrac{d}{dx}\left(\dfrac{\partial f}{\partial y'}\right) = 0$ which is a required Euler's equation.

Note: Other Forms of Euler's equation

1. $\dfrac{d}{dx} f(x, y, y') = \dfrac{\partial f}{\partial x}\dfrac{dx}{dx} + \dfrac{\partial f}{\partial y}\dfrac{dy}{dx} + \dfrac{\partial f}{\partial y'}\dfrac{dy'}{dx}$

or $$\frac{df}{dx} = \frac{\partial f}{\partial x} + \frac{\partial f}{\partial y}y' + \frac{\partial f}{\partial y'}y'' \quad ...(4)$$

But $$\frac{d}{dx}\left(y'\frac{\partial f}{\partial y'}\right) = y'\frac{d}{dx}\left(\frac{\partial f}{\partial y'}\right) + \frac{\partial f}{\partial y'}y'' \quad ...(5)$$

On subtracting (5) from (4), we have

$$\frac{df}{dx} - \frac{d}{dx}\left(y'\frac{\partial f}{\partial y'}\right) = \frac{\partial f}{\partial x} + \frac{\partial f}{\partial y}y' - y'\frac{d}{dx}\left(\frac{\partial f}{\partial y'}\right)$$

$$\frac{d}{dx}\left[f - y'\frac{\partial f}{\partial y'}\right] - \frac{\partial f}{\partial x} = y'\left[\frac{\partial f}{\partial y} - \frac{d}{dx}\frac{\partial f}{\partial y'}\right] = (y')(0) = 0 \qquad [Euler's equation]$$

Hence $$\mathbf{\frac{d}{dx}\left[f - y'\frac{\partial f}{\partial y'}\right] - \frac{\partial f}{\partial x} = 0} \quad ...(6)$$

Which is an another form of Euler's equation.

2. We know that $\dfrac{\partial f}{\partial y'}$ is also a function x, y, y' say $\phi(x, y, y')$.

$$\frac{d}{dx}\left(\frac{\partial f}{\partial y'}\right) = \frac{\partial \phi}{\partial x}\frac{dx}{dx} + \frac{\partial \phi}{\partial y}\frac{dy}{dx} + \frac{\partial \phi}{\partial y'}\frac{dy'}{dx} = \frac{\partial \phi}{\partial x} + \frac{\partial \phi}{\partial y}y' + \frac{\partial \phi}{\partial y'}y''$$

$$\frac{\partial}{\partial x}\left(\frac{\partial f}{\partial y'}\right) + \frac{\partial}{\partial y}\left(\frac{\partial f}{\partial y'}\right)y' + \frac{\partial}{\partial y'}\left(\frac{\partial f}{\partial y'}\right)y'' = \frac{\partial^2 f}{\partial x \partial y'} + y'\frac{\partial^2 f}{\partial y \partial y'} + y''\frac{\partial^2 f}{\partial y'^2}$$

Putting the value of $\dfrac{d}{dx}\left(\dfrac{\partial f}{\partial y'}\right)$ in Euler's equation, we get

$$\mathbf{\frac{\partial f}{\partial y} - \frac{\partial^2 f}{\partial x\,\partial y'} - y'\frac{\partial^2 f}{\partial y\,\partial y'} - y''\frac{\partial^2 f}{\partial y'^2} = 0} \quad ...(7)$$

This is the third form of Euler's equation.

54.5 EXTREMAL

Any function which satisfies Euler's equation is known as Extremal. Extremal is obtained by solving the Euler's equation.

Case 1. If f is independent of $x, i.e., \frac{\partial f}{\partial x} = 0.$

On substituting the value of $\frac{\partial f}{\partial x}$ in (6), we have $\frac{d}{dx}\left[f - y'\frac{\partial f}{\partial y'}\right] = 0$

Integrating, we get $f - y'\frac{\partial f}{\partial y'}$ = constant

Case 2. When f is independent of y, *i.e.*, $\frac{\partial f}{\partial y} = 0$.

Putting the value of $\frac{\partial f}{\partial y}$ in Euler's equation, we get

$\frac{d}{dx}\left(\frac{\partial f}{\partial y'}\right) = 0$, Integrating we get $\frac{\partial f}{\partial y'}$ = **constant**

Case 3. If f is an independent of $y', i.e., \frac{\partial f}{\partial y'} = 0$. On substituting the value of $\frac{\partial f}{\partial y'}$ in the Euler's equation, we get $\frac{\partial f}{\partial y} = 0$

This is the desired solution.

Case 4. If f is independent of x and y,

we have $\frac{\partial f}{\partial x} = 0$ and $\frac{\partial f}{\partial y} = 0$ or $\frac{\partial^2 f}{\partial x \partial y'} = 0$ and $\frac{\partial^2 f}{\partial y \partial y'} = 0$

Putting these value in Euler's equation (7), we have $y''\frac{\partial^2 f}{\partial y'^2} = 0$

If $\frac{\partial^2 f}{\partial y'^2} \neq 0$ then $y'' = 0$ whose solution is $\boldsymbol{y = ax+b}$.

Example 1. *Test for an extremum the functional*

$$I[y(x)] = \int_0^1 \left(xy + y^2 - 2y^2 y'\right) dx,\ y(0) = 1, y(1) = 2$$

Solution. Euler's equation

$$\frac{\partial f}{\partial y} - \frac{d}{dx}\left(\frac{\partial f}{\partial y'}\right) = 0 \qquad \text{...(1)}$$

Here $f = xy + y^2 - 2y^2 y'$

$$\frac{\partial f}{\partial y} = x + 2y - 4yy' \text{ and } \frac{\partial f}{\partial y'} = -2y^2$$

$$\frac{d}{dx}\left(\frac{\partial f}{\partial y'}\right) = \frac{d}{dx}\left(-2y^2\right) = -4yy'$$

Putting these values in (1), we get $x + 2y - 4yy' - (-4yy') = 0$

or

$x + 2y - 0$ *or* $y = -\frac{x}{2}$ At $x = 0, y = 0$; At $x = 1, y = -\frac{1}{2}$.

This extremal does not satisfy the boundary conditions $y(0) = 1, y(1) = 2$.
Hence there is no extremal. **Ans.**

Example 2. *Prove that the shortest distance between two points is along a straight line.*

Solution. Let $A\ (x_1, y_1)$ and $B\ (x_2, y_2)$ be the two given points and s the length of the arc joining these points.

Then $$s = \int_{x_1}^{x_2} ds = \int_{x_1}^{x_2} \sqrt{1 + \left(\frac{dy}{dx}\right)^2}\, dx = \int_{x_1}^{x_2} \sqrt{(1 + y'^2)}\, dx \qquad \ldots (1)$$

$$y(x_1) = y_1, \quad y(x_2) = y_2$$

If s satisfies the Euler's equation, then it will be minimum

$$\frac{\partial f}{\partial y} - \frac{d}{dx}\left(\frac{\partial f}{\partial y'}\right) = 0 \qquad \text{(Euler's equation)}$$

Here in (1), $$f = \sqrt{(1 + y'^2)}$$

f is independent of y, *i.e.*, $\frac{\partial f}{\partial y} = 0$

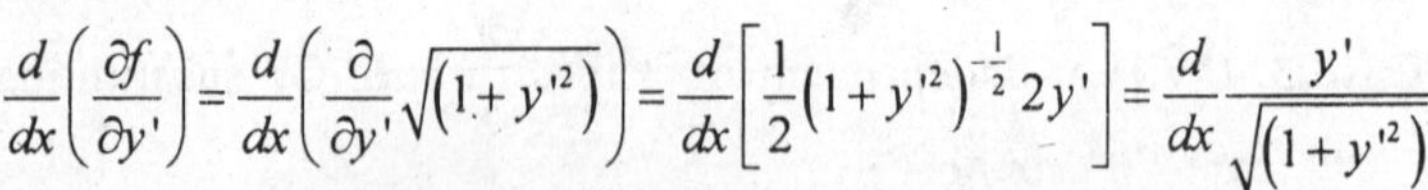

$$\frac{d}{dx}\left(\frac{\partial f}{\partial y'}\right) = \frac{d}{dx}\left(\frac{\partial}{\partial y'}\sqrt{(1 + y'^2)}\right) = \frac{d}{dx}\left[\frac{1}{2}(1 + y'^2)^{-\frac{1}{2}}\, 2y'\right] = \frac{d}{dx}\frac{y'}{\sqrt{(1 + y'^2)}}$$

Putting these values in Euler's Equation, we have

$$0 - \frac{d}{dx}\frac{y'}{\sqrt{(1 + y')}} = 0 \quad \text{or} \quad \frac{d}{dx}\frac{y'}{\sqrt{(1 + y'^2)}} = 0$$

On integrating $\frac{y'}{\sqrt{(1 + y'^2)}}$ constant (c), *i.e.*, $(y')^2 = c^2\ (1 + y'^2)$

or $$y'^2 (1 - c^2) = c^2 \text{ or } y'^2 = \frac{c^2}{1 - c^2} = m^2 \text{ or } y' = m \text{ or } \frac{dy}{dx} = m$$

Integrating $$y = mx + c \qquad \ldots (2)$$

which is a straight line. **Ans.**

Now $$y(x_1) = y_1 \quad \text{and} \quad y(x_2) = y_2$$

$$mx_1 + c = y_1 \text{ and } mx_2 + c = y_2 \qquad \ldots (3)$$

on subtracting, we get

or $$y_2 - y_1 = m(x_2 - x_1) \text{ or } m = \frac{y_2 - y_1}{x_2 - x_1}$$

Subtracting (3) from (2), we get

$$y - y_1 = m(x - x_1)$$

$$y - y_1 = \frac{y_2 - y_1}{x_2 - x_1}(x - x_1)$$ **Proved.**

Example 3. *Find the curve connecting the points (x_1, x_2) and (x_2, y_2) which when rotated about the x-axis gives a minimum surface.*

Find the extremal of the functional.

$$\int_{x_1}^{x_2} 2\pi y\, ds \text{ or } 2\pi \int_{x_1}^{x_2} y\sqrt{(1 + y'^2)}\, dx$$

Subject to $y\ (x_1) = y_1, y(x_2) = y_2$

Solution. 2π is constant so we have to find the extremal of

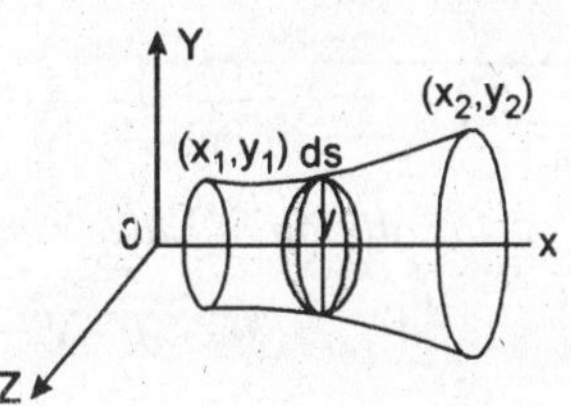

$$\int_{x_1}^{x_2} y\sqrt{(1+y'^2)}\, dx$$

Here $f = y\sqrt{(1+y'^2)}$ which is independent of x.

One form of Euler's equation is

$$\frac{d}{dx}\left[f - y'\frac{\partial f}{\partial y'}\right] - \frac{\partial f}{\partial x} = 0 \qquad \frac{d}{dx}\left[f - y'\frac{\partial f}{\partial y'}\right] = 0 \qquad \left(\frac{\partial f}{\partial x} = 0\right)$$

On integrating, we get, $f - y'\dfrac{\partial f}{\partial y'}$ = constant (c) ...(1)

Putting the value of f and $\dfrac{\partial f}{\partial y'}$ (1), we have

$$y = \sqrt{(1+y'^2)} - y'\frac{2y'}{2\sqrt{(1+y'^2)}}y = c$$

or $\quad y\sqrt{(1+y'^2)} - \dfrac{yy'^2}{\sqrt{(1+y'^2)}} = c$ or $y(1+y'^2) - yy'^2 = c\sqrt{(1+y'^2)}$

$$y = c\sqrt{(1+y'^2)} \quad or \quad y^2 = c^2(1+y'^2)$$

or $$y'^2 = \frac{y^2-c^2}{c^2} \quad or \quad y' = \frac{\sqrt{y^2-c^2}}{c} \quad or \quad \frac{dy}{dx} = \frac{\sqrt{y^2-c^2}}{c}$$

$$\frac{dy}{\sqrt{y^2-c^2}} = \frac{dx}{c} \Rightarrow \int\frac{dy}{\sqrt{y^2-c^2}} = \int\frac{dx}{c} \Rightarrow \cosh^{-1}\frac{y}{c} = \frac{x}{c} + b$$

$y = c\cosh\left(\dfrac{x}{c} + b\right)$ which is the equation of catenary. This is the required extremal. **Ans.**

Example 4. *Find the curve connecting two points (not on a vertical line), such that a particle sliding down this curve under gravity (in absence of resistance) from one point to another reaches in the shortest time. (Brachistochrone problem).*

Solution. Let the particle slide on the curve OA from O with zero velocity. Let OP = s and time taken from 0 to P = t. By the law of conservation of energy, we have

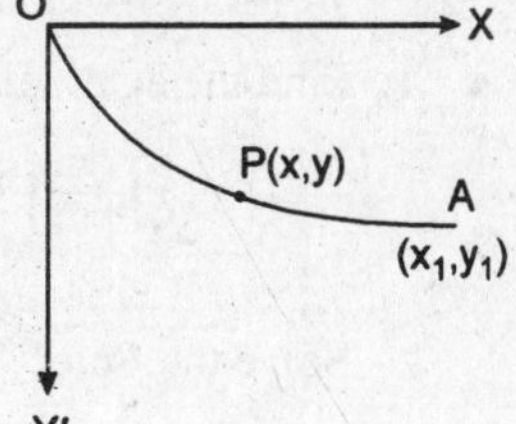

K.E. at P – K.E. at O = potential energy at P.

$$\frac{1}{2}mv^2 - 0 = mgh$$

or $\quad \dfrac{1}{2}m\left(\dfrac{ds}{dt}\right)^2 = mgh$ or $\dfrac{ds}{dt} = \sqrt{(2gy)}$

Time taken by the particle to move from O to A

$$T = \int_0^T dt = \int_0^{x_1}\frac{ds}{\sqrt{(2gy)}} = \frac{1}{\sqrt{(2g)}}\int_0^{x_1}\frac{ds}{\sqrt{y}} = \frac{1}{\sqrt{(2g)}}\int_0^{x_1}\frac{\sqrt{(1+y'^2)}}{\sqrt{y}}dx$$

Here, $f = \dfrac{\sqrt{(1+y'^2)}}{\sqrt{y}}$ *which is independent of x, i.e.,* $\dfrac{\partial f}{\partial x} = 0.$

and $$\frac{\partial f}{\partial y'} = \frac{1}{2\sqrt{y}} \frac{2y'}{\sqrt{(1+y'^2)}} = \frac{y'}{\sqrt{y}\sqrt{(1+y'^2)}}$$

Solution of Euler's equation is

$$f - y'\frac{\partial f}{\partial y'} = \text{constant } c$$

On substituting the values of f and $\frac{\partial f}{\partial y'}$, *We get*

$$\frac{\sqrt{(1+y'^2)}}{\sqrt{y}} - y'\frac{y'}{\sqrt{y}\sqrt{(1+y'^2)}} = c$$

$$\Rightarrow \quad \sqrt{1+y^2} - \frac{y^2}{\sqrt{(1+y^2)}} = c\sqrt{y} \quad \text{or} \quad 1+y'^2 - y'^2 = c\sqrt{(1+y'^2)}\sqrt{y}$$

$$\Rightarrow \quad 1 = c\sqrt{y(1+y'^2)} \quad \text{or} \quad 1+\left(\frac{dy}{dx}\right)^2 = \frac{1}{yc^2} \quad \text{or} \quad \frac{dy}{dx} = \frac{\sqrt{1-yc^2}}{yc^2}$$

$$\Rightarrow \quad \frac{dy}{dx} = \frac{\sqrt{1/c^2 - y}}{y} = \frac{\sqrt{a-y}}{y} \qquad \left(\frac{1}{c^2} = a\right)$$

$$dx = \sqrt{\frac{y}{a-y}}\, dy$$

$$\int_0^x x\, dx = \int_0^y \sqrt{\left(\frac{y}{a-y}\right)}\, dy$$

Put $y = a\sin^2\theta$

$dy = 2a\sin\theta\cos\theta\, d\theta$

$$x = \int_0^\theta \sqrt{\left(\frac{a\sin^2\theta}{a - a\sin^2\theta}\right)} 2a\sin\theta\cos\theta\, d\theta = \int_0^\theta \left(\frac{\sin\theta}{\cos\theta}\right) 2a\sin\theta\cos\theta d\theta = \int_0^\theta 2a\sin^2\theta d\theta$$

$$= a\int_0^\theta (1-\cos 2\theta)\, d\theta = a\left(\theta - \frac{\sin 2\theta}{2}\right)_0^\theta$$

$$\Rightarrow \quad x = \frac{a}{2}(2\theta - \sin 2\theta) \text{ and } y = a\sin^2\theta = \frac{a}{2}(1-\cos 2\theta)$$

On putting $\frac{a}{2} = A$ *and* $2\theta = \Theta$ $\quad \left.\begin{aligned} x &= A(\Theta - \sin\Theta) \\ y &= A(1-\cos\Theta)\end{aligned}\right]$ *which is a cycloid.*

EXERCISE 54.1

1. Find the external of the functional

$$I[y(x)] = \int_{x_0}^{x_1} \frac{1+y^2}{y'^2} dy$$

Ans. $y = \sinh(c_1 x + c^2)$

2. Solve the Euler's equations for $\int_{x_0}^{x_1} (x+y')y'\, dx.$ **Ans.** $y = -\frac{x^2}{4} + c_1 x + c_2$

3. **Solve the Euler's equation for** $\int_{x_0}^{x_1} (1+x^2y')y'\, dx$ **Ans.** $y = cx^{-1} + c_2$

Find the extremals of the functional and extremism value of the following:

4. $I[y(x)] = \int_{x_0}^{x_1} \frac{1+y^2}{y'^2} dx$ **Ans.** $y = \sinh(c_1 x + c_2)$

5. $I[y(x)] = \int_{\frac{1}{2}}^{1} x^2 y^2 dx$ subject to $y\left(\frac{1}{2}\right) = 1, y(2) = 4.$ **Ans.** $y = -\frac{c}{x} + d$, value = 1

6. $I[y(x)] = \int_0^2 (x - y')^2 dx$ subject to $y(0) = 0, y(2) = 4.$ **Ans.** $y = \frac{x^2}{2} + cx + d$, value = 2

7. $\int_0^{\frac{\pi}{2}} (y'^2 - y^2) dx$ subject to $y(0), y\left(\frac{\pi}{2}\right) = 1$ **Ans.** $y = \sin x$, value = 0

8. $\int_0^1 (y'^2 + 12xy) dx$ subject to $y(0) = 0, y(1) = 1$ **Ans.** $y = x^3$, value $= \frac{21}{5}$

9. $\int_1^2 \frac{\sqrt{(1+y'^2)}}{x} dx$ subject to $y(1) = 0, y(2) = 1.$ **Ans.** $y = x^3$

54.6 ISOPERIMETRIC PROBLEMS

The determination of the shape of a closed curve of the given perimeter enclosing maximum area is the example of isoperimetric problem. In certain problems it is necessary to make a given integral.

$$I = \int_{x_1}^{x_2} f(x, y, y')\, dx \qquad \text{....(1)}$$

maximum or minimum while keeping another integral

$$I = \int_{x_1}^{x_2} g(x, y, y')\, dx = K \text{ (Constant)} \qquad \text{... (2)}$$

Problems of this type are solved by Lagrange's multipliers method. We multiply (2) by λ and add to (1) to extremize (1)

$$I^* = \int_{x_1}^{x_2} f(x, y, y')\, dx + \lambda \int_{x_1}^{x_2} g(x, y, y')\, dx = \int_{x_1}^{x_2} F\, dx \text{ (say)}$$

Then by Euler's equation $\frac{\partial F}{\partial y} - \frac{d}{dx}\left(\frac{\partial F}{\partial y'}\right) l = 0$

Note. Isoperimetric problem. To find out possible curves having the same perimeter, the one which encloses the maximum area.

Example 5. *Find the shape of the curve of the given perimeter enclosing maximum area.*

Solution. Let P be the perimeter of the closed curve,

Then $$P = \int_{x_1}^{x_2} \sqrt{1 + y'^2}\, dx \qquad \text{...(1)}$$

The area enclosed by the curve, x-axis and two perpendicular lines is

$$A = \int_{x_1}^{x_2} y\, dx \qquad \text{....(2)}$$

We have to find the maximum value of (2) under the condition (1).
By Lagrange's multiplier method.

$$F = y + \lambda\sqrt{1 + y'^2}$$

For maximum or minimum value of A, F must satisfy Euler's equation

$$\frac{\partial F}{\partial y} - \frac{d}{dx}\left(\frac{\partial F}{\partial y'}\right) = 0$$

$$1-\lambda\frac{d}{dx}\left[\frac{1}{2}(1+y'^2)^{-\frac{1}{2}}(2y')\right]=0 \text{ or } 1-\lambda\frac{d}{dx}\left(\frac{y'}{\sqrt{1+y'^2}}\right)=0$$

Integraing w.r.t. 'x', we get $x-\dfrac{\lambda y'}{(1+y'^2)}=a$

or $$\frac{\lambda y'}{\sqrt{(1+y'^2)}}=x-a \text{ or } \lambda^2 y'^2=(1+y'^2)(x-a)^2$$

$$[\lambda^2-(x-a)^2]\,y'^2=(x-a)^2$$

or $$y'=\frac{x-a}{\sqrt{[\lambda^2-(x-a)^2]}} \quad \text{or} \quad \frac{dy}{dx}=\frac{x-a}{\sqrt{[\lambda^2-(x-a)^2]}}$$

Integrating w.r.t. (x), we obtain

$$y=-\sqrt{[\lambda^2-(x-a)^2]}+b$$

or $$y-b=-\sqrt{[\lambda^2-(x-a)^2]} \quad (y-b)^2=\lambda^2-(x-a)^2 \quad \text{or} \quad (x-a)^2+(y-b)^2=l^2$$

This is the equation of a circle whose centre is (a, b) and radius λ. **Ans.**

Example 6. *Find the extremal of the functional* $A=\int_{t_1}^{t_2}\frac{1}{2}(x\dot{y}-y\dot{x})dt$ *subject to the integral constraint* $\int_{t_1}^{t_2}\frac{1}{2}\sqrt{(\dot{x}^2-\dot{y}^2)}\,dt=l.$

Solution. Here $f=\frac{1}{2}(x\dot{y}-x\dot{y})$, $g=\sqrt{\dot{x}^2-\dot{y}^2}$

$$F=f+\lambda g$$
$$F=\frac{1}{2}(x\dot{y}-y\dot{x})+\lambda\sqrt{\dot{x}^2+\dot{y}^2}$$

For A to have extremal F must satisfy the Euler's equation

$$\frac{\partial F}{\partial x}-\frac{d}{dx}\left[\frac{\partial F}{\partial \dot{x}}\right]=0 \qquad ...(1)$$

$$\frac{\partial F}{\partial x}-\frac{d}{dx}\left[\frac{\partial F}{\partial \dot{x}}\right]=0 \qquad ...(1)$$

$$\frac{\partial F}{\partial x}-\frac{d}{dx}\left[\frac{\partial F}{\partial y}\right]=0 \qquad ...(2)$$

From (1)

$$\frac{1}{2}\dot{y}-\frac{d}{dt}\left(-\frac{y}{2}+\frac{\lambda 2\dot{x}}{2\sqrt{\dot{x}^2+\dot{y}^2}}\right)=0$$

$$\frac{d}{dt}\left(y-\frac{\lambda\dot{x}}{2\sqrt{\dot{x}^2+\dot{y}^2}}\right)=0 \qquad ...(3)$$

From (2)

$$-\frac{1}{2}\dot{x}-\frac{d}{dt}\left[\frac{x}{2}+\frac{\lambda y}{\sqrt{x^2+y^2}}\right]=0 \qquad ...(4)$$

$$\frac{d}{dt}\left[x-\frac{\lambda\dot{y}}{\sqrt{\dot{x}^2+\dot{y}^2}}\right]=0$$

Integrating (3) and (4), we have

$$y-\frac{\lambda\dot{x}}{\sqrt{\dot{x}^2+\dot{y}^2}}=c_1 \quad \text{or} \quad y-c_1=\frac{\lambda\dot{x}}{\sqrt{\dot{x}^2+\dot{y}^2}} \qquad ...(5)$$

$$x-\frac{\lambda\dot{y}}{\sqrt{\dot{x}^2+\dot{y}^2}}=c_1 \quad \text{or} \quad x-c_2=\frac{\lambda\dot{y}}{\sqrt{\dot{x}^2+\dot{y}^2}} \qquad ...(6)$$

Squaring (5) , (6) and adding, we get

$$(x-c_2)^2+(y-c_1)^2=\lambda^2\left(\frac{\dot{x}^2+\dot{y}^2}{\dot{x}^2+\dot{y}^2}\right)$$

$$(x-c_2)^2+(y-c_1)^2=\lambda^2$$

This is the equation of circle. **Ans.**

Example 7. *Find the solid of maximum volume formed by the revolution of a given surface area.*

Solution. Let the curve PA pass fthrough origion and it is rotated about the x - axis.

$$S=\int_0^a 2\pi y\,ds$$

$$S=\int_0^a 2\pi y\sqrt{(1+y^2)}\,dx \qquad ...(1)$$

$$V=\int_0^a \pi y^2\,dx \qquad ...(2)$$

Y, P (x,y), O, A, X

Here we have to extremize V with the given S.

Here $\qquad f=\pi y^2,\ g=2\pi y\sqrt{(1-y'^2)}$

$$F=f+\lambda g$$

$$F=\pi y^2+\lambda 2\pi y\sqrt{(1+y'^2)}$$

For maximum V, F must satisfy Euler's equation. But F does not contain x.

$$\therefore \qquad F-y'\frac{\partial F}{\partial y'}=C$$

or $$\pi y^2+\lambda 2\pi y\sqrt{(1+y'^2)}-y'\frac{1}{2}\frac{2\pi y\,\lambda 2y'}{\sqrt{(1+y'^2)}}=C$$

or $$\pi y^2+2\pi y\lambda\sqrt{(1+y'^2)}-\frac{2\pi\lambda yy'^2}{\sqrt{(1+y'^2)}}=C$$

or $$\pi y^2+\frac{2\pi y\lambda}{\sqrt{(1+y'^2)}}=C$$

As the curve passes through origin (0, 0), so $C=0$.

$$\pi y^2 + \frac{2\pi y\lambda}{\sqrt{(1+y'^2)}} = 0$$

or $$y + \frac{2\lambda}{\sqrt{(1+y'^2)}} = 0 \quad \text{or} \quad y\sqrt{(1+y'^2)} = -2\lambda$$

or $$1 + y'^2 = \frac{4\lambda^2}{y^2} \quad \text{or} \quad y'^2 = \frac{4\lambda^2}{y^2} - 1 = \frac{4\lambda^2 - y^2}{y^2}$$

or $$\frac{dy}{dx} = \frac{\sqrt{(4\lambda^2 - y^2)}}{y}$$

$$\int \frac{y\,dy}{\sqrt{(4\lambda^2 - y^2)}} = \int dx + C$$

$$-\sqrt{4\lambda^2 - y^2} = x + C \qquad \text{...(1)}$$

or $$\sqrt{4\lambda^2 - y^2} = -x - C$$

The curve passes through (0, 0). On putting $x = 0,\ y = 0$ in (1) we get

$$-C = 2\lambda$$

(1) becomes $$\sqrt{4\lambda^2 - y^2} = -x + 2\lambda$$

Squaring $$4\lambda^2 - y^2 = (x - 2\lambda)^2$$

or $$(x - 2\lambda)^2 + y^2 = 4\lambda^2$$

This is the equation of a circle.

Hence, on revolving the circle about x- axis, the solid formed is a sphere. **Ans.**

EXERCISE 54.2

1. Show that an isosceles triangle has the smallest perimeter for a given area and a given base.
2. Find the external in the isoperimetric problem of the extremum of

$$\int_0^1 (y'^2 + z'^2 - 4xz' - 4z)\,dx$$

subject to $\int_0^1 (y'^2 + xy' - z'^2)\,dx = 2,\ y(0) = 0,\ z(0) = 0,\ y(1) = 1,\ z(1) = 1.$

Ans. $y = \frac{-5x^2}{2} + \frac{7x}{2},\ z = x.$

3. Find the surface with the smallest area which encloses a given volume. **Ans.** Sphere
4. Find the external of the functional $\int_{t_1}^{t_2} \sqrt{x^2 + y^2 + z^2}\,dt$ subject to $x^2 + y^2 + z^2 = a^2$.

Ans. Arc of a great circle of a sphere.

5. Find the extremals of the isoperimetric problem $\int_{x_0}^{x_1} y'^2\,dx$ subject to $\int_{x_0}^{x_1} y\,dx = c.$ **Ans.** $y = x^2 + ax + b$

54.7 FUNCTIONALS OF SECOND ORDER DERIVATIVES

Let us consider the extremum of a functional.

$$\int_{x_1}^{x_2} [f(x, y, y', y'')]\,dx \qquad \text{...(1)}$$

The necessary condition for the above mentioned functional to be extremum is

$$\frac{\partial f}{\partial y} - \frac{d}{dx}\left(\frac{\partial f}{\partial y'}\right) + \frac{d^2}{dx^2}\left(\frac{\partial f}{\partial y''}\right) = 0$$

Proof. Let the boundary conditions be

$$y(x_1) = y_1,\ y(x_2) = y_2,\ y'(x_1) = y'_1,\ y'(x_2) = y'_2$$

Let α be a parameter and $\eta(x)$ is a differentiable function.

At the end points $\eta(x_1) = \eta(x_2) = 0$ and $\eta'(x_1) = \eta'(x_2) = 0$

Putting $y + \alpha\,\eta(x)$ for y in (1), we have

$$\int_{x_1}^{x_2} f[x, y+\alpha\,\eta(x), y'+\alpha\,\eta'(x), y''+\alpha\,\eta''(x)]\,dx$$

Writing $\int_{x_1}^{x_2} f[x, y+\alpha\,\eta(x), y'+\alpha\,\eta'(x), y''+\alpha\,\eta''(x)]\,dx = \int_{x_1}^{x_2} F dx = 1$

For extremum value of (1)

$$\frac{dI}{d\alpha} = 0$$

$$\frac{dI}{d\alpha} = \int_{x_1}^{x_2} \frac{\partial F}{\partial \alpha} dx$$

Differentiating under the sign of integral, we get

$$= \int_{x_1}^{x_2}\left(\frac{\partial F}{\partial y}\frac{\partial y}{\partial \alpha} + \frac{\partial F}{\partial y'}\frac{\partial y'}{\partial \alpha} + \frac{\partial F}{\partial y''}\frac{\partial y''}{\partial \alpha}\right) dx = \int_{x_1}^{x_2}\left(\frac{\partial F}{\partial y}\frac{\partial(\alpha\,n)}{\partial \alpha} + \frac{\partial F}{\partial y'}\frac{\partial(\alpha\,n')}{\partial \alpha} + \frac{\partial F}{\partial y''}\frac{\partial(\alpha\,n'')}{\partial \alpha}\right) dx$$

But $\frac{dI}{d\alpha} = 0$ when $\alpha = 0$

$$0 = \int_{x_1}^{x_2}\left[\frac{\partial f}{\partial y}\eta + \frac{\partial f}{\partial y}\eta' + \frac{\partial f}{\partial y}\eta''\right] dx \quad \text{or} \quad \int_{x_1}^{x_1}\frac{\partial f}{\partial y}\eta\,dx + \int_{x_1}^{x_2}\frac{\partial f}{\partial y'}\eta'\,dx + \int_{x_1}^{x_2}\frac{\partial f}{\partial y''}\eta''\,dx = 0$$

Integrating by parts, w.r.t. 'x', we have

$$\int_{x_1}^{x_2}\frac{\partial f}{\partial y}\eta\,dx + \left[\frac{\partial f}{\partial y}\eta - \int_{x_1}^{x_2}\frac{\partial}{\partial x}\left(\frac{\partial f}{\partial y'}\right).\eta\,dx\right]_{x_1}^{x_2} + \left(\frac{\partial f}{\partial y''}\eta' - \frac{d}{dx}\left(\frac{\partial f}{\partial y''}\right).\eta + \frac{d^2}{dx^2}\left(\frac{\partial f}{\partial y''}\right)\int_{x_1}^{x_2}\eta\,dx\right)_{x_1}^{x_2} = 0$$

But $n(x_1) = n(x_2) = 0$ and $\eta'(x_1) = \eta'(x_2) = 0$

so $\int_{x_1}^{x_2}\left[\frac{\partial f}{\partial y} - \frac{d}{dx}\left(\frac{\partial f}{\partial y'}\right) + \frac{d^2}{dx^2}\left(\frac{\partial f}{\partial y''}\right)\right]\eta(x)\,dx = 0 \Rightarrow \frac{\partial f}{\partial y} - \frac{d}{dx}\left(\frac{\partial f}{\partial y'}\right) + \frac{d^2}{dx^2}\left(\frac{\partial f}{\partial y''}\right) = 0$ **Proved.**

EXERCISE 54.3

1. Find the extremal of $\int_{x_0}^{x_1}(16y^2 - y''^2 + x^2)\,dx$. **Ans.** $y = c_1e^{2x} + c_2e^{-2x} + c_3\cos 2x + c_4\sin 2x$

2. Find the extremal of $\int_{-c}^{c}(ay + \frac{1}{2}by''^2)\,dx$ subject to $y(-c) = 0, y'(-c) = 0,$
$y(c) = 0, y'(c) = 0.$ **Ans.** $y = -\frac{a}{24b}(x^2 - c^2)^2$

3. Find the extremal of $\int_0^x y''^2\,dx$ subject to $\int_0^x y^2\,dx = 1, y(0) = y(p) = 0, y''(p) = 0.$
Ans. $y = a_1\sin x + a_2\sin 2x + \ldots\ldots$

4. Find the extremal of $\int_{x_0}^{x_1}(2xy + y'''^2)\,dx$. **Ans.** $y = \frac{x^7}{7!} + c_1x^5 + c_2x^4 + c_3x^3 + c_4x^2 + c_5x + c_6$

TENSOR ANALYSIS

55.1 INTRODUCTION

Scalars are specified by magnitude only, *vectors* have magnitude as well as direction. *Tensors* are associated with magnitude and two or more directions. For example, the stress of an elastic solid at a point depends upon two directions. One of the directions is given by the normal to the area, while the other is that of the force on it.

Tensors are similar to vectors. A vector can be specified by its components (Magnitude and direction). A tensor can be specified only by its components which depend upon the system of reference. The components of the same tensor will be different for two different sets of axes with different orientations.

Tensors analysis is suitable for mathematical formulation of natural laws in forms which are invariant with respect to different frames of reference. That is why Einstein used tensors for the formulation of his Theory of Relativity.

55.2 CO-ORDINATE TRANSFORMATION

If we have two systems of rectangular co-ordinate axes *OX, OY, OZ; OX', or, OZ'*; having same origin such that the direction cosines of the lines

OX', OY', OZ' relative to the system *OX YZ* are

$$l_1, m_1, n_1;\ l_2, m_2, n_2;\ l_3, m_3, n_3$$

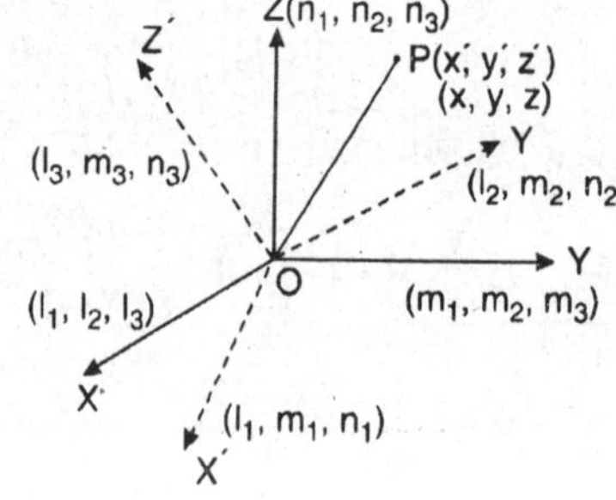

Two equivalent systems of transformation equations express x', y', z' in terms of x, y, z and vice versa.

$$\left.\begin{aligned} x' &= l_1 x + m_1 y + n_1 z \\ y' &= l_2 x + m_2 y + n_2 z \\ z' &= l_3 x + m_3 y + n_3 z \end{aligned}\right\} \quad \ldots (1)$$

$$\left.\begin{aligned} x &= l_1 x' + l_2 y' + l_3 z' \\ y &= m_1 x' + m_2 y' + m_3 z' \\ z &= n_1 x' + n_2 y' + n_3 z' \end{aligned}\right\} \quad \ldots (2)$$

where (x', y', z') and (x, y, z) are co-ordinates of point *P* relative to two systems of co-ordinate axes. System of transformation eq. shown above in (1) and (2) can be written as

	x	y	z
x'	l_1	m_1	n_1
y'	l_2	m_2	n_2
z'	l_3	m_3	n_3

55.3 SUMMATION CONVENTION

The sum of the following $a_1x_1 + a_2x_2 + a_3a_3 + ... + a_nx_n$ ...(1)

can be written in brief as $\sum_{i=1}^{i=n} a_i x_i$ ... (2)

More simplified and compact notation for (2) used by Einstein is $a_i x^i$. ...(3)

In (3) we have omitted Σ-sign.

$$a_i x^i = a_1 x^1 + a_2 x^2 + a_3 x_3^3 + ...a_n x^n$$

We write $x_1, x_2, x_3, ... x_n$ as $x^1, x^2, x^3, ... x^n$ in tenser analysis. These superscripts donot stand for powers of x but indicate different symbols. The power of x^i is written as $(x^i)^2, (x^i)^3$....

Example 1. *Write out* $a_{rs}x^s = b_r (r, s = 1, 2, 3....n)$ *in full:*

Solution. $a_{rs}\, x^s = b_r$

$a_{is}x^s + a_{2s}x^s + a_{3s}x^s + ... + a_{ns}x^n = b_1 + b_2 + b_3 + ... + b_n$ (r occurs 1 to n)

$(a_{11}x^1 + a_{12}x^2 + a_{13}x^3 + ... a_{1n}x^n) + (a_{21}x^1 + a_{22}x^2 + a_{23}x^3 + ... + a_{2n}x^n)$

$+ (a_{31}x^1 + a_{32}x^2 + a_{33}x^3 + ... +....a_{3n}x^n) + ... = b_1 + b_2 + b_3 +....+ b_n$

Example 2. If $f = f(x^1, x^2, x^3, ... x^n)$ then show that $df = \frac{\partial f}{dx^i} dx^i$

Solution. $df = \frac{\partial f}{dx^1} dx^1 + \frac{\partial f}{dx^2} dx^2 + ... + \frac{\partial f}{dx^n} dx^n = \frac{\partial f}{\partial x^i} dx^i$ **Proved.**

55.4 SUMMATION OF CO-ORDINATES

The equations of co-ordinates can be written in very compact form in terms of summation convention. We write (x_1, x_2, x_3) and $(\bar{x}_1, \bar{x}_2, \bar{x}_3)$ instead of (x, y, z) and (x', y', z') and denote the co-ordinate axes as OX_1, OX_2, OX_3 and $O\bar{X}_1, O\bar{X}_2, O\bar{X}_3$. Also we denote $x_i, \bar{x}_j$ as the co-ordinates of a point P relative to the two systems of axes; where $i = 1, 2, 3, j = 1, 2, 3$.

Let l_{ij} denote the cosines of the angles between $OX_i, O\bar{X}_j$. In general $l_{ij} \neq l_{ji}$

The eq. of co-ordinate transformation can be written as

$$\left.\begin{aligned} \bar{x}_1 &= l_{11}x_1 + l_{21}x_2 + l_{31}x_3 \\ \bar{x}_2 &= l_{12}x_1 + l_{22}x_2 + l_{32}x_3 \\ \bar{x}_3 &= l_{13}x_1 + l_{23}x_2 + l_{33}x_3 \end{aligned}\right\} \quad ...(1a)$$

$$\left.\begin{aligned} x_1 &= l_{11}\bar{x}_1 + l_{12}\bar{x}_2 + l_{13}\bar{x}_3 \\ x_2 &= l_{21}\bar{x}_1 + l_{22}\bar{x}_2 + l_{23}\bar{x}_3 \\ x_3 &= l_{31}\bar{x}_1 + l_{32}\bar{x}_2 + l_{33}\bar{x}_3 \end{aligned}\right\} \quad ...(1b)$$

These equations of co-ordinate transformation can be represented by means of a table form such that

	x_1	x_2	x_3
$\bar{x}_1$	l_{11}	l_{21}	l_{31}
$\bar{x}_2$	l_{12}	l_{22}	l_{32}
$\bar{x}_3$	l_{13}	l_{23}	l_{33}

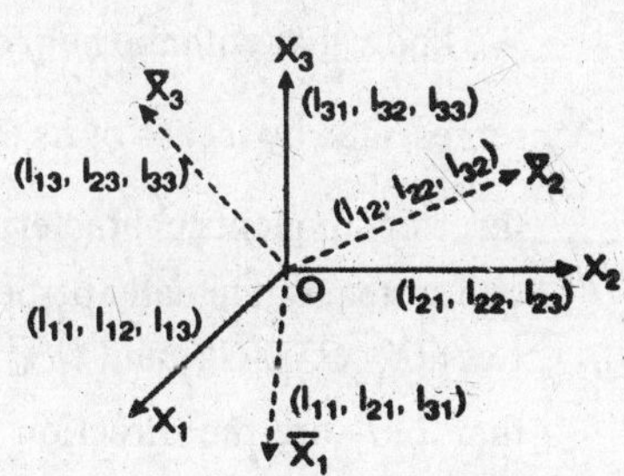

Adopting summation on convention *i.e.*,

$a_{11} + a_{22} + a_{33} = a_{ij}$

$a_{usip} b_{iq} = a_{lp} b_{1q} + a_{2p} b_{2q} + a_{3p} b_{3} q$ we re-write above equations as

$$\bar{x}_1 = l_{i1} x_i \qquad x_1 = l_{1j} \bar{x}_j$$
$$\bar{x}_2 = l_{i2} x_i \qquad x_2 = l_{2j} \bar{x}_j$$
$$\bar{x}_3 = l_{i3} x_i \qquad x_3 = l_{3j} \bar{x}_j$$

We can re-write these equations in single equation in the form.

$$\bar{x}_j = l_{ij} x_i, \; x_i = l_{ij} \bar{x}_j$$

which are complete equivalents of the equations of co-ordinate transformation from either system to another.

55.5 RELATION BETWEEN THE DIRECTION COSINES OF THREE MUTUALLY PERPENDICULAR STRAIGHT LINES

The direction cosines of any three mutually perpendicular straight lines $O\bar{X}_1, O\bar{X}_2, O\bar{X}_3$ relative to the system $OX_1, OX_2, 0X_3$ are $l_{11}, l_{21}, l_{31}, l_{12}, l_{22}, l_{32}, l_{23}, l_{33}$.

The relation between these direction cosines are

$l_{11} l_{11} + l_{21} l_{21} + l_{31} l_{31} = l_{j1} l_{j1} = 1 \qquad l_{12} l_{12} + l_{22} l_{22} + l_{32} l_{32} = l_{j2} l_{j2} = 1$

$l_{13} l_{13} + l_{23} l_{23} + l_{33} l_{33} = l_{j3} l_{j3} = 1.$

Similarly,

$l_{11} l_{12} + l_{21} l_{22} + l_{31} l_{32} = l_{j1} l_{j2} = 0 \qquad l_{12} l_{13} + l_{22} l_{23} + l_{32} l_{33} = l_{j2} l_{j3} = 0$

$l_{13} l_{11} + l_{23} l_{21} + l_{33} l_{31} = l_{j3} l_{j1} = 0$

Finally, we can write these equations by means of a single equation as

$$l_{ij} l_{kj} = \begin{cases} 1, \text{ when } i = k \\ 0, \text{ when } i \neq k \end{cases} \quad \text{or} \quad \delta_{ik} = \begin{cases} 1, \text{ when } i = k \\ 0, \text{ when } i \neq k \end{cases}$$

where δ_{ik} is the kronecker delta.

or $\qquad \delta_{ik} = l_{ij} l_{kj}$

Now, we know that $\bar{x}_j = l_{ij} x_i$

Multiplying both sides by l_{jk} then

or $\qquad l_{jk} \bar{x}_j = l_{ij} l_{jk} x_x \quad \Rightarrow \quad l_{jk} \bar{x}_j = \delta_{ik} x$

putting $i = k$ *i.e.*, $\delta_{ik} = 1$ when $i = k$

$$\delta_{kk} x_k = l_{jk} \bar{x}_j \Rightarrow x_k = l_{jk} \bar{x}_j$$

55.6 TRANSFORMATION OF VELOCITY COMPONENTS ON CHANGE FROM ONE SYSTEM OF RECTANGULAR AXES TO ANOTHER

We know that with the help of parallelogram law of velocities, that any given velocity can be represented by means of its three components along three mutually perpendicular lines and the three components characterise velocity completely. The components change as we pass from one system of mutually perpendicular lines to another.

Let OX_1, OX_2, OX_3 and $O\bar{X}_1, O\bar{X}_2, O\bar{X}_3$ are two systems of rectangular axes and suppose that $l_i, \bar{l}_j$ are the direction cosines of the line of action of the velocity and v, denote the magnitude of the velocity. Then

$$v_i = l v, \bar{v}_j = v\bar{l}_j \qquad \text{... (1)}$$

where v_i and $\bar{v}_j$, denotes the components of velocity relative to the two systems of axes.

By the equation of co-ordinate transformation, we have

$$\bar{l}_j = l_{ij} l_i, \; l_i = l_{ij}\bar{l}_j \qquad \text{... (2)}$$

From (1) and (2).

$$\frac{\bar{v}_j}{v} = l_{ij}\frac{v_i}{v}, \; \frac{v_i}{v} = l_{ij}\frac{\bar{v}_j}{v}. \qquad i.e., \; \bar{v}_j = l_{ij} v_i$$

Thus we see equation of transformation of velocity components are same as for the transformation of co-ordinate of points.

55.7 RANK OF A TENSOR

The rank of a tensor is the number (without counting an index which appears once as a subscript) of indices in the symbol representing a tensor. For example

Tensor	*Symbol*	*Rank*
Scalar	A	zero
Contravariant Tensor	B^i	1
Covariant Tensor	C_k	1
Covariant Tensor	D_y	2
Mixed Tensor	E^{il}_{jkl}	3

In an n-dimensional space, the number of components of a tensor of rank r is n^r.

55.8 FIRST ORDER TENSORS

Definition. Any entity representable by a set of three numbers (called components) relatively to a system of rectangular axes is called first order tensors, if its components a_i, a_j relatively to any two systems of rectangular axes OX_1, OX_2, OX_3, $O\bar{X}_1$, $O\bar{X}_2$, $O\bar{X}_3$ are connected by the relation,

$$\bar{a}_j = l_{ij} a_i \qquad \text{...(3)}$$

$$\Rightarrow \quad a_i = l_{ij}\,\bar{a}_j$$

l_{ij} being cosines of angle between OX_i and $O\bar{X}_j$. A tensor of first order is also called a *vector.*

Note. Consider any two tensors of first order and let $a_i, b_j, \bar{a}_p, \bar{b}_q$; be the components of the same relatively to two different systems of axes, we have

$$\bar{a}_p = l_{ip} a_i, \; \bar{b}_q = l_{jq} b_j$$

where l_{ip} and l_{jq} have their usual meanings. This gives

$$\bar{a}_p \bar{b}_q = l_{ip} a_i l_{jq} b_j = l_{ip} l_{jq} a_i b_j \qquad \text{... (1)}$$

The R.H.S. of (1) denotes the sum of 9 terms obtained by giving all possible pair of values to the dummy suffixes i,j so that each components of $\bar{a}_p \bar{b}_q$ is expressed as a linear combination of nine components of the set a_i, b_j; the coefficient being dependent only upon the positions of the two systems of axes relative to each other and not on the components of the sets $\bar{a}_p \bar{b}_q$, $a_i b_j$

55.9 SECOND ORDER TENSORS

Definition. Any entity representanle by a two suffixes set relatively to a system of rectangular axes is called a second order tensor, if the sets $a_{ij}, \bar{a}_{pq}$ representing the entity relative to any two systems of rectangular axes $OX_1, OX_2, OX_3, O\bar{X}_1, O\bar{X}_2, O\bar{X}_3$ are connected by relation.

$$\bar{a}_{pq} = l_{ip}\, l_{jp}\, a_{ij}$$

55.10 TENSORS OF ANY ORDER

Definition. Any entity representable by a set with m, suffixes relatively to a system of rectangular co-ordinate axes is called a tensor of order m, if the set $a_{ijkP}\ldots\ldots$, $\bar{a}_{pqrs}\ldots$ representing the entity relatively to any two systems of rectangular axes $OX_1, OX_2, OX_3, O\bar{X}_1, O\bar{X}_2, O\bar{X}_3$ are connected by the relation

$$\bar{a}_{pqrs}\ldots\ldots = l_{ip} l_{jq} l_{kr} l_{ls} \ldots a_{ijkl}\ldots$$

We say that $a_{ijkl}\ldots\ldots$ are the components of tensor relatively to the rectangular system of axes OX_1, OX_2, OX_3.

55.11 TENSOR OF ZERO ORDER

Definition. Any entity representable by a single number such that the same number represents the entity irrespective of any underlying system of axes is called a tensor of order zero. A tensor of order zero is also called a *scalar.*

55.12 ALGEBRAIC OPERATIONS ON TENSORS

Theorem. If $a_{ijkl}\ldots, b_{ijkl}\ldots$ are two tensors of the same order then

$$c_{ijkl}\ldots = a_{ijkl}\ldots + b_{ijkl}\ldots$$

is a tensor of the same order.

Proof. Let $a_{ijkl}\ldots, b_{ijkl}\ldots\ldots$ and $\bar{a}_{pqrs}\ldots\ldots, \bar{b}_{pqrs}\ldots$ be the components of the given tensors relatively to two systems $OX_1, OX_2, OX_3, O\bar{X}_1, O\bar{X}_2, O\bar{X}_3$.

We write

$$c_{ijkl}\ldots\ldots = a_{ijkl}\ldots\ldots + b_{ijkl}\ldots\ldots,$$

$$\bar{c}_{pqrs}\ldots\ldots = \bar{a}_{pqrs}\ldots\ldots + \bar{b}_{pqrs}\ldots\ldots$$

Let l_{ij} denote the cosine of the angle between OX_i and OX_j.

$$\bar{c}_{pqrs}\ldots\ldots = l_{ip} l_{iq} l_{kr}\ldots\ldots c_{ijkl}\ldots\ldots \qquad \ldots (1)$$

As $a_{ijkl}\ldots\ldots$ and $b_{ijkl}\ldots\ldots$ are tensors, we have

$$\bar{a}_{pqrs}\ldots\ldots = l_{ip} l_{jq} l_{kr} l_{ls}\ldots\ldots a_{ijkl}\ldots\ldots \qquad \ldots (2)$$

$$\bar{b}_{pqrs}\ldots\ldots = l_{ip} l_{jq} l_{kr} l_{ls}\ldots\ldots b_{ijkl}\ldots\ldots \qquad \ldots (3)$$

Adding (2) and (3), we obtain (1).

Hence the theorem

Similarly, we can show for difference

$$\bar{a}_{pqrs}\ldots\ldots - \bar{b}_{pqrs}\ldots\ldots = l_{ip} l_{jq} l_{kr} l_{ls}\ldots\ldots \left[a_{ijkl}\ldots - b_{ijkl}\ldots\right]$$

$$\bar{d}_{pqrs}\ldots\ldots = l_{ip} l_{jq} l_{kr} l_{ls}\ldots\ldots d_{ijkl}\ldots\ldots$$

55.13 PRODUCT OF TWO TENSORS

Theorem. If $a_{ijkl}\ldots, b_{pqrs}\ldots$ be two tensors of order α and β respectively, then $c_{ijkl\ldots pqrs}\ldots = a_{ijkl}\ldots b_{pqrs}\ldots$ is a tensor of order $\alpha + \beta$.

Proof. Let a_{ijkl} ....., b_{pqrs} ...... and $\overline{a}_{i_1j_1k_1l_1}$ ....., $\overline{b}_{p_1j_1k_1l_1}$ ....., $\overline{b}_{p_1q_1r_1s_1}$ ..... be the components of given tensor relatively to two systems $OX_1, OX_2, OX_3, O\overline{X}_1, O\overline{X}_2, O\overline{X}_3$ we write

$$c_{ijkl...pqrs} = a_{ijkl}.....b_{pqrs}$$

$$\overline{c}_{i_1j_1k_1l_1.......p_1q_1r_1s_1}..... = \overline{a}_{i_1j_1k_1l_1}.....\overline{b}_{p_1q_1r_1s_1}.....$$

Let l_{ij} be the direction cosines of the angle between OX_i and $O\overline{X}_j$, then

$$\overline{c}_{i_1j_1k_1l_1.......p_1q_1}..... = l_{ii_1}l_{jj_1}.....l_{pp_1}l_{qq_1}....c_{ijk...pqr.......} \quad ...(1)$$

As a_{ijkl} ..... and b_{pqrs} ...... are tensors we have

$$\overline{a}_{i_1j_1k_1l_1}..... = l_{ii_1}l_{jj_1}l_{kk_1}.....a_{ijkl}..... \quad ...(2)$$

$$\overline{b}_{p_1q_1r_1s_1}..... = l_{pp_1}l_{qq_1}l_{rr_1}.....b_{pqrs}.... \quad ...(3)$$

Multiplying (2) and (3) we get (1). The new tensor obtained is called product of the tensors.

55.14 QUOTIENT LAW OF TENSORS

Thorem. If there be an entity representable by a multisuffix set a_{ij} relatively to any given system of rectangular axes and if $a_{ij}b_i$ is a vector, where b_i is any arbitrary vector whatsoever then a_{ij} is a tensor of order two.

Proof. $a_{ij}b_i = c_j$ so that c_j is a vector. Let $a_{ij}b_i, c_j$ and $\overline{a}_{pq}, \overline{b}_p, \overline{c}_q$ be the components of the given entity and two vectors relatively to two systems of axes $OX_1, OX_2, OX_3, O\overline{X}_1, O\overline{X}_2, O\overline{X}_3$, then we have

$$a_{ij}b_i = c_j \quad ...(1)$$

$$\overline{a}_{pq}\overline{b}_p = \overline{c}_q \quad ...(2)$$

Also, b_i, c_j being vectors, we have

$$\overline{c}_q = l_{jq}c_j \quad ...(3)$$

$$b_i = l_{ip}\overline{b}_p \quad ...(4)$$

From these, we have

$$\overline{a}_{pq}\overline{b}_p = \overline{c}_q = l_{jq}c_j - l_{jq}a_{ij}b_i = l_{jq}a_{ij}l_{ip}\overline{b}_p = l_{ip}l_{jq}a_{ij}\overline{b}_p$$

i.e., $\left(\overline{a}_{pq} - l_{ip}l_{iq}a_{ij}\right)\overline{b}_p = 0$

As the vector $\overline{b}_p$ is arbitrary, we consider three vectors whose components relatively to $O\overline{X}_1, O\overline{X}_2, O\overline{X}_3$ are 1, 0, 0 ; 0, 1, 0 ; 0, 0, 1.

For these vectors, we have from (5)

$\overline{a}_{1q} - l_{il}l_{jq}a_{ij} = 0, \qquad \overline{a}_{2q} - l_{i2}l_{jq}a_{ij} = 0, \qquad \overline{a}_{3q} - l_{i3}l_{jq}a_{ij} = 0$

These are equivalent to

$$\overline{a}_{pq} - l_{ip}l_{jq}a_{ij} = 0$$

i.e., $\overline{a}_{pq} = l_{ip}l_{jq}a_{ij}$, [This shows that a_{ij} is of second order]

so that the components of the given entity obey the tensorial transformation laws. Hence the result.

55.15 CONTRACTION THEOREM

Theorem. If a_{ijkl}...... is a tensor of order m, then the set obtained on identifying any two suffixes is a tensor of order $(m-2)$.

Proof. Let a_{ijkl} ..... , $\overline{a}_{pqrs}$ ....., be the components of the given tensor relatively to two coordinate systems of axes $OX_1, OX_2, OX_3, O\overline{X}_1, O\overline{X}_2, O\overline{X}_3$, so that we have,

$$\bar{a}_{pqrs\ldots\ldots} = l_{ip}l_{jq}l_{kr}\ldots a_{ijkl\ldots\ldots} \qquad \ldots (1)$$

Let us identify q and s then,

$$\bar{a}_{pqrs\ldots\ldots} = l_{ip}l_{js}l_{kr}l_{ls}\ldots a_{ijkl\ldots\ldots}$$

$$\bar{a}_{pr\ldots\ldots} = l_{ip}l_{kr}\delta_{jl}\ldots a_{ijkl\ldots\ldots} \qquad \begin{bmatrix} \because \delta_{jl} = 0, & j \neq l \\ \delta_{jl} = 1, & j = l \end{bmatrix}$$

$$= l_{ip}l_{kr}\ldots a_{ilkl\ldots} = l_{ip}l_{kr}\ldots a_{ik\ldots}$$

This shows that the order of the tensor reduces by two.

Hence the theorem.

55.16 SYMMETRIC AND ANTISYMMETRIC TENSORS

If $\quad A^k_{rs} = A^k_{sr} \quad$ or $\quad \left(A^{r,s}_k = A^{s,r}_k\right)$

then A^k_{rs} (or A^k_{sr}) are said to be symmetric tensors.

If $\quad B^k_{rs} = -B^k_{sr} \quad$ or $\quad \left(B^{rs}_k = -B^{sr}_k\right)$

then B^k_{rs} (or B^{sr}_k) are known as antisymmetric tensors.

The symmetric (or antisymmetric) property is conserved under a transform of co-ordinates.

55.17 SYMMETRIC AND SKEW SYMMETRIC TENSORS

Invariance of the symmetric and skew-symmetric character of the sets of components of tensors

Theorem. Show that if $a_{ijkl\ldots\ldots}$ is symmetric (skew-symmetric) in any two suffixes, then so is also $\bar{a}_{pqrs\ldots\ldots}$ in the same suffixes.

Proof. Let $a_{ijkl\ldots\ldots}$, $\bar{a}_{pqrs\ldots\ldots}$ be the components of a tensor respectively to two systems of axes OX_1, OX_2, OX_3, $O\bar{X}_1, O\bar{X}_2, O\bar{X}_3$. Then we have

$$\bar{a}_{pqrs\ldots\ldots} = l_{ip}l_{jq}l_{kr}l_{ls}\ldots a_{ijkl\ldots\ldots} \qquad \ldots (1)$$

Now, suppose that $a_{ijkl\ldots\ldots}$ is symmetric in the second and fourth suffixes. Interchanging q and s on the two sides of (1) we obtain

$$\bar{a}_{psrq\ldots\ldots} = l_{ip}l_{js}l_{kr}l_{lq}\ldots a_{ijkl\ldots\ldots} \qquad \ldots (2)$$

As j and l are dummy, we can interchange them. Then interchanging j and l on the R.H.S. of (2) we get

$$\bar{a}_{psrq\ldots\ldots} = l_{ip}l_{ls}l_{kr}l_{jq}\ldots a_{ijkl\ldots\ldots} \qquad \ldots (3)$$

$$\bar{a}_{psrq\ldots\ldots} = l_{ip}l_{jq}l_{kr}l_{ls}\ldots a_{ijkl\ldots\ldots}$$

The set $a_{ijkl\ldots\ldots\ldots}$ is symmetric in the second and fourth suffixes.

Now from (1) and (3) we have

$$\bar{a}_{pqrs\ldots\ldots} = \bar{a}_{psrq\ldots\ldots\ldots}$$

Hence the result.

Definition. A tensor is said to be symmetric (skew-symmetric) in any two suffixes if its components relatively to every co-ordinate system are symmetric (skew-symmetric) in the two suffixes, in question.

A tensor is said to be symmetric (skew-symmetric) if it is so in every pair of suffixes, *e.g.*,

If u_i, v_j be any two vectors then the two second order tensors $u_iv_j + u_jv_i$, $u_iv_j - u_jv_i$ are respectively symmetric and skew-symmetric.

55.18 THEOREM

Every second order tensor can be expressed as the sum of a symmetric and a skew-symmetric tensor.

Proof. Let a_{ij} be any tensor of order 2. Now,

$$\bar{a}_{pq} = l_{ip} l_{jq} a_{ij}$$

$$= l_{jp} l_{iq} a_{ji} \qquad ...(1)$$

where we have interchanged the two dummy suffixes i and j. Then (1) shows that a_{ij} is also a tensor of order two.

Now, $$a_{ij} = \frac{1}{2}\left[a_{ij} + a_{ji}\right] + \frac{1}{2}\left[a_{ij} - a_{ji}\right]$$

= symmetric + skew-symmetric

Thus a_{ij} is the sum of symmetric and skew symmetric tensors.

55.19 A FUNDAMENTAL PROPERTY OF TENSORS

Theorem. If the components of a tensor relatively to any one system of co-ordinate axes are all zero, then the components relatively to every system of co-ordinate axes are also zero.

Proof. Consider a tensor whose components relatively to the systems of axes OX_1, OX_2, OX_3, $O\bar{X}_1, O\bar{X}_2, O\bar{X}_3$ are a_{ijkl} ...., $\bar{a}_{pqrs}$ ......and let a_{ijkl} ....= 0 for every system of values of i, j, k, l...... we have

$$\bar{a}_{pqrs\ldots\ldots} = l_{ip} l_{jq}\, l_{kr}\, l_{ls} \ldots a_{ijkl\ldots\ldots} = 0$$

for every system of values of p, q, r, s......

55.20 ZERO TENSOR

Def. A tensor whose components relatively to one co-ordinate system and, also relatively to every co-ordinate system are all zero is known as zero tensor.

A zero tensor of every order is denoted by 0.

EXERCISE 55.1

1. Write the following using summation convention:

   (a) $(x^1)^1 + (x^1)^2 + (x^1)^3 \ldots\ldots (x^1)^n$ **Ans.** $(x^1)^i$

   (b) $(x^1)^2 + (x^2)^2 + (x^3)^2 + \ldots (x^n)^2$ **Ans.** $(x^i)^2$

   (c) $\frac{df}{dt} = \frac{\partial f}{\partial x^1}\frac{dx^1}{dt} + \frac{\partial f}{\partial x^2}\frac{dx^2}{dt} + \ldots\ldots + \frac{\partial f}{\partial x^n}\frac{dx^n}{dt}$ **Ans.** $\frac{df}{dt} = \frac{\partial f}{\partial x^i}\frac{dx^i}{dt}$

2. Write out all the tensor in $S = a_{ij} x^i x^j$ taking $n = 3$.

   Ans. $S = (a_{11}x^1x^1 + a_{12}x^1x^2 + a_{13}x^1x^3) + (a_{21}x^2x^1 + a_{22}x^2x^2 + a_{23}x^2x^3) + (a_{31}x^3x^1 + a_{32}x^3x^2 + a_{33}x^3x^3)$

3. Write the tensor contained in x^{pq}, x_{qr} if $n = 2$

   Ans. $(x^{11} + x^{21})x_{11} + (x^{11} + x^{21})x_{12} + (x^{12} + x^{22})x_{21} + (x^{12} + x^{22})x_{22}$

4. How many equations in a four dimensional space are represented by $R^{\alpha}_{\beta\gamma\rho} = 0$ **Ans. 8**

5. Show that every tensor can be expressed as the sum of two tensors, one of which is symmetric and the other skew-symmetric in a pair of covariant or contravariant indices.

6. Show that the symmetric (or antisymmetric) property of a tensor is conserved under a transformation of co-ordinates.

7. If A^i and B_j are components of a contravariant and covariant tensor of rank one, then show that $C^i_j = A^i B_j$ are the components of a mixed tensor of rank 2.

8. Write down the laws of transformation for the tensors A^{ij}_k and B^{ij}_{klm}

Ans. $\overline{A}^{ij}_k = \frac{\partial \bar{x}^i}{\partial x^p}\frac{\partial \bar{x}^j}{\partial x^q}\frac{\partial x^r}{\partial \bar{x}^k}A^{pq}_r, \quad \overline{B}^{ij}_{klm} = \frac{\partial \bar{x}^i}{\partial x^p}\frac{\partial \bar{x}^j}{\partial x^q}\frac{\partial x^r}{\partial \bar{x}^k}\frac{\partial x^s}{\partial \bar{x}^l}\frac{\partial x^t}{\partial \bar{x}^m}B^{rq}_{rst}$

9. Evaluate (*a*) $\delta^i_j \delta^j_k$ (*b*) $\delta^i_j \delta^j_k \delta^k_l$ **Ans.** (*a*) δ^i_k (*b*) δ^i_j

10. Show that the velocity of a fluid at any point is a contravariant tensor of rank one.

55.21 TWO SPECIAL TENSORS

1. Alternate tensor

Consider an abstract entity of order 3 and dimension 3 such that its components relatively to every system of co-ordinate axes are the same and given by $\in_{ijk}$ where

$$\in_{ijk} = \begin{cases} 0 & \text{if any two of } ijk \text{ are equal} \\ 1 & \text{if } ijk \text{ is a cyclic permutation } 1, 2, 3 \\ -1 & \text{if } ijk \text{ is an anti cyclic permutation } 1, 2, 3 \end{cases}$$

Thus for unequal values of the suffixes, we have

$$\in_{123} = \in_{312} = \in_{231} = 1, \ \in_{132} = \in_{213} = \in_{321} = -1$$

Let $OX_1, OX_2, OX_3, O\overline{X}_1, O\overline{X}_2, O\overline{X}_3$ be two systems of rectangular axes. We want to show that $\in_{ijk}$ is a tensor of order three. Consider, now expression

$$l_{ip} l_{jq} l_{kr} \in_{ijk} \qquad \dots (1)$$

For any given system of values *p, q, r,* the expression (1) consists of a sum of $3^3 = 27$ terms of which 6 only are non-zero, for the other 21 terms corresponds to a case when atleast two of *i, j, k* are equal. The expression (1) can be written as in the form of determinant

$$\begin{vmatrix} l_{ip} & l_{2p} & l_{3p} \\ l_{1q} & l_{2q} & l_{3q} \\ l_{1r} & l_{2r} & l_{3r} \end{vmatrix}$$

From properties of determinants,

$$\text{Above determnant} = \begin{cases} 0 \text{ if any two of } p, q, r \text{ have equal value.} \\ 1 \text{ if } p, q, r \text{ is a cyclic permutation of } 1, 2, 3 \\ -1 \text{ if } p, q, r \text{ is a non cyclic permutation of } 1, 2, 3 \end{cases}$$

Thus we see that the components of the given entity in any two systems of rectangular axes satisfy the tensorial transformation equations so that the entity is a tensor. This tensor is known as *Alternate tensor.* Thus, we see alternate tensor is same as skew-symmetric tensor. $\in_{ijk}$, always denote the alternate tensor.

55.22 KRONECKER TENSOR

The symbol δ^k_i kronecker delta is defined as

$$\delta^k_i = 0 \text{ when } k \neq i$$

$$\delta^k_i = 1 \text{ when } k = i$$

It mean $\delta^1_1 = \delta^2_2 = \dots\dots = \delta^n_n = 1$ and $\delta^2_1 = \delta^3_1 = \delta^1_2 = \delta^3_2 = \dots\dots = 0$

In general $A_{ij}\delta^j_k = A_{i1}\delta^1_k = A_{i2}\delta^2_k + \dots + A_{ik}\delta^k_k + \dots + A_{1n}\delta^n_k$

$$= 0 + 0 + \dots + A_{ik}(1) + \dots + 0 = A_{ik}$$

Example 3. *If A^{ij} are the cofactors of a^{ij} in a deteminant Δ of order three, then show that*

$$a_{ij}A^{kj} = \Delta\delta_i^k$$

Solution. We know that

$$a_{11}A^{11} + a_{12}A^{12} + a_{13}A^{13} = \Delta \quad \text{... (1)}$$

$$a_{11}A^{21} + a_{12}A^{22} + a_{13}A^{23} = 0 \quad \text{... (2)}$$

$$a_{11}A^{31} + a_{12}A^{32} + a_{13}A^{33} = 0 \quad \text{... (3)}$$

These three equations can be written in brief as

$$a_{ij}A^{1j} = \Delta \quad \text{... (4)} \qquad a_{ij}A^{2j} = 0 \quad \text{... (5)} \qquad a_{ij}A^{3j} = 0 \quad \text{... (6)}$$

Using kronecker delta, equations (4), (5), (6) can be combined into a single equation:

$$a_{ij}A^{kj} = \Delta\delta_1^k \quad \text{... (7)}$$

Similarly six more equations are given by

$$a_{2j}A^{kj} = \Delta\delta_2^k \quad \text{.... (8)} \quad \text{and} \quad a_{3j}A^{kj} = \Delta\delta_3^k \quad \text{... (9)}$$

Equations (7), (8); (9) can be written as a single equation.

$$a_{ij}A^{kj} = \Delta\delta_i^k \quad \text{... (10)}$$

All the nine equations of the determinant are included in one equation (10).

55.23 ISOTROPIC TENSOR

A tensor which has the same set of components relatively to every system of co-ordinate axes is called an *Isotropic tensor.*

55.24 RELATION BETWEEN ALTERNATE AND KRONECKER TENSOR

Prove that $\in_{ijm}\in_{klm} = \delta_{ik}\delta_{jl} - \delta_{il}\delta_{jk}$. Here each side is a tensor of order 4 so that tensor equality is equivalent to set of 81 scalar equality. We have to prove that

$$\in_{ij1}\in_{kl1} + \in_{ij2}\in_{kl2} + \in_{ij3}\in_{kl3} = \delta_{ik}\delta_{jl} - \delta_{il}\delta_{jk}$$

Proof: Case I. When $i = j$ or $k = l$. There will be 45 such equations and for all these equations L.H.S. = 0 = R.H.S.

Case II. If the pair (i, j) such that $i \neq j$ is different from the pair (l, k), $l \neq k$ we see that there will be 24 such scalar equations for which L.H.S. = 0 = R.H.S.

Ex. $(i, j) = (1, 2), (j, k) = (1, 3), (3, 1), (2, 3), (3, 2)$

$(i, j) = (2, 1), (j, k) = (1, 3), (3, 1), (2, 3), (3, 2).$

Case III. Thus we are left to consider the possibility when i, j and k, l take the pairs of values,

$$(1, 2); (1, 3); (2, 3); (2, 1); (3, 1); (3, 2).$$

Consider the first case we have

$i = 1, j = 2, k = 1, l = 2$; $i = 1, j = 2, k = 2, l = 1$; $i = 2, j = 1, k = 1, l = 2$; $i = 2, j = 1, k = 2, l = 1$.

Each pair of (i, j) *i.e* (1, 2) gives two scalar equations. Thus 6 pairs of (i, j) give 12 such scalar equations. In these cases we have

L.H.S. = 1 = R.H.S., L.H.S. = –1 = R.H.S.

L.H.S. = – 1 = R.H.S., L.H.S. = 1 = R.H.S.

This result is also true for other five cases. Hence we have the result.

Example 4. *Prove that* $\in_{ilm}\in_{jlm} = 2\delta_{ij}$

Proof. We know $\in_{ilm}\in_{jkm} = \delta_{ij}\delta_{lk} - \delta_{ik}\delta_{lj}$

Taking $k = l$ we get

$$\in_{ilm} \in_{jlm} = \delta_{ij}\delta_{ll} - \delta_{il}\delta_{lj}$$

Now $$\delta_{ll} = \delta_{11} + \delta_{22} + \delta_{33} = 1+1+1 = 3$$

$$\delta_{il}\delta_{lj} = \delta_{ij} \qquad \because \delta_{ij}a_{im} = a_{jm}$$

$\therefore$ $$\in_{ilm} \in_{jlm} = 3\delta_{ij} - \delta_{ij} = 2\delta_{ij}$$ **Proved.**

Example 5. *Prove that* $\in_{ijk} \in_{ijk} = 6$

Proof. $\in_{ilm} \in_{jkm} = \delta_{ij}\delta_{lk} - \delta_{ik}\delta_{lj}$

Taking $k = l$, we get

$$\in_{ilm} \in_{jlm} = 2\delta_{ij}$$

Taking $i = j$ (Contraction)

$$\in_{ilm} \in_{ilm} = 2\delta_{ii} = 2 \times 3 = 6$$ **Proved.**

55.25 MATRICES AND TENSORS OF FIRST AND SECOND ORDER

Consider any vector. Its components a_i relatively to any system of axes may be written in the form of a row or a column matrix as

$$[a_1 a_2 a_3] \text{ or } \begin{bmatrix} a_1 \\ a_2 \\ a_3 \end{bmatrix}$$

We shall be writing $a_i = [a_1 a_2 a_3]$ or $a_i = \begin{bmatrix} a_1 \\ a_2 \\ a_3 \end{bmatrix}$

Now consider second order tensor. Its components a_{ij} relatively to any system of rectangular axes can be written as the form of matrix such that a_{ij} occurs at the intersection of the ith row and j^{th} column. Thus we shall write

$$\left[a_{ij}\right] = \begin{bmatrix} a_{11} & a_{12} & a_{13} \\ a_{21} & a_{22} & a_{23} \\ a_{31} & a_{32} & a_{33} \end{bmatrix}$$

A matrix obtained by interchanging rows and columns of a given matrix is called the transpose of the same. Transpose of $[a_{ij}]$ will be denoted by $[a_{ij}]'$.

Sum of two matrices of the same type is the matrix whose elements are the sums of the corresponding elements of two matrices.

55.26 SCALAR AND VECTOR PRODUCTS OF TWO VECTORS

Def. 1. *Scalar product.* The scalar $u_i\, v_i$ is called the scalar product of the two vectors u_i, v_j. Thus the scalar product $= u_1v_1 + u_2\, v_2 + u_3v_3$.

Def. 2. *Vector product* The vector $E_{ijk}u_iv_j$ is called vector product of two vectors u_i, v_j taken in this order. Components of these vectors are $u_2\, v_3 = u_3v_2$, $u_3v_1 = u_1v_3$, $u_1v_2 = u_2\, v_1$.

55.27 THE THREE SCALAR INVARIANTS OF A SECOND ORDER TENSOR

I. a_{ii} or $a_{11} + a_{22} + a_{33}$

II. $\frac{1}{2}(a_{ii}a_{jj} - a_{ij}a_{ji})$ or $a_{11}a_{22} + a_{22}a_{33} + a_{33}a_{11} - a_{12}a_{21} - a_{23}a_{32} - a_{31}a_{13}$

III. $|a_{ij}|$ or $\begin{vmatrix} a_{11} & a_{12} & a_{13} \\ a_{21} & a_{22} & a_{23} \\ a_{31} & a_{32} & a_{33} \end{vmatrix}$

Proof. I. Identifying i, j we see that a_{ii} is scalar. Thus

$$a_{ii} = a_{11} + a_{22} + a_{23} \qquad ...(1)$$

is invariant.

II. Consider now the tensor of 4th order, $a_{ij}a_{pq}$. Identifying i with q and j with p we see that $a_{ij}a_{ji}$ is scalar.

Thus $$a_{ij}a_{ji} = (a_{11})^2 + (a_{22})^2 + (a_{33})^2 + 2a_{12}a_{21} + 2a_{23}a_{32} + 2a_{31}a_{13} \qquad ... (2)$$

is invariant.Subtracting(2) from square of (1) and dividing by 2. We establish invariance of II.

(III) If a_{ij}, $\bar{a}_{pq}$ denote the components of tensor relatively to any two co-ordinate systems of axes, then is the usual notation, we have

$$\bar{a}_{pq} = l_{ip} l_{jq} a_{ij}$$

$$|\bar{a}_{pq}| = |l_{ip}||l_{jq}||a_{ij}|$$

since $$|l_{ip}| = |l_{jq}|$$

$\therefore$ $$|\bar{a}_{pq}| = |l_{ip}|^2 |a_{ij}| \qquad \text{but} \qquad |l_{ip}|^2 = 1$$

$\therefore$ $$|\bar{a}_{pq}| = |a_{ij}|$$

Hence it is an invariance.

55.28 SINGULAR AND NON-SINGULAR TENSORS OF SECOND ORDER

A tensor of second order is said to be singular or non-singular according as its determinant is zero or non zero.

55.29 RECIPROCAL OF A NON-SINGULAR TENSOR

Suppose a_{ij} be a second order tensor such that $|a_{ij}| \neq 0$

Lemma 1. We form another matrix

$$A_{ij} = \frac{\text{Cofactor of } a_{ij} \text{ in the determinant } a_{ij}}{|a_{ij}|}$$

Now, by theory of determinants, we know

$$A_{ki} a_{ij} = \delta_{kj} \qquad ...(1)$$

We shall now show that A_{ij} is a second order tensor, we can not do so, by using Quotient law, from equation (1) since a_{ij} is not an arbitrary tensor. Let c_j be an arbitrary vector, then

$$c_j a_{ij} = d_i \qquad ... (2)$$

So that d_i is also a vector. We shall prove that this is an arbitrary vector. Now (2) is equivalent to a set of 3 linear equations in the components of c_j and as the determinant of $a_{ij} \neq 0$, we may assign any arbitrary values to d_i and the resulting equations can be uniquely solved for the components of c_j. Thus d_i is an arbitrary vector. We now have

$$d_i A_{ki} = a_{ij} c_j A_{ki} = A_{ki} a_{ij} c_j = \delta_{kj} c_j = c_k$$

$$c_k = d_i a_{ki} \qquad ..(3)$$

Therefore by quotient law A_{ki} is a second order tensor.

Lemma 2. $e_{ij} = \dfrac{\text{Cofactor of } A_{ij} \text{ in the detterminant } A_{ij}}{|A_{ij}|}$

We know from the theory of determinants.

$$|A_{ij}||a_{ij}| = 1 \qquad \text{But} \qquad |a_{ij}| \neq 0$$

Hence determinant $|A_{ij}| \neq 0$

We shall now show that $e_{ij} = a_{ij}$

$$e_{ki}A_{ij} = \delta_{kj}$$

Take inner product with a_{ji}

$$e_{ki}A_{ij}a_{jl} = \delta_{ki}a_{jl}$$

$$e_{ki}\delta_{il} = a_{kl}$$

$$e_{kl} = a_{kl}$$

Def. Two second order non-singular tensors a_{ij} and A_{ij} are said to be conjugate (or reciprocal) tensors if they satisfy the equation

$$A_{ki}a_{ij} = \delta_{kj}$$

55.30 EIGEN VALUES AND EIGEN VECTORS OF A TENSOR OF SECOND ORDER

Def. A scalar, λ, is called an eigen value of second order tensor a_{ij}, if there exists a non-zero vector x, such that $a_{ij}x_j = \lambda x_i$. This equation is equivalent to

$$a_{ij}x_j = \lambda\delta_{ij}x_j$$

or $$(a_{ij} - \lambda\delta_{ij})x_j = 0 \qquad \ldots (1)$$

since $x_j \neq 0$, Hence $|a_{ij} - \lambda\delta_{ij}| = 0$ ...(2)

This is a necessary condition for λ, to be eigen value. Eq. (2) is cubic eq. in λ and therefore in general will give us three eigen values may not all be distinct corresponding to the tensor a_{ij}.

Consider now any system of co-ordinate axes OX_1, OX_2, OX_3, and let a_{ij} be the component of the given tensor in this system. Consider now a vector x_j, whose components relatively to OX_1, OX_2, OX_3 are given on solving (1). As the components of x_i are not zero relatively to one system OX_1, OX_2, OX_3, this vector can be zero vector *i.e.* its components relatively to any system of axes can not all be zero.

The tensor eq. (1) being true for one system OX_1, OX_2, OX_3 will be true for every system of axes.

Thus we see that every second order tensor possesses three eigen values, not necessarily all distinct. These eigen values are the roots of the cubic $|a_{ij} - \lambda\delta_{ij}| = 0$ in λ. Also to each eigen value corresponds an eigen vector. The vector x_i corresponding to eigen value λ is called an eigen vector.

55.31 THEOREM

Orthogonality of eigen vectors corresponding to distinct eigen values of a symmetric second order tensor.

Proof. Let a_{ij} be a symmetric second order tensor, and let x_i and y_i be the eigen vectors corresponding to the distinct eigen values λ_1 and λ_2 $(\lambda_1 \neq \lambda_2)$ we have

$$a_{ij}x_j = \lambda_1 x_i \qquad ...(1)$$

$$a_{ij}y_j = \lambda_2 y_i \qquad ...(2)$$

Now, $\lambda_1 x_i y_i = a_{ij}x_j y_i$

$= a_{ji}\, x_j y_i$ $\qquad [\because a_{ij} = a_{ji}]$

$= a_{ij}\, y_j x_i$ $\qquad$ (Interchanging dummy indices)

$\lambda_1 x_i y_i = a_{ij} y_j x_i$

$\therefore \quad \lambda_1 x_i y_i = \lambda_2 y_i x_i \qquad$ or $\qquad (\lambda_1 - \lambda_2)(x_i y_i) = 0$

since $\lambda_1 - \lambda_2 \neq 0 \quad \therefore \ x_i y_i = 0$

Thus x and y_i are orthogonal *i.e*. the eigen vectors are orthogonal.

55.32 REALITY OF THE EIGEN VALUES

Theorem. The eigen values of symmetric second order tensor are real

Proof. Let λ be any eigen value so that we have a relation

$$a_{ij}x_j = \lambda\, x_i \qquad ...(1)$$

Here the components of x_j cannot be assumed to be all real. Taking complex conjugate (denoted by bar) in (1), we get

$$\bar{a}_{ij}.\bar{x}_j = \bar{\lambda}\bar{x}_i$$

$$a_{ij}\bar{x}_j = \bar{\lambda}\bar{x}_i$$

$\left[\because a_{ij}\right.$ is symmetric $\therefore \ a_{ij} = a_{ji}$

$\left.\bar{a}_{ij} = a_{ij}\right.$ (all elements are real)

Take inner product by x_i

$$a_{ij}\bar{x}_j x_i = \bar{\lambda}\bar{x}_i x_i$$

$\bar{\lambda}\{\bar{x}_i x_i\} = a_{ji}(\bar{x}_j x_i)$ $\qquad$ by symmetry

$= a_{ij}\bar{x}_i x_j$ $\qquad$ [interchanging dummy indices]

$= a_{ij} x_i \bar{x}_j$ $\qquad$ $[\bar{x}_i x_i = \text{real}]$

$= \bar{a}_{ij} x_i \bar{x}_j = \text{real}$ $\qquad$ $[\because (a - ib)(a + ib) = a^2 + b^2$ which is real]

This shows that the right hand side is real. Hence $\bar{\lambda}$ is real. Thus λ is real *i.e.* eigen value are real. **Proved.**

55.33 ASSOCIATION OF A SKEW SYMMETRIC TENSORS OF ORDER TWO AND VECTORS

We associate the skew symmetric tensor of order two.

$$a_{ij} = \in_{ijk}\, a_k \qquad ...(1)$$

The tensor a_{ij} is skew symmetric for

$$a_{ji} = \in_{jik}\, a_k = - \in_{ijk}\, a_k = -a_{ij}$$

The relation (1) is equivalent to statements

$$a_{23} = a_1,\ a_{32} = -a_1;\ a_{31} = a_2,\ a_{13} = -a_2;\ a_{12} = a_3,\ a_{21} = -a_3;\ a_{11} = 0,\ a_{22} = 0;\ a_{33} = 0.$$

On the inner multiplication with $\in_{ijm}$ we obtain from (1)

$\in_{ijm}\, a_{ij} = \in_{ijm} \in_{ijk}\, a_k \qquad \in_{ijk} \in_{pqk} = \delta_{ip}\delta_{jq} - \delta_{iq}\delta_{jl}$

$= 2\delta_{mk} a_k \qquad \in_{ijk} \in_{pjk} = \delta_{ip}\delta_{jj} - \delta_{jp}\delta_{ij}$

$= 2a_m$ when $k = m \qquad = 3\delta_{ip} - \delta_{ip} = 2\delta_{ip}$

Hence $a_m = \dfrac{1}{2} \in_{ijm}\, a_{ij}$

This shows that association is one-one.

55.34 TENSOR FIELDS

A tensor field or a tensor point function is said to be defined when there is given a law which associates to each point of given region of space a tensor of the same order. Thus a tensor field a_{ij}...... of any order is defined if the components a_{ij}...... are functions of x_1, x_2, x_3.

55.35 GRADIENT OF TENSOR FIELDS:GRADIENT OF A SCALAR FUNCTION.

Let u be a scalar point function so that there is a value of u associated with each point of a given region of space. Thus if OX_1, OX_2, OX_3 and $O\overline{X}_1, O\overline{X}_2, O\overline{X}_3$ be any two systems, then u is a function of x_i and $\overline{x}_p$ which are co-ordinates of any point P relatively to the two systems of axes. For any point P, $x_i, \overline{x}_p$ are different but the values of u are same. Consider now two sets of first order

$$\frac{\partial u}{\partial x_i}, \frac{\partial u}{\partial \overline{x}_p} \qquad \text{we have } \frac{\partial u}{\partial x_i} = \frac{\partial u}{\partial \overline{x}_p}\frac{\partial \overline{x}_p}{\partial x_i}$$

We know that $\overline{x}_p = l_{ip}x_i$

$$\therefore \quad \frac{\partial \overline{x}_p}{\partial x_i} = l_{ip} \qquad \therefore \quad \frac{\partial u}{\partial x_i} = l_{ip}\frac{\partial u}{\partial \overline{x}_p}$$

Thus we see that $\frac{\partial u}{\partial x_i}$ is a tensor of order one *i.e.* a vector. This is usually denoted by u, i,

$$\text{grad } u = u, i$$

If components $\frac{\partial u}{\partial x_i}$ and $\frac{\partial u}{\partial \overline{x}_p}$ relatively to two systems of axes $OX_1, 0X_2, OX_3$ and $O\overline{X}_1$, $O\overline{X}_2, O\overline{X}_3$ obey the tensorial transformation law. This vector is called the *gradient of scalar u.*

55.36 GRADIENT OF VECTOR

Consider now any tensor field u_i of order one. If u_p, $\overline{u}_p$ be the components relatively to two systems of axes $OX_1, 0X_2, OX_3, O\overline{X}_1, O\overline{X}_2, O\overline{X}_3$ we have $\overline{u}_p = l_{ip}u_i$

$$\therefore \quad \frac{\partial \overline{u}_p}{\partial \overline{x}_j} = l_{ip}\frac{\partial u_i}{\partial \overline{x}_j} = l_{ip}\frac{\partial u_i}{dx_k}\cdot\frac{\partial x_k}{\partial \overline{x}_j} = l_{ip}l_{kj}\frac{\partial u_i}{\partial x_k} \qquad \left[\begin{array}{l} x_k = l_{kj}\overline{x}_j \\ \dfrac{\partial x_k}{\partial \overline{x}_j} = l_{kj} \end{array}\right.$$

We see $\frac{\partial u_i}{\partial x_k}$ is a tensor of second order. It is denoted by symbol $u_{i,j}$ and is called the *gradient of* $u_{i,j}$

55.37 DIVERGENCE OF VECTOR POINT FUNCTION

The scalar of the gradient of a vector point function is called the divergence of the point function. Thus if u_i is a vector point function so that

$$u_{i,j} = \frac{\partial u_i}{\partial x_j} \text{ is its gradient,then } u_{i,i} = \frac{\partial u_i}{\partial x_i} = \frac{\partial u_1}{\partial x_1} + \frac{\partial u_2}{\partial x_2} + \frac{\partial u_3}{\partial x_3} \text{ is called div } u_i$$

$$\text{div } u_i = u_{i,i}$$

55.38 CURL OF A VECTOR POINT FUNCTION

The vector of the gradient of a vector point function is called the curl of the point function.

Thus if u_i is a vector point function so that $u_{i,j}=\dfrac{\partial u_i}{\partial x_j}$ is its gradient, then the vector of a tensor *i.e.* the vector $\in_{jik}\, u_{i,j}$ is called the curl of u_i denoted by the symbol curl id, $u_i=\in_{jik}\, u_{i,j}$.

Example 6. *Prove the following results*

(*i*) grad $(\phi\psi)=\phi$ grad $\psi+\psi$ grad ϕ (*ii*) grad $(\vec{f}.\vec{g})=\vec{f}\times\text{curl}\,\vec{g}+\vec{g}\times\text{curl}\,\vec{f}+\vec{f}.\nabla\vec{g}+\vec{g}.\nabla\vec{f}$

(*iii*) div $(\phi\vec{f})=\phi$ div $\vec{f}+\vec{f}$ grad ϕ (*iv*) div $(\vec{f}\times\vec{g})=\vec{g}.$ curl $\vec{f}-\vec{f}.$curl $\vec{g}$

(*v*) curl $(\phi\vec{f})=$ grad $\phi\times\vec{f}+\phi$ curl $\vec{f}$ (*vi*) curl $(\vec{f}\times\vec{g})=\vec{f}\,\text{div}\,\vec{g}-\vec{g}\,\text{div}\,\vec{f}+\vec{g}.\nabla\vec{f}-\vec{f}.\nabla\vec{g}$

Proof. (*i*) grad $(\phi\psi)=(\phi\psi),\ i=\phi\psi,\ i+\psi\phi,\ i=\phi$ grad $\psi+\psi$ grad ϕ.

(*ii*) grad $(\vec{f}.\vec{g})=(f_i g_i)_{,j}$

$$=f g_{i,j}+g f_{i,j} \qquad \ldots(1)$$

Now $\vec{f}\times\text{curl}\,\vec{g}=\in_{pkm} f_p \in_{jik} g_{i,j}=-\in_{pmk}\in_{jik} f_p g_{i,j}$

$$=-\left[\delta_{pj}\delta_{mi}-\delta_{pi}\delta_{mj}\right]f_p g_{i,j}=-\delta_{pj}\delta_{mi}f_p g_{i,j}+\delta_{pi}\delta_{mj}f_p g_{i,j}$$

Identifying p, j and m, i in first part and p, i and m, j in second part, we get.

$$\vec{f}\times\text{curl}\,\vec{g}=-\delta_{pp}\delta_{mm}f_p g_{m,p}+\delta_{pp}\delta_{mm}f_p g_{p,m}=-f_p g_{m,p}+f_p g_{p,m}$$

$$=-\vec{f}.\nabla\vec{g}+f_p g_{p,m} \qquad \ldots(2)$$

similarly, $\vec{g}\times\text{curl}\,\vec{f}=-g.\nabla\vec{f}+g_p f_{p,m} \qquad \ldots(3)$

Substituting the value of $f_p g_{p,m} g_p f_{p,m}$ from (2) and (3) into (1), we get

$$\text{grad}(\vec{f}.\vec{g})=\vec{g}\times\text{curl}\,\vec{f}+\vec{g}.\nabla\vec{f}+\vec{f}\times\text{curl}\,\vec{g}+\vec{f}.\nabla\vec{g}$$

or $\text{grad}(\vec{f}.\vec{g})=\vec{g}\times\text{curl}\,\vec{f}+\vec{f}\times\text{curl}\,\vec{g}+\vec{g}.\nabla\vec{f}+\vec{f}.\nabla\vec{g}$

(*iii*) $\text{div}(\phi\vec{f})=(\phi f_i)_{,i}=\phi f_{i,i}+\phi_{,i}f_i=\phi\,\text{div}\vec{f}+f_i\phi_{,i}=\phi\,\text{div}\,\vec{f}+\vec{f}.\text{grad}\,\phi$

(*iv*) $\text{div}(\vec{f}\times\vec{g})=(\in_{ijk} f_i g_j)_{,k}=\in_{ijk}\{f_i g_{j,k}+g_j f_{i,k}\}=\in_{ijk} f_i g_{j,k}+\in_{ijk} g_j f_{i,k}$

$$=-\in_{kji} f_i g_{j,k}+\{\in_{kij} g_j f_{i,k}\}=-f_i\in_{kji} g_{j,k}+g_j\in_{kij} f_{i,k}$$

$$=-\vec{f}.(\text{curl}\,\vec{g})+\vec{g}.(\text{curl}\,\vec{f})=\vec{g}.(\text{curl}\,\vec{f})-\vec{f}.(\text{curl}\,\vec{g})$$

(*v*) $\text{curl}(\phi\vec{f})=\in_{jik}(\phi f_i)_{,j}=\in_{jik}\phi f_{i,j}+\in_{jik}\phi_{,j}f_i$

$$=\phi\in_{jik} f_{i,j}+\in_{jik} f_i\phi_{,j}=\phi(\text{curl}\,\vec{f})+(\text{grad}\,\phi)\times\vec{f}$$

(*vi*) curl $(\vec{f}\times\vec{g})=\in_{mkn}(\in_{ijk} f_i g_j)_{,m}=-\in_{mnk}\in_{ijk}[f_i g_{j,m}+g_j f_{i,m}]$

$$=-[\delta_{mi}\delta_{nj}-\delta_{mj}\delta_{nj}][f_i g_{j,m}+g_j f_{i,m}]=-[\delta_{mi}\delta_{nj}f_i g_{j,m}-\delta_{mj}\delta_{ni}f_i g_{j,m}+\delta_{mi}\delta_{nj}g_j f_{i,m}-\delta_{mj}\delta_{ni}g_j f_{i,m}]$$

$$=-\delta_{mm}\delta_{nn}f_m g_{n,m}+\delta_{mm}\delta_{nn}f_n g_{m,m}-\delta_{mm}\delta_{nn}g_n f_{m,m}+\delta_{mm}\delta_{nn}g_m f_{n,m}$$

$$=-f_m g_{n,m}+f_n g_{m,m}-g_n f_{m,m}+g_m f_{n,m}=-\vec{f}.\nabla\vec{g}+(\text{div}\,\vec{g})\vec{f}-(\text{div}\,\vec{f})\vec{g}+\nabla\vec{f}.\vec{g}$$

$$=\vec{f}\,\text{div}\vec{g}-\vec{g}\,\text{div}\,\vec{f}+\vec{g}.\nabla\vec{f}-\vec{f}.\nabla\vec{g}$$

55.39 SECOND ORDER DIFFERENTIAL OPERATORS

(*i*) div (grad ϕ) = $\nabla^2\phi$ (*ii*) curl (grad ϕ) = 0

(*iii*) div (curl $\vec{f}$) = 0 (*iv*) grad (div $\vec{f}$) = curl(curl $\vec{f}$) + $\nabla^2\vec{f}$

Proof. (I) div (grad ϕ) $= (\phi_{,i})_{,i} = \phi_{,ii} = \nabla^2\phi$

(*ii*) curl (grad ϕ) $= \in_{jik} (\phi_{,i})_{,i}$

$$= \in_{jik} \phi_{,ij} = I \text{ ,say} \quad \text{... (1)}$$

$$\phi_{,ij} = \phi_{,ji} \qquad \left[\frac{\partial^2\phi}{\partial x\partial y} = \frac{\partial^2\phi}{\partial y\partial x}\right]$$

Now, $I = \in_{jik} \phi_{,ji} = \in_{ijk} \phi_{,ij}$ ...(2)

From (1) and (2) $2I = (\in_{jik} + \in_{ijk})\phi_{,ij}$

$$2I = (-\in_{ijk} + \in_{ijk})\phi_{,ij} = 0$$

Hence, $I = 0$ or curl (grad ϕ) = 0

(*iii*) div (curl $\vec{f}$) $= (\in_{jik} f_{i,j})_{,k} = \in_{jik} f_{i,jk} = I$ (say) ... (1)

Because $f_{i,jk} = f_{i,kj}$

Then $I = \in_{jik} f_{i,kj} = \in_{kij} f_{i,jk}$ ...(2)

From (1) and (2) $2I = (\in_{jik} + \in_{kij}) f_{i,jk} = (\in_{jik} - \in_{jik}) f_{i,jk} = 0$

Hence, $I = 0$ or div (curl $\vec{f}$) = 0

(*iv*) grad (div $\vec{f}$) $= (f_{i,i})_{,j} = f_{i,ij}$ ... (1)

$$curl(curl\vec{f}) = \in_{mkn} (\in_{jik} f_{i,j})_{m} = \in_{nmk} \in_{jik} f_{i,jm} = (\delta_{nj}\delta_{mi} - \delta_{ni}\delta_{mj}) f_{i,jm}$$

$$= (\delta_{nn}\delta_{mm} f_{m,nm} - \delta_{nn}\delta_{mm} f_{n,mm}) = f_{m,nm} - f_{n,mm}$$

$$= f_{m,mn} - f_{n,mm} = \text{grad}(\text{div } \vec{f}) = \nabla^2\vec{f} \qquad \text{[From (1)]}$$

Thus, $\text{grad}(\text{div } \vec{f}) = curl(curl\vec{f}) + \nabla^2\vec{f}$

55.40 TENSORIAL FORM OF GAUSS'S AND STOKE'S THEOREM

Gauss's divergence theorem.

If $\vec{F}$, is a continuously differentiable vector point function and S is a closed surface enclosing a region V, then

$$\oint_s \vec{F}.\hat{n}ds = \int_V \text{div } \vec{F} dv \qquad \text{...(1)}$$

where $\hat{\eta}$ is a uunitt vector $\oint_s \vec{F}_i n_i ds = \int_V F_{i,i} dV = \int \frac{\partial F_i}{\partial x_i} dV$

55.41 STOKE'S THEOREM

If $\vec{F}$ is any continuously differentiable vector point function and S is a surface bounded by a curve c, then

$$\oint_c \vec{F}.dr = \int_S \text{curl } \vec{F}.\hat{\eta}ds \qquad \text{...(2)}$$

where $\hat{\eta}$ is a unit vector $\oint_c \vec{F}_i dr_i = \int_S (\in_{jik} F_{i,j}) n_k ds$

Example 7. *By means of divergence theorem of Gauss's, show that*

$$\oint_S \in_{qpi} n_p \in_{ijk} a_j x_k ds = 2a_q V$$

where V is the volume enclosed by the surface S, having the outward drawn normal n. The position vector to any point in V is x_i and a_i, *is an arbitrary constant vector.*

Proof. $\quad$ L.H.S. $= \oint_S \in_{ipq} \in_{ijk} n_p a_j x_k ds$

$$= \oint_S (\delta_{qj}\delta_{pk} - \delta_{qk}\delta_{pj}) n_p a_j x_k ds = \oint_S n_k a_q x_k ds - \oint_S n_j a_j x_q ds$$

$$= a_q \oint_S n_k x_k ds - a_j \oint_S n_j x_q ds = a_q \oint_V \frac{\partial x_k}{\partial x_k} dv - a_j \oint_V \frac{\partial x_q}{\partial x_j} dv = a_q \delta_{kk} v - a_j \delta_{qj} v$$

$$= 3a_q v - a_q v = 2a_q v$$ **Proved.**

Example 8. *If $\vec{q} = \vec{w} \times \vec{r}$, show that $2\vec{w} = \nabla \times \vec{q}$ using the index notation. The vector $\vec{w}$ is a constant.*

Solution. $\quad q_k = \in_{ijk} w_i x_j$ (given)

$$\left[\nabla \times \vec{q}\right]_m = \in_{lkm} q_{k,l} = \in_{lkm} \in_{ijk} (w_i x_j)_{,l}$$

$$\left[\nabla \times \vec{q}\right]_m = \in_{mlk} \in_{ijk} (w_i x_{j,l} + x_j w_{l,l}) = \in_{mlk} \in_{ijk} w_i x_{j,l}$$

Since $\vec{w}$ is a constant vector

$$\therefore \quad w_{i,l} = 0 = (\delta_{mi}\delta_{lj} - \delta_{mj}\delta_{li}) w_i x_{j,l} = w_m x_{l,l} - w_l x_{m,l} = w_m \delta_{ll} - w_i \delta_{ml} = 3w_m - w_m$$

$$\left[\nabla \times \vec{q}\right]_m w = 2w_m$$

Hence $\quad \nabla \times \vec{q} = 2\vec{w}.$ **Proved.**

55.42 RELATION BETWEEN ALTERNATE AND KRONECKER TENSOR

$$\in_{ijk} \in_{lmn} = \begin{vmatrix} l_{1i} & l_{2i} & l_{3i} \\ l_{1j} & l_{2j} & l_{3j} \\ l_{1k} & l_{2k} & l_{3k} \end{vmatrix} \times \begin{vmatrix} l_{1l} & l_{2l} & l_{3l} \\ l_{1m} & l_{2m} & l_{3m} \\ l_{1n} & l_{2n} & l_{3n} \end{vmatrix} = \begin{vmatrix} \delta_{il} & \delta_{im} & \delta_{in} \\ \delta_{jl} & \delta_{jm} & \delta_{jn} \\ \delta_{kl} & \delta_{km} & \delta_{kn} \end{vmatrix}$$

Identifying k and l, we get

$$\in_{ijk} \in_{kmn} = \begin{vmatrix} \delta_{ik} & \delta_{im} & \delta_{in} \\ \delta_{jk} & \delta_{jm} & \delta_{jn} \\ \delta_{kk} & \delta_{km} & \delta_{kn} \end{vmatrix} = \begin{vmatrix} \delta_{ik} & \delta_{im} & \delta_{in} \\ \delta_{jk} & \delta_{jm} & \delta_{jn} \\ 3 & \delta_{km} & \delta_{kn} \end{vmatrix}$$

$$\because \quad \delta_{kk} = \delta_{11} + \delta_{22} + \delta_{33} = 1 + 1 + 1 = 3.$$

Expanding the determinant, we have

$$\in_{ijk} \in_{kmn} = \delta_{ik} (\delta_{jm}\delta_{kn} - \delta_{jn}\delta_{km}) + \delta_{im} (3\delta_{jn} - \delta_{jk}\delta_{kn}) + \delta_{in} (\delta_{jk}\delta_{km} - 3\delta_{jm})$$

$$= \delta_{jm}\delta_{in} - \delta_{jn}\delta_{im} + 3\delta_{im}\delta_{jn} - \delta_{im}\delta_{jn} + \delta_{in}\delta_{jm} - 3\delta_{in}\delta_{jm}$$

$$\in_{ijk} \in_{kmn} = \delta_{im}\delta_{jn} - \delta_{in}\delta_{jm}$$

55.43 THE THREE SCALAR INVARIANTS OF A SECOND ORDER TENSOR

Let a_{ij} be a second order tensor

(*i*) a_{ii}

Proof. $\bar{a}_{pq} = l_{ip} l_{jq} a_{ij}$

Identifying p and q, we have $\bar{a}_{pp} = l_{ip} l_{jp} a_{ij}$

$\bar{a}_{pp} = \delta_{ij} a_{ij} = a_{ii}$ Hence a_{ii} is an invariant.

(*ii*) $\frac{1}{2} (a_{ii} a_{jj} - a_{ij} a_{ij})$

We know that a_{ii} and a_{jj} are invariants. Now we have to show that $a_{ij} a_{ji}$ is also an invariant. Then $(a_{ii} a_{jj} - a_{ij} a_{ij})$ will also be invariant.

Let $\quad \bar{a}_{pq} = l_{ip} l_{jp} a_{ij} \quad$ and $\quad \bar{a}_{rs} = l_{mr} l_{ns} a_{mn}$

Now consider the tensor of 4th order

$$\bar{a}_{pq}\ \bar{a}_{rs} = l_{ip}\, l_{jq}\, l_{mr}\, l_{ns}\, a_{ij}\, a_{mn}$$

First identifying r and q and then identifying p and s we have

$$\bar{a}_{pq}\ \bar{a}_{qp} = l_{is}\, l_{jr}\, l_{mr}\, l_{ns}\, a_{ij}\, a_{mn} = \delta_{in}\,\delta_{jm}\, a_{ij}\, a_{mn} = a_{ij}\, a_{ji}$$

Hence $a_{ij}\, a_{ji}$ is an invariant. Therefore $\frac{1}{2}(a_{ii}\, a_{jj} - a_{ij}\, a_{ji})$ is invariant

(*iii*) $|a_{ij}|$ **Proof.** $\bar{a}_{pq} = l_{ip}\, l_{jq}\, a_{ij} \quad \Rightarrow \quad |\bar{a}_{pq}| = |l_{ip}|\,|l_{jq}|\,|a_{ij}|$

We know by the property of determinants $|l_{ip}|\,|l_{jq}| = 1 \quad \Rightarrow \quad |\bar{a}_{pq}| = |a_{ij}|$

Hence $|a_{ij}|$ is an invariant.

55.44 TENSOR ANALYSIS

Example 9. *What is a mixed tensor of second rank? Prove that* $\delta_q^{\,p}$ *is a mixed tensor of the second rank*

Solution. The N^2 quantities $A_s^{\,q}$ are called components of a mixed tensor of the second rank if

$$\bar{A}_r^{\,p} = \frac{\partial \bar{x}^p}{\partial x^q}\frac{\partial x^s}{\partial \bar{x}^r} A_s^{\,q}$$

Now, if $\delta_s^{\,q}$ defined by $\delta_s^{\,q} = \begin{cases} 0 & \text{if } p \neq q \\ 1 & \text{if } p = q \end{cases}$

is a mixed tensor of second rank, it must transform according to the rule $\bar{\delta}_k^i = \frac{\partial \bar{x}^i}{\partial x^p}\frac{\partial x^q}{\partial \bar{x}^k}\delta_q^p$

The right side equals $\frac{\partial \bar{x}^j}{\partial x^p}\frac{\partial x^p}{\partial \bar{x}^k} = \delta_k^j$

since $\bar{\delta}_k^j = \delta_k^j = 1$ if $j = k$, and 0 if $j \neq k$, if follows that δ_r^p is a mixed tensor of rank two.

Example 10. *Evaluate* (*i*) $\delta_q^p\ A_s^{qr}$ (*ii*) $\delta_q^p\ \delta_r^p$

Solution. (*i*) $\delta_q^p\ A_s^{qr} = \delta_q^p\ A_s^{pr} = A_s^{pr}$

(*ii*) $\delta_q^p\ \delta_r^p = \delta_p^p\ \delta_r^p = \delta_r^p \qquad \because \delta_q^p = 1$

Example 11. *Show that every tensor can be expressed as the sum of two tensors one of which is symmetric and the other skew-symmetric in a pair of covariant or contravariant indices.*

Solution. Consider the tensor B^{pq}, we have

$$B^{pq} = \frac{1}{2}(B^{pq} + B^{qp}) + \frac{1}{2}(B^{pq} - B^{qp})$$

But $R^{pq} = \frac{1}{2}(B^{pq} + B^{qp}) = R^{qp}$ is symmetric and

$S^{pq} = \frac{1}{2}(B^{pq} - B^{qp}) = R^{qp}$ is skew-symm.

Thus B^{pq} = symm tensor + skew-symm tensor.

By similar reasoning the result is seen to be true for any tensor.

Example 12. *What is contraction as applied to tensors? Prove that the contraction of the tensor* A_q^p *is a scalar or invariant.*

Solution. *Contraction.* If one contravariant and one covariant index of a tensor are set equal, the result indicates that a summation over the equal indices is to be taken according to the summation convention. This resulting sum is a tensor of rank two less than that of the original tensor. The process is called contraction. For example, in the tensor of rank 3, B_q^{mp}, set $p = q$ we get $B_q^{mp} = C^m$, a tensor of rank 1.

To prove that contraction of A_q^p is a scalar or invariant.

we have
$$\bar{A}_k^j = \frac{\partial \bar{x}^j}{\partial x^p}\frac{\partial x^q}{\partial \bar{x}^k} A_q^p$$

putting $j = k$,
$$\bar{A}_j^j = \frac{\partial \bar{x}^j}{\partial x^p}\frac{\partial x^q}{\partial \bar{x}^j} A_q^p = \delta_p^q\, A_p^p = A_p^p$$

Then $A_j^j = A_q^p$ and it follows that A_p^p must be an invariant. Since A_q^p is a tensor of rank two and contraction with respect to a single index lowers the rank by two. Therefore, an invariant is a tensor of rank zero. **Proved.**

Example 13. *A covariant tensor has components xy, 2y – z², xz in rectangular co-ordinates. Find its covariant components in spherical co-ordinates.*

Solution. Let A_j denote the covariant component in rectangular co-ordinates

$$x^1 = x,\ x^2 = y,\ x^3 = z.$$

Then
$$A_1 = xy = x^1 x^2$$
$$A_2 = 2y - z^2 = 2x^2 - (x^3)^2$$
$$A_3 = xz = x^1 x^3$$

Let $\bar{A}_k$ denote the covariant component in spherical co-ordinates $\bar{x}^1 = r, \bar{x}^2 = \theta, \bar{x}^3 = \phi$

Then
$$\bar{A}_k = \frac{\partial \bar{x}^j}{\partial \bar{x}^k} A_j \quad \text{... (1)}$$

In spherical co-ordinates

$$x = r \sin\theta \cos\phi$$
or
$$x^1 = \bar{x}^1 \sin \bar{x}^2 \cos \bar{x}^3 \quad \text{...(2)}$$
$$y = r \sin\theta \sin\phi$$
or
$$x^2 = \bar{x}^1 \sin \bar{x}^2 \sin \bar{x}^3 \quad \text{...(3)}$$
$$z = r\cos\theta$$
or
$$x^3 = \bar{x}^1 \cos \bar{x}^2 \quad \text{...(4)}$$

Therefore equation (1) yields the covariant component.

$$\bar{A}_1 = \frac{\partial x^1}{\partial \bar{x}^1} A_1 + \frac{\partial x^2}{\partial \bar{x}^1} A_2 + \frac{\partial x^3}{\partial \bar{x}^1} A_3$$
$$= (\sin \bar{x}^2 \cos \bar{x}^3)(x^1 x^2) + (\sin \bar{x}^2 \sin \bar{x}^3) \times [(2x^2 - (x^3)^2)] + (\cos \bar{x}^2)(x^1 x^3)$$
$$= (\sin\theta\cos\phi))(r^2 \sin^2\theta \sin\phi\cos\phi) + (\sin\theta\sin\phi)(2r\sin\theta\sin\phi - r^2\cos^2\theta)$$
$$+ (\cos\theta)(r^2 \sin\theta\cos\theta\cos\phi)$$

$$\bar{A}_2 = \frac{\partial x^1}{\partial \bar{x}^2} A_1 + \frac{\partial x^2}{\partial \bar{x}^2} A_2 + \frac{\partial x^3}{\partial \bar{x}^2} A_3$$
$$= \left(\bar{x}^1 \cos \bar{x}^2 \cos \bar{x}^3\right)(x^1 x^2) + \left(\bar{x}^1 \cos \bar{x}^2 \sin \bar{x}^3\right)[(2x^2 - (x^3)^2)] + \bar{x}^1(-\sin \bar{x}^2)(x^1 x^2)$$

or $\bar{A}_2 = (r\cos\theta\cos\phi)(r^2\sin^2\theta\sin\phi\cos\phi) + (r\cos\theta\sin\phi)(2r\sin\theta\sin\phi - r^2\cos^2 0)$

$+ (-r\sin\theta)(r^2\sin\theta\cos\theta\cos\phi)$ $\bar{A}_3 = \frac{\partial x^1}{\partial \bar{x}^3} A_1 + \frac{\partial x^2}{\partial \bar{x}^3} A_2 + \frac{\partial x^3}{\partial \bar{x}^3} A_3$

$$= (-r\sin\theta\sin\phi)(r^2\sin^2\theta\sin\phi\cos\phi) + (r\sin\theta\cos\phi)(2r\sin\theta\sin\phi - r^2\cos^2\theta).$$ **Ans.**

Example 14. *Define symmetric and skew-symmetric tensors. Prove that a symmetric tensor of rank two has at most* $\frac{N(N+1)}{2}$ *different components in N-dimensional space* V_N.

Solution. Symmetric Tensor. A tensor is called symmetric with respect to two contravariant or two covariant indices if its components remain unaltered upon interchange of the indices.

Thus if $A_{qs}^{mpr} = A_{qs}^{pmr}$, the tensor is symmetric in m and p. If a tensor is symmetric with respect to any two contravariant and any two covariant indices, it is called symmetric.

Skew-symmetric. A tensor is called skew-symmetric with respect to two contravariant or two covariant indices if its component change sign upon interchange of the indices. Thus, if $A_{qs}^{mpr} = -A_{qs}^{pmr}$ the tensor is skew symmetric in m and p. If a tensor is skew-symmetric with respect to any two contravariant and any two covariant indices it is called skew-symmetric.

Let A_{pq} be a tensor of rank 2. The number of its all components in V_N is N^2.

The components of A_{pq} are

$$\begin{array}{l} A_{11} \quad A_{12} \quad A_{13} \ldots\ldots\ldots\ldots A_{1N} \\ A_{21} \quad A_{22} \quad A_{23} \ldots\ldots\ldots\ldots A_{2N} \\ \ldots\ldots\ldots\ldots\ldots\ldots\ldots\ldots\ldots\ldots\ldots\ldots\ldots\ldots \\ A_{N1} \quad A_{N2} \quad A_{N3} \ldots\ldots\ldots\ldots A_{NN} \end{array}$$

There are N independent components of the form

$$A_{11}, A_{22}, A_{33} \ldots\ldots\ldots\ldots A_{NN}$$

Hence number of components of the form $A_{12}, A_{23}, A_{34}, \ldots\ldots\ldots\ldots$ in which there are distinct subscripts will be $N^2 - N$. But these component are symmetric. *i.e.*, $A_{12} = A_{21}$ etc.

Hence number of different component of this form are $\frac{1}{2}(N^2 - N)$

$\therefore$ Total number of different components are

$$= \frac{1}{2}(N^2 - N) + N = \frac{N^2}{2} + \frac{N}{2} = \frac{N(N+1)}{2}$$

Example 15. *Define a metric or fundamental tensor. Determine the components of the fundamental tensor in cylindrical co-ordinates.*

Solution. Metric or Fundamental Tensor.

In rectangular coordinates (x, y, z) the differential of arc length ds is obtained from $ds^2 = dx^2 + dy^2 + dz^2$. By transforming to general curvilinear co-ordinates this becomes

$$ds^2 = \sum_{p=1}^{3} \sum_{q=1}^{3} g_{pq}\, du_p\, du_q$$

Such spaces are called three-dimensional Euclidean spaces. We define the line element ds in this space to be given by the quadratic form, called the metric form or metric,

$$ds^2 = \sum_{p=1}^{N} \sum_{q=1}^{N} g_{pq}\, dx^p\, dx^q \quad \ldots(1)$$

$$\Rightarrow \quad ds^2 = g_{pq}\, dx^p\, dx^q$$

The quantities g_{pq} are the components of a covariant tensor of rank 2 called the *metric* tensor or *fundamental* tensor.

We know that $ds^2 = dx^2 + dy^2 + dz^2$

In cylindrical co-ordinates,

$$x = r\cos\theta,\ y = r\sin\theta,\ z = z$$

$$\therefore \quad dx = -r\sin\theta\, d\theta + \cos\theta\, dr$$

$$dy = r\cos\theta\, d\theta + \sin\theta\, dr$$

$$dz = dz$$

$$ds^2 = dx^2 + dy^2 + dz^2$$

$$= (-r\sin\theta\, d\theta + \cos\theta\, dr)^2 + (r\cos\theta\, d\theta + \sin\theta\, dr)^2 + (dz)^2$$

or

$$ds^2 = (dr)^2 + r^2 (d\theta)^2 + (dz)^2$$

Also metric is given by

$$ds^2 = g_{pq}\, dx^p\, dx^q \qquad \dots (2)$$

If $x^1 = r,\ x^2 = \theta,\ x^3 = z$

then comparing (1) & (2), we have

$$g_{11} = 1,\ g_{22} = r^2,\ g_{33} = 1,\ g_{12} = g_{21} = 0,\ g_{13} = g_{31} = 0,\ g_{23} = g_{32} = 0.$$

Metric tensor is given by $\begin{pmatrix} g_{11} & g_{12} & g_{13} \\ g_{21} & g_{22} & g_{23} \\ g_{31} & g_{32} & g_{33} \end{pmatrix}$

Metric tensor in cylindrical co-ordinates = $\begin{pmatrix} 1 & 0 & 0 \\ 0 & r^2 & 0 \\ 0 & 0 & 1 \end{pmatrix}$ **Ans.**

Example 16. *Define what is meant by invariant? Show that the contraction of the outer product of the tensors* A^p and B_q *is an invariant.*

Solution. Scalar or Invariant. Suppose ϕ is a function of the co-ordinates x^k, and let ϕ denote the functional value under a transformation to a new set of co-ordinates $\bar{x}^k$. Then ϕ is called a *scalar* or *invariant* with respect to the co-ordinate transformation if $\phi = \bar{\phi}$.

Since A^P and B_q are tensors.

$$\bar{A}^j = \frac{\partial \bar{x}^j}{\partial x^p} A^p,\ \bar{B}_k = \frac{\partial x^q}{\partial \bar{x}^k} B_q$$

Then $$\bar{A}^j\,\bar{B}_k = \frac{\partial \bar{x}^j}{\partial x^p}\frac{\partial x^q}{\partial \bar{x}^k} A^p B_q$$

By contraction (putting $j = k$)

$$\bar{A}^j\,\bar{B}_j = \frac{\partial \bar{x}^j}{\partial x^p}\frac{\partial x^q}{\partial \bar{x}^j} A^p B_q = \delta^q_p A^p B_q = A^p B_p$$ **Proved.**

and so $A^P B_p$ is an invariant.

Example 17. *What do you understand by associated tensors ?*

Solution. Associated Tensors. Given a tensor we can derive other tensors by raising or lowering indices. For example, given the tensor A_{pq} we obtain by raising the index p, the tensor $A^p_{\cdot q}$,the dot indicating the original position of the moved index. By raising the index q also we obtain A^{pq}. We shall often write A^{pq}. These derived tensors can be obtained by forming inner products of the given tensor with the metric tensor g_{pq} or its conjugate g^{Pq}. Thus, for example

$$A_q^P = g^{rq} A_{rq}, \; A^{pq} = g^{rP} g^{sq} A_{rs}$$

All tensors obtained from a given tensor by forming inner products with the metric tensor and its conjugate are called *associated tensors* of the given tensor. For example; A^m and A_m are associated tensors, the first are contravariant and the second covariant components. The relation between them is given by

$$A_p = g^{pq} A^q \Rightarrow A^P = g^{pq} A_q.$$

For rectangular co-ordinates $g_{pq} = 1$ if $p = q$, and 0 if $p \neq q$, so that $A_p = A^P$, which explains why no distinction was made between contravariant and covariant components of a vector for rectangular co-ordinates.

55.45 CONJUGATE OR RECIPROCAL TENSORS

Let $g = |g_{pq}|$ denote the determinant with elements g_{pq} and suppose $g \neq 0$. Define g^{pq} by

$$g^{pq} = \frac{\text{cofactor of } g_{pq}}{g}$$

Then g^{pq} is a symmetric contravariant tensor of rank two called the conjugate or Reciprocal tensor of g_{pq}.

Also $\qquad g^{pq} g_{rq} = \delta_r^P$

55.46 CHRISTOFFEL SYMBOLS

The symbols $\qquad [pq, r] = \frac{1}{2}\left(\frac{\partial g_{pr}}{\partial x^q} + \frac{\partial g_{qr}}{\partial x^p} - \frac{\partial g_{pq}}{\partial x^r}\right); \begin{Bmatrix} s \\ pq \end{Bmatrix} = g^{sr} [pq, r]$

are called the Christoffel symbols of the first and second kind respectively.

Example 18. *Prove that* $[pq, r] = g_{rs} \begin{Bmatrix} s \\ pq \end{Bmatrix}$

Solution. $\qquad g_{ks} \begin{Bmatrix} s \\ pq \end{Bmatrix} = g_{ks}\, g^{sr} [pq, r] = \delta_k^r [pq, r] = [pq, k]$

$\Rightarrow \qquad [pq, k] = g_{ks} \begin{Bmatrix} s \\ pq \end{Bmatrix}$

ie. $\qquad [pq, r] = g_{rs} \begin{Bmatrix} s \\ pq \end{Bmatrix}$ **Proved.**

Example 19. *Prove that (i)* $[pq, r] = [qp, r]$ *(ii)* $\begin{Bmatrix} s \\ pq \end{Bmatrix} = \begin{Bmatrix} s \\ qp \end{Bmatrix}$.

Solution. (*i*) $\qquad [pq, r] = \frac{1}{2}\left(\frac{\partial g_{pr}}{\partial x^q} + \frac{\partial g_{qr}}{\partial x^p} - \frac{\partial g_{pq}}{\partial x^r}\right)$

$$= \frac{1}{2}\left(\frac{\partial g_{qr}}{\partial x^p} + \frac{\partial g_{pr}}{\partial x^q} - \frac{\partial g_{qp}}{\partial x^r}\right)$$

$[pq, r] = [qp, r]$

(*ii*) $\begin{Bmatrix} s \\ pq \end{Bmatrix} = g^{sr} [pq, r] = g^{sr} [qp, r] = \begin{Bmatrix} s \\ qp \end{Bmatrix}$ **Proved.**

Example 20. *Prove that* $\dfrac{\partial g_{pq}}{\partial x^m} = [pm, q] + [qm, p]$

Solution. $[pm, q] + [qm, p]$

$$= \frac{1}{2}\left(\frac{\partial g_{pq}}{\partial x^m} + \frac{\partial g_{mq}}{\partial x^p} - \frac{\partial g_{pm}}{\partial x^q}\right) + \frac{1}{2}\left(\frac{\partial g_{qp}}{\partial x^m} + \frac{\partial g_{mp}}{\partial x^q} - \frac{\partial g_{qm}}{\partial x^p}\right) = \frac{1}{2}\frac{\partial g_{pq}}{\partial x^m} + \frac{1}{2}\frac{\partial g_{qp}}{\partial x^m} = \frac{\partial g_{pq}}{\partial x^m}$$

Example 21. *Prove that* $\dfrac{\partial g^{pq}}{\partial x^m} = -g^{pn}\begin{Bmatrix} q \\ mn \end{Bmatrix} - g^{qn}\begin{Bmatrix} p \\ mn \end{Bmatrix}$

Solution. $\dfrac{\partial}{\partial x^m}(g^{jk} g_{ij}) = \dfrac{\partial}{\partial x^m}\left(\delta_i^k\right) = 0$

Then $g^{jk}\dfrac{\partial g_{ij}}{\partial x^m} + \dfrac{\partial g^{jk}}{\partial x^m} g_{ij} = 0 \quad \Rightarrow \quad g^{ij}\dfrac{\partial g^{jk}}{\partial x^m} = -g^{jk}\dfrac{\partial g_{ij}}{\partial x^m}$

Multiplying by g^{ir}

i.e. $$\delta_j^r \frac{\partial g^{jk}}{\partial x^m} = -g^{ir} g^{jk}[im, j] + [jm, i]$$

$\Rightarrow$ $$\frac{\partial g^{rk}}{\partial x^m} = -g^{ir}\begin{Bmatrix} k \\ im \end{Bmatrix} - g^{jk}\begin{Bmatrix} r \\ jm \end{Bmatrix}$$

Replace r, k, i, j by p, q, n, n, we get $\dfrac{\partial g^{pq}}{\partial x^m} = -g^{pn}\begin{Bmatrix} q \\ mn \end{Bmatrix} - g^{qn}\begin{Bmatrix} p \\ mn \end{Bmatrix}$ **Proved.**

Example 22. *Derive transformation laws for the christoffel symbols of the first and the second kind.*

Solution. Since $\bar{g}_{jk} = \dfrac{\partial x^p}{\partial \bar{x}^j}\dfrac{\partial x^q}{\partial \bar{x}^k} g_{pq}$

$$\therefore \quad \frac{\partial \bar{g}_{jk}}{\partial \bar{x}^m} = \frac{\partial x^p}{\partial \bar{x}^j}\frac{\partial x^q}{\partial \bar{x}^k}\frac{\partial g_{pq}}{\partial x^r}\frac{\partial x^r}{\partial \bar{x}^m} + \frac{\partial x^p}{\partial \bar{x}^j}\frac{\partial^2 x^q}{\partial \bar{x}^m \partial \bar{x}^k} g_{pq} + \frac{\partial^2 x^p}{\partial \bar{x}^m \partial \bar{x}^j}\frac{\partial x^q}{\partial \bar{x}^k} g_{pq} \quad \text{... (1)}$$

By cyclic permutation of indices n, k, m and p, q, r

$$\frac{\partial \bar{g}_{km}}{\partial \bar{x}^j} = \frac{\partial x^q}{\partial \bar{x}^k}\frac{\partial x^r}{\partial \bar{x}^m}\frac{\partial g_{qr}}{\partial x^p}\frac{\partial x^p}{\partial \bar{x}^j} + \frac{\partial x^q}{\partial \bar{x}^k}\frac{\partial^2 x^r}{\partial \bar{x}^j \partial \bar{x}^m} g_{qr} + \frac{\partial^2 x^q}{\partial \bar{x}^j \partial \bar{x}^k}\frac{\partial x^r}{\partial \bar{x}^m} g_{pq} \quad \text{...(2)}$$

$$\frac{\partial \bar{g}_{mj}}{\partial \bar{x}^k} = \frac{\partial x^r}{\partial \bar{x}^m}\frac{\partial x^p}{\partial \bar{x}^j}\frac{\partial g_{rp}}{\partial x^q}\frac{\partial x^q}{\partial \bar{x}^k} + \frac{\partial x^r}{\partial \bar{x}^m}\frac{\partial^2 x^p}{\partial \bar{x}^k \partial \bar{x}^j} g_{rp} + \frac{\partial^2 x^r}{\partial \bar{x}^k \partial \bar{x}^m}\frac{\partial x^p}{\partial \bar{x}^j} g_{rp} \quad \text{...(3)}$$

Subtracting (1) from the sum of (2) and (3) and multiplying by $\frac{1}{2}$, we obtain on using the definition of the Christoffel symbols of the first kind,

$$\overline{[jk, m]} = \frac{\partial x^p}{\partial \bar{x}^j}\frac{\partial x^q}{\partial \bar{x}^k}\frac{\partial x^r}{\partial \bar{x}^m}[pq, r] + \frac{\partial^2 x^p}{\partial \bar{x}^j \partial \bar{x}^k}\frac{\partial x^q}{\partial \bar{x}^m} g_{pq} \quad \text{... (4)}$$

55.47 TRANSFORMATION LAW FOR SECOND KIND

Multiplying (4) by $\bar{g}^{nm}$

$$\bar{g}^{nm} = \frac{\partial \bar{x}^n}{\partial x^s}\frac{\partial \bar{x}^m}{\partial x^t} g^{st} \qquad \text{we get}$$

$$\bar{g}^{nm}\,\overline{[jk, m]} = \frac{\partial x^p}{\partial \bar{x}^j}\frac{\partial x^q}{\partial \bar{x}^k}\frac{\partial x^r}{\partial \bar{x}^m}\frac{\partial \bar{x}^n}{\partial x^s}\frac{\partial \bar{x}^m}{\partial x^t} g^{st}[pq, r] + \frac{\partial^2 x^p}{\partial \bar{x}^j \partial \bar{x}^k}\frac{\partial x^q}{\partial \bar{x}^m}\frac{\partial \bar{x}^n}{\partial x^s}\frac{\partial \bar{x}^m}{\partial x^t} g^{st} g_{pq}$$

Then $$\overline{\begin{Bmatrix} n \\ jk \end{Bmatrix}} = \frac{\partial x^p}{\partial \bar{x}^j}\frac{\partial x^q}{\partial \bar{x}^k}\frac{\partial \bar{x}^n}{\partial x^s}\delta_t^r g^{st}[pq,r] + \frac{\partial^2 x^p}{\partial \bar{x}^j \partial \bar{x}^k}\frac{\partial \bar{x}^n}{\partial x^s}\delta_t^q g^{st} g_{pq}$$

$$\overline{\begin{Bmatrix} n \\ jk \end{Bmatrix}} = \frac{\partial x^p}{\partial \bar{x}^j}\frac{\partial x^q}{\partial \bar{x}^k}\frac{\partial \bar{x}^n}{\partial x^s}\begin{Bmatrix} s \\ pq \end{Bmatrix} + \frac{\partial^2 x^p}{\partial \bar{x}^j \partial \bar{x}^k}\frac{\partial \bar{x}^n}{\partial x^p} \qquad \text{... (5)}$$

(4) and (5) are required transformation laws.

Example 23. *If* $(ds)^2 = r^2 (d\theta)^2 + r^2 \sin^2\theta\, (d\phi)^2$, *find the value of*

(*a*) [22, 1] (*b*) [12, 2] (*c*) [1, 22] (*d*) [2, 12].

Solution. $(ds)^2 = r^2 (d\theta)^2 + r^2 \sin^2\theta\, (d\phi)^2$

$$g_{11} = r^2,\ g_{22} = r^2\sin^2\theta,\ g_{12} = 0 = g_{21}$$

$$g = \begin{vmatrix} g_{11} & g_{12} \\ g_{21} & g_{22} \end{vmatrix} = \begin{vmatrix} r^2 & 0 \\ 0 & r^2\sin^2\theta \end{vmatrix} = r^4\sin^2\theta$$

$$g^{11} = \frac{\text{cofactor of } g_{11}}{g} = \frac{r^2\sin^2\theta}{r^4\sin^2\theta} = \frac{1}{r^2}$$

$$g^{22} = \frac{\text{cofactor of } g_{22}}{g} = \frac{r^2}{r^4\sin^2\theta} = \frac{1}{r^2\sin^2\theta}$$

The christoffel symbols of first kind are

$$[ij,k] = \frac{1}{2}\left[\frac{\partial g_{jk}}{\partial x^i} + \frac{\partial g_{ik}}{\partial x^j} - \frac{\partial g_{ij}}{\partial x^k}\right]$$

(*a*) $$[22,1] = \frac{1}{2}\left[\frac{\partial g_{21}}{\partial x^2} + \frac{\partial g_{21}}{\partial x^2} - \frac{\partial g_{22}}{\partial x^1}\right] = \frac{1}{2}\left[\frac{\partial(0)}{\partial\phi} + \frac{\partial(0)}{\partial\phi} - \frac{\partial(r^2\sin^2\theta)}{\partial\theta}\right]$$

$$= r^2\sin\theta\cos\theta$$

(*b*) $$[12,2] = \frac{1}{2}\left[\frac{\partial g_{22}}{\partial x^1} + \frac{\partial g_{12}}{\partial x^2} - \frac{\partial g_{12}}{\partial x^2}\right] = \frac{1}{2}\left[\frac{\partial(r^2\sin^2\theta)}{\partial\theta} + \frac{\partial(0)}{\partial\phi} - \frac{\partial(0)}{\partial\phi}\right]$$

$$= r^2\sin\theta\cos\theta$$

(*c*) The christoffel symbols of second kind are

$$\begin{Bmatrix} k \\ ij \end{Bmatrix} = g^{kl}[ij,l]$$

$$\begin{Bmatrix} 1 \\ 22 \end{Bmatrix} = g^{1l}[22,l] = g^{11}[22,1]\, g^{12}[22,2]$$

$$= \frac{1}{r^2}[-r^2\sin\theta\cos\theta] + 0 \quad (g^{12} = 0) = -\sin\theta\cos\theta$$

(*d*) $$\begin{Bmatrix} 2 \\ 12 \end{Bmatrix} = g^{2l}[22,l] = g^{21}[12,1] + g^{22}[12,2]$$

$$= 0 + \frac{1}{r^2\sin^2\theta}[r^2\sin\theta\cos\theta] = \frac{\cos\theta}{\sin\theta} = \cot\theta = r^4\sin\theta\cos\theta$$ **Ans.**

Example 24. *If* $(ds)^2 = (dr)^2 + r^2 (d\theta)^2 + r^2\sin^2\theta\, (d\phi)^2$, *find the value of*

(*a*) [22, 1], (*b*) [33, 1], (*c*) [13, 3], (d) [23, 3],

(*e*) $\begin{Bmatrix} 1 \\ 22 \end{Bmatrix}$, (*f*) $\begin{Bmatrix} 1 \\ 33 \end{Bmatrix}$, (g) $\begin{Bmatrix} 3 \\ 13 \end{Bmatrix}$, (*h*) $\begin{Bmatrix} 3 \\ 23 \end{Bmatrix}$

Solution. $(ds)^2 = (dr)^2 + r^2 (d\theta)^2 + r^2 \sin^2 \theta (d\phi)^2$

$$x_1 = r,\ x_2 = \theta,\ x_3 = \phi$$
$$g_{11} = 1,\ g_{22} = r^2,\ g_{33} = r^2 \sin \theta$$
$$g_{12} = 0 = g_{13} = ...$$

$$g = \begin{vmatrix} g_{11} & g_{12} & g_{13} \\ g_{21} & g_{22} & g_{23} \\ g_{31} & g_{32} & g_{33} \end{vmatrix} = \begin{bmatrix} 1 & 0 & 0 \\ 0 & r^2 & 0 \\ 0 & 0 & r^2 \sin^2 \theta \end{bmatrix} = r^4 \sin^2 \theta$$

$$g^{11} = \frac{\text{cofactor of } g_{11}}{g} = \frac{\begin{vmatrix} r^2 & 0 \\ 0 & r^2 \sin^2 \theta \end{vmatrix}}{r^4 \sin^2 \theta} = \frac{r^4 \sin^2 \theta}{r^4 \sin^2 \theta} = 1$$

$$g^{22} = \frac{\text{cofactor of } g_{22}}{g} = \frac{\begin{vmatrix} 1 & 0 \\ 0 & r^2 \sin^2 \theta \end{vmatrix}}{r^4 \sin^2 \theta} = \frac{r^2 \sin^2 \theta}{r^4 \sin^2 \theta} = \frac{1}{r^2}$$

$$g^{33} = \frac{\text{cofactor of } g_{33}}{g} = \frac{\begin{vmatrix} 1 & 0 \\ 0 & r^2 \end{vmatrix}}{r^4 \sin^2 \theta} = \frac{r^2}{r^4 \sin^2 \theta} = \frac{1}{r^2 \sin^2 \theta}$$

The christoffel symbols of the first kind are

$$[ij, k] = \frac{1}{2}\left[\frac{\partial g_{jk}}{\partial x^i} + \frac{\partial g_{ik}}{\partial x^i} - \frac{\partial g_{ij}}{\partial x^k}\right]$$

(*a*) $[22, 1] = \frac{1}{2}\left[\frac{\partial g_{21}}{\partial x^2} + \frac{\partial g_{21}}{\partial x^2} - \frac{\partial g_{22}}{\partial x^1}\right] = \frac{1}{2}\left[\frac{\partial(0)}{\partial \theta} + \frac{\partial(0)}{\partial \theta} - \frac{\partial(r^2)}{\partial r}\right] = \frac{1}{2}(-2r) = -r$

(*b*) $[33, 1] = \frac{1}{2}\left[\frac{\partial g_{31}}{\partial x^3} + \frac{\partial g_{31}}{\partial x^3} - \frac{\partial g_{33}}{\partial x^1}\right] = \frac{1}{2}\left[\frac{\partial(0)}{\partial \phi} + \frac{\partial(0)}{\partial \phi} - \frac{\partial(r^2 \sin^2 \theta)}{\partial r}\right] = \frac{1}{2}(-2r \sin^2 \theta) = -r \sin^2 \theta$

(*c*) $[13, 3] = \frac{1}{2}\left[\frac{\partial g_{33}}{\partial x^1} + \frac{\partial g_{13}}{\partial x^3} - \frac{\partial g_{13}}{\partial x^3}\right] = \frac{1}{2}\frac{\partial g_{33}}{\partial x^1} = \frac{1}{2}\left[\frac{\partial(r^2 \sin^2 \theta)}{\partial r}\right] = r \sin^2 \theta$

(*d*) $[23, 3] = \frac{1}{2}\left[\frac{\partial g_{33}}{\partial x^2} + \frac{\partial g_{23}}{\partial x^3} - \frac{\partial g_{23}}{\partial x^3}\right] = \frac{1}{2}\frac{\partial g_{33}}{\partial x^2} = \frac{1}{2}\frac{\partial}{\partial \theta}(r^2 \sin^2 \theta) = r^2 \sin \theta \cos \theta$

The christoffel symbols of the second kind are

$$\begin{Bmatrix} k \\ ij \end{Bmatrix} = g^{kl}[ij, l]$$

(*e*) $\begin{Bmatrix} 1 \\ 22 \end{Bmatrix} = g^{1l}[22, l] = g^{11}[22, 1] + g^{12}[22, 2] + g^{13}[22, 3] = (1)(-r) + 0 + 0 = -r$

(*f*) $\begin{Bmatrix} 1 \\ 33 \end{Bmatrix} = g^{1l}[33, l] = g^{11}[33, 1] + g^{12}[33, 2] + g^{13}[33, 3] = (1)(-r \sin^2 \theta) + 0 + 0 = -r \sin^2 \theta$

(g) $\begin{Bmatrix} 3 \\ 13 \end{Bmatrix} = g^{3l}[13, l] = g^{31}[13, 1] + g^{32}[13, 2] + g^{33}[13, 3]$

$$= 0\,(r \sin^2 \theta) + 0\,[0] + \frac{1}{r^2 \sin^2 \theta}[r \sin^2 \theta] = \frac{1}{r}$$

(h) $\begin{Bmatrix} 3 \\ 23 \end{Bmatrix} = g^{3l}[23,l] = g^{31}[23, 1] + g^{32}[23, 2] + g^{33}[23, 3]$

$$= 0\,[0] + 0\,[0] + \frac{1}{r^2 \sin^2\theta}(r \sin\theta \cos\theta) = \cot\theta$$ **Ans.**

Example 25. *Prove that* $\begin{Bmatrix} p \\ pq \end{Bmatrix} = \frac{\partial}{\partial x^q} \log \sqrt{g}$

Solution. $g = g_{jk} G(j, k)$ (Sum over k only)

where $G(j, k)$ is the cofactor of g_{jk} in the determinant $g = |g_{jk}| \neq 0$ since $G(j,k)$ does not contain g_{jk} explicitly,

$$\frac{\partial g}{\partial g_{jr}} = G(j, r)$$

Then, summing over j and r

$$\frac{\partial g}{\partial x^m} = \frac{\partial g}{\partial g_{jr}} \frac{\partial g_{jr}}{\partial x^m} = G(j,r)\frac{\partial g_{jr}}{\partial x^m} = g\, g^{jr} \frac{\partial g_{jr}}{\partial x^m} = g\, g^{jr}([jm,r] + [rm,j])$$

$$\frac{\partial g}{\partial x^m} = g\left(\begin{Bmatrix} j \\ jm \end{Bmatrix} + \begin{Bmatrix} r \\ rm \end{Bmatrix}\right) = 2g\begin{Bmatrix} j \\ jm \end{Bmatrix}$$

Thus $$\frac{1}{2g}\frac{\partial g}{\partial x^m} = \begin{Bmatrix} j \\ jm \end{Bmatrix} \quad \text{or} \quad \begin{Bmatrix} j \\ jm \end{Bmatrix} = \frac{\partial g}{\partial x^m} \log \sqrt{g}$$

Replacing j by p and m by q, $\begin{Bmatrix} p \\ pq \end{Bmatrix} = \frac{\partial}{\partial x^q} \log \sqrt{g}$ **Proved.**

55.48 CONTRAVARIANT, COVARIANT AND MIXED TENSOR

If A_i be a set of n functions of the co-ordinates $x^1, x^2, x^3 \ldots x^n\,(x^i)$. They are transformed in another system of co-ordinates $\bar{x}^1, \bar{x}^2, \bar{x}^3 \ldots x^n$ according to $\bar{A}_i = \frac{\partial x^j}{\partial \bar{x}^i} A_j$

A_i are called the components of a covariant tensor.

If $\phi\,(x^1, x^2 \ldots x^n)$ be a scalar functions, then $\frac{\partial \phi}{\partial \bar{x}^i} = \frac{\partial \phi}{\partial x^1}\frac{\partial x^1}{\partial \bar{x}^i} + \frac{\partial \phi}{\partial x^2}\frac{\partial x^2}{\partial \bar{x}^i} + \ldots + \frac{\partial \phi}{\partial x^n}\frac{\partial x^n}{\partial \bar{x}^i}$

then $\frac{\partial \phi}{\partial x^1}, \frac{\partial \phi}{\partial x^2} \ldots \frac{\partial \phi}{\partial x^n}$ are the components of a covariant vector.

Since x is a function of x^i (*i.e.*, $x^1, x^2, \ldots x^n$)

so $$d\bar{x}^i = \frac{\partial \bar{x}^i}{\partial x^1} dx^1 + \frac{\partial \bar{x}^i}{\partial x^2} dx^2 + \ldots + \frac{\partial \bar{x}^i}{\partial x^n} dx^n \quad \ldots (2) \quad = \frac{\partial \bar{x}^i}{\partial x^1} dx^1$$

On comparing (1) and (2) we can say that $dx^1, dx^2, \ldots dx^n$ is an example of a contravariant tensor.

If q and s varry from 1 to n, then A^{qs} will be n^2 functions.

If N^2 quantities A^{qs} in a co-ordinate system $(x^1, x^2, \ldots x^N)$ are related to N^2 other quantities $\bar{A}^{pr}$ in another system $(\bar{x}^1, \bar{x}^2, \ldots \bar{x}^N)$ by the transformation equations

$$\bar{A}^{pr} = \sum_{s=1}^{N}\sum_{q=1}^{N} \frac{\partial \bar{x}^p}{\partial x^q}\frac{\partial \bar{x}^r}{\partial x^s} A^{qs} \quad \Rightarrow \quad \bar{A}^{pr} = \frac{\partial \bar{x}^p}{\partial x^q}\frac{\partial \bar{x}^r}{\partial x^s} A^{qs}$$

they are called *contravariant* components of a tensor of the second rank.

If the transformation law is $\bar{A}^{pr} = \frac{\partial x^q}{\partial \bar{x}^p}\frac{\partial x^s}{\partial \bar{x}^r} A_{qs,}$

then quantities A_{qs} are called components of *covariant* tensor of second rank.

The N^2 quantities A^q_s are called components of a *mixed* tensor of second rank if

$$\bar{A}^p_r = \frac{\partial \bar{x}^p}{\partial x^q}\frac{\partial x^s}{\partial \bar{x}^r} A^q_s$$

EXERCISE 55.2

1. If A^i are the components of an absolute contravariant tensor of rank one, show that $\frac{\partial A_i}{\partial x_j}$ are the components of a mixed tensor.

2. If ,A^{ji} and A_{ij} are reciprocal symmetric tensors and x_i are the components of a covariant tensor of rank one, show that $A_{ij}\ x^i\ x^j, = A^{ij}\ x_i\ x_j$ where $x^i = A^{i\alpha} x_\alpha$.

3. If the components of a tensor are zero in one co-ordinate system, then prove that the components are zero in all co-ordinate systems.

4. Show that the expression *A (i,j, k) is* a tensor if its inner product with an arbitrary tensor B^{jl}_k is a tensor.

5. A^{ij} is a contravariant tensor and B_i a covariant tensor. Show that $A^{ij}\ B_k$ is a tensor of rank three, but $A^{ij}\ B_j$ is a tensor of rank one.

6. If g_{ij} denotes the components of a covariant tensor of rank two, show that the product $g_{ij} dx^i\ dx^j$ is an invariant scalar.

7. Find g and g^{ij} corresponding to the metric

$$ds^2 = 5\ (dx^1)^2 + 3\ (dx^2)^2 + 4\ (dx^3)^2 - 6\ dx^1\ dx^2 + 4dx^2\ dx^3.$$

Ans. $g = 4$, $g^{11} = 2$, $g^{22} = 5$, $g^{33} = 1.5$, $g^{12} = 3$, $g^{23} = -2.5$, $g^{13} = -1.5$

8. Find the values of g and g^{ij}, if

$$ds^2 = \frac{dr^2}{1-\frac{r^2}{R^2}} + r^2(d\theta^2 + \sin^2\theta d\phi^2)\text{, where R is constant}$$

Ans. $g = \frac{r^4\sin^2\theta}{1-\frac{r^2}{R^2}}$; $g^{11} = 1-\frac{r^2}{R^2}$, $g^{22} = \frac{1}{r^2}$, $g^{33} = \frac{1}{r^2\sin^2\theta}$, $g^{ij} = 0\ (i \neq j)$

9. Prove that the angle $\theta_{12}, \theta_{23}, \theta_{31}$ between the co-ordinate curves in a three dimensional co-ordinate system are given by

$$\cos\theta_{12} = \frac{g_{12}}{\sqrt{g_{11}g_{22}}}, \cos\theta_{23} = \frac{g_{23}}{\sqrt{g_{22}\ g_{33}}}, \cos\theta_{31} = \frac{g_{31}}{\sqrt{g_{33}\ g_{11}}}$$

10. Prove that for an orthogonal co-ordinate system

(*a*) $g_{12} = g_{23} = g_{31} = 0$ (b) $g^{11} = \frac{1}{g_{11}}, g^{22} = \frac{1}{g_{22}}, g^{33} = \frac{1}{g_{33}}$

11. Surface of a sphere is a two dimensional Riemannian space. Find its fundamental metric tensor. If a be the fixed radius of the sphere.

Ans. $g_{11} = a^2$, $g_{22} = a^2\sin^2\theta$, $g = a^4\sin^2\theta$

$$g^{11} = \frac{1}{a^2}, g^{22} = \frac{1}{a^2\sin^2\theta} g^{12} = 0 = g^{21}.$$

LINEAR PROGRAMMING

56.1. INTRODUCTION

It will be of interest to know that linear programming had its origin during the second world war (1939-45). To fight the war man and material (resources) have to be maintained. There has to be efficient and safe land, sea and airtransport etc.

The government in England studied the problems during war particularly problems of armed forces, civil defence and navel strategy etc. The study for the solutions of the above problems resulted the linear programming.

Linear programming is the most popular mathematical technique which involve the limited resources in an optimal manner.

The term *programming* means planning to maximize profit or minimize cost or minimize loss or minimum use of resources or minimizing the time etc. Such problems are called *optimization Problem.* The term linear means that all equations or inequations involved are linear.

Example 1. *A manufacturer produces two types of toys i.e., A and B. Each toy of type A requires 4 hours of moulding and two hours of polishing whereas each toy of type B requires 3 hours of moulding 5 hours of polishing. Moulder works for 80 hours in a week and polisher works for 180 hours in a week. Profit on a toy of type A is Rs. 3 and on a toy of type B is Rs. 4. In what way the manufacturer allocates his production capacity for the two types of toys so that he may make the maximum profit per week.*

Solution.

Table. The above information can be written in tabular form as follows:

operation / *Toy*	*Moulding (in hours)*	*Polishing (in hours)*	*Profit (in Rs.)*
A	4	2	3
B	3	5	4
Time available (in hours)	80	180	

Let x be the number of toys of type A and y be the number of toys of type B produced per week.

Profit on one toy of type A = Rs. 3

Profit on x toys of type A = Rs. $3x$

Profit on one toy of type B = Rs. 4

Profit one y toys of type B = Rs. $4y$

Let Z be the weekly profit.

Then the weekly profit in Rs. is

$$Z = 3x + 4y$$

Here, Z is known as *objective function* which has to maximize/minimize.
One toy of type A on moulding requires = 4 hours.
x toys of type A on moulding requires = $4x$ hours
One toy of type B on moulding requires = 3 hours
y toys of type B on moulding requires = $3y$ hours
On moulding total time required = $4x + 3y$ hours
But moulder works for only 80 hours in a week.
So, $4x + 3y$ hours cannot exceed 80 hours.

$$\Rightarrow \qquad 4x+3y \leq 80$$

This is known as *first constraint*:
One toy of type A on polishing requires = 2 hours
x toys of type A on polishing requires = $2x$ hours
One toy of type B on polishing requires = 5 hours
y toys of type B on polishing requires = $5y$ hours
On polishing total time required = $2x + 5y$ hours
But polisher works for only 180 hours in a week.
So, $2x + 5y$ hours cannot exceed 180 hours.

$$\Rightarrow \qquad 2x+5y \leq 180$$

This is known as *second constraint.*
Since, the number of toys produced is non-negative.

$$\Rightarrow \qquad x \geq 0 \text{ and } y \geq 0$$

This is known as *third constraint.*
Under these three constraints (conditions) we have to plan the system to get the maximum profit.
Now, we summarize the above informations in mathematical form as follows :

To maximize $\qquad Z = 3x + 4y \qquad$ ... (1)

Subject to the constraints :

$$4x+3y \leq 80 \qquad \text{... (2)}$$

$$2x+5y \leq 180 \qquad \text{... (3)}$$

$$\left.\begin{array}{l} x \geq 0 \\ y \geq 0 \end{array}\right\} \qquad \text{... (4)}$$

The above mathematical expression is known as *mathematical formulation.*
From the above inequations we find out the values of x and y.
The values of x and y are substituted in the objective function $Z = 3x + 4y$.
The maximum/minimum value of the objective function is known as *optimal value.*

56.2. SOME DEFINITIONS

1. Linear Programming Problem

Here, we have to optimize the linear function Z subject to certain conditions. Such problems are called linear programming problems. As example 1 on page 613.

2. Objective functions

Objective function is a linear function of several variables, subject to the conditions that $Z = 3x + 4y$ in the previous example.

3. Optimal Value

Optimal value is a maximum or minimum value of a objective function to be calculated in a linear programming problem.

4. Non-negative Constants

Production of any item is always non-negative, so we write $x \geq 0, y \geq 0$.

5. Linear Relations

All the mathematical relations used in L.P.P. are linear relations.

6. Programming

Programming is the method of determining a particular programme.

7. Decision Variables

Decision variables are x and y which denote the required number of items/products.

8. Constraints

Constraints are linear inequalities or equations involved in linear programming problem. In the previous example (2), (3), (4) are constraints.

9. Optimization Problem

Optimization problem is a problem in which a objective function is to be maximize or minimize subject to the certain conditions. In the previous example we have to maxmize the profit, so it is optimization problem.

56.3. MATHEMATICAL FORMULATION OF LINEAR PROGRAMMING PROBLEMS

In the previous section we have defined certain technical terms of L.P.P. Conversion of the verbal description of L.P.P. into algebraic equations/inequations is known us Mathematical Formultion.

Working Rule to formulate the L.P.P.

Step 1. Identify the decision variables to be determined and expressed them as x, y etc.

Step 2. Identify all the limitations or constraints in the given problem and then express them as linear inequalities or equations in terms of x, y etc.

Step 3. Identify the objective function (Z) which is to optimize (maximize or minimize) and express Z in terms of x, y.

Procedure:

The solution of the given L.P.P. should be divided under the following heads:

1. Preparation of *table of information.*
2. Write down the *decision variables.*
3. Form the *objective function.*
4. Write down the *constraints.*
5. Write down Mathematical Formulation.

TYPE 1. TO MAXIMIZE OBJECTIVE FUNCTION

Example 2. *A company produces two types of ornaments A and B that require gold and silver. Each unit of type A requires 1 gram of silver and 2 grams of gold. Type B requires (each unit) 2 grams of silver and 1 gram of gold. The company has only 100 grams of silver and 80 grams of gold. Each unit of type A brings a profit of Rs. 500 and each unit of type B brings a profit of Rs. 400. Formulate the problem as a linear programming problem to maximize the profit.*

Solution. The above information is given in the following table:

1. Table:

Metal / *Ornaments*	*Silver (in grams)*	*Gold (in grams)*	*Profit (in Rs.)*
A	1	2	500
B	2	1	400
	100	80	

2. Decision variables. Decision variables are ornaments A and B.

Let the number of ornaments A be x and the number of ornaments B be y.

3. Objective function. To maximize profit.

Let Z be the objective function.

Profit of each unit of type A = Rs. 500

Profit of x units of type A = Rs. $500\ x$

Profit of each unit of type B = Rs. 400

Profit of y units of type B = Rs. $400\ y$

Total profit = $500\ x + 400\ y$

$\Rightarrow \qquad Z = 500\ x + 400\ y$

4. Constraint. (*i*) Company has only 100 grams of silver.

One unit of ornament A requires 1 gram of silver.

x units of ornament A require x grams of silver.

One unit of ornament B requires 2 grams of silver.

y units of ornament B require $2y$ grams of silver.

Total available quantity of silver for ornament A and B = 100 grams

$\Rightarrow \qquad x + 2y \leq 100$

Constraint. (*ii*) Company has only 80 grams of gold.

One unit of ornament A requires 2 grams of gold.

x units of ornament A require $2x$ grams of gold.

One unit of ornament B requires 1 gram of gold.

y units of ornament B require y grams of gold.

Total available quantity of gold = 80 grams

$\Rightarrow \qquad 2x + y \leq 80$

Constraint. (*iii*) Production of ornament A and ornament B cannot be negative.

$\Rightarrow \qquad x \geq 0$ and $y \geq 0$.

5. Mathematical Formulation. Hence, the linear programming problem for the given problem is as follows:

Maximize $\qquad Z = 500\ x + 400\ y \qquad$ ...(1)

Subject to the constraints:

$x + 2y \leq 100 \qquad$ ...(2)

$2x + y \leq 80 \qquad$ ...(3)

$x \geq 0 \qquad$ ...(4)

$y \geq 0 \qquad$ ...(5) **Ans.**

Example 3. *Two tailors A and B earn Rs. 150 and Rs. 200 per day respectively. A can stitch 6 shirts and 4 pants per day while B can stitch 10 shirts and 4 pants per day. Form a linear programming problem to minimise the labour cost to produce at least 60 shirts and 32 pants.* **[CBSE 2005]**

Solution. The given data can be put in the tabular form as:

1. Table:

Earning per day	*Rs. 150 per day*	*Rs. 200 per day*	*Minimum requirement*
Tailors / Stitch	A	B	
Shirts	6	10	60
Pants	4	4	32

2. Decision Variable. Let the tailors A and B work for x and y days respectively.

3. Objective function. The total labour cost for working x days of tailor A and y days of tailor B is Rs. $(150x + 200y)$.

Let Z denote the minimum labour cost, then
$Z = 150x + 200\,y$

4. Constraint (*i*). The minimum requirement of shirt is 60.

Tailor A stitches in one day = 6 shirts Tailor A stitches in x days = $6x$ shirts
Tailor B stitches in one day = 10 shirts Tailor B stitches in y days = $10y$ shirts

$\therefore \qquad 6x+10y \geq 60$

Constraint (*ii*). The minimum requirement of pants is 32.

Tailor A stitches in one day = 4 pants Tailor A stitches in x days = $4\,x$ pants
Tailor B stitches in one day = 4 pants Tailor B stitches in y days = $4y$ pants

$\Rightarrow \qquad 4x+4y \geq 32.$

Constraint (*iii*). The number of days worked by A or B is non negative.

$\therefore \qquad x \geq 0 \quad$ and $\quad y \geq 0.$

5. Mathematical formulation.

The mathematical formulation of given L.P.P. is as follows:

Minimize $Z = 150x + 200y$ ... (1)

subject to the constraints:

$$6x+10y \geq 60 \quad \text{... (2)}$$

$$4x+4y \geq 32 \quad \text{... (3)}$$

and $\qquad x, y \geq 0.$ ... (4)

Ans.

Example 4. *A manufacturer of leather belts makes three types of belts A, B and C which are processed on three machines M_1, M_2 and M_3. Belt A requires 2 hours on machine M_1 and 3 hours on machine M_3. Belt B requires 3 hours on machine M_1, 2 hours on machine M_2 and 2 hours on machine M_3 and belt C requires 5 hours on machine M_2 and 4 hours on machine M_3. There are 8 hours of time per day available on machine M_1, 10 hours of time per day available on machine M_2 and 15 hours of time per day available on machine M_3. The profit gained from belt A is Rs. 3.00 per unit, from belt B is Rs. 5.00 per unit and from belt C is Rs. 4.00 per unit. Formulate the L.P.P. to maximize the profit.*

Solution. The above information is given in the following Table:

1. Table:

Machines / Belts	M_1 (in hours)	M_2 (in hours)	M_3 (in hours)	Profit (in Rs.)
A	2	0	3	3
B	3	2	2	5
C	0	5	4	4
Available time (in hours)	8	10	15	

2. Decision Variables. The decision variables are the number of belt A, belt B and belt C. Let the number of belt A be x_1, the number of belt B be x_2, and the number of belt C be x_3.

3. Objective Function. To maximise the profit.

Profit on 1 belt A = Rs. 3 Profit on x_1 belts = Rs. $3\,x_1$
Profit on 1 belt B = Rs. 5 Profit on x_2 belts B = Rs. $5x_2$
Profit on 1 belt C = Rs. 4 Profit on x_3 belts C = Rs. $4x_3$
Total profit = $3x_1 + 5x_2 + 4x_3$ $\Rightarrow \qquad Z = 3x_1 + 5x_2 + 4x_3$

4. Constraint (*i*) The time available on machine M_1 = 8 hours per day.

Time required for 1 belt A on machine M_1 = 2 hours per day

Time required for x_1 belts A on machine M_1 = $2x_1$ hours per day

Time required for 1 belt B on machine M_1 = 3 hours per day

Time required for x_2 belts B on machine M_1 = $3x_2$ hours per day

$$2x_1 + 3x_2 \le 8$$

Constraint (*ii*) The time available on machine M_2 = 10 hours per day.

Time required for 1 belt B on machine M_2 = 2 hours per day

Time required for x_2 belts B on machine M_2 = $2x_2$ hours day

Time required for 1 belt C on machine M_2 = 5 hours per day

Time required for x_3 belts C on machine M_2 = $5x_3$ hours per day

$$2x_2 + 5x_3 \le 10$$

Constraint (*iii*) The time available on machine M_3 = 15 hour per day.

Time required for 1 belt A on machine M_3 = 3 hours per day

Time required for x_1 belts A on machine M_3 = $3x_1$ hours per day

Time required for 1 belt B on machine M_3 = 2 hours per day

Time required for x_2 belts B on machine M_3 = $2x_2$ hours per day

Time required for 1 belt C on machine M_3 = 4 hours per day

Time required for x_3 belts C on machine M_3 = $4x_3$ hours per day

$$3x_1 + 2x_2 + 4x_3 \le 15$$

Constraint (*iv*). The number of belt A, belt B and belt C are non-negative.

$$x_1 \ge 0, \qquad x_2 \ge 0 \qquad \text{and} \qquad x_3 \ge 0.$$

5. Mathematical Formulation.

The linear programming problem of the given problem is as follows

To maximise $\quad Z = 3x_1 + 5x_2 + 4x_3$ ...(1)

Subject to the constraints $\quad 2x_1 + 3x_2 \le 8$ ...(2)

$2x_2 + 5x_3 \le 10$ ...(3)

$3x_1 + 2x_2 + 4x_3 \le 15$ ...(4)

$x_1 \ge 0$ ...(5)

$x_2 \ge 0$ ...(6)

$x_3 \ge 0$ ...(7)

Ans.

TYPE II. DIET PROBLEMS

Example 5. *A dietician mixes together two kinds of food, say, X and Y in such a way that the mixture contains at least 6 units of vitamin A, 7 units of vitamin B, 12 units of vitamin C and 9 units of vitamin D. The vitamin contents of 1 kg of food X and 1 kg of food Y are given below:*

Cost	*Vitamin A*	*Vitamin B*	*Vitamin C*	*Vitamin D*
Food X	*1*	*1*	*1*	2
Food Y	2	*1*	3	*1*

One kg of food X costs Rs. 5, whereas one kg of food Y costs Rs. 8. Formulate the linear programming problem.

Solution.

1. **Decision Variables.** Decision Variables are units of food X and food Y. Let food X in the mixture be x kg. and food Y in the mixture be y kg.
2. **Objective Function.** To minimise the cost

   1 kg of food X costs Rs. 5
   x kg of food X costs Rs. $5x$
   1 kg of food Y costs Rs. 8
   y kg of food Y costs Rs. $8y$
   Total cost of food X and $Y = 5x + 8y$

   $$\Rightarrow \qquad Z = 5x + 8y$$

3. **Constraint (*i*)** The mixture contains atleast 6 units of vitamin A.

   1 kg of food X contains 1 unit of vitamin A.
   x kg of food X contains x units of vitamin A.
   1 kg of food Y contains 1 unit of vitamin A.
   y kg of food Y contains $2y$ units of vitamin A.

   $$\Rightarrow \qquad x + 2y \geq 6$$

   Constraint (*ii*) The mixture contains atleast 7 units of vitamin B.

   1 kg of food X contains 1 unit of vitamin B.
   x kg of food X contains x units of vitamin B.
   1 kg of food Y contains 1 unit of vitamin B.
   y kg of food Y contains y units of vitamin B.

   $$\Rightarrow \qquad x + y \geq 7$$

   Constraint (*iii*) The mixture contains at least 12 units of vitamin C.

   1 kg of food X contains 1 unit of vitamin C.
   x kg of food X contains x units of vitamin C.
   1 kg of food Y contains 3 units of vitamin C.
   y kg of food Y contains $3y$ units of vitamin C.

   $$\Rightarrow \qquad x + 3y \geq 12$$

   Constraint (*iv*) The mixture contains atleast 9 units of vitamin D.

   1 kg of food X contains 2 units of vitamin D.
   x kg of food X contains $2x$ units of vitamin D.
   1 kg of food Y contains 1 unit of vitamin D.
   y kg of food Y contains y units of vitamin D.

   $$\Rightarrow \qquad 2x + y \geq 9$$

   Constraint (*v*). The number of kg of food x and y is non-negative.

   $$x \geq 0$$
   $$y \geq 0$$

4. **Mathematical Formulation.** The linear programming problem of the given problem is as follows

   To minimise $\qquad Z = 5x + 8y \qquad$ ...(1)
   Subject to the constraints $\quad x + 2y \geq 6 \qquad$ ...(2)
   $x + y \geq 7 \qquad$ ...(3)
   $x + 3y \geq 12 \qquad$ ...(4)
   $2x + y \geq 9 \qquad$ ...(5)
   $x \geq 0 \qquad$ ...(6)
   $y \geq 0 \qquad$ ...(7)

Ans.

Example 6. *A firm is engaged in breeding goats. The goats are fed on various products grown on the farm. They need certain nutrients, named as X, Y, and Z. The goats are fed on two products A and B. One unit of product A contains 36 units of X, 3 units of Y and 20 units of Z, while one unit of product B contains 6 units of X, 12 units of Y and 10 units of Z. The minimum requirement of X, Y and Z is 108 units, 36 units and 100 units, respectively. Product A costs Rs. 20 per unit and product B costs Rs. 40 per unit. Formulate the L.P.P. to minimize the cost.*

Solution. The above information is give in the following table:

1. Table:

Nutrients / *Food*	*X*	*Y*	*Z*	*Cost (in Rs.)*
A	36	3	20	20
B	6	12	10	40
Minimum required units	108	36	100	

2. Decision variables. The decision variables are units of nutrients *X*, *Y* and *Z*. Let the units of food *A* be *x* the units of food *B* be *y*.

3. Objective function. To minimize the cost.

cost of 1 unit of food *A* = Rs. 20 — cost of *x* units of food *A* = Rs. 20 *x*

cost of 1 unit of food *B* = Rs. 40 — cost of *y* units of food *B* = Rs. 40 *y*

Total cost = 20 *x* + 40 *y* $\Rightarrow$ $Z = 20x + 40y$

4. Constraint (*i*). Minimum requirement of *X* nutrient = 108 units

One unit of food *A* contains 36 units of *X*.

x units of food *A* contains 36 *x* units of *X*.

One unit of food *B* contains 6 units of *X*.

y units of food *B* contains 6*y* units of *X*.

$\Rightarrow \quad 36x + 6y \geq 108$

Constraint (*ii*). The minimum requirement of *Y* nutrients = 36 units.

One unit of food *A* contains 3 units of nutrients *Y*.

x units of food *A* contains 3*x* units of nutrients *Y*.

One unit of food *B* contains 12 units of nutrients *Y*.

y units of food *B* contains 12*y* units of nutrients *Y*.

$\Rightarrow \quad 3x + 12y \geq 36$

Constraint (*iii*). The minimum requirement of nutrients *Z* = 100 units.

One unit of food *A* contains 20 units of nutrients *Z*.

x units of food *A* contains 20 *x* units of nutrients *Z*.

One unit of food *B* contains 10 units of nutrients *Z*.

y units of food *B* contains 10 *y* units of nutrients *Z*.

$\Rightarrow \quad 20x + 10y \geq 100$

Constraint (*iv*). Units of food *A* and *B* are non-negative.

$\therefore \quad x \geq 0$ and $y \geq 0$.

4. Mathematical Formulation.

To minimize, $Z = 20x + 40y$

Subject to the constraints: $36x + 6y \geq 108$

$3x + 12y \geq 36$

$20x + 10y \geq 100$

$x \geq 0,\ y \geq 0.$

TYPE III. TRANSPORTATION PROBLEM

Example 7. *There is a factory located at each of the two places P and Q. From these locations, a certain commodity is delivered to each of the three depots situated at A, B, and C. The weekly requirements of the depots are respectively 5, 5 and 4 units of the commodity while the production capacity of the factories at P and Q are 8 and 6 units respectively just sufficient for the requirement of depots. The cost of transportation per unit is given below:*

To / From	Cost (in Rs.) A	B	C
P	16	10	15
Q	10	12	10

How many units should be transported from each factory to each depot in order that the transportation cost is minimum. Formulate the above as a linear programming problem.

Solution. The given information is shown in the following figure

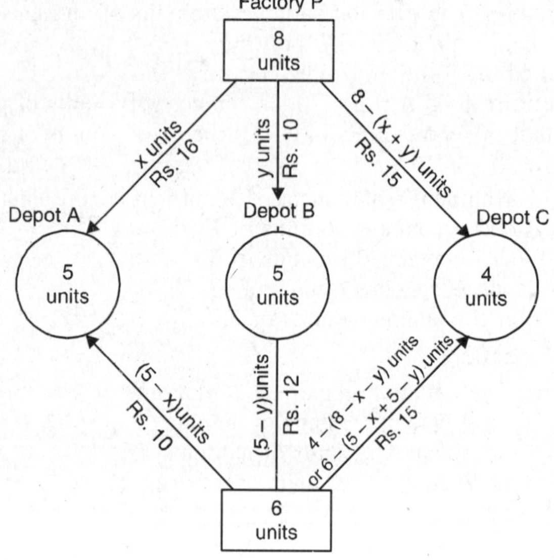

1. **Decision Variables.** Decision variables are units of commodity to be transported from the factories to the depots.
   Let the factory at P transport x units of commodity to depot at A and y units to depot at B so the remaining units at P, $8 - (x + y)$ will be transported to depot at C.
2. **Constraint (*i*).** From Factory at P, $(8 - x - y)$ units will be transported to the depot at C.

$$8 - x - y \geq 0 \quad \Rightarrow \quad x + y \leq 8 \qquad \text{... (1)}$$

Constraint (*ii*). x and y units are non-negative units.

$$x \geq 0 \qquad \text{... (2)}$$

$$y \geq 0 \qquad \text{... (3)}$$

Constraints (*iii*).

The remaining requirements $(5 - x)$ units are to be transported from the factory at Q to the depot at A. $\quad 5 - x \geq 0 \Rightarrow x \leq 5$ ... (4)

Constraints (*iv*).

The remaining requirements $(5 - y)$ units are to be transported from factory Q to the depot at B.

$$5 - y \geq 0 \Rightarrow y \leq 5 \qquad \text{... (5)}$$

Constraints (*v*).
The remaining requirements 4 – (5 – *x* + 5 – *y*) units of commodity will be transported from the factory at *Q* to the depot at *C*.

$$4-(5-x+5-y)\geq 0$$

$$\Rightarrow \quad x+y-4\geq 0 \quad \Rightarrow \quad x+y\geq 4 \quad \ldots (6)$$

3. Objective function. The transportation cost from the factory at *P* to the depots at *A, B* and *C* are respectively Rs. 16 *x*, 10 *y* and 15 (8 – *x* – *y*)
Total cost of transportation from factory at *P* = 16*x* + 10*y* + 15 (8 – *x* – *y*).
Similarly, the transportation cost from the factory at *Q* to the depots at *A, B* and *C* are Rs. $10\,(5-x), 12\,(5-y), 10\,(x+y-4)$ respectively.

Total transportation cost from factory $Q = 10\,(5-x)+12\,(5-y)+10\,(x+y-4)$
Grand total cost of transportation from both the factories at *P* and *Q* to all the depots

$$Z=16\,x+10\,y+15\,(8-x-y)+10\,(5-x)+12\,(5-y)+10\,(x+y-4)$$

$$\Rightarrow \quad Z=x-7y+190$$

4. Mathematical Formulation.
To minimize $Z = x - 7y + 190$
Subject to constraints

$$x+y\leq 8$$
$$x+y\geq 4$$
$$x\leq 5$$
$$y\leq 5$$
$$x\geq 0$$
$$y\geq 0.$$

Ans.

EXERCISE 56.1

1. A furniture dealer deals in only two items *viz.*, tables and chairs. He has Rs. 11,000 to invest and a space to store at most 40 pieces. A table costs him Rs. 500 and a chair Rs. 200. He can sell a table at a profit of Rs. 50 and a chair at a profit of Rs. 15. Assume that he can sell all the items that he buys. Formulate this problem as an L.P.P, so that he can maximize the profit.

Ans. Maximize, $Z = 50x + 15y$
Subject to the constraints :
$x+y\leq 40$
$500x + 200y \leq 11000$
$x\geq 0$
$y\geq 0$

2. A manufacturer produces nuts and bolts for industrial machinery. It takes 1 hour of work on machine *A* and 3 hours on machine *B* to produce a package of nuts; while it takes 3 hours on machine *A* and 1 hour on machine *B* to produce a package of bolts. He earns a profit of Rs. 2.50 per package on nuts and Re. 1 per package on bolts. Form a linear programming problem to maximize his profit, if he operates each machine for at the most 12 hours a day.

Ans. Maximize, $Z = 2.5x + y$
Subject to the constraints :
$x+3y\leq 12$
$3x+y\leq 12$
$x\geq 0$
$y\geq 0$

3. A person consumes two types of food, *A* and *B*, everyday to obtain 8 units of protein, 12 units of carbohydrates and 9 units of fat which is his daily minimum requirements. 1 kg of food *A* contains 2, 6, 1 units of protein, carbohydrates and fat, respectively. 1 kg of food *B* contains 1, 1 and 3 units of protein, carbohydrates and fat, respectively. Food *A* costs Rs. 8 per kg while food *B* costs Rs. 5 per kg. Form an LPP to find how many kg of each food should he buy daily to minimize his cost of food and still meet minimal nutritional requirements.

Ans. Minimize $Z = 5x + 8y$
Subject to the constraints :
$2x+y\geq 8$
$6x+y\geq 12$
$x+3y\geq 9$
$x\geq 0$
$y\geq 0$

4. A dietician wishes to mix two types of foods F_1 and F_2 in such a way that the vitamin contents of the mixture contains atleast 6 units of vitamin *A* and 8 units of vitamin *B*. Food F_1 contains 2 units/kg of vitamin A and 3 units/kg of vitamin B while food F_2 contains 3 units/kg of vitamin A and 4 units/kg of vitamin B. Food F_1 costs Rs. 50/kg and food F_2 costs Rs. 75/kg. Formulate the problem as a L.P.P. to minimize the cost of mixture.

Ans. Minimize $z = 50x + 75y$

Subject to the constraints :

$$2x + 3y \geq 6$$

$$3x + 4y \geq 8$$

$$x \geq 0, \quad y \geq 0$$

5. (Investment Problem) A retired person wants to invest an amount of upto Rs. 20,000. His broker recommends investing in two types of bonds A and B, bond A yielding 10% return on the amount invested and bond B yielding 15% return on the amount invested. After some consideration, he decides to invest at least Rs. 5000 in bond A and no more than Rs. 8000 in bond *B*. He also wants to invest at least as much in bond *A* as in bond *B*. What should his broker suggest if he wants to maximize his return on investments. Formulate LPP.

Ans. Minimize $Z = \frac{10}{100}x + \frac{15}{100}y$

$$x \geq 5000$$

$$y \leq 80,000$$

$$x \geq 0, \ y \geq 0$$

56.4 GRAPHICAL METHOD OF SOLVING LINEAR PROGRAMMING PROBLEMS

If a problem contains only two variables then we can solve the given problem by graphical method. There are two graphical method to solve a linear programming problem.

1. Corner point method **2.** Iso-profit or iso-cost method

56.5 CORNER POINT METHOD

This method is based on the fundamental extreme point theorem.

In previous class we have learnt how to formulate a system of linear inequalities involving two variables x and y mathematically.

Working Rule

Step 1. Formulate the given L.P.P. in mathematical form.

Step 2. The inequations are converted into equations.

In the equation on putting $y = 0$ we get x-coordinate on x-axis. Similarly, putting $x = 0$ we get y-coordinate on y-axis. Join these two points to get the graph of the equation.

Step 3. The inequation of a line divides the plane into two half planes, to choose the plane of the inequation we put $x = 0$ and $y = 0$ in the inequation. If origin satisfies the inequation then the region containing the origin is the region represented by the given inequation. Otherwise the half plane not containing the origin is the region represented by the given inequation.

Step 4. The region satisfying all the inequations is the feasible region.

Step 5. The vertices (corner points) of the required region are known as extreme points of the set of all feasible solutions of the L.P.P.

Step 6. By putting the values of x and y of each corner point in the objective function we get the values of the objective function at each of the vertices of the feasible region. Out of all the values of the objective function, we get a point at which the objective function is optimum (maximum or minimum).

Consider the following example:-

Example 8. *Solve the following L.P.P. graphically :*

Maximize, $Z = 3x + 2y$

Subject to the constraints $x + 2y \leq 10$

$$3x + y \leq 15$$

$$x \geq 0$$

$$y \geq 0$$

Solution. On changing the given inequations into equations, we have

$x+2y=10$ and $3x+y=15;$ $x=0$ and $y=0$

Region Represented by $x+2y \le 10$.

The line $x+2y=10$ meets the x-axis at the point $A(10, 0)$ and meets y-axis at the point B (0, 5). Join AB to obtain the graph of $x+2y=10$.

Here, origin ($x = 0$, $y = 0$) satisfies the inequation $x+2y \le 10$. So, the half plane containing the origin represents the solution set of the inequation $x+2y \le 10$.

Region Represented by $3x+y \le 15$.

The line $3x+y=15$ meets the x-axis at the point C (5, 0) and meets the y-axis at the point D (0, 15). Join CD to obtain the graph of the line $3x+y=15$. Here, origin $(x=0, y=0)$ satisfies the inequation $3x+y \le 15$. So, the half plane containing the origin represents the solution set of the inequation $3x+y \le 15$.

Region represented by $x \ge 0$ and $y \ge 0$.

All the points in the first quadrant satisfies $x \ge 0$ and $y \ge 0$. So, the first quadrant is the region represented by $x \ge 0$ and $y \ge 0$.

The shaded region $O\ C\ E\ B\ O$ represents the region satisfying all the inequations.

Each point of this region represents a feasible choice.

Every point of this region is called the feasible solution of the problem.

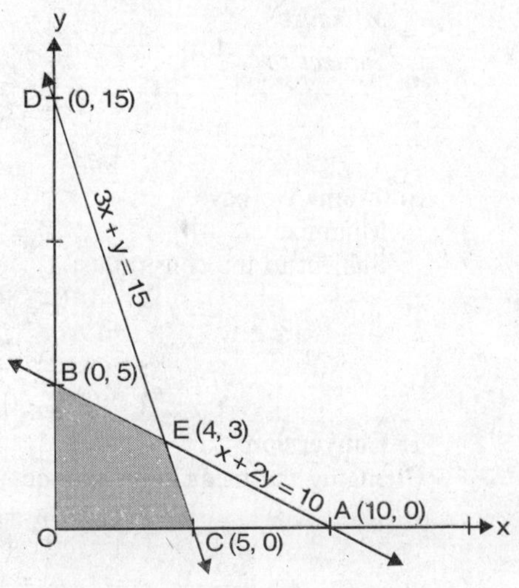

Feasible region. The shaded portion determined by all the constraints including non-negative constants of a linear programming problem is called the feasible region. In this example O BEC (shaded) is the feasible region for the problem.

Feasible Solution. Points within and on the boundary of the feasible region represent feasible solutions of the constraints.

Here, (5, 0), (4, 3) and (0, 5) are the feasible solutions. Any point outside the feasible region is called an infeasible solution. *For example*, (0, 15) and (10, 0) are infeasible solutions.

Optimal Solution. Any point in the feasible region that gives the optimal value (maximum or minimum) is called an optimal solution.

In the feasible region there are infinitely many points which satisfy all the constraints. It is not possible to check all the points for the maximum value of the objective function.

$$Z = 3x + 2y$$

For finding out the optimal solution we have to use the following theorems

Theorem 1. *The optimal value of objective function must occur at a corner point (vertex) of the feasible region.*

Theorem 2. *If the feasible region is bounded then the objective function Z has both maximum and minimum values on corner points of the bounded region.*

In the above example we have following table showing the value of the objective function at the corner points of the feasible region.

Vertex of the feasible region	*Corresponding value of Z = 3x + 2y (in Rs.)*
O (0, 0)	0
C (5, 0)	15
E (4, 3)	18 **Maximum**
B (0, 5)	10

We observe that the maximum value of the objective function is 18.

This method of solving linear programming problem is called *corner point method.*

Procedure: The solution of the given L.P.P. should be divided under the following heads:

1. Conversion of inequalities into equations.

2. Draw the graph of the lines and find regions represented by the inequations.

3. Apply Corner point method.

Example 9. *Solve the following linear programming problem graphically :*

Minimize $Z = -3x + 4y$

Subject to $x + 2y \leq 8$

$$3x + 2y \leq 12$$

$$x \geq 0, \quad y \geq 0.$$

Solution. We have,

Minimize $Z = -3x + 4y$ ...(1)

Subject to the constraints:

$$x + 2y \leq 8 \quad ...(2)$$

$$3x + 2y \leq 12 \quad ...(3)$$

$$x \geq 0, \; y \geq 0 \quad ...(4)$$

1. Conversion

Changing the inequations into equations, we have

$$x + 2y = 8$$
$$3x + 2y = 12$$
$$x = 0, \; y = 0$$

2. Drawing of Graphs of the lines:

Region represented by $x + 2y \leq 8$

$x + 2y = 8$

x	8	0
y	0	4
	A	B

On joining A and B, we get the graph of the equation $x + 2y = 8$.

Put $x = 0$, $y = 0$ in $x + 2y \leq 8$, then $0 + 0 \leq 8$, which is true. So the half plane containing origin represents the solution set of the inequation $x + 2y \leq 8$.

Region represented by $3x + 2y \leq 12$.

$3x + 2y = 12$

x	4	0
y	0	6
	C	D

On joining the points C and D, we get the graph of the equation $3x + 2y = 12$.

Put $x = 0$, $y = 0$ in the inequation $3x+2y \le 12$,

then $\quad 0+0 \le 12$ which is true.

So, the half plane containing origin represents the solution set of the inequation $3x+2y \le 12$.

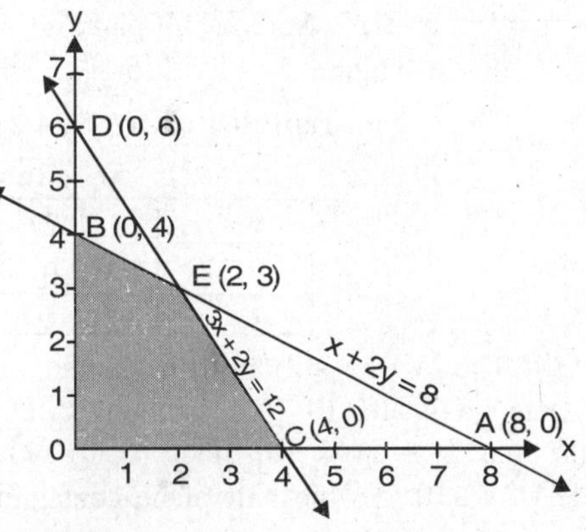

Region represented by $x \ge 0$ and $y \ge 0$.

First quadrant is the region of $x \ge 0$, $y \ge 0$.
The shaded region in figure is the feasible region $O\ C\ E\ B$ determined by the system of constraints (2) to (4), which is bounded. The coordinates of corner points are $C(4, 0)$, $E(2, 3)$ and $B\ (0, 4)$.

3. Corner Point Method

Now we evaluate $Z = -3x + 4y$ at these corner points.

Corner Point of the feasible region OCEBO	*Corresponding value of* $Z = -3x + 4y$
C (4, 0)	$-3(4) + 4(0) = -12$ **Minimum**
E (2, 3)	$-3(2) + 4(3) = 6$
B (0, 4)	$-3(0) + 4(4) = 16$

Hence, the minimum value of Z is -12 attained at the point C (4, 0). **Ans.**

Example 10. *Solve the following linear programming problem graphically:*

Maximise $\quad Z = 5x + 3y$

subject to $\quad 3x+5y \le 15$

$5x+2y \le 10$

$x \ge 0,\ y \ge 0.$

Solution. We have, Maximise $\quad Z = 5x + 3y$

subject to the constraints

$$3x+5y \le 15 \quad ...(1)$$
$$5x+2y \le 10 \quad ...(2)$$
$$x \ge 0,\ y \ge 0. \quad ...(3)$$

1. Conversion

On converting the above inequations, we have the following equations:

$$3x + 5y = 15$$
$$5x + 2y = 10$$
$$x = 0,\ y = 0$$

2. Drawing of the graph of the lines.

Region represented by the inequation $3x+5y \le 15$.

$3x + 5y = 15$

x	5	0
y	0	3
	A	B

The line $3x + 5y = 15$ meets the x-axis at A (5, 0) and meets y-axis at B (0, 3). On joining A, B we get the graph of the line $3x+5y=15$. Since (0, 0) satisfies the inequation

$3x+5y \le 15$. So, the half plane containing origin represents the region of the solution set of the inequation $3x+5y \le 15$.

The region represented by $5x+2y \le 10$.

$5x + 2y = 10$

x	2	0
y	0	5
	C	D

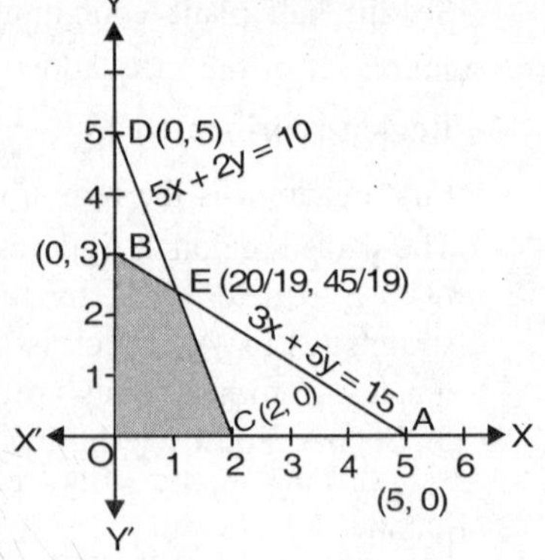

The line $5x + 2y = 10$ meets the x-axis at C (2, 0) and meets y-axis at D (0, 5). On joining C, D we get the graph of the line $5x + 2y = 10$. Since (0, 0) satisfies the inequation $5x+2y \le 10$. So the half plane containing origin represents the region of the solution set of the inequation $5x+2y \le 10$.

The region represented by $x \ge 0$ and $y \ge 0$.

$x \ge 0$ and $y \ge 0$ represent the first quadrant. The shaded region OCEBO bounded by the inequations (1) to (3) is the feasible region.

3. Corner Point Method. The coordinates of the vertices O, C, E, and B of the feasible region $OCEBO$ are O (0, 0), C (2, 0), E (20/19, 45/19) and B (0, 3) respectively.

Note that the coordinates of E are obtained by solving the equations $3x + 5y = 15$ and $5x + 2y = 10$. The value of the objective function at these points are given in the following table:

Corner point (x, y) of the feasible region OCEBO	*Value of the objective function* $Z = 5x + 3y$
O (0, 0)	$5(0) + 3(0) = 0$
C (2, 0)	$5(2) + 3(0) = 10$
E (20/19, 45/19)	$5\left(\frac{20}{19}\right)+3\left(\frac{45}{19}\right)=\frac{100}{19}+\frac{135}{19}=\frac{235}{19}$ **maximum**
B (0, 3)	$5(0) + 3(3) = 9$

Clearly, Z is maximum when $x=\frac{20}{19}, y=\frac{45}{19}$. Thus, $x=\frac{20}{19}$, $y=\frac{45}{19}$ is the optimal solution of the given L.P.P.

Hence the maximum value of Z is $\frac{235}{19}$ at $x=\frac{20}{19}, y=\frac{45}{19}$. **Ans.**

Example 11. *Solve graphically the following linear programming problem to minimise the cost $Z = 3x + 2y$ subject to the following constraints:*

$5x+y \ge 10$; $x+y \ge 6$; $x+4y \ge 12$; $x \ge 0, y \ge 0$.

Solution. We have,

Minimise $Z = 3x + 2y$ ... (1)

Subject to the constraints

$$5x+y \ge 10 \quad \text{... (2)}$$

$$x+y \ge 6 \quad \text{... (3)}$$

$$x+4y \ge 12 \quad \text{... (4)}$$

$$x \ge 0, y \ge 0 \quad \text{... (5)}$$

1. Conversion.

On converting the above inequations, we have the following equations

$$5x + y = 10$$
$$x + y = 6$$
$$x + 4y = 12$$
$$x = 0,\ y = 0$$

2. Drawing of graphs.

The region represented by $5x + y \geq 10$.

$$5x + y = 10$$

x	2	0
y	0	10
Point	*A*	*B*

On joining the points *A* and *B*, we get the graph of the line $5x + y = 10$. Put $x = 0, y = 0$ in $5x + y \geq 10$, then $0 + 0 \geq 10$ which is false. So half plane not containing origin represents the region of the solution set of the inequation $5x + y \geq 10$.

The region represented by $x + y \geq 6$

$$x + y = 6$$

x	6	0
y	0	6
Point	*C*	*D*

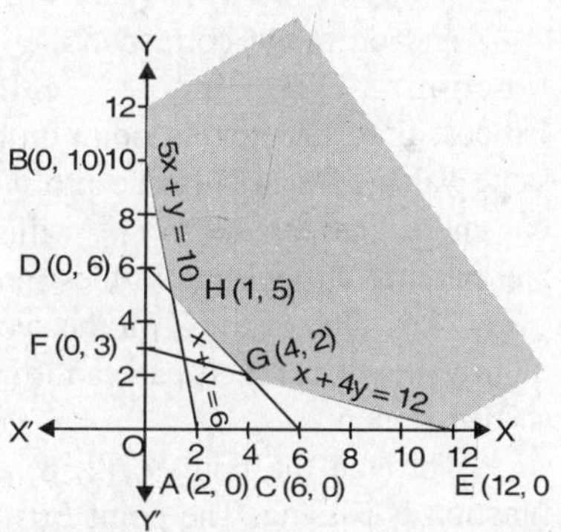

On joining the points *C* and *D*, we get the graph of the line $x + y = 6$.

Put $x = 0, y = 0$ in $x + y \geq 6$, then $0 + 0 \geq 6$ which is false. So the half plane not containing origin represents the region of the solution set of $x + y \geq 6$.

The region represented by $x + 4y \geq 12$

$$x + 4y = 12$$

x	12	0
y	0	3
Point	*E*	*F*

On joining the points *E* and *F* we get the graph of the line $x + 4y = 12$.

Put $x = 0, y = 0$ in $x + 4y \geq 12$, then $0 + 0 \geq 12$ which is false.

So, the half plane not containing origin represents the region of the solution set of the inequation $x + 4y \geq 12$.

The region represented by $x \geq 0$ and $y \geq 0$.

The first quadrant is represented by $x \geq 0$ and $y \geq 0$.

The shaded region is represented by the inequations (2) to (5).

3. Corner Point Method. The coordinates of the vertices of the feasible region are *E* (12, 0), *G* (4, 2), *H* (1, 5) and *B* (0, 10) respectively.

Note that the coordinates of *G* and *H* are obtained by solving the equations $x + y = 6$, $x + 4y = 12$ and $x + y = 6$, $5x + y = 10$ respectively.

The value of the objective function at these points are given in the following table :

Corner Point (x, y) of the feasible region EGHB	*Value of the objective function* $Z = 3x + 2y$
E, (12, 0)	$3 \times 12 + 2 \times 0 = 36$
G, (4, 2)	$3 \times 4 + 2 \times 2 = 12 + 4 = 16$
H, (1, 5)	$3 \times 1 + 2 \times 5 = 3 + 10 = 13$
B, (0, 10)	$3 \times 0 + 2 \times 10 = 0 + 20 = 20$

Clearly, H is minimum when $x = 1$, $y = 5$.

So, $x = 1$, $y = 5$ is the optimal solution of the given L.P.P.

Hence, Z is minimum when $x = 1$ and $y = 5$ and the minimum value of Z is Rs. 13.

56.6. ISO-PROFIT OR ISO-COST METHOD (MAXIMUM Z)

Let $Z = 3x + 2y$ be the objective function.

On putting any value of Z (say 12) in $Z = 3x + 2y$, we have

$$12 = 3x + 2y \qquad \ldots (1)$$

and draw the corresponding line of the objective function. This line is called Iso-profit or Iso-cost line, since every point on this line will give the same value of Z (12) (same profit or same cost). Draw one more line parallel to (1), within the feasible region and passing through the farthest point from the origin.

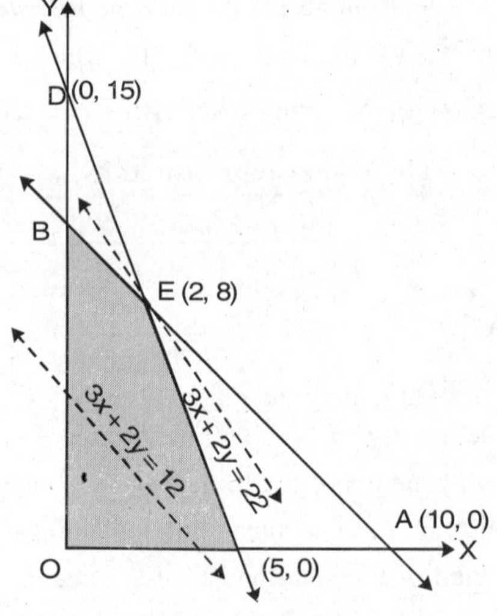

The value of the Z on the second line is the maximum value of the Z. It passes through one corner of the shaded region.

Through the point E (2, 8), the line of objective function is passing. The point E is on the shaded region and is vertex of the shaded region and the line $Z = 3x + 2y$ is farthest from the origin. Here the maximum value of $Z = 3\,(2) + 2\,(8) = 6 + 16 = 22$. So, the maximum value of the objective function is 22.

56.7 ISO-PROFIT OR ISO-COST METHOD (MINIMUM Z)

Let $Z = x + 2y$

Let us take three different values of Z; Z_1, Z_2 and Z_3 we get,

$$Z_1 = x + 2y \qquad \ldots (1)$$

$$Z_2 = x + 2y \qquad \ldots (2)$$

$$Z_3 = x + 2y \qquad \ldots (3)$$

The above three lines are parallel to each other having the same slope $\left(-\frac{1}{2}\right)$.

Out of these three lines, line (2) is the only line which passes through the point E (4, 2) nearest to the origin and passing through the feasible region.

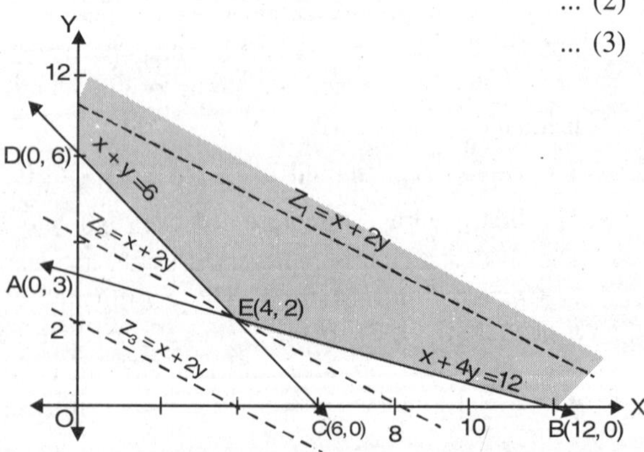

Thus, Z_2 is the minimum value of Z.

$x = 4$ and $y = 2$ gives the optimal solution.

$$Z_2 = 4 + 2(2) = 4 + 4 = 8$$

So, the minimum value of Z is 8.

Example 12. *Solve graphically the following linear programming problem:*

maximize $Z = 5x + 3y$

subject to the constraints :

$$x + y \leq 300$$
$$2x + y \leq 360$$
$$x \geq 0$$
$$y \geq 0$$

Solution. We have,

Maximum $Z = 5x + 3y$ ... (1)

Subject to the constraints

$x + y \leq 300$ ... (2)

$2x + y \leq 360$ ... (3)

$x \geq 0$ and $y \geq 0$ ... (4)

1. Conversion

On changing the inequations into equations we get the following equations.

$x + y = 300$ and $2x + y = 360$

2. Drawing of graphs

The region represented by $x + y \leq 300$

$x + y = 300$

x	300	0
y	0	300
Point	*A*	*B*

On joining the points *A* (300, 0) and *B* (0, 300) we get the graph of the line *AB*, $x + y = 300$.

Put $x = 0$, $y = 0$ in $x + y \leq 300$, then $0 + 0 \leq 300$, which is true. The half plane containing the origin is the region of the solution set of the inequation $x + y \leq 300$.

The region represented by $2x + y \leq 360$.

$2x + y = 360$

x	180	0
y	0	360
Point	*C*	*D*

On joining the points *C* (180, 0) and *D* (0, 360) we get the graph of the line $2x + y = 360$.

Put $x = 0$, $y = 0$ in $2x + y \leq 360$, then $0 + 0 \leq 360$, which is true.

The half plane containing the origin is the region of the solution set of the inequation $2x + y \leq 360$.

The region represented by $x \geq 0$ and $y \geq 0$.

The first quadrant is represented by $x \geq 0$, and $y \geq 0$.

The shaded region *OCEB* is the feasible solution bounded by the inequations (2) to (4).

On solving the equations (2) and (3), we get the point of intersection *E* (60, 240).

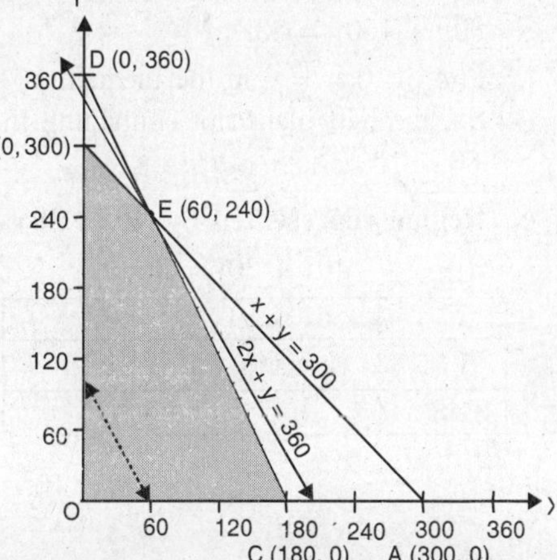

The coordinates of the corner points of the feasible region are C (180, 0), E (60, 240) and B (0, 300).

3. Iso-profit or Iso-cost Method.

Now, we take a constant value (say 300) i.e.,

(20 times the *l.c.m.* of 5 and 3).

On putting $Z = 300$ in objective function (1), we get $300 = 5x + 3y$ ... (5)

We draw the graph of the line $5x + 3y = 300$.

Draw one more line parallel to (5) which is the farthest from the origin and has atleast one point of the feasible region.

The parallel line to (5) passes through the farthest point E (60, 240) from the origin. Here, $x = 60$ and $y = 240$ will give the maximum value of Z.

The maximum value of Z is given by $Z = 5\,(60) + 3\,(240) = 300 + 720 = 1020$.

Hence, the maximum value of Z is 1020. **Ans.**

Example 13. *Solve graphically the following L.P.P.*

Minimize $Z = 5x + 4y$

subject to the constraints :

$$80x + 100y \geq 88$$
$$40x + 30y \geq 36$$
$$x \geq 0,\ y \geq 0$$

Solution. We have,

Minimize $Z = 5x + 4y$ ... (1)

Subject to the constraints

$$80x + 100y \geq 88 \qquad \text{... (2)}$$
$$40x + 30y \geq 36 \qquad \text{... (3)}$$
$$x \geq 0,\ y \geq 0 \qquad \text{... (4)}$$

1. Conversion

On converting the inequations into equations, we get

$$80x + 100y = 88$$
$$40x + 30y = 36$$

2. Drawing of graphs

Region represented by $80x + 100y \geq 88$.

$$80\,x + 100\,y = 88$$

x	1.1	0.0
y	0	0.83
Point	*A*	*B*

On joining the points A (1.1, 0) and B (0.0, 0.83) we get the graph of the line $80x + 100y = 88$.

Put $x = 0$, $y = 0$ in the inequation $80x + 100y \geq 88$ then $0 + 0 \geq 88$, which is false.

So, the half plane not containing the origin is the region of solution set of the inequation $80x + 100y \geq 88$

Region represented by $40x + 30y \geq 36$.

$$40x + 30y = 36$$

x	0.91	0
y	0	1.14
Point	*C*	*D*

On joining the points C (0.91, 0) and D (0, 1.14), we get the graph of the line $40x + 30y = 36$.

Put $x = 0, y = 0$ in the inequation $40x+30y \geq 36$, then $0+0 \geq 36$, which is false.

So, the half plane not containing the origin is the region of solution set of the inequation $40x+30y \geq 36$.

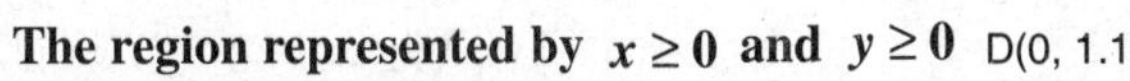

The region represented by $x \geq 0$ and $y \geq 0$

The first quadrant is represented by

$$x \geq 0 \text{ and } y \geq 0.$$

The shaded region AED represents the feasible solution bounded by the inequations (2) to (4). On solving (2) and (3), we get the point of intersection of the lines (2) and (3), E (0.6, 0.4).

3. Iso-profit or Iso-cost Method

Now, we take a constant = 20

On putting $Z = 20$ in (1), we get

$$5x + 4y = 20 \qquad \text{... (5)}$$

Now, we draw the graph of the line $5x + 4y = 20$.

Draw one more line parallel to (5), which is the nearest from the origin and has at least one point of the feasible region.

The parallel line to (5) passes through the nearest point E (0.6, 0.4) from the origin. Here, $x = 0.6$ and $y = 0.4$ will give the minimum value of Z. The minimum value of Z is given by

$$Z = 5\,(0.6) + 4\,(0.4) = 3.0 + 1.6 = 4.6 \qquad \textbf{Ans.}$$

Hence, the minimum value of Z is 4.6.

EXERCISE 56.2

Solve graphically each of the following linear programming problems :

1. Maximize $Z = 10x + 6y$ subject to the constraints

$$3x+y \leq 12$$
$$2x+5y \leq 34$$
$$x \geq 0,\ y \geq 0$$

Ans. Maximum : 56; $x = 2$, $y = 6$

2. Maximize $Z = 60x+15y$ subject to the constraints

$$x+y \leq 50$$
$$3x+y \leq 90$$
$$x \geq 0,\ y \geq 0$$

Ans. Maximum : 1650, $x = 20$, $y = 30$

3. Minimize $Z = 18x + 10y$ subject to the constraints

$$4x+y \geq 20$$
$$2x+3y \geq 30$$
$$x \geq 0,\ y \geq 0$$

Ans. Minimum : 134; $x = 3$, $y = 8$

4. Maximize $Z = 4x + 9y$ subject to the constraints

$$x+5y \leq 200$$
$$2x+3y \leq 134$$
$$x \geq 0,\ y \geq 0$$

Ans. Maximum : 382; $x = 10$, $y = 38$

5. Maximize $Z = 5x + 7y$ subject to the constraints

$$x+y \leq 4$$
$$3x+8y \leq 24$$
$$10x+7y \leq 35$$
$$x,\ y \geq 0$$

Ans. Maximum $Z = \frac{124}{5}$, $x = \frac{8}{5}$, $y = \frac{12}{5}$

6. Minimize $Z = 3x + 2y$ subject to the constraints :

$$x+y \geq 8$$
$$3x+5y \leq 15$$
$$x \geq 0,\ y \geq 0$$

Ans. No feasible region.

7. Minimize $Z = 3x + 5y$
subject to the constraints

$$x+3y \geq 3$$
$$x+y \geq 2$$
$$x, y \geq 0$$

Ans. Minimum $Z = 7, x = \frac{3}{2}, y = \frac{1}{2}$

8. Minimize $Z = 20x + 10y$
subject to the constraints

$$x+2y \leq 40$$
$$3x+y \geq 30$$

Ans. Minimum $Z = 200, x = 10, y = 0$

9. Minimize $Z = 30x + 20y$
subject to the constraints

$$x+y \leq 8$$
$$x+4y \geq 12$$
$$x, y \geq 0$$

Ans. Minimum $Z = 60, x = 0, y = 3$

10. Minimize $Z = x - 5y + 20$
subject to the constraints

$$x-y \geq 0$$
$$-x+2y \geq 2$$
$$x \geq 3$$
$$y \leq 4$$
$$x, y \geq 0$$

Ans. Minimum $Z = 4, x = 4, y = 4$

Find x and y for which $3x + 2y$ is minimum subject to these inequalities. Use a graphical method.

11. Find the minimum value of $3x + 5y$ subject to the constraints:

$$-2x+y \leq 4, \quad x+y \geq 3,$$
$$x-2y \leq 2, \quad x \geq 0, y \geq 0$$

Ans. Minimum $Z = -300, x = 6, y = 0$

12. Find the maximum value of $2x + y$ subject to the constraints :

$$x+3y \geq 6, \quad x-3y \leq 3, \quad 3x+4y \leq 24$$
$$-3x+2y \leq 6, \quad 5x+y \geq 5, \quad x, y \geq 0$$

Ans. Maximum $Z = \frac{43}{3}, x = \frac{84}{13}, y = \frac{15}{3}$.

56.8 SOLUTION OF LINEAR PROGRAMMING PROBLEMS

Here we will solve the linear programming problems.

Working Rule

Step 1. Define the problem mathematically.

Step 2. Graph the constraint inequalities by converting them into equations. Find out their respective intercept on both the axes and connect them by straight lines.

Step 3. Find out the vertices of the feasible region.

Step 4. Find out the value of the objective function on the vertices.

Step 5. Find out the optimum value of the objective function.

Procedure . The solution of the given LPP should be divided under the following heads:

1. Prepare a *table* of the data given in the problem. **2.** Write down the *decision variables.*

3. Form the *objective function.* **4.** Write down the constraints.

5. Mathematical formulation. **6.** Region represented by inequations.

7. Apply Corner point method/Iso-cost or iso-profit method.

Type I. To maximize the objective Function (Z)

Example 14. *If a young man rides his motor cycle at 25 km/hour he had to spend Rs. 2 per km on petrol. If he rides at a faster speed of 40 km/hour, the petrol cost increases at Rs. 5 per km. He has Rs. 100 to spend on petrol and wishes to find what is the maximum distance he can travel within one hour, express this as an L.P.P. and solve it graphically.*

Solution:

The above information are given in the following table :

1. Table

S.N.	*Speeds (km per hour)*	*Consumption of petrol per km.*	*Total amount Spent on petrol*
1.	25	Rs. 2	Rs. 100
2.	40	Rs. 5	

2. Decision Variables: Let the number of km riding motorcycle at the speed of 25k/h = x km

Let the number of km riding motor cycle at the speed of 40 km/hour = y km

3. Objective function

To maximize the distance of the journey.

$$Z = x + y$$

4. Constraint (*i*). The young man has Rs. 100 to spend on petrol.

When the speed is 25 km/hour cost of petrol for 1 km = Rs. 2

When the speed is 25 km/hour cost of petrol for x km = Rs. 2 x

When the speed is 40 km/hour cost of petrol for 1 km = Rs. 5

When the speed is 40 km/hour cost of petrol for y km = Rs. 5 y.

$$\therefore \quad 2x+5y \le 100$$

Constraint (*ii*) $\quad \text{Time} = \dfrac{\text{distance}}{\text{speed}}$

Time taken in the first journey $= \dfrac{x}{25}$

Time taken in the second journey $= \dfrac{y}{40}$

Total time given = 1 hour

$$\therefore \quad \frac{x}{25}+\frac{y}{40} \le 1$$

Constraint (*iii*) The distances in the journey are non-negative.

$$\therefore \quad x \ge 0 \text{ and } y \ge 0$$

5. Mathematical Formulation

To maximize, $\quad Z = x + y \quad$ ... (1)

Subject to the constraints :

$$2x+5y \le 100 \quad \text{... (2)}$$

$$\frac{x}{25}+\frac{y}{40} \le 1 \Rightarrow 8x+5y \le 200 \quad \text{... (3)}$$

$$x \ge 0,\ y \ge 0. \quad \text{... (4)}$$

6. Region represented by the contraints

$2x + 5y = 100$

x	50	0
y	0	20
Point	A	B

$8x + 5y = 200$

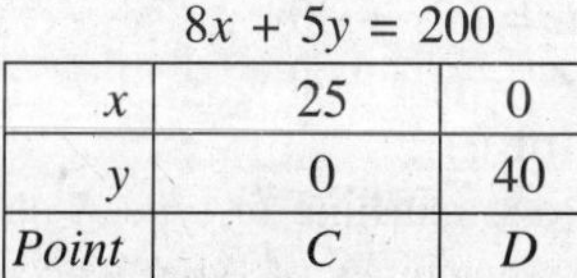

x	25	0
y	0	40
Point	C	D

We have drawn the graphs of the following lines :

$2x + 5y = 100$

$8x + 5y = 200$

$x = 0$

$y = 0$

Feasible region is represented by the shaded portion *OCEBO*.

7. Corner Point Method

The coordinates of the vertices of feasible region *OCEBO* are *O* (0, 0), *C* (25, 0), $E\left(\dfrac{50}{3}, \dfrac{40}{3}\right)$ and *B* (0, 20).

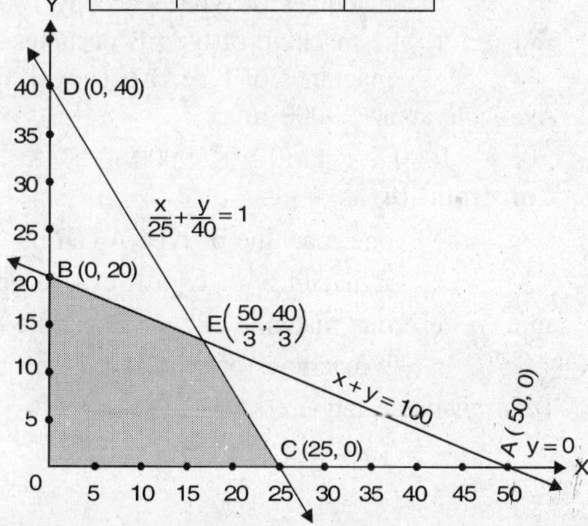

The values of the objective function at these points are as follows :

Corner Point (x, y) of the feasible region OCEBO	*Value of the objective function* $Z = x + y$
C (25, 0)	$25 + 0 = 25$
$E\left(\frac{50}{3}, \frac{40}{3}\right)$	$\frac{50}{3} + \frac{40}{3} = \frac{90}{3} = 30$ **Maximum**
B (0, 20)	$0 + 20 = 20$

Hence, $Z = 30$ is maximum when $x = \frac{50}{3}$ and $y = \frac{40}{3}$. **Ans.**

Example 15. *A factory owner purchased two types of machines, A and B for his factory. The requirements and the limitations for the machines are as follows:*

Machine	Area occupied	Labour force	Daily output (in units)
A	1000 m^2	12 men	60
B	1200 m^2	8 men	40

He has maximum area of 9000 m^2 available, and 72 skilled labourers who can operate both the machines. How many machines of each type should he buy to maximise the daily output ?

Solution.

1. Decision variables. Let x machines of type A and y machines of type B are bought to maximize the daily output.

2. Objective function. It is given that one machine of type A gives output 60 units so x machines of type A give output $60x$ units.

Similarly, y machines of type B give output $40y$ units.

Total output $= 60x + 40y$

$\Rightarrow \quad Z = 60x + 40y$

3. Constraint (*i*)

$\because$ one machine of type A occupies 1000 m^2 area

$\therefore$ x machines of type A occupy $1000\,x$ m^2 area

and $\because$ one machine of type B occupies 1200 m^2 area

$\therefore$ y machines of type B occupy $1200\,y$ m^2 area

Available area is 9000 m^2

$\therefore \quad 1000\,x + 1200\,y \le 9000.$

Constraint (*ii*)

$\because$ one machine of type A can be operated by 12 men

$\therefore$ x machines of type A can be operated by $12\,x$ men

and $\because$ one machine of type B can be operated by 8 men

$\therefore$ y machines of type B can be operated by $8y$ men

Total available labourers = 72.

$\therefore \quad 12\,x + 8\,y \le 72.$

Constraint (iii). Number of machines cannot be negative.

$\therefore \quad x \geq 0$ and $y \geq 0$

4. Mathematical Formulation

Thus the given L.P.P. is
Maximize $Z = 60x + 40y$

Subject to $\quad 1000\,x + 1200\,y \leq 9000$

$12\,x + 8\,y \leq 72$

$x \geq 0$ and $y \geq 0$

5. Region Represented by the inequations

To solve this L.P.P. we draw the lines

$1000\,x + 1200\,y = 9000.$ .... (1)

$12x + 8y = 72$ ... (2)

x	9	0
y	0	7.5
Points	A	B

x	6	0
y	0	9
Points	C	D

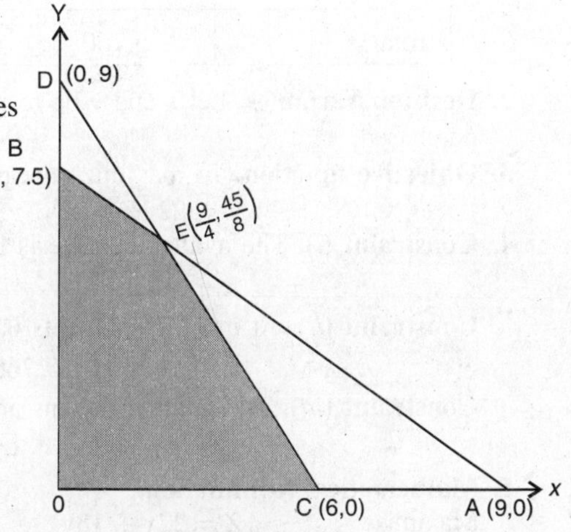

The feasible region of the L.P.P. is shaded in the adjoining figure.

6. Corner Point Method

On solving (1) and (2), we get the coordinates of $E\left(\frac{9}{4}, \frac{45}{8}\right)$.

The coordinates of corner points are C (6, 0), D (0, 7.5) and $E\left(\frac{9}{4}, \frac{45}{8}\right)$.

Now we evaluate $Z = 60\,x + 40\,y$ at the corner points.

Corner point of the feasible region	Corresponding value of $Z = 60x + 40y$
C (6, 0)	$Z = 60\,(6) + 0 =$ **360**
D (0, 7.5)	$Z = 60\,(0) + 40\,(7.5) = 300$
$E\left(\frac{9}{4}, \frac{45}{8}\right)$	$Z = 60\left(\frac{9}{4}\right) + 40\left(\frac{45}{8}\right) = \mathbf{360}$

Hence, $Z = 360$ is maximum when $x = 6$, $y = 0$ or $x = \frac{9}{4}, y = \frac{45}{8}$. **Ans.**

Example 16. *A dealer wishes to purchase a number of fans and sewing machines. He has only Rs. 5,760 to invest and has space for at most 20 items. A fan and sewing machine cost Rs. 360 and Rs. 240 respectively. He can sell a fan at a profit of Rs. 22 and sewing machine at a profit of Rs. 18. Assuming that he can sell whatever he buys, how should he invest his money in order to maximise his profit ? Translate the problem into LPP and solve it graphically.*

Solution. The above information are given in the following table :

1. **Table**

Items	*Cost (in Rs.)*	*Profit (in Rs.)*	*Space for total number of Items*
Fan	360	22	
Sewing machine	240	18	
Total	5,760		20

2. **Decision Variables.** Let x and y be respectively the number of fans and sewing machines purchased.
3. **Objective function.** To maximize the profit

$$Z = 22\,x + 18\,y$$

4. **Constraint (*i*).** The available space is for at most 20 items.

$\therefore$ $$x + y \le 20$$

Constraint (*ii*). At most investments is Rs. 5,760.

$\therefore$ $$360x + 240y \le 5{,}760 \quad \Rightarrow \quad 3x + 2y \le 48$$

Constraint (*iii*). The number of fans and sewing machines are non-negative.

$\therefore$ $$x \ge 0 \quad \text{and} \quad y \ge 0.$$

5. **Mathematical formulation.**

Maximize $$Z = 22x + 18y \qquad \text{... (1)}$$

Subject to the constraints

$$x + y \le 20 \qquad \text{... (2)}$$

$$3x + 2y \le 48 \qquad \text{... (3)}$$

$$x \ge 0,\; y \ge 0 \qquad \text{... (4)}$$

6. **Region represented by $x + y \le 20$ and $3x + 2y \le 48$**

$x + y = 20$

x	20	0
y	0	20
Point	*A*	*B*

$3x + 2y \le 48$

x	16	0
y	0	24
Point	*C*	*D*

7. **Corner Point Method**

The feasible region, *OCEBO*, is shaded. Here *C* (16, 0), *E* (8, 12) and *B* (0, 20) are the corner points.

Now the value of $Z = 22x + 18y$.

Corner point (x, y) of the feasible region	*Value of the objective function Z = 22x + 18y*
OCEBO	
C (16, 0)	22 (16) + 18 (0) = 352
E (8, 12)	22 (8) + 18 (12) = 392 **Maximum**
B (0, 20)	22 (0) + 18 (20) = 360

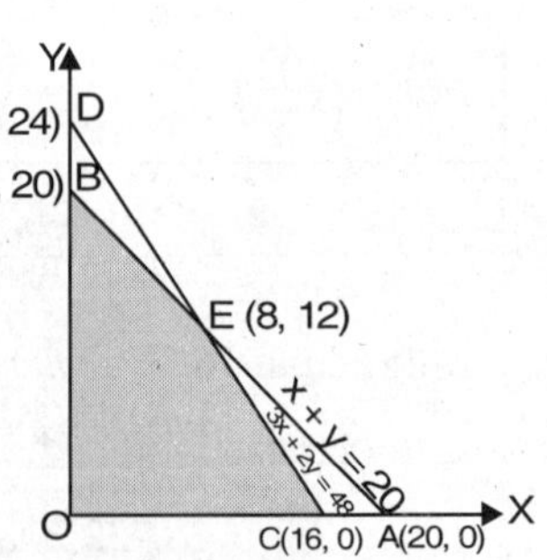

Thus the profit will be maximum, when 8 fans and 12 sewing machines are purchased and sold.

Also the maximum profit = Rs. 392. **Ans.**

Example 17. *Anil wants to invest at most Rs. 12,000 in Bonds A and B. According to the rules, he has to invest at least Rs. 2,000 in Bond A and at least Rs. 4,000 in Bond B. If the rate of interest on Bond A is 8% per annum and on Bond B is 10% per annum, how should he invest his money for maximum interest ? Solve the problem graphically.*

Solution. The above information are given in the following table:

1. Table

Bonds	*Interest*	*Investment in Rs.*
A	8%	At least 2,000
B	10%	At least 4,000
Total		12,000

2. Decision Variables. Let Anil invest Rs. x (more than Rs. 2,000) on bond *A*.
Let Anil invest Rs. y (more than Rs. 4,000) on bond *B*.

3. Objective function. To maximise the total interest

$$Z = \frac{8x}{100} + \frac{10y}{100}$$

$$\Rightarrow \quad Z = \frac{1}{50}(4x+5y)$$

4. Constraint (*i*). Total investment = 12,000

$$x + y \leq 12{,}000$$

Constraint (*ii*). Investment on Bond *A* = Minimum Rs. 2,000

$$x \geq 2{,}000$$

Constraint (*iii*). Investment on Bond *B* = Minimum Rs. 4,000.

$$y \geq 4{,}000.$$

5. Mathematical formulation

To maximise $\qquad Z = \frac{1}{50}(4x+5y) \qquad ... (1)$

Subject to the constraints

$$x + y \leq 12{,}000 \qquad ... (2)$$

$$x \geq 2{,}000 \qquad ... (3)$$

$$y \geq 4{,}000 \qquad ... (4)$$

6. Region represented by
$x + y = 12{,}000$

x	12,000	0
y	0	12,000
Points	*A*	*B*

$x = 2{,}000$ is represented by the line *CD*.
$y = 4{,}000$ is represented by the line *DE*.

7. Corner Point Method

The shaded bounded region *CDEC* is the feasible region of the given L.P.P., where the vertices are C (2,000; 10,000), *D* (2,000; 4,000) and *E* (8,000; 4,000).

Corner points of the feasible region CDE	*Value of the objective function* $Z = \frac{1}{50}(4x+5y)$
C (2,000; 10,000)	$\mathbf{\frac{1}{50}(4 \times 2{,}000 + 5 \times 10{,}000) = 1{,}160}$
D (2,000; 4,000)	$\frac{1}{50}(4 \times 2{,}000 + 5 \times 4{,}000) = 560$
E (8,000; 4,000)	$\frac{1}{50}(4 \times 8{,}000 + 5 \times 4{,}000) = 1{,}040$

Y, 12000, B, x = 2,000, C, E, y = 4,000, 4000, D, x + y = 12,000, O, 2000, A, 12000, X

∴ For earning maximum interest (Rs. 1,160), Anil should invest Rs. 2,000 in Bonds *A* and Rs. 10,000 in Bonds *B*. **Ans.**

Type II. To minimize the objective function (Z)

Example 18. *There are two types of fertilisers F_1 and F_2. F_1 consists of 10% nitrogen and 6% phosphoric acid and F_2 consists of 5% nitrogen and 10% phosphoric acid. After testing the soil conditions, a farmer finds that she needs atleast 14 kg. of nitrogen and 14 kg of phosphoric acid for her crop. If F_1 costs Rs. 6/kg and F_2 costs Rs. 5/kg, determine how much of each type of fertilizer should be used so that nutrient requirements are met at a minimum cost. What is the minimum cost?*

Solution. The above information are given in the table given below :

1. Table.

Fertilisers	*Nitrogen*	*Phosphoric acid*	*Cost per kg (in Rs.)*
F_1	10 %	6 %	6
F_2	5 %	10 %	5
Mixture	14 kg	14 kg	

2. Decision Variables.

Let the fertiliser F_1 in the mixture be x kg.
Let the fertiliser F_2 in the mixture be y kg.

3. Objective function.

To minimise the cost of the mixture of fertilisers F_1 and F_2.

$$Z = 6x + 5y$$

4. Constraint (*i*). Minimum requirement of Nitrogen = 14 kg

$$\frac{10}{100}x + \frac{5 \times y}{100} \geq 14 \quad \Rightarrow \quad 2x + y \geq 280$$

Constraint (*ii*). Minimum requirement of phosphoric acid = 14 kg

$$\frac{6}{100}x + \frac{10}{100}y \geq 14 \quad \Rightarrow \quad 3x + 5y \geq 700$$

Constraint (*iii*). Number of kg of fertiliser F_1 and F_2 be non-negative.

$x \geq 0$ and $y \geq 0$.

5. Mathematical Fermulation.

To minimise $\quad Z = 6x + 5y$

Subject to the constraints

$$2x + y \geq 280$$
$$3x + 5y \geq 700$$
$$x \geq 0, \quad y \geq 0.$$

6. Region represented by $2x+y\geq 280$, **and** $3x+y\geq 700$

$2x + y = 280$ $3x + 5y = 700$

x	140	0
y	0	280
Points	A	B

x	$\frac{700}{3}$	0
y	0	140
Points	C	D

7. Corner Point Method.

The feasible solution is the shaded portion.

The vertices of the feasible solution are B (0, 280), E (100, 80) and $C\left(\frac{700}{3}, 0\right)$

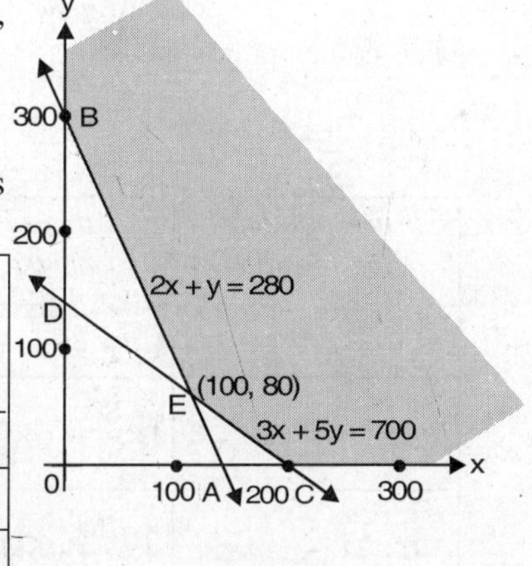

Let us calculate the value of Z at these corner points B, E and C.

Corner point (x, y) of the feasible region	*Value of the objective function* $Z = 6x + 5y$
B (0, 280)	$6\times 0+5\times 280=1400$
E (100, 80)	$\mathbf{6\times 100+5\times 80=1000}$
$C\left(\frac{700}{3}, 0\right)$	$6\times\frac{700}{3}+5\times 0=1400$

The minimum cost of the mixture of fertilisers F_1 and F_2 is Rs. 1000 if 100 kg of the fertiliser F_1 and 80 kg of fertiliser F_2 are mixed. **Ans.**

Example 19. *A diet is to contain at least 80 units of vitamin A and 100 units of minerals. Two foods F_1 and F_2 are available. Food F_1 costs Rs. 4 per unit and food F_2 costs Rs. 6 per unit. A unit of food F_1 contains at least 3 units of vitamin A and 4 units of minerals. A unit of food F_2 contains at least 6 units of vitamin A and 3 units of minerals. Formulate this as a linear programming problem. Find the minimum cost for diet that consists of mixture of these two foods and also meets the minimal nutritional requirements.*

Solution. The above information are given in the table below:

1. Table.

Food	*Vitamin A*	*Mineral (In units)*	*Cost per unit (in Rs.)*
F_1	At least 3 units	4	4
F_2	At least 6 units	3	6
Diet	At least 80 units	100	

2. Decision Variables. Let x units of food F_1 and y units of food F_2 be mixed in the diet.

3. Objective function. To \minimise the cost of the diet. $Z = 4x + 6y$

4. Constraint (*i*). Diet should contain at least 80 units of vitamin A. $3x+6y\geq 80$

Constraint (*ii*). Diet should contain at least 100 units of minerals $4x+3y\geq 100$

Constraint (*iii*). Number of units of food F_1 and F_2 are non-negative $x\geq 0$ and $y\geq 0$.

5. Region represented by $3x+6y\geq 80$ **and** $4x+3y\geq 100$.

$3x + 6y = 80$

x	$\frac{80}{3}$	0
y	0	$\frac{40}{3}$
Points	A	B

$4x + 3y = 100$

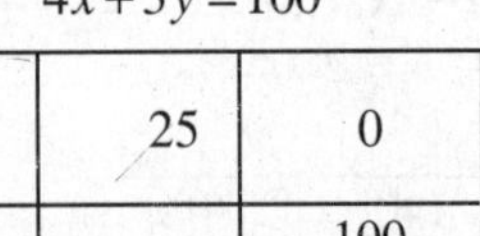

x	25	0
y	0	$\frac{100}{3}$
Points	C	D

7. Corner Point Method

The feasible region is the shaded portion.

The corner points are $A\left(\frac{80}{3}, 0\right)$

$E\left(24, \frac{4}{3}\right)$ and D (0, 30).

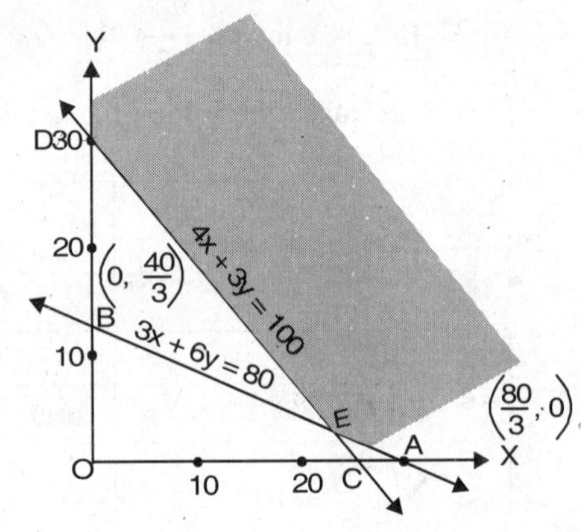

Corner points of the feasible region	*Value of the objective function* $Z = 4x + 6y$
$A\left(\frac{80}{3}, 0\right)$	$4 \times \frac{80}{3} + 6 \times 0 = \frac{320}{3}$
$\mathbf{E\left(24, \frac{4}{3}\right)}$	$\mathbf{4 \times \frac{24}{3} + 6 \times \frac{4}{3} = 40}$
$D(0, 30)$	$4 \times 0 + 6 \times 30 = 180$

Z is minimum for $x = 24$ and $y = \frac{4}{3}$.

The least cost of the mixture is Rs. 40 per unit. **Ans.**

Example 20. *A dietician wishes to mix together two kinds of food X and Y in such a way that the mixture contains atleast 10 units of vitamin A, 12 units of vitamin B and 8 units of vitamin C. The vitamin contents of one kg. food is given below:*

Food	*Vitamin A*	*Vitamin B*	*Vitamin C*	*Cost per kg.*
X	1	2	3	16
Y	2	2	1	20
Mixture	10	12	8	

One kg. of food X costs Rs. 16 and one kg. of food Y costs Rs. 20. Find the least cost of the mixture which will produce the required diet?

Solution.

1. Decision variables.

Let x kg. of food X and y kg of food Y are mixed together to make the mixture.

2. Objective function.

It is given that one kg. of food X costs Rs. 16 and one kg. of food Y cost Rs. 20. So, x kg. of food X and y kg of food Y will cost Rs. $(16x + 20y)$.

$\Rightarrow Z = 16x + 20y$

3. Constraint (i) Since one kg. of food X contains one unit of vitamin A

$\therefore$ x kg. of food X contains x units of vitamin A

Since one kg. of food Y contains 2 units of vitamin A

$\therefore$ y kg. of food Y contains $2y$ units of vitamin A

Therefore, the mixture contains $x + 2y$ units of vitamin A. But the mixture should contain at least 10 units of vitamin A.

$$\therefore \qquad x + 2y \geq 10$$

Constraint (*ii*).

Similarly, the mixture of x kg. of food X and Y kg. of food Y contains $(2x + 2y)$ units of vitamin B.

But the mixture should contain at least 12 units of vitamin B.

$$\therefore \qquad 2x + 2y \geq 12$$

Constraint (*iii*).

x kg. of food X and y kg. of food Y contains $3x + y$ units of vitamin C.

But the mixture should contain at least 8 units of vitamin C.

$$\therefore \qquad 3x + y \geq 8.$$

Constraint (*iv*).

Since the quantity of food X and food Y can not be negative:

$$x \geq 0 \qquad \text{and} \qquad y \geq 0.$$

4. Mathematical Formulation.

Thus the given L.P.P. is

Minimize $\qquad Z = 16x + 20y \qquad$ ...(1)

Subject to $\qquad x + 2y \geq 10 \qquad$ ...(2)

$2x + 2y \geq 12 \qquad$ ...(3)

$3x + y \geq 8 \qquad$ ...(4)

and $\qquad x \geq 0, \qquad y \geq 0 \qquad$ ...(5)

5. Region Represented by the inequations.

To solve this L.P.P. we draw the lines

$x + 2y = 10 \qquad 2x + 2y = 12 \qquad$ and $\; 3x + y = 8$

x	10	0
y	0	5
Points	A	B

x	6	0
y	0	6
Points	C	D

x	$\frac{8}{3}$	0
y	0	8
Points	E	F

The feasible region of the L.P.P. is shaded in the adjoining figure.

6. Corner Point Method

The coordinates of the corner points are $A(10, 0)$, Q (2, 4), $P(1, 5)$ and F (0, 8).

Now, we evaluate $Z = 16x + 20y$ at the corner points.

Corner points of the feasible region	*Corresponding values of $Z = 16x + 20y$*
A (10, 0)	$Z = 16(10) + 20(0) = 160$
Q (2, 4)	**$Z = 16(2) + 20(4) = 112$**
P (1, 5)	$Z = 16(1) + 20(5) = 116$
F (0, 8)	$Z = 16(0) + 20(8) = 160$

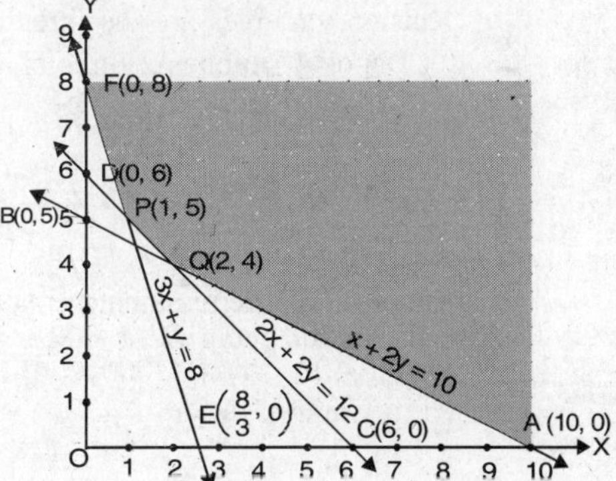

The minimum value of $Z = 112$ at the point $x = 2$, $y = 4$.

Hence, the least cost of the mixture is Rs. 112. **Ans.**

TYPE III. TRANSPORTATION PROBLEMS

Example 21. *An oil company has two depots A and B with capacities of 7000 l and 4000 l respectively. The company is to supply oil to three petrol pumps, D, E and F, whose requirements are 4500 l, 3000 l and 3500 l respectively. The distances (in km) between the depots and the petrol pumps is given in the following table:*

Distance in (km.)

From/To	*A*	*B*
D	7	3
E	6	4
F	3	2

Assuming that the transportation cost of 10 litres of oil is Re. 1 per km, how should the delivery be scheduled in order that the transportation cost is minimum ? What is the minimum cost?

Solution. The given information can be exhibited diagrammatically as follows:

1. **Table.**

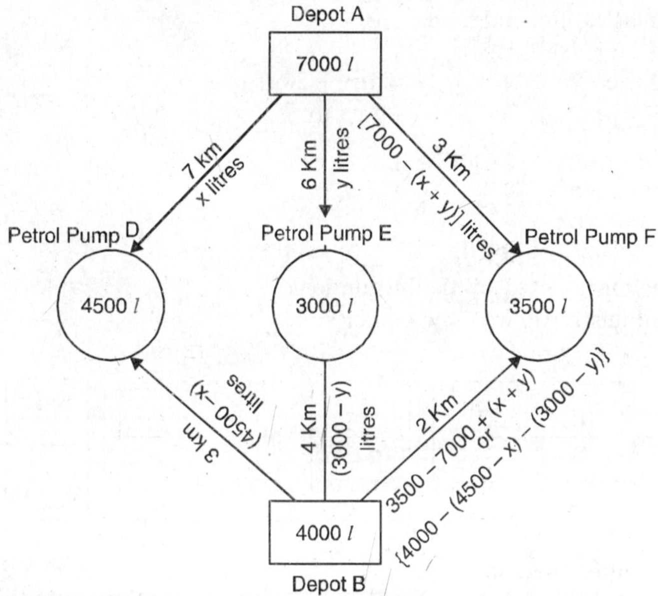

2. **Decision Variables.**

Decision variables are the litres of oil to be supplied from depots to the petrol pumps. Let Depot *A* supplies *x* litres of oil to petrol pump *D* and *y* litres of oil to petrol pump *E* and $[7000-(x+y)]$ litres of oil to petrol pump *F*.

Constraint (*i*). $7000-(x+y)\geq 0 \Rightarrow x+y\leq 7000$

$$x\geq 0,\ y\geq 0$$

Constraint (*ii*) The remaining requirement (4500 – *x*) litres of oil is supplied from the depot *B* to the petrol pump *D*, (3000 – *y*) litres of oil to petrol pump *E* and 4000 – [(4500 – *x*) + (3000 – *y*)] litres of oil to petrol pump *F*.

$$4500-x\geq 0 \quad\Rightarrow\quad x\leq 4500$$

$$3000-y\geq 0 \quad\Rightarrow\quad y\leq 3000$$

$$4000-4500+x-3000+y\geq 0 \quad\Rightarrow\quad x+y\geq 3500$$

3. Objective function.

The transportation cost from depot A to the petrol pump D, E and F is Rs. $\frac{x}{10}$, Rs. $\frac{6y}{10}$ and $Rs. \frac{3}{10}(7000 - x - y)$.

Similarly, the transportation cost from depot B to petrol pump D, E and F is $Rs.\ \frac{3}{10}(4500 - x)$, $Rs. \frac{4}{10}(3000 - y)$ and Rs. $\frac{2}{10}[x + y - 3500]$.

So, the total cost of transportation = Z

$$\Rightarrow \quad Z = \text{Rs.}\left(\frac{7x}{10} + \frac{6y}{10} + \frac{3}{10}(7000 - x - y) + \frac{3}{10}(4500 - x) + \frac{4}{10}(3000 - y) + \frac{2}{10}(x + y - 3500)\right)$$

$$\Rightarrow \quad Z = \left(\frac{7}{10} - \frac{3}{10} - \frac{3}{10} + \frac{2}{10}\right)x + \left(\frac{6}{10} - \frac{3}{10} - \frac{4}{10} + \frac{2}{10}\right)y + 2100 + 1350 + 1200 - 700$$

$$= \text{Rs.}\left(\frac{3}{10}x + \frac{1}{10}y + 3950\right)$$

4. Mathematical formulation.

Minimize $Z = \frac{3}{10}x + \frac{1}{10}y + 3950$ ...(1)

Subject to constraints

$x + y \leq 7000$ ...(2)

$x + y \geq 3500$ ...(3)

$x \leq 4500$ ...(4)

$y \leq 3000$ ...(5)

$x \geq 0$ ...(6)

$y \geq 0$ ...(7)

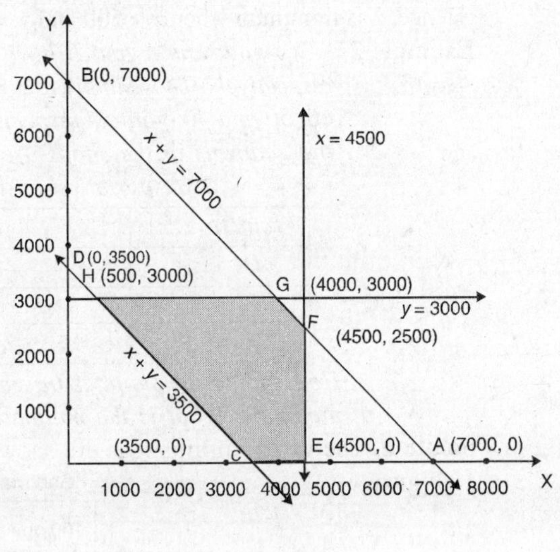

5. Region represented by the inequations.

$x + y = 7000$

x	7000	0
y	0	7000
Point	*A*	*B*

$x + y = 3500$

x	3500	0
y	0	3500
Point	*C*	*D*

The shaded region $C\ E\ F\ G\ H\ C$ represented by the inequations (2) to (7).

6. Corner Point method.

The coordinates of the vertices C, E, F, G, H and C of the feasible region $CEFGHC$ are $C(3500, 0)$, $E(4500, 0)$, $F(4500, 2500)$, $G(4000, 3000)$ and H (500, 3000) respectively. The coordinates of F are obtained by solving $x + y = 7000$ and $x = 4500$, coordinates of G are obtained by solving $x + y = 7000$ and $y = 3000$, the coordinates of H are obtained by solving $x + y = 3500$ and $y = 3000$. The value of the objective function at these points are given in the following table.

Corner points (x, y) of the feasible region C E F G H C	*Value of the objective function* $Z = \frac{3}{10}x + \frac{1}{10}y + 3950$
C (3500, 0)	$\frac{3}{10}(3500) + \frac{1}{10}(0) + 3950 = 5000$
E (4500, 0)	$\frac{3}{10}(4500) + \frac{1}{10}(0) + 3950 = 5300$
F (4500, 2500)	$\frac{3}{10}(4500) + \frac{1}{10}(2500) + 3950 = 5550$
G (4000, 3000)	$\frac{3}{10}(4000) + \frac{1}{10}(3000) + 3950 = 5450$
***H* (500, 3000)**	$\mathbf{\frac{3}{10}(500) + \frac{1}{10}(3000) + 3950 = 4400}$

Hence, Z is minimum when $x = 500$ and $y = 3000$; and the minimum value of Z is 4400. **Ans.**

Example 22. *Two godowns A and B have grain capacity of 100 quintals and 50 quintals respectively. They supply to 3 ration shops, D, E and F, whose requirements are 60, 50 and 40 quintals respectively. The cost of transportation per quintal from the godowns to the shops are given in the following table:*

Transportation cost per quintal (in Rs.)

From/To	*A*	*B*
D	*6*	*4*
E	*3*	*2*
F	*2.50*	*3*

How should the supplies be transported in order that the transportation cost is minimum. What is the minimum cost ?

Solution. The given informations are shown in the following figure:

1. Figure

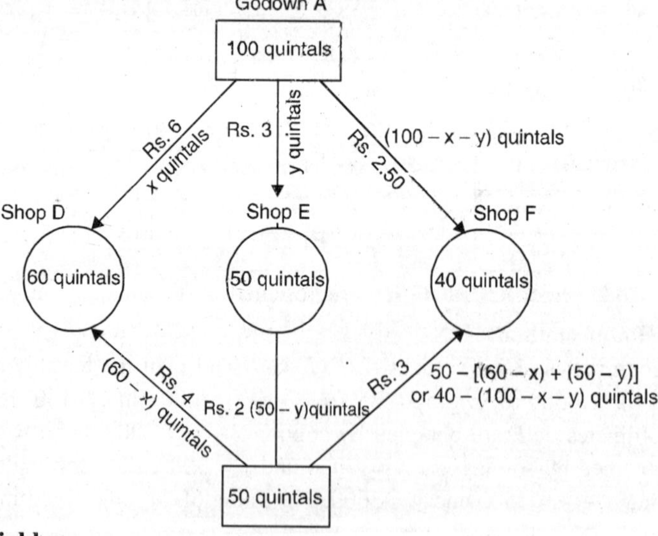

2. Decision variables.

Let the godown A transports x quintals grains to shop D and y quintals of grain to shop E, so the remaining quintals of grain at A, $100 - (x + y)$ quintals will be transported to the shop F etc.

3. Constraint (*i*).

$$100-(x-y)\geq 0 \quad \Rightarrow \quad x+y\leq 100$$
$$x\geq 0$$
$$y\geq 0.$$

Constraint (*ii*). The remaining requirements (60 – *x*), (50 – *y*), and $\{40-(100-x-y)\}$ quintals are to be transported from the godown *B* to the shops *D*, *E* and *F* respectively.

$$60-x\geq 0 \quad \Rightarrow \quad x\leq 60$$
$$50-y\geq 0 \quad \Rightarrow \quad y\leq 50$$
$$40-(100-x-y)\geq 0 \quad \Rightarrow \quad x+y-60\geq 0$$

or $\quad x+y\geq 60$

4. Objective function.

Cost of transportation from Godown *A* to shops *D*, *E* and *F* are Rs. 6*x*, Rs. 3*y* and Rs. 2.5 (100 – *x* – *y*). Similarly the cost of transportation from Godown *B* to shops *D*, *E* and *F* are Rs. 4(60 – *x*), Rs. 2 (50 – *y*) and Rs. 3 [40 – (100 – *x* – *y*)] respectively.

So total cost of transportation = *Z*

Z = Rs. $[6x + 3y + 2.5\,(100 - x - y) + 4\,(60 - x)\; 2\,(50 - 4) + 3\,(-60 + x + 4)]$

$= [(6 - 2.5 - 4 + 3)x + (3 - 2.5 - 2 + 3)\, y + 250 + 240 + 100 - 180]$

$= 2.5\, x + 1.5y + 410$

5. Mathematical Formulation

To minimize $\quad Z = 2.5x + 1.5y + 410$

$x+y\leq 100$ ...(1)

and

$x\geq 0$ ...(2)

$y\geq 0$ ...(3)

$x+y\geq 60$ ...(4)

and $\quad x\leq 60$ ...(5)

and $\quad y\leq 50$ ...(6)

6. (*i*) Region represented by $x+y\leq 100$

$x + y = 100$

x	100	0
y	0	100
Points	*A*	*B*

On joining *A* and *B* we get the graph of the line *x* + *y* = 100. Since, (0, 0) satisfies the inequation $0+0\leq 100$. So, the half plane containing origin represents the region of the inequation $x+y\leq 100$.

(*ii*) Region represented by $x+y\geq 60$.

$x + y = 60$

x	60	0
y	0	60
Points	*C*	*D*

On joining the points *C*(60, 0) and *D*(0, 60), we get the graph of the line represented by *x* + *y* = 60. Since, (0, 0) does not satisfy $x+y\geq 60$. So, the half plane not containing origin represents the region of $x+y\geq 60$.

(*iii*) Region represented by $x \geq 0$ and $y \geq 0$ is the first quadrant.

(*iv*) Region represented by $x \leq 60$ is the half plane containing the origin.

(*v*) Region represented by $y \leq 50$ is the half plane containing origin.

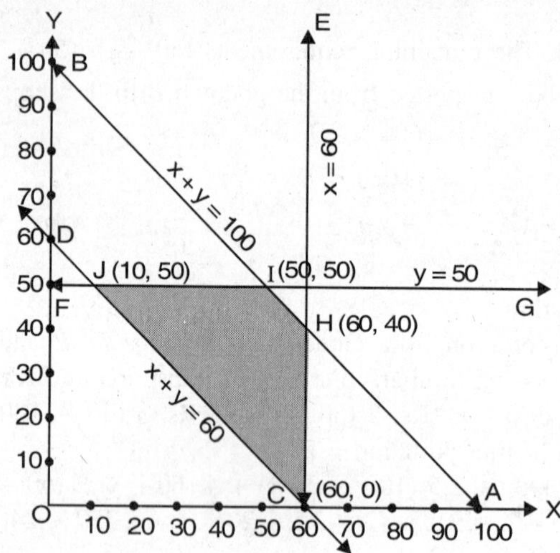

7. Corner point Method

The feasible region is the shaded portion *CHIJC*. Its vertices are *C*(60, 0), *H*(60, 40), *I*(50, 50) and *J*(10, 50).

The coordinates of *H* are calculated by solving the equations $x + y = 100$ and $x = 60$.

The coordinates of *I* are calculated by solving the equations $x + y = 100$ and $y = 50$.

The coordinates of *J* are calculated by solving the equations $x + y = 60$ and $y = 50$.

The values of the objective function at the corner points are given in the following table.

Corner point (x, y) of the feasible region	*Value of the objective function at the corner points in Rs.* $Z = 2.5x + 1.5y + 410$
C (60, 0)	2.5(60) + 1.5(0) + 410 = 560
H (60, 40)	2.5(60) + 1.5(40) + 410 = 620
I (50, 50)	2.5(50) + 1.5 (50) + 410 = 610
***J* (10, 50)**	**2.5 (10) + 1.5 (50) + 410 = 510**

Hence, *Z* is minimum when $x = 10$ and $y = 50$, the minimum cost of transportation in Rs. 510.

EXERCISE 56.3

1. A manufacturer produces two items X and Y. X needs two hours on machine A and two hours on machine B. Y needs 3 hours on machine A and 1 hour on machine B. If machine A can run for a maximum of 12 hours per day and machine B for 8 hours per day and profits from X and Y are Rs. 4 and Rs. 5 per items respectively. Formulate the problem as a linear programming problem and solve it graphically.

   Ans. Maximum profit = Rs. 22; Number of items X = 3; Number of items Y = 2

2. An aeroplane can carry a maximum of 200 passengers. A profit of Rs. 400 is made on each first class ticket and a profit of Rs. 300 is made on each economy class ticket. The airline reserves at least 20 seats for first class. However, at least 4 times as many passengers prefer to travel by economy class to the first class. Determine how many each type of tickets must be sold in order to maximize the profit for the airline. What is the maximum profit ?

   Ans. Maximum profit = Rs. 64000, First class tickets = 40. Economy class tickets = 160.

3. A manufacturer is trying to decide on the product quantities of two products-tables and chairs. There are 98 units of material and 80 labour-hours available. Each table requires 7 units of material and 10 labour-hours, while each chair requires 14 units of material and 8 labour-hours per chair. The profit on a table and a chair is Rs. 25 and Rs. 20, respectively. How many tables and chairs should be produced to have maximum profit?

(**Hint**: Use Iso-profit method).

Ans. Maximum profit = Rs. 200, Tables = 8, chairs = 0 or Tables = 4, chairs = 5.

4. A man owns a field of area 1000 sq. metre. He wants to plant trees in it. He has a sum of Rs. 1400 to purchase young trees. He has the choice of two types of trees. Type A requires 10 sq. metre of ground per tree and costs Rs. 20 per tree and type B requires 20 sq. metre of ground per tree and costs Rs. 25 per tree. When fully grown, type A produces an average of 20 kg. of fruits which can be sold at a profit of Rs. 2 per kg. and type B produces an average of 40 kg of fruits which can be sold at a profit of Rs. 1.50 per kg. How many trees of each type should be planted to achieve maximum profit when the trees are fully grown? What is the maximum profit ?

Ans. Maximum profit = Rs. 32,00, Type A = 20. Type B = 40.

5. A farm is engaged in breeding goats. The goats are fed on various products grown on the farm. They need certain nutrients, named as X, Y and Z. The goats are fed on two products A and B. One unit of product A contains 36 units of X, 3 units of Y and 20 units of Z, while one unit of product B contains 6 units of X, 12 units of Y and 10 units of Z. The minimum requirement of X, Y and Z is 108 units, 36 units and 100 units respectively. Product A costs Rs. 20 per unit and product B costs Rs. 40 per unit. How many units of each product must be taken to minimize the cost?

Ans. Minimum cost = Rs. 160, Product A = 4 units, Product B = 2 units.

6. A company producing soft drinks has a contract which requires that a minimum of 80 units of chemical A and 60 units of chemical B are to go in each bottle of the drink. The chemicals are available in a prepared mix from two different suppliers. Supplier X has a mix of 4 units of A and 2 units of B that costs Rs. 10 and the supplier *Y* has a mix of 1 unit of A and 1 unit of *B* that costs Rs. 4. How many mixes from X and Y should the company purchase to honour contract requirement and yet minimize the cost?

Ans. Minimum cost = Rs. 260. Mix of type A = 10 units, Mix of type B = 40 units.

7. To maintain one's health, a person must fulfil minimum daily requirements for the following three nutrients-calcium, protein and calories. His diet consists of only food items I and II whose prices and nutrient contents are shown below:

Price	*Food I* *Rs. 0.60 per unit*	*Food II* *Re. 1 per unit*	*Minimum requirements*
Calcium	10	4	20
Protein	5	5	20
Calories	2	6	12

Find the combination of food items so that the cost may be minimum.

Ans. Minimum cost = Rs. 2.80, Food I = 3 units, Food II = 1 unit.

8. A diet for a sick person must contain at least 4000 units of vitamin, 50 units of minerals and 1400 units of calories. Two foods A and B are available at a cost of Rs. 4 and Rs. 3 per unit respectively. If one unit of A contains 200 units of vitamin, 1 unit of mineral and 40 units of calories: one unit of food B contains 100 units of vitamin, 2 units of minerals and 40 units of calories, find what combination of foods should be used to have the least cost?

Ans. Minimum cost = Rs. 110, Food A = 5 units, Food B = 30 units.

9. Two godowns A and B have grain storage capacity of 100 quintals and 50 quintals respectively. They supply to three ration shopes D, E and F whose requirements are 60, 50 and 40 quintals respectively. The cost of transportation per quintal from the godown to the shopes are given in the following table:

From / To	Godown A	Godown B
D	6.00	4.00
E	3.00	2.00
F	2.50	3.00

How should the supplies we transported in order that the transportation cost is minimum.

Ans. From A : 10 quintals, 50 quintals and 40 quintals to D, E and F respectively
From B : 50 quintals, 0 quintal and 0 quintal to D, E and F respectively.

10. There is a factory located at each of the two places *P* and *Q*. From these locations, a certain commodity is delivered to each of these depots situated at *A*, *B* and *C*. The weekly requirements of the depots are respectively 5, 5 and 4 units of the commodity while the production capacity of the factories at *P* and *Q* are respectively 8 and 6 units. The cost of transportation per unit is given below:

To / From	Cost (in Rs.)		
	A	B	C
P	16	10	15
Q	10	12	10

How many units should be transported from each factory to each depot in order that the transportation cost is minimum. Formulate the above L.P.P mathematically and then solve it.

Ans. Minimum cost = Rs. 155,
From *P* : 0, 5, 3 units to depots at A, B, C respectively
From *Q* : 5, 0 and 1 units to depots at A, B and C respectively.

11. A brick manufacturer has two depots, A and B with stock of 30,000 and 20,000 bricks respectively. He receives orders from three builders *P*, *Q* and *R* for 15,000; 20,000 and 15,000 bricks respectively the cost in Rs. transporting 1000 bricks to the builders from the depots are given below:

From / To	*P*	*Q*	*R*
A	40	20	30
B	20	60	40

How should the manufacturer fulfill the orders so as to keep the cost of transportation minimum.

Ans. Minimum cost = Rs. 1200
From *A* : 0, 20 and 10 thousand bricks to builders P, Q and R.
From *B* : 15, 0 and 5 thousand bricks to builders P, Q and R.

12. A publisher sells a hard cover edition of a textbook for Rs. 72.00 and a paper back edition of the same text for Rs. 40.00. Costs to the publisher are Rs. 56.00 and Rs. 28.00 per book respectively in addition to weekly costs of Rs. 9600.00. Both types require 5 minutes of printing time, although hard cover requires 10 minutes binding time and the paper back requires only 2 minutes both the printing and binding operations have 4,800 minutes available each week. How many of each type of Book should be produced in order to maximize profit?

Ans. Max. profit = Rs. 2880, 360 hard cover edition, 600 paper back edition

56.9 THEORY OF SIMPLEX METHOD

The basis of the complex method consists of two fundamental conditions :-

(*i*) The feasibility condition

It ensures that the starting solution is basic feasible, only basic feasible solution will be obtained during computation.

(*ii*) Optimality condition

It guarantees only better solution (as compared to the current solution)

SOME IMPORTANT DEFINITIONS

Consider the following problem

Maximize $Z = c_1x_1 + c_2x_2 + \ldots\ldots\ldots\; c_nx_n,$ — Objective function

Subject to $a_{11}x_1 + a_{12}x_2 + \ldots\ldots\; a_{1n}x_n \le b_1,$ — Constraints

$a_{21}x_1 + a_{22}x_2 + \ldots\ldots\; a_{2n}x_n \le b_2,$

$a_{m1}x_1 + a_{m2}x_2 + \ldots\ldots, + a_{mn}x_n \le b_m,$

where $x_1, x_2 \ldots\ldots, x_n \ge 0.$

Introducing slack vairable $s, s_1, s_2, s_3 \ldots\ldots\ldots\; s_m$ in m constraint equations. It can be put in the following standard form:

Maximize $z = c_1x_1 + c_2x_2 + \ldots\ldots\; c_nx_n + s_1 + s_2 \ldots\ldots + s_m.$ .....(1)

Subject to $a_{11}x_1 + a_{12}x_2 + \ldots\ldots\ldots\; c_{1n}x_n + s_1 = b_1$

$a_{21}x_1 + a_{22}x_2 + \ldots\ldots\ldots\; a_{2n}x_n + s_2 = b_2$ .....(2)

$a_{m1}x_1 + a_{m2}x_2 + \ldots\ldots\; a_{mn}x_n + s_m = b_m$

where $x_1, x_2, \ldots\ldots\ldots, x_n, s_1, s_2 \ldots\ldots\; s_m \ge 0$ .....(3)

1. **Solution :** To find out the values of $x_1, x_2 \ldots\ldots\ldots\; x_n, s_1, \ldots\ldots\; s_m.$
2. **Feasible solution :** The values of $x_1, x_2 \ldots\ldots\ldots\; x_n, s_1, \ldots\ldots\; s_m$ is known as feasible solution if these values satisfy the equations (2) and (3)
3. **Basic solution :** By making any n variables out of $n + m$ equal to zero, the values of the remaining m variable is called the basic solution, if the determinant of the coefficients of these slack variables is non negative.
4. **Basic variables :** The variables of the basic solution are called basic variables (Some of them may be zero). The other n variables are called non-basic variables.
5. **Basic feasible solution :** The basic solution of non-negative variable is called basic feasible solution.
6. **Non- degenerate basic feasible solution :** If all m basic feasible variables non-negative, then the set of these variables known as non degenerate basic feasible solution.
7. **Degenerate basic feasible solution :** If one or more basic feasible variable is zero, then the solution known as degenerate basic feasible solution.
8. **Optimal basic feasible solution :** It contains basic feasible solution that optimize the objective function.
9. **Sets of points :** Linear equation in two variables represents a line and a linear equation in three variables represents a plane. Both of them are considered as set of points.

Example 23. *Solve by using simplex method* :

Maximize $z = 3x_1 + 4x_2$

Subject to $x_1 + x_2 \le 450$

$2x_1 + x_2 \le 600$

$x_1, x_2 \ge 0$

Solution.

Step 1. Express the the above objective function and inequations in the equation of standard form by adding slack variables.

Maximize $z = 3x_1 + 4x_2 + 0s_1 + 0s_2$

Subject to $x_1 + x_2 + s_1 = 450$

$2x_1 + x_2 + s_2 = 600$

where $x_1, x_2, s_1, s_2 \geq 0$

Step 2. Find initial basic feasible solution.

Putting $x_1 = x_2 = 0$, in the constraints, we get

$s_1 = 450$ and $s_2 = 600$ and $z = 0$

This is the initial basic feasible solution.

The above information can be expressed in the tabular form as follows.

Coefficient	C_J	3	4	0	0	
C_B	Basis	body	matrix	identity matrix		value
		x_1	s_2	s_1	s_2	b
0	s_1	1	1	1	0	450
0	s_2	2	1	0	1	600

Row, C_j represents the coefficients of the variables in the objective function.

Column, C_B represents the coefficients of the current basic variables. The basic variables are the slack variables s_1 and s_2.

Body matrix : The body matrix under non-basic variables x_1 and x_2 represents their coefficients in the constraints.

Indentity matrix : The identity matrix represents the coefficients of slack variables in the constraints.

Column b indicates the values of the basic variables, s_1 and s_2 in the initial basic feasible solution.

Step 3. Perform optimality test. Here, since $C_J - Z_J$ is positive under x_1 and x_2 columns so initial basic problem is not optimal and can be improved.

	C_j	3	4	0	0			
C_B	Basis	x_1	x_2	s_1	s_2	b	θ	
0	s_1	1	(1)	1	0	450	450	→ (Key row)
0	s_2	2	1	0	1	600	600	
	$Z_j = \Sigma(C_B a_{ij})$	0	0	0	0	0		
	$C_j - Z_j$	3	4	0	0			
			↑ K (Key column)					

Row $Z_j : Z_j = \Sigma\, C_B a_{ij}$

where a_{ij} are the matrix element in the i-th row and jth column. For example, $Z_1 = 0 \times 1 + 0 \times 2 = 0$

If the elements in the $C_j - Z_j$ row are negative or zero, then the current solution is optimal. Since here, two elements 3 and 4 under x_1 and x_2 variable columns are positive, so the solution is not optimal and we proceed to the next step.

Step 4. Iterate towards an optimal solution :

Selection of entering variable :

For this we look for maximum positive values in $C_j - Z_j$ row. If more than one vairable appears with the same maximum value then any one can be choosen arbitrarily.This variables is called entering variable. This column is knows as key column. Thus here we find x_2 as entering variable.

Selection of the leaving variable

Elements of column b are divided by the corresponding elements of the key column and row containing the minimum non-negative ratio is marked. This row is called the key row.

The element lying at the intersection of key column and key row is called key (or pivot) element and is enclosed in ( ). Here key element is 1.

Preparing new simplex table :

We replace s_1 by x_2 from the basis. Corresponding C_B coefficient is changed from 0 to 4. If Key element is not unity made it unity and other intersectional elements are made zeros by suitable row operations. Thus, we get the following table:

	C_j	3	4	0	0	
C_B	Basis	x_1	x_2	s_1	s_2	b
4	x_2	1	1	1	0	450
0	s_2	1	0	–1	1	150
	$Z_j = (\Sigma\, C_B a_{ij})$	4	4	4	0	1800
	$C_j - Z_j$	–1	0	–4	0	

Step 5. Since here all elements of $C_j - Z_j$ row are either zero or negative so the second feasible solution is optimal.

Hence the optimal solution is

$$x_1 = 0$$
$$x_2 = 450$$
$$Z_{max} = \text{Rs } 1800$$

Ans.

Example 24. *Use simplex method to solve the following problem.*

Maximize $Z = 2x_1 + 5x_2$

Subject to
$$x_1 + 4x_2 \le 24$$
$$3x_1 + x_2 \le 21$$
$$x_1 + x_2 \le 9$$
$$x_1, x_2 \ge 0$$

Solution.

Step 1. First we express the objective funciton and inequations in equations of standard form by adding slack variables s_1, s_2, and s_3.

Maximize $Z = 2x_1 + 5x_2 + 0s_1 + 0s_2 + 0s_3$.

Subject to
$$x_1 + 4x_2 + s_1 = 24$$
$$3x_1 + x_2 + s_2 = 21$$
$$x_1 + x_2 + s_3 = 9$$
$$x_1 + x_2 + s_3 = 9$$
$$x_1, x_2, s_1, s_2, s_3 \ge 0$$

Step 2. Find initial basic feasible solution.

Putting $x_1 = x_2 = 0$, in the constraints, we get

$$s_1 = 24,\ s_2 = 21 \text{ and } s_3 = 9 \text{ and } z = 0$$

This is initial basic feasible solution.

The above information can be expressed in matrix form as follow:

	Cj	2	5	0	0	0		
C_B	Basis	x_1	x_2	s_1	s_2	s_3	b	θ
0	s_1	1	(4)	1	0	0	24	6 → (Key row)
0	s_2	3	1	0	1	0	21	21
0	s_3	1	1	0	0	1	9	9
	Z_j	0	0	0	0	0	0	
	$C_j - Z_j$	2	5 K↑	0 key column	0	0	0	first feasible solution

Step 3. Perform optimality test :

Here, since $C_j - Z_j$ is positive under x_1 and x_2 - columns so initial basic feasible solution is not optimal and can be improved.

Step 4. Iterate towards an optimal solution.

First we make pivot element unity for this we divide the key row by 4.

	C_j	2	5	0	0	0		
C_B	Basis	x_1	x_2	s_1	s_2	s_3	b	
5	x_2	$\frac{1}{4}$	1	$\frac{1}{4}$	0	0	6	$R_1 \to \frac{1}{4} R_1$
0	s_2	3	1	0	1	0	21	
0	s_3	1	1	0	0	1	9	

Now we make the other elements of key column 0 by suitable following row operations.

	C_j	2	5	0	0	0		
C_B	Basis	x_1	x_2	s_1	s_2	s_3	b	
5	x_2	$\frac{1}{4}$	1	$\frac{1}{4}$	0	0	6	
0	s_2	$\frac{11}{4}$	0	$-\frac{1}{4}$	1	0	15	$R_2 \to R_2 - R_1$
0	s_3	$\frac{3}{4}$	0	$-\frac{1}{4}$	0	1	3	$R_3 \to R_3 - R_1$

	C_j	2	5	0	0	0		
C_B	Basis	x_1	x_2	s_1	s_2	s_3	b	θ
5	x_2	$\frac{1}{4}$	1	$\frac{1}{4}$	0	0	6	24
0	s_2	$\frac{11}{4}$	0	$-\frac{1}{4}$	1	0	15	$\frac{60}{4}$
0	s_3	$\left(\frac{3}{4}\right)$	0	$-\frac{1}{4}$	0	1	3	4 →

Z_j	$\frac{5}{4}$	5	$\frac{5}{4}$	0	0	30
$C_j - Z_j$	$\frac{3}{4}\uparrow$	0	$-\frac{5}{4}$	0	0	Second feasible solution

By the above table we observe that $\frac{3}{4}$ is the maximum positive value of $C_j - Z_j$ row so its column is the key column. Thus, here x_2 is entering variable.
Also 4 is the minimum ratio under θ, so row containing 4 is key row. Thus s_3 is leaving variable. At the intersection of key row and key column is $\frac{3}{4}$, so $\frac{3}{4}$ is pivot element.

C_B	Basis	x_1	x_2	s_1	s_2	s_3	b	
5	x_2	$\frac{1}{4}$	1	$\frac{1}{4}$	0	0	6	
0	s_2	$\frac{11}{4}$	0	$-\frac{1}{4}$	1	0	15	
0	x_1	1	0	$-\frac{1}{3}$	0	$\frac{4}{3}$	4	$R_3 \to \frac{4}{3} R_3$

C_B	Basis	x_1	x_2	s_1	s_2	s_3	b	
5	x_2	0	1	$\frac{1}{3}$	0	$-\frac{1}{3}$	5	$R_1 \to R_1 - \frac{1}{4} R_3$
0	s_2	0	0	$\frac{2}{3}$	1	$-\frac{11}{3}$	4	$R_2 \to R_2 - \frac{11}{4} R_3$
2	x_1	1	0	$-\frac{1}{3}$	0	$\frac{4}{3}$	4	

	C_j	2	5	0	0	0		
C_B	Basis	x_1	x_2	s_1	s_2	s_3	b	
5	x_2	0	1	$\frac{1}{3}$	0	$-\frac{1}{3}$	5	
0	s_2	0	0	$\frac{2}{3}$	1	$-\frac{11}{3}$	4	
2	x_1	1	0	$-\frac{1}{3}$	0	$\frac{4}{3}$	4	
	Z_j	2	5	1	0	1	33	Third feasible solution
	$C_j - Z_j$	0	0	−1	0	−1		

Since here all elements of $C_j - Z_j$ row are either zero or negative so the third feasible solution is optimal.

Hence, the optimal solution is

$$x_1 = 4$$
$$x_3 = 5$$
$$s_2 = 4$$
$$Z_{max} = 33$$

Ans.

Example 25. *Solve by simplex method the following L.P. problem.*

Minimize $\quad Z = x_1 - 3x_2 + 3x_3$

Subject to

$$3x_1 - x_2 + 2x_3 \le 7$$
$$2x_1 + 4x_2 \ge -12$$
$$-4x_1 + 3x_2 + 8x_3 \le 10$$
$$x_1, x_2, x_3 \ge 0$$

Solution.

Step 1. Right hand side of second constraint is made positive by multiplying –1.

Adding the slack variables s_1, s_2 and s_3 the above inequations can be written in standard form as follows :

Minimize $\quad Z = x_1 - 3x_2 + 3x_3 + 0s_1 + 0s_2 + 0s_3$

Subject to

$$3x_1 - x_2 + 2x_3 + s_1 = 7$$
$$-2x_1 - 4x_2 + s_2 = 12$$
$$-4x_1 + 3x_2 + 8x_3 + s_3 = 10$$
$$x_1, x_2, x_3, s_1, s_2, s_3 \ge 0$$

Step 2. To find initial basic feasible solution putting $x_1 = x_2 = x_3 = 0$ in the constraints, we get $s_1 = 7$, $s_2 = 12$, and $s_3 = 10$; $Z = 0$

The above information can be written in tabular form as follows:

	C_j	1	–3	3	0	0	0		$\frac{b}{\text{coefficient of } x_2}$	
C_B	Basis	x_1	x_2	x_3	s_1	s_2	s_3	b	θ	
0	s_1	3	–1	2	1	0	0	7	–7	
0	s_2	–2	–4	0	0	1	0	12	–3	
0	s_3	–4	(3)	8	0	0	1	10	$\frac{10}{3}$ →	Key Row
	Z_j	0	0	0	0	0	0	0		
	$C_j - Z_j$	1	–3	3	0	0	0			
			K↑ key column							

Step 3. **Perform optimality test:**

Here as we have to minimize Z, so if any $C_j - Z_j$ coefficient is negative the solution is not optimal

The column having most negative $C_j - Z_j$ value will be the key column.

Since, here element –3 under x_2 variable column is most negative, so the solution is not optimal and we proceed to the next step.

Step 4. Iterate towards an optimal solution. Variable s_3 is replaced by x_2 by performing the row operations in the usual ways. Solution will become optimal when all $C_j - Z_j$ coefficients become zero or positive.

	C_j	1	–3	3	0	0	0		$\frac{b}{\text{coefficient of } x_2}$
C_B	Basis	x_1	x_2	x_3	s_1	s_2	s_3	b	θ
5	s_1	$\left(\frac{5}{3}\right)$	0	$\frac{14}{3}$	1	0	$\frac{1}{3}$	$\frac{31}{3}$	$\frac{31}{5} \rightarrow$ Key Row
0	s_2	$\frac{-22}{3}$	0	$\frac{32}{3}$	0	1	$\frac{4}{3}$	$\frac{76}{3}$	$\frac{-38}{11}$
0	x_2	$\frac{-4}{3}$	1	$\frac{8}{3}$	0	0	$\frac{1}{3}$	$\frac{10}{3}$	$-\frac{5}{2}$
	Z_j	4	–3	–8	0	0	–1	–10	
	$C_j - Z_j$	–3 K↑	0	11	0	0	1		

Here further we have find a negative element –3 in $C_j - Z_j$ row, so we again proceed as above, with entering variable x_1 and leaving variable s_1.

	C_j	1	–3	3	0	0	0		
C_B	Basis	x_1	x_2	x_3	s_1	s_2	s_3	b	
1	x_1	1	0	$\frac{14}{5}$	$\frac{3}{5}$	0	$\frac{1}{5}$	$\frac{31}{3}$	
0	s_2	0	0	$\frac{156}{5}$	$\frac{22}{5}$	1	$\frac{14}{5}$	$\frac{354}{5}$	
–3	x_2	0	1	$\frac{32}{5}$	$\frac{4}{5}$	0	$\frac{3}{5}$	$\frac{58}{5}$	
	Z_j	1	–3	$-\frac{82}{5}$	$-\frac{9}{5}$	0	$-\frac{8}{5}$	$-\frac{143}{5}$	
	$C_j - Z_j$	0	0	$\frac{97}{5}$	$\frac{9}{5}$	0	$\frac{8}{5}$	optimal solution	

Since all elements of $C_j - Z_j$ row is positive or zero.
Hence, optimal solution is

$$x_1 = \frac{31}{5},\ x_2 = \frac{58}{5},\ x_3 = 0,\ Z_{min} = -\frac{143}{5}$$ **Ans.**

EXERCISE 56.4

Solve the following problems by the Simplex method.

1. Maximize $Z = 3x_1 + 2x_2$

Subject to $x_1 + x_2 \leq 4$

$x_1 - x_2 \leq 2$

$x_1, x_2 \geq 0$

Ans. $x_1 = 3,\ x_2 = 1$

$Z_{max.} = 11$

2. Maximize $Z = 4x_1 + 10x_2$

Subject to $2x_1 + x_2 \le 50$

$2x_1 + 5x_2 \le 100$

$2x_1 + 3x_2 \le 90$

$x_1, x_2 \ge 0$

Ans. $x_1 = 0, x_2 = 20$

$Z_{max.} = 200$

3. Maximize $Z = 2x_1 + x_2$

Subject to $x_1 + 2x_2 \le 10$

$x_1 + x_2 \le 6$

$x_1 - x_2 \le 2$

$x_1 - 2x_2 \le 1$

$x_1, x_2 \ge 0$

Ans. $x_1 = 4, x_2 = 2$

$Z_{max.} = 10$

4. Minimize $Z = x_1 - 3x_2 + 2x_3$

Subject to $3x_1 - x_2 + 2x_3 \ge 7$

$-2x_1 + 4x_2 \le 12$

$-4x_1 + 3x_2 + 8x_3 \le 10$

$x_1, x_2, x_3 \ge 0$

Ans. $x_1 = 4, x_2 = 5, x_3 = 0$

$Z_{min} = -11$

5. Minimize $Z = 4x_1 + x_2$

Subject to $3x_1 + x_2 = 3$

$4x_1 + 3x_2 \ge 6$

$x_1 + 2x_2 \le 3$

$x_1, x_2 \ge 0$

Ans. $x_1 = \frac{3}{5}, x_2 = \frac{6}{5}$

$Z_{min} = \frac{18}{5}$

Statistical Technique

(Moment, Moment Generating Function, Skewness, Kurtosis)

57.1 STATISTICS

Statistics is a branch of science dealing with the collection of data, organising, summarising, presenting and analysing data and drawing valid conclusions and thereafter making reasonable decisions on the basis of such analysis.

57.2 FREQUENCY DISTRIBUTION.

Frequency distribution is the arranged data, summarised by distributing it into classes or categories with their frequencies.

Wages of 100 workers

Wages in Rs.	0-10	10-20	20-30	30-40	40-50
Number of workers	12	23	35	20	10

57.3 GRAPHICAL REPRESENTATION

It is often useful to represent frequency distribution by means of a diagram. The different types of diagrams are

1. Histogram
2. Frequency polygon
3. Frequency curve
4. Cumulative frequency curve or Ogive
5. Bar chart
6. Circles or Pie diagrams.

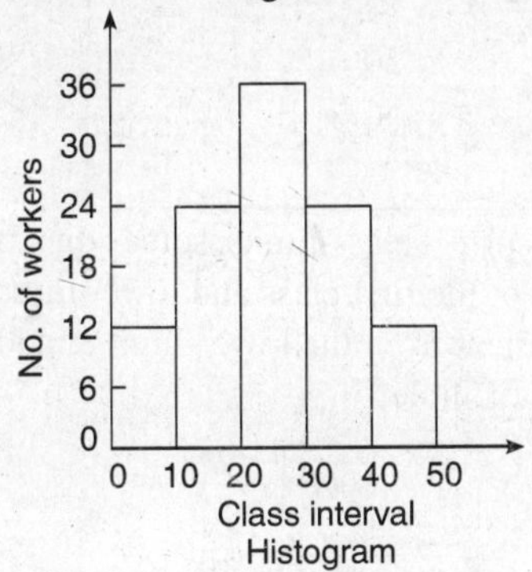

Histogram

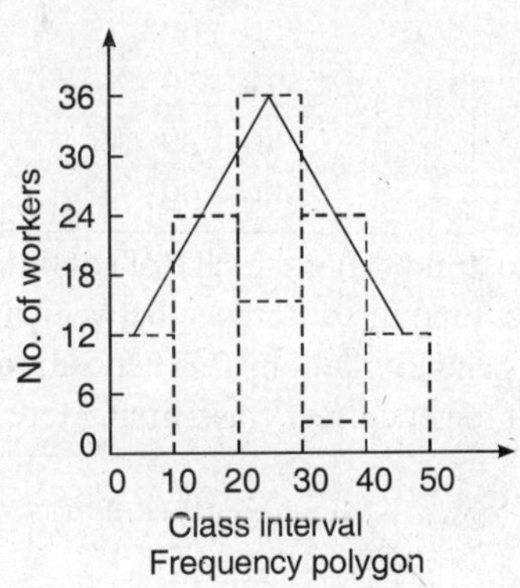

Frequency polygon

1. Histogram. Histogram consists of a set of rectangles having their heights proportional to the class-frequencies, for equal class-intervals. For unequal class-interval, the areas of rectangles are proportional to the frequencies.

2. Frequency Polygon. Frequency Polygon is a line graph of class-frequency plotted against class-mark. It can be obtained by connecting mid-points on the tops of the rectangles in the histogram.

3. Cumulative Frequency curve or the Ogive. If the various points are plotted according to the upper limit of the class as x co-ordinate and the cumulative frequency as y co-ordinate and these points are joined by a free hand smooth curve, the curve obtained is known as cumulative frequency curve or the Ogive.

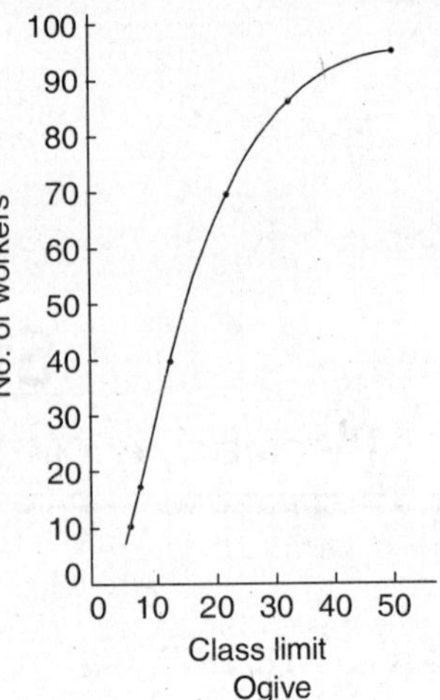

Ogive

57.4 EXCLUSIVE AND INCLUSIVE CLASS INTERVALS

Exclusive Class intervals: If the upper limit of a class is not included in its class interval, then this type of class interval is called exclusive, *e.g.*,

Income (Rs.)	*No. of workers*
40 – 50	20
50 – 60	30
60 – 70	40
70 – 80	22
80 – 90	18
90 – 100	10

In this method the upper limit of one class is the lower limit of next class. In the above example, in the class 40 – 50 there are twenty persons whose income is from Rs. 40 to 49.99. A person whose income is Rs. 50 is included in the class Rs. 50 – Rs. 60. Not in Rs. 40 – Rs. 50. These type of class-intervals are also known as overlapping class intervals.

Inclusive Class intervals: If the upper limit of a class is included in its class interval, then it is called *inclusive class interval.*

Income (Rs.)	*Frequency*
40 – 49	8
50 – 59	12
60 – 69	15
70 – 79	20
80 – 89	8
90 – 99	7

Adjustment: For the sake of continuity and to get correct class-limits, some adjustment is to be done. Find the difference between the upper limit of the first class and lower limit of the next class and divide it by 2. Then subtract this adjustment from the lower-limit & add it to upper limit to get correct class-limit. Here, in the above example,

$$\text{Adjustment} = \frac{50-49}{2} = 0.5$$

Subtract 0.5 from all the lower limit and add 0.5 to all the upper limits.

The adjusted class will be

Income (Rs.)	*Frequency*
39.5 – 49.5	8
49.5 – 59.5	12
59.5 – 69.5	15
69.5 – 79.5	20
79.5 – 89.5	8
89.5 – 99.5	7

57.5 AVERAGE OR MEASURES OF CENTRAL TENDENCY

An average is a value which is representative of a set of data. Average value may also be termed as measures of central tendency. There are five types of averages in common.

(*i*) Arithmetic average or mean (*ii*) Median (*iii*) Mode

(*iv*) Geometric Mean (*v*) Harmonic Mean

57.6 ARITHMETIC MEAN

(*a*) If $x_1, x_2, x_3, \ldots\ldots, x_n$ are *n* numbers, then their arithmetic mean (A.M.) is defined by

$$A.M. = \frac{x_1 + x_2 + x_3 + \ldots. + x_n}{n} = \frac{\sum x}{n}$$

If the number x_1 occurs f_1 times x_2 occurs f_2 times and so on, then

$$A.M. = \frac{f_1x_1 + f_2x_2 + f_3x_3 + \ldots. + f_nx_n}{f_1 + f_2 + \ldots.. + f_n} = \frac{\Sigma f x}{\Sigma f}$$

This is known as direct method.

Example 1. *Find the mean of 20, 22, 25, 28, 30.*

Solution. $A.M. = \frac{20+22+25+28+30}{5} = \frac{125}{5} = 25$ **Ans.**

Example 2. *Find the mean of the following*

Numbers	*8*	*10*	*15*	*20*
Frequency	*5*	*8*	*8*	*4*

Solution. (*a*) Direct method $\Sigma fx = 8 \times 5 + 10 \times 8 + 15 \times 8 + 20 \times 4$

$= 40 + 80 + 120 + 80 = 320$

$\Sigma f = 5 + 8 + 8 + 4 = 25$

$$A.M. = \frac{\Sigma fx}{\Sigma f} = \frac{320}{25} = 12.8$$ **Ans.**

(*b*) **Short cut method**

Let *a* be the assumed mean, *d* the deviation of the variate *x* from *a*. Then

$$\frac{\Sigma fd}{\Sigma f} = \frac{\Sigma f(x-a)}{\Sigma f} = \frac{\Sigma fx}{\Sigma f} - \frac{\Sigma fa}{\Sigma f} = A.M. - \frac{a\Sigma f}{\Sigma f} = A.M. - a$$

$$\therefore \quad \boxed{\mathbf{A.M.} = a + \frac{\Sigma fd}{\Sigma f}}$$

Example 3. *Find the arithmetic mean for the following distribution:*

Class	*0-10*	*10-20*	*20-30*	*30-40*	*40-50*
Frequency	*7*	*8*	*20*	*10*	*5*

Solution. Let assumed mean (a) = 25.

Class	*Class-mark* x	*Frequency* f	$x - 25 = d$	$f.d$
0 – 10	5	7	– 20	– 140
10 – 20	15	8	– 10	– 80
20 – 30	25	20	0	0
30 – 40	35	10	+ 10	+ 100
40 – 50	45	5	+ 20	+ 100
Total		$\Sigma f = 50$		$\Sigma fd = -20$

$$A.M. = a + \frac{\Sigma fd}{\Sigma f} = 25 + \frac{-20}{50} = 24.6$$ **Ans.**

(*c*) **Step deviation method**

Let a be the assumed mean, i the class length then

$$D = \frac{x-a}{i}, \quad A.M. = a + \frac{\Sigma f D}{\Sigma f} i$$

Example 4. *Find the arithmetic mean of the data given in example 3 by step deviation method.*

Solution. Let assumed mean (a) = 25, i = 10 (given)

Class	*class-mark* x	*Frequency* f	$D = \frac{x-a}{i}$	$f.D$
0 – 10	5	7	– 2	– 14
10 – 20	15	8	– 1	– 8
20 – 30	25	20	0	0
30 – 40	35	10	+ 1	+ 10
40 – 50	45	5	+ 2	+ 10
Total		$\Sigma f = 50$		$\Sigma fD = -2$

$$A.M. = a + \frac{\Sigma f D}{\Sigma f} . i = 25 + \frac{-2}{50} \times 10 = 24.6$$ **Ans.**

57.7 MEDIAN

Median is defined as the measure of the central item when they are arranged in ascending or descending order of magnitude.

(*a*) When the total number of the items is odd and equal to say n, then the value of $\frac{1}{2}(n+1)$th item gives the median.

(*b*) When the total number of the frequencies is even, say n, then there are two middle items, and so the mean of the values of $\frac{1}{2}n$th and $\left(\frac{1}{2}n+1\right)$th items is the median.

Example 5. *Find the median of 6, 8, 9, 10, 11, 12, 13.*

Solution. Total number of items = $n = 7$

Since, n is an odd number. Hence,

$$\text{The middle item} = \frac{1}{2}(7+1)^{th} = 4th$$

$$\text{Median} = \text{Value of the 4th item} = 10 \qquad \textbf{Ans.}$$

(*c*) For grouped data, $$\textbf{Median} = l + \frac{\frac{1}{2}N - C}{f} \cdot i$$

where l is the lower limit of the median class, f is the frequency of the class, i is the class-length, C is the cumulative frequency of the class preceding the median-class and N is the cumulative frequency of the data.

Example 6. *Find the value of Median from the following data:*

No. of days for which absent (less than)	*5*	*10*	*15*	*20*	*25*	*30*	*35*	*40*	*45*
No. of students	*29*	*224*	*465*	*582*	*634*	*644*	*650*	*653*	*655*

Solution. The given cumulative frequency distribution will first be converted into ordinary frequency as under

Class-Interval	*Cumulative frequency*	*Ordinary frequency*
0 – 5	29	29 = 29
5 – 10	224 = C	224 – 29 = 195
10 – 15	465	465 – 224 = 241 f
15 – 20	582	582 – 465 = 117
20 – 25	634	634 – 582 = 52
25 – 30	644	644 – 634 = 10
30 – 35	650	650 – 644 = 6
35 – 40	653	653 – 650 = 3
40 – 45	655 = N	655 – 653 = 2

Here, $$\frac{N}{2} = \frac{655}{2} = 327.5$$

Hence, Median class = class having *c.f.* just more than $\frac{N}{2}$ *i.e.* 327.5 = 10 – 15

Now, $$\text{Median} = l + \frac{\frac{N}{2} - C}{f} i$$

where l stands for lower limit of median class,

N stands for the total frequency,

C stands for the cumulative frequency of the class just preceding the median class,

i stands for width of class interval

f stands for frequency of the median class.

$$\text{Median} = 10 + \frac{\frac{655}{2} - 224}{241} \times 5 = 10 + \frac{103.5 \times 5}{241} = 10 + 2.15 = 12.15 \qquad \textbf{Ans.}$$

57.8 QUARTILES

Quartiles are the values of the variate which divide the total frequency into four equal parts. When the lower half before the median is divided into two equal parts, the value of the dividing variate is called **lower Quartile** and is denoted by Q_1. The value of the variate dividing the upper half into two equal parts is called the **upper Quartile** and is denoted by Q_3. Q_2 is the medium.

The formulae for computation of Q_1 and Q_3 are,

$$Q_1 = l + \frac{i}{f}\left(\frac{N}{4} - C\right) \Rightarrow Q_3 = l + \frac{i}{f}\left(\frac{3N}{4} - C\right).$$

57.9 DECILES

Deciles are those values of the variate which divide the total frequency into 10 equal parts.

$$D_1 = l + \frac{i}{f}\left(\frac{N}{10} - C\right) \Rightarrow D_2 = l + \frac{i}{f}\left(\frac{2N}{10} - C\right)$$

$$D_3 = l + \frac{i}{f}\left(\frac{3N}{10} - C\right)$$ and so on

D_5, the fifth decile is the median.

57.10 PERCENTILES

Percentiles are those values of the variate which divide the total frequency into 100 equal parts.

$$P_1 = l + \frac{i}{f}\left(\frac{N}{100} - C\right) \Rightarrow P_2 = l + \frac{i}{f}\left(\frac{2N}{100} - C\right)$$ and so on.

P_{50}, the 50th percentile is the median.

Note. In case of series where frequency is not given,

P_{10} = value of $\frac{10}{100}(n+1)^{th}$ observation, P_{50} = value of $\frac{50}{100}(n+1)^{th}$ observations etc.

57.11 MODE

Mode is defined to be the size of the variable which occurs most frequently.

In case of continuous frequency distribution.

$$\textbf{Mode} = l + \left(\frac{f - f_{-1}}{2f - f_{-1} - f_{+1}}\right) i$$

where l is lower limit, i is the class length, f is the frequeny of the modal class, f_{-1} and f_1 are the frequencies of the classes preceding and succeeding the modal class respectively.

The following points must be taken care of while calculating mode :

1. The values (or classes of values) of the variable must be in ascending order of magnitude.
2. If the classes are in inclusive form, then the actual limits of the modal class are to be taken for finding l and i.
3. The classes must be of equal width.

Example 7. *Find the mode of the following items : 0, 1, 6, 7, 2, 3, 7, 6, 6, 2, 6, 0, 5, 6, 0.*

Solution. 6 occurs 5 times and no other item occurs 5 or more than 5 times, hence the mode is 6. **Ans.**

Emperical formula

Mean – Mode = 3 [Mean – Median]

Example 8. *Find the mode from the following data:*

Age	0 – 6	6 – 12	12 – 18	18 – 24	24 – 30	30 – 36	36 – 42
Frequency	6	11	25	35	18	12	6

Solution.

Age	*Frequency*	*Cumulative frequency*
0 – 6	6	6
6 – 12	11	17
12 – 18	$25 = f_{-1}$	42
18 – 24	$35 = f$	77
24 – 30	$18 = f_1$	95
30 – 36	12	107
36 – 42	6	113

Here, max. frequency of any item is 35.

Hence modal class is 18.24

$$\text{Mode} = l + \frac{f - f_{-1}}{2f - f_{-1} - f_1} \times i = 18 + \frac{35-25}{70-25-18} \times 6$$

$$= 18 + \frac{60}{27} = 18 + 2.22 = 20.22 \qquad \textbf{Ans.}$$

57.12 GEOMETRIC MEAN

If $x_1, x_2, x_3, \ldots, x_n$ be n values of variates x, then the geometric mean

$$G = (x_1 \times x_2 \times x_3 \times x_4 \times \ldots \times x_n)^{1/n}$$

Example 9. *Find the geometric mean of 4, 8, 16.*

Solution. $G.M. = (4 \times 8 \times 16)^{1/3} = 8$ **Ans.**

57.13 HARMONIC MEAN

Harmonic mean of a series of values is defined as the reciprocal of the arithmetic mean of their reciprocals. Thus if H be the harmonic mean, then

$$\frac{1}{H} = \frac{1}{n}\left[\frac{1}{x_1} + \frac{1}{x_2} + \ldots + \frac{1}{x_n}\right]$$

Example 10. *Calculate the harmonic mean of 4, 8, 16.*

Solution. Let, $\frac{1}{H} = \frac{1}{3}\left[\frac{1}{4} + \frac{1}{8} + \frac{1}{16}\right] = \frac{7}{48}$, $H = \frac{48}{7} = 6.857$ **Ans.**

57.14 AVERAGE DEVIATION OR MEAN DEVIATION

It is the mean of the absolute values of the deviations of a given set of numbers from their arithmetic mean.

If $x_1, x_2, x_3, \ldots, x_n$ be a set of numbers with frequencies $f_1, f_2, \ldots f_n$ respectively. Let $\bar{x}$ be the arithmetic mean of the numbers $x_1, x_2, \ldots, x_n$, then

$$\text{Mean deviation} = \frac{\Sigma f_i \,|x_i - \bar{x}|}{\Sigma f_i}$$

Example 11. *Find the mean deviation of the following frequency distribution.*

Class	0 – 6	6 – 12	12 – 18	18 – 24	24 – 30
Frequency	8	10	12	9	5

Solution. $a = 15$

Class	*Mid-value* x	*Frequency* f	$d = x - a$	fd	$\lvert x - 14 \rvert$	$f \lvert x - 14 \rvert$
0 – 6	3	8	– 12	– 96	11	88
6 – 12	9	10	– 6	– 60	5	50
12 – 18	15	12	0	0	1	12
18 – 24	21	9	+ 6	54	7	63
24 – 30	27	5	+ 12	60	13	65
		$\Sigma f = 44$		$\Sigma fd = -42$		$\Sigma f \lvert x - 14 \rvert = 278$

$$\text{Mean} = \bar{x} = a + \frac{\Sigma f d}{\Sigma f} = 15 - \frac{42}{44} = 14 \text{ (approx.)}$$

$$\text{Mean or Average deviation} = \frac{\Sigma f \lvert x - \bar{x} \rvert}{\Sigma f} = \frac{278}{44} = 6.32$$ **Ans.**

57.15 STANDARD DEVIATION

Standard deviation is defined as the square root of the mean of the square of the deviation from the arithmetic mean.

$$S.D. = \sigma = \sqrt{\frac{\Sigma f (x - \bar{x})^2}{\Sigma f}}$$

Note. 1. The square of the standard deviation *i.e.*; σ^2 is called variance.

2. σ^2 is called the second moment about the mean and is denoted by μ_2 .

57.16 SHORTEST METHOD FOR CALCULATING STANDARD DEVIATION

We know that $$\sigma^2 = \frac{1}{N} \Sigma f (x - \bar{x})^2 = \frac{1}{N} \Sigma f (x - a - \overline{\bar{x} - a})^2$$

$$= \frac{1}{N} \Sigma f (d - \overline{\bar{x} - a})^2 \quad \text{where } x - a = d \quad = \frac{1}{N} \Sigma fd^2 - 2(\bar{x} - a) \frac{1}{N} \Sigma fd + (\bar{x} - a)^2 \frac{1}{N} \Sigma f$$

$$= \frac{1}{N} \Sigma fd^2 - 2(\bar{x} - a) \frac{1}{N} \Sigma fd + (\bar{x} - a)^2 \qquad [\because \Sigma f = N] \qquad \left[\bar{x} = a + \frac{\Sigma fd}{N} \text{ or } \bar{x} - a = \frac{\Sigma fd}{N} \right]$$

$$\sigma^2 = \frac{1}{N} \Sigma f d^2 - 2 \left(\frac{\Sigma f d}{N} \right) \left(\frac{\Sigma f d}{N} \right) + \left(\frac{\Sigma f d}{N} \right)^2 = \frac{1}{N} \Sigma f d^2 - 2 \left(\frac{\Sigma f d}{N} \right)^2 + \left(\frac{\Sigma f d}{N} \right)^2$$

$$\sigma^2 = \frac{1}{N} \Sigma fd^2 - \left(\frac{\Sigma fd}{N} \right)^2, \qquad S.D. = \sigma = \sqrt{\frac{\Sigma fd^2}{N} - \left(\frac{\Sigma fd}{N} \right)^2}$$

Note. Coefficient of variation $= \frac{\sigma}{x} \times 100$

Example 12. *Calculate the mean and standard deviation for the following table, given the age distribution of 542 members.*

Age in years	20 – 30	30 – 40	40 – 50	50 – 60	60 – 70	70 – 80	80 – 90
No. of members	3	61	132	153	140	51	2

Solution. Assumed mean = 55

Here, we take $d = \frac{x - a}{i} = \frac{x - 55}{10}$

Age grouped	*Mid value (x)*	*Frequency (f)*	$d = \dfrac{x-55}{10}$	fd	fd^2
20 – 30	25	3	– 3	– 9	27
30 – 40	35	61	– 2	– 122	244
40 – 50	45	132	– 1	– 132	132
50 – 60	55	153	0	0	0
60 – 70	65	140	1	140	140
70 – 80	75	51	2	102	204
80 – 90	85	2	3	6	18
		$\Sigma f = 542$		$\Sigma fd = -15$	$\Sigma fd^2 = 765$

$$\text{Mean} = \bar{x} = a + \frac{\Sigma fd}{\Sigma f}.i = 55 + \frac{(-15)\,10}{542} = 55 - 0.28 = 54.72$$

$$\text{Variance} = \sigma^2 = i^2\left[\frac{1}{N}\Sigma fd^2 - \left(\frac{\Sigma fd}{N}\right)^2\right] = 100\left[\frac{765}{542} - (0.028)^2\right] = 100 \times 1.4107 = 141.07$$

S.D. = σ = 11.9 years **Ans.**

57.17 SYMMETRY

A distribution is said to be symmetrical when its mean, median and mode are identical. *i.e.*;

Mean = Median = Mode.

In other words, a distribution is said to be symmetric when the frequencies are symmetrically distributed about the mean (or when the values of the variable are equidistant from the mean and have the same frequency).

Consider the following frequency distribution:

x	10	20	30	40	50	60	70
f	2	6	10	14	10	6	2

Mean
Median
Mode

$$\text{Mean} = \bar{x} = \frac{10\times2+20\times6+30\times10+40\times14+50\times10+60\times6+70\times2}{2+6+10+14+10+6+2} = \frac{2000}{50} = 40$$

In this distribution, we observe that the values 20 and 60 are equidistant from the mean, viz. 40 with the same frequency 6.

A symmetrical distribution when plotted on a graph will give a perfectly bell-shaped curve, which is known as normal curve.

57.18 SKEWNESS

Skewness denotes the opposite of symmetry. It is lack of symmetry,

Skew symmetrical Distribution (*U.P. III Semester, 2006*)

A distribution which is not symmetrical is said to be skew symmetrical distribution. In skew symmetrical distribution the left tail and the right tail are not of equal length. One tail will be longer than the other.

(*a*) **Negatively skew distribution.** In negatively skew distribution the left tail is longer than the right tail.

(*b*) **Positively skew distribution.** In positively skew distribution the right tail of the curve will be longer than the left.

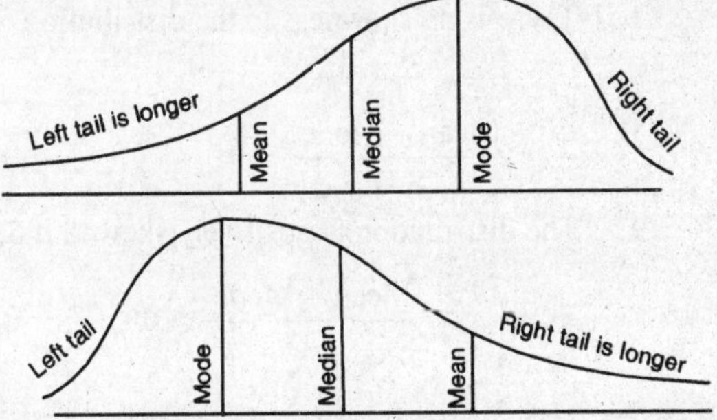

In skew distribution mean, median and mode are not equal.

57.19 TEST OF SKEWNESS

1. There is no skewness in the distribution if AM = Mode = Median
2. There is no skewness in the distribution if, Third quartile - Median = Median – First quartile.
3. There is no skewness if
   The sum of the frequencies which are less than Mode = Sum of the frequencies which are greater than Mode
4. There is no skewness if quartiles are equidistant from the median.
5. The distribution is negatively skewed if A.M. is less than Mode.
6. The curve is not symmetrical about the median if AM ≠ Median ≠ Mode.

57.20 USES OF SKEWNESS

1. It gives the nature of the curve.
2. It gives nature and concentration of observations about the mean.

57.21 TYPES OF DISTRIBUTION

1. Fairly symmetrical 2. Positively skewed 3. Negatively skewed.

57.22 MEASURE OF SKEWNESS

Measure of skewness is known as the measure of symmetry.

There are two types of measure of skewness.

1. Absolute measure: Absolute measure = Mean – Mode
2. Relative measure : These are four types of relative measure of skewness.
   (*i*) Karl Pearson's Coefficient of Skewness
   (*ii*) Bowley's Coefficient of Skewness.
   (*iii*) Kelly's Coefficient of Skewness and
   (*iv*) Measure of skewness based on the moments.
   (Mode = 3 Median – 2 Mode)

57.23 KARL PEARSON'S COEFFICIENT OF SKEWNESS:

$$\text{Karl Pearson's Coefficient of Skewness} = \boxed{\frac{\text{Mean} - \text{Mode}}{\text{Standard deviation}}}$$

$$= \frac{\text{Mean} - (3\,\text{Median} - 2\,\text{Mean})}{\text{Standard deviation}} = \boxed{\frac{3\,(\text{Mean} - \text{Median})}{\text{Standard deviation}}}$$

It generally lies between – 1 and 1.

If its value is zero then there is no skewness.

57.24 TYPES OF SKEWNESS IN TERMS OF MEAN AND MODE

1. There is no skewness in the distribution

$$S_k = 0$$

$$\Rightarrow \quad \frac{\text{Mean} - \text{Mode}}{\text{S.D.}} = 0 \quad \Rightarrow \quad \text{Mean} - \text{Mode} = 0 \quad \Rightarrow \quad \text{Mean} = \text{Mode}.$$

2. The distribution is positively skewed if $S_k > 0$.

$$\frac{\text{Mean} - \text{Mode}}{\text{S.D.}} > 0 \quad \Rightarrow \quad \text{Mean} - \text{Mode} > 0 \quad \Rightarrow \quad \text{Mean} > \text{Mode}.$$

3. The distribution is negatively skewed if

$$S_k < 0$$

$$\Rightarrow \quad \frac{\text{Mean} - \text{Mode}}{\text{S.D.}} < 0 \quad \Rightarrow \quad \text{Mean} - \text{Mode} < 0 \quad \Rightarrow \quad \text{Mean} < \text{Mode}$$

Example 13. *Compute the coefficient of Skewness from the following data:*

x	6	7	8	9	10	11	12
f	3	6	9	13	8	5	4

(U.P. III Semester, 2009-2010)

Solution. Let $a = 9$

x	f	$d = x - 9$	fd	fd^2	$c.f.$
6	3	− 3	-9	27	3
7	6	− 2	− 12	24	9
8	9	− 1	− 9	9	18
9	13	0	0	0	31
10	8	1	8	8	39
11	5	2	10	20	44
12	4	3	12	36	48
	$\Sigma f = 48$		$\Sigma fd = 0$	$\Sigma fd^2 = 124$	

$$\text{Mean} = a + \frac{\Sigma fd}{\Sigma f} = 9 + \frac{0}{48} = 9$$

Mode = Item of maximum frequency (13) = 9

$$\text{S.D.} = \sqrt{\frac{\Sigma fd^2}{\Sigma f} - \left(\frac{\Sigma fd}{\Sigma f}\right)^2} = \sqrt{\frac{124}{48} - \left(\frac{0}{48}\right)^2} = \sqrt{\frac{124}{48}} = 1.61$$

$$\text{Karl Pearson's Coefficient of Skewness} = \frac{\text{Mean} - \text{Mode}}{\text{S.D.}} = \frac{9-9}{1.61} = \frac{0}{1.61} = 0$$ **Ans.**

Example 14. *Calculate Karl Pearson's Coefficient of Skewness from the given data :*

Life time in months	*30-40*	*40-50*	*50-60*	*60-70*	*70-80*	*80-90*	*90-100*	*100-110*	*110-120*
No. of mobile	*4*	*6*	*8*	*26*	*28*	*12*	*8*	*5*	*3*

Solution. Let $a = 75$

Life time in months	No. of mobile (f)	Mid-value (x)	$d = x - 75$	fd	fd^2	*Cumulative frequency*
30 – 40	4	35	− 40	− 160	6400	4
40 – 50	6	45	− 30	− 180	5400	10
50 – 60	8	55	− 20	− 160	3200	18
60 – 70	26	65	− 10	− 260	2600	44
70 – 80	28	75	0	0	0	72
80 – 90	12	85	10	120	1200	84
90 – 100	8	95	20	160	3200	92
100 – 110	5	105	30	150	4500	97
110 – 120	3	115	40	120	4800	100
	$\Sigma f = 100$			$\Sigma fd = -210$	$\Sigma fd^2 = 31300$	

$l = 70$, $N = 100$, $cf = 40$, $\Sigma f = 100$, $i = 10$

$$\text{Mean} = a + \frac{\Sigma fd}{\Sigma f} = 75 + \frac{-210}{100} = 72.9$$

$$\text{Median} = l + \frac{\frac{N}{2} - c.f.}{f} i = 70 + \frac{\frac{100}{2} - 44}{28}.(10) = 70 + 2.143 = 72.143$$

$$\text{S.D.} = \sqrt{\frac{\Sigma fd^2}{\Sigma f} - \left(\frac{\Sigma fd}{\Sigma f}\right)^2} = \sqrt{\frac{31300}{100} - \left(\frac{-210}{100}\right)^2} = \sqrt{313 - 4.41}$$

$$= \sqrt{308.59} = 17.57$$

$$\text{Karl Pearson's Coefficient of Skewness} = \frac{3\,(\text{Mean} - \text{Median})}{\text{S.D.}} = \frac{3\,(72.9 - 72.143)}{17.57}$$

$$= \frac{3\,(0.757)}{17.57} = \frac{2.271}{17.57} = 0.1293$$ **Ans.**

EXERCISE 57.1

Calculate Karl Pearson's Coefficient of Skewness from the data given below:

1. S.D. = 6.5, AM = 29.6, mode = 27.52 **Ans.** $S_k = 0.32$
2. Mean = 100, Variance = 35, Median = 99.61. **Ans.** $S_k = 0.2$
3. AM = 45, Median = 48, S.D. = 22.5 **Ans.** $S_k = -0.4$
4. The sum of the 20 observation is 300 and sum of the squares of the observation is 5000, Median = 15. **Ans.** $S_k = 0$
5. Find the Karl Pearson's Coefficient of Skewness for the following

Years under	10	20	30	40	50	60
No. of persons	15	32	51	78	97	109

Ans. – 0.32

6. Calculate Karl Pearson's Coefficient of Skewness from the following data :

Cost per item (in Rs.)	4.5	5.5	6.5	7.5	8.5	9.5	10.5	11.5
No. of items	35	40	48	100	125	87	43	22

Ans. – 0.2445

7. From the following data calculate Karl Pearson's Coefficient of Skewness.

Scores	0	10	20	30	40	50	60	70	80
No. of players	150	140	100	80	80	70	30	14	0

Ans. – 0.462

8. The weekly wages in Rs. of the workers in a shoe factory are given below :

Weekly wages (in Rs.)	500 - 600	600 - 700	700 - 800	800 - 900	900 - 1000	1000-1100
No. of workers	8	12	4	2	1	1

Calculate Karl Pearson's Coefficient of Skewness. **Ans.** 0.34

9. Which of the following two series is symmetrical :
   Series (a) : Mean = 32, Median = 34, S.D. = 20
   Series (b) : Mean = 32, Median = 36, S.D. = 25 **Ans.** series (a) is more symmetrical than series (b)
10. Karl Pearson's Coefficient of Skewness of a distribution = 0.32, Standard deviation = 6.5
    A.M. = 29.6.
    From the above data find the mode and the median for the distribution.
    Ans. Mode = 27.52, Median = 28.91

57.25 BOWLEY'S COEFFICIENT OF SKEWNESS

Bowley's Coefficient of Skewness is based on the quartiles and Median. A distribution is symmetrical if the distance between the first quartile and median is equal to the distance between the median and third quartile i.e.

$$\text{Median} - Q_1 = Q_3 - \text{Median}$$

First Quartile Median Third Quartile

$Q_1 \longleftrightarrow M \longleftrightarrow Q_3$

It is not so *i.e.*

$$\text{Median} - Q_1 \neq Q_3 - \text{Median}$$

then the distribution is skewed.

57.26 MEASURE OF BOWLEY'S COEFFICIENT OF SKEWNESS

There are two ways to measure Bowley's Coefficient of Skewness.

1. Bowley's absolute measure of Skewness = $Q_3 + Q_1 - 2$ Median

2. Bowley's Relative Measure of Skewness

$$\text{Bowley's Coefficient of Skewness} = \frac{Q_3 + Q_1 - 2\text{ Median}}{Q_3 - Q_1}$$

This formula is also known as Quartile Coefficient of Skewness.

57.27 CHARACTERISTICS OF BOWLEY'S COEFFICIENT OF SKEWNESS

1. If the distribution is open of unequal class interval then Pearson's Coefficient of Skewness cannot be calculated and Bowley's Coefficient of Skewness can be calculated.

2. Bowley's Coefficient of Skewness lies between – 1 and + 1.

3. Bowley's measure is calculated only from the continuous distribution with exclusive classes.

57.28 LIMITATIONS OF BOWLEY'S COEFFICIENT OF SKEWNESS

1. It is based on the central 50% of the data and ignores the remaining 50% of the data on the extremes.

2. Bowley's formulae and Pearson's formulae cannot be compared. However if the distribu tion is symmetrical thus both coefficients are zero.

Example 15. *From the following data find Bowley's Coefficient of Skewness:*

Difference of quartiles = 80 *Mode = 60*

Sum of the quartiles = 120 *and Mean = 45*

Solution. Here, we have

$$Q_3 + Q_1 = 120$$
$$Q_3 - Q_1 = 80$$
$$\text{Mode} = 60$$
$$\text{Mean} = 45$$

We know that

$$\text{Mode} = 3\text{ Median} - 2\text{ Mean}$$

$$\Rightarrow \quad 60 = 3\text{ Median} - 2(45) \Rightarrow \text{Median} = 50$$

$$\text{Bowley's Coefficient of Skewness} = \frac{Q_3 + Q_1 - 2M}{Q_3 - Q_1} = \frac{120 - 2(50)}{80} = \frac{1}{4} = 0.25$$ **Ans.**

Example 16. *The sum of the upper quartile and lower quartile is 100 and the median is 55. The Bowley's Coefficient of Skewness is –0.6. Find the upper and lower quartiles.*

Solution. We know that

$$\text{Bowley's Coefficient of Skewness} = \frac{Q_3 + Q_1 - 2M}{Q_3 - Q_1} \qquad \ldots (1)$$

Here, $Q_1 + Q_3 = 100$, $M = 55$, $S_k = -0.6$,
Putting the values of $Q_1 + Q_3$, M and Coefficient of Skewness in (1), we get

$$-0.6 = \frac{100 - 2(55)}{Q_3 - Q_1} \Rightarrow -0.6 = \frac{100 - 110}{Q_3 - Q_1}$$

$$\Rightarrow \qquad -0.6 = \frac{-10}{Q_3 - Q_1} \Rightarrow Q_3 - Q_1 = \frac{50}{3}$$

But $\qquad Q_3 + Q_1 = 100 \quad \ldots (2) \qquad$ And $\quad Q_3 - Q_1 = \dfrac{50}{3} \qquad \ldots (3)$

On adding, we get $2Q_3 = \dfrac{500}{3} \Rightarrow Q_3 = \dfrac{250}{3}$

Putting the value of Q_3 in (2), we get

$$\frac{250}{3} + Q_1 = 100 \Rightarrow Q_1 = 100 - = \frac{250}{3} = \frac{50}{3}$$

Hence, $\qquad Q_1 = \dfrac{50}{3}$ and $Q_3 = \dfrac{250}{3}$ **Ans.**

Example 17. *Calculate Bowley's Coefficient of Skewness from the data given below.*

No. of houses	0	1	2	3	4	5	6
No. of Airconditioners	15	20	14	25	13	8	4

Solution.

No. of Houses (x)	*No. of Air conditioners (f)*	*Cumulative frequency*
0	15	15
1	20	35
2	14	49
3	25	74
4	13	87
5	8	95
6	4	99

Q_1 = size of $\dfrac{N+1}{4}$ = 25th item, Hence $Q_1 = 1$

Q_3 = size of $\dfrac{3(N+1)}{4}$ = 75th item. Hence $Q_3 = 4$

Median = size of $\dfrac{N+1}{2}$ = 50th item. Hence Mean = 3

$$\text{Bowley's Coefficient of Skewness} = \frac{Q_3 + Q_1 - 2M}{Q_3 - Q_1} = \frac{4 + 1 - 2(3)}{4 - 1} = -\frac{1}{3} = -0.33$$ **Ans.**

Example 18. *The following table shows the distances between the worker's residence and their office situated at Connaught Place, New Delhi.*

Distances	*0 – 10*	*10 – 20*	*20 – 30*	*30 – 40*	*40 – 50*	*50 – 60*	*60 – 70*
No. of workers	*2*	*5*	*10*	*15*	*10*	*4*	*1*

Calculate the Bowley's Coefficient of Skewness.

Solution.

Distance (*x*)	*No. of workers* (*f*)	*c.f*
0 – 10	2	2
10 – 20	5	7
20 – 30	10	17
30 – 40	15	32
40 – 50	10	42
50 – 60	4	46
60 – 70	1	47

$$N = 47, \quad \frac{N+1}{4} = 12, \quad \frac{N+1}{2} = 24, \quad \frac{3(N+1)}{4} = 36$$

The class 30 – 40 is the median class

$$l = 30, \ i = 10, \ f = 15, \ cf = 17, \ N = 47$$

$$\text{Median} = l + \frac{\frac{N+1}{2} - c.f.}{f}(i) = 30 + \frac{\frac{47+1}{2} - 17}{15}(10) = 30 + \frac{14}{3} = \frac{104}{3}$$

$$Q_1 = l + \frac{\frac{N+1}{4} - c.f.}{f} i = 20 + \frac{\frac{47+1}{4} - 7}{10}(10) = 20 + 5 = 25$$

$$Q_3 = l + \frac{3\left(\frac{N+1}{4}\right) - c.f.}{f} i = 40 + \frac{36-32}{10}(10) = 40 + 4 = 44$$

$$\text{Bowley's Coefficient of Skewness} = \frac{Q_3 + Q_1 - 2M}{Q_3 - Q_1} = \frac{44 + 25 - 2\left(\frac{104}{3}\right)}{44 - 25}$$

$$= \frac{69 - \frac{208}{3}}{19} = \frac{-\frac{1}{3}}{19} = -\frac{1}{57} = -0.0175 \qquad \textbf{Ans.}$$

EXERCISE 57.2

1. The data for a distribution is given below :

$Q_1 = 8.6$, Median = 12.3, $Q_3 = 14.04$

Calculate Bowley's Coefficient of Skewness. **Ans.** – 0.36

2 Calculate (*a*) Karl Pearson's Coefficient of Skewness

(*b*) Bowley's Coefficient of Skewness from the following data :

	City A	City B
A. M.	150	140
Median	142	155
S.D.	30	55
Q_3	195	260
Q_1	62	80

Ans.

	City A	City B
Karl Pearson's Coeff.	0.8	– 0.82
Bowley's Coeff.	– 0.203	0.167

3. Compute the quartile Coefficient of Skewness for the following distribution :

x	3 – 7	8 – 12	13 – 17	18 – 22	23 – 27	28 – 32	33 – 37	38 – 42
f	2	108	580	175	80	32	18	5

Ans. 0.119

4. Calculate Bowley's Coefficient of Skewness for the following data :

x	5	10	15	20	25	30	35	40	45
f	9	10	12	15	11	7	6	5	2

Ans. 0.33

5. A blood donation camp was held at Janakpuri. The distribution of the blood donar is given below :

Age in years	20 – 25	25 – 30	30 – 35	35 – 40	40 – 45	45 – 50	50 – 55	55 – 60
No. of donars	50	70	80	180	150	120	70	50

Calculate the Bowley's Coefficient of Skewness. **Ans.** 0.232

6. Calculate Bowley's Coefficient of Skewness for the following distribution.

Classes	1 – 5	6 – 10	11 – 15	16 – 20	21 – 25	26 – 30	31 – 35
Frequency	3	4	68	30	10	6	2

Hint: First change to exclusive distribution for taking real class limit as first class 0.5 – 5.5

Ans. 0.262

7. The participant of different ages took part in "Marathan" race at INDIA GATE on 2nd October 2008 as follows :

Age (in years)	10 – 20	20 – 30	30 – 40	40 – 50	50 – 60	60 – 70	70 – 80
No. of participants	358	2417	976	129	62	18	10

Calculate Bowley's Coefficient of Skewness. **Ans.** 0.131

57.29 KELLY'S COEFFICIENT OF SKEWNESS

Bowley's measure neglects the extreme data to measure skewness. The entire data should be taken into account in measuring skewness.

Kelly modified Bowley's measure of skewness by taking any two deciles equidistant from the median or any two percentiles equidistant from the median.

$$\text{Kelly's Coefficient of Skewness} = \frac{P_{10} + P_{90} - 2P_{50}}{P_{90} - P_{10}}$$

$$\text{Kelly's Coefficient of Skewness} = \frac{D_1 + D_9 - 2\,Median}{D_9 - D_1}$$

P denotes percentile and D denotes decile.

We know that, Median $= P_{50} = D_5$

Note : This method is only theoretical generally Karl Pearson's method is widely used.

Example 19. *Calculate percentile Coefficient of Skewness from the following :*

$P_{90} = 110, \qquad P_{10} = 30, \qquad P_{50} = 80$

Solution. Here, we have

$P_{90} = 110, \qquad P_{10} = 30, \qquad P_{50} = 80$

$$\text{Kelly's Coefficient of Skewness} = \frac{P_{90} + P_{10} - 2\,Median}{P_{90} - P_{10}}$$

$$= \frac{110 + 30 - 2\,(80)}{110 - 30} = \frac{140 - 160}{80} = \frac{-20}{80} = \frac{-1}{4} = -0.25 \qquad \textbf{Ans.}$$

Example 20. *Calculate Kelly's Coefficient of Skewness for the following data :*

$D_1 = 60$, $D_9 = 290$, Median = 165

Solution. Here, we have

$D_1 = 60$, $D_9 = 290$, Median = 165

$$\text{Kelly's Coefficient of Skewness} = \frac{D_1 + D_9 - 2\,Median}{D_9 - D_1} \quad \text{... (1)}$$

Putting the values of D_1, D_9 and the median in (1), we get

$$\text{Kelly's Coefficient of Skewness} = \frac{60 + 290 - 2\,(165)}{290 - 60}$$

$$= \frac{350 - 330}{230} = \frac{20}{230} = \frac{2}{23} = 0.087 \qquad \textbf{Ans.}$$

Example 21. *The weights in kg of 9 boys in a class are : 40, 42, 45, 48, 50, 52, 55, 56, 57. Calculate Kelly's Coefficient of Skewness based on percentiles.*

Solution. Here, we have 40, 42, 45, 48, 50, 52, 55, 56, 57

$n = 9$

P_{10} = Value of $\frac{10}{100}(n+1)^{\text{th}}$ observation = Value of $\frac{10}{100}(9+1)^{\text{th}}$ observation

= Value of first observation = 40

P_{50} = Value of $\frac{50}{100}(n+1)^{\text{th}}$ observation = Value of $\frac{50}{100}(9+1)$th observation

= Value of 5th observation = 50

P_{90} = Value of $\frac{90}{100}(n+1)$th observation = Value of $\frac{90}{100}(9+1)$th observation

= Value of 9th observation = 57

$$\text{Now, Kelly's Coefficient of Skewness} = \frac{P_{10} + P_{90} - 2\text{ Median}}{P_{90} - P_{10}} = \frac{40 + 57 - 2\,(50)}{57 - 40}$$

$$= \frac{97 - 100}{17} = -\frac{3}{17} = -0.176 \qquad \textbf{Ans.}$$

Example 22. *Calculate the Kelly's Coefficient of Skewness on the basis of percentiles for the following data.*

Marks obtained	20–30	30–40	40–50	50–60	60–70	70–80	80–90	90–100
No. of students	3	8	9	14	16	18	8	4

Solution.

Marks	*No. of students*	*c.f*
20 – 30	3	3
30 – 40	8	11
40 – 50	9	20
50 – 60	14	34
60 – 70	16	50
70 – 80	18	68
80 – 90	8	76
90 – 100	4	80

$$P_{10} = \text{Marks of } 10\left(\frac{N}{100}\right)\text{th student} = \text{Marks of } 10\left(\frac{80}{100}\right)\text{th student}$$

$$= \text{Marks of 8 th student.}$$

P_{10} lies in the class 30 – 40

$$P_{10} = l + \frac{10\left(\frac{N}{100}\right) - c.f}{f}(i) = 30 + \frac{10\left(\frac{80}{100}\right) - 3}{8}(10) = 30 + \frac{8-3}{8}(10)$$

$$= 30 + \frac{50}{8} = 36.25$$

Median = Marks of $\frac{80}{2}$ th student = Marks obtained by 40th student.

Median Class = 60 – 70.

$$\text{Median} = l + \frac{\frac{N}{2} - c.f.}{f}(i) = 60 + \frac{\frac{80}{2} - 34}{16}(10) = 60 + \frac{6}{16}(10) = 63.75$$

$$P_{90} = \text{Marks of } 90\left(\frac{N}{100}\right)\text{th student} = \text{Marks of } 90\left(\frac{80}{100}\right)\text{th student} = \text{Marks of 72nd student}$$

P_{90} class = 80 – 90

$$P_{90} = l + \frac{\frac{90}{100}(N) - c.f}{f}(i) = 80 + \frac{\frac{90}{100}(80) - 68}{8}(10) = 80 + \frac{72-68}{8}(10) = 80 + 5 = 85$$

$$\text{Kelly's Coefficient of Skewness} = \frac{P_{10} + P_{90} - 2\text{ Median}}{P_{90} - P_{10}}$$

$$= \frac{36.25 + 85 - 2(63.75)}{85 - 36.25} = \frac{-6.25}{48.75} = -0.128$$ **Ans.**

EXERCISE 57.3

Find Kelly's Coefficient of Skewness from the following table :

1. $P_{10} = 25$, $P_{90} = 200$, Median = 100 **Ans.** 0.143
2. $D_1 = 15.5$, $D_9 = 120.5$, Median = 70 **Ans.** – 0.038
3.

x	2.5	3.5	12.5	17.5
$c.f.$	7	18	25	30

On the basis of deciles. **Ans.** 0.866

4.

Rate (in Rs.)	Below 10	10 – 20	20 – 40	40 – 60	60 – 80	above 80
No. of workers	8	10	22	35	20	5

on the basis of deciles. **Ans.** – 0.07

5.

Wages (in Rs.)	800-900	900-1000	1000-1100	1100-1200	1200-1300	1300-1400	1400-1500
No. of workers	10	33	47	110	160	80	60

on the basis of percentiles. **Ans.** – 0.07

57.30 MOMENTS

The rth moment of a variable x about the mean $\bar{x}$ is usually denoted by μ_r is given by

$$\mu_r = \frac{1}{N}\Sigma f_i (x_i - \bar{x})^r, \qquad \Sigma f_i = N.$$

The rth moment of a variable x about any point a is defined by $\mu'_r = \frac{1}{N}\Sigma f_i (x_i - a)^r$

57.31 MOMENT ABOUT MEAN

Let $\overline{x}$ be the arithmetic mean, then

$$\mu_r = \frac{1}{N}\sum_{i=1}^{a} f_i (x_i - \overline{x})^r, \quad r = 0, 1, 3....... \qquad \text{where, } N = \sum_{i=1}^{n} f_i$$

If $r = 0$, $$\mu_0 = \frac{1}{N}\sum_{i=1}^{n} f_i (x_i - \overline{x})^0 = 1$$

If $r = 1$, $$\mu_1 = \frac{1}{N}\sum_{i=1}^{n} f_i (x_i - \overline{x}) = 0$$

If $r = 2$, $$\mu_2 = \frac{1}{N}\sum_{i=1}^{n} f_i (x_i - \overline{x})^2 \qquad [\mu_2 = \text{variance}]$$

If $r = 3$, $$\mu_3 = \frac{1}{N}\sum_{i=1}^{n} f_i (x_i - \overline{x})^3$$

If $r = 4$, $$\mu_4 = \frac{1}{N}\sum_{i=1}^{n} f_i (x_i - \overline{x})^4$$

57.32 MOMENTS ABOUT ANY NUMBER (RAW MOMENTS)

Let a be a arbitrary number then

$$\mu'_r = \frac{1}{N}\sum_{i=1}^{n} f_i (x_i - a)^r, \quad r = 0, 1, 2, ... \qquad \text{where } N = \sum_{i=1} f_i$$

If $r = 0$, $$\mu'_0 = \frac{1}{N}\sum_{i=1}^{n} f_i (x_i - a)^0 = 1$$

If $r = 1$, $$\mu'_1 = \frac{1}{N}\sum_{i=1}^{n} f_i (x_i - a)^1$$

$$= \frac{1}{N}\sum_{i=1}^{n} f_i x_i - \frac{a}{N}\sum_{i=1}^{n} f_i$$

$$= \overline{x} - a \qquad \left[\because \sum_{i=1}^{n} f_i = N\right]$$

If $r = 2$, $$\mu'_2 = \frac{1}{N}\sum_{i=1}^{n} f_i (x_i - a)^2$$

If $r = 3$, $$\mu'_3 = \frac{1}{N}\sum_{i=1}^{n} f_i (x_i - a)^3$$

If $r = 4$, $$\mu'_4 = \frac{1}{N}\sum_{i=1}^{n} f_i (x_i - a)^4$$

....

....

57.33 MOMENT ABOUT THE ORIGIN

$$v_r = \frac{1}{N} \sum_{i=1}^{n} f_i x_i^r$$

If $r = 0$, $$v_0 = \frac{1}{N} \sum_{i=1}^{n} f_i x_i^0 = 1$$

If $r = 1$, $$v_1 = \frac{1}{N} \sum_{i=1}^{n} f_i x_i = \bar{x}$$

If $r = 2$, $$v_2 = \frac{1}{N} \sum_{i=1}^{n} f_i x_i^2$$

If $r = 3$, $$v_3 = \frac{1}{N} \sum_{i=1}^{n} f_i x_i^3$$

If $r = 4$, $$v_4 = \frac{1}{N} \sum_{i=1}^{n} f_i x_i^4$$

....

....

57.34 RELATION BETWEEN μ_r AND μ'_r :

We have,

$$\mu_r = \frac{1}{N} \sum_{i=1}^{n} f_i (x_i - \bar{x})^r$$

$$= \frac{1}{N} \sum_{i=1}^{n} f_i [(x_i - a) - (\bar{x} - a)]^r$$

$$= \frac{1}{N} \sum_{i=1}^{n} f_i [(x_i - a) - \mu'_1]^r \qquad [\mu'_1 = \bar{x} - a]$$

On expanding by Binomial theorem,

$$= \frac{1}{N} \sum_{i=1}^{n} f_i [(x_i - a)^r - {}^rC_1 (x_i - a)^{r-1} \mu'_1 + {}^nC_2 (x_i - a)^{r-2} \mu'^2_1 - \ldots\ldots + (-1)^r \mu'^r_1]$$

$$\mu_r = \mu'_r - {}^rC_1 \mu'_{r-1} \mu'_1 + {}^rC_2 \mu'_{r-2} \mu'^2_1 - \ldots\ldots + (-1)^r \mu'^r_1$$

Putting $r = 2, 3, 4, \ldots\ldots$ we get

$$\mu_2 = \mu'_2 - 2\mu'^2_1 + \mu'^2_1 = \mu_2^1 - \mu'^2_1 \qquad [\because \mu_0^1 = 1]$$

$$\mu_3 = \mu'_3 - 3\mu'_2 \mu'_1 + 3\mu'^3_1 - \mu'^3_1$$

$$= \mu'_3 - 3\mu'_2\mu'_1 + 2\mu'^3_2$$

$$\mu_4 = \mu'_4 - 4\mu'_3\mu'_1 + 6\mu'_2\mu'^2_1 - 3\mu'^4_1$$

Thus, we have the following relations:

$$\boxed{\mu_1 = 0} \qquad \boxed{\mu_2 = \mu'_2 - \mu'^2_1}$$

$$\boxed{\mu_3 = \mu'_3 - 3\mu'_2\mu'_1 + 2\mu'^3_1} \qquad \boxed{\mu_4 = \mu'_4 - 4\mu'_3\mu'_1 + 6\mu'_2\mu'^2_1 - 3\mu'^4_1}$$

Note. 1. The sum of the coefficients of the various terms on the R.H.S. is zero.

2. The dimension of each term on R.H.S. is the same as that of terms on the L.H.S.

3. The number of terms on R.H.S. = order of the moment on L.H.S.

Conversely

$$\mu'_r = \frac{1}{N}\sum f_i(x_i - a)^r$$

$$= \frac{1}{N}\sum fi\,(\overline{x_i - \bar{x}} + \overline{\bar{x} - a})^r$$

$$= \frac{1}{N}\Big[\sum f_i(x_i - \bar{x})^r + {}^rC_1 \sum f_i(x_i - \bar{x})^{r-1}(\bar{x} - a) + {}^rC_2 \sum f_i(x_i - \bar{x})^{r-2}(\bar{x} - a)^2$$

$$\ldots + {}^rC_{r-1}\sum f_i(x_i - \bar{x})(\bar{x} - a)^{r-1} + \sum f_i(\bar{x} - a)^r\Big]$$

$$= \frac{1}{N}\sum f_i(x_i - \bar{x})^r + {}^rC_1\frac{1}{N}\sum f_i(x_i - \bar{x})^{r-1}(\bar{x} - a)$$

$$+ {}^rC_2\frac{1}{N}\sum f_i(x_i - \bar{x})^{r-2}(\bar{x} - a)^2 + \ldots + {}^rC_{r-1}\frac{1}{N}\sum f_i(x_i - \bar{x})(\bar{x} - a)^{r-1}$$

$$+ \frac{1}{N}\sum f_i(\bar{x} - a)^r$$

$$\Rightarrow \quad \mu'_r = \mu_r + r\,\mu_{r-1}\,\mu'_1 + \frac{r(r-1)}{2}\mu_{r-2}\,\mu'^2_1 + \ldots + r\,\mu_1\,\mu'^{r-1}_1 + \mu'^r_1$$

If $r = 1$ $\quad \mu'_1 = \bar{x} - a = \mu_1 - a$

If $r = 2$, $\quad \mu'_2 = \mu_2 + 2\mu_1\,\mu'_1 + \mu'^2_1$

$$= \mu_2 + 0 + \mu'^2_1 \qquad [\mu_1 = 0]$$

$$\Rightarrow \quad \mu'_2 = \mu_2 + \mu'^2_1$$

If $r = 3$ $\quad \mu'_3 = \mu_3 + 3\mu_2\,\mu'_1 + 3\mu_1\,\mu'^2_1 + \mu'^3_1$

$$= \mu_3 + 3\mu_2\,\mu'_1 + 0 + \mu'^3_1 \qquad [\mu_1 = 0]$$

$$\Rightarrow \quad \mu'_3 = \mu_3 + 3\mu_2\,\mu'_1 + \mu'^3_1$$

If $r = 4$, $\quad \mu'_4 = \mu_4 + 4\mu_3\,\mu'_1 + 6\mu_2\,\mu'^2_1 + 4\mu_1\,\mu'^3_1 + \mu'^4_1$

$$= \mu_4 + 4\mu_3\,\mu'_1 + 6\mu_2\,\mu'^2_1 + 0 + \mu'^4_1$$

$$\Rightarrow \quad \mu'_4 = \mu_4 + 4\mu_3\,\mu'_1 + 6\mu_2\,\mu'^2_1 + \mu'^4_1$$

$$\boxed{\mu'_1 = \mu_1 - a} \qquad \boxed{\mu'_2 = \mu_2 + \mu'^2_1}$$

$$\boxed{\mu'_3 = \mu_3 + 3\mu_2\,\mu'_1 + \mu'^3_1} \qquad \boxed{\mu'_4 = \mu_4 + 4\mu_3\,\mu'_1 + 6\mu_2\,\mu'^2_1 + \mu'^4_1}$$

57.35 RELATION BETWEEN v_r AND μ_r

$$v_r = \frac{1}{N}\sum_{i=1}^{n} f_i\,x_i^r\,; \qquad r = 0, 1, 2, \ldots$$

$$= \frac{1}{N}\sum_{i=1}^{n} f_i\,(x_i - a + a)^r$$

On expanding by binomial theorem, we have

$$= \frac{1}{N} \sum_{i=1}^{n} f_i \left[(x_i - a)^r + {}^rC_1 (x_i - a)^{r-1} a + \ldots.. + a^r\right]$$

$$= \mu'_r + {}^rC_1 \mu'_{r-1} a + \ldots.. + a^r$$

On putting $a = \bar{x}$, we get $\quad v_r = \mu_r + {}^rC_1 \mu_{r-1}\bar{x} + {}^rC_2 \mu_{r-2} \bar{x}^2 + \ldots.. + \bar{x}^r \qquad \ldots(1)$

On taking r = 1, 2, 3, 4 in (1), we get

$$v_1 = \mu_1 + \mu_0 \bar{x} = \bar{x} \qquad [\mu_1 = 0, \mu_0 = 1]$$

$$v_2 = \mu_2 + {}^2C_1 \mu_1 \bar{x} + {}^2C_2 \mu_0 \bar{x}^2 = \mu_2 + \bar{x}^2$$

$$v_3 = \mu_3 + {}^3C_1 \mu_2 \bar{x} + {}^3C_2 \mu_1 \bar{x}^2 + {}^3C_3 \mu_0 \bar{x}^3 = \mu_3 + 3\mu_2 \bar{x} + \bar{x}^3$$

$$v_4 = \mu_4 + {}^4C_1 \mu_3 \bar{x} + {}^4C_2 \mu_2 \bar{x}^2 + {}^4C_3 \mu_1 \bar{x}^3 + {}^4C_4 \mu_0 \bar{x}^4$$

$$= \mu_4 + 4\mu_3 \bar{x} + 6\mu_2 \bar{x}^2 + \bar{x}^4$$

$$\boxed{v_1 = \bar{x}} \qquad \boxed{v_2 = \mu_2 + \bar{x}^2}$$

$$\boxed{v_3 = \mu_3 + 3\mu_2 \bar{x} + \bar{x}^3} \qquad \boxed{v_4 = \mu_4 + 4\mu_3 \bar{x} + 6\mu_2 \bar{x}^2 + \bar{x}^4}$$

57.36 MEASURE OF SKEWNESS BASED ON MOMENT

1. Measure of skewness is given by β_1 where

$$\boxed{\beta_1 = \frac{\mu_3^2}{\mu_2^3}}$$

Karl Pearson's coefficient of skewness $= \pm \dfrac{\mu_3}{\mu_2^{3/2}}$

The sign of Karl Pearson's coefficient of skewness is determined from the sign of μ_3.

2. Measure of Kurtosis is given by β_2, where

$$\boxed{\beta_2 = \frac{\mu_4}{\mu_2^2}}$$

3. Gamma Coefficients

$$\boxed{\gamma_1 = \pm \sqrt{\beta_1}}$$

$$\boxed{\gamma_2 = \beta_2 - 3}$$

Example 23. *Find the relation between moment about the mean and moment about any arbitary point. The first four moments of a distribution about the value 4 of the variate are – 1.5, 17, – 30 and 108. Calculate the first four moments about the mean and find β_1 and β_2.* *(Uttarakhand, III Semester, 2008)*

Solution. We have,

$a = 4, \quad \mu_1' = -1.5, \quad \mu_2' = 17, \quad \mu_3' = -30$ and $\mu_4' = 108$

Moment about the mean

$$\mu_1 = 0$$

$$\mu_2 = \mu_2' - (\mu_1')^2 = 17 - (-1.5)^2 = 17 - 2.25 = 14.75$$

$$\mu_3 = \mu_3' - 3\mu_2' \mu_1' + 2(\mu_1')^3$$
$$= -30 - 3(17)(-1.5) + 2(-1.5)^3 = -30 + 76.5 - 6.75$$
$$= 39.75$$
$$\mu_4 = \mu_4' - 4\mu_3' \mu_1' + 6\mu_2' \mu_1'^2 - 3\mu_1'^4$$
$$= 108 - 4(-30)(-1.5) + 6(17)(-1.5)^2 - 3(-1.5)^4$$
$$= 108 - 180 + 229.5 - 15.19 = 142.31$$

$$\beta_1 = \frac{(\mu_3)^2}{(\mu_2)^3} = \frac{(39.75)^2}{(14.75)^3} = \frac{1580.06}{3209.05} = 0.4924$$

$$\beta_2 = \frac{\mu_4}{(\mu_2)^2} = \frac{142.31}{(14.75)^2} = \frac{142.31}{217.56} = 0.6541$$ **Ans.**

57.37 KURTOSIS *(U.P. III, Semester Dec. 2006)*

It measures the degree of peakedness of a distribution and is given by Measure of kurtosis

$$\beta_2 = \frac{\mu_4}{\mu_2^2}, \quad \mu_2 = \frac{\Sigma(x-\bar{x})^2}{N}, \quad \mu_4 = \frac{\Sigma(x-\bar{x})^4}{N}$$

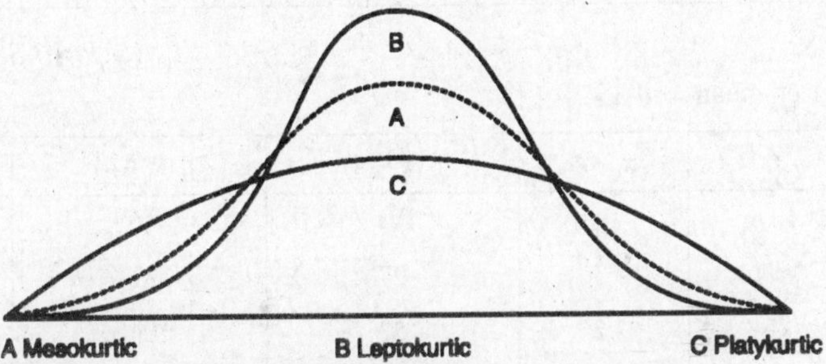

If $\beta_2 = 3$, the curve is normal or mesokurtic.

If $\beta_2 > 3$, the curve is peaked or leptokurtic.

If $\beta_2 < 3$, the curve is flat topped or platykurtic.

$$\gamma_2 = \beta_2 - 3$$

57.38 GROUPING ERROR AND ITS SHEPPARD'S CORRECTIONS (FOR MOMENTS)

If the distribution is not symmetrical and the number of class intervals is greater than $\frac{1}{20}$th of the range, then the computation of moments will have an error known as grouping error.

$$\mu_2\,(\text{corrected}) = \mu_2 - \frac{h^2}{12}$$

$$\mu_4\,(\text{corrected}) = \mu_4 - \frac{1}{2}h^2\mu_2 + \frac{7}{240}h^4$$

where h is the width of the class-interval while μ_1 and μ_3 require no correction. These formulae are known as **Sheppard's corrections.**

Example 24. *Find the corrected values of the following moments using Sheppard's correction. The width of classes in the distribution is 10 :*

$$\mu_2 = 210, \quad \mu_3 = 460, \quad \mu_4 = 96700, \quad h = 12$$

Solution. We have,

$$\mu_2 = 210, \quad \mu_3 = 460, \quad \mu_4 = 96700, \quad h = 12$$

$$\text{corrected } \mu_2 = \mu_2 - \frac{h^2}{12} = 210 - \frac{(12)^2}{12} = 210 - 12 = 198$$

$$\text{corrected } \mu_3 = 460$$

$$\text{corrected } \mu_4 = \mu_4 - \frac{1}{2} h^2 \mu_2 + \frac{7}{240} h^4$$

$$= 96700 - \frac{(12)^2}{2}(210) + \frac{7}{240}(12)^4$$

$$= 96700 - 15120 + 604.8 = 82184.8$$ **Ans.**

Example 25. *Calculate the variance and third central moment from the following data :*

x_i	0	1	2	3	4	5	6	7	8
f_i	1	9	26	59	72	52	29	7	1

(*U.P. III Semester Dec. 2005*)

Solution. Let mean = 4

x_i	f_i	$x_i - 4$	$f_i(x_i - 4)$	$f_i(x_i - 4)^2$	$f_i(x_i - 4)^3$
0	1	−4	−4	16	−64
1	9	−3	−27	81	−243
2	26	−2	−52	104	−208
3	59	−1	−59	59	−59
4	72	0	0	0	0
5	52	1	52	52	52
6	29	2	58	116	232
7	7	3	21	63	189
8	1	4	4	16	64
	$\Sigma f_i = 256$		$\Sigma fi(x_i - 4) = -7$	$\Sigma fi(x_i - 4)^2 = 507$	$\Sigma fi(x_i - 4)^3 = -37$

$$\mu'_1 = \frac{\Sigma f_i (x_i - 4)}{\Sigma f_i} = \frac{-7}{256}$$

$$\mu'_2 = \frac{\Sigma f_i (x_i - 4)^2}{\Sigma f_i} = \frac{507}{256}$$

$$\mu'_3 = \frac{\Sigma f_i (x_i - 4)^3}{\Sigma f_i} = \frac{-37}{256}$$

$$\mu_2 = \mu'_2 - \mu'^2_1 = \frac{507}{256} - \left(\frac{-7}{256}\right)^2 = 1.98047 - 0.00075 = 1.97972$$

$$\mu_2 = \mu'_2 - \mu'_1{}^2 = \frac{507}{256} - \left(\frac{-7}{256}\right)^2 = 1.98047 - 0.00075 = 1.97972$$

$$\mu_3 = \mu'_3 - 3\mu'_2 \,.\, \mu'_1 + 2(\mu'_1)^3$$

$$= \frac{-37}{256} - 3\left(\frac{507}{256}\right)\left(-\frac{7}{256}\right) + 2\left(-\frac{7}{256}\right)^3$$

$$= -0.14453 + 0.16246 - 0.00004$$

$$= 0.01789$$

Ans.

Example 26. *Calculate* $\mu_1, \mu_2, \mu_3, \mu_4$ *for the following frequency distribution :*

Marks	*0-10*	*10-20*	*20-30*	*30-40*	*40-50*	*50-60*
No. of students	*1*	*6*	*10*	*15*	*11*	*7*

Solution.

Marks	*No. of students* f	*Mid value* x	fx	$x-\bar{x}$	$f(x-\bar{x})$	$f(x-\bar{x})^2$	$f(x-\bar{x})^3$	$f(x-\bar{x})^4$
0-10	1	5	5	−30	−30	900	−27000	810000
10-20	6	15	90	−20	−120	2400	−48000	960000
20-30	10	25	250	−10	−100	1000	−10000	100000
30-40	15	35	525	0	0	0	0	0
40-50	11	45	495	10	110	1100	11000	110000
50-60	7	55	385	20	140	2800	56000	1120000
	$\Sigma f = 50$		$\Sigma fx = 1750$		$\Sigma f(x-\bar{x}) = 0$	$\Sigma f(x-\bar{x})^2 = 8200$	$\Sigma f(x-\bar{x})^3 = -18000$	$\Sigma f(x-\bar{x})^4 = 3100000$

$$\bar{x} = \frac{\Sigma fx}{\Sigma f} = \frac{1750}{50} = 35$$

$$\mu_1 = \frac{\Sigma f(x-\bar{x})}{\Sigma f} = \frac{0}{50} = 0$$

$$\mu_2 = \frac{\Sigma f(x-\bar{x})^2}{\Sigma f} = \frac{8200}{50} = 164$$

$$\mu_3 = \frac{\Sigma f(x-\bar{x})^3}{\Sigma f} = \frac{-18000}{50} = -360$$

$$\mu_4 = \frac{\Sigma f(x-\bar{x})^4}{\Sigma f} = \frac{3100000}{50} = 62000$$

Ans.

Example 27. *Find out the kurtosis of the data given below :*

Class-interval	*0-10*	*10-20*	*20-30*	*30-40*
Frequency	*1*	*3*	*4*	*2*

Solution. Let assumed mean be 25

Class	*Frequency* f_i	*Mid value* x_i	$x_i - 25$	$f_i(x_i - 25)$	$f_i(x_i - 25)^2$	$f_i(x_i - 25)^3$	$f_i(x_i - 25)^4$
0-10	1	5	−20	−20	400	−8000	160000
10-20	3	15	−10	−30	300	−3000	30000
20-30	4	25	0	0	0	0	0
30-40	2	35	10	20	200	2000	20000
	$\Sigma f = 10$			$\Sigma f_i(x_i - 25) = -30$	$\Sigma f_i(x_i - 25)^2 = 900$	$\Sigma f_i(x_i - 25)^3 = -9000$	$\Sigma f_i(x_i - 25)^4 = 210000$

$$\mu'_1 = \frac{\Sigma f_i(x_i - 25)}{\Sigma f_i} = \frac{-30}{10} = -3$$

$$\mu'_2 = \frac{\Sigma f_i(x_i - 25)^2}{\Sigma f_i} = \frac{900}{10} = 90$$

$$\mu'_3 = \frac{\Sigma f_i(x_i - 25)^3}{\Sigma f_i} = \frac{-9000}{10} = -900$$

$$\mu'_4 = \frac{\Sigma f_i(x_i - 25)^4}{\Sigma f_i} = \frac{210000}{10} = 21000$$

$$\mu_2 = \mu'_2 - {\mu'_1}^2 = 90 - (-3)^2 = 90 - 9 = 81$$ [Article 57.34 on page 1560]

$$\mu_4 = \mu'_4 - 4\,\mu'_3\,\mu'_1 + 6\mu'_2\,{\mu'_1}^2 - 3{\mu'_1}^4$$

$$= 21000 - 4(-900)(-3) + 6(90)(-3)^2 - 3(-3)^4$$

$$= 21000 - 10800 + 4860 - 243 = 14,817$$

$$\beta_2 = \frac{\mu_4}{\mu_2^2} = \frac{14,817}{(81)^2} = \frac{14,817}{6561} = 2.258$$

$$\gamma_2 = \beta_2 - 3 = 2.258 - 3 = -0.742$$ **Ans.**

Example 28. *Calculate the first four moments of the following data. Also make sheppard's correction.*

Values	*10-20*	*30-30*	*30-30*	*40-50*	*50-60*	*60-70*	*70-80*
Frequency	*1*	*20*	*69*	*108*	*78*	*22*	*2*

Solution. Calculation of Moments

Values	f	x	$d = x - 45$	$\frac{x-45}{10}$	$f_i\left(\frac{x-45}{10}\right)$	$f_i\left(\frac{x-45}{10}\right)^2$	$f_i\left(\frac{x-45}{10}\right)^3$	$f_i\left(\frac{x-45}{10}\right)^4$
10–20	1	15	– 30	– 3	– 3	9	– 27	81
20–30	20	25	– 20	– 2	– 40	80	– 160	320
30–40	69	35	– 10	– 1	– 69	69	– 69	69
40–50	108	45	0	0	0	0	0	0
50–60	78	55	10	1	78	78	78	78
60–70	22	65	20	2	44	88	176	352
70–80	2	75	30	3	6	18	54	162
	N = 300				$\Sigma f_i\left(\frac{x-45}{10}\right) = 16$	$\Sigma f_i\left(\frac{x-45}{10}\right)^2 = 342$	$\Sigma f_i\left(\frac{x-45}{10}\right)^3 = 52$	$\Sigma f_i\left(\frac{x-45}{10}\right)^4 = 1062$

Now
$$\mu'_1 = \frac{1}{N}\Sigma f_i\left(\frac{x-45}{10}\right).\ 10 = \left(\frac{16}{300}\right)10 = 0.53$$
$$\mu'_2 = \frac{1}{N}\Sigma f_i\left(\frac{x-45}{10}\right)^2.10^2 = \left(\frac{342}{300}\right)(10)^2 = 114$$
$$\mu'_3 = \frac{1}{N}\Sigma f_i\left(\frac{x-45}{10}\right)^3.10^3 = \left(\frac{52}{300}\right)(10)^3 = 173.33$$
$$\mu'_4 = \frac{1}{N}\Sigma f_i\left(\frac{x-45}{10}\right)^4.10^4 = \left(\frac{1062}{300}\right)(10)^4 = 35400.$$

Moments about mean

$$\mu_1 = 0 \text{ (always)}$$
$$\mu_2 = \mu'_2 - \mu'^2_1 = 114 - (0.53)^2 = 113.72$$
$$\mu_3 = \mu'_3 - 3\mu'_2\mu'_1 + 2\mu'^3_1 = 173.33 - 3(114)(0.53) + 2(0.53)^3 = -7.63$$
$$\mu_4 = \mu'_4 - 4\mu'_3\mu'_1 + 6\mu'_2\mu'^2_1 - 3\mu'^4_1$$ [Article 57.34 on page 1560]
$$= 35400 - 4(173.33)(0.53) + 6(114)(0.53)^2 - 3(0.53)^4 = 35224.44.$$

Sheppard's Correction, h = 10

$$\mu_1 \text{ (Corrected)} = \mu_1 = 0$$
$$\mu_2 \text{ (Corrected)} = \mu_2 - \frac{h^2}{12} = 113.72 - \frac{(10)^2}{12} = 105.39$$
$$\mu_3 \text{ (Corrected)} = \mu_3 = -7.63$$
$$\mu_4 \text{ (Corrected)} = \mu_4 - \frac{1}{2}h^2\mu_2 + \frac{7}{240}h^4$$
$$= 35224.44 - \frac{1}{2}(10)^2(113.72) + \frac{7}{240}(10)^4 = 29830.11$$ **Ans.**

Example 29. *The first four moments of a distribution about the value 4 of the variable arc – 1.5, 17, – 30 and 108. Find the moments about mean, β_1 and β_2. Find also the moments about (i) the origin, and (ii) the point $x = 2$.*

(U.P. III, Semester Dec. 2006)

Solution. In the usual notation, we are given assumed mean $a = 4$ and

$$\mu'_1 = -1.5, \quad \mu'_2 = 17, \quad \mu'_3 = -30 \quad \text{and} \quad \mu'_4 = 108.$$

(*a*) Moments about mean:

$$\mu_2 = \mu'_2 - \mu'^2_1 = 17 - (-1.5)^2 = 17 - 2.25 = 14.75$$

$$\mu_3 = \mu'_3 - 3\mu'_2\,\mu'_1 + 2\mu'^3_1$$ [Article 57.34 on page 1560]

$$= -30 - 3 \times (17) \times (-1.5) + 2(-1.5)^3$$

$$= -30 + 76.5 - 6.75 = 39.75$$

$$\mu_4 = \mu'_4 - 4\mu'_3\,\mu'_1 + 6\mu'_2\,\mu'^2_1 - 3\mu'^4_1$$

$$= 108 - 4(-30)(-1.5) + 6(17)(-1.5)^2 - 3(-1.5)^4$$

$$= 108 - 180 + 229.5 - 15.1875 = 142.3125$$

Hence $$\beta_1 = \frac{\mu_3^2}{\mu_2^3} = \frac{(39.75)^2}{(14.75)^3} = 0.4924$$

$$\beta_2 = \frac{\mu_4}{\mu_2^2} = \frac{142.3125}{(14.75)^2} = 0.6541$$

Also $$\bar{x} = a + \mu'_1 = 4 + (-1.5) = 2.5$$

(*b*) *Moments about origin.*

We have moments about $\bar{x}$

$\bar{x} = 2.5$, $\mu_2 = 14.75$, $\mu_3 = 39.75$ and $\mu_4 = 142.31$ (approx.)

We know that $\bar{x} = a + \mu'_1$, where μ'_1 is the first moment about the point $x = a$.

Taking $a = 0$, we get the first moment about origin as $\mu'_1 = \bar{x} =$ mean $= 2.5$.

[Using Article 57.35 of page 1561]

$$v_1 = x = 2.5$$

$$v_2 = \mu_2 + (\bar{x})^2 = 14.75 + (2.5)^2 = 14.75 + 6.25 = 21$$

$$v_3 = \mu_3 + 3\mu_2\,\bar{x} + (\bar{x})^3 = 39.75 + 3(14.75)(2.5) + (2.5)^3$$

$$= 39.75 + 110.625 + 15.625 = 166$$

$$v_4 = \mu_4 + 4\mu_3\,\bar{x} + 6\mu_2\,(\bar{x})^2 + (\bar{x})^4$$

$$= 142.3125 + 4(39.75)(2.5) + 6(14.75)(2.5)^2 + (2.5)^4$$

$$= 142.3125 + 397.5 + 553.125 + 39.0625$$

$$= 1132.$$

Moments about the point $x = 2$.

We have $\bar{x} = a + \mu'_1$. Taking $a = 2$, the first moment about the point $x = 2$ is

$$\mu'_1 = \bar{x} - 2 = 2.5 - 2 = 0.5$$ [Converse of Art. 57.34 on page 1560]

Hence, $$\mu'_2 = \mu_2 + \mu'^2_1 = 14.75 + 0.25 = 15$$

$$\mu'_3 = \mu_3 + 3\mu_2\,\mu'_1 + \mu'^3_1 = 39.75 + 3(14.75)(0.5) + (0.5)^3$$

$$= 39.75 + 22.125 + 0.125 = 62$$

$$\mu'_4 = \mu_4 + 4\mu_3 \mu'_1 + 6\mu_2 \mu'^2_1 + \mu'^4_1$$
$$= 142.3125 + 4(39.75)(0.5) + 6(14.75)(0.5)^2 + (0.5)^4$$
$$= 142.3125 + 79.5 + 22.125 + 0.0625 = 244$$ **Ans.**

Example 30. *The first three moments about the origin are given by*

$$\mu'_1 = \frac{n+1}{2},\ \mu'_2 = \frac{(n+1)(2n+1)}{6} \text{ and } \mu'_3 = \frac{n(n+1)^2}{4}$$

Examine the skewness of the data.

Solution.

$$\mu'_1 = \frac{n+1}{2},\ \mu'_2 = \frac{(n+1)(2n+1)}{6},\ \mu'_3 = \frac{n(n+1)^2}{4}$$

$$\mu_2 = \mu'_2 - \mu'^2_1$$
$$= \frac{(n+1)(2n+1)}{6} - \frac{(n+1)^2}{4}$$
$$= \frac{2(2n^2+3n+1) - 3(n^2+2n+1)}{12} = \frac{n^2-1}{12}$$

$$\mu_3 = \mu'_3 - 3\mu'_2 \mu'_1 + 2\mu'^3_1$$
$$= \frac{n(n+1)^2}{4} - \frac{3(n+1)(2n+1)}{6} \times \frac{(n+1)}{2} + 2\left(\frac{n+1}{2}\right)^3$$
$$= \frac{n^3+2n^2+n}{4} - \frac{2n^3+5n^2+4n+1}{4} + \frac{n^3+3n^2+3n+1}{4}$$
$$= 0$$

$$\text{Coefficient of skewness } = \gamma_1 = \frac{\mu_3}{\sqrt{\mu_2^3}} = \frac{0}{\sqrt{\mu_2^3}} = 0$$

The data is symmetrical. **Ans.**

Example 31. *The first three moments of a distribution, about the value '2' of the variable are 1, 16 and – 40. Show that the mean is 3, variance is 15 and* $\mu_3 = -86$.

Solution. We have

$$a = 2,\ \mu'_1 = 1,\ \mu'_2 = 16 \text{ and } \mu'_3 = -40$$

We know that $\mu'_1 = \bar{x} - a \Rightarrow \bar{x} = \mu'_1 + a = 1 + 2 = 3$

$$\text{Variance} = \mu_2 = \mu'_2 - \mu'^2_1 = 16 - (1)^2 = 15$$
$$\mu_3 = \mu'_3 - 3\mu'_2 \mu'_1 + 2\mu'^3_1 = -40 - 3(16)(1) + 2(1)^3 = -40 - 48 + 2 = -86.$$

Example 32. *The first four moments of a distribution, about the value '35' are – 1.8, 240, – 1020 and 144000. Find the values of* $\mu_1, \mu_2, \mu_3, \mu_4$.

Solution. We have,

$$\mu'_1 = -1.8,\quad \mu'_2 = 240,\quad \mu'_3 = -1020,\quad \mu'_4 = 144000$$
$$\mu_1 = 0$$
$$\mu_2 = \mu'_2 - \mu'^2_1 = 240 - (-1.8)^2 = 236.76$$ (Article 57.34 on page 1560]
$$\mu_3 = \mu'_3 - 3\mu'_2 \mu'_1 + 2\mu'^3_1 = -1020 - 3(240)(-1.8) + 2(-1.8)^3 = 264.36$$
$$\mu_4 = \mu'_4 - 4\mu'_3 \mu'_1 + 6\mu'_2 \mu'^2_1 - 3\mu'^4_1$$
$$= 144000 - 4(-1020)(-1.8) + 6(240)(-1.8)^2 - 3(-1.8)^4 = 141290.11.$$

Ans.

Example 33. *For a distribution, the mean is 10, variance is 16, γ_1 is 1, and β_2 is 4. Find the first four moments about the origin.*

Solution. We have,

$$\bar{x} = 10,\ \mu_2 = 16,\ \gamma_1 = 1, \beta_2 = 4$$

Now, $\gamma_1 = 1 \Rightarrow \sqrt{\beta_1} = 1 \Rightarrow \beta_1 = 1$

$\Rightarrow \quad \beta_1 = 1 \Rightarrow \dfrac{\mu_3^2}{\mu_2^3} = 1 \Rightarrow \mu_3^2 = \mu_2^3 = (16)^3 = (64)^2$

$\Rightarrow \quad \mu_3^2 = (64)^2 \Rightarrow \mu_3 = 64$

and $\quad \mu_2^3 = (16)^3 \Rightarrow \mu_2 = 16$

$\beta_2 = 4$

$\Rightarrow \quad \dfrac{\mu_4}{\mu_2^2} = 4 \Rightarrow \mu_4 = 4(16)^2 = 1024$

Moments about the origin

$v_1 = \bar{x} = 10$ [Using Article 57.35 of page 1560]

$v_2 = \mu_2 + \bar{x}^2 = 16 + 100 = 116$

$v_3 = \mu_3 + 3\mu_2\bar{x} + \bar{x}^3 = 64 + 3(16)(10) + (10)^3 = 64 + 480 + 1000 = 1544$

$v_4 = \mu_4 + 4\mu_3\bar{x} + 6\mu_2\bar{x}^2 + \bar{x}^4 = 1024 + 4(64)(10) + 6(16)(100) + (10)^4$

$= 1024 + 2560 + 9600 + 10000 = 23184$ **Ans.**

EXERCISE 57.4

1. Find the mean and s.d. of the following series:

Expenditure	No. of students
Below Rs. 5	6
" 10	16
" 15	28
" 20	38
" 25	46

Ans. Mean = Rs. 1293, S.D. = Rs. 6.41

2. Calculate the arithmetic mean and the standard deviation of the following values of the world's annual gold output (in millions of pounds) for 10 different years:

Year	94	95	96	93	87	79	73	69	68	67
Gold output	78	82	83	89	95	103	108	117	130	97

Ans. Mean = 80.50, standard deviation = 9.27

3. For a frequency distribution of marks in History of 200 candidates (grouped in intervals 0–5, 5–10, ...) the mean and standard deviation (s.d.) were found to be 40 and 15. Later it was discovered that the score 43 was misread as 53 in obtaining the frequency distribution. Find the corrected mean and s.d. corresponding to the corrected frequency distribution.

Ans. Mean = 39.95. Standard deviation = 14.975 approx.

4. A student while calculating the mean and the standard deviation on 25 observations, obtained the following values:

mean = 56 cms. : standard deviation = 2 cms.

It was later discovered at the time of checking that he had wrongly copied down an observation as 64. What is the mean and s.d. if correct value is omitted ?

Ans. Mean = 55.67 cms., S.D. = 1.18 cms. approx

5. Calculate first four moments about mean, for the following individual series:

5	5	5	5	5	5

Ans. $\mu_1 = \mu_2 = \mu_3 = \mu_4 = 0$

6. Calculate the first four moments about the mean for the following data. Also calculate β_1 and β_2.

x	1	2	3	4	5	6	7	8	9
f	1	6	13	25	30	22	9	5	2

Ans. $\mu'_1 = -0.09, \mu_2 = 2.4873, \mu_3 = 0.6789, \mu_4 = 18.3358$ approx., $\beta_1 = 0.0299$ approx. $\beta_2 = 2.9627$

7. Find the first four moments about mean for the following frequency distribution:

Marks	0–10	10–20	20–30	30–40	40–50
No. of Students	5	10	40	20	25

Ans. $\mu_1 = 0, \mu_2 = 125, \mu_3 = -300, \mu_4 = 37625$

8. The following table gives the monthly wages of workers in a factory. Compute the standard deviation, and skewness

Monthly wages (in Rs.)	*No. of workers*	*Monthly wages (in Rs.)*	*No. of workers*
125–175	2	375–425	4
175–225	22	425–475	6
225–275	19	475–525	1
275–325	14	525–575	1
325–375	3		

Ans. S.D. = Rs. 88.52, Skewness = 0.7

9. The first four moments of a distribution about x = 4 are 1, 4, 10, 45. Show that the mean is 5 and the variance is 3 and μ_3 and μ_4 are 0 and 26 respectively.

10. Calculate $\mu_1, \mu_2, \mu_3, \mu_4$ for the series : 4, 7, 10, 13, 16, 19, 22.

Ans. $\mu_1 = 0, \mu_2 = 36, \mu_3 = 0, \mu_4 = 2268$

11. If the first four moments of a distribution about the value 5 are equal to – 4, 22, – 117 and 560. Determine the corresponding moments:

(*i*) about the mean, and (*ii*) about zero **Ans.** (*i*) 0, 6, 83, 992 (*ii*) 1, 7, 102, 1361

12. In a certain distribution, the first four moments about the point $x = 4$ are – 1.5, 17, – 30 and 308. Calculate β_1 and β_2.

Ans. $\beta_1 = 0.4923, \beta_2 = 1.573$

13. Compute first four moments of the data 3, 5, 7, 9 about the mean. Also, compute the first four moments about the point 4.

Ans. $\mu_1 = 0, \mu_2 = 5, \mu_3 = 0, \mu_4 = 41$

$\mu'_1 = 2, \mu'_2 = 9, \mu'_3 = 38, \mu'_4 = 177$

14. The first four moments of distribution about the value 5 of the variable are 2, 20, 40 and 50. Calculate mean, $\mu_2, \mu_3, \mu_4, \beta_1$ and β_2. **Ans.** Mean = 7, $\mu_2 = 16, \mu_3 = -64, \mu_4 = 162, \beta_1 = 1$ and $\beta_2 = 0.63$

15. Show that, if the variable takes the values 0, 1, 2, ..., n with frequencies proportional to the binomial coefficients, 1, ${}^nC_1, {}^nC_2, \ldots {}^nC_n$ respectively then the mean of the distribution is $\frac{1}{2}n$, the mean square deviation about origin is $\frac{1}{4}n(n+1)$ and the variance is $\frac{1}{4}n$.

16. Show that, if the variable takes the values 0, 1, 2, ..., n with frequencies given by the terms of the binomial series $q^n, {}^nC_1 q^{n-1}p, {}^nC_2 q^{n-2}p^2, \ldots, p^n$ where p + q = 1, then the mean square deviation is $n^2p^2 + npq$ and the variance is npq.

57.39 MOMENT GENERATING FUNCTION

The moment generating function of the variate x about $x = a$ is defined as the expected value of $e^{t(x-a)}$ and is denoted by $M_a(t)$.

$$M_a(t) = \Sigma\, p_i\, e^{t(x_i - a)}$$

$$= \Sigma\, p_i \left[1 + t(x_i - a) + \frac{t^2}{2}(x_i - a)^2 + \ldots + \frac{t^r}{r!}(x_i - a)^r + \ldots\right]$$

$$= \Sigma\, p_i + t\,\Sigma\, p_i\,(x_i - a) + \frac{t^2}{2}\Sigma\, p_i\,(x_i - a)^2 + \ldots \frac{t^r}{r!}\Sigma\, p_i\,(x_i - a)^r + \ldots$$

$$= 1 + t\,\Sigma f_i\,(x_i - a) + \frac{t^2}{2}\Sigma f_i\,(x_i - a)^2 + \ldots + \frac{t^r}{r!}\Sigma f_i\,(x_i - a)^r + \ldots \qquad [p \approx f]$$

$$= \mu_0 + t\,\mu'_1 + \frac{t^2}{2!}\mu'_2 + \ldots + \frac{t^r}{r!}\mu'_r + \ldots \qquad [\Sigma\, p_i = 1 = \mu_0]$$

where μ'_r is the moment of order r about a.

Hence $\mu'_r =$ coefficient of $\frac{t^r}{r!}$ or $\mu'_r = \left[\frac{d^r}{dt^r}M_a(t)\right]_{t=0}$

Again $M_a(t) = \Sigma p_i\, e^{t(x_i - a)}$

$$= e^{-at}\,\Sigma\, p_i\, e^{tx_i}$$

$$= e^{-at}\, M_0(t)$$

Thus the moment generating function about the point $a = e^{-at}$ moment generating funtion about the origin.

Example 34. *Find the moment generating function of the discrete binomial distribution given by*

$$f(x) = {}^nC_x\, p^x\, q^{n-x}$$

Also, find the first and second moment about the mean and standard deviation.

Solution. Here, we have

$$f(x) = {}^nC_x\, p^x\, q^{n-x}$$

Moment generating function about the origin

$$= M_0(t) = \sum e^{tx} \,.\, {}^nC_x\, p^x\, q^{n-x}$$

$$= \sum {}^nC_x\, (pe^t)^x \,.\, q^{n-x}$$

$$= q^n + {}^nC_1\, q^{n-1}\, pe^t + {}^nC_2\, q^{n-2}\, (pe^t)^2 + \ldots.. \qquad \text{[By Binomial Theorem]}$$

$$= [q + pe^t]^n$$

$$v_1 = \left[\frac{d}{dt} M_0(t)\right]_{t=0} = [n\,(q + pe^t)^{n-1} \,.\, (pe^t)]_{t=0}$$

$$= n\,(q + p)^{n-1}\, p \qquad [q + p = 1]$$

$$= np$$

$$v_2 = \left[\frac{d^2}{dt^2} M_0(t)\right] = \frac{d}{dt}\,[n\,(q + pe^t)^{n-1}\,(pe^t)]_{t=0}$$

$$= [n\,(n-1)\,(q + pe^t)^{n-2}\,(pe^t)^2 + n\,(q + pe^t)^{n-1}\,(pe^t)]_{t=0}$$

$$= [n\,(n-1)\,(q + p)^{n-2}\,p^2 + n\,(q + p)^{n-1} \,.\, p]$$

$$= n\,(n-1)\,p^2 + np \qquad [q + p = 1]$$

$$= n\,p\,[(n-1)\,p + 1] = np\,[np + (1 - p)] = np\,[np + q]$$

$$= n^2p^2 + npq$$

$$\mu_1 = \bar{x} = v_1 = np$$

$$\mu_2 = \mu'_2 - \bar{x}^2 = v_2 - v_1^2 = (n^2p^2 + npq) - (np)^2$$

$$\Rightarrow \qquad \mu_2 = npq$$

Standard deviation $= \sqrt{npq}$

Mean $= np$ **Ans.**

Example 35. *Find the moment generating function of the discrete Poisson distribution given by*

$$f(x) = \frac{e^{-m} \,.\, m^x}{x!}$$

Also, find the first and second moments about mean and variance.

Solution. Here, we have

$$f(x) = \frac{e^{-m}\, m^x}{x!}$$

Moment generating function about the origin

$$= M_0(t) = \sum e^{tx} \frac{e^{-m}\, m^x}{x!}$$

$$= e^{-m} \sum \frac{(me^t)^x}{x!}$$

$$= e^{-m}\left[1 + me^t + \frac{(me^t)^2}{2!} + \frac{(me^t)^3}{3!} + \ldots.\right]$$

$$= e^{-m} . e^{me^t} = e^{m(e^t - 1)}$$

$$v_1 = \left[\frac{d}{dt} M_0(t)\right]_{t=0} = \left[\frac{d}{dt} e^{m(e^t-1)}\right]_{t=0} = \left[e^{m(e^t-1)} me^t\right]_{t=0} = e^{m(1-1)}.m = m$$

$$v_2 = \left[\frac{d^2}{dt^2} M_0(t)\right]_{t=0} = \frac{d}{dt}\left[e^{m(e^t-1)} me^t\right]_{t=0} = \left[e^{m(e^t-1)}(me^t)^2 + e^{m(e^t-1)}.me^t\right]_{t=0}$$

$$= \left[e^{m(1-1)} m^2 + e^{m(1-1)}.m\right] = m^2 + m$$

$$\mu_2 = v_2 - \bar{x}^2 = v_2 - v_1^2 = (m^2 + m) - m^2 = m$$

Hence, mean $= m$

Variance $= m$ **Ans.**

Example 36. *Find the moment generating function of the random variable whose moments are*

$$\mu_r' = (r+1)!\, 2^r$$

Solution. $M_x(t) = E(e^{tx}) = \sum_{x=0}^{\infty} e^{tx} P(X = x) = \sum_{r=0}^{\infty} \frac{t^r}{r!}\mu'_r = \sum_{r=0}^{\infty} \frac{t^r}{r!}(r+1)!.2^r = \sum_{r=0}^{\infty}(r+1)(2t)^r$

$$= 1 + 2.2t + 3.(2t)^2 + \ldots = (1-2t)^{-2}$$ **Ans.**

57.40 MOMENT GENERATING FUNCTION OF A FUNCTION OF CONTINUOUS VARIATE

The moment generating function of the continuous probability distribution about $x = a$ is given by

$$M_0(t) = \int_{-\infty}^{\infty} e^{t(x-a)} f(x)\, dx$$

57.41 PROPERTY OF MOMENT GENERATING FUNCTION

The moment generating function of the sum of two independent chance variables is the product of their respective moment generating functions.

Symbolically, $M_{x+y}(t) = M_x(t) \times M_y(t)$ provided that x and y are independent random variables.

Proof. Let x and y be two independent random variables so that $x + y$ is also a random variable. The m.g.f. of the sum $x + y$ w.r.t. origin is

$$M_{x+y}(t) = \Sigma p_i \{e^{t(x+y)}\} = \Sigma p_i \{e^{tx}.e^{ty}\} = \Sigma p_i (e^{tx}).\Sigma p_i (e^{ty})$$

Since x and y are independent variables and so are e^{tx} and e^{ty},

Hence, $M_{x+y}(t) = M_x(t) . M_y(t)$ **Proved.**

Example 37. *Find the moment generating function of the exponential distribution.*

$$f(x) = \frac{1}{c} e^{-x/c},\ 0 \le x \le \infty,\ c > 0$$ (*U.P. III Semester, Dec. 2005*)

Hence, find its mean and standard deviation.

Solution. The moment generating function about the origin is

$$M_0(t) = \int_0^{\infty} e^{tx}.f(x)\,dx = \int_0^{\infty} e^{tx}.\frac{1}{c} e^{-\frac{x}{c}}\,dx = \frac{1}{c}\int_0^{\infty} e^{\left(t-\frac{1}{c}\right)x}\,dx$$

$$= \frac{1}{c}\frac{1}{t-\frac{1}{c}}\left[e^{\left(t-\frac{1}{c}\right)x}\right]_0^{\infty}$$

$$= \frac{1}{c} \cdot \frac{1}{t - \frac{1}{c}} [0 - 1] = \frac{1}{ct - 1}(-1) = \frac{1}{1 - ct} = (1 - ct)^{-t}$$

$$= 1 + ct + c^2 t^2 + c^3 t^3 + \ldots.. \qquad \text{[Binomial Theorem]}$$

Moment about origin $= \left[\frac{d}{dt} M_0(t)\right]_{t=0} = \frac{d}{dt}[1 + ct + c^2t^2 + c^3t^3 + \ldots\ldots.]_{t=0}$

$$= [c + 2c^2 t + 3c^3 t^2 + \ \ldots..]_{t=0} = c$$

$$\mu_1 = \bar{x} = c$$

$$\mu'_2 = \left[\frac{d^2}{dt^2} M_0(t)\right] = \frac{d^2}{dt^2}[1 + ct + c^2t^2 + c^3t^3 + \ldots\ldots..]_{t=0}$$

$$= \frac{d}{dt}[c + 2c^2 t + 3c^3 t^2 + \ldots\ldots..]_{t=0} = [2c^2 + 6c^3 t + \ldots.]_{t=0}$$

$$\mu'_2 = 2c^2$$

$$\mu_2 = \mu'_2 - \mu_1^2 = 2c^2 - c^2 = c^2$$

Standard deviation $= \sqrt{\mu_2} = \sqrt{c^2} = c$ **Ans.**

Example 38. *Obtain the moment generating function of the random variable x having probability distribution*

$$f(x) = \begin{cases} x & \text{for } 0 < x < 1 \\ 2 - x & \text{for } 1 \le x < 2 \\ 0 & \text{elsewhere} \end{cases}$$ *Also determine* μ_1', μ_2' *and* μ_2.

Solution. $M_x(t) = \int e^{tx} f(x)\, dx$

$$= \int_0^1 x \,.\, e^{tx} dx + \int_1^2 (2 - x)\, e^{tx}\, dx + \int_2^\infty 0.e^{tx}\ dx = \left(\frac{xe^{tx}}{t} - \frac{e^{tx}}{t^2}\right)_0^1 + \left(\frac{2e^{tx}}{t} - \frac{xe^{tx}}{t} + \frac{e^{tx}}{t^2}\right)_1^2$$

$$= \frac{e^t}{t} - \frac{e^t}{t^2} + \frac{1}{t^2} + \left[\left(\frac{2e^{2t}}{t} - \frac{2e^{2t}}{t} + \frac{e^{2t}}{t^2}\right) - \left(\frac{2e^t}{t} - \frac{e^t}{t} + \frac{e^t}{t^2}\right)\right] = \frac{e^{2t} - 2e^t + 1}{t^2}$$

$$= \left(\frac{e^t - 1}{t}\right)^2 = \frac{1}{t^2}(e^{2t} - 2e^t + 1) = \frac{1}{t^2}\left[1 + \frac{2t}{1} + \frac{4t^2}{2} + \frac{8t^3}{6} + \frac{16t^4}{24} + \ldots \ -2\left(1 + t + \frac{t^2}{2} + \frac{t^3}{6} + \frac{t^4}{24} + \ldots\right) + 1\right]$$

$$= \frac{1}{t^2} + \frac{2}{t} + 2 + \frac{4t}{3} + \frac{2t^2}{3} + \ldots.. - \frac{2}{t^2} - \frac{2}{t} - 1 - \frac{t}{3} - \frac{t^2}{12} + \ldots. + 1$$

$$\mu_1' = \text{coefficient of } \frac{t}{1} = \frac{4}{3} - \frac{1}{3} = \frac{3}{3} = 1$$

$$\mu_2' = \text{coefficient of } \frac{t^2}{2!} = \frac{4}{3} - \frac{1}{6} = \frac{8 - 1}{6} = \frac{7}{6} \qquad \mu_2 = \mu_2' - (\mu_1')^2 = \frac{7}{8} - (1)^2 = \frac{1}{6}$$

Hence $\mu_1' = 1,\ \mu_2' = \frac{7}{6},\ \mu_2 = \frac{1}{6}$ **Ans.**

Example 39. *Find the moment generating function of the continuous normal distribution given by*

$$f(x) = \frac{1}{\sigma\sqrt{2\pi}}\, e^{-\frac{(x-\mu)^2}{2\sigma^2}}; -\infty < x < \infty$$

Solution. Here, we have

$$f(x) = \frac{1}{\sigma\sqrt{2\pi}} e^{-\frac{(x-\mu)^2}{2\sigma^2}}$$

Moment generating function about the origin

$$M_0(t) = \int_{-\infty}^{\infty} e^{tx} \frac{1}{\sigma\sqrt{2\pi}} e^{-\frac{(x-\mu)^2}{2\sigma^2}} dx \qquad \dots (1)$$

Putting $\frac{x-\mu}{\sigma} = z$ so that $dx = \sigma\, dz$ in (1), we get

$$M_0(t) = \frac{1}{\sigma\sqrt{2\pi}} \int_{-\infty}^{\infty} e^{t(\sigma z+\mu)} e^{-\frac{z^2}{2}} (\sigma\, dz) = \frac{e^{\mu t}}{\sqrt{2\pi}} \int_{-\infty}^{\infty} e^{t\sigma z - \frac{z^2}{2}} dz$$

$$= \frac{e^{\mu t + \frac{1}{2}t^2\sigma^2}}{\sqrt{2\pi}} \int_{-\infty}^{\infty} \left(e^{t\sigma z - \frac{1}{2}t^2\sigma^2 - \frac{z^2}{2}} \right) dz = \frac{1}{\sqrt{2\pi}} e^{\mu t + \frac{1}{2}t^2\sigma^2} \int_{-\infty}^{\infty} e^{-\frac{1}{2}(z^2 - 2t\sigma z + t^2\sigma^2)} dz$$

$$= \frac{1}{\sqrt{2\pi}} e^{\mu t + \frac{1}{2}t^2\sigma^2} \int_{-\infty}^{\infty} e^{-\frac{1}{2}(z-t\sigma)^2} dz \quad = e^{\mu t + \frac{1}{2}t^2\sigma^2} (1) \qquad \left[\int_0^{\infty} e^{-z^2} dz = \frac{\sqrt{\pi}}{2} \right]$$

$$= e^{\mu t + \frac{1}{2}t^2\sigma^2}$$

Ans.

Example 40. *The random variable X assuming only non-negative values has a gamma probability distribution if its probability distribution is given by*

$$f(x) = \begin{Bmatrix} \frac{\alpha^\beta}{\Gamma\beta} x^{\beta-1} e^{-\alpha x} \; ; x > 0, \; \alpha > 0, \; \beta > 1 \\ 0, \; elsewhere \end{Bmatrix}$$

Find the moment generating function of Gamma probability distribution.

Solution.

$$M_x(t) = \int_0^{\infty} e^{tx} f(x)\, dx = \int_0^{\infty} e^{tx} \cdot \frac{\alpha^\beta}{\Gamma\beta} \cdot x^{\beta-1} e^{\alpha}\, dx = \frac{\alpha^\beta}{\Gamma\beta} \int_0^{\infty} x^{\beta-1} e^{-x(\alpha-t)} dx$$

$$= \frac{\alpha^\beta}{(\alpha-t)^\beta\, \Gamma\beta} \int_0^{\infty} y^{\beta-1} e^{-y}\, dy \quad | \text{where } y = x(\alpha - t) \text{ so that } dy = (\alpha - t)\, dx$$

$$= \frac{1}{\left(1-\frac{t}{\alpha}\right)^\beta} \cdot \frac{1}{\Gamma\beta}\, \Gamma\beta = \left(1 - \frac{t}{\alpha}\right)^{-\beta} ; |t| < \alpha.$$

Ans.

Note:
$$M_x(t) = \begin{cases} \sum_x e^{tx}\, p(x), \text{ if } x \text{ is discrete} \\ \int_{-\infty}^{\infty} f_x(x)\, dx, \text{ if } x \text{ is continuous} \end{cases}$$

EXERCISE 57.5

Find the moment generating function of the following functions:

1. $f(x) = me^{-mx}; x, m > 0$ **Ans.** $\sum_{r=0}^{\infty} \left(\frac{t}{m}\right)^r, \; \mu'_r = \frac{r!}{m^r}$

2. $f(x) = e^{-x}(1 + e^{-x})^{-2}, -\infty < x < \infty$ **Ans.** $\beta(1 - t, \; 1 + t), \; 1 - t > 0$

   $\pi t \operatorname{cosec} \pi t, \; t < 1$

3. If x is a discrete random variable with probability function $f(x) = \frac{1}{k^x}, \; x = 1, \; 2\dots\dots,$ (k = constant)

Find its moment generating function. **Ans.** $\dfrac{e^t}{k-e^t}$

4. Find the moment generating function for the distribution where,

$$f(x) = \begin{cases} \dfrac{2}{3}, & \text{at } x = 1 \\ \dfrac{1}{3}, & \text{at } x = 2 \\ 0, & \text{otherwise} \end{cases}$$

Ans. $\dfrac{2}{3}.e^t + \dfrac{1}{3}e^{2t}$

5. Find the moment generating function of the random variable x having the probability density function:

$$f(x) = \begin{cases} \dfrac{1}{k}, & \text{for } -1 < x < 2 \\ 0, & \text{otherwise} \end{cases}$$

Ans. $\dfrac{1}{kt}\left(e^{2t} - e^{-t}\right),\ t \neq 0$; $\dfrac{3}{k},$ for $t = 0$

6. A random variable x has density function given by

$$f(x) = \begin{cases} 2e^{-2x}, & x \geq 0 \\ 0, & x < 0 \end{cases}$$

Obtain the moment generating function. **Ans.** $\dfrac{2}{2-t},$ if $t < 2$

7. Find the moment generating function, if it exists, given the probability distribution frequency :

$$f(x) = \begin{cases} xe^{-x}, & x > 0 \\ 0, & \text{otherwise} \end{cases}$$

Ans. $\dfrac{1}{1-t}$

8. Find the moment generating function for the given distribution

$$f(x) = \begin{cases} \dfrac{1}{b-a}, & \text{for } a < x < b \\ 0, & \text{otherwise} \end{cases}$$

Ans. $\dfrac{e^{bt} - e^{at}}{t(b-a)}$

9. A random variable x has the probability distribution function

$$f(x) = \begin{cases} \dfrac{1}{2x}, & \text{for } x = 1, 2, 3........ \\ 0, & \text{otherwise} \end{cases}$$

Ans. $\dfrac{e^t}{2-e^t}$

10. The first four moments of a distribution about the value '0' are –0.20, 1.76, –2.36 and 10.88. Find the moments about the mean and measure the kurtosis. (U.P., III Semester, Dec. 2009)

Hint: In the usual notation, we are given assumed mean $a = 0$ and

$$\mu_1' = -0.20, \quad \mu_2' = 1.76, \quad \mu_3' = -2.36, \quad \mu_4' = 10.88$$

Moment about mean

$$\mu_1 = \mu_1' = -0.20$$

$$\mu_2 = \mu_2' - \mu_1'^2 = 1.76 - (0.20)^2 = 1.76 - 0.04$$

$$\Rightarrow \quad \mu_2 = 1.72$$

$$\mu_3 = \mu_3' - 3\mu_2'\mu_1' + 22\mu_1'^3$$

$$= -2.36 - 3(1.76)(-0.20) + 2(-0.20)^3 = -2.36 + 1.056 - 0.016 = -1.32$$

$$\mu_4 = \mu_4' - 4\mu_3'\mu_1' + 6\mu_2'\mu_1'^2 - 3\mu_1'^4$$

$$= 10.88 - 4(-2.36)(-0.20) + 6(1.76)(-0.20)^2 - 3(-0.20)^4$$

$$= 10.88 - 1.888 + 0.4224 - 0.0048$$

$$= 9.4096$$

Kurtosis $= \beta_2 = \dfrac{\mu_4}{\mu_2^2} = \dfrac{9.4096}{(1.72)^2} = \dfrac{9.4096}{2.9584} = 3.180638188$

Here, $\beta_2 > 3$, so the curve is peaked or Leptokurtic. **Ans.**

METHOD OF LEAST SQUARES

58.1 PRINCIPLE OF LEAST SQUARE

The method of least squares is probably the most systematic procedure to fit a unique curve through the given points.

Let $y=f(x)$ be the equation of curve to be fitted to the given data (observed or experimental) points $(x_1, y_1), (x_2, y_2) .... (x_n, y_n)$. At $x = x_i$, the observed (or experimental) value of the ordinate is y_i and the corresponding value on the fitting curve is $N_1 M_1$, *i.e.*, $[f(x_1)]$. The difference of the observed and the expected (theoretical) value is

$$= P_1 M_1 - N_1 M_1 = P_1 N_1 = e_1.$$

This difference is called the error.

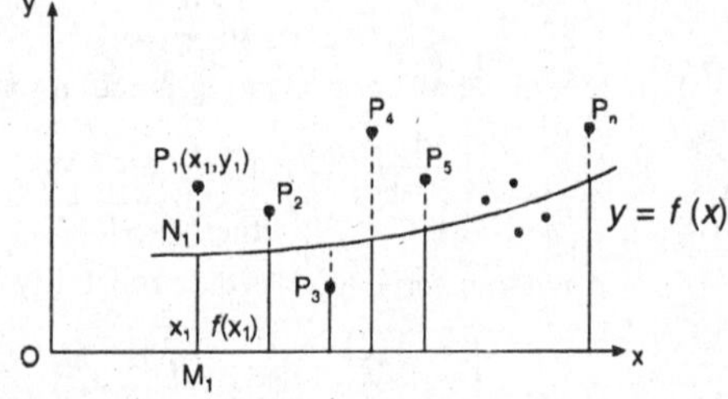

$$e_1 = y_1 - f(x_1)$$

Similarly, $$e_2 = y_2 - f(x_2)$$

$$e_3 = y_3 - f(x_3)$$

....

$$e_n = y_n - f(x_n)$$

Some of the errors $e_1, e_2, e_3, ... e_n$ will be positive and others negative.

In finding the total errors, errors are added. In addition, some negative and some positive errors may cancel and in some cases sum of all the errors may be zero, which leads to false result. To avoid such situation, we may make all the errors positive by squaring.

$$\text{Sum} = S = e_1^2 + e_2^2 + e_3^2 + ... + e_n^2$$

The curve of the best fit is that for which the sum of the squares of errors (S) is minimum. This is called the principle of least squares.

58.2 METHOD OF LEAST SQUARES

Let $$y = a + bx \qquad ...(1)$$

be the straight line to be fitted to the given data points $(x_1, y_1), (x_2, y_2) ...... (x_n, y_n)$.

Let y_{t_1} be the theoretical ordinate for x_1.

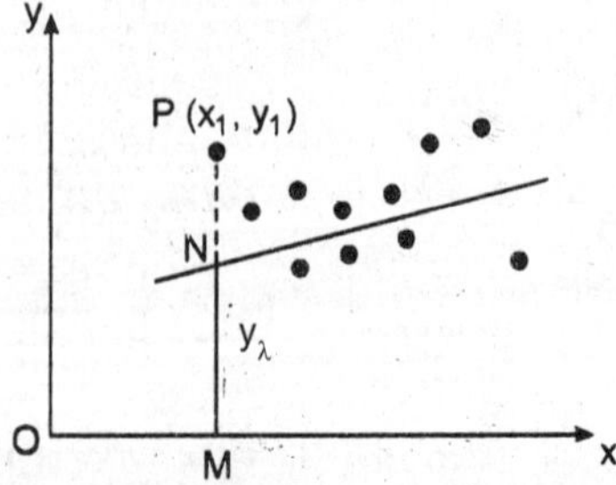

$$PM = y_1 \qquad NM = y_{t_1}$$

$$PN = PM - NM$$

Then $$e_1 = y_1 - y_{t_1} \qquad (PN = e_1)$$

$$e_1 = y_1 - (a + bx_1) \qquad (y_{t_1} = a + bx_1)$$

On squaring, we get $e_1^2 = (y_1 - a - bx_1)^2$

$$S = e_1^2 + e_2^2 + e_3^2 + ...e_n^2 = \Sigma e^2{}_i$$

$$S = \sum_{i=1}^{n} (y_i - a - bx_i)^2$$

For S to be minimum

$$\frac{\partial S}{\partial a} = \sum_{i=1}^{n} 2(y_i - a - bx_i)(-1) = 0 \text{ or } \Sigma (y - a - bx) = 0 \quad ...(2)$$

[To generalise y_i, y is written as y]

$$\frac{\partial S}{\partial b} = \sum_{i=1}^{n} 2 (y_i - a - bx_i)(-x_i) \text{ or } \Sigma (xy - ax - bx^2) = 0 \quad ...(3)$$

On simplification equations (2) and (3) become

$$\Sigma y = na + b \Sigma x \quad ...(4)$$

$$\Sigma xy = a \Sigma x + b \Sigma x^2 \quad ...(5)$$

The equations (4) and (5) are known as Normal equations.

On solving equations (4) and (5), we get the values of a and b.

On putting the values of a and b in (1), we get the equation of required line.

To Remember : The normal equations (4) and (5) are for

$$y = a + bx$$

(i) Equation (4) is obtained by putting Σ before all the terms on both sides of (1).

i.e., $\Sigma y = \Sigma a + \Sigma bx \Rightarrow \Sigma y = na + b\Sigma x$

(ii) Equation (5) is obtained on multiplying equation (1) by x and putting Σ before each obtained term on both the sides.

i.e.,

$$\Sigma xy = \Sigma ax + \Sigma bx^2$$

$$\Sigma xy = a \Sigma x + b \Sigma x^2$$

Example 1. *Find the best values of a and b so that $y = a + bx$ fits the data given in the table.*

x	0	1	2	3	4
y	1.0	2.9	4.8	6.7	8.6

Solution. $y = a + bx$ ...(1)

x	y	xy	x^2
0	1.0	0	0
1	2.9	2.9	1
2	4.8	9.6	4
3	6.7	20.1	9
4	8.6	34.4	16
$\Sigma x = 10$	$\Sigma y = 24.0$	$\Sigma xy = 67.0$	$\Sigma x^2 = 30$

Normal equations are

$$\Sigma y = na + b \Sigma x \quad ...(2)$$

$$\Sigma xy = a \Sigma x + b \Sigma x^2 \quad ...(3)$$

On putting the values of Σx, Σy, Σxy, Σx^2 in (2) and (3), we have

$$24 = 5a + 10\,b \quad ...(4)$$

$$67 = 10a + 30b \quad ...(5)$$

On solving (4) and (5), we get

$$a = 1, \quad b = 1.9$$

On substituting the values of a and b in (1), we get

$$y = 1 + 1.9\,x$$ **Ans.**

Example 2. *By the method of least squares, find the straight line that best fits the following data :*

x	*1*	*2*	*3*	*4*	*5*
y	*14*	*27*	*40*	*55*	*68*

Solution. Let the equation of the straight line best fit be $y = a + bx$ ... (1)

x	y	xy	x^2
1	14	14	1
2	27	54	4
3	40	120	9
4	55	220	16
5	68	340	25
$\Sigma x = 15$	$\Sigma y = 204$	$\Sigma xy = 748$	$\Sigma x^2 = 55$

Here, $n = 5$

Normal equations are

$$\Sigma y = na + b\,\Sigma x \quad \text{... (2)}$$

$$\Sigma xy = a\,\Sigma x + b\,\Sigma x^2 \quad \text{... (3)}$$

On putting the values of Σx, Σy, Σxy and Σx^2 in (2) and (3), we have

$$204 = 5a + 15\,b \quad \text{... (4)}$$

$$748 = 15a + 55b \quad \text{... (5)}$$

On solving equations (4) and (5), we get

$$a = 0, \quad b = 13.6$$

On substituting the values of a and b in (1), we get

$$y = 13.6\,x$$ **Ans..**

Example 3. *Use least-squares method to fit a curve of the form* $y = ae^{bx}$ *to the data :*

x	*1*	*2*	*3*	*4*	*5*	*6*
y	*7.209*	*5.265*	*3.846*	*2.809*	*2.052*	*1.499*

Solution.

$$y = ae^{bx} \quad \text{... (1)}$$

On taking log of both sides, we get

$$\log_e y = \log_e a + bx \quad \text{... (2)}$$

On putting $\log_e y = Y$, $\quad \log_e a = c \quad$ in (2), we get

$$Y = c + bx \quad \text{... (3)}$$

x	y	$Y = \log_e y$	xY	x^2
1	7.209	1.97533	1.97533	1
2	5.265	1.66108	3.32216	4
3	3.846	1.34703	4.04109	9
4	2.809	1.03283	4.13132	16
5	2.052	0.71881	3.59405	25
6	1.499	0.40480	2.4288	36
$\Sigma x = 21$		$\Sigma Y = 7.13988$	$\Sigma xY = 19.49275$	$\Sigma x^2 = 91$

Normal equations are $\Sigma Y = nc + b\Sigma x$ ... (4)

$$\Sigma xY = c\Sigma x + b\Sigma x^2 \quad ...(5)$$

On putting the values of n, Σx, ΣY, $\Sigma x Y$ and Σx^2 in equations (4) and (5), we get

$$7.13988 = 6c + 21b \quad ...(6)$$

$$19.49275 = 21c + 91b \quad ...(7)$$

On solving (6) and (7), we obtain $b = -0.3141$, $c = 2.28933$

$$c = \log_e a = 2.28933 \quad \Rightarrow \quad a = 9.86832$$

On substituting the values of a and b in (1), we get

$$y = 9.86832\, e^{-0.3141x}$$ **Ans.**

58.3 CHANGE OF ORIGIN AND SCALE

In some problems the magnitude of the variables in the given data is so large that the calculation becomes very tedious. The size of the data can be reduced by assuming some origin for x, y series.

The problem is further simplified by taking suitable scale for the values of x and y. If these values are equally spaced.

Let z be the width of the interval and (x_0, y_0) be taken as origin. Then putting

$$u = \frac{x - x_0}{h} \quad \text{and } v = y - y_0$$

Example 4. *Show that the line of fit to the following data is given by* $y = 0.7x + 11.285$:

x	0	5	10	15	20	25
y	12	15	17	22	24	30

Solution. Let $x_0 = 12.5$, $h = 2.5$, $y_0 = 20$ $\left[x_0 = \frac{\Sigma x}{N}\right.$, $\left.y_0 = \frac{\Sigma y}{N}\right]$

$$u = \frac{x - 12.5}{2.5} \quad \text{and } v = y - 20$$

$\Rightarrow$ The transformed equation is $v = a + bu$ ... (1)

x	y	$u = \frac{x-12.5}{2.5}$	$v = y - 20$	uv	u^2
0	12	– 5	– 8	40	25
5	15	– 3	– 5	15	9
10	17	– 1	– 3	3	1
15	22	1	2	2	1
20	24	3	4	12	9
25	30	5	10	50	25
		$\Sigma u = 0$	$\Sigma v = 0$	$\Sigma uv = 122$	$\Sigma u^2 = 70$

Normal equations are $\Sigma v = na + b\Sigma u$ ... (2)

$$\Sigma uv = a\Sigma u + b\Sigma u^2 \quad ...(3)$$

On putting the values of Σu, Σv, Σuv, Σu^2 in (2) and (3), we get

$$0 = 6a + 0 \quad \Rightarrow \quad a = 0$$

$$122 = a \times 0 + b \times 70 \quad \Rightarrow \quad b = \frac{122}{70} = 1.743$$

Putting the values of a and b in (1), we get

$$v = 1.743\,u \qquad \text{... (4)}$$

Putting $u = \dfrac{x-12.5}{2.5}$ and $v = y - 20$ in (4), we get

$$y - 20 = 1.743\left(\frac{x-12.5}{2.5}\right)$$

$$\Rightarrow \quad 2.5y - 50 = 1.743x - 1.743 \times 12.5$$

$$\Rightarrow \quad 2.5y = 1.743x - 21.7875 + 50$$

$$\Rightarrow \quad 2.5y = 1.743x + 28.2125$$

$$\Rightarrow \quad y = 0.7x + 11.285$$

Ans.

Example 5. *Fit a straight line to the following data :*

x	71	68	73	69	67	65	66	67
y	69	72	70	70	68	67	68	64

Solution. $\quad y = a + bx \qquad \text{... (1)}$

$$u = x - 69 \text{ and } v = y - 68$$

Transformed Equation is $\quad v = a + bu \qquad \text{... (2)}$

x	y	$u = x - 69$	$v = y - 68$	uv	u^2
71	69	2	1	2	4
68	72	– 1	4	– 4	1
73	70	4	2	8	16
69	70	0	2	0	0
67	68	– 2	0	0	4
65	67	– 4	– 1	4	16
66	68	– 3	0	0	9
67	64	– 2	– 4	8	4
		$\Sigma u = -6$	$\Sigma v = 4$	$\Sigma uv = 18$	$\Sigma u^2 = 54$

Normal equations are $\quad \Sigma v = na + b\,\Sigma u \qquad \text{... (3)}$

$$\Sigma uv = a\,\Sigma u + b\,\Sigma u^2 \qquad \text{... (4)}$$

On putting the values of Σu, Σv, Σuv, Σu^2 in (3) and (4), we get

$$4 = 8a + b(-6) \qquad \text{... (5)}$$

$$18 = -6a + 54b \qquad \text{... (6)}$$

On solving (5) and (6), we get

$$a = \frac{9}{11}, \quad b = \frac{14}{33}$$

On putting the values of a and b in (2), we get

$$v = \frac{9}{11} + \frac{14}{33}u \qquad \text{... (7)}$$

On putting $u = x - 69$ and $v = y - 68$ in (7), we get

$$y - 68 = \frac{9}{11} + \frac{14}{33}(x - 69)$$

$$y - 68 = 0.8182 + 0.4242\,x - 29.2727$$

$$y = 68.8182 - 29.2727 + 0.4242\,x$$

$$y = 39.5455 + 0.4242\,x$$

Ans.

EXERCISE 58.1

1. Find the linear least square polynomials based on data

x	-2	-1	0	1
y	6	3	2	2

is given. Find the least square straight line approximation to the data. **Ans.** $10y = 26 - 13x$

2. The following table shows the number of salesmen working for a certain concern :

Year	1998	1999	2000	2001	2002	2003
Number	28	38	46	40	56	60

Use the method of least squares to fit a straight line trend. **Ans.** $y = 5.9428x - 11843.9047$

3. If F is pull required to lift a load W by means of a pulley, fit a linear law $F = a - bw$ connecting F and W against the following data:

W	50	70	100	120
F	12	15	21	25

Ans. $F = 2.2785 - 0.1879W$

4. Determine the constants a and b by the least-squares method such that $y = ae^{bx}$ fits the following data :

x	1.0	1.2	1.4	1.6
y	40.170	73.196	133.372	243.02

Ans. $a = 2,\ b = 3$

5. Fit a least-square geometric curve $y = ax^b$ to the following data :

x	1	2	3	4	5
y	0.5	2	4.5	8	12.5

Ans. $a = 0.5012,\ b = 1.9977$

6. The pressure and volume of a gas are related by the equation $PV^r = k$, r and k being constants. Fit this equation to the following set of observations :

P (kg/cm^2)	0.5	1.0	1.5	2.0	2.5	3.0
V (litres)	1.62	1.00	0.75	0.62	0.52	0.46

Ans. $PV^{1.276} = 1.039$

7. The pressure of the gas corresponding to various volumes V is measured, given by the following data :

V (cm^3)	50	60	70	90	100
P (kg cm^{-2})	64.7	51.3	40.5	25.9	78

Fit the data to the equation $PV^r = C$ **Ans.** $PV^{0.28997} = 167.78765$

58.4 TO FIT UP THE PARABOLA

Let $$y = a + bx + cx^2 \qquad \text{... (1)}$$

be the equation of a parabola.

The following normal equations are obtained as in Art. 58.2

The normal equations are $$\Sigma y = na + b\,\Sigma x + c\,\Sigma x^2 \qquad \text{... (2)}$$

$$\Sigma xy = a\,\Sigma x + b\,\Sigma x^2 + c\,\Sigma x^3 \qquad \text{... (3)}$$

$$\Sigma x^2 y = a\,\Sigma x^2 + b\,\Sigma x^3 + c\,\Sigma x^4 \qquad \text{... (4)}$$

On solving these three normal equations, we get the values of a, b and c.

On putting the values of a, b and c in (1), we get the required equation of parabola.

To remember the normal equations (2), (3) and (4) for $y = a + bx + cx^2$.

(*i*) Equation (2) is obtained by putting Σ before each term on both sides of (1).

(*ii*) Equation (3) is obtained on multiplying (1) by x and putting Σ before each term on both sides of obtained equation.

(*iii*) Equation (4) is obtained on multiplying (1) by x^2 and putting Σ before each term on both sides of obtained equation.

Example 6. *Find least squares polynomial approximation of degree two to the data:*

x	0	1	2	3	4
y	– 4	– 1	4	11	20

Also compute the least error.

Solution. Let the equation of the polynomial be

$$y = a + bx + cx^2 \qquad \text{... (1)}$$

x	y	xy	x^2	x^2y	x^3	x^4
0	– 4	0	0	0	0	0
1	–1	–1	1	–1	1	1
2	4	8	4	16	8	16
3	11	33	9	99	27	81
4	20	80	16	320	64	256
$\Sigma x = 10$	$\Sigma y = 30$	$\Sigma xy = 120$	$\Sigma x^2 = 30$	$\Sigma x^2y = 434$	$\Sigma x^3 = 100$	$\Sigma x^4 = 354$

Normal equations are $\Sigma y = na + b\Sigma x + c\Sigma x^2$ ... (2)

$$\Sigma xy = a\Sigma x + b\Sigma x^2 + c\Sigma x^3 \qquad \text{... (3)}$$

$$\Sigma x^2 y = a\Sigma x^2 + b\Sigma x^3 + c\Sigma x^4 \qquad \text{... (4)}$$

On putting the values of Σx, Σy, Σxy, Σx^2, Σx^2y, Σx^3, Σx^4 in equations (2), (3), (4), we obtain

$$30 = 5a + 10b + 30c \qquad \text{... (5)}$$

$$120 = 10a + 30b + 100c \qquad \text{... (6)}$$

$$434 = 30a + 100b + 354c \qquad \text{... (7)}$$

On solving these equations, we get $a = -4$, $b = 2$, $c = 1$.

The required polynomial is

$$y = -4 + 2x + x^2, \quad \text{Error} = 0$$ **Ans.**

Example 7. *Employ the method of least squares to fit a parabola $y = a + bx + cx^2$ in the following data :*

(x, y) : (–1, 2), (0, 0), (0, 1), (1, 2)

Solution. Let the equation of the parabola be

$$y = a + bx + cx^2 \qquad \text{... (1)}$$

Here, $n = 4$

x	y	x^2	x^3	x^4	xy	x^2y
–1	2	1	– 1	1	– 2	2
0	0	0	0	0	0	0
0	1	0	0	0	0	0
1	2	1	1	1	2	2
$\Sigma x = 0$	$\Sigma y = 5$	$\Sigma x^2 = 2$	$\Sigma x^3 = 0$	$\Sigma x^4 = 2$	$\Sigma xy = 0$	$\Sigma x^2y = 4$

Normal equations are

$$\Sigma y = na + b\,\Sigma x + c\,\Sigma x^2 \qquad \text{... (2)}$$

$$\Sigma xy = a\,\Sigma x + b\,\Sigma x^2 + c\,\Sigma x^3 \qquad \text{... (3)}$$

$$\Sigma x^2y = a\,\Sigma x^2 + b\,\Sigma x^3 + c\,\Sigma x^4 \qquad \text{... (4)}$$

On putting the values of $\Sigma y, \Sigma xy, \Sigma x^2 y$ etc. in (2), (3) and (4), we get

$$5 = 4a + 0\,b + 2\,c \quad \Rightarrow \quad 5 = 4a + 2c \qquad \text{... (5)}$$

$$0 = 0a + 2\,b + 0\,c \quad \Rightarrow \quad 0 = 2b \qquad \text{... (6)}$$

$$4 = 2a + 0\,b + 2\,c \quad \Rightarrow \quad 4 = 2a + 2c \qquad \text{... (7)}$$

On solving (5), (6) and (7), we get

$$a = 0.5, \quad b = 0, \quad c = 1.5$$

On putting these values in (1), we get

$$y = 0.5 + 1.5x^2$$ **Ans.**

Example 8. *Fit a parabola $y = ax^2 + bx + c$ to the following data taking x as independent variable.*

x	1	2	3	5	7	11	13	17	19	23 .
y	2	3	5	7	11	13	17	19	23	29

(*U.P. III Semester, 2009-2010*)

Solution. Here, we have

$$y = ax^2 + bx + c \qquad \text{... (1)}$$

x	y	xy	x^2	x^2y	x^3	x^4
1	2	2	1	2	1	1
2	3	6	4	12	8	16
3	5	15	9	45	27	81
5	7	35	25	175	125	625
7	11	77	49	539	343	2401
11	13	143	121	1573	1331	14641
13	17	221	169	2873	2197	28561
17	19	323	289	5491	4913	83521
19	23	437	361	8303	6859	130321
23	29	667	529	15341	12167	279841
$\Sigma x = 101$	$\Sigma y = 129$	$\Sigma xy = 1926$	$\Sigma x^2 = 1557$	$\Sigma x^2y = 34354$	$\Sigma x^3 = 27971$	$\Sigma x^4 = 540009$

Normal equations are

$$\Sigma y = na + b\,\Sigma x + c\,\Sigma x^2 \qquad \text{... (2)}$$

$$\Sigma xy = a\,\Sigma x + b\,\Sigma x^2 + c\,\Sigma x^3 \qquad \text{... (3)}$$

$$\Sigma x^2y = a\,\Sigma x^2 + b\,\Sigma x^3 + c\,\Sigma x^4 \qquad \text{... (4)}$$

On putting the values of $\Sigma x, \Sigma y, \Sigma xy, \Sigma x^2, \Sigma x^2y, \Sigma x^3, \Sigma x^4$, in equations (2), (3), (4), we get

$$129 = 10a + 101\,b + 1557c \quad \text{... (5)}$$
$$1926 = 101a + 1557\,b + 27971c \quad \text{... (6)}$$
$$34354 = 1557a + 27971b + 540009c \quad \text{... (7)}$$

On solving (5), (6), (7), we get

$$a = 1.41259297$$
$$b = 1.089013957$$
$$c = 0.003136583595$$

Hence, the equation of the required parabola is

$$y = 1.41259297\,x^2 + 1.089013957x + 0.003136583595$$ **Ans.**

Example 9. *Fit a second degree parabola to the following :*

x	*1*	*2*	*3*	*4*	*5*
y	*1090*	*1220*	*1390*	*1625*	*1915*

(R.G.P.V., Bhopal, III Semester, Dec. 2003)

Solution. Let the equation of the parabola be

$$y = a + bx + cx^2 \quad \text{... (1)}$$

x	y	xy	x^2	x^2y	x^3	x^4
1	1090	1090	1	1090	1	1
2	1220	2440	4	4880	8	16
3	1390	4170	9	12510	27	81
4	1625	6500	16	26000	64	256
5	1915	9575	25	47875	125	625
$\Sigma x = 15$	$\Sigma y = 7240$	$\Sigma xy = 23775$	$\Sigma x^2 = 55$	$\Sigma x^2y = 92355$	$\Sigma x^3 = 225$	$\Sigma x^4 = 979$

Normal equations are
$$\Sigma y = na + b\,\Sigma x + c\,\Sigma x^2 \quad \text{... (2)}$$
$$\Sigma xy = a\,\Sigma x + b\,\Sigma x^2 + c\,\Sigma x^3 \quad \text{... (3)}$$
$$\Sigma x^2y = a\,\Sigma x^2 + b\,\Sigma x^3 + c\,\Sigma x^4 \quad \text{... (4)}$$

On putting the values of $n, \Sigma x, \Sigma x^2, \Sigma x^3, \Sigma x^4, \Sigma y, \Sigma xy, \Sigma x^2y$, in (3), (4) and (5), we get

$$7240 = 5a + 15b + 55c \quad \text{... (5)}$$
$$23775 = 15a + 55b + 225c \quad \text{... (6)}$$
$$92355 = 55a + 225b + 979c \quad \text{... (7)}$$

Steps for solution of (5), (6) and (7) are the following :-

3 (5), $\quad 21720 = 15a + 45b + 165c \quad$ ... (8)

(6) – (8), $\quad 2055 = 10b + 60c \quad$ ... (9)

11 (5), $\quad 79640 = 55a + 165b + 605c \quad$ ... (10)

(7) – (10), $\quad 12715 = 60b + 374c \quad$ ... (11)

6 (9), $\quad 12330 = 60b + 360c \quad$ ... (12)

(11) – (12), $\quad 385 = 14c \quad \Rightarrow \quad c = \dfrac{55}{2}$

From (9), $\quad 2055 = 10b + 60\left(\dfrac{55}{2}\right) \quad \Rightarrow \quad b = \dfrac{81}{2}$

From (5), $7240 = 5a + 15\left(\frac{81}{2}\right) + 55\left(\frac{55}{2}\right) \Rightarrow a = 1024$

On putting the values of a, b, c in (1), we get

$$y = 1024 + \frac{81}{2}x + \frac{55}{2}x^2$$

The equation of the required parabola is

$$2y = 2048 + 81x + 55x^2$$ **Ans.**

Example 10. *Fit a second degree parabola to the following data :*

x	10	15	20	25	30	35	40
y	11	13	16	20	27	34	41

(*R.G.P.V., Bhopal, III Semester, June 2005*)

Solution. Let the equation of the parabola be

$$y = a + bx + cx^2 \quad \text{... (1)}$$

x	y	xy	x^2	x^2y	x^3	x^4
10	11	110	100	1100	1000	10000
15	13	195	225	2925	3375	50625
20	16	320	400	6400	8000	160000
25	20	500	625	12500	15625	390625
30	27	810	900	24300	27000	810000
35	34	1190	1225	41650	42875	1500625
40	41	1640	1600	65600	64000	2560000
$\Sigma x = 175$	$\Sigma y = 162$	$\Sigma xy = 4765$	$\Sigma x^2 = 5075$	$\Sigma x^2y = 154475$	$\Sigma x^3 = 161875$	$\Sigma x^4 = 5481875$

Normal equations are

$$\Sigma y = na + b\,\Sigma x + c\,\Sigma x^2 \quad \text{... (2)}$$

$$\Sigma xy = a\,\Sigma x + b\,\Sigma x^2 + c\,\Sigma x^3 \quad \text{... (3)}$$

$$\Sigma x^2y = a\,\Sigma x^2 + b\,\Sigma x^3 + c\,\Sigma x^4 \quad \text{... (4)}$$

On putting the values of $n, \Sigma x, \Sigma x^2, \Sigma x^3, \Sigma x^4, \Sigma y, \Sigma xy, \Sigma x^2y$, in the equation (2), (3) and (4), we get

$$162 = 7a + 175b + 5075c \quad \text{... (5)}$$

$$4765 = 175a + 5075b + 161875c \quad \text{... (6)}$$

$$154475 = 5075a + 161875b + 5481875c \quad \text{... (7)}$$

$25 \times (5)$, $4050 = 175a + 4375b + 126875c$ ... (8)

$(6) - (8)$, $715 = 700b + 35000c$ ... (9)

$29 \times (6)$, $138185 = 5075a + 147175b + 4694375c$ ... (10)

$(7) - (10)$, $16290 = 14700b + 787500c$ ... (11)

$21 \times (9)$, $15015 = 14700b + 735000c$ ... (12)

$(11) - (12)$, $1275 = 52500c \Rightarrow c = \frac{17}{700}$

From (9), $715 = 700b + 35000\left(\frac{17}{700}\right) \Rightarrow b = -\frac{135}{700}$

From (5), $162 = 7a + 175\left(\frac{-135}{700}\right) + 5075\left(\frac{17}{700}\right)$

$$7a = 162 + \frac{135}{4} - \frac{493}{4} \Rightarrow a = \frac{145}{14}$$

On putting the values of a, b and c in (1), we get

$$y = \frac{145}{14} - \frac{135x}{700} + \frac{17x^2}{700}$$

Hence, the required parabola is

$$700y = 7250 - 135x + 17x^2$$ **Ans.**

58.5 CHANGE OF SCALE IN SECOND DEGREE EQUATIONS

If the data is of equal interval in large numbers then we change the scale as

$$u = \frac{x - x_0}{h} \text{ and } v = y - y_0.$$

Example 11. *Fit a second degree parabola to the following data by least squares method :*

x	1929	1930	1931	1932	1933	1934	1935	1936	1937
y	352	356	357	358	360	361	361	360	359

(U.P., II Semester, Summer 2001)

Solution. Taking $x_0 = 1933$, $y_0 = 357$

Again taking $u = x - x_0$, $v = y - y_0$

$u = x - 1933$, $v = y - 357$

The equation $y = a + bx + cx^2$ is transformed to $v = A + Bu + Cu^2$

x	$u = x - 1933$	y	$v = y - 357$	uv	u^2	u^2v	u^3	u^4
1929	– 4	352	– 5	20	16	– 80	– 64	256
1930	– 3	356	– 1	3	9	– 9	– 27	81
1931	– 2	357	0	0	4	0	– 8	16
1932	– 1	358	1	– 1	1	1	– 1	1
1933	0	360	3	0	0	0	0	0
1934	1	361	4	4	1	4	1	1
1935	2	361	4	8	4	16	8	16
1936	3	360	3	9	9	27	27	81
1937	4	359	2	8	16	32	64	256
Total	$\Sigma u = 0$		$\Sigma v = 11$	$\Sigma uv = 51$	$\Sigma u^2 = 60$	$\Sigma u^2v = -9$	$\Sigma u^3 = 0$	$\Sigma u^4 = 708$

Normal equations are

$$\Sigma v = nA + B\Sigma u + C\Sigma u^2 \Rightarrow 11 = 9A + 0B + 60C \Rightarrow 11 = 9A + 60C$$

$$\Sigma uv = A\Sigma u + B\Sigma u^2 + C\Sigma u^3 \Rightarrow 51 = 0A + 60B + 0C \Rightarrow 51 = 60B \Rightarrow B = \frac{17}{20}$$

$$\Sigma u^2v = A\Sigma u^2 + B\Sigma u^3 + C\Sigma u^4 \Rightarrow -9 = 60A + 0B + 708C \Rightarrow -9 = 60A + 708C$$

On solving these equations, we get

$$A = \frac{694}{231}, B = \frac{17}{20}, C = -\frac{247}{924}$$

$$v = \frac{694}{231} + \frac{17}{20}u - \frac{247}{924}u^2$$

Putting $v = y - 357$ and $u = x - 1933$, we get

$$y - 357 = \frac{694}{231} + \frac{17}{20}(x-1933) - \frac{247}{924}(x-1933)^2$$

$$\Rightarrow \quad y - 357 = \frac{694}{231} + \frac{17x}{20} - \frac{32861}{20} - \frac{247x^2}{924} - \frac{247}{924}(-3866x) - \frac{247}{924} \times (1933)^2$$

$$\Rightarrow \quad y - 357 = \frac{694}{231} - \frac{32861}{20} - \frac{247}{924}(1933)^2 + \frac{17}{20}x + \frac{247 \times 3866}{924}x - \frac{247}{924}x^2$$

$$\Rightarrow \quad \mathbf{y = 3 - 1643.05 - 998823.36 + 357 + 0.85x + 1033.44x - 0.267x^2}$$

$$y = -1000106.41 + 1034.29x - 0.267x^2$$ **Ans.**

Example 12. *Fit a parabolic curve of regression of y on x to the following data :*

x	1.0	1.5	2.0	2.5	3.0	3.5	4.0
y	1.1	1.3	1.6	2.0	2.7	3.4	4.1

Solution. Here, $n = 7$ (odd)

$$u = \frac{x-2.5}{0.5} = 2x-5$$

Let the equation be $y = a + bx + cx^2$

The transformed equation is

$$y = a + bu + cu^2 \quad \text{... (1)}$$

x	y	u	u^2	uy	u^2y	u^3	u^4
1.0	1.1	− 3	9	− 3.3	9.9	− 27	81
1.5	1.3	− 2	4	− 2.6	5.2	− 8	16
2.0	1.6	− 1	1	− 1.6	1.6	− 1	1
2.5	2.0	0	0	0	0	0	0
3.0	2.7	1	1	2.7	2.7	1	1
3.5	3.4	2	4	6.8	13.6	8	16
4.0	4.1	3	9	12.3	36.9	27	81
Total	$\Sigma y = 16.2$	$\Sigma u = 0$	$\Sigma u^2 = 28$	$\Sigma uy = 14.3$	$\Sigma u^2y = 69.9$	$\Sigma u^3 = 0$	$\Sigma u^4 = 196$

Normal equations are

$$\Sigma y = na + b\Sigma u + c\Sigma u^2 \quad \text{... (2)}$$

$$\Sigma uy = a\Sigma u + b\Sigma u^2 + c\Sigma u^3 \quad \text{... (3)}$$

$$\Sigma u^2y = a\Sigma u^2 + b\Sigma u^3 + c\Sigma u^4 \quad \text{... (4)}$$

On putting the values of Σu, Σy, Σuy etc. in (2), (3) and (4), we get

$$16.2 = 7a + 0 \times b + 28c \qquad 16.2 = 7a + 28c$$

$$14.3 = 0 \times a + 28b + 0 \times c \quad \Rightarrow \quad 14.3 = 28b$$

$$69.9 = 28a + 0 \times b + 196c \quad \Rightarrow \quad 69.9 = 28a + 196c$$

On solving the above equations, we get

$$a = 2.07, b = 0.511, c = 0.061$$

On putting the values of a, b, c in (1), we get

$$y = 2.07 + 0.511\,u + 0.061\,u^2 \quad \text{... (5)}$$

On putting the value of $u = 2x - 5$ in (5), we get

$$y = 2.07 + 0.511\,(2x - 5) + 0.061\,(2x - 5)^2$$

$$y = 2.07 + 1.022x - 2.555 + 0.061\,(4x^2 - 20x + 25)$$

$$\Rightarrow \quad y = 2.07 + 1.022x - 2.555 + 0.244x^2 - 1.22x + 1.525$$

$$\Rightarrow \quad y = 1.04 - 0.198\,x + 0.244\,x^2$$

Ans.

Example 13. *Fit a second degree parabola to the following data :*

x	*1*	*2*	*3*	*4*	*5*	*6*	*7*	*8*	*9*	*10*
y	*124*	*129*	*140*	*159*	*228*	*289*	*315*	*302*	*263*	*210*

(*U.P., III Semester, Dec. 2009*)

Solution. Taking $x_0 = 6$ and $y_0 = 228$

x	$u = x - 6$	y	$v = y - 228$	uv	u^2	$u^2 v$	u^3	u^4
1	– 5	124	– 104	520	25	– 2600	– 125	625
2	– 4	129	– 99	396	16	– 1584	– 64	256
3	– 3	140	– 88	264	9	– 792	– 27	81
4	– 2	159	– 69	138	4	– 276	– 8	16
5	– 1	228	0	0	1	0	– 1	1
6	0	289	61	0	0	0	0	0
7	1	315	87	87	1	87	1	1
8	2	302	74	148	4	296	8	16
9	3	263	35	105	9	315	27	81
10	4	210	– 18	– 72	16	– 288	64	256
Total	– 5		– 121	1586	85	– 4842	–125	1333

Let the equation of parabola be $a + bu + cu^2 = v$

Normal equations are

$$\Sigma v = na + b\Sigma u + c\Sigma u^2 \quad \Rightarrow \quad -121 = 10a - 5b + 85c \quad \text{... (1)}$$

$$\Sigma uv = a\Sigma u + b\Sigma u^2 + c\Sigma u^3 \quad \Rightarrow \quad 1586 = -5a + 85b - 125c \quad \text{... (2)}$$

$$\Sigma u^2 v = a\Sigma u^2 + b\Sigma u^3 + c\Sigma u^4 \quad \Rightarrow \quad -4842 = 85a - 125b + 1333c \quad \text{... (3)}$$

$$(1) + 2\,(2) \quad \Rightarrow \quad 3051 = 165b - 165c \quad \text{... (4)}$$

$$(3) + 17\,(2) \quad \Rightarrow \quad 22120 = 1320b - 792c \quad \text{... (5)}$$

$$8\,(4) - (5) \quad \Rightarrow \quad 2288 = -528c$$

$$\Rightarrow \quad c = -\frac{13}{3}$$

Putting the value of c in (4), we get

$$3051 = 165b - 165\left(-\frac{13}{3}\right)$$

$$\Rightarrow \quad 3051 = 165b + 715$$

$$\Rightarrow \quad 165b = 2336 \qquad \Rightarrow \quad b = \frac{2336}{165}$$

Putting the values of b and c in (1), we get

$$-121 = 10a - 5\left(\frac{2336}{165}\right) + 85\left(-\frac{13}{3}\right)$$

$$\Rightarrow \quad -121 = 10a - \frac{2336}{33} - \frac{1105}{3}$$

$$\Rightarrow \quad 10a = \frac{10498}{33} \qquad \Rightarrow a = \frac{5249}{165}$$

Putting the values of a, b and c in $a + bu + cu^2 = v$, we get

$$\frac{5249}{165} + \frac{2336}{165}u - \frac{13}{3}u^2 = v$$

Putting the values of $u = x - 6$ and $v = y - 228$, we get

$$\frac{5249}{165} + \frac{2336}{165}(x-6) - \frac{13}{3}(x-6)^2 = y - 228$$

$$\Rightarrow \quad \frac{5249}{165} + \frac{2336}{165}x - \frac{4672}{55} - \frac{13}{3}x^2 + 52x - 156 = y - 228$$

$$\Rightarrow \quad -\frac{13}{3}x^2 + \frac{10916}{165}x + \frac{283}{15} = y$$

$$\Rightarrow \quad y = -4.3333x^2 + 66.1576x + 18.8667 \text{ Ans.}$$

EXERCISE 58.2

1. Find the values of a, b, c so that $y = a + bx + cx^2$ is the best fit to the data :

x	0	1	2	3	4
y	1	0	3	10	21

Ans. $a = 1$, $b = -3$, $c = 2$

2. Fit a second degree parabola to the following data by Least squares method :

x	0	1	2	3	4
y	1	1.8	1.3	2.5	6.3

Ans. $y = 1.42 - 1.07x + 0.55x^2$

3. Fit a second degree parabola to the following data taking y as dependent variable :

x	1	2	3	4	5	6	7	8	9
y	2	6	7	8	10	11	11	10	9

Ans. $y = -1 + 3.55x - 0.27x^2$

4. Use the least-square method to obtain a parabola that approximates the data :

x	1.0	1.2	1.4	1.6	1.8	2
y	2.345	2.419	2.592	2.863	3.233	3.702

Ans. $y = 3.124 - 2.477x + 1.458x^2$

CHAPTER

59

CORRELATION AND REGRESSION

59.1 INTRODUCTION

A relationship may be obtained in two series. **For example**; two series relating to the heights and weights of a group of persons are given. It may be observed that weights increase with increase in heights - so that tall people are heavier than short sized people. We also know that the area A of circle of radius r is given by $A = \pi r^2$. It means larger radius will always have a larger area than a circle with smaller radius.

The intensity of light on the table decreases as the distance between source of light and table increases.

In this chapter we shall study the relationship of two series. Such a relationship is called statistical relationship.

59.2 TYPE OF DISTRIBUTION

There are two types of distributions.

(1) **Univariate Distribution.** A distribution in which there is only one variable, such as heights of students of a class.

(2) **Bivariate Distribution.** The distribution involving two variables such as heights and weights of the students of a class.

59.3 COVARIANCE

Let the corresponding values of two variables X and Y, given by ordered pairs

$(x_1, y_1), (x_2, y_2), (x_3, y_3), ... (x_n, y_n)$

Then the covariance between X and Y is denoted by cov. (X, Y).

It is defined as

$$\text{cov}\,(X, Y) = \frac{(x_1 - \bar{x})(y_1 - \bar{y}) + (x_2 - \bar{x})(y_2 - \bar{y}) + ... + (x_n - \bar{x})(y_n - \bar{y})}{n}$$

$$= \frac{1}{n}\sum_{n=1}^{n} (x_n - \bar{x})(y_n - \bar{y})$$

$\text{cov}\,(X, Y) = E\,(X\,Y) - E\,(X)\,E\,(Y)$

$E\,(XY)$, $E\,(X)$, $E\,(Y)$ are the corresponding means

Working Rule

Step I. Calculate the sums $\sum_{1}^{n} x_i$ and $\sum_{1}^{n} y_i$

Step II. Calculate the sum $\sum_{1}^{n} x_i\, y_i$ of the products of x_i and y_i.

Step III. Divide the values obtained in steps I, II by n to get $\dfrac{\Sigma x_i}{n}, \dfrac{\Sigma y_i}{n}, \dfrac{\Sigma x_i\, y_i}{n}$

Step IV. Obtain the difference $\sum_1^n \dfrac{x_i\, y_i}{n} - \left(\dfrac{\sum_i^n x_i}{n}\right)\cdot\left(\dfrac{\sum_i^n y_i}{n}\right)$ to get cov (X, Y).

Example 1. *Calculate the covariance of the following pairs of observations of two variates.*

(1, 4), (2, 2), (3, 4), (4, 8), (5, 9), (6, 12)

Solution.

$$\Sigma x_i = 1 + 2 + 3 + 4 + 5 + 6 = 21$$
$$\Sigma y_i = 4 + 2 + 4 + 8 + 9 + 12 = 39$$
$$\Sigma x_i y_i = (1 \times 4) + (2 \times 2) + (3 \times 4) + (4 \times 8) + (5 \times 9) + (6 \times 12)$$
$$= 4 + 4 + 12 + 32 + 45 + 72 = 169$$
$$\text{Cov}\,(X, Y) = \frac{\Sigma x_i\, y_i}{n} - \frac{\Sigma x_i}{n}\,\frac{\Sigma y_i}{n} = \left[\frac{169}{6} - \frac{21}{6} \times \frac{39}{6}\right] = \frac{169}{6} - \frac{91}{4} = \frac{65}{12}$$

Ans.

Example 2. *Find the covariance of the following pairs of observations of two variates :*

(10, 35) (15, 20) (20, 30) (25, 30) (30, 35)
(35, 38) (40, 42) (45, 30) (50, 40) (55, 70)

Solution.

$$\sum_{i=1}^n x_i = 10 + 15 + 20 + 25 + 30 + 35 + 40 + 45 + 50 + 55 = 325$$
$$\sum_{i=1}^n y_i = 35 + 20 + 30 + 30 + 35 + 38 + 42 + 30 + 40 + 70 = 370$$
$$\sum_{i=1}^n x_i\, y_i = (10 \times 35) + (15 \times 20) + (20 \times 30) + (25 \times 30) + (30 \times 35) + (35 \times 38)$$
$$+ (40 \times 42) + (45 \times 30) + (50 \times 40) + (55 \times 70)$$
$$= 350 + 300 + 600 + 750 + 1050 + 1330 + 1680 + 1350 + 2000 + 3850$$
$$= 13260$$
$$\text{Cov.}\,(X, Y) = \left[\frac{\Sigma x_i y_i}{n} - \frac{\Sigma x_i}{n} \cdot \frac{\Sigma y_i}{n}\right] = \left(\frac{13260}{10} - \frac{325}{10} \cdot \frac{370}{10}\right) = 1326 - 1202.5$$
$$= 123.5$$

Ans.

59.4 CORRELATION

Whenever two variables x and y are so related that an increase in the one is accompanied by an increase or decrease in the other, then the variables are said to be correlated.

For example, the yield of crop varies with the amount of rainfall.

59.5 TYPES OF CORRELATIONS

(1) Positive correlation

If an increase in the value of one variable X results in a corresponding increase in value of other variable Y on an average.

OR

If a decrease in the value of one variable X results in a corresponding decrease in value of other variable Y on an average.

The correlation is said to be positive.

(2) Negative correlation

If the increase in the values of one variable X results in a corresponding decrease in the values of other variable Y.

OR

If the decrease in the values of one variable X results in the increase to a corresponding values of *Y*.

The correlation between X and Y is said to be negative.

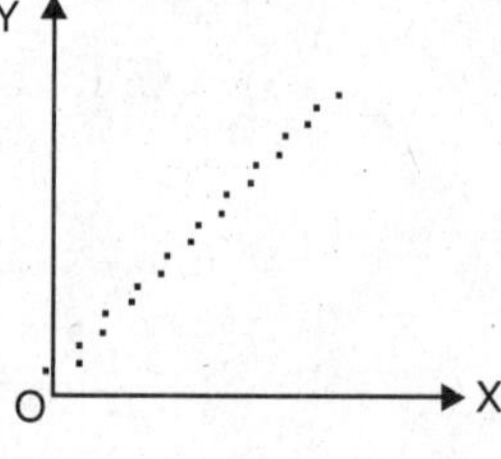

(3) Linear correlation

When all the ploted points lie approximately on a straight line, then the correlation is said to be linear correlation.

(4) Perfect correlation

If the deviation of one variable *X* is proportional to the deviation in other variable *Y*, then the correlation is said to be perfect.

In this case the plotted points on a graph lie exactly on a straight line.

4 (a) Positive perfect correlation

If increase in one variable *X* is proportional to the increase in the other variable *Y*. The graph will be exactly straight line.

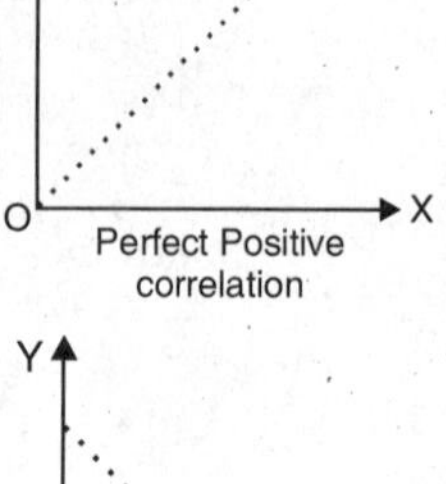

Perfect Positive correlation

4 (*b*) Negative perfect correlation

If increase in one variable is proportional to the decrease in the other variable. The graph will be exactly a straight line.

Perfect Correlation: If two variables vary in such a way that their ratio is always constant, then the correlation is said to be perfect.

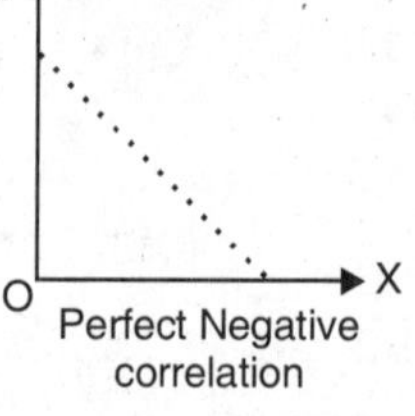

Perfect Negative correlation

59.6 SCATTER OR DOT-DIAGRAM

When we plot the corresponding values of two variables, taking one on x-axis and the other along y-axis, it shows a collection of dots.

This collection of dots is called a dot diagram or a scatter diagram.

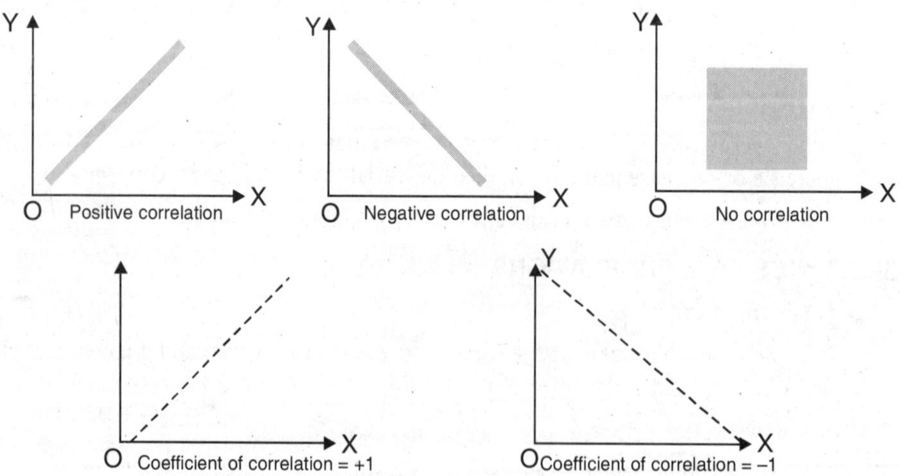

Methods of Determining Simple Correlation

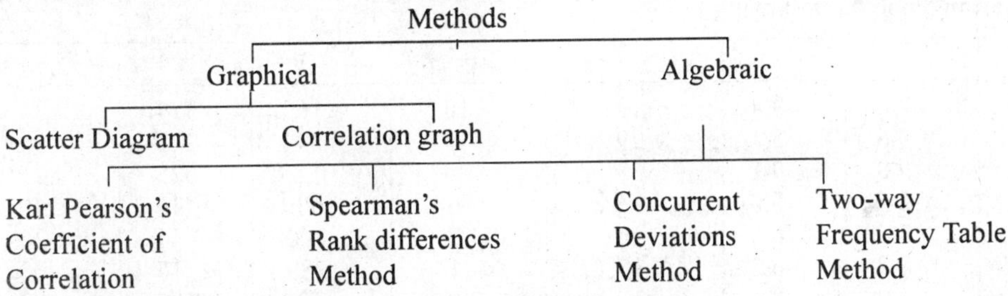

59.7 KARL PEARSON'S COEFFICIENT OF CORRELATION

r between two variables x and y is defined by the relation

$$r = \frac{\sum XY}{\sqrt{\left(\sum X^2\right)\left(\sum Y^2\right)}} = \frac{P}{\sigma_x \sigma_y} = \frac{Covariance(x, y)}{\sqrt{variance\, x}\sqrt{variance\, y}},$$

where $X = x - \bar{x}$, $Y = y - \bar{y}$

i.e. X, Y are the deviations measured from their respective means,

$$P = \left(\frac{\sum XY}{n}\right) = \text{co variance}$$

and σ_x, σ_y being the standard deviations of these series.

Example 3. *Calculate the coefficient of correlation between x and y series from the following data:*

$$\Sigma(x-\bar{x})^2 = 136, \qquad \Sigma(y-\bar{y})^2 = 138$$

$$\Sigma(x-\bar{x})(y-\bar{y}) = 122$$

Solution. Here, we have

$$\Sigma X^2 = \Sigma(x-\bar{x})^2 = 136$$

$$\Sigma Y^2 = \Sigma(y-\bar{y})^2 = 138$$

$$\Sigma XY = \Sigma(x-\bar{x})(y-\bar{y}) = 122$$

$$r = \frac{\Sigma XY}{\sqrt{\Sigma X^2}.\sqrt{\Sigma Y^2}} \qquad \ldots (1)$$

Putting the values of ΣXY, ΣX^2 & ΣY^2 in (1), we get

$$r = \frac{122}{\sqrt{136}\sqrt{138}} = \frac{122}{11.66 \times 11.75} = \frac{122}{137.005} = 0.89 \qquad \textbf{Ans.}$$

Example 4. *Ten students got the following percentage of marks in Economics and Statistics.*

Roll No.	*1*	*2*	*3*	*4*	*5*	*6*	*7*	*8*	*9*	*10*
Marks in Economics	*78*	*36*	*98*	*25*	*75*	*82*	*90*	*62*	*65*	*39*
Marks in Statistics	*84*	*51*	*91*	*60*	*68*	*62*	*86*	*58*	*53*	*47*

Calculate the coefficient of correlation.

Solution. Let the marks of two subjects be denoted by x and y respectively.

Then the mean for x marks $= \frac{650}{10} = 65$ and the mean of y marks $= \frac{660}{10} = 66$

If X and Y are deviations of x's and y's from their respective means, then the data may be arranged in the following form :

x	y	$X = x - 65$	$Y = y - 66$	X^2	Y^2	XY
78	84	13	18	169	324	234
36	51	– 29	– 15	841	225	435
98	91	33	25	10 89	625	825
25	60	– 40	– 6	1600	36	240
75	68	10	2	100	4	20
82	62	17	– 4	289	16	– 68
90	86	25	20	625	400	500
62	58	– 3	– 8	9	64	24
65	53	0	– 13	0	169	0
39	47	– 26	– 19	676	361	494
650	660	0	0	5 39 8	2224	2704

Here $\sum X^2 = 5398, \sum Y^2 = 2224, \qquad \sum XY = 2704$

$$\therefore \quad r = \frac{\sum XY}{\sqrt{\left(\sum X^2\right)\left(\sum Y^2\right)}} = \frac{2704}{\sqrt{5398 \times 2224}} = \frac{2704}{73.4 \times 47.1} = \frac{2704}{3457} = 0.78 \qquad \textbf{Ans.}$$

Example 5. *Calculate the coefficient of correlation between the marks obtained by 8 students in mathematics and statistics.*

Students	*A*	*B*	*C*	*D*	*E*	*F*	*G*	*H*
Mathematics	25	30	32	35	37	40	42	45
Statistics	08	10	15	17	20	23	24	25

(U.P. III Semester, 2009-2010)

Solution. Let the marks of two subjects be denoted by x and y respectively.

Let the assemed mean for x marks be 35 and that of for y be 17.

x	y	$X' = x - 35$	$Y' = y - 17$	X'^2	Y'^2	$X'Y'$
25	08	– 10	– 9	100	81	90
30	10	– 5	– 7	25	49	35
32	15	– 3	– 2	9	4	6
35	17	0	0	0	0	0
37	20	2	3	4	9	6
40	23	5	6	25	36	30
42	24	7	7	49	49	49
45	25	10	8	100	64	80
$N = 8$		$\Sigma X' = 6$	$\Sigma Y' = 6$	$\Sigma X'^2 = 312$	$\Sigma Y'^2 = 292$	$\Sigma X'Y' = 296$

We know that,
$$r = \frac{\frac{\Sigma X'Y'}{N} - \left(\frac{\Sigma X'}{N}\right)\left(\frac{\Sigma Y'}{N}\right)}{\sqrt{\left\{\frac{\Sigma X'^2}{N} - \left(\frac{\Sigma X'}{N}\right)^2\right\}\left\{\frac{\Sigma Y'^2}{N} - \left(\frac{\Sigma Y'}{N}\right)^2\right\}}} = \frac{\frac{296}{8} - \left(\frac{6}{8}\right)\left(\frac{6}{8}\right)}{\sqrt{\left[\left\{\frac{312}{8} - \left(\frac{6}{8}\right)^2\right\}\left\{\frac{292}{8} - \left(\frac{6}{8}\right)^2\right\}\right]}}$$

$$= \frac{\frac{583}{16}}{\sqrt{\left(\frac{615}{16}\right)\left(\frac{575}{16}\right)}} = \frac{583}{\sqrt{353625}} = \frac{583}{594.66} = 0.98039 \qquad \textbf{Ans.}$$

59.8 COEFFICIENT OF CORRELATION OF GROUPED DATA

$$r = \frac{\dfrac{\Sigma f X'Y'}{N} - \left(\dfrac{\Sigma f X'}{N}\right)\left(\dfrac{\Sigma f Y'}{N}\right)}{\sqrt{\left\{\dfrac{\Sigma f X'^2}{N} - \left(\dfrac{\Sigma f X'}{N}\right)^2\right\}}\sqrt{\left\{\dfrac{\Sigma f Y'^2}{N} - \left(\dfrac{\Sigma f Y'}{N}\right)^2\right\}}}$$

where r is the coefficient of correlation.

X' = Deviation from *assumed* mean of $x = x - a$

Y' = Deviation from *assumed* mean of $y = y - b$

N = Total number of items.

Example 6. *Find the coefficient of correlation between the age and the sum assured fron the following table.*

Sum assured in ₹.

Age – group	10,000	20,000	30,000	40,000	50,000	*No. of persons*
20 – 30	4	6	3	7	1	21
30 – 40	2	8	15	7	1	33
40 – 50	3	9	12	6	2	32
50 – 60	8	4	2	–	–	14
	17	27	32	20	4	100

Solution. Let the sum assured denoted by x and the age group by y.

$$x' = \frac{x - 30{,}000}{10{,}000}, \qquad y' = \frac{y - 45}{10}$$

$$r = \frac{\dfrac{\Sigma f X'Y'}{N} - \left(\dfrac{\Sigma f X'}{N}\right)\left(\dfrac{\Sigma f Y'}{N}\right)}{\sqrt{\left\{\dfrac{\Sigma f X'^2}{N} - \left(\dfrac{\Sigma f X'}{N}\right)^2\right\}}\sqrt{\left[\dfrac{\Sigma f Y'^2}{N} - \left(\dfrac{\Sigma f Y'}{N}\right)^2\right]}} \quad \ldots (1) \qquad [N = \Sigma f]$$

Putting the values in (1), we get

x			10,000		20,000		30,000		40,000		50,000		Row-wise			
		x'	–2		–1		0		1		2					
y		y'	f	$fX'Y'$	f	$fX'Y'$	f	$fX'Y'$	f	$fX'Y'$	f	$fX'Y'$	Σf	$f.Y'$	fY'^2	$fX'Y'$
20–30	25	–2	4	16	6	12	3	0	7	–14	1	–4	21	–42	84	+10
30–40	35	–1	2	4	8	8	15	0	7	–7	1	–2	33	–33	33	+3
40–50	45	0	3	0	9	0	12	0	6	0	2	0	32	0	0	0
50–60	55	1	8	–16	4	–4	2	0	–	0	–	0	14	14	14	–20
		Σf	17		27		32		20		4		N = 100	$\Sigma fY'$ = –61	$\Sigma fY'^2$ = 131	$\Sigma fX'Y'$ = –7
		fX'	–34		–27		0		20		8		$\Sigma fX'$ = – 33			
		fX'^2	68		27		0		20		16		$\Sigma fX'^2$ = 131			
		$fX'Y'$		4		16		0		–21		–6	$\Sigma fX'Y'$ = –7			

$$r=\frac{\dfrac{-7}{100}-\left(\dfrac{-33}{100}\right)\left(\dfrac{-61}{100}\right)}{\sqrt{\left\{\dfrac{131}{100}-\left(\dfrac{-33}{100}\right)^2\right\}}\sqrt{\left\{\dfrac{131}{100}-\left(\dfrac{-61}{100}\right)^2\right\}}}$$

Multiplying numerator and denominator by 10,000, we get

$$=\frac{100(-7)-(-33)(-61)}{\sqrt{100(131)-(-33)^2}\sqrt{100(131)-(-61)^2}}=\frac{-700-2013}{\sqrt{13100-1089}\sqrt{13100-3721}}$$

$$=\frac{-2713}{\sqrt{12011}\sqrt{9379}}=\frac{-2713}{109.59\times 96.85}=\frac{-2713}{10613.7915}=-0.2556$$

Hence, the age and sum assured are negatively correlated, *i.e.*, as age goes up the sum assured comes down. **Ans.**

Example 7. *Calculate the coefficient of correlation for the following table :*

x – age / *marks*	0 – 4	4 – 8	8 – 12	12 – 16	Total
0 – 5	7	—	—	—	7
5 – 10	6	8	—	—	14
10 – 15	—	5	3	—	8
15 – 20	—	7	2	—	9
20 – 25	—	—	—	9	9
Total	13	20	5	9	47

Solution.

		x	2		6		10		14		Row-wise			
		X'	-2		-1		0		1					
	y	Y'	f	$fX'Y'$	f	$fX'Y'$	f	$fX'Y'$	f	$fX'Y'$	Σf	fY'	fY'^2	$\Sigma fX'Y'$
0–5	2.5	-2	7	28							7	-14	28	28
5–10	7.5	-1	6	12	8	8					14	-14	14	20
10–15	12.5	0			5	0	3	0			8	0	0	0
15–20	17.5	1			7	-7	2	0			9	9	9	-7
20–25	22.5	2							9	18	9	18	36	18
	Σf		13		20		5		9		47	$\Sigma fY' = -1$	$\Sigma fY'^2 = 87$	$\Sigma fX'Y' = 59$
	fX'		-26		-20		0		9		$\Sigma fX' = -37$			
	fX'^2		52		20		0		9		$\Sigma fX'^2 = 81$			
	$fX'Y'$			40		1		0		18	$\Sigma fX'Y' = 59$			

Here. $\Sigma f X' = -37,\ \Sigma f X'^2 = 81,\ \Sigma f Y' = -1,\ \Sigma f Y'^2 = 87,\ \Sigma f X'Y' = 59$

$$r = \frac{\frac{\Sigma f X'Y'}{N} - \left(\frac{\Sigma f X'}{N}\right)\left(\frac{\Sigma f Y'}{N}\right)}{\sqrt{\frac{\Sigma f X'^2}{N} - \left(\frac{\Sigma f X'}{N}\right)^2}\sqrt{\frac{\Sigma f Y'^2}{N} - \left(\frac{\Sigma f Y'}{N}\right)^2}}$$

$$= \frac{\frac{59}{47} - \left(\frac{-37}{47}\right)\left(\frac{-1}{47}\right)}{\sqrt{\left\{\frac{81}{47} - \left(\frac{-37}{47}\right)^2\right\}}\sqrt{\left\{\frac{87}{47} - \left(\frac{-1}{47}\right)^2\right\}}} = \frac{1.255 - 0.017}{\sqrt{1.723 - 0.620}\sqrt{1.851 - 0.0005}}$$

$$= \frac{1.238}{\sqrt{1.103}\sqrt{1.8505}} = \frac{1.238}{1.05 \times 1.36} = \frac{1.238}{1.428} = 0.87$$ **Ans.**

Example 8. *A computer operator while calculating the coefficient between two variates x and y for 25 pairs of observations obtained the following constants :*

$n = 25,\ \Sigma x = 125,\ \Sigma x^2 = 650,\quad \Sigma y = 100,\ \Sigma y^2 = 460,\ \Sigma xy = 508$

It was however later discovered at the time of checking that he had copied down two pairs as (6, 14) and (8, 6) while the correct pairs were (8, 12) and (6, 8). Obtain the correct value of the correlation coefficient.

Solution. Here, corrected Σx = Incorrect $\Sigma x - (6 + 8) + (8 + 6) = 125 - 14 + 14 = 125$

Corrected Σy = Incorrect $\Sigma y - (14 + 6) + (12 + 8) = 100 - 20 + 20 = 100$

Corrected $\Sigma x^2 = 650 - (6^2 + 8^2) + (8^2 + 6^2) = 650 - 100 + 100 = 650$

Corrected $\Sigma y^2 = 460 - (14^2 + 6^2) + (12^2 + 8^2) = 460 - 232 + 208 = 436$

Corrected $\Sigma xy = 508 - [(6)(14) + (8)(6)] + (8)(12) + (6)(8)$

$= 508 - (84 + 48) + (96 + 48) = 508 - 132 + 144 = 520$

Corrected value of correlation coefficient is

$$r_{xy} = \frac{520 - \frac{125 \times 100}{25}}{\sqrt{\left[650 - \frac{(125)^2}{25}\right]\left[436 - \frac{(100)^2}{25}\right]}} = \frac{520 - 500}{\sqrt{(650 - 625)(436 - 400)}} = \frac{20}{\sqrt{25 \times 36}} = \frac{2}{3} = 0.67$$

Ans.

59. 9 SPEARMAN'S RANK CORRELATION

The coefficient of rank correlation is applied to the problems in which data cannot be measured quantitatively but qualitative assessment is possible such as beauty, honesty etc. In this case the best individual is given the rank no. 1 next rank no. 2 and so on.

59. 10 SPEARMAN'S RANK CORRELATION COEFFICIENT

$$r = 1 - \frac{6 \sum d^2}{n(n^2 - 1)}$$

Solution. Let $(x_1, y_1), (x_2, y_2) \ldots (x_n, y_n)$ be the ranks of n individuals corresponding to two characteristics.

Assuming nor two individuals are equal in either classification, each individual takes the values 1, 2, 3, ... n and hence their arithmetic means are, each

$$= \frac{\Sigma n}{n} = \frac{1}{n} \cdot \frac{n(n+1)}{2} = \frac{n+1}{2}$$

Let $x_1, x_2, x_3, \ldots x_n$ be the values of variable X and $y_1, y_2, y_3, \ldots y_n$ those of Y.

Then $$d = X - Y = \left(x - \frac{n+1}{2}\right) - \left(y - \frac{n+1}{2}\right) = x - y$$

where X and Y are deviations from the mean.

$$\sum X^2 = \sum\left(x - \frac{n+1}{2}\right)^2 = \sum x^2 - (n+1)\sum x + \sum\left(\frac{n+1}{2}\right)^2$$

$$= \frac{n(n+1)(2n+1)}{6} - \frac{(n+1)n(n+1)}{2} + n\left(\frac{n+1}{2}\right)^2 = \frac{n(n^2-1)}{12}$$

Clearly, $\sum X = \sum Y$ and $\sum X^2 = \sum Y^2$ $\therefore$ $\sum Y^2 = \frac{n(n^2-1)}{12}$

Hence $$\sum d^2 = \sum (x-y)^2 = \sum x^2 + \sum y^2 - 2\sum xy$$

$\therefore$ $$\sum XY = \frac{1}{2}\left[\frac{n(n^2-1)}{6} - \sum d^2\right] = \frac{1}{12}n(n^2-1) - \frac{1}{2}\sum d^2$$

Putting these values in $$r = \frac{\sum XY}{\sqrt{\sum X^2}\sqrt{\sum Y^2}} = \frac{\frac{1}{12}n(n^2-1) - \frac{1}{2}\sum d^2}{\frac{n(n^2-1)}{12}} = 1 - \frac{6\sum d^2}{n(n^2-1)}$$ **Ans.**

Working Rule

Step I. Assign ranks to each item of both series, if they are not given.

Step II. Calculate the difference D of ranks of X from the rank of Y and write it in a separate column.

Step III. Square the difference D and write D^2 in a separate column.

Step IV. Apply the formula to get the Rank correlation.

$$r = 1 - \frac{6\Sigma D^2}{n(n^2-1)}$$

where n is the total number of pairs of observations.

Example 9. *Compute Spearman's rank correlation coefficient r for the following data:*

Person	*A*	*B*	*C*	*D*	*E*	*F*	*G*	*H*	*I*	*J*
Rank in statistics	9	10	6	5	7	2	4	8	1	3
Rank in income	1	2	3	4	5	6	7	8	9	10

Solution.

Person	*Rank in statistics*	*Rank in income*	$d = R_1 - R_2$	d^2
A	9	1	8	64
B	10	2	8	64
C	6	3	3	9
D	5	4	1	1
E	7	5	2	4
F	2	6	–4	16
G	4	7	–3	9
H	8	8	0	0
I	1	9	–8	64
J	3	10	–7	49
				$\Sigma d^2 = 280$

$$r = 1 - \frac{6\Sigma d^2}{n(n^2-1)}, \quad r = 1 - \frac{6\times 280}{10(100-1)} = 1 - 1.697 = -0.697$$ **Ans.**

59.11 EQUAL RANKS

If there are more than one item with the same rank. The rank to the euqal items is assigned by average rank to each of these individuals.

For example; Suppose an item is repeated at the rank 5th (*i.e.*, the 5th and 6th items are having the same values then the common rank is assigned to 5th and 6th item is $\frac{5+6}{2} = 5.5$, which is the average of 5 and 6. The next rank assigned will be seven.

If an item is repeated thrice at rank 2, then the common rank assigned to each value will be $\frac{2+3+4}{3} = 3$ which is the arithmetic mean of 2, 3 and 4. Then next rank to be assigned would be 5.

To find the rank of correlation coefficient of repeated ranks, correlation factor is added to the Spearman's rank correlation formula.

59.12 CORRELATION FACTOR

In the formula of rank correlation coefficient, add the factor $\frac{m(m^2-1)}{12}$ to Σd^2, where m is the number of times an item (say a_1) is repeated. This factor is added for each repeated value in both the series.

The total number of observations is denoted by n.

The modified formula for the rank correlation coefficient is given below

$$r = 1 - \frac{6\left[\Sigma d^2 + \frac{1}{12}(m_1^3 - m_1) + \frac{1}{12}(m_2^3 - m_2) + \frac{1}{12}(m_3^3 - m_3) + \ldots\ldots\right]}{n(n^2-1)}$$

Example 10. *Obtain the ranks correlation coefficient for the following data :*

x	*68*	*64*	*75*	*50*	*64*	*80*	*75*	*40*	*55*	*64*
y	*62*	*58*	*68*	*45*	*81*	*60*	*68*	*48*	*50*	*70*

(Nagpur University, Summer 2002, Winter 2002)

Solution.

x	y	Rank in $x = x'$	Rank in $y = y'$	$d = x' - y'$	d^2
68	62	4	5	– 1	1
64	58	6	7	– 1	1
75	68	2.5	3.5	– 1	1
50	45	9	10	– 1	1
64	81	6	1	5	25
80	60	1	6	– 5	25
75	68	2.5	3.5	– 1	1
40	48	10	9	1	1
55	50	8	8	0	0
64	70	6	2	4	16
				Total	$\Sigma d^2 = 72$

Repeated Rank of x column	No. of times	Repeated Rank of y column	No. of times
2.5	$2 = m_1$	3.5	$2 = m_2$
6	$3 = m_3$		

Rank correlation coefficient

$$r = 1 - \frac{6\left[\Sigma d^2 + \frac{1}{12}(m_1^3 - m_1) + \frac{1}{12}(m_2^3 - m_2) + \frac{1}{12}(m_3^3 - m_3)\right]}{n(n^2-1)}$$

$$r = 1 - \frac{6\left[72 + \frac{1}{12}(2^3 - 2) + \frac{1}{12}(2^3 - 2) + \frac{1}{12}(3^3 - 3)\right]}{10(100-1)} = 1 - \frac{6(72 + 0.5 + 0.5 + 2)}{10(99)} = 0.545 \quad \textbf{Ans.}$$

Example 11. *Find rank correlation coefficient to the following data :*

x	65	63	67	64	68	62	70	66	68	67	69	71
y	68	66	68	65	69	66	68	65	71	67	68	70

(*Nagpur University, Summer 2005*)

Solution. Here we assign rank to the values of x and y and we have a table of the following form:

x	y	Rank in $x = x'$	Rank in $y = y'$	$d = x' - y'$	d^2
65	68	9	5.5	3.5	12.25
63	66	11	9.5	1.5	2.25
67	68	6.5	5.5	1	1
64	65	10	11.5	– 1.5	2.25
68	69	4.5	3	1.5	2.25
62	66	12	9.5	2.5	6.25
70	68	2	5.5	– 3.5	12.25
66	65	8	11.5	– 3.5	12.25
68	71	4.5	1	3.5	12.25
67	67	6.5	8	– 1.5	2.25
69	68	3	5.5	– 2.5	6.25
71	70	1	2	– 1	1
				Total	$\Sigma d^2 = 72.5$

Repeated Rank of x column	No. of times	Repeated Rank of y column	No. of times
4.5	$2 = m_1$	11.5	$2 = m_3$
6.5	$2 = m_2$	9.5	$2 = m_4$
		5.5	$4 = m_5$

The rank correlation coefficient r is given by

$$r = 1 - 6\left[\frac{\Sigma d^2 + \frac{1}{12}m_1(m_1^2 - 1) + \frac{1}{12}m_2(m_2^2 - 1) + \frac{1}{12}m_3(m_3^2 - 1) + \frac{1}{12}m_4(m_4^2 - 1) + \frac{1}{12}m_5(m_5^2 - 1)}{n(n^2-1)}\right]$$

Here $n = 12$, two x values are repeated twice so it is of the same rank. Two y values are repeated twice and one y values is repeated four times.

$$r = 1 - 6\frac{\left[72.5 + \frac{1}{12}2(2^2 - 1) + \frac{1}{12}2(2^2 - 1) + \frac{1}{12}2(2^2 - 1) + \frac{1}{12}2(2^2 - 1) + \frac{1}{12}4(4^2 - 1)\right]}{12(144-1)}$$

$$= 1 - \frac{6(72.5 + 2 + 5)}{12 \times 143} = 1 - 0.27797 = 0.722 \quad \textbf{Ans.}$$

Example 12. *Establish the formula* $\sigma_{x-y}^2 = \sigma_x^2 + \sigma_y^2 - 2r\sigma_x\sigma_y$
where r is the correlation coefficient between x and y.

Solution. We know that $\sigma_x^2 = \dfrac{\Sigma(x-\bar{x})^2}{n}$

$$\therefore \quad \sigma_{x-y}^2 = \frac{\Sigma[(x-y)-\overline{(x-y)}]^2}{n}$$

$\overline{x-y}$ = mean of $(x-y)$ series = mean of x– mean of $y = \bar{x} - \bar{y}$

$$\sigma_{x-y}^2 = \frac{\Sigma[(x-y)-(\bar{x}-\bar{y})]^2}{n} = \frac{\Sigma[(x-\bar{x})-(y-\bar{y})]^2}{n}$$

$$= \frac{\Sigma[(x-\bar{x})^2 + (y-\bar{y})^2 - 2(x-\bar{x})(y-\bar{y})]}{n}$$

$$= \frac{\Sigma(x-\bar{x})^2}{n} + \frac{\Sigma(y-\bar{y})^2}{n} - \frac{2\Sigma(x-\bar{x})(y-\bar{y})}{n} = \sigma_x^2 + \sigma_y^2 - \frac{2\Sigma(x-\bar{x})(y-\bar{y})}{n} \quad ...(1)$$

We know that $r = \dfrac{\Sigma(x-\bar{x})(y-\bar{y})}{n\sigma_x\sigma_y}$ or $\dfrac{\Sigma(x-\bar{x})(y-\bar{y})}{n} = r\sigma_x\sigma_y$

Putting this value in (1), we get

$$\sigma_{x-y}^2 = \sigma_x^2 + \sigma_y^2 - 2r\sigma_x\sigma_y$$ **Proved.**

Example 13. *If X and Y are uncorrelated random variables, find the coefficient of correlation between X + Y and X – Y.*

Solution.

Let $u = X + Y$ and $v = X - Y$

Then $r = \dfrac{\Sigma(u-\bar{u})(v-\bar{v})}{n\sigma_u\sigma_v}$

Now $u = X + Y, \bar{u} = \bar{X} + \bar{Y}$

Similarly $\bar{v} = \bar{X} - \bar{Y}$

Now $\Sigma(u-\bar{u})(v-\bar{v}) = \Sigma(X-\bar{X}+Y-\bar{Y})[(X-\bar{X})-(Y-\bar{Y})]$

$= \Sigma(x+y)(x-y) = \Sigma x^2 - \Sigma y^2 = n\sigma_x^2 - n\sigma_y^2$

Also $\sigma_u^2 = \dfrac{\Sigma(u-\bar{u})^2}{n} = \dfrac{1}{n}\Sigma[(X-\bar{X})+(Y-\bar{Y})]^2 = \dfrac{1}{n}\Sigma(x+y)^2 = \dfrac{1}{n}(\Sigma x^2 + \Sigma y^2 + 2\Sigma xy)$

$= \sigma_x^2 + \sigma_y^2$ (As X and Y are not correlated, we have $\Sigma xy = 0$)

Similarly $\sigma_v^2 = \sigma_x^2 + \sigma_y^2$

$$\therefore \quad r = \frac{\Sigma(u-\bar{u})(v-\bar{v})}{n\sigma_u\sigma_v} = \frac{n(\sigma_x^2 - \sigma_y^2)}{\sqrt{n(\sigma_x^2+\sigma_y^2)}\sqrt{n(\sigma_x^2+\sigma_y^2)}} = \frac{\sigma_x^2 - \sigma_y^2}{\sigma_x^2 + \sigma_y^2}$$ **Ans.**

EXERCISE 59.1

1. Calculate the coefficient of correlation from the data given below :

x	4	6	8	10	12
y	2	3	4	6	10

Ans. 0.95

3. Find the coefficient of correlation of the following data taking new origin of x at 70 and for y at 67:

x	67	68	64	68	72	70	69	70
y	65	66	67	67	68	69	71	73

(*A.M.I.E., Winter 2002*) **Ans.** 0.4 /2

3. Calculate Karl Pearson's coefficient of corrrelation from the following data, using 20 as working mean for price and 70 as working mean for demand.

Price	14	16	17	18	19	20	21	22	23
Demand	84	78	70	75	66	67	62	58	60

Ans. 1.044

4. The ranks of the same 16 students in two subjects A and B where as follows. Two numbers within brackets denote the ranks of the students in A and B respectively :

(1, 1), (2, 10), (3, 3), (4, 4), (5, 5), (6, 7), (7, 2), (8, 6), (9, 8), (10, 11), (11, 15), (12, 9), (13, 14), (14, 12), (15, 16), (16, 13).

Calculate the rank correlation for proficiencies of this group in subjects A and B. **Ans.** 0.8

5. Show that $E(x) = 0$, $E(x, y) = 0$ and hence deduce that the correlation between x and y is zero.

6. x and y are two random variables with the same standard deviation and correlation coefficient r. Show that the coefficient of correlation between x and $x + y$ is $\sqrt{\dfrac{1+r}{2}}$

7. Calculate the coefficients of correlation between x (Marks in Mathematics) and y (marks in Physics) given in this following data :

y \ x	10 – 40	40 – 70	70 – 100	Total
0 – 30	5	20	—	25
30 – 60	—	28	2	30
60 – 90	—	32	13	45
	5	80	15	100

Ans. 0.4517

8. Calculate from the data reproduced pertaining to 66 selected villages in Meerut district, the value of r, between 'total cultivated area' and 'the area under wheat'.

Area under wheat (in Bighas)	0 – 500	500 – 1000	1000 – 1500	1500 – 2000	2000 – 2500	Total
0 – 200	12	6	–	–	–	18
200 – 400	2	18	4	2	1	27
400 – 600	–	4	7	3	–	14
600 – 800	–	1	–	2	1	4
800 – 1000	–	–	–	1	2	3
Total	14	29	11	8	4	66

Ans. 0.749

9. Find the coefficient of correlation for the following data :

y \ x	16 – 18	18 – 20	20 – 22	22 – 24	Total
10 – 20	2	1	1	–	4
20 – 30	3	2	3	2	10
30 – 40	3	4	5	6	18
40 – 50	2	2	3	4	11
50 – 60	–	1	2	2	5
60 – 70	–	1	2	1	4
	10	11	16	15	52

Ans. 0.28

10. Two judges in a beauty contest rank the ten competitors in the following order :

6	4	3	1	2	7	9	8	10	5
4	1	6	7	5	8	10	9	3	2

Do the two judges appear to agree in their standard ? **Ans.** 0.224

59.13 REGRESSION

If the scatter diagram indicates some relationship between two variables x and y, then the dots of the scatter diagram will be concentrated round a curve. This curve is called the *curve of regression.*

Regression analysis is the method used for estimating the unknown values of one variable corresponding to the known value of another variable.

59.14 LINE OF REGRESSION

When the curve is a straight line, it is called a line of regression. A line of regression is the straight line which gives the best fit in the least square sense to the given frequency.

Regression will be called *non-linear* if there exists a relationship (parabola etc.) other than a straight line between the variables under consideration.

59.15 EQUATIONS TO THE LINES OF REGRESSION

Let $y = a + bx$ ... (1)

be the equation of the line of regression of y on x.

Let (x_r, y_r) be any point of dot.

From the figure

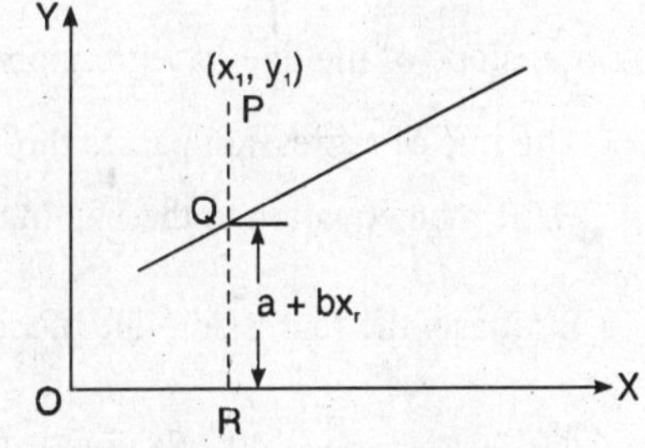

$$PR = y_r$$

$$QR = a + bx_r$$

$$PQ = PR - QR = y_r - a - bx_r$$

Let S be the sum of the squares of such distances, then

$$S = \sum (y - a - bx)^2$$

According to the principle of least squares, we have to choose a and b so that S is minimum. The method of least square gives the condition for minimum value of S.

$$\frac{\partial S}{\partial a} = -2\sum (y - a - bx), \qquad \frac{\partial S}{\partial b} = -2\sum (y - a - bx)x$$

$$\frac{\partial S}{\partial a} = 0, \qquad \frac{\partial S}{\partial b} = 0, \quad \text{for } S \text{ minimum}$$

i.e. $\sum (y - a - bx) = 0 \Rightarrow \sum y - na - b\sum x = 0 \Rightarrow \sum y = na + b\sum x$ ... (2)

and $\sum (xy - ax - bx^2) = 0 \Rightarrow \sum xy - a\sum x - b\sum x^2 = 0$

$\Rightarrow \sum xy = a\sum x + b\sum x^2$ ...(3)

Dividing (2) by n, we get

$$\frac{\Sigma y}{n} = a + b\frac{\Sigma x}{n} \qquad \left(\bar{y} = \frac{\sum y}{n}, \bar{x} = \frac{\sum x}{n}\right)$$

$$\bar{y} = a + b\bar{x}$$

where $\bar{x}$ and $\bar{y}$ are the means of x series and y series.

This shows that $(\bar{x}, \bar{y})$ lie on the line of regression (1), shifting the origin to $(\bar{x}, \bar{y})$, the equation (3) becomes

$$\sum(x-\bar{x})(y-\bar{y}) = a\sum(x-\bar{x}) + b\sum(x-\bar{x})^2$$

But $$\sum(x-\bar{x}) = 0$$

$\Rightarrow$ $$\sum(x-\bar{x})(y-\bar{y}) = b\sum(x-\bar{x})^2$$

or $$b = \frac{\sum(x-\bar{x})(y-\bar{y})}{\sum(x-\bar{x})^2} = \frac{\sum XY}{\sum X^2} \quad \ldots(4)$$

We know that $$r = \frac{\Sigma XY}{\sqrt{\Sigma X^2}\sqrt{\Sigma Y^2}} = \frac{\Sigma XY}{n\sqrt{\frac{\Sigma X^2}{n}}\sqrt{\frac{\Sigma Y^2}{n}}} = \frac{\Sigma XY}{n\sigma_x\sigma_y}$$

or $$\sum XY = nr\sigma_x\sigma_y$$

Putting the value of $\sum XY$ in (4), we get $b = \dfrac{nr\sigma_x\sigma_y}{\sum X^2} = \dfrac{r\sigma_x\sigma_y}{\frac{\sum X^2}{n}} = \dfrac{r\sigma_x\sigma_y}{\sigma_x^2} = \dfrac{r\sigma_y}{\sigma_x}$

i.e. slope of the line of regression $= b = r\dfrac{\sigma_y}{\sigma_x}$

The line of regression passes through $(\bar{x}, \bar{y})$.

Hence the equation to the line of regression is $y - \bar{y} = r\dfrac{\sigma_y}{\sigma_x}(x - \bar{x})$

Similarly the regression line of x on y is $x - \bar{x} = r\dfrac{\sigma_x}{\sigma_y}(y - \bar{y})$.

Note. $b_{yx} = r\dfrac{\sigma_y}{\sigma_x}$ and $b_{xy} = r\dfrac{\sigma_x}{\sigma_y}$ are known as the coefficients of regression.

$$b_{yx}.b_{xy} = \left(r\frac{\sigma_y}{\sigma_x}\right)\left(r\frac{\sigma_x}{\sigma_y}\right) = r^2$$

Example 14. *If θ be the acute angle between the two regression lines in the case of two variables x and y, show that*

$$\tan\theta = \frac{1-r^2}{r}\frac{\sigma_x\sigma_y}{\sigma_x^2+\sigma_y^2}$$

where r, σ_x, σ_y have their usual meanings. Explain the significance where $r = 0$ and $r = \pm 1$.

(*Nagpur University, Winter 2004, A.M.I.E., Winter 2001*)

Solution. Lines of regression are

$$y - \bar{y} = r\frac{\sigma_y}{\sigma_x}(x-\bar{x}) \quad \ldots(1) \qquad \therefore m_1 = r\frac{\sigma_y}{\sigma_x}$$

and $$x - \bar{x} = r\frac{\sigma_x}{\sigma_y}(y-\bar{y}) \quad \ldots(2) \qquad \therefore m_2 = \frac{1}{r}\frac{\sigma_y}{\sigma_x}$$

$$\tan\theta = \frac{m_2 - m_1}{1 + m_1 m_2} = \frac{\frac{1}{r}\frac{\sigma_y}{\sigma_x} - r\frac{\sigma_y}{\sigma_x}}{1 + r\frac{\sigma_y}{\sigma_x}\times\frac{1}{r}\frac{\sigma_y}{\sigma_x}} = \frac{\left(\frac{1}{r} - r\right)\frac{\sigma_y}{\sigma_x}}{1 + \frac{\sigma_y^2}{\sigma_x^2}}$$

$$\tan\theta = \frac{1-r^2}{r}\cdot\frac{\left(\frac{\sigma_y}{\sigma_x}\right)\sigma_x^2}{\sigma_x^2+\sigma_y^2} = \frac{1-r^2}{r}\,\frac{\sigma_x\sigma_y}{\sigma_x^2+\sigma_y^2} \quad ...(3)$$ **Proved.**

(*a*) If $r = 0$, then there is no relationship between the two variables and they are independent.

On putting the value of $r = 0$ in (3) we get $\tan\theta = \infty, \theta = \frac{\pi}{2}$. So the lines (1) and (2) are perpendicular.

(*b*) If $r = 1$ or -1

On putting these values of r in (3) we get, $\tan\theta = 0$ or $\theta = 0$

i.e. lines (1) and (2) coincide.

The correlation between the variables is perfect. **Ans.**

Example 15. *If the coefficient of correlation between two variables x and y is 0.5 and the acute angle between their lines of regression is.*

$\tan^{-1}\left(\frac{3}{5}\right)$, *show that* $\sigma_x = \frac{1}{2}\sigma_y$.

(*U.P. III Semester, June 2009*)

Solution. Here, we have

$$r = 0.5$$

$$\theta = \tan^{-1}\left(\frac{3}{5}\right) \Rightarrow \tan\theta = \frac{3}{5}$$

$$\tan\theta = \frac{1-r^2}{r}\cdot\frac{\sigma_x\sigma_y}{\sigma_x^2+\sigma_y^2} \quad ... (1) \quad \text{(From Example 14)}$$

Putting the values of r and $\tan\theta$ in (1), we get

$$\frac{3}{5} = \frac{1-\frac{1}{4}}{\frac{1}{2}}\,\frac{\sigma_x\sigma_y}{\sigma_x^2+\sigma_y^2}$$

$$\frac{3}{5} = \frac{3}{2}\,\frac{\sigma_x\sigma_y}{\sigma_x^2+\sigma_y^2}$$

$$2\sigma_x^2 + 2\sigma_y^2 - 5\sigma_x\sigma_y = 0$$

$$\Rightarrow \quad 2\sigma_x^2 - 5\sigma_x\sigma_y + 2\sigma_y^2 = 0$$

$$\Rightarrow \quad 2\sigma_x^2 - 4\sigma_x\sigma_y - \sigma_x\sigma_y + 2\sigma_y^2 = 0$$

$$\Rightarrow \quad 2\sigma_x(\sigma_x - 2\sigma_y) - \sigma_y(\sigma_x - 2\sigma_y) = 0$$

$$\Rightarrow \quad (2\sigma_x - \sigma_y)(\sigma_x - 2\sigma_y) = 0$$

$\Rightarrow$ Either $\sigma_x - 2\sigma_y = 0 \Rightarrow \sigma_x = 2\sigma_y$

(*Not desired*)

or $2\sigma_x - \sigma_y = 0 \Rightarrow \sigma_x = \frac{1}{2}\sigma_y$. **Proved.**

Example 16. *Find the correlation coefficient between x and y, when the lines of regression are:*

$$2x - 9y + 6 = 0 \text{ and } x - 2y + 1 = 0$$

Solution. Let the line of regression of x on y be $2x-9y+6=0$

Then, the line of regression of y on x is $x-2y+1=0$

$$\therefore \quad 2x-9y+6=0 \Rightarrow x=\frac{9}{2}y-3 \Rightarrow b_{xy}=\frac{9}{2}$$

$$\text{and} \quad x-2y+1=0 \Rightarrow y=\frac{1}{2}x+\frac{1}{2} \Rightarrow b_{yx}=\frac{1}{2}$$

$$r=\sqrt{b_{xy}\cdot b_{yx}}=\sqrt{\frac{9}{2}\times\frac{1}{2}}=\frac{3}{2}>1 \text{ which is not possible.}$$

So our choice of regression line is incorrect.

$\therefore$ The regression line of x on y is $x-2y+1=0$

And, the regression line of y on x is $2x-9y+6=0$

$$\therefore \quad x-2y+1=0 \Rightarrow x=2y-1 \Rightarrow b_{xy}=2$$

And $2x-9y+6=0 \Rightarrow y=\frac{2}{9}x+\frac{2}{3} \Rightarrow b_{yx}=\frac{2}{9}$

$$r=\sqrt{b_{xy}\cdot b_{yx}}=\sqrt{2\times\frac{2}{9}}=\frac{2}{3}$$

Hence, the correlation coefficient between x and y is $\frac{2}{3}$. **Ans.**

Example 17. *Two lines of regression are given by*

$5y-8x+17=0$ and $2y-5x+14=0$

If $\sigma_y^2=16$, find (i) the mean values of x and y (ii) σ_x^2 (iii) the coefficient of correlation between x and y.

Solution. We have, $5y-8x+17=0$ ... (1)

$$2y-5x+14=0$$

Since $(\bar{x},\bar{y})$ is a common point of the two lines of regression, we have

$$5\bar{y}-8\bar{x}+17=0 \quad \text{... (2)}$$

$$2\bar{y}-5\bar{x}+14=0 \quad \text{... (3)}$$

On solving (2) and (3) for $\bar{x}$ and $\bar{y}$, we have

$$\frac{\bar{x}}{17\times2-14\times5}=\frac{\bar{y}}{-8\times14-(-5\times17)}=\frac{1}{5\times(-5)-(-8)\times2}$$

$$\Rightarrow \frac{\bar{x}}{34-70}=\frac{\bar{y}}{-112+85}=\frac{1}{-25+16} \Rightarrow \frac{\bar{x}}{-36}=\frac{\bar{y}}{-27}=\frac{1}{-9} \Rightarrow \bar{x}=\frac{36}{9}=4 \text{ and } \bar{y}=\frac{27}{9}=3$$

The equations of line of regression can be written as

$$y=\frac{8}{5}x-\frac{17}{5} \text{ and } x=\frac{2}{5}y+\frac{14}{5}$$

$$r\frac{\sigma_y}{\sigma_x}=\frac{8}{5} \text{ and } r\frac{\sigma_x}{\sigma_y}=\frac{2}{5}$$

On multiplication of two equations, we get

$$\left(r\frac{\sigma_y}{\sigma_x}\right)\left(\frac{r\sigma_x}{\sigma_y}\right)=\frac{8}{5}\times\frac{2}{5}=\frac{16}{25} \Rightarrow r^2=\frac{16}{25} \Rightarrow r=\pm\frac{4}{5} \quad \text{... (4)}$$

Now we have to determine the sign of r *i.e.* + or –

as σ_x, σ_y are always +ve, so r is also +ve from (4). $r = \frac{4}{5}$

We are given $\sigma_y^2 = 16$ $\therefore \sigma_y = 4$

And $$r\frac{\sigma_y}{\sigma_x} = \frac{8}{5} \quad \text{... (5)}$$

On putting the values of r and σ_y in (5), we get

$$\left(\frac{4}{5}\right)\frac{4}{\sigma_x} = \frac{8}{5} \quad \Rightarrow \quad \frac{16}{5\sigma_x} = \frac{8}{5} \quad \Rightarrow \quad \sigma_x = 2$$

$$\Rightarrow \quad \sigma_x^2 = 4$$

Hence (*i*) $\bar{x} = 4,\ \bar{y} = 3$, (*ii*) $\sigma_x^2 = 4$, (*iii*) $r = \frac{4}{5}$ **Ans.**

Example 18. *In a partially destroyed laboratory record of an analysis of correlation data, the following results only are eligible:*

$\sigma_x^2 = 9$ *Regression equations :*

$$8x - 10y + 66 = 0$$
$$40x - 18y = 214$$

What were (a) the mean values of x and y (b) the standard deviation of y, (c) coefficient of correlation between x and y?

(U.P., III Semester, Dec. 2009, Nagpur University, Summer 2002)

Solution. Since both the lines of regression pass through the point $(\bar{x}, \bar{y})$, therefore, we have

$$8\bar{x} - 10\bar{y} + 66 = 0 \quad \text{... (1)}$$
$$40\bar{x} - 18\bar{y} - 214 = 0 \quad \text{... (2)}$$

On solving (1) and (2), by cross multiplication method, we have

$$\frac{\bar{x}}{(-10)(-214) - (66)(-18)} = \frac{\bar{y}}{(66)(40) - (8)(-214)} = \frac{1}{8(-18) - (-10)(40)}$$

$$\frac{\bar{x}}{2140 + 1188} = \frac{\bar{y}}{2640 + 1712} = \frac{1}{-144 + 400}$$

$$\Rightarrow \quad \frac{\bar{x}}{3328} = \frac{\bar{y}}{4352} = \frac{1}{256} \quad \Rightarrow \quad \bar{x} = \frac{3328}{256} = 13, \quad \bar{y} = \frac{4352}{256} = 17$$

Also given lines of regression can be written as $y = 0.8x + 6.6 : x = 0.45\,y + 5.35$

We get $r\frac{\sigma_y}{\sigma_x} = 0.8;$ $r\frac{\sigma_x}{\sigma_y} = 0.45$

$$\left(r\frac{\sigma_y}{\sigma_x}\right)\left(r.\frac{\sigma_x}{\sigma_y}\right) = (0.8)\ (0.45)$$

$$\Rightarrow \quad r^2 = 0.36 \quad \Rightarrow \quad r = 0.6$$

Ans. $$r\frac{\sigma_y}{\sigma_x} = 0.8 \quad \text{... (3)}$$

On putting the values of r and σ_x in (3), we get

$$(0.6)\frac{\sigma_y}{3} = 0.8 \quad \Rightarrow \quad \sigma_y = \frac{3(0.8)}{0.6} = 4$$

Hence (*a*) $\bar{x} = 13$, $\bar{y} = 17$, (*b*) $\sigma_y = 4$ (*c*) r = 0.6 **Ans.**

Example 19. *The two regression equations of the variables x and y are*

$$x = 19.13 - 0.87y \quad and \quad y = 11.64 - 0.50x.$$

Find (i) Mean of x's; (ii) Mean of y's; (iii) The correlation coefficient between x and y.

Solution.
$$x = 19.13 - 0.87y \qquad ...(1)$$
$$y = 11.64 - 0.50x \qquad ...(2)$$

As (1) and (2) pass through $(\bar{x}, \bar{y})$:

$$\bar{x} = 19.13 - 0.87\bar{y} \qquad ...(3)$$
$$\bar{y} = 11.64 - 0.50\bar{x} \qquad ...(4)$$

On solving (3) and (4) we get

$$\bar{x} = 15.937, \ \bar{y} = 3.67$$

From (1) $$r\frac{\sigma_x}{\sigma_y} = -0.87 \qquad ...(5)$$

From (2) $$r\frac{\sigma_y}{\sigma_x} = -0.50 \qquad ...(6)$$

As σ_x and σ_y are always positive, so r is negative.
Multiplying (5) and (6), we get

$$r\frac{\sigma_x}{\sigma_y} \cdot r\frac{\sigma_y}{\sigma_x} = -0.87 \times (-0.50)$$

$$r^2 = 0.435 \quad \Rightarrow \quad r = \pm\, 0.66$$ **Ans.**

Example 20. *The regression equations calculated from a given set of observations for two random variables are*

$$x = -0.4y + 6.4 \quad and \quad y = -0.6x + 4.6$$

Calculate $\bar{x}, \bar{y}$ and r.

Solution. The regression equations are

$$x = -\,0.4y + 6.4 \qquad ...(1)$$
$$y = -\,0.6x + 4.6 \qquad ...(2)$$

From (1) coefficient of regression of x on $y = r\frac{\sigma_x}{\sigma_y} = -\,0.4$ ...(3)

From (2) coefficient of regression of y on $x = r\frac{\sigma_y}{\sigma_x} = -\,0.6$ ..(4)

From (3) and (4), we have

$$\left(r\frac{\sigma_x}{\sigma_y}\right)\left(r\frac{\sigma_y}{\sigma_x}\right) = (-\,0.4)\,(-\,0.6)$$

$$\Rightarrow \quad r^2 = 0.24 \quad \Rightarrow \quad r = \pm\, 0.49$$

In (3) and (4), σ_x and σ_y are (always) negative so r is negative

$$r = -\,0.49$$

To find $\bar{x}$ and $\bar{y}$ we solve the equations (1) and (2) simultaneously. Their point of intersection is $(\bar{x}, \bar{y})$.

$$\bar{x} = 6, \qquad \bar{y} = 1$$ **Ans.**

Example 21. *Show that the geometric mean of the coefficients of regression is the coefficient of correlation.* *(AMIE, Summer 2001)*

Solution. The coefficients of regressions are $r\frac{\sigma_y}{\sigma_x}$ and $r\frac{\sigma_x}{\sigma}$

i.e.

$$G.M = \sqrt{r\frac{\sigma_y}{\sigma_x}.r\frac{\sigma_x}{\sigma_y}} = r$$

= coefficient of correlation. **Proved.**

Example 22. *The regression lines of y on x and of x on y are respectively y = ax + b and x = cy + d. Show that the means are* $\overline{x}$ *= (bc + d) (1 – ac) and* $\overline{y}$ *= (ad + b) (1 – ac) and correlation coefficient between x and y is* $\sqrt{ac}$. *Also, show that the ratio of the standard deviations of y and x is* $\sqrt{\frac{a}{c}}$.

Solution. Here, we have

The regression line of y on x is $y = ax + b$ ... (1)

The regression line of x on y is $x = cy + d$ ... (2)

As (1) and (2) pass through $(\overline{x}, \overline{y})$, so

$$\overline{y} = a\overline{x} + b \qquad \text{... (3)}$$

$$\overline{x} = c\overline{y} + d \qquad \text{... (4)}$$

Solving (3) and (4), we get

$$\overline{x} = \frac{bc + d}{1 - ac} \quad \Rightarrow \quad \overline{y} = \frac{ad + b}{1 - ac}$$

We know that $r\frac{\sigma_y}{\sigma_x} = a$ = slope of (1) ... (5)

and $r\frac{\sigma_x}{\sigma_y} = c$ ... (6)

Multiplying (5) and (6), we get

$$r\frac{\sigma_y}{\sigma_x}.r\frac{\sigma_x}{\sigma_y} = a.c \quad \Rightarrow \quad r^2 = ac \quad \Rightarrow \quad r = \sqrt{ac}$$ **Proved.**

Dividing (5) by (6), we get

$$\frac{r\frac{\sigma_y}{\sigma_x}}{r\frac{\sigma_x}{\sigma_y}} = \frac{a}{c} \quad \Rightarrow \quad \left(\frac{\sigma_y}{\sigma_x}\right)^2 = \frac{a}{c} \quad \Rightarrow \quad \frac{\sigma_y}{\sigma_x} = \sqrt{\frac{a}{c}}$$ **Proved.**

Example 23. *Prove that arithmetic mean of the coefficients of regression is greater than the coefficient of correlation.* (*A.M.I.E.T.E., Summer 2000*)

Solution. Coefficients of regression are $r\frac{\sigma_y}{\sigma_x}$ and $r\frac{\sigma_x}{\sigma_y}$

We have to prove that $A.M. > r$

$$\Rightarrow \quad \frac{1}{2}\left[r\frac{\sigma_y}{\sigma_x} + r\frac{\sigma_x}{\sigma_y}\right] > r \quad \Rightarrow \quad \frac{1}{2}\left[\frac{\sigma_y}{\sigma_x} + \frac{\sigma_x}{\sigma_y}\right] > 1$$

$$\Rightarrow \quad \frac{\sigma_y}{\sigma_x} + \frac{\sigma_x}{\sigma_y} - 2 > 0 \quad \Rightarrow \quad \frac{1}{\sigma_x\sigma_y}\left[\sigma_x^2 + \sigma_y^2 - 2\sigma_x\sigma_y\right] > 0$$

$$\Rightarrow \quad \frac{1}{\sigma_x\sigma_y}\left[\sigma_x - \sigma_y\right]^2 > 0$$ which is true. **Proved.**

Example 24. *In a study between the amount of rainfall and the quantity of air pollution removed the following data were collected.*

Daily Rainfall in 0.01 cm	*4.3*	*4.5*	*5.9*	*5.6*	*6.1*	*5.2*	*3.8*	*2.1*
Pollution Removed (mg/m³)	*12.6*	*12.1*	*11.6*	*11.8*	*11.4*	*11.8*	*13.2*	*14.1*

Find the regression line of y on x. (*A.M.I.E., Summer 2000*)

Solution.

S.N	*x (metre)*	*y*	*xy*	x^2
1	4.3	12.6	54.18	18.49
2	4.5	12.1	54.45	20.25
3	5.9	11.6	68.44	34.81
4	5.6	11.8	66.08	31.36
5	6.1	11.4	69.54	37.21
6	5.2	11.8	61.36	27.04
7	3.8	13.2	50.16	14.44
8	2.1	14.1	29.61	4.41
	37.5	98.6	453.82	188.01

Let $y = a + bx$ be the equation of the line of regression of y on x, where a and b are given by the following equations.

$$\Sigma y = na + b\,\Sigma x \quad \Rightarrow \quad 98.6 = 8a + 37.5b \qquad \text{...(1)}$$

$$\Sigma xy = a\,\Sigma x + b\,\Sigma x^2 \quad \Rightarrow \quad 453.82 = 37.5a + 188.01b \qquad \text{...(2)}$$

On solving (1) and (2), we get $a = 15.49$ and $b = -0.675$.

The equation of the line of regression is $y = 15.49 - 0.675x$ **Ans.**

Example 25. *Find the correlation coefficient and regression lines for the data :*

x	*1*	*2*	*3*	*4*	*5*
y	*2*	*5*	*3*	*8*	*7*

(*Nagpur University, Summer 2000*)

Solution. We have,

x	1	2	3	4	5	$\Sigma x = 15$
y	2	5	3	8	7	$\Sigma y = 25$

$$\bar{x} = \frac{1}{n}\Sigma x = \frac{1}{5} \times 15 = 3 \qquad \bar{y} = \frac{1}{n}\Sigma y = \frac{1}{5} \times 25 = 5$$

x	*y*	$X = x - 3$	$Y = y - 5$	X^2	Y^2	*XY*
1	2	− 2	− 3	4	9	6
2	5	−1	0	1	0	0
3	3	0	− 2	0	4	0
4	8	1	3	1	9	3
5	7	2	2	4	4	4
		Total		10	26	13

Correlation coefficient 'r' $= \dfrac{\Sigma XY}{\sqrt{\Sigma X^2}\sqrt{\Sigma Y^2}} = \dfrac{13}{\sqrt{10 \times 26}} = 0.8062$

and $r = \dfrac{\Sigma XY}{n\sigma_x \sigma_y} \Rightarrow \Sigma XY = n\,r\,\sigma_x\,\sigma_y$

Slope of regression line of y on $x = \frac{\Sigma XY}{\Sigma X^2} = \frac{13}{10}$

Slope of regression line of x on $y = \frac{\Sigma XY}{\Sigma Y^2} = \frac{13}{26} = \frac{1}{2}$

Equation of regression line of y on x is

$$y - \bar{y} = \frac{13}{10}(x - \bar{x}) \Rightarrow y - 5 = \frac{13}{10}(x - 3) \Rightarrow y = 1.3x + 1.1$$

Equation of regression line of x on y is

$$x - \bar{x} = \frac{1}{2}(y - \bar{y}) \Rightarrow x - 3 = 0.5\,(y - 5) \Rightarrow x = 0.5y + 0.5$$ **Ans.**

Example 26. *Find the coefficient of correlation and regression lines to the following data:*

x	*5*	*7*	*8*	*10*	*11*	*13*	*16*
y	*33*	*30*	*28*	*20*	*18*	*16*	*9*

(*Nagpur University, Winter 2003*)

Solution. Here $n = 7$, $\bar{x} = \frac{\Sigma x}{n} = \frac{70}{7} = 10$

$$\bar{y} = \frac{\Sigma y}{n} = \frac{154}{7} = 22$$

$\therefore$
$$X = x - \bar{x} = x - 10$$
$$Y = y - \bar{y} = y - 22$$

The various calculations are shown in the following table :

x	y	$X = x - 10$	$Y = y - 22$	XY	X^2	Y^2
5	33	− 5	11	− 55	25	121
7	30	− 3	8	− 24	9	64
8	28	− 2	6	− 12	4	36
10	20	0	− 2	0	0	4
11	18	1	− 4	− 4	1	16
13	16	3	− 6	− 18	9	36
16	9	6	− 13	− 78	36	169
$\Sigma x = 70$	$\Sigma y = 154$			$\Sigma XY = -191$	$\Sigma X^2 = 84$	$\Sigma Y^2 = 446$

Coefficient of correlation

Now, $$r = \frac{\Sigma XY}{\sqrt{\Sigma X^2 . \Sigma Y^2}} = \frac{-191}{\sqrt{84 \times 446}} = -0.9868$$

$$r\frac{\sigma_y}{\sigma_x} = r\sqrt{\frac{\Sigma Y^2}{\Sigma X^2}} = -0.9868\sqrt{\frac{446}{84}} = -2.2738$$

and $$r\frac{\sigma_x}{\sigma_y} = r\sqrt{\frac{\Sigma X^2}{\Sigma Y^2}} = -0.9868\sqrt{\frac{84}{446}} = -0.4283$$

$\therefore$ Equation of line of regression y on x is

$$y - \bar{y} = r\frac{\sigma_y}{\sigma_x}(x - \bar{x}), \qquad y - 22 = -2.2738\,(x - 10)$$

$$y - 22 = -2.2738x + 22.738, \qquad y = -2.2738x + 44.738$$

and equation of line of regression x on y is

$$x - \bar{x} = r\frac{\sigma_x}{\sigma_y}(y - \bar{y})$$

$\Rightarrow$ $x - 10 = -0.4283\,(y - 22)$

$\Rightarrow$ $x - 10 = -0.4283\,y + 9.4226$

$\Rightarrow$ $x = -0.4283\,y + 19.4226$

$x + 0.4283\,y = 19.4226$

Hence the equation of the line of regression of y on x is $y + 2.2738\,x = 44.738$

The equation of the line of regression of x on y is $x + 0.4283\,y = 19.4226$ **Ans.**

Example 27. *Find the coefficient of correlation and obtain the equation of the lines of regression for the data.*

x	6	2	10	4	8
y	9	11	5	8	7

(*Nagpur University, Winter 2000*)

Solution. Here, we have

x	6	2	10	4	8	$\Sigma x = 30$
y	9	11	5	8	7	$\Sigma y = 40$

$$\bar{x} = \frac{\Sigma x}{n} = \frac{30}{5} = 6, \quad \bar{y} = \frac{\Sigma y}{n} = \frac{40}{5} = 8$$

x	y	$X = x - 6$	$Y = y - 8$	X^2	Y^2	XY
6	9	0	1	0	1	0
2	11	– 4	3	16	9	– 12
10	5	4	– 3	16	9	– 12
4	8	– 2	0	4	0	0
8	7	2	– 1	4	1	– 2
				$\Sigma X^2 = 40$	$\Sigma Y^2 = 20$	$\Sigma XY = -26$

$$r = \frac{\Sigma XY}{\sqrt{\Sigma X^2 . \Sigma Y^2}} = \frac{-26}{\sqrt{40 \times 20}} = \frac{-26}{28.2842} = -0.919$$

The regression coefficient of y on x is

$$\frac{\Sigma XY}{\Sigma X^2} = -\frac{26}{40} = -0.65 \qquad \left(\frac{\Sigma XY}{\Sigma X^2} = r\frac{\sigma_y}{\sigma_x}\right)$$

The equation of line of regression of y on x is

$$y - \bar{y} = \frac{\Sigma XY}{\Sigma X^2}(x - \bar{x})$$

$\Rightarrow$ $y - 8 = -0.65\,(x - 6)$ $\Rightarrow$ $y = -0.65\,x + 11.9$

The regression coefficient of x on y is

$$\frac{\Sigma XY}{\Sigma Y^2} = -\frac{26}{20} = -1.3$$

The equation of line of regression of x on y is

$$x - \bar{x} = \frac{\Sigma XY}{\Sigma Y^2}(y - \bar{y}) \qquad \left(\frac{\Sigma XY}{\Sigma Y^2} = r\frac{\sigma_x}{\sigma_y}\right)$$

$\Rightarrow \qquad x - 6 = -1.3\,(y - 8) \quad \Rightarrow \quad x = -1.3y + 16.4$

Hence, the equation of the line of regression of y on x is

$$y = -0.65x + 11.9$$

The equation of the line of regression of x on y is

$$x = -1.3y + 16.4$$

Ans.

59.16 MULTIPLE REGRESSION

We know that the production of wheat depends not only on the amount of rain fall x_1 but also on the fertilizer x_2, pestisides x_3, quality of seeds x_4, quality of soil x_5 etc. In a multiple regression the dependent variable is a function of more than one independent variable.

Linear regression is a linear relationship between y and x_1, x_2, x_3.........

$y = a_0 + a_1 x_1 + a_2 x_2 + a_3 x_3 + \ldots\ldots\ldots$

In multiple non-linear regression equation is not linear; *for example*

$y = a_0 + a_1 x^{\alpha} + a_2 x^{\beta} + a_3 x^{\gamma} + \ldots\ldots$

NON LINEAR RELATIONSHIP

Example 28. *Fit a non linear relationship between the following data :*

x	1	2	3	4
y	1.7	1.8	2.3	3.2

(*U.P., III Semester, June 2009*)

Solution. Here, we have

	x	y	x^2	xy	x^3	x^4	x^2y
	1	1.7	1.	1.7	1	1	1.7
	2	1.8	4	3.6	8	16	7.2
	3	2.3	9	6.9	27	81	20.7
	4	3.2	16	12.8	64	256	51.2
Total	10	9.0	30	25.0	100	354	80.8

Let the non linear relationship is $y = a_0 + a_1 x + a_2 x^2$

Normal equations are

$$\Sigma y = na_0 + a_1\,\Sigma x + a_2\,\Sigma x^2$$

$$\Sigma xy = a_0\,\Sigma x + a_1 \Sigma x^2 + a_2 \Sigma x^3$$

$$\Sigma x^2 y = a_0\,\Sigma x^2 + a_1\,\Sigma x^3 + a_2\,\Sigma x^4$$

Substituting the values of Σy, n, Σx, Σx^2 etc. in these equations, we get

$$9 = 4a_0 + 10a_1 + 30a_2$$

$$25 = 10a_0 + 30a_1 + 100a_2$$

$$80.8 = 30a_0 + 100a_1 + 354a_2$$

Solving these equations, we get $a_0 = 2$, $a_1 = -0.5$ and $a_2 = 0.2$

Then the non-linear relationship is $y = 2 - 0.5x + 0.2x^2$ **Ans.**

59.17 ERROR OF PREDICTION

The deviation of the predicted value from the observed value is known as the standard error of prediction. It is given by

$$E_{yx} = \sqrt{\frac{\sum (y - y_r)^2}{n}}$$

where y is the actual value and y_r the predicted value.

Example 29. *Prove that* (*i*) $E_{yx} = \sigma_y . \sqrt{1-r^2}$ (*ii*) $E_{xy} = \sigma_x \sqrt{1-r^2}$

Solution. The equation of the line of regression of y on x is

$$y - \bar{y} = r\frac{\sigma_y}{\sigma_x}(x-\bar{x}), \quad y_r = \bar{y} + r\frac{\sigma_y}{\sigma_x}(x-\bar{x})$$

So, $$E_{yx} = \sqrt{\frac{\sum (y-y_r)^2}{n}} = \left[\frac{1}{n}\sum\left\{y-\bar{y}-r\frac{\sigma_y}{\sigma_x}(x-\bar{x})\right\}^2\right]^{1/2}$$

$$= \left[\frac{1}{n}\cdot\Sigma\left\{(y-\bar{y})^2 + \frac{r^2\sigma_y^2}{\sigma_x^2}(x-\bar{x})^2 - \frac{2r\sigma_y}{\sigma_x}(x-\bar{x})(y-\bar{y})\right\}\right]^{1/2}$$

$$= \left[\sum\frac{(y-\bar{y})^2}{n} + r^2\frac{\sigma_y^2}{\sigma_x^2}\sum\frac{(x-\bar{x})^2}{n} - 2r\frac{\sigma_y}{\sigma_x}\sum\frac{(x-\bar{x})(y-\bar{y})}{n}\right]^{1/2}$$

$$= \left[\sigma_y^2 + r^2\frac{\sigma_y^2}{\sigma_x^2}.\sigma_x^2 - 2r\frac{\sigma_y}{\sigma_x}r.\sigma_x.\sigma_y\right]^{1/2} = \left[\sigma_y^2 + r^2\sigma_y^2 - 2r^2\sigma_y^2\right]^{1/2} = \left[\sigma_y^2 - r^2\sigma_y^2\right]^{1/2}$$

$$= \sigma_y\sqrt{1-r^2}$$ **Proved.**

(*ii*) Similarly (*ii*) may be proved.

Example 30. *Find the standard error of estimate of y on x for the data given below:*

x	*1*	*3*	*4*	*6*	*8*	*9*	*11*	*14*
y	*1*	*2*	*4*	*4*	*5*	*7*	*8*	*9*

Solution. The equation of the line of regression of y on x is

$$y = \frac{7}{11}x + \frac{6}{11}. \quad \text{So } y_r = \frac{7x}{11} + \frac{6}{11}$$

S.No	x	y	y_r	$(y-y_r)$	$(y-y_r)^2$
1	1	1	$\frac{13}{11}$	$-\frac{2}{11}$	$\frac{4}{121}$
2	3	2	$\frac{27}{11}$	$-\frac{5}{11}$	$\frac{25}{121}$
3	4	4	$\frac{34}{11}$	$\frac{10}{11}$	$\frac{100}{121}$
4	6	4	$\frac{48}{11}$	$-\frac{4}{11}$	$\frac{16}{121}$
5	8	5	$\frac{62}{11}$	$-\frac{7}{11}$	$\frac{49}{121}$
6	9	7	$\frac{69}{11}$	$\frac{8}{11}$	$\frac{64}{121}$
7	11	8	$\frac{83}{11}$	$\frac{5}{11}$	$\frac{25}{121}$
8	14	9	$\frac{104}{11}$	$-\frac{5}{11}$	$\frac{25}{121}$
					$\Sigma(y-y_r)^2 = \frac{308}{121}$

$$E_{yx} = \sqrt{\frac{\sum(y-y_r)^2}{n}} = \sqrt{\frac{308}{121\times 8}} = \sqrt{\frac{7}{22}} = 0.564$$ **Ans.**

59.18 RELATION BETWEEN REGRESSION ANALYSIS AND CORRELATION ANALYSIS

Sr. No.	Correlation Analysis		Regression Analysis
1.	The relationship between two variables is given by the coefficient of correlation.	1.	In this case some points are stepped up and some are stepped down for making an average value.
2.	It is a measure of direction and degree of relationship between x and y.	2.	b_{xy} and b_{yx} are mathematical measure of average relationship between the two variables.
3.	Here, $r_{xy} = r_{yx}$	3.	$b_{xy} \neq b_{yx}$
4.	It does not reflect upon the nature of variable (dependent or independent variable).	4.	It indicates which is dependent variable and which is independent variable
5.	It does not imply cause and effect relationship between the variables.	5.	It indicates the cause and effect relationship between the variables.
6.	It is a relative measure and have no units.	6.	It is an absolute measure.
7.	It indicates the degree of association.	7.	It is used to forecast the nature of the dependent variable when the value of independent variable is given.
8.	It is confined to the study of linear relationship.	8.	It has not only application of linear relationship but non-linear relationship also.

EXERCISE 59.2

1. Find the regression line of y on x for the data :

x	1	4	2	3	5
y	3	1	2	5	4

Ans. $y = 2.7 + 0.1x$

2. Compute the regression line of x on y for the following data :

x	2	4	6	8	10
y	12	10	8	6	4

Ans. $x = 1 + 3yy$

3. Compute the regression line of y on x for the following data :

x	1	2	3	4	5	6
y	2	2	2	2	2	2

Ans. $y = 6 - x$

4. Find the regression lines of from the given data :

x	1	2	3	4	5	6	7	8	9	10
y	10	12	16	28	25	36	41	49	40	50

Ans. $x = 0.2y - 0.64, \quad y = 4.69x + 4.9$

5. Find the equations to the lines of regression and the coefficient of correlation for the following data:

x	2	4	5	6	8	11
y	18	12	10	8	7	5

Ans. $y - 10 = -1.34\,(x - 6),\ x - 6 = -0.632\,(y - 10),\ r = -0.92$

6. The following marks have been obtained by a class of students in statistics.

Paper I	80	45	55	56	58	60	65	68	70	75	85
Paper II	81	56	50	48	60	62	64	65	70	74	90

Compute the coefficient of correlation for the above data. Find the lines of regression.

Ans. $r = 0.918,\ y - 65.45 = 0.981\,(x - 65.18),\ x - 65.18 = 0.859\,(y - 65.45)$

7. The following results were obtained from records of age (x) and systolic blood pressure (y) of a group of 10 men :

	x	y
Mean	53	142
Variance	130	165

and $\Sigma(x-\bar{x})(y-\bar{y}) = 1220$

Find the appropriate regression eqation and use it to estimate the blood pressure of a man whose age is 45. **Ans.** $y = 0.94\,x + 92.26$, Blood pressure = 134.56

8. The following results were obtained from lineups in Applied Mechanics and Engineering Mathematics in an examination :

	Applied Mechanics(x)	*Engg. Maths.(y)*
Mean	47.5	39.5
Standard deviation	16.8	10.8

$r = 0.95$

Find both the regresssion equations. Also estimate the value of y for $x = 30$.

Ans. $y = 0.611x + 10.5$, $x = 1.478\,y - 1.143$, $y = 28.83$

9. If two regression coefficients are 0.8 and 0.2, what would be the value of coefficient of correlation? **Ans.** $r = 0.4$

10. The regression equation are : $7x - 16y + 9 = 0$, $5y - 4x - 3 = 0$, find $\bar{x}, \bar{y}$ and r.

(AMIE, Winter 2003) **Ans.** $\bar{x} = -\dfrac{3}{29}$, $\bar{y} = \dfrac{15}{29}$, $r = \dfrac{3}{4}$

11. The following regression equations and variances are obtained from a correlation table :

$20x - 9y - 107 = 0$, $4x - 5y + 33 = 0$, variance of $x = 9$.

Find (*i*) the mean values of x and y, (*ii*) the standard deviation of y. *(A.M.I.E., Winter 2000)*

Ans. $\bar{x} = 13, \bar{y} = 17, \sigma_y = 4$.

12. Two random variables have the least square regression lines with equation $3x + 2y = 26$ and $6x + y = 31$. Find mean values and correlation coefficient between x and y.

Ans. $\bar{x} = 4, \bar{y} = 7, r = 0.5$

13. The regression equations of two variables x and y are $x = 0.7y + 5.2$, $y = 0.3x + 2.8$. Find the means of the variables and the coefficient of correlation between them.

Ans. $r = 0.7395$, $\bar{x} = -0.1034$,

$\bar{y} = 0.5172$

14. Two lines of regression are given by $x + 2y = 5$ & $2x + 3y = 8$.

Calculate: (*i*) mean values of x and y

(*ii*) the coefficient of correlation

(*iii*) the ratio of the regression coefficients. **Ans.** $\bar{x} = 4$, $\bar{y} = 7$, $r = -0.5$

15. Fill in the blanks :

(*a*) Arithmetic mean of the coefficients of regression is ....than the coefficient of correlation.

(A.M.I.E., Summer 2000) **Ans.** greater

(*b*) If two regression lines concide then the coefficient of correlation is ....

(A.M.I.E., Winter 2000) **Ans.** ±1

CHAPTER

60

CORRELATION AND MULTIPLE REGRESSION ANALYSIS

60.1 INTRODUCTION

So for we have considered correlation between two variables only. But often it is necessary to find correlation between three or more variables.

For example; Crops are influenced not only by rainfall but of different fertilizers used.

Correlation between crops rainfall and fertilizer is the multiple correlation.

60.2 MULTIPLE CORRELATION

In multiple correlation we study three or more variables at a time.

In multiple correlation the effect of all the independent variables on a dependent variables is studied.

Let three variables be x_1, x_2 and x_3.

$R_{1.23}$ = Multiple correlation coefficient with x_1 as dependent variable and x_2 and x_3 as independent variables.

$R_{2.13}$ = Multiple correlation coefficient with x_2 as dependent variable and x_1 and x_3 as independent variables.

$R_{3.12}$ = Multiple correlation coefficient with x_3 as dependent variable and x_1 and x_2 a independent variables.

60.3 FORMULAE FOR THE CALCULATION OF MULTIPLE CORRELATION COEFFICIENT

$$R_{1.23} = \sqrt{\frac{r_{12}^2 + r_{13}^2 - 2r_{12}\, r_{13}\, r_{23}}{1 - r_{23}^2}}\,; \quad R_{2.13} = \sqrt{\frac{r_{21}^2 + r_{23}^2 - 2r_{12}\, r_{13}\, r_{23}}{1 - r_{13}^2}}\,; \quad R_{3.12} = \sqrt{\frac{r_{31}^2 + r_{32}^2 - 2r_{12}\, r_{13}\, r_{23}}{1 - r_{12}^2}}$$

60.4 PROPERTIES OF MULTIPLE CORRELATION

(1) Its value lies between 0 and 1. (2) If $R_{1.23} = 0$, then $r_{12} = 0$ and $r_{13} = 0$

(3) $R_{1.23} \geq r_{12}$ and $R_{1.23} \geq r_{13}$ Q (4) $R_{1.23} = R_{1.32}$

Coefficient of Multiple correlation between four Variables :

$$R_{1.234} = \sqrt{1 - \left(1 - r_{14}^2\right)\left(1 - r_{12.3}^2\right)\left(1 - r_{12.34}^2\right)}$$

Example 1. *A simple correlation coefficient between quantity of production of wheat* x_1, *fertilizer* (x_2) *and rainfall* (x_3) *are given*

$r_{12} = 0.4, \quad r_{13} = 0.5$ and $r_{23} = 0.6$

Find the coefficient of multiple correlation $R_{1.23}$.

Solution. Here, we have

$$r_{12} = 0.4, \quad r_{13} = 0.5 \quad \text{and } r_{23} = 0.6$$

We know that

$$R_{1.23} = \sqrt{\frac{r_{12}^2 + r_{13}^2 - 2r_{12}\, r_{13}\, r_{23}}{1 - r_{23}^2}} = \sqrt{\frac{(0.4)^2 + (0.5)^2 - 2(0.4)\,(0.5)\,(0.6)}{1-(0.6)^2}}$$

$$= \sqrt{\frac{0.16 + 0.25 - 0.24}{1 - 0.36}} = \sqrt{\frac{0.17}{0.64}} = \sqrt{0.2656} = 0.515$$ **Ans.**

Example 2. *If $r_{12} = 0.6$, $r_{23} = 0.35$ and $r_{31} = 0.4$ then find $R_{3.12}$.*

Solution. We have

$$r_{12} = 0.6, \quad r_{23} = 0.35, \quad \text{and } r_{31} = 0.4$$

We know that

$$R_{3.12} = \sqrt{\frac{r_{31}^2 + r_{32}^2 - 2r_{12}\, r_{13}\, r_{23}}{1 - r_{12}^2}} = \sqrt{\frac{0.4^2 + (0.35)^2 - 2(0.6)\,(0.4)\,(0.35)}{1-(0.6)^2}}$$

$$= \sqrt{\frac{0.16 + 0.1225 - 0.168}{1 - 0.36}} = \sqrt{\frac{0.1145}{0.64}} = \sqrt{0.1789} = 0.423$$ **Ans.**

Example 3. *If $r_{12} = 0.25$, $r_{13} = 0.35$ and $r_{23} = 0.45$ then find $R_{2.13}$.*

Solution. Here, we have

$$r_{12} = 0.25, \quad r_{13} = 0.35, \quad \text{and } r_{23} = 0.45$$

We know that

$$R_{2.13} = \sqrt{\frac{r_{21}^2 + r_{23}^2 - 2r_{12}\, r_{13}\, r_{23}}{1 - r_{13}^2}} = \sqrt{\frac{(0.25)^2 + (0.45)^2 - 2(0.25)\,(0.35)\,(0.45)}{1-(0.35)^2}}$$

$$= \sqrt{\frac{0.0625 + 0.2025 - 0.07875}{1 - 0.1225}} = \sqrt{\frac{0.18625}{0.8775}} = \sqrt{0.2123} = 0.461$$ **Ans.**

Example 4. *Given the following data :*

x_1	3	5	6	8	12	14
x_2	16	10	7	4	3	2
x_3	90	72	54	42	30	12

Compute the coefficient of linear multiple correlation of x_3 on x_1 and x_2.

Solution. Here $N = 6$; $\bar{x}_1 = \frac{\Sigma x_1}{6} = \frac{48}{6} = 8$, $\bar{x}_2 = \frac{\Sigma x_2}{6} = \frac{42}{6} = 7$, $\bar{x}_3 = \frac{\Sigma x_3}{6} = \frac{300}{6} = 50$

We have to compute the values of r_{13}, r_{23} and r_{12}.

$X_1 = x_1 - \bar{x}_1$			$X_2 = x_2 - \bar{x}_2$			$X_3 = x_3 - \bar{x}_3$					
x_1	X_1	X_1^2	x_2	X_2	X_2^2	x_3	X_3	X_3^2	X_1X_2	X_1X_3	X_2X_3
3	–5	25	16	+9	81	90	+40	1600	–45	–200	+360
5	–3	9	10	+3	9	72	+22	484	–9	–66	+66
6	–2	4	7	0	0	54	+4	16	0	–8	0
8	0	0	4	–3	9	42	–8	64	0	0	+24
12	+4	16	3	–4	16	30	–20	400	–16	–80	+80
14	+6	36	2	–5	25	12	–38	1444	–30	–228	+190
		$\Sigma X_1^2 = 90$			$\Sigma X_2^2 = 140$			$\Sigma X_3^2 = 4008$	$\Sigma X_1X_2 = -100$	$\Sigma X_1X_3 = -582$	$\Sigma X_2X_3 = 720$

Now, $r_{12} = \dfrac{\Sigma X_1X_2}{\sqrt{\Sigma X_1^2 \times \Sigma X_2^2}} = \dfrac{-100}{\sqrt{90 \times 140}} = \dfrac{-100}{\sqrt{12600}} = -0.89$

Also, $r_{13} = \dfrac{\Sigma X_1X_3}{\sqrt{\Sigma X_1^2 \times \Sigma X_3^2}} = \dfrac{-582}{\sqrt{90 \times 4008}} = -0.97$

Again $r_{23} = \dfrac{\Sigma X_2X_3}{\sqrt{\Sigma X_2^2 \times \Sigma X_3^2}} = \dfrac{720}{\sqrt{140 \times 4008}} = 0.96$

We know that

$$R_{3.12} = \sqrt{\frac{r_{13}^2 + r_{23}^2 - 2r_{12}\, r_{13}\, r_{23}}{1 - r_{12}^2}} \qquad \text{... (1)}$$

Substituting the values of r_{12}, r_{13} and r_{23} in (1), we get

$$R_{3.12} = \sqrt{\frac{(-0.97)^2 + (0.96)^2 - 2(-0.89)\,(-0.97)\,(0.96)}{1-(-0.89)^2}}$$

$$= \sqrt{\frac{0.9409 + 0.9216 - 1.66}{1 - 0.7921}} = \sqrt{\frac{0.2025}{0.2079}} = \sqrt{0.9740} = 0.987 \qquad \textbf{Ans.}$$

EXERCISE 60.1

1. If $r_{12} = 0.59$, $r_{13} = 0.46$, and $r_{23} = 0.77$ then find $R_{1.23}$. **Ans.** 0.416
2. If $r_{12} = 0.5$, $r_{13} = 0.6$, and $r_{23} = 0.7$ then find $R_{2.13}$. **Ans.** 0.707
3. If $r_{12} = 0.6$, $r_{13} = 0.7$, and $r_{23} = 0.65$ then find $R_{3.12}$. **Ans.** 0.757
4. If $r_{12} = 0.8$, $r_{13} = -0.5$, and $r_{23} = 0.40$ then prove that $R_{1.23} = R_{1.32}$
5. If $r_{12} = 0.45$, $r_{13} = 0.32$, and $r_{23} = 0.61$ then find $R_{1.23}$. **Ans.** 0.339

60.5 MULTIPLE REGRESSION ANALYSIS

We have considered two types of regression equations one of x on y and the other of y on x.

Multiple regression analysis represents an extension of two variables to three or more variables.

We take x_1 as dependent variable and x_2 and x_3 as independent variable.

60.6 PURPOSE OF MULTIPLE REGRESSION

(1) From the regression equation to find out the estimate of dependent variable from two or more independent variables.

(2) To find out the error in the estimate.

(3) To derive a measure of proportion of variance in the dependent variable from the independent variables.

60.7 REGRESSION EQUATION OF THREE VARIABLES

$$X_1 = a + b_{12.3}\, x_2 + b_{13.2}\, x_3 \qquad \text{... (A)}$$

Normal equation of multiple regression equation is

$$S = \Sigma\,(x_1 - X_1)^2 \qquad \text{... (B)}$$

Putting the value of X_1 from (A) in (B), we get

$$S = \Sigma\,(x_1 - a - b_{12.3}\, x_2 - b_{13.2}\, x_3)^2$$

Differentiating partially above equation w.r.t. a, $b_{12.3}$ and $b_{13.2}$, we get

$$\frac{\partial S}{\partial a} = \Sigma\,(x_1 - 1 - b_{12.3}\, x_2 - b_{13.2}\, x_3) = 0 \qquad \text{... (1)}$$

$$\frac{\partial S}{\partial b_{12.3}} = \Sigma\, x_2\,(x_1 - a - b_{12.3}\, x_2 - b_{13.2}\, x_3) = 0 \qquad \text{... (2)}$$

$$\frac{\partial S}{\partial b_{13.2}} = \Sigma\, x_3\,(x_1 - a - b_{12.3}\, x_2 - b_{13.2}\, x_3) = 0 \qquad \text{... (3)}$$

Equation (1) can be rewritten as

$$\Sigma x_1 - \Sigma a - b_{12.3}\, \Sigma x_2 - b_{13.2}\, \Sigma x_3 = 0$$

since $\Sigma x_1 = \Sigma(X_1 - \bar{X}_1) = 0$, $\Sigma x_2 = \Sigma(X_2 - \bar{X}_2) = 0$, $\Sigma x_3 = \Sigma(X_3 - \bar{X}_3) = 0$

[Sum of the deviations from the mean = 0]

Therefore from (1), $a = 0$

Putting the value of $a = 0$ in (2) and (3), we get

$$\Sigma x_1 x_2 - b_{12.3}\, \Sigma x_2^2 - b_{13.2}\, \Sigma x_2 x_3 = 0 \qquad \text{... (4)}$$

$$\Sigma x_1 x_3 - b_{12.3}\, \Sigma x_2 x_3 - b_{13.2}\, \Sigma x_3^2 = 0 \qquad \text{... (5)}$$

On solving (4) and (5), we get the values of $b_{12.3}$, and $b_{13.2}$.

On putting the values of a, $b_{12.3}$ and $b_{13.2}$ in (A), we get the required regression equation.

Similarly

$$x_2 = b_{21.3}\; x_1 + b_{23.1}\; x_3$$
$$x_3 = b_{31.2}\, x_1 + b_{32.1}\; x_2$$

Second Method :

On putting the values of $\Sigma x_1\, x_2$ etc. in (4) and (5), we get

$$r_{12} = \frac{\Sigma x_1 x_2}{\sigma_1 \sigma_2} \Rightarrow \Sigma x_1 x_2 = r_{12}\, \sigma_1 \sigma_2 \quad \text{etc.}$$

$$r_{12}\, \sigma_1 \sigma_2 = b_{12.3}\, r_{23}\, \sigma_2 \sigma_3 + b_{13.2}\, \sigma_3^2 \qquad \text{... (6)}$$

$$r_{13}\, \sigma_1 \sigma_3 = b_{12.3}\, r_{23}\, \sigma_2 \sigma_3 + b_{13.2}\, \sigma_3^2 \qquad \text{... (7)}$$

where r_{ij} = coefficient of correlation between x_i and x_j.

Solving (6) and (7), we get

$$b_{12.3} = \begin{vmatrix} r_{12}\sigma_1 & r_{23}\sigma_3 \\ r_{13}\sigma_1 & \sigma_3 \end{vmatrix} \div \begin{vmatrix} \sigma_2 & r_{23}\sigma_3 \\ r_{23}\sigma_2 & \sigma_3 \end{vmatrix} = \frac{\dfrac{-\sigma_1}{\sigma_2}\begin{vmatrix} r_{12} & r_{23} \\ r_{13} & 1 \end{vmatrix}}{\begin{vmatrix} 1 & r_{23} \\ r_{23} & 1 \end{vmatrix}} = -\frac{\sigma_1\, \Delta_{12}}{\sigma_2\, \Delta_{11}} \qquad \text{... (8)}$$

$$\text{and} \quad b_{13.2} = \frac{\dfrac{-\sigma_1}{\sigma_3}\begin{vmatrix} 1 & r_{12} \\ r_{23} & r_{13} \end{vmatrix}}{\begin{vmatrix} 1 & r_{23} \\ r_{23} & 1 \end{vmatrix}} = -\frac{\sigma_1}{\sigma_3}\frac{\Delta_{13}}{\Delta_{11}} \quad \text{... (9)}$$

Where Δ_{ij} is the co-factor of the element in the ith row and jth column in the determinant.

$$\Delta = \begin{vmatrix} 1 & r_{12} & r_{13} \\ r_{21} & 1 & r_{23} \\ r_{31} & r_{32} & 1 \end{vmatrix}$$

Hence, on substituting the values of $b_{12.3}$ and $b_{13.2}$ the equation to the regression plane of x_1 on x_2 and x_3 is

$$x_1 = \left[-\frac{\sigma_1}{\sigma_2}\frac{\Delta_{12}}{\Delta_{11}}\right]x_2 + \left[-\frac{\sigma_1}{\sigma_3}\frac{\Delta_{13}}{\Delta_{11}}\right]x_3 \quad \text{... (10)}$$

The above equation can also be written as :

$$x_1 = \frac{\sigma_1}{\sigma_2}\left[\frac{r_{12} - r_{13}\,r_{23}}{1 - r_{23}^2}\right]x_2 + \frac{\sigma_1}{\sigma_3}\left[\frac{r_{13} - r_{12}\,r_{23}}{1 - r_{23}^2}\right]x_3$$

Similarly,

$$x_2 = \frac{\sigma_2}{\sigma_3}\left[\frac{r_{23} - r_{12}\,r_{13}}{1 - (r_{13})^2}\right]x_3 + \frac{\sigma_2}{\sigma_1}\left[\frac{r_{12} - r_{23}.r_{13}}{1 - (r_{31})^2}\right]x_1, \quad x_3 = \frac{\sigma_3}{\sigma_2}\left[\frac{r_{23} - r_{12}\,r_{13}}{1 - (r_{12})^2}\right]x_2 + \frac{\sigma_3}{\sigma_1}\left[\frac{r_{13} - r_{23}\,r_{12}}{1 - (r_{12})^2}\right]x_1$$

Standard Error of the estimate for multiple regression and multiple correlation

The standard error of the estimate X_1, X_2 and X_3 is

$$S_{1.23} = \sqrt{\frac{\Sigma(X_1 - Y_1)^2}{N - 3}}$$

where $S_{1.23}$ is standard error of the estimate of X_1 on X_2 and X_3.

X_1 is the original value of X and Y_1 is the estimated value on the basis of the regression equation.

Standard error in terms of multiple correlation

$$S_{1.23} = \sigma_1\sqrt{\frac{1 - r_{12}^2 - r_{13}^2 - r_{23}^2 + 2r_{12}\,r_{13}\,r_{23}}{1 - r_{23}^2}}$$

Example 5. *If* $r_{12} = 0.6$, $r_{13} = 0.8$, $r_{23} = 0.3$

$\sigma_1 = 8$, $\sigma_2 = 9$, $\sigma_3 = 5$

Determine regression equation of x_1 *on* x_2 *and* x_3.

Solution. Let x_1, x_2 and x_3 be the respective deviations from the means of X_1, X_2 and X_3 series.

Regression equation of X_1 on X_2 and X_3 is

$$x_1 = b_{12.3}\,x_2 + b_{13.2}\,x_3 \quad \text{... (1)}$$

Now

$$b_{12.3} = \frac{\sigma_1}{\sigma_2} \times \left[\frac{r_{12} - r_{13}\,r_{23}}{1 - r_{23}^2}\right]$$

$$= \frac{8}{9}\left[\frac{0.6 - 0.8 \times 0.3}{1 - (0.3)^2}\right] = \frac{8}{9}\left[\frac{0.6 - 0.24}{1 - 0.09}\right] = \frac{8}{9}\left[\frac{0.36}{0.91}\right] = \frac{2.88}{8.19} = 0.35$$

$$b_{13.2} = \frac{\sigma_1}{\sigma_3} \times \left[\frac{r_{13} - r_{12}\,r_{23}}{1 - r_{23}^2}\right] = \frac{8}{5}\left[\frac{0.8 - 0.6 \times 0.3}{1 - (0.3)^2}\right]$$

$$= \frac{8}{5}\left[\frac{0.8-0.18}{1-0.09}\right] = \frac{8}{5}\left[\frac{0.62}{0.91}\right] = \frac{4.96}{4.55} = 1.09$$

Substituting the values of $b_{12.3}$ and $b_{13.2}$ in the equation (1), we get

$$x_1 = 0.35\, x_2 + 1.09\, x_3$$ **Ans.**

Example 6. *If* $r_{12} = 0.75,\ r_{13} = 0.65,\ r_{23} = 0.55,\ \sigma_1 = 9,\ \sigma_2 = 7,\ \sigma_3 = 4$

Determine the regression equation of X_2 *on* X_1 *and* X_3.

Solution. The regression equation of X_2 on X_1 and X_3 is given by

$$x_2 = b_{21.3}\, x_1 + b_{23.1}\, x_3 \qquad \ldots (1)$$

We know that

$$b_{21.3} = \frac{\sigma_2}{\sigma_1}\left[\frac{r_{12} - r_{23}\, r_{13}}{1 - r_{31}^2}\right] = \frac{7}{9}\left[\frac{0.75 - 0.55 \times 0.65}{1-(0.65)^2}\right] = \frac{7}{9}\left[\frac{0.75-0.3575}{1-0.4225}\right]$$

$$= \frac{7}{9}\left[\frac{0.3925}{0.5775}\right] = \frac{2.7475}{5.1975} = 0.5286$$

and

$$b_{23.1} = \frac{\sigma_2}{\sigma_3}\left[\frac{r_{23} - r_{12}\, r_{13}}{1 - r_{13}^2}\right] = \frac{7}{4}\left[\frac{0.55 - 0.75 \times 0.65}{1-(0.65)^2}\right] = \frac{7}{4}\left[\frac{0.55-0.4875}{1-0.4225}\right]$$

$$= \frac{7}{4}\left[\frac{0.0625}{0.5775}\right] = \frac{0.4375}{2.3100} = 0.1894$$

Substituting the values of $b_{12.3}$ and $b_{23.1}$ in (1), we get

$$x_2 = 0.5286\, x_1 + 0.1894\, x_3$$ **Ans.**

Example 7. *If* $\sigma_1 = 3,\quad \sigma_2 = 2.5,\quad \sigma_3 = 3.5$

$r_{12} = 0.3,\quad r_{13} = 0.5,\quad r_{23} = 0.4$

Find the regression equation of x_3 *on* x_1 *and* x_2.

Solution. Here, we have

$$\sigma_1 = 3,\quad \sigma_2 = 2.5,\quad \sigma_3 = 3.5$$

$$r_{12} = 0.3,\quad r_{13} = 0.5,\quad r_{23} = 0.4$$

The regression equation of x_3 on x_1 and x_2 is

$$x_3 = b_{31.2}\, x_1 + b_{32.1}\, x_2 \qquad \ldots (1)$$

We know that

$$b_{31.2} = \frac{\sigma_3}{\sigma_1}\left[\frac{r_{13} - r_{23}\, r_{12}}{1 - r_{12}^2}\right] = \frac{3.5}{3}\left[\frac{0.5-0.4\times 0.3}{1-(0.3)^2}\right]$$

$$= \frac{3.5}{3}\left[\frac{0.5-0.12}{1-0.09}\right] = \frac{3.5}{3}\left[\frac{0.38}{0.91}\right] = \frac{1.33}{2.73} = 0.487$$

$$b_{32.1} = \frac{\sigma_3}{\sigma_2}\left[\frac{r_{23} - r_{12}\, r_{13}}{1 - r_{12}^2}\right] = \frac{3.5}{2.5}\left[\frac{0.4-0.3\times 0.5}{1-(0.3)^2}\right]$$

$$= \frac{3.5}{2.5}\left[\frac{0.4-0.15}{1-0.09}\right] = \frac{3.5}{2.5}\left[\frac{0.25}{0.91}\right] = \frac{0.875}{2.275} = 0.385$$

Substituting the values of $b_{31.2}$ and $b_{32.1}$ in (1), we get

$$x_3 = 0.487\, x_1 + 0.385\, x_2$$ **Ans.**

Example 8. *Find the multiple regression equation of x_1 on x_2 and x_3 from the data given below :*

X_1	3	5	6	8	12	10
X_2	10	10	5	7	5	2
X_3	20	25	15	16	15	2

Solution. The regression equation of X_1 on X_2 and X_3 is given by

$$X_1 = a_{1.23} + b_{12.3}\ X_2 + b_{13.2}\ X_3 \quad \text{... (A)}$$

The three normal equations for getting the values of $a_{1.23}$, $b_{12.3}$ and $b_{13.2}$ are

$$\Sigma X_1 = N\ a_{1.23} + b_{12.3}\ \Sigma X_2 + b_{13.2}\ \Sigma X_3$$

$$\Sigma X_1 X_2 = a_{1.23}\ \Sigma X_2 + b_{12.3}\ \Sigma X_2^2 + b_{13.2}\ \Sigma X_2 X_3$$

$$\Sigma X_1 X_3 = a_{1.23}\ \Sigma X_3 + b_{12.3}\ \Sigma X_2 X_3 + b_{13.2}\ \Sigma X_3^2$$

X_1	X_2	X_3	X_1X_2	X_1X_3	X_2X_3	X_2^2	X_3^2
3	10	20	30	60	200	100	400
5	10	25	50	125	250	100	625
6	5	15	30	90	75	25	225
8	7	16	56	128	112	49	256
12	5	15	60	180	75	25	225
10	2	2	20	20	4	4	4
$\Sigma X_1 = 44$	$\Sigma X_2 = 39$	$\Sigma X_3 = 93$	$\Sigma X_1 X_2 = 246$	$\Sigma X_1 X_3 = 603$	$\Sigma X_2 X_3 = 716$	$\Sigma X_2^2 = 303$	$\Sigma X_3^2 = 1735$

Substituting the values in normal equations, we get

$$6\ a_{1.23} + 39\ b_{12.3} + 93\ b_{13.2} = 44 \quad \text{... (1)}$$

$$39\ a_{1.23} + 303\ b_{12.3} + 716\ b_{13.2} = 246 \quad \text{... (2)}$$

$$93\ a_{1.23} + 716\ b_{12.3} + 1735\ b_{13.2} = 603 \quad \text{... (3)}$$

Multiplying (1) by 13 and (2) by 2, we get

$$78\ a_{1.23} + 507\ b_{12.3} + 1209\ b_{13.2} = 572 \quad \text{... (4)}$$

$$78\ a_{1.23} + 606\ b_{12.3} + 1432\ b_{13.2} = 492 \quad \text{... (5)}$$

Subtracting (4) from (5), we get

$$99\ b_{12.3} + 223\ b_{13.2} = -\ 80 \quad \text{... (6)}$$

Multiplying (2) by 31 and (3) by 13, we get

$$1209\ a_{1.23} + 9393\ b_{12.3} + 22196\ b_{13.2} = 7626 \quad \text{... (7)}$$

$$1209\ a_{1.23} + 9308\ b_{12.3} + 22555\ b_{13.2} = 7839 \quad \text{... (8)}$$

Subtracting (8) from (7), we get

$$85\ b_{12.3} - 359\ b_{13.2} = -\ 213 \quad \text{... (9)}$$

Multiplying (6) by 85 and (9) by 99, we get

$$8415\ b_{12.3} + 18955\ b_{13.2} = -\ 6800 \quad \text{... (10)}$$

$$8415\ b_{12.3} - 35541\ b_{13.2} = -\ 21087 \quad \text{... (11)}$$

Subtracting (11) from (10), we get

$$54496\, b_{13.2} = 14287 \quad \Rightarrow \quad b_{13.2} = 0.2621$$

Putting the value of $b_{13.2}$ in (6), we get

$$99\, b_{12.3} + 223\,(0.2621) = -\,80 \quad \Rightarrow \quad 99\, b_{12.3} = -\,80 - 58.4483$$

$$\Rightarrow \quad b_{12.3} = \frac{-138.4483}{99} = -1.3984$$

Putting the values of $b_{12.3}$ and $b_{13.2}$ in (1), we get

$$6\, a_{1.23} + 39\,(-\,1.3984) + 93\,(0.2621) = 44$$

$$\Rightarrow \quad 6\, a_{1.23} - 54.5376 + 24.3753 = 44$$

$$\Rightarrow \quad 6\, a_{1.23} = 44 + 54.5493 - 24.3846 \quad \Rightarrow \quad a_{1.23} = \frac{74.1647}{6} = 12.3608$$

Substituting the values $a_{1.23} = 12.3608$, $b_{12.3} = -\,1.3984$ and $b_{13.2} = 0.2621$ in equation (A), we get

$$X_1 = 12.3608 - 1.3984\, X_2 + 0.2621\, X_3$$

which is the required regression equation of X_1 on X_2 and X_3. **Ans.**

Example 9. *If $r_{12} = 0.60$, $r_{13} = 0.70$, $r_{23} = 0.65$, and $\sigma_1 = 1.0$, find $S_{1.23}$.*

Solution. Here, we have

$$r_{12} = 0.60, \quad r_{13} = 0.70, \quad r_{23} = 0.65, \quad \text{and} \quad S_1 = 1.0$$

We know that

$$S_{1.23} = \sigma_1 \times \sqrt{\frac{1 - r_{12}^2 - r_{13}^2 - r_{23}^2 + 2 r_{12}\, r_{13}\, r_{23}}{1 - r_{23}^2}}$$

$$= 1.0 \times \sqrt{\frac{1 - (0.60)^2 - (0.70)^2 - (0.65)^2 + 2\,(0.60)\,(0.70)\,(0.65)}{1 - (0.65)^2}}$$

$$= 1.0 \times \sqrt{\frac{1 - 0.36 - 0.49 - 0.4225 + 0.546}{1 - 0.4225}}$$

$$= \sqrt{\frac{0.2735}{0.5775}} = \sqrt{0.4736} = 0.6882$$

Ans.

EXERCISE 60.2

1. If $r_{12} = 0.8$, $r_{13} = 0.7$, $r_{23} = 0.6$, $\sigma_1 = 10$, $\sigma_2 = 8$, $\sigma_3 = 5$, then find the regression eqaution of x_1 on x_2 and x_3. **Ans.** $x_1 = 0.742\, x_2 + 0.6875\, x_3$

2. If $\sigma_1 = 3$, $\sigma_2 = 4$, $\sigma_3 = 5$, $r_{12} = 0.7$, $r_{23} = 0.4$, $r_{31} = 0.6$, then determine the regression equation of x_1 on x_2 and x_3. **Ans.** $x_1 = 0.41\, x_2 + 0.229\, x_3$

3. If $r_{12} = 0.28$, $r_{23} = 0.49$, $r_{31} = 0.51$, $\sigma_1 = 2.7$, $\sigma_2 = 2.4$, $\sigma_3 = 2.7$, then find the regression eqaution of x_3 on x_1 and x_2. **Ans.** $x_3 = 0.405\, x_1 + 0.424\, x_2$

4. Find the multiple linear regression equation of x_1 on x_2 and x_3 from the data given below:

x_1	2	4	6	8
x_2	3	5	7	9
x_3	4	6	8	10

Ans. $x_1 = 2x_2 - x_3$

CHAPTER

61

PROBABILITY

61.1 PROBABILITY

Probability is a concept which numerically measure the degree of uncertainity and therefore, of certainity of the occurrence of events.

If an event A can happen in m ways, and fail in n ways, all these ways being equally likely to occur, then the probability of the happening of A is

$$= \frac{\text{Number of favourable cases}}{\text{Total number of mutually exclusive and equally likely cases}} = \frac{m}{m+n}$$

and that of its failing is defined as $\dfrac{n}{m+n}$

If the probability of the happening $= p$

and the probability of not happening $= q$

then $$p+q = \frac{m}{m+n} + \frac{n}{m+n} = \frac{m+n}{m+n} = 1 \text{ or } p+q = 1$$

For instance, on tossing a coin, the probability of getting a head is $\frac{1}{2}$.

61.2 DEFINITIONS

1. **Die :** It is a small cube. Dots are . .. ... :: ::. ::: marked on its faces. Plural of the die is dice. On throwing a die, the outcome is the number of dots on its upper face.
2. **Cards :** A pack of cards consists of four suits *i.e.* Spades, Hearts, Diamonds and Clubs. Each suit consists of 13 cards, nine cards numbered 2, 3, 4, ..., 10, and Ace, a King, a Queen and a Jack or Knave. Colour of Spades and Clubs is black and that of Hearts and Diamonds is red. Kings, Queens, and Jacks are known as *face* cards.
3. **Exhaustive Events or Sample Space :** The set of all possible outcomes of a single performance of an experiment is exhaustive events or sample space. Each outcome is called a sample point. In case of tossing a coin once, $S = (H, T)$ is the *sample space.* Two outcomes Head and Tail constitute an exhaustive event because no other outcome is possible.
4. **Random Experiment :** There are experiments, in which results may be altogether different, even though they are performed under identical conditions. They are known as random experiments. Tossing a coin or throwing a die is random experiment.
5. **Trail and Event :** Performing a random experiment is called a trial and outcome is termed as event. Tossing of a coin is a trial and the turning up of head or tail is an event.
6. **Equally likely events:** Two events are said to be '*equally likely*', if one of them cannot be expected in preference to the other. For instance, if we draw a card from well-shuffled pack, we may get any card, then the 52 different cases are equally likely.
7. **Independent event :** Two events may be *independent*, when the actual happening of one docs not influence in any way the probability of the happening of the other.
Example. The event of getting head on first coin and the event of getting tail on the second

coin in a simultaneous throw of two coins are independent.

8. **Mutually Exclusive events:** Two events are known as *mutually exclusive*, when the occurrence of one of them excludes the occurrence of the other. For example, on tossing of a coin, either we get head or tail, but not both.
9. **Compound Event :** When two or more events occur in composition with each other, the simultaneous occurrence is called a compound event. When a die is thrown, getting a 5 or 6 is a compound event.
10. **Favorable Events :** The events, which ensure the required happening, are said to be favourable events. For example, in throwing a die, to have the even numbers, 2, 4 and 6 are favourable cases.
11. **Conditional Probability : The probability of happening an event *A*, such that event B has already happened, is called the conditional probability of happening of *A* on the condition that *B* has already happened. It is usually denoted by *P (A/B)*.**
12. **Odds in favour of an event and odds against an event**
If number of favourable ways = m, number of not favourable events = n

(*i*) Odds in favour of the event = $\frac{m}{n}$, Odds against the event = $\frac{n}{m}$.

13. **Classical Definition of Probability.** If there are N equally likely, mutually, exclusive and exhaustive of events of an experiment and m of these are favourable, then the probability of the happening of the event is defined as $\frac{m}{N}$.
14. **Expected value.** If $p_1, p_2, p_3, \ldots, p_n$ of the probabilities of the events $x_1, x_2, x_3 \ldots x_n$ respectively the expected value

$$E(x) = p_1 x_1 + p_2 x_2 + p_3 x_3 + \ldots\ldots + p_n x_n = \sum_{r=1}^{n} p_r x_r$$

Example 1. *Find the probability of throwing*
(a) 5, (b) an even number with an ordinary six faced die.

Solution. (*a*) There are 6 possible ways in which the die can fall and there is only one way of throwing 5.

$$\text{Probability} = \frac{\text{Number of favourable ways}}{\text{Total number of equally likely ways}} = \frac{1}{6} \qquad \textbf{Ans.}$$

(*b*) Total number of ways of throwing a die = 6
Number of ways falling 2, 4, 6 = 3

$$\text{The required probability} = \frac{3}{6} = \frac{1}{2} \qquad \textbf{Ans.}$$

Example 2. *Find the probability of throwing 9 with two dice.*

Solution. Total number of possible ways of throwing two dice
= 6 × 6 = 36

Number of ways getting 9. *i.e.*, (3 + 6), (4 + 5), (5 + 4), (6 + 3) = 4.

$$\therefore \quad \text{The required probability} = \frac{4}{36} = \frac{1}{9}. \qquad \textbf{Ans.}$$

Example 3. *From a pack of 52 cards, one is drawn at random. Find the probability of getting a king.*

Solution. A king can be chosen in 4 ways.
But a card can be drawn in 52 ways.

$\therefore$ the required probability $= \frac{4}{52} = \frac{1}{13}$ **Ans.**

EXERCISE 61.1

1. In a class of 12 students, 5 are boys and the rest are girls. Find the probability that a student selected will be a girl. **Ans.** $\frac{7}{12}$
2. A bag contains 7 red and 8 black balls. Find the probability of drawing a red ball. **Ans.** $\frac{7}{15}$
3. Three of the six vertices of a regular hexagon are chosen at random. Find the probability that the triangle with three vertices is equilateral. **Ans.** $\frac{1}{10}$
4. What is the probability that a leap year, selected at random, will contain 53 Sundays. *(A.M.I.E., Dec. 2009, Summer 2001)* **Ans.** $\frac{2}{7}$

Fill in the blanks with appropriate correct answer

5. Chance of throwing 6 at least once in four throws with single dice is .... *(A.M.I.E., Summer 2000)* **Ans.** $\frac{671}{1296}$
6. A pair of fair dice is thrown and one die shows a four. The probability that the other die shows 5 is .... *(A.M.I.E., Summer 2000)* **Ans.** $\frac{1}{36}$

61.3 ADDITION LAW OF PROBABILITY

If A and B are two events associated with an experiment; then

$$P(A \cup B) = P(A) + P(B) - P(A \cap B)$$

Proof. Let m_1, m_2, and m be the number of favourable outcomes to the events A, B and $A \cap B$ respectively. The mutually exclusive outcomes in the sample space of the experiment be n.

$$P(A) = \frac{m_1}{n},\quad P(B) = \frac{m_2}{n},\quad P(A \cap B) = \frac{m}{n}$$

The favourable outcomes to the event A only $= m_1 - m$

The favourable outcomes to the event B only $= m_2 - m$

The favourable outcomes to the event $A \cap B$ only $= m$.

The favourable outcomes to the events A or B or both i.e.,

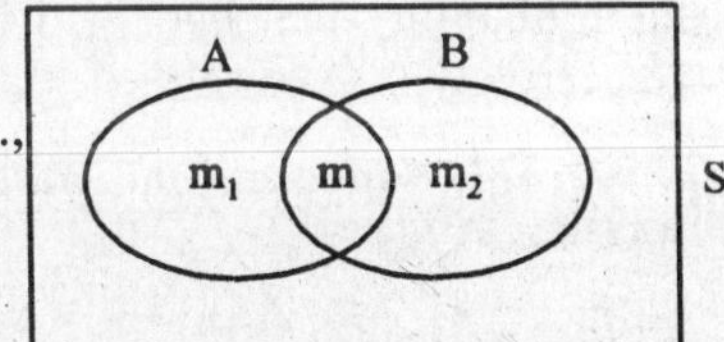

$$A \cup B = m_1 - m) + (m_2 - m) + m$$
$$= m_1 + m_2 - m$$

So, $$P(A \cup B) = \frac{m_1 + m_2 - m}{n}$$

$$= \frac{m_1}{n} + \frac{m_2}{n} - \frac{m}{n}$$
$$= P(A) + P(B) - P(A \cap B)$$

Theorem. If A and B are any two events prove that $P(A \cup B) = P(A) + P(B) - P(A \cap B)$ and hence prove that if A, B and C are any three events.

$$P(A \cup B \cup C) = P(A) + P(B) + P(C) - P(A \cap B) - P(B \cap C) - P(A \cap C) + P(A \cap B \cap C)$$

(AMIETE, June 2010)

Note: Mutually Exclusive Events

Consider the case where two events A and B are not mutually exclusive. The probability of the event that either A or B or both occur is given as

$$P(A \cup B) = P(A) + P(B) - P(A \cap B)$$

Example 4. *An urn contains 10 black and 10 white balls. Find the probability of drawing two balls of the same colour.*

Solution. Probability of drawing two black balls $= \dfrac{{}^{10}C_2}{{}^{20}C_2}$

$\therefore$ Probability of drawing two white balls $= \dfrac{{}^{10}C_2}{{}^{20}C_2}$

$\therefore$ Probability of drawing two balls of the same colour

$$= \frac{{}^{10}C_2}{{}^{20}C_2} + \frac{{}^{10}C_2}{{}^{20}C_2} = 2.\frac{{}^{10}C_2}{{}^{20}C_2} = 2.\frac{\dfrac{10\times 9}{2\times 1}}{\dfrac{20\times 19}{2\times 1}} = \frac{9}{19} \qquad \textbf{Ans.}$$

Example 5. *A bag contains four white and two black balls and a second bag contains three of each colour. A bag is selected at random, and a ball is then drawn at random from the bag chosen. What is the probability that the ball drawn is white ?.*

Solution. There are two mutually exclusive cases,
(*i*) when the first bag is chosen, (*ii*) when the second bag is chosen.

Now the chance of choosing the first bag is $\frac{1}{2}$ and if this bag is chosen, the probability of drawing a white ball is 4/6. Hence the probability of drawing a white ball from first bag is

$$\frac{1}{2} \times \frac{4}{6} = \frac{1}{3}$$

Similarly the probability of drawing a white ball from second bag is $\frac{1}{2} \times \frac{3}{6} = \frac{1}{4}$

Since the events are mutually exclusive the required probability $= \frac{1}{3} + \frac{1}{4} = \frac{7}{12}$ **Ans.**

61.4 MULTIPLICATION LAW OF PROBABILITY

If there are two independent events the respective probabilities of which are known, then the probability that both will happen is the product of the probabilities of their happening respectively.

$$P(AB) = P(A) \times P(B)$$

Proof. Suppose A and B are two independent events. Let A happen in m_1 ways and fail in n_1 ways.

$$\therefore \qquad P(A) = \frac{m_1}{m_1 + n_1}$$

Also let B happen in m_2 ways and fail in n_2 ways.

$$\therefore \qquad P(B) = \frac{m_2}{m_2 + n_2}$$

Now there are four possibilities

A and B both may happen, then the number of ways $= m_1 . m_2$.
A may happen and B may fail, then the number of ways $= m_1 . n_2$
A may fail and B may happen, then the number of ways $= n_1 . m_2$
A and B both may fail, then the number of ways $= n_1 . n_2$

Thus, the total number of ways $= m_1 m_2 + m_1 n_2 + n_1 m_2 + n_1 n_2 = (m_1 + n_1)(m_2 + n_2)$
Hence the probabilities of the happening of both A and B

$$P(AB) = \frac{m_1 m_2}{(m_1+n_1)(m_2+n_2)} = \frac{m_1}{m_1+n_1} \cdot \frac{m_2}{m_2+n_2} = P(A) \,.\, P(B)$$ **Proved.**

Example 6. *An article manufactured by a company consists of two parts A and B. In the process of manufacture of part A, 9 out of 100 are likely to be defective. Similarly, 5 out of 100 are likely to be defective in the manufacture of part B. Calculate the probability that the assembled article will not be defective (assuming that the events of finding the part A non-defective and that of B are independent).*

Solution. Probability that part A will be defective $= \frac{9}{100}$

Probability that part A will not be defective $= \left(1 - \frac{9}{100}\right) = \frac{91}{100}$

Probability that part B will be defective $= \frac{5}{100}$

Probability that part B will not be defective $= \left(1 - \frac{5}{100}\right) = \frac{95}{100}$

Probability that the assembled article will not be defective = (Probability that part A will not be defective)× (Probability that part B will not be defective)

$$= \left(\frac{91}{100}\right) \times \left(\frac{95}{100}\right) = 0.8645$$ **Ans.**

Example 7. *The probability that machine A will be performing an usual function in 5 years' time is $\frac{1}{4}$, while the probability that machine B will still be operating usefully at the end of the same period, is $\frac{1}{3}$*

Find the probability in the following cases that in 5 years time :

(i) Both machines will be performing an usual function.

(ii) Neither will be operating.

(iii) Only machine B will be operating.

(iv) At least one of the machines will be operating.

Solution. P (A operating usefully) $= \frac{1}{4}$, so $q\ (A) = 1 - \frac{1}{4} = \frac{3}{4}$

P (B operating usefully) $= \frac{1}{3}$, so $q\ (B) = 1 - \frac{1}{3} = \frac{2}{3}$

(*i*) P (Both *A* and *B* will operate usefully) $= P(A).\, P(B) = \left(\frac{1}{4}\right) \times \left(\frac{1}{3}\right) = \frac{1}{12}$

(*ii*) P (Neither will be operating) $= q(A)\,.\,q(B) = \left(\frac{3}{4}\right)\left(\frac{2}{3}\right) = \frac{1}{2}$

(*iii*) P (Only B will be operating) $= P(B) \times q(A) = \left(\frac{1}{3}\right) \times \left(\frac{3}{4}\right) = \frac{1}{4}$

(*iv*) P (At least one of the machines will be operating)

= 1 – P (none of them operates)

$$= 1 - \frac{1}{2} = \frac{1}{2}$$ **Ans.**

Example 8. *There are two groups of subjects one of which consists of 5 science and 3 engineering subjects and the other consists of 3 science and 5 engineering subjects. An unbiased* *die is cast. If number 3 or number 5 turns up, a subject is selected at random from the first group, otherwise the subject is selected at random from the second group. Find the probability that an engineering subject is selected ultimately.*

(A.M.I.E.T.E., Summer 2000)

Solution. Probability of turning up 3 or 5 $= \frac{2}{6} = \frac{1}{3}$

Probability of selecting engineering subject from first group $= \frac{3}{8}$

Now the probability of selecting engineering subject from first group on turning up 3 or 5

$$= \left(\frac{1}{3}\right) \times \left(\frac{3}{8}\right) = \frac{1}{8} \quad \dots (1)$$

Probability of not turning up 3 or 5 $= 1 - \frac{1}{3} = \frac{2}{3}$

Probability of selecting engineering subject from second group $= \frac{5}{8}$

Now probability of selecting engineering subject from second group on not turning up 3 or 5

$$= \frac{2}{3} \times \frac{5}{8} = \frac{5}{12} \quad \dots (2)$$

Probability of the selection of engineering subject $= \frac{1}{8} + \frac{5}{12}$ [From (1) and (2)]

$$= \frac{13}{24}$$ **Ans.**

Example 9. *An urn contains nine balls, two of which are red, three blue and four black. Three balls are drawn from the urn at random. What is the probability that*

(i) the three balls are of different colours?

(ii) the three balls are of the same colour? *(A.M.I.E., Summer 2000)*

Solution.

Urn contains 2 Red balls, 3 Blue balls and 4 Black balls.

(*i*) Three balls will be of different colours if one ball is red, one blue and one black ball are drawn.

$$\text{Required probability} = \frac{{}^2C_1 \times {}^3C_1 \times {}^4C_1}{{}^9C_3} = \frac{2 \times 3 \times 4}{84} = \frac{2}{7}$$ **Ans.**

(*ii*) Three balls will be of the same colour if either 3 blue balls or 3 black balls are drawn.

P (3 Blue balls or 3 Black balls) = P (3 Blue balls) + P (3 Black balls)

$$= \frac{{}^3C_3}{{}^9C_3} + \frac{{}^4C_3}{{}^9C_3} = \frac{1+4}{84} = \frac{5}{84}$$ **Ans.**

Example 10. *A bag contains 10 white and 15 black balls. Two balls are drawn in succession. What is the probability that first is white and second is black ?*

Solution. Probability of drawing one white ball $= \frac{10}{25}$

Probability of drawing one black ball without replacement = 15/24

Required probability of drawing first white ball and second black ball

$$= \frac{10}{25} \times \frac{15}{24} = \frac{1}{4}$$ **Ans.**

Example 11. *A committee is to be formed by choosing two boys and four girls out of a group of five boys and six girls. What is the probability that a particular boy named A and a particular girl named B are selected in the committee?*

Solution. Two boys are to be selected out of 5 boys. A particular boy A is to be included in the committee. It means that only 1 boy is to be selected out of 4 boys.

Number of ways of selection = ${}^4C_1$

Similarly a girl B is to be included in the committee.

Then only 3 girls are to be selected out of 5 girls.

Number of ways of selection = ${}^5C_3$

$$\text{Required probability} = \frac{{}^4C_1 \times {}^5C_3}{{}^5C_2 \times {}^6C_4} = \frac{4\times 10}{10\times 15} = \frac{4}{15}$$ **Ans.**

Example 12. *Three groups of children contain respectively 3 girls and 1 boy; 2 girls and 2 boys; 1 girl and 3 boys. One child is selected at random from each group. Find the chance of selecting 1 girl and 2 boys.*

Solution. There are three ways of selecting 1 girl and two boys.

I way : Girl is selected from first group, boy from second group and second boy from third group.

$$\text{Probability of the selection of (Girl + Boy + Boy)} = \frac{3}{4}\times\frac{2}{4}\times\frac{3}{4} = \frac{18}{64}$$

II way : Boy is selected from first group, girl from second group and second boy from third group.

$$\text{Probability of the selection of (Boy + Girl + Boy)} = \frac{1}{4}\times\frac{2}{4}\times\frac{3}{4} = \frac{6}{64}$$

III way : Boy is selected from first group, second boy from second group and the girl from the third group.

$$\text{Probability of selection of (Boy + Boy + Girl)} = \frac{1}{4}\times\frac{2}{4}\times\frac{1}{4} = \frac{2}{64}$$

$$\text{Total probability} = \frac{18}{64}+\frac{6}{64}+\frac{2}{64} = \frac{26}{64} = \frac{13}{32}$$ **Ans.**

Example 13. *The number of children in a family in a region are either 0, 1 or 2 with probability 0.2, 0.3 and 0.5 respectively. The probability of each child being a boy or girl 0.5. Find the probability that a family has no boy.*

Solution. Here there are three types of families

(*i*) Probability of zero child (boys) = 0.2

(*ii*)

Boy	Girl
0	1
1	0

Probability of zero boy in case II = 0.3 × 0.5 = 0.15

(*iii*)

Boy	Girl
0	2
1	1
2	0

In this case probability of zero boy = $0.5\times\frac{1}{3} = 0.167$

Considering all the three cases, the probability of zero boy

= 0.2 + 0.15 + 0.167 = 0.517 **Ans.**

Example 14. *A husband and wife appear in an interview for two vacancies in the same post. The probability of husband's selection is* $\frac{1}{7}$ *and that of wife's selection is* $\frac{1}{5}$. *What is the probability that*
(*i*) *both of them will be selected.* (*ii*) *only one of them will be selected, and* (*iii*) *none of them will be selected ?*

Solution. P (husband's selection) $= \frac{1}{7}$, P (wife's selection) $= \frac{1}{5}$

(*i*) P (both selected) $= \frac{1}{7} \times \frac{1}{5} = \frac{1}{35}$.

(*ii*) P (only one selected) = P (only husband's selection) + P (only wife's selection)\

$$= \frac{1}{7} \times \frac{4}{5} + \frac{1}{5} \times \frac{6}{7} = \frac{10}{35} = \frac{2}{7}$$

(*iii*) P (none of them will be selected) $= \frac{6}{7} \times \frac{4}{5} = \frac{24}{35}$ **Ans.**

Example 15. *A problem of statistics is given to three students A, B and C whose chances of solving it are* $\frac{1}{2}, \frac{3}{4}$ *and* $\frac{1}{4}$ *respectively. What is the probability that the problem will be solved?*

Solution. The probability that A can solve the problem $= \frac{1}{2}$

The probability that A cannot solve the problem $= 1 - \frac{1}{2}$.

Similarly the probability that B and C cannot solve the problem are

$$\left(1 - \frac{3}{4}\right) \text{ and } \left(1 - \frac{1}{4}\right)$$

$\therefore$ The probability that A, B, C cannot solve the problem

$$= \left(1 - \frac{1}{2}\right) \times \left(1 - \frac{3}{4}\right) \times \left(1 - \frac{1}{4}\right) = \frac{1}{2} \times \frac{1}{4} \times \frac{3}{4} = \frac{3}{32}$$

Hence, the probability that the problem can be solved $= 1 - \frac{3}{32} = \frac{29}{32}$ **Ans.**

Example 16. *A student takes his examination in four subjects α, β, γ, δ. He estimates his chances of passing in α as* $\frac{4}{5}$, *in β as* $\frac{3}{4}$, *in γ as* $\frac{5}{6}$ *and in δ as* $\frac{2}{3}$. *To qualify, he must pass in α and at least two other subjects. What is the probability that he qualifies?* (*AMIETE, Dec. 2010*)

Solution. $P(\alpha) = \frac{4}{5}, \quad P(\beta) = \frac{3}{4}, \quad P(\gamma) = \frac{5}{6}, \quad P(\delta) = \frac{2}{3}$

There are four possibilities of passing at least two subjects.

(*i*) Probability of passing β, γ and failing δ $= \frac{3}{4} \times \frac{5}{6} \times \left(1 - \frac{2}{3}\right) = \frac{3}{4} \times \frac{5}{6} \times \frac{1}{3} = \frac{5}{24}$

(*ii*) Probability of passing γ, δ and failing β $= \frac{5}{6} \times \frac{2}{3} \times \left(1 - \frac{3}{4}\right) = \frac{5}{6} \times \frac{2}{3} \times \frac{1}{4} = \frac{5}{36}$

(*iii*) Probability of passing δ, β and failing γ $= \frac{2}{3} \times \frac{3}{4} \times \left(1 - \frac{5}{6}\right) = \frac{2}{3} \times \frac{3}{4} \times \frac{1}{6} = \frac{1}{12}$

(*iv*) Probability of passing $\beta, \gamma, \delta = \frac{3}{4} \times \frac{5}{6} \times \frac{2}{3} = \frac{5}{12}$

Probability of passing at least two subjects $= \frac{5}{24} + \frac{5}{36} + \frac{1}{12} + \frac{5}{12} = \frac{61}{72}$

Probability of passing α and at least two subjects. $= \frac{4}{5} \times \frac{61}{72} = \frac{61}{90}$ **Ans.**

Example 17. *There are 6 positive and 8 negative numbers. Four numbers are chosen at random, without replacement, and multiplied. What is the probability that the product is a positive number ?*

Solution. To get from the product of four numbers, a positive number, the possible combinations are as follows :

S. No.	*Out of 6 Positive Numbers*	*Out of 8 Negative Numbers*	*Positive Numbers*
1.	4	0	${}^6C_4 \times {}^8C_0 = \frac{6 \times 5}{1 \times 2} \times 1 = 15$
2.	2	2	${}^6C_2 \times {}^8C_2 = \frac{6 \times 5}{1 \times 2} \times \frac{8 \times 7}{1 \times 2} = 420$
3.	0	4	${}^6C_0 \times {}^8C_4 = 1 \times \frac{8 \times 7 \times 6 \times 5}{1 \times 2 \times 3 \times 4} = 70$
			Total = 505

$$\text{Probability} = \frac{{}^6C_4 \times {}^8C_0 + {}^6C_2 \times {}^8C_2 + {}^6C_0 \times {}^8C_4}{{}^{14}C_4} = \frac{15 + 420 + 70}{\frac{14 \times 13 \times 12 \times 11}{1 \times 2 \times 3 \times 4}} = \frac{505 \times 4 \times 3 \times 2 \times 1}{14 \times 13 \times 12 \times 11} = \frac{505}{1001}$$ **Ans.**

Example 18. *A can hit a target 3 times in 5 shots, B 2 times in 5 shots and C three times in 4 shots. All of them fire one shot each simultaneously at the target. What is the probability that*
(i) 2 shots hit *(ii) At least two shots hit?*

Solution. Probability of A hitting the target $= \frac{3}{5}$

Probability of B hitting the target $= \frac{2}{5}$

Probability of C hitting the target $= \frac{3}{4}$

Probability that 2 shots hit the target

$$= P(A)\,P(B)\,q(C) + P(A)\,P(C)\,q(B) + P(B)\,P(C)\,q(A)$$

$$= \frac{3}{5} \times \frac{2}{5} \times \left(1 - \frac{3}{4}\right) + \frac{3}{5} \times \frac{3}{4} \times \left(1 - \frac{2}{5}\right) + \frac{2}{5} \times \frac{3}{4} \times \left(1 - \frac{3}{5}\right) = \frac{6}{25} \times \frac{1}{4} + \frac{9}{20} \times \frac{3}{5} + \frac{6}{20} \times \frac{2}{5}$$

$$= \frac{6 + 27 + 12}{100} = \frac{45}{100} = \frac{9}{20}$$ **Ans.**

(*ii*) Probability of at least two shots hitting the target

= Probability of 2 shots + probability of 3 shots hitting the target

$$= \frac{9}{20} + P(A)\,P(B)\,P(C) = \frac{9}{20} + \frac{3}{5} \times \frac{2}{5} \times \frac{3}{4} = \frac{63}{100}$$ **Ans.**

Example 19. *A and B take turns in throwing two dice, the first to throw 10 being awarded the prize. Show that if A has the first throw, their chances of winning are in the ratio 12:11.* *(AMIETE, Dec. 2009)*

Solution. The combinations of throwing 10 from two dice can be

$$(6 + 4), (4 + 6), (5 + 5).$$

The number of combinations is 3.

Total combinations from two dice = 6 × 6 = 36.

$\therefore$ The probability of throwing 10 = $p = \frac{3}{36} = \frac{1}{12}$

The probability of not getting 10 = $q = 1 - \left(\frac{1}{12}\right) = \frac{11}{12}$

If A is to win, he should throw 10 in either the first, the third, the fifth, ... throws.

Their respective probabilities are = $p,\ q^2 p,\ q^4 p,\ \ldots = \frac{1}{12},\ \left(\frac{11}{12}\right)^2 \frac{1}{12},\ \left(\frac{11}{12}\right)^4 \frac{1}{12} \ldots$

A's total probability of winning $= \frac{1}{12} + \left(\frac{11}{12}\right)^2 \cdot \frac{1}{12} + \left(\frac{11}{12}\right)^4 \cdot \frac{1}{12} + \ldots$

$$= \frac{\frac{1}{12}}{1 - \left(\frac{11}{12}\right)^2} = \frac{12}{23} \qquad \left[\text{This is infinte G.P. Its sum} = \frac{a}{1-r}\right]$$

B can win in either 2nd, 4th, 6th ... throws.

So B's total chance of winning $= qp + q^3 p + q^5 p + \ldots\ldots\ldots$

$$= \left(\frac{11}{12}\right)\left(\frac{1}{12}\right) + \left(\frac{11}{12}\right)^3 \left(\frac{1}{12}\right) + \left(\frac{11}{12}\right)^5 \left(\frac{1}{12}\right) + \ldots = \frac{\left(\frac{11}{12}\right)\left(\frac{1}{12}\right)}{1 - \left(\frac{11}{12}\right)^2} = \frac{11}{23}$$

Hence A's chance to B's chance $= \frac{12}{23} : \frac{11}{23} = 12 : 11$ **Proved.**

Example 20. *A and B throw alternatively a pair of dice. A wins if he throws 6 before B throws 7 and B wins if he throws 7 before A throws 6. Find their respective chances of winning, if A begins.*

Solution. Number of ways of throwing 6

i.e. $(1 + 5), (2 + 4), (3 + 3), (4 + 2), (5 + 1) = 5.$

Probability of throwing $6 = \frac{5}{36} = p_1,\ q_1 = \frac{31}{36}$

Number of ways of throwing 7

i.e.; $(1 + 6), (2 + 5), (3 + 4), (4 + 3), (5 + 2), (6 + 1) = 6$

Probability of throwing $6 = \frac{6}{36} = \frac{1}{6} = P_2,\ q_2 = \frac{5}{6}$

$$P(A) = p_1 + q_1 q_2 p_1 + q_1^2 q_2^2 p_1 + \ldots.$$

$$P(B) = q_1 p_2 + q_1^2 q_2 p_2 + q_1^3 q_2^2 p_2 + \ldots.$$

Probability of A's winning $= p_1 + q_1 q_2 p_1 + q_1^2 q_2^2 p_1 + \ldots.$

$$= \frac{p_1}{1-q_1q_2} = \frac{\frac{5}{36}}{1-\frac{31}{36}\times\frac{5}{6}} = \frac{5}{36}\times\frac{36\times 6}{61} = \frac{30}{61}$$

Probability of B's winning $= q_1 p_2 + q_1^2 q_2 p_2 + q_1^3 q_2^2 p_2 + \ldots\ldots$

$$= \frac{q_1 p_2}{1-q_1q_2} = \frac{\frac{31}{36}\times\frac{1}{6}}{1-\left(\frac{31}{36}\right)\left(\frac{5}{6}\right)} = \frac{31}{36\times 6}\times\frac{36\times 6}{61} = \frac{31}{61}$$ **Ans.**

EXERCISE 61.2

1. The probability that Nirmal will solve a problem is $\frac{2}{3}$ and the probability that Satyajit will solve it is $\frac{3}{4}$. What is the probability that (*a*) the problem will be solved (*b*) neither can solve it.
 Ans. (*a*) $\frac{11}{12}$, (*b*) $\frac{1}{12}$
2. An urn contains 13 balls numbering 1 to 13. Find the probability that a ball selected at random is a ball with number that is a multiple of 3 or 4. **Ans.** $\frac{6}{13}$
3. Four persons are chosen at random from a group containing 3 men, 2 women, and 4 children. Show that the probability that exactly two of them will be children is $\frac{10}{21}$.
4. A five digit number is formed by using the digits 0, 1, 2, 3, 4 and 5 without repetition. Find the probability that the number is divisible by 6. **Ans.** $\frac{4}{25}$
5. The chances that doctor A will diagnose a disease X correctly is 60%. The chances that a patient will die by his treatment after correct diagnosis is 40% and the chances of death by wrong diagnosis is 70%. A patient of doctor A, who had disease *X*, died, what is the chance that his disease was diagnosed correctly.
 Ans. $\frac{6}{13}$
6. An anti-aircraft gun can take a maximum of four shots on enemy's plane moving from it. The probabilities of hitting the plane at first, second, third and fourth shots are 0.4, 0.3, 0.2 and 0.1 respectively. Find the probability that the gun hits the plane. **Ans.** 0.6976.
7. An electronic component consists of three parts. Each part has probability 0.99 of performing satisfactorily. The component fails if two or more parts do not perform satisfactorily. Assuming that the parts perform independently, determine the probability that the component does not perform satisfactorily.
 Ans. 0.000298
8. The face cards are removed from a full pack. Out of the remaining 40 cards, 4 are drawn at random. What is the probability that they belong to different suits? **Ans.** $\frac{1000}{9139}$
9. Of the cigarette smoking population, 70% are men and 30% women, 10% of these men and 20% of these women smoke 'WILLS.' What is the probability that a person seen smoking a 'WILLS' will be a man.
 Ans. $\frac{7}{13}$
10. A machine contains a component *C* that is vital to its operation. The reliability of component *C* is 80%. To improve the reliability of a machine, a similar component is used in parallel to form a system S. The machine will work provided that one of these components functions correctly. Calculate the reliability of the system *S*. **Ans.** 96%

11. The odds that a book will be favourably reviewed by three independent critics are 5 to 2, 4 to 3, 3 to 4 respectively. What is the probability that of the three reviews, a majority will be favourable?

Ans. $\frac{209}{343}$

12. A man takes a step forward with probability 0.4 and backward with probability 0.6. Find the probability that at the end of 11 steps, he is just one step away from the starting point.

Ans. 0.5263

13. A candidate is selected for interview for three posts. For the first post there are three candidates, for the second there are 4, and for the third are 2. What is the chance of getting at least one post?

(*A.M.I.E., Summer 2001*) **Ans.** $\frac{3}{4}$

14. The chance of hitting a target by a bomb is 50% when 4 bombs are dropped, what is the probability of destroying the target, if one bomb is just sufficient to destroy it. (*A.M.I.E., Winter 2003*)

Ans. $\frac{15}{16}$

15. Tick √ the correct answer :

(*i*) *A*, *B*, *C* in order toss a coin, the first to throw a head wins. Assuming if A begins and the game continues indefinitely their respective chances of winning are:

(*a*) $\frac{4}{7}, \frac{2}{7}, \frac{1}{7}$ (*b*) $\frac{1}{7}, \frac{4}{7}, \frac{2}{7}$ (*c*) $\frac{2}{7}, \frac{4}{7}, \frac{1}{7}$ (*d*) None of these

(*A.M.I.E.,winter 2000*) **Ans.** (*a*)

(*ii*) An unbiased die with faces marked 1, 2, 3, 4, 5, 6 is rolled 4 times, out of four face values obtained, the probability that the minimum face value is not less than 2 and the maximum face value is not greater than 5 is then

(*a*) $\frac{16}{81}$ (*b*) $\frac{2}{9}$ (*c*) $\frac{80}{81}$ (*d*) $\frac{8}{9}$

(*A.M.I.E.T.E., Summer 2000*) **Ans.** (*a*)

(*iii*) India plays two matches each with England and Australia. In any match the probability of its getting points 0, 1 and 2 are 0.45, 0.05 and 0.5 respectively. Assuming the outcomes are independent, the probability that India gets at least seven points is

(*a*) 0.8750 (*b*) 0.0875 (*c*) 0.0625 (*d*) 0.0250

(*A.M.I.E.T.E., Summer 2001*) **Ans.** (*b*)

Fill up the blanks:

(*iv*) Probability of any event can not be greater than _____ and less than _____.

(*A.M.I.E.T.E, Winter 2001*) **Ans.** 0, 1

61.5 CONDITIONAL PROBABILITY

Let *A* and *B* be two events of a sample space *S* and let $P(B) \neq 0$. Then conditional probability of the event *A*, given *B*, denoted by $P(A/B)$, is defined by

$$P(A/B) = \frac{P(A \cap B)}{P(B)} \quad \text{... (1)}$$

Theorem. If the events *A* and *B* defined on a sample space S of a random experiment are independent, then

$$P(A/B) = P(A) \text{ and } P(B/A) = P(B)$$

Proof. A and B are given to be independent events,

$$P(A \text{ and } B) = P(A) \bullet P(B)$$

$$\Rightarrow \quad P(A/B) = \frac{P(A \cap B)}{P(B)} = \frac{P(A).P(B)}{P(B)} = P(A)$$

$$\Rightarrow \quad P(B/A) = \frac{P(B \cap A)}{P(A)} = \frac{P(B).P(A)}{P(A)} = P(B)$$

61.6 BAYE'S THEOREM

If $B_1, B_2, B_3, \ldots\ldots B_n$ are mutually exclusive events with $P(B_i) \neq 0$, $(i = 1, 2, \ldots n)$ of a random experiment then for any arbitrary event A of the sample space of the above experiment with $P(A) > 0$, we have

$$P(B_i / A) = \frac{P(B_i) P(A / B_i)}{\sum_{i=1}^{n} P(B_i) P(A / B_i)} \qquad \text{(for } n = 3\text{)}$$

$$P(B_2 / A) = \frac{P(B_2)P(A/B_2)}{P(B_1)P(A/B_1) + P(B_2)P(A/B_2) + P(B_3)P(A/B_3)}$$

Proof. Let S be the sample space of the random experiment.

The events $B_1, B_2, \ldots, B_n$ being exhaustive

$$S = B_1 \cup B_2 \cup \ldots . \cup B_n$$

$$[\because A \subset S]$$

$$\therefore \quad A = A \cap S = A \cap (B_1 \cup B_2 \cup \ldots . \cup B_n)$$

$$= (A \cap B_1) \cup (A \cap B_2) \cup \ldots . \cup (A \cap B_n) \quad \text{[Distributive Law]}$$

$$\Rightarrow \quad P(A) = P(A \cap B_1) + P(A \cap B_2) + \ldots . + P(A \cap B_n)$$

$$= P(B_1) P(A/B_1) + P(B_2) P(A/B_2) + \ldots . + P(B_n) P(A/B_n)$$

$$= \sum_{i=1}^{n} P(B_i) P(A/B_i) \qquad \ldots (1)$$

Now, $\quad P(A \cap B_i) = P(A) P(B_i / A)$

$$\Rightarrow \quad P(B_i / A) = \frac{P(A \cap B_i)}{P(A)} = \frac{P(B_i)P(A/B_i)}{\sum_{i=1}^{n} P(B_i)P(A/B_i)} \qquad \text{[Using (1)]}$$

Note. $P(B)$ is the probability of occurrence B. If we are told that the event A has already occurred. On knowing about the event A, $P(B)$ is changed to $P(B/A)$. With the help of Baye's theorem we can calculate $P(B / A)$.

Example 21. *An urn I contains 3 white and 4 red balls and an urn II contains 5 white and 6 red balls. One ball is drawn at random from one of the urns and is found to be white. Find the probability that it was drawn from urn I.*

Solution. Let U_1: the ball is drawn from urn I

U_2 : the ball is drawn from urn II

W : the ball is white.

We have to find $P(U_1/W)$

By Baye's Theorem

$$P(U_1/W) = \frac{P(U_1)P(W/U_1)}{P(U_1)P(W/U_1) + P(U_2)P(W/U_2)} \quad \dots (1)$$

Since two urns are equally likely to be selected, $P(U_1) = P(U_2) = \frac{1}{2}$

$P(W/U_1) = P$ (a white ball is drawn from urn I) $= \frac{3}{7}$

$P(W/U_2) = P$ (a white ball is drawn from urn II) $= \frac{5}{11}$

$\therefore$ From (1), $$P(U_1/W) = \frac{\frac{1}{2}\times\frac{3}{7}}{\frac{1}{2}\times\frac{3}{7}+\frac{1}{2}\times\frac{5}{11}} = \frac{33}{68}$$ **Ans.**

Example 22. *Three urns contains 6 red, 4 black; 4 red, 6 black; 5 red, 5 black balls respectively. One of the urns is selected at random and a ball is drawn from it. If the ball drawn is red find the probability that it is drawn from the first urn.*

Solution. Let U_1: the ball is drawn from U_1.

U_2: the ball is drawn from U_2.

U_3: the ball is drawn from U_3.

R : the ball is red.

We have to find $P(U_1/R)$.

By Baye's Theorem,

$$P(U_1/R) = \frac{P(U_1)P(R/U_1)}{P(U_1)P(R/U_1) + P(U_2)P(R/U_2) + P(U_3)P(R/U_3)} \quad \dots (1)$$

Since the three urns are equally likely to be selected $P(U_1) = P(U_2) = P(U_3) = \frac{1}{3}$

Also $P(R/U_1) = P$ (a red ball is drawn from urn I) $= \frac{6}{10}$

$P(R/U_2) = P$ (a red ball is drawn from urn II) $= \frac{4}{10}$

$P(R/U_3) = P$ (a red ball is drawn from urn III) $= \frac{5}{10}$

$\therefore$ From (1), we have $$P(U_1/R) = \frac{\frac{1}{3}\times\frac{6}{10}}{\frac{1}{3}\times\frac{6}{10}+\frac{1}{3}\times\frac{4}{10}+\frac{1}{3}\times\frac{5}{10}} = \frac{2}{5}$$ **Ans.**

Example 23. *In a bolt factory, machines A, B and C manufacture respectively 25%, 35% and 40% of the total. If their output 5, 4 and 2 per cent are defective bolts. A bolt is drawn at random from the product and is found to be defective. What is the probability that it was manufactured by machine B?*

Solution. A : bolt is manufactured by machine A.

B : bolt is manufactured by machine B.

C : bolt is manufactured by machine C.

$$P(A) = 0.25,\ P(B) = 0.35,\ P(C) = 0.40$$

The probability of drawing a defective bolt manufactured by machine A is $P(D/A) = 0.05$

Similarly, $P(D/B) = 0.04$ and $P(D/C) = 0.02$

By Baye's theorem

$$P(B/D) = \frac{P(B)P(D/B)}{P(A)P(D/A) + P(B)P(D/B) + P(C)P(D/C)}$$

$$= \frac{0.35 \times 0.04}{0.25 \times 0.05 + 0.35 \times 0.04 + 0.40 \times 0.02} = 0.41 \quad \textbf{Ans.}$$

Example 24. *An insurance company insured 2000 scooter drivers 4000 car drivers and 6000 truck drivers. The probability of accidents are 0.01, 0.03 and 0.15 respectively. One of the insured persons meets with an accident. What is the probability that he is a scooter driver?* *(AMIETE, Dec. 2009)*

Solution. Let E_1, E_2, E_3 and A be the events defined as follows :

E_1 = person chosen is a scooter driver

E_2 = person chosen is a car driver

E_3 = person chosen is a truck driver, and

A = person meets with an accident.

We have,

$n(E_1) = 2000$, $n(E_2) = 4000$, $n(E_3) = 6000$

Total number of persons = 2000 + 4000 + 6000 = 12000.

Therefore,

$$P(E_1) = \frac{2000}{12000} = \frac{1}{6} \text{ and } P(E_2) = \frac{4000}{12000} = \frac{1}{3} \text{ and } P(E_3) = \frac{6000}{12000} = \frac{1}{2}$$

It is given that

$P(A/E_1)$ = Probability that a person meets with an accident given that he is a scooter driver $= 0.01$.

Similarly, $P(A/E_2) = 0.03$ and $P(A/E_3) = 0.15$

We are required to find $P(E_1/A)$ *i.e.*, given that the person meets with an accident, what is the probability that he was a scooter driver.

By Bayes' rule, we have

$$P(E_1/A) = \frac{P(E_1)P(A/E_1)}{P(E_1)P(A/E_1) + P(E_2)P(A/E_2) + P(E_3)P(A/E_3)}$$

$$\Rightarrow \quad P(E_1/A) = \frac{\frac{1}{6} \times 0.01}{\frac{1}{6} \times 0.01 + \frac{1}{3} \times 0.03 + \frac{1}{2} \times 0.15} = \frac{1}{1 + 6 + 45} = \frac{1}{52} \quad \textbf{Ans.}$$

EXERCISE 61.3

A bag contains 3 coins of which one is two headed and the other two are normal & fair. A coin is selected at random and tossed 4 times in succession. If all the four times it appears to be head what is the probability that the two headed coin was selected.

(AMIETE, JUne 2010) **Ans.** $\frac{8}{9}$

2. An airline knows that 13% of the people who make reservations on a certain flight will not turn up. Consequently their policy is to sell 12 tickets which can accommodate only 10. What is the probability that everyone who turns up on a given day is accommodated? *(AMIETE, JUne 2010)*

SAMPLING METHODS

62.1 POPULATION (UNIVERSE)

Before giving the notion of sampling, we will first define *population.* The group of individuals under study is called *population* or *universe.* It may be finite or infinite.

62.2 SAMPLING

A part selected from the population is called *a sample.* The process of selection of a sample is called sampling. A *Random sample* is one in which each member of population has an equal chance of being included in it. There are ${}^{N}C_{n}$ different samples of size n that can be picked up from a population of size N.

62.3 PARAMETERS AND STATISTICS

The statistical constants of the population such as mean (μ), standard deviation (σ) are called parameters. Parameters are denoted by Greek letters.

The mean ($\bar{x}$), standard deviation $|S|$ of a sample are known as statistics. Statistics are denoted by Roman letters.

Symbols for Population and Samples

Characteristic	Population	Sample
	Parameter	Statistic
Symbols	population size = N population mean = μ population standard deviation = σ population proportion = p	sample size = n sample mean = $\bar{x}$ sample standard deviation = s sample proportion = $\tilde{p}$

62.4 AIMS OF A SAMPLE

The population parameters are not known generally. Then the sample characteristics are utilised to approximately determine or estimate of the population. Thus, static is an estimate of the parameter. To what extent can we depend on the sample estimates?

The estimate of mean and standard deviation of the population is a primary purpose of all scientific experimentation. The logic of the sampling theory is the logic of *induction.* In induction, we pass from a particular (sample) to general (population). This type of generalization here is known as *statistical inference.* The conclusion in the sampling studies are based not on certainties but on probabilities.

62.5 TYPES OF SAMPLING

Following types of sampling are common:

(1) Purposive sampling (2) Random sampling (3) Stratified sampling (4) Systematic sampling

62.6 SAMPLING DISTRIBUTION

From a population a number of samples are drawn of equal size n. Find out the mean of each sample. The means of samples are not equal. The means with their respective frequencies are grouped. The frequency distribution so formed is known as *sampling distribution of the mean.* Similarly, sampling distribution of standard deviation we can have.

62.7 STANDARD ERROR (S.E.)

is the standard deviation of the sampling distribution. For assessing the difference between the expected value and observed value, standard error is used. Reciprocal of standard error is known as *precision.*

62.8 SAMPLING DISTRIBUTION OF MEANS FROM INFINITE POPULATION

Let the population be infinitely large and having a population mean of μ and a population variance of σ^2. If x is a random variable denoting the measurement of the characteristic, then

Expected value of x, $E(x) = \mu$

Variance of x, Var $(x) = \sigma^2$

The sample mean $\bar{x}$ is the sum of n random variables, *viz.*, $x_1, x_2, \ldots, x_n$, each being divided by n. Here, $x_1, x_2, \ldots, x_n$ are independent random variables from the infinitely large population.

$$\therefore \quad E(x_1) = \mu \quad \text{and} \quad Var\,(x_1) = \sigma^2$$

$$E(x_2) = \mu \quad \text{and} \quad Var\,(x_2) = \sigma^2 \text{ and so on}$$

$$\text{Finally} \quad E(\bar{x}) = E\left[\frac{x_1 + x_2 + \ldots + x_n}{n}\right] = \frac{1}{n}Ex_1 + \frac{1}{n}E(x_2) + \ldots + \frac{1}{n}E(x_n) = \frac{1}{n}\mu + \frac{1}{n}\mu + \ldots + \frac{1}{n}\mu = \mu$$

$$\text{and} \quad Var(\bar{x}) = Var\left[\frac{x_1 + x_2 + \ldots + x_n}{n}\right] = Var\left(\frac{x_1}{n}\right) + Var\left(\frac{x_2}{n}\right) + \ldots + Var\left(\frac{x_n}{n}\right)$$

$$= \frac{1}{n^2}Var(x_1) + \frac{1}{n^2}Var(x_2) + \ldots + \frac{1}{n^2}Var(x_n) = \frac{1}{n^2}.\sigma^2 + \frac{1}{n^2}.\sigma^2 + \ldots + \frac{1}{n^2}.\sigma^2 = \frac{n\sigma^2}{n^2} = \frac{\sigma^2}{n}$$

The expected value of the sample mean is the same as population mean. The variance of the sample mean is the variance of the population divided by the sample size.

The average value of the sample tends to true population mean. If sample size *(n)* is increased then variance of $\bar{x}$, $\left(\frac{\sigma^2}{n}\right)$ gets reduced, by taking large value of n, the variance $\left(\frac{\sigma^2}{n}\right)$ of $\bar{x}$ can be made as small as desired. The standard deviation $\left(\frac{\sigma}{\sqrt{n}}\right)$ of $\bar{x}$ is also called **standard error of the mean.** It is denoted by $\sigma_{\bar{x}}$.

Sampling with Replacement

When the sampling is done with replacement, so that the population is back to the same form before the next sample member is picked up. We have

$$E(\bar{x}) = \mu$$

$$Var\,(\bar{x}) = \frac{\sigma^2}{n} \quad \text{or} \quad \sigma_{\bar{x}} = \frac{\sigma}{\sqrt{n}}$$

Sampling without replacement from Finite population

When a sample is picked up without replacement from a finite population, the probability distribution of second random variable depends on the outcome of the first pick up. n sample members do not remain independent. Now we have

$$E(\bar{x}) = \mu$$

and $$Var(\bar{x}) = \sigma_{\bar{x}}^2 = \frac{\sigma^2}{n}\cdot\frac{N-n}{N-1} \quad \text{or} \quad \sigma_{\bar{x}} = \frac{\sigma}{\sqrt{n}}\cdot\sqrt{\frac{N-n}{N-1}}$$

$$= \frac{\sigma}{\sqrt{n}} \text{ app} \qquad \left(\text{if } \frac{n}{N} \text{ is very small}\right)$$

Sampling from Normal Population

If $x \sim N\left(\mu, \sigma^2\right)$ then it follows that $\bar{x} \sim N\left(\mu, \frac{\sigma^2}{n}\right)$

Example 1. *The diameter of a component produced on a semi-automatic machine is known to be distributed normally with a mean of 10 mm and a standard deviation of* 0.1 *mm. If we pick up a random sample of size 5, what is the probability that the same mean will be between 9.95 and 10.05 mm?*

Solution. Let x be a random variable representing the diameter of one component picked up at random.

Here $x \sim N(10, 0.01)$, Therefore, $\bar{x} \sim N\left(10, \frac{0.01}{5}\right)$ $\qquad \left[\bar{x} = N\left(\bar{x}, \frac{\sigma^2}{n}\right)\right]$

$$Pr\{9.95 \le \bar{x} \le 10.05\} = 2 \times Pr\{10 \le \bar{x} \le 10.05\} \qquad \left\{z = \frac{\bar{x}-\mu}{\frac{\sigma}{\sqrt{n}}}\right\}$$

$$= 2 \times Pr\left\{\frac{10-\mu}{\frac{\sigma}{\sqrt{n}}} \le \frac{\bar{x}-\mu}{\frac{\sigma}{\sqrt{n}}} \le \frac{10.05-\mu}{\frac{\sigma}{\sqrt{n}}}\right\}$$

9.95 μ=10 10.05

$$= 2 \times Pr\left\{0 \le z \le \frac{10.05-10}{\frac{0.1}{\sqrt{5}}}\right\} = 2 \times Pr\{0 \le z \le 1.12\} = 2 \times 0.3686 = 0.7372 \qquad \textbf{Ans.}$$

Similar Question

A sample of size 25 is picked up at random from a population which is normally distributed with a mean 100 and a variance of 36. Calculate (*a*) $Pr\{\bar{x} \le 99\}$, (*b*) $Pr\{98 \le \bar{x} \le 100\}$

Ans. (*a*) 0.2023 (*b*) 0.4522

62.9 SAMPLING DISTRIBUTION OF THE VARIANCE

We use a sample statistic called the sample variance to estimate the population variance. The sample variance is usually denoted by s^2

$$s^2 = \frac{\sum_{i=1}^{n}(x_i - \bar{x})^2}{n-1}$$

62.10 TESTING A HYPOTHESIS

On the basis of sample information, we make certain decisions about the population. In taking such decisions we make certain assumptions. These assumptions are known as *statistical hypothesis.* These hypothesis are tested. Assuming the hypothesis correct we calculate the probability of getting the observed sample. If this probability is less than a certain assigned value, the hypothesis is to be rejected.

62.11 NULL HYPOTHESIS (H_0)

Null hypothesis is based for analysing the problem. Null hypothesis is the *hypothesis of no difference.* Thus, we shall persume that there is no significant difference between the observed value and expected value. Then, we shall test whether this hypothesis is satisfied by the data or not. If the hypothesis is not approved the difference is considered to be significant. If hypothesis is approved then the difference would be described as due to sampling fluctuation. Null hypothesis is denoted by H_0.

62.12 ERRORS

In sampling theory to draw valid inferences about the population parameter on the basis of the sample results.

We decide to accept or to reject the lot after examining a sample from it. As such, we are liable to commit the following two types of errors.

Type 1 Error. If H_0 is rejected while it should have been accepted.

Type II Error. If H_0 is accepted while it should have been rejected.

62.13 LEVEL OF SIGNIFICANCE

There are two critical regions which cover 5% and 1% areas of the normal curve. The shaded portions are the critical regions.

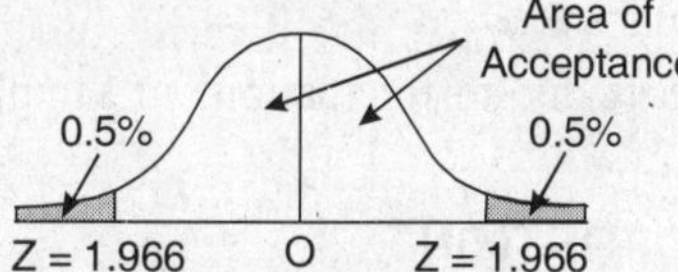

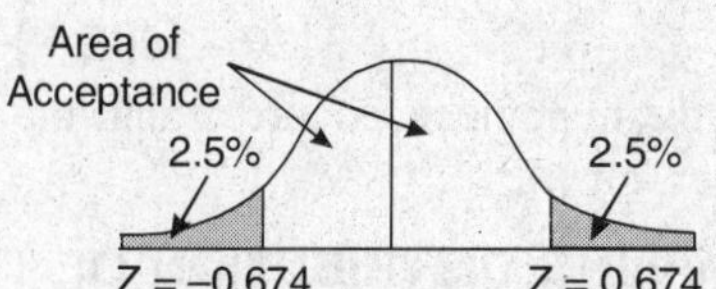

Thus, the probability of the value of the variate falling in the critical region is the level of significance. If the variate falls in the critical area, the hypothesis is to be rejected.

62.14 TEST OF SIGNIFICANCE

The tests which enables us to decide whether to accept or to reject the null hypothesis is called the tests of significance. If the difference between the sample values and the population values are so large (lies in critical area), it is to be rejected

62.15 CONFIDENCE LIMITS

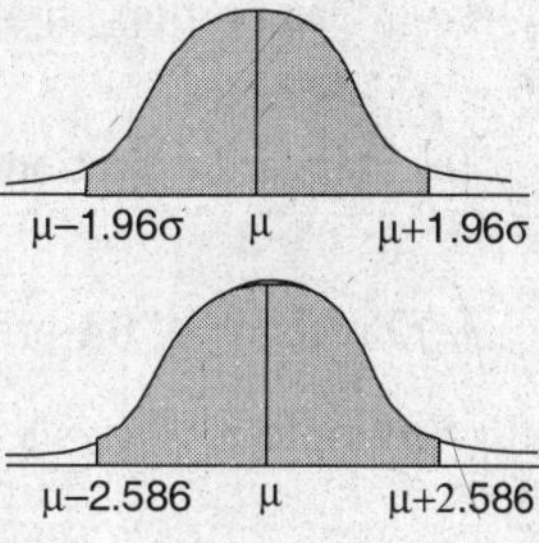

$\mu - 1.96\ \sigma, \mu + 1.96\sigma$ are 95% confidence limits as the area between $\mu - 1.96\sigma$ and $\mu + 1.96\sigma$ is 95%. If a sample statistics lies in the interval $\mu - 1.96\ \sigma$, $\mu + 1.96\sigma$, we call 95% confidence interval.

Similarly, $\mu - 2.58\sigma$, $\mu + 2.58\sigma$ is 99% confidence limits as the area between $\mu - 2.58\sigma$ and $\mu + 2.58\sigma$ is 99%. The numbers 1.96, 2.58 are called confidence coefficients.

62.16 TEST OF SIGNIFICANCE OF LARGE SAMPLES (*N* > 30)

Normal distribution is the limiting case of Binomial distribution when n is large enough. For normal distribution 5% of the items lie outside $\mu \pm 1.96\sigma$ while only 1% of the items lie outside $\mu \pm 2.586\ \sigma$.

$$z = \frac{x - \mu}{\sigma}$$

where z is the standard normal variate and x is the observed number of successes.

First we find the value of z. Test of significance depends upon the value of z.

(*i*) (*a*) If $|z| < 1.96$, difference between the observed and expected number of successes is not significant at the 5% level of significance.

(*b*) If $|z| > 1.96$, difference is significant at 5% level of significance.

(*ii*) (*a*) If $|z| < 2.58$, difference between the observed and expected number of successes is not significant at 1% level of significance.

(*b*) If $|z| > 2.58$, difference is significant at 1% level of significance.

Example 2. *A cubical die was thrown 9,000 times and 1 or 6 was obtained 3120 times. Can the deviation from expected value lie due to fluctuations of sampling?*

Solution. Let us consider the hypothesis that the die is an unbiased one and hence the probability of obtaining 1 or $6 = \frac{2}{6} = \frac{1}{3}$ *i.e.*, $p = \frac{1}{3}$, $q = \frac{2}{3}$

The expected value of the number of successes $= np = 9000 \times \frac{1}{3} = 3000$

Also $\sigma = \text{S.D.} = \sqrt{npq} = \sqrt{9000 \times \frac{1}{3} \times \frac{2}{3}} = \sqrt{2000} = 44.72$

$$3\sigma = 3 \times 44.72 = 134.16$$

Actual number of successes = 3120

Difference between the actual number of successes and expected number of successes

$$= 3120 - 3000 = 120 \text{ which is } < 3\sigma$$

Hence, the hypothesis is correct and the deviation is due to fluctuations of sampling due to randon causes. **Ans.**

62.17 SAMPLING DISTRIBUTION OF THE PROPORTION

A simple sample of n items is drawn from the population. It is same as a series of n independent trials with the probability p of success. The probabilities of 0, 1, 2, ..., n success are the terms in the binomial expansion of $(q + p)^n$.

Here mean $= np$ and standard deviation $= \sqrt{npq}$.

Let us consider the proportion of successes, then

(*a*) Mean proportion of successes $= \frac{np}{n} = p$

(*b*) Standard deviation (standard error) of proportion of successes $= \frac{\sqrt{npq}}{n} = \sqrt{\frac{pq}{n}}$

(*c*) Precision of the proportion of success $= \frac{1}{\text{S.E.}} = \sqrt{\frac{n}{pq}}$.

Example 3. *A group of scientific mens reported 1705 sons and 1527 daughters. Do these figures conform to the hypothesis that the sex ratio is* $\frac{1}{2}$.

Solution. The total number of observations = 1705 + 1527 = 3232

The number of sons = 1705

Theretore, the observed male ratio $= \frac{1705}{3232} = 0.5275$

On the given hypothesis the male ratio = 0.5000

Thus, the difference between the observed ratio and theoretical ratio

$$= 0.5275 - 0.5000$$
$$= 0.0275$$

The standard deviation of the proportion $= s = \sqrt{\dfrac{pq}{n}} = \sqrt{\dfrac{\frac{1}{2} \times \frac{1}{2}}{3232}} = 0.0088$

The difference is more than 3 times of standard deviation.

Hence, it can be definitely said that the figures given do not conform to the given hypothesis.

62.18 ESTIMATION OF THE PARAMETERS OF THE POPULATION

The mean, standard deviation etc. of the population are known as parameters. They are denoted by μ and σ. Their estimates are based on the sample values. The mean and standard deviation of a sample are denoted by $\bar{x}$ and s respectively. Thus, a static is an estimate of the parameter. There are two types of estimates.

(*i*) *Point estimation:* An estimate of a population parameter given by a single number is called a point estimation of the parameter. For example,

$$(\text{S.D.})^2 = \frac{\sum (x - \bar{x})^2}{n - 1}$$

(*ii*) *Interval estimation:* An interval in which population parameter may be expected to lie with a given degree of confidence. The intervals are

(*i*) $\bar{x} - \sigma_s$ to $\bar{x} + \sigma_s$ (68.27% confidence level)

(*ii*) $\bar{x} - 2\sigma_s$ to $\bar{x} + 2\sigma_s$ (95.45% confidence level)

(*iii*) $\bar{x} - 3\sigma_s$ to $\bar{x} + 3\sigma_s$ (99.73% confidence level)

$\bar{x}$ and σ_s are the mean and S.D. of the sample.

Similarly, $\bar{x} \pm 1.96\,\sigma_s$ and $\bar{x} \pm 2.58\,\sigma_s$ are 95% and 99% confidence of limits for

$\bar{x} \pm 1.96 \dfrac{\sigma}{\sqrt{n}}$ and $\bar{x} \pm 2.58 \dfrac{\sigma}{\sqrt{n}}$ are also the intervals as $\sigma_s = \dfrac{\sigma}{\sqrt{n}}$.

62.19 COMPARISON OF LARGE SAMPLES

Let two large samples of size n_1, n_2 be drawn from two populations of proportions of attributes A's as P_1, P_2 respectively.

(*i*) *Hypothesis:* As regards the attribute A, the two populations are similar. On combining the two samples

$$P = \frac{n_1 P_1 + n_2 P_2}{n_1 + n_2}$$

where p is the common proportion of attributes.

Let e_1, e_2 be the standard errors in the two samples, then

$$e_1^2 = \frac{pq}{n_1} \text{ and } e_2^2 = \frac{pq}{n_2}$$

If e be the standard error of the combined samples, then

$$e = P_1^2 + P_2^2 = \frac{pq}{n_1} + \frac{pq}{n_2} = pq\left[\frac{1}{n_1} + \frac{1}{n_2}\right]$$

$$z = \frac{P_1 - P_2}{e}$$

1. If $z > 3$, the difference between P_1 and P_2 is significant.

2. If $z < 2$, the difference may be due to fluctuations of sampling.

3. If $2 < z < 3$, the difference is significant at 5% level of significance.

(*ii*) *Hypothesis.* In the two populations, the proportions of attribute A are not the same, then standard error e of the difference $p_1 - p_2$ is

$$e^2 = p_1 + p_2$$

$$= \frac{P_1 - q_1}{n_1} + \frac{P_2 - q_2}{n_2}, z = \frac{P_1 - P_2}{e} < 3,$$

difference is due to fluctuations of samples.

Example 4. *In a sample of 600 men from a certain city, 450 are found smokers. In another sample of 900 men from another city, 450 are smokers. Do the data indicate that the cities are significantly different with respect to the habit of smoking among men.*

Solution. $n_1 = 600$ men, Number of smokers= 450, $P_1 = \frac{450}{600} = 0.75$

$n_2 = 900$ men, Number of smokers= 450, $P_2 = \frac{450}{900} = 0.5$

$$P = \frac{n_1 P_1 + n_2 P_2}{n_1 + n_2} = \frac{600 \times 0.75 + 900 \times 0.5}{600 + 900} = \frac{900}{1500} = 0.60$$

$$q = 1 - P = 1 - 0.6 = 0.4$$

$$e^2 = P_1^2 + P_2^2 = Pq\left(\frac{1}{n_1} + \frac{1}{n_2}\right)$$

$$e^2 = 0.6 \times 0.4\left(\frac{1}{600} + \frac{1}{900}\right) = 0.000667$$

$$e = 0.02582$$

$$z = \frac{P_1 - P_2}{e} = \frac{0.75 - 0.50}{0.02582} = 9.682$$

$z > 3$ so that the difference is significant. **Ans.**

Example 5. *One type of aircraft is found to develop engine trouble in 5 flights out of a total of 100 and another type in 7 flights out of a total of 200 flights. Is there a significant difference in the two types of aircrafts so far as engine defects are concerned.*

Solution. $n_1 = 100$ flights, Number of troubled flights = 5, $P_1 = \frac{5}{100} = \frac{1}{20}$

$n_2 = 200$ flights, Number of troubled flights = 7, $P_2 = \frac{7}{200}$

$$e^2 = \frac{P_1 q_1}{n_1} + \frac{P_2 q_2}{n_2} = \frac{0.05 \times 0.95}{100} + \frac{0.035 \times 0.965}{200}$$

$$= 0.000475 + 0.0001689 = 0.0006439$$

$$e = 0.0254$$

$$z = \frac{0.05 - 0.035}{0.0254} = 0.59$$

$z < 1$, Difference is not significant. **Ans.**

62.20 THE t-DISTRIBUTION (FOR SMALL SAMPLE)

The students distribution is used to test the significance of

(*i*) The mean of a small sample.

(*ii*) The difference between the means of two small samples or to compare two small samples.

(*iii*) The correlation coefficient.

Let $x_1, x_2, x_3, ..., x_n$, be the members of random sample drawn from a normal population with mean μ. If $\bar{x}$ be the mean of the sample then

$$t = \frac{\bar{x} - \mu}{\frac{s}{\sqrt{n}}} \quad \text{where } s^2 = \frac{\sum (x - \bar{x})^2}{n - 1}$$

Example 6. *A machine which produces mica insulating washers for use in electric device to turn out washers having a thickness of 10 mm. A sample of 10 washers has an average thickness 9.52 mm with a standard deviation of 0.6 mm. Find out t.*

Solution. $\bar{x} = 9.52, M = 10, S' = 0.6, n = 10$

$$t = \frac{\bar{x} - M}{\frac{s}{\sqrt{n}}} = \frac{9.52 - 10}{\frac{0.6}{\sqrt{10}}} = -\frac{0.48\sqrt{10}}{0.6} = -\frac{4}{5}\sqrt{10}$$

$$= -0.8 \times 3.16 = -2.528$$ **Ans.**

62.21 WORKING RULE

To calculate significance of sample mean at 5% level.

Calculate $t = \frac{\bar{x} - \mu}{s}\sqrt{n}$ and compare it to the value of t with $(n-1)$ degrees of freedom at 5% level, obtained from the table. Let this tabulated value of t be t_1.

If $t < t_1$, then we accept the hypothesis *i.e.*, we say that the sample is drawn from the population.

If $t > t_1$, we compare it with the tabulated value of t at 1% level of significance for $(n - 1)$ degrees of freedom. Denote it by t_2. If $t_1 < t < t_2$ then we say that the value of t is significant.

If $t > t_1$, we reject the hypothesis and the sample is not drawn from the population.

Example 7. *A manufacturer intends that his electric bulbs have a life of 1000 hours. He tests a sample of 20 bulbs, drawn at random from a batch and discovers that the mean life of the sample bulbs is 990 hours with a S.D of 22 hours. Does this signify that the batch is not up to the standard?*

[Given: *The table value of t at 1% level is significance with 19 degrees of freedom is 2.539*]

Solution. $\bar{x} = 990, \sigma = 22, x = 1000$

$$t = \frac{\bar{x} - \mu}{\frac{s}{\sqrt{n}}} = \frac{990 - 1000}{\frac{22}{\sqrt{20}}} = -\frac{10\sqrt{20}}{22} = -\frac{22.36}{11} = -2.033$$

0 2.033 2.539

Since the calculated value of t (2.032) is less than the value of t (2.539) from the table. Hence, it is not correct to say that this batch is not upto this standard. **Ans.**

Example 8. *Ten individuals are chosen at random from a population and their heights are found to be in inches 63, 63, 64, 65, 66, 69, 69, 70, 70, 71. Discuss the suggestion that the Mean height of universe is 65.*

For 9 degree of freedom t at 5% level of significance = 2.262.

Solution.

x	$x-67$	$(x-67)^2$
63	–4	16
63	–4	16
64	–3	9
65	–2	4
66	–1	1
69	+2	4
69	+2	4
70	+3	9
70	+3	9
71	+4	16
$\sum x = 670$		$\sum (x-\bar{x})^2 = 88$

$$\bar{x} = \frac{\sum x}{n} = \frac{670}{10} = 67,$$

$$s = \sqrt{\frac{\sum (x-\bar{x})^2}{n-1}} = \sqrt{\frac{88}{9}} = 3.13$$

$$t = \frac{\bar{x}-\mu}{\frac{s}{\sqrt{n}}} = \frac{67-65}{\frac{3.13}{\sqrt{10}}} = \frac{2\sqrt{10}}{3.13} = 2.02$$

$$2.02 < 2.262$$

Region of acceptance

0 2.02 2.262

Calculated value of t (2.02) is less than the table value of t (2.262). The hypothesis is accepted the mean height of universe is 65 inches. **Ans.**

Example 9. *The mean life time of sample of 100 fluorescent light bulbs produced by a company is computed to be 1570 hours with a standard deviation of 120 hours. The company claims that the average life of the bulbs produced by it is 1600 hours. Using the level of significance of 0.05, is the claim acceptable?*

Solution. $\bar{x} = 1570$, $s = 120$, $n = 100$, $\mu = 1600$

$$t = \frac{\bar{x}-\mu}{\frac{s}{\sqrt{n}}} = \frac{1570-1600}{\frac{120}{\sqrt{100}}} = \frac{1570-1600}{12} = 2.5$$

At 0.05 the level of significance, $t = 1.96$

Calculated value of $t >$ Table value of t.

$$2.5 > 1.96$$

Hence the claim is to be rejected. **Ans.**

Example 10. *A sample of 6 persons in an office revealed an average daily smoking of 10, 12, 8, 9, 16, 5 cigarettes. The average level of smoking in the whole office has to be estimated at 90% level of confidence.*

$t = 2.015$ for 5 degree of freedom

Solution.

x	$x-10$	$(x-10)^2$
10	0	0
12	2	4
8	–2	4
9	–1	1
16	+6	36
5	–5	25
Total	0	$\sum (x-10)^2 = 70$

$$\text{Mean} = a + \frac{\Sigma fd}{\Sigma f} = 10 + \frac{0}{6} = 10$$

$$s = \sqrt{\frac{\sum (x-\bar{x})^2}{n-1}} = \sqrt{\frac{70}{5}} = 3.74$$

At 90% level of confidence, $t = \pm 2.015$.

$$t = \frac{\bar{x}-\mu}{\frac{s}{\sqrt{n}}} \Rightarrow \pm 2.015 = \frac{10-\mu}{\frac{3.74}{\sqrt{6}}}$$

$$\Rightarrow \mu = 2.015 \times \frac{3.74}{\sqrt{6}} + 10 = 6.92, 13.08$$ **Ans.**

Example 11. *A fertiliser mixing machine is set to give 12 kg of nitrate for quintal bag of fertiliser: Ten 100 kg bags are examined The percentages of nitrate per bag are as follows:*

11, 14, 13, 12, 13, 12, 13, 14, 11, 12

Is there any reason to believe that the machine is defective? Value of t for 9 degrees of freedom is 2.262.

Solution. The calculation of $\bar{x}$ and s is given in the following table:

x	$d = x - 12$	d^2
11	−1	1
14	2	4
13	1	1
12	0	0
13	1	1
12	0	0
13	1	1
14	2	4
11	−1	1
12	0	0
$\sum x = 125$	$\sum d = 5$	$\sum d^2 = 13$

$$\mu = 12 \text{ kg}, \; n = 10, \bar{x} = \frac{\sum x}{n} = \frac{125}{10} = 12.5$$

$$s^2 = \frac{\sum d^2}{n} - \left(\frac{\sum d}{n}\right)^2 = \frac{13}{10} - \left(\frac{5}{10}\right)^2 = \frac{13}{10} - \frac{1}{4} = \frac{21}{20} = \frac{105}{100}$$

$$s = 1.024$$

Value of t for 9 degrees of freedom = 2.262

Also $t = \dfrac{\bar{x} - \mu}{s}\sqrt{n} = \dfrac{12.5 - 12}{1.024}\sqrt{10} = 1.54$

Since the value of t is less than 2.262, there in no reason to believe that machine is defective. **Ans.**

Example 12. *A random sample of size 16 values from a normal population showed a mean of 53 and a sum of squares of deviation from the mean equals to 150. Can this sample be regarded as taken from the population having 56 as mean? Obtain 95% and 99% confidence limits of the mean of the population.*

$$\gamma = 15, \; \alpha = 0.05, t = 2.131$$
$$\alpha = 0.01, t = 2.947$$

Solution. $\mu = 56, n = 16, \; \bar{x} = 53, \; \sum (x - \bar{x})^2 = 150$

$$s^2 = \frac{\sum (x - \bar{x})^2}{n - 1} = \frac{150}{15} = 10$$

$$s = \sqrt{10}$$

$$t = \frac{\bar{x} - \mu}{\frac{s}{\sqrt{n}}} = \frac{53 - 56}{\frac{\sqrt{10}}{\sqrt{16}}} = \frac{-3 \times 4}{\sqrt{10}} = -3.79$$

Region of rejection

−3.79 −2.131 O

$$t = 3.79$$

When α = 0.5 then 3.79 > 2.131

When α = 0.01 then 3.79 > 2.947

Thus, the sample cannot be regarded as taken from the population. **Ans.**

62.22 TESTING FOR DIFFERENCE BETWEEN MEANS OF TWO SMALL SAMPLES

Let the mean and variance of the first population be μ_1 and σ_1^2 and μ_2 . σ_2^2 be the mean and variance of the second population.

Let $\bar{x}_1$ be the mean of small sample of size n_1 from first population and $\bar{x}_2$ the mean of a sample of size n_2 from second population.

We know that

$$E(\bar{x}_1) = \mu_1 \text{ and } Var\ (\bar{x}_1) = \frac{\sigma_1^2}{n_1}$$

$$E(\bar{x}_2) = \mu_2 \text{ and } Var(\bar{x}_2) = \frac{\sigma_2^2}{n_2}$$

If the samples are independent, then $(\bar{x}_1)$ and $(\bar{x}_2)$ are also independent.

$$E(\bar{x}_1 - \bar{x}_2) = \mu_1 - \mu_2 \text{ and } Var(\bar{x}_1 - \bar{x}_2) = \frac{\sigma_1^2}{n_1} + \frac{\sigma_2^2}{n_2}$$

$$\bar{x}_1 \sim N\left(\mu_1, \frac{\sigma_1^2}{n_1}\right) \text{ and } \bar{x}_2 \sim N\left(\mu_2, \frac{\sigma_2^2}{n_2}\right) \text{ then } (\bar{x}_1 - \bar{x}_2) \sim N\left(\mu_1 - \mu_2, \frac{\sigma_1^2}{n_1} + \frac{\sigma_2^2}{n_2}\right)$$

$$t = \frac{(\bar{x}_1 - \bar{x}_2) - (\mu_1 - \mu_2)}{\sqrt{\frac{\sigma_1^2}{n_1} + \frac{\sigma_2^2}{n_2}}}$$

If the population is the same then

$$t = \frac{\bar{x}_1 - \bar{x}_2}{\sigma\sqrt{\frac{1}{n_1} + \frac{1}{n_2}}} \qquad (\mu_1 - \mu_2 = \mu_1 - \mu_1 = 0)$$

Example 13. *Two independent samples of 8 and 7 items respectively had the following values of the variable (weight in ounces):*

Sample 1: 9 11 13 11 15 9 12 14

Sample 2: 10 12 10 14 9 8 10

Is the difference between the means of the sample significant?

[Given for V = 13, $t_{0.05}$ = 2.16]

Solution.

Assumed mean of x = 12, Assumed mean of y = 10

x	$(x-12)$	$(x-12)^2$	y	$(y-10)$	$(y-10)^2$
9	–3	9	10	0	0
11	–1	1	12	2	4
13	1	1	10	0	0
11	–1	1	14	4	16
15	3	9	9	–1	1
9	–3	9	8	–2	4
12	0	0	10	0	0
14	2	4	–	–	–
94	–2	34	73	3	25

$$\bar{x} = \frac{\sum x}{n} = \frac{94}{8} = 11.75$$

$$\sigma_x^2 = \frac{\Sigma(x-12)^2}{n} - \left(\frac{\Sigma(x-12)}{n}\right)^2 = \frac{34}{8} - \left(\frac{-2}{8}\right)^2 = 4.1875$$

$$\bar{y} = \frac{\sum y}{n} = \frac{73}{7} = 10.43$$

$$\sigma_y^2 = \frac{\sum(y-10)^2}{n} - \left[\frac{\Sigma(y-10)}{n}\right]^2 = \frac{25}{7} - \left(\frac{3}{7}\right)^2 = 3.388$$

$$s = \sqrt{\frac{(x-\bar{x})^2 + \sum(y-\bar{y})^2}{n_1 + n_2 - 2}} = \sqrt{\frac{34+25}{8+7-2}} = \sqrt{\frac{59}{13}} = \sqrt{4.54} = 2.13$$

$$t=\frac{\bar{x}-\bar{y}}{s\sqrt{\frac{1}{n_1}+\frac{1}{n_2}}}=\frac{11.75-10.43}{2.13\sqrt{\frac{1}{8}+\frac{1}{7}}}=\frac{1.32}{2.13\sqrt{0.268}}=\frac{1.32}{2.13\times 0.518}=\frac{1.32}{1.103}=1.2$$

Thus, 5% value of t for 13 degree of freedom is given to be 2.16. Since calculated value of t is 1.2 is less than 2.16, the difference between the means of samples is not significant. **Ans.**

EXERCISE 62.1

1. A random sample of six steel beams has mean compressive strength of 58.392 psi (pounds per square inch) with a standard deviation of s = 648 psi. Test the null hypothesis $H_0 = \mu = 58{,}000$ psi against the alternative hypothesis H_1: $\mu > 58{,}000$ psi at 5% level of significance (value for t at 5 degree of freedom and 5% significance level is 2.0157). Here μ denotes the population mean. *(A.M.I.E., Summer 2000)*
2. A certain cubical the was thrown 96 times and shows 2 upwards 184 times. Is the the biased?
   Ans. die is biased.
3. In a sample of 100 residents of a colony 60 are found to be wheat eaters and 40 rice eaters. Can we assume that both food articles are equally popular?
4. Out of 400 children, 150 are found to be under weight. Assuming the conditions of simple sampling, estimate the percentage of children who are underweight in, and assign limits within which the per-centage probably lies.
   Ans. 37.5% approx. Limits = 37.5 ± 3 (2.4)
5. 500 eggs are taken at random from a large consignment, and 50 are found to be bad. Estimate the percentage of bad eggs in the consignment and assign limits within which the percentage probably lies.
   Ans. 10%, 10 ± 3.9
6. A machine puts out 16 imprefect articles in a sample of 500. After the machine is repaired, puts out 3 imprefect articles in a batch of 100. Has the machine been improved?
   Ans. The machine has not been improved.
7. In a city *A*, 20% of a random sample of 900 school boys had a certain slight physical defect. In another city *B*, 18.5% of a random sample of 1600 school boys had the same defect. Is the difference between the proportions significant?
   Ans. z = 0.37, Difference between proportions is significant.
8. In two large populations there are 30% and 25% respectively of fair haired people. Is this difference likely to be hidden in samples of 1200 and 900 respectively from the two populations?
   Ans. z = 2.5, not hidden at 5% level of significance.
9. One thousand articles from a factory are examined and found to be three percent defective. Fifteen hundred similar articles from a second factory are found to be only 2 percent defective. Can it reasonably be concluded that the product of the first factory is inferior to the second?
   Ans. It cannot be reasonable concluded that the product of the first factory is inferior to that of the second.
10. A manufacturing company claims 90% assurance that the capacitors manufactured by them will show a tolerance of better than 5%. The capacitors are packaged and sold in lots of 10. Show that about 26% of his customers ought to complain that capacitors do not reach the specified standard.

62.23 CHI SQUARE TEST

The Chi-square distribution is one of the most extensively used distribution function in statistics. It was first discovered by Helmert in 1875 and later on Karl Pearson's in 1900.

62.24 CHI-SQUARE VARIATES

The square of a standard normal variate is known as Chi-square variate (χ^2) with one degree function :

$$z = \frac{x-\mu}{\sigma} \text{ is a normal variate.}$$

Hence $\left(\frac{x-\mu}{\alpha}\right)^2$ is a Chi-square variate.

If x be a normaly distributed variate and $x_1, x_2, \ldots\ldots\ldots, x_n$ be a random sample of n-values from this population then $w = x_1^2 + x_2^2 + \ldots\ldots + x_n^2$ has χ^2 distribution with n-degree of freedom.

62.25 CONDITIONS FOR CHI-SQUARE TEST

There are some conditions which are necessary for Chi square test.

1. The sample under study must be large and may be total of cell frequency should not be less than 50.
2. The member of the cells should be independent.
3. The cell frequency of each cell should be greater than 5. If any cell has frequency less than 5 then it should be combined with the next or preceding cell until the total frequency exceeds 5.
4. If there are any constraint on the cell frequencies they should be linear
i.e.; $\alpha_1 x_1 + \alpha_2 x_2 + \alpha_3 x_3 + \ldots\ldots + \alpha_i x_i + \ldots\ldots\ldots\ldots + \alpha_n x_n = \lambda$

Note: Cell frequency should not involve any logarithmic, exponential or trigonometric relation.

62.26 CHI-SQUARE (χ^2) IS USED AS:

(1) Test of independence (2) Test of goodness of fit

(3) To test if the hypothetical value of the population variate is σ^2

(4) To test the homogeneity of independent estimate of the population variance.

We shall mainly use the first two test

62.27 CHI-SQUARE TEST OF GOODNESS OF FIT

This test is used to test significance of the discrepancy between theory and experiment. It helps us to find if the deviation of the experiment from the theory is just by chance or it is due to the inadequacy of the theory to fit the observed data.

The theoretical frequencies for various classes are calculated from the assumption of the population. The significant deviation between the observed and theoretical frequencies is tested by means of this test.

χ is calculated by means of the following formula

$$\chi^2 = \sum_{i=1}^{k} \frac{(O_i - E_i)^2}{E_i} \quad \text{and } \Sigma O_i = \Sigma E_i = N$$

where O_i is the observed frequency E_i is the expected (Theoretical) frequency of the cell.

62.28 WORKING RULE TO CALCULATE χ^2:

Step 1. Calculate the expected frequencies.

Step 2. Calculate the difference between each observed frequency O_i and the corresponding expected frequency E_i for each class i.e.; to find $O_i - E_i$

Step 3. Square the difference obtained in step 2 for each value i.e.; Calculate $(O_i - E_i)^2$.

Step 4. Divide $(O_i - E_i)^2$ by the expected frequency E to get $\frac{(O_i - E_i)^2}{E_i}$

Step 5. Add all these quotients obtained in step 4. Then $\chi^2 = \frac{\sum_{i=1}^{n}(O_i - E_i)^2}{E_i}$

It is to be noted

(1) The value of χ^2 is always positive. (2) χ^2 will be zero if each pair is zero.

(3) The value of χ^2 lies between 0 and ∞.

62.29 DEGREE OF FREEDOM

Case I. If the data is given in the form of a series of variables in a row or column then the degree of freedom = (No. of items in the series) – 1

Case 2. When the number of frequencies are put in cells in a contingency table.

The degree of freedom = (R – 1) (C – 1)

where R is number of rows and C is the number of columns.

Example 14. *A survey of 320 families with 5 children is given below :*

No. of boys	*5*	*4*	*5*	*2*	*1*	*0*	*Total*
No. of girls	*0*	*1*	*2*	*3*	*4*	*5*	
No. of families	*14*	*56*	*110*	*88*	*40*	*12*	*320*

Is this result consistent with hypothesis i.e.; the male and female birth are equally possible.

Solution. Null Hypothesis H_0.

(1) Male and Female birth are equally probable.

Alternate Hypothesis H_1: Male and female birth are not equally probable.

Calculation of expected frequencies $(q + p)^n$

Probability of female birth $= p = \frac{1}{2}$

Probability of male birth $= q = \frac{1}{2}$

$(q+p)^n = q^n + {}^nC_1\, p\, q^{n-1} + {}^nC_2\, p^2\, q^{n-2} + {}^nC_3\, p^3\, q^{n-3} + \ldots\ldots + p^n$

$$\left(\frac{1}{2}+\frac{1}{2}\right)^5 = \left(\frac{1}{2}\right)^5 + 5\left(\frac{1}{2}\right)^1\left(\frac{1}{2}\right)^4 + 10\left(\frac{1}{2}\right)^2\left(\frac{1}{2}\right)^3 + 10\left(\frac{1}{2}\right)^3\left(\frac{1}{2}\right)^2 + 5\left(\frac{1}{2}\right)^4\left(\frac{1}{2}\right) + \left(\frac{1}{2}\right)^5$$

$$\text{No. of girls} = 320\left[\frac{1}{32}+\frac{5}{32}+\frac{10}{32}+\frac{10}{32}+\frac{5}{32}+\frac{1}{32}\right]$$

$$= 320\times\frac{1}{32} + 320\times\frac{5}{32} + 320\times\frac{10}{32} + 320\times\frac{10}{32} + 320\times\frac{5}{32} + 320\times\frac{1}{32}$$

$= 10 + 50 + 100 + 100 + 50 + 10$

These are the expected frequencies of the female births.

O	E	$O-E$	$(O-E)^2$	$\frac{(O-E)^2}{E}$
14	10	4	16	1.60
56	50	6	36	0.72
110	100	10	100	1.00
88	100	− 12	144	1.44
40	50	− 10	100	2.00
12	10	2	4	0.40
			Total	7.16

Level of significance Let $\alpha = 0.05$

Critical value. The table value of χ^2 at $\alpha = 0.05$ for $(6-1)(2-1) = 5$ degree of freedom is 11.07

Decision Since the calculated value of χ^2 (7.16) < Table value of χ^2 at level of significance 0.05 for $5df = 11.07$

Hence, the null hypothesis is accepted i.e.; the male and female birth is equally probable.

Example 15. *The table below give the number of air craft accidents that occured during the various days of the week. Test whether the accidents are uniformly distributed over the week.*

Days	*Mon.*	*Tue.*	*Wed.*	*Thu.*	*Fri*	*Sat*	*Sun*	*Total no. accidents*
No. of accidents	*14*	*18*	*12*	*11*	*15*	*14*	*14*	*98*

Solution. H_0 : Null Hypothesis : The accidents are uniformly distributed over the week.

The expected frequencies of the accidents on each day $= \frac{98}{7} = 14$

O	E	$O-E$	$(O-E)^2$	$\frac{(O-E)^2}{E}$
14	14	0	0	0
18	14	4	16	1.14
12	14	− 2	4	0.29
11	14	− 3	9	0.64
15	14	1	1	0.07
14	14	0	0	0
14	14	0	0	0
98			Total	2.14

Level of significance :

Let $\alpha = 0.05$

Critical value: The table value of χ^2 at $\alpha = 0.05$ is for $(7-1)(2-1)$ i.e.; 6 degree is $\chi^2 = 12.592$

Since the calculated value of χ^2 (7.16) < Table value of χ^2 at level of significance 0.05 for six degree = 12.592

Hence, the null Hypothesis is accepted i.e.; the air craft accident are uniformly distributed over the week.

Ans.

62.30 CHI-SQUARE TEST AS A TEST OF INDEPENDENCE

$$\text{Expected Frequency} = \frac{\text{Row total} \times \text{Column total}}{\text{Grand total}}$$

Example 16. *In an investigation into the health and nutrition of two groups of children of different social status the following results are obtained.*

Social Status / *Health*	*Poor*	*Rich*	*Total*
Below Normal	*130*	*20*	*150*
Normal	*102*	*108*	*210*
Above normal	*24*	*96*	*120*
Total	*256*	*224*	*480*

Discuss the relation between the health and their social status.

Solution. H_0. Null Hypothesis : There is no association between health and social status.

H_1. Alternate Hypothesis: there an no association between health and social status.

$$\text{Expected frequency} = \frac{\text{Row total} \times \text{Column total}}{\text{Grand total}}$$

Social Status / *Health*	*Poor*	*Rich*	*Total*
Below Normal	$\frac{256 \times 150}{480} = 80$	$\frac{224 \times 150}{480} = 70$	150
Normal	$\frac{256 \times 210}{480} = 112$	$\frac{224 \times 210}{480} = 98$	210
Above normal	$\frac{256 \times 120}{480} - 64$	$\frac{224 \times 120}{480} = 56$	120
Total	256	224	480

Total number of observed frequencies = Total number of expected frequencies = 480

Degree of freedom = (3 – 1) (2 – 1) = 2

Level of significance, take $\alpha = 0.5$

Critical value the table value of χ^2 at $\alpha = 0.05$ for degree of freedom 2 is 5.99.

Decision. Since the calculated value of χ^2 (122.44) > table value of χ^2 at level of significance 0.05 for two $2 d.f. = 5.991$.

Hence, the null hypothesis is rejected i.e., social status and health are associated (Dependent). **Ans.**

Calculation of Chi-square

Observed value (O)	Expected Value (E)	$O-E$	$(O-E)^2$	$\frac{(O-E)^2}{E}$
130	80	50	2500	$\frac{2500}{80} = 31.25$
102	12	– 10	100	$\frac{100}{112} = 0.89$
24	64	– 40	1600	$\frac{1600}{64} = 25$
20	70	– 50	2500	$\frac{2500}{70} = 35.71$
108	98	10	100	$\frac{100}{98} = 1.02$
96	56	40	1600	$\frac{1600}{56} = 28.57$
			Total	122.44

Example 17. *The I.Q. and economic condition of home of 1000 students of an engineering college, Delhi were noted as given in the table :*

I.Q. / *Economic con.*	*High*	*Low*	*Total*
Rich	*100*	*300*	*400*
Poor	*350*	*250*	*600*
Total	*450*	*550*	*1,000*

Find out whether there is any association between economic condition at home and I.Q. of the students.

Given for 1 d.f., χ^2 *at the level of significance 0.05 is 3.84.*

Solution.

Null Hypothesis H_0: There is no association between economic condition at home and I.Q.

Alternative hypothesis H_1: There is an association between economic condition at home and *I . Q.*

$$\text{Expected frequency E} = \frac{\text{Row total} \times \text{Column total}}{\text{Grand total}}$$

I.Q. / *Economic cond.*	*High*	*Low*	*Total*
Rich	$\frac{400 \times 450}{1000} = 180$	$\frac{400 \times 550}{1000} = 220$	400
Poor	$\frac{600 \times 450}{1000} = 270$	$\frac{600 \times 550}{1000} = 330$	600
Total	450	550	1000

Calculation of Chi-square

Observed value (O)	Expected Value (E)	$O-E$	$(O-E)^2$	$\frac{(O-E)^2}{E}$
100	180	– 80	6400	35.5
350	270	80	6400	23.7
300	220	80	6400	29.1
250	330	– 80	6400	19.4
			Total	107.7

Degree of freedom = (R – 1) (C – 1) = (2 – 1) (2 – 1) = 1 given for *d.f.* = 1,
χ^2 at the level of significance 0.05 = 3.84

Decision. The calculated value of χ^2 is greater than Table value of χ^2. Hence, the hypothesis is rejected and the alternative hypothesis is accepted.

Hence, there is an association between economic condition at home and I.Q. **Ans.**

Example 18. *To test the effectiveness of inoculation against cholera, the following table was obtained.*

	Attached	*Not attached*	*Total*
Inoculated	30	160	190
Notinoculated	140	460	600
Total	170	620	790

(The figures represent the number of persons)

Use χ^2 *- test to defend or refute the statement. The inoculation prevents attack from cholera.* *(U.P. III Semester Dec. 2009)*

Solution. **H_0 Null Hypothesis:** No inoculation prevents attack from cholera.

H_1 Alternate Hypothesis:

The inoculation prevents attack from cholera.

$$\text{Expected frequency } E = \frac{\text{Row total} \times \text{Column Total}}{\text{Grand total}}$$

	Attacked	Not Attacked	Total
Inoculated	$\frac{170 \times 190}{790} = 40.9$	$\frac{620 \times 190}{790} = 149.1$	190
Not inoculated	$\frac{170 \times 600}{790} = 129.1$	$\frac{620 \times 600}{790} = 470.9$	600
Total	170	620	790

Total number of observed frequencies
= Total number of expected frequencies = 790

Calculation of Chi-square

Observed value (O)	Expected value (E)	(O – E)	$(O-E)^2$	$\frac{(O-E)^2}{E}$
30	40.9	– 10.9	118.81	2.904
140	129.1	10.9	118.81	0.920
160	149.1	10.9	118.81	0.797
460	470.9	– 10.9	118.81	0.252
			Total	4.873

Degree of freedom = (R – 1) (C – 1) = (2 – 1) (2 – 1) = 1

The critical value of the table value of χ^2 at $\alpha = 0.05$ for 1d.f. is 3.841.

Decision: Since the calculated value of χ^2 (4.873) is greater than the table value (3.841). Thus, the hypothesis is rejected and the alternative hypothesis is accepted. Hence, the inoculation prevents attack from cholera. **Ans.**

EXERCISE 62.2

1. In an experiment immunization of cattle from a disease, the following results are obtain:

	Affected	Unaffected	Total
Inoculated	12	28	40
No Inoculated	13	7	20
Total	25	35	60

Examine the effect of vaccin in controlling the incidence of the disease. **Ans.** Not independent

2. In the contigency table given below use Chi-square test to test for independence of hair colour and eye colour of persons:

Hair colour / Eye colour	Light	Dark	Total
Blue	26	9	35
Brown	7	18	25
Total	33	27	60

Ans. Hair colour and eye colour are associated

3. A survey amongst women was conducted to study the family life. The observations are as follows:

Family life

	Happy	Not Happy	Total
Educated	70	30	100
Not educated	60	40	100
Total	130	70	200

Test whether there is any association between family life and education.

Ans. there is no association between family life and education.

4. A certain drug was administrated to 500 people out of a total of 800 included in a sample to test its efficiency against typhoid, the results are given below :

	Typhoid	No Typhoid	Total
Drug	200	300	500
No Drug	280	20	300
	48	320	800

On the basis of the data, can we say that drug is effective in preventing Typhoid.

Ans. $\chi^2 = 222.22$ drug is effective

5. The following table gives the number of person's whose eye sight is attacked and an injection of macugen is injected by Prof. Atul.

	Eye sight Improved	Eye sight not improved	total
Injected	216	145	361
Not injected	105	234	339
Total	321	379	700

Do you think macugen injection can improve the eye sight. **Ans.** By injection, eye sight has improved

6. From the table given below, whether the colour of the sons eyes is associated with that of father's eye.

Eyes colour in sons

		Not light	light	
Eyes colour in fathers	Not light	230	148	378
	light	151	471	622
		381	619	1000

There is an association between the colour of eyes of sons and colours of eye's of fathers.

Ans. Null hypothesis in rejected

7. The following table gives the classification of 500 plants according to the nature of leaf and flower colour.

	Blue flowers	White flower	Total
Flat leaf	329	121	450
Creppled leaf	78	32	110
Total	407	153	560

Test whether they have any association between them. **Ans.** No association between them

8. The table below gives the data obtained from a hospital of sugar patients:

	Cured	Not cured	Total
Inoculated	31	469	500
Not inoculated	185	1,315	1500
	216	1784	2000

Test the effectiveness of inoculation in preventing the sugar disease.

Ans. Inoculation is effective in preventing the attack of sugar

9. The following table give the no. of good and defective parts produced by each of these shifts in a factory. Test whether the shift has any association with good or defective parts.

	Day shift	Evening shift	Night shift	Total
Good parts	960	940	950	2850
Defective parts	40	50	45	135
	1000	990	995	2985

Ans. $\chi^2 = 0.3485$, Null Hypothesis is accepted

10. The following table shows the results of drug against B.P.

	Not attacked	Attacked	Total
Drug	267	37	304
No Drug	757	155	912
Total	1024	192	1216

Find out whether there is any significance association between drug and attack.

Ans. Drug prevents the attack of B.P.

CHAPTER

63

Binomial Distribution

63.1 BINOMIAL DISTRIBUTION $P(r) = {}^nC_r\ p^r \cdot q^{n-r}$

To find the probability of the happening of an event once, twice, thrice, ... r times ...... exactly in n trials.

Let the probability of the happening of an event A in one trial be p and its probability of not happening be $1 - p = q$.

We assume that there are n trials and the happening of the event A is r times and its not happening is $n - r$ times.

$$\underbrace{A\ A .... A}_{r \text{ times}} \qquad \underbrace{\bar{A}.\bar{A}......\bar{A}}_{n-r \text{ times}} \qquad ...(1)$$

A indicates its happening, $\bar{A}$ its failure and $P(A) = p$ and $P(\bar{A}) = q$.

We see that (1) has the probability

$$\underbrace{pp .... p}_{r \text{ times}}\ \underbrace{q \cdot q .... q}_{n-r \text{ times}} = p^r \cdot q^{n-r} = p^r q^{n-r} \qquad ...(2)$$

Clearly (1) is merely one order of arranging r A's.

The probability of (1) $= p^r q^{n-r} \times$ Number of different arrangements of A's and $(n - r)\ \bar{A}$'s.

The number of different arrangements of r A's and $(n - r)\ \bar{A}$'s $= {}^nC_r$

$\therefore$ **Probability of the happening of an event r times $= {}^nC_r\ p^r \cdot q^{n-r}$.**

$$P(r) = {}^nC_r\ p^r \cdot q^{n-r} \quad (r = 0, 1, 2, ..., n).$$
$$= (r + 1)\text{th term of } (q + p)^n$$

If $r = 0$, probability of happening of an event 0 times $= {}^nC_0\ q^n p^0 = q^n$

If $r = 1$, probability of happening of an event 1 time $= {}^nC_1\ q^{n-1}\ p$

If $r = 2$, probability of happening of an event 2 times $= {}^nC_2\ q^{n-2}\ p^2$

If $r = 3$, probability of happening of an event 3 times $= {}^nC_3\ q^{n-3}\ p^3$ and so on.

These terms are clearly the successive terms in the expansion of $(q + p)^n$.

Hence it is called Binomial Distribution.

Example 1. *Find the probability of getting 4 heads in 6 tosses of a fair coin.*

Solution. $p = \dfrac{1}{2}, q = \dfrac{1}{2}, n = 6, r = 4$.

We know that $\quad P(r) = {}^nC_r\ q^{n-r}\ p^r$

$$P(4) = {}^6C_4\ q^{6-4}\ p^4 = \frac{6 \times 5}{1 \times 2}\left(\frac{1}{2}\right)^2\left(\frac{1}{2}\right)^4 = 15 \times \left(\frac{1}{2}\right)^6 = \frac{15}{64} \qquad \textbf{Ans.}$$

Example 2. *If on an average one ship in every ten is wrecked, find the probability that out of 5 ships expected to arrive, 4 at least will arrive safely.*

Solution. Out of 10 ships, one ship is wrecked.

i.e., Nine ships out of ten ships are safe.

$$p\text{ (safety)} = \frac{9}{10}$$

P (At least 4 ships out of 5 are safe) $= P$ (4 or 5) $= P$ (4) + P (5)

$$= {}^5C_4\, p^4 q^{5-4} + {}^5C_5\, p^5 q^0 = 5\left(\frac{9}{10}\right)^4\left(\frac{1}{10}\right) + \left(\frac{9}{10}\right)^5 = \left(\frac{9}{10}\right)^4\left(\frac{5}{10}+\frac{9}{10}\right) = \frac{7}{5}\left(\frac{9}{10}\right)^4 \qquad \textbf{Ans.}$$

Example 3. *The overall percentage of failures in a certain examination is 20. If six candidates appear in the examination, what is the probability that at least five pass the examination?*

Solution. Probability of failures $= 20\% = \dfrac{20}{100} = \dfrac{1}{5}$

Probability of pass $(P) = 1 - \dfrac{1}{5} = \dfrac{4}{5}$

Probability of at least five pass $= P$ (5 or 6)

$$= P(5) + P(6) = {}^6C_5\, p^5 q + {}^6C_6 p^6 q^0 = 6\left(\frac{4}{5}\right)^5\left(\frac{1}{5}\right) + \left(\frac{4}{5}\right)^6 = \left(\frac{4}{5}\right)^5\left[\frac{6}{5}+\frac{4}{5}\right] = 2\left(\frac{4}{5}\right)^5 = \frac{2048}{3125}$$

$= 0.65536$ **Ans.**

Example 4. *Ten percent of screws produced in a certain factory turn out to be defective. Find the probability that in a sample of 10 screws chosen at random, exactly two will be defective.*

Solution. $p = \dfrac{1}{10}, \quad q = \dfrac{9}{10}, n = 10, r = 2$

$$P(r) = {}^nC_r\; p^r\, q^{n-r}$$

$$P(2) = {}^{10}C_2\left(\frac{1}{10}\right)^2\left(\frac{9}{10}\right)^{10-2} = \frac{10\times 9}{1\times 2}\left(\frac{1}{10}\right)^2\left(\frac{9}{10}\right)^8 = \frac{1}{2}\cdot\left(\frac{9}{10}\right)^9 = 0.1937 \qquad \textbf{Ans.}$$

Example 5. *The probability that a man aged 60 will live to be 70 is 0.65. What is the probability that out of 10 men, now 60, at least 7 will live to be 70 ?*

Solution. The probability that a man aged 60 will live to be 70 $= p = 0.65$

$$q = 1 - p = 1 - 0.65 = 0.35$$

Number of men $= n = 10$

Probability that at least 7 men (7 or 8 or 9 or 10) will live to 70

$$= P\ (7) + P\,(8) + P\,(9) + P\,(10) = {}^{10}C_7\, q^3\, p^7 + {}^{10}C_8\, q^2 p^8 + {}^{10}C_9\, q\, p^9 + p^{10}$$

$$= \frac{10\times 9\times 8}{1\times 2\times 3}(0.35)^3(0.65)^7 + \frac{10\times 9}{1\times 2}(0.35)^2(0.65)^8 + 10(0.35)(0.65)^9 + (0.65)^{10}$$

$$= (0.65)^7\,[\,120\ (0.35)^3 + 45\,(0.35)^2\,(0.65) + 10\,(0.35)\,(0.65)^2 + (0.65)^3]$$

$$= (0.65)^7 \times 125\,[120 \times (0.07)^3 + 45 \times (0.07)^2\,(0.13) + 10\,(0.07)\,(0.13)^2 + (0.13)^3]$$

$$= 0.04902 \times 125\ [0.04 + 0.028665 + 0.011830 + 0.002197]$$

$= 6.1275 \times 0.082692 = 0.5067$ **Ans.**

Example 6. *If 10 % of bolts produced by a machine are defective. Determine the probability that out of 10 bolts, chosen at random (i) 1 (ii) none (iii) at most 2 bolts will be defective.*

Solution. Probability of defective bolts $= p = 10\% = 0.1$

Probability of not defective bolts $= q = 1 - p = 1 - 0.1 = 0.9$

Total number of bolts = n = 10

(i) Probability of 1 defective bolt $= {}^{10}C_1 (0.1)^1 (0.9)^9 = 0.3874$

(ii) Probability that none is defective = Probability of 0 defective bolt

$= P(0) = {}^{10}C_0 (0.1)^0 (0.9)^{10} = 0.3487$

(iii) Probability of 2 defective $= {}^{10}C_2 (0.1)^2 (0.9)^8 = 0.1937$

Probability of at most 2 defective $= P(0 \text{ or } 1 \text{ or } 2)$

$= P(0) + P(1) + P(2) = 0.3487 + 0.3874 + 0.1937$

$= 0.9298$ **Ans.**

Example 7. *An underground mine has 5 pumps installed for pumping out storm water, the probability of any one of the pumps failing during the storm is* $\frac{1}{8}$. *What is the probability that (i) at least 2 pumps will be working; (ii) all the pumps will be working during a particular storm ?*

Solution. (i) Probability of pump failing $= \frac{1}{8}$

Probability of pump working $= 1 - \frac{1}{8} = \frac{7}{8}, \quad P = \frac{7}{8}, \quad q = \frac{1}{8}, \quad n = 5$

(i) P (At least 2 pumps working) $= P$ (2 or 3 or 4 or 5 pumps working)

$= P(2) + P(3) + P(4) + P(5) = {}^5C_2\, p^2 q^3 + {}^5C_3\, p^3 q^2 + {}^5C_4\, p^4 q + {}^5C_5\, p^5 q^0$

$$= 10\left(\frac{7}{8}\right)^2\left(\frac{1}{8}\right)^3 + 10\left(\frac{7}{8}\right)^3\left(\frac{1}{8}\right)^2 + 5\left(\frac{7}{8}\right)^4\left(\frac{1}{8}\right) + \left(\frac{7}{8}\right)^5$$

$$= \frac{1}{8^5}[10 \times 49 + 10 \times 343 + 5 \times 2401 + 16807]$$

$$= \frac{1}{8^5}[490 + 3430 + 12005 + 16807] = \frac{32732}{8^5} = \frac{8183}{8192}$$

(ii) P (All the 5 pumps working) $= P(5) = {}^5C_5 p^5 q^0 = \left(\frac{7}{8}\right)^5 = \frac{16807}{32768}$ **Ans.** (i) $\frac{8183}{8192}$ (ii) $\frac{16807}{32768}$

Example 8. *Write two-three areas where binomial distribution is applied. The probability of entering student in chartered accountant will be graduate 0.5. Determine the probability that out of 10 students (i) none (ii) one or (iii) at least one will graduate.*

(R.G.P.V., Bhopal, Dec., 2003)

Solution. Given, the probability of an entering student in chartered accountant will graduate is $p = 0.5$

$\therefore$ The probability of an entering student in charactered accountant will not graduate is $q = 0.5$.

Therefore

(i) The probability of none will graduate out of 10 students

$P(0) = {}^{10}C_0\, p^0 q^{10} = {}^{10}C_0 (0.5)^0 (0.5)^{10} = 9.765625 \times 10^{-4}$ **Ans.**

(ii) The probability of exactly one student will graduate out of 10 students.

$P(1) = {}^{10}C_1 (0.5)^1 (0.5)^9 = 10 \times 0.5 \times (0.5)^9 = 9.765625 \times 10^{-3}$ **Ans.**

(iii) The probability of at least one will graduate out of 10 students

P (At least one) = 1 – (probability of none will graduats)

$= 1 - 9.765625 \times 10^{-4} = 0.99$ **Ans.**

Example 9. *The probability that a bomb dropped from a plane will strike the target is* $\frac{1}{5}$. *If six bombs are dropped, find the probability that:*

(i) Exactly two will strike the target.

(ii) At least two will strike the target. *(R.G.P.V., Bhopal, II Semester, Feb. 2006)*

Solution. Here, $p = \frac{1}{5}$, $q = 1 - \frac{1}{5} = \frac{4}{5}$, $n = 6$

We know that $P(r) = {}^nC_r p^r q^{n-r}$

$$P(2) = {}^6C_2 \left(\frac{1}{5}\right)^2 \left(\frac{4}{5}\right)^{6-2} = 15\left(\frac{256}{15625}\right) = \frac{768}{3125} = 0.24576$$

$$\begin{aligned} P(\text{at least } 2) &= P(2, 3, 4, 5, 6) = P(2) + P(3) + P(4) + P(5) + P(6) \\ &= P(0) + P(1) + P(2) + P(3) + P(4) + P(5) + P(6) - P(0) - P(1) \\ &= 1 - [P(0) + P(1)] \\ &= 1 - \left[{}^6C_0 \left(\frac{1}{5}\right)^0 \left(\frac{4}{5}\right)^6 + {}^6C_1 \left(\frac{1}{5}\right)\left(\frac{4}{5}\right)^5\right] = 1 - \left[\frac{4096}{15625} + 6\left(\frac{1024}{15625}\right)\right] \\ &= 1 - \frac{10240}{15625} = \frac{5385}{15625} = \frac{1077}{3125} = 0.34464 \end{aligned}$$

Hence (*i*) $P = 0.24576$ (*ii*) $P = 0.34464$ **Ans.**

EXERCISE 63.1

1. If 20% of the bolts produced by a machine are defective, determine the probability that out of 4 bolts chosen at random
(*a*) 1 (*b*) 0 (*c*) At most 2
bolts will be defective. **Ans.** (*a*) 0.4096, (*b*) 0.4096, (*c*) 0.9728.

2. Six dice are thrown 729 times. How many times do you expect at least three dice to show a five or a six ? **Ans.** 233

3. Find the probability of getting a total of 7 at least once in 4 tosses of a pair of fair dice.?
(*A.M.I.E., Winter 2002*) **Ans.** $\frac{671}{1296}$

4. If the chance that any one of the 10 telephone lines is busy at any instant is 0.2, what is the chance that 5 of the lines are busy ? What is the probability that all the lines are busy ?
Ans. ${}^{10}C_5\ (0.2)^5\ (0.8)^5$, $(0.2)^{10}$

5. An insurance salesman sells policies to 5 men, all of identical age in good health. According to the actuarial tables the probability that a man of this particular age will be alive 30 years hence is $\frac{2}{3}$. Find the probability that in 30 years.
(*a*) All 5 men (*b*) At least 3 men (*c*) Only 2 men (*d*) At least 1 man will be alive.
Ans. (*a*) $\frac{32}{243}$ (*b*) $\frac{192}{243}$ (*c*) $\frac{40}{243}$ (*d*) $\frac{242}{243}$

6. Consider an urn in which 4 balls have been placed by the following scheme : A fair coin is tossed; if the coin falls head, a white ball is placed in the urn, and if the coin falls tail, a red ball is placed in urn. (*i*) What is the probability that the urn will contain exactly 3 white balls ? (*ii*) What is the probability that the urn will contain exactly 3 red balls, given that the first ball placed was red?
Ans. (*i*) $\frac{1}{8}$, (*ii*) $\frac{3}{8}$

7. A box contains 10 screws, 3 of which are defective. Two screws are drawn at random without replacement. Find the probability that none of the two screws is defective. **Ans.** $\frac{7}{15}$

8. Out of 800 families with four children each, how many families would be expected to have :
(*i*) 2 boys and 2 girls; (*ii*) at least one boy; (*iii*) no girl; (*iv*) at most two girls ?
Assume equal probabilities for boys and girls. **Ans.** (*i*) 300, (*ii*) 750, (*iii*) 50, (*iv*) 550.

9. In a hurdle race, a player has to cross 10 hurdles. The probability that he will clear each hurdle is 5/6. What is the probability that he will knock down less than 2 hurdles ? **Ans.** $\frac{8}{3}\left(\frac{5}{6}\right)^9$

10. An electronic component consists of three parts. Each part has probability 0.99 of performing satisfactorily. The component fails if 2 or more parts do not perform satisfactorily. Assuming that the parts perform independently, determine the probability that the component does not perform satisfactorily.

Ans. 0.000298

11. The incidence of occupational disease in an industry is such that the workers have 20% chance of suffering from it. What is the probability that out of 6 workers 4 or more will catch the disease ?

(*A.M.I.E., Winter 2005*) **Ans.** $\dfrac{53}{2125}$

12. Among 10,000 random digits, find the probability p that the digit 3 appears at most 950 times.

(*A.M.I.E., Summer 2003*) Ans. $\sum_{n=0}^{950} 1000\ C_r \left(\frac{1}{10}\right)^r \left(\frac{9}{10}\right)^{1000-r}$

13. A fair coin is tossed 400 times. Using normal approximation to the binomial, find the probability that a head will occur (*a*) more than 180 times and (*b*) less than 195 times.

(*A.M.I.E. Winter 2004*) **Ans.** (*a*) $1-\left(\frac{1}{2}\right)^{221}$ (*b*) $1-\left(\frac{1}{2}\right)^{115}$

14. In a bombing action there is 50% chance that any bomb will strike the target. Two direct hits are needed to destroy the target completely. How many bombs are required to be dropped to give a 99% chance or better of completely destroying the target. (*R.G.P.V., Bhopal, June 2008*) **Ans.** 11

63.2 MEAN OF BINOMIAL DISTRIBUTION

(*GBTU, Dec. 2012, AMIETE, Winter 2002, Summer 2000, A.M.I.E., Winter 2002*)

$$(q+p)^n = q^n + {}^nC_1\, q^{n-1}p^1 + {}^nC_2\, q^{n-2}p^2 + {}^nC_3\, q^{n-3}p^3 + \ldots\ldots + {}^nC_r\, q^{n-r}p^r + \ldots\ldots + p^n$$

Successes r	*Frequency f*	*Product rf*
0	q^n	0
1	$n\,q^{n-1}p$	$n\,q^{n-1}p$
2	$\frac{n(n-1)}{2}q^{n-2}p^2$	$n(n-1)\,q^{n-2}p^2$
3	$\frac{n(n-1)(n-2)}{6}q^{n-3}p^3$	$\frac{n(n-1)(n-2)}{2}q^{n-3}p^3$
....	......	....
n	p^n	np^n

$$\Sigma fr = nq^{n-1}p + n(n-1)q^{n-2}p^2 + \frac{n(n-1)(n-2)}{2}q^{n-3}p^3 + \ldots + n\,p^n$$

$$= np\left[q^{n-1} + \frac{(n-1)}{1!}q^{n-2}p + \frac{(n-1)(n-2)}{2}q^{n-3}p^2 + \ldots + p^{n-1}\right]$$

$$= n\,p\,(q+p)^{n-1} = np \qquad \text{(since } q+p=1\text{)}$$

$$\Sigma f = q^n + n\,q^{n-1}p + \frac{n(n-1)}{2}q^{n-2}p^2 + \ldots\ldots + p^n$$

$$= (q+p)^n = 1 \qquad \text{(since } q+p=1\text{)}$$

Hence, $\quad \text{Mean} = \dfrac{\Sigma f r}{\Sigma f} = \dfrac{np}{1}$ **Ans.**

63.3 STANDARD DEVIATION OF BINOMIAL DISTRIBUTION

(*A. M. I. E.T. E., Winter 2002, A.M.I.E., Winter 2002*)

Successes r	*Frequency f*	*Product* $r^2 f$
0	q^n	0
1	$n\,q^{n-1}p$	$n\,q^{n-1}p$
2	$\frac{n(n-1)}{2}q^{n-2}p^2$	$2\,n(n-1)\,q^{n-2}p^2$
3	$\frac{n(n-1)(n-2)}{6}q^{n-3}p^3$	$\frac{3n(n-1)(n-2)}{2}q^{n-3}p^3$
.....	.....	......
n	p^n	n^2p^n

We know that $$\sigma^2 = \frac{\Sigma f r^2}{\Sigma f} - \left(\frac{\Sigma f r}{\Sigma f}\right)^2 \qquad ...(1)$$

r is the deviation of items (successes) from 0.

$$\Sigma f = 1, \;\; \Sigma fr = np$$

$$\Sigma fr^2 = 0 + nq^{n-1}p + 2n(n-1)q^{n-2}p^2 + \frac{3n(n-1)(n-2)}{2}q^{n-3}p^3 + ... + n^2p^n$$

$$= np[q^{n-1} + \frac{2(n-1)}{1!}q^{n-2}p + \frac{3(n-1)(n-2)}{2!}q^{n-3}p^2 + ... + np^{n-1}]$$

$$= np[q^{n-1} + \frac{(n-1)q^{n-2}p}{1!} + \frac{(n-1)(n-2)}{2!}q^{n-3}p^2 + ... + p^{n-1}$$

$$+ \frac{(n-1)q^{n-2}p}{1!} + \frac{2(n-1)(n-2)}{2!}q^{n-3}p^2 + ... + (n-1)\,p^{n-1}\Big]$$

$$= n\,p\left[q^{n-1} + (n-1)q^{n-2}p + \frac{(n-1)(n-2)}{2!}q^{n-3}p^2 + ... + p^{n-1}\right.$$

$$\left. + (n-1)\,p\left\{q^{n-2} + (n-2)\,q^{n-3}p + \frac{(n-2)(n-3)}{2!}q^{n-4}p^2 + ... + p^{n-2}\right\}\right]$$

$$= np\,[(q+p)^{n-1} + (n-1)\,p\,(q+p)^{n-2}] = np\,[1 + (n-1)\,p]$$

$$= np\,[np + (1-p)] = np[np + q] = n^2p^2 + npq$$

Putting these values in (1), we have

$$\text{Variance} = \sigma^2 = \frac{n^2p^2 + n\,pq}{1} - \left(\frac{np}{1}\right)^2 = npq\,,$$

$$S.D. = \sigma = \sqrt{n\,pq}$$

Hence for the binomial distribution,

$$\text{Mean} = np\,,\;\; \mu_2 = \sigma^2 = n\,p\,q$$

Example 10. *Find the first four moments of the binomial distribution.* (*AMIETE, Summer 2000*)

Solution. First moment about the origin

$$\mu_1' = \sum_{r=0}^{n} {}^nC_r\, p^r q^{n-r}.r = \sum_{r=0}^{n} r.\frac{n(n-1)(n-2).......(n-r+1)}{r!}p^r q^{n-r}$$

$$= n\sum_{r=1}^{n} \frac{(n-1)(n-2).......(n-r+1)}{(n-r)!}p^r q^{n-r} = np\sum_{r=1}^{n} {}^{n-1}C_{r-1}\,p^{r-1}q^{n-r}$$

$$= np\,(q+p)^{n-1} = np$$

Thus, the mean of the Binomial distribution is np.

Second moment about the origin

$$\mu_2' = \sum_{r=0}^{n} {}^nC_r\, p^r q^{n-r}.r^2 \qquad [r^2 = r(r-1) + r]$$

$$= \sum_{r=0}^{n}\{r(r-1)+r\}\, {}^nC_r\, p^r q^{n-r} = \sum_{r=0}^{n} r(r-1)\, {}^nC_r\, p^r q^{n-r} + \sum_{r=0}^{n} r\cdot {}^nC_r\, p^r q^{n-r}$$

$$= \sum_{r=0}^{n} \frac{r(r-1)\,n(n-1)(n-2)\,.......(n-r+1)}{r!}\, p^r q^{n-r}$$

$$\sum_{r=0}^{n} \frac{r\,n(n-1)(n-2)\,.......(n-r+1)}{r!}\, p^r q^{n-r}$$

$$= n(n-1)p^2 \sum_{r=2}^{n} \frac{(n-2)(n-3)\,.......(n-r+1)}{(r-2)!}\, p^{r-2} q^{n-r}$$

$$+ np\,\Sigma \frac{(n-1)(n-2)\,.......(n-r+1)}{(r-1)!}\, p^{r-1} q^{n-r}$$

$$= n\,(n-1)\,p^2\,(q+p)^{n-2} + np\,(q+p)^{n-1} = n\,(n-1)\,p^2 + np$$

Third moment about the origin

$$\mu_3' = \sum_{r=0}^{n} {}^nC_r\, p^r q^{n-r}\cdot r^3$$

[Let $r^3 = Ar(r-1)(r-2) + Br(r-1) + Cr$
By putting $r = 1, 2, 3,$ we get A = 1, B = 3, C = 1]

$$\mu_3' = \sum_{r=0}^{n}\{r(r-1)(r-2)+3r(r-1)+r\}\, {}^nC_r\, p^r q^{n-r}$$

$$= \sum_{r=0}^{n} r(r-1)(r-2)\, {}^nC_r\, p^r q^{n-r} + 3\sum_{r=0}^{n} r(r-1)\cdot {}^nC_r\, p^r q^{n-r} + \sum_{r=0}^{n} r\cdot {}^nC_r\, p^r q^{n-r}$$

$$= \sum_{r=0}^{n} \frac{r(r-1)(r-2)\cdot n(n-1)\,.......(n-r+1)}{r!}\, p^r q^{n-r}$$

$$+ 3\sum_{r=0}^{n} \frac{r(r-1)\cdot n\,(n-1)\,.......(n-r+1)}{r!}\, p^r q^{n-r} + \sum_{r=0}^{n} r\,\frac{n(n-1)\,.......(n-r+1)}{r!}\, p^r q^{n-r}$$

$$= \sum_{r=3}^{n} \frac{n(n-1)(n-2)(n-3)\,.......(n-r+1)}{(r-3)!}\, p^r q^{n-r}$$

$$+ 3\sum_{r=2}^{n} \frac{n(n-1)(n-2)(n-3)\,.......(n-r+1)}{(r-2)!}\, p^r q^{n-r}$$

$$+ \sum_{r=1}^{n} \frac{n(n-1)(n-2)\,.......(n-r+1)}{(r-1)!}\, p^r q^{n-r}$$

$$= n(n-1)(n-2)p^3 \sum_{r=3}^{n} {}^{n-3}C_{r-3}\, p^{r-3} q^{n-3} + 3n(n-1)p^2 \sum_{r=2}^{n} {}^{n-2}C_{r-2}\, p^{r-2} q^{n-2}$$

$$+ np \sum_{r=1}^{n} {}^{(n-1)}C_{r-1}\, p^{r-1} q^{n-1}$$

$$= n\,(n-1)\,(n-2)\,p^3 (q+p)^{n-3} + 3\,n\,(n-1)\,p^2 (q+p)^{n-2} + np\,(q+p)^{n-1}$$

$$= n\,(n-1)\,(n-2)\,p^3 + 3\,n\,(n-1)\,p^2 + np$$

Fourth Moment

$$\mu_4' = \sum_{r=0}^{n} {}^nC_r\, p^r q^{n-r} \cdot r^4$$

[Let $r^4 = Ar(r-1)(r-2)(r-3) + Br(r-1)(r-2) + Cr(r-1) + Dr$
By putting r = 1, 2, 3, 4, we get A = 1, B = 6, C = 7, D = 1]

$$\mu_4' = \sum_{r=0}^{n} r(r-1)(r-2)(r-3) \cdot {}^nC_r\, p^r q^{n-r} + \sum_{r=0}^{n} 6r(r-1)(r-2) \cdot {}^nC_r\, p^r q^{n-r}$$

$$+ \sum_{r=0}^{n} 7r(r-1) \cdot {}^nC_r\, p^r q^{n-r} + \sum_{r=0}^{n} r \cdot {}^nC_r\, p^r q^{n-r}$$

$$= \sum_{r=0}^{n} \frac{r(r-1)(r-2)(r-3) \cdot n(n-1).......(n-r+1)}{r!} p^r q^{n-r}$$

$$+ 6 \sum_{r=0}^{n} \frac{r(r-1)(r-2) \cdot n(n-1).......(n-r+1)}{r!} p^r q^{n-r}$$

$$+7 \sum_{r=0}^{n} \frac{r(r-1) \cdot n(n-1).......(n-r+1)}{r!} p^r q^{n-r} + \sum_{r=0}^{n} \frac{r \cdot n(n-1).......(n-r+1)}{r!} p^r q^{n-r}$$

$$= \sum_{r=4}^{n} \frac{n(n-1)(n-2)(n-3)(n-4)...(n-r+1)}{(r-4)!} p^r q^{n-r}$$

$$+ 6 \sum_{r=3}^{n} \frac{n(n-1)(n-2)(n-3)...(n-r+1)}{(r-3)!} p^r q^{n-r}$$

$$+ 7 \sum_{r=2}^{n} \frac{n(n-1)(n-2)...(n-r+1)}{(r-2)!} p^r q^{n-r} + \Sigma \frac{n(n-1)...(n-r+1)}{(r-1)!} p^r q^{n-r}$$

$$= n(n-1)(n-2)(n-3) \sum_{r=4}^{n} {}^{n-4}C_{r-4}\, p^r q^{n-r} + 6n(n-1)(n-2) \sum_{r=3}^{n} {}^{n-3}C_{r-3} \cdot p^r q^{n-r}$$

$$+ 7n(n-1) \sum_{r=2}^{n} {}^{n-2}C_{r-2} \cdot p^r q^{n-r} + n \sum_{r=1}^{n} {}^{n-1}C_{r-1} \cdot p^r q^{n-r}$$

$$= n(n-1)(n-2)(n-3)\, p^4 (q+p)^{n-4} + 6n(n-1)(n-2)\, p^3 (q+p)^{n-3}$$
$$+ 7n(n-1)\, p^2 (q+p)^{n-2} + np(q+p)^{n-1}$$

$$= n(n-1)(n-2)(n-3)\, p^4 + 6n(n-1)(n-2)\, p^3 + 7n(n-1)\, p^2 + np$$

63.4 CENTRAL MOMENTS : (Moments about the mean)

Now, the first four central moments are obtained as follows:

Second Central Moment

$\mu_2 = \mu_2' - \mu_1' = [n(n-1)p^2 + np] - n^2 p^2 = np[(n-1)p + 1 - np] = np(1-p) = npq$

Variance of Binomial distribution is npq

Third Central Moment

$$\mu_3 = \mu_3' - 3\mu_2' \mu_1' + 2\mu_1^{3'}$$
$$= \{n(n-1)(n-2)p^3 + 3n(n-1)p^2 + np\} - 3\{n^2p^2 + npq)np\} + 2n^3p^3$$
$$= np[-3np^2 + 3np + 2p^2 - 3p + 1 - 3npq]$$
$$= np[3np(1-p) + 2p^2 - 3p + 1 - 3npq]$$
$$= np[3npq + 2p^2 - 3p + 1 - 3npq] = np[2p^2 - 3p + 1] = np[2p^2 - 2p + q]$$
$$= np[-2p(1-p) + q] = np(-2pq + q) = npq(1-2p) = npq(q-p)$$

Fourth Central Moment

$$\mu_4 = \mu_4' - 4\mu_3'\mu_1' + 6\mu_2'\mu_1'^2 - 3\mu_1'^4$$

$$= n(n-1)(n-2)(n-3)p^4 + 6n(n-1)(n-2)p^3 + 7n(n-1)p^2 + np - 4[n(n-1)(n-2)p^3 + 3n(n-1)p^2 + np]\,np + 6[n(n-1)p^2 + np]n^2p^2 - 3n^4p^4$$

$$= np[(n-1)(n-2)(n-3)p^3 + 6(n-1)(n-2)p^2 + 7(n-1)p + 1 - 4\{n(n-1)(n-2)p^3 + 3n(n-1)p^2 + np\} + 6\{n(n-1)p^2 + np\}np - 3n^3p^3]$$

$$= np[(n^3 - 6n^2 + 11n - 6)p^3 + (6n^2 - 18n + 12)p^2 + 7np - 7p + 1 + \{(-4n^3 + 12n^2 - 8n)p^3 - 4(3n^2 - 3n)p^2 - 4np\} + \{(6n^3 - 6n^2)p^3 + 6n^2p^2\} - 3n^3p^3]$$

$$= np[(n^3 - 6n^2 + 11n - 6 - 4n^3 + 12n^2 - 8n + 6n^3 - 6n^2 - 3n^3)p^3 + (6n^2 - 18n + 12 - 12n^2 + 12n + 6n^2)p^2 + (7n - 7 - 4n)p + 1]$$

$$= np[(3n-6)p^3 + (-6n+12)p^2 + (3n-7)p + 1]$$

$$= np[3np^3 - 6p^3 - 6np^2 + 12p^2 + 3np - 7p + 1]$$

$$= np[3np^3 - 3np^2 - 6p^3 + 6p^2 - 3np^2 + 3np + 6p^2 - 6p - p + 1]$$

$$= np[-3np^2(1-p) + 6p^2(1-p) + 3np(1-p) - 6p(1-p) + (1-p)]$$

$$= np[-3np^2q + 6p^2q + 3npq - 6pq + q] = npq[-3np^2 + 6p^2 + 3np - 6p + 1]$$

$$= npq[3np(1-p) - 6p(1-p) + 1] = npq[3npq - 6pq + 1]$$

$$= npq[1 + 3(n-2)pq]$$ **Ans.**

63.5 MOMENT GENERATING FUNCTIONS OF BINOMIAL DISTRIBUTION ABOUT ORIGIN

$$M_0(t) = E(e^{tx}) = \Sigma\, {}^nC_x p^x q^{n-x} \cdot e^{tx}$$

$$= \Sigma\, {}^nC_x (pe^t)^x q^{n-x} = (q + pe^t)^n$$

Differentiating w.r.t. 't' we get $Ma'(t) = n(q + pe^t)^{n-1} p \cdot e^t$

On putting $t = 0$, we get $\mu_1' = n(q+p)^{n-1}p$

$$\mu_1' = np$$

Since $\qquad M_a(t) = e^{-at}M_0(t)$

Moment generating function of the Binomial distribution about its mean $(m) = np$ is given by

$$M_m(t) = e^{-npt}M_0(t)$$

$$M_m(t) = e^{-npt}(q + pe^t)^n = (qe^{-pt} + pe^{-pt} + t)^n = (qe^{-pt} + pe^{(1-p)t})^n$$

$$= \left[q\left(1 - pt + \frac{p^2t^2}{2!} - \frac{p^3t^2}{2!} + \frac{p^4t^4}{4!} + \ldots\right) + p\left(1 + qt + \frac{q^2t^2}{2!} + \frac{q^3t^3}{3!} + \frac{p^4t^4}{4!} + \ldots\right)\right]^n$$

$$= \left[1 + pq\frac{t^2}{2!} + pq(q^2 - p^2)\frac{t^3}{3!} + pq(q^3 + p^3)\frac{t^4}{4!} + \ldots\right]^n$$

$$1 + \mu_1 t + \mu_2\frac{t^2}{2!} + \mu_3\frac{t^3}{3!} + \mu_4\frac{t^4}{4!} + \ldots$$

$$= 1 + npq\frac{t^2}{2!} + npq(q-p)\frac{t^3}{3!} + npq[1 + 3(n-2)pq]\frac{t^4}{4!} + \ldots$$

Equating the coefficients of like powers of t on both sides, we get

$$\mu_2 = npq, \quad \mu_3 = npq(q-p), \ \mu_4 = npq[1 + 3(n-2)pq]$$

Hence the moment coefficient of skewness is

$$\beta_1 = \frac{\mu_3^2}{\mu_2^3} = \frac{[npq\,(q-p)]^2}{(npq)^3} = \frac{(q-p)^2}{npq} \quad ; \quad \gamma_1 = \sqrt{\beta_1} = \frac{q-p}{\sqrt{npq}}$$

Coefficient of Kurtosis is given by

$$\beta_2 = \frac{\mu_4}{\mu_2^2} = \frac{npq\,[1+3pq\,(n-2)]}{(npq)^2} = 3 + \frac{1-6pq}{npq} \;;\; \gamma_2 = \beta_2 - 3 = \frac{1-6pq}{npq}$$

Example 11. *If the probability of a defective bolt is 0.1, find (a) the mean (b) the standard deviation for the distribution bolts in a total of 400.*

Solution. $n = 400,\ p = 0.1,\ \text{Mean} = np = 400 \times 0.1 = 40$

$$\text{Standard deviation} = \sqrt{npq} = \sqrt{400 \times 0.1(1-0.1)}$$

$$= \sqrt{400 \times 0.1 \times 0.9} = 20 \times 0.3 = 6$$ **Ans.**

Example 12. *A die is tossed thrice. A success is getting 1 or 6 on a toss. Find the mean and variance of the number of successes.* *(AMIETE, Dec. 2010)*

Solution. $$n = 3,\ p = \frac{1}{3},\ q = \frac{2}{3}$$

$$\text{Mean} = np = 3 \times \frac{1}{3} = 1$$

$$\text{Variance} = npq = 3 \times \frac{1}{3} \times \frac{2}{3} = \frac{2}{3}$$ **Ans.**

Example 13. *If mean and variance of a binomial distribution are 4 and 2 respectively, find the probability of (i) exactly 2 successes (ii) less than 2 successes (iii) at least 2 successes.* *(R.G.P.V., Bhopal, II Semester, June 2005)*

Solution. Mean = 4 $\Rightarrow$ $np = 4$ ... (1)

Variance = 2 $\quad npq = 2$ ... (2)

Dividing (2) by (1), we get

$$\frac{npq}{np} = \frac{2}{4} \quad \Rightarrow \quad q = \frac{1}{2}$$

$$p = 1 - q = 1 - \frac{1}{2} = \frac{1}{2}$$

Putting the value of p in (1), we get

$$n\left(\frac{1}{2}\right) = 4 \quad \Rightarrow \quad n = 8$$

(*i*) Probability of r successes $= {}^nC_r\, p^r q^{n-r}$

$$P(2) = {}^8C_2 \left(\frac{1}{2}\right)^2 \left(\frac{1}{2}\right)^{8-2} = {}^8C_2 \left(\frac{1}{2}\right)^8 = \frac{8 \times 7}{2}\,\frac{1}{256} = \frac{7}{64}$$

(*ii*) P (less than 2 successes) $= P(0) + P(1) = {}^8C_0\, p^0 q^8 + {}^8C_1\, p^1 q^7$

$$= \frac{1}{256} + 8\,\frac{1}{2}\left(\frac{1}{2}\right)^7 = \frac{9}{256}$$

(*iii*) P (at least 2 successes) $= P(2) + P(3) + \ldots + P(8)$

$$= P(0) + P(1) + P(2) + P(3) + \ldots + P(8) - P(0) - P(1)$$

$$= 1 - P(0) - P(1) = 1 - [P(0) + P(1)] = 1 - \frac{9}{256} = \frac{247}{256}$$ **Ans.**

Example 14. *Fit a Binomial distribution for the following data and compare the theoretical frequencies with actual ones :*

x	0	1	2	3	4	5
y	2	14	20	34	22	8

(*R.G.P.V., Bhopal, II Semester, June 2006*)

Solution.

x	$y = f$	fx	$P = {}^5C_r p^r q^{5-r}$	Theoretical Frequency
0	2	0	${}^5C_0 (0.568)^0 (0.432)^5 = 0.015$	$100 \times 0.015 = 1.5$
1	14	14	${}^5C_1 (0.568)^1 (0.432)^4 = 0.099$	$100 \times 0.099 = 9.9$
2	20	40	${}^5C_2 (0.568)^2 (0.432)^3 = 0.260$	$100 \times 0.260 = 26.0$
3	34	102	${}^5C_3 (0.568)^3 (0.432)^2 = 0.342$	$100 \times 0.342 = 34.2$
4	22	88	${}^5C_4 (0.568)^4 (0.432)^1 = 0.225$	$100 \times 0.225 = 22.5$
5	8	40	${}^5C_5 (0.568)^5 (0.432)^0 = 0.0591$	$100 \times 0.0591 = 5.91$
	100	284		

$$\Sigma f = 100, \qquad \Sigma f x = 284$$

$$\text{Mean} = \frac{\Sigma f x}{\Sigma f} = \frac{284}{100} = 2.84$$

$$\text{Mean} = np = 2.84$$

$$5p = 2.84 \quad \Rightarrow \quad p = \frac{2.84}{5} = 0.568$$

$$q = 1 - p = 1 - 0.568 = 0.432$$

Binomial Distribution $= 100\,(0.432 + 0.568)^5$ **Ans.**

63.6 RECURRENCE RELATION FOR THE BINOMIAL DISTRIBUTION

By Binomial distribution, $P(r) = {}^nC_r\; p^r\, q^{n-r}$ ...(1) (*A.M.I.E., Summer 2002*)

$$P(r+1) = {}^nC_{r+1}\; p^{r+1}\, q^{n-r-1} \qquad ...(2)$$

On dividing (2) by (1), we get

$$\frac{P(r+1)}{P(r)} = \frac{{}^nC_{r+1}}{{}^nC_r}\;\frac{p^{r+1} q^{n-r-1}}{p^r q^{n-r}}$$

$$= \frac{n(n-1)(n-2).....(n-r)}{(r+1)!}\;\frac{r!}{n(n-1)(n-2).....(n-r+1)}\;\frac{p}{q}$$

$$\frac{P(r+1)}{P(r)} = \frac{n-r}{r+1}\,\frac{p}{q} \quad \text{or} \quad P(r+1) = \frac{n-r}{r+1}\,\frac{p}{q}\,P(r)$$ **Ans.**

EXERCISE 63.2

1. Fit a binomial distribution to the following frequency data:

x	0	1	3	4
f	28	62	10	4

(*U. P. III Sem. Dec. 2004*)

Ans. $P(r) = {}^{104}C_r (0.00999)^r (0.99111)^{104-r}$

2. Four coins were tossed 200 times. The number of tosses showing 0, 1, 2, 3 and 4 heads were found to be as under. Fit a binomial distribution to these observed results. Find the expected frequencies.

No. of heads:,	0	1	2	3	4
No. of tosses:	15	35	90	40	20

(A.M.I.E. Winter 2004)

3. Fill in the blanks :

(*a*) If three persons selected at random are stopped on a street, then the probability that all of them were born on Sunday is _______. *(A.M.I.E., Winter 2001)* **Ans.** $\frac{1}{343}$

(*b*) The mean, standard deviation and skewness of binomial distribution are ______, ____ and_____.

(A.M.I.E., Summer 2001) **Ans.** $np, \sqrt{npq}$

4. Tick √ the correct answer :

(*a*) The variance for a Binomial distribution is :

(*i*) np (*ii*) $\sqrt{np}$ (iii) npq (*iv*) $\sqrt{npq}$

(R.G.P.V., Bhopal, II Semester, June 2007) **Ans.** (*iii*)

(*b*) For the Binomial distribution $(p + q)^n$, he relation of mean and variance is :

(*i*) means = variance (*ii*) means < variance

(*iii*) mean > variance (*iv*) $(\text{mean})^2$ = variance

(R.G.P.V., Bhopal, II Semester, June 2006) **Ans.** (*iii*)

(*c*) In usual notation, for Binomial distribution, npq, is

(*i*) $< np$ (*ii*) np (*iii*) $> np$ (*iv*) None of the above

(A.M.I.E., Winter 2005) **Ans.** (*i*)

POISSON DISTRIBUTION

64.1 POISSON DISTRIBUTION

Poisson distribution is a particular limiting form of the Binomial distribution when p (or q) is very small and n is large enough.

Poisson distribution is

$$\boxed{P(r)=\frac{m^r e^{-m}}{r!}}$$

where m is the mean of the distribution.

Proof. In Binomial distribution.

$$P(r) = {}^nC_r q^{n-r} p^r = {}^nC_r (1-p)^{n-r} p^r$$

$$\left(\text{since mean} = m = np,\ p=\frac{m}{n}\right)$$

$$= {}^nC_r \left(1-\frac{m}{n}\right)^{n-r}\left(\frac{m}{n}\right)^r \qquad (m \text{ is constant})$$

$$= \frac{n(n-1)(n-2)\ldots(n-\overline{r-1})}{r!}\left(\frac{m}{n}\right)^r\left(1-\frac{m}{n}\right)^{n-r}$$

$$= \frac{\frac{n}{n}\left(\frac{n}{n}-\frac{1}{n}\right)\left(\frac{n}{n}-\frac{2}{n}\right)\ldots\left(\frac{n}{n}-\frac{r-1}{n}\right)m^r\left(1-\frac{m}{n}\right)^n}{r!\quad\left(1-\frac{m}{n}\right)^r}$$

$$= \frac{1\left(1-\frac{1}{n}\right)\left(1-\frac{2}{n}\right)\ldots\left(1-\frac{r-1}{n}\right)m^r\left(1-\frac{m}{n}\right)^n}{r!\quad\left(1-\frac{m}{n}\right)^r}$$

Taking limits, when n tends to infinity

$$\lim_{n\to\infty}\left(1-\frac{m}{n}\right)^n = \lim_{n\to\infty}\left[\left(1-\frac{m}{n}\right)^{-\frac{n}{m}}\right]^{-m} = e^{-m} \text{ and } \lim_{n\to\infty}\left(1-\frac{m}{n}\right)^r = 1$$

$$P(r)=\frac{m^r}{r!}e^{-m}$$

$$\boxed{P(r)=\frac{e^{-m}.m^r}{r!}}$$

64.2 MEAN OF POISSON DISTRIBUTION

$$P(r) = \frac{e^{-m} . m^r}{r!}$$ (*A.M.I.E.T.E., Summer 2004, 2002*)

Successes r	*Frequency f*	*f.r*
0	$\frac{e^{-m} m^0}{0!}$	0
1	$\frac{e^{-m} m^1}{1!}$	$e^{-m}. m$
2	$\frac{e^{-m} m^2}{2!}$	$e^{-m}. m^2$
3	$\frac{e^{-m} m^3}{3!}$	$\frac{e^{-m} . m^3}{2!}$
...	...	...
r	$\frac{e^{-m} m^r}{r!}$	$\frac{e^{-m} .m^r}{(r-1)!}$
...	...	...

$$\sum f\, r = 0 + e^{-m}.m + e^{-m}.m^2 + e^{-m}.\frac{m^3}{2!} + ... + e^{-m}\frac{m^r}{(r-1)!} + ... = e^{-m}.m\left[1 + \frac{m}{1!} + \frac{m^2}{2!} + ... + \frac{m^{r-1}}{(r-1)!} + ...\right]$$

$= m . e^{-m}. [e^m] = m$

$$\text{Mean} = \frac{\sum fr}{\sum f} = \frac{m}{1},$$

Mean = *m*. **Ans.**

64.3 STANDARD DEVIATION OF POISSON DISTRIBUTION

$$P(r) = \frac{e^{-m} m^r}{r!}$$ (*A.M.I.E.T.E., Summer 2002*)

Successes r	*Frequency* f	*Product* rf	*Product* r^2f
0	$\frac{e^{-m} m^0}{0!}$	0	0
1	$\frac{e^{-m} m^1}{1!}$	$e^{-m}. m$	$e^{-m}.m$
2	$\frac{e^{-m} m^2}{2!}$	$e^{-m}. m^2$	$2e^{-m}. m^2$
3	$\frac{e^{-m} m^3}{3!}$	$\frac{e^{-m} .m^3}{2!}$	$3e^{-m}.\frac{m^3}{2!}$
....	....	....	....
r	$\frac{e^{-m} m^r}{r!}$	$\frac{e^{-m} .m^r}{(r-1)!}$	$\frac{re^{-m} .m^r}{(r-1)!}$
....	....	....	....

$\Sigma f = 1,$ $\Sigma fr = m$

$$\sum f\, r^2 = 0 + e^{-m}.m + 2e^{-m}. m^2 + 3. e^{-m}.\frac{m^3}{2} + ... + \frac{r\, e^{-m}.m^r}{(r-1)!} + ..$$

$$= m.e^{-m}\left[1+2m+\frac{3m^2}{2!}+\frac{4m^3}{3!}+\ldots\frac{r.m^{r-1}}{(r-1)!}+..\right]$$

$$= m.e^{-m}\left[1+m+\frac{m^2}{2!}+\frac{m^3}{3!}+\ldots\frac{m^{r-1}}{(r-1)!}+\ldots+ m+2\frac{m^2}{2!}+\frac{3m^3}{3!}+\ldots+\frac{(r-1)m^{r-1}}{(r-1)!}+\ldots\right]$$

$$= m.e^{-m}\left[\left\{1+m+\frac{m^2}{2!}+\frac{m^3}{3!}+\ldots+\frac{m^{r-1}}{(r-1)!}+\ldots\right\} +m\left\{1+\frac{m}{1!}+\frac{m^2}{2!}+\ldots+\frac{m^{r-2}}{(r-2)!}+\ldots\right\}\right]$$

$$= m.e^{-m}\,[e^m + m\,e^m] = m + m^2$$

$$\sigma^2 = \frac{\Sigma fr^2}{\Sigma f}-\left(\frac{\Sigma fr}{\Sigma f}\right)^2 = \frac{m+m^2}{1}-(m)^2 = m \quad \text{or} \quad \sigma=\sqrt{m}$$

S. D. = $\sqrt{m}$

Hence mean and variance of a Poisson distribution are each equal to m. Similarly we can obtain,

$$\mu_3 = m, \qquad \mu_4 = 3m^2 + m$$

$$\beta_1 = \frac{1}{m}, \qquad \beta_2 = 3+\frac{1}{m}$$

$$\gamma_1 = \frac{1}{\sqrt{m}}, \qquad \gamma_2 = \frac{1}{m}$$

64.4 MEAN DEVIATION

Show that in a Poisson distribution with unit mean, and the mean deviation about the mean is $\left(\frac{2}{e}\right)$ times the standard deviation.

Solution. $P(r) = \frac{m^r}{r!}e^{-m}$ But mean = 1 *i.e.* $m = 1$ and S.D. = $\sqrt{m} = 1$

Hence, $P(r) = \frac{e^{-m}}{r!} = \frac{e^{-1}}{r!} = \frac{1}{e}.\frac{1}{r!}$

r	$P(r)$	$\|r-1\|$	$P(r)\,\|r-1\|$
0	$\frac{1}{e}$	1	$\frac{1}{e}$
1	$\frac{1}{e}$	0	0
2	$\frac{1}{e}\frac{1}{2!}$	1	$\frac{1}{e}\frac{1}{2!}$
3	$\frac{1}{e}\frac{1}{3!}$	2	$\frac{1}{e}\frac{2}{3!}$
4	$\frac{1}{e}\frac{1}{4!}$	3	$\frac{1}{e}\frac{3}{4!}$
....	....	....	....
r	$\frac{1}{e}\frac{1}{r!}$	$r-1$	$\frac{1}{e}\frac{r-1}{r!}$

$$\Sigma P(r)\,|r-1| = \frac{1}{e}+0+\frac{1}{e}\frac{1}{2!}+\frac{1}{e}\frac{2}{3!}+\frac{1}{e}\frac{3}{4!}+\ldots+\frac{1}{e}\frac{r-1}{r!}+\ldots = \frac{1}{e}\left[1+0+\frac{1}{2!}+\frac{2}{3!}+\frac{3}{4!}+\ldots+\frac{r-1}{r!}+\ldots\right]$$

$$= \frac{1}{e}\left[1+\left(\frac{1}{1!}-\frac{1}{1!}\right)+\left(\frac{2}{2!}-\frac{1}{2!}\right)+\left(\frac{3}{3!}-\frac{1}{3!}\right)+\left(\frac{4}{4!}-\frac{1}{4!}\right)+\ldots+\left(\frac{r}{r!}-\frac{1}{r!}\right)+\ldots\right]$$

$$= \frac{1}{e}\left[1+\frac{1}{1!}+\frac{2}{2!}+\frac{3}{3!}+\frac{4}{4!}\times\ldots+\frac{r}{r!}+\ldots-\frac{1}{1!}-\frac{1}{2!}-\frac{1}{3!}-\frac{1}{4!}\ldots-\frac{1}{r!}-\ldots\right]$$

$$= \frac{1}{e}\left[1+\left\{1+\frac{1}{1!}+\frac{1}{2!}+\frac{1}{3!}+\ldots+\frac{1}{(r-1)!}+\ldots\right\}-\left\{1+\frac{1}{1!}+\frac{1}{2!}+\frac{1}{3!}+\frac{1}{4!}+\ldots+\frac{1}{r!}\ldots\right\}+1\right]$$

$$= \frac{1}{e}[1+e-e+1] = \frac{2}{e} = \frac{2}{e}(1) = \frac{2}{e}\text{ S.D.}$$ **Proved.**

64.5 MOMENT GENERATING FUNCTION OF POISSON DISTRIBUTION

(A.M.I.E., Summer 2000)

Solution. $$P(r) = \frac{e^{-m}m^r}{r!}$$

Let $M_x(t)$ be the moment generating function, then

$$M_x(t) = \sum_{r=0}^{\infty} e^{tr}\frac{e^{-m}.m^r}{r!} = \sum_{r=0}^{\infty} e^{-m}.\frac{(me^t)^r}{r!} = e^{-m}\left[1+me^t+\frac{(me^t)^2}{2!}+\frac{(me^t)^3}{3!}+\ldots\right] = e^{-m}.e^{me^t} = e^{m(e^t-1)}$$

64.6 CUMULANTS

The cumulant generating function $K_x(t)$ is given by

$$K_x(t) = \log_e M_x(t) = \log_e e^{m(e^t-1)} = m(e^t-1)\log_e e$$

$$= m(e^t-1) = m\left[1+t+\frac{t^2}{2!}+\frac{t^3}{3!}+\ldots.+\frac{t^r}{r!}+\ldots-1\right] = m\left[t+\frac{t^2}{2!}+\frac{t^3}{3!}+\ldots+\frac{t^r}{r!}+\ldots\right]$$

Now $K_r = r$ th cumulant = coefficient of $\frac{t^r}{r!}$ in $K(t) = m$

i.e., $k_r = m,$ where $r = 1, 2, 3, \ldots.$

Hence, all the cumulants of the Poisson distribution are equal. In particular, we have

Mean $= K_1 = m,\ \mu_2 = K_2 = m,\ \mu_3 = K_3 = m$

$$\mu_4 = K_4 + 3K_2^2 = m + 3m^2$$

$$\beta_1 = \frac{\mu_3^2}{\mu_2^3} = \frac{m^2}{m^3} = \frac{1}{m},\quad \beta_2 = \frac{\mu_4}{\mu_2^2} = \frac{m+3m^2}{m^2} = \frac{1}{m}+3$$

64.7 RECURRENCE FORMULA FOR POISSON DISTRIBUTION

Solution. By Poisson distribution

$$P(r) = \frac{e^{-m}.m^r}{r!} \quad \ldots (1)$$

$$P(r+1) = \frac{e^{-m}m^{r+1}}{(r+1)!} \quad \ldots (2)$$

By dividing (2) by (1), we get

$$\frac{P(r+1)}{P(r)} = \frac{e^{-m}m^{r+1}}{(r+1)!}\frac{r!}{e^{-m}.m^r} = \frac{m}{r+1}$$

$$P(r+1) = \frac{m}{r+1}P(r)$$ **Ans.**

Example 1. *If the variance of the Poisson distribution is 2, find the probabilities for $r = 1, 2, 3, 4$ from the recurrence relation of the Poisson distribution. Also find $P(r \geq 4)$.*

Solution. Variance $= m = 2;$
Mean $= 2$

$$P(r+1) = \frac{m}{r+1}P(r) \quad \text{[Recurrence relation]}$$

Now $$P(r+1) = \frac{2}{r+1} P(r) \qquad (m = 2)$$

If $r = 0,\ P(1) = \frac{2}{0+1} P(0) = \frac{2}{0+1}(0.1353) = 0.2706 \qquad P(0) = e^{-m} = e^{-2} = 0.1353$

If $r = 1,\ P(2) = \frac{2}{1+1} P(1) = \frac{2}{2}(0.2706) = 0.2706$

If $r = 2,\ P(3) = \frac{2}{2+1} P(2) = \frac{2}{3}(0.2706) = 0.1804$

If $r = 3,\ P(4) = \frac{2}{3+1} P(3) = \frac{1}{2}(0.1804) = 0.0902$

$$P(r \geq 4) = P(4) + P(5) + P(6) + \ldots$$
$$= 1 - [P(0) + P(1) + P(2) + P(3)]$$
$$= 1 - [0.1353 + 0.2706 + 0.2706 + 0.1804]$$
$$= 1 - 0.8569 = 0.1431$$ **Ans.**

Example 2. *Assume that the probability of an individual coal miner being killed in a mine accident during a year is $\frac{1}{2400}$. Use appropriate statistical distribution to calculate the probability that in a mine employing 200 miners, there will be at least one fatal accident in a year.* (*A.M.I.E.T.E., Summer 2001*)

Solution. $P = \frac{1}{2400}, \quad n = 200$

$$m = np = \frac{200}{2400} = \frac{1}{12}$$

P (At least one) $= P(1 \text{ or } 2 \text{ or } 3 \text{ or } \ldots\ldots \text{ or } 200) = P(1) + P(2) + P(3) + \ldots + P(200)$

$$= 1 - P(0) = 1 - \frac{e^{-m} . m^0}{0!} = 1 - e^{-\frac{1}{12}} = 1 - 0.92 = 0.08$$ **Ans.**

Example 3. *Suppose 3% of bolts made by a machine are defective, the defects occurring at random during production. If bolts are packaged 50 per box, find*
(a) exact probability and
(b) Poisson approximation to it, that a given box will contain 5 defectives.

Solution. $$p = \frac{3}{100} = 0.03$$

(*a*) $$q = 1 - p = 1 - 0.03 = 0.97$$

Hence the probability for 5 defective bolts in a lot of 50

$$= {}^{50}C_5 (0.03)^5 (0.97)^{45} = 0.013074 \quad \text{(Binomial Distribution)}$$

(*b*) To get Poisson approximation $m = n\,p = 50 \times \frac{3}{100} = \frac{3}{2} = 1.5$

Required Poisson approximation $= \frac{m^r e^{-m}}{r!} = \frac{(1.5)^5 e^{-1.5}}{5!} = 0.01412$ **Ans.**

Example 4. *The number of arrivals of customers during any day follows Poisson distribution with a mean of 5. What is the probability that the total number of customers on two days selected at random is less than 2?*

Solution. $$m = 5$$

$$P(r) = \frac{e^{-m} m^r}{r!}, \quad P(r) = \frac{e^{-5}(5)^r}{r!}$$

If the number of customers on two days < 2 = 1 or 0

First day	*Second Day*	*Total*
0	0	0
0	1	1
1	0	1

$$\text{Required probability} = P(0)\,P(0) + P(0)\,P(1) + P(1)\,P(0)$$

$$= \frac{e^{-5}(5)^0}{0!}\cdot\frac{e^{-5}(5)^0}{0!} + \frac{e^{-5}(5)^0}{0!}\cdot\frac{e^{-5}(5)^1}{1!} + \frac{e^{-5}(5)^1}{1!}\,\frac{e^{-5}(5)^0}{0!}$$

$$= e^{-5}\cdot e^{-5} + e^{-5}\cdot e^{-5}\cdot 5 + e^{-5}\cdot 5\cdot e^{-5}$$

$$= e^{-10}[1 + 5 + 5] = 11e^{-10} = 11 \times 4.54 \times 10^{-5}$$

$$= 4.994 \times 10^{-4}$$

Ans.

Example 5. *In a certain factory producing cycle tyres, there is a small chance of 1 in 500 tyres to be defective. The tyres are supplied in lots of 10. Using Poisson distribution, calculate the approximate number of lots containing no defective, one defective and two defective tyres, respectively, in a consignment of 10,000 lots.*

Solution. $p = \frac{1}{500},\quad n = 10$

$$m = np = 10\,.\,\frac{1}{500} = \frac{1}{50} = 0.02, \quad P(r) = \frac{e^{-m}.m^r}{r!}$$

S.No.	Proability of defective	Number of lots containing defective
1	$P(0) = \frac{e^{-0.02}(0.02)^0}{0!} = e^{-0.02} = 0.9802$	$10{,}000 \times 0.9802 = 9802$ lots
2	$P(1) = \frac{e^{-0.02}(0.02)^1}{1!}$	$10{,}000 \times 0.019604 = 196$ lots
	$= 0.9802 \times 0.02 = 0.019604$	
3.	$P(2) = \frac{e^{-0.02}(0.02)^2}{2!}$	$10{,}000 \times 0.000196 = 2$ lots

$$= 0.9802 \times 0.0002 = 0.00019604$$

Ans.

Example 6. *A car hire firm has two cars which it hires out day by day. The number of demands for a car on each day is distributed as a Poisson distribution with mean 1.5. Calculate the number of days in a year on which*

(i) neither car is on demand $(e^{-1.5} = 0.2231)$

(ii) a car demand is refused.

(MDU, Dec. 2010, A.M.I.E., Summer 2004 Winter 2001, June 2009)

Solution. $m = 1.5$

(*i*) If the car is not used, then demand $(r) = 0$

$$P(r) = \frac{e^{-m}.m^r}{r!}, \quad P(0) = \frac{e^{-1.5}(1.5)^0}{0!} = e^{-1.5} = 0.2231$$

Number of days in a year when the demand is zero = 365 × 0.2231 = 81.4315 **Ans.** 81 days

(*ii*) Some demand is refused if the number of demands is more than two *i.e.* $r > 2$.

$$P(r > 2) = P(3) + P(4) + \ldots = 1 - [P(0) + P(1) + P(2)]$$

$$= 1 - \left[\frac{e^{-1.5}(1.5)^0}{0!} + \frac{e^{-1.5}(1.5)^1}{1!} + \frac{e^{-1.5}(1.5)^2}{2!}\right]$$

$$= 1 - [e^{-1.5} + e^{-1.5} \times 1.5 + e^{-1.5} \times 1.125] \quad = 1 - e^{-1.5}[1 + 1.5 + 1.125] = 1 - e^{-1.5} \times 3.625$$

$$= 1 - 0.2231 \times 3.625 = 1 - 0.8087375 \quad = 0.1912625$$ **Ans.**

Number of days in a year when some demand of car is refused

$$= 365 \times 0.1912625 = 69.81 = 70 \text{ days}$$ **Ans.**

Example 7. *If the probability that an individual suffers a bad reaction from a certain injection is 0.001, determine the probability that out of 2000 individuals*

(a) exactly 3 (b) more than 2 individuals (c) None (d) More than one individual will suffer a bad reaction. *(A.M.I.E.T.E., Winter, 2002, 2000)*

Solution. $p = 0.001, \quad n = 2000$

$$m = np = 2000 \times 0.001 = 2$$

$$\therefore \quad P(r) = \frac{e^{-m} m^r}{r!} = e^{-2}\frac{2^r}{r!} = \frac{1}{e^2} \times \frac{2^r}{r!}$$

(*a*) P (exactly 3) $\quad P(3) = \frac{1}{e^2} \cdot \frac{2^3}{3!} = \frac{1}{(2.718)^2} \times \frac{8}{6} = (0.135) \times \frac{4}{3} = 0.18$

(*b*) P (more than 2) $= P(3) + P(4) + P(5) + \ldots + P(2000)$

$$= 1 - [P(0) + P(1) + P(2)] = 1 - \left[\frac{e^{-2}(2)^0}{0!} + \frac{e^{-2}(2)^1}{1!} + \frac{e^{-2}(2)^2}{2!}\right]$$

$$= 1 - e^{-2}[1 + 2 + 2] = 1 - \frac{5}{e^2} = 1 - 5 \times 0.135 = 1 - 0.675 = 0.325$$ **Ans.**

(*c*) $\quad P$ (none) $= P(0) = \frac{e^{-2}(2)^0}{0!} = 0.135$

(*d*) P (more than 1) $= P(2) + P(3) + P(4) + \ldots + P(2000) = 1 - [P(0) + P(1)]$

$$= 1 - \left[\frac{e^{-2}(2)^0}{0!} + \frac{e^{-2}(2)^1}{1!}\right] = 1 - 3e^{-2} = 1 - 3 \times 0.135 = 1 - 0.405 = 0.595$$ **Ans.**

Example 8. *In a certain factory turning out razor blades, there is a small chance of 0.002 for any blade to be defective. The blades are supplied in packets of 10. Use appropriate and suitable distribution to calculate the approximate number of packets containing no defective, one defective and two defective blades respectively in a consignment of 50000 packets.* *(R.G.P.V., Bhopal, II Semester, June 2006)*

Solution. Here, $p = 0.002, \quad n = 10$

$$m = np \quad \Rightarrow \quad m = 10 \times 0.002 = 0.020$$

$$P(r) = \frac{e^{-m} \cdot (m)^r}{r!}$$

r	$p(r) = \frac{e^{-0.02}(0.02)^r}{r!}$	Number of packets = 50000 p
0	$p(0) = \frac{e^{-0.02}(0.02)^0}{0!} = 0.980$	$50000 \times (0.980) = 4900$
1	$p(1) = \frac{e^{-0.02}(0.02)^1}{1!} = 0.0196$	$50000 \times (0.0196) = 980$
2	$p(2) = \frac{e^{-0.02}(0.02)^2}{2!} = 0.000196$	$50000 \times (0.000196) = 9.8$

Hence, number of packets containing no defective razor blades = 49000.
Number of packets containing one defective razor blade = 980
Number of packets containing two defective razor blase = 9.8

Example 9. *If there are 3 misprints in a book of 1000 pages find the probability that a given page will contain*
(i) no misprint (ii) more than 2 misprints. *(U.P., III Semester, Dec. 2009)*

Solution. Total number of pages = 1000
No. of misprints =3

$$P = \frac{3}{1000} = 0.003, \qquad n = 1, \qquad m = nP = 1 \times 0.003 = 0.003$$

Poisson distribution

$$P(r) = \frac{e^{-m} \cdot m^r}{r!}, \qquad P(0) = \frac{e^{-0.003}(0.003)^0}{0!} = e^{0.003} = 0.997$$

$$P(r > 2) = P(3) = \frac{e^{-0.003}(0.003)^3}{3!} = 0.0000000045$$

Hence (*i*) the probability that a page will contain no error = 0.997
(*ii*) the probability that a page will contain more than two misprints = 0.0000000045 **Ans.**

Example 10. *A manufacturer knows that the condensers he makes contain on an average 1% of defectives. He packs them in boxes of 100. What is the probability that a box picked out at random will contain 4 or more faulty condensers?*

Solution. $P = 1\% = 0.01, n = 100, \quad m = np = 100 \times 0.01 = 1$

$$P(r) = \frac{e^{-m}.(m)^r}{r!} = \frac{e^{-1}(1)^r}{r!} = \frac{e^{-1}}{r!}$$

$$P(4 \text{ or more faulty condensers}) = P(4) + P(5) + \ldots + P(100) = 1 - [P(0) + P(1) + P(2) + P(3)]$$

$$= 1 - \left[\frac{e^{-1}}{0!} + \frac{e^{-1}}{1!} + \frac{e^{-1}}{2!} + \frac{e^{-1}}{3!}\right] = 1 - e^{-1}[1 + 1 + \frac{1}{2} + \frac{1}{6}] = 1 - \frac{8}{3e} = 1 - 0.981 = 0.019 \qquad \textbf{Ans.}$$

Example 11. *An insurance company found that only 0.01% of the population is involved in a certain type of accident each year. If its 1000 policy holders were randomly selected from the population, what is the probability that not more than two of its clients are involved in such an accident next year? (given that $e^{-0.1} = 0.9048$)*

Solution.

$$P = 0.01\% = \frac{1}{100} \times \frac{1}{100} = \frac{1}{10000}, \qquad n = 1000$$

$$m = np = (1000) \times \frac{1}{10000} = \frac{1}{10} = 0.1$$

$$P(r) = \frac{e^{-m} m^r}{r!}$$

$$P(\text{not more than } 2) = P(0, 1 \text{ and } 2) = P(0) + P(1) + P(2)$$

$$= \frac{e^{-0.1}(0.1)^0}{0!} + \frac{e^{-0.1}(0.1)^1}{1!} + \frac{e^{-0.1}(0.1)^2}{2!} = e^{-0.1}\left(1 + 0.1 + \frac{0.01}{2}\right)$$

$$= 0.9048 \times 1.105 = 0.9998 \qquad \textbf{Ans.}$$

Example 12. *Fit a Poisson distribution to the set of observations :*

x	0	1	2	3	4
f	122	60	15	2	1

(R.G.P.V., Bhopal, II Semester, Dec. 2007, June 2007)

Solution. The mean number = $\dfrac{\Sigma f \cdot x}{\Sigma f}$.

x	f	fx
0	122	0
1	60	60
2	15	30
3	2	6
4	1	4
Total	200	100

$$\text{Mean} = \frac{\Sigma f x}{\Sigma f} = \frac{100}{200} = \frac{1}{2}$$

x	$P(x) = \dfrac{e^{-1/2}(1/2)^x}{x!}$	Theoretical frequency	Given frequency
0	$P(0) = \dfrac{e^{-\frac{1}{2}}\left(\frac{1}{2}\right)^0}{0!} = 0.6065$	$0.6065 \times 200 = 121.3$	121
1	$P(1) = \dfrac{e^{-\frac{1}{2}}\left(\frac{1}{2}\right)^1}{1!} = \dfrac{0.6065}{2} = 0.3033$	$0.3033 \times 200 = 60.7$	61
2	$P(2) = \dfrac{e^{-\frac{1}{2}}\left(\frac{1}{2}\right)^2}{2!} = \dfrac{0.6065}{8} = 0.0758$	$0.0758 \times 200 = 15.2$	15
3	$P(3) = \dfrac{e^{-\frac{1}{2}}\left(\frac{1}{2}\right)^3}{3!} = \dfrac{0.6065}{48} = 0.0126$	$0.0126 \times 200 = 2.5$	2
4	$P(4) = \dfrac{e^{-\frac{1}{2}}\left(\frac{1}{2}\right)^4}{4!} = \dfrac{0.6065}{384} = 0.0016$	$0.0016 \times 200 = 0.32$	1

Ans.

EXERCISE 64.1

1. Find the probability that at most 5 defective fuses will be found in a box of 200 fuses if experience shows that 2 per cent of such fuses are defective. **Ans.** 0.785

2. The number of accidents during a year in a factory has the Poisson distribution with mean 1.5. The accidents during different years are assumed independent. Find the probability that only 2 accidents take place during 2 years time. **Ans.** 0.224

3. A manufacturer of cotter pins knows that 5% of his product is defective. If he sells cotter pins in boxes of 100 and guarantee that not more than 10 pins will be defective, what is the approximate probability that a box will fail to meet the guaranteed quality. [$e^{-5} = 0.006738$] **Ans.** 0.0136875

4. Suppose the number of telephone calls on an operator received from 9.00 to 9.05 follow a Poisson distribution with mean 3. Find the probability that

   (*i*) the operator will receive no calls in that time interval tomorrow,

   (*ii*) in the next three days the operator will receive a total of 1 call in that time interval.
   [$e^{-3} = 0.04978$] **Ans.** (*i*) e^{-3} (*ii*) $3 \times (e^{-3})^2 (e^{-3}.\, 3)$

5. On the basis of past record it has been found that there is 70% chance of power cut in a city on any particular day. What is the probability that from the first to the 10th date of the month, there are 5 or more days without power cut. (*A.M.I.E.T.E., Summer 2001*)

   Ans. $\left[\dfrac{3^5}{5!} + \dfrac{3^6}{6!} + \dfrac{3^7}{7!} + \dfrac{3^8}{8!} + \dfrac{3^9}{9!} + \dfrac{3^{10}}{10!}\right] e^{-3}$

6. The distribution of typing mistakes committed by a typist is given below. Assuming a Poisson model, find out the expected frequencies:

Mistakes per pages	0	1	2	3	4	5
No. of pages	142	156	69	27	5	1

Ans. 147, 147, 74, 25, 6, 1 pages.

7. Let x be the number of cars per minute passing a certain crossing of roads between 5.00 P.M. and 7.00 P.M. on a holiday. Assume x has a Poisson distribution with mean 4. Find the probability of observing atmost 3 cars during any given minute between 5.00 P.M. and 7 P.M. (given $e^{-4} = 0.0183$) **Ans.** 0.4331

8. Number of customers arriving at a service counter during a day has a Poisson distribution with mean 100. Find the probability that at least one customer will arrive on each day during a period of five days. Also find the probability that exactly 3 customers will arrive during two days.

Ans. $(1-e^{-100})^5,\ e^{-200} \times \dfrac{4(100)^3}{3}$

9. In a normal summer, a truck driver gets on an average one puncture in 1000 km. Applying Poisson distribution, find the probability that he will have
(*i*) no puncture, (*ii*) two punctures in a journey of 3000 kms. **Ans.** (*i*) e^{-3} (*ii*) $4.5\ e^{-3}$

10. Wireless sets are manufactured with 25 soldered joints each. On the average, 1 joint in 500 is defective. How many sets can be expected to be free from defective joints in a consignment of 10000 sets ?

Ans. 9512

11. In a certain factory turning out razor blades, there is small chance $\dfrac{1}{500}$ for any blade to be defective. The blades are supplied in packets of 10. Using Poisson's distribution, calculate the approximate number of packets containing (i) no defective (ii) one defective and (iii) two defective blades respectively in a

12. If m and μ_r denote by the mean and central rth moment of a Poisson distribution, then prove that

$$\mu_{r+1} = r\,m\,\mu_{r-1} + m\frac{d\mu_r}{dm}.\qquad \left[\textbf{Hint. } \mu_r = \sum_{r=0}^{\infty}(x-m)^r\frac{e^{-m}m^x}{x!},\ \text{find } \frac{d\mu_r}{dm}\right]$$

13. A certain screw-making machine produces an average 2 defective screws out of 100, and pack them in boxes of 500. Find the probability that a box contains 15 defective screws.

(*A.M.I.E., Winter 2005*) **Ans.** 0.0347

14. The distribution of the number of road accidents per day in a city is Poisson with mean 4. Find the number of days out of 100 days when there will be:
(*i*) no accident (*ii*) at least 2 accidents (*iii*) at most 3 accidents (*iv*) between 2 and 5 accidents

Ans. (*i*) 2 days (*ii*) 91 days (*iii*) 43 days (*iv*) 39 days

15. Fill in the blanks.

(*a*) If x has a Poisson distribution such that $P(x=k) = P(x=k+1)$ for some positive integer k then mean of x is .... **Ans.** $k+1$

16. Choose the correct answer:

(*a*) In the Poisson distribution if $P(x=k) = P(x=k+1)$, then the mean is :
(*i*) k (*ii*) 2k (*iii*) k + 1 (*iv*) k – 1 (*R.G.P.V. Bhopal, II Semester, June 2007*) **Ans.** (*iii*)

(*b*) The value of measure of skewness of Poisson distribution is :
(*i*) m (*ii*) $\sqrt{m}$ (*iii*) $\dfrac{1}{m}$ (*iv*) $\dfrac{1}{\sqrt{m}}$ (*R.G.P.V., Bhopal, II Semester, June 2006*) **Ans.** (*iii*)

(*g*) Poisson distribution with unit mean, mean-deviation about the mean is :
(*i*) $\dfrac{1}{e}$ (*ii*) $\dfrac{\sigma}{e}$ (*iii*) $\dfrac{2\sigma}{e}$ (*iv*) $\dfrac{2}{e}$ (*R.G.P.V., Bhopal, II Semester, Feb 2006*) **Ans.** (*iv*)

(h) In the Poisson distribution if $2P(x=1) = P(x=2)$, then the variance is :
(*i*) 0 (*ii*) –1 (*iii*) 4 (*iv*) 2 (*R.G.P.V., Bhopal, II Semester, June 2007*) **Ans.** (*iii*)

(*i*) In the Poisson distribution if $2p(x=1) = p(x=2)$
Then mean is
(*i*) 0 (*ii*) – 1 (*iii*) 4 (*iv*) 2 (*R.G.P.V., Bhopal, II Semester, June 2007*) **Ans.** (*iii*)

NORMAL DISTRIBUTION

65.1 CONTINUOUS DISTRIBUTION

So far we have dealt with discrete distributions where the variate takes only the integral values. But the variates like temperature, heights and weights can take all values in a given interval. Such variables are called continuous variables.

Let $f(x)$ be a continuous function, then Mean $= \int_{-\infty}^{+\infty} x . f(x)\, dx$

$$\text{Variance} = \int_{-\infty}^{+\infty} (x - \bar{x})^2 . f(x)\, dx \qquad (\bar{x} = \text{mean})$$

Note. $f(x)$ is called probability density function if

(1) $f(x) \geq 0$ for every value of x. (2) $\int_{-\infty}^{+\infty} f(x)\, dx = 1$

(3) $\int_a^b f(x)\, dx = P, \qquad (a < x < b)$

Example 1. *The probability density function $f(x)$ of a continuous random variable x is defined by*

$$f(x) = \begin{cases} \dfrac{A}{x^3}, & 5 \leq x \leq 10 \\ 0, & otherwise \end{cases} \qquad \textit{Find the value of } A.$$

Solution. Here, $f(x) = \dfrac{A}{x^3}, \qquad 5 \leq x \leq 10$

Since $f(x)$ is probability density function, so

$$\int_5^{10} \frac{A}{x^3}\, dx = 1 \quad \Rightarrow \quad \left[-\frac{A}{2x^2}\right]_5^{10} = 1$$

$$\frac{A}{2}\left[-\frac{1}{100} + \frac{1}{25}\right] = 1$$

$$\frac{A}{2}\left(\frac{3}{100}\right) = 1 \quad \Rightarrow \quad A = \frac{200}{3}$$ **Ans.**

Example 2. *A function $f(x)$ is defined as follows*

$$f(x) = \begin{cases} 0, & x < 2 \\ \dfrac{1}{18}(2x+3), & 2 \leq x \leq 4 \\ 0, & x > 4 \end{cases}$$

Show that it is a probability density function.

Solution.
$$f(x) = \begin{cases} 0, & x < 2 \\ \frac{1}{18}(2x+3), & 2 \le x \le 4 \\ 0, & x > 4 \end{cases}$$

If f(x) is a probability density function, then

(*i*) $\int_{-\infty}^{\infty} f(x)\,dx = 1$

Here $\int_2^4 \frac{1}{18}(2x+3)dx = \frac{1}{18}[x^2+3x]_2^4 = \frac{1}{18}(16+12-4-6) = 1$

(*ii*) $f(x) > 0$ for $2 \le x \le 4$

Hence, the given function is a probability density function. **Proved.**

Example 3. *The diameter of an electric cable is assumed to be continuous random variate with probability density function:*

$f(x) = 6x(1-x),\ 0 \le x \le 1$

(*i*) *verify that above is a p.d.f.* (*ii*) *find the mean and variance.*

Solution. (*i*) $\int_{-\infty}^{\infty} f(x)\,dx = \int_0^1 6x(1-x)dx = \int_0^1 (6x-6x^2)dx$

$$= (3x^2-2x^3)_0^1 = 3-2 = 1$$

Secondly $f(x) > 0$ for $0 \le x \le 1$.

Hence the given function is a probability density function.

(*ii*) Mean $= \int_{-\infty}^{\infty} x.f(x)\,dx = \int_0^1 x.6x(1-x)dx$

$$= \int_0^1 (6x^2-6x^3)dx = \left(2x^3 - \frac{3}{2}x^4\right)_0^1 = 2-\frac{3}{2} = \frac{1}{2}$$ **Ans.**

Variance $= \int_{-\infty}^{\infty} (x-\bar{x})^2.f(x)\,dx = \int_0^1 \left(x-\frac{1}{2}\right)^2.6x(1-x)\,dx$

$$= \int_0^1 \left(x^2-x+\frac{1}{4}\right)(6x-6x^2)dx = \int_0^1 \left(12x^3-6x^4-\frac{15}{2}x^2+\frac{3}{2}x\right)dx$$

$$= \left(3x^4-\frac{6}{5}x^5-\frac{5}{2}x^3+\frac{3x^2}{4}\right)_0^1 = \left(3-\frac{6}{5}-\frac{5}{2}+\frac{3}{4}\right) = \frac{1}{20}$$ **Ans.**

Example 4. *A continuous random variable has p.d.f.*

$$f(x) = ke^{-\frac{x}{5}},\ x \ge 0,$$
$$= 0, \text{ else where}$$

then the value of k is...... (*A.M.I.E., Winter 2002*)

Solution. $\int_{-\infty}^{\infty} p.d.f(x)\,dx = 1$

$\Rightarrow \int_{-\infty}^{0} p.d.f(x)\,dx + \int_0^{\infty} p.d.f(x)\,dx = 1$

$$0 + \int_0^{\infty} ke^{-\frac{x}{5}}\,dx = 1, \Rightarrow k\left[\frac{e^{-\frac{x}{5}}}{-\frac{1}{5}}\right]_0^{\infty} = 1 \Rightarrow -5k\left[\frac{1}{e^{\infty}}-1\right] = 1 \Rightarrow 5k = 1 \Rightarrow k = \frac{1}{5}$$ **Ans.**

Example 5. *If the probability density function of a random variable x is*

$$f(x) = \begin{cases} kx^{\alpha-1}(1-x)^{\beta-1}, & \text{for } 0 < x < 1, \quad \alpha > 0, \quad \beta > 0 \\ 0, & \text{otherwise} \end{cases}$$

Find k and mean of x.

Solution. If $f(x)$ is a probability density function,

Then $\int_{-\infty}^{\infty} x.f(x)\,dx = 1$

Here $\int_0^1 k\,x^{\alpha-1}(1-x)^{\beta-1}dx = 1$ [$f(x)$ is a Beta function.]

$$\Rightarrow \qquad k\frac{\overline{|\alpha}\,\overline{|\beta}}{\overline{|\alpha+\beta}} = 1 \quad \Rightarrow \quad k = \frac{\overline{|\alpha+\beta}}{\overline{|\alpha}\,\overline{|\beta}} \qquad \textbf{Ans.}$$

$$\text{Mean} = \int_{-\infty}^{\infty} x.f(x)\,dx = \int_0^1 x.k\,x^{\alpha-1}(1-x)^{\beta-1}dx$$

$$= k\int_0^1 x^{\alpha+1-1}(1-x)^{\beta-1}dx \qquad \dots (1)$$

[$f(x)$ is Beta function]

Putting the value of k and the integral in (1), we get

$$\text{Mean} = \frac{\overline{|\alpha+\beta}}{\overline{|\alpha}\,\overline{|\beta}} \cdot \frac{\overline{|\alpha+1}\,\overline{|\beta}}{\overline{|\alpha+\beta+1}} = \frac{\overline{|\alpha+\beta}}{\overline{|\alpha}\,\overline{|\beta}} \frac{\alpha\overline{|\alpha}\,\overline{|\beta}}{(\alpha+\beta)\overline{|\alpha+\beta}} = \frac{\alpha}{\alpha+\beta} \qquad \textbf{Ans.}$$

EXERCISE 65.1

1. The probability density $p(x)$ of a continuous random variable is given by

   $p(x) = y_0\, e^{-|x|}, -\infty < x < \infty$

   Prove that $y_0 = 1/2$. Find mean and variance of the distribution. **Ans.** 0.2

2. If $f(x) = \begin{cases} \frac{1}{2}(x+1), & -1 < x < 1 \\ 0, & \text{elsewhere} \end{cases}$ **Ans.** $\frac{1}{3}, \frac{2}{9}$

   Find the mean and variance.

3. X is a random variable giving time (in minutes) during which a certain electrical equipment is used at maximum load in a specified time period. If the *pdf* is given by

   $$f(x) = \begin{cases} \dfrac{x}{(1500)^2}, & 0 \le x \le 1500 \\ -\dfrac{(x-3000)}{(1500)^2}, & 1500 \le x \le 3000 \\ 0, & \text{elsewhere,} \end{cases}$$

   represents the density of a variable x, find $E(x)$ and $Var(x)$

   find the expected value of X. **Ans.** 1500, 375000

4. A function is defined as under :

   $$f(x) = \frac{1}{k}, \; x_1 \le x \le x_2 \; = 0, \text{ elsewhere.}$$

   Find the cumulative distribution of the variate x when k satisfies the requirements for $f(x)$ to be a density function.

   Ans. $f(x) = 0, \; x < x_1; \quad (x - x_1)/(x_2 - x_1), \quad x_1 \le x \le x_2, \; 1; x \ge x_2.$

5. A continuous distribution of a variable x in the range $(-3, 3)$ is defined as

$$f(x)=\begin{cases}\frac{1}{16}(3+x)^2, & -3\le x<-1\\ =\frac{1}{16}(6-2x^2), & -1\le x<1\\ =\frac{1}{16}(3-x)^2, & 1\le x\le 3.\end{cases}$$

Verify that the area under the curve is unity. Show that the mean is zero.

65.2 MOMENT GENERATING FUNCTION OF THE CONTINUOUS PROBABILITY DISTRIBUTION ABOUT $x = a$ is given by

$$M_a(t) = \int_{-\infty}^{\infty} e^{t(x-a)} f(x)\,dx \qquad \text{where } f(x) \text{ is p.d.f.}$$

Example 6. *Find the moment generating function of the exponential distribution*

$$f(x) = \frac{1}{c}e^{-x/c} \qquad 0 \le x \le \infty,\ c>0$$

Hence find its mean and S.D.

Solution. The moment generating function about origin is

$$M_0(t) = \int_0^{\infty} e^{tx}\frac{1}{c}e^{-x/c}\,dx = \frac{1}{c}\int_0^{\infty} e^{(t-1/c)x}\,dx = \frac{1}{c}\left[\frac{e^{(t-1/c)x}}{t-\frac{1}{c}}\right]_0^{\infty} = \frac{1}{c}\left[-\frac{1}{t-\frac{1}{c}}\right] = \frac{1}{1-ct} = (1-ct)^{-1}$$

$$= 1 + ct + c^2t^2 + c^3t^3 + c^4t^4 + \ldots$$

$$\mu_1' = \frac{d}{dt}[M_0(t)]_{t=0} = [c + 2c^2t + 3c^3t^2 + 4c^4t^3 + \ldots]_{t=0} = c$$

$$\mu_2' = \frac{d^2}{dt^2}[M_0(t)]_{t=0} = [2c^2 + 6c^3t + 12c^4t^2 + \ldots]_{t=0} = 2c^2$$

$$\mu_2 = \mu_2' - (\mu_1')^2 = 2c^2 - c^2 = c^2$$

S.D. = c

Hence, Mean $= c$, S.D. $= c$ **Ans.**

65.3 NORMAL DISTRIBUTION

(*A. M. I. E., Summer 2002*)

Normal distribution is a continuous distribution. It is derived as the limiting form of the Binomial distribution for large values of n and p and q are not very small.

The normal distribution is given by the equation

$$f(x) = \frac{1}{\sigma\sqrt{(2\pi)}}e^{-\frac{(x-\mu)^2}{2\sigma^2}} \qquad \ldots(1)$$

where μ = mean, σ = standard deviation, $\pi = 3.14159\ldots$, $e = 2.71828\ldots$

$$P(x_1 < x < x_2) = \int_{x_1}^{x_2} \frac{1}{\sigma\sqrt{(2\pi)}}e^{-\frac{(x-\mu)^2}{2\sigma^2}}\,dx$$

On substitution $z = \dfrac{x-\mu}{\sigma}$ in (1), we get $f(z) = \dfrac{1}{\sqrt{2\pi}}e^{-\frac{1}{2}z^2}$ ...(2)

Here mean = 0, standard deviation = 1.

(2) is known as standard form of normal distribution.

Theorem. To derive normal distribution as a limiting case of Binomial distribution where $p \ne q$ but $p \approx q$. (*U.P. III Semester, Dec. 2006; R.G.P.V., Dec. 2001*)

Statement. The limiting case of binomial distribution $(p+q)^n$, as $n \to \infty$ and neither p nor q are very small, generates the normal distribution.

Proof. The frequency for r and $(r + 1)$ successes in binomial distribution are

$$f(r) = N.{}^nC_r\, p^r\, q^{n-r} \text{ and } f(r+1) = N.{}^nC_{r+1}\, p^{r+1}\, q^{n-(r+1)}$$

The frequency of r successes > frequency of $(r + 1)$ successes if

$$f(r) > f(r+1) \Rightarrow \frac{f(r)}{f(r+1)} > 1$$

$$\Rightarrow \frac{N.{}^nC_r\, p^r\, q^{n-r}}{N.{}^nC_{r+1}\, p^{r+1}\, q^{n-r-1}} > 1 \quad \Rightarrow \quad \frac{\dfrac{n!}{r!(n-r)!}.p^r.q^{n-r}}{\dfrac{n!}{(r+1)!(n-r-1)!}p^{r+1}\, q^{n-r-1}} > 1$$

$$\Rightarrow \frac{n!\, p^r.q^{n-r}\,(r+1)!(n-r-1)!}{r!(n-r)!.n!\, p^{r+1}.q^{n-r-1}} > 1 \quad \Rightarrow \quad \frac{q.(r+1)}{(n-r)\,p} > 1 \quad \Rightarrow \quad qr + q > np - pr$$

$$\Rightarrow \quad q > np - r(p+q)$$

$$\Rightarrow \quad r > np - q \qquad \text{... (1)}$$

Again, similarly the frequency of r successes > the frequency of (r – 1) successes if

$$f(r) > f(r-1) \Rightarrow \frac{f(r)}{f(r-1)} > 1$$

$$\Rightarrow \frac{N.{}^nC_r\, p^r\, q^{n-r}}{N.{}^nC_{r-1}\, p^{r-1}\, q^{n-(r-1)}} > 1 \quad \Rightarrow \quad \frac{\dfrac{n!}{r!(n-r)!}.p^r.q^{n-r}}{\dfrac{n!}{(r-1)!(n-r+1)!}p^{r-1}\, q^{n-r+1}} > 1$$

$$\Rightarrow \frac{n!\, p^r q^{n-r}\,(r-1)!(n-r+1)!}{r!(n-r)!n!\, p^{r-1}\, q^{n-r+1}} > 1 \quad \Rightarrow \quad \frac{p(n-r+1)}{rq} > 1$$

$$\Rightarrow \quad pn - pr + p > rq \quad \Rightarrow \quad pn + p > pr + qr$$

$$\Rightarrow \quad pn + p > r(p+q) \quad \Rightarrow \quad pn + p > r \qquad \text{... (2)}$$

$$[\because p + q = 1]$$

From (1) and (2), we have

$$pn + p > r > np - q$$

$$pn + p + q > r > np$$

$$np + 1 > r > np$$

Since a possible value of r is np, therefore, without loss of generality we can assume that np is an integer as $n \to \infty$. Hence the frequency of np successes can be assumed to be maximum frequency. Let y_0 be the frequency of np sccesses and y_x be the frequency of $(np + x)$ successes.

Then

$$y_0 = f(np) = N.{}^nC_{np}\, p^{np}\, q^{n-np} \qquad \text{[From (1), for } r = np]$$

$$= N\frac{n!}{(np)!(n-np)!}\, p^{np}\, q^{n-np} = N\frac{n!}{(np)!(nq)!}\, p^{np}\, q^{nq} \quad \text{... (3)} \quad [\because q = 1 - p]$$

and
$$y_x = N.\frac{n!}{(np+x)!(nq-x)!}\, p^{np+x}\, q^{nq-x} \qquad \text{... (4)}$$

Diving (4) by (3), we get $\dfrac{y_x}{y_0} = \dfrac{(np)!(nq)!}{(np+x)!(nq-x)!}\, p^x\, q^{-x}$ ... (5)

If n be large, then according to James Stirling, we have

$$n! = e^{-n}\, h^{n+\frac{1}{2}}\, \sqrt{(2\pi)},$$

From (5) $\dfrac{y_x}{y_0} = \dfrac{e^{-np}(np)^{np+1/2}\sqrt{2\pi}\, e^{-nq}(nq)^{nq+1/2}\sqrt{2\pi}\; p^x q^{-x}}{e^{-(np+x)}(np+x)^{np+x+1/2}\sqrt{2\pi}\, e^{-(nq-x)}(nq-x)^{nq-x+1/2}\sqrt{2\pi}}$

$$= \frac{(np)^{np+1/2}(nq)^{nq+1/2}(nq/nq)^x}{(np)^{np+x+1/2}\left\{1+\dfrac{x}{np}\right\}^{np+x+1/2}(np)^{nq-x+1/2}\left\{1-\dfrac{x}{nq}\right\}^{nq-x+1/2}}$$

$$= \frac{1}{\left\{1+\dfrac{x}{np}\right\}^{np+x+1/2}\left\{1-\dfrac{x}{nq}\right\}^{np+x+1/2}}$$

$\therefore$ $$\log\frac{y_x}{y_0} = -\left(np+x+\frac{2}{2}\right)\log\left(1+\frac{x}{np}\right)-\left(nq-x+\frac{1}{2}\right)\log\left(1-\frac{x}{nq}\right)$$

$$= -\left(np+x+\frac{1}{2}\right)\left(\frac{x}{nq}-\frac{x^2}{2n^2p^2}+\frac{x^3}{3p^3q^3}-\ldots\right)+\left(nq-x+\frac{1}{2}\right)\left(\frac{x}{nq}+\frac{x^2}{2n^2q^2}+\frac{x^3}{3n^3q^3}+\ldots\right)$$

$$= x\left(1-\frac{1}{2np}+1+\frac{1}{2np}\right)+x^2\left(\frac{1}{2np}-\frac{1}{np}+\frac{1}{4n^2p^2}+\frac{1}{2nq}-\frac{1}{nq}+\frac{1}{4n^2q^2}\right)$$

$$+x^3\left(\frac{1}{3n^2q^2}+\frac{1}{6n^3q^3}-\frac{1}{2n^2q^2}-\frac{1}{2n^2p^2}-\frac{1}{3n^2q^2}+\frac{1}{6n^3p^3}\right)-\ldots$$

$$= \frac{p-q}{2npq}x+\frac{p^2+q^2}{4n^2p^2q^2}x^2-\frac{x^2}{2npq}+\ldots + \text{terms of higher orders.}$$

Neglecting terms containing $1/n^2$, we have

$\therefore$ $$\log\frac{y_x}{y_0} = -\frac{q-p}{2npq}x-\frac{x^2}{2npq}$$

Since p < 1, q < 1 and so q – p is very samll as compared with n. Therefore I[st] term may be neglected. $(q-p=0)$

$\therefore$ $$\log\frac{y_x}{y_0} = -\frac{x^2}{2npq} = -\frac{x^2}{2\sigma^2}$$ [$\because \sigma^2 = npq$, the variance of Binomial distribution]

$\Rightarrow$ $$y_x = y_0\, e^{-x^2/2\sigma^2}$$

65.4 NORMAL CURVE

Let us show binomial distribution graphically.

The probabilities of heads in 1 tosses are

${}^{10}C_0 q^{10} p^0$, ${}^{10}C_1 q^9 p^1$, ${}^{10}C_2 q^8 p^2$,

${}^{10}C_3 q^7 p^3$, ${}^{10}C_4 q^6 p^4$, ${}^{10}C_5 q^5 p^5$,

${}^{10}C_6 q^4 p^6$, ${}^{10}C_7 q^3 p^7$, ${}^{10}C_8 q^2 p^8$,

${}^{10}C_9 q^1 p^9$, ${}^{10}C_{10} q^0 p^{10}$.

$p = \frac{1}{2}$, $q = \frac{1}{2}$. It is shown in the figure given.

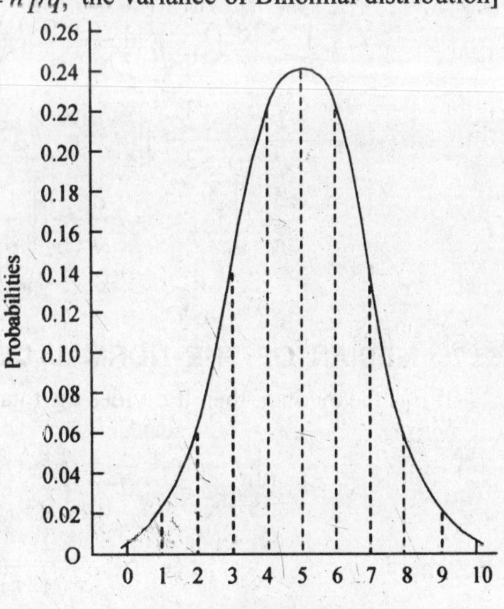

If the variates (heads here) are treated as if they were continuous, the required probability curve will be a *normal curve* as shown in the above figure by dotted lines.

Properties of the normal curve. $y = y_0 e^{-\frac{x^2}{2\sigma^2}}$

1. The curve is symmetrical about the y-axis. The mean, median and mode coincide at the origin.
2. The curve is drawn, if mean (origin of x) and standard deviation are given. The value of y_0 can be calculated from the fact that the area of the curve must be equal to the total number of observations.
3. y decreases rapidly as x increases numerically. The curve extends to infinity on either side of the origin.
4. (*a*) $P(\mu - \sigma < x < \mu + \sigma) = 68\%$
   (*b*) $P(\mu - 2\sigma < x < \mu + 2\sigma) = 95.5\%$
   (*c*) $P(\mu - 3\sigma < x < \mu + 3\sigma) = 99.7\%$

Hence (*a*) About $\frac{2}{3}$ of the values will lie between $(\mu - \sigma)$ and $\mu + \sigma$

(*b*) About 95% of the values will lie between $(\mu - 2\sigma)$ and $(\mu + 2\sigma)$.

(*c* About 99.7 % of the values will lie between (m – 3 s) and (m + 3s

65.5 MEAN FOR NORMAL DISTRIBUTION

$$\text{Mean} = \int_{-\infty}^{+\infty} \frac{1}{\sigma\sqrt{(2\pi)}} e^{-\frac{x^2}{2\sigma^2}} . x\,dx = \frac{1}{\sigma\sqrt{(2\pi)}} \int_{-\infty}^{+\infty} e^{-\frac{t^2}{2}} (t\sigma)(\sigma dt) \qquad \left[\text{putting } \frac{x}{\sigma} = t\right]$$

$$= \frac{\sigma}{\sqrt{(2\pi)}} \int_{-\infty}^{+\infty} t\, e^{-\frac{t^2}{2}} dt = \frac{\sigma}{\sqrt{2\pi}} \left[e^{-\frac{t^2}{2}}\right]_{-\infty}^{+\infty} = \frac{\sigma}{\sqrt{(2\pi)}} [0] = 0$$

65.6 STANDARD DEVIATION FOR NORMAL DISTRIBUTION

$$\mu_2' = \int x^2 . f(x)\, dx \quad \text{or} \quad \mu_2' = \int_{-\infty}^{+\infty} x^2 . \frac{1}{\sigma\sqrt{(2\pi)}} e^{-\frac{x^2}{2\sigma^2}} . dx$$

Put $\qquad \frac{x^2}{2\sigma^2} = t \Rightarrow x = \sqrt{2}\,\sigma t^{1/2}. \qquad \Rightarrow \qquad dx = \frac{\sqrt{2}\,\sigma}{2t^{1/2}} dt$

$$\mu_2' = \int_{-\infty}^{+\infty} (2\sigma^2 t) \frac{1}{\sigma\sqrt{(2\pi)}} e^{-\frac{2\sigma^2 t}{2\sigma^2}} . \frac{\sqrt{2}\sigma}{2t^{\frac{1}{2}}} dt = \int_{-\infty}^{+\infty} (2\sigma^2 t) \frac{1}{\sigma\sqrt{(2\pi)}} e^{-t} \left(\frac{\sqrt{2}\sigma}{2t^{1/2}}\right) dt$$

$$= \frac{2\sigma^2}{\sigma\sqrt{(2\pi)}} \frac{\sqrt{2}\sigma}{2} \int_{-\infty}^{+\infty} t^{\frac{3}{2}-1} e^{-t} dt, = \frac{\sigma^2}{\sqrt{\pi}} . 2 \int_0^{+\infty} t^{\frac{3}{2}-1} e^{-t} dt \qquad \left[\int_0^{+\infty} x^{n-1} e^{-x} dx = \overline{|n}\right]$$

$$= \frac{\sigma^2}{\sqrt{\pi}} . 2 \overline{\left|\frac{3}{2}\right.} = 2\frac{\sigma^2}{\sqrt{\pi}} . \frac{1}{2} \overline{\left|\frac{1}{2}\right.} = \frac{\sigma^2}{\sqrt{\pi}} \sqrt{\pi} = \sigma^2$$

$$\mu_2 = \mu_2' - (\mu_1')^2 = \sigma^2 - (0)^2 = \sigma^2$$

$$S.D. = \sigma$$ **Ans.**

65.7 MEDIAN OF THE NORMAL DISTRIBUTION

If a is the median, then it divides the total area into two equal halves so that

$$\int_{\infty}^{a} f(x)\, dx = \frac{1}{2} = \int_{a}^{\infty} f(x)\, dx$$

where $\qquad f(x) = \frac{1}{\sigma\sqrt{(2\pi)}} e^{-\frac{(x-\mu)^2}{2\sigma^2}}$

Suppose Median a > mean, μ then

$$\int_{-\infty}^{\mu} f(x)\,dx + \int_{\mu}^{a} f(x)\,dx = \frac{1}{2} \qquad \left[\text{But } \int_{-\infty}^{\mu} f(x)\,dx = \frac{1}{2}\right]$$

$$\frac{1}{2} + \int_{\mu}^{a} f(x)\,dx = \frac{1}{2} \qquad (\mu = \text{mean})$$

$$\int_{\mu}^{a} f(x)\,dx = 0$$

Thus $a = \mu$

Similarly, when a < mean, we have $a = \mu$.

Thus, median = mean = μ

65.8 MEAN DEVIATION ABOUT THE MEAN μ

(*U.P. III Semester, Dec. 2009*)

Mean deviation $= E\,|x - \mu|$

$$= \int_{-\infty}^{\infty} |x-\mu| \frac{1}{\sigma\sqrt{(2\pi)}} e^{-\frac{(x-\mu)^2}{2\sigma^2}}\,dx$$

$$= \Rightarrow dz = \frac{1}{\sqrt{2}} t^{-\frac{1}{2}}\,dt \quad \text{where } z = \frac{x-\mu}{\sigma} \quad \Rightarrow \quad dz = \frac{dx}{\sigma}$$

$$= \sigma \frac{1}{\sqrt{(2\pi)}}\left[\int_{-\infty}^{0} -z e^{-\frac{z^2}{2}}\,dz + \int_{0}^{\infty} z e^{-\frac{z^2}{2}}\,dz\right]$$

$$= \frac{2\sigma}{\sqrt{(2\pi)}} \int_{0}^{\infty} z e^{-\frac{z^2}{2}}\,dz = \sigma\sqrt{\frac{2}{\pi}}\left[-e^{-\frac{z^2}{2}}\right]_0^{\infty} \quad \text{(as the function is even)}$$

$$= \sigma\sqrt{\frac{2}{\pi}}[-0+1] = \sigma\sqrt{\frac{2}{\pi}} = \frac{4}{5}\sigma \text{ approximately.}$$

65.9 MODE OF THE NORMAL DISTRIBUTION

We know that mode is the value of the variate x for which $f(x)$ is maximum. Thus, by differential calculus $f(x)$ is maximum if $f'(x) = 0$ and $f''(x) < 0$

where $$f(x) = \frac{1}{\sigma\sqrt{(2\pi)}} e^{-\frac{(x-\mu)^2}{2\sigma^2}}$$

Clearly $f(x)$ will be maximum when the exponent will be maximum which will be the case

$$\frac{(x-\mu)}{2\sigma^2} = 0 \quad \Rightarrow \quad (x-\mu)^2 = 0 \quad \Rightarrow \quad x = \mu$$

Thus mode is μ, and modal ordinate $= \dfrac{1}{\sigma\sqrt{(2\pi)}}$

65.10 MOMENT OF NORMAL DISTRIBUTION

$$\mu_{2n+1} = \int_{-\infty}^{\infty} (x-\mu)^{2n+1} f(x)\,dx = \frac{1}{\sigma\sqrt{(2\pi)}} \int_{-\infty}^{\infty} (x-\mu)^{2n+1} e^{-\frac{(x-\mu)^2}{2\sigma^2}}\,dx$$

$$= \frac{1}{\sqrt{(2\pi)}} \int_{-\infty}^{\infty} (\sigma z)^{2n+1} e^{-\frac{z^2}{2}}\,dz \qquad \left[z = \frac{x-\mu}{\sigma}\right]$$

$$= \frac{\sigma^{2n+1}}{\sqrt{(2\pi)}} \int_{-\infty}^{\infty} z^{2n+1} e^{-\frac{z^2}{2}} dz = 0 \qquad \left(\text{since } z^{2n+1} e^{-\frac{z^2}{2}} \text{ is an odd function}\right)$$

$$\mu_{2n} = \int_{-\infty}^{\infty} (x-\mu)^{2n} f(x)\, dx = \frac{1}{\sqrt{(2\pi)}} \int_{-\infty}^{\infty} (\sigma z)^{2n} e^{-\frac{z^2}{2}} dz = \frac{\sigma^{2n}}{\sqrt{2\pi}} \int_{-\infty}^{\infty} z^{2n} e^{-\frac{z^2}{2}} dz$$

$$= \frac{2\sigma^{2n}}{\sqrt{(2\pi)}} \int_0^{\infty} z^{2n} e^{-\frac{z^2}{2}} dz = \frac{2\sigma^{2n}}{\sqrt{(2\pi)}} \int_0^{\infty} (2t)^n e^{-t} \frac{1}{\sqrt{2}} t^{-\frac{1}{2}} dt$$

$$\left[z^{2n} \cdot e^{-\frac{z^2}{2}} \text{ is an even function}\right]$$

$$= \frac{2^n \sigma^{2n}}{\sqrt{\pi}} \int_0^{\infty} t^{\left(n+\frac{1}{2}-1\right)} e^{-t\, dt} = \frac{2^n \sigma^{2n}}{\sqrt{\pi}} \int_0^{\infty} e^{-t} t^{\left(n-\frac{1}{2}\right)} dt \qquad \left[\frac{z^2}{2} = t \Rightarrow dz = \frac{1}{\sqrt{2}} t^{-\frac{1}{2}} dt\right]$$

$$= \frac{2^n \sigma^{2n}}{\sqrt{\pi}} \left\lceil n+\frac{1}{2} \right.$$

Changing n to $(n-1)$, we get

$$\mu_{2n-2} = \frac{2^{n-1} \sigma^{2n-2}}{\sqrt{\pi}} \left\lceil n-\frac{1}{2} \right.$$

On dividing, we get

$$\frac{\mu_{2n}}{\mu_{2n-2}} = 2\sigma^2 \frac{\left\lceil n+\frac{1}{2}\right.}{\left\lceil n-\frac{1}{2}\right.} = \frac{2\sigma^2\left(n-\frac{1}{2}\right)\left\lceil n-\frac{1}{2}\right.}{\left\lceil n-\frac{1}{2}\right.} = 2\sigma^2\left(n-\frac{1}{2}\right)$$

$$\mu_{2n} = \sigma^2 (2n-1)\mu_{2n-2}$$

which gives the recurrence relation for the moments of normal distribution

$$\mu_{2n} = [(2n-1)\sigma^2]\,[(2n-3)\sigma^2]\mu_{2n-4}$$

$$= [(2n-1)\sigma^2]\,[(2n-3)\sigma^2][(2n-5)\sigma^2]\mu_{2n-6}$$

$$= [(2n-1)\sigma^2]\,[(2n-3)\sigma^2][(2n-5)\sigma^2] \ldots\ldots (3\sigma^2)\ (1\cdot\sigma^2)\mu_0$$

$$= (2n-1)\,(2n-3)\,(2n-5) - \ldots\ldots 1\cdot\sigma^{2n} \qquad (\mu_0 = 1)$$

$$= 1.\,3.\,5.\,7\ldots\ldots (2n-5)\,(2n-3)\,(2n-1)\,\sigma^{2n}$$

65.11 MOMENT GENERATING FUNCTION OF NORMAL DISTRIBUTION

Normal distribution function is given by

$$f(x) = \frac{1}{\sigma\sqrt{2\pi}} e^{-\frac{(x-\mu)^2}{2\sigma^2}}; \qquad -\infty < x < \infty$$

Moment generating function about the origin $= M_0'(t)$

$$= \int_{-\infty}^{\infty} e^{tx}\, \sigma \frac{1}{\sqrt{2\pi}} e^{-\frac{1}{2}\left(\frac{x-\mu}{\sigma}\right)^2} dx \qquad \ldots(1)$$

On putting $\frac{x-\mu}{\sigma} = z$ so that $dx = \sigma\, dz$ in (1), we get

$$M_0(t) = \frac{1}{\sigma\sqrt{2\pi}} \int_{-\infty}^{\infty} e^{t(\sigma z+\mu)} e^{-\frac{z^2}{2}} \sigma\, dz = \frac{\sigma e^{\mu t}}{\sigma\sqrt{2\pi}} \int_{-\infty}^{\infty} e^{t\sigma z} . e^{-\frac{z^2}{2}} dz = \frac{e^{\mu t}}{\sqrt{2\pi}} \int_{-\infty}^{\infty} e^{-\frac{1}{2}(z^2-2t\sigma z)} dz$$

$$= \frac{e^{\mu t}}{\sqrt{2\pi}} \int_{-\infty}^{\infty} e^{-\frac{1}{2}[z^2-2t\sigma z+t^2\sigma^2]+\frac{1}{2}t^2\sigma^2} dz = \frac{e^{\mu t+\frac{1}{2}t^2\sigma^2}}{\sqrt{2\pi}} \int_{-\infty}^{\infty} e^{-\frac{1}{2}(z-t\sigma)^2} dz \qquad \ldots(2)$$

On putting $\frac{1}{2}\;(z-t\sigma)^2 = y^2$ in (2) so that $(z-t\,\sigma)\,dz \quad = 2y\,dy$

i.e. $\Rightarrow$ $\sqrt{2}\;y\,dz\; = 2\,y\,dy$

$$\Rightarrow \qquad dz = \sqrt{2}\,dy \quad M_0(t) = \frac{e^{\mu t + \frac{1}{2}t^2\sigma^2}}{\sqrt{2\pi}} \int_{\infty}^{\infty} e^{-y^2}\sqrt{2}\,dy = \frac{e^{\mu t + \frac{1}{2}t^2\sigma^2}}{\sqrt{\pi}} \int_{-\infty}^{\infty} e^{-y^2}\,dy$$

$$= e^{\mu t + \frac{1}{2}t^2\sigma^2}\frac{1}{\sqrt{\pi}}(\sqrt{\pi}) = e^{\mu t + \frac{1}{2}t^2\sigma^2} \qquad \left[\because \int_{-\infty}^{\infty} e^{-y^2}\,dy = \sqrt{\pi}\right]$$

65.12 AREA UNDER THE NORMAL CURVE

By taking $z = \dfrac{x-\bar{x}}{\sigma}$; the standard normal curve is formed.

The total area under this curve is 1. The area under the curve is divided into two equal parts by z = 0. Left hand side area and right hand side area to $z = 0$ is 0.5. The area between the ordinate $z = 0$ and any other ordinate can be noted from the table.

Example 7. *On a final examination in mathematics, the mean was 72, and the standard deviation was 15. Determine the standard scores of students receiving graders.*

(*a*) 60 (*b*) 93 (*c*) 72

Solution.

(*a*) $z = \dfrac{x-\bar{x}}{\sigma} = \dfrac{60-72}{15} = -0.8$ (*b*) $z = \dfrac{93-72}{15} = +1.4$ (*c*) $z = \dfrac{72-72}{15} = 0$ **Ans.**

Example 8. *Find the area under the normal curve in each of the cases*

(*a*) *z = 0 and z = 1.2;* (*b*) *z = – 0.68 and z = 0;*

(*c*) *z = – 0.46 and z = 2.21;* (*d*) *z = 0.81 and z = 1.94;*

(*e*) *To the left of z = 0.6;* (*f*) *Right of z = –1.28.*

Solution.

(*a*) Area between $z = 0$ and $z = 1.2$ (*b*) Area between $z = 0$ and $z = -0.68$

$= .3849$ $= 0.2518$

(*a*)

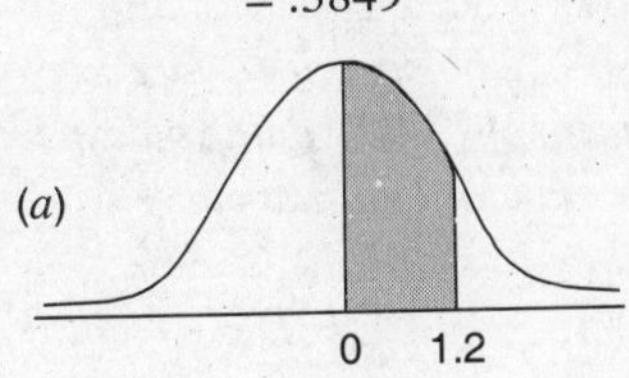

(*b*)

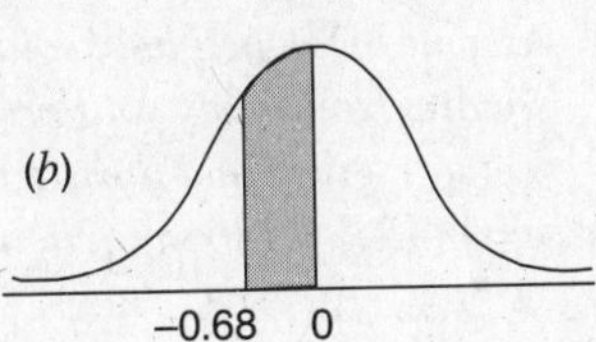

(*c*) Required area = (Area between $z = 0$ and $z = 2.21$)

+ (Area between $z = 0$ and $z = -0.46$)

= (Area between $z = 0$ and $z = 2.21$)

+ (Area between $z = 0$ and $z = 0.46$)

$= 0.4865 + 0.1772 = 0.6637.$

(*c*)

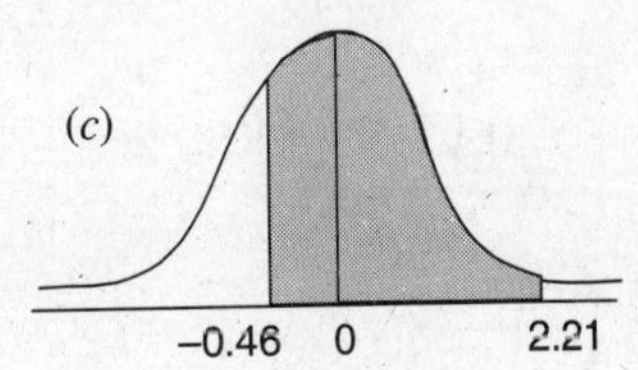

(*d*) Required area = (Area between $z = 0$ and

$z = 1.94$) – (Area between $z = 0$ and $z = 0.81$)

$= 0.4738 - 0.2910 = 0.1828$

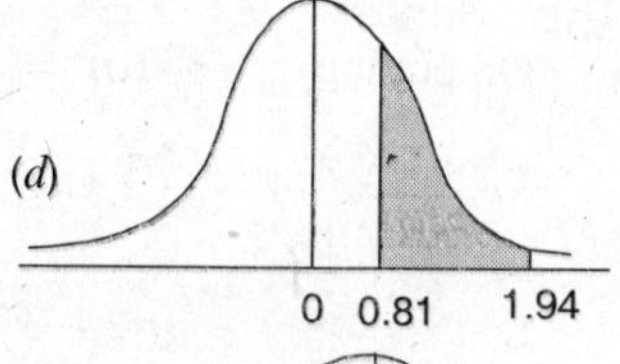

(*e*) Required area $= 0.5 -$ (Area between $z = 0$ and $z = 0.6$)

$= 0.5 - 0.2257 = 0.2743$

(*f*) Required area = (Area between $z = 0$ and $z = -1.28$) + 0.5

$= 0.3997 + 0.5$

$= 0.8997.$

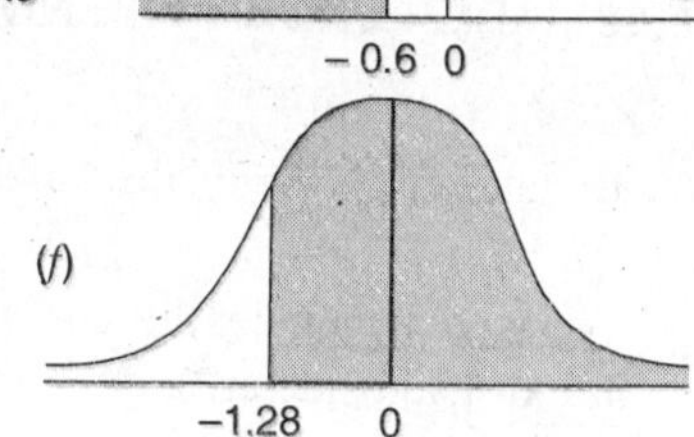

Example 9. *Find the value of z in each of the cases*

(*a*) Area between 0 and z is 0.3770

(*b*) Area to the left of z is 0.8621

Solution.

(*a*) $z = \pm 1.16$

(*b*) Since the area is greater than 0.5.

Area between 0 and z.

$= 0.8621 - 0.5 = 0.3621$

from which $z = 1.09$ **Ans.**

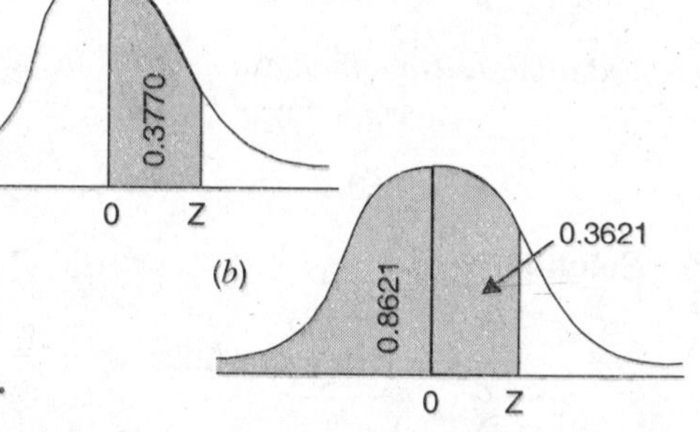

Example 10. *Students of a class were given an aptitude test Their marks were found to be normally distributed with mean 60 and standard deviation 5. What percentage of students scored more than 60 marks ?*

Solution. $x = 60,\ \bar{x} = 60,\ \sigma = 5$

$$z = \frac{x - \bar{x}}{\sigma} = \frac{60 - 60}{5} = 0$$

if $x > 60$ then $z > 0$

Area lying to the right of $z = 0$ is 0.5.

The percentage of students getting more than 60 marks = 50 % **Ans.**

Example 11. *Assume mean height of soldiers to be 68.22 inches with a variance of 10.8 inches square. How many soldiers in a regiment of 1,000 would you expect to be over 6 feet tall, given that the area under the standard normal curve between x = 0 and x = 0.35 is 0.1368 and between x = 0 and x = 1.15 is 0.3746.*

(*U.P. III Semester Dec. 2001*)

Solution. Mean $= \bar{x} = 68.22$ inches

variance $= \sigma^2 = 10.8$ inches squares

If $x = 72$ inches then $z = \frac{x - \mu}{\sigma} = \frac{72 - 68.22}{\sqrt{10.8}} = 1.15$

$$P(x > 72) = P(z > 1.15)$$

$$= 0.5 - P(0 \le z \le 1.15) = 0.5 - 0.3746 = 0.1254$$

Number of soldiers $= 1000 \times 0.1254 = 125.4 \approx 125$ (app.) **Ans.**

$$z = \frac{x-\mu}{\sigma} \quad \Rightarrow \quad z = \frac{x-64.5}{3.3} \qquad \dots (1)$$

Area between 0 and $z = 0.99 - 0.5 = 0.49$

From the table, z for area 0.49 is 2.327.

Putting the value of z in (1), we get

$$\Rightarrow \quad \frac{x-64.5}{3.3} = 2.327 \quad \Rightarrow x - 64.5 = 3.3 \times 2.327$$

$$x - 64.5 = 7.68$$

$$\Rightarrow \quad x = 7.68 + 64.5 = 72.18 \text{ inches}$$

Hence 99% students are of height less than 72.18 inches. **Ans.**

Example 12. *A sample of 100 dry battery cells tested to find the length of life produced the following results:*

$\bar{x}$ = 12 hours, σ = 3 hours

Assuming the data to be normally distributed, what percentage of battery cells are expected to have life

(i) more than 15 hours *(ii) less than 6 hours*

(iii) between 10 and 14 hours ? *(U.P. III Semester Dec. 2003)*

Solution. Here, Mean $= \bar{x} = 12$ hours

and Standard deviation $= \sigma = 3$ hours

x denotes the length of life of dry battery cells.

$$z = \frac{x-\bar{x}}{\sigma}$$

(*i*) When $x = 15$, then $z = \frac{15-12}{3} = 1$

$$\therefore \quad P(x > 15) = P(z > 1)$$

$$= P(0 < z < \infty) - P(0 < z < 1)$$

$$= 0.5 - 0.3413 = 0.1587 = 15.87\%$$

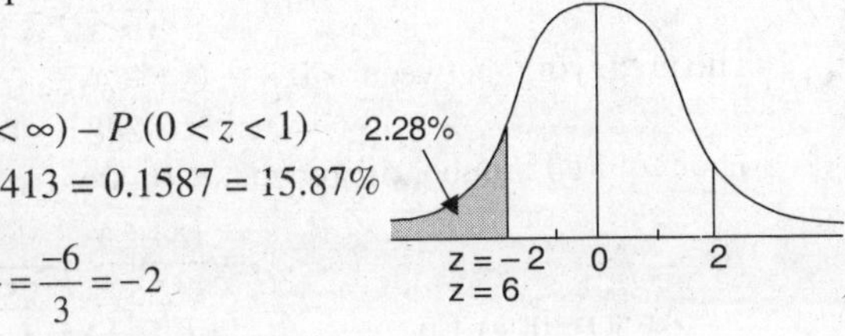

(*ii*) When x = 6, then $z = \frac{6-12}{3} = \frac{-6}{3} = -2$

$$P(x < 6) = P(z < -2)$$

$$= P(z > 2) = 0.5 - P(0 < z < 2)$$

$$= 0.5 - 0.4772 = 0.0228 = 2.28\%$$

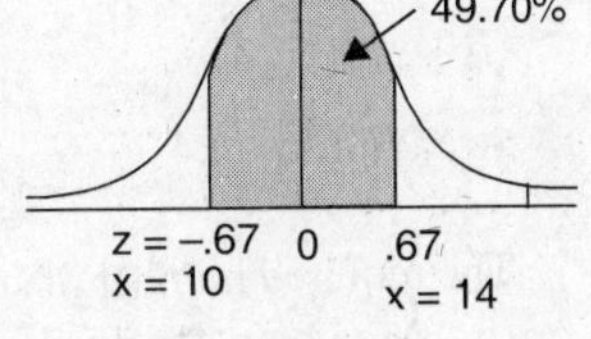

(*iii*) When $x = 10$, then $z = \frac{10-12}{3} = \frac{-2}{3} = -0.67$

When $x = 14$, then $z = \frac{14-12}{3} = \frac{2}{3} = 0.67$

$$P(10 < x < 14) = P(-0.67 < z < 0.67)$$

$$= 2\,P(0 < z < 0.67) = 2 \times 0.2485 = 0.4970 = 49.70\%$$ **Ans.**

Example 13. *The mean yield per plot of a crop is 17 kg and standard deviation is 3 kg. If distribution of yield per plot is normal, find the percentage of plots giving yields:*

(i) Between 15.5 kg and 20 kg ; and

(ii) More than 20 kg. *[U.P. (MBA) 2005)]*

Solution. Mean $= \mu = 17$ kg

S.D. $= \sigma = 3$ kg

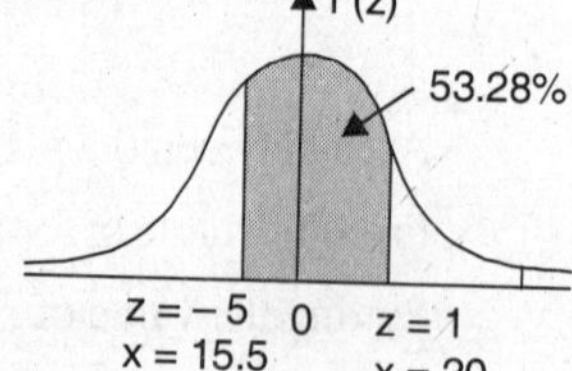

Standard Normal variable $z = \dfrac{x-\mu}{\sigma}$

(*i*) When $x_1 = 15.5$, $\quad z_1 = \dfrac{x_1-\mu}{\sigma} = \dfrac{15.5-17}{3} = -0.5$

When $x_2 = 20$, $\quad z_2 = \dfrac{x_2-\mu}{\sigma} = \dfrac{20-17}{3} = 1$

$\therefore \quad P(15.5 < x < 20) = P(-0.5 < z < 1)$

$= P(0 < z < -0.5) + P(0 < z < 1) = 0.1915 + 0.3413 = 0.5328$

$\therefore$ Required percentage of plots $= 53.28\%$

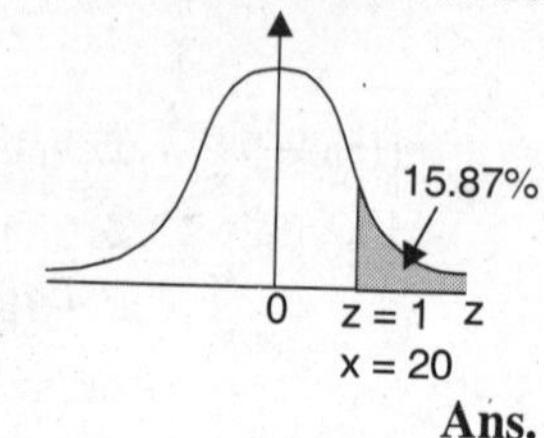

(*ii*) When x = 20, $z = \dfrac{20-17}{3} = 1$

$P(x > 20) = P(z > 1)$

$= 0.5 - P(0 < z < 1) = 0.5 - 0.3413$

$= 0.1587$

Ans.

Example 14. *In a sample of 1000 cases, the mean of a certain test is 14 and standard deviation is 2.5. Assuming the distribution to be normal, find*

(i) how many students score between 12 and 15 ?

(ii) how many score above 18 ? *(iii) how many score below 8 ?*

(iv) how many score 16 ?

Solution. $n = 1000, \ \bar{x} = 14, \sigma = 2.5$

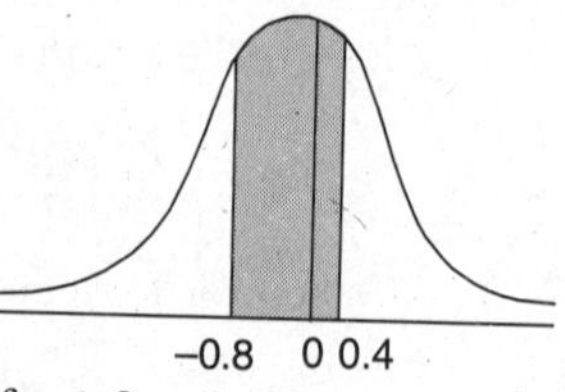

(*i*) $\quad z_1 = \dfrac{x-\bar{x}}{\sigma} = \dfrac{12-14}{2.5} = -0.8$

$z_2 = \dfrac{15-14}{2.5} = \dfrac{1}{2.5} = 0.4$

The area lying between -0.8 to 0.4 = Area from 0 to -0.8 + area from 0 to 0.4

$= 0.2881 + 0.1554 = 0.4435$

The required number of students $= 1000 \times 0.4435 = 443.5 = 444$ (say)

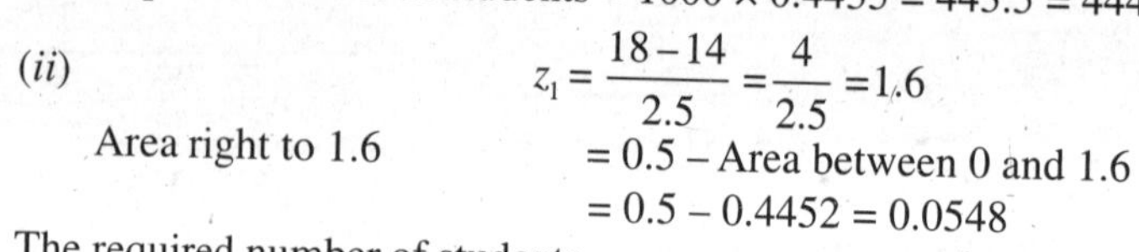

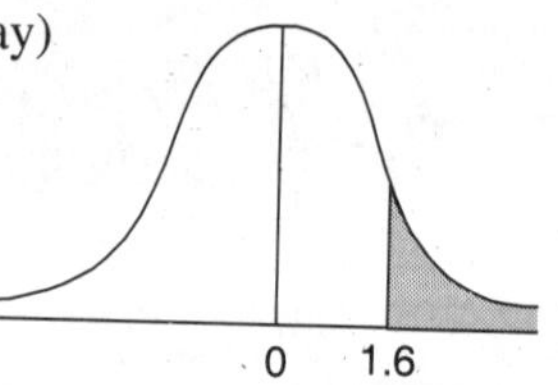

(*ii*) $\quad z_1 = \dfrac{18-14}{2.5} = \dfrac{4}{2.5} = 1.6$

Area right to 1.6 $\quad = 0.5$ – Area between 0 and 1.6

$= 0.5 - 0.4452 = 0.0548$

The required number of students

$= 1000 \times 0.0548 = 54.8 = 55$ (say)

(*iii*) $\quad z = \dfrac{8-14}{2.5} = -\dfrac{6}{2.5} = -2.4$

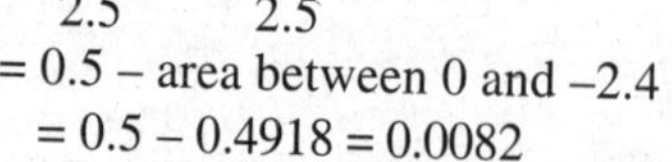

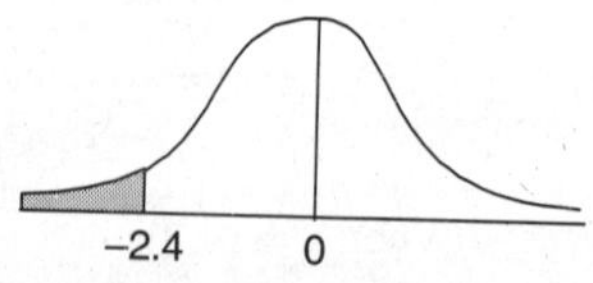

Area left to -2.4 $\quad = 0.5$ – area between 0 and -2.4

$= 0.5 - 0.4918 = 0.0082$

The required number of students $= 1000 \times 0.0082 = 8.2 = 8$ (say)

(*iv*) Area between 15.5 and 16.5

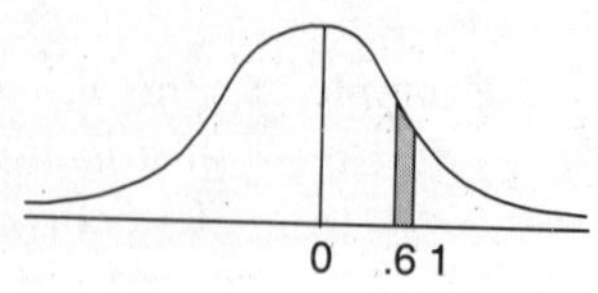

$z_1 = \dfrac{15.5-14}{2.5} = 0.6$

$z_2 = \dfrac{16.5-14}{2.5} = 1$

Area between 0.6 and 1 $\quad = 0.3413 - 0.2257 = 0.1156$

The required number of students $= 0.1156 \times 1000 = 115.6 = 116$ say

Ans.

Example 15. *The mean inside diameter of a sample of 200 washers produced by a machine is 0.502 cm and the standard deviation is 0.005 cm. The purpose for which these washers are intended allows a maximum tolerance in the diameter of 0.496 to 0.508 cm, otherwise the washers are considered defective. Determine the percentage of defective washers produced by the machine, assuming the diameters are normally distributed* *(A.M.I.E., Summer 2001)*

Solution. $z_1 = \dfrac{x-\bar{x}}{\sigma} = \dfrac{0.496-0.502}{0.005} = -1.2$

$z_2 = \dfrac{x-\bar{x}}{\sigma} = \dfrac{0.508-0.502}{0.005} = +1.2$

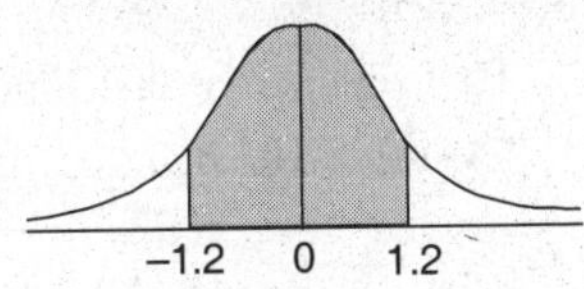

Area for non-defective washers = Area between $z = -1.2$ and $z = +1.2$

$= 2$ Area between $z = 0$ and $z = 1.2$.

$= 2 \times (0.3849) = 0.7698 = 76.98\%$

Percentage of defective washers $= 100 - 76.98$

$= 23.02\%$ **Ans.**

Example 16. *A manufacturer of envelopes knows that the weight of the envelopes is normally distributed with mean 1.9 gm and variance* 0.01 *gm. Find how many envelopes weighing (i) 2 gm or more, (ii) 2.1 gm or more, can be expected in a given packet of1000 envelopes.*

[Given : if t is the normal variable, then $\phi\,(0 \le t \le 1) = 0.3413$ *and* $\phi\,(0 \le t \le 2) = 0.4772$*]*

Solution. $\mu = 1.9$ gm, Variance = 0.01 gm $\Rightarrow \sigma = 0.1$

(*i*) $x = 2$ gms or more

$$z = \frac{x-\mu}{\sigma} = \frac{2-1.9}{0.1} = \frac{0.1}{0.1} = 1$$

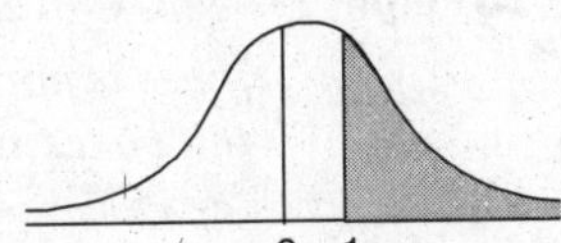

$P\,(z > 1) =$ Area right to $z = 1$

$= 0.5 - 0.3413 = 0.1587$

Number of envelopes heavier than 2 gm in a lot of 1000

$= 1000 \times 0.1587 = 158.7 = 159$ (app)

(*ii*) $$z = \frac{2.1-1.9}{0.1} = \frac{0.2}{0.1} = 2$$

$P\,(z > 2) =$ Area right to $z = 2$

$= 0.5 - 0.4772 = 0.0228$

0 2

Number of envelopes heavier than 2.1 gm in a lot of 1000

$= 1000 \times 0.0228 = 22.8 = 23$ (app) **Ans.** (*i*) 159 (*ii*) 23

Example 17. *The life of army shoes is 'normally' distributed with mean 8 months and standard deviation 2 months. If 5000 pairs are issued how many pairs would be expected to need replacement after 12 months?*

[Given that $P\,(z \ge 2) = 0.0228$ *and* $z = \dfrac{(x-\mu)}{\sigma}$*]*

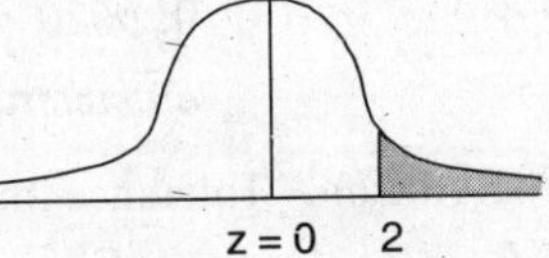

Solution. Mean $(\mu) = 8$

Standard deviation $(\sigma) = 2$

Number of pairs of shoes = 5000

Total months $(x) = 12$

$$\text{When } z = \frac{x-\mu}{\sigma} = \frac{12-8}{2} = 2$$

Area when $(z \geq 2) = 0.0228$

Number of pairs whose life is more than 12 months $(z > 2)$

$$= 5000 \times 0.0228 = 114$$

Replacement after 12 months = 5000 – 114 = 4886 pairs of shoes

Example 18. *In a male population of 1000, the mean height is 68.16 inches and standard deviation is 3.2 inches. How many men may be more than 6 feet (72 inches) ?*
[ϕ (1.15) = *0.8749,* ϕ (1.2) = 0.8849, ϕ (1.25) = *0.8944*]
where the argument is the standard normal variable.

Solution. Male population = 1000

Mean height = 68.16 inches

Standard deviation = 3.2 inches

Men more than 72 inches = ?

0 1.2

ϕ (1.15) = 0.8749, ϕ (1.2) = 0.8849

ϕ (1.25) = 0.8944

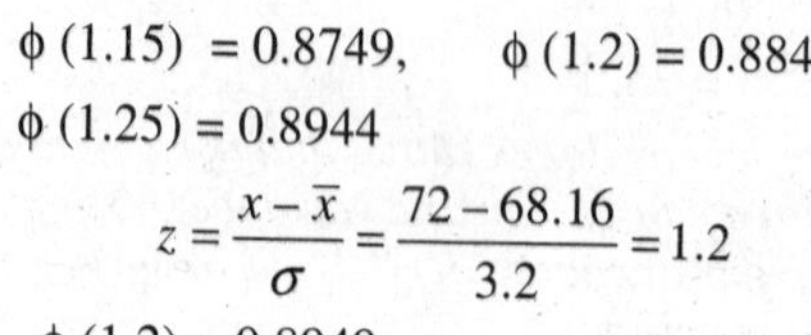

$$z = \frac{x-\bar{x}}{\sigma} = \frac{72-68.16}{3.2} = 1.2$$

ϕ (1.2) = 0.8949

ϕ for more than 1.2 = 1 – 0.8849 = 0.1151

Men more than 72 inches = 1000 × 0.1151 = 115.1 = 115 (say) **Ans.**

Example 19. *Pipes for tobacco are being packed in fancy plastic boxes. The length of the pipes is normally distributed with* μ = *5" and a = 0.1". The internal length of the boxes is 5.2". What is the probability that the box would be small for the pipe?*

[*given that* ϕ *(1.8)* = *0.9641,* ϕ(2) = *0.9772,* ϕ (2.5) = 0.9938]

Solution. $\mu = 5''$, $\sigma = 0.1''$, $x = 5.2''$

ϕ (1.8) = 0.9641, ϕ (2) = 0.9722, ϕ (2.5) = 0.9938

$$z = \frac{x-\mu}{\sigma} = \frac{5.2-5}{0.1} = 2$$

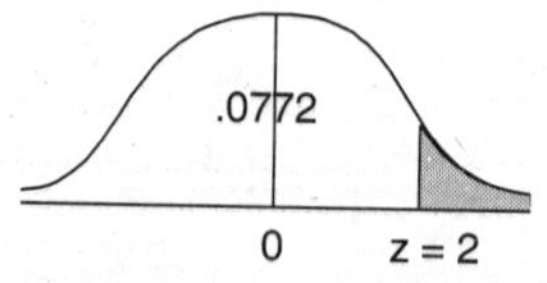

ϕ (2) = 0.9772

ϕ (z > 2) = 1– 0.9772 = 0.0228

The box will be small if the length of the pipe is more than 5.2" $(z = 2)$.

Hence the probability is 0.0228 **Ans.**

Example 20 *Assuming that the diameters of 1,000 brass plugs taken consecutively from a machine form a normal distribution with mean* 0.7515 *cm and standard deviation* 0. *0020 cm, how many of the plugs are likely to be rejected if the approved diameter is* 0.752 ± *.004 cm ?*

Solution. Tolerance limits of the diameter of non-defective plugs are

0.752 – 0.004 = 0.748 cm and

0.752 + 0.004 = 0.756 cm

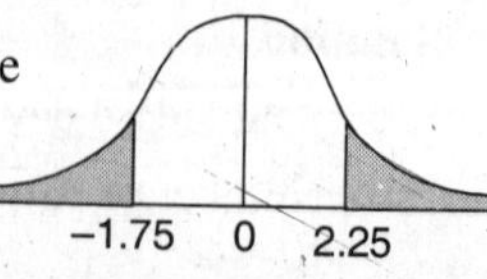

$$z = \frac{x-\mu}{\sigma}$$

If $x_1 = 0.748$, $z_1 = \frac{0.748 - 0.7515}{0.002} = -1.75$

If $x_2 = 0.756$, $z_2 = \frac{0.756 - 0.7515}{0.002} = 2.25$

Area under $z_1 = -1.75$ to $z_2 = 2.25$

$= (\text{Area from } z = 0 \text{ to } z_1 = -1.75) + (\text{Area from } z = 0 \text{ to } z_2 = 2.25)$

$= 0.4599 + 0.4878 = 0.9477$

Number of plugs likely to be rejected

$= 1000\,(1 - 0.9477) = 1000 \times .0523 = 52.3$

Approximately 52 plugs are likely to be rejected. **Ans.**

Example 21. *A manufacturer knows from experience that the resistance of resistors he produces is normal with mean $\mu = 100$ ohms and standard deviation $\sigma = 2$ ohms. What percentage of resistors will have resistance between 98 ohms and 102 ohms ?*

Solution. $\mu = 100$ ohms, $\sigma = 2$ ohms

$x_1 = 98, \quad x_2 = 102$

$z = \frac{x-\mu}{\sigma}, \quad z_1 = \frac{98-100}{2} = -1$

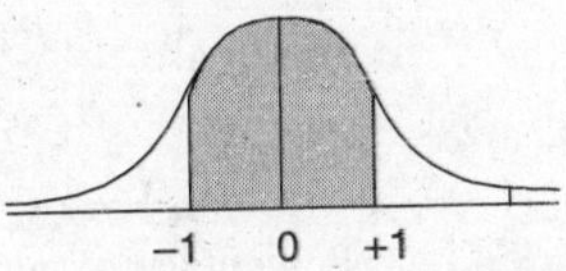

$z_2 = \frac{102-100}{2} = +1$

–1 0 +1

Area between $z_1 = -1$ and $z_2 = +1$

$= (\text{Area between } z = 0 \text{ and } z = -1)$

$+ (\text{Area between } z = 0 \text{ and } z = +1)$

$= 2\,(\text{Area between } z = 0 \text{ and } z = +1) = 2 \times 0.3413 = 0.6826$

Percentage of resistors having resistance between 98 ohms and 102 ohms = 68.26 **Ans.**

Example 22 *In a normal distribution, 31% of the items are under 45 and 8% are over 64. Find the mean and standard deviation of the distribution.* (AMIETE, Dec. 2010)

Solution. Let μ. be the mean and σ the S.D.

It $x = 45$, $z = \frac{45-\mu}{\sigma}$

If $x = 64$, $z = \frac{64-\mu}{\sigma}$

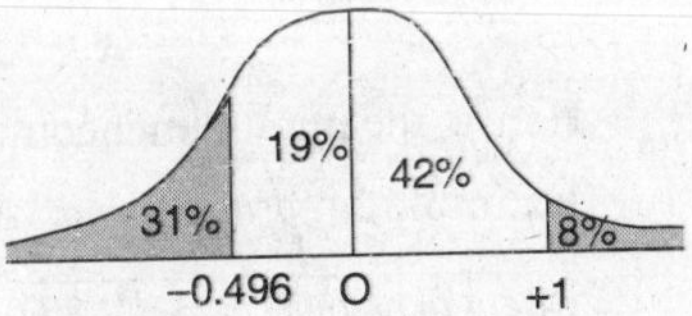

19% 42% 31% 8%
–0.496 O +1

Area between 0 and $z = \frac{45-\mu}{\sigma} = 0.50 - .31 = 0.19$

[From the table, for the area 0.19, $z = 0.496$]

$$\frac{45-\mu}{\sigma} = -0.496 \qquad ...(1)$$

Area between $z = 0$ and $z = \frac{64-\mu}{\sigma} = 0.5 - 0.08 = 0.42.$

(From the table, for area 0.42, $z = 1.405$)

$$\frac{64-\mu}{\sigma} = 1.405 \qquad ...(2)$$

Solving (1) and (2) we get $\mu = 50$, $\sigma = 10$. **Ans.**

Example 23. *The income of a group of 10,000 persons was found to be normally distributed with mean Rs. 750 p.m. and standard deviation of Rs. 50. Show that, of this group, about 95% had income exceeding Rs. 668 and only 5% had income exceeding Rs. 832. Also find the lowest income among the richest 100.*

(U.P. III Semester Dec. 2004)

Solution. Mean $= \mu = 750$

Standard deviation $= \sigma = 50$

and $z = \dfrac{x - \mu}{\sigma}$

(i) If $x_1 = 668$, then $z_1 = \dfrac{668 - 750}{50} = -1.64$

$P(x_1 > 668) = P(z_1 < -1.64)$

$= 0.5 + P(-1.64 \le z \le 0) = 0.5 + P(0 \le z \le 1.64) = 0.5 + 0.4495 = 0.9495$

$\therefore$ Percentage of persons having income exceeding Rs. 668 = 94.95% $\approx$ 95% (approx.)

(*ii*) If $x = 832$, then $z = \dfrac{832 - 750}{50} = 1.64$

$P(x_2 > 832) = P(z_2 > 1.64)$

$= 0.5 - 0.4495$

$= 0.0505$

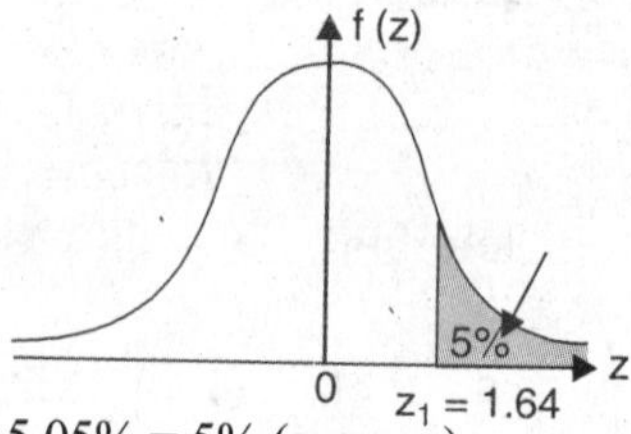

$\therefore$ Percentage of persons having income exceeding Rs. 832 = 5.05% = 5% (approx.)

(*iii*) Let x be the lowest income among the richest 100 persons.

100 persons = 1% of 10,000

100 persons represents 1% area under the curve on the right hand side.

Thus the area between 0 and z

$= 0.5 - 0.01 = 0.49$

From the table z for area 0.49 is 2.33

$z = \dfrac{x - \mu}{\sigma}$

$\Rightarrow \quad 2.33 = \dfrac{x - 750}{50} \quad \Rightarrow x - 750 = 50 \times 2.33$

$\Rightarrow \quad x - 750 = 116.5 \quad \Rightarrow x = 866.5$

Hence, the minimum income among the 100 richest persons is equal to Rs. 866.5. **Ans.**

Example 24. *Fit a normal curve to the following data :*

Length of line (in cm)	8.60	8.59	8.58	8.57	8.56	8.55	8.54	8.53	8.52
Frequency	2	3	4	9	10	8	4	1	1

Solution. Let assumed mean = 8.56 cm

x_i	f_i	$x_i - 8.56$	$f_i(x_i - 8.56)$	$f_i(x_i - 8.56)^2$
8.60	2	.04	.08	.0032
8.59	3	.03	.09	.0027
8.58	4	.02	.08	.0016
8.57	9	.01	.09	.0009

8.56	10	0	0	0
8.55	8	– .01	– .08	.0008
8.54	4	– .02	– .08	.0016
8.53	1	– .03	– .03	.0009
8.52	1	– .04	– .04	.0016
	$\Sigma f_i = 42$		$\Sigma f_i (x_i - 8.56) = 0.11$	$\Sigma f_i (x_i - 8.56)^2 = 0.0133$

$$\text{Mean} = a + \frac{\Sigma f_i (x_i - 8.56)}{\Sigma f_i} = 8.56 + \frac{0.11}{42} = 8.56 + 0.00262 = 8.56262 \quad \textbf{Ans.}$$

$$\text{Standard deviation} = \sqrt{\frac{\Sigma f_i (x_i - 8.56)^2}{\Sigma f_i} - \left(\frac{\Sigma f_i (x_i - 8.56)}{\Sigma f_i}\right)^2} = \sqrt{\frac{0.0133}{42} - \left(\frac{0.11}{42}\right)^2}$$

$$= \sqrt{0.000316666 - 0.000006859} = \sqrt{0.00030980} = 0.0176$$

Hence, the equation of the normal curve fitted to the given data is

$$P(x) = \frac{N}{\sigma\sqrt{2\pi}} e^{\frac{-(x-\mu)^2}{2\sigma}}$$

where, $\mu = 8.56262$

$\sigma = 0.0176$ and $N = 42$ **Ans.**

EXERCISE 65.2

1. In a regiment of 1000, the mean height of the soldiers is 68.12 units and the standard deviation is 3.374 units. Assuming a normal distribution, how many soldiers could be expected to be more than 72 units? It is given that

$P(z = 1.00) = 0.3413$, $P(z = 1.15) = 0.3749$ and

$P(z = 1.25) = 0.3944$, where z is the standard normal variable. **Ans.** 125

2. The lifetime of radio tubes manufactured in a factory is known to have an average value of 10 years. Find the probability that the lifetime of a tube taken randomly (*i*) exceeds 15 years, (*ii*) is less than 5 years, assuming that the exponential probability law is followed. **Ans.** (*i*) 0.2231, (*ii*) 0.3935.

3. The breaking strength X of a cotton fabric is normally distributed with $E(x) = 16$ and $\sigma(x) = 1$. The fabric is said to be good if $X \geq 14$. What is the probability that a fabric chosen at random is good. Given that $\phi(2) = 0.9772$ **Ans.** 0.9772

4. A manufacturer knows from experience that the resistance of resistors he produces is normal with mean it $\mu = 140\ \Omega$ and standard deviation $\sigma = 5\Omega$. Find the percentage of the resistors that will have resistance between 138 Ω and 142 Ω. (given $\phi(0.4) = 0.6554$, where z is the standard normal variate). **Ans.** 31.08%

5. A manufacturing company packs pencils in fancy plastic boxes. The length of the pencils is normally distributed with $\mu = 6''$ and $\sigma = 0.2''$. The internal length of the boxes is 6.4″. What is the probability that the box would be too small for the pencils (Given that a value of the standardized normal distribution function is $\phi(2) = 0.9772$). **Ans.** 0.0228.

6. A manufacturer produces airmail envelopes, whose weight is normal with mean $\mu = 1.95$ gm and standard deviation $\sigma = 0.05$ gm. The envelopes are sold in lots of 1000. How many envelopes in a lot will be heavier than 2 gm? Use the fact that $\frac{1}{\sqrt{2\pi}} \int_{0}^{1} \exp\left(\frac{-x^2}{2}\right) dx = 0.3413$ **Ans.** 159

7. The mean height of 500 students is 151 cm and the standard deviation is 15 cm. Assuming that the heights are normally distributed, find how many student's height lie between 120 and 155 cm. **Ans.** 294

8. A large number of measurement is normally distributed with a mean of 65.5″ and S.D. of 6.2″. Find the percentage of measurements that fall between 54.8 and 68.8″. **Ans.** 66.01%

9. Find the mean and variance of the density function $f(x) = \lambda e^{-\lambda x}$ **Ans.** $\frac{1}{\lambda}, \frac{1}{\lambda^2}$

10. If x is normally distributed with mean 1 and variance 4,

(*i*) Find $Pr(-3 \leq x \leq 3)$; (*ii*) Obtain k if $Pr(x \leq k) = 0.90$ **Ans.** (*i*) 0.8185, (*ii*) 3.56

11. A normal variable x has mean 1 and variance 4. Find the probability that $x \geq 3$. (Given: z is the standard normal variable and $\phi(0) = 0.5$, $\phi(0.5) = 0.6915$, $\phi(1) = 0.8413$, $\phi(1.5) = 0.9332$) **Ans.** 0.1587

12. The random variable x is normally distributed with $E(x) = 2$ and variance $V(x) = 4$. Find a number p (approximately), such that $P(x > p) = 2P(x \leq p)$. [The values of the standard normal distribution are $0(-0.43) = 0.3336$, and $0(-0.44) = 0.3300$]. **Ans.** 1.13834

If X ~ N (10, 4) find Pr [|X| ≥ 5]. **Ans.** $\frac{1}{e}, \frac{1}{5\sqrt{2\pi}} e^{-\frac{(x-75)^2}{2(0.5)}}, 0.062$

13. The continuous random variable x is normally distributed with $E(x) = \mu$ and $V(x) = \mu^2$. If $Y = cx + d$, then find $V(Y)$. **Ans.** $c^2 \mu^2$

14. The pdf of X is given by $f(X) = \lambda e^{-\lambda x}$ $x \geq 0, \lambda \geq 0$. Calculate $Pr[X > E(X)]$.

If $X \sim N(75, 25)$, find $Pr[X > 80/X > 77]$ (*A.M.I.E., Winter 2001*)

If $X \sim N(10, 4)$, find $Pr[|X| \geq 5$ **Ans.** $\frac{1}{e}, \frac{1}{5\sqrt{2\pi}} e^{-\frac{x-75)^2}{2(0.5)}}, 0.062$

15. A random variable x has a standard normal distribution ϕ. Prove : $Pr(1|X| > k) = 2[1 - \phi(k)]$

16. The random variable x has the probability density function $f(x) = kx$ if $0 \leq x \leq 2$

Find k. Find x such that

(*i*) $Pr(X \leq x) = 0.1$ (*ii*) $Pr(X \leq x) = 0.95$ **Ans.** $k = \frac{1}{2}$ (*i*) $x = 0.632$ (*ii*) $x = 1.949$

17. For a normal curve, show that $\mu_{2n+1} = 0$ and $\mu_{2n} = (2n-1)\sigma^2 \mu_{2n-2}$.

18. In a mathematics examination, the average grade was 82 and the standard deviation was 5. All the students with grades from 88 to 94 received a grade B. If the grades are normally distributed and 8 students received a B grade, find how many students took the examination. Given

$\frac{x}{\sigma}$	1.20	2.00	2.40	2.45
A	0.3849	0.4772	0.4918	0.4929

(*A.M.I.E.T.E., Winter 2001*) **Ans.** 75 students

19. Explain the characteristics and importance of a normal distribution. (*A.M.I.E., Summer 2004*)

20. The life time of a certain component has a mean life of 400 hours and standard deviation of 50 hours. Assuming normal distribution for the life time of 1000 components, determine approximately the number of components whose life time lies between 340 to 465 hours. You may use the following data? Where symbols have their usual meanings. (*A.M.I.E., Winter 2002*) **Ans.** 788

21. For standard normal variate mean μ. is (*A.M.I.E., Winter 2005*)

(*a*) 1 (*b*) 0 (*c*) 6 (*d*) none of the above **Ans.** (*b*)

22. Fill in the blanks :

(*f*) The mean, median and mode of a normal distribution are........... (*A.M.I.E., Summer 2000*) **Ans.** zero

(*h*) The probability density function of Beta distribution with $\alpha = 1$, $\beta = 4$ is $f(x)$=......

(*AMIE., Summer 2000*) **Ans.** $4(1-x)^3$

TABLE – 1

AREA UNDER STANDARD NORMAL CURVE FROM $z = 0$ T0 $z = \frac{x - \mu}{\sigma}$

An entry in the table is the proportion under the entire curve which is between $z = 0$ and a positive value of z. Area for negative values of z are obtained by symmetry.

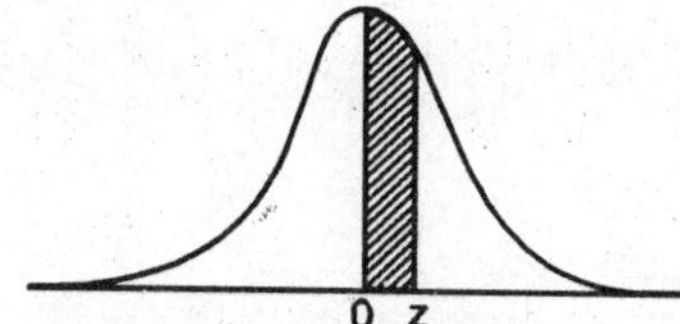

z	.00	.01	.02	.03	.04	.05	.06	.07	.08	.09
0.0	.0000	.0040	.0080	.0120	.0160	.0199	.0239	.0279	.0319	.0359
0.1	.0398	.0438	.0478	.0517	.0557	.0596	.0636	.0675	.0714	.0753
0.2	.0793	.8832	.0871	.0910	.0948	.0987	.1026	.1064	.1103	.1141
0.3	.1179	.1217	.1255	.1293	.1331	.1368	.1406	.1443	.1480	.1517
0.4	.1554	1591	.1628	.1664	.1700	.1736	.1772	.1808	.1844	.1879
0.5	.1915	.1950	.1985	.2019	.2054	.2088	.2123	.2157	.2190	.2224
0.6	.2257	.2291	.2324	.2357	.2389	.2422	.2454	.2486	.2517	.2549
0.7	.2580	.2611	.2642	.2673	.2703	.2734	.2764	.2794	.2823	.2852
0.8	.2881	.2910	.2939	.2967	.2995	.3023	.3051	.3078	.3106	.3133
0.9	.3159	.3186	.3212	.3238	.3264	.3289	.3315	.3340	.3365	.3389
1.0	.3413	.3438	.3461	.3485	.3508	.3531	.3554	.3577	.3599	.3621
1.1	.3643	.3665	.3686	.3708	.3729	.3749	.3770	.3790	.3810	.3830
1.2	.3849	.3869	.3888	.3907	.3925	.3944	.3962	.3980	.3997	.4015
1.3	.4032	.4049	.4066	.4082	.4099	.4115	.4131	.4147	.4162	.4177
1.4	.4192	.4207	.4222	.4236	.4251	.4265	.4279	.4292	.4306	.4319
1.5	.4332	.4345	.4357	.4370	.4382	.4394	.4406	.4418	.4429	.4441
1.6	.4452	.4463	.4474	.4484	.4495	.4505	.4415	.4525	.4535	.4545
1.7	.4554	.4564	.4573	.4582	.4591	.4599	.4608	.4616	.4625	.4633
1.8	.4641	.4649	.4656	.4664	.4671	.4678	.4686	.4693	.4699	.4706
1.9	.4713	.4719	.4726	.4732	.4738	.4744	.4750	.4756	.4761	.4767
2.0	.4772	.4778	.4783	.4788	.4793	.4798	.4803	.4808	.4812	.4817
2.1	.4821	.4826	.4830	.4834	.4838	.4842	.4846	.4850	.4854	.4857
2.2	.4861	.4864	.4868	.4871	.4875	.4878	.4841	.4884	.4887	.4890
2.3	.4893	.4896	.4898	.4901	.4904	.4906	.4909	.4911	.4913	.4916
2.4	.4918	.4920	.4922	.4925	.4927	.4929	.4931	.4932	.4934	.4936
2.5	.4938	.4940	.4941	.4943	.4945	.4946	.4948	.4949	.4951	.4952
2.6	.4953	.4955	.4956	.4957	.4959	.4960	.4961	.4962	.4963	.4964
2.7	.4965	.4966	.4967	.4968	.4969	.4970	.4971	.4972	.4973	.4974
2.8	.4974	.4975	.4976	.4977	.4977	.4978	.4979	.4979	.4980	.4981
2.9	.4981	.4982	.4982	.4983	.4984	.4984	.4985	.4985	.4986	.4986
3.0	.4987	.4987	.4987	.4988	.4988	.4989	.4989	.4989	.4990	.4990

CHAPTER

66

FUZZY SET

66.1 INTRODUCTION

If a doctor asks a patient "How are you" patient replies almost O.K."

The word "almost" is a vague term and not mathematical i.e. It means neither "yes" nor "No" but between them.

But the word "almost" gives lots of information to the doctor & the doctor decides the further future treatment of the patient.

For this reason a mathematical modelling of vague knowledge is necessary. To convey such an information "Fuzzy set" is introduced. L.A. Zadeh in 1965 introduced this concept on the basis of membership function defined as:

$$\mu : X \to \{0,1\}$$

Here

$\mu(x) = 1$, means full membership

$\mu(x) = 0$, means non-membership and

$0 < \mu(x) < 1$, means intermediate membership.

Due to Zadeh's work, a theory of vagueness (fuzziness) is now fully developed. The concept of 'Fuzzy set" & membership degree were introduced to form a mathematical model of vagueness.

A set is collection of well-defined distinct objects. A set of intelligent students is not a set. Because the criteria to be intelligent is not well-defined. We cannot say whether a particular student belongs to or not. The belonginess is not clear but vague. In fuzzy set theory, we assume that all students are members, all belong to the set upto certain extent.

Let Among Anil, Rajiv and Suresh, Anil got 10 marks, Rajiv got 40 marks and Suresh get 90 marks out of 100. Hence in comparison to Anil Rajiv is intelligent but if compared to Suresh he is not intelligent. All the three can be said to be intelligent to some extent and hence all the three are the members of this set of intelligent students.

For example: We write

A = {0.5 Rita, 0.9 Kusum, 0.4 Suresh, 0.6 John, 0.2 Latif} for the set of rich people.

It indicates that Rita has 0.5 degree of membership in A, Kusum has 0.9 degree of membership in A. Suresh has a 0.4 degree of membership in A, John has a 0.6 degree of membership in A and Latif has a 0.2 degree of membership in A. Thus, Kusum is the richest and Latif is the poorest of these people.

66.2 FUZZY SET

Definition. Let X be a non-zero set. A fuzzy set A of this set X is defined by the following set of pairs.

$$A = \{(x, \mu_A (x))\}: x \in X$$

Where, $\mu_A : X \to [0, 1]$

is a function called as the membership function of A & μ_A (x) is the grade of membership or degree of belonginess or degree of membership of $x \in X$ in A.

Thus a fuzzy set is a set of pairs consisting of a particular element of the universe and its degree of membership.

A can also be written as

$$A = \{(x_1, \mu_A(x_1)), (x_2, \mu_A(x_2)), \ldots .(x_n \mu_A(x_n))\}$$

Symbolically we write

$$A = \left\{ \frac{x_1}{\mu_A(x_1)}, \frac{x_2}{\mu_A(x_2)} \ldots . \frac{x_n}{\mu_A(x_n)} \right\}$$

66.3 EQUALITY OF TWO FUZZY SETS:

Example 1. Let $X = \{2, 3, 4\}$

Consider the three fuzzy sets A, B, C of X as given below

$$A = \left\{ \frac{2}{5}, \frac{3}{6}, \frac{4}{1} \right\}$$

$$B = \left\{ \frac{2}{7}, \frac{4}{8}, \frac{3}{9} \right\}$$

$$C = \left\{ \frac{4}{1}, \frac{2}{5}, \frac{3}{6} \right\}$$

Here in set 'A' and set "C" members and their degrees are same.

$\therefore$ $A = C$, But in set "A" & set "B" members are the same but their degrees of membership are not the same.

Hence $A \neq B$.

66.4 COMPLEMENT OF A 'FUZZY SET'

The component of a fuzzy set A is the set $\overline{A}$ with degree of the membership of an element in $\overline{A}$ is equal to one minus the degree of the membership of this element in A.

Example. The set "A" is written as

A = [0.9 Rama, 0.4 Manju, 0.8 Neera, 0.1 Jyoti] for the set of beautiful girls.

Thus $\overline{A}$ = {0.1 Rama, 0.6 Manju, 0.2 Neera, 0.9 Jyoti} for the set of girls who are not beautiful.

66.5 UNION OF TWO FUZZY SETS

The union of two fuzzy sets A and B is the fuzzy set $A \cup B$, where the degree of membership of an element in $A \cup B$ is the maximum of the degrees of membership of this element in A and in B.

Example. We write

A = {0.3 Radha, 0.9 Pawan, 0.6 Mahesh, 0.4 Kunal} for the set of rich people.

B = {0.4 Radha, 0.8 Pawan, 0.2 Mahesh, 0.7 Kunal} for the set of famous people.

Here the degree in $A \cup B$ of each element is the maximum of degrees of membership of this element in A & in B.

$A \cup B$ = {0.4 Radha, 0.9 Pawan, 0.6 Mahesh, 0.7 Kunal}

66.6 INTERSECTION OF TWO FUZZY SETS

The intersection of two fuzzy sets A and B is the fuzzy set $A \cap B$, where the degree of membership of an element in $A \cap B$ is the minimum of the degrees of membership of this element in A and in B.

Example: Let A = [0.5 Pushpa, 0.1 Suman, 0.8 Rani, 0.4 Kailash] for the set of fat people.

B = [0.3 Pushpa, 0.6 Suman, 0.2 Rani, 0.7 Kailash } for the set of tall people.

Here the degree of membership in $A \cap B$ of each element is the minimum of degree of membership of this element in A and in B.

$A \cap B$ = {0.3 Pushpa, 0.1 Suman, 0.2 Rani, 0.4 Kailash}

66.7 TRUTH VALUE (R.G.P.V. Bhopal I/II Sem. Summer 2004)

***(i)** Truth value of the negation of a proposition*

The truth value of the negation of a proposition in fuzzy logic is 1 minus the truth value of the proposition.

Example. If the truth value "Vimla is happy" is 0.9.

Then the truth value of the statement "Vimla is not happy" is 1–0.9 = 0.1

Example. If the truth value of the statement "Devendra is smart" is 0.8.

Then the truth value of the statement that Devendra is not smart" is 1–0.8 = 0.2

***(ii)** Truth value of the conjuction of two prepositions.*

The truth value of the conjunction of two prepositions in the fuzzy logic is the minimum of the truth values of the two prepositions.

Example. If the truth value of the statement "Khan is brave" is 0.7 And the truth value of the statement "Kamal is brave". is 0.8.

Then the truth value of the statement "Khan and Kamal are brave is 0.7 (minimum of the two).

And the truth value of "neither Khan nor Kamal is brave" is 0.2 (As truth value of the negation of 1st statement is 1–0.7 = 0.3 and that of second statement is 1–0.8 = 0.2 & minimum of 0.3 and 0.2 is 0.2)

***(iii)** The truth value of the disjunction of two prepositions.*

The truth value of the disjunction of two prepositions in fuzzy logic is the maximum of the truth values of the two prepositions.

Example. If the truth value of the statement "Neera is intelligent" is 0.9 and the truth value of the statement "Rekha is intelligent" is 0.6

Then the truth values of the statements

"Neera is intelligent, or "Rekha is intelligent" is 0.9 (Maximum of the two).

The truth values of the statements "Neera is not intelligent" or "Rekha is not intelligent" is 0.4 (Maximum of 1 – 0.9 = 0.1 and 1 –0.6 = 0.4).

EXERCISE 66.1

1. Interpret the following:

(*i*) The set A = {0.7 Anu, 0.9 Rasika, 0.2 Sarita, 0.5 Kartik} for the set of honest people.

(*ii*) The set of B = [0.2 John, 0.4 Charu, 0.9 Medha, 0.8 Gagan] for the set of brave people.

2. What are the constituents of the pair in a fuzzy set.

3. Which of the two fuzzy sets are equal of the following:

A = [0.3 Sonu, 0.8 Renu, 0.9 Paul, 0.5 Kunal]

B = [0.6 Kunal, 0.9 Paul, 0.7 Renu, 0.3 Sonu]

C = [0.8 Renu, 0.9 Paul, 0.3 Sonu, 0.5 Kunal]

4. Write down complement set of A, if

A = [0.3 Krishna, 0.8 Kamal, 0.7 Rajnish, 0.6 Surendra]

5. Write down $A \cup B$ in fuzzy sets.

If fuzzy set = A [0.5 x_1, 0.3 x_2, 0.7 x_3, 0.8 x_4]

Fuzzy set B = [0.6 x_1, 0.4 x_2, 0.9 x_3, 0.1 x_4]

6. Write down $A \cap B$ in fuzzy sets.

If fuzzy set A = [0.4 P, 0.7 Q, 0.2 R, 0.5 S]

fuzzy set B = [0.8 P, 0.6 Q, 0.1 R, 0.4 S]

7. Find the truth value of the negation of the following prepositions.

(*i*) The truth value of "A is happy" is 0.8.

(*ii*) The truth value of "B is rich" is 0.7.

(*iii*) The truth value of "Kamla is beautiful" is 0.9.

8. Find the truth value of the conjunction of the two prepositions. If the truth value of the statements "Ranjeet is a good driver" is 0.7.

Latif is a good driver is 0.6.

9. Give the truth value of the disjunction of the two prepositions

If the truth value "Sarla has a good health" is 0.6.

And the truth value "Vijay possesses a good health" is 0.8.

10. Write short notes on "Fuzzy sets".

ANSWERS

1. (*i*) Rasika is the most honest and Sarita is the least honest.

(*ii*) Medha is the bravest girl and John is the least.

2. Members and its degree of membership.

3. $A = C$

4. $\overline{A}$ = [0.7 Krishna, 0.2 Kamal, 0.3 Rajnish, 0.4 Surendra]

5. $A \cup B$ = [0.6 x_1, 0.4 x_2, 0.9 x_3, 0.8 x_4]

6. $A \cap B$ = [0.4 P, 0.6 Q, 0.1 R, 0.4 S]
7. (*i*) The truth value of "*A* is not happy" is = 1– 0.8 = 0.2.
   (*ii*) The truth value of "*B* is not rich" is 1 – 0.7 = 0.3.
   (*iii*) The truth value of "Kamla is not beautiful" is 1 – 0.9 = 0.1
8. The truth value of the conjunction, Ranjeet & Latif are good drivers is 0.6 (minimum of the two).
9. The truth value "Sarla has a good health" or "Vijay possesses a good health" is 0.8 (Maximum of the two).

66.8 APPLICATIONS

All engineering disciplines have already been affected to various degrees by new methodological possibilities opened by fuzzy sets, fuzzy measures.

(*i*) **Electrical Engineering**

By developing fuzzy controllers, electrical engineering was first engineering discipline within which the utility of fuzzy sets and fuzzy logic was recongised. Fuzzy image processing, electronic circuits for fuzzy logic or robbotics is also developed in electrical engineering.

(*ii*) **Civil Engineering**

In civil engineering, some initial ideas regarding the application of fuzzy sets emerged in 1970. There is the uncertainity in applying theoretical solution to civil engineering projects, designing at large. Designer deals with the uncertainity, in safety which is required in the construction of bridges; buildings, dams etc. Fuzzy set theory has already proven useful, consists of problems of assessing or evaluating existing constructions.

(*iii*) **Mechanical Engineering**

It was realised around mid-1980s that fuzzy set theory is eminently suited for mechanical engineering design.

A wide range of material might be used in mechanical engineering and the membership function is expressed in terms of corrosion, thermal expansion or some other measurable material property. A combination of several properties including the cost of different materials, may also be used.

(*iv*) **Industrial Engineering**

Two well-developed areas of fuzzy set theory that are directly relevant to industrial engineering are fuzzy control and fuzzy decision making.

Numerous their applications of fuzzy set theory in industrial engineering have also been explored to various degrees, Fuzzy set are convenient for estimating the service life of a given piece of equipment for various conditions under which it operates.

In industrial environment, fuzzy sets are also applied in designing built-in tests for industrial systems.

(*v*) **Computer Engineering**

In mid 1980s, when the utility of fuzzy controllers became increasingly visible, the need for computer hardware to implement the various operations involved in fuzzy logic and approximate reasoning has been recognised. All inference rules of a complex fuzzy inferences engine are processed in parallel. This increases efficiency tremendously and extends the scope of applicability of fuzzy controllers, and potentially, other fuzzy expert systems. In digital mode, fuzzy sets are represented as vectors of numbers (0, 1). Analog fuzzy hardware is characterised by high speed and good compatibility with sensors, it is thus suitable for complex on-line fuzzy controllers.

(*vi*) **Reliability theory**

The classical theory of reliability is developed in world war II on the following assumptions.

(*a*) **Assumption of dichotomous states.** At any given time, the engineering products is either in functioning state or in failed state.

(*b*) **Probability assumption.** The behaviour of the engineering product with respect to the two critical states (functioning and failed) can adquately be characterised in terms of probability theory.

An alternative reliability theory, rooted in fuzzy sets and probability.

(*c*) **Assumption of fuzzy sets.** At any time the engineering products may be in functioning states to some degree and in failed state to another degree.

(*d*) **Possibility assumption.** The behaviour of the engineering product with respect to the two critical fuzzy states (fuzzy functioning state and fuzzy failed state) can adequately be characterised in terms of possibility theory, while second theory based on fuzzy sets is more meaningful.

(*vii*) **Robotics**

The fuzzy set theory that is relevant to robotics include approximate reasoning, fuzzy controllers and other kind of fuzzy systems, fuzzy pattern recognition and image processing, fuzzy data bases.

EXERCISE 66.2

1. Write short note on the following:
Fuzzy logic affects many disciplines. (Rajiv Gandhi University, M.P. Summer 2001)

2. Define with example
Fuzzy graph, fuzzy relations. (Ravi Shanker Uni. I semester 2003)

INDEX

E

F

G

H

I

J

K

L

M

N

O

P

R

S

T

U

V

W

Z

INDEX

A

B

C

D

E

F

G

H

I